6th Asian/Australian Rotorcraft Forum & Heli Japan 2017

Advanced Rotorcraft Technology and Safety Operations

Kanazawa, Japan
7 - 9 November 2017

ISBN: 978-1-5108-7002-4

Printed from e-media with permission by:

Curran Associates, Inc.
57 Morehouse Lane
Red Hook, NY 12571

Some format issues inherent in the e-media version may also appear in this print version.

Printed by Curran Associates, Inc. (2018)

Vertical Flight Society
2701 Prosperity Ave, Suite 210
Fairfax, VA 22031
USA

Phone: (703) 684-6777
Fax: (703) 739-9279

www.vtol.org

Additional copies of this publication are available from:

Curran Associates, Inc.
57 Morehouse Lane
Red Hook, NY 12571 USA
Phone: 845-758-0400
Fax: 845-758-2633
Email: curran@proceedings.com
Web: www.proceedings.com

FINAL PROGRAM (as of November 16, 2017)

The 6th Asian/Australian Rotorcraft Forum & Heli Japan 2017, Kanazawa, Japan	
Monday, November 6th, 2017	
15:00~18:00	**Registration Open [Hall B, 3F]**
15:00~18:00	**Welcome Reception [Hall B, 3F]**

Tuesday, November 7th, 2017	
8:00	**Registration Open [3F]**
9:00~9:10	**Opening Ceremony [Hall A, 3F]**
	Opening Address: Prof. Kawachi, Mr. Katayama, Dr. Aoyama
9:10~10:50	**Keynote Speeches I [Hall A, 3F]** **Chair: Noboru Kobiki (JAXA)**
	KN-I 9:10~9:30 :ARF6-101, AHS International Executive Director, Mike Hirschberg, "The Future of Vertical Flight".....1
	KN-II 9:30~9:50 : ARF6-105, MHI, Japan, VP and GM, Mr. Takashi Katayama, "MHI Helicopter Overview".....N/A
	KN-III 9:50~10:10 : ARF6-106, KHI, Japan, Mr. Noriaki Katayama, "Helicopter Activities in Kawasaki Heavy Industries, Ltd.".....19
	KN-IV 10:10~10:30 : ARF6-107, SUBARU, Japan, VP and GM, Mr. Hiroshi Wakai, "SUBARU CORPORATION Helicopter Overview".....N/A
	KN-V 10:30~10:50 : ARF6-085, Pusan Univ & KAI, Korea, Prof. Sejong Oh and Dr. Jongho Choi, "Introduction to Korean Rotorcraft Development Programs"....N/A
10:50~11:00	**Break**
11:00~12:20	**Keynote Speeches II [Hall A, 3F] (Simultaneous Interpretation)** **Chair: Takashi Aoyama (JAXA)**
	KN-VI 11:00~11:20 : ARF6-102, AVICopter, China, Dr. Huang Chuanyue, "A Glance of Helicopter Industry in China".....N/A
	KN-VII 11:20~11:40 : ARF6-103, IIS, India, Prof. Dineshkumar Harursampath, "Helicopter Ecosystem in India".....26
	KN-VIII 11:40~12:00 : ARF6-104, DoD, Australia, Dr. Arvind K. Sinha, "Australian Rotary Wing Imported Design Technologies – Issues and Challenges".....N/A
	KN-IX 12:00~12:20 : ARF6-094, METU, Turkey, Prof. Ilkay Yavrucuk, "A Brief Look at Helicopter Related Activities in Turkey".....N/A

12:20~13:30	**Luncheon meeting sponsored by Weather News [Hall A] and by SUBARU [Room C]**				
13:30~14:30	**Special Lectures I [Hall A, 3F] (Simultaneous Interpretation)** **Chair: Dock Joo Lee (KAIST)**				
	SL-I 13:30~14:00 : ARF6-108, HemNet, Mr. Wataru Nishikawa, "Doctor-Heli in Japan and Air Ambulance in Europe and US".....45				
	SL-II 14:00~14:30 : ARF6-109, ACSL, Prof. Kenzo Nonami, "Recent Technologies and Future Prospective of ACSL Drone".....N/A				
14:30~14:40	**Break**				
14:40~16:20	**Hall A, 3F** **Safety Operations I** **(Simultaneous Interpretation)** **Chair: Maki Nagao (Aero Asahi)**	**Hall B, 3F** **Aerodynamics I** **Chair: Sejong Oh (Pusan Univ)**	**Room C, 3F** **UAV I** **Chair: JVR Prasad (GIT)**	**Room D, 3F** **Design I** **Chair: Yasutada Tanabe (JAXA)**	**Room E, 2F** **Dynamics** **Chair: Hyeonsoo Yeo (US Army ADD)**
	A101 14:40~15:00 ARF6-042 JHSTによる安全活動について (Safety Activities of JHST) Osamu Washida (Aero Asahi) 48	B101 14:40~15:00 ARF6-002 Physics Based Models for Helicopters in Maneuvering Flight Lakshmi N Sankar (GIT) 71	C101 14:40~15:00 ARF6-035 Interim report of study about requirements of UAV sense and avoid system 112 Akitoshi Sakaguchi (SJAC)	D101 14:40~15:00 ARF6-084 Optimal Aerodynamic Design of Main Rotor Blade for a High-Speed Compound Helicopter 140 Masahiko Sugiura (JAXA)	E101 14:40~15:00 ARF6-069 Vibration Suppression of Rotating Beams Through Piezoelectric Shunt Circuits 188 Jinwei Shen (Univ Alabama)
	A102 15:00~15:20 ARF6-048 小型航空機に係る総合的な安全対策の推進 (General Aviation Safety Improvement) Tomofumi Chida (JCAB) 51	B102 15:00~15:20 ARF6-074 A Study of Rotor/Wing Aerodynamic Interaction at High Speed Flight on a Compound Helicopter Hideaki Sugawara (Ryoyu Systems) 82	C102 15:00~15:20 ARF6-001 Simulation and evaluation of linear velocity control by Cellular Air Traffic System (CATS) for high-density logistical VTOL-UAS traffic Yu Ito (Yamato Holdings) 115	D102 15:00~15:20 ARF6-037 Performance Optimization for an Optimum Rotor Speed Helicopter Jiayi Xie (Tsinghua Univ) 150	E102 15:00~15:20 ARF6-092 Performance and Loads Analyses of Compound Helicopters using a Rigid Coaxial Rotor 201 Sang-Hyun Chae (KARI)
	A103 15:20~15:40 ARF6-003 改正航空法の現状と今後の制度設計の方向性 (The Amended Civil Aeronautical Act for Safety and Harmonization between Aircraft and UAV) Kohei Hase (JCAB) 56	B103 15:20~15:40 ARF6-075 Improvement of actuator surface method for various solidity rotor computation Jaeyoung Park (Pusan Univ) 94	C103 15:20~15:40 ARF6-028 Design of a Ducted-Fan Micro-Rotorcraft with Structural Battery Dineshkumar Harursampath (IIS) 126	D103 15:20~15:40 ARF6-034 Active Flow Control of Rotor via Combinational Strategy Yiyang Ma (NUAA) 159	E103 15:20~15:40 ARF6-057 Reduction of Helicopter Rotor Vibratory Loads Through Robust and Hybrid Optimization Muhammed Emre Bilen (METU) 211

	A104 15:40~16:00 ARF6-089 CARATSにおける小型航空機用RNAV施策の紹介 (Introduction of the RNAV Measures for Small Aircraft in CARATS) Syuro Yamano (JCAB) 60	B104 15:40~16:00 ARF6-022 Numerical Simulations of the Flowfield of Tiltrotor Aircrafts in Ground Effect Based on N-S Equations Jiang Shuang (NUAA) 99	C104 15:40~16:00 ARF6-059 Development of VTOL with movable legs Sho Goto (Tokushima Univ) 136	D104 15:40~16:00 ARF6-030 The Design Technology of All-electric Autogyro Liang Lei (NUAA) 168	E104 15:40~16:00 ARF6-013 Blade Loads Analysis of a Variable Speed Helicopter Rotor Dong Han (NUAA) 232
	A105 16:00~16:20 ARF6-093 安全運航の為の航空気象観測インフラと気象予測について (Aviation Meteorological Observation Infrastructure and Weather Prediction for Safety Operation) Mizuki Nishimura (Weathernews) 64	B105 16:00~16:20 ARF6-038 Development of Nonlinear Vortex Lattice Method and Its Application to Aerodynamic Analysis of Wind turbine Hakjin Lee (KAIST) 107	C105 16:00~16:20	D105 16:00~16:20 ARF6-008 Human Factors Issues of Glass Cockpit Automation Cigdem Gunes (TAI) 177	E105 16:00~16:20 Special Presentation #1 Individual Blade Control of a 5-bladed Rotor Using the Multiple Swashplate System Berend van der Wall (DLR) 243
16:20~16:30	**Break**				
	Hall A, 3F Safety Operations II (Simultaneous Interpretation) Chair: Osamu Washida (Aero Asahi)	**Hall B, 3F Acoustics I Chair: Schwarz Thorsten (DLR)**	**Room C, 3F UAV II Chair: Shigeru Sunada (Nagoya Univ)**	**Room D, 3F Test & Evaluation I Chair: Noboru Kobiki (JAXA)**	**Room E, 2F Flight Controls Chair: Berend van der Wall (DLR)**
	A106 16:30~16:50 ARF6-010 Lesson Learnt from Data Exchange Between Manned and Unmanned Aerial Vehicles and Accommodating 12 Different UAS Operations Hiroko Nakamura (Univ Tokyo) 260	B106 16:30~16:50 ARF6-009 Tackling the HARTII test case: A joint effort by DLR and JAXA Gunther Wilke (DLR) 281	C106 16:30~16:50 ARF6-090 Prediction of Rotor Flow for Unmanned Aerial System Using Nonlinear Vortex Lattice Method Yeongmin Jo (KAIST) 311	D106 16:30~16:50 ARF6-014 Separation Control using Rail Plasma Actuator on a VR-12 Airfoil Jayant Sirohi (UT Austin) 343	E106 16:30~16:50 ARF6-004 Comprehensive Flight Control Law Design Method for Helicopters Wei Wu (NUAA) 377
	A107 16:50~17:10 ARF6-072 高い高度でホバリングするヘリコプタロータのダウンウォッシュが地表面付近に及ぼす影響(The influence of the downwash of a helicopter rotor hovering at high altitude on the ground surface) Kenta Nakashima (NDA) 264	B107 16:50~17:10 ARF6-040 Development of bio-inspired low-noise propeller for drone Ryusuke Noda (Chiba Univ) 293	C107 16:50~17:10 ARF6-018 Design and Analysis Tools for New Rotorcraft Concepts and Emerging Markets Christopher Silva (Aeronautic Concepts) 321	D107 16:50~17:10 ARF6-053 An experimental study of aerodynamics of a compound helicopter Koichi Yonezawa (CRIEPI) 356	E107 16:50~17:10 ARF6-073 Usage of a Driving Force of Wheels for Shortening a Take-off Distance of Small Airplanes Emi Matsui (Osaka Pref Univ) 389
16:30~18:10	A108 17:10~17:30 ARF6-099 Helicopter downwash measurement using Doppler lidar Naoki Matayoshi (JAXA) 268	B108 17:10~17:30 ARF6-045 A Study on High-Order Numerical Methods for Aerodynamic Noise Prediction of Helicopter Blades in Hovering Flight Ki Ro Kim (Konkuk Univ) 297	C108 17:10~17:30 ARF6-091 Development of Integrated Optimization Design Framework for a Small-scale Multirotor UAV Hyeongseok Kim (Seoul Univ) 326	D108 17:10~17:30 ARF6-078 Endurance Test of the Revised SNUF blade with Active Trailing-Edge Flap SangJoon Shin (Seoul Univ) 362	E108 17:10~17:30 ARF6-024 State-space Identification of Unmanned Helicopter Dynamics using Invasive Weed Optimization Algorithm on Flight Data S.N Omkar (IIS) 394
	A109 17:30~17:50 ARF6-100 病院屋上ヘリポートの実情と課題 (Current Situation and Issues for Roof-Top Heliports on Hospitals) Motomi Kinoshita (Aero Facility) 274	B109 17:30~17:50 ARF6-046 The research of rotor noise radiation characteristics from an in-flight helicopter based on acoustic mapping Fei Wang (NUAA) 301	C109 17:30~17:50 ARF6-068 Robust Design of Type-II Small-Scale Unmanned Helicopter for Hover Performance using Taguchi Method Dineshkumar Harursampath (IIS) 334	D109 17:30~17:50 ARF6-031 An Experimental Investigation of Ground Effect on Rotorcraft in the presence of Side walls M Rohith (IIT Madras) 370	E109 17:30~17:50 ARF6-061 Obstacle avoidance system for multicopter Oda Yasuto (Tokushima Univ) 400
	A110 17:50~18:10 ARF6-079 石川県消防防災航空隊の活動について (Activities of the Ishikawa Air Rescue Team) Kazuhiro Nishijima (Ishikawa Pref) 277	B110 17:50~18:10	C110 17:50~18:10 ARF6-060 Development of Tailsitter VTOL using Multi copter Kenta Kodera (Tokushima Univ) 340	D110 17:50~18:10	E110 17:50~18:10
18:10~19:00	**Break**				
19:00~21:00	**Banquet [Hall A & B, 3F]**				

Wednesday, November 8th, 2017					
8:15	**Registration Open**				
9:00~10:00	**Special Lectures II [Hall A, 3F] (Simultaneous Interpretation)** **Chair: Qijun Zhao (NUAA)**				
	SL-III 9:00~9:30 : ARF6-110, Leonardo, Dr. James Wang, "What has happened around the world since Project Zero Started the Electric Revolution? When will Electric Vertical Flight become a Reality and What are the Challenges?".....N/A				
	SL-IV 9:30~10:00 : ARF6-095, Airbus Helicopters, Joeh Leo, "Helicopter Safety – A review of world-wide efforts and focus for our futures".....N/A				
10:00~10:10	**Break**				
10:10~11:50	**Hall A, 3F** **Safety Operations III** **(Simultaneous Interpretation)** **Chair: Naoki Matayoshi (JAXA)**	**Hall B, 3F** **Aerodyanmics II** **Chair: Lakshmi N Sankar (GIT)**	**Room C, 3F** **UAV III** **Chair: Takeshi Akasaka (KIT)**	**Room D, 3F** **Design II** **Chair: Masahiko Sugiura (JAXA)**	**Room E, 2F** **Structures & Materials** **Chair: Daisuke Matsuura (MHI)**
	A201 10:10~10:30 ARF6-086 Operation and Prospect of EMS (Emergency Medical Service) Helicopter in Korea 403 Myeang Hyen Bae (Konkuk Univ)	B201 10:10~10:30 ARF6-082 Effect of Blade Deformation on Air and Structure Loads of LCH Rotor Younghyun You (Konkuk Univ) 447	C201 10:10~10:30 ARF6-081 Influences of Upper and Side Walls on the Performance of a Multiple Rotor Drone Yasutada Tanabe (JAXA) 497	D201 10:10~10:30 ARF6-087 Structural Optimization of Large Scale Wind Turbine Cross-Sections using Mixed Beam Theory 534 Jae Seong Bae (Konkuk Univ)	E201 10:10~10:30 ARF6-049 Generalized Timoshenko Theory for Functionally Graded Beams with Three-dimensional Warping 566 Manoj Kumar Dhadwal (Konkuk Univ)
	A202 10:30~10:50 ARF6-083 Construction of AIR-TRAFFIC-SYSTEM for all Weather Operations of Helicopter Takeshi Tomio (Helicopter Flight Safety) 409	B202 10:30~10:50 ARF6-066 Computational Analysis on Aerodynamic Interaction of Quad Tilt Rotor during Forward Flight Seonggi Lee (Pusan Univ) 455	C202 10:30~10:50 ARF6-096 Experimental study on fluid field and interference at downstream of quadrotor Sida Yang (Nagoya Univ) 506	D202 10:30~10:50 ARF6-012 Performance Analysis of Variable Speed Tail Rotors with Gurney Flaps Dong Han (NUAA) 543	E202 10:30~10:50 ARF6-097 Significant Effects Obtained from Health and Usage Monitoring System (HUMS) on JASDF Rescue Helicopter 578 Kosuke Watanabe (MHI)
	A203 10:50~11:10 ARF6-017 ALB Operation in a Helicopter - Its Application for Infrastructure Maintenance and Disaster Prevention Hiroshi Isobe (Aero Asahi) 414	B203 10:50~11:10 ARF6-054 A New Ground Effect Factor based on Image Vortices Cibin Joseph (IIT Madras) 461	C203 10:50~11:10 ARF6-039 Flow Visualization of Wake of a Quad-copter in Ground Effect Hikaru Otsuka (Tohoku Univ) 513	D203 10:50~11:10 ARF6-055 Analysis on the influence of the variable speed and variable diameter strategies on aerodynamic performance of the tiltrotor Bo Wang (NUAA) 550	E203 10:50~11:10 ARF6-011 Nonlinear Analysis of Pretwisted Anisotropic Beam with Multiple Delaminations 584 Dineshkumar Harursampath (IIS)
	A204 11:10~11:30 ARF6-019 Design light weight emergency cot with enhanced spinal immobilization capability 419 Musaddiq Al Ali (Hiroshima Univ)	B204 11:10~11:30 ARF6-033 An Evaluation of Ground Effect Modelling for Rotors in Hover Salini S. Nair (IIT) 467	C204 11:10~11:30 ARF6-050 Effects of the size and the position of the center of gravity of a multirotor UAV on hovering flight characteristics 523 Shuichi Umezaki (Osaka Pref Univ)	D204 11:10~11:30 ARF6-007 Affective Human Factors in Aviation Cigdem Gunes (TAI) 560	E204 11:10~11:30
	A205 11:30~11:50 ARF6-113 VTOL GoFly Prize (Special Presentation) Gwen Lighter 430	B205 11:30~11:50 ARF6-006 Helicopter Rotor Trim and Blade Motion Perturbations Caused by Wake Vortices of Wind Turbines and Fixed Wing Aircraft 476 Berend G. van der Wall (DLR)	C205 11:30~11:50	D205 11:30~11:50	E205 11:30~11:50
11:50~13:00	**Luncheon meeting sponsored by NAVICOM AVIATION CORPORATION [Hall A] and by SUBARU [Room C]**				

	Hall A, 3F Test & Evaluation II **Chair: Jayant Sirohi (UT Austin)**	Hall B, 3F Aerodynamics III **Chair: Qijun Zhao (NUAA)**	Room C, 3F UAV IV **Chair: Hiroshi Tokutake (Kanazawa Univ)**	Room D, 3F Acoustics II **Chair: Blanche Demaret (ONERA)**	Room E, 2F
13:00~14:40	A206 13:00~13:20 ARF6-098 Helicopter Ship Interface Testing - Innovative SHOL-X predictive software tool 592 Alrik Hoencamp (AeroMath | Aeronautical Engineering Services)	B206 13:00~13:20 ARF6-058 Numerical Study on Ducted Fan Tail Rotor in Forward Flight Conditions Nahyeon Roh (Pusan Univ) 645	C206 13:00~13:20 ARF6-047 Development of a ducted rotor for multicopters Koichi Yonezawa (CRIEPI) 681	D206 13:00~13:20 ARF6-036 Numerical Investigations on Aerodynamic and Acoustic Characteristics of a Rigid Coaxial Rotor in Hover Qijun Zhao (NUAA) 712	E206 13:00~13:20
	A207 13:20~13:40 ARF6-016 Deformation Measurement and Modal Identification of an Extremely Flexible Rotor Blade 603 Daiju Uehara (UT Austin)	B207 13:20~13:40 ARF6-023 CFD/CSD coupling investigations on aeroelastic characteristics of rotor with different blade-tip planforms Li Ma (NUAA) 649	C207 13:20~13:40 ARF6-015 Modeling of Aerodynamic Characteristics of Drone and Improvement of Gust Response 687 Raijo Torita (Kanazawa Univ)	D207 13:20~13:40 ARF6-076 Triple Swept Rotor Design for Helicopter BVI Noise Reduction Sanghyun Chae (KARI) 720	E207 13:20~13:40
	A208 13:40~14:00 ARF6-063 Experimental Investigation of the Rotor-Wing Aerodynamic Interaction on a Compound Helicopter in Hover Keiichiro Makino (KIT) 616	B208 13:40~14:00 ARF6-027 Aerodynamic Performance Analysis for Rotor with Ice Accretion in Forward Flight Xi Chen (NUAA) 657	C208 13:40~14:00 ARF6-080 Development of High-Fidelity Flight Dynamic Models for Unmanned Rotorcrafts and multicopters with Applications 695 Seong Han Lee (Konkuk Univ)	D208 13:40~14:00 ARF6-020 Development and validation of a new hybrid method for predicting helicopter rotor impulsive noise Liangquan Wang (NUAA) 726	E208 13:40~14:00
	A209 14:00~14:20 ARF6-064 Experimentall Investigation of the Rotor-Wing Aerodynamic Interaction on a Compound Helicopter in High Advance Ratio Yusuke Hamamoto (KIT) 621	B209 14:00~14:20 ARF6-041 CFD Analysis on Aerodynamic Characteristics of Helicopter Ducted Tail Rotor Based on Sliding Mesh Chenkai Cao (NUAA) 666	C209 14:00~14:20 ARF6-044 Low Speed Performance Analysis of a Compound Rotorcraft with Tandem Ducted Fan and Pusher Wonik Seo (INHA Univ) 701	D209 14:00~14:20 ARF6-021 Suppression on Rotor BVI Noise by IBC Active Control Methodology Qijun Zhao (NUAA) 739	E209 14:00~14:20
	A210 14:20~14:40 Special Presentation #2 From ERATO Basic Research to the Blue Edge Rotor Blade *(EREA Best Paper Award 2016, AHS2016 T&E Session Best Paper)* Berend van der Wall (DLR) 626	B210 14:20~14:40 ARF6-029 Numerical research on the aerodynamic impact of ship superstructure for helicopter operations Yi Xu (NUAA) 674	C210 14:20~14:40 ARF6-077 Design, Manufacturing, and Experimental Investigation of Composite Rotor Blades for Small-Scale Unmanned Helicopters with Flight Test Verification 706 Dineshkumar Harursampath (IIS)	D210 14:20~14:40	E210 14:20~14:40

14:40~14:50	**Break**
14:50~16:50	**Special Lectures III [Hall A, 3F] (Simultaneous Interpretation)** **Chair: Arvind Sinha (DoD Australia)**
	SL-V 14:50~15:30 : ARF6-111, US Army ADD, Dr. Hyeonsoo Yeo, "Recent Advances in High Fidelity Rotorcraft Aeromechanics Analysis".....749
	SL-VI 15:30~16:10 :ARF6-005, ONERA, Dr. Blanche Demaret, "ONERA Recent Contribution to Reduce Environmental Signature for Rotorcraft, and to Provide Tools for Flight and Operation Safety".....765
	SL-VII 16:10~16:50 : ARF6-112, GIT, Prof. JVR Prasad, "Envelope Protection for Carefree Maneuvering".....780
16:50~17:10	**Closing Remarks, Best Paper Awards Ceremony [Hall A, 3F]**

Thursday, November 9th, 2017	
	Technical Bus Tour (From/Return: Kanazawa Railway Station)
8:30 ~ 12:30	**Cosmo Isle Hakui (Space Science Museum); Chrihama (Thousand Mile Sand Beach)**
12:30 ~ 13:30	**Lunch at Nikko (JAL) Hotel Kanazawa**
13:30 ~ 16:00	**Hakko-kan (Haku-Ichi) (Gold Leaf Art Museum)**

6th Asian-Australian Rotorcraft Forum
&
Heli Japan 2017

先進回転翼技術と安全運航に関する国際会議

Advanced Rotorcraft Technology and Safety Operations

2017.11.7~9

in KKR Hotel Kanazawa, Ishikawa, Japan

Sponsored by **The Japan Helicopter Society**
AHS International

Co-sponsored by **The Japan Society for Aeronautical and Space Sciences**

Final Program and Abstracts

6th Asian-Australian Rotorcraft Forum (ARF) & Heli Japan 2017

Technical Meeting on Advanced Rotorcraft Technology and Safety Operations

November 7-9, 2017
KKR Hotel Kanazawa, Ishikawa, JAPAN

This meeting will be held at the **KKR Hotel Kanazawa**, located neighbor to the historic Kanazawa Castle Park, in front of the Ote-Mon Gate, about 1.2 km from JR Kanazawa Station

KKR Hotel Kanazawa
Ote-Machi 2-32, Kanazawa, Ishikawa, 920-0912, JAPAN
Tel. (+81) 76-264-3261

KKR ホテル金沢
〒920-0912 石川県金沢市大手町 2-32
TEL.076-264-3261

Access

□Walk from Kanazawa Station --- About 25 minutes

□Taxi from Kanazawa Station --- About 10 minutes, taxi fee about 1,000JPY

アクセス

□徒歩　　・・・JR 金沢駅より２５分。
東口出口を直進し（向かって右側の道路を歩く）、名鉄エムザまで進み、むさし交差点を渡り尾張町交差点まで進む、尾張長交差点を右折、大手掘方面へ徒歩５分

□タクシー・・・JR 金沢駅から約 10 分（約 1,000 円）
□バス　　・・・JR 金沢駅東口７番乗り場より橋場町経由バス（11･12･16 番）で 10 分（料金 200 円）
※城下町かなざわ周遊バス・ライトアップバスは尾張町には停車致しません。
・尾張町バス停で下車、尾張町交差点を右折、大手掘方面へ徒歩５分

For more information, please visit
http://www.helijapan.org/ARFHeliJapan2017.html

Japan Helicopter Society, Japan Chapter of AHS International
国際会議ヘリジャパン2017実行委員会

国際会議ヘリジャパン 2017 実行委員会（Heli Japan 2017 Executive Committee）

議　長	河内　啓二	東京大学名誉教授
General Chairperson	Keiji Kawachi	The University of Tokyo, Emeritus Professor
実行委員長	青山　剛史	国立研究開発法人 宇宙航空研究開発機構
Executive Chairman	Takashi Aoyama	Japan Aerospace Exploration Agency (JAXA)

総務委員会（General Affairs Committee）

委員長	小曳　昇	国立研究開発法人 宇宙航空研究開発機構
Chairman	Noboru Kobiki	JAXA
委員	澤田　実宏	三菱重工業(株)
Member	Mitsuhiro Sawada	Mitsubishi Heavy Industries, Ltd
委員	坂口　晴治	川崎重工業(株)
Member	Seiji Sakaguchi	Kawasaki Heavy Industries, Ltd.
委員	赤坂　剛史	金沢工業大学
Member	Takeshi Akasaka	Kanazawa Institute of Technology
委員	小坂　充	(株)SUBARU
Member	Mitsuru Kosaka	SUBARU Corporation

プログラム委員会 (Program Committee)

委員長	田辺　安忠	国立研究開発法人 宇宙航空研究開発機構
Chairman	Yasutada Tanabe	JAXA
委員	杉浦　正彦	国立研究開発法人 宇宙航空研究開発機構
Member	Masahiko Sugiura	JAXA
委員	長尾　牧	朝日航洋(株)
Member	Maki Nagao	Aero Asahi Co.
委員	饗庭　昌行	防衛装備庁　航空装備研究所
Member	Masayuki Aiba	Aerial Systems Research Center, ATLA

Asian/Australian Rotorcraft Forum (ARF) International Executive Committee Members

「ヘリジャパン2017」へようこそ

ヘリジャパン2017議長
河内啓二
Conference General Chairperson
Keiji Kawachi

ここ金沢において「第6回アジア／オーストラリア回転翼フォーラム＆ヘリジャパン2017」を開催し、多くの国から優れたエンジニアのご参加を頂いたことをとてもうれしく思います。私たちは皆様を心より歓迎致します。金沢は歴史的史跡に恵まれた魅力的な街です。会議の合間にこの街も大いに楽しんで頂きたいと思います。

近年、ヘリコプタ、ティルトロータ、ドローンなどに代表される回転翼航空機はますます社会の注目を集めています。深刻な地震や火事、交通事故などの救助救援活動において、回転翼機はその有効性や利便性を示し続けて来ました。この会議において、皆様の貴重な情報を交換し相互理解を進めて、この重要な回転翼機の一層の発展に資することを期待しております。

会議を開催するにあたって、多くの方々から暖かいご支援を頂きました。厚く御礼申し上げます。

Welcome to Heli Japan 2017

It is my great pleasure to hold the 6th Asian/Australian Rotorcraft Forum (ARF) & Heli Japan 2017 here in Kanazawa and to see excellent researchers and engineers from many countries. We welcome very much all these participants. Kanazawa is an attractive town having many historical sites. I hope that you will enjoy this town in addition to the professional activities.

Rotorcraft, such as helicopter, tilt-rotor and UAV is gathering more and more interests from public these days. We always found effectiveness and convenience of rotorcraft through serious earthquakes, large fires, traffic accidents etc. I hope that all of you exchange valuable information and obtain mutual understanding through this conference. I believe these will make it possible to develop an excellent new rotorcraft in the future.

I appreciate many supports from various organizations. These supports are useful and valuable to realize this conference.

ご挨拶

日本ヘリコプタ協会会長
片山　健
President of Japan Helicopter Society
Takeshi Katayama

皆さんこんにちは、そしてようこそ Heli Japan2017 へ。片山健と申します。2016 年 7 月から日本ヘリコプタ協会の会長をさせていただいております。本協会は、1989 年に設立され、ヘリコプタに関する技術的観点だけでなく、運用面に関しても取り組んでまいりました。

国際会議 Heli Japan は、今回で 5 回目の開催となり、初めて Asian/Australian Rotorcraft Forum (ARF)と連名での開催となります。Heli Japan を主催し、ここ金沢に世界中からたくさんのヘリコプタの専門家に参加いただけることは、大変喜ばしい限りです。Heli Japan/ARF が全ての出席者にとって貴重で価値のある機会になると信じております。

最後に、ご支援していただきました諸団体及び企業の皆様に、厚く御礼申し上げます。

Greetings

Hello everyone and welcome to Heli Japan2017, my name is Takeshi Katayama. I am in charge of Japan Helicopter Society (JHS) as a president since July 2016. JHS was established in 1989 and has worked on helicopter not only technical perspective but also operational perspective.

This is the fifth Heli Japan international conference and the first to joint Asian/Australian Rotorcraft Forum (ARF). It is my great pleasure to host Heli Japan and have many helicopter experts from worldwide courtiers at here in Kanazawa. I believe this Heli Japan/ARF becomes precious and valuable opportunity for all attendees.

Finally, I would like to thank many organizations and companies for their great support.

ご挨拶

Heli Japan 2017 実行委員長
青山　剛史
Conference Executive Chairman
Takashi Aoyama

第6回Asian/Australian Rotorcraft Forum (ARF) & Heli Japan 2017へようこそ！ARFを構成するアジア/オーストラリアの各国はもとより、米国や欧州など世界の各国からご参加をいただき、ここ金沢で国際会議として開催できましたことは大変喜ばしく、実行委員会を代表して厚く御礼申し上げます。Heli Japan は今回で5 回目となりますが、ARFとしての開催は初めてで、今後の発展がますます期待されるところです。従来からHeli Japanでは、自然災害が多い我が国におけるヘリコプタ活用の特徴に鑑みて、ヘリコプタの先進技術のみならず防災や安全運航の話題提供にも力を入れてきたところですが、今回もその伝統を引き継いでおります。会議では、特別講演7件、キーノートスピーチ9件、一般講演約90件（内、海外約60件）の発表が予定されています。特別講演におきましては、ドクターヘリ、ドローン、電動化、IHSTの取り組み、最先端技術など、時代の流れを映し出す講演が聞けるものと期待しております。

本国際会議の開催に際しまして、地元の石川県、金沢市のご後援をいただくとともに、多くの関係機関、諸団体ならびに各企業、大学のご理解とご協力、重厚なるご支援を賜りました。ここに改めて厚くお礼申し上げます。

Greetings

Welcome to the 6th Asian/Australian Rotorcraft Forum (ARF) & Heli Japan 2017! On behalf of the Heli Japan 2017 Executive Committee, I am pleased to invite you to Kanazawa from around the world such as Asian/Australian countries, the United Sates, and European countries. This fifth Heli Japan will be the first to be held as ARF and further development is expected. In this conference, discussions on not only advanced technologies but also disaster relief and safety operations will be provided considering the special circumstances of the helicopter usage in Japan where natural disasters frequently occur. We have 7 special lectures, 9 keynote speeches and about 90 general lectures including about 60 from abroad. The topics which mirror the trend of the times such as EMS, Drone, electric vertical flight, IHST activities, and advanced technologies are expected in the special lectures.

Finally, I would like to thank everyone who supports us, especially Kanazawa City, Ishikawa pref., many helicopter-related organizations, companies and universities that take part in this event. I hope you will have an amazing time in this 6th Asian/Australian Rotorcraft Forum & Heli Japan 2017.

■ 主 催　HOSTS

日本ヘリコプタ協会　The Japan Helicopter Society
The AHS International Japan Chapter

Asian/Australian Rotorcraft Forum International
Executive Committee

■ 共 催　CO-HOST

一般社団法人　日本航空宇宙学会
The Japan Society for Aeronautical and Space Sciences

■ 後 援　WITH SUPPORT OF

石川県
Ishikawa Prefecture

金沢市
Kanazawa-city

一般社団法人　全日本航空事業連合会
All Japan Air Transport and Service Association

認定NPO法人　救急ヘリ病院ネットワーク
Emergency Medical Network of Helicopter and Hospital

一般社団法人　日本航空宇宙工業会
The Society of Japanese Aerospace Companies

一般財団法人　航空振興財団
Japan Civil Aviation Promotion Foundation

■ 協 賛　SPONSORS

三菱重工業(株)
Mitsubishi Heavy Industries, Ltd.

(株)SUBARU
Subaru Corporation

川崎重工業(株)
Kawasaki Heavy Industries, Ltd.

朝日航洋(株)
Aero Asahi Co.

日本エアロスペース(株)
Japan Aerospace Corporation

東京航空計器(株)
Tokyo Aircraft Instrument Co. Ltd.

MHIエアロスペースシステムズ(株)
MHI Aerospace Systems Corp.

シンフォニアテクノロジー(株)
SINFONIA TECHNOLOGY CO., LTD.

現在（いま）、最もご採用いただいている旬なピエゾアクチュエータ

Piezo Actuators LineUPs

薄型・扁平形状により、装置の小型化に貢献するシリーズ

MTD & MTDD　コンパクトサイズで大きな変位を発生

MTD　印加電圧の上昇とともに縮む

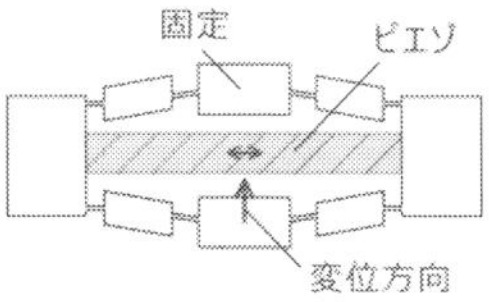

MTDD　印加電圧の上昇とともに伸びる

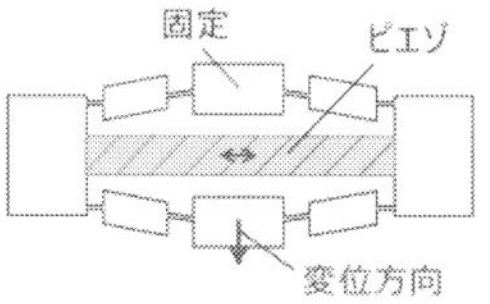

特長
- 1mmに近い（750μm）ほどの大変位を出力可能
- 扁平かつ薄型により装置の小型化に寄与
- 圧電素子の変位方向と90°直行方向に駆動

用途例
- 露光装置用精密ステージ
- オートフォーカス機構
- 精密ポンプ

MTK & MTKK　大きな変位を発生かつ共振周波数が高い

MTK　印加電圧の上昇とともに縮む

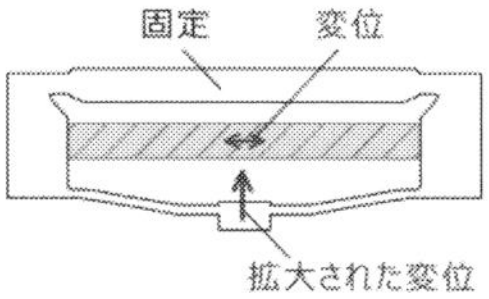

MTKK　印加電圧の上昇とともに伸びる

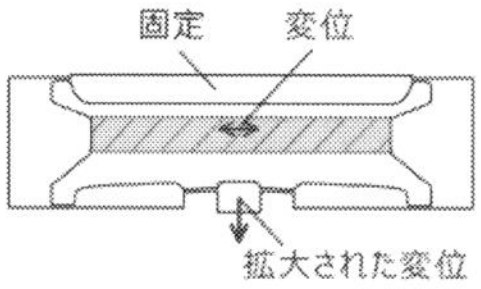

特長
- 数キロHzの共振周波数を維持し、180μm発生変位を実現
- 扁平かつ薄型により装置の小型化に寄与
- 圧電素子の変位方向と90°直行方向に駆動

用途例
- ハプティック用アクチュエータ
- 露光装置用精密ステージ
- オートフォーカス機構
- マスフロー

ピエゾ駆動用電源・制御システム

power supply & control unit

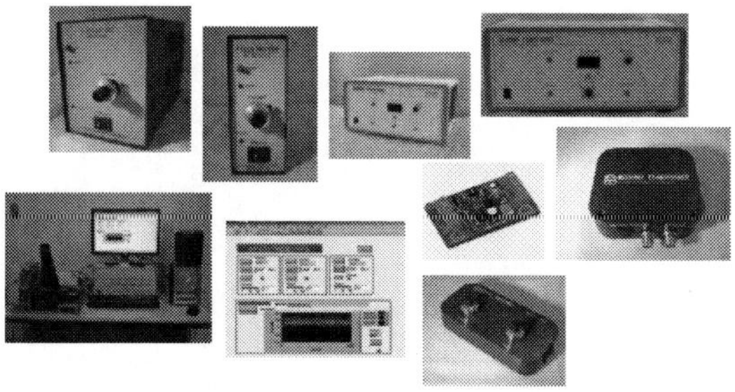

MT Intelligence

世界で初めて、積層圧電素子の
商用製品化を達成したメンバーは
これからも挑戦を続けて参ります。

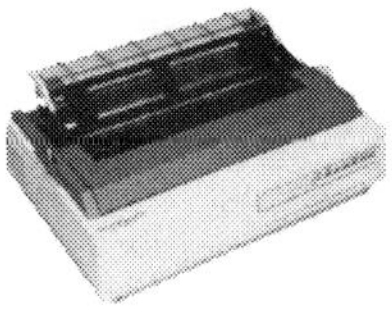

NM-5020 ミニエース漢字ライタ

TEL 03-5835-0108　FAX 03-5835-0109　info@mechano-transformer.com
〒101-0032　東京都千代田区岩本町２丁目７番12号ビルックス No.3 4階

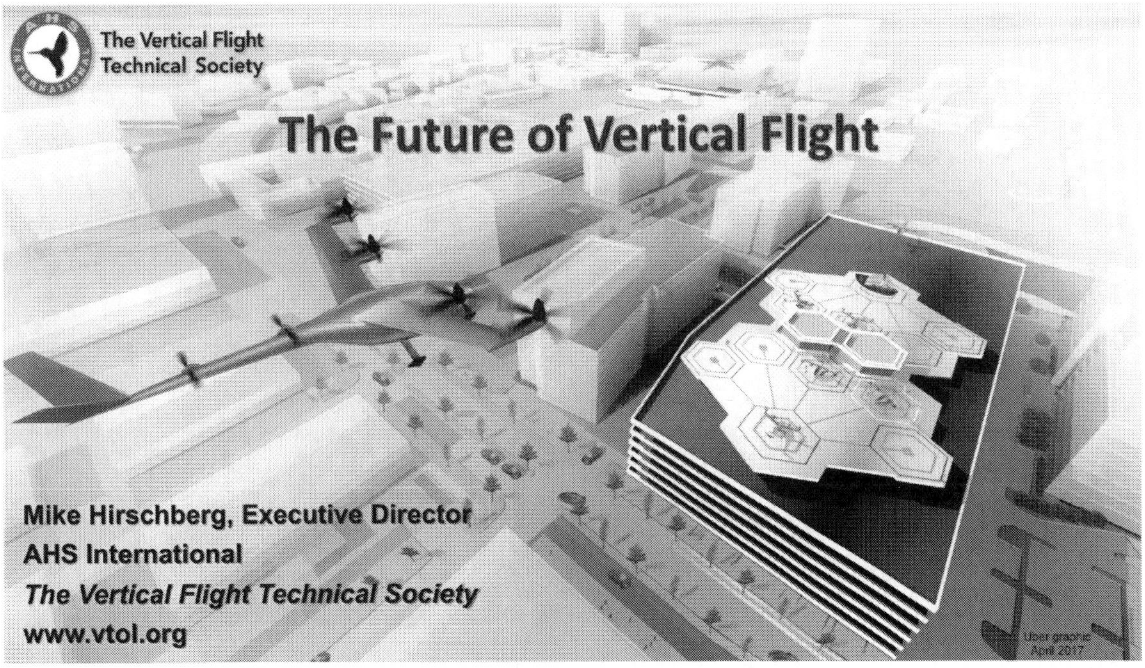

The Vertical Flight Technical Society
www.vtol.org

What is AHS International?

- The global **Vertical Flight Technical Society**
  - Founded in 1943 as the **American Helicopter Society**
  - Everything from VTOL **MAVs/UAS** to **helicopters** and **eVTOL** to **STOVL** (everything vertical except rockets)
- **Expands knowledge** about vertical flight technology and promotes its application around the world
- Advances **safety and acceptability**
- Advocates for vertical flight **R&D funding**
- Helps **educate and support** today's and tomorrow's vertical flight engineers and leaders

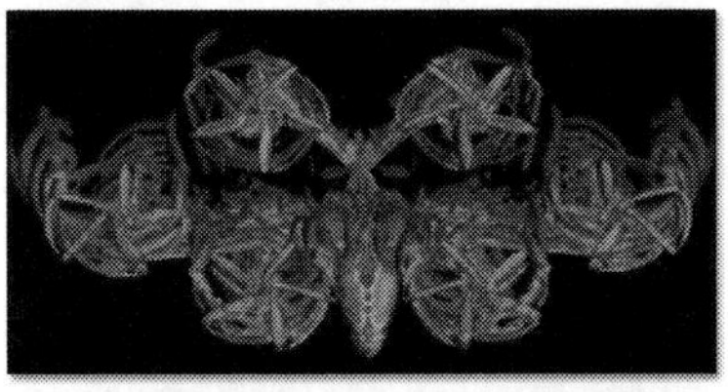

CFD of Joby S4, Aug 2015

VFF Scholarship Winners at AHS Forum 71, May 2015

2

The Vertical Flight Technical Society
www.vtol.org

FORUM 74 AHS INTERNATIONAL
MAY 14-17, 2018 PHOENIX, ARIZONA

The Vertical Flight Technical Society
www.vtol.org

The Future of Vertical Flight

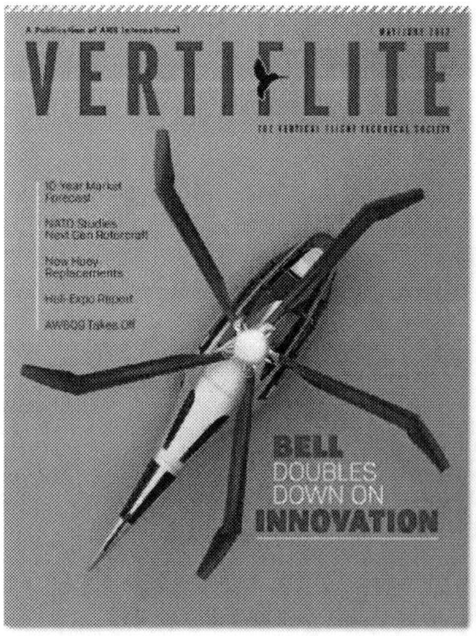

4

The Vertical Flight Technical Society
www.vtol.org

OEM Perspective

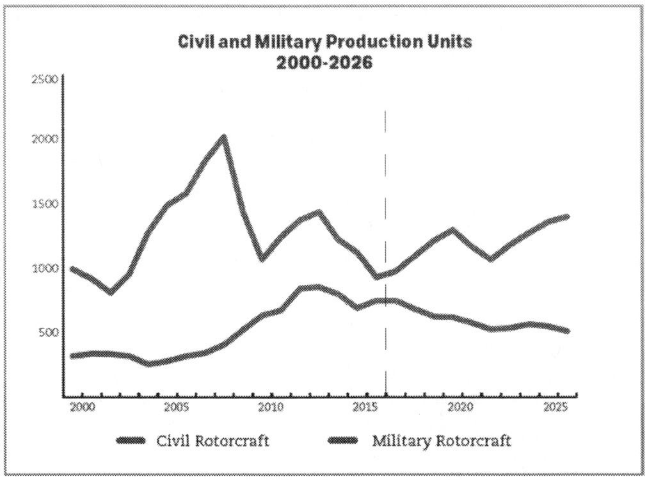

- Global downturn in helicopter production units began in 2012
- Rising oil and gas prices will boost civil helicopter sales
- Overall global military units continue to decline
- OEMs investing in new products and capabilities

5

The Vertical Flight Technical Society
www.vtol.org

Bell 525 Relentless & 505 Jet Ranger X

The Vertical Flight Technical Society
www.vtol.org

Bell FC-X Concept

The Vertical Flight Technical Society
www.vtol.org

Leonardo's AW Family

The Vertical Flight Technical Society
www.vtol.org

AW609 Civil Tiltrotor

The Vertical Flight Technical Society
www.vtol.org

Next Gen Civil Tiltrotor (NGCTR)

Clean Sky 2

The Vertical Flight Technical Society
www.vtol.org

Airbus H175 / AC352

The Vertical Flight Technical Society
www.vtol.org

Airbus H160 Medium Twin

The Vertical Flight Technical Society
www.vtol.org

Airbus X6 (2020s)

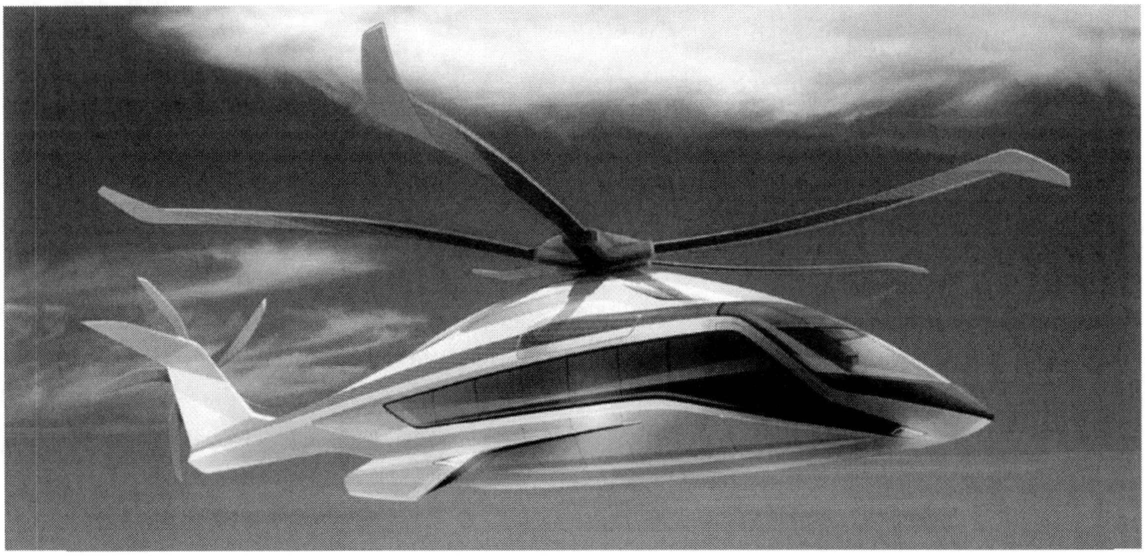

The Vertical Flight Technical Society
www.vtol.org

Clean Sky 2: Airbus RACER

Clean Sky 2

The Vertical Flight Technical Society
www.vtol.org

US Military: VTOL Capability Gaps

- **Performance shortfalls**
  - Speed, range, payload, endurance, altitude
- **Unexploited autonomy/collaboration**
  - Significantly increased mission effectiveness remains untapped
- **Unacceptable survivability & situational awareness shortfalls**
  - Safety and threat losses, no common picture
- **Costly sustainment**
  - Supportability, maintainability, reliability and availability

16 years of conflict and DoD studies reveal significant VTOL mission capability gaps

www.vtol.org/FVL

15

The Vertical Flight Technical Society
www.vtol.org

Future Vertical Lift (FVL) Concepts for Capability Set 3

www.vtol.org/FVL

16

The Vertical Flight Technical Society
www.vtol.org

Joint Multi-Role (JMR) Demonstrators Under Construction

Propulsion System Test Bed
August 2017

www.vtol.org/FVL

17

The Vertical Flight Technical Society
www.vtol.org

Bell V-280 Valor FVL Concept

The Vertical Flight Technical Society
www.vtol.org

Bell V-280 Valor

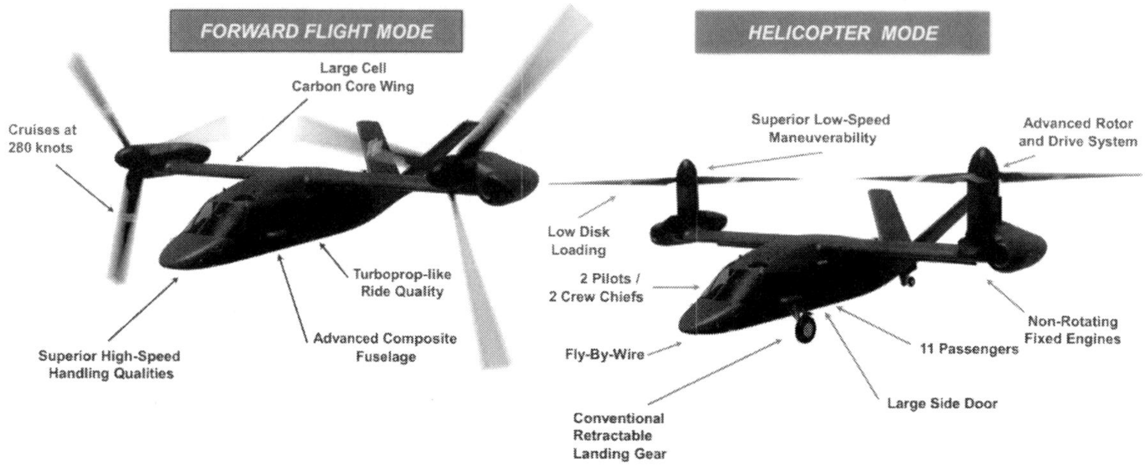

www.vtol.org/FVL

19

The Vertical Flight Technical Society
www.vtol.org

Sikorsky-Boeing SB>1 Defiant

The Vertical Flight Technical Society
www.vtol.org

Sikorsky-Boeing SB>1 Defiant

X2™ Technology

www.vtol.org/FVL

The Vertical Flight Technical Society
www.vtol.org

Sikorsky S-97 Raider

- Announced October 2010 (LTH)
- Construction begun October 2012
- First Flight 22 May 2015; crashed in 2017
- 2nd aircraft under construction
- Advancing Blade Concept (ABC) compound
- 0-2 crew + 6 troops
- 34 ft (10 m) rotor diameter
- Speed: 220 kt (cruise) – 240 kt (dash)

www.vtol.org/FVL

The Vertical Flight Technical Society
www.vtol.org

Compounds & Tiltrotors

Sikorsky-Boeing SB>1 Defiant™ (2018)

30,000 lb (13.6 t) class

Sikorsky S-97 Raider™ (2015)

11,000 lb (5 t)

Sikorsky X2 Technology™ **Demonstrator** (2008)

5,500 lb (2.5 t)

Bell Helicopter V-280 Valor (2017)

30,000 lb (13.6 t) class

Leonardo (with Bell) AW609 (2003)

16,800 lb (7.6 t)

Bell Boeing V-22 Osprey (1989)

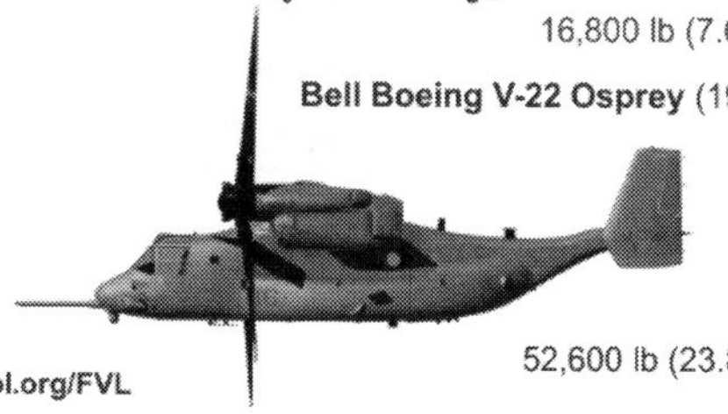

52,600 lb (23.8 t)

www.vtol.org/FVL

AHS graphic

23

The Vertical Flight Technical Society
www.vtol.org

Uber Elevate

- Uber Elevate
  - Unveiled at AHS Workshop in Sep 2016
  - White Paper in Oct 2016
  - Summit in April 2017
- Developing an "Ecosystem"
  - Partnerships with cities, real estate companies, aircraft manufacturers, EV charger manufacturers and cities
  - Connecting innovators, investors, regulators, technical experts, media
- Smaller aircraft, but higher barriers
  - Technical, regulatory, environmental, economic, infrastructural and cultural

- Started **www.eVTOL.news** website
  - 30+ aircraft concept detailed
  - Many missions beyond Uber's Elevate

www.eVTOL.news

24

The Vertical Flight Technical Society
www.vtol.org

Aurora eVTOL

- 8 VTOL electric props
- 1 pusher electric prop
- 97 kt (180 km/h)
- First Flight 2018

Graphics courtesy of Aurora
Manassas, Virginia, USA

www.eVTOL.news

25

The Vertical Flight Technical Society
www.vtol.org

Joby Aviation

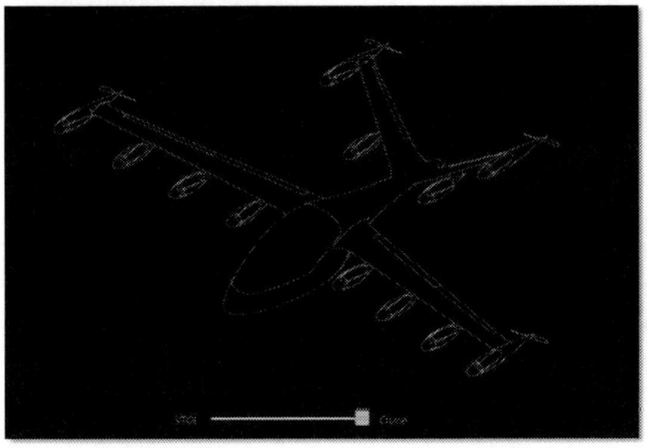

Original 2-seat Joby S2
12 lift/cruise propellers + 4 cruise propellers
All electric

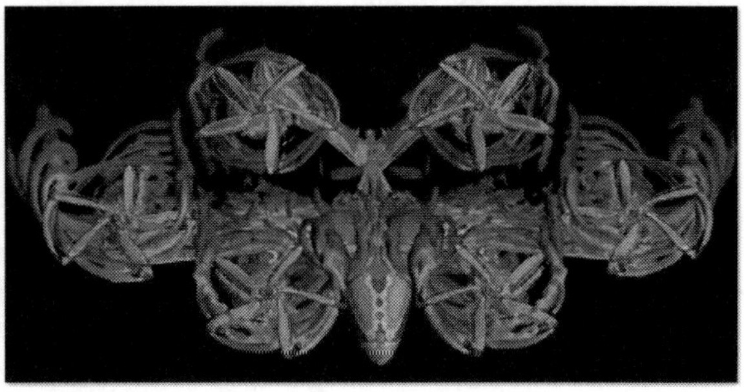

Current 4-seat Joby S4
6 lift/cruise propellers
All electric

Graphics courtesy of Joby Aviation
Santa Cruz, California, USA

www.eVTOL.news

26

The Vertical Flight Technical Society
www.vtol.org

e-volo 2x Multicopter
Now in pre-production

Photos courtesy of e-volo GmbH

www.eVTOL.news

27

The Vertical Flight Technical Society
www.vtol.org

Lilium Aviation

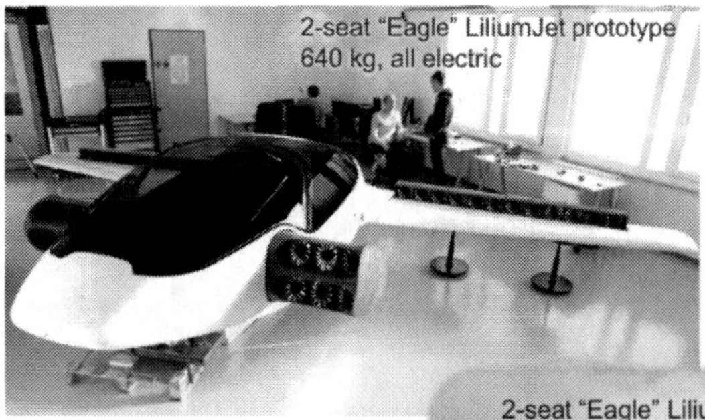

- 36 electric fans
  - 24 on wings
  - 12 on canards
- 160 kt (300 km/h)
- "Eagle" first flight April 2017

Graphics courtesy of Lilium Aviation
Garching, Germany

www.eVTOL.news

28

The Vertical Flight Technical Society
www.vtol.org

A^3 by Airbus: Vahana

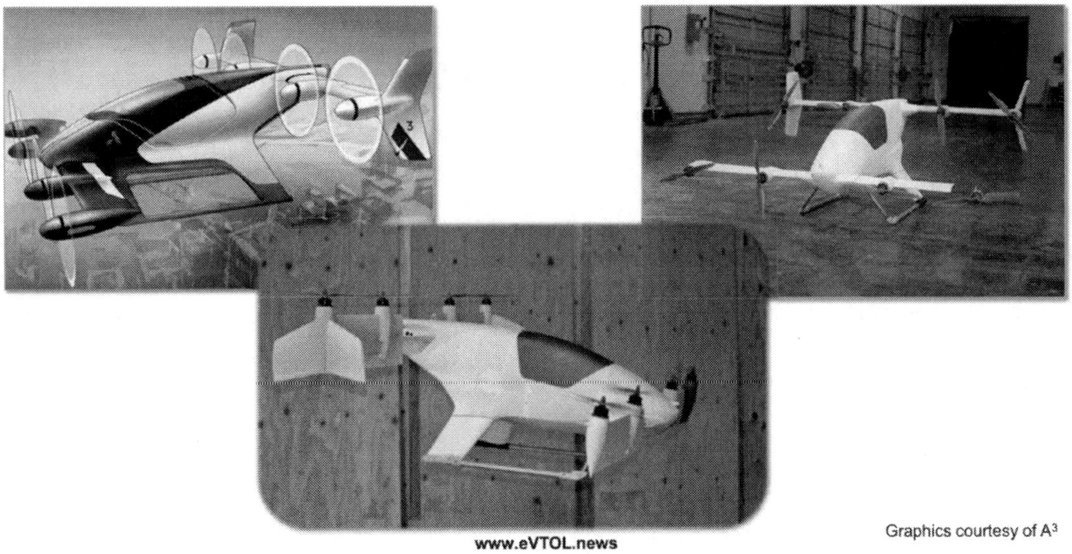

www.eVTOL.news

Graphics courtesy of A^3

30

The Vertical Flight Technical Society
www.vtol.org

Ehang 184

www.eVTOL.news

Graphics courtesy of EHang

31

The Vertical Flight Technical Society
www.vtol.org

Cartivator SkyDrive

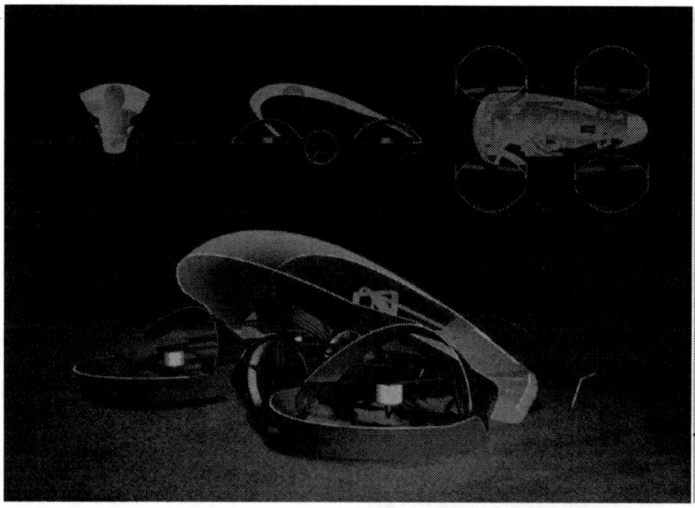

www.eVTOL.news

32

The Vertical Flight Technical Society
www.vtol.org

XTI Aircraft: TriFan 600

Hybrid-electric concept

www.eVTOL.news

33

The Vertical Flight Technical Society
www.vtol.org

Aurora XV-24A LightningStrike

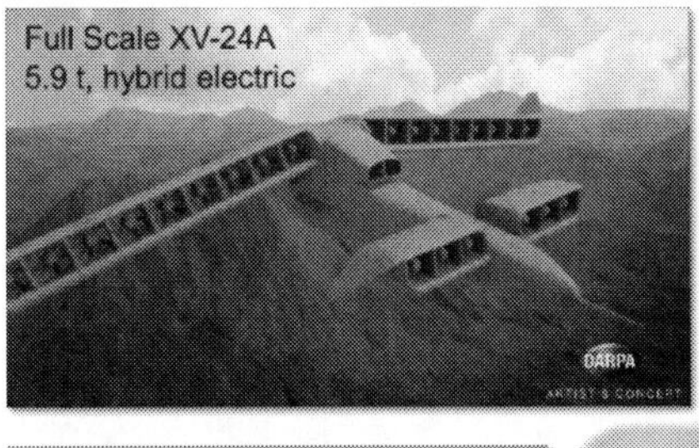

- 18 + 6 electric fans
- 3x 1 MW Generators
- 1 RR Liberty engine
  - 6k shp (4.5 MW)
- 300 kt (555 km/h)
- First Flight 2018

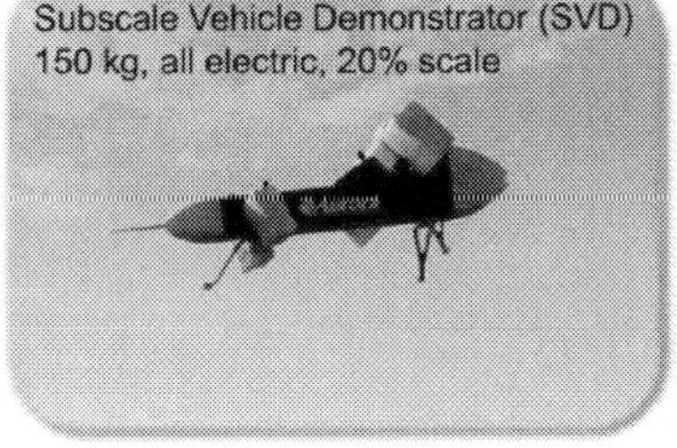

Graphics courtesy of Aurora Flight Sciences
Manassas, Virginia, USA

www.eVTOL.news

34

The Vertical Flight Technical Society
www.vtol.org

eVTOL Online Resources

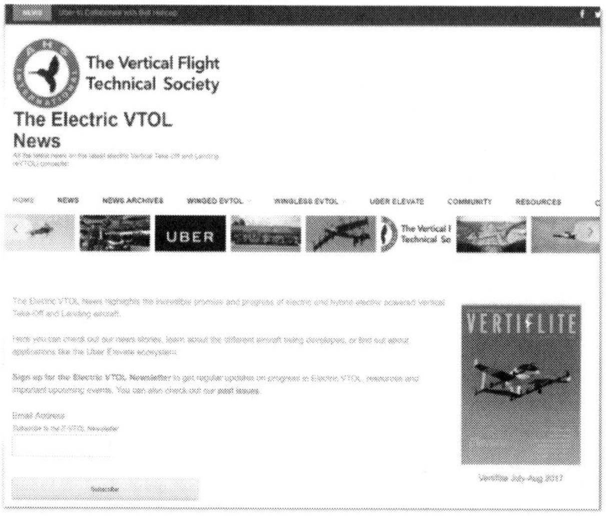

- Electric VTOL News
  - www.eVTOL.news
  - www.facebook.com/electricVTOL
  - www.twitter.com/electricVTOL
- Also
  - Email newsletter
  - eVTOL News videos
  - Comprehensive analytical report (Dec 2017)

www.eVTOL.news

35

The Vertical Flight Technical Society
www.vtol.org

Summary

FORUM 74
AHS INTERNATIONAL
MAY 14-17, 2018 PHOENIX, ARIZONA

- AHS is the global Vertical Flight Technical Society
  - If you aren't a member, join today at **www.vtol.org!**
  - 74th Annual Forum is May 14-17 in Phoenix, Arizona
- Significant European & N. American rotorcraft developments underway
  - State-of-the-art convention helicopter developments and AW609 tiltrotor
  - Clean Sky 2: advanced compound and tiltrotor
  - FVL/JMR: advanced compound and tiltrotor — see **www.vtol.org/FVL**
- Significant funds being invested in electric VTOL (>$250M)
  - 30+ companies investing heavily in electric and hybrid/electric VTOL aircraft
  - The explosive interest in drones may be repeated with manned eVTOL
  - For more info, see **www.eVTOL.news**

www.vtol.org

36

6th Asian/Australian Rotorcraft Forum & HeliJapan 2017

Helicopter activities in Kawasaki Heavy Industries, Ltd.

2017/11/7

Noriaki Katayama
Aerospace Company
Kawasaki Heavy Industries, Ltd.

Kawasaki
Powering your potential

KHI Helicopter Projects

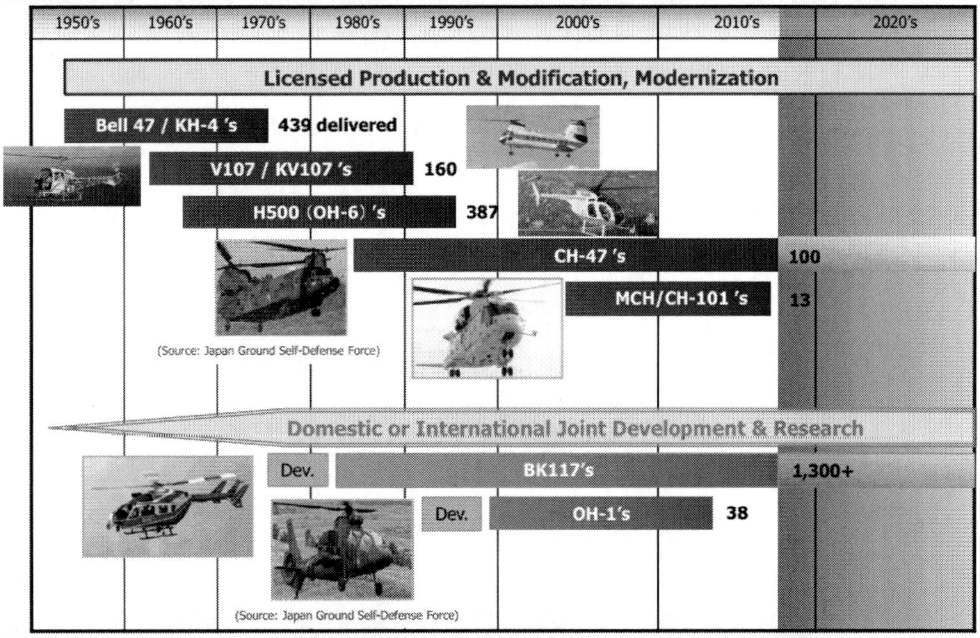

Kawasaki
Powering your potential

Bell Model 47 to KH-4

- Kawasaki Heavy Industries, Ltd.(KHI) started production of licensed Bell 47 helicopters in 1952.
- KHI has redesigned 3 seated 47G3 to 4 seated KH-4. This was the first step toward the development of original design helicopter.

Bell 47

KH-4

Bell 47/KH-4 ⇒ KHR-1

- In 1966, KHI built KHR-1 research helicopter from KH-4, which had so called rigid rotor, that is "stiff-in-plane hingeless rotor system".

Demonstration of Large Hub Moment

Demonstration of Stability

KHR-1

- The rotor system had steel hub plate, and had no lead-lag damper because it was stiff-in-plane rotor.
- You can see this helicopter in Kakamigahara Aerospace Museum in Gifu Pref.

(Source: http://oh1ninja.la.coocan.jp/)

V107 ⇒ KV107IIA

- V107 was developed by Vertol Corporation, and adopted by US Marine Corps as CH-46.
- The manufacturing right for the civil market was completely transferred to KHI.
- In 1968, TC of power up version, KV107IIA was obtained with both FAA and JCAB in cooperation with Boeing Vertol.

KV107IIA

BK117

- In 1977, MBB(now, Airbus Helicopters) and KHI had started international co-development of BK117.
- KHI work share was the Main transmission, Fuselage, Control System, Power Plant, Fuel System, Electric System, etc.

First Flight in 1979

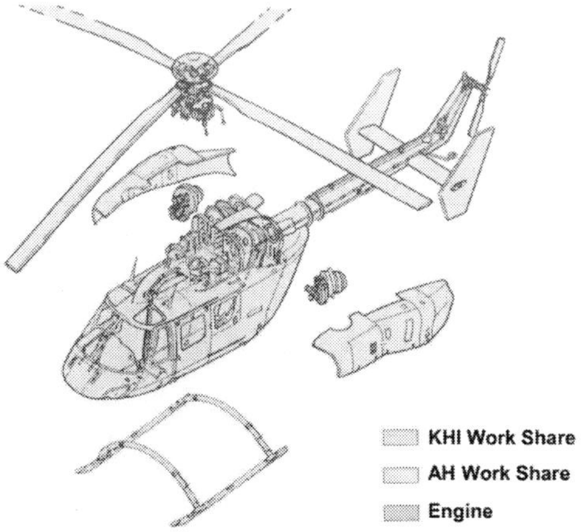

H500 ⇒ FRP Blade Research

- The research of FRP blade was conducted using H500 helicopter from 1980.
- The construction of the blade was unique as shown below.

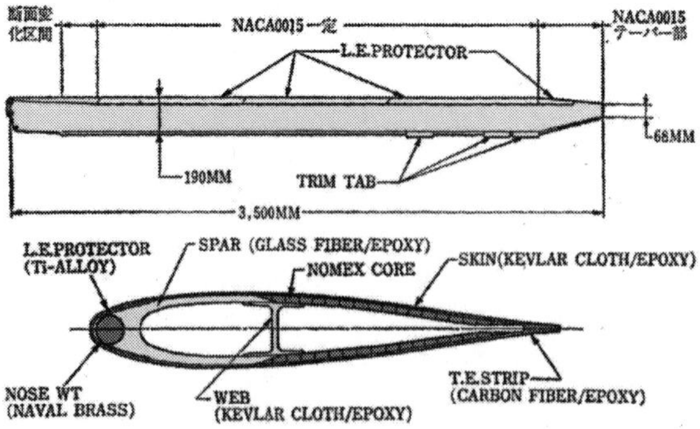

FRP Rotor Hub Research

- The research of the bearingless rotor hub started in 1980, and continued until 1989.

(Source: Niwa, Y. et al., A Study of Bearingless Rotor-Hub System for XOH-1, AHS 54th Forum, 1998)

Reflect Experiences to New Helicopter

- These researches have lead to the pure domestic development of OH-1.

(Source: Niwa, Y. et al., A Study of Bearingless Rotor-Hub System for XOH-1, AHS 54th Forum, 1998)

(Source: Niwa, Y. et al., The Development of The New Observation Helicopter (XOH-1), AHS 54th Forum, 1998)

OH-1 Observation Helicopter

- OH-1 is the pure domestically developed helicopter.
- OH-1 has achieved simultaneously excellent maneuverability and handleability with all-composite hingeless rotor system.

(Source: Japan Ground Self-Defense Force)

(Source: http://oh1ninja.la.coocan.jp/)

Maneuverability
(0g maneuver)

(Source: Aiba, M. et al., The Flight Test Results of The New Observation Helicopter (XOH-1), AHS 56th Forum, 2000)

Handleability
(Smooth landing to 15 degree slope)

Research Activities toward Development

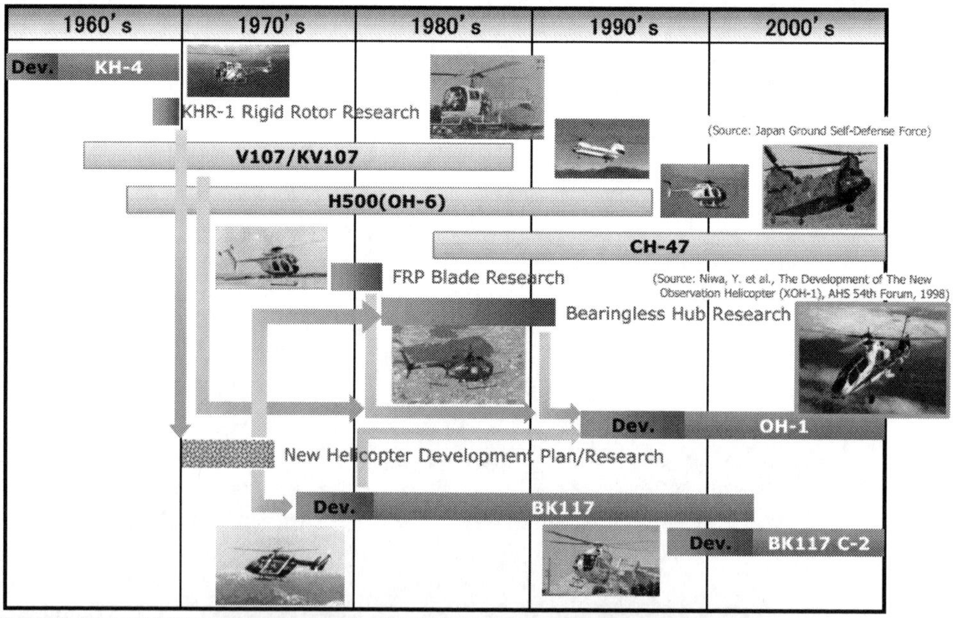

(Source: Japan Ground Self-Defense Force)

(Source: Niwa, Y. et al., The Development of The New Observation Helicopter (XOH-1), AHS 54th Forum, 1998)

Research for the Future

- KHI is researching key technologies in the area of Cost, Safety, Performance and Operability.

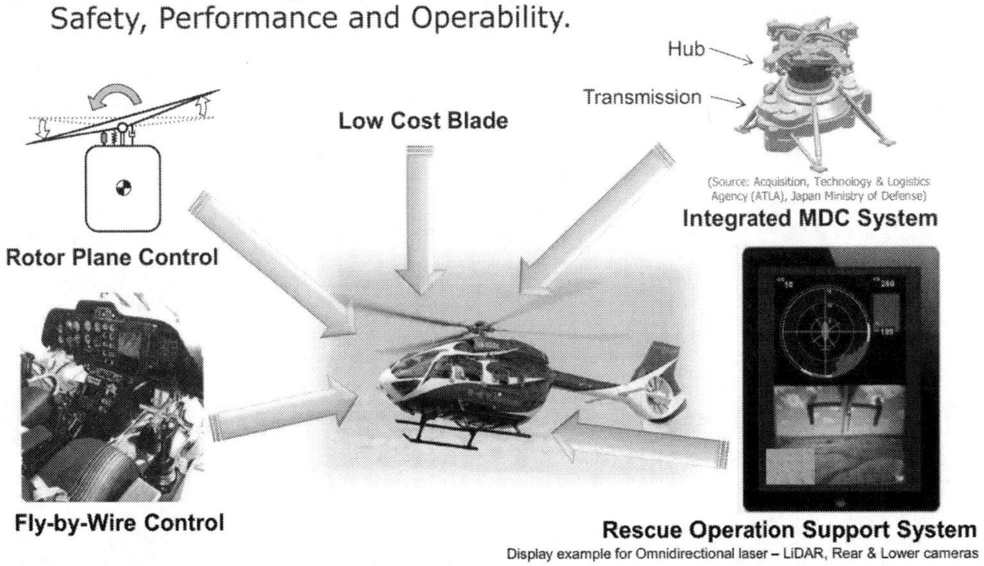

Kawasaki Powering your potential

HELICOPTER ECOSYSTEM IN INDIA
Prof. Dineshkumar Harursampath PhD (Georgia Tech)
NMCAD Lab, Dept. of Aerospace Engg.
Indian Institute of Science
भारतीय विज्ञान संस्थान
IISc
NMCAD
6th ARF & Heli Japan 2017
Technical Meeting on Advanced Rotorcraft
Technology and Safety Operations
November 7-9, 2017 Kanazawa, Ishikawa, Japan
AHS
INTERNATIONAL

PROFILE OF INDIAN HELICOPTER INDUSTRY
Bell Helicopter
A Textron Company
DYNAMATIC TECHNOLOGIES LIMITED
BOEING
MAKE IN INDIA
LOCKHEED MARTIN
TATA
TATA ADVANCED SYSTEMS
HAL
Hindustan Aeronautics Limited

Breaking News!

Saab-Adani Group JV to make drones & **military choppers!**

Sep'17: HAL gets $1b order: 41 ALH

Aug'17: LCH Production Launch
RWR&DC-designed 5.8ton LCH
-cleared by CEMILAC
15-unit order already!
Twin-engine,
narrow FL &
tandem config. for pilot & copilot/weapon system operator
IISc
NMCAD

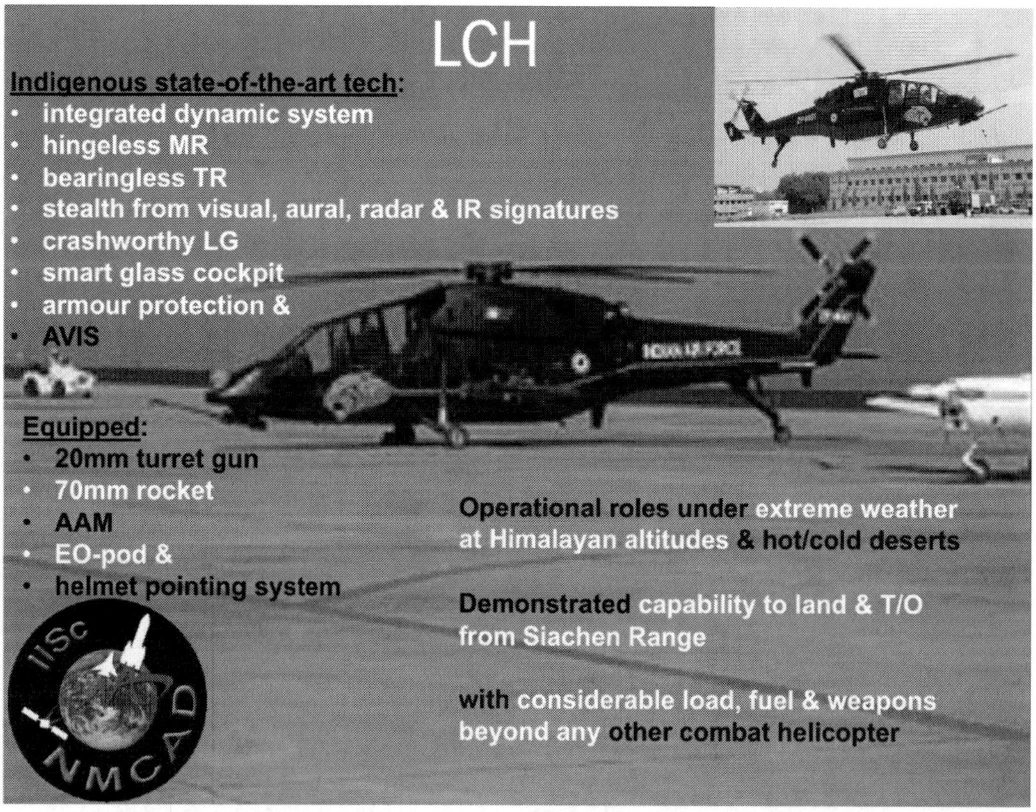
LCH
Indigenous state-of-the-art tech:
• integrated dynamic system
• hingeless MR
• bearingless TR
• stealth from visual, aural, radar & IR signatures
• crashworthy LG
• smart glass cockpit
• armour protection &
• AVIS
Equipped:
• 20mm turret gun
• 70mm rocket
• AAM
• EO-pod &
• helmet pointing system
Operational roles under extreme weather at Himalayan altitudes & hot/cold deserts
Demonstrated capability to land & T/O from Siachen Range
with considerable load, fuel & weapons beyond any other combat helicopter
IISc
NMCAD

May'17: LUH (PT-2) Maiden Flight

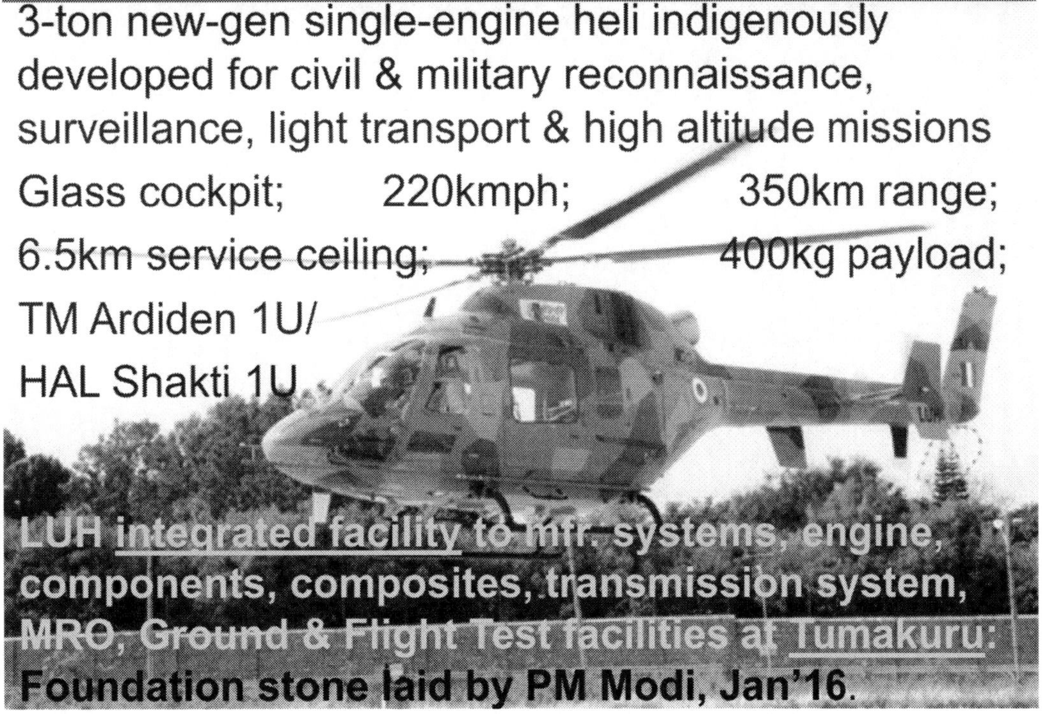

3-ton new-gen single-engine heli indigenously developed for civil & military reconnaissance, surveillance, light transport & high altitude missions

Glass cockpit; 220kmph; 350km range;

6.5km service ceiling; 400kg payload;

TM Ardiden 1U/
HAL Shakti 1U

LUH integrated facility to mfr. systems, engine, components, composites, transmission system, MRO, Ground & Flight Test facilities at Tumakuru: Foundation stone laid by PM Modi, Jan'16.

IMRH: Preliminary design of 10-ton twin-engine Indian Multi Role Helicopter for Indian Defence

Blade folding option for ship deck operations

Roles: support air assault, air transport, combat logistics, combat S&R, casualty evacuation

HAL

Hindustan Aeronautics Ltd.

Setup: 1940

Founders: Seth Walchand & Mysore Maharaja

Taken over by GoI: 1942

Turned public ltd. co.: 1964

Helicopter Division estb. 1970

Mfr. single (>600) & twin engine (>100) heli:

- Rudra, Dhruv, Cheetah, Chetak, Cheetal & Lancer

Supported by co-located

- RWR&DC, MRO division & Rotory Wing Academy to train pilots

www.hal-india.com

Seth Walchand Hirachand

Chetak (Alouette III)

HAL began heli mfr: 1962
with Sud-Aviation (Eurocopter)
1st delivery: 1965

Cheetah (LAMA SA 315-B)
IISc
NMCAD
License agreement to produce Cheetah with SNIAS (Eurocopter): 1970
1st delivery: 1976

Derivatives: Lancer & Cheetal
IISc
NMCAD
LAH (Lancer) (1990s):
Basic str. from reliable & proven Cheetah;
Bullet-proof front panels;
12.5mm Gun cum 70mm Rocket pods;
Optical sighting for accurate firing.
Cheetal:
Re-engined variant of Cheetah set
2006 world record of highest landing
at 'Saser Kangri' of Himalayas

The ALH "Dhruv"
Self-reliance in design & mfr. of twin-engine Dhruv: multi-role, multi-mission all-weather 5.5-ton heli
Started series production: 2001-02
Military & civil transport, disaster relief, S&R operations

ALH Dhruv variants
•Mk I: Initial config. (conventional cockpit, electro-mechanical gauges, turboshaft engines, designed in collaboration with MBB)
INDIAN AIR FORCE
•Mk II: Glass cockpit with 4 Multi-Functional Displays
•Mk III: Improved version (glass cockpit, higher power engines & mission systems [Electro-Optic Pod, Helmet Pointing System, Electronic Warfare Suite, IR suppressor & Engine Air Particle Separator])
•Mk IV: Weaponised platform designed for armed roles, "Rudra"

Rudra (certified:2013)

ALH & Acrobatics

Sikorsky in India

LOCKHEED MARTIN

- In June 2009, Sikorsky entered into JV with Tata Advanced Systems Limited (TASL) to mfr. S-92 in India for export & domestic markets
- $200m mfr. plant operational in Hyderabad from 2010
- Initially, TASL supplying **complete** S-92 cabins
- 1st cabin delivered: Nov'10
- >120 cabins delivered

IISc NMCAD

Bell Helicopter in India

Bell Helicopter
A Textron Company

- 1995: India liaison office for customer support
  dedicated Customer Service Engineers
  & authorized Customer Service Facility
- 2004: Textron engg. & tech. center, Bengaluru
  > 500 engineers for various programs
- 2016: TASL & Bell join forces to develop
  commercial & govt. (incl. military)
  Light Utility & Reconnaissance markets
- Sales office, New Delhi

IISc NMCAD

Bell's Indian market growing fast

After taking 52 yrs to sell 1st 100 choppers in India, Bell expects to sell next 100 in < 5yrs!

"India: our fastest growing market in whole world" - Bell's director of communications, Greg Hubbard

1st Bell 429 delivered to India: 2011

1st Bell 407GXP delivered to India: 2016

MRO centers being set up

407 to be made in India

BOEING

Boeing – AH 6i Vertical Stabilizer Assembly
Successfully Delivered 11 Shipsets On and Ahead of Schedule with meeting 100% quality standards

TATA

TATA ADVANCED SYSTEMS

Programs under Execution

Boeing AH64 Apache Fuselage
TATA and Boeing have entered into a Joint Venture to fabricate and assemble the fuselage for the AH64 Apache helicopter

Boeing AH64 Apache Secondary Structures
TATA Advanced Systems have also been contracted for the AH64 Apache secondary structures and the first delivery planned by Q4, 2017

Boeing CH47 ChinookAft. Fuselage Panels
TATA Advanced Systems have been contracted to deliver the Aft Fuselage Panels for Boeing CH47 Chinook Helicopter. First delivery expected from Q4 of 2016.

PROFILE OF THE INDIAN CIVIL HELICOPTER OPERATION

THE ROTARY WING SOCIETY OF INDIA

Estd.1998

INDIAN ROTORCRAFT OPERATION PROFILE

1st Civil Heli Flight Capt. Rustam H Captain in Hiller UH 12B in Nov '53

- **271 civil & 600+ military turbine heli**
- **Military heli count to double in 5 yrs, but civil fleet declining!**
- **HAL mfr. commercial heli eg., Cheetah, Chetak & Dhruv**
- **Now, few pvt. companies mfr. components for major OEMs eg., Bell, Boeing & Sikorsky**

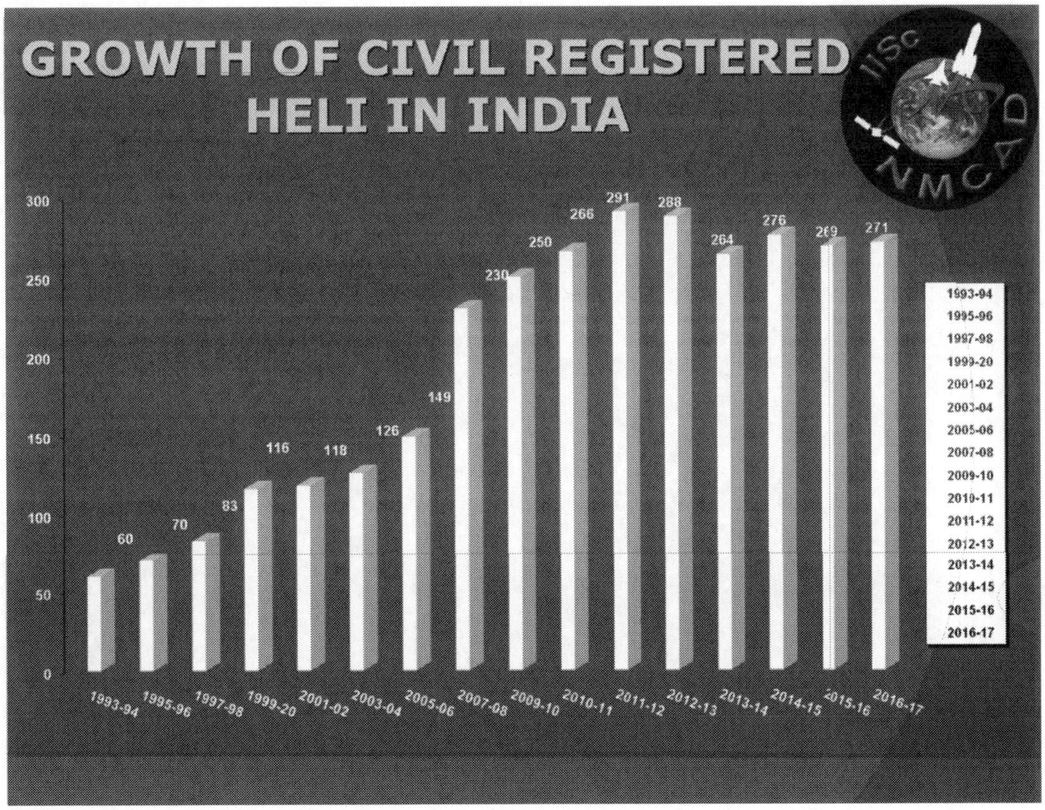
GROWTH OF CIVIL REGISTERED HELI IN INDIA
IISc
NMCAD
300
250
200
150
100
50
0
60
70
83
116
118
126
149
230
250
266
291
288
264
276
269
271
1993-94
1995-96
1997-98
1999-20
2001-02
2003-04
2005-06
2007-08
2009-10
2010-11
2011-12
2012-13
2013-14
2014-15
2015-16
2016-17

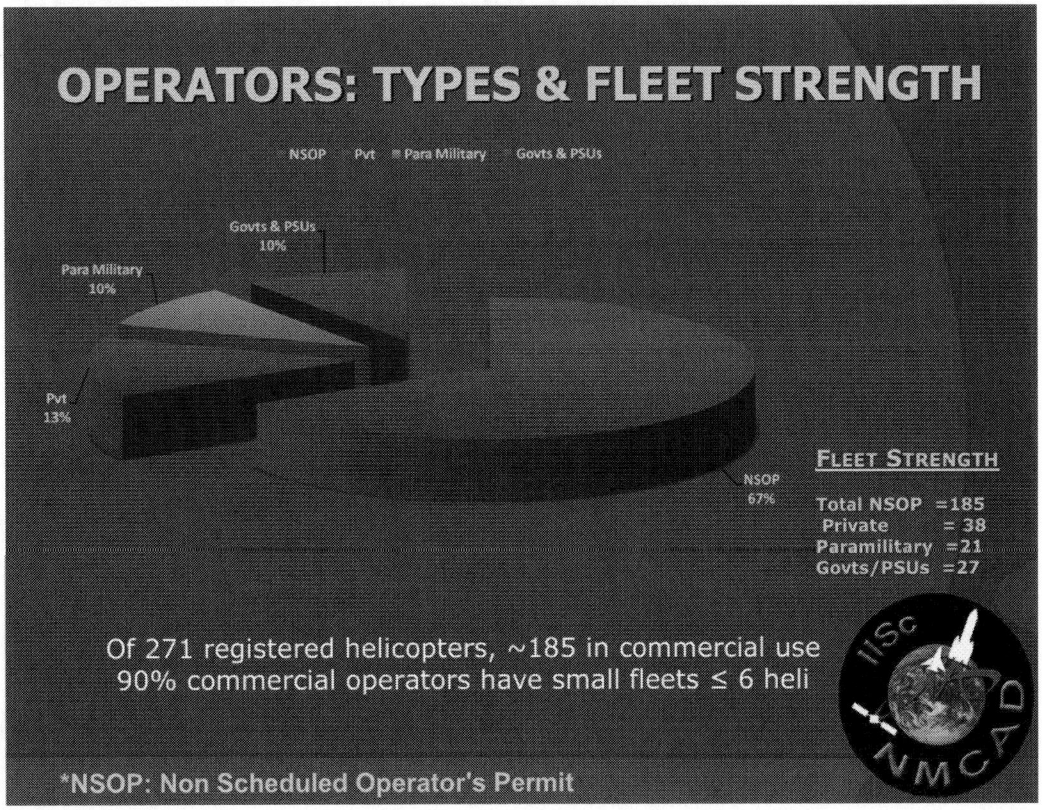
OPERATORS: TYPES & FLEET STRENGTH
NSOP
Pvt
Para Military
Govts & PSUs
Govts & PSUs 10%
Para Military 10%
Pvt 13%
NSOP 67%
FLEET STRENGTH
Total NSOP =185
Private = 38
Paramilitary =21
Govts/PSUs =27
Of 271 registered helicopters, ~185 in commercial use
90% commercial operators have small fleets ≤ 6 heli
IISc
NMCAD
*NSOP: Non Scheduled Operator's Permit

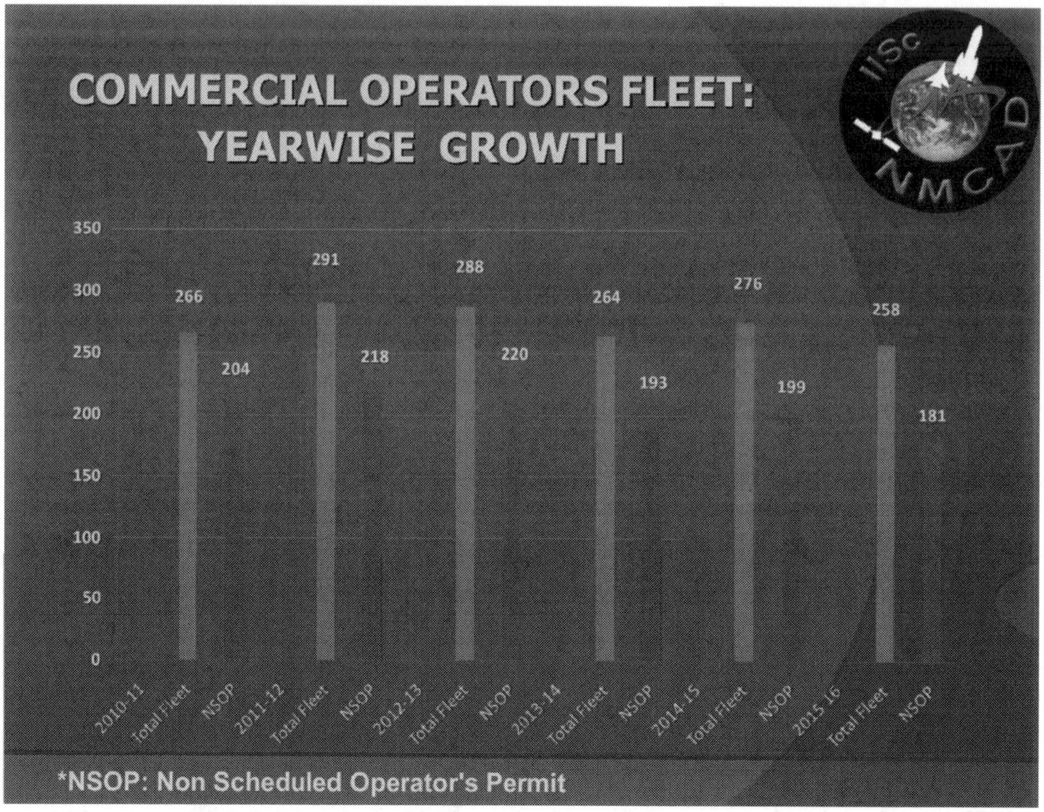
COMMERCIAL OPERATORS FLEET: YEARWISE GROWTH
IISc
NMCAD
350
300
250
200
150
100
50
0
266
204
291
218
288
220
264
193
276
199
258
181
2010-11
Total Fleet
NSOP
2011-12
Total Fleet
NSOP
2012-13
Total Fleet
NSOP
2013-14
Total Fleet
NSOP
2014-15
Total Fleet
NSOP
2015-16
Total Fleet
NSOP
*NSOP: Non Scheduled Operator's Permit

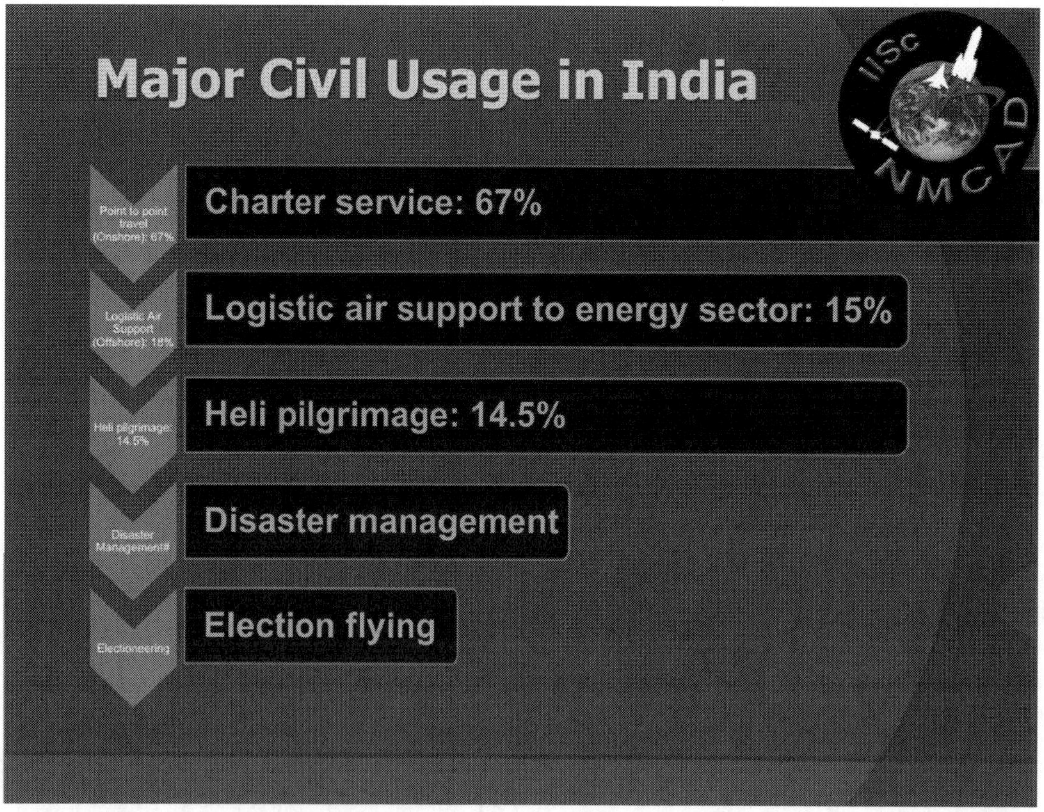
Major Civil Usage in India
IISc
NMCAD
Point to point travel (Onshore): 67%
Charter service: 67%
Logistic Air Support (Offshore): 18%
Logistic air support to energy sector: 15%
Heli pilgrimage: 14.5%
Heli pilgrimage: 14.5%
Disaster Management#
Disaster management
Electioneering
Election flying

COMMERCIAL HELICOPTER BRANDS IN INDIA

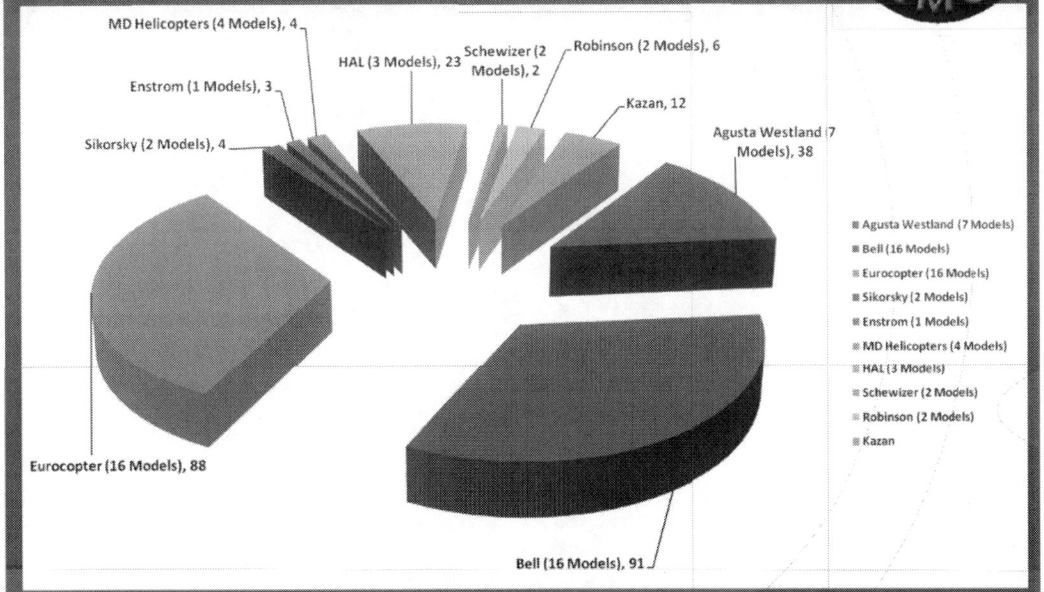

PRESENT USAGE & CHALLENGES

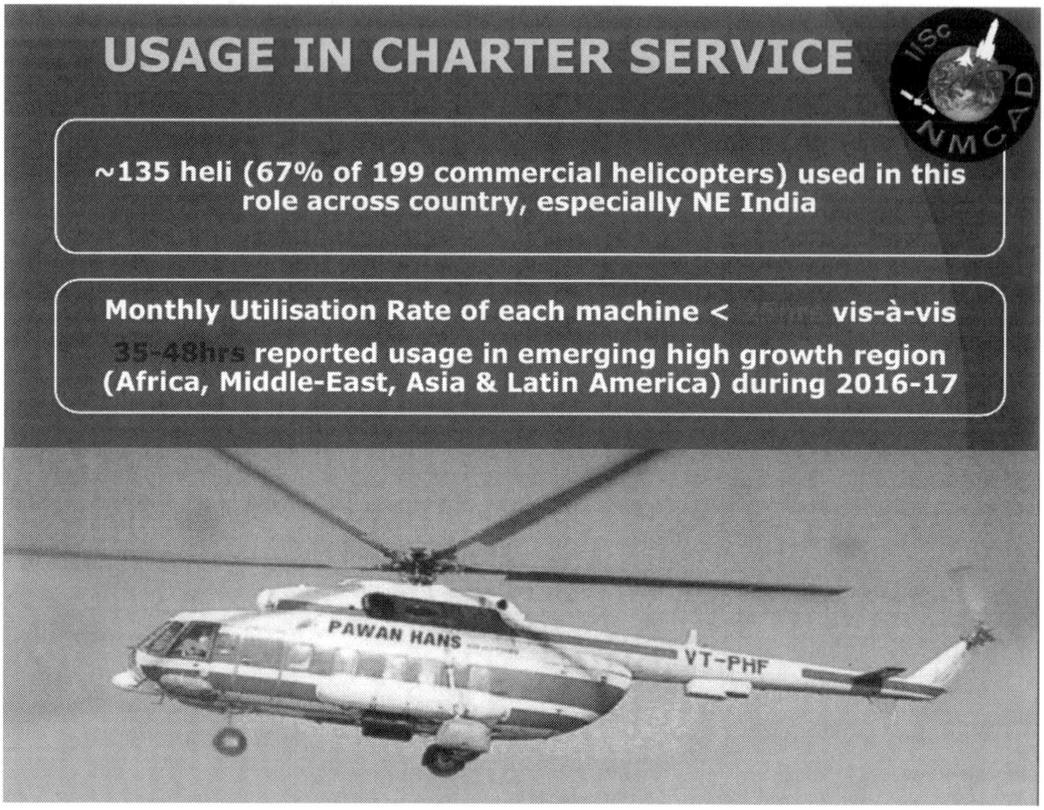

ENERGY SECTOR

AVG HELICOPTERS	SECTOR
1 (AW 139)	Crew Change Bundle Services
4 (AW 139)	Crew Change ONGC
2+3	Crew Change ONGC
3+6	Production ONGC
2 -3	Crew Change & Prod BGEPIL
2	Crew Change & Prod RIL
2	Crew Change & Prod CAIRN
1	Crew Change & Prod GSPC
1	Crew Change Bundle Services
2-3	Seasonal Seismic Survey
29-31 Helicopters are Used in offshore	**Total Flying per year-25,000 HRS**

HELICOPTER PILGRIMAGE ROLE

AVG HELICOPTERS	SECTOR
#10	KEDARNATH (UTTARAKHAND)
2	HEMKUND (UTTARAKHAND)
#5	CHARDHAM (UTTARAKHAND)
2(+2 RES)	VAISHNO DEVI (J & K)
6 (+2/3 RES)	AMARHNATH (J & K)
2(+1 RES)	MACHAIL MATA (J & K)
2	MANIMAHESH (HP)
29 Helicopters are used in Heli Pilgrimage (14.5%)	**Flying Effort 18500-21000hs per year**

Poor present Usage in following roles:

With ↑ aspirations of people, use of heli for public good will ↑ ... subject to?

Civil Helicopter Accident Rate Trend in India

[Yearly Accident Count]

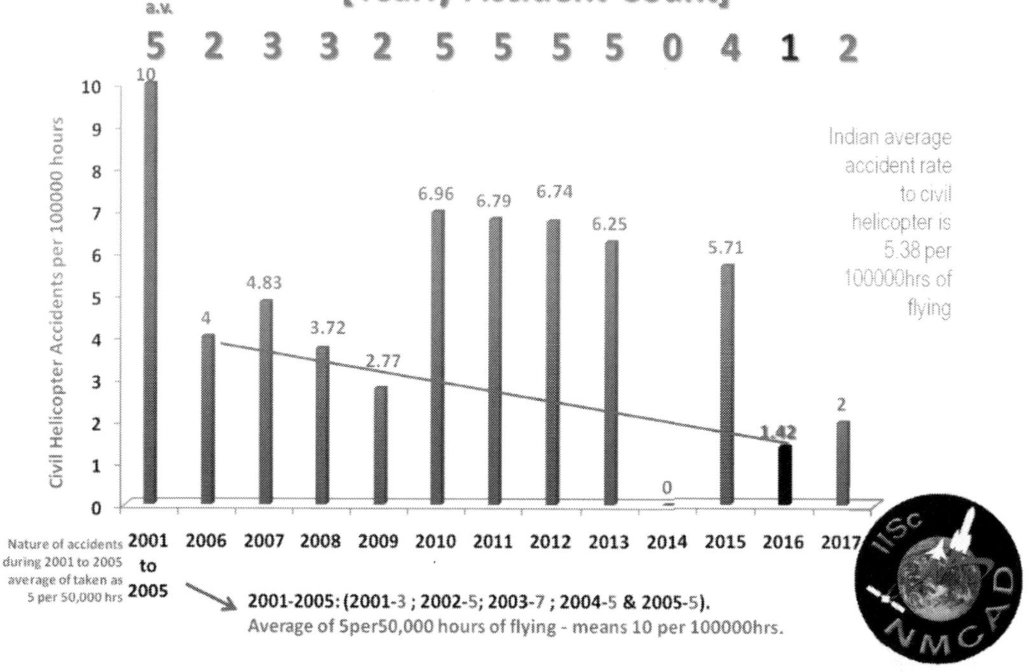

Conclusions

- Heli R&D in India: enormous strides in past 2 decades
- HAL's pioneering work resulted in design & mfr. of Dhruv, sophisticated & modern heli in its class
- Other major heli initiatives underway at HAL
- Infrastructure growth in helicopter-related components & suppliers as well as training of engg. designers
- Surge in basic academic helicopter engg. research in last decade

Conclusions

- However, academic research concentrated at 2 institutions (IISc+IIT-K) needs to spread to more institutes/univs.
- Most research addressed heli dynamics & control. Need to address heli aerodynamics & aeroacoustics in basic univ. research, with enhanced *international collaboration*
- Heli MNCs now consider India as mfr. as well as R&D base for advanced heli development
- Considerable investment on simulators, pilot training & MRO facilities needed to spread heli usage & increase # of operators

Acknowledgements

- Mr. N. Sheshadri, Executive Director, RWR&DC, HAL
- AVM K Sreedharan VM (G) Retd., President, RWSI
- Prof. Venkatesan, IIT-K
- Prof. R. Ganguli, IISc

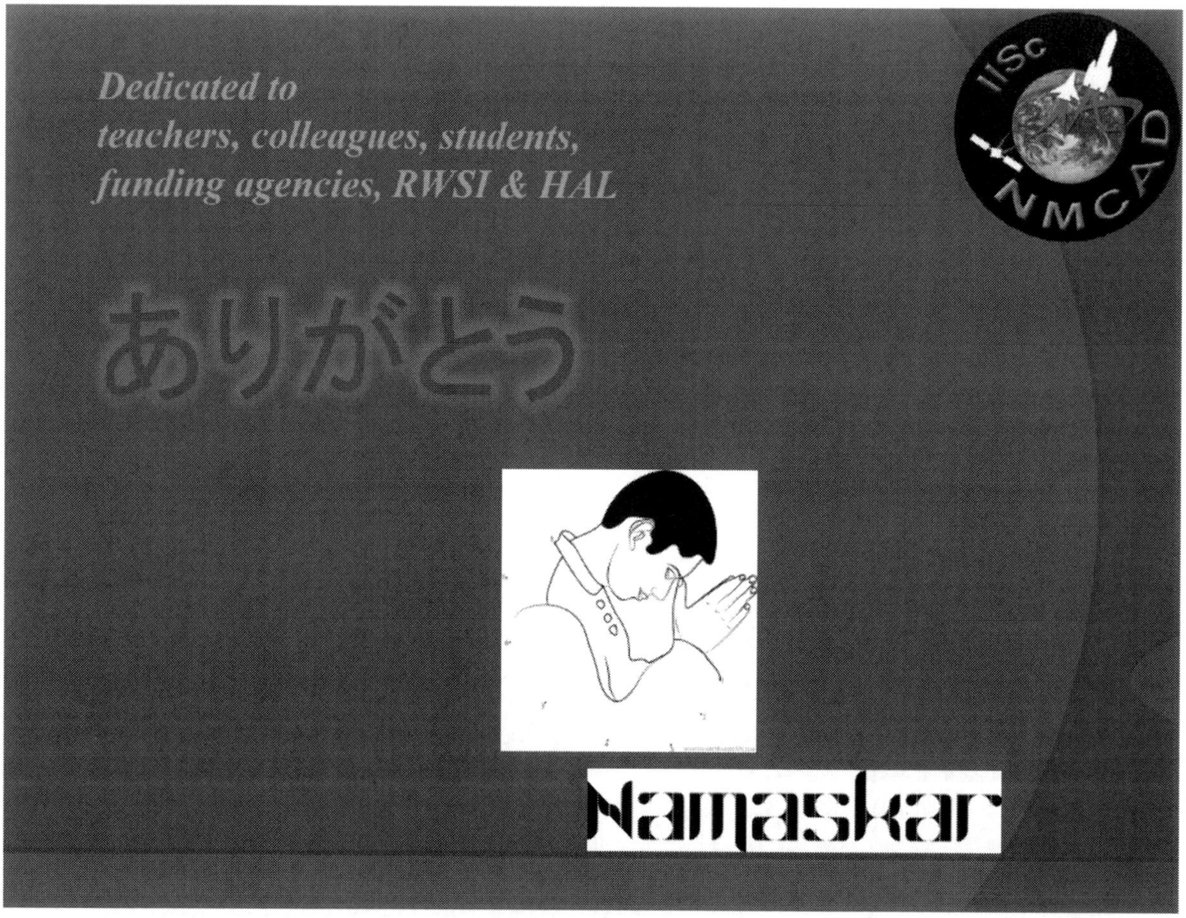
Dedicated to
teachers, colleagues, students,
funding agencies, RWSI & HAL
IISc
NMCAD
ありがとう
Namaskar

Rotorcraft Forum & Heli Japan
November 6th, 2017

Doctor-Heli in Japan

Nishikawa, Wataru
former Chairman, AHS Japan Chapter
Director, HEM-Net
(Helicopter Emergency Medical Network)

51 Doctor-Heli Stations

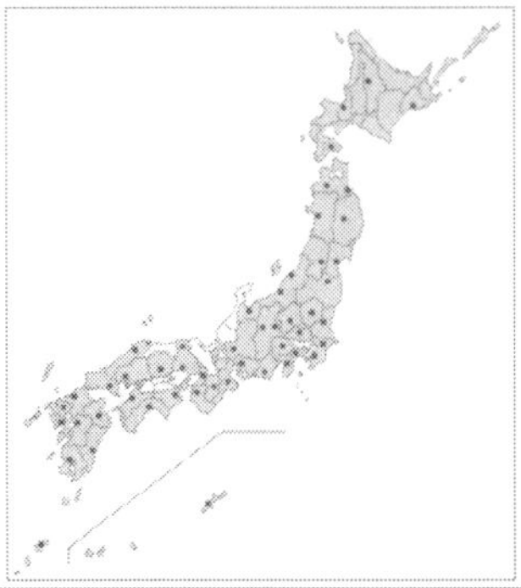

Except Tokyo and other 2 prefectures

2015 Response Time
Tokyo : 51.4min.
National Average : 39.4min.

From Start to Now

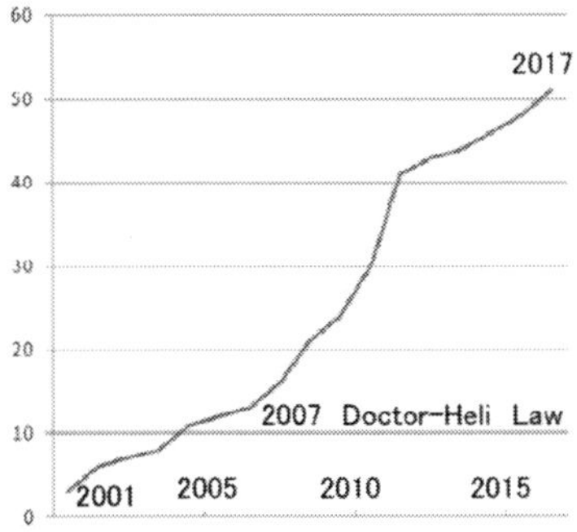

Doctor-Heli Committee

- The Doctor-Heli Investigation Committee in the Internal Affairs Office of the Japanese Cabinet (August, 1999～June, 2000)
- Subjects:
  - Operation Cost
  - Medical Effects
  - Safety

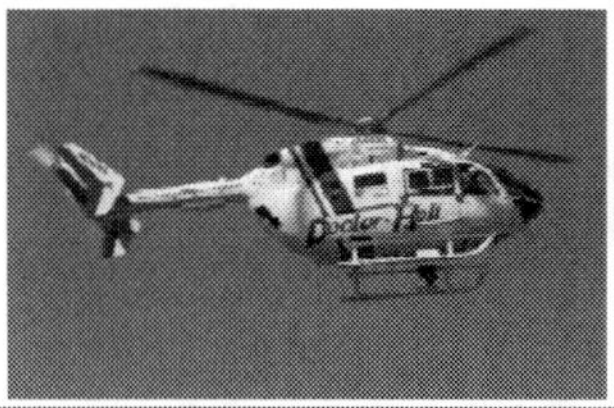

Medical Effects

- The Results of Trial Operation by Ministry of Health (Oct.1999～Sept.2000)

	Death	Severe	Recovery	Moderate	Total
Actual Results	70	53	203	140	466
Estimated by car	116	112	98	140	466

- Mortality reduction by almost half
- Two-fold increase in patients returning to normal activities.
- Results : Doctor-Heli started officially (April, 2001)

Operation System

- Operated by commercial helicopter operators
- Twin-turbine helicopter with medical equipment
- Standing-by at big general hospital with life-saving emergency center
- Flight Crew : pilot, mechanical engineer, doctor, nurse
- Mechanical engineer works as co-pilot in flight.
- CS(Communication Specialist) receives the mission request from ambulance center, and dispatches the mission flight within a few minutes.

—Daytime Standing-by & Scramble Starting—

Helicopter Models for Doctor-Heli

- H-135 22
- BK117......19
- MD900......4
- AW109......4
- Bell429......2

- Total.........51
  (Operated by 11 operators)

Activities at Scenes

- Landing at pre-designated "Rendezvous Point".
- Average 500 Rendezvous Points per helicopter station. Maximum 1,900 points per station.
- At rendezvous point, the doctor meets the patient transported there by an ambulance car and starts medical treatment immediately.
- After primary treatment, the patient transported to hospital by Doctor-Heli.

Number of Missions

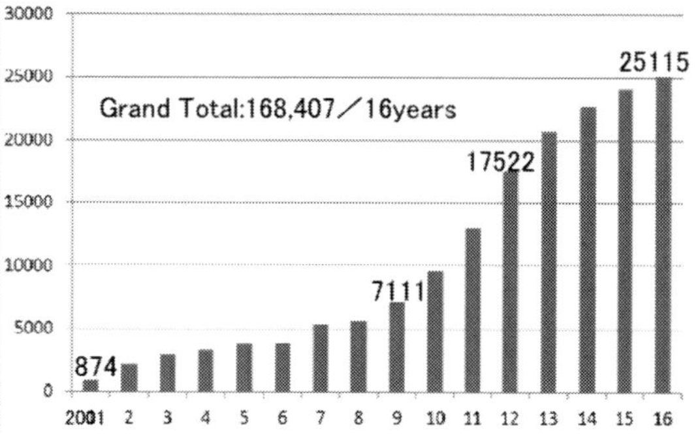

Cost Performance

- Standard cost per station : ¥250mil
- Born by :
  - Ministry of Health : 50%
  - Ministry of Internal Affairs : 10 ～40%
  - Local Prefectures : remaining
- Number of missions : 25,115 in 2016
- Average 500 missions per station
- Average 500,000 yen per mission
- No Charge to patient

World Helicopter Ambulance

Country	Start	Station	Major Operator
Germany	1970	78	ADAC, DRF etc
US	1972	846	Civil helicopter operators
Swiss	1973	13	REGA
Austria	1983	28	OEAMTC
France	1983	30	SAMU
UK	1987	34	Civil helicopter operators
Italy	1987	51	Civil helicopter operators
Spain	1989	25	Civil helicopter operators
Nederland	1995	4	Civil helicopter operators
Japan	2001	51	Civil helicopter operators
合 計	—	1,160	—

11

Adverse Teacher (US)

- Number of ambulance helicopter accidents (2006 – 2015, NTSB)
  - Total Accident : 92
  - Fatal Accident : 34
  - Dead: 97
- Charge to patient and insurance company : about $30,000 per transport
- Because of free competition among many operators in the same area

Risk Prevention

- Helicopter Ambulance should be
  - Day scene flights only
  - Land only at pre-designated LZs
  - Twin-engine helicopter
  - Two pilots---reduce the likelihood of pilot error

Dr. Ira Blumen,
University of Chicago Medical Center, 2011

these are "Necessary conditions" and not enough "Sufficient condition"

Deregulation

- Airline Deregulation, 1978
- Many operators in the same area
- Excessive competition
- Operation by force
  - In adverse weather
  - Pilots fly overwork day and night
- Accident and cost increased
- Helicopter ambulance should be an exception of the deregulation principle.

Troubles in Doctor-Heli

- Accident : 1(no fatal and no injured)
- Incidents : 2
  - landed safely with one engine failure
- Small incidents
  - Cabin door opened in flight
  - Cowling opened in flight

Words of a Patient

helped miraculously by Doctor-Heli

"I hope my granddaughter to be a bride of Flight Doctor"

END

Safety Activities of JHST

Osamu Washida
General Manager
Aero Asahi Corporation
Tokyo, Japan

Shigekichi Takeuchi
Airbus Helicopters Japan
Kobe, Japan

Takashi Hisaizumi
Airbus Helicopters Japan
Kobe, Japan

PLEASE NOTE: DO NOT INCLUDE EMAIL ON 1ST PAGE

ABSTRACT

Japan Helicopter Society (JHS) re-started safety activities as JHST (Japan Helicopter Safety Team) in 2015. JHS had been visited members of Asia Pacific Helicopter Safety Team in Sep/2015, and decided to do it again. This activity is the part of IHST (International Helicopter Safety Team), it's formed similar teams in the world. This time, we JHST make presentations what we did, and what we will do for the future of safety of the air.

INTRODUCTION

The JHST is not an independent activity in Japan right now. It's formed in the JHS as part of flight safety activities. We three has been assigned as board member of the JHST, and are going how do we proceed it. This time, the presentation of our activities shows intermittent report on the way to goal. But no goal is for the safety. It's long, long way to the future.

HISTORY OF JHST

Starting activity

The history of the safety activity for helicopters has been existed for a long time. But one of important thing was happened in Sep/2005. AHS (American Helicopter Society) and HAI (Helicopter Association International) held the IHSS (International Helicopter Safety Symposium) to reduce accidents of helicopters. Many people joined it from operators, manufactures, authorities and investigators.

"Reducing 80% of the accidents rates in 10years", it was created as target of the meeting in that time.

IHST is consisted several local teams, first at US (USHST) and Europe (EHEST), then spread worldwide, and more than ten teams join it.

In Japan, Japan IHST committee was formed in JHS in 2008, and they established Japan JHSAT (Japan Helicopter Safety Analysis Team). The team included some operators, manufactures, academies and authority. And analyzed some case of accidents, then reported it. But it stopped 2010 then in the rest for some years.

In 2015, as mentioned, Mr. Leo Jeoh, Chair of AeHST and Mr. Klein, Head of Safety, AirbusHelicopters visited us and talking about HST activity in Japan. They would like to proceed safety activity in Asia region, and propose us to do it. In 2016, starting of safety activity was approved at the yearly meeting of the JHS, then we're working together to do it.

After establishment of JHST

After starting, we discuss following committees and authorities.

- JCAB (Japan Civil Aviation Bureau)
- JTSB (Japan Transport Safety Bureau)
- Some organizations regarding helicopters, operators, manufacturers, and other committees.

They agreed our policy, but we haven't showed them actuary process and structure yet.

All other committees and authorities have their own directions and we have to show them how should we coordinate regarding the helicopter safety including our position.

Our Activities

We would like to show you our activities base on the presentation at JA2016 last year.

Our main activities are;

- Japanese Safety Leaflets
- Safety Road Show

These two activities are linked to Airbus Helicopters safety activity.

Safety Leaflet in Japanese

EHEST (now RSC/Rotor Sectorial Committee in EASA) issued some leaflet in some language. Airbus Helicopter Japan translates them into Japanese. 11 leaflets are already translated and put in our home page as follows.

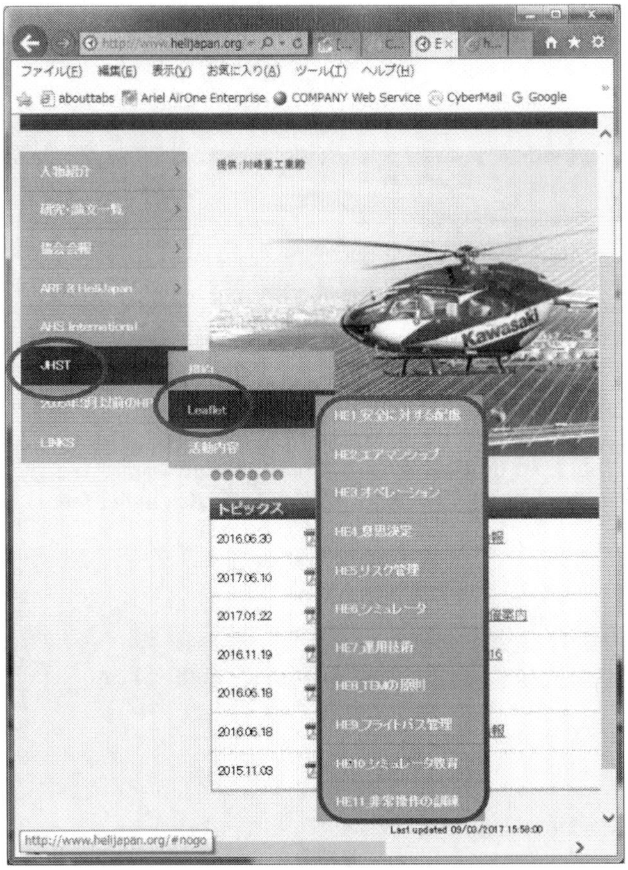

Figure-1 Website of Japan Helicopter Society

Titles and summary are;

HE1 Safety considerations for helicopter pilots

Improving safety by sharing good practices.

HE2 Airman ship

The good consistent use of good judgement and well-developed knowledge.

HE3 Helicopter Off Airfield Landing Site Operation

Notice for helicopter operation outside of airfield.

HE4 Single Pilot Decision making

Preventing human error during decision making.

HE5 Risk management in training

Tools and methods to improve risk management in training.

HE6 Advantages of simulators in Helicopter Flight Training

Highlight the helicopter flight simulation training devices and review additional training.

HE7 Techniques for Helicopter Operations in Hilly and Mountainous Terrain

Basic principal of operation in those area.

HE8 The Principles of Thread and Error Management for Helicopter pilots, instructors and training organization

Introduce thread and error management.

HE9 Automation and Flight Path Management

Identify current best practice on automation and flight path management.

HE10 Teaching and Testing in Flight Simulation Training Devices (FSTD)

How to conduct aircrew training and testing in FSTD.

HE11 Training and Testing of Emergency and Abnormal Procedures in Helicopters

Guidance and safety tips for pilots, instructors and examiners for such flight.

HE12 and 13 are under translation.

HE12 Helicopter Performance

How to use performance data of helicopters.

HE13 Weather Threat for VMC Flights

Correct decision making about weather before/in flight situation.

These leaflets are available at; http://www.helijapan.org/

You can see it on the left column.

Original of these leaflet are published by EHEST, based on their analysis of more than hundred accident last decade. Almost accidents included many factors and they are put in the leaflet to prevent accidents.

Original version (in English) from ;

http://www.ihst.org/Default.aspx?tabid=3089&language=en-US

Safety Road Show

The Safety Road Show is created by Airbus Helicopters, and their HQ and local blanches spread it in the world wide. Some part of the world, for example Latin America, AH repeated their show many times in the region, finally, the accidents reduced approx. 80% during 2009 to 2014. Before that, this region recorded the worst rate of the helicopter accidents, but that was reduced. Of course it's not only by the show, however it exactly affected about safety culture of people there. In Japan, the show was started May/2016, and repeated many times at JCAB, JTSB, commercial operators, para-public operators, and institutes. Also the show for the public, including private pilots will be held next October in Tokyo and Osaka. Few years before, JCAB decided to introduce "SMS (Safety Management System)" based on decision of ICAO, but it was very complex and every people had a problem for introducing it. The show also present how to do it, and it helps almost person to understanding how should they do it.

Following figure is the sample of items in workshop for SMS at Safety Road show. Point of SMS are;

・Continuous improvement is necessary to mature SMS.

・Activity and efficiency should be checked to keep function of SMS.

・Continuous improvements and keep it in the organization, is the one of the factor of successful SMS.

課：

企業名／組織名：

記入日：

実行内容	責任者	完了日
1. Safety Road Show によって安全意識を持つ。		
2. 安全オペレーション・グループの立ち上げ、および運航／業務安全管理者「SM」（能力および適切な技能を有すること）の任命を行う。		
3. 安全オペレーション・グループおよび運航／業務安全のトップが果たす役割を定義する。また、企業安全方針および運航／業務安全目標を定義する。		
4. パイロットおよび技術者に対し訓練が必要かを分析する。		
5. パイロットおよび技術者に対する年間訓練計画を立てる。		
6. CAMO タイプの整備支援システムを使用し、ヘリコプターの耐空性を維持する。		
7.『飛行データ管理』タイプの装置、VISION 1000 を使用する。		
8. トップ・マネジメントの確約を確実に得る。トップ・マネジメントが確実に方針、運航/業務安全目標および SMS の実施計画を承認する。		
適切なリソースを割り当て、これら安全措置の支援を行う。		

Figure-2 Example of work shop items at the Show

Road map of JHST

Since starting of JHST, we talk to others to establish JSHT, however not yet. One of the points is gathering people in other organization. Already in Japan, some organizations are existed and all of them are working to improve safety in their part, like pilots, mechanics, authorities, manufacturers and so on. We would like gather them under safety flag. And JCAB started their activity "Safety committee for General Aviation" regarding improvement of safety in general aviation last year, because we had some catastrophic accidents not only for the helicopters. We JHST is invited it and working together right now. I hope our activity and it will be joined in the future like Europe. (EHEST was already merged by EASA at beginning of 2017, now they are working as RSC, Rotorcraft Sectorial Committee, and it works under EASA Rotorcraft Department.) The committee in Japan is going to be such style, I think it's better and all other organizations and industries join it easier.

The following organization chart which we're considering now. Key point is a "Working Group", it'll work to develop some actions what will be better for helicopter operations in Japan.

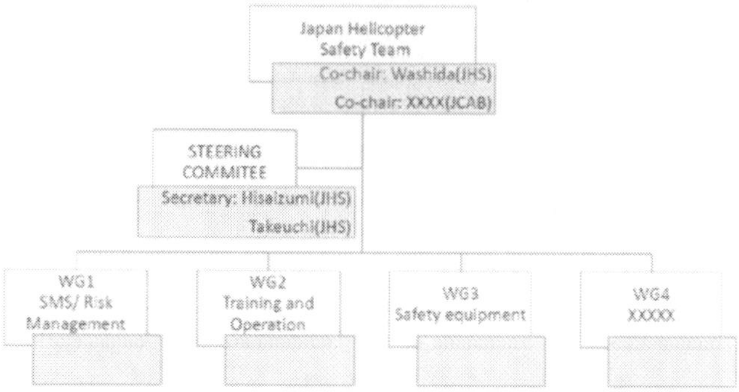

Figure-3 Organization chart of JHST

Every people in this world think about safety, and how should we do that, but it's not completely same. We JHST believe that it will be one of result to do that in the future.

REFERENCES

EHEST (RSC) Home page;
https://essi.easa.europa.eu/ehest/index.html%3Fcat=6.html

IHST Home Page;
http://www.ihst.org/

JHS Home Page;
http://www.helijapan.org/#nogo

The 6th Asian/Australian Rotorcraft Forum
& Heli Japan 2017, in Kanazawa, Japan

小型航空機に係る総合的な安全対策の推進
General Aviation Safety Improvement

国土交通省航空局安全部運航安全課
小型航空機安全対策官
千田　知史

CHIDA Tomofumi
Director for Safety of General Aviation
Flight Standards Division
Civil Aviation Bureau
Ministry of Land, Infrastructure, Transport and Tourism

国土交通省
Ministry of Land, Infrastructure, Transport and Tourism

近年の小型航空機による事故状況概要
Recent Small Aircraft Accidents in Japan

国土交通省

- Since 2015, the number of small aircraft accidents such as General Aviation (GA) have increased significantly, including a fatal accident of a small aircraft that crashed in an urban area.
- To ensure safety, drastic improvement measures of small aircraft are necessary.

○ 平成27年から自家用等の小型航空機の事故が頻発し、空港周辺で小型航空機が墜落し死者を含む被害も発生
○ 国民の安全・安心を確保するべく、小型航空機の更なる安全確保に向けた抜本的な対策が必要

近年の小型航空機による航空事故件数

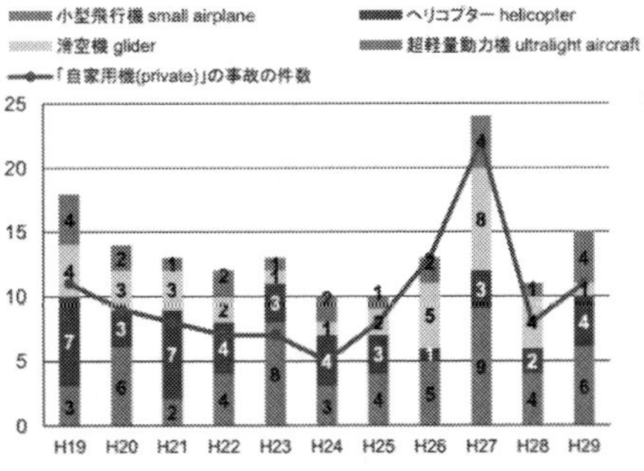

※ H29は9月10日現在

近年の小型航空機による死亡事故等

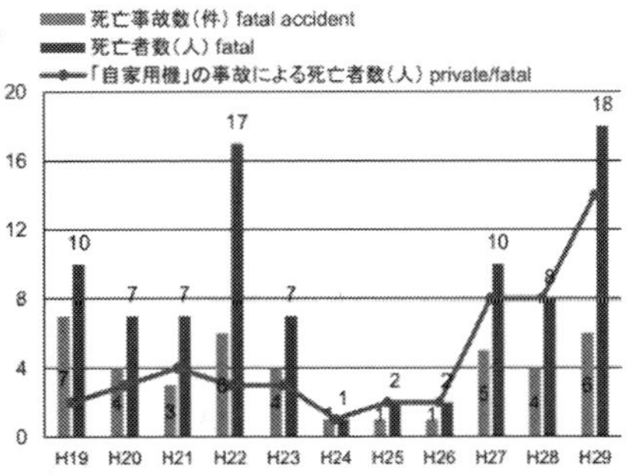

※ H29は9月10日現在

個人に係る航空事故発生率の日米比較

	日本 JAPAN	米国 USA
H24	6.62	7.05
H25	16.23	6.26
H26	16.77	6.23
H27	31.10	5.85

※10万運航時間あたりの航空事故発生件数。統計方法が異なる可能性があるため、参考値。

安全意識の徹底 ／ 操縦技能・知識の向上 ／ 事故等の未然・再発防止　が喫緊の課題

今年発生した主な航空事故

Small Aircraft Major Accidents In 2017

国土交通省

Between January 1st and September 10, 2017, there have been 15 accidents.
The summary of the major accidents is as follows;

平成29年1月1日から9月10日の間、小型航空機*の航空事故が15件発生

主な事故については、以下のとおり。

小型航空機* (Small aircraft) = 小型飛行機 (small airplane)、回転翼航空機 (helicopter)、滑空機 (glider)、超軽量動力機 (ultralight aircraft)

○ 3月5日、長野県消防防災航空センター所属の回転翼航空機が松本空港を離陸し、飛行中、長野県の鉢伏山付近に墜落し、搭乗者9名が死亡

(インターネット記事より)

○ 5月14日、山梨県警航空隊所属の回転翼航空機が山梨県の丹波山村付近で救助活動中、樹木等が救助対象者に接触し、当該救助対象者が死亡

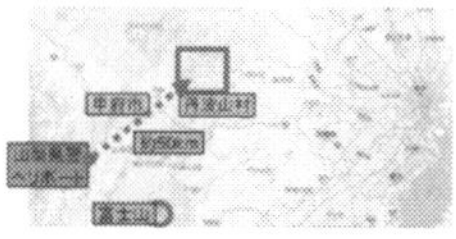

○ 8月14日、個人所属の小型飛行機が八尾空港を離陸し、飛行中、奈良県の山中に墜落し、搭乗者2名が死亡

(インターネット記事より)

○ 5月3日、個人所属の超軽量動力機が山形県内の場外離着陸場を離陸し、場周経路を飛行中、墜落し、搭乗者1名が死亡

(機体メーカーのカタログより)

○ 6月3日、事業者所属の小型飛行機が富山空港を離陸し、飛行中、富山県の山中に墜落し、搭乗者4名が死亡

(インターネット記事より)

○ 8月27日、個人所属の滑空機が福島県内の場外離着陸場を離陸し、飛行中、福島県の山中に墜落し、搭乗者1名が死亡

(インターネット記事より)

これまでの安全対策

Present safety measures

国土交通省

- Conventional Measures
  - ・Pilot Flight Review to verify pilot performance every 2 years (since 2014)
- Additional Measures since 2015 when the accident rates increased conspicuously
  - ・Safety seminars for pilot ・Safety seminars Engineers/Mechanics
  - ・Promotion of subscription to Aviation insurance for private aircraft etc.

自家用操縦士の技量維持に係る制度

これまでの主な追加対策(平成27年より幅広い対策を推進)

1. 小型航空機の操縦士に対する安全確保を徹底

・各地区において安全講習会を開催

・定期的な技能審査を受ける際に操縦技能審査員から指導
(パンフレットを作成・配布)

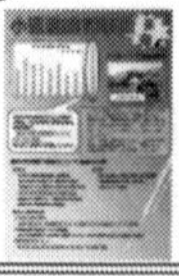

2. 小型航空機の整備士に対する確実な整備の実施を徹底

小型航空機を整備する整備士等を対象とした講習会を新たに開催)

3. 自家用機の航空保険加入を促進

・小型航空機の操縦士に適切な保険加入を奨励・指導
(パンフレットを作成・配布)
・自家用機が国管理空港等を使用する場合には、保険加入を条件とし、国以外が管理する空港等にも同様の措置を講じるよう指導

4. 事業許可の適切な取得の徹底

事業許可取得に関する啓発ペーパーにより、事業許可取得の必要性や無許可事業の通報窓口を自家用操縦士等に周知

小型航空機等に係る安全推進委員会の開催

国土交通省

"Small Aircraft Safety Improvement Committee"

- ☐ It is necessary to investigate many items including foreign measures, discuss the results with experts and related groups, and implement them.
- ☐ Regularly hold "Small Aircraft Safety Improvement Committee" meetings, where further improvement of small aircraft safety measure will be investigated, considered and promoted.

今後の小型航空機の安全対策の構築に係る調査・検討を有識者や関係団体等の意見を踏まえながら行うため、「小型航空機等に係る安全推進委員会」を定期的に開催し、新たに安全対策や安全啓発活動等の取組を推進

委員等構成

○委員(敬称略)　有識者(大学教授、 航空評論家、 研究機関など)

○オブザーバ　小型航空機の操縦者関係団体、小型航空機の運送事業者団体、小型航空機に係る製造者団体、小型航空機に係る装備品/整備関連事業者、関係省庁

○事務局　国土交通省航空局安全部運航安全課

検討事項

○近年の事故・重大インシデント発生の背景の分析

○海外における小型航空機の安全対策の状況

○安全性向上のために活用可能な技術に係る検討

○安全啓発活動のあり方　等

※　上記の他、今後の小型航空機の安全対策を構築のため必要と考えられる事項を適宜検討事項に追加するものとする。

スケジュール

○平成28年12月13日：　第1回委員会

○平成29年3月28日　：　第2回委員会

○平成29年9月25日　：　第3回委員会

○その後も定期的に開催(年2回程度)

東京都調布市における小型機墜落事故に関する調査報告書への対応(概要)

国土交通省

Response of small aircraft accidents report (Safety Recommendation Report) in an urban area (summary)

- ☐ Distribute Brochure to inform the following measures for pilots of private small aircraft,
  ・Confirm takeoff weight on performance before departure, etc.
- ☐ Announcement of ways to make the maximum use of the runway length for aerodrome operators.

1. 事故の概要及び推定原因

※事故の詳細については、以下の運輸安全委員会のホームページに掲載の事故報告書を参照
http://jtsb.mlit.go.jp/jtsb/aircraft/detail.php?id=2131

発生日時：　平成27年7月26日　午前10時58分頃

場　　所：　東京都調布市

概　　要：　自家用小型機が調布飛行場から離陸した直後、住宅に墜落。同機は大破し炎上、機長等2名死亡、同乗者3名重傷。住宅1棟が全焼、周辺住宅3棟にも火災等被害、住民1名死亡、2名軽傷。

原　　因：　離陸重量限界を超過した状態で、低速で離陸し、過度な機首上げ姿勢を継続したことで、離陸上昇中、速度低下により、失速し墜落したと推定

(運輸安全委員会ホームページより)

2. 運輸安全委員会による勧告内容及び当該勧告に対する国土交通省航空局の対応(案)

航空局では事故直後から、離陸重量等の出発前の確認手順の再点検を含む注意喚起文書の発出、航空安全講習会の開催、安全啓発リーフレットの配布等、再発防止の取組みを実施中。

運輸安全委員会による勧告

(1)　自家用小型機の操縦士に対し、
- ・飛行前に性能上の最大重量を確認する重要性の理解促進
- ・離陸中に性能低下が発生した時に再接地する等の対処方法の確認について指導強化

(2)　空港の設置・管理者に対し、
- ・滑走路長を最大限に利用するための事例(誘導路の接続方法等)の周知

国土交通省航空局による対応(案)

(1)　自家用小型機の操縦士に対する理解促進・指導強化のため、

① 改めて、小型航空機の運航者や関係団体に対し注意喚起文書を発出

② 勧告内容等を踏まえたリーフレットを今後作成・配布し、定期的な技能審査の機会に理解を確認

③ 事故直後から全国主要空港において開催している航空安全講習会を引き続き実施

④ 国土交通省のホームページを通じて更なる安全啓発

(2)　空港の安全性向上に寄与すると考えられる方法として、ターニングパッドや取付誘導路の配置によって既存の滑走路長を最大限に利用している事例を取りまとめ、空港の設置・管理者に対し周知文書を発出

今後の取り組み

国土交通省

Further improvement of safety measures

In addition to conducting analysis of accidents and serious incidents, we will considered further safety measures such as utilization of new technologies for safety improvement, strengthening of relations between CAB and pilot.

これまでの「小型航空機等に係る安全推進委員会」における検討を踏まえ、引き続き事故等の分析を行うとともに、先進的な技術の活用、国と操縦士の連携強化などの更なる安全対策について検討。

先進的な技術の活用について

委員会において先進的な技術の事例紹介

FDM
(Flight Data Monitoring)

エアバッグ付
シートベルト

AOA (Angle of Attack)
表示システム

等

これらの技術の活用可能性について今後検討

操縦技能・知識の向上 / **事故等の未然・再発防止**

国と操縦士の連携強化

1. 定期的な技能審査の機会を捉え、操縦士のメールアドレスを収集
2. 安全情報の発信強化
   （1）スレッド・アンド・エラー・マネジメント（TEM）に関するポスター配布
   （2）安全性向上につながる動画を今後作成
   （3）小型機安全ポータルサイトの構築の検討

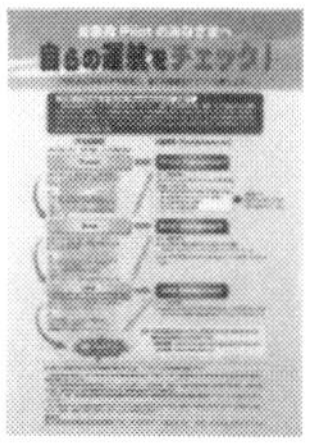

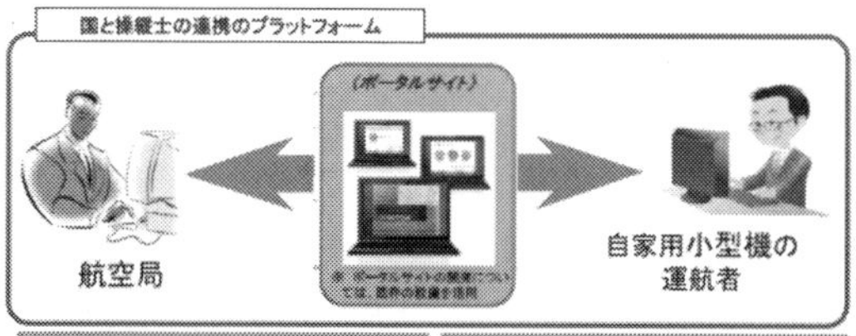

操縦技能・知識の向上 / **安全意識の徹底**

ヘリコプター操縦士の養成・確保に向けた取組

国土交通省

Efforts to foster and secure helicopter pilot

- ☐ Fostering/Securing a sufficient number of competent pilots for helicopter air ambulance(HAA)/helicopter emergency medical service(HEMS) and fire /disaster relief helicopters has been recognized as a pressing issue in Japan.
- ☐ Relevant ministries jointly discussed on this matter and delivered a set of counter measures in July 2015.

ドクターヘリ、消防・防災ヘリ操縦士に係る課題

<ドクターヘリ操縦士>

○ドクターヘリの操縦士に対しては2000時間以上の飛行経歴（業界基準）が求められており、操縦士の高齢化が進んでいる

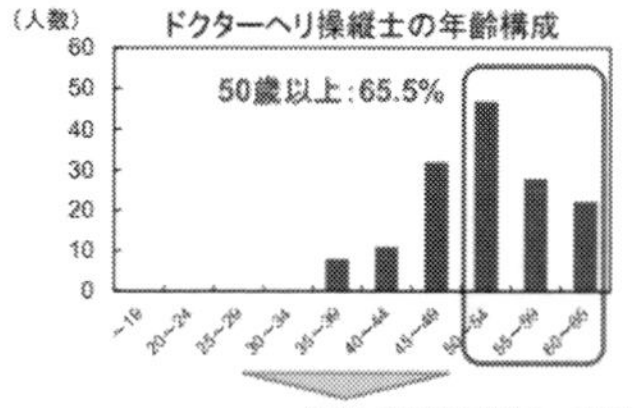

○一方、若手操縦士が経験を積んでいた農業散布等の業務が減少

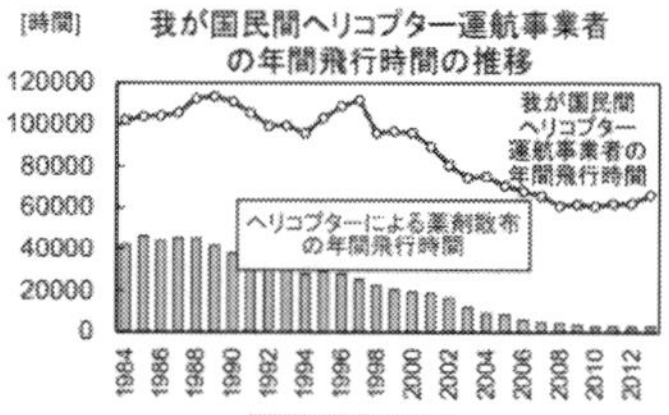

今後必要な技量・経験のある操縦士の確保が困難に

<消防・防災ヘリ>

○ドクターヘリの場合と同様、一定の飛行経験を求めることが一般的

○多くの自治体で操縦士不足を背景として24時間運航体制が確保できていない

○一部の自治体では操縦士不足から運航を制限している例もある

ドクターヘリ・消防防災ヘリに対応可能な技量・経験のあるヘリコプター操縦士の確保が課題

ヘリ操縦士の養成・確保に関する関係省庁連絡会議取りまとめ（平成27年7月）（抜粋）

【民間事業者等と関係省庁で連携しつつ、以下を推進】

ドクターヘリ及び消防・防災ヘリ操縦士の訓練プログラム等の開発
- ・シミュレータ等を活用した訓練プログラムの策定　等

ドクターヘリ及び消防・防災ヘリ操縦士の乗務要件見直し
- ・飛行経歴2000時間等の乗務要件（業界標準）の見直し

ヘリコプター操縦士のキャリアパス確保
- ・若手操縦士への効果的な業務の割り当て、同乗訓練の実施
- ・新規業務の開拓促進　等

ドクターヘリ、消防・防災ヘリに対応可能な技量・経験のある操縦士を確保

シミュレータの活用
- ・高額な費用を要する訓練の合理化　等

若年定年退職自衛官の活用
- ・50歳代半ばで若年定年退職する自衛官の活用方策の検討

ヘリコプター操縦士の養成・確保に向けた取組

国土交通省

Efforts to foster and secure helicopter pilot

- To implement the counter measures, committee composed of academic expert, industry, the relevant ministries/agencies was established in Dec 2016.The Committee investigated the training requirements and crew experience requirements of each Japanese operator as well as those in Europe/US.
- The committee concluded that flight crew requirement for HAA/HEMS should be 1000 hours of flight experience as PIC, instead of 2000 flight hours (regardless of PIC or Co-pilot), which used to be the industry voluntary requirement. Pilots are also required to receive the dedicated training program focusing on the characteristics of HAA/HEMS operation and undergo annual competency check.
- JCAB has amended the regulation on the flight crew requirements for HAA/HEMS to incorporate the above new requirements in June 2017.

■ **ドクターヘリ、消防・防災ヘリ操縦士の乗務要件及び訓練プログラムに関する検討委員会**

- ドクターヘリ、消防・防災ヘリの操縦士に係る乗務要件の見直し及び訓練プログラム(シミュレータの活用も含む)の開発を目的として、学識経験者、業界関係者等から構成される検討委員会を設置し、平成28年12月14日から平成29年3月16日までの間、4回に亘り開催。
- 欧米での乗務要件についての調査や国内で実際に業務を行っている操縦士に対する調査等を踏まえつつ、検討を行った。

■ **検討委員会におけるドクターヘリ、消防・防災ヘリ操縦士の乗務要件及び訓練プログラムの検討結果**

- ドクヘリ操縦士の現在の乗務要件である「2,000時間以上の飛行時間」(業界自主基準)を、「1,000時間以上の機長時間」とする。
- 上記乗務要件の見直しにあたり、ドクヘリ運航の特性に着目した訓練プログラム(狭隘地や傾斜地での離着陸、離着陸時の土煙等による視界不良への対応等)の設定と実施、事業者における十分な操縦士の能力確認を求める。
- なお、離着陸時の土煙等による視界不良への対応など、安全性の観点から実機で実施することが比較的困難な訓練項目については、シミュレータで実施することを推奨する。
- 消防・防災ヘリ操縦士については、ドクヘリ同様の乗務要件(1,000機長時間)等とするが、夜間飛行、吊下揚収及び消防防災活動も行うことから、これらに対応した経験要件及び訓練項目を追加する。

(検討結果を受けた当局の対応)

- 従来は業界自主基準として設定されていたドクヘリ操縦士の乗務要件等について、国の基準として明確化した(6/19(月)付けで新しい基準を公布)。
- 消防・防災ヘリ操縦士の乗務要件等については、検討委員会においてとりまとめられた内容を6/26(月)付けで消防庁あてに通知。

The Amended Civil Aeronautical Act for Safety and Harmonization between Aircraft and UAV

7 November 2017
Ministry of Land, Infrastructure, Transport, and Tourism
Civil Aviation Bureau

国土交通省
Ministry of Land, Infrastructure, Transport and Tourism

Outline of Amendments to the Civil Aeronautical Act

国土交通省

Background

- Recently, the use of unmanned aerial vehicle (UAV) has grown at a rapid rate, specifically the use of UAV is expanding to various fields such as photography, pesticide spraying, and infrastructure inspection.
- It is expected that UAV will contribute to the creation of new industries and services, and they will significantly contribute to the convenience and quality of people's lives through a variety of new applications.
- On the other hand, we must confront safety issues regarding UAV accidents and incidents.

It is necessary to establish basic rules for airspaces that will allow UAV to operate and operational procedures that can be used as an emergency measure with consideration to the current international situation.

Outline

(1) Types of airspace that require permission to operate UAV※

※Any airplane and rotorcraft which cannot accommodate any person on board and can be remotely or automatically piloted (Excluding super light weight ones)

It is prohibited to operate UAV in the following airspace without permission from the Ministry of Land, Infrastructure, Transport, and Tourism.

※Permission to operate UAV, within these airspace types, is issued to operators who have established a safety system.

○ The airspace around airports, which might impact the safe aircraft operation【A and B in the figure below】
○ Any airspace above densely inhabited or districts【C in the figure below】

(2) Operational procedures of UAV

Any person who intends to operate UAV is required to adhere to the following procedures unless otherwise approved by the Ministry of Land, Infrastructure, Transport, and Tourism

※More flexible operations can be approved subject to safety systems.

○ Operation in the daytime only.
○ Visual surveillance of surroundings at all times.
○ Maintain a minimum of 30m distance between the UA and any other people or property.

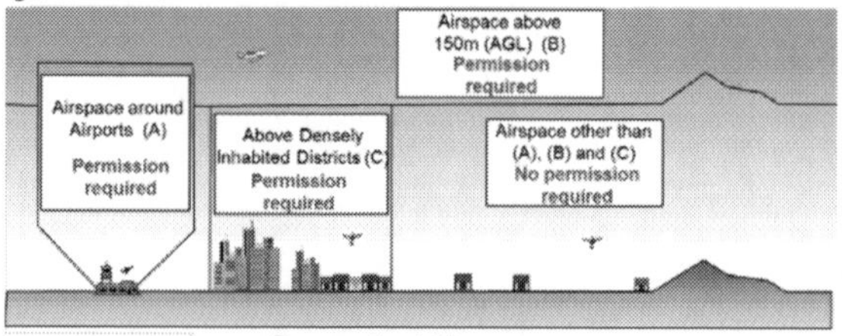

Conceptual Airspace

1

Recent Trends in the Number of Applications for UAV Operation

国土交通省

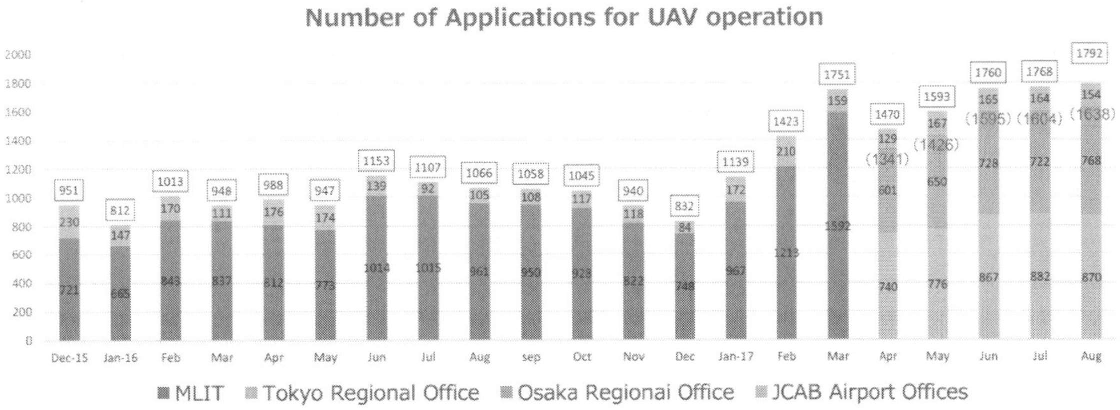

※Counting of respective airport offices

- During the period from December 2015 to February 2017. Months are tracked starting the 10th of the current month through the 9th of the following month.
- As for March 2017, counted from the 9th to 31st of March 2017.
- Note: MLIT transitioned application processing over to JCAB Regional Offices effective April 2017

2

The Public-Private Sector Conference on Improving the Environment for UAVs

国土交通省

The 2nd "Public-Private Dialogue towards Investment for the Future"(5 Nov. 2015) Prime Minister Abe stated

"We will aim to make parcel delivery by drones a reality, as soon as three years from now. For this purpose, the government will immediately establish a Public-Private Council, in which users and the relevant ministries and agencies will discuss the specific structural and systemic requirements. A policy to improve the system should be established by such a council by summer next year."

The public-Private Sector Conference on Improving the Environment for UAVs

○ **Establishment of The public-Private Sector Conference with members of related departments, agencies, manufacturers, users. (Held 6 times from 7 Dec 2015)**

○ **The public-Private Sector Conference reviewed roadmap about technological development on April, also reviewed direction of institutional design on July in 2016.**
And then they revised roadmap about technological development, made new roadmap for the Aerial Industrial Revolution on May in2017.

Summary of Public/Private Sector Conference

1.Institutional design for safety of small UAVs
2.Grasp the operation of the revised air law and systematize and share safety measures
3.Improvement of environment for promotion of business · business utilizing by small UAVs
4.Verification of voluntary efforts to ensure safety of small UAVs
5.Improvement of the environment to realize "the Aerial Industrial Revolution"

Member of Public/Private Sector Conference

Cabinet Secretariat

Affiliated government agencies

Manager of related ministries and agencies participate
Cabinet Secretariat (Deputy Minister's Office, Countermeasure Crisis management office, IT Comprehensive Strategy Office, Japan Revitalization Comprehensive Office, Cabinet Cyber Security Center, Regional Creation Promotion Office) The Metropolitan Police Department, Consumer Affairs Agency, Ministry of Public Management , Fire and Disaster Management Agency , Department of Justice , Ministry of Education Culture Sports Science and Technology , Ministry of Health Labor and Welfare, Ministry of Agriculture Forestry and Fisheries, Ministry of Economy Trade and Industry, Ministry of Land, Infrastructure, Transport and Tourism

Affiliated organizations

- **Comprehensive organization consisting of manufacturers and users of small UAVs**
- **Association of makers of small UAVs**
- **User organization in a specific field**
- **Aviation organization**
- **Economic organization etc.** **34 organizations · 10 companies**

3

※Established "Meeting for system design for further safety of UAVs usage"

Road map for the Aerial Industrial Revolution | Technology development and environment improvement for safe utilization of small UAVs

19 May 2017
The Public-Private Sector Conference on Improving the Environment for UAVs

国土交通省

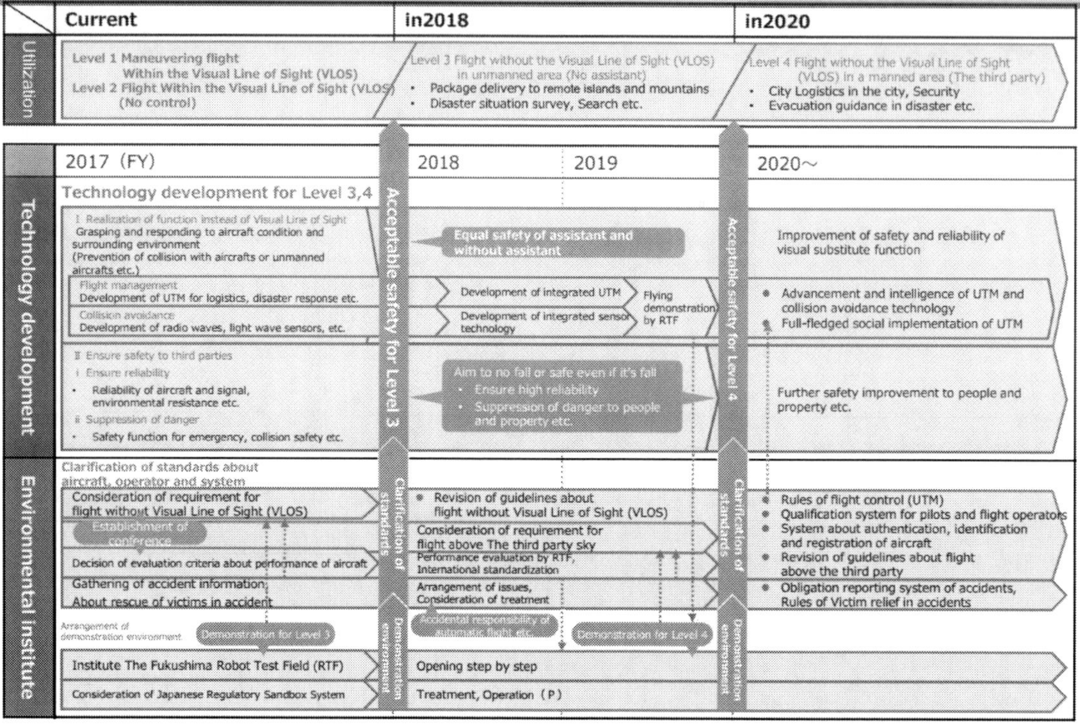

4

Basic policy of system design toward securing further safety of UAV < Summary >

国土交通省

Basic stance

- **Through operation of the amended Civil Aeronautical Act enforced on 10 December 2015, requirements concerning airframe, operator and operation control system have become more concrete ones and formulation of comprehensive rules including guidelines and activities of private associations has also progressed**
- **In order to correspond to rapidly progressing societal implementation of new technology and diversification of its usage, rules have been made or modified with expedition and flexibility but in a phased manner from where applicable**

Basic policy of system design

＜Basic flight rules＞

- **Enlighten as to prohibition of operation while drinking and pre-flight check, and prepare rules based on verification result of the effect**
- **Study to formulate a mandatory reporting system of accidents, a voluntary reporting system of small incidents/hazards, and gathering and analysis system of accident information**

<Secure further safety of airframe, pilot and operation control system>

- **Pilot training and flight manuals prepared by private associations are to be put on MLIT website provided that they comply with certain standards, and if they are actually utilized, review procedures will be streamlined partly.**
- **System will be introduced to fully provide goods delivery services in remote islands or mountain areas around 2018.**
- **In order to fully provide goods delivery services in urban areas in 2020s, type certification and pilot license system will promptly be studied and prepared.**
- **Even in case that permissions or approvals are not required, safety shall be enhanced through utilization of pilot training and flight manuals.**

＜Safety and harmonization between UA and aircraft＞

- **Investigative commission with participation of operators of UAV and manned aircraft will promptly be launched to prepare rules for collision avoidance between manned aircraft and unmanned aircraft, and amongst UAVs as well, by the end of FY 2016.**
- **Rules and countermeasures will be studied to prevent risks due to malfunction and operation mistake around airports.**
- **System to share flight information between operators of manned aircraft and unmanned aircraft will be established. NOTAM will be improved.**

＜Others＞

- **Prerequisite insurances shall be continued and safety awareness shall be maintained or improved.**
- **Publicity of guidelines and formulation of voluntary rules will be promoted with regard to privacy protection and flight over the land of the third party.**
- **Voluntary efforts to identify owners of UAV will be encouraged.**
- **Proper frequency system to support flight out of the Visual Line of Sight will be studied.**

5

Safety and Harmonization between Aircraft and Unmanned Aerial Vehicle

国土交通省

○Further safety securing is required. There were no accidents such as death or injury to third parties by UAVs, but there was near miss case with the aircraft.

○They made system design for further safety of UAVs usage on July in 2016.

○Consideration about collision avoidance between manned aircraft and unmanned aircraft at Investigative commission on MLIT. Interim report on 30 Mar 2017

Case of approach to aircraft

●Date：31 Jun 2016

●Place：Chiba Prefecture (Airspace of No need permit)

Small UAVs (radio controlled aircraft) and a doctor helicopter approached. According to a report from a doctor helicopter, radio controlled machine dropped almost vertically on the front left side of the doctor helicopter about 15 to 25 m and passed through.

Public-Private Sector Conference

○ Establishment of The public-Private Sector Conference with members of related departments, agencies, manufacturers, users on December in 2015.

○ They made system design for further safety of UAVs usage on July in 2016.

Direction of Safety and Harmonization between Aircraft and Unmanned Aerial Vehicle (Outline)

- Begin meetings soon with participation of operators for small UAVs and airline operators, aim to establish rules by end of FY2016. To build a new system to avoid collision of UAs with each other.

Installation of Investigative commission on MLIT

Member

- UAVs operation organization
- Aircraft operation organization
- The study organization
- Affiliated government agencies

Main Considerations

- Collision avoidance between UAVs and aircraft
- Collision avoidance between UAVs and UAVs
- Measures to improve visibility of UAVs
- System of sharing information or contents between aircraft and UAVs etc.

Schedule

8 Nov 2016 1st Meeting

9 Feb 2017 2nd Meeting

30 Mar 2017 3rd Meeting (Interim report)

Report to the Public and Private Council

After 2017 Holding as necessary

6

Study Panel Interim Report on Safety and Harmonization between Aircraft and Unmanned Aircraft (Outline)

国土交通省

Basic Concept of Collision Avoidance

- Avoid proximity between aircraft and unmanned aircraft (UA) beforehand by sharing flight information among parties concerned
- Avoid collision by complying with flight rules to be developed, in case proximity is unavoidable
- Strengthen research and development efforts toward practical application of collision avoidance technology and operating control system through government –private cooperation as well as international coordination
- Basic rules shall be incorporated in the guidelines for safety flight of UA and be requirements for approval or permission pertaining to UA operation as well

Measures to avoid collision of aircraft and UA with each other

＜Information sharing before operation＞

To make available to parties concerned flight information on medical helicopter obtained by private businesses utilizing satellite communications. Further, any person who intends to operate UA shall confirm such flight information prior to the operation

＜Rules before operation＞

Not to operate UA when cruising aircraft is visually recognized

＜Rules during operation＞

When cruising aircraft is visually recognized, take immediate avoidance actions such as emergency landing of UA at a secure place

Measures to avoid collision of UAs with each other

＜Information sharing before operation＞

To build a new system to make available to parties concerned flight information of UA. Further, any person who intends to operate UA shall share and confirm such flight information prior to the operation.

＜Rules before operation＞

When other flying UA is visually recognized, planned route, altitude and time of flight shall be coordinated with parties concerned

＜Rules during operation＞

When other flying UA is visually recognized, take immediate avoidance actions such as emergency landing of UA at a secure place and the route, altitude and time of flight shall be coordinated with parties concerned

Others

＜Improvement of visibility of fuselage＞

It is required to obtain weather information prior to operation and not to operate UA in the clouds or fog where sufficient visibility range is not secured. Also, technology investigation on improvement of visibility of fuselage will be conducted.

＜Tightening of regulations around airports＞

Various issues such as utilization of geo-fence, introduction of stricter rules to prevent proximity to or collision with aircraft and educational campaign will be taken up for study

＜Educational campaign on the significance and function of flight information＞

Undertake educational campaigns through website to any person intending to operate UK, on the significance and function of flight information and its proper usage

7

Introduction of the RNAV measures for small aircraft in CARATS

Yamano Shuro
Japan Civil Aviation Bureau
Tokyo, Japan

PLEASE NOTE: DO NOT INCLUDE EMAIL ON 1ST PAGE

ABSTRACT

Civil Aviation Bureau, Ministry of Land, Infrastructure, Transport and Tourism install "The committee for promoting renovation of the air traffic system" in order to meet the growing demand in air traffic and other needs appropriately. All stakeholders such as person with relevant knowledge and experience, operator, research institution, manufacturer, the related government ministries and agency have developed CARATS, the long-term vision for air traffic systems, through collaborative decision-making and have been making efforts to push forward the reforms.

The goals to be achieved by 2025 within the framework of CARATS(Collaborative Actions for Renovation of Air Traffic Systems) had been set the measures such as meet growing demand in air traffic, improve safety, improve user-friendliness(punctuality, low cancellation rates, speedy service) and improve operational efficiency. Also within framework of CARATS, the measures for 8 types of reforms have been developed. JCAB prepared roadmaps to judge the implementation of the measures, and works for decision making of the implementation and beginning operation every year.

I introduce the progress of consideration of the RNAV measures (establishment of low altitude RNAV route, establishment of SID, STAR and instrument approach procedure) for small aircraft that activity is expected for search and rescue, disaster relief operation and lifesaving.

Air traffic system

Necessity of long-term vision – Present situation and issue

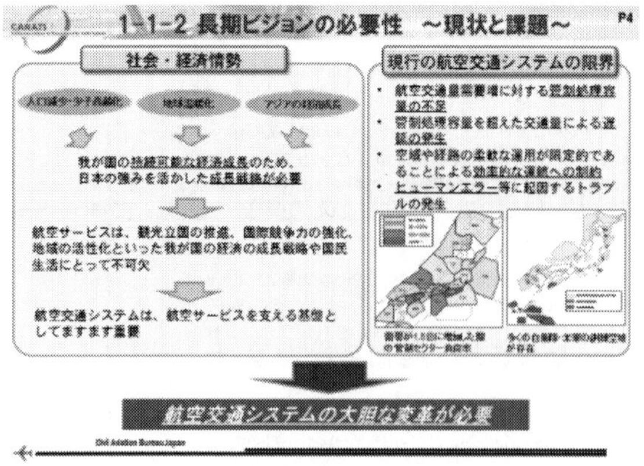

Long-term vision for air traffic systems – CARATS

CARATS roadmap and implementation system

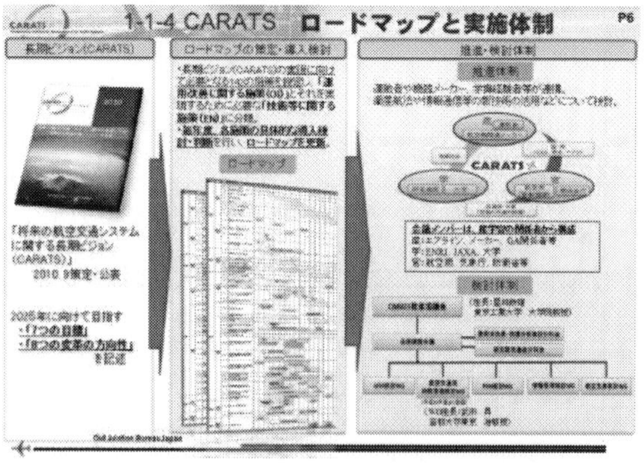

Structure of long-term vision (CARATS)

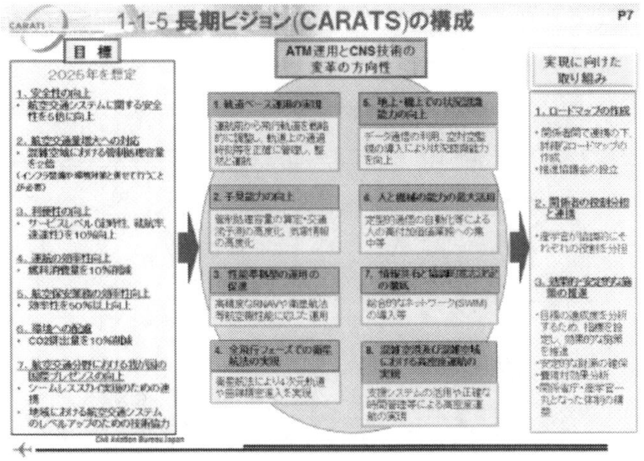

Direction of the renovation

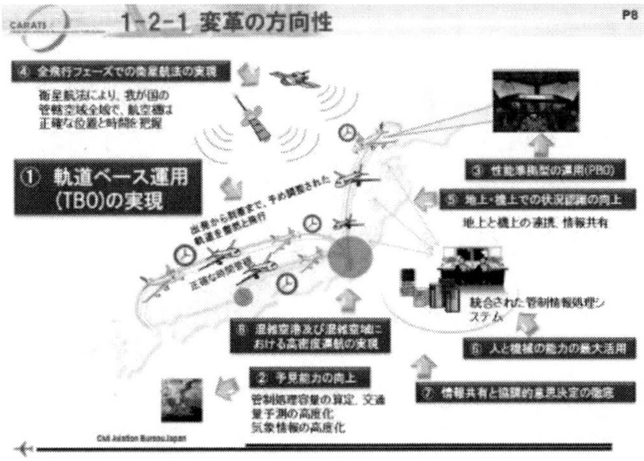

Direction of the renovation 1

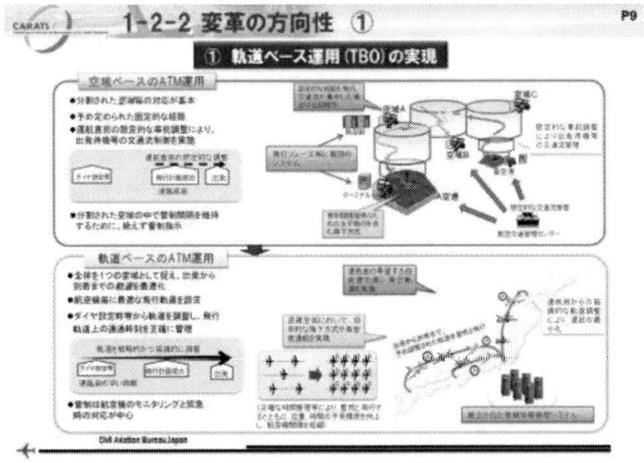

Direction of the renovation 2・3

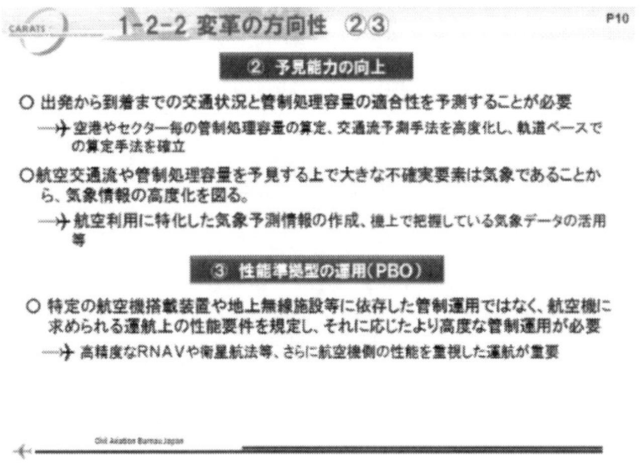

Direction of the renovation 4

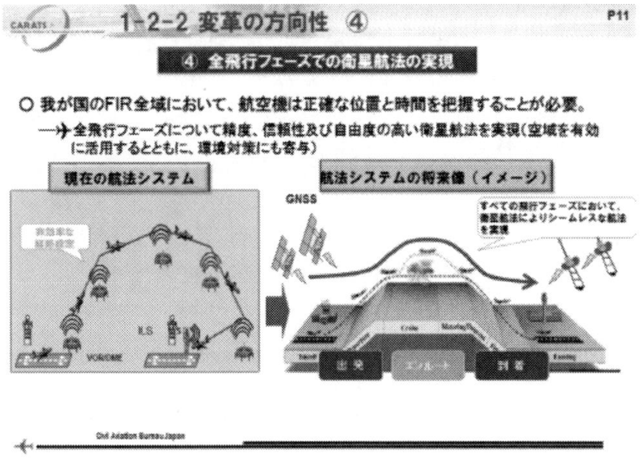

Direction of the renovation 5

Direction of the renovation 6

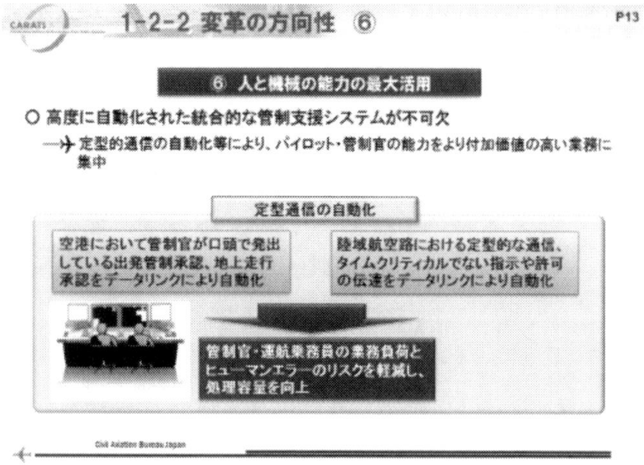

Direction of the renovation 7・8

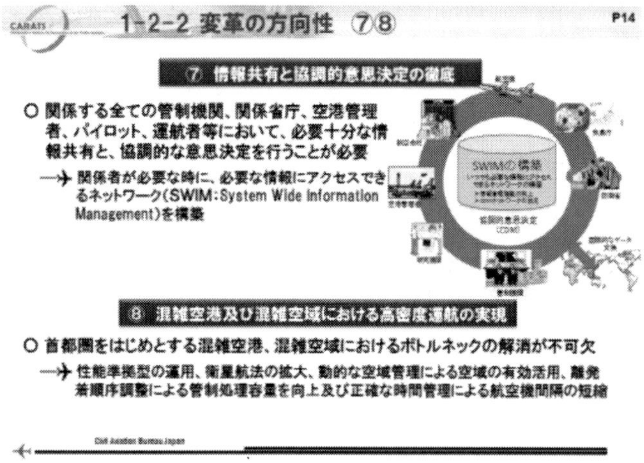

Overview of the measures for small aircraft

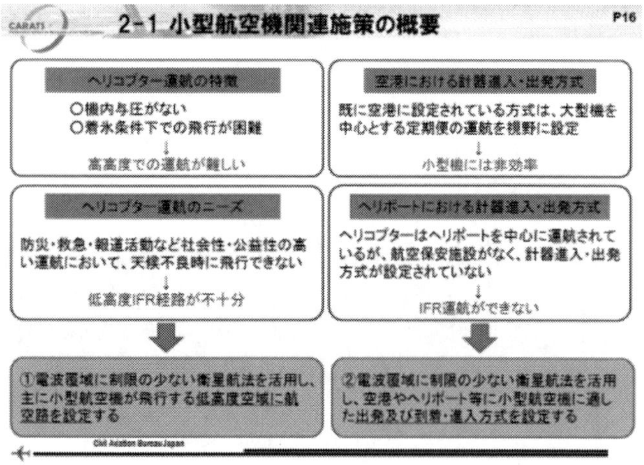

Establishment of low altitude RNAV route - Overview of the measures

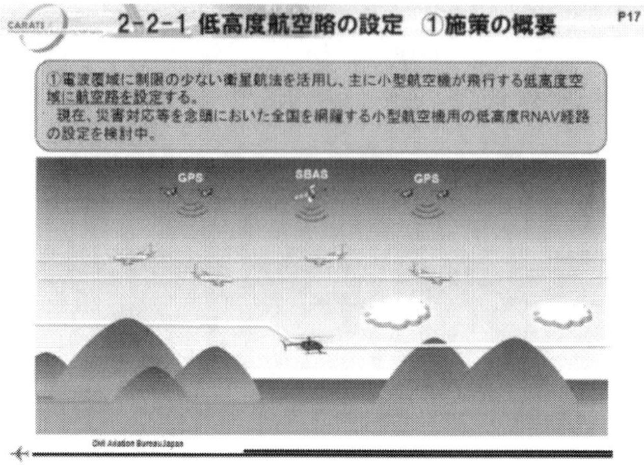

Establishment of low altitude RNAV route - Current status

Establishment of SID, STAR and instrument approach procedure - Overview of the measures

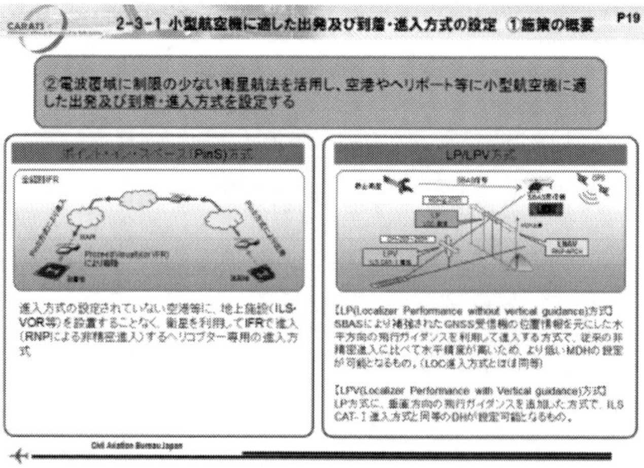

Establishment of SID, STAR and instrument approach procedure - Current status

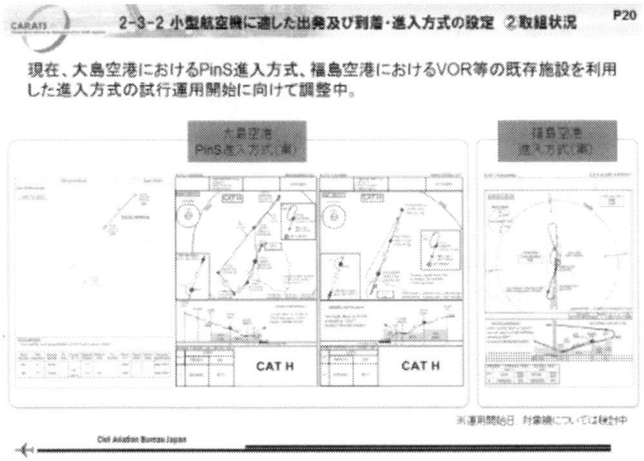

安全運航の為の航空気象観測インフラと気象予測について

Aviation Meteorological Observation Infrastructure and Weather Prediction for Safe Operations

西村　瑞紀 (Mizuki Nishimura)

SKY Content Service

Weathernews Inc.

Japan

高森　美枝 (Mie Takamori)

SKY Content Service General manager

Weathernews Inc.

Japan

川畑　貴義 (Takayoshi Kawahata)

SKY Content Service Marketing leader

Weathernews Inc.

Japan

要旨

航空機の安全運航を行う上で、気象条件は非常に大きな要素となっている。株式会社ウェザーニューズ（以下、ウェザーニューズ）では航空気象サービスとしてエアライン、使用事業者向けに約 30 年間に渡り航空気象サービスの提供を行っている。本書では、近年、各地で発生している局所的な気象現象に対し、ウェザーニューズが独自に整備・観測している観測インフラ（独自レーダー、ライブカメラ等）データについて、運航上有効となった事例について報告を行う。

Weather is one of the most important factors for safe aircraft operations. Weathernews Inc. has experience in providing aviation weather solutions to commercial and general aviation companies for the past 30 years. This paper introduced case studies of the effective use of the Weathernews proprietary observational infrastructure (original radars, web cameras, etc.) and databases in effectively providing risk information regarding localized weather phenomena.

はじめに

この 10 年を振り返ってみると、航空事故や重大インシデントと認定された事例は 239 件[1]に上り、2017 年に入っても、3 月の長野県鉢伏山のフライト訓練中での事故、6 月の北アルプス、8 月の奈良県山添村での事故など航空機事故とりわけ小型航空機の事故が後を絶たない。運輸安全委員会によると、過去 10 年に発生した航空事故の約 5 割が小型機とヘリコプターに占められる。様々な要因が考えられるが、時には気象も航空事故・重大インシデントをもたらす一つの原因になりかねない。それゆえ航空機の安全運航を行う上で、気象条件は非常に大きな要素となっており、ウェザーニューズでは 30 年来、空の安全を守ることを使命に、航空気象サービスの提供を行っている。そこでは気象情報の提供だけではなく、独自に観測インフラを開発・設置し、局地的な気象現象の捕捉・解析・予測に取り組んでいる。本書では更なる安全運航支援を実現するための気象観測インフラ・独自インフラの有効性について報告する。

航空気象観測インフラ

ウェザーニューズでは、従来の気象観測器では捉えることのできない気象現象や運航上の課題に対して、新たな視点で様々なインフラを整備している。その中

[1]運輸安全委員会 報告書検索より。

から航空機の運航を行う上で有効と考えられる気象観測インフラについて、いくつか取り上げる。

独自観測レーダーによる取り組み

ウェザーニューズでは、ゲリラ雷雨をはじめとする局地的な気象現象を捉えるために航空機レーダーを応用した超小型レーダー「WITH レーダー」を独自開発し、2017 年 8 月時点で 80 基を設置・運用している（図 1）。

図 1 WITH レーダー外観

表 1 に示す WITH レーダーの性能のとおり、WITH レーダーの特徴として、6 秒に 1 回の高頻度観測が可能な点、水平方向観測以外に鉛直方向観測による雨雲の雲長高度・反射強度から発達・衰弱などの傾向をリアルタイムに観測できる点などが挙げられる。

表 1 WITH レーダーの性能

	特性	WITH レーダー	従来レーダー
	観測高度	0km 以上	2km 以上
	観測範囲	半径 50km	半径 250km
	観測頻度	6 秒	5 分
	解像度	150m	1km
	観測方向	水平・鉛直	水平のみ
観測要素	降水強度	○	○
	移動速度	○	×
	移動方向	○	×
	降水層判定	○	×

また、起伏の激しい地形では、従来レーダーのブラインドエリアとなることがあり、実際に雨雲が発生していてもレーダーに映らない難点を抱えている（図 2）。山の峠や平地であっても山影にあたる地域などに WITH レーダーを設置することで、雨雲の見える化が実現され、従来レーダーのブラインドエリアを補うことができる。従来レーダーと WITH レーダーの組み合わせによって、ゲリラ雷雨の捕捉率は今や 90%を上回る。

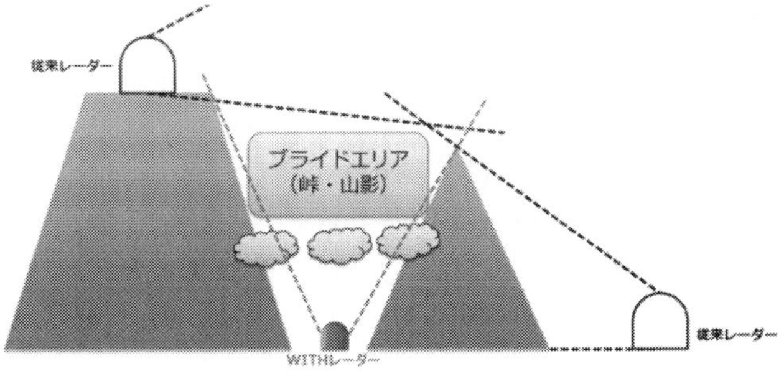

図 2 ブライドエリアに対する WITH レーダーの効果

更に、WITH レーダーではドップラー速度を活用することによって雨雲の移動速度や移動方向も観測することができる。この特性を活かして竜巻を捉えた事例を以下に示す。2010 年 7 月、横浜付近で局地的に雨雲が急発達し、WITH レーダーで観測を行った。図 3 に WITH レーダーで捉えた雨雲の様子を示す。横浜付近に強い降水強度を表す黄色や赤色のエコーが観測され、局地的に雨雲が発達している状況がうかがえる。

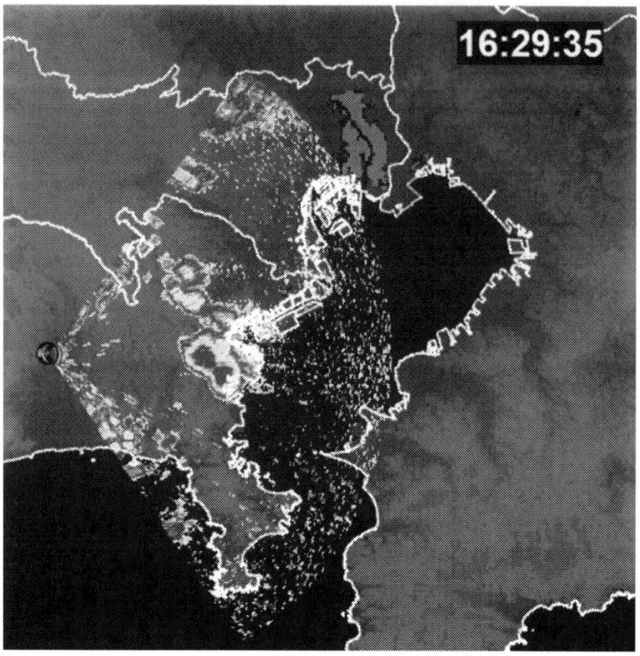

図 3 WITH レーダーで捉えた発達した雨雲の様子（赤ほど強い降水強度を表す）

ドップラー速度に切り替えて雨雲の内部構造を映したものが図 4 である。ちょうど発達した雨雲と同じ位置に赤と青が合わさっていることが見て取れる。ドップラー速度では、遠ざかる風の動きは赤色で、近づく風の動きは青色で表現されるため、小規模ながら渦を巻いている様子が確認できた。

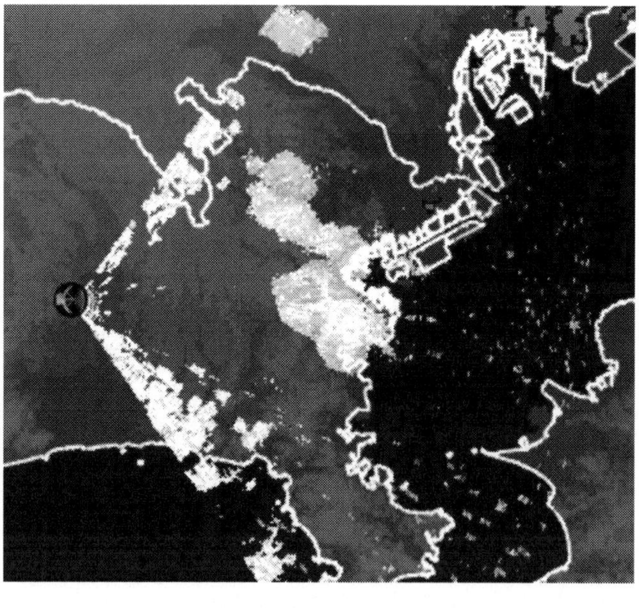

図 4 WITH レーダードップラー速度の様子（遠ざかる風の動きは赤色、近づく風の動きは青色）

当時の状況は、ウェザーニューズの会員から送られてくる天気リポートでも「小さな竜巻が近くで発生しました！しばらくして二つになって北から南の方へ移動しました。」と報告があった。天気リポート写真を図 5 に示す。

図 5 ウェザーニューズ会員の天気リポート写真

この竜巻は、南風と東風の強い収束線上で発生した竜巻と考えられ、上空には南から乾燥域の先端が接近していたため、大気の状態が不安定な場であった。当初はマルチセル型の構造だったが、途中からメソサイクロンを伴うスーパーセル型に変化し、20 分程度持続した。メソサイクロンは極めて小規模な低気圧性の循環構造のため、事前の予測が難しい。このような状況下でも 6 秒に 1 回の高頻度観測が可能な WITH レーダーで雨雲の発生・発達の様子を監視し、一様な色の中に異なる色が渦巻き構造で現れた場合の竜巻発生の可能性を検知することがでる。WITH レーダーを活用した空港やヘリポート付近での実況監視は、竜巻・突風・降雹の注意喚起に極めて有効であると考える。

続いて、冬場の短時間強雪を捉えた事例[2]を示す。2015 年 12 月、北海道石狩湾を小低気圧が南下し、札幌や岩見沢を中心に活発な雪雲が流れ込んだ。小低気圧周辺では雪雲が発達し、時間 5cm 前後の強い降雪によって視程が悪化した。WITH レーダーによる観測の結果、小低気圧に伴う発達した雪雲が小樽方面から接近し 5 時に小樽上空で観測された雪雲の雲頂高度は

[2] ふゆトピア研究発表会（2017 年）

2,000m 以上に発達していた（図 6）。また、その一時間後には降水強度を示すエコーの色が黄色になり、雪雲が更に発達傾向であることが見て取れる（図 7）。その後、8 時頃になると札幌～小樽間でエコーに隙間が出始め（図 8）、9 時には札幌～小樽付近に流れ込む雪雲がほとんどなく（図 9）降雪のピークは越え天気の回復が早いことが見て取れた。

図 6 WITH レーダー観測事例（12 月 7 日 5 時：雲頂高度）

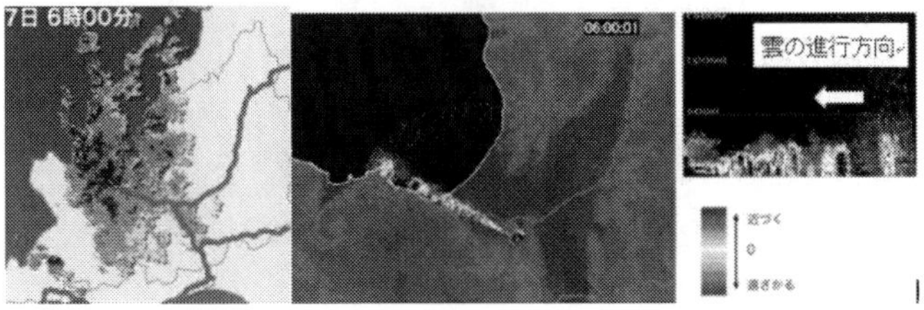

図 7 WITH レーダー観測事例（12 月 7 日 6 時：移動速度）

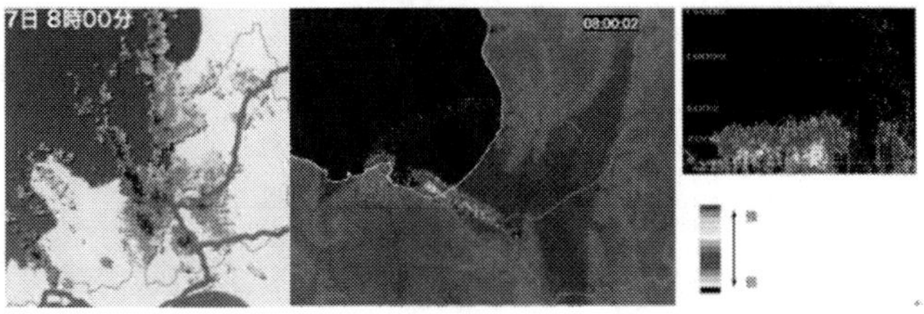

図 8 WITH レーダー観測事例（12 月 7 日 8 時：雲頂高度）

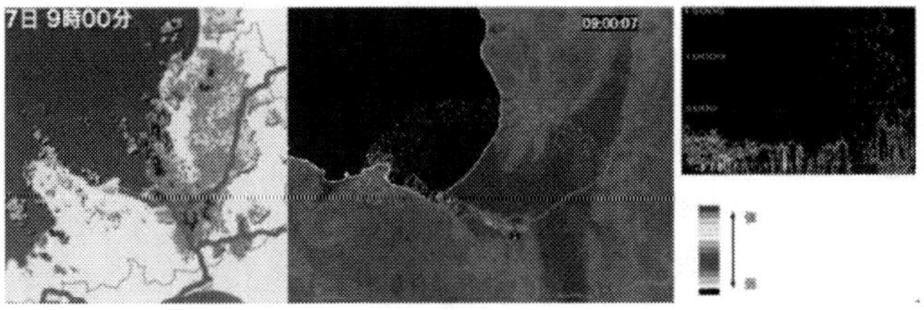

図 9 WITH レーダー観測事例（12 月 7 日 9 時：雲頂高度）

WITH レーダーを活用することによって、短時間強雪の可能性や回復のタイミングを見極めることができ、航空機を安全に運航する上での判断材料の一つとして有効性が確認できた。

航空用ライブカメラの取り組み

国土の約 7 割[3]が山地や丘陵地である日本において、低高度を飛行する小型航空機にとっては山間部の飛行や峠越えは避けられない。一方、そのような地域では気象観測器の配備が限定的で情報量が乏しいことが現状である。ウェザーニューズでは、空港付近や小型航空機の主要ルート沿いに独自のライブカメラを設置し、「峠の茶屋」と題し運用を行っている（図 10）。峠の茶屋は、航空機を運航する視点で画角を設定し、リアルタイムで現地の様子を確認することができるため、運航判断を行う上での重要な判断材料となっている。

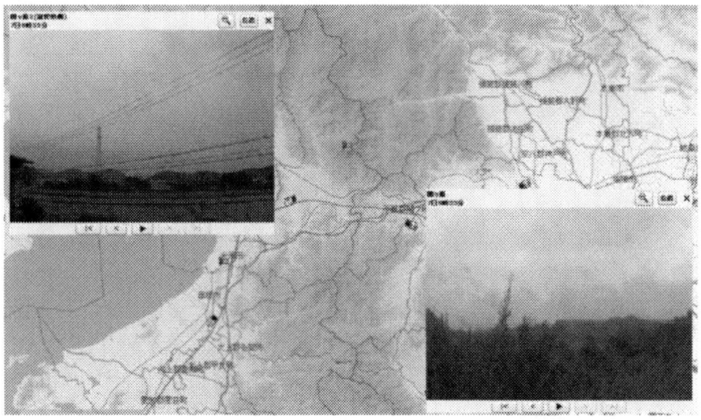

図 10 峠の茶屋（航空用ライブカメラ）

ウェザーニューズでは峠の茶屋を全国約 110 箇所に設置し、一般用のライブカメラと合わせると 1,000 台に上る。また、特定の峠の茶屋に対して、独自の画像解析を施し、悪天のアラートや予測の実況補正に活用し精度向上を図っている。画像解析の仕組みとしては、画角内の対象物に対してコントラスト比判定によるエッジ検出技術を応用し、画角内の対象物が見えるのか、見えないのかを自動判定している（図 11・12）。

---

[3]国土技術研究センター

図 11 晴天時の峠の茶屋

図 12 悪天時の峠の茶屋（中央の対象物の画像解析によって悪天を検知）

その判定結果をもとに、図 13 に示す悪天のアラートや図 14 のように予測への反映を行っており、観測データのない地域での実況監視や実況補正に有効であると言える。

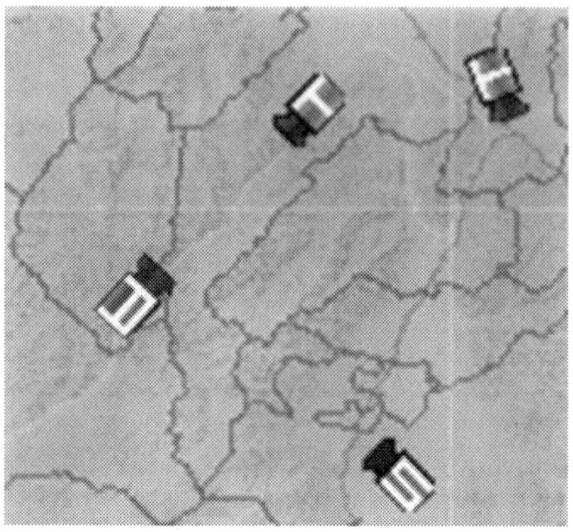

図 13 画像解析による悪天アラート（峠の茶屋アイコンを赤で示す）

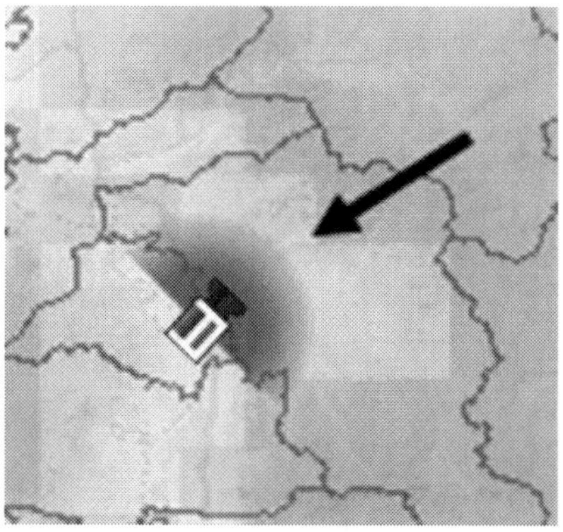

図 14 画像解析による実況補正（峠の茶屋周辺の予報を悪化修正）

小型動態監視システムの取り組み

かねてより航空機用の機体動態監視システムは存在していたものの、小型航空機への導入においては費用・工数の面からも導入数に限りがあった。一方で、航空機の安全運航を支援する上で、気象情報だけではなく機体が今どこを飛行しているのかを把握することが求められた。そこで、ウェザーニューズでは、2012 年にイリジウム衛星通信による持ち込み型の小型動態監視システム FOSTER-CoPilot を開発した（図 15）。

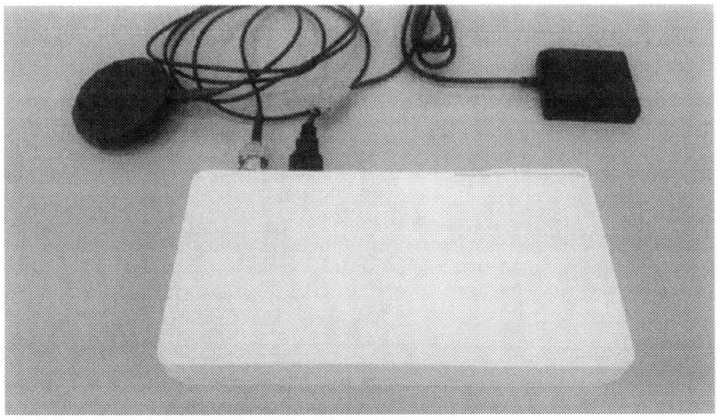

図 15 FOSTER-CoPilot 外観

小型航空機での実用化の条件として、①小型で軽量かつ機体の修理改造が不要であること。②通信コストを最小限に抑えること。③山間部や洋上でも通信可能であること。④運航上必要とする航空気象情報、航空情報、位置情報との一元管理が可能であること。に着目した。合わせて、航空機で使用するために極めて重要となる電波法と航空法の使用条件を満たしたすことを最も重要視し開発を行った。使用条件は以下のとおりである。

- 電波法（総務省電波衛星移動通信課と確認）
  - 技術基準適合証明を取得したイリジウム通信モジュールとアンテナを使用。
  - 電波法で定める技術基準適合証明ならびに電気通信端末機器として適合認定を取得した設備で、電波法に基づく無線免許を取っていると認められた端末。
- 航空法（航空局安全部空港安全・保安対策課と確認）
  - 航空法施行規則第 164 条の 15 第 4 号「航空機の運航の安全に支障を及ぼすおそれのある電子機器」ではない事を証明するために RTCA（Radio Technical Committee for Aeronautics：米国航空無線通信委員会）に準ずる 3 つの試験①DO-160 Section21（放射妨害試験）、②RTCA DO-294 強電界試験、③RTCA DO-294 弱電界試験を実施（図 16）し「運航の安全に支障を及ぼすおそれがないため、航空法施行規則第 164 条の 15 第 4 項に規定する電子機器に該当しない」ことを確認。

航法装置	最小受信感度(dBm)	Citation 560受信端電力(dBm)	AS350B受信端電力(dBm)
VHF	-93	-109.3	-97.8
VOR/LOC	-86	-137.6	-112.7
GS	-87	-120.1	-
DME	-83	-140.7	-110.5
Transponder	-75	-121.6	-133.9
TCAS	-74	-116.3	-

図 16 電磁干渉試験の概要

FOSTER-CoPilot の導入により、図 17 に示すとおり航空機の位置情報を把握できるようになり、加えて、運航判断に欠かすことのできない、雨雲レーダーやライブカメラなどの航空気象コンテンツと位置情報を一元的に監視することで、より安全な運航判断の実現につながっている。

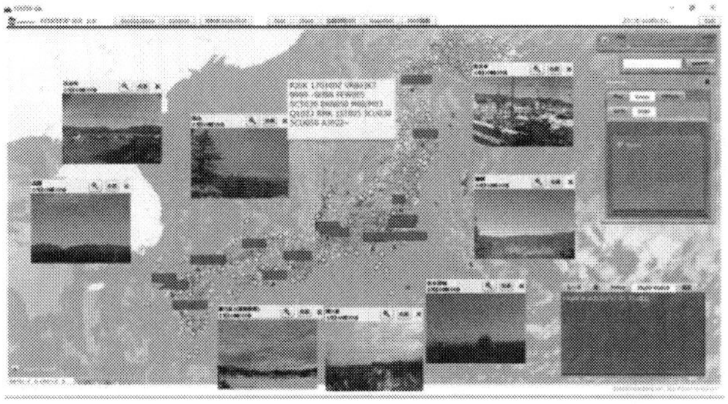

図 17 航空気象コンテンツと航空機の位置情報（赤の帯が機体の表示機番名）

FOSTER-CoPilot を導入することによって位置情報を地上側でリアルタイムに把握することができるようになった一方で、利用者からのニーズとして、①無線の不感地帯、着陸後の圏外エリアにおける地上と機上のコミュニケーション対策。②機上におけるナビゲーションと気象情報の組み合わせ。が顕在化した。ウェザーニューズでは、FOSTER-CoPilot を活用して、地上と機上のコミュニケーション、および機上での自機位置の把握、目的地設定を可能とするアプリケーション FOSTER-NAV（図 18）を開発し、2016 年 8 月に運用試験を行った。

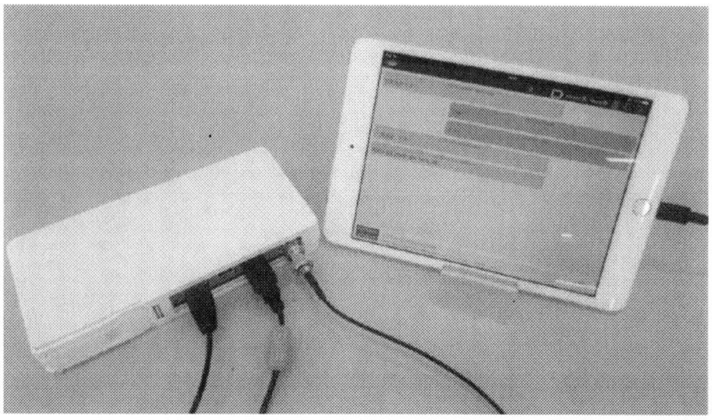

図 18 FOSTER-CoPilot（左）と FOSTER-NAV（右）

FOSTER-NAV とは、従来の FOSTER-CoPilot にディスプレイ用のタブレットを接続し運用するアプリケーションである。主な機能しては、①図 19 に示すイリジウム衛星通信を介した地上と機上のメッセージ送受信。②GPS モジュールで測位した自機位置のタブレット表示。③自機位置から目的地（ランデブーポイント）までの方向・距離・所要時間の表示。④地上のインターネット環境化で取得した気象コンテンツの保存・再表示である。

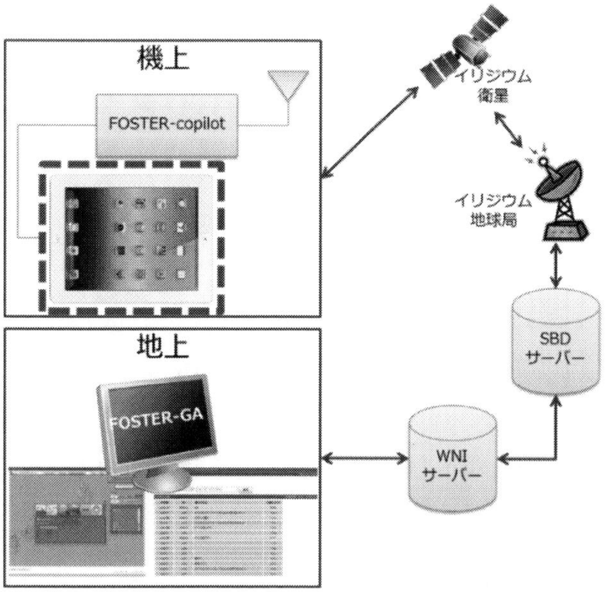

図 19 FOSTER-NAV 構成図

運用試験の結果、①離陸直前の気象情報の確認が可能な点。②機上にて詳細地図が確認できる点。③無線の届かいないエリアにおいてテキスト通信を用いたコミュニケーションの実現。という点で有効性が確認できた。一方で、持ち込み型であるが故の機内での簡易設置方法や気象情報の充実が課題に上がった。

結論

航空機の安全運航において、気象は極めて重要な要素である。ゲリラ雷雨・竜巻や短時間強雪など局地的で激甚な気象現象が多数発生する昨今、従来の気象観測データだけでは不十分な場合がある。局地的な現象や従来レーダーのブラインドエリア内の現象を捕捉することを得意とする WITH レーダーや航空機の運航視点で設置した峠の茶屋などの気象観測インフラを駆使しつつ、持ち込み型の小型動態監視システム FOSTER-CoPilot によって航空機の位置情報を把握することで、更なる安全運航が実現できると考える。

Author contact: Mizuki Nishimura nish-m@wni.com

Physics Based Models for Helicopters in Maneuvering Flight

Ritu Marpu Eshcol
Siemens PLM
Orlando, FL, USA

Lakshmi N Sankar
Regents Professor
Georgia Institute of Technology
Atlanta, GA, USA

ABSTRACT

The aeromechanical loads developed during selected revolutions of severe diving turn maneuver 11679 is studied. The diving turn maneuver (11680), ranked the most severe out of all the flight conditions recorded in the UH-60A flight database, experiences a peak load factor of nearly 1.8 during the maneuver and is characterized by the most severe torsional and pitch-link loads. Time histories of aerodynamic sectional normal loads and pitching moments, blade structural loads, and pitch-link loads are presented. The rotor experiences extensive occurrence of stall including an advancing blade stall phenomenon for this maneuver. The fundamental mechanism behind this stall occurrence is explored.

INTRODUCTION

Maneuvers are critical aspects of rotor and hub design. The strength and durability of rotor hub components is dependent on the accurate prediction of peak-to-peak structural loads at the blade root. Accurate knowledge of control loads is important for sizing the expensive swash-plate components, for fatigue. The component size is generally dictated by large loads that occur during extreme flight conditions. Rotor components need to be redesigned if the loads occurring during extreme maneuvers are outside of the design load spectrum, or else a shorter lifespan must be accepted for those components.

Incorrect estimation of the fatigue load spectrum for rotor components can be detrimental to an aircraft design program. The rotor oscillatory loads define the fatigue loading on the rotating system components. Oscillatory loads are defined as 1P and higher in the rotating system. They are also the major source of airframe vibratory loads. The vibratory loads that are passed on from the rotating system to the fixed system cause fatigue damage for fixed-system components.

The controls loads are primarily dependent on the accuracy of torsional response which is the most difficult to be captured by existing, state-of-the-art predictive tools. The peak-to-peak loads in the fixed system are dependent on the harmonic content of the rotating system loads. For example, for a 4-bladed rotor, the 4P swash-plate servo loads in the fixed system are dependent upon the 3-5P pushrod loads in the rotating system. Although accurate prediction of the peak-to-peak pushrod loads is adequate for sizing the pushrod, the harmonic content of pushrod loads must be accurately predicted to capture the 4P servo loads.

Presented at the 6th Asian/Austrailian Rotorcraft Forum / Heli Japan 2017, Kanazawa, Japan, November 7-9, 2017.

The rotor aeromechanics prediction tools must therefore be equipped to accurately predict mean, peak-to-peak and higher harmonic content of rotating system loads. This forms the motivation for the current work.

RESEARCH OBJECTIVES

The thrust of this research is to investigate important physical phenomena that cause severe loads on the rotor in steady and maneuvering flight conditions. The secondary goal is to improve the prediction quality of the harmonic content of the rotating system loads by accurately predicting key aerodynamic phenomena.

UH-60A FLIGHT TEST PROGRAM

The NASA-Army UH-60A Airloads Program [1] has provided the highly researched and most extensive rotor flight test database, which includes pilot input, vehicle attitudes, blade airloads and structural loads. A number of steady and maneuvering flight counters were flown and studied as a part of this program. A number of maneuvering flight conditions were carried out including level bank-turns, diving bank-angle turns, symmetric pull-ups, rolling pullouts and pushovers, UTTAS maneuvers among the others [1], [2]. A ranking system based on a subset of six measurements was developed to rank the maneuvers based on severity of the loads developed. The six measurements are: the pitch-link load, torsional moment at 0.3R, and the flap and chord bending moment measurements at 0.113R and 0.6R. During the diving turn 11680, the helicopter developed the most severe push-rod loads, torsion moment at 0.3R and chord bending moment at 0.113R. The second most severe maneuver is the longitudinal pull-up (11029) which had the highest flapwise-bending moment at 0.113R.

COMPUTATIONAL MTHODOLOGY

The CFD methodology used in this study is GT-Hybrid [3-9]. This analysis utilizes a hybrid methodology where the flow field near the blade is resolved through the Navier-Stokes solution, whereas the influence of the other blades and of the trailing vorticity in the far field wake are accounted for by modeling them as a collection of piece-wise linear bound and trailing vortex elements. The near wake is inherently captured in the Navier-Stokes analysis.

A multi-body dynamics analysis, DYMORE [10] is used for the structural analysis (CSD). The rotor blades are modeled as elastic beams with geometrically exact composite beam finite element formulation. DYMORE includes an internal lifting line based aerodynamic model and auto-pilot algorithm which can be used to perform a fully trimmed aeroelastic simulation of an isolated rotor configuration. A four bladed UH-60A rotor model is used for the simulations. The model includes blade, lag damper and push rod, rotor hub and swash-plate. A lifting line based 2-D table lookup aerodynamics with dynamic wake model is used for computing airloads internally at the start of the simulations. The airloads from the table-look up are replaced with the CFD loads (applied as "delta" externally applied forces) during the second and subsequent iterations. Typically 5 to 10 coupled iterations are needed for the aerodynamic effects from the CFD side, and the elastic deformations and trim adjustments from the comprehensive analysis, to converge. In the present study, these coupled aeroelastic simulations were computed over 10 hours of wall clock time on an 8 node Linux cluster.

RESULTS AND DISCUSSION

Calculations have been done for two dive turn maneuvers shown in figure 1 below, and a pull-up maneuver using the hybrid methodology above. For brevity, this paper only discusses the 11679 dive turn maneuver.

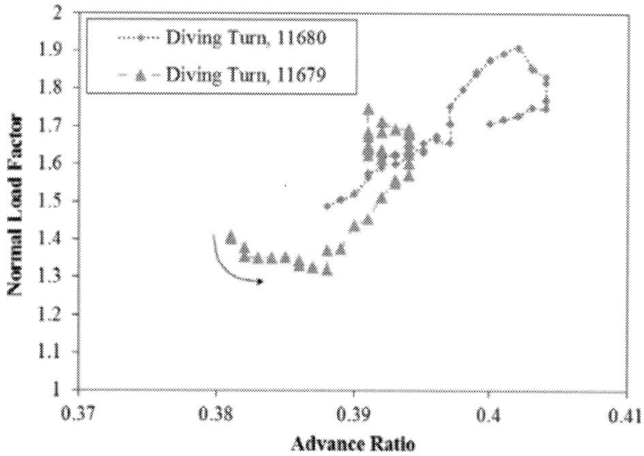

Figure 1 Advance ratio vs load factor for 11680 and 11679

Figure 2 shows the time history of angular rates for the dive turn maneuver C11679 based on rotor revolutions. While the yaw rate is relatively constant throughout the maneuver, the pitch and roll rates change considerably. At the start of the maneuver, the helicopter has a pitch rate of 6.26 degrees/s. The vehicle experiences a peak pitch rate at Rev 15 with a peak average pitch rate of 9.4 degrees/s. The load factor changes relatively slowly over the first 10 revolutions as shown in Figure 2 (b). The vehicle attitudes, angle of attack and sideslip angle are shown in Figures 2 (c) and (d). The vehicle pitch attitude does not coincide with the vehicle angle of attack, the difference being the flight path angle for this maneuver. The sideslip is nominally zero throughout the maneuver with the vehicle roll attitude varying between a minimum of 55 degrees and maximum of 60 degrees. The alternating pitch-link loads (half peak-to-peak) during the dive turn maneuver C11679 gradually increase reaching a peak at Rev 28, as shown in Figure 3. Lowest half peak-to-peak pitch-link loads are found at Rev 10. The slight variation in rotor RPM was accounted for in the structural and aerodynamic models.

SECTIONAL AIRLOADS, C11679

Blade sectional airloads consisting of sectional normal loads and pitching moments for the first 7 revolutions are shown in Figure 4 - Figure 7. The mean loads as well as the waveform are captured reasonably well for this extreme maneuver. The stall characteristics for Revs 1-7 remain the same except for differences in magnitude of the loads. For Maneuver 11680, the starting revolutions are characterized by a high load factor of nearly 1.9, which causes the rotor to operate in complex aerodynamic environment starting from maneuver initiation. Unlike the first few revolutions of Maneuver 11680, there is no advancing transonic stall observed for the starting revolutions of Maneuver 11679, since the normal load factor is less than 1.5. It is observed that although the inboard stations give accurate loading results at the 0 and 180 (nose) azimuth positions, stations from 0.675R to the tip have a phase difference between simulated and flight test data. Although mean loads are predicted reasonably well, the peak loads especially at the 0.675R and 0.775R are not adequately captured. Lift stall is observed beginning in the fourth quadrant, continuing into the first quadrant at radial locations between 0.675R and 0.865R, as seen in Figure 5.28. This stall has been predicted with a phase lag of about 12 degrees compared to flight test data.

The pitching moment data without removing the means have been plotted in Figures 6 and 7. In Figure 6, at the inboard stations 0.225R and 0.4R, negative pitching moments are observed at 170 degrees azimuth over the nose of the vehicle likely due to increased angle of attack resulting from flow over the fuselage. This effect is not accurately captured in the predictions, since the fuselage is not being simulated. Moment stall for this flight condition is observed at almost all the radial stations. At 0.225R, moment stall occurs between 140 and 160 degrees as indicated by flight test data. At 0.4R moment stall is observed at about 170 degrees in both measured and predicted data. At the outboard locations of 0.675R to 0.775R, moment stall associated with first dynamic stall cycle is seen in the flight test data near 210 deg. The first dynamic stall event is three-dimensional in nature and sets in at 0.775R and becomes more pronounced as one goes outboard in the span-wise direction. A second stall cycle occurs between 330 and 360 deg.

Rev 20 of Maneuver 11679 is characterized by peak-load factor of 1.7g, and an advance ratio 0.3 for Maneuver 11679. Therefore, Rev 20 has been studied in detail. The sectional normal loads and pitching moments at selected stations for this revolution are presented in Figure 8. In this revolution, the advancing side stall event is seen in addition to stall cycles on the retreating side. At 0.775R the first stall vortex sets in at about 220 degrees azimuth. The sectional airloads for this revolution are very similar to the aerodynamic loading observed in Rev 12, Counter 11680. Multiple and distinct localized stall cycles are observed toward the blade tip region. Over the retreating side, there is a large region of separated flow, which extends to the first quadrant, following which there is advancing side stall seen at about 45 deg azimuth, similar to that found in Rev 12, Counter 11680.

For a high thrust flight condition (C9017) of the UH-60A consisting of two stall cycles on the retreating side, a study by Datta et al. [11] showed that the first retreating blade stall was caused by high trim settings (collective and longitudinal control) which led to high local angles of attack on the retreating blade. If the trim state is correctly computed, the first stall event may be captured. His study also showed that the second stall event which occurs a few degrees of azimuth behind the first stall event is due to the 5P elastic twist deformation. In order to capture the second stall event, it is important to employ accurate aerodynamic and turbulence models so that the flow reattachment and separation is captured adequately. Very often, the flow may not reattach leading to the merging of first and second stall events indistinguishable from each other, spreading over a large region on the retreating side. The third stall event in the diving maneuvers is associated with supercritical flow over the front of the leading edge [12]. Accurate prediction of the retreating stall cycles is important for capturing the advancing side stall which is affected by a right combination of control angles, elastic twist as well as effective angle of attack.

STRUCTURAL LOADS, C11679

Normal bending moments at 0.3R and 0.5R blade sections for the first 6 revs have been plotted in Figure 9. The first thing that is observed is that the load behavior between one revolution and the other are repetitive showing periodicity in solution. The peak-to-peak and mean values of the normal bending moment have been captured reasonably well. The time histories of normal bending moments are captured well in magnitude, however there is a phase difference observed. Similar to the previous maneuver, there is a spike observed in normal bending moment time history at 0.3R location at the beginning of every rotor revolution. This may possibly be due to over-prediction in lift coefficient at this azimuthal location (ψ=0) or an artifact of solution restart for every revolution. The waveform and peak-to-peak normal bending moments at 0.5R are captured well and the prediction results are consistent with flight test data at this radial station. There is an over-prediction of compressive push-rod loads in the third quadrant between ψ = 180° and ψ = 225°.

SUMMARY OF RESULTS

1. The mean and waveform of sectional normal loads and pitching moments were captured moderately well despite the severity of the maneuver. In particular, the peak-to-peak sectional airloads at inboard locations are modeled reasonably accurately.

2. Due to the combination of high advance ratio (0.4) and high normal load factor (1.8), there is extensive occurrence of dynamic stall phenomena. The interaction of dynamic stall vortex with supercritical flow is critical for this maneuver, and crucial for the prediction of the advancing side third stall event at 0.675R to 0.775R radial locations. This was not captured by the current simulations.

3. The peak-to-peak structural loads and the mean loads are accurately predicted. The harmonic content of structural loads with harmonics 1P to 3P is well-captured.

4. The prediction quality of harmonic content of structural loads varies with radial location. For instance, for the torsional moments, the 4P harmonic 4P is well predicted at 0.5R location, while it is poorly predicted at 0.3R location.

5. The peak-to-peak pitch-link loads were reasonably well-predicted and the time history of pitch-link loads correlates well with the time history of blade torsional moment at 0.3R. However, there is an inconsistency in the prediction of the pitch-link loads at ψ = 180°.

6. Harmonics 4P to 6P of structural loads were not accurately captured. However, the higher harmonic content (greater than 7 per rev) of pitch-link loads was adequately captured.

CONCLUSIONS

Based on the studies performed, the following conclusions may be drawn:

1. The hybrid methodology in conjunction with the multibody dynamics code provide an economical means of modeling maneuvers.

2. For the maneuvers studied in this work, the loose coupling approach was found to be adequate to model all the necessary physical phenomena.

3. A 3rd stall event occurring in a region of super-critical flow is observed in the first quadrant at 92% radial location on the rotor blade, during rotor revolutions characterized by peak load factor of both

Maneuver 11679. The current CFD/CSD methodology is able to capture this event.

4. The mean and the peak-to-peak structural loads occurring in Maneuver 11679 were well-predicted. However, the retreating blade wave-form was not captured well. The structural model will further need to be improved in order to better capture the retreating side events.

Author contact: Lakshmi N Sankar lsankar@ae.gatech.edu

ACKNOWLEDGMENTS

This work was funded by the Vertical Lift Consortium, formerly the Center for Rotorcraft Innovation and the National Rotorcraft Technology Center (NRTC), U.S. Army Aviation and Missile Research, Development and Engineering Center (AMRDEC) under Technology Investment Agreement W911W6-06-2-0002, entitled National Rotorcraft Technology Center Research Program. The authors would like to acknowledge that this research and development was accomplished with the support and guidance of the NRTC and VLC. The views and conclusions contained in this document are those of the authors and should not be interpreted as representing the official policies, either expressed or implied, of the AMRDEC or the U.S. Government. The U.S. Government is authorized to reproduce and distribute reprints for Government purposes notwithstanding any copyright notation thereon.

REFERENCES

1. Bousman, W. G., and Kufeld, R. M. UH-60A airloads catalog: National Aeronautics and Space Administration, Ames Research Center, 2005.
2. Kufeld, R. M., and Bousman, W. G. "High load conditions measured on a UH-60A in maneuvering flight," Journal of the American Helicopter Society Vol. 43, No. 3, 1998, pp. 202-211. http://dx.doi.org/10.4050/JAHS.43.202.
3. Rajmohan N., Sankar L.N., Bauchau O., Makinen S.M., Egolf T.A. and Charles B.D., "Application of Hybrid Methodology to Rotors in Steady and Maneuvering Flight," AHS 64th Annual Forum, Montreal, Canada, 2008.
4. Rajmohan, N., Manivannan, V., Sankar, L.N., Costello. M., "Development of a Methodology for Coupling Rotorcraft Aeromechanics and Vehicle Dynamics to study Helicopters in Maneuvering Flight," American Helicopter Society Annual Forum 09, Dallas, TX, May 27 – 29, 2009.
5. Rajmohan, N., "Application of Hybrid Methodology to Rotors in Steady and Maneuvering Flight", Ph.D. Dissertation, Georgia Institute of Technology, 2010.
6. Rajmohan N., Sankar L.N., Costello M., "Effect of Inflow Model on Coupling between Aeromechanics and Flight Mechanics", 49th AIAA Aerospace Sciences Meeting, Orlando, Florida, Jan 2011.
7. Rajmohan N., Marpu R., Sankar L.N., Baeder J.D., Egolf T. A., "Improved Prediction of Rotor Maneuvering Loads using a Hybrid Methodology," AHS 67th Annual Forum, Virginia Beach, VA, 2011.
8. Marpu R., Sankar L.N., Stephen Makinen., Baeder J.D., "Computational Modeling of Diving-turn Maneuvers using Hybrid Methodology," AHS 68th Annual Forum, Fort Worth, TX, 2012
9. Marpu R., Sankar L.N., Makinen Stephen, Baeder J.D., Egolf T. A., Wasikowski M., "Physics Based Modeling of Maneuver Loads for Rotor and Hub Design", AHS Specialist's Conference Design, San Francisco, CA, Jan. 23-25, 2012.
10. Bauchau, O. A., "Computational Schemes for Flexible, Nonlinear Multi-body Systems", Multibody System Dynamics, Vol. 2, 1998, pp. 169-225.
11. Datta, A., and Chopra, I. "Validation and understanding of UH-60A vibratory loads in steady level flight," Journal of the American Helicopter Society Vol. 49, No. 3, 2004, pp. 271-287. http://dx.doi.org/10.4050/JAHS.49.271.
12. Bousman, W. G. "A qualitative examination of dynamic stall from flight test data," *Journal of the American Helicopter Society* Vol. 43, No. 4, 1998, pp. 279-295. http://dx.doi.org/10.4050/JAHS.43.279

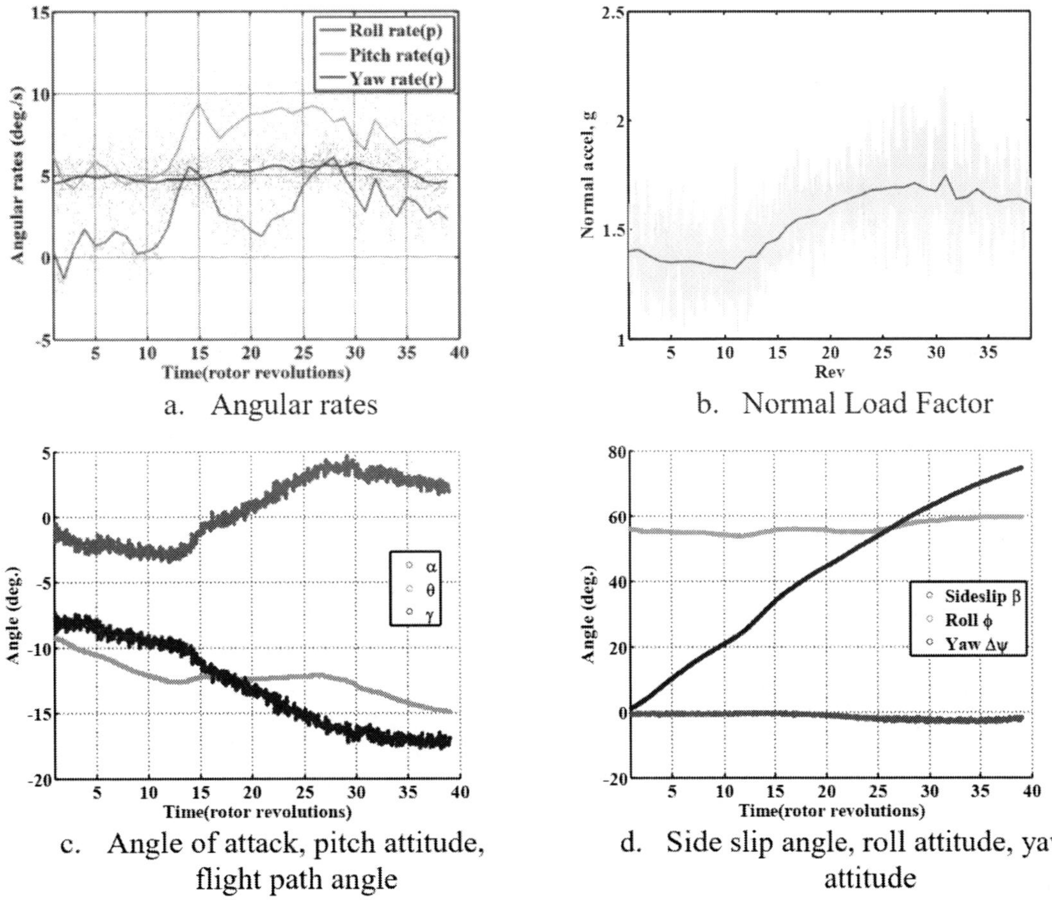

a. Angular rates

b. Normal Load Factor

c. Angle of attack, pitch attitude, flight path angle

d. Side slip angle, roll attitude, yaw attitude

Figure 2 Vehicle attitudes in the inertial frame, for 11679.

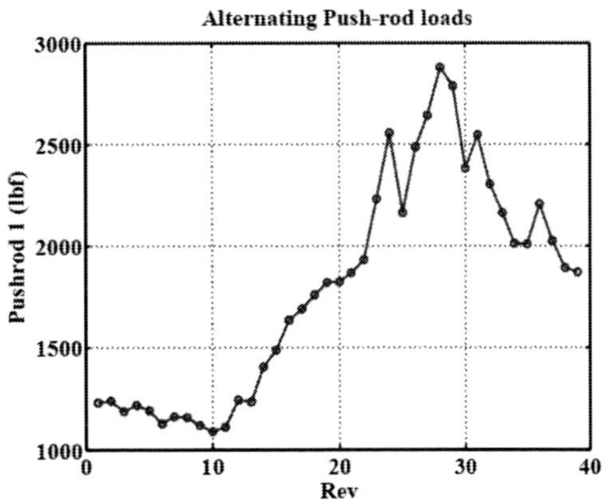

Figure 3. Alternating push-rod loads, for 11679.

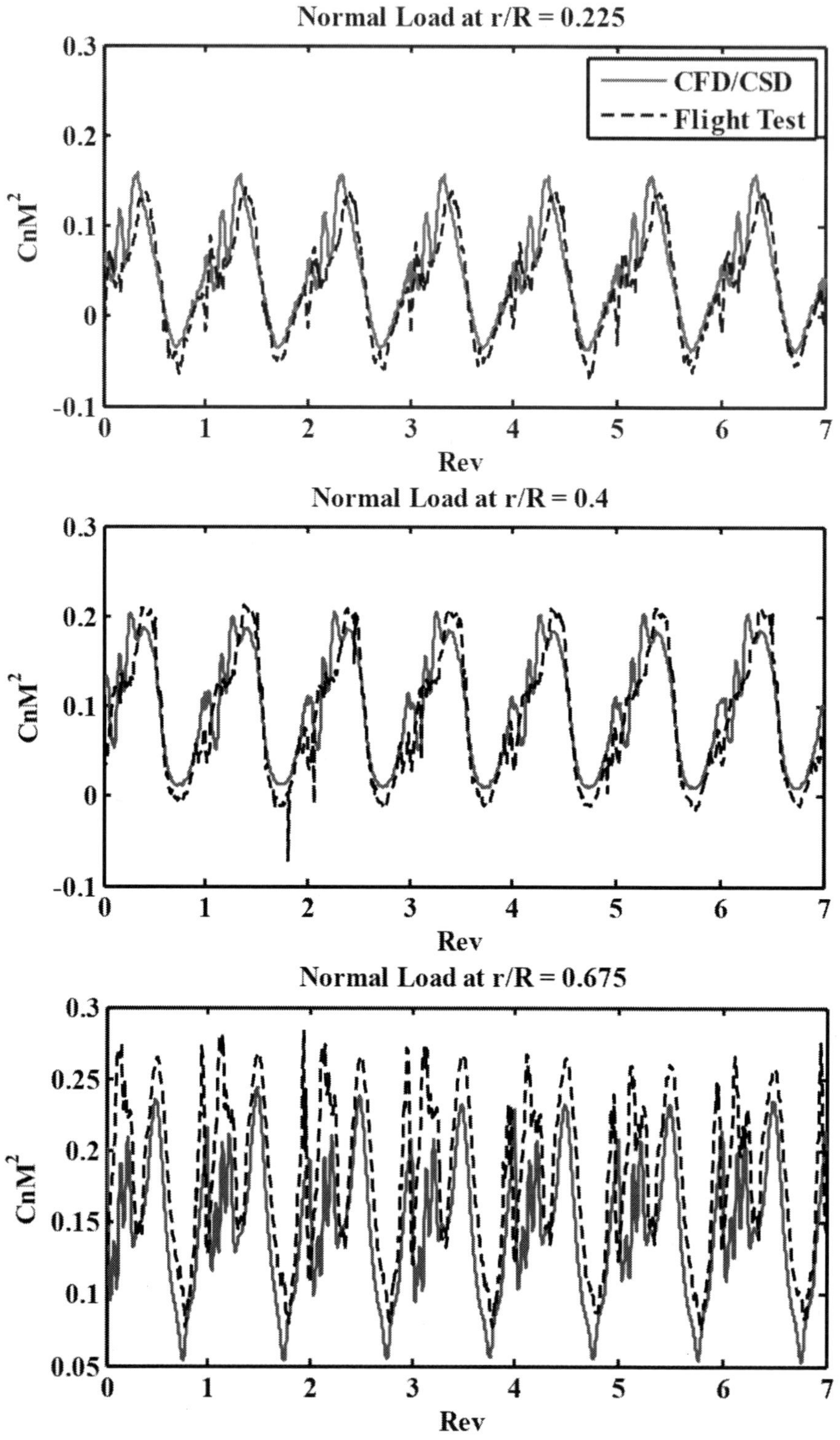

Figure 4. Sectional Normal Loads for Revs 1-7 of C11679, at 0.225R, 0.4R and 0.675R respectively.

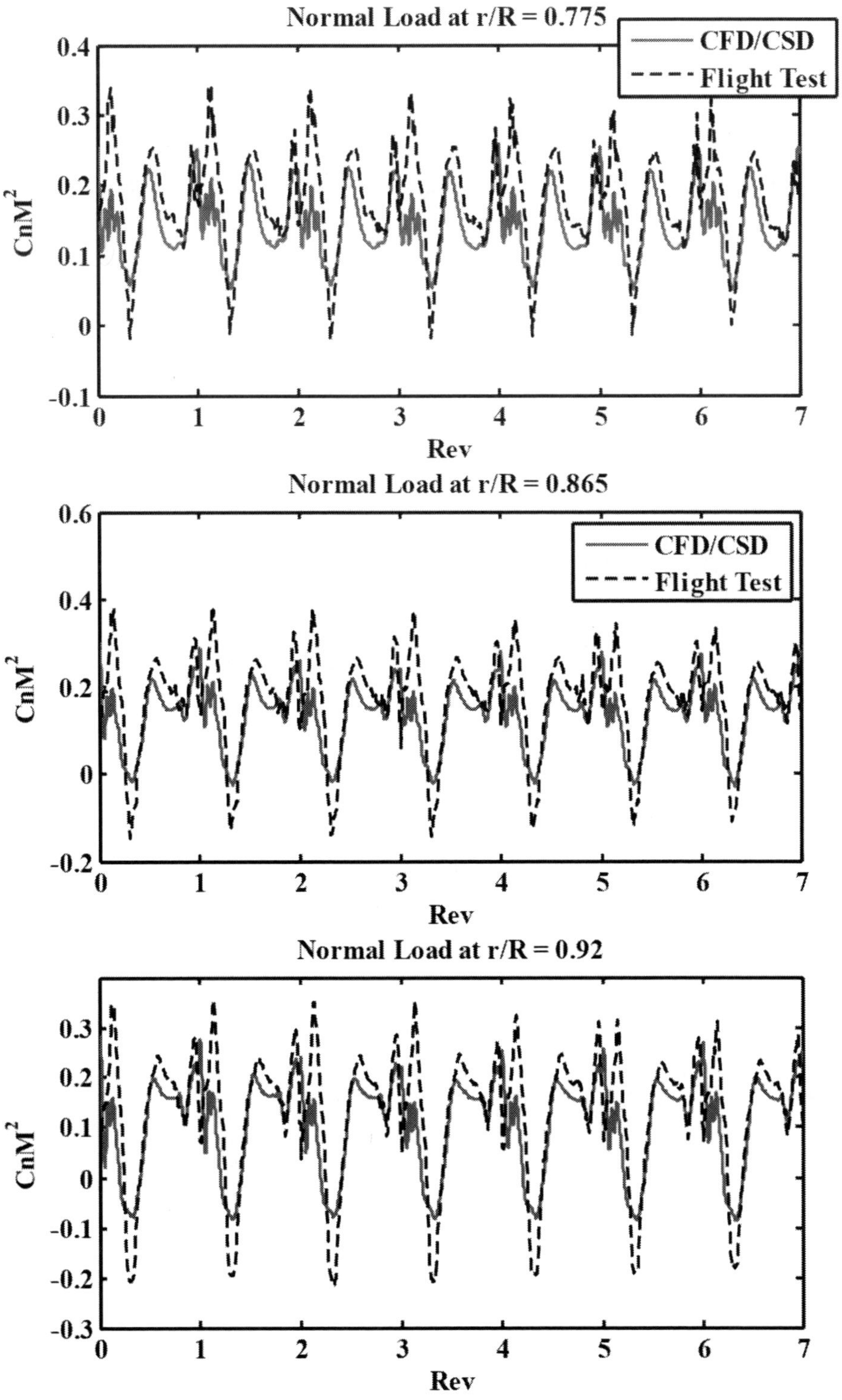

Figure 0. Sectional Normal Loads for Revs 1-7 of C11679, at 0.775R, 0.865R and 0.92R respectively.

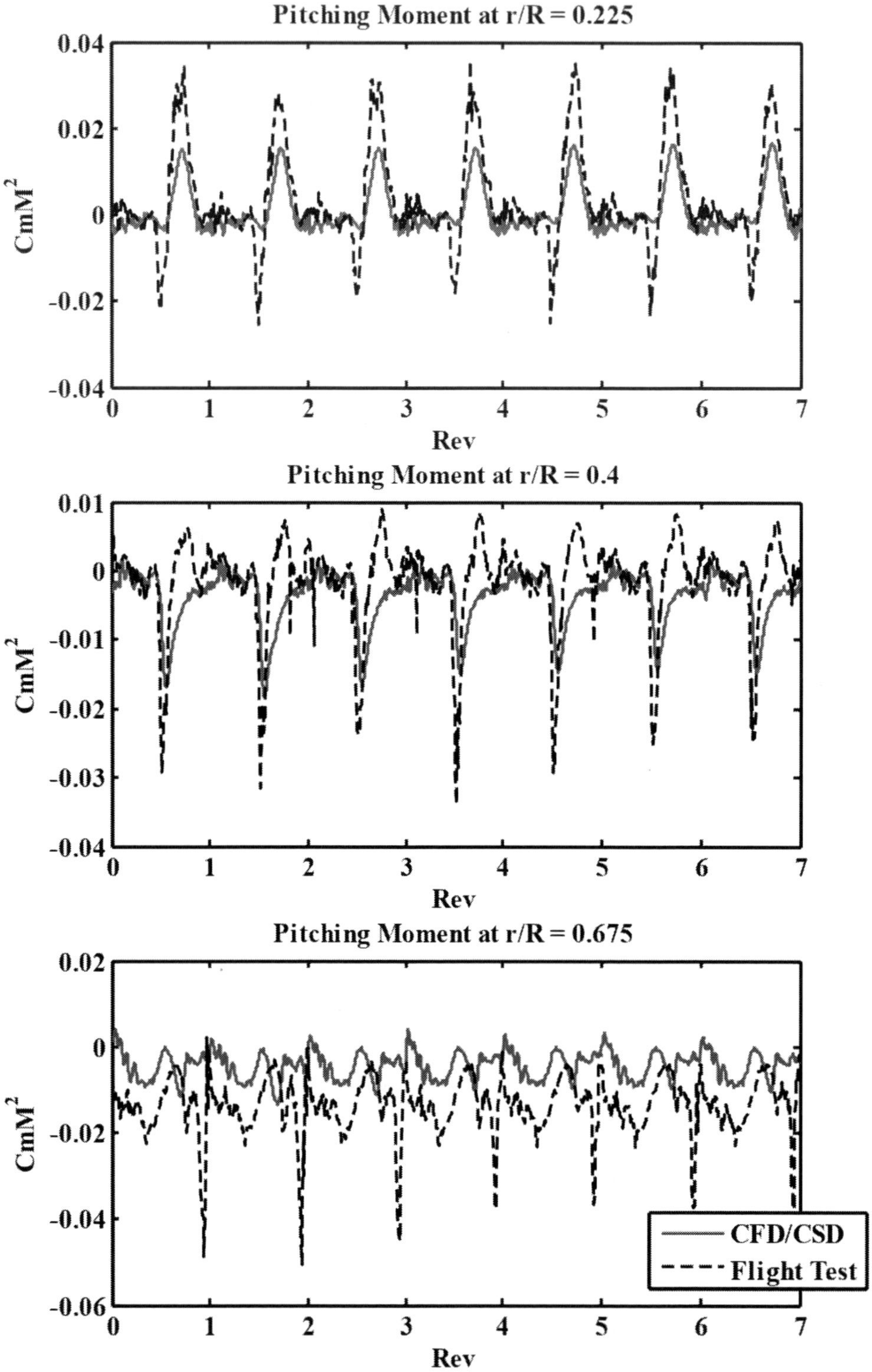

Figure 6. Sectional Pitching Moments for Revs 1-7 of C11679, at 0.225R, 0.4R and 0.675R respectively.

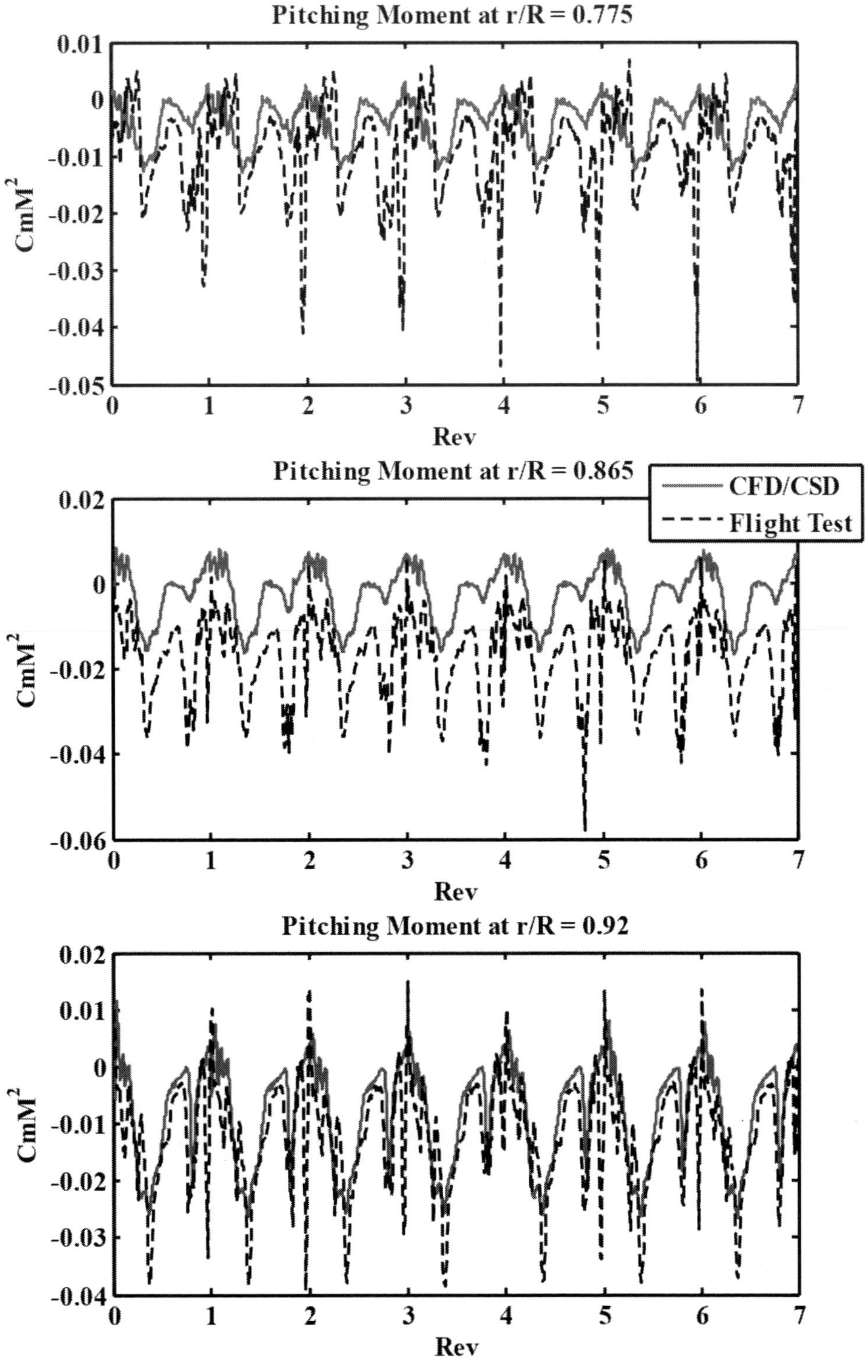

Figure 7. Sectional Pitching Moments for Revs 1-7 of C11679.

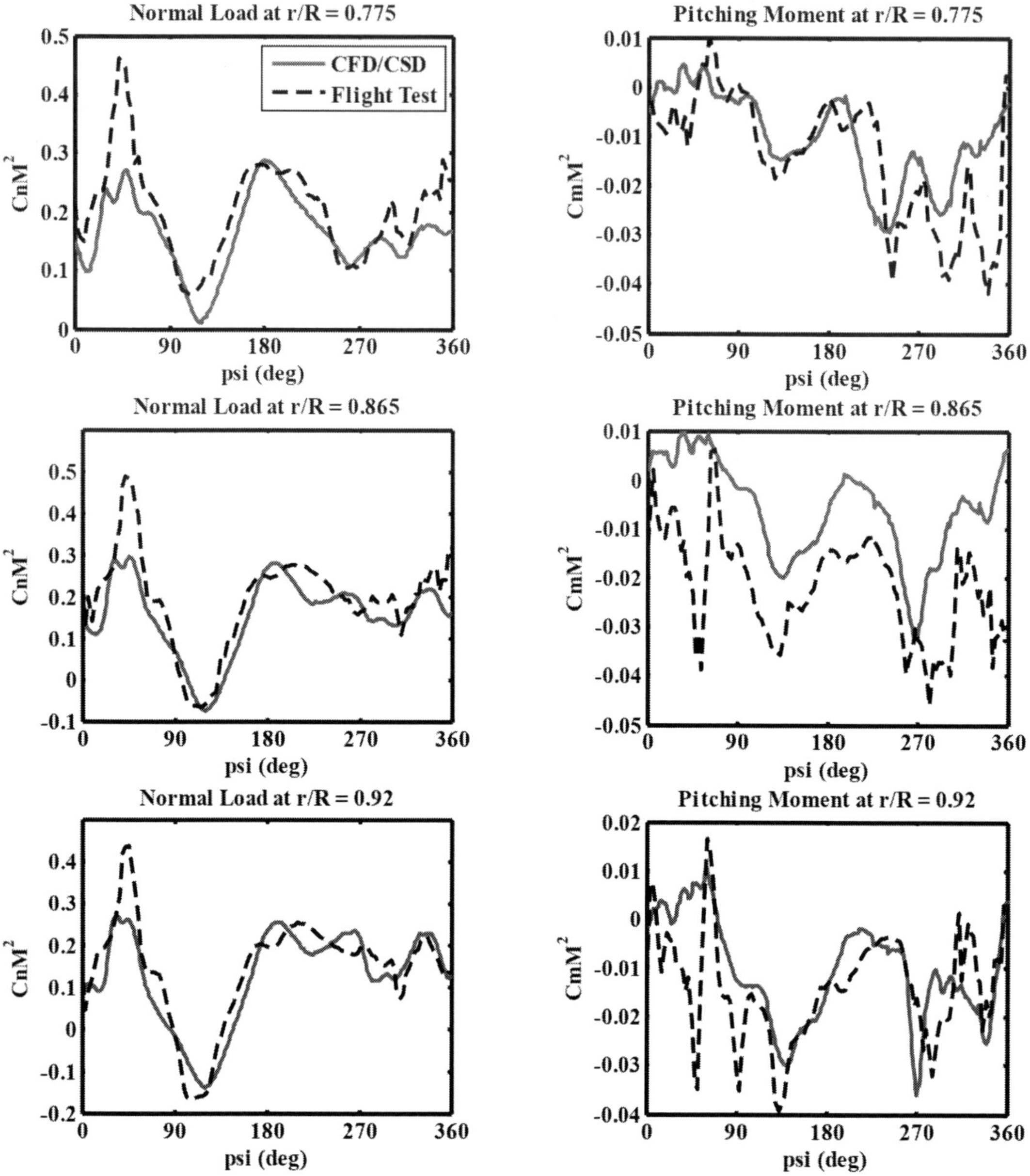

Figure 8. Sectional normal loads and pitching moments for Rev 20 characterized by normal load factor of 1.7, 11679

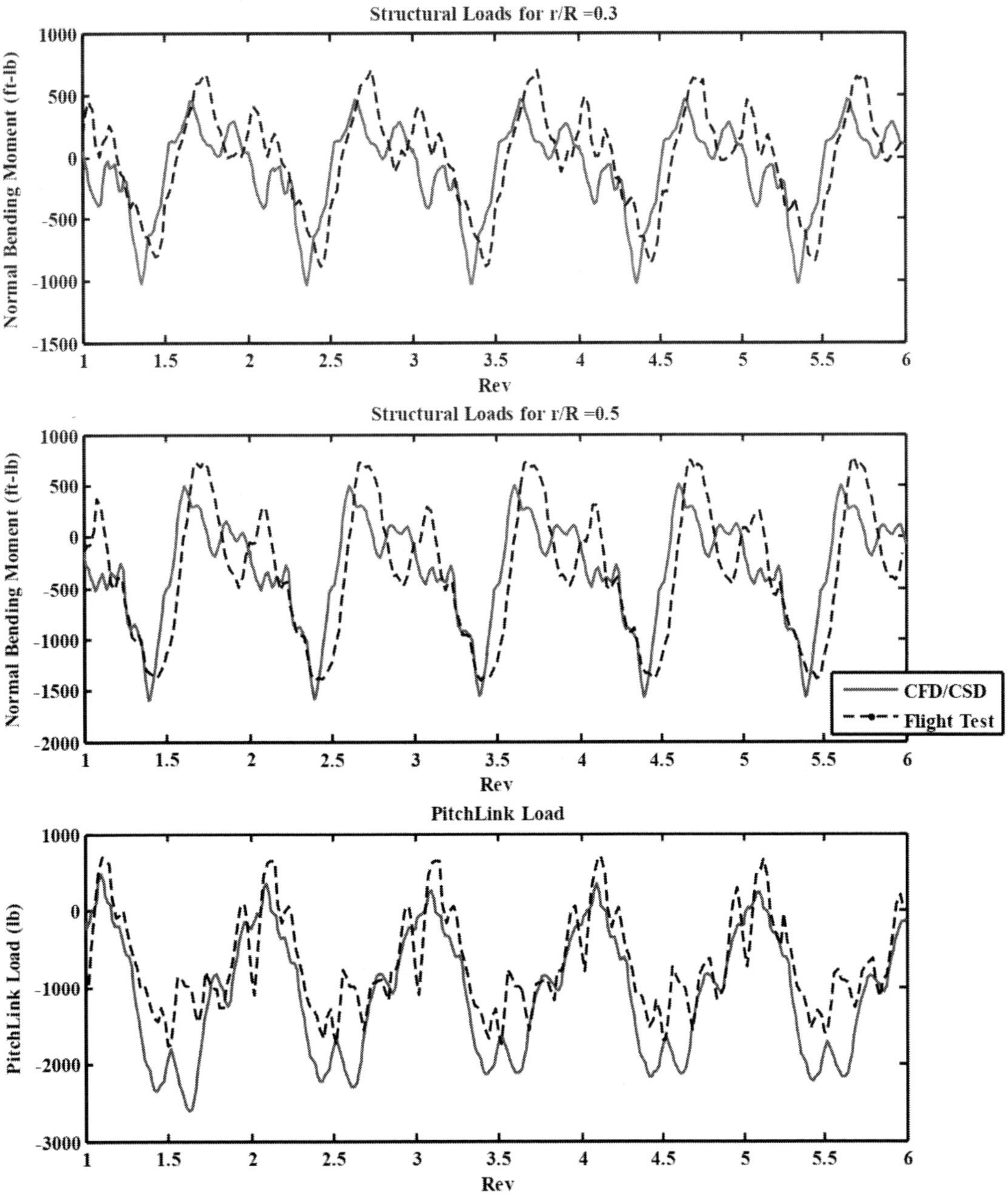

Figure 9. Blade structural loads for Revs 1-6 of C11679.

A Study of Rotor/Wing Aerodynamic Interaction at High Speed Flight on a Compound Helicopter

Hideaki Sugawara
Research Engineer
Ryoyu Systems Co., Ltd.
Nagoya, Aichi, JAPAN

Yasutada Tanabe
Associate Senior Researcher
Japan Aerospace Exploration Agency
Mitaka, Tokyo, JAPAN

ABSTRACT

Rotor/wing aerodynamic interaction is considered one of the key technology issues for compound helicopter design. Understanding of the influence on the rotor and the wing aerodynamic characteristics by the aerodynamic interaction is significantly important. This paper describes about the investigation of rotor/wing interaction through a numerical simulation using a rotorcraft CFD solver, rFlow3D, developed at JAXA. A simplified computational model with a rotor and a wing is considered. A rectangular wing model is added to the UH-60A helicopter with the wing size for the original maximum takeoff weight. To investigate the influence of rotor/wing aerodynamic interaction, the isolated rotor and the isolated wing configurations are firstly simulated and then compared to the rotor/wing combined configuration. Two flight conditions at the cruising speed and at the maximum speed are computed. The rotor speed is reduced to 75% RPM at high speed flight condition. The numerical simulation result indicates that the total effective drag of the rotor/wing configuration is increased than the simple summation of the isolated models. The wing lift is reduced by rotor/wing aerodynamic interaction, and the required rotor thrust is increased to keep the same total lift. The sectional normal force on the wing has an asymmetric distribution. The sectional lift is significantly reduced and the sectional drag is increased on the wing side below the rotor advancing side. On the wing side below the blade retreating side, the rotor has much slight effects on the sectional lift and drag on the wing.

NOTATION

ρ_∞	air density
a_∞	speed of sound
U_∞	freestream velocity
R	rotor radius
Ω	rotor rotational speed
c	blade chord
A	rotor disk area, πR^2
σ	rotor solidity
μ	advance ratio
θ_0	blade collective pitch angle
θ_{1c}	blade lateral cyclic pitch angle
θ_{1s}	blade longitudinal cyclic pitch angle
c_w	wing chord
b	wing span
S	wing area
C_T	rotor thrust coefficient, $T/\rho_\infty A(\Omega R)^2$
C_Q	rotor torque coefficient, $Q/\rho_\infty A(\Omega R)^2$
C_D	rotor drag coefficient, $D/\rho_\infty A(\Omega R)^2$
C_{DE}	rotor effective drag coefficient, $C_Q/\mu + C_D$
C_X	X-force coefficient, $F_X/\rho_\infty A(\Omega R)^2$
C_Y	Y-force coefficient, $F_Y/\rho_\infty A(\Omega R)^2$
C_Z	Z-force coefficient, $F_Z/\rho_\infty A(\Omega R)^2$
C_{MX}	rolling moment coefficient, $M_X/\rho_\infty A(\Omega R)^2 R$
C_{MY}	pitching moment coefficient, $M_Y/\rho_\infty A(\Omega R)^2 R$
C_{MZ}	yawing moment coefficient, $M_Z/\rho_\infty A(\Omega R)^2 R$
C_{DET}	total effective drag coefficient, $C_{DE} + C_{X_{wing}}$
C_nM^2	sectional normal force coefficient, $f_z/(1/2\rho_\infty a_\infty^2 c)$
c_l	sectional lift coefficient
c_d	sectional drag coefficient
c_x	sectional X-force coefficient, $f_x/(1/2\rho_\infty U_\infty^2 c_w)$
c_z	sectional Z-force coefficient, $f_z/(1/2\rho_\infty U_\infty^2 c_w)$

Presented at the 6th Asian/Austrailian Rotorcraft Forum / Heli Japan 2017, Kanazawa, Japan, November 7-9, 2017.

INTRODUCTION

A compound helicopter with a single main rotor and a wing is considered one of the high-speed rotorcraft concepts. Recently, Japan Aerospace Exploration Agency (JAXA) proposed a compound helicopter concept as shown in Figure 1. This rotorcraft is constructed with a single main rotor, a main wing and multiple propellers. Tanabe (Ref. 1) suggested that there are several key technology issues for this type of compound helicopter. Within them, rotor/wing aerodynamic interaction is one of the key technology issues. The wing download is caused by the rotor downwash because the wing is generally placed beneath the rotor. Therefore, understanding the change of the aerodynamic characteristics by rotor/wing aerodynamic interaction is necessary for the design of compound helicopter.

Figure 1. Conceptual drawing of the compound helicopter.

There are several researches about rotor/wing aerodynamic interactions (Refs. 2-3) so far, but the influence of aerodynamic characteristics by rotor/wing interaction in forward flight condition has been little reported. It is important to understand the rotor/wing aerodynamic interaction in forward flight, especially at high speed flight.

In JAXA, a rotorcraft CFD solver, rFlow3D, has been developed (Refs. 4-5). The authors validated the aerodynamic prediction accuracy of rotor performance at high advance ratio based on full-scale UH-60A rotor (Ref. 6). The prediction results showed good correlation with the test data especially for the effective drag of the rotor. The authors also applied rFlow3D to simulate the rotor/wing aerodynamic interaction in hover based on Felker's experiment (Ref. 3, 7). The computed wing download was overestimated, but the trends were predicted qualitatively agreeable with experimental data.

In this paper, simulation of rotor/wing aerodynamic interaction at high speed flight using rFlow3D is discussed. The computational model is based on UH-60A with simplified configuration of the rotor and the wing. Wing sizing is carried out based on the maximum takeoff weight of the UH-60A for this numerical simulation. The change of the aerodynamics of the rotor and the wing caused by the rotor/wing aerodynamic interaction is described.

COMPUTATIONAL MODEL

UH-60A rotor

The blade geometry of the full-scale UH-60A is created referring Refs. 8 and 9. The blade geometry is shown in Figure 2. In this blade, there is sweepback of 20 degrees at blade tip, and a trim tab. The trim tab angle is assumed of 0 degrees in the numerical simulation. The number of blades is 4, the rotor radius is 8.18 m, the nominal chord is 0.527 m, the blade twist is nonlinear distribution and the equivalent blade twist is -18 degrees. The blade airfoils are SC1095 and SC1095R8. The nominal rotor speed is 258 RPM, and the hovering tip Mach number is 0.650. Table 1 summarizes the UH-60A rotor parameters.

Mach number on the blade around the tip at the advancing side approaches the sound speed in high speed flight. Thus, the reduction of rotor speed is required at high speed flight condition. In this paper, target maximum speed is set to 500 km/h, and cruising speed is assumed about 420 km/h. Then, the rotor speed is reduced to 75% RPM at high speed flight which results in a high advance ratio about 0.703 at the cruising speed and 0.837 at the maximum speed. The advancing tip Mach number at the cruising and maximum speeds are 0.830 and 0.896, respectively. Table 2 shows the assumed conditions in high speed flights.

Figure 2. UH-60A blade geometry.

Table 1. UH-60A Rotor Parameters

Number of blades	4
Radius	8.18 m
Nominal chord	0.527 m
Equivalent blade twist	-18 deg.
Solidity	0.0826
Airfoil	SC1095/SC1095R8
Rotational speed	258 RPM

Table 2. Assumed Conditions in High Speed Flights

Rotor speed	193.5 RPM (75% RPM)
Flight speed	116.7 m/s (cruising speed) 138.9 m/s (maximum speed)
Advancing blade tip Mach number	0.830 (cruising speed) 0.896 (maximum speed)

Wing

Wing sizing is performed by reference to UH-60A information in Prouty's book (Ref. 10). Maximum takeoff weight is assumed of 97860.8 N as the same as UH-60A. The ratio of lift share between the rotor and the wing is supposed 1:1 at the cruising speed. Assuming the design lift coefficient as 0.5, the required wing area is determined as 12 m^2 from above assumptions. The wing chord length is determined as 1.048 m assuming the wing planform is rectangular and the wing span length is 0.7 times rotor diameter, and then the aspect ratio is 10.9. The wing airfoil is selected as NACA23015. The wing vertical installing position is determined as 0.25 times rotor radius beneath the rotor that is referred from the drawing in Prouty's book. The mounting angle of the wing is determined by the results of α-sweep thorough the numerical simulations. Figure 3 and Figure 4 show the lift coefficient and the lift to drag ratio variations, respectively. The lift coefficient of 0.538 at the angle of attack of 5 degrees is obtained, which is close to the design lift coefficient, and the lift to drag ratio can be seen at the maximum value within the computational results. Therefore, the mounting angle is selected as 5 degrees. Table 3 summarizes the wing specifications. Figure 5 illustrates the computational model of the rotor and wing configuration. Figure 6 shows the estimation of the lift share ratio between the rotor and the wing using the wing lift coefficient of the numerical simulation result at the angle of attack of 5 degrees assuming the rotor shaft angle is zero degrees. The cruising Mach number is about 0.34 and the maximum Mach number is about 0.41. The lift share ratios of the rotor to the wing are about 0.45 : 0.55 at the cruising speed, and about 0.22 : 0.78 at the maximum speed.

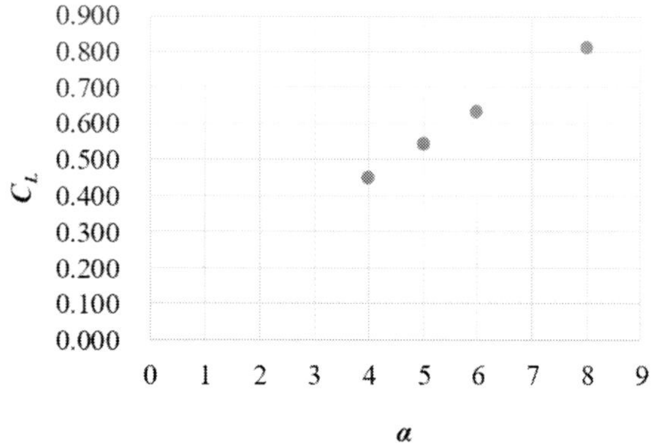

Figure 3. Lift coefficient vs. angle of attack.

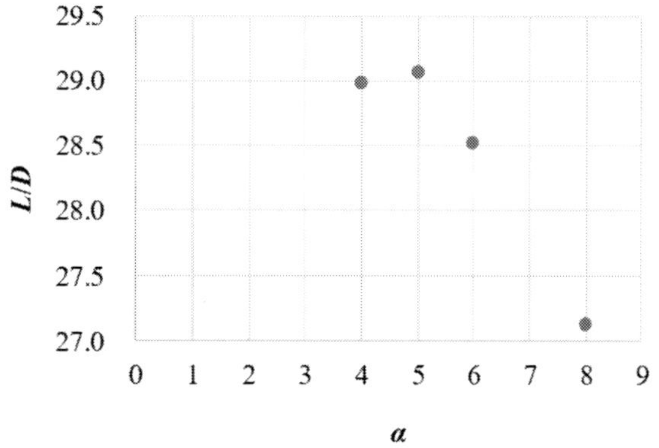

Figure 4. Lift to drag ratio vs. angle of attack.

Table 3. Wing Specifications

Span length	11.452 m
Chord length	1.048 m
Aspect ratio	10.9 m
Area	12.0 m²
Planform	Rectangle
Twist	0 deg.
Wing position	Mid wing
Mounting angle	5 deg.

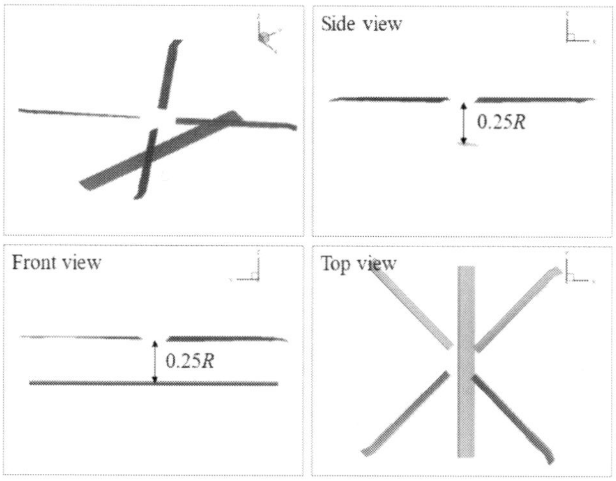

Figure 5. Computational model of rotor/wing configuration.

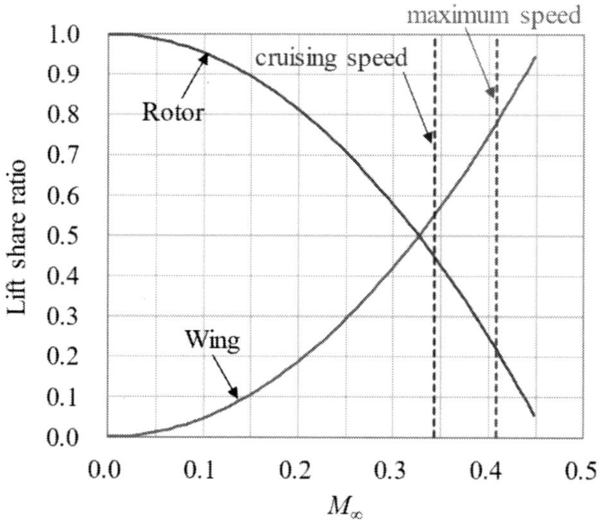

Figure 6. Lift share ratio between the rotor and wing.

NUMERICAL METHODS AND CONDITIONS

Rotorcraft flow solver, rFlow3D

rFlow3D is a structured grid solver based on moving overlapped grid method. The governing equations are the three-dimensional compressible Navier-Stokes equations discretized with Finite-Volume Method (FVM). The inviscid flux is computed using mSLAU (modified SLAU) (Ref. 4) and viscous flux is computed using a second order central difference scheme. Time integration methods are provided by explicit four-stage Runge-Kutta (R-K) integration for the background grids and LU-SGS/DP-LUR (Ref. 11) implicit unsteady method extended by dual time-stepping method (DTS) for the body fitted grids. A Fourth-order Compact MUSCL TVD (FCMT) interpolation method (Ref. 12) is used for high-order spatial accuracy. Several common turbulence models were implemented such as Spalart-Allmaras model and *k-ω* two equation model. Tri-linear interpolation method is used for interpolation between grids. Utilized computational methods in this paper are shown in Table 4.

Table 4. Computational Method

	Background grids	Blade/Wing grids
Governing eqns.	3-D compressible Navier-Stokes eqns.	
Spatial discretization	Cell-vertex FVM	Cell-centered FVM
Time integration	4 stages R-K	DTS/LU-SGS
Numerical flux	mSLAU	
Reconstruction	FCMT	
Viscous flux	2nd order central difference	
Turbulence model	Menter k-ω SST 2003 model	

Trim analysis

Trim adjustment is based on the partial derivatives of the rotor thrust and moments to the rotor controls (Ref. 5). The partial derivatives are numerically obtained based on the Blade Element Theory (BET). Several inflow models on the BET can be selected, Pitt & Peters model is used in this paper. The sectional aerodynamic forces are computed using the aerodynamic table such as in C81 format. To improve the computational stability and convergence, the trim adjustment is added with a relaxation factor multiplied.

To investigate the rotor/wing aerodynamic interaction, three configurations are examined. They are the isolated rotor and the isolated wing, and the combined rotor and wing configurations. The isolated rotor and the isolated wing are computed as the state without the effect of rotor/wing aerodynamic interaction, and the rotor/wing configuration is computed for its effect. The trim adjustment for full configuration is considered only for the vertical trim which is the summation of the rotor thrust and the wing normal force. The trim for rolling and pitching moments are adjusted by the blade control angle using only the rotor forces without the wing forces. In the isolated rotor and wing, the target rotor thrust is determined by difference between the max takeoff weight and the normal force of the isolated wing. In the rotor/wing configuration, the blade control angle is computed by summation of the rotor thrust and the wing normal force.

Numerical conditions

The flight conditions are the cruising flight and the maximum speed flight, and the attitude angle of the fuselage is zero degrees in the level flight. The rotor shaft angle is also zero degrees. The rotor tip Mach number is set to 0.48750 as 75% RPM as described above. The advance ratio of the rotor at the cruising and maximum speed are 0.703 and 0.837, respectively. Reynolds number based on the blade chord length is about 10.9 million at the cruising flight and about 11.7 million at the maximum speed. The target thrust and the target rolling and pitching moments are defined as the trim condition. The target thrust is 97860.8 N as the maximum takeoff weight and the target rotor rolling and pitching moments are set to 0.0 Nm. Where, the target thrust is the summation of rotor thrust and wing normal force as previously described. Summarized computational conditions are shown in Table 5. The blade flap and lag motions are not considered in this numerical simulation.

Table 5. Computational Conditions

	Cruise	Maximum
Flight condition	Level flight	
Freestream Mach No.	0.34284	0.40814
Rotor tip Mach No.	0.48750 (75%RPM)	
Rotor shaft angle	0 deg.	
Advance ratio	0.703	0.837
Target thrust	97860.80 N	
Target roll/pitch moments	0.0 Nm	

The overlapping computational grids are consisted of the outer background grid, the inner background grid, the blade grid and the wing grid. The grid system is shown in Figure 7. The spatial grid resolution in the inner background grid is 20% of the blade tip chord length of 0.561 m. The outer and inner background grids are Cartesian grids. The blade and wing grids are body fitted grids. The grid topology of the body fitted grids is in SOH type as a single block grid, where a singular line is formed at the blade tip and blade root, O-type around the blade and H-type in the body-wise. The first grid spacing from the wall corresponds to a y-plus of less than 1.0.

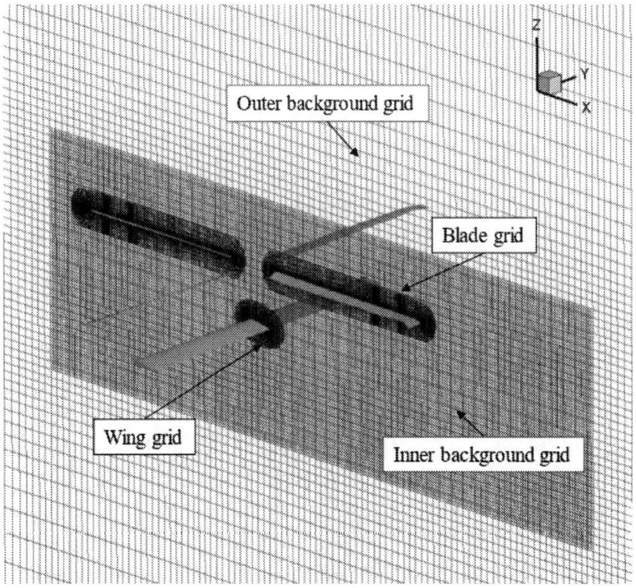

Figure 7. Overlapping grids for rotor and wing model.

RESULTS

Influence on total aerodynamic forces and moments

The definition of total aerodynamic forces and moments are shown in Figure 8. The moment center is defined at the rotor center. The force and moment are nondimensionalized by the air density, the rotor disk area, the blade tip speed and the rotor radius as the same as the non-dimensional coefficients of the rotor. Where, the X-force, Y-force and Z-force indicates the drag, side force and thrust, respectively.

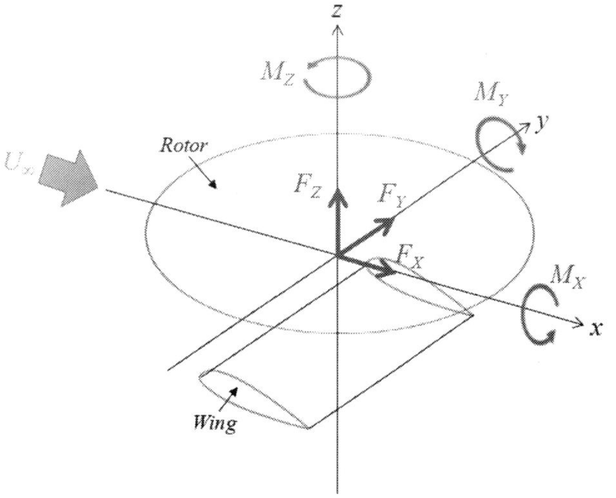

Figure 8. Definition of total aerodynamic forces and moments.

Figure 9 shows the total thrust of the rotor and the wing at the cruising speed and at the maximum speed. The total thrust is compared between the isolated configurations (the isolated rotor and the isolated wing) and the rotor/wing configuration in this figure. The orange bar shows the rotor thrust, and the blue bar shows the wing normal force. The error of the total thrust is less than 0.3% as compared with the target thrust. In comparison with the isolated configuration in each condition, the wing normal force is decreased and the required rotor thrust is increased in the rotor/wing configuration. The reduction ratio of the wing normal force is 12.0% at the cruising speed and 6.9% at the maximum speed. Table 6 compares the lift share ratio between the design value and the computational results in each flight condition. The lift share ratio in the isolated configuration is obtained as designed. At the maximum speed, the different lift share ratios between the design value and the isolated configuration are considered due to Mach number effect and Reynolds number effect. The computed lift coefficient of the wing is about 0.552 at the maximum speed. Calculating the wing lift using this lift coefficient, the lift share ratio corresponds with the simulation result. From the rotor/wing configuration results, the lift ratio of the wing is reduced from 0.55 to 0.48 at the cruising speed and 0.80 to 0.75 at the maximum speed. The reduction ratios of the wing normal forces caused by the rotor are different and can be interpreted that decreasing wing lift associates the rotor thrust.

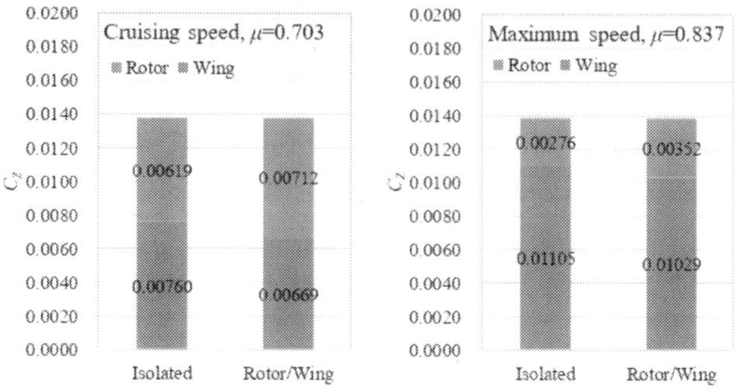

Figure 9. Total thrust of the rotor and wing at the cruising speed and at the maximum speed.

Table 6. Comparison of Lift Share Ratios

	Design		Isolated		Rotor/Wing	
	Rotor	Wing	Rotor	Wing	Rotor	Wing
Cruise	0.45	0.55	0.45	0.55	0.52	0.48
Maximum	0.22	0.78	0.20	0.80	0.25	0.75

Figure 10 shows the comparison of the blade control angle in each condition. The collective and cyclic pitch angle in the rotor/wing configuration are increased by the increasing required rotor thrust as compared with the isolated configuration. The maximum difference is 3.9 degrees in the collective pitch angle, 1.0 degrees in the lateral cyclic pitch angle and 4.4 degrees in the longitudinal cyclic pitch angle. The variation of the lateral cyclic pitch angle is small as compared with the collective pitch and longitudinal cyclic pitch angles. Therefore, the result indicates that the rotor aerodynamics by varying rotor thrust at the high advance ratio has significant impacts to the rotor pitching moment than the rotor rolling moment.

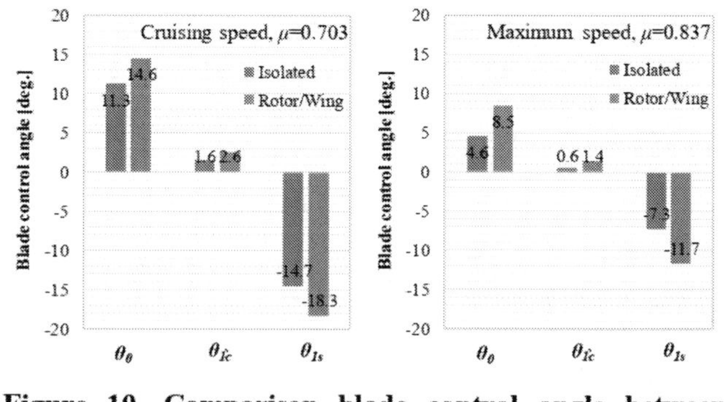

Figure 10. Comparison blade control angle between isolated and rotor/wing configuration.

The comparison of the total drag in each condition are shown in Figure 11. The drag of the rotor and the wing is increased in the rotor/wing configuration than the isolated configurations. The wing drag is increased by about 19.0% at the cruising speed and about 11.7% at the maximum speed. The wing aerodynamic performance is degraded by rotor/wing aerodynamic interaction. The increased rotor thrust has caused the rotor drag to increase, the rotor drag is increased by 13.7% at the cruising speed and 30.2% at the maximum speed. The rotor drag is sensitive for the rotor thrust variation at the high advance ratio as shown in Reference 6. The drag of the rotor accounts about 80% of the total drag. Please notice that the rotor of current UH-60A has a large blade twist and is not designed for high advance ratio flight. Optimal design to reduce the rotor drag at high speed is required.

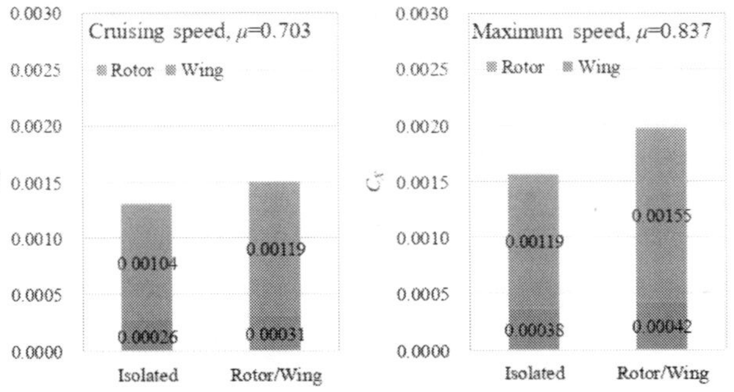

Figure 11. Total drag of the rotor and wing at the cruising speed and at the maximum speed.

Figure 12 compares the total effective drag coefficient of two configurations in each condition. The total effective drag in the rotor/wing model is increased by 23.9% at the cruising speed and 19.6% at the maximum speed than the summation of the isolated models. The effective drag of the rotor accounts about 88% for the total effective drag.

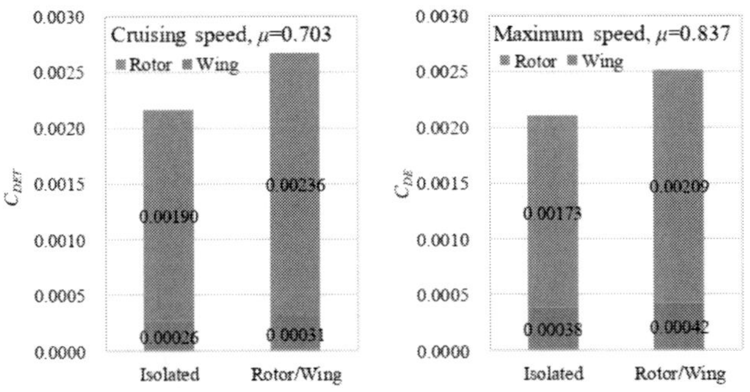

Figure 12. Total effective drag of the rotor and wing at the cruising speed and at the maximum speed.

Figure 13 shows the total side force in each condition. There is very little influence on the total side force by rotor/wing aerodynamic interaction. The side force occurs in the rotor/wing configuration but much smaller than the total drag force as seen in Figure 11.

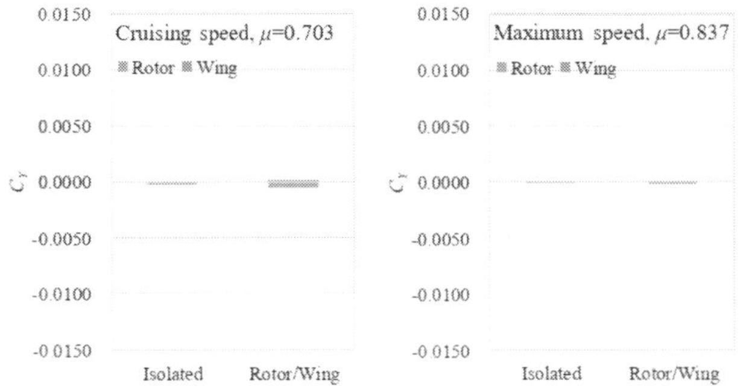

Figure 13. Total side force of the rotor and wing at the cruising speed and at the maximum speed.

Figure 14 to 16 show the comparison of the total rolling, pitching and yawing moments at each condition. The rolling and pitching moments of the rotor are trimmed to about zero by the rotor trim adjustment. The yawing moment of the rotor indicates opposite sign to the rotor torque. The rolling moment on the wing arises to the negative direction in the rotor/wing configuration as shown in Figure 14, but this is a favorable trend. In general, the rolling moment of the rotor occurs to the positive direction. Thus, the required blade cyclic pitch angle can be reduced partially by this trend. The same tendency is shown for the pitching moment as shown in Figure 15. The yawing moment on the wing occurs slightly as shown in Figure 16 but much smaller than the rotor yawing moment. The detail of these phenomena will be described in the next section.

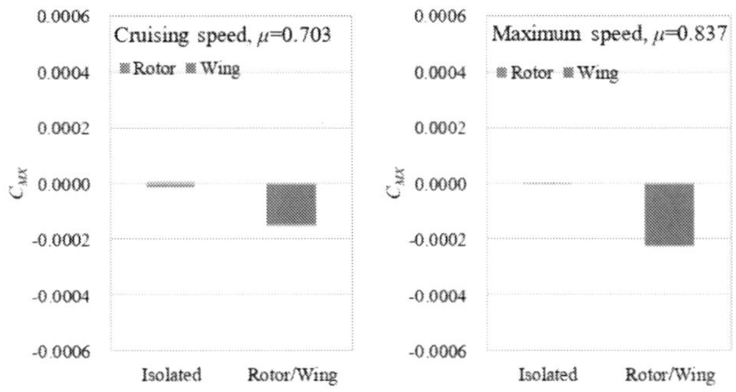

Figure 14. Total rolling moment of the rotor and wing at the cruising speed and at the maximum speed.

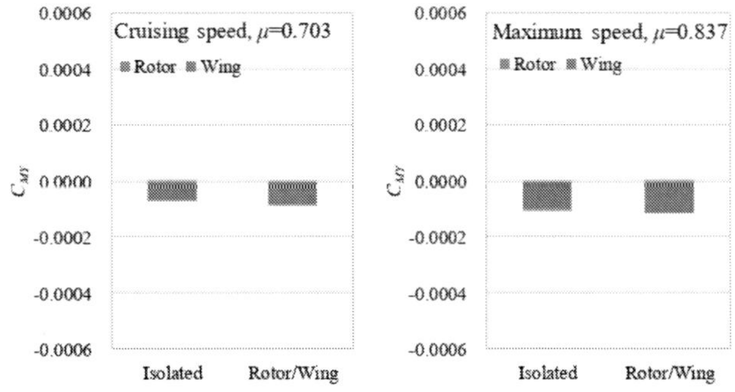

Figure 15. Total pitching moment of the rotor and wing at the cruising speed and at the maximum speed.

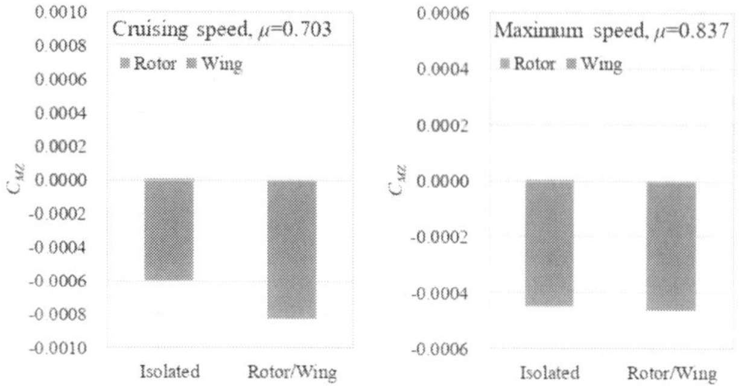

Figure 16. Total yawing moment of the rotor and wing at the cruising speed and at the maximum speed.

Effect on aerodynamics around wing

The average sectional normal force on the wing of the two configurations in each condition are shown in Figure 17 and Figure 18. Where, the sectional Z-force indicates the sectional normal force of the wing. From these figures, the sectional normal force on the wing is observed in asymmetric distribution in the rotor/wing configuration. The sectional normal force in the rotor/wing configuration is reduced significantly than the isolated wing at the wing starboard below the rotor advancing side and has minor difference at the wing port below the rotor retreating side. The wing rolling moment in Figure 14 is caused by this distribution. Similarly, the average sectional drag is also in asymmetric distribution as shown in Figure 19 and 20. Where, the sectional X-force indicates the sectional drag of the wing. The sectional drag is increased at the wing starboard and shows slight change at the wing port. This trend is associated with the yawing moment as shown in Figure 16, but the wing yawing moment is small as compared with the rotor torque. Thus, this effect is considered small on the airframe aerodynamics. The existence of the rotor has a significant impact on the wing force on the wing starboard below the rotor advancing side.

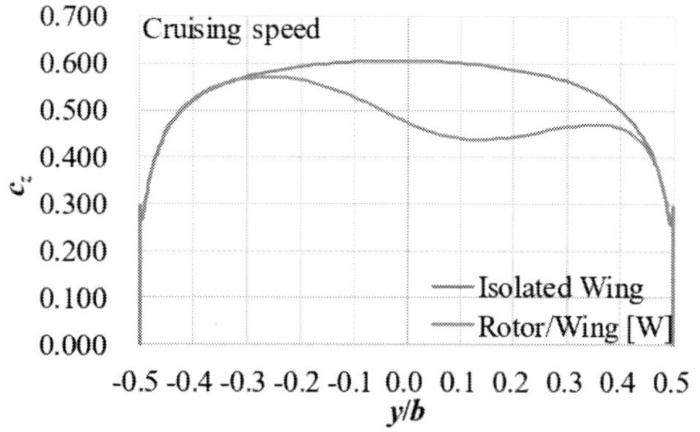

Figure 17. Comparison average sectional normal force on the wing between the isolated wing and the rotor/wing configuration at the cruising speed.

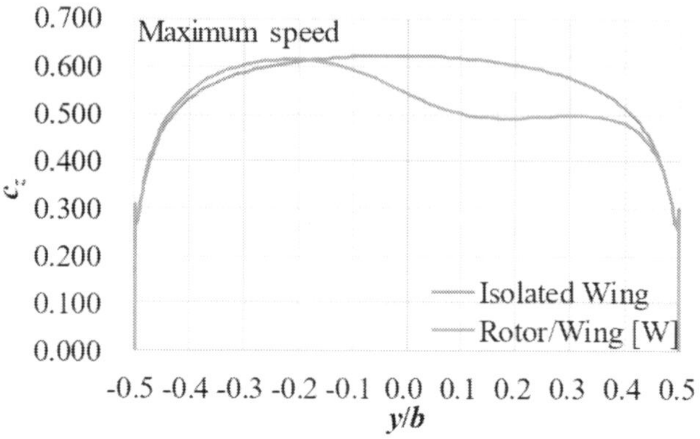

Figure 18. Comparison average sectional normal force on the wing between the isolated wing and the rotor/wing configuration at the maximum speed.

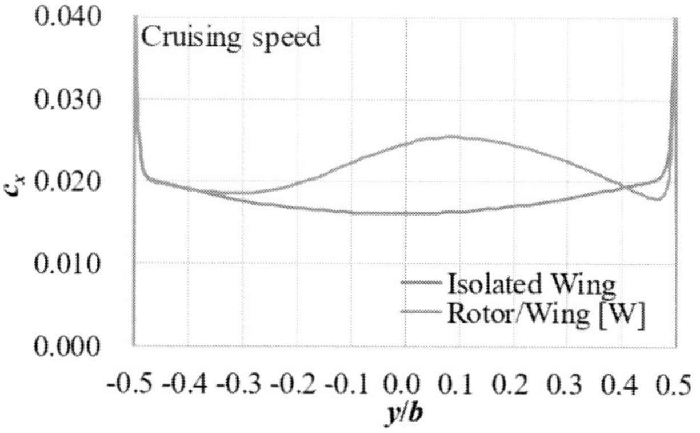

Figure 19. Comparison average sectional drag on the wing between the isolated wing and the rotor/wing configuration at the cruising speed.

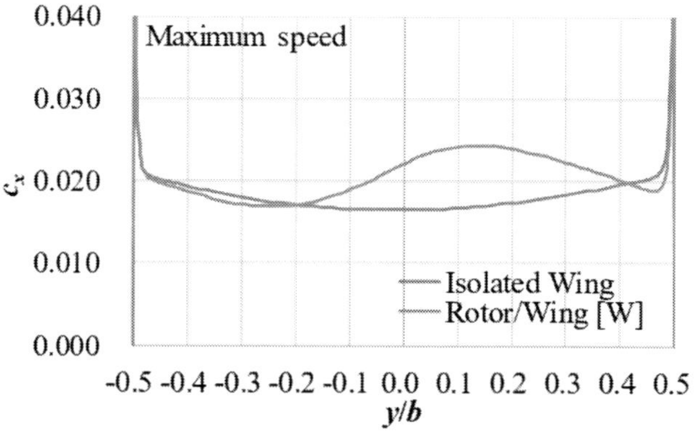

Figure 20. Comparison average sectional drag on the wing between the isolated wing and the rotor/wing configuration at the maximum speed.

To investigate the influence on the aerodynamic characteristics of the wing by the rotor, the sectional normal force on the blade at each azimuth angle is examined. Figure 21 and Figure 22 shows the distribution of the sectional normal force coefficient on the rotor blade at each azimuth angle at the cruising speed and at the maximum speed, respectively. The rectangle of the solid line shows the wing and the dash line indicates the reverse flow region on the rotor, with a diameter of μR. The wing port exists inside the reverse flow region form these figures. The high normal force region is observed at the rotor front on the right in the both conditions. This region corresponds with the reduction region of the normal force on the wing, it can be presumed that the induced velocity from the rotor interacts with the flowfield around the wing. The negative sectional normal force is shown on the reverse flow region, and it interprets that the influence on the wing aerodynamics is scarcely affected. The sectional normal force distribution shows that the rotor thrust is obtained at the rotor front and aft area on the rotor advancing side. This distribution explains furthermore that the rotor pitching moment is sensitive than the rotor rolling moment at high advance ratio as previously described about the blade control.

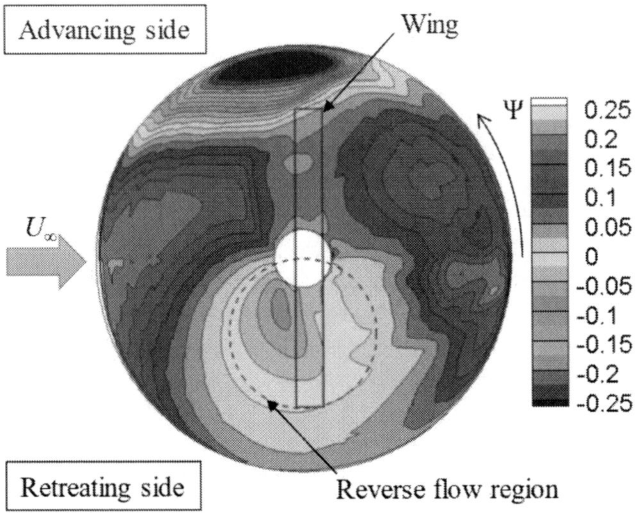

Figure 21. Sectional blade normal force distribution at the cruising speed.

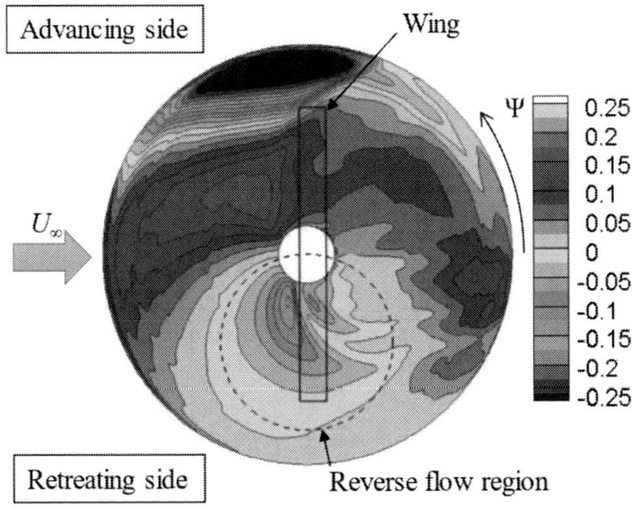

Figure 22. Sectional blade normal force distribution at the maximum speed.

The sectional flowfields around the rotor and wing at azimuth angle of 0 degrees and at 45 degrees at the cruising speed are visualized in Figure 23 and 24. The contour color shows the perpendicular velocity vector as nondimensionalized by the speed of sound. The red color indicates the zero or the upwash, and the other colors show the downwash. The section positions are shown at -0.25b and 0.25b. The velocity distribution illustrates that the rotor induced velocity interacts on the flowfield around the wing starboard and slightly impacts around the wing port. At the wing starboard below the rotor advancing side, the downwash is observed around the wing leading edge. At the wing port of the rotor retreating side, the velocity distribution around the wing leading edge is showing almost freestream velocity.

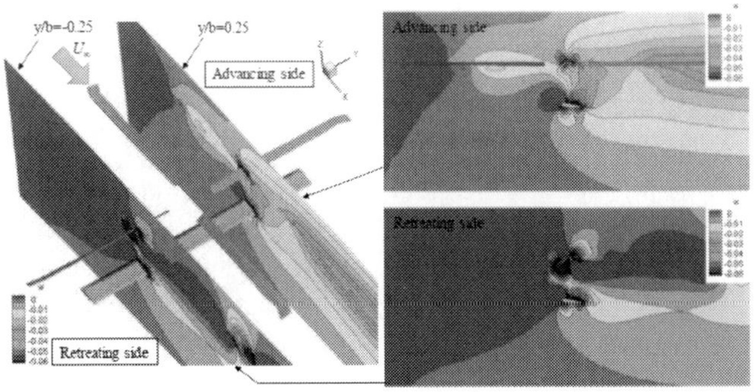

Figure 23. Sectional normal velocity distribution at the rotor azimuth angle of 0 degrees at the cruising speed.

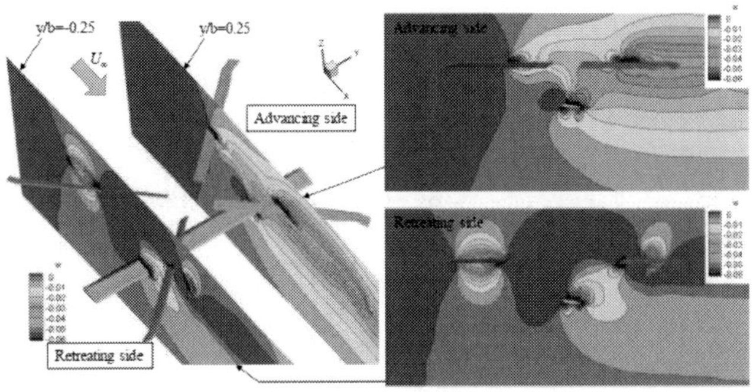

Figure 24. Sectional normal velocity distribution at the rotor azimuth angle of 45 degrees at the cruising speed.

To understand more detail of the wing aerodynamics in rotor/wing aerodynamic interaction, the estimation of the sectional normal and drag forces are calculated based on the blade element theory. The wing sectional aerodynamics is illustrated in Figure 25. Where, u_t is the tangential velocity vector to the wing, u_p is the perpendicular velocity vector to the wing, U is the resultant velocity, θ is the installing angle of the wing, ϕ is the inflow angle and α is the effective angle of attack. The tangential velocity is assumed to be the freestream velocity. The perpendicular velocity is presumed to construct by the induced velocity from the wing and the rotor. In this estimation, the induced velocity from the wing is assumed based on the simple wing theory. The induced velocity around the wing is supposed double at the downstream in the simple wing theory as shown in Figure 26. Where, v_i is the induced velocity around the wing. The perpendicular velocity is extracted as the induced velocity of the wing from the isolated wing simulation result. The extraction position is about $9c_w$ backward from the wing trailing edge, and the perpendicular velocity is approximately constant as shown in Figure 26. The induced velocity from the rotor is extracted as the perpendicular velocity at the wing quarter chord position from the isolated rotor simulation result. A sample of the extracted perpendicular velocity distribution at the cruising speed is shown in Figure 27. The velocity is nondimensionalized by the speed of sound. The perpendicular velocity of the isolated rotor is in asymmetric distribution, and large downwash is observed at the wing starboard below the rotor advancing side. The red line indicates the summation of the wing induced velocity and the induced velocity from the rotor. This distribution is used as the perpendicular velocity for the blade element theory. Note that the velocity around the wing tip is less accurate by the existence of the wing tip vortex. However, this estimate is focused on the trend of the sectional force distribution, and this method can be expected to obtain the force distribution excluding around the wing tip. The effective angle of attack is different locally about 1.5 degrees at the maximum as compared with the isolated wing. The sectional lift and drag coefficients are based on the aerodynamic data of the NACA23015 airfoil for the computation of the sectional force.

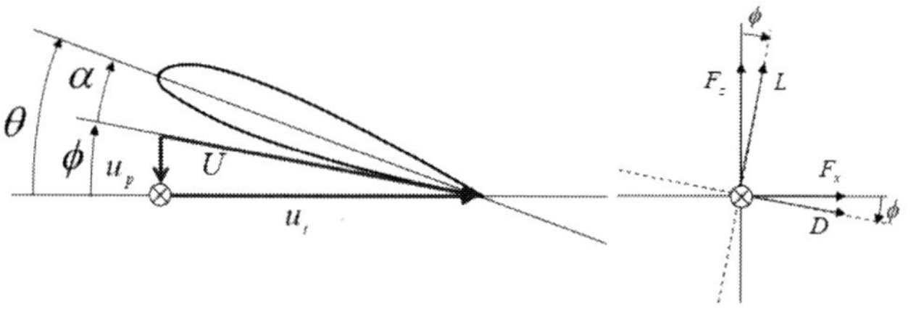

Figure 25. Wing sectional aerodynamics.

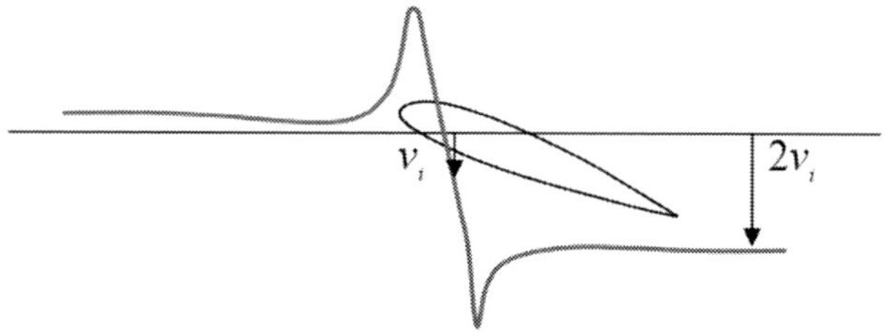

Figure 26. Induced velocity around wing.

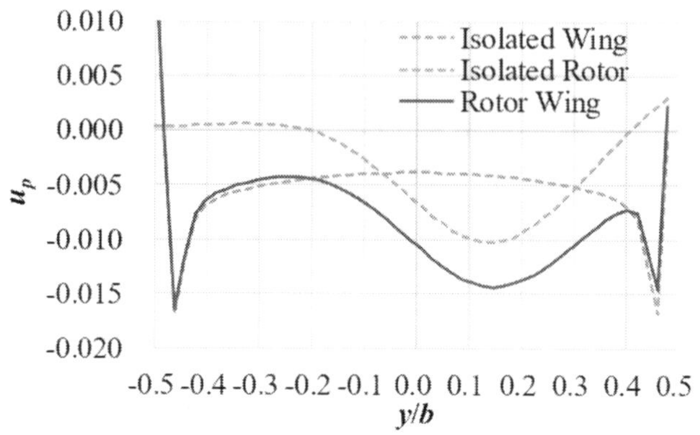

Figure 27. Sample extracted perpendicular velocity distribution.

Figure 28 to Figure 31 compares the sectional normal and drag forces distribution between the numerical simulation results and the estimated results based on the blade element theory. The estimated results are shown in good correlation with the numerical simulation results except around the wing tip. The inflow angle at the wing starboard is increased by the rotor induced velocity. Thus, the wing normal force is decreased and the wing drag is increased as shown in Figure 25. The estimated result explains the association between the rotor induced velocity and the wing sectional aerodynamic forces. Therefore, the induced velocity of the rotor interacts with the flowfield around the wing and is the cause of the asymmetric force distribution on the wing.

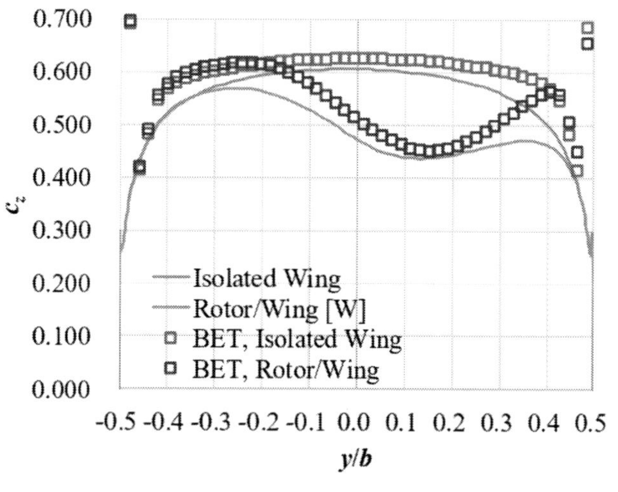

Figure 28. Comparison of the sectional normal force distribution on the wing at the cruising speed.

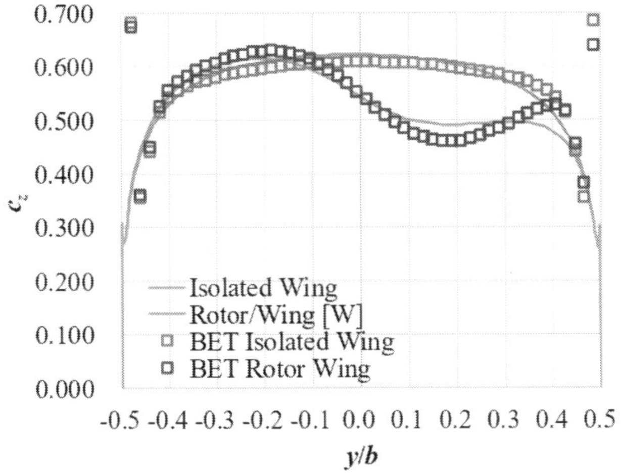

Figure 29. Comparison of the sectional normal force distribution on the wing at the maximum speed.

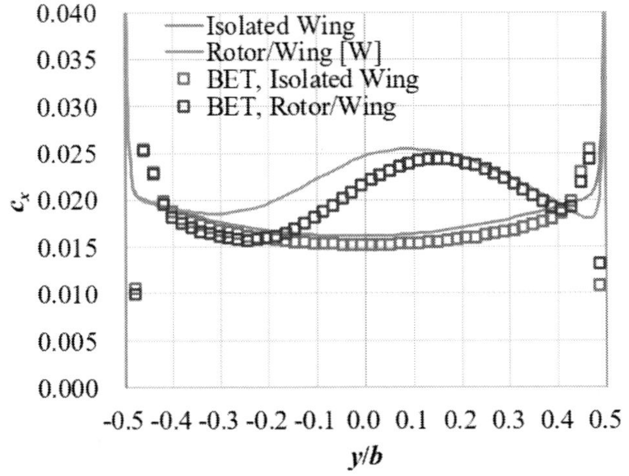

Figure 30. Comparison of the sectional drag distribution on the wing at the cruising speed.

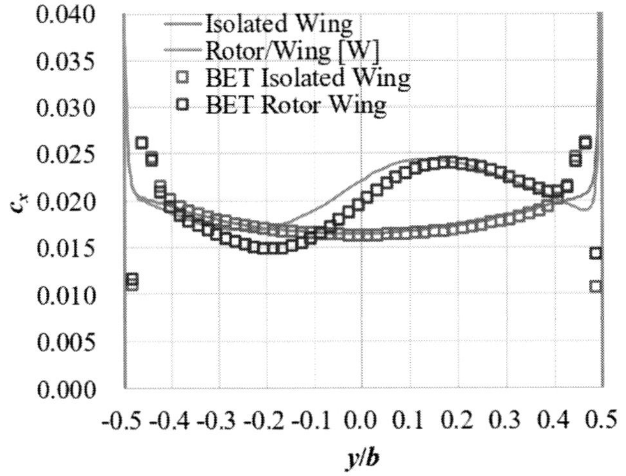

Figure 31. Comparison of the sectional drag distribution on the wing at the maximum speed.

Figure 32 and Figure 33 show the sectional normal force distribution on the wing at some rotor azimuth angles at the cruising and maximum speed flight. The variation of the normal force on the wing by the blade passage is observed. Similarly, Figure 34 and Figure 35 show the sectional drag distribution. The variation of the sectional drag on the wing by the blade passage is also observed. The large variation is seen on the wing port (rotor retreating side) than the wing starboard (rotor advancing side). Therefore, the steady effect is dominant on the wing starboard, and the unsteady effect is prominent on the wing port. The amplitude of the fluctuations seems to be proportional to the rotor thrust oscillations where at the cruising speed condition larger rotor thrust is generated than at the maximum speed condition.

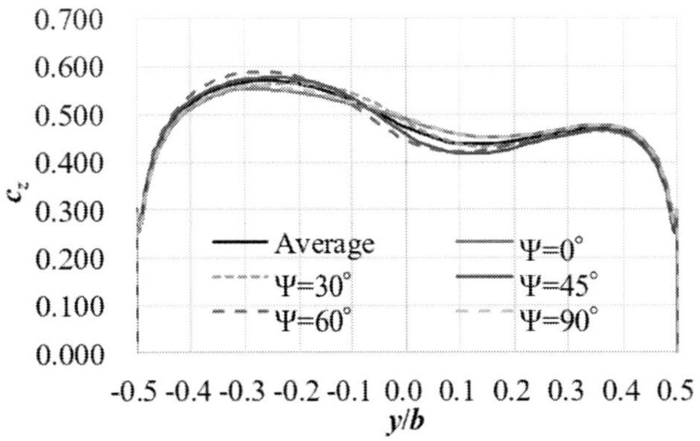

Figure 32. Sectional normal force distribution on the wing at each rotor azimuth angle at the cruising speed.

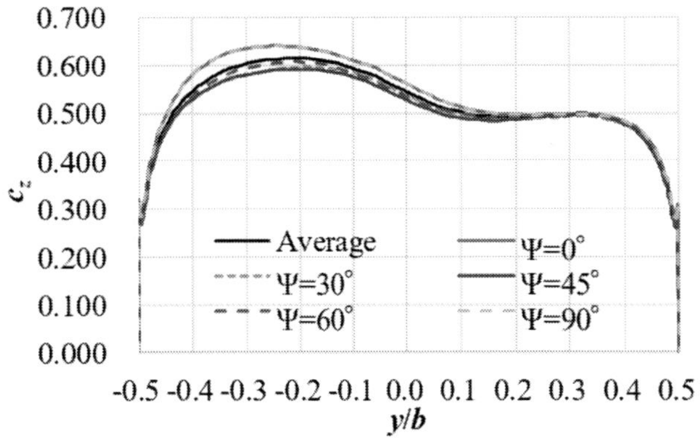

Figure 33. Sectional normal velocity force on the wing at each rotor azimuth angle at the maximum speed.

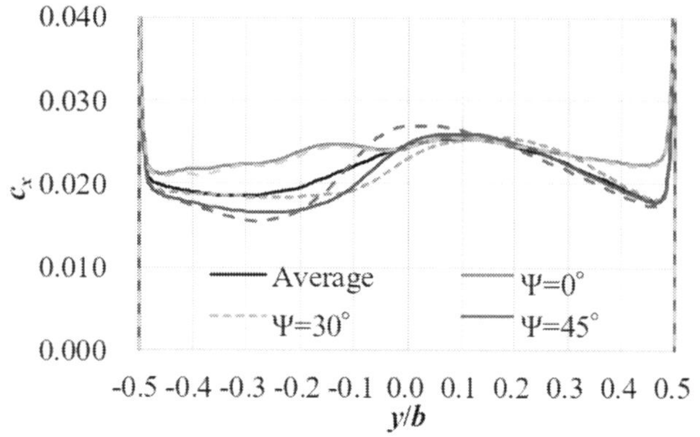

Figure 34. Sectional drag distribution on the wing at each rotor azimuth angle at the cruising speed.

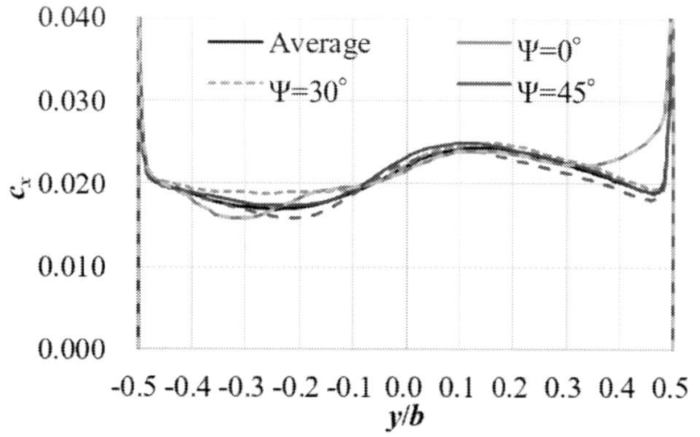

Figure 35. Sectional drag distribution on the wing at each rotor azimuth angle at the maximum speed.

CONCLUSIONS

A study of rotor/wing aerodynamic interaction at high advance ratio through the numerical simulation using the rotorcraft CFD solver, rFlow3D, developed at JAXA is carried out. The computational models utilized simplified rotor and wing configurations. The wing model is assumed to be installed on UH-60A helicopter, and the wing sizing is performed based on its maximum takeoff weight. To investigate the influence of rotor/wing aerodynamic interaction, the isolated rotor and the isolated wing configurations and the rotor/wing configuration are computed. Two flight conditions are simulated at the cruising speed and at the maximum speed. The rotor speed is assumed to reduce to 75% RPM at high speed flight condition.

In comparison with the isolated configurations, the wing normal force is reduced by the rotor, and the required rotor thrust has to be increased in the rotor/wing configuration. The total drag and effective drag are increased by rotor/wing aerodynamic interaction. The rotor drag is shown to be dominant in each drag and accounts for 80 and 88%. The wing rolling moment is caused by the asymmetric rotor induced velocity, with the direction being opposite to the rotor rolling moment. The pitching moment shows a similar trend to the rolling moment. There is very little influence on the total side force and the yawing moment. The numerical simulation results show that the total forces and moments of the rotor and wing configuration are not simple summation of the isolated rotor and the isolated wing. Therefore, the corrections of the aerodynamic forces and moments are required with consideration of rotor/wing aerodynamic interaction in the design of a compound helicopter.

The sectional force distribution on the wing is shown in asymmetric distribution, especially, on the wing starboard below the rotor advancing side. The wing normal force is decreased and the drag is increased on the wing starboard. This trend is in good correlation with the results estimated by the blade element theory using the flowfield of the numerical simulation data of the isolated models. The reason of this tendency is explained by the result of blade element theory that the rotor induced velocity interacts with the flowfield around the wing and is the cause of the asymmetric force distribution on the wing.

The variation of the sectional forces on the wing by the blade passage is observed by the numerical simulation result. The fluctuation of the sectional forces is dominant at the wing port below the rotor retreating side. This result indicates that the steady effect is dominant on the wing starboard below the rotor advancing side and the unsteady effect is prominent on the wing port below the rotor retreating side.

A further study of the influence on the wing aerodynamics at different rotor advance ratio condition and with variation of the lift share ratio between the rotor and wing should be conducted. Detailed experiments about rotor/wing aerodynamic interaction are also necessary for the understanding its phenomenon and validation of the numerical simulations.

Author contact:
Hideaki Sugawara hideaki8@chofu.jaxa.jp
Yasutada Tanabe tan@chofu.jaxa.jp

REFERENCES

[1]Tanabe, Y., Aoyama, T., Kobiki, N., Sugiura, M., Miyasita, R., Sunada, S., Kawachi, K., and Nagao, M., "A Conceptual Study of High Speed Rotorcraft", 40th European Rotorcraft Forum, Amsterdam, Netherlands, 2014.

[2]Makofski, R. A. and Menkick, G. F., "Investigation of Vertical Drag and Periodic Airloads Acting on Flat Panels", NACA-TN 3900, December 1956.

[3]Felker, F. F. and Light, J. S., "Aerodynamic Interaction Between a Rotor and Wing in Hover", Journal of the American Helicopter Society, April 1988.

[4]Tanabe, Y. and Saito, S., "Significance of All-Speed Scheme in Application to Rotorcraft CFD Simulations", 3rd

International Basic Research Conference on Rotorcraft Technology, Nanjing, China, October 14-16, 2009.

[5]Tanabe Y., Saito, S. and Sugawara, H., "Construction and Validation of an Analysis Tool Chain for Rotorcraft Active Noise Reduction", 38th European Rotorcraft Forum, Amsterdam, Netherlands, 2012.

[6]Tanabe, Y., Sugawara, H., "Aerodynamic Validation of rFlow3D Code with UH-60A Data Including High Advance Ratios", the 41st European Rotorcraft Forum, Munich, Germany, Sept. 1-4, 2015.

[7]Tanabe, Y, and Sugawara, H., "Numerical Simulation of Aerodynamic Interaction Between a Rotor and a Wing", the Asia-Australian Rotorcraft Forum, November 17-18, 2016.

[8]Bousman, W. G., "Aerodynamic Characteristics of SC1095 and SC1094 R8 Airfoils", NASA/T2003-212265, AFDD/TR-04-003, December 2003.

[9]Davis, S. J., "Predesign Study for a Modern 4-Blade Rotor for the RSRA", NASA Contractor Report 166155, March 1981.

[10]Prouty, R. W., *Helicopter Performance, Stability, and Control*, Krieger Publishing Company, Malabar, Florida, 2005, pp. 698.

[11]Zhang, L.P. and Wang, Z.J., "A Block LU-SGS Implicit Dual Time-Stepping Algorithm for Hybrid Dynamic Meshes," Computers & Fluids, Vol.33, pp.891–916, 2004.

[12]Yamamoto S. and Daiguji H., "Higher-Order-Accurate Upwind Schemes for Solving the Compressible Euler and Navier-Stokes Equations," Computers & Fluids, Vol.22, No.2/3, pp.259-270, 1993.

Improvement of actuator surface method for various solidity rotor computation

Jaeyoung Park	Sejong Oh	Donghun Park	Myungsik Tai
Integrated PhD candidate	Professor	Assistant Professor	MS candidate
Pusan National University	Pusan National University	Pusan National University	Pusan National University
Busan, Korea	Busan, Korea	Busan, Korea	Busan, Korea

ABSTRACT

Actuator based CFD solvers has practical advantages to predict performance of rotorcraft. Especially, the actuator surface method(ASM) can describe motions and physical phenomena of each blade as pressure discontinuity surface. The force distribution on actuator surface, calculated using velocity of reference line, determines characteristic of each blade. In case of high solidity rotor computation, effects of nearby blades should be concerned. In the present paper, methodologies of the force distribution using source-doublet panel method and reference line velocity correction considering effects of front and rear blade are proposed. The numerical simulations are conducted with open source code, OpenFOAM. Validation of developed ASM code is carried out for the case of ONERA7A rotor.

INTRODUCTION

In rotorcraft design, it is essential to analyze aerodynamic performance using various parameters of each blade. One of important parameter, Solidity defines as the area ratio of blades to that of rotor. Actually, it is well-known that lower solidity rotor has higher value of figure of merit. However, to get same thrust, low solidity rotor needs high angle of attack of blade and encounters blade stall. On the other hand, in case of high solidity rotor, interaction of each blade has influences on rotor performance. Therefore, it is necessary to analyze co-relations between solidity and aerodynamic characteristics.

Generally, Navier-Stokes equation based on computational fluid dynamics(CFD) is utilized for aerodynamic performance analysis. Although CFD is reliable, it is computationally expensive and requires much time and efforts. As an alternative, actuator based blade modeling methods, such as actuator surface method(ASM), actuator disk method(ADM), are employed. In contrast to Full CFD, actuator based method utilizes pressure discontinuity to represent blade of rotor disk regions. Especially, ASM can consider each blade effects as volume force distributions on blade region. These techniques can reduce computational time and grid generation efforts, but these method needs physical modeling to get reliable results. There are many analyses applying actuator concept. Kim performed analysis of helicopter rotor using ASM. (Ref.1) C.Sibuet and S. P. Breton applied actuator concept to wind turbine. (Ref.2) I Dobrev and F Massouh suggested actuator surface hybrid model with pressure jump boundary condition of blade region. (Ref.3)

In present paper, to improve ASM for the case of various solidity rotor problem, those two approaches are considered.

1) Apply chord-wise volume force distribution method to represent reliable pressure distribution.

2) Consider interaction effects of neighboring blades in case of high solidity rotor problem.

Validations of suggested method are conducted in cases of hovering rotor and aligned rectangular wing.

ANALYSIS METHODS

Actuator model

To conduct numerical simulation, the ASM is adopted. ASM replaces each blade as pressure discontinuity of blade region. Pressure discontinuity is considered as source term on momentum equation of CFD.

Presented at the 6th Asian/Austrailian Rotorcraft Forum / Heli Japan 2017, Kanazawa, Japan, November 7-9, 2017.

$$\frac{\partial U}{\partial t} + (U \cdot \nabla)U - \nabla \cdot (\nu \nabla U) = s - \frac{1}{\rho}\nabla p \qquad (1)$$

Overall ASM calculation procedures are present in Figure 1.

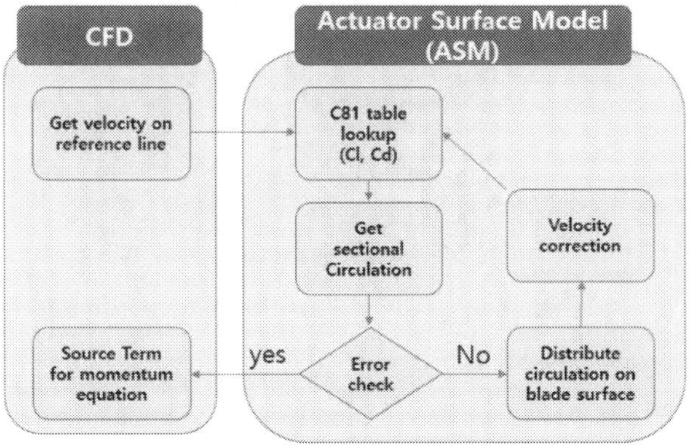

Figure 1. Flow chart of ASM

Procedures of ASM start from getting velocity on reference line. Then characteristic of blade is determined from table lookup and velocity correction. In this analysis, open source code, OpenFOAM, is utilized as CFD. ASM developed based on IASM that Kim developed in his analysis. (Ref.1)

Chord-wise volume force distribution

Determination of pressure distribution is important for reliable description of volume force distribution on blade surface. Preceding researches of ASM adopted fixed chord-wise volume force distribution like Ref.1 and Ref.3 or empirical formulation of pressure distribution as Ref.4. In present analysis, Source-doublet panel method(SDM) is used to get chord-wise pressure distribution of blade. SDM is potential methodology using the distribution of source-doublet elements. This method is appropriate to consider thickness effect of blade. Using SDM allows treatment of realistic pressure distribution on blade section. In ASM, velocity data from reference line is used as freestream velocity of SDM.

Effectiveness of front and rear blades

ASM concept uses reference line to determine characteristics of blade section. Velocity data of reference line includes various effects of flow field. Thus, in high solidity case, effectiveness of front and rear blade should be concerned. The analysis of Ref.1 used Biot-Savart law and corrected reference velocity with concerning one blade.

In this research, terms of front and rear blade effectiveness are included in correction equation. Circulations of each blade describe influences of neighboring blade. Circulation effect can be described as a function of distance and strength. Distances and circulation strength can be expressed as Figure 2. Here, $\boldsymbol{\Gamma}$ is bound circulation of blade section and $\mathrm{d}\boldsymbol{\Gamma}$ is shed vortex driven from different circulation along span-wise direction.

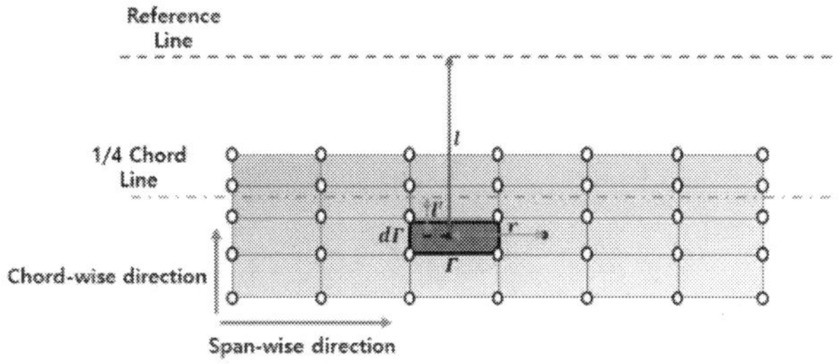

Figure 2. Definition distance and strength of a blade

Blade circulation effect on reference line can be expressed as Biot-Savart law.

$$\boldsymbol{w}_i = \frac{\Gamma_{i,j}}{4\pi}\int_{\tan^{-1}\frac{l_{i,(i,j)}}{r_{i,i-h}}}^{\tan^{-1}\frac{l_{i,(i,j)}}{r_{i,i+h}}} \frac{\sin\theta}{l_{i,(i,j)}} d\theta \;+\; \frac{d\Gamma_{di,dj}}{4\pi}\int_{\tan^{-1}\frac{r_{i,di}}{l_{i,(i,j)}}}^{\pi} \frac{\sin\theta}{r_{i,di}} d\theta \quad (2)$$

Here, i and j are span-wise and chord-wise blade panel numbering respectively. $\boldsymbol{w}$ is velocity induced by blade circulation on 1/4 chord line. First term and second term are substituted as $f_1(l,\boldsymbol{\Gamma})$ and $f_2(l,r,\mathrm{d}\boldsymbol{\Gamma})$ for convenience. Then blade circulation effect becomes equation (3).

$$\boldsymbol{w} = f_1(l,\boldsymbol{\Gamma}) + f_2(l,r,d\boldsymbol{\Gamma}) \qquad (3)$$

The effectiveness of front and rear blade circulation can be concerned in reference line velocity correction. (Figure 3.)

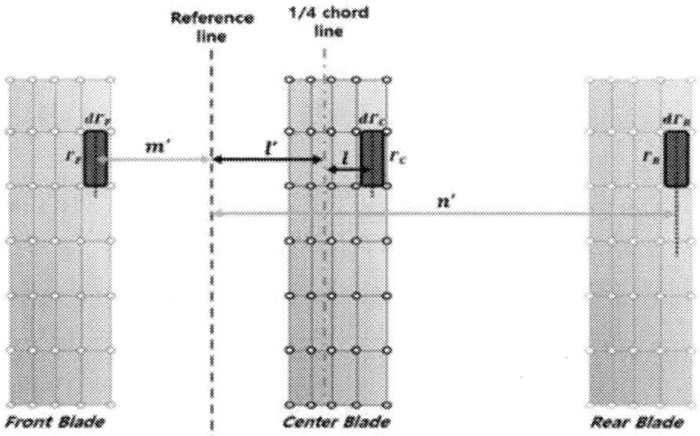

Figure 3. Definition distance and strength front, center and rear blade

Front and rear blade contributions can be included center blade reference velocity correction as equation (4)

$$\begin{aligned}\boldsymbol{w} = {} & w'_F + w'_C + w'_R + f_1(l,\Gamma_C) + f_2(l,r,d\Gamma_C) \\ & - f_1(l',\Gamma_C) - f_2(l',r,d\Gamma_C) - f_1(m',\Gamma_F) - f_2(m',r,d\Gamma_F) \\ & - f_1(n',\Gamma_R) - f_2(n',r,d\Gamma_R) \qquad (4)\end{aligned}$$

Here, w' is velocity induced on reference line. Right hand side of equation (4) can be expressed as four terms.

$$\boldsymbol{w} = \boldsymbol{w}' + \boldsymbol{F_C} + \boldsymbol{F_F} + \boldsymbol{F_R} \qquad \textbf{(5)}$$

Where

$$w' = w'_F + w'_C + w'_R$$

$$F_C = f_1(l, \Gamma_C) + f_2(l, r, d\Gamma_C) - f_1(l', \Gamma_C) - f_2(l', r, d\Gamma_C)$$

$$F_F = -f_1(m', \Gamma_F) - f_2(m', r, d\Gamma_F)$$

$$F_R = -f_1(n', \Gamma_R) - f_2(n', r, d\Gamma_R)$$

In equation (5), first term represents velocity induced on reference line by front, center and rear blades. Second term and fourth term means front and rear blade effectiveness on reference line of center blade, respectively. Third term includes center blade influences on reference line and 1/4 chord line. This equation is applied as reference velocity correction equation in ASM.

Geometry of rectangular wing cases

To validate ASM, span-wise lift coefficient distribution of rectangular wing compared with experiment and CFD. (Ref.5) For single rectangular wing cases, results of ASM are compared with those of CFD and experiment. Calculations of three aligned rectangular wings are conducted to validate reference line velocity correction equation. The Reynolds number of flow field is 1.5×10^6 . Rectangular wing has NACA0012 as wing sections and aspect ratio is 4.96. The ASM and CFD grid for single wing is shown in Figure4.

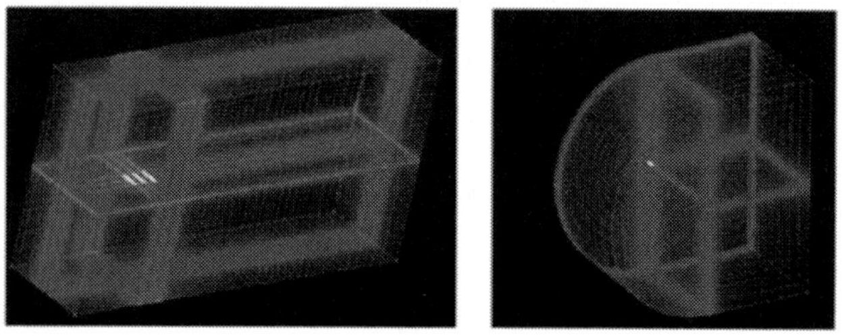

Figure 4. Grid geometry of ASM(left) and CFD(right) for rectangular wing

Total Grid number is 1.2 million for ASM and 2.9 million for CFD. The Grid of ASM also used for three aligned wings case. In case of CFD grid, symmetry boundary condition applied on half wing grid. For three aligned wing case, gap distance is defined as the distance from trailing edge of front wing to leading edge of rear wing. (Figure 5.) Gap distance of CFD and ASM grid is twice of wing chord length. CFD grid of three aligned wings is shown in Figure6. The number of grid is 6 million and symmetry boundary condition applied on half wing grid. This validation conducted to prove capability of ASM code to predict accurate performance of blade.

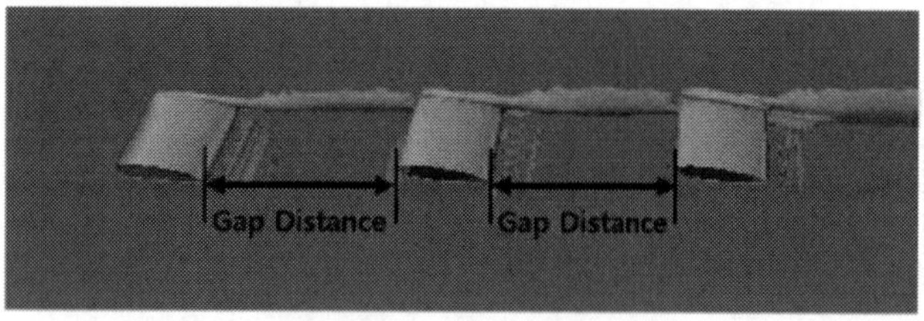

Figure 5. Definition of gap distance

Figure 6. CFD grid for three aligned wing

Geometry of Rotor cases

In rotor case, ONERA 7A model rotor is adopted to validate ASM. ONERA 7A rotor was tested in DNW. (Ref.6) For hovering condition, collective pitch is $\theta_{0.7} = 7.5°$ with the nominal thrust coefficient $C_T/\sigma = 0.08$ and tip Mach number is $M_{tip} = 0.6612$. Aspect ratio of blade is 15 and blade consists of OA213 and OA209 airfoil. Rotor case ASM grid shown in Figure 7. Total number of rotor ASM grid is 3.8 million.

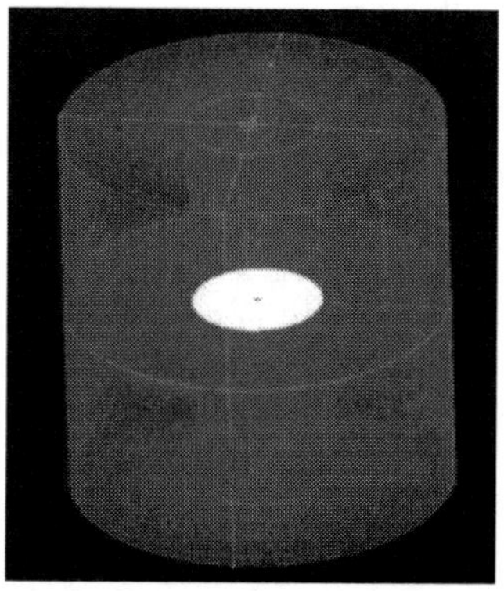

Figure 7. ASM grid for ONERA 7A

The Grid using ONERA 7A validation is also utilized for application of various solidity case. In the various solidity analysis, just the number of ONERA 7A blade changed.

RESULT OF ANALYSIS

In this section, results of ASM are compared with experimental data and CFD. For rectangular wing, data of span-wise lift coefficient distribution is presented. In rotor cases, results including sectional normal force coefficient and total thrust coefficient are compared.

Single rectangular wing

The analysis of rectangular wing is carried out in various angle of attack, $\alpha = 4°, 8°, 12°$. The results of fixed Gaussian and SDM force distribution are compared with experimental data in figure8.

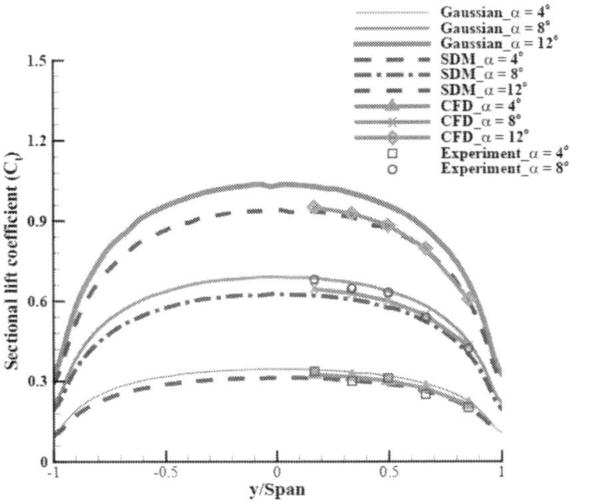

Figure 8. Sectional lift coefficient of single rectangular wing

The maximum differences of sectional lift coefficient of CFD and experimental data is 4% for $\alpha = 4°$ and 6% for $\alpha = 8°$. In case of ASM, SDM force distribution method has average 3% difference compared with CFD result, while fixed Gaussian has average 6% difference. Thus, ASM with SDM force distribution method is more accurate than that with fixed force distribution.

Three aligned wings

The case of three aligned wings is conducted to validate center blade reference velocity correction equation. Gap distance of each wing is 2 chord length and angle of attack fixed as $\alpha = 8°$. The result of ASM compared with CFD is presented in Figure9.

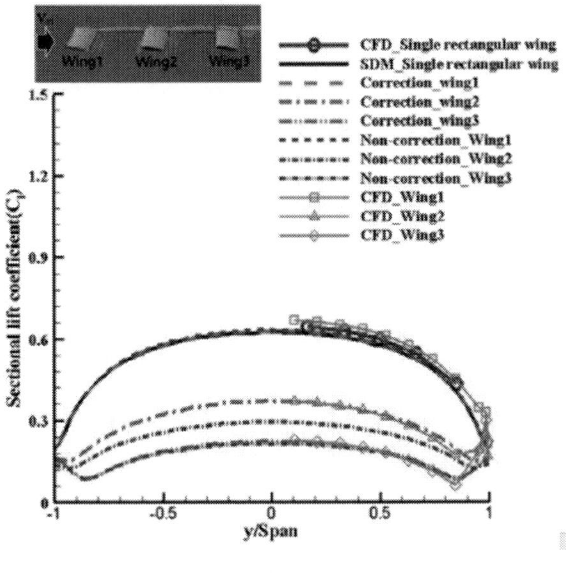

Figure 9.Sectional lift coefficient of three aligned wings at $\alpha = 8°$

In three aligned wings case, front wing, center wing and rear wing correspond to wing1, wing2 and wing3, respectively. The influence of reference velocity correction equation is clearly shown in wing2. The maximum difference of non-correction and CFD is 20% in wing2. However, after applying correction equation to ASM, sectional lift coefficient agrees with that of CFD with 0.2% difference. In wing3, the maximum difference is 5% and 3% for non-correction and correction respectively. The result of wing1 shows insignificant influence of correction equation. It is because of relatively small influence of rear wing(wing2). The sectional lift coefficient of wing2 is much smaller than that of wing1 and comparatively long distance from reference line of wing1 to panel of wing2 causes ineffectiveness of correction equation.

ONERA 7A hovering rotor

To confirm accuracy of ASM in rotor problem, ONERA 7A hovering rotor case is adopted. Figure10. illustrates normal force coefficient distribution along blade radial position. Here, ASM with SDM force distribution compared with experimental data and computed result of EROS. (Ref.7) The EROS is three-dimensional Euler-equation code using Chimera method. Although there are some differences near tip regions, overall normal force coefficient distributions of ASM agree with those of experiment and EROS. The nominal coefficient of ASM is $C_T/\sigma = 0.078977$ with 1% error.

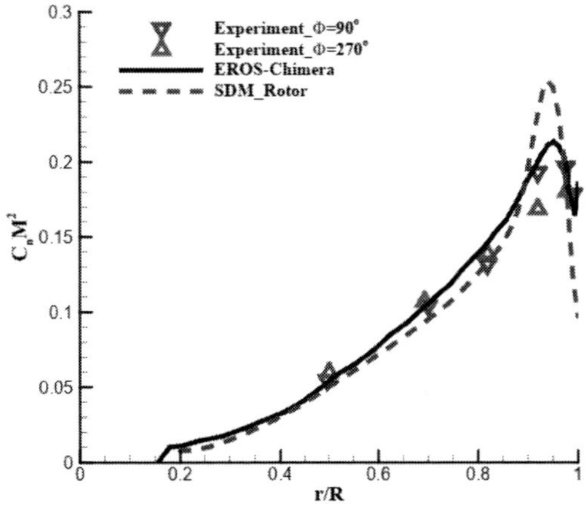

Figure 10. ONERA 7A hovering rotor case: Normal force coefficient distributions along blade radial position

Various solidity rotor

To confirm influence of correction equation varying solidity, single rotor hovering case is computed varying the number of ONERA 7A blade. Blade number vary from 4 to 10. Total thrust coefficient of each rotor shows in Figure 11.

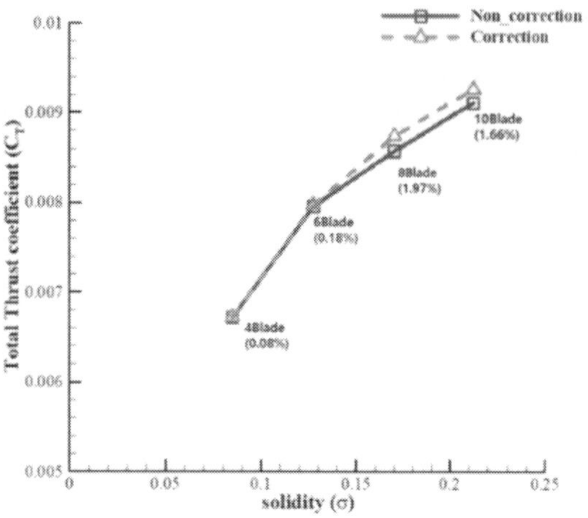

Figure 11. Total Thrust coefficient trend of various solidity to confirm effectiveness of correction equation

As the solidity of rotor increases, difference between correction and non-correction appears. For the case of 4 blade and 6blade, difference between them is less than 1%. However, 8 blade and 10 blade cases have almost 2% of thrust coefficient difference. This indicates that the influence of neighboring blades important feature for ASM in high solidity rotor problem.

CONCLUSIONS

The main purpose of this research is improvement of actuator surface model(ASM) for computing various solidity rotor problem. The reference line velocity correction equation is derived from Biot-Savart law and applied to ASM calculations with rectangular wing and hovering rotor cases. Through comparisons of improved ASM results with experiment and CFD data, the validation is carried out. In various solidity rotor problem, consideration of neighboring blades had influence on prediction of rotor performance.

Author contact: Jaeyoung Park wodud051@pusan.ac.kr,
Sejong Oh tazo@pusan.ac.kr,
Donghun Park parkdh@pusan.ac.kr
Myungsik Tai tms3636@pusan.ac.kr

ACKNOWLEDGMENTS

This work was conducted at High-Speed Compound Unmanned Rotorcraft (HCUR) research laboratory with the support of Agency for Defense Development (ADD).

REFERENCES

[1] Kim, T., Oh, S., & Yee, K. "Novel Actuator Surface Method for Helicopter Rotor Analysis" *Journal of Aircraft* Vol. 53, No. 6 (2016), pp. 1947-1952.

doi:10.2514/1.C033666

[2] Sibuet Watters, C., Breton, S. P., & Masson, C. "Application of the actuator surface concept to wind turbine rotor aerodynamics" *Wind Energy*, *13*(5), 433-447.

[3] Dobrev, I., Massouh, F., & Rapin, M. "Actuator surface hybrid model" In *Journal of Physics: Conference Series* (Vol. 75, No. 1, p. 012019). IOP Publishing.

doi:10.1088/1742-6596/75/1/012019

[4] Shen, W. Z., Zhang, J. H., & Sørensen, J. N. "The actuator surface model: a new Navier–Stokes based model for rotor computations" *Journal of Solar Energy Engineering*, *131*(1), 011002.

doi:10.1115/1.3027502

[5] Khodadoust, A. "*An Experimental Study of the Flowfield on a Semispan Rectangular Wing With a Simulated Glaze Ice Accretion*" phD Thesis, University of Illinois at urbana-Champaign. 1993

[6] Schultz, K. J., Splettstoesser, W., Junker, B., Wagner, W., Schoell, E., Mercker, E., ... & Fertis, D. "A parametric windtunnel test on rotorcraft aerodynamics and aeroacoustics (Helishape)—test procedures and representative results" *The Aeronautical Journal*, *101*(1004), 143-154.

[7] Renzoni, P., D'Alascio, A., Kroll, N., Peshkin, D., Hounjet, M. H., Boniface, J. C., ... & Schöll, E. "EROS—a common European Euler code for the analysis of the helicopter rotor flowfield" *Progress in aerospace sciences*, *36*(5), 437-485.

Numerical Simulations of the Flowfield of Tiltrotor Aircrafts in Ground Effect Based on N-S Equations

Jiang Shuang Master student
Zhao Qi-Jun Professor
Zhang Hang Master student
National Key Laboratory of Science and Technology Rotorcraft Aeromechanics
Nanjing University of Aeronautics and Astronautics
Nanjing, China

ABSTRACT

Focusing on the interference flowfield of a tiltrotor aircraft in ground effect (IGE), a computational fluid dynamics (CFD) method based on hybrid grid technique is established in order to capture the viscous flow in the boundary layer which affects the aerodynamic performance of the tiltrotor aircraft. In this method, the Navier-Stokes equations have been taken as governing equations, the Spalart-Allmaras model is used as turbulence model, the discretization of convective fluxes and the time marching are completed by Jameson scheme and five-step Runge-Kutta iteration respectively. The influence of the tiltrotor on its flowfield is modeled as a distribution of momentum source. The flowfield characters of V-22 tiltrotor aircraft have been calculated at different altitudes in hovering and forward flight at low-speed in the IGE. In hover, the rotor wakes form spanwise flows along the ground, which meet underneath the centerline of the fuselage and form a fountain flow in a similar fashion to what happens on the wing's upper surface. The interference flowfield of tiltrotors is simulated at different forward flight speed, and the ground vortexes are captured at low speed. Furthermore, the flowfield of the tiltrotor aircraft is calculated. As the speed increases, the high-pressure zone on the wing's upper surface diminishes and moves to the wing tip fast.

INTRODUCTION

The tiltrotor aircraft has the ability of hovering in the sky and cruising in high speed, which couples the advantages of the helicopter and fixed-wing aircraft. Compared with traditional helicopter and fixed-wing aircraft, the tiltrotor aircraft has wider flight envelope. In hover and small-speed forward flight, the rotor is in the helicopter-rotor mode providing vertical tension which reduces requirements for the landing plane and the length taxiway. At the high-speed forward flight, the rotor leans forward until it reaches the fixed-wing propeller mode to achieve higher cruising speed. Therefore, both in military and civilian areas, it has a very wide range of applications.

Though the tiltrotor aircraft has the advantage of both helicopter and fixed-wing aircraft, it bring more aerodynamic problems at the same time, that affects aerodynamic performance. One of the most important factors affecting the aerodynamic performance of the tiltrotor aircraft is aerodynamic interference between the rotor and the wing in hover. It is important to carry out aerodynamic interference analysis of tiltrotor aircraft. And while on duty, experiencing nap-of-the-earth flight is a must. The flowfield IGE is more complex.

At present, the main research methods of tiltrotor interference flowfield are experimental methods[1-4]and computational methods[5-10]. However, due to the unsteady characteristics of the rotor in the transient state of the rotor, the study of the interference flowfield of the tiltrotor aircraft is mainly in hover state and forward flight state. The researches[11-16] show that in hover, as the ground height increases, the load on the tilt rotor decreases as the height decreases further, and even becomes an upward force. This has the net effect of reducing the download produced on the fuselage IGE.

Considering the computational efficiency and computational accuracy, a high-precision CFD method is applied to the simulation of the aerodynamic interference field of the tiltrotor aircraft IGE and momentum source is adopted to simulate the effect of the rotor on the flowfield.

NUMERICAL METHOD

V-22 is adopted to simulate the interference flowfield. In order to capture the viscous flow in boundary layer which affects the aerodynamic performance of tilt-rotor aircraft, considering multi-flight conditions of tilt-rotor aircraft, a hybrid mesh method (as shown in Fig. 1) is proposed in this paper, which is composed of prismatic (near boundary layer) and tetrahedral mesh (other fields). 8L×7L×3.6L(L is the length of fuselage) is adopted as the computation zone.

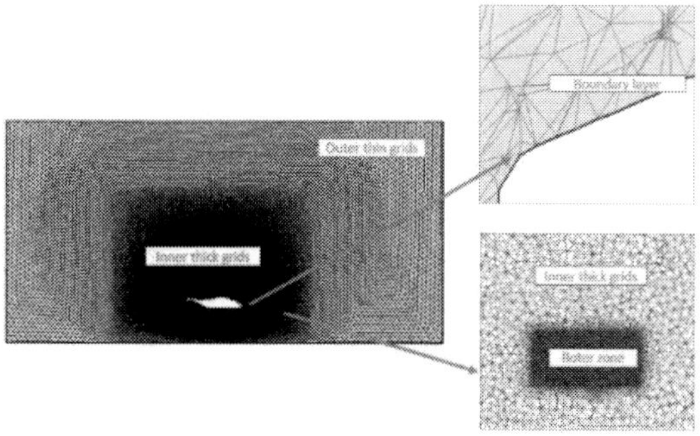

Figure 1. Grids for flowfield computation

N-S equations are adopted as governing equations, a CFD method is established for predicting the interaction flowfield of helicopter fuselage/rotor/ground, and in this method, the effects of rotor on flowfield are simulated by a momentum source model, the viscosity effect is calculated by one equation S-A turbulence model. Jameson second-order central difference scheme is employed in spatial discretization, and the five-step Runge-Kutta method is adopted in temporal discretization.

N-S equations containing momentum source is shown below:

$$\frac{\partial}{\partial t}\int_{\Omega} \boldsymbol{W} d\Omega + \oiint_{\partial S} (\boldsymbol{F}_c - \boldsymbol{F}_v) \bullet \boldsymbol{n} dS = \mathbf{Q} \quad (1)$$

$$\boldsymbol{W} = \begin{Bmatrix} \rho \\ \rho u \\ \rho v \\ \rho w \\ \rho E \end{Bmatrix}, \quad \boldsymbol{F}_c = \begin{Bmatrix} \rho \boldsymbol{V} \\ \rho u \boldsymbol{V} + p\vec{i} \\ \rho v \boldsymbol{V} + p\vec{j} \\ \rho w \boldsymbol{V} + p\vec{k} \\ \rho h \boldsymbol{V} \end{Bmatrix},$$

$$\boldsymbol{F}_v = \begin{Bmatrix} 0 \\ \tau_{xx}\vec{i} + \tau_{xy}\vec{j} + \tau_{xz}\vec{k} \\ \tau_{yx}\vec{i} + \tau_{yy}\vec{j} + \tau_{yz}\vec{k} \\ \tau_{zx}\vec{i} + \tau_{zy}\vec{j} + \tau_{zz}\vec{k} \\ \Theta_x\vec{i} + \Theta_y\vec{j} + \Theta_z\vec{k} \end{Bmatrix}, \quad Q = \begin{bmatrix} 0 \\ S_x \\ S_y \\ S_z \\ 0 \end{bmatrix}。$$

Where Ω is the volume of cell, t is time, ρ is density, $\boldsymbol{V} = (u, v, w)^T$ is absolute velocity, $\boldsymbol{V}_r = (u_r, v_r, w_r)^T$ is relative velocity, $\boldsymbol{V}_t = (u_t, v_t, w_t)^T$ is convected velocity, $\vec{i}$、$\vec{j}$、$\vec{k}$ is a unit vector in three directions in x、y、z directions, $\tau_{(\bullet)}$ $\Theta_{(\bullet)}$ are viscosity dependent terms. E and h are total energy and total enthalpy. Momentum source is adopted to simulate effect of the rotor on the flowfield. Fig.2 shows the blade element in azimuth angle ψ and radius r. Inflow angle

$$\beta_* = \arctan(V_Z / V_O) = \arctan\left(W_n / V_\theta\right) .$$

Supposing that the pitch angle is φ, then angle of attack is $\alpha = \varphi - \beta_*$.

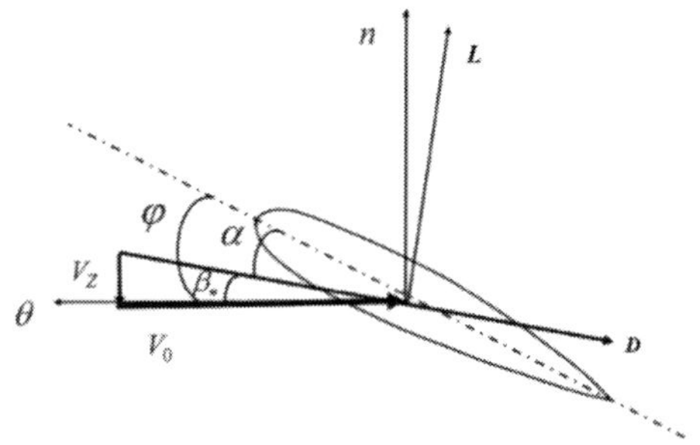

Figure 2. Sketch of flow and aerodynamics of blade element

Supposing that the chord length is b, C_l and C_d can be obtained. Then lift and drag are shown below:

$$\begin{cases} \boldsymbol{L} = \dfrac{1}{2}\rho V^2 C_l b dr \\ \boldsymbol{D} = \dfrac{1}{2}\rho V^2 C_d b dr \end{cases} \quad (2)$$

Where $V = \sqrt{U_r^2 + V_\theta^2 + W_n^2}$ is resultant velocity.

$$\begin{cases} d\boldsymbol{T} = \boldsymbol{L}\cos\beta_* - \boldsymbol{D}\sin\beta_* \\ d\boldsymbol{Q} = -\boldsymbol{L}\sin\beta_* - \boldsymbol{D}\cos\beta_* \end{cases} \quad (3)$$

Where dT and dQ are lift and revolving resistence

$dF = dT + dQ$ represents resultant force of flowfield on the blade element. Without considering the radial force and the effect of the blade on the flow field is $-dF$ based on the Newton's third law.

$$F = N \cdot (-dF) \cdot \frac{\Delta\psi}{2\pi} \tag{4}$$

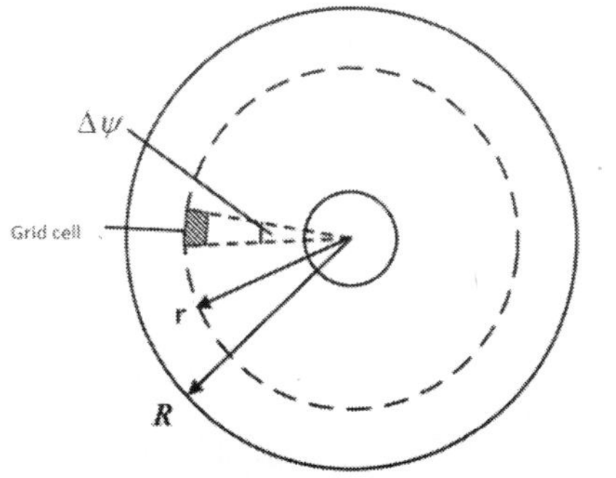

Figure 3. Momentum source in the rotor disc

In order to verify the validity of the momentum source method in simulating the rotor flowfield, the rotor flowfield in hover state is calculated by taking the rotor model as an example, the one can be compared with the experimental results in the literature (Ref. 17). Figure 3 shows the distributions of calculated and tested pressure values of different places. The calculated results have good agreements with the experimental datas in Ref. 17, which shows the accuracy of the method.

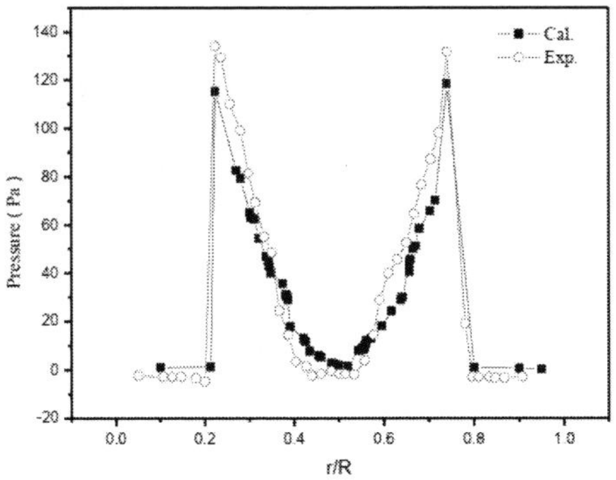

(a) 0.125R below the rotor disc

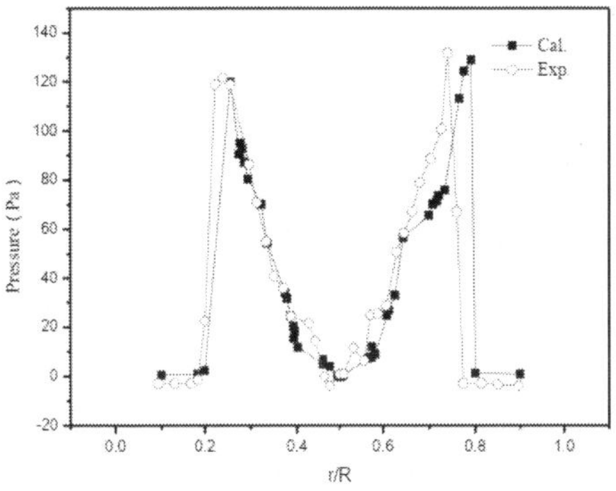

(b) 0.325R below the rotor disc

Figure 4. Comparison of calculated and experimental values of rotor pressure

RESULTS

Hover IGE

Figure 4 shows the interference flowfield around the tiltrotor aircraft, from which the whole flowfield can be obtained. In hover, the wings remain stationary, and the rotor down washes hits the surface of the wing, increasing the download. In hover, the rotor wakes form spanwise flows along the ground, which meet underneath the centerline of the fuselage and form a fountain flow in an similar fashion to "Fountain Effect" happening on the wing's upper surface, that makes the flowfield around the aircraft more complicated.

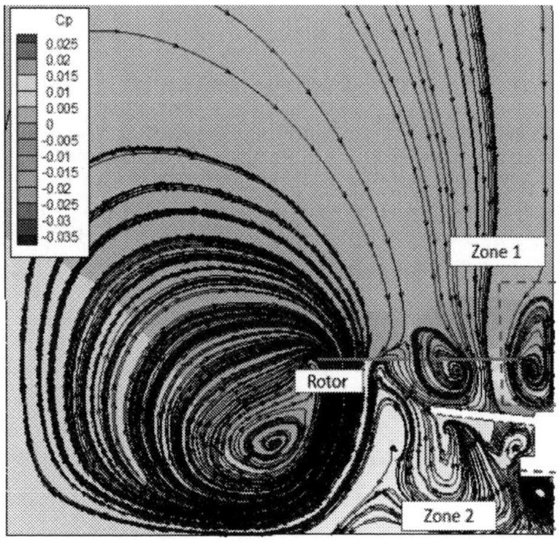

Figure 4. Flowfield diagram of rotor center section IGE

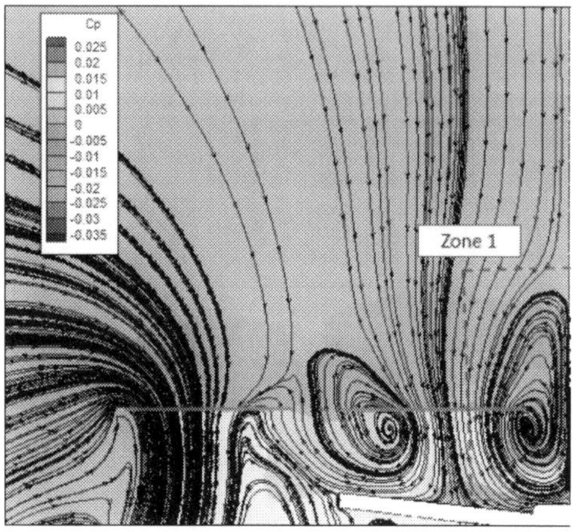

(a) Enlargement of Zone 1

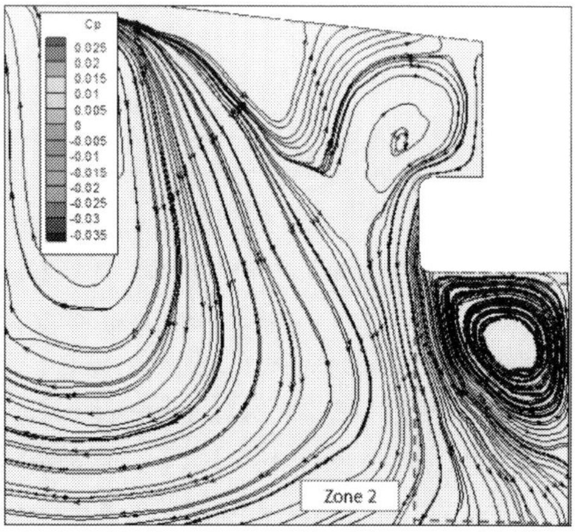

(b) Enlargement of Zone 2

Figure 5. Enlargement of part of the flowfield

Figure 6 shows the flowfield in different height (0.1R、0.5R、1.0R、1.5R、2.0R). With the increase of ground height, the downwash of the rotor to the middle of the fuselage decreases, and the ground effect decreases gradually. As the body is closer to the ground, the two airflows that meet in the middle of the fuselage turn upward due to ground obstruction. However, with the increase of the ground height, the obstruction of the ground to the airflow decreases or even disappears, so that the airflows can move downward freely, and the ground effect on the flowfield will weaken.

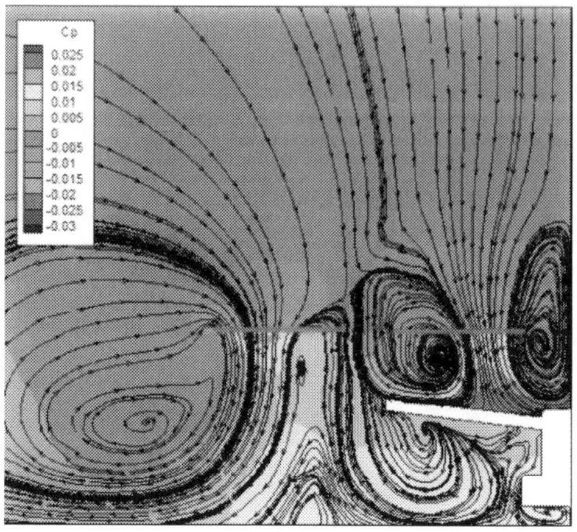

(a) h/R=0.1

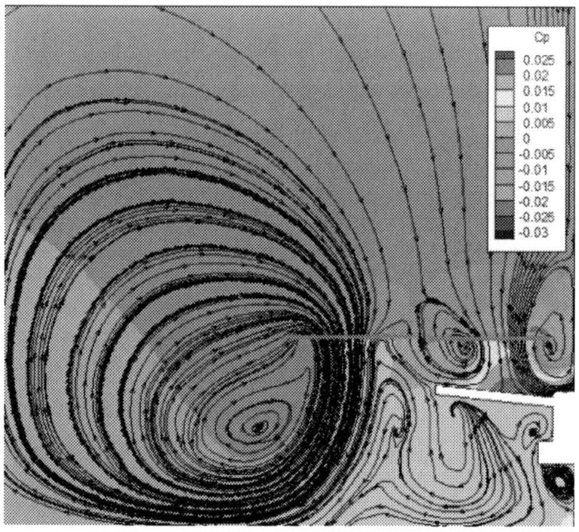

(b) h/R=0.5

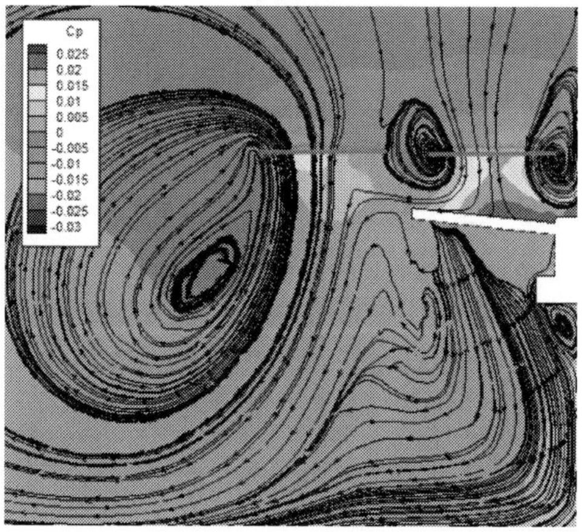

(c) h/R=1.0

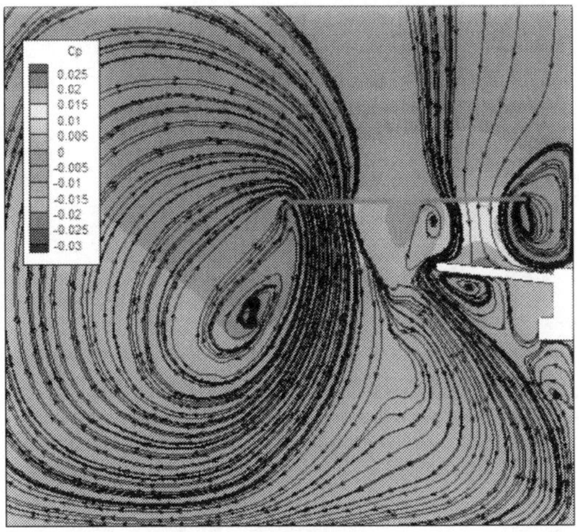

(d) h/R=1.5

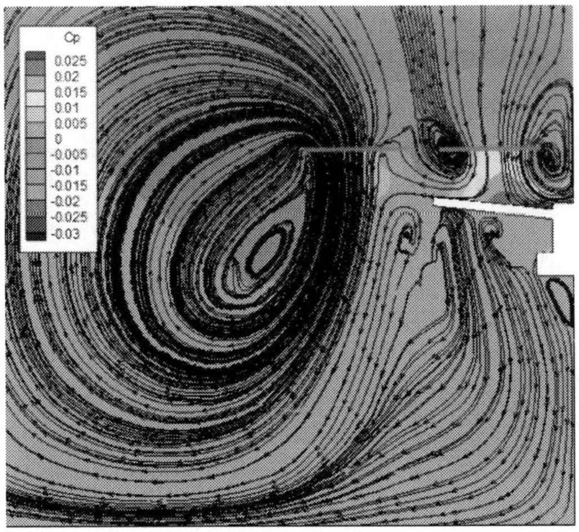

(e) h/R=2.0

Figure 6. Flowfield in different ground height

Figure 7 shows the comparisons of pressure coefficient between IGE and OGE (h/R=0.1). The bottom pressure IGE is higher than that OGE in the inner side because of the effect of fuselage which increases the ground effect to some degree. In the middle of the wing where is not close to the body, the bottom pressure IGE is lower than OGE.

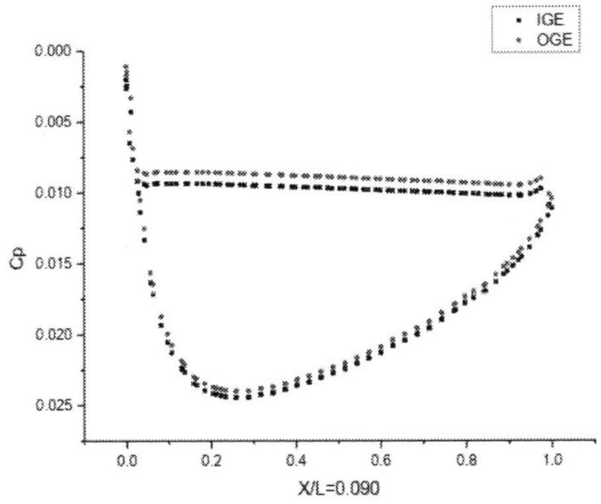

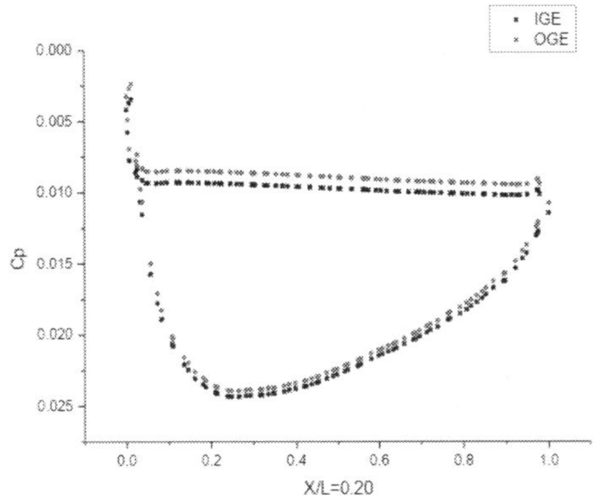

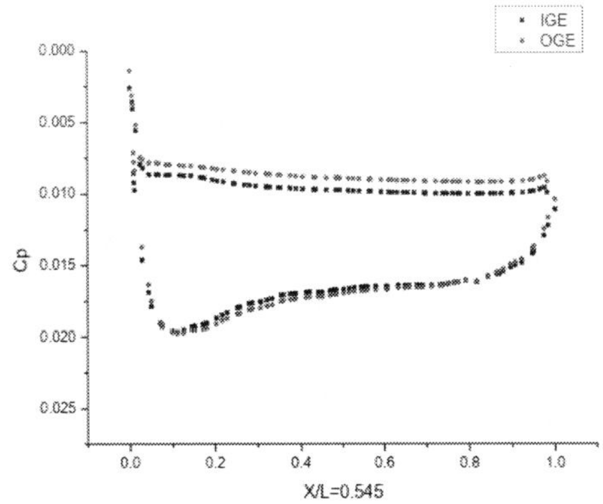

Figure 7. Comparisons of pressure coefficient between IGE and OGE along the spanwise of wing

Low speed flight IGE

Figure 8 shows the interference flowfield between the rotors and ground, which indicates that as μ increases gradually, the ground vortex gradually appears, and move along the direction of the inflow stream. As μ continues to increase, the ground vortex continues to diffuse and weaken until it disappears. At the same time, with the increasement of μ, the deflection effect of the rotor on the flow decreases gradually.

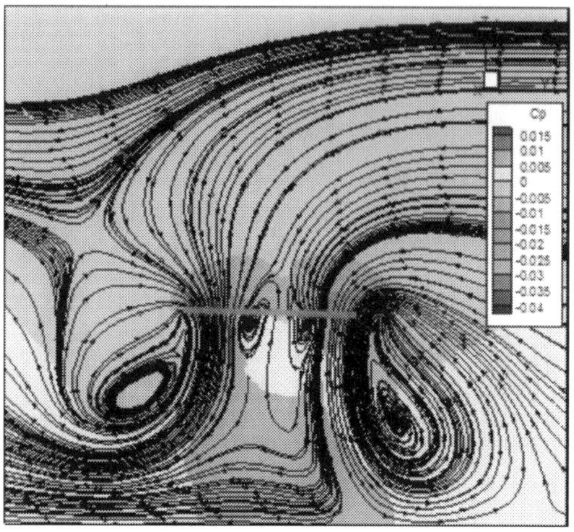

(a) μ=0.01

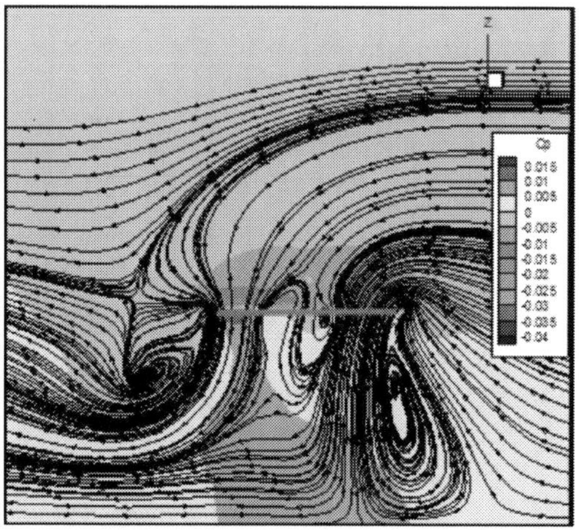

(b) μ=0.02

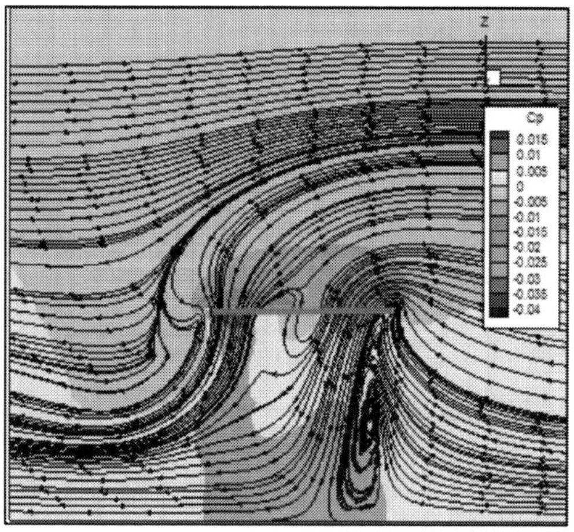

(c) μ=0.03

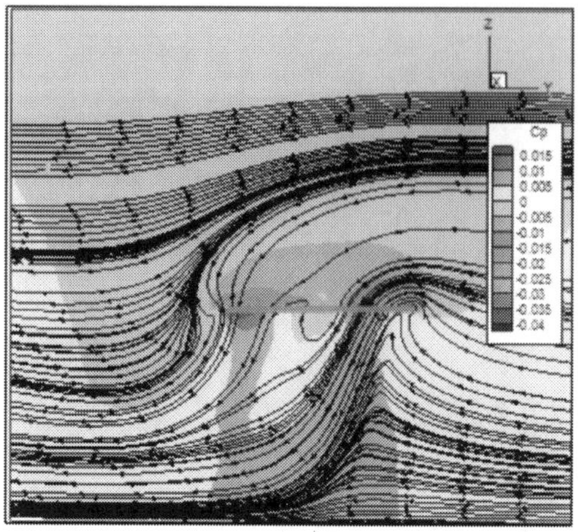

(d) μ=0.04

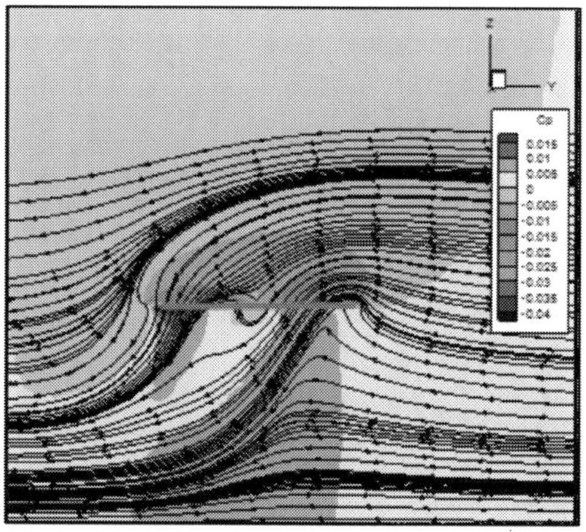

(e) μ=0.05

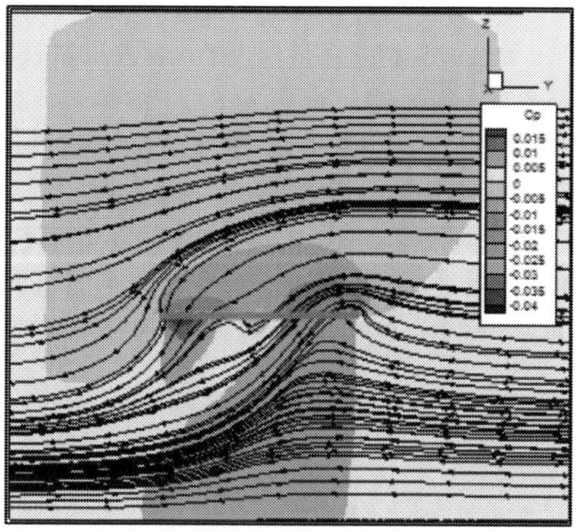

(f) μ=0.06

Figure 8. Flowfield in h/R=1.0

Figure 9 shows the distribution of pressure coefficient. When the forward speed is low, there is a high-pressure zone on the upper surface of the wing which indicates lift loss of the rotor because the downwash hits the upper surface directly and vertically. As the speed increases, the high-pressure zone is rapidly becoming smaller with the position moving to the tip.

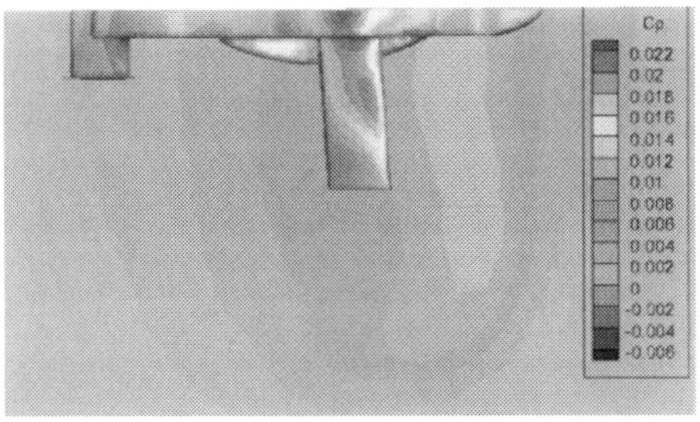

(a) μ=0.02

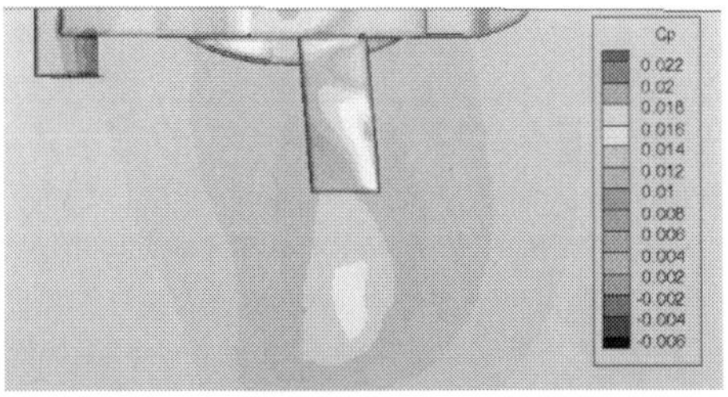

(b) μ=0.04

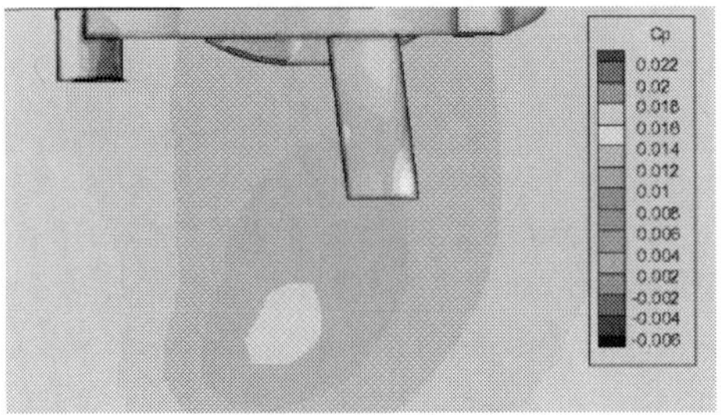

(c) μ=0.06

Figure 9. Distribution of pressure coefficient

SUMMARY AND CONCLUSIONS

Focusing on the interference flowfield of the tiltrotor in ground effect. A high-precision CFD method coupled with hybrid grids is applied to the simulation of the aerodynamic interference field of the tiltrotor aircraft IGE and momentum source is adopted to simulate the effect of the rotor on the flowfield. Some conclusions are obtained:

(1) With the increasement of ground height, the downwash of the rotor to the middle of the fuselage decreases, and the ground effect decreases gradually.

(2) Ground vortex appears first in low speed forward flight. As μ continues to increase, the ground vortex continues to diffuse and weaken until it disappears.

(3) The downwash hits the upper surface vertically which results in a high pressure zone there. As the speed increases, the high-pressure zone is rapidly becoming smaller with the position moving to the tip.

REFERENCES

[1]Felker F F, Maisel M D, Betzina M D. Full-scale tiltrotor hover performance. Journal of the American Helicopter Society. 1986, 31(2): 10-18.

[2]Felker F F, Signor D B, Young L A, et al. Performance and loads data from a hover test of a 0.658-scale V-22 rotor and wing. NASA TM-8419, 1987.

[3]Johnson W, Lau B H, Bowles J V. Calculated performance, stability, and maneuverability of high-speed tilting-prop-rotor aircraft. NASA TM-88349, 1986.

[4]Mcveigh M. The V-22 Tilt-rotor large-scale rotor performance/wing download test and comparison with theory. Proceedings of the 11th European Rotorcraft Forum, London, England: 1986.

[5]Meakin R L. Moving body overset grid methods for complete aircraft tiltrotor simulations. AIAA Paper 1993-3350., 1993.

[6]Ruith M R. Unstructured, multiplex rotor source model with thrust and moment trimming–Fluent's VBM model. Proceedings of 23rd AIAA applied aerodynamics, Toronto: 2005: 6-9.

[7]Yeo H, Johnson W. Aeromechanics analysis of a heavy lift slowed-rotor compound helicopter. Journal of Aircraft. 2007, 44(2): 501-508.

[8]Sheng C, Narramore J C. Computational simulation and analysis of bell boeing quad tiltrotor aero interaction. Journal of the American Helicopter Society. 2009, 54(4): 42002.

[9]Wissink A, Potsdam M, Sankaran V, et al. A coupled unstructured-adaptive Cartesian CFD approach for hover prediction. Proceeding of the 66th Annual Forum of American Helicopter Society, Alexandria: 2010: 1300-1317.

[10]Abras J, Narducci R. Analysis of CFD modeling techniques over the MV-22 tiltrotor. Proceeding of the 66th Annual Forum of American Helicopter Society, Alexandria: 2010: 11-13.

[11]Marr R L, Ford D G, Ferguson S W. Analysis of the wind tunnel test of a tilt rotor power force model[J]. 1974.

[12]Tomassoni J E, Taylor R B, Delarm L N, et al. Wind Tunnel Test of a Powered Tilt-Rotor Dynamic Model on a Simulated Free Flight Suspension System. Volume 6[R]. BOEING VERTOL CO PHILADELPHIA PA, 1971.

[13]Fradenburgh E A. Aerodynamic factors influencing

overall hover performance[J]. Aerodynamics of Rotary Wings, 1972, 7(1).

[14]Wilson J C, Mineck R E, Freeman C E. Aerodynamic characteristics of a powered tilt-proprotor wind tunnel model[J]. 1976.

[15]Desopper A, Routhieau V, Roth G. Study of the low speed characteristics of a tiltrotor[J]. 2002.

[16]HEUZÉ O, DIAZ S, DESOPPER A, et al. Simplified models for tiltrotor aerodynamic phenomena in hover and low speed flight[C]. The Aerodynamics Research Conference, CEAS Aerospace, Cambridge, UK. 2002.

[17]McKee J W, Naeseth R L. Experimental investigation of the drag of flat plates and cylinders in the slipstream of a hovering rotor, NACA TN 4239, 1958.

Development of Nonlinear Vortex Lattice Method and Its Application to Aerodynamic Analysis of Wind turbine

Hakjin Lee
Ph.D. student
Korea Advanced Institute of Science and Technology
Daejeon, Republic of Korea

Yeongmin Jo
Ph.D. student
Korea Advanced Institute of Science and Technology
Daejeon, Republic of Korea

Duck-Joo Lee
Professor
Korea Advanced Institute of Science and Technology
Daejeon, Republic of Korea

ABSTRACT

The accurate aerodynamic models for the wind turbine is needed to design a more efficient and reliable rotor blades of wind turbine. The vortex method is one of the most suitable models, because of its affordable computational cost and reasonably good results. However, it cannot capture the nonlinear aerodynamic behaviors that occur pre-stall region in low Reynolds number and stalled flow. In this research, nonlinear vortex lattice method has been developed to overcome the drawbacks of vortex lattice method by using table loo-up and vortex strength correction process. The proposed method has been utilized to predict aerodynamic performances of NREL Phase VI rotor models that exposed to both attached and stalled flow in operating conditions. The results show that the nonlinear vortex lattice method can yield the accurate results in both axial and yawed flow conditions.

INTRODUCTION

Wind turbine is becoming a pressing issue in industry and academia because it is one of the most promising renewable energy technology. The wind turbine industry and the market has been continuously growing in recent years. The rotor blades of the wind turbine convert the kinetic energy of wind into aerodynamic loadings that will be translated into electrical power. In order to design more efficient and cost-effective blade, it is important to predict accurately the aerodynamic performance of wind turbine system and to model the detailed flow characteristic around rotor blades.

Several aerodynamic models for the wind turbine like Blade Element Momentum Theory (BEMT), Computational Fluid Dynamics (CFD), and Vortex method have been develop to accurately predict the aerodynamic loads on the rotor blade of wind turbine. The BEMT has simple structures and easy to implementation. However, several correction model is need to consider three-dimensional effect tip loss, yawed or tilted flow, dynamic inflow, because it inherently has some assumptions. CFD based methods can capture the unsteady and transient aerodynamic behaviors that occurs around the rotor blade, and accurately describe the flow fields. However, the problems related to the turbulence modeling and unexpected numerical dissipation error due to discretized domain still remain. Although the computing performance of computer has been considerably improving in recent years, the CFD method still requires the high computational costs and resources. The vortex method is one of popular models to predict an aerodynamic performance of rotary systems such as helicopter, wind turbine and propeller, because of its affordable cost and reasonably good results. It can give more physical solution than BEMT, and give more cost-effective solution than CFD.

The vortex method is composed of various blade modeling, such as lifting line, vortex lattice (lifting surface), and panel method. Among them, the lifting line method has been extensively used for aerodynamic analysis of wind turbine rotor by coupling with free-wake model (Refs. 1-2). This approaches cannot correctly calculate the strength of bound vortex when the wind turbine is exposed to stalled flow condition, because the vortex method inherently cannot consider the viscous effects like boundary layer separation. Therefore, there have been attempts to consider the aerodynamic behaviors in post-stall region. A. van Garrel (Ref. 3) suggested that lifting line method with nonlinear vortex strength computation. In this method, the bound vortex strengths are determined by iteratively matching the lift force from Kutta-Joukowski theorem with the lift force that is obtained from table look-up with the local velocity and effective angle of attack. However, the camber effect cannot consider, because the rotor blade is modeled with only one chordwise vortex panel. Kim et al. (Ref. 4) developed the vortex lattice method with nonlinear vortex correction method to take account of the effects of separated flow on bound vortex strength. In order to find the corrected bound vortex strength of each blade section, a nonlinear system of equation has to be constructed and solved by using optimization algorithm. The location of control points where the effective velocity and angle of attack were calculated is mathematically ambiguous, and the bound vortex strength of each chordwise vortex lattice elements will be adjusted by the same amount with regardless of their location.

The purpose of this research is to develop the nonlinear vortex lattice method that can consider the camber effect and nonlinear aerodynamic behaviors in the post-stall region.

Presented at the 6th Asian/Austrailian Rotorcraft Forum / Heli Japan 2017, Kanazawa, Japan, November 7-9, 2017.

Our co-author suggested and mathematically proved the location of control points where the lift force is calculated by table look-up for nonlinear vortex lattice method. In the proposed method, the bound vortex strength of chordwise elements will be corrected depending on their location. The nonlinear vortex lattice method is used to simulate the NREL Phase VI rotor model which is subjected to both attached and separated flow conditions. Compared to experimentally measured data, it was observed that the proposed nonlinear vortex lattice method can give more accurate solutions.

NUMERICAL METHOD

Vortex Lattice Method

Among the vortex methods, the vortex lattice method is used to calculate the aerodynamic loads generated by the rotor blades of wind turbine. Compared to lifting line method, the main advantage of using this method is that the boundary conditions are imposed on the actual lifting body, which can have camber and various planform shapes. The rotor blade is represented by a lifting surface with zero thickness and discretized into a number of quadrilateral vortex ring elements that consists of four concentrated (lumped) vortex lines with constant strength along the vortex lines. The leading edge segment of each vortex ring element is placed on the quarter chord line, and the collocation point where the flow tangency boundary condition will be imposed is at the center of three-quarter chord line. The unknown strength of bound vortex is determined by solving a linear system of equations, which derived from the general integral solution of Laplace equation with no normal flow boundary condition. The aerodynamic loads can be computed by unsteady Kutta-Joukowski theorem as following:

$$\vec{F}_{ij} = \rho\left(\vec{V}_{ij}\times\vec{\Gamma}_{ij}\Delta l_{ij} + \frac{\partial}{\partial t}\vec{\Gamma}_{ij}\Delta S_{ij}\right) \tag{1}$$
$$= \rho\left(\vec{V}_{ij}\times\vec{\Gamma}_{\tau_i}\Delta b_{ij} + \vec{V}_{ij}\times\vec{\Gamma}_{\tau_j}\Delta c_{ij} + \frac{\partial}{\partial t}\vec{\Gamma}_{ij}\Delta S_{ij}\right)$$

$$dL_j = \sum_{i=1}^{N}\vec{F}_{ij}\cdot\vec{n}_{ij} \tag{2}$$

where $\vec{V}$ is a total velocity including freestream, rotational velocity, and induced velocity from rotor blade and wake at each collocation point, $\vec{\Gamma}_{\tau_i}$ and $\vec{\Gamma}_{\tau_j}$ are the circulation strength of vortex lines along the chordwise and spanwise direction, respectively. dL_j is sectional lift force and N is a number chordwise vortex lattice elements.

Vortex Particle Method

The wake structures shed from rotor blades are represented as number of vortex particles instead of vortex filaments. The strength of vortex particles has been already determined at the previous time step by imposing Kutta condition at the trailing edge vortex panels. These vortex particles mutually affect each other and is allowed to freely distort and move to the downstream.

The vorticity field in downstream can be represented by asset of S number of Lagrangian vector-valued particles.

$$\vec{\omega}(\vec{x},t) = \nabla\times\vec{u} = \sum_{i=1}^{N}\vec{\alpha}_i(t)\xi_\sigma\left(\vec{x}-\vec{x}_i(t)\right) \tag{3}$$
$$= \sum_{i=1}^{N}\vec{\omega}_i(t)V_i\xi_\sigma\left(\vec{x}-\vec{x}_i(t)\right)$$

During time-marching step, the vortex particles will be convected with the local velocity. The induced velocity by the vortex particles is calculate as following equation.

$$\vec{u}(\vec{x},t) = \sum_{i=1}^{S}K_\sigma\left(\vec{x}-\vec{x}_i(t)\right)\times\vec{\alpha}_i(t) \tag{4}$$
$$= \sum_{i=1}^{S}\frac{-g_\sigma(\vec{x})}{|\vec{x}|^3}\left(\vec{x}-\vec{x}_i(t)\right)\times\vec{\alpha}_i(t)$$

where $K_\sigma(\vec{x})$ is the regularized Biot-Savart kernel. For more detailed description of vortex particle method, refer to the references (Refs. 5-8)

Nonlinear Vortex Lattice Method

In post-stall region, vortex lattice method tend to excessively overpredict the lift force that results from failing to properly calculate the circulation strength of vortex ring elements, because it cannot inherently deal with an influence of stalled flow. In the present research, the nonlinear vortex lattice method coupled with aerodynamic table look-up and vortex strength correction process has been suggested to overcome this limitation

In this study, the control points where the local sectional lift forces are calculated from table look-up are specified at the center of half camber line. A problem that is associated with the location of control points has been mathematically solved by our co-author. It turned out that the most suitable location of control point for nonlinear vortex lattice method is half camber line (Ref. 9). The main advantage of using vortex lattice method and control points on the camber line is that the camber effect can be included.

The lift forces of each blade section depend upon the local inflow velocity and effective angle of attack. The local inflow velocity at control point is composed of time-dependent kinematic velocity components due to freestream and the motion of rotor blades, and the vorticity contribution of rotor blades and wake. Therefore, the inflow velocity included the velocity components induced by the wake vortices. It is known at each time step, and is used to estimate the local inflow angle at control point as following:

$$\vec{V}_{inflow} = \vec{V}_{wind} + \vec{V}_{motion} + \vec{V}_{vorticity} \tag{5}$$

$$\alpha_{inflow} = \tan^{-1}\left(\frac{\vec{V}_{cp} \cdot \vec{a}_3}{\vec{V}_{cp} \cdot \vec{a}_1}\right) \tag{6}$$

where $\vec{V}_{cp}$ and α_{cp} are local inflow velocity and inflow angle at control point. The inflow velocity included the e $\vec{a}_1$ and $\vec{a}_3$ are the unit vectors along parallel and normal to rotating plane, respectively.

The effective angle of attack at control points is defined by

$$\alpha_{effective} = \alpha_{inflow} - \theta_{geometric} \tag{7}$$

where $\theta_{geoemetric}$ is local geometric pitch angle including local twist angle and collective pitch angle. The lift and drag coefficients corresponding to local inflow velocity and effective angle of attack can be obtained by table look-up. The lift force generated by the blade section is computed by

$$dL_j = \frac{1}{2}\rho\left|\vec{V}_{inflow,j}\right|^2 C_{l,j} dA_j \tag{8}$$

The effective angle of attack of blade could exceed a stall onset angle when wind turbine is subject to high speed wind. The vortex lattice method cannot consider a nonlinear aerodynamic behavior that occurs in stalled flow, and has a tendency to give over-predicted aerodynamic forces and circulation strength on the blade and wake. If the aerodynamic table contains the lift coefficients over stall onset angle, the sectional lift force in the post-stall region can be approximately calculated. However, the solution might be inaccurate, because the vortex strength is used to compute the local inflow velocity and inflow angle. An influence of stalled flow on the vortex strength should be included to obtain more accurate solution. In present research, the circulation strength of vortex ring elements on the blade were corrected by matching two lift forces that are computed from unsteady Kutta-Joukowski theorem and table look-up. Figure 1 describes the nonlinear vortex strength correction process.

1. Find the initial circulation strength of vortex ring elements on the blade by solving a set of linear equations Γ_{ij}

2. For each blade section, define the ratio of the strength of chordwise elements to the strength of the vortex element placed at the trailing edge, and averaged circulation strength

$$\lambda_{ij} = \frac{\Gamma_{ij}}{\Gamma_{T.E.j}}, \quad \Gamma_{avg,j} = \sum_{i=1}^{N} \Gamma_{ij} / N$$

3. Compute the inflow velocity, inflow angle and effective angle of attack at control points for each blade section

4. Obtain the aerodynamic coefficients for each blade section through table look-up, and compute the corresponding lift forces

5. Compute a new circulation strength for each blade section by matching Equation (2) and (8) $\Gamma^{k}_{table,j}$

6. For each blade section, update the circulation strength using the difference between the new and current values with under-relaxation factor γ

$$\Gamma_j^{k+1} = \Gamma^{k-1}_{table,j} + \gamma(\Gamma^{k}_{table,j} - \Gamma^{k-1}_{table,j}) = \Gamma^{k-1}_{table,j} + \gamma\Delta\Gamma^{k}_{table,j}$$

7. Apply the updated strength of each blade section into total $N \times M$ vortex ring elements using vortex strength ratio

$$\Gamma_{ij} = \Gamma_j^{k+1}\lambda_{ij}$$

8. It the strength difference between previous and current time step is large than some convergence criterion go to Step 2, else go to Step 8

9. Apply the wake rollup for time-marching wake convection with updated circulation strength

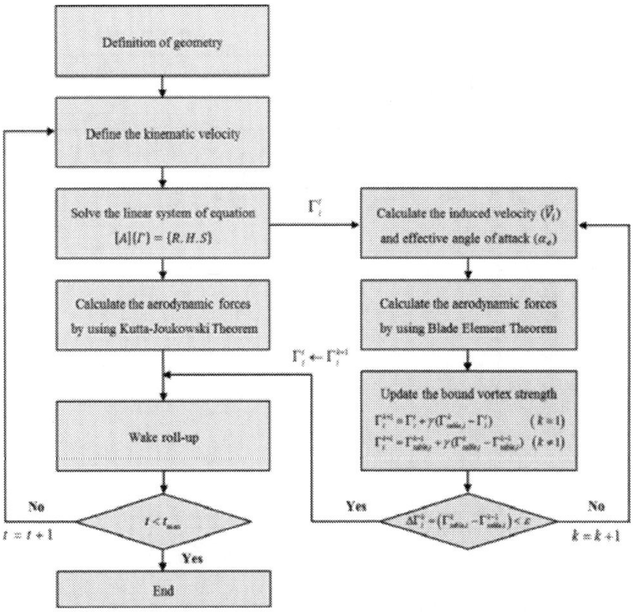

Figure 1. Flow chart for nonlinear vortex lattice method

NUMERICAL RESULTS

NREL Phase VI rotor model

The measurements of wind turbine were conducted in the natural environment where the direction and magnitude of wind continuously and randomly change with a high uncertainty. However, the data obtained by this field test cannot directly compare with results predicted by the numerical simulation due to an uncontrolled wind condition. In order to overcome this problem, NREL Phase VI experiment with a dimeter of 10m rotor blade that consisted of S809 airfoil with tapered-twisted planform shape was conducted in 24.4m × 36.6m NASA-Ames wind tunnel facility in 2000 (Ref. 10). They provide measured data including surface pressure, normal and tangential force distribution along radial direction in both axial and yawed flow conditions. This measurements have extensively been used to validate an accuracy of aerodynamic models for wind turbine.

Axial flow condition

NREL Phase VI rotor model is subject to various wind speed corresponding to both attached and stalled flow condition. During wind tunnel experiment, the rotor speed remains constant. Flow conditions are listed in Table 1.

Table 1. Axial flow conditions

Run	Wind speed [m/s]	Rotating speed [RPM]	Air density [kg/m^3]
S050000	5.0	71.7	1.244
S070000	7.0	71.9	1.246
S100000	10.0	72.1	1.246
S130000	13.0	72.1	1.227
S150000	15.1	72.1	1.224
S200000	20.1	72.0	1.221
S250000	25.1	72.1	1.220

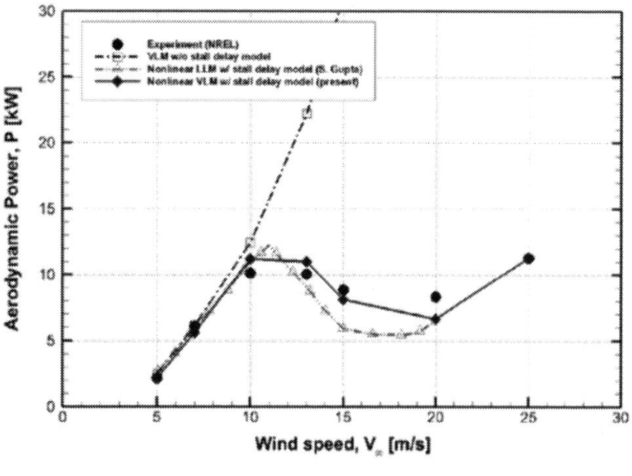

Figure 2 Comparison of the measured and predicted aerodynamic power for NREL Phase VI rotor in axial flow condition

Results and discussion

Figure 2 shows the comparison of experimentally measured data (Ref. 10) and numerically predicted aerodynamic power depending on wind speed. When the wind speed increased over 10m/s, the separated flow start to cover the rotor blade from inboard to tip section that result in decreased power generation. The massive flow separation on the rotor blade is taking place for wind speed above 20m/s. However, the vortex lattice method cannot completely capture a tendency of power regulation that occurs due to separated flow. The nonlinear lifting line method suggested by S. Gupta (Ref. 11) over-predicted the stall onset of wind speed and the excessively under-predicted the aerodynamic power in stalled flow condition, even with the inclusion of stall delay model. The numerical simulation results show the proposed nonlinear vortex lattice method with Raj and Selig stall delay model (Ref. 12) yields the most accurate results. The reason for improved results could be the circulation strength correction on vortex lattice elements and the camber effect of rotor blade. The effect of camber of rotor blade on quality of solution will be examined in the future research.

CONCLUSIONS

Predicting an accurate rotor performance of wind turbine in attached and separated flow condition is needed to design more cost-effective rotor blade. In the present research, the nonlinear vortex lattice method was developed to for aerodynamic analysis of wind turbine. The main advantage of this method is that both camber effect of blade geometry and nonlinear aerodynamic characteristics that occurs in stalled flow can be considered. It was used to compute the aerodynamic power generated by NREL Phase VI rotor. The numerical simulation results showed that the proposed method can give the reasonable solutions at the non-stalled and stalled flow condition. The nonlinear vortex lattice method will be applied to predict the rotor performance of highly cambered blade to confirm the camber effect. It will also be utilized for rotor simulation in yawed flow condition.

Author contact:
Hakjin Lee, hakin@kaist.ac.kr
Yeongmin Jo, yminjo@kaist.ac.kr
Duck-Joo Lee djlee@kaist.ac.kr

ACKNOWLEDGMENTS

This research was supported by the Climate Change Research Hub of KAIST (Grant No. N11170061). This work was also supported by the Korea Institute of Energy Technology Evaluation and Planning (KETEP) and the Ministry of Trade, Industry & Energy (MOTIE) of the Republic of Korea (No. 20168520021200)

REFERENCES

[1]Gupta, S., and Leishman, J. G., "Performance Predictions of the NREL Phase VI Combined Experiment Rotor using a Free-Vortex Wake Model," Paper AIAA 2006-390, 44th AIAA Aerospace Sciences Meeting and Exhibit, Reno, Nevada, January 9-12, 2006.

[2]Dueque, E. P. N., Burklund, M. D., and Johnson, W., "Navier-Stokes and Comprehensive Analysis Performance Predictions of the NREL Phase VI Experiment," *Journal of Solar Energy Engineering*, Vol. 125, (4), Nov. 2003, pp. 457-467.
doi: 10.1115/1.1624088

[3]Garrel van A., "Development of a Wind Turbine Aerodynamics Simulation Module," ECN-C-03-079, 2003.

[4]Kim, H., Lee, S., and Lee, S., "Numerical Analysis on The Aerodynamics of HAWTs using Nonlinear Vortex Strength Correction," *Current Applied Physics*, Vol. 10, (2) Mar. 2010, pp. 311-315.
doi: 10.1016/j.cap.2009.11.033

[5]Winckelmans, G. S., "Topics in Vortex Methods for the Computation of Three- and Two-Dimensional Incompressible Fluid," Ph. D. dissertation, California Institute of Technology, 1989.

[6]Lee, D. J. and Na, S. U., "Numerical Simulations of Wake Structure Generated by Rotating Blades using a Time Marching, Free Vortex Blob Method," *European Journal of Mechanics – B/Fluids*, Vol. 18 (1), 1999, pp.147-159.
doi: 10.1016/S0997-7546(99)80011-9

[7]Park, S. H., Chu, Y. B., and Lee, D. J., "Investigation of Main Rotor-Tail Rotor Interaction in Hover Flight using Vortex Particle Method," 5th Asian-Australian Rotorcraft Forum, Singapore, November 17-18, 2016.

[8]Park, S. H., Chu, Y. B., Lee, D. J., and Jang, J. S., "Trimmed Simulation of a Complete Helicopter Configuration in Hover Flight," AHS 72nd Annual Forum, West Palm Beach, Florida, USA, May 17-19, 2016.

[9]Jo, Y. M., Lee, H. J., and Lee D. J., "Prediction of Rotor Flow for Unmanned Aerial System Using Nonlinear Vortex Lattice Method," Presented at the 6th Asian/Austrailian Rotorcraft Forum / Heli Japan 2017, Kanazawa, Japan, November 7-9, 2017.

[10]Hand, M. M., Simms, D. A., Fingersh, L. J., Jager, D. W., Cotrell, J. R., Schreck, S., and Larwood S. M., "Unsteady Aerodynamics Experiment Phase VI: Wind Tunnel Test Configurations and Available Data Campaigns," NREL/TP-500-29955, 2001.

[11]Gupta, S., "Development of a Time-Accurate Viscous Lagrangian Vortex Wake Model for Wind Turbine Applications," Ph.D dissertation, University of Maryland, 2006.

[12]Raj, N. V., "An Improved Semi-Empirical Model for 3-D Post-Stall Effects in Horizontal Axis wind Turbines," Master's thesis, University of Illinois at Urbana Champaign, 2000.

Interim report of study about requirements of UAV sense and avoid system

Akitoshi SAKAGUCHI
Chairman of the UAS Committee
The Society of Japanese Aerospace Companies(SJAC)
Minato-ku, Tokyo, JAPAN

ABSTRACT

Market of Unmanned Aerial Vehicle (UAV) is rapidly growing, and risk of collision between UAV and aircrafts or ground objects is more serious. Especially in Japan, the risk of collision between helicopters operated in low altitude and VTOL UAVs (helicopter type or multi rotor type) will be main issue, because it is difficult for existing air traffic control system to cover all UAV flights and low altitude and VFR operation of helicopters in Japan. The Society of Japanese Aerospace Companies (SJAC) is studying technical requirements for collision avoidance system responding to the situation.

Collision avoidance system is classified Air Borne Sense and Avoid (ABSAA) and Ground Based Sense and Avoid(GBSAA). Due to incompleteness of data link between UAV and ground control station, UAV should be have any ABSAA to prepare the loss of data link. But ABSAA may be too large or too heavy for small UAV to cover enough range. So GBSAA should be alternative or back up for small UAV's ABSAA.

This paper is the interim report of SJAC activity developing basic requirements of ABSAA and GBSAA based on some scenarios of operation, and suggesting ABSAA system combined with radar, camera and ADS-B and GBSAA using radar.

INTRODUCTION

Currently, various approaches have been taken jointly with the public and private sectors concerning industrial development utilizing UAV called "the Industrial Revolution in the Sky" in JAPAN. As an example, in May 2017, for the realization of level 4 operation(Beyond Line of Sight flight over third parties) of UAV around 2020 , "the Industrial Revolution in the Sky road map" (ref.1) was announced by the public and private committee for the improvement of the environment related to the small unmanned aircraft.

In response to these developments, SJAC established "the Unmanned Aircraft Systems Committee" (UAS Committee) in April 2016, in order to disseminate knowledge and opinions from Japan's aerospace manufacturing industries.

In this paper, the UAS Committee shows an interim report of study about the requirement of Sense and Avoid (SAA) between manned aircraft and UAVs, which we studied in FY 2016.

Presented at the 6th Asian/Austrailian Rotorcraft Forum / Heli Japan 2017, Kanazawa, Japan, November 7-9, 2017.

RESULT AND DISCUSSION

Currently, the rapid spread of small multi-rotor type UAVs is exploring the possibility of various industrial uses, but on the other hand, accompanying the rapid increase in the operational opportunities of UAVs in various situations, risk of collision with aircrafts and/or ground objects is becoming serious. Particularly in Japan, we consider it is the most imminent problem that helicopters with VFR such as doctor helicopters flying at low altitude have serious risk of a collision or near miss with UAVs. It is difficult for low altitude VFR aircraft to obtain sufficient support from air traffic control in Japan, but on the other hand, as the performance of UAV is increased, so flight of UAV in the same airspace with manned aircrafts is real problem faced with.

Also, in the operation of UAVs, the means to replace the "Pilot's Obligation for Keeping Watch" specified in the Civil Aeronautics Act, that is, the realization of SAA is an important issue for UAV. Though many countries have also taken various developments, there is no practical system which can withstand actual operation, either standardized internationally at the present time.

Therefore, SJAC assumed that it is the most recent challenge in Japan that VFR flight of a manned helicopter at low altitude and flight of VTOL type UAV which may coexist in a same airspace. We are studying the necessary requirements of the SAA system for such situation.

Organization of prerequisites

As the SAA requirements review is diverse, the UAS Committee assumed scenarios to serve as the basis for consideration based on the following preconditions.

(Precondition 1)

A manned aircraft has right of way to UAV. (Therefore, avoidance of collision is the duty of the UAV side) (ref.2)

(Precondition 2)

Japan's existing air traffic control system can't adequately monitor low altitude / VFR manned aircrafts and UAVs.

(Precondition 3)

In order to realize collision avoidance, a three-level hierarchy such as Planning, Control and Self Separation is necessary. UAV Traffic Management (UTM) corresponds to the Planning and Control phase, and SAA is Self Separation system. (ref.3)

(Precondition 4)

Some studies of a collision avoidance system for UAV are also conducted abroad, however, those predicting airplanes cruising at relatively high altitudes with constant course are mainstream and can't be directly applied to low altitude VTOL aircraft and UAV.

Based on the above four assumptions, the conditions for collision avoidance considered by the UAS committee are shown in Table 1, and the assumed scenarios are shown in Table 2.

Item	Selected Condition	Reason of selection
Size of UAS	Below Aircraft without pilot (Small Aircraft)	Japanese Regulation divides UAS and Aircraft without pilot which is same size of manned A/C
Flight Rule	VFR	It can't rely air traffic controller.
Target of SAA	Manned Helicopter	In Japan, helicopters are main aircrafts access to low altitude in which UAS fly.
Weather Condition	VMC	Because of VFR flight
ATC	N/A	Because of VFR flight
Airspace	Class E & G	There is high risk to collision or near miss due to No ATC support.
Operation of UAS	Beyond Line of Sight RPAS	Higher requirement of regulation than LOS operation. ICAO requirement

Table1. conditions for collision avoidance

No	Scenario	UAS	Target of SAA	Altitude	Day/Ngt	Visibility	Max Num UAS	[illegible]
1	Flying for disaster Observation and Helicopter coming to land	VTOL UAS Velocity 0km/h	Helicopter Velocity 50km/h	Below 50m	Day	1500m	About 100	Tempolaly
2	Flying for transportation and Helicopters coming across	VTOL UAS Velocity 100km/h	Helicopter Velocity 200km/h	About 150m	Day & Night	5000m	About 10	Permanent
3	Flying for inspection of infrastructure	VTOL UAS Velocity 50km/h	Power lines Bridge Building Mountains (above UAS)	Below 50m	Day	1500m	About 5	Tempo/ Permanent
4	Bad Weather	VTOL UAS Velocity 100km/h	Cloud	About 150m	Day	Surface can be insight	About 10	Permanent

Table2. assumed scenario

Based on the above criteria and scenarios, we studied ABSAA, GBSAA, and collision avoidance rules.

Study of ABSAA

As data link between UAV and ground control station is practically difficult to ensure complete reliability, ABSAA which is mounted on UAV and makes UAV independently achieve collision avoiding , is considered a mandatory system. Assuming collision avoidance between a manned helicopter (flight velocity 200 km / h) and a large unmanned helicopter (flight velocity 100 km / h) as the scenario, we made a conclusion that a combination of radar, optical sensor and ADS - B receiver is effective as ABSAA. As for the optical sensor for ABSAA, the maximum detection range is assumed to be about 500 m even with a 4K camera. So it is assumed that range of radar for ABSAA is required about 5 km to 10 km.

Therefore, it is considered that ABSAA system using radar as the main sensor and complementing the optical sensor with identification of objects that are difficult to grasp with radar, such as weather conditions or obstacles on the ground is effective.

It is possible to grasp the position information of the target by the ADS - B receiver installed in UAV. However, since there are some aircraft without ADS-B, it seems to be insufficient as ABSAA only by the ADS - B receiver. Other moving objects management systems (such as JAXA's DNET etc.)(ref.4) can also be used, but there are some constraints like data update rate, they should be used as supplementary for ABSAA.

Study of GBSAA

ABSAA assumed in the previous paragraph may be difficult to be installed in all in small scale UAV, due to size, weight or electric power supply. For some small UAVs that can't mount radars for ABSAA, it is necessary to detect objects far from by the equipment on the ground as alternative. Especially, assuming aircrafts without ADS - B or any position identifying device dedicated to UAV, GBSAA should have surveillance radar on the ground. By our estimation, as radar for GBSAA, a system that secures a position resolution of about 100 m at a detection distance of about 10 km is assumed to be a realistic performance. As for the GBSAA radar, it is desirable to be integrated as a function of UTM.

Study of collision avoidance rules

There is no international or domestic common rule at present for avoiding collision maneuver when UAV detects a conflict object. In particular, there is no study worldwide on collision avoidance rules between manned helicopters of low altitude and unmanned helicopter or multi-rotor (VTOL aircraft).

At present, we consider it is one of option that avoidance maneuver should be implemented for minimum interval of Near Mid - Air Collision distance (horizontal radius 150 m / vertical ± 30 m) prescribed by FAA. However, considering the influence of manned helicopter's downwash and wake turbulence on small unmanned aerial vehicles, the near miss distance of manned aircraft and UAV should be set in the future. The timing to start avoidance maneuvering can't be determined at the present time due to uncertain of the avoidance maneuver performance of UAVs, such as acceleration, deceleration, climb, descent, turning, but in the other example, around 30 seconds is assumed as one index. Since there are various restrictions such as flight performance, flight altitude, detection range of sensor, downwash of manned helicopter etc. upward and downward avoidance maneuvers assumed by the usual TCAS may not be suitable in this scenario of VTOL. We may consider unique maneuvers, including hovering in the air. Also, assuming that manned aircraft have priority right of way to UAV, UAVs change the course for manned aircrafts. Regarding avoidance between UAV and UAV, it is necessary to review the tasks and set unique rules as necessary, while referring to the flight rules of manned aircraft.

Policy of activities from FY 2017

Demonstration of collision avoidance system is planned in Ministry of Economy, Trade and Industry (METI)/ New Energy and Industrial Technology Development Organization (NEDO) "Realization project of energy saving society where robot drone is active". In addition, at the "Study Group on UAV flight beyond line of sight and over the Third Party" held jointly by METI, Trade and Industry and the Ministry of Land, Infrastructure, Transport and Tourism (MLIT), will discuss about requirements examination including avoidance of collision with manned aircraft .Based on the study of the collision avoidance system by SJAC, we can expect to contribute to the realization of the collision avoidance system in Japan by using the above framework.

CONCLUSIONS

The UAV Committee of SJAC assumes collision avoidance between a manned helicopter and a VTOL type UAV carrying out low altitude VFR flight is supposed to be the most recent issue in Japan. ABSAA, GBSAA and collision avoidance rules have been examined for the scenario. In the future, we will continue to work together with related government agencies and organizations to develop collision avoidance systems and to realize the development of operational rules.

REFERENCES

[1]METI, *Industrial Revolution in the SKY Roadmap* http://www.meti.go.jp/policy/mono_info_service/mono/robot/drone.html, May 2017

[2]JCAB, *About Measures of collision avoidance between aircraft and unmanned aircraft* http://www.mlit.go.jp/common/001172562.pdf Feb.2017

[3]ICAO, *Manual on Remotely Piloted Aircraft System" Doc10019* 2015

[4]Keiji KOBAYASHI et al, "*Development of Disaster Relief Aircraft Management System Network (D-NET)," Aeronautical and Space Sciences JapanVol. 64 (2016) No. 1* ,2016

Simulation and evaluation of linear velocity control by Cellular Air Traffic System (CATS) on high-density logistical VTOL-UAS traffic

Yu Ito
Corporate researcher
Yamato Holdings Co., Ltd.
Tokyo, Japan

ABSTRACT

For a realization of delivery or logistical VTOL-UAS transport in the urban area, management over high-density traffic is required. Conventional practices and studies on air traffic control (ATC) having problems for this situation, a new method of flight control is required. Adapting and adjusting the signaling block system for railroad control into a two-dimensional manner, the needed concept of ATC for high-density traffic will be constructed, given its name as Cellular Air Traffic System (CATS). In CATS, the only function needed for ground navigation computers is to receive GPS location from each vehicle, reflect them on the virtual map, and send the information of the map to each vehicle. The ability of enhancing speed planning and course plot on each onboard computer will help the craft travel more efficiently, the improvement will be up to each carrier keeping the basic cost for running the infrastructure low. In this study, how CATS will actually act like in high-density traffic, and what detailed configuration of this system will be the most efficient for the practice is evaluated by cellular automata simulation using spreadsheet software.

INTRODUCTION

As technology for inexpensive civilian UAS been rapidly developed since the mid-2010s, various business ideas to use them as a completely new airborne platform have emerged. In response to the future vision that a swarm of "drones" should fly in business areas, how to control them and keep both ground and airspace free from capable accidents have been one of the major concerns in the research of this field.

Existing studies have faced this issue on a regulative and comprehensive approach. The concept of UTM (Unmanned Aircraft System Traffic Management) proposed by Kopardekar et al. (Ref. 1), the de facto most-dominant project towards this area in 2017, is aimed to make itself the public protocol to control all kinds of unmanned air vehicles, including small multicopters for surveillance, delivery drones, or fixed wing UASs. Though this kind of common traffic management law is essential to avoid a hazardous and chaotic airspace crowded with Hobbes-like unregulated drones, conceiving all kinds of unmanned aircrafts regardless of their propulsion, configuration, and purpose will not necessarily provide maximum efficiency to each of the systems; aside from public regulating protocols, each business entity has also to establish flight management systems in order to pursuit maximum efficiency regarding its own business and crafts. The concept of UTM itself also does not deny the act of each business, and assumes them to become external linked system of the entire UTM structure.

Presented at the 6th Asian/Australian Rotorcraft Forum / Heli Japan 2017, Kanazawa, Japan, November 7-9, 2017.

Based on this awareness, this study is going to construct a flight control system specially designed for urban autonomous VTOL logistics aircrafts which is named CATS (Cellular Air Traffic System) from its method of decision applying the idea of cellular automata. First, the motive and philosophy of the system will be explained. How this private system can be connected to existing and conceived public aviation environment (e.g. ATC or UTM) to form an ecosystem as a part will also discussed. Finally, simulation of CATS and its evaluation will be conducted using a simplified model using spreadsheet software.

MOTIVE

The subject service of the air traffic system to be discussed is heavier-than-air aircrafts capable of autonomous navigation and equipped with VTOL ability, whose main purpose of flight is carrying cargo from a point to another inside urban area. This type of aircrafts has three operational features that discriminate themselves from other flying vehicles.

First, the biggest virtue of aircrafts as a transport is that they can fly almost straightly to the destination free from terrain or artificial structure. This will even increase their advantage to ground transports of their service speed, in addition to that of their high cruise speed. Especially at cities in growth, locations that need to be linked by transports tend to scatter regardless of road configuration, and thus, are rarely connected by strait roads. As reference 3 suggests, the course should be set above the urban skyline to avoid buildings in order to realize this strength of urban flight.

Second, thanks to their VTOL ability, they can not only land on or take-off from limited spaces, but also change velocity and course, and even stop (hover) without losing altitude. Moreover, because they don't carry passengers like the aerial taxi concept conceived in Reference 2, and therefore rideability does not have to be considered, larger momentum is acceptable for their payload. This allows more crowded traffic than that with fixed-wing aircrafts, since they can safely avoid collision and share narrower space with others on the same altitude level as automobiles on ground do. However, it should be reminded that this ability differs by the craft's configuration (e.g. whether it is an unmanned helicopter, multicopter, tiltrotor, tiltwing, lift-fan craft, or a tail sitter), total mass, and power.

Finally, the service fleet has to be in large scale and dense. Because of their high service speed and relatively light payload, compared to ground transports, one of the considerable business models will be to use multiple crafts to maximize their efficiency and cover up the lacking payload of each craft, and therefore, the density of crafts in an area should be high. This almost necessarily means that both the craft and the fleet has to be navigated by computers, because human control on such complicated operation would be impossible, or even if not, cost the service provider a tremendous labor cost.

The most suitable system for navigation has to rationalize all these characteristics. One of the conceivable solutions for high density traffic, linearly derived from the current and existing transport system, is to make virtual highways that resembles to the Skyways project proposed in Reference 4, in the sky that cross over each other in levels, so that the aircraft can avoid mid-air collision by adjusting the distance between the aircraft ahead, as automobiles do. However, in reality, this solution will not be simple as in above. Primarily, there is only a limited height to construct multi-crossing layers that is decided from the height of the buildings in the city as its minimum altitude to the boundary with the piloted aircrafts' traffic as its maximum altitude (see Figure 1). Secondly, regardless of the aim to make the system simple, every craft has to be provided with a high-level sense-and-avoid system, because there is still danger of mid-air collision left such as when merging into another virtual highway, danger of one craft to fail maintain its designated altitude and stray into another traffic, or a swarm of bird flies into its path[1]. Regarding such assumption, it will kill the advantage of a simple, therefore power-and-payload-friendly autopilot configuration to be loaded. Finally, the number of lines connecting two among the multiple available loading points in urban and suburban area will be enormous, and are almost impossible to register them all. Even if to combine resembling paths and rationalize them into determinate number of important routes, cities are always changing organically, and location of economic activities are to change them time by time. Therefore, to maintain the system rational means that the routes have to be rearranged constantly, that will need an organizational structure for such practice.

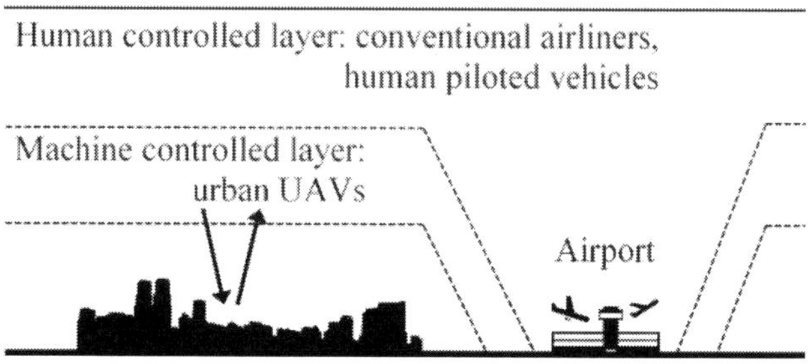

Figure 1 Concept of urban UAV flight layer

Another solution will be to oppositely install all the craft a high-level autopilot on each craft and let fly while avoiding collision by themselves, without any special infrastructure for the system. References 5 and 6 suggest installing such computer on manned aircraft to alternate the current complicated ATC made by human operators. However, this solution also has two major problems when to apply on unmanned urban VTOL aircrafts. One is that it requires any but all aircrafts to install a designated fast and highly reliable computing autopilots, that will consume precious power (and probably payload), and moreover increase the cost of each craft whose great number is needed for operation. Secondly, the process of decision making for a high-density VTOL traffic will become even more complicated than just to consider existing fixed-wing airliners, because it has to consider multiple crafts, more than two, to come across in the crowded traffic. In fact, Reference 5 mainly considers avoidance path planning for two crafts to come across, and withholds the solution for three or more crafts to become extremely complicated.

Therefore, a completely new method to use neither both of the solutions referred above should be established for an urban logistics mission.

PHILOSOPHY

In order to fulfill the requirement for a completely new guiding method for urban logistics, traffic control method of neither automobiles nor fixed-wing aircrafts but railroad is referred in this study. Trains which only move linearly (or, in a one-dimensional manner) are controlled by signals given to separated track sections with electrical existence detection called blocks in relation with the other train traveling just ahead of one: red light shows the next block is occupied by another train and therefore suggests to stop; yellow indicates train on near and orders caution

[1] In Luc Besson's film "The Fifth Element" (1997), failure in such aerial system of the 23rd century earth is comically illustrated, firstly as the hero nearly collides with another aerial car just as he starts his taxi from his garage, and secondly his taxi is caught in an accident and falls to an altitude that contains the traffic opposite to his original direction, triggering a chain of chaotic events on the traffic.

and slow speed; green means the track is free of other trains for a distance and allow maximum speed to enter the next section. When the Shinkansen was developed in Japan as a high-speed track standard, technicians developed a new way of exclusive control to alternate conventional signal in fear of drivers missing the signal in high velocity and colliding into the train ahead. The newly introduced system was to transfer the signaling information to onboard devices and apply brakes automatically and designedly without overrun (Ref. 7). The point of this system for the Shinkansen was that location of trains are recognized by pre-divided blocks free from observation error, and trains can be controlled safely however simply in relation with just the occupation status of the blocks directly given to the onboard device. Adapting and adjusting this idea into a two-dimensional (not three because of the lack of height of the traffic layer) manner, the needed concept of clearance control for the UAS will be constructed. In this concept, the linear blocks on railroad track will become cells on maps, that gives itself provisional name in this study as, Cellular Air Traffic System (CATS). The configuration of CATS will be stated as following.

Virtual map

Air space will be divided into cells of (several-) hundred-meters square on the virtual map. The status of each cell reflects its occupation and speed limitation subjectively to a craft as:

Green: The cell is reserved exclusively for the vehicle. Entering the cell in maximum velocity is allowed. The status is shown as red to other crafts.

Yellow: Several crafts are allowed to enter the cell. Speed inside this cell is restricted to a proper level that VDs can stop suddenly for collision avoidance. When the cell is occupied with maximum number of crafts allowed, status turns red for other VDs.

Red: Entering the cell is prohibited while it is red.

White: No status registered. The cell can be reserved into green or yellow for entering now and after.

Black: Area which flight control by CATS is prohibited, e.g. high mountains, airports, military and government facilities, or nuclear power stations. The cell cannot be entered nor reserved.

Figure 2 shows the image of this virtual map applied to central Tokyo. The status is only shown as white or green, however, the upper-left section that contains important public facilities, such as government offices, the Metropolitan Police Department, the US Embassy, and the Imperial Palace is registered as black to prohibit UAS for good.

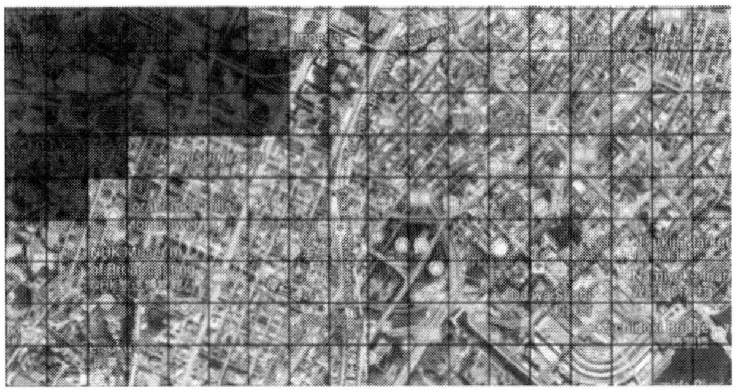

Figure 2 Image of the virtual map applied to central Tokyo area (image originates in Google Maps)

White cells neighboring the cell to be passed by one craft exclusively in maximum speed (green cell for the vehicle) will turn yellow for other crafts. The definition of "neighborhood" is adopted from Moore's study on cellular automata (Ref. 8) that indicates eight surrounding cells including those in northeast, southeast, southwest, and northwest location to its own. The occupation area of the craft on the map will be ranged in about a hundred-meters diameter from the center of the airframe. Vehicles with this existence area to enter whether entirely or partially will be regarded as "inside" the cells (Figure 3).

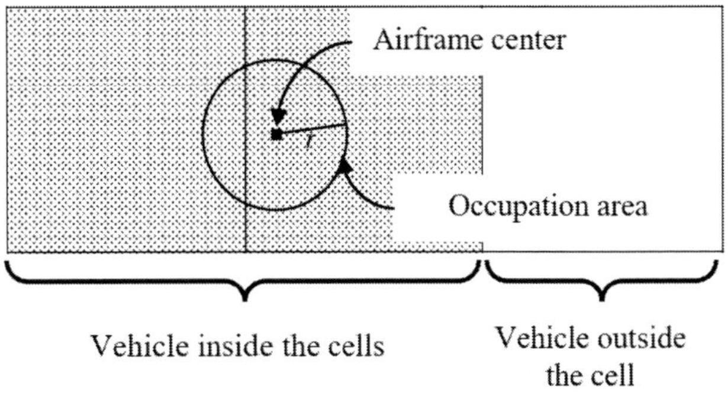

Figure 3 Definition of the location of the vehicle and occupation of cells

System configuration

The system consists of ground control computer and onboard computers on each craft. Regarding the name of the system, the former will be named "Mother" and the latter "Kitten". Each craft will be equipped with a GPS locator, broad band communication, Kitten computer, and sense-and-avoid systems to be used only in low velocity. Mother communicates with multiple Kittens and gives them instruction whether to enter the next cell on their planned path.

The conversation between Mother and Kitten is illustrated as Figure 4. The function needed for Mother system is to receive GPS location from each vehicle, reflect them on the virtual map, and send the clearance information to each Kitten. On the other hand, Kitten is designed simply to react to the instruction from Mother and guide the aircraft as told. Thanks to the existence of the virtual map, decision making of both Mother and Kitten is made very simple.

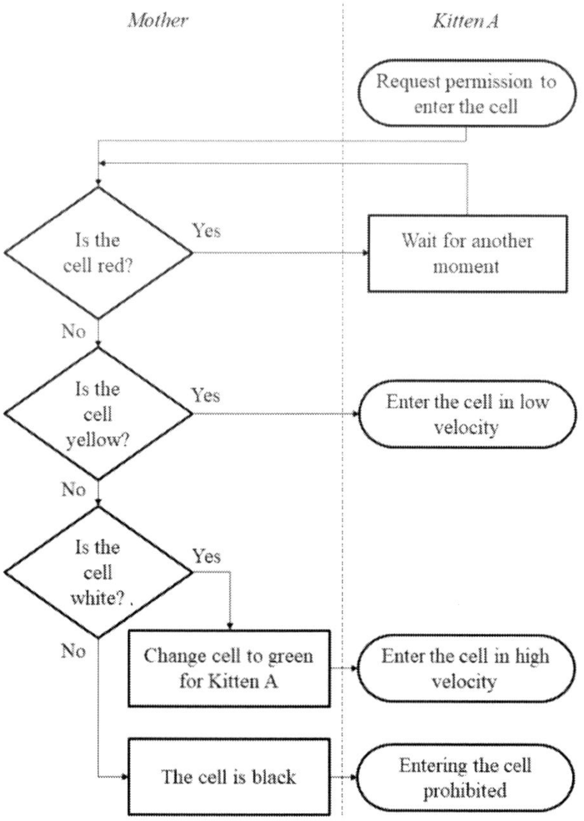

Figure 4 Conversations between Mother and Kitten

The function of Kitten can be considered as in multiple levels. The simplest assumption is that they only set course at the initial phase and adjust velocity as told from the Mother. In a more sophisticated configuration, Kitten will avoid the location where the traffic is stuck, and set new course towards its destination. The ideal phase of the Kittens would be those capable of set and change course by themselves regarding the wind condition at the point. Whatever the version of the Kitten is, Mother can accept them and form a system as a whole. This Internet-like thought of liberty should allow a constant and gradual transformation on the function of Kitten, and therefore the total CATS, without major service stop.

At its simplest function, each route has to be designed in advance in the control center referred later; if not, two crafts traveling straightly towards each other, or crafts to connect two points over blacked cells will be stuck on their course for ever. Even when manually designing route for the simple Kitten, the whole balance of configuration does not have to be considered, and chaotic setting of course is acceptable; the only two simple rules that have to be obeyed is; (1) keep right [or, left] when facing straightly another vehicle, and (2) avoid black cells. The course can be at the shortest if it is to fulfill these two rules.

Operation

In a two-dimensional manner, relation among multiple vehicles have to be considered, while regarding only the single relation between the train ahead is required on signaling block. This causes the difference that CATS have to show the status of the cells to each Kitten individually, while railway signals show objective status, which becomes a little more complicated in the system. However, by following this regulation, each Kitten can simply adjust speed and course obeying the indication given as the status of each cell while following essentially the initial flight plan without multi-level crossing.

Though the system is designed to minimize human involvement into the service, the operation of CATS will require some form of control center to give additional information in time to the system. When to consider the simplest function of Kittens, course for the crafts as described in the former section will also be designed in this control center manually. However, this operation is very easy and does not require re-designing of the whole system.

For example, all needed to add a landing zone into the CATS is just registering the status of the cells on and near the location of the new vertiport to yellow permanently to limit the airspeed to avoid collision of arriving and departing flights, at the same time allowing multiple aircrafts to enter the zone. By connecting flight information of ATC or monitoring the situation on each cell, external information such as approach of a piloted aircraft or a swarm of birds can be taken into the system for collision avoidance. This maneuverability will also enhance the social importance, because it can provide safe air space for observation, search, rescue, ambulance, or emergency logistics for vehicles under CATS control instantly by changing the cell status to yellow or red in occasion of disaster, such as large magnitude fire, earthquake, tsunami, or terrorism. If Kittens do not have the function to plot course automatically, the task to redesign course after the additional information reflected to CATS will be also required for the control center.

Ecosystem

CATS will not stand by itself, and will interact with other structure that consists the air space traffic. The position of CATS in the whole aerospace traffic management structure is to become one of the locally optimized solutions for UAS traffic. This means that the concept needs to be designed referring the dynamically changing policy of UTM. Aerodynamical character of crafts also have to be considered whether they can join this system or not, since their ability to slow down or stop (hover) in the air differ among each other. Therefore, the system defines requirement for the crafts to join.

By considering ATC as a system of communication for the presence information of flying objects, the ecosystem to surround CATS becomes even obvious. Regarding the OSI reference model (Ref. 9) for computer communication, an abstract layer model for aircraft communication can be illustrated as Figure 5. The basic idea of abstract layer is that information is transferred between systems containing the common protocol in each layer. Each user (pilot/autopilot for this study) is placed on the top of the structure, and the lower the layer goes, the wider the inclusion of the upper system (usually) becomes. This structure in the Internet system allows various types and versions of computers to be connected for information exchange, which the same is required for a safe air traffic management.

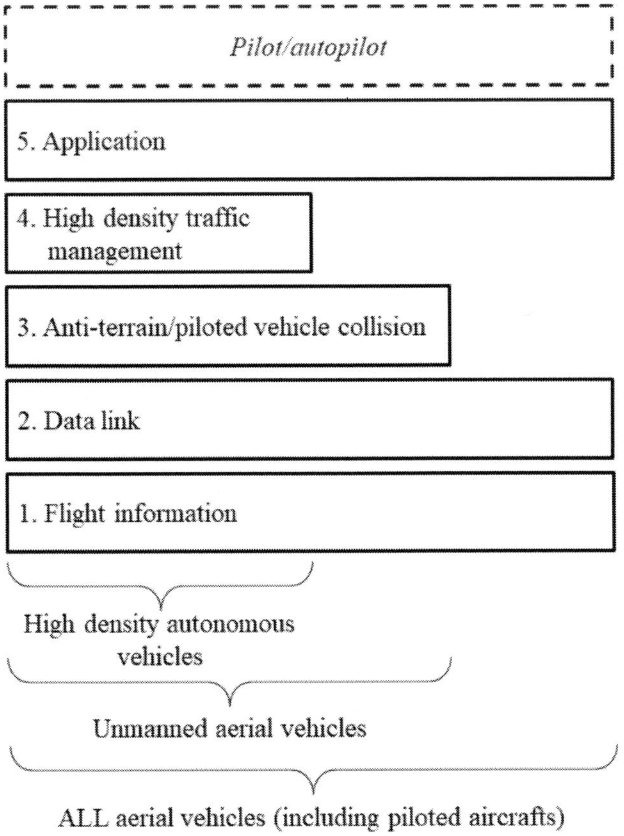

Figure 5 An abstract layer model for aircraft communication

CATS locate itself over the fourth and fifth layer: Mother is to be located on the fourth layer, "high density traffic management", while Kitten will be on the fifth "application" layer. Common UAS traffic control architecture including UTM should be related to the third, "anti-terrain/piloted vehicle collision" layer. This means that CATS will conversate with the lower layer for information of CATS controlled vehicles and non-CATS aircrafts (including piloted vehicles), while aircrafts both under CATS control will exchange their presence information exchange simply on the platform of CATS.

SIMULATION

In order to verify that the concept of CATS actually works and does not contain impossibilities in its philosophy, simulation using spread sheet software was conducted. Microsoft Excel 2016 was used for applicability for future extended studies. For the configuration of the simulator, see Appendix. The simulation itself was successful of showing vehicles guided by Kittens can avoid each other on a cellular divided airspace following an algorithm set in Mother.

As an initial study, the simulation held here assumed the function of Kitten to be at the simplest state, or in other words, only to adjust velocity of the craft. Therefore, Kitten were given almost no function except to plot course to connect their indicated start and goal, and tell Mother their current position. In further studies, how to make Kittens smarter so that they can re-plot course have to be considered.

Three crafts were placed in an area filled by ten-by-ten cells. The findings from the trial process of the construction of the simulator are stated as follows.

Findings

The most important findings from the construction of the simulator was that Mother had to keep virtual maps somewhat as many as the number of Kittens. Because the status of cells is shown to each craft subjectively, each cell has to contain its information related to every Kitten. Though these multiple maps could be piled up and ordered visually, the number of information to be contained in the system does not change essentially. Because of the nature of each map of the Mother to relate to each Kitten one-by-one, this map will be called "Nipple" further in this study.

This means that the number of Kittens that can be operated by one Mother is limited at her number of Nipples originally designed. When considering reality, controlling one city with a single Mother will lack expansibility in a long run, because it is almost impossible to predict the growth or reduction of an active city, and therefore has the danger to yield a massive lack or surplus of the computing resource for a long period of time. To optimize the use of CATS throughout a long time, it would be better to assign Mother to each subdivided sector and control the whole city by a higher rank computer that relates Kitten moving to the neighboring sector and its controlling Mother's Nipple. (Of course, this upper ranked computer should be named "Grandmother.") This configuration shall be studied in further research.

Because of this same number of Kittens and Nipples in total, essentially, there are no difference in the whole computing process between whether a central computer (Mother) proceeds the traffic management, or is it negotiated between nodes (Kittens) without central processor (Peer-to-Peer; P2P). However, regarding the communication channel between computers that will be source-limited wireless communication, the existence of Mother is very important to utilize communication sessions: when the total number of Kittens is n, the number of edges (channels) needed for CATS control is n, while P2P negotiation will need $(n-1)!$ (see Figure 6).

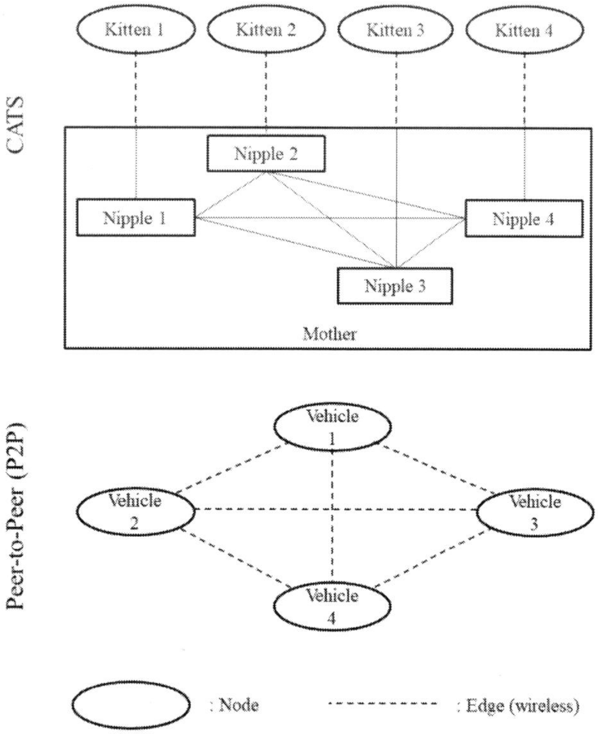

Figure 6 CATS vs P2P negotiation

CONCLUSIONS

The basic idea of controlling traffic both by onboard and ground computers using a cellular divided airspace was verified by a simulation using spread sheet software. Throughout the construction of this simulator, it was found that the number of onboard computers to be related to a ground computer should be somehow limited to the original design of the ground computer. Thus, in order to optimize the operation, higher rank computer was found to be needed. How to make onboard computer smarter so that they can re-plot their course while at flight, and the configuration of the elder rank computer(s) remained to be studied at further research.

Author contact: Yu Ito yuito@kuronekoyamato.co.jp

APPENDIX: CATS SIMULATOR CONFIGURATION

The spread sheet consists of 6 pages: "Cockpit & Recorder", "Mother (0)", "Mother (n)", "Kitten1", "Kitten2", and "Kitten3". As an appendix to this study, configuration of each sheet and their interaction will be noted. Software for implementation and running that took place in this study is Microsoft Excel 2016. For formula spread to multiple cells, formula in the upper-and-left-most cell of the concerned area is shown for representation. Value of cells except referred particularly bellow are those who don't have any dynamic formula.

Sheets "Kitten1", "Kitten2", and "Kitten3"

Each onboard computer named "Kitten" simply plans the occupation of cells that contain straight path from the start to the goal of each craft. All the three sheets have the configuration and formula in common, such as shown in Figure 7.

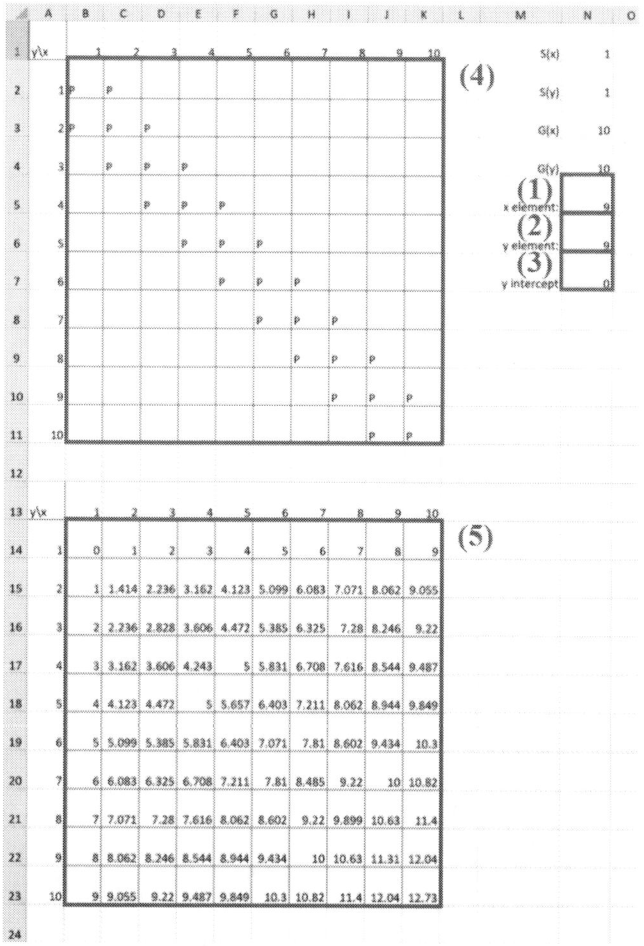

y\x	1	2	3	4	5	6	7	8	9	10
1	P	P								
2	P	P	P							
3		P	P	P						
4			P	P	P					
5				P	P	P				
6					P	P	P			
7						P	P	P		
8							P	P	P	
9								P	P	P
10									P	P

y\x	1	2	3	4	5	6	7	8	9	10
1	0	1	2	3	4	5	6	7	8	9
2	1	1.414	2.236	3.162	4.123	5.099	6.083	7.071	8.062	9.055
3	2	2.236	2.828	3.606	4.472	5.385	6.325	7.28	8.246	9.22
4	3	3.162	3.606	4.243	5	5.831	6.708	7.616	8.544	9.487
5	4	4.123	4.472	5	5.657	6.403	7.211	8.062	8.944	9.849
6	5	5.099	5.385	5.831	6.403	7.071	7.81	8.602	9.434	10.3
7	6	6.083	6.325	6.708	7.211	7.81	8.485	9.22	10	10.82
8	7	7.071	7.28	7.616	8.062	8.602	9.22	9.899	10.63	11.4
9	8	8.062	8.246	8.544	8.944	9.434	10	10.63	11.31	12.04
10	9	9.055	9.22	9.487	9.849	10.3	10.82	11.4	12.04	12.73

Figure 7 Configuration of sheets named "Kitten"

Cells N1:N5 designate the coordinates of the point of start (S) and goal (G). Sections (4) and (5) in each of the three sheets ("Kitten1", "Kitten2", and "Kitten3") shown in Figure refer to the same map, while (4) indicates the path of each craft, and (5) show each cell from the start distance, which equals to the linear concentration time in assumption that the basic velocity of the craft is 1 cell side/time. If needed, experimenter can multiply the value to change the assumed speed of the craft. X coordinates are decided from left to right, and y coordinates are from above to bellow.

Regardless to the complicated formula in these sheets, these parts are out of the velocity control algorithm. Thus, alternative path deciding process (including manual input of “P” to the cells to be designated to contain the arbitrary course) for non-linear route are adaptable instead of the formula in (4) for further studies. This design also reflects the basic concept that each of the on-board computer does not have to be standardized, and also shall be free from central anti-collision decision making.

Each numbered cell and section in Figure contains each formula as described in the following:

(1) N5 (x element of the path vector)

```
=$N$3-$N$1
```

(2) N6 (y element of the path vector)

```
=$N$4-$N$2
```

(3) N7 (y intercept of the primary function that contains the track)

```
=($N$2-0.5)-($N$6/$N$5)*($N$1-0.5)
```

The value “0.5” in the first and last sets of parentheses originate in the assumption that the point of start and goal is on the center of the cell, so that their position should be adjusted from x/y coordinate numbers which indicates the farthest distance from the origin.

(4) B2:K11

```
=IF($N$1=$N$3,IF(AND(B$1=$N$1,
OR(AND($N$2<=$A2,$A2<=$N$4),
AND($N$4<=$A2,$A2<=$N$2))),"P",""),
IF(AND(OR(AND($A2>=B$1*$N$6/$N$5+$N$7,
$A2-1<=B$1*$N$6/$N$5+$N$7),
AND($A2>=(B$1-1)*$N$6/$N$5+$N$7,
$A2-1<=(B$1-1)*$N$6/$N$5+$N$7),
AND($A2<=B$1*$N$6/$N$5+$N$7,
$A2-1>=B$1*$N$6/$N$5+$N$7),
AND($A2<=(B$1-1)*$N$6/$N$5+$N$7,
$A2-1>=(B$1-1)*$N$6/$N$5+$N$7),
AND(B$1-1>=($A2-$N$7)*$N$5/$N$6,
B$1<=($A2-$N$7)*$N$5/$N$6),
AND(B$1-1<=($A2-$N$7)*$N$5/$N$6,
B$1>=($A2-$N$7)*$N$5/$N$6)),
OR(AND($N$1<=B$1,B$1<=$N$3),
AND($N$3<=B$1,B$1<=$N$1)),
OR(AND($N$2<=$A2,$A2<=$N$4),
AND($N$4<=$A2,$A2<=$N$2))),"P",""))
```

(5) B14:K23

```
=((B$13-$N$1)^2+($A14-$N$2)^2)^(1/2)
```

Sheet “Mother (0)”

This sheet, whose configuration is shown as in Figure 8, shows the cognition of the ground controller “Mother” at time 0. Simulation held in this study assumes that there is no overcrossing in time 0, which makes the formula in this sheet simple compared to that in “Mother (n)” explained after. Three maps (Nipples) in the upper column (“Green sections (0)” that indicates (6), (7), and (8)) show the green section of each craft subjectively, and those in the lower (“Occupying station (0)” which is to (9-1), (9-2), and (9-3)) show the yellow and red cells to the other crafts objectively. Kitten1, for instance, is given (6), (9-2), and (9-3) for its approach clearance.

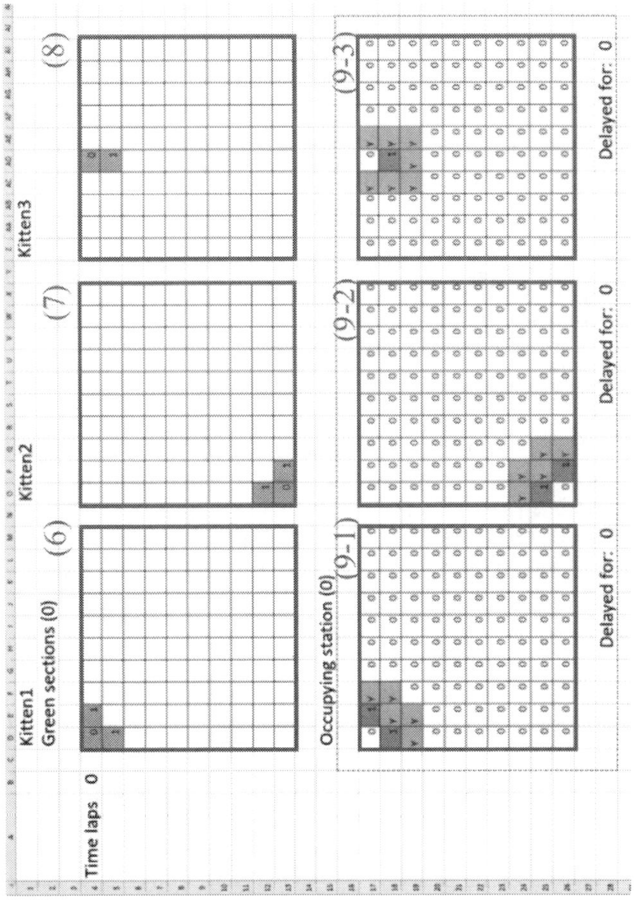

Figure 8 Configuration of Sheet “Kitten (0)”

Each numbered section in Figure contains each formula as described in the following. (9-1), (9-2), and (9-3) have the same formula in common, thus the explanation is unified as (9). Notice that green sections ((6), (7), and

(8)) respectively have structures in common. Simply extending the same structure for additional Kittens will do fine, when to increase the number of crafts for further simulations.

Each numbered cell and section in Figure contains each formula as described in the following:

(6) D4:M13

```
=IF(AND(Kitten1!B2="P",$B$4<=Kitten1
!B14,Kitten1!B14<=$B$4+1),Kitten1!B14,"
")
```

(7) O4:X13

```
=IF(AND(Kitten2!B2="P",$B$4<=Kitten2
!B14,Kitten2!B14<=$B$4+1),Kitten2!B14,"
")
```

(8) Z4:AI13

```
=IF(AND(Kitten3!B2="P",$B$4<=Kitten3
!B14,Kitten3!B14<=$B$4+1),Kitten3!B14,"
")
```

(9) D17:L26, O17:X26, Z17:AI26

```
=IF(D4="",IF(AND(C3="",D3="",E3="",
E4="",E5="",D5="",C5="",C4=""),0,"Y"),D
4)
```

Cells M28, X28, and AI28 don't have any function in "Kitten (0)" solely, but play an important role in the whole velocity control process.

Sheet "Mother (n)"

This sheet whose configuration is shown as in Figure 9, shows the cognition of the ground controller "Mother" at time n (n=1, 2, ...). As the status of time n-1 is given to area C3:AI15 ("Occupying station (n-1)") as value by the process of the VBA as stated afterwards, this sheet decides the approach clearance for each Kitten at time n.

As a craft asks for permition to enter a cell that is collored yellow by other craft(s), ground controller Mother is designed to tell the craft to reduce the craft's velocity and to stop before entering the red section ahead. This function is implemented in this sheet by returning "D" (for *delay*) for the correspondent cells in the Green sections at time n ((10), (11), and (12)), deciding to delay the craft if there are one or more "D" in the field ((13)), and feeding the result to the decision at n+1 ((10), (11), and (12)).

Though the basic policy for this infrastructure is "the fastest craft to reach the cell gets the path", priority for occupation when multiple Kittens claim for the same cell simultaneously is decided as: Kitten1, Kitten2, Kitten3, ... This idea is reflected to the differences of the bolded signs of equality and inequality.

Each numbered cell and section in Figure contains each formula as described in the following:

(10) D18:M27

```
=IF(AND(Kitten1!B2="P",
$B$4-$M$15<=Kitten1!B14,
Kitten1!B14<$B$4-$M$15+2),
IF(AND(NOT(D4=SMALL($D$4:$M$13,
COUNTIF($D$4:$M$13,0)+1)),
OR(AND(O4="Y",Kitten1!B14+$M$15>Kitten2
!B14+$X$15),
AND(Z4="Y",Kitten1!B14+$M$15>Kitten3!B1
4+$AI$15))),
"D",Kitten1!B14+$M$15),"")
```

(11) O18:X27

```
=IF(AND(Kitten2!B2="P",
$B$4-$X$15<=Kitten2!B14,
Kitten2!B14<$B$4-$X$15+2),
IF(AND(NOT(O4=SMALL($O$4:$X$13,
COUNTIF($O$4:$X$13,0)+1)),
OR(AND(D4="Y",Kitten2!B14+$X$15>=Kitten
1!B14+$M$15),
AND(Z4="Y",Kitten2!B14+$X$15>Kitten3!B1
4+$AI$15))),
"D",Kitten2!B14+$X$15),"")
```

(12) Z18:AI27

```
=IF(AND(Kitten3!B2="P",
$B$4-$AI$15<=Kitten3!B14,
Kitten3!B14<$B$4-$AI$15+2),
IF(AND(Z4=SMALL($Z$4:$AI$13,
COUNTIF($Z$4:$AI$13,0)+1),
OR(AND(D4="Y",Kitten3!B14+$AI$15>=Kitte
n1!B14+$M$15),
AND(O4="Y",Kitten3!B14+$AI$15>=Kitten2!
B14+$X$15))),
"D",Kitten3!B14+$AI$15),"")
```

(13) M29, Z29, AI29

```
=IF(COUNTIF(D18:M27,"D")>0,
M15+1,M15)
```

(14) D32:M41

```
=IF(D18="",IF(AND(C17="",D17="",E17=
"",E18="",E19="",D19="",C19="",C18=""),
0,IF(Kitten1!B2="","Y",
IF(Kitten1!B14+$M$29<$B$4+2,
Kitten1!B14+$M$29,
IF(Kitten1!B14+$M$29<=$B$4+3,"Y",0)))),
Kitten1!B14+$M$29)
```

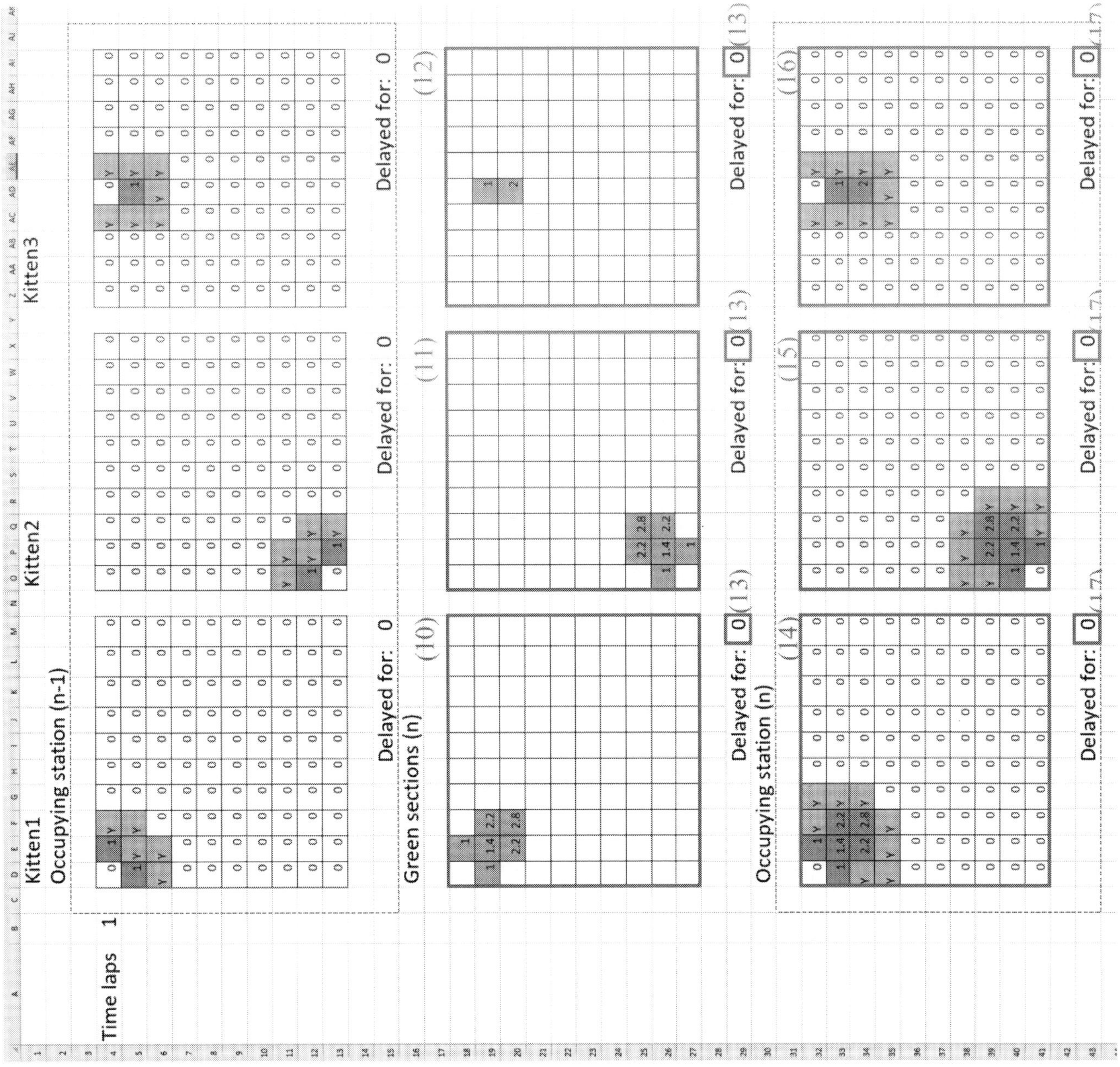

Figure 9 Configuration of sheet "Mother (n)"

(15) O32:X41

```
=IF(O18="",IF(AND(N17="",O17="",
P17="",P18="",P19="",O19="",N19="",
N18=""),0,IF(Kitten2!B2="","Y",
IF(Kitten2!B14+$X$29<$B$4+2,
Kitten2!B14+$X$29,
IF(Kitten2!B14+$X$29<=$B$4+3,"Y",0)))),
Kitten2!B14+$X$29)
```

(16) Z32:AI41

```
=IF(Z18="",IF(AND(Y17="",Z17="",
AA17="",AA18="",AA19="",Z19="",Y19="",
Y18=""),0,IF(Kitten3!B2="","Y",
IF(Kitten3!B14+$AI$29<$B$4+2,
Kitten3!B14+$AI$29,
IF(Kitten3!B14+$AI$29<=$B$4+3,"Y",0))))
,Kitten3!B14+$AI$29)
```

(17) M43, X43, AI43

```
=M29
```

Sheet "Cockpit & Recorder"

This sheet, shown as in Figure 10, consists of cockpit, which adjusts the situation of the simulation, and recorder, which records the whole result of the experiment.

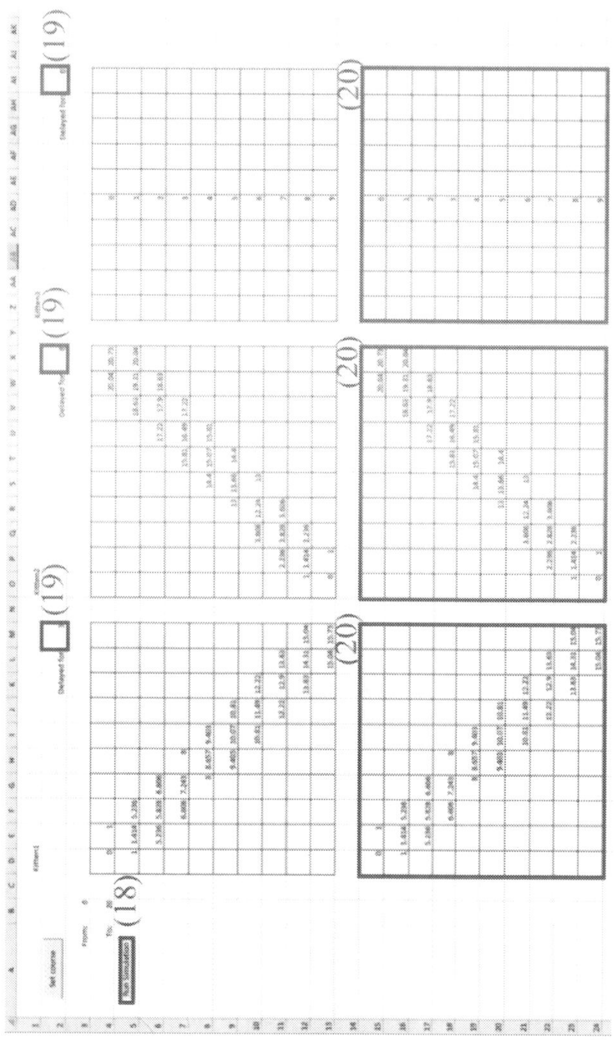

Figure 10 Configuration of sheet "Cockpit & Recorder"

As for the latter, this program, whose script will be explained afterwards, is to run simulation from time 0 to that designated in cell B4 (though the start time is also shown in cell B3, this study does not accept figures except for 0). This simple program repeats copying and pasting proper areas while advancing time so that the formulae written in the sheets can work as intended to decide the process regarding as it was in the time one step before.

The recorder function is designed to overwrite the path information regarding the situation of sheets "Kitten (0)" or "Kitten (n)". As a result, the whole "contrail" of each craft is recorded. By looking into the figure of each cell that suggests the time of passing of the cell, where and how the traffic was delayed can be read.

Each numbered button and cell and section in Figure contains each macro or formula as described in the following:

(18) Button "Run Simulation" (written in VBA)

```
Sub RunSimulation()

Sheets("Mother (n)").
Range("D4:AI13,M15,X15,AI15").
ClearContents

Sheets("Cockpit & Recorder").
Range("D4:AI13").ClearContents

Dim Value1 As String

Dim Value2 As String

Value1 = Sheets("Cockpit & Recorder").
Range("B3")

Value2 = Sheets("Cockpit & Recorder").
Range("B4")

Dim i As Integer

For i = Value1 To Value2

If i = 0 Then

Sheets("Mother (0)").
Range("D17:AI26").Copy

Sheets("Mother (n)").Range("D4").
PasteSpecial (xlPasteValues)

Sheets("Mother (0)").
Range("D4:AI13").Copy

Sheets("Cockpit & Recorder").
Range("D4").PasteSpecial
(xlPasteValues)

Sheets("Mother (n)").Range("B4") = 1

End If

If i = Value1 And Value1 >= 1 Then

Sheets("Mother (n)").Range("D18:AI27").
Copy

Sheets("Cockpit & Recorder").
Range("D4").PasteSpecial
(xlPasteValues)

End If

Sheets("Cockpit & Recorder").
Range("D15:AI24").Copy

Sheets("Cockpit & Recorder").
Range("D4").PasteSpecial
(xlPasteValues)
```

```
Sheets("Mother (n)").
Range("D32:AI43").Copy

Sheets("Mother (n)").
Range("D4").PasteSpecial
(xlPasteValues)

Sheets("Mother (n)").Range("B4")
 = i + 1

Next

End Sub
```

(19) M2, X2, AI2

```
= 'Mother (n)'!M15
```

(20) D15:M24, O15:X24, Z15:AI24

```
=IF(AND(NOT(D4=""),NOT(D4="D"),
OR(D4>='Mother (n)'!D18,
'Mother (n)'!D18="")),D4,
'Mother (n)'!D18)
```

REFERENCES

[1]Kopardekar, P., Rios, J., Prevot, T., Johnson, M., Jung, J., and Robinson, J. E., "Unmanned Aircraft System Traffic Management (UTM) Concept of Operations," Paper AIAA 2016–3293, 16th AIAA Aviation Technology, Integration, and Operations Conference Proceedings, Washington, D. C., June 13–17, 2016.

[2]Uber Elevate, "Fast-Forwarding to a Future of On-Demand Urban Air Transportation," Uber, *available online* https://www.uber.com/elevate.pdf, 2016.

[3]Atkins, E. M., Khalsa, A. D., Groden, M. D., "Commercial Low-Altitude UAS Operations in Population Centers," Paper AIAA 2090-7070, 9th AIAA Aviation Technology, Integration, and Operations Conference (ATIO) Proceedings, Hilton Head, South Carolina, September 21–23, 2009.

[4]Airbus Group, "A New Perspective on Urban Mobility," *Forum*, vol. 88, 2016, pp. 6–13.

[5]Tomlin, C., Pappas, G. J., Sastry, S., "Conflict resolution for air traffic management: a study in multiagent hybrid systems," *IEEE Transactions on Automatic Control*, Vol. 43, Issue 4, 1998, pp. 509–521. doi: 10.1109/9.664154

[6]Wei, Z., Li, J., "Design of air traffic control operation system using multi agent technology and simulation," *The Open Fuels & Energy Science Journal*, Vol. 8, 2015, pp. 368–374.

[7]Takashige, T., "Signalling systems for safe railway transport," *Japan Railway & Transport Review*, Vol. 21, 1999, pp. 44–50.

[8]Moore, E. F., "Machine models of self-reproduction," *Proceedings of Symposia in Applied Mathematics*, vol. 14, 1962, pp. 17–33.

[9]Zimmermann, H., "OSI Reference Model - The ISO Model of Architecture for Open Systems Interconnection," *IEEE Transactions on Communications*, Vol. COM-28, No.4, 1980, pp. 425–432.

Design of a Ducted-Fan Micro-Rotorcraft with Structural Battery

Chetan Gupta
Intern
Indian Institute of Science
India

Dineshkumar Harursampath
Asst. Professor
Indian Institute of Science
India

Ramesh Gupta
Asst. Professor
Shiv Nadar University
India

ABSTRACT

Low flight endurance is a pitfall commonly identified in most modern-day battery powered micro aerial vehicles. This paper presents the design and analysis of a ducted fan micro aerial vehicle(MAV) having a multifunctional, modular structure in order to address the above problem. The ducted fan rotorcraft has been designed particularly to meet the safety and hover efficiency needs for a prolonged indoor surveillance mission. A detailed literature survey on rotorcraft MAVs and multifunctional structural batteries has been reported and discussed. The paper involves detailed conceptual, preliminary and final design stages of a structurally and aerodynamically optimized axial fan that successfully caters to lower power consumption as compared to similar off the shelf propellers available in the market. The MAV incorporates multifunctionality by merging its structural and energy elements into a single sub-system (PAN-based carbon fiber batteries) thereby enhancing flight endurance.

INTRODUCTION

The concept of micro-sized Unmanned Aerial Vehicles or MAVs has gained increasing interest over the past decade with the principal aim of carrying out surveillance missions. This class of aircrafts were under research for decades after the world wars, however, most concepts and prototypes were not easily materialised. With the proliferation of the digitization and miniaturization that took the world stage in the 1990s, the concept of MAVs soon became palpable. Development of MAVs boomed there on and has now transformed into a full-fledged market within 2 decades and caters to not only aircraft companies and the militaries around the world, but has also found its way to the layman aircraft enthusiast. There have been many conflicts over the maximum size of a MAV- according to DARPA, a MAV has to be smaller than 15cm, while in books like 'Unmanned Rotorcraft Systems' by Guowei Cai (Ref.1) claim that rotorcrafts under 0.7m are also MAVs.

The study of multifunctional entities has long been part of the aerospace domain. The need for reducing weight, maximizing mechanical and overall efficiencies, minimizing drag are all primal to the design of an aircraft. These varied problems have compelled researchers to come up with subsystems that could perform the complete or partial function of two or more subsystems, thus becoming multifunctional in nature. The concept of multifunctional structural batteries originates from the basic solid-state thin-film rechargeable battery technology was first invented at Oak Ridge National Laboratory, TN, USA, in the 1990s and has evolved drastically.

Presented at the 6th Asian/Austrailian Rotorcraft Forum / Heli Japan 2017, Kanazawa, Japan, November 7-9, 2017.

In 2000, ITN Energy Systems and Global Solar Energy developed copper-indium-gallium-selenium (CIGS) photovoltaic material a flexible copper polymide printed circuit and a solid state thin film Li battery (Ref.2). The system was developed for space satellites and had a projected specific power of 1000W/kg. To improve on the poor efficiency of the CIGS system, ITN Energy systems came out with PowerFibers (Ref.3) - solid-state thin-film rechargeable batteries on fibre substrates. Large length-to-diameter ratio permits small bending radii and substantial flexing and an inherently larger surface area allows them to store more power. (see Figure 1)

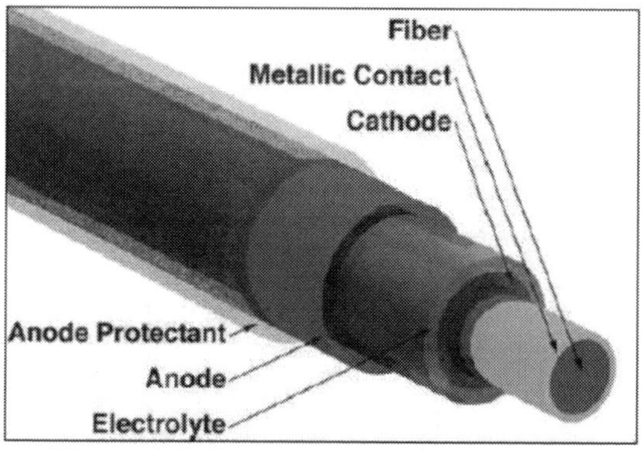

Figure 1. Schematic cross-sectional view

Expanding on their existing research ITN went on to make PowerComposites. It used PowerFibres as the core and added an adhesive matrix to impart a mechanical/structural component to the battery fibres. (Figure 2).

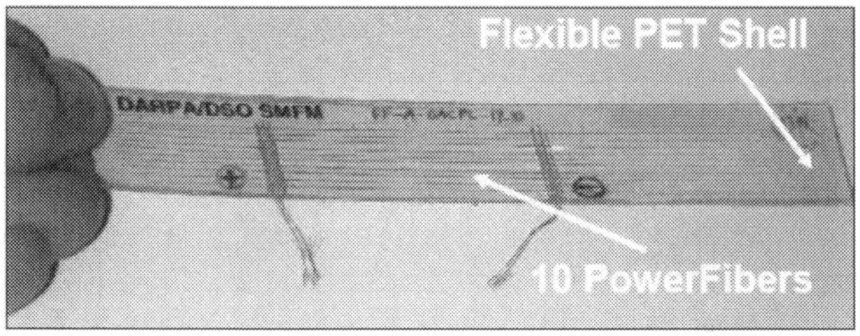

Figure 4. PowerComposite electrically connected in parallel by auxiliary Cu wires

Meanwhile, studies (Ref.4) by the US Naval Research Laboratory focused on developing laminated polymer lithium ion (PLI) bicell materials with barrier-layer packaging. The PLI bicell uses $LiCoO_2$ particles in the cathode and graphite particles in the anode. The anode and cathode particles are held in layer form by a PVDF+HFP polymer matrix "glue". The bicell is thin (~0.5 mm thick) and capable of arbitrary planar shaping and stacking. Experiments were done on their practicality by implementing these cells onto fixed wing MAVs like Wasp I and Wasp II. Research (Ref.5) from the University of California, LA brings Carbon Fiber Reinforced Polymers(CFRP) into the multifunctional structure-battery domain. This study proposes embedding solid state thin film Li-ion energy cells in CFRP in order to harvest energy in addition to providing structural support and power. The study demonstrates the ability of embedded Li-ion energy cells to withstand pressure and temperature required during the manufacturing of CFRP composites. Simultaneously, the study confirms that the charge/discharge rates of embedded energy cells did not deviate much from the previous standard charge/discharge baseline parameters.

The latest revolution in MSB came through the research (Ref.6) from KTH Royal Institute of technology gives evidence that shows carbon fibers can be used as the active negative material and current collector in structural batteries. Several grades of commercially-available polyacrylonitrile (PAN)-based carbon fibers were studied and the results show how fibers with intermediate modulus have better electrochemical properties than high modulus materials and fibers with similar modulus and tensile strength display similar electrochemical performance. A number of studies surrounding this concept of using carbon fibers as electrodes have since developed. One such research (Ref.7) from KTH provided important conclusions about various interactions between the electric and mechanical subsystems. Two specimens of PAN-based carbon fiber (sized T800 and IMS65) identified as most suitable for multifunctional batteries were observed to have a benign effect on their tensile properties on lithium ion intercalation. Neither the Lithium -ion intercalation nor the cycle count affected the tensile strength of the fibers. On the contrary after the first cycle, each lithiation causes a drop in the ultimate tensile load that is recovered for the delithiation that follows which supports the claim that PAN based carbon fibers are optimal for use as electrodes. Also, the outer surface of the carbon fiber electrode itself did not either appear to be damaged by the cycling on FE-SEM images after 1000 cycles. (Figure 4)

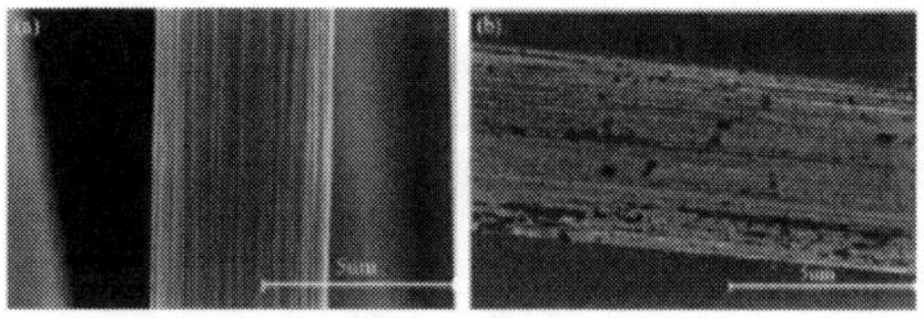

Figure 10. (a) Desized IMS65 carbon fibre before cycling (b) lithiated after 1000 cycles

Another research study (Ref.8) conducted by professors from Sweden, London and Austria has reported a method to deposit a thin solid polymer electrolyte (SPE) coating around individual carbon fibers for the realization of novel battery designs. In this study, an electrocoating method is used to coat methacrylate-based solid polymer electrolytes onto carbon fibers. By this approach a dense uniform, apparently pinhole-free, poly (methoxy polyethylene glycol (350) monomethacrylate) coating with an average coating thickness of 470 nm was deposited around carbon fibers. The Li-ion content in the solid polymer coating was found to be sufficiently high for battery applications. A research paper (Ref.9) does an evaluation of commercial polyacrylonitrile-based carbon fibers in terms of capacity and coulombic efficiency, as well as a microstructural and surface evaluation. Using high precision coulometry, they found a capacity of around 250–350 mAh/g and a very high coulombic efficiency of over 99.9% after ten cycles, which is even higher than a commercial state-of-the art graphitic electrode evaluated as reference. The high coulombic efficiency is attributed to the very low surface area of the carbon fibers, resulting in a small and stable solid-electrolyte interface layer, thus reinstating that PAN-based intermediate modulus fibers are well suited for structural-battery application. Below is a table of the modern rotorcrafts MAVs, conforming to the 0.7m size restriction, that have incorporated novel techniques to optimize different aircraft subsystems and have been used to benchmark our design.

Table 3. Signle/Twin Rotor MAV

MAV Name	Power Source	Max. Dimension (mm)	Time of Flight (min)	Gross Weight (g)	Max. Payload (g)
MICOR	Electric	152	17	100	10
iSTAR	IC Engine	226	60	1500	
RP1	Electric	260	_	982	300
RQ-16A T-Hawk	IC Engine	368	40	7800	_
Lama V4	Electric	410	12	230	_
T-REX 250	Electric	430	13	260	_

[a] RP – Research Prototype

CONCEPTUAL DESIGN

The design strategy adopted in developing the MAV revolves around four major criteria – endurance, scale-ability, power source and application. Flight Endurance, highlighted in the literature survey as the foremost MAV parameter that needs improvement, was set at achieving 10-15 minutes above the average time of flight of 20 minutes (current-day average for MAV). Scale-ability is a feature that refers to a MAV design that can be scaled up or scaled down in size with minimal design changes so that the MAV shows multi-mission adaptability. Further, the MAV should incorporate modularity to its design to ensure ease in manufacturing, assembly and maintenance. The literature survey revealed the absence of an MAV solely designed for indoor environments. An indoor environment poses simpler dynamics as compared to an outdoor setting due to the absence of environmental factors like crosswinds. Thence, a ducted fan MAV design powered by conventional battery powered systems and a multi-functional structural battery has been proposed.

Ducted-Fan MAV

The MAV hosts a compact shrouded design as the duct surrounds the rotating propeller thereby ensuring safety to the people/objects in the indoor environment. (see Figure 1) Ducted fan UAVs maintain a higher payload to stored volume ratio (compared to non-ducted counterpart) and are thus are more efficient than multi-copters of similar footprint. Additionally, the duct imparts a gradient in lift force using the Coandă effect when tilted, which in turn generates a moment (about its COM) which self-stabilizes the MAV for hover as seen in Figure 2. The presence of a duct around the rotating propeller reduces vortices generated at the blade tips thereby reducing noise, which can be of paramount importance in the case of a reconnaissance mission.

Figure 1. MAV Conceptual Design

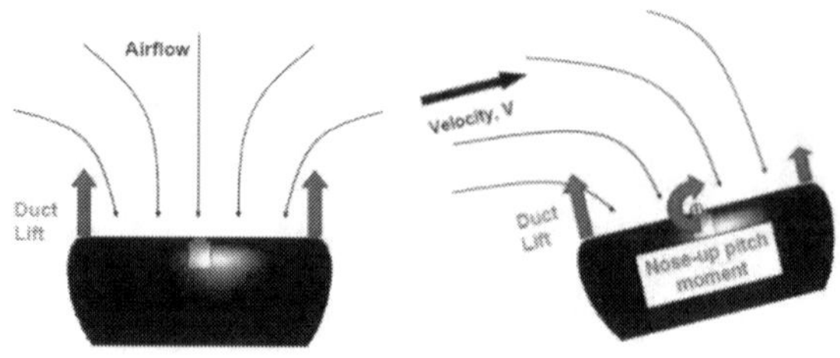

Figure 2. Self-stabilizing Duct Aerodynamics

Taking a factory shop floor as a surveillance target and based on the literature study, we proposed an MAV with a maximum dimension of 11 inches weighing (gross) in at approximately 1.5kg with a payload capacity of 200g-250g. (see Figure 3 and Table 1)

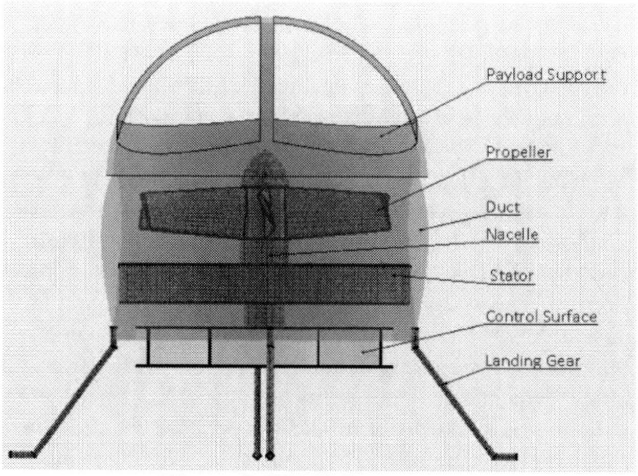

Figure 3. MAV Components

Table 1: MAV Structural Weight Approximation

Part	Qty	Material	Density(kg/m^3)	Weight(g)
Duct	1	MSB	1550	~400
Stator	1	PVC	1300	230
Axial Fan	1	PVC	1300	70
Landing Gear	1	PVC	1300	30
Control Vane	4	PVC	1300	80
Nacelle	1	PVC	1300	110
Top Cone	1	PVC	1300	30
Payload Support	1	PVC	1300	100

[a] Total Structural Weight: 1050g

Multi-Functional Body

Carbon fibers have been shown to intercalate lithium ions (Ref.6), suggesting that combined with their high specific mechanical properties they can be used as electrode-reinforcement in structural battery composite materials. One of these fiber types, the intermediate modulus fiber IMS65 by Toho Tenax has particularly shown better performance (Ref.7) over other carbon fibers with respect to a combination of energy density and tensile strength. The duct of the MAV has been especially designed to cover a large surface area to house the multifunctional structural battery such that a large amount of electrical energy can be stored within the structure as a secondary power source. The energy stored within the duct is channeled with the help of wiring through the stator towards the nacelle. The nacelle encases the electric motors that run the MAV control system through an Arduino chip. (see Figure 4) Additionally, the MSB can be used to power surveillance systems placed at the top of the payload support. Such an electrical model puts lesser load on the primary Li-ion battery which is then used to power the propulsion system alone.

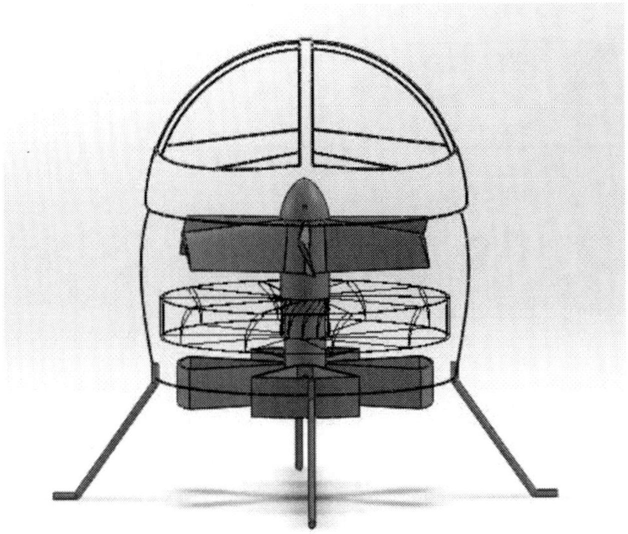

Figure 4. Multi-functional Body

PRELIMINARY DESIGN

The propulsion system chosen for the MAV was an axial fan. An axial fan is a type of fan whose flow is axially in and axially out, linearly. Such fans generate large volume flow rates of air that is crucial for a rotorcraft MAV. Below are the design parameters analysed to structurally and aerodynamically optimize the fan.

Fan Hub-to-Tip Ratio

Axial fans show varying performance as the ratio of the hub diameter and the fan diameter are varied. According to research (Ref.10) from Bu-Ali Sina University, the hub to tip ratio of an axial fan affects its axial flow speed and rate as well as the static pressure developed in the fan. As the hub to tip ratio increases (up to a limit), there is corresponding increase of axial flow speed, rate and static pressure. This results in high thrust which is essential to the ducted fan MAV. A hub to tip ratio of 0.3 and 0.4 is considered for the preliminary design of the axial fan. The fan blades are also give a twist from 60° (to the plane of the fan) at the hub to 30° (to the plane of the fan) at the blade tip. Also, the fan tip clearance from the duct can play a mojor role in determining the hover efficiency of the MAV. Research (Ref.11) shows that adding a duct with a tip clearance of 3.08% of the radius of the fan, increases hover efficiencies by 38% and reducing this tip clearance to 1.71% of the fan radius further increases hover efficiency by 17.85%. In this model, a fan tip clearance of 2.5% is suggested as the duct would contain MSB which need to be in safe limits from rotating parts to avoid possible heating up of the MSB and thereby lowering battery performance.

Blade Solidity

Blade solidity plays a role in determining how much mechanical energy of the rotating blade is transferred to the airflow, thus determining the efficiency of the fan. The solidity is a measure of the resistance offered by the blades to the flow of air and is an important design variable. The pressure drop across a turbine is proportional to the axial force acting on the blades. An increase of solidity increases the axial force and likewise the pressure drop (Ref.12). This increase in pressure difference across the fan is important for the ducted fan to generate thrust, however, is restricted by the fan's aerodynamics. Larger number of blades cause interference of airflow over the airfoil shaped blades. This causes high turbulence and irregular lift on the blades thus leading to a very inefficient axial fan design. This is unlike the case of turbines where greater number of vanes (high balde solidity) are designed to improve the momentum transfer and guidance provided to fluid molecules to flow effectively through. Hence the preliminary axial fan design is considered for a 6 blade and 8 blade configurations. (Below is the calculation for solidity of a 6-blade fan)

$s_{hub} = 0.35$ and $s_{tip} = 0.51$

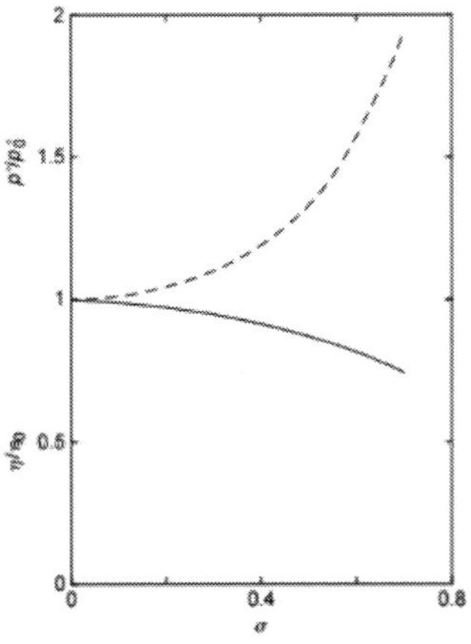

Figure 5. Variation of peak efficiency and non-dimensional pressure drop vs. solidity

Axial Fan Speed

The fan speed was a design parameter that was benchmarked to existing commercially available axial fan designs to understand the efiiciency improvement achieved in our design. Vaneaxial fans of similar size developed by Aerovent Inc. were studied with respect to their volume flow rate and axial outflow velocity. This indoor MAV would spend maximum flight time in the hovering state; thereby the MAV required a 15N thrust which corresponds to an axial fan outflow velocity of 17.7m/s and a volume flow rate of 0.721m^3/s. These parameters intersect with the VPBD vaneaxial fan by Aerovent Inc. giving us an approximate the axial outflow velocity of 4000rpm.

Blade Airfoil Selection

In order to ensure that the MAV has the ability to exude the best possible performance, a detailed study was performed on the shape of the cross section of each blade of the fan. Instead of limiting the fan to a regular thin profile blade, seen in many axial fans commercially, the axial fan for the MAV was chosen to have an airfoil shape that would help improve the fan's performance and efficiency, however, bearing a higher cost of designing and manufacturing. Below is the calculation for the Reynold's number of the blade from hub to tip. Using the relation,

$$\text{Reynold's number, } Re = (\rho * v * c)/\mu,$$

we calculate the Reynolds number at three points on the blade keeping the atmospheric temperature to be 25°C. Re_{hub} = 50,000; $Re_{mid\text{-}span}$ = 76,000; Re_{tip} = 102,000. Based on the calculations, 50,000<Re<110,00 is the range of Reynold's numbers experienced by the blade (MAV hover condition) due to different linear velocities at different cross sections of the blade. What is most important is to note the fact that this range of Re corresponds to a very low Reynold's number. In relation to the airfoil boundary layer, important areas of concern are the separated regions which occur near the leading and/or trailing edges and transition from laminar to turbulent flow if it occurs. (Ref.13) Separation and transition are highly sensitive to Reynolds number, pressure gradient, and the disturbance environment. Transition and separation play a critical role in determining the development of the boundary layer which, in turn, affects the overall performance of the airfoil and finally the whole aircraft. At Reynolds numbers above 50,000 and below 200,000 extensive laminar flow can be obtained. Although the laminar separation bubble may still present a problem for a particular airfoil, as the Reynolds number increases, the performance of the airfoil improves. In order to maximize the performance of the rotorcraft, only airfoils that have an exceptional performance at low Reynolds numbers were considered. In the case of rotorcrafts, pitching moment, plays an important role in selection of airfoils. For a rotating blade, the airfoil pitching moment coefficient, Cm is required to be within the range: -0.04<Cm<0.04 in order to keep the pitching moment at a minimum to ensure static and longitudinal stability of the aircraft. Bearing this in mind, below is a table (see Table 2) of all airfoils (from the UIUC airfoil database) conforming to these parameters put through a 2D airfoil analysis. (see Fig.6-9).

Table 2. Selection of Airfoil for Axial Fan Model

S. no	Airfoil Name	Airfoil Status	Remarks
1	M06-13-128	Rejected	High C_m for given Re no. range
2	DAE-51	Rejected	High C_m for given Re no. range
3	Clark Y	Rejected	High C_m for given Re no. range
4	S6063	Selected	_
5	LA2573A	Selected	_
6	NACA23012	Selected	_
7	NACA 4702	Rejected	Insufficient data
8	NACA 64(3)-418	Selected	_
9	NACA 63-415	Rejected	High C_m for given Re no. range
10	FX63-137	Rejected	High C_m for given Re no. range
11	NACA 2204	Selected	_

PRELIMNARY AXIAL FAN ANALYSIS

Blade Airfoil Analysis

Of the 5 airfoils that showed exceptional performance at low Reynold's numbers of flow, airfoils S6063 and NACA 23012 were chosen for the final design of the fan due to their relatively stable values of C_m over a range of angle of attack (α). S6063 has a maximum C_l/C_d = 45 at a Reynold's number of 120,000 (Figure 6) and a C_m within specified limits up to a positive angle of attack of 10° (figure 7).

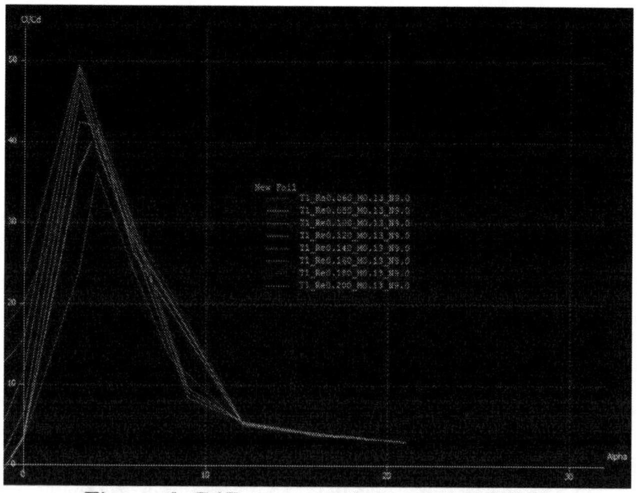

Figure 6. C_l/C_d vs α graph for airfoil S6063

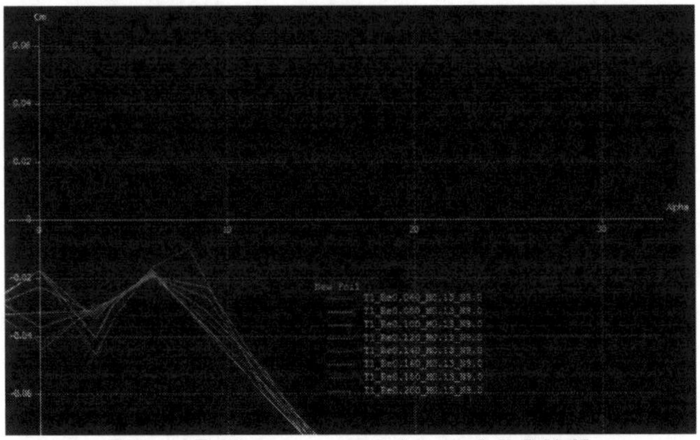

Figure 7. C_m vs α graph for airfoil S6063

NACA23012 has a maximum C_l/C_d = 35 at a Reynold's number of 120,000 (Figure 8) and a C_m within specified limits up to a positive angle of attack of 15° (figure 9).

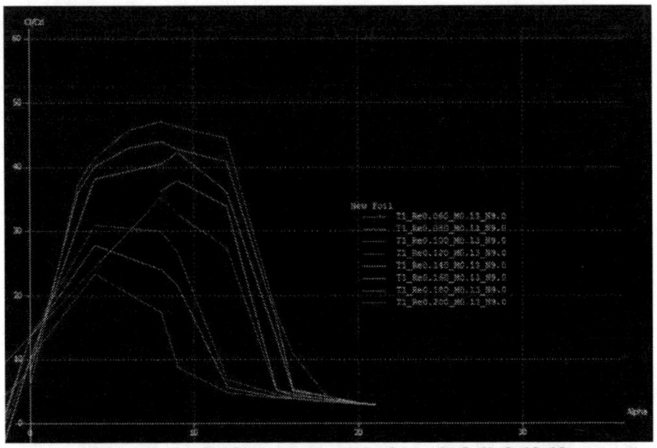

Figure 8. C_l/C_d vs α graph for airfoil S6063

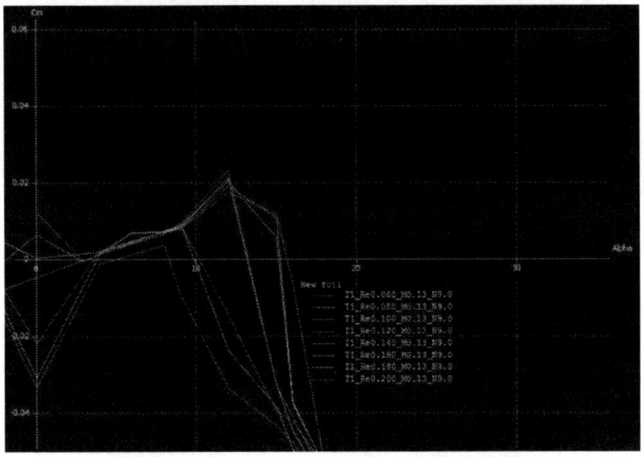

Figure 9. C_m vs α graph for airfoil S6063

Blade Twist Selection and CFD Modelling

The preliminary axial fan design uses a pre-set twist angle at the root of the blade (hub end) of 60°, a value obtained from literature (Ref.10). This twist of the axial fan was designed in two ways with respect to the geometry of the leading edge of the fan blades. First, taking the mid-point of the chord of the blade airfoil as the center of rotation, the blade was twisted from a root angle of 60° to a blade tip angle of 20°,30°, 40° and 50° forming 4 different configurations. (see Figure 10)

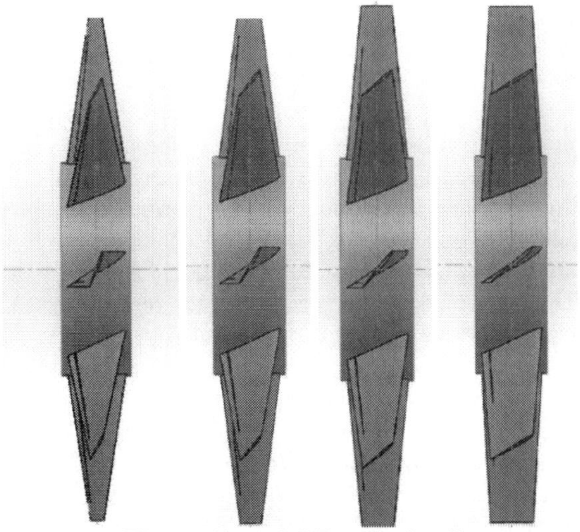

Figure 10. Axial Fan Design I

Second, taking the leading edge of the blade airfoil as the center of rotation, the blade was twisted from a root angle of 60° to a blade tip angle of 20°,30°, 40° and 50° forming 4 different configurations. (see Fig 11)

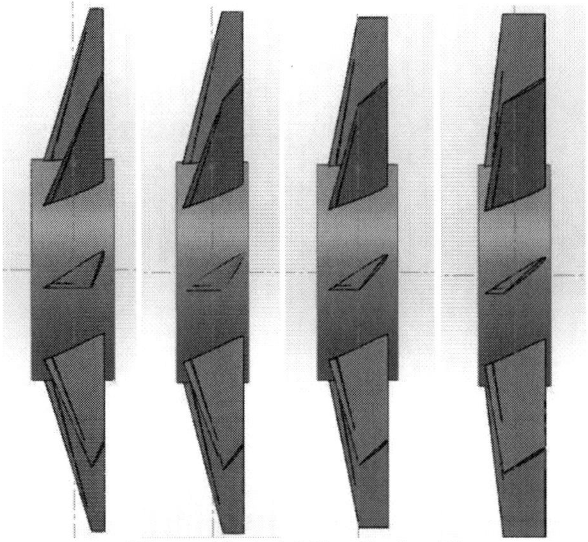

Figure 10. Axial Fan Design II

The alternative of twisting the blade along the quarter-chord or the aerodynamic center, that lies at approximately 0.23 times the chord length from the leading edge, was also considered. This type of twist however involved a complex design due to the continuously varying chord length of the blade from the hub to the tip resulting in manufacturing difficulties and was hence avoided.

All designs above were simulated using ANSYS Fluent to understand the flow pattern and velocity contours generated on rotating the fan at the design speed of 4000rpm. Further, each model above had two counterparts, one having a smaller hub to tip ratio of 0.3 and the other having a larger hub to tip ratio of 0.415. This was done in order to understand the pressure contours developed before and aft the axial fan and to study the variation in mass flow rate of the propelled air with varying hub to tip ratios.

Multi-Reference Frame(MRF) modelling is a steady-state approximation in which individual cell zones move at different rotational and/or translational speeds. The flow in each moving cell zone is solved using the moving reference frame equations. At the interfaces between cell zones, a local reference frame transformation is performed to enable flow variables in one zone to be used to calculate fluxes at the boundary of the adjacent zone. It should be noted that the MRF approach does not account for the relative motion of a moving zone with respect to adjacent zones (which may be moving or stationary); the grid remains fixed for the computation. This is analogous to freezing the motion of the moving part in a specific position and observing the instantaneous flow field with the rotor in that position. Hence, the MRF is often referred to as the "frozen rotor approach."

Below (Figure 11) is the design with the inlet volume, outlet volume (shaped like a cone) and MRF volume containing the axial fan used for the CFD analysis.

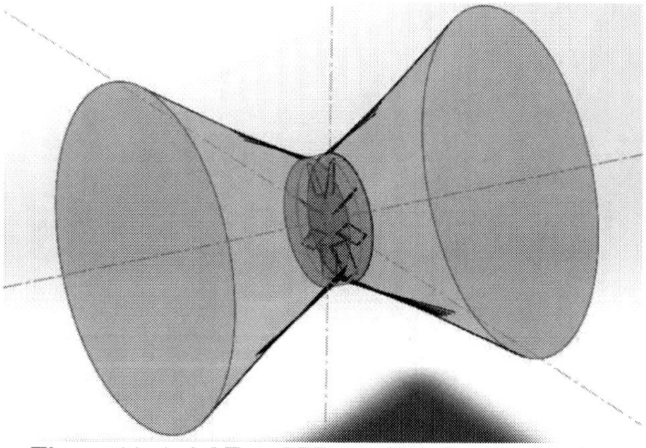

Figure 11. Axial Fan CFD Analysis Pre-Processing Model

In order to display the CFD velocity/pressure contours, an iso-surface running across the mid-plane of the geometry is used. Simulations using design I showed non-axial components of flow at the outlet as seen in Fig 12.

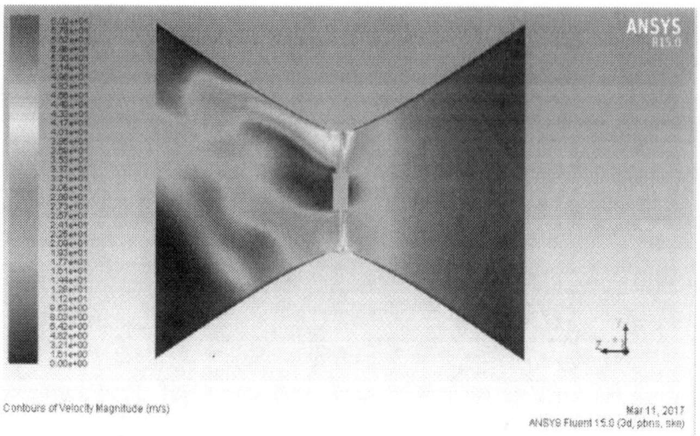

Figure 12. Axial Fan CFD Simulations – Design I (Velocity Contours)

The non-axial components of thrust generated by the axial fan (design I) make it unfit for the application as a propeller and thereby we consider design II (6 blades) which shows highly axial flows for all the 8 variations (Figure 13). Another important point to note is the velocity and volume flow rate that is generated using this design of the axial fan. Initial calculations in the conceptual design phase showed that an axial outlet velocity of 17.7m/s and a volume flow rate of 0.721m3/s would be required for the MAV to hover. At the design speed that was approximated to 4000rpm in the preliminary design phase of the axial fan, the 8 variations of design II show much higher axial outlet velocity and mass flow rate.

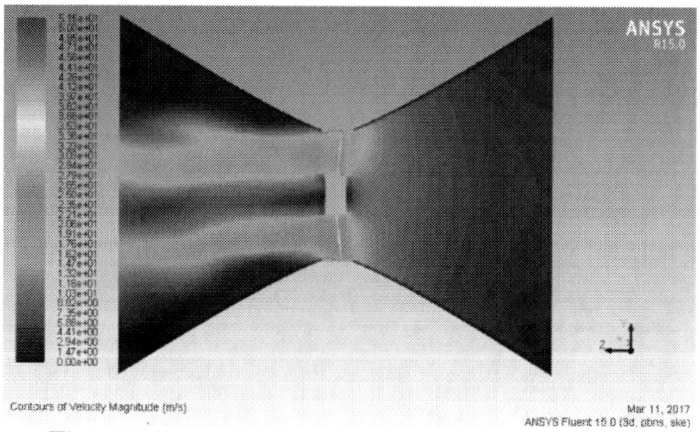

Figure 13. Axial Fan CFD Simulations – Design II (Velocity Contours)

AXIAL FAN FINAL DESIGN & ANALYSIS

Blade Element Theory

The Disc Actuator Theory developed by W.J.M. Rankine describes the efficiency of propellers to be directly proportional to the rotor diameter and inversely proportional to the rotor speed. Since the rotor diameter is fixed due to MAV size constraints, we optimize the axial fan for generating optimal thrust at lowest rotational speeds possible using the Blade Element Theory. (Figure 14) It involves breaking a blade down into several small parts then determining the forces on each of these small blade elements. These forces are then integrated along the entire blade and over one rotor revolution in order to obtain the forces and moments produced by the entire propeller or rotor. One of the key difficulties lies in modelling the induced velocity on the rotor disk.

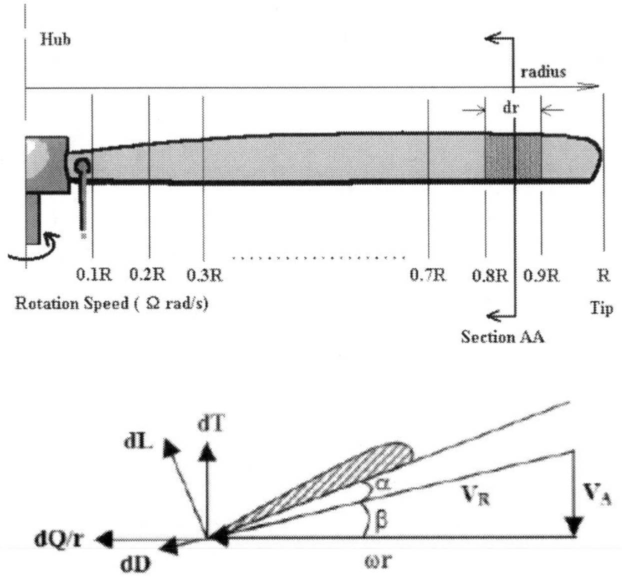

Figure 14. Blade Element Theory

$$dL = \tfrac{1}{2}\, CL\, \rho dr\, V_R^2$$

$$dD = \tfrac{1}{2}\, CD\, \rho dr\, V_R^2$$

$$dT/z = dL\cos\beta - dD\sin\beta$$

$$dT/z = dL\cos\beta[1 - \tan\gamma.\tan\beta] \text{ (where } \tan\gamma = dD/dL)$$

$$dQ/rz = dL\sin\beta + dD\cos\beta$$

$$dQ/rz = dL\cos\beta\ [\ \tan\beta + \tan\gamma]$$

$$\eta = (VA/r\omega)\ [\tan\beta/\tan(\beta+\gamma)]$$

However, the pitch angle β increases to β_I or induced pitch angle due to the increment in axial velocity of air on crossing the propeller plane and on incorporating the induced components of axial and rotational velocities, we get an efficiency of

$$\eta = [(1 - a')/(1+a)]\ \tan\beta_I/\ \tan\ (\beta_I + \gamma)$$

From the propeller theory above and our 2D airfoil study we design an axial fan whose blades have been twisted from the root(hub) to the tip at a gradient to ensure maximum lift at each section of the blade. The section of the blade at the hub is inclined at 50° and 36° at the tip with respect to the horizontal plane. Bearing an airfoil shape of S6063, the blade is optimized with respect to high lift to drag ratios and limited pitching moment throughout its length. An MRF analysis of the above design produced highly axial flows with mass weighted average axial outlet velocity of 26m/s and a volume flow rate of 0.9m3/s. (Figure 15)

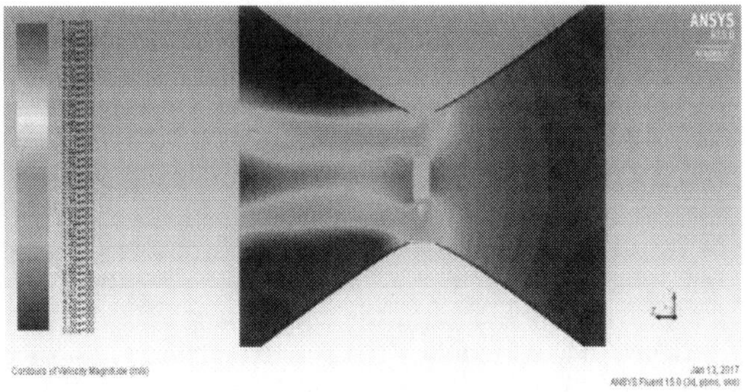

Figure 15. Axial Fan CFD Simulations – Final Design (Velocity Contours)

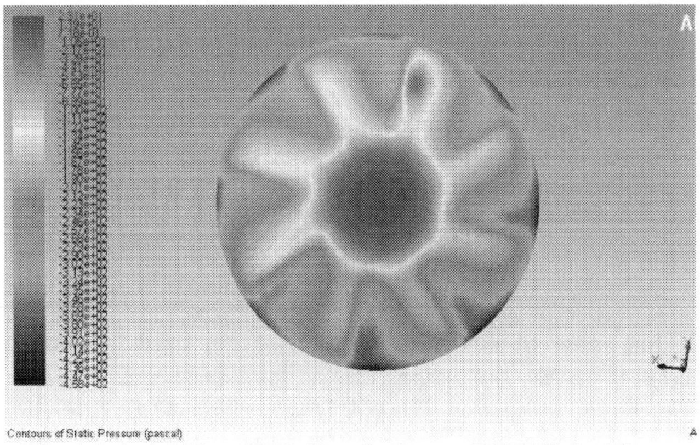

Fig 16. Axial Fan CFD Simulations – MRF Inlet Plane (Pressure Contours)

Figure 16 describes the pressure contours on a plane right before the air enters the rotational MRF zone. We see clear green contours in a radial pattern which correspond to suction gauge pressures of up to -350 Pa, while the central red coloured zone represents the propeller hub area where the gauge pressure remains close to 0 Pa.

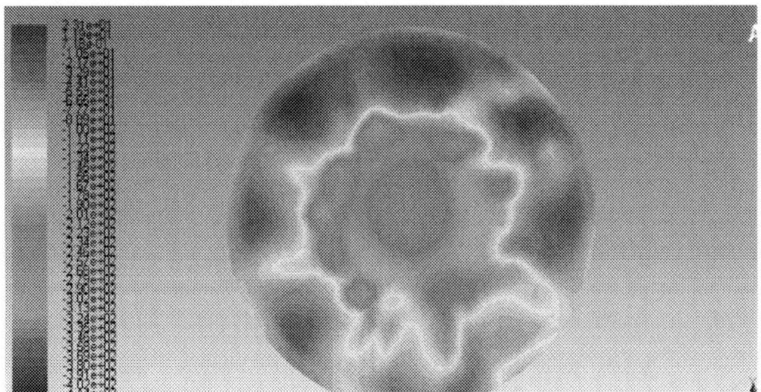

Figure 17. Axial Fan CFD Simulations – MRF Outlet Plane (Pressure Contours)

The figure 17 clearly describes the pressure contours on a plane right after the rotational MRF zone. High airflow is indicated in the central green coloured zone which is representative of axial flow. Along the periphery, the flow appears to be less and hence coloured red (close to atmospheric pressure).

ELECTRICAL PERFORMANCE

Axial Fan Performance

The axial fan used for the benchmarking the design speed of the MAV propulsion system is a terrestrial, robust, belt driven blower which has a much higher power requirement than what is feasible on a micro aerial vehicle. In order to bring up a meaningful comparison, we looked into lightweight, carbon fiber, UAV propellers developed by a company called T-Motor. Loading test data with respect to different motors and their corresponding propellers performed by T-Motors were studied. It was observed that propellers of similar size as our axial fan (ϕ11 inch) were significantly lighter. Thence, larger propellers (ϕ20 inch) with similar values of moment of inertia as our axial fan were studied to approximate the power consumed at ~3000 rpm (see Table 3)

Table 3. Power Requirement for Axial Fan

Motor	Rotor Diameter (ϕ inch)	Rotor Speed (rpm)	Power I/P (W)
U7 V2.0	20	2992	115
MN601-S	20	3089	106

This value of power input was in turn used to compare propellers of the same size as our axial fan for thrust generated as seen in Table 4. It was observed that our axial fan produced 40% greater thrust (neglecting mechanical friction) in comparison to off-the shelf rotors in the market.

Table 4. Thrust Comparison

Motor	Rotor Diameter (ϕ inch)	Rotor Speed (rpm)	Power I/P (W)	Thrust (g)
U3	11	7500	121	870
MN3110	11	7100	111	860
MN3508	12	6200	102	870
MN4010	12	6200	111	860
MN2510	12	6600	122	930
Axial Fan	*11*	*3000*	*~110*	*1230*

Multi-functional Structural Body

The proposed use of the multi-functional structural battery is for fabricating the duct, stator and payload support of the MAV, which has a total surface of 0.565 m^2 and weighs in at 0.87 kg. (thickness of MSB = 1mm, Ref.14) Since, it is only feasible to have the duct (surface area = 0.225 m^2; weight = 0.348 kg) made of MSB (MAV weight constraint), the forecasted power stored (Ref.15) on the MAV is 70W.

CONCLUSIONS

1. The concept of using a ducted fan MAV in an indoor environment is critically discussed and the MAV design strategy is centered around the environmental factors and ergonomics.

2. Incorporation airfoil shaped blades in the axial fan in place of conventional planar twisted blades and aerodynamic optimization of the fan with respect to specified hover conditions shows significantly higher thrust generating capacity of approximately 40% higher (neglecting mechanical friction).

3. The proposed multifunctional structural body is capable of storing 70W of power at a single charge, sufficient to run basic control systems such as an Arduino chip. Further possible uses of this energy could be – powering the stepper motors attached to the control vanes, backup power system for the surveillance payload.

Author contact: Chetan Gupta cgupta@student.tudelft.nl
Dineshkumar Harursampath dinesh@aero.iisc.ernet.in
Ramesh Gupta ramesh.gupta@snu.edu.in

ACKNOWLEDGMENTS

Thanks to Mr. Sreedhara Harlanayaka, Scientist, National Aerospace Laboratories, India for his guidance on multifunctional battery technology and Dr. Nitin Banker, Asst. Prof., Shiv Nadar University, India for his technical insight on concepts of Computational Fluid Dynamics.

REFERENCES

[1]Cai, Guowei, Chen, Ben M., Lee, Tong Heng, *Unmanned Rotorcraft Systems*, Advances in Industrial Control, Springer-Verlag London Limited 2011, Chapter 1.

[2] Clark, C., Summers, J., "Innovative flexible lightweight thin-film power generation and storage for space applications" Energy Conversion Engineering Conference and Exhibit, (IECEC) 35th Intersociety, Las Vegas, USA, 2000.

[3]ITN Energy Systems, Inc., "PowerFibers: Thin-Film Batteries on Fiber Substrate", ITN Energy Systems, Inc., Littleton, CO 80127, USA, 2003.

[4]Thomas, J., Keennon, M., T., DuPasquier, A., Qidwai, M., A., Matic, P., "Multifunctional Structure-Battery Materials for Enhanced Performance in Small Unmanned Air Vehicles" ASME International Mechanical Engineering Congress and Exposition, Washington, DC, USA, November 15–21, 2003.

[5]Pereira, T., Guo, Z., Nieh, S., Arias, J., H., Hahn, T., "Embedding thin-film lithium energy cells in structural composites" *Composites Science and Technology,* Vol. 68, pp 1935–194 ,2008.
doi: 10.1016/j.compscitech.2008.02.019

[6]Kjell, M., H., Jacques, E., Zenkert, D., Behm, M., Lindbergha, G., "PAN-Based Carbon Fiber Negative Electrodes for Structural Lithium-Ion Batteries", *Journal of The Electrochemical Society*, volume 158, issue 12, A1455-A1460, 2011.
doi: 10.1149/2.053112jes

[7]Kjell, M., H., Jacques, E., Zenkert, D., Behm, M., Lindbergh, G., Willgert, M., "Impact of electrochemical cycling on the tensile properties of carbon fibres for structural lithium-ion composite batteries", *Composites Science and Technology*, Volume 72, Issue 7, pp. 792-798 13 April 2012.
doi: 10.1016/j.compscitech.2012.02.006

[8]Leijonmarck, S., Carlson, T., Lindbergh, G., Asp, L., E., Maples, H., Bismarck, A., "Solid polymer electrolyte-coated carbon fibres for structural and novel micro batteries", *Composites Science and Technology*, Volume 89, pp. 149–157, 13 December 2013.
doi: 10.1016/j.compscitech.2013.09.026

[9]Leijonmarck, S., Hagberg, J., Lindbergh, G., "High Precision Coulometry of Commercial PAN-Based Carbon Fibers as Electrodes in Structural Batteries", *Journal of The Electrochemical Society*, volume 163, issue 8, A1790-A1797, 2016, doi: 10.1149/2.0041609jes.

[10]Izadi, M., J. and Falahat, A., "Effect of Blade Angle of Attack and Hub to Tip Ratio on Mass Flow Rate in an Axial Fan at a Fixed Rotational Speed", ASME Proceedings - 8th Symposium on Applications in Computational Fluid Dynamics, Paper No. FEDSM2008-55179, pp. 903-913, Florida, USA, August 10–14, 2008.

[11]Akturk, A. and Camci, C., "Tip Clearance Investigation of a Ducted Fan Used in VTOL Unmanned Aerial Vehicles—Part I: Baseline Experiments and Computational Validation", *Journal of Turbomachinery*, Volume 136, Issue 2, Sep 26, 2013.
doi: 10.1115/1.4023468

[12]Dixon, S., L. and Hall, C., A., *Fluid Mechanics and Thermodynamics of Turbomachinery,* Butterworth-Heinemann, 30-March-2005, Pages 143-181, Chapter 5.

[13]Thomas J. Mueller, "Aerodynamic Measurements at Low Reynolds Numbers for Fixed Wing Micro-Air Vehicles", University of Notre Dame, USA, 13-17 September 1999.

[14]Asp, L., E. and Greenhalgh, E., S., "Structural power composites", *Composites Science and Technology,* Volume 101, pp. 41-61, 12 September 2014.
doi: 10.1016/j.compscitech.2014.06.020

[15]Asp, L., E., Leijonmarck, S., Carlson, T., Lindbergh, G., "Realisation of structural battery composite materials", 20th International Conference on Composite Materials, Copenhagen Denmark, 2015.

Development of VTOL with movable legs

Sho Goto
1st grade in MA
Tokushima University
Tokushima, Tokushima, Japan

Masafumi Miwa
professor
Tokushima University
Tokushima, Tokushima, Japan

ABSTRACT

In this research, we propose VTOL which transitions from vertical takeoff to a horizontal flight with the movable leg equipped with a propulsion device facing the back of the aircraft. We conducted experiments on vertical take-off and landing and hovering by a test airframe using a fixed leg equipped with ducted fans as propulsion device.

I. Introduction

At present, in the aeronautical survey at the time of natural disaster, airborne photography is getting to be used a small radio controlled UAV (unmanned aerial vehicle). Therefore, research on autonomous control of UAV is being conducted. As a kind of UAV, research on Vertical Take and Landing (VTOL) machines combining the maneuverability of rotary wing aircraft and the cruise distance of fixed wing aircraft has been conducted.

In the previous research, we proposed an inverted model of the aircraft and aimed to realize a propulsive type VTOL machine. The propulsion formula is characterized in that the center of gravity is at the top of the thrust generator. Further, by securing a certain amount of the area of the intake portion of the thrust generating device, it is possible to provide the load, the structure of the body, and the work space on the upper portion of the thrust generating device, thereby enlarging the use applications. This is because the influence of the air current on the upper part is smaller than the lower part of the thrust generator, due to the characteristics of the propeller. On the other hand, since the propulsive force exerts an unstable force on the attitude, automatic control of the three axis attitude is required. Although we succeeded in vertical takeoff and landing and hovering in the previous research, transition to the horizontal state which is the ultimate goal of VTOL machine research has not been achieved yet. However, in an inverted flying object, when the body tilts 90 degrees at the time of transition to the horizontal state, a gimbal lock phenomenon occurs and it is necessary to perform attitude control using quaternion, which complicates control.

Therefore, in this study, we propose a VTOL machine that performs vertical takeoff and landing on a movable leg with built-in thrust generator and hovering, and at the time of hovering, move the movable leg itself horizontally toward the rear of the fuselage to simplify control. Figure 1 shows a diagram of the aircraft going to the horizontal state. As an introduction, a test airframe named FLFO (Fixed Legs Flying Object) having a fixed leg provided with a thrust generator on fixed wing aircraft was manufactured, and stable vertical takeoff and landing and hovering were attempted.

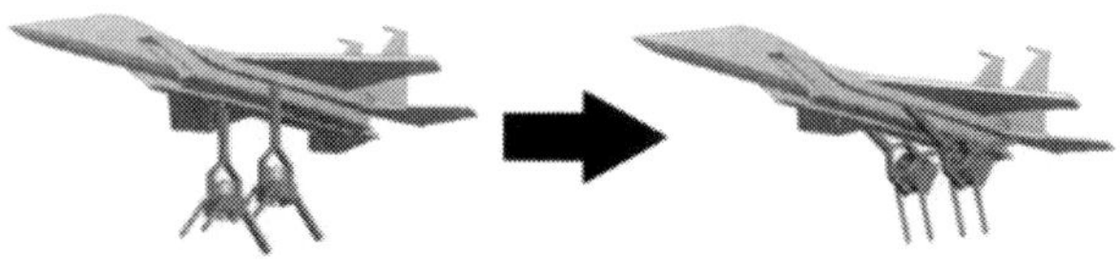

Fig. 1 Flying Object

II. Transition to horizontal flight

Consider the case of moving horizontally during vertical flight. A simplified diagram is shown in figure 2. The velocity at this time is V, the fixed blade area is S, the atmospheric density is ρ, and the lift coefficient is C_L. The generated lift L is expressed by the following equation.

$$L = \frac{1}{2}\rho C_L S V^2 \quad (1)$$

When the leg is further tilted proportional to generated lift force, the horizontal force increases and the speed increases. Since the lift is proportional to the square of the speed, the lift increases according to the speed. Lifting force and weight of the aircraft will soon be balanced, and the transition to level flight is completed. Actually, the influence of the movement of the center of gravity position appears, but the thrust control can cancel it.

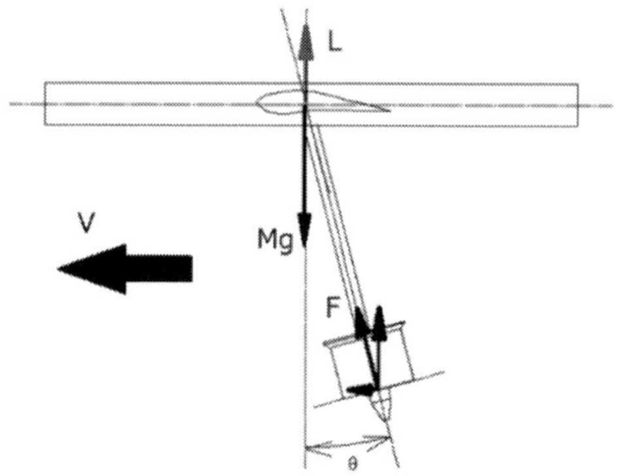

Fig. 2 Level flight

III. OVERVIEW OF AIRCRAFT MODEL

As a first step we aim for stable vertical take-off and landing and hovering. As a testing airframe, we designed FLFO with fixed leg with thrust generator on fixed wing aircraft. As a thrust generator, a pair of duct fans of forward and reverse rotation is used. In the duct fan, a plastic propeller is attached to the rotating shaft of the motor, and the propeller is covered by the duct. Therefore, it does not come into contact with the propeller from the outside, and there is less danger than an air vehicle using a general rotor. It also has the feature that its shape is very compact. A two-dimensional deflection nozzle is used for deflection of the duct fan in the thrust direction. Since the two-dimensional deflection nozzle can control the thrust vector by operating at a constant rotation speed and squeezing it with a deflecting plate, it is advantageous that the response is faster than a gimbal deflection mechanism or the like.

IV. DEVELOPMENT OF FIXED LEGS

In this research, the final goal is to make the transition to horizontal flight with movable legs. When the duct fan and the servomotor which have relatively large mass move to the rear of the aircraft, the aircraft attitude is affected. This is caused by reaction torque and movement of the center of gravity position. Therefore, in order to reduce the influence on the attitude of the aircraft, we want to keep the mass as close as possible to the moving part of the movable leg. In this section, we investigate the performance difference concerning the thrust control by the distance between the thrust generator and the deflection plate, using two kinds of legs with different distances between the thrust generator and the deflector. Based on this experiment, we will develop fixed legs to be attached to fixed wing aircraft. Duct fans and the servomotors are attached to the legs, and the deflector plate is operated by the link rod to deflect the thrust. Distances between the thrust generator and the deflector were 0 mm and 40 mm, respectively. A leg portion having an offset of 0 mm is referred to as a C (Close) type, and a leg portion of 40 mm is referred to as an S (Separate) type. Figure 3 shows the appearance of the legs made for comparative experiments. As the linear thrust change from the previous research, it is known that the deflection plate length which is comparatively easy to control the thrust is 20 mm, but this time we measure the performance difference concerning the thrust control by the distance between the thrust generator and the deflection plate Therefore, in order to make the change of thrust during control larger, the experiment was carried out with the deflection plate length being 50 mm.

Fig. 3 Legs type C, type S

A. Thrust Measurement

The magnitude of the duct fan thrust in the experimental leg was measured. The measurement results are shown in Figure 4. Red line represents the thrust of the duct fan alone, blue line represents the thrust from the leg, the plot shows the measured value, and the solid line shows the result of approximating the measured value.

B. Measurement of Thrust Change by Throttle

A constant rotational speed of duct fan was output downward and the magnitude of the thrust changed was measured by manipulating the angle of the deflector plate in the direction to narrow the nozzle from 0 ° to 30 ° in steps of 5 °. The measurement results are shown in Figure 5. Red line indicates S-shaped leg portion, blue line indicates C-shaped leg portion, and green dotted line indicates theoretical value.

C. Deflection Thrust Measurement

A constant rotational speed of duct fan was output and the angle of the deflector was operated in the same direction from 0 ° to 30 ° in steps of 5 ° and the magnitude of thrust generated downward was measured and compared. The measurement results are shown in Figure 6. Red line indicates S-shaped leg portion, blue line indicates C-shaped leg portion, and green dotted line indicates theoretical value.

D. Experimental Results and Discussion

In the S type which the distance between the duct fan and the deflecting plate is 40 mm, the thrust reduction ratio remains around 5% compared to the C type and can be said to be operational. Regarding throttling and deflection control, it was found that the S type having a longer distance between the duct fan and the deflection plate has better performance. This is because the C type is considered to be due to the difference in the front projected area of the deflector on which the jet actually strikes differs from the theoretical formula.

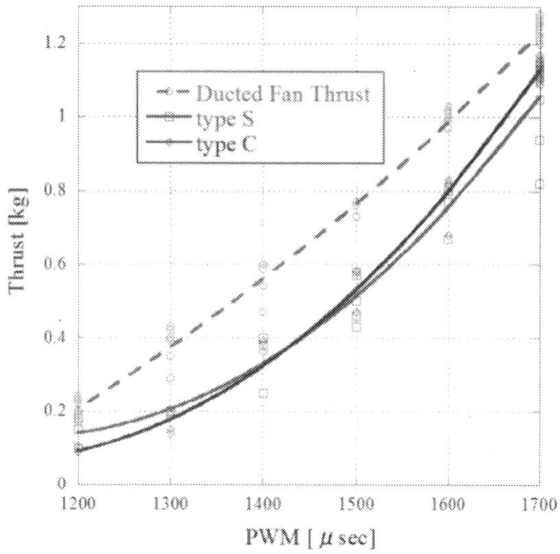

Fig. 4 PWM and Thrust

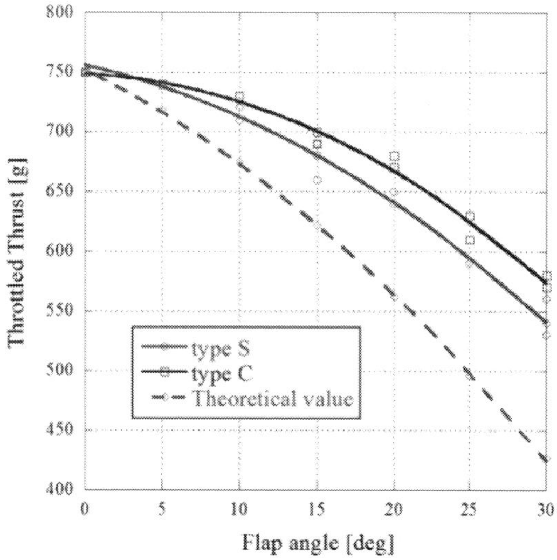

Fig. 5 PWM and Throttled Thrust

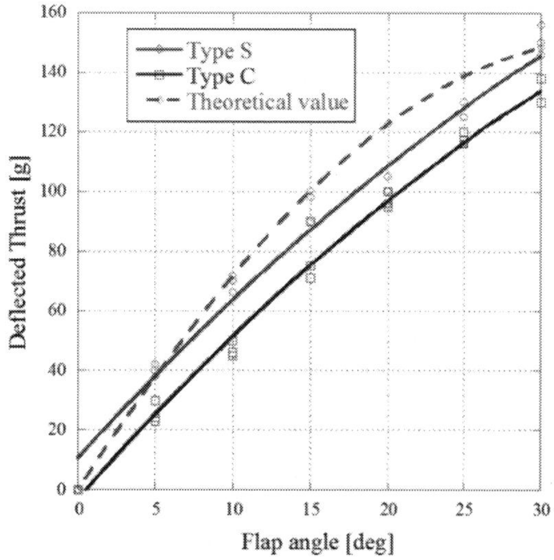

Fig. 6 PWM and Deflected Thrust

V. DEVELOPMENT OF FLFO

New fixed legs were manufactured, and the aircraft model FLFO was manufactured. Figure 7 shows the FLFO. The basic structure of the legs, the thrust generator and the deflector are the same as the S type experimental legs used in Chapter 3, but in order to secure the payload, they were not installed in the fixed wing aircraft, and Its weight is reduced by 100 g per leg. Also, the deflection plate length was shortened and the amount of change in the front projected area of the deflector plate was reduced. As a result, the thrust change becomes linear, making it easier to control thrust. The program for control was produced with reference to the tilt type bicoper library[5]. The control gain was set in real time from the aircraft information during flight.

Fig. 7 FLFO

A. Hovering Experiment of FLFO

Vertical takeoff and landing using a two dimensional deflection nozzle, hovering, and attitude manipulation were done to see the stability of attitude.

B. Response to Step Input

We investigated the step response of FLFO. Figure 8 shows the experimental result. It follows that it follows the target attitude instantly without divergence.

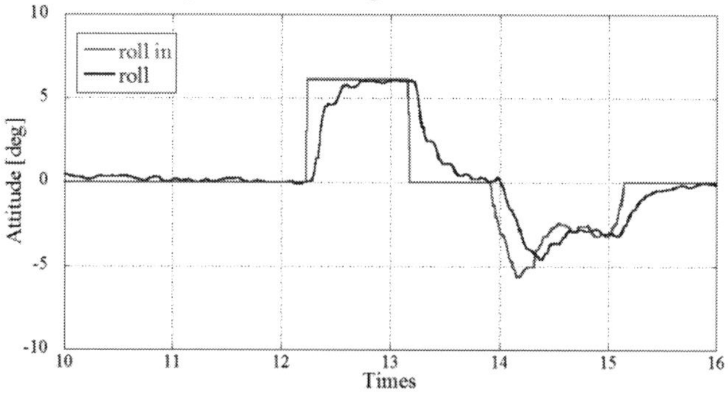

Fig. 8 Response to step input

C. Response to Input of Disturbance

We investigated the response to the disturbance of the FLFO pushed by the finger during hovering. The experimental landscape is shown in Figure 9 and the experimental results are shown in Figure 10. Granted three large disturbance inputs, both the roll axis and the pitch axis instantly stabilized without divergence.

Fig. 9 Response to disturbance: image

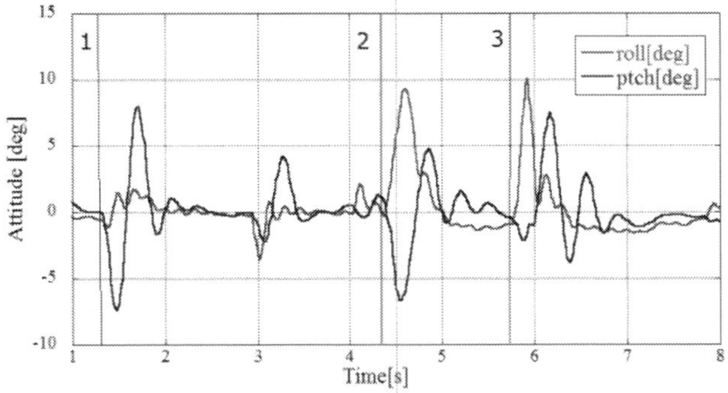

Fig. 10 Response to disturbance

VI. Conclusion

Experimental results show that the attitude control by the two-dimensional deflection nozzle is useful because it is fast response and resistant to disturbance. Regarding the legs, we succeeded in weight saving, but the thrust declined due to the structure of the suction part, and the payload fell down rather. For good control performance, from experience it is ideal to keep the aircraft's body weight to 70% or less of the maximum thrust 3). However, the current FLFO has a weight of 74% against the maximum thrust.

In the future, we will review the design of the aircraft including the legs and secure the payload. Then, we developed a new testing machine MLFO (Movable Legs Flying Object) using movable legs, aiming to shift to horizontal flight by tilting the movable leg. Fig. 11 shows the engine section of the MLFO manufactured. We also make a test machine in the state after the transition and investigate the performance as a fixed wing machine.

Fig. 11 MLFO

Acknowledgment

This work was supported by JSPS KAKENHI Grant Number16K01330.

References

[1] Masafumi Miwa, Yuki Shigematsu, and Takashi Yamashita, "Control of Ducted Fan Flying Object Using Thrust Vectoring", Journal of System Design and Dynamics, Vol. 6, No. 3, 2012.

[2] Yamashita Takeshi, Heisei 24th master's thesis Control of a duct fan aircraft equipped with a variable nozzle, the University of Tokushima Graduate School February 12, 2012.

[3] Masafumi Miwa and Shinya Marubashi: Ducted Fan Flying Object with Normal and Reverse Ducted Fan Units, International Journal Robotics and Mechatronics, Vol. 1, No. 1, pp. 8 - 15, 2014.

[4] Masafumi Miwa, Yuki Shigematsu, and Takashi Yamashita, "Control of Ducted Fan Flying Object Using Thrust Vectoring", Journal of System Design and Dynamics, Vol.6, No. 3, (2012), pp. 322-334.

[5] Masafumi Miwa, Atsushi Mitsumata, Shinji Uemura, "Ducted fan Bicopter type flight unit", System Integration Meeting 2014, 3M1-6.

Optimal Aerodynamic Design of Main Rotor Blade for a High-Speed Compound Helicopter

Masahiko Sugiura
Research Scientist
JAXA
Mitaka, Tokyo, Japan

Yasutada Tanabe
Associate Senior Research Scientist
JAXA
Mitaka, Tokyo, Japan

Hideaki Sugawara
Research Engineer
Ryoyu Systems
Nagoya, Aichi, Japan

ABSTRACT

This paper proposes one of the optimal main rotor blade geometries for a high-speed compound helicopter. Recently, compound helicopter or tilt-rotor has been developed to achieve high-speed while preserving VTOL capabilities. In the compound helicopter, asymmetric flow field on the main rotor increases as the flight speed increases. In other words, reverse flow region on the retreating side spreads. In such a flow field, different blade geometries are deemed to be suited than conventional main rotor blades. Optimization methods are applied in this study to propose new blade geometries. In particular, design parameters of blade geometry are defined as twist angle, chord length, and span position of root cutout. As a result, a new blade shape which is remarkably different from the conventional helicopter is obtained. The optimal blade's root cutout position is approximately forty percent of the span, but it is a reasonable result since the effect of the reverse flow region can be reduced with that. The comparison with UH-60 on rotor performance in hover and high-speed forward flights are also conducted. The optimal blade shows considerably low effective drag coefficient in high-speed flight compared to UH-60.

INTRODUCTION

Helicopters were put to practical use in 1940s. After that, they have been utilized in many situations such as firefighting, emergency medical service (EMS), broadcasting and transportation since they have excellent VTOL/hovering performance. However, helicopter's maximum flight speed is lower than that of fixed wing aircraft. Developments of high-speed helicopters have been attempted since 1950s (Refs. 1-3). There are mainly two approaches to achieve high-speed flight. One is tilting the rotary wing forward depending on flight speed. Tilt rotor (Ref. 4) and tilt wing (Ref. 5) are representative examples. These rotorcraft change flight modes from helicopter to fixed wing as flight speed increases. Osprey V-22 is in practical use, though its mechanics tend to be complicated (Ref. 6). The other approach is called compound helicopter, which adds main wing and propeller to a conventional helicopter while maintaining a main rotor of a helicopter (Ref. 1). Since 1950s, there has been several research prototypes, and the compound helicopters are in product development process now, following the successes of technology demonstrators such as Sikorsky's X2 (Ref. 7) and Airbus Helicopters' X3 (Ref. 8).

On the other hand, Japan Aerospace Exploration Agency (JAXA) has been conducting a conceptual study of high-speed helicopters, considering it is technologically efficient to start to develop lightweight high-speed helicopter, then to heavyweight sequentially. The first target is a light helicopter such as an EMS or a rescue helicopter. And the next target would be medium weight and heavy weight in the near future. Compound helicopter is selected since its hovering performance is higher than tilt rotor which uses prop rotors both in hover and forward flights. This hovering performance which arise from using helicopter's main rotor is effective in the supposed EMS or rescue operations. Moreover, anti-torque generation by thrust difference between two electric propellers at the main wing tips is proposed, taking into account the recent technology development of electric aircraft. Main wing's position can be freely designed with this electric anti-torque system. Required thrust in high-speed forward flight is mainly provided by a pusher propeller at the empennage which is directly driven by the engine. The redundant thrust of propellers at the both sides of the main wing is also utilized for forward flight. Delinking pusher propeller at the empennage enables effective hover and low speed forward flights.

Figures 1 and 2 show a conceptual design of high-speed HEMS (helicopter emergency medical service) proposed by JAXA and its major technical issues, respectively. There are some technical issues in the compound helicopter configuration such as developing aerodynamically low drag fuselage as shown in Fig. 2. Among them, optimal rotor blade shape for high-speed flight is supposed to be one of the most important technical issues since flow field around a rotor drastically varies with advance ratio (Fig. 3). Figure 4 shows reverse flow region at high advance ratio, in which flow's direction for a blade is opposite to the other region.

Presented at the 6th Asian/Australian Rotorcraft Forum / Heli Japan 2017, Kanazawa, Japan, November 7-9, 2017.

At these high advance ratios, thrust reversal is a well-known phenomenon (Fig. 5, Ref. 12). In spite of positive collective pitch angle, the total thrust becomes negative at high advance ratios due to negative lift generation in the reverse flow region. Considering these characteristics at high advance ratio, this paper proposes a brand new optimal blade shape through multi-optimization of high-speed and hover flights.

NUMERICAL METHODS

For the modification of blade geometry, optimization technique is combined with blade element theory (BET). In the present research, BET technique to get aerodynamic solution and optimization technique to obtain optimized blade geometry are weakly coupled. Blade geometry for BET analysis is modified using design variables which are suggested by the optimization method. The results of BET analysis are utilized to evaluate the objective functions for optimization process. In this section, optimization procedure and blade element theory are elaborated.

Figure 1. Conceptual figure of high-speed HEMS (helicopter emergency medical service) proposed by JAXA.

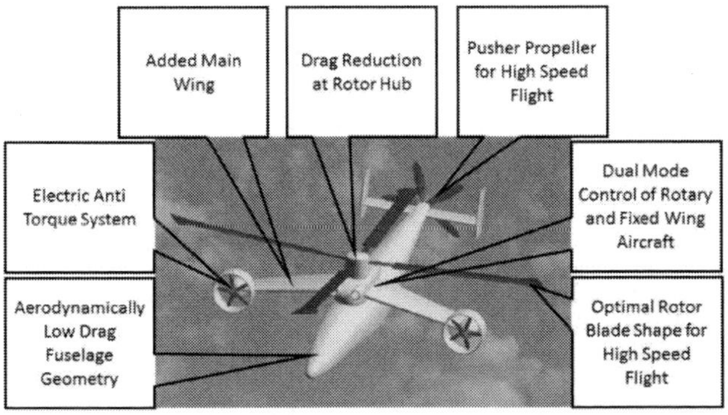

Figure 2. Technical issues of compound helicopter.

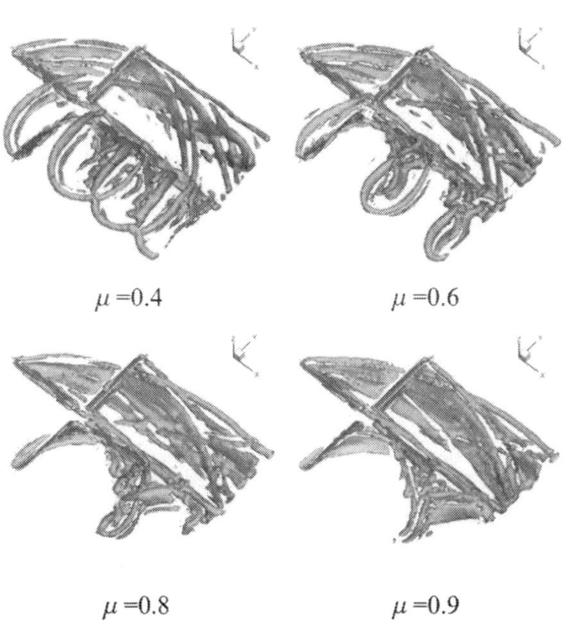

Figure 3. Flow filed variations with advance ratio.

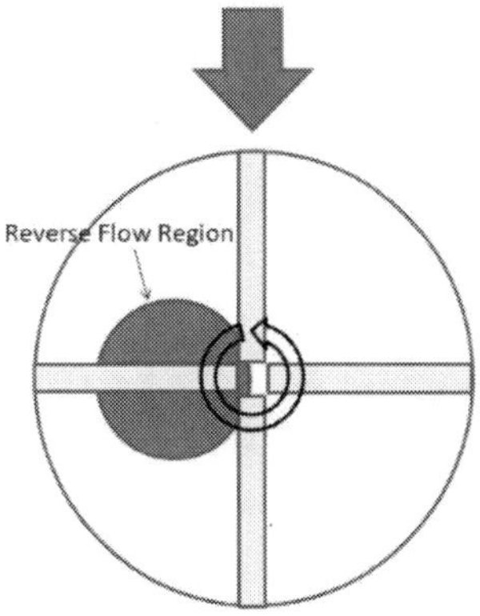

Figure 4. Reverse flow region at high advance ratio.

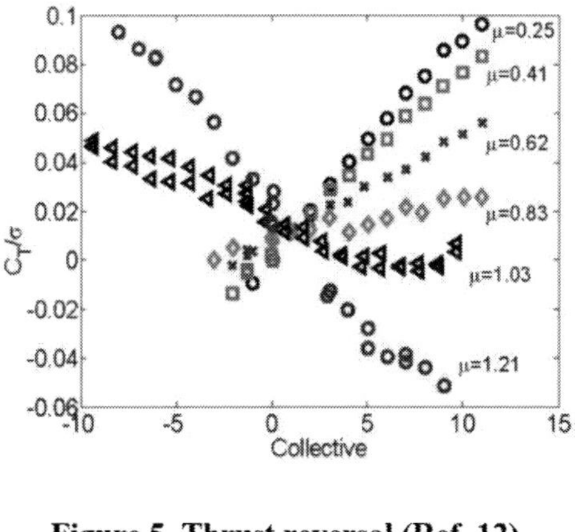

Figure 5. Thrust reversal (Ref. 12).

Optimization

A Kriging model based genetic algorithm (GA) optimization as shown in Fig. 6 is adopted. Once the GA optimization is over, the validity of the search region is examined using Kriging model. The sample points are interpolated with the Gaussian random function as the correlation function to estimate the trend of the stochastic processes. Expected improvement (EI) of objective function is directly used as fitness values in the optimization. GA maximizes EIs of objective functions to find the non-dominated solutions about EIs and several points are selected from the non-dominated solutions to update the Kriging model. The details of GA and Kriging model are described in Refs. 13 and 14.

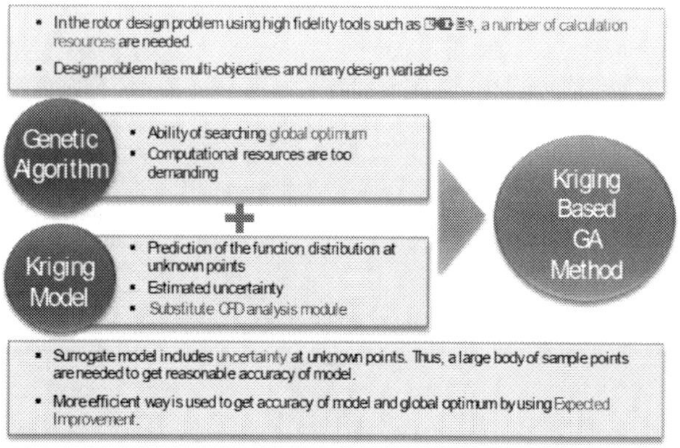

Figure 6. Kriging based GA optimization method combined with rotor BET code.

Blade Element Theory

JAXA has been developing and validating CFD based rotorcraft comprehensive analysis tool <rFlow3D> (Ref. 15). Validation for high advance ratio rotor also has been conducted thus far (Ref. 16). However, blade element theory is employed in this study since the effect of various kinds of parameters must be estimated at the conceptual design stage. Aerodynamic coefficients of a blade element are interpolated using look-up tables of airfoil, which covers angle of attack from 0 to 360 deg. and includes Mach number effect. Linear inflow model of Pitt & Peters is applied as induced velocity model (Ref. 17).

OPTIMIZATION

Conditions

In this paper, the compound helicopter's weight is assumed as four-ton class since helicopters in that category is often utilized as EMS in Japan (Ref. 11). Considering four-ton class helicopter like H145 (BK117, Fig. 7), flight conditions of hover and high-speed forward flights are set as Table 1 and 2. In high-speed forward flight, the target speed is 463km/h (250kt), which is approximately twice of a conventional helicopter. Each fifty percent of the maximum takeoff weight is generated by the main rotor and main wing, respectively. Rotor revolutions are reduced to 73% to avoid Mach divergence in the advancing side.

Figure 8 shows control points of blade design variables and constraints of design variables at each control points are described in Table 3. Point 1 controls root cutout position. Chord length is constant and the blade is linearly twisted between point 1 and 2. At point 3, twist angle can be varied from +10 deg. to -10 deg. and chord length is also changed. At point 4 (blade tip), twist angle and chord length are design variables, but chord length is determined by keeping weighted blade area constant. Weighted blade area is calculated by integrating products of a local spanwise position length from the rotor hub center and a local chord length. Baseline rotor blade properties are also shown in Table 4. In terms of objective functions in optimization, figure of merit (FM) and coefficient of effective drag (C_{DE}) are selected in hover and forward flights.

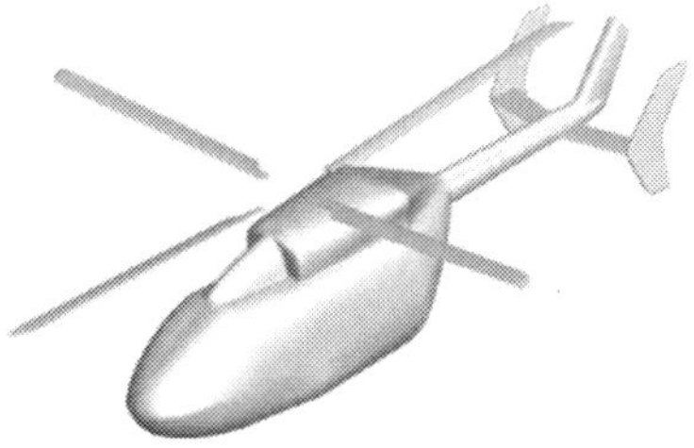

Figure 7. H145 (BK117) like helicopter.

Table 1. Flight Condition of Hover Flight.

Trim Condition Lift, kg	4,100 (Maximum Takeoff Weight) $Mx=My=0$
Rotor Revolutions, rpm	383.36
Tip Mach Number $Mtip$	0.650
Blade loading, C_T/σ	0.075

Table 2. Flight Condition of High-Speed Forward Flight.

Forward Flight Speed, km/h	463
Trim Condition Lift, kg	2,050 (50% of Maximum Takeoff Weight) $Mx=My=0$
Rotor Revolutions, rpm	279.12 (73%)
Advance Ratio μ	0.8
Tip Mach Number $Mtip$	0.850
Blade loading, C_T/σ	0.070

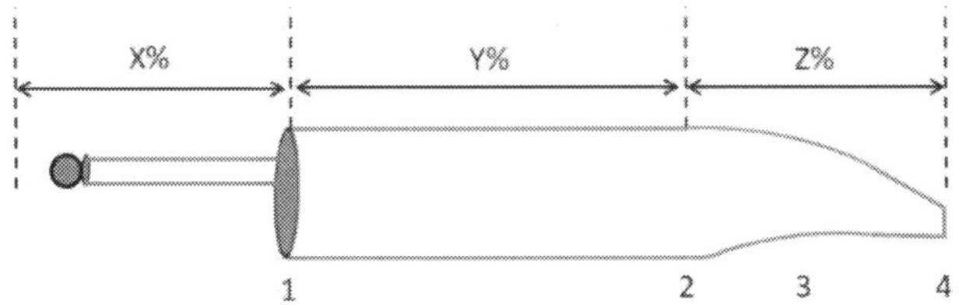

Figure 8. Control points of blade design variables.

Table 3. Constraints of blade design variables at each control points.

Control Points	Span Position	Constraints Twist	Chord Length
1	0~0.6r/R	-	-
2	0~1.0r/R	-	-
3	0~1.0r/R	-10° ~+10°	0.5c~2.0c
4	1.0r/R	-10° ~+10°	Weighted Blade Area Constant

Table 4. Rotor Properties of the Baseline.

Number of Blades	4
Airfoil	NACA23012
Rotor Radius	5.5 m
Chord Length	0.41 m
Root Cutout	0.30 (/R)
Twist Angle	-8.0 deg

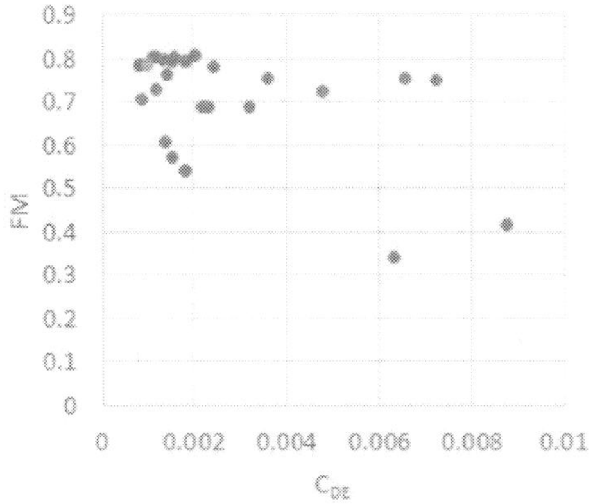

Figure 9. Sampling data points in multi-point optimization (red: optimal point, green: baseline, blue: other sampling points).

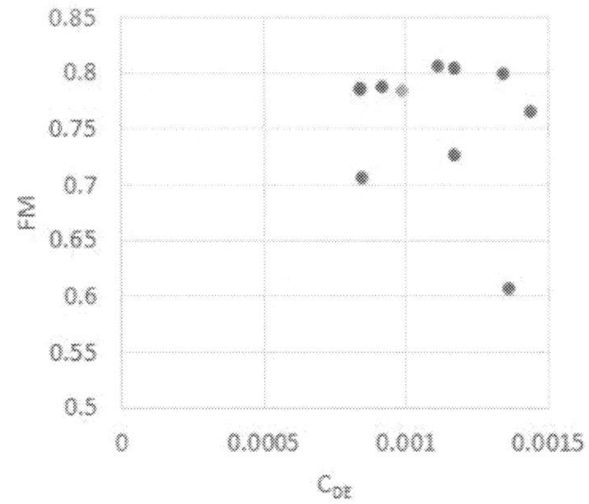

Figure 10. Enlarged drawing of area around the optimal point and the baseline in Fig. 9.

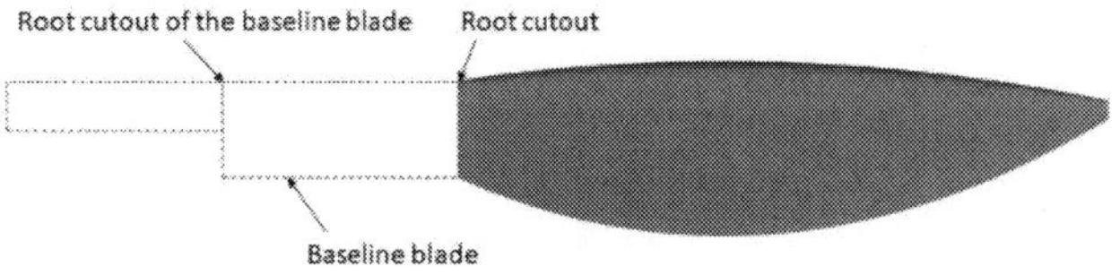

Figure 11. Top view of Opti-Rotor blade shape.

Results

Figure 9 shows sampling data points in multi-point optimization. It's a trade-off between hovering and forward flight efficiencies, but a blade shape which has high FM and low C_{DE} is a favorable solution. Red point is the optimum, green one is the baseline, and blue ones are other sampling points. Each objective function varies depending on the design variables. The optimal point and the baseline looks very close in Fig. 9, but area around them are enlarged in Fig. 10. Coefficient of effective drag of the optimal (Opti-Rotor) blade is approximately 15% lower than the baseline, FMs are approximately same though.

Figure 11 shows a top view of Opti-Rotor blade shape. Root cutout position of forty percent and propeller like blade shape are different from conventional rotor blades. However, this kind of blade shape is reasonable since it can avoid reverse flow regions at high advance ratio. The exact values of chord length and twist angle are shown in Fig. 12.

ROTOR PERFORMANCE

In this session, rotor performances of Opti-Rotor blade in hover and high-speed forward flights are compared with UH-60 blade (Refs. 18 and 19), which is commonly utilized as a reference. Figure 13 illustrates a top view of UH-60 blade shape. And the chord length and twist angle distributions along spanwise position are shown in Fig. 14. Table 4 summarizes the rotor properties of Opti-Rotor blade and UH-60. Opti-Rotor blade is expanded proportionally to correspond with the rotor radius of UH-60. Twist angle of UH-60 is approximately twice of Opti-Rotor.

Hover Flight

Computational conditions of hover flight are determined based on Ref. 20. Table 5 summarizes the computational

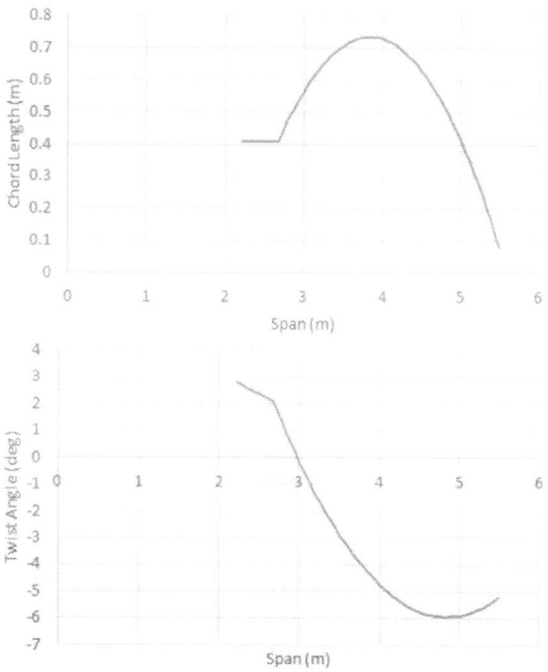

Figure 12. Chord length and twist angle distributions along spanwise position.

Figure 13. Top view of UH-60 blade shape.

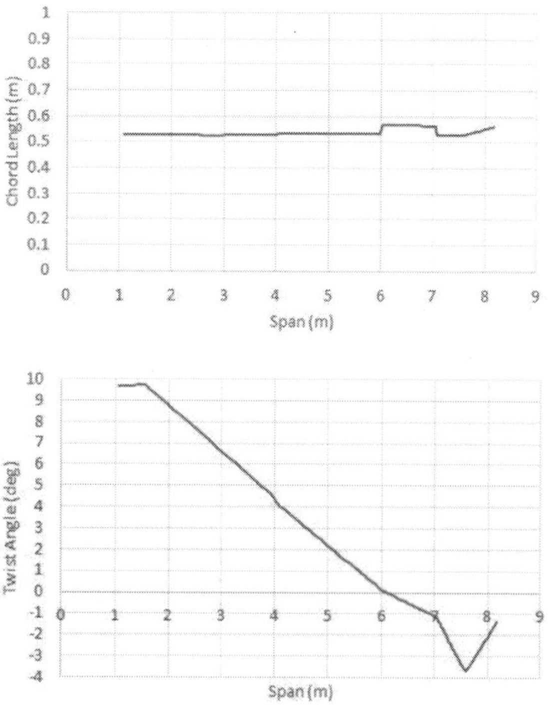

Figure 14. Chord length and twist angle distributions along spanwise position (UH-60).

Table 4. Rotor Properties of the Optimum and UH-60.

	Optimum	UH-60
Number of Blades	4	4
Airfoil	NACA23012	SC1095
Rotor Radius	8.18 m	8.18 m
Chord Length	0.61 m	0.53 m
Root Cutout	0.40 (/R)	0.13 (/R)
Twist Angle	-8.0 deg	-16.0 deg
Solidity	0.095	0.082

conditions. Hover performances of Opti-Rotor blade and UH-60 are compared in terms of FM in Fig. 15. In this figure, experimental data of Sikorsky WT1/WT2 and DNW are model rotor results. UH-60 indicates its computational results. From this figure, it is seen that FMs of Opti-Rotor and UH-60 are identical, but Opti-Rotor shows slightly higher FM in high blade loading. When it comes to the difference between BET and experiment, BET results show higher FMs than the experiments in general. Figure 16 shows the comparison of power loading. The power loading of Opti-Rotor corresponds with UH-60, but is slightly higher. With respect to power loading, the difference between BET and experiment is indistinctive.

Figures 17-22 describe CnM^2 distributions along spanwise position in each blade loadings. Blade loadings of UH-60 are added in the captions of these figures. Overall, CnM^2 of UH-60 increases approximately linearly along spanwise position, but there are dimples around a blade tip. This trend is supposed to be due to steep twist angle dimple around the blade tip (Fig. 14). On the other hand, there are dimples around 80% of spanwise position in CnM^2 distributions of Opti-Rotor.

Table 5. Computational Conditions of Hover Flight.

Trim Condition Thrust, kg	21817, 44673, 64412, 83112, 93501, 103891 $Mx=My=0$
Blade loading (UH-60), C_T/σ	0.021, 0.043, 0.062, 0.080, 0.090, 0.100
Rotor Revolutions, rpm	258.00
Tip Mach Number $Mtip$	0.650

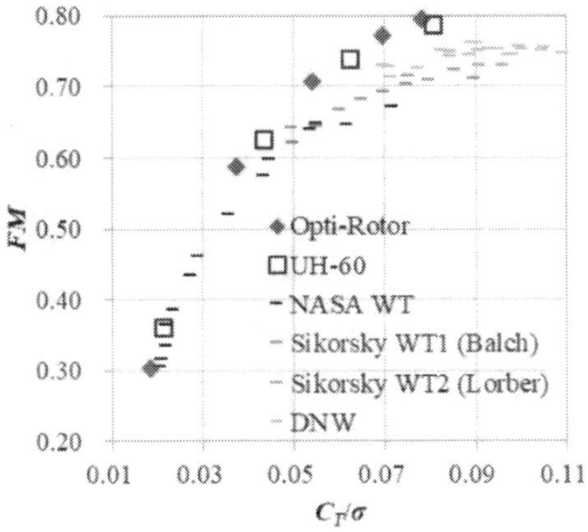

Figure 15. Blade loading vs. figure of merit.

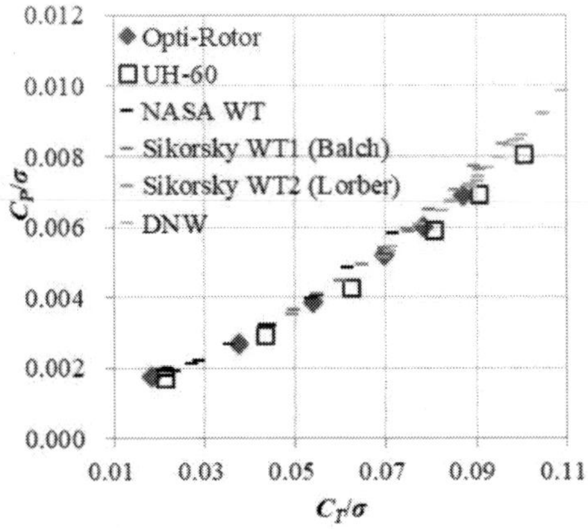

Figure 16. Blade loading vs. power loading.

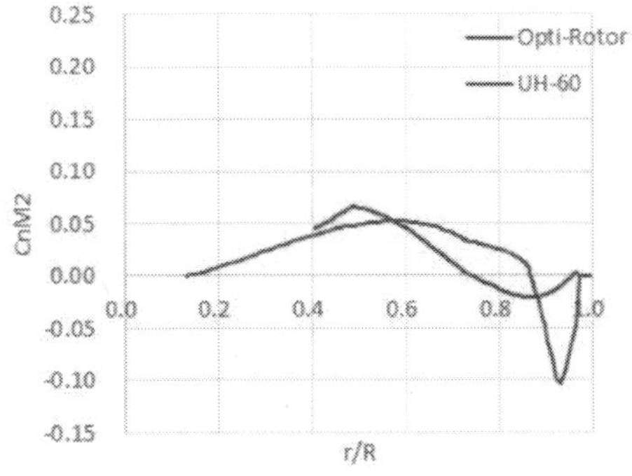

Figure 17. CnM² distributions along spanwise position (C_T/σ=0.021).

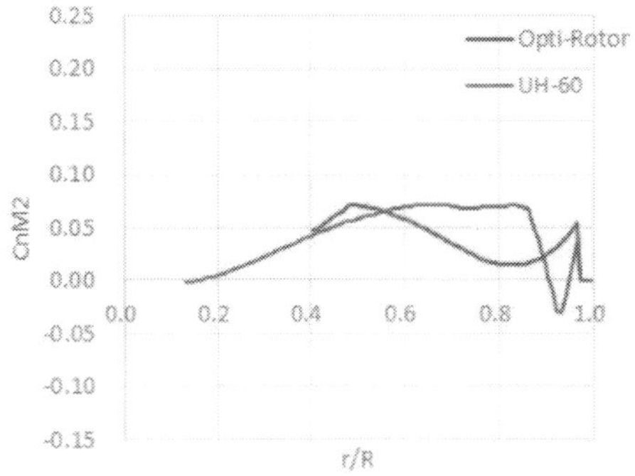

Figure 18. CnM² distributions along spanwise position (C_T/σ=0.043).

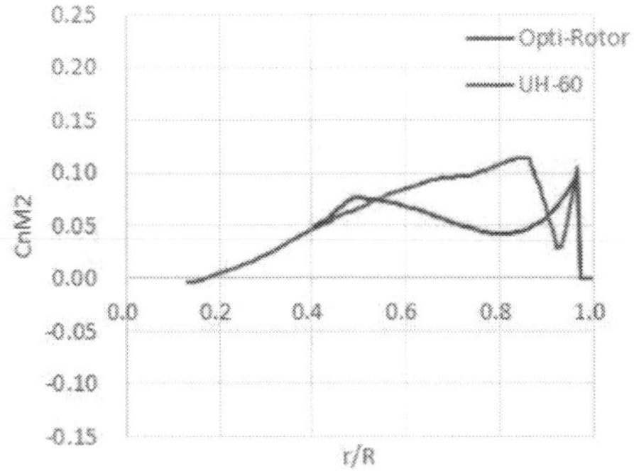

Figure 19. CnM² distributions along spanwise position (C_T/σ=0.062).

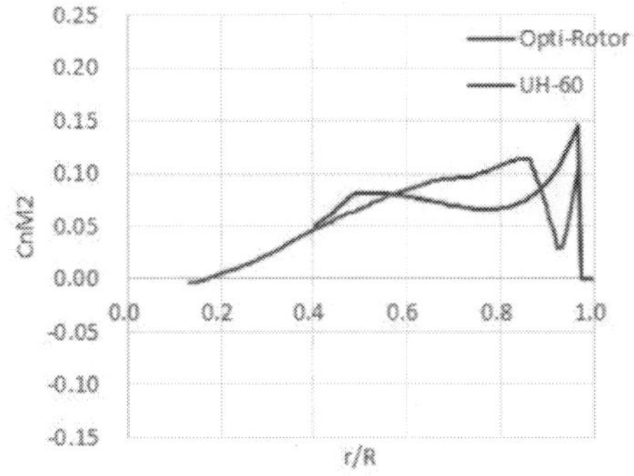

Figure 20. CnM² distributions along spanwise position (C_T/σ=0.080).

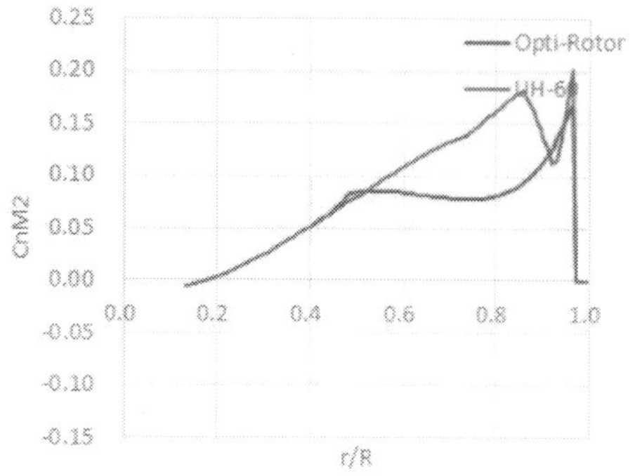

Figure 21. CnM^2 distributions along spanwise position (C_T/σ=0.090).

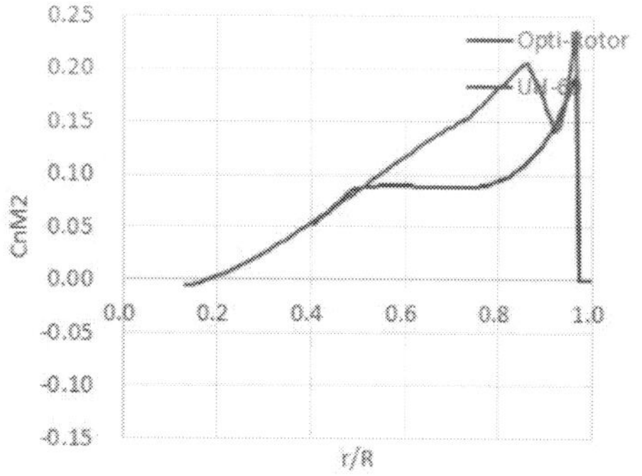

Figure 22. CnM^2 distributions along spanwise position (C_T/σ=0.100).

High-Speed Forward Flight

Computational conditions of high-speed forward flight are set based on Refs. 21-23. Table 6 and Fig. 23 summarize the computational conditions, which are simulated by Potsdam et. al. (Refs. 22 and 23). The tip Macn number 0.26 is 40% of the nominal rotor revolutions. And the shaft angle is assumed as zero deg. The computations are conducted at advance ratios 0.4, 0.6, 0.8, and 0.9.

Effective drags are compared between Opti-Rotor and UH-60 in Figs. 24-27. At advance ratios 0.4 and 0.6, the effective drags of Opti-Rotor are slightly higher than UH-60 especially when the blade loadings are low. However, at advance ratios 0.8 and 0.9, the effective drags of Opti-Rotor are lower than UH-60 in all blade loadings. This difference is supposed to be larger at the optimal design point whose advance ratio and blade loading are 0.8 and 0.070, respectively.

Figures 28-35 show CnM^2 contours on a rotor plane at each computational condition. CnM^2 contours of Opti-Rotor are considerably different from those of UH-60. At the advancing side in all conditions, positive CnM^2 area around the blade root is seen in a wider azimuth angle in Opti-Rotor. On the other hand, it is also seen that smaller negative values of CnM^2 are generated at the blade tips of Opti-Rotor.

Table 6. Computational Conditions of High-Speed Forward Flight.

Tip Mach Number *Mtip*	0.26
Shaft Angle, α_s	0

μ	M_∞	C_T/σ	T, N
0.4	0.104	0.0131	2156.78
0.4	0.104	0.0347	5717.85
0.4	0.104	0.0447	7373.14
0.4	0.104	0.0547	9028.27
0.4	0.104	0.0723	11937.36
0.4	0.104	0.0885	14595.79
0.6	0.156	0.0198	3260.25
0.6	0.156	0.0325	5366.86
0.6	0.156	0.0374	6169.34
0.6	0.156	0.0435	7172.48
0.6	0.156	0.0544	8978.10
0.6	0.156	0.0623	10282.23
0.8	0.208	0.0201	3310.41
0.8	0.208	0.0274	4514.22
0.8	0.208	0.0313	5166.20
0.8	0.208	0.0325	5366.86
0.8	0.208	0.0337	5567.52
0.8	0.208	0.0404	6670.99
0.8	0.208	0.0444	7322.98
0.9	0.234	0.0204	3360.58
0.9	0.234	0.0231	3811.90
0.9	0.234	0.0258	4263.39
0.9	0.234	0.0283	4664.55
0.9	0.234	0.0258	4263.39

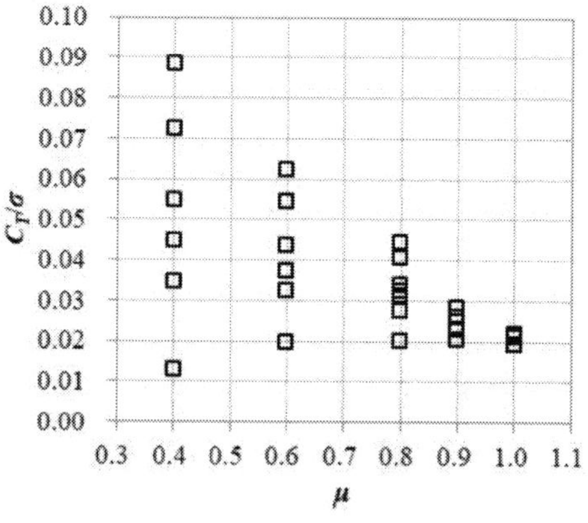

Figure 23. Computational conditions of forward flight (advance ratio vs. blade loading, UH-60).

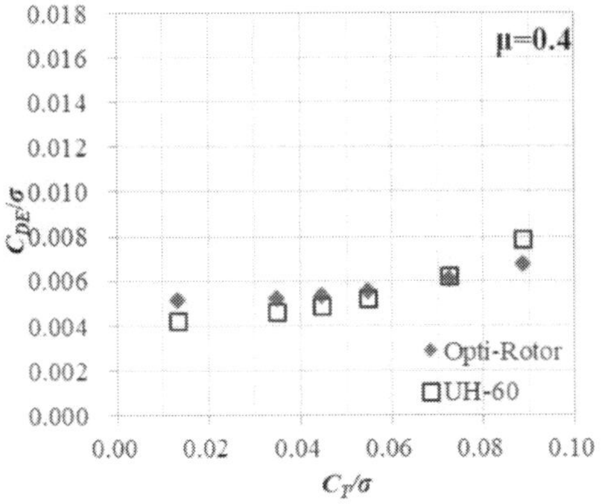

Figure 24. Thrust vs. effective drag (μ =0.4).

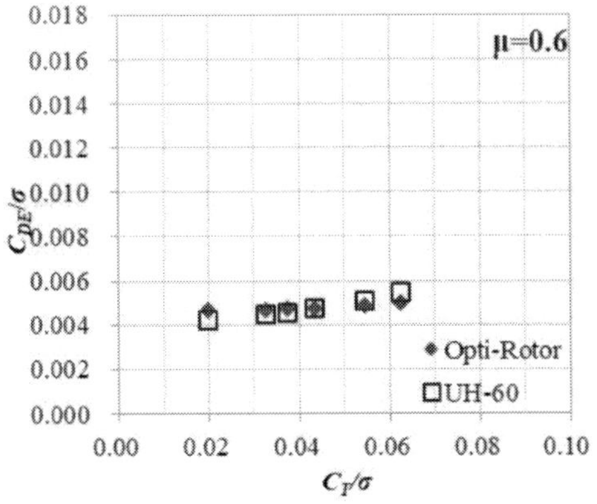

Figure 25. Thrust vs. effective drag (μ =0.6).

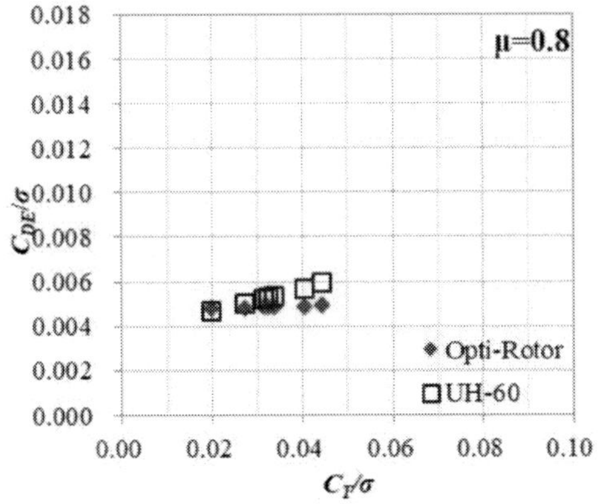

Figure 26. Thrust vs. effective drag (μ =0.8).

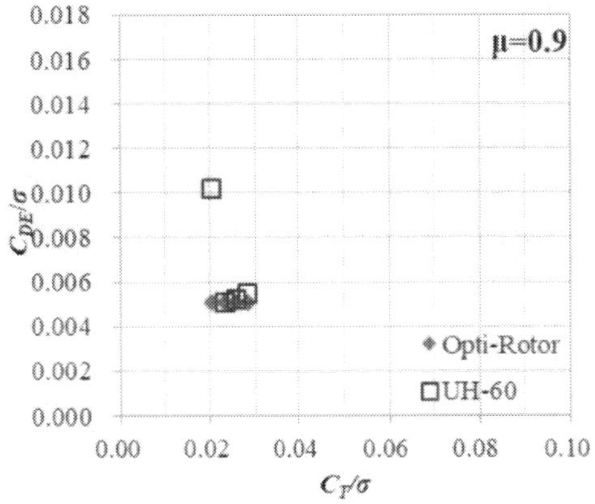

Figure 27. Thrust vs. effective drag (μ =0.9).

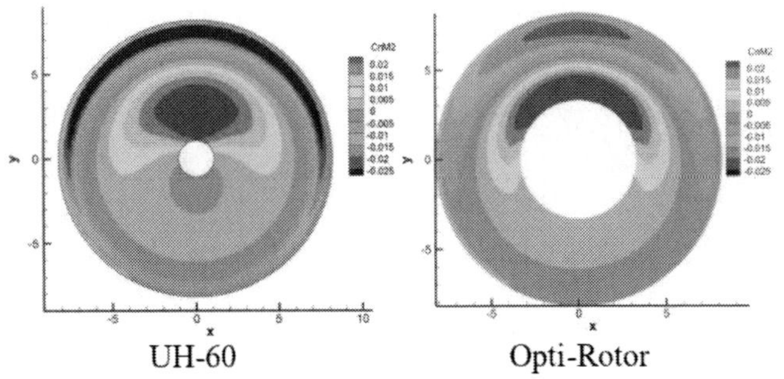

Figure 28. CnM² contour on a rotor plane (μ =0.40, C_T/σ =0.013).

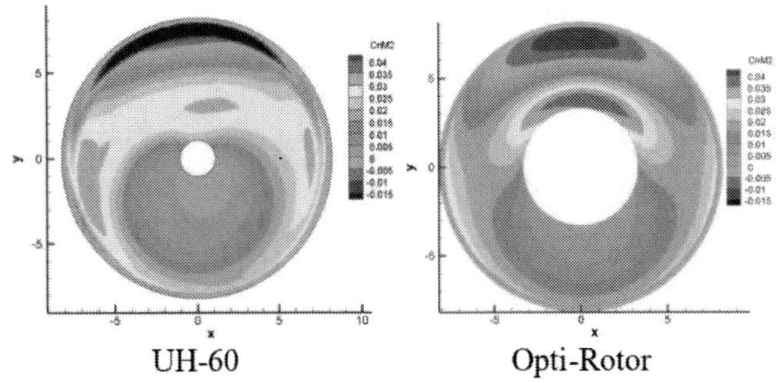

Figure 29. CnM² contour on a rotor plane (μ =0.40, C_T/σ =0.089).

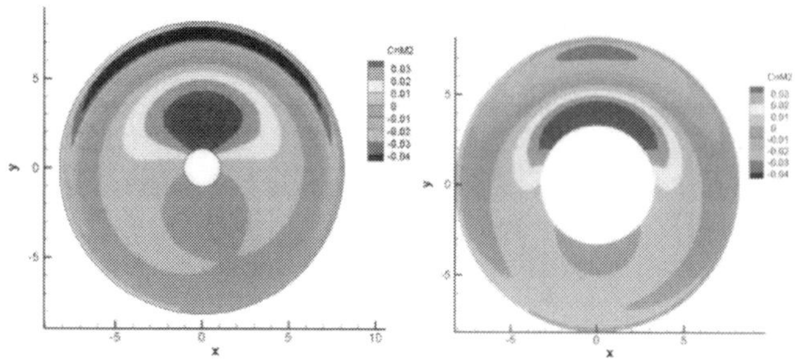

UH-60 Opti-Rotor

Figure 30. CnM² contour on a rotor plane (μ=0.60, C_T/σ=0.019).

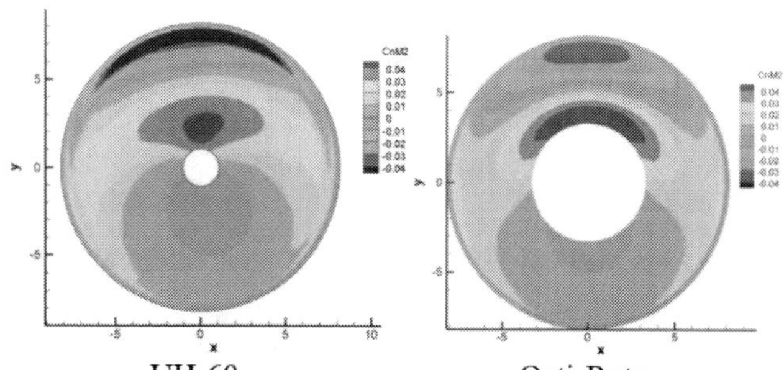

UH-60 Opti-Rotor

Figure 31. CnM² contour on a rotor plane (μ=0.60, C_T/σ=0.062).

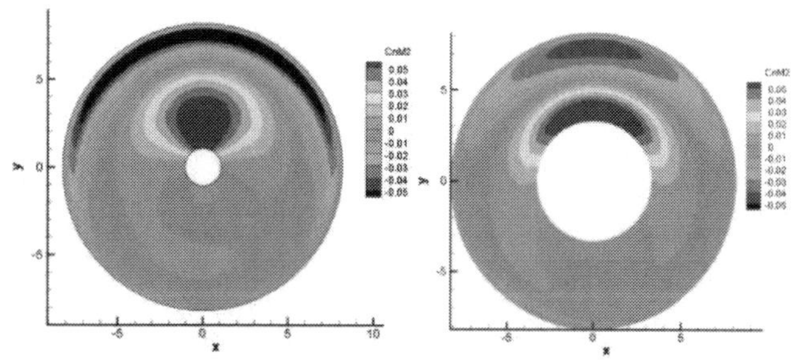

UH-60 Opti-Rotor

Figure 32. CnM² contour on a rotor plane (μ=0.80, C_T/σ=0.020).

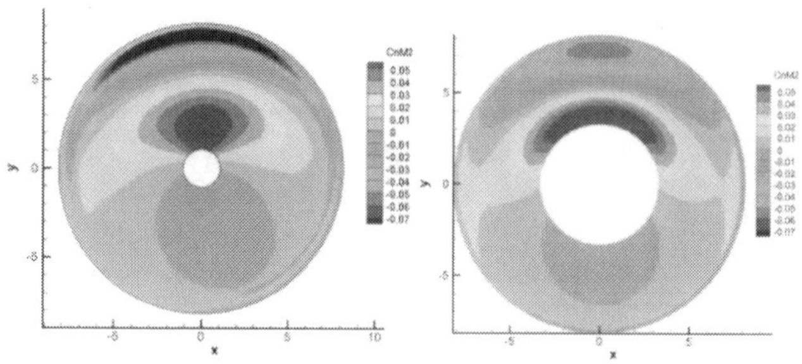

UH-60 Opti-Rotor

Figure 33. CnM² contour on a rotor plane (μ=0.80, C_T/σ=0.044).

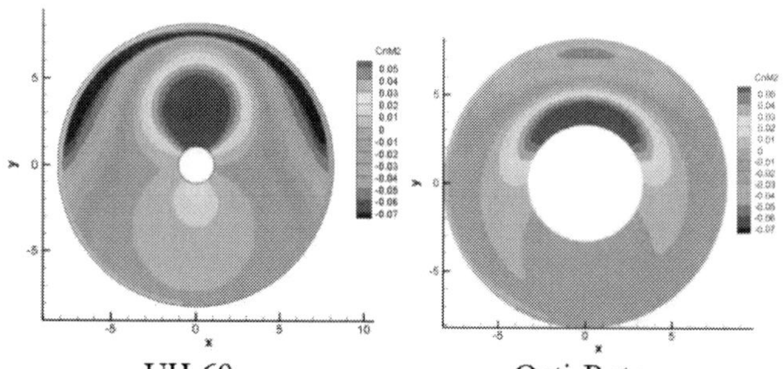

UH-60 Opti-Rotor

Figure 34. CnM² contour on a rotor plane (μ=0.90, C_T/σ=0.020).

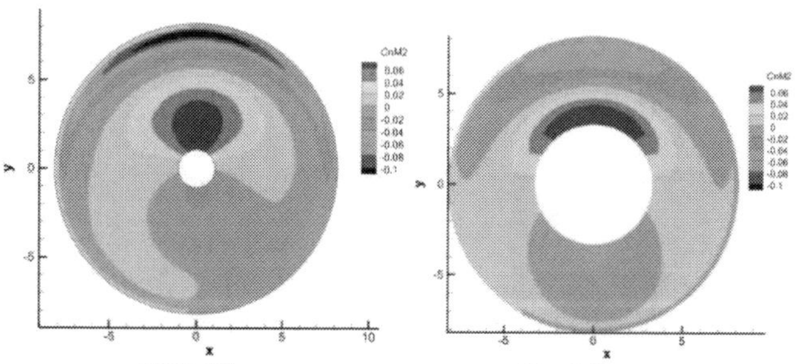

UH-60 Opti-Rotor

Figure 35. CnM² contour on a rotor plane (μ=0.90, CT/s=0.028).

CONCLUSIONS

In this study, a novel main rotor blade shape for a high-speed compound helicopter is proposed. Forty percent root cutout position is found to be aerodynamically efficient through optimization. Opti-Rotor's effective drag is 15 % lower than the baseline blade, while maintaining hover performance of the baseline. Performance of Opti-Rotor is also investigated in various conditions of blade loading and advance ratio by comparing with UH-60. In hover condition, Opti-Rotor shows approximately the same performance as UH-60. On the other hand, the drag of Opti-Rotor becomes lower as advance ratio increases in high-speed flight condition, compared to UH-60. In this paper, comparisons between Opti-Rotor and UH-60 are conducted in the conditions of existing experimental data, referring other papers, though. Comparison in the conceptual design points of high-speed forward flight in EMS, whose advance ratio and blade loading are 0.8 and 0.070, respectively, will be added in the near future. In addition, optimization with CFD adding design parameters of swept and dihedral angles will be planned.

Author contact: Masahiko Sugiura sugiura.masahiko@jaxa.jp, Yasutada Tanabe tan@chofu.jaxa.jp, Hideaki Sugawara hideaki8@chofu.jaxa.jp.

REFERENCES

[1]Harris, F. D., "An Overview of Autogyros and the McDonnell XV-1 Convertiplance," NASA/CR-2003-212799, 2003.

[2]Harris, F. D., "Rotor Performance at High Advance Ratio; Theory versus Test," NASA/CR-2008-215370, 2008.

[3]Prouty, R. W., and Yackle, A. R., "The Lockheed AH-56 Cheyenne - Lessons Learned," AIAA Paper 92-4278, AIAA Aircraft Design Systems Meeting, Hilton Head, SC, August 24-26, 1992.

[4]XV-3, http://www.aviastar.org/helicopters_eng/bell_xv-3.php

[5]Markman, S. and Bill H., "Tilt-Wing VTOL Systems". Straight Up: A History of Vertical Flight. Schiffer Publishing, 2000. ISBN 0-7643-1204-9.

[6]Whittle, R., "Osprey Shows Its Mettle", American Helicopter Society / VERTIFLITE, Vol. 61, No. 3, page 23-26, May/June 2015.

[7]Walsh, D., Weiner, S., Arifian, K., Lawrence, T., Wilson, M., Millott, T. and Blackwell, R., "High Airspeed Testing of the X2 Technology Demonstrator," American Helicopter Society 67th Annual Forum, Virginia Beach, VA, U.S., May 9-11, 2011.

[8]Nelms, D., "Aviation Week Flies Eurocopters X3," Aviation Week & Space Technology, Vol. 174, No. 24, Jul. 9, 2012.

[9]Brannen, Kate. "U.S. Army Eyes Joint Multirole Helo". Defense News, 19 July 2010.

[10]http://www.airbushelicopters.com/website/en/press/Airbus-Helicopters-to-design-new-compound-rotorcraft-demonstrator-in-the-frame-of-Clean-Sky-2-program_1550.html

[11]Tanabe, Y., Aoyama, T., Kobiki, N., Sugiura, M., Miyashita, R., Sunada, S., Kawachi, K. and Nagao, M., "A Conceptual Study of High Speed Rotorcraft," 40th European Rotorcraft Forum, Southampton, UK, September 2-5, 2014.

[12]Berry, B. and Chopra, I., "Slowed Rotor Wind Tunnel Testing of an Instrumented Rotor at High Advance Ratio," 40th European Rotorcraft Forum, Southampton, UK, 2-5 September 2014.

[13]Kanazaki, M., Yokokawa, Y., Murayama, M., Ito, T., Jeong, S. and Yamamoto, K., "Nacelle Chine Installation Based on Wind-Tunnel Test Using Efficient Global Optimization," Transactions of the Japan Society for Aeronautical and Space Sciences, Vol. 51, No. 173, pp. 146-150, 2008.

[14]Kanazaki, M., Jeong, S. and Yamamoto, K., "High-Lift System Optimization Based on Kriging Model Using High Fidelity Flow Solver," Transactions of Japan Society for Aeronautical and Space Science, Vol. 49, No. 165, pp. 169-174, 2006.

[15]Tanabe, Y., Saito S. and Sugawara, H., "Construction and Validation of an Analysis Tool Chain for Rotorcraft Active Noise Reduction," 38th European Rotorcraft Forum, Amsterdam, NL, September 4-7, 2012.

[16]Tanabe, Y. and Sugawara, H., "Aerodynamic Validation of rFlow3D Code with UH-60A Data Including High Advance Ratios," 41st European Rotorcraft Forum, Munich, Germany, September 1-4, 2015.

[17]Leishman, J. G., *Principles of Helicopter Aerodynamics*, Second Edition, Cambridge Aerospace Series, Cambridge University Press, New York, NY, 2006, Chapter 3.

[18]Davis, S. J., "Predesign Study for a Modern 4-Bladed Rotor for the RSRA," NASA CR 166155, 1981.

[19]Bousman, W. G., "Aerodynamic Characteristics of SC1095 and SC1094 R8 Airfoils," NASA/TP–2003-212265, AFDD/TR-04-003, 2003.

[20]Shinoda P., "Rotor Performance of a UH-60 Rotor System in the NASA Ames 80- by 120-Foot Wind Tunnel," American Helicopter Society 58th Annual Forum, Montreal, Canada. June 11-13, 2002.

[21]Datta, A., Yeo, H. and Norman, T. R., "Experimental Investigation and Fundamental Understanding of a Slowed UH-60A Rotor at High Advance Ratios," American Helicopter Society 67th Annual Forum, Virginia Beach, VA, U.S., May 3-5, 2011.

[22]Potsdam, M., Datta, A. and Jayaraman, B., "Computational Investigation and Fundamental Understanding of a Slowed UH-60A Rotor at High Advance Ratios," American Helicopter Society 68th Annual Forum, Fort Worth, TX, U.S., May 1-3, 2012.

[23]Potsdam, M., Yeo, H. and Ormiston, R., "Performance and Loads Predictions of a Slowed UH-60A Rotor at High Advance Ratios," 39th European Rotorcraft Forum, Moscow, Russia, 2013.

Performance Optimization for an Optimum Rotor Speed Helicopter

Jiayi XIE
PhD candidate
Institute for Aero-Engine, School of Aerospace Engineering,
Tsinghua University
Beijing, China

Ming ZHOU
Professor
Institute for Aero-Engine, School of Aerospace Engineering,
Tsinghua University
Beijing, China

Zhifeng XIE
Assistant Professor
Institute for Aero-Engine, School of Aerospace Engineering,
Tsinghua University
Beijing, China

PLEASE NOTE: DO NOT INCLUDE EMAIL ON 1ST PAGE

ABSTRACT

Aerodynamic design of rotor blade is challenging and crucial for the development of helicopter, it is difficult to design an optimal blade across the flight envelope. The current work discusses the rotor speed optimization problem. This study proposes two kinds of optimization objective for the optimum rotor speed helicopter: a bi-objective optimization problem including aerodynamic performance measurement and mechanical performance measurement; and a piecewise objective function optimization problem. The Pareto solution for the bi-objective optimization problem is implemented, and the optimum rotor speed for the piecewise objective function optimization problem is carried out.

INTRODUCTION

Over the last decades, there are many researches focusing on variable rotor speed helicopter, because of its significant optimization in rotor performance as discussed in Ref.1, which includes helicopter efficiency improvement, energy saving and flight envelope extension.

A variable-speed rotor applies a wide-speed-range turboshaft engine or variable speed transmission to change rotor revolution, making it operate optimally in different flight conditions, which is demonstrated to be beneficial to improve helicopter performance. An optimum rotor system (ORS) showed in Ref.2 was developed by Boeing in last century, and made it configured to the A160/A160T Hummingbird UAV (unmanned aerial vehicle). The maximum flight speed could achieve 263 km/h and its endurance was up to 24 hours at that time. Guo (Ref 3) optimized the rotor speed of UH-60A at different forward conditions. The rotor power could achieve 85% of the baseline through an optimal rotor speed with respect to flight speed. While there existed an increase of shaft torque, about 14% of the baseline. Han and Barakos (Ref 4) studied a variable speed tail rotor. The maximum power improvement could be 30% of the baseline.

It is known that the most helicopters are designed to fixed its rotor speed in flight envelope. One of the causes is about the rotor dynamics. The rotor works near a resonance point, if its revolution is allowed to change a lot, there may be more resonance states encountered, and rotorcraft engineers have to face a more challenging task to design rotor structure. Another aspect to limit a variable rotor speed helicopter is the engine. The turbo-shaft engine applied in a medium- or heavy helicopter has a narrow power turbine speed. Once the work point is off design, the power turbine efficient drops down quickly. So it is difficult to change the rotor speed by varying power turbine revolution. Some designs applied a variable-ratio transmission, however it makes the drive system more complicated and heavier, which is discussed in Ref.5. Those mentioned above are potential difficulties in applying a variable-speed rotor in industry. Sela and Rosen (Ref 6) studied the effects of inter-blade connections and variable rotor speed on the aeromechanical stability of a helicopter and concluded that well-designed inter-blade connections may be more efficient to stability problems. Han, Wang, Smith and Lesieutre (Ref 7) explored an embedded chordwise damper to reduce transient lag bending moments. Guo (Ref 3), Guo and Horn (Ref 8) also investigated the rotor stability in a variable-rotor speed UH-60A. Considering an integrated performance including engine, drive system and the helicopter rotor, Feng (Ref 5) showed that the face gear and piston engine could be an alternative solution to variable rotor speed rotorcraft in the future.

Since there is a trade-off relationship between the hovering and forward flight performances of a helicopter, it is difficult to design an optimal blade across the flight envelope, although for the fixed rotor speed helicopter. The two reasons causing the compromise problem lies in: first, the induced power and profile power behave different characteristics across the flight envelope; second, there are opposite parameter effects on the two power consumptions. In recent years, researches on solving the performance trade-off problem through advanced optimization algorithms and up-graded modelling methods have been published

Presented at the 6th Asian/Austrailian Rotorcraft Forum / Heli Japan 2017, Kanazawa, Japan, November 7-9, 2017.

continuously. Le Pape (Ref 9) applied three strategies in 7A rotor and ERATO rotor optimizations, namely a constraint method, a weighting of objective functions (WOF) method, and a genetic algorithm. The constraint method and WOF were both implemented to balance hover performance and rotor torque coefficient in $\mu = 0.4$ in the study. The genetic method optimized the ERATO rotor aerodynamic shape in three design points simultaneously: hover, $\mu = 0.2$ and $\mu = 0.4$. Léon, Pape, Costes, Désidéri and Alfano (Ref 10) applied a Nash game method to balance hover FM and forward flight power requirement. Imiela (Ref 11) coupled the Computational Fluid Dynamics (CFD) and Computational Structural Dynamics (CSD) to build a high-fidelity optimization framework. The weighted sum of hover FM and torque coefficient in $\mu = 0.4$ was optimized in the study. One of the limitations of a high-fidelity modeling is the expensive time cost problem, which causes it a difficult task to apply across the flight envelope. Advanced optimization algorithms also have to face the similar problem. Besides, the optimal blade design considering a combination of limited design points are hardly to say it the optimum across the flight envelope. And the WOF approach usually causes some arbitrariness in weights setting as mentioned in Ref.10. A traditional solution in practice to compromise hover and forward flight performances is to design a rotor of optimum hover FM, with acceptable performance penalty in forward flight referred to Ref.12.

The main purpose of current work is to extend a novel and efficient method called GOE to bi-objective optimization for the blade parameter for a variable-rotor-speed helicopter. The method is applied and validated in the blade span optimization in this paper. While considering the applications of the ORS-helicopter, a rigid rotor assumption is likely to be reasonable in the current work. Rohit and Hyeonsoo (Ref 13) have demonstrated the stiffer rotor to have a better performance than a softer one. The current work mainly contributes to extend an efficient and original solution, GOE, to achieve blade parameter and rotor revolution optimizations for a variable-rotor-speed rotorcraft.

MODELING

The main rotor provides a lift surface, propulsion surface and control surface for a rotorcraft, therefore its accurate modeling makes a significant effect on the entire vehicle flight simulation. Since in the most level flight cases, the small AOAs on a rotor blade will be observed, a linear aerodynamics and small incidence assumption are often applied in flight dynamics modeling. However in a reduced rotor speed situation, the two assumptions may be broken down. Because of compensating a lower air dynamic pressure, the rotor requires a higher blade pitch along with increased AOAs.

This study models the blade dynamics in a hypothesis of the rigid flap, and solves its periodic motion in a rational hub coordinate by the General-Alpha method. The blade flap can be modeled by a synthesized acting of moment balance.

$$M_a + M_{cf} + M_f + M_{cor} + M_s + M_r = 0 \qquad (1)$$

The moment components include centrifugal moment M_{cf}, blade inertia moment M_f, Coriolis acceleration moment M_{cor}, restrain moment at the flap hinge M_s, helicopter body induced moment M_r and the aerodynamic moment M_a.

Substitutes the moment components above and reconstructs Eq. (1) as a standard 2nd order ordinary differential equation,

$$m\ddot{\beta} + c\dot{\beta} + k\beta = f\left(\beta, \dot{\beta}, t\right) \qquad (2)$$

where

$m = I_\beta$, $c = 0$, $k = K_\beta + \omega^2\left(I_\beta + eM_\beta\right)$,

$$f\left(\beta,\dot{\beta},t\right) = M_a\left(\beta,\dot{\beta},t\right) + 2\omega\left(I_\beta + eM_\beta\right)\left(p\cos\psi - q\sin\psi\right) + I_\beta\left(\dot{p}\sin\psi + \dot{q}\cos\psi\right) + M_\beta\left(\dot{w} - uq + pv\right)$$

[primary] showed that in an assumption of the small AOAs and linear aerodynamics, the Eq.(2) can be simplified to a 2nd differential-equations when neglecting high order responses. As mentioned, this study directly solves the nonlinear Eq.(2) in a rotational coordinate. The blade element aerodynamic modeling is obtained by an interpolation of airfoil C81 table.

Especially for a teetered rotor in a level flight trim, Eq.(2) can be further simplified as

$$\begin{aligned} &2I_\beta\ddot{\beta} + \left(2K_\beta + 2\omega^2 I_\beta\right)\beta \\ &= M_a\left(\beta,\dot{\beta},t\right) - M_a\left(-\beta,-\dot{\beta},t+\pi/\omega\right) \end{aligned} \qquad (3)$$

The rotor flap coefficients can be obtained till the Eq.(2) or Eq.(3) converges to a periodic solution.

In order to capture the blade parameter effects on aerodynamic performance, a nonlinear inflow model should be applied. This study uses a prescribed wake method to compute the induced velocity, which the method describes the rotor tip wake by a series of equations[Beddoes]. The interested age makes at least one circle completed wake convect beyond the rotor disc, that is to say at least $(\mu\pi)^{-1}$ circle.

This study uses both the simple Prandtl's momentum theory and Beddoes' prescribed wake method to calculate the rotor inflow as Figure 1 shows. Within the trim iteration, the simpler momentum theory will be applied to speed up computation, while in the aerodynamics update out of trim iteration, the prescribed wake will be applied to obtain a more accurate solution. The Jacobian matrix used in a trim iteration actually provides a convergence path to the equilibrium solution, therefore an approximating Jacobian matrix has no effect on the final trim state. The prescribed wake method with an updated wake geometry guarantees the solution precision.

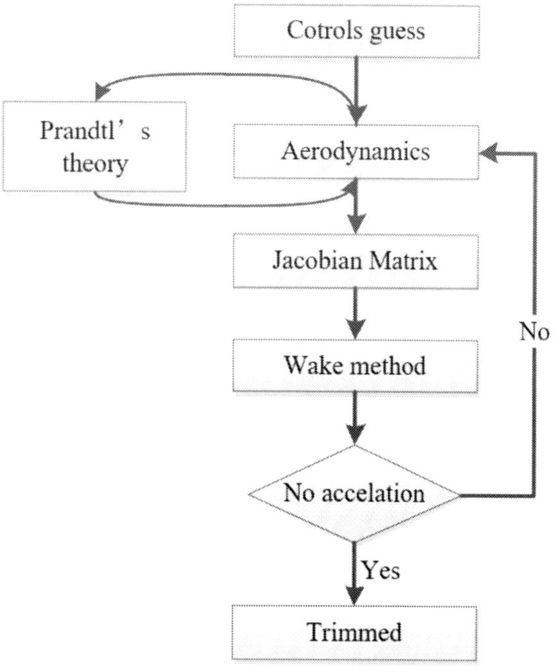

Figure 1. Induced flow computation procedure.

This study extended a three-step design method from single objective multi-points optimization to bi-objective multi-points optimization. As Ref.14 showed that, the design procedure includes a global optimization step, allowing parameters to vary along with flight conditions; and a secondary-optimization determining the blade parameters, which their corresponding aerodynamic performances are the closest to the global optimizations. This study extends the design thinking to a bi-objective optimization problem. The framework of bi-objective multi-points optimization framework is shown in Figure 2.

In this study, the main rotor power requirement is developed by blade element theory that may provide a better insight into the power causes.

$$dQ = \tilde{r}\tan\varphi dF_z + \tilde{r}\frac{dD}{\cos\varphi} = dQ_i + dQ_o \tag{4}$$

where dF_z, dD are blade section normal and drag forces respectively, φ is the HP (hub plane) inflow angle, and $\tilde{r}$ is the effective rational radius for a certain blade section.

The Eq.(4) shows that an incremental blade torque comes from the blade section normal and drag forces. The blade backward force can be dealt in the same say.

$$\begin{aligned} dH &= \left(\tan\varphi\sin\psi - \sin\beta\cos\psi\right)dF_z + \frac{\sin\psi}{\cos\varphi}dD \\ &= dH_i + dH_o \end{aligned} \tag{5}$$

where ψ, β are the azimuth and blade flap in the rational reference system.

To understand the rotor power causes better, we compute the quantity $R\mu dH_i + dQ_i$ mentioned by Johnson (Ref 15) in a hub wind coordinate, where in this coordinate any y-axis terms can be eliminated that may simplify its development.

$$\begin{aligned} dQ_1 &= R\mu dH_i + dQ_i \\ &= \left(R\mu\sin\psi + \tilde{r}\right)\tan\varphi dF_z - R\mu\sin\beta\cos\psi dF_z \end{aligned} \tag{6}$$

where R is the rotor radius and μ is advance ratio in HP.

Derive Eq.(6) further by introducing the blade section velocity formula,

$$dQ_1 = \frac{\dot{\beta}}{\omega}\left(r - \varepsilon R\right)dF_z + \lambda_h R\cos\beta dF_z \tag{7}$$

where $\dot{\beta}$ is the flap velocity and λ_h expresses HP inflow ratio.

The first part must be zero if integrating within one rotor revolution, since there is no net work during a circle. While if the rotor speed varies, there will be more or less power consumption to accelerate or decelerate the rotor speed.

By introducing the relationship $\lambda_h = \lambda_i - \lambda_\mu$, where λ_i is the induced ratio acting on blade section, λ_μ expresses the inflow ratio from freestream in HP, we can have the incremental torque dQ like,

$$\begin{aligned} dQ &= \lambda_i RdT + \left(R\mu dH_o + dQ_o\right) \\ &- R\mu dH - \lambda_\mu RdT \\ &+ \frac{\dot{\beta}}{\omega}\left(r - \varepsilon R\right)dF_z \end{aligned} \tag{8}$$

It can be seen that the dQ is separated into four causes: the component one $\lambda_i RdT$ required to provide rotor thrust, defined induced power P_i; the component two $\left(R\mu dH_o + dQ_o\right)$ to overcome air viscous, defined as profile power P_o; the component three $-\left(R\mu dH+\lambda_\mu RdT\right)$ relates with trim solution, defined as P_f+P_c, named parasite power and "climb" power respectively; and the component four $\frac{\dot{\beta}}{\omega}\left(r-\varepsilon R\right)dF_z$ determined by the rotor dynamics. Once given a flight condition and trimmed orientations, the term $-\left(R\mu dH+\lambda_\mu RdT\right)$ is completely determined to overcome a fuselage equivalent drag. Thus it may not be a must consideration when to implement a blade optimization design.

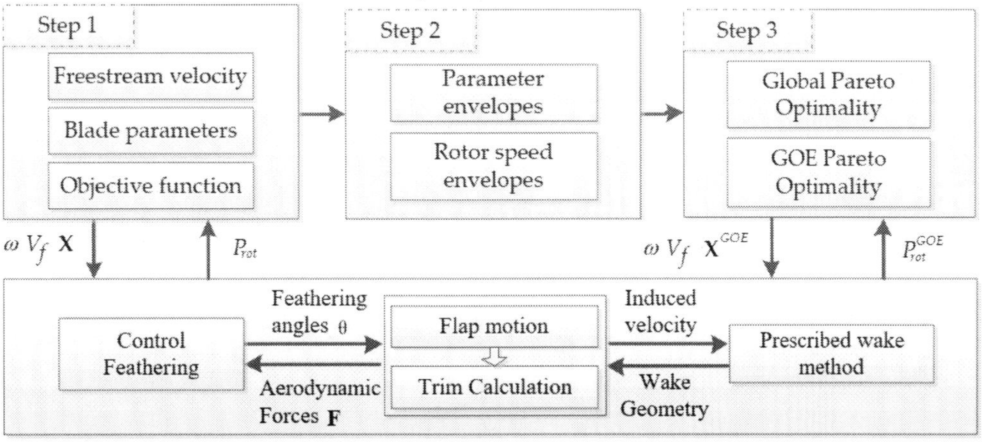

Figure 2. Blade parameter design procedure for the variable-speed rotor.

VALIDATION

The wake method in the present simulation is validated by a 2MRTS (2-meter rotor test system) wake experiment implemented by Ghee, Berry, Zori and Elliott (Ref 16). The rotor parameter and experiment condition are listed in Table 1 and Table 2. Figure 3 shows the wake position trailed from different azimuths. It can be seen that the prescribed wake method can capture wake distortion. In the advancing and retreating blades, the wake roll-up can be simulated. In both side view and top view we can see that the vertical position is slightly within the disc, which may induce blade-vortex interaction (BVI), aerodynamic noise and significant vibration problem.

Table 1. 2MRTS rotor parameters.

Parameter	Value	Parameter	Value
Blade	4	Planform	rectangle
radius (m)	0.861	Airfoil	NACA0012
Aspect ratio	12.98	Rpm	2000
Twist(°)	-8	Blade solidity	0.0977
Flap offset(m)	0.0508	Lag offset(m)	0.0508

Table 2. Robin experiment condition.

Parameter	Value	Parameter	Value
Advance ratio	0.15	Long. Cyclic pitch(°)	1.99
radius (m)	0.861	Later. Cyclic pitch(°)	1.39
Rotor shaft(°)	-3	Coll. Pitch(°)	6.55
Precone (°)	1.5		

An ultra-light helicopter is used to validate the in-house rotorcraft simulation tool and as a baseline for the further bi-objective optimization. The vehicle and rotor parameters are listed in Table 2.

Table 3. Main parameters of an ultra-light helicopter.

Vehicle Parameters	
Gross weight	414 kg
CG position	19 mm in front of hub center; 1285 mm beneath the hub
Parasite drag area ratio	0.0315
Rotor Parameters	
Blade number	2
Rotor type	Teetering; rectangular blade
Nominal speed	525 rpm
Rotor diameter	7010 mm
Chord length	170 mm
Twist angle	−8 deg
Pre-cone angle	3 deg
Shaft tilt	3 deg
Solidity	0.0309
Lock number	5.01
Airfoil	Boeing VR-7

The flap coefficients and power required plots in Fig demonstrate a reasonable result comparing in-house rotorcraft simulation tool with CAMRAD and UMARC. There can be observed that an increasing difference of β_s between present simulation and UMARC results. As Shen mentioned it may be caused by the inflow model, which UMARC applied an averaged inflow model, while the present and CARAD applied the wake method in this case.

As for rotor power loss, minimal consumption is achieved with the medium advance ratio, due to a reduction of induced power P_i and an increase of profile power P_o as flight speed rises. With flight speed increases, P_i is lower than in hover, which is because there is more mass flux contributed by freestream in forward flight, a bit of induced inflow can produce adequate thrust for the entire helicopter. Therefore, it can be seen that P_o is larger than P_i in forward flight, and becomes the bottleneck in performance improvement.

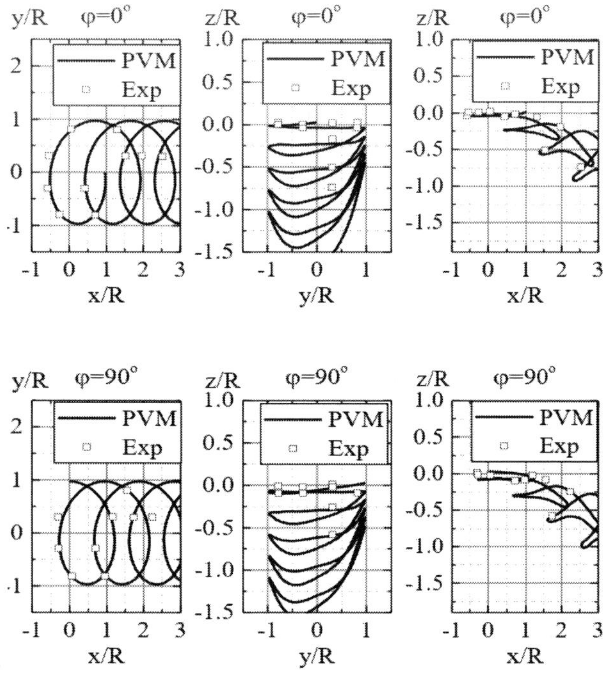

(a)

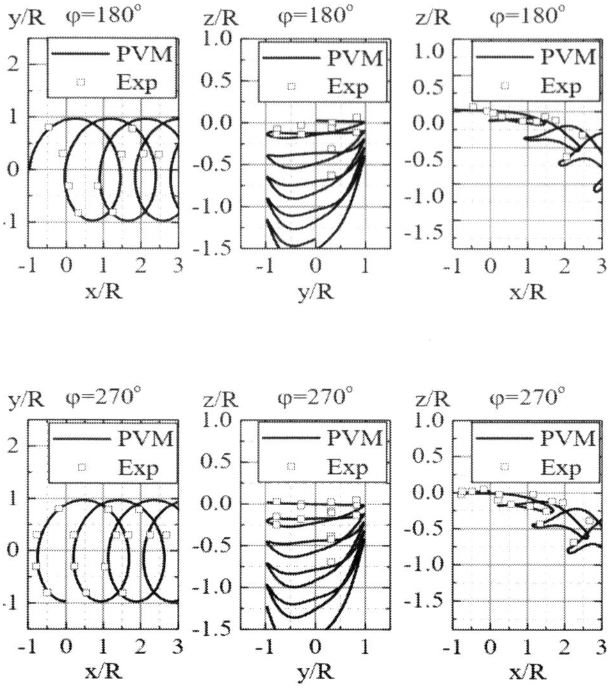

(b)

Figure 3. Wake position validation.

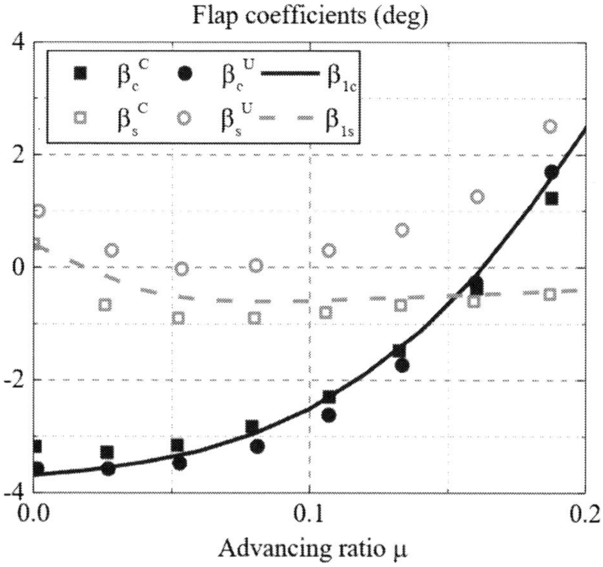

Figure 4. Flap coefficients

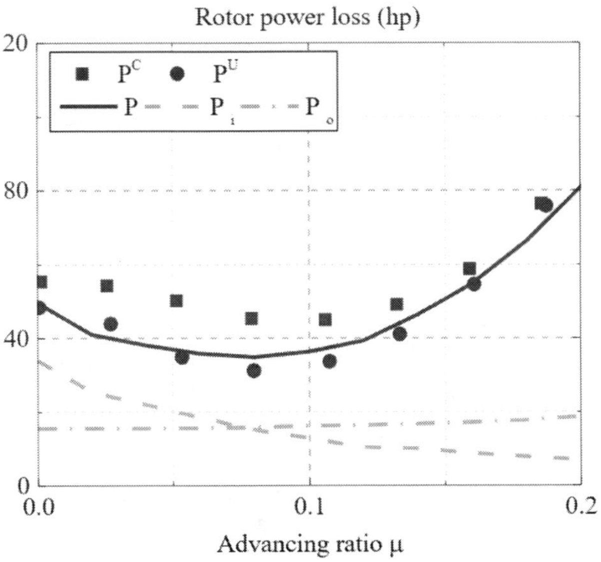

Figure 5. Power required.

RESULTS AND DISCUSSION

Rotor Speed Effects

The current work studies the rotor speed effects on rotor aerodynamic performance (required power) and structure dynamics (shaft torque). Figure 6 and Figure 7 respectively show the required power contour and shaft torque contour, with their corresponding optimum rotor speed. It can be seen that there are significant impacts on both aerodynamic performance and structure dynamics. On one hand, the dynamic pressure reduces with the rotor speed reduces, which makes profile power decreases. On the other hand, it is required a higher blade pitch angle to maintain rotor thrust in a reduced rotor speed condition. Because the flow separation is likely to happen, large airfoil drag coefficients may appear. This reflects the fact that profile power improvement is affected by two opposite impacts: decrease in dynamic pressure and increase in attack of angle (AOA). (1) When dynamic pressure reduction is more prominent, profile power loss decreases with rotor speed decreases; (2) when drag coefficients increase due to a higher feathering angle, higher AOA, profile power loss increases dramatically. With the reduction of rotor revolution, the total required power decreases first and increases later. When closing to the stall boundary, the minimal required power is observed.

There is a relationship among the rotor speed, required power and the shaft torque.

$$T = P/\omega \tag{9}$$

Figure 7 depicts the shaft torque contour and its corresponding optimum rotor speed. It can be seen that the its relationship with rotor speed and flight condition is different from the required power, and the optimum rotor speed as well. There is a larger optimum rotor speed than the one of minimizing required power. Figure 8 shows the four components of required power, which $P_*^{r}\left(*=i,o,c,f\right)$ is defined as the component power ratio to the total required power in the same flight condition. The induced power and profile power are affected by the rotor speed more significantly as mentioned above. The term P_f and P_c are both unchanged with rotor speed. The shaft torque contour is depicted in Figure 9. It can be seen from Eq. (9) that the all four components relate to rotor speed variation. Since the terms P_f and P_c are unchanged, their corresponding shaft torques are inverse proportion to rotor speed variation. In hover, the induced power dominates in rotor required power. Since a reduced rotor speed makes less improvement in induced power, there is a relative higher optimum rotor speed in hover. In fast flight, the improvement in profile power benefits to rotor aerodynamic performance. However in the fast flight, the parasite power P_f and "climb" power P_c increase and dominate in the total required power, a reduced rotor speed requires a higher mechanical performance of the rotor shaft.

Based on the observation, we can propose two rotor optimization problems. One is a bi-objective optimization including aerodynamic performance and mechanical performance,

$$\begin{cases} f_1\left(V_f\right)=\min\limits_{\omega}\dfrac{P_i\left(V_f\right)+P_o\left(V_f\right)}{P_{i0}\left(V_f\right)+P_{o0}\left(V_f\right)} \\ f_2\left(V_f\right)=\min\limits_{\omega}\dfrac{T_i\left(V_f\right)+T_o\left(V_f\right)+T_f\left(V_f\right)+T_c\left(V_f\right)}{T_{i0}\left(V_f\right)+T_{o0}\left(V_f\right)+T_{f0}\left(V_f\right)+T_{c0}\left(V_f\right)} \end{cases} \tag{10}$$

The other one is to optimize a piecewise objective function. Considering that the rotor shaft torque is far below to its limit (at least the value in maximum flight speed) in medium advance ratio, we could maintain as much improvement of aerodynamic performance as possible by reducing rotor speed.

$$f\left(V_f\right)=\begin{cases} \min\limits_{\omega}\dfrac{T\left(V_f\right)}{T_0\left(V_f\right)}, \textit{fast advance ratio} \\ \min\limits_{\omega}\dfrac{P_i\left(V_f\right)+P_o\left(V_f\right)}{P_{i0}\left(V_f\right)+P_{o0}\left(V_f\right)}, \textit{otherwise} \end{cases} \tag{11}$$

where $T_*\left(*=i,o,c,f\right)$ is the rotor shaft torque corresponding to four required power components, and subscript 0 means the baseline value in the same flight conditions.

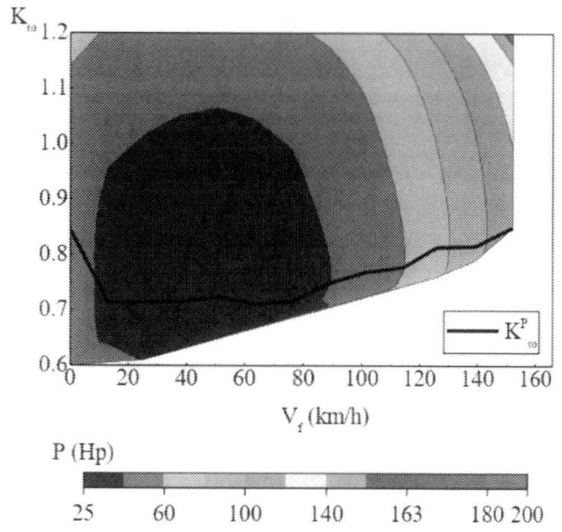

Figure 6. Power contour and minimal power-corresponding optimum rotor speed.

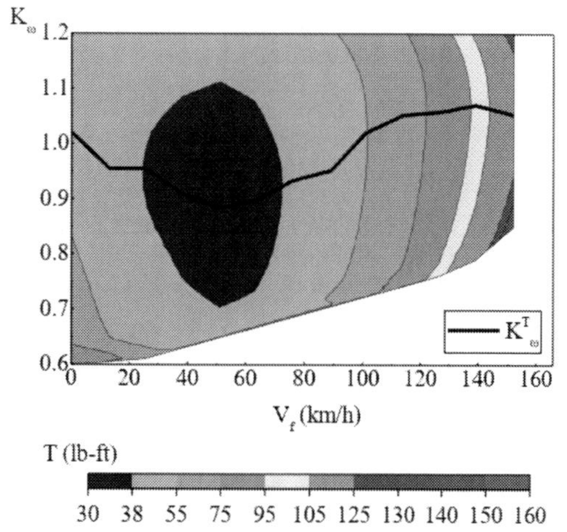

Figure 7. Torque contour and minimal torque-corresponding optimum rotor speed.

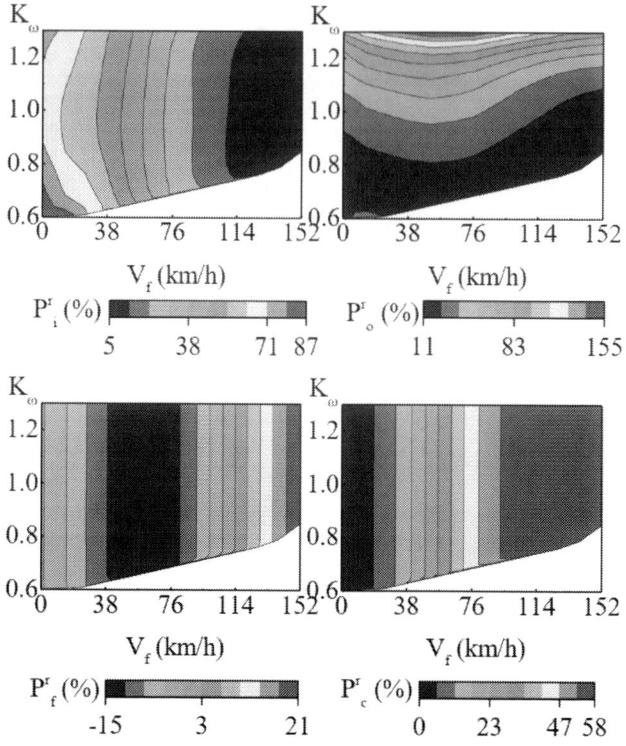

Figure 8. Power components contour.

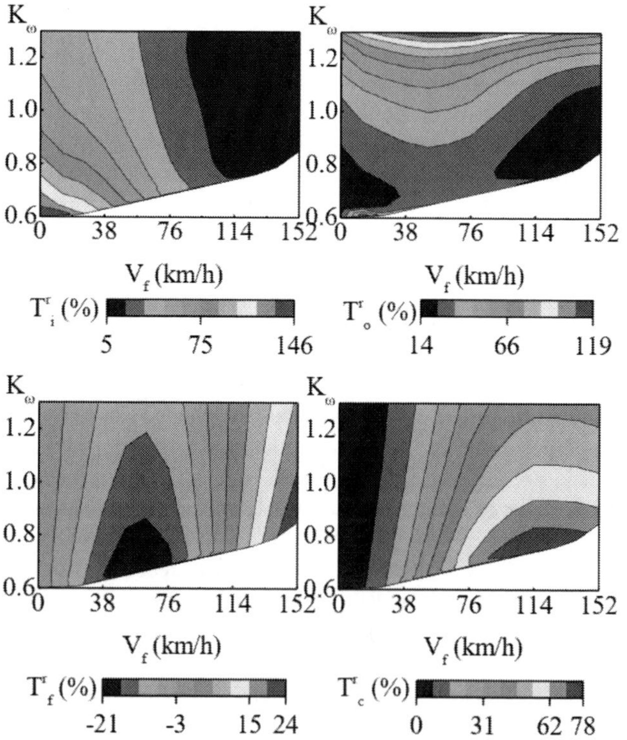

Figure 9. Torque components contour.

Figure 10 depicts the Pareto solution of the bi-objectives. It can be seen that there are three different relationships between aerodynamic performance and mechanical performance along the rotor speed decreases: firstly the two performances improves with the rotor speed decreases, and then the aerodynamic performance improves but the structural performance start to become worse. In the end, both aerodynamic and mechanical performances become worse with the rotor speed reduces. Figure 11 gives the piecewise objective solution. For the baseline in the current work, we choose the 140km/h flight speed as the break point.

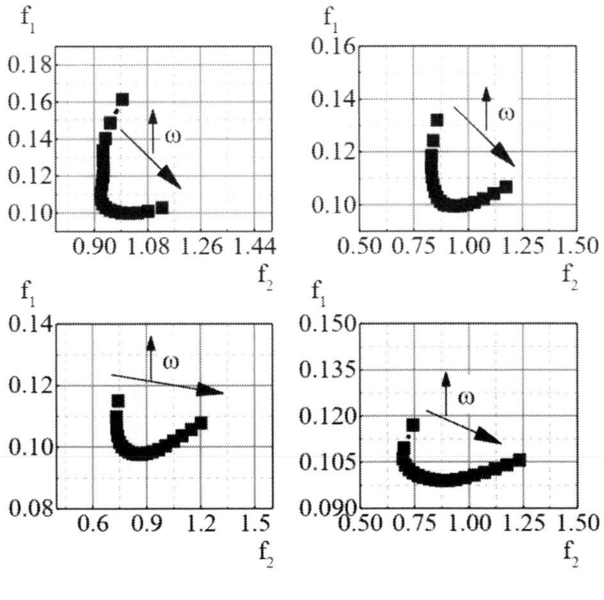

Figure 10. Pareto solution in different flight conditions.

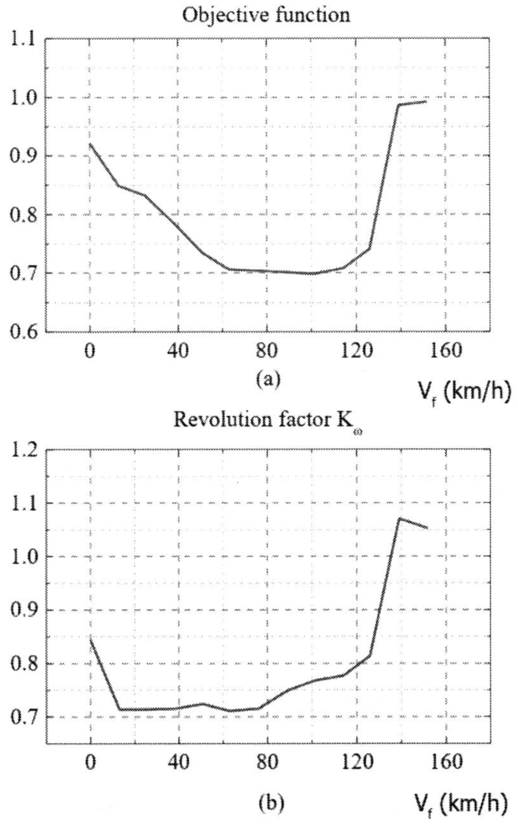

Figure 11. Optimum rotor speed.

CONCLUSIONS

This paper discusses the optimum rotor speed problem, and proposes two kinds of optimization objective-a bi-objective optimization problem including aerodynamic performance measurement and mechanical performance measurement; and a piecewise objective function optimization problem-and discusses the rotor speed effects on rotor performances. The work on performance optimization for an optimum rotor speed helicopter is undergoing, and a more accurate rotor modeling including structure dynamics and free-wake method will be implemented in the future.

Author contact: Zhifeng XIE xzhf@tsinghua.edu.cn

REFERENCES

[1]J. H. Steiner. "An investigation of performance benefits and trim requirements of a variable speed helicopter rotor." Vol. Master, The Pennsylvania State University, 2008.

[2]A. E. Karem. "Optimum speed rotor." Google Patents, 1999.

[3]W. Guo. "Flight control design for rotorcraft with variable rotor speed," *Aerospace Engineering*. Vol. Ph.D., The Pennsylvania State University, 2009.

[4]D. Han and G. N. Barakos. "Variable-speed tail rotors for helicopters with variable-speed main rotors," *The Aeronautical Journal* Vol. 121, No. 1238, 2017, pp. 433-448. doi: https://doi.org/10.1017/aer.2017.4

[5]G. Feng. "Study on Key Technologies of Heavy Fuel Helicopter Powertrain." Vol. Ph.D., Tsinghua University, 2016.

[6]N. M. Sela and A. Rosen. "Modeling the influence of inter-blade connections and variable rotor speed on the aeromechanical stability of a helicopter," *Aerospace science and technology* Vol. 4, No. 3, 2000, pp. 173-188.

[7]D. Han, J. Wang, E. C. Smith and G. A. Lesieutre. "Transient loads control of a variable speed rotor during lagwise resonance crossing," *AIAA journal*, 2013.

[8]W. Guo and J. Horn. "Rotor State Feedback Control for Rotorcraft with Variable Rotor Speed," *AIAA Guidance, Navigation, and Control Conference*. American Institute of Aeronautics and Astronautics, 2009.

[9]A. Le Pape. "Numerical aerodynamic optimization of helicopter rotors: multi-objective optimization in hover and forward flight conditions," *31st European Rotorcraft Forum*, Florence, Italy,13-15 September 2015, 2005.

[10]E. R. Léon, A. L. Pape, M. Costes, J.-A. Désidéri and D. Alfano. "Concurrent Aerodynamic Optimization of Rotor Blades Using a Nash Game Method," *Journal of the American Helicopter Society* Vol. 61, No. 2, 2016, pp. 1-13. doi: 10.4050/jahs.61.022009

[11]M. Imiela. "High-fidelity optimization framework for helicopter rotors," *Aerospace Science and Technology* Vol. 23, No. 1, 2012, pp. 2-16. doi: 10.1016/j.ast.2011.12.011

[12]J. L. Walsh, G. J. Bingham and M. F. Riley. "Optimization methods applied to the aerodynamic design of helicopter

rotor blades," *Journal of the American Helicopter Society* Vol. 32, No. 4, 1987, pp. 39-44.

[13]J. Rohit and Y. Hyeonsoo. "Effects of torsion frequencies on rotor performance and structural loads with trailing edge flap," *Smart Materials and Structures* Vol. 21, No. 8, 2012, p. 085026.

[14]J. Xie, Z. Xie, M. Zhou and J. Qiu. "Multidisciplinary Aerodynamic Design of a Rotor Blade for an Optimum Rotor Speed Helicopter," *Applied Sciences* Vol. 7, No. 6, 2017, p. 639.

[15]W. Johnson. *Helicopter theory*: Courier Corporation, 2012.

[16]T. A. Ghee, J. D. Berry, L. A. Zori and J. W. Elliott. "Wake geometry measurements and analytical calculations on a small-scale rotor model,"NASA-TP-3584 1996.

Active Flow Control of Rotor via Combinational Strategy

Ma Yi-Yang Ph.D. student

Zhao Qi-Jun Professor

Wang Bo Lecturer

National Key Laboratory of Science and Technology Rotorcraft Aeromechanics
Nanjing University of Aeronautics and Astronautics
Nanjing, China

ABSTRACT

A new active flow control strategy by placing a synthetic jet actuator (SJA) and a trailing-edge flap (TEF) has been proposed, and its control effects on mitigation of large negative pitching moments and drag caused by rotor dynamic stall are numerically investigated by CFD method. A moving-embedded grid method and an unsteady Reynolds averaged Navier–Stokes (URANS) solver are established for predicting the complex flowfields of rotor. Calculated results indicate that TEF and SJ can suppress the formation of dynamic stall vortex and postpone flow separation over rotor, resulting in much lower Cdmax and CQmax comparing to the baseline state. At last, simulations are conducted for the flow control of rotor by the combinational method, and the numerical results indicate that aerodynamic characteristics of rotor could be further improved by the new control method comparing to individual control method, the TEF can better improve the thrust coefficient of rotor and synthetic jet further increases the aerodynamic characteristics of rotor.

NOTATION

$\mathbf{W}$	=the conservative variable
$\mathbf{F}_c$	=the convective fluxes
$\mathbf{F}_v$	=the viscous fluxes
τ	=the pseudo time
t	=the real time
Ω	=the volume of grid cell
S	=the boundary of grid cell
δ	=the deflection angle of the trailing-edge
δ_0	=the mean deflection angle
δ_m	=the angular amplitude of TEF
φ_0	=the phase difference
k	=the reduce frequency of rotor airfoil
k^*	=the non-dimensional frequency with respect to k
ξ	=the chordwise directions of airfoil
η	=the normal directions of airfoil
U_0	=the steady component of jet velocity
U_m	=the amplitude value of the oscillatory component
ω_{jet}	=the angular frequency
$\mathbf{d}_{jet}$	=the blowing direction of synthetic jet
F^+	=the non-dimensional frequency
C_u	=the momentum blowing coefficient
U^*	=the the RMS velocity of the period synthetic jet
U_∞	=the velocity of free steam
c	=the chord length of airfoil
h	=the width of jet orifice
$\omega_{airfoil}$	=the angular frequency of the oscillatory airfoil

Presented at the 6th Asian/Austrailian Rotorcraft Forum / Heli Japan 2017, Kanazawa, Japan, November 7-9, 2017.

INTRODUCTION

Due to combination of freestream velocity and rotation velocity of rotor, helicopter rotor encounters different flow conditions at advancing (transonic) and retreating (low-speed) sides respectively (Ref. 1). The periodic variation of angle of attack (AoA) makes rotor undergo dynamic stall on retreating blade, leading to large negative pitching moments which might bring on excessive vibration load and flutter of rotor, and further limits the operational envelope of helicopter (Ref. 2).

In recent years, the rotor technology has tended toward methods of enhancing performance over a wide range of operating conditions. Passive designs involving blade geometry are limited by the conflicting requirements created by advancing blade compressibility and retreating blade stall, hence concepts such as fixed and dynamic slats, variable droop leading-edge airfoils, synthetic jet control, and trailing-edge flaps (TEFs) have been considered for active aerodynamic flow control. Of these, TEFs appears attractive since its control system is expected to be a compact, lightweight, low-actuation-power, and high-bandwidth device that can be used for multifunctional roles such as vibration and noise suppression, rotor performance enhancement, and aeromechanical stability improvement.

Mechanically a trailing edge flap (TEF) is easy to design and implement, and is a viable alternative because of its high control authority, low actuation power, and low aerodynamic drag (Ref. 3).The capabilities of TEFs for vibration control and enhancing rotor performance using higher-harmonic inputs have been studied extensively, and is well documented in literature – see Refs. (Ref. 4-6). TEFs have analytically been shown to be capable of reducing power requirement through lower harmonic inputs (Ref. 7-8), with suitably applied deflections to alter the aerodynamic characteristics of the blade section and target performance benefits especially at high speeds, where the disk loading is highly nonuniform, and making small redistributions of the loads is beneficial to performance. A wind-tunnel testing of the Boeing SMART (Smart Material Actuated Rotor Technology) rotor, which has an on-blade piezoelectric-actuated trailing-edge flap on each blade of a full-scale, five-bladed MD 900 helicopter rotor, demonstrated significant rotor-induced vibration and BVI noise reductions (Ref. 9).

The TEF strategy may encounter a complex dynamic stall phenomenon with varied Mach numbers along the retreating blade of rotor, which would limit the control effects of TEF. As another promising active flow control method, synthetic jet (SJ) has large application potential on controlling static and dynamic stall of rotor airfoil (Ref. 10-12). Different from the TEF, synthetic jet could delay or prevent the formation of dynamic stall vortex of an oscillatory airfoil by introducing periodic disturbance into the separated flow, at the same time, it does not change the aerodynamic shape of airfoil, which is helpful to maintain a low level of radar area of rotor airfoil (Ref. 13). Hassan numerically analyzed the effectiveness of synthetic jet on dynamic stall control for VR-7 airfoil (Ref. 14), and the benefits of synthetic jet for improving the aerodynamic characteristics of airfoil with onset of dynamic stall were well validated. Woo el al. further carried out wind-tunnel tests on synthetic jet control of NACA4415 airfoil dynamic stall (Ref. 15), it showed that synthetic jet could significantly reduce the nose-down moment and increase lift coefficient of airfoil at the same time.

Though the drag and moment coefficient of rotor can be maintained at a low level with active control of TEF, the main control effects of TEF are to increase the lift coefficient of rotor. Additionally, due to the low jet velocity of current actuators, the applications of synthetic jet on flow control of rotor are still limited. By considering both the advantages and shortcomings of TEF and SJ on dynamic stall control, a combinational active control strategy is developed. The new strategy is proposed by combining the synthetic jet control and the TEF control, which has the potentiality to suppress the divergences of drag and moment coefficient of rotor with onset of dynamic stall and enhance the lift coefficient at the same time.

In this paper, numerical simulations are conducted on active flow control of rotor by using the new active flow control strategy. Firstly, solution of rotor flowfield is obtained by solving the unsteady RANS equations with k-ω SST turbulence model, and moving-embedded grid method is employed. Surface points of airfoil with a TEF are reconstructed by CST method and grids around rotor are quickly regenerated by solving Poisson equations. Additionally, a dimensionless velocity boundary condition is established to simulate flow control effects of synthetic jet. As validations, control cases of rotor by employing TEF and SJ are calculated and compared to the test data respectively. Furthermore, control effects of the new combination control method on rotor are investigated. Based on the numerical results, simulations are conducted for active flow control of rotor by the combinational method, and some valuable conclusions are achieved.

AERODYNAMIC METHODS

A high-efficiency and high-precision CFD method for predicting the unsteady flow characteristics of rotor and airfoil with TEF and SJ control has been developed.

CFD Method

The CFD methods (Ref. 16-17) based on Reynolds averaged Navier-Stokes (URANS) equations are employed, and improved to simulate the unsteady characteristics of flowfield around rotor with active flow control.

$$\frac{\partial}{\partial \tau}\int \mathbf{W}d\Omega + \frac{\partial}{\partial t}\int \mathbf{W}d\Omega + \oiint (\mathbf{F}_c - \mathbf{F}_v)dS = 0 \qquad (1)$$

where $\mathbf{W}$ are conservation variables, $\mathbf{F}_v$ are the viscous fluxes, and $\mathbf{F}_c$ are convective fluxes including the motion of the dynamic grid system. τ and t are pseudo and real time respectively, Ω is the volume of grid cell, and S is the boundary of grid cell.

The dual time method is used to well simulate the unsteady characteristics of rotor flowfield, and the sub-iteration is fulfilled by the implicit LU-SGS method (Ref. 18) which can be implemented easily on vector and parallel computers. To accurately predicting the non-linear and unsteady characteristics of vortical flowflield of rotor, the third-order Roe scheme together with MUSCL approach (Ref. 19) are employed for the discretization of convective fluxes. Aiming at well simulating flow separation over surface of rotor and airfoil, the k-ω SST turbulence model (Ref. 20) is employed to calculate turbulence viscosity.

The unsteady aerodynamic characteristics of the AH-1G helicopter rotor in forward flight (Ref. 21) have been simulated. Figure 1 shows the distributions of normal forces coefficients at different blade sections and azimuthal angles. The calculated results have good agreements with the experimental data in Ref. 21, which well indicate the accuracy of the present CFD method.

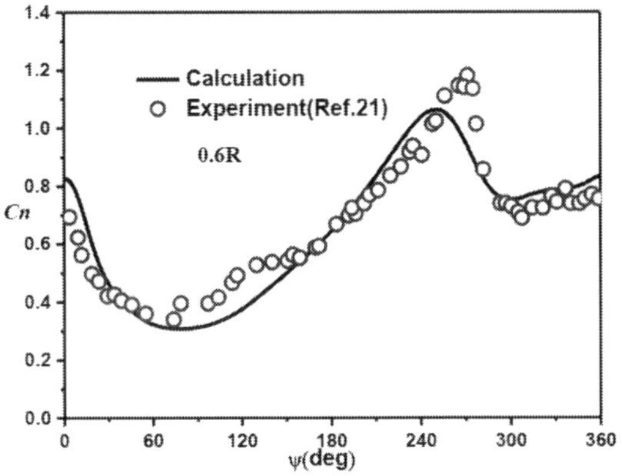

(a) r=0.6R

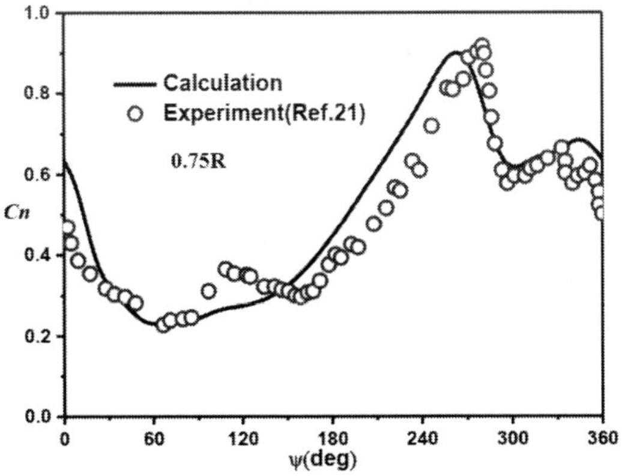

(b) r=0.75R

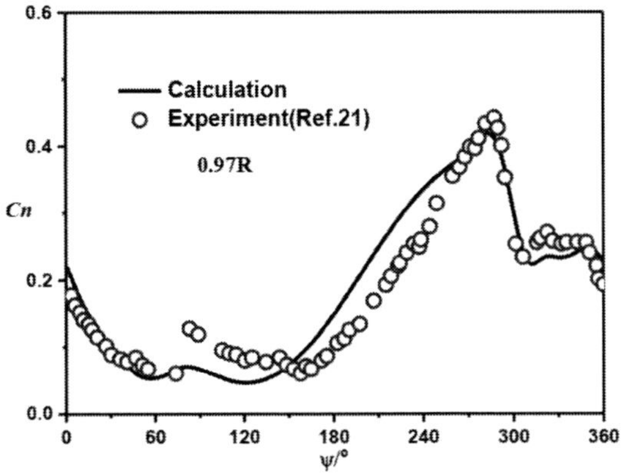

(c) r=0.97R

Figure 1. Comparisons of rotor normal forces between calculated results and the experimental data.

TEF Method

To simulate the control effects of TEF on airfoil dynamic stall, the trailing-edge is set to deflect around a grid point on the surface of airfoil according to

$$\delta = \delta_0 + \delta_m \sin(2k^* kt - \varphi_0) \quad (2)$$

where δ is the deflection angle of the trailing-edge, δ_0 and δ_m are the mean deflection angle and angular amplitude of TEF respectively, φ_0 is the phase difference, k^* is the non-dimensional frequency of the TEF with respect to reduce frequency k of rotor airfoil. As configuration of airfoil is changed, the distributions of grid points over airfoil are reconstructed based on CST method (Ref. 22). Figure 2 shows the sketch of oscillatory center of airfoil and TEF and Figure 3 gives the regenerated girds around an airfoil with TEF.

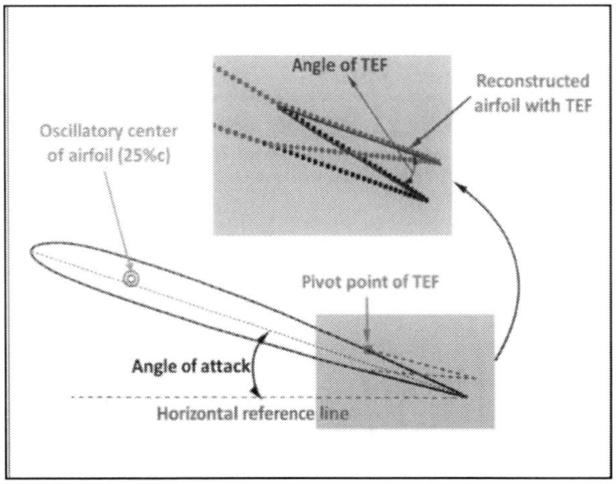

Figure 2. sketch of oscillatory center of airfoil and TEF.

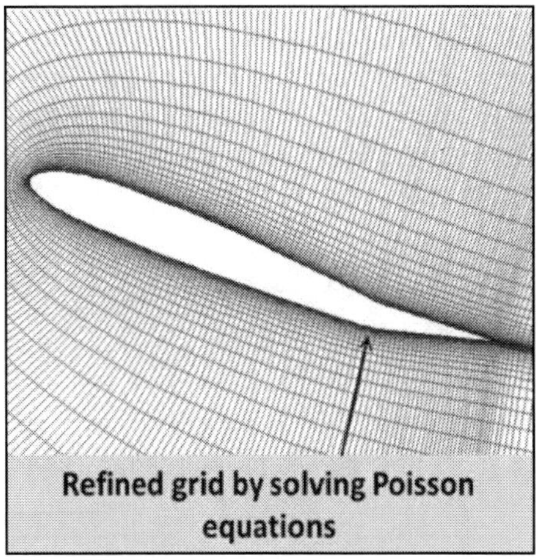

Figure 3. Regeneration of grid around airfoil with TEF by solving Poisson equations.

SJ Method

The contribution of synthetic jet to the surrounded fluid is modeled by a suction/blowing type boundary condition (Ref. 23). Taking 2-D case for example, the perturbation of synthetic jet to the unsteady flowfield is established by specifying the sinusoidal velocity over the orifice of jet actuator as

$$\mathbf{U}(\xi, \eta = 0, t) = [U_0 + U_m \sin(\omega_{jet} t)] f(\xi) \mathbf{d}_{jet} \quad (3)$$

where ξ , η denote chordwise and normal directions of airfoil respectively. U_0 is steady component of jet velocity, U_m is the amplitude value of the oscillatory component of jet velocity, ω_{jet} is the angular frequency, and $\mathbf{d}_{jet}$ is the blowing direction of synthetic jet. $f(\xi, \eta)$ determines spatial variation of the jet over the actuator orifice, as shown in Figure 4.

Considering relationship between periodic jet and periodic motion of rotor airfoil, the non-dimensional frequency F^+ and momentum blowing coefficient C_u are defined, which are shown as

$$C_{um} = 2\frac{h}{c}(\frac{U^*}{U_\infty})^2, C_{u0} = 2\frac{h}{c}(\frac{U_0}{U_\infty})^2, F^+ = \frac{\omega_{jet}}{\omega_{airfoil}}$$

where U^* is the RMS (Root Mean Square) velocity of the period synthetic jet, U_∞ is the velocity of free steam, c is the chord length of airfoil, h is the width of jet orifice in chordwise direction of airfoil, and $\omega_{airfoil}$ is the angular frequency of the oscillatory airfoil.

As a result, the dimensionless expression of the jet boundary condition is written as

$$U(\xi,\eta=0,t) = \sqrt{\frac{c}{2h}}U_\infty[\sqrt{C_{u0}} + \sqrt{C_{um}}\sin(2F^+kt)] \quad (4)$$

Figure 5 shows lift coefficient of VR-7B airfoil w/o synthetic jet control. As can be seen from the figure, the variation trends of the numerical results correlate well with the experimental data (Ref. 24). Additionally, the lift coefficient of airfoil could be significantly improved under synthetic jet control, which indicates that synthetic jet could efficiently delay dynamic stall of rotor airfoil.

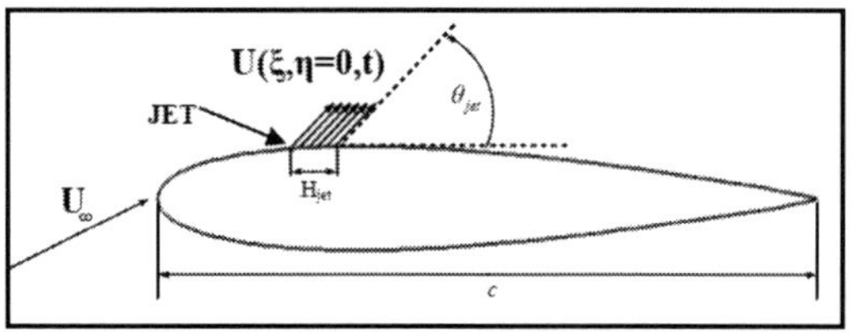

Figure 4. Jet blowing/suction boundary condition on airfoil.

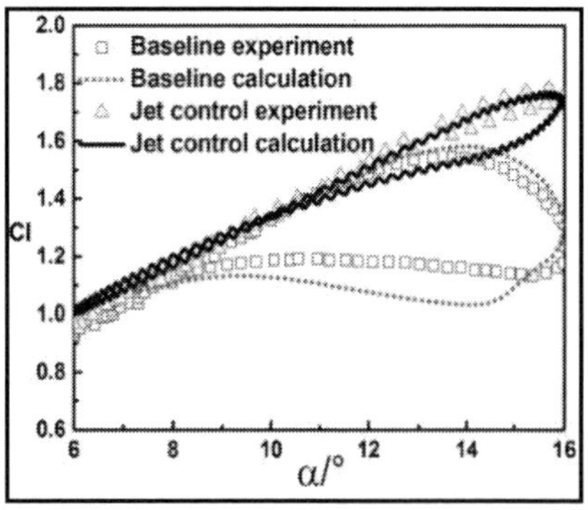

Figure 5. Lift coefficient of VR-7B airfoil w/o synthetic jet control.

RESULTS

Active flow control on rotor via SJ

The C-T rotor is chosen as the analysis object in this paper. There are four jet actuators at different chordwise and spanwise positions of the blade. The actuators (A1, A3) are placed at 10%c on the suction surface of blade while the other ones (A2, A4) are at 50%c on the upper surface, and the width of jet orifices is 1%c, as indicated in the table 1.

Table 1. The position of actuators on the surface of blade.

Actuator	A1	A2	A3	A4
Spanwise position	50%R-70%R	50%R-70%R	70%R-90%R	70%R-90%R
Chordwise position	10%c (1%c)	50%c (1%c)	10%c (1%c)	50%c (1%c)

The grid of blade has a resolution of 343×40×131, there are 137 points on the lower and upper surfaces of the blade section, such as in the Figure 6-7. In which, there are 11 points over the jet orifice.

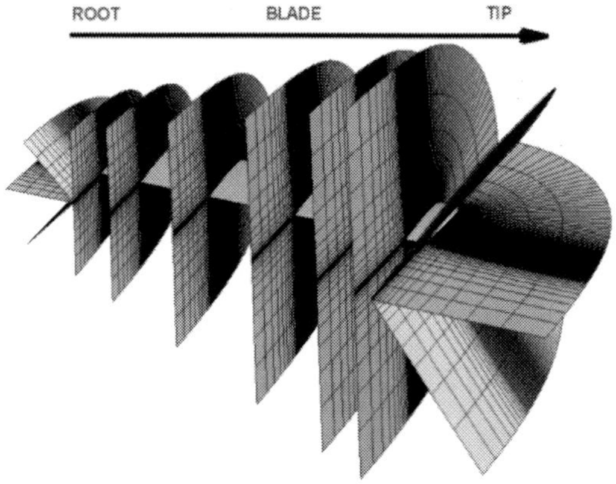

Figure 6. Body-fitted grids around rotor airfoil and blade.

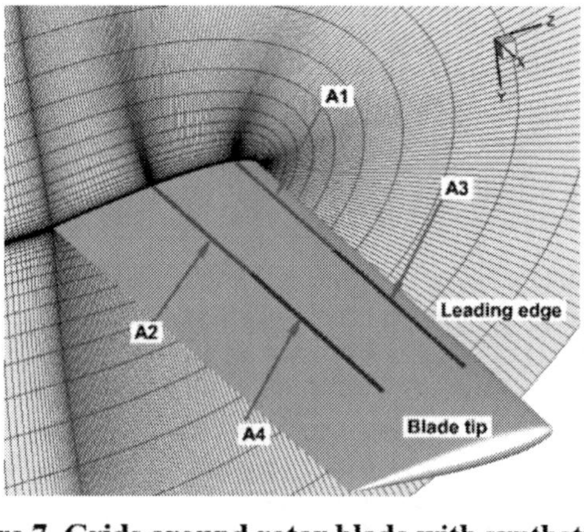

Figure 7. Grids around rotor blade with synthetic jet actuators.

The different control effects of the synthetic jet on aerodynamic performance of blades due to the location of the actuators are investigated (M_{tip}=0.628 , μ=0.3) and the excitation frequency is F^{+}=10.

The comparisons of the control effects of synthetic jet on the normal force coefficient by the four actuators (A1, A2, A3, A4) respectively are illustrated in Figure 8. The collective pitch is 8°, and the jet control condition of two actuators is the same: momentum coefficient is C_{μ}=0.004 and jet angle is θ_{jet}=25°. It can be seen immediately, the synthetic jet has a significant effect on the sectional normal force coefficient of blades, and the synthetic jets induced by actuator A2 and A4 are more efficient on increase of normal force coefficient than actuator A1and A3. The main reason is that the control effect of the actuator closer to the middle of chord is better when there is no flow separation on the blade surface. At the same time, the relative velocity is smaller in the retreating blade, and the ratio of the jet velocity to the relative velocity (the definition of jet momentmum coefficient in the forward flight) is bigger. So there is no significant influence of synthetic jet on the normal force coefficient in the advancing blade, and that is bigger in the retreating blade.

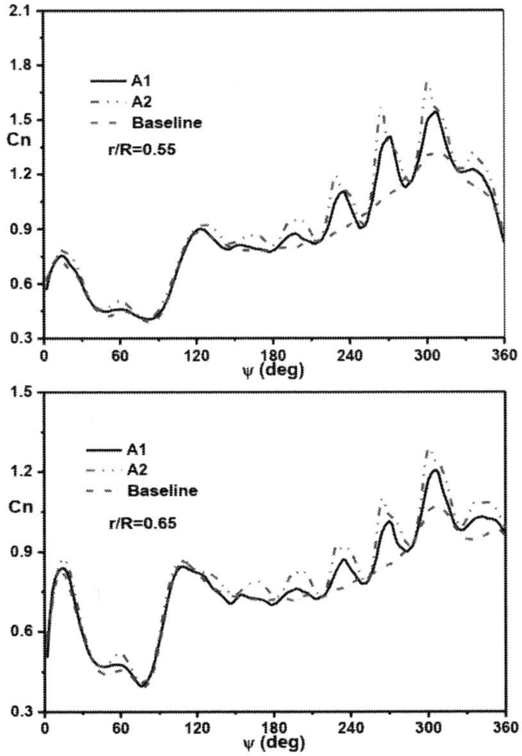

(a) A1 and A2

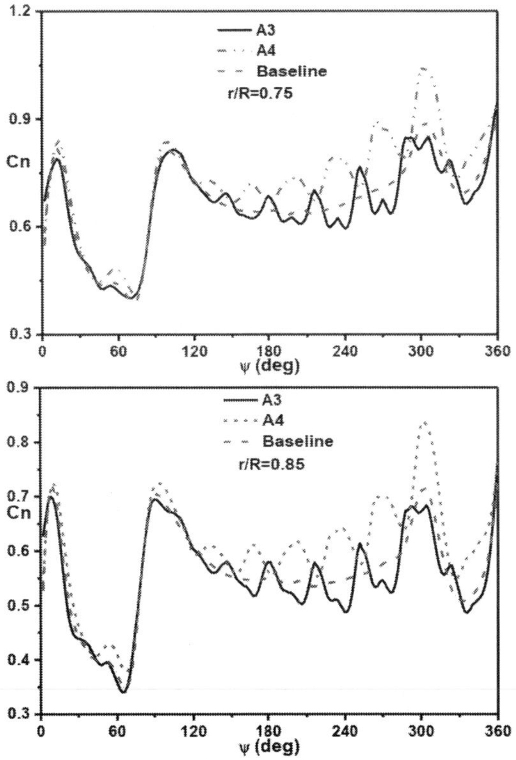

(b) A3 and A4

Figure 8. Comparisons of control effects for different actuators on sectional normal force.

The streamlines over the r=0.8R blade section at the 90° azimuthal angel is illustrated in Figure 9. There is a large flow separation at the upper surface due to the increase of adverse pressure gradient by the shock wave. Since 0.8R section isn't in the control area of actuator A1 or A2, the phenomenon of flow separation remains. Under the control of actuator A3 which is located upstream of the separation point, the synthetic jet effectively decreases the area of flow separation region. Since actuator A4 is located in the flow separation region, the synthetic jet induced by this actuator has a best effect on preventing the flow separation, and the area of flow separation region significantly decreases. In addition, a better control effect on preventing flow separation can be got if all actuators work.

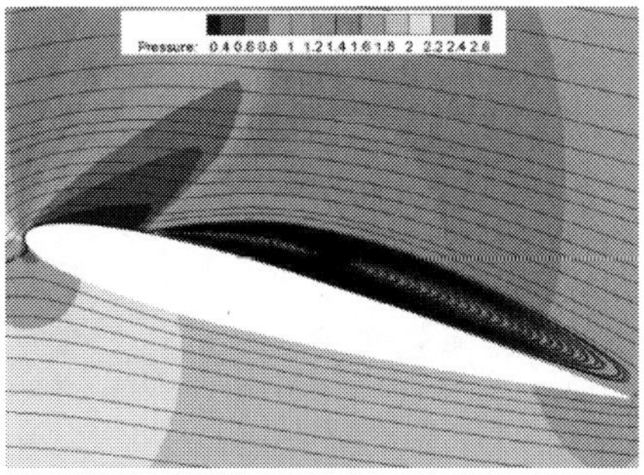

(a) baseline

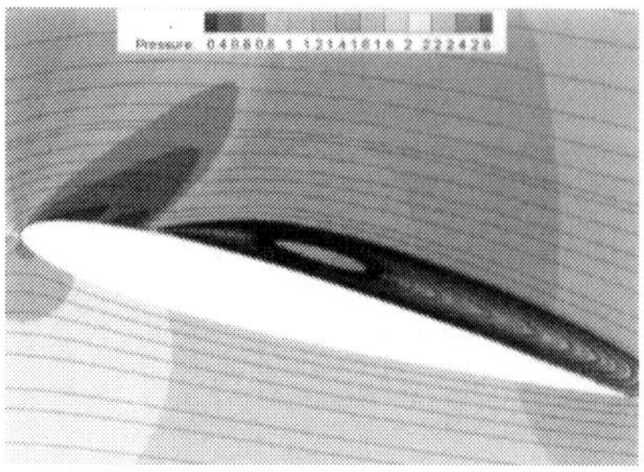

(b) actuator 1

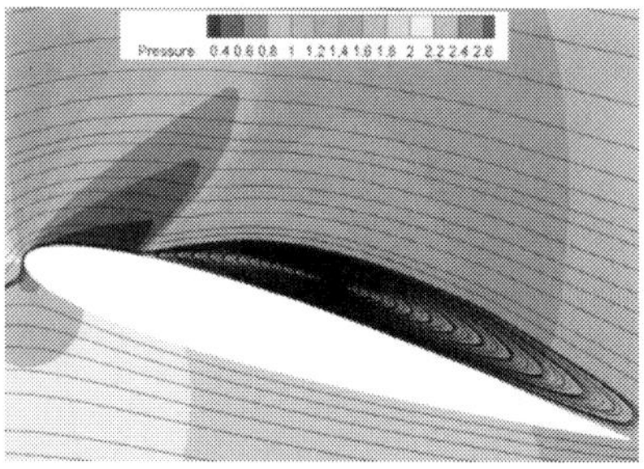

(c) actuator 2

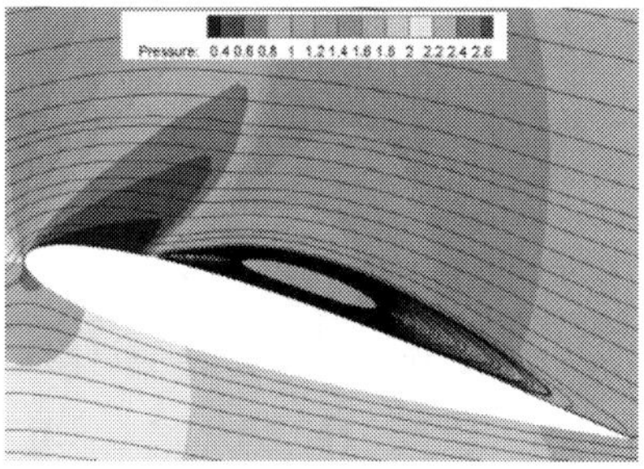

(d) actuator 3

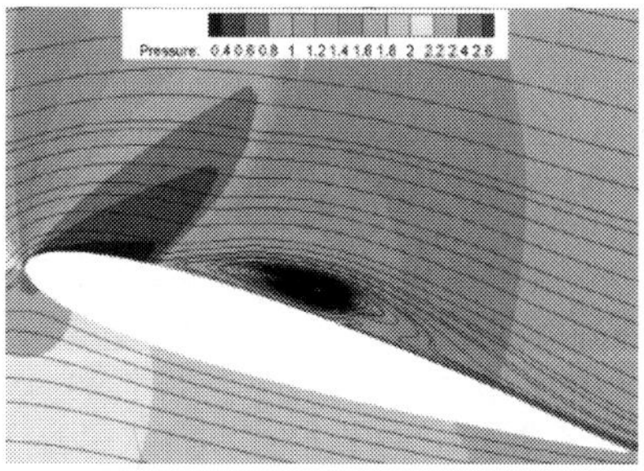

(e) actuator

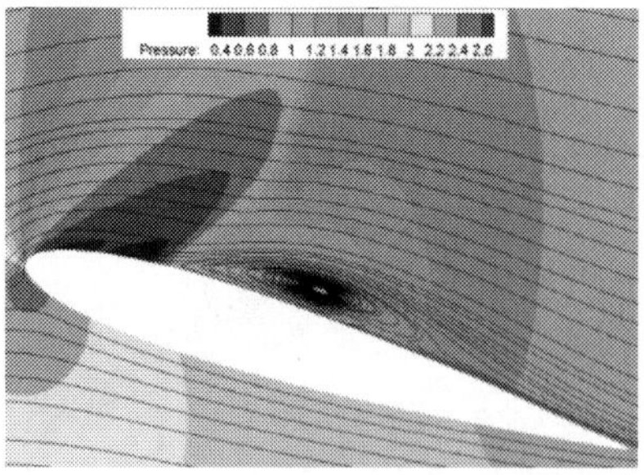

(f) all actuators

Figure 9. Streamlines over the section of blade(r=0.8R) under the control of different actuators at 90° azimuthal angle.

Active flow control on rotor via TEF

Based upon the simulations of dynamic stall control of rotor airfoil by TEF method, the flow control of rotor is further investigated. The simulated rotor consists of VR-12 airfoil with an aspect ratio 15. The 25%c TEF is located from 0.5R to 0.8R of the blade. Figure 10 shows the schematic diagram of body-fitted grids around the rotor blade with TEF.

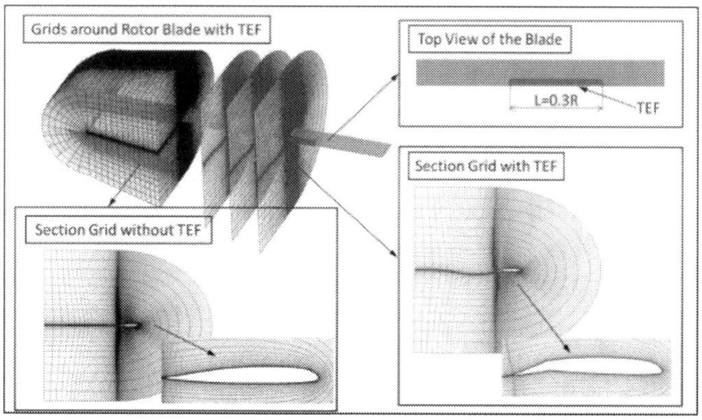

Figure 10 schematic diagram of body-fitted grids around the rotor blade with TEF.

Figure 11 shows the aerodynamic characteristics of the rotor. As δm of TEF increases, caused by the deflecting of the trailing-edge, the lift coefficient of rotor increases and the lift coefficient of rotor decreases. The torque coefficient of blade decreases obviously at the retreating blade side with increases of δm, while it is less affected by the TEF at the advancing side. Overall, trailing-edge flap has the capability in improving aerodynamic characteristics of rotor, especially in increasing the lift of rotor. In addition, trailing-edge flap has a better control effect on torque coefficient in retreating blade than that in advancing blade, and this will facilitate the control on dynamic stall in retreating blade.

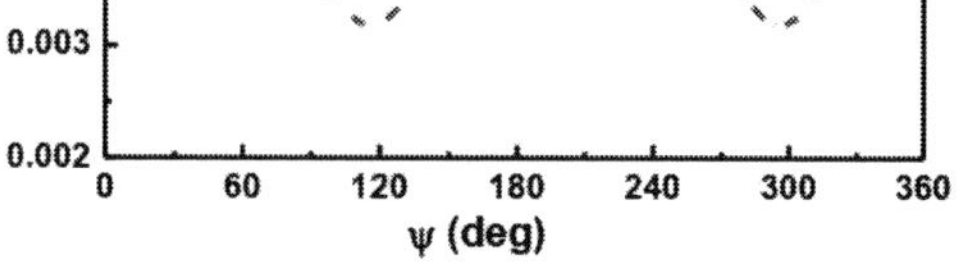

(a)Lift coefficient

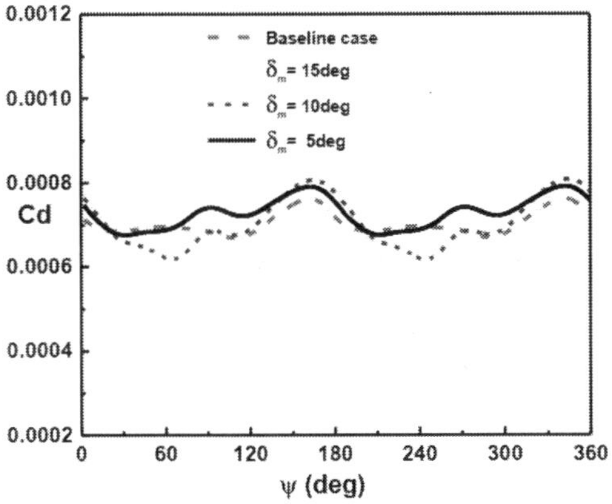

(b) Drag coefficient

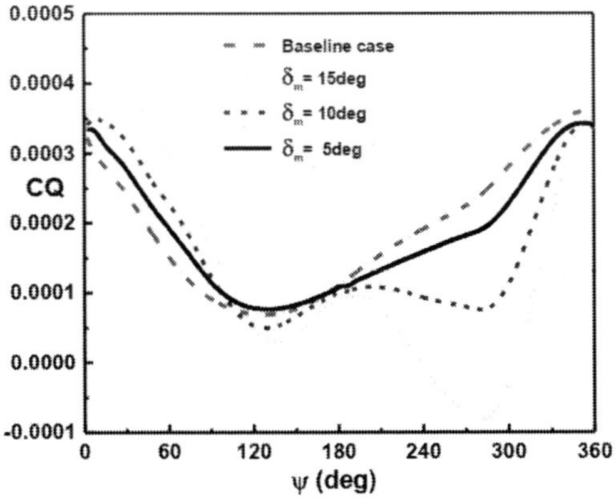

(c) Torque coefficient

Figure 11. Aerodynamic characteristics of rotor for baseline case and different TEF control cases.

Combinational control strategy on rotor

The combination control strategy is further taken into account, and its control effects on rotor consists of VR-12 airfoil with an aspect ratio 15. The 25%c TEF is located from 0.5R to 0.8R of the blade. The motion of the trailing-edge flap is the same as above section and the oscillation of the TEF is

$$\delta = -10^\circ \sin(\omega \times t)$$

The actuators are placed at 25%c on the suction surface of blade and the width of jet orifices is 1%c. Figure 12 illustrates the aerodynamic characteristics of the rotor with jet with $C_\mu = 2.7\times10^{-4}, F^+ = 10, \theta_{jet} = 30^\circ$.

The combinational control strategy can further improve aerodynamic characteristics of rotor under the unsteady condition comparing to the control case by using TEF only and SJ only. The thrust coefficient of rotor could be significantly improved under TEF control, while dynamic stall of rotor airfoil could not be suppressed. The SJ control could efficiently delay dynamic stall of rotor. The combinational control strategy could improve the lift coefficient and efficiently delay dynamic stall of rotor at the same time.

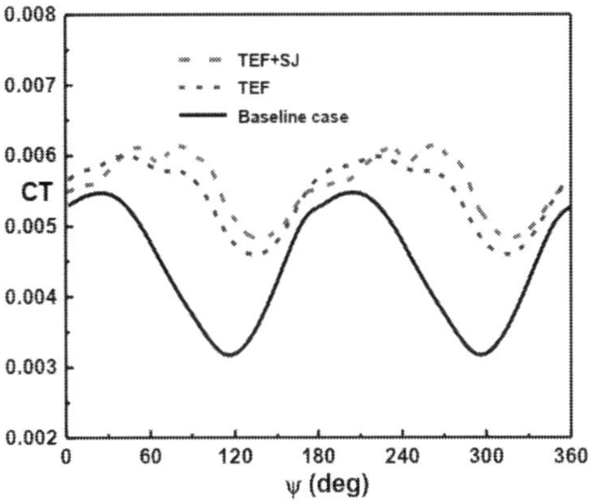

(a) Lift coefficient

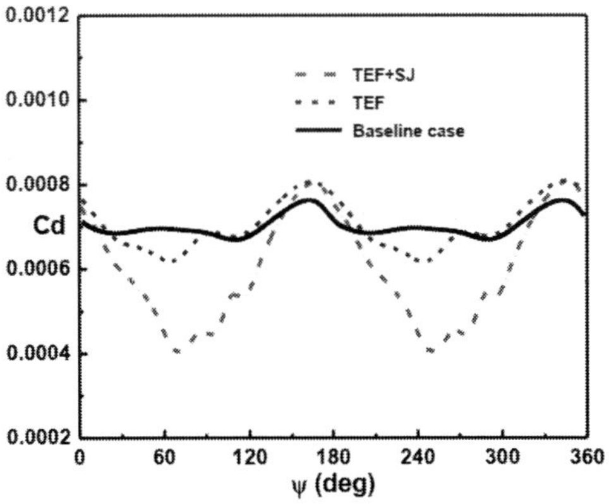

(b) Drag coefficient

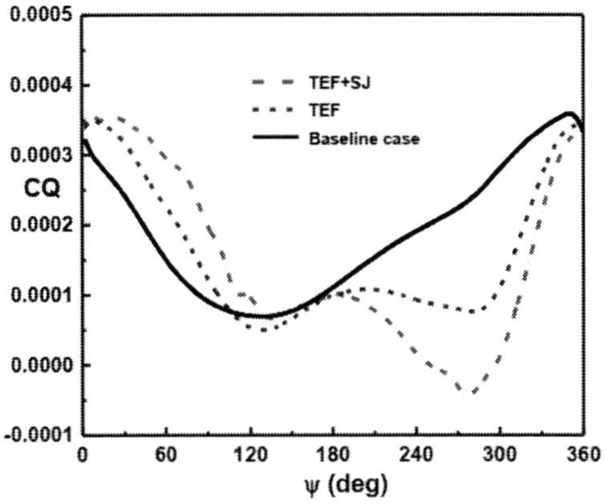

(c) Torque coefficient

Figure 12. Aerodynamic force coefficients of rotor with different control strategy.

SUMMARY AND CONCLUSIONS

Aiming at investigations of dynamic stall control via the new active flow control method with the combination of TEF and SJ, an unsteady flowfield solving method is established based on the grid system regeneration around a rotor. The control effects of the TEF and SJ on aerodynamic forces of rotor rotor are effectively simulated. Finally, the combinational control strategy applied on rotor is simulated. Some valuable conclusions are obtained as follows.

(1) With control of SJ, the dynamic stall vortex is well suppressed at large AoAs of rotor, as results, the peaks of drag and torque coefficients are significantly reduced and delayed, meanwhile, the thrust coefficient of rotor is maintained at a high level.

(2) The TEF control could improve the aerodynamic characteristics of rotor in forward flight, especially increasing the thrust coefficient of rotor, and it has a better control effect on torque coefficient in retreating blade than that in advancing blade. The combinational control strategy can further improve aerodynamic characteristics of rotor.

(3) The thrust coefficient of rotor could be significantly improved under TEF control, while dynamic stall of rotor airfoil could not be suppressed. The SJ control could efficiently delay dynamic stall of rotor. The combinational control strategy could improve the lift coefficient and efficiently suppress dynamic stall of rotor at the same time.

Author contact:

MA Yi-Yang mayiyang@nuaa.edu.cn
Zhao Qi-Jun zhaoqijun@nuaa.edu.cn
Wang Bo wangbo@nuaa.edu.cn

REFERENCES

[1]Conlisk, A. T., "Modern helicopter rotor aerodynamics". *Progress in Aerospace Sciences*, Vol. 37, (5), 2002, pp. 417-476.

[2]Yu, Y. H., Lee, S., McAlister, K. W., Tung, C., Wang, C. M., "Dynamic Stall Control for Advanced Rotorcraft Application". *AIAA Journal*, Vol. 33, (2), 1995, pp. 289-295.

[3]Viswamurthy, S. R., and Ganguli, R., "Effect of Piezoelectric Hysteresis on Helicoptor Vibration Control Using Trailing-Edge Flaps," *Journal of Guidance, Control, and Dynamics*, Vol. 29, No. 5, September-October 2006, pp. 1201–1209.

[4]Shen, J., and Chopra, I., "Aeroelastic Stability of Trailing-Edge Flap Helicopter Rotors," *Journal of the American Helicopter Society*, Vol. 48, No. 1, October 2003, pp. 236 – 243.

[5]Johnson, W., "Rotorcraft Aerodynamics Models for a Comprehensive Analysis," American Helicopter Society 54th Annual National Forum, Washington, DC, May 20-22 1998.

[6]Straub, F. K., and Charles, B. D., "Aeroelastic Analysis of Rotors with Trailing Edge Flaps Using Comprehensive Codes," 55th Annual Forum of the American Helicopter Society International, Montr'eal, Canada, May 25–27 1999.

[7]Jain, R., Yeo, H., and Chopra, I., "Computational Fluid Dynamics—Computational Structural Dynamics Analysis of Active Control of Helicopter Rotor for Performance Improvement," *Journal of the American Helicopter Society*, 55, 42004 (2010).

[8]Ravichandran, K., Falls, J., Ananthan, S., Chopra, I., and Hein, B., "Active Rotor Controls for Vibration Reduction and Performance Enhancement," Proceedings of the AHS Specialists' Conference on Aeromechanics, San Francisco, CA, January 20–22, 2010.

[9]Sim, B. W., Janakiram, R. D., Barbely, N. L., and Solis, E., "Reduced In-Plane, Low Frequency Noise of An Active Flap Rotor," American Helicopter Society 65th Annual Forum Proceedings, Grapevine, TX, May 27-29, 2009.

[10]Zhao Guoqing, Zhao Qijun, "Parametric analyses for synthetic jet control on separation and stall over rotor airfoil". *Chinese Journal of Aeronautics*, Vol. 27, (5), 2014, pp. 1051-1061.

[11]Tang, H., Salunkhe, P., Zheng, Y., et al, "On the use of synthetic jet actuator arrays for active flow separation control". *Experimental Thermal and Fluid Science*, 2014, Vol. 57, pp. 1-10.

[12]Jee, S., Mejia, O. D. L., Moser, R. D., et al, "Simulation of rapidly maneuvering airfoils with synthetic jet actuators". *AIAA Journal*, Vol. 51, (8), 2013, pp. 1883-1897.

[13]Melton, L. P., Hannon, J., Yao, C. S., et al., "Active flow control at low Reynolds numbers on a NACA 0015 airfoil". AIAA Paper 2008-6407, 2008.

[14]Hassan A. A., Nagib H., Wygnanski I., "Oscillatory jets-benefits and numerical modeling issues". AHS International, 58 th Annual Forum Proceedings, 2002, 1:711-731.

[15]Woo, G. T. K., Glezer, A., "Transient control of separating flow over a dynamically-pitching airfoil". AIAA Paper 2010-861, 2010.

[16]Zhao Qijun, Xu Guohua, Zhao Jinggen, "New hybrid method for predicting the flowfields of helicopter rotors". *Journal of Aircraft*, Vol. 43, (2), 2006, pp. 372-380.

[17]Zhao Qijun, Xu Guohua, Zhao Jinggen, "Numerical simulations of the unsteady flowfield of helicopter rotors on moving embedded grids". *Aerospace Science and Technology*, Vol. 9, (2), 2005, pp. 117–124.

[18]Sharov D., Nakahashi K., "Low speed preconditioning and LU-SGS scheme for 3D viscous flow computations on unstructured grids". AIAA Paper 98-0614, 1998.

[19]Edwards J. R., Franklin R. K., Liou M. S., "Low-diffusion flux-splitting methods for real fluid flows with phase transitions". *AIAA Journal*, Vol. 38, (9), 2000, pp. 1624-1633.

[20]Menter F. R., "Two-equation eddy-viscosity turbulence models for engineering applications". *AIAA Journal*, Vol. 32, (8), 1994, pp.1598-1605.

[21]Kim, J. W., Park, S. H., Yu, Y. H., "Euler and Navier-Stokes simulations of helicopter rotor blade in forward flight using an overlapped grid solver". AIAA Paper 2009-4268, 2009.

[22]Ceze, M., Hayashi, M., Volpe, E., "A study of the CST parameterization characteristics". AIAA Paper 2009-3767, 2009.

[23]Donovan, J. F., Kral, L. D., Cary, A. W., "Active flow control applied to an airfoil". AIAA Paper 1998-16119, 1998.

[24]Nagib, H., Kiedaisch, J., Greenblatt, D., Wygnanski, I., Hassan, A., "Effective Flow Control for Rotorcraft Applications at Flight Mach Numbers". AIAA Paper 2001-2974, 2001.

The Design Technology of All-electric Autogyro

Liang Lei master
Qinghua Zhu professor
Nanjing University of Aeronautics and Astronautics
Nanjing City, Jiangsu Province, China

ABSTRACT

This paper presents an all-electric autogyro, whose powerplant, prerotating mechanism and control system are all powered by electric energy. In this paper, some key technologies of all-electric autogyro are described firstly. Then a design method of energy and power for all-electric autogyro is proposed by calculating the required power and energy consumption for prerotation, takeoff, climb and cruise respectively. Next the primary design and flight performance of a single all-electric autogyro are presented. Several methods for increasing endurance and flying range are proposed, and an improved design scheme is given. Finally, compared with autogyro driven by fuel engine, the characteristics of all-electric autogyro are explained. This paper is intended to provide a reference for the research of all-electric autogyro.

NOTATION

a_o = acceleration of running (m/s^2)
b = blade chord length (m)
C_d = resistance coefficient of fuselage
D = resistance (N)
D_e = energy density of battery (W·h/kg)
E = energy required (Wh)
F_D = ground friction (N)
F_{xFc} = horizontal component of fuselage air resistance when cruising (N)
F_{xFo} = horizontal component of fuselage air resistance when running (N)
F_{xFu} = horizontal component of fuselage air resistance when climbing (N)
F_{xMRc} = horizontal component of rotor aerodynamic force when cruising (N)
F_{xMRo} = horizontal component of rotor aerodynamic force when running (N)
F_{xMRu} = horizontal component of rotor aerodynamic force when climbing (N)
F_{yFu} = vertical component of fuselage air resistance when climbing (N)
F_{yMRu} = vertical component of rotor aerodynamic force when climbing (N)
G_w = gross weight of autogyro (N)
k = blade number
L = lift (N)
M = required torque of rotor (N·m)
M_e = weight of battery (kg)
M_w = total weight of autogyro (kg)
M_e = empty weight of autogyro (kg)
M_u = available weight of autogyro (kg)
$\overline{M}$ = ratio of available weight and total weight
$\overline{M}_e$ = relative weight of battery
n = rotating speed of rotor (rpm)
$N_e^{(0)}$ = power rating of electromotor (kW)
p = disk load (kg/m^2)
P = required power (kW)
P_c = required power for cruise (kW)
P_e = power of battery (kW)
P_o = required power for takeoff (kW)
P_p = required power for prerotation (kW)
P_u = required power for climb (kW)
q = power load (N/kW)
R = rotor radius (m)
R_f = flying range (km)
S_f = drag area of fuselage (m^2)
t = endurance (h)
T_c = thrust of propeller when cruising (N)
T_o = thrust of propeller when running (N)
Tr = rotor lift (N)
V_c = cruising speed (m/s)
V_o = running speed (m/s)
V_{xu} = horizontal component of climb speed (m/s)
V_{yu} = vertical component of climb speed (m/s)
η = power utilization ratio
η_1 = power utilization ratio of electromotor
η_2 = power utilization ratio of propeller
θ = inclination angle of rotor shaft (degree)
ρ = air density (kg/m^3)
σ = rotor solidity

INTRODUCTION

Development of general aviation industry is inseparable from advances in design and production technology of aircrafts. Among many aircrafts, autogyro is considered as the safest. Because of simple structure and convenient operation, autogyro has been concerned widely. Autogyros now almost are powered by fuel engines, which cause lots of air pollution, vibration and noise. With development of the electric power industry and battery technology, as well as

---

Presented at the 6th Asian/Austrailian Rotorcraft Forum / Heli Japan 2017, Kanazawa, Japan, November 7-9, 2017.

the application of electric drive technology in automobiles, many research institutes and aircraft manufacturers are exploring a road for the application of electric drive technology in aviation industry. This paper presents an all-electric autogyro, which is a new form. The powerplant, prerotating mechanism and control system of the all-electric autogyro are all powered by electric energy.

The key technologies of an all-electric autogyro mainly involve three aspects: electric drive technology, autogyro technology and the combination of them. The first two aspects have already been studied and discussed widely by many experts and scholars, there is no longer too much discussion about them in this paper. This paper is going to discuss the last one, that is, some problems involved in applying electric drive technology to an autogyro. This paper is intended to provide a reference for the research of all-electric autogyro.

PRELIMINARY CALCULATIONS OF ENERGY AND POWER

Battery is the energy source of an all-electric autogyro. First of all, it is necessary to determine the amount of energy required by the entire flight. If the energy density of the battery is known, the amount of energy required would be figured out.

$$M_e=E/D_e \quad (1)$$

For an all-electric autogyro, the total weight remains constant during the flight, and the energy consumed could be divided into two parts, that is, the energy consumed by the power system and the energy consumed by other systems.

$$E=E_1+E_2 \quad (2)$$

E_1 is the energy consumed by the power system, E_2 is the energy consumed by other systems.

Furthermore, the energy consumed by the power system could be roughly divided into four, that is the energy consumed by the power system when prerotating, running, climbing and cruising. (Due to the descent and landing of an autogyro can be completed in the case of closing the power system, there is no consideration here.)

$$E_1=E_{11}+E_{12}+E_{13}+E_{14} \quad (3)$$

E_{11}, E_{12}, E_{13} and E_{14} represent the energy consumed by the power system when prerotating, running, climbing and cruising respectively.

According to

$$\eta E=Pt \quad (4)$$

$$\eta = \eta_1\eta_2 \quad (5)$$

In order to know the energy consumed by the power system when prerotating, running, climbing and cruising respectively, it is necessary to know required power at these stages.

Required power for prerotation

When prerotating, only need to rotate the rotor of autogyro and speed it up to a certain speed, so the required power can be expressed as

$$P_p=2n\pi M/60000 \quad (6)$$

Required power for takeoff

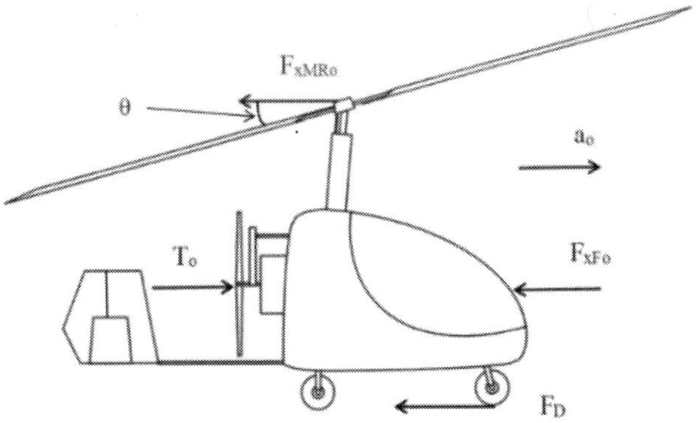

Figure 1. Horizontal force analysis when running

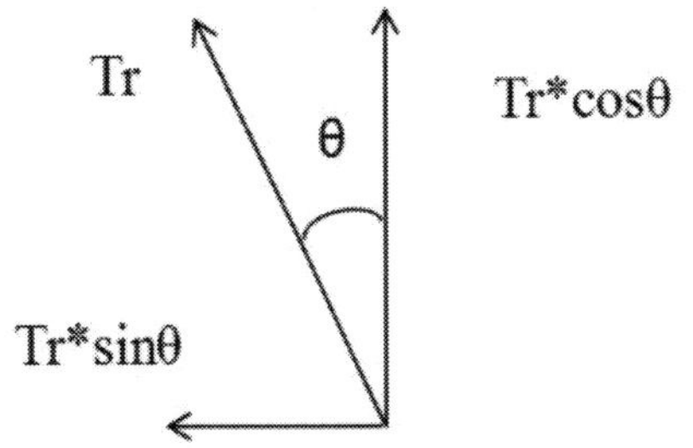

Figure 2. Horizontal and vertical decomposition of the rotor lift

The take-off of the autogyro is a horizontal accelerated motion, the required power can be expressed as

$$P_o=T_oV_o \quad (7)$$

$$T_o=F_D+F_{xMRo}+F_{xFo}+M_wa_o \quad (8)$$

$$F_D=0.02(M_wg-Tr\cos\theta) \quad (9)$$

$$F_{xFo}=0.5\rho C_dS_fV_o^2 \quad (10)$$

$$F_{xMRo}=Tr\sin\theta \quad (11)$$

Required power for climb

The climb of autogyro is approximately uniform motion in a oblique line, it can be considered as combination of uniform linear motion in horizontal and vertical directions, so the required power for climb can be considered as two parts

$$P_u=P_{xu}+P_{yu} \tag{12}$$

Its horizontal component can be expressed as

$$P_{xu}=T_{xu}V_{xu} \tag{13}$$

$$T_{xu}=F_{xMRu}+F_{xFu} \tag{14}$$

$$F_{xMRu}=Tr\sin\theta=M_w g\tan\theta \tag{15}$$

$$F_{xFu}=0.5\rho C_d S_f V_{xu}^2 \tag{16}$$

Its vertical component can be expressed as

$$P_{yu}=T_{yu}V_{yu} \tag{17}$$

$$T_{yu}=F_{yMRu}+F_{yFu} \tag{18}$$

$$F_{yMRu}=Tr\cos\theta=M_w g \tag{19}$$

$$F_{yFu}=0.5\rho C_d' S_f' V_{yu}^2 \tag{20}$$

Required power for cruise

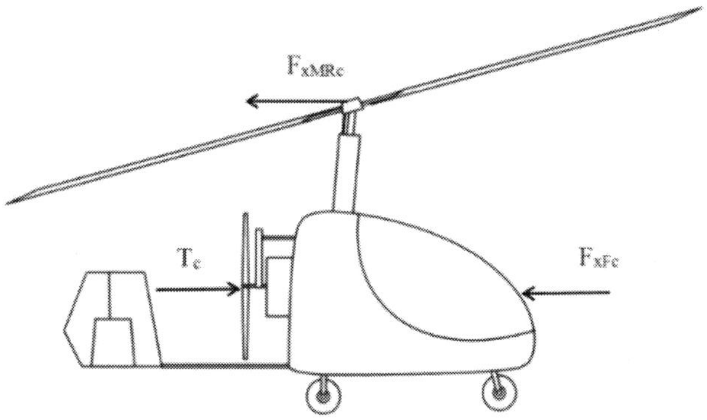

Figure 3. Horizontal force analysis when cruising

The cruise of an all-electric autogyro is approximately a uniform rectilinear motion, the required power can be expressed as

$$P_c=T_cV_c \tag{21}$$

The whole machine is in the state of force balance when cruising.

$$T_c=F_{xMRc}+F_{xFc} \tag{22}$$

$$F_{xMRc}=M_w g\tan\theta \tag{23}$$

$$F_{xFc}=0.5\rho C_d S_f V_c^2 \tag{24}$$

When relevant datas are known, the required power of each stage and the total energy consumption can be calculated by the above method, and then the suitable battery and electromotor for an all-electric autogyro can be selected. But in the preliminary development stage of an all-electric autogyro, much of the data used in calculations is often unknown. Generally, firstly, the electromotor and battery are selected according to available range of the power load and the desired time of flight. The power rating of the electromotor can be preliminarily determined by the following formula.

$$N_e^{(0)}=\frac{G_w}{q} \tag{25}$$

Then according to the expected endurance, the required battery size can be preliminarily determined.

$$E=N_e^{(0)}t \tag{26}$$

Next, weight of the required battery can be determined according to formula (1).

In the subsequent development, the weight of battery and the power of electromotor should be modified repeatedly by above methods, so as to select suitable battery and electromotor for the all-electric autogyro. The battery needs to meet the total energy consumption and the motor needs to meet the required power of each stage.

DESIGN OF A SINGLE ALL-ELECTRIC AUTOGYRO

In this section, a single all-electric autogyro is taken as an example, the main design parameters and flight performance are calculated to verify feasibility of the design scheme about the all-electric autogyro.

Primary Design

Set a single all-electric autogyro with a total weight of 300kg, whose empty weight is 200kg and available weight is 100kg, then the ratio of available weight and total weight is about 0.33, and most of autogyros in use are similar. Take some autogyros existing as reference, the weight distribution of this single all-electric autogyro is preliminary listed in Table 1.

$$M_w=M_e+M_u \tag{27}$$

$$\overline{M}=\frac{M_u}{M_w} \tag{28}$$

Table 1. Weight Distribution of Empty Machine

NAME	Weight(kg)	Percentage(%)
Power system	110	55.0
Rotor system	23	11.5

Control system	7	3.5
Body structure	35	17.5
Landing gear	20	10.0
Prerotating mechanism	5	2.5
Total	200	100

The rotor radius is selected as 3m, then it can be considered that the disk load is about 10.6kg/m^2 (generally less than 12kg/m^2).

$$p = \frac{M_w}{\pi R^2} \tag{29}$$

The single all-electric autogyro uses a 2-blade hanging teetering rotor, the blade chord length is taken as 0.15m, and then it can be considered that the rotor solidity is about 0.032 (generally 0.03~0.04).

$$\sigma = \frac{kb}{\pi R} \tag{30}$$

According to the method introduced earlier, the electromotor with power rating of 40kW is preliminary selected for this all-electric autogyro. According to the formula (33), the power load is 73.5N/kW (generally 60~80N/kW). If a lithium battery with weight of 100kg and energy density of 200W·h/kg is used as the energy source, the endurance of this all-electric autogyro is about 0.5h by the formula (34). Because its flying range is related to specific energy consumption characteristics of the electromotor and the cruise speed, there is no detailed calculation here.

Table 2. Preliminary Results of Gross Parameters

NAME	Value
Total weight	300kg
Empty weight	200kg
Available weight	100kg
Rotor radius	3m
Disk load	10.6kg/m^2
Blade number	2
Blade chord length	0.15m
Rotor solidity	0.032
Drag area of fuselage	2.3m^2
Resistance coefficient of fuselage	0.25
Weight of battery	100kg
Energy density of battery	200W·h/kg
Power rating	40kW
Power load	73.5N/kW
Endurance	0.5h

According to calculations, if traditional layout and design scheme are adopted, the single all-electric autogyro cannot obtain sufficient endurance and flying range. In order to improve its flight performance, it is necessary to carry out an improved design. Prior to the improved design, factors affecting endurance and flying range are examined.

Factors Affecting Endurance and Flying Range

First, the endurance can be expressed as

$$t = \frac{D_e M_e}{P_e} \tag{31}$$

If $\overline{M}_e$ is used to represent the relative weight of the battery, then

$$\overline{M}_e = \frac{M_e}{M_w} \tag{32}$$

The power load is represented by q.

$$q = \frac{G_w}{N_e^{(0)}} \tag{33}$$

The formula for calculating endurance can be rewritten as

$$t = \frac{D_e \overline{M}_e q \eta_1}{g} \tag{34}$$

From this formula, it can be seen the influence of the energy density and relative weight of battery, the power load and the power utilization ratio of electromotor on endurance.

The flying range of an all-electric autogyro mainly depends on its cruising speed and time, its cruising speed refers to the speed at which the energy consumption of motor is minimum. If V_c is used to indicate the cruising speed, the flying range of an all-electric autogyro can preliminary be expressed as

$$R_f = 3.6 V_c t = 3.6 V_c \frac{D_e \overline{M}_e q \eta_1}{g} \tag{35}$$

Improvements

Based on the above analysis, endurance of the all-electric autogyro can be improved by increasing the energy density and relative quality of battery and the power load.

<u>Increase the energy density of battery</u>

The energy density of battery is a major factor that restricts the flight performance of the all-electric autogyro. The energy density of lithium battery is 150~220W·h/kg now. With efforts of researchers all over the world, the energy density is still increasing.

When the power rating of electromotor and the weight of battery are constant, if the energy density of lithium battery can reach 500Wh/kg, the variation trend of endurance is shown in figure 4.

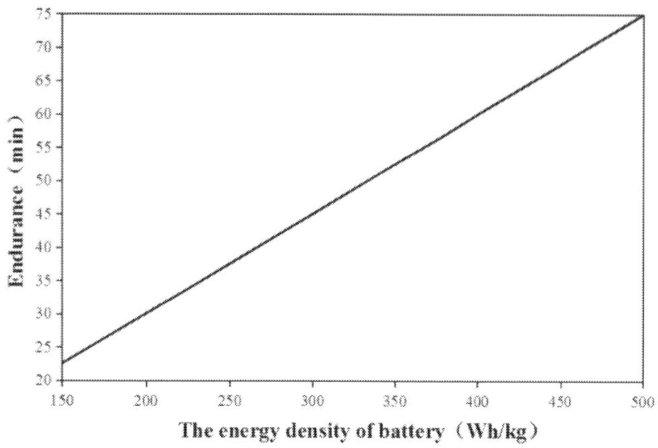

Figure 4. The curve of endurance changing with the energy density of battery

Increase the relative weight of battery

Generally, the relative weight of battery can only be in a small range. When the value is too small, sufficient endurance and flying range cannot be obtained. The excessive value would debase the flight performance and load efficiency. When the weight of an autogyro is determined, the weight of battery is relatively determined.

If the power rating of electromotor and the energy density of battery are constant, the weight of battery is only changed, the variation trend of endurance is shown in figure 5.

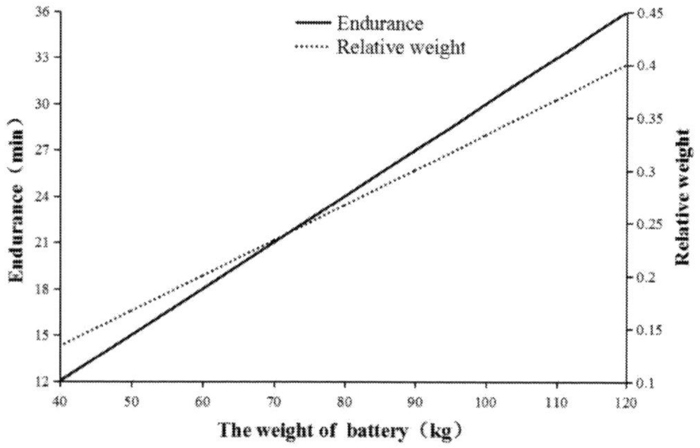

Figure 5. The curve of endurance changing with the weight of battery

Increase the power load

According to formula (33), if the total weight is unchanged, it is necessary to reduce the power rating of electromotor so as to increase the power load. And selection of the power rating of electromotor is based on the required power for each state. Most of time, an all-electric autogyro is in cruising state. The relationship between the required power for cruise and the power rating of electromotor can be expressed as:

$$\eta_2 N_c^{(0)} = P_c \tag{36}$$

Therefore, in order to reduce the power rating of electromotor, on the one hand, the power utilization ratio of propeller can be increased, on the other hand, the required power for cruise can be reduced.

According to formula(21), the required power for cruise is mainly related to the cruise speed and the thrust of propeller.

Because cruise of an all-electric autogyro is approximately a uniform rectilinear motion, the whole machine is in the state of force balance. Take formula (22), (23) and(24) into (21):

$$P_c = T_c V_c = (F_{xMRc} + F_{xFc}) V_c = (M_w g \tan\theta + 0.5\rho C_d S_f V_c^2) V_c \tag{37}$$

It can be seen from the formula (37) that factors associated with the required power for cruise are total weight, inclination angle of rotor shaft, air density, cruising speed, resistance coefficient and drag area of fuselage. But inclination angle of rotor shaft should be obtained by trimming. The required power for cruise is calculated by following empirical formulas here.

$$P_c = P_f + P_x + P_i \tag{38}$$

$$P_f = 0.5\rho S_f C_d V_c^3 \tag{39}$$

$$P_x + P_i = R_d V_c \tag{40}$$

$$R_d = M_w g(-0.000022 a_r^2 + 0.0079 a_r + 0.08 + 0.25 C_t) \tag{41}$$

$$C_t = \frac{M_w g}{0.5 \rho V_c^2 \pi R^2} \tag{42}$$

$$a_r = \frac{C_t}{\sigma} \tag{43}$$

Take formula (39), (40), (41), (42) and (43) into (38):

$$P_c = 0.5\rho S_f C_d V_c^3 + M_w g V_c[-0.000022(\frac{M_w g}{0.5\rho kbRV_c^2})^2 + 0.0079\frac{M_w g}{0.5\rho kbRV_c^2} + 0.25\frac{M_w g}{0.5\rho \pi R^2 V_c^2} + 0.08] \tag{44}$$

According to formula (44), if air density, total weight and cruising speed are constant, main factors that influence the required power for cruise are resistance coefficient and drag area of fuselage, rotor radius, number and chord length of blades. According to variable-controlling approach, curves of the required power changing with these parameters can be obtained, as shown in figure 6.

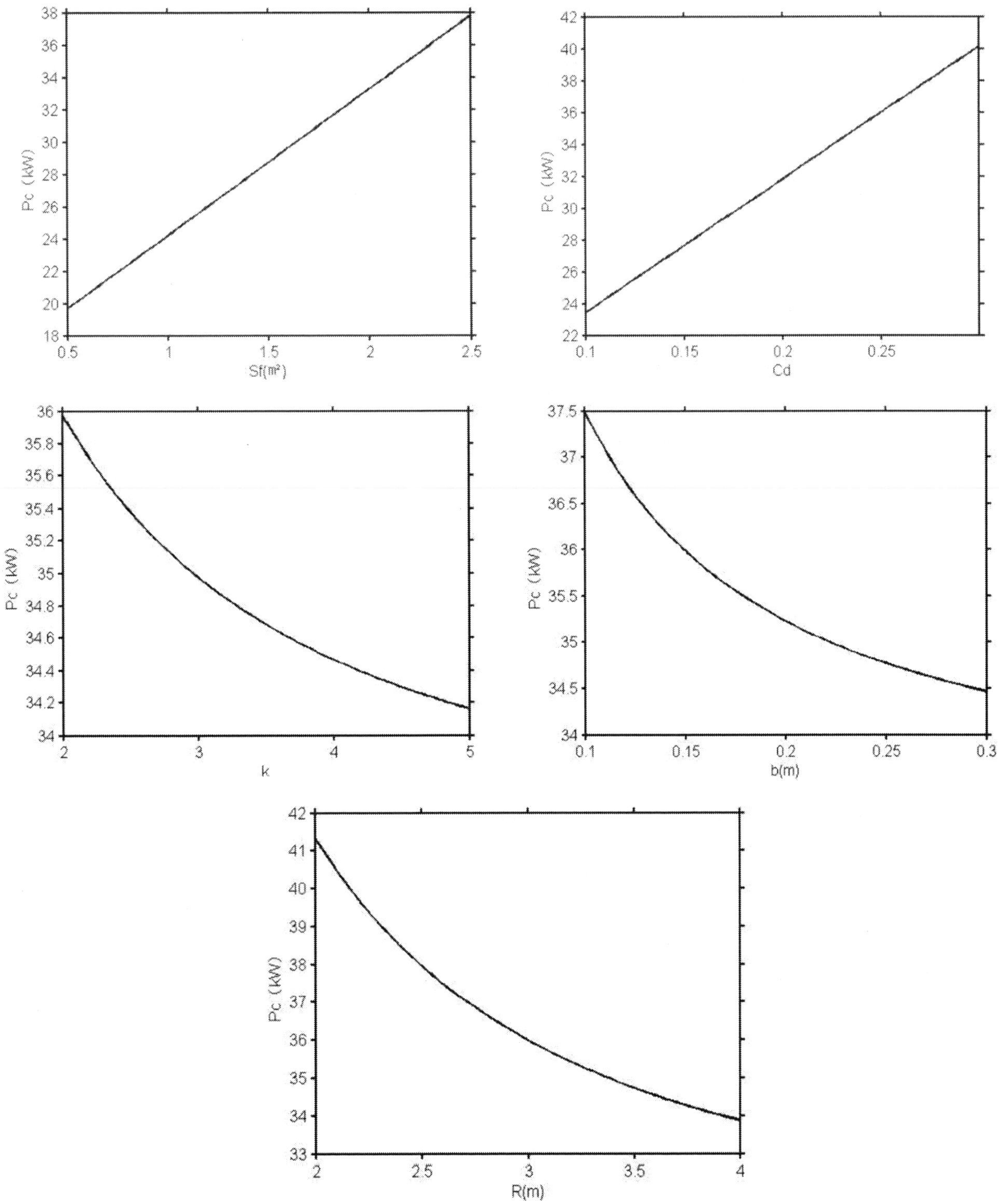

Figure 6. Curves of the required power changing with some parameters

As can be seen from the figure, the required power for cruise is directly proportional to the resistance coefficient and drag area of fuselage, and it is inversely proportional to rotor radius, number and chord length of blades. Therefore, in the premise of keeping total weight and cruising speed unchanged, resistance coefficient and drag area of fuselage should be reduced, and rotor radius, number and chord length of blades should be increased.

This law can also be verified in formula (37).

On the one hand, the horizontal component of rotor aerodynamic force is related to the total weight and the inclination angle of rotor shaft. Since the total weight is unchanged, in order to reduce the horizontal component force of rotor aerodynamic force, it is necessary to reduce the inclination angle of rotor shaft. When the inclination angle of rotor shaft is decreased, the flow of air will also decrease in the forward flight. In order to get enough lift, it is necessary to increase rotor radius, number and chord length of blades.

On the other hand, the air resistance of fuselage is mainly related to air density, cruising speed, resistance coefficient and drag area of fuselage. If air density and cruising speed is unchanged, in order to reduce the air resistance of fuselage, it is necessary to reduce the resistance coefficient and drag area of fuselage.

When the required power is reduced, the power rating of electromotor can be reduced accordingly, and the power load is increased. When the battery is constant but the power rating of electromotor is changed, variation trend of endurance is shown in figure 7.

Keeping the weight of battery constant, for different energy densities of battery, variation trend of endurance is shown in figure 8.

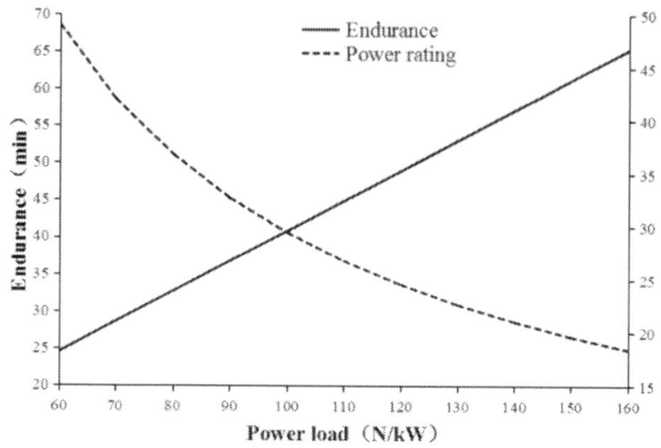

Figure 7. The curve of endurance changing with the power load

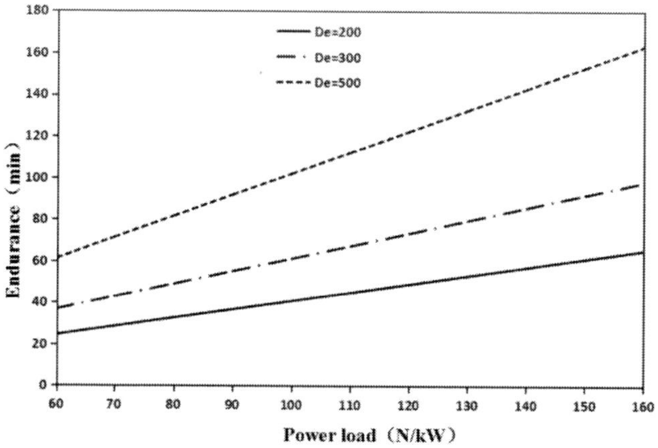

Figure 8. The curve of endurance changing with the power load for different energy densities of battery

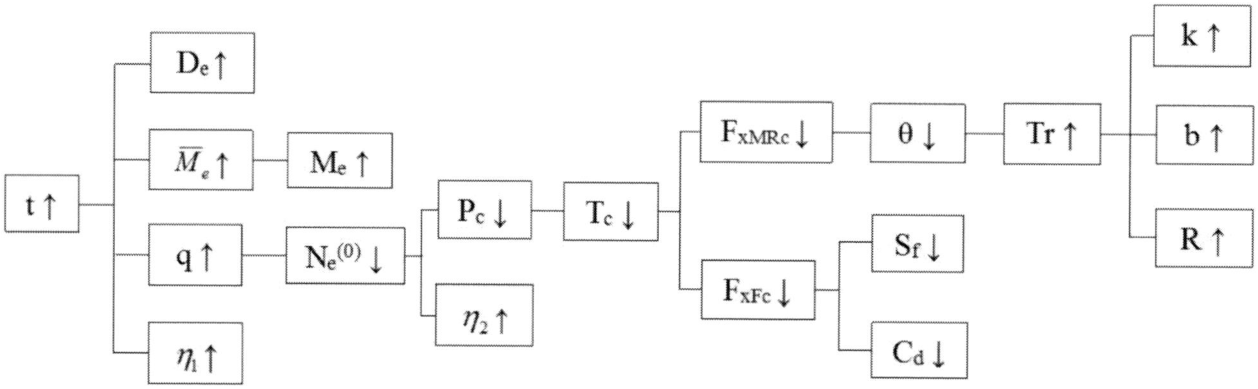

Figure 9. Measures to increase endurance

Summary

According to above analysis, main measures to increase the endurance of all-electric autogyro are:

a. Increase the energy density of battery;

b. Increase the relative weight of battery;

c. Reduce the power rating of electromotor;

d. Increase rotor radius, number and chord length of blades;

e. Reduce the resistance coefficient and drag area of fuselage;

f. Increase the power utilization ratio.

Among them, the d and e are rectifications of the all-electric autogyro itself, this is similar to increasing flying range and endurance by increasing lift-to-drag ratio for all electric aircraft.

For an all-electric autogyro cruising:

$$D=F_{xMRc}+F_{xFc}=M_{w}g\tan\theta+0.5\rho C_{d}S_{f}V_{c}^{2} \quad (45)$$

$$L=M_{w}g \quad (46)$$

The formula for calculating endurance can be rewritten as:

$$t=\frac{L}{D}\cdot\frac{\eta_{1}D_{e}\overline{M}_{e}}{g}\cdot\frac{1}{V_{c}} \quad (47)$$

The formula for calculating flying range can be rewritten as:

$$R_{f}=V_{c}t=\frac{L}{D}\cdot\frac{\eta_{1}D_{e}\overline{M}_{e}}{g} \quad (48)$$

It can be seen that the increase of lift-to-drag ratio could contribute to the increase of flying range and endurance.

Improved Design

According to improvements in the previous section, keeping the total weight and cruising speed unchanged, an improved design scheme is put forward to increase flying range and endurance.

On the one hand, a streamlined fuselage is used to reduce the resistance coefficient and drag area, as shown in figure 10.

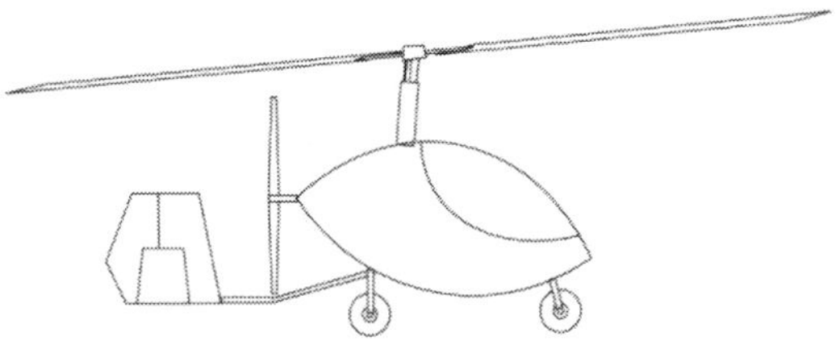

Figure 10. An all-electric autogyro with a streamlined fuselage

On the other hand, the gross parameters and weight distribution are recarried out, the results are shown in the following tables.

Table 3. Weight Distribution of Empty Machine after Improved Design

NAME	Weight(kg)	Percentage(%)
Power system	107	53.5
Rotor system	32	16
Control system	7	3.5
Body structure	29	14.5
Landing gear	20	10.0
Prerotating mechanism	5	2.5
Total	200	100

Table 4. Gross Parameters after Improved Design

NAME	Value
Total weight	300kg
Empty weight	200kg
Available weight	100kg
Rotor radius	3.5m
Disk load	7.8kg/m^2
Blade number	2
Blade chord length	0.2m
Rotor solidity	0.036
Drag area of fuselage	1.5m^2
Resistance coefficient of fuselage	0.2
Weight of battery	100kg
Energy density of battery	300W·h/kg
Power rating	30kW
Power load	98N/kW
Endurance	1h

With the total weight and cruising speed unchanged, the endurance is doubled after improved design. According to the formula (35), the flying range will also be doubled.

CONCLUSIONS

Compared with autogyro driven by fuel engine, all-electric autogyro has following characteristics:

(1) Larger lift-to-drag ratio, power load and rotor solidity, smaller disc load and inclination angle of rotor shaft, streamlined fuselage with smaller resistance coefficient and drag area.

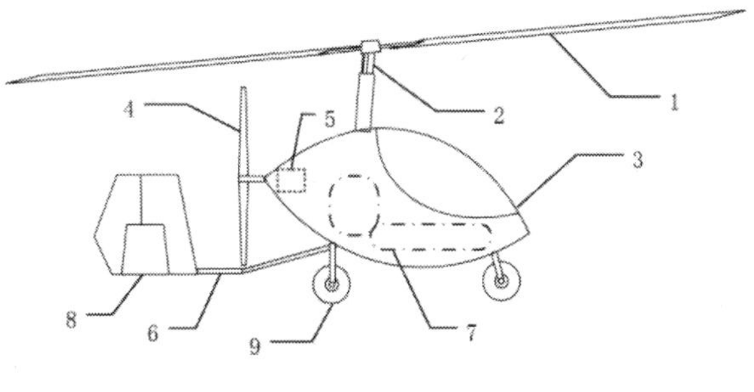

Figure 11. The overall layout of an all-electric autogyro

(1 is the rotor, 2 is the rack, 3 is the fuselage, 4 is the propeller, 5 is the electromotor, 6 is the tail boom, 7 is the battery, 8 is the tail, 9 is the landing gear)

(2) Small size and flexible layout: An all-electric autogyro is powered by an electromotor, and the weight of battery does not change during the flight, so there is no destabilizing moment. Therefore, the position of battery can be very flexible, for example, it can be arranged at the bottom of

cabin or under the seat (as shown in Figure 11), so as to save space and reduce volume of the whole machine.

(3) Better balance, stability and maneuverability: Since the weight of battery does not vary with time, distance and angle of flight, an all-electric autogyro has better balance, stability and maneuverability.

(4) More convenient in assembly, disassembly and maintenance: For an all-electric autogyro, the revolution speed of motor and propeller can be controlled by adjusting the output current of battery. As shown in Figure 12, the driver can operate the controller through the stick to adjust the output voltage and current of battery and the revolution speed of motor, so as to change the thrust of the propeller, and then the all-electric autogyro could run and fly. The electric drive system is more convenient in assembly, disassembly and maintenance.

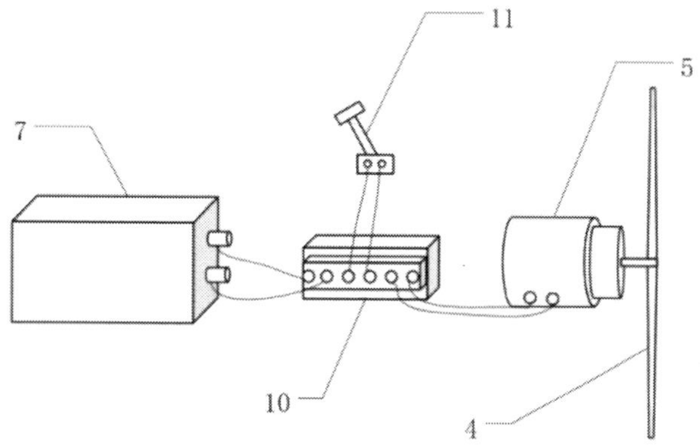

Figure 12. Assembly relations of electric drive systems

(4 is the propeller, 5 is the electromotor, 7 is the battery, 10 is the controller, 11 is the stick)

(5) Higher load efficiency and better maneuverability: Cyclic pitch and yaw of an all-electric autogyro can be achieved by controlling the steering gear. The brakes can aslo be in electric form. The reducer and the oil system can be removed, the weight of an autogyro would be greatly reduced, so an all-electric autogyro has higher load efficiency and better maneuverability.

(6)More environmentally-friendly and comfortable.

Author contact: Liang Lei 1044918155@qq.com
Qinghua Zhu zhuqinghua@nuaa.edu.cn

REFERENCES

[1]Wang, S.C., *Helicopter Aerodynamics*, Aviation Professional Material Editing Group, Beijing, 1985, pp. 18-29.

[2]Zhang, C.L., Guo, C.G, *Helicopter Preliminary Design*, Defense Industry Press, Beijing, 2006, pp. 40-49.

[3]Martin, H., *Modern Gyroplane Design*, Monterey, 1992, pp. 9-14.

[4]Federal Aviation Administration, *Rotorcraft Flying Handbook*, FAA-H-8083-21, Chapter15.

[5]Huang, J., Yang, F.T., "Development and challenges of electric aircraft with new energies," *Acta Aeronautica et Astronautica Sinica*, Vol. 37, (1), 2016, pp. 57-68.

[6]Nie, Z., Chen, M., "Calculation and analysis of electric-powered helicopter flight performance," *Journal of Beijing University of Aeronautics and Astronautics*, Vol. 38, (9), Sept. 2012, pp. 1139-1143. DOI:10.13700/j.bh.1001-5965.2012.09.029

[7]Liu, B., Ma, X.P., Wang, H.P., Zhou, K.S., "Design Analysis Methodology for Electric-Powered Mini-UAV, " *Journal of Northwestern Polytechnical University*, Vol. 23, (3), Jun. 2005, pp. 396-400.

[8]Leishman, J.G., "Development of the Autogiro: a Technical Perspective," *Journal of Aircraft*, Vol. 41, (4), 2004, pp. 765-781.

[9]Coton, F.N., Smrcek, L., Patek, Z., "Aerodynamic characteristics of a gyroplane configuration," *Journal of Aircraft*, Vol. 35, (2), 1998, pp. 274-279.

[10]Zhu, Q.H., Li, J.B., Ni, X.P., Zhang, C.L., "Several Aspects of Gyroplane Design," Chinese Helicopter Society 22th Annual Forum Proceedings, Baoding, Hebei, Aug 2006.

Human Factors Issues of Glass Cockpit Automation

Çiğdem GÜNEŞ
Human Factors Specialist, MSc
Turkish Aerospace Industries, Inc. (TAI)
Ankara, TURKEY

Serkan GÜNEŞ
Assoc. Prof. Dr.
Gazi University, Industrial Design Department
Ankara, TURKEY

ABSTRACT

With the advances in technology, clutter of mechanical indicators in the aircraft cockpit is replaced with digital displays. This revolution does not make only visual changes, but also changes the interaction between the pilot and the automated system. Cockpit automation has changed cockpit design philosophy with many promised benefits such as improvements in the precision, improved system safety, efficiency of operations, less workload and less human error. However, to achieve perfect design has not been fulfilled yet. Despite providing innovation and easiness, cockpit automation brings about some 'Human Factors' issues because of lack of support of human-machine interaction and cooperation. In literature, eleven human factors issues which are the consequences of glass cockpit automation are detected as changing role of the pilots, willingness to use automation, cognitive conflicts between pilot and the automation behavior, information overload, workload, lack of situation awareness, insufficient feedback, pilot skill degradation, cockpit design based problems, amount of automation and change in human error. This paper is a comprehensive review of the related literature that describes these eleven human factors issues.

INTRODUCTION

A glass cockpit is an aircraft cockpit that contains electronic instrument displays with highly automated systems. Parasuraman and Riley (Ref. 30, pp.231) defined automation as "the execution by a machine agent (usually a computer) of a function that was previously carried out by human". In first place, automation takeovers the time-consuming and labor-intensive work from human and will continue with the activities that requires human thinking and problem solving. Despite automated systems enable to decrease the need of human, it is now unfeasible to free human operator completely in advanced systems. Therefore, while automation is implemented, the human operator and its limitations, biases and potentials should be taken into consideration. For this reason, Wickens and Hollands (Ref. 43, p.531) emphasized that automation should be used to compensate the limitation or inability of the operator to perform functions.

Traditional cockpits had cluttered design as all the information had to be displayed at all the time, fairly small text sizes and very little if any use of color or graphics. Contrarily, glass cockpit has a visually minimalist design relative to traditional cockpit as shown in Figure 1.

Figure 1. Revolution of Cockpit Design from DC-9 to A380.

In aviation, automation is applied by some promises such as; increased safety, situation awareness, accuracy, and decreased human workload, cost, and complexity. Although automation was applied with good intentions and high

Presented at the 6th Asian/Australian Rotorcraft Forum / Heli Japan 2017, Kanazawa, Japan, November 7-9, 2017.

expectations, and being successful in terms of some aspects such as accuracy and economy of operations, unfortunately also produces some interaction problems between the user and the system (Ref. 1). Automated systems changed the nature of the work, so the cognitive action of the human operator. Therefore, this created complex work environments inevitably. One of major reasons of this situation is the belief of designers how much the human activity is decreased: the more successful the system (Ref. 7). However, it should not be ignored that the system is still used by a human as well as designed by a human (Ref. 33).

When the fatal accidents are analyzed (Table 1), it is seen that total pilot error based on different causes is the main reason of the accidents.

Table 1. Causes of Fatal Accidents by Decade-percentage (Ref. 35).

Cause	1950s	1960s	1970s	1980s	1990s	2000s	Avg
Pilot Error	40	32	24	25	27	25	29
Pilot Error (weather related)	11	18	14	17	21	17	16
Pilot Error (mechanical related)	7	5	4	2	4	3	4
Total Pilot Error	**58**	**57**	**42**	**44**	**53**	**45**	**50**
Other Human Error	0	8	9	6	8	9	7
Weather	16	10	13	15	9	8	12
Mechanical Failure	21	20	23	21	21	28	22
Sabotage	5	5	11	13	10	9	9
Other Cause	0	2	2	1	0	1	1

HUMAN FACTORS ISSUES OF AIRCRAFT AUTOMATION

Human factors problems of automated systems have been pointed out by many Human Factors researchers such as Wiener, Curry, Woods, Wickens, Parasuraman, and other scholars. Each researcher pointed out a different aspect of automation issues.

Some of the issues in literature are selected and similar issues are combined in terms of the aim of this study. Some of the issues are ignored such as software related problems because they are not the concern of this study. The reasons of these accidents are grouped into 11 issues including role of the pilot, cognitive conflicts in the mental model, excessive information, workload extremes (too much or too little), loss of situation awareness, insufficient feedback, loss of skills, design problems, human error, automation extremes and trust problems(too much or too little) (Ref. 15). These eleven human factors issues will be discussed in detail in the remaining of the paper.

New Roles for the Pilot

With rapid growth in microprocessor technology, designers have built very complex systems based on the rationale that the systems could operate automatically without human assistance. Wiener (Ref. 47) confutes this argument with two contrary aspects; all systems need a human to set up, feedback about what to do and monitoring, and they always need a human to take over the system when it fails.

Automation decreases the routine and time-consuming activities, however inherit human abilities such as cognitive flexibility, creative thinking and decision making activities cannot be supported by the automation.

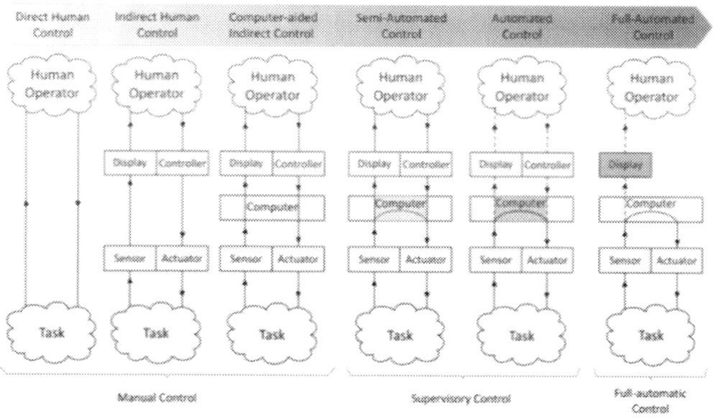

Figure 2: From Direct Human Control to Full-Automated Control.

In highly automated systems human operators are no longer need to perform the activity in person; they only monitor the ongoing processes. Although the system can operate by itself, attentional factors still affect system performance. Because the human operator must follow the system in case a fault cannot be detected by the automated system or an emergency case is occurred. When needed human must always be in the loop (Ref. 29).

Just monitoring and giving feedback when required seems to be a very easy mission at first glance; however it is more complicated than estimated. While the human role is changed from active controller to passive monitor, the ability of the human observer to maintain his/her focus on the continuing process for long periods of time gets more important. However human beings are not good at monitoring. Studies have shown that human cannot maintain visual attention toward a source of information for more than about half an hour even if he/she is highly motivated (Ref. 1). Wiener (Ref. 45) determined that the role of attention and vigilance is more complex and difficult than that is before the automation.

Eastern Flight 401 has given an example for vigilance problem by Wiener and Curry (Ref. 48). In this crash, autopilot descended the aircraft from 2000 feet by some reasons. The descent was not recognized by the crew members (despite the indicators and alarms), as they were absorbed in the disassembly of a warning light bulb. Soon, the aircraft crashed into Florida Everglades. The crash was a result of the flight crew members' failure to monitor the flight instruments properly.

Parasuraman et al. (Ref. 32) also claims that even operators behave equally in automated and manual systems when they are focused on a single-task, they are poor at monitoring in automated systems when they are engaged in other tasks compared with the manual systems.

Out-of-the-Loop Unfamiliarity (Ref. 42) is another problem of the change in human role. When the human operator is removed from the control loop, awareness of the operator may be reduced (Ref. 18). Because of the highly automated systems over-reliance, complacency and skill degradation problems may occur (These problems will be explained in the following chapters). Wickens (Ref. 42) claimed that these costs of high-level automation -reduced situation awareness, complacency, and skill degradation- might collectively cause the "out-of-the-loop" unfamiliarity.

Although highly automated systems work precisely and generally very well, there will always be possibility of the automation to make an undesired action or fail. In case such conditions, the human operator is needed to take over the manual function and cope with the situation. However, the "Out-of-the-Loop Unfamiliarity" may prevent the operator from settling the current situation of the aircraft successfully in time (Ref. 31).

Automation Use / Disuse/ Misuse

The use of the automation depends on the positive or negative attitudes of automation. Generally, these attitudes are affected by the reliability or accuracy of the automation. For instance, while automatic breaking systems have been welcomed and used efficiently, smoke detectors are disliked by many people because of the false alarms (Ref. 30).

During interaction with the automated system, pilot develops a strategy about using it. Although an appropriate usage is desired by designers, this strategy sometimes appears as disusing (underutilization of automation) or sometimes misusing (overreliance on automation) the automation Parasuraman and Riley (Ref. 30).

Although Parasuraman and Riley (Ref. 30) based on the automation usage to reliability or accuracy, Dzindolet, et al. (Ref. 10) claimed that it is not such simple. They created a framework of automation use claims that automation usage depends on some processes that are originated from the causes of productivity loss in groups model of Mullen et al. (Ref. 24). They believe that this model can be applied to human-computer teams, as well. Several researchers also support this thought that the human-computer system can be accepted as a team in which one member is not human (Ref. 6; Ref. 51). This model includes the cognitive processes, social processes and motivational processes of the human operator.

Cognitive Sources: Automation bias is one of the factors that influence the automation usage of the human operator. Mosier and Skitka (Ref. 23) defined automation bias as "the tendency to use automated cues as a heuristic replacement for vigilant information seeking and processing" (p.205). The decision reached by the automated aid before decision making may lead the operator to rely on the automation in a heuristic manner. Even though, this strategy is generally appropriate, sometimes this reliance may be inappropriate and causes misuse.

Dzindolet et al. (Ref. 9) conducted a study with students about the reliability of the aid was manipulated in a Visual Detection Task. Students are asked to mark the slides that a soldier is camouflaged in a terrain over two hundred slides. A group of subjects was informed that a computer program would help them in detecting targets and computer gave its result before the decision making of the participants. For the other groups of participants, automation bias is eliminated and computer gave the results only after the participant chose her/his own decision. In conclusion, it is emerged from the survey that when the automated aid's decisions is given first, that make automation bias to occur, participants tended to misuse rather than disuse their automated aids, however when automation bias is eliminated participants accept their own decisions, and they tended to disuse rather than misuse. In this survey, Dzindolet et al. (Ref. 9) found that, automation reliability did not affect the automation bias.

Social Sources: Most of the researchers propose that trust is the one of the major variables that affects the automation use (Ref. 30; Ref. 21; Ref. 22). "Trust can be defined as the attitude that an agent will help achieve an individual's goal in a situation characterized by uncertainty and vulnerability" (Ref. 22, p.51).

Pilot may show different reaction across the automation. The role of trust is mediating between human and automated systems. Pilots need to trust automation in some degree in order to use the automation. However, for a healthy interaction between automation and pilot, it should be neither too much nor too less.

If over-reliance on automation is occurred then the pilot gives up the tracking the automation that prevent the recognition of system malfunctions (Ref. 30). "When it works well, it usually works very well indeed – so well in fact that we sometimes that it more than we should. Yet on the rare occasions when it does fail, those failures may often be more catastrophic, less forgiving, or at least more frustrating than would have been the corresponding failures of a human in the same circumstance" (Ref. 44, p.493). Pilots tend not to use cues available from the displays or controls when they trusted the automated system too much as Eastern Flight 401 case, which was mentioned before.

If the human operator does not trust the automation, he/she will not use it. As mentioned before, during World War II, improved radar systems are thought to be a great advantage towards the enemies by United States. However, radar misidentified Spaniel fish vessels and large whales as enemy submarines. Increased misidentification caused human operator's trust to automated system's alerts and to ignore critical signals (Ref. 29).

As Lee and Moray (Ref. 21) said that, the main goal is to achieve calibrated trust that means ensuring correspondence between human operators' trust to the automation and the automation capability. Good calibration of the automation relative to the trust is represented by the diagonal line in the Figure 3 that means above this line is over-trust and below is distrust (Ref. 22).

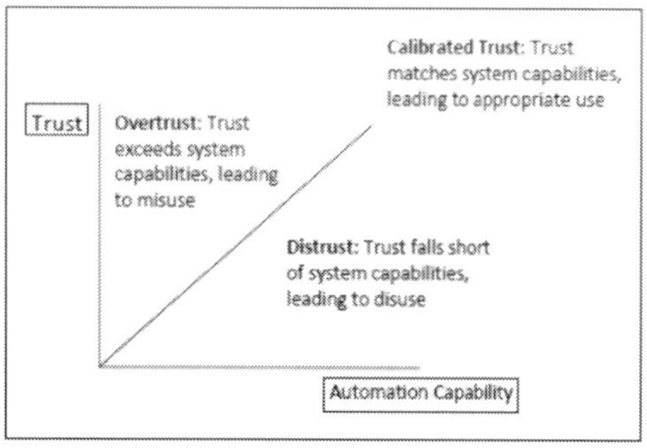

Figure 3: Automation Capability (trustworthiness) (Lee and See, 2004, pp.55)

Dzindolet et al. (Ref. 10, p.8) claims that trust is "outcome of a comparison process between the perceived reliability of the automated aid (trust in aid) and the perceived reliability of manual control (trust in self)". However, some errors may be occurred in comparison, which causes misuse or disuse problems. One of these errors is that humans tend to over-estimate their own performance, and the other is if information is given to the human operators about the reliability of the automated aids then the automation bias gets high.

Motivational Sources: Dzindolet et al. (Ref. 11) determined that the third variable of the automation use is the motivational processes of the human operator. Humans tend to feel less responsible when working with automated systems so they are likely less motivated to make effort as regards working alone. This phenomenon is called "social loafing" or "free riding" in social psychology by Latane et al (Ref. 20). According to the Shepperd's Expectancy Value Theory, there are three factors that trigger the motivation; expectancy, instrumentality, and outcome value (Ref. 19).

Expectancy is the feeling of human being necessary for a group to be successful. If the human operator feels he/she is dispensable for the system, then the bias to the automated aid will be high. However, in some instances this tendency may cause automation misuses.

Instrumentality is another factor of motivational process. If the human operator believes that the overall outcome contributes positive values to the job, then he/she will be willing to being in cooperation with the automated system.

The value of the outcome may also affect the motivation of the human. Dzindolet et al. (Ref. 11) defines the outcome value as the difference between the outcome and the cost it requires. Number of tasks, interest in task, fatigue, and cognitive overhead are the four contributor of cost. If the number of the tasks so the workload is increased, then the cost of performing that task increases. Boring tasks' cost is higher than the interesting tasks. Fatigue is another cost of the outcome, if the operator feels fatigue then misuse may increase. Finally, if the cost of transferring from manual operation to automation or vice versa is high then the operator maintains the condition. This may cause complacency or misuses.

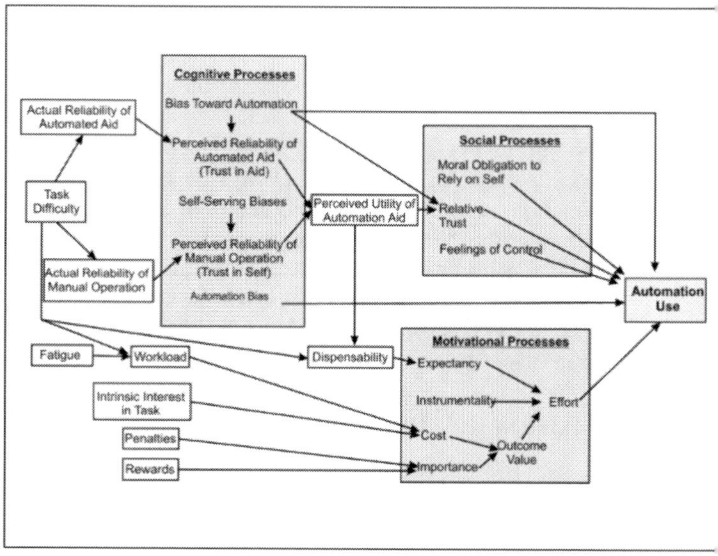

Figure 4: Framework of Automation Use (Ref. 10, pp.18)

In conclusion, Dzindolet et al.'s (Ref. 10) framework of automation use indicates that the cognitive, motivational and social process of the human cause misuse, disuse and appropriate automation use which is illustrated in Figure 4. Designers should consider the interaction of these domains attentively to ensure the appropriate usage of automation.

Cognitive Conflicts

Cognitive conflict between human mental model and the operation of automation is one of the consequences of the goal of removing human from the loop. Cognitive conflict in a human computer interaction is the mismatch between the mental model that human create to define the current situation and the real behavior of the machine under control. Therefore, in order to understand the mental conflict first how the human does understand the world / situations should be detected.

People produce schemas about the usage of a system with which they interact. Norman (Ref. 26) called these schemas the 'system image' (Figure 5). Human perceives the real action of the automation as how it is coded in his/her system image and reaction is given according to this mental process outcome. The success of the human-computer interaction of a glass cockpit is provided by calibrating the mental model of the pilot with the working model of the glass cockpit. If there exists an interaction gap between them, this often shows itself as a surprise on the operator (Ref. 3). If this interaction cannot be performed properly, it is inevitable to face with accidents related with automation.

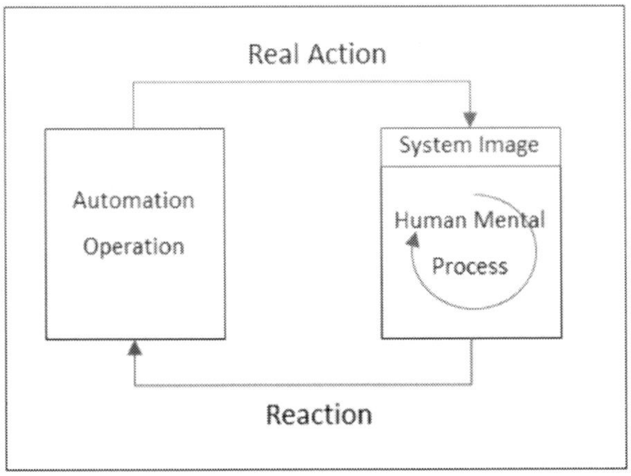

Figure 5: Human Mentality of Automation

Sarter and Woods (Ref. 37) declared that this mental model discrepancy could only be solved by better design, better displays, and better training.

Besnard and Baxter (Ref. 3) claims that the conflicts between human mental model and the system's operation are because of some factors or a combination of them such as high-tempo-dynamic system, low predictability of machine's behavior, occurrence of an undetected error and erroneous continuation of an initial plan.

The Cali accident, which is accepted as a milestone in glass cockpit automation and examined by most of the human factors researchers (ref. 13; Ref. 3; Ref. 14; Ref. 32; Ref. 34; Ref. 10; Ref. 49), illustrates the results of a combination of all these factors. In 1995, a Boeing B757 crashed into a 12,000ft mountain near Cali. In this accident, pilots tried to implement a direct approach suggested by air traffic control. While pilots reprogramming the Flight Management System (FMS) for an approach starting from ROZO4 beacon, because of writing 'RO' and accepting the first proposed beacon they caused the aircraft to change the route to ROMEO, which is 132 miles far from Cali. When they noticed that the aircraft was turning to a wrong heading and corrected the beacon as ROZO4, the aircraft fell into a fatal course and crashed into the mountain (Ref. 3). This accident shows that what a flight crew without a clear picture of the situation of the aircraft can cause.

Information Overload

Modern automated glass cockpits are capable of presenting vast amounts of data to the pilots. This information is presented to the pilots in text, numeric, or graphic form via displays. Some of the information is displayed automatically when needed (e.g. Calculated Air Release Point (CARP) or a message belong to a waypoint) or in an emergency (e.g. warnings), and some of the information is needed pilot to access (information in the Control and Display Unit (CDU) pages). Glass cockpit technology is feasible to implement more and more information. Increasing of information load causes the system become more complex and sophisticated. As Woods et al. (Ref. 50, Ref. 52) emphasized that this situation is paradoxical although more and more data is available in principle; the ability of human operator to process this information has not been improved. Therefore, information overload may create errors in the system that may cause human error.

Wiener (Ref. 47) concerns that if the complexity of the systems cannot be restrained, then the flight crews cannot perform their duties effectively and in an occurrence of failure, they cannot take over the control. Because information overload may cause obscuring needed information among that too much information presented in the cockpit (Ref. 4).

Woods et al. (Ref. 52) determined three characterization of information overload; (1) clutter problem, (2) workload bottleneck, (3) significance of data. The first way of characterizing data overload is that too much information is displayed to the human operators. Designers tend to present all the information available technically to the operators. This brings human to an impasse, as he/she has to know where to look or what to look for over all these information. If the cockpit environment is considered, there may be more than two hundred pages full of information. Ironically, although automation allows accessing any data, designers have to make a choice and eliminate some of the data in order to cope with the complexity. Secondly, workload bottleneck occurs in settings where access to the data has increased quickly. Pilots have to do too much work simultaneously with time constraints. Finally, large amounts of data create obscurity in significance of information and may stress the operator's cognitive activity of focusing on the relevant data for a particular situation. Woods et al. (Ref. 52) indicate that when human face with excessive available data, he/she feels like a baby seeing the world as a "great blooming".

Workload Dilemma

The word "workload" generally refers to the mental processes (cognitive and affective) of the human operator rather than the physical workload in aviation. Workload is defined by Hart and Staveland (Ref. 16) as "the perceived relationship between the amount of mental processing capability or resources and the amount required by the task". Huey and Wickens (Ref. 17) determined four factors that influence the workload; **(1) Imposed task demands:** If the task's difficulty, number, rate, or complexity increases then the workload is expected to increase. **(2) The level of operator's ability:** If the errors of the human operator increase then the workload is assumed to increase. **(3) The mental and physical effort the operator exerts**: Workload reflects the operator's response to a task rather than task demand directly. **(4) Operator's perceptions:** Operator's workload can also be affected by the feeling of being effortful and over loaded even though his/her performance has not changed. In addition to these factors fatigue, stress, training, knowledge, crew coordination, environmental factors should also be considered while mentioning about the workload.

One of the promises of the cockpit automation is to reduce workload, by helping the pilot in his intractable missions. However, this expectation has not been achieved completely as some researches show that workload has not been decreased in practice, instead it affect pilot workload adversely. Wiener (Ref. 46) pointed that although automated systems support pilots in routine flight phases, which require low workload, they increase the workload in time-critical phases when help is needed most. This phenomenon is named as 'clumsy automation'. According to his researches, introduction of glass cockpit has a little effect on overall pilot workload and he believes that automation tends to redistribute instead of reducing pilot workload. The major source of this increment in workload is because of the requirement to reprogram FMS when the conditions changed (e.g., changing the route, changing the mission, etc.). Therefore, pilots feel that the overall workload has not been decreased.

In addition to Wiener, Kirlik (Ref. 55) also claims that automation which has difficulties in initiating and engaging, increases the operator workload, or produce an unbalanced pattern of workload. In this point, it should be underlined that excessive workload (too much or too little) levels may cause faults in the interaction. Pilots need a proper workload to maintain their attention on a particular task. If the workload is too little then the pilot may get bored and lose vigilance, and if the workload is too much then pilot may lose his/her motivation. Therefore, system design should have some challenge. Some human factors researchers believe that workload and performance have a complex relationship that means human performance is coherent under moderate workload that does not change suddenly or unpredictably (Ref. 56). According to the Yerkes - Dodson law, some arousal helps the increment in the performance up to a point, but when the optimal point is exceeded then the performance decreases.

The majority of workload is imposed on the pilot who interacts with automation. Pilot-non-flying has to do more work in comparison with the pilot flying, as he/she has to set up the automated system. The feedback that the automated systems need is provided by this pilot such as changing the route or selecting an approach type. Therefore, this may cause a "programming nightmare" on the pilot-non-flying (Ref. 38).

According to a phenomenon called 'cognitive tunneling' (Ref. 41) if the workload is very high, then the pilot tends to focus on only some particular cues and ignore the other information sources. Thus, pilots may fail to recognize other errors of automated system and so may be late to take over the control. As in the Eastern Flight 401, pilots cannot recognize the decent into the terrain while focusing on a burned lamp (Ref. 4).

Situation Awareness (SA)

Situation Awareness is described by Endsley (Ref. 57) as "the perception of the elements in the environment within a volume of time and space, the comprehension of their meaning, and the projection of their status in the near future". It can be defined as intuitively knowing what is going on in an existing environment.

Situation awareness consists of three levels; Level - 1: Perception, Level - 2: Comprehension and Level - 3: Projection (Ref. 58). The first level refers to perceiving critical factors in the environment (Level 1 SA), second level refers to analyze those factors in accordance to the operator's goals (Level 2 SA) and finally the third level means understanding what will happen in near future (Level 3 SA). These levels of situation awareness allow the human operator to take action in a timely and effective manner even in very complicated and challenging tasks (Ref. 13).

Wickens (Ref. 58) indicates that each of these levels point to different perceptual/cognitive operations and have different consequences in design. For instance, while a Level-1 SA requires designing alerts more observable, Level-3 SA may require predictive displays.

Pilot needs to have good situation awareness about the aircrafts state, potential threats, what automation is doing, and what it is going to do to direct a flight mission well. However, sometimes they may not create a good mental model about what is going on, as he/she has delivered the control to the automation. Wiener (Ref. 46) says pilots in the glass cockpits often ask these questions: What is it (the auto flight system) doing? Why did it do that? What is it going to do next?

When human has to make critical decisions in the operation of complex systems like aviation, generally with time constraints, most of the errors are occurred because of the lack of situation awareness (Endsley, 2001). As the aviation is fragile, sometimes results may be very catastrophic with many lives lost.

Furthermore, Sarter and Woods (Ref. 37) pointed to the 'mode confusion' of autopilot which means that pilots have insufficient awareness of how the modes functioning and changing. Mode confusion problem may also occur when autopilot changes its mode from one to another. Pilots may not understand which mode the auto-flight system performing in some cases. Aircraft will behave different in each of the modes and if the pilot needs to get the authority, there is not much clue which mode the autopilot is.

In advanced glass cockpits, as pilot's monitoring ability the automation is correlated with his/her expectations from the automated system, this may cause the pilot to canalize his/her attention on a particular situation. Moreover, there is a potential risk that he/she may miss an indication from the automation (Ref. 37). This may create 'Automation Surprise' when he/she finally detects the result of this discrepancy between expected and actual (Ref. 36).

On the other hand, some researchers propose that the automation increases the situation awareness by assisting the pilot. Curry and Ephrath (Ref. 8) stated that pilots who work with an automated system are better on monitoring than the pilots working with manual systems. They attributed this improvement to the monitors that helps the pilots in detecting system failures. Pilots were able to allocate their attention in automated system whereas; they needed to devote their attention on the manual systems that has a narrower range of cues (Ref. 12).

Of course, a proper design should increase the situation awareness with its highly safe systems. For example, some automated systems display the route of the aircraft, current position, previous and next waypoints of the aircraft, by the help of the ground proximity warning system (GPWS), radar, terrain alerting warning system (TAWS), and traffic collision avoidance system (TCAS) and warns the pilot against the weather threats, terrain threats or intruders around the aircraft.

To sum up, an appropriate design is required to increase the situation awareness. Endsley and Kiris (Ref. 12) declared that situation awareness of the pilot may be increased with automated systems by providing superior, integrated information to the pilots.

Insufficient Feedback

Designers tend to automate everything that is feasible and remove the human operator out of the loop. Most of the human factors issues are emerged from this design approach that cause inadequate feedback of automation's actions and intentions (Ref. 27).

As Norman (Ref. 27) says, automation isolates the pilot mentally from the operations of the aircraft with inadequate feedback of the moment-to-moment activities of the aircraft. Appropriate feedback prevents the cognitive conflicts between the pilot and the automated system, and many situation awareness problems so the human errors.

In 1985, a China Airlines 747 crashed because of lack of feedback of autopilot. Autopilot was compensated the loss of power from its outer right engine and did not inform the pilots sufficiently. When autopilot reached the limit of its compensatory abilities and could not keep the aircraft stable, it was too late for the pilots to take over the control. If the automatic pilot had informed the pilots that it was recovering the balance because of a problem, so this would make pilots warned. When people work with automated systems, feedback is crucial for the effectual monitoring the activities of aircraft that may help to detect the errors. Feedback is also essential, to make human-in-the-loop by updating the mental model of the pilot and provide him to take control easily when needed. Autopilots work by physically, so pilot has the subtle information from the physical movement of the control wheels. In this case, feedback was available; however, it was insufficient to attend pilot's attention properly.

On the other hand, excessive rate of feedback may cause representing too much information that can cause detracting attention, which is explained in the previous issue; information overload. Norman (Ref. 27) indicates that automation gives feedback but it will not give information about what it is actually doing. Therefore, it leaves the pilot uninformed again and lumbers the pilot in obscurity.

Low observability is another problem of the automation. Woods et al. (Ref. 50) point to that problem is not based on only the inadequate feedback; it is actually the required effort of the pilot's to monitor and process the information.

As automation does not function correctly all the times, pilots need to be taking over the control. In order to control successfully to unexpected or erroneous system behavior of the automation, pilot must determine the source of the problem quickly. This may be achieved only by appropriate feedback.

Pilot Skill Loss

Skill is not only the combination of operators' knowledge and performance it is also a fundamental aspect of human cognition (Ref. 25). Wickens (Ref. 41) pointed out higher degrees of automation limits the pilots' activities by doing most of the operations without pilots' participation. This changed the human role from active participant to passive observer as described previously. Therefore, this change in human role also causes the degradation in manual skills when automation is used for a long time. Pilots' abilities may be rasped. Likewise, in the course of time, improving manual skills with time may not be enabled experienced as in traditional cockpits. Just like Norman (Ref. 28) concerns technology can make us smart or dumb.

The major difference between an expert and non-expert is the time. An expert has lots of time to perform a task whereas a non-expert barely able to cope with the situation (Ref. 25). An expert performs the minimum steps required for a certain mission while a non-expert follows indirect methods and performs more steps than required to complete the same mission. To become an expert, a non-expert needs time and practice. Unfortunately, as Bainbridge (Ref. 1) indicated physical skills degrade in course of time when they are not used. Loss in pilot skill is the result of automation and it is the source of human error.

Glass Cockpit Design Concept

Another alteration derived from the automation is the new environment of the cockpit. There are many advantages of glass cockpit design about usability. New design allows to display colored graphics, to present only the information needed for a specific operation, enlarge text sizes so increase the readability and to make color coding. Contrarily, traditional cockpits has cluttered design as all the information has to be displayed at all the time, fairly in small text sizes and very little if any use of color or graphics.

Glass cockpit allows unnecessary information to be disappeared temporally. Before, each data had to be displayed during the flight mission continually, with the advanced technology it is now enough to display a data when it exceeds a particular value or when it is needed. For example, there is no need to display the instruments of Instrument Flight Rules (IFR) during a Visual Flight Rules (VFR) flight. This simplification of flight deck, allows pilot to focus on the most needed information during a specific mission. The new cockpit design allows pilots to reconfigure the displays according to their needs. This flexibility of displays serves several possibilities in the information presentation. Pilots can select the information, which should be displayed in a display. They can both display primary flight controls and navigational information at the same time.

However, new cockpit philosophy may cause some problems at the same time. Glass cockpit needs feedback from the pilot via the (CDU). Again, the pilot needs to know the exact place where the information is implemented in the architecture of the CDU pages (it should beneficial to be remembered that CDU may have more than 200 pages).

As Wiener (Ref. 46) concerns after long period times of human factors researches and application, there are more than two or three keyboard types, so different layouts, in a glass cockpit that causes human error. The un-standard design may cause serious problems in automated systems. Every system has a different interface and this may lead to number of errors as result of the pilots moving from one aircraft to another (Ref. 40). Standardization of hardware is also desired to reduce training and maintenance costs (Ref. 46). Wiener (Ref. 47) points out that the keyboards are highly vulnerable to finger error. These problems does not have to be seen when they are created, also they can lie dormant in the system for hours after they created until they become active. For instance, in the Cali accident, typing ROMEO instead of ROZO4 caused a catastrophic error.

Moreover, it is difficult for pilot to accomplish other tasks while pressing buttons or keys. The pilot has to take the risk to face with division of attention and remaining head down while accomplishing keyboard inputs, which may cause missing the situation awareness of the external flight environment (Ref. 2).

Amount of Automation

With the rapid growth in microprocessor technology, some designers tend to design complex systems based on the rationale that systems can operate without human contribution (As it is discussed previously, this design goal cannot be actualized as the automated system needs human feedback and also needs human to substitute in case of malfunctions). If this complexity cannot be restrained then the pilots may have difficulties in performing their duties. On the other hand, some designers became aware of the sophistication of automated systems so simplification arises as an alternative to automation (Ref. 47). According to that approach, simplification rather than complicated automation may also achieve the same goal.

Lee and See (2004) claim that an appropriate communication between pilots and glass cockpit depends on how well the operation capabilities of the automated system are expressed to the user and this can only be done by simplifying the automation or making its operations clear.

The potential difficulty of over-automation is that it degrades the situation awareness of the pilots with complicated systems and insufficient feedback. Increases boredom (Ref. 53) so reduces the job satisfaction of the pilots and reduces monitoring (Ref. 31). Norman (Ref. 27) on the other hand, stresses that the problem in glass cockpit is not the over-automation, but the lack of feedback. He proposes that it is possible to reduce the problems of automation through appropriate design considerations such as giving feedback to operator and ensuring interaction with the operator in an effective manner whatever the automation level.

According to a survey of Yvette et al. (Ref. 54) pilots flying with A-320, MD-11 and B747-400 did not complained about the over-automation problem even they claimed that they appreciate more automation. However, instead of a fully autonomous level of automation, they indicated that they would be more comfortable with a shared pilot automation performance.

In addition, Sarter, Woods and Billings (Ref. 38) also claim that the unexpected consequences of the automation is not the result of over-automation or human error, the responsible agent is the design which allows for interaction problems between human and the automated system. Billings (Ref. 5) supports this argument and he says that the development in the automation of aircraft cockpits may have gone too far too quickly, without understanding its possible effects on pilots.

These two sides of approaches are also found supporter in the aircraft manufacture. Wiener (Ref. 47) compares the two aircraft manufacturer; The Douglas and the Boeing. He declares that whereas the design philosophy of Douglas is to remove the pilot from the loop (automation takes the decisions automatically and does not ask the pilot's approval), Boeing's approach is to share the responsibility (automation informs the pilot and gives the eventual decision to the pilots).

Change in Human Error

Change in human error is the consequences of all other factors that have just been explained in this paper. This is the major ironies (Ref. 1) of the automation of the systems.

Although the major aim in the glass cockpit is to eliminate the human from the loop, human is still in the system from designing to operating the automated system. Civil aviation authorities stated that while accidents caused by machine have decreased over the years, the number of accidents caused by humans has increased comparatively (Ref. 39).

Automation is hoped to remove human error by replacing the human who makes errors inevitably with inerrable machines. However, Wiener (Ref. 47) claims that automated systems try to prevent the causes of the error instead of preventing the original error and change the nature of the error. Briefly, automation may create more serious problems while eliminating small errors occurred in traditional cockpits (Ref. 47).

Designer anticipates an interface design according to his/her mental model. Barriers are created to obstruct the faults on the extent of the interface designer's anticipations. However it is impossible to estimate all the situations in a system in which human is in the loop. In addition to that, designers have implemented more information onto displays and more options into FMS. This hamper the pilots to master the whole system so with lack of knowledge pilots may make mistakes (Ref. 49).

Moreover, new system creates new human errors. As automation interactively works with the pilots, it needs the pilot to give input to the FMS. Depending on these inputs, FMS makes calculations and performs its mission. Thus, this system is open to human errors. In order to prevent this types of errors, system designers creates some barriers, however this design barriers are also depended to the prescience of another human 'the designer'.

CONCLUSIONS

Glass cockpit automation has been implemented with best intends of designers such as increasing the safety, increasing accuracy of systems, increasing situation awareness of pilots, decreasing workload and decreasing the costs of the flight. It has actually provided undeniable contributions to aviation in precision, efficiency, and safety; unfortunately, it has also come along side effects. These side effects are grouped into eleven human factors issues in this paper. These eleven issues include design-based problems that shall be taken into consideration while designing.

Looking all these problems one can think that automation should not be implemented to cockpits, but it not like that at all. Contrarily, implementation is indeed encouraged as long as these problems are considered during design process. Although glass cockpits have experienced several accidents and many incidents related to pilot interface problems, this does not mean these aircrafts are unsafe. Actually, they generally work very well in accuracy and efficiency.

Author contact:

Çiğdem GÜNEŞ cigdem.gunes@tai.com.tr
Serkan GÜNEŞ serkangunes@gazi.edu.tr

REFERENCES

[1]Bainbridge, L., "Ironies of Automation," *Automatica*, Vol. 19, (6), 1983, pp. 775-779.

[2]Baron, S., "Pilot Control," *Human Factors in Aviation,* edited by E. L. Wiener and D. C. Nagel, Academic Press, San Diego, 1988, pp.347-383.

[3]Besnard, D. and Baxter, G., "Cognitive Conflicts in Aviation: Managing Computerised Critical Environments," *Proceedings of the Fifth DIRC Research Conference,* edited by Jo Mackie and M. Rouncefield, Lancaster University Press, Edinburgh, 2005, pp. 119-124.

[4]Billings, C. E., "Human-Centered Aircraft Automation; a Concept and Guidelines," NASA Tech. Memorandum 103885, NASA Ames Research Center, California, 1991.

[5]Billings, C. E., *Aviation Automation: The Search for a Human Centered Approach*, Mahwah, Lawrence Erlbaum Associates, NJ, 1997.

[6]Bowers, C. A., Oser, R. L., Salas, E., and Cannon-Bowers, J. A., "Team Performance in Automated Systems," R. Parasuraman and M. Mouloua, *Automation and Human Performance: Theory and Applications*, Lawrence Erlbaum Associates, NJ, 1996, pp. 243-263.

[7]Christoffersen, K. and Woods, D. D., "How to Make Automated Systems Team Players," edited by E. Salas, *Advances in Human Performance and Cognitive Engineering Research*, Vol. 2, JAI Press, Elsevier, Oxford, 2002, , pp. 1-12.

[8]Curry, R. E., and Ephrath, A. R., "Monitoring and Control of Unreliable Systems," edited by T. B. Sheridan and G. Johannsen, *Monitoring Behavior and Supervisory Control*, Plenum, New York, 1997, pp. 193-203.

[9]Dzindolet, M. T., Pierce, L. G., Beck, H. P., Dawe, L. A., "Misuse and Disuse of Automated AIDS" *Human Factors and Ergonomics Society Annual Meeting Proceedings, Cognitive Engineering and Decision Making*, Human Factors and Ergonomics Society, California, 1999, pp. 339-343

[10]Dzindolet, M. T., Dawe, L. A., Beck, H. P., and Pierce, L. G., *A Framework of Automation Use*, Army Research Laboratory, Maryland, 2001

[11]Dzindolet, M. T. and Pierce, L. G., "A Computerized Text Analysis Can Detect Deception," *Proceedings of the Human Factors and Ergonomics Society 48th Annual Meeting*, Human Factors and Ergonomics Society, Santa Monica, California, 2004.

[12]Endsley Mica R. and Kiris, E. O., "The Out-of-the-loop Performance Problem and Level of Control in Automation," *Human Factors*, Vol.37, 1995, pp. 381-394.

[13]Endsley, M. R., "Designing for Situation Awareness in Complex Systems," *Proceedings of the Second International Workshop on Symbiosis of Humans, Artifacts and Environment*. Kyoto, Japan, 2001.

[14]Funk K., Lyall B., Wilson, J., Vint, R., Niemczyk M., Suroteguh C. and Owen G., "Flight Deck Automation Issues," *The International Journal of Aviation Psychology*, 1999, Vol. 9 (2), pp. 109-123.

[15]Güneş, Ç., "Human Factors Issues of Glass Cockpit Automation," Presented in Partial Fulfillment of the Requirements for the Master of Science in the Graduate School of Natural end Applied Sciences, Middle East Technical University, Ankara, 2010.

[16]Hart, S. G., and Staveland, L. E., "Development of A Multi-Dimensional Workload Rating Scale: Results Of Empirical And Theoretical Research," edited by P. A. Hancock and N. Meshkati, *Human Mental Workload*, Elsevier, Amsterdam, Netherlands, 1988, pp.139-183

[17]Huey, B. M., Wickens, C. D., *Workload Transition-Implications for Individual and Team Performance*, National Academy Press, Washington DC, 1993

[18]Kaber, D. B. and Endsley, M. R., "Out-of-the-Loop Performance Problems and the Use of Intermediate Levels of Automation for Improved Control System Functioning and Safety," *Process Safety Progress*, Vol. 16 (3), 1997, pp. 126-131.

[19]Kerr, N. L., and Bruun, S. E., "Dispensability of Member Effort and Group Motivation Losses: Free Rider Effects," *Journal of Personality and Social Psychology*, Vol. 44, 1983, pp. 78-94.

[20]Latane, B., Williams, K., and Harkins, S., "Many Hands Make Light The Work: The Causes and Consequences of Social Loafing," *Journal of Personality and Social Psychology*, Vol. 37, 1979, pp. 822-832.

[21]Lee, J. D. and Moray, N., "Trust, Control Strategies and Allocation of Function in Human-Machine Systems," *Ergonomics*, Vol. 35 (10), 1992, pp. 1243-1270.

[22]Lee, J. D. and See, K. A., "Trust in Automation: Designing for Appropriate Reliance," *Human Factors: The Journal of the Human Factors and Ergonomics Society*, Vol. 46, 2004, pp. 50-80.

[23]Moiser, K. L. and Skitka L. J., "Human Decision Makers and Automated Aids: Made for Each Other?," edited by R. Parasuraman and M. Mouloua, *Automation of Human Performance: Theory and Applications,* Lawrence Erlbaum Associates, Inc, New Jersey, 1996, pp. 201-220.

[24]Mullen, B., Johnson, C. and Salas, E., "Productivity Loss in Brainstorming Groups: A Meta-Analytic Integration," *Basic and Applied Social Psychology*, Vol. 12, 1991, pp. 3-23.

[25]Norman, D. A., *Errors in Human Performance*, Center for Human Information Processing, San Diego, California, 1980.

[26]Norman, D. A., "Cognitive Engineering," edited by D. A. Norman, and S. W. Draper, *User Centered System Design: New Perspectives on Human-Computer Interaction*, Lawrence Erlbaum Associates, Hillsdale, NJ, 1986, pp. 32-65.

[27]Norman, D. A., "The 'Problem' with Automation: Inappropriate Feedback and Interaction, Not 'Over Automation'," *Philosophical Transactions of the Royal Society of London, Series B. Bilological Sciences*. Vol. 327 (1241), 1990, pp. 585-593.

[28]Norman, D. A. *Things That Make Us Smart: Defending Human Attributes In The Age Of The Machine*, Addison&Wesley, New York, 1993.

[29]Parasuraman, R., "Human Computer Monitoring," *Human Factors*, Vol. 29 (6), 1987, pp.695-706.

[30]Parasuraman, R. and Riley, V., "Humans and Automation: Use, Misuse, Disuse, Abuse," *Human Factors*, Vol. 39 (2), 1997, pp. 230-253.

[31]Parasuraman, R., "Designing Automation for Human Use: Empirical Studies and Quantitative Models," *Ergonomics*, Vol. 43, (7), 2000, pp. 931-951.

[32]Parasuraman, R., Molloy, R., Mouloua, M., and Hilburn, B. "Monitoring of Automated Systems," edited by R. Parasuraman and M. Mouloua, *Automation and Human Performance: Theory and Applications*, Lawrence Erlbaum Associates, Mahwah, NJ, 1996, pp. 91-115.

[33]Parasuraman, R. and Wickens, C. D. "Humans: Still Vital After All These Years of Automation," *Human Factors*, Vol. 50 (3), 2008, pp.511-520.

[34]Phillips, R. O., "Investigation of Controlled Flight into Terrain: Descriptions of Flight Paths for Selected Controlled Flight into Terrain (CFIT) Aircraft Accidents 1985-1987," Project No. DOT-TSC-FA9DI-99-01, Department of Transportation, Washington, DC, 1999.

[35]Plane Crash Info Website, http://www.planecrashinfo.com/cause.htm, last visited on September, 2017

[36]Sarter, N. B., "Strong, Silent, and 'Out-Of-The-Loop': Properties of Advanced (Cockpit) Automation and Their Impact on Human-Automation Interaction," Presented in Partial Fulfillment of the Requirements for the Degree Doctor of Philosophy in the Graduate School of the Ohio State University, 1994.

[37]Sarter, N. B. and Woods D. D., "Team Play with a Powerful and Independent Agent: A Corpus of Operational Experiences and Automation Surprises on the Airbus A-320," *Human Factors*, Vol.39 (4), 1997, pp. 553-569.

[38]Sarter, N.B., Woods, D. D. and Billings, C. E., "Automation Surprises," edited by G. Salvendy, *Handbook of Human Factors & Ergonomics*, Second Edition, Wiley, New York, 1997, pp.1926-1943.

[39]Sheridan, T. B., *Telerobotics, Automation and Human Supervisory Control*, MIT Press, Cambridge, MA, 1992.

[40]Singer, G. "Filling the Gaps in the Human Factors Certification," edited by S. Dekker and E. Hollnagel, *Coping with Computers in the Cockpit*, Ashgate Publishing, Aldershot, UK 1999, pp. 87-107.

[41]Wickens, C. D. "Attention in Perception And Display Space," edited by C.D. Wickens and J. G. Hollands, *Engineering Psychology and Human Performance*, Harper-Collins, New York, 1992, pp.74-115.

[42]Wickens, C. D., "Designing for Situation Awareness and Trust in Automation," *Proc. IFAC Conference Integrated Systems Engineering*. Baden, Germany, 1994.

[43]Wickens, C. D. and Hollands, J., *Engineering Psychology and Human Performance,* Prentice Hall, New Jersey, 2000.

[44]Wickens, C. D., Gordon, S. E. and Liu, Y., *An Introduction to Human Factors Engineering*, Longman, New York, 1998.

[45]Wiener, E. L., "Vigilance and Inspection," edited by J. S. Warm, *Sustained Attention in Human Performance,* Wiley, Chichester, UK, 1984, pp. 207-246.

[46]Wiener, E. L., "Human Factors of Advanced Technology ("Glass Cockpit") Transport Aircraft", NASA Technical Report No. 177528, NASA, Ames Research Center, California, 1989.

[47]Wiener, E. L., "Intervention Strategies for the Management of Human Error," NASA, Ames Research Center, California, 1993.

[48]Wiener, E. L. and Curry, R. E., "Flight-Deck Automation: Promises and Problems," *Ergonomics*, Vol.23 (10), 1980, pp. 995-1011.

[49]Wiener, E. L., Chute, R. D., Moses, J. H., *Transition to Glass: Pilot Training for High-technology Transport Aircraft*. Flight Safety Foundation, Flight Safety Digest. Vol.18 (6-8), 1999, pp.1-136.

[50]Woods, D. D., Johannesen, L. J., Cook, R. I., Sarter, N. B., *Behind Human Error: Cognitive Systems, Computers and Hindsight*, Crew Systems Ergonomics Information Analysis Center, Ohio, 1994.

[51]Woods, D. D., "Decompositing Automation: Appearant Simplicity, Real Complexity," edited by R. Parasuraman and M. Mouloula, *Automation Technology and Human Performance*, Erlbaum, 1996, pp. 3-17

[52]Woods, D. D., Patterson, E. S. and Roth, E. M., "Can We Ever Escape from Data Overload? A Cognitive Systems Diagnosis," *Cognition, Technology & Work*, Vol. 4, 2002, pp. 22-36.

[53]Woods, D. D., Cook R. I., Billings, C. E., "The Impact of Technology on Physician Cognition and Performance," *Journal of Clinical Monitoring and Computing*, Vol. 11 (1), 1995, pp. 5-8.

[54]Yvette J. T., William H. R., and Richard W. P., "Pilot Opinions on High Level Flight Deck Automation Issues," NASA Contractor Report 4669, BBN Systems and Technologies, Massachusetts, 1995.

[55]Kirlik, A., "Modeling Strategic Behavior in Human-Automation Interaction: Why an Aid Can (and Should) Go Unused," *Human Factors*, Vol. 35, 1993, pp. 221-242.

[56]Kantowitz, B.H. and Casper, P.A., "Human workload in aviation," edited by E. L. Wiener and D. C. Nagel, *Human Factors in Aviation*, Academic Press, San Diego, 1988, pp. 157-188.

[57]Endsley, M. R., "Situation Awareness Global Assessment Technique (SAGAT)," Proceedings of the National Aerospace and Electronics Conference (NAECON), New York, 1988, pp. 789-795.

[58]Wickens, C. D., "Situation Awareness: Review of Mica Endsley's 1995 Articles on Situation Awareness Theory and Measurement," *The Journal of the Human Factors and Ergonomics Society*, Vol: 50, 2008, pp. 397- 403.

Vibration Suppression of Rotating Beams Through Piezoelectric Shunt Circuits

Preston Powell
Graduate Student
The University of Alabama
Tuscaloosa, AL, USA

Jinwei Shen
Associate Professor
The University of Alabama
Tuscaloosa, AL, USA

ABSTRACT

This paper analytically investigates the feasibility of passive vibration damping of a rotating beam-like structure, such as a helicopter rotor, through the use of piezoelectric materials. When piezoelectric materials are mounted to a structure undergoing bending stresses, such as a cantilever beam, there is an exchange of mechanical and electrical energy between the beam and the piezoelectric material. This electrical energy can be used to power small electronics such as onboard data transmitters or be dissipated through electrical shunt circuits. Electrical components in a shunt circuit (resistors, capacitors, and inductors) release energy from the system as heat. Energy dissipation corresponds to a vibration damping effect in the electromechanical system. Numerous configurations of electrical components and mechanical structures are explored. First, the Rayleigh-Ritz method of assumed modes is adopted for a rotating uniform single degree-of-freedom cantilever beam. Both in-plane (lag) and out-of-plane (flap) bending directions are considered. The beam model is modified to include piezoelectric elements and electrical shunt circuits. Two types of shunt circuits are considered: one with a single resistive element and one with a resistor and inductor in series. Various resistances are used in finding the frequency and impulse responses of the rotating beam with a shunt circuit. The change in damping potential between resistors is evaluated for each electromechanical system. The effects of the number of modes assumed when modeling the beam are also highlighted. The settling times of the impulse responses are used as the figure of merit to assess energy dissipation in the systems. Successful vibration damping of rotating cantilever beams is predicted through the piezoelectric shunt circuits.

NOTATION

C_p Piezoelectric Capacitance

F External force

I Current

K Stiffness matrix

L Inductance

M Mass matrix

q Piezoelectric charge

R Resistance, Beam length

T Kinetic energy

U Potential energy

υ Voltage

v Lag displacement

V Modal lag displacement

w Flap displacement

W Modal flap displacement

W_e Electrical work energy

δW Virtual work

Θ Electromechanical coupling matrix

Ψ Shape function

Ω Rotational speed

Subscript

b Relating to beam structure

flap Relating to flap bending

lag Relating to lag bending

p Relating to piezoelectric element

sh Shunt

v Relating to lag displacement

w Relating to flap displacement

Superscript

T Transpose

$(\dot{\ })$ First derivative with respect to time

$(\ddot{\ })$ Second derivative with respect to time

INTRODUCTION

Vibration reduction of mechanical structures is ideal in nearly every field of engineering. In aerospace applications, vibrations lead to noise, instability, inefficiency, and deterioration of important mechanical parts. Helicopter rotors, in particular, experience significant vibrations during flight. Due to the complexities of helicopter blade motion and design, external damping methods are difficult to implement on rotorcraft. Active methods such as that in Ref. 1, which use a power source and/or controller, have been shown to successfully dampen vibrations in non-rotating structures such as fixed wing aircraft and robotic appendages. Passive methods, which operate independently without an external battery, usually achieve less damping but are significantly simpler to integrate into dynamically complex systems, as shown in Ref. 2.

Piezoelectric materials have been researched for their potential use in both active and passive vibration suppression methods. These materials are unique, in that, they produce an electrical charge under the presence of mechanical stresses. Conversely, they generate mechanical stresses under applied electrical loads, as demonstrated in Ref 3. When mounted to a structure undergoing bending stresses, such as a cantilever beam with an impulse load at the tip, there is an exchange of mechanical and electrical energy between the beam and the piezoelectric material. Active methods operate by converting an electrical load input to an applied mechanical load such as a bending moment. This bending moment counteracts the ambient vibrations of the structure and a net damping effect is achieved. In many cases, active methods use multiple piezoelectric elements that act as either sensors or actuators. An effective control strategy is required to employ such a method.

Contrary to active methods, passive damping methods involving piezoelectric materials suppress vibrations by dissipating energy from the overall system. Ref. 4 explored the most common passive method which incorporates a piezoelectric element in an electrical circuit. This circuit, referred to as a shunting circuit, consists of various electrical elements (resistors, inductors, and capacitors) designed to remove energy from the system. This energy dissipation corresponds to a vibration damping effect in the electromechanical system. Although much research has been reported on piezoelectric shunt circuits for damping in vibrating structures, few studies have aimed to apply such a method to rapidly rotating members such as helicopter rotors. In addition to the passive nature and simple implementation of this technique, the piezoelectric material and shunt circuit have minimal mass and stiffness contributions to the mechanical structure.

These characteristics are ideal for rotor blades. Accordingly, this research aims to assess and confirm the potential damping capabilities of passive piezoelectric shunt circuits on a rotating beam. First, the Rayleigh-Ritz method of assumed modes is adopted for a rotating uniform single degree-of-freedom beam with cantilever boundary conditions. The beam model is modified to include piezoelectric elements and electrical shunt circuits. The figure of merit used to measure energy dissipation is the impulse response settling time. Changes in damping potential between various electrical components are highlighted to help choose which resistance and/or inductance to use in the shunt circuit. Ultimately, successful vibration damping of rotating beams is predicted through the piezoelectric shunt circuits.

APPROACH

As previously mentioned, a rotating beam with cantilever boundary conditions is be modeled using the Rayleigh-Ritz method of assumed modes. Although the beam displacement is restricted to a single degree-of-freedom, two types of bending are considered in this study: in-plane (lag) and out-of-plane (flap). The equations of motion for the two bending directions are slightly different due to the centrifugal force acting in the plane of displacement for lag bending. No inherent structural damping is included in the model so that the damping seen in the impulse responses is the result of the shunt circuit alone.

In addition to two bending directions, two different shunt circuit designs will be investigated, one with a single resistor

Presented at the 6th Asian/Austrailian Rotorcraft Forum / Heli Japan 2017, Kanazawa, Japan, November 7-9, 2017.

and one with both a resistor and inductor, each of which has its own advantages. A resistive shunt circuit (R-Shunt) dissipates energy for a wide range of frequencies while a resistive-inductive shunt circuit (RL-Shunt) is tuned to a target frequency to maximize energy dissipation at a particular mode of vibration. The mathematical beam/piezo model will now be shown.

Electromechanical Model

A rotating beam with a single piezoelectric element mounted near the root is modeled using energy principles. The electrical elements in the shunt circuit are incorporated into the model in the next section and are, therefore, not included in the following electromechanical model. Figure 1 and Figure 2 show the schematics of a beam displacing in both flap and lag directions. w is the flap displacement, v is the lag displacement, and Ω is the rotational speed.

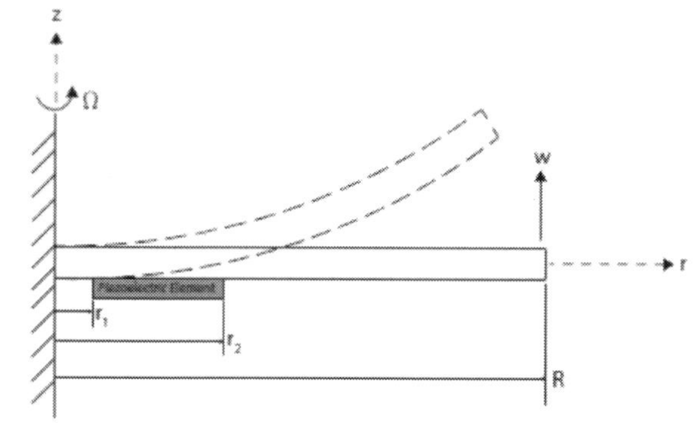

Figure 1. Schematic of a Rotating Beam/Piezo Undergoing Flap Bending

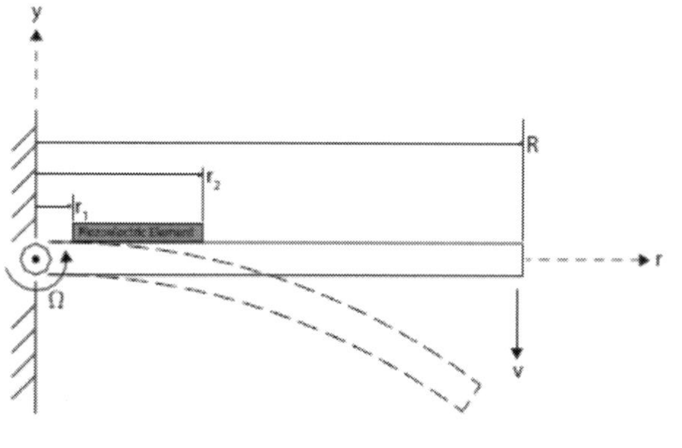

Figure 2. Schematic of a Rotating Beam/Piezo Undergoing Lag Bending

The method of assumed modes separates displacements into admissible spatial shape functions and temporal modal coordinates.

$$w(r,t) = \sum_{i=1}^{N} \psi_{w,i}(r) W_i(t) = \boldsymbol{\Psi}_w(r)\boldsymbol{W}(t) \tag{1}$$

$$v(r,t) = \sum_{i=1}^{N} \psi_{v,i}(r) V_i(t) = \boldsymbol{\Psi}_v(r)\boldsymbol{V}(t) \tag{2}$$

In Eqs. (1) and (2), $\boldsymbol{\Psi}_w$ and $\boldsymbol{\Psi}_v$ are the vectors of shape functions and $\boldsymbol{W}$ and $\boldsymbol{V}$ are the vectors of modal coordinates for flap and lag bending, respectively. This separation of variables allows the equations of motion to be derived using Hamilton's principle.

$$V.I. = \int_{t_1}^{t_2} [\delta(T - U + W_e) + \delta W] dt \tag{3}$$

Minimizing the variational indicator in Eq. (3) leads to two differential equations that describe the mechanical and electrical behavior of the system. In the equation above, T is the kinetic energy, U is the potential energy, W_e is the electrical work energy, and δW is the virtual work. The equations of motion for flap and lag take a similar form but differ in the formulation of their stiffness matrices. For flap bending, the final electromechanical equations of motion are derived as

$$\left(M_{b,flap} + M_{p,flap}\right)\ddot{\boldsymbol{W}} + \left(K_{b,flap} + K_{p,flap}\right)\boldsymbol{W} - \Theta_{flap} v(t) = F_z \tag{4}$$

$$\Theta_{flap}^T \boldsymbol{W} + C_p v(t) = q(t) \tag{5}$$

Similarly, the lag bending electromechanical equations of motion are found to be

$$\left(M_{b,lag} + M_{p,lag}\right)\ddot{\boldsymbol{V}} + \left(K_{b,lag} + K_{p,lag}\right)\boldsymbol{V} - \Theta_{lag} v(t) = F_x \tag{6}$$

$$\Theta_{lag}^T \boldsymbol{V} + C_p v(t) = q(t) \tag{7}$$

Eqs. (4)-(7) show the exchange of energy between the beam structure and piezoelectric element, where Θ is the electromechanical coupling matrix, C_p is the inherent piezoelectric capacitance, $v(t)$ is the voltage across the piezoelectric element, and $q(t)$ is the electric charge on the piezoelectric electrodes. In these equations, M, K, and Θ are all determined by the material and dimensional properties of the beam and piezoelectric element. Therefore, our state variables are the charge $q(t)$, the voltage $v(t)$, and the modal coordinates $\boldsymbol{W}$ and $\boldsymbol{V}$.

At this point, however, there is no energy dissipation in the system. A simulation of the above equations of motion will simply show the piezoelectric element acting as a transducer. The shunt circuit elements are now introduced to account for the energy released as heat through a resistor.

Shunt Circuit Equations

The two types of shunt circuit designs, R-Shunt and RL-Shunt, are relatively simple to include in the electromechanical equations of motion found in the previous section. Ohm's law is used to describe the drop in voltage across electrical elements and the resulting differential equations are substituted into Eqs. (4)-(7) to determine the final equations of motion. We will treat the two circuit designs separately.

1. R-Shunt

Figure 3 and Figure 4 shows the circuit designs for a piezoelectric element connected to an R-Shunt. The current I is equal to the first time derivative of the charge, q. As the current flows across the resistor, energy is released and a dampening effect is seen in the vibrating beam.

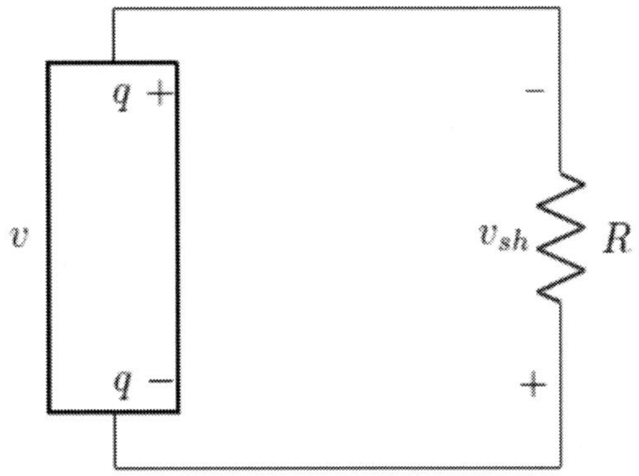

Figure 3. R-Shunt Circuit Design: Voltage Source Representation

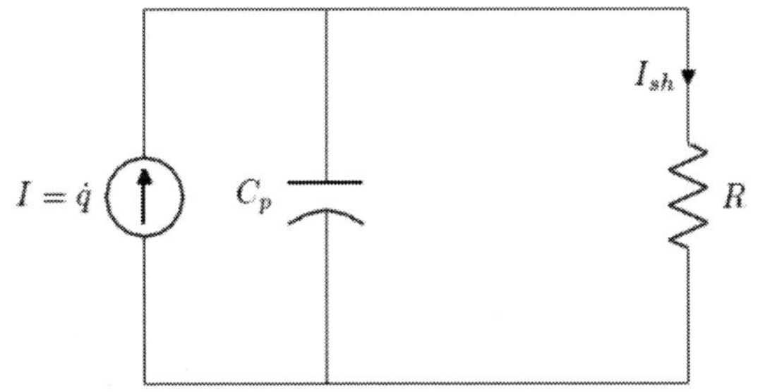

Figure 4. R-Shunt Circuit Design: Current Source Representation

According to Ohm's law, the change in voltage across the resistive element in the above circuits can be expressed by

$$v(t) = -RI(t) = -R\dot{q}(t) \tag{8}$$

Rearranging Eq. (5) and substituting into Eqs. (4) and (8), we find the final equations of motion for flap bending with an R-Shunt to be

$$(M_b + M_p)\ddot{\boldsymbol{W}} + (K_b + K_p)\boldsymbol{W} - \Theta\left[\frac{1}{C_p}q - \frac{\Theta^T}{C_p}\boldsymbol{W}\right] = F_z \tag{9}$$

$$\dot{q} = \frac{\Theta^T}{RC_p}\boldsymbol{W} - \frac{1}{RC_p}q \tag{10}$$

Similarly, rearranging Eq. (7) and substituting into Eqs. (6) and (8) gives the lag bending equations for an R-Shunt as

$$(M_b + M_p)\ddot{\boldsymbol{V}} + (K_b + K_p)\boldsymbol{V} - \Theta\left[\frac{1}{C_p}q - \frac{\Theta^T}{C_p}\boldsymbol{V}\right] = F_x \tag{11}$$

$$\dot{q} = \frac{\Theta^T}{RC_p}\boldsymbol{V} - \frac{1}{RC_p}q \tag{12}$$

1. RL-Shunt

Figure 5 and Figure 6 show the circuit designs for an RL-Shunt with a resistor and inductor in series. Again, the current flows across the resistor and energy is released as heat. However, the introduction of an inductor results in an electrical frequency. If the beam vibrates at a frequency equal to the shunt circuit electrical frequency, current through the electrical elements will be maximized. However, if the frequencies do not match, less current will flow through the circuit and a significant reduction in energy dissipation will be apparent.

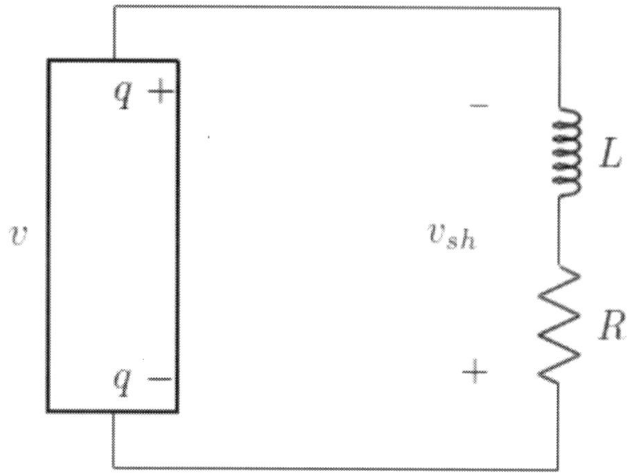

Figure 5. RL-Shunt Circuit Design: Voltage Source Representation

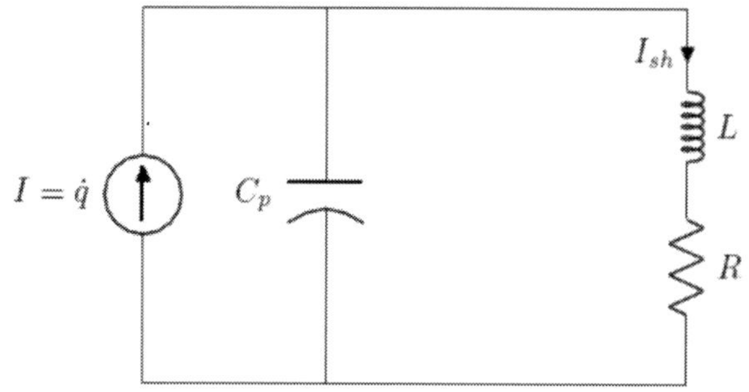

Figure 6. RL-Shunt Circuit Design: Current Source Representation

Ohm's law shows that the drop in voltage across the electrical elements in the circuit is

$$v(t) = -L\dot{I}(t) - RI(t) = -L\ddot{q}(t) - R\dot{q}(t) \tag{13}$$

Rearranging Eq. (5) and substituting into Eqs. (4) and (13), we find the final equations of motion for flap bending with an RL-Shunt to be

$$\begin{gathered}(M_b + M_p)\ddot{\boldsymbol{W}} + (K_b + K_p)\boldsymbol{W} - \Theta\left[\frac{1}{C_p}q - \frac{\Theta^T}{C_p}\boldsymbol{W}\right] \\ = F_z\end{gathered} \tag{14}$$

$$\ddot{q} = \frac{\Theta^T}{LC_p}\boldsymbol{W} - \frac{1}{LC_p}q - \frac{R}{L}\dot{q} \tag{15}$$

Similarly, rearranging Eq. (7) and substituting into Eqs. (6) and (13) gives the lag bending equations for an RL-Shunt as

$$\begin{gathered}(M_b + M_p)\ddot{\boldsymbol{V}} + (K_b + K_p)\boldsymbol{V} - \Theta\left[\frac{1}{C_p}q - \frac{\Theta^T}{C_p}\boldsymbol{V}\right] \\ = F_z\end{gathered} \tag{16}$$

$$\ddot{q} = \frac{\Theta^T}{LC_p}\boldsymbol{V} - \frac{1}{LC_p}q - \frac{R}{L}\dot{q} \tag{17}$$

We now have four sets of equations of motion that can be simulated for various resistances and inductances.

RESULTS

The impulse response of each of the four systems is simulated for flap and lag bending from a tip load of 1 *N*. Each system was modeled using both single and multiple mode assumptions. The resistances are chosen by investigating the transfer response of each system for various sized resistors. The figure of merit used to assess each simulation is settling time, t_s . The beam and piezoelectric material properties are shown in Table 1.

Table 1. Beam and Piezoelectric Material Properties

m_b	0.95	Beam Mass per Unit Length [*kg/m*]
R	2	Beam Length [*m*]
b_b	0.121	Beam Width [*m*]
t_b	0.0265	Beam Thickness [*m*]
c_b	1.33×10^9	Beam Modulus [*Pa*]
r_1	0.1	PZT Patch Starting Coordinate [*m*]
r_2	0.3	PZT Patch Ending Coordinate [*m*]
ρ_p	7800	PZT Density [*kg/m³*]
c_p	62×10^9	PZT Modulus [*Pa*]
b_p	0.121	PZT Width [*m*]
t_p	8×10^{-4}	PZT Thickness [*m*]
d_{31}	-320×10^{-12}	PZT Piezoelectric Constant [*m/V*]

g_{31}	-9.5×10^{-3}	PZT Piezoelectric Voltage Constant [Vm/N]
K_3^T	3800	PZT Dielectric Constant
C_p	2×10^{-7}	PZT Capacitance [F]

Single Mode

1. R-Shunt

The first shunt to be evaluated is a simple resistive network shunt, as shown in Figure 3. This shunt circuit system behaves most like a viscoelastically damped system. As with all of the following results, the first step in assessing the damping potential of the shunt system is deciding which resistance to use. This is done by investigating the frequency response of the system for various resistances. It is to be noted that these values are chosen rather arbitrarily and do not represent the "optimum" resistances for vibration damping.

Figure 7 and Figure 8 show the frequency responses for the single mode R-Shunt system undergoing flap and lag bending, respectively. Clearly, the flap bending system achieves the most damping through the 30000 Ω resistor while the lag bending system achieves the most through the 50000 Ω resistor.

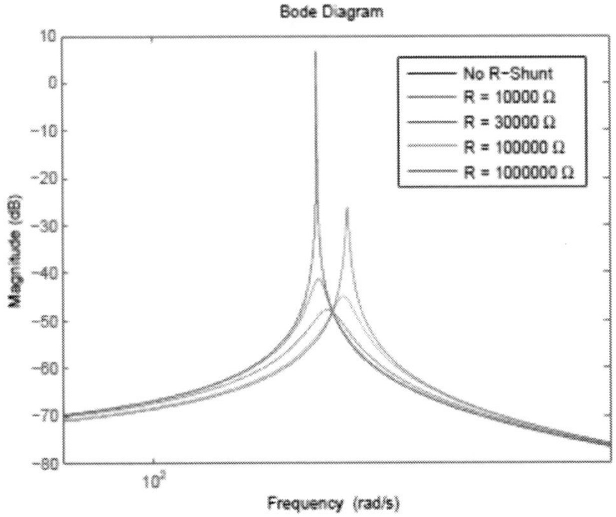

Figure 7. Flap Bending Frequency Response of R-Shunt System for Various Resistor Values

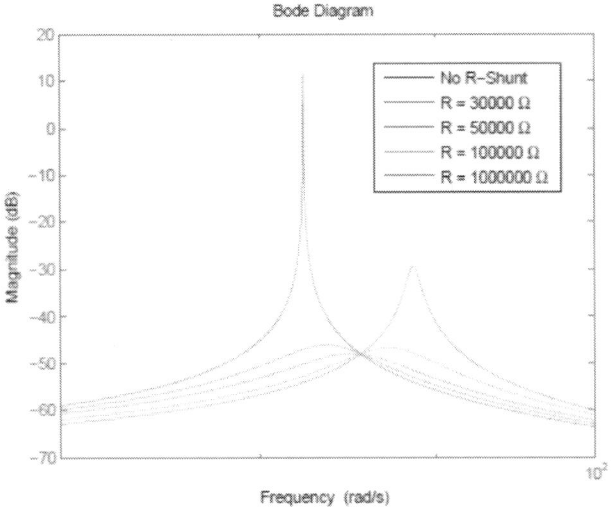

Figure 8. Lag Bending Frequency Response of R-Shunt System for Various Resistor Values

2. RL-Shunt

The resistive-inductive shunt shown in Figure 5 will now be evaluated for various resistances. Figure 9 and Figure 10 show the frequency responses for the single mode R-Shunt system undergoing flap and lag bending, respectively.

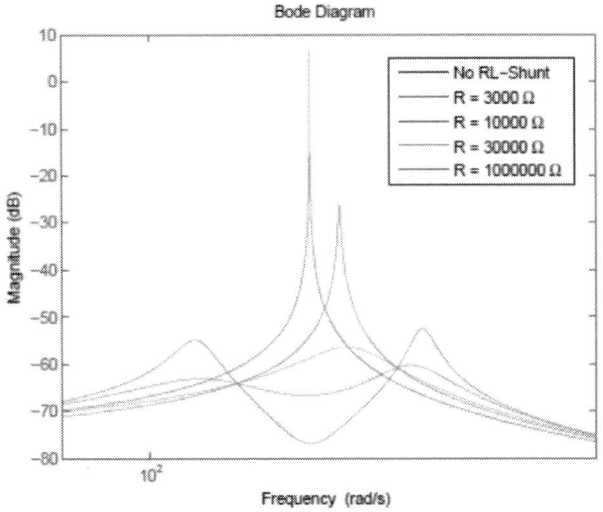

Figure 9. Flap Bending Frequency Response of RL-Shunt System for Various Resistor Values

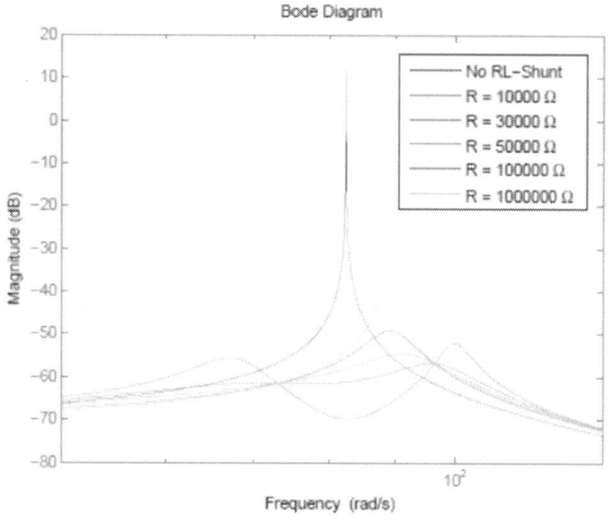

Figure 10. Lag Bending Frequency Response of RL-Shunt System for Various Resistor Values

According the figures above, the flap bending system should employ a 10000 Ω resistor and the lag bending system should use a 30000 Ω. However, for comparison, the impulse responses will be found using the resistances chosen for the R-Shunt systems shown previously. The settling times for the impulse response of both flap and lag bending of an R-Shunt and an RL-Shunt are shown in Table 2.

3. Settling Times

Table 2. Summary of Shunt Circuit Resistances and Settling Times for Single Mode Impulse Responses

Bending	Modes	Shunt Type	Resistor	Settling Time
Flap	1	R-Shunt	30000 Ω	1.84
Flap	1	RL-Shunt	30000 Ω	0.64
Lag	1	R-Shunt	50000 Ω	1.24
Lag	1	RL-Shunt	50000 Ω	0.57

For comparison purposes, the same resistances were used for the two shunt types for both flap and lag bending. It is clear that using the RL-shunt that is tuned to the natural frequency of the electromechanical system achieves a significant reduction in settling time. The flap bending settling time is reduced by 65% with the addition of an inductor while that of the lag bending is reduced by 54%.

A single mode approximation for the equations of motion of a rotating cantilever beam is helpful but undoubtedly less accurate than a model that accounts for more than one mode of vibration. Energy is clearly being dissipated through the shunt circuits, but the simplicity of a one-mode model may be neglecting some behavior that the beam will exhibit. Accordingly, the same responses are determined for the beam model including multiple modes.

Multiple Modes

1. R-Shunt Flap and Lag Bending

Once again, a simple resistive shunt circuit will be added to the rotating beam/piezo system. The advantage to this method, especially for multiple mode approximations, is that there will be energy dissipation no matter what vibration frequency the beam is exhibiting. However, choosing the optimal resistor becomes more difficult when multiple modes are present. The resistor that most dampens the first mode may not be as effective for the second or third. Accordingly, it is ideal to find the dominant mode and choose the resistor according to that mode's frequency response.

Figure 11 shows the flap bending frequency response of the multiple mode R-Shunt at each resonant frequency. It is apparent that the first mode optimal resistance is 30000 Ω while the second and third modes show the most damping with a 15000 Ω resistor. Therefore, the impulse responses for both resistances will be found.

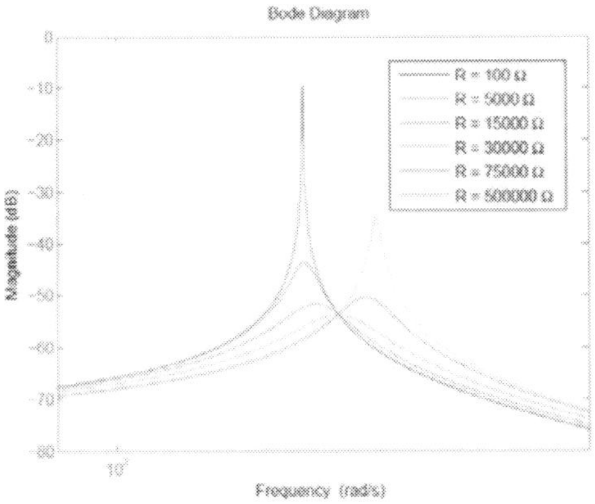

(a) First Mode

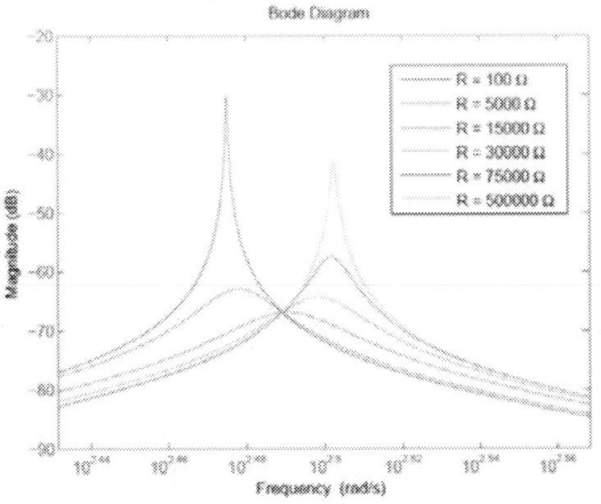

(b) Second Mode

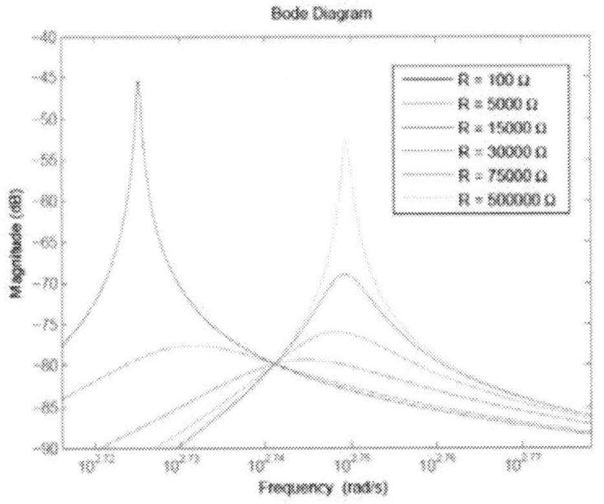

(c) Third Mode

Figure 11. Flap Bending Frequency Response at Resonant Frequencies of Multiple Mode R-Shunt for Various Resistor Values

The same process will be carried out for the lag bending system. Figure 12 shows the lag bending frequency response of the multiple mode R-Shunt at each resonant frequency. In the case of lag bending, each mode has a different optimum resistance. The resistors that provide the greatest reduction in vibration amplitude are 50000 Ω for the first mode, 10000 Ω for the second mode, and 5000 Ω for the third mode. Accordingly, all three resistances will be simulated.

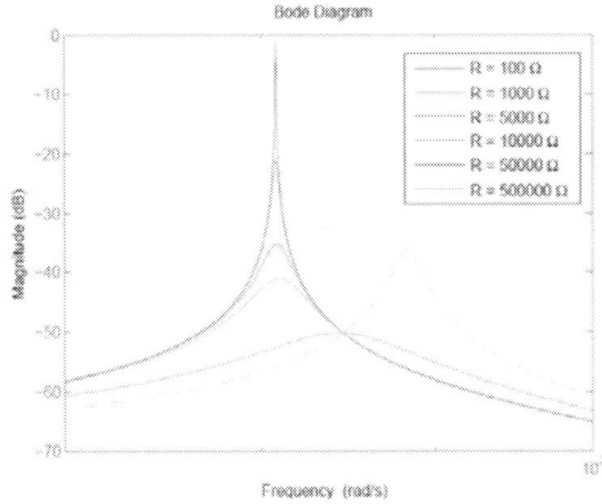

(a) First Mode

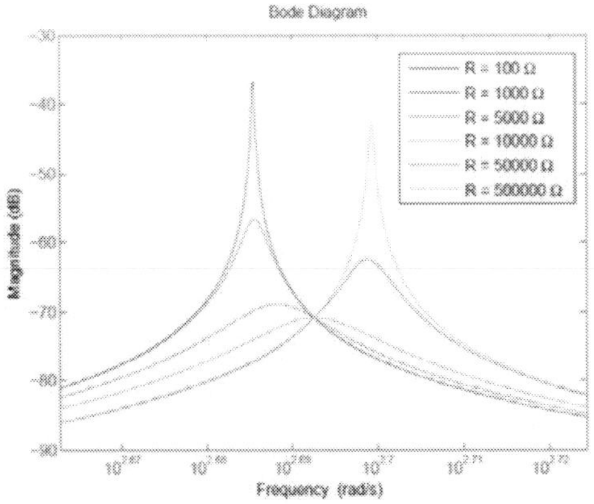

(b) Second Mode

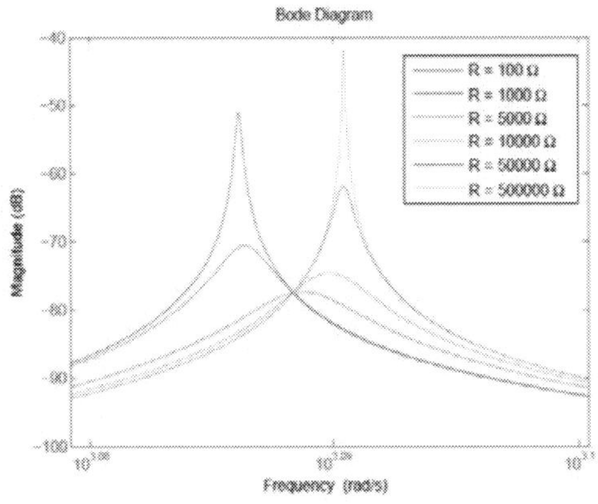

(c) Third Mode

Figure 12. Lag Bending Frequency Response at Resonant Frequencies of Multiple Mode R-Shunt for Various Resistor Values

Table 3. Summary of R-Shunt Circuit Resistances and Settling Times for Multiple Mode Impulse Responses

Bending	Modes	Shunt Type	Resistor	Settling Time
Flap	3	R-Shunt	30000 Ω	0.95
Flap	3	R-Shunt	15000 Ω	1.23
Lag	3	R-Shunt	50000 Ω	1.31
Lag	3	R-Shunt	10000 Ω	2.7
Lag	3	R-Shunt	5000 Ω	5.3

Table 3 summarizes the settling times for each impulse response and resistance. It is clear that, for both flap and lag motion, there is always going to be some amount of energy dissipation provided by a simple resistive shunt. Any amount of deflection will cause a strain on the piezoelectric element resulting in a charge on the electrodes of the PZT. This will always generate a current through the shunt circuit, and therefore, the resistor. We will now add an inductor to the shunt circuit to tune the electrical frequency to a target mode.

2. RL-Shunt Flap Bending

As previously discussed, it is advantageous to include an inductor in the shunt circuit if we want to target a specific mode or frequency of vibration for energy dissipation. For the single mode simulations, the RL-Shunt system provided significantly more damping than the purely resistive shunt circuit. This is because there were no other modes present to interfere with the mechanical vibration frequency. Because we assumed that the beam vibrated at or near a single frequency, the shunt circuit could be tuned such that it would allow the maximum amount of current to ow through it. If multiple modes are introduced to the simulation, we cannot assume that the shunt circuit will always remove energy from the system. This concept will be investigated for both flap and lag motion.

To begin, the first mode flap bending frequency will be targeted. Therefore, the frequency we will tune the RL-Shunt to is 117 rad/s. The inductance required to tune the circuit to that frequency is 365 H. Figure 13 shows the flap bending frequency responses of the RL-Shunt. The resistances that show the most vibration reduction for the first and second mode are 25000 Ω and 120000 Ω, respectively.

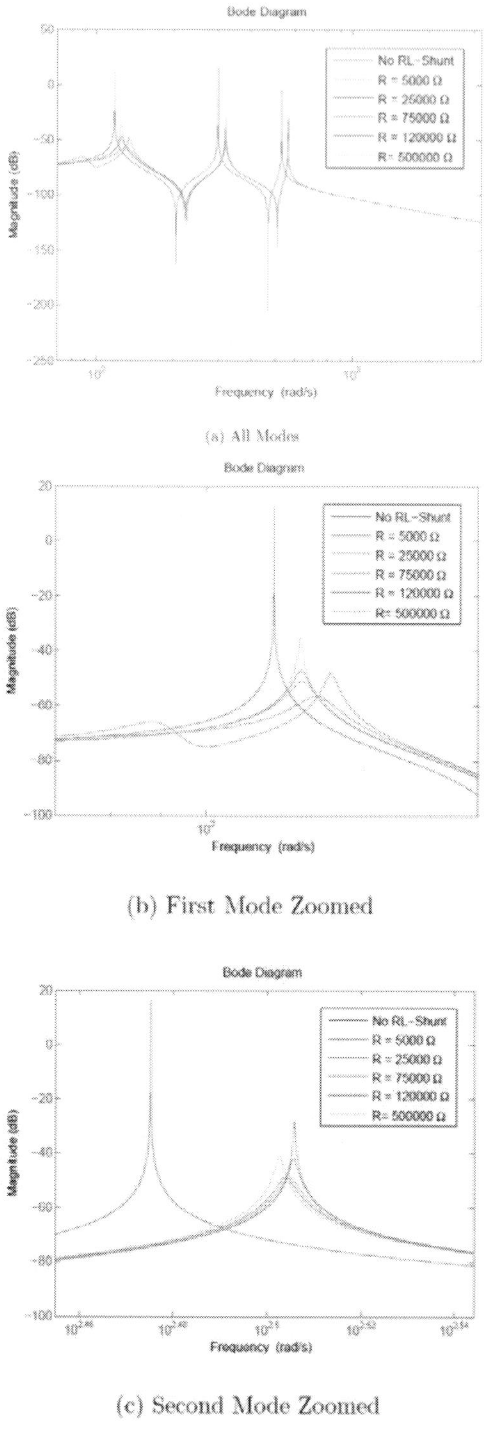

(a) All Modes

(b) First Mode Zoomed

(c) Second Mode Zoomed

Figure 13. Flap Bending Frequency Response of First Mode Tuned RL-Shunt System for Various Resistors

We will now tune the RL-Shunt to the second mode of vibration, which corresponds to a frequency of 399 rad/s. Therefore, the inductance required for this shunt system is 55.9 H. Figure 14 shows the lag bending frequency responses of the RL-Shunt. The resistors that provide the greatest reduction in vibration amplitude are 30000 Ω for the first mode and 10000 Ω for the second mode.

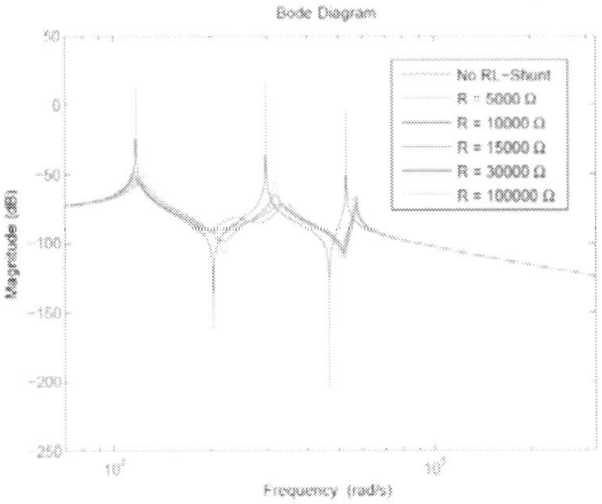

(a) All Modes

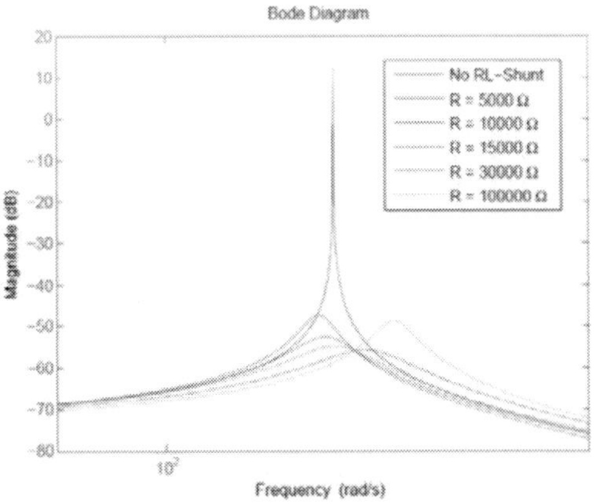

(b) First Mode Zoomed

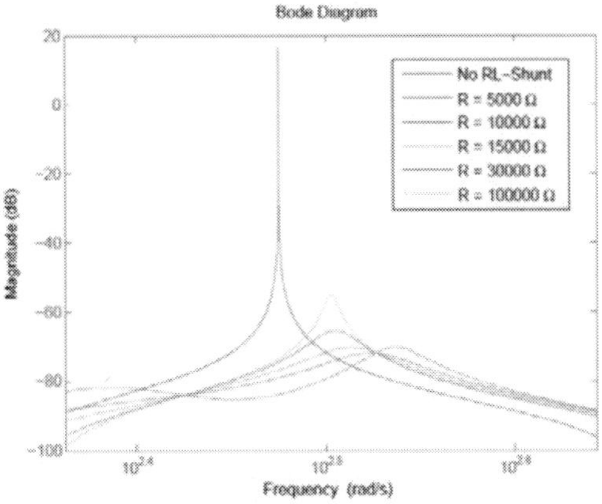

(c) Second Mode Zoomed

Figure 14. Flap Bending Frequency Response of Second Mode Tuned RL-Shunt System for Various Resistors

The impulse responses for the resistances and systems shown above are simulated. The results are shown in Table 4.

Table 4. Summary of Flap Bending RL-Shunt Circuit Resistances and Settling Times for Multiple Mode Impulse Responses

Bending	Modes	Shunt Type	Resistor	Settling Time
Flap	3	RL-Shunt (Tuned to 1st Mode)	25000 Ω	0.95
Flap	3	RL-Shunt (Tuned to 1st Mode)	120000 Ω	1.23
Flap	3	RL-Shunt (Tuned to 2nd Mode)	30000 Ω	1.31
Flap	3	RL-Shunt (Tuned to 2nd Mode)	10000 Ω	2.7

2. RL-Shunt Lag Bending

As we did for flap bending, we will start by targeting the lag bending first mode frequency of 80.8 rad/s. Therefore, an inductance of 766 H is required to tune the RL-Shunt. According to Figure 15, which shows the frequency responses of this system, the addition of the RL-Shunt has little damping effect on the second and third modes. A 35000 Ω resistor will achieve the most damping for the first mode. The variation in resistance does not make a significant difference in energy dissipation at the other modes.

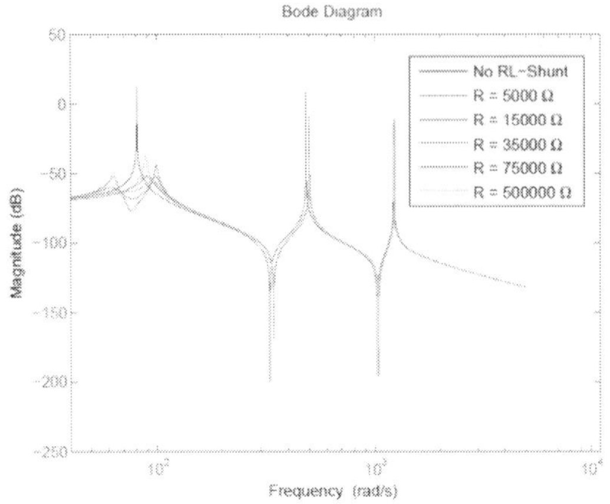

(a) All Modes

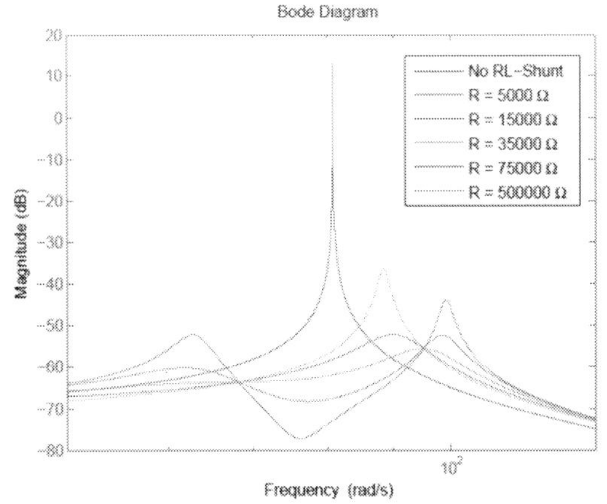

(b) First Mode Zoomed

Figure 15. Lag Bending Frequency Response of First Mode Tuned RL-Shunt System for Various Resistors

The impulse response of the system with a 35000 Ω resistor, shown in Figure 16, illustrates that energy is dissipated at the beginning of the simulation. However, as the frequency response showed, there is almost no damping of the beam for the other modes. The first mode vibrations are dampened out quickly but the vibration amplitude is near constant after 0.5 seconds. This is a major disadvantage of using an RL-Shunt for a system that vibrates at a wide range of frequencies.

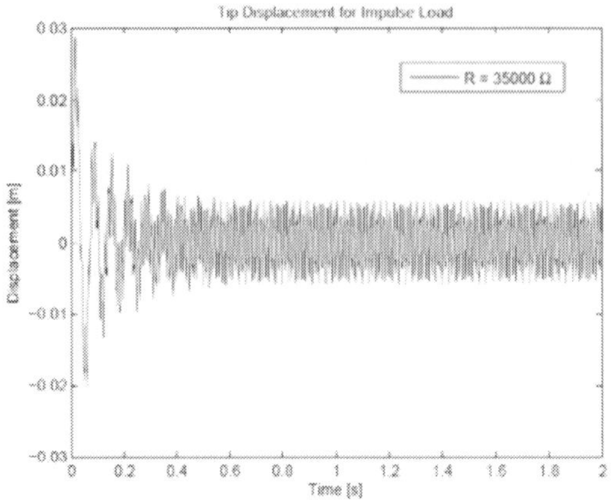

Figure 16. Lag Bending Impulse Response with RL-Shunt and 35000 Ω Resistor

The second mode frequency of lag bending is 485 rad/s. Accordingly, an inductance of 21.3 H is needed to tune the RL-Shunt circuit. The frequency response of the new system in Figure 17 shows that changes in resistor affect not only the second mode to which the RL-Shunt is tuned, but also the first mode. The resistors that provide the greatest reduction in vibration amplitude are 35000 Ω for the first mode and 5000 Ω for the second mode.

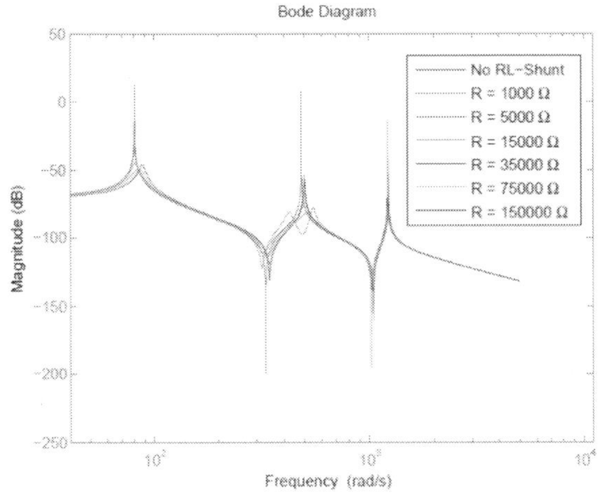

(a) All Modes

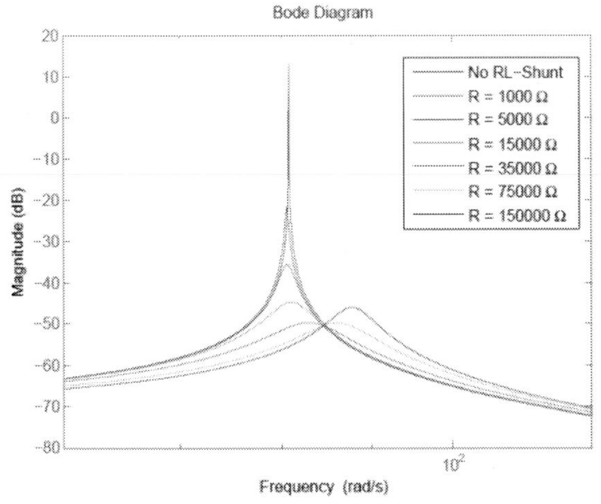

(b) First Mode Zoomed

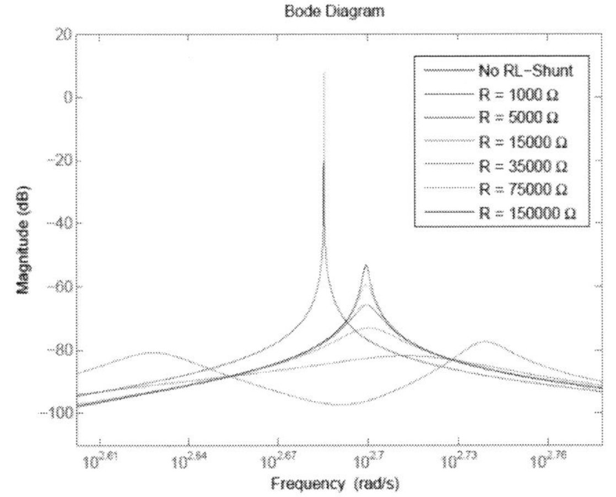

(c) Second Mode Zoomed

Figure 17. Lag Bending Frequency Response of Second Mode Tuned RL-Shunt System for Various Resistors

The impulse responses for the resistances and systems shown above are simulated. The results are shown in Table 5.

Table 5. Summary of Lag Bending RL-Shunt Circuit Resistances and Settling Times for Multiple Mode Impulse Responses

Bending	Modes	Shunt Type	Resistor	Settling Time
Lag	3	RL-Shunt (Tuned to 1st Mode)	35000 Ω	N/A
Lag	3	RL-Shunt (Tuned to 2nd Mode)	35000 Ω	1.27
Lag	3	RL-Shunt (Tuned to 2nd Mode)	5000 Ω	5.36

SUMMARY AND CONCLUSIONS

The previous section explored the energy dissipation capabilities of two different shunt circuits applied to a rotating cantilever beam undergoing flap or lag bending. Both single-mode and multi-mode approximations were included which resulted in a number of electromechanical systems. The results are summarized in Table 6.

Table 6. Summary of Shunt Circuit Resistances and Settling Times for Impulse Responses of Flap and Lag Bending

Bending	Modes	Shunt Type	Resistor	Settling Time
Flap	1	R-Shunt	30000 Ω	1.84
Flap	1	RL-Shunt	30000 Ω	0.64
Flap	3	R-Shunt	30000 Ω	0.95
Flap	3	RL-Shunt (1st Mode Tuning)	25000 Ω	13.0
Flap	3	RL-Shunt (1st Mode Tuning)	120000 Ω	3.43
Flap	3	RL-Shunt (2nd Mode Tuning)	10000 Ω	1.15
Flap	3	RL-Shunt (2nd Mode Tuning)	30000 Ω	0.77
Lag	1	R-Shunt	50000 Ω	1.24
Lag	1	RL-Shunt	50000 Ω	0.57
Lag	3	R-Shunt	50000 Ω	1.31
Lag	3	RL-Shunt (1st Mode Tuning)	35000 Ω	N/A
Lag	3	RL-Shunt (2nd Mode Tuning)	5000 Ω	5.36
Lag	3	RL-Shunt (2nd Mode Tuning)	35000 Ω	1.27

While the single mode simulations provided insight to the fundamental behavior of our piezo/beam system, the multiple mode approximations gave a more realistic prediction of the dynamics. The multiple mode simulations accounted for behavior that was neglected by the single mode approximations. Therefore, the general increase in settling time is found when more modes are introduced.

The most notable conclusion is that we could drastically reduce the settling time of the same exact beam/piezo system solely by adjusting the electrical components in the shunt circuit. We reduced the multiple mode settling time by 12.23 seconds and 4.09 seconds for flap and lag bending, respectively. It is also important to recognize that the inductances required to tune the RL-Shunts to the second mode are much smaller than those for the first mode. The 766 H inductor for the first lag mode is unrealistically large compared to the second mode inductance of 21.3 H. It would be impractical to install a 766 H inductor on a complex structure like a helicopter rotor.

A possible explanation for the reduced energy dissipation of the lag bending simulations involves the geometry of the beam. While the energy expressions differ between the two bending directions, the areas of the bending surfaces to which the piezoelectric element is attached differ greatly between flap and lag displacements. The beam parameters in these simulations are similar to those of an actual rotor blade. The surface areas on top and bottom of a rotor are much larger than those of the leading and trailing edges. This reduced area corresponds directly to a smaller electromechanical coupling matrix. Accordingly, less current is fed through the circuit and, therefore, the damping potential drops. In all cases, however, there is clearly energy dissipation in the system and the damping capabilities of a piezoelectric shunt circuit are confirmed.

The results show that energy dissipation of a rotating cantilever beam is possible through piezoelectric shunt circuits. Variations in electrical components are strongly related to the damping capabilities of the shunts. For both flap and lag bending, the second mode tuned RL-Shunts had smaller settling times than the R-Shunts. This is because the second mode is dominant during the impulse response. The various resistance values used in the simulations were chosen somewhat arbitrarily. Although we chose an "optimal" resistor for each system by looking at the frequency response, the true "best case" value for achieving the most energy dissipation is most likely near our optimal resistance but slightly higher or lower. The range of whole number resistances was chosen to simplify the comparison of different systems and to emphasize the effects on the frequency responses. Energy dissipation found in the impulse responses confirms that passive vibration damping of a rotating cantilever beam has been predicted through piezoelectric shunt circuits.

Author contact:

Preston Powell preston.c.powell@gmail.com
Jinwei Shen jinwei.shen@eng.ua.edu

REFERENCES

[1]Baz, A. and Ro, J., "Vibration control of plates with active constrained layer damping," *Smart Materials and Structures*, Vol. 5, 1996, pp. 272-280.

[2]Hagood, N.W. and von Flotow, A., "Damping of structural vibrations with piezoelectric materials and passive electrical networks," *Journal of Sound and Vibration*, Vol. 146, (2), 1991, pp. 243-268.

[3]Cady, W.G., *Piezoelectricity*, McGraw-Hill, New York, NY, 1946.

[4]Park, C.H., "Dynamics modelling of beams with shunted piezoelectric elements," *Journal of Sound and Vibration*, Vol. 268, (2), 2003, pp. 115-129.

Performance and Loads Analyses of Compound Helicopters Using a Rigid Coaxial Rotor

Jeong-In Go[1]
Researcher
Agency for Defense Development
Daejeon, Korea

Do-Hyung Kim[2]
Senior Researcher
Korea Aerospace Research Institute
Daejeon, Korea

Jae-Sang Park[3]
Assistant Professor
Chungnam National University
Daejeon, Korea

Sang-Hyun Chae[4]
Researcher
Korea Aerospace Research Institute
Daejeon, Korea

Seong-Yong Wie[5]
Researcher
Korea Aerospace Research Institute
Daejeon, Korea

ABSTRACT

The present study validates the performance, vibration, and load analyses of the XH-59A compound helicopter to establish the analysis techniques for lift-offset compound helicopters using a rigid coaxial rotor. For the performance analysis, this work uses two different analysis codes such as CAMRAD II (Comprehensive Analytical method of Rotorcraft Aerodynamics and Dynamics II) and NDARC (NASA Design and Analysis of Rotorcraft), and for the vibration and load analyses, CAMRAD II is used. Performance analyses using CAMRAD II and NDARC are compared well to each other, and they are also correlated nicely with the flight test results. The 3/rev hub moments and blade structural loads of the XH-59A helicopter with auxiliary propulsions are analyzed using CAMRAD II in forward flight and validated reasonably well to the flight test results. Since the present analysis results are in moderate or good agreement with the flight test data, the techniques for performance, vibration, and structural load analyses for lift-offset compound helicopters are established appropriately. Furthermore, the blade section lift forces are investigated for the XH-59A with auxiliary propulsions in forward flight to study the unique characteristics of the ABC™ (Advancing Blade Concept™) rotor.

INTRODUCTION

Conventional helicopters using a single main rotor and a single tail rotor have the unique capabilities such as vertical take-off/landing (VTOL) and hovering. However, their maximum flight speed is slower than the fixed-wing aircrafts. Therefore, compound helicopters that can flight in high-speed while maintaining VTOL and hovering capabilities have been developed recently to solve this problem of conventional helicopters.

Not only rotors but also wings or auxiliary propulsions are used for the compound helicopters. Particularly, the XH-59A (Figure 1), X2 technical demonstrators, and S-97 Raider developed by Sikorsky are the lift-offset compound helicopters using the ABC™ (Advancing Blade Concept™, [1]). The ABC™ uses a counter-rotating rigid coaxial rotor for high-speed forward flight of helicopters. The ABC™ is named since it produces lift significantly on the advancing blades (Figure 2, [2]). Alternately, the ABC™ is explained as the lift-offset (LOS), which is calculated by dividing the rolling moment of each rotor by its thrust. Compound helicopters using the ABC™ have the unique advantages as follows. Since the rolling moments for the upper and lower rotors are the same in magnitudes but are opposite directions, the performance loss of the rotor because of the trim for rolling moment can be reduced. In addition, the dynamic stall on the retreating side of the rotor can be avoid. Furthermore, the lift-to-drag ratio of the ABC™ helicopter can be improved, as compared to that of conventional helicopters. Thus, the rotor speed can be reduced; therefore, high-speed forward flight is possible [2]. However, a significant disadvantage of the lift-offset helicopter is a serious vibration during high-speed flight. In case of the XH-59A helicopter, severe 3/rev cockpit vibration during high-speed forward flight was observed in its flight test because of the use of the rigid coaxial rotor and the absence of a vibration control system [3]. Thus, both performance and vibration predictions are important when compound rotorcrafts using the ABC™ rotor are developed. However, few works using analysis codes have been conducted for the predictions of performance and vibration, in spite of the development of lift-offset compound helicopters such as the XH-59A, X2 technical demonstrators, and S-97 Raider.

Presented at the 6th Asian/Austrailian Rotorcraft Forum / Heli Japan 2017, Kanazawa, Japan, November 7-9, 2017.

(a) Pure helicopter configuration

(b) Compound helicopter configuration

Figure 1. XH-59A technical demonstrator.

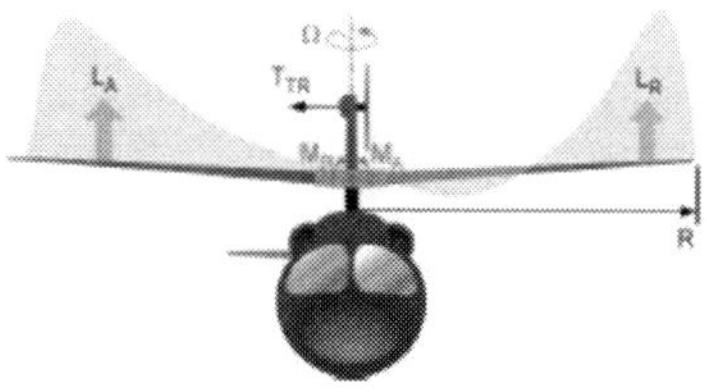

(a) Single main rotor

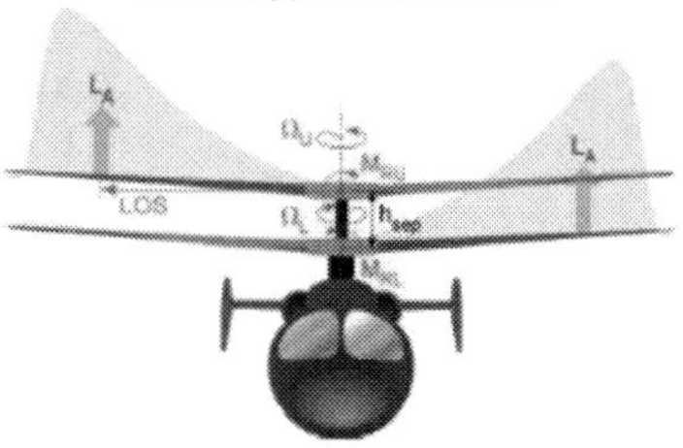

(b) Rigid coaxial rotor with ABC™

Figure 2. Lift and moment characteristics.

The performance analyses [4, 5] of an XH-59A lift-offset helicopter were conducted using NDARC (NASA Design and Analysis of Rotorcraft, [4]) and CAMRAD II (Comprehensive Analytical Method of Rotorcraft Aerodynamics and Dynamics II, [6]). In Ref. [5], the free wake model was used only for the performance analyses in the hover condition; however, the prescribed wake model, which is relatively simpler than the free wake model was used for the performance analyses in the forward flight condition. Moreover, the load and vibration of the XH-59A helicopter were not studied in that work [5]. Recently, RCAS (Rotorcraft Comprehensive Analysis System, [7]) was used in the work for performance, loads, and vibration of the XH-59A helicopter in high-speed forward flight [8]. But, most of results in this study [8] were obtained using a finite-state dynamic inflow model, which is simpler than the free wake model; therefore the validation for the 3/rev hub pitch moment was not good. Also, the analysis of the hover performance was not conducted.

The goal of this research is to validate the predictions of performance, vibration, and load of the ABC™ helicopter, XH-59A, in both the hover and forward flight conditions. As prediction tools, two different codes are used. For the performance analyses, CAMRAD II using a general free wake model and NDARC are used. Unlike the previous research [4] using NDARC, the performance analyses using NDARC in this paper are conducted for the aircraft model obtained from the conceptual design [9]. For the vibration and loads analyses, CAMRAD II is used. Most of the present prediction results are compared to flight test results as well as to each other. Through the present study, techniques for analyses of performance and vibration for the ABC™ helicopter are established appropriately. Finally, the blade section lifts of the XH-59A helicopter rotor are studied although these predictions are not correlated with the flight test results.

ANALYSIS MODEL

In this study, the XH-59A helicopter is considered as a model for the performance, vibration, and load analyses of the lift-offset compound helicopter. The XH-59A helicopter is an ABC™ technology demonstrator for high-speed flight while maintaining hovering and VTOL capabilities of conventional helicopters. The XH-59A helicopter was developed initially as a pure helicopter configuration (Figure 1(a), [2]) in 1964. After its successful flight tests in pure helicopter mode in 1973, and two auxiliary propulsions were added to the aircraft for transformation into a compound helicopter (Figure 1(b), [10]). The XH-59A in compound helicopter configuration reached a maximum level flight speed of 240 knots [3]. The general properties of XH-59A helicopter are summarized in Table 1 [3, 11, 12].

Table 1. Properties of XH-59A helicopter.

Hub type	Hingeless rotor
Radius, R [ft]	18
Number of rotors	2
Number of blades	3
Total solidity	0.127
Tip speed [ft/sec]	
Pure helicopter	650
Compound helicopter	450
Horizontal tail	
Area [ft^2]	60
Span [ft]	15.50
Vertical tail	
Area [ft^2]	30
Span [ft]	12
Fuselage	
Length [ft]	40.5
Rotor separation [ft]	2.5
Power plants	
Lift	PT6T-3 turboshaft engine
Thrust	J60-P-3A turbojet engine

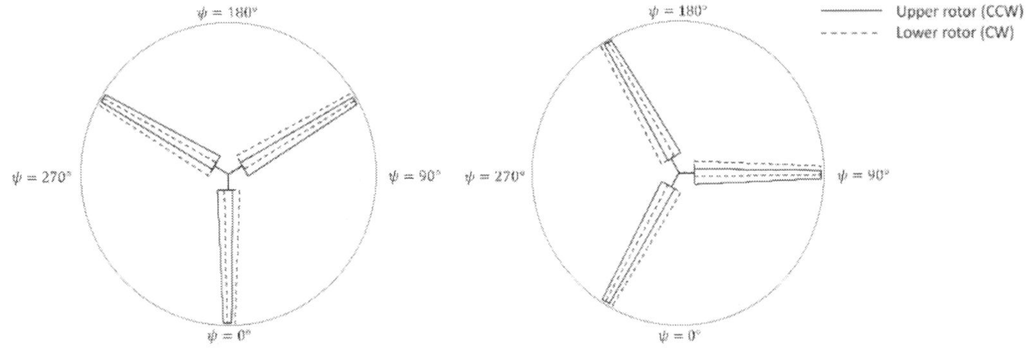

(a) Cross-over angle of 0° (b) Cross-over angle of 90°

Figure 3. Definition of cross-over angles for a coaxial rotor.

Although the XH-59 helicopter in compound configuration solved the low-speed problem of conventional helicopters, it had serious problems such as severe vibration and technical limitation associated with the reduction of rotor speed [2]. The XH-59A compound helicopter not only uses very stiff rigid blades but also has no vibration control system. As a result, the XH-59A helicopter experienced severe vibration problems especially in high-speed flight conditions. The cross-over angle is defined as the azimuth angle where the upper and lower blades of a coaxial rotor cross over each other, as given in Figure 3 [13]. It is interesting that the vibration characteristics of the XH-59A in compound helicopter configuration were definitely different for the different cross-over angles of 90° and 0° [11]. For an example, the 3/rev hub pitch moment is more dominant as compared to the 3/rev hub roll moment for a cross-over angle of 0°. However, the characteristics of the 3/rev hub loads are opposite for a cross-over angle of 90°. The 3/rev pitch or roll moment has a strong effect on the vibration of the XH-59A helicopter [11]. In addition, the 2/rev component of loads on the rotating blade is the main source for the 3/rev hub vibratory loads of the XH-59A helicopter [13, 14].

In the present work, the flight test data [5, 11, 15] are used for the correlation between the present analyses and the measured data.

COMPUTATIONAL METHODS

NDARC

For the performance analyses of the XH-59A helicopter, this study uses the NDARC, which is a rotorcraft design and analysis tool, developed by NASA. NDARC can conduct rapidly the conceptual design and performance analysis for various rotorcrafts as well as conventional helicopters. However, it is difficult to establish the modeling and analysis techniques using NDARC because a number of input values are required for design and analysis using NDARC [16]. The tasks and functions of the NDARC are briefly represented in Figure 4 [4].

In this study using NDARC, performance analyses of the XH-59A in pure and compound helicopter configurations are conducted. The XH-59A aircraft model for analyses is obtained from the previous work [9]. In addition, the lift-offset value is assumed as 0.25, which is the recommended value in the reference [5].

In the NDARC, the rotor power of each of the upper and lower rotors is calculated as the sum of the induced power (P_i), profile power (P_o), and parasite power (P_p), as shown in Eq. (1). In addition, various parameters such as induced power factor, and mean drag coefficient are used to calculate the performance of the rotor. These performance parameters in this study are obtained from Ref. [4] and modified appropriately. The lift-to-drag ratio of aircraft (L/D), and effective lift-to-drag ratio of the rotor (L/D_e) can be calculated from the Eqs. (2) and (3), respectively. Rotor power of the coaxial rotor (P_{coaxial}) is the sum of the upper and lower rotor powers.

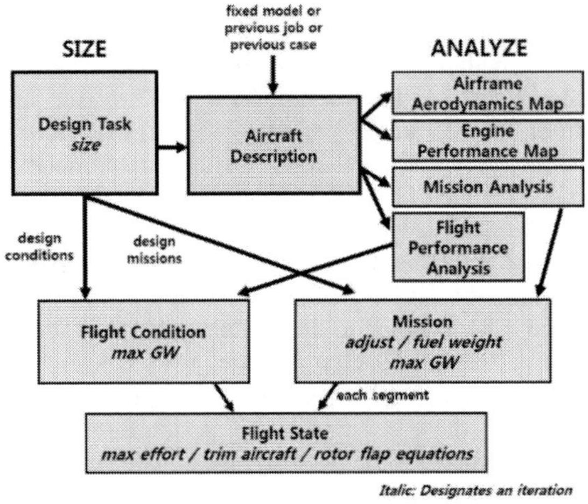

Figure 4. Outline of NDARC tasks.

$$P = P_i + P_o + P_p \tag{1}$$

$$\frac{L}{D} = \frac{WV}{P_{\text{coaxial}}} \tag{2}$$

$$\frac{L}{D_e} = \frac{L}{(\frac{P_{\text{coaxial}}}{V} + X)} \tag{3}$$

CAMRAD II

CAMRAD II [6] is a comprehensive analysis code for performance, loads, and aeroelasticity stability of rotorcrafts. CAMRAD II has been used widely for the comprehensive analyses of rotorcrafts because it can be used for not only conventional helicopters but also rotorcrafts with various configurations. CAMRAD II has nonlinear finite elements, multibody dynamics, and rotor unsteady aerodynamics based on the lifting-line theory along with various inflow models and sophisticated wake models [17]. However, it is not easy to construct an analysis model using CAMRAD II since many input and empirical parameters are required.

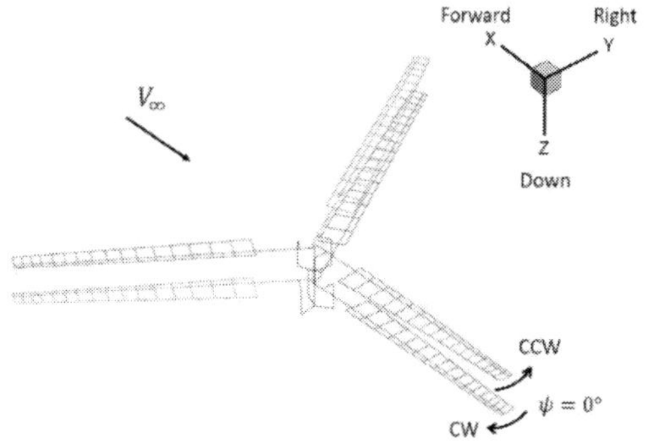

Figure 5. CAMRAD II model of XH-59A rotor.

This study conducts the performance, vibration, and load analyses using CAMRAD II for the XH-59A helicopter. To compare with the present authors' performance analysis results [9] using NDARC, the value of lift-offset (LOS) is assumed as 0.25 in the present CAMRAD II analysis. As represented in Figure 5, the XH-59A helicopter model with a cross-over angle of 0° is used for the present analyses.

Since the C81 tables for the actual airfoils of the XH-59A rotor blade are not available in public domain, this study uses the C81 tables for the airfoils similar to the actual airfoils of the XH-59A rotor blade, as shown in Figure 6. This method was used successfully in the previous works [5, 8]. However, the C81 airfoil tables used in this study cannot accurately represent the aerodynamics characteristics of the XH-59A rotor blade, for an example, Reynolds number. Therefore, the drag coefficients are adjusted in order to obtain better performance analysis results, similar to the method used in Ref. [5].

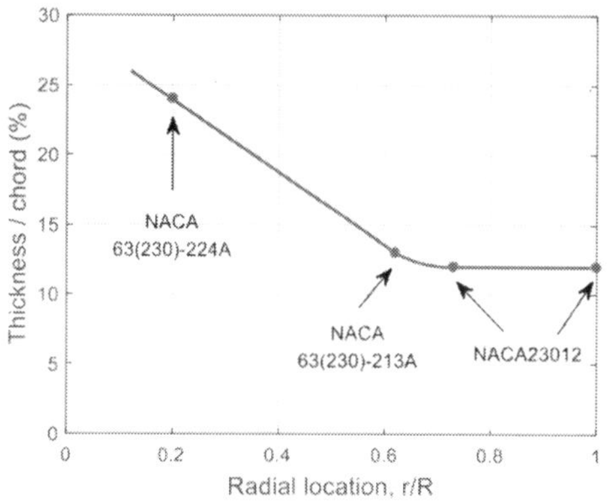

(a) Actual airfoils

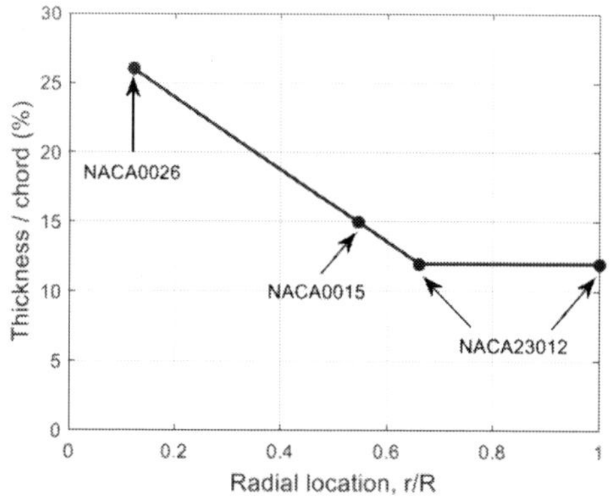

(b) Present airfoil model

Figure 6. Blade thickness and airfoil distribution.

Each blade is modeled as 16 aerodynamic panels, and the general free wake model is used for forward flight condition as well as for hover condition. The initial size of the vortex core of the free wake model is modeled as 50% of the chord length at the tip [5]. In addition, each blade is modeled as 7 nonlinear finite beam elements. The cross-sectional properties of the XH-59A rotor blade in Ref. [18] are used appropriately for the present blade modeling. In addition, the pitch hinge of the XH-59A rotor is modeled to be located at 5%R, and the rotor control system such as pitch links, pitch horns, and swashplates is modeled sophisticatedly.

The aerodynamic coefficients of the XH-59A fuselage are obtained from Ref. [15]. The trim analysis at the forward flight condition is performed such that the 6 components of loads acting on the aircraft are zero. The mean collective pitch angle, and lateral and longitudinal cyclic pitch angles of each rotor are used as trim variables for lift-offset helicopter. The control phase angle to control the lift-offset rotor is represented as the combination of the lateral and longitudinal cyclic pitches [5]. Alternatively, the trim analysis can be conducted using 6 rotor controls of the upper and lower rotors when the pitch angle of the aircraft is fixed as 0°, which gives the best performance for the lift-offset helicopter [11]. In this approach, trim targets are considered as the vertical force, the torque offset of the upper and lower rotors, and the rolling and pitching moments of the upper and lower rotors. The lift-offset can be considered as the differential rolling moment.

To validate the structural dynamics of the present CAMRAD II model, a fan plot analysis for an XH-59A rotor blade is performed [19] and the results are shown in Figure 7. The predicted non-rotating frequencies by CAMRAD II are correlated well with the measured values [15]. In addition, the present frequencies in lower modes are also compared well with the previous predictions [11, 15]. However, for the first torsional frequency (T1) at nonrotating condition, the present result is better than the previous analyses [11, 15], as compared to the measured data. Therefore, it is considered that the present structural dynamics model using CAMRAD II for XH-59A is successfully validated.

In the performance analysis using CAMRAD II, the rotor power and lift-to-drag ratio are calculated using Eqs. (1) to (3), as given previously. For the vibration analysis of the XH-59A helicopter, the 3/rev hub moment (M_{3P}) is calculated using Eq. (4). In Eq. (4), superscripts, *upper* and *lower*, denote upper and lower rotors, respectively, subscripts, *c* and *s*, mean the cosine and sine components of the hub loads, respectively [8].

$$M_{3P} = \sqrt{[M_{3Pc}^{upper} + M_{3Pc}^{lower}]^2 + [M_{3Ps}^{upper} + M_{3Ps}^{lower}]^2} \qquad (4)$$

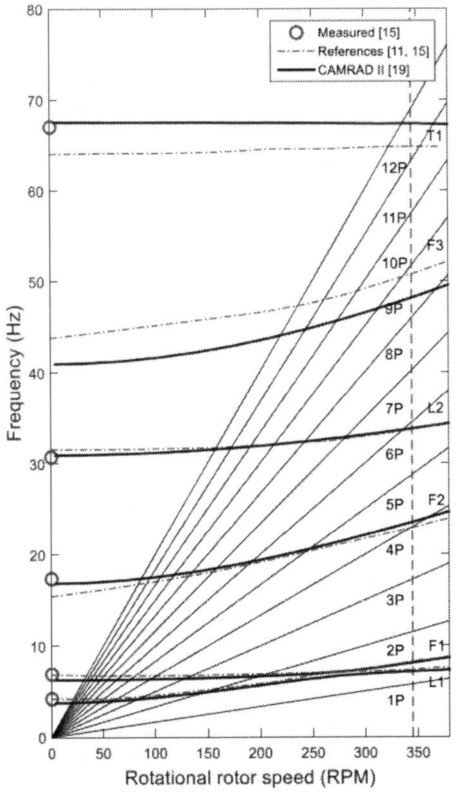

Figure 7. Validation of the fan plot analysis.

RESULTS AND DISCUSSIONS

Performance analyses

In this section, the performance are studied for the XH-59A helicopter with and without auxiliary propulsions in hover and forward flight conditions. For the correlation, the analysis results using CAMRAD II and NDARC are compared to each other as well as flight test data [5]. Some of prediction results using CAMRAD II and NDARC in this section are obtained from the present authors' recent works [9, 19] for the comparison and validation.

Figure 8 shows the correlation results of the figure of merit for the XH-59A without the auxiliary propulsion in hover. As seen in the figure, two prediction results by CAMRAD II and NDARC are quite similar to each other, although they use different analytical models and analysis techniques. They are also compared well with the flight test data. In addition, the figure of merit of the XH-59A helicopter (between 0.75 and 0.8) is better than that of conventional helicopter (between 0.6 and 0.75, [20]) since the wake of the upper rotor is reduced before it reaches the lower rotor in case of a coaxial rotor system [5].

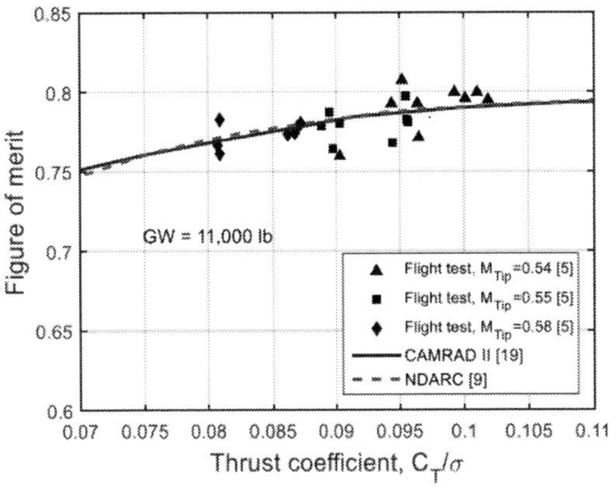

Figure 8. Figure of merit in hover.

Figures 9 and 10 show the results of performance analyses using CAMRAD II and NDARC for the XH-59A without auxiliary propulsions in forward flight. As given in the figures, the rotor power and the aircraft lift-to-drag ratio both are predicted well, as compared to the flight test data. In addition, CAMRAD II and NDARC predictions are quite similar to each other; although CAMRAD II over-predicts slightly the aircraft lift-to-drag ratio than the NDARC result at 80-160 knots in Figure 10 because the rotor power using CAMRAD II is estimated to be lower slightly than the result by NDARC, as shown in Figure 9.

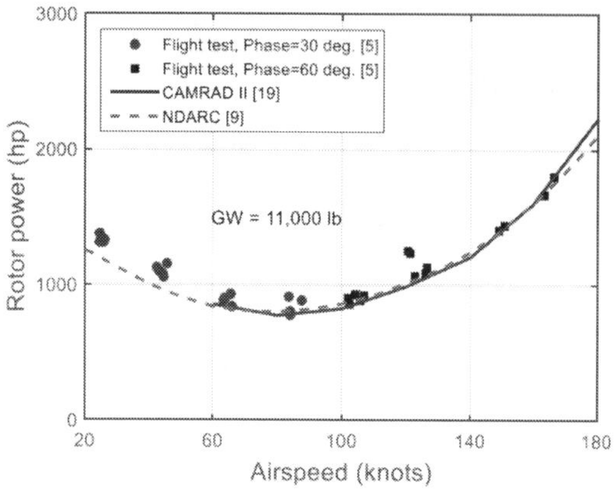

Figure 9. Validation of rotor power in pure helicopter mode.

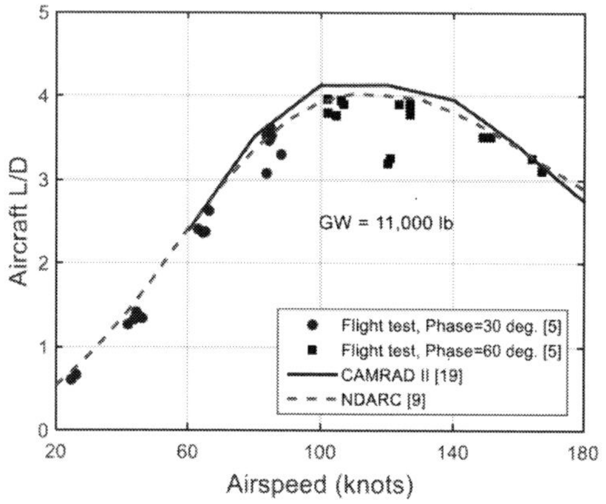

Figure 10. Validation of aircraft lift-to-drag ratio in pure helicopter mode.

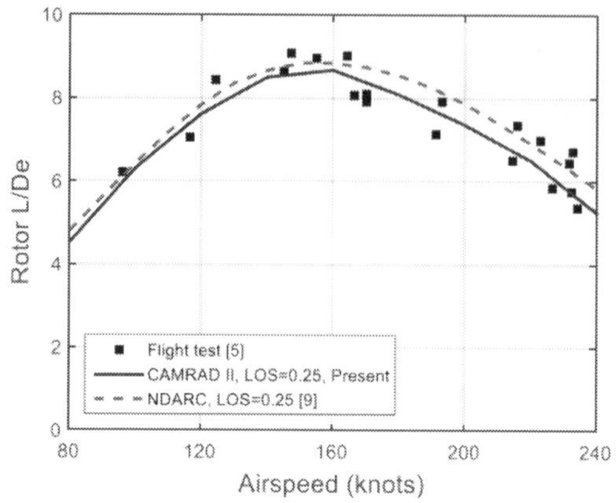

Figure 11. Validation of the rotor effective lift-to-drag ratio in compound helicopter mode.

The effective lift-to-drag ratios of the rotor for XH-59A using auxiliary propulsions are validated in Figure 11. Note that the flight test results include all the test data for the gross weight ranging from 11,000 to 13,000 lb, but CAMRAD II and NDARC results in this figure are for 13,000 and 11,000 lb, respectively. In addition, although the lift-offset value in the flight test varies with the flight speed, the lift-offset value is fixed as 0.25 for both CAMRAD II and NDARC analyses. As shown in Figure 11, the analysis results using CAMRAD II and NDARC are all within the upper and lower bounds of the flight test results. When the flight speed is above 160 knots, the NDARC result is slightly overpredicted, as compared to the prediction by CAMRAD II. Again, two predictions are quite similar to each other, although different analysis models and techniques are used.

Vibration and load analyses

In this section, the 3/rev hub vibrations and blade loads of the XH-59A compound helicopter in forward flight are analyzed using CAMRAD II and the results are validated to the flight test results. For the CAMRAD II analyses given in this section, the compound helicopter configuration using auxiliary propulsions are considered and the lift-offset value is fixed as 0.25.

Figure 12 validates the 3/rev hub vibratory loads in forward flight. The CAMRAD II analysis and flight test [11] results both show that the 3/rev hub pitch moment is much higher than the 3/rev hub roll moment for the cross-over angle of 0°. In addition, the 3/rev hub moments increase as the flight speed increases in both the analysis and flight test. Although there are only two flight test data that can be compared, the present CAMRAD II predictions are correlated nicely with the measured data [11]. Therefore, the analysis method of the hub vibratory loads for the XH-59A compound helicopter is appropriately established.

Since the 2/rev load components of the rotating blade have a strong effect on the 3/rev hub vibratory loads for the XH-59A helicopter, the 2/rev flap bending moments of the upper and lower blades at 10%R are investigated in Figure 13. The present analysis results are not validated because there is no flight test data for these. As shown in Figure 13, the 2/rev flap bending moments of upper and lower rotors increase with the increase of the flight speed, which is a similar trend to the relation between the 3/rev hub loads and flight speed, as given previously in Figure 12.

Figures 14 shows the 1/2 peak-to-peak values of the flap bending moments of upper and lower blades at 10%R for the XH-59A helicopter in compound mode since the flap bending moment is dominant at the blade root. For the lower rotor blade, the present CAMRAD II shows a reasonable prediction as compared with the flight test, although the down-up behavior at 200-220 knots is not clearly predicted in this analysis. For the upper rotor blade, the CAMRAD II result is over-predicted, as compared to the flight test data. In addition, the measured 1/2 peak-to-peak value for the upper rotor blade is lower than that for the lower rotor blade

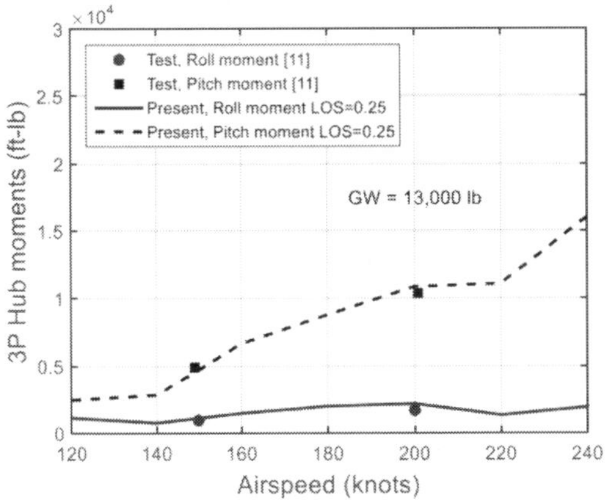

Figure 12. Validation of 3/rev hub moments.

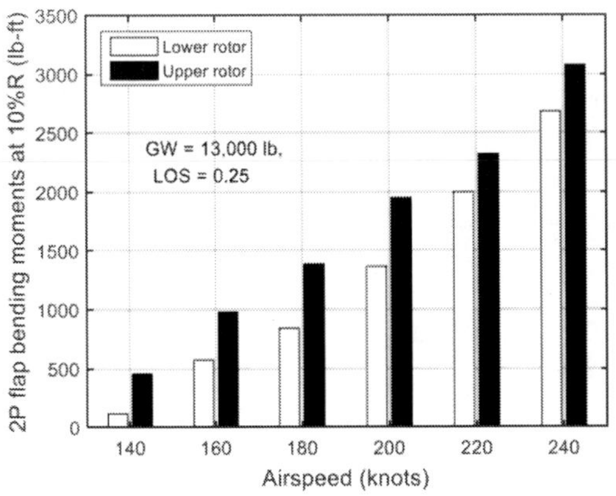

Figure 13. 2/rev flap bending moments at 10%R.

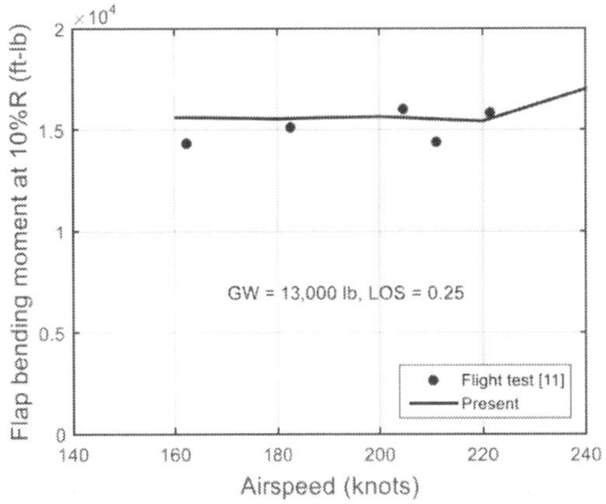

(a) Lower rotor

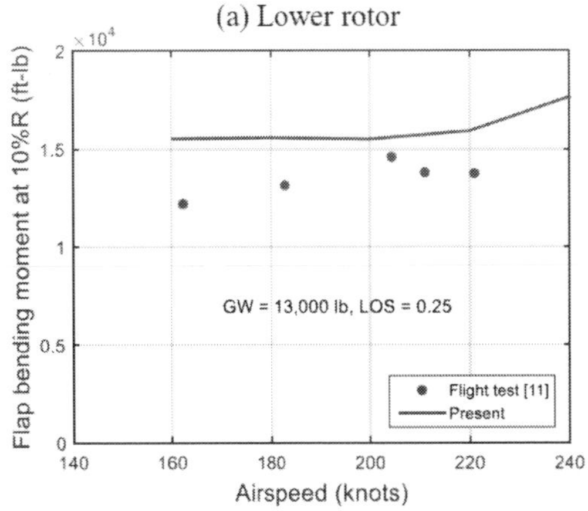

(b) Upper rotor

Figure 14. Validation of blade structural loads at 10%R (1/2 peak-to-peak values).

in the flight test; however, the present predictions for upper and lower rotor blades are quite similar to each other. The reasons for this discrepancy between the analysis and test data are as follows. First, the trim condition used in this CAMRAD II analysis is different from the trim condition for the flight test [11]. Second, the lift-offset value is fixed as 0.25 in the present analysis while the lift-offset value in the flight test changes as the flight speed increases. Third, it is not easy to calculate the blade structural loads accurately using the rotorcraft comprehensive analysis.

Figure 15 shows the lift distributions for the XH-59A helicopter in compound mode at 140 and 240 knots. The flight speeds of 140 and 240 knots are for the best effective lift-to-drag ratio and maximum forward flight speed conditions, respectively. Because the flight test for the XH-59A helicopter did not measure the rotor airloads, the correlation between the measured data and the analysis results is not possible for the rotor airloads. As given in the figures, most lift is produced on the advancing side of each of the upper and lower rotors, which is an inherent characteristic of the ABC™ rotor. In addition, the lift distributions of the upper and lower rotors are almost similar to each other. Particularly, the negative tip loadings on the advancing side are observed clearly at 240 knots than the results at 140 knots.

Figures 16 and 17 predict the blade section lift forces at 140 and 240 knots, respectively. In the present predictions, three blade stations (23.5, 50.0, and 98.5% R) of the upper and lower rotor blades are considered. It should be noted that the azimuth angles in these figures are defined based on the rotation direction of an upper rotor (counter-clockwise direction). As shown in these figures, the amplitude and variation of the section lift forces of the upper and lower rotor blades are similar to each other at three blade locations at both 140 and 240 knots. The lift fluctuations are predicted at 98.5%R of both upper and lower rotor blades at 140 knots, while the fluctuations of section lift forces are definitely observed at 50.0 and 98.5%R of upper and lower rotor blades at 240 knots. Particularly, the amplitude of fluctuations of the lower rotor blade is larger than that of the upper rotor blade because the wake of the upper rotor may affect the lower rotor. In addition, the section lift forces at 98.5%R at 240 knots in Figure 17 clearly exhibits negative loading behaviors in the first quadrants, however this is not shown in the prediction at 140 knots in Figure 16.

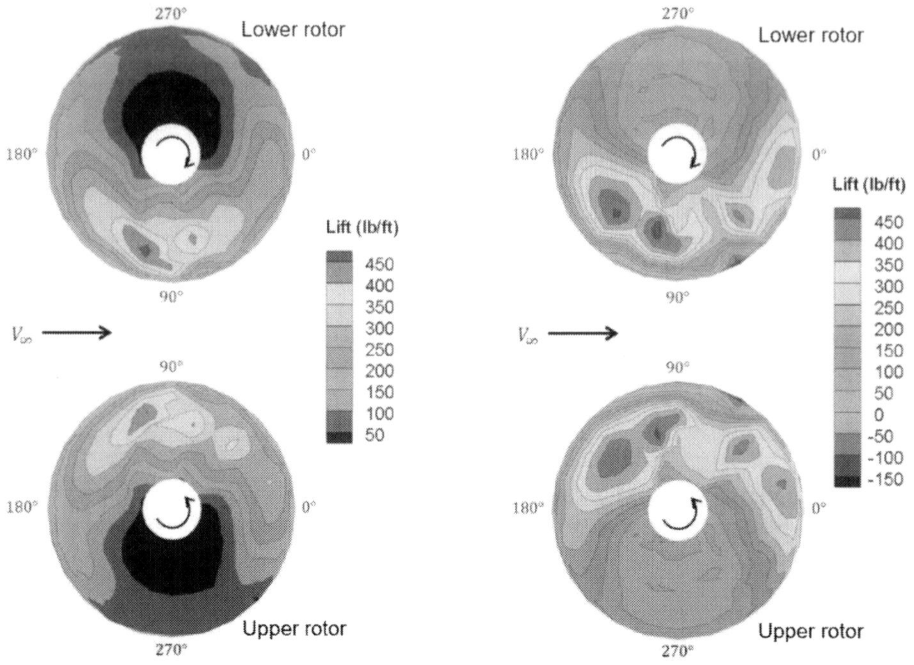

(a) 140 knots (b) 240 knots

Figure 15. Lift distributions in compound helicopter mode (LOS=0.25).

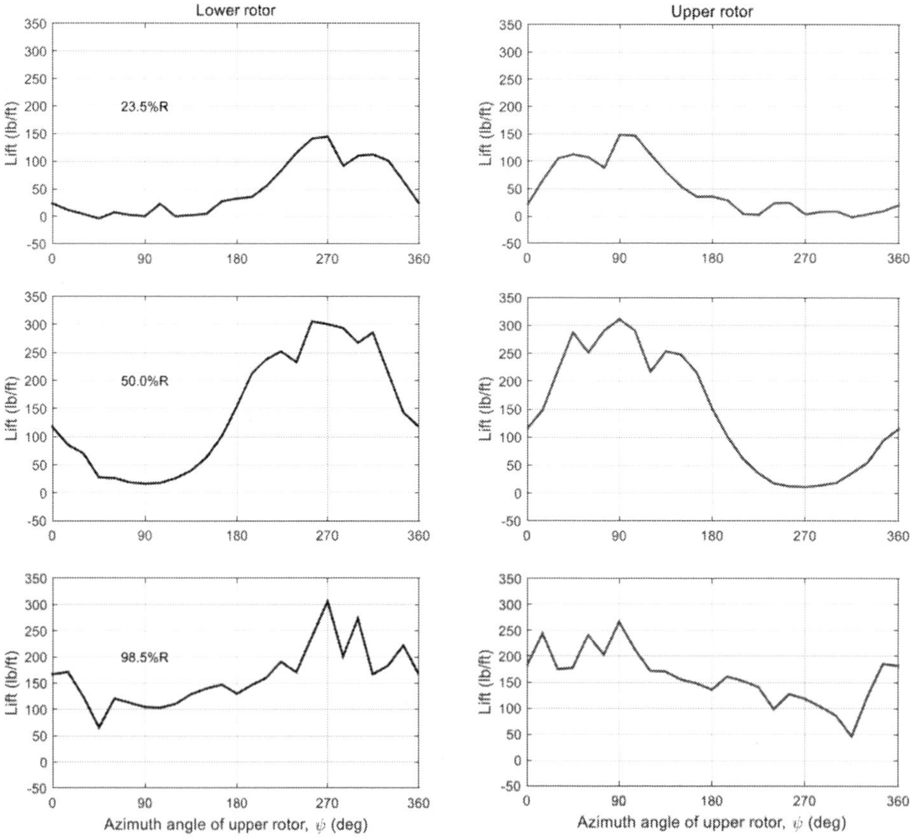

Figure 16. Blade section lifts at 140 knots in compound helicopter mode (LOS=0.25).

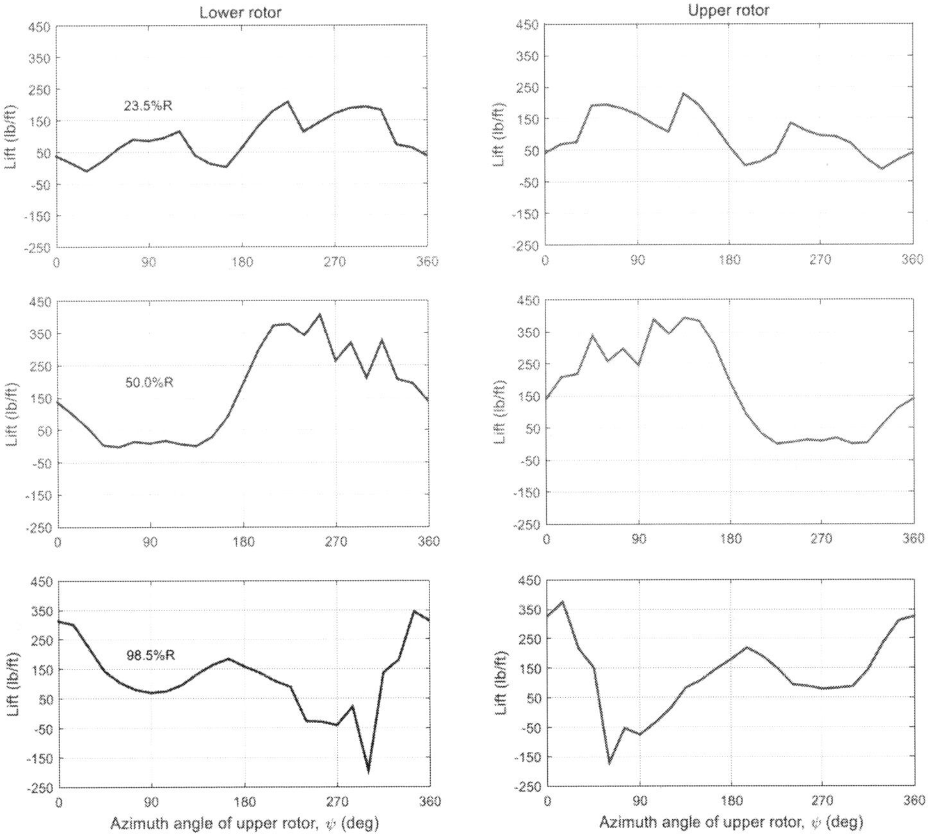

Figure 17. Blade section lifts at 240 knots in compound helicopter mode (LOS=0.25).

CONCLUSIONS

This work conducted the validation of the performance, vibration, and load analyses of the XH-59A compound helicopter using a rigid coaxial rotor. For the performance analyses in various flight conditions, two analysis codes, CAMRAD II and NDARC, were used. The prediction results by CAMRAD II and NDARC were quite similar to each other, and also they were correlated well with the flight test data.

For the hub vibratory loads and the blade loads in forward flight, CAMRAD II analyses using the lift-offset value of 0.25 was conducted for the XH-59A in compound helicopter mode. The predicted 3/rev hub loads were compared excellently with the flight test data. But, the validation of the 1/2 peak-to-peak values for the flap bending moments was moderate. In addition, the lifts of the XH-59A rotor in compound helicopter mode in forward flight were calculated in order to investigate a unique characteristic of the ABC™ rotor, although they were not validated to the flight test results.

Through the present study, the analysis techniques of the performance, vibration, and load for the ABC™ helicopter using a rigid coaxial rotor were established appropriately.

Author contact: Jeong-In Go[1] jigo@add.re.kr
Do-Hyung Kim[2] dhkim@kari.re.kr
Jae-Sang Park[3] aerotor@cnu.ac.kr
Sang-Hyun Chae[4] nyugnas@kari.re.kr
Seong-Yong Wie[5] wie@kari.re.kr

ACKNOWLEDGMENTS

This research was supported by Basic Science Research Program through the National Research Foundation of Korea (NRF) funded by the Ministry of Science, ICT and Future Planning (NRF-2016R1C1B1007199). This work was supported by research fund of Korea Aerospace Research Institute. This work was conducted at High-Speed Compound Unmanned Rotorcraft (HCUR) research laboratory with the support of Agency for Defense Development (ADD).

REFERENCES

[1]Ruddell, A. J., "Advancing Blade Concept (ABC™) Development," Proceedings of the American Helicopter Society International 32nd Annual Forum, 1976.

[2]Bagai, A., "Aerodynamic Design of the X2 Technology Demonstrator™ Main Rotor Blades," Proceedings of the American Helicopter Society International 64th Annual Forum, 2008.

[3]Blackwell, R., and Millott, T., "Dynamics Design Characteristics of the Sikorsky X2 Technology™ Demonstrator Aircraft," Proceedings of the American Helicopter Society 64th Annual Forum, 2008.

[4]Johnson, W., "NDARC – NASA Design and Analysis of Rotorcraft Validation and Demonstration," Proceedings of the American Helicopter Society Aeromechanics Specialists' Conference, 2010.

[5]Johnson, W., "Influence of Lift Offset on Rotorcraft Performance," Proceedings of the American Helicopter Society 64th Annual Forum, 2008.

[6]Johnson, W., CAMRAD II: Comprehensive Analytical Method of Rotorcraft Aerodynamics and Dynamics, Johnson Aeronautics, Palo alto, Calif, USA, 2012.

[7]Saberi, H., and Khoshlahjeh, M., "Overview of RCAS and Application to Advanced Rotorcraft Pproblems," Proceedings of the American Helicopter Society 4th Decennial Specialist's Conference on Aeromechanics, 2004.

[8]Jacobellis, G., Gandhi, F., and Floros, M., "A Physics-Based Approach to Trim Optimization of Coaxial Helicopters in High-Speed Flight," Proceedings of the American Helicopter Society International 71st Annual Forum, 2015.

[9]Go, J. -I., Park, J.-S., and Choi, J. S., "Validation on Conceptual Design and Performance Analyses for Compound Rotorcrafts Considering Lift-offset," International Journal of Aeronautical and Space Sciences, Vol. 18, No. 1, 2017, pp. 154-164.

[10]Coleman, C. P., "A Survey of Theoretical and Experimental Coaxial Rotor Aerodynamic Research," NASA Technical Paper TP-3675, 1997.

[11]Ruddell, A. J., "Advancing Blade Concept (ABC) technology demonstrator," USAAVRADCOM-TR-81-D-5, 1981.

[12]Bauchspies, J. S., Bryant Jr., W. R., and Simpson, W. E., "Trade Off Analysis of Technology Needs for Public Service Helicopters," NASA Contractor Report 3927 ORI TR-2459, 1958.

[13]Jenney, D. S., "ABC™ aircraft development status," Proceedings of the European Rotorcraft and Powered Lift Aircraft Forum, 1980.

[14]O'Leary, J., and Miao, W., "Design of Higher Harmonic Control for the ABC™," Proceedings of the American Helicopter Society Mideast Region National Specialists Meeting on Rotor System Design, 1980.

[15]Felker III, F. F., "Performance and Loads Data from a Wind Tunnel Test of a Full-Scale, Coaxial, Hingeless Rotor Helicopter," NASA-TM-81329, 1981.

[16]Kim, S. B., and Choi, J. S., "A Study on Establishment of the Helicopter Initial Design Model using the Modified Weight Estimation Equations," Journal of the Korean Society for Aeronautical and Space Sciences, Vol. 43, No. 3, 2015, pp. 213-223.

[17]Johnson, W., Moodie, A. M., and Yeo, H., "Design and Performance of Lift-Offset Rotorcraft for Short-Haul Missions," Proceeding of the American Helicopter Society Future Vertical Lift Aircraft Design Conference, 2012.

[18]Pleasants, W. A., "A Rotor Technology Assessment of the Advancing Blade Concept," AVRADCCM Technical Report 82-A-18, 1983.

[19]Go, J. I., Kim, D. H., and Park, J.-S., "Performance and Vibration Analyses of Lift-Offset Helicopters," International Journal of Aerospace Engineering, Vol. 2017, Article ID 1865751, 2017.

[20]Simon, N., The Foundations Helicopter Flight, Butterworth-Heinemann, 1st edition, 1994.

Reduction of Helicopter Rotor Vibratory Loads Through Robust and Hybrid Optimization

M. Emre Bilen
Graduate Student

Ender Cigeroglu
Assoc. Prof. Dr.

H. Nevzat Özgüven
Prof. Dr.

Department of Mechanical Engineering
Middle East Technical University
Ankara, Turkey

ABSTRACT

Vibration during helicopter operation causes many problems like flight crew and passenger discomfort, reduced fatigue life of helicopter components and even instabilities leading to catastrophic failure. This work focuses on the reduction of vibratory loads of helicopter rotors through optimization of rotor geometry and blade cross-section. For optimization, Particle Swarm Optimization and Nelder-Mead Optimization algorithms have been used in conjunction. The study is divided into four parts. In the first part, the robust and hybrid optimization algorithm with parallel processing capability has been explained and comparisons have been made with other search algorithms using a number of benchmark problems. On the second part, the rotor models to be optimized are presented. The rotors with conventional hub-to-blade and inter-blade damper designs are modeled in multibody dynamics code Dymore. In the third part, preliminary analytic analyses are conducted for a better understanding of the rotor design parameters on vibratory loads and using this information, the targets and the constraints are determined. Finally, the optimization results are presented. Comparisons between the initial and the final states of the problem in terms of flight dynamic loads and vibration levels have been given.

NOTATION

B	Particle Best
c	Chord Length, (m)
c_1, c_2	Global Best and Particle Best Trust Coefficient
CL	Centrifugal Load, (N)
f	Objective Function
F	Force, (N)
EA	Axial Stiffness, (N)
EI_{yy}	Flapwise Bending Stiffness, (Nm^2)
EI_{zz}	Chordwise Bending Stiffness, (Nm^2)
g	Constraint Function
G	Global Best
GJ	Torsional Stiffness, (Nm^2)
k	Iteration Number
m	Mass, (kg)
M	Moment, (Nm)
m_0	Blade Sectional Mass-per-Length, (kg/m)
m_{xx}	Blade Sectional Polar Inertia, (kgm)
n	Design Space Dimension
N_s	Swarm Size
p	Penalty or Target Function
r, R	Rotor Station and Rotor Radius, (mm)
r_1, r_2	Random Numbers Between 0 and 1
v	Particle Velocity
x	Design Point
α	Skin Lay-up Angle, (°)
γ	Lock Number, Contraction Coefficient
ϵ	Strain, (mm/mm)
ζ	Blade Lag Motion
η	Non-dimensional Rotor Station
ρ	Reflection Coefficient
χ	Expansion Coefficient
ω	Natural Frequency, (rad/s)
Ω	Rotor Rotation Speed, (rpm)
Δt	Time Step
$\Delta\theta$	Blade Torsional Deformation, (°)

Presented at the 6th Asian/Australian Rotorcraft Forum / Heli Japan 2017, Kanazawa, Japan, November 7-9, 2017.

INTRODUCTION

Helicopters are notorious for their high vibration levels. For years, engineers have tried to solve this problem either by reducing vibration at its source, that is, mainly the rotors or by isolating/absorbing the vibration. However, the latter choice generally targets the vibration in the specific locations on the airframe and leaves other locations susceptible to vibration; hence, reducing the life and strength of components exposed to high vibration.

In this study, the problem of reducing vibration created by the main rotor blades is addressed. During a steady flight, the rotor motion and loadings are periodic, so the blades are exposed to excitation at integer multiples of rotor rotation speed. In the case where one of the blade natural frequencies and excitation frequency coalesce, resonance occurs and the vibrations of the blades are greatly amplified. This leads to many adverse effects which include, but not limited to, reduced life and strength of the blades requiring heavier blades to compensate, increased vibration transferred to the airframe, passenger and pilot discomfort and even stability issues.

In earlier studies to minimize the vibration levels at the source, Friedmann constrained the blade natural frequencies to be in preassigned upper and lower bounds. Flapwise, chordwise and blade mass at 11 spanwise sections were selected as the design variables. It was possible to reduce 4 Ω harmonic of hub shear loading by 77 % through targeting the blade mass and 4 Ω harmonic of hub shear of a 4-bladed rotor (Ref. 1). Similar to Friedmann, Adelman bounded the blade natural frequencies in a predefined range. An upper limit for blade response amplitude is imposed and vertical vibratory shear load transmitted from the rotor to airframe was selected as the objective function. It was possible to reduce the vibration by around 70 % (Ref. 2).

As the technology and the production methods are advanced, the approach for this problem has evolved as well. Lee and Hajelat applied Genetic Algorithm to minimize the weighted sum of the rotor hub shear force and bending moments for a hingeless rotor (Ref. 3). As constraints, helicopter performance, blade mass, and composite structure failure criteria were considered. It was pointed that the design space for rotor optimization is non-convex, i.e., the objective function contains many local minima and non-gradient optimization methods are very suitable for such problems. Due to the non-convex nature of this problem, the selection of optimization method also plays a great role in finding the best possible solution. The gradient-based search algorithms lack the exploration aspect and many times stuck in the nearest local minimum.

The popularity of non-gradient optimization algorithms has increased over time. The popular methods are, but not limited to, Genetic Algorithms, Simulated Annealing, Particle Swarm Optimization and Nelder-Mead optimization.

Reduction of vibratory loads is investigated through two different objective functions for two different fully articulated rotor types. One of the objective functions tackles vibration through frequency separation and the other targets minimization maximum oscillatory loads along the blade span, separately. However, each objective function shares same constraints during optimization like torsional deformation of the blade etc. The studied rotors are characterized by the attachment type of the lag damper; one with inter-blade attachment, other with hub-to-blade attachment. A total of four cases are investigated and optimized.

HYBRID OPTIMIZATION

Consider the optimization problem in the following form,

$$\underset{x}{\text{minimize}} \quad f(x)$$

$$\text{subject to} \quad g(x) \leq 0$$

$$l_b \leq x \leq u_b$$

where $f: \mathbb{R}^n \to \mathbb{R}$, $g(x) = (g_1(x), \dots, g_m(x))$. Here f can be discontinuous, non-differentiable, non-convex meaning with many local minima in the design space, and $g_i(x)$ are the constraints applied during optimization. Here l_b and/or u_b are lower and upper bounds and can be infinite to represent unbounded design variables. The purpose of optimization algorithm is to find the global minima subjected to given constraints and design space limits.

In this study, to construct by an efficient and effective optimization algorithm, Particle Swarm Optimization and Nelder-Mead Optimization methods are selected and combined. The advantage of both methods is that they do not require any gradient information; hence, lowering the computation effort. Particle Swarm Optimization provides a wide design space exploration, whereas, Nelder-Mead Optimization provides its convergence properties to the hybrid algorithm. The details of each algorithm and their combination are discussed in the following sections.

Particle Swarm Optimization

Particle Swarm Optimization (PSO) algorithm, developed by Kennedy et al., is based on the interpretations of the movement behaviors of swarms (Ref. 4). The synchronous behavior of large organism groups like a sudden change of direction, scattering and regrouping for the optimum outcome is the underlying principle of PSO. Similar to swarms of organisms, PSO tries to obtain the optimum outcome of a problem exploring the design space and converging to the best solution.

At each iteration, the function values corresponding to each particle is calculated. Next, the position of each particle is adjusted according to three influence parameters; global best, particles best, and its inertia. Global best, G^k, is defined as the best objective value obtained by any particle during optimization up to iteration k; whereas, particle best, B_i^k, is the position corresponding to the best objective value obtained by a specific particle up to iteration k. The inertia of a particle can be described as the effect of its velocity from the last iteration. The effect of inertia, the global best and the particle best are multiplied by inertia weight, w, the global best trust coefficient, c_1, and the particle best trust coefficient, c_2, respectively. To add some randomness to the algorithm, the trust coefficients are then multiplied by random numbers both ranging between 0 to 1, respectively. The velocity of each particle is calculated as follows

$$v_i^k = wv_i^{k+1} + c_1 r_1 G^k + c_2 r_2 B_i^k. \quad (1)$$

Once the velocity is calculated, the positions of each particle are updated by adding velocities to current positions. The positions are calculated with the following equation

$$x_i^{k+1} = x_i^k + v_i^k \Delta t. \quad (2)$$

The algorithm of PSO used in this study is given in Algorithm 1.

Algorithm 1 Particle Swarm Optimization

1. Calculate swarm size with $N_s = 25n$ if $n \geq 3$, else $N_s = 50$. Distribute particles with random positions and velocities to design space. Set $k = 1$.
2. Start iteration k. Calculate the function value for each particle.
3. Determine particle best B_i^k by comparing current particle function value with its previous function values. Set the position of the best function value as the particle best. Repeat for each particle.
4. Compare the best of each particle and select the position of the particle best with best function value as the global best, G^k.
5. Update the position of each particle by calculating its velocity by using Eq. (1) and adding it to its current position according to Eq. (2).
6. Set $k = k + 1$. If $k < k_{max}$, terminate iteration k and go to step 2. If $k = k_{max}$, terminate algorithm.

Nelder-Mead Optimization

Published in 1965 by Nelder and Mead (Ref. 5), Nelder-Mead Optimization (NMO) has become one of the most popular search algorithms. In this paper, the algorithm is constructed based on Ref. 6. Consider a design space with n-dimensions. A simplex is defined as the convex hull of $n + 1$ distinct points so that in two-dimensional space it forms a triangle and in three-dimensional space a tetrahedron and so forth. Each point of the simplex represents a design choice and there are corresponding function values of each point. NMO tries to improve the worst function value with simplex operations namely; *reflection, expansion, contraction, shrinkage.*

At beginning of each iteration, the design points are sorted in ascending order according to their function values. Then, the centroid $\bar{x}$ of first n points is calculated. This point is used as a reference point for simplex operations. After calculation of the centroid, the simplex operations are performed. The objective of the simplex operations is to improve the function value of the worst point. NMO algorithm is expressed in Algorithm 2. The simplex operations are depicted in Figure 1 for a two-dimensional case, where $f(x_1) < f(x_2) < f(x_3)$.

Algorithm 2 Nelder-Mead Optimization

1. For n-dimensional space, form a simplex using $n + 1$ distinct points. Set $k = 1$.
2. Start iteration k. Sort the points $x_1^k, \dots, x_{n+1}^k$ such that $f_1^k < f_2^k \dots < f_{n+1}^k$ where $f_i^k = f(x_i^k)$.
3. Calculate centroid of first n points from

$$\bar{x} = \sum_{i=1}^{n} x_i^k / n. \quad (3)$$

4. *Reflection*: Calculate reflection point x_r as follows

$$x_r = \bar{x} + \rho(\bar{x} - x_{n+1}^k). \quad (4)$$

Calculate its function value with $f_r = f(x_r)$. If $f_1 \leq f_r$, assign $x_{n+1}^{k+1} = x_r$ and $x_i^{k+1} = x_i^k, i = 1, \dots, n$, then proceed to step 8.

5. *Expansion*: If $f_r < f_1^k$, calculate expansion point x_e as follows

$$x_e = \bar{x} + \chi(x_r - \bar{x}). \quad (5)$$

Evaluate its function value with $f_e = f(x_e)$. If $f_e < f_r$, assign $x_{n+1}^{k+1} = x_e$ and $x_i^{k+1} = x_i^k, i = 1, \dots, n$, then proceed to step 8, else assign $x_{n+1}^{k+1} = x_r$, $x_i^{k+1} = x_i^k, i = 1, \dots, n$, then proceed to step 8.

6. Contraction: If $f_r \geq f_n$, and,
   a. if $f_n \leq f_r < f_{n+1}$, calculate outside contraction point x_{oc} as follows

   $$x_{oc} = \bar{x} + \gamma(x_r - \bar{x}). \quad (6)$$

   Calculate its function value with $f_{oc} = f(x_{oc})$. If $f_{oc} \leq f_r$, assign $x_{n+1}^{k+1} = x_{oc}$ and $x_i^{k+1} = x_i^k, i = 1, \dots, n$, then proceed to step 8.

   b. if $f_r \geq f_{n+1}$, calculate inside contraction point x_{ic} as follows

   $$x_{ic} = \bar{x} - \gamma(x_r - \bar{x}). \quad (7)$$

   Evaluate its function value with $f_{ic} = f(x_{ic})$. If $f_{ic} \leq f_r$, assign $x_{n+1}^{k+1} = x_{ic}$ and $x_i^{k+1} = x_i^k, i = 1, \dots, n$, then proceed to step 8.

7. *Shrinkage*: If steps 4 to 6 do not provide any improvement on x_{n+1}^k, perform shrinkage, evaluate function values for $x_i^{k+1} = x_1 + \sigma(x_i - x_1)$, $i = 2, \dots, n + 1$.

8. If $abs(x_{n+1}^k - x_1^k) < 10^{-6}$ and $abs(f_{n+1}^k - f_1^k) < 10^{-6}$, terminate algorithm, else set $k = k + 1$, terminate iteration and go to step 2.

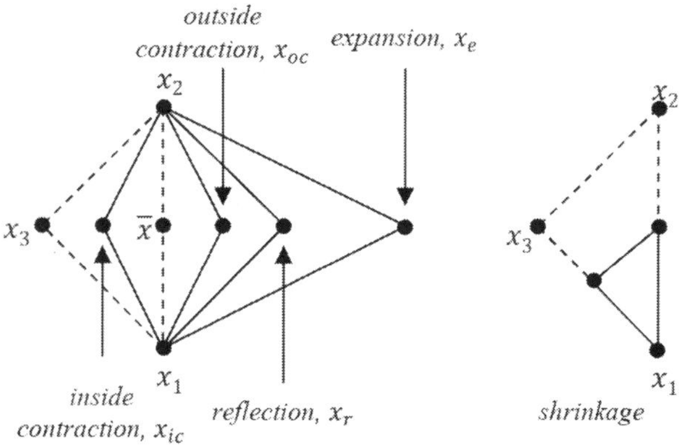

Figure 1. Simplex operations for a two-dimensional design space

Hybrid Algorithm

The convergence speed of PSO is very slow and requires a large number of iterations for pinpointing and converging to the global minima. Moreover, the capability of NMO for finding the global minima in non-convex problems is very weak and largely depends on the initial guesses. Hence, by combining PSO and NMO, an efficient and effective algorithm, which is referred as Hybrid Optimization (HO) can be constructed and aforementioned weaknesses can be eliminated.

In this study, two algorithms are used in conjunction. The optimization starts with PSO for design space exploration. After Algorithm 1 is performed, HO feeds the best solution of PSO to NMO. Since the fed best solution constructs only one point in NMO, remaining n points are randomly selected in the design space. Hence, a simplex with $n + 1$ points is constructed. Using this initial simplex, Algorithm 2 is performed and the resulting best point of NMO is recorded as the result of the hybrid optimization.

Since PSO is selected for its capability of design space exploration, the coefficients, c_1, c_2, w, and Δt are selected not for convergence but for exploration. According to Ref. 7, it is suggested that $c_1 = c_2 = 1.494$, $w = 0.5 + r/2$ where r is a random number between 0 and 1, and $\Delta t = 1$ for best exploration and convergence of the algorithm. However, in this study, it is desired that the exploration capability of PSO is come into prominence. A series of analyses are conducted with several benchmark problems and it is observed that $c_1 = 0.8$, $c_2 = 1.2$, $w = 0.9$, and $\Delta t = 1$ provide best exploration capability for minimum number of iterations.

In Ref. 6, the suggested parameters for simplex operations are given as $\rho = 1, \chi = 2, \gamma = 0.5$, and $\sigma = 0.5$. However, to increase the convergence rate, the coefficient for shrinking operation is halved. As a result, the parameters for reflection, expansion, contraction, and shrinkage are selected as $\rho = 1$, $\chi = 2$, $\gamma = 0.5$, and $\sigma = 0.25$ respectively.

Since there are some elements of randomness embedded in HO, the robustness of the code is tested. In this study, the robustness of an optimization algorithm is defined as the probability of finding global minima in each run. The more successful the algorithm in finding the global minima, the more robust it is. Robustness values of HO, Genetic Algorithm (GA), Simulated Annealing (SA) are tested and compared. In order to assess the hybridization capabilities of GA and SA, they are also hybridized with `fmincon` function of MATLAB® which is one of the gradient-based search algorithms of Optimization Toolbox in MATLAB®. In these hybrid optimizations first GA or SA finishes their algorithms then they feed their best outcome to `fmincon`.

Seven benchmark problems with a different number of variables are investigated. The problems are selected for their multimodalities. The explicit formulations of benchmark problems are given in Appendix A. Results are tabulated in Table 1. It is observed that, even though GA and SA are hybridized, the proposed hybrid function has outperformed both in the majority of the problems in terms of the robustness and the total number of function calls. Moreover, comparing the average results, it is clear that hybrid algorithm is more successful in finding the global minima than other algorithms.

Table 1. Robustness of algorithms

Problem	Algorithm	Robustness	Avg. f Call
	HO	0.992	4673
Ackley	GA	0.122	5849
	SA	0.068	2820
	HO	0.495	3087
Tripod	GA	0.527	6791
	SA	0.229	1500
	HO	0.650	3078
Mishra 1	GA	0.039	5475
	SA	0.144	1593
	HO	0.805	3062
Mishra 2	GA	0.141	5488
	SA	0.385	1617
	HO	0.730	6285
Mishra 5	GA	1.000	9943
	SA	0.677	3888
	HO	0.603	3056
Periodic	GA	0.228	3658
	SA	0.737	1638
	HO	0.403	6660
Rosenbrock	GA	0.795	18212
	SA	0.921	5567
	HO	0.668	4271
AVERAGE	GA	0.407	7916
	SA	0.405	2660

The upper and lower bounds for variables are handled by enforcing each variable to reside in the predefined interval at the end of each iteration for both PSO and NMO. The boundary enforcement is conducted as follows

$$x_i^k = \begin{cases} x_i^k, & x_i^k < x_i^{ub} \\ x_i^{ub}, & x_i^k \geq x_i^{ub} \end{cases}, \quad (8)$$

$$x_i^k = \begin{cases} x_i^k, & x_i^k > x_i^{lb} \\ x_i^{lb}, & x_i^k \leq x_i^{lb} \end{cases}. \quad (9)$$

Eq. (8) and Eq. (9) are performed for each variable of each design point.

ROTOR MODELS

In this study, two 4-bladed, fully articulated rotors are investigated; one with inter-blade lag damper (IBR) and one with hub-to-blade lag damper (HBR). Apart from the lag damper configurations, rotors are modeled identically at the initial state. The purpose of this investigation is to understand the requirements of these rotors in terms of optimization.

Both rotors are assumed to be wind tunnel rotors and are fixed to the ground with rigid connections. The Wind is introduced to the systems as far field free stream flows.

Rotor Topology

Both rotors are equipped with non-rotating actuators, swash plate, pitch link, and pitch control lever. Swashplate rotation is provided with scissors attached to the mast. Articulation for both models is provided with spherical bearings which connect the blade to the hub with the help of tension links. Since articulation is provided by a spherical joint, lag, flap and pitch hinges are coincident. The blades are modeled as one-dimensional beams. For IBR, lag damper resides in between two consecutive blades and is connected to the tension links of these blades. On the other hand, for HBR, the damper is located between the hub and the blade and it is connected to the pitch control lever on the blade side. Simple schematics of lag damper configurations, shown with gray, are provided in Figure 2.

The models are developed with Dymore, a finite element based multibody dynamics code for the comprehensive modeling of flexible and nonlinear multibody systems (Ref. 8). Dymore utilizes three-dimensional geometrically correct beam theory for modeling the beams. Detailed description of Dymore and its beam formulations can be found in Ref. 8 and Ref. 9. The aerodynamic loads are calculated with built-in simple aerodynamics code which employs Peters unsteady aerodynamics and the inflows models (Ref. 10).

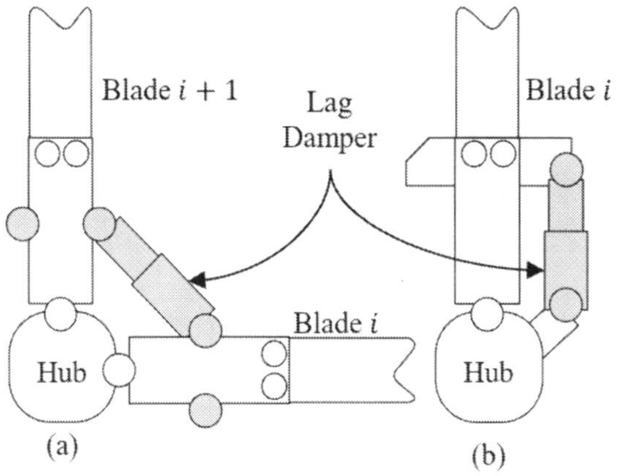

Figure 2. (a) Inter-blade damper configuration (b) Hub-to-blade damper configuration

In Table 2, basic rotor parameters are provided. Apart from the lag damper configuration, both IBR and HBR share the same rotor parameters. The non-dimensional rotor span is defined as $\eta = r/R$, where r is dimensional rotor span.

Table 2. Rotor parameters for IBR and HBR

Number of Blades	4
Rotor Speed, Ω	$360\ rpm$
Rotor Radius, R	$6000\ mm$
Hinge Offset	$0.05\ \eta$
Root Cut-out	$0.25\ \eta$
Airfoil Profile	NACA0012
Torque Offset	-
Pitch-Flap Coupling	11.9°, pitch up nose down
Solidity	0.0849
Chord, c	$0.4\ m$, constant
Twist Rate	$2°/m$, linear, nose down
Tip Loss Factor	0.97
Damper Length	$350\ mm$

The reference axis system of the rotors are as follows; x-axis parallel to far field velocity, positive downstream; y-axis perpendicular to far field stream, positive towards right; z-axis parallel to rotor masts, positive up.

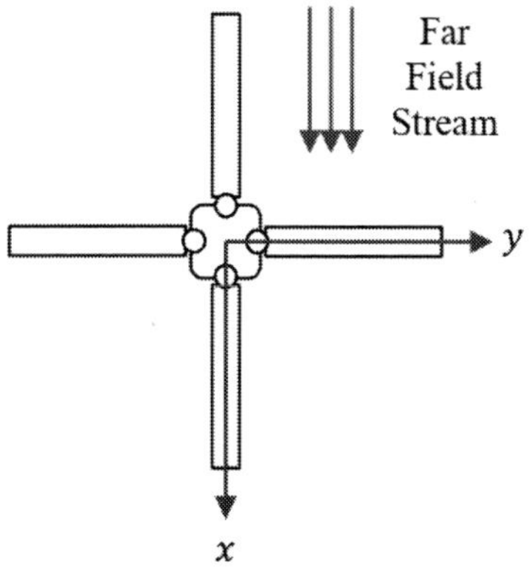

Figure 3. Reference axis systems for the rotors

For the rotor blades, x-axis is coincident with feathering axis and positive towards blade tip; y-axis is coincident with chord line and positive towards leading edge; z-axis is positive up. Reference axis for the rotors and the blades are given in Figure 4.

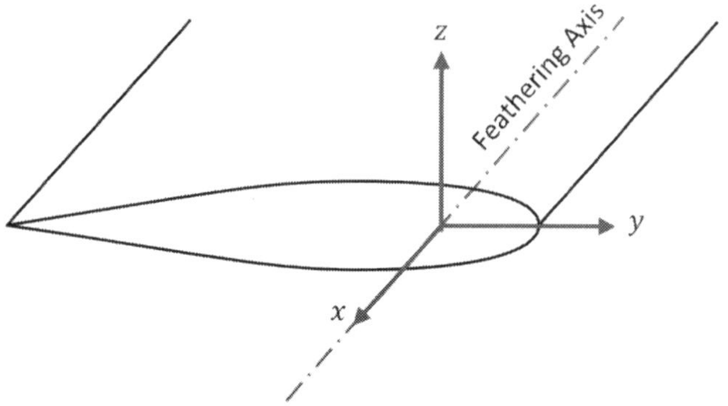

Figure 4. Reference axis systems for the blades

Blade Cross-Section

This study focuses on modeling and tailoring of composite rotor blades. Blade cross-sections are modeled with PreVABS and analyzed using VABS. PreVABS is a design-driven pre-processing computer program which can effectively generate high-resolution finite element modeling data for VABS by directly using design parameters such as simple geometric parameters and both the spanwisely and chordwisely varying composite laminate lay-up schema for rotor blade (Ref. 11). Developed by Cesnik, Hodges and Yu et al, VABS can perform classical analysis for inhomogeneous, anisotropic beams with initial twist and curvature having arbitrary reference and material properties yielding stiffness and mass matrices. Moreover, the three-dimensional stress and strain fields can be recovered using VABS (Ref. 12).

The blade structure is divided into three parts; root region, transition region, and functional region. In order keep the model simplistic, tip sweep is not included. The schematic of blade regions is given in Figure 5.

The cross-section properties of the root region are taken stiff and during optimization process it is untouched. The transition region is not modeled and it is handled in Dymore by linear interpolation. This study mainly deals with the structure of the functional region.

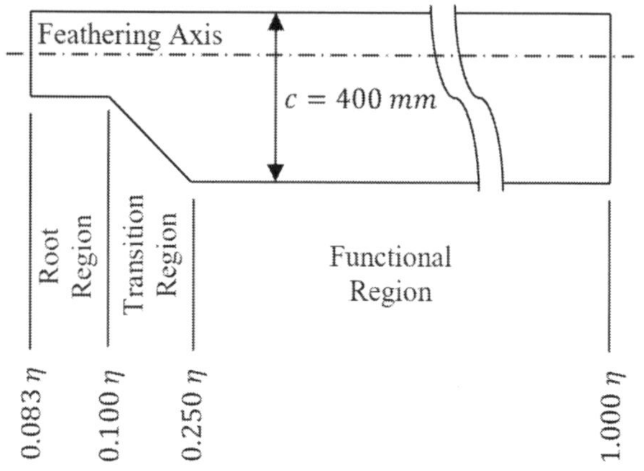

Figure 5. Blade structural regions, viewed from top

The cross-section of the functional region mainly consists of carbon-fiber (CF), glass-fiber (GF), honeycomb (HC), film adhesives and steel. 0° lay-up angle is taken as the blade span axis. Carbon-fiber lay-ups are used for skin and its lay-up configuration is $[\pm45/\pm\alpha/\pm45]$. The angle α is used as an optimization parameter. In tailoring the torsional stiffness, the lay-up angles of the skin has significant effect. Glass-fiber lay-ups are used for spar and inner/outer wraps. Lay-up configuration for spar is $[0_8]$. Inner wrap and outer wrap lay-up configurations are $[\pm45]$ and $[\pm45]_s$ glass-fiber, respectively. Moreover, two walls are placed in between inner and outer wraps for fortifying the cross-section. Wall lay-up configurations are $[\pm45/0_2/\mp45]$ glass-fiber. For avoiding crushing in the trailing edge zone, honeycomb is utilized. Film adhesives are used in between the lay-ups. Steel erosion shield is implemented into the cross-section for protection from sand and dust abrasion. Nose block consists of 0° glass-fiber material. The lay-up configurations for upper and lower surfaces are identical to eliminate any coupling effects. A simple schematic and composite lay-up configuration of the cross-section is given in Figure 6. The composite material properties and their thicknesses are provided in Table 3.

Table 3. Material properties used in blade design

Parameter	GF	CF	Steel	Film A.	HC
$E_{11}\ (GPa)$	48	14	200	1.0	0.00
$E_{22}\ (GPa)$	14	10	200	1.0	0.00
$E_{33}\ (GPa)$	14	10	200	1.0	0.01
$G_{12}\ (GPa)$	4.9	6.0	78	0.50	0.04
$G_{13}\ (GPa)$	4.9	6.0	78	0.50	0.04
$G_{23}\ (GPa)$	5.1	4.0	78	0.50	0.03
ν_{12}	0.30	0.20	0.30	0.25	0.01
ν_{13}	0.30	0.20	0.30	0.25	0.01
ν_{23}	0.30	0.20	0.30	0.25	0.01
$\rho\ (g/cm^3)$	1800	1500	8000	1200	50
$t\ (mm)$	0.226	0.133	0.750	0.180	-

For the evaluation of the cross-section structural integrity, two approaches exist namely; maximum stress and maximum strain. In their study, Li et al used stress approach for evaluating the structural integrity of the cross-sections. For isotropic materials, Von Mises criterion is employed and for composite materials, Tsai-Wu criterion is utilized (Ref. 13). The second approach for checking failure criteria is to evaluate the strain levels on the cross-section. In Ref. 14, Isik and Kayran used a maximum strain criterion for optimizing a blade cross-section.

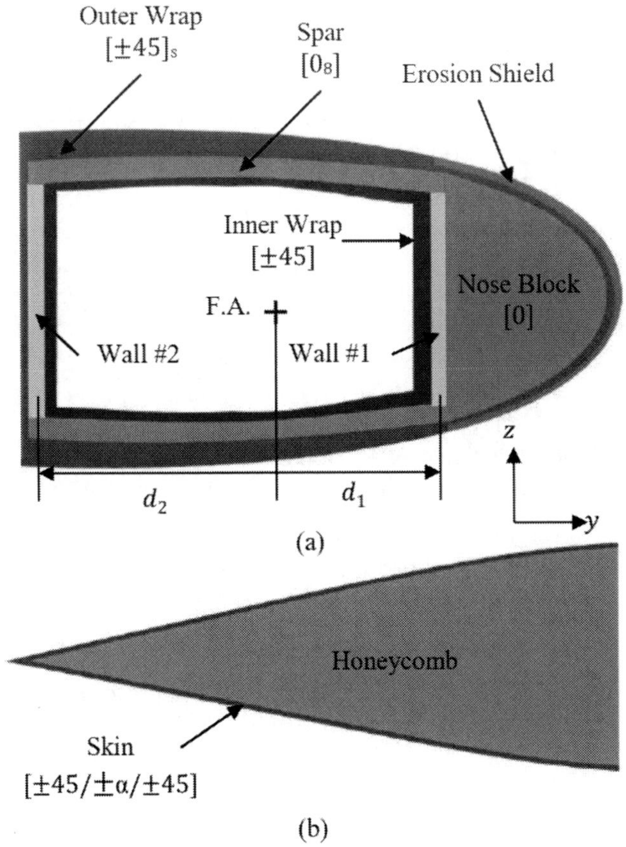

Figure 6. Schematic of the functional region cross-section, (a) leading edge region, (b) trailing edge region

In this study, maximum strain criterion is used for strength evaluations. For calculation of the strain, only the centrifugal loading is used. Since functional region spans from $0.25\,\eta$ to $1.00\,\eta$ and since functional region has uniform blade cross-section, the maximum centrifugal loading (CL) occurs at $0.25\,\eta$ and can be calculated as follows;

$$CL = m_f \Omega^2 R_f. \quad (10)$$

where, m_f is the mass of the functional region and R_f is the spanwise center-of-gravity position of the functional region measured from hub center. For IBR and HBR, R_f is constant and $R_f = 3750\,mm$. The mass of the functional region as a function of mass-per-length of the cross-section can be calculated using the following equation

$$m_f = m_0 L_f. \quad (11)$$

where, m_0 is the mass-per-length value of the cross-section and L_f is the length of the functional region which is equal to $4500\,mm$.

In order to compensate for manufacturing defects, hot/wet conditions, material impurities, combined loading cases where every force and moment terms are applied to the cross-section, limit load considerations and fatigue strength, during the calculation of Eq. (10), the rotational speed is taken as 125 % higher than the rotor rotational speed in Table 2. Moreover, the resulting CL is multiplied by a safety factor of 1.5 for the final loading applied on the blade cross-section. Resulting applied load is as follows

$$CL_\epsilon \cong 2.34\,CL. \quad (12)$$

OPTIMIZATION CONSTRAINTS AND TARGETS

In order to optimize the rotors, some physical and behavioral constraints and targets are integrated into the optimization problem. To properly add the penalty and target functions, each function is normalized with factors defined along the penalty and target functions. After normalization, these functions range between 0 and approximately 1. Hence, the contributions of each function to the objective function are in the same order of magnitude.

Frequency Separation

One of the major targets in helicopter rotor optimizations is to adjust the natural frequencies of the blades so that they do not coalesce with excitation frequencies. During a steady flight, rotor blades are exposed to excitations at frequencies corresponding to integer multiple of the rotor rotational speed. Hence, the natural frequencies of the blade should be away from these excitation frequencies to avoid resonance.

During an operational flight, the external excitations are generally dominated by the first two harmonics. The contributions of the remaining harmonics are much smaller. However, during a high altitude, high-speed flight, the angle of attack may reach stall caused by increased collective angle requirements leading to higher contributions from higher harmonics. Figure 7 shows the contributions of each harmonic to total lift-per-length at $0.75\,\eta$. The simulation is based on a wind tunnel model. Two flight cases are investigated; low collective and high collective. For low collective case 8° collective and 7° cyclic input and for high collective case is 14° collective and 10° cyclic input are given as pilot inputs. Both cases are run with 72 *knots* wind to simulate high speed flight. High collective case reaches stall region since for NACA0012 the static stall angle is around 16° (Ref. 15). It can be observed that the amplitudes of 3rd, 4th, and 5th harmonics for high collective case are approximately twice of the amplitudes of the same harmonics of low collective case. However, the contribution of each harmonic to total vibratory load depends heavily on flight characteristics.

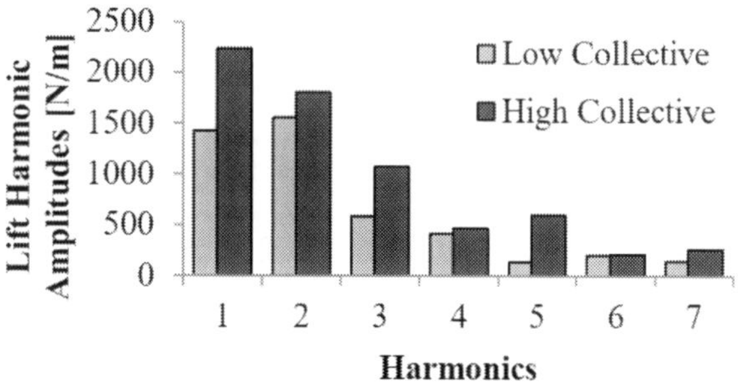

Figure 7. Harmonics for high and low collective wind tunnel simulations

In order to assess the effect of natural frequency and excitation frequency interaction, a natural frequency of the second flap, the second lag, and the first torsion modes are varied independently. To do so, sectional stiffnesses are altered. During analysis 12° collective and 10° longitudinal cyclic is given to the rotor as pilot input. Moreover, 120 *knots* constant wind is provided as far field stream.

The second flapwise mode is swept from 2.5 Ω to 3.5 Ω and maximum load amplification is recorded at 0.45 η. The amplification of 3 Ω blade sectional flapwise bending moment at 0.45 η with second flapwise mode frequency is graphed in Figure 8. Harmonic amplitude is normalized with maximum oscillatory flapwise bending moment on the blade for each analysis point. It is observed that near 3 Ω, the excitation and natural frequencies coalesce and the load is increased by 1.5 times compared to response at 2.8 Ω and 4.0 times compared to response at 2.6 Ω. Moreover, the contribution of 3 Ω to total oscillatory load is nearly 45 %, when coalescence occurs.

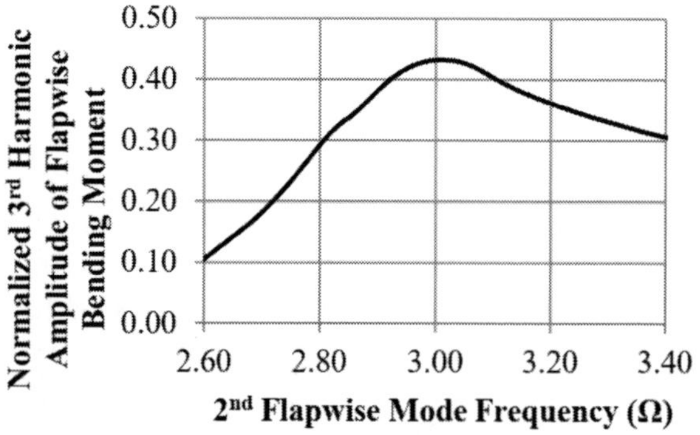

Figure 8. Change of 3rd harmonic of flapwise moment with 2nd flapwise mode frequency at 0.45 η

The second chordwise mode is swept from 5.5 Ω to 6.5 Ω and maximum load amplification is observed at 0.45 η similar to flapwise mode. In Figure 9, the response of 6 Ω blade sectional chordwise bending load at 0.45 η with second chordwise mode frequency is given. For each analysis point, 6th harmonic amplitude is normalized with maximum oscillatory chordwise bending load on the blade. In the vicinity of 6 Ω, the excitation frequency and natural frequency coalesce and the load is amplified by 1.3 times compared to response at 5.8 Ω and 2.2 times compared to response at 5.6 Ω. Moreover, 3 Ω harmonic of chordwise bending load contributes nearly 50 % to total oscillatory chordwise bending load.

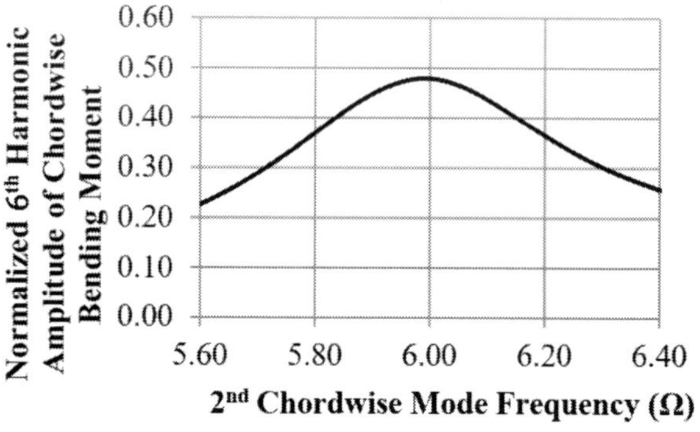

Figure 9. Change of 6 Ω harmonic of chordwise moment with 2nd chordwise mode frequency at 0.45 η

The first torsion mode is swept from 3.5 Ω to 4.5 Ω and at blade root maximum load amplification is observed. The amplification of 4 Ω blade sectional torsional moment at 0.08 η with first torsion mode frequency is graphed in Figure 10. Harmonic amplitude is normalized with maximum oscillatory torsional moment on the blade for each analysis point. It is observed that near 4 Ω, the excitation and natural frequencies coalesce and the load is amplified by 1.05 times compared to response at 3.8 Ω and 1.15 times compared to response at 3.6 Ω. In addition, the contribution of 4 Ω to total oscillatory load is nearly 25 % at coalescence.

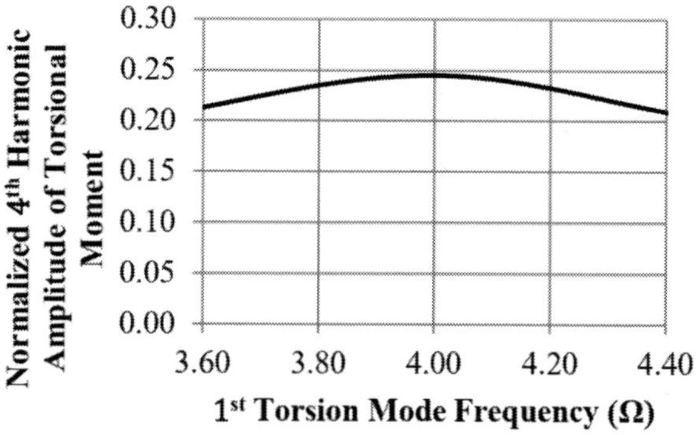

Figure 10. Change of 4th harmonic amplitude of torsional moment with 1st torsion mode frequency at 0.08 η

Maximum load amplification is observed with flapwise mode. Since flapwise bending stiffness is much lower than chordwise bending stiffness, the flapwise mode frequencies tend to be lower than chordwise mode frequencies leading to higher amplifications. Moreover, lag dampers dampens the response at resonance regions. The distribution of harmonic amplitudes throughout the blade can be found in Appendix B for all sweep studies.

In ideal conditions where each natural frequency can be tailored independently, it is desirable to place each natural frequency of the blade under operational speed 0.5 Ω away from the closest excitation frequency for minimum amplification. However, since frequency of each mode is indirectly coupled through structural design of the blade and there are numerous other constraints on the blade design, it is very unlikely to achieve this target. Many rotor optimization approaches target frequency separation to be between 0.15 Ω and 0.20 Ω from excitation frequencies. In this study, the frequency separation is targeted to be $\Delta\omega_{trgt} = 0.20\,\Omega$. The penalty function for frequency separation can be expressed as follows;

$$p_f = \begin{cases} \dfrac{|\Delta\omega_i - \Delta\omega_{trgt}|}{\Delta\omega_{trgt}}, & |\Delta\omega_i| < \Delta\omega_{trgt} \\ 0, & |\Delta\omega_i| \geq \Delta\omega_{trgt} \end{cases}. \qquad (13)$$

Blade Torsional Stiffness

One of the most critical characteristics of rotor blades is the torsional deformation of the blade. The tennis racket effect, tension-torsion couplings of composite blades and aerodynamic moment generated by cambered airfoil sections are mainly the causes of this deformation. These effects tend to deform the blade torsionally in nose down direction. The effect of torsional deformation can be investigated in two parts: mean deformation and 1 Ω deformation. Consider the pilot input and the deformed blade response in Figure 11. Under aforementioned effects, blade collective and cyclic inputs are degraded, decreasing the effectivity of pilot input over the blade. The collective loss leads to unnecessary increase in pilot collective input range for accounting the deformation. This approach is also applicable for cyclic inputs which are affected by 1 Ω deformations.

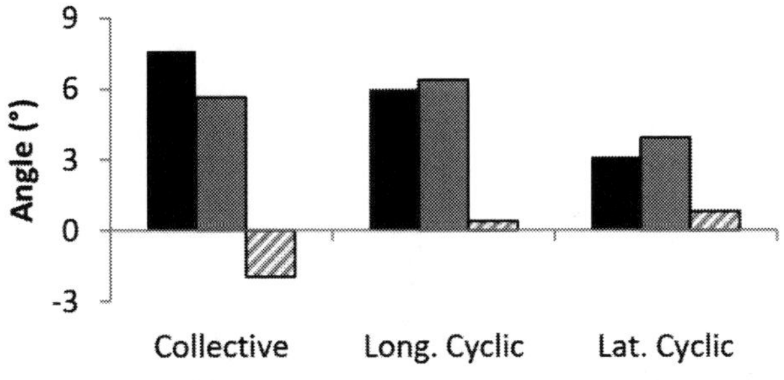

Figure 11. Pilot input, blade response, and control loss

It is crucial for the blade to be tailored good enough to compensate for this deformation yet be lightweight as well. In this study, it is targeted that the mean and 1 Ω deformations of the blade does not exceed $\Delta\theta_0^{max} = \Delta\theta_1^{max} = 1.5°$. The penalty functions for mean and 1 Ω deformations are expressed as follows respectively:

$$p_{\theta 0} = \begin{cases} 0, & |\Delta\theta_0| < \Delta\theta_0^{max} \\ |\,\Delta\theta_0/\Delta\theta_0^{max}\,| - 1, & |\Delta\theta_0| \geq \Delta\theta_0^{max} \end{cases}, \qquad (14)$$

$$p_{\theta 1} = \begin{cases} 0, & \Delta\theta_1 < \Delta\theta_1^{max} \\ \Delta\theta_1/\Delta\theta_1^{max} - 1, & \Delta\theta_1 \geq \Delta\theta_1^{max} \end{cases}. \qquad (15)$$

Flutter

One of the instabilities rotor blades encounter is flutter. Flutter can be avoided by mass balancing the blades such that the center-of-gravity (CG) of the blade is in front of the aerodynamic center and towards the leading edge (Ref. 16). For this purpose, it is targeted that CG is in front of the aerodynamic center and shear center (SC) is as close to the aerodynamic center as possible. It is assumed that the aerodynamic center is on the feathering axis. CG and SC are measured from feathering axis. The penalty functions for each center are given as shown below

$$p_{CG} = \begin{cases} |y_{CG}|/(c/4), & y_{CG} < 0 \\ 0, & y_{CG} \geq 0' \end{cases} \qquad (16)$$

$$p_{SC} = |y_{SC}|/(c/4). \qquad (17)$$

Tension Center

For cross-sections with anisotropic materials, it is likely that the tension center (TC) and CG do not coincide. The centrifugal loading acts on the CG and the offset between TC and CG creates a bending moment. Since centrifugal loading during helicopter operation is constant, this bending moment acts on the blade steadily. In order to reduce the mean bending moment acting on the blade, it is targeted to minimize the distance between CG and TC. Because of vertical symmetry of the blade functional region cross-section, the offset along chord line is considered only. The penalty function is set as

$$p_{TC} = |y_{CG} - y_{TC}|/(c/4). \qquad (18)$$

Blade Structural Integrity

It is vital to consider structural integrity of the blade while optimizing the rotor. In this study, a maximum strain level is imposed as a constraint. The penalty function for strain is taken as

$$p_{\epsilon} = \begin{cases} 0, & \epsilon < \epsilon^{max} \\ \epsilon/\epsilon^{max} - 1, & \epsilon \geq \epsilon^{max} \end{cases}. \qquad (19)$$

The maximum strain is imposed by carbon-fibers since carbon epoxy has the lowest ultimate tensile strength among other materials used in the blade cross-sectional modeling (Ref. 17). It was suggested by Isik and Kayran that for carbon epoxy, the maximum longitudinal strain should be lower than $\epsilon^{max} = 5400\ \mu\epsilon$ [14].

First Cyclic Inplane Frequency

Frequency of the lag mode is driven by the position of the hinge, blade second mass moment of inertia about the hinge offset, lag damper stiffness and rotor speed. For an articulated rotor system, the first cyclic inplane mode has a characteristic natural frequency less than 1 Ω and, hence, the rotor system is prone to ground resonance instability which in turn imposes additional damping requirements for the system (Ref. 18). Typically, a lag damper in the rotor and some airframe damping provided through landing gears are needed to maintain stability. Analysis indicates that higher lag frequencies tend to lower system damping requirements (Ref. 16).

As shown in Figure 12, as the first inplane frequency increases, the necessary damping ratio for stability decreases. First cyclic inplane frequency is reflected to the optimization problem as a target and $\omega_\zeta^{target} = 0.40\ \Omega$ to minimize airframe damping requirements while keeping lag damper damping between 15 % and 20 %. The penalty function for the first inplane frequency is provided in Eq. (20). It is assumed that during optimization first inplane frequency does not go above 0.8 Ω. Hence, for normalization $\omega_\zeta^{max} = 0.4\ \Omega$ is used.

$$p_{\omega_\zeta}^{cyc} = \left| \frac{\omega_\zeta^{cyc} - \omega_\zeta^{target}}{\omega_\zeta^{max}} \right|. \quad (20)$$

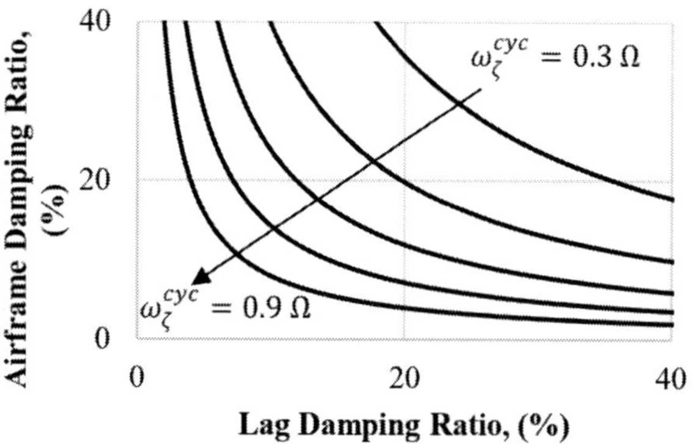

Figure 12. Damping requirement as a function of lag frequency

Lag Damper Effectivity of IBR

The inter-blade lag dampers work on the principle of relative lag motion between blades. When the blades lead-lag with the same phase angle, the lag damper motion is very close zero. This leads to very low damping for collective modes as the lag dampers are not excited. However, it is possible to alleviate this problem by carefully orienting the damper.

Inter-blade damper works best for scissor inplane modes. Hence, in ideal conditions, the collective mode frequency; ω_ζ^{coll}, is equal to scissor mode frequencies, ω_ζ^{sci} since in scissors mode the lag dampers are utilized most effectively. Therefore, the effectiveness of inter-blade damper for the collective mode can be increased by targeting the collective inplane mode frequency to the scissor inplane mode frequency. The penalty function for this case is taken as

$$p_{\omega_\zeta}^{coll} = \left| \frac{\omega_\zeta^{coll} - \omega_\zeta^{sci}}{\omega_\zeta^{sci}} \right|. \quad (21)$$

Lock Number

Determined by the balance of the centrifugal and aerodynamic flap moments, rotor coning angle is directly proportional to Lock number, γ (Ref. 18). Although, smaller γ implies lighter blade, it also causes more coning angle. In order to limit the coning angle in suitable range, it is targeted that γ is smaller than $\gamma^{max} = 10$. The penalty function is defined as

$$p_\gamma = \begin{cases} 0, & \gamma < \gamma^{max} \\ \gamma/\gamma^{max} - 1, & \gamma \geq \gamma^{max}. \end{cases} \quad (22)$$

Blade Mass

In order to increase the available useful payload for helicopter operations, it is desirable to keep the blade mass at minimum. Therefore, it is targeted to minimize the blade mass. The target function for blade mass minimization is expressed as

$$p_{mb} = m_b / m_b^{max}. \quad (23)$$

For normalization, $m_b^{max} = 60\ kg$ is used.

Blade Oscillatory Load

Reducing the vibratory loads through blade span is another target in this study. The reduction of blade vibration is achieved by the minimization of the maximum oscillatory load occurring along the blade span. Here, oscillatory load refers to the half-peak amplitude of the resulting forces and moments on the blade. In order to express the moments along with forces, the resulting moment is normalized with rotor radius R. The oscillatory force along x-axis is disregarded as it will be difficult to control the oscillatory force created along centrifugal loading direction. The resulting oscillatory force and moment combination is expressed as follows

$$F_{res}^{osc} = \sqrt{F_y^{osc\,2} + F_z^{osc\,2}} + \sqrt{M_x^{osc\,2} + M_y^{osc\,2} + M_z^{osc\,2}}/R. \quad (24)$$

The target function for blade oscillatory load minimization is provided below

$$p_{osc} = F_{res}^{osc} / F_{res}^{osc,max}. \quad (25)$$

For normalization, $F_{res}^{osc,max} = 2000\ N$ is used.

PROBLEM DEFINITION

The optimization of IBR and HBR is handled with two different objective functions. The first objective function includes the target function to reduce the vibratory load through frequency separation; whereas, the second objective function uses the blade oscillatory load target. Since the effectivity of the lag dampers of HBR does not depend on the type of inplane mode, the first collective inplane target function is not involved in HBR optimization objective functions. There are two types of rotors and two objective functions, a total of 4 cases are investigated. The list of target and penalty functions used in each case is summarized in Table 4.

Table 4. Target and penalty functions employed in each case

Functions	HBR		IBR	
	Case 1	Case 2	Case 3	Case 4
p_f	✓	×	✓	×
p_{osc}	×	✓	×	✓
$p_{\theta 0}$	✓	✓	✓	✓
$p_{\theta 1}$	✓	✓	✓	✓
p_{CG}	✓	✓	✓	✓
p_{SC}	✓	✓	✓	✓
p_{TC}	✓	✓	✓	✓
p_{ϵ}	✓	✓	✓	✓
$p_{\omega_\zeta}^{cyc}$	✓	✓	✓	✓
$p_{\omega_\zeta}^{coll}$	×	×	✓	✓
p_{γ}	✓	✓	✓	✓
p_{mb}	✓	✓	✓	✓

All cases optimizations are performed with a wind tunnel rotor and are exposed to 140 *knots* far field wind. The environmental conditions are selected as ISA 0, (Ref. 19). For pilot inputs, 12° collective, 6° longitudinal forward cyclic and 3° lateral left cyclic are given for real flight simulation. The pilot inputs are kept constant for all design points.

The objective function for each case is constructed through summation of target and constraint functions applicable for each case as follows

$$f(x) = \sum p_i. \quad (26)$$

For HBR optimization, a total of 6 design variables are utilized; the amount and the spanwise location of concentrated mass placed on the blade, the positions of Wall #1 and Wall #2, the lay-up angle of skin, and lag damper spring coefficient. These variables are labeled as $x_1, \ldots, x_6$, respectively. In addition to 6 design variables used in HBR optimization, 3 more design variables are employed for IBR optimization. The added design variables are to optimize the position of the inter-blade lag damper to increase its effectiveness in the first collective inplane mode. Hence, a total of 9 design variables are employed. These variables are labeled as $x_1, \ldots, x_9$ in given order. The additional 3 design variables the distance of the lag damper center point from hub center, the orientation angle of the lag damper, the angle of the center point of the lag damper from feathering axis. In Figure 14, a sketch is provided for the additional design variables for IBR optimizations.

For each case, a concentrated mass is utilized for center-of-gravity and frequency adjustments. To move the center-of-gravity towards leading edge, the mass is located at the nose. Schematic for the concentrated mass is given in Figure 13.

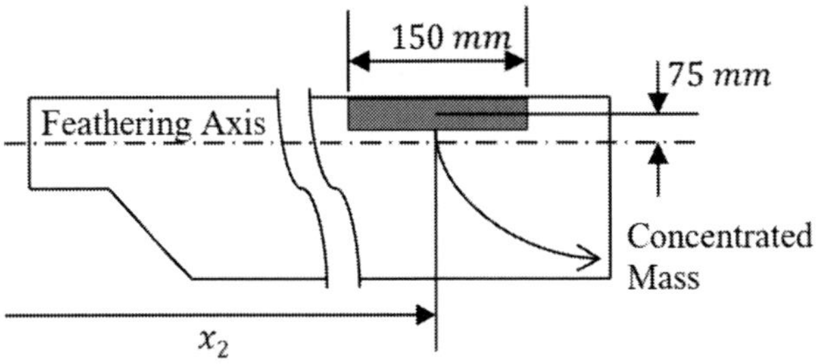

Figure 13. Concentrated mass placed on the blade

The physical lower and upper bounds of design variables are given in Table 5.

Table 5. The lower and upper bounds used for IBR and HBR optimization

Design Variable	Lower Bound	Upper Bound
x_1	$0\ kg$	$2\ kg$
x_2	$1000\ mm$	$5500\ mm$
x_3	$30\ mm$	$90\ mm$
x_4	$130\ mm$	$30\ mm$
x_5	$0°$	$90°$
x_6	$10^5\ N/m$	$10^7\ N/m$
x_7	$200\ mm$	$500\ mm$
x_8	$30°$	$60°$
x_9	$30°$	$60°$

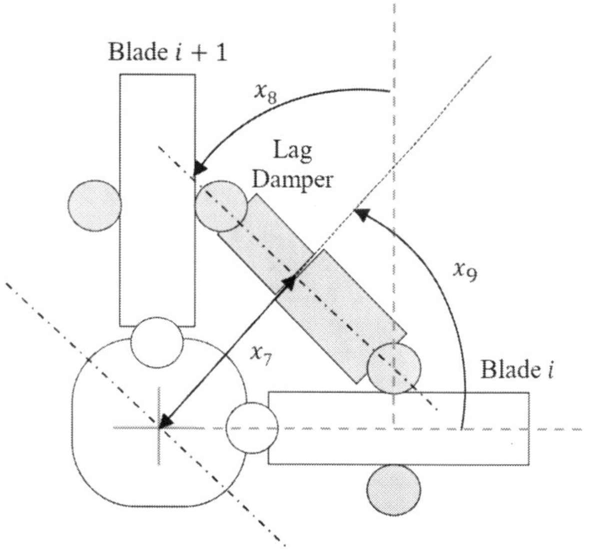

Figure 14. Design variables for inter-blade damper

The first six variables are same for both rotor models. The last three variables are used for IBR only.

RESULTS

Four cases are optimized with HO. For each case, optimization runs are repeated for 10 times to ensure that the global minimum is attained. The initial and minimum achieved objective function and the corresponding design parameter values are provided in Table 6 and Table 7 for HBR and IBR, respectively.

Table 6. Design parameters before and after optimization for HBR

Param.	Initial Conf.	Case 1	Case 2	Unit
x_1	2	1.546	2	kg
x_2	5500	5500	5500	mm
x_3	90	76.2	45.7	mm
x_4	70	85.7	130	mm
x_5	45	0	0	°
x_6	2	1.47	1.78	$10^6\ N/m$
$f(x)$	5.58	2.39	2.39	-

Comparing Case 1 and Case 2 used for optimizing HBR, the optimization results are remarkably different. Although two approaches favors 0° lay-up angle for skin, the wall locations are dissimilar. For Case 1, the walls are closer to leading edge; whereas, for Case 2, optimum wall locations are found to be towards trailing edge. Moreover, the total distance between the walls is higher for Case 2 leading to larger inner and outer wrap areas. In Case 2, the location of Wall #2 reaches the upper boundary. Similarly, minimization of objective function for Case 2 utilizes the maximum amount of concentrated mass at the blade tip. The damper spring coefficient is slightly higher for Case 2 than Case 1.

Table 7. Design parameters before and after optimization for IBR

Param.	Initial Conf.	Case 3	Case 4	Unit
x_1	2	0.469	0.019	kg
x_2	5500	5086	5494	mm
x_3	90	56.9	45.1	mm
x_4	70	96.4	130	mm
x_5	45	0	0	°
x_6	2	4.35	4.39	$10^6\ N/m$
x_7	300	260	300	mm
x_8	45	30.0	59.4	°
x_9	45	38.1	30.3	°
$f(x)$	5.45	2.10	2.65	-

For the optimization of IBR, similar results are obtained. The locations of walls are closer to leading edge for Case 3 than Case 4. Similar to HBR cases, Case 4 favors for higher distance between the walls resulting in larger inner and outer wrap areas. Both cases uses 0° skin lay-up angle. Minimization of objective functions for Case 3 and Case 4 require less amount of concentrated mass compared to Case 1 and Case 2. The resulting damper spring coefficients are nearly the same for IBR optimization; however, when compared to HBR optimization they are significantly higher indicating that to achieve first inplane frequency target, IBR requires considerably larger spring coefficients. Optimization for IBR favors the damper line-of-action to be parallel to one of the neighboring blades in order to increase the effectiveness of lag damper on the collective inplane mode.

Using the design variables in Table 6, the blade cross-sectional properties and rotor parameters for HBR are tabulated in Table 8.

Table 8. Blade cross-sectional properties and rotor parameters for HBR

Param.	Initial Conf.	Case 1	Case 2	Unit
m_0	4.87	5.15	7.14	kg/m
m_{xx}	47.3	51.6	62.0	$10^{-3}\ kgm$
EA	87.2	124	167	$10^6\ N$
GJ	42.7	37.8	41.5	$10^3\ Nm^2$
EI_{yy}	24.4	32.5	38.1	$10^3\ Nm^2$
EI_{zz}	0.60	1.27	1.51	$10^6\ Nm^2$
y_{CG}	9.93	4.41	1.05	mm
y_{SC}	20.3	11.5	7.02	mm
y_{TC}	16.0	5.95	0	mm
m_b	36.1	38.9	47.5	kg
γ	10.1	9.41	7.28	-
ϵ	5534	3359	2708	$\mu\epsilon$
$\Delta\theta_0$	1.63	1.48	1.47	°
$\Delta\theta_1$	1.72	1.51	1.50	°
ω_ζ^{cyc}	0.43	0.40	0.40	Ω
ω_ζ^{coll}	0.43	0.40	0.40	Ω
ω_ζ^{sci}	0.43	0.40	0.40	Ω

Design parameter values in Table 7 result in the blade cross-sectional properties and rotor parameters given in Table 9 for IBR.

Table 9. Blade cross-sectional properties and rotor parameters for IBR

Param.	Initial Conf.	Case 3	Case 4	Unit
m_0	4.87	6.17	7.17	kg/m
m_{xx}	47.3	54.9	62.1	$10^{-3}\ kgm$
EA	87.2	141	168	$10^6\ N$
GJ	42.7	38.9	41.5	$10^3\ Nm^2$
EI_{yy}	24.4	34.4	38.2	$10^3\ Nm^2$
EI_{zz}	0.60	1.36	1.52	$10^6\ Nm^2$
y_{CG}	9.93	3.78	-0.63	mm
y_{SC}	20.3	10.7	7.01	mm
y_{TC}	16.0	-0.51	0.31	mm
m_b	36.1	41.1	45.65	kg
γ	10.1	9.17	8.13	-
ϵ	5534	2577	2488	$\mu\epsilon$
$\Delta\theta_0$	1.63	1.49	1.46	°
$\Delta\theta_1$	1.71	1.51	1.51	°
ω_ζ^{cyc}	0.37	0.40	0.40	Ω
ω_ζ^{coll}	0.29	0.39	0.38	Ω
ω_ζ^{sci}	0.43	0.40	0.41	Ω

For HBR, Case 2 results in significantly heavier blade design compared to Case 1. Although the centrifugal loading on the blade increases with total blade mass, the strain at 0.25 η is lower in Case 2 since the total distance between the walls are wider for Case 2 leading to increased axial stiffness of the blade, hence, reducing the strain. Larger wall distance also helps to increase the torsional stiffness of the blade which reduces the torsional deformations which are within the given constraints. Moreover, since the spars, inner and outer wraps and walls are closer to trailing edge, center-of-gravity, shear center and tension centers are closer to feathering axis as well, since the spars, inner and outer wraps and walls are closer to trailing edge.

The resulting blade cross-sectional properties and rotor parameters are similar to HBR for IBR optimizations. Case 4 requires heavier blade for optimum design but wider wall distance reduces the strain at 0.25 η. The center-of-gravity, shear center and tension center are as well closer to feathering axis for Case 4 showing similar behavior to HBR optimization. The orientation of the lag damper increases its effectiveness over the collective inplane mode and the resulting collective, cyclic and scissor mode frequencies are closer to each other compared to the initial configuration.

The frequency placements are given in Table 10 and Table 11 for HBR and IBR, respectively. Case 1 and Case 3 target to reduce the vibration by frequency separation. Except for the 3rd flap mode frequency of Case 3, the natural frequencies are away from the excitation frequencies by at least 0.20 Ω. Case 2 and Case 4 does not impose any frequency separation; however, it is observed that the 2nd flap mode is 0.20 Ω away from 3 Ω excitation frequency at the minimum indicating that one of the major contributions to total oscillatory load is due to 3 Ω excitation.

Table 10. Blade natural frequencies for HBR

Mode	Frequency (Ω)		
	Initial Conf.	Case 1	Case 2
1st Lag	0.43	0.40	0.40
1st Flap	1.04	1.03	1.04
2nd Flap	2.80	2.80	2.80
3rd Flap	4.73	5.32	5.28
1st Torsion	5.38	6.20	6.05
2nd Lag	7.78	7.20	6.84
4th Flap	8.92	9.20	9.06

Table 11. Blade natural frequencies for IBR

Mode	Frequency (Ω)		
	Initial Conf.	Case 3	Case 4
1st Lag	0.37	0.40	0.40
1st Flap	1.04	1.04	1.04
2nd Flap	2.80	2.74	2.70
3rd Flap	4.71	5.16	5.08
1st Torsion	5.35	6.24	6.20
2nd Lag	7.78	7.20	7.02
4th Flap	8.91	8.80	8.71

Harmonic excitations created by the rotor are transferred to the airframe at the frequencies that are integer multiples of blade number. Figure 15 shows the 4th harmonic amplitudes of forces transferred to airframe for HBR initial configuration, Case 1 and Case 2. The reductions of forces are larger for Case 2 than Case 1 for inplane shears and for vertical shear. For rotor moments, the reduction does not present similar trend as shown in Figure 16. The y component of moment is decreased slightly for Case 1; however, y component of moment increases compared to initial HBR configuration for Case 2. The reduction attained for the 4th harmonic rotor torque, i.e. in z direction, is significant. Reduction of rotor forces are superior for Case 2 than Case 1 indicating that reducing blade oscillatory loads rather than separating the frequencies results in lower rotor inplane and vertical shears.

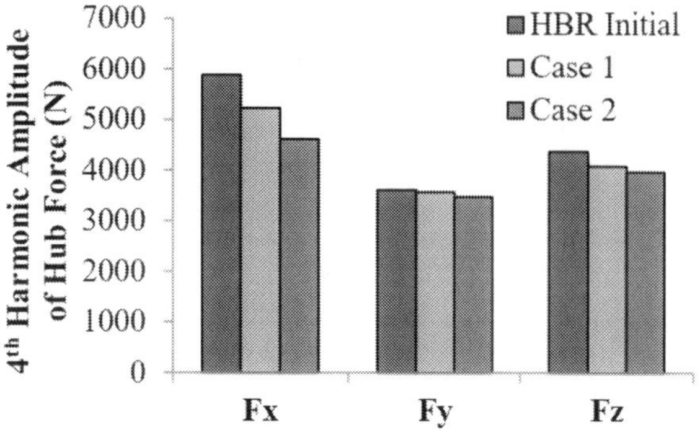

Figure 15. The 4th harmonic amplitudes of rotor inplane and vertical shears for HBR optimization

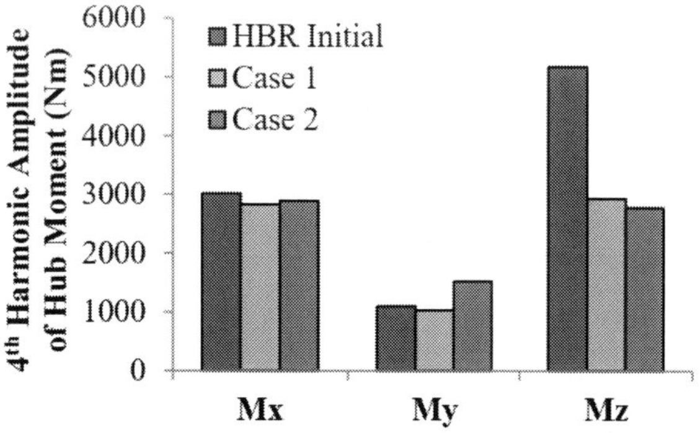

Figure 16. The 4th harmonic amplitudes of rotor moments and torque for HBR optimization

IBR optimization results show similar outcomes in terms of rotor forces as presented in Figure 17. However, the reduction in vertical shear is much larger for both Case 3 and Case 4 compared to Case 1 and Case 2. The lateral inplane force does not present substantial reduction between the initial state and the final states for both rotor types. In Figure 18, the rotor moments are provided for IBR. Unlike HBR optimization results, the reduction in rotor moments attained through optimization is greater for IBR optimizations. Case 4 which directly targets blade oscillatory loads gives the best reduction in overall in terms of rotor loads except for M_z which is the rotor torque.

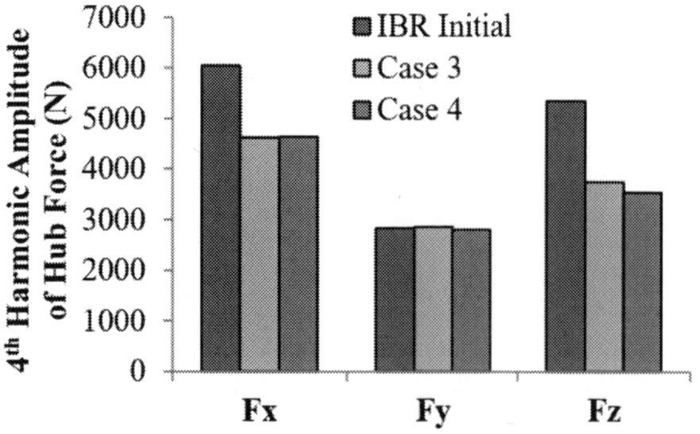

Figure 17. The 4th harmonic amplitudes of rotor inplane and vertical shears for IBR optimization

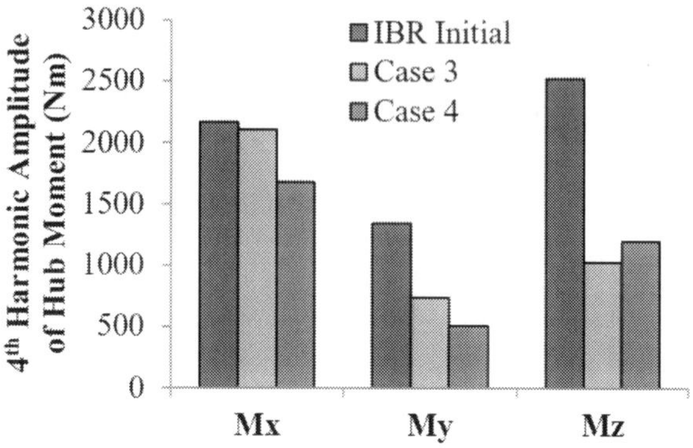

Figure 18. The 4th harmonic amplitudes of rotor moments and torque for IBR optimization

Figure 19 provides the blade mean and oscillatory force distributions for HBR optimizations. Since Case 2 results is heavier blade, the mean axial force distribution is larger than Case 1 and initial configuration. Apart from the mean axial load, reduction in mean and oscillatory loads is achieved. Specifically in the root region, reduction in the blade oscillatory loads is more pronounced. Although the blade design comes out to be heavier for Case 2, the decrease in oscillatory forces is larger for this case. Similar trends are observed for blade moment distributions, which are given in Figure 20. Blade mean torsional moment increases for both cases compared to the initial HBR design. For oscillatory moment distributions Case 2 provides the best reduction.

Blade mean and oscillatory force distributions for IBR optimizations are given in Figure 21. Case 3 and Case 4 favors for high weight blade design resulting in an increased mean axial force on the blade. However, comparing the oscillatory forces F_y and F_z, both cases give similar force reductions and they are lower than forces created by the initial configuration. Comparing the optimization results with the initial configuration in terms of oscillatory and mean moments show significant reductions in oscillatory moments for both cases. Case 4 is more successful in reducing of oscillatory moments compared to Case 3.

For both HBR and IBR, the oscillatory loads are reduced as a result of the optimizations. For each rotor, the reduction of vibratory loads by targeting the maximum blade oscillatory load yields better results compared to targeting frequency separation.

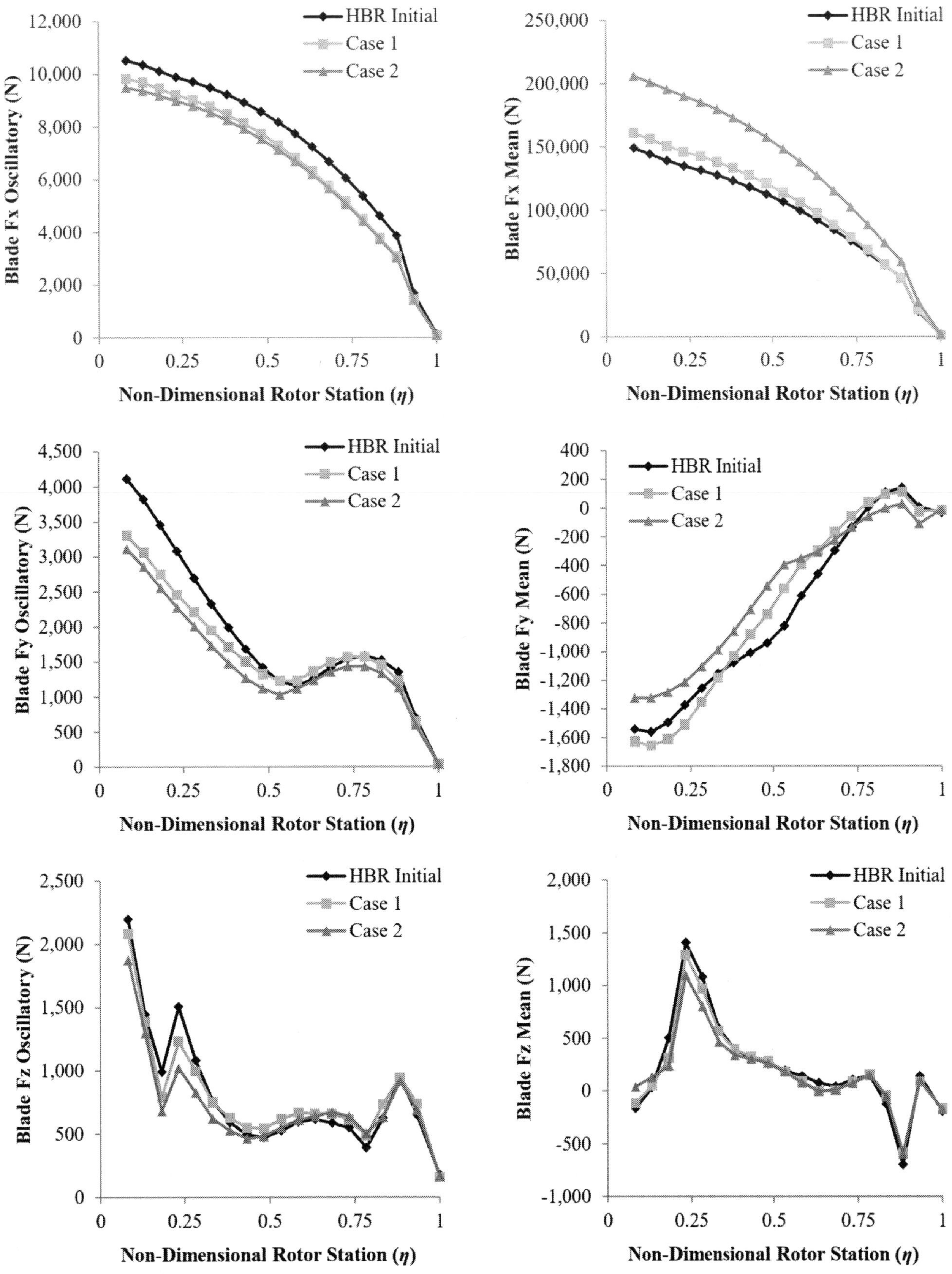

Figure 19. Blade oscillatory and mean force distributions for HBR optimizations

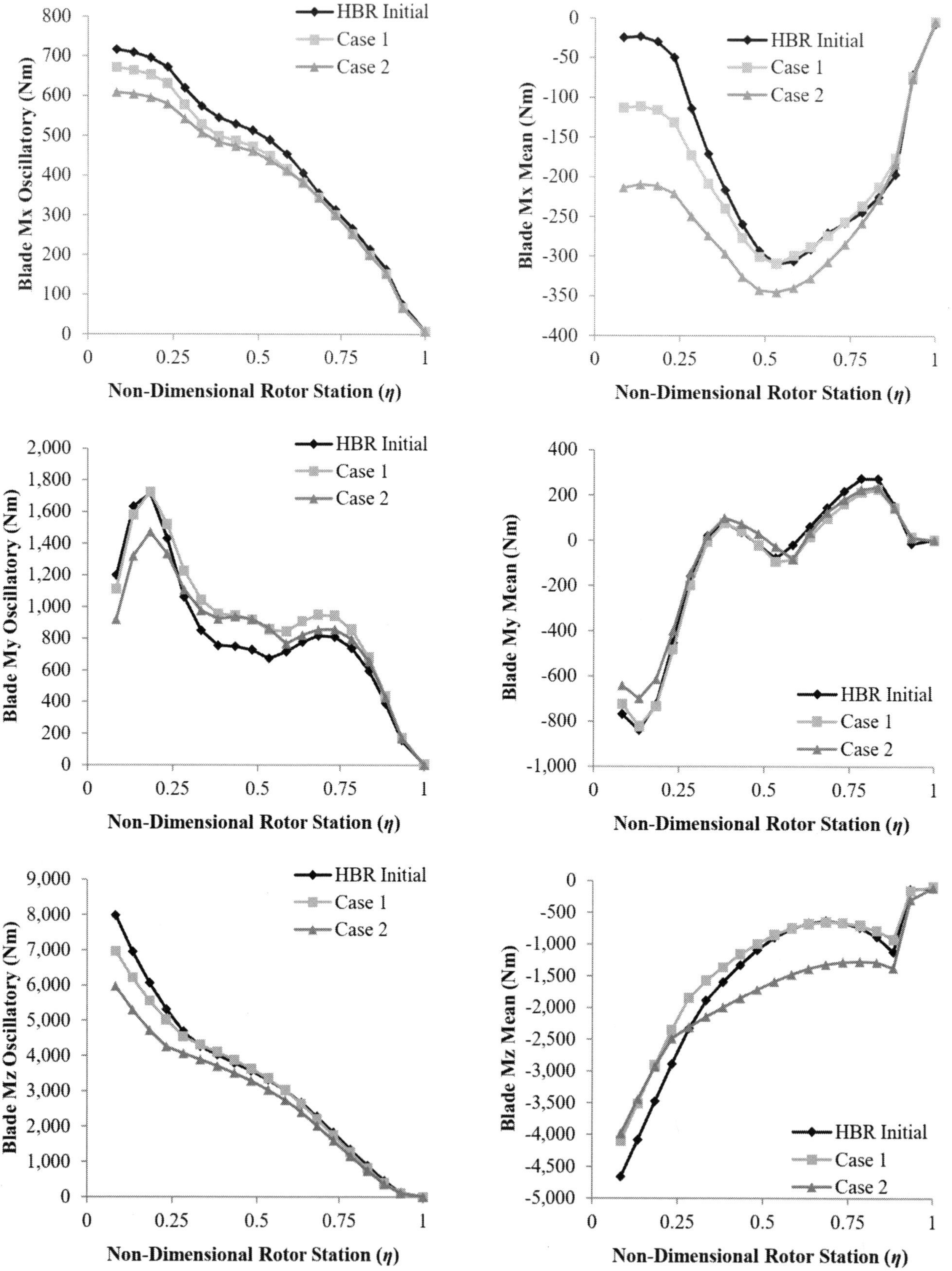

Figure 20. Blade oscillatory and mean moment distributions for HBR optimizations

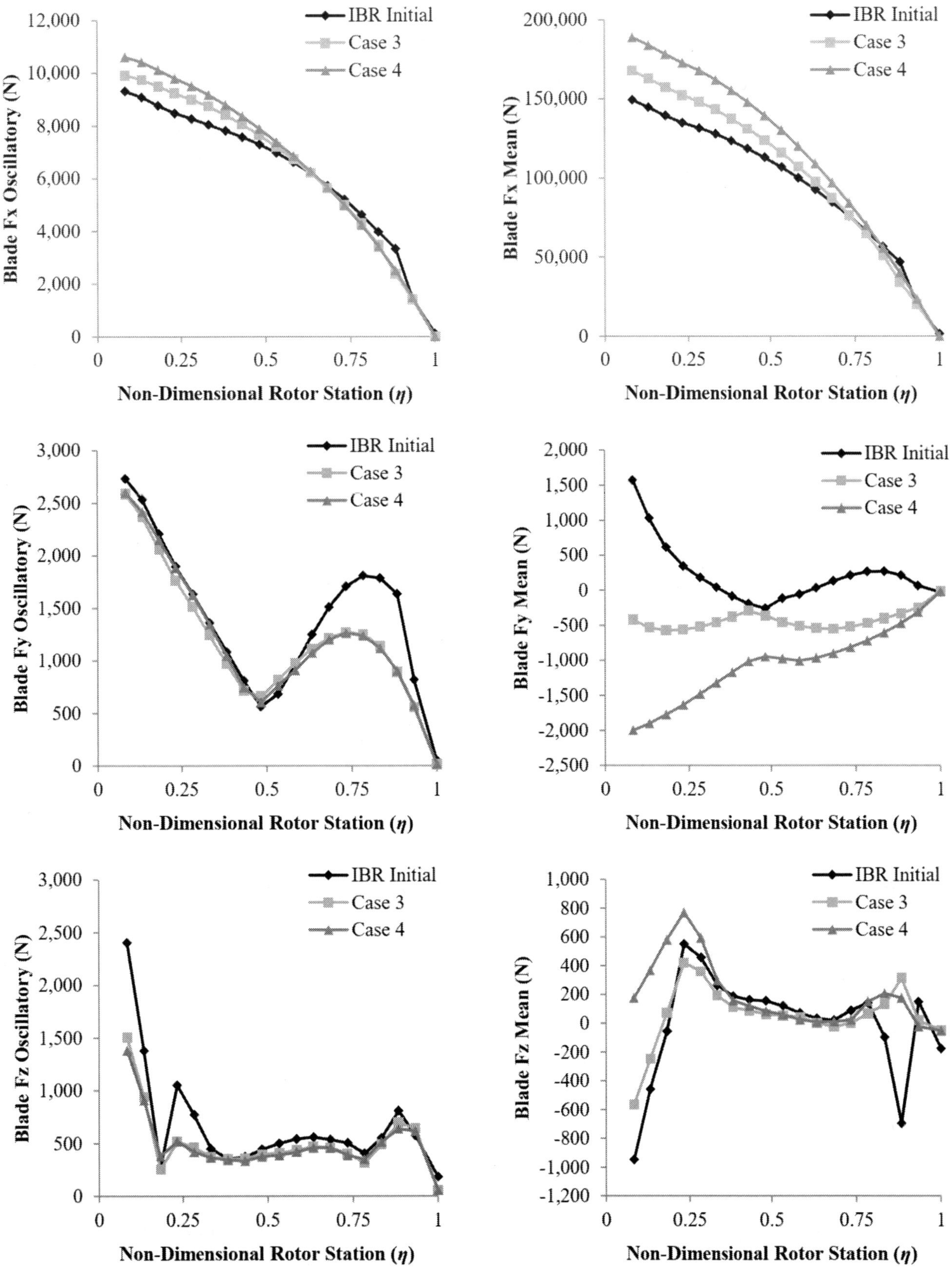

Figure 21. Blade oscillatory and mean force distributions for IBR optimizations

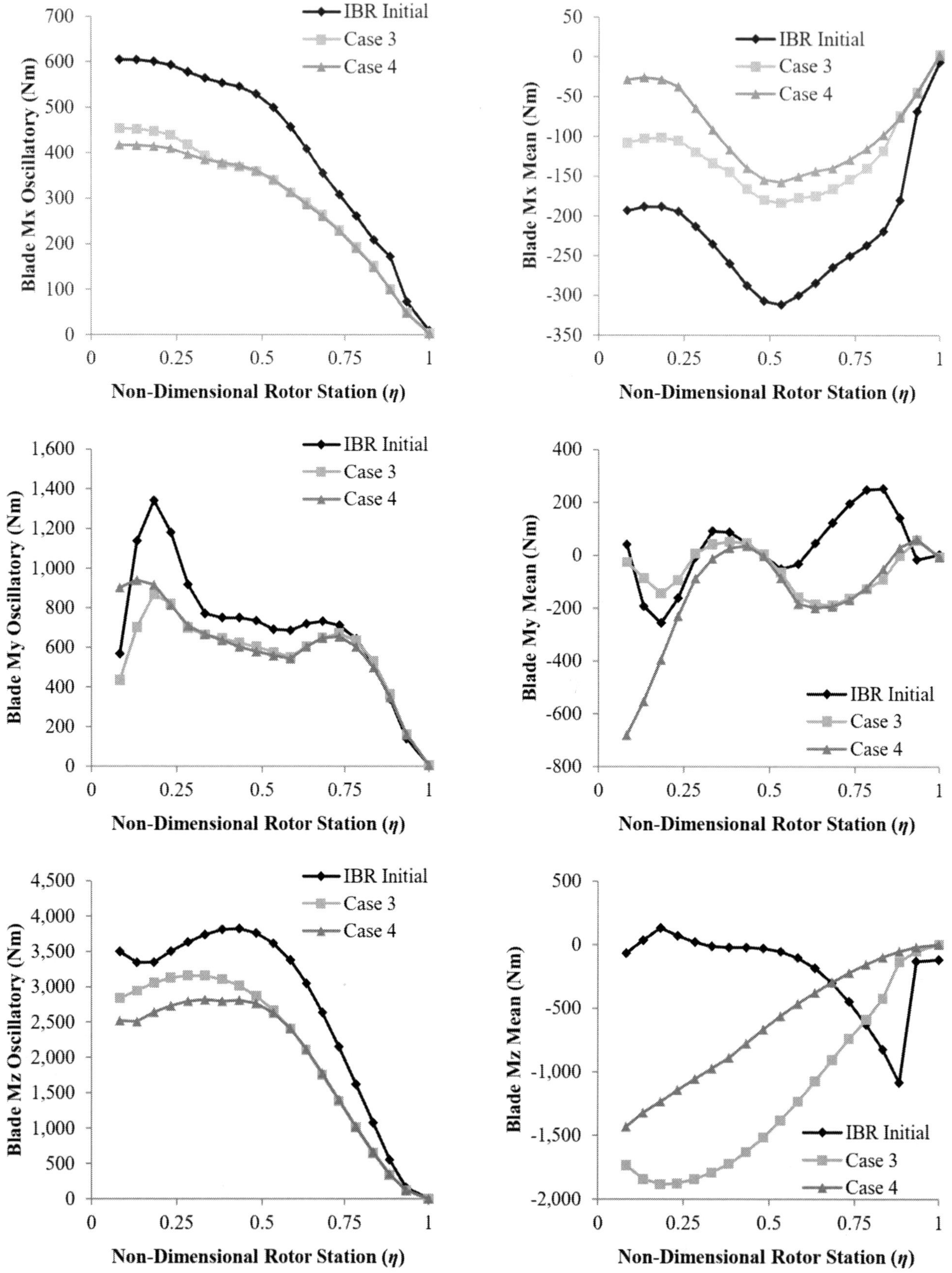

Figure 22. Blade oscillatory and mean moment distributions for IBR optimizations

CONCLUSIONS

In this study, a robust and hybrid optimization method has been developed by using PSO and NMO in conjunction. Utilizing the hybrid optimization method, two rotors were optimized for minimum vibratory loads on the blades. Reductions of blade oscillatory loads were investigated with two main approaches; frequency separation and direct blade oscillatory load minimization. For frequency separation, the effect of resonance on the blades was demonstrated and minimum separation limit has been defined. In blade oscillatory load minimization, the maximum oscillatory forces and moments occurring on the blade were reduced to single resultant force and the objective function was defined based on the minimization of the resultant force.

Besides from targeting vibration reduction, other blade and rotor related behaviors have been considered. In order to limit control efficiency loss, blade torsional deformation limit were defined. To avoid flutter, shear center was targeted to be in front of the feathering axis and center-of gravity was required to be as close to feathering axis as possible. Blade structural integrity was ensured by constraining the maximum strain on the functional region. Lock number was required to be less than 10. In order to keep the blade mass as low as possible, targets functions were defined. For both rotors, the required damping for ground resonance stability was kept in reasonable limits by targeting a cyclic inplane mode frequency. For IBR only, the increase in lag damper effectiveness for collective inplane mode was targeted through decreasing the frequency differences between collective inplane mode frequency and scissor inplane frequency.

A total of 4 cases have been investigated. For each case hybrid optimization was employed and optimization routines were repeated for 10 times to ensure global minima attainment.

As a result of optimizations, following conclusions were drawn:

1. The developed hybrid optimization outperformed GA and SA when compared in terms of their robustness in finding the global minima.
2. The effect of resonance on the blade sectional loads indicates significant load amplifications. Due manufacturing limitations, coupling between modes and design constraints, it is not possible to tweak separate all to modes to be 0.5 Ω away from closest excitation frequency. Hence, it is suggested that 0.2 Ω separation the best compromise for load reduction.
3. The results for IBR and HBR optimization revealed that reducing the vibratory load through reduction of maximum oscillatory loads on the blade results in better outcomes. Although it is possible to reduce the rotor loads transferred to airframe and the blade oscillatory loads through frequency separation, more successful results are obtained by directly targeting maximum blade oscillatory loads.
4. It is noted that the excitation characteristics depend heavily on flight scenario and the rotors should be optimized considering the harsh excitation scenarios.
5. Albeit the rotor configurations are the same except for the damper connection type for IBR and HBR, the optimum designs require different blade designs for minimum load.

Author contact:
M. Emre Bilen, emre.bilen@metu.edu.tr,
Ender Cigeroglu, ender@metu.edu.tr,
H. Nevzat Özgüven, ozguven@metu.edu.tr.

APPENDIX A

Benchmark problems tabulated in Table 1 are listed below.

Ackley's Function

$$f(x) = -20e^{-0.02\sqrt{n^{-1}\sum_{i=1}^{n} x_i^2}} - e^{n^{-1}\sum_{i=1}^{n}\cos(2\pi x_i)} + 20 + e. \tag{27}$$

The function, retrieved from Ref. 20, is subjected to $-35 \leq x_i \leq 35$ and $n = 3$. The global minimum and corresponding function value is $f(0,0,0) = 0$.

Tripod Function

$$f(x) = p(x_2)\big(1 + p(x_1)\big) + \big|x_1 + 50p(x_2)\big(1 - 2p(x_1)\big)\big| + \big|x_2 + 50\big(1 - 2p(x_2)\big)\big|. \tag{28}$$

The function, retrieved from Ref. 20, is subjected to $-100 \leq x_i \leq 100$ and $p(x) = 1$ for $x \geq 0$ otherwise $p(x) = 0$. The global minimum and the corresponding function value is $f(0, -50) = 0$.

Mishra 1 Function

$$f(x) = \sqrt{\left|\cos\left|x_1^2 + x_2\right|\right|} + (x_1 + x_2)/100. \tag{29}$$

The function, retrieved from Ref. 21, is subjected to $-10 \leq x_i \leq 10$ The global minimum and the corresponding function value is $f(-8.4666, -9.9988) = -0.18466$.

Mishra 2 Function

$$f(x) = \sqrt{|\sin|x_1^2 + x_2||} + (x_1 + x_2)/100. \qquad (30)$$

The function, retrieved from Ref. 21, is subjected to $-10 \leq x_i \leq 10$ The global minimum and the corresponding function value is $f(-9.94112, -9.99952) = -0.199441$.

Mishra 5 Function

$$f(x) = \sum_{i=1}^{n} |x_i^5 - 3x_i^4 + 4x_i^3 + 2x_i^2 - 10x_i - 4|. \qquad (31)$$

The function, retrieved from Ref. 21, is subjected to $-10 \leq x_i \leq 10$ and $n = 4$. The global minima are located at $x_i = -1$ *or* 2 and the corresponding function value is $f(x_i) = 0$.

Periodic Function

$$f(x) = 1 + \sin^2(x_1) + \sin^2(x_2) - 0.1e^{-(x_1^2+x_2^2)}. \qquad (32)$$

The function, retrieved from Ref. 20, is subjected to $-10 \leq x_i \leq 10$. The global minimum and the corresponding function value is $f(0,0) = 0.9$.

Rosenbrock Function

$$f(x) = \sum_{i=1}^{n-1} [100(x_{i+1} - x_i^2)^2 + (x_i - 1)^2]. \qquad (33)$$

The function, retrieved from Ref. 20, is subjected to $-5 \leq x_i \leq 5$ and $n = 6$. The global minimum and the corresponding function value is $f(1, \ldots, 1) = 0$.

APPENDIX B

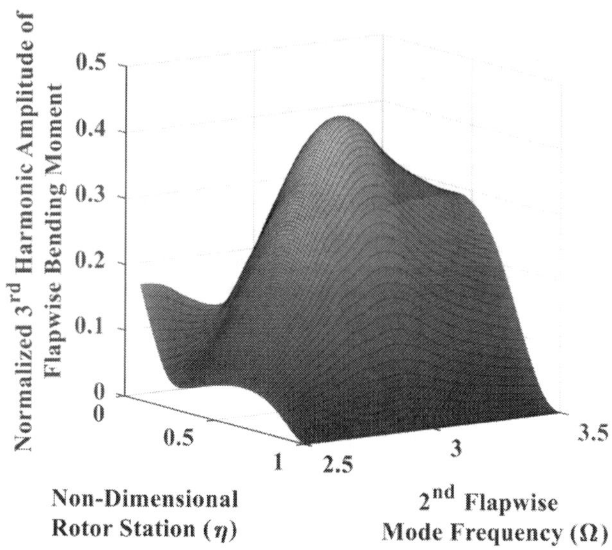

Figure 23. Change of the 3rd harmonic amplitude of flapwise bending moment distribution with the 2nd flap mode frequency

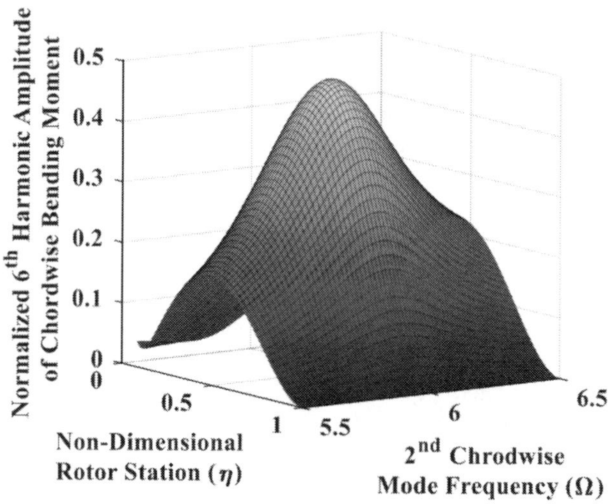

Figure 24. Change of the 6th harmonic amplitude of chordwise bending moment distribution with the 2nd chordwise mode frequency

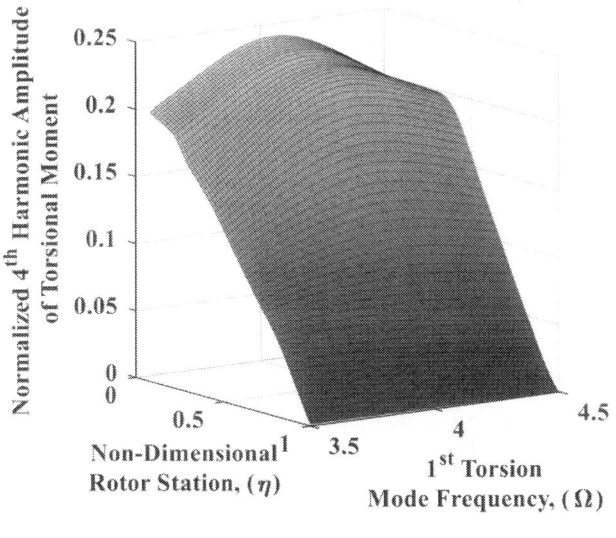

Figure 25. Change of the 4th harmonic amplitude of torsional moment distribution with the 1st torsion mode frequency

ACKNOWLEDGMENTS

The authors would like to thank Alperen Ayberk Isik for his guidance in VABS modeling.

REFERENCES

[1]P. P. Friedmann, "Helicopter Vibration Reduction Using Structural Optimization with Aeroelastic / Multidisciplinary Constraints - A Survey," vol. 28, no. 1, 1990.

[2]H. M. Adelman and W. R. Mantay, "Integrated Multidisciplinary Design Optimization of Rotorcraft," *J. Aircr.*, vol. 28, no. 1, pp. 22–28, 1991.

[3]J. Lee and P. Hajelat, "Parallel Genetic Algorithm Implementation in Multidisciplinary Rotor Blade Design," *J. Aircr.*, vol. 33, no. 5, pp. 962–969, 1996.

[4]J. Kennedy and R. Eberhart, "Particle swarm optimization," *Proc. ICNN'95 - Int. Conf. Neural Networks*, pp. 1942–1948, 1995.

[5]J. A. Nelder and R. Mead, "A Simplex Method for Function Minimization," *Comput. J.*, vol. 7, no. 4, pp. 308–313, Jan. 1965.

[6]J. C. Lagarias, J. A. Reeds, M. H. Wright, and P. E. Wright, "Convergence Properties of the Nelder--Mead Simplex Method in Low Dimensions," *SIAM J. Optim.*, vol. 9, no. 1, pp. 112–147, 1998.

[7]R. C. Eberhart and Yuhui Shi, "Particle swarm optimization: developments, applications and resources," *Proc. 2001 Congr. Evol. Comput. (IEEE Cat. No.01TH8546)*, vol. 1, pp. 81–86, 2001.

[8]O. A. Bauchau, "DYMORE: a finite element based tool for the analysis of nonlinear flexible multibody systems," *Georg. Inst. Technol.*, pp. 1–17, 2001.

[9]O. A. Bauchau, "Computational Schemes for Flexible, Nonlinear Multi-Body Systems," *Multibody Syst. Dyn.*, vol. 2, pp. 169–225, 1998.

[10]D. A. Peters and C. J. He, "Comparison of Measured Induced Velocities with Results from a Closed-Form Finite State Wake Model in Forward Flight." 1989.

[11]T. Hu, "A Validation and Comparison About VABS-IDE and VABS-GUI," 2012.

[12]W. Yu, V. Volovoi, D. H. Hodges, and X. Hong, "Validation of the variational asymptotic beam sectional analysis," *AIAA J.*, vol. 40, no. 10, pp. 2105–2112, 2002.

[13]L. Li, V. V. Volovoi, and D. H. Hodges, "Cross-Sectional Design of Composite Rotor Blades," *J. Am. Helicopter Soc.*, vol. 53, no. September 2016, p. 56, 2008.

[14]A. A. Isik and A. Kayran, "Structural Optimization of Composite Helicopter Rotor Blades," *Am. Soc. Compos. Thirty-First Tech. Conf.*, 2016.

[15]S. Mittal and P. Saxena, "Prediction of Hysteresis Associated with the Static Stall of an Airfoil," *AIAA J.*, vol. 38, no. 5, pp. 933–935, 2000.

[16]A. R. S. Bramwell, G. Done, and D. Balmford, *Bramwell's Helicopter Dynamics*, 2nd ed. 2000.

[17]D. D. Samborsky, T. J. Wilson, P. Agastra, and J. F. Mandell, "Delamination at Thick Ply Drops in Carbon and Glass Fiber Laminates Under Fatigue Loading," *J. Sol. Energy Eng.*, vol. 130, no. 3, p. 22, 2008.

[18]W. Johnson, *Rotorcraft Aeromechanics*. Cambridge University Press, 2013.

[19]M. Cavcar, "The International Standard Atmosphere (ISA)," *Anadolu Univ. Turkey*, pp. 1–7, 2000.

[20]M. Jamil and X.-S. Yang, "A Literature Survey of Benchmark Functions For Global Optimization Problems," pp. 1–47, 2013.

[21]S. K. Mishra, *Performance of Repulsive Particle Swarm Method in Global Optimization of Some Important Test Functions: A Fortran Program*. 2006.

Blade Loads Analysis of a Variable Speed Helicopter Rotor

Dong Han*
Ph.D
Nanjing University of Aeronautics and Astronautics
Nanjing, China

ABSTRACT

To investigate the influence of blade stiffness and rotor speed on the flapwise, lagwise and torsional loads of variable speed helicopter rotors, the aeroelastic modeling of the coupled rotor and fuselage system is derived based on the generalized force formulation. The loads analysis concentrates on the 1/rev and 2/rev loads in the blade flapwise, lagwise and torsional directions. Reducing rotor speed generally causes the increase of blade flapwise, lagwise and torsional loads. Lager flapwise stiffness is preferred for the 1/rev flapwise, 1/rev and 2/rev lagwise, and 1/rev and 2/rev torsional loads. The 2/rev lagwise resonance has significant influence on the loads in the three directions, especially the 2/rev lagwise load. The rotor should avoid working near these resonance rotor speeds. The torsional stiffness has substantially small influence on the lower harmonic loads.

NOMENCLATURE

A	blade cross-sectional area
C_D	fuselage drag coefficient
C_H	rotor drag coefficient
C_{M_X}	rotor rolling moment coefficient
$C_{M_{X_F}}$	fuselage rolling moment coefficient
C_{M_Y}	rotor pitching moment coefficient
$C_{M_{Y_F}}$	fuselage pitching moment coefficient
C_T	thrust coefficient
C_W	weight coefficient
C_Y	rotor side force coefficient
C_{Y_F}	fuselage side force coefficient
F	force
h	vertical distance from helicopter mass center to hub center
I_β	moment inertia about flapping hinge
$I_{\beta\zeta}$	inertial flap-lag coupling
I_ζ	moment inertia about lag hinge
k_P	radius of gyration of blade cross section
l	blade length
M	moment
n	number of generalized degrees
Q	generalized force
q	generalized coordinate
r	axial position
R	rotor radius
S	section elastic constant
$\mathbf{T}$	transformation matrix
t	time
u	axial displacement
u_e	axial elastic displacement
v	lagwise elastic displacement / velocity
w	flapwise elastic displacement
x	axial coordinate
x_{CG}, y_{CG}	longitudianl and lateral distance from helicopter center to rotor shaft
α_s	longitudinal tilt of rotor shaft
β	blade flapping angle
δ	variational symbol
ζ	blade lag angle
η, ζ	section coordinates in deformed coordinate system
θ_0	collective pitch angle
θ_{1c}	lateral cyclic pitch angle
θ_{1s}	longitudinal cyclic pitch angle
ν_β	non-dimensional fundamental flap frequency
ν_ζ	non-dimensional fundamental lag frquency
κ	moment strain curvature
ρ	blade mass density
ϕ	elastic rotation
ϕ_s	lateral tilt of rotor shaft

Subscript

i, j	matrix or vector index

Superscript

$()^A$	contribution from aerodynamics
$()^E$	contribution from elastic potential energy
$()^T$	contribution from kinetic energy
$'$	$\partial/\partial x$
$\dot{}$	$\partial/\partial t$
$\bar{()}$	$()/R$

INTRODUCTION

Varying rotor speed following the variation of helicopter flight states provides an effective and feasible means to improve helicopter performance [1, 2], which has been applied to A160, V-22 and X2 rotorcrafts. However, varying helicopter rotor speed may introduce some troublesome dynamic problems. First, the variation of rotor speed causes the change of excitation frequencies of rotor

---

Presented at the 6th Asian/Austrailian Rotorcraft Forum / Heli Japan 2017, Kanazawa, Japan, November 7-9, 2017.

blades. Second, varying rotor speed causes the change of blade frequency ratios. With decreasing rotor speed, the frequency ratios of the fundamental frequencies increase and the blade turns to be more 'stiff'. Third, the reduction of rotor speed can worsen the asymmetry of aerodynamic environment between the advancing and retreating sides. The advance ratio and reverse flow region increase, which can cause the increase of the region of dynamic stall. These effects may have paramount influence on blade loads of variable speed rotors.

Han et al. investigated the pitch link loads of a variable speed rotor [3]. The investigation indicated that the rotor speed had significant influence on the pitch link loads and the lagwise resonance had a critical influence on the pitch link loads. Bowen-Davies and Chopra investigated the aeromechanics of a variable speed rotor system [4]. The results indicated that the hub loads were sensitive to the reduced rotor speed. Berry and Chopra conducted a wind tunnel test of a Mach-scale variable speed rotor [5]. The test showed that reducing rotor speed could lower the fixed frame hub vibratory loads. Han et al. utilized the fluidlastic dampers embedded at the blade tip to reduce the high transient loads of variable speed rotors during resonance crossing [6]. The results indicated the effectiveness of that method to control the transient loads. Han and Smith investigated the steady and transient lagwise loads of a variable speed rotor [7]. The flap motion had vital contribution to the high steady and transient lagwise loads, and increasing blade lag damping could effectively reduce the transient lagwise loads. Variable speed rotors usually utilize the design of rigid rotor, which has much larger blade stiffness than conventional articulated or hingeless rotors. How the rotor speed and blade stiffness affecting blade loads needs to be investigated.

This work concentrates on the blade loads analysis of a variable speed rotor. The system modeling is derived based on the generalized force formulation, which includes a rotor model, a fuselage model and a propulsive trim method. The blade loads will be analyzed to explore the effect of rotor speed and blade stiffness on the flapwise, lagwise and torsional loads.

MODEL

The system modeling primarily includes a rotor model, a fuselage model and a propulsive trim method.

ROTOR MODEL

This rotor modeling includes a moderate deflection composite beam model, blade kinematics, rotor aerodynamics and an induced velocity model.

COMPOSITE BEAM MODEL

Composite materials are widely applied in manufacturing helicopter rotor blades. A sophisticated moderate deflection composite beam model is utilized to describe the anisotropy characteristics of composite beams [8]. This model is based on the assumptions of small strain and moderate rotations of beam cross section. The variation of the strain energy can be written as

$$\delta U = \sum_{i=1}^{n} Q_i^E \delta q_i = \int_l \left[F_x \delta u_e' + M_x \delta\phi' + M_y \delta\kappa_y + M_z \delta\kappa_z\right] dl \tag{1}$$

The axial linear strain and the second-order moment strain measure are

$$u_e' = u' + \frac{1}{2}\left(v'^2 + w'^2\right) \tag{2}$$

$$\boldsymbol{\kappa} = \begin{Bmatrix} \kappa_x \\ \kappa_y \\ \kappa_z \end{Bmatrix} = \begin{Bmatrix} \phi' \\ -C_\theta w'' + S_\theta v'' \\ S_\theta w'' + C_\theta v'' \end{Bmatrix} \tag{3}$$

where, C_θ denotes $\cos\theta$, and S_θ is $\sin\theta$. θ equals the elastic twist angle plus the pretwist angle. With the nonlinear coupling terms between extension and torsion, the section loads are

$$\begin{Bmatrix} F_x \\ M_x \\ M_y \\ M_z \end{Bmatrix} = \begin{bmatrix} S_{uu} & S_{u\phi} + \frac{1}{2}\phi' S_{uu} k_P^2 & S_{uw} & S_{uv} \\ S_{\phi u} + \phi' S_{uu} k_P^2 & S_{\phi\phi} & S_{\phi w} & S_{\phi v} \\ S_{wu} & S_{w\phi} & S_{ww} & S_{wv} \\ S_{vu} & S_{v\phi} & S_{vw} & S_{vv} \end{bmatrix} \begin{Bmatrix} u_e' \\ \kappa_x \\ \kappa_y \\ \kappa_z \end{Bmatrix} \tag{4}$$

The stiffness matrix of the cross section is calculated by explicit expressions or some beam cross section analysis software. The position vector of an arbitrary point on the blade can be written as

$$\begin{Bmatrix} \mathbf{R_0} \\ 1 \end{Bmatrix} = \mathbf{T_0 r_0} = \begin{bmatrix} & \mathbf{C} & & \begin{matrix} x+u \\ v \\ w \end{matrix} \\ 0 & 0 & 0 & 1 \end{bmatrix} \begin{Bmatrix} 0 \\ \eta \\ \zeta \\ 1 \end{Bmatrix} \tag{5}$$

where, **C** denotes the transformation matrix from the deformed coordinate frame to the undeformed coordinate frame.

KINETIC ENERGY

For the convenience of the description of the kinematics of rigid motions, the rotations are usually defined in the local rotating frame. The rotations about rotor shaft, and flap, lag and feathering hinges are described as generalized coordinates. The 4×4 transformation matrix is utilized to describe the rigid rotation [9]. If a blade has several rigid rotations (for example, m), the position vector after the elastic deformations and rigid rotations can be written as

$$\begin{Bmatrix}\mathbf{R}\\1\end{Bmatrix} = \begin{bmatrix}\mathbf{H} & \mathbf{d}\\ \mathbf{0}_{3\times1}^{\mathbf{T}} & 1\end{bmatrix}\mathbf{r}_0 = \mathbf{T}_m\mathbf{T}_{m-1}\cdots\mathbf{T}_2\mathbf{T}_1\mathbf{T}_0\cdot\mathbf{r}_0 \tag{6}$$

The transformation matrix $\mathbf{T}_i$ denotes a rigid rotation ($i > 0$).

The methodology in Ref. 10 and Ref. 11 to calculate the kinetic energy is employed to describe the nonlinear coupling characteristics between elastic deflections and rigid rotations. The variation of the kinetic energy of the blade is

$$\delta T = \sum_{i=1}^{n} Q_i^T\delta q_i = \sum_{i=1}^{n}\int_l\iint_A -\rho\ddot{\mathbf{R}}\cdot\frac{\partial\mathbf{R}}{\partial q_i}dA\,dl\,\delta q_i \tag{7}$$

The tangent mass, damping, and stiffness matrices introduced by the kinetic energy can be derived by the partial derivative of the variation of the kinetic energy to the generalized acceleration, velocity and displacement, respectively.

AERODYNAMICS

The relative velocity of an arbitrary point on the blade pitch axis to the local air flow is determined by the air flow, the induced velocity over the rotor disk and the blade motion. According to the angle of attack and the resultant velocity, the lift coefficient C_l, drag coefficient C_d and pitching moment coefficient C_m of the airfoil segment is determined by Leishman-Beddoes aerodynamics with dynamic stall [12,13]. Leishman-Beddoes aerodynamic model is a semi-empirical model, which can capture the unsteady lift, drag and pitching moment of an airfoil undergoing dynamic stall. The trailing edge separation and vortex shedding phenomena in dynamic stall are considered based on the modeling of the attached flow model. Under attached flow conditions, the normal force coefficient and pitching moment coefficient includes three components: 1) the circulatory component due to angle of attack and pitch rate; 2) the non-circulatory component due to angle of attack; 3) the non-circulatory component due to pitch rate. The two coefficients are

$$C_n = C_n^c + C_{n_\alpha}^i + C_{n_q}^i \tag{8}$$

$$C_m = C_m^c + C_{m_\alpha}^i + C_{m_q}^i \tag{9}$$

The circulatory component due to angle of attack and pitch rate is modified to consider the trailing edge separation. For the vortex shedding phenomenon in dynamic stall, the extra terms C_n^v and C_m^v are added to the normal force and pitching moment coefficients.

The variation of the work done by the aerodynamics is

$$\delta W^A = \sum_{i=1}^{n} Q_i^A\delta q_i = \int_l\left(\mathbf{F}_A\cdot\frac{\partial\mathbf{R}}{\partial q_i} + \mathbf{M}_A\cdot\frac{\partial\mathbf{a}_s}{\partial q_i}\right)\delta q_i dl \tag{10}$$

It should be noted that the aerodynamic force vector $\mathbf{F}_A$, aerodynamic moment vector $\mathbf{M}_A$ and angle vector $\mathbf{a}_s$ are defined in the inertial coordinate frame. The Pitt-Peters inflow model is utilized to determine the distribution of induced velocity over the rotor disk [14].

EQUATIONS OF MOTION

By using Hamilton's principle, the implicit nonlinear dynamic equations of motion based on the generalized force formulation include four parts: blade elastic potential energy, blade kinetic energy and work done by the aerodynamics. The equations of motion are

$$Q_i^E(\mathbf{q}) + Q_i^T(\mathbf{q},\dot{\mathbf{q}},\ddot{\mathbf{q}},t) + Q_i^A(\mathbf{q},\dot{\mathbf{q}},t) = 0 \quad (i = 1,\cdots,n) \tag{11}$$

The Newmark integration method is utilized to calculate the periodic responses [15]. This unconditionally stable implicit scheme permits the use of large time integration step. This step is determined by the considerations of accuracy.

PROPULSIVE TRIM

For the propulsive trim analysis, the fuselage is treated as a rigid body with aerodynamic forces and moments. Three pitch controls and two rotor attitude angles are the input variables to solve the three force and two moment equilibrium equations [16], as seen in Figure 1. The control input vector is $\mathbf{x} = \{\theta_0\ \ \theta_{1c}\ \ \theta_{1s}\ \ \alpha_s\ \ \phi_s\}^T$, and the output or target vector is $\mathbf{y} = \{y_1\ \ y_2\ \ y_3\ \ y_4\ \ y_5\}^T = \mathbf{0}$. The equilibrium equations in the non-dimensional form are

$$\begin{cases} y_1 = C_T - C_W \\ y_2 = C_D + C_H - C_T\alpha_s \\ y_3 = C_Y + C_{Y_F} + C_T\phi_s \\ y_4 = C_{M_Y} + C_{M_{Y_F}} + C_W(\bar{h}\alpha_s - \bar{x}_{CG}) - \bar{h}C_D \\ y_5 = C_{M_X} + C_{M_{X_F}} + C_W(\bar{h}\phi_s - \bar{y}_{CG}) + \bar{h}C_{Y_F} \end{cases} \tag{12}$$

Newton–Raphson method is utilized to calculate the trimmed values in hover and forward flight.

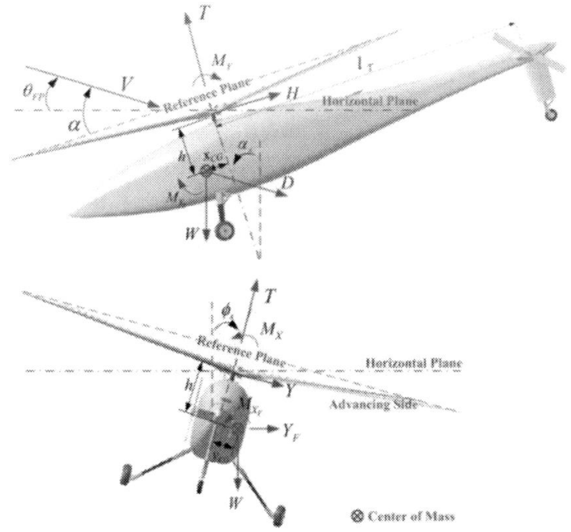

Fig. 1 Forces and moments on the rotor and fuselage.

AEROELASTIC ANALYSIS

The aeroelastic analysis of a variable speed rotor is presented. Comprehensive parametric analysis is conducted to investigate the influence of rotor speed and blade stiffness on the blade loads. Since the 1/rev and 2/rev loads are usually the major components of blade loads, this work concentrates on the analysis of these lower harmonic blade loads.

BASELINE PARAMETERS

The weight of the baseline helicopter is 2000 kg. In the fuselage model, only aerodynamic drag is considered in the aeroelastic analysis and the ratio of the equivalent flat plate area to the rotor disk area is 0.004. The vertical length from the rotor hub center to the rotor shaft of the helicopter is 1.1 m. The longitudinal distance from the hub center of the tail rotor to the mass centre of the helicopter is 8.25 m. The parameters of the rotor are shown in Table 1. The baseline blade has uniform mass and stiffness distribution. The fundamental frequency ratios in the flapwise, lagwise and torsional directions at the rotor speed 350 rpm are 1.30, 1.40 and 8.57 respectively. The rotor blade frequencies with rotor speed are shown in Figure 2. 'F' denotes the frequency in the flapwise direction. 'L' denotes the frequency in the lagwise direction. 'T' denotes the frequency in the torsional direction. From the figure, it is obvious that the fundamental lagwise frequency ratio crosses the 2/rev resonant region at the rotor speed 238 rpm. Figure 3 shows the rotor power with the forward speed and rotor speed. It is obvious that reducing rotor speed can significantly reduce the rotor power. It can be concluded that decreasing rotor speed is an effective means to reduce helicopter power.

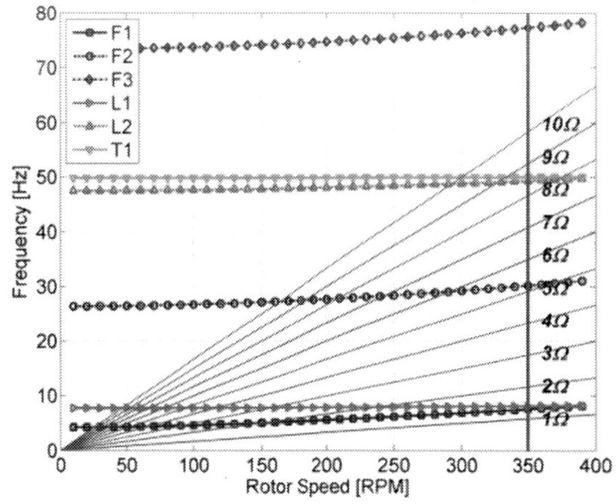

Fig. 2 Rotor blade frequencies.

Table 1 Parameters of the rotor

Parameter	Value
Rotor Radius	5.5 m
Chord Length	0.35 m
Number of Blades	4
Airfoil	NACA0012
Rotor Speed Range	250 - 350 rpm
Blade Pretwist	-12°
Blade Mass	23.1 kg/m

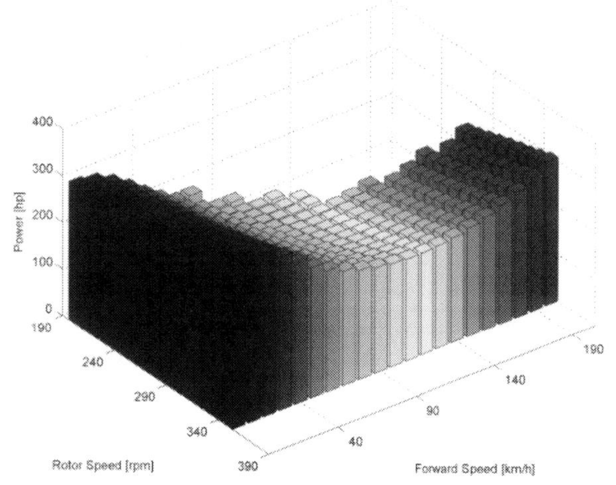

Fig. 3 Rotor power with forward speed and rotor speed

INFLUENCE OF FLAPWISE STIFFNESS

In the following analysis, we concentrate on the investigation of blade root loads, since usually the location of the maximum loads occur at the blade root. Varying rotor speed can achieve better performance improvement at low to medium forward speed flight. In the following analysis, the typical forward speed 100 km/h is adopted. The blade flapwise stiffness is adjusted to make the fundamental flapwise frequency ratio be 1.10, 1.20, 1.30 and 1.40 at the rotor speed 350 rpm respectively. Figure 4 shows the influence of the blade flapwise stiffness and rotor speed on the 1/rev flapwise root bending moment. The 1/rev flapwise load decreases with the rotor speed and then increases. The variation of this load after 300 rpm is limited. With the increase of the flapwise stiffness, the 1/rev flapwise load decreases significantly. The frequency ratio away from 1.0 causes the frequency away from the resonance frequency 1Ω, then the 1/rev flapwise load decreases. Figure 5 shows the influence of the blade flapwise stiffness and rotor speed on the 2/rev flapwise root bending moment. When the frequency ratio is 1.10, the variation of the 2/rev flapwise load is substantially small. When the ratio is 1.40, the 2/rev flapwise load is much larger than the 1/rev load, and decreases with the blade stiffness. With increasing blade flapwise stiffness, the 2/rev flapwise load increases significantly. For the consideration of flapwise loads, higher flapwise stiffness is preferred for the 1/rev flapwise load and lower flapwise stiffness is preferred for the 2/rev load.

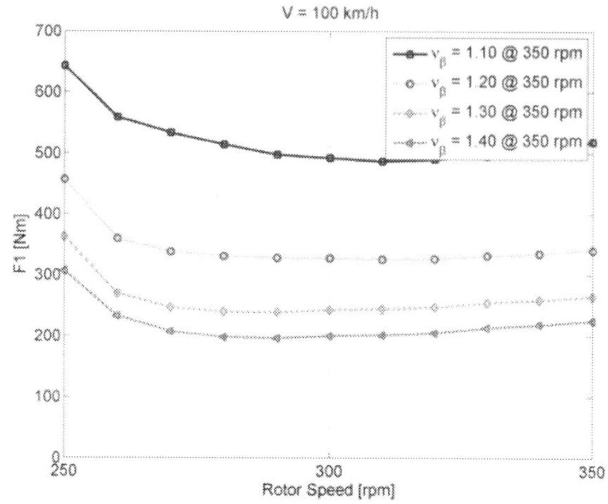

Fig. 4 The first harmonic flapwise load with rotor speed and flapwise stiffness.

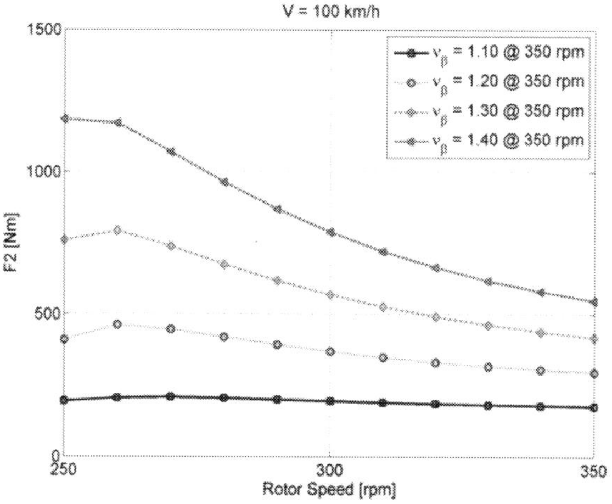

Fig. 5 The second harmonic flapwise load with rotor speed and flapwise stiffness.

Figure 6 shows the influence of the blade flapwise stiffness and rotor speed on the 1/rev lagwise root bending moment. It is obvious that the blade with larger flapwise stiffness has much smaller 1/rev lagwise load. The amplitudes increase by 6.7% and 17.1% for the cases with the frequency ratios 1.10 and 1.20, when the rotor speed changes from 250 rpm to 350 rpm. The amplitudes decrease by 8.9% and 29.8% for the cases with the frequency ratios 1.30 and 1.40. The variation of the 1/rev lagwise load is limited. Figure 7 shows the influence of the blade flapwise stiffness and rotor speed on the 2/rev lagwise root bending moment. With the increase of the blade flapwise stiffness, the 2/rev lagwise load decreases. The amplitude decreases by 43.7% at the rotor speed 350 rpm, when the frequency ratio changes from 1.10 to 1.40. This value is 26.8% at the rotor speed 250 rpm. With the reduction of the rotor speed, the 2/rev lagwise load increases significantly, and the amplitude is much larger than the other harmonic loads. This is due to the rotor speed approaching the lagwise resonance rotor speed 238 rpm. Increasing the flapwise stiffness can decrease the high lagwise loads near lagwise resonance rotor speed.

The coupled flap and lag equations of motion [17] for a rigid blade are

$$I_\beta \ddot{\beta} + I_\beta \Omega^2 \upsilon_\beta^2 \beta - 2I_{\beta\zeta} \Omega \beta \dot{\zeta} = F_\beta \tag{13}$$

$$I_\zeta \ddot{\zeta} + I_\zeta \Omega^2 \upsilon_\zeta^2 \zeta + 2I_{\beta\zeta} \Omega \beta \dot{\beta} = F_\zeta \tag{14}$$

It is obvious that the blade flap motion can excite the lag motion due to the Coriolis force. Large flap motion ($\beta\dot{\beta}$) can introduce large lag motion. This indicates the blade with smaller stiffness in the flapwise direction can excite larger lag motion and larger lagwise loads than the blade with larger stiffness. For the consideration of lagwise loads, larger flapwise stiffness is preferred to reduce the coupling.

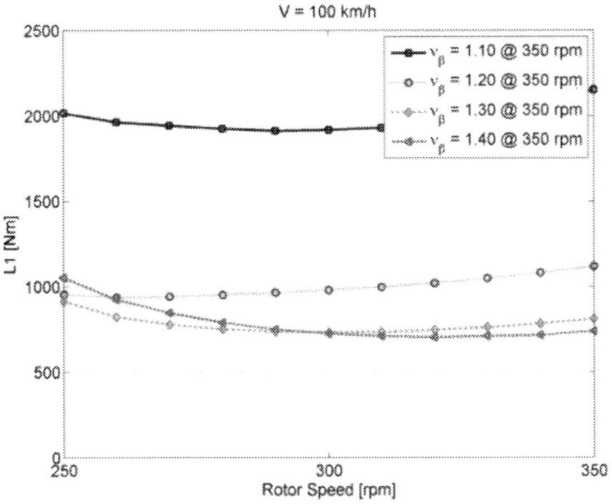

Fig. 6 The first harmonic lagwise load with rotor speed and flapwise stiffness.

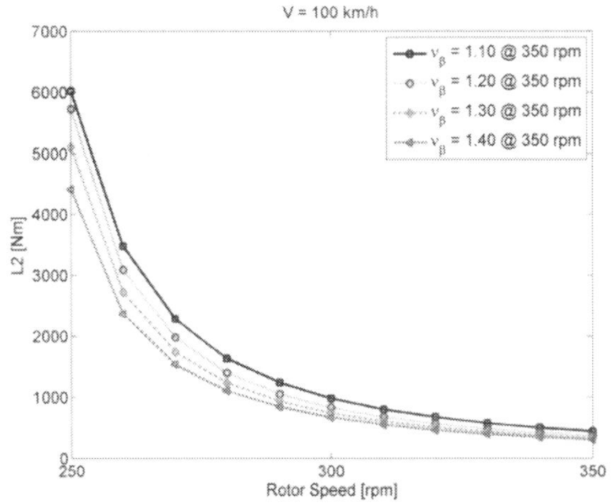

Fig. 7 The second harmonic lagwise load with rotor speed and flapwise stiffness.

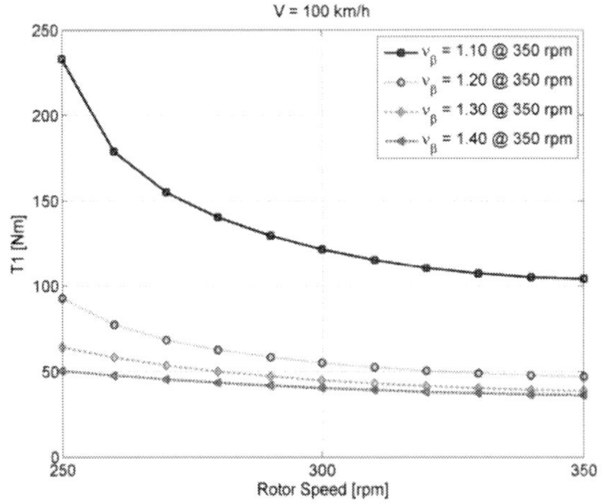

Fig. 8 The first harmonic torsional load with rotor speed and flapwise stiffness.

Figure 8 shows the influence of the blade flapwise stiffness and rotor speed on the 1/rev torsional load. The 1/rev torsional load decreases with the rotor speed. The amplitude with smaller flapwise stiffness reduces much more than that with larger flapwise stiffness. With increasing flapwise stiffness, the 1/rev torsional load decreases. The amplitude decreases by 78.3% at the rotor speed 250 rpm when the frequency ratio changes from 1.10 to 1.40. This value is 65.2% at the rotor speed 350 rpm. Figure 9 shows the influence of the blade flapwise stiffness and rotor speed on the 2/rev torsional load. It is obvious the trend is similar with Figure 8. At the rotor speed 250 rpm, the 2/rev torsional load is much larger than the 1/rev torsional load. Increasing flapwise blade stiffness can significantly reduce the 2/rev torsional load. For the consideration of the torsional loads, larger flapwise stiffness is preferred.

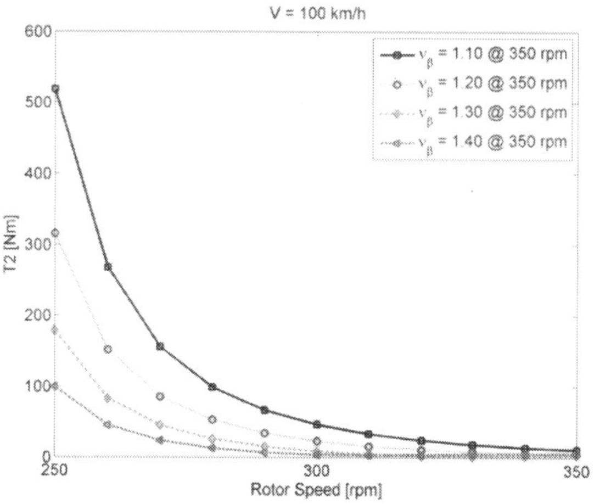

Fig. 9 The second harmonic torsional load with rotor speed and flapwise stiffness.

INFLUENCE OF LAGWISE STIFFNESS

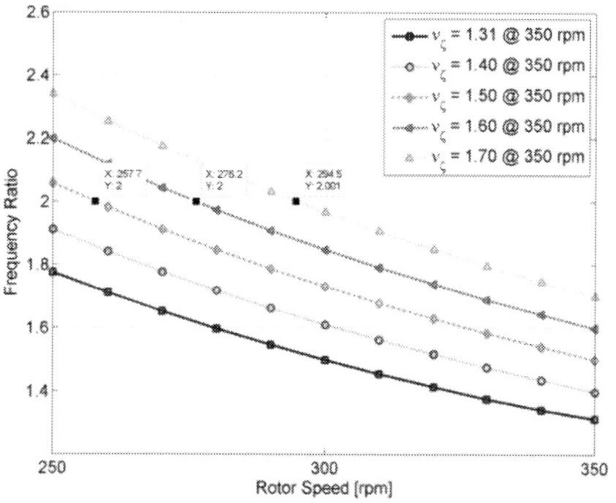

Fig. 10 Lagwise frequency ratio with rotor speed.

The blade lagwise stiffness is adjusted to make the fundamental flapwise frequency ratio be 1.31, 1.40, 1.50, 1.60 and 1.70 at the rotor speed 350 rpm respectively. The fundamental lagwise frequency ratios with the rotor speed are shown in Figure 10. With increasing rotor speed, the frequency ratio decreases almost linearly. The 2/rev resonance rotor speed changes to 258 rpm, 276 rpm and 295 rpm for the lagwise frequency ratios 1.50, 1.60 and 1.70, respectively. The resonance rotor speed increases with blade lagwise stiffness.

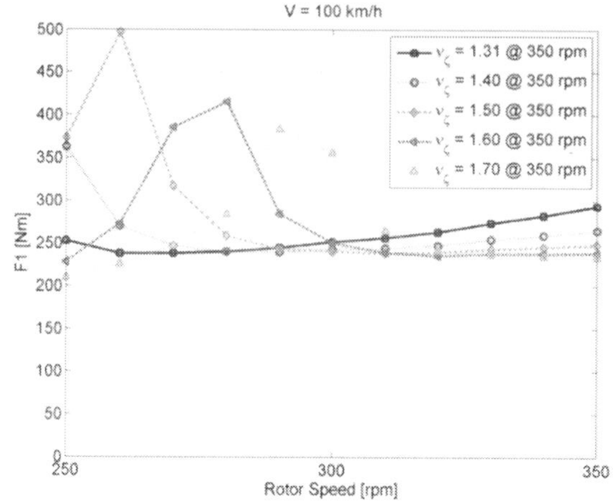

Fig. 11 The first harmonic flapwise load with rotor speed and lagwise stiffness.

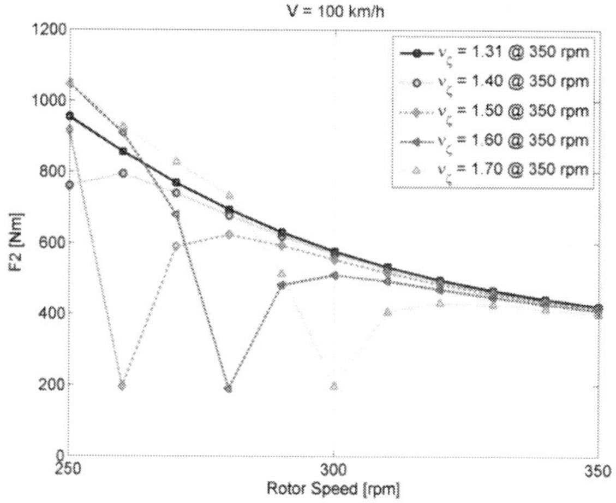

Fig. 12 The second harmonic flapwise load with rotor speed and lagwise stiffness.

Figure 11 shows the influence of the blade lagwise stiffness and rotor speed on the 1/rev flapwise load. The 1/rev flapwise load increases sharply near the lagwise resonance rotor speed. At higher rotor speeds, for example larger than 320 rpm, the load decreases with increasing blade lagwise stiffness. The variation is limited. Figure 11 shows the influence of the blade lagwise stiffness and rotor speed on the 2/rev flapwise load. The 2/rev flapwise load decreases sharply near the lagwise resonance rotor speed. At higher rotor speeds, the variation of the load with blade lagwise stiffness is little. The lagwise resonance has very strong influence on the flapwise loads.

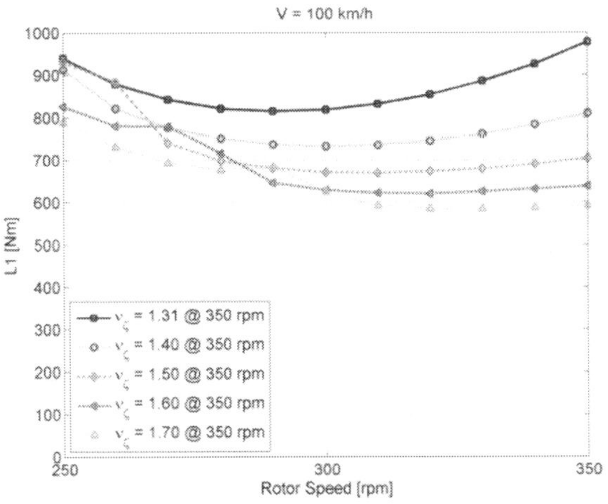

Fig. 13 The first harmonic lagwise load with rotor speed and lagwise stiffness.

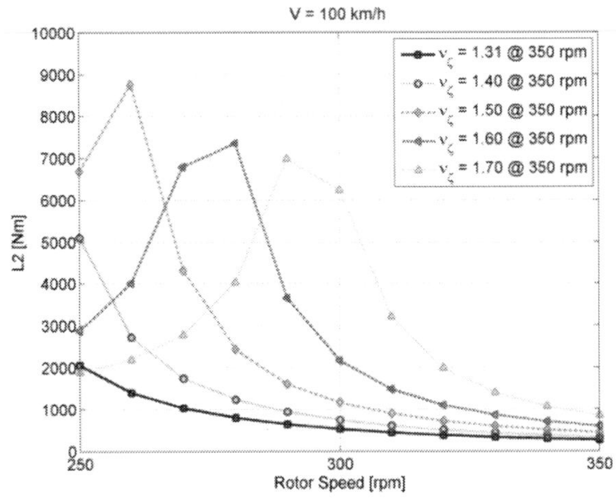

Fig. 14 The second harmonic lagwise load with rotor speed and lagwise stiffness.

Figure 13 shows the influence of the blade lagwise stiffness and rotor speed on the 1/rev lagwise load. Generally, the 1/rev lagwise load decreases with increasing blade lagwise stiffness, which is due to the deviation of the fundamental lagwise frequency away the 1Ω rotor speed. At higher or lower rotor speeds, the 1/rev load is usually larger than the load at medium rotor speeds, which means at higher or lower advance ratios, the 1/rev lagwise load can be larger. Figure 14 shows the influence of the blade lagwise stiffness and rotor speed on the 2/rev lagwise load. It is obvious that the 2/rev load increases sharply near the resonance rotor speed. The amplitude is much larger than that at other rotor speeds. The rotor should avoid working near lagwise resonance rotor speeds. The peak resonance lagwise load decreases with increasing rotor speed, which means the resonance occurring at lower rotor speed can cause larger lagwise load.

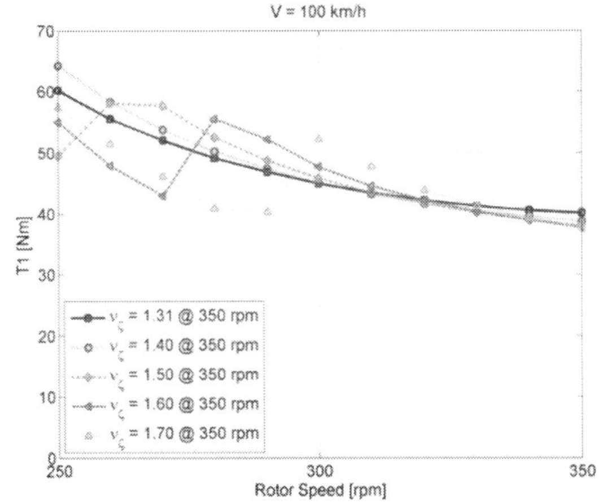

Fig. 15 The first harmonic torsional load with rotor speed and lagwise stiffness.

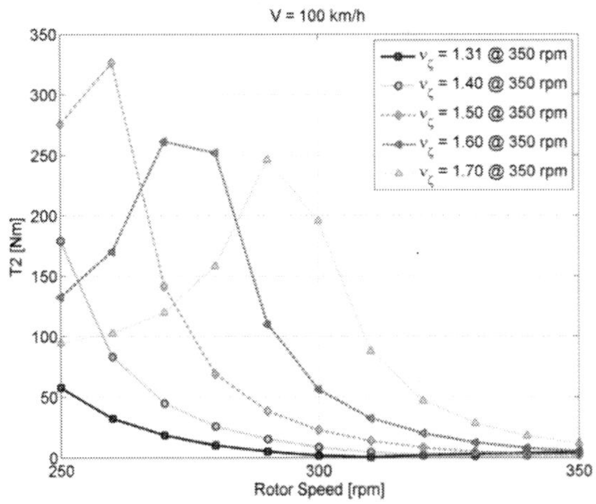

Fig. 16 The second harmonic torsional load with rotor speed and lagwise stiffness.

Figure 15 shows the influence of the blade lagwise stiffness and rotor speed on the 1/rev torsional load. The variation occurs near the resonance rotor speeds. At higher rotor speeds, the variation of the 1/rev torsional load is substantially small. The 1/rev torsional load decreases with increasing rotor speed for the cases with frequency ratio 1.31 and 1.40. Figure 16 shows the influence of the blade lagwise stiffness and rotor speed on the 2/rev torsional load. It is obvious that the 2/rev load increases sharply near the resonance rotor speed. The amplitude is much larger than that at other rotor speeds. The 2/rev torsional load is much smaller than the 1/rev load at higher rotor speeds. But it is much larger near the resonance rotor speeds. The rotor should avoid working near lagwise resonance rotor speeds. The peak resonance torsional load decreases with increasing rotor speed, which means the resonance occurring at lower rotor speed can cause larger torsional load.

INFLUENCE OF TORSIONAL STIFFNESS

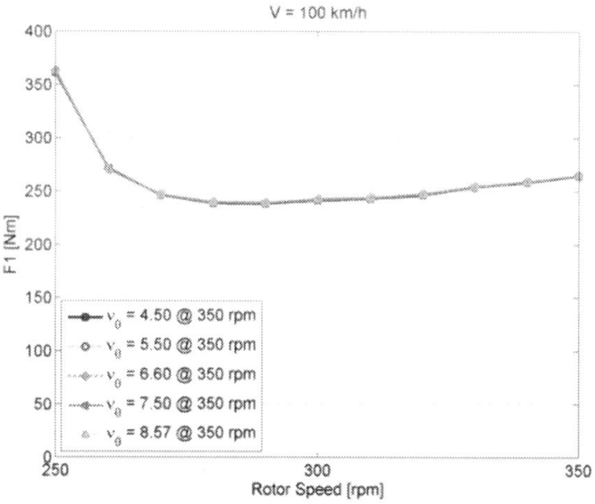

Fig. 17 The first harmonic flapwise load with rotor speed and torsional stiffness.

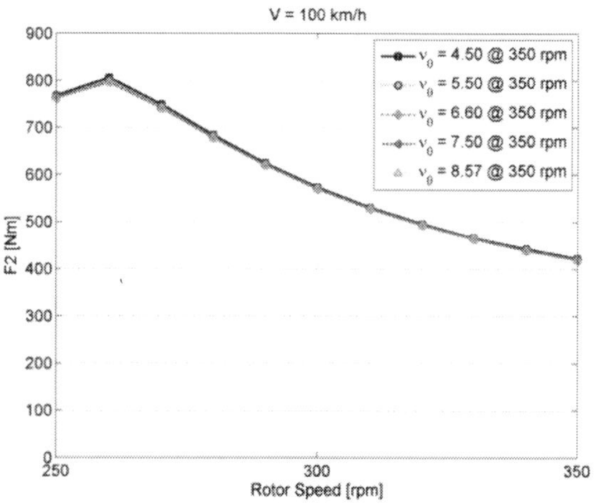

Fig. 18 The second harmonic flapwise load with rotor speed and torsional stiffness.

The blade torsional stiffness is adjusted to make the fundamental torsional frequency ratio be 4.50, 5.50, 6.50, 7.50 and 8.57 at the rotor speed 350 rpm respectively. Figure 17 and Figure 18 show the influence of the blade torsional stiffness and rotor speed on the 1/rev and 2/rev flapwise loads. It is obvious that the torsional stiffness has substantially small influence on the 1/rev and 2/rev flapwise loads.

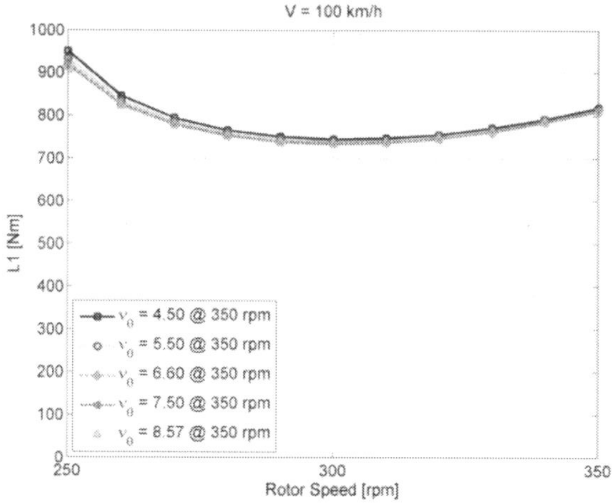

Fig. 19 The first harmonic lagwise load with rotor speed and torsional stiffness.

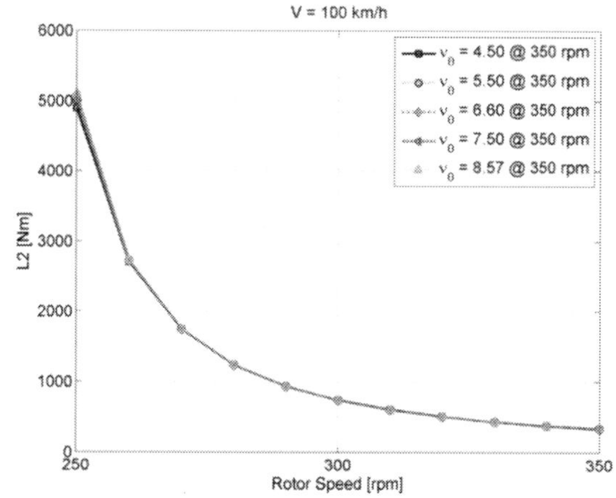

Fig. 20 The second harmonic lagwise load with rotor speed and torsional stiffness.

Figure 19 and Figure 20 show the influence of the blade torsional stiffness and rotor speed on the 1/rev and 2/rev lagwise loads. It is obvious that the torsional stiffness has substantially small influence on the 1/rev and 2/rev lagwise loads. At lower rotor speeds, the effect is very small. The 1/rev lagwise load decreases with increasing blade torsional stiffness.

Figure 21 and Figure 22 show the influence of the blade torsional stiffness and rotor speed on the 1/rev and 2/rev torsional loads. It is obvious that the torsional stiffness has substantially small influence on the 1/rev and 2/rev torsional loads. The 1/rev and 2/rev torsional load decreases significantly with increasing rotor speed. It can be concluded that the torsional stiffness has substantially small influence on the 1/rev and 2/rev flapwise, lagwise and torsional loads. The rotor working at lower rotor speeds will survive large torsional loads.

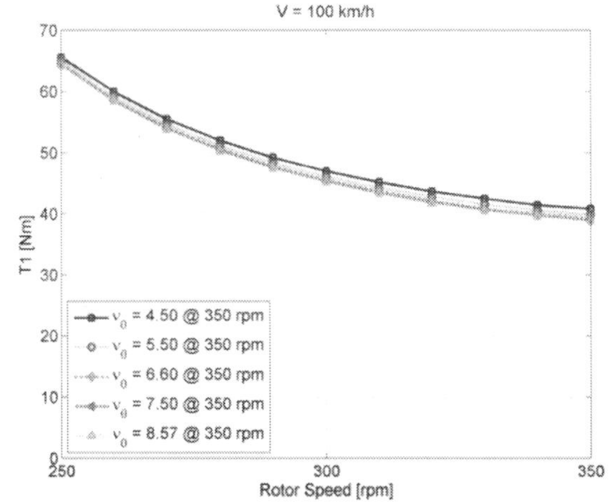

Fig. 21 The first harmonic torsional load with rotor speed and torsional stiffness.

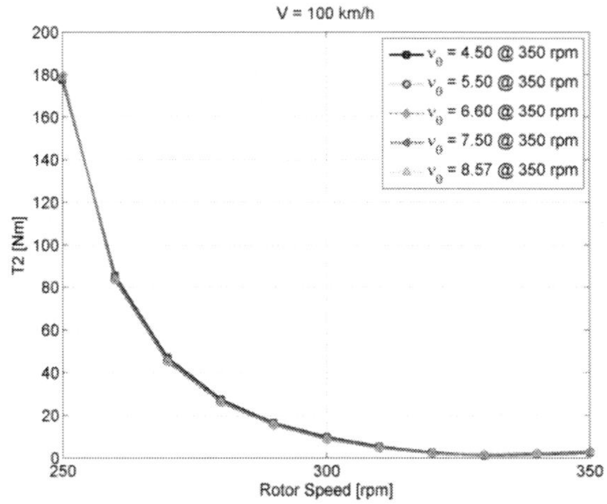

Fig. 22 The second harmonic torsional load with rotor speed and torsional stiffness.

CONCLUSIONS

To investigate the influence of the blade stiffness and rotor speed on the flapwise, lagwise and torsional loads of variable speed rotors, the rotor blade is treated as moderate deflection beam model undergoing coupled flap, lag and torsion deflections. An unsteady aerodynamic model with dynamic stall is adopted to calculate the blade airfoil aerodynamics. A propulsive trim method is utilized to simulate the real flight state. The equations of motion of the system are derived based on the generalized force formulation. The loads analysis concentrates on the 1/rev and 2/rev loads in the blade flapwise, lagwise and torsional directions. The investigation yielded the following conclusions:

1) Reducing rotor speed generally causes the increase of blade flapwise, lagwise and torsional loads.

2) The 2/rev lagwise resonance has significant influence on the flapwise, lagwise and torsional loads, especially the 2/rev lagwise load. The rotor should avoid working near these resonance rotor speeds.

3) Lager flapwise stiffness is preferred for the 1/rev flapwise, 1/rev and 2/rev lagwise, and 1/rev and 2/rev torsional loads. Increasing blade flapwise stiffness can increase the 2/rev flapwise load.

4) Lagwise stiffness has limited influence on the 1/rev and 2/rev flapwise and torsional loads, except the rotor working near lagwise resonance rotor speed. Increasing lagwise stiffness can decrease the 1/rev lagwise load and increase the 2/rev load.

5) The torsional stiffness has substantially small influence on the 1/rev and 2/rev flapwise, lagwise and torsional loads. The rotor working lower rotor speeds will survive large torsional loads.

Author contact: Dong Han donghan@nuaa.edu.cn

ACKNOWLEDGMENTS

This work is supported from the Natural Nature Science Foundation of China (11472129), Aeronautical Science Foundation of China (20165752048) and Science and Technology on Rotorcraft Aeromechanics Laboratory (61422200504162200002).

REFERENCES

[1] Prouty, R. W., "Should We Consider Variable Rotor Speeds?" *Vertiflite*, Vol. 50, No. 4, 2004, pp. 24-27.
[2] DiOttavio, J. and Friedmann, D., "Operational Benefits of an Optimal, Widely Variable Speed Rotor," 66th Annual Forum Proceedings of the American Helicopter Society, Phoenix, AZ, May 11 - 13, 2010.
[3] Han, D., Smith, E. C. and Rahn, C. D., "Pitch Link Loads Analysis of a Variable Speed Rotor," 67th Annual Forum Proceedings of the American Helicopter Society, Virginia Beach, Virginia, May 3-5, 2011.
[4] Bowen-Davies, G. and Chopra, I., "Aeromechanics of a Variable-Speed Rotor," 67th Annual Forum Proceedings of the American Helicopter Society, Virginia Beach, VA, May 3 - May 5, 2011.
[5] Berry, B. and Chopra, I., "Wind Tunnel Testing for Performance and Vibratory Loads of a Variable-Speed Mach-Scale Rotor," 67th Annual Forum Proceedings of the

American Helicopter Society, Virginia Beach, VA, May 3 - May 5, 2011.

[6] Han, D., Wang, J., Smith, E. C. and Lesieutre, G. A., "Transient Loads Control of a Variable Speed Rotor during Resonance Crossing," *AIAA Journal*, Vol. 51, No. 1, 2013, pp. 20-29.

[7] Han, D. and Smith, E. C., "Lagwise Dynamic Analysis of a Variable Speed Rotor," *Aerospace Science and Technology*, Vol. 29. No. 1, 2013, pp. 277-286.

[8] Johnson, W., "Rotorcraft Dynamics Models for a Comprehensive Analysis," Helicopter Society 54th Annual Forum, Washington, DC, May 20–22, 1998.

[9] Shabana, A. A., Dynamics of Multibody Systems, 2nd ed., Cambridge University Press, Cambridge, England, UK, 1998, pp. 28-88.

[10] Zheng, Z. C., Ren, G., and Cheng, Y. M., "Aeroelastic Response of a Coupled Rotor/Fuselage System in Hovering and Forward Flight," *Archive of Applied Mechanics*, Vol. 69, No. 1, 1999, pp. 68-82.

[11] Han, D. and Smith, E. C., "Lagwise Loads Analysis of a Rotor Blade with an Embedded Chordwise Absorber," *Journal of Aircraft*, Vol. 46, No. 4, 2009, pp. 1280-1290.

[12] Leishman, J. G. and Beddoes, T. S., "A Semi-Empirical Model for Dynamic Stall," *Journal of the American Helicopter Society*, Vol. 34, No. 3, 1989, pp. 3-17.

[13] Leishman, J. G., "Validation of Approximate Indicial Aerodynamic Functions for Two-Dimensional Subsonic Flow," *Journal of Aircraft*, Vol. 25, No. 10, 1988, pp. 914-922.

[14] Peters, D. A., and HaQuang, N., "Dynamic Inflow for Practical Application," *Journal of the American Helicopter Society*, Vol. 33, No. 4, 1988, pp. 64–68.

[15] Owen, D. R. J. and Hinton, E., Finite Elements in Plasticity: Theory and Practice, Pineridge Press, Swansea, U.K., 1980, pp. 431-436.

[16] Leishman, J. G., Principles of Helicopter Aerodynamics, 2nd ed., Cambridge Univ. Press, Cambridge, New York, 2006.

[17] Johnson, W., Helicopter Theory, Princeton University Press, Princeton, New Jersey, 1980: chapter 9.

Individual Blade Control of a 5-bladed Rotor Using the Multiple Swashplate System

Philip Küfmann
philip.kuefmann@dlr.de
project leader

Rainer Bartels
rainer.bartels@dlr.de
research engineer

Berend G. van der Wall
berend.vanderwall@dlr.de
senior scientist

Oliver Schneider
oliver.schneider@dlr.de
research engineer

German Aerospace Center (DLR), Institute of Flight Systems, Braunschweig, Germany

Hermann Holthusen
hermann.holthusen@dnw.aero
project aerodynamicist

Jos Postma
jos.postma@dnw.aero
R&D engineer

German-Dutch Wind Tunnels (DNW), Marknesse, Netherlands

Abstract

After its first wind tunnel test in 2015, the multiple swashplate system (META) as well as the DLR's rotor test rig were modified and upgraded extensively to allow IBC operation on a five-bladed rotor system. In late 2016 a second wind tunnel test was performed on a Mach-scaled, five-bladed model rotor with the goal to reduce vibration, noise and required rotor power on a five-bladed rotor in different flight conditions using proven IBC strategies. Highlights of the test matrix were 2/rev sweeps and the test of different localized pitch control (LPC) strategies for reduction of noise and required rotor power. In simulated high-speed flight, the required rotor power was successfully reduced using a 2/rev input with an amplitude of $1°$. In descent flight condition, significant BVI noise reductions relative to the baseline case were achieved on both sides of the rotor disk through the application of 2/rev HHC as well as a localized pitch control (LPC) schedule. In addition the 5/rev hub vibration levels were reduced significantly both through the application of 3/rev IBC and a vibration controller using a 4-6/rev multi-harmonic IBC signal. The highest reduction was achieved for the vertical 5/rev force, nearby eliminating the most prominent component of rotor hub vibrations

Notation

C_T	thrust coefficient
$F_{x,y,z}$	hub forces (non-rotating frame)
$I_{BM,5}$	index for blade bending moments with influence on 5/rev hub loads
$M_{x,y,z}$	hub moments (non-rotating frame), Nm
N_b	number of blades
p	pressure, kPa
r	radial coordinate, m
R	rotor radius, m
VI_5	weighted 5/rev vibration intrusion index
$y_{tip,el}$	elastic lag displacement at the blade tip
$z_{tip,el}$	elastic flap displacement at the blade tip
α_S	rotor shaft angle, deg
β_{PC}	pre-cone angle
γ	flight path angle of the model, deg
$\Theta_{0,C,S}$	collective and cyclic control angles, deg
Θ_n	n/rev HHC amplitude, deg
σ	solidity
ϕ_n	phase of elastic n/rev blade torsion, deg
φ_n	n/rev HHC phase, deg
ϑ	individual blade pitch angle, deg
μ	advance ratio
ψ	azimuth angle, deg
Ω_{ref}	rotor rotational frequency, rad/s, Hz

Abbreviations

ADC	Analog-Digital Converter
BL	Baseline
DLR	Deutsches Zentrum für Luft- und Raumfahrt e.V. (German Aerospace Center)
DNW	Deutsch-Niederländischer Windkanal (German-Dutch Wind Tunnels)
FTK	Fortschrittliche Taumelscheibenkonzepte (advanced swashplate-concepts)
HART	HHC Aeroacoustics Rotor Test
HHC	Higher Harmonic Control
h.p.	High-Pass filtered
IBC	Individual Blade Control
LLF	Large Low-Speed Facility
LPC	Localized Pitch Control
META	MEhrfach-TAumelscheibe (Multiple Swashplate Control System)
MN, MV	Minimum Noise, Minimum Vibration
SKAT	Skalierung und Risikominimierung von Technologie bei innovativem Design (Scaling and risk-minimization for technology with innovative design)
SPL	Sound Pressure Level
SPL.rel	Sound Pressure Level relative to max. level of the respective frequency range
SPR	Stereo Pattern Recognition

Originally presented at the 43rd European Rotorcraft Forum, Milano, Italy, Sept. 12-15, 2017.

1. Introduction

Helicopters, while having advantages making them indispensable in many applications, are prone to high noise and vibration levels as well as comparatively high power consumption in high-speed flight. In addition to passive measures – such as optimized fuselage and rotor blade designs – the field of active rotor control offers a variety of approaches to address these specific problems directly at their source [1–3], many of which have been the subject of research at DLR [4–7]. In 2008, the DLR introduced and patented a novel active rotor control system for achieving complete individual blade control (IBC) on rotors with more than three blades without using actuators in the rotating frame [8–11]. The Multiple Swashplate System (META) has since been integrated into the DLR's rotor test rig and its IBC capabilities have been successfully demonstrated on a four-bladed rotor both in hover conditions [12] and in the wind tunnel [13, 14] within the project FTK-META.

As part of the follow-up project SKAT (scalability and risk-minimization of technology with innovative design) together with Airbus Helicopter Germany, the Multiple Swashplate System and the DLR's rotor test rig have since been equipped with a five-bladed rotor hub and modified for 5-blade IBC operation. After extensive updates to both soft- and hardware and pretests at DLR's own rotor test facility, the system entered the DNW's large low speed facility (LLF) in the Netherlands for nine days of testing. The goal of this wind tunnel test was to research the transferability of established IBC strategies for performance enhancement as well as noise and vibration reduction from a four-bladed to a five-bladed rotor system.

The first part of this paper outlines the changes and updates of the META system, the test setup and the measurement systems used in the wind tunnel. The second part of the paper then discusses selected results of the wind tunnel test with regards to changes in noise emissions, vibration levels and rotor performance and in comparison to results from the earlier FTK-META test.

2. Test setup

2.1. Text conditions and test matrix

In order to maintain comparability between the results, the test hardware as well as the principal test conditions and procedures closely resembled those of the first wind tunnel test of the META system in the project FTK-META, which are extensively documented in [13] and [14]. The following section therefore only outlines the changes and upgrades made for accurate and safe 5-blade IBC operation of the META system as well as changes in the measurement setup at the DNW LLF.

As in the FTK-META test, IBC tests were performed for three different simulated flight conditions – descent flight, cruise and high-speed level flight. The respective trim conditions are summarized in table 1.

Table 1: Forward flight trim conditions overview

Condition	V_∞, m/s	$100 \cdot C_T$	γ	α_S
Descent	33	0.600	$7.0°$	$+6.7°$
Cruise	66	0.604	$0.0°$	$-5.6°$
High-speed	76	0.605	$0.0°$	$-7.5°$

Besides single frequency HHC with frequencies from 2/rev to 6/rev, multi-harmonic control inputs were tested using a T-matrix based vibration control algorithm. Based on random multi-harmonic inputs and measured 5/rev vibrations, this algorithm first identifies a linear model of the rotor (the so called T-matrix) and subsequently uses a quadratic programming approach to determine optimized HHC signals for vibration reduction while adhering to user-set limits. The details of the control algorithm are documented in [15]. Additionally, the effects of localized pitch control inputs which have been shown to be potentially beneficial for BVI-noise reduction [16] and rotor performance enhancement [17–20], were evaluated. An overview of the test matrix is given in table 2.

Table 2: Test matrix overview

IBC input / phase sweeps	V_∞, m/s 33	66	76
2/rev HHC $\Theta_2 = 0.4°$	✓		
2/rev HHC $\Theta_2 = 0.8°$, $\Theta_2 = 1.0°$		✓	✓
3/rev $\Theta_3 = 0.3°$	✓		
3/rev $\Theta_3 = 0.5°$, $\Theta_2 = 0.8°$		✓	✓
4/rev $\Theta_4 = 0.2°$	✓	✓	✓
4/rev $\Theta_4 = 0.4°$	✓		✓
4/rev $\Theta_4 = 0.55°$	✓	✓	
5/rev $\Theta_5 = 0.2°$	✓		✓
5/rev $\Theta_5 = 0.35°$	✓	✓	✓
5/rev $\Theta_5 = 0.45°$	✓	✓	
6/rev $\Theta_6 = 0.2°$	✓		✓
6/rev $\Theta_6 = 0.3°$	✓	✓	✓
6/rev $\Theta_6 = 0.4°$	✓	✓	
Vibration control 4-6/rev	✓	✓	✓
LPC $\Theta_{dip} = \pm 0.3°$, $\Theta_{dip} = \pm 0.6°$	✓	✓	
LPC $\Theta_{dip} = \pm 0.75°$	✓	✓	✓
LPC $\Theta_{dip} = \pm 1.1°$		✓	✓

The LPC function tested in the wind tunnel consists of one so-called 'dip', which increases or decreases the pitch angle of the rotor blade in a certain, predefined sector of the rotor disk. To make a realization of this kind of control signal possible with the META system, the change in pitch was not defined as a step function, but as a function with a semi-sinusoidal lead-in and lead-out. The details regarding the definition of the LPC-function can be found in [20].

Both the amplitude Θ_{dip} and the azimuthal location (or phase) φ_{dip} were modified during the LPC-tests documented in this paper. The width of the LPC functions remained constant at $90°$ with a symmetrical lead-in and lead-out ($\Phi_1 = \Phi_2 = 45°$) and no plateau ($\Phi_H = 0°$). An example of the tested LPC function with a pitch increase of $\Delta\vartheta_{LPC} = 0.75°$ at the azimuthal position $\varphi_{dip} = 90°$ is shown in Fig. 1.

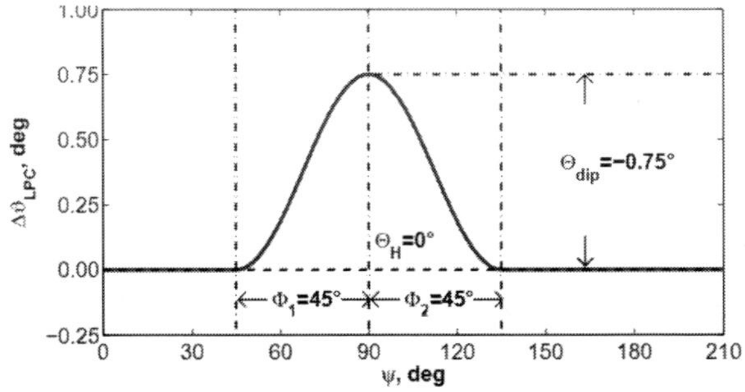

Figure 1: Basic form of LPC-function (dip)

2.2. Model rotor and META

In order to support IBC operation with 5 blades, the DLR's rotor test rig was equipped with a five-bladed rotor hub designed on the basis of the preexisting Bo105 model rotor hub, see Fig. 2.

Figure 2: CAD-drawing of 5-bladed Bo105-type hub

Furthermore, the multiple swashplate system originally designed for 4-bladed model rotors was modified in both hard- and software. Both swashplates were redesigned to accommodate a fifth control rod, resulting in an asymmetric distribution of control rods on the swashplates. The rotor is split up into two independent subrotors with 2 blades on the outer swashplate and 3 blades on the inner swashplate. A schematic depiction of the META's two subrotors for five-bladed operation is shown in Fig. 3.

The META's control software for IBC and LPC-operation was modified accordingly. Due to the asym-

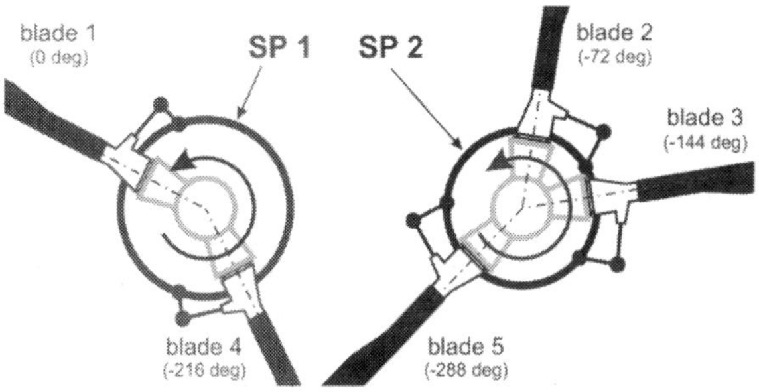

Figure 3: Subrotors of the 5-bladed META system

metric arrangement of rotor blades on the respective swashplates, 2/rev and 3/rev control in the rotating frame result in multi-frequency control signals for the actuators in the non-rotating frame. Besides limiting control authority for these frequencies, this made the control system much more susceptible to small differences in the dynamic behaviour of the hydraulic actuators themselves as well as signal runtime differences between actuators, resulting in inacceptably high blade-to-blade differences in both amplitude and phase of HHC signals.

As a remedy, the controller for the META's hydraulic actuators was upgraded to a notch-filter based system and the rotor hub was equipped with redundant pitch angle sensors (2 per blade). On top of the acuator strokes computed in real time using a kinematic model, the signals from the pitch angle sensors are fed back into a PI-controller with limited autority to ensure the correct IBC amplitude and phase for each blade. With the upgraded control system, control signal deviations and blade-to-blade differences were effectively reduced to $\Delta\Theta_n \leq 0.05°$ in amplitude and $\Delta\varphi_n \leq 1°$ in phase for each frequency.

The resulting control frequencies for the actuators of the 5-bladed META system for each n/rev HHC input (2-6/rev) are summarized in table 3.

Table 3: Control modes on 5-bladed META system

IBC input	Actuator control frequencies				
	1/rev	2/rev	3/rev	4/rev	5/rev
2/rev	✓	✓	✓		
3/rev		✓	✓	✓	
4/rev					✓
5/rev					✓
6/rev					✓

2.3. Measurement equipment

The measurement equipment and the measurement procedures were mostly identical to the FKT-test as documented in [13,14]. The following sections therefore contain only a short description of the measurements systems, as well as upgrades and deviations from the FTK-setup.

2.3.1. Acoustics

The microphone systems used to measure the acoustic characteristic of the SKAT rotor are practically identical to the setup which was applied for the META test in 2016 [13]. Both systems and their position relative to the wind tunnel model are shown in Fig. 4.

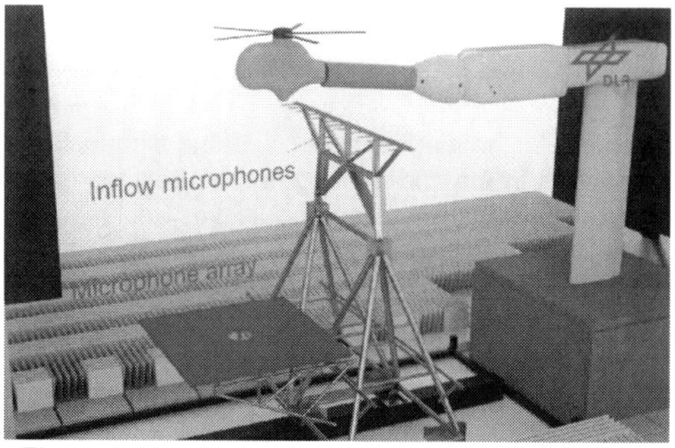

Figure 4: Wind tunnel model and acoustic systems

With the first system the flyover noise field below the rotor plane was measured in order to produce so called 'noise-carpet' plots. The system consists of 13 microphones installed on a traversable in-flow wing. The microphone signals were measured in parallel by two of DNW's 16bit-ADC Viper acquisition systems. One data set of signals was measured in phase-locked mode with 2048 acquisition steps per rotor revolution and the other data set was measured with a fixed sampling frequency of $102.4\ kHz$. For both acquisition types the continuous sampling duration was at least $10\ s$ ($\approx$175 rotor revolutions).

The second acoustic system with a phased microphone array of $4x4\ m^2$ aperture and 140 microphones was used for noise source localization within the rotor disk. The array was placed out-of the flow onto the traverse support for the in-flow microphones of the first acoustic system. The signals from the array were measured with DNW's 24bit-ADC Viper HDR system with a fixed sampling frequency of $102.4\ kHz$ and a sampling duration of at least $10\ s$.

2.3.2. SPR-system

The same SPR-system as in the FTK-test [13] was used for stereoscopic tracking of blade-mounted markers and to determine the contribution of the individual modes in flapping, lead-lag, and torsion after post-processing.

The markers were detected at 30 equally spaced azimuthal positions and the number of averaged samples to determine the final marker positions was reduced from 10 (FTK-project) to 5, which proved sufficient in terms of tracking accuracy while reducing acquistion times considerably. Twelve strobe lights were mounted on a ring-shaped truss illuminating the rotor from above with light pulses of $0.5\ \mu s$ duration. The cameras were equipped with band pass filters to pass only the fluorescence light and block out stray light interfering with the measurement, thus ensuring good contrast of the markers detected in the images.

To obtain optical pitch angle measurements (see also 'Blade deformations' section), two additional markers per blade were applied to the aluminum blade mounting bracket, directly adjacent to the two resistive pitch angle sensors mounted on the pitch arms of rotor hub. Fig. 5 shows the different marker positions on the wind tunnel model mounted in the test section of the DNW LLF.

Figure 5: Marker positions on model rotor blades, blade roots and rotor hub

2.3.3. Blade instrumentation

For this wind tunnel test, the FTK-blades from the first wind tunnel test [13, 14] were re-used. In addition, a fifth FTK-blade was manufactured and equipped with several dynamic pressure sensors for the detection of BVI-events. The positions of the pressure sensors on the 5th FTK-blade are shown in Fig. 6.

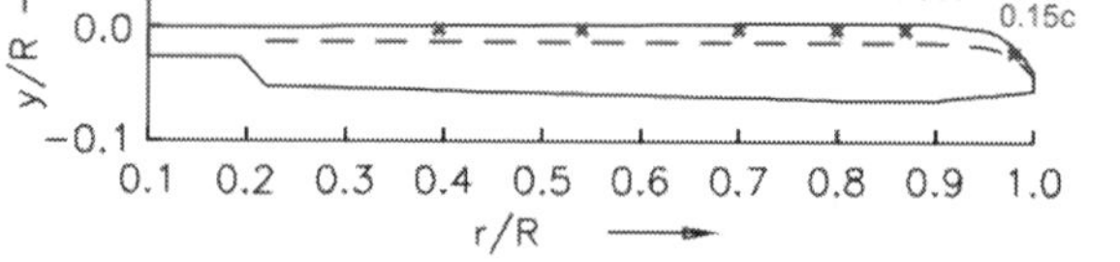

Figure 6: Pressure sensor positions on 5th FTK-blade

2.4. Test procedure

In general the test procedure as well as the procedures for SPR blade tracking and acoustic measurements were identical to the procedures applied during the FTK-test. Details on the test and measurement procedures as well as the applied post-processing techniques can be found in [13].

The model rotor was manually trimmed for zero hub moments ($M_X \stackrel{!}{=} M_Y \stackrel{!}{=} 0$) and the respective thrust level for the chosen flight condition. If IBC was applied, the model was retrimmed before measuring and taking a datapoint.

For acoustic measurements, the microphone array system was used to quickly and efficiently determine maximum and minimum BVI-noise cases as well as the noise source locations within the rotor disk. In a second step, measurements with the microphone traverse were taken for specific test cases which were selected based on the results from the array system. In addition, online results of the blade pressure measurements were used to verify the validity of the array results for the determination of the specific test point parameters.

2.5. Power corrections

Due to the dynamic behaviour of the wind tunnel model slight changes of thrust, propulsive force and hub moments occured and their effects on the measured, effective rotor power P_{eff} had to be corrected during post-processing, as outlined in [13]. During the wind tunnel test with a five-bladed rotor system, additional trim measurements were obtained to determine the effect of hub moment changes ($M_X \neq 0$, $M_Y \neq 0$) on the effective rotor power P_{eff}. As a result, a correction based on changes in hub moments was added:

(1) $$\Delta P_M = c_x \cdot M_X + c_y \cdot M_Y$$

The coefficients c_x and c_y were determined from additional trim measurements by applying the method of least squares:

(2)
$$c_x = \frac{-\sum M_Y^2 \cdot \sum M_X \Delta P + \sum M_X M_Y \cdot \sum M_Y \Delta P}{\sum M_X^2 \cdot \sum M_Y^2}$$
$$c_y = \frac{-\sum M_X^2 \cdot \sum M_Y \Delta P + \sum M_X M_Y \cdot \sum M_X \Delta P}{\sum M_X^2 \cdot \sum M_Y^2}$$

However, compared to the power corrections due to changes in thrust and propulsive force, ΔP_M was found to be negligible. As a result, measurements from FTK-META remain comparable to those obtained with the 5-bladed rotor system.

3. Results

3.1. Rotor performance

Several HHC inputs were tested for rotor performance enhancement, with 2/rev being the most effective approach. Fig. 8 shows the change in effective rotor power P_{eff} over the 2/rev phase input in high-speed flight condition ($\mu = 0.345$) or different values of Θ_2.

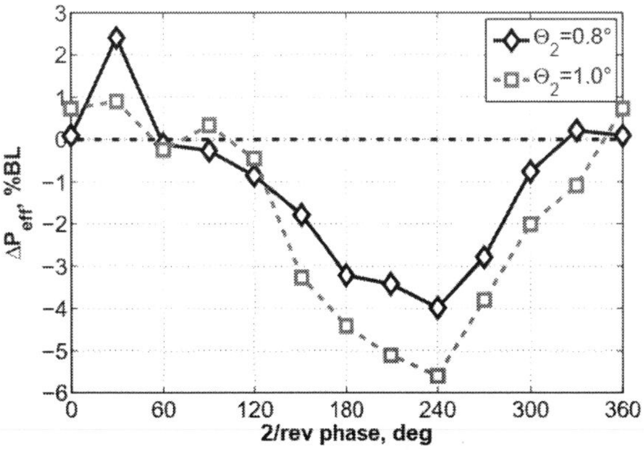

Figure 7: Change of P_{eff} during 2/rev phase sweep

With 2/rev HHC, the maximum reduction in effective rotor power of $\Delta P_{eff} = -5.6\%$ was measured for an amplitude of $\Theta_2 = 1.0°$ at a phase setting of $\varphi_2 = 240°$. Both curves covering the phase sweeps for the respective amplitudes are near sinusoidal. The optimum phase angle corresponds well with results from the FTK-META-WK test, where the maximum power reductions using 2/rev HHC were found for phase angles of $184° \leq \varphi_2 \leq 244°$ [13, 14]. The increase of the measured power reduction with Θ_2 suggests that even higher power savings can be reached with higher 2/rev amplitudes.

The effects of 3/rev HHC on effective rotor power are shown in Fig. 8 for two different amplitudes. Here, the

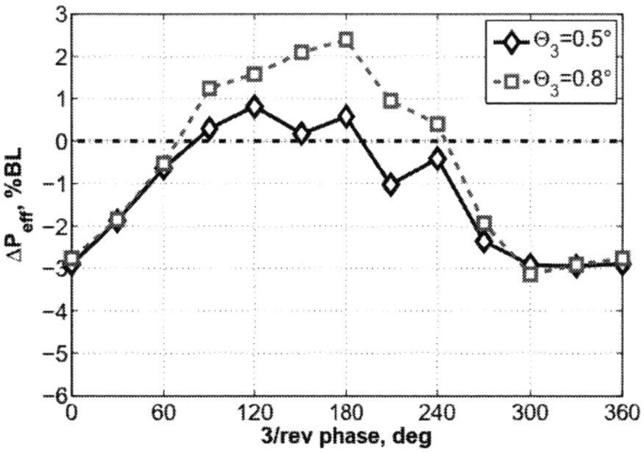

Figure 8: Change of P_{eff} during 3/rev phase sweep

effective rotor power was reduced by a maximum of $\Delta P_{eff} = -2.8\%$ at zero phase angle ($\varphi_3 = 0$) and an amplitude of $\Theta_3 = 0.5°$. While a higher 3/rev amplitude seems to lead to a higher power increase for $90° \leq \varphi_3 \leq 210°$, this is not the case for power reductions where a saturation seems to be reached at $\Theta_3 = 0.5°$.

3.2. Vibration reduction

3.2.1. Single frequency phase sweeps

To assess the vibration level of the model a vibration intrusion index VI_5 was defined in accordance with [21, 22] as a weighted combination of 5/rev hub forces and moments using the virtual weight W_0 of the scaled model and rotor radius R to obtain non-dimensional values.

(3)

$$VI_5 = \frac{1}{W_0}\sqrt{(0.5 \cdot F_{X,5})^2 + (0.67 \cdot F_{Y,5})^2 + F_{Z,5}^2} + \frac{1}{R \cdot W_0}\sqrt{M_{X,5}^2 + M_{Y,5}^2}$$

Fig. 9 shows the averaged baseline case vibration index of the five bladed SKAT rotor over the test conditions. The rotor exhibits vibrations which almost linearly increase with the flight speed. At high speed flight ($76\ m/s$, $\mu = 0.345$) the vibration level is about three times higher than in descent flight at $33\ m/s$ ($\mu = 0.15$).

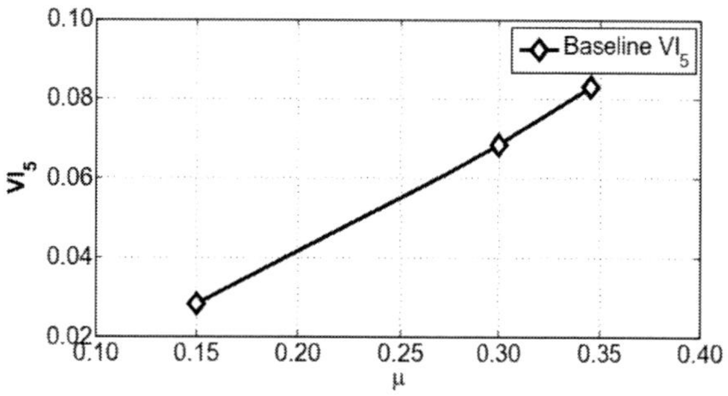

Figure 9: Vibration intrusion index VI_5 at different flight conditions.

3/rev to 6/rev blade pitch inputs with different control amplitudes and phases were tested to analyse the vibration characteristics of the five bladed rotor. For reasons of consistency and comparability of results, the HHC amplitudes for 3/rev and 4/rev HHC were as close as possible to the amplitudes used during the FTK-test [13].

However, the amplitudes for 4/rev to 6/rev HHC (see table 2) proved too high to achieve reductions of the vibration intrusion index VI_5, which was increased at all flight speeds and phase settings for these HHC frequencies. The main reason for this behaviour lies within the kinematics of the multiple swashplate system. Since 4–6/rev HHC in the rotating frame is realized by (collective and cyclic) 5/rev swashplate movements, the META's hydraulic actuators mounted on the baseplate directly affect the 5/rev forces and moments measured by the rotor balance and offset the value for the intrusion index VI_5 considerably.

In contrast, 3/rev HHC inputs on the 5-bladed rotor system are realized with multicyclic movements of the two swashplates with frequencies between 2/rev and 4/rev. Without the 5/rev actuation forces acting on the rotor balance, 3/rev HHC led to suitable vibration reductions. In the high speed case at $\mu = 0.345$, the vibration level was decreased by 60% with an amplitude of $\Theta_3 = 0.8°$ and a phase angle of $\varphi_3 = 0°$, see Fig. 10.

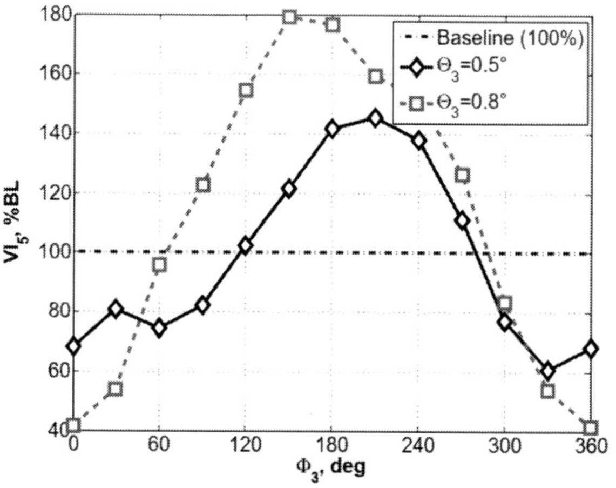

Figure 10: Effect of 3/rev HHC on VI_5 at high speed flight ($\mu = 0.345$).

The highest reduction in effective rotor power had been measured for the same control settings (see Fig. 8), showing the potential of 3/rev HHC for simultaneous performance enhancement and vibration reduction on a 5-bladed rotor system.

In order to make a qualitative assessment of changes in vibration levels possible without the influence of 5/rev actuator forces, a new index was defined using the 4/rev to 6/rev out-of-plane and in-plane blade bending moments from $N_b = 5$ blades, see Eq. (4).

(4)

$$I_{BM,5} = N_b\sqrt{M_{f,5}^2} + \frac{N_b}{2}\sqrt{M_{f,4}^2 + M_{f,6}^2} + \frac{N_b}{2}\sqrt{M_{l,4}^2 + M_{l,6}^2}$$

The corresponding forces in the rotating frame are transformed into 5/rev loads in the non-rotating frame and thus can be used as an indicator for changes in 5/rev vibration levels. For the 3/rev phase sweep, the curves for $I_{BM,5}$ qualitatively match the curves for the vibration intrusion index VI_5 and the maximum reduction in both cases occurs for an amplitude of $\Theta_3 = 0.8°$ at a phase setting of $\varphi_3 = 0°$, see Fig. 10 and Fig. 11.

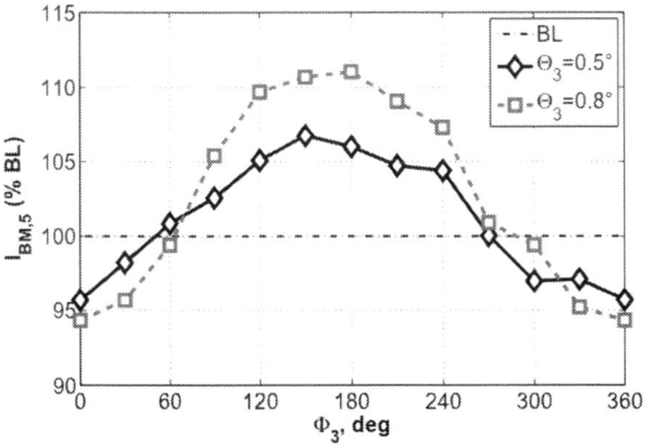

Figure 11: Effect of 3/rev HHC on $I_{BM,5}$ at high speed flight ($\mu = 0.345$).

Fig. 12 shows the development of $I_{BM,5}$ for the 4-6/rev phase sweeps with different amplitudes. With 3/rev control the maximum reduction of $I_{BM,5}$ was measured around 5% and corresponded to a reduction of the vibration intrusion index of $\Delta VI_5 \approx 60\%$. Here, reductions of up to $\Delta I_{BM,5} \approx 12\%$ were reached using a 4/rev input with an amplitude of $\Theta_4 = 0.4°$ at a phase setting of $\varphi_4 = 360°$. Additionally, the maximum reduction of $I_{BM,5}$ occured for phase settings between $\varphi_n = 300°$ and $\varphi_n = 360°$ for all control frequencies from 4/rev to 6/rev.

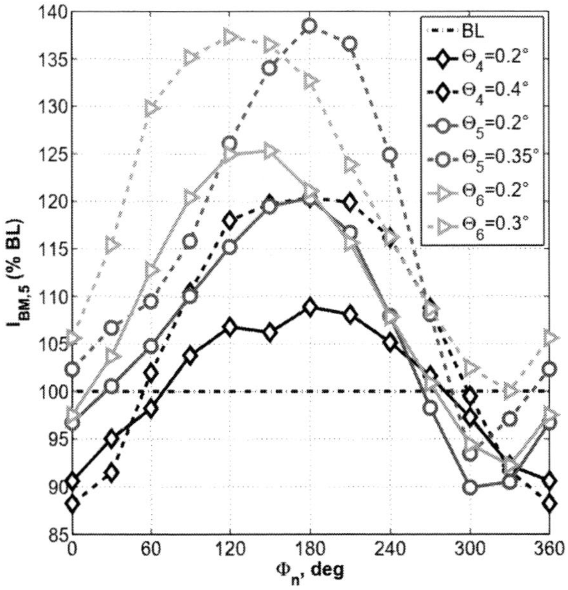

Figure 12: Effect of 4-6/rev HHC on $I_{BM,5}$ at high speed flight ($\mu = 0.345$).

Therefore, when compared to the results of the 3/rev phase sweep, a much higher reduction or even a near complete elimination of 5/rev vibratory hub loads (without taking actuator loads into account) is deemed possible by using single- or multiharmonic control inputs with frequencies of 4-6/rev.

3.2.2. Multi-harmonic vibration control

During the project FTK-META-WK, a vibration control algorithm had been integrated into the META control software [15]. This control algorithm combines the well-known T-matrix contol approach [23] with methods for optimizing control solutions while still adhering to user-set limits (e.g. limits of the used actuation system) [24]. Following the integration process and first tests of the controller on a 4-bladed model rotor in hover conditions, the algorithm was modified for 5-bladed operation, targeting 5/rev hub vibrations using multicyclic 4–6/rev HHC inputs.

The test of this vibration controller in the wind tunnel demonstrated its effectiveness for all test conditions. Depending on the test condition, the controller was able to reduce the vibration intrusion index by up to 83% (high-speed flight), see Fig. 13. Due to the 5/rev actuator forces acting on the rotor balance during 4–6/rev HHC operation, the optimum HHC amplitudes determined by the control algorithm were low compared to the earlier open-loop tests, see table 4.

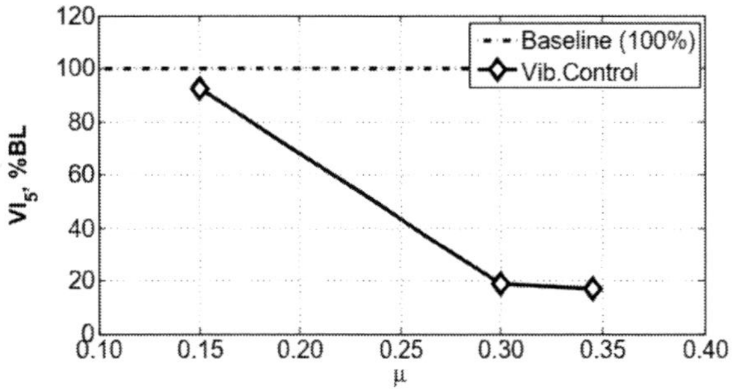

Figure 13: Vibration controller results using 4/rev–6/rev HHC at all tested flight conditions.

Table 4: Vibration controller results

condition	descent $33\ m/s$	cruise $66\ m/s$	high-speed $76\ m/s$
4/rev Θ_4	$0.02°$	$0.0°$	$0.02°$
5/rev Θ_5	$0.0°$	$0.05°$	$0.05°$
6/rev Θ_6	$0.04°$	$0.05°$	$0.07°$
ΔVI_5	7%	81%	83%

Fig. 14 shows the 5/rev magnitudes of the hub forces and moments for the baseline case as well as two subsequent controller inputs at $76\ m/s$ ($\mu = 0.345$). The first contoller solution ('step 1') reduced mainly the vertical and lateral 5/rev forces ($F_{Z,5}$ and $F_{Y,5}$) while the 5/rev roll moment ($M_{X,5}$) remained nearly unchanged. In a second step, further significant load reductions compared to the first control solution were reached in all axes, and the previously unaffected roll moment $M_{X,5}$ was reduced to 56% of its baseline value. The highest single reduction was achieved for the vertical 5/rev force $F_{Z,5}$ with nearly 93%.

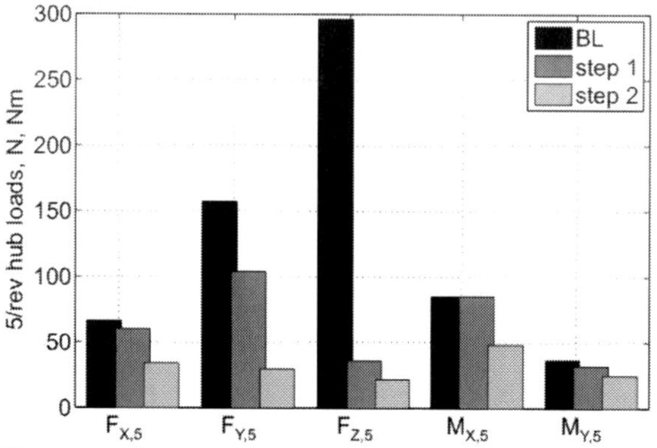

Figure 14: Changes in 5/rev hub loads during vibration controller operation

While the results presented here prove the effectiveness of the control algorithm used, it must be noted that due to the mechanical setup of the META system, a significant portion of the achieved 5/rev vibration reduction was reached due to 5/rev actuator forces counteracting the 5/rev baseline vibrations rather than aerodynamic effects in the rotating frame.

3.3. Noise investigations

3.3.1. Finding the maximum BVI condition

In a first step, the maximum BVI condition in simulated descent flight at $33\ m/s$ was determined by varying the rotor shaft angle α_S and analyzing measurements taken by the microphone array system. Fig. 15 shows the narrow band spectra of the relative sound pressure level (SPL.rel) measured within the rotor plane during the α_S-variation.

As can be seen in Fig. 15, the highest noise level was recorded for a rotor shaft angle of $\alpha_S = 7°$ with the highest overall peak near $630\ Hz$. The maximum BVI case was also confirmed by the analysis of leading edge pressure distributions which was performed in parallel to the microphone array analysis. For this purpose the high frequency content ≥10/rev was analyzed for radial sections of 54, 70, 80 and 98% radius within the regions of interest, i.e., the azimuth ranges on advancing side ($30° \leq \psi \leq 90°$) and on the retreating side ($270° \leq \psi \leq 330°$).

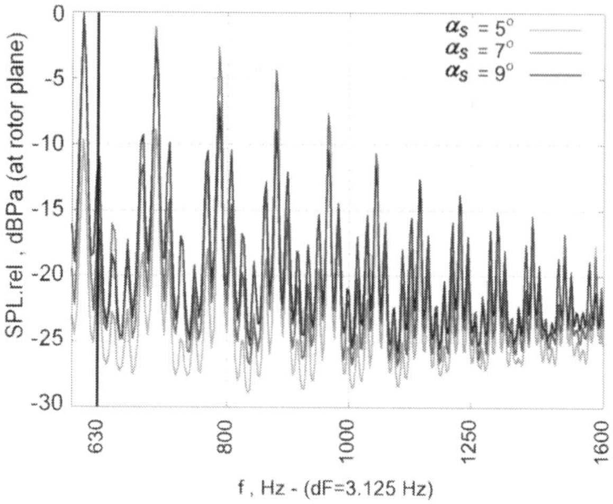

Figure 15: Narrow-band spectra of SPL.rel measured in the rotor plane

Instead of time histories, time derivatives may provide a clearer impression of the BVI-intensity, because the derivative is the physical source of noise radiation. Examples are given in Fig. 16 (adv. side) and Fig. 17 (ret. side) for the identification of the maximum noise case at 33 m/s wind speed for the range of shaft angle $5° \leq \alpha_S \leq 9°$. Those BVI that are most parallel to the blade leading edge are located within the grey shaded area. Large pressure gradients (positive and negative) over a large radial extension (especially at the blade tip where the highest Mach numbers occur) and within the grey shaded areas (representing the blade parallel BVI zones) can be seen as an indicator for strong BVI noise. The strongest events on both the advancing and retreating side can be found at 7° shaft angle of attack.

For the same cases the qualitative contours of BVI locations and their intensity are shown in Fig. 18. The 5° shaft angle (top graph) shows advancing side BVI more upstream (to the left) and the 9° (lower graph) more downstream (to the right) than at 7° (middle graph). The same is found on the retreating side. It is obvious that the shaft angle setting of $\alpha_S = 7°$ is the one with highest BVI intensity and thus the maximum noise radiation condition, which confirms the analysis of microphone array measurements shown before in Fig. 15. Therefore, this case was selected as the baseline (BL) case for all active control parameter variations.

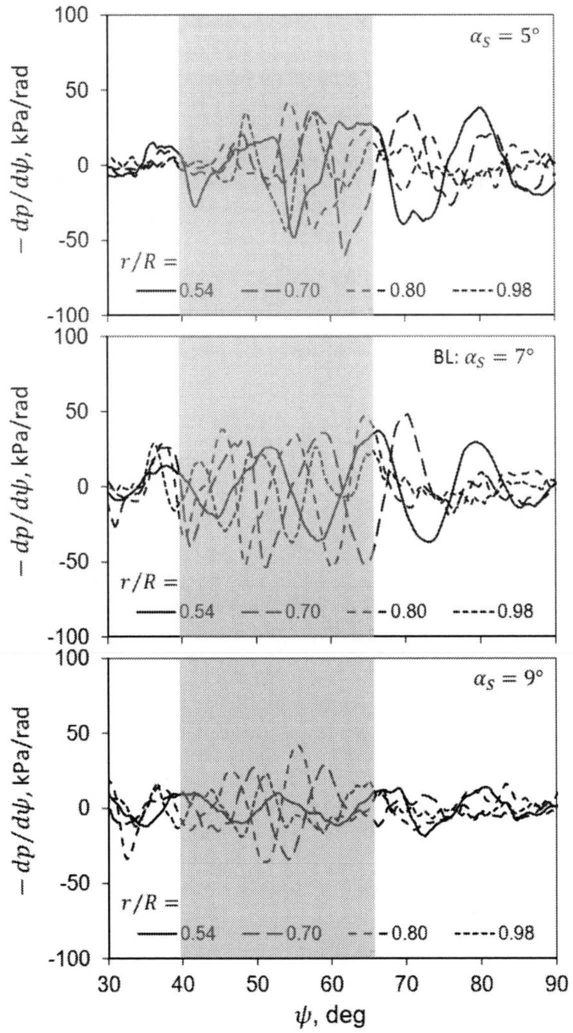

Figure 16: Time derivative of pressure signals during sweep of angle of attack, h.p. filtered at 10/rev, advancing side, 33 m/s.

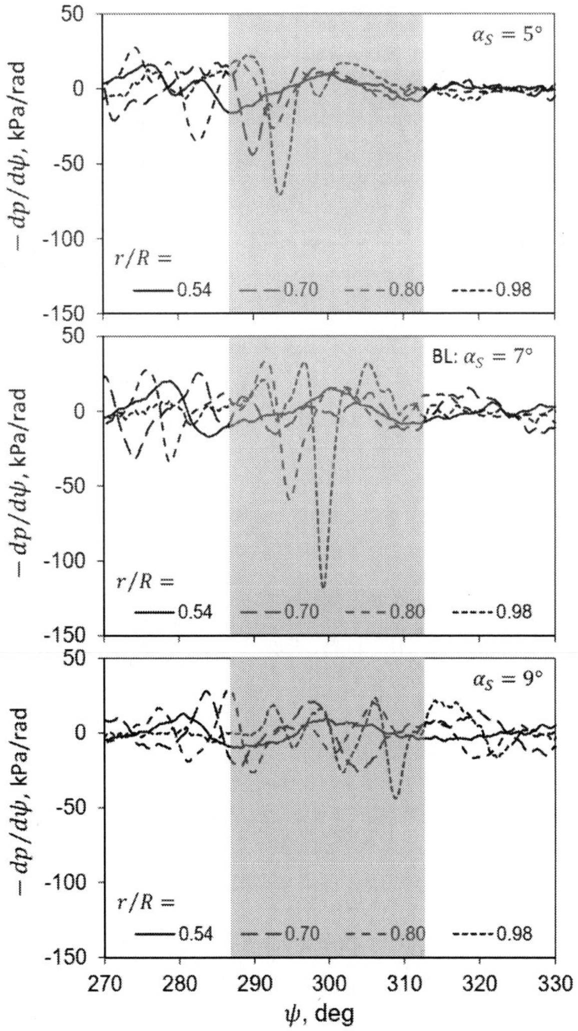

Figure 17: Time derivative of pressure signals during sweep of angle of attack, h.p. filtered at 10/rev, retreating side, 33 m/s.

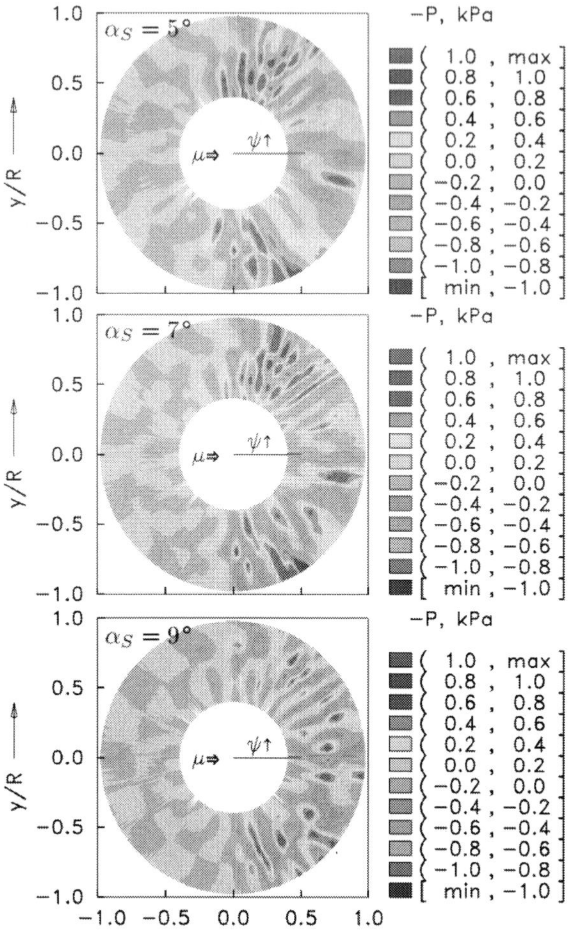

Figure 18: BVI in pressure signals during sweep of angle of attack, h.p. filtered at 10/rev, $33\ m/s$.

3.3.2. Noise reduction through HHC and LPC

Once the BL case was defined, HHC phase sweeps at 2 to 6/rev were performed with three different amplitudes. In addition, LPC (single 'dip' with $45°$ slopes) was also employed with a phase sweep, different amplitudes and different sign. The most promising noise reductions were identified in the same manner as the maximum BVI condition during the sweep of angle of attack, but now the minimum BVI was searched for. From all the variations, the following conditions were found most promising for a minimum BVI noise radiation: 2/rev HHC with an amplitude of $\Theta_2 = 1°$ and a phase setting of $\varphi_2 = 270°$ and also LPC with a pitch increase of $\Delta\vartheta_{LPC} = 0.75°$ at $\varphi_{dip} = 140°$ rotor azimuth.

For the analysis of array source plots, which show the sources of BVI noise within the rotor plane, the 3rd octave band around $630\ Hz$ was chosen, since at this frequency the highest peak within the narrow-band noise spectrum occurred in the baseline case (see Fig. 15). Fig. 19 shows such array source plots with a dynamic range of $20\ dB$ for the baseline case ($33\ m/s$, $\alpha_S = 7°$) as well as the minimum noise cases for both 2/rev HHC and LPC.

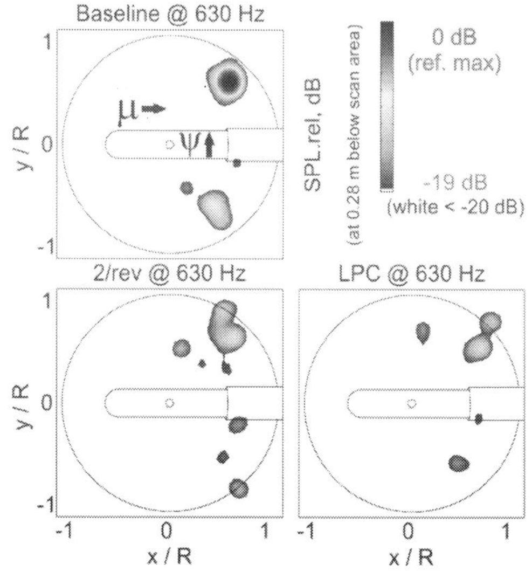

Figure 19: Array source plots for baseline (BL) and minimum noise (MN) cases

Clearly both 2/rev HHC and single dip LPC exhibit significant reductions in this frequency band within the rotor plane – on the advancing side, the intensity is greatly reduced and on the retreating side of the rotor disk the noise source is almost eliminated in both cases. When the respective narrow-band spectra are compared, the differences between minimum noise 2/rev HHC and LPC become more clear. Fig. 20 shows a comparison between the baseline case and minimum noise 2/rev HHC, as well as between both minimum noise cases.

As can be seen in Fig. 20 (top), minimum noise 2/rev HHC significantly reduces noise levels for frequencies $\geq 630\ Hz$ but increases noise for lower frequencies. Compared to the noise levels obtained with the 2/rev HHC input, the minimum noise LPC exhibits lower noise levels over the whole frequency range, see Fig. 20 (bottom).

To maintain comparability of results from the HART and HART II tests [5, 6, 25], the range between the 6th and 40th blade passage frequency ($N_b \cdot \Omega_{ref}$) was chosen for the evaluation of BVI noise levels via noise carpet plots. On a five-bladed rotor system with a nominal rotor speed of $\Omega_{ref} = 110\ rad/s = 17.5\ Hz$, this covers a range from $525\ Hz$ to $3500\ Hz$.

For the baseline case and both minimum noise cases such carpet plots were compiled covering an area of $6x6\ m$ centered around the rotor hub using data

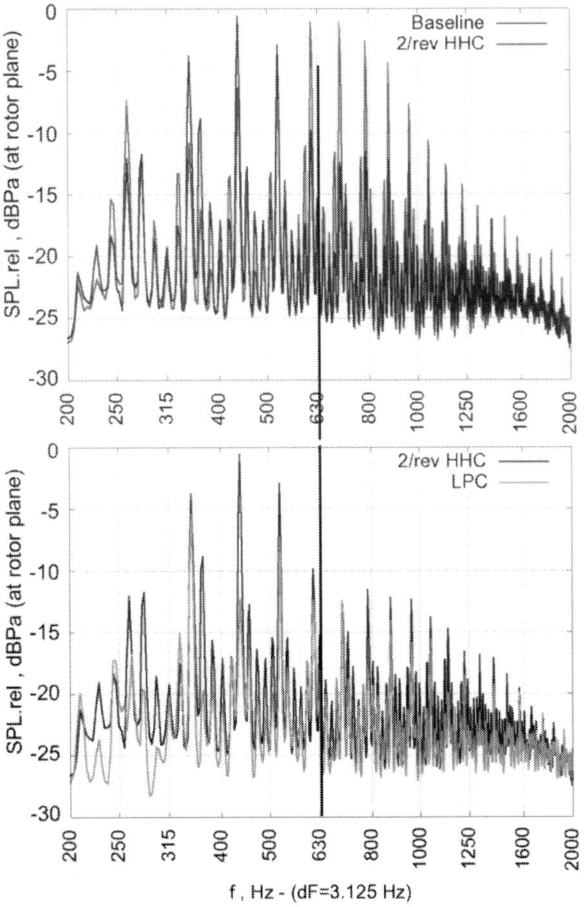

Figure 20: Narrow-band spectra for the baseline case and both MN cases in comparison.

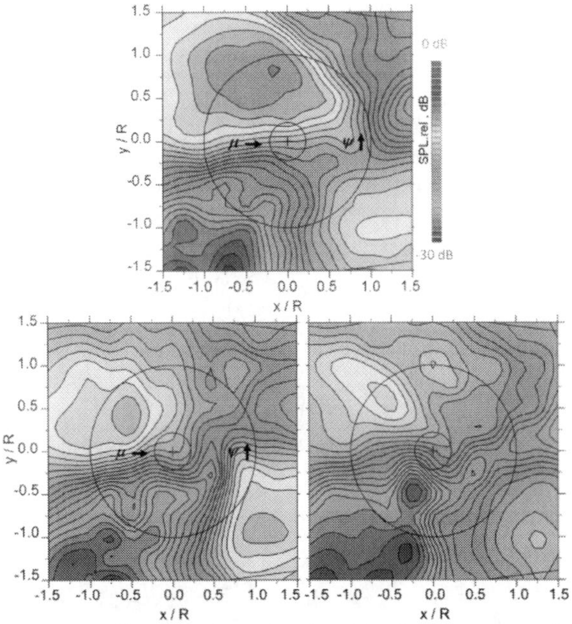

Figure 21: Carpet plots for the baseline case (top), minimum noise 2/rev HHC (bottom left) and minimum noise LPC (bottom right)

from the microphone traverse measurements. Fig. 21 shows all three cases in comparison.

In both minimum noise cases, a significant reduction of the measured noise level is visible, especially on the advancing side of the rotor, where the highest noise levels appeared for the baseline case. For 2/rev HHC (bottom left plot), the peak value on the advancing side of the rotor disk is reduced by $1.5\ dB$ and the area of high-noise emissions is reduced considerably. However, in the fourth quadrant on the retreating side of the rotor disk, noise levels were increased considerably by 2/rev control.

In comparison, the minimum noise LPC input yielded an even higher reduction of BVI noise on the advancing side of the rotor disk (peak value reduced by 3 dB) as well as a slight decrease on the retreating side (fourth quadrant) relative to the baseline. These LPC-results also correspond fairly well with findings from Malovrh et al. [16], where similar inputs in the second quadrant of the rotor disc were found to be highly effective in reducing advancing side BVI noise.

It must be noted that due to the asymmetric distribution of rotor blades among the two swashplates and the resulting multi-harmonic actuator control frequencies (see 'Model rotor and META' section) the control authority for 2/rev (and 3/rev) inputs was severely limited compared to the previous test of the META system on a four-bladed rotor [13]. Therefore it cannot be ruled out that with higher 2/rev control authority ($\Theta_2 > 1.0°$) further reduction of BVI noise could be achieved on the five-bladed rotor.

3.3.3. Pressure measurements for minimum noise HHC and LPC

The physical phenomena leading to a reduction of BVI-noise through 2/rev control and LPC can be understood by a closer look at the blade loading and the high-frequency pressure changes as a result of the blade-vortex interactions. Both control schemes alter the low frequency content of blade aerodynamic loading both radially and azimuthally, and with it the intensity of trailed tip vortex strength. In addition, the distribution of induced velocities within the rotor disk will be modified accordingly, which in return strongly affects the tip vortex trajectory when travelling across the rotor disk. The third physical consequence is a modified blade dynamic response due to the change of airloads mainly in flap and torsion. All together result in a different blade-vortex separation distance at azimuth locations critical for noise radiation. The resulting noise intensity can be amplified or reduced, depending on the HHC or LPC parameters, and those parameters for a minimum of BVI noise radiation were identified.

In the low frequency (0-9/rev) pressure time histories corresponding to the baseline case and the two minimum noise cases, shown in Fig. 22, it can be seen that the 2/rev control (middle) causes an increase of loading at $\psi = 135°$ and at $315°$, where the blade pitch has the maximum 2/rev upwards deflection, as well as a reduction of loading at $\psi = 45°$ and $225°$, where it is minimum.

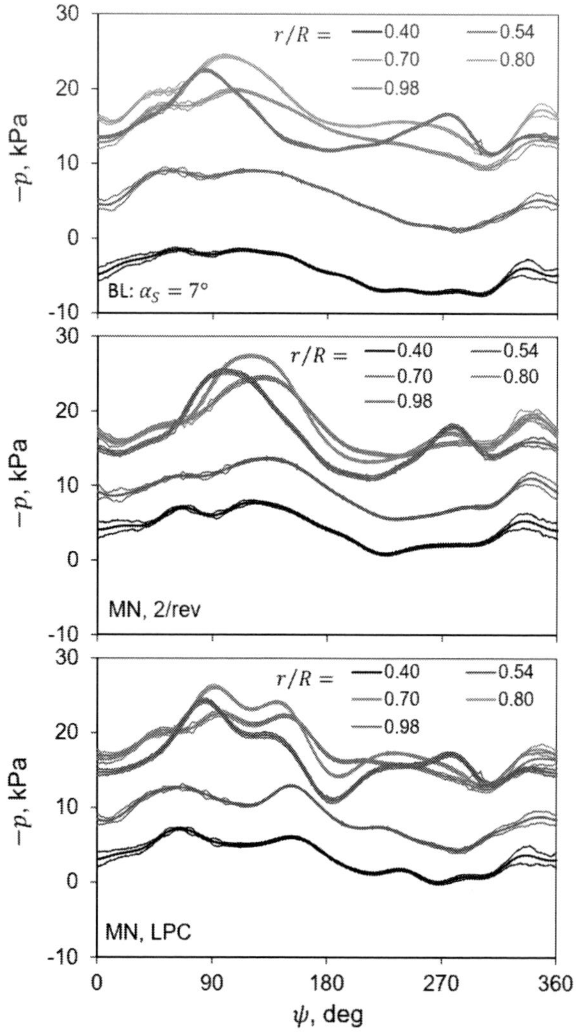

Figure 22: Low frequency pressure signals for BL and MN cases, 0-9/rev, advancing side, $33\ m/s$, $\alpha_S = 7°$.

LPC acts only within a limited range of azimuth, but the blade dynamic response to that excitation extends for a longer range of the rotor revolution (until post-pulse oscillations die out). In Fig. 22 (bottom) the loading exhibits an extra hump relative to the baseline case at the azimuthal position of $\psi = 140°$, and the torsion response following that persists almost until the end of the revolution. The thin lines around the thicker ones indicate the range of standard deviation of local unfiltered pressure data, computed from 32 successive revolutions measured. The time derivatives for the most BVI-relevant azimuthal regions in comparison to the BL cases are given in Fig. 23 and Fig. 24.

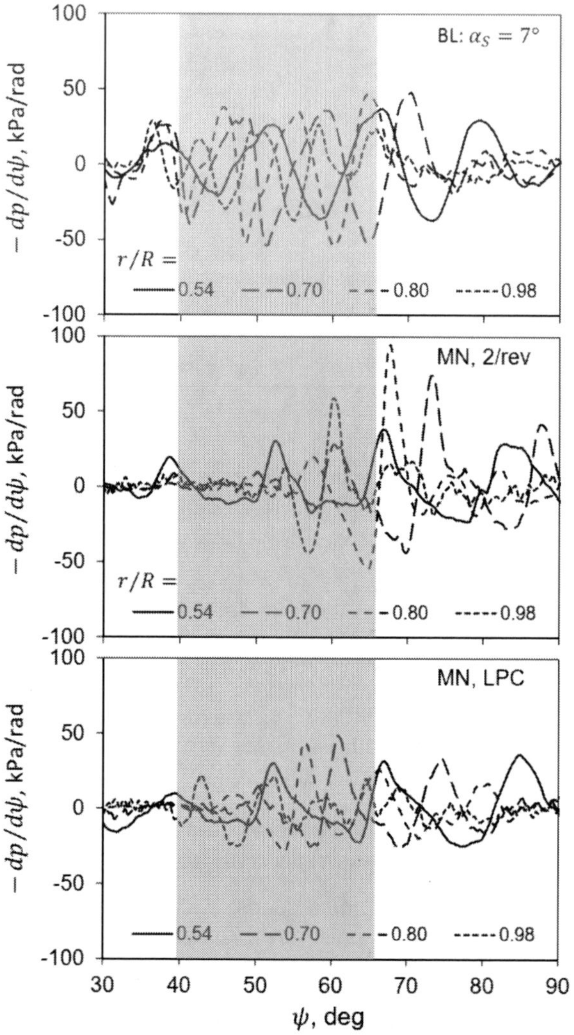

Figure 23: Time derivative of pressure signals for BL and MN cases, h.p. filtered at 10/rev, advancing side, $33\ m/s$, $\alpha_S = 7°$.

The 2/rev control reduces the BVI intensity on both the advancing and the retreating side, see Fig. 23 (middle) and Fig. 24 (middle), each one compared to the BL case (top). The physics behind are an increased blade lift in the second quadrant of the disk, which results in an increased downwash in that region, causing the tip vortices to more quickly convect downwards than in the BL case and to increase the blade-vortex miss-distance in the first quadrant. On the retreating side the opposite is happening: the reduced blade lift in the third quadrant gener-

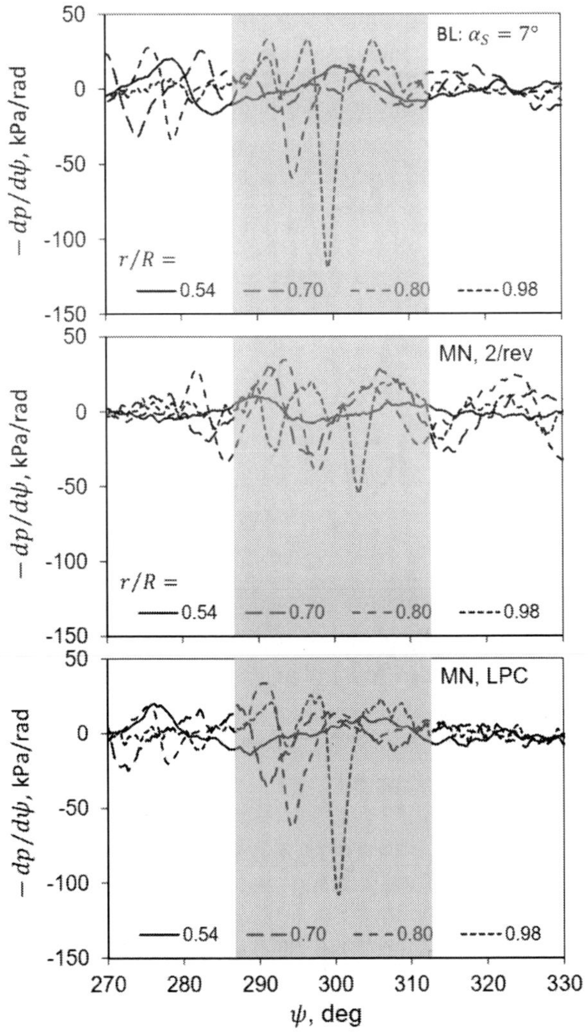

Figure 24: Time derivative of pressure signals for BL and MN cases, h.p. filtered at 10/rev, retreating side, $33\ m/s$, $\alpha_S = 7°$.

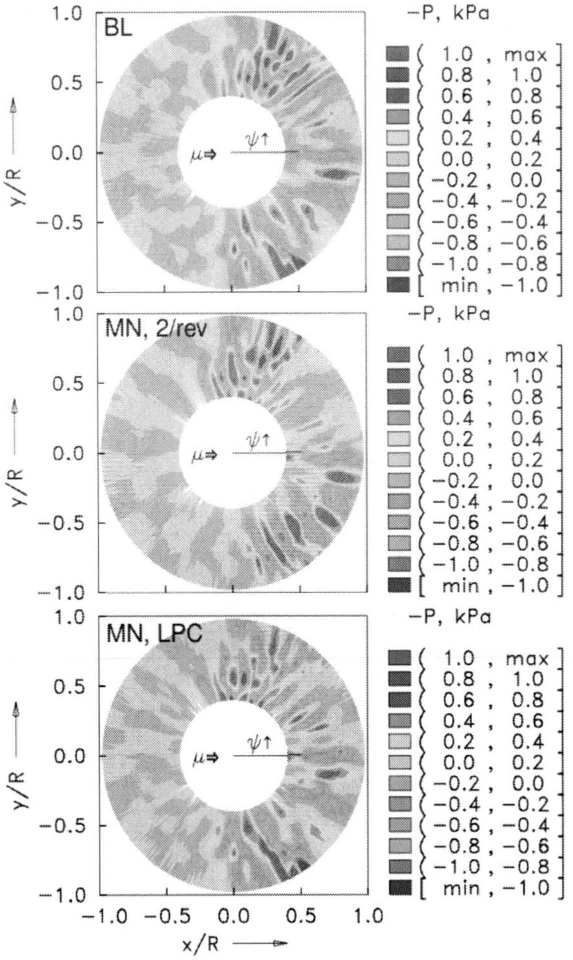

Figure 25: BVI in pressure signals for BL and MN cases, h.p. filtered at 10/rev, $33\ m/s$, $\alpha_S = 7°$.

ates weaker vortices than in the BL case and less induced inflow due to less lift, causing tip vortices to a slower downwards convection and to interact with rotor blades later in the fourth quadrant.

Any HHC input changes the blades' pitch all around the revolution, while LPC is affecting only a small azimuthal range without any control outside. Only elastic blade response to it will persist for some successive fraction of the revolution. Therefore the major effect on the advancing side of the disk is about the same as for the 2/rev HHC with the same physics behind, while the retreating side appears only marginal affected relative to the BL case. LPC results are given in the bottom plots of the respective figures, to be compared to the BL case (top).

Fig. 25 shows the pressure fluctuations in a more qualitative form with the contours in the rotor disk. The baseline case (upper graph) clearly shows several strong BVI on the advancing side ($40° \leq \psi \leq 65°$). On the retreating side one strong BVI appears at the blade tip at $\psi = 300°$, plus several more inboard BVI in the range $280 \leq \psi \leq 315°$. A 2/rev minimum noise HHC (middle graph) shifts the advancing side BVI to more upstream locations ($60 \leq \psi \leq 80°$) outside the range of blade-parallel BVI, thus reducing BVI noise. The strong BVI at the blade tip on the retreating side is almost eliminated, while the BVI intensity more inboard of the blade appears increased relative to the baseline case. However, these are confined to a small radial extension, and at these locations the Mach numbers are smaller than at the tip. Finally, the LPC for minimum noise (bottom graph) shows modifications of BVI on the advancing side similar to those of the 2/rev HHC, but with less remaining BVI intensity. The retreating side BVI of the LPC control appears similar to the baseline case, but the inboard BVI appear to be shifted upstream to $\psi = 285°$, almost eliminating the BVI following it. The presented analysis of blade pressure measurements thus confirms the results gathered from both microphone array (source plots, Fig. 19) and traverse measurements (Fig. 21).

3.4. Blade deformations

To localize the rotor hub center position and for measuring model movements the circular hub fairing was equipped with 11 SPR markers of $20\ mm$ diameter (Fig. 5). By using these hub markers it was possible to determine the rotor hub center location as well as the current azimuth position of the rotor for calculation of the lead-lag motion of the rotor blades and to determine the model roll and pitch angle. Measurements were done with azimuth increments of $\Delta\psi = 12°$, respectively 30 azimuth positions per revolution such that the analysis allows to synthesize the lower harmonics from 0-6/rev from the time history of the blade motion with good confidence. On every azimuthal position 5 images were taken for averaging purposes in order to get smooth data with reduced errors and eliminated non-harmonic vibrations.

The calculation of the blade motion parameters for flap and lead-lag was be done using the quarter chord line of the rotor blade as described in [26] and [27]. The elastic blade lead-lag deflection y_{el} is given by the distance between the radial position of the quarter chord line and a straight line defined by the current azimuth position of the blade (lag positive). The elastic blade flap deflection z_{el} (positive up) is given by the distance between the quarter chord line and a straight line defined by the pre-cone angle of the rotor head ($\beta_{PC} = 1.85°$). The elastic pitch deformation (θ_{el}, positive nose up) can be calculated by the distance between the z-coordinates of the front and rear blade marker, the associated pitch control angle, the pre-twist angle and the pitch offset in z-direction due to the different distance of the front and rear blade markers to the quarter chord line.

3.4.1. Validation of SPR pitch measurements

The blade root was equipped with two SPR markers to measure the blade pitch angle at the innermost radius at $r/R = 0.1$ without blade torsion. Since the potentiometer based sensors for blade pitch angle measurement are mounted in the same location, a direct comparison between the two sensor systems / approaches was made to test and validate the results computed from the SPR pitch measurements. An example is given in Fig. 26 for the minimum noise LPC control input in descent flight with an amplitude of $\Theta_{dip} = 0.75°$ at an azimuthal location of $\varphi_{dip} = 140°$. The collective and cyclic pitch control angles are subtracted. The comparison shows a remarkable congruence between the SPR results and the pitch angle sensor(s). Both amplitude and phase angle match very well confirming the high accuracy of the SPR system.

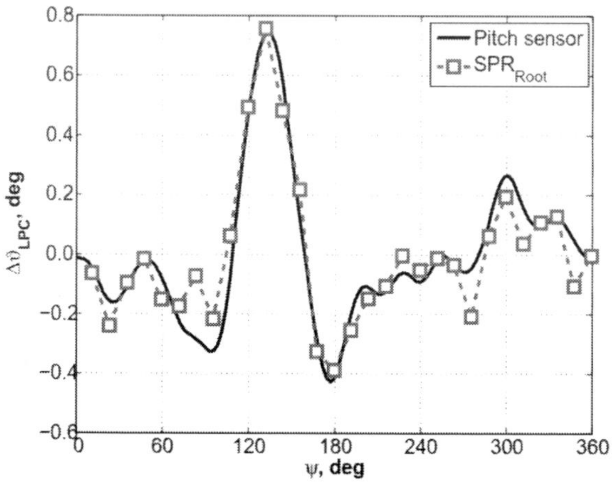

Figure 26: Comparison of measured pitch angle for Dip1 (control angle subtracted)

3.4.2. Hover results

Since in the FTK-META and SKAT wind tunnel tests the same blade loading and identical blades were used, a comparison of a hover thrust sweep can be made. Thrust sweeps were performed between $0N$ and $7000\ N$ in FTK-META and between $-1500\ N$ and $7700\ N$ in SKAT. In Fig. 27 the elastic tip deflection in lag (averaged over all rotor blades) is given for both rotors. For $C_T/\sigma < 0.05$ the blade tips are in a lead position (partly due to a built-in pre-lead angle), above $C_T/\sigma = 0.05$ the blades are in lag position.

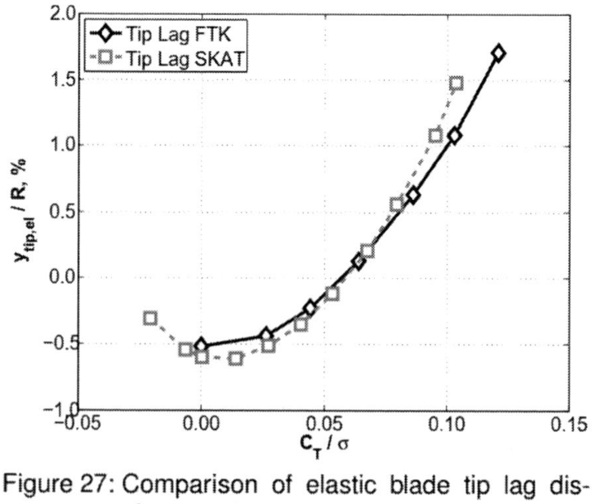

Figure 27: Comparison of elastic blade tip lag displacement

The comparison of the elastic blade tip deflection in flap direction is shown in Fig. 28. The gradient of both curves is very similar but an offset of about 12 mm can be found for the SKAT test compared to the flap displacement measured during the FTK-test. This offset results from the different pre-cone angles of both rotor heads ($\beta_{pc,FTK} = 2.5°$, $\beta_{pc,SKAT} = 1.85°$).

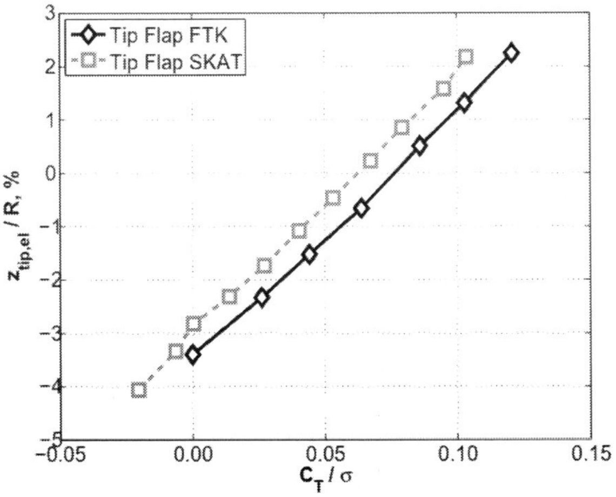

Figure 28: Comparison of elastic flap displacement at the blade tip

3.4.3. HHC results

During the SKAT test HHC inputs with frequencies from 2/rev to 6/rev were executed. The results of n/rev SPR measurements for the descent flight condition ($\mu = 0.15$) are presented in Fig. 29.

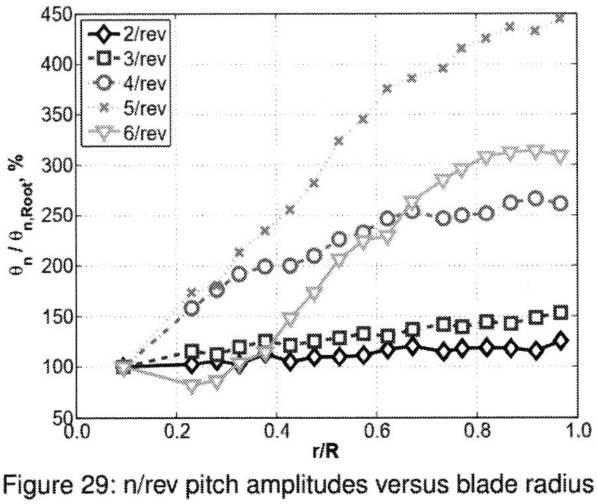

Figure 29: n/rev pitch amplitudes versus blade radius

The changes in n/rev amplitude are plotted oder the blade radius relative to the amplitude of the control input at the blade root. In all n/rev cases the control input amplitude is amplified with increasing blade radius up to the tip ranging from 125% for 2/rev input up to 450% for 5/rev HHC. Since the first torsion frequency of the FTK blades is about 4.7/rev, the maximum amplitude amplification can be found for 4/rev and 5/rev control. The maximum phase delay was found for the 6/rev input at $\Delta\varphi_6 \approx 135°$, as shown in Fig. 30 for all n/rev HHC inputs.

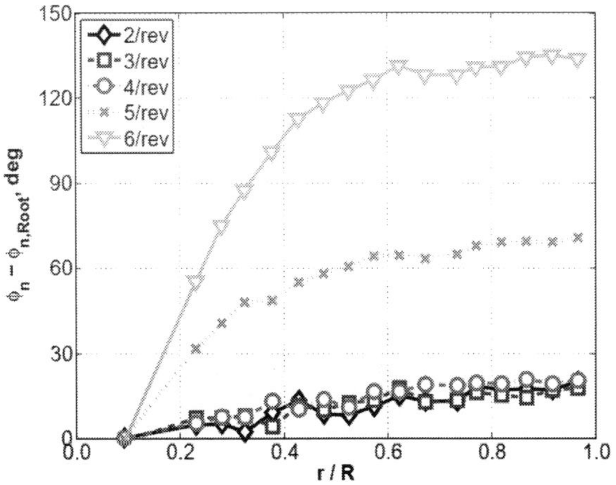

Figure 30: n/rev pitch phases versus blade radius

4. Conclusions

Following earlier wind tunnel tests with 4 rotor blades [13] and as part of a joint research project with Airbus Helicopters Germany, the DLR's Multiple Swashplate System was modified for 5-bladed IBC operation and tested in the Large Low-Speed Facility of the DNW. During those tests, several different IBC strategies were employed to increase rotor performance and to reduce vibration levels and noise emissions. The main conclusions of the test are:

- Using 2/rev HHC with an amplitude of $\Theta_2 = 1.0°$ the required rotor power in high-speed flight condition ($\mu = 0.345$) was reduced by a maximum of 5.6% compared to the baseline case. The optimum 2/rev phase of $\varphi_2 = 240°$ corresponds well with results from earlier 4-bladed wind tunnel tests.
- In high-speed flight ($\mu = 0.345$) 3/rev control yielded a maximum power reduction of $\Delta P_{eff} = -2.8\%$ with an amplitude of $\Theta_3 = 0.8°$ and an optimum phase setting of $\varphi_3 = 0°$.
- The optimum 3/rev HHC settings for performance enhancement in high-speed flight also yielded the lowest 5/rev vibration instrusion index VI_5, which was reduced to 40% of its baseline value. Thus, 3/rev HHC shows potential to reduce both required rotor power and vibration simultaneously on a 5-bladed, hingeless rotor system.

- Due to the actuators of the META system excerting 5/rev forces on the rotor balance during 4-6/rev IBC operation, similar reductions of vibrational hub loads could not be measured for these control frequencies. However, the analysis of blade bending moments indicates that significant vibration reductions can be achieved with 4-6/rev IBC on a five-bladed rotor.
- A T-matrix based vibration controller was adapted for operation on a five-bladed rotor system and its effectiveness successfully demonstrated in the wind tunnel. However, the reduction of the weighted vibration intrusion index VI_5 by 83% in high speed flight ($\mu = 0.345$), was mostly a result of 5/rev actuator forces counteracting the 5/rev vibrational loads measured by the rotor balance.
- The maximum BVI-case in descent flight ($\mu = 0.15$) was found for a shaft angle of $\alpha_S = 7°$ using measurements from the DNW's microphone array and confirmed by on-blade pressure measurements. This shaft angle setting was used as baseline case for further noise investigations.
- Both 2/rev HHC and LPC were effectively used to reduce BVI-noise levels in simulated descent flight at $33\ m/s$ ($\mu = 0.15$).
- In the frequency range of 6-40th BPF, which was used to ensure comparability with the earlier HART and HART II tests, single-'dip' LPC yielded significantly higher noise reductions than 2/rev HHC. High-noise areas were reduced considerably in both size and magnitude, with a reduction in peak SPL of $3.0\ dB$.
- Pressure measurements both confirmed the results from the acoustic measurements and served to identify changes in tip-vortex strength and vortex trajectories as the physical phaenomena resulting in BVI-noise reductions.
- Using the proven and enhanced SPR system of DNW-LLF, the blade motion of all five rotor blades as well as the model pitch and roll angles were analyzed. A comparison of the SPR results with the conventionally measured blade pitch angles shows very good congruence.

Acknowledgments

The tests in the DNW wind tunnel were conducted within the framework of the national research project SKAT (Scaling and risk-minimization for technology with innovative design) funded by the German Federal Ministry of Economic Affairs and Energy (BMWi). The authors would like to express their gratitude towards everyone involved at DLR and DNW during the planning, preparation and execution of the SKAT wind tunnel test and the subsequent data analysis.

Supported by:

Federal Ministry
for Economic Affairs
and Energy

on the basis of a decision
by the German Bundestag

References

[1] C. Kessler, "Active Rotor Control for Helicopters: Motivation and Survey on Higher Harmonic Control," *CEAS Aeronautical Journal*, vol. 1, no. 1-4, pp. 3–22(20), 2011.

[2] C. Kessler, "Active Rotor Control for Helicopters: Individual Blade Control and Swashplateless Rotor Designs," *CEAS Aeronautical Journal*, vol. 1, no. 1-4, pp. 23–54(22), 2011.

[3] P. P. Friedmann, "On-blade control of rotor vibration, noise, and performance: Just around the corner?," *Journal of the American Helicopter Society*, vol. 4, no. 59, pp. 1–37(37), 2014.

[4] Y. H. Yu, B. Gmelin, H. Heller, J. J. Philippe, E. Mercker, and J. S. Preisser, "HHC Aeroacoustics Rotor Test at the DNW - The Joint German / French / US HART Project," in *20th European Rotorcraft Forum*, (Amsterdam, The Netherlands), Oct. 4-7, 1994.

[5] W. R. Splettstoesser, R. Kube, W. Wagner, U. Seelhorst, A. Boutier, F. Micheli, E. Mercker, and K. Pengel, "Key Results From a Higher Harmonic Control Aeroacoustic Rotor Test (HART)," *Journal of the American Helicopter Society*, vol. 42, no. 1, pp. 58–78(21), 1997.

[6] B. G. van der Wall, B. Junker, C. L. Burley, T. F. Brooks, Y. H. Yu, C. Tung, M. Raffel, H. Richard, W. Wagner, E. Mercker, K. Pengel, H. Holthusen, P. Beaumier, and Y. Delrieux, "The HART II Test in the LLF of the DNW - a major step towards rotor wake understanding," in *28th European Rotorcraft Forum*, (Bristol, UK), Sept. 17-20, 2002.

[7] B. G. van der Wall, J. W. Lim, M. J. Smith, S. N. Jung, J. Bailly, J. D. Baeder, and D. D. Boyd, "The HART II International Workshop: An Assessment of the State-of-the-Art in Comprehensive Code Prediction," *CEAS Aeronautical Journal*, vol. 4, no. 3, pp. 223–252(30), 2013.

[8] B. G. van der Wall and R. Bartels, "Patent für eine Hubschrauber-Rotorsteuereinrichtung."

Pat.-Nr.: DE-10-2006-030-089-D, Deutsches Patent- und Markenamt, Jan. 2008.

[9] P. Küfmann, "Patent für ein Verfahren zum Ermitteln von Stellgrößen." Pat.-Nr.: DE-10-2010-024-089-B4, Deutsches Patent- und Markenamt, Jan. 2013.

[10] R. Bartels, P. Küfmann, and C. Kessler, "Novel Concept for Realizing Individual Blade Control (IBC) for Helicopters," in *36th European Rotorcraft Forum*, (Paris, Frankreich), Sept. 7-9, 2010.

[11] P. Küfmann, R. Bartels, C. Kessler, and B. G. van der Wall, "On the Design and Development of a Multiple-Swashplate Control System for the Realization of Individual Blade Control for Helicopters," in *67th Annual Forum of the American Helicopter Society*, (Virginia Beach, VA, U.S.A.), May 3-5, 2011.

[12] P. Küfmann, R. Bartels, and O. Schneider, "DLR's Multiple Swashplate Control System: Operation and Preliminary Testing," in *38th European Rotorcraft Forum*, (Amsterdam, The Netherlands), Sept. 4-7, 2012.

[13] P. Küfmann, R. Bartels, B. G. van der Wall, O. Schneider, H. Holthusen, J. Gomes, and J. Postma, "The First Wind-Tunnel Test of the DLR ' s Multiple Swashplate System : Test Procedure and Preliminary Results," in *72nd Annual Forum of the American Helicopter Society*, (West Palm Beach, FL, U.S.A.), May 17-19, 2016.

[14] P. Küfmann, R. Bartels, B. G. van der Wall, O. Schneider, H. Holthusen, J. Gomes, and J. Postma, "The First Wind Tunnel Test of the Multiple Swashplate System: Test Procedure and Principal Results," *Journal of the American Helicopter Society*, accepted for publication May 4th 2017.

[15] P. Küfmann and C. Brillante, "Implementation and Test of a Semi-Closed Loop HHC-Algorithm with the DLR's Multiple Swashplate System," in *41st European Rotorcraft Forum*, (Munich, Germany), Sept. 1-4, 2015.

[16] B. Malovrh and F. Gandhi, "Localized Individual Blade Root Pitch Control for Helicopter Blade-Vortex Interaction Noise Reduction," *Journal of the American Helicopter Society*, vol. 55, no. 3, pp. 032007–1–032007–12(12), 2010.

[17] F. Kody, E. Corle, M. D. Maughmer, and S. Schmitz, "Non-Harmonic Deployment of Trailing-Edge Flaps for Rotor-Performance Enhancement and Vibration Reduction," in *5th Decennial AHS Aeromechanics Specialists' Conference*, (San Francisco, CA, U.S.A.), American Helicopter Society Inc., Jan. 22-24, 2014.

[18] P. Küfmann, R. Bartels, and B. G. Wall, "Localized Blade-Root IBC for Rotor-Performance Enhancement and Vibration Reduction," in *40th European Rotorcraft Forum*, (Southampton, UK), Sept. 2-5, 2014.

[19] F. Kody, E. Corle, M. D. Maughmer, and S. Schmitz, "Higher-Harmonic Deployment of Trailing-Edge Flaps for Rotor-Performance Enhancement and Vibration Reduction," *Journal of Aircraft*, vol. 53, no. 2, pp. 1–10(11), 2015.

[20] P. Küfmann, R. Bartels, and B. G. van der Wall, "Rotor performance enhancement via localized pitch control and its effects on hub vibrations and pitch link loads," *CEAS Aeronautical Journal*, vol. 8, no. 1, pp. 181–196(16), 2017.

[21] S. Crews and B. Hamilton, "Army helicopter crew seat vibration – past performance, future requirements," in *American Helicopter Society North East Region National Specialists' Meeting on Helicopter Vibration*, November 1981.

[22] AMCOM, "ADS-27A-SP - requirements for rotorcraft vibration specifications, modeling and testing," 2006.

[23] W. Johnson, "Self-Tuning Regulators for Multicyclic Control of Helicopter Vibration," NASA Technical Paper TP-1996 (A-8719), Scientific and Technical Information Branch, Moffett Field, CA, U.S.A., March 1982.

[24] R. M. Morales, M. C. Turner, P. Court, and C. Hutchin, "Actuator Constraints Handling in Higher Harmonic Control Algorithms for Vibration Reduction," in *Proceedings of the 40th European Rotorcraft Forum*, (Southampton, UK), Sept. 2-5, 2014.

[25] W. R. Splettstoesser, K.-J. Schultz, R. Kube, T. F. Brooks, E. Booth, G. Niesl, and O. Streby, "A Higher Harmonic Control Test in the DNW to Reduce Impulsive BVI Noise," *Journal of the American Helicopter Society*, vol. 39, no. 4, pp. 3–13(11), 1994.

[26] O. Schneider, B. G. van der Wall, and K. Pengel, "HART II Blade Motion Measured by Stereo Pattern Recognition (SPR)," in *Proceedings of the 59th Annual Forum of the American Helicopter Society*, (Phoenix, AZ, U.S.A.), May 6-8, 2003.

[27] O. Schneider, "Analysis of SPR measurements from HART II," *Aerospace Science and Technology*, vol. 9, no. 5, pp. 409–420(12), 2005.

Lesson Learnt from Data Exchange between Manned and Unmanned Aerial Vehicles and Accommodating 12 Different UAS Operations in a Airspace

Hiroko Nakamura
Project Associate Professor
University of Tokyo
Tokyo, Japan

Yoshihiko Matsumoto
Master Candidate
University of Tokyo
Tokyo, Japan

Shinji Suzuki
Professor
University of Tokyo
Tokyo, Japan

ABSTRACT

This manuscripts addresses the importance of establishing stable communications systems among UA and carefully designed operation processes among manned aircrafts and UAS flying in a same airspace, based on the authors' recent experiments in Japan.

INTRODUCTION

How to accommodate (relatively small) UAS safely into the national airspace, especially in urban environments is recognized as a important issue at both the governmental and private level in Japan. A study group that considered the collision risk among manned aircrafts and UASs at the airspace under 150 meters was formed by the Japan Civil Aviation Bureau (JCAB) with members from public and private entities and three meetings have been held since November 2016 as of August 2017. Development of UTM is appeared at the roadmap for safe application of UAS made by the Ministry of Economy, Trade and Industry (METI) (http://www.meti.go.jp/policy/mono_info_service/mono/robot/kaiteiroadmap.pptx (in Japanese)). And in July 2016, 11 companies, institute and university launched Japan Unmanned System Traffic & Radio Management Consortium (JUTM) for discussing and developing environment and technology necessary for safe operation and wide introduction of unmanned systems (Table 1). JUTM activities are not limited to aerial but also ocean and ground vehicles and the number of members reached more than 120 as of July 2017.

Table 1. JUTM Founding Members

Industry	ANA Holdings, NTT Docomo, Japan Post, Hitachi, Fuji Heavy Industry, Yamato Holdings
Academia	University of Tokyo
National Institute	National Institute of Advanced Industrial Science and Technology (AIST), National Institute of Information and Communications Technology (NICT), Japan Aerospace Exploration Agency (JAXA), Electric Navigation Research Institute (ENRI)
Municipality	Fukushima Prefecture (Observer)

Information sharing among operators of both manned and unmanned aircrafts will support the situation awareness of operators. And there should be a proper information or alerts that can help UAS operators to make a effective decision to avoid any collision risks. In order to investigate what kind of information should be effective, how such information can be gathered and how such information can be effectively shared among operators, two distinctive experiments were conducted in Japan, University of Tokyo (UTokyo) Amakusa experiment in December 2016 and JUTM Fukushima experiment in March 2017.

Next section describes the overview of UTokyo Amakusa experiment and there are three highlights. One is the position information acquisition systems, namely D-NET systems, available for manned aircrafts and installed part of Fire and Emergency helicopters (FE helicopter) operated by prefectures in Japan. While the D-NET systems were only available systems to obtain the real-time position of the helicopter at the experiment, several problems were discovered through the experiment. Second is options to obtain the position data of different UAS deployed for the experiment. And third is necessary of further researches on designing efficient alarms and designing operation process for UAS when UAS received a alarm of a manned aircraft approach.

Thrid section describes the overview of JUTM Fukushima experiment and there are two highlights. One is the difficulty to manage multiple operators which use the same airspace without communication platforms. The UTM system used for the experiment is still at a immatured level and a lot of misunderstanding happened. Another is also the off nominal situation caused by loss of LTE and unexpected bad weather.

Last section summaries lesson learnt from the experiments.

Presented at the 6th Asian/Austrailian Rotorcraft Forum / Heli Japan 2017, Kanazawa, Japan, November 7-9, 2017.

UTOKYO AMAKUSA DATA EXCHANGE DEMONSTRATION BETWEEN A MANNED HELICOPTER AND TWO DIFFERENT TYPE OF SMALL UAS

The authors conducted a work to build and test a concept to share information among operators of manned aircrafts and sUASs using the same low altitude airspace. This work is to support sUAS operators in making decisions and actions that avoid collisions. The concept is to share the positions of manned aircraft and sUASs on a web-based UAS Traffic Management (UTM) platform and to provide alarms at the timing operators of sUASs so that they can land before a manned aircraft reaches the minimum distance to separate. Such a platform to share information among different operators of manned aircraft flying at Visual Flight Rules (VFR) and sUASs at low altitude airspace does not yet exist today comprehensively, at least, in Japan.

The web system to is made up of databases and websites. Each operator sends the position information of their aircraft or sUAS to the database in real time by Internet connection via the radio waves of their smartphone (Fig.1).

The relative positions are calculated from the position information stored in the database so that an alert is issued to the browser of each operator if it gets closer than the prescribed distance.

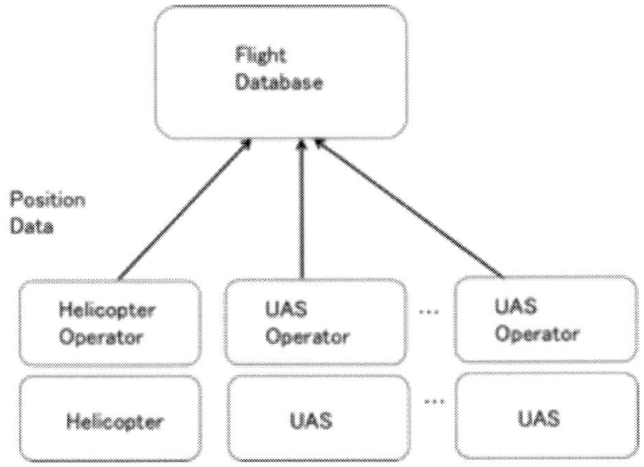

Figure 1. Overview of the web system used at UTokyo Amakusa demonstrateon

Position data of different aircrafts were acquired in different manner.

Position of the helicopter

Amakusa-city, Kumamoto prefecture and All Nippon Airways supported the experiment. Thanks to their great efforts, the FE helicopter of the Disaster and Fire Aviation Center of Kumamoto-prefecture joined to the experiment. D-NET system which uses Iridium satellite to send the GPS data (Fig.2) to a server is equipped to 56 helicopters, including the one of Kumamoto prefecture, out of 76 FE helicopters in Japan, as of 2016. D-NET system was the only and handy way to obtain the position information of the FE helicopter.

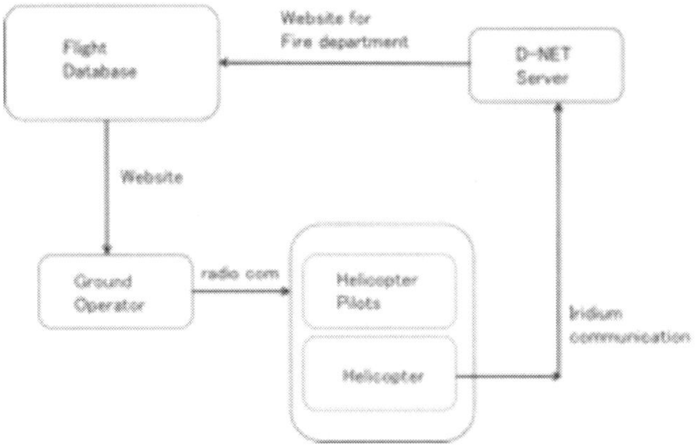

Figure 2. Overview of the flow of information for the Hibari operation

Position of UAS

This experiment uses two UAS. Due to the restrictions of the Radio Law, position information should have been uploaded to the database via the ground control system (Fig. 3). Different Ground Control Stations are used for each UAS (Fig. 4). For DJI Phantom 4, the position information was extracted, using the Software Development Kit (SDK) from DJI's application, and uploaded from a transmitter to the database via a smart phone. For a UAS made by authors with MicroKopter kits, the GPS information of the hobby sUAS was sent to the ground station PC by the Xbee module.

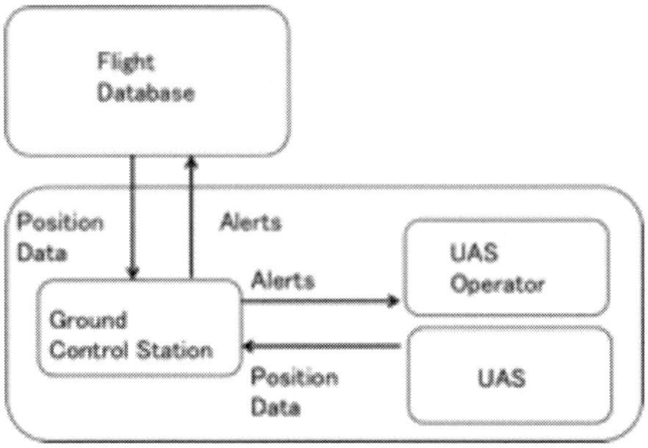

Figure 3. Overview of flow of information for the sUAS operation

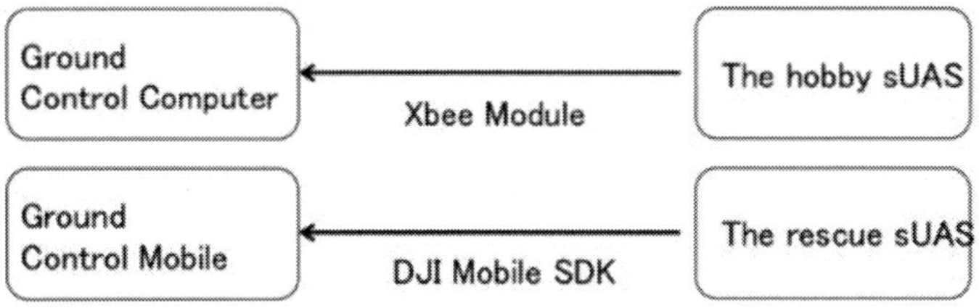

Figure 4. Methods for sending the position of the sUAS to the ground station

Figure 5. Pictures from the UTokyo Amakusa experiment

JUTM FUKUSHIMA DEMONSTRATION WITH 12 DIFFERENT DRONE OPERATIONS IN A AIRSPACE DATA EXCHANGE DEMONSTRATION BETWEEN A MANNED HELICOPTER AND TWO DIFFERENT TYPE OF SMALL UAS

JUTM to which the first and third authors are belong to conducted a demonstration of basic concept of UTM, that is airspace management by flight plan and real-time tracking, and prototype UTM core system in 15-16 March 2017. 9 companies and 2 industries operated 12 UAS for different missions (Table 2, Fig. 6). The experiment was also supported by companies and institute listed in Table 3 and has Fukushima prefecture and Minami-soma city as Associate and Cabinet Office, Ministry of Internal Affairs and Communications, Ministry of Economy Trade and Industry, Ministry of Land, Infrastructure, Transport, and Tourism, and National Institute of Advanced Industrial Science and Technology. The prototype system has nothing related with the system described in the previous section.

Four scenarios (Fig. 7-9) were prepared in order to demonstrate following basic functions of UTM;

- Warning when a UA flies away from the allocated airspace
- Warning when a UA flies into restricted area
- Geo-fence to the area where and a helicopter planed to land
- Landing control at a “drone-port”
- Collision avoidance instruction in a shared airspace with a dummy UA

As the UTM prototype is still not matured, some of functions have been made not spontaneously or automatically. For example, flight plan management was made long time before the experiment. Final approve of taking off or landing and the management were made not through the system but through voice transceiver.
In Japan, use of LTE network not from ground is prohibited at the current Radio Law but with some conditions experiments to use the LTE are allowed. Some of UA carried LTE smartphone and sent their information directly to the UTM.

Despite the long preparation, the weather was very bad at the week so that the detail of scenario had to be modified at the field.

Figure 6. Pictures from the JUTM Fukushima experiment

Table 2. JUTM Fukushima demonstration operators

Operator	Mission	UA Type	Comm
Japan Post	Postal delivery	MR	LTE
Yamato holdings	Parcel delivery	MR	LTE
enRoute/ enRouteM’s	Crop-dusting	MR	
Watami	Remote sensing agriculture	MR	
Hitachi Const. Machinery	Construction surveying	MR	
OFF line	Missing person investigation	MR	
Sompo Japan Nipponkoa	Emergency warning & Damage investigation	MR	
AIST	Disaster Monitoring	MR	
Yamaha Motor	Delivery of relief supplies	Helicopter	sat.com

TV-U Fukushima	Broadcasting	MR	
NICT	Radio monitoring	FW	

[a](MR: Multi roters, FW: Fixed wing)

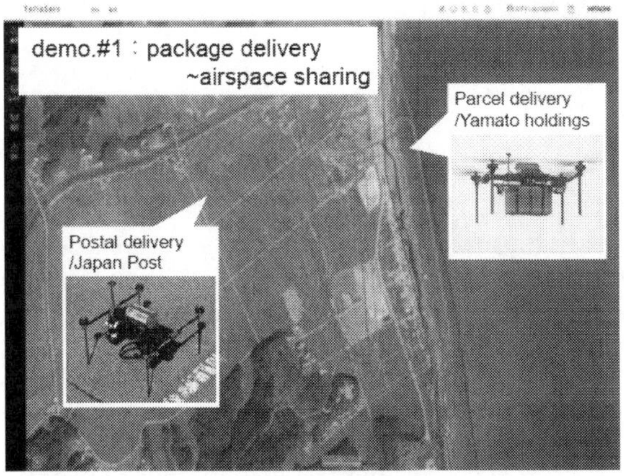

Figure 6. Scenario 1

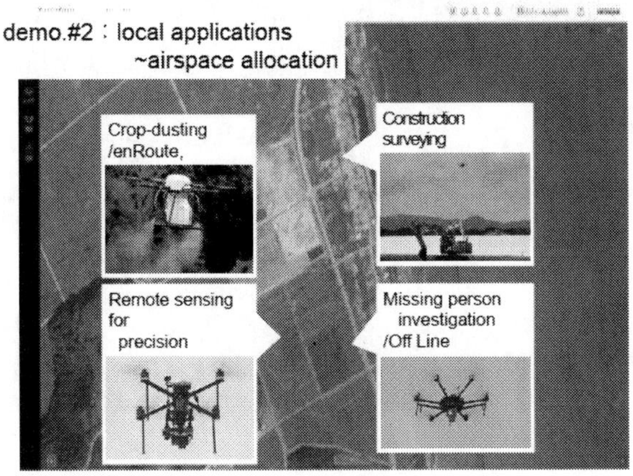

Figure 7. Scenario 2

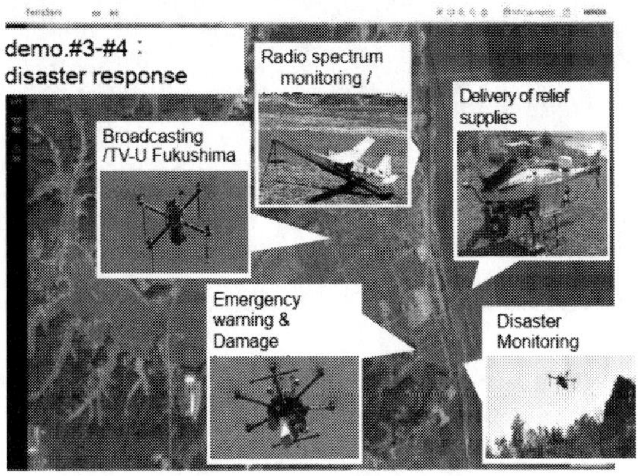

Figure 8. Scenario 3 and 4

LESSON LEARNT

Infrastructure requirement

Position information of a manned aircraft which may fly in the same airspace is required for UAS if UAS is expected to fly away or land before the manned aircraft reached at certain distance, for example, 600 meter. While we need more investigation on appropriate separation distance, use of satellite system such as D-NET system has some problems. Due to the communication cost and vulnerability in connection especially in mountains, the data was available infrequently, with average of 60 seconds time interval. And there are about 47 seconds of delay for a data to be informed at the server. These delay and interval made difficult for designing alarm timing and decision timing for UAS operator to land the UA.

For UAS, there are various options to obtain the position data and if the UTM is web-system, obtaining methods are not critical. If the position data can be sent from air to server, it will reduce the time delay. Use of LTE onboard could be useful but we experienced lost of connection, too.

Security consideration

As there were concerns about allowing the real-time position information of FE helicopter to be known to people outside of the operator, the position of some aircrafts can be shared with some uncertainty. And the uncertainty of the real positioning could be solved by incursion of time buffer in a alarm.

Other

When a UTM uses a function to have UAS to land, landing procedure should be agreed among stakeholders. Some may land immediately to the closest position from the point that UA were when the operator received the alert. Some may land to the closest position from the operator. There were confusions at the Amakusa experiment.

The prototype systems both in Amakusa and Fukushima experiments, the real-time status of aircraft, such as landing, ready for fly, flying, were not clear and we found the information is very useful and important, especially in unexpected bad weather.

Author contact: Hiroko Nakamura techhn@mail.ecc.u-tokyo.ac.jp

ACKNOWLEDGMENTS

We would like to express our gratitude to Amakusa city, Minamisoma city, Fukushima prefecture, the Disaster and Fire Aviation Center of Kumamoto-prefecture, Ama-usa Interjurisdictional Center, All Nippon Airways and all the member of JUTM for sup-porting this demonstrations. This work was also partly supported by JSPS KAKENHI Grant Number 16K17161.

The Influence of the Downwash of a Helicopter Rotor Hovering at High Altitude on the Ground Surface

Kenta NAKASHIMA Graduate Student
Noriaki ITOGA Professor
National Defense Academy
Yokosuka, JAPAN

ABSTRACT

In this study, the horizontal flow velocity near the ground surface caused by the downwash of a helicopter rotor hovering at high altitude is experimentally measured by using a Pitot tube. The experimental parameters are the horizontal distance between the rotor and a Pitot tube, the vertical distance between the ground plate and a Pitot tube, and the rotor thrust coefficient. Experimental results show that the influence of the rotor thrust coefficient does not appear in the velocity ratio of the horizontal flow velocity to the induced velocity at a rotor disk obtained by simple momentum theory. At even a position of 3.5 times the rotor radius away from the hovering point, the horizontal flow velocity is the same magnitude as the induced velocity at a rotor disk.

INTRODUCTION

There are many researches on the downwash of a helicopter rotor, but most of them are cases that a helicopter hovers at low altitude[1-5]. Therefore, in this study, the influence of the downwash of a helicopter rotor hovering at high altitude is clarified by measuring the horizontal flow velocity near the ground surface using a Pitot tube under the condition of constant rotor thrust coefficient.

EXPERIMENTAL EQUIPMENT

Figure 1 shows a general view of the experimental equipment. The specifications of the model rotor are shown in Table 1. The model rotor, which suspended from the top so as not to disturb the rotor wake, is articulated type. A load detector for measuring thrust and torque of the rotor is built in the base of the model rotor. In addition, the model rotor has a swash plate, and it is possible to change the collective and cyclic pitch angles every 0.01°, respectively. The rotor rotational direction is clockwise looking from the top.

The ground plate is made of polystyrene foam. The thickness of the ground plate is 30mm. The ground plate is a rectangle with a long side of 5,747mm and a short side of 3,585mm.

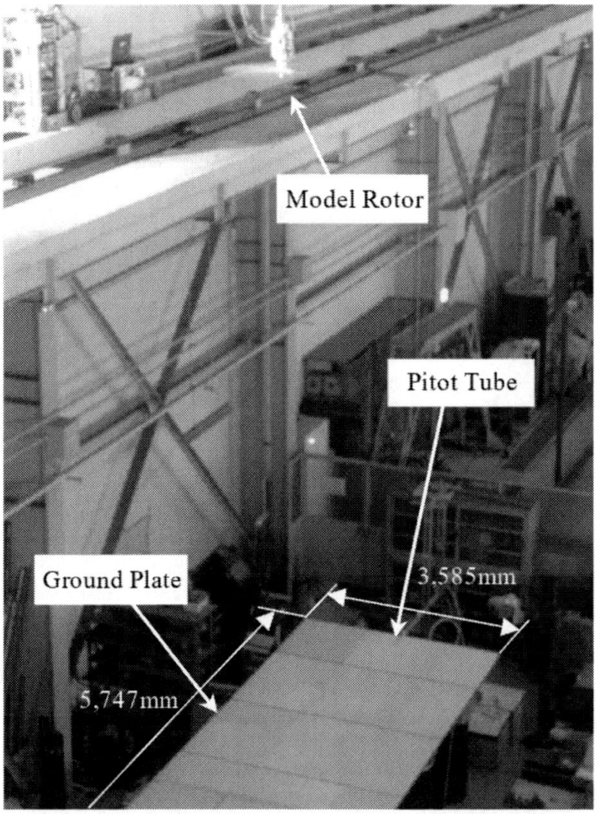

Figure 1. General View of the Experimental Equipment.

Table 1. Major Parameters of the Model Rotor.

Airfoil	NACA0015
Built-in Twist	NA
Number of Blades	2
Rotor Radius: R	569mm
Blade Chord	60mm
Flapping Hinge Offset	17mm
Root Cutout	113mm

EXPERIMENTAL PARAMETERS

Definitions of experimental parameters are shown in Figure 2. The Parameters are the vertical distance between the rotor plane and the ground plate, the horizontal distance from the rotor axis to a Pitot tube, and the vertical distance from the

Presented at the 6th Asian/Australian Rotorcraft Forum / Heli Japan 2017, Kanazawa, Japan, November 7-9, 2017.

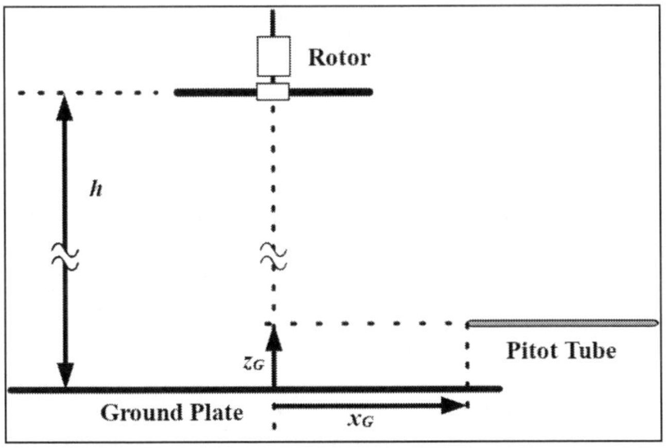

Figure 2. Definitions of Experimental Parameters.

Table 2. Range of Experimental Parameters.

Rotor Height: h	12.0
Pitot Tube Distance: x_G	$1.25 \le x_G \le 5.0$
Pitot Tube Height: z_G	$0.01 \le z_G \le 0.5$
Rotor Thrust Coefficient: C_T	6.71×10^{-3}, 4.00×10^{-3}

ground plate to a Pitot tube. These parameters are normalized by the rotor radius R. Normalized parameters are called rotor height h, Pitot tube distance x_G, Pitot tube height z_G, respectively. The range of them and the rotor thrust coefficient C_T are indicated in Table 2.

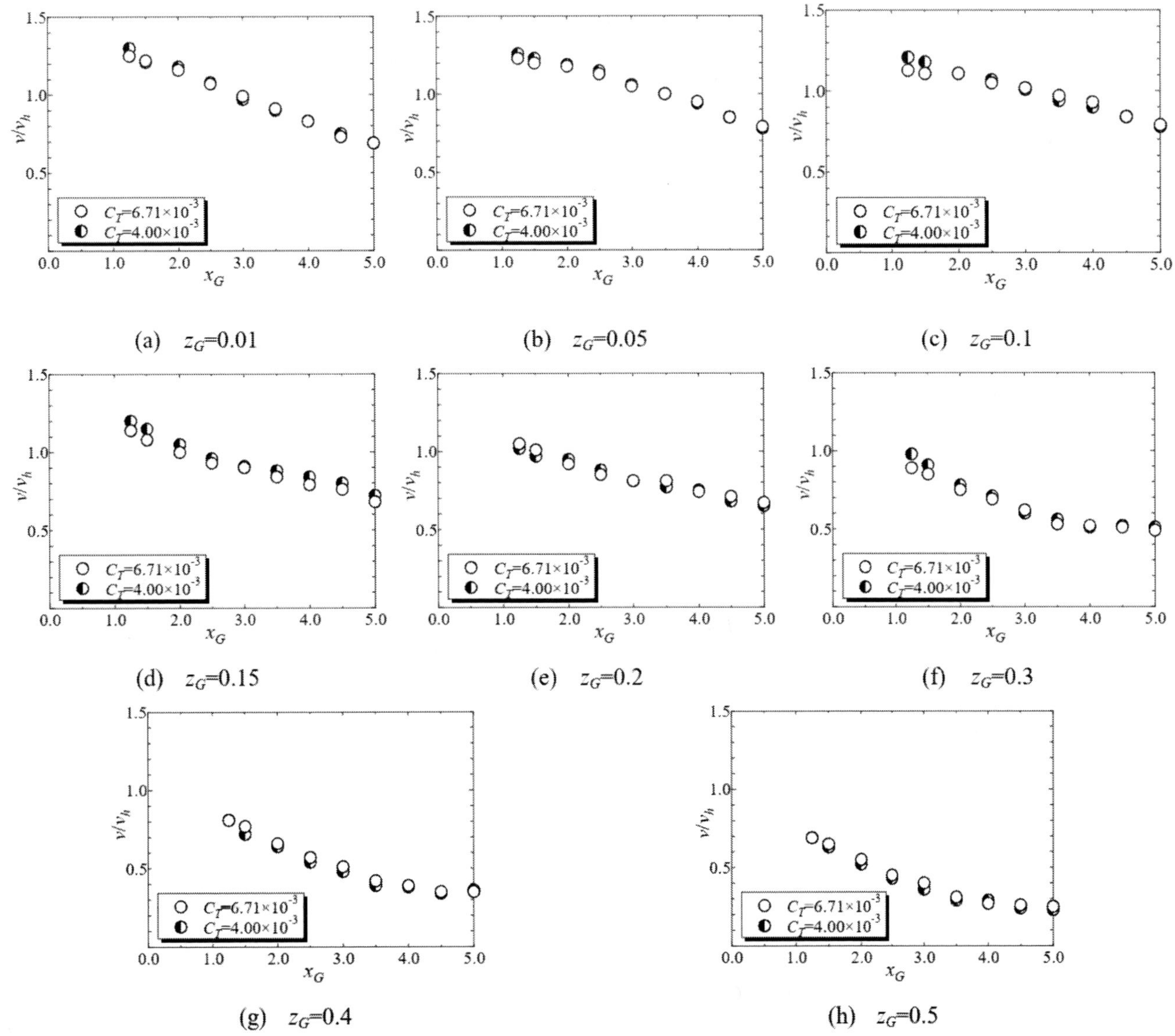

Figure 3. Velocity Profiles at Pitot Tube Height. $\left(C_T = 6.71\times10^{-3}, 4.00\times10^{-3}\right)$

EXPERIMENTAL METHOD

In this experiment, the collective pitch angle is adjusted so that the rotor thrust coefficient becomes constant within an accuracy of ±0.25% of the standard value each time Pitot tube distance x_G and Pitot tube height z_G are changed. The horizontal velocity v near the ground surface is measured using a Pitot tube. A number of rotor revolutions is constant at 1,020 rpm. In order to improve the measurement accuracy, only the total pressure is measured with a Pitot tube, and the static pressure is measured in a box isolated from the surrounding air flow. v is obtained from the steady component of differential pressure processed by FFT analysis.

RESULTS AND DISCUSSIONS

The dimensionless quantity used in this study is the velocity ratio v/v_h, which is defined as the equation (1).

$$\frac{v}{v_h} = \frac{v}{(R\Omega)\sqrt{C_T/2}} \tag{1}$$

R is the rotor radius, Ω is the rotational angular velocity of the rotor, and C_T is the rotor thrust coefficient.

Figure 3 shows the velocity ratio v/v_h versus Pitot tube distance x_G for each Pitot tube height z_G. Circular marks show results in case of the rotor thrust coefficient $C_T = 6.71\times10^{-3}$, and semicircular marks show results in case of $C_T = 4.00\times10^{-3}$. Each result shows a good agreement each other at any x_G and z_G. That is, there is no influence of C_T on v/v_h. Therefore, in the following discussion, results of $C_T = 6.71\times10^{-3}$ alone are shown.

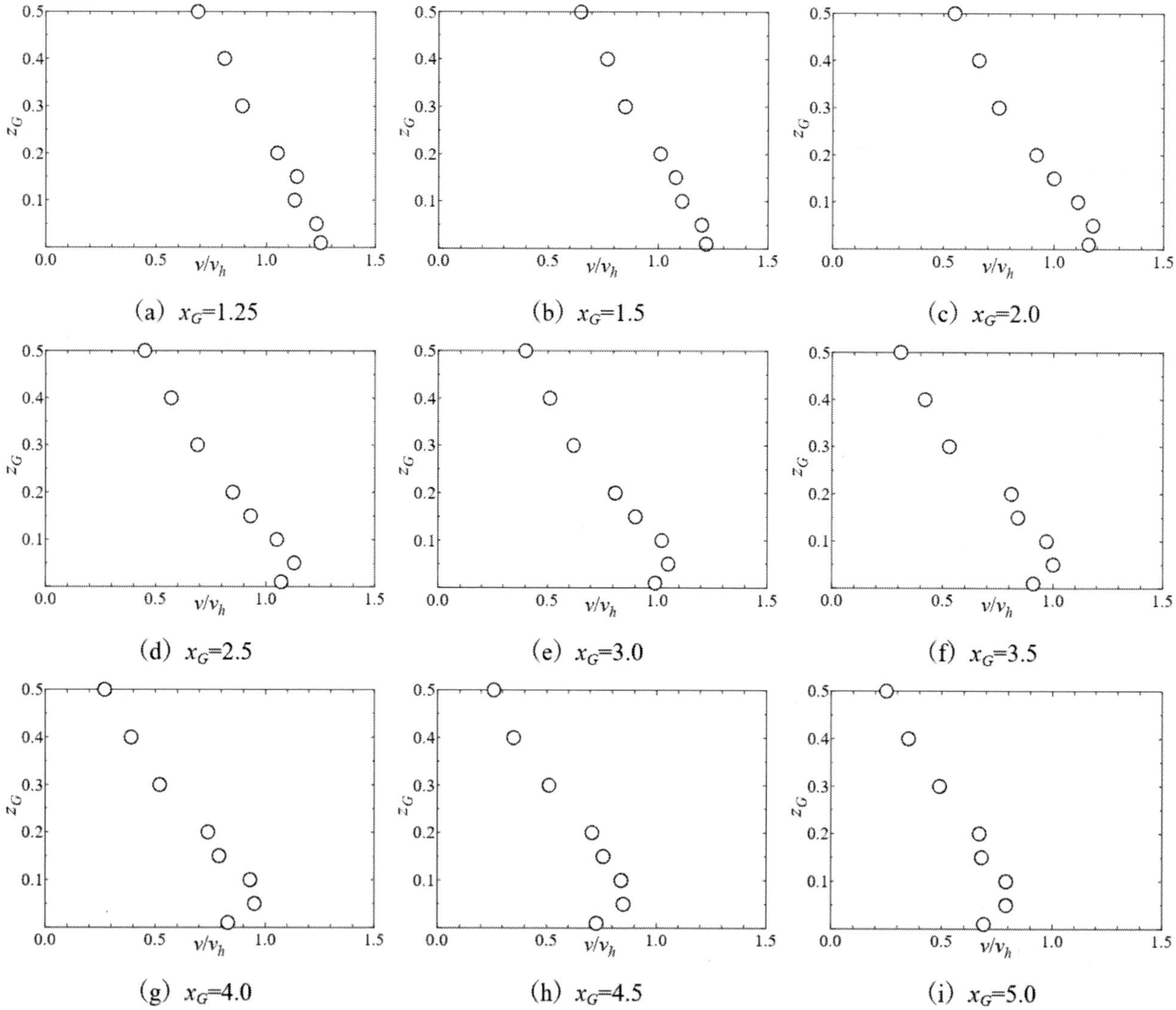

(a) x_G=1.25 (b) x_G=1.5 (c) x_G=2.0
(d) x_G=2.5 (e) x_G=3.0 (f) x_G=3.5
(g) x_G=4.0 (h) x_G=4.5 (i) x_G=5.0

Figure 4. Velocity Profiles at Various Pitot Tube Distance. (C_T=6.71×10^{-3})

Figure 4 shows the velocity ratio v/v_h versus Pitot tube height z_G for each Pitot tube distance x_G. The behavior of v/v_h with respect to z_G changes at x_G=1.5. v/v_h in the range of $x_G \leq 1.5$ take the maximum value at z_G=0.01, which is the closest to the ground plate, and decrease substantially as z_G increases. On the other hand, v/v_h in the range of $x_G \geq 2.0$ tend to increase once as z_G increases and then decrease after taking the maximum value. These states mean the development of the boundary layer. It is also understood that the maximum value of v/v_h decreases as x_G increase. The maximum value is about 1.25 at $x_G = 1.25$, while it is about 0.8 at $x_G = 5.0$.

Figure 5 shows the velocity ratio v/v_h with respect to Pitot tube distance x_G. It is understood that v/v_h gradually decreases as x_G increases. In the range of $x_G \leq 3.5$, it can be seen that the value of v/v_h is 1.0 or more, that is, the horizontal flow velocity near the ground is over 10~15m/s. At even a position of x_G=5.0, the horizontal flow velocity is about 8~12 m/s. Comparing the flight test measurements[6] indicated by the solid line with the experimental ones, they are roughly in agreement.

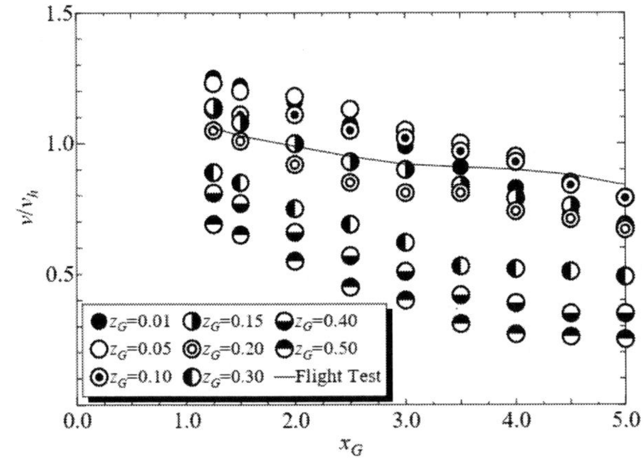

Figure 5. Velocity Profiles at Pitot Tube Height. (Experimental Results, Flight Test Measurements)

CONCLUDING REMARKES

The findings obtained by this research are summarized as follows.

- The rotor thrust coefficient has no influence on the velocity ratio of the horizontal flow velocity to the induced velocity at a rotor disk obtained by simple momentum theory.
- At even a position of 3.5 times the rotor radius away from the hovering point, the horizontal flow velocity is the same magnitude as the induced velocity at a rotor disk.

REFERENCES

[1]Leese, G. V. and Knight, I. T., Jr., " Helicopter Downwash Data," Army Engineer Waterways Experiment Station, 1974, pp.1-40.

[2]Lee, T. E., Leishman, J. G. and Ramasamy, M., "Fluid Dynamics of Interacting Blade Tip Vortices with a Ground Plane," *Journal of the American Helicopter Society*, Vol.55, (2), 2010, pp.22005.

[3]Cheeseman I.C. and Bennett W.E., "The effect of the ground on a helicopter rotor," R&M 3021, ARC, 1955, pp.1-10.

[4]Light, J. S., "Tip Vortex Geometry of a Hovering Helicopter Rotor in Ground Effect," *Journal of the American Helicopter Society*, Vol.38, (2), 1993, pp.34-42.

[5]Bolanovich, M. and Marks, M. D., " Experimental Downwash Velocity, Static Pressure, and Temperature Distributions in Ground Effect for a 75 - Foot Jet Driven Rotor," *Journal of the American Helicopter Society*, Vol.4, (2), 1959, pp.22-36.

[6]Nagao, M., Matayoshi, N., Nakashima, K., and Itoga, N., " Herikoputa no daunnwhossyu ga chijyo ni oyobosu eikyou ni tsuite (The Influence of the Downwash of a Helicopter on the ground)," Heisei 29 nenndo koukuu ninngenn kougaku bukai, 2017 (in Japanese).

Helicopter downwash measurement using Doppler lidar

Naoki Matayoshi
Senior Researcher
Japan Aerospace Exploration Agency
Mitaka, Tokyo, Japan

Hirokazu Ishii
Associate Senior Researcher
Japan Aerospace Exploration Agency
Mitaka, Tokyo, Japan

Eiichi Yoshikawa
Researcher
Japan Aerospace Exploration Agency
Mitaka, Tokyo, Japan

Yasutada Tanabe
Senior Researcher
Japan Aerospace Exploration Agency
Mitaka, Tokyo, Japan

Masahiko Sugiura
Researcher
Japan Aerospace Exploration Agency
Mitaka, Tokyo, Japan

Maki Nagao
Aero Asahi Corporation
Koto, Tokyo, Japan

ABSTRACT

Helicopter downwash becomes strong at low airspeed region and can be hazardous for the following aircraft and surrounding personnel. The Japan Aerospace Exploration Agency has conducted several flight tests and CFD analysis to clarify the characteristics of helicopter downwash especially in ground effect. In the flight tests, Doppler lidar and ultrasonic anemometers were used to measure downwash and observed data indicate that downwash would roll-up into vortices like fixed wing aircraft when helicopter's airspeed is over 15kt and such vortices might linger long according to surrounding wind condition. The acquired downwash data will be utilized to improve CFD analysis capability and be submitted to ICAO to establish new wake turbulence separation minima for helicopters.

INTRODUCTION

Helicopter downwash (wake) becomes strong at low airspeed region and can be hazardous for the following aircraft and surrounding personnel. In the UK, there occurred several crashes of small aircraft, which are thought to be caused by helicopter downwash (Ref. 1). The International Civil Aviation Organization (ICAO) is therefore considering helicopter's own wake turbulence separation minima. It is known that helicopter downwash rolls-up into a pair of vortices like a fixed wing aircraft at airspeed of over 40 kt and several measurements clarified its circulation strength (Ref. 2). However, there are still little measurement data for helicopter downwash at low airspeed region near ground such as taxing. The Japan Aerospace Exploration Agency (JAXA) is therefore conducting helicopter downwash measurement and CFD analysis to clarify the wake formation process at low airspeed region (Ref. 3). This paper describes the measurement of helicopter downwash at low airspeed region using Doppler lidar and ultrasonic anemometers. This measurement was conducted by the JAXA and the Aero Asahi Corporation.

DOWNWASH MEASUREMENT METHOD

Helicopters to be measured

The helicopters whose downwash was measured were the BK117-C2 owned by the JAXA and the AS332L owned by the Aero Asahi Corporation. The flight test condition and appearance of both helicopters are shown in Table 1 and Fig. 1, respectively.

Presented at the 6th Asian/Australian Rotorcraft Forum / Heli Japan 2017, Kanazawa, Japan, November 7-9, 2017.

Table 1. Flight test condition.

	BK117-C2	AS332L
Rotor diameter	11.0m	16.2m
Weight at flight test	2750–3100kg	5900–6400kg
Disc loading at flight test	29–33kg/m^2	29–31kg/m^2

(a) BK117-C2 (JAXA research helicopter).

(b) AS332L (Aero Asahi Corporation).

Figure 1. Helicopters whose downwash was measured.

Measurement equipment

The JAXA's Doppler lidar and ultrasonic anemometers were used for downwash measurements.

Doppler lidar: A Doppler lidar emits laser light in the air, receives laser light scattered by small particles (aërosol) in the atmosphere, and measures movement of aerosol, that is, wind from the Doppler shift component of received light. The WINDCUBE200S manufactured by the Leosphere (Table 2, Fig. 2) was used for this measurement. The JAXA has used this lidar for wake vortex measurement of fixed wing aircraft (Ref. 4).

Ultrasonic anemometer: The CYG-81000 manufactured by the Young (Table 3, Fig. 3) was used for this measurement. The ultrasonic anemometer has good responsiveness enough to capture wake behaviors. We measured the 3-axis components of wind at a measurement rate of 10Hz.

Table 2. Major specifications of WINDCUBE200S.

Item	Specification
Lidar type	Pulsed Doppler lidar
Laser wave length	1.54μm
Laser power	700mW
Measurement item	Wind speed (laser irradiation direction component), spectrum/ power of received signal
Measurement rate	0.1–30Hz
Measurement range	0–6km (at measurement rate of 1Hz)
Range resolution	Physical resolution (laser pulse length): 25–100m Output resolution: 1m (minimum)
Wind measurement range	0–30m/s
Scan pattern	FIX, PPI, RHI, DBS
Size and weight	1.0×0.8×1.5m, 230kg

Figure 2. Doppler lidar WINDCUBE200S.

Table 3. Major specifications of CYG-81000.

Item	Specification
Measurement item	3-axis component of wind speed
Measurement rate	1–10Hz
Wind measurement rage	0–40m/s
Size and weight	0.57×0.32×0.32m, 1.7kg

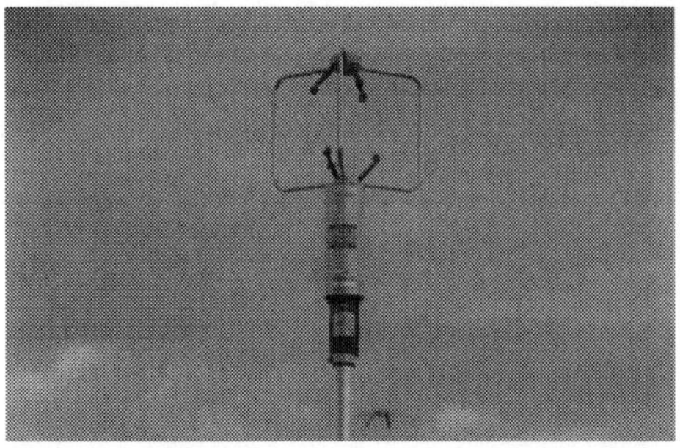

Figure 3. Ultrasonic anemometer CYG-81000.

Measurement setup

Table 4 shows the location and date of the downwash measurements, and Fig. 4 shows the arrangement of the measurement equipment. The helicopter passed over runway centerline, and the ultrasonic anemometers measured 3-axis components of downwash at 2.5m height above ground. The Doppler lidar was set about 200m away from the runway, and measured the wind distribution on the vertical section orthogonal to the runway by RHI (Range-Height Indicator) scan that changes elevation angle with fixed azimuth angle. Please note that the Doppler lidar can measure only the wind speed component in the laser irradiation direction and its spatial resolution is not isotropic. On the runway, the Doppler lidar's spatial resolution in the vertical direction is about 1m, corresponding to 0.2degree resolution in elevation angle. On the other hand, the lidar's spatial resolution in the horizontal direction (laser irradiation direction) is about 50m, corresponding to laser pulse length, too coarse for downwash measurement. We therefore decided to measure only the vertical distribution of the maximum/minimum wind speeds of horizontal wind (component orthogonal to the runway) in the range up to about 20m above the runway. The horizontal size of the measurement region was about 50m left and right of the runway centerline.

Table 4. Location and date of downwash measurement.

	BK117-C2	AS332L
Location	Taiki aviation park, Hokkaido, Japan	Shikabe aerodrome, Hokkaido, Japan
Date	October 23-26, 2014	March 7-10, 2016

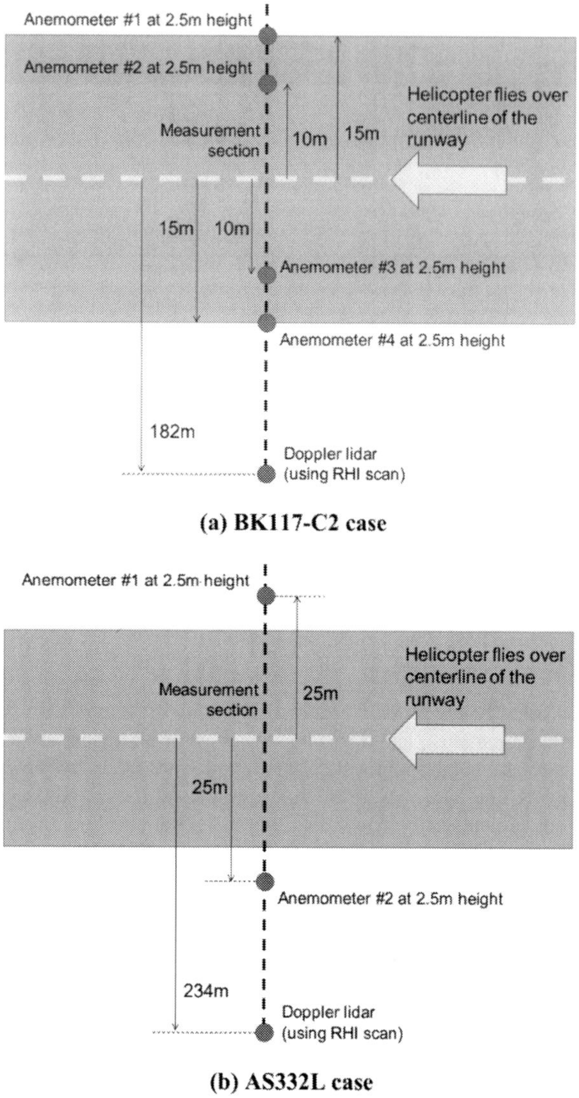

Figure 4. Equipment arrangement of downwash measurement.

DOWNWASH MEASUREMENT RESULT

Measurement condition

Figure 5 shows ground wind condition at the measurement. The ground wind speeds were below 5kt for both BK117-C2 and AS332L cases. It indicates that the measurements were carried out under weak wind conditions, which facilitated the downwash measurement. Figure 6 shows the helicopter's airspeed and height above ground when the helicopters crossed the measurement section which the Doppler lidar monitored. The airspeeds were mostly below 40kt and the heights were below 80ft, and meets the target measurement condition. The airspeed of the helicopter was calculated by correcting the ground speed of the helicopter, measured by the GPS receiver mounted on the helicopter, with the ground wind speed measured by the ultrasonic anemometer.

Downwash measurement result by ultrasonic anemometer

Downwash wind speed over the runway: Figure 7 shows the AS332L's downwash wind speed measured by the ultrasonic anemometer next to the runway. In the measurement example with a helicopter airspeed of 18kt, a sharp rising edge of the wind speed suggests that the downwash rolls-up into a pair of vortices. On the contrary, such sharp rising edge cannot be seen in the measurement example with a helicopter airspeed of 11kt.

The results of comparing the maximum downwash wind speeds of the BK117-C2 and the AS332L are shown in Fig. 8. Since the disc loadings of two helicopters were almost the same as shown in Table 1, the maximum downwash wind speeds of the two helicopters are nearly the same. Although there are variations, as the airspeed increases, the downwash tends to weaken. In addition, vertical wind of up to 20kt was measured, still it is weaker than the horizontal wind speed because it is close to the ground.

Downwash duration time over the runway: Figure 9 shows the period during which a downwash of 10kt or more was measured by the ultrasonic anemometer next to the runway. The time when the helicopter passed through the measurement section is set to be zero. The longest time during which the downwash of 10kt or more was observed was about 30 seconds after the passage of the helicopter. The downwash was also measured even before the passage of the helicopter at low airspeeds of 10kt or less. However, the downwash duration can be longer than we observed when the downwash rolls-up into vortex form. As well known, the wake vortex moves so as to separate in the left and right direction near the ground by its own induced winds. Please note that the wake vortex may linger on the runway under conditions where the crosswind is balanced with the moving speed of the wake vortex.

Downwash measurement result by Doppler lidar

Although the Doppler lidar measurement was carried out in both the BK117-C2 and the AS332L cases, the valid measurement data could not be extracted from the AS332L measurement because the signal-to-noise ratio of the received signal was too low. There is a possibility that the density of the aerosol which is the scattering source of the laser light is low at the time of measurement. Therefore, only the measurement result of the BK117-C2 is shown.

Downwash wind speed over the runway: Figure 10 shows maximum downwash wind speed of the BK117-C2 measured by the Doppler lidar. The maximum wind speed is increased by 5 to 10kt compared to the measurement result of the ultrasonic anemometer shown in Fig. 8. The Doppler lidar measures the wind speed up to about 20m above the ground where the influence of friction on the ground surface is weak, so the measured wind speed is faster than the speed near the ground.

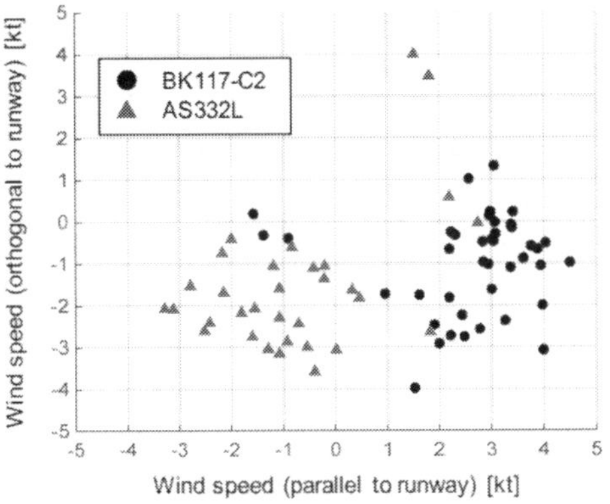

Figure 5. Ground wind conditions at measurement.

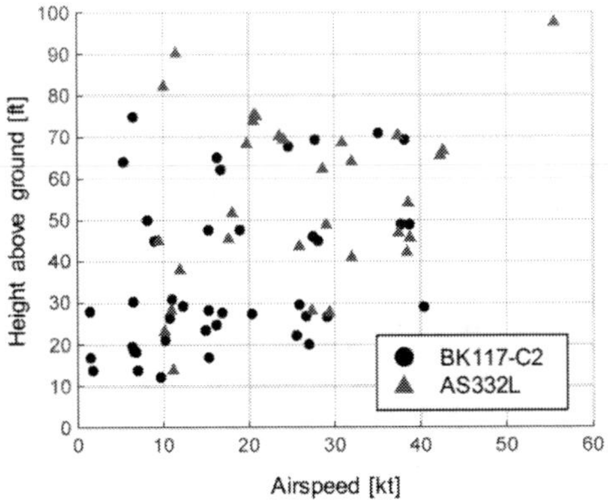

Figure 6. Helicopter's airspeed and height above ground when crossing the measurement section.

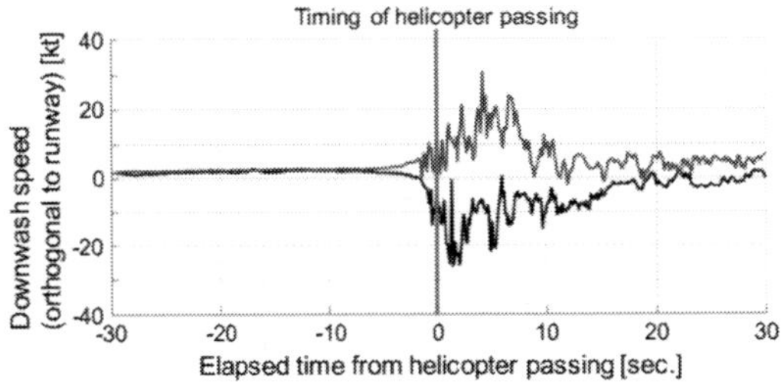

(a) Helicopter airspeed of 11kt

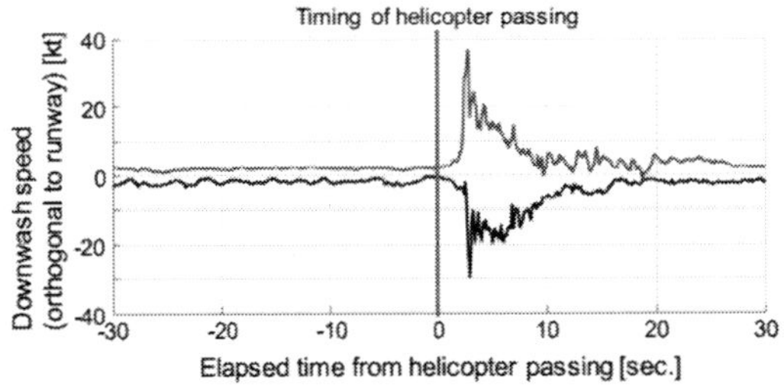

(b) Helicopter airspeed of 18kt

Figure 7. AS332L's downwash measured by ultrasonic anemometer.

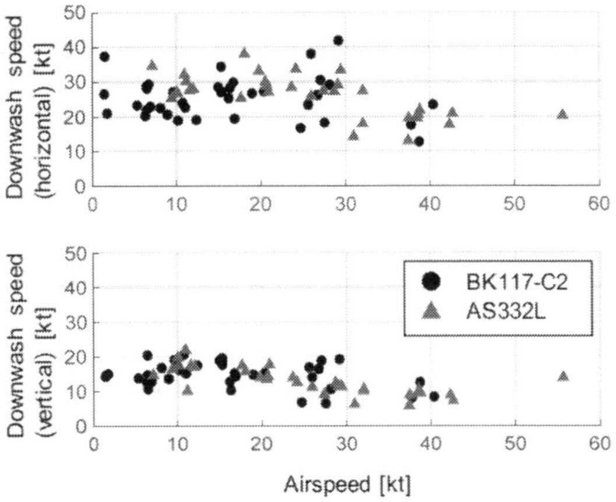

Figure 8. Maximum downwash wind speed measured by ultrasonic anemometer.

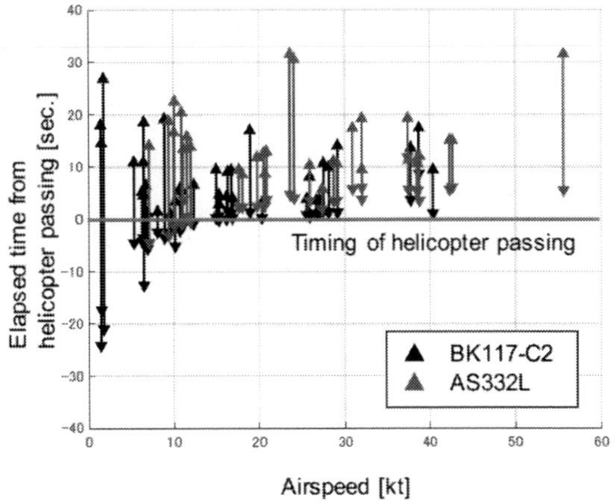

Figure 9. Period during which downwash of 10kt or more measured by ultrasonic anemometer.

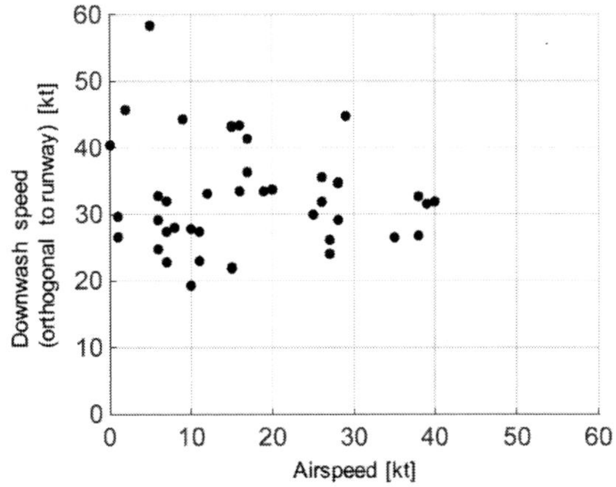

Figure 10. Maximum downwash wind speed of BK117-C2 measured by Doppler lidar.

Wake vortex formation: Figure 11 illustrates the results where the Doppler lidar measures the vertical distribution of the downwash of the BK117-C2. We compare cases where the helicopter passes the measurement section at 12kt and 40kt airspeeds. In the figure, the intensity of the received signal of the lidar is shown in color, the vertical axis is height above ground, and the horizontal axis is wind speed component in the laser irradiation direction. A region where the signal intensity is higher than the noise level of the lidar, that is, the wind speed range where the wind was actually observed is shown in green to red color. In the case of the airspeed of 40kt, the maximum and minimum peak values of the wind speed are slightly displaced in the vertical direction, three peaks are seen, and it suggests that the two vortices are displaced in the vertical direction. On the other hand, in the case of the airspeed of 12kt, such vortex structure cannot be seen. The downwash simply diffuses near the ground. Figure 12 shows an example where the helicopter downwash was visualized by smoke at an airspeed of about 20kt. The downwash clearly rolls-up into vortex form.

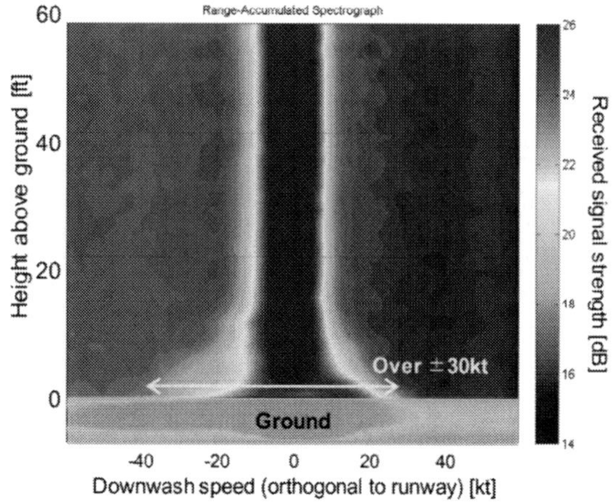

(a) Helicopter airspeed of 12kt.

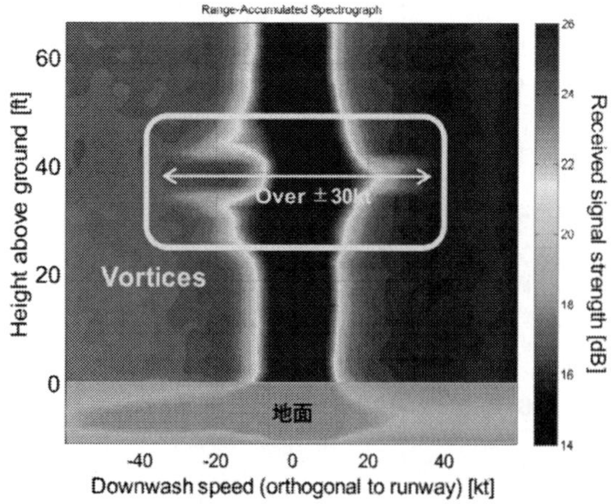

(b) Helicopter airspeed of 40kt.

Figure 11. Vertical distribution of BK117-C2's downwash measured by Doppler lidar.

Fig. 13 shows the result of estimating the circulation intensity of the helicopter wake vortex by reading the peak wind speed and the core radius of the vortex from the lidar measurement. The vortex-type downwash was observed only at airspeed of over 15kt. Theoretical circulation intensity based on simple momentum theory (equation 1) is also shown in Fig. 13.

$$\Gamma = mg/\rho/b/V \qquad (1)$$

Γ: circulation intensity, m: aircraft mass, g: gravity acceleration, ρ: air density, b: separation between two vortices (rotor diameter multiplied by π/4 is used this time), V: aircraft airspeed

The circulation intensity tends to decrease as the airspeed increases, corresponding to the theoretical formula. Please note that there is a possibility that the errors of estimation based on the lidar measurement is large because the measuring resolution of the lidar is not fine enough to estimate core radius.

Figure 12. Visualized BK117-C2's downwash at airspeed of 20kt.

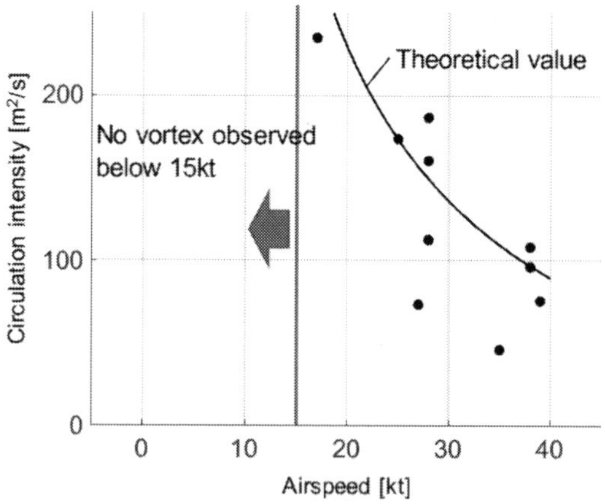

Figure 13. Circulation intensity estimation of BK117-C2's wake vortex based on lidar measurement result.

CONCLUSIONS

This paper reported the downwash measurement results of the BK117-C2 and the AS332L helicopters. The downwash parameters such as maximum/minimum wind speed, duration time and circulation intensity at low airspeed region were successfully measured by the Doppler lidar and the ultrasonic anemometers. The measurement result well corresponds to the CFD analysis result and would contribute to improve the CFD analysis accuracy (Ref. 3). The JAXA regularly reports the measurement data and the CFD analysis results to the ICAO Wake Turbulence Working Group and continuously support the revision work of the wake turbulence separation minima.

ACKNOWLEGEMENTS

The authors express their deepest appreciation to the people of Taiki Aviation Park and Shikabe Aerodrome who kindly supported this downwash measurement.

Author contact: Naoki Matayoshi, matayoshi.naoki@jaxa.jp

REFERENCES

[1] CAA, Aircraft accident report 1/93. Technical report, Civil Aviation Authority, 1993.

[2] Teager, S. A., et. al., "Flight Test Investigation of Rotorcraft Wake Vortices in Forward Flight", DOT/FAA/CT-94/117, 1996.

[3] Masahiko S., et. al., "Numerical Simulations and Measurements of the Helicopter Wake in Ground Effect", *Journal of Aircraft*, Vol. 54, No. 1 (2017), pp. 209-219.

[4] Yoshikawa E., et. al., "Aircraft Wake Vortex Retrieval Method on Lidar Lateral Range-Height Indicator Observation", *AIAA Journal*, Vol. 55, No. 7 (2017), pp. 2269-2278.

"Current Situation and Issues for Roof-Top Heliports on Hospitals"

Mr. Motomi Kinoshita

President

Aero Facility Co., Ltd.

Shimbashi Plaza Bldg. 15F,

4-9-1 Shimbashi, Minato-ku, Tokyo,

105-0004 Japan.

SUMMARY

About hospital rooftop heliports:

1) "In recent years, the number of EMS helicopters have been landing on Fire Fighting Helipads on the roofs of hospitals with increasing frequency which are not supposed to be used for repetitive landing."
2) "Fire Fighting Helipads are not strong enough for repetitive landing, therefore this practice could cause a serious crisis in 30, 40 years."
3) "Hospital heliports should have the same construction design and strength as Transportation (Non-public) Heliports and should be categorized as the Temporary (Non-airfield Take Off and Landing Site) Helipads, which require relatively simple application processes."
4) "The goal is to expand the use of safe and low cost heliports."

INTRODUCTION □

"In recent years, the number of EMS helicopters have been repetitively landing on Fire Fighting Helipads on the roof of hospitals with increasing frequency which are not supported to be used for repetitive landing", says Motomi Kinoshita, CEO of Aero Facility Co., Ltd. located in Minato-ku, Tokyo, which undertakes the designs and constructions of heliports. He spreads the warning that "if repetitive landings continue on such heliports that are not strong enough, there would be potential danger in 30 or 40 years. Therefore, he insists "hospital heliports should have the same construction design and strength as Transportation Heliports, and should be categorized as the Temporary Helipads which require relatively simple application processes." As Motomi Kinoshita proclaims, "The goal is to expand the use of safe and low cost heliports". The interview with him has been held about what hospital heliports should be.

MAIN BODY - MAJOR HEADING

Rooftop heliports are mainly divided into three types, 1. "Emergency take off and landing sites (Fire Fighting Helipads)" where the Fire Department confirms them during construction

---

Presented at the 6th Asian/Austrailian Rotorcraft Forum / Heli Japan 2017, Kanazawa, Japan, November 7-9, 2017.

and landing is accepted exclusively in Fire Emergencies. 2. "Non-public Heliports (Transportation Heliports)" which are strong enough for repetitive landings. 3. "Non-airfield take off and landing sites (Temporary Helipads)", for which allowance is necessary for each landing. (Please refer to Figure 1 and Figure 2 for their structures.) The Fire Fighting Helipads are the system based on a notification issued by the former Housing Bureau of the Ministry of Construction and Fire Defense Division and Fire Prevention Division of the Fire and Disaster Management Agency in 1990, which means it is not supposed to be used for transportation from the beginning.

Actually, "Fire Fighting Helipads" can not be landed on

There are two symbols for Fire Fighting Helipads, "H" known as Emergency take off and landing sites and "R" for Emergency Rescue Space, for which the Fire Departments give guidance and certification, however, there is no unified standard or designation in the whole country.

Helipads "H" can be landed exclusively for fire extinguishing and rescue operations and have sufficient strength for such landings. However, installation costs are tens of millions yen and there are many supplementary facilities. Helipads "R" are meant for fire extinguishing and rescue operations during hovering and do not have sufficient strength for landing. However, installation costs are millions yen and a low amount of supplementary facilities are required. "Therefore, helipads "R" are essentially destined to be destructed at the end of life of the building without a helicopter landing at all", mentioned by Motomi Kinoshita.

"Transportation Heliports" require higher costs and longer time for installation

On the other hand, Transportation Heliports are installed under the guidance and supervision of JCAB, with the permission of the Minister of Land, Infrastructure and Transport (MLIT). There are only 4 Transportation Heliports in Tokyo, Tokyo Metropolitan Police Department Headquarters, Ministry of Internal Affairs, Ark Hills and the Asahi Shimbun. There are various requirements applied, such as JCAB instructions from the design stage and various actions such as environmental assessment clearance and practice in a public hearing. Additionally, installation costs are about 100 million yen and there are still such risks to be forced to change the flight routes or of the termination of the approval that even after it has been once obtained, for example when a new building is built around.

The third Temporary Helipads are such locations where the takeoff and landing was specially permitted. They are mainly used for pesticide spraying on fields or golf courses. There are no strength checks on the takeoff and landing sites as authorizations are issued to the operating company.

The question "why" to heliports with different purpose of use

But why have heliports with different purposes been installed on the roofs of hospitals? The background is the operation of EMS helicopters starting in 2001 and the hospital heliports have increased in number accordingly. Motomi Kinoshita explains that, "There is a vicious circle that heliports with different purpose of use have been installed because the authorizations for heliport installations and for operation of heliports come from different authorities, and also due to the lack of knowledge of the designers and constructors and furthermore, the landing permissions have still been granted.

Now, which of the three types of heliports are appropriate for hospital rooftop heliports?

Motomi Kinoshita explains, "it is recommended by the Japan Civil Aviation Bureau for hospitals to install a Transportation Heliport as doctors and patients will be boarding on to and off from helicopters. However, Transportation Heliports require a long time and high cost for installation, and also risks are high even after the operation has started. That's why, at Aero Facility, we suggest hospital heliports should have the same construction design and strength as Transportation Heliports, and operating procedures shall be as simple as Temporary Helipads". This creates a safe and low cost heliport for hospitals.

Another issue is the red "H" mark in the center of the white cross on the landing site which is an international standard indication for hospital heliports. In Japan, however, this mark is not allowed to be painted on fire fighting or transportation heliports even in hospitals. Regarding this matter, Motomi Kinoshita insists "Japan should follow with the standardization in order to accomplish fast and safe operations".

FIGURES AND TABLES

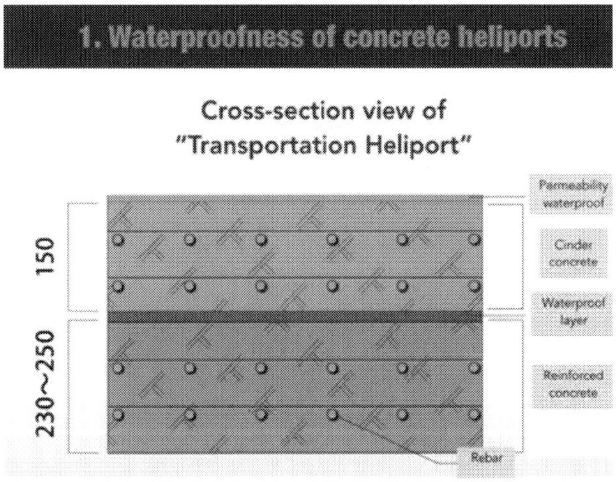

Figure 1. Cross-section view of "Transportation Heliport"

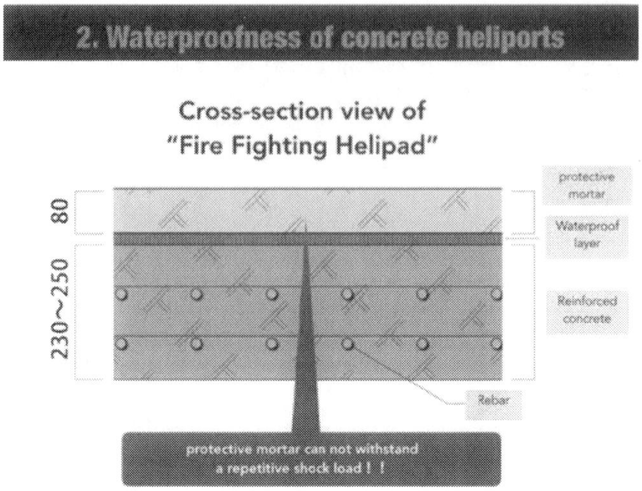

Figure 2. Cross-section view of "Fire Fighting Helipad"

CONCLUSIONS

Summary of Rooftop Heliports

The concrete structure of Temporary Helipads (Non-airfield Take Off and Landing Sites) should have the same construction designs as Transportation (Non-public Heliports) Heliports that have been made with sufficient strength for repetitive takeoff and landing.

However, due to the false knowledge there are many cases that Temporary Helipads where helicopters take off and land repetitively have been installed with the construction design of Fire Fighting Helipads (Emergency Take Off and Landing Sites) which are exclusive for fire fighting usage and not sufficient for repetitive landing.

This is dangerous because there is a possibility that the structure of concrete floor could be damaged due to aged degradation in 30, 40 years and it might cause a serious aviation incident.

Author contact: Mr. motomi kinoshita

石川県消防防災航空隊の活動について

Kazuhiro Nishijima
Captain
Ishikawa Air Rescue Team
Komatsu, Ishikawa, Japan

ABSTRACT

石川県消防防災航空隊は、石川県の消防職員である航空隊員 8 名と、民間航空会社に業務委託された運航クルー4 名で構成される。山岳地帯での救助活動などのため、「はくさん」の名称で中型ヘリコプターのベル 412EP を 1997 年から運航開始している。「はくさん」の装備や救助・救急資器材、運航基準、活動内容、救助・火災防御活動について紹介した。

石川県の概要

石川県は日本のほぼ中心部に位置し、面積は約４，２００km²。人口は、約１１５万人で県庁所在地の金沢市をはじめ、１９の市町で構成されている。

地形は、東西に約１００km、南北に約２００kmで、南北に細長い形状をしており東側は富山県との県境に沿って山々が連なり、西側は、日本海に面して海岸線が続いている。

南東部には県内最高峰で標高 2,702ｍの白山がそびえ、北側は日本海に突き出た能登半島で、沖合約５０㎞には舳倉島が位置している。

石川県の気候

石川県の気象は、平均気温が約１５度で夏場は気温が３５℃を超えて猛暑日が続く一方、冬は冷え込みが厳しく、氷点下になる日もある。

年間の降水日数は約１６０日で、雨が降りやすく、天候は変わりやすい。特に冬場は、北西からの季節風が続くため降水量が多く、山間部は豪雪地帯となる。

また、雷が年間約４０日発生することから、ヘリコプターの運航に厳しい気象になる日が多い地域である。

航空隊の体制

石川県消防防災航空隊は、グループリーダー1 名、航空隊員は隊長、副隊長２名、隊員５名の計８名で、運航クルーは、操縦士１名、整備士２名、運航管理１名の計４名で、航空隊員は県内の消防職員が、運航は民間の航空会社に業務を委託している。

飛行方式は、有視界飛行方式を採用し、運航時間は、日の出から日没となっている。

県内各地への所要時間

小松空港は、石川県の南寄りに位置し、航空自衛隊、民間航空会社が利用する空港で、県内各地への飛行時間は、金沢市までは約１０分、北端の舳倉島へは約５０分を要する（図 1）。

ヘリコプター「はくさん」の紹介

「はくさん」の機体は、山岳地帯での救助活動に適し、機内での作業性に優れた能力を有する中型ヘリコプター「ベル式４１２ＥＰ型」で１９９７年４月から運航を開始し、現在までの飛行時間は約５，５００時間である。

Presented at the 6th Asian/Austrailian Rotorcraft Forum / Heli Japan 2017, Kanazawa, Japan, November 7-9, 2017.

機体のカラーは、消防のイメージ色である赤を基調とし、「白」は白山に積もる雪を表すデザインで、愛称は県民に分かりやすく親しまれている白山に因み、「はくさん」と名づけられた。

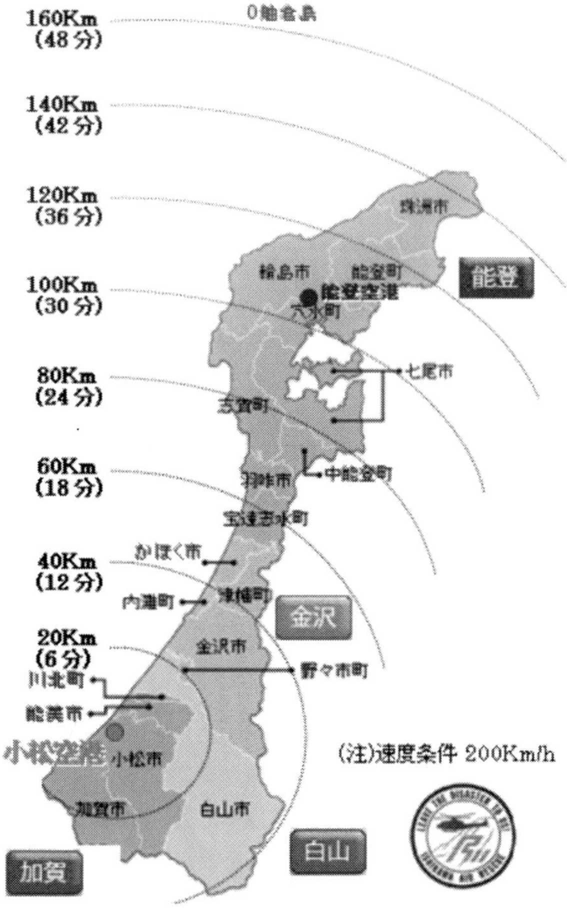

図 1. 石川県内各地への所要時間

図 2. ヘリコプター「はくさん」

「はくさん」の主な装備

救助活動等を行うための、主な装備は次のとおりとなっている。

● 動態管理システム

衛星通信を利用して航空隊事務所や県庁において、ヘリコプターの位置情報などをリアルタイムで把握する。

● ホイスト装置

隊員を降下させたり、傷病者を吊り上げて救出する。

● ドロップタンク

機体の下面に取り付け、ダム湖や河川から水を吸い揚げ、山林火災等で消火活動を行う。

● 救急医療用ベッド

救急処置を行いながら、傷病者を搬送する。

● 機外拡声器

災害時の避難誘導や広報活動に活用する。

● カーゴフック

救援物資や機内に搭載できない物を吊り上げ輸送する。

主な救助資器材

クイックストラップ、エバックハーネス、バーティカルストレッチャー、水難救助資器材一式等

主な救急資器材

心電図モニター、スクープストレッチャーAED、救急医療用ベッド、救急バック、携帯酸素等

緊急運航基準

災害等が発生し、消防からの要請受け、次の要件を満たす場合に出動する。

- 公共性：災害等から生命、身体及び財産を保護し被害の軽減を図る。
- 緊急性：差し迫った必要性がある場合。
- 非代替性：消防防災ヘリコプター以外に適切な手段がない場合。

主な活動内容

- 救助活動

海難事故、河川等での水難事故及び山岳等での遭難事故で、捜索又は救助活動が必要と認められる場合等

- 救急活動

遠距離の救急患者搬送及び高度医療機関への転院搬送

傷病者発生場所への医師の搬送、医薬品等の輸送移植のための臓器の搬送

- 火災防御活動

林野火災等における空中からの消火活動

火災発生時の、被害調査及び情報収集等

- 災害応急活動

地震、津波等災害発生時の被害調査及び情報収集活動

- 広域応援活動

東日本大震災等の緊急消防援助における活動及び富山県、福井県への応援出動等

災害出動状況

現在までの災害総出動件数は９２３件で年平均約５０件の出動をしている。

出動の内訳は、救助件数が５２８件（５７％）、救急件数が１９６件（２１％）、広域応援活動が１０７件（１２％）、災害応急活動が５９件（７％）火災防御活動が３２件（４％）で、救出人員は３９１名、搬送人員は３９７人となっている。

昨年度の災害出動件数は３８件で、出動の内訳は、救助件数が３０件、救急件数が６件、火災防御活動、広域応援活動が１件で、救出人員は１３名、救急搬送人員は１９名となっている。

災害事例（映像で紹介）

救助活動

火災防御活動

図 3. 火災防御活動

まとめ

石川県消防防災航空隊は、今後もヘリコプターのスピード、機動性、ホバーリング性能等ヘリコプターの特性を最大限に活用して、安全に救助、救急、消火等の活動を行い「県民の安全と安心」を空から守っていきたいと思います。

Tackling the HARTII test case: A joint effort by DLR and JAXA

Gunther Wilke, Annika Möller-Länger
research engineers
German Aerospace Center (DLR)
Braunschweig, Germany

Yasutada Tanabe, Masahiko Sugiura
research engineers
Japan Aerospace Agency
Tokyo, Japan

Hideaki Sugawara
research engineer
Ryoyu Systems Co., Ltd.
Tokyo, Japan

ABSTRACT

The paper focuses on the validation and verification of the high-fidelity aero-mechanical tools of DLR and JAXA. The CFD based tools are validated against the HART II test campaign, for which blade deformations, airloads and noise carpets have been measured. The validation is split into three steps. The first step compares the different blade motion modelling approaches for the baseline case, here either fully or partially utilizing the measured blade motion from the experiment or fully simulating the blade motion. The second step compares the fully elastic modelling between the DLR and JAXA codes for the three common HARTII test cases (baseline, minimum noise, and minimum vibration). In the last step, the resolution of the blade-vortex interaction (BVI) is investigated. The effect on the noise carpets of the applied higher order schemes of DLR and JAXA is analyzed for different grid sizes. From this research, it is observed that the blade motion can be simulated in a good agreement with the experiment. For capturing BVI correctly, a higher order method along with a fine mesh resolution in the flow computation is required.

NOTATION

c_nM^2 – Mach scaled normal force coefficient

Θo – Collective pitch control angle, degrees

Θc – Lateral pitch control angle, degrees

Θs – Longitudinal pitch control angle, degrees

INTRODUCTION

The design of the helicopter rotor is a challenging task. The unsteady nature of the airflow around a helicopter rotor blade forces it to bend and flap, while the helicopter itself must be considered in a forces and moments equilibrium for correct comparison between different blade designs. Therefore, high-fidelity simulations are necessary to correctly predict its performance as well as acoustics. The simulation itself requires taking into account multiple disciplines at once: aerodynamics, structural dynamics, flight mechanics and aero-acoustics.

One design-driving flight-condition is the descent flight for the helicopter. The blade tip vortices trailing off are met again by blades in the rotor plane causing blade-vortex interaction (BVI) noise making this distinguished 'wop-wop' sound. Due to its very present nature, it has become a certification requirement to stay within specific noise level limits for this flight condition. Therefore, the need to correctly predict it during the rotor blade design phase is of major importance.

This flight condition has been investigated during the HART II campaign (Ref. 1), where a model Bo 105 rotor has been used. Besides the regular operation of the blade, additional high harmonic controls (HHC) for either noise or vibration reduction have been assessed. While this has been a demonstration of the potential of HHC, the (economical) risk of putting it on a mass produced helicopter is still high. In order to be able to predict its worthiness better, accurate simulation tools are necessary. (Ref. 2)

The greatest challenge of simulating the descent flight condition of a helicopter rotor is the correct prediction of the vortex flow field in and around the rotor plane. The blade element theory coupled with a wake model can lead to good results as already shown by van der Wall et al. (Ref. 3), yet many empirical and semi-empirical factors govern the solution.

The hope is through the application of computational fluid dynamics (CFD) that the dependency on such factors is reduced. Smith et al. (Ref. 4) demonstrated the general applicability of CFD to the HART II test case. From their review of different methods, it becomes clear that for a good prediction of the BVI phenomenon, it is necessary to apply higher order schemes as well as a large amount of grid points to resolve this efficiently.

DLR and JAXA have been working in a cooperation to advance their rotor modelling approaches. In this paper, the partners DLR and JAXA present their approaches to simulate the HARTII test case. This is done in three steps:

Presented at the 6th Asian/Australian Rotorcraft Forum / Heli Japan 2017, Kanazawa, Japan, November 7-9, 2017.

1. In the first phase, the different blade motion modelling approaches are tried between the partners to identify the most suitable method for further studies. This is done for the baseline test case.

2. With the best motion modelling approach established, the other two test cases (minimum noise and minimum vibration) are additionally computed to see how well the impact of the HHC is captured.

3. Lastly, the effect of the here applied higher order methods is investigated with respect to their accuracy in predicting the higher-frequency airloads as well as noise carpets for the baseline test case.

METHODOLOGY

The general methodology of the simulation of BVI noise for a helicopter rotor used within this work is the following:

1. Comprehensive code: It computes an initial trim state of the helicopter rotor using simplified aerodynamics based on the blade element theory (BET), a linear-elastic beam model and a trim procedure.

2. Flow solver: with the trim state and deformation of the rotor given, a more accurate prediction of the aerodynamic loads is obtained by utilizing CFD. These corrected airloads are coupled back into the comprehensive code through the delta airloads approach (Ref. 5) to gradually achieve equilibrium of the blade deformation with the high-fidelity airloads.

3. Aero-acoustic extrapolation: In this context, the Ffowcs-Williams Hawkins (FW-H) equations are solved in the Farassat 1 & 1A formulation (Ref. 6) for helicopter rotor blades. As an input, the aerodynamic pressure of the rotor blades is fed into the FW-H code, which then transports this pressure onto the microphone array below the helicopter.

In the following, these tools are explained for each partner.

Comprehensive Codes

For the fluid-structural coupling, the Institute of Aerodynamics and Flow Technology of DLR relies on the Helicopter Overall Simulation Tool (HOST, Ref 7) originally developed by Eurocopter France (now Airbus Helicopters France). As stated above, the aerodynamics of this comprehensive code are obtained using the BET, which relies on 2D airfoil lookup tables. The elastic model is based on an extended version of the Beam model. It is linearized by the assumption of small deformations in flap, lead-lag bending and torsion. The offset between neutral axes and center of gravity is accounted for. The blade is then split up into a finite-amount of fictive hinges, for which the Lagrange equations are solved taking into account the aerodynamic, inertial and structural forces. For increased efficiency, a modal approach is utilized by HOST, which decouples time and space. The equilibrium of forces is found by a Newton-Raphson procedure.

JAXA has developed an analysis tool chain for the rotorcraft aeroelasticity and aeroacoustics as described in (Ref. 8). The rotor blade is treated as a rotating pre-twisted nonuniform beam based on the Houbolt and Brooks' equations (Ref. 9). The coupled flapping and lead-lagging bending natural frequencies and mode shapes together with the uncoupled torsional frequencies and mode shapes are calculated using an in-house code rMode from the blade properties and blade root constraints, using the method given by Isakson and Eisley (Ref. 10) for the coupled bending and the general Myklestad method for the uncoupled torsion. Output of rMode is then used as input to the CFD/CSD/Trim coupling analysis code rFlow3D. The elastic deformations of the rotor blades are expressed as a summation of N natural modes for bending and M natural modes for torsion with time varying amplitudes based on the modified Galerkin procedure. The periodic amplitude variation for each mode is then approximated with a Fourier series up to 6th harmonics and being used in the CFD calculations.

The trim adjustment is based on the partial derivatives of the rotor thrust and moments to the rotor controls for the JAXA simulations. The partial derivatives are numerically obtained based on the simple Blade Element Theory where the inflow is uniform and depending on the target thrust only. To improve the computational stability and convergence, the elastic motion changes and the trim adjustment are added with a relaxation factor multiplied.

Flow Solvers

DLR employs two different flow solvers for the here presented studies. On the one hand there is the structured legacy code FLOWer (Ref. 11), while on the other side the novel unstructured code TAU (Ref. 12) is used.

The FLOWer code is a cell-centered block-structured finite-volume code based on the 2nd order Jameson-Schmidt Turkel (JST) central scheme (Ref. 13) in space and an explicit five-step Runge-Kutta scheme in time. For acceleration it utilizes multi-grid and implicit residual smoothing. Enk (Ref. 14) enhanced FLOWer with a 4th order accurate line-implicit finite difference scheme, the Pade scheme as presented by Lele (Ref. 15). The weak code coupling with HOST code has been validated by Dietz et al. (Ref. 16).

In a recent work it has been extended to be used in a mixed spatial scheme environment for helicopters, e.g. by using the Chimera approach (Ref. 17): the rotor blades and fuselage are computed with the JST scheme, while the background mesh is computed with the Pade scheme to exploit the better vortex conservation, see Figure 1.

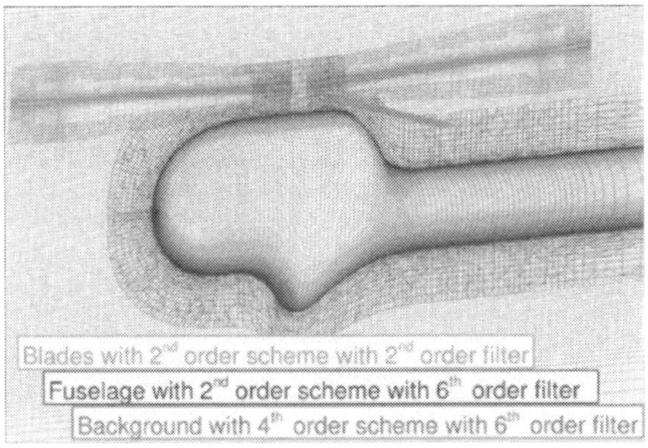

Figure 1. HART II simulation setup with DLR's FLOWer code

Opposing the FLOWer code, the TAU code uses an explicit three-step Runge-Kutta scheme. It is also a finite volume code, yet utilizes a central scheme and is vertex-centered. With the central scheme of TAU, it is also possible to utilize cell types different from structured hexahedral cells, such as tetrahedral cells or prisms. For convergence acceleration, multigrid may also be employed. The fluid-structural coupling of TAU with HOST has been validated by Wendisch and Raddatz (Ref. 18).

Both DLR codes feature a rich variety of turbulence models. In this work, the Wilcox k-ω turbulence model (Ref. 19) is used for the FLOWer simulations, while for TAU the Spallart-Allmaras turbulence model (Ref. 20) is employed.

On the JAXA side, the Navier-Stokes equations in an Arbitrary Lagrangian Eulerian (ALE) form are solved using Finite Volume Method (FVM). The all-speed numerical scheme mSLAU adopted in rFlow3D and the moving overlapped grid method are described in detail in (Ref. 21). An example of the grid system used with rFlow3D code is shown in Figure 2. In Figure 3, the fuselage is in a realistic complex geometry, which is solved with the JANUS code (Ref. 22).

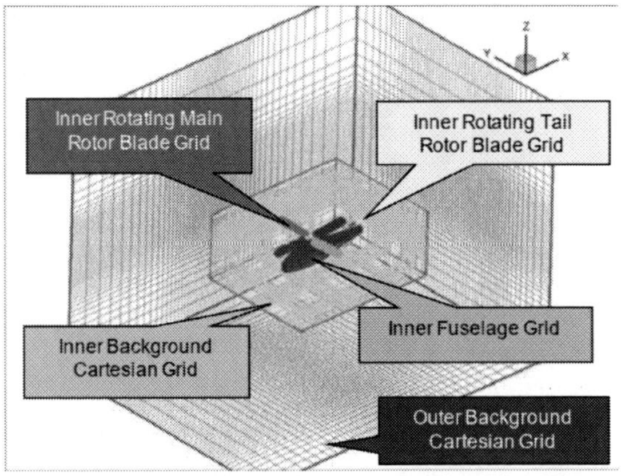

Figure 2: Overlapped grids used in rFlow3D

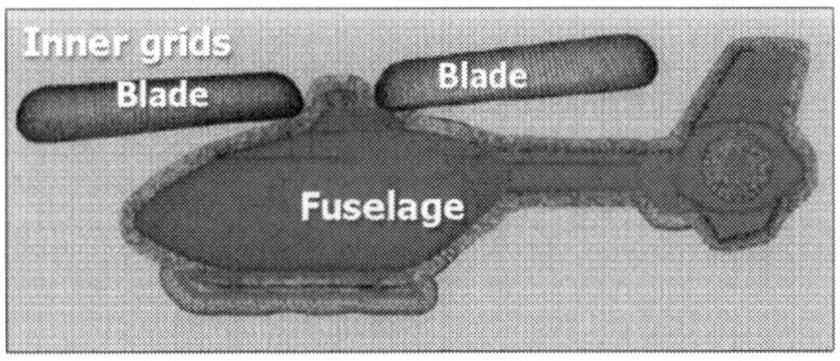

Figure 3: Grid system for JANUS code

The numerical scheme mSLAU adopted in rFlow3D is a modified version of SLAU (Ref. 23) with the extension to the three dimensional moving grid. It is found very suitable for the flow around a rotary wing, where the local flow speed varies from very low to transonic velocities. Combining with a Fourth-order Compact MUSCL TVD (FCMT) interpolation method (Ref. 24) for reconstruction, 4th order spatial accuracy is obtained in the smooth areas, which improves the capturing of blade tip vortices.

Acoustic Analogy Codes

DLR's FW-H code is APSIM (Ref. 25), which evaluates the blade surface pressure to predict the pressure at the microphones. It neglects the quadrupole terms, but includes surface dilatation (Ref. 26). The blade motion is obtained by differentiating the surface grid obtained from the flow solver with a 2nd order accurate central stencil. This relieves the user from specifying the blade motion explicitly and no further coordinate transformations are required; however it comes at the price of numerical rounding errors. The Latter has been found to be of negligible influence.

After the converged periodic pressure distributions on the blade surface are obtained for the JAXA simulations, rotor noise at the observing position is calculated using the Farassat Formulation 1 (Ref. 27) based on Ffowcs Williams-Hawkings equation. The most time consuming part in the noise prediction is to find the source cells that satisfies the retarded time criteria and while the blade is rotating and deforming, a lot of coordinate transforming is required. Because the blade rotating, pitching, lead-lagging, flapping and also the elastic deforming are expressed as functions of the azimuth angle, smooth retard time searching is realized. The surface pressure on the cell surface is interpolated. It must be notified that because the velocities and accelerations of the surface cells are not required in the Farassat Formulation 1, it is more suitable to be used combined with the CFD results compared to the Farassat Formulation 1A.

STEP 1: BLADE MOTION MODELLING

In this first phase, different blade motion modelling approaches are studied for the baseline test case. DLR and JAXA have two each. DLR's first approach takes the recoded blade motion from the experiment for each blade and models it individually with the recorded blade pitch control angles. Opposing this, JAXA utilizes the averaged measured blade deformations, but re-determines the trim control angles numerically. At last, both partners compute the whole blade motion numerically, assuming that each blade features the same structural properties (stiffness, mass distribution, center of gravity and neutral axes). One DLR simulation included the wind tunnel fuselage as it was seen in (Ref. 28) that it allows achieving trim control angles matching better with the experiment. For comparison, DLR also provided a simulation without a fuselage. The different approaches are summed up in Table 1.

DLR conducted this research using the FLOWer code with a mesh setup consisting of roughly 13 million mesh points and a time step equivalent of 0.5° rotor azimuth. The Pade scheme is employed for better vortex conservation. More details on the setups can be found in (Ref. 29).

JAXA used two layers of background grid as shown in Figure 4. The grid resolution in the inner background layer has been changed so that the results on a standard grid resolution of 1/10 chord length (about 43 million cells) and a fine grid resolution of 75/1000 chord length (about 77 million cells) can be compared. The blade grid size is 121 points in spanwise, 101 points in chordwise and 21 points in normal directions. The mesh size in normal direction is adjusted according to the resolution in the inner background layer so that sufficient overlapping between the blade grid and inner background grid is secured taking into account the 7 points stencil of the current 4th order scheme adopted. At this stage, only the isolated rotor configuration has been compared to the experiment.

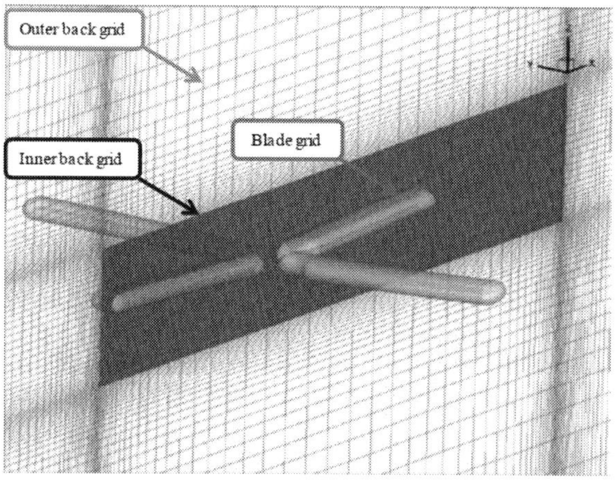

Figure 4: JAXA overlapped grids for HART-II isolated rotor simulations

Table 1. Different motion modelling approaches by DLR and JAXA

Motion type	DLR	JAXA
Full experimental	✓	
Exp. deformation with num. pitch		✓
Fully simulated / isolated rotor	✓	✓
Fully simulated / rotor + fuselage	✓	

First, the obtained airloads at the 87% radial cut are presented in Figure 5. The mean values have been removed in accordance to the CFD/CSM comparison paper by Smith et al. (Ref. 4) and are listed in Figure 5. Here it is seen that directly imposing the experimental data onto the CFD simulation does not yield the appropriate result. Two potential reasons are identified for this. The first reason might be that despite the added complexity of the fuselage, it is still necessary to include more parts of the wind tunnel setup into the simulation, such as wind tunnel nozzle, collector and likely the wind tunnel walls. The second reason may be the inherent difficulty in measuring the exact location of the blades in a rotating system. This becomes more obvious when looking at the partially experimental motion, where the trim control angles have been re-trimmed. Here it is clearly perceived that the results improve in contrast to the purely experimental motion, and the mean c_nM^2 value is reduced. Going further to the isolated rotor simulation, where the blade motion is fully determined through the simulation, the mean c_nM^2 is underpredicted for DLR as well as JAXA. However, the removed mean airloads are in better agreement with the experiment. This can only be further enhanced by including the fuselage. For this case, the best agreement of the trim control angles is observed for the simulated motion test cases, check Table 3. Despite the fact, that the fully simulated approaches achieve a better airloads correlation, this comes from a cancellation of effects. When looking at the occurring torsional deformation, Figure 6, it is observed that the fully simulated results by DLR are below the experimental results by 0.64° on average, which is in accordance with the results by ONERA, who also utilize the HOST code (Ref. 30). The opposite effect is observed for JAXA, where the blade torsion is overpredicted by 1.06°. A strong dependency on the elastic blade modelling is thus concluded. However, when the rotor control angles are trimmed, the effect of false structural dynamics is compensated, which yields better results for the airloads and ultimately for the noise prediction instead of directly feeding the experimental motion to the flow solvers.

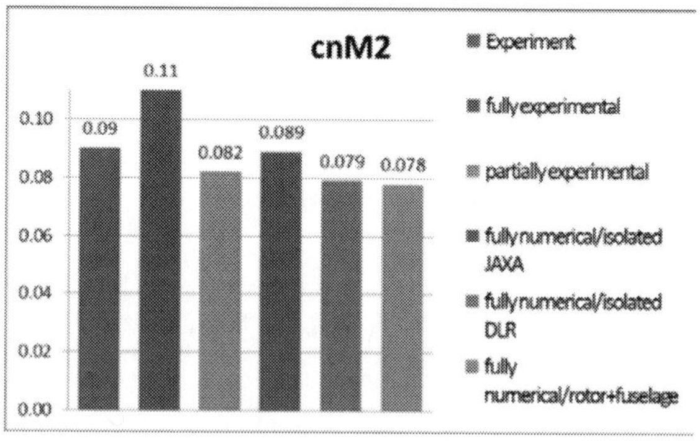

Figure 5: mean cnM2 for STEP 1

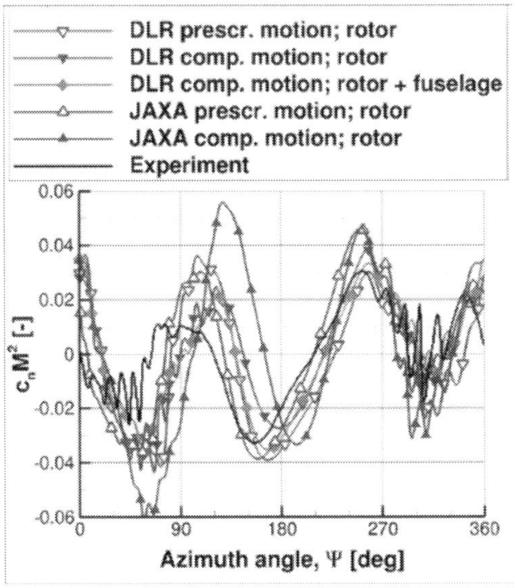

Figure 6. Comparison of different motion modelling approaches (mean removed)

Table 2. Trim control angles for different motion modeling approaches

Pitch angles	Collective Θ_0	Lateral Θ_c	Longitudinal Θ_s
Experiment	3.80	1.92	-1.34
Experimental deformation, and num. pitch	3.11	1.47	-1.31
Fully Simulated / isolated rotor			
JAXA	2.27	1.76	-0.94
DLR	3.77	1.63	-0.85
rotor + fuselage DLR	3.72	1.87	-0.98

STEP 2: FULLY SIMULATED MOTION MODELLING FOR ALL HART II CASES

In this part, all HART II motion test cases are compared against each other: baseline, minimum noise and minimum vibration. These minimum noise and minimum vibration test cases feature additional higher harmonic controls (HHC) to induce additional 3/rev pitching oscillations.

DLR's isolated rotor simulations are carried out using the TAU code with a mesh setup of 9.7 million grid points in total. The corresponding FLOWer results are also based on the same blade meshes, with a slightly different background grid, yet same number of grid points. Further details on these computations can be found in (Ref. 31).

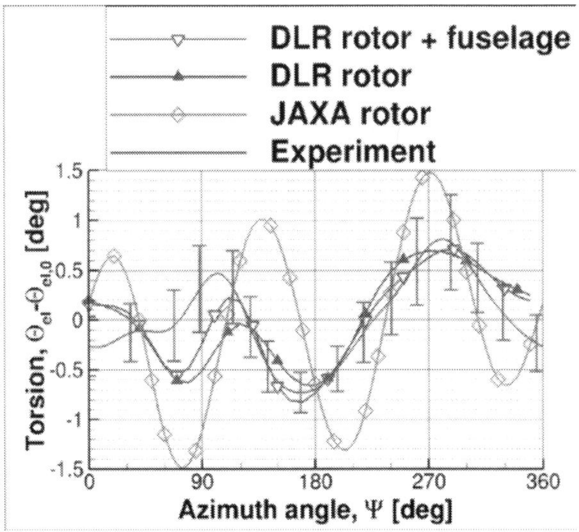

Figure 7. Torsional deformation for different motion modelling approaches.

In contrast to changing the setup, JAXA's simulations are performed on the same grid as before consisting of 43 million grid cells.

The obtained trim control angles of each case are listed in Table 4. It is observed that the collective pitch angle Θ_0 is under predicted in all case by all simulations, with the minimum noise case being mostly off. For the lateral control angle Θ_c, a similar behavior is noticed, yet the greatest offset is found for the minimum vibration case. The longitudinal control angle Θ_s features the greatest offset to the experiment in general, again with minimum vibration being the case with the largest difference.

These offsets are caused by a) partial representation of the wind tunnel and assumption of a mean inflow change throughout the test section and b) by a differently modelled blade deformation. The different flapwise deformations are shown in Figure 7, while the torsional deformations are shown in Figure 8. For all flapping motions, a fair correlation is perceived with the baseline being the best. However, the greatest offset is noticed on the advancing side between Ψ = 90-180°. Comparing this with the torsional distribution, further discrepancies with the experiment are found. This is argued with a lack of correct flap-torsion coupling being eminent in both structural codes. However, the HOST code agrees slightly better in the torsional deformation, especially in the minimum noise case where the strong 3/rev signal from the HHC is seen. The largest difference to the experiment is observed for the minimum vibration test case, where DLR underpredicts the amplitude, while the JAXA result shows a strong phase shift.

Looking at the airloads in Figure 9, it is observed that the best agreement with the experiment is achieved for the minimum noise case. Here, the strong 3/rev signal is mostly well captured, as already observed for the torsional deformation. On the advancing as well as retreating side the force is slightly below the experiment. This is also a trend found for the baseline case, where JAXA as well as DLR are below the experimentally measured airloads in the first and last quadrant. An overshoot is observed just past Ψ = 90°. From the paper by Lim et al. (Ref. 28) it is deduced that the effect of including the fuselage may bring improvement to this. Again, the minimum vibration case is captured the worst by the simulations.

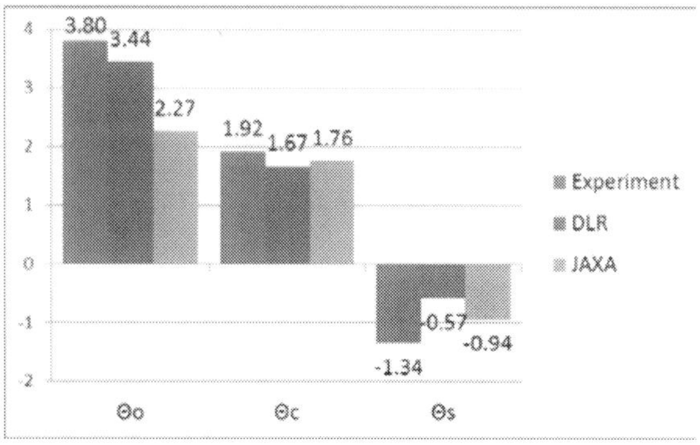

a) baseline

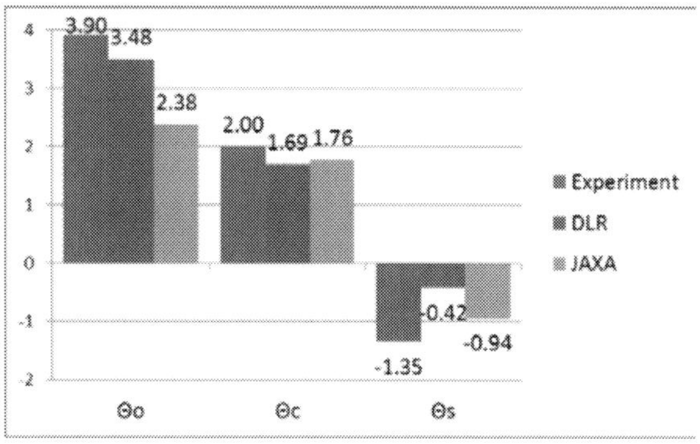

b) minimum noise

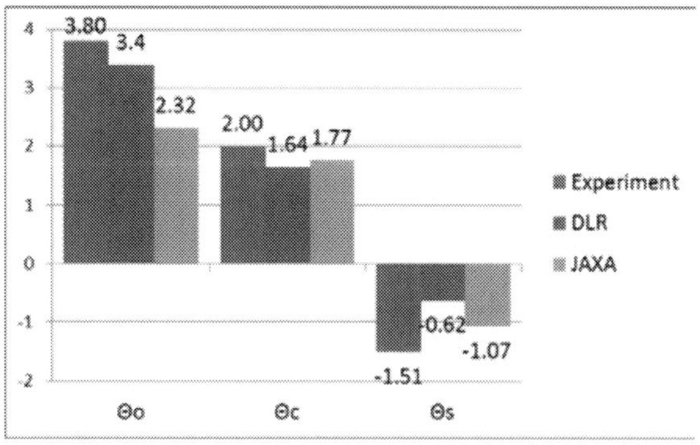

c) minimum vibration

Figure 8: Trim control angles for STEP 2 verification

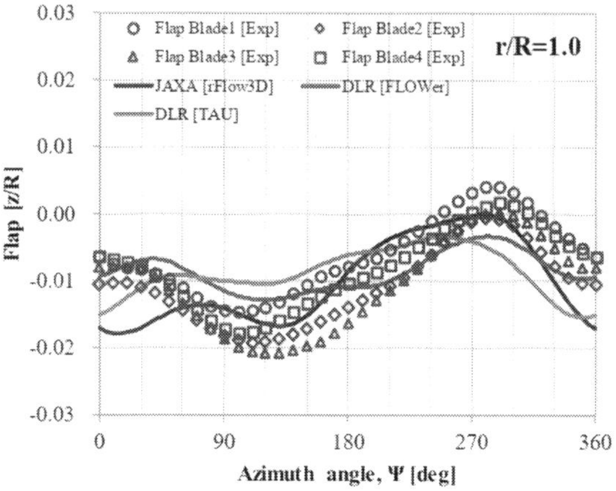

a) Baseline

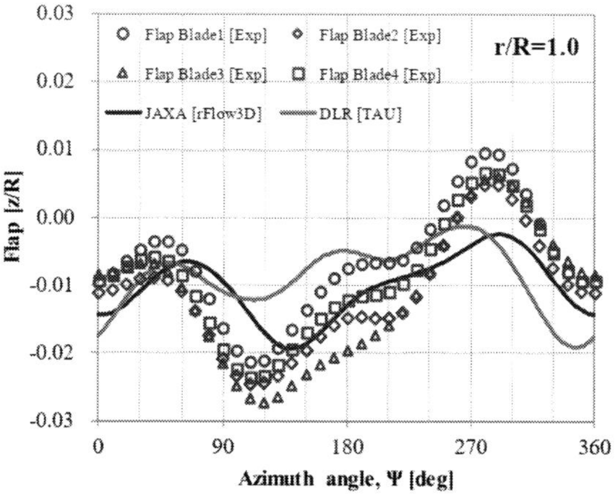

b) minimum noise

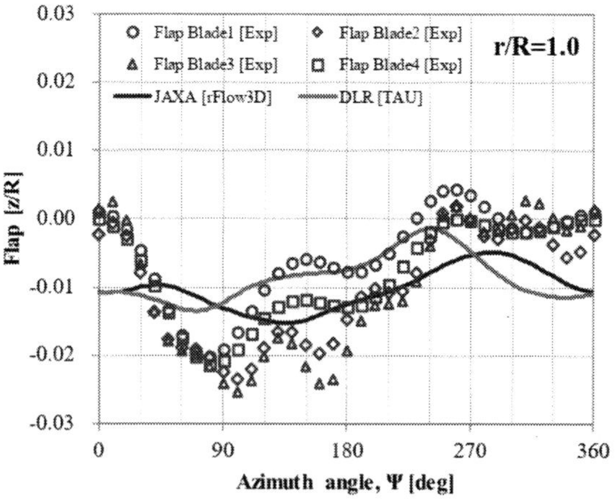

c) minimum vibration

Figure 9. Flapwise deformations for STEP 2 verification

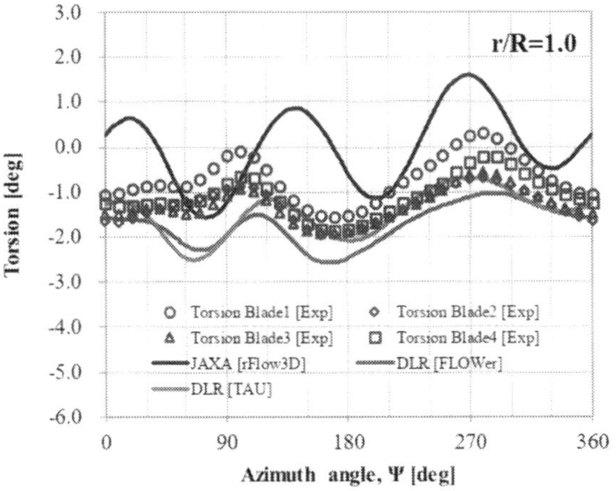

a) baseline

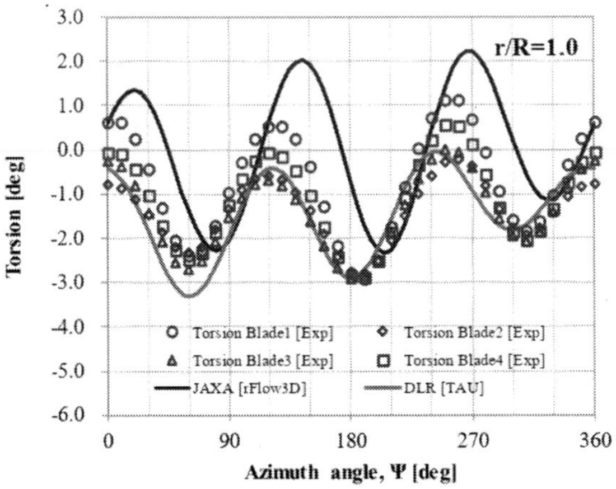

b) minimum noise

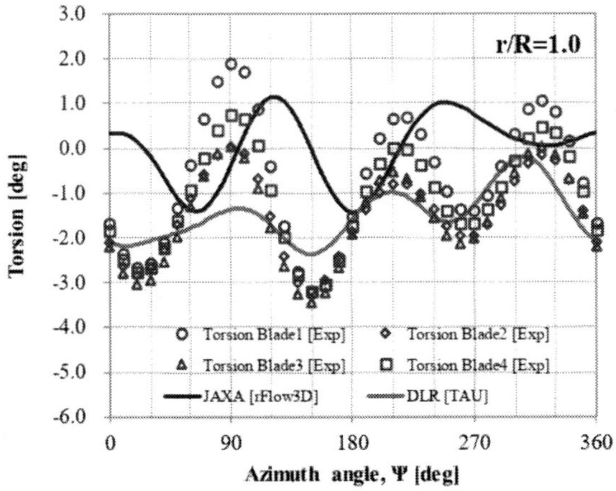

c) minimum vibration

Figure 10. Torsional deformation for STEP 2 verification

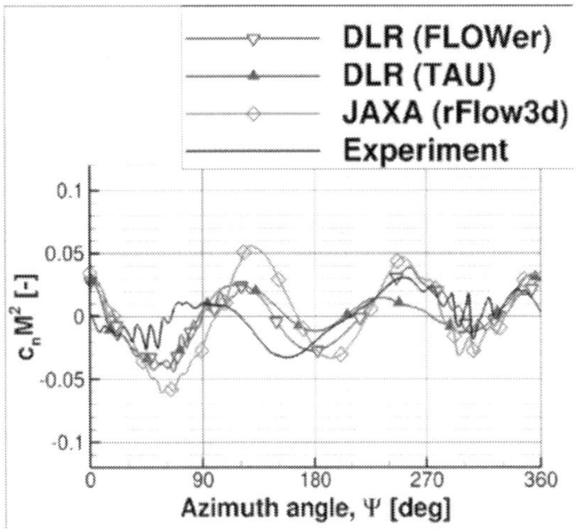

a) baseline

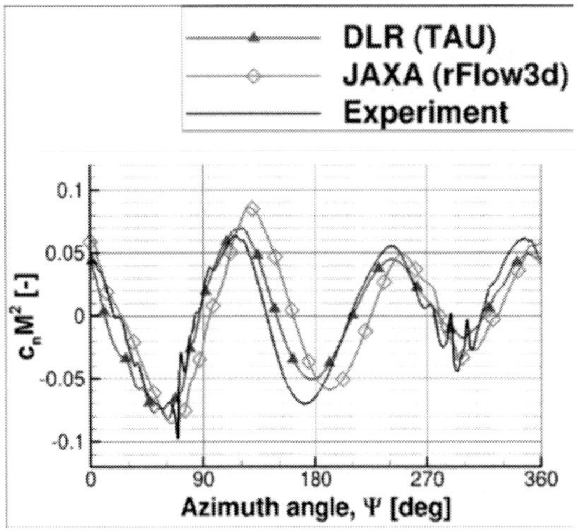

b) minimum noise

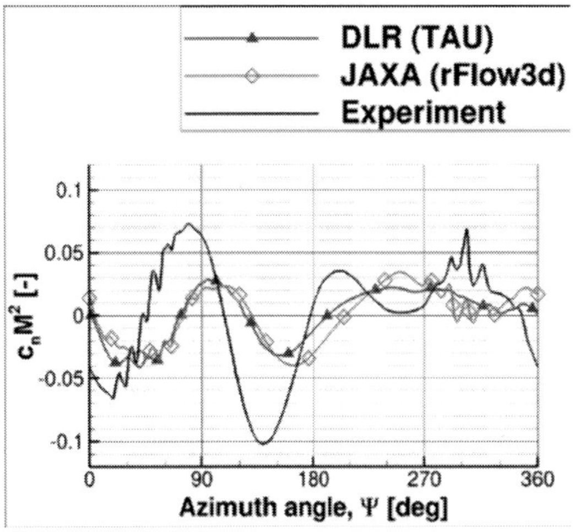

c) minimum vibration

Figure 11. Airloads at 87% r/R for STEP 2 verification

STEP 3: ADVANCED AERODYNAMIC MODELLING WITH HIGHER ORDER METHODS

For this section, the higher order schemes available to JAXA and DLR are utilized with a high grid resolution to see how well the high-frequency airloads can be captured by the given approaches and ultimately the noise carpets. On the DLR side, a mesh setup consisting of 103 million mesh points is employed featuring an inner background mesh resolution equivalent to 1/10 chord length. This setup is tested with the JST as well as the hybrid Pade scheme to see how large the effect of the increased order is. Opposing this, JAXA investigates two different grid resolutions in the rotor plane, one also with 1/10 chord spacing and one with 75/1000 chord spacing using the SLAU scheme.

In Figure 10 is seen that by exchanging the spatial scheme, the vortex conservation is strongly improved. Whereas the JST scheme shows a vague representation of the vortex flow field, the Pade scheme resolves the downwash and the turbulent wake with many details. Comparing this with the JAXA solution on similar grid spacing, it is seen that the horse shoe vortices have increased more in core radius than for the DLR simulation. While the Pade scheme shows slightly better vortex conservation within the rotor plane, DLRs vortex tube of the retreating blade stays red for about 45° aft of the blade and for JAXA about 20°, the effect of the growing core radius is likely to be attributed to a different effect. The JAXA mesh already becomes coarse right after the rotor plane, while DLR maintains the high grid resolution of 1/10 chord length spacing for two more rotor radii behind the rotor plane yielding a better resolution of the complete wake behind the rotor. With decreasing the grid spacing in the JAXA simulation, the vortex preservation properties in the rotor plane are about the same for the simulations by DLR and JAXA.

For the given simulations, only minor changes in the deformations, as well as control angles are seen relative to the previous simulations and are thus not further addressed in this paper. However, the higher frequency content of the airloads has changed and is therefore shown in Figure 11. In the airloads plot, the integral behavior of DLR's solutions is closer to the experiment than the JAXA solutions. This has already been observed in the first step and is mostly attributed to the inclusion of the fuselage in the DLR simulation. Moving onto the temporal derivative of the airloads, the high-frequency content is more clearly perceived and it is seen that DLR's higher order result even overpredicts the amplitude of the wiggles on the advancing side. The result of the JST scheme also captures the locations of the wiggles correctly; however, the amplitudes are too small in comparison with the experiment, which is similar to JAXA's solution on the coarser mesh. The overshoot of DLR's higher order results on the advancing side is associated with the discrepancies seen in the motion modelling. On the retreating side, both DLR and JAXA are in good alignment with the phase of the BVI oscillations in the loads, yet the strong negative spike is not captured any partner.

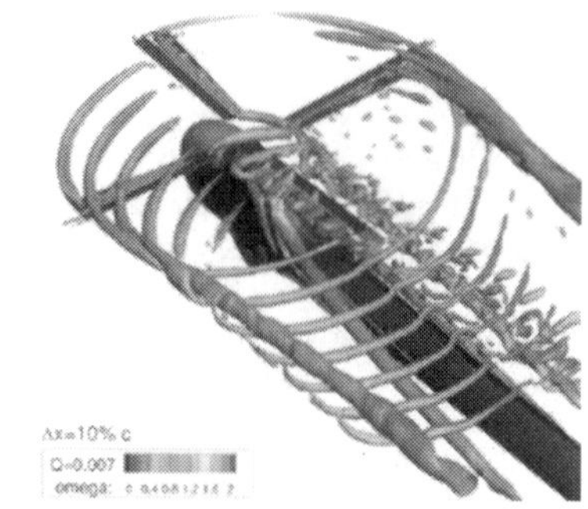

a) DLR simulation JST scheme

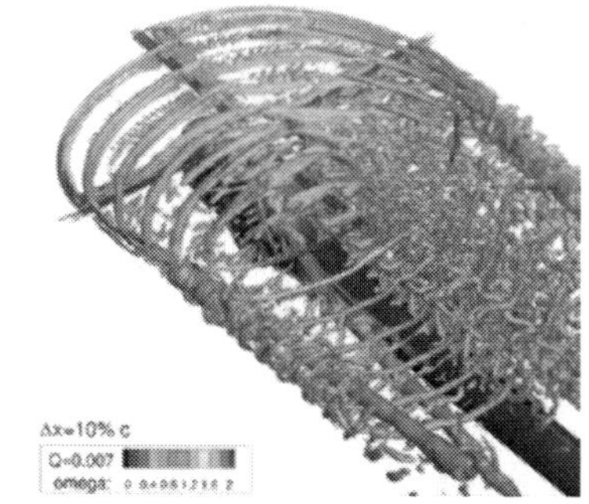

b) DLR simulation with hybrid Pade Scheme

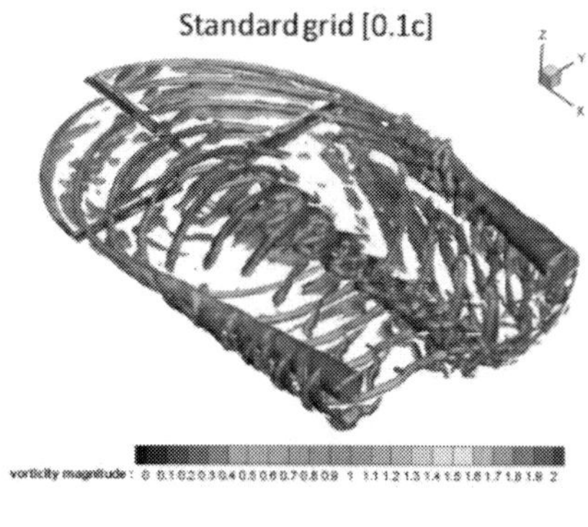

c) JAXA simulation with 1/10 chord spacing

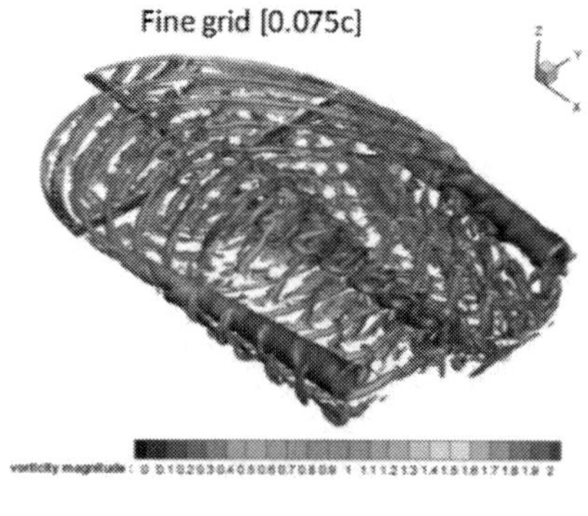

d) JAXA simulation with 75/1000 chord spacing

Figure 12. Vorticity plots of the STEP 3 verification. (NOTE: same coloring scales used)

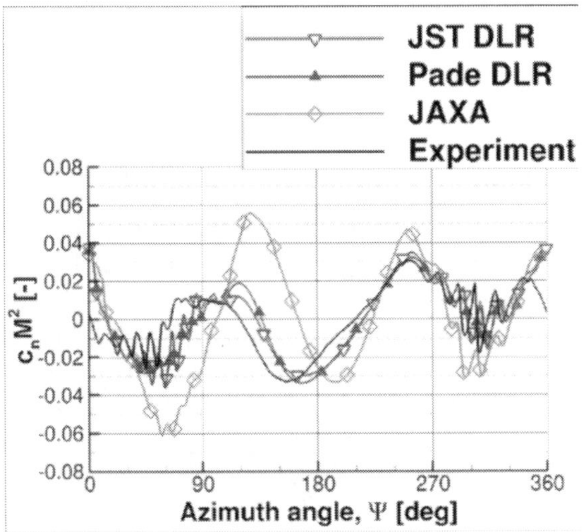

a) Airloads

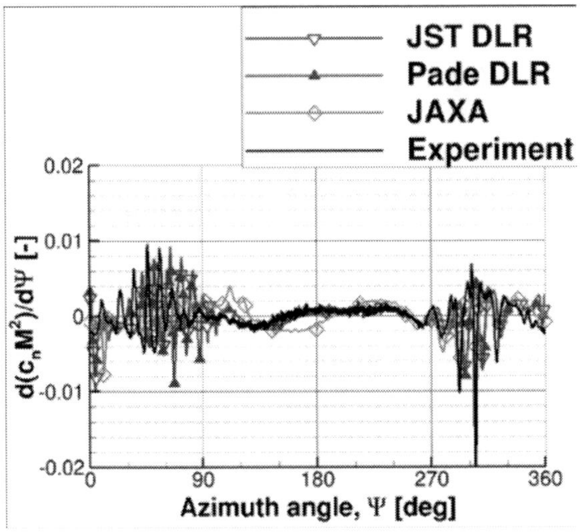

b) Derivative of airloads

Figure 13. Normal forces at 87 % r/R for STEP 3 verification

As the final results, the noise carpets of the different simulations are compared against each other along with the experimentally measured noise carpet in Figure 12. The best match with the experiment is obtained with the DLR simulation using the hybrid Pade scheme, followed by JAXA's fine mesh solution. The least correlation is given for DLR's JST solution, where the overall noise level is clearly to low and three, instead of two hotspots are identified. A connection with the airloads at 87% r/R is also observed. While both partners are able to capture the advancing side well, the retreating side is too quiet. Since DLR slightly overpredicts the loads on the advancing side with the Pade scheme, the resulting noise carpet peak is louder by about 2 dB and the area above 113 dB is larger than for the experiment. On the JAXA side, for the finer resolution with 75/1000 chord spacing, the retreating side is slightly louder than the advancing side, which is reasoned with the different motion obtained in contrast to the experiment already highlighted in STEP 1 and 2.

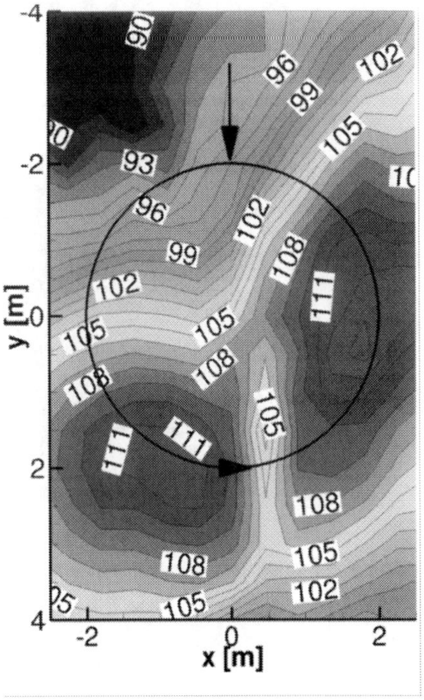

a) Experimental

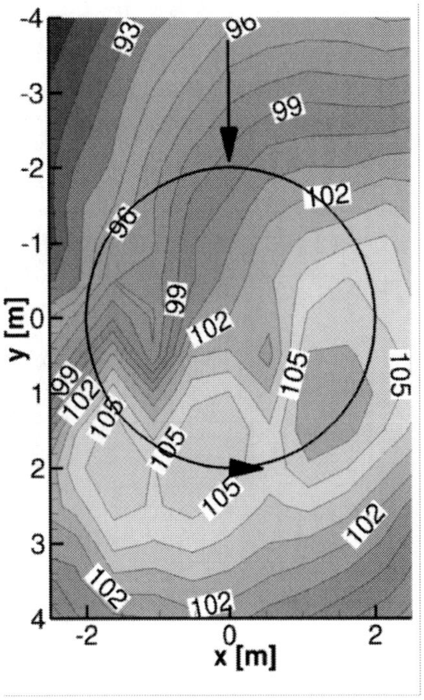

b) DLR solution with JST scheme

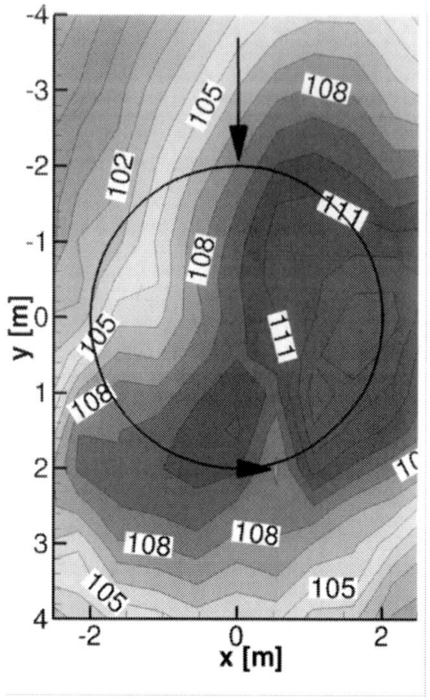

c) DLR solution with hybrid Pade scheme

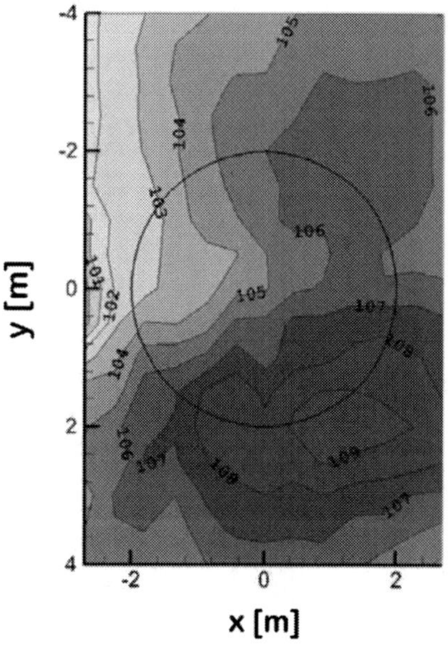

d) JAXA standard grid (0.1c) solution

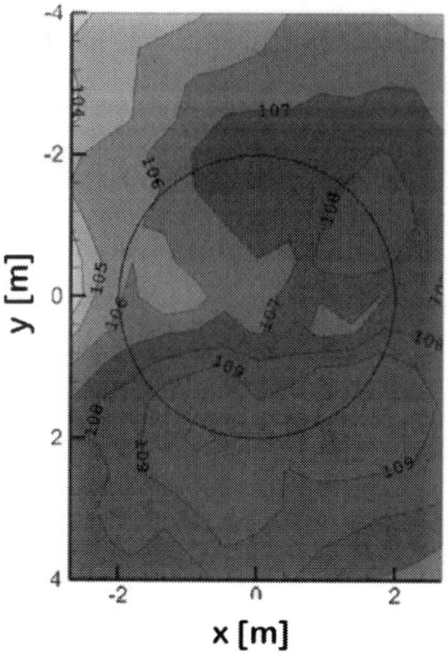

e) JAXA fine grid (0.075c) solution

Figure 14. Noise carpets for the HART II baseline case in STEP 3 verifcation

CONCLUDING REMARKS

In this paper, the tool set of DLR and JAXA for modelling BVI noise have been introduced. Within the DLR/JAXA cooperation, the HART II test case has been agreed as the common research database. The tool set has then been validated in three phases:

1. Blade motion modelling for the baseline case:

   In this step, different approaches have been tried, from purely utilizing the measured blade motion and deformation up to fully simulating it. From this research it has been seen that the best results in terms of airloads correlation are achieved by fully simulating the motion, despite the fact that a different motion than the experiment is obtained.

2. Simulating motion modelling for all HART II cases:

   From the result of the previous step, the other two HART II cases (minimum noise and minimum vibration) have also been simulated with a fully fluid-structural coupling. From this investigation it was seen that the strong 3/rev signal of the minimum noise case is well replicated seen in the torsional deformation as well as the airloads. In contrast to this, the minimum vibration case had the least correlation with the experiment.

3. Application of higher order schemes and high grid resolutions:

The lessons learned from this step is that the application of a higher order scheme allows for an improved aero-acoustic prediction, despite the fact that the blade motion modelling still needs to be further refined. On DLR's side, the going from a 2nd to a 4th order accurate scheme in the background mesh, increased the peak level in noise by 8 dB and even overshot the experimental result. JAXA showed that not only a higher order scheme is necessary, but also a higher grid resolution: 1 db has been gained by decreasing the background grid spacing from 1/10 chord length to 75/1000 chord length.

Concluding from these studies, it becomes clear that modelling BVI noise is a non-trivial task and high-fidelity codes on the aerodynamical, but also on the structural side are necessary. Further improvements of the current status may be achieved by

a) Increase of the aerodynamic modelling complexity, adding the wind tunnel nozzle as well collector may be beneficial, but also the more trivial but resource demanding approach of increasing resolution will improve on the general results.

b) Employing a structural model based on finite-element analysis in the time domain may further improve the correct capturing of flap-torsion deformation coupling and thus the blade motion modelling.

Author contact:

Gunther Wilke gunther.wilke@dlr.de

Annika Länger Möller annika.laenger@dlr.de

Yasudata Tanabe tan@chofu.jaxa.jp

Masahiko Sugiura sugiura.masahiko@jaxa.jp

Hideaki Sugawara hideaki8@chofu.jaxa.jp

ACKNOWLEDGMENTS

-

REFERENCES

[1]B. G. van der Wall, "A Comprehensive Rotary-Wing Data Base for Code Validation: The HART II International Workshop," The Aeronautical Journal, vol. 115, pp. 91-102, 2011.

[2]F. H. Schmitz, "The Challenges and Possibilities of a Truly Quiet Helicopter 29th Alexander A. Nikolsky Honorary Lecture," Journal of the American Helicopter Society, vol. 61, pp. 1-33, 2016.

[3]B. G. van der Wall, S. N. Jung, J. W. Lim, M. J. Smith, J. Bailly, J. D. Baeder and D. D. Boyd Jr., "An Assessment of Comprehensive Code Prediction State-of-the-Art Using the HART II International Workshop Data," in Annual Forum 68 of the American Helicopters Society, Ft. Worth, TX, 2012.

[4]M. J. Smith, J. W. Lim, B. G. van der Wall, J. D. Baeder, R. T. Biedron, D. D. Boyd Jr., B. Jayaraman, S. N. Jung and B.-Y. Min, "An assessment of CFD/CSD prediction state-of-the-art using the HART II international workshop data," in 68th Annual Forum of the American Helicopter Society, Ft. Worth, TX, 2012.

[5]M. Potsdam, H. Yeo and W. Johnson, "Rotor Airloads Prediction Using Loose Aerodynamic/Structural Coupling," in American Helicopter Society 60th Annual Forum, Baltimore, MD, 2004.

[6]K. S. Brentner and F. Farassat, "Modeling aerodynamically generated sound of helicopter rotors," Progress in Aerospace Sciences, Vols. 39-2, pp. 83-120, 2003.

[7]B. Benoit, A.-M. Dequin, K. Kampa, W. Grünhagen, P.-M. Basset and B. Gimonet, "HOST, a General Helicopter Simulation Tool for Germany and France," in 56th Annual Forum of the American Helicopters Society, 2000.

[8]Y. Tanabe, S. Saito and H. Sugawara, "Construction and Validation of an Analysis Tool Chain for Rotorcraft Active Noise Reduction," in 38th European Rotorcraft Forum, Amsterdam, Netherlands, 2012.

[9]J. C. Houbolt and G. W. Brooks, "Differential Equations of Motion for Combined Flapwise Bending, Chordwise Bending, and Torsion of Twisted Nonuniform Rotor Blades," NASA TN-3905, 1957.

[10]G. Isakson and G. Eisley, "Natural Frequencies in coupled bending and torsion of twisted rotating and non-rotating blades," NASA CR-65, 1964.

[11]N. Kroll, P. Aumann, W. Bartelheimer, H. Bleecke, B. Eisfeld, J. Fassbender, M. Kuntz, J. Lieser, E. Monsen, R. Heinrich, M. Mauss, J. Raddatz, U. Reisch, B. Roll and T. Schwarz, "FLOWer Installation and User Handbook, Release 116," 2008.

[12]D. Schwamborn, T. Gerhold and R. Heinrich, "The DLR TAU-Code: Recent Applications in Research and Industry," in ECCOMAS CFD 2006 Conference, 2006.

[13]A. Jameson, W. Schmidt and E. Turkel, "Numerical Solution of the Euler Equations by Finite Volume Methods Using Runge-Kutta Time-Stepping Schemes," in 14th AIAA Fluid and Plasma Dynamics Conference, 1981.

[14]S. Enk, "Zellzentriertes Padeverfahren für DNS und LES," PhD Thesis, TU Braunschweig, Germany, 2015.

[15]S. K. Lele, "Compact finite difference schemes with spectral-like resolution," Journal of Computational Physics, vol. 103, pp. 16-42, 1992.

[16]M. Dietz, W. Khier, B. Knutzen, S. Wagner and E. Krämer, "Numerical Simulation of a Full Helicopter Configuration Using Weak Fluid-Structure Coupling," in American Institute of Aeronautics and Astronautics, 2007.

[17]T. Schwarz, "Ein blockstrukturiertes Verfahren zur Simulation der Umströmung komplexer Konfigurationen," 2005.

[18]J.-H. Wendisch and J. Raddatz, "Validation of an unstructured CFD solver for complete helicopter configurations with loose CSD-Trim coupling," in 40th European Rotorcraft Forum, 2014.

[19]D. C. Wilcox, Turbulence Modeling for CFD, 2nd ed., C. A. La Canada, Ed., DCW Industries, Inc, 1998.

[20]P. R. Spalart and S. R. Allmaras, "A One-Equation Turbulence Model for Aerodynamic Flows," in 30th Aerospace Sciences Meeting and Exhibit, 1992.

[21]Y. Tanabe and S. Saito, "Significance of All-Speed Scheme in Application to Rotorcraft CFD Simulations," in The 3rd International Basic Research Conference on Rotorcraft Technology, Nanjing, China, 2009.

[22]Y. Tanabe, S. Saito, O. Takayama, D. Sasaki and K. Nakahashi, "A New Hybrid Method of Overlapping Structured Grids Combined with Unstructured Fuselage Grids for Rotorcraft Analysis," in 36th European Rotorcraft Forum, Paris, France, 2010.

[23]E. Shima and K. Kitamura, "On New Simple Low-Dissipation Scheme of AUSM-Family for All Speeds," in 47th AIAA Aerospace Sciences Meeting, Orlando, FA, 2009.

[24]S. Yamamoto and H. Daiguji, "Higher- Order-Accurate Upwind Schemes for Solving the Compressible Euler and Navier-Stokes Equations," Computers & Fluids, vol. 22, no. 2/3, pp. 259-270, 1993.

[25]J. Yin and J. Delfs, "Improvement of DLR Rotor Aeroacoustic Code (APSIM) and its Validation with Analytic Solution," in 29th European Rotorcraft Forum, 2003.

[26]C. C. Hennes and K. S. Brentner, "The Effect of Blade Deformation on Rotorcraft Acoustics," in 31st European Rotorcraft Forum, 2005.

[27]F. Farassat, "Derivation of Formulation 1 and 1A of Farassat," NASA TM-2007-214853, 2007.

[28]J. W. Lim and A. C. B. Dimanlig, "The Effect of Fuselage and Rotor Hub on Blade-Vortex Interaction Airloads and Rotor Wakes," in 36th European Rotorcraft Forum, 2010.

[29]G. Wilke, "Efficient Aero-Acoustic Simulation of the HART II Rotor with the Compact Pade Scheme," in 42nd European Rotorcraft Forum, 2016.

[30]B. Rodriguez, R. Boisard and J. Mayeur, "Eulerian-Lagrangian Coupling for Helicopter Rotor Aerodynamics," in 38th European Rotorcraft Forum, Amsterdam, Netherlands, 2012.

[31]A. Länger, "New Results in Numerical and Experimental Fluid Mechanics X Notes on Numerical Fluid Mechanics and Multidisciplinary Design," A. a. H. G. a. K. E. a. W. C. Dillmann and C. Breitsamter, Eds., Springer International Publishing Switzerland, 2016, pp. 327-337.

Development of bio-inspired low-noise propeller for drone

Ryusuke Noda
Dr
Chiba University
Chiba, Japan

Toshiyuki Nakata
Dr
Chiba University
Chiba, Japan

Kenta Ishibashi
Mr
Chiba University
Chiba, Japan

Teruaki Ikeda
Mr
Chiba University
Chiba, Japan

Chen Di
Mr
Chiba University
Chiba, Japan

Yuma Yoshinaga
Mr
Chiba University
Chiba, Japan

Hao Liu
Prof.
Chiba University
Chiba, Japan

ABSTRACT

In recent years, multicopter drone has been attracting attention because of its great potentials for various missions such as surveillance, investigation and delivery. For application of drone, however, its noise will be a serious issue especially in urban areas, and, to our knowledge, it has not been resolved yet. In this study, inspired by our previous studies of bio-inspired design system focusing on unique wing structures of insects and birds, we have developed new low-noise type propellers for drone. The various bio-inspired attachments such as the serrations on the leading-edge, velvet-like surface and fringe on the trailing edge were tested and noise levels and aerodynamic performance were evaluated experimentally and numerically. Our results show that an attachment at the trailing edge can suppress the noise level while maintaining the aerodynamic efficiency at the close level to the basic propeller.

INTRODUCTION

Along with the development of various miniature devices such as a motor, battery and video camera, multicopter typed unmanned aerial vehicles (UAVs), called "drones", has become widespread during the past few years as reported in Ref. 1-2. Many commercial drones have already achieved the stable hovering and maneuver flights and it is expected to apply to a variety of fields as discussed in Ref. 1. However, considering the use of drones in an urban or disaster area, its high noise level can be a problem as reported in Ref. 3. In this paper, we developed the low-noise propeller based on the bio-inspired technologies.

Recently, bio-inspired technologies has been attracting wide attention due to the various impressive abilities of animals in nature and it has been started to apply to drones. From the view point of software, inspired by a honeybee flight, Sabo et al (Ref. 4) developed the bio-inspired navigation system by using the optic flow. Keshavan et al (Ref. 5) present the bio-inspired wide field processing of optic flow for enhancing the flight stability under the various flight conditions. In the development of bio-inspired hardware for drones, Mintchev et al (Ref. 6) apply the elastic materials to the frame and present the collision-tolerance strategies inspired by flexible wings of insects. For enabling the various flight modes and enhancing the maneuverability, Henderson et al (Ref. 7) suggest the shape deformable structure for drones. However, few studies have focused on the bio-inspired propellers due to the difference between the inherent flight of animals and the flight of multicopter.

Owl's flight is widely known for silent flight as reported in Ref. 8 due to their unique wing morphologies such as leading edge serrations, trailing edge fringe and velvet like surfaces as shown in Figure 1. In this study, inspired by their unique morphologies, we attached various micro/macro structures to the propeller and then evaluated the noise level experimentally and the aerodynamic performances numerically under the hovering flight condition of the drone.

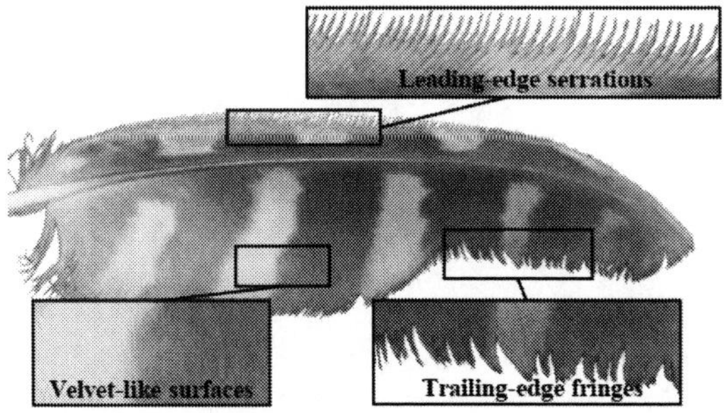

Figure 1. Microstructures on primary feather of ural owl (*Strix uralensis japonica*).

Presented at the 6th Asian/Austrailian Rotorcraft Forum / Heli Japan 2017, Kanazawa, Japan, November 7-9, 2017.

METHODOLOGY

Propeller model

Propeller of Phantom 3 (DJI Ltd.) was employed as the basic propeller. We attached various structures to this propeller and its performance was evaluated. The three-dimensional shape of this propeller was reconstructed by using the laser scanner (Laser ScanArm V2, FARO Technologies Inc.) as shown in Figure 2. Note that all reconstructed points were smoothed by the filter for removing the measurement errors.

Noise level measurement

For measuring the noise level under the hovering condition, the precision sound level meter (NL-52, RION Ltd.) was located at 1.5 m vertical and 4.0 m horizontal position from in-flight drone. This measurement was conducted indoor and the noise was measured for 10 seconds.

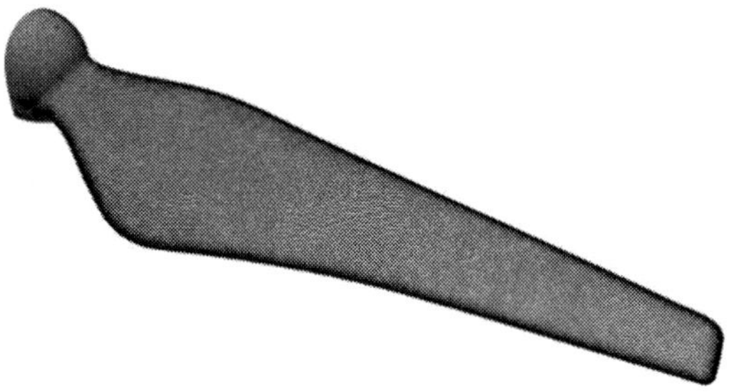

Figure 2. Reconstructed shape of the basic propeller.

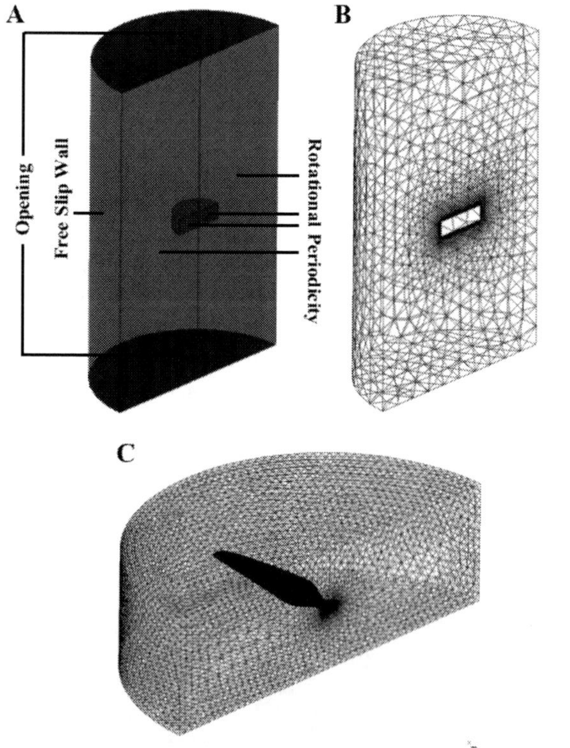

Figure 3. (A) Grid systems and boundary conditions. Note that two domains colored red and blue are set to be rotational and stational domains, respectively. (B) and (C) express the meshes for two domains.

Numerical simulation

For investigating the flow fields around the propeller and estimating aerodynamic forces, the numerical simulation for a single propeller was conducted through ANSYS CFX (ANSYS Inc.). The LES WALE model was adopted as the turbulence model for performing the transition from laminar flow to developed turbulent flow. Considering the total weight of Phantom 3, the rotational speed was set to be 4500 rpm which is almost the same value under the hovering flight condition. The total time step was set to be 1800 step and it can reach from rotational angle = 0° to 180°. Note that the steady analysis was performed before the time transient analysis and its result was given as the initial flow condition for the transient analysis. Figure 3 shows the grid system and boundary conditions. The total mesh number for the analysis of the basic propeller is about 10 million. In this study, the lift and drag forces coefficients C_L and C_D on a propeller are defined in the same way as in Ref. 9:

$$C_L = L/0.5\rho U^2 S_2, \tag{1}$$

$$C_D = Q/0.5\rho U^2 S_3, \tag{2}$$

where L is the lift force and Q is the torque about the rotational axis, ρ is the density of air, U is the wing tip speed, S_2 and S_3 are the second and third moment of wing area, respectively. For evaluating the efficiency with consideration of the differences of lift forces due to the additional structures in tested models, the figure of merit, FM for a propeller is defined as

$$\mathrm{FM} = P_{\mathrm{RF}}/P_{\mathrm{CFD}}, \tag{3}$$

where P_{RF} is the minimum power for generating the resultant lift force obtained from numerical analysis derived by Rankin-Frounde momentum theory (Ref. 10):

$$P_{\mathrm{RF}} = L\sqrt{L/2\rho A_0}, \tag{4}$$

where A_0 is the area of actuator disk.

RESULTS AND DISCUSSION

Tested propeller models are shown in Figure 4. Compared to the basic propeller, the serration, winglet and velvet surface models (Figure 4A-C) show the decrease of lift forces and FMs at 4,500 rpm, while the noise levels are also decreased in all cases. The measured noise characteristics when the drone hovers with these models are almost the same level. In this paper, we mainly discuss the results of the Figure 4D which show the most preferable performances for the development of the low-noise propeller.

Figure 4D shows the low-noise propeller which has the additional structure of the 15 × 30 mm aluminum plate at the trailing edge. The measured noise characteristics of the hovering drone with the basic and the low-noise propeller are illustrated in Figure 5. It is seen that the sound pressure level of the low-noise propeller model was uniformly suppressed within the range of frequency from 200 to 20000 Hz. The overall values called Z-weighted sound pressure level of the basic and low-noise propellers are 72.5 and 70.1 dB respectively. It is confirmed that the additional structure at trailing edge can suppress about 2.4 dB with the same amount of lift generation.

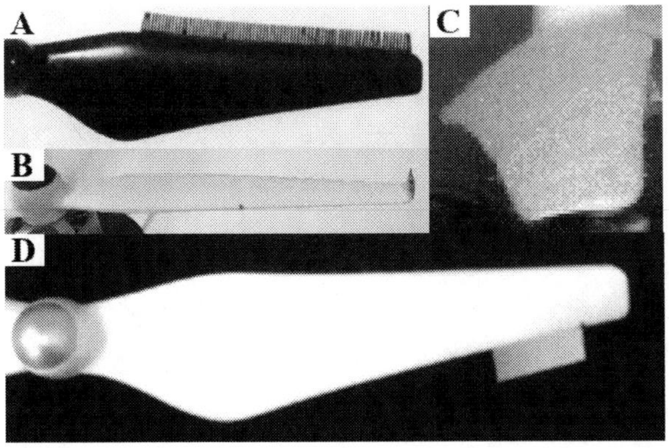

Figure 4. Tested propeller models. Various structures such as (A) serrations, (B) winglet, (C) velvet surface and (D) aluminum plate at trailing edge were attached.

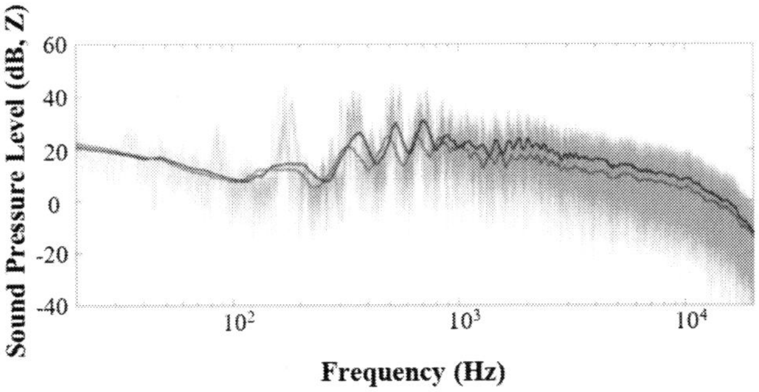

Figure 5. Measurement noise from hovering drone with basic and low-noise propeller.

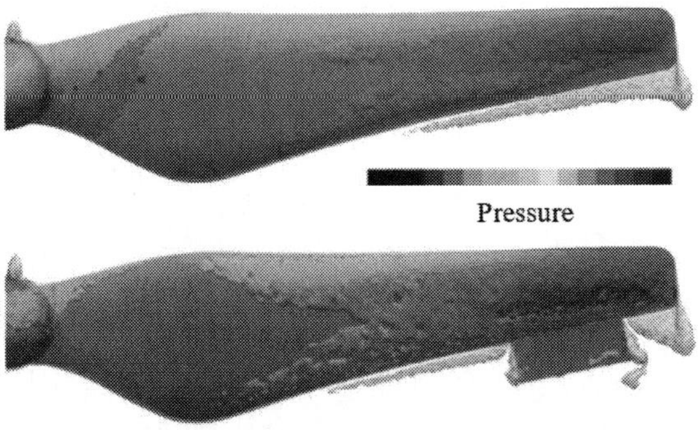

Figure 6. Pressure distribution and iso-surface of *Q*.

Table 1. Geometry parameters for the basic and low noise propellers.

	Base	Low-noise
Wingspan	0.24 m	0.24 m
Wing area	5.2×10^{-3} m^2	5.6×10^{-3} m^2
Second moment of wing area	9.5×10^{-6} m^4	11.5×10^{-6} m^4
Third moment of wing area	7.9×10^{-7} m^5	9.9×10^{-7} m^5

Table 2. Aerodynamic performances of basic and low noise propellers.

	Base	Low-noise
Lift	1.13 N	1.34 N
Torque	0.015 Nm	0.021 Nm
Lift coefficient	0.435	0.437
Drag coefficient	0.140	0.150
Lift to drag ratio	3.10	2.92
Power	7.299 W	9.775 W
Figure of merit	0.1747	0.1745

Table 1 and 2 show the parameters of the morphologies of the basic and low-noise propellers, and the results of the numerical simulations, respectively. It is seen that the lift and torque of the low-noise propeller are equally increased, which is thought to be due to the increase of the second and third moment of wing area, while the lift and drag force coefficients are maintained at the similar level. It is also confirmed that the lift-to-drag ratio of the low-noise propeller is slightly decreased while the FM is maintain ed at the similar level. The power consumptions of the basic and low-noise propellers during the hovering flight are confirmed to be nearly identical from these results. The noise suppression in the measurement is thought to be caused mainly by the reduction of the rotational speed with the increase in wing area of the low-noise propeller. Figure 6 shows the pressure distribution on the upper surface and iso-surface of Q value. It can be seen that they have the different structures of the shedding tip and trailing edge vortices and, in this case, the decrease of the shedding trailing edge vortices of the low-noise propeller may effectively work for the suppression of noise level further, especially in a higher frequency level.

CONCLUSIONS

In this study, we developed the low-noise propeller inspired by the unique wing morphologies of an owl. By comparing the performance of the basic and low-noise propellers, it is found that the aluminum plate attached at the trailing edge can suppress the noise level more than 2 dB while the power consumption is maintained at the similar level when it is attached to the hovering drone. This is thought to be due to the reduction of the rotational speed by the increase of the wing area, and the changes in the structures of shedding vortices.

Author contact: Hao Liu hliu@faculty.chiba-u.jp

ACKNOWLEDGMENTS

This work was supported by Impulsing Paradigm Change through Disruptive Technologies (ImPACT) program of Japan Science and Technology (JST) Agency.

REFERENCES

[1]Nonami, K., "Drone technology, cutting-edge drone business, and future prospects," *Journal of Robotics and Mechatronics*, Vol. 28, 2016, pp. 262–272.

[2]Rothstein, A., *Drone (Object Lessons)*, Bllomsbury Publishing, New York, NY, 2015, Chapter 1.

[3]Christian, A. and Cabell, R., "Initial investigation into the psychoacoustic properties of small unmanned aerial system noise," AIAA Aviation Technology, Integration, and Operations Conference (AVIATION 2017), Denver, CO, June 5-9, 2017.

[4]Sabo, C., Cope, A., Gurny, K., Vasilaki, E. and Marshall, J.A., "Bio-inspired visual navigation for a quadcopter using optic flow," AIAA infotech @ Aerospace, San Diego, CA, January 4-8, 2016.

[5]Keshavan, J., Gremillion, G., Escobar - Alravez, H. and Humbert, J. S., "A μ analysis-based controller-synthesis framework for robust bioinspired visual navigation in less-structured environments," *Bioinspiration and Biomimetics*, Vol. 9, 2014, pp. 25011.

[6]Mintichev, S., Rivaz, S. and Floreano, D., "Insect-inspired mechanical resilience for multicopters," *IEEE Robotics and automation letters*, Vol. 2, 2017, pp. 1248-1255.

[7]Henderson, L., Glaser, T. and Kuester, F., "Towards bio-inspired structural design of a 3D printable, ballistically deployable, multi-rotor UAV," IEEE Aerospace conference, Big Sky, MT, March 4-11, 2017.

[8]Wagner, H., Weger, M., Klaas, M. and Schro, W., "Features of owl wings that promote silent flight," Interface Focus, Vol. 7, 2017, pp. 20160078.

[9]Usherwood, J. R. and Ellington, C. P., "The aerodynamics of revolving wings. I Model hawkmoth wings," *J. Exp. Biol.*, Vol. 205, 2002, pp. 1547–1564.

[10]Ellington, C. P., "The aerodynamics of hovering insect flight. V. a vortex theory," *Phil. Trans. R. Soc. B*, Vol. 305, 1984, pp. 115–144.

A Study on High-Order Numerical Methods for Aerodynamic Noise Prediction of Helicopter Blades in Hovering Flight

Ki Ro Kim 1
Ph.D. Candidate
Konkuk University
Seoul, Republic of Korea

Soo Hyung Park 1
Professor
Konkuk University
Seoul, Republic of Korea

Min Jun Park 2
Ph.D. Candidate
Korea Advanced Institute of Science and Technology
Deajeon, Republic of Korea

Duck Joo Lee 2
Professor
Korea Advanced Institute of Science and Technology
Deajeon, Republic of Korea

Nam Eun Park 3
Ph.D.
Senior Research Engineer
Sacheon-city, Republic of Korea

Dong Kyun Im 3
Ph.D.
Senior Research Engineer
Sacheon-city, Republic of Korea

ABSTRACT

Aero-acoustics in far-field with the permeable methodology was applied to predict the aerodynamic noise according to the numerical schemes. A grid system was constructed by using overlapped grid method. The present high-order numerical method was coupled with the acoustic analogy using the Kirchhoff approach to Ffowcs Williams and Hawkings equations. Multi permeable surfaces surrounding the blades were constructed to include the flow noise generated from the shock waves and the tip vortices. The high speed impulsive noise on UH-1H rotor blades in hovering flight was predicted and compared with the experimental data.

INTRODUCTION

The flow field around the rotating rotor blades has unsteady aerodynamic characteristics due to the blades-vortex interaction and fluid-structure interaction at the blade tip and has a very complicated flow structure. Therefore, a technology is necessary that can simulate numerically accurately the aerodynamic characteristics of helicopter blades. In recent years, the aerodynamic analysis using CFD has attracted attention and studies using the Chimera grid methodologies are actively underway as in Ref. 1~3.

Helicopters have a wide range of applications not only for military purpose but also for civilian purpose. However, helicopters are severely limited in their mission performance due to the complexity of the rotor system and the aerodynamic noise during the flight. As a result, noise regulations are being strengthened globally, and researches are underway to more accurately predict noise generated from the rotor blades. Helicopter noise can be categorized into BVI (Blade-Vortex Interaction) noise and HSI (High Speed Impulsive) noise. BVI noise can be predicted by far-field noise using acoustic pressure information on the blade surface which is discussed in Ref. 4, and HSI noise can be predicted by far-field noise by volume integration of acoustic pressure information surrounding the flow field as discussed in Ref. 5.

Presented at the 6th Asian/Austrailian Rotorcraft Forum / Heli Japan 2017, Kanazawa, Japan, November 7-9, 2017.

In this study, the aerodynamic analysis is performed by applying the three-dimensional Euler's equations, the high-order numerical methods and a Chimera grid method for UH-1H rotor blades in hovering flight. Also, permeable surfaces surrounding the physical blades were constructed to include the flow noise generated from the shock waves and the tip vortices. The acoustic analogy using the Kirchhoff approach to FW-H (Ffowcs Williams and Hawkings) equations is used. Far-field noise results computed with all noise sources are predicted and validated. The effect of spatial accuracy is investigated by applying high-order spatial methods.

NUMERICAL SOLUTION PROCEDURE

In this chapter, the aerodynamic analysis and the noise analysis according to the high-order numerical methods for UH-1H rotor blades in hovering are discussed.

Chimera grid system and Boundary conditions

In this paper, the noise analysis is performed by applying FW-H equations after computing the aerodynamic analysis using UH-1H rotor blades shape as shown in Table 1.

Table 1. Characteristics of UH-1H rotor blades

Airfoil	Planform	Number of Blades	Aspect ratio
NACA0012	Rectangular	2	13.713

To perform the aerodynamic analysis, a grid system using a Chimera grid method is constructed. A Chimera grid method overlaps each grid system (main grid and sub-grids) and transmits information between the grids through interpolation points which is discussed in Ref. 6. Therefore, in the analysis around the object with relative motion, it is easy to analyze the aerodynamic characteristics around the rotating blade. Figure 1 shows the Chimera grid system used for the aerodynamic analysis and a permeable surface concept used for the noise analysis.

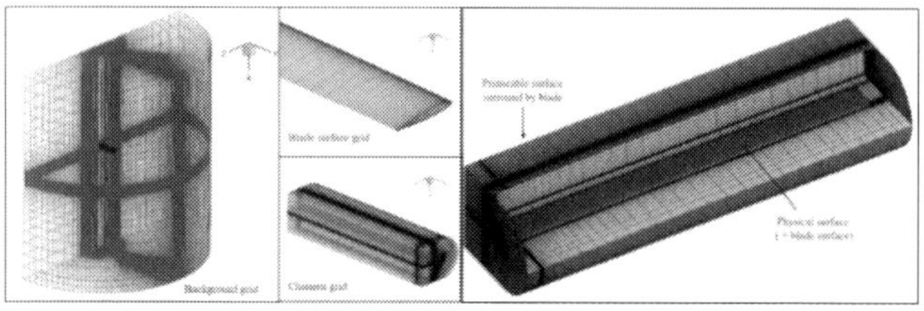

Figure 1. Chimera grid system and concept of a permeable surface.

The permeable surface constructed of the arbitrary virtual permeable surface to include the physical blade surface and the volume in the flow field where the noise source is located. The acoustic pressure at a fixed point is computed from flow solutions obtained at each node point of the multi-permeable surfaces.

Inviscid wall condition is applied to the solid wall surface and the characteristic boundary conditions using the Riemann invariant are applied at the far-field. Since the UH-1H rotor blades are composed of two blades, a sub-grid is constructed for one blade in this analysis. Thus, periodic interface condition is applied as a boundary condition for the rotation axis interface and the convergence of numerical results is improved through steady-state aerodynamic analysis.

Numerical Methods

A three-dimensional Euler's equations are used to analyze the flow field around the rotor blade. The finite volume method is used for spatial discretization, and Roe's FDS (Flux Difference Splitting) scheme in Ref. 7 and 5th order WENO (Weighted Essentially Non-Oscillatory) scheme in Ref. 8 are applied for the analysis. The aerodynamic analysis is computed by DADI (Diagonalized Alternate Directional Implicit Method) which in discussed Ref. 9 is applied as a time advance method and 3rd order TVD (Total Variation Diminishing) interpolation method as a limiter which in discussed Ref. 10. And the data of the steady-state aerodynamic analysis for the permeable surface is transformed to the results of the unsteady analysis according to the time advance.

As a numerical method for the noise analysis, FW-H equations with permeable boundary condition is used by Farassat formulation 1A which in discussed Ref. 11. Acoustic analogy using Kirchhoff approach provides aeroacoustics from all noise sources including thickness noise, loading noise and flow noise (Monopole, Dipole, Quadrupole) only by using surface integration without volume integration. All noise sources are predicted at the 3.09R (Blade radius) position of the observer from the center of the rotation plane. The flow information needed to calculate the acoustic pressure at every moment is generated at dt = 0.152307e-04 intervals corresponding to 0.25 degrees in azimuth angle.

Numerical Results

The collective pitch angle is set to 0° for non-lifting hovering flight and the aerodynamic analysis is performed with respect to tip Mach number 0.8, 0.85 and 0.88. Pressure distribution at the r/R = 0.95 position of the blade using WENO scheme is shown as Figure 2. A strong shock wave is captured at the blade tip.

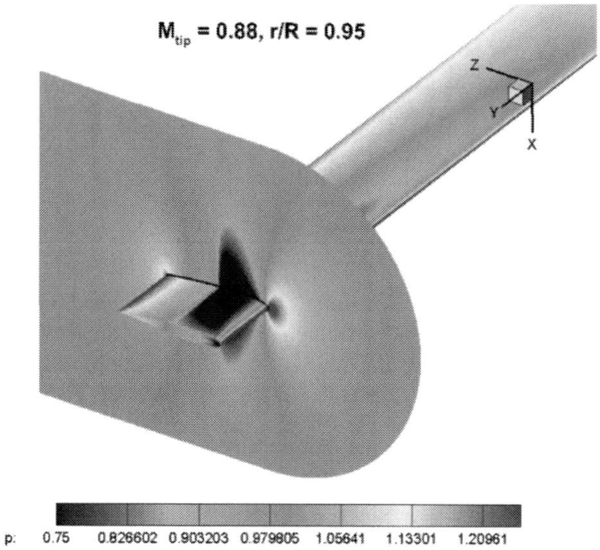

Figure 2 Pressure distribution on the blade using WENO scheme.

The noise analysis is computed based on CFD results from the Roe's FDS scheme and the WENO scheme according to the permeable surface locations. For the shape of multi permeable surfaces, C-H type grids are generated. Each permeable surface has different locations according to the chord length to height from the blade surface. The numerical noise results are compared and validated with those of Baeder (Ref. 12), Brentner (Ref. 13) and Experimental data (Ref. 14)

Figure 3 and Figure 4 show the results of the far-field noise analysis at tip Mach numbers 0.8 and 0.85, respectively, when the permeable surface locations are z/c = 0.5 and 1.5. In case of tip Mach number of 0.8, there is no shock waves on the blade surface. The results of the noise analysis according to the numerical schemes and the permeable surface locations are very similar to those of Baeder (Ref. 12).

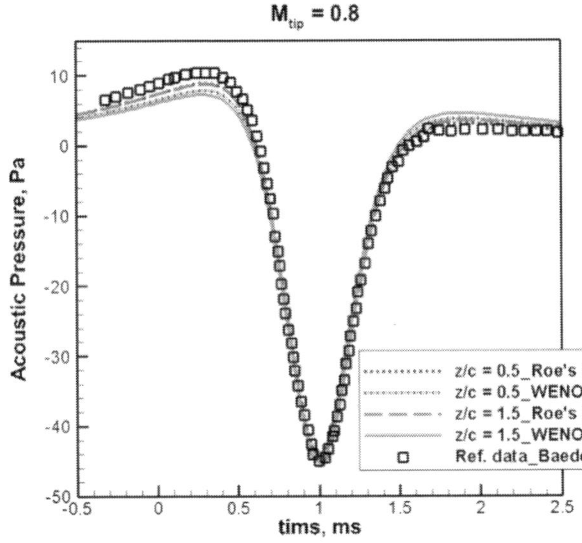

Figure 3 Comparison of rotor noise at tip Mach 0.8 (Ref. 12).

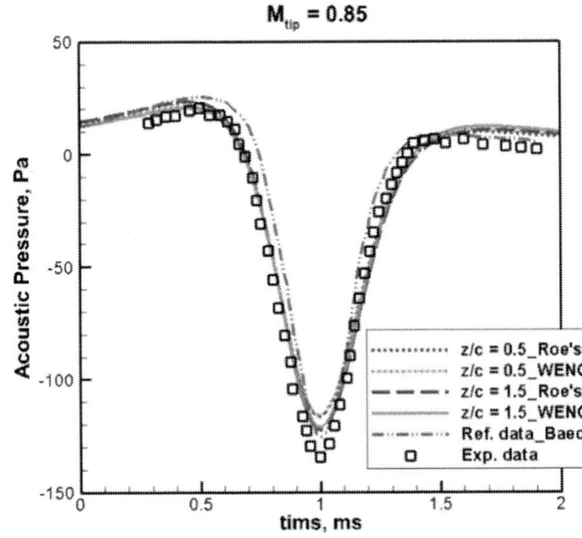

Figure 4 Comparison of rotor noise at tip Mach 0.85 (Refs. 12 and 14).

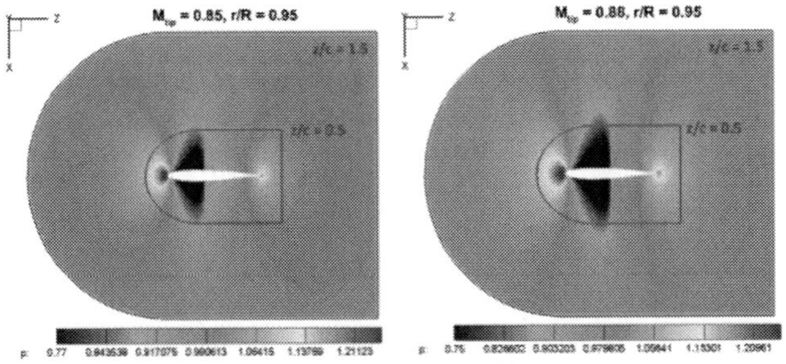

Figure 5 Permeable surface locations with pressure distribution.

However, as shown in Figure 5, in case of tip Mach number 0.85, it can be seen that there is no difference in the peak value of the acoustic pressure according to the numerical schemes but has a difference with respect to the permeable surface locations only. It is considered that the permeable surface of z/c = 0.5 does not completely contain the shock waves generated from the blade tip as shown in Figure 5.

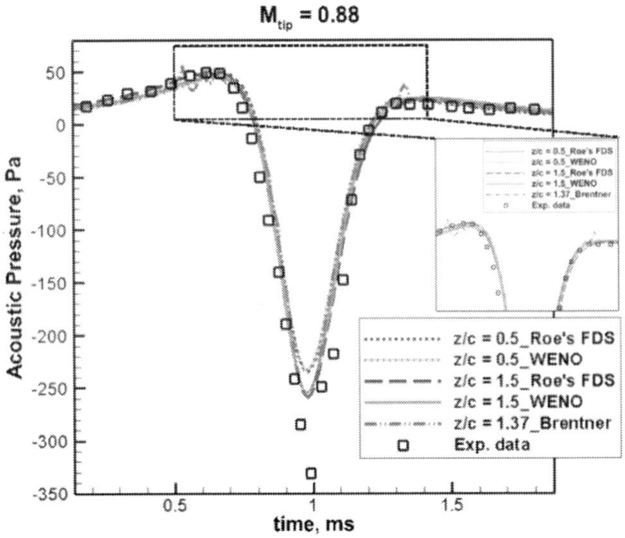

Figure 6 Comparison of rotor noise at tip Mach 0.88 (Refs. 13 and 14).

The results of the noise analysis at tip Mach number of 0.88 are shown in Figure 6. Similarly, when the permeable surface location is z/c = 0.5, there is a difference in the acoustic pressure because the shock waves generated from the blade tip is not completely contained. When the permeable surface location is z/c = 1.5, the peak values obtained from both Roe's FDS scheme and WENO scheme are closer to the experimental data. It implies that the noise source caused by the shock waves generated on the blade surface is a dominant factor in the far-field noise prediction. As a result, the permeable surface should completely contain the shock waves to help predict more accurate volume noise sources.

It is interesting that, in Brentner (Ref. 13) case, there is a small oscillation of the acoustic pressure at time = 0.5~0.7 and 1.25~1.4 as shown in Figure 6. In contrast, the numerical results of present study yielded linear acoustic pressure results without any oscillation. It should be noted that the noise analysis is performed with steady-state flow solution. The fully converged flow solution has much smaller error than that produced from the unsteady aerodynamic analysis.

CONCLUSIONS

In this study, Roe's FDS scheme and 5th order WENO scheme as a high-order finite volume spatial scheme are applied to the aerodynamic analysis for UH-1H rotor blades in hovering flight. Permeable surface methodology is used and coupled with an acoustic analogy code using Kirchhoff approach to FW-H equations. To accurately predict far-field acoustic noise, the permeable surface should be constructed to surround the shock waves completely. Results show that it is important to use fully-converged flow field solutions with high-order spatial schemes in hovering flight.

Author contact:

Ki Ro Kim kornk2r@naver.com
Soo Hyung Park pish@konkuk.ac.kr
Nam Eun Park nameunpakraero@koreaaero.com
Dong Kyun Im dongkyun.im@koreaaero.com

ACKNOWLEDGMENTS

This work was supported by the KEIT Research Grant of 2017 ('10053155').

REFERENCES

[1]Ahmad, J., and Duque, E. P. N., "Helicopter rotor blade computation in unsteady flows using moving overset grids," *Journal of Aircraft*, Vol. 33, (1), 1996, pp. 54-60.

[2]Kim, J. W., Park, S. H., Yu, Y. H., Kim, Y. J., and Kwon, J. H., "Verification of Hovering Rotor Analysis Code Using Overlapped Grid," *Journal of The Korean Society for Aeronautical and Space Sciences*, Vol. 36, (8), 2008, pp. 719-727.

[3]Kim, J. W., Park S. H., and Yu Y. H., "Euler and Navier-Stokes simulations of helicopter rotor blade in forward flight using an overlapped grid solver," 19th AIAA CFD Conference, 2009

[4]Wie, S. Y., Im, D. K., Kwon, J. H., Lee, D. J., Chung, K. H., and Kim, S. B., "Helicopter rotor noise in the merged tip-vortex and blade interaction condition," *International Journal of Aeroacoustics*, Vol. 10, (4), 2011, pp. 427-442.

[5]Farassat, F. "Quadrupole source in prediction of the noise of rotating blades-A new source description," American Institute of Aeronautics and Astronautics Conference, 1987,

[6]Im, D. K., Wie, S. Y., Kim, E., Kwon, J. H., Lee, D. J., Park, S. H., Chung, K. H., and Kim, S. B., "Unsteady Aerodynamic Analysis for Helicopter Rotor in Hovering and Forward Flight Using Overlapped Grid," *Journal of The Korean Society for Aeronautical and Space Sciences*, Vol. 37, (3), 2009, pp. 215-223.

[7]Roe, P. L., "Approximate Riemann Solvers, Parameter Vectors, and Difference Schemes," *Journal of computational physics*, Vol. 2, (43), 1981, pp. 357-372.

[8]Jiang, G. S., and Shu, C. W., "Efficient implementation of weighted ENO schemes," *Journal of computational physics*, Vol. 126, (1), 1996, pp. 202-228.

[9]Park, S. H., and Kwon. J. H., "Implementation of k-w Turbulence Models in an Implicit Multigrid Method," *AIAA journal*, Vol. 7, (42), 2004, pp. 1348-1357.

[10]Kim, K. H., and Kim, C., "Accurate, Efficient and Monotonic Numerical methods for Multi-dimensional Compressible Flows Part II: Multi-dimensional Limiting Process," *Journal of Computational Physics*, Vol. 208, 2005, pp. 570-615.

[11]Farassat, F., and George P. S., "The prediction of helicopter rotor discrete frequency noise," 38th American Helicopter Society Annual Forum, Proceedings, (A82-40505 20-01), 1982, pp. 497-507.

[12]Baeder, J. D., Judith M. G., and Yu, Y. H., "A computational study of the aeroacoustics of rotors in hover." *Journal of the American Helicopter Society*, Vol. 42, (1), 1997, pp. 39-53.

[13]Brentner, K. S., and Farassat, F., "Analytical comparison of the acoustic analogy and Kirchhoff formulation for moving surfaces," *AIAA journal*, Vol. 36, (8), 1998, pp. 1379-1386

[14]Purcell T. W., "CFD and transonic helicopter sound," 14th European Rotorcraft Forum, 1988, pp. 17.

The research of rotor noise radiation characteristics from an in-flight helicopter based on acoustic mapping

Fei Wang
Phd student
NANJING UNIVERSITY OF AERONAUTICS AND ASTRONAUTICS
Nanjing, China

Guohua Xu
Professor
NANJING UNIVERSITY OF AERONAUTICS AND ASTRONAUTICS
Nanjing, China

Yongjie Shi
Assist Professor
NANJING UNIVERSITY OF AERONAUTICS AND ASTRONAUTICS
Nanjing, China

ABSTRACT

Based on the acoustic mapping, a prediction model for the ground noise radiated from an in-flight helicopter is established. For the enhancement of calculation efficiency, a high-efficiency second-level acoustic radiation model is developed by the combination of the point-source idea and the feature of rotor noise radiation. Rotor free-wake analysis method and Ffowcs Williams-Hawkings (FW-H) equation are applied for the aerodynamics and noise prediction. A database of noise spheres is established by the helicopter trim model together with a parametric modeling approach. Furthermore, based on acoustic mapping, a method of rapid simulation for the ground noise radiated from an in-flight helicopter is developed. The noise footprint for OLS rotor is then calculated and the influence of some parameters including advance ratio and flight path angle on ground noise is deeply analyzed The results suggest that with the increase of advance ratio and flight path angle, the peak noise levels on the ground first increase and then decrease, in the meantime, the maximum SEL noise on the ground shifts toward the advancing side of rotor.

INTRODUCTION

Although helicopters with the unique ability to take off and land vertically, hover as well as flexibly maneuver have been widely used, the high level of noise radiated by their rotors has drawn more and more attention. In particular, the unique Blade-Vortex Interaction (BVI) noise of helicopters in descent flight has a severe influence on ground objects. In the past, a lot of work on rotor noise analysis methods[1-3], noise-generating mechanism[4] as well as acoustic properties[5] have been done. However, researches on the rotor noise footprint prediction method and corresponding rotor noise radiation characteristics when the helicopter is in flight have barely been carried out.

Commonly, the boundary integral method[1-3] and Computational Aero Acoustics (CAA) method[6] are used for the calculation of rotor noise. Given a unique spot of the helicopter, both methods are mainly used to calculate the rotor noise directly and generally cannot take the influence of atmosphere absorption into account. Whereas, for the prediction of ground noise radiated from a helicopter in flight, the calculation time will significantly increase and the demand of high computation efficiency ensues. This is because the flight path will contain lots of track points rather than one[7], and for a better understanding of ground noise characteristics, an observer plane instead of one or several observers is used, which consists of numerous orderly distributed observers. Besides, impacts of atmosphere absorp-tion[8] on rotor noise have to be considered in the propagation of rotor noise. There will be excessive computations when applying the two aforementioned methods directly to the prediction of ground noise from an in-flight helicopter directly. Specifically, the CAA method itself not only has such problems as numerical dissipation but also need to compute the noise throughout the entire computation domain, which extends to ground observers, and the amount of calculation will increase dramatically with the increase of computation domain. Relative to the CAA method, the boundary integral method has a higher computation efficiency, where sound sources are computed using aerodynamic analysis models, and then near-field noise is propagated to far-field by the acoustic equation. However, the boundary integral method is not suitable for the calculation of dynamic ground noise from an in-flight helicopter due to the time-consuming computation of sound sources and radiation of noise radiated from rotating rotors in one period by solving the retarded time equation repeatedly[9].

Conner[7], Fleming[10] et al conducted researches where an experiment based method was used to predict the noise footprint of a helicopter in flight. In the research, the measurement of ground noise of a helicopter in different flight conditions came first and then the database of radiation spheres was built, which projects inversely ground noise to the near field of rotors by the Acoustic Repropagation Technique (ART). Then in simulation of ground noise, the noise footprint was obtained by the projection of the corresponding radiation spheres extracted from the database when the helicopter "flied" along the flight trajectory rather than calculated from analysis models directly. This is what makes the method efficient. However,

Presented at the 6th Asian/Austrailian Rotorcraft Forum / Heli Japan 2017, Kanazawa, Japan, November 7-9, 2017.

since the method is constrained by test conditions, the experiment based ground noise prediction technically and economically demands more, which is especially embodied in outfield noise tests where the ever-changing flight conditions and ambient atmosphere data need to be recorded continuously.

In view of this, this paper attempts to develop a high-efficiency method for the simulation of noise footprint from an in-flight helicopter by employing the idea of radiation spheres. At first, a second-level acoustic radiation model capable of accounting for atmosphere absorption by the combination of the idea of point source and rotor noise radiation properties is built. The efficiency is improved by substituting a single stationary radiation point on the surface of the second-level radiation sphere in this model for surface radiation points rotating with blades. Secondly, the helicopter trim model is derived and the parametric modeling is performed. Furthermore, a database of radiation spheres taking advance ratio and Tip-Path-Plane (TPP) angle as characteristic parameters is generated. Then a no-table improvement in computational efficiency is obtained by mapping the noise on the radiation sphere surface to ground observers in the ground noise simulation. In addition, the descent flight is taken as a numerical example, and the effects of different advance ratios and flight path angles on the ground noise are analyzed by using the method developed. On the basis of the analysis, meaningful results for descent trajectory with low ground noise are obtained.

EFFICIENT PREDICTION MODEL OF GROUND NOISE

In order to predict the different time-variant characteristics and sound pressure level at various directions, a second-level acoustic radiation model that can account for spherical spreading and atmosphere absorption is built referring to the point-source theory[11]. On the basis of parametric expression of rotor noise derived from helicopter trim model, a high-efficient ground noise prediction model based on acoustic mapping is established.

Second-level acoustic radiation model

In the prediction of ground noise from an in-flight helicopter, flight trajectory is discretized first. In this paper, the trajectory is discretized according to the distance traveled by helicopter over one rotor revolution, namely, $2\pi\mu R$ (μ is advance ratio, and R is rotor radius) . The comparison between second-level acoustic radiation model and direct ground noise prediction method is sketched in Figure 1. In general, the computation of ground noise is composed of the rotor aerodynamics prediction, rotor noise radiation as well as post-processing of ground noise. The efficient ground noise method is achieved by inserting the second-level acoustic radiation sphere between phase ① and phase ② in Figure 1, from which the rotor noise is propagated. The Sound Exposure Level (SEL) in the figure is the weighted Sound Pressure Level (SPL) taking the effect of time exposure into account. In the figure, means the maximum bound circulation over a blade.

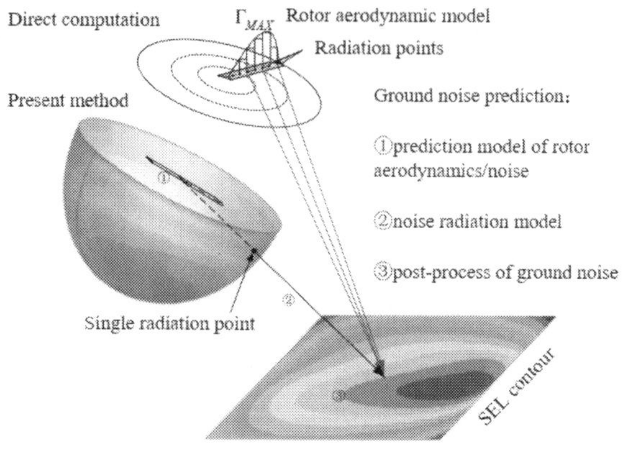

Figure 1 Comparison of two ground noise prediction methods

Corresponding to the discretization period of flight path, the mapping surface of acoustic radiation spheres representing acoustic energy radiated by a helicopter rotor over one rotor revolution is calculated. The zero degree in elevation angle direction is located in the tip-path-plane and the elevation angle will be negative blow the tip-path-plane. The zero degree in azimuth direction points toward the rear of helicopter. The radiation points on the radiation sphere surface are distributed with an interval of 5 ° in both the elevation angle and azimuth angle. The SPL is computed at every radiation point. The noise characteristics at different radiation points represent the noise radiation property in the corresponding direction. The mapping surface is located at 25R from the rotor hub. The distance insures that radiation sphere only captures far-field acoustic radiation (namely it insures that the attenuation of noise is linear with the increase of radiation distance without near-field noise). In the process of noise propagation, a line connecting the sphere center (namely rotor hub) to the ground observer represents the ray along which noise propagates from the helicopter rotor to the observer. The point of the intersection between sphere surface and the line represents the second-level sound source, whose noise will be mapped to the ground observer along the ray. The SPL at the ground observer can be expressed as:

$$L(r) = L(r_0) + A_{\text{geometry}} + A_{\text{atmosphere}} \tag{1}$$

where $L(r)$ is the SPL at the ground observer location r, $L(r_0)$ the SPL of the second-level radiation point, and A_{geometry} represents spherical spreading loss, $A_{\text{atmosphere}}$ the atmospheric losses[8]。

As described in introduction, this paper focuses on the establishment of the efficient method and the research of rotor noise radiation characteristics of an in-flight helicopter. The observers are set to be on the ground. So, in the process of noise propagation, the effects of terrain conditions on noise are not considered. This is a conventional method that has been adopted by other researchers in Refs.[12-14].

The noise calculation of radiation points on mapping surface includes air-loading and noise radiation calculation. The free-wake method with rotor trimming is used to predict the aerodynamics. A schematic of rotor wake model15 is shown in Figure 2. The rotor wake model is composed of near-wake and far-wake. The near-wake includes trailed vortices and shed vortices resulting from the differences of bound vortex strengths in spanwise and azimuthal direction respectively. The far-wake comprises only a single tip vortex filament which evolves from a roll-up of the near-wake detached from the trailing edge of blade after a revolve of 30 °. The motion equation of a vortex filament can be written as:

$$\frac{\mathrm{d}\boldsymbol{r}(\psi,\zeta)}{\mathrm{d}t}=\boldsymbol{V}\left(\boldsymbol{r}(\psi,\zeta)\right) \tag{2}$$

where $\boldsymbol{r}(\psi,\zeta)$ defines the positions of control points, ψ the azimuth at which the blade located, ζ the wake age angle, t the time and $\boldsymbol{V}\left(\boldsymbol{r}(\psi,\zeta)\right)$ is the velocity of control points including free-stream and induced velocities from bound vortex as well as the other vortex filaments. The induced velocities are computed by Biot-Savart law with Scully vortex core model[16] and an initial vortex core radius of 0.2*c* (*c* is blade chord).

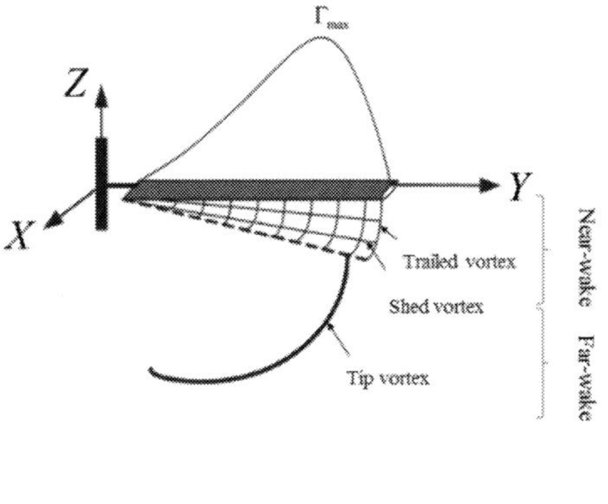

Figure 2 Schematic of rotor free-wake model

To eliminate rotor pitching and rolling moments and match the calculated rotor thrust coefficient to the desired level, a rotor trimming model is presented, in which the difference correction of control angles is calculated using Newton-Rhapson iterative methods. The expression is as follows:

$$\begin{Bmatrix}\Delta\theta_0\\ \Delta\theta_{1c}\\ \Delta\theta_{1s}\end{Bmatrix}=\begin{bmatrix}\frac{\partial C_T}{\partial\theta_0} & \frac{\partial C_T}{\partial\theta_{1c}} & \frac{\partial C_T}{\partial\theta_{1s}}\\ \frac{\partial C_{My}}{\partial\theta_0} & \frac{\partial C_{My}}{\partial\theta_{1c}} & \frac{\partial C_{My}}{\partial\theta_{1s}}\\ \frac{\partial C_{Mx}}{\partial\theta_0} & \frac{\partial C_{Mx}}{\partial\theta_{1c}} & \frac{\partial C_{Mx}}{\partial\theta_{1s}}\end{bmatrix}^{(-1)}\times\begin{Bmatrix}C_T^{desire}-C_T\\ -C_{Mx}\\ -C_{My}\end{Bmatrix} \tag{3}$$

where, $\left(C_T^{desire},0,0\right)$ are the desired trim state. $\left(C_T,C_{Mx},C_{My}\right)$ are the calculated thrust and moment coefficients. $\left(\Delta\theta_0,\Delta\theta_{1c},\Delta\theta_{1s}\right)$ represents the adjustment of rotor collective and cyclic pitch angle.

The rotor noise prediction is conducted using the well-known Farassat 1A equation[9] as follows:

$$p'(\boldsymbol{x},t)=p'_{\mathrm{T}}(\boldsymbol{x},t)+p'_{\mathrm{L}}(\boldsymbol{x},t) \tag{4}$$

where p'_{T} and p'_{L} are the thickness and loading noise respectively, $\boldsymbol{x}$ the observation location vector, and t is the observation time.

High-efficiency prediction model for ground noise

Previous studies[17] reveal that under trimmed steady-state flights, the BVI noise is governed by four parameters: rotor thrust coefficient C_T, hover tip Mach number M_{ht}, rotor tip-path-plane angle α_{TPP} and advance ratio μ. Figure 3 shows the forces exerted on a helicopter under steady-state flight in the longitudinal plane.

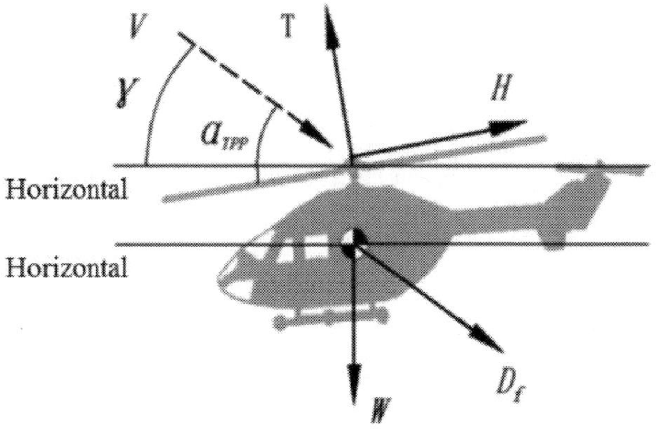

Figure 3 Schematic of forces exerted on helicopter

Equilibrium equations of the helicopter in the longitudinal plane are given in Eq. (5).

$$\begin{cases}\sum F_x: & T\sin(-\alpha_{\mathrm{TPP}})-D_{\mathrm{f}}-H\cos(\alpha_{\mathrm{TPP}})-W\sin\gamma=0\\ \sum F_y: & T\cos(-\alpha_{\mathrm{TPP}})+H\sin(-\alpha_{\mathrm{TPP}})-W\cos\gamma=0\end{cases} \tag{5}$$

where T and H are rotor thrust and drag force respectively, D_f and W the fuselage drag and helicopter weight, and γ denotes helicopter path angle.

In descent, the H-force is negligible[18]. The rotor thrust is assumed to be equal to helicopter weight. Solving Eq.(5) yields the following expression for tip-path-plane angle:

$$\alpha_{TPP} = -\frac{D_f}{W} - \gamma \qquad D_f = \frac{1}{2}\rho V^2 f \tag{6}$$

where, f represents equivalent flat plate area, ρ the air density, and V is helicopter velocity.

In descent, rotor thrust and hover tip Mach number are assumed to be constant in general and thus the governing parameters will be the advance ratio and tip-path-plane angle. A database of radiation spheres characterizing advance ratio and tip-path-plane angle is then prebuilt. The database is shown in the right-hand side of Figure 4. In the computation of ground noise, the radiation sphere at every flight path spot will be selected from the database according to the flight state (advance ration and tip-path-plane angle). Then the ground noise is calculated by projecting the noise on the sphere to ground observers rather than calculated in real-time with analysis models. This is what makes the method efficient further.

The flow chart of the ground noise calculation using the present method is depicted in Figure 4. The calculation procedure can be divided into three steps: First of all, the flight trajectory is discretized and the corresponding flight parameters (μ, α_{TPP}) are calculated at each discretized point. Secondly, the radiation sphere is selected and oriented by rotating an angle $\gamma + \alpha_{TPP}$. The radiation point is ascertained by connecting the discretized point to the ground observer and the corresponding noise is propagated to the ground. Finally, the post-processing of ground noise is conducted.

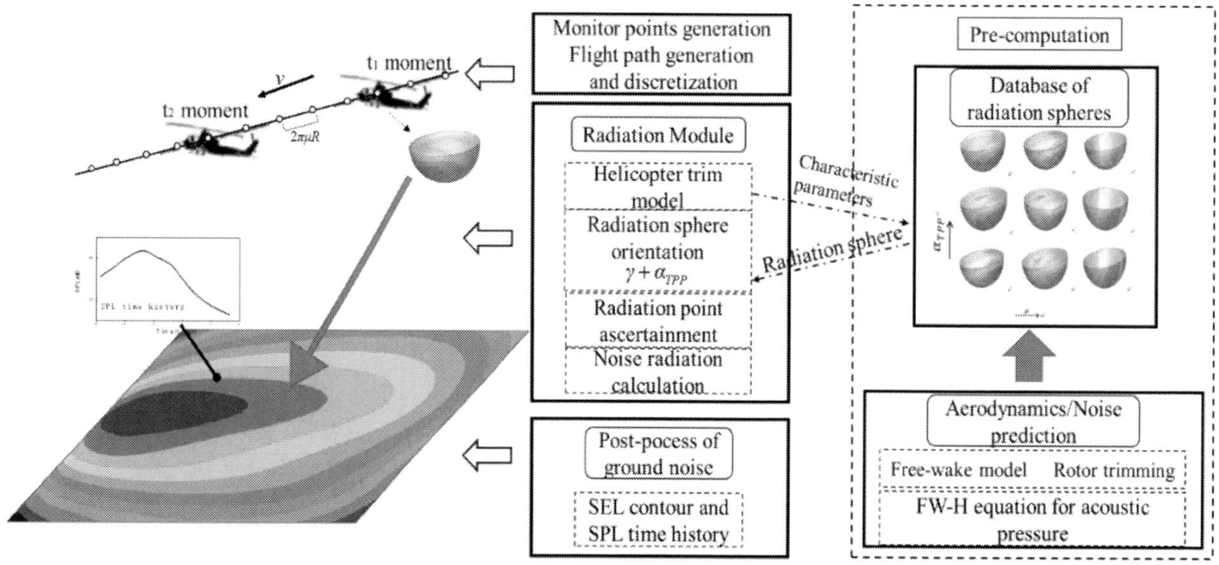

Figure 4 Flow chart of high-efficiency ground noise computation

VALIDATION OF CALCULATION METHODS

In order to validate the capability of the present wake models, Elliott et al's experiment rotor[19] is first taken as a numerical example and induced velocities are compared. Then the BVI blade loads and noise are calculated on the well-known Operational Loads Survey OLS rotor and compared with available data[17]. By comparing the calculation time between the second-level acoustic radiation model and direct calculation method of noise, the high computation efficiency of the present model is verified.

Validation of free-wake model

To verify the capacity of present free-wake in inflow prediction. The experiment data[19] measured at the conditions with C_T=0.008 , μ=0.15 and μ=0.23 are used in this section. In the experiment, the inflow one chord above the rotor disk was measured and non-dimensioned with respect to the blade tip speed. The four-bladed rotor has a forward shaft tilt angle of −3 °. The blade has a linear twist of −8° from the root to tip, a radius of 0.86 m, and the aspect ratio is 13.

Figure 5 gives the comparisons of inflow ratio among the experimental data, the results by time-marching free-wake method in reference 20 and the present calculation was given. As is shown, the present results demonstrate better agreements with the experimental data when compared with the results by the time-marching method. The discrepancies of inflow in central regions may be caused by the presence of fuselage and rotor hub in experiment since the blockage effects of fuselage and rotor hub have not been simulated in the present method. Besides, the influence of "vortex-induced" upwash at the front of the rotor disk is also precisely predicted with the present method.

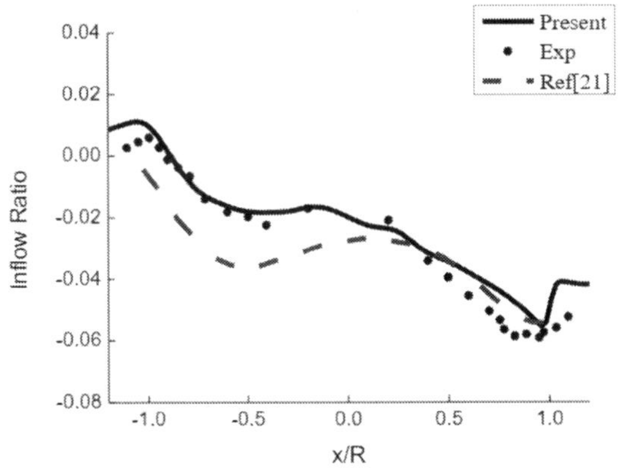

(a) μ=0.15

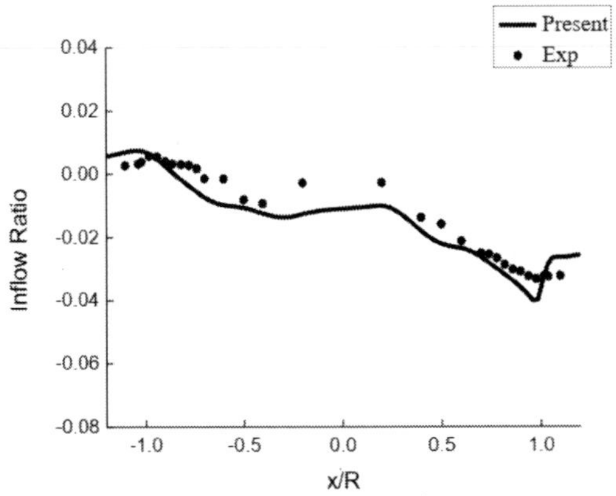

(b) μ=0.23

Figure 5 Comparisons of inflow ratio across disk at different advance ratios

The rotor used for BVI loads and noise validation is the two-bladed OLS rotor-- a 1/7 scale model of AH-1 helicopter main rotor. The blade has an aspect ratio of 9.22 and a linear twist of –8.2°. The simulation condition is as follows: an advance ratio of 0.164, thrust coefficient of 0.0054 and hover tip Mach number of 0.664. Table 1 gives control angles calculated by the present method, experiment17 and Ref 20.

Table 1. Control angles of OLS rotor

Blade Motion	θ_0	θ_{1s}	θ_{1c}	β_0	β_{1s}	β_{1c}
Experiment	6.14	-1.39	0.9	0.5	0.0	-1.0
Ref.[20]	6.07	-2.37	1.2	0.5	0.0	-1.0
Present	6.51	-2.28	1.5	0.5	0.0	-1.0

Figure 6 shows the comparison of the predicted blade loads between the present method and other three methods (AFDD, DLR, ONERA) from Ref. [21] at 0.91*R* and 0.95*R* spanwise locations. It is shown that the predicted value by present method is intermediate compared with other results and has good agreements with them. In addition, the present method captures several BVI interactions on the advancing side, while the other methods only capture a single strong interaction.

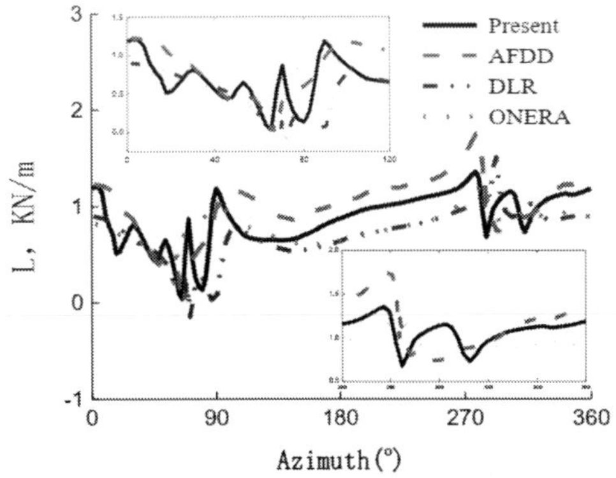

(a) r/R=0.91

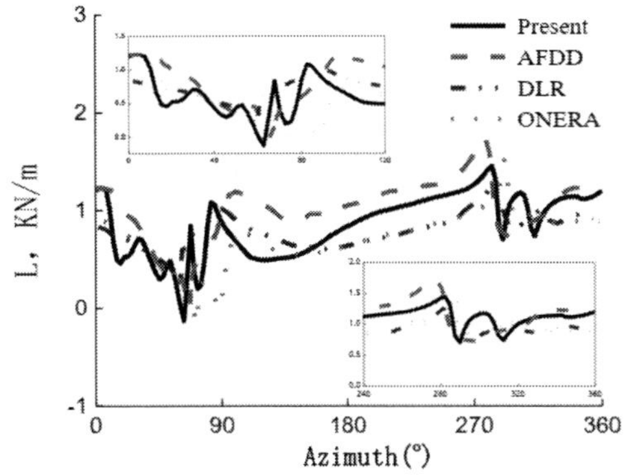

(b) r/R=0.955

Figure 6 Comparisons of blade loadings at several spanwise locations

Validation of noise prediction method

The noise signal of OLS rotor at two observers are calculated. Both observers are 3.44*R* away from the rotor hub, 30° below the rotor plane. Laterally, observers No.3 and No.9 are located at $\psi = 180^\circ$ and $\psi = 210^\circ$ respectively.

The comparisons of sound pressure time histories at two different observations between predicted values, experimental data are presented in Figure 7. It can be seen that the present method has the capability of capturing two peaks of noise signals and a better agreement with the experiment data. But the method shows a little discrepancy with experimental results at $t^* = 0.2$, where the peak drop is not captured well.

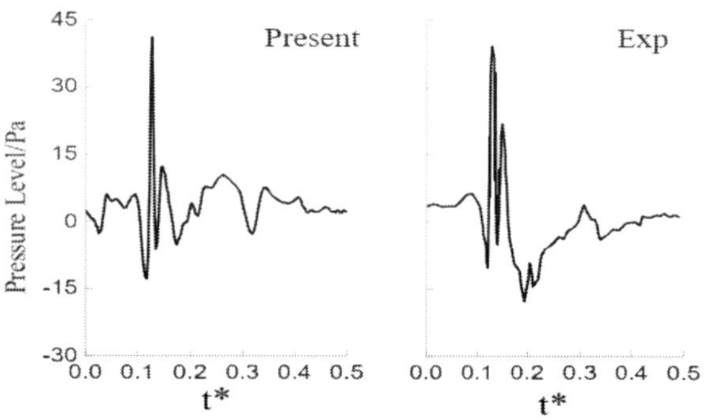

(a) Sound pressure time history at #3

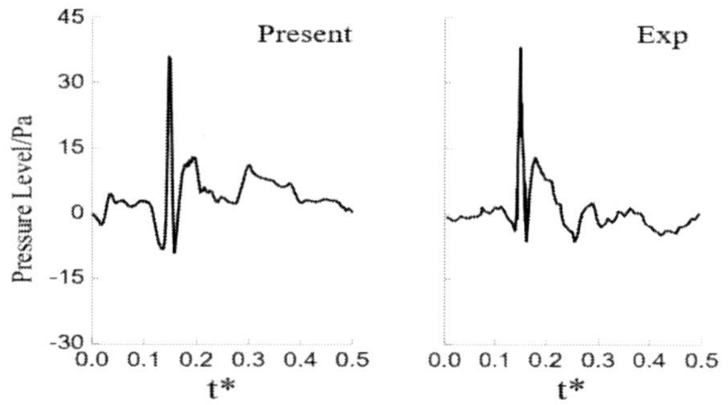

(b) Sound pressure time history at #9

Figure 7 Comparisons of sound pressure time history at two observations

Computational efficiency validation of second-level acoustic radiation model

The comparison of calculation time between two methods when the helicopter is located at a specified spot is listed in Table 2. In this case, the number of the sound sources on blade surface is 18×28×201 (spanwise segments × chordwise segments × the number of points of pressure signature). It is necessary to calculate the noise radiated from all the sound sources with the direct calculation method of rotor noise. As a contrast, only one second-level sound source on the mapping surface of radiation sphere is needed to be calculated with the present second-level acoustic radiation method. Namely, noise of every observer was radiated from only one radiation point for the latter method, while noise of every observer was radiated from 101304 radiation points for the former method. As seen in Table 2, it takes only five minutes to calculate the ground noise with one hundred thousand observers compared to three hours by direct calculation method (operated on PC with Core i7、3.6GHz). It is concluded that the present method is high-efficiency when compared with direct computation method. It should be pointed out that, although the efficiency can be greatly improved with second-level acoustic radiation method, the present method is not suitable for the computation of near-field observer. This is because that the radiation sphere will capture both near-field and far-field acoustic radiation, and the attenuation of noise will be nonlinear with the increase of radiation distance. The direct calculation method should be applied under the situation.

Table 2 Comparison of computation time between two different methods

Methods	Present	Direct
Number of radiation points	1	101304
Number of observers	100,000	100,000
Computation time	5min	3.2h

PARAMETRIC STUDIES AND DISCUSSIONS

In this section, the ground noise is simulated using the developed method and the effects of flight velocity and flight path angle on ground noise are analyzed on the AH-1 helicopter.

The schematic of ground observer plane and flight trajectory is depicted in Figure 8. The observer plane which has observers with an interval of 2.5 m in both directions extends to 60 m from the origin to either side of *y* axis and is 190 m in length along the *x*-direction. Flight trajectory is right above the centerline (*y*=0). To simulate the final stages of the approach, the helicopter will fly another 20 m. Three specified observers locating at the advancing/retreating side of rotor and below the flight trajectory are set to examine SPL time histories in different directions. In the subsequent parametric studies, both SEL contours on the ground and SPL time histories at three specified observers are calculated. In Figure 9, the observer time is negative because the source emission time is stipulated to be zero when the helicopter arrives at the landing point (see Figure 8).

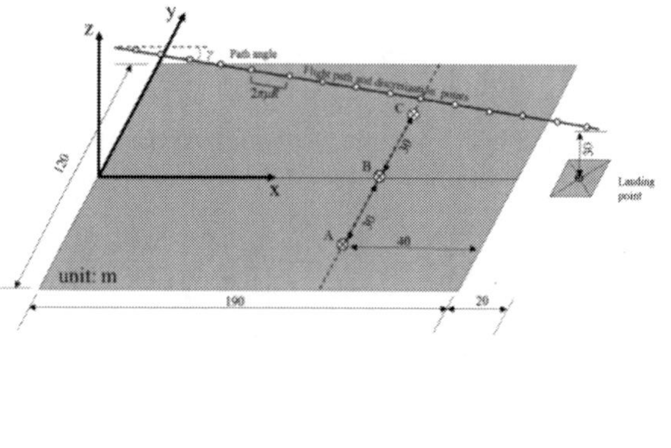

Figure 8 Schematic of helicopter trajectory and observer plane

Effects of flight path angle

The effect of flight path angle on ground noise is investigated and the results are plotted in Figure 9. The SEL contours and SPL time histories are calculated for the approach flight with advance ratio of 0.164 as well as flight path angles of 0°, −6° and −9°. For all the cases shown in Figure 9, it can be seen that: (1) The same noise variation trend at three observers is exhibited, where the SPL increases with the approach of helicopter and decreases with the departure of the helicopter. Observer B has the highest SPL value because of a closer proximity to the trajectory and thus a lower attenuation of atmospheric absorption as well as spherical spreading loss. Observer A has a higher SPL value than Observer C because the advancing side is the preferred propagated direction of BVI noise. (2) The SEL increases along the centerline (y=0) as the helicopter flies forward. The max SEL increases first and then decreases as the flight path angle increases, meanwhile the SEL gradually turns asymmetric about the ground centerline, namely, the max SEL contour shifts from the centerline to the advancing side (bottom-half of observer plane).

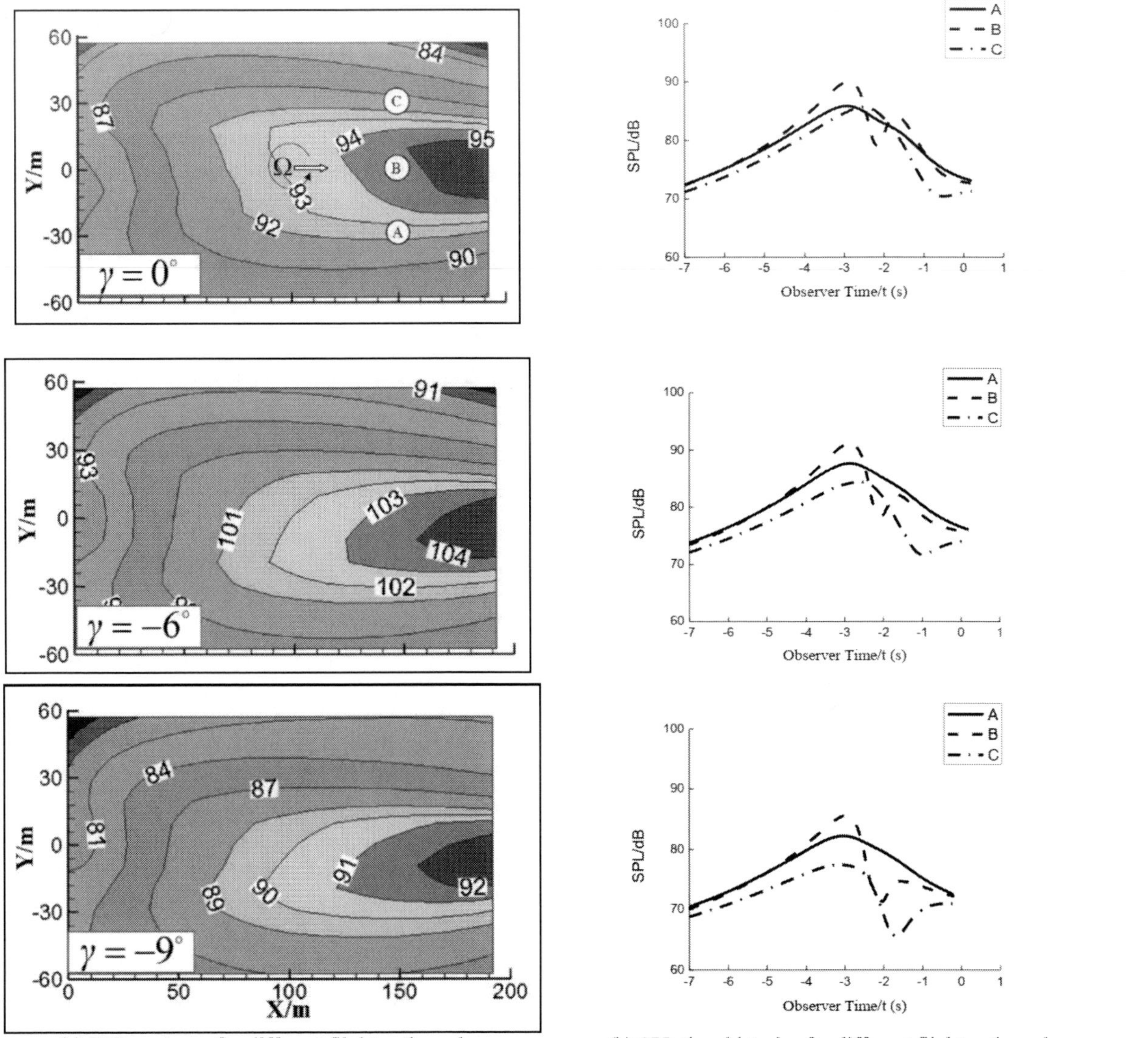

(a) SEL contours for different flight path angles (b) SPL time histories for different flight path angles

Figure 9 Ground noise for different flight path angles

The inflow as a function of the advance ratio for several flight path angles is depicted in Figure 10. The inflow and advance ratio have been non-dimensioned by hover induced velocity of rotor. It can be seen that in level flight the inflow across the tip-path-plane is negative and thus rotor wake is blown downward. So, the miss-distance between blade and vortex is relatively large and the BVI noise is low. However, the inflow ratio (which is negative) is nearly zero with the steeper of flight path angle and the wake is not blown downward as strongly as level flight. At the same time, a steeper flight path tends to tilt the tip-path-plane in the nose-up direction. Both mentioned factors make the miss-distance decrease and thus the maximum BVI noise increase dramatically. Meanwhile, with the tilting back of the tip-path-plane, interactions on the rear side of the rotor disk are strengthened and the max SEL are radiated to the advancing side of the main rotor. But, the inflow across the tip-path-plane turns to outflow as the rotor disk tilts back further (flight path angle of $-9°$) and then the rotor wakes are blew toward the upper side of the tip-path-plane, leading to the increase of miss-distance and the reduction of BVI noise.

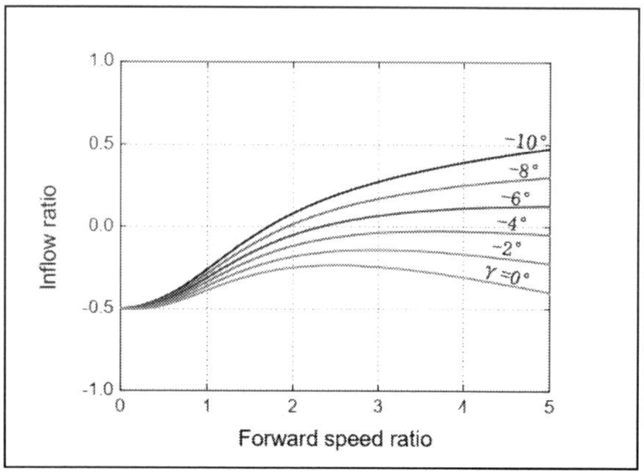

Figure 10 Inflow in different flight conditions

Effects of advance ratio

SEL contours on ground and SPL time histories at three specified observers for the approach flight with a flight path angle of $-6°$ and several advance ratios are presented in Figure 11. It shows the same trend that the value of SEL increases along the centerline as the helicopter flies. The figures also indicate that the peak SPL exhibits a trend of increasing first and then decreasing with the increase of tip-path-plane angles. Meanwhile, the location of maximum SEL contours is changed with the increase of advance ratio from location y=0 to location y=−5 then to location y=−10. It means the maximum SEL shifts from ahead of rotor to the advancing side of the rotor. This is mainly because that the rotor wakes move to the rear of the rotor disk faster as the forward velocity is increased. Accordingly, the interactions at the rear of rotor disk are strengthened and thus, the BVI noise is radiated more toward the advancing side. However, with the further increase of forward velocity, the inflow turns positive gradually and the wake will move above the tip-path-plane, which is below the tip-path-plane initially. Therefore, the miss-distances will decrease first and then increase in an opposite direction, which makes the peak BVI noise exhibit a reverse trend.

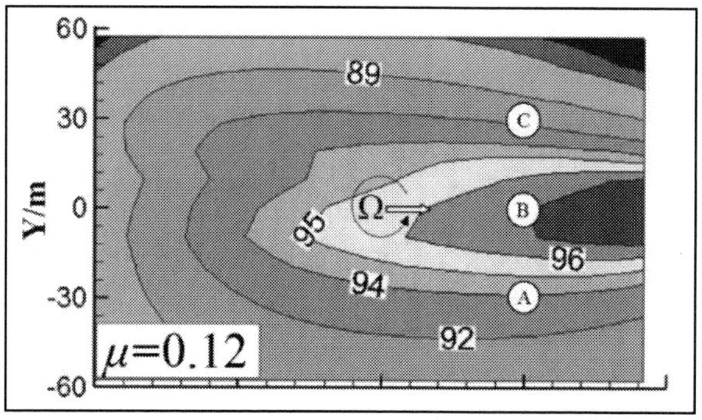

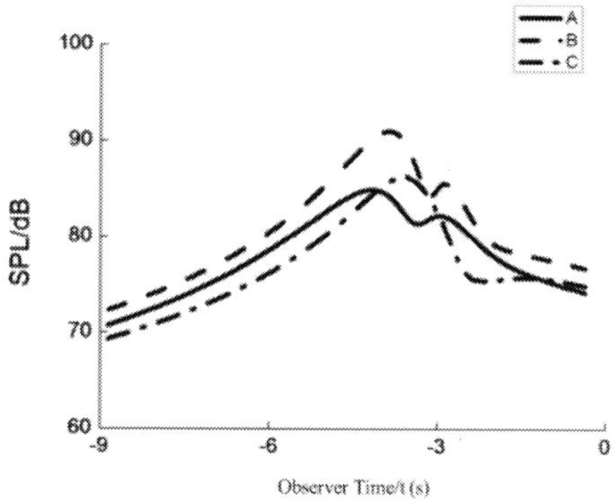

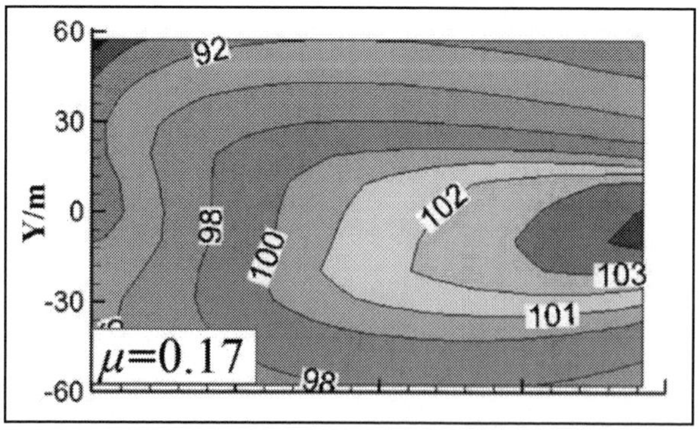

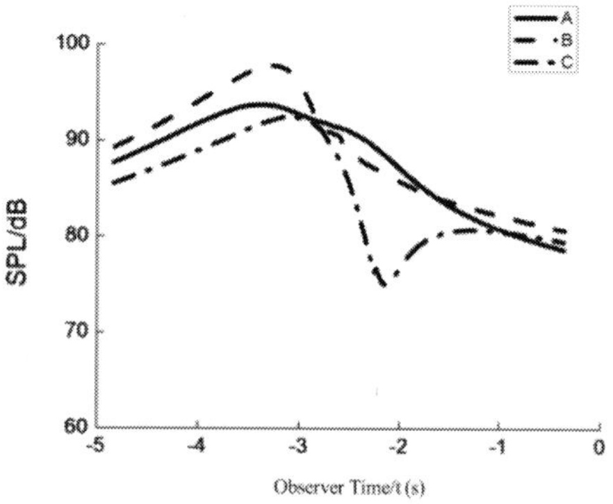

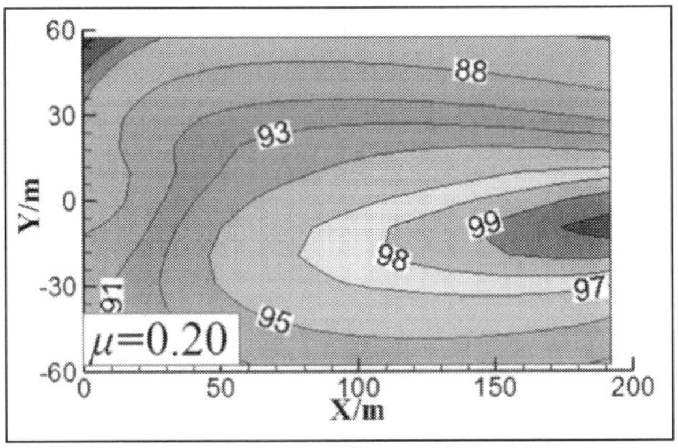

(a) SEL contours for different advance ratios

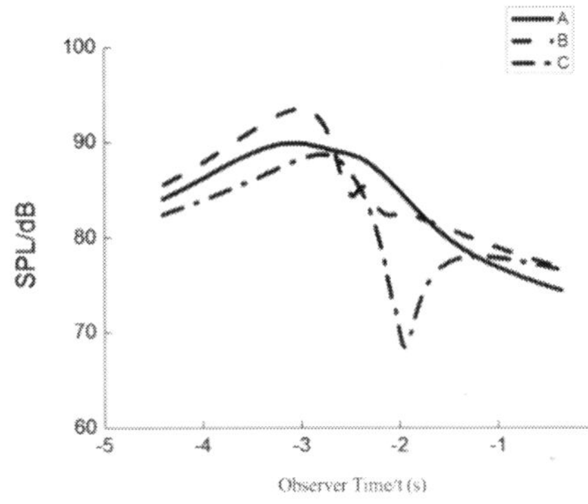

(b) SPL time histories for different advance ratios

Figure 11 Ground noise for different advance ratios

CONCLUSION

A high-efficiency method for the calculation of ground noise radiated from an in-flight helicopter is proposed. And the ground noise characteristics are analyzed for different advancing ratios and flight path angles by the method. Based on the analysis, effects of longitudinal forces (*X*-force and *Z*-force) on miss-distances are studied. The effects of *Z*-forces on BVI noise are calculated. The conclusions are as follows:

(1) A significant enhancement in computational efficiency is obtained by two models. The first is second-level radiation sphere model, by which the noise from only one sound source is radiated. The second is the prebuilt database of radiation spheres. In the simulation of noise footprint, the radiation sphere is selected or interpolated from the database. The computational efficiency is improved further.

(2) The peak SEL contours exhibit a trend of increasing first and then decreasing with the increase of both flight path angle and advancing ratio. Meanwhile, the peak SEL gradually shifts to the advancing side of the rotor disk because of the variation of miss-distance.

(3) The method established in this paper can be used to study the rotor noise radiation characteristics of an in-flight helicopter. In the following study, the effects of terrain conditions on ground noise will be considered.

REFERENCES

[1]Boyd, D. D., "HART-II acoustic Predictions Using A Coupled CFD/CSD Method," *American helicopter society 65th annual forum*; 2009 May 27-29; Texas, United States.2009.p.1-19.

[2]Sugiura, M., Tanabe, Y., Sugawara, H., "Development of A Hybrid Method of CFD And Prescribed Wake Model For Helicopter BVI Noise Prediction," *Transactions of the Japan Society for Aeronautical & Space Sciences* 2012, 27(56):343-50.

[3]Fan, F., Shi, Y.J., Xu, G.H., "Computational Research on Aerodynamic and Aeroacoustic Characteristic of Scissor Tail-Rotor In Hover," *Acta Aeronautica et Astronautica Sinica* 2013, 36(1):135-38.

[4]Edwards, B., Cox, C., Earl R., "Revolutionary Concepts for Helicopter Noise Reduction: SILENT Program," Washington, DC.:NASA;2002. Report No.: NASA-2002-211650.

[5]Sakoswky, P.C., Charles, B.D., "Noise measurement test results of AH-1G operational loads survey," Fort Worth :Bell Helicopter Co; 1976. Report No.: 299-099-831.

[6]Long, L.N., Morris, P.J., Ahuja, V., "Several aerospace applications of computational aero acoustics," *Proceedings of the ASME noise control and acoustics division*, 1998, Anaheim, CA, 25:13-20.

[7]Conner, D.A., Burley, C.L., Smith, C.D., "Flight acoustic testing and data acquisition for the rotor noise model," *Proceedings of the 62nd annual forum of the American helicopter society*; 2006 May 9-11; Phoenix, United States.2006.p.1-17.

[8]Weir, D.S., Jumper, S.J., Burley, C.L., Golub, R. A., "Aircraft noise prediction program theoretical manual: Rotorcraft system noise prediction system," Washington, D.C.:NASA;1995.Report No.: NASA-TM-83199.

[9]Farassat, F., Succi, G.P., "The prediction of helicopter rotor discrete frequency noise," *Vertica* 1983, 7(4): 309-20.

[10]Fleming, G.G., and Rickley, E.j., "HNM heliport noise model, Version 2.2, User's Guide," Dot/FAA/EE/94-01, OT-VNTSC-FAA-94-3, February 1994.

[11]Dowling, A. P., Williams, J.E.F., Wright, W. M., "Sound and sources of sound," *American Journal of Physics* 1985; 53(6):320.

[12]Schmitz, F.H., Stepniewski, W.Z., "Reduction of VTOL operational noise through flight trajectory management," *Journal of Aircraft* 1973; 10(7): 385–94.

[13]Tsuchiya, T., Ikaida, H., Ishii, H., Gomi, H., "Optimal flight for ground noise reduction in helicopter landing approach," *The Japan Society of Mechanical Engineers* 2007; 169 (50):209-17.

[14]Tsuchiya, T., Ishii, H., Uchida, J., Ikaida, H, Gomi, H., Matayoshi, N., "Flight trajectory optimization to minimize ground noise in helicopter landing approach," *Journal of Guidance Control & Dynamics* 2009, 32(2):605-15.

[15]Bagai, A., Leishman, J.G., "Rotor free-wake modeling using a relaxation technique –including comparisons with experimental data," *Journal of the American Helicopter Society* 1995; 40(3):29-41.

[16]Johnson, W., *Helicopter theory*. 1st ed. Princeton: Princeton University Press 1980. p.145-60.

[17]Boxwell, D.A., Schmitz, F.H., Splettstoesser, W.R., Schultz, K.J., H"elicopter model rotor-blade vortex interaction impulsive noise: scability and parameteric variations," *Journal of the American Helicopter Society* 1987; 32(1):3-12.

[18]Leishman, J. G., *Principles of Helicopter Aerodynamics*, Cambridge University Press, New York, NY, 2000, p.580-605.

[19]Elliot, J.W., Althoff, S.L., Sailey, R.H., "Inflow measurements made with a laser velocimeter on a helicopter model in forward flight," Vol. II and Vol. III. Washington, D.C.:NASA;1988. Report No.: NASA TM-100542, TM-100543.

[20]Chung. K.J., Hwang, C.J., Park, Y.M,, Jeon, W.J., Lee, D., "Numerical predictions of rotorcraft unsteady air-loading and BVI noise by using a time-marching free-wake and acoustic analogy," *31th European Rotorcraft Forum* 2005, 1071-5.

[21]Yu, Y.H., Tung, C., Gallman, J., Schultz, K.J., Wall, BVD., Spiegel, P., "Aerodynamics and acoustics of rotor blade-vortex interaction," *Journal of Aircraft* 1995; 32(5):970-7.

Prediction of Rotor Flow for Unmanned Aerial System Using Nonlinear Vortex Lattice Method

Yeongmin Jo
Ph.D. Candidate
Korea Advanced Institute of Science and Technology
Daejeon, South Korea

Hakjin Lee
Ph.D. Candidate
Korea Advanced Institute of Science and Technology
Daejeon, South Korea

Duck-Joo Lee
Professor
Korea Advanced Institute of Science and Technology
Daejeon, South Korea

ABSTRACT

Small scale of unmanned aerial vehicle (UAV) is becoming more important subject for next generation. In order to investigate the aerodynamics of the UAV, very low Reynolds number and slow flight velocity corresponding to its small scale should be considered. There were attempts to include the low Reynolds number effect into the conventional vortex lattice method (VLM) using table look-up, but location for induced velocity for the table look-up was ambiguous. The main purpose of the current study is to solve the ambiguity by introducing an equivalent lattice approach. An appropriate location for the approach is analyzed, and a nonlinear VLM with that approach is developed. Finally, it is validated in two flow cases of a fixed wing and a UAV propeller.

INTRODUCTION

Recently, small scale of unmanned aerial vehicle (UAV) or system (UAS) with rotor system is becoming more important subject for next generation owing to its high mobility and usability. Accordingly, there have been many research attempts to create the innovative systems or improve performances of the current system. In that situation, the aerodynamics of the rotor flow is a very important issue, and experimental measurements or computer simulations are essential. The major difference between the conventional helicopter and the small scale UAV in terms of the aerodynamics is that very low Reynolds number flows under 1×10^5 are generated near the UAV system because of its relatively slow flight speed and its small scale. Thus, the low Reynolds number flow corresponding to the viscous effects should be considered.

Linear potential flow based solvers such as lifting line method (LLM), vortex lattice method (VLM), and panel method have been widely used during past decades because of their computational efficiencies. Furthermore, highly vortical flows such as rotor flow can be accurately simulated by the solvers with appropriate wake models. However, they may inaccurate when the viscous and nonlinear flow effects are not negligible. Thus, the solution accuracy is not guaranteed when the solvers are used to simulate very low Reynolds number flows.

There have been attempts to include the viscous effect into the potential flow based solvers. Owen (Ref. 1) used the Weissinger's LLM with a table look-up technique for 2-D viscous flow around the sectional airfoil (nonlinear LLM). The fundamental is that the sectional circulation strength calculated by the conventional LLM is replaced with the value found by the table look-up. For that, induced velocity at the control (or center) point at each span-wise lattice is computed, and effective angle of attack (AoA) corresponding to the geometric angle and the induced velocity is calculated. Using the effective AoA and sectional flow speed, the lift and drag forces for the 2-D viscous flow are found in the pre-constructed look-up table. This is mathematically simple and widely used, but it cannot consider geometric camber because it uses one lattice along the chord-wise direction. Kim et al. (Ref. 2) suggested the VLM with the table look-up, but location of the control point to calculate the induced velocity is ambiguous because several control points are located along the chord-wise direction, whereas it is clear in the nonlinear LLM.

The main purpose of the current study is to develop a nonlinear VLM without the ambiguity, and validate it in low Reynolds number flows. An equivalent lattice approach is suggested, and appropriateness of the approach is validated using asymptotic analysis.

This paper is organized as follow. Concept of the equivalent lattice approach is introduced, and asymptotic analysis is performed to designate the location of the control point in the following section. Next, a nonlinear VLM corresponding to the equivalent lattice approach is developed, and the developed methodology is validated in two low

Presented at the 6th Asian/Austrailian Rotorcraft Forum / Heli Japan 2017, Kanazawa, Japan, November 7-9, 2017.

Reynolds number flow cases of a fixed wing and a UAV propeller. Finally, this paper is concluded in the final section.

EQUIVALENT LATTICE APPROACH

In this section, the equivalent lattice approach is introduced. Fundamental equations to designate an appropriate location of the control point for this approach are derived. Asymptotic analysis is performed with the equations in cases of two-dimensional airfoil with and without a camber, and it is directly applied to three-dimensional flow cases.

Conceptual Model: Equivalent Lattice Approach

In this equivalent lattice approach, several chord-wise straight-line vortex lattices placed on the mean camber line are considered as an equivalent vortex lattice curved along the mean camber line. The schematic description is shown in Figure 1. Then, a control point of the lattice is placed at the mid chord of the camber line, and the bound circulation strength is set as an average value of the chord-wise bound circulation strengths. Finally, the induced velocity is calculated at the control point as the same as the Weissinger's LLM (Ref. 1). It should be noted that the equivalent lattice approach is a conceptual model, and it does not generate any new lattices. In addition, this approach is only applied when the induced velocity is computed.

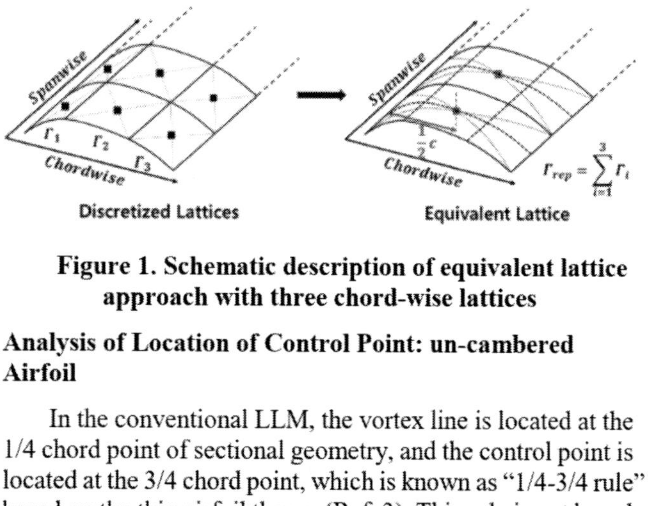

Figure 1. Schematic description of equivalent lattice approach with three chord-wise lattices

Analysis of Location of Control Point: un-cambered Airfoil

In the conventional LLM, the vortex line is located at the 1/4 chord point of sectional geometry, and the control point is located at the 3/4 chord point, which is known as "1/4-3/4 rule" based on the thin airfoil theory (Ref. 3). This rule is not based on theoretical law, but it worked well and it became a rule of thumb for the LLM. It was also used to the VLM that the rule is applied to each lattice discretized along span-/chord-wise directions. In this current study, this original rule is extended for the equivalent lattice approach. At first, the fundamental equations for a two-dimensional un-cambered airfoil are derived, and a location of the control point is designated by an asymptotic analysis. After that, the same procedure is applied to a cambered airfoil geometry.

For the analysis for the un-cambered airfoil case, a flat plate is considered and it is schematically described in Figure 2.

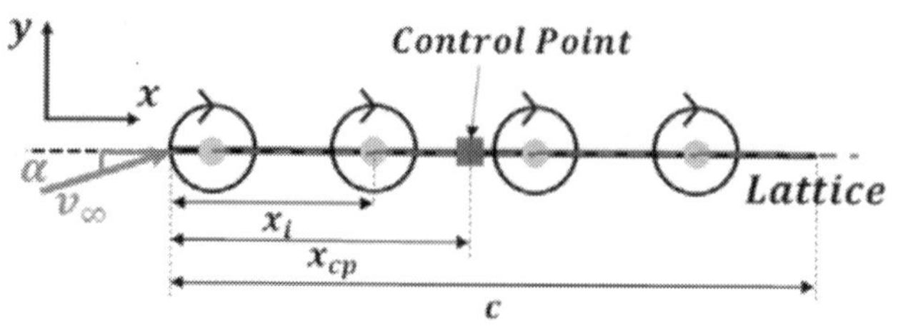

Figure 2. Notation for analysis of location of control point: un-cambered airfoil case

where x_i and x_{cp} are the x-directional (chord-wise) location of the point vortex element for i-th lattice and the representative control point for equivalent lattice, respectively, and c is the chord length. In this analysis, the airfoil chord line is discretized by n number of lattices with same length of $^{c}/_{n}$, thus the point vortex elements are equally spaced. In addition, all points are located on the flat plate and the y-directional components are zero. Then, the downward induced velocity at the control point by the vortex elements can be expressed as follows:

$$v_{cp} = -\frac{1}{2\pi}\sum_{i=1}^{n}\frac{\Gamma_i}{r_i} = -\frac{1}{2\pi}\sum_{i=1}^{n}\frac{\Gamma_i}{x_{cp}-x_i} \tag{1}$$

where Γ_i is the vortex strength for i-th lattice. In addition, upward normal free-stream velocity at the control point can be expressed as follows with the small angle approximation:

$$v_{\infty}^{nor} = v_{\infty}cos\left(\frac{dy_c}{dx}-\alpha\right) \simeq v_{\infty}\alpha \tag{2}$$

where v_{∞} is the free-stream velocity, $\frac{dy_c}{dx}$ is the derivative of mean camber line y_c along x-direction, which is zero for the flat plate, and α is the flow angle of attack (AoA). Then, sum of the downward induced velocity and the upward normal free-stream velocity should be equal for zero normal boundary condition:

$$-\frac{1}{2\pi}\sum_{i=1}^{n}\frac{\Gamma_i}{x_{cp}-x_i} = -v_{\infty}\alpha \tag{3}$$

On the other hand, the Kutta-Joukowsky theorem can be explained together with the thin airfoil theory as follows:

$$L = \rho_{\infty}v_{\infty}\Gamma_{rep} = \frac{1}{2}\rho_{\infty}v_{\infty}^{2}c2\pi\alpha \tag{4}$$

where L is the lift force, ρ_{∞} is the fluid density, Γ_{rep} is the representative vortex strength, which is the averaged value for the equivalent lattice. By equating Eq. (3) and Eq. (4) about α, the following equation is derived, and it can be simplified as Eq. (6).

$$\alpha = \frac{2\rho_\infty v_\infty \Gamma_{rep}}{\rho_\infty v_\infty^2 c 2\pi} = \frac{\Gamma_{rep}}{v_\infty c\pi} = \frac{1}{2\pi v_\infty}\sum_{i=1}^{n}\frac{\Gamma_i}{x_{cp}-x_i} \quad (5)$$

$$\Gamma_{rep} = \frac{1}{2}\sum_{i=1}^{n}\frac{\Gamma_i}{{x_{cp}}/{c} - {x_i}/{c}} \quad (6)$$

As the same as the conventional VLM, each point vortex is located at 1/4 chord point on the lattice, and the unit chord-wise location can be expressed as follows:

$$\frac{x_i}{c} = \frac{i-1}{n} + \frac{1}{4n} = \frac{1}{n}\left(i - \frac{3}{4}\right) \quad (7)$$

Finally, Eq. (6) can be summarized as follows:

$$\Gamma_{rep} = \frac{1}{2}\sum_{i=1}^{n}\frac{\Gamma_i}{{x_{cp}}/{c} - {x_i}/{c}} = \frac{1}{2}\sum_{i=1}^{n}\frac{\Gamma_i}{{x_{cp}}/{c} - {1}/{n}\left(i - {3}/{4}\right)} \quad (8)$$

The above equation is sum of the n terms which can be expressed as n-th polynomial equation. Thus, it has n number of solutions about ${x_{cp}}/{c}$ term, and the solutions are found numerically in the current study, because it is difficult to obtain the analytic solutions. For the vortex strength Γ_i, three types of chord-wise distributions of constant ($\Gamma_i = 1$), linear increase ($\Gamma_i = i$) and linear decrease ($\Gamma_i = n-(i-1)$) were considered, and the solutions were obtained by using a root finding method in Wolfram Alpha (Ref. 4), and the most consistent solution about the number of lattices among the numerical solutions was selected in each case.

Figure 3 shows the solution convergence of each case along the number of lattices, where the solutions are represented as distance between the solution and the first point vortex. In the constant distribution (red line) case, the distance was exactly 0.5 regardless of the number of lattices. The other two cases of the linear increase (green line) and decrease (blue line) showed a little bit lower and upper values of 0.5 respectively, but they converged to that value asymptotically. In consideration of the rule of thumb, it is reasonable to place the control point for the equivalent lattice at the mid point on the chord line.

Analysis of Location of Control Point: cambered Airfoil

As the second, the cambered airfoil case is described. In the current study, a parabolic camber model is considered which is shown in Figure 4. The mean camber line and its derivative are expressed as follows:

$$y_c = 4\delta\left(\frac{x}{c}\right)(c-x) \rightarrow \frac{dy_c}{dx} = 4\delta\left[1 - 2\left(\frac{x}{c}\right)\right] \quad (9)$$

where δ is the maximum camber. In this case, the control point and the point vortex elements are placed on the mean camber line, and their y-directional locations of y_{cp} and y_i can be derived as follow:

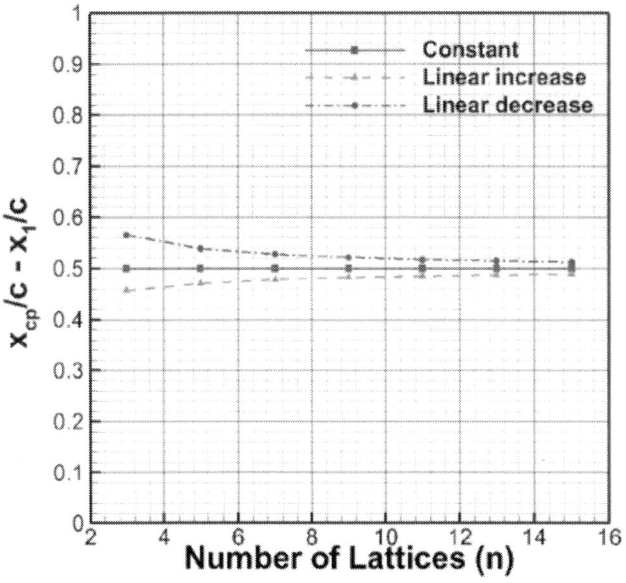

Figure 3. Distance between solution (x_{cp}/c) and first point vortex (x_1/c) along number of lattices: un-cambered airfoil case

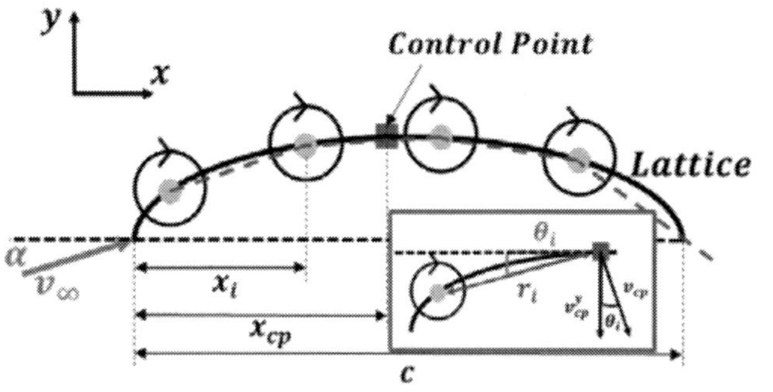

Figure 4. Notation for analysis of location of control point: cambered airfoil case

$$y_{cp} = 4c\delta\left(\frac{x_{cp}}{c}\right)\left(1 - \frac{x_{cp}}{c}\right) \ at\ x = x_{cp} \quad (10)$$

$$y_i = 4c\delta\left(i - \frac{3}{4}\right)\left[1 - \frac{1}{n}\left(i - \frac{3}{4}\right)\right] \ at\ x_i = \frac{c}{n}\left(i - \frac{3}{4}\right) \quad (11)$$

Then, distance r_i between the control point and i-th point vortex and corresponding $cos(\theta_i)$ depicted in Figure 4 can be expressed as Eq. (12) and Eq. (13), respectively. And, the y-directional downward induced velocity at the control point by the point vortex elements and the corresponding velocity normal to the camber line can be expressed as follows:

$$v_{cp}^{y} = -\frac{1}{2\pi}\sum_{i=1}^{n}\frac{\Gamma_i}{r_i}cos(\theta_i) = -\frac{1}{2\pi}\sum_{i=1}^{n}\frac{\Gamma_i}{r_i}\frac{c\left({x_{cp}}/{c} - {x_i}/{c}\right)}{r_i}$$

$$= -\frac{1}{2\pi c}\sum_{i=1}^{n}\frac{\Gamma_i\left({x_{cp}}/{c} - {x_i}/{c}\right)}{A^2} \simeq v_{cp}^{nor} \quad (14)$$

In addition, the upward free-stream velocity normal to the camber line at the control point can be expressed as follows with the small angle approximation:

$$r_i = \sqrt{(x_{cp} - x_i)^2 + (y_{cp} - y_i)^2} = \sqrt{\left[x_{cp} - \frac{c}{n}\left(i - \frac{3}{4}\right)\right]^2 + \left\{4c\delta\left(\frac{x_{cp}}{c}\right)\left(1 - \frac{x_{cp}}{c}\right) - \frac{4c\delta}{n}\left(i - \frac{3}{4}\right)\left[1 - \frac{1}{n}\left(i - \frac{3}{4}\right)\right]\right\}^2}$$

$$= c\sqrt{\left[\frac{x_{cp}}{c} - \frac{1}{n}\left(i - \frac{3}{4}\right)\right]^2 + 16\delta^2\left\{\left(\frac{x_{cp}}{c}\right)\left(1 - \frac{x_{cp}}{c}\right) - \frac{1}{n}\left(i - \frac{3}{4}\right)\left[1 - \frac{1}{n}\left(i - \frac{3}{4}\right)\right]\right\}^2} = c\boldsymbol{A} \quad (12)$$

$$\cos(\theta_i) = \frac{x_{cp} - x_i}{r_i} = \frac{c\left({}^{x_{cp}}/_{c} - {}^{x_i}/_{c}\right)}{r_i} \quad (13)$$

$$v_\infty^{nor} = v_\infty \cos\left(\frac{dy_c}{dx} - \alpha\right) = v_\infty\left\{4\delta\left[1 - 2\left(\frac{x_{cp}}{c}\right)\right] - \alpha\right\} \quad (15)$$

Then, the two velocities should be equal for the zero normal boundary condition as follows:

$$-\frac{1}{2\pi c}\sum_{i=1}^{n}\frac{\Gamma_i\left({}^{x_{cp}}/_{c} - {}^{x_i}/_{c}\right)}{\boldsymbol{A}^2} = v_\infty\left\{4\delta\left[1 - 2\left(\frac{x_{cp}}{c}\right)\right] - \alpha\right\} \quad (16)$$

In addition, the Kutta-Joukowsky theorem and the thin airfoil theory for the cambered airfoil should be equated as follows:

$$L = \rho_\infty v_\infty \Gamma_{rep} = \frac{1}{2}\rho_\infty {v_\infty}^2 c 2\pi(\alpha + 2\delta) \quad (17)$$

Then, Eq. (16) and Eq. (17) can be connected together about the free-stream velocity as follows:

$$v_\infty = \frac{\Gamma_{ref}}{c\pi(\alpha + 2\delta)}$$

$$= -\frac{1}{2\pi c\left\{4\delta\left[1 - 2\left(\frac{x}{c}\right)\right] - \alpha\right\}}\sum_{i=1}^{n}\frac{\Gamma_i\left({}^{x_{cp}}/_{c} - {}^{x_i}/_{c}\right)}{\boldsymbol{A}^2} \quad (18)$$

The independency about α should be satisfied in the inverse term of Eq. (18), and the final derivation of the relation is done as follows:

$$\Gamma_{rep} = \frac{1}{2}\sum_{i=1}^{n}\frac{\Gamma_i\left[\frac{x_{cp}}{c} - \frac{1}{n}\left(i - \frac{3}{4}\right)\right]}{\boldsymbol{A}^2} \quad (19)$$

For the validation, as the same as the un-cambered airfoil case, three types of the vortex strength distributions were considered. the maximum camber was set as 0.2 which is similar to NACA 25XX. The solutions were obtained numerically, and the most consistent solution about the number of lattices among the solutions were selected in each case.

Figure 5 shows the solution convergences along the number of lattices, where the solution values are represented as distance between the value and the first point vortex,

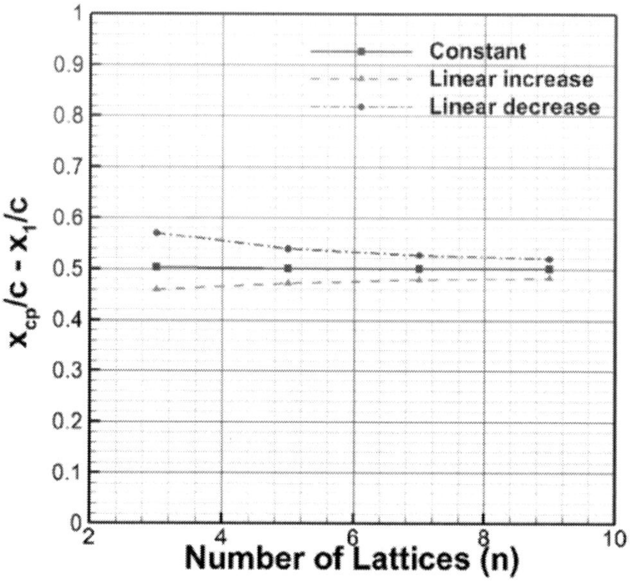

Figure 5. Distance between solution (x_{cp}/c) and first point vortex (x_1/c) along number of lattices: cambered airfoil case ($\delta = 0.2$)

respectively. As the same as the un-cambered airfoil case, all the solutions converged to 0.5 asymptotically. Finally, it can be concluded that the placement of the control point for the equivalent lattice at the mid chord on the mean camber line is reasonable for both cases with and without the camber.

NONLINEAR VORTEX LATTICE METHOD (NONLINEAR VLM)

In this section, fundamental of the nonlinear VLM is described. First of all, a vortex particle method for the wake modeling is formulated. After that, the table look-up and corresponding bound circulation update procedure is described. Finally, the overall computation procedure of the nonlinear VLM is explained.

Vortex Particle Method for Wake Modeling

A vortex filament method has been widely used for wake modeling by coupling it with the linear potential solvers. However, it has limitations in that the connectivity of the filaments should be retained, and empirical values such as vortex decay factor are required. Considering those aspects, a vortex particle method (VPM) (Ref. 5, 6) is used in the current study.

In the VPM, the vorticity field induced by the wake flows is represented by a set of N number of Lagrangian vector-valued particles as follows:

$$\vec{\omega}(\vec{x},t) = \nabla \times \vec{v} = \sum_{i=1}^{N} \vec{\alpha}_i(t)\zeta(\vec{x} - \vec{x}_i(t))$$

$$= \sum_{i=1}^{N} \vec{\omega}_i(t) V_i \zeta(\vec{x} - \vec{x}_i(t)) \quad (20)$$

where $\vec{x}$ is an arbitrary location in the flow field, t is the physical time, $\vec{x}_i$ is the position of i-th particle, and $\vec{\alpha}_i$ is the total vorticity vector of i-th particle in volume V_i. and $\zeta = \zeta\left(\vec{x}/\sigma\right)/\sigma^3$ is the regularized vorticity distribution function within the core of the particle where σ is the core size.

In the current study, the high-order algebraic smoothing function in Ref. 7 is used. Then, the induced velocity by the particle is represented as a curl of the stream function which solves $\nabla^2\psi(\vec{x},t) = -\omega(\vec{x},t)$. The stream function and the velocity can be expressed by the Green's function $G(\vec{x})$ and the regularized Biot-Savart kernel $K_\sigma(\vec{x}) = -g_\sigma(\vec{x})/|\vec{x}|^3$ respectively, as follow:

$$\psi(\vec{x},t) = -G(\vec{x}) \times \vec{\omega}(\vec{x},t) \quad (21)$$

$$\vec{v}(\vec{x},t) = \nabla \times \psi(\vec{x},t) = \sum_{i=1}^{N} K_\sigma(\vec{x} - \vec{x}_i(t)) \times \sigma_i(t) \quad (22)$$

The VPM cooperates with wake lattices convected from the trailing edge of wings or rotor blades. After a few time steps, the wake lattices are split into several vortex particles and each particle is convected by its local flow velocity.

Table Look-up and Bound Circulation Update

The induced velocity $\vec{v}_{ind}$ at the control point of the equivalent lattice is computed by the classical Biot-Savart law and Eq. 22, respectively. Then, the induced AoA can be calculated as follows:

$$\alpha_{ind} = arctan\left(\frac{v_{y,ind}}{v_\infty + v_{x,ind}}\right) \quad (23)$$

where $v_{x,ind}$ and $v_{y,ind}$ are the x- and y-directional components of the induced velocity, respectively. Finally, resultant effective AoA is calculated as following relation:

$$\alpha_{eff} = \alpha_{geo} + \alpha_{ind} \quad (24)$$

where α_{geo} is the geometric AoA such as the twist, incidence, pitch angles. This relation is depicted in Figure 6.

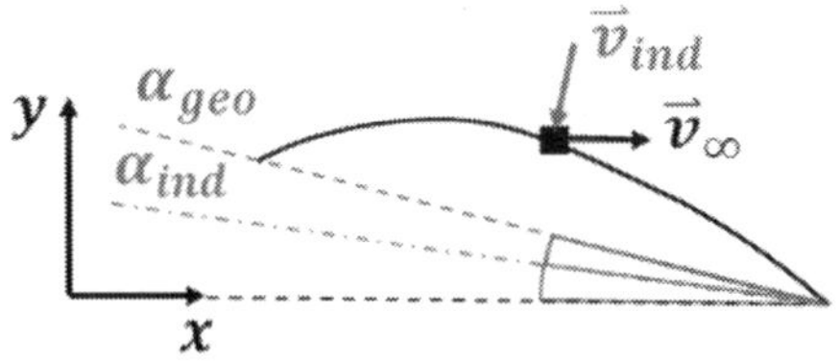

Figure 6. Schematic description for relation of geometric AoA (α_{geo}) and induced AoA (α_{ind})

After calculating the effective AoA and corresponding Reynolds number at each section, the lift and drag coefficients can be easily found by linear interpolation of those values in a pre-constructed look-up table, respectively. Then, the bound circulation strength Γ for each section can be obtained by the Kutta-Joukowsky theorem as follows:

$$\Gamma = \frac{1}{2} v_{eff} c_l c \quad (25)$$

where $v_{eff} = v_\infty + v_{x,ind}$ is the effective velocity, c_l is the lift coefficient found by the table look-up, and c is the sectional chord length. Finally, the representative bound circulation for the equivalent lattice is updated to the calculated circulation strength.

Computation Procedure

The computation procedure of the nonlinear VLM is depicted in Figure 7. After initializing the solver, the influence matrix between the lattices is constructed and the corresponding Laplace's equation is solved by a linear solver, which is the same as the conventional VLM. After that, the linear solution is used as an initial solution of the bound vortex update procedure.

In the update procedure, the induced velocity at the control point on each equivalent lattice is computed, and corresponding effective AoA and the Reynolds number are calculated. Then, the sectional lift and drag coefficients are found by the table look-up. The sectional lift coefficient is converted to the circulation strength, and the bound circulation for the equivalent lattice is updated with under relaxation for numerical stability as follows:

$$\Gamma^{k+1} = \Gamma^k + 0.05(\Gamma^k - \Gamma^{k-1}) \quad (26)$$

Then, the induced velocity is computed again, and this iteration is repeated until the all bound circulation strengths along the span-wise direction are sufficiently converged.

The surface force and moments are calculated based on the finally updated bound circulation, and the wake lattices and corresponding the vortex particles are convected. This overall procedure is terminated when the surface forces are sufficiently converged.

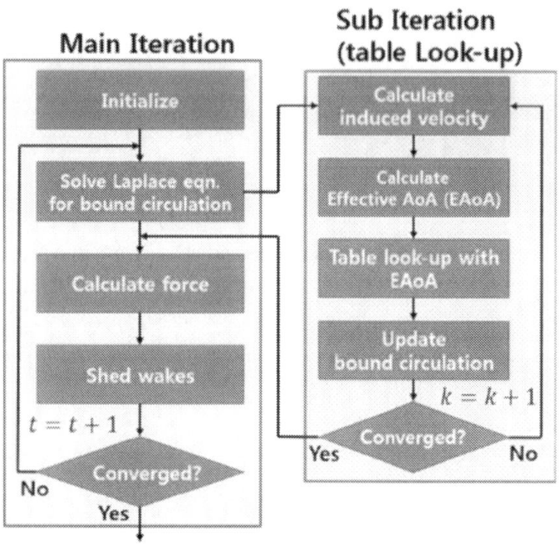

Figure 7. Computation procedure of nonlinear VLM with table look-up

VALIDATION OF NONLINEAR VLM

In this section, the nonlinear VLM described in the previous section is validated in two flow cases: a fixed wing and a propeller for the UAV.

Case1: Wortmann FX63-137 Rectangular Wing

In Ref. 8, aerodynamic characteristics of FX63-137 rectangular wing with various aspect ratios were measured at low Reynolds numbers of $8x10^4$ and $20x10^4$. The FX63-137 airfoil is highly cambered airfoil. In that case, the conventional LLM and the nonlinear LLM may not accurately compute the aerodynamic performances caused by the camber effect. In addition, the flow condition is in very low Reynolds number range, and the conventional VLM may not consider the low Reynolds number effect. Thus, it is thought that this case is good to validate the nonlinear VLM.

In the current study, Reynolds number of $20x10^4$ is considered which corresponds 19.685 m/s of free-stream velocity for chord length of 0.1524 m. Aspect ratio of 4 is considered, and the geometric information, flow condition, and numerical setup are summarized in Table 1 and depicted in Figure 8. The number of chord-wise and span-wise lattices were determined in internal validation test. Each simulation stops when the lift coefficient is sufficiently converged.

In order to make a look-up table for both the nonlinear LLM and VLM, two-dimensional experimental data in Ref. 8 were digitized and used. For this validation, the LLM and VLM without (conventional) and with (nonlinear) the table look-up procedure were used to compute the lift coefficients along with increasing AoA.

Figure 9 shows C_L-AoA curves for the 2-D (black dots) and 3-D experiments (black solid line) in Ref. 8 and the computation results. The conventional LLM (red solid line)

Table 1. Geometric Information, Flow Condition, and Numerical Setup for FX63-137 Rectangular Wing

Section Airfoil	FX63-137
Chord Length	0.1524 m (6 in.)
Aspect Ratio w.r.t. full span	4
Reynolds Number	200,000
Free-stream Velocity	19.685 m/s
Angle of Attack	-2~20 deg.
# of Chord-wise Lattices	21
# of Span-wise Lattices	28
# of Wake Lattice Strip	3
Time Step	0.001 s

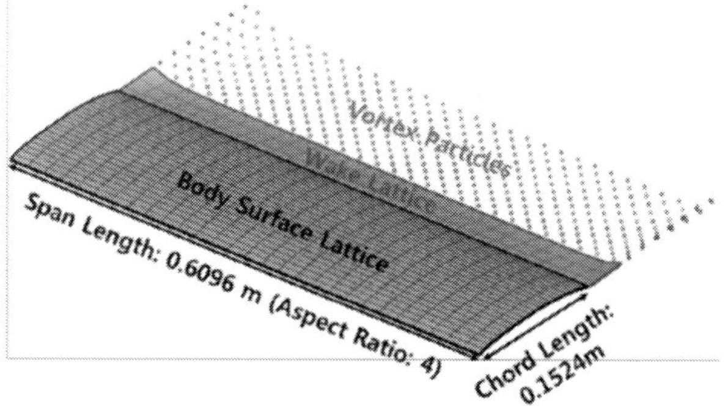

Figure 8. Surface (21x28) and wake (3x28) lattices

under-predicted the lift coefficient at all the AoA because of the absence of the camber effect. That effect can be also shown that the 2-D and 3-D experimental data that the lift coefficients are over 0.6 and 0.4 at the zero AoA, respectively. The large discrepancy between the computation result and the experimental data became small by the table look-up procedure (nonlinear LLM, red dotted line), but the nonlinear LLM rather over-predicted the lift coefficient. On the other hand, the conventional VLM (green solid line) with the camber effect over-predicted it at all the AoA, which is because it did not include the low Reynolds number effect. On the contrary, the nonlinear VLM with the table look-up (green dotted line) showed excellent agreement with the 3-D experimental data. Figure 10 shows the downstream tip vortex flow computed by the nonlinear VLM at AoA of 6°. The helical shape of the vortex described by the vortex particles seems to be well simulated.

Case2: DJI 9443 Phantom2 UAV Propeller

DJI Phantom series is a representative small-scale UAV product of DJI company in typical consumer market. The fourth version of the UAV released recently, but the second version is considered in the current study. The DJI Phantom 2 has four propellers with two blades named DJI 9443 propeller which has diameter of 9.45 in. (0.24 m). The geometric information such as section airfoil shape, twist and chord distributions along radial direction is not open to the public, and they were obtained by slicing the propeller along the radial direction and scanning the cross-sectional geometries. The CAD geometry restored by the geometric information is shown in Figure 11.

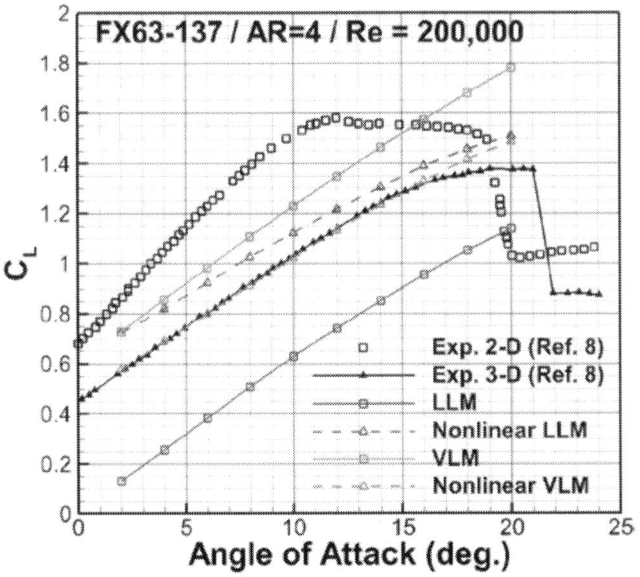

Figure 9. Comparison of C_L-AoA curve

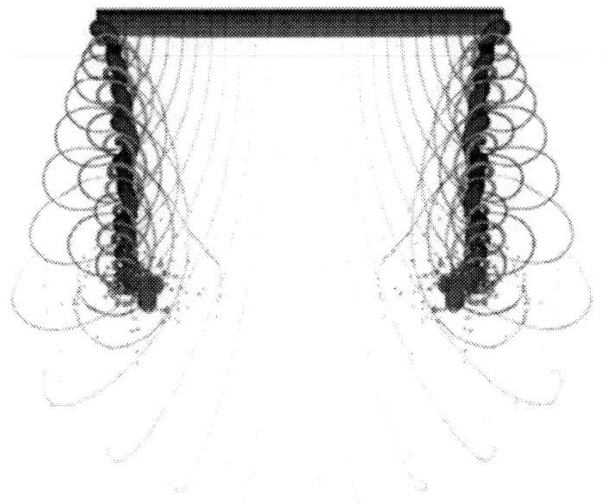

Figure 10. Downstream tip vortex flow (Nonlinear VLM, AoA= 6 deg.)

Rotating speed of the propeller for hovering (static) condition is about 5,500~6,000 revolution per minute (RPM) which corresponds the Reynolds number range from about 5,000 for the root section to about 60,000 for 0.75R section (75% of radius). This range of the very low Reynolds number implies that the nonlinear computation with the table look-up procedure should be used for accurate prediction of the aerodynamic characteristics.

As there is no references for aerodynamic data for the sectional airfoil, CFD simulation for the airfoil was conducted. In the current study, SU2 (Stanford University Unstructured) open source CFD solver (Ref. 9) was used to solve Reynolds-averaged Navier-Stokes (RANS) equations. This solver is based on finite volume method with unstructured grid system. ROE's 2nd order upwind scheme was used for spatial discretization. Considering computation time with various flow conditions, steady simulations were conducted using Euler implicit scheme with lower-upper symmetric-gauss-seidel (LU-SGS) for temporal integration. For the RANS computation, Spalart-Allmaras (SA) one-equation method was used, and a correlation based algebraic transition model

Figure 11. CAD geometry of DJI 9443 propeller

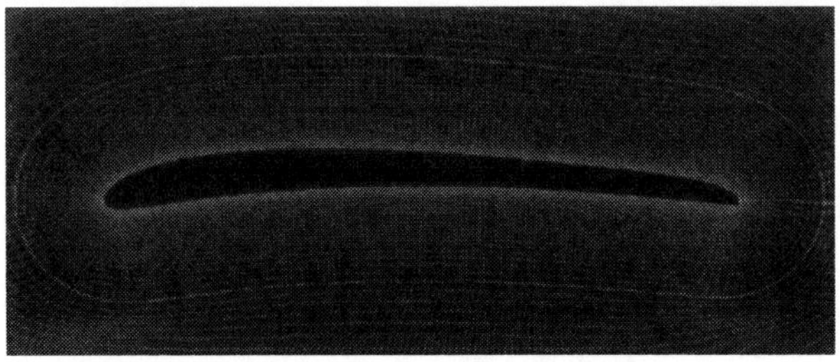

Figure 12. Airfoil shape and boundary layer elements

of Bas-Cakmakcioglu was used to simulate transitional Laminar-turbulent flow.

A computational grid was generated by using Pointwise®. In order to capture the boundary layer flow, 45 layers of rectangular elements were generated near the airfoil which is shown in Figure 12. The first layer thickness was set as $y^+ \leq 1$ for all considered Reynolds numbers. Including the boundary layer elements, a total of 42,000 rectangular elements were generated in a computational domain as far as 150 times the airfoil chord length.

In order to construct a look-up table, a total of 198 flow cases with different Reynolds number and AoA were simulated. Table 2 describes the minimum and maximum values of the Reynolds number, AoA, and Mach number. Most of the flow cases did not converged to a steady solution, but oscillated. Figure 13 shows computation histories at Reynolds number of 40,000 and AoA of 0, 2, and 4 deg.. In this case, even the solution at AoA of 0 deg. (red line) oscillated irregularly, whereas the solution at AoA of 4 deg. (blue line) converged well. This infers that the flow is highly unstable because of the very low Reynolds number. Thus, unsteady simulation may be needed for more accurate simulation. But averaged values during a certain iterations in steady computations were used in the current study. Figure 14 and Figure 15 show plots of the constructed look-up tables for lift and drag coefficients, respectively. In comparison of different lift slopes at low and high Reynolds numbers, it is expected that the low Reynolds number effect will be a dominant factor of the flow.

Table 3 describes the geometric information, flow condition, and numerical setup for the nonlinear VLM. The rotating speeds were set as the same as Ref. 10 to for comparison. The time-step was set as the azimuthal angle of 5 deg. at any rotating speed. In order to prevent unrealistic flow phenomena

Table 2. Minimum and Maximum Values of Flow Condition for Look-up Table

Flow Condition	Minimum	Maximum
Reynolds Number	3333.3	73333.3
Angle of Attack	-6 deg.	10 deg.
Mach Number	0.009	0.198

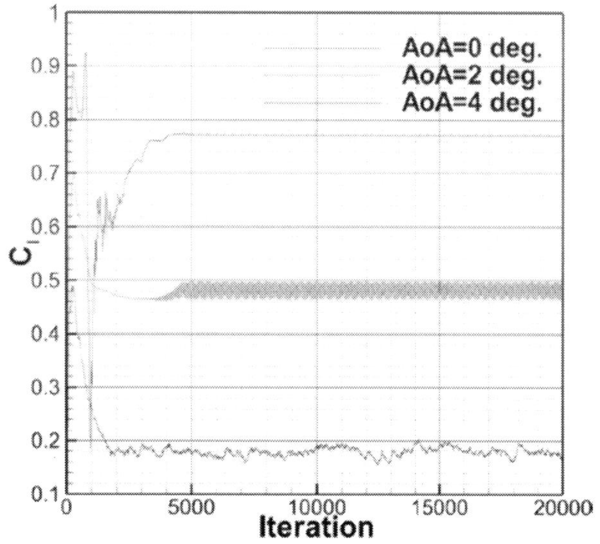

Figure 13. Computation history of 2-D flow at Re= 40,000

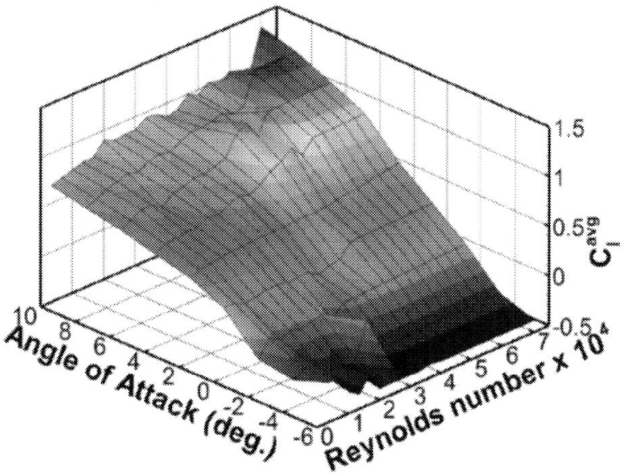

Figure 14. Look-up table for C_l along Reynolds number and AoA

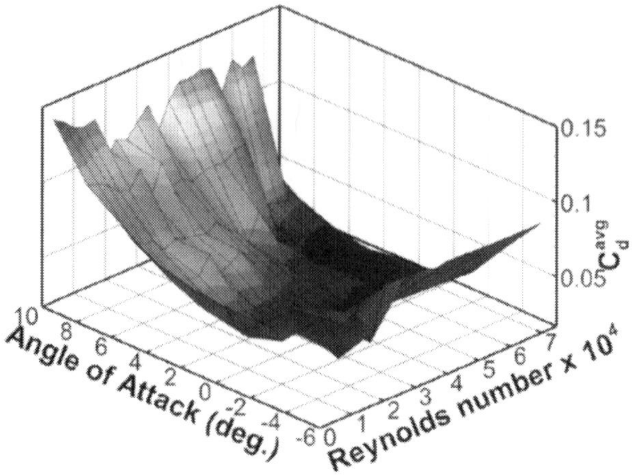

Figure 15. Look-up table for C_d along Reynolds number and AoA

Table 3. Geometric Information, Flow Condition, and Numerical Setup for DJI Phantom2 UAV Propeller

Propeller	DJI 9443 Prop.
Diameter	0.24 m (9.45 in.)
Rotating Speed	4,800, 5,400, 6,000 RPM
Free-stream Velocity	0 m/s (static)
# of Chord-wise Lattices	21
# of Span-wise Lattices	28
# of Wake Lattice Strip	3
Root cut	7% of Radius
Time Step	$\Delta\varphi$ =5 deg.

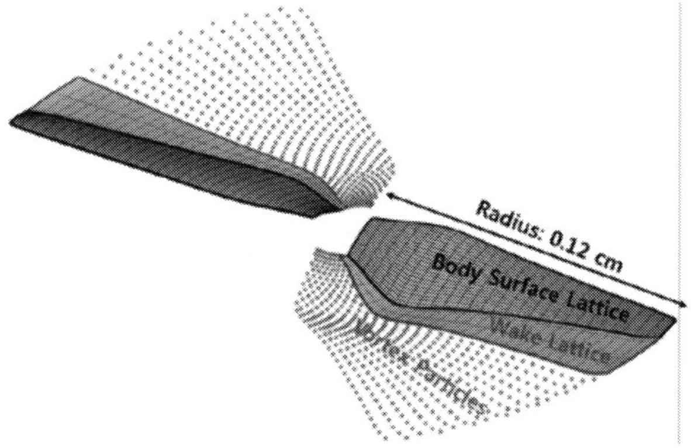

Figure 16. Surface (21x28) and wake (3x28) lattices

by impulsive rotating flow, the slow-starting approach (Ref. 5) was applied during initial first revolution. Figure 16 shows the body surface and wake lattices and corresponding vortex particles.

Figure 17 shows the history of thrust coefficient computed by each method and measured by Structural Acoustic Loads and Transmission (SALT) anechoic chamber facility at NASA Langley Research Center (blue solid line) and computed by detached eddy simulation (DES) of OVERFLOW2 CFD solver (pink triangle) in Ref. 10. In Ref. 10, the CFD solution at the fifth revolution was only available.

As similar with the previous fixed wing case, the conventional LLM (red solid line) under-predicted the thrust compared to the experimental data. In case of the nonlinear LLM (red dotted line), it over-predicted the thrust, and the solution was not converged, but highly fluctuated. This may be non-physical phenomena as other computation results converged well. The conventional VLM (green solid line) over-predicted the thrust, whereas the nonlinear VLM (green dotted line) under-predicted it. Among the computation results in the current study, the nonlinear VLM was most accurate and the error was about 6.06%. This is a little bit high compared to the error of the OVERFLOW2 as 2.99%, but it is acceptable in engineering sense.

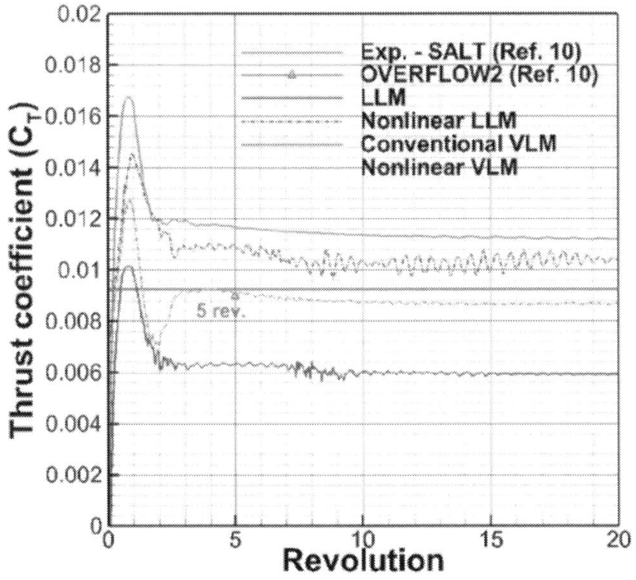

Figure 17. Computation history of thrust coefficient (5,400 RPM)

In terms of the computation time, the DES simulation of OVERFLOW2 in Ref. 10 took approximately 30 hours to simulate one revolution of the rotor using 1,008 CPUs, and converged solution was obtained at the fifth revolution, which corresponds 151,200 CPU hours. In case of the nonlinear VLM, excluding the time to construct the look-up table, it took approximately 24 hours to simulate 20 revolutions using 40 CPUs, which corresponds 960 CPU hours, and this is about 160 times faster than the DES simulation even the nonlinear VLM took the revolution of the rotor 4 times more than the DES simulation. Including the time for the look-up table construction of 32,000 CPU hours, the nonlinear VLM was about 4.6 times faster than the DES simulation. Typically, a design optimization of a planform of wing or rotor does not consider modification of the sectional airfoil shape, and a look-up table is constructed once and is used repeatedly in the design process. Thus, the nonlinear VLM will be very efficient in a design optimization.

Figure 18 shows the downstream vortex flows depicted by the vorticity contour with the vortex particles. It seems that the tip vortex flow was captured well, and the overall flow was simulated reasonably.

Figure 19 shows the thrust coefficients along with the rotating speed. In Figure 19, both the OVERFLOW2 and the nonlinear VLM under-predicted the thrust at all the rotating speed, and the nonlinear VLM had more errors compared to the DES simulation of OVERFLOW2. The errors can be reduced by using unsteady simulations for the construction of the look-up table, and by using some correction methodologies such as stall delay model for rotating flow, which will be considered in future study. But, the overall trend was well predicted.

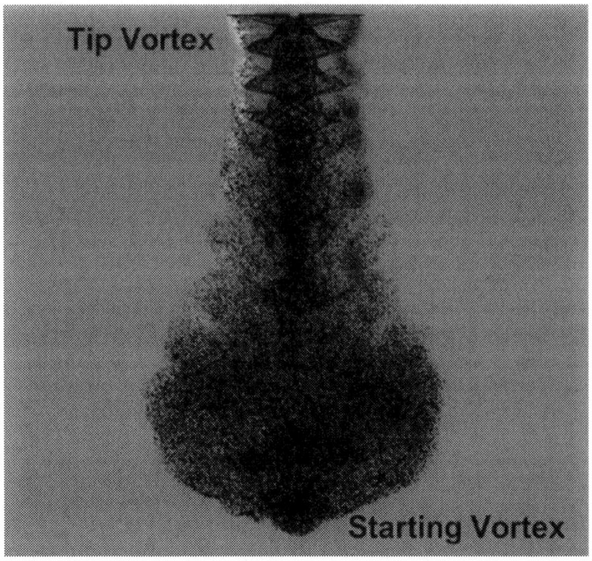

Figure 18. Downstream Vorticity contour (Nonlinear VLM / 20 rev. / 5,400 RPM)

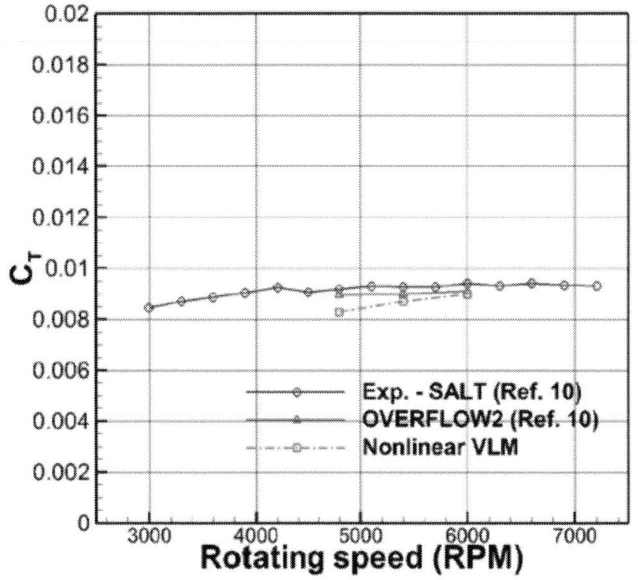

Figure 19. Comparison of thrust coefficient along rotating speed

From the results in the two cases, it is thought that the nonlinear VLM developed in the current study is applicable very low Reynolds number flows with highly cambered geometries.

CONCLUSION AND FUTURE WORK

In the current study, the nonlinear VLM with the table look-up procedure was developed in order to include the camber and low Reynolds number effects into the rotor flow simulations. The nonlinear VLM was validated in the two cases of the fixed wing and the UAV propeller. As the validation results, the nonlinear VLM had excellent agreement with the experimental data in the fixed wing case. In the UAV propeller case, it had acceptable accuracy, and the computation time was considerably short compared to the unsteady CFD solution.

As future works, the UAV propeller case will be re-considered, and the 2-D unsteady flow simulation and the 3-D rotating flow correction will be included to improve the solution accuracy. Furthermore, noise characteristics of the propeller will be predicted and compared with experimental data.

Author contact:
Yeongmin Jo yminjo@kaist.ac.kr
Hakjin Lee hakjin@kaist.ac.kr
Duck-Joo Lee djlee@kaist.ac.kr

ACKNOWLEDGMENTS

This research was supported by Development of Design Technologies for High Efficiency, Low Noise Propellers of Multi-copter UAVs through the National Research Foundation of Korea (NRF), funded by the Ministry of Science, ICT & Future Planning (No. 2016937799). It was also supported by the Korea Institute of Energy Technology Evaluation and Planning(KETEP) and the Ministry of Trade, Industry & Energy(MOTIE) of the Republic of Korea (No. 20168520021200). Finally, it was supported by the KEIT Research Grant of 2017 (No. 10053155).

REFERENCE

[1]Owens, D. B., "Weissinger's Model of The Nonlinear Lifting-Line Method for Aircraft Design," Paper AIAA 98-0597, AIAA Aerospace Sciences Meeting and Exhibit, Reno, NV, Jan. 12-15, 1998.

[2]Kim, H., Lee, S., and Lee, S., "Numerical Analysis on The Aerodynamics of HAWTs using Nonlinear Vortex Strength Correction," *Current Applied Physics*, Vol. 10, 2010, pp. 311-315.

[3] McDaniel, D. R., Cummings, R. M., Morton, S. A., and Mason, W. H., "Applied Computational Aerodynamics," Cambridge Unversity Press, 2015.

[4]"Wolfram Alpha: Computational Knowledge Engine," url: "www.alframalpha.com"

[5]Na, S. U. and Lee, D. J., "Predictions of Airloads and Wake Geometry for Slowly Starting Rotor Blades in Hovering Flight by using Time Marching Free Vortex Blob Method," American Helicopter Society 52nd Annual Forum Proceedings, Washington, D. C., June 4, 1996.

[6]Park, S. H., Chu, Y. B., and Lee, D. J., "Investigation of Main Rotor-Tail Rotor Interaction in Hover Flight using Vortex Particle Method," 5th Asian-Australian Rotorcraft Forum, Singapore, Nov. 17-18, 2016.

[7]Winckelmans, G. S., "Topics in Vortex Methods for the Computation of Three- and Two-Dimensional Incompressible Fluid," Ph.D. Dissertation, California Institute of Technology, 1989.

[8]Bastedo, W. G. and Mueller T. J., "Spanwise Variation of Laminar Separation Bubbles on Wings at Low Reynolds Numbers," Journal of Aircraft, Vol. 23, (9), 1986, pp. 687-694.

[9]Economon, T. D., Palacios, F., Copeland, S. R., Lukaczyk, T. W., and Alonso, J. J., "SU2: An Open-Source Suite for Multiphysics Simulation and Design," AIAA Journal, Vol. 54, (3), 2016, pp. 828-846.

[10]Zawodny, N. S., Boyd, D. D., and Burley, L. C., "Acoustic Characterization and Prediction of Representative, Small-Scale Rotary-Wing Unmanned Aircraft System Components," American Helicopter Society 72nd Annual Forum Proceedings, West Palm Beach, FL, May 17-19, 2016.

Design and Analysis Tools for New Rotorcraft Concepts and Emerging Markets

Christopher Silva **Jeffrey D. Sinsay**
Aeronautic Concepts, LLC
Livermore, CA, USA

Wayne Johnson
Johnson Aeronautics
Palo Alto, CA, USA

ABSTRACT

New and expanded markets for vertical takeoff and landing aircraft are emerging around the world, providing numerous opportunities for businesses to thrive by using advanced unmanned and manned aircraft. Opportunities exist to provide dramatically better performance and economics than those aircraft which have made early entries into these markets. Example cases are documented for two of these potential markets: unmanned package delivery and autonomous air taxi. A sensitivity analysis indicates that improving speed and vehicle reliability are keys to making a viable economic case for aircraft in these new markets. Application of aeromechanics tools originally developed and validated for conventional rotorcraft to these new types of vehicles should result in improved productivity and reduced operating costs. In some cases, it is indeed necessary to use tools such as comprehensive analysis, structural analysis, and computational fluid dynamics to capture the relevant phenomena and therefore design the best vehicle for the missions.

INTRODUCTION

New market opportunities are being envisioned for vertical takeoff aircraft which may be manned or unmanned, or may operate as either. In order for an entrepreneur or investor to assess whether or not there is a feasible business case employing VTOL aircraft, the purchase and operating economics of the flight vehicle to be used must be quantified. Harris and Scully (Ref. 1) have previously identified rotorcraft costs as problematic, and others (Ref. 2) continue to identify the purchase and operating economics of traditional rotorcraft as not compatible with business cases for these markets. The two potential markets considered for the present study are (1) Unmanned Package Delivery (UPD) (Ref. 3); and (2) on-demand Autonomous Air Taxi (Ref. 4).

In the present study, we consider the drivers for operating economics, and whether there are opportunities to apply aeromechanic design and analysis tools to reduce those costs. In order to perform an economic analysis, factors affecting both the costs and productivity need to be considered. For the purposes of this study, absolute values of the cost components are unknown because there are no current examples of vehicles operating in the emerging markets of unmanned pckage delivery or autonomous air taxi. Two of the main drivers of traditional operating cost are fuel (energy) consumption and maintenance due to vibratory loads (Ref. 5). These factors are important for both traditional and new types of aircraft which may be different in both their physical configuration and their manner of operation.

For semi-autonomous or remotely piloted aircraft, there are additional costs associated with maintaining communications between the control (or safety) pilot and the air vehicle. These additional costs are directly related to utilization tend to put a premium on vehicle speed in order to minimize time in flight.

METHODOLOGY

In order to estimate the economics of operating a fleet of aircraft, simple models are employed for operating and support costs and for operating revenue. The cost models of Harris and Scully (Ref. 1) are used to capture basic flyaway and operating costs for Autonomous Air Taxi vehicles. The present study is focused on the aspects of cost and revenue which may be tightly linked to aeromechanic features and design approaches. For instance, no demand curve for price vs. supply has been assumed, and profits are not considered. Readers may find an investigation of AAT demand sensitivity with varying supply and cost presented in Narkus-Kramer (Ref. 6). If desired, the methods of the present study can be used to augment the cost and productivity factors assumed in other economic models in order to establish market feasibility. Magnitudes of the cost components and their relative importance may be skewed for these new types of vehicles as compared to the conventional aircraft which were used to generate the empirical relations.

6th Annual Asian-Australia Rotorcraft Forum
HeliJapan 2017, Kanazawa, Japan
7-9 November 2017

Cost Formulation

The elements of operating cost are as identified by Harris (Ref. 7) and implemented in NDARC (Ref. 8), supplemented by specialized and new terms for the following components (Table 1).

Table 1. Modified and new cost parameters

Parameter	Definition	Units
C_{batt}	Cost of battery replacement	\$/FH
C_{comm}	Cost of communications	\$/FH
C_{crew}	Cost of crew (split among aircraft)	\$/FH
C_{ins}	Cost of insurance	\$/FH

Aeromechanics Tools

CAMRAD II (Ref. 9) has been used to perform the performance and loads analysis on the rotor systems of the quadrotor vehicles in the present study. Recent results (Ref. 10) have validated the ability of CAMRAD II to provide accurate results for this class of vehicle, and at small sizes in addition to the large vehicles for which the software has been validated in the past. NDARC (Ref. 8) has been used to size the aircraft and perform basic calculations of acquistion and operating cost.

UNMANNED PACKAGE DELIVERY (UPD)

One emerging rotorcraft market, which has attracted significant interest, is the potential use of VTOL UAVs to provide last-mile delivery of small parcels. The market for unmanned aerial package delivery has been selected for the present study, with the particular missions as defined in the 2015 AHS International Student Design Competition (Ref. 3) and overviewed in Figure 1. For the present study, it has been assumed that the maintenance burden of UPD vehicles is about an order of magnitude lower in terms of maintenance man hours per flight hour than is the case for manned aviation today.

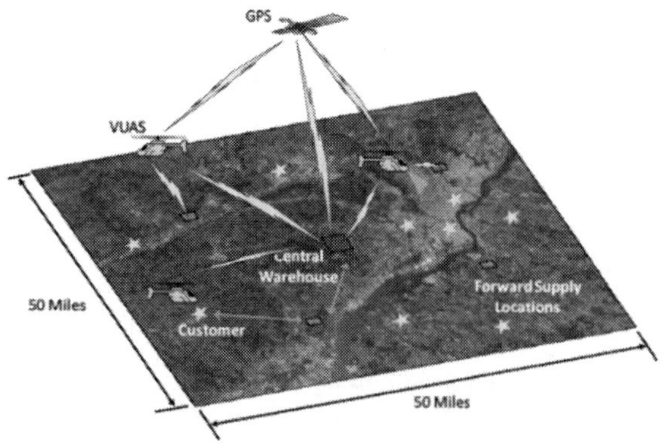

Fig. 1. Overview of Parcel Delivery UAV System (Ref. 3)

AUTONOMOUS AIR TAXI (AAT)

An more ambitious new market is for autonomous air taxis, which would be available for on-demand transport between a pickup vertiport and a landing vertiport in response to a user's request. For this market to be viable, the additional costs relative to other modes of transportation are offset by the ability to rapidly transit, especially

BASELINE MULTIROTOR DESIGN

An electric quadrotor type of aircraft has been designed for both the UPD and AAT missions of the general arrangement shown in Figure 2. Excursions examining fixed pitch versus collective pitch, rpm versus collective pitch control, and hingeless versus flapping rotors were performed. The baseline mission is designed to carry 250 lb of payload 20 mi with 20 minutes of reserve flight at V_{be}. Rotor diameter for all three designs was fixed at 12 ft, resulting in disk loadings consistent with existing light helicopters. Summary results for three designs are showing in Table 2. Initial results show that for quad rotor vehicles of this size, collective pitch and hub load relief from flapping are desirable. Twist of the rotor system has been varied to reduce overall vehicle size.

Fig. 2. A possible multicopter designed for package delivery

SENSITIVITIES AND REMEDIES

A very high portion of the operating cost for new vehicle types is that of insurance. A survey of publicly-available information from various drone insurance providers has found that, due to a dearth of data on the safety of the vehicles and therefore the loss rates, premiums are very high (placing an extraordinarily high value on the speed of performing a mission). For a near-term application of UPD, the insurance cost will likely be based on an abundance of caution on the part of the underwriters. Two mitigation approaches are: (1) to reduce the exposure to this per-flight-hour cost by reducing the flight hours required to perform a delivery; and (2) reduce the insurance premiums by convincing the insurer that the liability exposure is less than they presently fear. The present study considered a simple evaluation of costs based on varying block speed, and number of pilots per operation, with results shown in Fig. 3. With the assumptions of the present study it is clear the point of diminishing returns for autonomy has been reached at about 20 aircraft/pilot, and a look at the components of direct operating cost per operation for a particular set of assumptions (Fig. 4) reveals that the insurance cost per flight hour dominates all others.

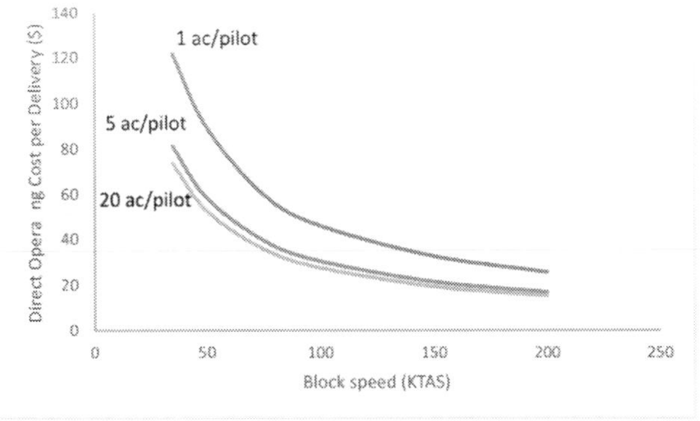

Fig. 3. Variation in DOC with speed and autonomy

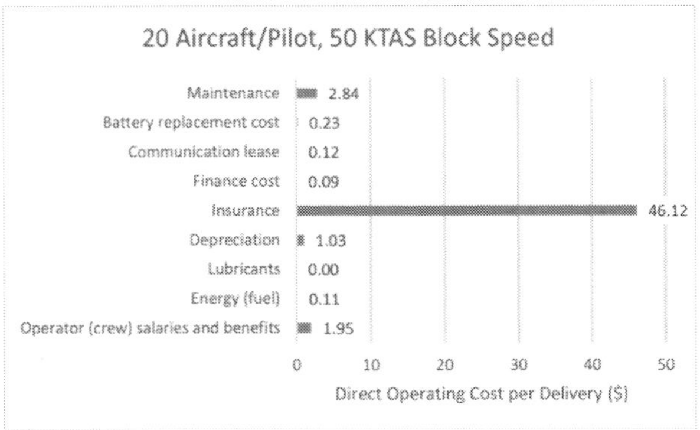

Fig. 4. Insurance cost dominates DOC

The speed sensitivity indicates that aircraft which have a higher inherent speed will be vastly superior, even if they cost a little more to acquire and maintain. Apart from different types of vehicles, a quadrotor aircraft with an improved rotor system can achieve better speed with enhancements to the rotor system.

For the AAT vehicle, the present study considered vehicles with three different types of rotor: (1) fixed-pitch hingless propeller; (2) collective- and cyclic-pitch hingless rotor; (3) collective- and cyclic-pitch flapping rotor. For the larger vehicles of the AAT market, loads are significant, and maintenance costs, which should be more in line with the maintenance costs of existing aircraft, will be a more significant part of the total. For the present study, maintenance man-hours per flight hour were estimated at 0.2 MMH/FH for the flapping rotor aircraft, which is in line with general aviation aircraft, and reducing engine-related maintenance by 1/2. For the hingeless rotor aircraft, maintenance is estimated to be Maintenance costs in rotating machinery are often driven by cyclic loads, and component lives may be directly related to cyclic loads (Ref. 11). CAMRAD II analysis of a selection of rotor types for the AAT vehicle revealed significantly different loads for different controls on the rotor system. A comparison of root 1/2 peak-to-peak flap loads is presented in Fig. 5, and it is readily seen that introducing a flapping hinge (in this case with 45 degrees of delta-3) reduces the loading by a factor of 2 or more.

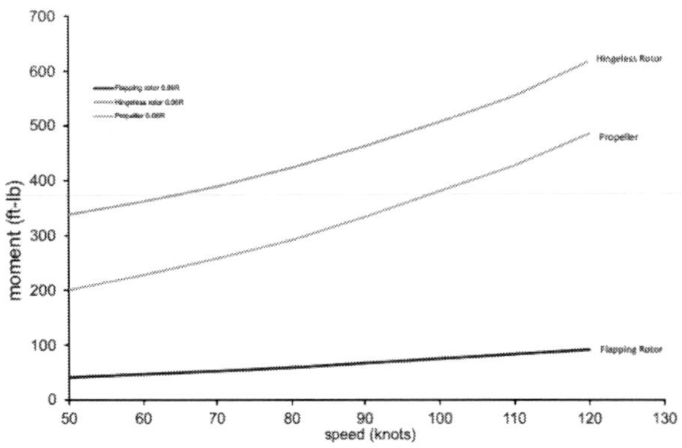

Fig. 5. A comparison of peak-to-peak root flap loads

CONCLUSIONS

Two significant impediments to commercial viability in the Unmanned Package Delivery and Autonomous Air Taxi markets have been identified. In the UPD market, insurance rates drive faster aircraft. In the AAT market, maintaining low rotor loads is important for maintaining manageably low maintenance costs. Legacy rotorcraft system design and comprehensive analysis tools can help in the design of these vehicles. Further design and optimization work is required to identify if these vehicle might approach the level of affordability desired by the emerging AAT and UPD markets.

REFERENCES

[1]Harris, F. D. and Scully, M. P., "Rotorcraft Cost Too Much." Journal of the American Helicopter Society, Vol. 43, No. 1, 1998.

[2]Dudley, M. R. E., "Second Annual Transformative Vertical Flight Concepts Workshop: Enabling New Flight Concepts Through Novel Propulsion and Energy Architectures," NASA/CP-2016-219141, 2016.

[3]AHS International and Boeing, "Request for Proposal for Distributed Logistics in an Urban Setting Using Small Unmanned Aerial Vehicles," 32nd Annual AHS International Student Design Competition, 2015.

[4]Moore, M. D., "AviationFrontiers: On-DemandAircraft," 10th AIAA Aviation Technology, Integration, and Operations (ATIO) Conference, 2010.

[5]Devin, R. M., "Influencing Direct Operating Costs: When and How," 39th Annual American Helicopter Society Forum, 1983.

[6]Narkus-Kramer, M. *et al.*, "Net Present Value, Tradespace, and Feasibility of On-Demand Aircraft," 16th AIAA Aviation Technology, Integration, and Operations (ATIO) Conference, 2016.

[7]Harris, F. D., "An Economic Model of U.S. Airline Operating Expenses," NASA CR 2005-213476, 2005.

[8]Johnson, W. R., "NDARC NASA Design and Analysis of Rotorcraft," NASA/TP-2015-218751, 2015.

[9]Johnson, W. R., "CAMRAD II References," retrieved from http://johnson-aeronautics.com/documents/CAMRADreferences.pdf, 2017.

[10]Russell, C. and Sekula, M., "Comprehensive Analysis Modeling of Small-Scale UAS Rotors," AHS International 73rd Annual Forum, May 2017.

[11]Sun, F.-B. and Kececioglu, D., "Fatigue reliability evaluation and preventive maintenance policy for a structural element under random vibration," AIAA/ASME/ASCE/AHS/ASC Structures, Structural Dynamics and Materials Conference and Exhibit, AIAA-1996-1563, 1996.

Table 2. Comparision of three quad rotor designs. Showing impact of variable vs. fixed rpm operation and rigid vs. flapping hinge configuration.

		Design 1	Design 2	Design 3
RPM Type		Fixed	Variable	Variable
Collective		Yes	No	No
Hub Type		Flapping	Flapping	Rigid
Disk Loading	psf	3.1	3.2	4.1
Solidity (TW)		0.0798	0.0556	0.0712
Nominal Tip Speed	fps	450	550	550
Operating Tip Speed	fps	450	850-550	850-550
Maximum Speed	kt	67	65	65
Design Gross Weight	lb	1,396	1,454	1,861
Weight Empty	lb	1,141	1,199	1,606
Rotor Group Weight	lb	136.9	122.1	150.5
Propulsion Group Weight	lb	494.7	586.4	887.6
Motor Power	hp	4x25	4x38	4x60
Flight Energy	MJ	65	82	123
Reserve Energy	MJ	54	66	104
Battery Size	Ah	7,880	9,770	15,000
Flyaway Cost	$M	1.696	1.892	2.409
Maintenance Hours per Flight Hour		0.61	0.63	0.74
Mission DOC	$	63.70	71.06	89.39

Development of Integrated Optimization Design Framework for a Small-scale Multirotor UAV

Hyeongseok Kim M.S. Candidate

Daejin Lim M.S. Candidate

Kwanjung Yee Professor

Department of Mechanical and Aerospace Engineering, Seoul National University

Seoul, Republic of Korea

ABSTRACT

As the applicable field of multirotor UAVs becomes wider, it is necessary that applicability should be reflected in the design process to satisfy the specific mission requirements. To establish a systematic design framework for the UAVs, conceptual layout optimization of the universal drone system was developed with multidisciplinary analysis modules which consist of attitude, aerodynamics, electric system, and structure analysis. Analyzing steady level flight, blade element momentum theory, Kirchhoff's rule, and 1-D Euler beam theory were used to calculate the required electrical power to perform the mission. The framework could calculate variable motor efficiency using a motor speed constant, which considers each required electrical power in various mission segments. This paper presents the design framework to obtain required electrical power from a mission segment, and the quadrotor design were carried out to evaluate the applicability to perform the mission.

NOTATION

B	tip loss factor
C_d	sectional drag coefficient
C_l	sectional lift coefficient
C_P	rotor power coefficient
C_T	rotor thrust coefficient
dC_{P_o}	sectional rotor profile power coefficient
dC_{P_i}	sectional rotor induced power coefficient
dC_T	sectional rotor thrust coefficient
GW	gross weight
I_0	motor no load current
I_M	motor drive current
I_{zz}	moment of inertia about z-axis
Kv	motor speed constant
P_{Co}	copper loss
P_E	required electrical power
P_{Ir}	iron loss
P_{Co}	motor no load current
P_{Mech}	required mechanical power
P_{ML}	mechanical loss
P_S	stray loss
r	non-dimensional radial distance
R_B	battery inner resistance
R_E	ESC inner resistance
R_M	motor internal resistance
W_M	motor weight
X	root cut out ratio
x	beam sectional distance
κ	rotor induced power factor
λ	rotor inflow ratio through disk
λ_C	climb inflow ratio
σ	rotor solidity
$\sigma(x,y)$	bending stress
σ_{abw}	allowable bending stress
σ_{max}	maximum bending stress
σ_{ut}	ultimate bending stress

INTRODUCTION

Multirotor unmanned aerial vehicle (UAV) has three advantages: Manufacturability, maneuverability, and applicability. Manufacturability is that the UAV could be made by the simple combination of several commercial products such as propellers, brushless DC (BLDC) motors, electric speed controllers (ESC), a flight control computer, batteries, and frame. Maneuverability gets better by advanced control techniques which has been adapted for the drone. Numerous research has been conducted for the better maneuverability of the multirotor UAV (Refs. 1-3). Current trend shows an increase in drone applications such as photography, agriculture, leisure, sports, and delivery. However, the applicability is still immature in terms of design despite its mature maneuverability and manufacturability It is

---

Presented at the 6th Asian/Austrailian Rotorcraft Forum / Heli Japan 2017, Kanazawa, Japan, November 7-9, 2017.

necessary that the applicability be reflected in conceptual and preliminary design level. In order to satisfy the specific mission requirements, the mission-intensive design is essential.

For the performance analysis of the multirotor UAV, Justin, W. (Ref. 4) showed individual components weights of the multirotor UAV were estimated using commercial products and required mechanical power was analyzed using blade element momentum theory (BEMT). But required electrical power and flight time could not be considered by only mechanical power analysis. Bershadsky, D. (Ref. 5) developed overall sizing and performance analysis tool of the multirotor UAV, Electric Multirotor Sizing Tool (EMST). However, there was a limitation in estimating the flight time using the constant value of motor efficiency without varying the motor drive current. It was inappropriate to treat several mission segments needed a different motor drive current. Open-source web-based multirotor UAV analysis tool, eCalc, using performance of commercial products (Ref. 6). This tool referred to the performance of the commercially available products such as battery, ESC, motor, and propeller. It analyzed and calculated the overall performance of the multirotor UAV such as hovering RPM, required mechanical and electrical power, motor efficiency and flight time. As these analysis methods and tools were unable to account for the design that components of the multirotor UAV were undetermined, they were also inadequate for the mission-intensive design process.

To establish a systematic and mission-intensive design process of the multirotor UAV starting with mission requirements, conceptual layout optimization of the universal drone system (CLOUDS) was developed in this study. CLOUDS was composed of one estimation module and four analysis modules to calculate the overall performance of the multirotor UAV that satisfies the mission requirements.

The analysis modules were validated with experimental data or commercial product data. The design result was compared with the analysis result of the eCalc. Not only were optimized design solutions calculated for the maximum hovering time and endurance, but also a proper design solution for a specific mission profile was calculated by the CLOUDS. Then, the numerical design solution was substituted for commercial products with similar specifications, and the commercial products were analyzed as inputs in CLOUDS to check the performance gap between the numerical solutions and the commercial products.

METHODOLOGY

The multirotor UAV multidisciplinary design optimization framework, CLOUDS, consisted of a weight estimation module and four analysis modules: 1) attitude analysis, 2) aerodynamic analysis, 3) electric system analysis, and 4) structure analysis. The overall design process was divided into three hierarchical design levels: overall design process, mission segment analysis process, and analysis process in each analysis modules.

Design Level 1: Overall process

A mission profile, one of the inputs of CLOUDS, was broken down to several mission segments. Through the analysis modules, total required capacity for performing the mission profile was calculated. One of the design convergence criteria was comparing total required capacity with the battery capacity of design variables. The other was comparing maximum bending stress with allowable bending stress of the rotor-frame support bar (Figure 1).

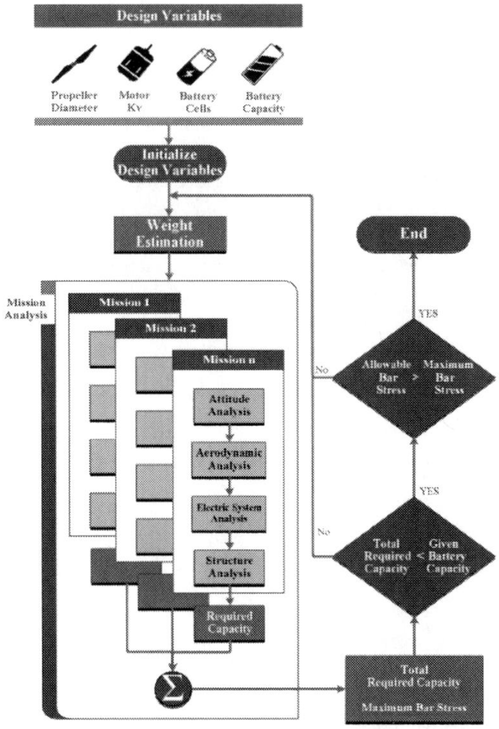

Fig. 1. Overall design process.

Optimization to achieve maximizing hovering time and endurance or minimizing the gross weight with representative design variables such as rotor radius, motor speed constant (Kv), battery voltage (3.7V per a cell), and battery capacity were carried out. Sequential least square programming (SLSQP) was used for optimization (Ref. 7).

Most component weights were estimated by curve fitting result of Ref. 5 except for the motor weight. The motor weight was estimated from 1179 commercial products in Ref. 6. Although the motors had various manufacturing features and different dimension at the same Kv, weight had a trend whose R-square was 0.69 in 14-pole motors (Figure 2). Since many commercial motors were 14-pole and the trend was acceptable in conceptual and preliminary design. Thus, 14 poles motor was used in the whole design process of this study. The curve fitting result of the motor was:

$$W_M = 323{,}392\ \mathrm{Kv}^{-1.192},\ \mathrm{g} \quad (1)$$

Design Level 2: Mission segment analysis

In each mission segment, required electric capacity was calculated through multidisciplinary analysis which were aerodynamics, electric system, and structure (Figure 3). In contrast with fuel-consuming propulsion such as helicopters, electric energy of the battery was consumed for the multirotor UAV to perform the mission without changing the battery weight. Therefore, once design variables were determined, gross weight was constant due to the electric motor propulsion system.

Attitude analysis module received the gross weight as an input and calculated required thrust at each rotor. This was used as an input for the aerodynamic analysis module, then rotor RPM and required mechanical power were calculated accordingly. Simultaneously, required thrust was the input for the structure analysis module and maximum bar stress was calculated. After that, motor efficiency and required electric capacity were analyzed by electric system analysis module with inputs of the mechanical power and RPM. This level 2 process was proceeded in every mission segments summarizing the electric capacities, which was the total electric capacity.

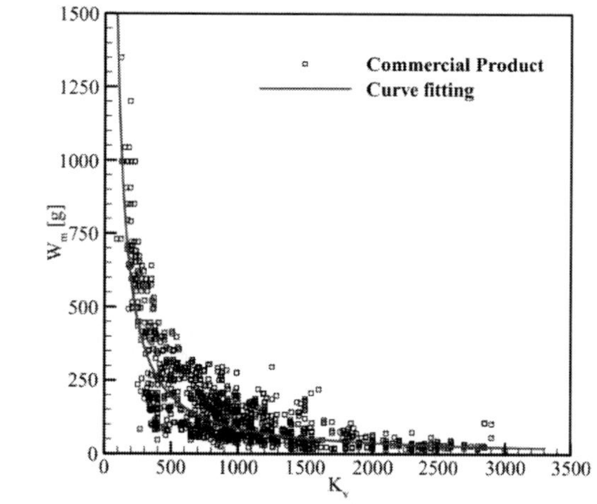

Fig. 2. Motor weight estimation.

Design Level 3: Analysis Modules

Attitude analysis follows steady level flight assumption that all rotors generated equivalent thrust and torque. Also, frame drag coefficient was modeled referring to EMST (Ref. 5). Forward flight was assumed to vertical flight whose vertical speed was geometrically calculated. This assumption might be acceptable in conceptual and preliminary design level to the multirotor UAV which generated higher inflow ratio and larger tilt angle than conventional helicopters. Properly required thrust, vertical speed, and tilt angle were identified for each mission type (Table 1).

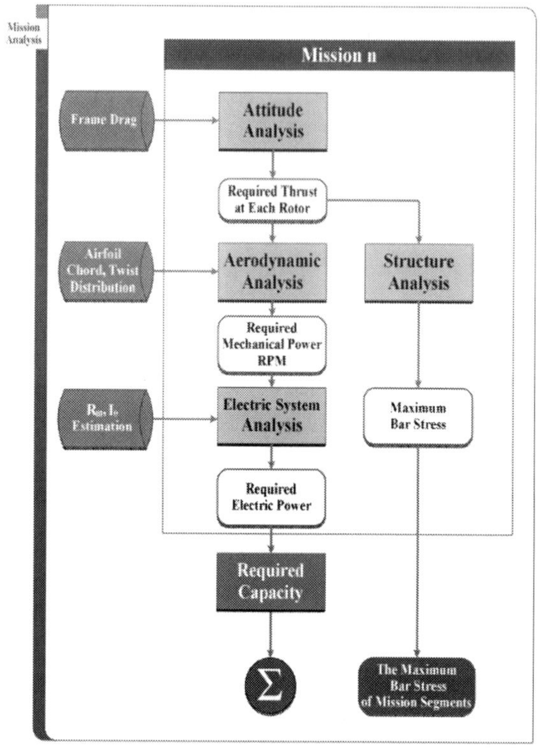

Fig. 3. Mission segment analysis.

Table 1. Attitude analysis.

Mission type	Required thrust	Vertical speed	Tilt angle
Hovering	$\frac{\mathrm{GW}}{\#\ \mathrm{of\ Rotor}}$	0	0
Climbing	$\frac{\mathrm{GW} + \mathrm{Frame\ Drag}}{\#\ \mathrm{of\ Rotor}}$	V_∞	0
Descent	$\frac{\mathrm{GW} - \mathrm{Frame\ Drag}}{\#\ \mathrm{of\ Rotor}}$	$-V_\infty$	0
Forward flight	$\frac{\mathrm{GW}}{\#\ \mathrm{of\ Rotor} \times \cos\theta}$	$V_\infty \sin\theta$	$\tan^{-1}\left(\frac{\mathrm{Frame\ Drag}}{\mathrm{GW}}\right)$

After the required thrust of a rotor was calculated, rotor RPM and required mechanical power were analyzed by the aerodynamic analysis module. Blade element momentum theory (BEMT) was used with several assumptions (Refs. 8, 9).

$$dC_{T,BET} = \frac{1}{2}\sigma C_l r^2 dr \quad (2)$$

$$dC_{T,MT} = 4\lambda(\lambda - \lambda_c) r dr \quad (3)$$

$$C_T = \int_X^B dC_T dr \quad (4)$$

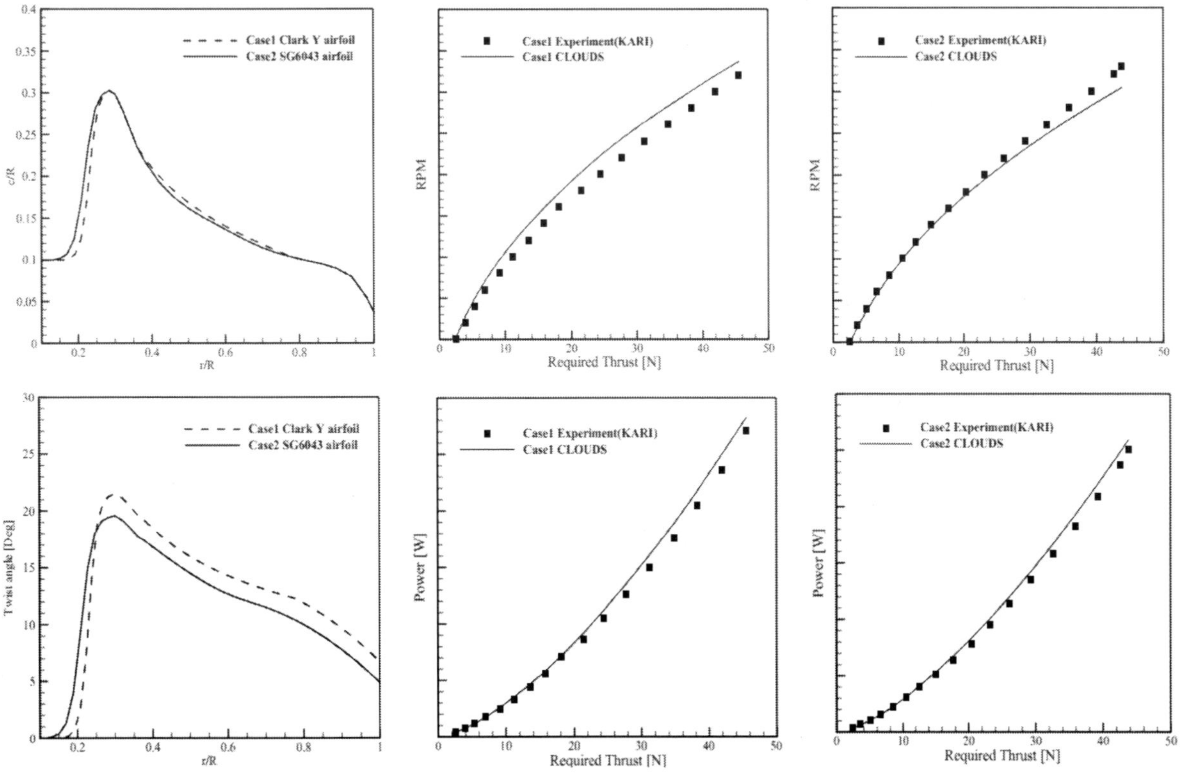

Fig. 4. Validation of the aerodynamic module.

$$C_P = \int_X^B dC_{P_i} + \int_X^1 dC_{P_o} \qquad (5)$$
$$= \int_X^B \kappa dC_T + \int_X^1 \frac{1}{2}\sigma C_d r^3 dr$$

There were two assumptions in the aerodynamic analysis: 1) tip loss factor was 0.88 with 0.15 root cut out ratio, and 2) induced power factor was 1.25, which was induced from experimental data of Korea Aerospace Research Institute (KARI).

For the validation of the aerodynamic analysis module, rotor RPM and mechanical power were compared with KARI's two propeller experimental data. Case1 was a propeller whose airfoil was Clark Y and Case2 was a propeller whose airfoil was SG6043. 2D airfoil aerodynamic characteristics were used by Xfoil tool (Ref. 10). and geometry distributions which were chord-radius distribution and twist-radius distribution along the span was shown in Figure 4.

The aerodynamic analysis result was well-matched with the experimental data for both cases. The maximum 5% error was calculated in RPM at about 29 N of required thrust in case 1. Case1 geometry distributions and airfoil were used to the overall design.

To calculate the required electrical power of a motor, the mechanical power, and battery voltage, motor speed constant was used in the electric system analysis module. With assuming that battery voltage was 3.7V per a cell and battery internal resistance was 0.003Ω per a cell referred to Ref. 5-6, the motor drive current was analyzed with motor internal resistance and no load current in Figure 5.

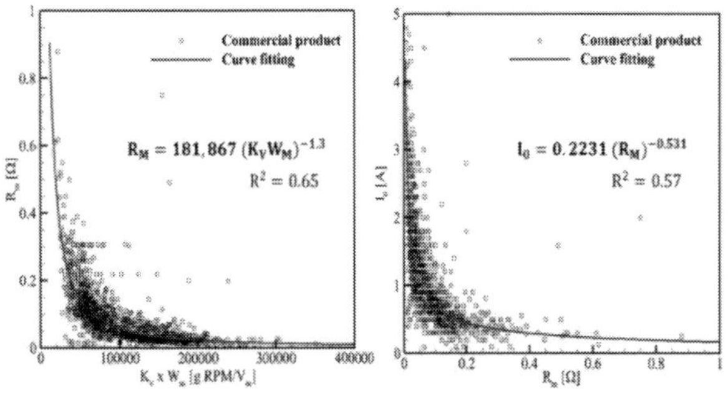

Fig. 5. Motor properties estimation.

As the motor weight was estimated by (1)., internal resistance and no load current of the motor was also estimated by the motor speed constant. Using estimated motor properties, drive voltage, and current of all motors required electrical power and motor efficiency were calculated considering not only the required mechanical power, but also copper, iron loss, mechanical loss, and stray loss (Ref. 11, Figure 6).

$$V_M = V_B - I_M(R_E + R_B N) \qquad (6)$$

$$P_E = P_{Mech} + P_{Co} + P_{Ir} + P_{ML} + P_S \quad (7)$$

$$P_{Co} = I_M^2 R_M\,,\ P_{Ir} = (V_M - I_M R_M) I_0 \quad (8)$$

$$\eta_M = \frac{P_{Mech}}{P_E} \quad (9)$$

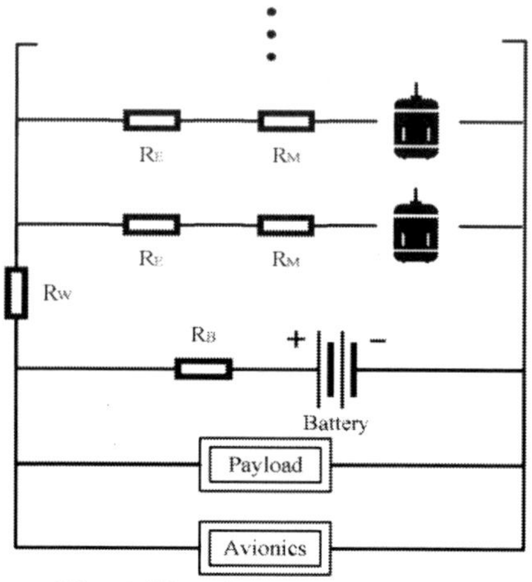

Fig. 6. Electric system schematics.

The required electrical power directly related to the overall performance was compared with eCalc based on the quadrotor composed to commercial products. Using constant motor efficiency concerning the motor drive current, EMST was also compared in Table 2 and Figure 7.

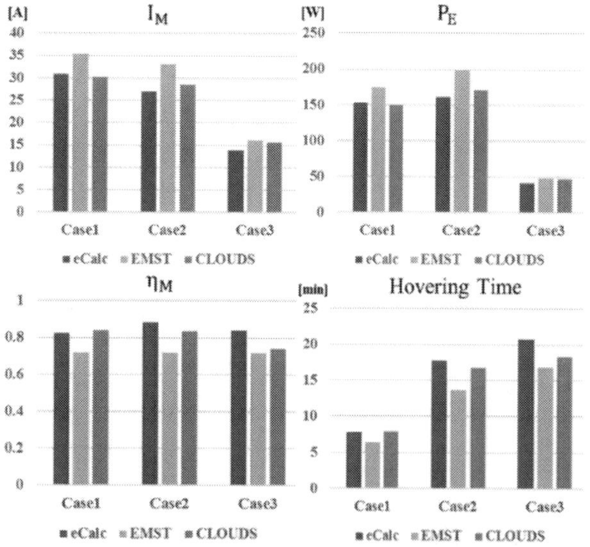

Fig. 7. Comparison of electric system analysis.

It was shown that the electric system analysis considering variable motor efficiency was more comparable with eCalc analysis based on commercial products than 72% constant motor efficiency method. The difference was directly reflected on the reliability of the flight time analysis. Because of the variable motor drive current for performing each mission segments, it was necessary to estimate the motor efficiency in the design procedure. At the same ESC, mechanical power and motor voltage, the electric system analysis of CLOUDS was well-matched with the result of the eCalc which was combined with commercial products in different Kv.

Table 2. Validation cases of electric system analysis

Specification		Case 1	Case 2	Case 3
Rotor	Diameter [in]	15	15	11
	Model		GWS	
Motor	Kv	420	520	870
	Model	EMAX GT4030-06	Sunny Sky X3525-520	Sunny Sky V2814-870
Battery	Capacity [Ah]	5	10	6
	Cell	6	6	3
	Model	LiPo 5Ah 65/100C	LiPo 10Ah 65/100C	LiPo 6Ah 65/100C
ESC	Max Amp.[A]		50	
	Model		Suppo 50A	

Structure analysis was primarily fulfilled to the motor-frame support bar with 1-D Euler beam theory. The geometrical assumptions were defined as in Figure 8.

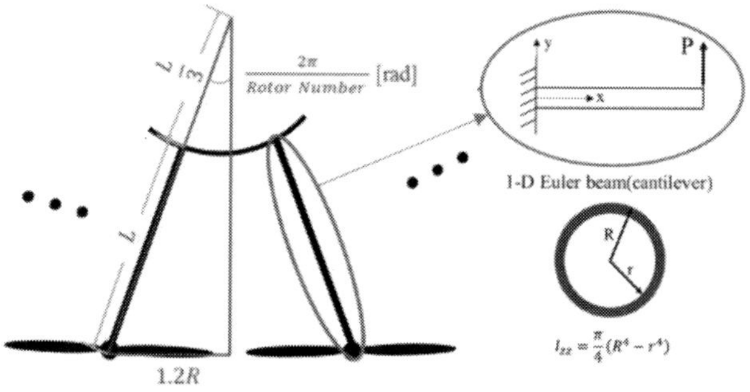

Fig. 8. Geometric assumptions.

The rotor generated thrust at the edge of the bar and the stress of the cantilever bar was described:

$$\sigma(x,y) = \frac{g_f\, GW\,(x-L)}{4I_{zz}} \quad (10)$$

$$\sigma_{max} = \pm\frac{g_f\, GW\, L}{4I_{zz}} \quad (11)$$

$$\sigma_{alow} = \sigma_{ut}\, n_{safe} \quad (12)$$

Carbon fiber is the widely used material for the support bar and its ultimate stress is 2.068 Gpa (Ref. 12). Accelerate factor, g_f was 1.5 for universal multirotor UAV and safety factor, n_{safe} was 1.5 in the whole design process. Comparing maximum stress and allowable stress, CLOUDS checked the structural breakdown of the support bar. To verify the structural analysis, 10kg quadrotor UAV with 22-inch propeller and 10mm x 8mm circular carbon fiber bar was analyzed in high maneuvering condition (g_f, n_{safe} = 2). As a result, 37.6Mpa maximum stress and 1,034Mpa allowable stress was calculated, which was the same result as Ref. 13.

All analysis modules were modeled based on physics and validated with experimental data or commercial products. For the mission segment, required electrical power was calculated through these multidisciplinary analysis modules, and it could be analyzed to overall performance including flight time, endurance, vertical rate of climb, and cruise speed.

DESIGN RESULT

Rotor diameter, motor speed constant, battery cell, and battery capacity were used to design variables, which could determine overall size and flight performance. With regard to the minimized gross weight as the most proper design solution to a specific mission profile, quadrotor UAVs were designed in 2 cases. Case 1 checked the validation of design result indirectly. Case 2 was gross weight optimization for a photograph mission profile. The boundary of all design variables was based on the purchasable product range as shown in Table 3.

Table 3. Boundary of the design variables.

Design variables	Lower bound	Upper bound	Baseline
Rotor diameter [in]	7	30	24
Motor Kv	150	2000	1800
Battery cell	2	6	6
Battery capacity [Ah]	1	30	10

All design cases followed these assumptions:

- Rotors have the same geometry ratio of radius (c/R, θ/R) distribution along a rotor span with Case 1 airfoil in Figure 4.
- All flights were regarded as steady level flight.
- Frame weight is 19%, wiring weight is 5% of gross weight, and avionics consumes 0.5 ampere (Ref. 5).
- Nylon reinforced plastic rotor is used.
- P_{ML} is 1% and P_S is 0.5% of mechanical power for the electric system analysis in Equation 7 (Ref. 11).
- Payload weight is 500g and consumes electrical power in forward flight, and the battery had a limitation of 85% discharge.

Design case 1: validation of design solution result

For a given hovering time, a combination of design variables, and numerical optimized design solution, was deduced from CLOUDS. Then, the numerical solution was converted to commercial products which had analogous properties. The commercial products were placed in eCalc and hovering time was calculated. By comparing initial required hovering time and the calculated hovering time, CLOUDS solution could be indirectly clarified.

From required 15-minute hovering time in design case 1, the optimized design solution was Table 4.

Table 4. Case 1 design solution.

Design variables	CLOUDS	Commercial products (Model)
Rotor diameter [in]	10.91	11 (Assumption 1)
Motor Kv	2000	2000 (T-motor MN2206)
Battery cells	2.98	3 (LiPo 3300mAh, 55/80C)
Battery capacity [Ah]	3.25	3.3

CLOUDS calculated 15.02-minute hovering time with inputs of the commercial products, which occurred under the 1% hovering time gap between numerical solution and commercial products solution. Also, the eCalc showed 14.2-minute hovering time with the same inputs, which was 5% error. Considering the ±15% analysis reliability of eCalc (Ref. 6), it might be a valid crosscheck procedure that CLOUDS started design with 15-minute hovering time requirement and 14.2-minute hovering time was analyzed by eCalc with the solution of CLOUDS.

Design case 2: proper design solution to a mission profile

For a photograph mission profile in Figure 9, it was accomplished to proper quadrotor design solution which was optimized to minimize gross weight. The mission profile could be separated into 7 mission segments and total mission flight time was 22-minute. Filming mission included a specific camera that consumed 3.8W electrical power for recording in forward flight.

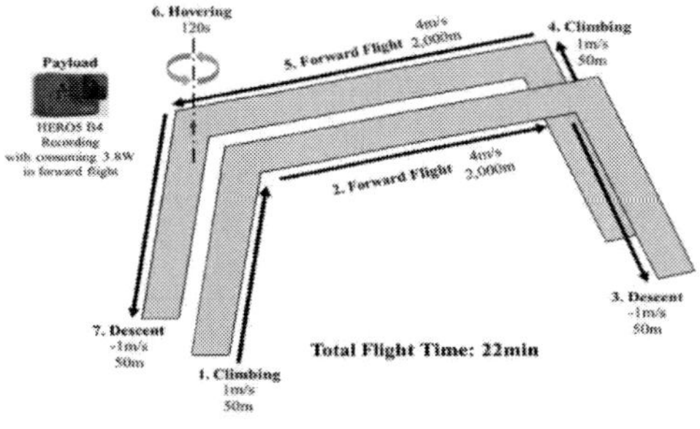

Figure 9. Filming mission profile.

The 1,585g gross weight of quadrotor solution was calculated by CLOUDS and commercial products in Table 5 were chosen for substituting numerical solution.

Table 5. Case 2 design solution.

Design variables	CLOUDS	Commercial products (Model)
Rotor diameter [in]	13.77	14 (Assumption 1)
Motor Kv	1920	1900 (EMAX MT2206)
Battery cells	3.598	4 (LiPo 4600mAh, 55/80C)
Battery capacity [Ah]	4.596	4.6

Gross weight of the quadrotor consisting of a set of the commercial products was 1,638g, which had 3% difference due to the substitution from the numerical solution to the set. Each mission segment required different electrical power as shown in Figure 5. Along with the variation of motor drive current, the motor efficiency and electrical power were calculated by electric system analysis. This is the reason why it should be considered to the variation of motor efficiency in design procedure from different mission segments.

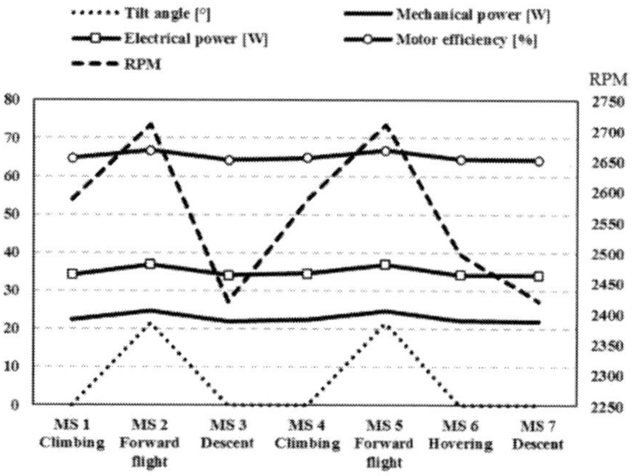

Figure 10. Performance analysis in mission segments.

CONCLUSION

A systematic mission-intensive design framework for multirotor UAVs, CLOUDS, was developed. From a single mission segment to a particular mission profile, CLOUDS calculates overall performance using multidisciplinary analysis and determined the design variables proper to the mission. The following conclusions have been made.

1. The novel design process was developed for the electric propulsion system of multirotor UAVs, which was distinct from fuel propulsion based design. The design process treated a specific mission profile, and a required battery capacity was calculated by attitude, aerodynamic, and electric system analysis in each mission segment.

2. All analysis modules in CLOUDS were modeled based on physics and validated with experimental data or commercially available products. Aerodynamic analysis used blade element momentum theory with 0.88 tip loss factor and 1.25 induced power factor, which was deduced from experimental data of KARI. The forward flight was considered as a vertical flight due to the large tilt angle and rotor inflow ratio than conventional helicopters. At aerodynamic analysis in the design process, a specific airfoil and geometry ratios, c/R and θ/R, were needed.

3. By estimating inner resistance and no load current of a motor, it could be analyzed to the motor drive current, electrical power and motor efficiency. 14-pole motor had an inner resistance and no load current trend with motor speed constant, and it was used in whole design process.

4. It was necessary to consider variable motor efficiency along with the motor drive current, and the efficiency was mainly compared with eCalc which was based on commercial product properties. Developed electric system analysis was well-matched with the eCalc analysis, which was acceptable for the conceptual design level.

5. Minimizing gross weight was considered as the proper solution to a particular mission. For validation of the design solution, the solution of CLOUDS was cross-checked with eCalc, trading the design requirement and outputs.

6. A 22-minute photograph mission was given for finding the proper quadrotor design solution. About 3% gross weight gap was generated by converting numerical solution to commercial products. Also, rotor RPM, electric power, and tilt angle were calculated in each mission segments.

ACKNOWLEDGMENTS

This research was supported by National R&D Program through the National Research Foundation of Korea(NRF) funded by the Ministry of Science & ICT (NRF-2016M1B3A1A03937680)

REFERENCES

[1]Mahony, R., Vijay K., and Peter C., "Multirotor aerial vehicles," *IEEE Robotics and Automation*, Vol. 20, (32), Sep 2012.

[2]Bangura, M., Lim, H., Kim, H. J., and Mahony, R., "Aerodynamic power control for multirotor aerial vehicles," *Robotics and Automation (ICRA), 2014 IEEE International Conference*, May 2014, pp.529-536.

[3]Hoffmann, G. M., Huang, H., Waslander, S. L., and Tomlin, C. J., "Quadrotor helicopter flight dynamics and control: Theory and experiment," *AIAA Guidance, Navigation, and Control Conference*, Vol. 2, (4), Aug 2007.

[4]Winslow, J., Hrishikeshavan, V., and Chopra, I., "Design Methodology for Small Scale Unmanned Quadrotors," *55th AIAA Aerospace Sciences Meeting,* Grapevine, Texas, Jan 2017, pp.1-15.

[5]Bershadsky, D., Haviland, S., and Johnson, E. N., "Electric Multirotor Propulsion System Sizing for Performance Prediction and Design Optimization," *AIAA SciTech 2016*, AIAA, San Diego, CA, 2016.

[6]eCalc xcopter, http://www.ecalc.ch/xcoptercalc.php

[7]Dieter K., "A Software Package for Sequential Quadratic Programming," DFVLR-FB, 1988.

[8]Leishman, J. G., *Principles of Helicopter Aerodynamics*, Cambridge University Press, New York, NY, 2000, Chapter 10.

[9]Johnson, W., *Helicopter Theory*, Princeton University Press, Princeton, NJ, 1980, pp. 808–813.

[10]Drela M., "XFOIL: an analysis and design system for low Reynolds number airfoils," *Low Reynolds Number Airfoil Aerodynamics*, Notre Dame, Indiana, 1989.

[11]Ilka, R., Tilaki, A. R., Alamdari, H. A., and Baghipour, R., "Design Optimization of Permanent Magnet-Brushless DC Motor using Elitist Genetic Algorithm with Minimum loss and Maximum Power Density," *Int. J. Mecha. Elecr. Comput. Technol*, vol.4, (10), 2014, pp. 1169-1185.

[12]CLEARWATER COMPOSITES, LLC URL: http://www.amesweb.info/StructuralBeamDeflection/CantileverBeamConcentratedLoad.aspx

[13]Advanced Mechanical Engineering Solution (AMES) beam calculator, URL: http://www.amesweb.info/StructuralBeamDeflection/CantileverBeamConcentratedLoad.aspx

Robust Design of Type-II Small-Scale Unmanned Helicopter for Hover Performance using Taguchi Method

D. R. Abhiram
PhD Student
Indian Institute of Science, Bangalore, India.

Ranjan Ganguli
Professor
Indian Institute of Science, Bangalore, India.

Dineshkumar Harursampath
Assistant Professor
Indian Institute of Science, Bangalore, India.

ABSTRACT

A simple approach for robust design of a small-scale unmanned helicopter main rotor using Taguchi method is presented. Multiple parameters of the main rotor such as, blade length, blade planform variation, blade pre-twist and rotor speed are considered with the objective of improving the hover performance of the helicopter. The value of each design parameter is considered at three distinct levels. A numerical scheme based on blade element momentum theory is used for performance analysis of the helicopter. The Taguchi method is used to obtain the optimal combination of design parameter values that result in a helicopter design with improved hover performance. Robustness of the design is facilitated by using practical tolerance values for the design variables as noise factors. The resulting robust design displays improved hover efficiency of the helicopter, with minimum variability in presence of reasonable variations in design parameters.

INTRODUCTION

Interest in development of unmanned aerial vehicles (UAVs) has been increasing. Developed initially for military purposes, UAVs are used for civilian applications as well. UAVs can be broadly classified, based on the mechanism of lift generation, as fixed wing UAVs, rotary wing UAVs, and flapping wing UAVs. This paper focuses on rotary wing UAVs.

Rotary wing UAVs, or unmanned rotorcraft, use rotors to generate thrust and propulsive forces. They are classified based on their size and characteristics, as Type-I, Type-II and Type-III unmanned rotorcraft (Ref. 1).

Type-I unmanned rotorcraft are relatively large. Their rotor diameter is in the range of three to four meters. They are powered by internal combustion engines, and are suitable for autonomous or remotely controlled unmanned missions that require longer endurance (30 minutes to a couple of hours) and heavier payload (20 to 30 kg). Type-I unmanned rotorcraft are dominated by the traditional main-rotor-tail-rotor helicopter configurations having two to four main-rotor blades. These unmanned vehicles are typically made of composite materials, aluminium and stainless steel.

Type-II unmanned rotorcraft are smaller than their Type-I counterparts. Their rotor diameter is in the range of one to two meters. They are usually powered by on-board batteries. Most of the remote controlled helicopters used for recreational and hobby purposes fall under Type-II category. These are made of carbon fiber composites and plastics.

Type-III are the smallest, they are almost entirely made of plastics and are powered by small batteries. Small remote controlled single and multiple rotor helicopters, coaxial rotor helicopters, most of the quadcopters, etc. fall in this category. Their small size and light weight makes them attractive for applications involving remote surveillance and stealth monitoring but at the same time makes them unsuitable for applications that require substantial flight time or payload carrying capacity.

Type-I and Type-II unmanned helicopters are attractive for practical applications that require significant payload carrying capability and longer endurance. In addition to some of the tasks that can be performed by fixed-wing UAVs, these unmanned helicopters can use their unique abilities such as hover, vertical flight, and slow speed flight in forward, backward and other directions, to perform tasks in a variety of situations. This versatility makes them suitable for both military and civilian applications. The MQ-8B Fire Scout developed by Northrop Grumman is used for naval reconnaissance (Ref. 2). Yamaha's R-50 and R-MAX unmanned helicopters are used for spraying agrochemicals over farms and for investigating vegetation. The R-MAX has also been used for monitoring volcanic activity, surveying of areas affected by earthquake, monitoring radiation levels during nuclear emergencies, and for detecting damage in bridges and highways

Presented at the 6th Asian/Austrailian Rotorcraft Forum / Heli Japan 2017, Kanazawa, Japan, November 7 – 9, 2017.

(Ref. 3). A similar unmanned helicopter, the X-Copter, is currently under development for agricultural and industrial applications (Ref. 4). The Sky Surveyor, an unmanned helicopter developed at Chiba University, is used to inspect power transmission lines (Ref. 5). The use of such unmanned helicopters can also be extended to many other civilian applications such as aerial surveillance in urban areas, traffic monitoring, broadcasting sports events, aiding and assisting workers and engineers at construction sites, etc.

A large potential for unmanned helicopters exists since they are suitable for a host of practical applications. Despite their potential, the use of unmanned helicopters in civilian applications has been limited, due to the high costs associated with the development and operation of such unmanned helicopters for civilian applications (Ref. 4).There is also a surprising lack of available literature on design and analysis of unmanned helicopters that are suitable for many practical civilian applications. Most of the research on unmanned rotorcraft in this category, has focussed on their autonomous navigation and control aspects.

The main rotor is the most important component of a helicopter, and its design is critical for ensuring satisfactory performance. This paper considers several design parameters affecting the rotor of a typical Type-II unmanned helicopter to obtain a robust design employing the Taguchi method.

The Taguchi method was initially developed to improve the quality of manufactured products. It is now used in many different engineering fields for robust system design (Refs. 6–8). It is a fractional factorial design of experiments approach, which use orthogonal arrays of design parameters to reduce the variation in system performance measure while determining the combination of design parameter values that will lead to enhanced performance. It involves use of noise values to account for variations in the design variables. The use of orthogonal arrays in Taguchi method of design provides a systematic approach for reducing the number of experiments or numerical simulations required for the analysis and at the same time provides an opportunity to evaluate the effect of each design parameter on the design.

In the current paper the Taguchi method is used to generate a robust design of a small-scale 2-bladed unmanned helicopter by considering 4 design parameters of the main rotor: blade length, blade planform variation (i.e. blade taper), span-wise distribution of blade twist and the rotational speed. The performance of the resulting robust helicopter design is compared with initial design and is found to have better hover efficiency. The baseline helicopter is modeled to simulate the autonomous mini-helicopter being developed at IIT Kanpur, India (Ref. 9). The objective is to reduce the power required in hover while ensuring that the design is sufficiently robust so as to be insensitive to reasonable variation in design parameter values.

PERFORMANCE ANALYSIS OF HELICOPTER IN HOVER

The main rotor governs three important functions: 1) Thrust generation in hover and vertical flight; 2) Generation of horizontal thrust component needed for forward flight; 3) Production of forces and moments (jointly with the tail rotor) for attitude and position control. Therefore, estimation of main rotor performance is essential for good rotor design. A numerical implementation of the blade element momentum theory (BEMT) is used in this study for predicting the hover performance since, this model is suitable for representing radial variations in rotor blade twist and taper. The BEMT combines the momentum theory and the blade element theory (BET), which are treated in detail in (Ref. 10). The non-dimensional hover power measure, C_P is given as,

$$C_P = \sum_{n=1}^{N} [\lambda_n (\Delta C_T)_n] + \sum_{n=1}^{N} (\Delta C_{P_o})_n \tag{1}$$

$$P = \rho A (\Omega r)^3 C_P \tag{2}$$

$$FM = \frac{C_T^{3/2}}{C_P} \tag{3}$$

where N is the number of blade elements, λ_n is the inflow ratio at the nth blade element, C_T is the thrust coefficient, C_{P_o} is the profile power coefficient, C_P is power coefficient, P is the rotor power, ρ is the density of air, A is the rotor disk area, Ω is the rotor speed, r is the non-dimensional radial position and FM is the figure of merit.

BASELINE HELICOPTER DESIGN

The baseline design is modeled to simulate the autonomous mini-helicopter (Ref. 9), which is a Type-II unmanned helicopter. Table 1 gives overall configuration of the mini-helicopter.

TAGUCHI METHOD OF DESIGN

The Taguchi method of design is a fractional factorial design of experiments approach. The Taguchi method can be used to determine the combination of design parameters that lead to an enhanced performance of a system, with minimum variation in its performance. It was initially applied to solve industrial problems to reduce product variation by considering the source of the variation at the design stage itself and making the product insensitive to these variations and thereby improve its quality. It is now used in many engineering fields for robust system design (Ref. 11).

In Taguchi method, design variables that primarily influence the system performance are called as control factors since, their values can be easily set. The factors that represent the sources of variation in the system performance are called noise factors. Noise factors often represent environmental variables such as humidity, temperature, etc. and may also represent other sources of variations such as variation in material properties, which are not easily controllable. Taguchi method leads to robust design which is insensitive to the noise factors. This is achieved by using system performance measure that combines the mean response and variability and is obtained from designs based on Taguchi orthogonal arrays (OAs).

OAs are a unique set of tables developed for design of experiments (DoE); they represent the smallest fractional factorials which can be used for experiment designs. The orthogonal array used for the control factors is called inner array and that used for noise factors is called outer array. For each design experiment in the inner array a set of design experiments are conducted based on the outer array, to account for the effect of noise factors (Ref. 12).

The response of the system at each design point in the inner array is represented by a performance measure called as, signal-to-noise (S/N) ratio. The S/N ratio used depends on the objective of the problem. Three commonly used S/N ratio are:

1. Larger-the-better: used when the goal is to maximize the system response.
2. Nominal-the-better: used when the goal is to make the system response achieve a specific target value.
3. Smaller-the-better: used when the goal is to minimize the system response.

In the present study, the smaller-the-better S/N ratio is used since, the objective is to minimize the hover power of the helicopter. The smaller-the-better S/N ratio is defines as follows,

$$\frac{S}{N} = -10\log_{10}\left(\frac{1}{k}\left[\sum_{i=1}^{k}(y_i)^2\right]\right) \quad (4)$$

where y_i is the system response of the ith trial in the outer array and k is the total number of trials in the outer array.

The combination of design parameters for which the S/N ratio is highest corresponds the robust design system configuration. This robust design will have minimum variation in performance with given noise factors. Higher values of S/N ratio implies that the system is less sensitive to noise factors. The robust design of a 2-bladed unmanned helicopter using Taguchi method is described in the following section.

ROBUST DESIGN USING TAGUCHI METHOD

Based on the expression for the power required for a hovering helicopter, four main-rotor parameters are selected as design variables: rotor blade radius, chord length at blade tip (along with blade taper, this captures the blade planform variation), blade pre-twist, and rotor speed. Three levels of the design variables are considered in this study to explore the design space. The chord length at the blade tip is also used for determining the rotor blade planform shape. At level 1, blade has constant chord, whereas levels 2 and 3 correspond to linear taper.

Noise values are introduced for each design variable to analyze the sensitivity of performance to noise factors. Realistic tolerance of the design parameters have been assumed as noise values (Ref. 6). The value of design variables at three levels along with the noise levels are given in Table 2.

In the present work, four design variables at three levels are used as control factors and two levels are considered for the four noise factors for sensitivity analysis. Therefore, the standard $L_9(3^4)$ orthogonal array is used as the inner array for the control factors and L_8 array is used for the noise factors. The orthogonal arrays, $L_9(3^4)$ and L_8 (corresponding to 4 design variables), are given in Table 3 and Table 4, respectively. For the inner array, '0', '1', and '2' represents the control factors at level 1, level 2, and level 3, respectively. For the outer array, '2' and '1' indicate the high and low levels of the noise factors, respectively.

Table 5 shows nine design of experiments for the inner array. Eight trials are conducted in the outer array corresponding to nine DoEs in inner array (in L_9 OA). Therefore, a total of $(9 \times 8 =)$ 72 simulations are carried out. The robust design with the best hover performance will correspond to the combination of design parameters that will lead to the highest S/N ratio. The results of all the runs are summarized in Table 6, with the mean power required to hover and the corresponding S/N ratio for each design in the inner array.

RESULTS AND DISCUSSIONS

In this study, numerical results are obtained for different configurations of an unmanned helicopter based on orthogonal array design of experiments. Baseline values for the four design variables, rotor radius R, Tip chord c_{tip}, Blade pre-twist θ_{tw} and rotor speed Ω are: $0.710m$, $0.065m$, $0°$ and 1400 RPM, respectively. The bounds on the design space are determined by the physical constraints.

The results of the Taguchi method, summarized in Table 6, indicates that the design parameter combination that has the highest S/N ratio is the combination from row 1. This design leads to minimum power consumption in hover and the highest S/N ratio.

$$\begin{aligned} \text{Rotor radius, } R &= 0.8\text{m} \\ \text{Blade tip chord, } c_{tip} &= 0.050\text{m} \\ \text{Blade pre-twist, } \theta_{tw} &= -5° \\ \text{Rotor speed, } \Omega &= 1300\text{RPM} \end{aligned} \quad (5)$$

Table 7 gives a comparison of the baseline helicopter and the robust helicopter design obtained using the Taguchi method. The robust helicopter configuration, when compared with the baseline helicopter, shows a 11.3% reduction in power required to hover.

CONCLUSION

This study utilizes the Taguchi method of design to obtain a robust design of main rotor blade of a Type-II unmanned helicopter, with an aim to improve its hover performance. The robust design obtained indicates that the design configuration with rotor radius $0.8m$, tip chord $0.050m$, blade pre-twist $-5°$ and rotor speed equal to 1300 RPM corresponds to the design with the minimum requirement of hover power. The obtained robust design shows 11% hover power reduction compared to baseline. Taguchi method of design uses S/N ratio which combines mean and variation measures to arrive at a robust design configuration. The robustness of the Taguchi design implies that the helicopter hover performance is insensitive to reasonable variations in the blade design variables. This study will help helicopter designers to perform preliminary performance calculations to study the rotor performance in hover and assist them in initial sizing of the rotor without the use of computationally expensive high fidelity analysis tools.

Author contact
1) Mr. D. R. Abhiram: abhiram@aero.iisc.ernet.in
2) Dr Ranjan Ganguli: ganguli@aero.iisc.ernet.in
3) Dr. Dineshkumar Harursampath: dinesh@aero.iisc.ernet.in

REFERENCES

[1]Cai, G., Chen, B. M., and Lee, T. H., "An overview on development of miniature unmanned rotorcraft systems," *Frontiers of Electrical and Electronic Engineering in China*, Vol. 5, (1), 2010, pp. 1–14.

[2]Grumman, N., "MQ-8B Fire Scout," Accessed on: 19-08-2014, 2011.

[3]Sato, A., "The RMAX helicopter UAV," Technical report, DTIC Document, 2003.

[4]Shim, D. H., Han, J.-S., and Yeo, H.-T., "A development of unmanned helicopters for industrial applications," *Unmanned Aircraft Systems*, Springer, 2009, pp. 407–421.

[5]Nonami, K., "Prospect and recent research & development for civil use autonomous unmanned aircraft as UAV and MAV," *Journal of System Design and Dynamics*, Vol. 1, (2), 2007, pp. 120–128.

[6]Hu, Y. and Rao, S. S., "Robust Design of Horizontal Axis Wind Turbines Using Taguchi Method," *Journal of Mechanical Design*, Vol. 133, (11), 2011, pp. 111009.

[7]Mallick, R., Ganguli, R., and Bhat, M. S., "Robust design of trailing edge flap with orthogonal array inspired response surface for helicopter vibration reduction," Proceedings of the 39th European Rotorcraft Forum, Moscow, Russia, 2013.

[8]Bhadra, S. and Ganguli, R., "Aeroelastic optimization of a helicopter rotor using orthogonal array-based metamodels," *AIAA Journal*, Vol. 44, (9), 2006, pp. 1941–1951.

[9]Venkatesan, C., Swaroop, B., Haritha, P., and Gupta, R., "Development of autonomous mini helicopter: Challenges faced," *Journal of Aerospace Sciences and Technologies*, Vol. 65, (1A), 2013.

[10]Leishman, J. G., *Principles of Helicopter Aerodynamics*, Cambridge University Press, second edition, 2006.

[11]Chen, W.-C., Nguyen, M.-H., Chiu, W.-H., Chen, T.-N., and Tai, P.-H., "Optimization of the plastic injection molding process using the Taguchi method, RSM, and hybrid GA-PSO," *The International Journal of Advanced Manufacturing Technology*, 2015, pp. 1–14.

[12]Roy, R. K., *A primer on the Taguchi method*, Society of Manufacturing Engineers, Dearborn, MI, second edition, 2010.

Table 1. Baseline helicopter (Ref. 9)

Specification	Value
Main rotor radius (m)	0.9
Main rotor chord (m)	0.065
Main rotor RPM	1400
Number of main rotor blades	2
Total weight (kg)	10

Table 2. Design parameters, levels and noise values

	Control factor	Units	Level-0	Level-1	Level-2	Noise factor	
x_1	Rotor radius	m	0.8	0.9	1.0	z_1	± 0.005
x_2	Tip chord	m	0.050	0.065	0.035	z_2	± 0.005
x_3	Blade linear twist	degrees	-5	-7.5	-10	z_3	± 1
x_4	Rotor speed	RPM	1300	1400	1500	z_4	± 5

Table 3. Standard $L_9(3^4)$ orthogonal array (inner array)

Experiment Number	Control factors x_1	x_2	x_3	x_4
1	0	0	0	0
2	0	1	1	1
3	0	2	2	2
4	1	0	1	2
5	1	1	2	0
6	1	2	0	1
7	2	0	2	1
8	2	1	0	2
9	2	2	1	0

Table 4. L_8 orthogonal array (outer array)

Simulation	Noise factors z_1	z_2	z_3	z_4
1	1	1	1	1
2	1	1	1	2
3	1	2	2	1
4	1	2	2	2
5	2	1	2	1
6	2	1	2	2
7	2	2	1	1
8	2	2	1	2

Table 5. Design parameter matrix: Inner array $L_9(3^4)$

Experiment Number	Rotor radius x_1 (m)	Tip chord x_2 (m)	Blade linear twist x_3 (degrees)	Rotor speed x_4 (RPM)
1	0.8	0.050	-5	1300
2	0.8	0.065	-7.5	1400
3	0.8	0.035	-10	1500
4	0.9	0.050	-7.5	1500
5	0.9	0.065	-10	1300
6	0.9	0.035	-5	1400
7	1.0	0.050	-10	1400
8	1.0	0.065	-5	1500
9	1.0	0.035	-7.5	1300

Table 6. Results of nine design of experiments (L_9 OA)

Experiment No.	Mean Power (W)	S/N Ratio
1	690.62	−56.79
2	700.99	−56.92
3	732.90	−57.31
4	920.99	−59.29
5	703.73	−56.95
6	771.73	−57.76
7	1006.73	−60.07
8	1062.73	−60.54
9	798.73	−58.06

Table 7. Comparison of baseline helicopter design with robust design

Design parameter	Rotor radius	Tip chord	Blade linear twist	Rotor speed	Hover power
Symbol	R	c_{tip}	θ_{tw}	Ω	P
Units	m	m	degrees	RPM	W
Baseline	0.710	0.065	0	1400	778.6
Robust	0.8	0.050	−5	1300	690.62

Development of Tailsitter VTOL using Multi copter

Kenta Kodera
Graduate student
Tokushima University
Tokushima, Japan

Masafumi Miwa
Associate professor
Tokushima University
Tokushima, Japan

ABSTRACT

In recent years, Unmanned Aerial Vehicle(UAV) is used for various things such as disaster investigation and Aerial photography. UAV is classified as fixed wing aircraft and rotary wing aircraft. Rotary wing aircraft can perform vertical take-off and landing(VTOL), and low-speed flight. Vertical takeoff and landing used in this study is a UAV with the advantages of fixed wing aircraft and rotary wing aircraft. Tail sitter is one of VTOL. It has a simple structure compared to other VTOLs and, it can perform vertical takeoff and landing like a rotary wing aircraft, also it can travel long distances at high speed like a fixed wing aircraft. In this study, we developed the test aircraft with four rotor helicopters installed near the center of gravity of all wing aircraft and selectively use rotary wing aircraft and fixed wing aircraft depending on the situation.

INTRODUCTION

In recent years, Unmanned Aerial Vehicle(UAV) is used for various things such as disaster investigation and Aerial photography. UAV is classified as fixed wing aircraft and rotary wing aircraft. Rotary wing aircraft can vertical take-off and landing, and low-speed flight by hovering is possible. Rotary wing aircraft are used for many applications such as surveying, aerial photographing and the like based on its flight characteristics. On the other hand, the fixed wing aircraft has an advantage that the cruising time, cruising distance, and stability during flight are superior to those of a rotary wing aircraft. Vertical takeoff and landing(VTOL) is a UAV with the advantages of fixed wing aircraft and rotary wing aircraft. Tail sitter is one of VTOL aircraft. Tail slitter can perform maneuvering as shown in Figure 1[(1)]. Also, tail slitter has advantages. It has few complicated drive parts like a tilt rotor, it is possible to take off and landing like a helicopter or a rotary wing aircraft in a vertical take-off and landing, and in a horizontal flight, a long distance as a fixed wing aircraft which is possible to fly. On the other hand, as a disadvantage of tail sitter, there is a problem that the viewpoint of the pilot is upward when landing and it is not possible to see the landing surface, but it is resolved because it is an unmanned aircraft. In the previous study, the wing length of the airframe was longer, the more susceptible to the influence of the disturbance, the stable hovering could not be done[2)]. In this study, we developed the test aircraft with four rotor helicopters installed near the center of gravity of all wing aircraft and selectively use rotary wing aircraft and fixed wing aircraft depending on the situation.

TEST AIRCRAFT

In this research, we used a test aircraft with four rotor helicopters installed near the center of gravity of all wing aircraft. The test aircraft specification is shown in Figure 2 and TABLE 1. We created a control program based on ArduCopter of open source project "ArduPilot".

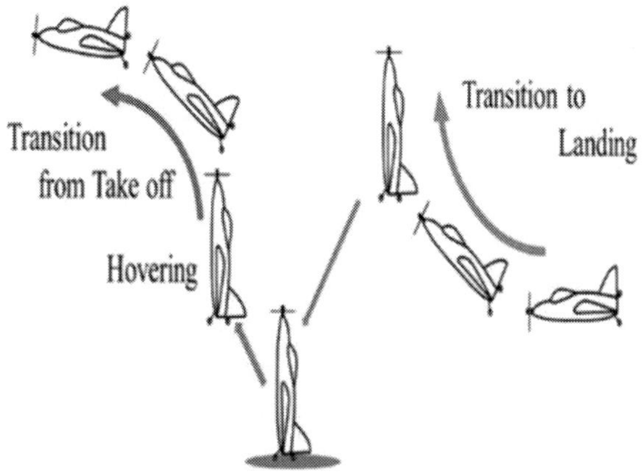

Fig. 1 Takeoff and landing of Tail sitter VTOL

Fig. 2 Test aircraft

Presented at the 6th Asian/Austrailian Rotorcraft Forum / Heli Japan 2017, Kanazawa, Japan, November 7-9, 2017.

TABLE I TEST AIRCRAFT SPECIFICATION

Weight (include battery)	738[g]
Weight (not include battery)	566[g]
Radius of rotor	65[mm]
Distance of center of rotor	115[mm]
Wing span	720[mm]

AIRCRAFT DESIGN

Thrust measurement

In order to design and make the wing attached to the fuselage, thrust measurement was performed first. The motor was attached to the thrust force measurement table, the thrust measurement table was placed on the scale, thrust was generated downward, and the magnitude of the thrust was measured. As a result of measuring the thrust of four motors, average value of the maximum thrust was 3.34 N. From this, 2.04 N which is 60% of maximum thrust was regarded as the thrust of a propeller. Since the test aircraft equipped four motors, we designed the wing with the total thrust 8.17 N.

Wing design

First of all, we selected the airfoil as main wing. There is a camber to generate a large lift in ordinary airplanes, but in the test aircraft, unnecessary force is generated at hovering due to the camber, so NACA0015 of symmetrical wing without camber was selected. Next, the main wing area was calculated. The main wing area can be expressed by equation (1), where L is the lift force required for horizontal flight, ρ is the air density, V is the body speed, S is the main wing area, and C_L is the lift coefficient. The lift force shall be obtained by subtracting the vertical component of the thrust measured from the weight of the aircraft. The lift coefficient shown in Figure3.

$$S = \frac{2L}{\rho V^2 S C_L} \tag{1}$$

Main wing area was calculated as 0.177m^2 by equation (1) and Figure 3. Assuming that the lift force is 8.17N, the air density is 0.119 kgs^2/m^4, the speed is 6 m/s, and the angle of attack is 15 °. In the previous study, the wing shape was made into a long and narrow rectangle, so it became affected by disturbance, so in this research we adopted a receding wing. In the previous experiment, the moment to down the nose due to the lift generated by the wing during the transition from the horizontal flight to the vertical hovering became larger than the moment to up the nose caused by the differenced thrust of multirotor. Since the test aircraft could not up the nose and could not return the posture, we mounted the elevon behind the main wing in order to create a momentum to assist transition in case of aircraft transition. Finally, main wing is 0.183 m^2 in total for wings and it meets the necessary area.

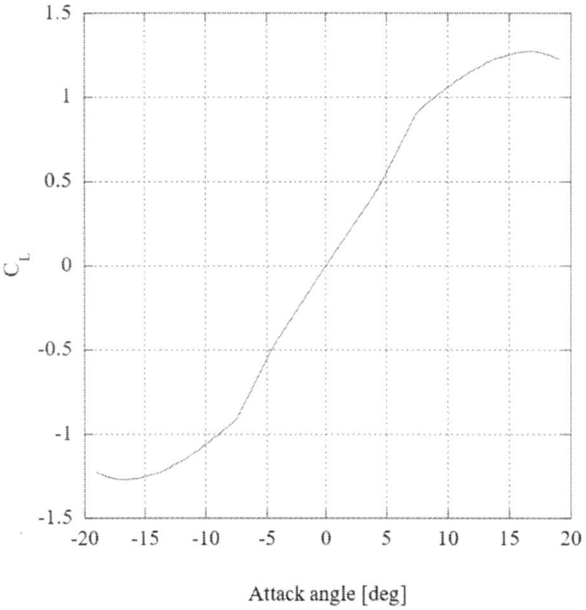

Fig. 3 Lift coefficient

Aircraft assembling

We assembled airframe using balsa wood. Finally, the weight of the main wing is 102g.The aerodynamic center of this test aircraft is located 283 mm from the ground and the center of gravity is 325.3 mm from the ground. If the force which up the nose by the air current during horizontal flight is affected and the aerodynamic center is below the center of gravity, the attack angle and the lift increase. If the aerodynamic center is at the same position as the center of gravity, the lift cannot cancel the force. However, if the aerodynamic center is below the center of gravity, the lift will generate moment to down the nose, and it cancels the force due to disturbance and make stable flying.

HOVERING TEST

Hovering test procedure

Hovering experiments were carried out using the test aircraft. In this study, "Stabilize" mode was used for attitude control at hovering. After taking off, we sent a signal of the target posture value from the transmitter to the airframe. It was confirmed that aircraft followed the target angle. Furthermore, we investigated the aircraft characteristics during hovering.

Results and Discussion

The hovering experiment was carried out in the riverbed of the Yoshino River in Tokushima. The scene of hovering experiment is shown in Figure 4. Figure 5 and Figure 6 show the results of the pitch and roll angle response of the experiment. The red line in Figure 5 shows the target value of the pitch angle of the aircraft, and the blue line shows the attitude pitch angle of the actual aircraft. In addition, the red line in Figure 6 indicates the target value of the Roll angle of the aircraft, and the blue line indicates the attitude roll angle

Fig. 4 Test landscape

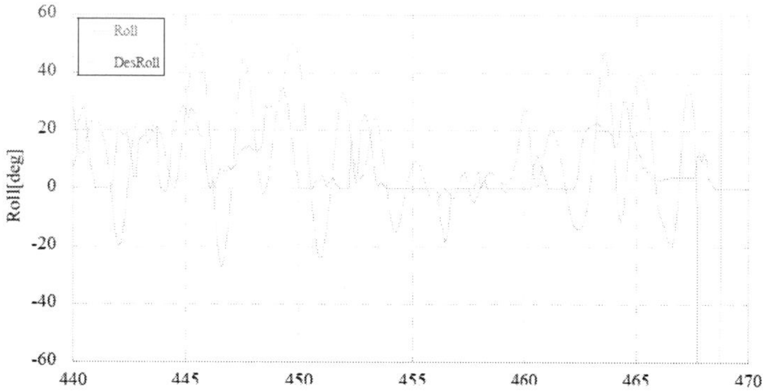

Fig. 5 Roll of Test aircraft

of the actual aircraft. From Figure 5 and Figure 6, the posture during hovering follows the target value. However, because the experiment was conducted outdoors, the aircraft was affected by the wind and it was not able to stabilize the posture. In addition, this time we mounted the elevon to control attitude of stabilizing the posture at horizontal flight, but since the propeller and the elevon did not overlap, propeller slipstream did not hit the elevon and could not control attitude during transition. Based on these results, we thought that it is necessary to stabilize the flight during hovering by increasing the power of the aircraft.

CONCLUSION

We developed a tail sitter type test VTOL driven with multi-copter, and tried hovering. Test aircraft was able to perform hovering. But it was difficult to stabilize during hovering because it was influenced by the wind. In order to solve this problem, we will increase the power of the aircraft body and stabilize the attitude of the aircraft by enlarging the rotor. Also, it was necessary to extend the elevon to the bottom of the propeller, apply the wind from the propeller, and generate force.

ACKNOWLEDGMENTS

This work was supported by JSPS KAKENHI Grant Number16K01330.

REFERENCES

[1]Matsumoto, T., Kita, K., Suzuki, R., Oosedo, A., Go, K., Hoshino, at all. (2010, May). A hovering control strategy for a tail-sitter VTOL UAV that increases stability against large disturbance. In Robotics and Automation (ICRA), 2010 IEEE International Conference on (pp. 54-59). IEEE2 .

[2]SAKAI Yoshihiro, and MIWA Masahumi. "2A2-F06 Development of Tailsitter VTOL using multi rotor system." JSME annual Conference on Robotics and Mechatronics (Robomec) 2015 (2015).

Separation Control using Rail Plasma Actuator on a VR-12 Airfoil

Young-Joon Choi
Graduate Research Assistant
University of Texas at Austin
Austin, Texas, United States

Jayant Sirohi
Associate Professor
University of Texas at Austin
Austin, Texas, United States

ABSTRACT

An experimental study was conducted to alleviate static stall using a rail plasma actuator (RailPAc). This device consists of parallel rails flush mounted to a VR-12 wing with a high-current ($\sim$ 1.3 kA) arc bridging the gap between the rails. The actuator transfers momentum to the flow by generating Lorentz force ($\sim$0.3 N lasting $\sim$1 ms) in the arc. Experiments were conducted in a low speed wind tunnel at two different Reynolds numbers (Re $\sim$ 45,000 and 90,000) and various angles of attack (up to $\sim$ 30°). Particle image velocimetry (PIV) visualized the flow over the passive and actuated airfoil and improvement in lift coefficient was measured using a force balance. The results showed that the RailPAc can improve the aerodynamic performance of a wing. Flow reattachment was observed and demonstrated the actuator's ability to suppress stall over a wide range of angles of attack. In addition, a single RailPAc operation resulted in $\sim$ 40% improvement in post-stall lift and $\sim$ 4° increase in stall angle compared to passive airfoil with a RailPAc. The results provide insight into the actuation mechanism behind a RailPAc and quantified the benefits in post-stall regimes.

INTRODUCTION

The next generation of helicopters will feature added range, speed, and payload. These improvements can be realized with active flow control technologies that rely on unsteady flow disturbance to improve the aerodynamic performance of a rotorcraft (Ref. 1). Past studies have shown that active actuators can suppress stall (Ref. 2), reduce noise (Refs. 3, 4), and control vibrations (Ref. 5) by manipulating the boundary layer, shock, and vortex interactions. In case of static stall, active control on a rotor blade can increase its post stall lift and improve the payload capacity of a helicopter (Ref. 6).

While several actuators demonstrated stall alleviation in academic research, their practical implementation on a full-scale helicopter remains a challenge. An ideal actuator for a rotorcraft application must have large control authority (defined as momentum delivered to the flow) and high bandwidth (frequency of actuation) to control the flow over a wide range of operating envelope of a helicopter. In addition, actuator must preserve the passive performance of a rotorcraft by minimizing its footprint (i. e. compact volume, low mass, and no modifications to the profile of a body). Unfortunately, studies based on mechanical (Ref. 7), fluidic (Ref. 8), combustion (Ref. 9), and plasma (Ref. 10) actuators have yet to produce a device that can deliver substantial aerodynamic benefits with minimal cost to the passive performance of a blade. Mechanical, fluidic, and combustion devices require moving parts and external thermal/fluid source that add complexity as well as mass to the blade structure. In addition, the trade-off between actuator stroke and bandwidth, as well as the time scales associated with the thermal and fluid cycles, limit their actuation frequency. While higher bandwidth is possible with plasma actuators, studies have shown that the actuators based on electrohydrodyanamic forcing are ineffective for controlling external flow with velocities in excess of $\sim$10 m/s (Refs. 11–13). The deficiencies of past concepts demonstrate a clear need for a new actuator suited for rotorcraft application.

A new plasma actuator based on magnetohydrodynamic forcing has been developed at the University of Texas at Austin to address the drawbacks of past actuator concepts. A Rail Plasma Actuator (RailPAc), first documented by Pafford et. al. (Ref. 14), possesses all the required attributes of an ideal actuator. It is a railgun based device that mediates momentum transfer through an electrical discharge in the flow. The absence of moving parts allows this type of actuator to operate with a bandwidth on the order of several $\sim$kHz. In addition, the actuator can be mounted conformal to an existing aerodynamic body and requires no structural modification or performance degradation on the passive configuration.

The device consists of parallel rails flush mounted to the aerodynamic surface and a thermal arc that bridges the gap between the rails (Figure 1). The current through the rails generates a magnetic field ($\vec{B}$) that couples with the arc current density ($\vec{j}$) to impart a Lorentz force ($\vec{F_B}$). The arc accelerates along the rails and imparts momentum to the surrounding air through compression and entrainment effects.

Presented at the 6th Asian/Australian Rotorcraft Forum / Heli Japan 2017, Kanazawa, Japan, November 7-9, 2017.

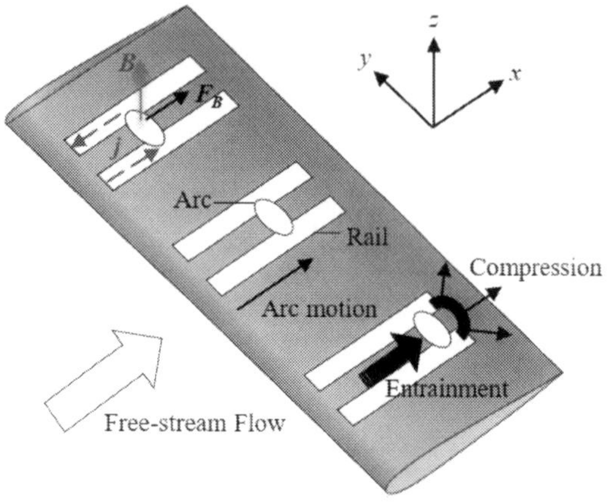

Fig. 1: A schematic of a RailPAc mounted on an aerodynamic surface.

The objective of this paper is to demonstrate static stall alleviation with a RailPAc on a rotorcraft blade. This is achieved by mounting a prototype RailPAc on a VR-12 wing and conducting the experiment at various Reynolds numbers (*Re*) and angles of attack (α). As the operating condition is varied from pre-stall to post-stall regimes, various measurements, including electrical, high-speed imaging, and PIV, are performed to give insight into the RailPAc actuation. Lift measurements are conducted to quantify the improvement in the aerodynamic performance of a VR-12 wing.

EXPERIMENTAL SETUP

The experiments were conducted in an open test section and a closed circuit wind tunnel (Figure 2) at the University of Texas at Austin. The maximum free-stream velocity (U_∞) is 20 m/s and the test section measures 0.56 m by 0.91 m. Two mounting arms with force balances position the wing at the center of the test chamber. The force balances are constructed out of strain gauges mounted on brass flexures that are only sensitive to the force applied in the lift direction ($\hat{e}_z$-direction). Note that the two side plates (polycarbonate sheets) at the ends of the wing prevent formation of wing tip vortices. The side plates are structurally decoupled from the wing, such that the aerodynamic forces acting on the side plates are not measured by the force balances.

The test article is a VR-12 wing (chord c = 79 mm and span s = 1020 mm) comprised of a foam core with carbon fiber layers (Figure 3). Note that the three RailPAcs are mounted along the span direction ($\hat{e}_y$-direction) in a notch built into the mid-span location. This array covers $\sim 7\%$ of the wing span. Each RailPAc consists of copper anode and Elkonite cathode rails (width h = 2 mm with 2 mm spacing between rails) flush mounted to the suction side of the wing. The rails permit the arc to transit from a location close to the leading edge ($y/c \sim$ 10 %) all the way to the trailing edge. A ceramic insert is present between the rails to electrically isolate the anode from the cathode. The PEEK inserts separate each RailPAc from one another with a 25.4 mm gap. This distance was chosen to prevent the three separate arcs in each RailPAc from merging into a single arc. Electrical wires run underneath the PEEK inserts to connect the cathode of one RailPAc to the anode of the adjacent RailPAc (i. e., the RailPAcs are connected electrically in series). A notch built into the foam core allows the RailPAc power supply to connect to the RailPAc array from the mounting end of the wing. Figure 3a shows the top view and side view of the current flow.

The free-stream velocity was measured using a pitot-static probe. The Reynolds number based on chord were 45,000 and 90,000, corresponding to the free-stream velocities of 10 m/s and 20 m/s, respectively. The angle of attack (α), measured with an inclinometer attached to the wing spar, can be varied from 0° to 30°.

Flow Visualization

Both qualitative and quantitative flow visualizations were performed by introducing Glycol seed particles (1μm diameter and $\rho/\rho_a \sim 1.1$) into the wind tunnel. The motion of the seed particles was captured in the $\hat{e}_x\hat{e}_z$-plane using a Phantom Miro M-310 camera fitted with 100 mm lens. The framing rate was 4000 fps and viewing window was 2 by 1.5 chord lengths. A 2D particle imaging velocimetry (PIV) was performed by projecting a $\sim$ 2 mm thick laser sheet on the top surface of the wing with the laser sheet centered on the mid-span RailPAc gap (Fig. 2). The sheet was formed by redirecting a 29 mJ/pulse Nd/YLF laser beam from the ground to the top of the windtunnel via system of lenses and mirrors. The laser was pulsed synchronously with the camera frames.

The span-wise effect of a RailPAc was measured by moving the laser sheet to different positions along the $\hat{e}_y$-direction. The sheet location was moved at an increment of 25.4 mm from 46.1 mm ($y/s \sim 5\%$) to 160.4 mm ($y/s \sim 16\%$) away from the center of the RailPAc array. A 2D PIV was performed at each location and the results of three RailPAc operations were averaged using the arc initiation as phase reference.

RailPAc Circuit

The RailPAc circuit (Figure 4) is comprised of a single electrolytic capacitor (3.5 mF), trigger circuit, and a RailPAc array. A Xantrex XHR 600 DC power supply charges the capacitor to 300 V. The trigger circuit, constructed with an automobile ignition system (HyFire), momentarily provides a -12 kV pulse to one of the cathode of the RailPAc array, initiating the arc along the graphite trail (Fig. 3b). A high voltage diode, situated between trigger circuit and capacitor, prevents the -12 kV pulse from discharging into the ground. The pulse creates a breakdown along the graphite trail on the outermost RailPAc gap, eventually leading to a sequential break down of all RailPAc gaps in the array. A large current ($\sim$ kA) conducts across the break down region and a thermal arc is formed as a result.

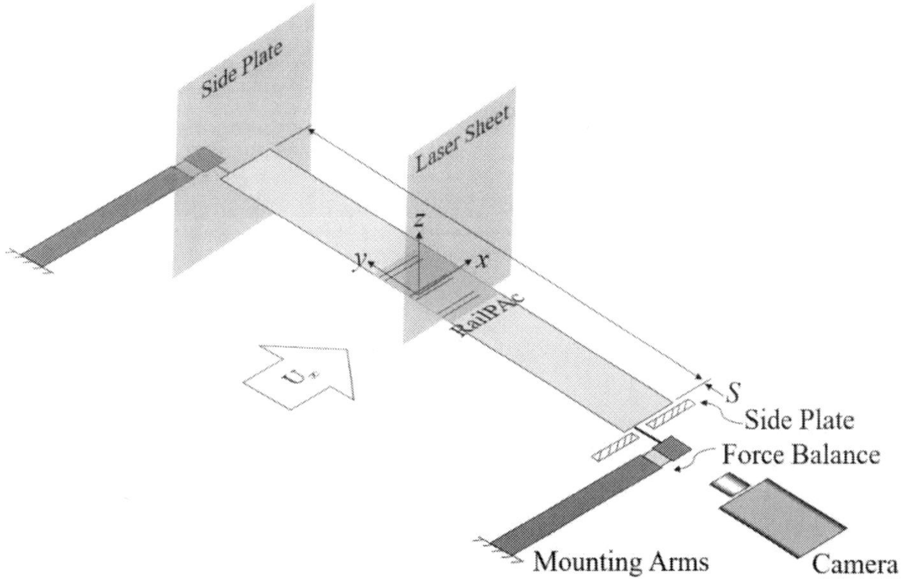

Fig. 2: A wind tunnel schematic showing the VR-12 test article, PIV, and force balance setup.

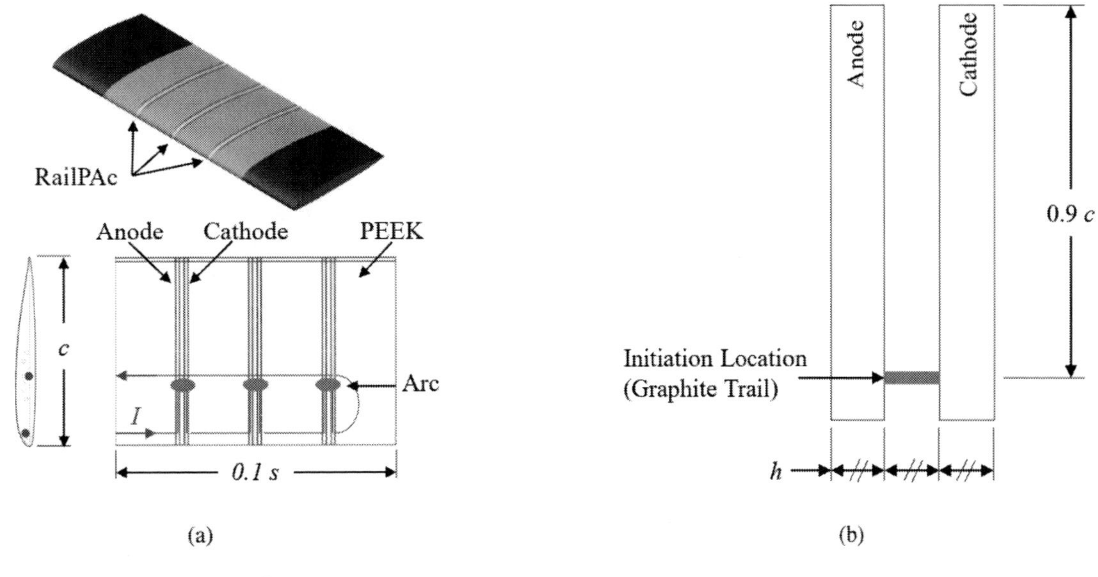

Fig. 3: (a) A schematic of the RailPAc array mounted on a VR-12 wing and (b) rail configuration of a single RailPAc.

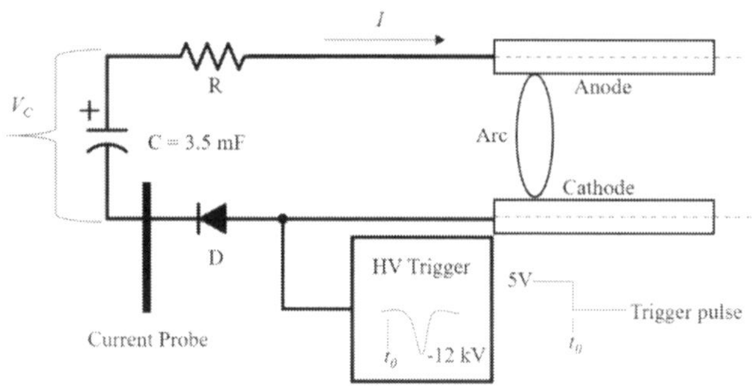

Fig. 4: A Schematic of a RailPAc power supply.

Electrical measurements were conducted with a current probe (Pearson 110A) and high voltage probes to measure the arc current and arc voltage, respectively.

Transit Video

The arc motion was captured from above using a Phantom Miro M-310 camera. A high f-stop ($f/30$) lens with 100 mm focal length and small exposure time (1μs) prevented image saturation from the high intensity arc. A band-pass optical filter centered at the wavelength of 650 nm filtered out the PIV laser (527 nm). The camera frame rate was 20,000 fps, and the camera was oriented such that all three RailPAcs could be captured simultaneously. The arc position as a function of transit time was found by locating the region of highest intensity along the rail gap. The results characterized the arc velocity as well as changes in the arc shape.

RESULTS

The experimental results were obtained by operating the wind tunnel at two different velocities. Table 1 summarizes the tunnel operating conditions. The momentum coefficient (C_μ) was defined as,

$$C_\mu = \frac{|F_B|_{max}}{\rho U_\infty^2 hc} = \frac{L' I_{max}^2}{\rho U_\infty^2 A} \quad (1)$$

The actuator area was estimated as $3hc$. The expression for the Lorentz force F_B was obtained from measurements conducted by Choi *et al.* (Ref. 15). Their studies on a bench-top RailPAc showed that the Lorentz force can be modeled as the product of the inductance gradient (a measure of the induced magnetic field) and square of the arc current. Inductance gradient (0.33 μH/m) was obtained from the published value for a similar rail geometry (Ref. 18). A peak current (I_{max} ~1.3 kA) was measured by the current probe (Rogowski coil). These values gave a peak force ($|F_B|_{max}$ ~0.3 N), corresponding to a peak C_μ of 4.8 and 1.2 at Re of 45,000 and 90,000, respectively.

The test matrix (Table 2) summarizes the free-stream velocity and angles of attack explored in this study.

Table 1: Summary of the experimental conditions.

Parameter	Re ~45,000	Re ~90,000
U_∞	10 m/s	20 m/s
T_∞	20°C	22°C
P_∞	106.7 kPa	106.2 kPa
C_μ	4.8	1.2

Table 2: Summary of the free-stream velocity and Reynolds numbers.

α	Re ~45,000	Re ~90,000
0°	X	X
4°	X	X
8°	X	X
10°	X	X
12°	X	X
14°		X
17°	X	X
22°	X	X
26°	X	X
30°	X	

Transit Video

The arc structure and motion were characterized using high-speed images of the arc transit (Figure 5). The arc structure can be described as a luminous filament bridging the gap between the rails. The luminosity is not uniform across the gap with highest luminosity found near the electrode attachment points (arc roots). Note that the arc is not confined within the actuator width (~ 6 mm) and extends along the span-wise direction to ~18 mm. This structure reveals that the arc interacts with a large volume of air, even over the wing area not directly covered by the RailPAc.

The initial transit of the arc ($t \leq 0.16$ ms) can be described as a coherent motion of a single filament arc. The end of the transit is marked by deformation of the arc into distinct cathode and anode roots. As shown in the snapshot at t = 0.32 ms

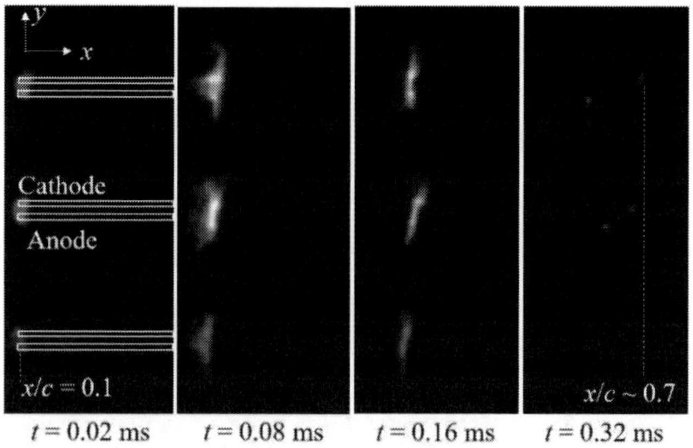

Fig. 5: High-speed images of the arc transit. Time t is referenced from the arc initiation time at t = 0 ms.

of Figure 5, the anode root is anchored at a upstream location, while the cathode root transits smoothly to a downstream location. Choi *et al.* (Ref. 15) also observed this phenomenon in their RailPAc transient force measurement and showed that the anchoring of anode root reduces the Lorentz force. Therefore, the ability of the arc to transfer momentum to the flow is highest in the initial transit stage, and is reduced towards the end of the transit.

The cathode root is observed to transit upto $x/c \sim 0.7$, such that the arc transit covers $\sim 60\%$ of the wing chord. The transit velocity was obtained by tracking the position of highest arc luminosity along the rail. The average arc transit velocities were 54 ± 1 m/s and 58 ± 1 m/s for U_∞ = 10 m/s and 20 m/s, respectively. These average velocities were obtained with 90 RailPAc operations for each free-stream velocity. The transit velocity did not vary with respect to the angle of attack, revealing that the arc transit is insensitive to the surface pressure distribution of the wing.

Electrical Measurements

The electrical measurements (Figure 6) show that the arc current is delivered via a short duration pulse, with the peak current occurring approximately 1 ms after initiation. The peak current is 1.3 kA at the initial capacitor voltage of 300 V. The instantaneous power can be computed by multiplying the capacitor voltage with the arc current. The peak electrical power is 350 kW, however, the total energy discharged is 22 J per RailPAc per pulse due to the small duration of the current pulse. Although the initial capacitor voltage is constant throughout the experiment, the discharge voltage and current varies between each run. In particular, the variation between experiments becomes more significant towards the end of the arc transit, in comparison to the time shortly after the arc initiation. It can be concluded that the electrical characteristics of the arc are repeatable if the arc motion is coherent (shown by initial transit stages in Figure 5), while the measurement is not repeatable if the arc exhibits anchoring.

Flow Visualization

The high luminosity and temperature (causing vaporization of seed particles) of the arc precludes measurement of velocity vectors during the arc transit using PIV. However, the effect of a moving arc on the separated flow can be qualitatively assessed from the seed particle images. The passive wing (Figure 7a) is in deep stall with separation occurring at the leading edge. A shear layer forms between the free-stream region and the separated region. The RailPAc generates an arc that extends to a significant vertical distance ($\sim$20 mm) above the wing surface (Figure 7b). The high temperature of the arc vaporizes the seed particles causing dark patches to appear above the wing (Figure 7c). These patches advect at the local convection velocity and move upward due to thermal convection. Free-stream fluid is entrained behind the patches and the separation over the entire suction surface of the wing is suppressed (Figure 7d). This process is similar to

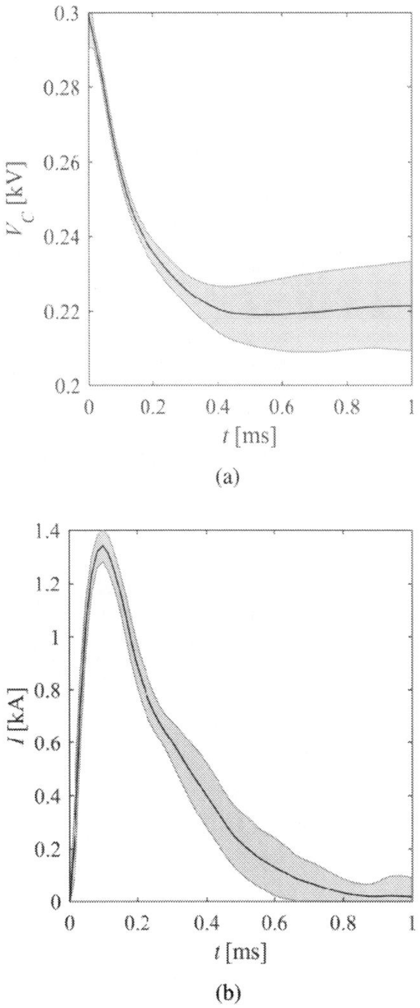

Fig. 6: Mean electrical measurement on a RailPAc power supply showing (a) capacitor voltage (V_C) and (b) discharge current (I) as a function of time. The uncertainty bounds represent one standard deviation of mean measurement.

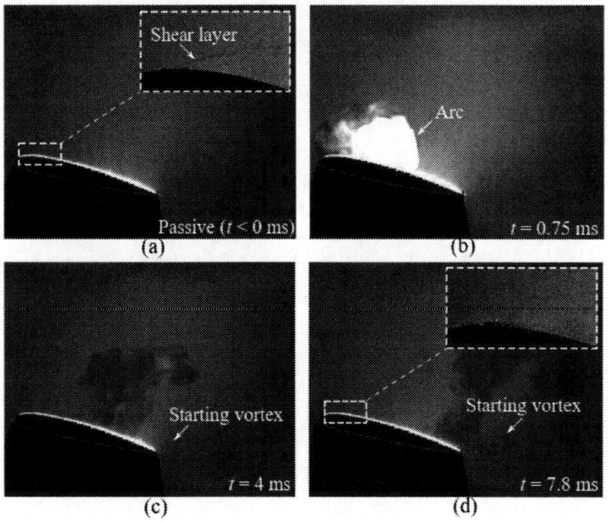

Fig. 7: Side view of the arc transit and seed particles (Re = 45,000 and α = 12°)

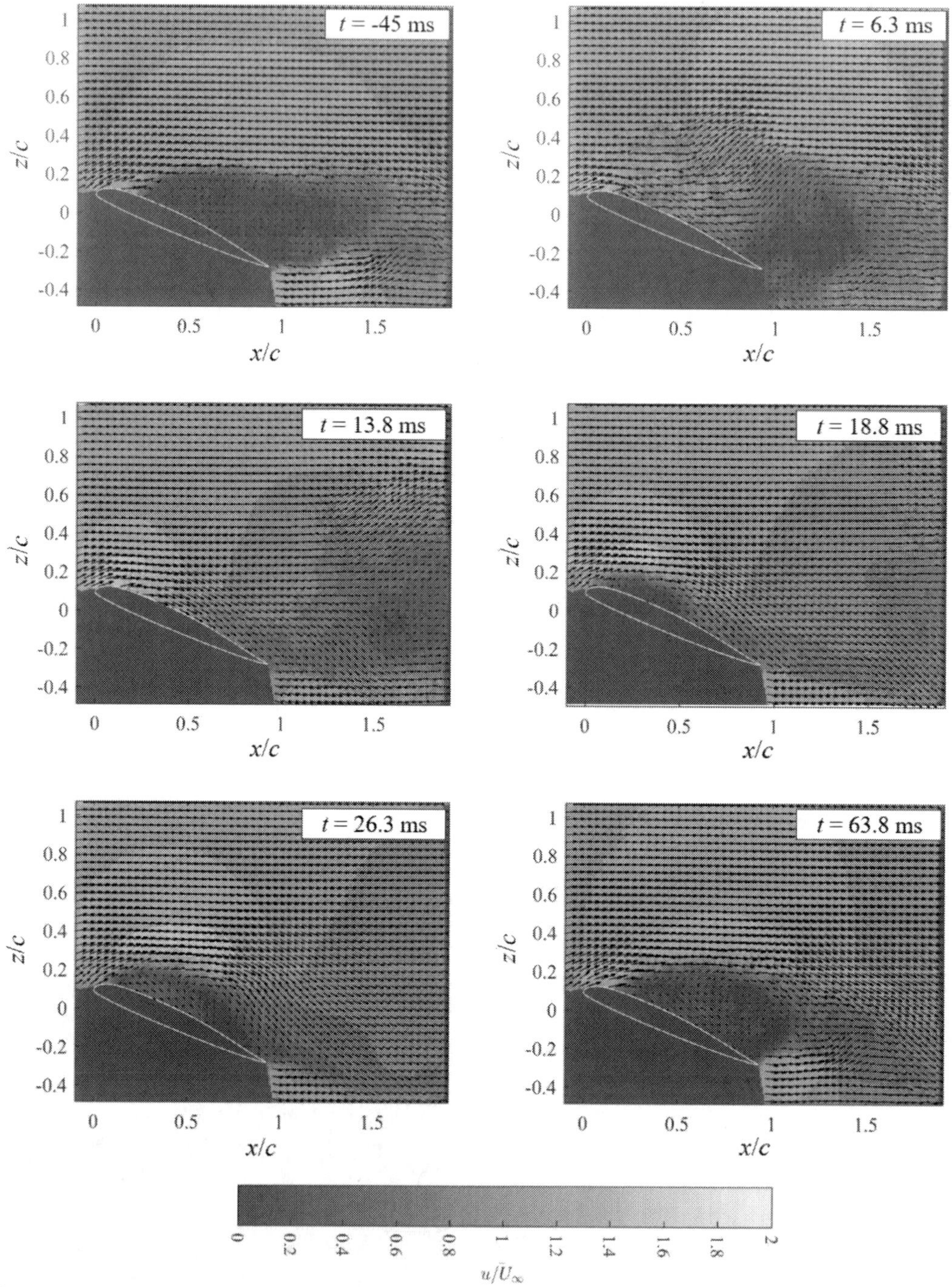

Fig. 8: Measured velocity fields for U_∞ = 10 m/s and α=22° obtained with moving averages of 10 successive frames.

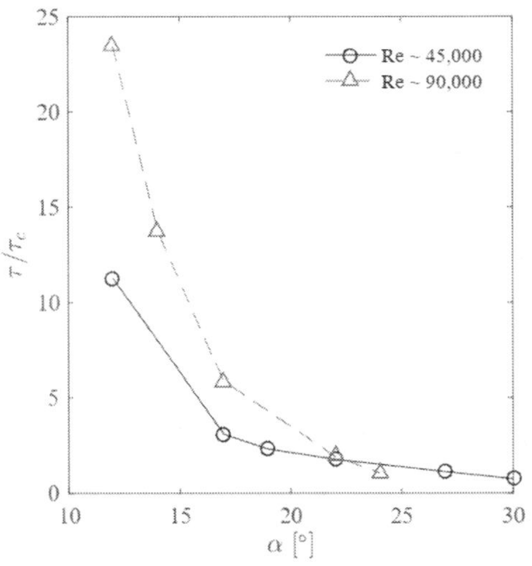

Fig. 9: Reattachment time (τ) normalized by convection time (τ_c) for different Reynolds numbers (*Re*) and angles of attack (α).

the dynamic reattachment observed by Greenblatt and Wygnanski (Ref. 17). They proposed that active flow actuators reattach flow by removing low momentum fluid from the surface and replacing the lost fluid with the free-stream air. In the case of RailPAc, the entrainment of free-stream is due to the Lorentz force acting on the arc. This process energizes the boundary layer and promotes reattachment. The formation of a starting vortex signifies the recovery of the bound circulation.

The motion of the arc on the wing surface induces a large flow-field disturbance as a result of flow separation and reattachment. The change in the velocity field over the wing is captured using a high-speed PIV, before the arc initiation as well as after the arc transit. As shown in Figure 8, the wing is initially in deep stall with the flow separating near the leading edge. The separation region is characterized by a large vortex above the trailing edge and a reverse flow that extend over the upper surface. Shortly after the actuation (t = 6.3 ms), the free-stream flow is entrained into the separated region, and the reattachment occurs. At t = 13.8 ms, the flow is attached over the wing and a large increase in measured velocity is observed at the location of maximum thickness. A vortex near the leading edge begins to grow and continue promote flow reattachment at the trailing end of the wing. The vortex eventually separates from the leading edge, convects down stream, and the wing re-enters deep stall. The unsteady flow persists for $\sim$ 100 ms in the form of large scale vortices that shed at the trailing-edge of the wing. These structures lead to wake oscillations that decrease over time, and the wing resumes its passive condition. These remarkable phenomena are observed at all post-stall angles of attack, albeit at different time scales.

The flow over the wing remains reattached for a finite duration (referred to as reattachment time, τ), that can be measured by examining the changes in the shear layer (Figure 7a). The passive wing develops a shear layer with a particular angle with the wing surface. After actuation, the angle decreases until the shear layer is suppressed due to flow reattachment. After some time, the flow separation causes the angle to increase until the shear layer resumes its passive condition. The duration between the RailPAc initiation and the instant at which the shear layer returns to its passive condition is defined as the reattachment time. These times were measured using high-speed images of the seed particles at various angles of attack and Reynolds numbers (Figure 9).

The reattachment time was found to decrease with increasing angle of attack; however, it is insensitive to the Reynolds number. If the reattachment time is normalized by the convection time over the airfoil chord (Figure 9), the normalized time is smaller for lower Reynolds number than for the higher Reynolds number. In fact, the normalized time for $Re \sim 45{,}000$ is half that of the $Re \sim 90{,}000$ due to the two-fold difference in the convection time scales. These observations reveal that the reattachment time is strongly influenced by the strength of the adverse pressure gradient, and insensitive to the free-stream condition for low Reynolds number flows. Note that the strength of adverse pressure gradient rises dramatically with increasing angle of attack. Hence, the RailPAc actuation depends strongly on the pressure distribution over the wing surface in the Reynolds number regimes considered in this study.

The RailPAc is able to achieve flow reattachment at various angles of attack without exerting adverse effects on the passive wing. Figures 10 and 11 show the flow over the passive and actuated (after one convection time scale) wing at different angles of attack for $Re \sim 45{,}000$ and $Re \sim 90{,}000$, respectively. At pre-stall angles of attack, the high-temperature fluid introduced by the moving arc momentarily disturbs the wing wake. The flow over the wing remains attached albeit with small reduction in the velocity magnitude. This velocity change is attributed to the rail vaporization that ejects large particles into the flow. A bias error from the non-uniform sizes of the seed particles manifests as a small change in the measured velocity before and after actuation at pre-stall conditions.

The RailPAc can reattach flow at high angles of attack, although full reattachment of the flow over the entire wing surface is possible only for angles of attack upto $\alpha \sim 17°$. At higher angles of attack, the flow reattachment occurs over a portion of the wing before the re-occurrence of stall. At all angles of attack, a significant increase in the flow velocity over the wing is observed in addition to the reduction in the size of the wake. These observations reveal that the RailPAc actuation at post-stall angles of attack results in lift enhancement and drag reduction.

Span-wise Flow Visualization

The span-wise effect of a RailPAc was examined by performing PIV at different locations along the $\hat{e}_y$-direction away from the actuator. The flow-field at different span-wise locations is shown in Figure 12, with measurements taken at $Re \sim 45{,}000$ and $\alpha \sim 12°$. The suction surface of the wing and the outlines of the rails are provided as a reference. Before the arc

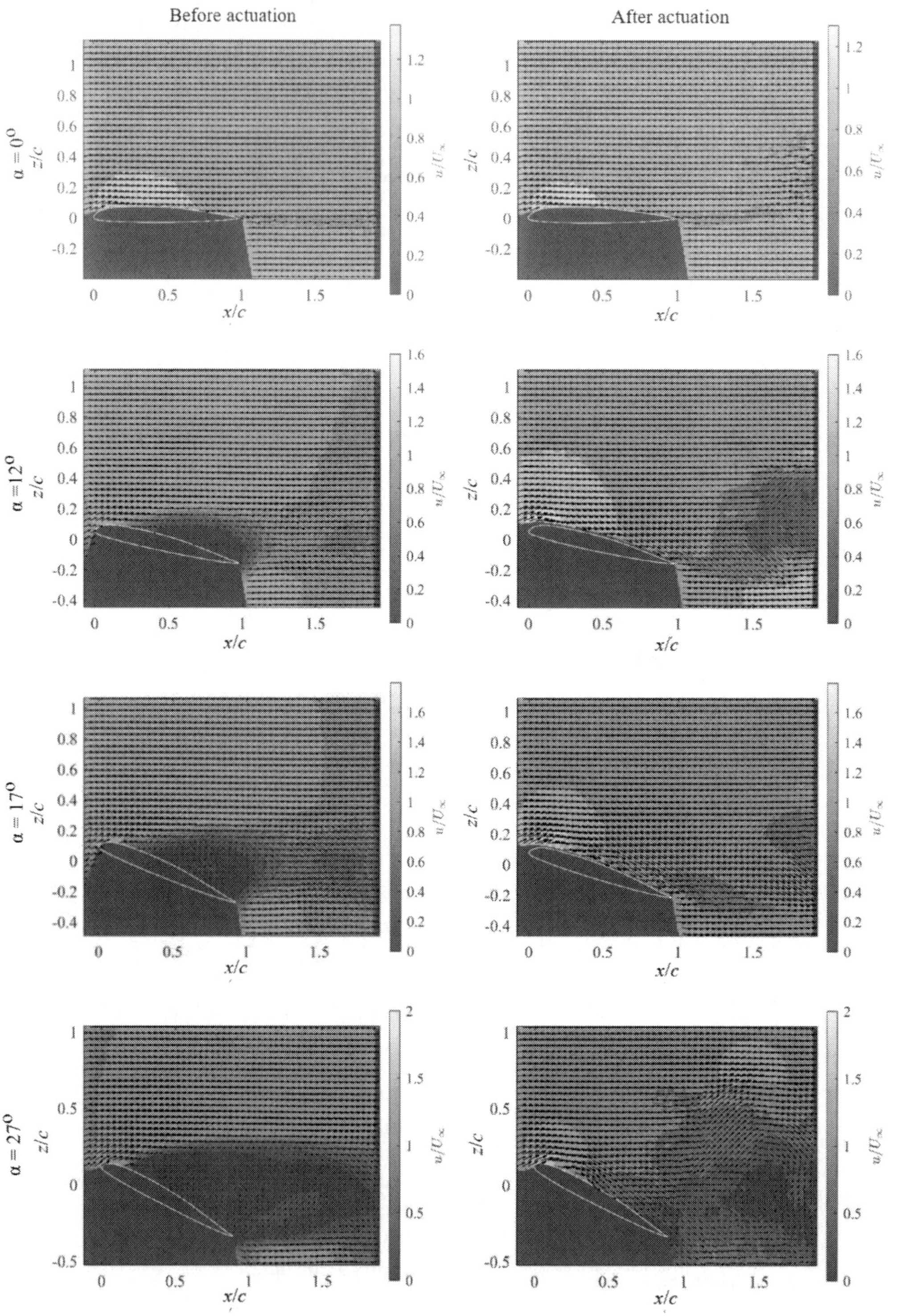

Fig. 10: PIV measurement before actuation and one convection time scale after actuation at $Re \sim 45{,}000$.

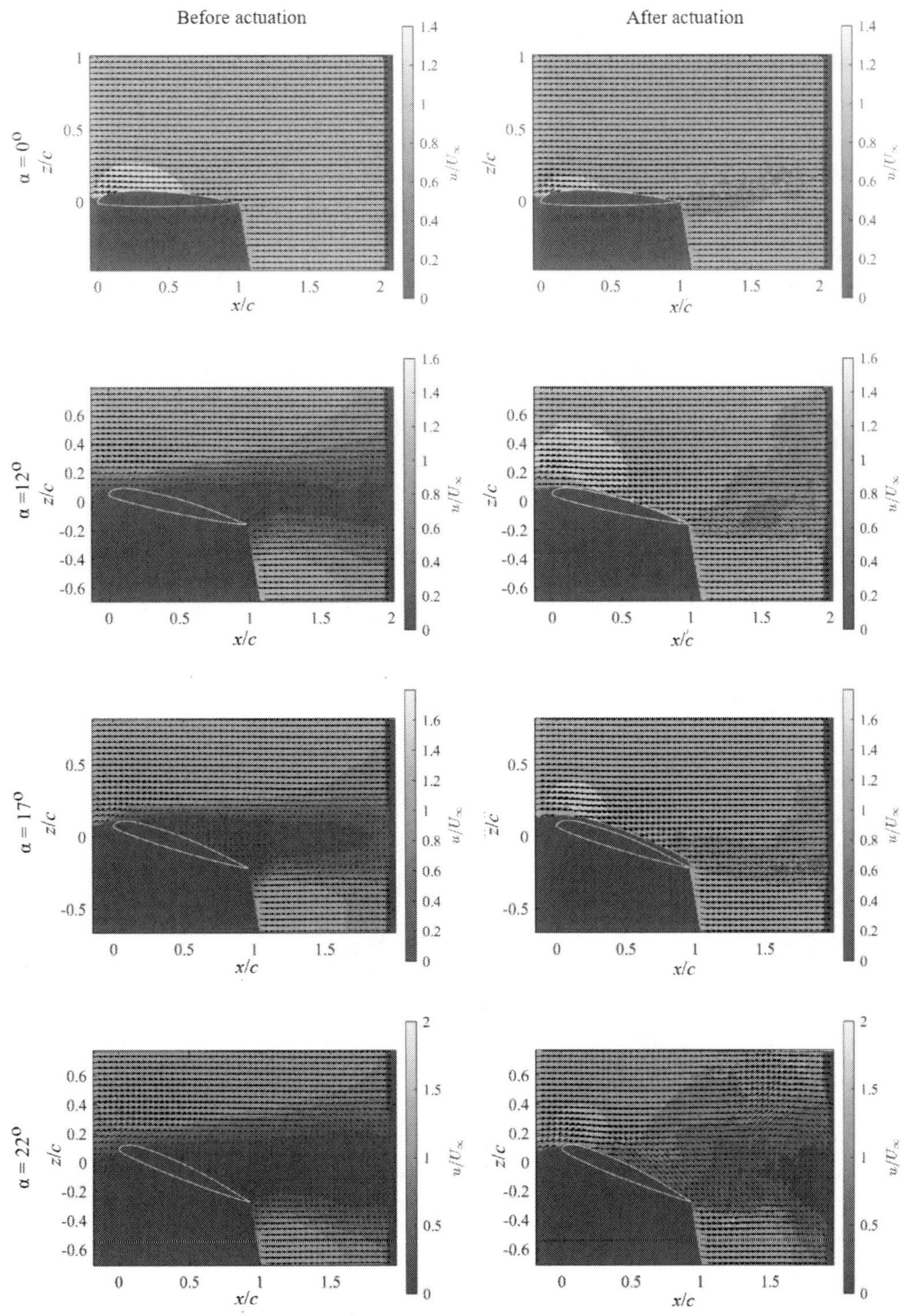

Fig. 11: PIV measurement before actuation and one convection time scale after actuation at $Re \sim 90{,}000$

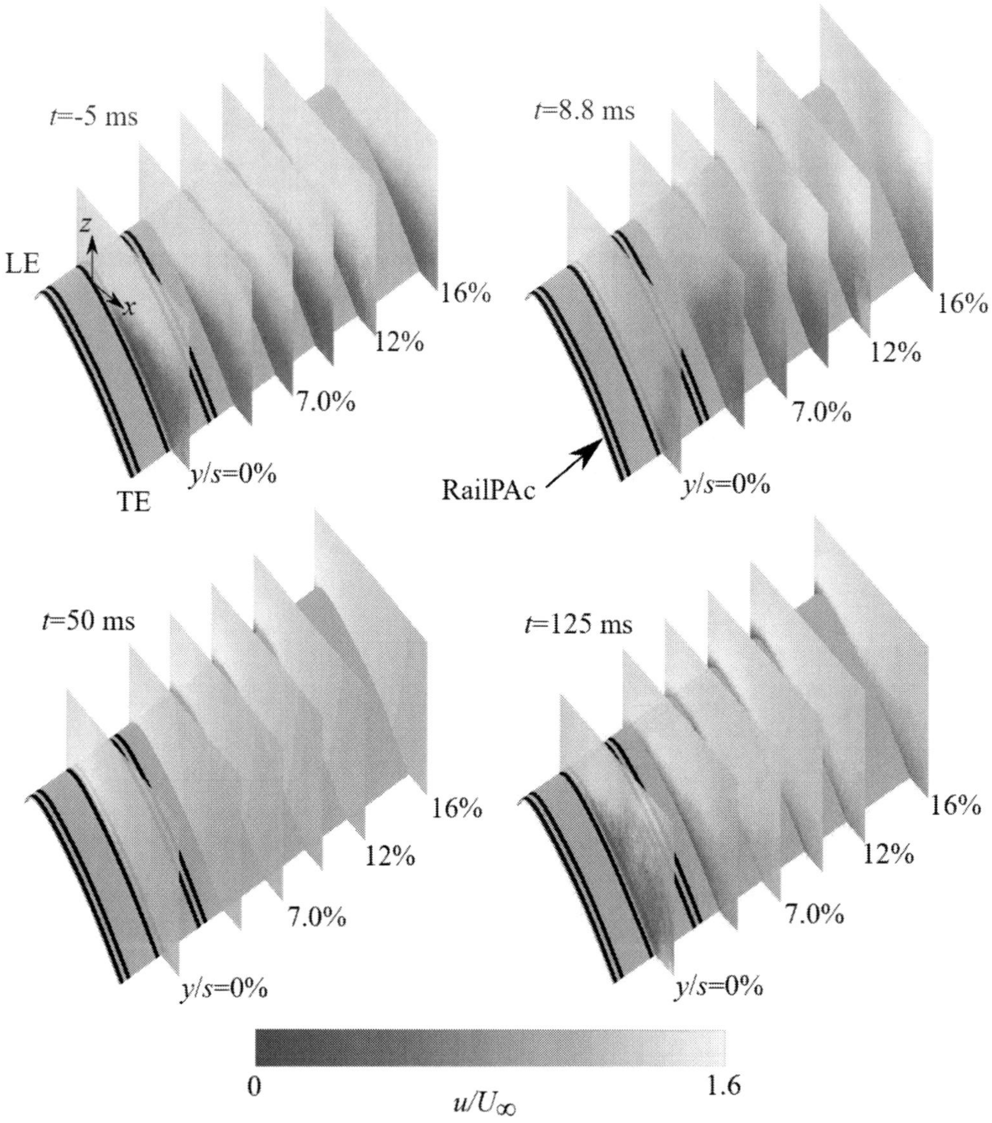

Fig. 12: Span-wise PIV measurements at $Re \sim 45,000$ and α=12°. Time (t) is referenced from the arc initiation. Outline of the rails is indicated.

initiation (t =-5 ms), the wing is in static stall, with separated flow observed at all span-wise locations. After the arc passage, the flow begins to reattach over all the observed locations, albeit at different time scales. For example, at $t = 8.8$ ms, the flow over the center of the RailPAc array (y/s =0%) is fully attached to the trailing edge, while all other locations are partially attached upto the mid-chord location. Thus, the RailPAc attaches flow directly over the arc passage first, and the reattachment progressively continues along the span-wise direction.

The induced velocity measurements performed by Choi *et al.* (Ref. 18) on a high-speed arc provide physical reason behind this delay. Their measurements revealed that the entrainment of the surrounding air into the arc wake is delayed by the thermal effect from the high arc temperature. When the arc travels at high speeds, a stratified flow is created behind the arc that separates the surrounding air from the high-temperature wake. The surrounding air experiences temporary reverse flow against the direction of the arc motion and entrainment is delayed until some time after the arc passage. Hence, the flow over the RailPAc is attached first, while the outboard locations experience delayed reattachment.

The flow over the wing is fully attached at t=50 ms at all the observed locations (upto ~16%). This is remarkable given that the RailPAc array covers only ~ 7% of the span. The result shows that the actuator's effect is highly three-dimensional and the flow disturbance caused by a RailPAc is not confined to the area over the actuator array.

As shown at t =125 ms of Figure 12, the flow separation begins from the actuator location. This is attributed to the installation of the rails on the top surface of the wing. Due to the manufacturing tolerance, a small ledge was present between the wing surface and the rails and prompted an earlier separation of the boundary layer. While stall clearly occurs over the RailPAc array first, the sequential order of flow separation along the span-wise direction away from the actuator could not be discerned from the measurements.

Lift Measurement

The maximum change in the loadcell signal ($|F_R|_{max}$) was found for each angle of attack and Reynolds number. The change in the coefficient of lift is defined as,

$$\Delta C_L = \frac{|F_R|_{max}}{0.5\rho U_0^2 cs} \tag{2}$$

This value was added to the passive coefficient of lift to quantify the increase in aerodynamic performance due to RailPAc actuation. Figure 13 shows both passive and actuated lift characteristics of a RailPAc modified wing for different Reynolds numbers.

The measured lift coefficients reveal transition to stall at angles of 8° and 10° for $Re \sim 45,000$ and $90,000$, respectively. At the angles of attack below the stall angle, the lift characteristics of the actuated wing do not vary significantly from those of the passive configuration. Thus, the thermal disturbance accompanying the arc passage has negligible effect on the subsequent lift characteristics of the wing. In contrast, a remarkable improvement in lift characteristics at post-stall angles of attack can be observed. The stall angles for the actuated wing are 12° and 14° for $Re \sim 45,000$ and $90,000$, respectively. These values represent 4° increase over the passive wing due to RailPAc actuation. The maximum C_L for the actuated configuration are ~ 1.3 and ~ 1.4 for $Re \sim 45,000$ and $90,000$, respectively. Thus, ~ 40% increase over the passive wing with a RailPAc are observed. Lift improvement continues beyond the stall angles, with RailPAc delivering a remarkable performance improvement at all angles of attack tested in this study.

SUMMARY AND CONCLUSIONS

In summary, a prototype RailPAc was embedded in a VR-12 airfoil operating at $Re \sim 45{,}000$ and 90,000. The actuator was characterized with high-speed imaging and electrical measurements. The images revealed an arc lasting ~ 1 ms traveling at a velocity of ~ 54 m/s and 58 m/s at the two Reynolds numbers tested. The transit velocity was insensitive to the angle of attack. The electrical measurements captured a peak current of 1.3 kA with each actuator consuming ~22 J per operation.

Particle image velocimetry (PIV) captured the changes in the flow over the airfoil before and after the actuation. The images of the seed particles captured the formation and convection of a high-temperature wake behind the arc, entrainment of free-stream fluid, and (for post-stall angles of attack) reattachment of the flow over the airfoil. These observations demonstrate that the flow separation is suppressed by the entrainment effect due to Lorentz force and thermal effect that removes the low momentum flow away from the airfoil surface. The comparison of PIV results before and after actuation showed that the RailPAc is capable of suppressing flow separation at different angles of attack. A full reattachment of the flow was possible at angles of attack up to ~17°, while partial reattachment was observed at higher values. The reattachment time was found to be sensitive to the angle of attack, but insensitive to the Reynolds number.

The PIV images quantified the flow response before and after the RailPAc actuation, while the measurements from the force balance quantified the improvement in the lift performance. At both Reynolds numbers, the stall angle was increased by 4° compared to the passive airfoil with RailPAc. The RailPAc increased the maximum lift coefficient by ~ 40% compared to the passive case. The results obtained demonstrate that the RailPAc was effective in suppressing static stall at high angles of attack.

ACKNOWLEDGMENTS

This material is based upon work supported by, or in part by, the U. S. Army Research Laboratory and the U. S. Army Research Office under contract/grant number W911NF1410226.

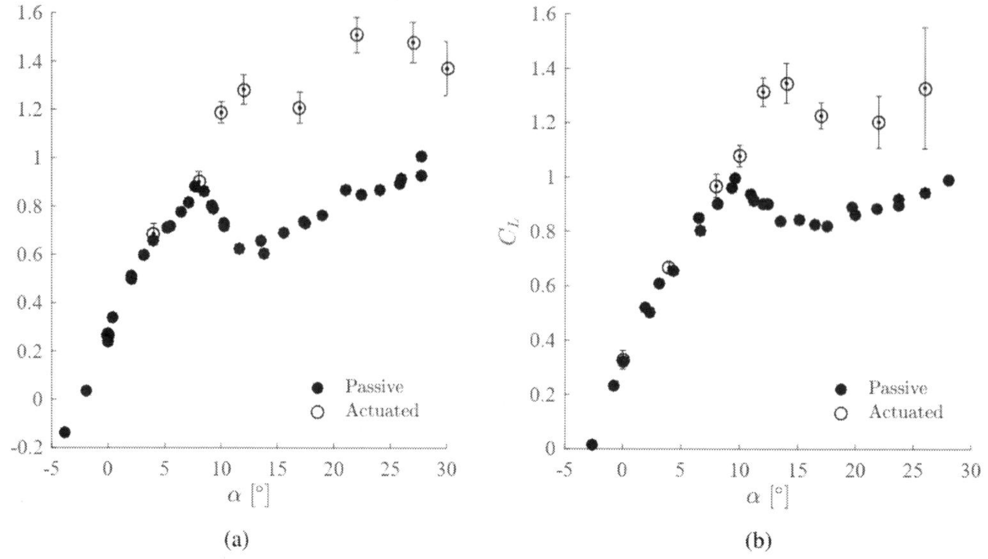

Fig. 13: The passive and actuated lift characteristics of a VR-12 wing with a RailPAc operating at (a) $Re \sim 45{,}000$ and (b) $Re \sim 90{,}000$.

REFERENCES

[1]Cattafesta, L. and Sheplak, M., "Actuators for Active Flow Control," *Annu. Rev. Fluid Mech.*, Vol. 43, 2011, pp. 247–272.

[2]Greenblatt, D. and Wygnanski, I., "Use of Periodic Excitation to Enhance Airfoil Performance at Low Reynolds Numbers," *Journal of Aircraft*, Vol. 38, (1), 2001, pp. 190–192.

[3]Yu, Y. H., Gmelin, B., Splettstoesser, W., Philippe, J. J., Prieur, J., and Brooks, T. F., "Reduction of Helicopter Blade-Vortex Interaction Noise by Active Rotor Control Technology," *Progress in Aerospace Sciences*, Vol. 33, (9), 1997, pp. 647–687.

[4]Smith, A. N., Babinsky, H., Fulker, J. L., and Ashill, P. R., "Shock-Wave/Boundary Layer Interaction Control using Streamwise Slots in Transonic Flows," *Journal of Aircraft*, Vol. 41, (3), 2004, pp. 540–546.

[5]Nixon, M. W., Kvaternik, R. G., and Settle, T. B., "Tiltrotor Vibration Reduction Through Higher Harmonic Control," *Journal of the American Helicopter Society*, Vol. 43, (3), 1998, pp. 235–245.

[6]Hassan, A. A., McVeigh, M. A., and Wygnanski, I., "Flow Control for Rotorcraft Applications," *Fundamentals and Applications of Modern Flow Control*, edited by R. D. Joslin and D. N. Miller, Vol. 231, Chap. 13, Progress in Astronautics and Aeronautics, Reston, VA, 2009, pp. 403–441.

[7]Lee, T. and Chopra, I., "Design of Piezostack-Driven Trailing-Edge Flap Actuator for Helicopter Rotors," *Smart Mater. Struct.*, Vol. 10, (1), 2001, pp. 15–24.

[8]Chen, C., Seele, R., and Wygnanski, I., "Flow Control on a Thick Airfoil Using Suction Compared to Blowing," *AIAA Journal*, Vol. 51, (6), 2013, pp. 1462–1472.

[9]Crittenden, T., Glezer, A., Funk, R., and Parekh, D., "Combustion-Driven Jet Actuators for Flow Control," Paper AIAA 2000-2768, 15th AIAA Computational Fluid Dynamic Conference, Anaheim, CA, June 11–14, 2001.

[10]Caruana, D., "Plasmas for Aerodynamic Control," *Plasma Phys. Control. Fusion*, Vol. 52, (12), 2010, pp. 124045.

[11]Klimov, A., Moralev, I., Bityurin, V., Kazansky, P., and Chertov, D., "Flow Separation Control Near NACA23012 Airfoil by a Capacity Coupled Surface HF Discharge," Paper AIAA 2011-3733, 42nd AIAA Plasma Dynamics and Lasers Conference, Honolulu, HI, June 27–30, 2011.

[12]Opaits, D. F., Edwards, M. R., Zaidi, S. H., Likhanskii, A. V., and Macheret, S. O., "Surface Plasma Induced Wall Jets," Paper AIAA 2010-469, 48th AIAA Aerospace Sciences Meeting, Orlando, FL, Jan 4–7, 2010.

[13]Likhanskii, A. V., Schneider, M. N., Opaits, D. F., Miles, R. B., and Macheret, S. O., "Limitations of the DBD Effects on the External Flow," Paper AIAA 2006-470, 44th AIAA Aerospace Sciences Meeting, Orlando, FL, Jan 4–7, 2010.

[14]Pafford, B., Sirohi, J., and Raja, L. L., "Propagating-arc Magnetohydrodynamic Plasma Actuator for Directional High-Authority Flow Control in Atmospheric Air," *J. Phys. D: Appl. Phys*, Vol. 46, (48), 2013, pp. 485208.

[15]Choi, Y., Gray, M., Sirohi, J., and Raja, L. L., "Measurement of Transient Force Produced by a Propagating Arc Magnetohydrodynamic Plasma Actuator in Quiescent Atmospheric Air," *J. Phys. D: Appl. Phys*, Vol. 48, (42), 2015, pp. 425204.

[16]Choi, Y., Gray, M., Sirohi, J., and Raja, L. L., "Effects of Anchoring and Arc Structure on the Control Authority of a

Rail Plasma Actuator," *J. Phys. D: Appl. Phys*, Vol. 50, (35), 2017, pp. 355203.

[17]Greenblatt, D. and Wygnanski, I., "The Control of Flow Separation by Periodic Excitation," *Prog. Aerosp. Sci.*, Vol. 36, (7), 2000, pp. 487–545.

[18]Choi, Y., Gray, M., Sirohi, J., and Raja, L. L., "Measurement of Velocity Induced by a Propagating Arc Magnetohydrodynamic Plasma Actuator," Paper AIAA 2007-0156, 55th AIAA Aerospace Sciences Meeting, Grapevine, TX, January 9–13, 2017.

An Experimental Study of Aerodynamics of a Compound Helicopter

Koichi Yonezawa
Project researcher
Central Research Institute
of Electric Power Industry
Abiko, Japan

Tomoki Kagayama
Graduate Student
Osaka University
Toyonaka, Japan

Shigeru Sunada
Professor
Nagoya University
Nagoya, Japan

Kazuyasu Sugiyama
Professor
Osaka University
Toyonaka, Japan

Noboru Kobiki
Associate senior Researcher
Japan Aerospace Exploration Agency
Tokyo, Japan

Yasutada Tanabe
Associate senior Researcher
Japan Aerospace Exploration Agency
Tokyo, Japan

Masahiko Sugiura
Researcher
Japan Aerospace Exploration Agency
Tokyo, Japan

Takashi Aoyama
Associate Principal Researcher
Japan Aerospace Exploration Agency
Tokyo, Japan

ABSTRACT

Aerodynamic characteristics of a small-scale concept model of a compound helicopter are examined. Experiments are carried out with the side propellers and the pusher propeller without the main rotor. Influences of aerodynamic interactions of the fuselage, the main wing and the propellers are examined using various configurations. Results shows that the wake of the side propellers increases the thrust of the pusher propeller and the lift force on the main wing. The pusher propeller does not affect the aerodynamic characteristics of the airframe and the side propellers significantly.

NOTATION

CD: Drag coefficient, $CD = \left(\text{Drag force}\right) / \frac{1}{2}\rho U^2 S$

CL: Lift coefficient, $CL = \left(\text{Lift force}\right) / \frac{1}{2}\rho U^2 S$

CP: Propeller collective pitch angle

d: Propeller diameter, 0.262 m

J: Advance ratio, $J=U/nd$

n: Propeller rotational speed [1/s]

S: Main wing area, 0.06615 m^2

U: Forward velocity [m/s]

α: Angle of attack of main wing

ρ: Density of fluid, 1.22 kg/m3

INTRODUCTION

Helicopters are quite useful aircraft because of capabilities of the vertical take-off and landing and the hovering. Various attempts have been made to increase the horizontal flight speed of helicopters as fast as airplanes. One of the successful VTOL concepts of the airplane is a tiltrotor aircraft represented by V-22 Osprey. The tiltrotor technology of V-22 Osprey has realized the VTOL as well as the high-speed cruising with two rotors. The benefit of the tiltrotor concept is less number of rotors. As a concept of the high-speed helicopter, on the other hand compound helicopters are developing, which have the main wing and the propeller as well as the main rotor. Japan aerospace exploration agency, JAXA is carrying out a conceptual study of the compound helicopter and developed a small-scale concept model as shown in Fig. 1 [1]. The compound helicopter developed by JAXA has a main rotor, two side propellers, and a pusher propeller. The side propellers are put at the main wing tip and produce the anti-torque against the main rotor. Since the compound helicopter has both the vertical rotor and the horizontal propellers, aerodynamic interactions occur [2,3]. There are possibilities that the aerodynamic interaction can be advantages[4], which improve the aerodynamic performances of the compound helicopter.

In the present study, the aerodynamic forces have been measured in a wind tunnel and effects of aerodynamic interactions have been discussed. In the wind tunnel test, the main rotor was not equipped and the interactions between the

Presented at the 6th Asian/Austrailian Rotorcraft Forum / Heli Japan 2017, Kanazawa, Japan, November 7-9, 2017.

Fig. 1 Small-scale concept model of compound helicopter developed by JAXA.

Table 1 Specifications of compound helicopter model

Fuselage and main rotor	
Length of fuselage	1.28 m
Span length of main wing	0.98 m
Main wing chord length	0.12 m (tip) 0.15 m (root)
Main wing airfoil	NACA4416 (tip) NACA4420 (root)
Diameter of main rotor	1.5 m [Not used]
Side/Pusher propeller	
Diameter	0.262 m
Number of blade	3
Airfoil	NACA0009

side propellers, the pusher propeller and the main wing were examined.

EXPERIMENTAL SETUP

Two kinds of airframes were used as shown in Fig. 2. The conventional helicopter model does not have the main wings but has tail wing. The compound helicopter model has the main and the tail wings. The shapes of fuselage are identical to each other. Principal dimensions of the compound helicopter model are shown in Table 1. The specifications of the side and the pusher propellers are identical. Variable corrective pitch mechanisms are built in the side and the pusher propellers.

Experiments were carried out using a wind tunnel at Osaka University. The cross section of the test section is a chamfered rectangle with 1.8 m of the height and width. The turbulence intensity of the wind is less than 0.5 %. The experimental apparatus is presented in Fig. 3. Three components of the forces and three components of moments, which are the difference between the thrust and the drag force, the side force, the lift force, the pitching moment, the yawing moment and the rolling moments, were measured using a load cell. Each rotational speed of the side and the pusher propellers was measured using optical tachometers and fixed at 8000 rev/min. The rotational direction of the side propeller is opposite direction to the trailing vortex of the main wing. The main flow speed of the wind tunnel was measured using a pitot tube and was fixed at 10 m/s. Four kinds of configurations were tested without the main rotor blades as shown in Table 2. The configuration FULL is with both the side and the pusher propellers. The configuration "P" means the pusher propeller is equipped without the side propeller and the configuration "S" means the side propeller is equipped without the pusher propeller. The configuration "AF" means the airframe is used without propellers.

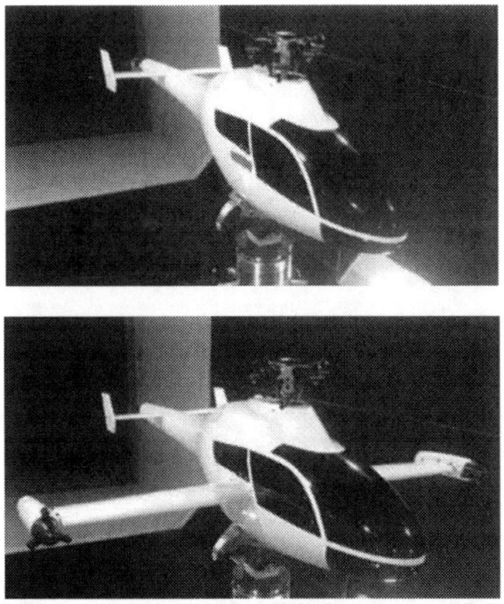

Fig. 2 Airframes of conventional helicopter model (top) and compound helicopter model (bottom) without rotor and propeller blades.

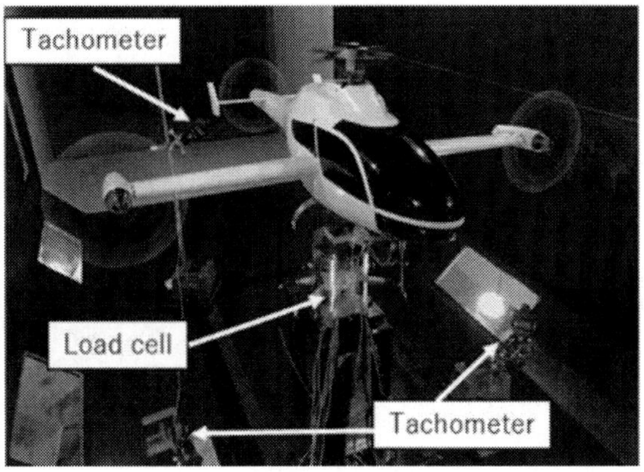

Fig. 3 Experimental apparatus

Table 2 Configuration of compound helicopter model

Configuration	Side prop.	Pusher prop.
"FULL"	Yes	Yes
"P"	No	Yes
"S"	Yes	No
"AF"	No	No

RESULTS AND DISCUSSIONS

Aerodynamic forces are examined at various angles of attack without propeller blades at first. The lift force and the

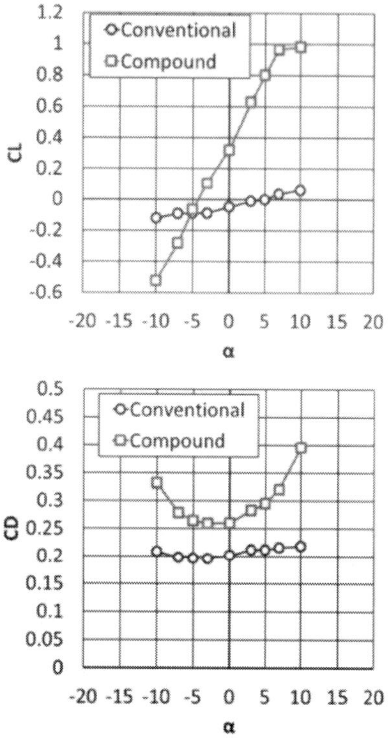

Fig. 4 Aerodynamic force measurements of airframes of the conventional and the compound helicopter models without propeller blades. (Configuration "AF")

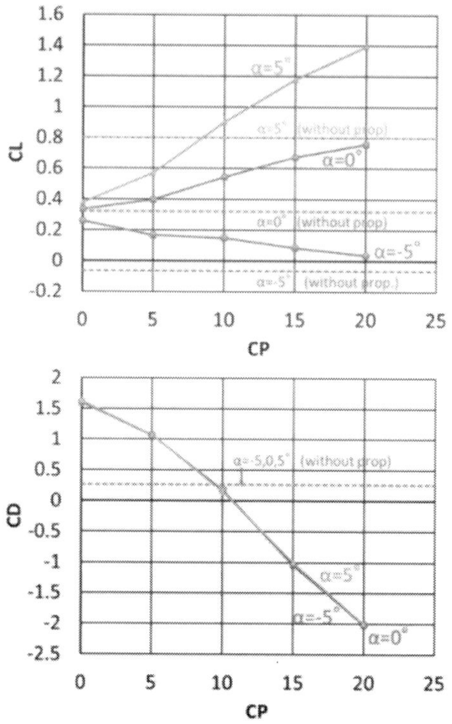

Fig. 5 Aerodynamic force measurements with side propeller blades without pusher propeller blades. (Configuration "S")

drag force coefficients, *CL* and *CD* versus the angle of attack α are shown in Fig. 4. The *CL* of the conventional helicopter model is small and the lift force is negative at $\alpha = 0°$. The *CD* of the compound helicopter model is larger than that of the conventional helicopter model. The most of the large drag force is due to the drive devices of the side propellers at the main wing tip.

The aerodynamic interaction between the side propeller and the main wing is examined without the pusher propeller. The *CL* and *CD* are plotted as functions of the propeller corrective pitch, *CP* in Fig. 5. Aerodynamic forces were measured at three kinds of angles of attack of the main wing, α = -5°, 0°, 5°. The *CL* and the *CD* of the airframe without propellers are plotted as the horizontal dashed lines. The *CLs* with the side propellers are larger than those without the side propellers at all αs examined. This is because the wake of the side propeller suppresses the trailing vortex of the main wing and the downwash is decreased. As the *CP* increases the *CL* increases at $\alpha = 0°$ and $\alpha = 5°$. This is because the upwash becomes stronger as the *CP* increases. The *CL* decreases as the *CP* increases at $\alpha = -5°$ because the vertical component of the thrust is negative at negative α. The *CDs* decrease as the *CP* increases at all αs because the direction of the propeller thrust is opposite to the drag force direction. The propeller causes the large drag forces at small *CP* condition. The influence of the propeller is so significant that the influence of the angle of attack is small.

The aerodynamic interaction between the pusher propeller and the airframe is examined without the side propeller. The *CL* and the *CD* are plotted as functions of the *CP* in Fig. 6. The difference of the *CL*s between with and without propeller can be seen in Fig. 6. The *CL* increases as the *CP* increases at $\alpha = 0°$ and $\alpha = 5°$ with the pusher propeller. However, the influence of the pusher propeller on the *CL* is not so significant. The *CD*s show similar values at all αs. These suggest that the flows induced by the pusher propeller does not affect the aerodynamic characteristics of the fuselage and the main wing significantly.

The *CL* and *CD* with both the side propellers and the pusher propeller are plotted as functions of the propeller corrective pitch, *CP* in Fig. 7. Although the values are different from the results shown in Fig. 5, similar tendency can be seen. The *CL*s are larger than those without the propellers at $\alpha = 0°$ and 5°. The *CD*s are almost identical at all αs examined. According to these results, the induced flows by the side and the pusher propellers affect the *CL*. However, it is unclear the influences of aerodynamic interactions between the side propellers and the pusher propeller. Then the aerodynamic forces of the propellers caused by each interaction are estimated from the combination of results of two configurations.

The CL and CD of the pusher propeller with the influence of the side propellers can be estimated from the results of configurations FULL and S as follows:

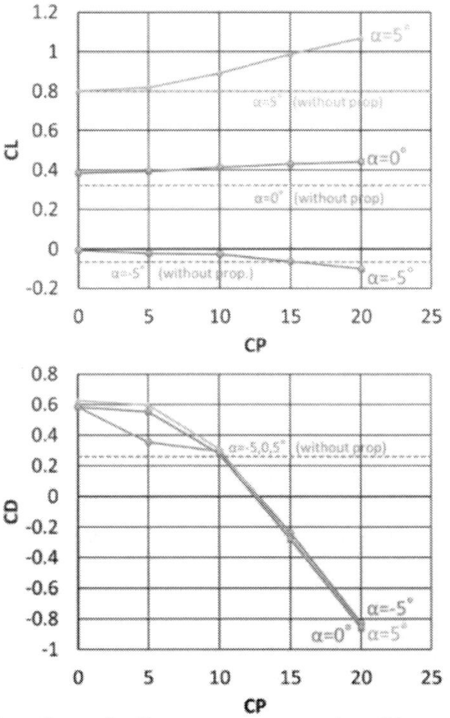

Fig. 6 Aerodynamic force measurements with pusher propeller blades without side propeller blades. (Configuration "P")

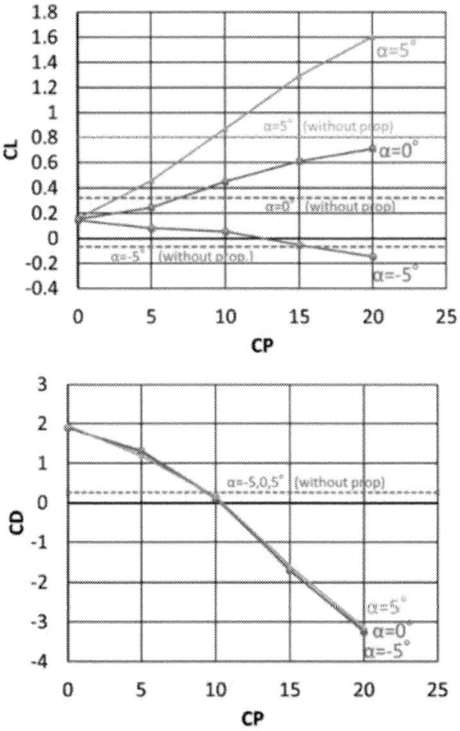

Fig. 7 Aerodynamic force measurements with side propellers and pusher propeller. (Configuration "FULL")

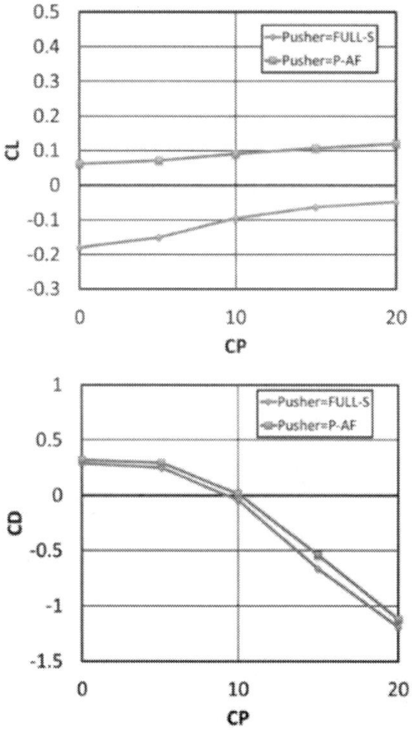

Fig. 8 Influence of aerodynamic interaction of side propellers on pusher propeller characteristics.

$$\begin{pmatrix} CL \\ CD \end{pmatrix}_{\text{Pusher prop. affected by side props.}} = \begin{pmatrix} CL \\ CD \end{pmatrix}_{\text{FULL}} - \begin{pmatrix} CL \\ CD \end{pmatrix}_{\text{S}} \quad (1).$$

The CL and CD of the pusher propeller without the influence of the side propellers are:

$$\begin{pmatrix} CL \\ CD \end{pmatrix}_{\text{Pusher prop. without influence of side props.}} = \begin{pmatrix} CL \\ CD \end{pmatrix}_{\text{P}} - \begin{pmatrix} CL \\ CD \end{pmatrix}_{\text{AF}} \quad (2).$$

Since all *CL* and *CD* values measured with configuration FULL, S, P and AF include the influence of the airframe, the both estimated values, $(CL, CD)_{\text{Pusher prop. affected by side prop.}}$ and $(CL, CD)_{\text{Pusher prop. without influence side prop.}}$ are considered not to include the influence of the airframe. The influence of the side propellers on the pusher propeller at $\alpha = 0°$ can be observed in Fig. 8. When the influence of the side propeller is not included, $CL_P - CL_{AF}$ is positive, in other words, the pusher propeller increases the *CL*. With the influence of the side propellers, however, $CL_{FULL} - CL_S$ is negative, which means the pusher propeller decreases the *CL*. The *CD* of the pusher

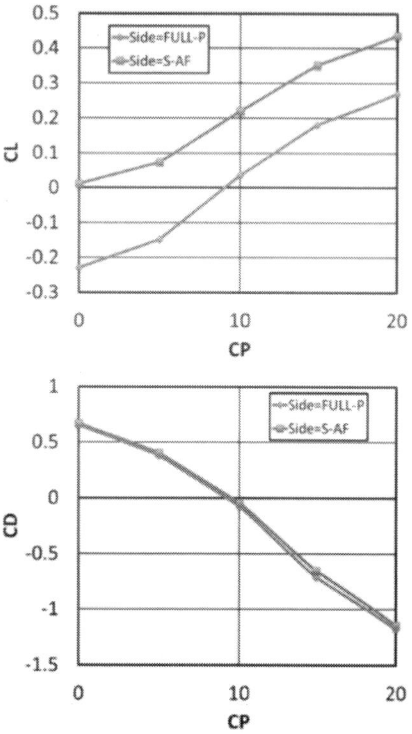

Fig. 9 Influence of aerodynamic interaction of pusher propellers on pusher propeller characteristics.

propeller is also affected by the side propellers. CD_{FULL} - CD_{S} is smaller than CD_{P} - CD_{AF} as can be seen in the bottom of Fig. 8. This suggests that the thrust of the pusher propeller is increased as a result of the influence of the side propellers.

The *CL* and *CD* of the side propellers with the influence of the pusher propellers can be estimated from the results of configurations FULL and P as follows:

$$\begin{pmatrix} CL \\ CD \end{pmatrix}_{\text{Side props. affected by pusher prop.}} = \begin{pmatrix} CL \\ CD \end{pmatrix}_{\mathrm{FULL}} - \begin{pmatrix} CL \\ CD \end{pmatrix}_{\mathrm{P}} \quad (3).$$

The *CL* and *CD* of the side propellers without the influence of the pusher propeller are:

$$\begin{pmatrix} CL \\ CD \end{pmatrix}_{\text{Side props. without influence of pusher prop.}} = \begin{pmatrix} CL \\ CD \end{pmatrix}_{\mathrm{S}} - \begin{pmatrix} CL \\ CD \end{pmatrix}_{\mathrm{AF}} \quad (4).$$

The influence of the pusher propeller on the side propellers at $\alpha = 0°$ can be observed in Fig. 9. With the influence of the pusher propeller, the side propellers increase the *CL* at larger α. Without the pusher propeller, however, the side propellers increase the *CL* furthermore. The results shown in Figs. 8 and 9 suggest that the interaction between the side and the pusher propeller decrease the *CL*. On the other hand, the *CD* values with the side propellers are not affected by the pusher propeller.

According to the results obtained by the present study, the effects of the aerodynamic interactions between the airframe, the side propellers and the pusher propeller are summarized as follows. The *CL* is affected by both the side propeller and the pusher propeller. This suggests that the induced flows by both the side propellers and the pusher propeller affect the flow pattern around the airframe. The influence of the side propellers is more significant than that of the pusher propeller. This may be because the side propeller is located upstream the main wing. It is worth noting that the suction of the pusher propeller far downstream of the main wing affects the *CL* at a noticeable amount. The interaction between the side propellers and the pusher propellers gives a negative effect on the side propeller effect on the *CL*. The detailed mechanisms have not been clarified. There are also effects of the aerodynamic interaction on the propeller thrust. The wakes of the side propellers affect the pusher propeller thrust though the pusher propeller does not affect the thrusts of the side propellers.

CONCLUSION

In the present study, the aerodynamic forces of a small-scale concept model of a compound helicopter have been measured with various configurations without the main rotor. Effects of aerodynamic interactions between the airframe, the side propellers, and the pusher propeller are discussed. The lift coefficient is affected by both the side propeller and the pusher propeller. The influence of the side propellers is more significant than that of the pusher propeller. The pusher propeller thrust is affected by the side propellers but the pusher propeller does not affect the thrust of the side propellers. Although the influences of the main rotor are not taken into account, the aerodynamic interaction of the propellers seems to be beneficial for the flight of the compound helicopter. It can be considered that further investigations including the main rotor is valuable to develop the compound helicopter.

REFERENCES

[1]Sugiura, M., Tanabe, Y., Kobiki, N., and Aoyama, T., "Flight Test of a Model Size Compound Helicopter," 5th Asian Australian Rotorcraft Forum, Singapore, November, 2016.

[2]Tanabe, Y. and Sugawara, H., "Numerical Simulation of Aerodynamic Interaction Between a Rotor and a Wing," 5th Asian Australian Rotorcraft Forum, Singapore, November, 2016.

[3]Inaba M., Akasaka T., Furuta T., Kobiki, N, Tanabe, Y., "Experimental Investigation of the Rotor-Wing Aerodynamic Interaction on a Compound Helicopter in Hover," 5th Asian Australian Rotorcraft Forum, Singapore, November, 2016.

[4]Nakai, G., Yoshida, N., Matsumoto, S., Yonezawa, K., Sugiyama, K., Sunada, S., "Propeller-Wing Interaction in Low Reynolds Number Flows," 13th Asian International Conference on Fluid Machinery, Tokyo, Japan, 2015.

Endurance Test of the Revised SNUF blade with Active Trailing-Edge Flap

WonJong Eun
Graduate student
Seoul National University
Seoul, Republic of Korea

ByungWook Lim
Graduate student
Seoul National University
Seoul, Republic of Korea

SangJoon Shin
Professor
Seoul National University
Seoul, Republic of Korea

ABSTRACT

SNUF blades equipped with active trailing-edge flaps has been suggested and revised in order to reduce the oscillatory loads in rotary wing hubs over the past few years. The flap hinge moment was predicted by the unsteady CFD analysis and the flap driving components were designed based on the relevant prediction. In addition, blade planform and section design were conducted to maximize its vibration reduction capability by multi-body dynamics analysis. In this paper, results of the endurance experiment on the flap driving mechanism will be presented. In this experiment, the present driving components except the actuator will be operated, and the integrity of the designed flap driving components will be examined. Then, a load equivalent to the centrifugal load due to the rotation will be applied on the flap. Finally, the present SNUF blades will be fabricated and tensile experiment will be performed. In the future, a ground rotation test will be conducted.

NOTATION

c	Blade chord length
$F_{\max}$	Maximum blocked force of an actuator
$F_{X,4P}$	4P hub forward vibratory shear load
$F_{Y,4P}$	4P hub sideward vibratory shear load
$F_{Z,4P}$	4P hub vertical vibratory shear load
H	Hinge moment
k	Torsional spring coefficient
L	Moment arm length
$M_{X,4P}$	4P hub vibratory chord bending moment
$M_{Y,4P}$	4P hub vibratory flap bending moment
R	Blade radius
V	Input voltage
V_{index}	Vibration index for 4P hub vibratory loads
$V_{\max}$	Maximum blocked force of an actuator
$x_{\max}$	Maximum stroke of an actuator
θ	Flap deflection
θ_0	Initial flap setting

INTRODUCTION

Active trailing-edge flap (ATF) is an effective method to diminish high vibratory loads and noise that are induced by complex unsteady aerodynamics in a rotorcraft. A variety of piezoelectric material actuators that provide high bandwidth actuation and exhibit high energy densities were introduced during the recent decade, and ATF was suggested as an effective vibration control methodology that did not require a high actuation power and weight penalty (Ref. 1).

One of the important considerations in the flap-driving mechanism of the trailing-edge flap is that the components mounted on the blade receive excessive centrifugal load. The flap-driving components consist of various parts and joints, and there is a concern that flap deflection may be reduced by an increment in friction due to the centrifugal load. Sufficient flap deflection is required to adjust the vibratory loads using the trailing-edge flap. Therefore, various studies have been investigated on the flap-driving mechanism that generated the flap deflection exceeding a target value in a rotating condition.

Bimorph piezoelectric actuators were implemented in early servo-flap experiments targeting the Boeing CH-47D tandem helicopter (Ref. 2). A 1/5-scale stationary-model with a 10% chord flap was wind-tunnel tested for various advance ratios between zero and 0.21, and at frequencies up to 13/rev. Significant flap deflection, lift, and pitching moments were obtained, but the values were smaller than the theoretical predictions. An improved design using a multilayer bender actuator, solid state flexural hinges, and

---

Presented at the 6th Asian/Australian Rotorcraft Forum / Heli Japan 2017, Kanazawa, Japan, November 7-9, 2017.

impedance matching principles, was subsequently conceptualized, manufactured, and tested (Ref. 3), with improved bench-test results (±11° flap deflection over the 0–90 Hz bandwidth). However, due to inherent limitations in the bimorph excitation concept (low mass efficiency and need for leading edge balance weights) such investigation was not continued; attention was refocused on piezoelectric stacks (Ref. 4). Walz and Chopra (Ref. 5), and Koratkar and Chopra (Refs. 6, 7) also used bimorph piezoelectric actuator for rotor blade flaps. Koratkar and Chopra (Ref. 6) used a 4% span flap actuated by a 4-layer piezoelectric bimorph actuator. A Hall sensor was incorporated into the blade to measure flap deflection during rotating blade testing. The testing was performed at various collective pitch values up to 4° collective pitch. Flap deflections of ±8° were measured by the Hall sensor for the Froude scaled operating speed 900 rpm. Further tests were performed on a Mach-scaled rotor. For 10 Hz excitation, flap deflection from ±10° (non-rotating) to ±5.7° (rotating at 1,800 rpm) were obtained. The flap deflection was also shown to increase with frequency from ±6° at 1/rev to ±10° at 4/rev. Fulton and Ormiston (Refs. 8, 9) also used bimorph flap actuation to control a 10% chord, 12% span elevon mounted on a 1.143 m (7.5 ft) radius two-blade rotor. Elevon deflections of up to ±10°, for non-rotating conditions, and ±5°, at 760 rpm, were demonstrated at frequencies up to 4/rev. Reduction in the blade vibratory loads in forward flight was demonstrated.

Prechtl and Hall (Ref. 4) built a mechanically amplified X-frame actuator using a pair of EDOCorp. EC-98 PMN-PT piezo-ceramic stacks in 140 layers. In a 1/6 Mach scale CH-47D rotor blade, the X-frame actuator occupied the leading edge part of the airfoil (Ref. 10). In the scaled rotor blade, deflection amplitudes of approximately -7° to +10° sustained for increasing frequencies up to 150 Hz, were recorded. Lee and Chopra (Refs. 11, 12) also reported using piezostacks for servo-flap actuation of a scaled rotor blade. A two-stage (L-L) displacement amplification mechanism and Physik Instrumente P-804.10 piezostacks were used. The L-L actuator was installed in the spar of a blade section and mechanically connected to a trailing edge flap. Wind tunnel test at speed up to 120 ft/sec and angle of attack of up to 12° demonstrated peak-to-peak deflection of 13° for 1/rev, 19° for 2/rev, 17.5° for 3/rev, and 8° for 4/rev. A federally funded program for full-scale implementation of smart materials actuation rotor technology (SMART) was conducted by Boeing Mesa, Arizona. In a gradual approach, the full-scale proof-of-concept demonstration was planned on the MD 900 bearingless rotor (Ref. 13). A prototype flap actuator with a two-stage amplification and bi-axial operation was constructed and tested (Ref. 14). The actuator was also successfully subjected to environmental qualification tests consisting of shake and spin testing. Since then, the double X-frame (2-X) concept was adopted, fabricated, and tested. The 2-X prototype demonstrated 1 mm stroke and successfully sustained environmental and spin tests (Ref. 15). When incorporated into the 2-X actuator, such piezo-stacks were predicted to exceed the 2°. Design specification (Ref. 14). Up to 3° flap-authority was predicted with high voltage stacks, and 4° with single-crystal stacks.

In the last few years, the present authors have been steadily studying to reduce the vibratory loads, which were generated by the rotating blade blades, using the active trailing-edge flap (Refs. 16 – 20). In the early stage, various flap-driving mechanisms have been fabricated and tested to achieve a maximum deflection of the flap of ±4°. However, none of these flap-driving mechanisms achieved the desired flap deflection. These results were static bench tests except for the aerodynamic and loads to apply in the flap. When the aerodynamic loads are applied to the flap, it is obvious that the flap deflection will be reduced. In addition, the Seoul National University Flap (SNUF) blade is a Mach-scaled blade with a rotation speed of 1,300 RPM. Therefore, degradation in the flap deflection may occur due to an increment in friction force at high rotational speed.

In this paper, an experiment was conducted to verify the operability of the revised flap-driving component. The revised flap-driving component was designed considering the governing equation of the flap-driving mechanism and the unsteady flow analysis results that estimated the flap hinge moments. Such flap-driving component was fabricated and the flap deflection was measured under the expected centrifugal loading condition using the dummy weights. This experiment was performed for one hour to ensure that the designed and fabricated flap-driving components maintains consistent performance over the operating time. In the future, tensile test will be carried out on the blade including the flap-driving component, and the ground rotation test will be performed using the whirl tower facility in Korea Aerospace Research Institute.

SNUF BLADE IMPROVEMENTS

Hinge Moment Estimation

In order to move the flap up to ±4° even under the aerodynamic loading condition, it was necessary to consider the external aerodynamic load in the design process of the flap mechanism. Therefore, the hinge moment acting on the flap was considered as a torsional spring as shown in Figure 1. The governing equation of the flap-driving mechanism can be expressed as Eq. (1) which includes the flap hinge moment as the torsional spring constant.

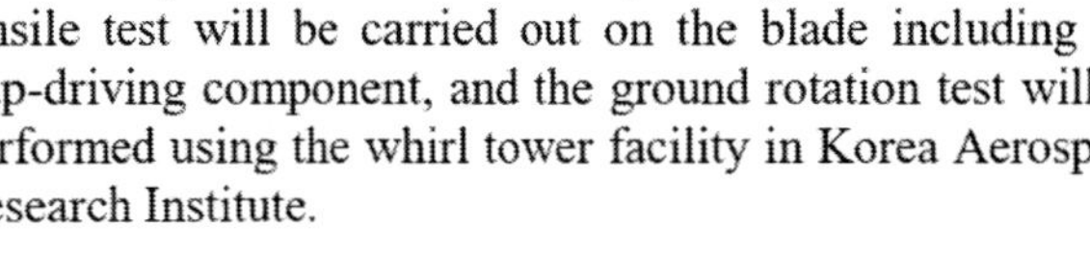

Figure 1. Flap-driving mechanism schematic diagram.

$$V = -V_{max}\left(\frac{k}{LF_{max}} + \frac{L}{x_{max}}\right)\theta + \frac{LV_{max}}{x_{max}}\theta_0 \quad (1)$$

The unsteady flow analysis using ANSYS Fluent was performed to predict the aerodynamic forces precisely (Ref. 19). Figure 2 illustrates the hinge moment results according to the flap deflection. Assuming that the relationship between the flap deflection and the hinge moment was linear, the slope k had the same unit as the coefficient of the torsional spring. By this process, the Cedrat APA1000L was selected as the piezoelectric actuator candidate and the flap-driving mechanism was fabricated.

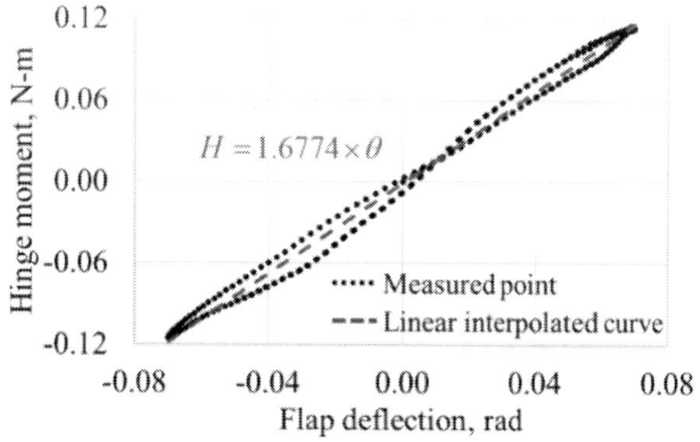

Figure 2. Hinge moment vs. Flap deflection.

Blade Planform Selection

The configuration of the SNUF blade is shown in Figure 3. The detailed design of the special section was created considering the effect of the flap center location, length and chord length on the hub vibratory load reduction. The vibration reduction according to the flap dimension and location was estimated by Dymore (Ref. 21). As shown in Eq. (2) the hub vibratory loads were compared using the vibration index.

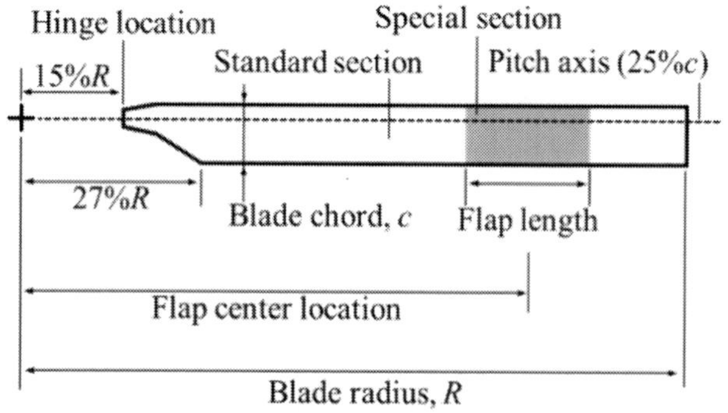

Figure 3. Planform of the SNUF blade.

$$V_{index} = \sqrt{\frac{F_{X,4P}^2 + F_{Y,4P}^2 + F_{Z,4P}^2}{\text{Thrust}} + \frac{M_{X,4P}^2 + M_{Y,4P}^2}{\text{Torque}}} \quad (2)$$

Figure 4 shows the vibration index values according to the flap shape parameters, and the flap parameter was determined as shown in Table 1.

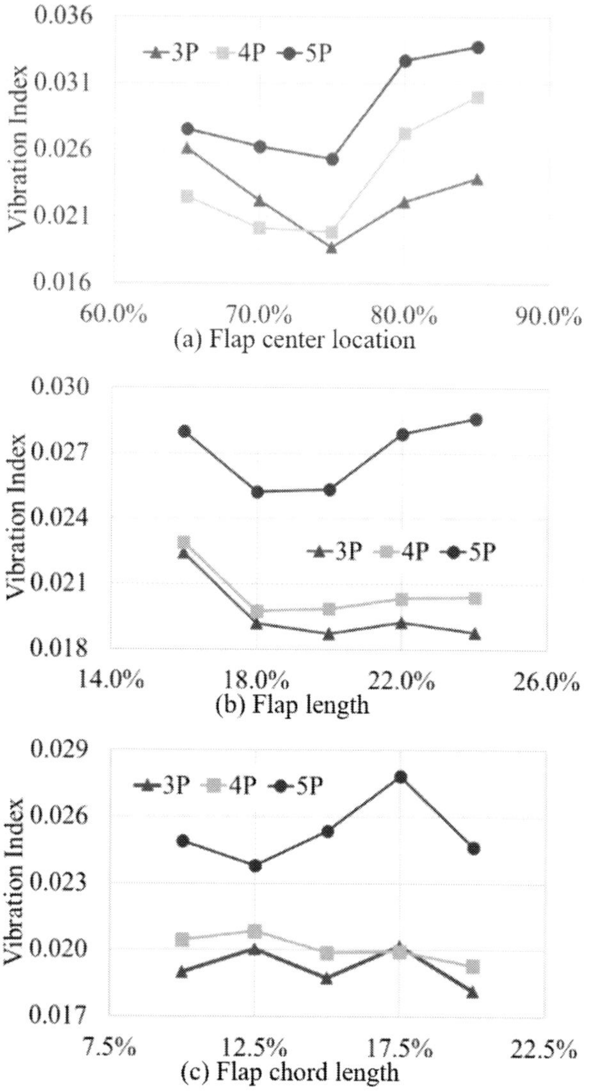

Figure 4. Hinge moment vs. Flap deflection.

Table 1. Flap shape parameter design results.

Flap shape parameter	Value
Flap center location	75%*R*
Flap length	20%*R*
Flap chord length	15%*R*

Cross Section Optimization

In this paper, the cross-sectional design optimization framework, which based on the ATR optimization framework (Refs. 22, 23), was established to perform the cross-section optimization of SNUF blade. Figure 5 shows the flow chart of the present optimization framework. The objective function used in the current optimization was established to minimize the weight of the blade in the range that satisfied the defined Lock number range, thereby reducing the 1st torsional frequency.

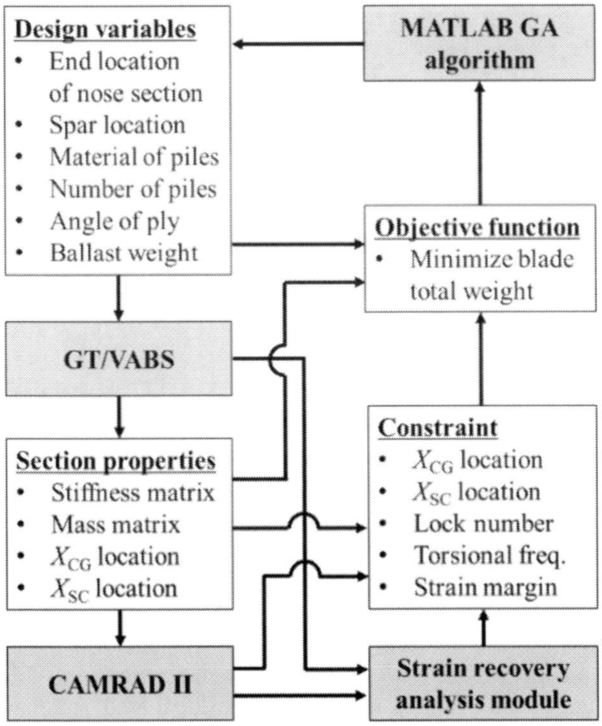

Figure 5. Flow chart of the SNUF blade design optimization (Ref. 22).

The optimized cross-sectional design configurations are illustrated in Figure 6. With respect to the present optimization, the tail position of the front section and the spar position was determined as 15%c and 42%c, respectively. The total weight of the optimized SNUF blade corresponded to 1.016 kg, which was 10% smaller than the weight of the previous SNUF blade (1.130 kg). The detailed optimized cross sectional design results are given in Ref. 20.

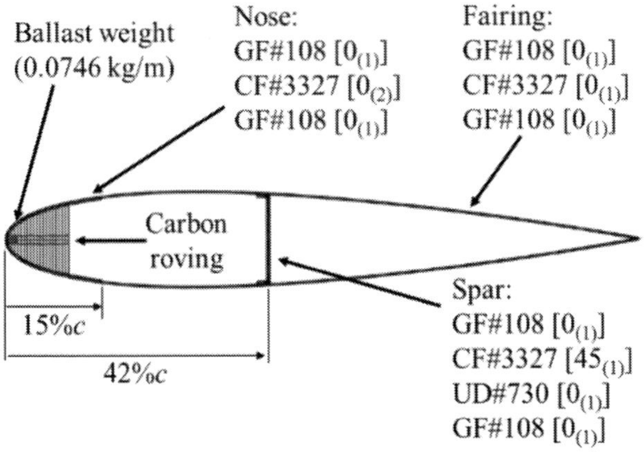

Figure 6. Final cross-sectional configuration of the SNUF blade.

Three Dimensional Static Structural Analysis

A detailed three-dimensional analysis that considered the contact situation was required for a better understanding of structural behavior due to the additional components, such as the piezoelectric actuator and flap-driving components. The three-dimensional finite element model and contact analysis was implemented by MSC.NASTRAN. The detailed flap actuation components and composite blade skins and spars were incorporated into the three-dimensional finite element analysis. The external aerodynamic loads obtained from the CAMRAD II (Ref. 24) trim analysis were imposed. Centrifugal loads and gravity were also considered to validate the effectiveness of the stress within the contact surfaces and attachment points.

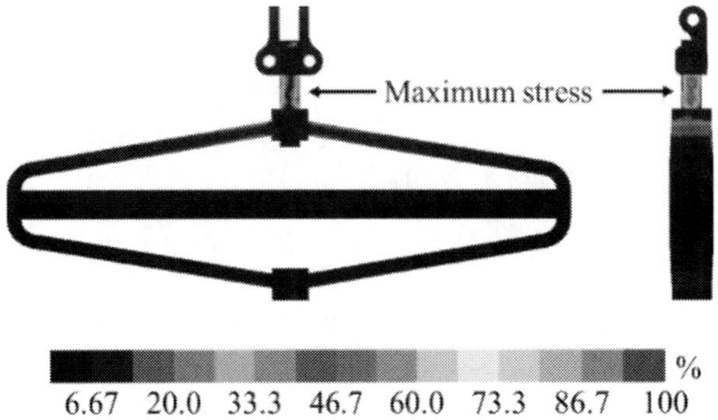

Figure 7. Margin of safety on the actuator rod.

Maximum stress value was observed to exist between the actuator rod and the guide, which corresponded to the contact point as shown in Figure 7 and represented 53% of the allowable stress. The von Mises stress results corresponding to the blade skin was also examined, and the maximum stress values corresponded to 68% of the permissible stress and were observed at the part at which the guide and blade skin were connected. These results are presented in Ref. 20.

EXPERIMENTAL SET-UP

Flap-driving Components Fabrication and Bench Test

A three dimensional CAD drawing of the flap-driving components with the selected APA1000L actuator and designed flap-driving mechanism, is shown in Figure 8. Based on the design shown in Figure 8, the practical flap-driving components were fabricated as shown in Figure 9.

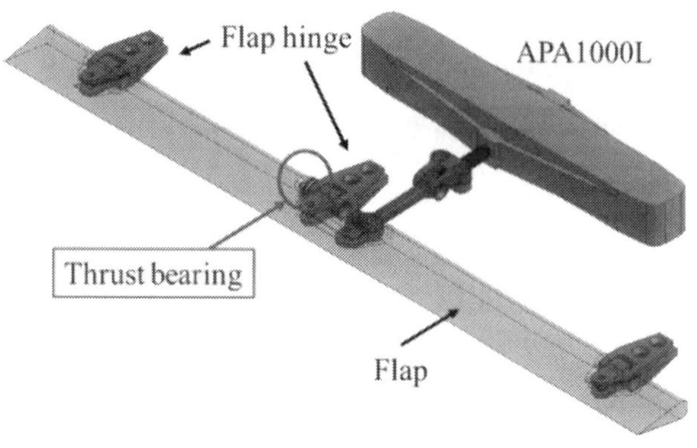

Figure 8. Flap-driving components design.

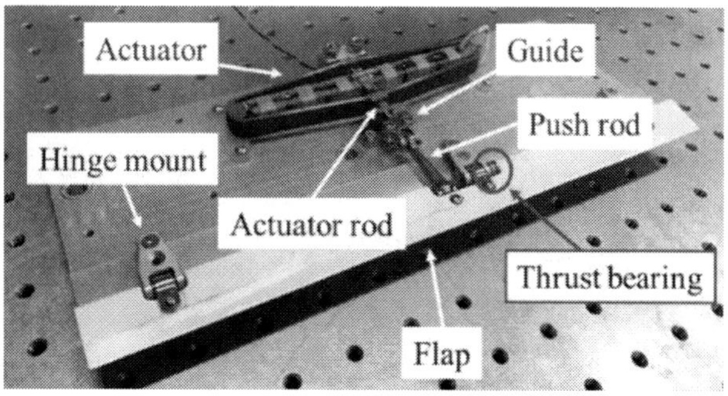

Figure 9. Fabricated flap-driving components.

First, in order to examine the operability of the fabricated flap-driving component, the static bench experiments were performed without centrifugal loading. The static bench test results of present flap-driving components was found to be much higher than the flap deflection target, ±4° as shown in Figure 10.

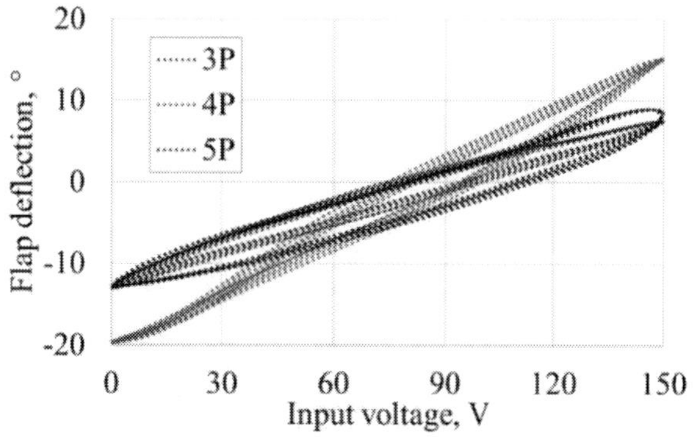

Figure 10. Flap deflection measurement without centrifugal loading.

Endurance Experiments with Centrifugal Load

In order to investigate whether the designed and fabricated flap-driving components were capable of producing the target flap deflection under the operating condition, the endurance experiment was devised to take into account the following two aspects.

- The flap deflection under the centrifugal loading condition
- Durability of the flap-driving component which consist of various bearings and hinges

There is a concern that the flap deflection may degrade due to the centrifugal load on the flap in the rotating condition of the blade. Therefore, as shown in Figure 11, an experiment considering the centrifugal load acting on the flap was devised. The test bed equipped with the flap-driving components was fixed to the vibration absorbing table, and the dummy weight was suspended on the flap. In the setup, the dummy weights were connected to the five holes in the flap by fishing lines so that the virtual centrifugal loading might be distributed on the flap.

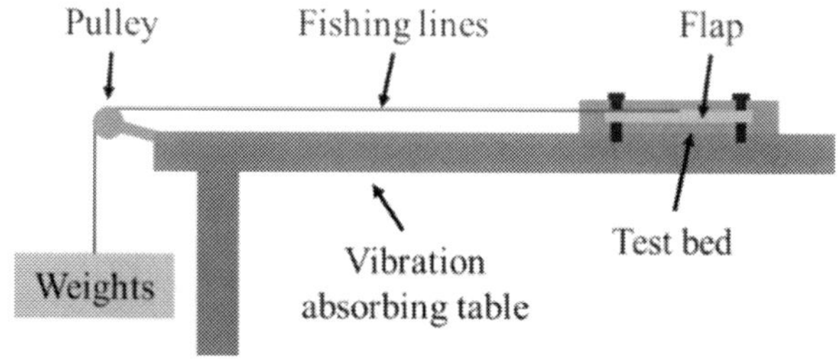

Figure 11. Schematic of the centrifugal loading condition.

When the flap-driving component is mounted on the blade and operated, fatigue accumulates on the component, such as hinges and bearings, due to the centrifugal load, and the flap deflection may be degraded. Therefore, the centrifugal load experiment shown in Figure 11 was carried out during the target operating time, one hour. The conditions of the endurance experiment are summarized in Table 2.

Table 2. Endurance experiment conditions

Condition	Value
Dummy weights hanging on the flap	20 kg
Operational time	3600 sec (1 hour)

In the endurance experiment, instead of the piezoelectric actuator APA1000L, a screw ball type electric cylinder and a servo motor were implemented to produce the same stroke of the actuator. The screw ball type electric cylinder transfers the rotational motion to the rectilinear motion by its screw. Thus, the motions and torques of the motor were converted into rectilinear stroke motions and loads with little loss. Figure 12 shows a test bed equipped with the electric cylinder and the servo motor.

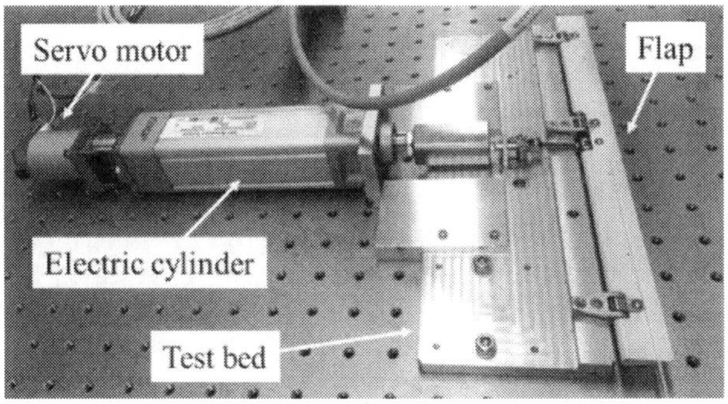

Figure 12. Test bed with the ball screw type electric cylinder and the servo motor.

In order to control the servo motor, the 332CPR encoder was installed and motor control was performed using NI DAQ. The endurance experiment was conducted at a frequency of 10 Hz, considering the bandwidth of the motor. In this experiment, as shown in Figure 13, two laser displacement sensor was employed. Sensor A measured the flap deflection, and Sensor B measured the flap drift displacement in the direction of the centrifugal loads.

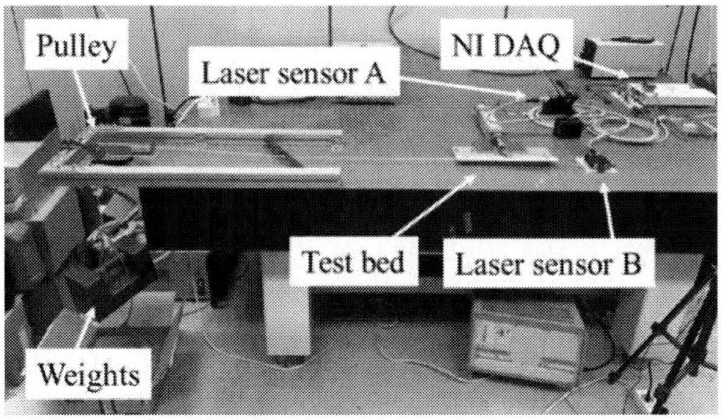

Figure 13. Endurance experiment set-up.

EXPERIMENTAL RESULTS

Flap Deflection Measurement

The endurance experiment was conducted for 3,600 seconds. Figure 14 shows the flap deflection measured by the laser sensor A. It is observed that the flap deflection at the initial stage (59 ~ 60 sec) and the final stage (3599 ~ 3600 sec) show similar trends. Figure 15 shows the results of FFT to analyze the measured flap deflection in the frequency domain.

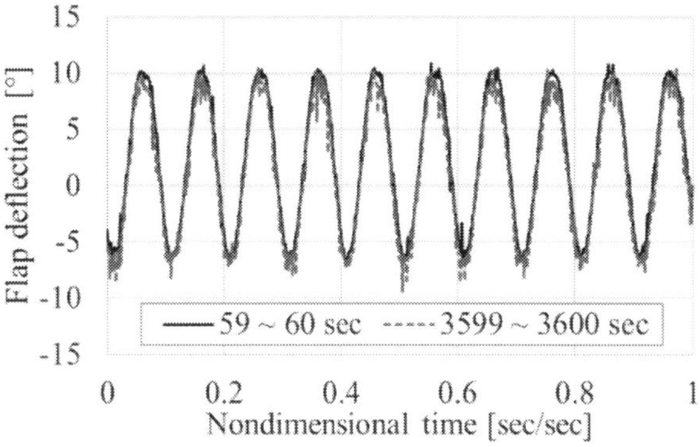

Figure 14. Flap deflection measurement.

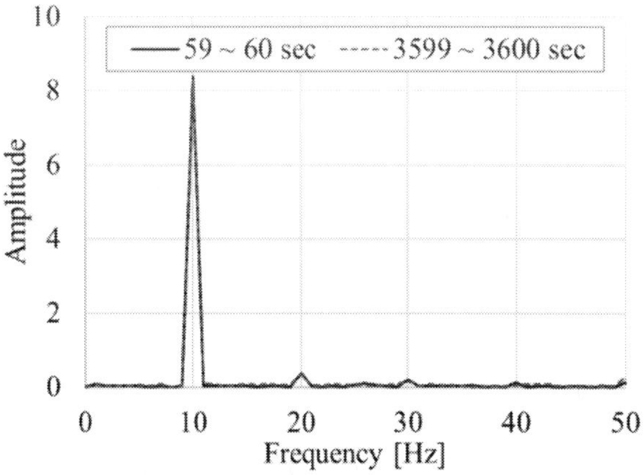

Figure 15. FFT results of the flap deflection.

As shown in Figure 14, the flap deflection reached -8° to +10°, exceeding the target value of ±4°, even under the centrifugal load of 20 kg acting on the flap. The flap deflections were maintained throughout the entire actuation, and its amplitude were observed to remain constant during one hour operation. As a result of FFT analysis of the flap deflection, the largest values were found at the operating frequency of 10 Hz in both initial and final periods, and the magnitudes were also similar.

In addition, the results of the flap drift displacement measurements in the centrifugal direction and the FFT results of these values are shown in Figure 16, Figure 17, respectively. The maximum flap drift value was 0.016 mm, which indicated that the effect of centrifugal loading on the flap drift displacement was insignificant. According to the FFT results of the flap drift measurements, only values corresponding to N times the operating frequency (10 Hz) were obtained.

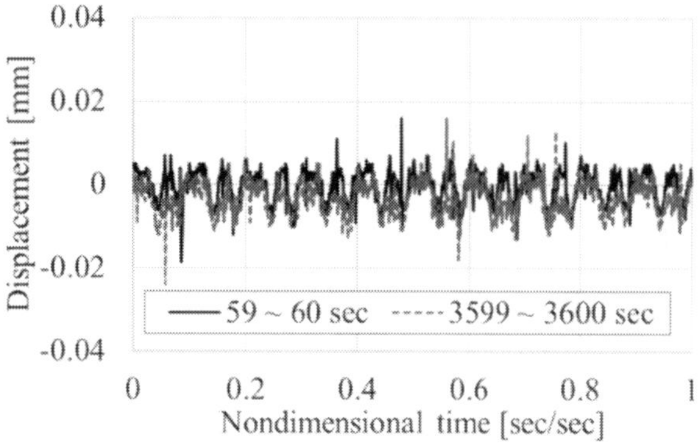

Figure 16. Flap drift measurement.

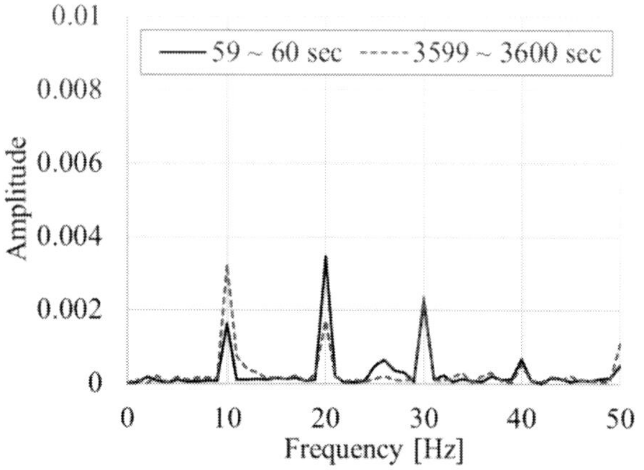

Figure 17. FFT results of the flap drift.

It was verified that the amplitude and frequency of the flap deflection of the present flap-driving components were maintained during the operating time (3,600 sec) under the expected centrifugal loading condition.

Plans for Further Experiments

In the future, a prototype blade including a flap-driving components will be fabricated in accordance with a designed cross-section laminated pattern, and a tensile experiment will be performed. After verifying the structural integrity of the composite blade through tensile experiment, the operational test at the hover situation will be carried out using the whirl tower in Korea Aerospace Research Institute, as shown in Figure 18.

Figure 18. Whirl tower facility of the Korea Aerospace Research Institute.

CONCLUSION

In this paper, flap-driving components of SNUF blades designed for vibration reduction were fabricated and the flap deflections under centrifugal loading condition were measured during the expected operating time.

The flap-driving components were designed through previous studies, and thrust bearings were added to handle the centrifugal load acting on the flap. The flap-driving components were fabricated and the static bench experiment was performed and the flap deflections, sufficiently larger than the target value of ±4° under the 3, 4 and 5P actuation conditions, were obtained.

In order to measure the flap deflection of the flap-driving components in the situation where the centrifugal load imposed on the flap, a virtual centrifugal loading experiment using pulley and weight was devised. Also, this experiment was performed for one hour, and it was verified that the present flap-driving components maintained a constant capability during a desired operation time.

As a result of the endurance experiment, the flap deflection reached +10° to -8° under centrifugal load of 20 kg. Also, it was observed that there was no change in the magnitude of the flap deflection even when the flap-driving component was operated for one hour.

In the future, the tensile test of the designed blade will be carried out, and the operability of the rotor system under the hover condition will be verified by using the whirl stand in Korea Aerospace Research Institute.

Author contact: WonJong Eun eun0914@gmail.com
ByungWook Lim goody147@snu.ac.kr
SangJoon Shin ssjoon@snu.ac.kr

ACLNOWLEDGMENTS

This work was conducted at High-Speed Compound Unmanned Rotorcraft (HCUR) research laboratory with the support of Agency for Defense Development (ADD).

REFERENCE

[1]Ham, N. D., "Helicopter Individual Blade Control and Its Applications," American Helicopter Society 38th Annual Forum Proceedings, St. Louis, MO, 1983.

[2]Spangler, R. L. Jr. and Hall, S. R., "Piezoelectric Actuators for Helicopter Rotor Control," Report # SSL 1-89, SERC 14-90, MIT Space Engineering Research Center, Massachusetts Institute of Technology, 1989.

[3]Hall, S. R. and Prechl, E. F., "Development of a Piezoelectric Servoflap for Helicopter Rotor Control," Journal of Smart Materials and Structures, Vol. 5, pp. 26–34, 1996.

[4]Prechl, E. F. and Hall, S. R., "Design of a High Efficiency Discrete Servo-Flap Actuator for Helicopter Rotor Control," SPIE Vol. 3041, pp. 158–182, 1997.

[5]Walz, C. and Chopra, I., "Design, Fabrication, and Testing of a Helicopter Rotor Model with Smart Trailing Edge Flaps," Proceedings of the AIAA/ASME/ASCE/AHS/ASC 35th Structures, Structural Dynamics, and Materials Conference - Adaptive Structures Forum, Hilton Head, SC, April 21–22, paper AIAA-94-1767-CP, pp. 298–319, 1994.

[6]Koratkar, N. A. and Chopra, I., "Analysis and Testing of a Mach Scaled Helicopter Rotor in Hover with Piezoelectric Bender Actuated Trailing-Edge Flaps," SPIE's 1998 Symposium on Smart Structures and Materials, San Diego, CA, March, SPIE Vol. 3329, pp. 321–332, 1998.

[7]Koratkar, N. A. and Chopra, I., "Hover Testing of a Mach-Scaled Rotor with Piezoelectric Bender Actuated Trailing Edge Flaps," in Smart Structures and Materials 2000: Smart Structures and Integrated Systems, N. W. Wereley (Ed.), SPIE Vol. 3985, pp. 2–10, 2000.

[8]Fulton, M. and Ormiston, R. A., "Hover Testing of a Small-Scale Rotor with On-Blade Elevons," Proceedings of the 53rd Forum of the American Helicopter Society, Virginia Beach, VA, 1997.

[9]Fulton, M. and Ormiston, R. A., "Small-Scale Rotor Experiments with On-Blade Elevons to Reduce Blade Vibratory Loads in Forward Flight," Proceedings of the 54th Forum of the American Helicopter Society, Washington DC, 1998.

[10]Prechtl, E. F. and Hall, S. R., "An X-frame Actuator Servo-Flap Actuation System for Rotor Control," SPIE Vol. 3329, pp. 309–319, 1998.

[11]Lee, T. and Chopra, I., "Design and Static Testing of a Trailing-Edge Flap Actuator with Piezostacks for a Rotor Blade," SPIE Vol. 3329, pp. 321–332, 1998.

[12]Lee, T. and Chopra, I., "Development of a Smart Trailing-Edge Flap Actuator with Multi-Stage Stroke Amplifier for a Rotor Blade," in Smart Structures and Materials 2000: Smart Structures and Integrated Systems, N. W. Wereley (Ed.), SPIE Vol. 3985, pp. 11–25, 2000.

[13]Straub, F. K. and King, R. J., "Application of Smart Materials to Control of a Helicopter Rotor," SPIE Symposium on Smart Structures and Materials, San Diego, CA, 1996.

[14]Straub, F. K., "Development of a Full Scale Smart Rotor System," 8th ARO Workshop on Aeroelasticity of Rotorcraft Systems, Penn State University, 1999.

[15]Hall, S. R., Tzianetopolou, T., Straub, F. and Ngo, H., "Design and Testing of a Double X-Frame Piezoelectric Actuator," in Smart Structures and Materials 2000: Smart Structures and Integrated Systems, N. W. Wereley (Ed.), SPIE Vol. 3985, pp. 26–37, 2000.

[16]Lee, J. H., Natarajan, B., Eun, W. J., R., V. S., Park, J. S., Kim, T. S., and Shin, S. J., "Structural and Mechanism Design of an Active Trailing Edge Flap Blade," *Journal of Mechanical Science and Technology*, Vol. 29, (9), pp. 2605–2617, 2013.

[17]Natarajan, B., Eun, W. J., Lee, J. H., and Shin, S.-J., "Structural Design of an Active Trailing Edge Flap Blade for Helicopter Vibration Control," 53rd AIAA/ASME/ASCE/AHS/ASC Structures, Structural Dynamics, and Material Conference and Exhibit Proceedings, Honolulu, HI, 2012.

[18]Kang, J. P., Eun, W. J., Lim, J. H., Visconti, U., and Shin, S.-J., "Design Improvements of Smart Active Trailing-Edge Flap for Rotating Test," 56th AIAA/ASME/ASCE/AHS/ASC Structures, Structural Dynamics, and Material Conference and Exhibit Proceedings, Kissimmee, FL, 2015.

[19]Visconti, U., Eun, W. J., Sim, J. S., Lee, S. W. and Shin, S. J., "Design Improvements and Flap Deflection Evaluations with Considering Centrifugal Load on Active Trailing Edge Flap," Aircraft Engineering and Aerospace Technology, Accepted, 2017.

[20]Eun, W. J., Sim, J. S., Lee, S. W. and Shin, S. J., "Further Improvements in the SNUF Blade Design by Numerical Design Optimization Framework," ASCE's Journal of Aerospace Engineering, Accepted, 2017.

[21]Bauchau, O. A., Bottasso, C. L., and Nikishkov, Y. G., "Modeling Rotorcraft Dynamics with Finite Element Multibody Procedures," *Mathematical and Computer Modeling*, Vol. 33, 2001, pp.1113–1137.

[22]Cesnik, C. E. S., Mok, J. W., Parikh, A. S., and Shin, S. J., "Optimization Design Framework for Integrally Twisted Helicopter Blades," 45th AIAA/ASME/ASCE/AHS/ASC Structures, Structural Dynamics, and Material Conference and Exhibit Proceedings, Palm Springs, CA, 2004.

[23]Shin, S.-J., Cesnik, C. E. S., and Hall, S. R., "Closed-Loop Control Test of NASA/ARMY/MIT Active Twist Rotor for Vibration Reduction," *Journal of the American Helicopter Society*, Vol. 50, (2), pp. 178–194.

[24]Johnson, W. "Technology Drivers in the Development of CAMRAD II," American Helicopter Society 38th Annual Forum Proceedings, San Francisco, CA, 1994.

An Experimental Investigation of Ground Effect on Rotorcraft in the Presence of Side Wall

Rohith M
Research Scholar
Aerospace Engineering
Indian Institute of Technology
Madras, India

Cibin Joseph
Doctoral Student
Aerospace Engineering
Indian Institute of Technology
Madras, India

Ranjith Mohan
Assistant Professor
Aerospace Engineering
Indian Institute of Technology
Madras, India

ABSTRACT

Rotorcraft flying in close proximity to ground are known to generate a larger thrust due to the shed wake being obstructed by the ground. The effect has been well researched and empirical corrections are available in literature. However, studies on the influence of side walls on ground effect is still lacking. Such studies may be applied to full scale rotorcraft hovering near walls, skyscrapers and similar structures, rotor-based UAVs flying in a constrained space and in corrections for wall interference in wind tunnel testing. Analytical models are presently available for modelling the effect of wind tunnel wall interferences in specific wind tunnels, however they are rarely prescribed in a form applicable to generalized situations like a helicopter approaching a single side wall in ground effect. This work aims to study the influence of side walls in the presence of ground using experiments conducted on a model rotor. Force and Torque measurements are obtained using a six-axis load cell.

NOMENCLATURE

A	Rotor disk area
C_T	Thrust Coefficient, $T/\rho A V_{TIP}^2$
H	Elevation from Ground Level
IGE	In Ground Effect condition
N_b	Number of Blades
OGE	Out of Ground Effect condition
R	Rotor radius
T_g	Rotor thrust in ground effect
T_∞	Rotor thrust out of ground effect
V_h	Inflow Velocity in Hover
W	Distance from wall
$(X;Y;Z)$	Reference Frame
c	Blade Chord
h	Non-dimensional elevation from ground, H/R
kG	Thrust Augmentation factor
v_{ig}	Induced velocity in ground effect
$v_{i\infty}$	Induced velocity out of ground effect
w	Non-dimensional distance from wall, W/R
Ω	Rotor RPM
θ	Blade Pitch angle

INTRODUCTION

Rotorcraft often operate in scenarios associated with confined environment and their ability to hover is used extensively in many operational roles. In confined environment applications the rotorcraft operates near to the ground with other vertical obstacles present nearby. A UAV or MAV operating inside a building or a helicopter landing on a ship deck are examples to such scenarios. One of the major areas of interest in these scenarios is the interaction of the rotor wake with structures nearby (ground and walls). One of the popular theoretical studies on interaction of ground and rotor wake is by Betz (Ref. 1), based on an image sink method. The study was carried out by replacing the rotor with a sink and the ground is represented using an image sink placed below the ground level at a depth of rotor height (H) from ground. But the experimental studies (Ref. 1) showed opposite trends from the analytical predictions as the flow pattern below the rotor resembles a source instead of a sink as considered.

The effect of ground on a helicopter in forward flight was studied later by Cheeseman and Bennet (Ref. 2) using image source method. Further a number of numerical and computational investigations on the topic have been produced indicating some similar interpretations (Refs. 3–6). Analytical methods proposed by Knight and Hefner (Refs. 7, 8) consider the effect of ground as variation on the induced power and also have experimental validation supporting the same. Later experiments by Zbrozek (Ref. 9) proposed the dependence of rotor height on the improvement of thrust due to ground effect.

The rotorcraft is said to be in ground effect (IGE) when the rotor height from the ground is less than one rotor diameter. The aerodynamics of rotor near to ground (IGE) is

Presented at the 6th Asian/Australian Rotorcraft Forum/Heli Japan 2017, Kanazawa, Japan, November 7-9, 2017.

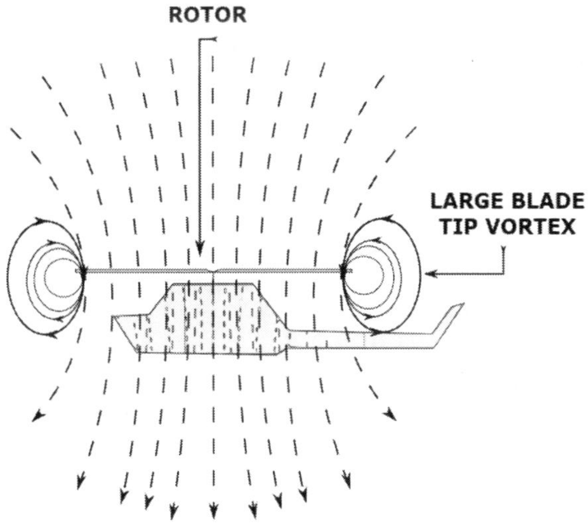

Fig. 1: Schematic of rotor wake in hover: OGE

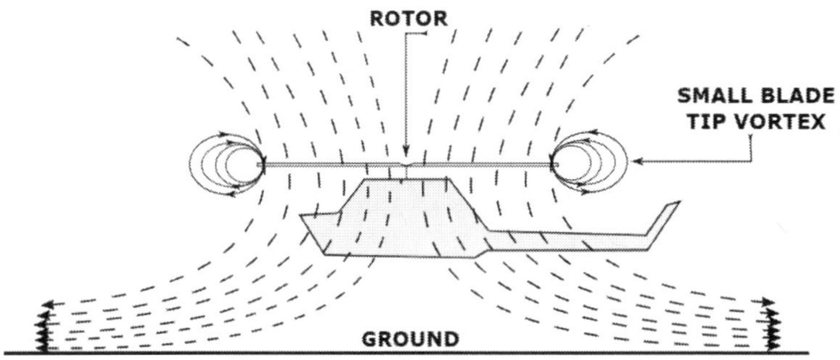

Fig. 2: Schematic of rotor wake in hover: IGE

different when compared to out of ground (OGE) scenario. When rotor shed wake is obstructed by the ground, an increase in the rotor thrust and reduction in inflow velocity is observed (Refs. 10, 11). The contracting nature of the rotor slipstream is altered to a more expanding wake type in the presence of ground (Figs.1,2). This aerodynamic interaction between the rotor induced wake and surrounding obstructions is quantified conventionally using the ratio of thrust in ground effect to that out of ground effect represented as k_G.

The thrust augmentation factor represents the ratio of thrust produced in ground effect to that of out of ground effect at constant power, or the power required in ground effect to that in out of ground effect at constant thrust. Three more commonly used empirical formulas based on experimental results are those proposed by Cheeseman (Ref. 2), Newman (Ref. 12) and Hayden (Ref. 13). They are as given in Eqns. 1, 2 and 3 respectively and are plotted in Fig. 3. Recent work (Ref. 14) also attempts to improve such theoretical factors in the lower elevation ranges.

$$kG = \frac{1}{1-\left(\frac{1}{4h}\right)^2} \tag{1}$$

$$kG = \frac{1}{1-\exp-2.75h} \tag{2}$$

$$kG = .9926 + .03794\left(\frac{2}{h}\right)^2 \tag{3}$$

A similar phenomenon concerned with constrained environment flight of rotorcraft is the effect of vertical obstructions or wall on the performance of the system. There are a few numerical studies on the topic, specifically, on the close proximity of helicopter to ship superstructures while landing. Some of the recent experiments using Particle Image Velocimetry(PIV) and pressure sensitive paint measurements for studying the behaviour of the rotor wake and outwash are

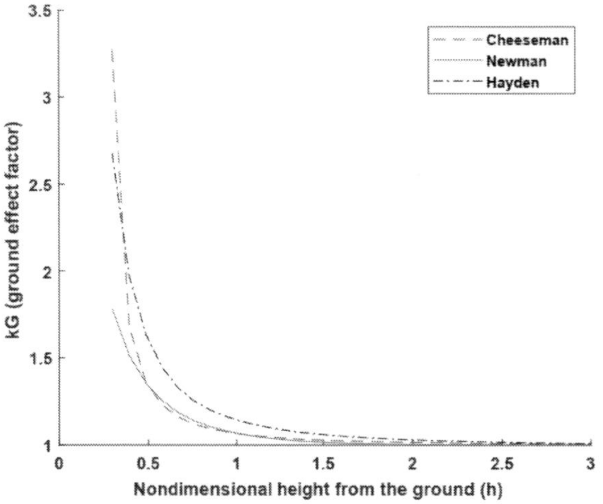

Fig. 3: kG factor plotted with respect to non dimensional height from the ground

noteworthy (Refs. 15–18)). However, a detailed experimental study depicting the effect of side walls on rotorcraft in ground effect is lacking. A closely related development in recent literature is the interaction between the rotor wake and a cubic obstacle using Laser Doppler Anemometry (LDA) and Stereoscopic Particle Image Velocimetry(SPIV) (Ref. 19).

Alhough there are factors and empirical relations to account for the interference of wind tunnel side walls on test rotors, they are often confined to certain wind tunnel geometries. CFD studies was performed on such scenarios in (Ref. 20) wherein the inflow pattern and moments were analysed for varying distances from the wall and the ground. The results in (Ref. 20) show that there is a considerable variation of roll away moment caused by the presence of wall.

Some of the applications of this study on hovering helicopters near buildings close to ground, landing and take-off of

Parameter	Value
Number of blades Nb	2
Blade radius (mm) R	485
Chord (mm) c	42
Ω (rpm) - without side wall	460
Ω (rpm) - with side wall	460
Blade collective pitch θ	8^0

Table 1: Parameters of the experimental setup

rotorcraft from ships(here the ship superstructure acts as the obstacle to the wake), Unmanned Aerial Vehicles (UAVs) flying inside a building with confined corridors and rotor experiments in wind tunnels with the interference of side wall.

In this paper, we study the forces and moments experienced by rotor while hovering near ground and side wall.

METHODOLOGY

Test Rig

The experimental setup consists of a rotor fixed on a high torque DC motor. Main features of rotor system are reported in the Table 1. Rotor blades are connected to the blade grip through a pitch hinge and pitch adjustment plate. Collective pitch of the blades can be set to a fixed value within a range of $\pm 16^0$. Rotor hub is connected directly to the shaft of high torque motor using a coupler. Motor is fixed to the force/torque sensor using four connection rods and it is secured to the top of a stand, the height of which is adjustable. Stand has a heavy base and two concentric tubes, the smaller diameter tube moves inside the outer tube and can be locked at different elevations thus enabling the variation of height from the ground. Relative position of rotor with the side wall is varied by changing the location of the stand base to record the readings.

Rotor blades are fixed with a negative pitch angle, which creates a rotor wake directed upwards and the ceiling acts as the ground surface. This arrangement is followed to minimize the obstruction of wake due to the experimental setup. Effectively the distance between the ceiling and model is varied as an alternative of the height of rotor from ground.

The schematic of the experimental setup is given in Fig. 4 where the variable height stand, load cell, DC motor and the rotor is shown. Rotor reaction measurements are carried out using an ATI gamma 9230 transducer. The transducer is a six axis load cell which monitors all three axis forces and moments on the rotor with an accuracy of .75% and input range of $\pm 32N$(lateral axis) and $\pm 100N$ (vertical axis) for forces and $\pm 2.5Nm$ for moments. Alignment of load cell with respect to wall and the axis orientation is shown in Figs. 5 and 6.

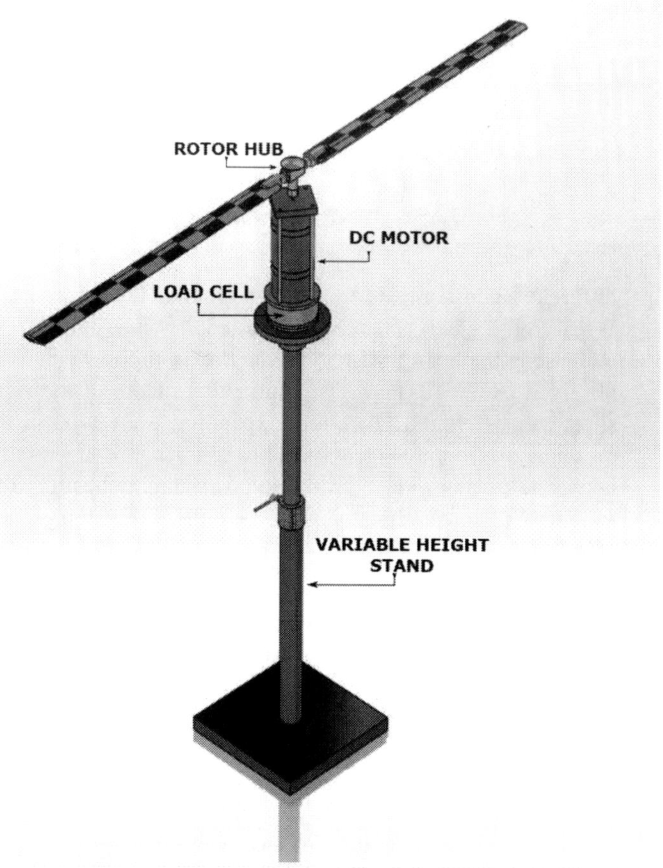

Fig. 4: Experimental Test Rig

Test Matrix Each test is conducted by keeping the setup in a defined position with respect to the wall and ground. Figure 7 indicates the parameters of setup in various tests. Height measurements are taken from the rotor disk and wall distance measurements are recorded from the blade tip.

Procedure The experiment position of rotor is shown in the test matrix (figure 1). Rotor test points are varied manually and kept to an accuracy of $1mm$ in both linear dimensions (height and distance from wall). The rotational speed of the rotor throughout the experiment, that is for all cases, is kept a constant (within 0.5%).

At each position, measurements are obtained for about 10 seconds for near to wall and near to ground points. The data

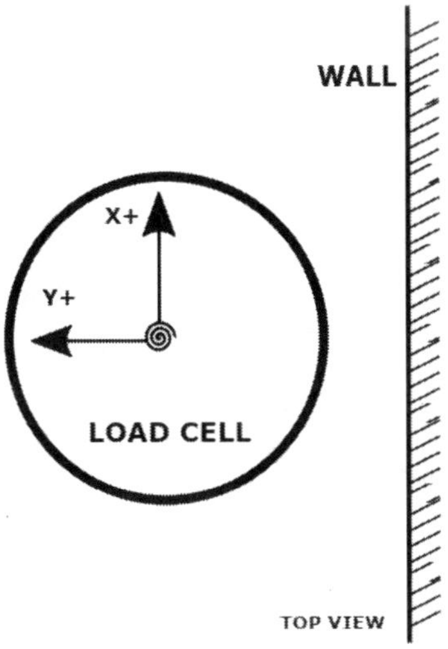

Fig. 5: Orientation of experimental setup : Loadcell w.r.t wall

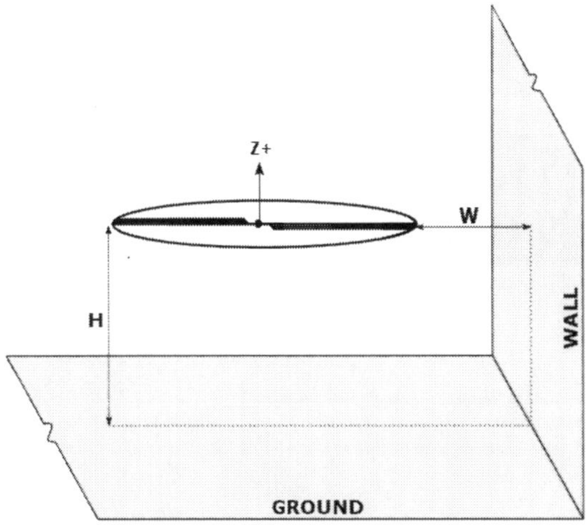

Fig. 6: Orientation of experimental setup : Rotor reference and measuring system

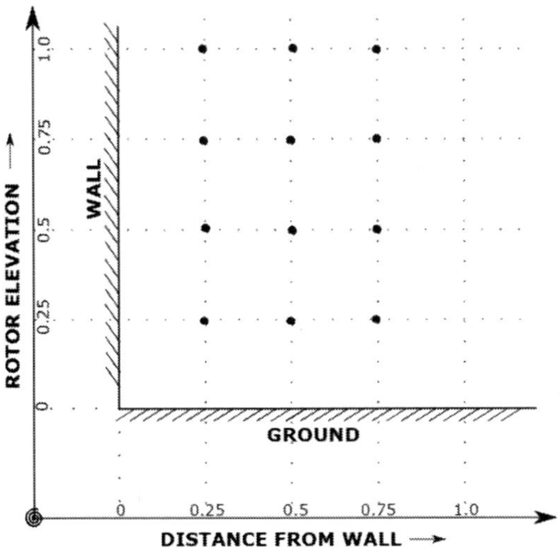

Fig. 7: Test matrix

sampling rate is set to be 150 samples per second effectively collecting 1500 samples corresponding to each position of the rotor. The high baud rate and small sampling time is adopted to capture the transients in the readings and to avoid the data drifting of the load-cell respectively. Data has been logged using ATI interface and analysed using MATLAB. The noise levels in the readings are found to be negligible and hence no further filters are used on the data obtained. Average force/moment values are reported here.

RESULTS AND DISCUSSION

In this section we present the main results and output of the experiment. All the force values obtained from the experiment are normalised using the rated thrust in OGE. Also the moments and linear measurements (height/distance from wall) are non-dimensionalised using the rotor torque and rotor radius(R) respectively.

Ground Effect without side wall

Augmentation of Thrust The augmentation in thrust due to effect of ground is illustrated using a plot of normalised rotor thrust against non-dimensionalised rotor elevation. The plot clearly discriminates the region of IGE influence and OGE region. Figure 8 indicates the augmentation of thrust obtained from the experiments. The results are found to be in agreement with the computational results.

Torque variation Figure 9 shows the plot of normalised IGE torque against normalised rotor elevation. Rotor torque does not show any considerable variation with rotor elevation.

Ground Effect with side wall

Thrust Variation The increase in thrust due to the ground effect remains unchanged in the presence of a side wall while the variation with side wall distance appear to be within 5% of the OGE thrust as seen in Figure 10

Torque variation As the rotor approaches the ground, the torque decreases by about 4% due to ground effect while there appears to be only slight variations due to the side wall as visible in Figure 11

Moment Variation The moment about the X-axis, is referred to as a roll moment for ease of reporting and about the Y-axis as a pitch moment.

A general trend for the rotor to roll away from the wall was observed in the presence of the side wall as shown in Figure 12. Contrarily, the test case at an elevation of 1R appear to stand out with a roll in moment for the rotor. However, such a roll -in moment is also reported in computational studies (Ref. 20). The magnitude of the rolling moment also appears to be significant with a maximum of around 20% the rotor OGE torque occurring at very close distances to the side wall.

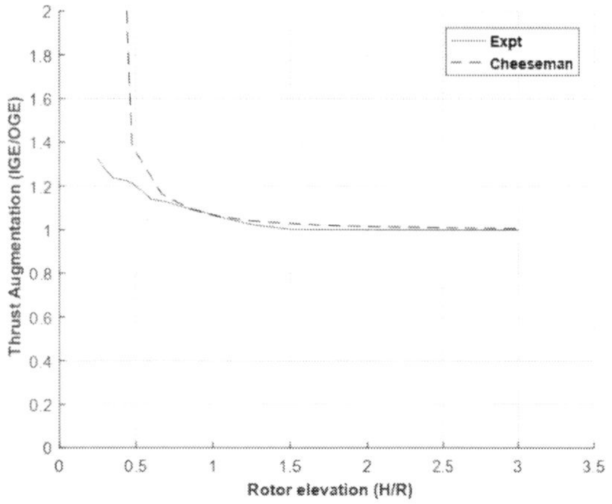

Fig. 8: Thrust augmentation factor for rotor in ground effect at constant power

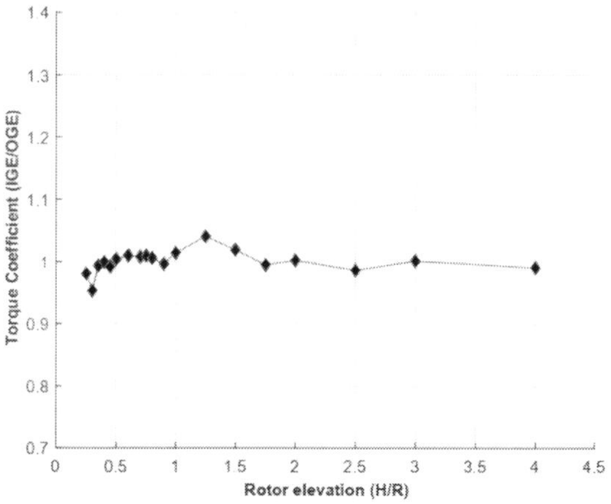

Fig. 9: Variation of Torque coefficient of rotor in ground effect from experiment

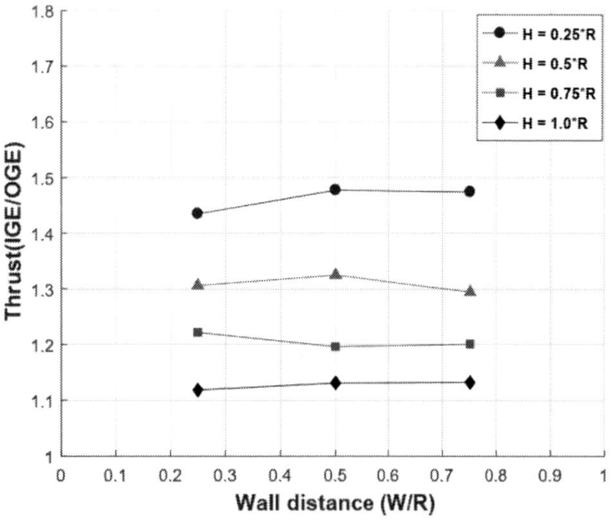

Fig. 10: Variation in thrust of the rotor in ground effect due to side wall at various elevations

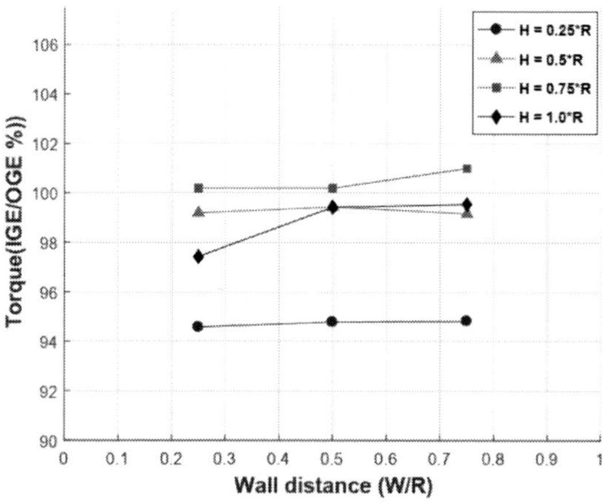

Fig. 11: Variation in torque of the rotor in ground effect due to side wall at various elevations

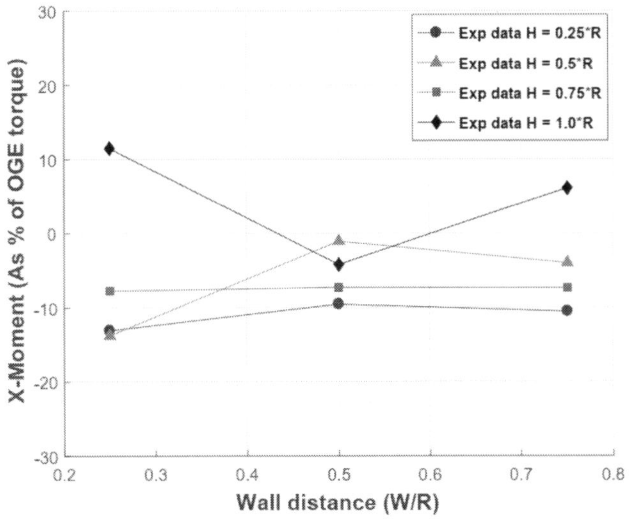

Fig. 12: Variation in rolling moment of the rotor in ground effect due to side wall at various elevations

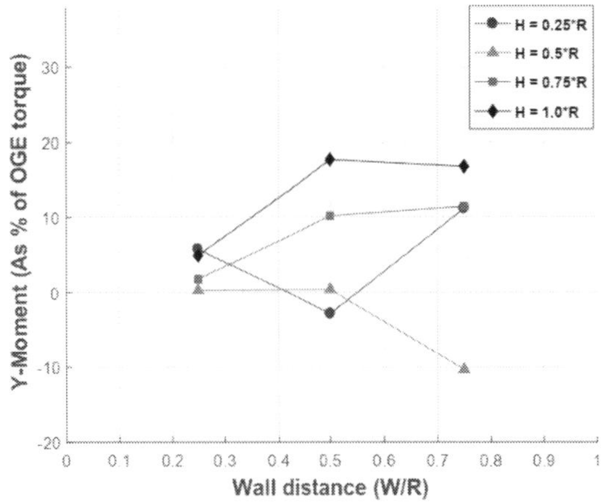

Fig. 13: Variation in pitching moment of the rotor in ground effect due to side wall at various elevations

The pitching moments appeared to follow a zeroing out trend with a largely positive moment. The magnitudes were found to be of the order of 20% the rotor torque as observed from Figure 13

CONCLUSIONS

From the experiments conducted on a constant pitch rotor in ground effect and side wall influence, it was concluded that:

1. Thrust variation is similar to that observed in ground effect.
2. Torque decreases as the rotor nears the ground with no considerable changes as the side wall distance decreases.
3. A rolling away moment was observed with the presence of the side wall in ground effect, while for larger elevations, this appeared to change to a roll in moment with magnitudes of around 20% rotor torque in OGE.
4. A pitching moment was also observed in the rotor of similar magnitudes to that of the rolling moment. However, a trend was unobservable.

Author contact:
Rohith M rohithm92@gmail.com
Cibin Joseph cibinjoseph92@gmail.com
Ranjith Mohan ranjith.m@iitm.ac.in

REFERENCES

[1]Betz, A., "The ground effect on lifting propellers," Technical report, National Advisory Committee for Aeronautics, 1937.

[2]Cheeseman, I. and Bennett, W., "The effect of ground on a helicopter rotor in forward flight," , 1955.

[3]Kutz, B. M., Kowarsch, U., Keßler, M., and Krämer, E., "Numerical investigation of helicopter rotors in ground effect," 30th AIAA Applied Aerodynamics Conference, Vol. 7, 2012.

[4]Brown, R. E. and Whitehouse, G. R., "Modelling rotor wakes in ground effect," *Journal of the American Helicopter Society*, Vol. 49, (3), 2004, pp. 238–249.

[5]Griffiths, D. A., Ananthan, S., and Leishman, J. G., "Predictions of Rotor Performance in Ground Effect Using a Free-Vortex Wake Model," *Journal of the American Helicopter Society*, Vol. 50, (4), 2005, pp. 302–314.

[6]Kalra, T. S., Lakshminarayan, V. K., and Baeder, J. D., "Cfd validation of micro hovering rotor in ground effect," American Helicopter Society 66th Annual Forum Proceedings, 2010.

[7]Knight, M. and Hefner, R. A., "Static Thrust Analysis of the Lifting Airscrew," NACA Technical Notes 626, 1937.

[8]Knight, M. and Hegner, R. A., "Analysis of ground effect on the lifting airscrew," , 1941.

[9]Zbrozek, J., "Ground Effect on the Lifting Rotor," ARC Reports and Memoranda 2347, July 1947.

[10]Light, J. S., "Tip vortex geometry of a hovering helicopter rotor in ground effect," *Journal of the American helicopter society*, Vol. 38, (2), 1993, pp. 34–42.

[11]Johnson, W., *Rotorcraft aeromechanics*, Vol. 36, Cambridge University Press, 2013.

[12]Newman, E. M., "A New Approach to the Calculation of the Effect of the Ground on the Performance of Rotary Wing Aircraft," Technical report, Naval Air Systems Command Washington DC Professional Development Center, 1971.

[13]Hayden, J. S., "The effect of the ground on helicopter hovering power required," 32nd annual forum of the American helicopter society, Washington, DC, 1976.

[14]Joseph, C. and Mohan, R., "A New Ground Effect Factor based on Image Vortices," 6th Asian/ Australian Rotorcraft Form, 2017.

[15]Tanner, P. E., Overmeyer, A. D., Jenkins, L. N., Yao, C.-S., and Bartram, S. M., "Experimental Investigation of Rotorcraft Outwash in Ground Effect," , 2015.

[16]Nathan, N. and Green, R., "Measurements of a rotor flow in ground effect and visualization of the brown-out phenomenon," , 2008.

[17]Nathan, N. and Green, R., "Flow visualisation of the helicopter brown-out phenomenon," *The Aeronautical Journal*, Vol. 113, (1145), 2009, pp. 467–478.

[18]Nathan, N. and Green, R., "Wind tunnel investigation of flow around a rotor in ground effect," , 2010.

[19]Gibertini, G., Clavel, C., Grassi, D., Parolini, C., Zagaglia, D., and Zanotti, A., "An Experimental Setup for the Study of Helicopter and Building Aerodynamic Interaction," 40th European Rotorcraft Forum, South Hampton, United Kingdom, September 2014.

[20]Joseph, C. and Mohan, R., "A Numerical Investigation of Ground Effect on Rotorcraft in the presence of a Side wall," 43rd European Rotorcraft Fotum, Milan, Italy, 2017.

Comprehensive Flight Control Law Design Method for Helicopters

Wu Wei 1
Associate Professor
Nanjing University of Aeronautics and Astronautics
Nanjing, China

ABSTRACT

A comprehensive flight control law design method based on adaptive control is presented in this paper. The proposed method consists of three basic modules – model decoupling, online system identification and adaptive pole placement. The model decoupling module decouples the helicopter flight dynamics model based on dynamic inversion technique. This procedure helps to reduce the difficulties in online system identification and adaptive control design. In online system identification module, a recursive extended least squares algorithm is established to identify the augmented linear flight dynamics model which is composed of helicopter model and unideal noise model. The helicopter model parameters and the noise parameters are identified simultaneously which improves the identification accuracy as well as robustness compared with conventional method. Pole placement is implemented in the last module to make a helicopter keep tracking the idea poles which can be designed according to ADS-33E-PRF. The adaptive rule in this step is designed based on eigenvalue analysis of the model to remove all unnecessary oscillations of the control parameters. An adaptive controller is designed according to the developed method for the UH-60 helicopter based on a nonlinear simulation program. Typical response types are also implemented. The simulation results show that the designed adaptive controller has high performance as well as robustness in both hover and forward flight.

INTRODUCTION

The helicopter is a special aircraft which can perform hovering, vertical takeoff and landing as well as low speed maneuvers, and it has already been widely used in civil and military domain. However, the flying quality of a helicopter is poor due to many reasons, such as a helicopter is highly coupled, it is quite unstable and the vibration and noise level of a helicopter is high etc. The poor flying quality brings many problems in actual flight, for example, the pilot need to make compensation controls to eliminate undesirable cross coupling responses at any time during the flight which means the workload is very high. Moreover, in some severe condition, such as during aggressive maneuver procedure, the poor flying quality will cause safety problems. Therefore, the operational capability and the mission effectiveness of a helicopter are limited for these reasons. In order to solve these problems, the flying quality of a helicopter must be improved. A proper flying quality specification is required for flying quality design. However, since the dynamic characteristics of a helicopter are very complex, the helicopter flying quality specification is also experienced a long time of evolution [1]. In the first helicopter flying quality specification, the MIL-H-8501 and its revised version (the MIL-H-8501A), the primary requirements only consisted of simple time domain parameters. However, in the current specification, the ADS-33E-PRF [2], the coverage of design requirements is expanded significantly. When we have the specification on hand, the rest problem is how to improve the flying quality of a helicopter based on certain specification. Basically, there are two different ways to deal with this problem. The first way is to optimize some of the design parameters of a helicopter such as rotor diameter, rotor height, rotor flapping hinge offset, position and area of vertical tail or horizontal tail etc. However, since the helicopter is highly coupled and the dynamic behavior of a helicopter is very complex, adjusts one certain design parameter may increase the flying quality level of some requirements defined in the specification but decrease the others. Therefore, the optimization results of all design parameters will always be a compromised solution. The second way is to design a proper flight control system with high performance, and it has already been proven that this kind of approach is much more efficient and effective than the first way.

However, the complexity of a helicopter also brings many difficulties in flight control system design [3-4]. Although most of the current controllers used in helicopters are designed based on classical PID control method, the disadvantages of PID controllers in helicopter flight control are also very obvious. For example, a helicopter has strong nonlinearity, which means the dynamic characteristics of a helicopter have large differences in different flight states. If a PID controller is designed in one flight state, such as hover, the performance of the controller will be decreased quite a lot in other flight states, forward flight for instance. Therefore, the overall performance of a PID controller is

Presented at the 6th Asian/Austrailian Rotorcraft Forum / Heli Japan 2017, Kanazawa, Japan, November 7-9, 2017.

limited. In order to overcome the shortages of PID controllers, there are a lot of researches concentrate on development of advanced flight control design tools based on modern control theory. Optimal control and model-following control method have already been used to design the flight control laws for helicopters that equipped with fly-by-wire or flight-by-light control system for decades [5-7]. Flight test results indicate the performance of such controllers is increased significantly compared with PID controllers. Robust control is another advanced tool for helicopter flight control law design [8-9]. In this kind of method, the influences of model uncertainties caused by nonlinearity, measurement noise and environment disturbances etc. are considered during the design process, so the controllers designed by this approach will be much more robust than PID controllers. Adaptive control, which is even more powerful than robust control, is a perfect tool for flight control law design of complex aircraft including helicopters. There are plenty of papers addressed their achievements in adaptive control research, and these studies proposed many excellent ideas in neural adaptive control and model reference adaptive control for fixed wings and helicopters [10-16]. New theoretical achievements for adaptive control design were also obtained in recent years [17]. However, it is still a challenge of designing a stable adaptive controller with instant control accuracy at different flight states for a helicopter today due to its complicated dynamic characteristics.

A comprehensive adaptive control design method is developed in this paper for improving flying quality of a helicopter. The flight control law is designed using combined adaptive and dynamic inversion technique. An improved online system identification algorithm is also established that is able to consider the influence of measurement noise. The adaptive strategy is selected to ensure the controller keeps high performance from hover to maximum forward flight speed, while the feedback coefficients have minimum changing frequency. Application of the developed method to a UH-60 helicopter shows the proposed adaptive control design method is effective, efficient and robust.

GENERAL DESCRIPTION OF THE DESIGN METHOD

A general controller structure based on the developed design method can be found in Fig. 1. Basically, there are two control loops in the controller. The inner loop is used to decouple the helicopter, and the outer loop is designed to make the helicopter reach required flying quality level according to the ADS-33E-PRF. Since the identification of full coupled flight dynamics model is a difficult task especially in real time, the helicopter is decoupled before the identification module to simplify the identification problem by a dynamic inversion controller. Then in the system identification module, only several decoupled or weak coupled models are need to be identified. In the pole placement controller module, ideal poles in each control channel are designed for different purpose according to the ADS-33E-PRF, and a pole placement algorithm is applied to the identified flight dynamics model to ensure the poles of final closed-loop model are identical to ideal poles. Because the sampling frequency of airborne sensors is high, so the differences among each identified models in contiguous sampling points are very small. In some cases, such as in steady hover or steady forward flight, the identified models are nearly identical all the time. Therefore, there is no need to do pole placement at each sampling points. The adaptive rule is set then based on model analysis to avoid unnecessary calculation of feedback matrix. As a consequence, the efficiency and the stability of the controller are increased.

Details of the three basic modules, i.e. inner loop decoupling controller, online system identification and adaptive pole placement will be discussed in the following sections. A nonlinear model of UH-60 helicopter which can be found in Ref. [23] is used in this paper to do numerical simulation and validation of the designed adaptive controller.

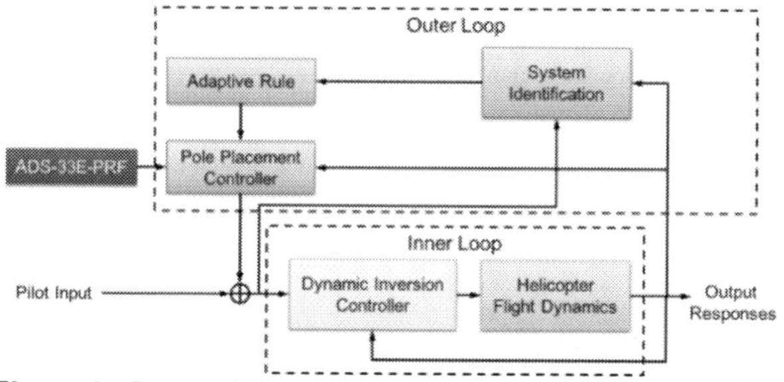

Figure1. General Structure of the Adaptive Controller.

INNER LOOP DECOUPLING CONTROLLER

In order to decouple a helicopter, an inner loop controller is designed based on dynamic inversion technique. Because the helicopter flight dynamics in Fig. 1 is nonlinear, so a nonlinear dynamic inversion solver is the best tool to deal with this problem. However, since the structure of helicopter flight dynamics model is very complex, the nonlinear dynamic inversion problem of a helicopter is also very complicated. Besides, the nonlinear dynamic inversion calculation is very time consuming, it is difficult to use such solver in real time. Therefore, a linear dynamic inversion solver is applied in this paper, while a linear state space flight dynamics model is used in dynamic inversion calculation.

Fig. 2 shows the basic principle of dynamic inversion controller. A linear state space flight dynamics model of a helicopter is represented as Eq. (1). The feedback matrix $\boldsymbol{K}_{inv}$ and feedforward matrix $\boldsymbol{L}_{inv}$ can be determined according to current stability matrix $\boldsymbol{A}$ and control matrix $\boldsymbol{B}$ of a helicopter.

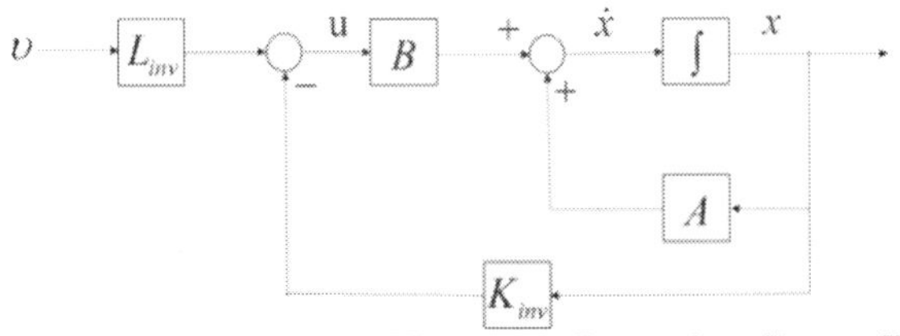

Figure2. Basic Scheme of Dynamic Inversion Controller.

$$\dot{\boldsymbol{x}} = \boldsymbol{A}\boldsymbol{x} + \boldsymbol{B}\boldsymbol{u} \tag{1}$$

A closed-loop model is easily obtained according to Fig.2, which can be written as Eq. (2). It is obviously that the closed-loop model can be decoupled by selecting proper $\boldsymbol{K}_{inv}$ and $\boldsymbol{L}_{inv}$ matrix. In order to obtain the solution, matrix inverse calculation is required. However, the control matrix $\boldsymbol{B}$ is not a square matrix, which means the normal inverse matrix does not exist.

$$\dot{\boldsymbol{x}} = (\boldsymbol{A} - \boldsymbol{B}\boldsymbol{K}_{inv})\boldsymbol{x} + \boldsymbol{B}\boldsymbol{L}_{inv}\boldsymbol{\upsilon} \tag{2}$$

Considering there are both fast and slow state variables in x, the whole system will be stable if the fast state is stabilized. Therefore, Eq. (2) can be replaced by Eq. (3) for solving the dynamic inversion problem.

$$\dot{\boldsymbol{x}}_{inv} = (\boldsymbol{A}_{inv} - \boldsymbol{B}_{inv}\boldsymbol{K}_{inv})\boldsymbol{x} + \boldsymbol{B}_{inv}\boldsymbol{L}_{inv}\boldsymbol{\upsilon} \tag{3}$$

In which, $\boldsymbol{x}_{inv} = [w, p, q, r]^T$,

$\boldsymbol{x} = [u, v, w, p, q, r, \phi, \theta, \psi]^T$, $\boldsymbol{\upsilon} = \left[\delta_{long}, \delta_{lat}, \delta_{col}, \delta_{ped}\right]^T$, $\boldsymbol{A}_{inv}$ is a 4×9 matrix, $\boldsymbol{B}_{inv}$ is a 4×4 matrix.

The next step is to define ideal decoupled model structure. Weak decoupled stability matrix $\boldsymbol{A}_{exp}$ and control matrix $\boldsymbol{B}_{exp}$ are shown in Eq. (4) and Eq. (5) respectively. The reason for not using fully decoupled model is to keep lateral-directional oscillation modes in the model. In ADS-33E-PRF, special requirements are defined for these modes such as Dutch roll. Therefore, these modes should be remained and optimized in pole placement module.

$$\boldsymbol{A}_{exp} = \begin{bmatrix} 0 & 0 & Z_w & 0 & 0 & 0 & 0 & 0 & 0 \\ 0 & 0 & 0 & L_p & 0 & L_r & 0 & 0 & 0 \\ 0 & 0 & 0 & 0 & M_q & 0 & 0 & 0 & 0 \\ 0 & 0 & 0 & N_p & 0 & N_r & 0 & 0 & 0 \end{bmatrix} \tag{4}$$

$$\boldsymbol{B}_{exp} = \begin{bmatrix} 0 & 0 & Z_{\delta_{col}} & 0 \\ 0 & L_{\delta_{lat}} & 0 & L_{\delta_{ped}} \\ M_{\delta_{long}} & 0 & 0 & 0 \\ 0 & N_{\delta_{lat}} & 0 & N_{\delta_{ped}} \end{bmatrix} \tag{5}$$

The solution for $\boldsymbol{K}_{inv}$ and $\boldsymbol{L}_{inv}$ matrix can be easily obtained by combing Eq. (3), Eq. (4) and Eq. (5), as shown in Eq. (6) and Eq. (7).

$$\boldsymbol{L}_{inv} = \boldsymbol{B}_{inv}^{-1}\boldsymbol{B}_{exp} \tag{6}$$

$$\boldsymbol{K}_{inv} = (\boldsymbol{A}_{inv} - \boldsymbol{A}_{exp})\boldsymbol{B}_{inv}^{-1} \tag{7}$$

A decoupling controller is designed for a UH-60 helicopter in hover condition based on the preceding method. The solution can be found in Eq. (8) and Eq. (9).

$$\boldsymbol{L}_{inv} = \begin{bmatrix} 0.9993 & 0.0106 & 0.0218 & -0.0443 \\ -0.0337 & 0.997 & 0.0422 & 0.0046 \\ -0.0172 & 0 & 0.9919 & 0.0733 \\ 0.0015 & 0 & -0.1055 & 0.9922 \end{bmatrix} \tag{8}$$

$$\boldsymbol{K}_{inv} = \begin{bmatrix} 0.0105 & 0.0394 & 0.0062 & 0.9375 & -0.0014 & -0.0092 & -0.0036 & 0.0078 & 0 \\ 0.0073 & -0.0214 & 0.0015 & -0.0315 & -2.0038 & 0.0019 & -0.0070 & 0.0151 & 0 \\ -0.0028 & 0.0017 & 0 & -0.0149 & -0.0804 & 0.0377 & -0.1638 & 0.3542 & 0 \\ 0.0035 & 0.0169 & -0.0033 & 0.0012 & -0.4632 & -0.0040 & 0.0174 & -0.0377 & 0 \end{bmatrix} \tag{9}$$

Step inputs of longitudinal cyclic, lateral cyclic, collective and pedal are applied to the decoupled model respectively. Simulation results are shown in Fig. 3 ~ Fig. 6. It is obviously to see that the longitudinal channel and vertical channel are fully decoupled, while the lateral channel and directional channel are weakly coupled. Moreover, the performance of the controller is kept well with flight states changing.

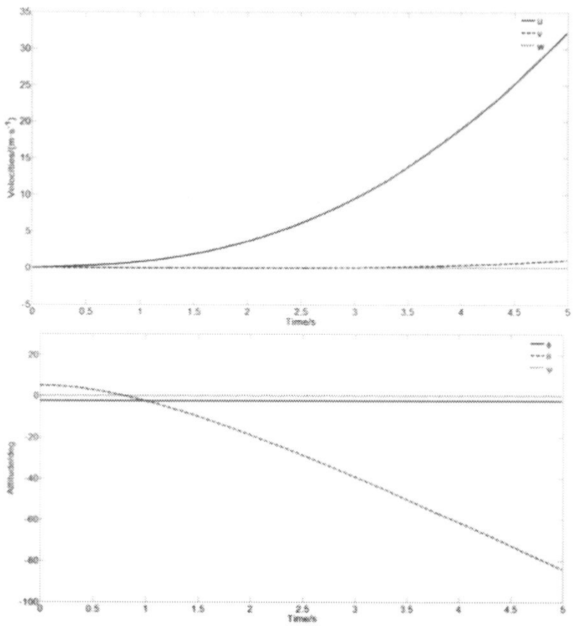

Figure 3: Decoupled Helicopter Responses with Longitudinal Cyclic Step Input.

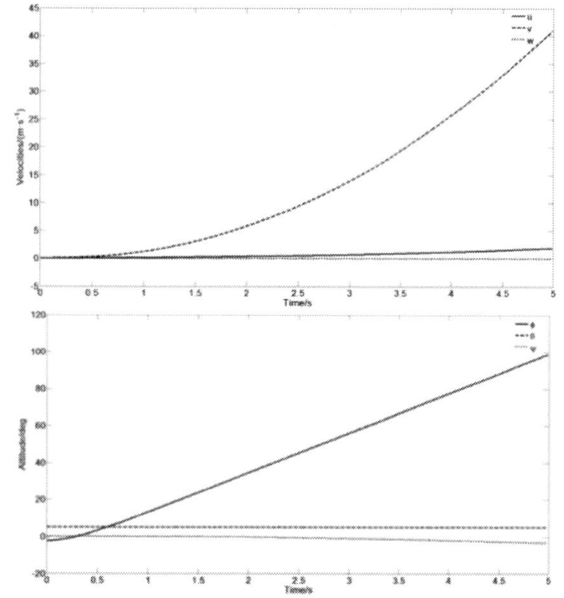

Figure 4: Decoupled Helicopter Responses with Lateral Cyclic Step Input.

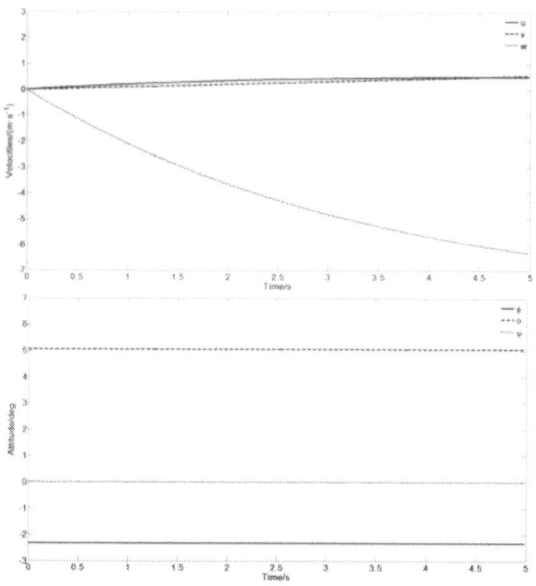

Figure 5: Decoupled Helicopter Responses with Collective Step Input.

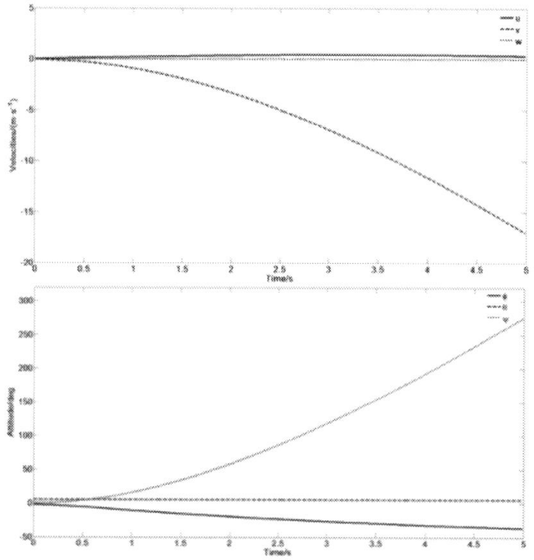

Figure 6: Decoupled Helicopter Responses with Pedal Step Input.

ONLINE SYSTEM IDENTIFICATION

Online system identification is a very important module in adaptive controller, since the accuracy of identified model has large influences on controller's performance. However, the identification problem of a helicopter is very complex. Nearly 5 decades have been past from the first attempt in applying system identification into helicopter flight dynamics modeling, and plenty of methods have been developed both in time domain and frequency domain [18-22]. However, most of these methods are used for offline identification purpose, and it is not possible to integrate these comprehensive methods into an adaptive controller. The most frequently used identification algorithm in an adaptive controller currently is recursive least squares (RLS) method. This kind of method is easy to implement and efficient, but it is not robust to noise. A helicopter has high level of vibration and measurement noise during the flight, which means the measured responses are contaminated. Although there are low pass filters in flight control system, the influences of measurement noise to measured data are still existed. Theoretically speaking, the least squares estimator is unbiased and consistent only when there is white noise or no noise in measured data. Therefore, in most of the current online identification methods, measurement noise is neglected or simplified to white noise. But unfortunately, the measurement noise of a helicopter will always be colored, and the white noise assumption will bring additional identification errors.

In this paper, a recursive extended least squares algorithm is established. Because the helicopter is decoupled before identification, full coupled flight dynamics model identification problem can be replaced by identifying 3 decoupled and weak coupled models, as shown in Eq. (10) ~ Eq. (12).

$$\begin{bmatrix} \dot{u} \\ \dot{q} \\ \dot{\theta} \end{bmatrix} = \begin{bmatrix} X_u & X_q & -g\cos\theta_0 \\ M_u & M_q & 0 \\ 0 & 1 & 0 \end{bmatrix} \begin{bmatrix} u \\ q \\ \theta \end{bmatrix} + \begin{bmatrix} X_{\delta_{long}} \\ M_{\delta_{long}} \\ 0 \end{bmatrix} \delta_{long} \tag{10}$$

$$\dot{w} = Z_w \cdot w + Z_{\delta_{col}} \cdot \delta_{col} \tag{11}$$

$$\begin{bmatrix} \dot{v} \\ \dot{p} \\ \dot{r} \\ \dot{\phi} \\ \dot{\psi} \end{bmatrix} = \begin{bmatrix} Y_v & Y_p + w_0 & Y_r - u_0 & g\cos\theta_0 cos\phi_0 & 0 \\ L_v & L_p & L_r & 0 & 0 \\ N_v & N_p & N_r & 0 & 0 \\ 0 & 1 & \tan\theta_0 & 0 & 0 \\ 0 & 0 & 1/\cos\theta_0 & 0 & 0 \end{bmatrix} \begin{bmatrix} v \\ p \\ r \\ \phi \\ \psi \end{bmatrix} + \begin{bmatrix} Y_{\delta_{lat}} & Y_{\delta_{ped}} \\ L_{\delta_{lat}} & L_{\delta_{ped}} \\ N_{\delta_{lat}} & N_{\delta_{ped}} \\ 0 & 0 \\ 0 & 0 \end{bmatrix} \begin{bmatrix} \delta_{lat} \\ \delta_{ped} \end{bmatrix} \tag{12}$$

In order to consider the influences of unideal measurement noise, a second order noise model is established as shown in Eq. (13).

$$\boldsymbol{e}(t) = \boldsymbol{\xi}(t) + \boldsymbol{d}_1\boldsymbol{\xi}(t-\Delta t) + \boldsymbol{d}_2\boldsymbol{\xi}(t-2\Delta t) \tag{13}$$

In which, $\boldsymbol{e}$ is unideal noise vector, t is time variable, Δt is sampling time interval, d_1 and d_2 are noise model parameters vector, ξ is noise description vector which can be approximated by model prediction error.

Substitutes Eq. (13) into Eq. (10) ~ Eq. (12), and after some mathematical manipulation, a standard discrete least squares identification model can be obtained as shown in Eq. (14). Detailed structure of each component in Eq. (14) can be found in Eq. (15) ~ Eq. (17) for different models.

$$\boldsymbol{y}(k) = \boldsymbol{h}^T(k)\boldsymbol{\theta} + \overline{\boldsymbol{e}}(k) \tag{14}$$

Longitudinal model:

$$\begin{cases} \boldsymbol{y}^{long}(k) = \left[a_x(k), \dot{q}(k)\right]^T \\ \boldsymbol{\theta}^{long} = \left[X_u, X_q, X_{\delta_{long}}, d_1^{long1}, d_2^{long1} M_u, M_q, M_{\delta_e}, d_1^{long2}, d_2^{long2}\right]^T \\ \boldsymbol{h}^{long}(k) = \begin{bmatrix} u(k) & q(k) & \delta_{long}(k) & \xi_1^{long}(k-1) & \xi_1^{long}(k-2) & 0 & 0 & 0 & 0 & 0 \\ 0 & 0 & 0 & 0 & 0 & u(k) & q(k) & \delta_{long}(k) & \xi_2^{long}(k-1) & \xi_2^{long}(k-2) \end{bmatrix}^T \\ \overline{\boldsymbol{e}}^{long}(k) = \left[\xi_1^{long}(k), \xi_2^{long}(k)\right]^T \end{cases} \tag{15}$$

Vertical model:

$$\begin{cases} \boldsymbol{y}^{ver}(k) = \left[a_z(k)\right]^T \\ \boldsymbol{\theta}^{ver} = \left[Z_w, Z_{\delta_{col}}, d_1^{ver}, d_2^{ver}\right]^T \\ \boldsymbol{h}^{ver}(k) = \left[w(k) \quad \delta_{col}(k) \quad \xi_1^{ver}(k-1) \quad \xi_1^{ver}(k-2)\right]^T \\ \overline{e}^{ver}(k) = \xi_1^{ver}(k) \end{cases} \tag{16}$$

Lateral-directional model:

$$
\begin{cases}
\boldsymbol{y}^{latyaw}(k)=\left[a_y(k),\dot{p}(k),\dot{r}(k)\right]^T \\
\boldsymbol{\theta}^{latyaw}=\left[Y_v,Y_p,Y_r,Y_{\delta_{lat}},Y_{\delta_{ped}},d_1^{latyaw1},d_2^{latyaw1},L_v,L_p,L_r,L_{\delta_{lat}},L_{\delta_{ped}},d_1^{latyaw2},d_2^{latyaw2},N_v,N_p,N_r,N_{\delta_{lat}},N_{\delta_{ped}},d_1^{latyaw3},d_2^{latyaw3}\right]^T \\
\boldsymbol{h}^{latyaw}=\begin{bmatrix} v(k) & p(k) & r(k) & \delta_{lat}(k) & \delta_{ped}(k) & \xi_1^{latyaw}(k-1) & \xi_1^{latyaw}(k-2) & 0 & 0 & 0 & 0 & 0 & 0 & 0 & 0 & 0 & 0 & 0 & 0 & 0 & 0 \\ 0 & 0 & 0 & 0 & 0 & 0 & 0 & v(k) & p(k) & r(k) & \delta_{lat}(k) & \delta_{ped}(k) & \xi_2^{latyaw}(k-1) & \xi_2^{latyaw}(k-2) & 0 & 0 & 0 & 0 & 0 & 0 & 0 \\ 0 & 0 & 0 & 0 & 0 & 0 & 0 & 0 & 0 & 0 & 0 & 0 & 0 & 0 & v(k) & p(k) & r(k) & \delta_{lat}(k) & \delta_{ped}(k) & \xi_3^{latyaw}(k-1) & \xi_3^{latyaw}(k-2) \end{bmatrix}^T \\
\bar{\boldsymbol{e}}^{latyaw}(k)=\left[\xi_1^{latyaw}(k),\xi_2^{latyaw}(k),\xi_3^{latyaw}(k)\right]^T
\end{cases}
\quad (17)
$$

Standard recursive least squares iterations, as shown in Eq. (18), can be applied directly to the above extended models. Data saturation problem is avoided by adding a window with length L in the identification process, when the number of data used in identification reaches L, the covariance matrix $\boldsymbol{P}$ will be reset according to current states.

$$
\begin{cases}
\hat{\boldsymbol{\theta}}(k)=\hat{\boldsymbol{\theta}}(k-1)+\boldsymbol{K}(k)[\boldsymbol{y}(k)-\boldsymbol{h}^T(k)\hat{\boldsymbol{\theta}}(k-1)] \\
\boldsymbol{K}(k)=\boldsymbol{P}(k-1)\boldsymbol{h}(k)\left[\boldsymbol{h}^T(k)\boldsymbol{P}(k-1)\boldsymbol{h}(k)+1\right]^{-1} \\
\boldsymbol{P}(k)=\left[\boldsymbol{I}(k)-\boldsymbol{K}(k)\boldsymbol{h}^T(k)\right]\boldsymbol{P}(k-1)
\end{cases}
\quad (18)
$$

Identification results of a UH-60 helicopter in hover condition are shown in Table1 ~ Table 3 and Fig. 7 ~ Fig. 8, while comparative study between the developed method in this paper with standard RLS algorithm is also conducted. 3-2-1-1 input signal, which is different with sweep excitation used in identification procedure, is applied to verify the identified models. The verification results are shown in Fig. 9 ~ Fig. 12. It is apparently that the identification accuracy of the developed method is increased significantly compared with standard RLS method, and this is especially true when the signal to noise ratio is low (Fig. 11 and Fig. 12). This is because in standard RLS algorithm, noise is treated as white random signal which has zero mean value and constant variance. However, the real noise is colored that has non-zero mean value and time variant variance. So there exists a bias term between the ideal noise and the real noise, and biased estimation of model parameters is obtained. In the method developed in this paper, the bias term is represented as colored noise model. Therefore the bias in helicopter model parameter identification is eliminated. Fig. 13 shows the noise prediction error, and good prediction capability of the noise model is proven.

Table 1. Identification Results of Longitudinal Model.

Model Parameter	Identified in This Paper	Identified by Standard RLS	True Value
X_u	-0.0236	-0.0265	-0.02349
X_q	2.8395	2.8827	2.809
$X_{\delta_{long}}$	-1.6589	-1.6347	-1.659
d_1^{long1}	0.4870	-	0.5
d_2^{long1}	0.1662	-	0.2
M_u	0.0036	0.0038	0.003554
M_q	-0.8199	-0.8996	-0.8161
$M_{\delta_{long}}$	0.3348	0.3219	0.3346
d_1^{long2}	0.4996	-	0.5
d_2^{long2}	0.2433	-	0.2

Table 2. Identification Results of Vertical Model.

Model Parameter	Identified in This Paper	Identified by Standard RLS	True Value
Z_w	-0.2923	-0.2804	-0.2931
$Z_{\delta_{col}}$	-7.9217	-7.8761	-7.921
d_1^{ver1}	0.5211	-	0.5
d_2^{ver1}	0.1985	-	0.2

Table 3. Identification Results of Lateral-directional Model.

Model Parameter	Identified in This Paper	Identified by Standard RLS	True Value
Y_v	-0.0474	-0.0501	-0.0473
Y_p	-1.7377	-1.8916	-1.723
Y_r	0.6428	0.6806	0.6383
$Y_{\delta_{lat}}$	0.9384	0.7833	0.942
$Y_{\delta_{ped}}$	-1.4885	-1.4674	-1.486
$d_1^{latyaw1}$	0.5197	-	0.5
$d_2^{latyaw1}$	0.2239	-	0.2
L_v	-0.0415	-0.04874	-0.04124
L_p	-3.5429	-3.5992	-3.551
L_r	0.0807	0.0560	0.07467
$L_{\delta_{lat}}$	1.3243	1.2362	1.334
$L_{\delta_{ped}}$	-0.8341	-0.8043	-0.8406
$d_1^{latyaw2}$	0.5321	-	0.5
$d_2^{latyaw2}$	0.2434	-	0.2
N_v	0.0096	0.0100	0.00976
N_p	-0.0981	-0.1295	-0.1013
N_r	-0.3312	-0.3596	-0.3342
$N_{\delta_{lat}}$	0.0259	0.0371	0.02734
$N_{\delta_{ped}}$	0.6042	0.6507	0.604
$d_1^{latyaw3}$	0.4997	-	0.5
$d_2^{latyaw3}$	0.1910	-	0.2

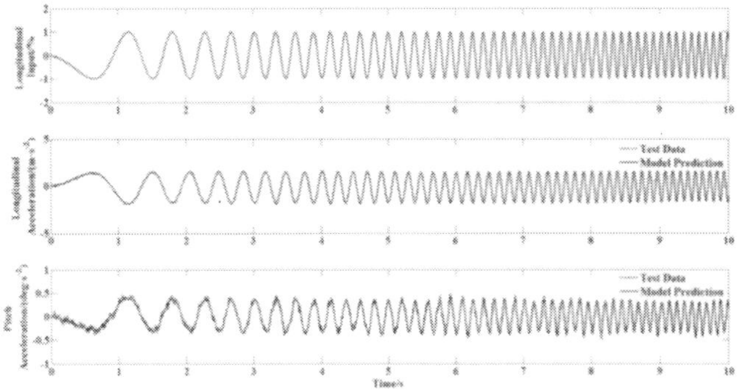

Figure 7: Longitudinal decoupled model identification result.

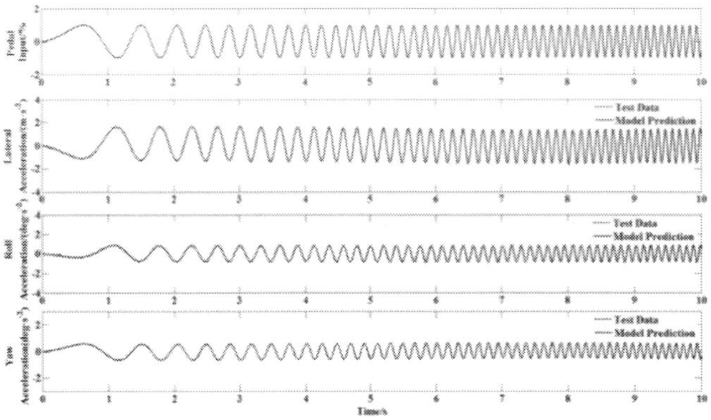

Figure 8: Lateral-directional weak coupled model identification result.

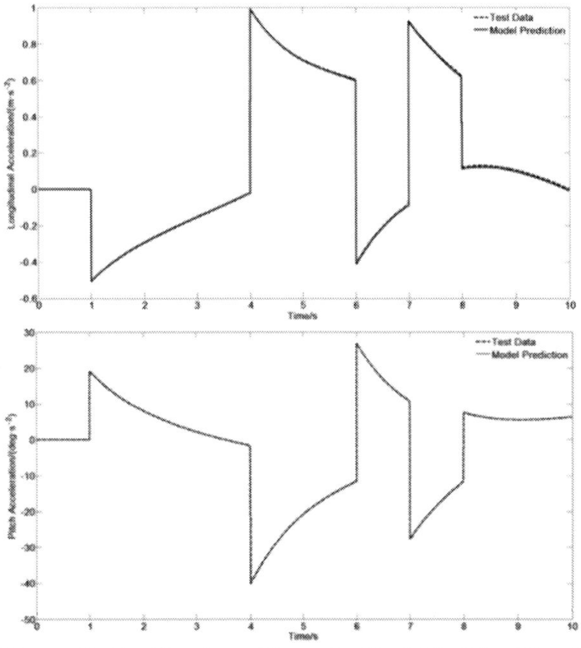

Figure 9: Verification of Longitudinal Model without Noise Contamination.

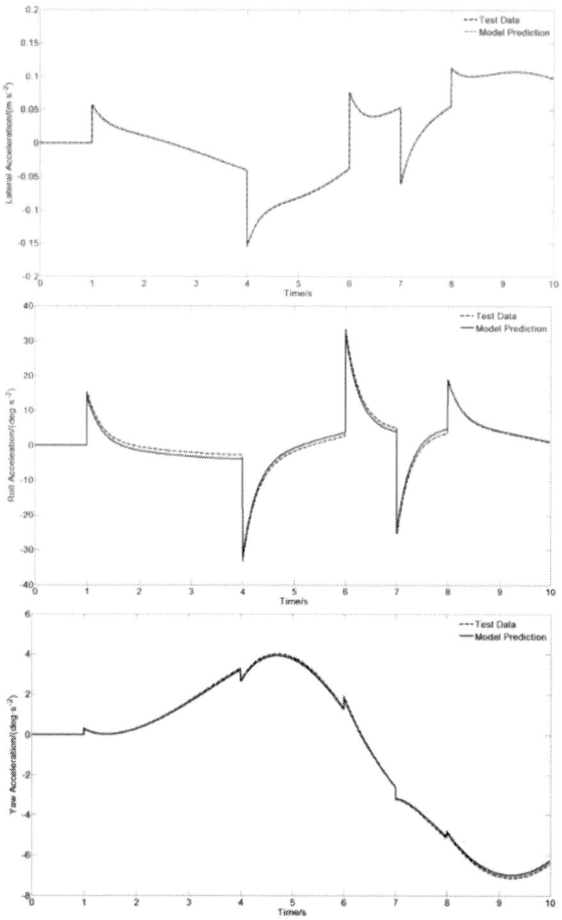

Figure 10: Verification of Lateral-directional Model without Noise Contamination.

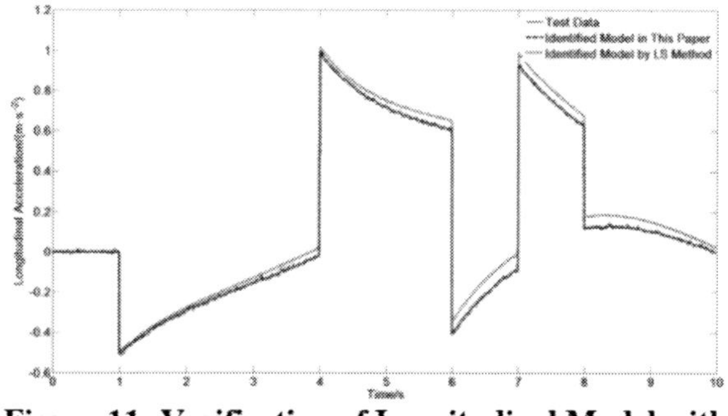

Figure 11: Verification of Longitudinal Model with Colored Noise Contamination.

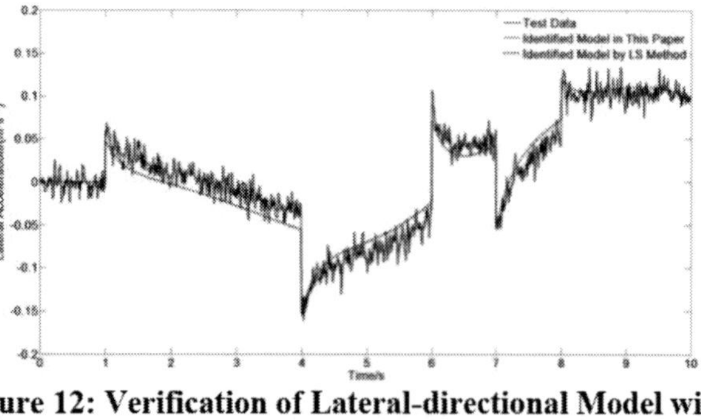

Figure 12: Verification of Lateral-directional Model with Colored Noise Contamination.

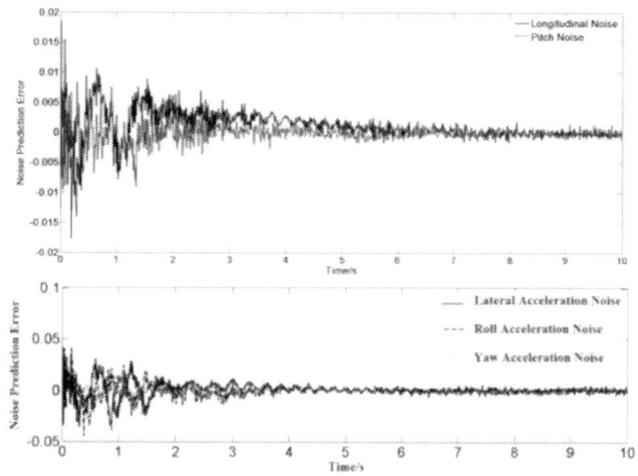

Figure 13: Verification of Colored Noise Model.

ADAPTIVE POLE PLACEMENT

Pole placement is a powerful tool in modern control theory, and it is very suitable for flying quality design. This is because in ADS-33E-PRF, many of the flying quality requirements can be represented by poles or eigenvalues. Therefore, it is very convenient to use ideal poles or eigenvalues to implement flying quality design. If the closed loop model of a helicopter has the designed poles or eigenvalues, then this helicopter will reach the expected flying quality level. Since different requirements are defined for hover/low speed flight and forward flight in ADS-33E-PRF, ideal poles should also be designed separately in different flight states.

The flight dynamics model used in adaptive pole placement is 6 degrees of freedom state space model, the same as Eq. (1). However, since the model is decoupled, stability matrix A and control matrix B in Eq. (1) now have very simple structure, as shown in Eq. (19) and Eq. (20).

$$\boldsymbol{A}=\begin{bmatrix} X_u & 0 & 0 & 0 & X_q & 0 & 0 & -g\cos\theta_0 & 0 \\ 0 & Y_v & 0 & Y_p & 0 & Y_r & g\cos\theta_0\cos\phi_0 & 0 & 0 \\ 0 & 0 & Z_w & 0 & 0 & 0 & 0 & 0 & 0 \\ 0 & L_v & 0 & L_p & 0 & L_r & 0 & 0 & 0 \\ M_u & 0 & 0 & 0 & M_q & 0 & 0 & 0 & 0 \\ 0 & N_v & 0 & N_p & 0 & N_r & 0 & 0 & 0 \\ 0 & 0 & 0 & 1 & 0 & \tan\theta_0 & 0 & 0 & 0 \\ 0 & 0 & 0 & 0 & 1 & 0 & 0 & 0 & 0 \\ 0 & 0 & 0 & 0 & 0 & 1/\cos\theta_0 & 0 & 0 & 0 \end{bmatrix} \tag{19}$$

$$\boldsymbol{B}=\begin{bmatrix} X_{\delta_{long}} & 0 & 0 & 0 \\ 0 & Y_{\delta_{lat}} & 0 & Y_{\delta_{ped}} \\ 0 & 0 & Z_{\delta_{col}} & 0 \\ 0 & L_{\delta_{lat}} & 0 & L_{\delta_{ped}} \\ M_{\delta_{long}} & 0 & 0 & 0 \\ 0 & N_{\delta_{lat}} & 0 & N_{\delta_{ped}} \\ 0 & 0 & 0 & 0 \\ 0 & 0 & 0 & 0 \\ 0 & 0 & 0 & 0 \end{bmatrix} \tag{20}$$

Since Eq. (19) and Eq. (20) are nearly decoupled, the flying quality level of interaxis coupling of the helicopter may reach level 1 now, and this can be proven by Fig. 3 ~ Fig. 6. However, the other flying quality requirements such as stability, control bandwidth etc. are still not improved. Therefore, ideal poles should be designed according to these flying quality requirements. In this paper, the ideal poles in hover and low speed flight states are designed based on small-amplitude attitude change (short-term and mid-term responses to control inputs) and response to collective controller. In forward flight state, besides the above requirements, lateral-directional stability is also considered in ideal poles design.

The ideal poles in longitudinal channel include 1 negative real pole and 2 conjugate complex poles. The negative real pole represents longitudinal damping, and the complex poles represent longitudinal dynamic mode of a helicopter, such as hover oscillation or phugoid in forward flight. The values of ideal poles can be determined by numeric optimization method. The original longitudinal damping of a helicopter can be used as initial value of negative real pole, initial guess of complex poles can be determined easily according to Fig. 14 by selecting an arbitrary point in Level 1 area. Then these poles can be optimized by newton algorithm to ensure the bandwidth of a helicopter reaches desired Level 1 value according to ADS-33E-PRF.

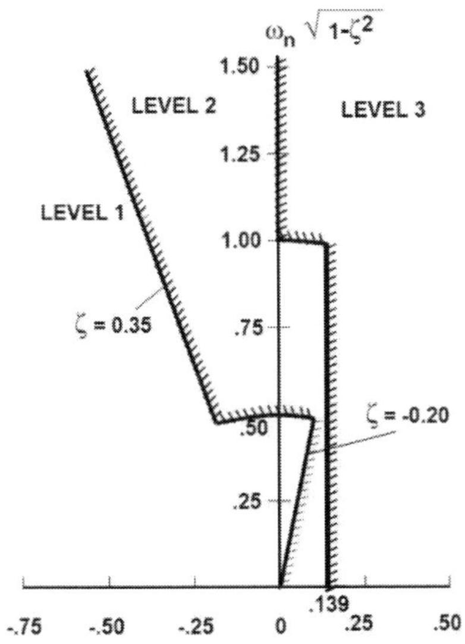

Figure 14: Limits on pitch (roll) oscillations in hover and low speed.

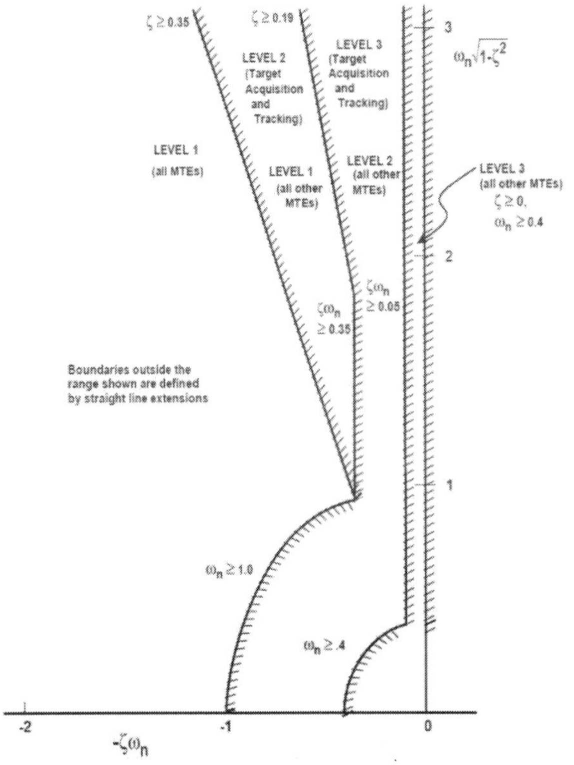

Figure 15: Lateral-directional oscillatory requirements.

The ideal poles in lateral-directional channel include 3 negative real poles and 2 conjugate complex poles. The negative real poles represent lateral damping, yaw damping and spiral mode, and the complex poles represent lateral hover oscillation or Dutch roll mode in forward flight. The values of lateral ideal poles can be determined using the same method as in longitudinal channel. Initial guess of complex poles can be determined according to Fig. 14 in hover and low speed flight and Fig. 15 in forward flight state.

Vertical idea pole can be set based on height response characteristics requirement. In ADS-33E-PRF, this requirement only need vertical rate response have a qualitative first-order appearance for at least 5 seconds following a step collective input. Therefore, vertical ideal pole can be set as a proper negative real number directly.

When all the ideal poles are determined, the next step is to solve the pole placement problem, as defined in Eq. (21). A sufficient and necessary condition for Eq. (21) has a solution is (***A***, ***B***) is fully controllable. It is obviously that all state variables of a helicopter can be controlled by pilot, i.e. (***A***, ***B***) is fully controllable. There are many comprehensive solvers to calculate the feedback control matrix Kpole. In this paper, the same algorithm as "place" function in MATLAB software is used to obtain a solution of Eq. (21).

$$\lambda_i(\boldsymbol{A}-\boldsymbol{B}\boldsymbol{K}_{pole})=\lambda_i^*,\quad i=1,2,\cdots,n \tag{21}$$

The closed-loop poles in Eq. (21) will be exactly the same value as desired ideal poles. Moreover, if the open-loop poles of A matrix in Eq. (21) have small changes, the closed-loop poles will be very close to the desired ideal poles by using the same feedback matrix $\boldsymbol{K}_{pole}$. This indicates the performance of the controller will maintain well by using one constant $\boldsymbol{K}_{pole}$ matrix in a reasonable flight state range. Therefore, there is no need to do pole placement calculation at each sampling time. In this paper, an adaptive strategy for pole placement is developed which can be illustrated by Fig. 16. In each sampling time, the open-loop eigenvalues of matrix A in Eq. (19) are calculated. Then these eigenvalues will be compared with baseline eigenvalues. If the relative variation of any eigenvalues is greater than 10%, the value of cost function J will be 1. Otherwise, the cost function J will be -1. If the cost function has a positive value, then the pole placement is carried out, and the baseline eigenvalues will be replaced by current open-loop eigenvalues. On the contrary, if the cost function has a negative value, no calculation and no update are required. The baseline eigenvalues can be initialized by a zero vector, and then the pole placement will surely be carried out at the first sampling point.

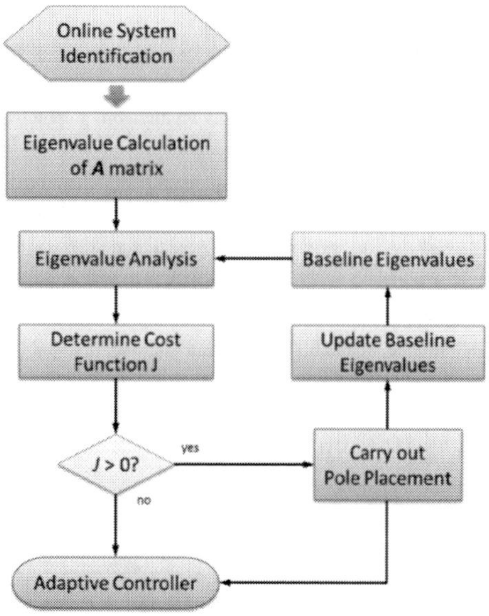

Figure 16: Adaptive Pole Placement Strategy.

Based on the developed adaptive pole placement controller, Attitude Command Attitude Hold (ACAH), Height Hold (HH) and Translational Rate Command (TRC) response types defined in ADS-33E-PRF are implemented for a UH-60 helicopter. Fig. 17 ~ Fig. 19 shows the verification of these 3 response types. It can be found that all 3 response types meet the requirements in flying quality specification. Moreover, the performance of the controller is maintained well from hover to high speed forward flight state. This indicates the adaptive control law designed in this paper is effectiveness.

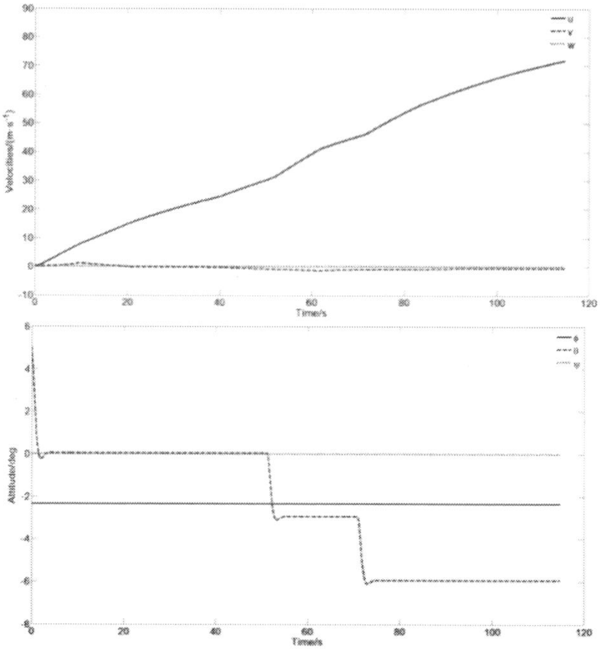

Figure 17: Verification of ACAH response types.

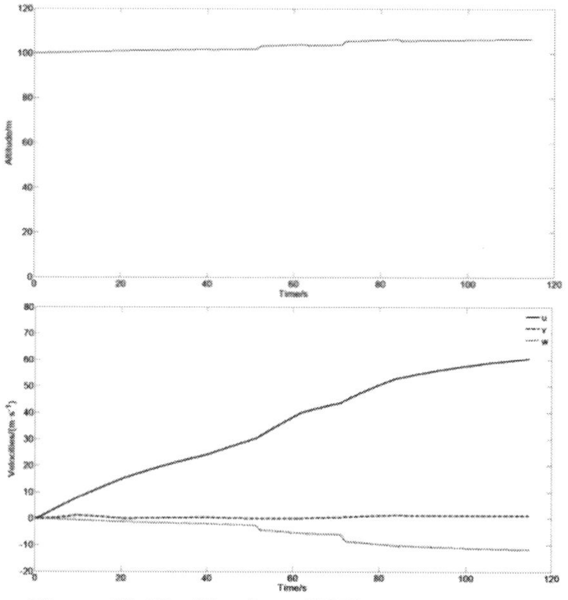

Figure 18: Verification of HH response types.

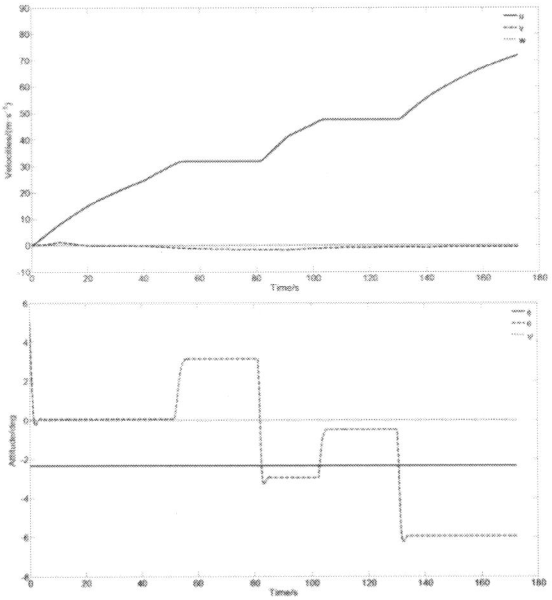

Figure 19: Verification of TRC response types.

CONCLUSIONS

A comprehensive adaptive flight controller design method for improving helicopter flying quality is developed in this paper. Main achievements and improvements can be summarized as follows:

1) An inner-loop decoupling controller based on dynamic inversion technique is very useful to decrease the difficulties in designing outer-loop adaptive controller. Weak coupled model has very simple model structure, which is very helpful to increase the efficiency in online system identification and pole placement calculation.

2) An improved online system identification method is developed. The introduction of noise model helps to solve the biased estimation problem caused by colored measurement noise, so that the overall identification accuracy is increased. The established recursive extended least squares algorithm with a fixed length window is proven to be very effective and efficient in real time system identification of helicopter flight dynamics model.

3) Outer-loop adaptive controller is designed based on real time pole placement algorithm. It is very convenient to use such controller to improve flying quality since many flying quality requirements can be represented by ideal poles. The adaptive strategy based on eigenvalue analysis is easy to implement. This adaptive rule eliminates all unnecessary adjustments of feedback control matrix, which helps to increase the stability and efficiency of the adaptive controller.

Author contact: Wu Wei 1 scorpio_nuaa@nuaa.edu.cn

ACKNOWLEDGMENTS

This work was co-supported by the National Natural Science Foundation of China (No. 61503183) and the Aeronautical Science Foundation of China (No. 2015ZA52002).

REFERENCES

[1]David G. Mitchell, David B. Doman, David L. Key, David H. Klyde, David B. Legget, David J. Moorhouse et al, " Evolution, Revolution, and Challenges of Handling Qualities, " *Journal of Guidance, Control and Dynamics*; Vol. 27, No. 1, 2004, pp. 12-28.
doi: 10.2514/1.3252

[2]Aeronautical Design Standard, Handling Qualities Requirements for Military Rotorcraft. U.S. Army Aviation and Missile Command, Redstone Arsenal, AL; 2000. Standard No.: ADS-33E-PRF.

[3]Robert T.N. Chen, William S. Hindson, "Analytical and Flight Investigation of the Influence of Rotor and Other High-Order Dynamics on Helicopter Flight-Control System Bandwidth," NASA -TM-86696, 1985.

[4]C. Fielding, P. K. Flux, "Non-linearities in flight control systems," *The Aeronautical Journal*, 2003, pp. 673-696.

[5]K. H. Landis, S. I. Glusman, "Development of ADOCS controllers and control laws. Volume 1- Executive summary," NASA CR-177339, 1985.

[6]Chad R. Frost, William S. Hindson, Ernesto Moralez, George E. Tucker, James B. Dryfoos, "Design and Testing of Flight Control Laws on The RASCAL Research Helicopter," AIAA Modeling and Simulation Technologies Conference and Exhibit, Monterey, California, 2002.
doi: 10.2514/6.2002-4869

[7]Mark B. Tischler, Christopher L. Blanken, Kenny K. Cheung, Sean S. M. Swei et al, "Modernized Control Laws for UH-60 BLACK HAWK Optimization and Flight-Test Results," *Journal of Guidance, Control and Dynamics*, Vol. 28, No. 5, 2005, pp. 964-978.
doi: 10.2514/1.8021

[8]Peter J. Gorder, Ronald A. Hess, "Robust Rotorcraft flight Control System Design Including Rotor Degrees of Freedom," AIAA Guidance, Navigation, and Control Conference, Scottsdale, AZ, U.S.A, 1994.
doi: 10.2514/6.1994-3692

[9]J. Gadewadikar, F.L Lewis, Kamesh Subbarao, Kemao Peng, Ben Chen, "H-Infinity Static Output-Feedback Control for Rotorcraft," AIAA Guidance, Navigation, and Control Conference and Exhibit, Keystone, Colorado, 2006.
doi: 10.2514/6.2006-6238

[10]S. Suresh, S. N. Omkar, V. Mani, N. Sundararajan, "Nonlinear Adaptive Neural Controller for Unstable

Aircraft," *Journal of Guidance, Control and Dynamics*, Vol. 28, No. 6, 2005, pp. 1103-1111.
doi: 10.2514/1.12974

[11]Zheng Qu, Anuradha M. Annaswamy, "Adaptive Output-Feedback Control with Closed-Loop Reference Models for Very Flexible Aircraft," *Journal of Guidance, Control and Dynamics*, Vol. 39, No. 4, 2016, pp. 873-888.
doi: 10.2514/1.G001282

[12]S. Suresh, N. Sundararajan, P. Saratchandran, "Neural Adaptive Flight Controllers for Helicopters," 44th AIAA Aerospace Sciences Meeting and Exhibit, Reno, Nevada, 2006.
doi: 10.2514/6.2006-1476

[13]Keng Peng Tee, Shuzhi Sam Ge, Francis E. H. Tay, "Adaptive Neural Network Control for Helicopters in Vertical Flight," *IEEE Transactions on Control Systems Technology*, Vol. 16, No. 4, 2008, pp. 753-762.
doi: 10.1109/TCST.2007.912242

[14]Nathan D. Richards, Richard J. Adams, David H. Klyde, Bruce Cogan, "Flight-Test Evaluation of an Adaptive Controller for Flying Qualities Specification and Protection," *Journal of Guidance, Control and Dynamics*, Vol. 38, No. 12, 2015, pp. 2241-2256.
doi: 10.2514/1.G000873

[15]Geethalakshmi S. Lakshmikanth, Radhakant Padhi, John M. Watkins, James E. Steck, "Adaptive Flight-Control Design Using Neural-Network-Aided Optimal Nonlinear Dynamic Inversion," *Journal of Aerospace Information Systems*, Vol. 11, No. 11, 2014, pp. 785-806.
doi: 10.2514/1.I010165

[16]Vahram Stepanyan, Kalmanje Krishnakumar, "Adaptive Control with Reference Model Modification," *Journal of Guidance, Control and Dynamics*, Vol. 35, No. 4, 2012, pp. 1370-1374.
doi: 10.2514/1.55756

[17]L. Hoecht, T. Bierling, F. Holzapfel, "Ultimate Boundedness Theorem for Model Reference Adaptive Control Systems," *Journal of Guidance, Control and Dynamics*, Vol. 37, No. 5, 2014, pp. 1595-1603.
doi: 10.2514/1.G000324

[18]J. Grauer, J. Conroy, J. Humbert, D. Pines, "System Identification of a Miniature Helicopter," *Journal of Aircraft*, Vol. 46, No. 4, 2009, pp. 1260-1269.
doi: 102514/1.40561

[19]M. Elshafei, S. Akhtar, M. S. Ahamed, "Parametric Models for Helicopter Identification Using ANN," *IEEE Transactions on Aerospace and Electronic Systems*, Vol. 36, No. 4, 2000, pp. 1242-1252.

[20]Ravindra Jategaonkar, Dietrich Fischenberg, Wolfgang von Gruenhagen, "Aerodynamic Modeling and System Identification from Flight Data—Recent Applications at DLR," *Journal of Aircraft*, Vol. 41, No. 4, 2004, pp. 681-691.
doi: 10.2514/1.3165

[21]Tischler, M., Remple, R., *Aircraft and Rotorcraft System Identification: Engineering Methods with Flight test Examples 2nd Edition*, AIAA Education Series, AIAA, Reston, VA, 2012, pp. 1-20.

[22]Wu Wei, Chen Renliang, "Set-Membership Identification Method for Helicopter Flight Dynamics Modeling," *Journal of Aircraft*, Vol. 52, No. 2, 2015, pp. 553-560.
doi: 10.2514/1.C032774

[23]Li Pan, Chen Renliang, "A mathematical model for helicopter comprehensive analysis," *Chinese Journal of Aeronautics*, Vol. 23, No. 3, 2010, pp. 320-326.

Usage of a Driving Force of Wheels for Shortening a Take-off Distance of Small Airplanes

Emi Matsui
Graduate Student
Osaka Prefecture University
Osaka, Japan

Takakage Arai
Professor
Osaka Prefecture University
Osaka, Japan

Shigeru Sunada
Professor
Nagoya University
Aichi, Japan

ABSTRACT

Kobayashi and Nishizawa (2007) suggested that the take-off distance of an electric airplane, which can be used personally, can be shorten by using both a propeller thrust and a wheel driving force. Recently flying cars such as Aeromobil 3, which can also be used personally, have been developed. A short take-off distance will contribute to increase the possibility that a flying car will be used in our daily life. The purpose of the present study is to make clear how a take-off distance of an existing commercial car with a flying ability can be decreased when both the propeller thrust and the wheels driving force can be used. The study about a flying car will be a reference of developing an electric airplane with high performance.

NOTATION

A	=	Aspect ratio	[-]
b	=	Number of blades	[-]
C_{D_f}	=	Body drag coefficient	[-]
C_{Di}	=	Induced drag coefficient	[-]
C_{D0}	=	Profile drag coefficient	[-]
C_L	=	Wing lift coefficient	[-]
c	=	Blade chord length	[m]
D_{car}	=	Total drag	[N]
D_f	=	Parasite drag	[N]
D_i	=	Wing induced drag	[N]
D_p	=	Wing profile drag	[N]
D_{wheel}	=	Rolling drag	[N]
e	=	Oswald efficiency factor	[-]
M	=	Mach number at propeller blade tip	[-]
m_{car}	=	Mass of the car	[kg]
m_p	=	Mass of the propeller	[kg]
m_w	=	Mass of the main wing	[kg]
N	=	Normal force	[N]
P_a	=	Available power	[W]
P_e	=	Engine power	[W]
P_p	=	Power consumed by a propeller	[W]
P_{wheel}	=	Power consumed by wheels	[W]
R	=	Propeller radius	[m]
S	=	Area of the main wing	[m^2]
S_f	=	Frontal projected area	[m^2]
T_{engine}	=	Wheel driving force by engine	[N]
$T_{friction}$	=	Wheel driving force due to a friction force	[N]
T_0	=	Propeller thrust at $V = V_s$	[N]
T_p	=	Propeller thrust	[N]
T_{wheel}	=	Wheel driving force	[N]
V	=	Forward velocity	[m/s]
V_0	=	Take-off velocity	[m/s]
V_s	=	Minimum forward velocity at the usage of T_p	[m/s]
η	=	Efficiency	[-]
λ	=	Advance ratio	[-]
μ_f	=	Friction coefficient	[-]
μ_r	=	Rolling drag coefficient	[-]
ρ	=	Air density	[kg/m^3]
ω	=	Angular velocity	[rad/s]

INTRODUCTION

Recently, cars with flying ability, which is a flying car, such as Aeromobil 3 and Terrafugia have been developed. Technologies for making a car lighter make possible these cars with a flying ability.

A car has an advantage against other ways of transportation that it enables us to move from a door to a door. When a car has a flying ability, the area we can move from a door to door will be much wider. Moreover, a car with a flying ability will have a new value which a car without a flying ability does not have. For example, three-dimensional information can be obtained during a flight, while two-dimensional information is obtained during a drive.

Presented at the 6th Asian/Austrailian Rotorcraft Forum / Heli Japan 2017, Kanazawa, Japan, November 7-9, 2017.

In the present analysis, a take-off distance of a flying car will be analyzed. This is because a short take-off distance is required so that the flying car is used in our daily life. Furthermore, the analysis will be made about a existing commercial car where a flying ability is added. This is because a running characteristics of the car are required to remain when a flying ability is added to the car.

MODEL OF A FLYING CAR

Figure 1 shows the model of the flying car we discuss in the present analysis. A take-off distance of this model will be investigated and the take-off speed is 150km/h. Though tail wings are shown in Figure 1, the aerodynamic forces acting on tail wings are neglected in the present analysis.

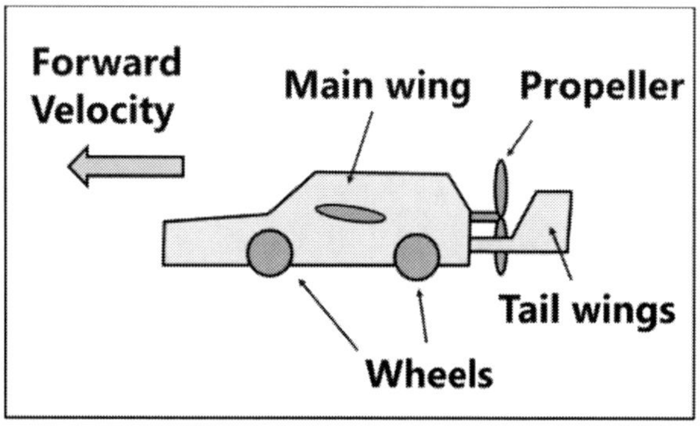

(a)

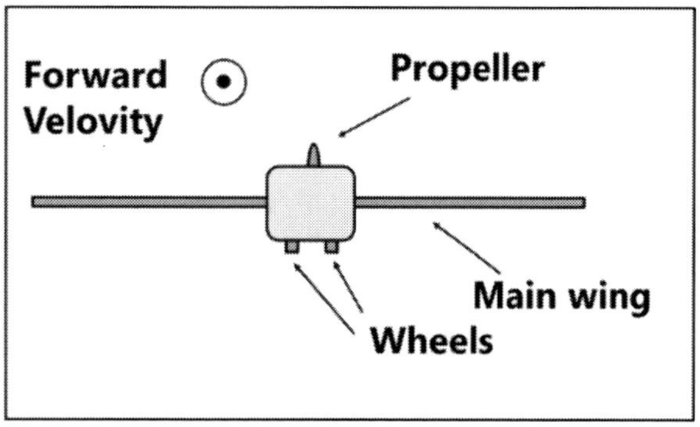

(b)

Figure 1. Model of a Flying Car.

(a)Side View (b)Front View

Car Body

A flying car in the present study is composed of a main wing, a propeller and an existing commercial car, Civic VTi (HONDA)[1]. Table 1 shows the specifications of the car. In this table, the values except for the mass and the frontal projected area are assumed for the present study.

Table 1. Characteristics of Civic VTi.

Characteristic	Symbol	Value
Mass of car[1]	m_{car}	1020 kg
Friction coefficient	μ_f	0.85
Rolling drag coefficient	μ_r	0.01
Frontal projected area[1]	S_f	2.33 m^2
Body drag coefficient	C_{D_f}	0.3

Engine

An engine with its power, $P_e = 81.7$kW (= 111 PS) is mounted on the flying car for driving the wheels and the propeller. The engine power is close to that of Civic VTi (HONDA). The efficiency η, which is a ratio between an engine power and an available power for the wheels and the propeller P_a, is set to be 0.9. Then,

$$P_a = \eta P_e = 73.5 \text{ kW } (= 100 \text{ PS}) \tag{1}$$

Main Wing

Table 2 shows the values about a main wing. Planform of the main wing is rectangular. The airfoil is NACA0009. The wing mass was obtained by the equation indicated in Ref.2.

Table 2. Characteristics of a Main Wing.

Characteristic	Symbol	Value
Area	S	9.0 m^2
Aspect ratio	A	10.7
Mass	m_w	34.9 kg
Oswald efficiency factor	e	0.9

Propeller

Table 3 shows the values about the propeller. The airfoil is NACA0009. Values such as the mass, chord distribution and twist angle distribution will be described later.

Table 3. Characteristics of a Propeller.

Characteristic	Symbol	Value
Number of blades	b	2
Propeller diameter	R	0.8 m

AERODYNAMIC FORCES ON A FLYING CAR

Drag

About the main wing, profile drag D_w and induced drag D_i are given by Eq.(2) and (3). About the car body, parasite drag D_f is given by Eq.(4) and about the wheels, rolling drag is given by Eq.(5).

$$D_p = \frac{1}{2}\rho S V^2 C_{D0} \tag{2}$$

$$D_i = \frac{1}{2}\rho S V^2 C_{Di} = \frac{1}{2}\rho S V^2 \frac{C_L^2}{e\pi A} \tag{3}$$

$$D_f = \frac{1}{2}\rho S_f V^2 C_{Df} \tag{4}$$

$$D_{wheel} = \mu_r N \tag{5}$$

$$= \mu_r \left\{(m_{car} + m_w + m_p)g - \frac{1}{2}\rho S V^2 C_L\right\}$$

Total drag acting on a flying car is

$$D_{car} = D_p + D_i + D_f + D_{wheel} \tag{6}$$

Wheel Driving Force T_{wheel}

Ozaki[3] showed that thrust by wheels is equal to the smaller value between a driving force with an engine T_{engine} and a driving force due to the friction between wheels and the ground $T_{friction}$. Then,

$$T_{wheel} = \min(T_{engine}, T_{friction}) \tag{7}$$

T_{engine} is given by

$$T_{engine} = \frac{P_{wheel}}{V} \tag{8}$$

$T_{friction}$ is given by

$$T_{friction} = \mu_f N$$
$$= \mu_f \left\{(m_{car} + m_w + m_p)g - \frac{1}{2}\rho S V^2 C_L\right\} \tag{9}$$

Figure 2 shows the value of T_{wheel} with T_{engine} and $T_{friction}$, when P_{wheel}=P_a. Figure 2 shows the values when C_L=+0.5 and C_L= -0.5. The comparison of T_{wheel} between these figures, the differences are negligibly small, though Eq. (10) indicates that a negative C_L can enhance the T_{wheel}. Then, hereafter, C_L =+0.5 will be assumed because C_L is required to be positive during a flight.

A total mass of the present flying car is about 1100kg. The lift coefficient is required to be 1.1 for a take-off at V=150km/h. To decrease the induced drag, it is assumed that the flying car runs with C_L=+0.5 and the C_L is increased by the increase of the angle of attack and by using a flap just before the take-off.

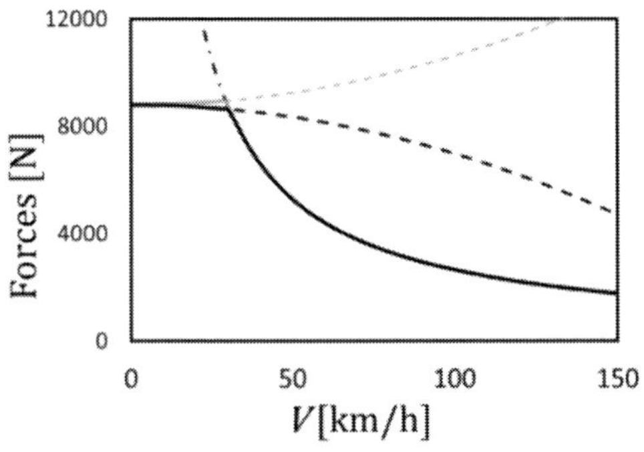

Figure 2. Wheel Driving Force.

(—·— : T_{engine},
– – – : $T_{friction}$ when C_L=+0.5,
–·–·– : $T_{friction}$ when C_L=-0.5,
—— : T_{wheel} when C_L=+0.5,
—— : T_{wheel} when C_L=-0.5)

Propeller Thrust T_p

Propeller thrust T_p was calculated using Adkins and Liebeck Method, the details of which are stated in Ref. 4. Under the condition that the propeller consumes $P_p(V$=150 km/h) is equal to the P_a, the shape of propeller, that is the span-wise distributions of chord length and twist angle were obtained as Figures 3 and 4, respectively. Advance ratio $\lambda = \frac{V_0}{\Omega_0 R}$ is 0.2 and the blade cut-off radius is $0.2R$. This advance ratio is determined so that tip Mach number is close to about 0.6. The tip Mach number will be shown in Figure 5. The thrust T_0 for V_0 is 1345 N and propeller efficiency $\frac{T_0 V_0}{P_a}$ is 0.76.

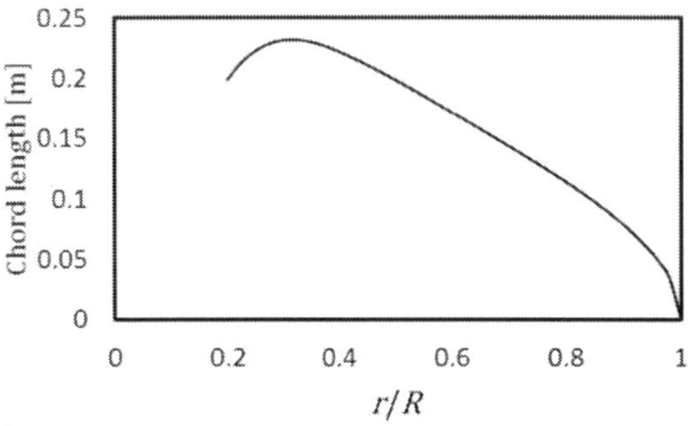

Figure 3. Span-wise Distribution of Chord Length.

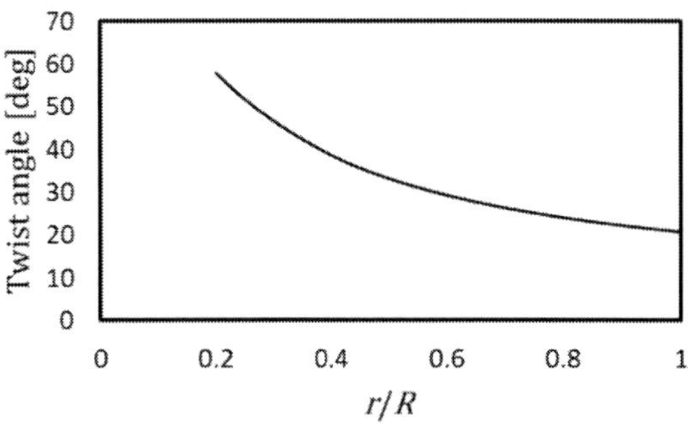

Figure 4. Span-wise Distribution of Twist Angle.

About the designed propeller, we estimated the thrust at other forward speeds V as follows: Advance ratio $\lambda = \frac{V}{\Omega R}$ is independent of V and its value is 0.2. The effect of Reynolds number on an airfoil characteristics was ignored. The thrust T_p and power P_p for V are given by

$$T_p = \left(\frac{V}{V_0}\right)^2 T_0 \tag{10}$$

$$P_p = \left(\frac{V}{V_0}\right)^3 P_a \tag{11}$$

Figure 5 shows the values of T_p/T_0 and P_p/P_a and M, which is the Mach number at the blade tip. Because the Mach number at the blade tip is 0.61, the possibility of drag divergence does not exist.

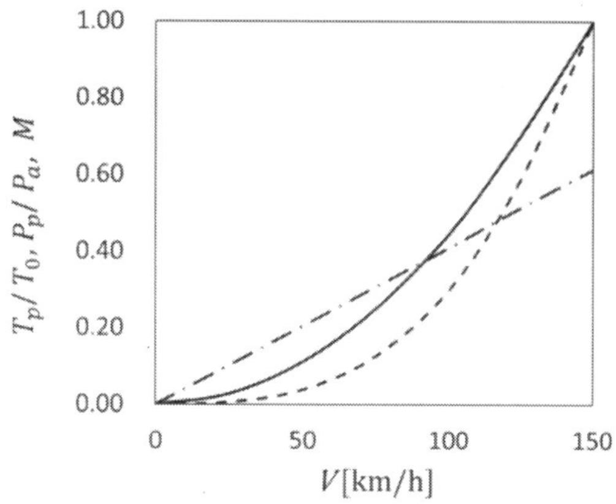

Figure 5. Thrust and Power by the Propeller and Tip Mach Number.

(—— : T_p/T_0, - - - : P_p/P_a, —·— : M)

Comparison between drag, wheel driving force, and propeller thrust

Figure 6 shows the wheels driving force T_{wheel}, propeller thrust T_p, and drag D_{car}. This figure shows that at any forward velocity between 0 km/h and 150 km/h, a wheel driving force is larger than propeller thrust. Then, a take-off distance when the T_{wheel} is used at any forward velocity is minimum. Also, it shows that the propeller thrust T_p is larger than the drag D_{car} when V>71 km/h.

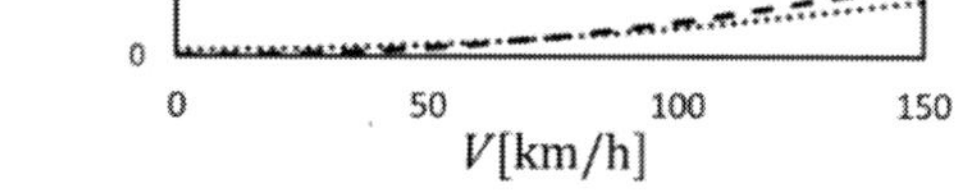

Figure 6. Wheel Driving Force T_{wheel}, Propeller Thrust T_p and Drag D_{car}.

(—— : T_{wheel}, - - - : T_p, ········· : D_{car})

DISCUSSION ON REDCUTION OF TAKE–OFF DISTANCE

Hereafter, we discuss how a take-off distance can be reduced by the combination of T_{wheel} and T_p. As stated above, when the T_{wheel} is used at any forward velocity till take-off, the take-off distance can be minimum. However, a flying car needs to propulsive with a propeller after its take-off. Therefore, it is better that the generation of propeller thrust starts before the take-off for a smooth switch between the T_{wheel} and the T_p.

We compare some take-off distances between the five cases shown in Table 4. In the compared cases, the followings are different.

・Forward velocity at the beginning of the generation of T_p, V_s

(Case75: V_s=75 km/h, Case120: V_s=120 km/h, Case150: V_s=150 km/h)

・After the beginning the usage of T_p, just the T_p is used or both the T_p and the T_{wheel} are also used. For the latter case, the total power consumed for the T_p and the T_{wheel} is equal to P_a =100 PS. Then, P_{wheel} is $P_a - P_p$ = 100 PS $- P_p$. Therefore,

$$T_{wheel} = \min\{(100\text{ PS} - P_p)/V, T_{friction}\} \quad (12)$$

As shown in Figure 2, when P_{wheel}=P_a, the T_{wheel} is equal to T_{engine} except for small V (V<31 km/h). And, T_{engine} when P_{wheel} = 100 PS $- P_p$ is smaller than that when P_{wheel}=P_a. $T_{friction}$ is independent of P_{wheel}. Then, when P_{wheel}=100 PS $- P_p$, T_{wheel} is equal to T_{engine}, except for small V.

Table 4. Cases for the comparison.

	V_s [km/h]	Thrust at V>V_s
Case75P	75	Just T_p
Case75PW	75	Both T_{wheel} and T_p
Case120P	120	Just T_p
Case120PW	120	Both T_{wheel} and T_p
Case150W	150	Just T_p

A take-off distance was calculated by

$$(m_{car} + m_w + m_p)\frac{dV}{dt} = T_p + T_{wheel} - D_{car} \quad (13)$$

Figure 7 shows the comparison between the five cases indicated in Table 4.

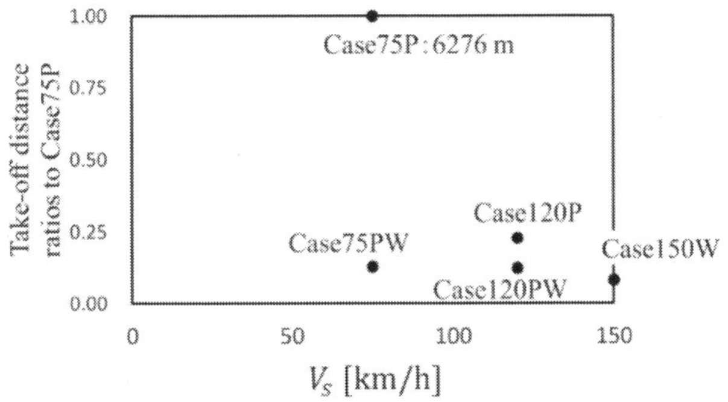

Figure 7. Comparison of Take-off Distances between the Cases in Table 4.

Figure 7 shows take-off distances in Case75P is 6276 m and this distance is maximum. This figure shows take off-distances in Case75PW and Case120PW are close that in Case150W, which is minimum. These distances are less than 1 km. When a wheel driving force till take-off is utilized, the take-off distance is close to its minimum value independently of V_s. The flying car can take off at a small airport such as Yao airport.

These small differences in take-off distances between Case 75PW, Case 120PW and Case150W can be explained as follows: Figure 8 shows the variations of T_p, $T_{wheel} + T_p$ for P_{wheel}=$P_a - P_p$ and T_{wheel} for $P_{wheel} = P_a$, when V is varied. The line of T_{wheel} for $P_{wheel} = P_a$ corresponds to the solid line in Figure 2 Table 5 shows the variations of the total thrust in Figure 8 about the five cases. Table 5 and Figure 8 show that the variations of the total thrust ($T_p + T_{wheel}$) in Case75PW and Case120PW are close to that T_{wheel} of Case150W. This causes the small differences in take-off distances in Figure 7 between Case75PW, Case120PW and Case150W.

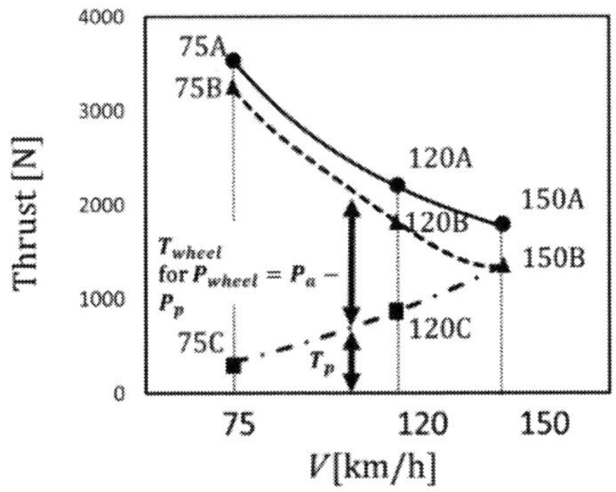

Figure 8. Variations of T_p, $T_{wheel} + T_p$ for P_{wheel}=$P_a - P_p$ and T_{wheel} for P_a.

(——— : T_{wheel} for P_a
– – – : $T_{wheel} + T_p$ for P_{wheel}=$P_a - P_p$
– · – · : T_p)

Table 5. Variations of thrust in Figure 8.

Case75P	V[km/h]	75C → 120C → 150B
	Thrust[N]	337 → 862 → 1347
Case75PW	V[km/h]	75B → 120B → 150B
	Thrust[N]	3223 → 1828 → 1347
Case120P	V[km/h]	75A → 120A → 120C → 150B
	Thrust[N]	3530 → 2207 → 862 → 1347
Case120PW	V[km/h]	75A → 120A → 120B → 150B
	Thrust[N]	3530 → 2207 → 1828 → 1347
Case150W	V[km/h]	75A → 120A → 150A
	Thrust[N]	3530 → 2207 → 1765

For a smooth operation of the flying car, the generation of the propeller thrust should start before the take-off.

CONCLUSIONS

(1) The effect of a negative lift on main wing on the wheel driving force is negligibly small. Then, a pitch angle of a main wing should be fixed at a positive value for a take-off.
(2) A wheel driving force should be utilized till a take-off. For smooth take-off, a propeller thrust should be also used before the take-off.
(3) The required take-off distance is about 1km and an infrastructure such as a small airport will be required.

REFERENCES

[1]Honda Motor Co., Ltd, "CIVIC 1995.9", Honda, http://www.honda.co.jp/factbook/auto/CIVIC/19950904/cv95-038.html,(accessed 2017-08-21).

[2]Rinoie, K., *Method of Designing Aircraft*, 1st ed., CORONA PUBLISHING CO.,LTD.,2011,298p (in Japanese).

[3]Ozaki, T., *Automotive Engineering*, 2nd ed., Morikita Publishing Co., Ltd., 1979, 259p (in Japanese).

[4]Adkins, C; Liebeck, R. "Design of optimum propellers," Journal of propulsion and power, 1994, Vol.10, No 5, pp.67

State-Space Identification of Unmanned Helicopter Dynamics using Invasive Weed Optimization Algorithm on Flight Data

Navaneethkrishnan B Research Assistant
Pranjal Biswas Project Intern
Saumya Kumaar Research Assistant
Gautham Anand Research Assistant
S N Omkar Chief Research Scientist
Indian Institute of Science
Bengaluru,INDIA

ABSTRACT

In order to achieve a good level of autonomy in unmanned helicopters, an accurate replication of vehicle dynamics is required, which is achievable through precise mathematical modeling. This paper aims to identify a parametric state-space system for an unmanned helicopter to a good level of accuracy using Invasive Weed Optimization (IWO) algorithm. The flight data of Align TREX 550 flybarless helicopter is used in the identification process. The rigid-body dynamics of the helicopter is modeled in a state-space form that has 40 parameters, which serve as control variables for the IWO algorithm. The results after 1000 iterations were compared with the traditionally used Prediction Error Minimization (PEM) method and also with Genetic Algorithm (GA), which serve as references. Results show a better level of correlation of the actual and estimated responses of the system identified using IWO to that of PEM and GA.

NOTATION

- A_b, B_a : Flapping Angle Coupling Constants
- B_d, A_c : Electronic Stabilizer bar-rotor coupling derivatives
- A_{lat}, A_{lon}, B_{lat}, B_{lon} : Flapping Angle Control Derivatives
- C_{lon}, D_{lat} : Intercoupling constants of the stabilizer bar
- K_r, K_{rfb} : Tail rotor damping coefficients
- L_u, L_v, L_b, L_w : Rolling Moment Derivatives
- M_u, M_v, M_w, M_a, M_{col} : Pitching Moment Derivatives
- N_v, N_p, N_w, N_r, N_{rfb}, N_{ped}, N_{col} : Directional Moment Derivatives
- X_u, X_a : Lateral Force Derivatives
- Y_v, Y_b, Y_{ped} : Longitudinal Force Derivatives
- Z_a, Z_b, Z_w, Z_r, Z_{col} : Heave Force Derivatives
- a, b : Lateral and longitudinal Flapping angle
- c, d: Lateral and longitudinal electronic stabilizer bar feedback
- $iter_{max}$, $iter$: Maximum and current iteration
- n: Nonlinear modulation index
- p, q, r: Roll, Pitch and Yaw Rate
- r_{fb}: Yaw rate feedback
- u, v, w: Lateral, longitudinal and vertical translational velocity
- ϕ, θ: Roll and Pitch angles
- δ_{lat}, δ_{lon}, δ_{ped}, δ_{col} : Lateral, longitudinal, pedal and collective inputs
- μ_A, μ_B : Mean of sets A and B
- σ_A, σ_B, $\sigma_{initial}$, σ_{final}, σ_{iter}: Standard deviations of sets A, B, the initial iteration, the final iteration and the current iteration respectively
- τ_f, τ_s, : Bare rotor and stabilizer bar time constants

Presented at the 6th Asian/Australian Rotorcraft Forum / Heli Japan 2017, Kanazawa, Japan, November 7-9, 2017.

INTRODUCTION

The use of unmanned helicopters have been increasing steadily in the past decade not only for military applications but also in civilian applications like crop-dusting, seed-bombing, search and rescue etc. (Ref. 1-4). The popularity stems from the vertical take-off and landing (VTOL) capabilities of helicopters. For a given amount of battery power, helicopters provide longer service range and hence can cover larger area when compared to commonly used quad-copters (Ref. 5) However, autonomous operation is favorable to improve the efficiency of the task or even render it feasible (for example, in a case where the vehicle needs to fly out of communication range of the radio transmitter). Design of controllers to enable autonomy in unmanned helicopters proves to be an arduous task due to the complexity involved in obtaining the mathematical model of the same. This difficulty in modeling arises due to the non-linear dynamics of helicopters and a high degree of inter axis coupling. There have been

Fig. 1: Align TREX 550L - Helicopter used for data collection

many attempts at modeling both small scaled and its full sized counterpart. One of the most notable pieces of research on this topic was carried out by Mettler *et al.* (Ref. 6-8). The research uses a frequency domain identification tool called CIFER (Comprehensive Identification from Frequency Responses), which was developed by the U.S Army and NASA, to identify the state-space parameters of a Yamaha R-50 helicopter. However, due to the high price and limited applicability of the CIFER tool, we decided to use an optimization algorithm to estimate the parameters of the state-space model.

Many studies have been conducted evaluating the effectiveness of optimizing algorithms like Prediction Error Minimization (PEM) (Ref. 9), Subspace Method (Ref. 10) and Recursive Least squares (Ref.11, 12) in identifying dynamic systems. However, these traditionally used algorithms work well only for simple linear dynamic models because of their tendency to get stuck in local optima. An exhaustive study has been conducted on the potency of meta-heuristic algorithms on global optimization problems and has been presented in (Ref. 13). The unassuming nature of the algorithms towards the differentiability of the cost-function and their dynamic nature enable it to jump out of any local minima. The effectiveness of meta-heuristic algorithm in identifying the yaw-heave dynamics of a small scaled helicopter has been demonstrated by Tang et. al. (Ref 14).

Invasive Weed Optimization (IWO) algorithm is a population based random search algorithm which was proposed by Mehrabian et.al. (Ref. 15) inspired by the colonizing behavior of invasive weed in agricultural farmlands. The algorithm is structurally simple, robust and has lesser hyper-parameters as compared to its counterparts. The algorithm has found use in a wide range of application such as Neural Network training (Ref. 16), combustion optimization (Ref. 17), PID controller design (Ref 18), trajectory planning (Ref. 19), auto-disturbance rejection controller tuning (Ref. 20) etc.

This paper aims to obtain a mathematical model of an unmanned helicopter (Fig 1.) in the state-space form using IWO algorithm and compare its effectiveness against the traditionally used PEM and a commonly used meta-heuristic tool like Genetic Algorithm (GA). The following section describes the physical characteristics and the onboard instrumentation of the helicopter used to collect the flight data which is followed by the methodology used to acquire flight data including the flight test procedure. This is followed by the description of the state-space model used and then succeeded by the system identification methodology using the IWO algorithm. The paper concludes by tabulating the results and a discussion on the same.

DESCRIPTION OF HELICOPTER UAV

In order to be able to model a vehicle successfully, we need to record the outputs from the actual vehicle to be modeled. The small-scaled helicopter used for modeling is an Align TREX 550L which is a hobby grade vehicle capable of 3-D flight. Unlike most commercially available hobby helicopters, the TREX 550L does not possess a flybar. A flybar is a mechanical device that, using the principles of Coriolis force, provides a negative feedback to the longitudinal and lateral motions. The negative feedback for the motions in the TREX 550L is provided by a Micro Electro-Mechanical System (MEMS) Gyroscope which serves as a replacement for the mechanical flybar. The advantage of using an electronic flybar is that the amplitude of the negative feedback can be tuned to the pilots comfort. It may be tuned to provide high stability or to provide high input sensitivity. Figure 2 describes the physical dimensions of the helicopter.

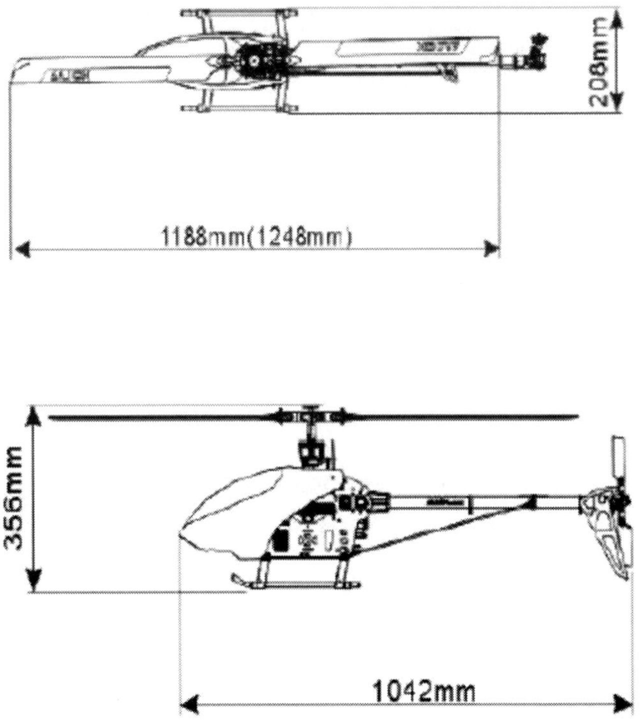

Fig. 2: Dimensions of the helicopter used for flight data acquisition

Helicopter Instrumentation

The helicopter used is equipped with an array of state-of-the-art sensors and flight control board that record high-quality flight data. The flight control board used is Pixhawk, an ARM based flight controller. This arrangement records the velocities (u, v and w), angular rates (p, q, and r) and blade flapping angles (a and b, that are functions of swash-plate servo arm positions). The data is recorded onto a micro SD card on the Pixhawk at the rate of 100 Hz. The data is filtered for sensor noise using a Butterworth filter at half-power frequency. Figure 3 depicts the instrumentation mounted atop the unmanned helicopter.

FLIGHT DATA ACQUISITION

A helicopter has a wide range of dynamic responses inclusive of responses in the nonlinear range. However, missions requiring autonomy rarely call for operations in the nonlinear domain. Hence, we linearize the system and model it in the state-space form. In order to obtain an accurate dynamic model, the maximum dynamic range of the vehicle should be excited while keeping it from entering the nonlinear domain. One of the well-established techniques to perform this excitation is the Frequency Sweep method (Ref. 21). In this

Fig. 3: Instrumentation of TREX 550L

method, the vehicle is given sinusoidal inputs of constant amplitude and increasing frequency in both lateral and longitudinal direction. Frequency Sweep can be executed either using a signal-generator or manually. To maintain the safety aspect of the flight test and considering the expertise of our pilot, we decided to manually perform the frequency sweep using the '3-2-1-1' input sequence. In a '3-2-1-1' sequence, a positive deflection of the control stick is made for the duration of 3 seconds and then a negative deflection is given for 2 seconds and the stick is deflected in the positive direction again for 1 second and then back to negative for 1 second. This sequence is repeated for both lateral and longitudinal directions. The pedal input was given only for directional corrections. The tests were conducted at constant rpm and hover conditions. In order to keep the external perturbations to the minimum, the flight tests were conducted during the period where ambient wind speed was lesser than 1.029 m/s (2 knots) as high winds can degrade the accuracy of the identified dynamic model.

STATE SPACE MODEL

The parametric state-space form of the unmanned helicopter was adopted from Mettler *et al.* (Ref. 6), which has been derived from first principles. The state space form is defined as shown in Eq.1. Where x (Eq.2) is the state matrix and u (Eq.3) is the input matrix.

$$\dot{\mathbf{x}} = A\mathbf{x} + B\mathbf{u} \tag{1}$$

Where,

$$\mathbf{x} = \begin{bmatrix} u & v & p & q & \phi & \theta & a & b & w & r & r_{fb} & c & d \end{bmatrix}^T \tag{2}$$

$$u = \begin{bmatrix} \delta_{lat} & \delta_{lon} & \delta_{ped} & \delta_{col} \end{bmatrix}^T \tag{3}$$

$$A = \begin{bmatrix}
X_u & 0 & 0 & 0 & 0 & -g & X_a & 0 & 0 & 0 & 0 & 0 & 0 \\
0 & Y_v & 0 & 0 & g & 0 & 0 & Y_b & 0 & 0 & 0 & 0 & 0 \\
L_u & L_v & 0 & 0 & 0 & 0 & 0 & L_b & L_w & 0 & 0 & 0 & 0 \\
M_u & M_v & 0 & 0 & 0 & 0 & M_a & 0 & M_w & 0 & 0 & 0 & 0 \\
0 & 0 & 1 & 0 & 0 & 0 & 0 & 0 & 0 & 0 & 0 & 0 & 0 \\
0 & 0 & 0 & 1 & 0 & 0 & 0 & 0 & 0 & 0 & 0 & 0 & 0 \\
0 & 0 & 0 & -\tau_f & 0 & 0 & 1 & A_b & 0 & 0 & 0 & A_c & 0 \\
0 & 0 & -\tau_f & 0 & 0 & 0 & B_a & -1 & 0 & 0 & 0 & 0 & B_d \\
0 & 0 & 0 & 0 & 0 & 0 & Z_a & Z_b & Z_w & Z_r & 0 & 0 & 0 \\
0 & N_v & N_p & 0 & 0 & 0 & 0 & 0 & N_w & N_r & N_{rfb} & 0 & 0 \\
0 & 0 & 0 & 0 & 0 & 0 & 0 & 0 & 0 & K_r & K_{rfb} & 0 & 0 \\
0 & 0 & 0 & -\tau_s & 0 & 0 & 0 & 0 & 0 & 0 & 0 & -1 & 0 \\
0 & 0 & -\tau_s & 0 & 0 & 0 & 0 & 0 & 0 & 0 & 0 & 0 & -1
\end{bmatrix} \tag{4}$$

$$B = \begin{bmatrix}
0 & 0 & 0 & 0 \\
0 & 0 & Y_{ped} & 0 \\
0 & 0 & 0 & 0 \\
0 & 0 & 0 & M_{col} \\
0 & 0 & 0 & 0 \\
0 & 0 & 0 & 0 \\
A_{lat} & A_{lon} & 0 & 0 \\
B_{lat} & B_{lon} & 0 & 0 \\
0 & 0 & 0 & Z_{col} \\
0 & 0 & N_{ped} & N_{col} \\
0 & 0 & 0 & 0 \\
0 & C_{lon} & 0 & 0 \\
D_{lat} & 0 & 0 & 0
\end{bmatrix} \tag{5}$$

The A and B matrices (Eq.4 and Eq.5) together have 40 parameters which require to be estimated for successful identification of the system dynamics.

SYSTEM IDENTIFICATION

The IWO algorithm is used to estimate these 40 parameters such that the model replicates the responses as close to the recorded training data as possible. The algorithm scatters seeds in a 40 dimensional search space and tries to converge at a global optimum. Which essentially means that the algorithm tries to find the best set of values for the state-space matrices. The steps involved in the identification is explained as follows.

Step 1 - Initialization: A finite number of seeds (solutions) are deposited randomly over a search space.

Step 2 - Cost Function: Cost/Fitness of each seed is determined by running it through a cost function.The Population Pearson Coefficient is used to define a cost function for this algorithm. The Population Pearson correlation coefficient of two sets of data is a measure of their frequency correlation or rather their shape fitness. Consider data sets A and B have N scalar observations each, then the Pearson correlation coefficient is defined as shown in Eq.6

$$\rho(A,B) = \frac{1}{N-1} \sum_{n=1}^{N} \left(\frac{\overline{A_l - \mu_a}}{\sigma_a} \right) \left(\frac{\overline{B_l - \mu_b}}{\sigma_b} \right) \tag{6}$$

From this we define the cost function as depicted in Eq.7

$$C = \sum_{i=1}^{13} (1 - \rho_i)^2 \tag{7}$$

Step 3 - Reproduction: Every seed grows into a flowering plant which in turn deposits its own seeds. But the number of seeds each flowering plant can deposit linearly varies from the minimum possible seeds to the maximum possible seeds based on its fitness value. In other words, the plant with a higher fitness score can deposit the maximum number of seeds.

Step 4 - Spatial Dispersal: The spawned seeds are distributed randomly over the search space by normally distributed random number with zero mean and varying variance. The variance varies as shown in Eq.8

$$\sigma_{iter} = \frac{(iter_{max} - iter)^n}{iter_{max}} (\sigma_{initial} - \sigma_{final}) + \sigma_{final} \quad (8)$$

This ensures that the probability of dropping a seed in a distant area decreases nonlinearly with each iteration due to which fitter plants are grouped and the weak ones are eliminated.

Step 5 - Competitive Exclusion: Initially all the seeds are allowed to grow unchecked till it reaches the population limit. Once the population limit is reached, a function to eliminate plants with poor fitness gets called. This function ranks all the seeds in the population with their parents' ranks and eliminates weeds with lower fitness and allows newer seeds to fill the population. This way plants with low fitness scores are allowed to survive if their offspring gives a high fitness score.

The above steps are iterated 1000 times in order to obtain the highest fitness. The maximum iterations are set at 1000 because the cost plateaued out at around 1000 iterations during tuning. 10 trials of these iterations are conducted so that the confidence interval of 95% can be calculated. A 95% confidence interval means that the probability of the exact solution falling within that interval is 0.95 or 95%. The solution with the best fitness or the lowest cost is used to generate the state space model. The validation of this model is done using another dataset from the same helicopter.

RESULTS

Of the 13 identified states, we perform time domain verification only on p, q, ϕ and θ as they are the only states crucial for controller design. The results from the verification are tabulated in Table 1. Further the identified parameters with their 95% confidence interval has been tabulated in Table 2. It can be observed from Table 1 that PEM performs an inferior job of identifying the model as compared to the metaheuristic algorithms. This, as discussed in the introduction of this paper, is due to its inability to handle complex systems like helicopter dynamics due to its tendency to get stuck in a local optima. Among IWO and GA, we notice that IWO performs noticeably better than its counterpart.

We can observe that a few values in Table 2 are set 0 by the algorithm. Those parameters come into play on while in forward flight and not during hover. This corroborates the values which was identified using CIFER by Mettler *et al.* (Ref. 6) for hover condition. Figures 4, 5 and 6 helps in visualizing the time domain verification performed for models identified using PEM, GA and IWO respectively.

Table 1: Performance Comparison

States	PEM	GA	IWO
p	0.5535	0.8385	0.9575
q	0.5329	0.7691	0.9089
ϕ	0.4946	0.6765	0.9084
θ	0.0336	0.6576	0.8722

Table 2: Parameters Identified using IWO

		95% Confidence Interval	
Parameter	Value	Upper Limit	Lower Limit
X_u	-0.32066	-0.212	-0.32658
X_a	40.21598	40.43	-5.36256
Y_v	-0.93658	-0.745	-0.92882
Y_b	-16.1151	18.05	-28.9708
L_u	-0.00121	-0.0024	-0.07221
L_v	-0.47665	-0.31	-0.65863
L_b	133.6111	183.6	145.5951
L_w	0	0	0
M_u	0.1	0.10005	0.099401
M_v	-0.09822	-0.021	-0.07774
M_a	104.9063	87.47161	73.76749
M_w	0	0	0
T_f	0.093851	0.176949	0.09222
A_b	-0.19213	-0.12113	-0.43625
A_c	0.061597	0.351488	0.057038
B_a	0.083523	0.125623	-0.0954
B_d	0.984168	0.778003	0.32217
Z_a	8.166105	8.446461	-4.93675
Z_b	1.028478	8.586446	-87.4542
Z_w	0.045724	0.038181	0.017925
Z_r	-1.39101	0.85415	-0.28432
N_v	0.009652	0.008206	0.003685
N_p	-8.23373	-2.13475	-6.88379
N_w	0	0	0
N_r	-8.69927	0.18654	-4.02513
N_{rfb}	42.69381	18.9659	-34.6882
K_r	2.350899	2.220768	-0.31141
K_{rfb}	-14.5913	-11.0303	-15.1889
τ_s	0.134939	0.212613	0.08313
Y_{ped}	0	0	0
M_{col}	0	0	0
A_{lat}	-0.09993	-0.01799	-0.09167
A_{lon}	0.701979	0.766162	0.083986
B_{lat}	-0.07779	0.510326	-0.15717
B_{lon}	-0.09942	-0.0212	-0.09096
Z_{col}	-6.05944	40.69521	-18.7476
N_{ped}	-27.4672	4.452297	-36.3945
N_{col}	-3.22316	3.742138	-2.04491
C_{lon}	-0.09815	0.495883	-0.09044
D_{lat}	0.793573	0.671539	0.025611

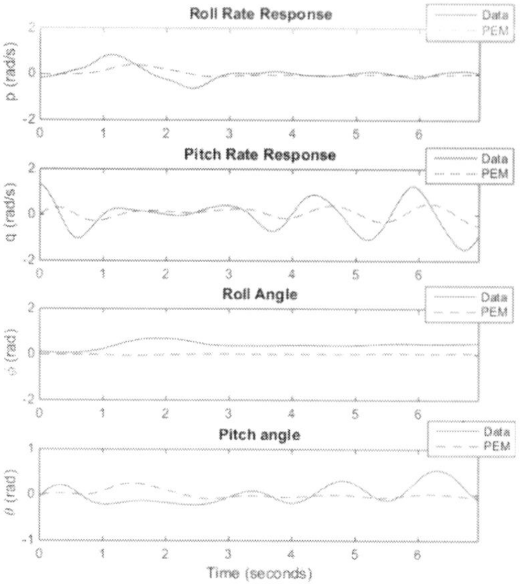

Fig. 4: Time domain verification of validation data using PEM

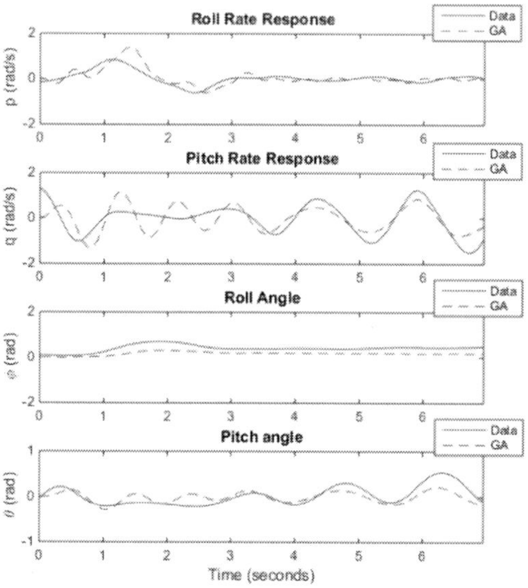

Fig. 5: Time domain verification of validation data using GA

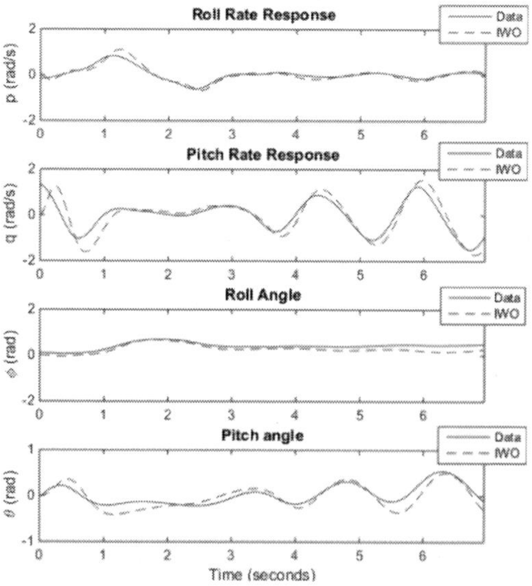

Fig. 6: Time domain verification of validation data using IWO algorithm

CONCLUSIONS

A highly accurate state space model of the helicopter dynamics was identified using the IWO algorithm. From this we can conclude that the IWO algorithm is a powerful tool for identifying helicopter dynamics and possibly even other complex systems. This work could be taken further by using the identified model to design controllers to enable helicopter autonomy. Apart from that, there is a big scope in comparing the effectiveness of metaheuristic algorithms on state space and other nonlinear models like Nonlinear Autoregressive with Exogenous inputs (NARX) and Hammerstein-Wiener to have a better understanding of the algorithms.

Author Contact :
Navaneethkrishnan B, navaneeth@aero.iisc.ernet.in
Pranjal Biswas, pranjalbiswas27@gmail.com
Saumya Kumaar, saumya@aero.iisc.ernet.in
Gautham Anand, gauthama.anand@gmail.com
S N Omkar, omkar@aero.iisc.ernet.in

REFERENCES

[1]Valavanis, Kimon P., ed. *Advances in unmanned aerial vehicles: state of the art and the road to autonomy* Vol. 33. Springer Science &Business Media, Berlin, Germany, 2008, Chapter 1.

[2]Wong, K.C., "Survey of Regional Developments: Civil Applications," *UAV Australia Conference*, Melbourne, Australia, 8-9 February 2001 pp. 8-9.

[3]R. Sugiura, N. Noguchi, and K. Ishii, "Remote-sensing technology for vegetation monitoring using an unmanned helicopter," *Biosyst. Eng.*,vol. 90, no. 4, pp. 369379, Apr. 2005.

[4]M. Bernard, K. Kondak, I. Maza, and A. Ollero, "Autonomous transportation and deployment with aerial robots for search and rescue missions," *Journal of Field Robotics*, vol. 28, no. 6, pp. 914-931, 2011.

[5]Bart Theys, Which one has the highest range: VTOL, helicopter, multi-rotor (or flying squirrel?), *DIY Drones*, accessed 25th May 2017,<http://diydrones.com/profiles/blogs/what-has-the-best-range-vtol-helicopter-multi-rotor-or-flying>

[6]Mettler Bernard, Takeo Kanade, and Mark Brian Tischler. *System identification modeling of a model-scale helicopter*. Carnegie Mellon University, The Robotics Institute, 2000.

[7]Mettler Bernard, *Identification modeling and characteristics of miniature rotorcraft*. Springer Science & Business Media, Berlin, Germany, 2013.

[8]Tischler M.B., Remple R.K., "Aircraft and Rotorcraft System Identification," *Engineering Methods with Flight Test Examples*, American Institute of Aeronautics and Astronautics, Virginia, USA, 2006.

[9]Shim, D. Hyunchul, Hyoun Jin Kim, and Shankar Sastry. "Control system design for rotorcraft-based unmanned aerial vehicles using time-domain system identification." In *Control Applications, 2000. Proceedings of the 2000 IEEE International Conference* on, pp. 808-813. IEEE, 2000.

[10]Li, Ping, Ian Postlethwaite, and Matthew C. Turner. "Subspace-based System Identification for Helicopter Dynamic Modelling." *American Helicopter Society 63rd Annual Forum*, Virginia Beach, VA, May 1-3, 2007.

[11]Raptis, Ioannis A., Kimon P. Valavanis, and Wilfrido A. Moreno, "System identification and discrete nonlinear control of miniature helicopters using backstepping," *Journal of intelligent & robotic systems*, vol. 55, no. 2, pp. 223-243, 2009.

[12]Raptis, Ioannis A., Kimon P. Valavanis, and Wilfrido A. Moreno, "System Identification for a Miniature Helicopter at Hover Using Fuzzy Models, *Journal of intelligent & robotic systems*, vol. 55, no. 2, pp. 347-360, 2009.

[13]Khajehzadeh M, Taha MR, El-Shafie A, Eslami M, "A survey on metaheuristic global optimization algorithms," *Research Journal of Applied Sciences, Engineering and Technology*, vol. 3, no. 6, pp. 569-78, 2011.

[14]Tang, Shuai, Xinye Zhao, Li Zhang, and ZhiQiang Zheng. "System identification of heave-yaw dynamics for small-scale unmanned helicopter using genetic algorithm." In *32nd Chinese Control Conference* (CCC), pp. 1797-1802. IEEE, 2013

[15]Mehrabian, A.R. and Lucas, C., 2006. A novel numerical optimization algorithm inspired from weed colonization. *Ecological informatics*, 1(4), pp.355-366.

[16]Giri, Ritwik, Aritra Chowdhury, Arnob Ghosh, Swagatam Das, Ajith Abraham, and Vaclav Snasel. "A modified invasive weed optimization algorithm for training of feed-forward neural networks." *IEEE International Conference on Systems Man and Cybernetics (SMC)*, 2010, pp. 3166-3173. IEEE, 2010.

[17]Zhao, Huan, Pei-hong Wang, Xianyong Peng, Jin Qian, and Quan Wang. "Constrained optimization of combustion at a coal-fired utility boiler using hybrid particle swarm optimization with invasive weed." In *Energy and Environment Technology*, 2009. ICEET'09. International Conference on, vol. 1, pp. 564-567. IEEE, 2009.

[18]Kundu, Debarati, Kaushik Suresh, Sayan Ghosh, and Swagatam Das. "Designing Fractional-order PI D controller using a modified invasive Weed Optimization algortihm." In *World Congress on Nature & Biologically Inspired Computing*, 2009. NaBIC 2009., pp. 1315-1320. IEEE, 2009.

[19]Pahlavani, P., Delavar, M.R. and Frank, A.U., 2012. Using a modified invasive weed optimization algorithm for a personalized urban multi-criteria path optimization problem. *International Journal of Applied Earth Observation and Geoinformation*, 18, pp.313-328.

[20]Chen, Z., Wang, S., Deng, Z. and Zhang, X., 2011, September. Tuning of auto-disturbance rejection controller based on the invasive weed optimization. *Sixth International Conference on Bio-Inspired Computing: Theories and Applications (BIC-TA)*, 2011 (pp. 314-318). IEEE.

[21]Tischler Mark B., Jay W. Fletcher, Vernon L. Diekmann, Robert A. Williams, and Randall W. Cason. *Demonstration of frequency-sweep testing technique using a Bell 214-ST helicopter*. No. NASA-A-87073. National Aeronautics and Space Administration, Moffett Feild CA Ames Research Centre, 1987.

Obstacle avoidance system for multicopter

Yasuto Oda
Graduate Student
Tokushima University
Tokushima, Japan

Masafumi Miwa
Associate Professor
Tokushima University
Tokushima, Japan

ABSTRACT

Recently, Unmanned Aerial Vehicle is used in many purpose such as wall inspection and rescue activities. When UAV moves remotely, a pilot uses First Person View (FPV) to get circumstance information. It is useful but it is difficult to operate UAV using only FPV stably, because it cannot detect a bit motion either slant of airframe of multicopter. In this study LIDARs are equipped on a multicopter to detect distance toward obstacle and avoid it.

INTRODUCTION

Recently, because Unmanned Aerial Vehicle is easy to operate and can be utilized at danger place (Ref. 1), UAV is used in many purpose such as aerial photograph, wall inspection and rescue activities. In particular, UAV is operated where disaster happens and people cannot enter in. For example, in 2014 August, Hiroshima had heavy rain and landside disaster, so people could not enter the disaster side. In that case, UAV was operated and it examined the damage situation. It was in the spotlight, because it was shown on TV (Ref. 2).

To operate UAV remotely, sometimes operator uses camera to know surrounding condition of UAV, and this method is called FPV (First Person View). FPV is hard to operate stably, because the image from camera is bent and operator cannot know a bit movement of the airframe of multicopter (Ref. 3).

To support FPV operating, we present an obstacle avoidance system uses LIDAR (LIght Detection And Ranging).

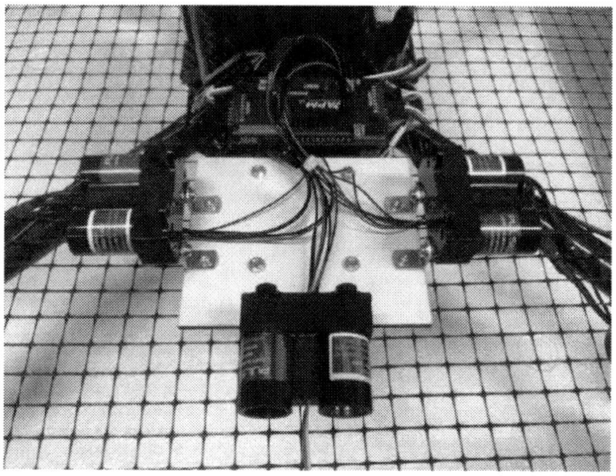

Figure 1. Experimental Equipment

EXPERIMENTAL SETUP

Experimental equipment

Figure 1 shows experimental setup of obstacle avoidance equipment and Figure 2 shows the experimental airframe. ZION-PG560 (enRoute) was used as the experimental airframe. It has two APM2.6 (3D Robotics). One is used as FC (Flight Controller) with Arducopter3.2.1 and the other is programmed as OAS (Obstacle Avoidance System). Three LIDAR-Lite (Garmin) sensors are put on the multicopter, and they measure forward, right side and left side distance.

Figure 2. ZION PG560

Presented at the 6th Asian/Austrailian Rotorcraft Forum / Heli Japan 2017, Kanazawa, Japan, November 7-9, 2017.

Program for obstacle avoidance

Radio receiver on the airframe transmits signals channel 1 to channel 8 from radio controller to OAS by Pulse-width Modulation signal. Channel 1 is aileron signal, channel 2 is elevator signal, channel 3 is throttle signal, and channel 4 is rudder signal. Those signals show input value by changing their pulse width 1100 microseconds to 1900 microseconds. This study uses channel 1 and channel 2. Considering channel 1 signal, when that pulse width is larger than 1520 microseconds, the airframe moves right, and when it is smaller than 1520 microseconds, the airframe moves left. Similar to channel 2 signal, when its pulse width is larger than 1520 microseconds, the airframe moves forward, and when it is smaller than 1520 microseconds, the airframe moves back.

LIDARs measure distance of three directions (forward, left side and right side) toward obstacles. When the airframe is approaching obstacles, OAS cancels channel 1 and 2 input signal, and transmits aileron and elevator PWM (Pulse-Width Modulation) signals related to distance to obstacles.

Figure 3 shows the three LIDARs equipped on the experimental airframe. They were positioned 0.25m from the left and right side edge of the airframe, and 0.3m from frontward edge of the airframe. The distance that collision avoidance system operates is 1m, because the airframe avoids colliding to people.

The obstacle avoidance control is selected to 3 types with the transmitter's toggle switch. They are called mode A, mode B and mode C, respectively. In mode A, obstacle avoidance is only activated in the forward direction. In mode B, obstacle avoidance motion is activated in the left side and right side directions. In mode C, obstacle avoidance movement is activated in the forward, left side and right side directions.

EXPERIMENT

Experiment of obstacle avoidance movement

In this experiments, an obstacle approached to the test airframe to the position 0.5m from the left side. This experiment measures and logs signal that OAS received and FC received from OAS of the channel 1. The state of the experiment is shown in Figure 4.

EXPERIMENTAL RESULT AND CONSIDARATION

The results of obstacle avoidance experiments are shown in Figures 5 and 6. Figure 5 shows the roll angle of the aircraft and distance to the obstacle, and Figure 6 shows input signal to OAS and output signal by OAS. 17 seconds and 40 seconds after the start of the experiment, the obstacles were brought close to the left of the aircraft.

When the obstacle approaches to the airframe, a part of the output signal linearly changes according to the distance in Figure 6. When the obstacle is not approaching, the output signal follows the input signal.

In Figure 5 and 6, 46 seconds after the start of the experiment, although the obstacle is approaching, the signal was disturbed. It is because LIDAR measured the distance to the wall rather than the distance to the obstacle shown as Figure 7. At that time, disturbance of this signal occurs. This problem can be solved by attaching a gimbal to the device and adjusting the LIDAR so that it always keeps horizontally.

CONCLUSIONS

This research proposes the operation of LIDAR as an aid to FPV maneuvering etc. of small unmanned aerial vehicles, and we assembled a device to detect obstacles and avoid collision. The device was mounted on a multicopter aircraft, and this study measured input and output signals during obstacle avoidance operation.

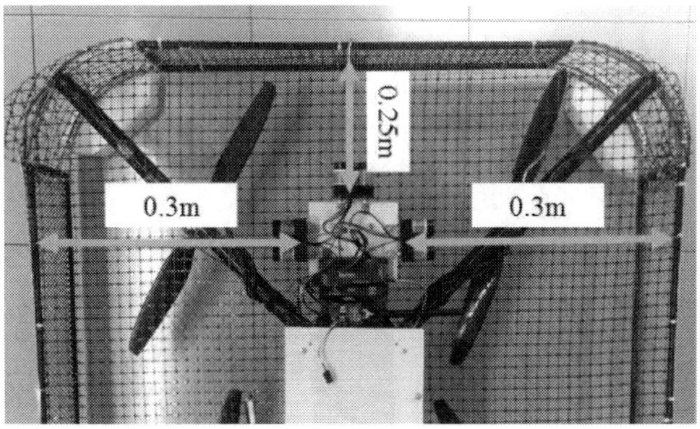

Figure 3. Setting LIDARs

Figure 4. Scene of Avoiding Obstacles

Laser rangefinder (LRF) is considered to be introduced as shown in Figure 8. Using LRF, it is possible to measure the distance in the range of 240 degrees from the front at once (yellow area in Figure 9). Moreover, in order to use a small unmanned aerial vehicle in various situations, not only obstacle avoidance device but also many other devices must be mounted on the airframe. In this study, PG560 never suffered from lack of thrust. But in order to make space to mount other device and to reduce the weight, it is necessary to introduce the OAS into the FC.

Author contact: Yasuto Oda c501732057@tokushima-u.ac.jp

ACKNOWLEDGMENTS

This work was supported by JSPS KAKENHI Grant Number16K01330.

REFERENCES

[1]Demizu Akira, "Development of safety inspection system for dam embankment using unmanned aircraft," Public benefit support project report of Kyushu Regional Planning Association in 2012.

[2]Kitano Yukihiro, "Aerial shooting using UAV and its use," Construction Technology Development Exchange Meeting in Chugoku Region in 2015.

[3]Chatani Naoki, "FPV pilot for flight rescue robot," Graduation thesis of the University of Tokushima.

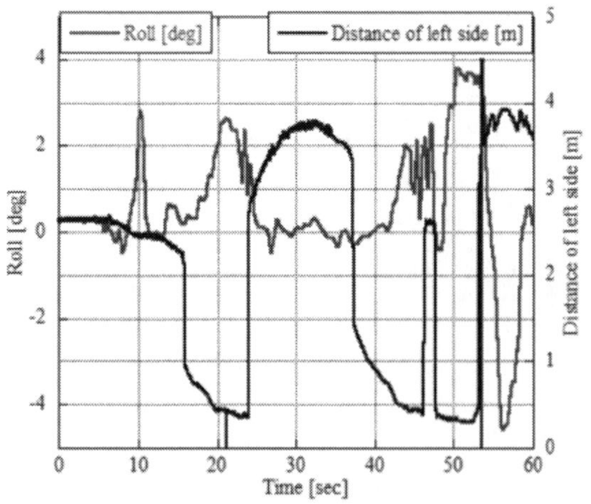

Figure 5. Roll Angle and Distance of the Left Side

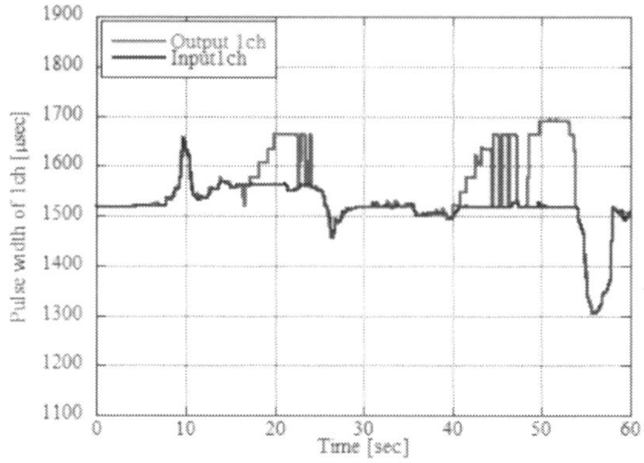

Figure 6. Pulse Width of Input and Output on OAS

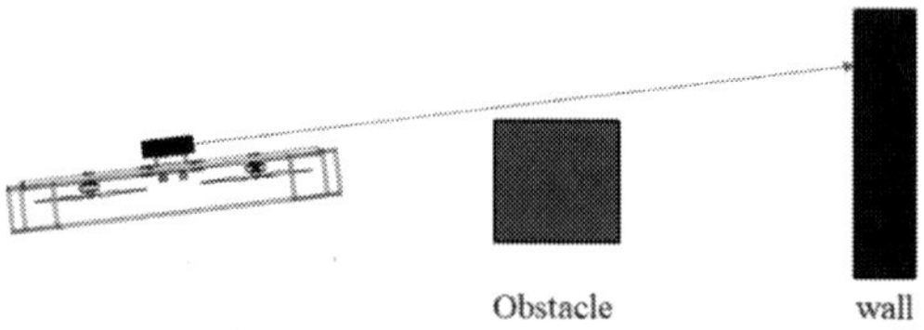

Figure 7. Incorrect Measurement

Figure 8. Laser Rangefinder

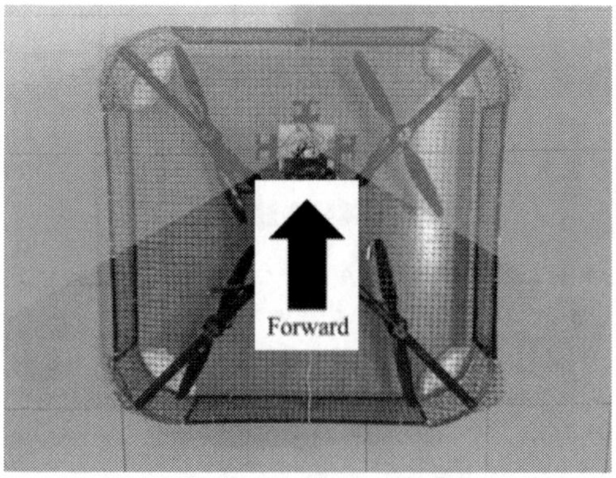

Figure 9. Area where LRF measures

Operation and Prospect of Helicopter Emergency Medical Service in Republic of Korea

Myeang-Hyen Bae
Former Commander of Aviation of
Operations Command, Republic of Korea Army
Research Associate
Korea Advanced Institute of
Science and Technology
Daejeon, South Korea

Daejeon, Korea
+82-42-350-3798
baemyh@kaist.ac.kr

Duck-Joo Lee
Professor
Korea Advanced Institute of
Science and Technology
Daejeon, South Korea

Daejeon, Korea
+82-42-350-3716
djlee@kaist.ac.kr

ABSTRACT

Germany has had 40 years of history of helicopter emergency medical service (HEMS) since it began in 1970. The first permanent civil air ambulance helicopter, Christoph 1 started at the hospital of Harlaching, Munich, Germany, in November 1970. The HEMS in South Korea started with three helicopters in 2011, and nine helicopters are operating in six cities now. In the near future, South Korea will purchase more emergence medial service (EMS) helicopters to accomplish the missions. Now, the HEMS has three models of the helicopters of AW-109, BC-135, and AW-169, which more models will be added. The HEMS is mainly operated in the provinces and metropolitan areas facing coastal and island areas, which is 88% of the operations. Even with the short history of the HEMS of about six years in South Korea, it saved lives almost 3,953 times. The HEMS should be further supplemented and developed to cover the whole region of South Korea in terms of institutional aspects such as addition of the helicopters for emergency rescue and the safety regulation of aircraft operation.

Keywords

HEMS(Helicopter Emergency Medical Service), Operation, Prospect, korea

INTRODUCTION

The helicopter emergency medical service (HEMS), which also called 'Air Ambulance' or 'Doctor Helicopter', in South Korea started in 2011, and it has achieved the emergency transportation of seriously injured patients over 5,000 times until 2017.

This HEMS (Helicopter Emergency Medical Service) have been operating to transport rapidly for a serious illness emergency patient on mountainous region and island area in the Korean peninsula. The Air Ambulance system has a area emergency medical treatment center. When the emergency patient is happened the up-to-date medical treatment equipment and medical professional doctor team move to patient is happened place. The Korean Air Ambulance system (Doctor Helicopter) is operating in 6 areas that is INCHON Metropolitan City, JEON-NAM, KANGWON, KEONG-BUK, CHOONG-NAM, JEON-BUK province etc.

This HEMS contributes to compensate the biased emergency medical treatment to the around Capital area and Metropolitan area for 6 years, and to save seriously injured emergency patient. This is to keep necessary golden time to save the patient. The government announced that The Air Ambulance (Doctor Helicopter) contributes to rapid transportation to save the emergency patient and to improve the survival rate of emergency patients. In the future, the operation of the doctor helicopter will be expanded from 6 areas to 11 areas in South Korea. In addition, the central government will coordinate and reinforce the local governments to arrange hospitals and collaborated systems for another medical evacuation helicopter operation agency.

GENERAL OPERATIONAL SITUATION OF HEMS IN SOUTH KOREA

Medical Support Equipment of HEMS in South Korea

The Korean peninsula consists of 70% mountainous regions and many islands. Therefore, when emergency patients are occurred in that area, we need medical evacuation helicopter to operate the HEMS. How to operation about HEMS?

First, the HEMS in South Korea is operated expeically in vulnerable area of mountainous area and island area. a medical professional doctor team get boards the helicopter to care the seriously injured patients. Thus, the HEMS resolves gap of the medical benefit, and contributes to reduce the patient's death and disability.

Procedure of HEMS Operation

First of all, we support mountainous area and island area the transportation procedure is simple. The operation period is from sunrise to sunset. The crewmember is pilot, co-pilot, emergency professional medical doctor, emergency nurse, and emergency medical technician.

Type and Characteristics of Helicopters for HEMS

Now, we use three types of of EC-135, AW109, and AW-169 for six mountainous and island area. Thus, we have total of nine helicopters in 2017. EC-135 for KANGWON and KYONG-BUK, AW-109 for CHOONK-NAM, JEON-BUK, and AW-169 for INCHON and JEON-NAM are operated by the government of Ministry of Health and Welfare.

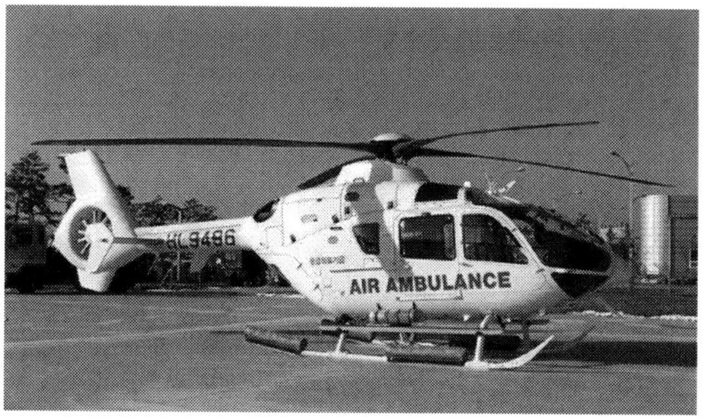

Figure 1 EC-135 for KANGWON and KANG-BUK: 8 people of crew and passenger, and flight distance of 220 km

Figure 2 AW-109 for CHOONG-NAM and JEON-BUK: 8 people of cre and passenger, and flight distance of 260 km

Presented at the 6th Asian/Austrailian Rotorcraft Forum / Heli Japan 2017, Kanazawa, Japan, November 7-9, 2017.

Figure 3 AW-169 for INCEON and JEON-NAM: 9 people of crew and passenger, and flight distance of 400 km

Area and Time of HEMS operation

The Korean HEMS has been operated since 2011 in INCHON city and JEON-NAM Province. It has been operated since 2013 in KANGWON, KEONG-BUK Provinces and since 2016 in CHOONK-NAM and JEON-BUK Provinces.

The operational are 6 regions now which are INCHON city, JEON-NAM, KANGWON, KEONG-BUK, CHOONG-NAM, and JEON-BUK Provinces. First of all, we think that the operational regions are mountainous area and island area, etc. for people in vulnerable area.

Figure 4 Operational area for HEMS in South Korea

IMPLEMENTATION PERFORMANCE OF HEMS (SEP. 2011 – DEC. 2016) IN SOUTH KOREA

Movement Performance of HEMS (SEP . 2011~DEC. 2016)

SECTION	REQUEST	NO. OF MOVEMENT		REJECT
		TRANS(PATIENT)	STOP	
TOTAL	4,266	3,953(3,968)	313	1,591
INCHON	709	661(667)	48	361
JEON-NAM	1,111	1,018(1,022)	93	510
KANGWON	841	771(773)	70	227
KEONG-BUK	1,290	1,197(1,200)	93	409
CHOONG-NAM	242	237(237)	5	64
JEON-BUK	73	69(69)	4	20

Figure 5 Movement performance of HEMS

Transportation Trend of HEMS

SECTION	SUM	2011	2012	2013	2014	2015	2016
TRANS	3,953	75	319	483	947	935	1,194
DAILY TRANS	1.88	0.75	0.87	1.25	2.6	2.56	3.27

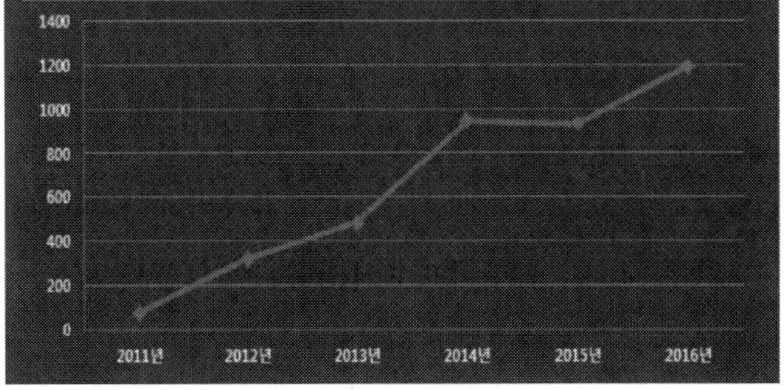

Figure 6 Transportation trend of HEMS

The transportation trend of HEMS is gradually increasing from 2011 to 2016. Therefore, total number of transportation is 3,953 people. The number of transportation in 2011 was 75 people, and the number of transportation in 2016 was 1,194 people.

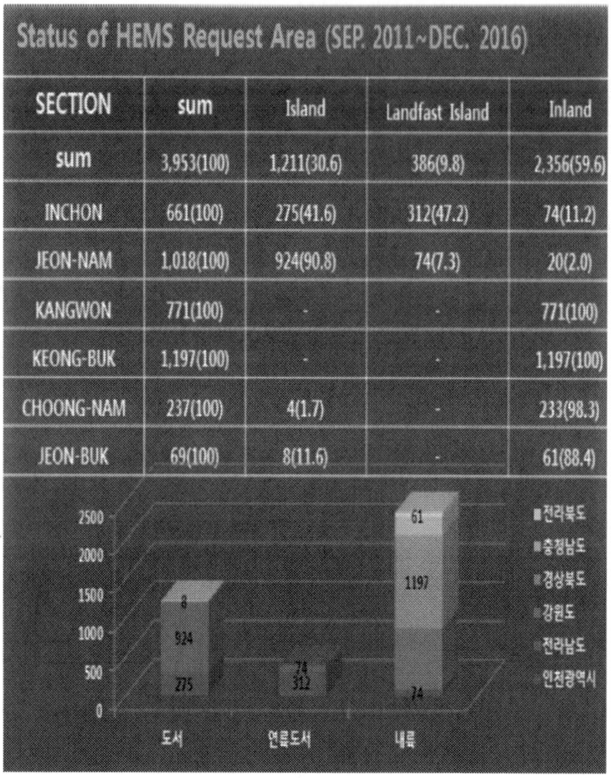

SECTION	sum	Island	Landfast Island	Inland
sum	3,953(100)	1,211(30.6)	386(9.8)	2,356(59.6)
INCHON	661(100)	275(41.6)	312(47.2)	74(11.2)
JEON-NAM	1,018(100)	924(90.8)	74(7.3)	20(2.0)
KANGWON	771(100)	-	-	771(100)
KEONG-BUK	1,197(100)	-	-	1,197(100)
CHOONG-NAM	237(100)	4(1.7)	-	233(98.3)
JEON-BUK	69(100)	8(11.6)	-	61(88.4)

Figure 7 Status of HEMS request area (Sep. 2011 – Dec. 2016)

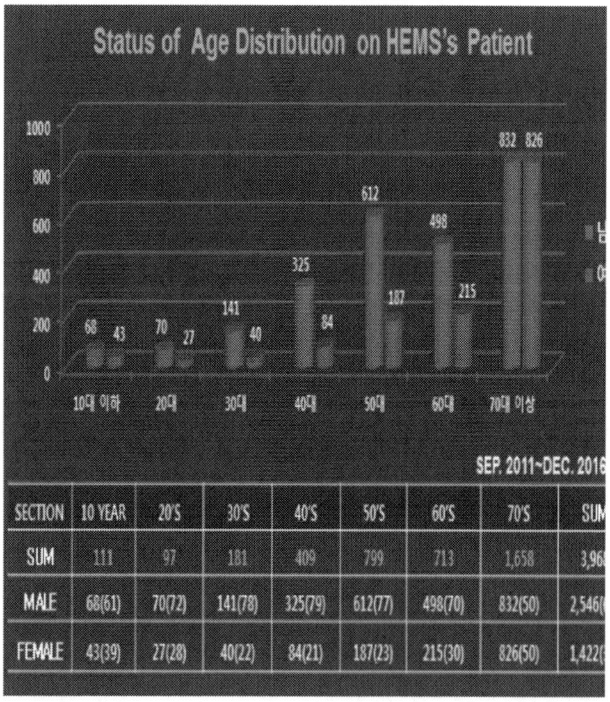

SECTION	10 YEAR	20'S	30'S	40'S	50'S	60'S	70'S	SU
SUM	111	97	181	409	799	713	1,658	3,96
MALE	68(61)	70(72)	141(78)	325(79)	612(77)	498(70)	832(50)	2,546(
FEMALE	43(39)	27(28)	40(22)	84(21)	187(23)	215(30)	826(50)	1,422(

Figure 8 Status of age distribution on HEMS's patient

The male patient is more than female patient. That is male patient 64% and female patient 36%. And all of them male or female patient are gradually increasing from 40 years to 70 years etc.

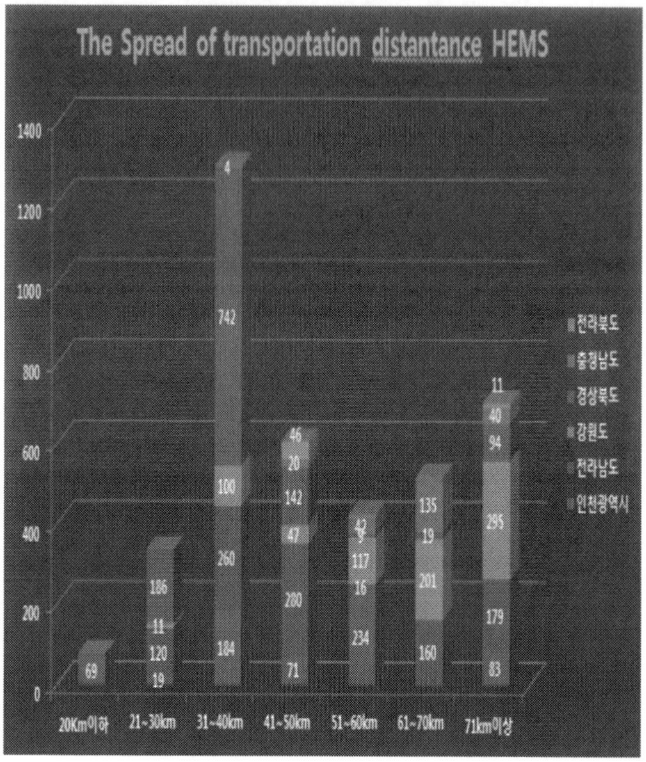

Figure 9 The spread of transportation distance HEMS

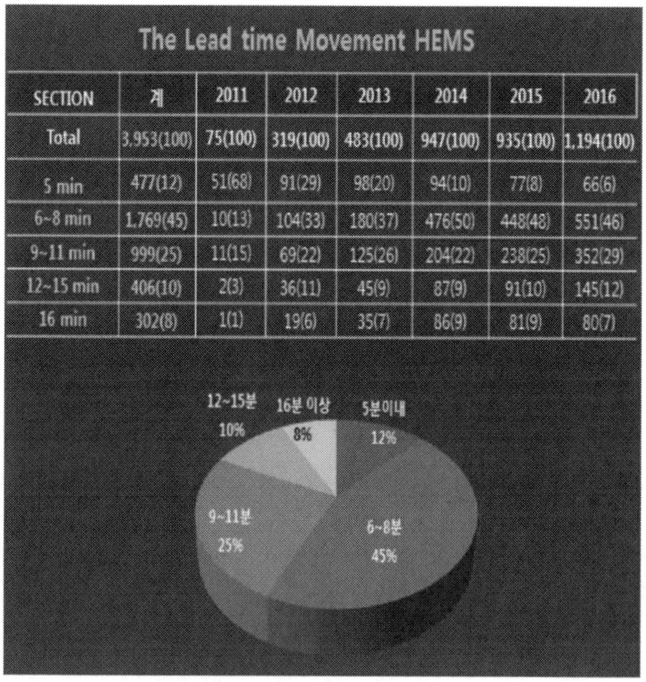

SECTION	계	2011	2012	2013	2014	2015	2016
Total	3,953(100)	75(100)	319(100)	483(100)	947(100)	935(100)	1,194(100)
5 min	477(12)	51(68)	91(29)	98(20)	94(10)	77(8)	66(6)
6~8 min	1,769(45)	10(13)	104(33)	180(37)	476(50)	448(48)	551(46)
9~11 min	999(25)	11(15)	69(22)	125(26)	204(22)	238(25)	352(29)
12~15 min	406(10)	2(3)	36(11)	45(9)	87(9)	91(10)	145(12)
16 min	302(8)	1(1)	19(6)	35(7)	86(9)	81(9)	80(7)

Figure 10 Lead time movement HEMS

The Lead time Movement HEMS is usually the most from 6 minutes to 8 minutes as a 45%. Next from 9 minutes to 11 minutes as a 25%.

AVIATION SAFETY OF HEMS IN SOUTH KOREA

The Korean Helicopter Emergency Medical Service set up to September 2011 in Inchon Metropolitan City. Gradually, The HEMS is increased up to 2017. So we are operating to 9 helicopters for 6 regions island and mountainous area peoples. This Doctor Helicopter transport to transportation for emergency patients rapidly. So that is contributed for big improvement survival rate. In the future Korean Government will expand from 6 Metropolitan city and Province area to 11 Metropolitan city and Province area. Also Korean Government will reinforce coordination system with local government to hospital and fire defense and coast guard etc..

Now we need to prepare for aviation safety to aircraft, aircrew member education for professional doctor especially, pilot flight training and mechanic maintenance training, air traffic control training etc. Also we need to necessary for aviation safety to human factor, weather, maintenance etc. Therefore we need to management system for aviation safety. In this management system have to coordinate with pilot and flight doctor and flight company and all crew members. Because aviation safety is very important to prepare for aviation accident prevention.

DEVELOPMENT OF HEMS IN SOUTH KOREA

South Korea developed the HEMS with SURION helicopter of Korea Aerospace Industries (KAI), which the first flight was done in Jan. 2016. The helicopter can pick up 6 patients with emergency medical treatments in medevac mission under adverse weather conditions. It has exterior parts of weather radar and hoist, and interior cockpit with traffic collision avoidance system with PDA, TRS, and wireless-ICS, and interior cabin with medical support equipment, detachable aux. fuel tank and active vibration control system.

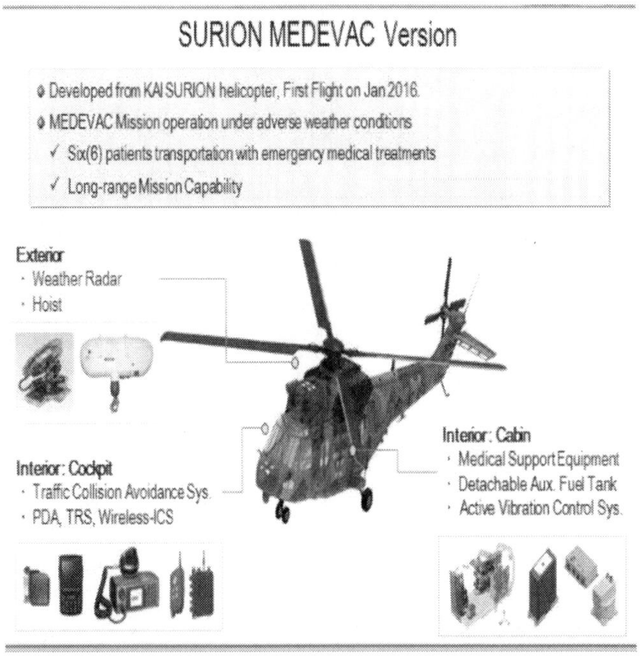

Figure 11 MEDEVAC version of SURION

Medical Support Equipment of HEMS in South Korea

- Medical crew's swivel and movable crash-worthy seat
- Electric powered, easy patient loading and litter height adjustable litter lift system (Max. 6 patients)
- Easily removable, folding type and TSO-C127a compliant ambulatory patient's seat
- Lights to provide for visualization, examination, and treatment of all patients
- Commercial medical equipment and power outlets for charging battery
- Defibrillator, ventilator, patient monitor, IV pump, portable refrigerator, etc.

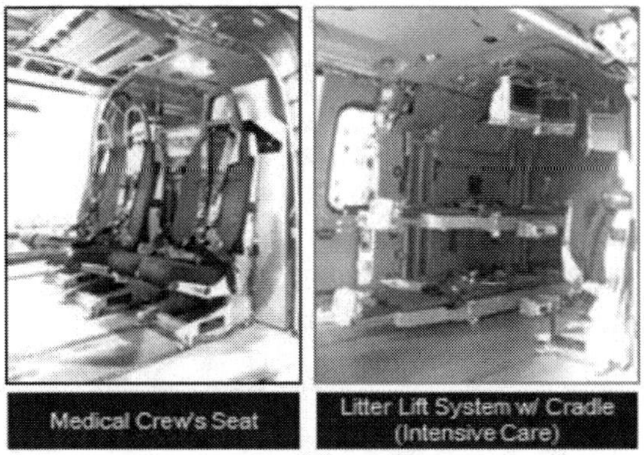

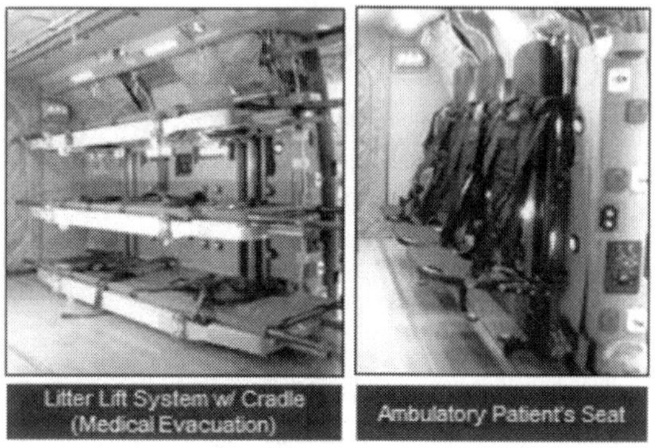

Figure 12 Medical support equipment of MEDEVAC version of SURION

Various Mission Configuration for HEMS in South Korea

- Advanced Life Support: intensive care for more than two patients
- Medical Evacuation Six littered or ambulatory patients
- Aeromedical helicopter service is highly demanding on needs in military and civil markets.
- KAI will provide the customized export version of the helicopter to achieve the MEDEVAC mission.

Figure 13 Various mission configurations of MEDEVAC version of SURION

CONCLUSION

What is the HEMS in South Korea? We think that it is to reduce the time for transportation for saving life in the future. Today, the government announced that the doctor helicopter achieved to save over 5,000 people by transportation of patient from 2011 to 2016 during 6 years. In this period, the doctor helicopter operated in island and mountainous areas in Korean Peninsula. It transported seriously injured patient, patient with cerebrovascular disease, and patient with cardiovascular disorder. The HEMS's crew was educated to crew members education in South Korea already. In addition, pilot and maintenance mechanic had the job training.

In the near future, the HEMS will be expanded to operational area from 6 regions to 11 regions by the government. We need nigh fly for emergency patients living in island and mountainous areas. In addition we need to prepare aviation safety to manage the system for crew members. In the future, we want the bigger helicopter to manage the emergency patients comfortably inside the cabin.

Now, South Korea is developing the HEMS with SURION of KAI. We have various mission configurations of medevac helicopter for HEMS. We also have a Advanced Life Support. It is intensive care for more than two patients and medical evacuation six littered or ambulatory patients. KAI will provides customized export version of helicopter to achieve the medical evacuation mission.

Author contact:
Myeang-Hyen Bae baemyh@kaist.ac.kr
Duck-Joo Lee djlee@kaist.ac.kr

REFERENCE

[1]Reason, J. T., "Managing the Risks of Organizational Accidents," Gower House Croft Road Aldershot, England, 1997.

[2]Oster, C. V., Strong, J. S., and Zorn, K., "Why Airplanes Crash: Aviation Safety in a Changing World," 200 Madison Avenue, NY, 1992.

[3]National Emergency Medical Center, Situation and Performance of Korean HEMS, Seoul, South Korea, 2017.

[4]"The history of HEMS System," Yonsei University Wonju Medical College EMS, Wonju, South Korea, 2016.

[5]"The Enhance of HEMS ," The 5th Autumn Symposium of Korean Association of Aero Emergency Medical Service, Mokpo Korean Hospital, Mokpo, South Korea 2015.

[6]"The Doctor Helicopter: The New Start and The More Development," The 6th Autumn Symposium of Korean Association of Aero Emergency Medical Service, Dan-kuk University, Hospital, ChonAn, South Korea, 2016.

Construction of AIR-TRAFFIC-SYSTEM for all Weather Operations of Helicopter

== for keeping the Safe and Secure Society by the Improvement of Helicopter-Flight-Safety ==

Takeshi Tomio , Ph.D

Secretary

a Society for the Research on Helicopter-Flight-Safety

(acting in JAPAN CIVIL AVIATION PROMOTION FOUNDATION)

Ooyaku Tokyo

ABSTRACT

The helicopter-Operation in Japan is now expected to come into the All-Weather-Operations on 365 days and 24 hours, in response to the frequently occurrence of huge Disasters , and responding to the request from the citizen for a construction and keeping of the Safe and Secure Society. The helicopters are now making every efforts for conducting many kinds of the missions for Disaster-Relief , however, these missions sometimes face to the difficulties coming from the condition of the weather and/or nighttime. For surmounting such various difficulties, the construction of AIR-TRAFFIC-SYSTEM for all Weather Operations of Helicopter is absolutely necessary and is now being conducted so as to be able to fulfill the qualities necessary for matching to the helicopter's specific flight-quality and flight-mode.This paper outlines some topics of the subjected air-traffic-system.

NOTATION

CARATS	Collaborative Actions for Renovation of Air Traffic Systems
FDMA	Fire and Disaster Management Agency
JASDF	Japan Self Defense Forces
JAXA	Japan Aerospace Exploration Agency
JCAB	Japan Civil Aviation Bureau
MLIT	Ministry of Land, Infrastructure, Transport and Tourism

INTRODUCTION

At PM14:46 on March11 2011, huge earthquake attacked the Tohoku area (East-North Area) of Japan with magnitude 9.0. Many kinds (or by other words, every considerable kinds) of disasters, such as Collapse of houses, buildings, bridges, roads, cliffs,and many facilities for Life-Line,etc., and Tsunami Fire, and others. As shown on Figure 1,many helicopters, of which maximum total numbers was 300 , went in action coming from all over country of Japan. The missions had continued on a long time duration, as sample on the FDMA-helicopter, as sample, 80 days (Miyagi Prefecture), and 50 days (in Fukushima Prefecture). These missions were successfully conducted with many fruits fortunately without critical accident of helicopter. On the other hand, it is necessary to remark that the above mentioned fruits were gotten fortunately under many personal efforts or Resistance for overcoming of the the difficulties which may come from the bad condition of the weather and/ or nighttime.

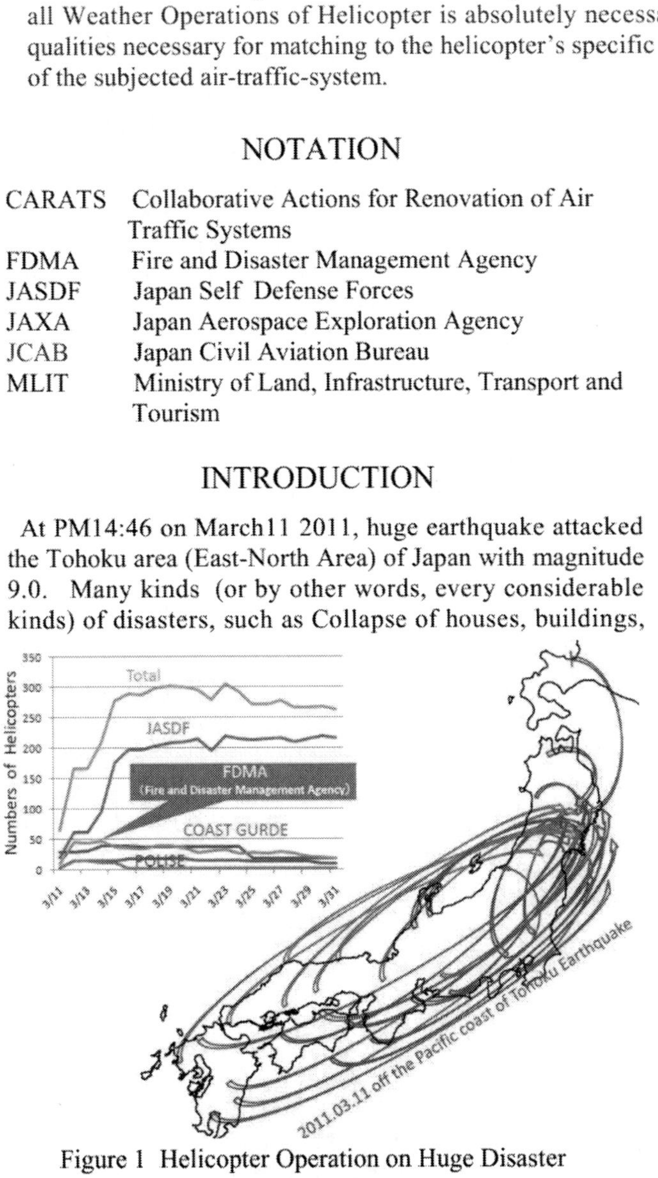

Figure 1 Helicopter Operation on Huge Disaster

Presented at the 6th Asian/Austrailian Rotorcraft Forum / Heli Japan 2017, Kanazawa, Japan, November 7-9, 2017.

I think that the infrastructure for the helicopter operations under such bad conditions is still not constructed now, and so,the helicopters still have the risk for keeping the Flight-Safety, as shown on Figure2. We, the Society for the Research on Helicopter Flight Safety have been conducting the R&D mainly for relaxation of such risk. On the base of the fruits of the R&D of our society, some topics on the matters titled above are presented and discussed in the following sections.

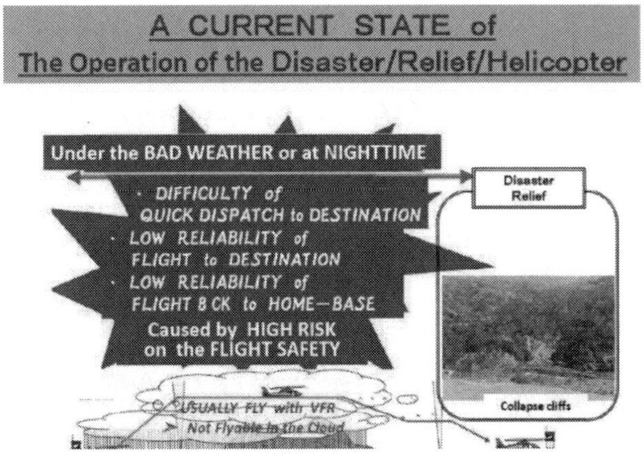

Figure 2 Current State of Helicopter Operation on Disaster Relief

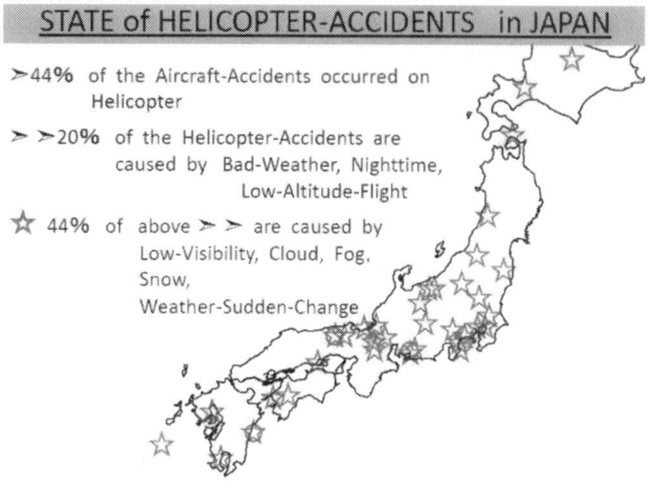

Figure 3 State of Helicopter

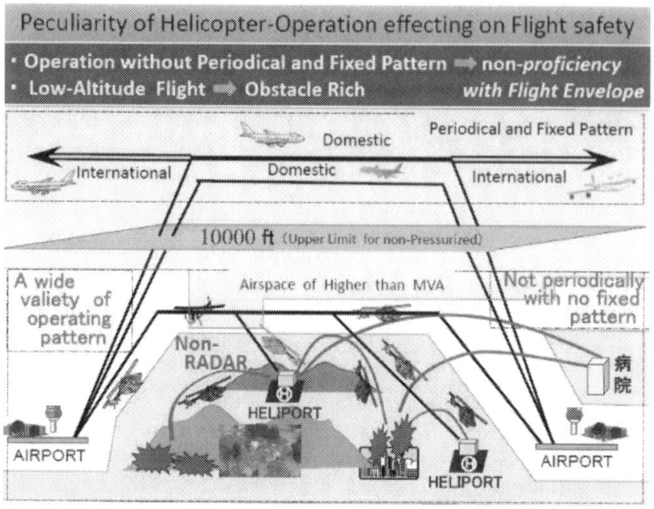

Figure 4 Peculiarity of Helicopter-Operation

PRESENT STATE OF THE HELICOPTER FLIGHT SAFETY

The disaster-relief-helicopters are now operating with every efforts to overcome the risk on Flight-Safety. For reducing the risk and promoting of the disaster-relief-operation as practical as best, a state of the helicopter safety flight should be reviewed, and, after then, a method for removal of the risk should be developed and applied.

State of the Accidents occurred on the Helicopter

A state of the occurrence of the Helicopter-Accidents is shown on the above Figure 3. The summary is as follows.

➢1 44% of the Aircraft-Accidents occurred on Helicopter (432 on helicopter / 983 on aircraft)
(983=432+551 of airplane (Large165 and small386))

➢2 20% of above ➢1 (Helicopter-Accidents) was caused by ***Bad-Weather, Nighttime, and Low-Altitude-Flight***

➢3 44%of above ➢2 was caused by ***Low-Visibility due to Cloud/Fog/Snow, Wind, Combi-nation of them,*** *and* ***Sudden Change of Weather,etc.***
(Accidents caused by only of wind or snow is not counted)

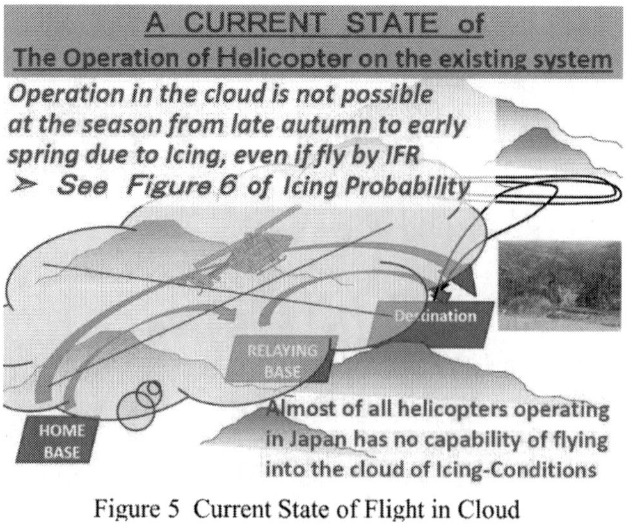

Figure 5 Current State of Flight in Cloud

Factors affecting on the Helicopter-Accidents

The helicopter operates with the Peculiarities shown on the Figure 4. Major factors affecting on the flight safety are,

➢ Operation without Periodical-Schedule
➢ Operation without Fixed Pattern
(Operation with the wide variety of flight pattern)
➢ Operation with the wide variety of task
➢➢ Mastering by repeating of the pattern is not expectable
➢ Low-Altitude-Flightaaa
➢➢Obstacle Rich, and so, Higher Risk of the collision
➢ Weak to bad weather and sudden change of it as shown in left column
➢ Almost of all helicopters operating in Japan has no capability of flying into the cloud of Icing-Conditions

Figure 5 and Figure 6 show the typical peculiarities which should be take into account on the development of subjected system.

INDISPENSABLE REQUIREMENT for all Weather Helicopter-Operation-System

For conducting all weather and 365days and 24 hrs operations, with a reliability as higher as practicable and with a quick-response, the air-traffic-system should have the full quality to be able to accept all of the above Peculiarities.

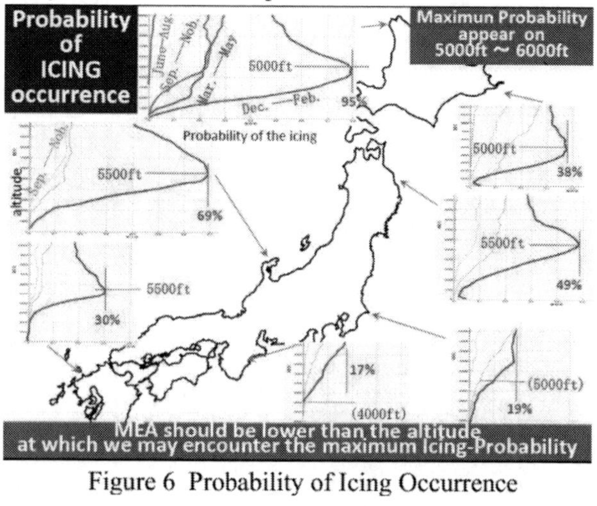

Figure 6 Probability of Icing Occurrence

CONSTRUCTION of AIR-TRAFFIC-SYSTEM for all Weather Operations of Helicopter

An air traffic system for all weather operations of the helicopter is now under the construction as the major project of the CARATS promoting by JCAB. The priority of this construction is putting on to give the system a usefulness for Disaster-relief as much as practicable. Our Society for the Research on Helicopter Flight Safety is now working as one of the members of CARATS, and expect that the system is build up as earlier as possible.

The system consists with two major parts of a Enroute-Network and a Network of Bases where the helicopter can do a IFR Departure and Landing as shown on Figure 7.

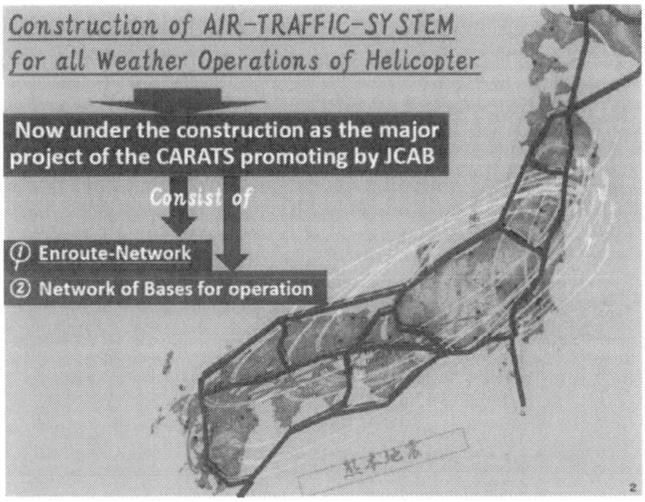

Figure 7 Consist of Air-Traffic-System

Enroute-Network (First Phase)

The network is now being constructed on CARATS with the properties mentioned below.

- ➢RNAV5
- ➢GNSS (SBAS and/or ABAS)
- ➢Communication Required
- ➢Radar-Surveillance Required
  (both for the Normal Control and for the Radar-Vectoring as the contingency procedure)
- ➢MEA(Minimum Enroute Altitude) should be set on the altitude on which the above mentioned features could be fulfilled.

NOTE

1. The lowest MEA of this Enroute has been estimated so as to fulfill the requirement of MOC of 2000ft, by CARATS. This lowest MEA is 3000ft and the highest MEA is 7000ft as shown on Figure 8.
2. In addition to above factor (MOC), especially for helicopter operation, it is needed that the MEA should be set as lowest as possible to minimize the risk of encountering of Icing-Condition, so, our Society evaluate an acceptable maximum MEA as shown on Figure 9. This altitude is based on the Probability of occurrence of Icing Weather shown on Figure 6.

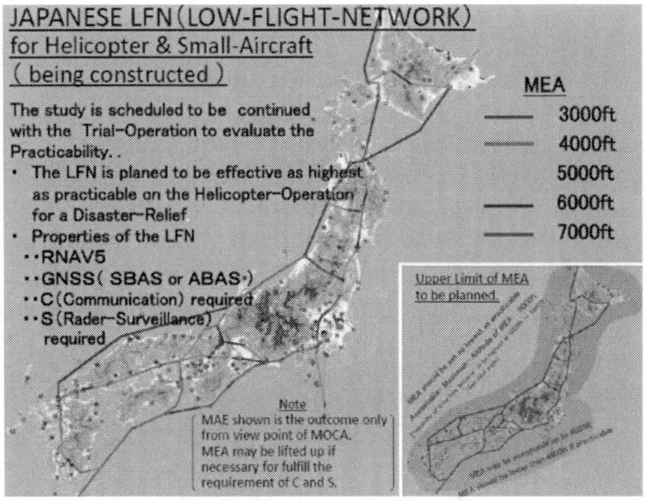

Figure 8 The lowest MEA estimated to fulfill the Minimum Obstacle Clearance on the Japanese LFN

Enroute-Network (Second Phase and after)

== to be improved in the future ==

On the above Enroute-Network(First Phase) that is based on the specification of RNAV5, it is possible that the MEA could be pulled up, on the some portions of the Network, higher than the required MEA shown on Figure 9.

It is afraid that the MEA higher than Figure 9 makes the Network unusable for IFR-Operation in the cloud at the season from late autumn to early spring due to Icing. For elimination of this problem, it needs to replace the RNAV5 system with RNP2 and/or RNP0.3 for resetting MEA as lower as possible within range of altitude shown on Figure 9.

In addition to, more improvement to pull down of MEA should be incorporated by using RNP0.3 if necessary.

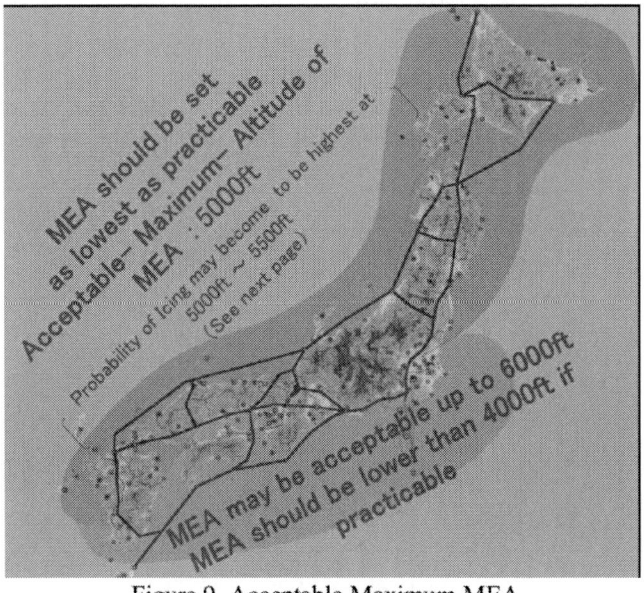

Figure 9 Acceptable Maximum MEA

NETWORK of HELIPORTs/AIRPORTs for APPROACH and DEPARTURE

The Integration of the NETWORK of HELIPORTs/ AIRPORTS with the ENROUTE-NETWORK is now being constructed on CARATS.

A Base of the Helicopter-Operation is not only Airports but also Heliports. Figure 10 outlines the state of operating Bases of the Disaster-Relief-Helicopter of FDMA, as a typical sample. From the view point of the preparation of the Infrastructure for the All-Weather-Operations on 365 days and 24 hours, It may be most important how to overcome the complexities and difficulties that the incorporation of the Heliports into the total air system may confront.

With this recognition, our Society is now doing the R & D. This paper presents a outline of them.

Enabler for incorporation of Heliport into Total System

A Helicopter point-in-space (PinS) departures and a Point-in-space approach procedures (PinS RNP APCH) for helicopters are now become to be available. These procedure is developed by ICAO and published in Sixth edition of Doc 8168 OPS/611 - 2014. These procedure is developed mainly for operation on the Visual Heliport specified in Annex14 Volume II Heliports revised as Fourth Edition July 2013.

A Figure 11 shows only of the Key-Point of most usable part of PinS that is the segment from MAPt (Mist Approach Point) to the heliport called as Fly-Visual-Segment (IFR category). This segment is also flyable under the VFR.

Case Study of application on the Heliport

A Figure 12 shows the sample of the case study for leaning how to able to apply the Pins with good practicability so as to meet the airspace conditions surrounding the heliport.

Integration of Airport and Heliport into the system as the Base for relaying on Long Range Cruising.

An integration subjected is scheduled to do on the Project of CARATS. A Figure 13 is the contents of the study that our Society now try.

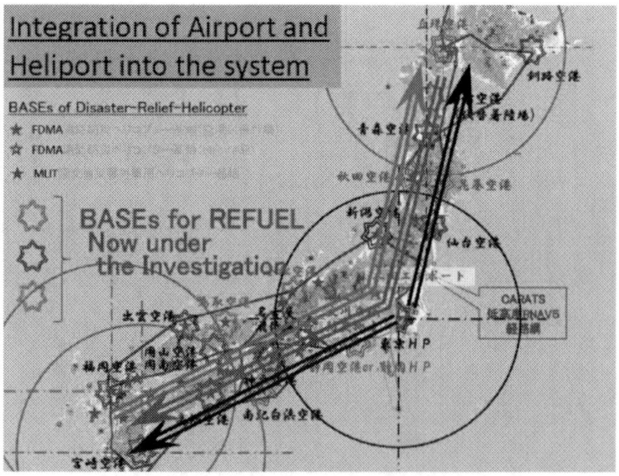

Figure 13 Integration of Airport and Heliport

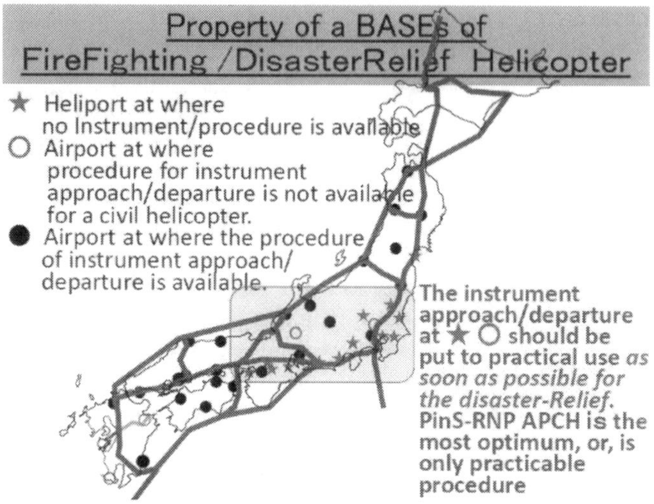

Figure 10 Property of the Bases 0f FDMA Helicopter

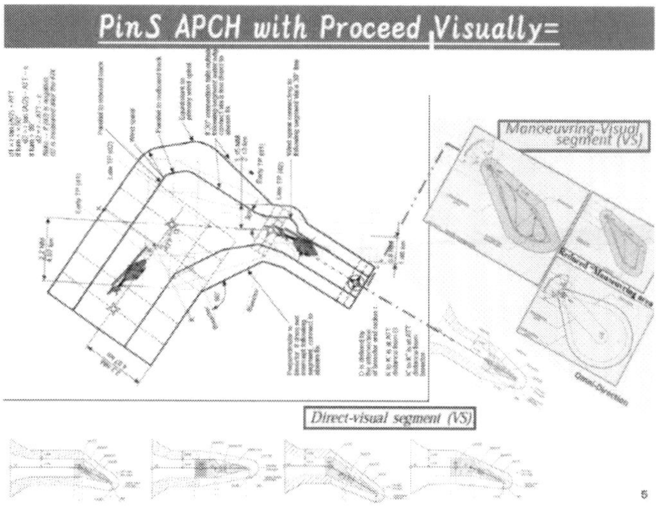

Figure 11 Key-Point of most usable part of PinS

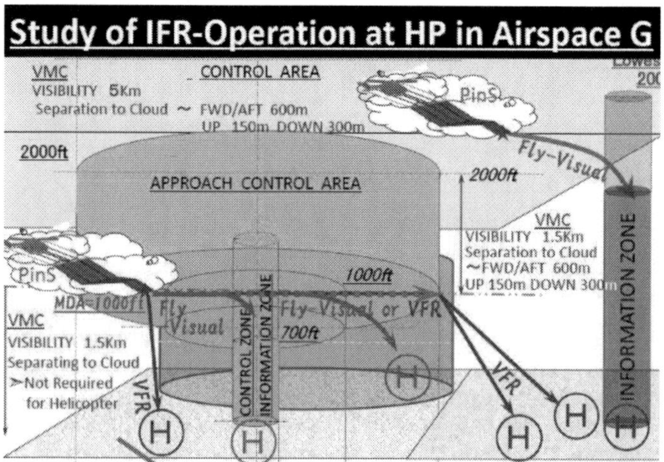

Figure 12 sample of the case study to apply the Pins to the Heliport located in Airspace Class G

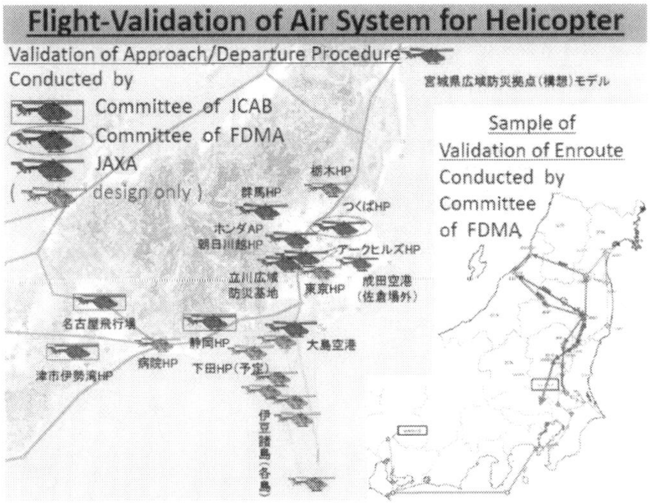

Figure 14 Flight-Validation of Air System has been Conducting in Japan

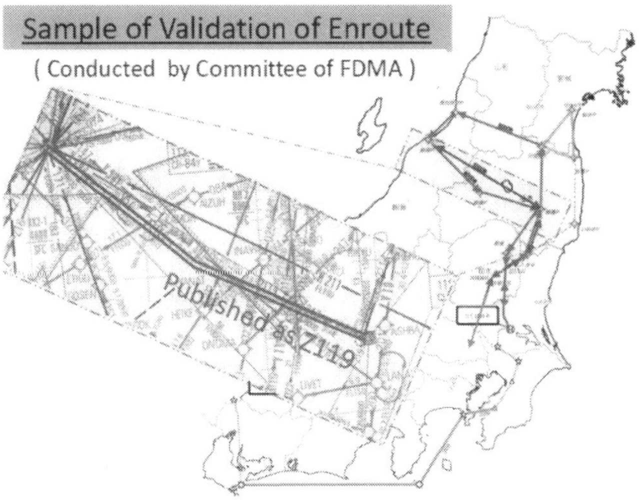

Figure 15 Sample of the Flight-Validation of ENROUTE

FLIGHT-VALIDATION of AIR SYSTEM

On the enablers for construction of Helicopter Air system, the Flight-Validations for a Practicability and a Probability has been conducing more than 20-times on many Helicopter-Bases more than 10-Heliport/Airport,as shown on Figure 14.

Validation of Enroute

Figure 15 is the Flight-Validation conducted in the committee of FDMA. The fruits of this Flight-Validation promoted the publication of Z119 which is first GNSS-RNAV-ENROUTE for HELICOPTER in Japan.

Validation of PinS Approach Procedures

The Practicability of PinS, and, the Probability to adopt Pins at Japanese-Heliport has been evaluating, Figure 16 is the sample of them.

These Flight-Validation lets us know the importance of a establishment of the measures to set MDA as lower as practicable, especially for Heliport located in the Airspace-Category-G as shown on Figure 12.

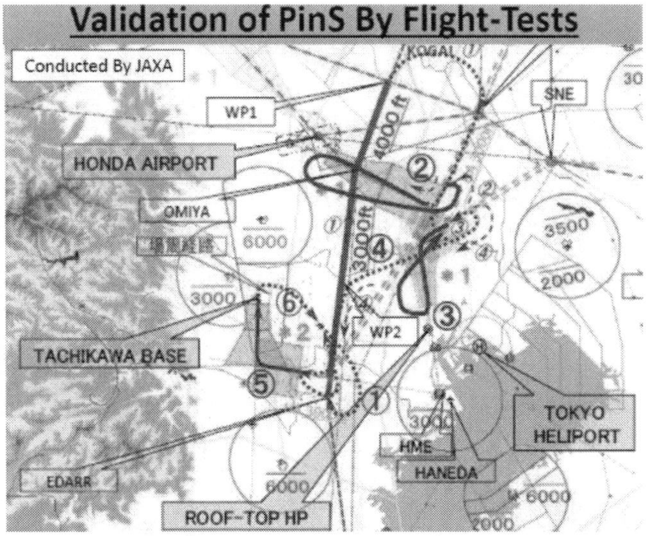

Figure 16 Sample of the Flight-Validation of PinS

Validation expected to be continued in the Future

As mentioned above, the All-Weather-Operations of the Helicopter on 365 days and 24 hours is essential function for a construction and keeping of the Safe and Secure Society, and it is requested to do as quick as possible and need to do with the high reliability.

For make it realize with high qualities, the incorporation and integration of the Enablers shown on Figure 17 is absolutely necessary.

However, almost of the Enablers is now under the evaluation of adaptability on the Japanese Environments.

So, our Society think that the validations should be continued without discontinuity.

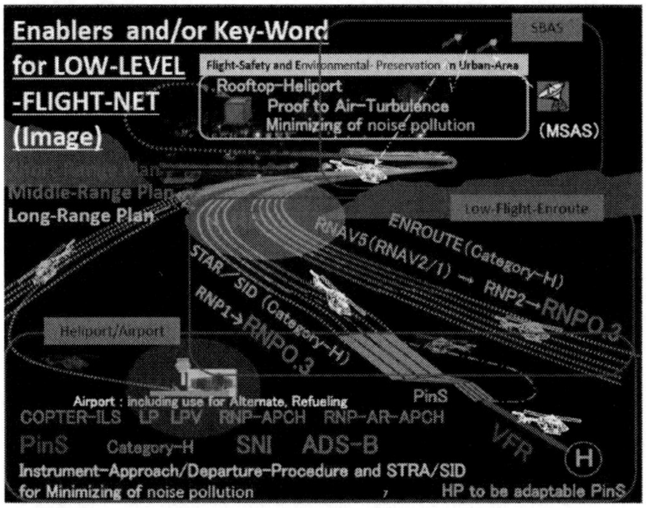

Figure 17 Enabler and/or Key-Word for Low-Level Flight-Network

CONCLUSION

The following outcomes is expected as the fruits of the construction of the subjected system.

Construction of the Safe and Secure Society

The helicopter-Operation in Japan is now coming into the All-Weather-Operations on 365 days and 24 hours, and the helicopters come to be able to fly for conducting the every kinds of missions for the Disaster-Relief as shown on Figure 18, to every places and at all times, when the Disaster attack the Citizen.

Improvement of the Flight Safety of the Helicopter

The Network of Enroute of the subjected system may act as the Safety-Network for reducing the possibility of accidents of the helicopter. As shown on Figure 19.

Growth of the Helicopter Business

The helicopter is operating, until now, mostly under the VFR category, and so, the helicopter often face to the the difficulties of the Safe-Flight due to the bad weather or the nighttime.

Figure 20 is a typical sample of the business that could not continue in Japan. On the business category of the Passenger-Transport, many Lines had started or had tried to start in, however, almost of them was discontinued by the causes of Less rate of trafficability (lower than 80%), or Fatal accidents affected by the bad weather,nighttime, or sudden change of the weather.

By the improvement the trafficability and/or improvement of the Flight-Safety mentioned above, the growth of this business of Passenger-Transportation can be expected.

These Fruits mentioned on the left column can not come true without the efforts for overcoming of all difficulties, and we can hand such overcoming only by the collaborating activity of many organization with a specific knowledge of them.

I like to conclude this paper with proposing of doing the above mentioned collaborating activity.

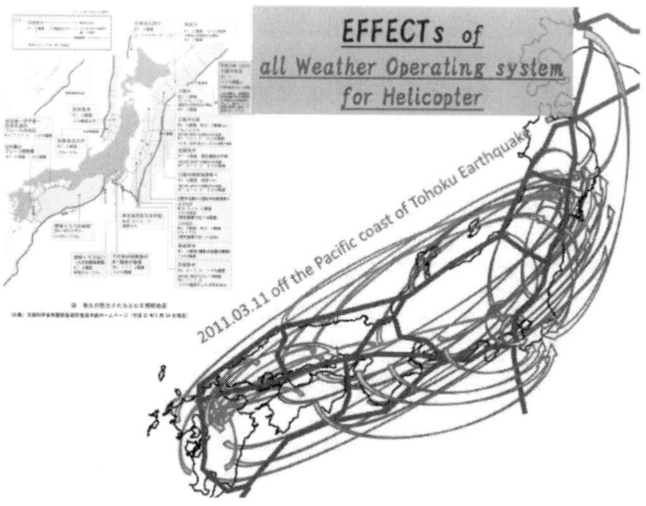

Figure 18 Effect expected on the Disaster-Relief

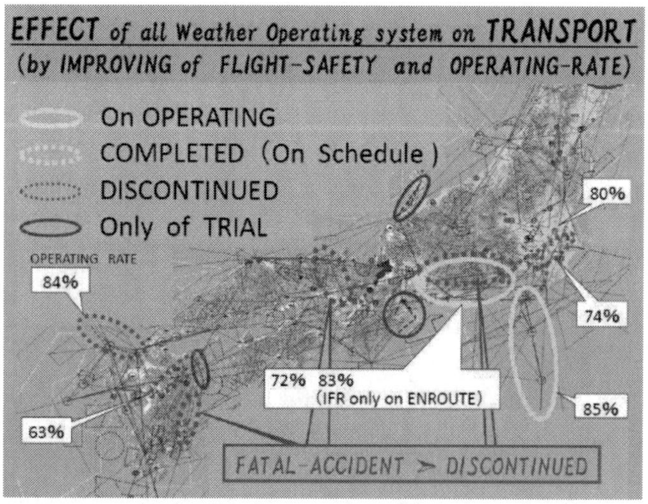

Figure 20 Effect expected on the growth of the Helicopter Business

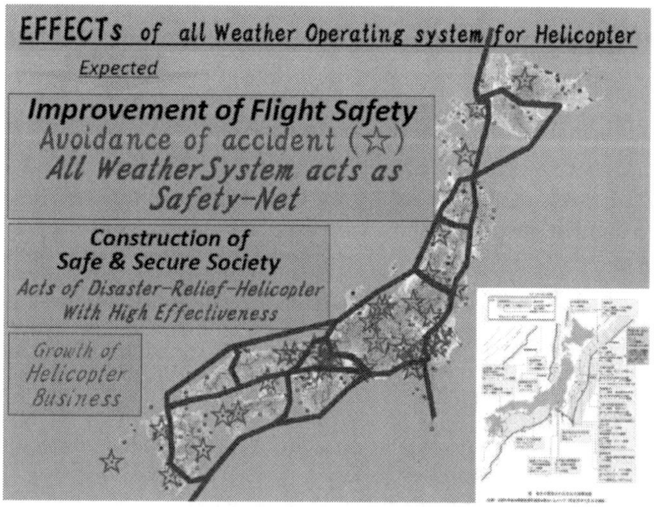

Figure 19 Effect expected on the Improvement of the Flight-Safety

This page intentionally left blank.

ヘリコプタ運用におけるALBの有用性と社会インフラ維持管理への応用

磯部寛
朝日航洋株式会社　商品化推進室

ALB Operation in a Helicopter - Its Application for Infrastructure Maintenance and Disaster Prevention, by Aero Asahi

Hiroshi Isobe
Airborne Operations, Aero Asahi Corporation, Kawagoe, Japan

Abstract

Much of Japan's infrastructure such as roads, railways, ports, river ways, and power lines, were designed and built in the high-growth period between the late 60's and mid 70's. This infrastructure is now aging and deteriorated. In addition to these factors, the infrastructures also need to be protected from natural disasters such as flooding rivers during the rainy season and earthquakes. Aero Asahi Corporation's mission is to protect life and property from any disaster or incident by using the latest technologies including hardware and software and thus make an important social contribution by performing "early deployment" survey. We, as a Japan's biggest helicopter operation company, operate ALB with advantages of helicopter mobility to survey in various geographical conditions and thus make an social contribution. In this presentation I will introduce types of work we perform using ALB and describe its advantage, issues, and feasibility in order to fulfill our mission.

1. はじめに

社会インフラストラクチャには、主に道路、鉄道、港湾、河川、送電がある。これらの多くは、日本の高度成長期（1960年代後半～1970年代）に建築され、今では老朽化による劣化が多く目立つようになった。加えて、地震を含む自然災害直後には、初動派遣（アーリー・ディプロイメント）を基本とした計測支援が求められる。

ここでは最新のALB(Airborne Lidar Bathymetry)技術を使って、ヘリの機動性を活かすことで、インフラ設備の維持管理、災害に効率よく対応し、国民の生命と財産を守るために朝日航洋として貢献できる技術とその取り組みについていくつか紹介する。

2. 航空レーザー測深機とは

ALBとは、航空レーザー測深機のことであり、Airborne Lidar Bathymetryの省略である。技術的には1980年代から海外（ロシア、アメリカ、スウエーデン）で研究開発され、主に北米やヨーロッパなどで利用されてきた。日本国内では、海上保安庁においてH15年にカナダOPTECH社のSHOALS1000を導入し、海図の補正や各種調査に使用している。

海上保安庁のSHOALS1000は朝日航洋が納入し、以降保守を継続してきた。2013年には同じOPTECH社製CZMILを後継機として同じく納入し、以降保守を行っている。

近年、測深性能の向上や機材の小型化が進み、2011年頃から国内の民間測量会社で国内に機材を持ち込んで実験が行われた。スウエーデン、Leica Geosystems製キロプテラⅡは、2015年に国内で初号機が導入され、弊社は2016年に導入した。近赤外レーザとグリーンレーザの両方を使用して海面を推定し、水中を透過するグリーンレーザで屈折補正と高速度補正を行い、より正確な水面下の地形（楕円体）計測を実現している。

弊社で運用するレーザー測深機は、近赤外レーザは波長1064nm、グリーンレーザは波長515nmであり、レーザー発射数（PRF）はそれぞれ500,000発/sec、35,000発/secであり、測深能力では透明度の約1.5倍の測深が可能である。レーザーのスキャンパターンはこれまでの近赤外レーザーにあったいわゆるzigzagではなく、オブリークパターンである。これにより構造物の壁面をより的確に計測することが可能となっている。

3. レーザー測深機の装備と運用

本システムの装備は、当社AS350-B3型回転翼機の機外テールブーム下方に近赤外レーザー発振部、グリーンレーザー発振部、80MPデジタルカメラ（CIR+RGB）が専用のハウジングに収納され（Fig.1、

Fig. 2）、機体キャビン後方 RH にレーザーコントローラ、データストレージ、慣性計測装置、電源等が組み込まれたコントロールラック及びカメラコントロールユニットがそれぞれラックに収納され取り付けられており、これらは光ケーブルを含む各種通信ケーブルで連結されている。（Fig. 3）　また、垂直尾翼上部に GNSS アンテナが取り付けられている。また、直下及び斜め下 45 度の向きで機体のクロスチューブにカメラを装備し、オペレータがこれらをモニタリングすることにより特に送電線の接近木レーザー計測における運航の安全と効率化に役立てている。

通常の乗員配置は、オペレータ１名（Aft-LH）、運用補助１名（Co-Pilot 席）、機長１名（Pilot 席）の計３名である。

Fig. 1 AS350-B3 Helicopter

Fig. 2 Laser Housing Installation

Fig. 3 Rack Installation

4. 計測におけるヘリコプタ運用のメリット（vs. 固定翼）

1）例えば河川のように曲がりくねった場所でもヘリの機動性を活かし計測コースとスピードを柔軟に地形に対し合わせることで、運用時間の短縮、コストの削減がはかれる。

2）ヘリでは飛行高度をある程度フレキシブルに変更できる。結果急峻な山岳地域など地形に沿った等対地飛行高度が維持でき、データの安定した精度が保たれる。(Fig. 4)

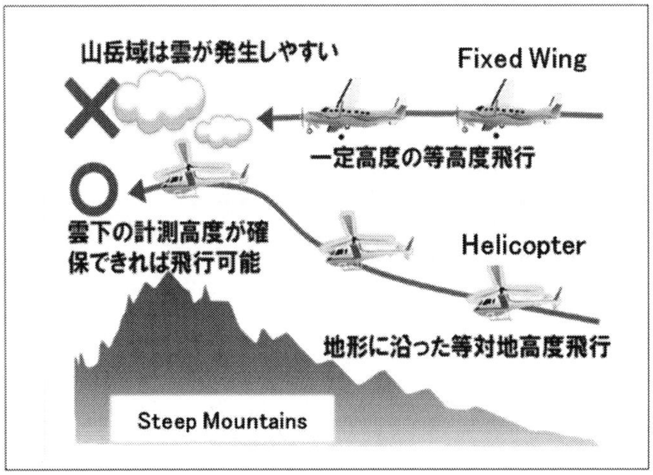

Fig.4 Comparison between Fixed Wing and Rotary Wing

3）場外ヘリポートの利用により計測地域の天候が正確にわかる。

4）水部測量の場合、ボート等に装備した音響測深機で測量を実施するため浅瀬域ではボートの進入が非常に難しい。ヘリではこういった障害が無く、且つ面的にスピーディに計測が可能である。

5）以下に回転翼機と固定翼機の運用比較サマリを記す。

	Advantage	Disadvantage
HELICOPTER 	✦VARIABLE AIRSPEED ✦HIGH DENSITY POINT CLOUD ✦FLEXIBLE BASE LOCATION ✦SHORTER TURNING TIME ✦NOE (Nap of the Earth)	✦LIMITED FLIGHT TIME ✦NOISE POLLUTION ✦FLIGHT RESTRICTION IN DENSED AREA - Altitude (City) ✦NO PAV INSTALL
FIXED WING 	✤LONG RANGE ✤QUIET ✤LARGER CABIN ✤LESS FLIGHT COST	✤MINIMUM SPEED LIMIT (Stall) ✤LOW DENSITY (due to higher speed) ✤INCONVENIENT BASE LOCATION ✤LONGER TURNING TIME ✤CREW EXHAUSTION (due to longer flight)

Table 1 Advantage and Disadvantage on Aircrafts

5. 震災後の利活用-航路啓開調査

2011年3月に東日本を襲った震災では多くの人命が失われた。また海岸や港湾における設備の被害も大きく、その際港湾内外に流れ出た流木やデブリによって被災した方々への医療品や食料の海からの供給が困難となった。（Fig.5）同時にあらゆる物資の供給が途絶えることで、工場等へのサプライチェーンが円滑に機能しなくなることにより日本経済に大きな影響を与える結果となった。

Fig. 5 Debris in Port after Tsunami

航空レーザー測深機を利用したデブリの調査は、オランダのNPO団体であるThe Ocean CleanupによるGreat Pacific Garbage Patch（いわゆるオーシャンデブリ）における調査活動でも知られるように海外でも積極的に進められている。一方、我が国の災害時においては当該装置を利活用することで海岸や港湾内外のデブリのボリューム、位置、更に流れ等を調査することで、物資補給のルート（航路啓開）を確保することの手助けができる。

6. 河川維持管理－定期縦横断測量の代替への適用

現在我が国では、河川の維持管理が定期的に実施されており、豪雨時の河川氾濫といった甚大な災害を最小限に食い止めるべく取組みがなされている。具体的には、主に国土交通省の直轄河川では定期縦横断測量を5年周期で200mピッチ（横断測量）で実施しており、河道全体の地形及び変状をモニタリングすることにより災害につながらないための河川維持管理を行っている。

これは従前の実測作業を行っており、膨大なマンパワーと時間を要しているのが実状である。また、場合によっては人が川の中に入り測深を行っており、安全面でも課題がある。

新技術のレーザー測深機を用い計測を実施することにより、スピーディ、安全、広域を面で計測することが可能となり、現在代替が適用できるかのフィージビリティのための検証が行われており、これが実現できれば極めて画期的な新技術の導入となる。

当社では様々な地形における河川で航空レーザー測深機を用いてデータを蓄積し、過去の実測値や音響測深機で取得した値と比較した上で実現性の検証を進めている。（Fig. 6）

但し、濁度が酷い場所、また一定以上の水深では機材の特性上、計測が不可能ゆえ、その代替案についても検討を進めている。

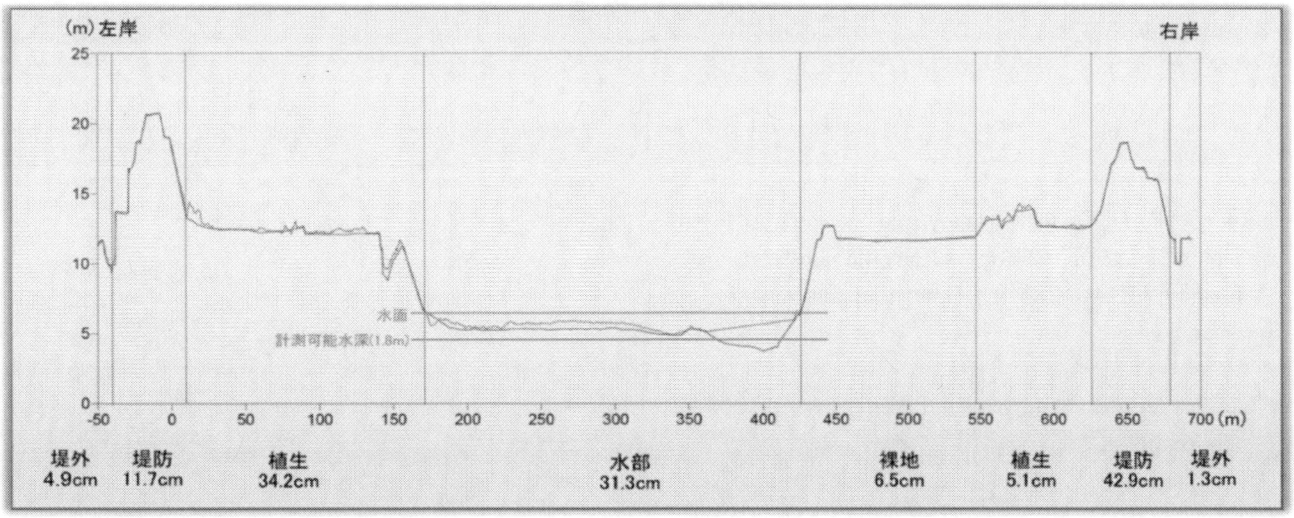

Fig. 6 Comparison between ALB data and actual data surveyed in river (at cross section)

（最下部の数値 cm は Std. Deviation）

7. 港湾・河川維持管理－ヘリ+ 無人測深機（水中点検ロボット）の活用

先術のように、ALB には弱点がある。水部の濁度が悪い場所、白波が立つ所、特定の水質（例えばフミン酸質）が多くある水域ではデータ取得が非常に困難である。そこで、弊社では無人測深機「水中点検ロボット」（Fig. 7）を開発し搭載された音響測深機により水底の地形を計測し、更に水中カメラで矢板護岸や橋脚を撮影し変状抽出を行い設備のモニタリングを行う。

運用のシナリオとして、①まず ALB で概査計測を実施し、②データ検出できなかった箇所を水中ロボを用い精査計測を実施する。音響測深機は濁度、水質の影響を受けないため、これにより水域全体の計測が可能となる。（Fig. 8）

Fig. 7 Under Water Inspector

Fig. 8 Data combined with ALB and Under Water Inspector

8. まとめ

我が国は世界に類を見ない地震大国であり、台風も多く、そして近年は地球温暖化の影響もあり不安定な気象に遭遇している。高度経済成長時に建築された多くのインフラ設備は劣化し始めており、いかにマンパワーをかけずに安全にスピーディにそれらを維持管理することが急務である。今回導入した「ヘリ ALB」の新技術は、これまで見えなかった水部の形状を把握しインフラ維持管理の要求に応えることが可能である。今後はデータの蓄積を積極的に行い、検証を進めることで少しでも国民の生命と財産を守ることに寄与できることが期待される。

謝　辞

ALB 実証実験において、テストフィールドをご提供下さいました国土交通省各地方整備局並びに河川事務所のご関係者の皆様、また、さまざまなご助言をいただきましたライカジオシステムズ様にはこの場をお借りし御礼申し上げます。更に、ご協力をいただきました弊社空間情報事業本部へリレーザープロジェクト関係者、同航空事業本部の皆様に記して御礼申し上げます。

文　献

LaRocque, P., G. Tuell, T. Craney, and J.Y. Park (2011) A Status Update of the Coastal Zone Mapping and Imaging LIDAR(CZMIL), Optech Inc. Presentation Document

Viktor Feygels (2016) 46 years in Laser Bathymetry: Myths and Reality

Koji Kawai (2015) Overview of the new LIDAR “CZMIL”, Report of Hydrographic and Oceanographic Researches No.52 March , 2015

Airborne Hydrography AB (2014) User Manual Chiroptera II Operation Manual

Design light weight emergency cot with enhanced spinal immobilization capability

Musaddiq Al ALi 1	Muazez Al ALi 2	Amjad Sahib 3
Mr.	Dr.	Mr.
Hiroshima University	University of Baghdad	University of Wasit
Higashihiroshima, Japan	Baghdad, Iraq	Wasit, Iraq

ABSTRACT

Spinal Injuries are common for traffic ascendants. Securing spinal position and eliminate unnecessary moving is vital during emergency transport. The purpose of this paper is to introduce enhance the design for a cot that handles spinal position securely, and with lowest cot weight as possible. This will be done by design and optimization composite based cot. The main objective is secure displacement of the Cot due to the static load, i.e. patient weight, and dynamically induced displacement due to transportation vibration. The system consists of two parts, first, a mobile cot to transfer the patient to and from the ambulance helicopter, and the second part is the shelf like support. to maximize strength to weight ratio with minimizing cost, hybrid composite materials (carbon fiber - epoxy and glass fiber -epoxy) are used for the mobile cot. Genetic algorithm gave preliminary design, i.e. the stacking sequence and ply angle. Displacement is less than 2 mm. Shape optimization extremum thickness of ply for shifting the range of natural modes from working vibration range. Topology optimization is used to designed stiffener. The processes shifted the natural mode with 35%. To secure the patient. The polymeric cellular pillow is designed using inverse homogenization. Aluminum is used as the material design for fixed bed. The supporting parts designed using two topology optimizations strategy (conformal lattice structure CLS, and free topology optimization) targeting von mises stress minimization. free topology optimization shows better results 38% compared to CLS topology optimization. The main bed structure is stiffened with topology optimization based stiffener design. It showed 59% improvement in vibrational characteristics.

NOTATION

C compliance (Pa^{-1})

d the distance of neighbor elements to the central one

E modulus of elasticity (Pa)

F force (N)

K finite element Stiffness matrix (N/m)

mf mass fraction

N number of elements

r checkboard filter radius

S stiffness matrix (Pa)

t linkage criterion

u element displacement (m)

V domain volume (m^3)

V_d volume fraction

W plate lateral displacement (m)

x element density

Y local coordinates

ε strain field

σ stress field (Pa)

ω frequency (Hz)

λ adjoint operator

Γ Global domain

Ψ objective function

Ω local domain

INTRODUCTION

Spinal injuries are common for automotive accidents and falling incidents. In certain cases, such as remote areas. patient transportations can be only done by using helicopter ambulance. As an aerial vehicle, optimum weight handling is a vital issue. In order to improve weight capacity, the need for designing light emergency cot is needed. In other hands, spinal injuries have certain specifications to be met, otherwise, it will be fatal for the patient. First, spin should be secured both statically and dynamically. i.e., body inertia reaction with acceleration should be eliminated. Cot lateral and normal displacement should be limited as much as possible. In the case of dynamic and vibrational induced motion, a cot should withstand deformation, and the occurrence of cot natural frequencies should be shifted beyond working vibration spectrum. Secondary considerations should also be taken into consideration, such as handling in and out of the ambulance space, and optimal space handling of possible attached specialized emergency equipment. The design is the activity in which engineers accomplish the preceding task. The designer task is to create set of specification for making or manufacturing. Optimization is a technique for improving or increasing the value of some numerical quantity that improves or satisfy certain conditions. The current work is an attempt to introduce design using shape, and topology optimization associated with a genetic algorithm, as multistage design for a specialized cot to meet spinal injury requirements. Targeted materials are hybrid fiber composites (carbon fiber-epoxy, and glass fiber -epoxy), ABS polymer, and Aluminum. Fiber-reinforced composite materials are continuing to replace the conventional metals in primary and secondary aerospace structural elements owing to their good mechanical properties to weight ratio. Material selection is based on the availability, ease of manufacturability, and light weight. Getting optimal design is being studied intensively since last century. To achieve feasible mechanical design for the certain application, practically there are several conflicting aspects to be considered. For example, global minimum compliance of a design domain is not associated with minimum stress field distribution. To locate a solution for which, the design variables and constraint, satisfy the design criterion in the best way of minimum triage of all needed or desired aspects; multi objective optimization gives such opportunity. Heuristic and meta heuristic methods were intensively used. Metaheuristic methods, such as genetic algorithm (GA) and particle swarm optimization try to mimic biological activities. They prove their competitiveness with another optimality criterion. Yet, they lose their edge with increasing design variables. The direct use of GA (Ref. 1) is questionable for topology optimization such that, practically, topography to be optimized are complicated in nature, making finding a valid design with the orthodox method a challenging task. Multi-objective optimization (Ref. 2) in general form can be GA optimization target. The genetic algorithm has been used in design composite structures. Riche and Haftka studied stacking sequence optimization of the thin composite plate (Ref. 3). Liu et al studied stacking sequences of predetermining fiber orientations for maximizing buckling load (Ref. 4). Park et al, used Finite element analysis as a numerical solution to address stacking sequence and orientation considering moderate thick plates (considering first order shear deformation theory), optimizing the plate using a genetic algorithm (Ref. 5). Farshi and Herasati implemented Tsai-Hill criterion, to develop optimality criterion based optimization technique for layer thickness and orientation (Ref. 6). Topology optimization has been developed rapidly in last decades, and still a considerable attractive topic to be addressed due to free computer design. It based on based auto design in order to find the optimal shape of the designed part based on objective subjected to the constraint. Topology optimization (TO) be designed method based on a mathematical scheme to introduce non-existence material within design space to get designer properties such as light weight or certain mechanical properties. Topology optimization was been developed from Maxwell dilemma of optimal weight truss design, by Michell (Ref. 7). Himp (Ref. 8) (1) studied strain field by adopting slip line, which at that time was regarded as an important aspect of plastic deformation implementations. Drucker et al (Ref. 9) applied constant dissipation per unit volume as their study to stress strain fields and strain energy. Chan (Ref. 10) study the optimization of static stability of truss structure by developing a technique to determine topographic based strain filed. Charrett and Rozvany (Ref. 11) adopted Prager – shield implementation in order to find optimal design criterion considering rigid-perfectly plastic systems under multiple loading. Rozvany and Prager (Ref.12) studied optimal design of grillage like continua. Their approach was spatial distribution within confined grillage units. Rossow and Taylor (Ref.13) used finite element method as a numerical solution to find the optimum thickness of variable thickness sheets. Potential energy for the elastic sheet in plane stress assumption was addressed. By introducing holes into plate structure, this work founded shape optimization. Cheng and Olhoff (Ref.14) implement finite element method as a numerical solution to optimize the thickness of annual plate with stiffened like approach. Bendsoe (Ref.15) studied Solid Isotropic Material with Penalization (SIMP) based on Taylors work.

GENETIC ALGORITHM OPTIMIZATION

Optimization is a field of applied mathematics consisting of a collection of principles and methods used for the solution of quantitative problems in many disciplines, including physics, biology, engineering, economics, and business, evolutionary methods, and control. The methods that used in this research are divided into two main types, first, approximation methods, and that achieved by using linear programming with random equations sets, and the

---

Presented at the 6th Asian/Austrailian Rotorcraft Forum / Heli Japan 2017, Kanazawa, Japan, November 7-9, 2017.

second type represents in Knowledge-based information systems or Evolutionary computing algorithms are designed to mimic the performance of biological systems, and that the Genetic Algorithms. the genetic algorithm (GA) use techniques derived from biology and rely on the application of Darwin's principle of survival of the fittest. When a population of biological creatures is allowed to evolve generations, individual characteristics that are useful for survival tend to be passed on to future generations, because individuals carrying them get more chances to breed. In biological populations, these characteristics are stored in what called Chromosomal Strings. The mechanics of natural genetics is based on operations that result in the structured yet randomized exchange of genetic information between the chromosomal strings of the reproduction parents and consists of Reproduction, Crossover, occasional Mutation and Inversion of the chromosomal strings.

SOLID ISOTROPIC MATERIAL WITH PENALIZATION

Solid Isotropic Material with Penalization (SIMP) is a scheme apply for discretized design domain to find smooth optimal structure.in which, material properties set to be constant for the discretized domain, however, the existence of building block which is set in what so-called density function composed of material existence ρ of conditioned power multiply by mechanical properties.

Figure 1. Topology optimization Problem

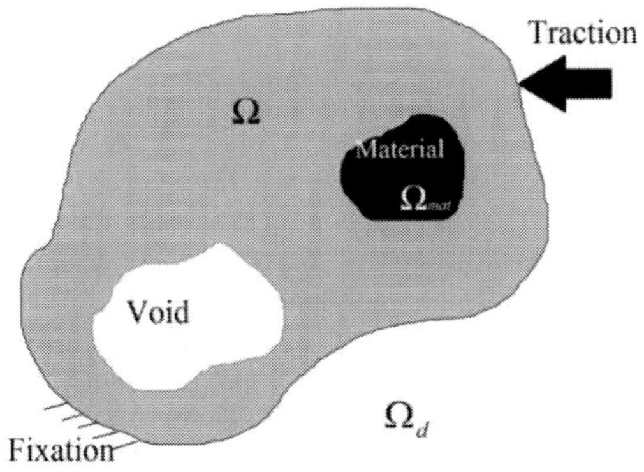

$$\lim_{volf.\to 0} x\{\Gamma(\Omega \cap \Omega_{mat} = 0 \ )\} \qquad \ldots 1$$

$$Extermum., \Psi(\rho)\,|\,E_{\min\to 0} + x^q(E_{\min\to 0} + E_{mat\to 1})$$

$$s.t. \quad \int_{\Omega_d} \rho dx \le V_d\,,\ 0 < x_{\min} < x_e < 1 \quad \forall x \in \Omega_d$$

Where $\Psi(\rho)$ is objective function,

Power (q)that satisfy condition of two dimensions (Ref.16) set to be within

$$q \in \left]15\frac{1-v}{7-5v}, \frac{3-3v}{2+4v}\right] \qquad \ldots 2$$

The solution in the scope of current scheme face some challenges as checkerboards and Nonexistence. The latter is Mesh dependency problem so Nonexistence tends to introduce non-existence element, severely, to satisfy solution with decreasing value of objective function. To solve it, relaxation principal is introduced, modifying density function with what so-called gray region (Ref.17) ($0 < \rho < 1$). As well as gray area, heuristic searching with sufficient constraint will be a remedy. Checkerboards is also a byproduct of discretization. The solution in it tends to be non-applicable due to mesh dependency. One of an effective solution uses high order elements, however, computation resources and time will easily be consumed under pressure of real life complexity of design domain. Another solution is filtered independence filtering process. One of which, Sigmund (Ref.18) was introduced

$$\frac{\partial f'(\rho)}{\partial x_e} = \frac{1}{x_e \sum_{n=1}^{N}(r_{\min} - d(e,n))} \sum_{n=1}^{N}(r_{\min} - d(e,r))x_n \frac{\partial f(\rho)}{\partial x_e} \qquad \ldots 3$$

Where $d_{(e,n)}$ is distance between central element x_e and each neighbor elements $x_{(n=1 \to N)}$. The produced modified sensitivity will update the designated Optimally criteria.

HOMOGENIZATION

Homogenization is a mathematical technique used intensively in the last decades to calculate accurate effective properties of a composite structure. The Early shade of such capability might be found in Hill (Ref.19). In his work, he adopted energy approach. Sanchez et al (Ref.20) followed by Bensoussan et al (Ref.21), successfully established the applicable fundamentals to use homogenization for composite materials. Within topology optimization topic, Bendson and Kikuch (Ref.22) studied successfully the use of homogenization method as topology optimization scheme. Finite element method played a vital role in such development. The homogenization method is based on study local unit cell such as Here the use of periodic base cells (PBCs) to be generalized to the structure (Ref. 23, and 24). Such generalization force to adopt periodic boundary condition to ease the calculation of field within the structure. The microscopic effective tensor of properties can be presented periodically as(Ref.25)

$$E^H_{ijkl} = \frac{1}{|V|}\int E_{pqrs}(\varepsilon^{0(ij)}_{pq} - \varepsilon^{ij}_{pq})(\varepsilon^{0(kl)}_{rs} - \varepsilon^{kl}_{rs})dV \qquad \ldots 4$$

Here, V is unit cell volume. E_{pqrs}, is the cell property tensor. $\varepsilon^{0(ij)}_{pq}$, is the external known field to be applied. ε^{ij}_{pq}, is the varying associated field of finding. Representative Volumetric Element (RVE), is increasingly used, associated with homogenization to calculate effective mechanical properties of composites (Ref. 26). RVE is the minimum unit cell in which static fields are sufficiently representative. The word "sufficient" means mathematically

$$\sum_{e=1}^{N}\frac{\left(\sigma^{*}-\langle\sigma\rangle\right)^{2}}{\langle\sigma\rangle} \quad \ldots 5$$

Which is the Chi square (χ^2) test. Volumetric unit calculations can be ease according to a specific application such as for fiber reinforced composites.

Here σ^*, and $\langle\sigma\rangle$ are the normalized average value of the stress in the current unit cell and an average of it respectively. The topology optimization of the microstructure can be performed in terms of minimization of potential energy.

$$Extermum\sum_{ij=1}^{2} r_{ij}(\overline{E}_{ijkl}^{H})|\ E_{ijkl}^{H}=\frac{1}{|Y|}\sum_{e=1}^{N}(u_{e}^{ij})^{T}K_{e}u_{e}^{kl}$$

$$Y=]0,Y_1^o[\times]0,Y_2^o[\times]0,Y_3^0[, \quad ..6$$

$$s.t.\ \int_{\Omega}\rho^{e}d\Omega=V_{d}.$$

Here, r_{ij} denotes to waiting for a factor to balance the objective function. $\overline{E}_{ijkl}^{H}$, is the effective properties formulation according to bounding formulas such as Wiener bounds, and Hashin-Strikman bonds (Ref. 27-29).

CONFORMAL LATTICE STRUCTURES OPTIMIZATION

Additive manufacturing of polymeric based materials gives theoretically, thermal stress-free structures, with good surface finish. this will ease manufacturing process and extra-design consideration for overcoming such problems, especially with complicated shapes such as inverse homogenization material design. However, here is an impressive advance in metallic additive manufacturing; still, surface finish within complicated lattice is challenging. Truss like structure can be a solution. The capability to cleaning the structure will increase the expected fatigue life. Truss like lattice structure can be formulated with using ground structure method. It states that the structure can be optimally built by adding or vanishing chosen links linking between an appropriate fixed set of nodal points. Such as in Conformal Lattice Structures (CLS) (Ref. 30).

Figure 2. conformal lattice structure problem scheme

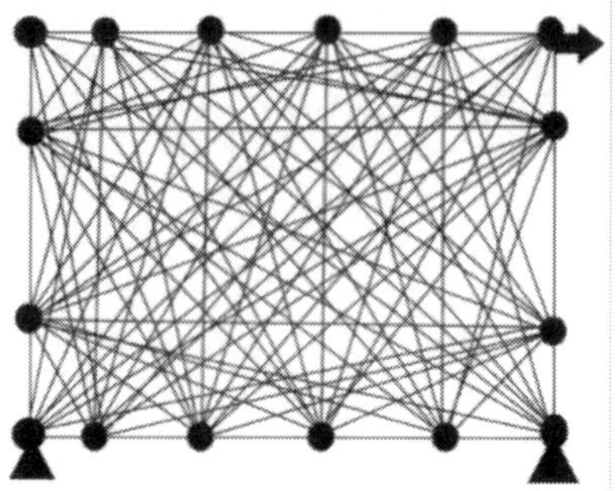

The basic optimization scheme is

$$\max_{t>0}\ \min_{u},\frac{1}{2}u^{T}(\sum_{e=1}^{N}t_{e}K_{e})u \quad \ldots 7$$

$$s.t.\ \sum_{i=1}^{m}t_{i}=V$$

SENSITIVITY ANALYSIS

Topology optimization in terms of sensitivity analysis approaches needs derivative to apply an iterative Lagrangian multiplier. As Heuristic and meta-heuristic methods are used to extremum functions for the field of optimization, heuristic methods, as Lagrangian is implemented in many topology optimization types of research, showing its effectiveness and simplicity to apply and control. Adding to that, metaheuristics showed in some techniques weakness toward increasing design variables number. Some research's value advancing of quantum advancing in quantum-based searching approaches and advancing in metaheuristic hyperdilation, but in the scope of near future, heuristic searching proved to be as a keystone in global optimization of many variables as topology optimization. Global search algorithms need sensitivity analysis. In such case, derivative might contain certain difficulties, especially, the derivative should be in term of design variables. First order sensitivity analysis is required be perform each iteration. The adjoint variable method is used to develop a unified formulation for representing response variation in term of variation design. Aggregation objective function which subjected to a constraint with respect to design variable x. finite element based sensitivity analysis using adjoint method gives

$$\frac{d\Psi}{dx_e}=\lambda^{T}(\frac{dF}{dx_e}-\frac{dK}{dx_e})+\frac{\partial\Psi}{\partial x_e} \quad \ldots 8$$

Here, Ψ is a density related function to be derivative.

DESIGN PROCESS

The design process is divided into three major steps. Steps are categorized according to part to be designed. The first step is designing the light weight bed, which should be mobile to carry the patient from and to the aerial ambulance. The second step is designing the fixture metallic base, the third step is designing auxiliary parts such as the securing techniques for patient and cot itself and evaluate the completed parts.

1- **Step 1 Designing the light weight bed (Composite structure)**

Uni-directional carbon fiber – epoxy system will be the main body of cot that directly withstands the patient weight. It should be satisfying the following purposes: its normal displacement should as minimum as possible, add to that, the natural frequency should be controlled to avoid undesired noise during unloading transportation and to avoid interlaminar failure due to the harmonic motion. In general, the design is being considered as a thin plate, subjected to a normal load. Preliminary design of static worst-case scenario is being adopted to design ply orientations. Worst case scenario was chosen to be point load on the simply supported plate.

$$\min.\ W_{mn} \mid S^{-1}F \qquad \text{s.t.}\ \theta \in [0,180] \qquad \ldots 9$$

Where W_{mn} is normal displacement of the plate which comes from the following?

$$\begin{bmatrix} \hat{S}_{11} & \hat{S}_{12} & 0 & \hat{S}_{14} & \hat{S}_{15} \\ \hat{S}_{12} & \hat{S}_{22} & 0 & \hat{S}_{24} & \hat{S}_{25} \\ 0 & 0 & \hat{S}_{33} & \hat{S}_{34} & \hat{S}_{35} \\ \hat{S}_{14} & \hat{S}_{24} & \hat{S}_{34} & \hat{S}_{44} & \hat{S}_{45} \\ \hat{S}_{15} & \hat{S}_{25} & \hat{S}_{35} & \hat{S}_{45} & \hat{S}_{55} \end{bmatrix} \begin{bmatrix} U_{mn} \\ V_{mn} \\ W_{mn} \\ X_{mn} \\ Y_{mn} \end{bmatrix} = \begin{bmatrix} 0 \\ 0 \\ Q_{mn} \\ 0 \\ 0 \end{bmatrix} \qquad \ldots 10$$

Real coding GA is used to identify ply angles. An analytical solution is being adopted, using Navier's solution. Topology optimization is used for the plate that results from GA step in order to optimize ply thickness for maximizing natural frequency, and minimizing plate displacement, with decreasing weight as much as possible. as a multi objective optimization problem, it is computational power consumer, add to the that the difficulties of optimality criterion. For the case of weight and frequency, the required range can be determined, so minimizing displacement with setting bounds for the other desired characteristics is suffice. Optimization scheme can be summarized in eq

$$\text{find} \quad x \rightarrow x^{q} \mid q = 3$$

$$\min.,\ W_{mn} \mid E_{mat\rightarrow 1} + x^{p}(E_{\min\rightarrow 0} + E_{mat\rightarrow 1})$$

$$\text{s.t.} \begin{cases} \int_{\Omega_d} \rho dx \leq mf_d\ ,\ 0 < x_{\min} < x_e < 1 \quad \forall x \in \Omega_d \\ \omega . \in]\omega^{u}, \omega^{l}] \end{cases} \qquad \ldots 11$$

to ensure the structural rigidity of the plate, stiffeners are used. The optimal location of stiffeners and stiffener topography itself can be deduced using topology optimization. Finite element based modal analysis was performed to identify

2- **Step 2 Designing the fixed metallic support**

To limit the patient mobility and secure it within ambulance space, a fixed metallic bed is being introduced. Two important characteristics are being targeted, modal analysis shifting to the safe operating vibrational range and reducing weight with minimizing stress concentration in order to maximize, the structure lifespan.

3- **Step 3 Auxiliary systems spongy back support**

The basic shape of the spine is a combination of curves, vary in gradually in thickness and radius. To give more support to the spine in transportation phase, a spongy back sport is designed to withstand back weight and adopted with its shape. The design objective is to minimize weight, with maximizing bulk modulus. To this goal, inverse homogenization based topology optimization technique is used. Homogenization technique is used for calculating effective properties in a wide spectrum of physics application, such as astrophysics, and mechanical engineering. This work uses the implementation of homogenization in composite materials calculations.

APPLIED DESIGN PROCESS

Materials in use

The mechanical characteristics of composite is summarizing in Table 1 and 2

Table 1. Mechanical properties and cost of carbon and glass fiber - epoxy system

Orthotropic Properties	Glass Fibers	Carbon Fibers
Ex	22.3e9 Pa	70.4e9 Pa
Ey	6.8e9 Pa	10.3e9 Pa
Ez	6.8e9 Pa	10.3e9 Pa
υ12	.35	.3
υ23	.4	.29
υ13	.35	.28
G12	2.8e9 Pa	7.17e9 Pa
G23	2.72e9 Pa	7.17e9 Pa
G13	2.8e9 Pa	10.1e9 Pa
Density	2.5e3 Kg/m³	1.6e3 Kg/m³
cost	10 $	105 $

Table 2. Mechanical properties of secondary systems materials

Isotropic Properties	Aluminum	ABS
E	70e9 Pa	44e9 Pa
υ	.3	.34
Density	2.8e3 Kg/m³	1.04e3 Kg/m³

Design mobile cot part

The design problem is consisting of three parts, mobile cot made of a composite plate of dimensions (2 by .6 m). it consists of 8 plies of thickness 2 mm for ply. Ply has the possibility to be a unidirectional carbon fiber epoxy system or unidirectional glass fiber epoxy system. Orientation angle is varying between 0 and π angle. The cost of manufacturing ply of glass fiber composite is 10 $ while it is more than 10 times that for carbon fiber epoxy composite **(Table 1)**. A genetic algorithm was used to give the preliminary design of the mobile cot. the preliminary design boundary conditions are simply supported plate with the eccentric concentrated load.

Figure 3. Initial design problem of the mobile hybrid composite cot.

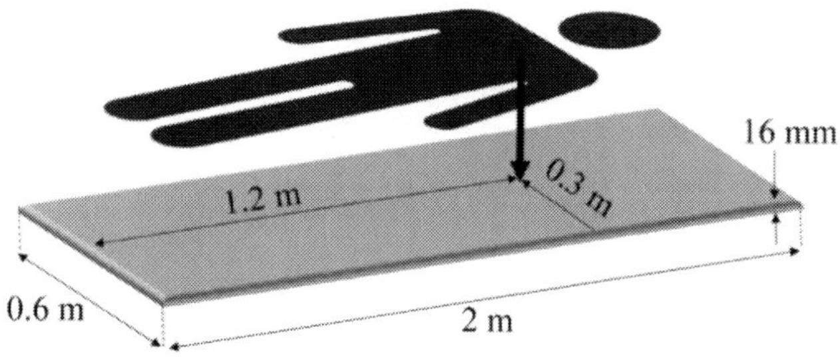

Weighted sum function is used as Multi-objective optimization function. Selection of weights for is vital for global optimality to be satisfied. GA is adopted as global optimization solver to select the best weighting for sum multi-objective function. It can be selected using an evolutionary algorithm with pre-determined sets by initialization of GA. The real coding genetic algorithm was used. GA is used successfully for the more complex problem, yet reasonable constraint numbers. Designing scheme is

$$\min, w_1(\Im) + w_2(\Re) + w_3(\aleph)$$
$$s.t. \quad ply \in [1,2] \qquad \dots 13$$
$$\theta \in]0, \pi[$$

Here, $\Im$, is the weight of the designed plates, $\Re$, is the price of the design plates, and $\aleph$, is the maximum lateral displacement of the plates. Genetic algorithm designed plate is shown in Table 3.

Table 3. Generated optimal Preliminary mobile cot design using GA

Ply number	Composite system	Angle of orientation (degree)
1	Carbon epoxy	161
2	Glass epoxy	22
3	Glass epoxy	40
4	Carbon epoxy	129
5	Carbon epoxy	117
6	Glass epoxy	153
7	Carbon epoxy	23
8	Carbon epoxy	173

A model using finite element method is built to using ANSYS to identify natural frequencies within working vibration range. Suitable working vibration range has been adopted for between the range [0, 250] (Ref. 31-33). It shows the range of natural modes as in Table

Table 3. Modal analysis of GA design plate.

Mode	Frequency (Hz)
1	132.32
2	172.55
3	238.49

Shape optimization was used to multi-stage optimization of the hybrid composite plate. The first step is reducing the weight with maximizing lateral displacement with actual body load by free resizing of ply thickness. Topology optimization performed using OptiStruct solver. The optimization scheme of ply thickness is done by minimizing overall compliance of plate, with displacement and volume fraction as a constraint.

$$find \quad h \qquad \dots 14$$

$$\min., F^T U \mid E_{mat\to 1} + x^p (E_{\min\to 0} + E_{mat\to 1})$$

$$s.t. \begin{cases} \int_{\Omega_d} \rho dx \le V_d., \ 0 < x_{\min} < x_e < 1 \quad \forall x \in \Omega_d \\ \lim W \to 0 \end{cases}$$

Objective function history is shown in figure 4

Figure 4. Shape Optimization objective function.

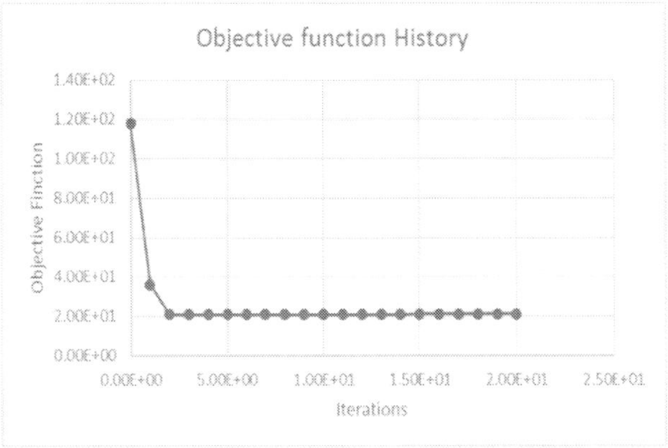

The designed ply thickness is shown in figure 5

Figure 5. Shape optimization ply design.

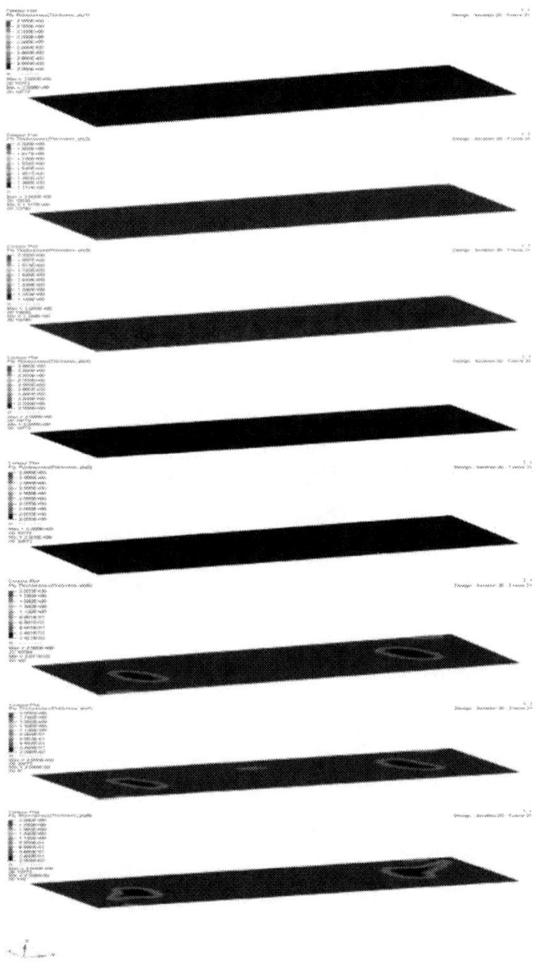

The results of modal analysis of newly generated plate are

Table 4. Modal analysis of shape optimization design plate.

Mode	Frequency (Hz)
1	144.6812
2	197.4815

Designing a stiffener combination is preferable to support the plate.

$$\min . \sqrt[2]{\left(\sum_{e=1}^{M} w_c \frac{C_e - C^{\min}}{C^{\max} - C^{\min}}\right)^2 + \left(w_\omega \frac{\omega^{\max} - \omega_e}{\omega^{\max} - \omega^{\min}}\right)^2} \quad \ldots 15$$

$$s.t.\ s.t.\quad \int_{\Omega_d} \rho dx \le V_d ,\ 0 < x_{\min} < x_e < 1 \quad \forall x \in \Omega_d$$

Here, C_e, $C^{\min}$, and $C^{\max}$ are structure compliance, lower bound, upper bound respectively. And the same for natural frequency.

Figure 6. Shape optimization based stiffener design, (a) optimization thickness results, (b) the designed stiffener based on scheme a

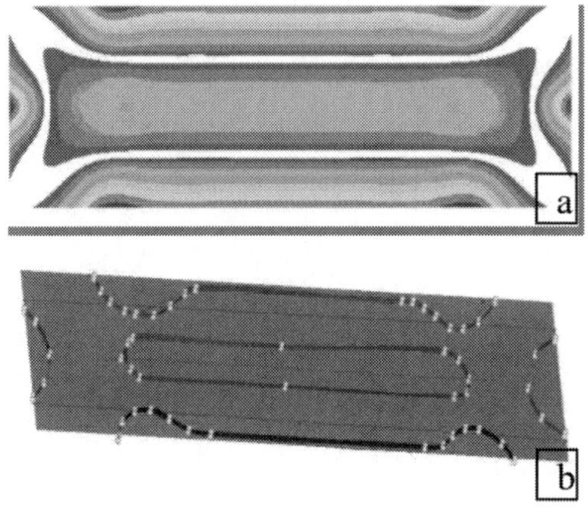

The modal analysis of final designed mobile cot is listed in table 5

Table 5. Stiffened mobile cot modal analysis

Mode	Frequency (Hz)
1	225.773
2	234.435
3	249.846

DESIGN METALLIC FIXED BED

The Aluminum bed was boundary conditions is as shown in Figure (7-a). The assumed design for which the boundary condition was taken is a plate supported by two cantilevers on the ends of the plate. Table 6 shows the modal analysis of the Aluminum bed

Table 6. Modal analysis of Aluminum Plate with block rigid support

Mode	Frequency (Hz)
1	41.28567
2	55.1153
3	114.2948
4	156.2364
5	178.224

The range of the first mode could be modified by introducing stiffener to the plate. To design optimal stiffener, design scheme in using equation 15 is been adopted. Results are shown in Table 7 (Figure 7-b).

Table 7. Modal analysis of Aluminum Plate with stiffener

Mode	Frequency (Hz)
1	100.005
2	118.7942
3	150.5186
4	150.5202

Stress base CLS and free stress based - shape topology optimization was adopted to decrease support weight with maintaining feasibility of the support structure. Stress that targeted is von mises stress. Adjoint sensitivity was used for optimality criterion Stress results and displacements promote free shape topology optimization as shown in Table (8).

Table 8. Topology optimized of support results with 0.3 volume fraction design

Mode	Displacement (mm)	Stress (N/mm^2)
CLS	9.293e-2	1.933
Free shape	5.767e-2	1.188

Figure 7. (a) designed problem, (b) stiffened metallic bed, (c) optimized support using CLS, (d) optimized support with Free shape optimization.

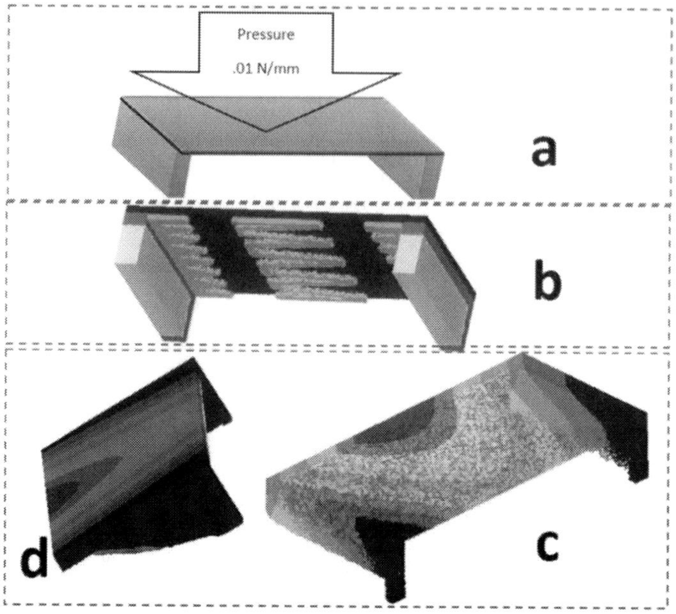

DESIGN SUPPORT PILLOW (MICRO STRUCTURE EXTREMUM WITH TOPOLOGY OPTIMIZATION)

Securing spine needs active and passive systems. Active systems are the strips and belts that secure the body in order to limit inertia action of the human body. In most cases, it is possible to take advantage of the spinal creature.

Figure (8) Targeted Spinal Curvature for pillow fixation.

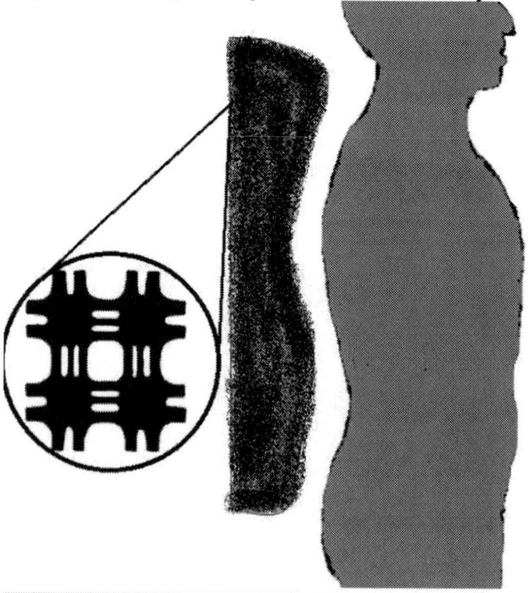

Pillow is design optimized for minimum weight and maximum bulk modulus of elasticity. Cellular composites

are being considered here as an optimization strategy. It consists of the self-repeated optimized unit cell. Hashin-Strikman bounds are used to estimate the local bulk modulus. The optimized unit cell is being shown in Figure 9

Figure 9. Unit Cell optimization

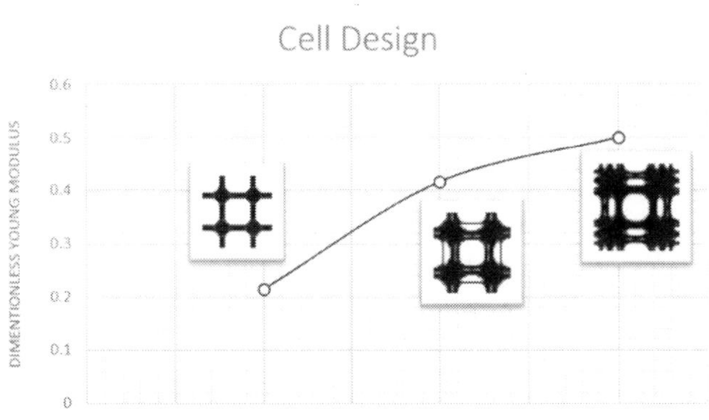

Optimization algorithm is given in equation 16

$$\sigma = C : \varepsilon$$

$$\min(C_{11} + C_{22} + C_{12})|$$

$$C^{H}_{ijkl} = \frac{1}{|Y|}\sum_{e=Y}(d_0^{e(ij)} - d^{e(ij)})^T k^e(\rho^e)(d_0^{e(kl)} - d^{e(kl)})$$

$$K(\rho^e)d^{(ij)} = f^{(ij)} \qquad \dots 16$$

Young modulus of elasticity can be found from equation (17)

$$E = \frac{(1+\upsilon)(1-2\upsilon)}{(1-\upsilon)}C_{11} \qquad \dots 17$$

CONCLUSIONS

In this paper, the superposition way of thinking is being adopted in the design process. Composite plate optimization was performed maximizing modal shape vibration. The results improved 8.5 % compared to initial, non-optimized plate. Designing stiffener improved vibrational response by 35% comparing to shape optimized ply design (Figure 10). The suggested stationary Aluminum bed will secure the patient with in the Ambulance designated space. Weight to performance ratio was approach by adopting stress based topology optimization objective function. 70% volume reduction as a constraint. Free shape topology optimization shows better results than CLS topology optimization with 38%. In order to shift the range of natural modes, stiffener design scheme was used, as same as for the mobile hybrid composite cot. The designed stiffeners show 59% improvement compare to non-stiffener one (Figure 11).

Figure 10. Free vibration analysis of mobile cot designs

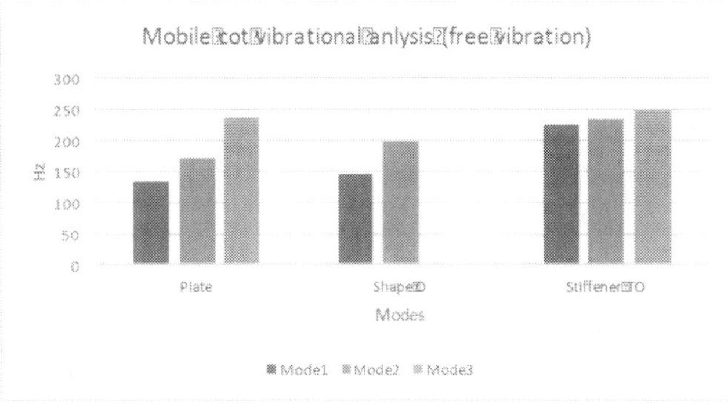

Figure 11. Free vibrational analysis of fixed bed designs.

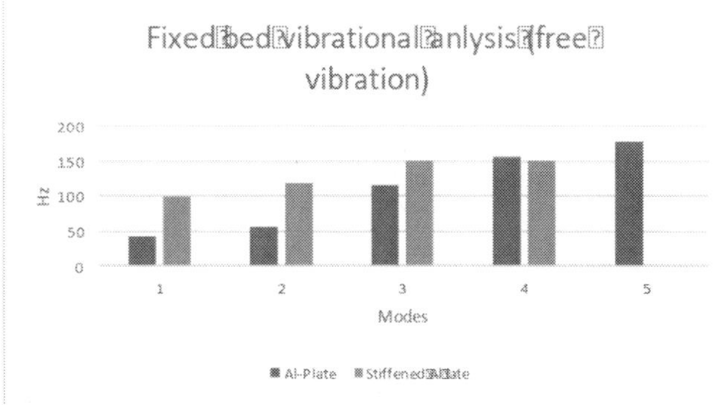

In aeronautic designs, weight reduction is required. Pillow weight reduction was achieved by adopting inverse homogenization based topology optimization. Bulk modulus was chosen to be maximized.

Author contact:

Musaddiq Al Ali 1 mosadeq007@yahoo.com
Muazez Al ALi 2 drmuazez@gmail.com
Amjad Yusif 3 3dphotonic@gmail.com

ACKNOWLEDGMENTS

The authors would like to thank Dr. Abdil Khaliq Maleh Kadem, for his valuable information and knowledge sharing in spinal injury handling subject.

REFERENCES

[1]Blanco A., Delgado M., Pegalajar M. C. , "A real-coded genetic algorithm for training recurrent neural networks. Neural networks", 2001;14(1):93-105.

[2]Marler R. T., Arora J. S., "The weighted sum method for multi-objective optimization: new insights", Structural and multidisciplinary optimization. 2010;41(6):853-62.

[3]Le Riche R., Haftka R. T., "Optimization of laminate stacking sequence for buckling load maximization by genetic algorithm", AIAA journal. 1993;31(5):951-6.

[4]Liu B., Haftka R. T., Akgün M. A., Todoroki A., "Permutation genetic algorithm for stacking sequence design of composite laminates", Computer methods in applied mechanics and engineering. 2000;186(2):357-72.

[5]Park J., Hwang J., Lee C., Hwang W.,"Stacking sequence design of composite laminates for maximum strength using genetic algorithms",Composite Structures. 2001;52(2):217-31.

[6]Farshi B., Herasati S.,"Optimum weight design of fiber composite plates in flexure based on a two level strategy", Composite structures. 2006;73(4):495-504.

[7]Michell A. G. M.,"The limits of economy of material in frame-structures", The London, Edinburgh, and Dublin Philosophical Magazine and Journal of Science. 1904;8(47):589-97.

[8]Hemp W. S.,"Theory of structural design", College of Aeronautics Cranfield; 1958.

[9]Drucker D. C., Shield R.,"Design for Minimum Weight", BROWN UNIV PROVIDENCE RI; 1956.

[10]Chan A.,"The design of Michell optimum structures", College of Aeronautics Cranfield; 1960.

[11]Charrett D., Rozvany G.,"Extensions of the Prager-Shield theory of optimal plastic design", International Journal of Non-Linear Mechanics. 1972;7(1):51-64.

[12]Rozvany G., Prager W.,"Optimal design of partially discretized grillages", Journal of the Mechanics and Physics of Solids. 1976;24(2-3):125-36.

[13]Rossow M., Taylor J.,"A finite element method for the optimal design of variable thickness sheets", Aiaa Journal. 1973;11(11):1566-9.

[14]Cheng K. T., Olhoff N.,"An investigation concerning optimal design of solid elastic plates", International Journal of Solids and Structures. 1981;17(3):305-23.

[15]Bendsøe M. P.,"Optimal shape design as a material distribution problem", Structural and multidisciplinary optimization. 1989;1(4):193-202.

[16]Bendsoe M., Sigmund O.,"Topology Optimization-Theory, Methods and Applications", Springer; 2003.

[17]Sigmund O., Petersson J.,"Numerical instabilities in topology optimization: a survey on procedures dealing with checkerboards, mesh-dependencies and local minima", Structural and Multidisciplinary Optimization. 1998;16(1):68-75.

[18]Sigmund O.,"A 99 line topology optimization code written in Matlab", Structural and multidisciplinary optimization. 2001;21(2):120-7.

[19]Hill R.,"Elastic properties of reinforced solids: some theoretical principles", Journal of the Mechanics and Physics of Solids. 1963;11(5):357-72.

[20]Sanchez-Palencia E.," Homogenization method for the study of composite media", Asymptotic analysis II. 1983;985:192-214.

[21]Bensoussan A., Lions J. L., Papanicolaou G.,"Asymptotic analysis for periodic structures", North-Holland Publishing Company Amsterdam; 1978.

[22]Bendsøe M. P., Kikuchi N.,"Generating optimal topologies in structural design using a homogenization method", Computer methods in applied mechanics and engineering. 1988;71(2):197-224.

[23]Sigmund O.,"Materials with prescribed constitutive parameters: an inverse homogenization problem", International Journal of Solids and Structures. 1994;31(17):2313-29.

[24]Cadman J. E., Zhou S., Chen Y., Li Q.,"On design of multi-functional microstructural materials", Journal of Materials Science. 2013;48(1):51-66.

[25]Andreassen E., Andreasen C. S.,"How to determine composite material properties using numerical homogenization", Computational Materials Science. 2014;83:488-95.

[26]Gitman I. M., Askes H., Sluys L. J., Lloberas O.,"The concept of representative volume for elastic, hardening and softening materials", Proceedings of the XXXII International Summer School-Conference Advance Problems in Mechanics, Saint Petersburg (Repino), Russia; 2004.

[27]Hashin Z.,"Analysis of composite materials", J appl Mech. 1983;50(2):481-505.

[28]Hashin Z., Shtrikman S.,"A variational approach to the theory of the elastic behaviour of multiphase materials", Journal of the Mechanics and Physics of Solids. 1963;11(2):127-40.

[29]Castañeda P. P., Telega J. J., Gambin B.,"Nonlinear Homogenization and its Applications to Composites, Polycrystals and Smart Materials: Proceedings of the NATO

Advanced Research Workshop", held in Warsaw, Poland, 23-26 June 2003: Springer Science & Business Media; 2006.

[30]Choi K., Kuhn J. L., Ciarelli M. J., Goldstein S. A.,"The elastic moduli of human subchondral, trabecular, and cortical bone tissue and the size-dependency of cortical bone modulus", Journal of biomechanics. 1990;23(11):1103-13.

[31]Clevenson S. A., Leatherwood J. D., Hollenbaugh D. D., "Interior noise and vibration measurements on operational military helicopters and comparisons with various ride quality criteria". 1983.

[32]Konstanzer P., Enenkl B., Aubourg P., Cranga P., editors. "Recent advances in Eurocopter's passive and active vibration control",. annual forum proceedings-american helicopter society; 2008: american helicopter society, inc.

[33]Bongers P., Hulshof C., Dljkstra L., Boshuizen H., Groenhout H., Valken E.," Back pain and exposure to whole body vibration in helicopter pilots", Ergonomics. 1990;33(8):1007-26.

Join us in making the impossible, possible.
GO FLY
PRIZE

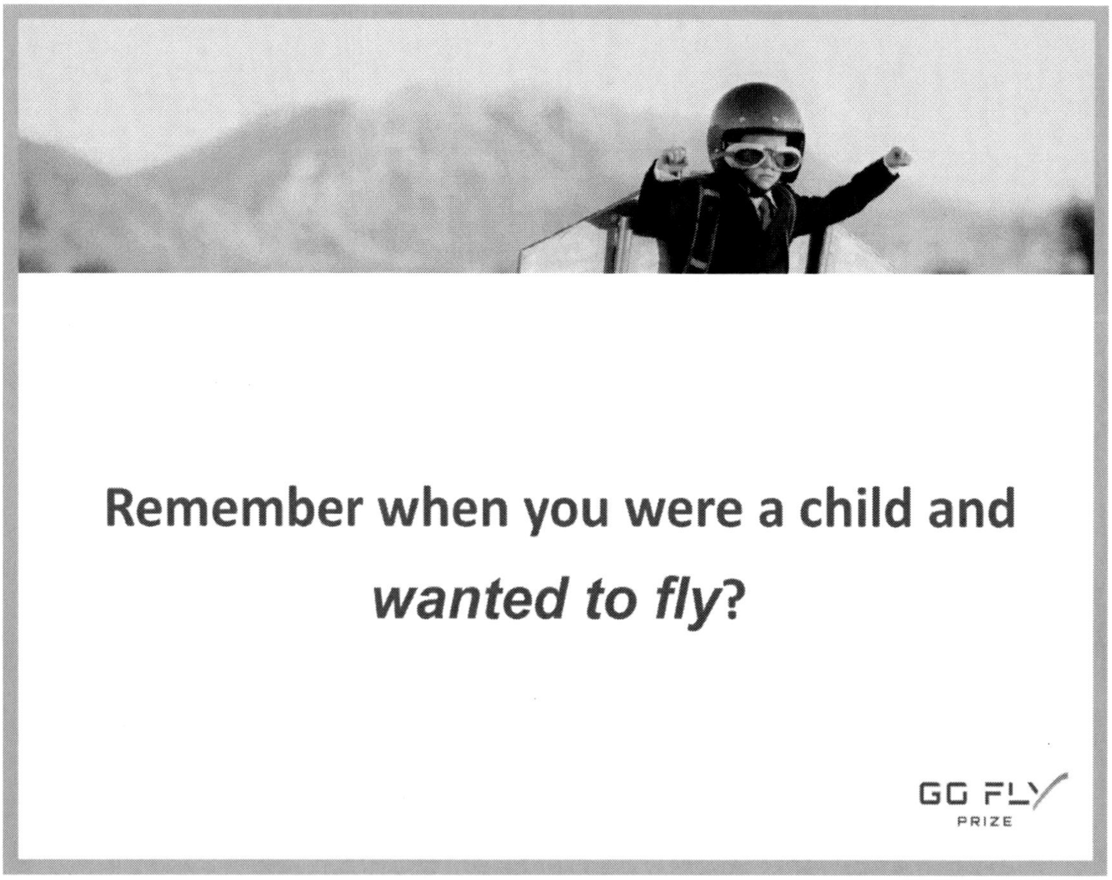
Remember when you were a child and
wanted to fly?
GO FLY
PRIZE

We are going to make that dream a reality.

Together, we are going to make humans fly.

A Special Thank You

Eric Allison
Richard Pat Anderson
Ashish Bagai
Eric Bartsch
Gregg Beeber
Paul Bevilaqua
Joe Ben Bevirt
Rob Bulaga
Carlos Cesnik
Mark Costello
John Del Frate
Mark Drela
Mike Duffy
Sean Elliot
Steve Ericson
Mark Forger
Brian German
Tony Ginn
Martin Gomez
Mike Hirschberg
Douglas Jarmusz
David Josephson
Jeff King
Ilan Kroo
Earl Lawrence
Dimitri Mavris
Mag Dean Mirmirani
Jerry Montague
Mark Moore
Dan Newman
Rich Ouellette
Mark Page
Bob Pearce
Darryll Pines
Blaine Rawdon
Stephen Rizzi
Sky Sartorius
Scott Schultz
Tony Tether
Ben Tigner
John Tracy
Mike Ward
James Wong
Chip Yates
Hal Youngren

Objective

To build a **safe, ultra-compact, quiet, socially acceptable, urban-compatible, near-VTOL** individual flying device capable of carrying one 200 lb. person a distance of 20 miles without refueling or recharging.

GG FLY PRIZE

GG FLY PRIZE

Vision: Anyone, Anywhere

CHALLENGE LANGUAGE:

"What the device looks like or how it works is up to you. We welcome revolutionary design. The device should function safely in both crowded cities and rural areas; it should be lightweight enough and maneuverable enough so that anyone can move it around, and it should be quiet not only for the user, but also for the general public. We are propulsion agnostic, but like all great inventions the device should be user-friendly, almost an extension of the user's body, and provide the thrill of flight."

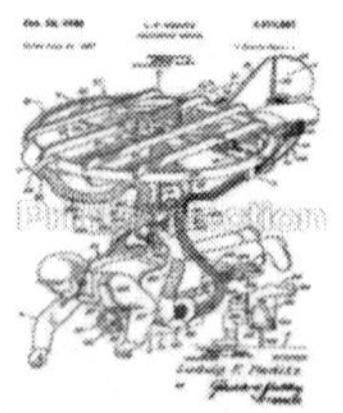

The GoFly Ecosystem

	HUB TO HUB	HUB TO HUB	DOOR TO DOOR
SKY	BOEING 747 	UBER AIR SHUTTLE 	GOFLY GO FLY PRIZE
LAND	BUS 	CARPOOL VANS/SHUTTLES 	MOPED, MOTORCYCLE, BIKE SCOOTER, SEGWAY, SKATEBOARD 

Demographic Basis for Vision

- Urbanization
- Failing Infrastructure
- Commuting Statistics

Urban Migration 1950-2050

http://www.unicef.org/sowc2012/urbanmap/

Failing Infrastructure

- American Society of Civil Engineers estimates the US needs to spend $2 Trillion over the next five years to make necessary infrastructure repairs--- and that is just repairs, not modernization.
- 25% of US bridges are structurally deficient or functionally obsolete.
- By 2023, 1 in 4 bridges, which are typically designed to serve for 50 years, will be more than 65 years old.

Commuting 85% of world travels *20 miles* or less to get to work

Worldwide we spend a year of our lives commuting.

Region	Vehicles on Road in millions	Number of Commuters in millions	Average Commute Time in minutes	Commute Distance in miles	% of Car Commuters
USA	253	128.3	25.4	25	75.7
UK	35	11.3	54	9.3	60

Region	Vehicles on Road (mil)	Average Commute Time	% of Car Commuters
Germany	53.6	21	58
China	240	39	15
Japan	75	45	24
India	72.7	29	67
Brazil	80	33	20*
Russia	48	52	40
Worldwide	**1000**	**37**	**64**

Now is the Time

Convergence of 4 Breakthrough Technologies:

1. **Stability and Control Systems**:
   - Commercial and consumer driven access to autopilots and control systems make otherwise unflyable concepts easy to operate.
2. **Electrics:**
   - Advances in electric components (batteries, capacitors, motors, controllers) open the door to new hybrid and all-electric concepts.
3. **Materials and structures innovations**
4. **Advanced manufacturing**
   - Offers quick and low-cost prototyping.

Why People Participate?

1. The Glory
2. The Prize Money
3. The Post-Prize Opportunities

Opens up a whole new world to winners and creates a new future.

Prizes Create the Future

Orteig Prize ushers in the age of Airmail and the Commercial Airline Industry

BLUE ORIGIN

GALACTIC

SPACEX

Ansari XPRIZE (space) creates the Commercial Space Industry. (Virgin bought the winner)

DARPA GRAND CHALLENGE

DARPA Grand Challenge creates Autonomous Cars. (Google bought the winner)

Strategy:

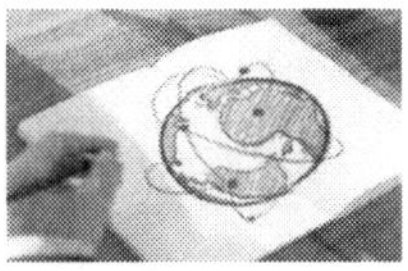

GLOBAL INCENTIVE COMPETITION

The GoFly Prize is a $2M, two year incentive challenge to create personal flying devices, providing global exposure and large financial rewards.

ACCELERATORS INCUBATORS

INNOVATION INCUBATOR MODEL

GoFly empowers innovator teams, giving them access to experienced mentors in design, engineering, finance, law, branding, and marketing.

STEM

SCIENCE, TECHNOLOGY, ENGINEERING & MATHEMATICS

SCHOOL STEM OUTREACH

GoFly offers high school and middle school students the opportunity to connect STEM to the real world by participating in our programs.

The Incubator - Mentor Program

Technical Mentors

- Renowned experts from leading aerospace corporations and universities provide guidance in areas of VTOL, propulsion, noise mitigation, electrics, and others.

Funding Mentors

- GoFly Network of financial professionals work with teams to introduce them to funding sources.
- Services may include business plan development, strategic consulting, presentation skills.

Legal Mentors

- Mentors may assist with IP filing, corporate formation documents, insurance procurement.

Branding & Marketing Mentors

- Experts help with messaging and pitch books.

In-Kind Product Acquisition

- Procurement of corporate in-kind product for all teams.

Challenge Details

Time Frame:	24 Month Competition: Completion Summer 2019
3 Stages:	Stage 1: Technical Specs with ten $20k prizes Stage 2: Prototype with four $50k prizes Stage 3: Fly-Off with one $1.5M prize Innovation Prize: $100,000
Innovators:	4 Types: Universities, Backyard Engineers, Small and Medium Sized Companies

Attributes of a Good Rules Package

- State everything once and only once.
- No requirement can be ignored and the entry remain viable.
- No strong incentives for unsafe design or operation.
- Nothing that unfairly favors teams with abnormally skilled or brave pilots.
- Nothing existing can achieve a prize-eligible score.
- Design space not overly restricted by requirements that have traditionally burdened aerospace.

Judging Criteria

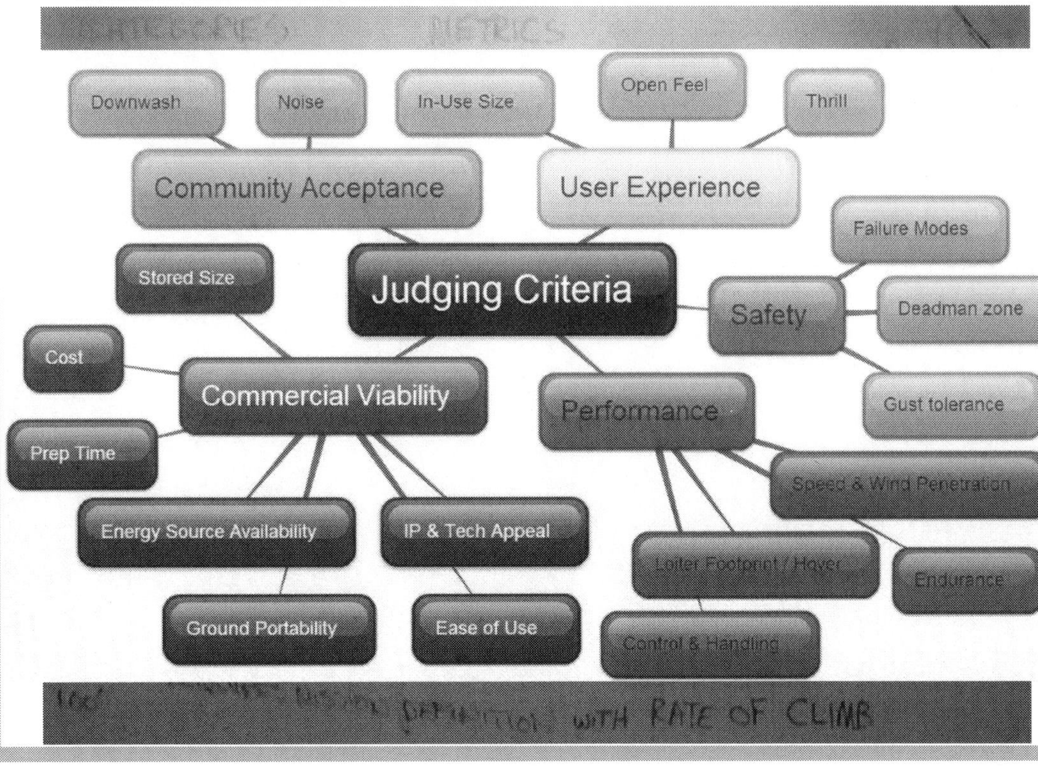

The Journey to the Rules

The Process:
Close to 100 SME interviews.
15+ Drafts of Rules and Guidelines.
Over 50 private one-on-one technical expert commentaries.
Multiple full day wire-brushing sessions. *Listening.*
Revisions, comments, revisions, comments, revisions, comments.
Thank you, thank you, thank you.

Obstacles we have overcome (and some we are still dealing with):

- Fighting against other visions and the efficiency debate.
- Understanding and capturing noise.
- Incentivizing safety without "owning" it.
- Allowing unmanned; equating manned and unmanned for scoring.
- Regulatory hurdles.

And the Winner Is...

FINAL SCORE = $\theta_{Size} + \theta_{Noise} + \theta_{Speed}$

Fly-Off Tasks & Scoring Measurements

Size Measurement → Scoring Parameter θ_{Size}

Threshold: 8.5 ft diameter *(to be finalized)*

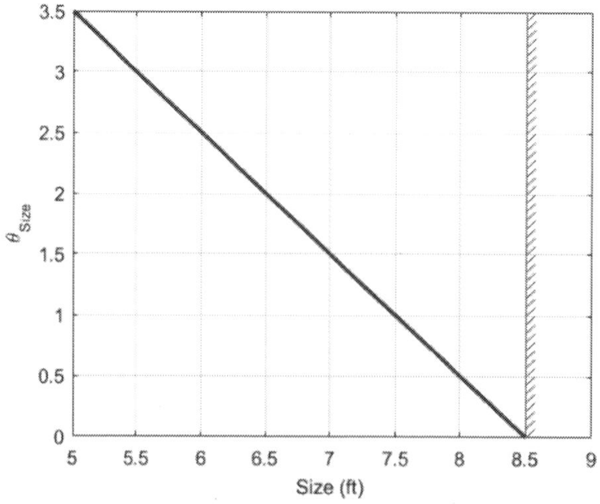

Fly-Off Tasks & Scoring Measurements

Noise Measurement → Scoring Parameter θ_{Noise}

Measured 50' from envelope center for takeoff and landing

Threshold: 87 dBA

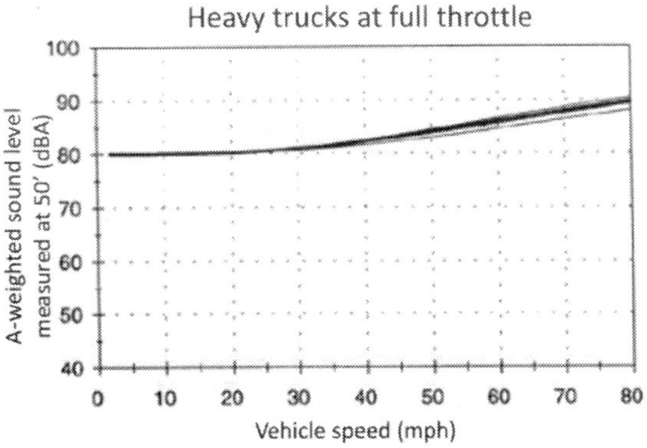

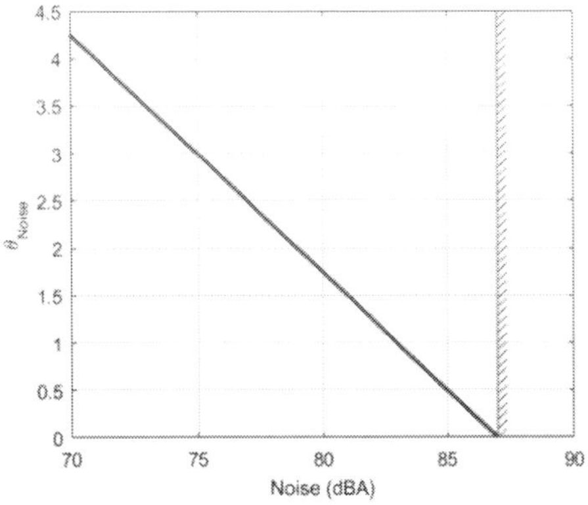

Fly-Off Tasks & Scoring Measurements

Speed Measurement → Scoring Parameter θ_{Speed}

Threshold: 30 kts average

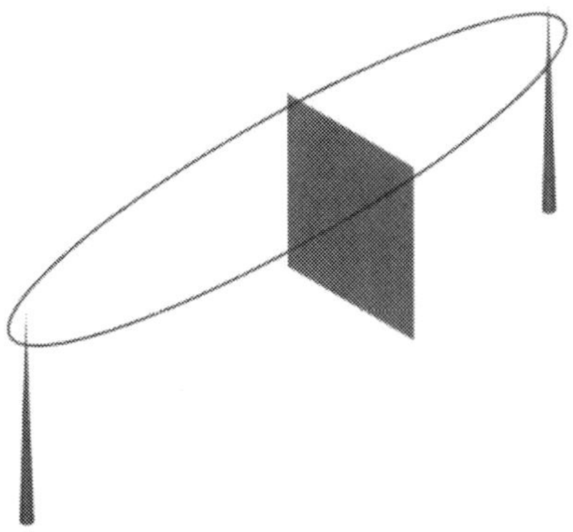

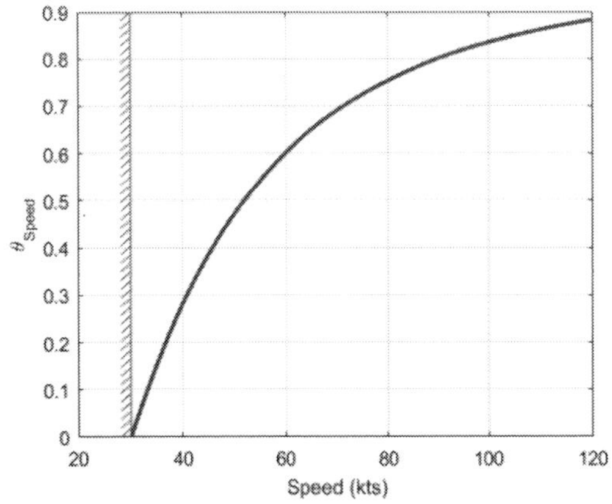

Fly-Off Tasks & Scoring Measurements

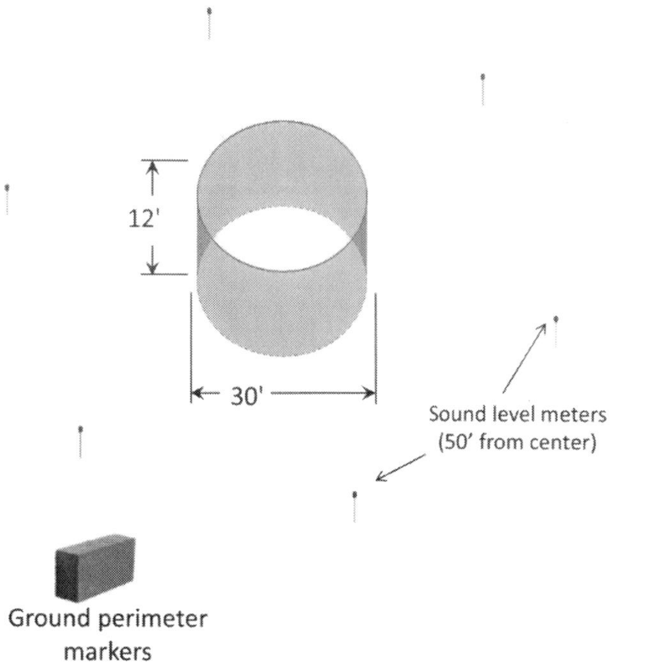

Flight Demonstration

1. Takeoff and climb without violating the takeoff/landing envelope.
2. Conduct a speed run of six laps around a 1 nmi course.
3. Demonstrate the capability to abort a landing by performing a touch & go without violating the takeoff/landing envelope.
4. After loitering such that total flight demonstration endurance is greater than **20 minutes**, descend and land without violating the takeoff/landing envelope.

Fly-Off Tasks & Scoring Measurements

Sightlines – enforce 90° field of view
- Incentive for open, thrilling flying experience
- Disincentive for overhead open rotors

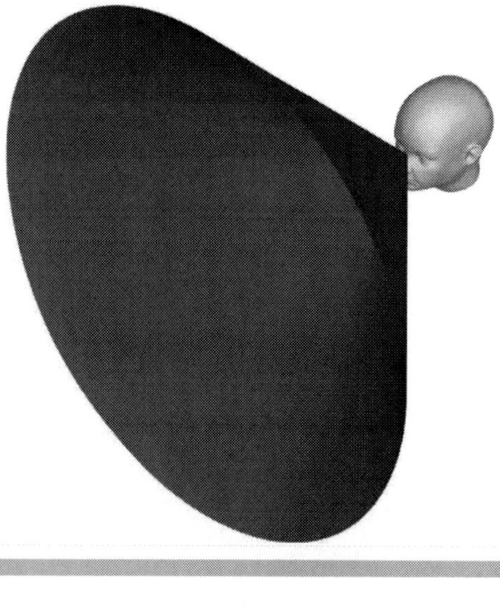

Safety

Safety is ultimately the responsibility of the teams.

However, the intent is to develop technologies that can be rated for people to safely fly in the future. Consequently, the competition requires that all device architectures are, at the fundamental concept level, "human-ratable," i.e.,

A device that has no systems or components, except for primary structure, in which a single point failure could result in loss of an operator's life or limb.

Reporting and documentation of safety aspects *are* required.

Unmanned Option

Autonomous or remotely piloted with "dummy" operator.

Creates extra regulatory and operational challenges for both team practice and the fly-off.

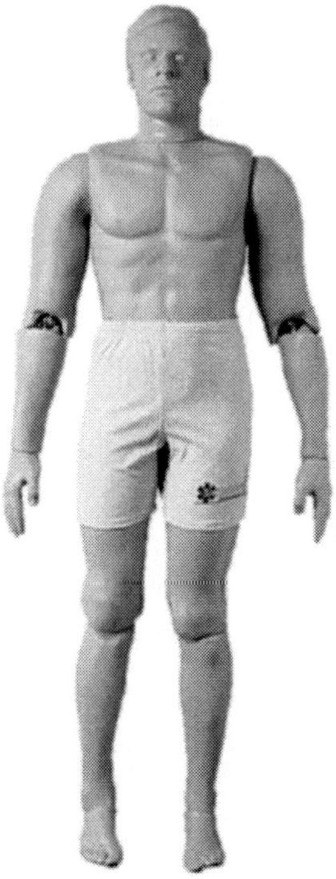

Where We Are Now

Almost Ready to Launch...

We would welcome your expertise and guidance.
Any and all comments and criticisms are welcome.
... and of course as you are the brightest minds in VTOL, we would love for you to join us.

1. Come help us as an Advisor to the GoFly Prize.
2. Be one of our Incubator Technical Mentors helping the teams create.
3. Form a Team and win $2,000,000 in prizes.

lighter@goflyup.com

GO FLY PRIZE

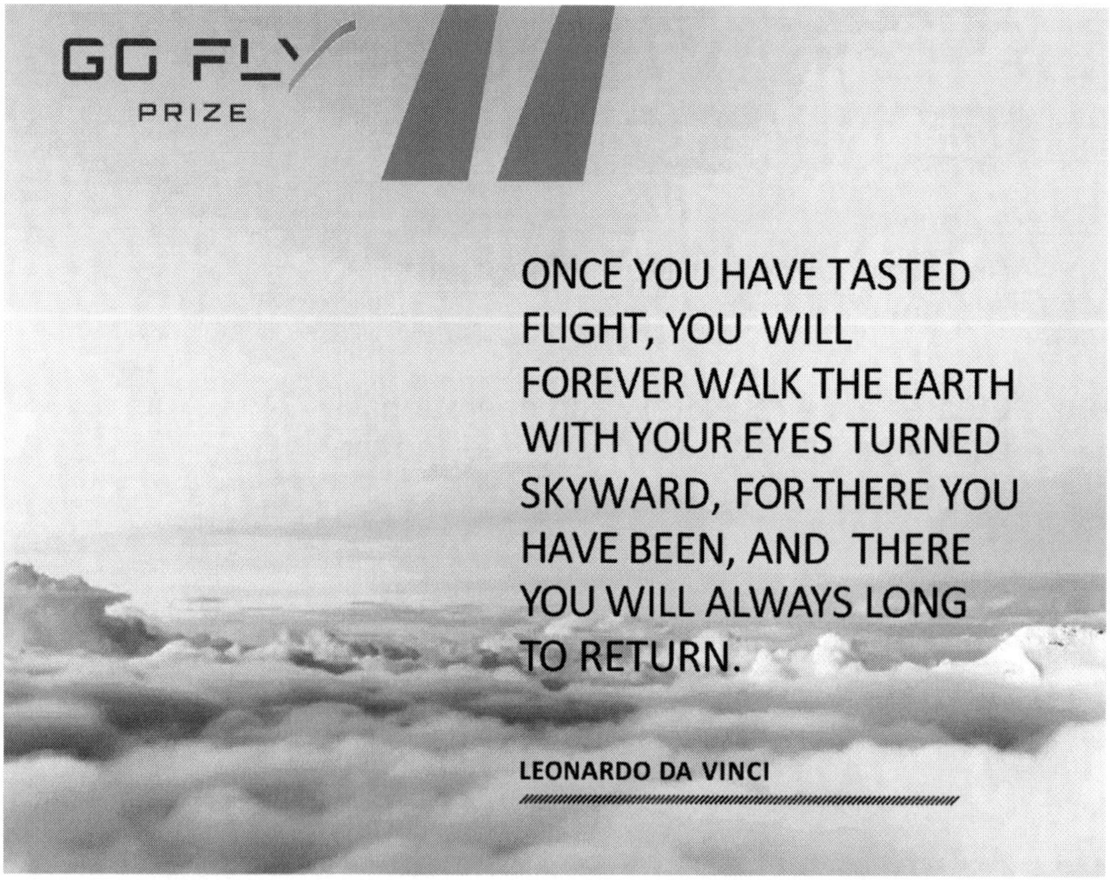

Join us in making the impossible, possible.
GO FLY
PRIZE

GO FLY
PRIZE

Scoring

Add vs. multiply

Product-based schemes carry the danger that two teams with very close scoring parameters in one area can get wildly different scores, overpowering much more significant differences in another scoring parameter

Which team should win, red or blue?

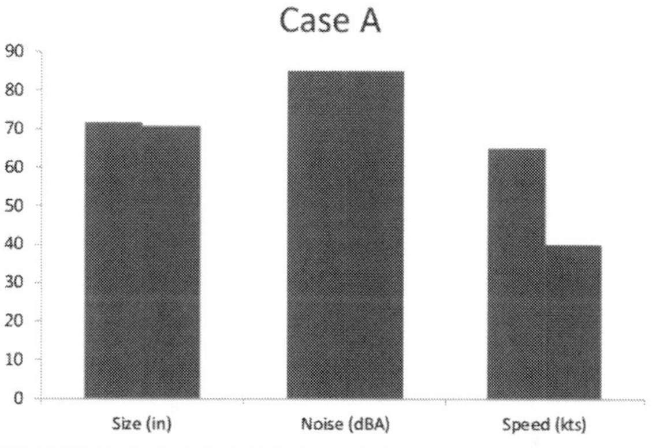

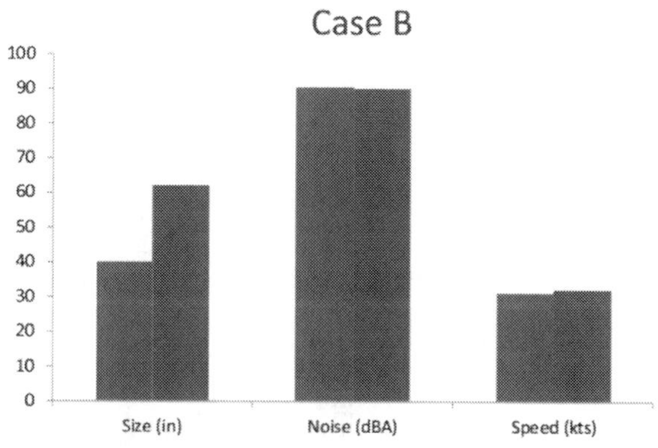

Effect of Blade Deformation on Air and Structural Loads of LCH Rotor

Younghyun You
Post-Doctor
Konkuk University
Seoul, Republic of Korea

Deokhwan Na
Graduate Research Assistant
Konkuk University
Seoul, Republic of Korea

Sung N. Jung
Professor
Konkuk University
Seoul, Republic of Korea

Kiro Kim
Graduate Research Assistant
Konkuk University
Seoul, Republic of Korea

Soo Hyung Park
Professor
Konkuk University
Seoul, Republic of Korea

Gu Hwan Park
Senior Research Engineer
Korea Aerospace Industries
Sacheon, Republic of Korea

ABSTRACT

In this work, a CFD-CSD loose coupling strategy is applied to predict airloads, blade motions, and structural loads of main rotor of Light Civil Helicopter (LCH) which is under development at KAI (Korea Aerospace Industries). A computational structural dynamics (CSD) analysis code CAMRAD II is coupled with a computational fluid dynamics (CFD) code KFLOW for this purpose. The flight condition includes a high-speed forward flight where the formation of shock along with the negative loading near the tip of the advancing side is of special concern. The convergence behavior of the coupled analysis is investigated first by tracing the iteration histories of section airloads, trim control angles, and blade deformations. CFD-CSD coupling results are then correlated against those by CFD analysis with rigid blade motion and CSD analysis as well. The influence of blade elastic motions on airloads and structural loads of LCH rotor are examined. The flow physics affecting the aeroelastic predictions of the rotor is also investigated.

INTRODUCTION

During the past decades, the planform shapes of rotor blades and their internal structures become more complicated because of the introduction of the new design and manufacturing technologies. Either a simple lifting-line based rotorcraft comprehensive analysis or a CFD analysis taking into account only the rigid blade motion is often insufficient to estimate the accurate airloads and structural loads during the new rotorcraft design process. Now, the general consensus reached is that more reliable prediction tools are crucial for modern helicopter design. Even with advanced computational power and state-of-the-art solution technologies, however, it is still challenging tasks due to complex flow environment around the rotor disk. The tip regions in the advancing side are under high dynamic pressure and transonic effect while the retreating side is confronted with low dynamic pressure and stall. The Computational Fluid Dynamics (CFD) / Computational Structural Dynamics (CSD) coupling technique allows a first-principal based approach and offers improvements on air and structural loads predictions because of the capability to model the complex aerodynamic environment around the rotor disk and its strong interaction with the blade motions in a refined fashion. Recent studies using CFD-CSD coupling (Ref. 1) have shown remarkable accomplishments in the prediction of loads and vibrations of various helicopter rotors in diverse flight conditions.

The CFD-CSD coupling in rotorcraft applications is pioneered by Tung et al. (Ref. 2) for predicting rotor aerodynamic loadings in high speed conditions to take advantage of the first principle-based flow representation by CFD approach in the aeromechanics analysis. The algorithm has been evolved during various validation studies such as the UH-60A rotor (Ref. 3-4). Even though much heavier computational resources are required, a CFD-CSD coupled approach is conceived to be desired particularly for airloads and structural loads prediction of a helicopter rotor.

The LCH program is to develop a new 4.5 ton-class, twin-engined, multi-purpose light civil helicopter at KAI in South Korea. The rotor blade of LCH has an extended chord at the blade outboard region with swept, tapered and anhedral tip. The objectives of the present work are 1) to model LCH rotor for the analysis of either CSD or CFD level; 2) to perform a loose coupling between KFLOW (CFD) and CAMRAD II (CSD) to estimate the airloads and structural loads of LCH rotor in a high speed flight condition; and 3) to identify the effect of blade elastic deformations on the aeromechanics analysis solutions.

Presented at the 6th Asia/Australian Rotorcraft Forum / Heli Japan 2017, Kanazawa, Japan, November 7-9, 2017.

ANALYSIS METHODS

The analytical approaches adopted in the present study are described in the following subsections.

Rotorcraft aeromechanics analysis

A rotorcraft dynamics analysis code CAMRAD II (Ref. 5) is used to analyze the rotor. The code is characterized by multibody dynamics, nonlinear finite elements, and various levels of rotorcraft aerodynamic models. For the structural analysis, the blade motion is expressed as the summation of the rigid body motion and the elastic motion. The rigid body motion describes the motion of an end of a beam element, and the elastic motion is measured relative to the rigid body motion. The beam elements are represented by three translations (axial, flap, and lead-lag) and three rotational degrees of freedom (DOF) resulting in a 15 DOF per beam element. The blade structure is discretized into several nonlinear beam finite elements discretized along the blade span. The aerodynamic model of CAMRAD II is based on the lifting-line theory combined with the airfoil table look-up as well as various levels of the vortex wake representations. The ONERA-EDLIN unsteady aerodynamic theory along with C81 airfoil table look-up is used for the aerodynamic loads acting on the blade. In addition, a rolled-up free wake model is introduced to the non-uniform induced inflow of the rotor. The rolled-up wake model is based on the feature that a tip vortex forms at the blade tip and convects to the flow field behind the rotor.

Computational fluid dynamics

A three-dimensional compressible RANS (Reynolds Averaged Navier-Stokes) flow solver KFLOW (Ref. 6-7) is used for the CFD analysis. The KFLOW is capable of computing time-accurate moving body problems by employing a Chimera overlapped grid system. The overset grid methodology needs an auxiliary algorithm to interchange information between grids. In the present Chimera overset method, the cut-and-paste algorithm is used to compose a cross section that allows exchange of information between grids. A second-order accurate, dual-time stepping scheme combined with a diagonalized alternating-directional implicit (DADI) method is used to compute the unsteady flow fields around a rotor. The dual-time stepping also eliminates factorization and linearization errors by marching the solutions in pseudotime. The inviscid fluxes are calculated using the fifth-order WENO scheme, while the central differencing technique is used to the viscous fluxes. The k-ω Wilkox-Durbin (WD+) scheme is employed for the turbulence model. The characteristic boundary conditions using the Riemann invariant are applied at the far-field boundary, whereas a no-slip condition is used at the solid wall of the blade.

In the aeroelastic simulation of a rotor system, the blade elastic deformations lead to changes in grid cell volumes, necessitating rigorous enforcement of the geometric conservation law (GCL), which states that the increase of a moving cell volume should be equal to the sum of the variations along the surface that encloses the volume. The GCL is applied to meet the conservation relation for the surfaces and volumes of the control cells in a moving grid system. The elastic blade motions at each physical time step are considered for the CFD solution using a modified volume calculated from the GCL.

Coupling method

A loose coupling between CAMRAD II and KFLOW codes is used for the analysis. The basic principle of the coupling is to exchange information between CSD-computed blade motions and CFD-computed airloads, per revolution base, to benefit the strength of the other code. The coupling iteration begins with CSD analysis using a built-in aerodynamic model. The resulting blade motions along with trim control angles are transferred to the CFD code to update the aerodynamic forces and moments. The difference in airloads between the two codes (i.e. delta airloads) is calculated and superposed to the CSD airloads for the updated blade motions and trim controls for the subsequent iteration stage. This process continues until the airloads and trim control angles indicate little or no noticeable difference compared to the previous iteration steps. It is remarked that the CFD results have a timewise interval of 0.2° ($\Delta\psi$ = 0.2°) while CSD having 5° resolutions. Due to the varying resolutions in both results, appropriate data regression schemes are required to transfer data between CFD and CSD analyses. For blade motions, the interpolation in the radial direction is represented using a polynomial with the sixth order while the timewise domain is interpolated using a Fourier series containing up to the ninth components. The airloads in the spanwise direction are interpolated using a cubic spline fit whereas a random data selection at every 5° azimuth angle is applied for the timewise airloads data.

VALIDATION STUDY OF COUPLED ANALYSIS

The present CFD-CSD coupled analysis is validated by evaluating the accuracy of predictions via measured HART I rotor data as is presented in this section.

HART I test

The HART (Higher harmonic Aeroacoustic Rotor Test) I rotor is tested at the German-Dutch wind tunnel (DNW) in 1994 (Ref. 8-9). A 40% scaled BO-105 rotor model along with a fuselage is used for the test. The measured data set of HART I rotor exhibits a wide spectrum – the noise level, blade surface pressure, tip vortices, blade motions, and structural moments with and without the application of HHC (Higher Harmonic Control) pitch control inputs. The rotor has a four-bladed hingeless configuration with a solidity 0.077 and blade radius 2 m. The baseline case of HART I in low-speed descent flight condition is considered for the validation study. The target trim values specified are 3100 N, -11.2 N-m, and -20 N-m respectively for the thrust, hub roll and pitching moments. The positive signs are defined respectively when the advancing side moves down and a pitch up is induced for the moments.

Numerical modeling

Fig. 1 shows the computational grid systems for the HART rotor. The blade grids extend 1.5 times of a chord length (c) in the normal direction, measured from the blade surface. The cell spacing for the first grid point from the wall boundary used is $1.0\times10^{-5}c$. The off-body grids consist of an inner region which extends 4c upward, 3c below from the blade, and 1.5c away from the blade tip. The far field boundary is stretched up to 5R (blade radius), centered at the rotor hub. The cell spacing is 0.1c for the Cartesian off-body grids. The CFD computational grids consist of 6.4M (million) cells for the blade grid, 29.1M for the off-body grid, and another 2.5M for the fuselage grid, leading to a total of about 37.5M grids.

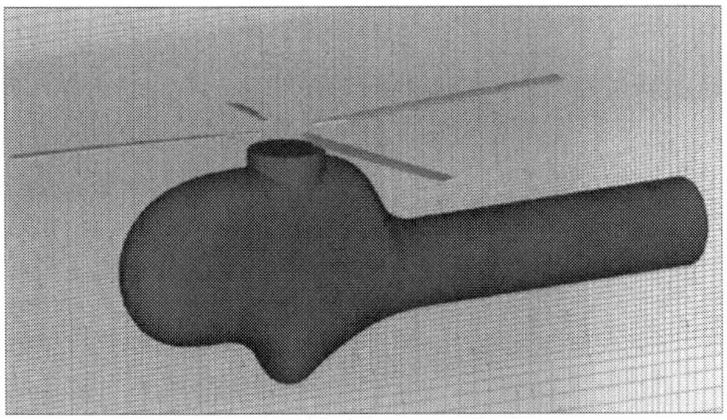

Figure 1. CFD model for HART I.

For the CSD analysis, the blade structure is modeled using 15 beam finite elements while the airfoil blade region is divided into 17 non-uniform aerodynamic panels with finer segments toward the blade tip. Specifically, the centers of each aerodynamic panel are aligned to coincide with the measured airloads stations to minimize the discretization error.

Validation results

Fig. 2 shows the variation of CFD trim forces and moments (denoted as continuous lines with symbols) against the measured target values (dashed lines) with the coupling iterations. Since the trim process is managed by the CSD code, the trim results are post-processed and computed using the CFD code at each iteration step. The resulting CFD trim values indicate a clear convergence after about 6 iteration cycles while slightly underestimating the trim forces and moments, as compared with the measured results.

In Fig. 3, the trim control angles (collective, lateral and longitudinal cyclic angles) predicted using a CSD approach and CFD-CSD couplings are compared against the measured HART I data. All predictions are obtained after the trim is converged. The CSD predictions overestimate the collective pitch by about 12.3% whereas CFD-CSD coupled results correlate reasonably (maximum up to 1.8%) with the measured data.

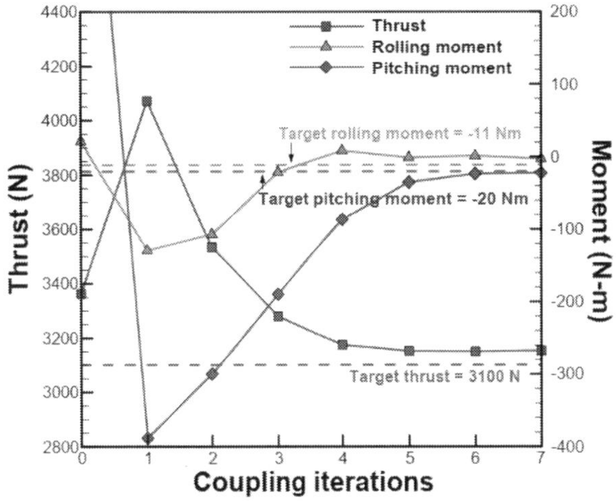

Figure 2. Convergence behaviors of target trim values with coupling iterations.

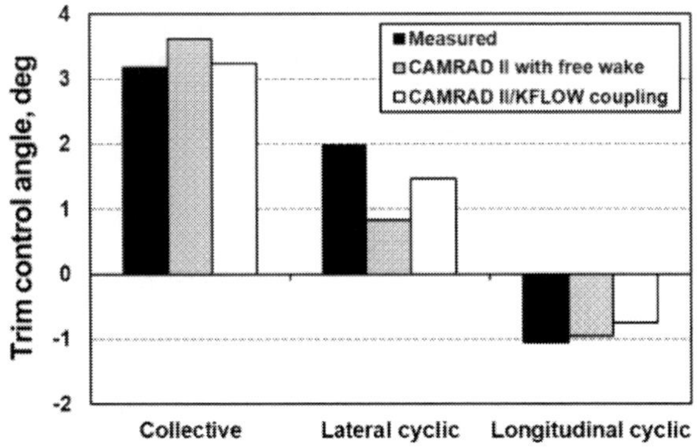

Figure 3. Comparison of control trim angles.

Figs. 4 show the predicted pressure distributions over the upper and lower surfaces of the blade for the 97% radial station as compared with those of the measured data (denoted as hollow circles). The azimuth angles are 30° and 120°, respectively. A total of 44 pressure sensors are distributed at the radial stations of 0.97R. The output signals from each pressure sensor are sampled at a rate of 2,048/rev and averaged for 64 rotor revolutions. Good correlation is obtained at the specified azimuth angles. CFD results predict pressure fluctuations near the trailing-edge tab with a finite thickness of 0.9 mm. However, the measured data show no such indication since pressure sensors are not installed to verify the pressure variations because of the limited space allowed in this region of the blade.

Fig. 5 shows the comparison of the spanwise distribution of section airloads M^2C_n predicted using CSD analysis and CFD-CSD coupling against the measured data. The CSD analysis predicts the mean airloads response reasonably, however, much of the oscillatory peaks present either in the first or the fourth quadrant of the measured airloads are missed and a down-up pulse near the front edge of the rotor (around the azimuth of 180°) is not captured as well. The highly oscillatory peaks are due to a close encounter between the blades and the nearby vortices formed by the preceding blade motions. This phenomenon is

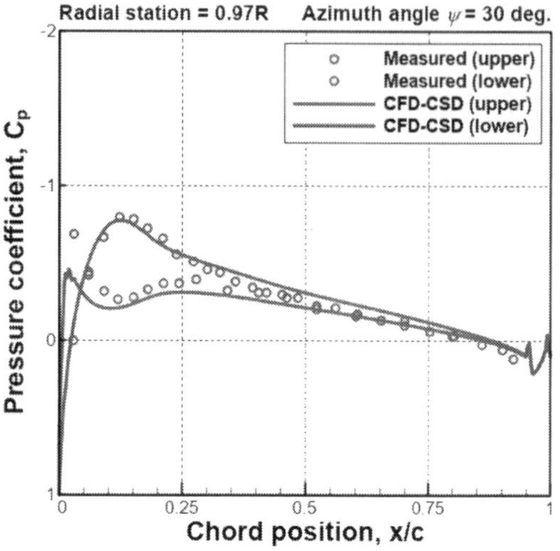

(a) $\psi = 30^{\circ}$

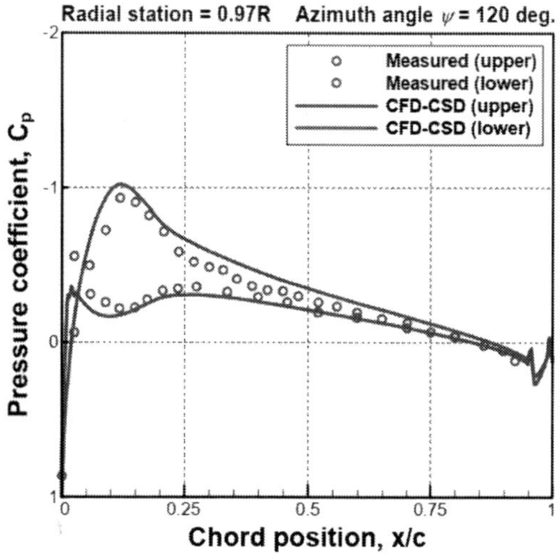

(b) $\psi = 120^{\circ}$

Figure 4. Comparison of chordwise pressure distributions at spanwise location $r/R = 0.97$.

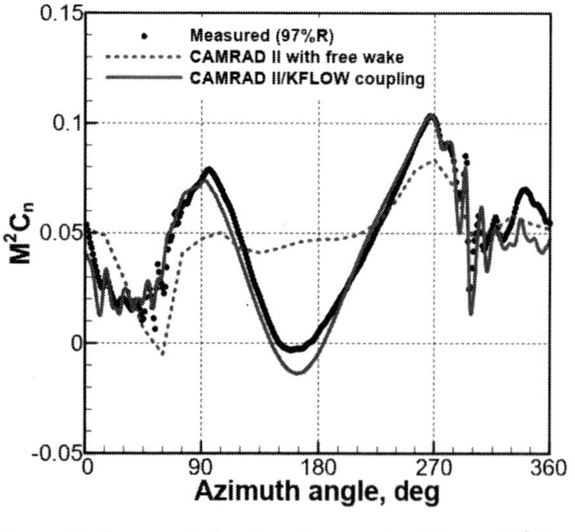

Figure 5. Comparison of section normal force M^2C_n at radial station $r/R = 0.97$.

known as BVI leading to severe vibration and noise (Ref. 10). The CFD-CSD coupled approach improves the correlation drastically, especially for BVI response apparent in both the first and the fourth quadrants, overall 2/rev (two per revolution) waveform, and the down-up pattern near the front edge noticed in the measured airloads signals. The retreating side signals including the highest peak around the azimuth angle of 300° at $0.97R$ are accurately predicted. Some of the advancing side signals are underestimated slightly with the present CFD-CSD approach. This is related with the aging of tip vortices and more refinements in CFD grids may be necessary to capture the wake behavior.

Fig. 6 presents the iso-surfaces of Q-criterion colored by vorticity when the reference blade is positioned at 25° (measured from the rear). It is noted that the vorticity is clearly visible for over one and half revolutions of the rotor and the possible BVI encounters during the period are reflected accordingly in the airloads prediction results shown earlier.

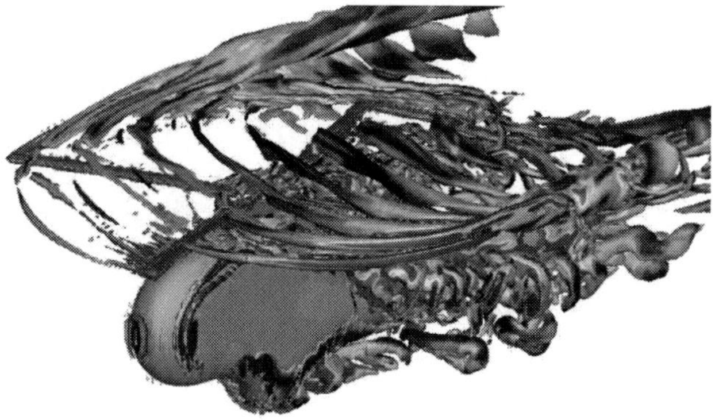

Figure 6. Iso-surfaces of Q-criterion plots colored by vorticity.

In Fig. 7, the predicted blade elastic deformations at the tip using CFD-CSD as well as CSD with a rolled-up wake model are compared against the measured data. It is seen that all predicted results capture the general trends of the measured data. Among other cases, a 2/rev waveform apparent in the measured flap motion is correctly predicted using the CFD-CSD coupling. The comparison of tip elastic torsion results shows an improved predictions on magnitudes and phases of the elastic torsion are met with the CFD-CSD coupling, even though the mean is underestimated significantly as compared with the measured data. The discrepancy in the mean is induced by the fact that the structural properties of HART I blades are not measured using the instrumented (blade No. 1) blade but from one (blade No. 3) of the blades (Ref. 11). Furthermore, the instrumented blade is heavier by about 6% compared to the other blades and the rotor encounters substantial blade-to-blade dissimilarities. For lag mode deformation results, a 4/rev behavior present in the measured signals is not captured in any of the analyses. The source of the discrepancy is unclear at this moment.

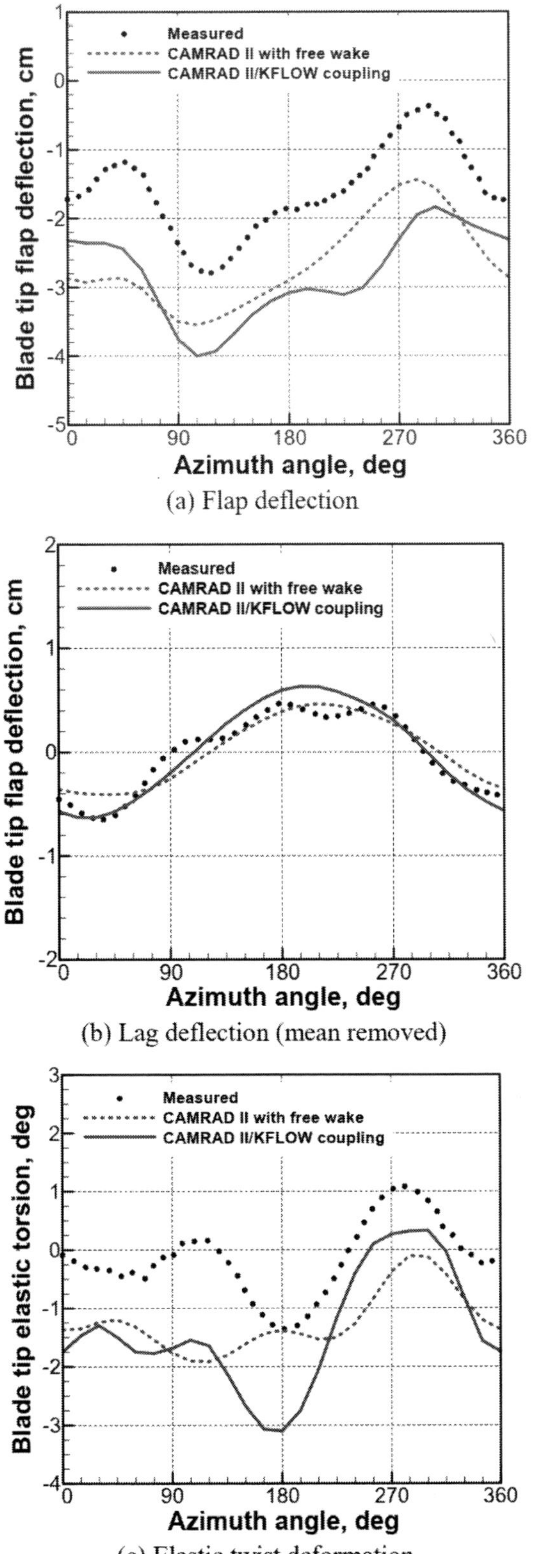

(a) Flap deflection

(b) Lag deflection (mean removed)

(c) Elastic twist deformation

Figure 7. Comparison of elastic deformations at the blade tip.

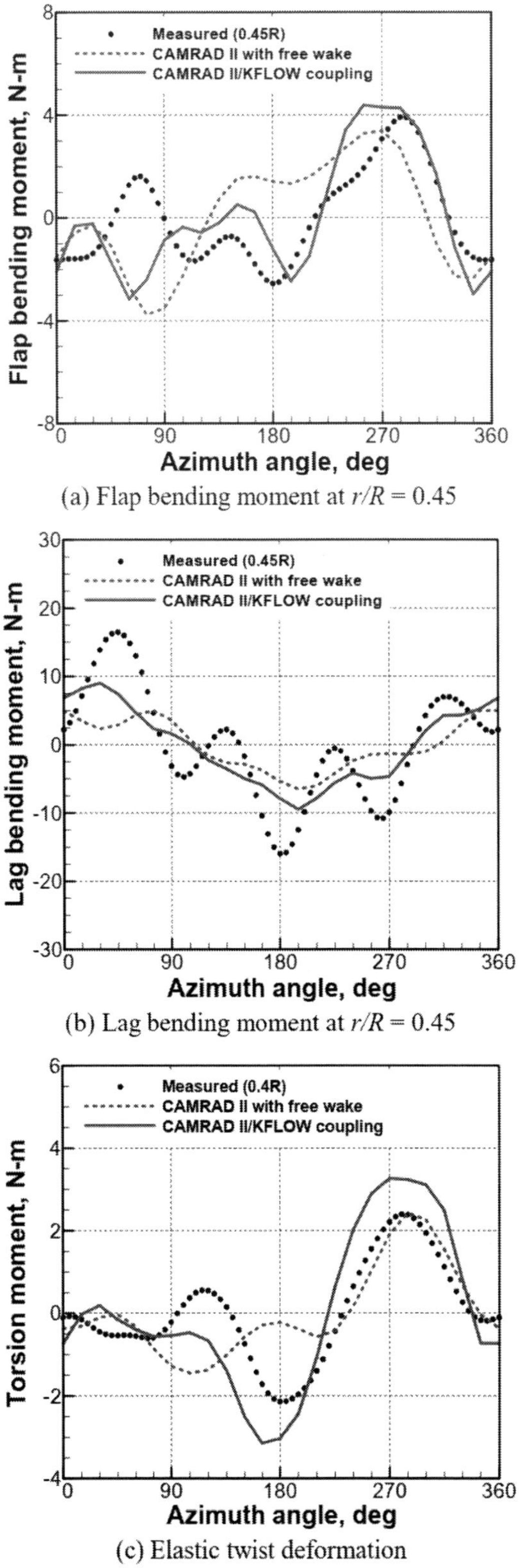

(a) Flap bending moment at $r/R = 0.45$

(b) Lag bending moment at $r/R = 0.45$

(c) Elastic twist deformation

Figure 8. Comparison of oscillatory structural moments at blade outboard stations.

Finally, the predicted flap bending, lag bending, and torsion moments obtained at an outboard blade station (around $r/R = 0.45$) are validated against the measured data in Fig. 8. The mean is removed for a convenience. As expected, the CSD predictions with rolled-up free wake model show only marginal correlations with the measured data. The oscillatory patterns in the measured flap bending are correlated poorly with the CSD analysis while predicting opposite phases in the second quadrant of the measured torsion moments. The CSD-CFD coupled approaches improve the correlation significantly. The mismatched waveforms in the second quadrant of the disk are corrected somewhat for both flap bending and torsion moments. However, the first peak in the measured flap bending along with a 4/rev harmonic behavior in the measured lag bending are not captured by any of the present analysis models. Some degree of uncertainties in the blade structural properties (in association with the instrumented blade) may cause the discrepancies.

In summary, the comparison results of the chordwise pressure distribution, section airloads, blade deformations and structural moments indicate that the quality of the CFD-CSD prediction result appears excellent and deserves to exploit further for any analysis system.

COUPLED ANALYSIS OF LCH ROTOR

In this section, the airloads and structural loads of LCH main rotor in forward flight condition (μ = 0.28) is evaluated using a loose CFD-CSD coupled approach.

Computational modeling

The 5-bladed main rotor system of LCH is analyzed using a CSD code CAMRAD II. The blade structure is discretized using 11 beam finite elements while the airfoil blade region is modeled using 20 unequal spaced aerodynamic panels. Note that finer aerodynamic segments are placed toward the blade tip. The inter-blade damper configuration of LCH rotor is also implemented.

Figure 9 shows a perspective view of the computational grid system for CFD analysis. Only blade region of LCH main rotor is modeled. A moving overlapped Chimera grid system with different types of grids (background grids and blade grids) is used for the present computation. In this case, a three-dimensional incompressible government equation is used for the CFD analysis. The inviscid fluxes are calculated using the ROE scheme in this work. The total number of grid cells is about 14.4 million. Each blade grid has 0.44 million cells and the background grid has 12 million cells. The inner region of background grid has a uniform spacing (15% of chord length) in all directions to capture the vortex.

Figure 9. CFD model for LCH rotor.

Numerical results and discussion

The CSD trim analysis is performed for the forward flight condition by solves the controls and aircraft attitudes that balance the forces and moments of LCH helicopter. Table 1 represents blade pitch control angles and flapping angles. The amplitude of cyclic angles of flapping motion is less than 1 deg. The calculated collective and cyclic pitch angles are rigid motion of blade twist deformation. And the rigid flap deformation is obtained using the conning angle and cyclic flapping angles. The rigid lead-lag deformation is assumed zero.

Table 1. LCH main rotor blade rigid motion.

	CSD with free wake
Collective pitch angle (θ_0)	5.99
Lateral cyclic angle (θ_{1c})	0.95
Longitudinal cyclic angle (θ_{1s})	-4.21
Conning angle (β_0)	3.23
Longitudinal flapping angle (β_{1c})	-0.53
Lateral flapping angle (β_{1s})	-0.92

Figure 10 shows the elastic flap, lead-lag, and twist deformations obtained using a CSD analysis with a free vortex wake model. It is defined as positive when the blade undergoes flap up, lag back, and nose up deformations, respectively. The elastic flap deflections are obtained by removing the vertical displacements produced by conning angle and cyclic flapping angles. The elastic twist deformation is obtained by subtracting the collective and cyclic pitch angles and pretwist angles from the total geometric pitch angles. It is observed that the positive peak of the elastic flap deformation (Fig. 10a) is produced at near 270 deg azimuth angle and the negative peak is occurred at near 0 deg azimuth angle. In the elastic twist angles, the 3/rev harmonic wave forms are represented.

Using the blade deformation data predicted by CSD analysis, CFD analysis is performed. Two different blade deformation data are considered in the present study; one is used rigid blade deformations only and another used elastic blade deformations adding the rigid blade motion. Figure 11 illustrates the predicted contour plots obtained for the section normal forces (M^2C_n) using CSD with free wake model and CFD with blade deformations, denoted over the entire rotor disk. The similar trends are observed that the peak magnitudes at the rotor disk aft outside region. In the CFD analysis results, a sudden increase of section normal force are observed after 80% span location due to the extended chord of the LCH blade.

The high lift at near the front and outboard region in the CFD prediction with rigid blade motion become reduced significantly in the CFD calculation with elastic blade deformation considered. It is affected by decrease of the elastic twist at near 180 deg azimuth angle. However, the

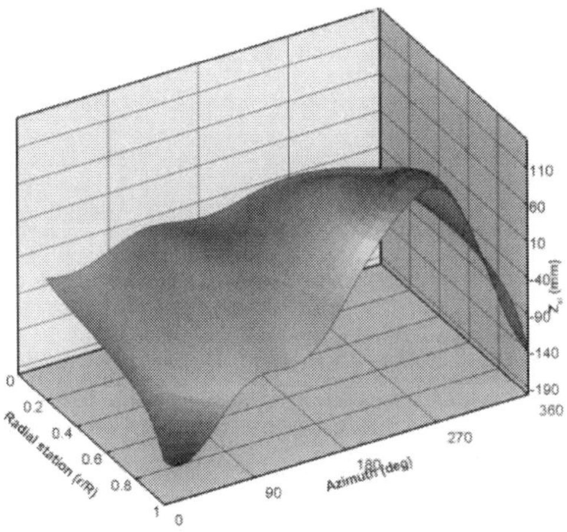

(a) Flap deformation

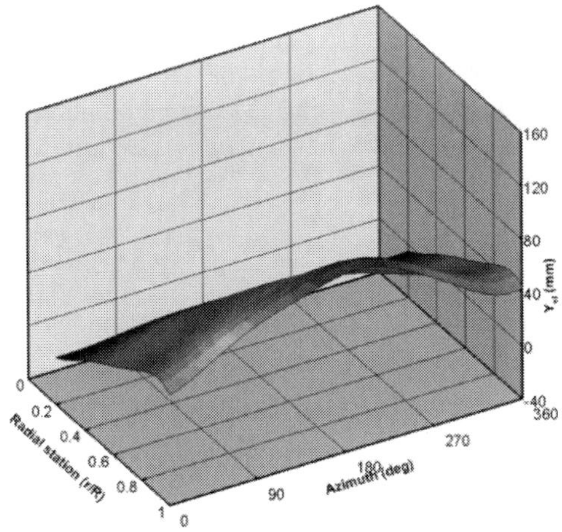

(b) Lag deformation

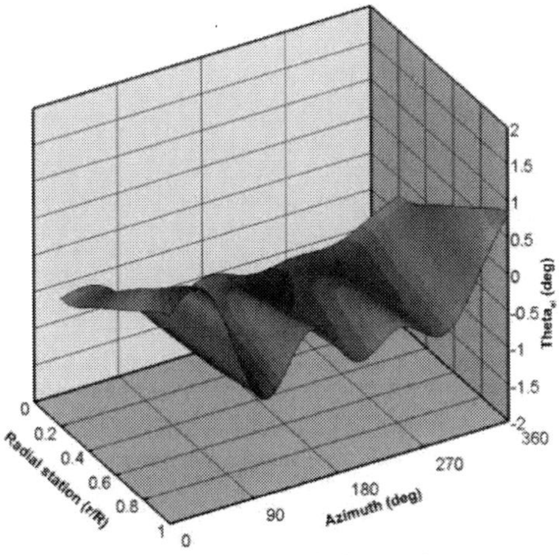

(c) Elastic torsion deformation

Figure 10. LCH main rotor blade elastic deformation.

increase of the elastic twist at near 0 deg azimuth angle increase the section normal force at the aft and outboard region of the rotor disk.

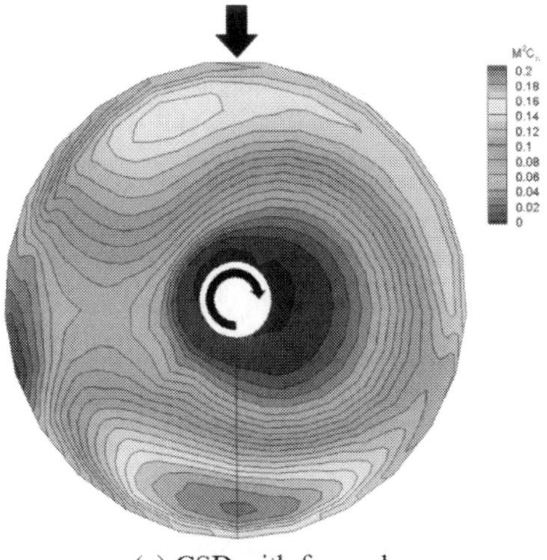

(a) CSD with freewake

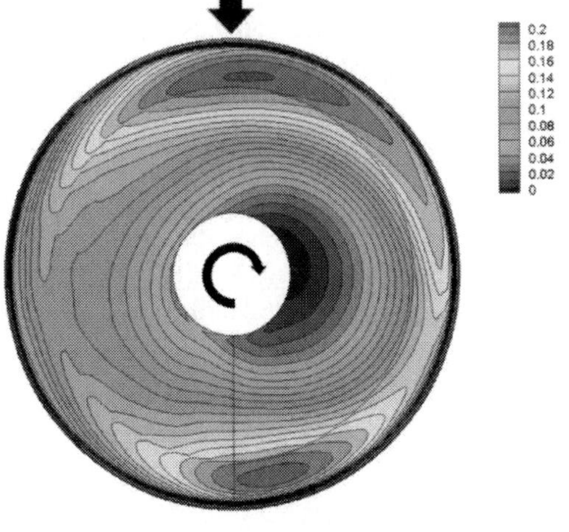

(b) CFD with rigid blade control

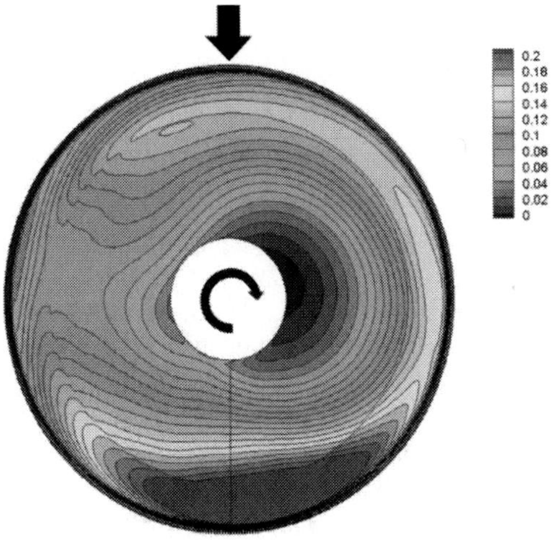

(c) CFD with elastic blade deformation

Figure 11. Comparison of section normal forces of LCH rotor.

CONCLUDING REMARKS

In this work, a CFD-CSD loose coupling strategy is applied to predict airloads, blade motions, and structural loads of LCH rotor in forward flight condition. The CFD-CSD coupling strategy is validated using HART I rotor test data in low-speed descending flight condition. The predicted results show close agreements with the measured data. The blade tip motions of HART I rotor predicted using coupling analysis indicate reasonable accuracy solutions while showing some offsets in the mean as compared with the measured data. A dissimilarity in the structural properties between the blades is probable source of the deviation in the mean. The preliminary prediction of airloads and structural loads using CFD-CSD coupled analysis is successfully conducted in forward flight condition of LCH rotor under developing. The effects of elastic blade deformation are identified by illustrating the section airloads computed over the rotor disk. The alleviation of peak loads in the front side and outboard region of the rotor disk is reached due to the distribution of the elastic twist deformation.

Author contact: Younghyun You 1 john82@konkuk.ac.kr
Deokhwan Na 2 ladug12@naver.com
Sung Nam Jung 3 snjung@konkuk.ac.kr
Kiro Kim 4 kornk2r@naver.com
Soo Hyung Park 5 pish@konkuk.ac.kr
GuHwan Park 6 ilypark@koreaaero.com

ACKNOWLEDGMENTS

This work was supported by the KEIT Research Grant of 2016 (No. 10053155). We thank HART team for the test data.

REFERENCES

[1]You, Y. H., Sa, J. H., Park, J. S., Park, S. H., and Jung, S. N., "Modern Computational Fluid Dynamics/Structural Dynamics Simulation for a Helicopter in Descent," *Journal of Aircraft*, Vol. 50, (5), 2013, pp. 1450-1464.

[2]Tung, C., Caradonna, F. X., and Johnson, W., "The Prediction of Transonic Flows on an Advancing Rotor," *Journal of the American Helicopter Society*, Vol. 31, (3), 1986, pp. 4-9.

[3]Potsdam, M., Yeo, H. S., and Johnson, W., "Rotor Airloads Prediction using Loose Aerodynamic/Structural Coupling," *Journal of Aircraft*, Vol. 43, (3), 2006, pp. 732-742.

[4]Datta, A., Nixon, M., and Chopra, I., "Review of Rotor Loads Prediction with the Emergence of Rotorcraft CFD," *Journal of the American Helicopter Society*, Vol. 52, (4), 2007, pp. 287-317.

[5]Johnson, W., *CAMRAD II*, Comprehensive Analytical Model of Rotorcraft Aerodynamics and Dynamics, Johnson Aeronautics, 1992.

[6]Kim, J. W., Park, S. H., and Yu, Y. H., "Euler and Navier-Stokes Simulations of Helicopter Rotor Blade in Forward Flight using an Overlapped Grid Solver," Proceedings of the 19th AIAA computational fluid dynamics conference, San Antonio, TX, 2009.

[7]Sa, J. H., Kim, J. W., Park, S. H., You, Y. H., Park, J. S., Jung, S. N., and Yu, Y. H., "Prediction of HART II Airloads Considering Fuselage Effect and Elastic Blade Deformation," Proceedings of the Heli Japan 2010, Sitama, Japan, 2010.

[8]Yu, Y. H., Gmelin, B., Heller, H., Philippe, J. J, Mercker, E., and Preisser, J. S., "HHC Aeroacoustics Rotor Test at the DNW - the Joint German/ French/US HART Project," Proceedings of the 20th European Rotorcraft Forum, 1994.

[9]Lim, J. W., Nygaard, T. A., Strawn, R., and Potsdam, M., "Blade-Vortex Interaction Airloads Prediction using Coupled Computational Fluid and Structural Dynamics," *Journal of the American Helicopter Society*, Vol. 52, (4), 2007, pp. 318-328.

[10]Yu, Y. H. "Rotor Blade-Vortex Interaction Noise," *Progress of Aerospace Sciences*, Vol. 36, (2), 2000, pp. 97-115.

[11]Jung, S. N., You, Y. H., Lau, B. H., Johnson, W., and Lim, J. W., "Evaluation of Rotor Structural and Aerodynamic Loads using Measured Blade Properties," *Journal of the American Helicopter Society*, Vol. 58, (4), 2013, pp. 1-12.

Computational Analysis on Aerodynamic Interaction of Quad Tilt Rotor during Forward Flight

Seonggi Lee
MSc candidate
Pusan National University
Busan, Korea

Donghun Park
Assistant Professor
Pusan National University
Busan, Korea

Sejong Oh
Professor
Pusan National University
Busan, Korea

ABSTRACT

Aerodynamic interaction of Quad Tilt Rotor (QTR) during forward flight is studied using hybrid CFD method developed in the open source CFD code (OpenFOAM). Rotor blades effects are considered as source term and inserted to the momentum equation. The rotors fixed to the wing tip nacelles rotate upward from lower to upper wing surface inside the wings. The front and rear rotors have interference effects on front and rear wings. When the rotors rotate upward direction in in-board wing, up-wash occurs near the rotor blade tip due to rotating blade and down wash occurs in the vicinity of nacelle due to the wake of rotor blade tip vortex. High thrust is locally generated at rear rotors due to the wake of front nacelles.

INTRODUCTION

The ability of vertical take-off and cruising with high speed is unique advantage of tilt rotor aircraft. It is performed by tilting its nacelles which is fixed at wing tip. The Quad Tilt Rotor (QTR) has four rotors, while general tilt rotor has only two rotors. Thus, QTR has the higher payload compared to typical tilt rotor.

There have been several researches to analyze complex aerodynamic characteristics of QTR. Gupta and Baeder studied QTR aerodynamics in forward flight and grounded effect using actuator disk method (Ref. 1, 2). Sheng and Narramore simulated Bell Boeing QTR vehicle using high-fidelity CFD solver (Ref. 3).

In this paper, numerical simulations are performed using hybrid CFD method. Aerodynamic interactions of QTR during forward flight are analyzed. The effects of front and rear rotors are evaluated separately by comparing four cases, unpowered, front rotors in operation, rear rotors in operation, quad rotors in operation. Sectional lift coefficient over the wings and sectional thrust coefficient over rotors are compared.

METHODOLOGY

Numerical setup

Improved Actuator Surface Method (IASM) is employed to simulate flow fields around QTR (Ref. 4). It is hybrid three-dimensional incompressible transient Reynolds-averaged Navier-Stokes solver developed for the rotor simulations. In this solver, rotor blades are modeled as source term of momentum equation. The source term is determined by Blade Element Method (BEM). The PIMPLE algorithm in OpenFOAM, which combines PISO and SIMPLE, is used for calculation of flow fields (Ref. 5). Details of the IASM are described in Ref. 4. In Ref. 6, the validity and efficiency of the IASM is verified about rotorcraft simulation.

Temporal integration is performed using the second order Euler backward scheme. And the second order upwind scheme is used for spatial discretization. The k- ω SST turbulent model is employed. The rotors proceed 1degree for each time step, which require approximately 4.8-6.3 hours to complete one rotor revolution using Intel Xeon E5-2604v4 3.4GHz CPU 20 Processors.

Grid topology

The geometry of QTR, including rotor information, is estimated from previous researches and general configuration of QTR. The length of fuselage of QTR is 3m. The rear wing has the longer span than that of front wing. The incidence angle of front and rear wing are 4.5 degree and 2 degree. The diameter of each rotor is 1.4m.

Unstructured hybrid mesh is generated for simulation. Hexahedrons and prisms are layered around the QTR body. And then, pyramids and tetrahedrons fill remaining computational domain. The grid of rotor blades is not generated and replaced by disk plane because all rotor blades are treated as momentum source. Total number of cell is 14.75 million. Figure 1 shows the grid system used in this study. In Figure 1-(a), the cells colored in red represent virtual blade cells.

Presented at the 6th Asian/Austrailian Rotorcraft Forum / Heli Japan 2017, Kanazawa, Japan, November 7-9, 2017.

(a) Surface mesh

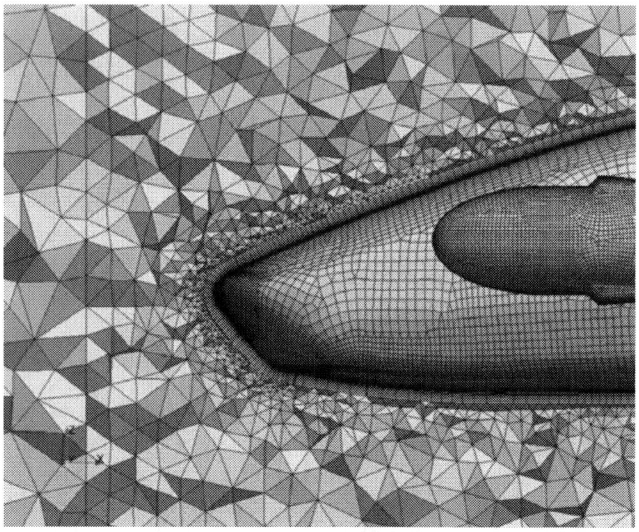

(b) Volume mesh

Figure 1. Grid system used for analysis of QTR

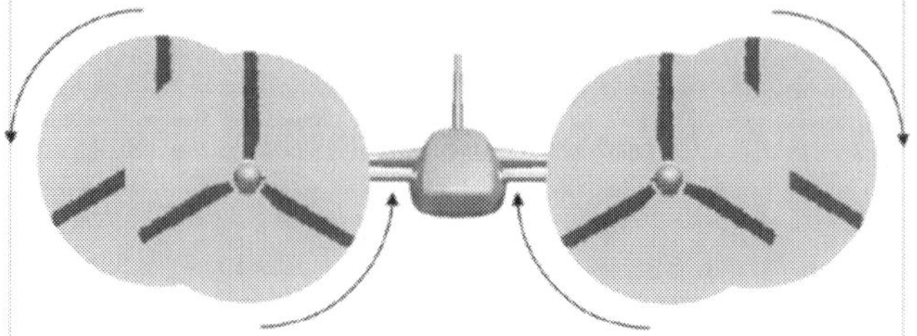

Figure 2. Rotation direction of rotors

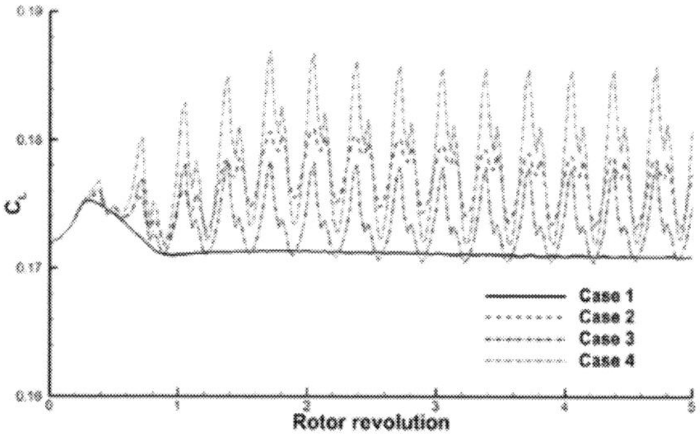

(a) Lift coefficient

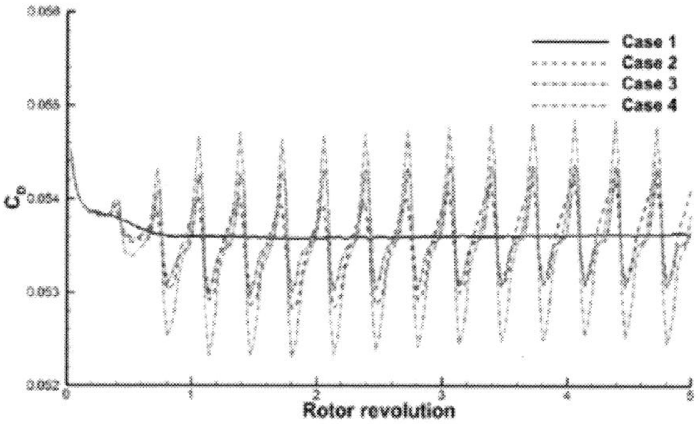

(b) Drag coefficient

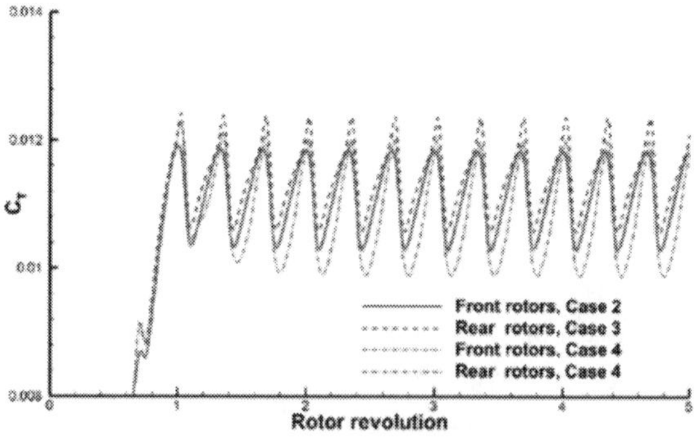

(c) Thrust coefficient

Figure 3. Force coefficients history

RESULTS AND DISCUSSION

In this study, four cases are investigated to separately evaluate interaction effects of front and rear rotors. Case 1 means the numerical result of unpowered model which account no rotor effect. Case 2 and Case 3 mean analysis with only front and rear twin rotor. Case 4 result includes all four rotors (Table 1). Rotors at each port wing rotate in

Table 1. Conditions of simulation

Case	Front rotors	Rear rotors	Runtime for 1 revolution (h)
1	X	X	4.2
2	O	X	5.0
3	X	O	4.8
4	O	O	6.3

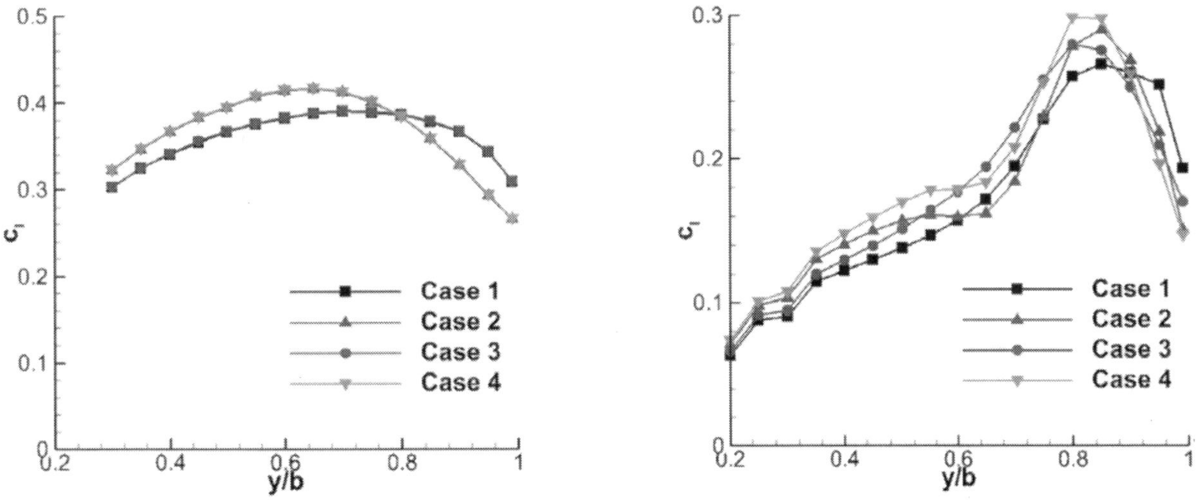

(a) Front wing **(b) Rear wing**

Figure 4. Sectional lift coefficient along the wing span at 0 deg. azimuth angle

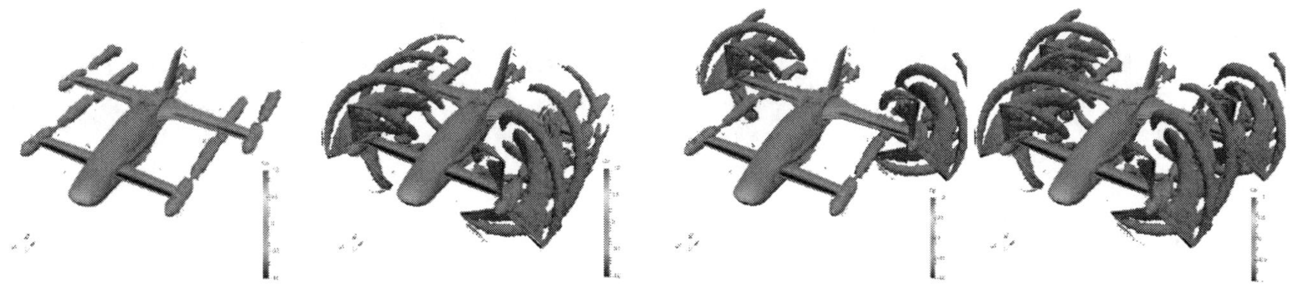

(a) Case 1 **(b) Case 2** **(c) Case 3** **(d) Case 4**

Figure 5. Q Criterion with C_p contour at 0 deg. azimuth angle

clock-wise when observed at front of QTR. Rotors at each starboard wing rotate in opposite direction (Figure 2).

Force coefficients history

Lift, drag and thrust coefficient history are shown in Figure 3. For the simulations including rotors, 5 revolutions are required for the force coefficients to reach the periodic state. The advance ratio, defined as ratio of free stream velocity and rotor tip velocity, is 0.57. It is as high as the wakes of rotors quickly convect away. Thus, Periodic results are obtained with relatively few rotor revolutions.

Spanwise lift distribution

Sectional lift coefficient distributions and Q criterion with pressure coefficient contours at 0-degree and 60-degree azimuth angle are presented in Figure 4 ~ Figure 7. Azimuth angle is 0-degree when one of the three rotor blades is pointing upward. Spanwise position is normalized by half span length (b). It is observed that the effect of rear rotors on front wing is negligible. When the front rotor blades rotate in the vicinity of front wing in-board, it causes up-wash near the blade tip. This up-wash creates a high effective angle of attack, and the higher lift is generated compared to wing without rotors. Down-wash is also observed near the nacelle caused by blade tip vortex.

Interference effects at rear wing is more complex compared to that at front wing. In the vicinity of wing root, the higher lift is generated when front rotor is in operation (Case 2, Case 4). Interaction effects of rotor and wing, described above, are also observed at rear wing when rear rotor is in operation (Case 3, Case 4). When both front and rear rotors work, the phenomena are added up. Thus, the higher lift is generated at wing root due to front rotor and at wing tip due to interaction of rear rotor and rear wing. Lift forces on rear wings are compared in Figure 8. Lift are normalized by lift on rear wings in Case 1.

Thrust distribution

Sectional thrust coefficient distribution at front rotor and rear rotor are presented in Figure 9. All contours are

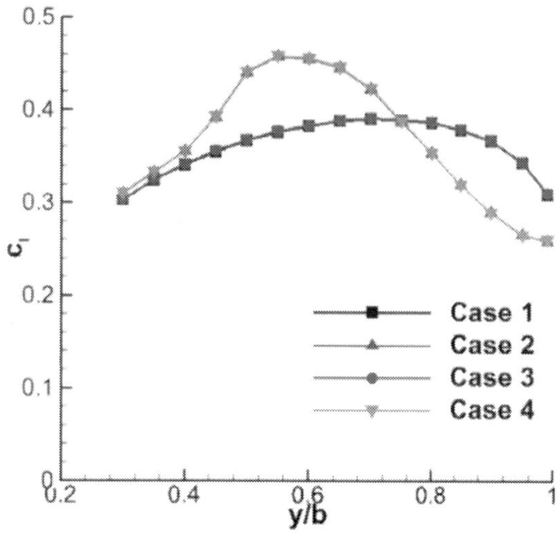

(a) Front wing

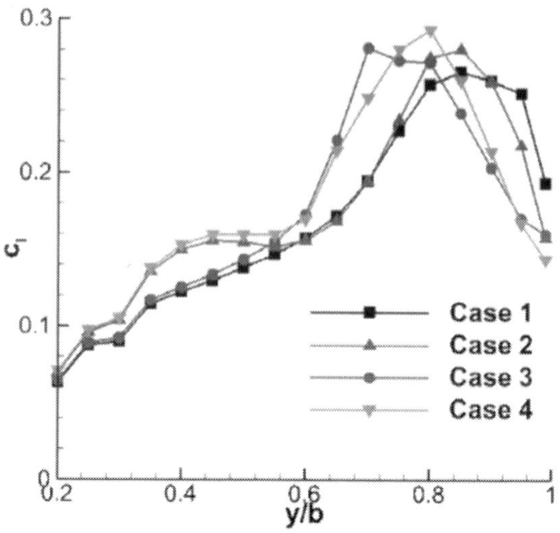

(b) Rear wing

Figure 6. Sectional lift coefficient along the wing span at 60 deg. azimuth angle

(a) Case 1

(b) Case 2

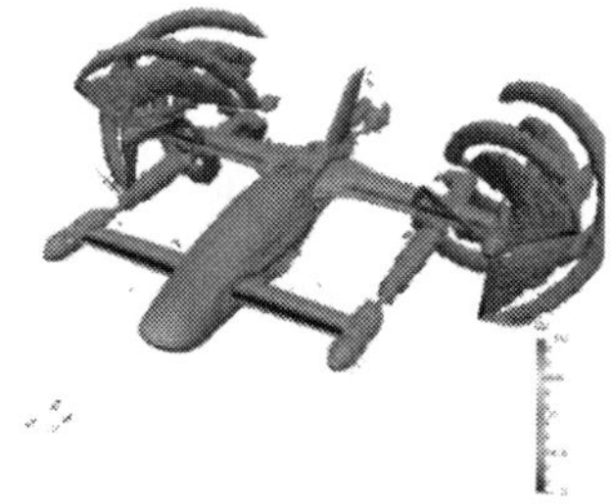

(c) Case 3

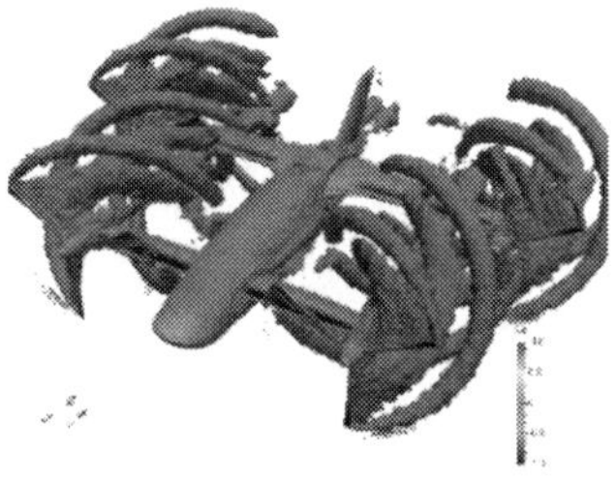

(d) Case 4

Figure 7. Q Criterion with C_p contour at 60 deg. azimuth angle

measured at rotors that is fixed at port wings. There is no significant difference between Thrust distribution at front rotor of Case 2 and that of Case 4. Interference effects of rear wing and rotor on front rotor are not considerable. Comparing Figure 9 (c) and (d), It is observed that the higher thrust is generated about the rotor tip, and the lower thrust is generated about the rotor root, when front rotor is in operation. Thus, the variation from the lowest value of sectional thrust coefficient to the highest value of that is larger in Case 4. The thrust at rear rotor locally increase due to the wake of the front nacelle (Figure 10).

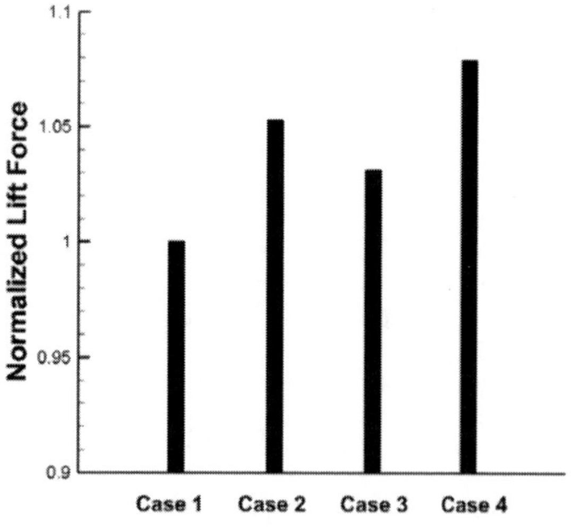

Figure 8. Normalized lift force on rear wings

CONCLUSIONS

Aerodynamic interaction among wings and rotors of QTR were investigated using hybrid CFD method, IASM. The IASM efficiently presented solution of complex flow fields around QTR. The lift of wings consisting of QTR increased when rotors are in operation. This is due to interaction between wing and rotor. The rotors that rotates upward near the wing make up-wash. Thrust of front rotor is not affected by rear rotor. However, thrust variation at rear rotor is large when front rotor is in operation. There is region which yield locally high thrust at rear rotor due to the wake of the front nacelle.

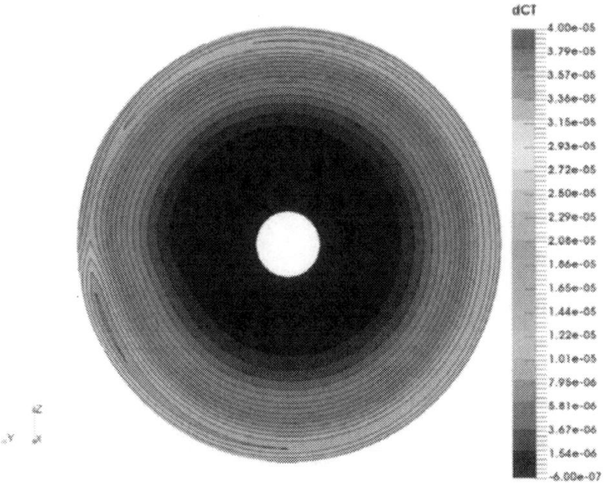

(a) Front rotor, Case 2

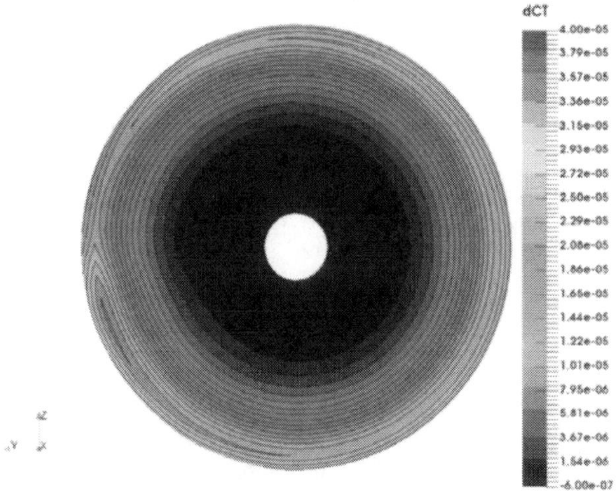

(b) Front rotor, Case 4

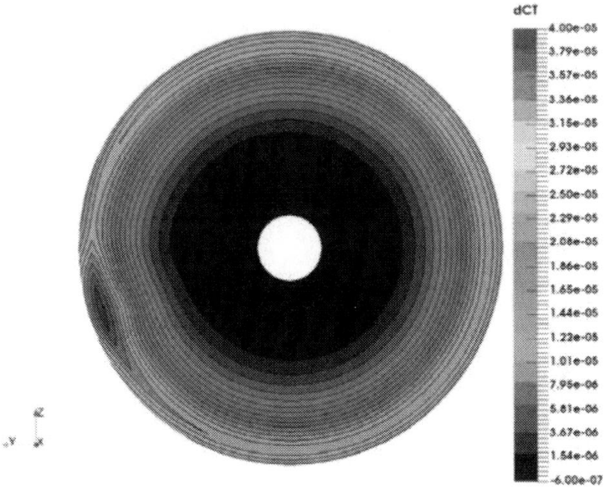

(c) Rear rotor, Case 3

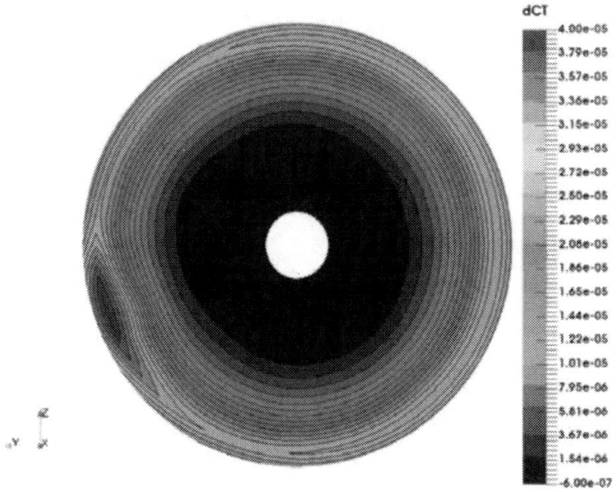

(c) Rear rotor, Case 4

Figure 9. Sectional thrust coefficient distribution over rotors

As a future work, detail flow features of QTR would be investigated. In this study, only forward flight condition of QTR was treated. Flow fields of QTR during transient flight would also be studied.

Author contact: Seonggi Lee zacwhee@pusan.ac.kr
Donghun Park parkdh@pusan.ac.kr
Sejong Oh. tazo@pusan.ac.kr

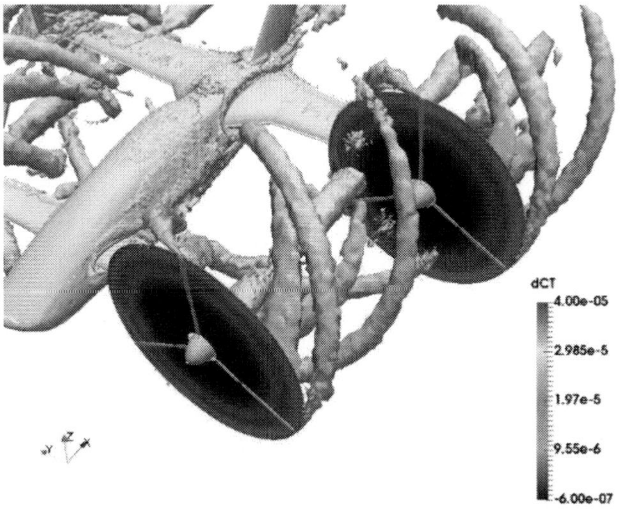

Figure 10. Q-criterion with sectional thrust coefficient

REFERENCES

[1]Gupta, V., and Baeder, J., "Investigation of Quad Tiltrotor Aerodynamics in forward Flight Using CFD," 20th AIAA Applied Aerodynamics Conference, St. Louis, MO, June 24-26, 2002.

[2]Gupta, V., and Baeder, J., "Quad Tiltrotor

Aerodynamics in Ground Effect," American Helicopter Society 58th Annual Forum Proceedings, Montreal, Canada, June 11-13, 2002.

[3]Sheng, C. and Narramore, J., "Computational Simulation and Anlaysis of Bell Boeing Quad Tiltrotor Aero Interaction," *Journal of the American Helicopter Society*, Vol. 54, (4), Oct. 2009, pp. 42002-42002.
doi: 10.4050/JAHS.54.042002

[4]Kim, T., Oh, S., and Yee, K., "Improved Actuator Surface Method for Wind Turbine Application," *Renewable Energy*, Vol. 76, April 2015, pp. 16-26.
doi: 10.1016/j.renene.2014.11.002

[5]Jasak, H., "Error Analysis and Estimation for the Finite Volume Method with Applications to Fluid Flows," Ph.D. Dissertation, Imperial College, London, 1996.

[6]Kim, T., Oh, S., and Yee, K., "Novel Actuator Surface Method for Helicopter Rotor Analysis," Journal of Aircraft, Vol. 53, (6), 2016, pp. 1947-1952.
doi: 10.2514/1.C033666

A New Ground Effect Factor Based on Image Vortices

Cibin Joseph
Doctoral Student
Aerospace Engineering
Indian Institute of Technology
Chennai, India

Ranjith Mohan
Assistant Professor
Aerospace Engineering
Indian Institute of Technology
Chennai, India

ABSTRACT

Cheeseman's factor for ground effect—based on image sources—has widely been used to account for the increase in thrust experienced by a lifting rotor in ground effect. However, for elevations lesser than half the rotor radius, the factor is known to drastically over predict this thrust augmentation. Even though these elevations may not be viable to full scale rotorcraft due to the fuselage height, they are quite commonly encountered in UAVs and MAVs. A new factor is proposed based on image vortices that is able to predict the thrust augmentation more accurately. Results are compared with multiple experiments from literature and are found to match exceptionally well.

NOTATION

A	Rotor disk area
C_T	Thrust Coefficient, $T/\rho A V_{TIP}^2$
IGE	In Ground Effect condition
OGE	Out of Ground Effect condition
R	Rotor radius
r_o	Reference radial location
$\bar{r_o}$	Non-dimensional reference radial location, r_o/R
T_g	Rotor thrust in ground effect
T_∞	Rotor thrust out of ground effect
v_{ig}	Induced velocity in ground effect
$v_{i\infty}$	Induced velocity out of ground effect
z	Elevation from ground level
$\bar{z}$	Non-dimensional elevation from ground, z/R
δv_i	Velocity induced by image rotor on actual rotor
Γ	Strength of point vortex

INTRODUCTION

Over the course of history, there have been several attempts at formulating a theoretical factor to predict the thrust augmentation experienced by a lifting rotor in the presence of a ground plane. The ground effect phenomenon, caused by a ground plane obstructing the wake shed from a lifting rotor was first examined theoretically by Betz (Ref. 1) and later experimentally by Zbrozek (Ref. 2). Cheeseman and Bennett (Ref. 3), Hayden (Ref. 4), Knight and Hefner (Refs. 5, 6) and Light (Ref. 7) are a few others who have investigated the phenomenon theoretically and/or experimentally later on.

Betz approached the rotor as a mechanism that creates suction, modelling it as a sink and the ground as an image sink of equal strength placed at a distance twice the actual rotor elevation. A substantial improvement in results from Betz's formulation was obtained when Cheeseman and Bennett represented the rotor as a constant strength source and the ground as an image source of equal strength. The analysis also utilized blade element theory to obtain the augmentation to rotor thrust at constant power due to the presence of the ground plane in both hover and forward flight. The theoretical investigation by Knight and Hefner utilized a set of cylindrical vortex sheets for the rotor and a corresponding set for the ground. Although the correlation obtained is good, the formulation requires numerical integration and binomial expansion of Legendre's polynomials. Cheeseman's factor, in this regard, is favourable as the expression for thrust augmentation is simple. However, the factor approaches infinity near elevations of 0.25R and is only practically applicable to elevations at 0.5R and above. This limitation is of little significance to full scale rotorcraft as the fuselage height restricts these elevations. On the other hand, for UAVs, MAVs and other smaller rotorcraft, these elevations are frequently encountered.

From the perspective of a theoretical formulation, an analysis using the method of images is promising as the resulting expressions are compact and the relationships between parameters are easily ascertained. On that note, the method of images is applied here to a set of 2D vortices in a manner similar to that used by Cheeseman to obtain a prediction for the thrust augmentation on a hovering rotor in ground effect at elevations of 0.1R and above. The results obtained are compared with Cheeseman's and Hayden's factor (Ref. 4) and other ground effect experiments from literature (Refs. 6–9).

Presented at the 6th Asian/Austrailian Rotorcraft Forum/Heli Japan 2017, Kanazawa, Japan, November 7-9, 2017.

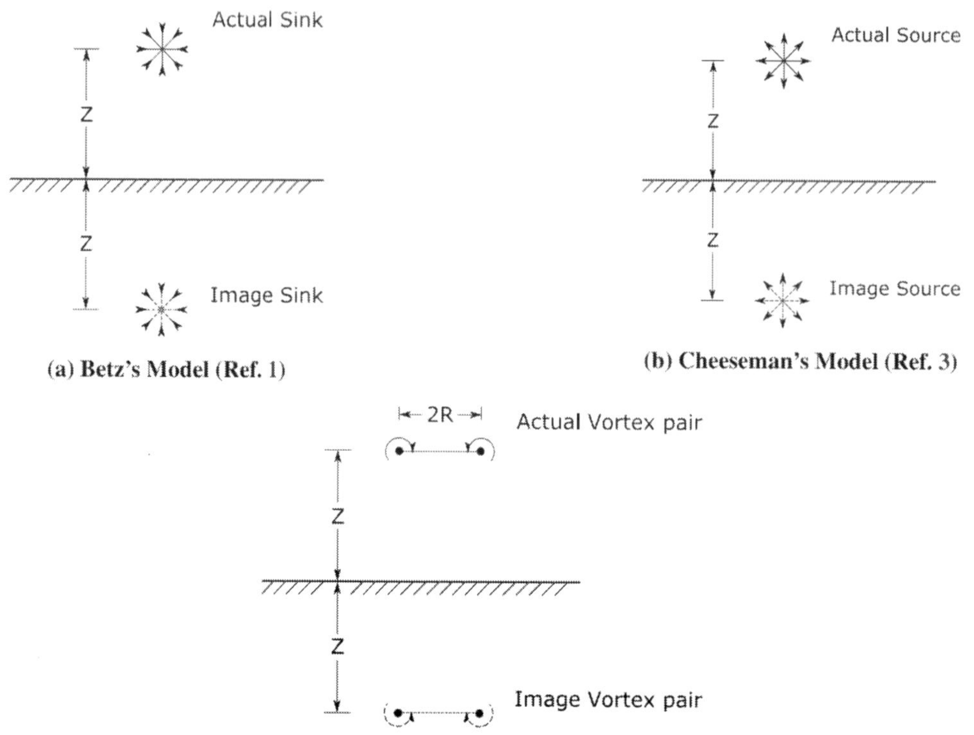

(a) Betz's Model (Ref. 1)

(b) Cheeseman's Model (Ref. 3)

(c) Proposed Model

Fig. 1: Schematic illustrating previous and proposed models

METHODOLOGY

The method of images is widely used in fluid mechanics to model potential fields containing a 'no-flux' boundary (Ref. 10). The inherent symmetry in the potential field due to the image establishes the impenetrable boundary condition required to be modelled. Shown in Figure 1 is an illustration of the arrangement used by Betz, Cheeseman and the authors for comparison. The change in induced velocity at the rotor disk due to ground effect is modelled as the velocity induced by the image rotor on the actual rotor. A constant power assumption is introduced to utilize this change in induced velocity in an easy manner.

FORMULATION

Assumptions

1. The lifting rotor is in hover.
2. As the elevation changes, the rotor maintains constant power.
3. The ground plane extends to infinity on all sides and is a no-flux boundary.
4. Induced velocity distribution across the rotor disk is assumed to be uniform.

Thrust Augmentation Factor Derivation

At constant power and uniform induced velocity distribution,

$$T_g \, v_{ig} = T_\infty \, v_{i\infty} \tag{1}$$

Hence,

$$\frac{T_g}{T_\infty} = \frac{v_{i\infty}}{v_{ig}} \tag{2}$$

If the velocity induced by the image rotor on the actual rotor is δv_i, then

$$v_{ig} = v_{i\infty} - \delta v_i \tag{3}$$

Substituting this to determine the thrust augmentation factor, we obtain from Equation 2

$$\frac{T_g}{T_\infty} = \frac{v_{i\infty}}{v_{i\infty} - \delta v_i} \tag{4}$$

$$= \frac{1}{1 - \frac{\delta v_i}{v_{i\infty}}} \tag{5}$$

An expression for $\delta v_i / v_{i\infty}$ may be substituted in Equation 5 to obtain the thrust augmentation factor.

Strength of vortices

Two point vortices are used to model the rotor flowfield in 2D. These vortices may be assumed to be the tip vortices shed by the rotor. The subsequent vortices shed are neglected to reduce the complexity to the formulation. This assumption is not vastly inconsistent as the rotor wake in the presence of the ground and side walls is highly diffusive due to turbulence/viscous effects (Ref. 11). The formulation is retained in two dimensions since a 3D ring vortex requires evaluation of complex integrals similar to those used by Knight and Hefner.

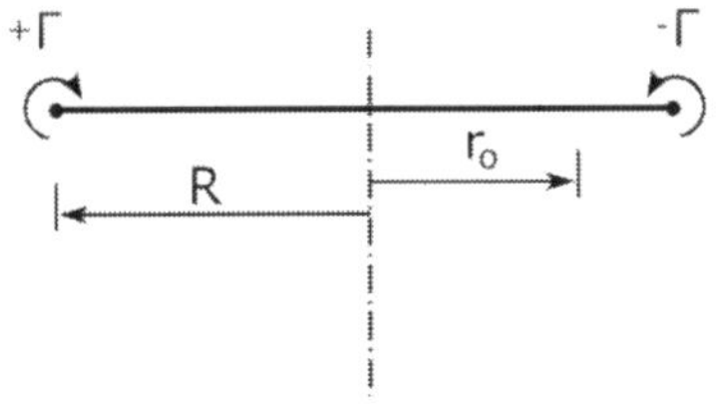

Fig. 2: Modeling rotor using 2 vortices

To determine the strength of the vortices, a reference radial location is chosen at which the velocity induced by the vortices is assumed to be equal to that across the actual rotor. Flow symmetry dictates that the vortices are of equal and opposite strengths.
If r_o is the reference radial location,
then induced velocity at r_o,

$$v_i(r_o) = \frac{\Gamma}{2\pi(R+r_o)} + \frac{\Gamma}{2\pi(R-r_o)} \tag{6}$$

$$= \frac{\Gamma}{\pi R(1-(r_o/R)^2)} \tag{7}$$

$$= \frac{\Gamma}{\pi R(1-\bar{r_o}^2)} \tag{8}$$

$$\Rightarrow \quad \Gamma = \pi R(1-\bar{r_o}^2)\, v_i \tag{9}$$

The non-dimensional quantity $\bar{r_o} = r_o/R$ is introduced here.

Expression for perturbation velocity, δv_i

We proceed to obtain an expression for the velocity induced by the image rotor across the actual rotor, δv_i, by summing up the vertical components of the velocities induced by each of the image vortices. Note that the strengths of the individual vortices for the image rotor is inverted as shown in Figure 3.

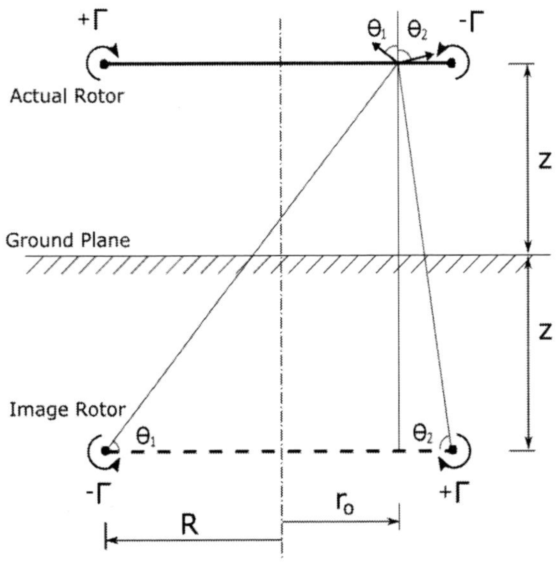

Fig. 3: Ground effect schematic

$$\delta v_i = \frac{\Gamma}{2\pi}\frac{\cos\theta_1}{\sqrt{(2z)^2+(R-r_o)^2}} + \frac{\Gamma}{2\pi}\frac{\cos\theta_2}{\sqrt{(2z)^2+(R+r_o)^2}} \tag{10}$$

$$= \frac{\Gamma}{2\pi}\frac{(R-r_o)}{((2z)^2+(R-r_o)^2)} + \frac{\Gamma}{2\pi}\frac{(R+r_o)}{((2z)^2+(R+r_o)^2)} \tag{11}$$

Where, from trigonometry,

$$\cos\theta_1 = \frac{R+r_o}{((2z)^2+(R+r_o)^2)} \tag{12}$$

$$\cos\theta_2 = \frac{R-r_o}{((2z)^2+(R-r_o)^2)} \tag{13}$$

Substituting for Γ from Equation 9, we obtain

$$\frac{\delta v_i}{v_i} = \frac{(1-\bar{r_0}^2)}{2}\left\{\frac{(1-\bar{r_o})}{(4\bar{z}^2+(1-\bar{r_o})^2)} + \frac{(1+\bar{r_o})}{(4\bar{z}^2+(1+\bar{r_o})^2)}\right\} \tag{14}$$

Note that the elevation here is written using the non-dimensional quantity $\bar{z} = z/R$.

Value of reference location r_o

To settle on a value for r_o, the inflow distribution for the physical rotor is assumed to be cubic. The reference location r_o may be obtained by equating the mass flow through an actuator disk with uniform inflow and one with the cubic inflow distribution.

$$\pi R^2 v_i = \int_0^R v_i(r)\, 2\pi r dr \tag{15}$$

$$= \int_0^R v_o r^3\, 2\pi r dr \tag{16}$$

$$= 2\pi v_o \frac{R^5}{5} \tag{17}$$

$$\Rightarrow \quad v_i = \frac{2}{5} v_o R^3 \tag{18}$$

For this uniform induced velocity, the reference location is,

$$v_i = \frac{2}{5} v_o R^3 = v_o r_o^3 \tag{19}$$

$$\Rightarrow \quad r_o = \sqrt[3]{\frac{2}{5}}R = 0.7368R \tag{20}$$

This value of r_o was also found to match acceptably with results of ground effect experiments as discussed in the following section. On substituting the value of r_o from Equation 20 in Equation 14, the expression for perturbation velocity simplifies to,

$$\frac{\delta v_i}{v_i} = 0.2286\left\{\frac{0.2632}{(0.0693+4\bar{z}^2)} + \frac{1.7368}{(3.0165+4\bar{z}^2)}\right\} \tag{21}$$

and further from Equation 5, the thrust augmentation factor reduces to the form,

$$\frac{T_g}{T_\infty} = \frac{1}{1-0.2286\left\{\frac{0.2632}{(0.0693+4\bar{z}^2)}+\frac{1.7368}{(3.0165+4\bar{z}^2)}\right\}} \tag{22}$$

$$= \frac{0.0131-0.7714\,\bar{z}^2+\bar{z}^4}{0.6572\,\bar{z}^2+\bar{z}^4} \tag{23}$$

or in a form similar to that used by Cheeseman's as

$$\frac{T_g}{T_\infty} = 1+\frac{0.0199}{\bar{z}^2}+\frac{0.0944}{0.6572+\bar{z}^2} \tag{24}$$

It is to be noted that assuming a different inflow distribution would produce a different value for r_o. In fact, it may easily be found out for an nth degree distribution, $v_i(r) = v_o r^n$, r_o is given by

$$r_o = \left(\frac{2}{n+2}\right)^{\frac{1}{n}} \tag{25}$$

and hence for a quadratic distribution, r_o is found to be 0.7071 and for a 4th order distribution, 0.7598 and so on. It was found that values of r_o in the range 0.70 - 0.76 agree well with experimental results.

Also, the inflow distribution between the two vortices that model the rotor is neither proportional to $1/r$ nor cubic (from the assumption used to find r_o). The vortices only serve the purpose of modelling the flowfield around the rotor and the perturbation in the induced velocity at r_o by the image vortices. Effectively, in the absence of the image vortices, the induced velocity at the rotor is assumed to be uniform with a magnitude equal to that induced by the vortices at r_o. This is a consequence of the uniform inflow distribution assumption that lead to equation 1.

RESULTS

Figure 4 shows the proposed thrust augmentation factor for a constant power lifting rotor plotted alongside results from experiments (Refs. 6–9) and popular ground effect factors (Refs. 3,4).

The root-mean-square error (RMSE) is a frequently used measure of the deviation in predicted and observed values. To offer a quick comparison between the proposed model and Cheeseman's and Hayden's models, the RMSE of the relative errors for each experimental dataset are provided in a tabular form in Table 1. The RMSE of the relative errors is calculated as

$$RMSE = \sqrt{\frac{(\sum_{i=1}^{n}\varepsilon_i^2)}{n}} \tag{26}$$

where, the relative error is calculated as,

$$\varepsilon_i = \frac{(\hat{y}_i - y_i)}{y_i} \tag{27}$$

$\hat{y}_i$ being the predicted value and y_i being the actual value obtained from experiments. The RMSE values are provided for an elevation range 0.1R to 4.0R. From the plot shown in Figure 4, the proposed factor appears to be closer to experimental values for an elevation less than 1.0R and hence RMSE values of relative errors for $z < 1.0R$ and $z > 1.0R$ are also provided separately in Table 1. Note that the datasets that were used from experiments may not have equal number of samples between experiments by different authors.

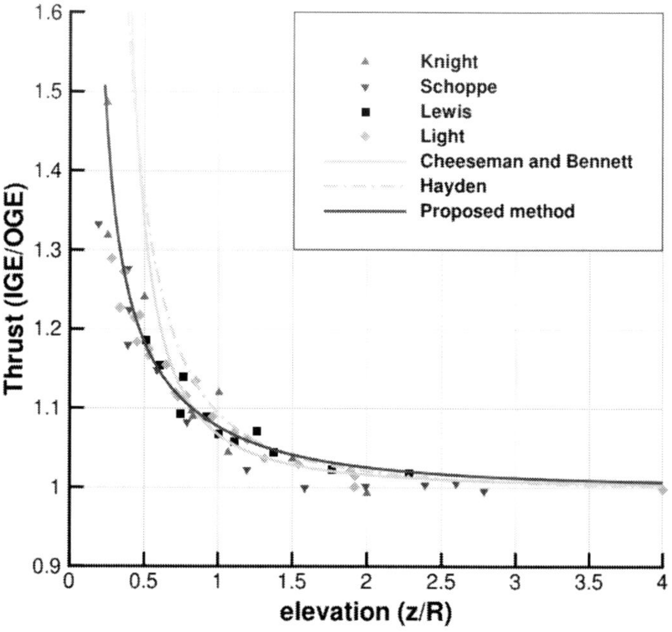

Fig. 4: Thrust augmentation factor for a lifting rotor at constant power in ground effect

Reference location r_o as a fitting parameter

Although in Equation 20, the value of the reference location r_o has been derived, the parameter may be changed as prescribed in Equation 25 to obtain slightly different curves for the thrust augmentation factor as shown in Figure 5. The value of r_o may in fact be considered as a fitting parameter and optimized to obtain a better correlation with specific experimental data or flight conditions. As an example, the Schoppe dataset—which happened to have the largest RMSE value for the proposed method—was used to find a better estimate to r_o. This may be posed as an optimization problem where the objective is to minimize the error in the values predicted by the proposed method by varying r_o. Standard algorithms like the simplex search method (Ref. 12) may be used to solve this single variable unconstrained non-linear optimisation problem. For the present analysis, the MATLAB function, *fminsearch* was used to find the minimum RMSE value to be 3.84 at $r_o = 0.8R$ as opposed to 8.20 at $r_o =$

Model	Light		Lewis		Schoppe		Knight	
Cheeseman	82.8	0.960	5.60	1.24	86.5	1.66	4.32	2.79
	66.0		**3.89**		**61.2**		**3.53**	
Hayden	24.2	1.08	8.24	1.31	55.6	2.46	6.82	2.59
	19.3		**5.64**		**39.4**		**4.87**	
Proposed	2.84	1.27	1.50	0.863	11.3	2.58	2.72	2.84
	2.39		**1.19**		**8.20**		**2.79**	

Table 1: RMSE values of relative error(%) in prediction for proposed and existing models over the range z=0.1R to 4.0R(shown in bold). The left and right values above that are RMSE values of relative error(%) calculated for the range z<1R and z>1R respectively.

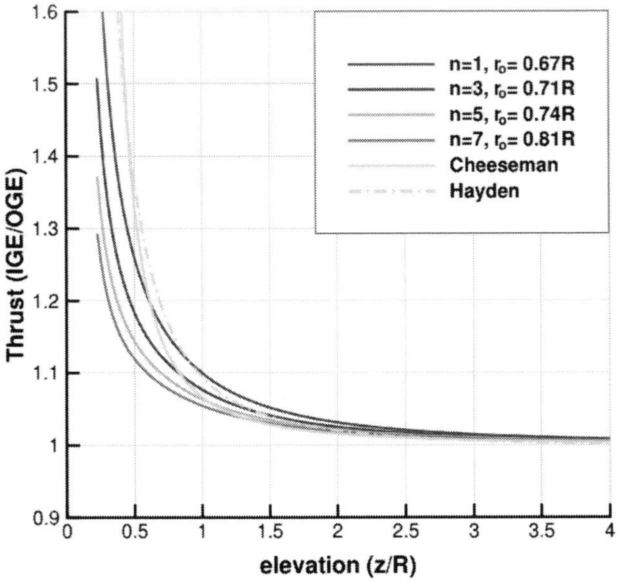

Fig. 5: Thrust augmentation factor for a lifting rotor at constant power in ground effect for various degrees(n) of approximating polynomial for inflow

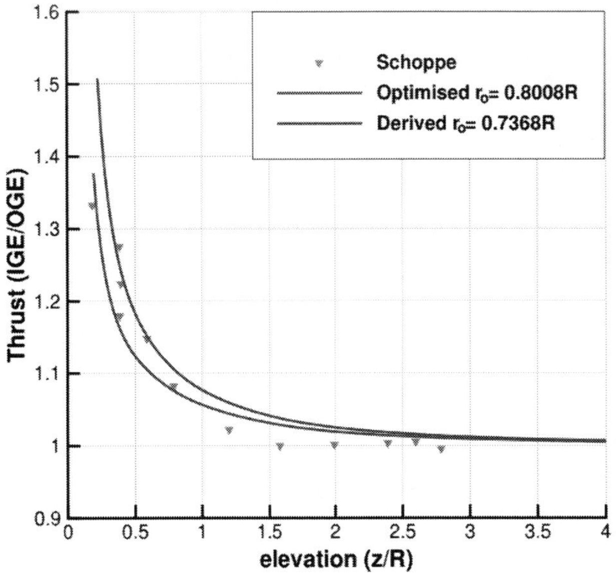

Fig. 6: Thrust augmentation factor for a lifting rotor at constant power in ground effect compared with derived and optimised models for the Schoppe dataset

$0.7368R$. Both the actual and optimised expressions along with the Schoppe dataset are plotted in Figure 6.

CONCLUSIONS

The proposed model based on method of images using point vortices is able to predict the thrust augmentation factor with much less error compared to existing models for ground effect. In the elevation range of one rotor radius and below, the results were found to match exceptionally well, while for larger elevations from the ground the variation was found to be similar to existing expressions in literature. The reference location r_o used to derive the expression for the thrust augmentation factor may also be used as a fitting parameter for specific experimental data or flight conditions in ground effect.

Author contact:
Cibin Joseph cibinjoseph92@gmail.com
Ranjith Mohan ranjith.m@iitm.ac.in

REFERENCES

[1]Betz, A., "The Ground Effect on Lifting Propellers," NACA Technical Memorandum 836, 1937.

[2]Zbrozek, J., "Ground Effect on the Lifting Rotor," ARC Reports and Memoranda 2347, July 1947.

[3]Cheeseman, I. and Bennett, W., "The Effect of the Ground on a Helicopter Rotor in Forward Flight," Aeronautical Research Council Reports and Memoranda 3021, 1955.

[4]Hayden, J., "The Effect of the Ground on Helicopter Hovering Power Required," 32nd Annual Forum of the

American Helicopter Society, Washington, D.C. USA, May 1976.

[5]Knight, M. and Hefner, R. A., "Static Thrust Analysis of the Lifting Airscrew," NACA Technical Notes 626, 1937.

[6]Knight, M. and Hefner, R. A., "Analysis of Ground Effect on the Lifting Airscrew," NACA Technical Notes 835, 1941.

[7]Light, J., "Tip Vortex Geometry of a Hovering Helicopter Rotor in Ground Effect," 45th Annual Forum of the American Helicopter Society, Boston, Massachusetts, USA, 1989.

[8]Schoppe, G., "Standschubprfungen mit der Modell-Hubschraube und Vorversuche mit der Modell-Hubschraube (Static Thrust Tests of the Model Helicopter Rotor and Preliminary Researchof the Model Helicopter Rotor)," Forschungsbericht FB275/1, Zentrale fr Technisch-wissen-schaftliches Berichtswesen ber Luftfahrtforschung, translated by J. Vanier, NACA US AMC Report F-TS-1003-RE, September 1946.

[9]Lewis, I. and Richard, B., "Army Helicopter Performance Trends," *Journal of the American Helicopter Society, April, 1972*.

[10]Kundu, P., Cohen, I., and Dowling, D., *Fluid Mechanics*, Elsevier Science, 2015.

[11]Milluzzo, J. I. and Leishman, J. G., "Vortical Sheet Behavior in the Wake of a Rotor in Ground Effect," *AIAA Journal*, Vol. 55, January 2017, pp. 24–35.

[12]Lagarias, J. C., Reeds, J. A., Wright, M. H., and Wright, P. E., "Convergence properties of the Nelder–Mead simplex method in low dimensions," *SIAM Journal on optimization*, Vol. 9, (1), 1998, pp. 112–147.

An Evaluation of Ground Effect Modelling for Rotors in Hover

Salini S. Nair
Doctoral Student
Aerospace Engineering
Indian Institute of Technology
Madras, India

Cibin Joseph
Doctoral Student
Aerospace Engineering
Indian Institute of Technology
Madras, India

Ranjith Mohan
Assistant Professor
Aerospace Engineering
Indian Institute of Technology
Madras, India

ABSTRACT

A well-known phenomenon in rotorcraft, the ground effect, occurs when a rotor is at close proximity to the ground creating an obstruction to the trailing wake from the rotor. This results in an increase in thrust and a decreased inflow through the rotor disk, in turn affecting handling qualities. Until recently, analytical modelling of ground effect involved using empirical factors or method of images to account for the variation in thrust as a function of the elevation. The Morillo-Peters dynamic inflow model has now been known to satisfactorily predict the variations in rotor thrust for a rotor in ground effect. Although the variation in thrust seem to agree with empirical formulations, the model needs to be evaluated with experimental or computational simulations by comparing the variation of inflow over the rotor disk. An investigation on how closely the model predicts the inflow distribution is beneficial to future developments, considering the possibility of extending the model to other dynamic environments involving ground effect. This paper draws a comparison between the inflow models of Morillo, Ke Yu and Peters, Nowak and He, and a CFD package, *RotCFD*, for a hovering rotor in ground effect.

NOMENCLATURE

A	Area of the rotor
$[A]$	Transform matrix
C_T	Thrust coefficient, $T/\rho AV_{TIP}^2$
D	Damping matrix
G	Coordinates relation matrix
H	Elevation from ground
IGE	In Ground Effect condition
$[L]$	Influence Coefficient matrix
$[M]$	Mass matrix
MD	Morillo-Duffy parameters
NH	Nowak-He parameters
OGE	Out of Ground Effect condition
P	Pressure
$P(\nu), Q(i\eta)$	Normalized associated Legendre functions of first and second kind respectively
R	Rotor radius
T	Rotor Thrust
V_m	Mass flow matrix
V_{TIP}	Blade tip velocity
a, α	Main rotor inflow coefficients
c, β	Ground rotor inflow coefficients
h	Non-dimensional elevation from ground, H/R
k_G	Ground effect factor/Thrust augmentation factor
m, r	Index for harmonics
n, j	Index for radial shape function
$\bar{r}$	Non dimensional radial location (r/R)
t	time
v, w	Inflow in the axial direction
Φ	Pressure potential
λ	Non dimensional inflow $(\frac{v}{\Omega R})$
ν, η, ψ	Ellipsoidal coordinates for rotor disc
ρ	Density
σ	Rotor solidity
τ	Main rotor pressure coefficients
τ	Collective pitch angle
ω	Ground rotor pressure coefficients

Presented at the 6th Asian/Australian Rotorcraft Forum/Heli Japan 2017, Kanazawa, Japan, November 7-9, 2017.

INTRODUCTION

Ground effect, characterised by an increase in thrust and reduction in inflow velocity across the rotor disk as a lifting rotor approaches the ground (Ref. 1), has been experimentally and theoretically investigated in literature using various techniques (Refs. 2–5).

Cheeseman (Ref. 3) analysed ground effect using the method of images applied to a pair of 3D flow sources and provided a compact expression for the increase in thrust at constant power as the rotor approaches the ground. Hayden (Ref. 6) developed an expression for this increase in thrust using experimental data obtained from various rotors. The expression for Cheeseman's k_G factor is provided in Eqn.1

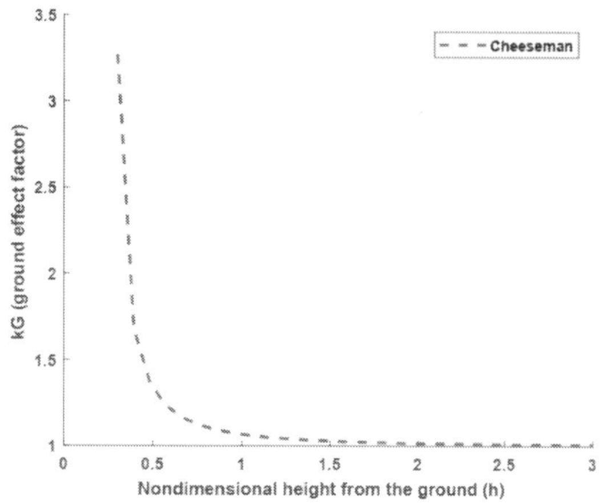

Fig. 1: Variation of Cheeseman's k_G factor with distance from the ground

and plotted in Fig.1.

$$k_G = \frac{1}{1-\left(\frac{1}{4h}\right)^2} \tag{1}$$

Here h is the distance from the ground non-dimensionalised with respect to the rotor radius. As observed from the figure, the effect of ground on thrust is relatively insignificant for elevations above $1.5R$ to $2R$ from the ground and is generally classified as the out of ground effect regime(OGE). However, experiments on ground effect (Ref. 5) suggest that Cheeseman's factor and Hayden's formulations drastically over-predict the thrust at small elevations like $0.25R$. In (Ref. 7), a modified form of image vortex method is used to theoretically formulate k_G factor and it has been found to have better correlation with experimental results.

Quite recently, finite state inflow models have been developed for modelling ground effect (Ref. 8). They appear to hold promise since both steady and dynamic responses (Refs. 9,10) to the ground and forward speed may be modelled using finite state models.

The focus of this paper is on two different techniques, finite state dynamic inflow models and using a CFD package, *RotCFD*. The results obtained are validated using experimental results available in literature (Ref. 5).

Ground effect modelling using finite state dynamic inflow model

Finite state dynamic inflow models are used for computation of inflow through the rotor both above and below the rotor disc. The model accounts for the variation of induced flow in the radial and azimuthal direction as well as time (the unsteady effect). The most widely accepted finite state model is the one proposed by Peters-He (Ref. 11), which can be used for computing the inflow anywhere on the rotor disc. The model is derived from the basic potential flow equations.

The pressure distribution at the rotor disk is given by Eqn.2 where ν, η, ψ are ellipsoidal coordinates. The pressure potential Φ is a solution to Laplace equation and satisfies the condition of pressure discontinuity across the rotor disc due to the nature of $P(\nu)$ and $Q(i\eta)$, which are normalized associated Legendre functions of first and second kind. In Eqn. 2, τ represents the rotor pressure coefficients computed from loading on the rotor blade.

$$\Phi(\nu,\eta,\psi,t) = -\frac{1}{2}\sum_{m=0}^{\infty}\sum_{n=m+1,m+3..}^{\infty} \bar{P}_n^m(\nu)\bar{Q}_n^m(i\eta) \left[\tau_n^{mc}(t)\cos(m\psi) + \tau_n^{ms}(t)\sin(m\psi)\right] \tag{2}$$

The inflow in the axial direction through the rotor is computed using Eqn.3 where α represent the inflow coefficients.

$$w = \sum_{r=0}^{\infty}\sum_{j=r+1,r+3..}^{\infty} \bar{P}_j^r(\nu)\bar{Q}_j^r(i\eta)\left[\alpha_j^{rc}(t)\cos(r\psi) + \alpha_j^{rs}(t)\sin(r\psi)\right] \tag{3}$$

Here the Legendre functions describe the radial variation of inflow and trigonometric functions the azimuthal variation. The model is in the form of a set of ordinary differential equations (Eqn.4) where M and L represent the mass and influence coefficient matrices respectively and have closed form representations, V_m represents the mean flow through the rotor.

$$[M]\{\dot{\alpha}_j^r\} + V_m[L]^{-1}\{\alpha_j^r\} = \left\{\frac{\tau_n^m}{2}\right\} \tag{4}$$

The model however cannot be used for computation of inflow outside the disc region and also cannot account for ground effect. The simplest way to account for ground effect in the dynamic inflow model is to use the empirical factor k_G to account for the change in mass flow through the rotor by using $V_m k_G$ instead of V_m in Eqn. 4. This implies all the inflow coefficients α are reduced to account for ground effect. The accuracy of the results depend on the accuracy of the k_G factor used.

As given in Eqn.2, only $m+n = odd$ terms, which accounts for the discontinuity across the rotor disc were included in Peters-He model. Further advances were made on Peters-He model to include the effect of mass source terms by incorporating $m+n = even$ terms as well into the expansion in (Ref. 12).The model is of the form given in Eqn.5, where all the matrices have closed form representations and $j+r = even$ as well as $j+r = odd$ terms are included with the exclusion of $j = r$ terms in the velocity expansion. The model converged slowly but could be used for computing inflow on or above the rotor plane and paved the way for using dynamic inflow models in ground effect analysis.

$$[M]\{\dot{\alpha}_j^r\} + [D]V_m\{\alpha_j^r\} = \{\tau_n^m\} \tag{5}$$

In (Ref. 13), which is based on Peters-He dynamic inflow model, the no penetration condition at the ground was implemented by equating the pressure due to the main rotor (Eqn. 6) on the ground to an equivalent pressure at the ground. The ground pressure (Eqn. 7) was represented as mass source perturbations and hence $r+j = even$ terms alone are taken to represent the ground pressure potential. By equating the two, the ground pressure states ω_j^{rc} and ω_j^{rs} can be determined from the main rotor pressure states τ_n^{mc} and τ_n^{ms} using appropriate coordinate transformations.

$$P_{main} = -\sum_{m=0}^{\infty} \sum_{n=m+1,m+3...}^{\infty} \left[\tau_n^{mc}\phi_n^{mc} + \tau_n^{ms}\phi_n^{ms}\right], \tag{6}$$
$$\phi_n^{mc} = \bar{P_n^m}(\nu)\bar{Q_n^m}(i\eta)\cos(m\psi),$$
$$\phi_n^{ms} = \bar{P_n^m}(\nu)\bar{Q_n^m}(i\eta)\sin(m\psi)$$

$$P_{grnd} = -\sum_{r=0}^{\infty} \sum_{j=r,r+2..}^{\infty} \left[\omega_j^{rc}\phi_n^{mc} + \omega_j^{rs}\phi_n^{ms}\right] \tag{7}$$

In this model, the unsteady part of the ground pressure perturbation is neglected and using potential theory equations, the ground interference velocity states are directly computed from the ground pressure states and the pressure equivalence relation. The model was able to get fairly good results for complete, partial and dynamic ground effect. However, the implementation involved the computation of several integrals, which is time consuming.

The representation of ground effect based on the finite state model from (Ref. 12) was developed in (Ref. 10) by incorporating a ground rotor into the system. The ground rotor generates pressure at the ground rotor disc equal to the pressure due to the main rotor on the ground thereby satisfying the no-penetration condition. The ground rotor does not generate lift and is assumed to have infinite number of blades. The model was an extension of the model proposed in (Ref. 12) by including the $m = n$ terms in the pressure expansion of the ground rotor. Thus, a set of differential equations, one representing the main rotor (Eqn. 8), and the other representing the ground rotor (Eqn. 9) were developed.

$$[M]\{\dot{a}_n^m\} + [D][V]\{a_n^m\} = [D]\{\tau_n^m\} \tag{8}$$

$$[M]\{\dot{c}_n^m\} + [D][V]\{c_n^m\} = [D]\{\omega_n^m\} \tag{9}$$

The model follows the pressure equivalence approach to compute the ground rotor pressure states ω from the main rotor pressure states τ (Eqn.10). The model can be used to simulate complete and partial ground effect and can be used to represent ground surface of any shape by calculating appropriate G_{jn}^{rm} matrix.

$$\{\omega_j^r\} = \left[G_{jn}^{rm}\right]\{\tau_n^m\}, \tag{10}$$
$$n = m+1, m+3.., j = r, r+2..$$

The inflow at the main rotor plane due to main rotor is calculated from Eqn. 11 and the inflow at the main rotor due to the ground rotor, which in effect accounts for the ground effect, is determined from Eqn. 12. The net inflow through the rotor in the presence of ground is given by Eqn.13.

$$v_{main} = \sum_{m=0}^{\infty} \sum_{n=m+1}^{\infty} (a_n^{mc}\phi_n^{mc} + a_n^{ms}\phi_n^{ms}) \tag{11}$$

$$v_{grnd} = \sum_{m=0}^{\infty} \sum_{n=m+1}^{\infty} (c_n^{mc}\phi_n^{mc} + c_n^{ms}\phi_n^{ms}) \tag{12}$$

$$v = v_{main} + v_{grnd} \tag{13}$$

However, the model was observed to under predict the effect of ground and a correction factor was required to be applied in the pressure equivalence relation to obtain better correlation with the actual solution.

Further work was done on the model in (Ref. 10) by including $m = n$ terms in the velocity expansion in (Ref. 14). This eliminated the requirement of the correction factor and gave fairly good correlation with the actual solutions obtained using convolution.

The advanced models compared to the basic model in (Ref. 11) were not well conditioned and the nonlinear models had convergence issues. An improved model developed in (Ref. 15) combines modified form of He model (Ref. 11), referred to as Nowak-He model, and the model from (Ref. 14), referred to as the Morillo Duffy model, using a blending function. The blending function is such that on the rotor disc the results will be similar to He model while at distances quite far from the rotor disc Morillo-Duffy model dominates. The fundamental aim behind developing such a model is to compute the inflow in the rotor wake. Here the model is used to compute the inflow when the rotor is in ground effect.

The model is as given in Eqn. 14, where $[M_{new}]$ is the new mass matrix. Similar formulation is obtained for the ground rotor with the ground rotor pressure states and the main rotor pressure states related as in Eqn.10. The Morillo-Duffy variables $\left\{a_j^m\right\}$ can be obtained from the Nowak-He variables $\{\alpha_n^m\}$ using the transformation given in Eqn. 15, where the matrix $\left[A_{nj}^m\right]$ has closed from representation.

$$[M_{new}]\{\dot{\alpha}_n^m\} + [D]V_m[L]^{-1}[M_{new}]\{\alpha_n^m\} = [D]\{\tau_n^m\} \tag{14}$$

$$\{a_j^m\} = [A_{nj}^m]^{-1} \{\alpha_n^m\} \quad (15)$$

The inflow is computed from two different equations, giving the Morillo-Duffy component V_{MD} (Eqn.16) and the Nowak-He component V_{NH} (Eqn.17). The blended inflow is obtained as in Eqn. 18, where b is a blending factor and $h = \nu - \varepsilon$, where ε is a parameter used to ensure that quite close to the rotor disc Morillo-Duffy component is not used.

$$V_{MD} = \sum_{m=0}^{\infty} \sum_{n=m}^{\infty} a_n^{mc} \phi_n^{mc} + a_n^{ms} \phi_n^{ms} \quad (16)$$

$$\begin{aligned} V_{NH} = &\sum_{m=0}^{\infty} \sum_{n=m+1,m+3..}^{\infty} \alpha_n^{mc} \frac{1}{\nu} \phi_n^{mc} + \alpha_n^{ms} \frac{1}{\nu} \phi_n^{ms} \\ &+ \sum_{m=0}^{\infty} \sum_{n=m,m+2..}^{\infty} a_n^{mc} \phi_n^{mc} + a_n^{ms} \phi_n^{ms} \end{aligned} \quad (17)$$

$$V_{BL} = \frac{1}{1+bh} V_{NH} + \frac{bh}{1+bh} V_{MD} \quad (18)$$

The approach to simulate ground effect is similar to previous models, that is pressure equivalence. A similar equation as Eqn.14, given by Eqn.19 is used for the ground rotor states and the relation given in Eqn.10 is used to compute the ground rotor pressure states.

$$[M_{new}] \left\{ \dot{\beta}_n^m \right\} + [D] V_m [L]^{-1} [M_{new}] \{\beta_n^m\} = [D] \{\omega_n^m\} \quad (19)$$

METHODOLOGY

The focus in this paper is on obtaining the coefficient of thrust, k_G factor and radial distribution of inflow and loading from both CFD simulation using *RotCFD* and dynamic inflow models. The parameters used for modeling are obtained from (Ref. 5) and is given in Table 1. For OGE case, two different collective pitch angles, $\theta_0 = 4^0$ and $\theta_0 = 8^0$ are considered while for the IGE case , $\theta_0 = 8^0$ is considered. Distances from the ground to the rotor are varied from $0.25R$ to $3R$. Hover condition alone is considered. The lift coefficient corresponding to different angles of attack for the NACA 0012 airfoil include static stall and the effect of Mach number.

Dynamic Inflow Model

Three different dynamic inflow models are considered in this paper, He model (Ref. 11), Morillo-Duffy model (Ref. 14), Blended model (Ref. 15). The nonlinear inflow models are used to obtain the coefficient of thrust (C_T), the radial variation of inflow and loading for different distances of the rotor from the ground. The differential equations corresponding to the different models are solved in Matlab until convergence is obtained for C_T.

He model The differential equations corresponding to the model are as in Eqn.4, with the difference that the model is nonlinear. This implies the mass flow matrix V_m is not a constant but dependent on the inflow states as given in Eqn.20 (Ref. 11). To incorporate the effect of ground, V_m is replaced by $V_m k_G$, where Hayden's formula for thrust augmentation given in (Ref. 6) is used.

$$V_m = \begin{bmatrix} \bar{\nu} & 0 & 0 & \dots \\ 0 & 2\bar{\nu} & 0 & \dots \\ \vdots & \vdots & \ddots & \\ 0 & \dots & \dots & 2\bar{\nu} \end{bmatrix}, \bar{\nu} = \frac{2}{\sqrt{3}} [1 \ 0 \ 0 \ \dots \ 0] [L]^{-1} \{\alpha\} \quad (20)$$

Morillo-Duffy model The differential equations used are as given in Eqns.8 and 9, with $m = n$ terms included. The $[V]$ matrix for nonlinear formulation is given in Eqn.21.

$$[V] = \begin{bmatrix} \bar{\nu} & 0 & 0 & \dots \\ 0 & 2\bar{\nu} & 0 & \dots \\ \vdots & \vdots & \ddots & \\ 0 & \dots & \dots & 2\bar{\nu} \end{bmatrix}, \bar{\nu} = \frac{2}{\sqrt{3}} a_1^0 \quad (21)$$

Blended model The differential equations given in Eqn.14 are used. For the computation of inflow at the main rotor due to the main rotor, the Nowak-He component alone is used, implying $h = 0$ in Eqn.18. For computing the inflow due to the ground rotor on the main rotor, Eqn. 19 is used. Here the value of the blending parameter b is taken as 20 and that of ε as .01.

CFD Solver

RotCFD, the flow solver used in this investigation is a software package capable of simulating a complete rotorcraft and rotor aerodynamic interactions with other aircraft or bodies. Of the various modules provided by *RotCFD*, this research utilizes RotUNS, the fluid solver module that uses unstructured octree type meshing. The rotor is modelled as a distribution of momentum sources, the strengths of which are determined from flow-field properties, rotor geometry and aerodynamic characteristics of the blade cross-section (Refs. 16–19). The realizable $k-\varepsilon$ turbulence model was used for all simulations.

For simulating ground effect, the boundary of the domain was assigned a viscous wall boundary condition. The fluid properties were set to ambient conditions as shown in Table 2.

Convergence was ensured by monitoring thrust and torque coefficients alongside residuals of the fluid equations. On an average, the simulations required to be run to around 200 rotor revolutions for convergence with an azimuthal resolution of 5° per timestep (iteration).

Parameter	Value
Number of blades N_b	2
Blade radius (m) R	.762
Chord (non dimensional) c/R	.067
Airfoil	NACA0012
Solidity σ	.0426
Hinge offset (non dimensional) e/R	.034
Root cutout (non dimensional) rc/R	.167
Omega (rpm)	960
Density (kg/m^3)	1.28

Table 1: Parameters used for modeling (Ref. 5)

Parameter	Metric
Tip Mach number	0.23
Tip Reynolds number[a]	2.78×10^5
Static Density	1.28 kg/m^3
Static Pressure	103351.5 bar
Static Temperature	279.65 K
Dynamic Viscosity	1.8×10^{-5} kg/ms

[a]chord-based

Table 2: Flow Properties

Grid System

Unstructured 3D grids with tetrahedral elements, generated using an in-built octree-type method (Ref. 20) were used for all simulations. Grid refinement was provided at the rotor and in the region the wake impinges the ground. The domain extents chosen and a representative cross-section of the grid used is shown in Fig. 2.

RESULTS AND DISCUSSION

Out of Ground Effect (OGE)

Two different collectives are considered in this analysis, 4^0 and 8^0. The converged values of C_T obtained using the different inflow models and *RotCFD* is shown along with the experimental results from (Ref. 5) in Table 3. The results show that C_T values obtained using the different methods compares fairly well with each other and with the experimental result.

Now the radial variation of inflow and loading is considered. Figures 3 and 4 shows the variation of inflow and loading on the blade along the radius, the horizontal axis is the radial distance from the root non dimensional with respect to the rotor radius. The inflow variation obtained using blade element momentum theory (BEMT) is also plotted in the inflow plot (Fig.3) for comparison (indicated + sign in the figure). The BEMT formula for inflow ratio λ is given in Eqn.22 where σ is the solidity, a the lift curve coefficient, θ_0 the collective and $\bar{r}$ is the non dimensional radial distance. In the calculation here, the lift curve coefficient has been assumed to have a constant value 2π.

$$\lambda = \frac{\sigma a}{16}\left[\sqrt{1+\frac{32}{\sigma a}\theta_0 \bar{r}}-1\right] \tag{22}$$

The plot shows reasonable agreement between the dynamic inflow model, BEMT and the *RotCFD* results in the region on the blade ranging from .45R to .75R, which is the major region on the blade contributing to thrust and hence agreement in the C_T values as given in Table 3. There is considerable difference in inflow in the root and the tip regions (Fig.3), which is reflected in the loading plots (Fig.4). This should be due to the nature of the radial shape functions (Legendre polynomials) used to describe the inflow and loading variation in the radial direction in the dynamic inflow model.

Similar results are observed for collective $\theta_0 = 4^0$ as shown in Figs.5 and 6.

In Ground Effect (IGE)

In IGE analysis, collective $\theta_0 = 8^0$ alone is considered and the varying distances from the ground ranging from .25R to 3R is considered. Figure 7 shows the variation of C_T with distance from the ground h (non dimensional with rotor radius R). It can be observed that the C_T values obtained using dynamic inflow models are quite close to the values from experiment (Ref. 5) for smaller distances from the ground (say, .3R to 1R), while the results are over predicted beyond 1R. The results obtained using *RotCFD* are slightly higher than the values obtained using dynamic inflow models in region ranging from .3R to 1R and slightly lower in the region beyond 1R.

Figure 8 shows the variation of k_G factor with distance from the ground h. It is observed that the dynamic inflow models compared with experimental values under predict the k_G factor while *RotCFD* results show fairly good correlation.

Figures 9 and 10 shows the radial variation of inflow and loading obtained using different methods when the rotor

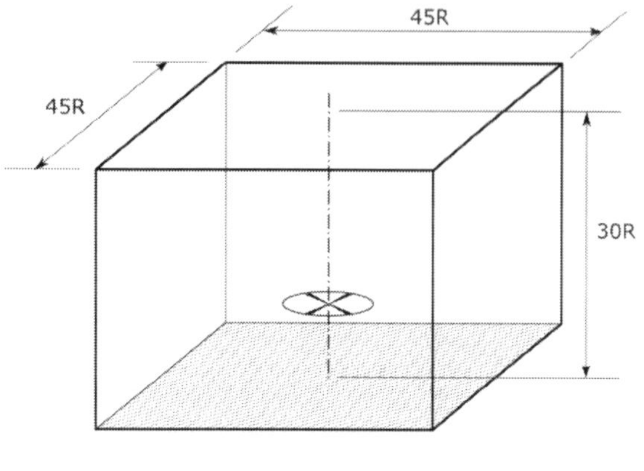

(a) Domain extents

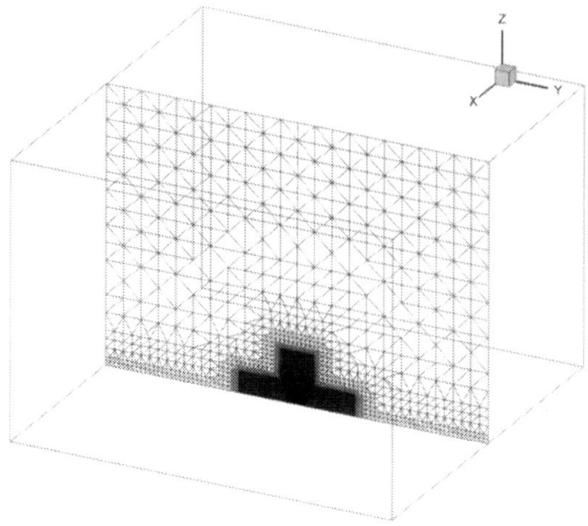

(b) Representative mesh

Fig. 2: Ground Effect: Domain extents and representative mesh

Table 3: Converged C_T values obtained using different models compared with available experimental results from (Ref. 5)

Collective	He model	Morillo-Duffy model	Blended model	*RotCFD*	Experimental Results (Ref. 5)
4^0	.0013	.0013	.0013	.001221	
8^0	.0031	.0032	.0032	.003106	0.00292136

is at $h = .5R$ while Figs. 11 and 12 show the same plots for $h = .25R$. It is observed that while the loading plots show fairly good correlation between *RotCFD* results and dynamic inflow model results, the inflow plots widely differ between the two. This again could be attributed to the nature of Legendre polynomials and also to the fact that dynamic inflow models under predict the effect of ground (Fig. 8).

This can be clearly observed from Fig.13, which shows $\frac{\lambda^{OGE} - \lambda^{IGE}}{\sqrt{C_T^{OGE}/2}}$ plotted with respect to non dimensional radial distance. Here, λ^{OGE} refers to the non dimensional inflow ($\lambda = \frac{v_z}{\Omega R}$) in out of ground effect while λ^{IGE} refers to the non dimensional inflow in ground effect when $h = .25R$. C_T^{OGE} refers to the value of C_T in out of ground effect condition (Table 3). The plot shows that upwash or difference in inflow predicted by dynamic inflow models is much lower than that predicted by *RotCFD*. This is the reason behind the difference in the inflow plots in Figs. 9 and 11.

CONCLUSIONS

When a rotor is hovering close to the close to the ground, it experiences an upwash or reduction in inflow through the rotor disc. This results in an increase in thrust or, in other words, more thrust is produced at the same power as distance of the rotor from the ground decreases. Several methods have been proposed in literature to quantify this effect and experimental analysis was also done from which empirical formulas were developed to quantify this effect.

In this paper, three different dynamic inflow models and a CFD package called *RotCFD* were used to simulate rotor in ground effect. Th coefficient of thrust and k_G factor obtained from these analysis were compared with available experimental results from literature. The radial variation of inflow , loading and upwash obtained using dynamic inflow models and *RotCFD* were also compared.

In out of ground effect analysis, it was observed that the C_T values obtained from dynamic inflow model analysis, *RotCFD* analysis and experiment matches quite well. The radial variation of inflow and loading obtained using dynamic inflow models and *RotCFD* were quite comparable.

In ground effect analysis, it was observed for smaller distances from the ground, the results obtained showed a fairly good correlation with experimental results, while at distances from the ground beyond $1R$, the analysis values of C_T were higher than those obtained from experiments. The k_G factor from *RotCFD* analysis match quite well with those obtained from experiments and the dynamic inflow models were found to under-predict the effect of ground. That is, a higher upwash is predicted by *RotCFD* compared to dynamic inflow models.

Author contact:

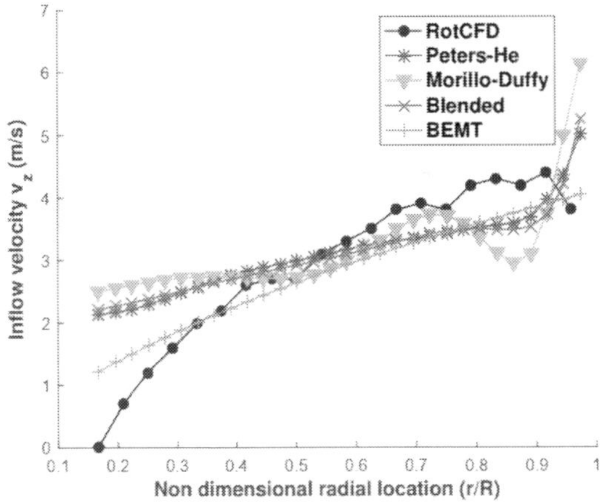

Fig. 3: Variation of inflow velocity (v_z) along the radius using *RotCFD*, BEMT and different dynamic inflow models ($\theta_0 = 8^0$)

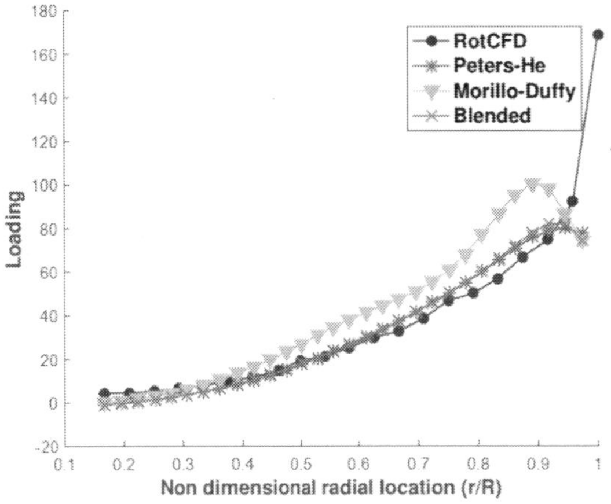

Fig. 4: Variation of loading along the radius using *RotCFD* and different dynamic inflow models ($\theta_0 = 8^0$)

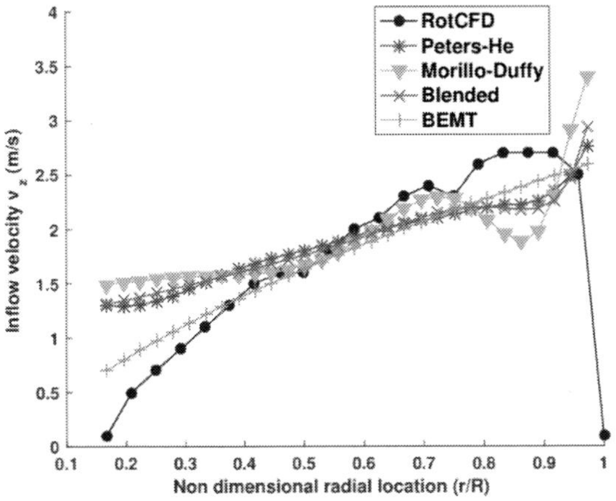

Fig. 5: Variation of inflow velocity (v_z) along the radius using *RotCFD*, BEMT and different dynamic inflow models ($\theta_0 = 4^0$)

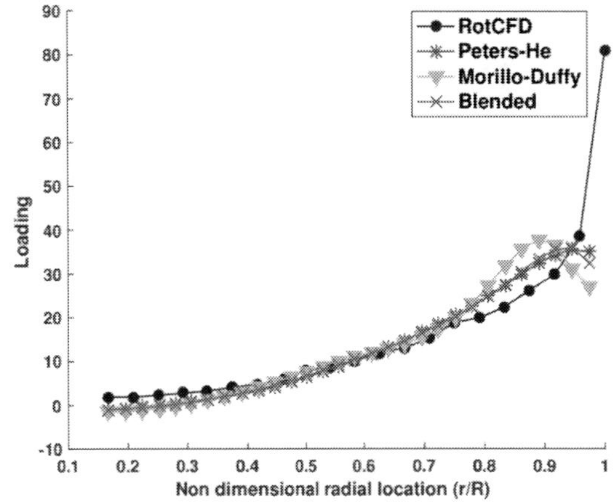

Fig. 6: Variation of loading along the radius using *RotCFD* and different dynamic inflow models ($\theta_0 = 4^0$)

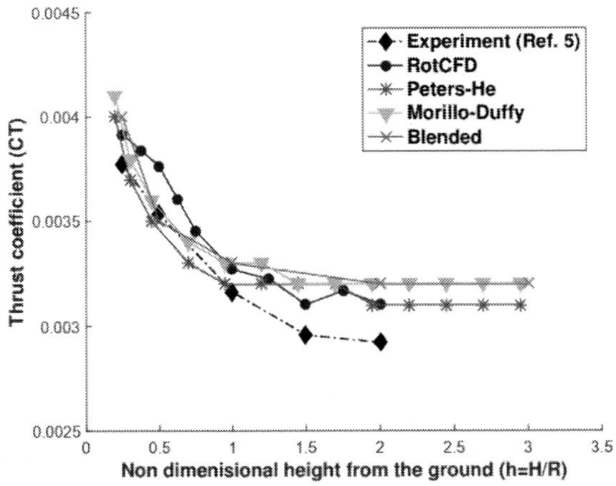

Fig. 7: Variation of C_T obtained with non dimensional height from the ground ($\theta_0 = 8^0$)

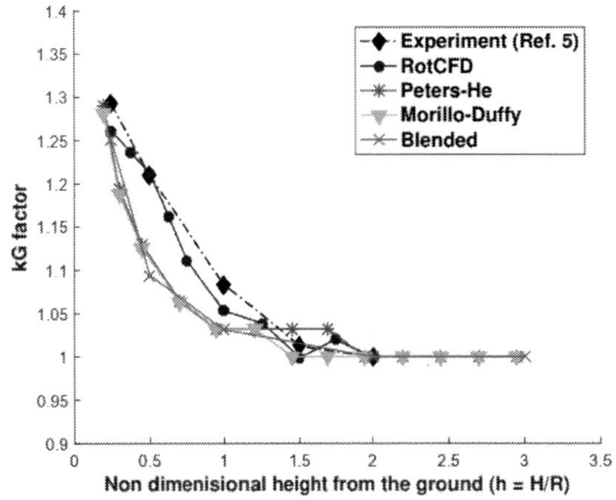

Fig. 8: Variation of k_G with non dimensional height from the ground ($\theta_0 = 8^0$)

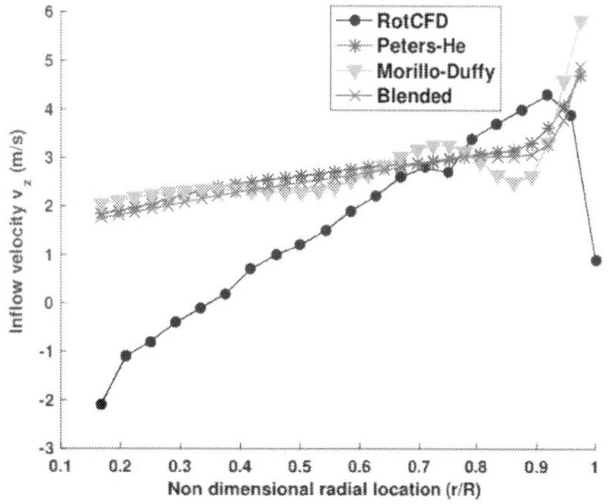

Fig. 9: Variation of inflow velocity (v_z) along the radius using *RotCFD* and different dynamic inflow models at $h = .5R$ ($\theta_0 = 8^0$)

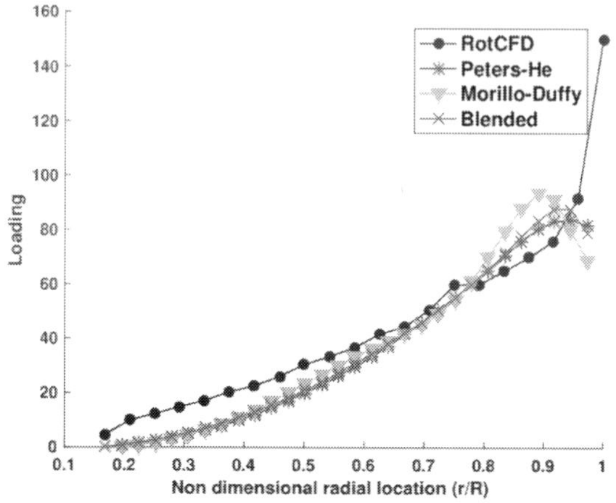

Fig. 10: Variation of loading along the radius using *RotCFD* and different dynamic inflow models at $h = .5R$ ($\theta_0 = 8^0$)

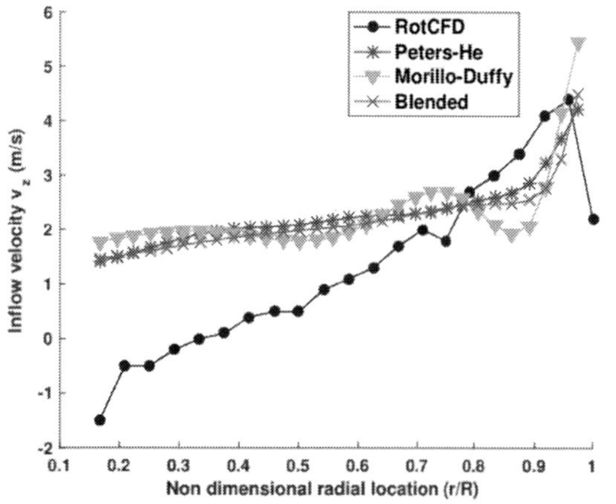

Fig. 11: Variation of inflow velocity (v_z) along the radius using *RotCFD* and different dynamic inflow models at $h = .25R(\theta_0 = 8^0)$

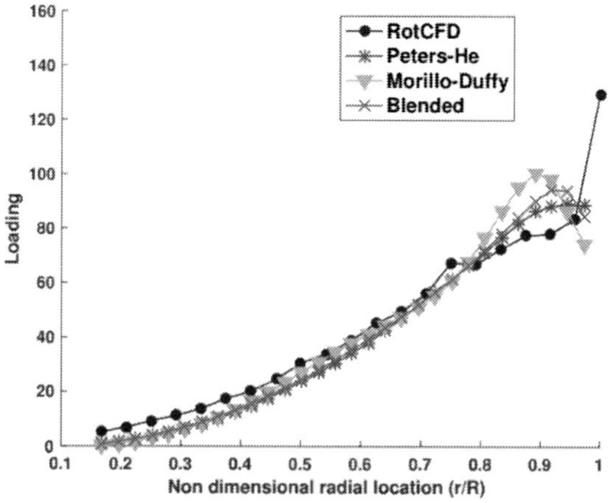

Fig. 12: Variation of loading along the radius using *RotCFD* and different dynamic inflow models at $h = .25R$ ($\theta_0 = 8^0$)

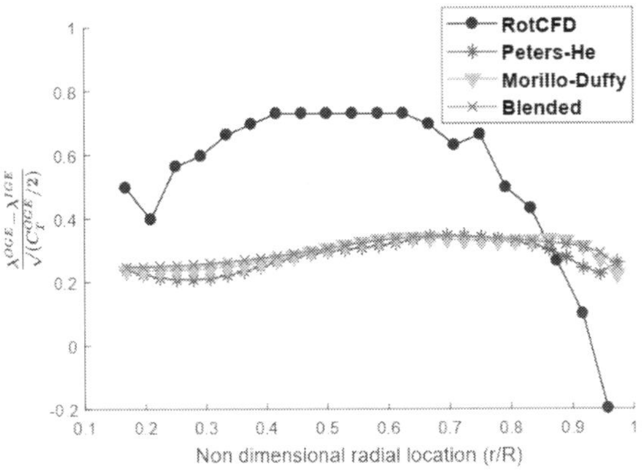

Fig. 13: Variation of difference in inflow (non dimensional) caused by ground effect along the radius at $h = .25R$ ($\theta_0 = 8^0$)

Salini S. Nair snairsalini@gmail.com
Cibin Joseph cibinjoseph92@gmail.com
Ranjith Mohan ranjith.m@iitm.ac.in

REFERENCES

[1]Johnson, W., *Rotorcraft aeromechanics*, Vol. 36, Cambridge University Press, 2013.

[2]Betz, A., "The ground effect on lifting propellers," Technical report, National Advisory Committee for Aeronautics, 1937.

[3]Cheeseman, I. and Bennett, W., "The effect of ground on a helicopter rotor in forward flight," , 1955.

[4]Knight, M. and Hegner, R. A., "Analysis of ground effect on the lifting airscrew," , 1941.

[5]Light, J. S., "Tip vortex geometry of a hovering helicopter rotor in ground effect," *Journal of the American helicopter society*, Vol. 38, (2), 1993, pp. 34–42.

[6]Hayden, J. S., "The effect of the ground on helicopter hovering power required," 32nd annual forum of the American helicopter society, Washington, DC, 1976.

[7]Joseph, C. and Mohan, R., "A Numerical Investigation of Ground Effect on Rotorcraft in the presence of a Side wall," 43rd European Rotorcraft Fotum, Milan, Italy, 2017.

[8]Peters, D. A., "Review of dynamic wake models for application to dynamics and stability of rotorcraft," Technical report, WASHINGTON UNIV ST LOUIS MO DEPT OF MECHANICAL ENGINEERING, 2003.

[9]Gaonkar, G. H. and Peters, D. A., "Effectiveness of Current Dynamic-Inflow Models in Hover and Forward Flight," *Journal of the American Helicopter Society*, Vol. 31, (2), 1986, pp. 47–57.

[10]Yu, K. and Peters, D. A., "Nonlinear State-Space Modeling of Dynamic Ground Effect," *Journal of the American Helicopter Society*, Vol. 50, (3), 2005, pp. 259–268.

[11]Peters, D. A., Boyd, D. D., and He, C. J., "Finite-State Induced-Flow Model for Rotors in Hover and Forward Flight," *Journal of the American Helicopter Society*, Vol. 34, (4), 1989, pp. 5–17.

[12]Peters, D. A., Morillo, J. A., and Nelson, A. M., "New developments in dynamic wake modeling for dynamics applications," *Journal of the American Helicopter Society*, Vol. 48, (2), 2003, pp. 120–127.

[13]Xin, H., *Development and validation of a generalized ground effect model for lifting rotors*, Ph.D. thesis, Georgia Institute of Technology, 1999.

[14]Garcia-Duffy, C., Hsieh, A., and Peters, D. A., "A COMPLETE, NONLINEAR INDUCED FLOW THEORY FOR ROTORS IN INCOMPRESSIBLE FLOW," Proceedings of the 11th pan American congress of applied mechanics–PACAM XI, 2009.

[15]Huang, J., Peters, D. A., and Prasad, J., "Converged velocity field for rotors by a blended potential flow method," *GSTF Journal on Aviation Technology (JAT)*, Vol. 1, (2), 2015.

[16]Rajagopalan, R. G. and Mathur, S., "Three Dimensional Analysis of a Rotor in Forward Flight," *Journal of the American Helicopter Society*, May 1991.

[17]Rajagopalan, R. G., Baskaran, V., Hollingsworth, A., Lestari, D., A. Garrick, Solis, E., and Hagerty, B., "RotCFD - A Tool for Aerodynamic Interference of Rotors: Validation and Capabilities," Future Vertical Lift Aircraft Design Conference, January 2012.

[18]Mark, A. and Strawn, R. C., 58th Annual Forum of The American Helicopter Society.

[19]Guntupalli, K. and Rajagopalan, R. G., "Momentum Source Model for Discrete Blades," International Powered Lift Conference, American Helicopter Society, 2010.

[20]Ochs, S. S., "An Adaptively-Refined Quadtree Grid Method for Incompressible Flows," *PhS Thesis*, 1998.

Helicopter Rotor Trim and Blade Motion Perturbations Caused by Wake Vortices of Wind Turbines and Fixed Wing Aircraft

Berend G. van der Wall
berend.vanderwall@dlr.de

Paul H. Lehmann
paul.lehmann@dlr.de

German Aerospace Center (DLR), Institute of Flight Systems
Lilienthalplatz 7, 38108 Braunschweig, Germany

Abstract

With increasing numbers of large wind turbines on-shore and off-shore, interactions of their wake with helicopters become interesting. In this paper the wake is modeled as a tip vortex helix with a vortex strength estimated from the wind turbine thrust. Helicopter rotors of different size are subjected to the wake and the collective and cyclic controls required to keep the trim are compared to the maximum available controls of the rotorcraft. In addition the blade flapping response due to the vortex influence without pilot action is computed and compared to maximum flapping angles allowed. It is found that typical onshore wind turbines could be a hazard for ultralight helicopters, but not for larger helicopters. Large offshore wind turbines, however, could even be a danger for small helicopters that may be used for maintenance. In addition the results are compared to fixed-wing wake vortex interaction with a helicopter as given in the literature.

NOMENCLATURE

A, B	Non-dimensional effective begin and end of rotor blade, referenced to R
c, c_{eq}	Airfoil chord and equivalent chord, m
$C_{l\alpha}$	Lift curve slope
C_T	Thrust coefficient of the helicopter rotor
c_n	Radial integral coefficients, $n = 0,1,2,...$
$L, \overline{L}$	Blade lift, N; non-dimensional blade lift
$M_\beta, \overline{M_\beta}$	Aerodynamic flap moment about the flapping hinge, Nm; non-dimensional flap moment
N_b	Number of rotor blades
r	Non-dimensional radial coordinate, referenced to R
R_c, r_c	Vortex core radius, m; non-dimensional core radius, referenced to R
R, R_{WT}	Helicopter and wind turbine rotor radius, m
RCR	Rotor control ratio
RFR	Rotor flapping ratio
u, v, w	Velocity components in x, y, z directions, m/s
U	Rotor blade tip speed, m/s
v_{iV}	Wake vortex induced velocity, m/s
V_T, V_P	Non-dimensional velocities acting tangential and normal at the blade element, referenced to U
V_W	Wind speed, m/s
V_∞	Helicopter flight speed, m/s
x, y, z	Non-dimensional hub-fixed coordinates of the helicopter rotor, x pos. downstream, y pos. starboard, z pos. up, referenced to R
y_0	Vortex position within the rotor disk; non-dimensional, referenced to R
$\beta, \beta_0, \beta_C, \beta_S$	Rotor blade flapping angle, mean, and cyclic components, deg
β_V	Core radius shape factor
Γ, Γ_{eq}	Wind turbine tip vortex and equivalent straight line vortex circulation strength, m²/s
Δ	Perturbation of a variable
$\Theta, \Theta_0, \Theta_C, \Theta_S$	Rotor blade pitch angle, collective, lateral and longitudinal control angle, deg
λ_i	Thrust-induced inflow velocity normal to the rotor disk, non-dimensionalized by U
λ_V, λ_{V0}	Wind turbine wake vortex induced inflow ratio and its amplitude, normal to the rotor disk, non-dimensionalized by U
μ	Rotor advance ratio, $\mu = V_\infty/U$
ν_β	Natural frequency of flapping, non-dimensionalized by Ω
ρ	Air density, kg/m³
σ	Rotor solidity, $\sigma = N_b c/(\pi R)$
ψ	Rotor blade azimuth, deg
ψ_V	Wake age in terms of azimuth behind the blade, deg
Ω, Ω_{WT}	Rotor rotational speed of helicopter rotor and wind turbine, rad/s

1. INTRODUCTION

In past decades the effects of rotorcraft encountering the wake vortex of large fixed-wing aircraft as sketched in Fig. 1 have been a subject of interest. This may happen in military operations with large transport aircraft and utility helicopters, but also heliports in the vicinity of airport runways may be subjected to large civil aircraft wake vortices. In the 1980s, NASA [1] and US Army [2] investigated a UH-1H helicopter trimmed at 60 kts flying through the wake vortices of a Douglas C-54 airplane at varying distances. Over the range of 0.42 nm to 6.64 nm, maximum rotor blade structural loads, helicopter attitude response and tail rotor flapping were measured, the helicopter reactions were recognized but did not constitute safety hazards.

(a) Sketch of a fixed wing wake with a helicopter

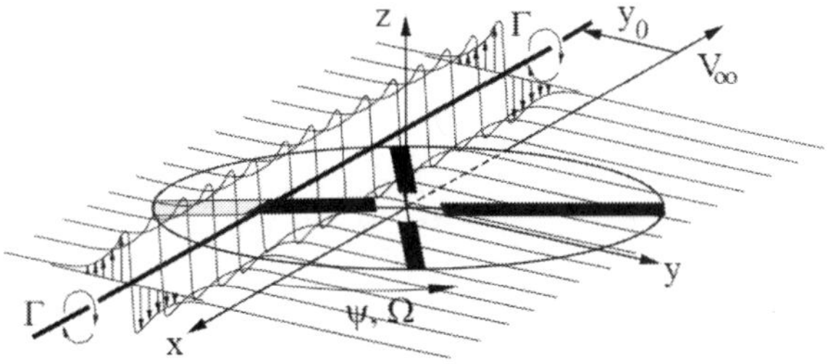

(b) Helicopter rotor with a single straight vortex

Fig. 1: Fixed wing vortex - rotor interaction.

Numerical simulation was used in [3] and [4] to investigate the effects from a pair of trailing vortices of a preceding large Boeing 747 (B-747) airplane on the flight dynamics of a fixed and a rotary wing aircraft. The responses of airplane and helicopter are described as different in a way that the helicopter reacts more damped to the disturbance by the tip vortex. More recently, work has been conducted by the University of Liverpool and QinetiQ, [5]-[7]. They investigated the influence of an active runway from an international airport to helicopter operations from a nearby approach and takeoff area. For their work they used flight mechanics simulation tools to examine the effects when a helicopter encounters the shed tip vortex from a large aircraft.

A model of the vortex velocity profile was established by the use of LIDAR measurement data from the airport. Several calculations for the Lynx helicopter in forward speeds from hover to 80 kts were conducted. The results showed that the helicopter reaction is primarily dependent on the rotor position relative to the vortex center. In some combinations, hazardous helicopter reactions were recognized. The main question is, whether a rotorcraft which meets handling performance standards is able to recover the disturbed flight attitude after encountering the vortex.

The rotor blade and helicopter response to vortex encounters was numerically investigated in [8] and [9], emphasizing the mutuality of the interaction. It was concluded that the interacting vortex deformation under rotor airloads significantly reduces the impact on rotor and helicopter response, compared to a rigid vortex and an unidirectional interaction.

Today another source of rotorcraft-vortex encounter is emerging: more than 27000 on-shore wind turbines (WT) in Germany (end of 2016) with up to 7.5 MW power often are in the vicinity of the numerous glider and sport aircraft airfields. Ultralight autogyros are flying in large numbers in Germany since 2009 and ultralight helicopters regulations were established in Dec. 2016 and are to be expected also to enter the market in large numbers. Off-shore wind farms consist of WT having 5 to 8 MW power and larger ones being planned. For maintenance small helicopters are used that will encounter the wakes of such WT. In addition, the interaction of ultralight rotorcraft with WT wake vortices as sketched in Fig. 2 is an emerging subject of interest as well. A first paper on the subject was presented by the authors in [10].

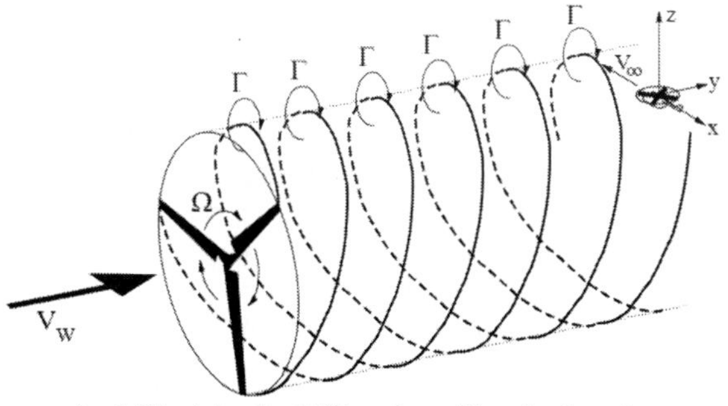

(a) Sketch of a WT wake with a helicopter

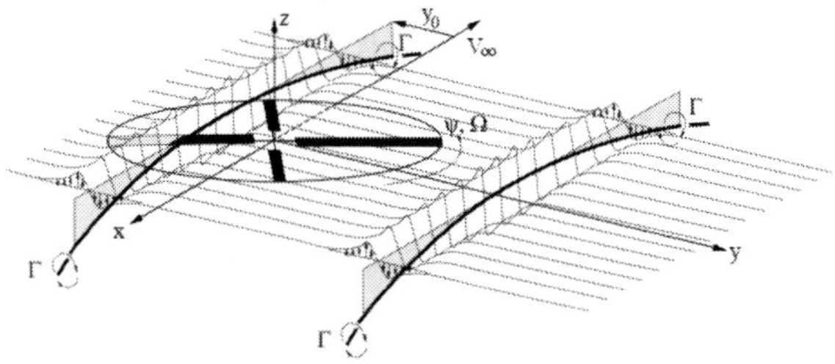

(b) Helicopter rotor in a curved WT wake with multiple vortices

Fig. 2: Wind turbine vortex - rotor interaction.

Such horizontal axis WTs have a downstream wake which is characterized by a spiral helix of usually three blade tip vortices on the surface of the wake tube, as the number of blades is three for the overwhelming majority of the installed systems, see Fig. 2 (a).

The tip vortex spiral on the surface of the tube is fed with circulation from the WT blades. Around each of these vortices induced swirl velocities are generated that are oriented towards the turbine inside the tube – resulting in a global wind deficit – and adding on the wind velocity outside the tube. Thus, inside the tube a global "wind deficit" is present that manifests itself as a loss of air momentum. Crossing the tube horizontally at half of its height will therefore generate a side-slip angle for the aircraft when penetrating the wake tube boundary on one side. This side-slip angle vanishes again when penetrating the wake tube boundary on the opposite side. When entering the boundary of the wake large horizontal vortex swirl velocities are encountered that change their sign at the boundary itself, representing a dual lateral pulse for the aircraft.

The situation is very different when crossing the wake tube at its upper or lower boundary, i.e. in the immediate vicinity of the center of these vortices as indicated in Fig. 2 (a), showing the staggered vortices of the wake spiral and a helicopter approaching it from the right. The interactional problem of a helicopter rotor with four individual rotor blades passing the upper end of the wake spiral in almost normal direction to it is shown in Fig. 2 (b). In general this problem is unsteady because the rotating blades enter and pass the WT tip vortex induced velocity field periodically during their revolution.

The curvature of the WT wake spiral within the rotor is very small and can be ignored, because the WT radius is more than 10 times of the ultralight rotor radius, which reduces the problem to that shown in Fig. 1 (b) with a straight vortex. An analytical closed-form estimate to compute rotor controls required for rejection of the vortex disturbance by means of a simplified blade element method was derived in [11] for the first time in the literature. In this paper it will be extended to the rotor blade flapping response when no pilot action is employed.

2. ANALYTICAL TREATMENT

2.1. Problem statement

For an analytical solution of the problem sketched in Fig. 1 (b) the WT vortex within the rotor disk is replaced by an equivalent infinitely long straight line vortex with an associated equivalent circulation Γ_{eq} as sketched by the dashed line in Fig. 3.

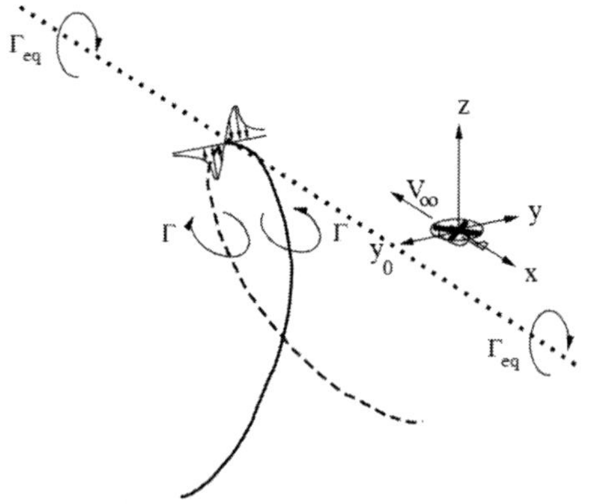

Fig. 3: Replacement of the WT spiral vortex by an infinite long straight line vortex.

The swirl velocity profile includes a core radius r_c, a lateral position within the rotor disk y_0 and the swirl velocity magnitude depends on the vortex circulation strength Γ. Coordinates x, y, z and lengths such as the core radius and vortex location within the rotor disk are made non-dimensional by the helicopter rotor radius R, velocities are referred to the tip speed $U = \Omega R$ to provide the wake-induced inflow ratio λ_V as fraction of the helicopter blade tip speed. The circulation is made non-dimensional by division through UR.

This straight line vortex is modeled with a core radius model of Burnham-Hallock [13], which is a special case of the Vatistas' model [14]. The equivalent circulation Γ_{eq}, or the inflow ratio λ_{V0}, are then estimated based on the computed wake-induced velocity profiles. It turns out that for the cases investigated here this equivalent circulation is about half of the value of the WT spiral vortex strength.

$$\text{(1)}\quad \begin{aligned} \lambda_V &= \frac{v_{iV}}{U} = \frac{\Gamma_{eq}}{2\pi UR}\frac{y-y_0}{(y-y_0)^2+r_c^2} \\ &= \lambda_{V0}\frac{r\sin\psi - y_0}{(r\sin\psi - y_0)^2+r_c^2}; \quad \lambda_{V0} = \frac{\Gamma_{eq}}{2\pi UR} \end{aligned}$$

The peak value of the inflow ratio is obtained at the core radius itself.

$$\text{(2)}\quad \lambda_{V,max} = \frac{\lambda_{V0}}{2r_c} = \frac{\Gamma_{eq}}{4\pi UR_c} = \frac{V_c}{U}$$

From Eqs. (1) and (2) it can already be concluded that the smaller the helicopter rotor radius R, or the smaller its tip speed U, the larger will be the vortex-induced inflow ratio. Therefore, because small ultralight helicopters and autogyros have both smaller radius and smaller tip speed than small to medium size helicopters, these rotorcraft will respond stronger to such vortex encounters than the larger helicopters.

Natural diffusion is represented by time-dependent decay (or aging) functions for both the circulation strength (which reduces asymptotically to zero for long time) and the vortex core radius (which widens with time) in the manner following [13] and [15].

$$\text{(3)}\quad \begin{aligned} r_c &= \frac{R_{c0}}{R}\sqrt{1+\frac{5\cdot 10^{-6}\, rad/s}{(R_{c0}/R_{WT})^2\,\Omega_{WT}}\psi_v} \\ \Gamma &= \Gamma_0 e^{-0.001932\psi_V} \end{aligned}$$

The initial core radius R_{c0} is set to 5% of the WT blade chord length at 93% radius (i.e., the reference chord at the blade tip area). Both decay factors of the core radius and the circulation aging functions are empirical, based on measurements.

Based on the sketch of Fig. 1 (b) the velocities acting on a rotor blade element tangential (in the rotational plane, normal to the radial axis) and normal to the rotational plane can be established. It is assumed that:

- the helicopter's flight path is parallel to the WT's wake vortex axis,
- the vortex center lies in the plane of the rotor disk,
- the rotor is horizontal.

Then, as indicated in the sketch, only vertical vortex-induced velocities are acting on the blade elements and they are a function of the lateral coordinate y only. All velocities can be split into those components present in an isolated rotor (index 0) and those components due to the WT wake λ_V from Eq. (1) that are considered as perturbations (Δ values). The in-plane velocity V_T acting at a blade element consists of the circumferential velocity expressed by r and the contribution of the advance ratio μ. Contributions to the normal velocity come from the inflow ratio due to forward flight μ_z, the thrust-induced inflow ratio λ_i, the WT vortex-induced inflow ratio λ_V, and due to rotor blade flapping position β as well as due to flapping velocity $d\beta/d\psi$.

(4)
$$\begin{aligned} V_T &= r + \mu \sin\psi = V_{T0} + \Delta V_T ; \quad \Delta V_T = 0 \\ V_P &= \mu_z + \lambda_{i0} + \Delta\lambda_i + \lambda_V + \mu\beta\cos\psi + r\frac{d\beta}{d\psi} \\ &= V_{P0} + \Delta\lambda_i + \Delta V_{PV} + \Delta V_{P\beta} \end{aligned}$$

A trim of the rotor requires collective control Θ_0, longitudinal Θ_S and lateral cyclic control angles Θ_C in order to establish the required thrust, propulsive and lateral forces, as well as the hub moments needed for a steady flight. Any perturbations of the velocities acting at the blades therefore require perturbations in the controls $\Delta\Theta$ in order to maintain the trim. Rotor blade flapping can also be considered to consist of a flapping motion resulting in trim β and a perturbation $\Delta\beta$ caused by the vortex.

(5)
$$\begin{aligned} \Theta &= \Theta_0 + \Theta_S \sin\psi + \Theta_C \cos\psi \\ \Delta\Theta &= \Delta\Theta_0 + \Delta\Theta_S \sin\psi + \Delta\Theta_C \cos\psi \\ \beta &= \beta_0 + \beta_S \sin\psi + \beta_C \cos\psi \\ \Delta\beta &= \Delta\beta_0 + \Delta\beta_S \sin\psi + \Delta\beta_C \cos\psi \end{aligned}$$

For simplification a perturbation parallel to the x-axis is assumed (see Fig. 1 (b)), which causes only lateral unbalance of vortex-induced inflow. In longitudinal direction the disturbances are always balanced fore and aft of the rotor hub, thus $\Delta\Theta_C = 0$. The governing equation of rotor blade flapping for trim and perturbations is given by Eq. (6), including the natural frequency of flapping ν_β and the blade Lock number γ. Like the flapping angle, the aerodynamic moment $\overline{M_\beta}$ again can be split into a trim value plus perturbations in the mean and 1/rev components.

(6)
$$\begin{aligned} \frac{d^2\beta}{d\psi^2} + \nu_\beta^2 \beta &= \gamma \overline{M_\beta} \\ \frac{d^2\Delta\beta}{d\psi^2} + \nu_\beta^2 \Delta\beta &= \gamma\, \Delta\overline{M_\beta} \end{aligned}$$

The blade section angle of attack is defined by three contributions: first, the pitch angle Θ, which is needed for the helicopter trim in undisturbed atmosphere. Second, a perturbation $\Delta\Theta$ (Eq. (5)) to compensate the trim disturbance caused by the third contribution, namely the WT vortex-induced velocities given in Eq. (4). First, this results into an equation for the steady (mean) lift perturbation to be zero. Second, the $1/\mathrm{rev}$ aerodynamic flapping moment perturbation about the hub must be zero as well. Both of these values require a radial integration of the section lift distribution from the effective begin of the airfoiled part of the blade A to the effective end of it B. Lift and moment are referenced to $\rho(\Omega R)^2 cRC_{l\alpha}$ and $\rho(\Omega R)^2 cR^2 C_{l\alpha}$, respectively.

(7)
$$\begin{aligned} \Delta\overline{L} &= \int_A^B \Delta\overline{dL} \approx \frac{1}{2}\int_A^B V_T^2 \left(\Delta\Theta - \frac{\Delta V_P}{V_T}\right) dr \\ &= \Delta\Theta_0 \frac{1}{2}\int_A^B (r + \mu\sin\psi)^2\, dr \\ &+ \Delta\Theta_S \frac{1}{2}\int_A^B (r + \mu\sin\psi)^2 \sin\psi\, dr \\ &- \Delta\lambda_i \frac{1}{2}\int_A^B (r + \mu\sin\psi)\, dr \\ &- \lambda_{V0} \frac{1}{2}\int_A^B \frac{(r + \mu\sin\psi)(r\sin\psi - y_0)}{(r\sin\psi - y_0)^2 + r_c^2}\, dr \end{aligned}$$

The equation for the moment is:

(8)
$$\Delta\overline{M_\beta} = \int_A^B r\Delta\overline{dL}$$

To solve the wake integral of the perturbation lift a Fourier series is needed for ease of further processing. However, a broken rational function with periodic terms in both nominator and denominator is present. A Fourier analysis transforms this into the desired form, [10].

(9)
$$\begin{aligned} &\frac{(r + \mu\sin\psi)(r\sin\psi - y_0)}{(r\sin\psi - y_0)^2 + r_c^2} \\ &= a_{V0} + \sum_{i=1}^{\infty} \left(a_{Vi}\cos i\psi + b_{Vi}\sin i\psi \right) \end{aligned}$$

The mean value a_{V0} is needed for keeping the mean value of the lift perturbation zero ($\Delta\overline{L_0} = 0$) and the $1/\mathrm{rev}$ sine part b_{V1} is needed for keeping the rotor roll moment perturbation zero (in the rotating frame that is $\Delta\overline{M_{\beta S}} = 0$), all higher harmonics are of no interest for this study since they do not affect the rotor trim. A rotor pitching moment perturbation is not generated due to the vortex axis being parallel to the rotor y-axis (in the rotating frame: $\Delta\overline{M_{\beta C}} = 0$). The problem consists in: first, the computation of the Fourier coefficients of Eq. (9); second, solving the integrals occurring in the last term of Eq. (7) and Eq. (8):

$$\frac{\Delta\overline{L}}{\lambda_{V0}} = -\frac{1}{2}\int_A^B a_{V0}(r,\mu,y_0,r_c)\,dr$$
(10)
$$\frac{\Delta\overline{M_{\beta 0}}}{\lambda_{V0}} = -\frac{1}{2}\int_A^B r a_{V0}(r,\mu,y_0,r_c)\,dr$$
$$\frac{\Delta\overline{M_{\beta 0}}}{\lambda_{V0}} = -\frac{1}{2}\int_A^B r b_{V1}(r,\mu,y_0,r_c)\,dr$$

2.2. Solution

2.2.1. Controls required for constant trim

Constant trim means to keep rotor thrust and hub moments constant by applying collective and cyclic control perturbations. Constant thrust yields $\Delta\lambda_i = 0$ and is expressed by Eq. (11), derived from Eq. (7).

(11)
$$\begin{aligned} 0 &= \Delta\Theta_0\frac{1}{2}\int_A^B\left(r^2+\frac{\mu^2}{2}\right)dr + \Delta\Theta_S\frac{1}{2}\int_A^B \mu r\,dr \\ &-\lambda_{V0}\frac{1}{2}\int_A^B a_{V0}\,dr \\ &= \Delta\Theta_0\frac{2c_3+c_1\mu^2}{4} + \Delta\Theta_S\frac{\mu c_2}{2} - \frac{\lambda_{V0}}{2}\int_A^B a_{V0}\,dr \\ c_i &= \int_A^B r^{i-1}dr = \frac{B^i - A^i}{i}\end{aligned}$$

Constant hub moments require the 1/rev aerodynamic moments about the hub in the rotating frame to be constant, and these imply that the 1/rev flapping remains constant as well, thus $\Delta\beta_C = \Delta\beta_S = 0$. Three governing equations result: one for the mean lift Eq. (11) and two for the Cosine and Sine part of the moment.

(12)
$$\begin{aligned} 0 &= \Delta\Theta_0\frac{1}{2}\int_A^B 2\mu r^2\,dr - \lambda_{V0}\frac{1}{2}\int_A^B r b_{V1}\,dr \\ &+\Delta\Theta_S\frac{1}{2}\int_A^B\left(r^3+\frac{3\mu^2}{4}r\right)dr \\ &= \Delta\Theta_0\mu c_3 + \Delta\Theta_S\frac{4c_4+3c_2\mu^2}{8} - \frac{\lambda_{V0}}{2}\int_A^B r b_{V1}\,dr\end{aligned}$$

With a vortex oriented parallel to the rotor x-axis no Cosine moments can develop and thus no controls needed to counteract them, hence $\Delta\Theta_C = 0$, reducing the number of equations to two. In addition, $\Delta\Theta_C$ does not affect the rotor thrust. From the 1/rev Sine part of the aerodynamic moment the second equation, Eq. (12), is derived from Eq. (8). The vortex-related integrals in Eqs. (11) and (12) were first given in [10] and their results are given in the Appendix A1 and A2.

2.2.2. Blade flapping for fixed controls

This case represents a rotor trimmed in undisturbed atmosphere first and then the control angles are kept fixed such that all control perturbations are zero $\Delta\Theta = 0$. Then the vortex is introduced into the rotor, which will cause blade flapping. The perturbation part of Eq. (6) provides the governing equation for the mean flapping (coning), the longitudinal and the lateral flapping. Recall that now the rotor thrust will vary and thus $\Delta\lambda_i \neq 0$. For the case of zero rotor angle of attack as considered here a linearization about the trim condition can be done:

(13)
$$\Delta\lambda_i = \frac{d\lambda_{i0}}{dC_T}\Delta C_T;\quad \lambda_{i0} = \sqrt{\sqrt{\frac{C_T^2}{4}+\frac{\mu^4}{4}}-\frac{\mu^2}{2}}$$
$$\frac{d\lambda_{i0}}{dC_T} = \frac{C_T}{\sqrt{8}\sqrt{\sqrt{C_T^2+\mu^4}-\mu^2}\,\sqrt{C_T^2+\mu^4}}$$

The aerodynamic moment Eq. (8) consists of three parts: one that involved the flapping motion only, one due to perturbations in rotor thrust and one due to the vortex. For fixed controls:

(14)
$$\begin{aligned}\Delta\overline{M_\beta} &= -\frac{1}{2}\int_A^B rV_T\Delta V_P\,dr \\ &= -\frac{1}{2}\int_A^B rV_T\left(\mu\Delta\beta\cos\psi + r\frac{d\Delta\beta}{d\psi}\right)dr \\ &-\frac{1}{2}\int_A^B rV_T\Delta\lambda_i\,dr - \frac{1}{2}\int_A^B rV_T\lambda_V\,dr\end{aligned}$$

First, the part due to rotor blade flapping is considered and all higher harmonics ignored, because the trim is defined by the constant and 1/rev parts.

(15)
$$\begin{aligned}\Delta\overline{M_\beta}(\Delta\beta) &= -\frac{1}{2}\int_A^B rV_T\left(\mu\Delta\beta\cos\psi + r\frac{d\Delta\beta}{d\psi}\right)dr \\ &= -\frac{\mu^2c_2-4c_4}{8}\Delta\beta_C\sin\psi \\ &-\left(\frac{\mu c_3}{2}\Delta\beta_0 + \frac{\mu^2c_2+4c_4}{8}\Delta\beta_S\right)\cos\psi\end{aligned}$$

The second part is due to rotor thrust perturbations.

(16)
$$\overline{\Delta M_\beta}(\Delta C_T) = -\frac{1}{2}\int_A^B rV_T \frac{d\lambda_{i0}}{dC_T}\Delta C_T\, dr = -\frac{c_3 + \mu c_2 \sin\psi}{2}\frac{d\lambda_{i0}}{dC_T}\Delta C_T$$

This requires a computation of the change in thrust coefficient due to the vortex influence.

(17)
$$\Delta C_T = -\frac{\sigma C_{l\alpha}}{2}\frac{1}{2\pi}\int_0^{2\pi}\int_A^B V_T\lambda_V\, dr\, d\psi = -\frac{\sigma C_{l\alpha}}{2}\lambda_{V0}\int_A^B a_{V0}\, dr$$

Finally, the impact of the vortex on the flapping moment is needed. Again only the mean and 1/rev part is retained.

(18)
$$\overline{\Delta M_\beta}(\lambda_V) = -\frac{1}{2}\int_A^B rV_T\lambda_V dr = -\frac{\lambda_{V0}}{2}\left(\int_A^B ra_{V0}\, dr + \int_A^B rb_{V1}\, dr \sin\psi\right)$$

The solution of the first integral in Eq. (18) is given in Appendix A3, the second integral and that occurring in Eq. (17) were already given before. Inserting the flapping perturbations of Eq. (5) into Eq. (6) and the aerodynamic moments of Eqs. (15) to (18) on the right side the harmonic balance method can be applied, leading to a system of linear equations for the flapping coefficients. The change of thrust and induced velocity is automatically obtained.

2.2.3. Rotor control and flapping ratios

A rotor control ratio (RCR) can be established that relates the required maximum control angle to the maximum control angle that is available for mitigating vortex effects, $\Delta\Theta_{max}$. Similarly, a rotor flapping ratio (RFR) can be defined, relating the maximum flapping perturbation angle to the maximum allowed flapping angle $\Delta\beta_{max}$. That angle may be defined by mast bumping, fuselage collision, or blade collisions in a coaxial rotor design.

(19)
$$RCR = \frac{|\Delta\Theta_0| + \sqrt{\Delta\Theta_S^2 + \Delta\Theta_C^2}}{\Delta\Theta_{max}}$$
$$RFR = \frac{|\Delta\beta_0| + \sqrt{\Delta\beta_S^2 + \Delta\beta_C^2}}{\Delta\beta_{max}}$$

3. RESULTS OF THE ANALYTICAL MODEL

Realistic rotor blades have a root cutout of about $0.22R$ such that with blade tip and root losses the effective begin and end of the aerodynamic lifting part of the blade can be considered as $A = 0.25$ and $B = 0.97$. For analysis of blade flapping response the natural frequency of flapping is assumed to be $\nu_\beta = 1$ (central flapping hinge) and a representative Lock number of $\gamma = 8$ is taken.

Although in hover a space-fixed vortex is unrealistic because the strong rotor downwash will blow it away, the results provide a deep insight into the physics of such a vortex-rotor interaction. In forward flight an advance ratio of $\mu = 0.3$ is considered here, which is close to the cruise speed of most helicopters. Then, the underlying assumption of a space-fixed vortex is much better because the rotor downwash is significantly smaller than in hover and its impact on the vortex is thus small as well.

3.1. Controls required for constant trim

Collective control angle: Results are given in Fig. 4 (a) for the hovering case and in (b) for forward flight. In hover, for vortex positions in the left half of the rotor ($y_0 < 0$) a positive collective control angle is needed to compensate the vortex induced downwash that dominates over the majority of the rotor disk. When the vortex core is at the rotor hub ($y_0 = 0$) the collective control angle is nearly zero because the downwash on the left side equals the upwash on the right side. Vortex positions on the right side of the rotor disk require negative collective control angle because the vortex-induced upwash dominates over the rotor disk. The flat part of the line for $r_c = 0$ for vortex positions around the hub is due to the root cutout. This effect vanishes for zero cutout and for core radii larger than the cutout.

The smaller the core radius, the larger the collective control angle required to compensate vortex effects on thrust. This is due to larger peak values of the vortex-induced inflow. Maximum controls are required when the vortex has a small core radius and is just inboard of the rotor disk, i.e., when $|y_0|$ is just little less than 1. In these cases the maximum of upwash or downwash affects the rotor's lift. For $r_c = 0.5$ the maximum control is required for vortex positions $|y_0| = 1$ and for larger core radii the maximum collective control angle is required when $|y_0| \approx 1.3$. Note that in hover the mean blade lift and thus rotor thrust is not affected by the longitudinal cyclic control angle, see Eq. (7).

In forward flight, shown in Fig. 4 (b), an asymmetry of the curves is developing, which is caused by the different dynamic pressure on advancing and retreating side of the rotor. This also involves the longitudinal control angle into the thrust development, following Eq. (7). Although the general appearance of the curves is similar to that in hover, some remarkable differences exist. First, the required collective control angle is larger than in hover for vortex positions on the retreating side. This is because in forward flight the retreating side has less dynamic pressure than in hover and thus the same induced velocities of the vortex generate larger modifications in the angle of attack.

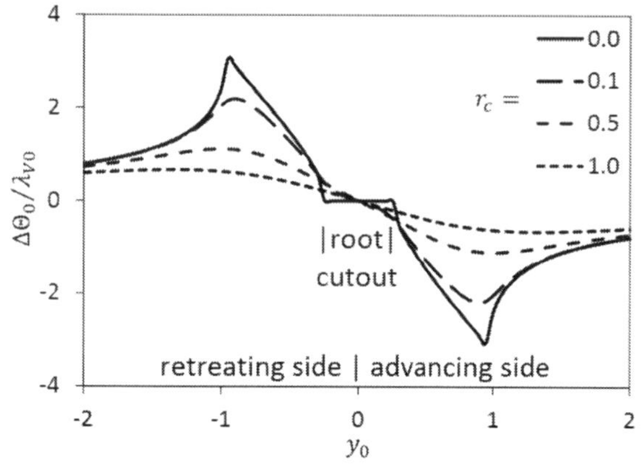

(a) Hover, $\mu = 0.0$

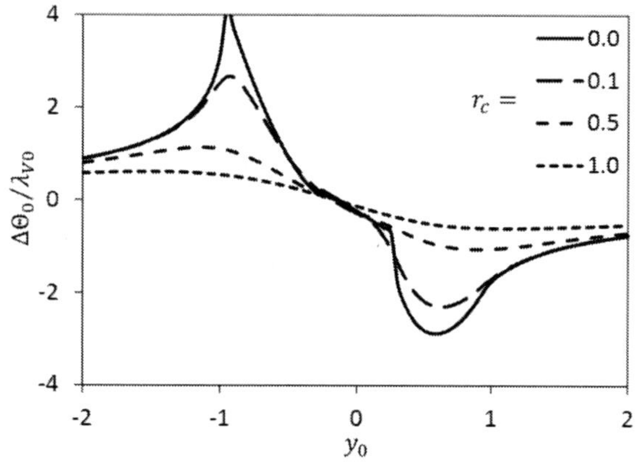

(b) Forward flight, $\mu = 0.3$

Fig. 4: Collective control angle.

Second, for vortex positions on the advancing side in forward flight, Fig. 4 (b), the magnitude of collective control angle required is similar to that in hover shown in (a), but the vortex positions for maximum control angles are more inboard than in hover. Also, the curve around the maximum appears more rounded. Both effects are due to the larger dynamic pressure on the advancing side. Additionally, the diminishing vortex-induced inflow on the retreating side affects the aerodynamics there more than in hover because of less dynamic pressure. The combination of all these effects leads to the curves shown.

Longitudinal cyclic control angle: Results are given in Fig. 5 (a) for the hovering case and in (b) for forward flight. Due to the vortex orientation relative to the rotor the resulting aerodynamic roll moments cause changes in the longitudinal control angle to cancel the vortex impact on trim. The plot shows the biggest changes when the vortex is located at $y_0 = -1, 0$ and $+1$. A vortex position at the rotor hub ($y_0 = 0$) generates as much downwash on the right side of the rotor as upwash on the left side. Then, a positive aerodynamic roll moment appears and needs to be compensated by a positive longitudinal control angle.

When the vortex is located at either end of the rotor disk the longitudinal control angle needed to compensate its influence is negative in both cases. This is due to the vortex-induced velocity gradients within the rotor disk. For $y_0 = -1$ the entire disk is immersed in downwash of the vortex, with largest values on the left side of the disk, thus a positive gradient from left to right. For $y_0 = +1$ the entire disk is immersed in upwash of the vortex, with largest values on the right side of the disk, thus again a positive gradient from left to right. Therefore, in both cases a negative longitudinal control is needed to compensate this gradient.

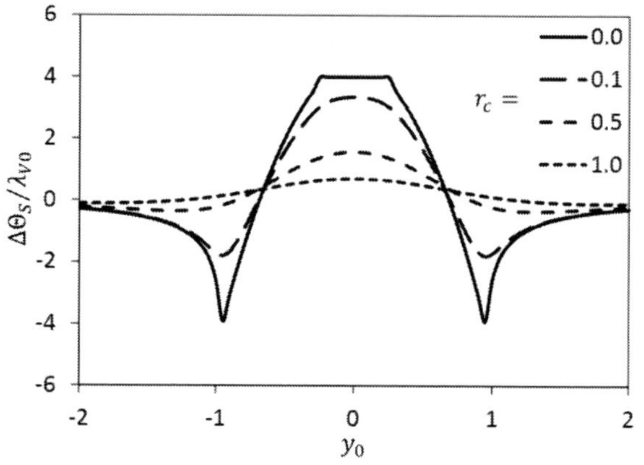

(a) Hover, $\mu = 0.0$

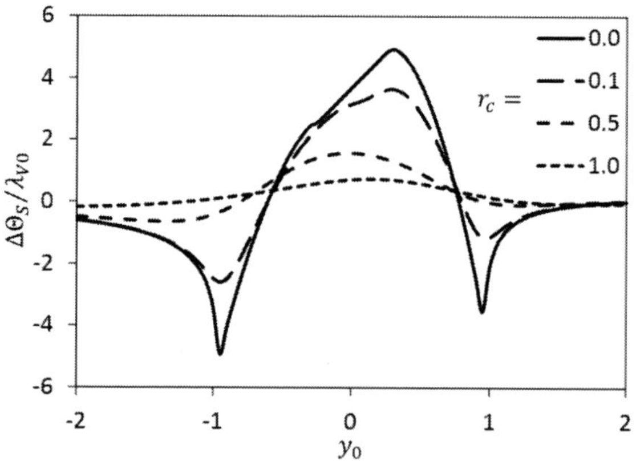

(b) Forward flight, $\mu = 0.3$

Fig. 5: Longitudinal cyclic control angle.

In forward flight, shown in Fig. 5 (b), a similar behavior is seen, but larger maximum cyclic control angles are needed for vortex positions on the retreating side, compared to those in hover. In contrast, the peak values for vortex positions in the vicinity of $y_0 = 1$ are less in forward flight than in hover. Finally, the control angle maximum for vortex positions around the hub center ($y_0 = 0$) in forward flight is increasing and shifted to positions more on the advancing side, compared to hover. All these effects can be traced down to the increase of dynamic pressure on the advancing side and its reduction on the retreating side.

Collective flapping (coning) angle: Results are given in Fig. 6 (a) for the hovering case and in (b) for forward flight. Although a re-trim to the same thrust as in undisturbed atmosphere is done, requiring collective control angles as given in Fig. 4, some collective flapping angle will still develop. This is due to a radial re-distribution of the mean blade section lift, when the vortex is affecting the aerodynamics.

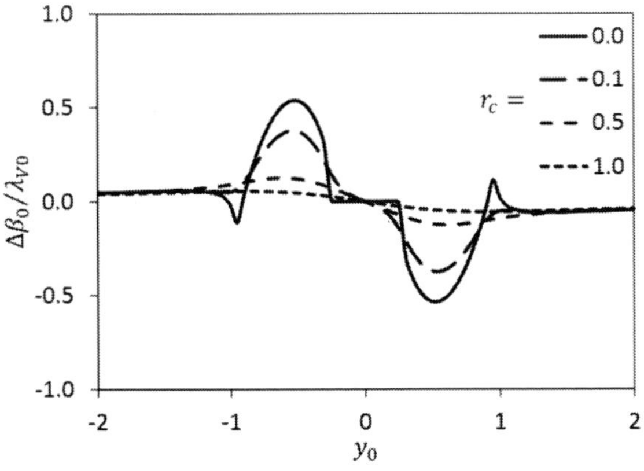

(a) Hover, $\mu = 0.0$

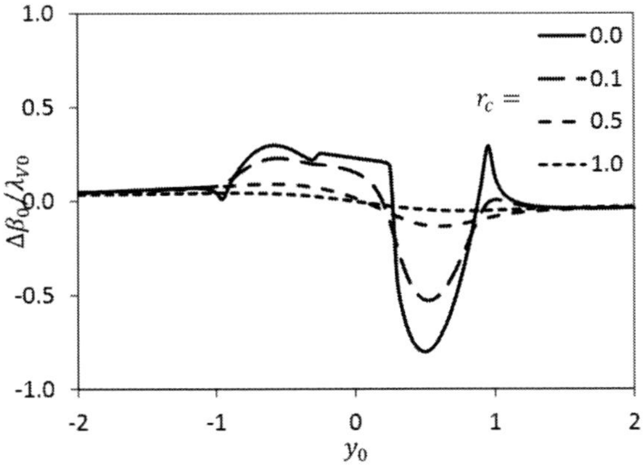

(b) Forward flight, $\mu = 0.3$

Fig. 6: Collective flapping (coning) angle.

In hover, Fig. 6 (a), when the vortex is located around half of the radius on the retreating side, its upwash is affecting the blade tip region and its downwash the root area. This leads to a re-distribution of sectional lift with more lift at the blade tip area and less at the root area. The overall lift remains the same, but the aerodynamic flapping moment is larger than without the vortex and thus the blade has a larger upwards flapping. For vortex positions around the half radius on the advancing side the opposite happens.

In forward flight, due to the difference of dynamic pressure acting on advancing and retreating side, the vortex effect on steady flapping is exaggerated for positions on the advancing side and less on the retreating side, compared to hover.

3.2. Blade flapping for fixed controls

In this section the rotor blade flapping as caused by the vortex influence is investigated. For this purpose the rotor is first trimmed in undisturbed atmosphere, then the control angles are fixed, the vortex introduced at different lateral positions and the blade flapping evaluated without re-trimming the rotor.

Collective flapping (coning) angle: Results are given in Fig. 7 (a) for the hovering case and in (b) for forward flight.

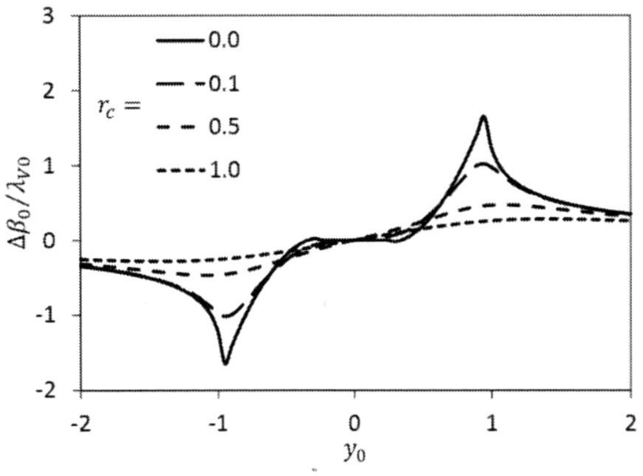

(a) Hover, $\mu = 0.0$

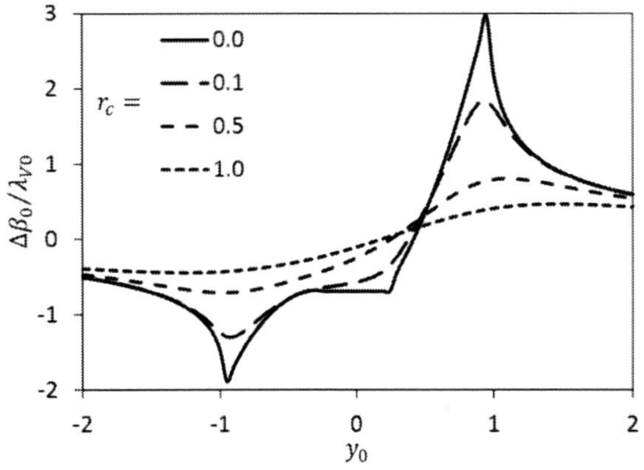

(b) Forward flight, $\mu = 0.3$

Fig. 7: Collective flapping (coning) angle.

For the clockwise sense of vortex rotation as used here, its position on the retreating side in- or outboard the rotor disk generates downwash over the majority of the disk and thus reduces the rotor thrust. As a consequence of less thrust the blades statically flap down relative to the trim. For vortex positions on the advancing side of the disk the opposite happens because in that case the vortex-induced upwash is dominating over the disk, leading to increased thrust and thus an increased upwards flap deflection.

In forward flight, Fig. 7 (b), the asymmetry of dynamic pressure on advancing and retreating side increases the lift more for vortex positions on the advancing side to almost twice the value of hovering. It reduces the lift more for vortex positions on the retreating side than in hover to a much less extent.

Longitudinal and lateral cyclic flapping angles: Results are given in Fig. 8 for an advance ratio of $\mu = 0.3$ only. In (a) results are given for the longitudinal flapping angle and in (b) for the lateral flapping angle.

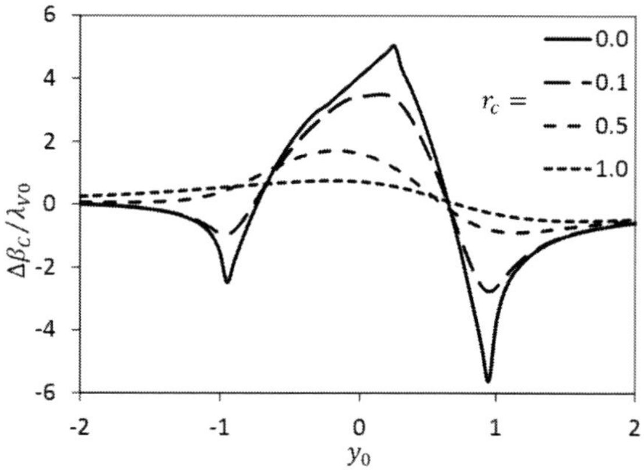

(a) Longitudinal cyclic flapping angle

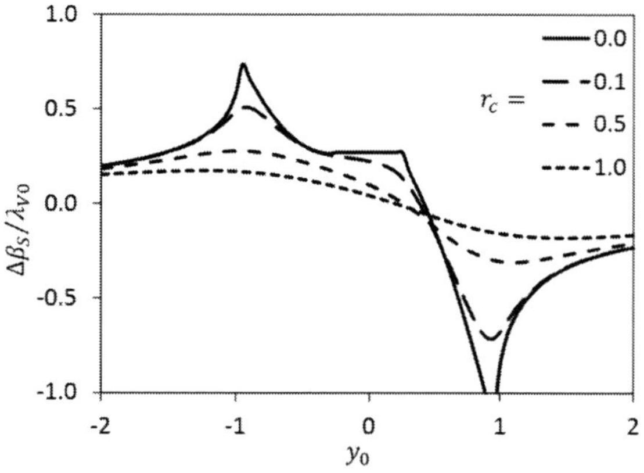

(b) Lateral cyclic flapping angle

Fig. 8: Cyclic flapping (disk tilt) angles in forward flight, $\mu = 0.3$.

In hover only longitudinal flapping develops due to the natural frequency of flapping chosen ($\nu_\beta = 1$) and due to the Lock number chosen ($\gamma = 8$). Then the longitudinal flapping angle developing in hover is identical to the longitudinal control angle required to mitigate the vortex effects shown in Fig. 5 (a), such that in that figure Θ_S only has to be replaced by β_C. For the same reasons the lateral flapping angle in hover is $\beta_S = 0$, because $\Theta_C = 0$.

In forward flight the aerodynamic moment includes both longitudinal and lateral flapping due to the coupling terms proportional to the advance ratio in Eq. (15). The main response still is longitudinal flapping angle, which is about six times larger than the lateral flapping angle. Again, the higher dynamic pressure on the advancing side causes a larger blade response for vortex positions on that side of the disk, while the response for vortex positions on the retreating side is less than in hover.

4. EXEMPLARY RESULTS

4.1. Rotor simulation codes

4.1.1. HOST

For flight mechanics investigations the Institute of Flight Systems at DLR uses Airbus Helicopters "Helicopter Overall Simulation Tool" (HOST, [16]) developed for desktop simulation. It is a modular tool that has the ability to simulate any type of helicopter and to calculate trim, time domain response and other. For the results obtained here HOST was used in a special configuration with an isolated rotor trim instead of a complete helicopter trim. HOST includes an "Atmospheric Environment Submodule" (AES). This submodule is connected to HOST's flight mechanics model, calculates disturbed air and provides all relevant helicopter parts with the related turbulence parameters. It includes a "big size aircraft vortex wake model" and it is assumed that this model is able to represent an idealized vortex in the wake of a WT.

4.1.2. DLR's S4 rotor simulation code

S4 is a high-resolution isolated rotor comprehensive simulation code described in detail in [20]. A Bo105 full-scale elastic rotor blade model is used with two flapping and one lag modes. Nonlinear steady aerodynamics and constant inflow are used for this study for comparison with HOST and with the analytical rotor model described in the following section. The rotor blade is discretized into 40 blade elements and 2 deg azimuth steps are used for time-marching numerical integration of the modal equations of motion to solve the dynamic response problem.

A finite length straight line vortex that can be arbitrarily oriented with respect to the rotor was implemented into the S4 code. To compute all three components of the vortex-induced velocities at all blade elements instantaneous positions a core radius and the Biot-Savart law is used. As shown in Fig. 9 this vortex model essentially represents a Vatistas type of velocity profile.

4.2. Rotors investigated and method applied

Within this study an isolated rotor trim is the basis of analysis. The purpose is to compare HOST results with the analytical model estimates and with DLR's S4 rotor code in order to investigate the differences between these methods.

Various rotors are investigated with respect to the controls required to mitigate the vortex effects and to compute the flapping perturbations when no controls are applied. With respect to on-shore WT interactions, sports ultralight rotorcrafts (UL) are of interest such as autogyros (AG) and coaxial helicopters (COAX). Regarding off-shore WT interaction helicopters of the Bo105 size are of interest. The the Bo105 is a light utility rotorcraft with a mass of 2200 kg representative for those used in maintenance of offshore wind farms. The HOST Bo105 model assumes an articulated rotor with rigid blades flapping about an effective hinge offset with a spring and the Meier-Drees model [19] is used for the induced velocity calculation due to rotor thrust.

An interaction with a vortex of a wide body aircraft like a B-747 is considered for these rotorcraft and larger ones such as an UH-1D and a CH-53D. Parameters used are given in Table 1; the values for the AG and the UL helicopter are considered as representative.

Table 1: Properties of the rotor blades investigated.

	AG	COAX	Bo105	UH-1D	CH-53D
type	see-saw	see-saw	hinge-less	see-saw	articu-lated
R, m	4.22	3.25	4.91	7.32	11.0
U, m/s	155	153	218	248	213
γ, Lock	4.84	6.22	8.00	6.53	8.91
ν_β, /rev	1.00	1.00	1.12	1.00	1.09
$\Delta\Theta_{max}$	-	8°	8°	8°	8°
$\Delta\beta_{max}$	7°	5.73°	15°	12°	14°

To determine the perturbation in main rotor control angles due to the aircraft vortex, the rotor is trimmed at prescribed forward speed in the influence area of the vortex. The rotor is positioned in a way that the vortex is in the same height as the rotor hub and is varied in position from $y = -2R$ to $+2R$ in steps of $0.25R$. The perturbation control angles are obtained by subtracting the trim controls of the undisturbed atmosphere (without a vortex) from the trim controls including the vortex.

When the rotor is always trimmed to constant thrust, then λ_i is constant and thus $\Delta\lambda_i = 0$. This also implies trimming for constant hub moments which yields rotor blade flapping perturbations in 1/rev will be zero. However, the mean flapping (coning) – despite constant thrust – will slightly vary because the steady lift radial distribution will vary depending on the vortex position and its properties. Otherwise, when keeping the controls of trim in undisturbed atmosphere fixed and then introducing the vortex, the thrust will change depending on the vortex influence and hence $\Delta\lambda_i \neq 0$ and $\Delta\beta \neq 0$.

4.3. The wind turbines and the wake model

The investigations of this paper focus on two WT of different power class: a representative on-shore 3 MW turbine and a representative off-shore 7 MW turbine. The reference chord at 93% radius is used to define the initial tip vortex core radius, while the equivalent solidity of the WT rotors is based on the thrust-weighted chord distribution. Data for the WT are given in Table 2. The "worst case" scenario is of interest, which is the operational condition of the WT with maximum tip vortex circulation strength which can be estimated from the rotor thrust coefficient.

The computation of the maximum circulation strength is outlined in [12] and not repeated here. It must be noted that the rotational speed as function of the wind speed is not revealed by the manufacturers and thus had to be estimated by the authors. Also, the thrust curve of the 7 and 10 MW WT is unknown and therefore the same thrust coefficient as for the 3 MW turbine was used instead. The WT wake-induced velocities are computed numerically at distances up to four WT rotor radii behind the turbine, and around distances centered at 500 m behind it.

Table 2: Vortex generator data (10 MW estimated).

Generator:	3 MW	7 MW	10 MW	B-747
R, m	56.5	77.0	95.0	29.8
Γ_0, m²/s	63.7	98.6	130	660
$c(0.9R)$,m	1.000	1.363	1.682	5.070
R_{c0}, m	0.050	0.068	0.084	0.253

The wake is represented by eight revolutions of each blade's wake (beginning at the turbine in the first case, centered about the mean distance in the second case). Every revolution is discretized by 72 straight line vortex elements, each one representing a 5 deg increment. Induced velocities of these finite-length straight vortex elements, including a core radius model, are computed numerically.

While the Vatistas model is used in the analytical isolated rotor investigations HOST makes use of its "big size aircraft wake vortex model". The influence of the helicopter's thrust-induced velocity on the aircraft vortex motion is not modelled.

All calculations performed here were conducted with the flight mechanics model of the BO105 helicopter and the so-called "Big Size Aircraft Vortex Wake Model". Therein, the vortices are described following the Lamb-Oseen formula [17], [18]. Its equivalent circulation strength or swirl velocity V_c, a core radius R_c and a decay factor β_V must be provided; they are estimated from the computed WT wake-induced velocity field and given in Table 3. At a distance $y = 0$ from the core center this formula has a numerical singularity, but it can be proven that the analytical boundary value results in $v_V = 0$, because the expression in brackets of Eq. (20) approaches the zero faster than the division by y.

(20)
$$\lambda_V = \frac{v_{iV}}{U} = \frac{V_c}{U}\frac{R_c}{y}\left[1 - e^{-\beta_V\left(\frac{y}{R_c}\right)^2}\right]$$
$$\lambda_{V\,\max} = 0.638\frac{V_c}{U}; \quad y\left(\lambda_{V\,\max}\right) = 1.1209 R_c$$

Note that for the value used of $\beta_V = 1.0$ the maximum swirl velocity is $v_{iV,max} \approx 0.638 V_c$, obtained at $y \approx 1.12 R_c$.

Three cases A, B and C representing a vortex of a 3, 7 or a 10 MW WT in a distance of 100 m are investigated, larger distances are considered less important due to vortex aging. Therefore the HOST airplane wake model is provided with the data given in Table 3 for these cases. The fourth case D represents a vortex trailed by a B-747 aircraft 2 km behind it, as used in [4], also considered a young and thus most intense vortex. For these cases the vortex-induced velocity distributions for both models based on Eq. (1) and Eq. (20) are shown in Fig. 9.

Table 3: Properties of the HOST big size aircraft vortex wake model and for the Vatistas' model.

	WT,	Dist.,	HOST		Vatistas	
	MW	km	R_c, m	V_c, m/s	R_c, m	V_c, m/s
A	3	0.1	0.415	10.3	0.393	6.18
B	7	0.1	0.568	11.6	0.542	7.00
C	10	0.1	-	-	0.646	7.76
D	B-747	2.0	3.345	26.5	3.280	16.0

Although the formulae are quite different, the resulting velocity profiles are rather close to each other and allow a direct comparison of HOST results with the analytical model. The rotor radii of various rotorcraft vehicles are indicated in the figure as well in order to give an impression of the vortex size relative to the radius of the encountering vehicle. It is also obvious from Fig. 9 that the WT vortices seem unimportant for large helicopters like the CH-53, but already important for UL-helicopters and autogyros, especially considering the smaller tip speed of those compared to larger helicopters. The B-747 vortex appears huge for the smaller rotorcraft, and for the UL helicopter represents almost a linear variation of normal velocity across its disk with about twice of its mean thrust-induced velocity at the core radius.

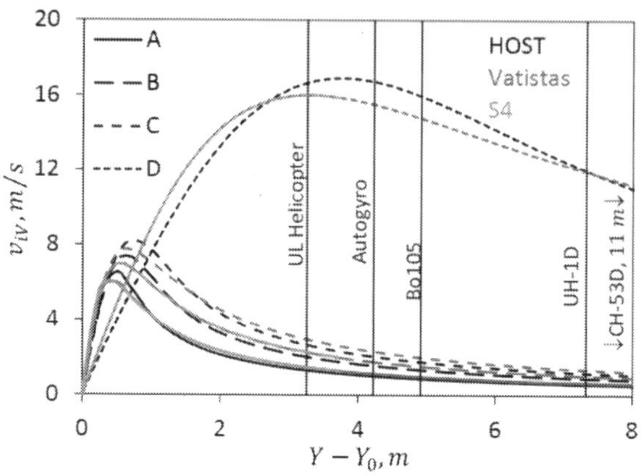

Fig. 9: Induced velocity distributions for cases A-D of Table 3 in relation to various helicopter rotor radii.

4.4. Bo105: controls required to maintain trim

The following results were calculated with the helicopter rotor model described in Section 4.1. In HOST computations the vortex axis has a remaining 5 deg orientation misalignment with respect to the x-axis, thus generating a slight aerodynamic pitching moment. HOST results are compared with the analytical rotor model and with S4 results for the cases A, B and D of Table 3, based on the velocity profiles shown in Fig. 9.

Fig. 10 shows the perturbations of the main rotor trim control angles caused by the vortex. Fig. 10 (a) gives results for the 3 MW WT, (b) for the 7 MW WT and (c) for the B-747 vortex. The vortex location relative to the rotor hub ranges from $y_0 = -2$ to $+2$, i.e. from the far left side to the far right side. The agreement of HOST, S4 and the analytical model is excellent.

Collective control: For vortex positions in the left half of the rotor ($y_0 < 0$) a positive collective control angle is needed to compensate the vortex induced downwash that dominates over the rotor disk. When the vortex core is at the rotor hub ($y_0 = 0$) the collective control angle is nearly zero because the downwash on the left side equals the upwash on the right side. Vortex positions on the right side of the rotor require negative collective control angle because the vortex-induced upwash dominates over the rotor disk. Comparing Fig. 10 (a), (b) and (c) the 7 MW turbine causes larger perturbations in control angles than the 3 MW turbine and the B-747 vortex requires about 4 times more control angle employed than in case of the 7 MW WT. This is due to the increasing circulation strength from case A to D.

Longitudinal rotor control angle $\Delta\theta_S$: Due to the vortex orientation relative to the rotor the resulting aerodynamic roll moments cause changes in the longitudinal control angle to cancel the vortex impact on trim. The plot shows the biggest changes when the vortex is located at $y_0 = -1$, 0 and +1. A vortex position at the rotor hub ($y_0 = 0$) generates as much downwash on the right side of the rotor as upwash on the left side. Then, a positive aerodynamic roll moment appears and needs to be compensated by a positive longitudinal control angle.

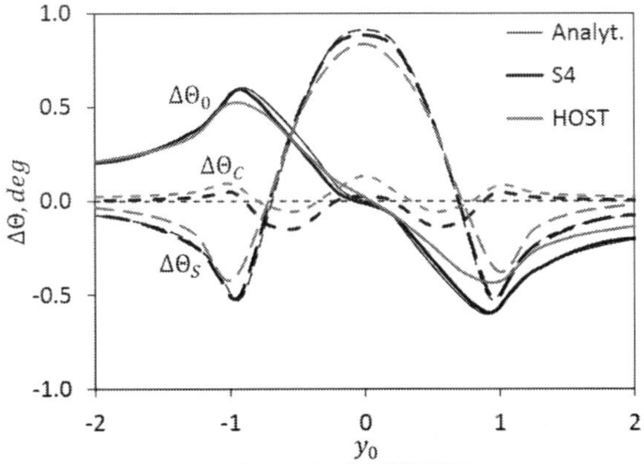

(a) Case A: 3 MW WT

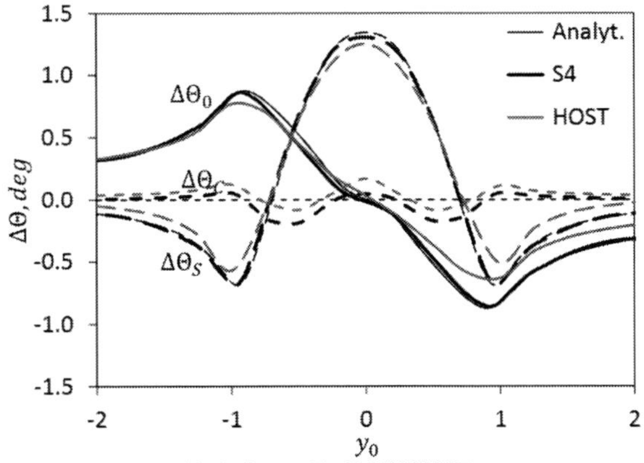

(b) Case B: 7 MW WT

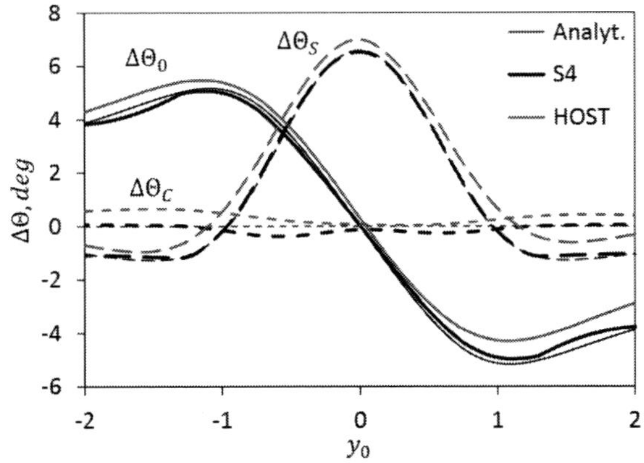

(c) Case D: B-747

Fig. 10: Vortex impact on rotor control angles, hover.

When the vortex is located at either end of the rotor disk the longitudinal control angle needed to compensate its influence is negative in both cases. This is due to the vortex-induced velocity gradients within the rotor disk. For $y_0 = -1$ the entire disk is immersed in downwash of the vortex, with largest values on the left side of the disk, thus a positive gradient from left to right. For $y_0 = +1$ the entire disk is immersed in upwash of the vortex, with largest values on the right side of the disk, thus again a positive gradient from left to right. Therefore, in both cases a negative longitudinal control is needed to compensate this gradient.

Lateral control angle $\Delta\theta_C$: From simple considerations due to the setup of the problem no perturbations are expected, since the vortex does not introduce a longitudinal velocity gradient within the rotor disk. This is also the result of the analytical model. However, the non-linear simulation codes HOST and S4 both compute a small lateral control angle. All controls shown in Fig. 10 that are required to compensate the WT wake vortex effects are small to moderate, compared to an available control bandwidth of approximately 8 deg. For the cases of Fig. 10 these ratios are given in Fig. 11.

As long as the vortex center is outside to the right or left of the rotor disk its influence on blade flapping is small due to small vortex-induced velocities. For small vortex core radii relative to the rotor radius, which is in effect for cases A and B, the maximum RCR is obtained when the vortex axis is located close to the right and left end of the disk, see Fig. 11 (a) and (b), because the largest influence on collective is combined with a large influence on cyclic control, as seen in Fig. 10 (a) and (b).

Moving the vortex center more inboard reduces its influence on the collective and the cyclic control angles quickly approaches zero, thus the RCR correspondingly becomes smaller. When the vortex is centered in the disk, its influence on collective control is zero, but the largest influence on cyclic control is present, again leading to a local maximum RCR.

Although the general behavior is similar for large vortex core radius relative to the rotor radius as is present in case D, the induced velocity gradients becomes smaller, see Fig. 9. This reduces the vortex influence on cyclic control for positions close to the right and left end of the disk as shown in Fig. 10 (c), while the influence on cyclic flapping for central positions within the disk remains large. In this case, Fig. 10 indicates an almost linear vortex-induced velocity across the disk. Therefore, the maximum RCR for large vortex core radii is found for vortex positions close to the center of the rotor disk and not at the right or left end of it, see Fig. 11 (c).

Finally, the change in rotor power is of interest, because it must be expected that such an aerodynamic disturbance causes a large variety of angles of attack and thus will modify both the induced drag and the airfoil drag. Results are given in Fig. 12 for the same cases as before. Due to re-trimming the rotor for constant thrust the changes in thrust and thrust-induced velocity are zero. Only in HOST a slight deviation of thrust-induced velocity is resulting; the reason for this has to be investigated.

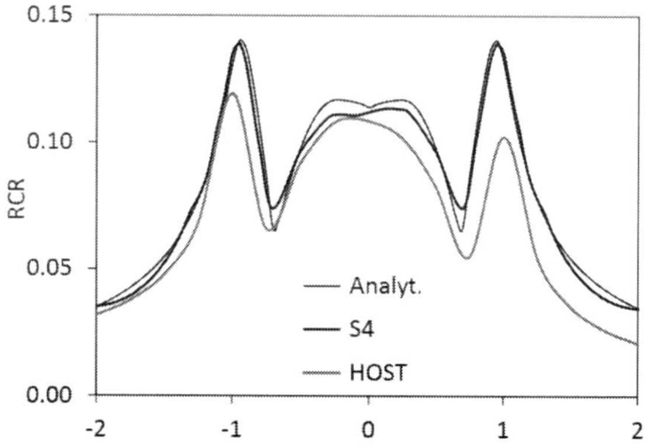

(a) Case A: 3 MW WT

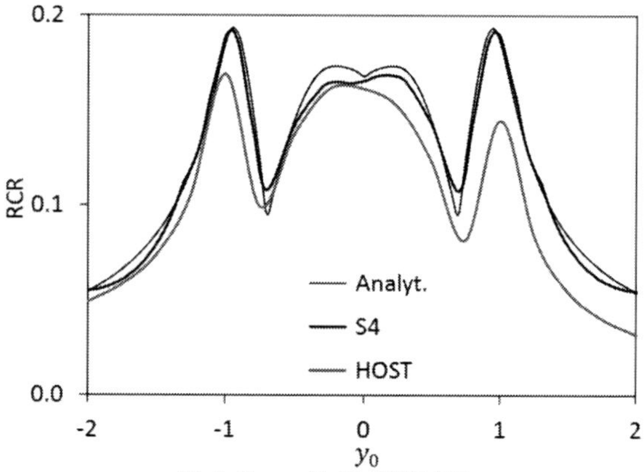

(b) Case B: 7 MW WT

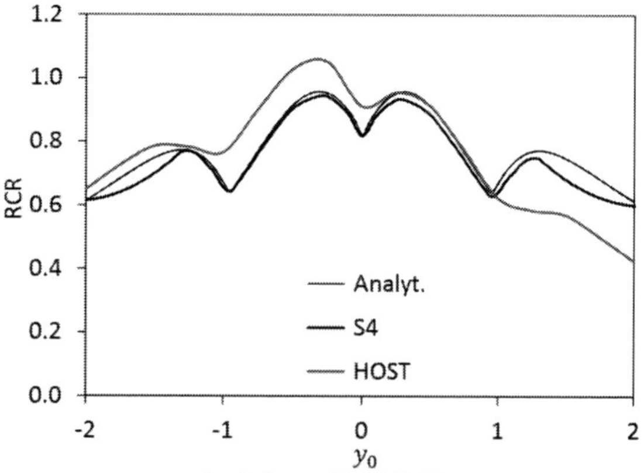

(c) Case D: B-747

Fig. 11: Rotor control ratio in hover.

When the vortex is on the left of the rotor the induced downwash increases and with it the power required, as if it is in a climb. At positions on the right side of the rotor vortex-induced upwash acts like a descent flight and thus reduces the power required. It may be somewhat surprising that vortex positions around the rotor center $-0.5 < y_0 < -0.3$ cause a reduction of rotor power for cases A and B (vortex core radius small relative to rotor radius, Fig. 12 (a) and (b), while large core radii as in Fig. 12 (c) do not show an influence on the power for a center position of the vortex.

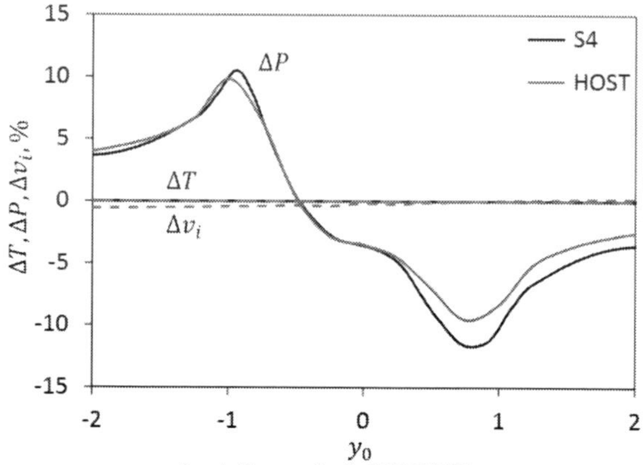

(a) Case A: 3 MW WT

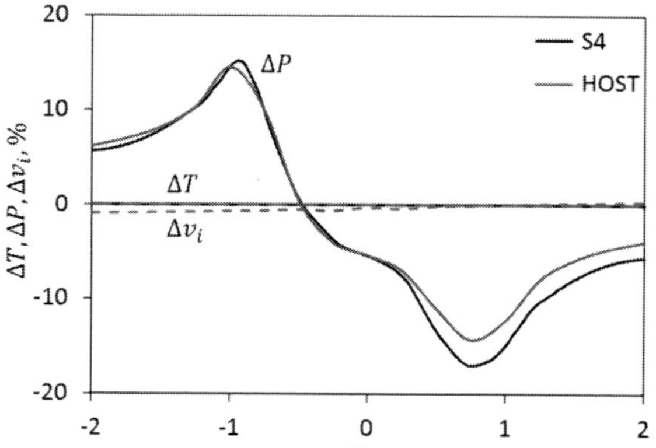

(b) Case B: 7 MW WT

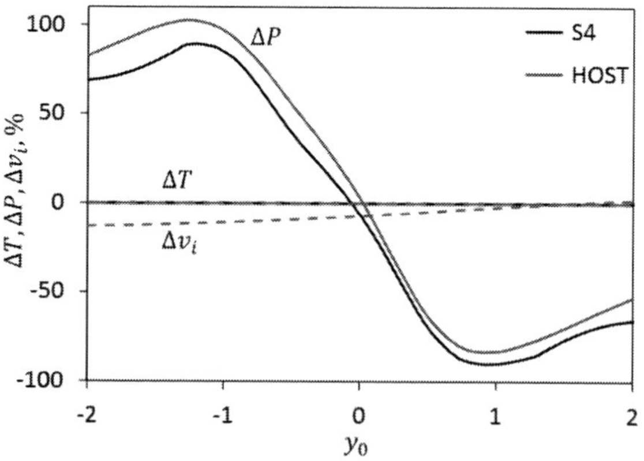

(c) Case D: B-747

Fig. 12: Vortex impact on rotor power in hover.

This effect may be explained with the highly nonlinear velocity profile within the rotor disk of the small core radius vortices, see Fig. 9. Obviously the induced drag reduction on the upwash side of the vortex core "outperforms" the increased induced drag on the downwash side of it. This effect is not present for large vortex core radius relative to the rotor radius as seen in Fig. 12 (c). The vortex-induced velocity profile for a central vortex position within the rotor disk is almost linear across it, thus the reduction of induced drag on the upwash side is balanced by the increase of it on the downwash side and the integral effect on power is zero.

4.5. Bo105: vortex-induced blade flapping

The rotor first is trimmed in undisturbed atmosphere, then the controls are kept fixed and thereafter the vortex is introduced without re-trimming the rotor. In this case rotor blade flapping perturbations will develop in coning and 1/rev. Higher harmonics are not considered here, but will emerge as well. Cases A, B and D as before are investigated.

The hingeless rotor type of the Bo105 with a natural frequency of flapping of $\nu_\beta = 1.12/rev$ has a phase lag of about 72 deg in 1/rev flapping, in contrast to a centrally hinged teetering rotor with $\nu_\beta = 1.0/rev$ and 90 deg phase lag. Therefore, the blade response to a pure Sine excitation of the vortex will develop both a major Cosine and a smaller Sine component in blade flapping. This is clearly visible in Fig. 13, and the longitudinal and lateral flapping angles follow the behavior of the longitudinal control shown in Fig. 10 with about the same magnitude. The arcus tangent of the ratio of longitudinal and lateral flapping yields a phase of 72 deg as expected and the results of the analytical model, the isolated rotor code S4 and of the complete helicopter simulation – used in isolated rotor mode – are in surprisingly good agreement.

The blade coning correlates well with the collective of Fig. 12, but with less magnitude. This is due to the balance of the centrifugal forces due to blade mass and the blade lift. Here the blade Lock number γ has a great influence, representing the ratio of blade inertia to aerodynamic moments. Light-weight blades will have a larger coning because the centrifugal forces are less, and heavy blades will have less coning due to the opposite.

As observed before in the control angles required and the RCR, a large vortex core radius relative to the rotor radius such as in case D shown in Fig. 13 (c) has less impact on blade flapping for vortex positions near the edges of the disk. For the same reasons outlined in description of Fig. 10 the largest influence of the vortex on cyclic flapping is found for the central position within the disk, while the largest coning is developing when the vortex center is outside the disk such that the maximum amount of upwash or downwash affects the blade lift.

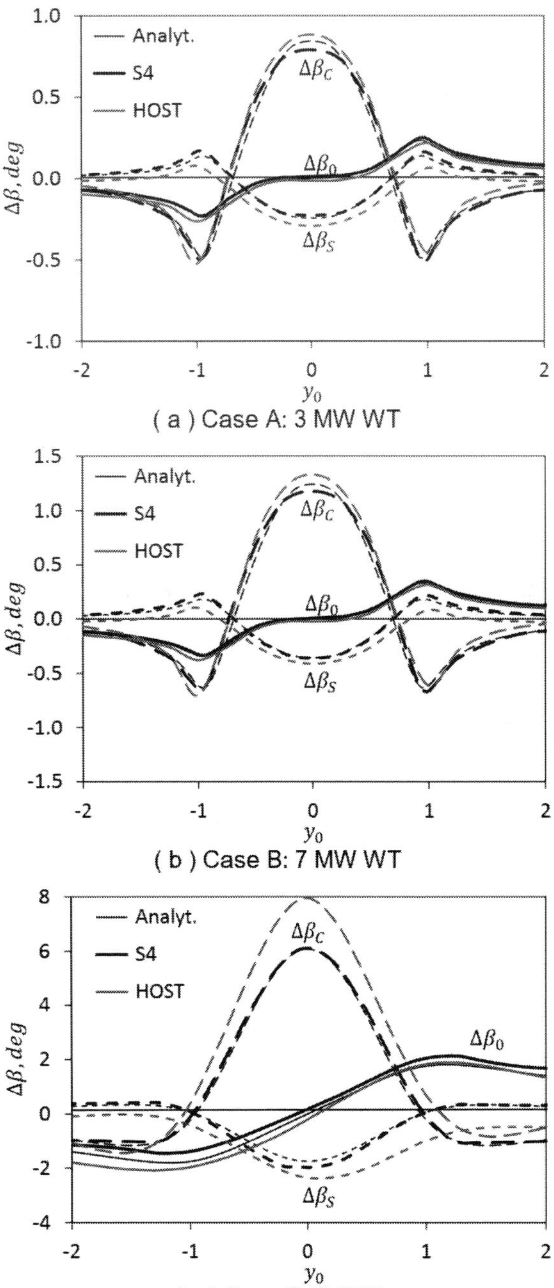

(a) Case A: 3 MW WT

(b) Case B: 7 MW WT

(c) Case D: B-747

Fig. 13: Vortex impact on flapping angles, hover.

In parallel to the rotor control ratio (RCR) that reflects the ability of the pilot to counteract the vortex impact on trim, a rotor flapping ratio (RFR) reflects how much flapping evolves before the blades will collide with the tail boom or the blade stops at the mast. The Bo105 has a relative large amount of allowed flapping of about 15 deg, before the rotor blades collide with the tail boom or the end plates of the horizontal stabilizer. The results for the RFR are given in Fig. 14. Due to the large amount of allowed flapping the ratios are small and considered as uncritical.

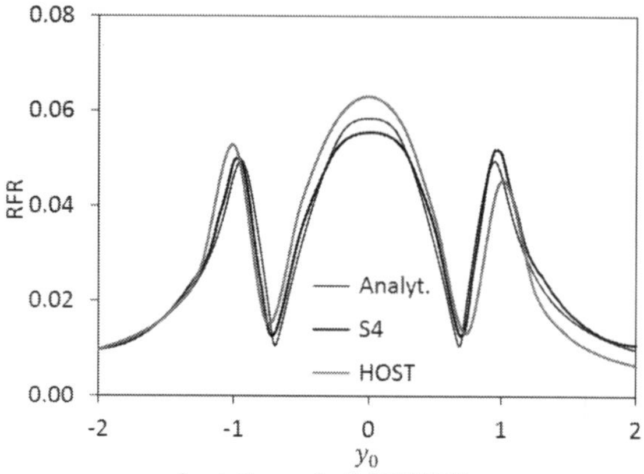

(a) Case A: 3 MW WT

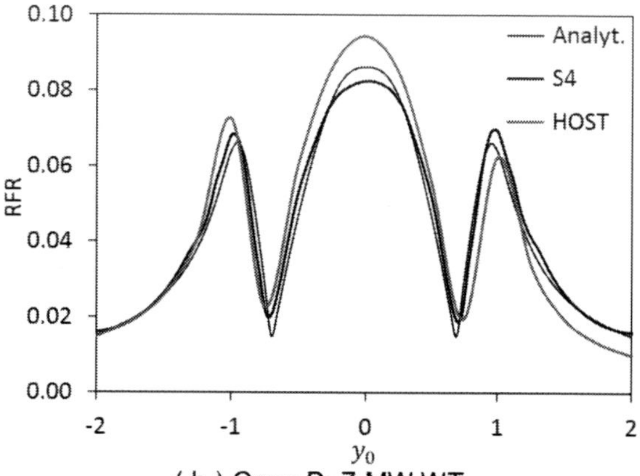

(b) Case B: 7 MW WT

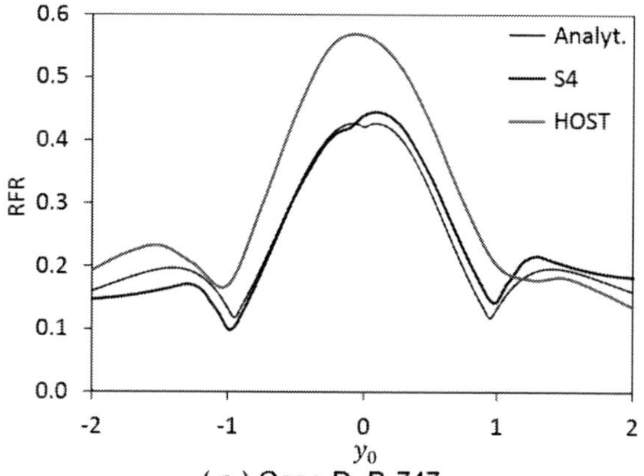

(c) Case D: B-747

Fig. 14: Rotor flapping ratio in hover.

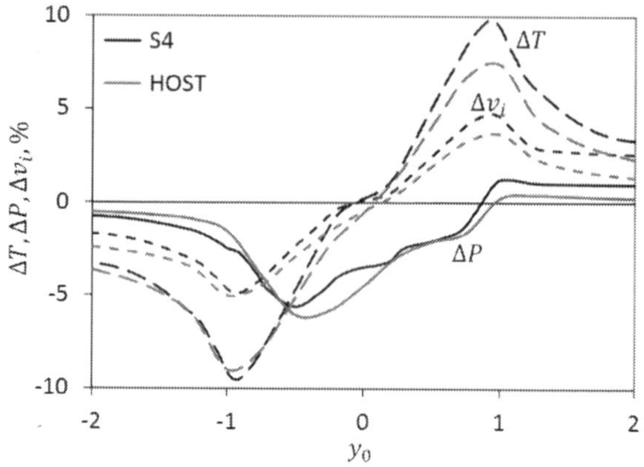

(a) Case A: 3 MW WT

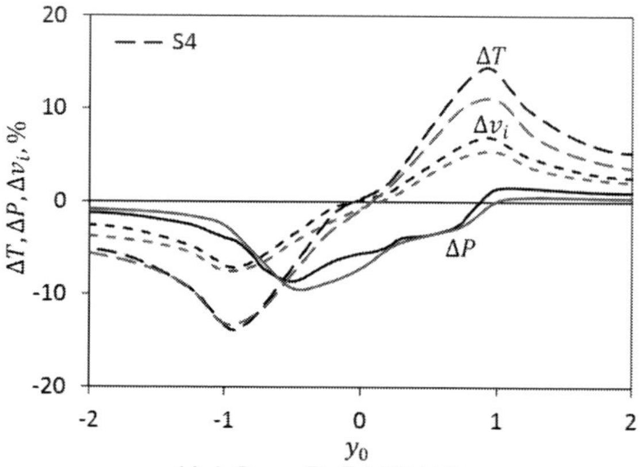

(b) Case B: 7 MW WT

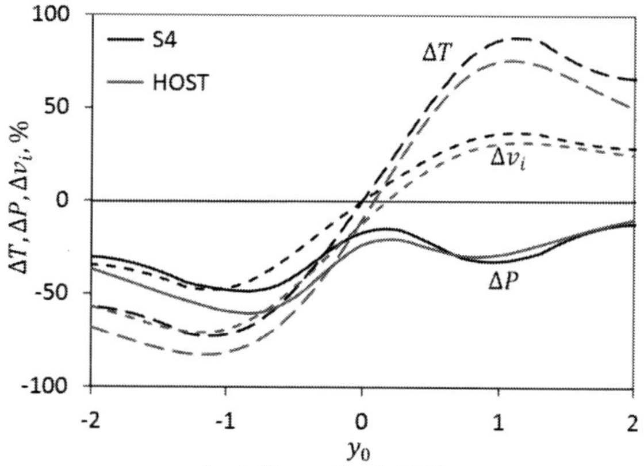

(c) Case D: B-747

Fig. 15: Vortex impact on rotor power in hover.

Not re-trimming the rotor causes the rotor thrust and moments to vary from the undisturbed trim, and therefore induced velocities and power will vary accordingly. This is shown in Fig. 15, to be compared to the case of re-trimming the rotor in Fig. 12. Thrust and with it the thrust-induced velocity varies with the vortex-induced mean velocity acting on the rotor disk. The behavior of the power, however, is very different from the re-trimmed case. A power reduction is found for almost all vortex positions.

Vortex positions on the left side of the rotor it mainly induces downwash, reducing the rotor thrust, the mean thrust-induced velocity and thus the major part of rotor induced power. When the vortex is on the right side of the rotor disk it mainly induces upwash, increasing rotor thrust and with it the mean thrust-induced velocity. In this case the vortex-induced upwash and the thrust-induced downwash are acting against each other. When the vortex is outside to the right side of the rotor disk both influences cancel each other and the power remains virtually unchanged.

4.6. Analytical model applied to other rotors

In the foregoing investigations the analytical model is found in excellent agreement with the more sophisticated rotor codes.

Therefore (and because the further rotorcraft listed in Table 1 are not modeled in S4 or in HOST), the rotor controls required to compensate vortex effects on trim (in case of helicopters only) and the vortex influence on blade flapping (including the autogyro) are computed for cases A-D. For the small UL rotorcraft with teetering rotor systems it is of interest whether mast bumping can happen due to excessive flapping (especially for AGs and 2-bladed single rotor helicopters), or blade collisions (in case of coaxial rotors).

In the following only the RCR and the RFR are shown for the various types of rotors listed in Table 1, exposed to the cases of Table 3. The effect of vortices generated by WTs of 3, 7, and 10 MW power (cases A to C) and that of a large fixed-wing aircraft (case D) on the rotor control ratio in dependence on the lateral vortex position within the rotor disk is shown in Fig. 16. It is obvious that for increasing size and weight of the rotorcraft the amount of control to mitigate the vortex effects on rotor trim will be reduced.

The most critical RCR are thus found for the smallest vehicle, the ultralight COAX helicopter, Fig. 16 (a). Even the large WT vortices require close to 50% of the available (assumed) control margin, and the large aircraft vortex exceeds it, causing uncontrollable reactions of the vehicle.

For the next larger helicopter, the Bo105 shown in Fig. 16 (b), the effects are reduced, but the large aircraft vortex requires almost the full amount of available control to keep the rotor trim constant. The much larger UH-1D and the even larger CH-53D are both less affected, but still the large aircraft vortex requires up to 80% of the available control margin Fig. 16 (c) and (d). The WT vortices appear to be no problem for all helicopters of size Bo105 and up, requiring up to 25% (Bo105) to 15% (CH-53D) of the control margin.

The influence of the same vortices on the rotor blade flapping in terms of the RFR of the same helicopters as in Fig. 16, and additionally on the ultralight autogyro, are given next in Fig. 17. Note that every rotorcraft has different allowed range of flapping, in contrast to the control margin of the helicopters, which was assumed to 8 deg for all (Table 1). In single rotor teetering blade articulation the maximum allowed flapping may be limited by either mast bumping, which is a catastrophic event with usually rotor separation as the consequence of it, or by blade collision with the tail boom or other elements of the fuselage or empennage.

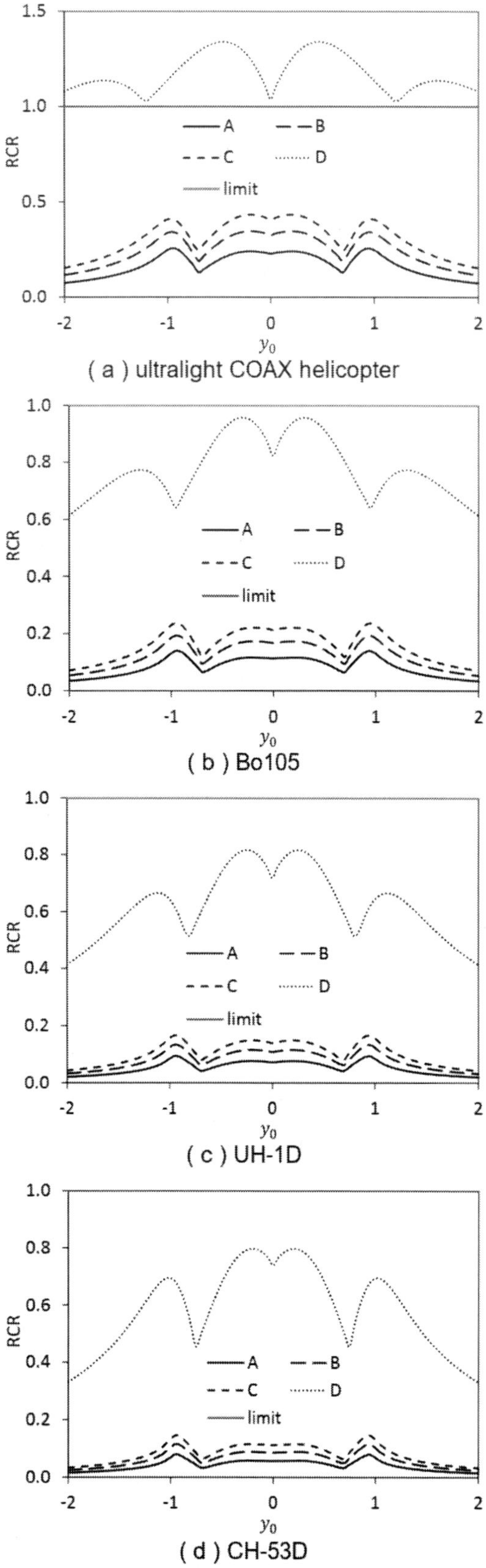

Fig. 16: Rotor control ratios for different helicopters exposed to WT and fixed-wing vortices.

The major problem of coaxial rotors is a blade collision occurring when blades are flapping in opposite directions. This is especially the case in vortex-rotor interactions, because – for example – the vortex upwash will act on the advancing side of one of the rotors and simultaneously on the retreating side of the other, causing differential harmonic flapping motion. Such differential flapping is the cause of the vertical rotor separation distance, which typically amounts to 20% of the rotor radius.

Mast bumping – or fuselage/tail boom/empennage contact – may be a limit of flapping only in cases where both rotor disks tilt parallel to each other. It must be kept in mind that in the following results it is assumed that the trim condition does not have any 1/rev flapping. If this is not the case the flapping margin is reduced by the amount of flapping already present in the trim state. In trimmed steady state flight the blade flapping of an autogyro can already amount up to 3.5 deg, and even more at lower RPM or during a maneuver [21]. RFR results for the autogyro are given in Fig. 17 (a). Mast bumping is occurring only during interactions with the fixed wing vortex at central positions of it. WT vortices cause blade flapping of 20 to 40% of the available margin (7 deg, see Table 1). However, taking into account the flapping during normal steady state flight the remaining margin for flapping motion becomes only half of that value, doubling the RCR.

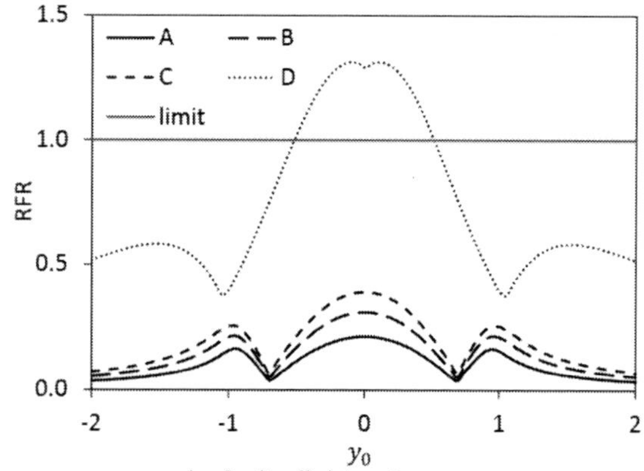

(a) ultralight autogyro

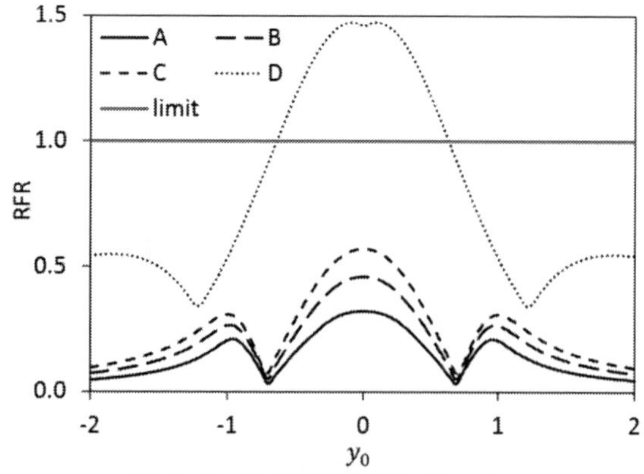

(b) ultralight COAX helicopter

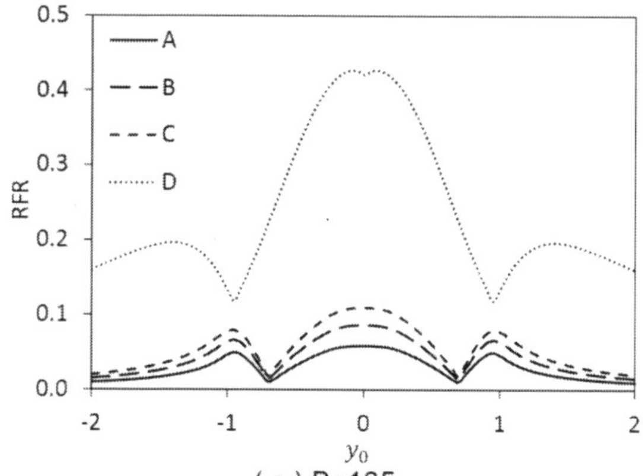

(c) Bo105

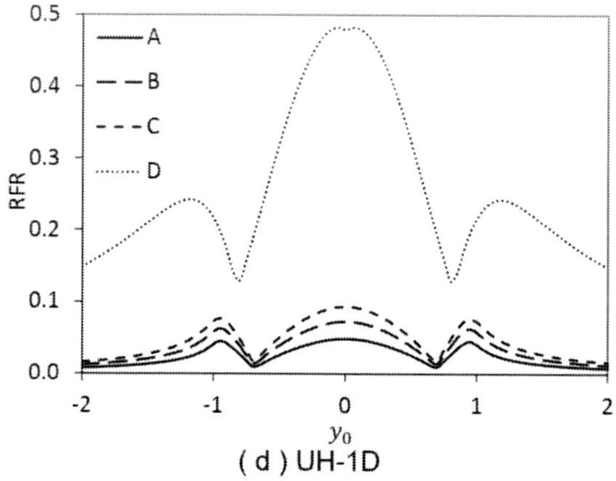

(d) UH-1D

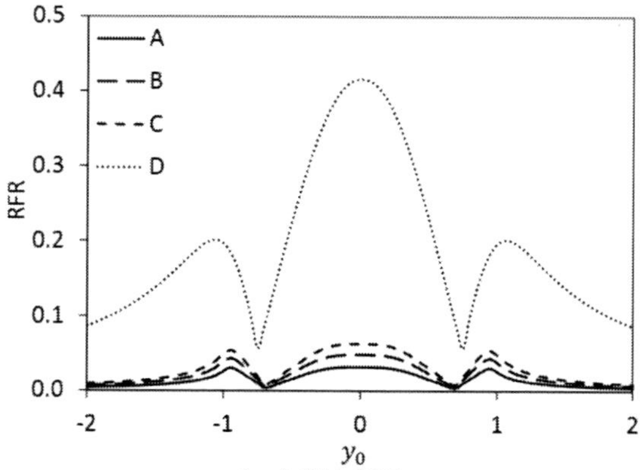

(e) CH-53D

Fig. 17: Rotor flapping ratios for different helicopters exposed to WT and fixed-wing vortices.

In addition, operating the rotor at lower RPM or during a maneuver the remaining margin for flapping is even further reduced and then even WT vortices may become a hazard for autogyros. In contrast to autogyros the RPM of helicopters is fixed within rather small bounds. The RFR results for the ultralight COAX helicopter are given in Fig. 17 (b). It may be surprising that the RFR are generally larger than for the autogyro shown in (a). This is partly caused by the smaller rotor radius and thus larger relative influence of the vortex, see Fig. 9, but mainly due to a smaller flapping margin for the critical case of blade collision, see Table 1.

Again the fixed wing vortex is the most severe hazard, causing blade collisions for a wide range of vortex positions inside the rotor disk. WT vortices cause RFR between 0.3 and almost 0.6. This may not be critical as such, but some differential flapping will occur during normal flight as well, reducing the available margin and increasing further the RFR. For all other larger helicopters such as the Bo105, the UH-1D and the CH-53D results are shown in Fig. 17 (c)-(e), respectively. These rotorcraft have a significantly higher blade tip speed and larger margins of blade flapping, see Table 1, and thus less influence of the WT or fixed wing vortices, as was already demonstrated in Fig. 10. Consequently the RFR is significantly less than for the ultralight rotorcraft discussed before. The maximum RFR of almost 0.5 is found for the UH-1D in Fig. 17 (d), and less for the Bo105 and the CH-53D. The RFR due to WT vortices range around 0.1 and are considered as not of importance.

4.7. Bo105: forward flight

The effect of forward flight at an advance ratio (for zero shaft angle of attack) of μ = 0.3 is investigated next at the example of the Bo105 rotor and for case B only. Again results of the analytical model are compared with those of the S4 and HOST codes. First the vortex effect on the rotor thrust and hub moments is compensated by rotor controls, Fig. 18. In Fig. 18 (a) the general form of collective and cyclic controls as a function of lateral vortex position is shown, to be compared with those in hover given in Fig. 10 (b). The overall appearance is similar as in hover, but asymmetric in forward flight.

Vortex position at the edge of the rotor disk on the retreating side cause larger disturbances in terms of angle of attack due to smaller tangential velocities at the blade elements, thus larger cyclic and with it larger collective controls than in hover are required. Contrary, vortex positions at the edge on the advancing side are less disturbing due to the larger tangential velocities there, compared to hover. Inboard positions on the advancing side again have a slightly larger impact on controls than in hover.

The consequences of these are reflected in the RCR given in Fig. 18 (b), to be compared with Fig. 11 (b): an asymmetry is clearly visible and the maximum RCR are larger than in hover. Fig. 18 (c) shows the impact on power required, to be compared with Fig. 12 (b). Again, the aforementioned asymmetry is found and the maximum power increase is 30% (in hover: 15%) of the undisturbed trim power. The maximum decrease is about -35% (versus -17% in hover). This is caused by vortex upwash or downwash, felt by the rotor as an equivalent to a climb or descent situation, respectively.

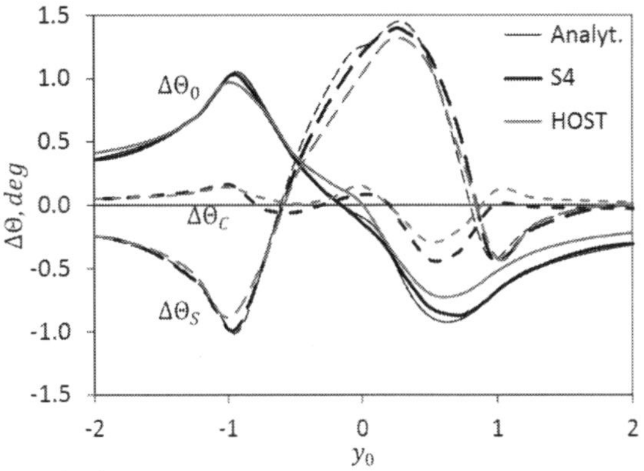

(a) Vortex impact on rotor control angles

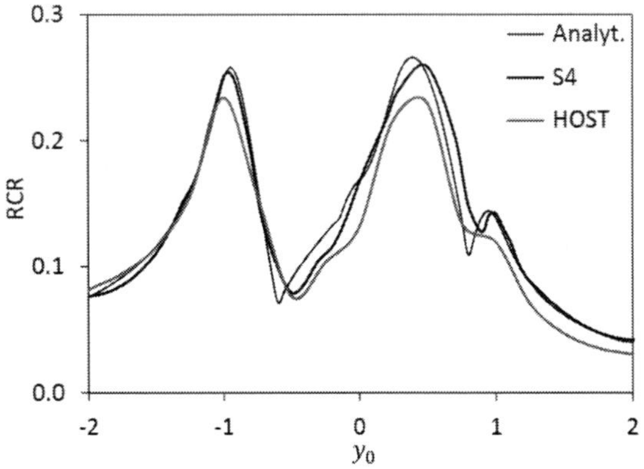

(b) Rotor control ratio

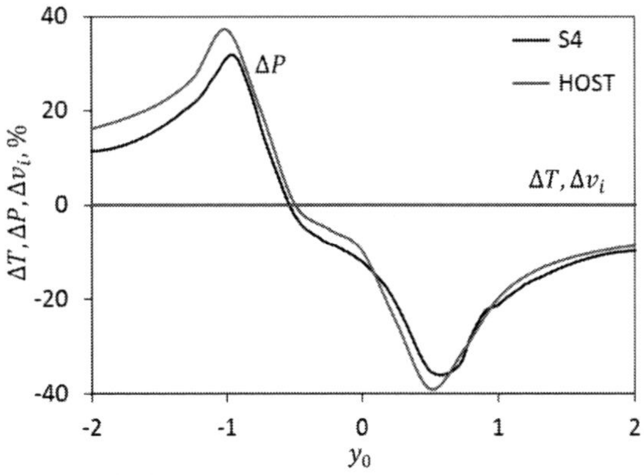

(c) Vortex impact on rotor power

Fig. 18: Rotor controls required for vortex disturbance rejection and power in forward flight, case B.

The vortex impact on a rotor without re-trimming is shown next in Fig. 19 (a) with respect to rotor blade flapping, to be compared with Fig. 13 (b) for the hovering case. The asymmetry already found in the controls is found again, but with larger flapping response for vortex positions on the advancing side and less on the retreating side. This is due to the larger dynamic pressure and associated aerodynamic forces acting on the blade, compared to those on the retreating side.

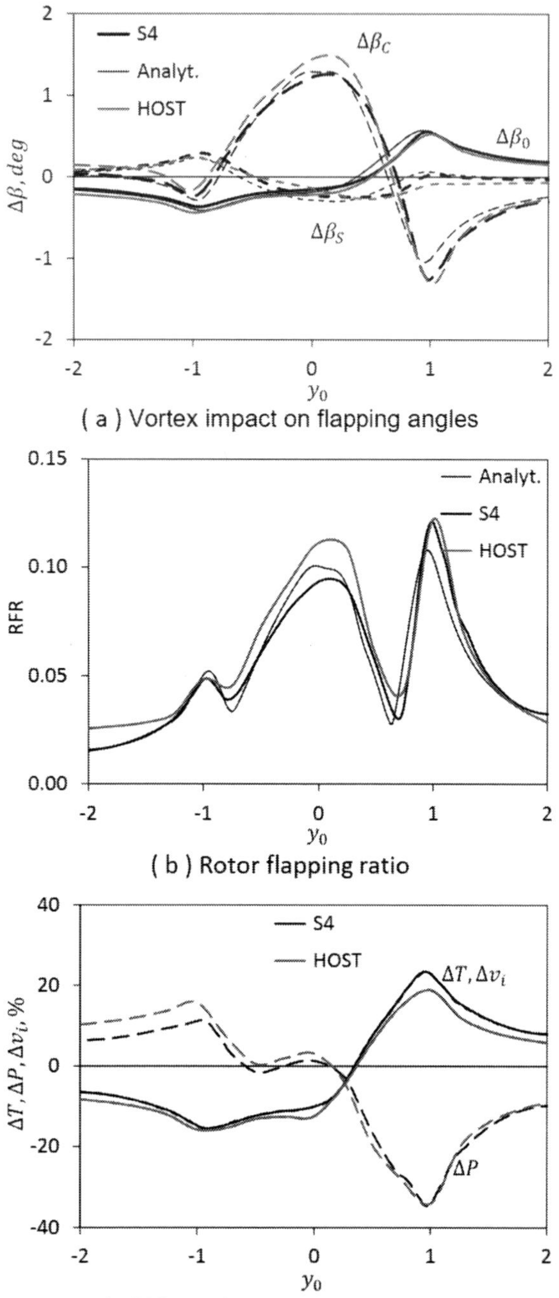

(a) Vortex impact on flapping angles

(b) Rotor flapping ratio

(c) Vortex impact on rotor power

Fig. 19: Vortex-induced blade flapping and influence on power and thrust in forward flight, case B.

Consequently the RFR in Fig. 19 (b) is largest for vortex positions on the advancing side edge of the disk, while in hover it is largest for central vortex positions. Finally the impact on thrust, thrust-induced velocity and rotor power is shown in Fig. 19 (c), to be compared with Fig. 15 (b) in hover. The percentage change in power and induced velocity is virtually the same, and amounts to an increase of up to 23% thrust and induced velocity for vortex positions at the advancing side edge of the disk, versus 15% in hover. The loss of thrust for positions on the retreating side edge of the disk is about the same as in hover.

However, the resulting change of power required shown in Fig. 19 (c) is different from the hovering case Fig. 15 (b). The vortex positions on the retreating side of the disk, especially close to the edge of it and outside, now cause a power increase versus a decrease in hover. In this forward flight condition the vortex-induced peak velocities are larger than the thrust-induced velocities, which appear vice-versa in the hovering case, and therefore causing the reversal of the impact on power. Overall, the analytical model results are very close to those obtained with more sophisticated codes.

5. CONCLUSIONS

In this paper the effect of a young wake vortex generated by wind turbines (WT) 100 m behind them and by a large fixed-wing aircraft (2000 m behind it) on the rotor control ratio (RCR) as well as on the rotor flapping ratio (RFR) is presented. Rotors considered range from ultralight rotorcraft (autogyro and coaxial helicopter), to small, medium and large helicopters. Results from the Bo105 helicopter rotor simulated by HOST are compared with those obtained by a simplified analytical treatment of an isolated rotor, and those of DLR's rotor code S4.

The major conclusions are:

- The simplified isolated rotor analysis generates results very close to those of the nonlinear sophisticated rotor or complete helicopter codes for this kind of investigation, despite the simplifications made.
- For autogyros WT vortices may become hazardous, because the blade flapping developing in addition to that of the flapping due to the flight condition may result in mast bumping. Large aircraft vortices are always dangerous for autogyros.
- Small (Bo105 size), medium and large helicopters are more limited by the available control margin than by the margin of flapping. A large fixed wing aircraft vortex can result in RCR of 1 for the small helicopter, and up to 0.8 for the larger ones, but the RFR for all remains below 0.5 due to a larger flapping margin, compared to the control margin. In any case, the WT vortices are not causing hazardous RCR or RFR for these helicopter sizes.
- Ultralight coaxial helicopters may also be in danger due to the potential of blade collisions, because a vortex encounter results in differential blade flapping that cannot be compensated by pilot control. Big aircraft vortex encounters exceed a RCR of 1, and WT vortices cause a

RCR of almost 0.5. The limited amount of differential flapping becomes dangerous in case of a large fixed wing vortex encounter, and the RFR generated by WT vortices may reach 0.6.

- In hover the impact of the vortex on rotor controls, flapping and power are symmetric, while in forward flight they become asymmetric and larger than in hover. In forward flight larger controls are required for re-trimming when the vortex is on the retreating side of the rotor, and larger flapping develops when it is on the advancing side.
- RCR and RFR are larger in forward flight than in hover. However, for the Bo105 example they remain uncritical for the 7 MW WT investigated.
- With no pilot action the vortex-induced velocities modify the small helicopter (Bo105) rotor thrust by up to ±90% in hover, which will cause a significant influence on flight mechanics of the helicopter. In forward flight the change of thrust is much less (±20%).
- The rotor power in these cases is always reduced by up to 50% in hover. In forward flight a power increase up to +20% is found when the vortex generates downwash on the retreating side, and a power decrease of -30% for upwash on the advancing side.
- When re-trimming for constant thrust the rotor power is varying by ±100% in hover and ±40% in forward flight. In the latter a power increase is found when the vortex generates downwash on the retreating side, and a power decrease for upwash on the advancing side.
- Despite re-trimming for constant thrust and zero 1/rev flapping the mean flapping angle varies for vortex positions within the rotor disk due to a radial re-distribution of steady lift.
- All of these effects can be explained with the relation of the magnitude of rotor thrust-induced velocities to the vortex-induced velocities. In hover, the rotor thrust-induced velocities are larger, while in forward flight the thrust-induced velocities quickly become smaller than the vortex-induced velocities that are independent of the helicopter's flight speed.

Future investigations will focus on the complete helicopter trim and to the flight dynamics response.

6. REFERENCES

[1] Dunham, R.E., Holbrook, G.T., Mantay, W.R., Campbell, R.L., Van Gunst, R.W., "Flight-Test Experience of a Helicopter Encountering an Airplane Trailing Vortex", 32nd Annual National V/STOL Forum of the American Helicopter Society, Washington, DC, May 10-12, 1976.

[2] Mantay, W.R., Holbrook, G.T., Campbell, R.L. and Tamaine, R.L., "Helicopter response of an airplane's trailing vortex," *Journal of Aircraft*, **14**(4), 1977, pp. 357-363.

[3] Saito, S., Azuma, A., Kawachi, K., Okuno, Y., Hasegawa, T., "Numerical simulations of dynamic response of fixed and rotary wing aircraft to a large airplane wake," 13th European Rotorcraft Forum, Arles, France, Sept. 8-11, 1987.

[4] Azuma, A., Saito, S., Kawachi, K., "Response of a helicopter penetrating the tip vortices of a large airplane," *VERTICA*, **11**(1), 1987, pp. 65-76.

[5] Turner, G.P., Padfield, G.D., Harris, M., "Encounters with Aircraft Vortex Wakes; The Impact on Helicopter Handling Qualities," *Journal of Aircraft*, **39**(5), 2002, pp. 839–849.

[6] Padfield, G.D., Manimala, B., Turner, G.P., "A Severity Analysis for Rotorcraft Encounters with Vortex Wakes," *Journal of the American Helicopter Society*, **49**(4), 2004, pp. 445–456.

[7] Lawrence, B., Padfield, G.D., "Wake Vortex Encounter Severity for Rotorcraft in Approach and Landing," 31st European Rotorcraft Forum, Florence, Italy, Sept. 13-15, 2005.

[8] Whitehouse, G.R., Brown, R.E., "Modeling the Mutual Distortions of Interacting Helicopter and Aircraft Wakes," *Journal of Aircraft*, **40**(3), 2003, pp. 440-449.

[9] Whitehouse, G.R., Brown, R.E., "Modelling a helicopter rotor's response to wake encounters," *The Aeronautical Journal*, **108**(1079), 2004, pp. 15-26.

[10] van der Wall, B.G., Fischenberg, D., Lehmann, P.H., van der Wall, L.B., "Impact of Wind Energy Rotor Wakes on Fixed-Wing Aircraft and Helicopters," 42nd European Rotorcraft Forum, Lille, France, Sept. 5-8, 2016

[11] van der Wall, B.G., van der Wall, L.B., "Analytical Estimate of Rotor Controls Required for a Straight Vortex Disturbance Rejection," *Journal of the American Helicopter Society*, **62**(1), 2017, pp. 015001-1-4

[12] Leishman, J.G., *Principles of Helicopter Aerodynamics*, Cambridge: Cambridge University Press, UK, 2000.

[13] Burnham, D.C., Hallock, J.N., "Chicago Monostatic Acoustic Vortex Sensing System," U.S. Department of Transportation, DOT-TSC-FAA-79-103, 1982.

[14] Vatistas, G.H., Kozel, V., Mih, W.C., „A Simpler Model for Concentrated Vortices," *Experiments in Fluids*, **11**(1), 1991, pp. 73-76.

[15] Wang, X., White, M., Barakos, G., Wind Turbine Wake Encounter Study. Liverpool: University of Liverpool, UK, 2015.

[16] Benoit, B., Dequin, A.-M., Kampa, K., von Grünhagen, W., Basset, P.-M., Gimonet, B., „HOST, a General Helicopter Simulation Tool for Germany and France“, 56th Annual Forum of the American Helicopter Society, Virginia Beach, VA, May 2-4, 2000.

[17] Oseen, C.W., "Über Wirbelbewegung in einer reibenden Flüssigkeit," *Arkiv för Matematik, Astronomi och Fysik*, **7**, 1912, pp. 14-21.

[18] Lamb, H., *Hydrodynamics*, Cambridge: Cambridge University Press, UK, 1932, pp. 592-593, 668-669.

[19] Meijer-Drees, J., "A Theory of Airflow Through Rotors and its Application to Some Helicopter Problems," *Journal of the Helicopter Association of Great Britain*, **3**(2), 1949, pp. 79-104

[20] van der Wall, B.G., Lim, J.W., Smith, M., Jung, S.N., Bailly, J., Baeder, J.D., Boyd, D.D., "The HART II International Workshop: An Assessment of the State-of-the-Art in Comprehensive Code Prediction, *CEAS Aeronautical Journal*, **4**(3), 2013, pp. 223-252

[21] Duda, H., Seewald, J., *Flugphysik der Tragschrauber*, Berlin: Springer Vieweg, 2016

APPENDIX

The integral in Eq. (18), based on the kernel function of Eq. (7), is given here in A3, together with the integrals of Eqs. (11) and (12) that have already been presented in [10]. To simplify the expressions of the results the following abbreviations are used.

(21) $$\xi = r^2 - |y_0|^2 + |r_c|^2 \qquad \eta = 2|y_0||r_c|$$

A1. Radial integration of the coefficient a_{V0}

(22) $$\int a_{V0}\,dr = \mu \ln\sqrt{\sqrt{\zeta^2+\eta^2}+\zeta+2y_0^2} + \mu\ln\left(1+\frac{\sqrt{2}|r_c|}{\sqrt{\sqrt{\zeta^2+\eta^2}+\zeta}}\right) + y_0\frac{\sqrt{2}|r_c|}{\sqrt{\sqrt{\zeta^2+\eta^2}+\zeta}}$$

A2. Radial integration of the product rb_{V1}

(23) $$\begin{aligned}\int rb_{V1}\,dr &= 2\mu\,\mathrm{sgn}(y_0)|r_c|\arctan\frac{\sqrt{2}|y_0|}{\sqrt{\zeta^2+\eta^2}+\zeta}\\ &+2\mu y_0\left[\ln\sqrt{\sqrt{\zeta^2+\eta^2}+\zeta+2y_0^2}+\ln\left(1+\frac{\sqrt{2}|r_c|}{\sqrt{\sqrt{\zeta^2+\eta^2}+\zeta}}\right)\right]\\ &+r^2-\sqrt{2}\left(|r_c|\sqrt{\sqrt{\zeta^2+\eta^2}+\zeta}-|y_0|\sqrt{\sqrt{\zeta^2+\eta^2}-\zeta}\right)\end{aligned}$$

A3. Radial integration of the product ra_0

(24) $$\begin{aligned}\int ra_{V0}\,dr &= \mu r+\frac{y_0|r_c|}{\sqrt{2}}\frac{\sqrt{\sqrt{\zeta^2+\eta^2}+\zeta}}{\sqrt{\zeta^2+\eta^2}+\zeta}r\\ &-\frac{y_0+\mu}{2}|r_c|\ln\left[\frac{r\sqrt{2}\sqrt{\sqrt{\zeta^2+\eta^2}+\zeta}+\sqrt{\zeta^2+\eta^2}+r^2}{\sqrt{\left(y_0^2-r_c^2\right)^2+\eta^2}}\right]\\ &-\left(\mathrm{sgn}(y_0)\frac{y_0^2-r_c^2}{2}+\mu|y_0|\right)\arctan\left(\frac{r\sqrt{2}\sqrt{\sqrt{\zeta^2+\eta^2}+\zeta}+\sqrt{\zeta^2+\eta^2}+\zeta}{\eta}\right)\end{aligned}$$

Influences of Upper and Side Walls on the Performance of a Multiple Rotor Drone

Yasutada Tanabe
Associate Senior Researcher
JAXA
Mitaka, Tokyo, Japan

Masahiko Sugiura
Researcher
JAXA
Mitaka, Tokyo, Japan

Takashi Aoyama
Principal Researcher
JAXA
Mitaka, Tokyo, Japan

Hideaki Sugawara
Research Engineer
Ryoyu Systems, Co. Ltd
Tokyo, Japan

Shigeru Sunada
Professor
Nagoya Univ.
Nagoya, Japan

Koichi Yonezawa
Project Researcher
CRIEPI
Chiba, Japan

Hiroshi Tokutake
Associate Professor
Kanazawa Univ.
Kanazawa, Japan

ABSTRACT

Influences of the walls on the performance of the multiple rotor type drones are numerically simulated. With current wide spreading applications of the autonomously flyable UAVs, there are potential needs to use the drones for inspections and observation near various structures, such as buildings and bridges. The flowfields around multiple rotors are significantly influenced with the existence of an upper or a side wall, so that the thrust and required torque and other forces and moments are changed. It is found that when a rotor is approaching an upper wall, as the distance between the rotor and the upper wall is less than a diameter of the rotor, the thrust suddenly increases that causes the rotors to collide onto the upper wall. When an isolated rotor is operated near a side wall, the thrust decreases and a rolling moment appears to tilt the rotor toward the wall as the distance becomes shorter. For a multiple rotor drone near a side wall, the rotors have different distances from the wall, which causes the whole aircraft tilts toward the wall. From a view of safety operations, the multi-rotor drone should be kept away from both the upper wall and the side wall at least 1.5 times of the rotor diameter to prevent unexpected motions of the aircraft caused by the wall during hovering flight.

INTRODUCTION

Multi-rotor type drones (will be referred as multi-rotors in this paper) have been applied to various fields including observations and inspections near infrastructures, such as bridges and buildings. In such cases, flying near a wall is often required.

When a multi-rotor is flying near a upper wall, absorbed onto the wall is often experienced. Also, when a multi-rotor is flying near a side wall, it tends to tilt to the wall and stick to the wall. The mechanism of the causes of the absorption of the multi-rotors to the wall is not well understood yet. From a view of safety operations, the relationships of the aerodynamic forces and moments of a multi-rotor with the distance from the wall need to be clarified and built into the flight controls of the multi-rotors.

The objective of this study is to understand the cause of the wall absorption phenomena and quantitatively clarify from what distances such phenomena start to appear. A multi-rotor model based on a variable pitch controlled hexa-rotor drone is numerically simulated.

NUMERICAL METHODOLOGIES

Studies of the aerodynamic interactions between multiple rotors with an upper or a side wall are carried out through numerical simulations using a CFD code, rFlow3D, developed in JAXA for rotorcraft. In the rFlow3D code,[1, 2] a three dimensional moving overlapping grid method is used. This code is originally developed to treat the conventional helicopters where a main rotor and a tail rotor are considered. It has been extended so that any number of rotors can be handled. [3] Any number of inner background grids can also be placed for efficient capturing of the rotor near wakes.

The full N-S equation in ALE form is shown in Eq. 1, where **U** is the conservative flow variable vector and $\mathbf{F}^i$ and $\mathbf{F}^v$ are inviscid and viscous flux vector respectively. $V(t)$ is the time-varying cell volume and $S(t)$ is the time-varying cell boundary. **n** is the normal vector to the cell surface.

Presented at the 6th Asian-Australian Rotorcraft Forum & Heli Japan 2017, Kanazawa, Japan, November 7-9, 2017.

(1) $$\frac{\partial}{\partial t}\int_{V(t)}\mathbf{U}dV + \oint_{S(t)}\left(\mathbf{F}^{i} - \mathbf{F}^{v}\right)\cdot \mathbf{n}dS = 0$$

where,

$$\mathbf{U} = \begin{bmatrix} \rho \\ \rho\mathbf{v} \\ \rho e \end{bmatrix},$$

$$\mathbf{F}^{i}\cdot\mathbf{n} = \begin{bmatrix} (\mathbf{v}-\dot{\mathbf{x}})\cdot\mathbf{n}\rho \\ (\mathbf{v}-\dot{\mathbf{x}})\cdot\mathbf{n}\rho\mathbf{v} + p\mathbf{n} \\ (\mathbf{v}-\dot{\mathbf{x}})\cdot\mathbf{n}\rho e + p\mathbf{v}\cdot\mathbf{n} \end{bmatrix},$$

$$\mathbf{F}^{v}\cdot\mathbf{n} = \begin{bmatrix} 0 \\ \boldsymbol{\tau}\cdot\mathbf{n} \\ (\boldsymbol{\tau}\cdot\mathbf{v}-\mathbf{q})\cdot\mathbf{n} \end{bmatrix}$$

An all-speed numerical scheme SLAU (Simple Low-dissipation AUSM) [4] with extension to three dimensional moving grids (referred as mSLAU) is adopted. [5] It is very suitable for the flow calculation around a rotary wing, where the local flow speed may vary from very low on the root area to transonic at the tip. Combining SLAU with a Fourth-order Compact MUSCL TVD (FCMT) interpolation scheme, [6] fourth-order spatial accuracy is obtained in shock free regions. Implicit LU-SGS and Dual-Time-Stepping method [7] are used for time integration on blade grids. Yet for the background grids, explicit four stages Runge-Kutta time integration method [8] is used. In this study, considering the relative low Reynolds numbers on the rotors, no turbulence model is adopted.

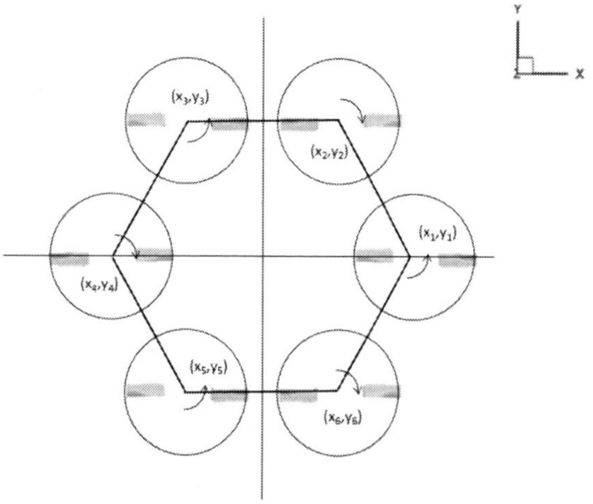

Figure 1. Definitions of multi-rotor coordinates.

The coordinate definitions of a hexa-copter are shown in Figure 1. The rotors are counter-rotating regarding to their neighbours. The aerodynamic forces and moments acting on the centre of a multi-rotor can be calculated from Eq. (2) and (3).

(2)
$$X_{sum} = \sum_{i=1}^{Nr} X_i$$
$$Y_{sum} = \sum_{i=1}^{Nr} Y_i$$
$$Z_{sum} = \sum_{i=1}^{Nr} Z_i$$

(3)
$$Mx_{sum} = \sum_{i=1}^{Nr} Mx_i + \sum_{i=1}^{Nr} Z_i y_i$$
$$My_{sum} = \sum_{i=1}^{Nr} My_i + \sum_{i=1}^{Nr} -Z_i x_i$$
$$Mz_{sum} = \sum_{i=1}^{Nr} Mz_i + \sum_{i=1}^{Nr} \{-X_i y_i + Y_i x_i\}$$

The rotor forces and moments are non-dimensionalized as in Eq. (4), where the torque and power relationship holds only for a single rotor, remembering that neighbouring rotors are counter-rotating to each other. Also the Figure of Merit (*FM*) of a rotor is defined as the ratio of ideal induced power to the actually required power, the nearer to the value of 1, the better rotor performance in hover.

(4)
$$C_X = X/\left(\rho\pi R^2 V_{tip}^2\right)$$
$$C_Y = Y/\left(\rho\pi R^2 V_{tip}^2\right)$$
$$C_Z = C_T = Z/\left(\rho\pi R^2 V_{tip}^2\right)$$
$$C_{MX} = M_X/\left(\rho\pi R^2 V_{tip}^2 R\right)$$
$$C_{MY} = M_Y/\left(\rho\pi R^2 V_{tip}^2 R\right)$$
$$C_{MZ} = -C_Q = M_Z/\left(\rho\pi R^2 V_{tip}^2 R\right)$$
$$C_P = P/\left(\rho\pi R^2 V_{tip}^3\right) = C_Q$$
$$FM(\text{Figure of Merit}) = \frac{(C_T)^{3/2}/\sqrt{2}}{C_Q}$$

MULTI-ROTOR MODELLING

A prototype drone with 6 pitch-controlled rotors as shown in Figure 2 is manufactured as a part of a research of the ImPACT program funded by the Japanese government and managed by the Japan Science and Technology Agency (JST). [9] The aim of the adoption of the pitch-control for the rotors is to attain fast attitude change of the drone so that it will be more robust during flight in gusty wind. Also, stability of the drone can be maintained during descent flight before landing. A thin airfoil (NACA0009) is chosen and the rotor blade has no twist, allowing negative pitch angle during flight control and upside-down flight.

The layout of the hexa-rotor drone is shown in Figure 3 and the rotor blade is shown in Figure 4. The design parameters for this test vehicle MR-Imp #1 is tabulated in Table 1. Design gross weight of this drone is 6 kg, so that each rotor carries 1 kgf during hovering flight.

Figure 2. A pitch-control multi-rotor test vehicle, MR-Imp #1.

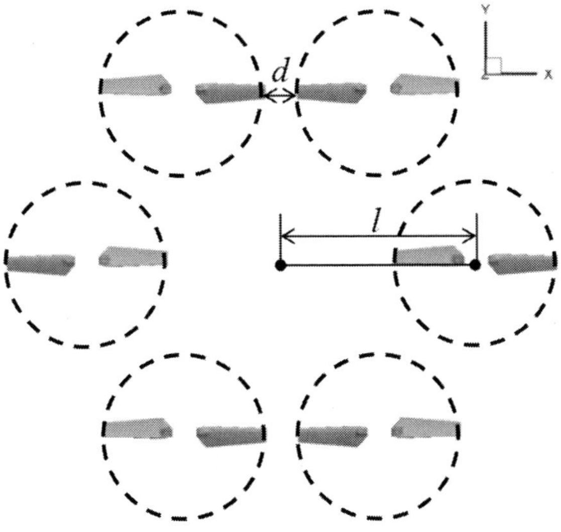

Figure 3. Layout of the hexa-rotor drone, MR-Imp #1.

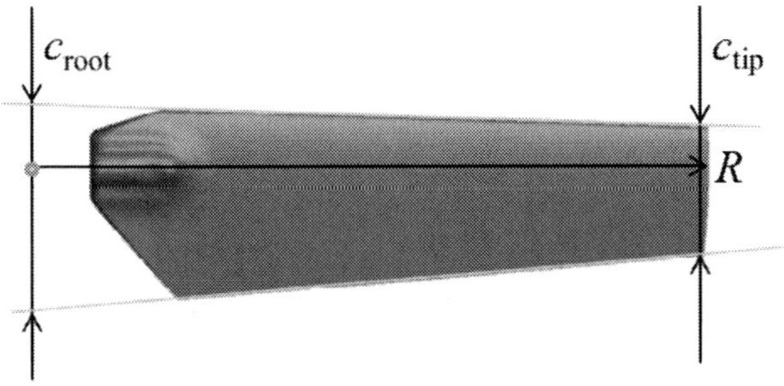

Figure 4. The rotor blade of MR-Imp #1.

Table 1. Design parameters of the hexa-rotor drone

Hexac-Rotor Design Parameters	
Rotor diameter, D	0.330 m
Rotor radius, R	0.165 m
Chord length (c_{root}, c_{tip})	0.048 m, 0.029 m
Blade planform	Tapered
Blade twist angle	0 deg
Blade airfoil	NACA0009
Number of rotors, Nr	6
Number of blades per rotor, Nb	2
Tip Mach number, M_{tip}	0.29412 (100 m/s)
Reynolds number based on chord length	1.99×10^5
Distance between rotor, d	0.063 m (0.38R)
Distance between rotor center and aircraft center, l	0.393 m

ROTORS WITH AN UPPER WALL

Firstly, a single rotor below a wall is studied. Then a hexa-rotor drone is simulated.

A single rotor with an upper wall

A single isolated rotor with two blades of the test vehicle MR-Imp #1 is placed under a wall (ceiling) as shown in Figure 5. The distance between the rotor and the upper wall is varied as shown in Table 2. A case without a wall is also simulated for comparison. The pitch angle of the rotor blade is fixed at 10 degrees which produce nearly the desired thrust during hovering flight without a wall.

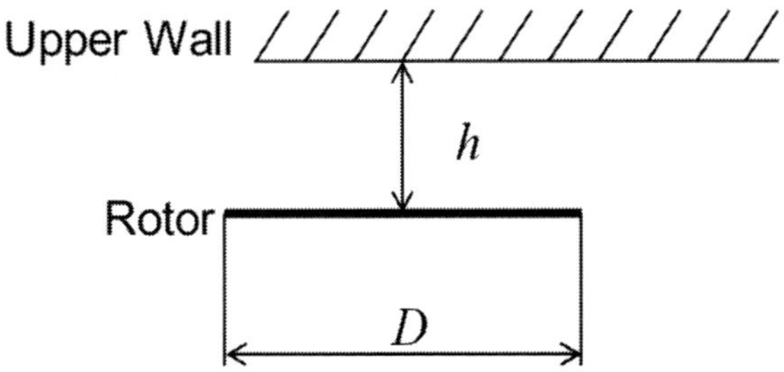

Figure 5. Notations of a single rotor and an upper wall.

Table 2. Test cases for a rotor with an upper wall

Case No.	Pitch Angle θ_0	Distance from Upper Wall, *h*
1	10°	*0.2D*
2		*0.3D*
3		*0.5D*
4		*1.0D*
5		*1.5D*
6		*2.0D*
7		*3.0D*
8		Without Wall

The overlapping grid used for the simulations of a rotor and an upper wall is shown in Figure 6. The top surface of the background grid is set to a solid wall condition. The blade grid vertical position is changed to simulated the distance from the upper wall. The coverage of the background grid is wide enough as of 20 times of the rotor radius to simulate the infinite area of the upper wall. The blade grid is shown in Figure 7, where a SOH-type topology (singular lines at both blade ends, O-type grid in chordwise direction and H-type grid in the bodywise direction) is adopted to model the blade shape with high fidelity. The point numbers and minimum spacing of the grids are shown in Table 3.

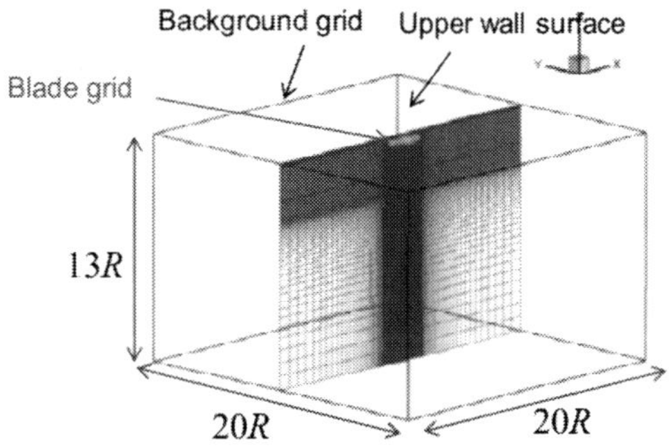

Figure 6. CFD grid for simulation of a rotor with an upper wall.

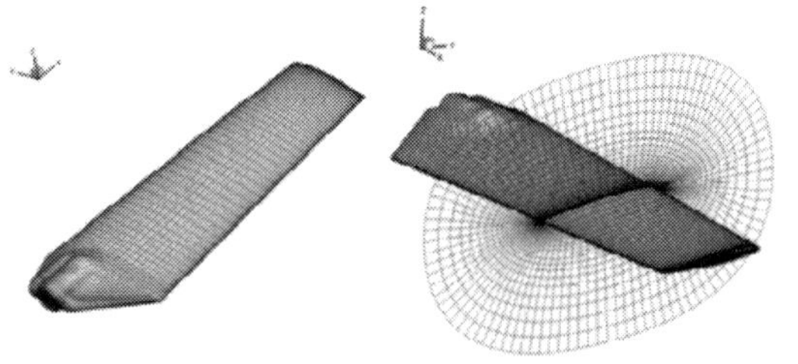

Figure 7. Blade grid topology

Table 3. Grid points for simulation of a rotor with an upper wall.

Grid	Grid points
Background grid for upper wall	153 x 153 x 147 (X x Y x Z) Wall minimum spacing: 1.0×10^{-3}
Blade grid	91 x 101 x 51 (span x chord x normal) Re : 1.99×10^5 Minimum spacing: 1.0×10^{-5}

The thrust coefficient of the rotor changes with the distance from the upper wall as shown in Figure 8. When the distance is less than 1 diameter of the rotor, the thrust increases gradually and when is distance is less than 0.5 diameter (1 radius) the thrust increases abruptly at the fixed blade pitch angle. The required rotor torque also increases as shown in Figure 9. However, the Figure of Merit of the rotor increases as the rotor gets nearer to the upper wall, meaning that even with same power, the rotor produces more thrust. When a rotor is getting near to the upper wall, it will be absorbed to the wall in an accelerating rate when the rotating speed and the pitch angle of the rotor is fixed. It must be mentioned that the rotor starts to feel the existence of the upper wall from a much longer distance about 3 diameters. A drop of the thrust at 1 diameter position before abrupt rise is observed.

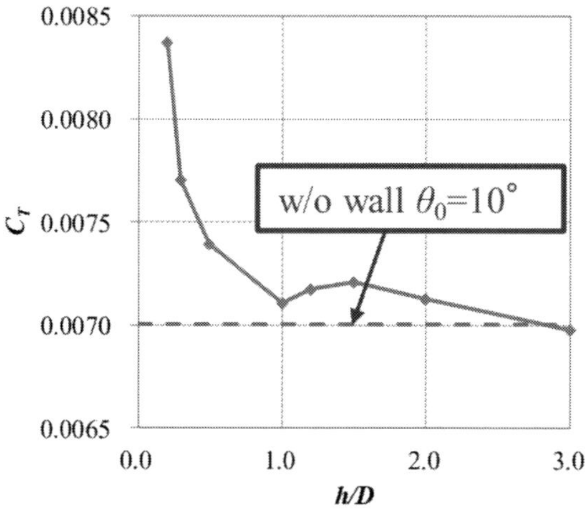

Figure 8. Thrust coefficient change with distance from the upper wall for a single rotor.

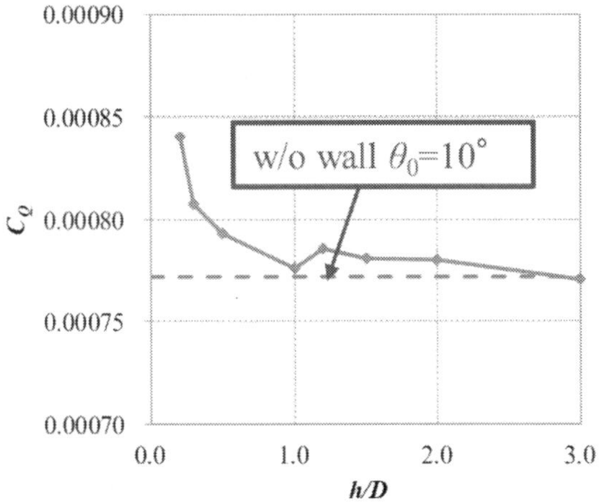

Figure 9. Torque coefficient change with distance from the upper wall for a single rotor.

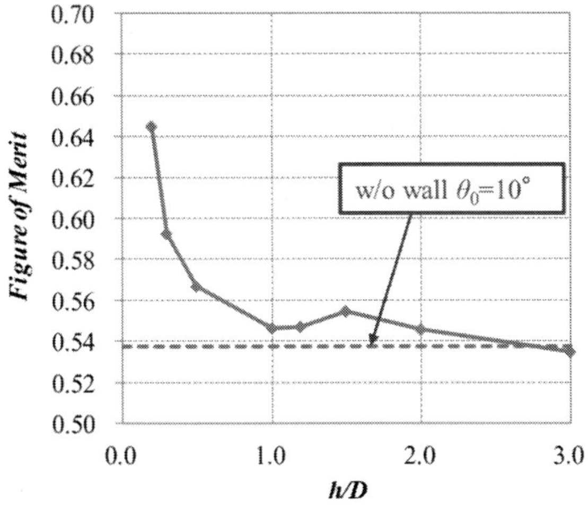

Figure 10. Figure of merit change of a single rotor with the upper wall distance

Hexa-rotors with an upper wall

For a drone with hexa-rotors, as shown in Figure 11, the thrust decreases about 2% and has a minimum value at distance about 1.5 diameters and then increases about 4% until the distance of 0.5 diameters (1 radius) and then increases abruptly. The Figure of Merit of each rotor is also showing similar tendency.

From a view of safety operations, this hexa-rotor drone should be kept at least more than 1.5 diameter away from the upper wall to prevent unexpected movement of the drone.

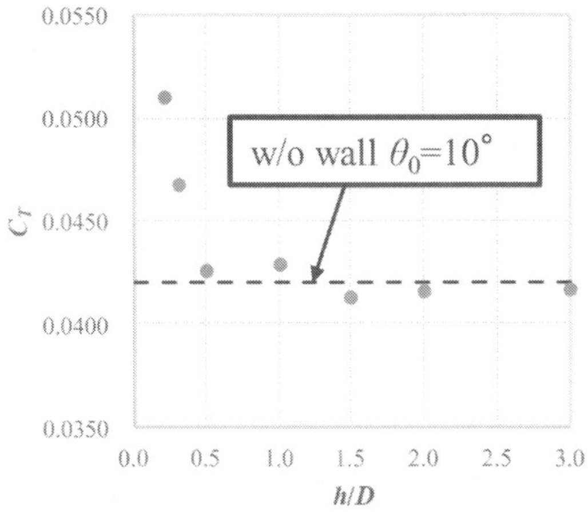

Figure 11. Total thrust coefficient changes of hexa-rotors with upper wall distance.

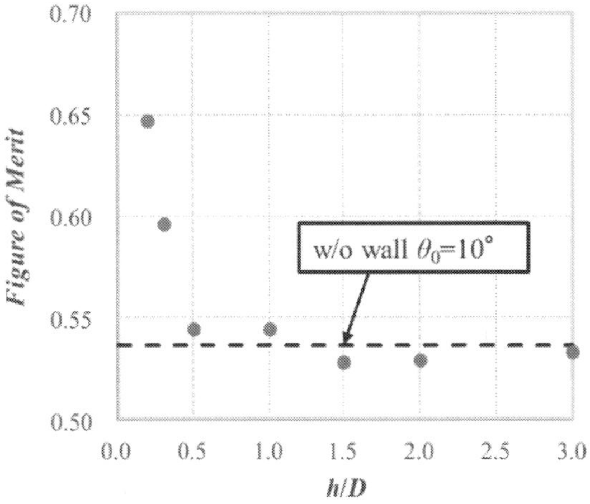

Figure 12. Figure of merit change for hexa-rotors with the upper wall distance.

ROTORS WITH A SIDE WALL

Similarly, a single isolated rotor with a side wall is first studied and then hexa-rotors are simulated.

A single rotor with a side wall

The distance between a rotor and a side wall is defined as shown in Figure 13. The gap between the rotor blade tip with wall, *g*, is 0.5 diameter less than the distance between the rotor center and the side wall, *l*. The rotor blade pitch is fixed at 10 degrees also and the distance of the rotor from wall is changed as shown in Table 4.

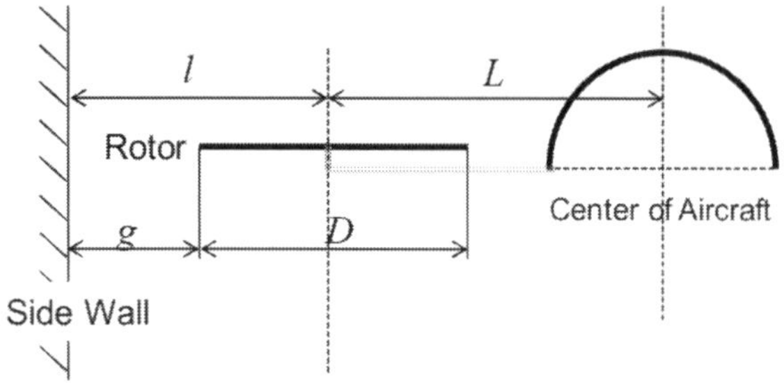

Figure 13. Notations of a rotor with a side wall.

Table 4. Test cases for a rotor with a side wall.

Case No.21	Pitch Angle, θ_0	Distance from Side Wall, l	Gap between rotor tip and side wall, g
1	10°	0.65D	0.15D
2		0.70D	0.20D
3		0.80D	0.30D
4		0.90D	0.40D
5		1.15D	0.65D
6		1.20D	0.70D
7		1.30D	0.80D
8		1.40D	0.90D
9		1.50D	1.0D
10		2.00D	1.5D

The overlapped grid used for the simulations of a rotor with a side wall is shown in Figure 5 and the grid parameters are shown in Table 5.

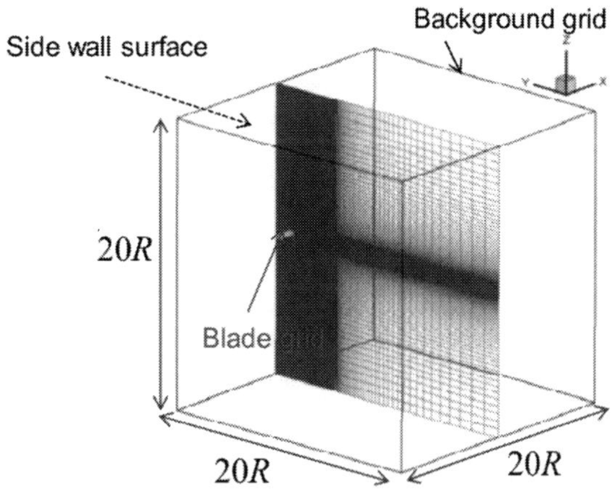

Figure 14. CFD grid for simulation of a rotor with a side wall.

Table 5. Overlapped grid parameters for simulations of a rotor with a side wall.

Grid	Grid points
Background grid for side wall	199 x 245 x 143 (X x Y x Z) Wall minimum spacing: 1.0×10^{-3}
Blade grid	91 x 101 x 51 (span x chord x normal) Re : 1.99×10^5 Minimum spacing: 1.0×10^{-5}

The coordinates of the rotors with a side wall is illustrated in Figure 15, where the x-axis is defined parallel to the wall.

The thrust of a single rotor only change a small fraction when the distance between the rotor and wall is nearer as shown in Figure 16. When the gap between the rotor tip with the wall is smaller than 1 diameter, decrease of thrust start to appear. And the rolling moment of the rotor is observed with the direction to tilt the rotor toward the wall. The Figure of Merit of the rotor also only decreases in less than 1%.

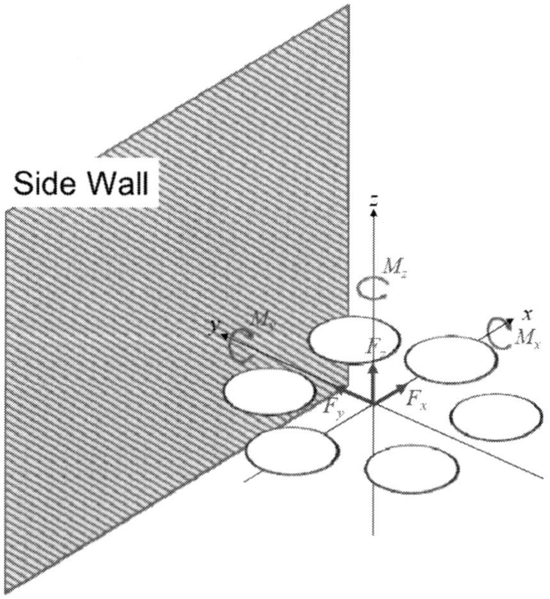

Figure 15. Coordinates for hexa-rotors with a side wall.

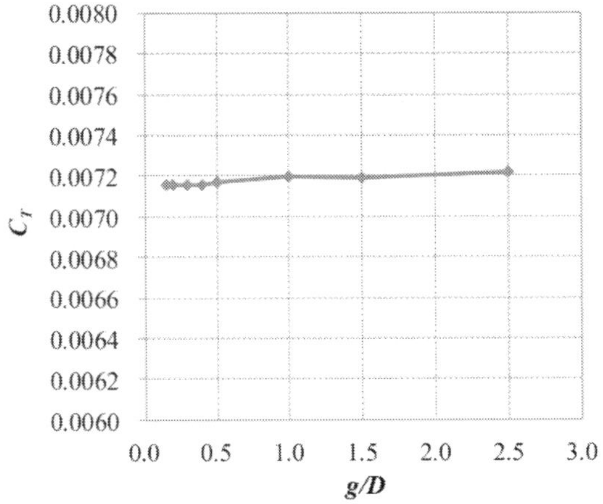

Figure 16. Thrust coefficient change with the distance from a side wall.

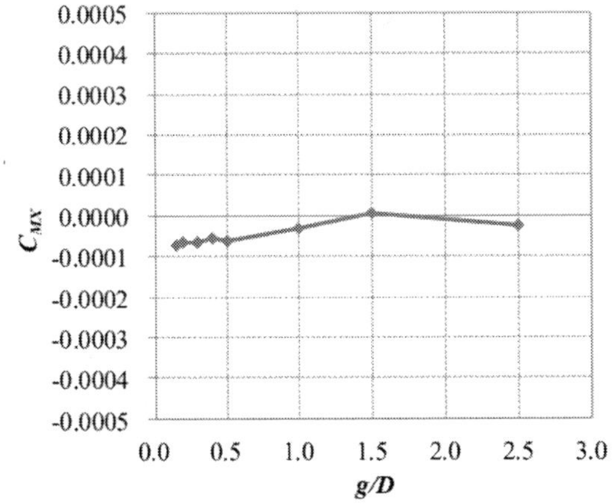

Figure 17. Rolling moment change of a single rotor with distance from a side wall

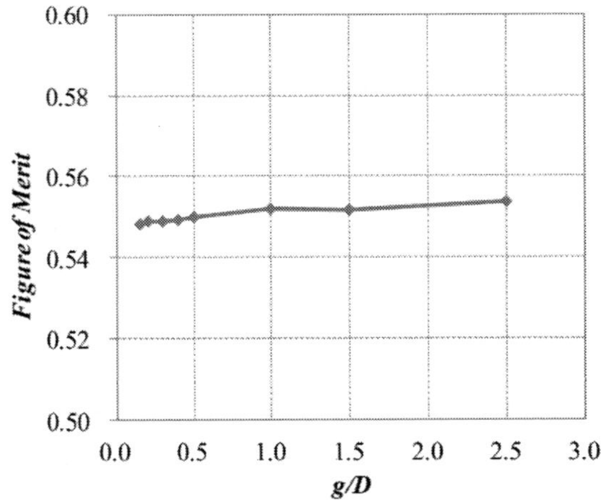

Figure 18. Figure of merit change of a single rotor with distance from a side wall.

Hexa-rotors with a side wall

As shown in Figure 19, in a hexa-rotor layout, each rotor has different distances from the side wall. The rotor 2 and rotor 3 have small distance to the wall, and as shown in Figure 20, relatively large decreases of thrust is observed. The individual rolling moments are shown in Figure 21 and as a total, which do not contribute much to the whole rolling moment of the drone. The Figure of Merit of rotor 2 and rotor 3 also decreases when the drone is approaching the wall as shown in Figure 22.

The summed thrust of the drone decreases gradually when the distance is shorter as shown in Figure 23 only with a small amount. However, the rolling moment of the drone as shown in Figure 24 decreases (increases in the direction to tilt the whole drone toward the wall) significantly almost linearly with the distance decreases when the gap of the rotor with the wall is less than 1.0 diameter. Which produce an effect to cause the drone fly toward the wall in an accelerating rate.

From a view of safety operation of the drone, the rotor tip should be kept away from the wall more than 1.5 diameter of the rotor to prevent unexpected movement of the drone absorbed toward the side wall.

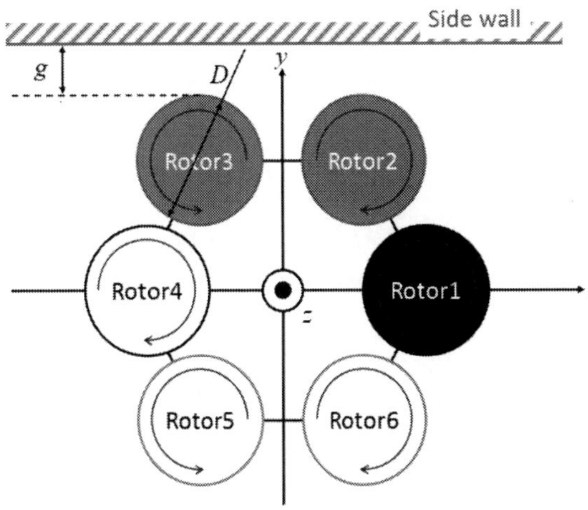

Figure 19. Hexa-rotors layout relative to a side wall.

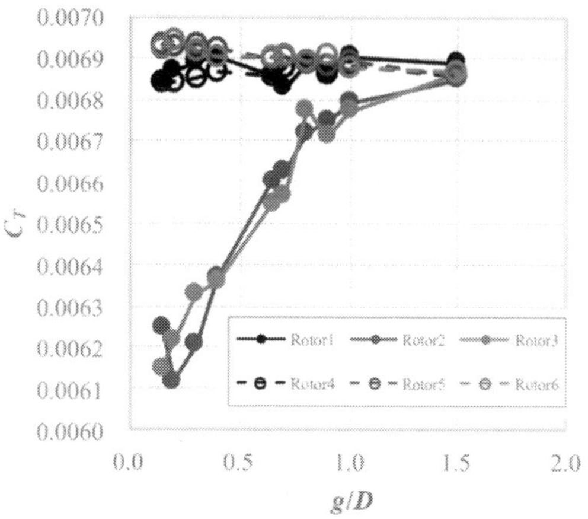

Figure 20. Thrust coefficients for each rotor with distance between the center of hexa-rotors and a wll.

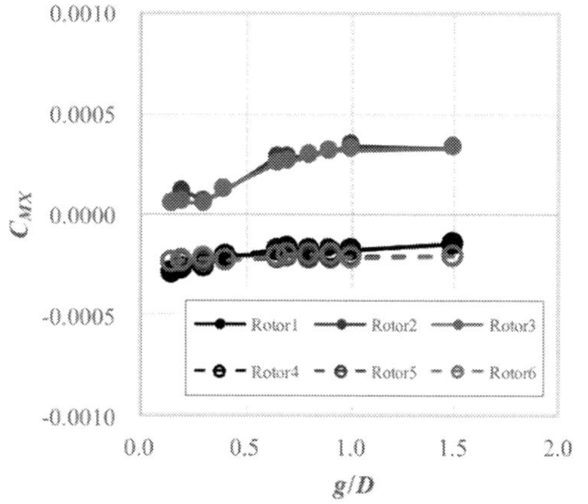

Figure 21. Rolling moment coefficients of hexa-rotors with the distance between the drone center and the side wall.

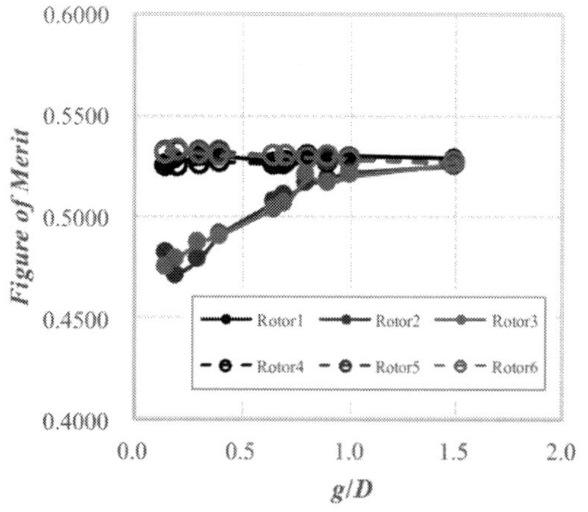

Figure 22. Figure of merit of each rotor with the distance from the side wall.

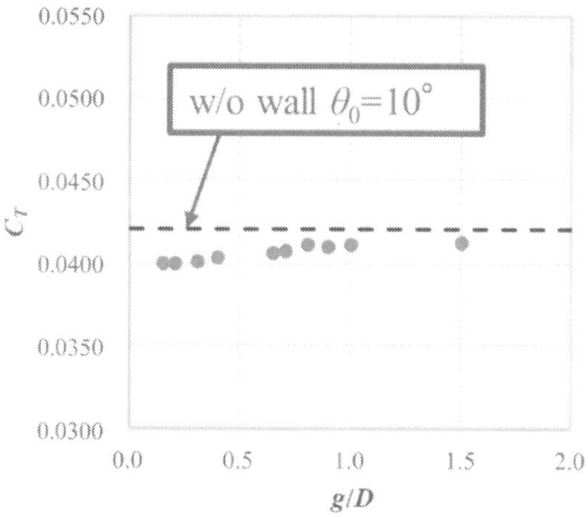

Figure 23. Total thrust coefficients of hexa-rotors with the distance from a side wall.

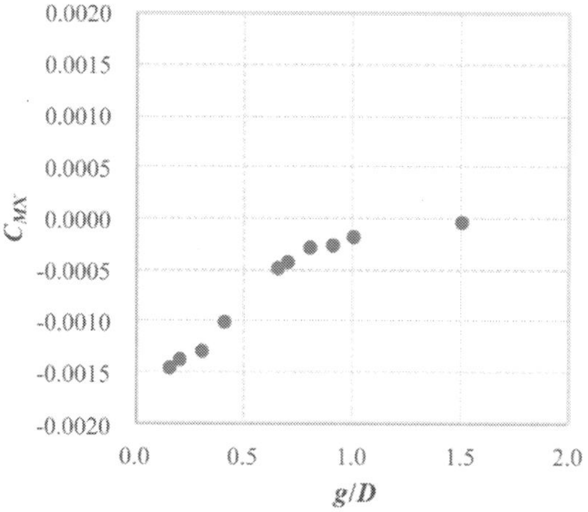

Figure 24. Rolling moment of hexa-rotors with the distance from a side wall.

CONCLUSIONS

It is found that when a rotor is approaching an upper wall, as the distance between the rotor and the upper wall is less than a diameter of the rotor, the thrust suddenly increases that causes the rotors to collide onto the upper wall. When an isolated rotor is operated near a side wall, the thrust decreases and a rolling moment appears to tilt the rotor toward the wall as the distance becomes shorter than 1 diameter of a rotor. For a multiple rotor drone near a side wall at a distance less a rotor diameter, the rotors have different distances from the wall, which causes the whole aircraft tilt toward the wall significantly.

From a view of safety operations of multi-rotor drones, the distance from the upper wall and the gap between the rotor tip and the side wall should be both larger than 1.5 times of the diameter of the rotor to prevent the multi-rotor drone to move unexpectedly during hovering flight near the walls.

Author contact: Yasutada Tanabe, tan@chofu.jaxa.jp;
Masahiko Sugiura, sugiura.masahiko@jaxa.jp;
Takashi Aoyama, aoyama.takashi@jaxa.jp;
Hideaki Sugawara, hideaki8@chofu.jaxa.jp;
Shigeru Sunada, sunada@aero.osakafu-u.ac.jp;
Koichi Yonezawa, koichi-y@criepi.denken.or.jp;
Hiroshi Tokutaka, tokutake@se.kanazawa-u.ac.jp.

ACKNOWLEDGMENTS

This research was funded by ImPACT Program of Council for Science, Technology and Innovation (Cabinet Office, Government of Japan).

REFERENCES

[1] Tanabe, Y., Saito S. and Sugawara, H., Construction and Validation of an Analysis Tool Chain for Rotorcraft Active Noise Reduction, 38th ERF, Amsterdam, NL, Sept. 4-7, 2012.

[2] Tanabe, Y., Saito, S., Takayama, O., Sasaki, D. and Nakahashi, K., A New Hybrid Method of Overlapping Structured Grids Combined with Unstructured Fuselage Grids for Rotorcraft Analysis, 36th European Rotorcraft Forum, Paris, France, September 9-11, 2010.

[3] Tanabe, Y., Aoyama, T., Sugiura, M., Sugawara, H., Sunada, S., Yonezawa, K. and Tokutake, H., "Numerical Simulations of Aerodynamic Interactions Between Multiple Rotors," 42nd European Rotorcraft Forum, Lille, France, Sept. 6-9, 2016.

[4] Shima, E., and Kitamura, K., On New Simple Low-Dissipation Scheme of AUSM-Family for All Speeds," 47th AIAA Aerospace Sciences Meeting, Orlando, FA, January 5-8 2009, AIAA Paper 2009-136.

[5] Tanabe, Y. and Saito, S., "Significance of All-Speed Scheme in Application to Rotorcraft CFD Simulations," The 3rd International Basic Research Conference on Rotorcraft Technology, Nanjing, China. October, 2009.

[6] Yamamoto, S. & Daiguji, H., "Higher- Order-Accurate Upwind Schemes for Solving the Compressible Euler and Navier-Stokes Equations," Computers & Fluids, Vol.22, No.2/3, pp.259-270, 1993.

[7] Zhang, L.P. and Wang, Z.J., "A Block LU-SGS Implicit Dual Time-Stepping Algorithm for Hybrid Dynamic Meshes," Computers & Fluids, Vol.33, pp.891–916, 2004.

[8] Arnone, A., Liou, M.S. and Povinelli, L. A., "Multigrid Time-Accurate Integration of Navier-Stokes Equations," NASA TM-106373, November 1993.

[9] The Japan Science and Technology Agency (JST), ImPACT: Impulsing Paradigm Change through Disruptive Technologies Program, Tough Robotics Challenge (TRC), http://www.jst.go.jp/impact/en/program/07.html, accessed on Oct. 14, 2016.

Experimental Study on Flow Field and Interference at Downstream of Quadrotor Helicopter

Sida Yang
Master degree candidate
Department of Aerospace Engineering,
Nagoya University
Nagoya, Japan

Koichi Mori
Associate professor
Department of Aerospace Engineering,
Nagoya University
Nagoya, Japan

ABSTRACT

The final objective of this study was investigate the flow field and interference at downstream of a quadrotor helicopter via theoretical calculation and experiment. However, in this paper just the theoretical and experimental study at single rotor case was discussed. First, the flow velocity distribution at the cases of RPM from 5000 to 10000, at downstream of quadrotor was analyzed based on three dimensional circular jet flow theory; second, the experiment was conducted to measure the flow velocity distribution at the same RPM cases and downstream area of quadrotor; finally, the result of calculation and experiment was comparatively analyzed. As a result, it was found that the flow velocity which was measured in experiment was agrees well with theoretical velocity merely at low radial distance from propeller axis. However, at high radial distance from propeller axis the flow velocity measured in experiment was higher than that of theory. Furthermore, at the center point of quadrotor, experimental flow velocity was double times of theoretical flow velocity at far downstream. It is demonstrated that the three dimensional circular jet flow theory cannot be completely applied to analyze the flow field at downstream of quadrotor, it probably because rotation and tip vortex of propeller caused more complex flow field compare with jet hole.

Key Words: Quadrotor, Interference, Velocity Distribution, Jet Flow

NOTATION

d: distance from experiment or calculation position to center of propeller disc [m]

J: momentum of the jet flow [kg·m/s]

K: $= J/\rho$

$\dot{m}$: mass flow rate [kg/s]

n: number of propeller

r: radius of propeller [m]

T: propeller thrust [N]

u: flow velocity at downstream [m/s]

v: induced velocity of propeller [m/s]

u_{exp}: measurement velocity [m/s]

u_{the}: theory velocity [m/s]

y: coordinate at y axis [m]

z: coordinate at z axis [m]

ε: swirl viscosity coefficient

η: diffusion coefficient

ρ: density of the air [kg/m^3]

ω: flow velocity at far downstream [m/s]

INTRODUCTION

Quadrotor helicopter is a kind of unmanned aerial vehicle equipped with four rotors. This is supposed to can be applied for high risk area, such as aerial photography, military reconnaissance or disaster relief.

As can be seen in the Figure 1, the neighboring rotors rotate in opposite directions to each other, and the opposite side rotors rotate in the same direction. Thus, it is easy to change yaw direction just by increasing or decreasing RPM of opposite side rotors, without interfere roll and pitch attitude. Furthermore, it also has excellent stability performance at hovering case. Because of these features, quadrotor helicopter is easy to operated and competent to various flight missions. This is the main reason why a lot of

Presented at the 6th Asian/Austrailian Rotorcraft Forum / Heli Japan 2017, Kanazawa, Japan, November 7-9, 2017.

enterprises and researchers engage in the research and development of quadrotor helicopter.

However, the research about quadrotor helicopter is mainly focused on the control and automation, and its aerodynamic characteristics almost have not been studied. As one of the few exceptions, [3]Hidemasa Yasuda conducted numerical analysis on flow field of quadrotor at no wind and side wind case. But in many occasions, CFD is also facing the problem like low accuracy, and lack of clear flow field model, etc.

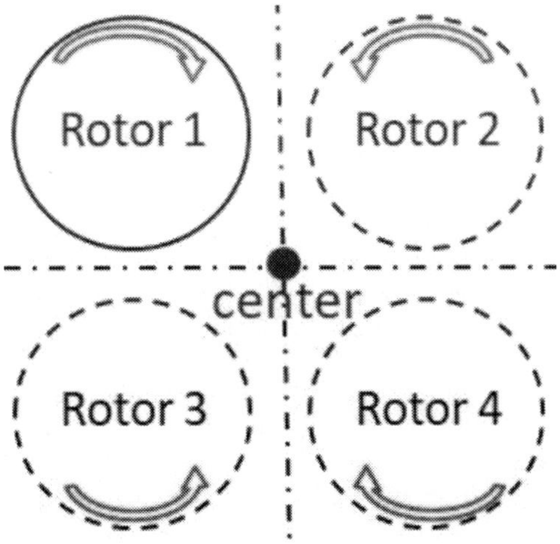

Figure.1 Simplified diagram of quadrotor

Therefore, at the first step of this study, flow velocity distribution of downstream at RPM from 5000 to 10000 cases was calculated based on three dimensional circular jet flow theory. At the second step, the experiment was conducted at the same cases with first step. In this paper, the flow field at single rotor cases (merely rotated rotor 1) was investigated.

CALCULATION AND EXPERIMENT

Calculation method

The calculation model was based on the three dimensional circular jet flow theory (see Figure.2). In this theory, flow velocity at each coordinate point can be calculated by equation (1)~(4):

$$u = 3K/[(8\pi\varepsilon_0 z)(1+\eta^2/4)^2] \quad (1)$$

$$\eta = [3K/(16\pi\varepsilon_0^2)]^{0.5}(d/z) \quad (2)$$

$$\varepsilon_0 = 0.0161K^{0.5} \quad (3)$$

$$K = J/\rho = 2\pi\int_0^R u^2 r\, dr \quad (4)$$

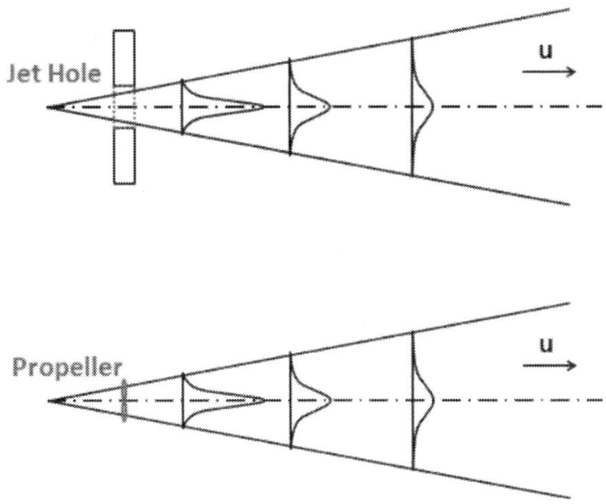

Figure.2 Three dimensional circular jet flow model

The propeller was seen as the jet hole, which jet flow velocity equal to induced velocity itself. As can be seen in the figure.3, the calculation area was rectangle area which range was z direction from 0.2m to 1.0m and y direction from -0.24m to 0.24m at cross-section which includes rotation axis of adjacent rotors.

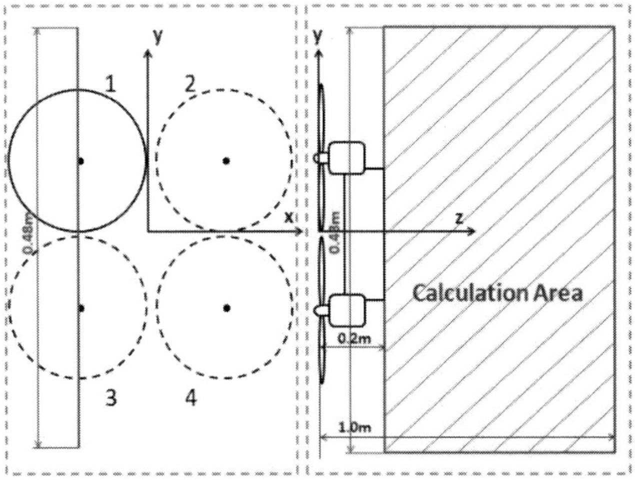

Figure.3 Calculation area

Experiment model

Figure.4 shows the experimental model, and Table.1 shows its specifications.

Figure.4 Experimental model

Table.1 Specifications of experimental model

Characteristic	Value
Distance between neighboring rotors [cm]	13.0
Rotor radius r [cm]	12.7
Motor RPM range	4000~12000

Experiment method

In the Figure.5, the experimental rectangle area was shown. Velocity distribution was investigated at the cross-section which the same with that of calculation, and coordinates of measurement points were also z direction from 0.2m to 1.0m, and y direction from -0.24m to 0.24m.

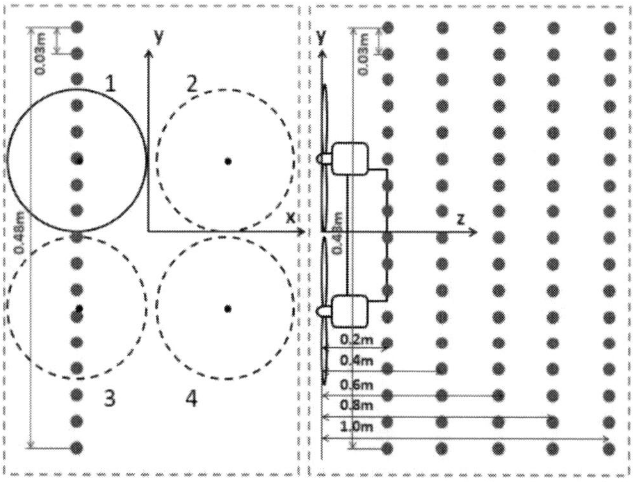

Figure.5 Experiment area

The pitot tube (see Figure.6) was used to measure the dynamic pressure P_d, and the differential pressure meter (see Figure.7) was used to display and read P_d. The flow velocity was formulated as:

$$u = \sqrt{2P_d/\rho} \quad (5)$$

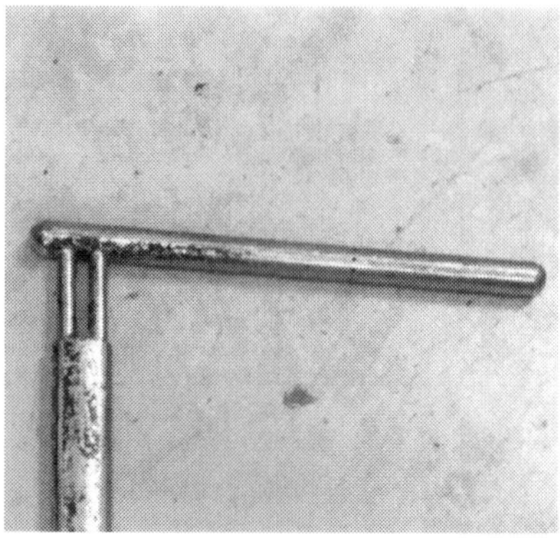

Figure.6 Pitot tube

Figure.7 Differential pressure meter

RESULTS AND DISCUSSIONS

Velocity distribution at the cross-section which includes rotation axis of adjacent rotors

(a) RPM5000

In this experiment model, rotation axis of rotor at y=0.06m position.

Figure.8 shows the theoretical flow velocity distribution at downstream of rotor at RPM5000 case. At this cross-section, the velocity at z=0.4m and y=0.06m (behind of rotor) was about 5m/s, the highest in this downstream area. This was because the closer to the rotor, its induced velocity should be higher and it meets common sense. Moreover the velocity distribution was symmetrical along the rotation axis.

Figure.9 shows the experimental flow velocity distribution at downstream of rotor at RPM5000 case. The velocity at z=0.4m and y=0.06m (behind of rotor) was also about 5m/s but a little higher than theoretical velocity. Furthermore, the high velocity area of experiment was wider than that on theory especially at high radial distance from the propeller axis. In addition, because of inevitable measurement error, the measurement velocity distribution was not completely symmetric along the rotation axis.

In the Figure.10, the theoretical data was indicated by solid lines and experimental data was indicated by points (the same below). As seen this figure, at the low radial distance from propeller axis (around y=0.06m), experimental velocity at z=0.6m and z=1.0m was almost the same with theoretical velocity. However, the rotor supporter interfered the flow field at downstream, especially the area near to the supporter (see Figure.11). It was considered that the point at z=0.2m in the interfered field. Therefore its experimental velocity was lower than theoretical velocity at y=0.06m position. But, at high radial distance from propeller axis, experimental velocity was higher than theoretical velocity a little bit. The reason was considered that this downstream area was interfered by tip vortex of propeller.

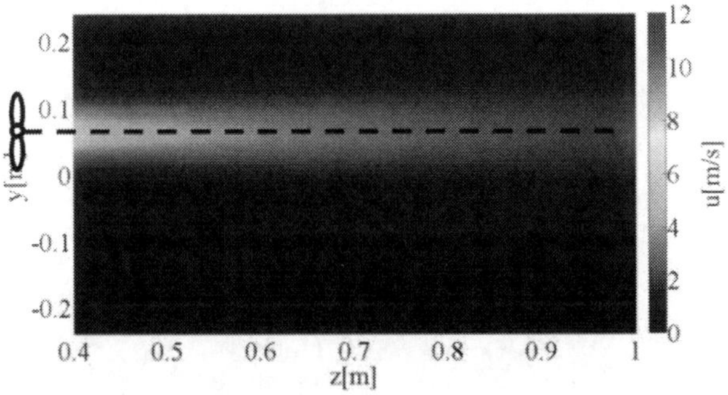

Figure.8 Velocity distribution in RPM5000 (theory)

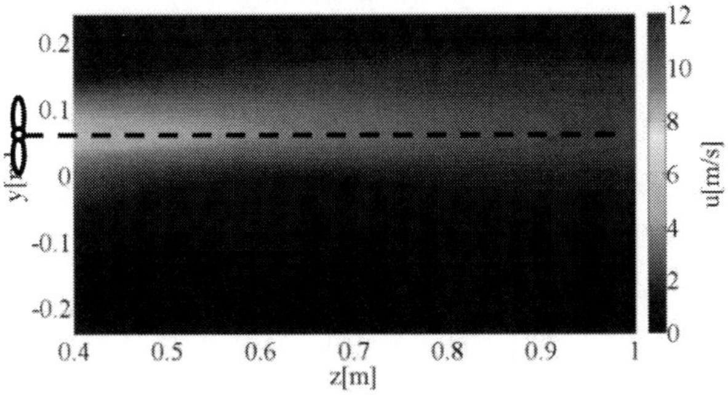

Figure.9 Velocity distribution in RPM5000 (experiment)

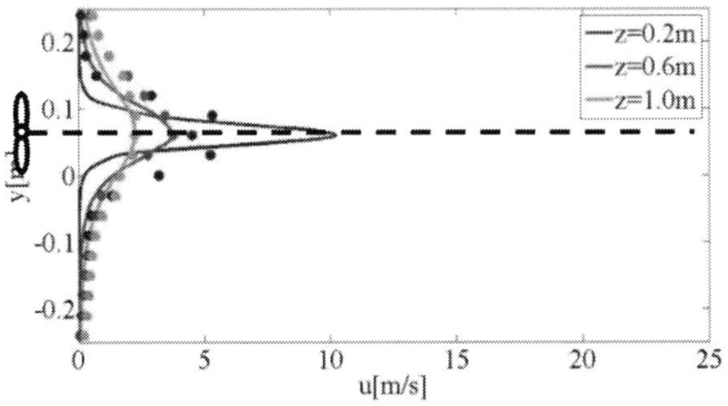

Figure.10 Contradistinction between theory and experiment at RPM5000

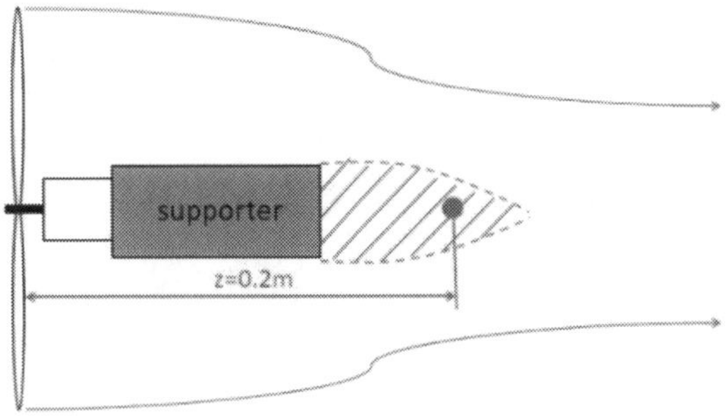

Figure.11 Interference behind the supporter

(b) RPM7000

Figure.12 shows the theoretical velocity distribution at downstream of rotor at RPM7000 case. The velocity at z=0.4m and y=0.06m (behind of rotor) was about 8m/s, the highest in this downstream area. The velocity distribution was symmetrical along the rotation axis.

Figure.13 shows the experimental velocity distribution at downstream of rotor at RPM7000 case. The velocity at z=0.4m and y=0.06m (behind of rotor) was about 7m/s, a bit lower than theoretical velocity. Furthermore, the high velocity area of experiment was wider than that on theory. The measurement velocity distribution was also asymmetric along the rotation axis.

As seen in Figure.14, at the axis y=0.06m, experimental velocity at z=0.6m and z=1.0m was almost the same with theoretical velocity. Yet at z=0.2m, experimental velocity was lower than theoretical velocity because it in the interfered field at behind of rotor supporter. However, at y≠ 0.06m experimental velocity was slightly higher than theoretical velocity.

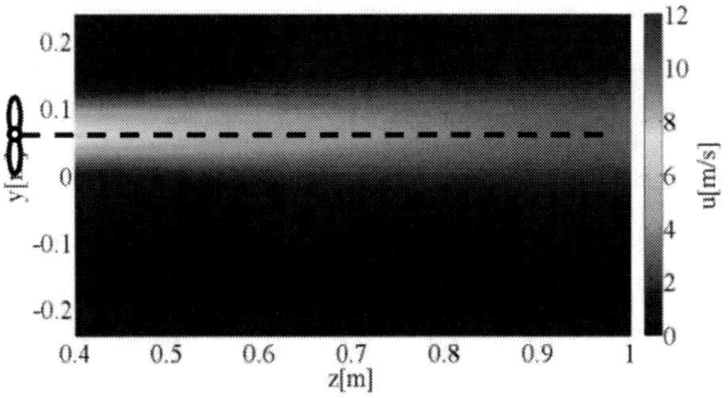

Figure.12 Velocity distribution in RPM7000 (theory)

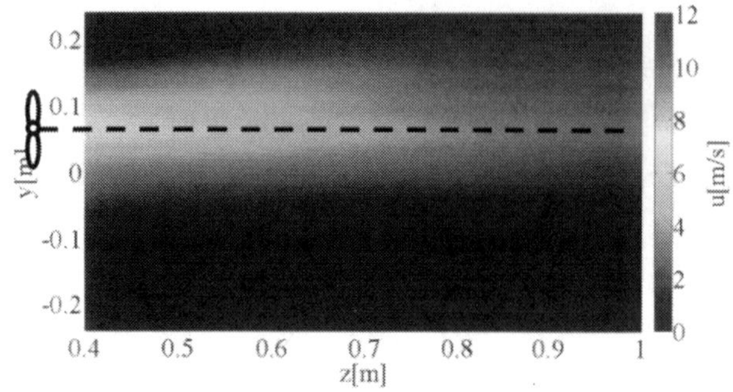

Figure.13 Velocity distribution in RPM7000 (experiment)

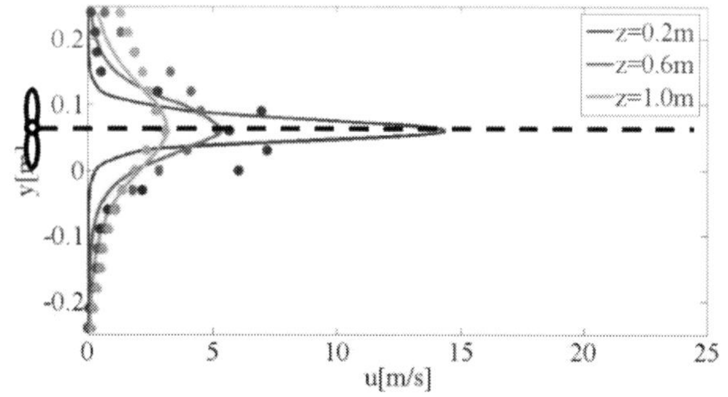

Figure.14 Contradistinction between theory and experiment at RPM7000

(c) RPM9000

Figure.15 shows the theoretical velocity distribution at downstream of rotor at RPM9000 case. The velocity at z=0.4m and y=0.06m (behind of rotor) was about10m/s, the highest in this downstream area. The velocity distribution was symmetrical along the rotation axis.

Figure.16 shows the experimental velocity distribution at downstream of rotor at RPM9000 case. The velocity at z=0.4m and y=0.06m (behind of rotor) was about 9m/s, slightly lower than theoretical velocity. In addition, the high velocity area of experiment was wider than that on theory. The measurement velocity distribution was not absolutely symmetric along the rotation axis.

As seen in Figure.17, at the axis y=0.06m, experimental velocity at z=0.6m was almost the same with theoretical velocity, and at z=1.0m experimental velocity was a bit higher than theoretical velocity. At z=0.2m, experimental velocity was lower than theoretical velocity because it in the interfered field at behind of rotor supporter. However, at y≠ 0.06m experimental velocity was higher than theoretical velocity.

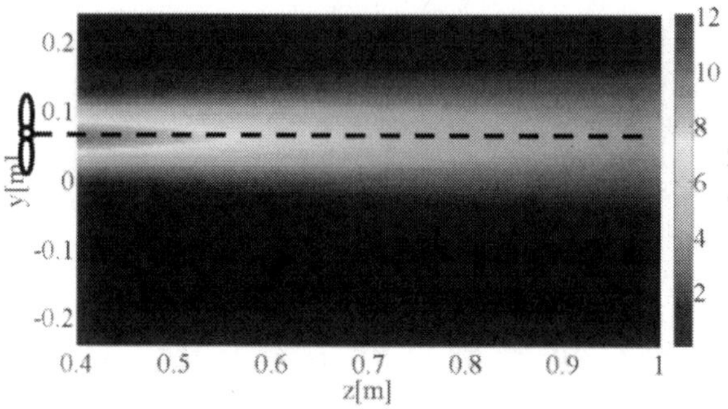

Figure.15 Velocity distribution in RPM9000 (theory)

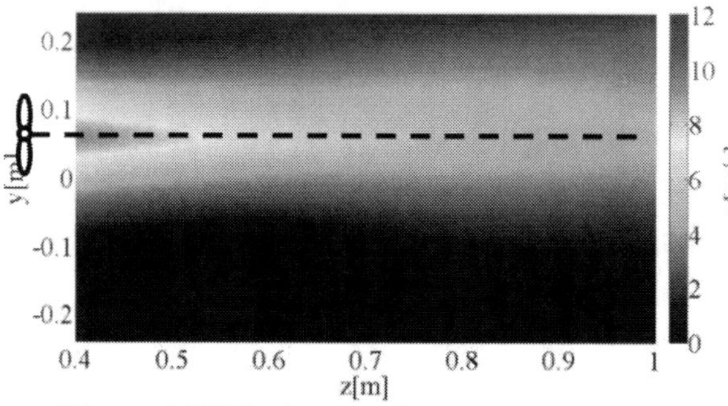

Figure.16 Velocity distribution in RPM9000 (experiment)

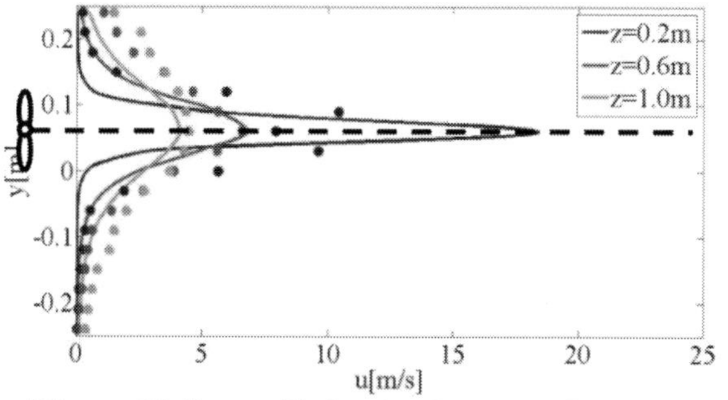

Figure.17 Contradistinction between theory and experiment at RPM9000

Flow velocity at center of quadrotor

In order to investigate the interference at downstream of the center of quadrotor (see Figure.1), theoretical calculation and experimental test was conducted. In this study, the investigated cases were RPM was from 5000 to 10000 and z was from 0.2m to 1.0m.

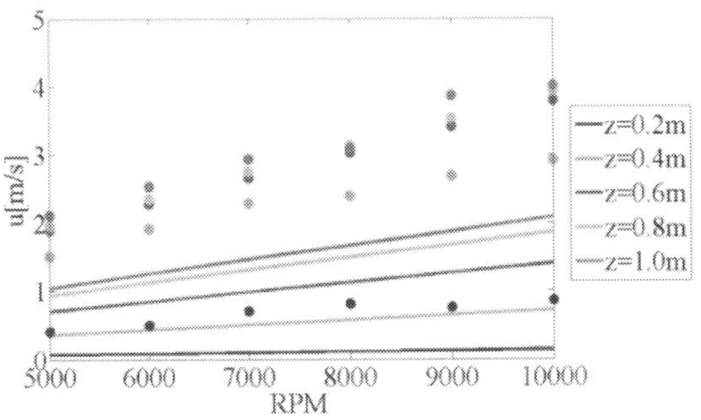

Figure.18 Velocity at center of quadrotor

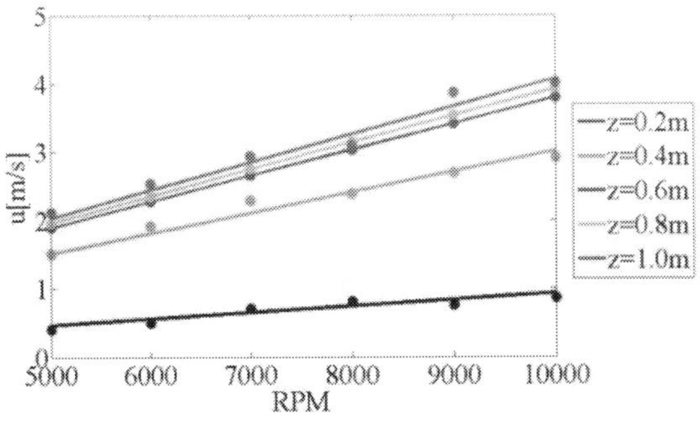

Figure.19 Velocity at center of quadrotor

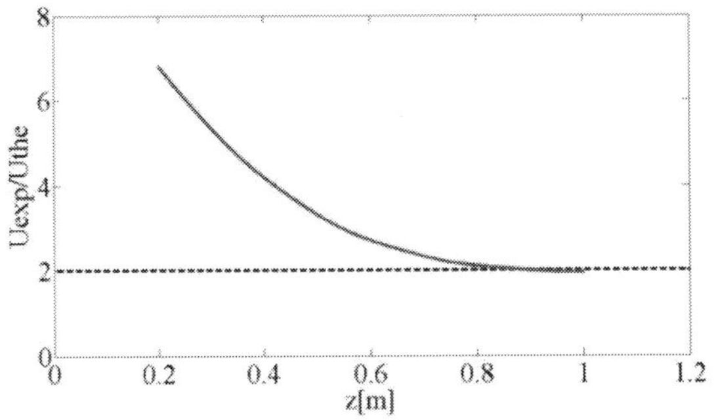

Figure.20 *f(n)*-z

Figure.18 shows the summary of theoretical flow velocity and experimental flow velocity distribution at downstream of the quadrotor center. As can be seen, at the same position (identical z), experimental velocity was higher than theoretical velocity.

In order to find the relation between them, the theoretical velocity at each position was multiplied by the formula *f(n)*. At z=0.2m, if *f(n)* was set to 6.8; at z=0.4m, if *f(n)* was set to 4.2; at z=0.6m, if *f(n)* was set to 2.7; at z=0.8m, if *f(n)* was set to 2.1; at z=1.0m, if *f(n)* was set to 1.95, the experimental velocity was almost agrees with the theoretical velocity which was shown in Figure.19, and *f(n)*-z relation was shown in Figure.20.

As the distance z increased, the value of *f(n)* was gradually approaching to 2. Therefore, it was guessed that at far downstream of quadrotor center, experimental flow velocity was double times of theoretical flow velocity.

CONCLUSIONS

The flow velocity at low radial distance from the propeller axis (value of y was around 0.06m) can be calculated by three dimensional circular jet flow theory. However, at nearby the rotor supporter (value of z was small), actual flow velocity was lower than theoretical flow velocity because of the interference from supporter.

Experimental flow velocity was higher than theoretical flow velocity at high radial distance from the propeller axis. The reason was considered to be tip vortex of propeller resulting in complicated flow field at this downstream area and lead to actual flow velocity higher than theoretical.

The practical flow velocity at center of the quadrotor downstream was higher than theoretical flow velocity, and actual flow velocity was double times of theoretical flow velocity at far downstream.

ACKNOWLEDGMENTS

A sincerely thank is to Mr. Mori for his instruction and supports in doing this test.

REFERENCES

[1] Gabriel M. Hoffmann, Haomiao Huang, Steven L. Waslander, Claire J. Tomlin, "Precision flight control for a multi-vehicle quadrotor helicopter testbed," *Control Engineering Practice*, Volume 19, Issue 9 (2011).

[2] Ho ffmann, G. M., Huang, H., Waslander, S. L. and Tomlin, C.J., "Quadrotor Helicopter Flight Dynamics and Control: Theory and Experiment," *AIAA Paper 2007-6461*, 2007.

[3] Hidemasa YASUDA, Keiichi KITAMURA, et al, "Numerical Analysis of Flow Field and Aerodynamic Characteristics of a Quadrotor," *Trans, JSASS Aerospace Tech. Japan*, Vol. 11, pp. 61-70, 2013.

[4] Marcoc. DE Simone, Fleming J. and Ng W, "Influence of Aerodynamics on Quadrotor Dynamics," *Recent Researches in Mechanical and Transportation Systems*, NV, 2016.

[5] Mark Cutler, Nazim-Kemal Ure, Bernard Michini, and Jonathan How, "Comparison of Fixed and Variable Pitch Actuators for Agile Quadrotors," *AIAA Guidance, Navigation, and Control Conference*, August 2011.

[6] Mohd Khan, "Quadcopter Flight Dynamics," International Journal of Scientific & Technology Rearch Volume 3, ISSUE 8, AUGUST 2014.

[7] N. Abas, A. Legowo and R. Akmeliawati, "Parameter identification of an autonomous quadrotor," *2011 4th International Conference on Mechatronics (ICOM)*, Kuala Lumpur, 2011, pp. 1-8.

[8] Osgar John Ohanian III, Paul A. Gelhausen, et al., "A

Compact Method for Modeling the Aerodynamics of Ducted Fan Vehicles," *48th AIAA Aerospace Sciences Meeting Including the New Horizons Forum and Aerospace Exposition*, 2010.

[9] Schaefer, Jr., C. G., Baskett, L. J. "Goldeneye: The Clandestine UAV," *2nd AIAA "Unmanned Unlimited" Systems, Technologies, and Operations*, San Diego, CA, September 2003.

[10] Seddon, J., and Newman, S., *Basic Helicopter Aerodynamics*, John Wiley & Sons, Ltd, 2011, pp. 133-136.

[11] Waslander, S., Wang, C., "Wind disturbance estimation and rejection for quadrotor position control", *AIAA Infotech@Aerospace conference*, Seattle, Washington, April 2009.

[12] Wang Bo, Guo Zheng, et al., "Investigation of the Aerodynamic Characters of Ducted Fan System," *World Academy of Science, Engineering and Technology International Journal of Mechanical*, Aerospace, Industrial, Mechatronic and Manufacturing Engineering Vol:6, No:9, 2012.

[13] Weir, Robert J, "Aerodynamic Design Considerations for a Free-Flying Ducted Propeller," *AIAA Atmospheric Flight Mechanics Conference*, Paper No.88-4377-CP, Washington DC, 1988.

[14]Yang, C., Aoyama, T., Kondo, N. and Saito, S, "Aerodynamic/ Acoustic Analysis for Main Rotor and Tail Rotor of Helicopter," *Trans. Japan Soc. Aero. Space Sci.*, **51** (2008), pp. 28-36.

[15] Yang, C. and Aoyama, T, "Comprehensive Study on Interpolation Methods for Helicopter Simulation: Interpolation between Cartesian and Curvilinear Grids in Parallel Computation," *JAXA Research and Development Report*, JAXA-RR-07-024E, 2008.

[16] Yogianandh Naidoo, Riaan Stopforth, et al,"Rotor Aerodynamic Analysis of a Quadrotor for Thrust Critical Applications," *Presented at the 4th Robotics and Mechatronics Conference of South Africa (ROBMECH 2011)*, 23-25 November 2011, CSIR Pretoria South Africa.

[17]Yoon, S. and Kwak, D, " Implicit Navier-Stokes Solver for Three-Dimensional Compressible Flows," *AIAA Journal, 30 (1992)*, pp. 2653-2659.

[18] Zar ovy, S., Costello, M., Mehta, A., Flynn, A., Gremillion, G., Miller, D., Ranganathan, B. and Humbert, J. S , "Experimental Study of Gust Effects on Micro Air Vehicles, " *AIAA Paper 2010-7818*, AIAA

Flow Visualization of Wake of a Quad-copter in Ground Effect

Hikaru Otsuka	Masayoshi Kohno	Keiji Nagatani
Graduate Student	Graduate Student	Associate Professor
Tohoku University	Tohoku University	Tohoku University
Sendai, Japan	Sendai, Japan	Sendai, Japan

ABSTRACT

The wake flow of a small quadcopter unmanned aerial vehicle generates a ground effect during take-off and landing. The ground effect affects the rotor thrust and posture stability. However, the structure of the wake with plural rotors has not been unveiled experimentally. In this study, the wake flow structure of a quadcopter with flow interaction between rotors was examined, and the wakes of hovering were visualized using a laser light sheet. In addition, particle image velocimetry measurements were conducted, changing ground off height and the distance between rotors. The visualization results in several planes suggested a merging wake between rotors and pulled wake at the center of the quadcopter. Furthermore, the particle image velocimetry results showed the development of a soaring flow at the center of the quadcopter in the ground effect operation, and the velocity of the soaring flow decelerated with closing rotor distance. The experimental results indicate that the soaring flow pushes the quadcopter; this flow affects the lifting force on the quadcopter during the ground effect.

NOTATION

A area of the rotor disk, πR^2, m^2
Ct rotor thrust coefficient, $T/\rho A \Omega^2 R^2$
Ω rotational speed of the rotor, rad/s
ρ air density, kg/m^3
z height of rotor above ground plane, m
r radius of blade, m
l distance between coincided rotors, m

INTRODUCTION

Small multirotor unmanned aerial vehicles (UAVs) are used for applications such as surveillance of disaster site, inspecting wall cracks of old bridges, and patrol drones. Similar to other vertical take-off and landing aircrafts, rotors of the small multirotor UAVs are affected by the aerodynamic interference effect produced by their wake development. Particularly, the aerodynamic interaction effect is significant when the rotors operate near the ground plane; this is known as the ground effect [1, 2]. The ground effect changes the rotors performance at take-off and landing, and causes body posture instability. However, the detailed aerodynamics of wake in multicopters have been unclear compared with the isolated rotor wake in the ground effect (IGE). The wake flow seems to contain an effect of flow interaction between rotors. We presumed that the wakes of plural rotors are different from the isolated rotor wake, and the wake structures depend on the placement of rotors. In our preliminary experiments, we confirmed that thrust of a quadcopter IGE depends on the distance between rotors. Thus, the ground effect to the rotor performance is expected to be changed with rotor configuration. Therefore, understanding the IGE wake structure is important to explain the mechanism of thrust change affected by flow interaction in the ground effect and to realize stable take-off and landing. Experimental visualization is profitable to examine rotor wake structures. It gives us hints into improving rotor placement design to reduce ground effect on the rotors and validate numerical analysis results by comparing with experimental results.

Studies concerning isolated rotor wake have conducted numerical and experimental analyses [3,4,5,6,7,8]. Komerath [9] provided an overview of the progress of investigations on rotor wake. For experimental wake flow visualization, photograph investigations using dust particles [3], shadowgraph [10,11] and particle image velocimetry (PIV) measurements [11,12] were conducted. One of the motivations of rotor wake visualization is the wake blowing flow of helicopter IGE [13,14,15]. The blowing flow causes sight loss of helicopter pilots in desert areas. Thus, the understanding of the rotor wake is important in the design of helicopters. PIV measurements were conducted intensively on rotor wake flow during IGE operations conducted intensively by the groups of researchers of in the University of Maryland led by Professor Leishman [12,11,16,17]. Lee, Leishman, and Ramasamy [17] conducted flow visualization and PIV measurement on a small isolated rotor with 86 mm radios near a ground plane. They clarified a detailed structure of the wake in and out of the ground effect. Moreover, they clarified tip vortex structures in the flow development near the ground plane with clear images. In addition to experimental investigations, numerical analyses were conducted using computational fluid dynamics (CFD) and other methods [5,18,19]. Although numerical analysis clarifies the detailed structures of wake flows, validation of

---

Presented at the 6th Asian/Australian Rotorcraft Forum / Heli Japan 2017, Kanazawa, Japan, November 7-9, 2017.

the accuracy of calculation results is indispensable because CFD calculations on the rotor consider unsteady flow. Moreover, calculations on moving objects as rotor blades require a massive calculation cost. For multicopters wakes, flow interaction is anticipated between the wakes. Therefore, experimental visualization is indispensable to examine structures of rotor wake and is important for CFD results validation. In addition, most of the wake-flow studies focus on isolated rotor and not on plural rotors. For quad tiltrotor aircrafts, the flow interactions were assessed using CFD calculations [20,21].

The objective of the current study is to examine the wake flow structure of a quadcopter UAV in the ground effect. The wake flow affects the flow interaction between the rotors. We conducted flow visualization and PIV measurement on a quadcopter model. The experimental examination of wake structures of plural rotor configuration is important for understanding the mechanism of the ground effect and validating numerical analysis results. We selected the quadcopter configuration because it is the most popular in multicopter usages, and is the most basic configuration of numerical analysis of multicopters flow.

METHOD

We conducted flow visualization and PIV measurement on a quadcopter model. Characteristics of the flow structure were analyzed through a momentum static image during flow visualization. The averaged velocity vector maps of the wake in foul target planes were measured using the PIV test to clarify the movement of wake in different rotor configurations. The detailed method is described as follows.

Quadrotor model setup

The experiments were conducted using a quadcopter model. The model was composed of support frames, four rotors, and brushless motors as shown in Fig. 1. There was not an object, such as the body, among the rotors to examine the flow structures for avoiding the effect of flow interaction between the body and the rotor wake. The rotors were attached at the end of the support pillars to avoid flow interaction. Figure 2 shows the rotor used in the experiment. The rotor is used in commercial small multicopter UAVs. The rotor diameter is 239 mm, and the rotor speed was set at 6,000 rpm. The Reynolds number of the rotor based on 75% span chord length was 54,700. Furthermore, the chord length at 75% of the blade span was 20 mm, and the rotor solidity was 0.105. The rotor speed was controlled using an electronic speed controller. The thrust coefficient of the rotor out of the ground effect (OGE) was 0.11 at 6,000 rpm.

The experiment was conducted in a large room with sufficient space to avoid flow circulation in the room from the rotor wake. Figure 3 shows the experimental set up used in the two experiments. Except for the method of inserting smoke, most of the experimental setup was common for both the experiments. The flat floor of the setup was square, and the side length was 3,600 mm, which is 30 times that of the rotor radius (R) to avoid the effect of floor roughness.

In the experiments, we changed the distance between the rotors and rotor height from the ground plane. The rotor distance was set at 2.1 and 2.7 R; 2.1 R was close to the minimum rotor shaft distance of quadcopter configuration. The ground off height was set to 4.5 R for OGE and 0.5 R for IGE for comparison. The thrust at 4.5 R is almost the same as that out of ground plane. Thus, the value of 4.5 R was assumed as the OGE height in the experiment.

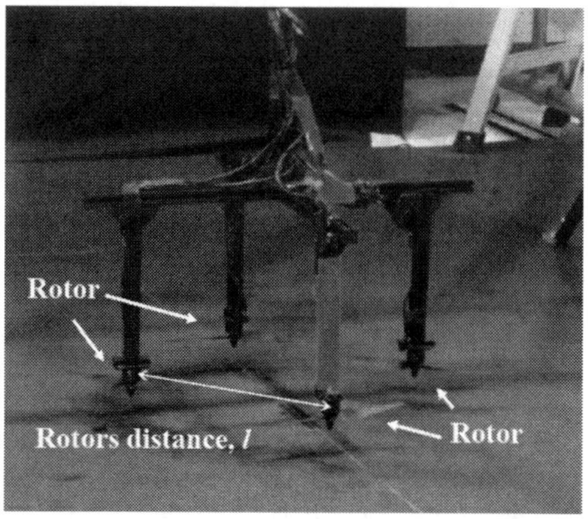

Figure 1 Experimental setup for the PIV measurement.

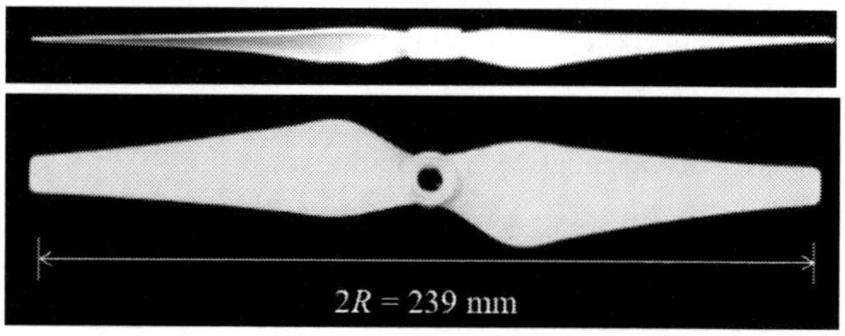

Figure 2. Experimental rotor.

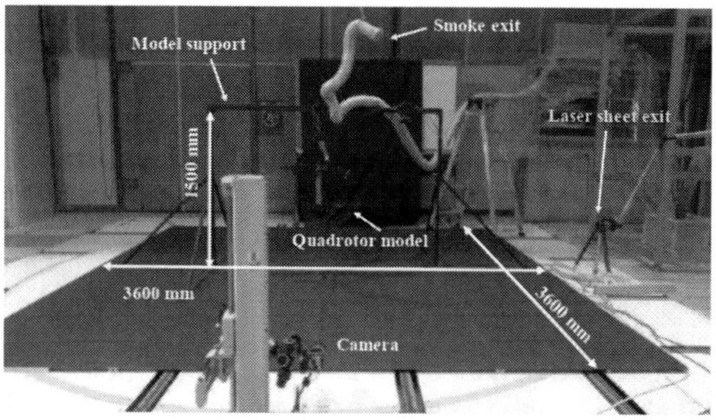

Figure 3. Experimental setup for flow visualizations and PIV measurements.

The visualization planes were set at foul patterns, as shown in Fig. 4. Planes #1 and #2 contain a rotor each and were 3 cm offset from the centers of the rotors. Planes #3 and #4 are at the midpoints of the rotors, implying that there is no rotor at the planes. For the quadcopter configuration, the rotation direction of adjoined rotors differs. Therefore, planes #3 and #4 differ in rotor blade movement near the plane. In plane #3, rotor blades move toward the center of the quadcopter, while in plane #4, they move toward the outer side of the quadcopter. Table 1 summarizes the conditions for rotor measurements. Smoke was generated from the smoke generator and released from the outlet, which was set above the setup. Nd: Yttrium aluminium garnet (YAG) laser was the light source of the laser sheet. The laser illuminated target planes from the side of the quadcopter. The width of the sheet was 2 mm. Figure 5 shows the setup illuminated using a laser sheet. The results of flow visualization and PIV measurements were assembled with pictures of different areas at different moments in the same conditions because the camera viewing area was very small to realize a moderate resolution. Furthermore, moderate density smoke spreads in a narrow area, implying that it could not cover the whole target area simultaneously. Therefore, a linear guides system was used to linearly move the camera position to the target plane, and images were captured in different areas at the same plane. Moreover, PIV vector maps were assembled with each target plane because of the camera view area limitation.

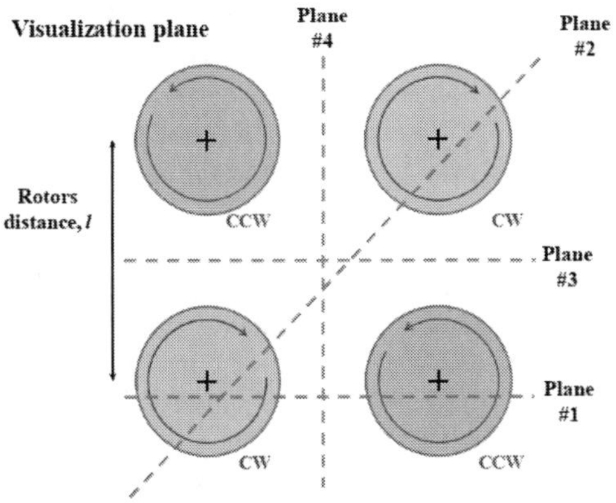

Figure 4. Position of the visualization planes.

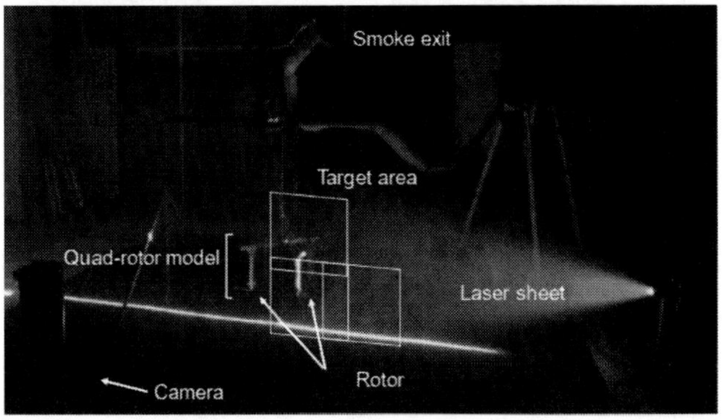

Figure 5. Setup with the laser sheet.

Flow visualization

A flow visualization was conducted to examine the structural characteristics of the quadrotor wake. Pictures were taken from the side by using a 4-megapixel resolution CMOS camera. The frame rate was set at 10 Hz. We recorded pictures of the wake, and selected pictures to present wake features. The pictures were taken over 50 s. To capture the wake structures, the smoke was generated intermittently to adjust the density. Flow visualizations were conducted only for plane #2 in Fig. 4, which is the slice view of two diagonal rotors.

PIV measurements

PIV measurements were conducted to examine the flow direction of the wake at different heights and rotors distances. The frame rate was 10 Hz, and the interval between a pair of pictures to calculate the flow field vector was 400 μs. We measured the vector map of wake 1000 times and averaged them. The seed particle of the smoke was generated using a heat exchanger from fog fluid. Diameter of the seed particle was 1.5 μm. The room used in the experiment was filled with smoke once. After stopping smoke generation, we initiated visualization by uniformly spreading a moderate density of particles in the room.

Table 1. Conditions of flow visualization and PIV measurements.

z/R	Rotors distance	Test conditions	
		Flow visualization plane	PIV plane
4.5	2.7	#2	#1, #2, #3, #4
4.5	2.1	Not conducted	#2
0.5	2.7	#2	#1, #2, #3, #4
0.5	2.1	#2	#2

RESULTS

The flow field of the quadcopter wake was visualized using a laser sheet at plane #2, and PIV measurements were conducted for planes #1–4 at different ground off heights and different distances between the rotors. The visualization images show wakes of the right-side rotor.

Flow visualization

OGE operation with the wider rotor distance. Figure 6 shows the flow visualization result with ground off height of 4.5 R and l = 2.7 R. The figure shows that the wake collides with the ground plane and spreads to outer side. The wake contains vortex rings near the inner and outer boundaries. The vortex rings merge to the flow while moving down to the ground plane. We could not confirm vortex rings near the ground plane. After reaching the ground plane from the rotors, the wake soared in the upper direction at the center of the body. The soaring flow reached the height of the rotor plane and entered the rotor again. Outside the body, the wake spread in the outer direction. Furthermore, the wake boundary was slightly bent toward the center of the body.

IGE operation with the wider rotor distance. Figure 7 shows the flow visualization result with ground off height of 0.5 R and l = 2.7 R. Here, the wake contains tip vortex rings the same as those in the isolated rotor wake. However, the movement of the rings along the wake flow was asymmetrical to the inner and outer sides of the quadrotor. The wake is shrunken compared with the wake at $z/R = 4.5$. Therefore, the vortex rings reached the ground plane and spread. Inside the quadrotor, the vortex rings moved in the upper direction with the soaring flow from the wake. The soaring flow passed the height of the rotor plane and width of the flow boundary shrunk. Outside the quadrotor, the vortex rings moved horizontally along the ground plane. The wake between the rotor and ground plane was difficult to observe in the visualized flow with smoke because of the unsteady and fast flow. The flow is generated from the upper side of the rotor shrink as it gets close to the rotor. The right edge of the smoke entrained into the rotor was approximately 2 R away from the rotor shaft at $z/R = 2$.

IGE operation with the closer rotor distance. Figure 8 shows the flow visualization result with the ground off height of 0.5 R and $l = 2.1$ R. Here, the wake contains tip vortexes similar to those in the case of IGE operation with wider rotor distance. The tip vortexes in the rotor wake at the side farther from the center of the quadcopter spread to the outer side of the quadcopter and moved along the ground plane. The tip vortexes at the inner side of the quadrotor soared at the center of the quadrotor. However, the observed vortexes were thinner and the width of soaring flow was smaller than that in the case of the wider rotor distance with IGE. Moreover, smoke entrained into the rotor coming from the upper right side area of the rotor was confirmed.

PIV measurement

The PIV measurement results were expressed using a vector map. The counter color and size of arrow expressed the velocity magnitude of each vector. The evaluation of the exact speed at the measured points needs discussion because flow speed evaluated in the PIV measurements differs from the true speed of the flow. We used the results to examine the flow path, direction, and features of the velocity map.

OGE operation with the wider rotor distance. Figure 9 shows the averaged vector map from PIV measurements with ground off height of 4.5 R and $l = 2.7$ R. The features sketch of the wake is illustrated at bottom part of the figure. For plane #1 in Figs. 9 (a) and (e), the rotor wake is almost symmetrical to the center of the rotor at a height farther from the ground plane. However, at the ground plane, the velocity of the wake outside the quadrotor is faster than that inside. Furthermore, a strong soaring flow was not observed between the right and left rotors, and a weak circulation remained at the blade tip inside the rotor. Moreover, tip vortices were observed at both edges of the wake in moment velocity map as sketched.

For plane #2 in Figs. 9 (b) and (f), the rotor wake was asymmetrical to the center of the rotor. The outer wake velocity was faster than the inner near the ground plane. In addition, the horizontal velocity in the wake vectors moved toward the inside of the quadrotor. Therefore, the wake flow direction slanted toward the inside of the quadrotor. The speed of the wake core region was over 10 m/s. A soaring flow developed at the center of the quadrotor, and reached the height of the rotor plane. The flow entrained the rotor again and formed a circulation flow. In this plane, tip vortexes were confirmed at wake edges in momentum flow images, as sketched in Fig. 9 (f).

For plane #3 in Figs. 9 (c) and (g), soaring flow was confirmed at the center of the quadrotor. The center area was the same for the soaring flows of planes #3 and #4. The plane does not contain rotor; thus, the area under the rotor was not exactly under the rotor center. At this area, the flow emanated from the outside. At midheight between the rotor and ground plane, the flow appeared to move downward. Near the ground plane, a large part of the downwash spread to the outside, and part of the downwash moved toward the soaring flow at the center of the quadrotor. The top end of the soaring flow was near the inside of the rotor blade tip, and a circulation flow was formed, as sketched in Fig. b1(g).

For plane #4 in Figs. 9 (d) and (h), the basic structure of the flow was the same as that in plane #3. Soaring flow and downwash were confirmed. However, velocity of the downwash under the rotor was faster than that in plane #3. Moreover, the amount of flow entering the soaring flow from the downwash was lesser than that in plane #3.

IGE operation with the wider rotor distance. Figure 10 shows the averaged vector map of PIV measurements with ground off height of 0.5 R and $l = 2.7$ R. For plane #1 in Figs. 10 (a) and (e), wake shrinking was observed and the wake spread toward the inside and outside of the quadrotor along the ground plane. A strong circulation was observed at the blade tip near the center of the quadcopter. Moreover, the wake sored between the rotors and entered the rotor again. The flow entering into the rotor disk emanated from the upper parts of the rotor planes, and soaring flow was not confirmed except for the circulation at blade tip. The same as in higher rotor case in Fig. 9 (e), vortex rings were observed at both sides of the wake edges. Vortexes that were formed inside the quadrotor disappeared promptly compared with those formed outside.

Plane #2 in Figs. 10 (b) and (f) contains the center of the quadrotor. At the center of the quadrotor, a strong soaring flow was developed. The flow reached the top support frame. The velocity of the soaring flow was approximately 11m/s. Moreover, a strong circulation was formed similar to that in plane #1. Outside the rotor, the flow entrained in the rotor emanated from the upper area of the rotor, and the wake of the rotor spread along the ground plane. In addition, vortexes were observed at both sides of wake edges, as sketched in Fig. 9(f). The visualized flow of this condition is shown in Fig. 7.

OGE and IGE operation with closer rotor distance. Fig. 11 shows the averaged vector map of PIV measurements at ground off height of 0.5 and 4.5 R and l = 2.1 R. For z/R = 4.5 at plane #2 in Figs. 11 (a) and (b), the rotor wake is asymmetrical, the same as that of plane #2 of the wider rotor in Figs. b1 (a) and (e). The rotor wake was pulled to the center of the quadrotor, and the wake flow velocity outside the quadcopter was faster than that inside. Therefore, the vortex sheet was slanted toward the center of the body. The wake flow pulled toward the center merged with the opposite side rotor wake at midheight from the ground plane. At the ground plane, the wake flow moved toward the outside of the quadcopter along the plane. In this condition, a soaring flow was not observed at the center of the quadcopter, although it was observed at the wider rotor distance case. Near the blade tip inside the quadcopter, a small soaring flow area was observed. This rising flow formed circulation around the rotor blade tips at the center of the quadcopter, as sketched in Fig. 11 (b). In this condition, the flow above the rotors moved lower and entered the rotor disk. Therefore, a rising flow was not observed above the rotor height. Blade tip vortexes were confirmed at the wake boundaries.

For z/R = 0.5 at plane #2 in Figs. 11 (c) and (d), the wake was compressed, the same as in the other cases when z/R = 0.5. Outside the quadrotor, the wake spread toward the outside along the ground plane. Inside the quadcopter, the wake reached the ground plane and sored at the center of the quadcopter. we could not determine the exact velocity because the front rotor overlapped the captured image. Nevertheless, the soaring flow speed was approximately 10 m/s at the height of 2 R. At both side edges of the wake, blade tip vortexes were observed. At the inside of the quadcopter, vortexes were transported toward the upper side of the rotor along the soaring flow.

DISUCUSSIONS

Comparison between isolated rotor and quadrotor

The wake structure of the quadrotor in our experiments is different from the isolated rotor wake shown by Lee, Leishman, and Ramasamy [17]. The wake of the isolated rotor is symmetrical to the center of the rotor, while the wake of a rotor of the quadrotor is asymmetrical to the rotor shaft and interfered with the other three rotor wakes. Outside the quadcopter, the wake spread toward the outside. However, the wake flow inside the quadcopter moved toward the center at the ground plane and resulted in a rising flow. A soaring flow did not appear in the wake of the isolated rotor. Thus, the occurrence of the soaring flow and asymmetry of rotor wake to rotor shaft axis are differences of the quadrotor wake to isolated rotor wake.

Comparison between OGE and IGE with the wider rotor distance.

The comparison of the ground heights with a rotor distance of l = 2.7 R in Figs. 9 and 10 shows that the soaring flows at the center of the quadcopter are different. At OGE operation, the soaring flow is weak and the velocity is approximately 3 m/s; the flow reaches a height of the rotor plane. In contrast, at IGE operation, the soaring flow is strong and velocity is approximately 11 m/s; the flow passes the rotor plane and moves to top of the target view area. Therefore, a soaring flow develops with the lowering height of the rotor planes at the wider rotor distance. Moreover, the wake flow of the rotor spreading toward the outside of the quadcopter moved horizontally at z/R = 4.5. However, for z/R =0.5, wake flow of the rotor spreading toward the outside soared from the ground plane at planes #3 and #4. Thus, at IGE operations, a blowing flow to soar was developed from the rotors clearance and the rotors edge outside of the quadcopter.

Comparison between OGE and IGE with the closer rotor

The comparison of the ground off heights with rotors distance l = 2.1 R in Figs. 11 (a) to (d) shows that the soaring flow development at the center of the quadcopter are different. The basic feature is the same as that of the wider rotor distance operations. At z/R = 4.5, soaring flow is not observed. As the rotor got close to the ground plane, a soaring flow appeared at the center of the quadcopter.

Comparison of rotors distance OGE operations

The comparison of the rotors distances at OGE operations in Figs. 9, 11 (a), and 11 (b) shows that the soaring flow at the center of quadrotor is different. In case of a small distance between rotors, a soaring flow was not observed. In contrast, soaring flow appeared for wide rotor distances. Therefore, the rising flow disappeared with closing rotor distance. We presumed that the closing rotor distance eliminates rotor clearance at the height of the rotor plane for the passing flow. Therefore, it is easier for the flow to spread outside than to converge in the center of the quadcopter, and the soaring flow disappears.

Comparison of rotors distance IGE operations

The comparison of the rotors distances at IGE operations in Figs. 10, 11 (c), and 11 (d) shows that the basic characteristics of the flow structure are similar. Soaring flow develops at the center of the quadcopter at both rotor distances. However, velocity of the rising flow with l = 2.7 R is faster than that with l = 2.1 R. The velocities with l = 2.7 and 2.1 R were 11 and 10 m/s, respectively. The width of the rising flow with the close rotor distance is smaller than that of the flow with the wide rotor distance. The closing rotor distance seems to impede the development of soaring flow.

Effect of the soaring flow.

In quadcopters, the body is located at the center of four rotors. Therefore, the body seems to be pushed by the soaring flow refracting from the ground plane. We speculate that this mechanism reinforces the thrust increase in the ground effect. Thus, it is expected that changing the distance between the rotors will affect rotor thrust in the ground effect. Furthermore, a blockage plate between rotors can eliminate soaring flow. The dependency of thrust change in the ground effect on the rotor distance can be weakened. To examine the effect of soaring flow in the thrust in the ground effect, we must examine the relation of thrust and rotor distances. In addition, a quadcopter controller contains the atmospheric pressure sensor, which is used as height sensor in a number of cases. The effect of unstable soaring flow on the pressure sensor vibration is a topic of concern. For stable take-off and landing of multicopter UAVs, more studies are required on the wake of the multicopters.

CONCLUSIONS

To examine the structure of wake of the quadrotor experimentally, we conducted flow visualization and PIV measurement at different ground off heights and different rotor distances. We draw the following conclusions from this study.

1) In the quadcopter wake, a soaring flow reflected from the ground plane develops at the center of the quadcopter at a height in the ground effect. The velocity of the soaring flow depends on the rotor clearance at the center of the quadcopter. With closing rotor distance, the wake is weakened. At a rotor distance of 2 R, the soaring flow does not appear.

2) The rotor wake of the quadcopter is asymmetrical to rotor shaft axis, which is not as similar as that of the isolated rotor wake. The wake is pulled to the center of the quadcopter. Thus, the rotor wake is slanted toward the inside of the quadcopter. The vertical speed of the wake outside the quadcopter is faster than the speed inside.

3) The velocity and direction of the spreading flow from the quadcopter at the ground plane are different in the visualization planes. The spreading flow has a directional difference on the velocity. The speed depends on whether the plane contains the rotor wake.

We experimentally analyzed the wake structures of the quadcopter configuration and verified the existence of the soaring flow, which is unique for the wake of plural rotor configuration. The flow can affect the stability of the body posture in the ground effect. Moreover, the effect on thrust increase with the ground effect is a topic for concern. To realize stable take-off and landing, further studies are required with experiments and numerical analysis validations.

Author contact: Hikaru Otsuka h-otsuka@frl.mech.tohoku.ac.jp, Masayoshi Kohno masayoshi@frl.mech.tohoku.ac.jp, Keiji Nagatani keiji@ieee.org

ACKNOWLEDGMENTS

This work was supported by the Grand-in Aid for the JSPS Research Fellow Grant, Number 16J02686. The authors would like to thank technical assistants of the low-speed wind tunnel at the Institute of Fluid Science in Tohoku university for the technical support of the experiments

REFERENCES

[1]Johnson, W., *Helicopter Theory*, Princeton University Press, Princeton, NJ, 1980, pp. 122–124.

[2]Leishman, J. G., *Principles of Helicopter Aerodynamics*, Cambridge University Press, New York, NY, 2000, Chapter 5.8.

[3]Taylor, M. K., "A Balsa-dust Technique for Air-flow Visualization and Its Application to Flow through Model Helicopter Rotors in Static Thrust," *NACA Technical Note*, 2220, Langley Aeronautical laboratory, Nov. 1950.

[4]Caradonna, F. X., and Tung. C., "Experimental and Analytical Studies of a Model Helicopter Rotor in Hover," NASA TM81232, 1981.

[5]Betz, A., "The Ground Effect on Lifting Propellers," NACA TM No. 836, 1937.

[6]Itoga, N., Nagashima, T., Yoshizawa, Y., and PRASAD, J. V. R., "Numerical Method for Predicting I. G. E. Hover Performance of a Lifting Rotor," *Transaction of the Japan society for Aeronautical and Space Sciences*, Vol. 43, (144), pp. 122–129, 2000.

[7]Curtiss, H. C., Sun, M., Putman, W. F., and Hanker, E. J., "Rotor Aerodynamics in Ground Effect at low Advance Ratios," the *Journal of the American Helicopter Society*, Vol. 29, (1), Jan. 1984, pp. 48–55.
doi: 10.4050/JAHS.29.48

[8]Koo, J., and Oka, T., "Experimental Study on the Ground Effect of a Model Rotor in Hovering," *Technical Report of NAL*, TR-113, National Aerospace Lab., Tokyo, Japan, 1966. (In Japanese) (Available in English as NASA TT F-13, 938, 1971.)

[9]Komerath, N. M., Smith, M. J., and Tung, C., "A Review of Rotor Wake Physics and Modeling," *Journal of the American Helicopter Society*, Vol. 56, (2), Apr. 2011, pp. 22006.
doi: 10.4050/JAHS.56.022006

[10]Light, J. S., "Tip Vortex Geometry of a Hovering Helicopter Rotor in Ground Effect," *Journal of the American helicopter Society*, Vol. 38, (2), Apr. 1993, pp. 34–42. Doi: https://doi.org/10.4050/JAHS.38.34

[11]Moedersheim, E., Daghir, M., and Leishman, J. G., "Flow Visualization of Rotor Wakes using Shadowgraph and Schlieren Techniques," *19th European Rotorcraft Forum*, Amsterdam, Oct. 1994, pp. 31.

[12]Ramasamy, M., and Leishman, J. G., "Benchmarking Particle Velocimetry with Laser Doppler Velocimetry for Rotor Wake Measurements," *AIAA Journal*, Vol. 45, (11), Nov. 2007, pp. 2622–2633.
doi: 10.2514/1.28130

[13]Rauleder, J., and Leishman, J. G., "Interactions of Vortical Flow with Dispersed Particles below a Rotor," *AIAA Journal*, Vol. 54, (11), Nov. 2016, pp. 3361–3373.
doi: 10.2514/1.J053823

[14]Phillips, C., and Brown, R. E., "Eulerian Simulation of the Fluid Dynamics of Helicopter Brownout," *Journal of Aircraft*, Vol. 46, (4), Jul. – Aug. 2009, pp. 1416–1429.
doi: 10.2514/1.41999

[15]Milluzzo, J., and Leishman, J. G., "Assessment of Rotorcraft Brownout Severity in Terms of Rotor Design

Parameters," *Journal of the American Helicopter Society*, Vol. 55, (3), Jul. 2010, pp. 32009.
doi: 10.4050/JAHS.55.032009

[16]Milluzzo, J., and Leishman, J. G., "Vortical Sheet Behavior in the Wake of a Rotor in Ground Effect," *AIAA Journal*, Vol. 55, (1), Jan. 2017, pp. 24–35.
doi: 10.2514/1.J054498

[17]Lee, T. E., Leishman, J. G., and Ramasamy. M., "Fluid Dynamics of Interacting Blade Tip Vortices with a Ground Plane," *Journal of the American Helicopter Society*, Vol. 55, (2), Apr. 2010, pp. 22005.
doi: 10.450/JAHS.55.022005

[18]Lakshminarayan, V. K., and Baeder. J. D., "Computational Investigation of Micro Hovering," *Journal of the American Helicopter Society*, Vol. 55, (2), Apr. 2010, pp. 22001.
doi: 10.4050/JAHS.55.022001

[19]Duraisamy, K., and Baeder, J. D., "High Resolution Wake Capturing Methodology for Hovering Rotors," *Journal of the American Helicopter Society*, Vol. 52, (2), Apr. 2007, pp. 110–122.
doi: 10.4050/JAHS.52.110

[20]Gupta, V., and Baeder, J. D., "Investigation of Quad Tiltrotor Aerodynamics in Forward Flight Using CFD," *20th AIAA Applied Aerodynamics Conference*, St. Louis, Missouri, USA, AIAA paper 2002–2812.

[21]Yoon, S., Lee, H. C., and Pulliam, T. H., "Computational Analysis of Multi-Rotor Flows," *AIAA SciTech Forum, 54th AIAA Aerospace Science Meeting*, Jan. 2016, San Diego, USA, AIAA paper 2016-0812.

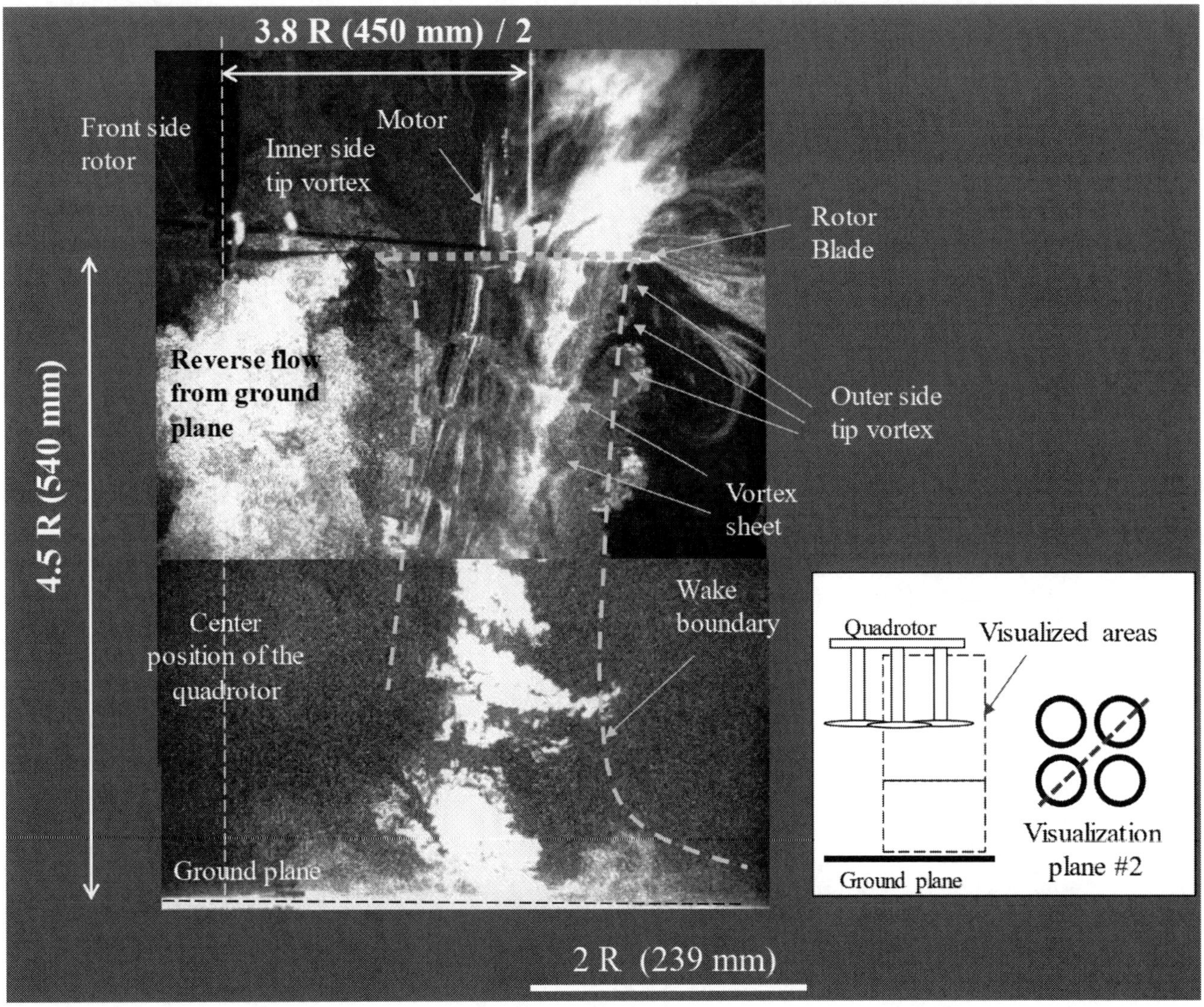

Figure 6 Visualized rotor wake of IGE operations at close rotor distances.

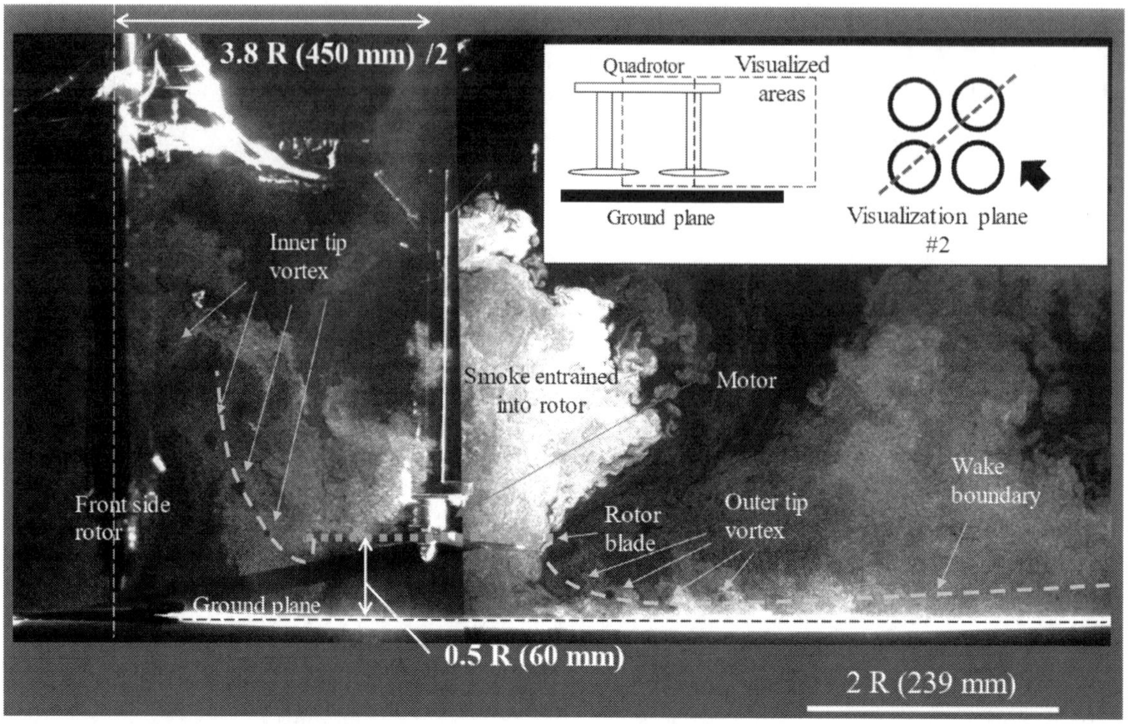

Figure 7. Visualized rotor wake of IGE operation at wide rotor distances.

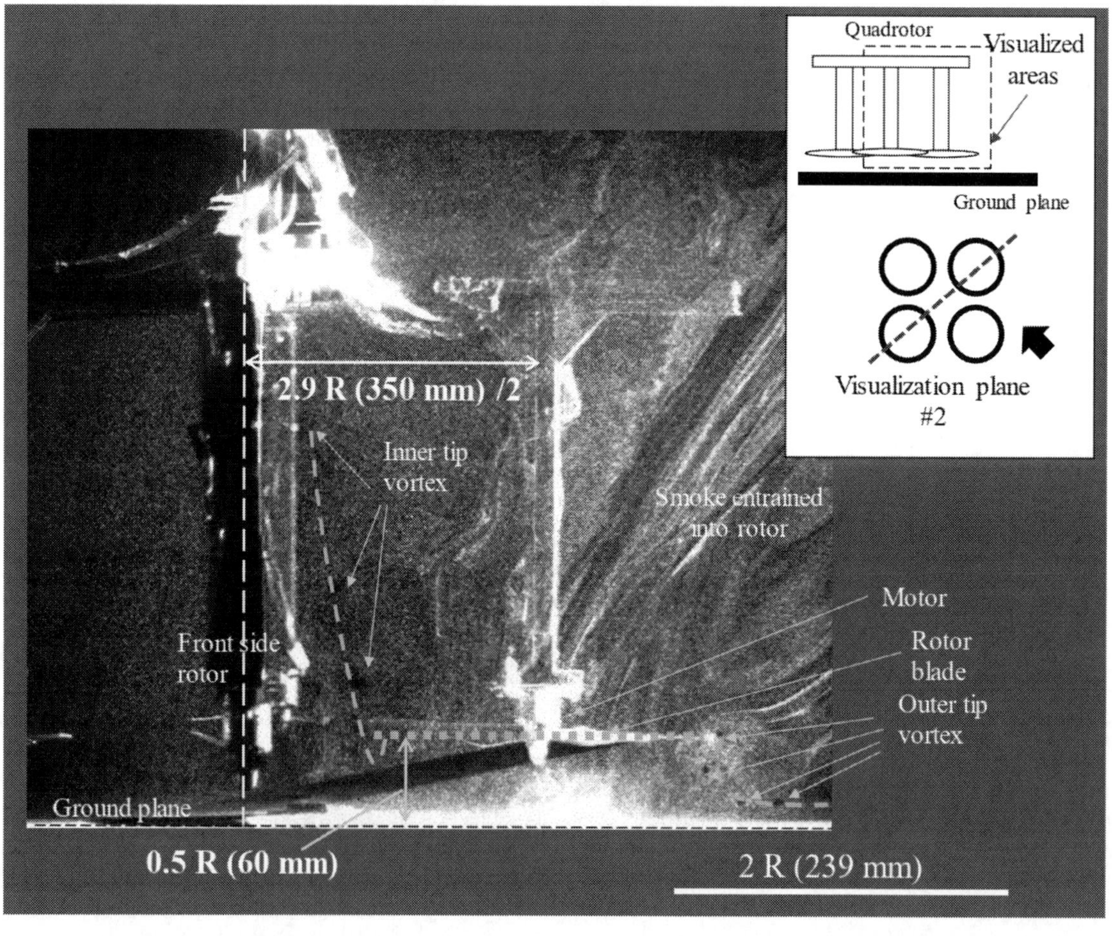

Figure 8 Visualized rotor wake of IGE operation at close rotor distances.

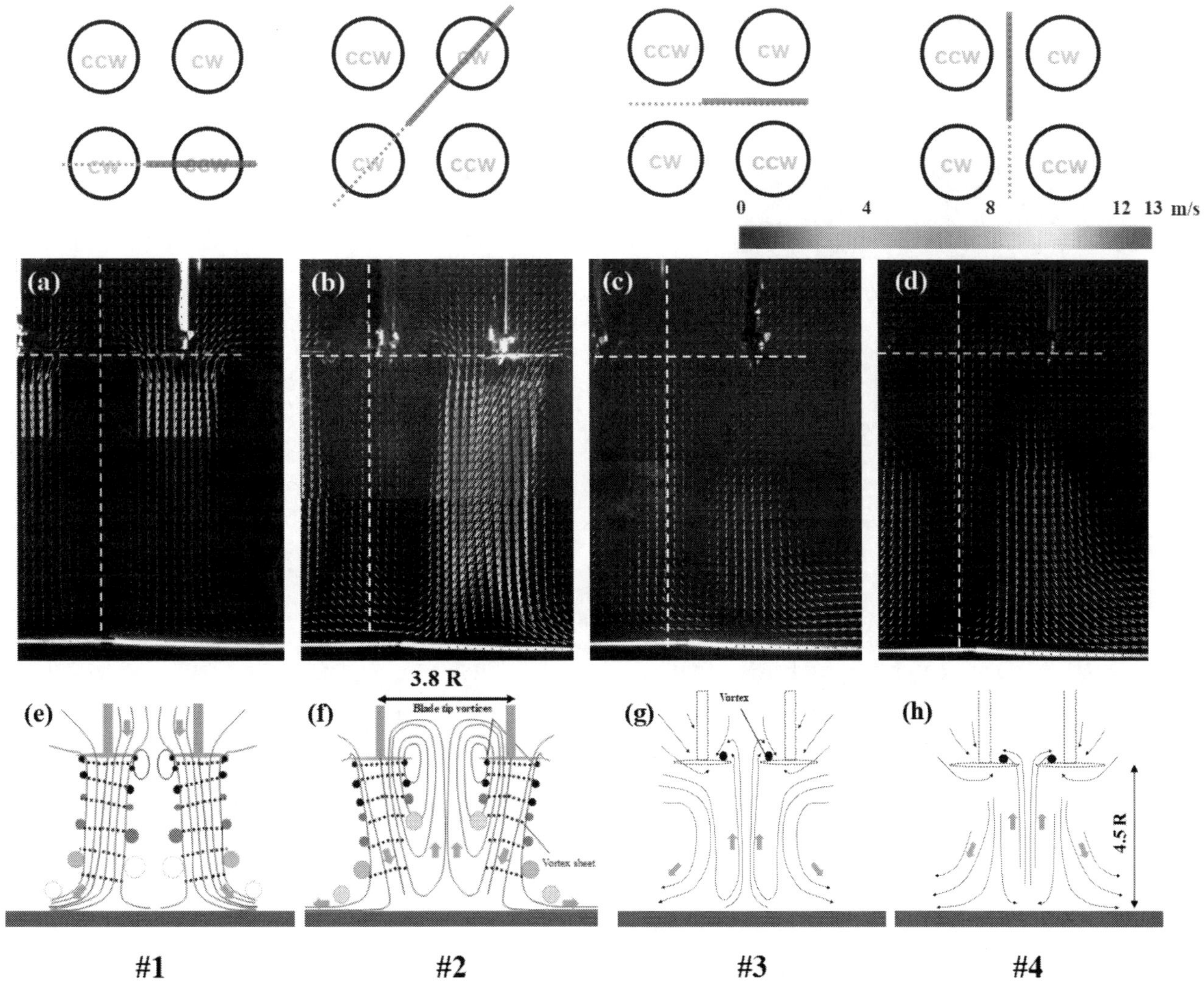

Figure 9 Wake of OGE operation at wide rotor distances.

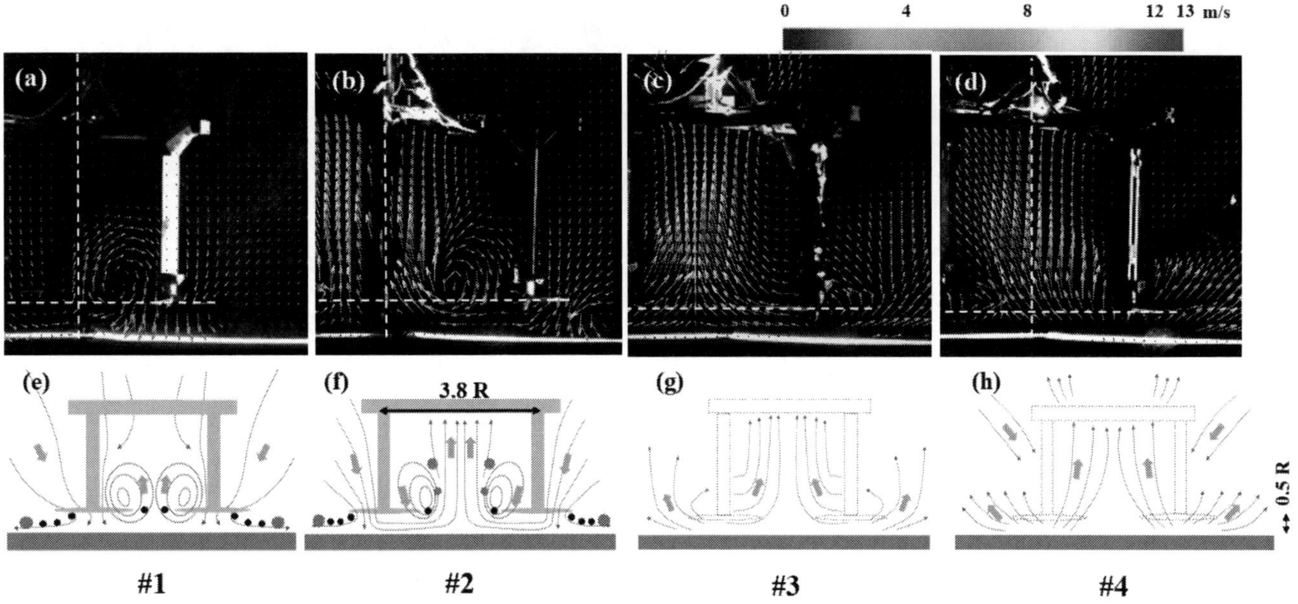

Fig. 10 Wake of IGE operation at wide rotor distances.

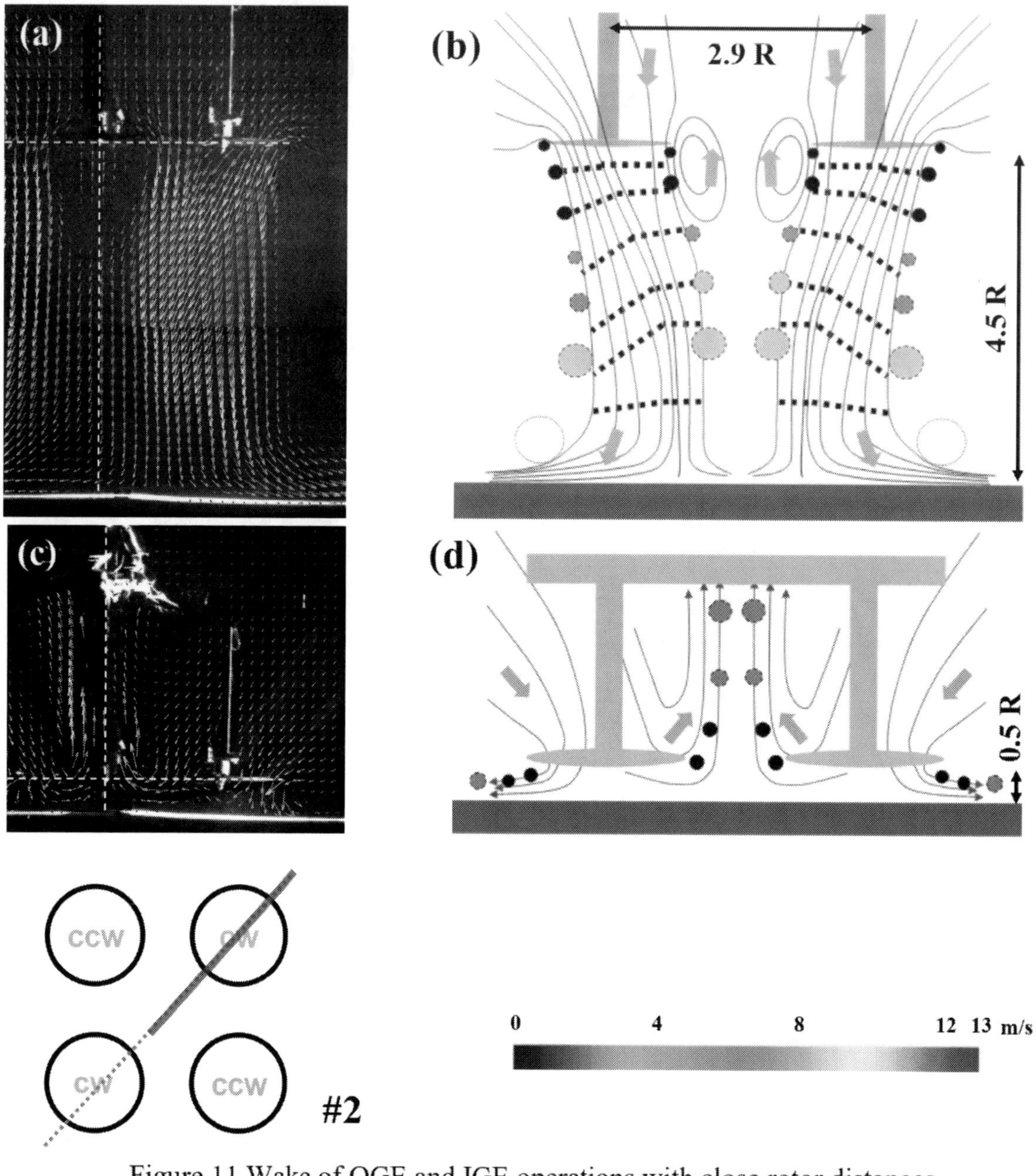

Figure 11 Wake of OGE and IGE operations with close rotor distances.

Effects of the size and the position of the center of gravity of a multirotor UAV on hovering flight characteristics.

Shuichi Umezaki
Mr.
Osaka Prefecture University
Osaka, Japan

Shigeru Sunada
Professor
Nagoya University
Aichi, Japan

Yasutada Tanabe
Senior Researcher
JAXA
Tokyo, Japan

Koichi Yonezawa
Doctor
Central Research Institute of Electric Power Industry
Chiba, Japan

Hiroshi Tokutake
Professor
Kanazawa University
Ishikawa, Japan

Takakage Arai
Professor
Osaka Prefecture University
Osaka, Japan

ABSTRACT

There have been various kinds of the multirotor UAVs (Unmanned Aerial Vehicles) recently. Their scale and their position of the center of gravity are various. In this paper, the effects of the scale and the position of the center of gravity of a multirotor UAV which has six rotors on hovering flight characteristics are investigated. The flight characteristics are discussed by using the eigenvectors obtained by solving linear equations of motion. In the scale analysis, the mass and the moment of inertia are varied proportionally to the third and fifth powers of the size. The variations of aerodynamic characteristics of rotors due to that of Reynolds number are also considered. In the center of gravity position analysis, the position of center of gravity was moved up and downward vertically. The present analysis is a fundamental study which can be expected to contribute to developments of multirotor UAVs in the future.

INTRODUCTION

It is well known that the use of multirotor UAVs (Unmanned Aerial Vehicles) has been rapidly expanding in recent years, for example, UAVs for aerial shoots, carrying payloads and spreading pesticides. They have various sizes and numbers of rotors. Moreover, it is expected that more and more kinds of multirotor UAVs will be developed. However, little attention has been paid to aerodynamics of a multirotor UAV in comparison with that of a single rotor helicopter. To design a multirotor UAV which can be operated more safely, it is required to analyze its dynamics not only from control engineering but also from aerodynamics.

The purpose of this study is to clarify the scale effects of a multirotor UAV on its inherent stability. As a calculation model, a hexarotor UAV developed by the authors' group was adopted. First, the stability derivatives at hovering state were obtained analytically. Second, its inherent stability was investigated by using the stability derivatives. Finally, the effects of 1) its scale and 2) its center of gravity position(CGP) on its inherent stability were investigated. In analyses of the scale effect, it was assumed that the mass and the moment of inertia of a multirotor UAV were changed proportionally to the third and fifth powers of the representative length. In analyses of the CGP, it was assumed that the position of center of gravity of a multirotor UAV was moved up and downward vertically while the variation of its moment of inertia was ignored. The present analysis is a fundamental study which can be expected to contribute to the developments of multirotor UAVs in the future.

METHOD & MATERIALS

Analysis model

Fig.1 shows the analysis model of a multirotor UAV, which was developed by the authors' group. Tab.1 shows its parameters. In analysis, the body fixed coordinate system is used (shown in Fig.2). Its origin coincides with the center of gravity of the multirotor UAV.

Presented at the 6th Asian/Austrailian Rotorcraft Forum / Heli Japan 2017, Kanazawa, Japan, November 7-9, 2017.

Tab.1 The parameters of the analysis model

	symbol	value	unit
mass	m	4.99	kg
distance between z-axis and the rotational axis of each rotor	R_A	0.393	m
number of revolution in hovering	Ω	628	rad/s
distance between the center of gravity and the rotational plane of rotors	h	0.0675	m
number of blades of each rotor	b	2	-
rotor radius	R	0.16	m
blade chord length	c	0.0385	m
collective pitch angle	θ_0	0.105	rad
lift slope of a blade	a	5.73	
drag coefficient of a blade	C_d	0.015	
induced velocity	v_i	6.86	m/s
moment of inertia (around x-axis)	I_{xx}	0.207	kg m^2
moment of inertia (around y-axis)	I_{yy}	0.203	kg m^2
moment of inertia (around z-axis)	I_{zz}	0.373	kg m^2

Fig.1 The analysis model of a multirotor UAV

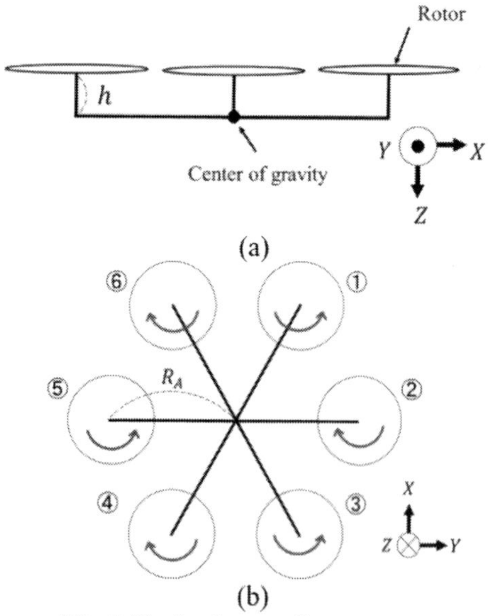

Fig.2 The body coordinate system

Tab.2 the rotational direction and the coordination of rotors

number		Coordination
①	CCW	$(\frac{\sqrt{3}}{2}R_A,\ \frac{1}{2}R_A,\ -h)$
②	CW	$(0,\ R_A,\ -h)$
③	CCW	$(-\frac{\sqrt{3}}{2}R_A,\ \frac{1}{2}R_A,\ -h)$
④	CW	$(-\frac{\sqrt{3}}{2}R_A,\ -\frac{1}{2}R_A,\ -h)$
⑤	CCW	$(0,\ -R_A,\ -h)$
⑥	CW	$(\frac{\sqrt{3}}{2}R_A,\ -\frac{1}{2}R_A,\ -h)$

Equations of motion

When small disturbances around a hovering are considered, the equations of motion of the multirotor UAV is expressed as:

$$m\ddot{x} = -mg\Theta + X \quad (1)$$

$$m\ddot{y} = mg\Phi + Y \quad (2)$$

$$m\ddot{z} = Z \quad (3)$$

$$I_{xx}\dot{p} - I_{xz}\dot{\bar{r}} = L \quad (4)$$

$$I_{yy}\dot{q} = M \quad (5)$$

$$-I_{xz}\dot{p} + I_{zz}\dot{\bar{r}} = N \quad (6)$$

$$\dot{\Phi} = p \quad (7)$$

$$\dot{\Theta} = q \quad (8)$$

$$\dot{\Psi} = \bar{r} \quad (9)$$

where x, y, z are the coordinates of the center of gravity of the multirotor UAV, $p, q, \bar{r}$ are angular velocities, Φ, Θ, Ψ are Euler angles and X, Y, Z, L, M, N are aerodynamic forces and moments.

To make the linear equations of motion, aerodynamic forces and moments are Taylor-expanded around their equilibrium. For example, the x-axis aerodynamic force X is expressed as:

$$X = \frac{\partial X}{\partial u}u + \frac{\partial X}{\partial v}v + \frac{\partial X}{\partial w}w + \frac{\partial X}{\partial p}p + \frac{\partial X}{\partial q}q + \frac{\partial X}{\partial \bar{r}}\bar{r} \quad (10)$$

where u, v, w are velocites of the multirotor UAV.

Blade element theory is used to calculate aerodynamic forces and moments on the rotors. The geometry of a rotor blade is shown in Fig.3. The blade element has the collective pitch angle θ_0, the local inflow angle ξ, and the azimuth angle ψ. The lift and drag acting on the blade element are defined as:

$$dL = \frac{1}{2}\rho V^2 C_L c dr \quad (11)$$

$$dD = \frac{1}{2}\rho V^2 C_D c dr \quad (12)$$

where ρ is the density of air, V is the inflow velocity the blade element.

The aerodynamic forces on the blade element are:

$$dX = (-dL\sin\xi - dD\cos\xi)\sin\psi \quad (13)$$

$$dY = (-dL\sin\xi - dD\cos\xi)\cos\psi \quad (14)$$

$$dZ = -dL\cos\xi + dD\sin\xi \quad (15)$$

The velocity of the blade element is expressed as below by using the tangential component of velocity V_T and the radial component V_P:

$$V = \sqrt{V_P^2 + V_T^2} \quad (16)$$

$$V_T = r\Omega \quad (17)$$

$$V_P = v_i \quad (18)$$

The induced velocity v_i is derived from the simple momentum theory:

$$v_i = \sqrt{\frac{mg}{12\rho\pi R^2}} \quad (19)$$

Assuming $V_T \gg V_P$, the local inflow angle and is the velocity of the blade element can be written as:

$$\sin\xi \cong \xi \quad \cos\xi \cong 1 \quad (20)$$

$$V^2 = V_P^2 + V_T^2 \cong V_T^2 \quad (21)$$

$$\xi = \tan^{-1}\frac{V_P}{V_T} \cong \frac{V_P}{V_T} \quad (22)$$

Then, the lift coefficient of the blade element can be written as:

$$C_L = a\alpha = a(\theta_0 - \xi) \cong a\left(\theta_0 - \frac{V_P}{V_T}\right) \quad (23)$$

The aerodynamic forces of a blade are:

$$X = \frac{b}{2\pi}\int_0^{2\pi}\int_0^R \frac{dX}{dr} dr d\psi \quad (24)$$

$$Y = \frac{b}{2\pi}\int_0^{2\pi}\int_0^R \frac{dY}{dr} dr d\psi \quad (25)$$

$$Z = \frac{b}{2\pi}\int_0^{2\pi}\int_0^R \frac{dZ}{dr} dr d\psi \quad (26)$$

When a small disturbance u in the x-axis occurs on the counterclockwise rotational blade, the tangential component of velocity V_T and the radial component V_P are expressed as below (shown in Fig.3):

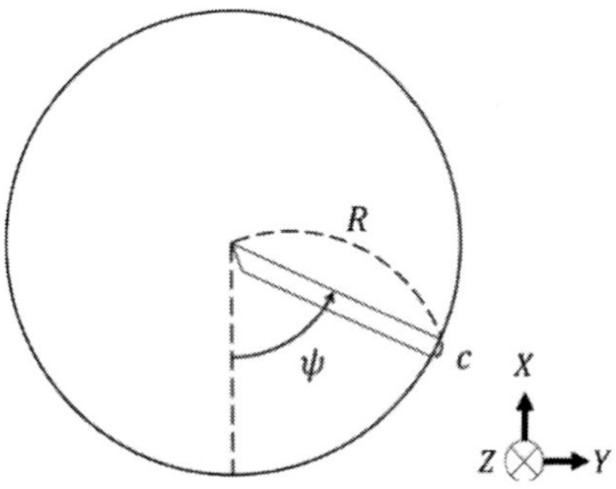

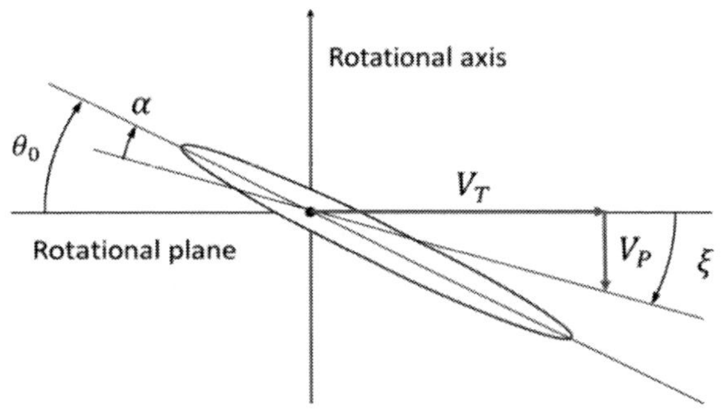

Fig.3 The geometry of a rotor blade

$$V_T = r\Omega + u \sin\psi \tag{27}$$

$$V_P = v_i \tag{28}$$

Substituting them to the equations (1), (2) and (13), the aerodynamic force on a blade in the x-axis is:

$$X|_u = -\frac{\rho bcR}{4}(av_i\theta_0 + C_D R\Omega)u \tag{29}$$

Similarly, the $X|_u$ in Eq. (29) can be obtained about the clockwise rotational blade.:

$$\psi' = -\psi \tag{30}$$

$$d\psi = -d\psi'$$

ψ	$0 \rightarrow 2\pi$
ψ'	$0 \rightarrow -2\pi$

So, the total aerodynamic force in the x-axis caused by a disturbance u is:

$$X_{total} = 6X = -\frac{3}{2}\rho bcR(av_i\theta_0 + C_D R\Omega)u \tag{31}$$

The partial differential coefficient $\partial X/\partial u$ and the stability derivative are:

$$\frac{\partial X_{total}}{\partial u} = -\frac{3}{2}\rho bcR(av_i\theta_0 + C_D R\Omega) \tag{32}$$

$$X_u = -\frac{3\rho bcR}{2m}(av_i\theta_0 + C_D R\Omega) \tag{33}$$

Other partial differential coefficient can be also derived in a similar way.

A small disturbance v in the y-axis:

$$V_T = r\Omega + v \cos\psi \tag{34}$$

$$V_P = v_i \tag{35}$$

A small disturbance w in the z-axis:

$$V_T = r\Omega \tag{36}$$

$$V_P = v_i - w \tag{37}$$

When a small disturbance p around the x-axis occurs, the rotors in one side of x-axis have upwash while the rotors in the other side have downwash. Moreover, each rotor's tangential component of velocity are varied because of h: the distance between the center of gravity position and the rotational plane of rotors. Then the tangential component of velocity V_T and the radial component V_P on the i-th rotor are:

$$V_T = r\Omega + hp \cos\psi \tag{38}$$

$$V_P = v_i - (y_i + r \sin\psi)p \tag{39}$$

When a small disturbance q around the y-axis occurs, the rotors in one side of y-axis have upwash while the rotors in the other side have downwash. In addition, each rotor's tangential component of velocity are also varied. Then the components of velocity on the i-th rotor are:

$$V_T = r\Omega - hq \sin\psi \tag{40}$$

$$V_P = v_i + (x_i - r\cos\psi)q \tag{41}$$

When a small disturbance $\bar{r}$ around the z-axis occurs, it gives each rotor a different influence. However, apparently, the total effect about the six rotors averaged during one rotational period is 0.

By using six kinds of velocity components described above, the partial differential coefficient (stability derivatives) of aerodynamic forces can be calculated.

$$X_v = 0 \tag{42}$$

$$X_w = 0 \tag{43}$$

$$X_P = -\frac{\rho abcR^2(\theta_0\Omega R - 3v_i)}{2m} \tag{44}$$

$$X_q = \frac{3}{2m}\rho bchR(av_i\theta_0 + C_D R\Omega) \tag{45}$$

$$X_{\bar{r}} = 0 \tag{46}$$

$$Y_u = 0 \tag{47}$$

$$Y_v = -\frac{3\rho bcR}{2m}(av_i\theta_0 + C_D R\Omega) \tag{48}$$

$$Y_w = 0 \tag{49}$$

$$Y_p = -\frac{3}{2m}\rho bchR(av_i\theta_0 + C_D R\Omega)p \tag{50}$$

$$Y_q = \frac{\rho abcR^2(\theta_0\Omega R - 3v_i)}{2m} \tag{51}$$

$$Y_{\bar{r}} = 0 \tag{52}$$

$$Z_u = 0 \tag{53}$$

$$Z_v = 0 \tag{54}$$

$$Z_w = -\frac{3}{2m}\rho bcR^2\Omega(a + C_D) \tag{55}$$

$$Z_p = 0 \tag{56}$$

$$Z_q = 0 \tag{57}$$

$$Z_{\bar{r}} = 0 \tag{58}$$

The aerodynamic moments are expressed as below:

$$L = \sum_{i=1}^{6} (-z_i Y_i + y_i Z_i) \tag{59}$$

$$M = \sum_{i=1}^{6} (z_i X_i - x_i Z_i) \tag{60}$$

$$N = \sum_{i=1}^{6} (-y_i X_i + x_i Y_i) \tag{61}$$

By substituting the equations (14) - (16) in the equations (59) - (61) with six kinds of velocity components in a disturbance, the partial differential coefficient (stability derivatives) of aerodynamic moments can be calculated.

$$L_u = 0 \tag{62}$$

$$L_v = -\frac{3\rho bchR(av_i\theta_0 + C_D R\Omega)}{2I_{xx}} \tag{63}$$

$$L_w = 0 \tag{64}$$

$$L_p = -\frac{3\rho bch^2R(av_i\theta_0 + C_D R\Omega)}{2I_{xx}} - \frac{3\rho bcR^2{R_A}^2\Omega(a + C_D)}{4I_{xx}} \tag{65}$$

$$L|_q = \frac{1}{2}\rho abchR^2(\theta_0\Omega R - 3v_i)q \tag{66}$$

$$L_{\bar{r}} = 0 \tag{67}$$

$$M_u = \frac{3\rho bchR(av_i\theta_0 + C_D R\Omega)}{2I_{yy}} \tag{68}$$

$$M_v = 0 \tag{69}$$

$$M_w = 0 \tag{70}$$

$$M_p = \frac{\rho abchR^2(\theta_0\Omega R - 3v_i)}{2I_{yy}} \tag{71}$$

$$M_q = -\frac{3\rho bch^2R(av_i\theta_0 + C_D R\Omega)}{2I_{yy}} - \frac{3\rho bcR^2{R_A}^2\Omega(a + C_D)}{4I_{yy}} \tag{72}$$

$$M_{\bar{r}} = 0 \tag{73}$$

$$N_* = 0 \quad (* = u, v, w, p, q, \bar{r}) \tag{74}$$

The linear equations of motion is expressed as:

$$\dot{\boldsymbol{x}} = \boldsymbol{A}\boldsymbol{x} \tag{75}$$

$$\boldsymbol{x} = [u \quad v \quad w \quad p \quad q \quad \bar{r} \quad \Phi \quad \Theta \quad \Psi]^T \tag{76}$$

$$\boldsymbol{A} = \begin{bmatrix} X_u & 0 & 0 & X_p & X_q & 0 & 0 & -g & 0 \\ 0 & Y_v & 0 & Y_p & Y_q & 0 & g & 0 & 0 \\ 0 & 0 & Z_w & 0 & 0 & 0 & 0 & 0 & 0 \\ 0 & L'_v & 0 & L'_p & L'_q & 0 & 0 & 0 & 0 \\ M_u & 0 & 0 & M_p & M_q & 0 & 0 & 0 & 0 \\ 0 & N'_v & 0 & N'_p & N'_q & 0 & 0 & 0 & 0 \\ 0 & 0 & 0 & 1 & 0 & 0 & 0 & 0 & 0 \\ 0 & 0 & 0 & 0 & 1 & 0 & 0 & 0 & 0 \\ 0 & 0 & 0 & 0 & 0 & 1 & 0 & 0 & 0 \end{bmatrix} \tag{77}$$

$$L'_i = \frac{L_i + (I_{xz}/I_{xx})N_i}{1 - \{I_{xz}^2/(I_{xx}I_{zz})\}} \quad (i = p, q, \bar{r}) \tag{78}$$

$$N'_i = \frac{N_i + (I_{xz}/I_{zz})L_i}{1 - \{I_{xz}^2/(I_{xx}I_{zz})\}} \quad (i = p, q, \bar{r}) \tag{79}$$

To make the non-dimensional equation, let R represent the characteristic length. The non-dimensional time and velocity are:

$$\tau = \frac{10R\Omega}{g} \tag{80}$$

$$(u, v, w) = \frac{R}{\tau}(u^*, v^*, w^*) \tag{81}$$

Then, the equation of motion:

$$\dot{\boldsymbol{x}}^* = \boldsymbol{A}^*\boldsymbol{x}^* \tag{82}$$

$$\boldsymbol{x}^* = [u^* \quad v^* \quad w^* \quad p^* \quad q^* \quad \bar{r}^* \quad \Phi^* \quad \Theta^* \quad \Psi^*]^T \tag{83}$$

$$\boldsymbol{A}^* = \begin{bmatrix} X_u^* & 0 & 0 & X_p^* & X_q^* & 0 & 0 & -\frac{\tau^2 g}{R} & 0 \\ 0 & Y_v^* & 0 & Y_p^* & Y_q^* & 0 & -\frac{\tau^2 g}{R} & 0 & 0 \\ 0 & 0 & Z_w^* & 0 & 0 & 0 & 0 & 0 & 0 \\ 0 & {L'_v}^* & 0 & {L'_p}^* & {L'_q}^* & 0 & 0 & 0 & 0 \\ M_u^* & 0 & 0 & M_p^* & M_q^* & 0 & 0 & 0 & 0 \\ 0 & {N'_v}^* & 0 & {N'_p}^* & {N'_q}^* & 0 & 0 & 0 & 0 \\ 0 & 0 & 0 & 1 & 0 & 0 & 0 & 0 & 0 \\ 0 & 0 & 0 & 0 & 1 & 0 & 0 & 0 & 0 \\ 0 & 0 & 0 & 0 & 0 & 1 & 0 & 0 & 0 \end{bmatrix} \tag{84}$$

Analysis method

To make the effect of the scale on the hovering flight clear, a scale ratio k is defined. The analysis model under consideration has the blade length Rk, the chord length ck, the total mass mk^3, the moment of inertia Ik^5, the distance between z-axis and the rotational axis of each rotor $R_A k$ and the distance between the center of gravity position and the rotational plane of rotors hk. By using these parameters, the induced velocity and the angular velocity in hovering flight are expressed as $v_i\sqrt{k}$ and $\Omega/\sqrt{k}$, respectively. Then, the effect of the scale ratio on the hovering flight is investigated by calculating stability derivatives and eigenvalues.

To investigate the effect of the center of gravity position on a hovering flight, stability derivatives and eigenvalues are calculated on the condition that the distance between the center of gravity position and the rotational plane of rotors h is changed. $h > 0$ corresponds to that the center of gravity is lower than the rotational plane of rotors, and $h < 0$ corresponds to that it is higher than the rotational plane of rotors.

RESULT & DISCUSSION

As shown in Fig.4, when the scale ratio gets larger, the stability derivatives X_p , X_q , Y_p and Y_q increase while the others decrease. The stability derivatives X_p , X_q, Y_p and Y_q are proportional to $\sqrt{k}$, the stability derivatives L'_v , M_u and N'_v are proportional to $1/\sqrt{k^3}$ and others are proportional to $1/\sqrt{k}$. The stability derivatives N_* ($* = v, p, q$) are of the order of $10^{-8} \sim 10^{-6}$, which are much smaller than the others. So, yawing motion is not caused by a disturbance, v, p, q. This is because the two rotors rotate clockwise and the other two rotors do counterclockwise. And the yawing moment due to a disturbance cancel between the four rotors.

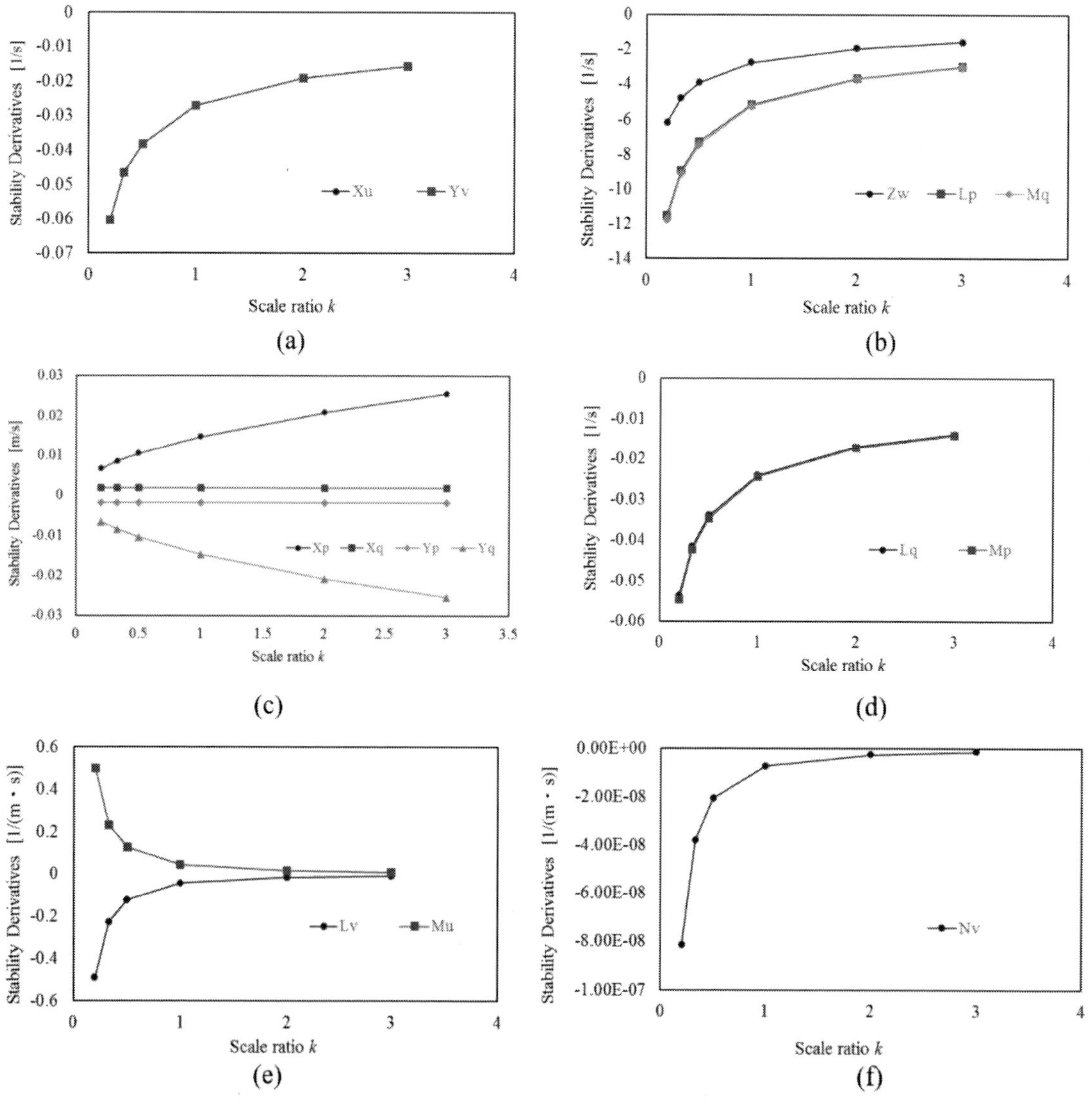

Fig.4 The stability derivatives in the scale analysis (1/2)

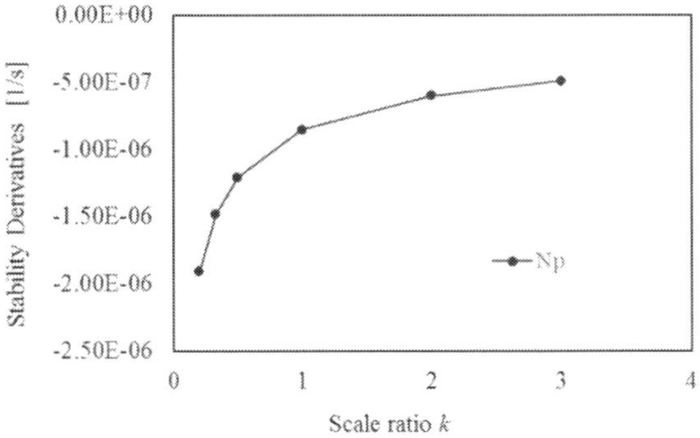

(g)

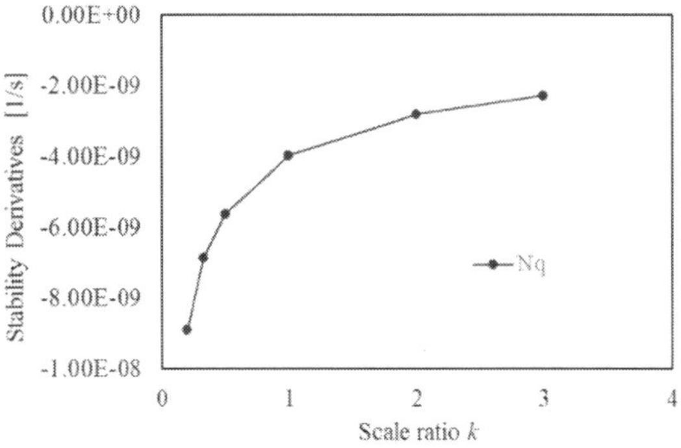

(h)

Fig.4 The stability derivatives in the scale analysis (2/2)

When the scale ratio $k = 1$, the eigenvalues of the equation (78) are -5.27, -5.16, -2.76, 0 (multiple root), $-0.00545 \pm 0.288i$, $-0.00543 \pm 0.289i$. The motion of the multirotor UAV is stable because all the eigenvalues, in the complex plane.

Fig.5 show the magnitudes of components of each eigenvector. They are the ratio of each component to the component u. When the component u is 0, other suitable component is 0 instead of u. In Fig.5 (a), the components u and q are much higher than the others. So, the eigenvalue of -5.27 corresponds to a pitching and translational motion. When other eigenvectors are considered in the same manner, the eigenvalue of -5.16 corresponds to a rolling and a translational motion. The eigenvalue of 0 corresponds to a yawing motion. The yawing motion must be controlled by changing the rotors' angular velocity. The eigenvalue of -2.76 corresponds to a vertical translational motion and the eigenvalue of 0 corresponds to a yawing motion. The eigenvalues of $-0.00545 \pm 0.288i$ have negative real numbers and their eigenvector component w are 0. Therefore, these eigenvalues correspond to oscillational motions in x-y plane. Tab.4 shows the phases of the eigenvalues of $-0.00545 \pm 0.288i$ and $-0.00543 \pm 0.289i$. About these eigenvalues, the phase of the component u are close to that of the component q while the phase difference between the v and p is close to

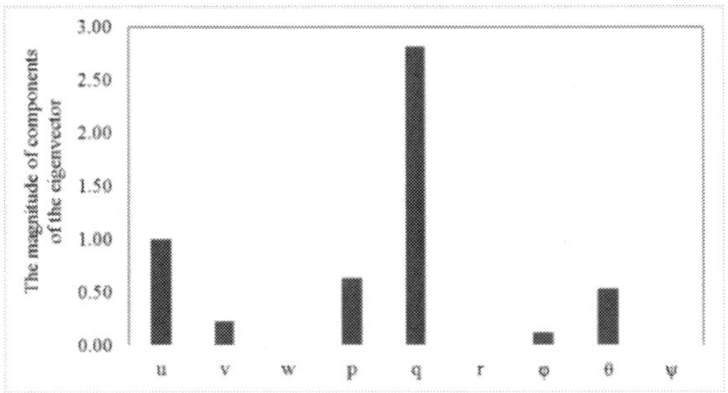

$\sigma = -5.27$

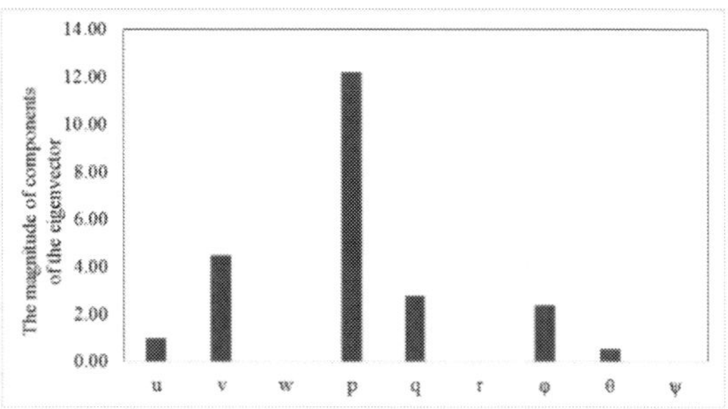

$\sigma = -5.16$

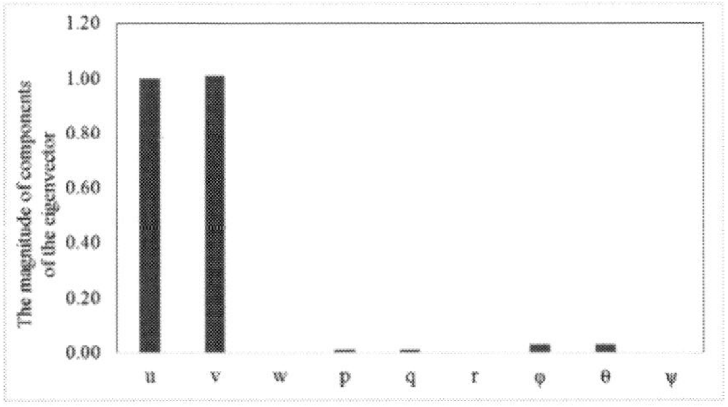

$\sigma = -0.00545 \pm 0.288i$

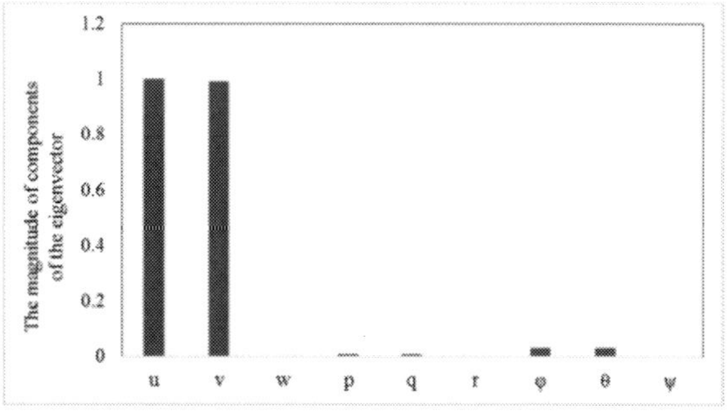

$\sigma = -0.00543 \pm 0.289i$

Fig.5 The magnitude of components of each eigenvector (1/2).

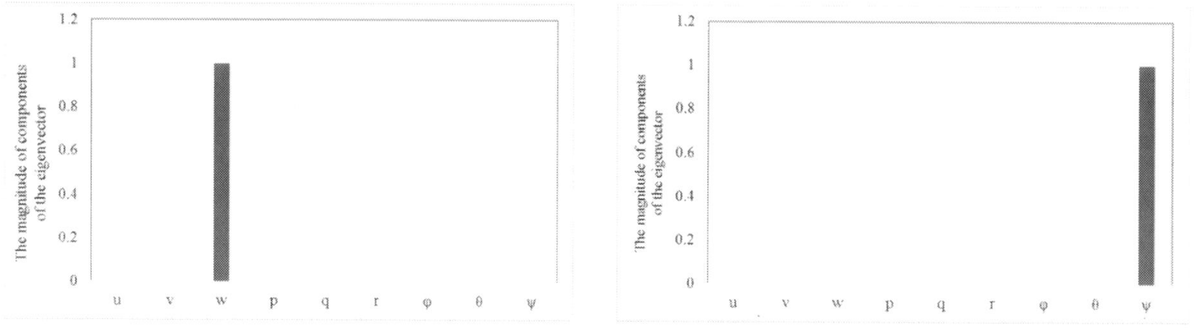

$\sigma = -2.76$ $\sigma = 0$

Fig.5 The magnitude of components of each eigenvector (2/2)

Tab.3 The phases of each component of eigenvectors of $\sigma = -0.00545 \pm 0.288i$ and $-0.00543 \pm 0.289i$

eigenvalue	$-0.00545+0.288i$	$-0.00545-0.288i$	$-0.00543+0.289i$	$-0.00543-0.289i$
component	Phase [deg]	Phase [deg]	Phase [deg]	Phase [deg]
u	-6.70	6.70	0.00	0.00
v	180	180	-6.69	6.69
w	0.00	0.00	0.00	0.00
p	-3.00	3.00	170	-170
q	-9.80	9.80	-3.19	3.19
$\bar{r}$	-3.00	3.00	170	-170
Φ	-94.2	94.2	79.0	-79.0
Θ	-101	101	-94.2	94.2
Ψ	-94.2	94.2	79.0	-79.0

180deg. These eigenvalues correspond to a swing motion shown in Fig.7. Therefore, the multirotor UAV in hovering flight has five kinds of motions: pitching, rolling, yawing, vertical translational and swing motions.

When the cases for scale ratios k are considered, the eigenvalues changed but the components of each eigenvector do not change. Furthermore, as shown in Fig.6, the scale of all the eigenvalues decrease while scale ratio k increases. When the size of the multirotor UAV is larger, more time is required for the attenuation of disturbed motions shown in Figs. 6(a) and (b). And the period of the oscillation shown in Fig. 6(b) is larger

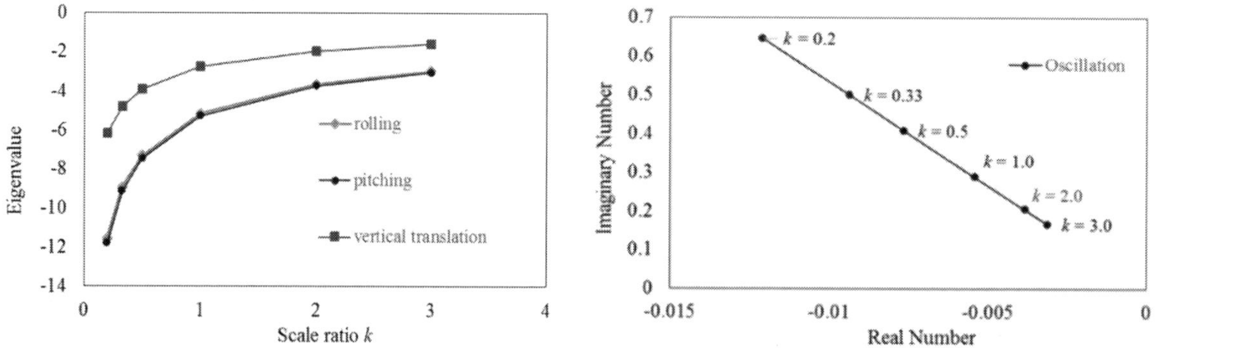

(a) (b)

Fig.6 The eigenvalues in the scale analysis

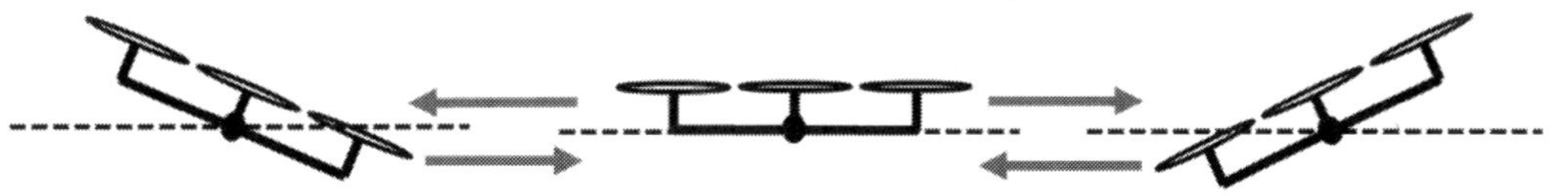

Fig.7 The swing motion

Fig.8 shows the curve $h-$ each of the stability derivatives on the condition of scale ratio $k = 1$. The stability derivatives X_q, Y_p, L_v, L_q, M_u and M_p are linear functions of h. The L_p, M_q, N_p are quadratic functions of h. The others are independent of h

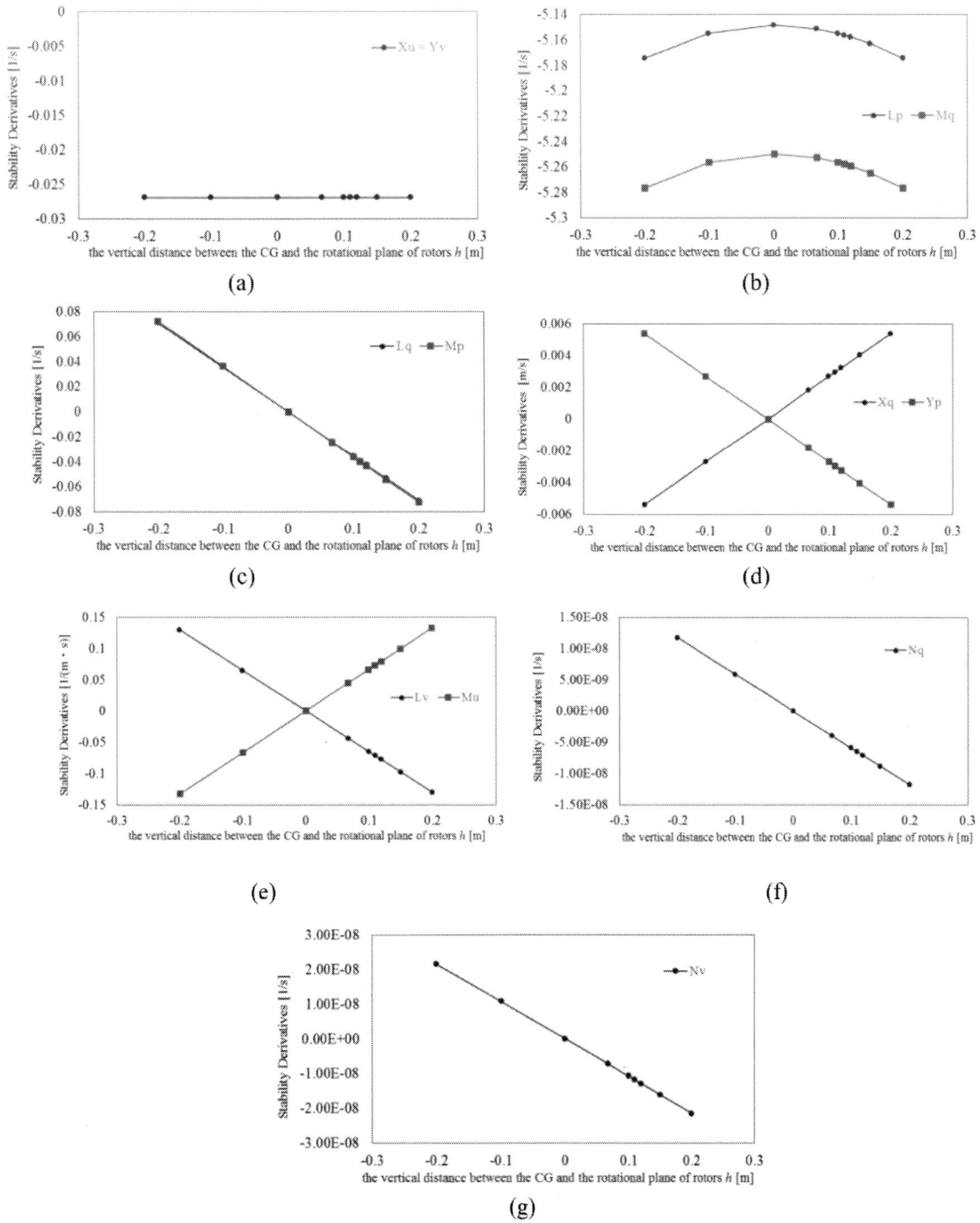

Fig.8 The stability derivatives of the center of gravity position analysis

In Fig.9, the eigenvalue of pithing motion is a convex function of h and that of rolling motion is a downward-convex function of h. These differences are caused by the asymmetric arrangement of rotors around x-axis and y-axis.

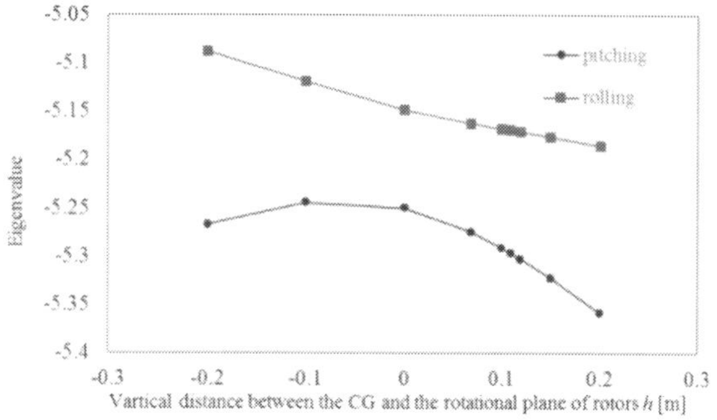

(a)

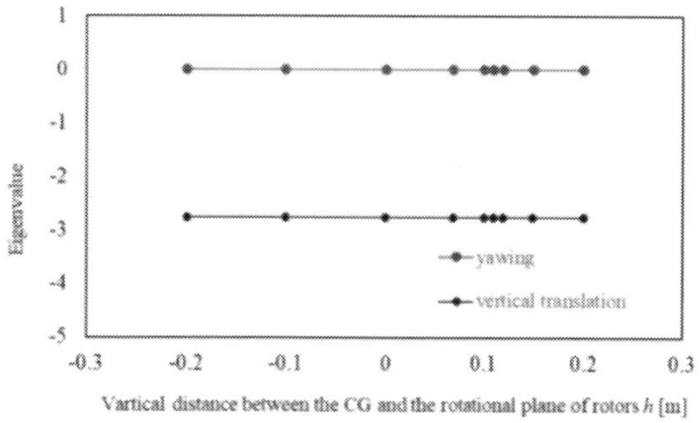

(b)

Fig.9 The eigenvalues of pitching, rolling, yawing and vertical translational motions

Fig.10 shows the eigenvalues of each swing motions in the complex plane. They become real numbers when $h < 0$. The eigenvalue of one swing motion on the condition of $h < 0$ is negative while that of the other swing motion is positive. Moreover, when h is larger than about 0.11, the real part of them becomes positive. Therefore, the motion of the multirotor UAV is unstable when the center of gravity position is 1) higher or 2) much lower than the rotational plane of the rotors. In a combined motion of a translational motion and a pitching motion around hovering of a single rotor helicopter[1], the h should be positive and small for making the flight stable. The result about of the h in the multirotor UAV, which was stated above, is similar to this fact about a single rotor helicopter.

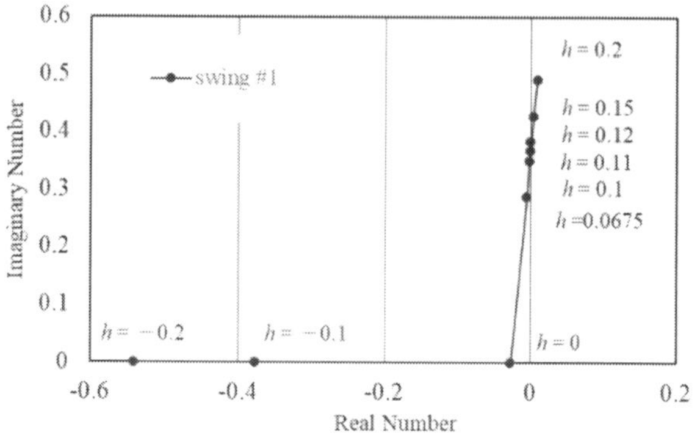

(a)

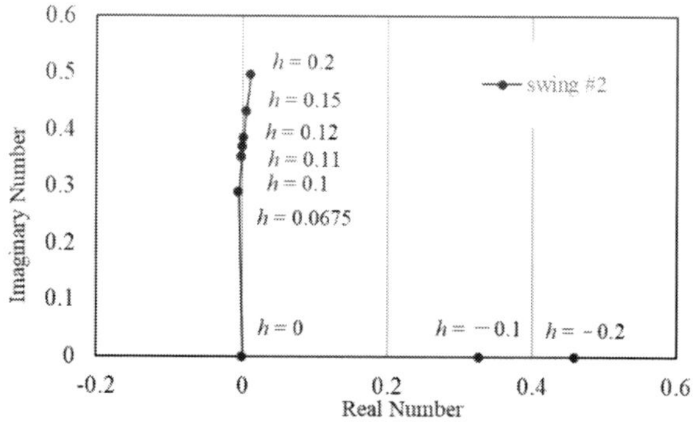

(b)

Fig10. The eigenvalues of swing motions

CONCLUSIONS

This study investigated the effects of scale and the center of gravity position of the multirotor UAV, and the following results were obtained.

- The multirotor UAV with six rotors has six kinds of motion: pitching, rolling, yawing, vertical translational and swing motions.
- The magnitude of eigenvalues of these motions except for yawing decrease when the scale ratio k increases. All the motions of the multirotor UAV except for the yawing motion are inherently stable because their real parts are negative. However, the yawing motion must be controlled because its eigenvalue is 0.
- When the center of position of the multirotor UAV is higher or much lower than its rotational plane of rotors, its motion becomes unstable inherently.

ACKNOWLEDGMENTS

This research is supported by ImPACT (Impulsing Paradigm Change through Disruptive Technologies program) of Council for Science, Technology and Innovation, Cabinet Office, Government of Japan.

REFERENCES

[1]Kato, K. and Imanaga,I., Introduction of helicopter, Tokyo University Press, Tokyo, 1985 (in Japasnese).

Author contact: Shuichi Umezaki sxb03015@edu.osakafu-u.ac.jp

Structural Optimization of Large-scale Wind Turbine Blade Cross-sections via an Analytical Beam Approach

Jae Seong Bae
Graduate Research Assistant
Konkuk University
Seoul, Republic of Korea

Manoj Kumar Dhadwal
Postdoctoral Researcher
Konkuk University
Seoul, Republic of Korea

Sung Nam Jung
Professor
Konkuk University
Seoul, Republic of Korea

ABSTRACT

In this work, the structural optimization of cross-sections of a large-scale wind turbine blade is performed using an analytic thin-walled beam theory with the support of genetic algorithm (GA). The Reissner's mixed variational theorem is applied to derive 7×7 stiffness constants that characterize the beam in Timoshenko level for bending and Vlasov level for torsion. Recovery of beam stresses along with cross-sectional offsets are incorporated for the optimum design of cross-sections. The present beam formulation is validated against other existing beam approaches and detailed finite element (FE) analysis results. These include cross-sectional stiffness constants, rotating natural frequencies, and sectional shear components for thin-walled composite beams. A reasonable correlation is obtained for the cases considered. A total of 7 wind turbine cross-sections are considered in the optimization to reduce structural weight while keeping structural strength within the allowable limit and varying thicknesses over the cross-sections. With the optimization framework, all cross-sections show reductions in weights from 5% to 20% based from the reference configurations while meeting constraint conditions.

INTRODUCTION

Wind energy is expected to become one of the important sources of renewable energy to substitute fossil fuels and solutions to reduce emissions of green-house gases in near future. As an example, European Wind Energy Association (EWEA) (Ref. 1) expects that in 2020, it is expected that installed capacity for wind turbine exceeds more than 41% compared to the one in 2013. The capacity anticipated in 2020 grows more than 12.8% of the total energy demand in European Union.

To meet the increasing demands for wind energy, it is required not only to increase the number of wind farms but also to enlarge the size of individual wind turbine blades. Recently, the wind turbine rotor blades are designed at preliminary design stage through the upscaling process (Refs. 2–4), which enlarges the radius of rotor blades by reflecting the historical trends of the pasts. Upscaling affects the important design parameters, such as mass, power, and length of wind turbine blades.

Increasing the length of wind turbine blades through upscaling process is challenging on blade mass, which is proportional approximately to the cube of blade radius based on historical trends (Refs. 2, 5). Considering the future prospect on wind energy, which requires bigger, and more number of wind turbine rotors with less cost of energy (COE), it is necessary to make efficient design methodology for wind turbine blades. There have been studies on optimizing rotor blades which can be applicable on helicopter and wind turbine rotor blades (Refs. 6–14). Optimization of wind turbine rotors can be categorized into global and local levels (Refs. 10, 12). In Global level optimization, targets are set to find target stiffness, performance, vibration reduction, and stability. In the local level optimization, the internal structures of blades are changed to meet the target stiffness found by global level optimization, and to find structurally robust designs to failure, fatigue life, and so on.

In local level optimization, the two-dimensional (2D) and the three-dimensional (3D) FE model, the cross-sectional analysis, or the both methods are used to design the blade internal structures. Although 2D and 3D FE models (Refs. 11, 14) can give the detailed stress levels of the blade, the computational costs are heavy using this method in the optimization process. The cross-sectional analyis (Refs. 6–10) decomposes the 3D beam structure into the 2D cross-sectional problem and the one-dimensional (1D) problem. This method only extracts the cross-sectional properties that is used in the 1D beam analysis. Therefore, this method takes less computation time than the 2D or 3D FE analysis, which is the better option for the optimization. The wind turbine blade is a kind of thin-walled beams. Therfore, the 1D contour-based analytical thin-walled beam theories (Refs. 15, 16) are preferred to the 2D FE based cross-sectional analysis methods (Refs. 17, 18) because the cross-sectional geometry has to be remeshed at every iteration, which requires the additional schemes, such as the paremeterizations to model the cross-sections (Refs. 6, 9, 12).

Presented at the 6th Asian/Australian Rotorcraft Forum / Heli Japan 2017, Kanazawa, Japan, November 7-9, 2017.

The present study adopts an analytical beam approach to optimize large scale wind turbine blade cross-sections. The thin-walled beam theory developed by Jung and Park (Ref. 19) is extended to compute the 7×7 cross-sectional stiffness constants that take into account the transverse shear deformation and contrained warping effect associated with torsion. The recovery of sectional stress components and centroidal offset properties are newly incorporated in the formulation. The maximum strain criterion is combined with the beam analysis to compose the failure analysis. The cross-sectional analysis method is integrated with MATLAB GA optimizer to minimize the blade weights over seven different cross-sections chosen arbitrarily for a reference wind turbine blade.

CROSS-SECTIONAL ANALYSIS

Beam Kinematics

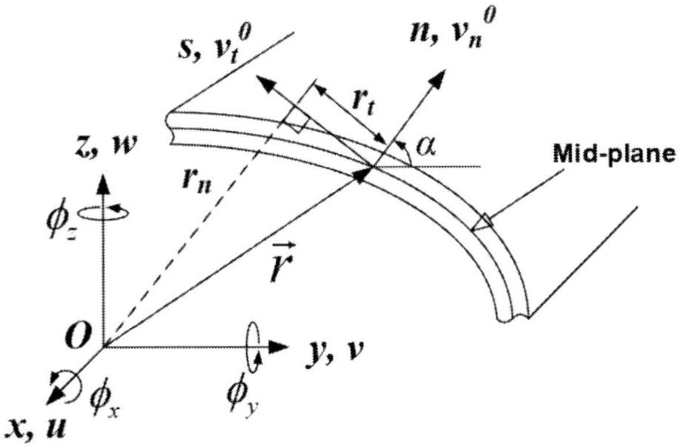

Figure 1. Coordinate systems on beam cross-section

Figure 1 shows the coordinate systems used for thin-walled beam cross-section. The x, y and z are coordinates of global reference frame which are used to describe beam's three-dimensional motion. The n, s, and x forms local shell coordinate system. The orientation angle, α is used to measure the angle between the local and global reference coordinate systems.

Following the work of Jung and Park (Ref. 19), the mid-plane strains are expressed as

$$\varepsilon_{xx} = u^0_{,x} = U_{,x} + z\beta_{y,x} + y\beta_{z,x} - \bar{\omega}\phi_{,xx} \quad (1)$$

$$\gamma_{xs} = u^0_{,s} + v^0_{t,x} = u^0_{,s} + V_{,x}y_{,s} + W_{,x}z_{,s} + r_n\phi_{,x} \quad (2)$$

where ε_{xx} and γ_{xs} are the normal and shear strains at the mid-plane location, respectively. U, V, and W are the displacements of the cross-sectional reference origin to the axial, chordwise, and flapwise directions. ϕ, β_y, and β_z are the rotation angles due to twist, flapwise and chordwise bending respectively. $\bar{\omega}$ is the torsional warping function and r_n is the vertical distance to the arbitrary midplane material point. $()_{,s}$, $()_{,x}$ denote the derivatives with respect to s and x, respectively.

Constitutive Relations

Assuming negligable hoop stress, $N_{ss} = 0$, the constitutive relations of the shell wall can be expresses by

$$\begin{Bmatrix} N_{xx} \\ N_{xs} \end{Bmatrix} = \begin{bmatrix} A'_{11} & A'_{16} \\ A'_{16} & A'_{66} \end{bmatrix} \begin{Bmatrix} \varepsilon_{xx} \\ \gamma_{xs} \end{Bmatrix} \quad (3)$$

where

$$A'_{11} = A_{11} - \frac{A^2_{12}}{A_{22}},\ A'_{16} = A_{16} - \frac{A_{12}A_{26}}{A_{22}},\ A'_{66} = A_{66} - \frac{A^2_{26}}{A_{22}} \quad (4)$$

A_{ij} are the laminate stiffness constants in the classical lamination plate theory (Ref. 27). In this work, both displacements and shear stress are set as the unknows together, the constitutive relations is semi-inverted as

$$\begin{Bmatrix} N_{xx} \\ \gamma_{xs} \end{Bmatrix} = \begin{bmatrix} A_{n\varepsilon} & A_{n\gamma} \\ -A_{n\gamma} & A_{\gamma\gamma} \end{bmatrix} \begin{Bmatrix} \varepsilon_{xx} \\ N_{xs} \end{Bmatrix} \quad (5)$$

where

$$A_{n\varepsilon} = A'_{11} - \frac{{A'_{16}}^2}{A'_{66}}, \quad A_{n\gamma} = \frac{A'_{16}}{A'_{66}}, \quad A_{\gamma\gamma} = \frac{1}{A'_{66}} \quad (6)$$

Governing Equations

We can derive the variational statement of equilibrium equation of a beam slice by using Reissner's mixed variational theorem (Ref. 25) as follows.

$$\begin{aligned} \delta\Pi &= \delta U - \delta W \\ &= \int_C [N_{xx}\delta\varepsilon_{xx} + N_{xs}\delta(u^0_{,s} + v^0_{t,x}) + \delta N_{xs}(u^0_{,s} + v^0_{t,x} - \gamma_{xs})]ds \\ &- \int_A (\delta\mathbf{u}^{\mathrm{T}}\mathbf{t})' dA = 0 \end{aligned} \quad (7)$$

where $\mathbf{u}$ and $\mathbf{t}$ mean the displacement vector and surface tractions applied on the beam's cross-section, respectively. In Eq. (7), δN_{xs} acts as the Lagrange multiplier enforcing the compatibility of shear strains, which implies the shear strains derived from Eqs. (2) and (5) are compatible with each other. Using Eqs. (1), (2), (5) and (7), we can derive the equilibrium equations at the membrane level

$$\frac{\partial N_{xx}}{\partial x} + \frac{\partial N_{xs}}{\partial s} = 0 \quad (8)$$

$$\frac{\partial N_{xs}}{\partial x} = 0 \quad (9)$$

along with the constraint equations as

$$\gamma_{xs} - u^0_{,s} - v^0_{t,x} = 0 \quad (10)$$

From Eqs. (8), (9), and (5), we can compute N_{xs} as follows.

$$N_{xs} = N_{xs_0} - \int_{s_0}^{s} N_{xx,x} ds = N_{xs_0} - \int_{s_0}^{s} A_{n\varepsilon}\varepsilon_{xx,x} ds \quad (11)$$

N_{xs} has the constant part, $N_{xs,0}$, and the part that varies along contour coordinate s. The constant term must be determined by using the strain compatibility condition, Eq. (10). Enforcing Eq. (10) to be satisfied over each cell comprising of the beam section can be stated in mathematical form, using Eqs. (1), (2) and (5). The above relation is written as

$$\oint A_{\gamma\gamma} N_{xs} ds = \oint A_{n\gamma} \varepsilon_{xx} ds + \oint r_n ds \phi \tag{12}$$

Inserting Eq. (11) into Eq. (12), and solving for N_{xs_0} gives

$$N_{xs_0} = \mathbf{f}\mathbf{q}_b + \mathbf{F}\mathbf{q}_{b,x} \tag{13}$$

where

$$\mathbf{q}_b = \lfloor \ \bar{u}' \quad \phi_{,x} \quad \beta_y' \quad \beta_z' \quad \phi_{,xx} \ \rfloor^{\mathrm{T}} \tag{14}$$

Inserting Eq. (13) into Eq. (11) gives

$$N_{xs} = \mathbf{f}\mathbf{q}_b + \mathbf{F}\mathbf{q}_{b,x} - \mathbf{S}\mathbf{q}_{b,x}' \tag{15}$$

where

$$\mathbf{S} = \int_{s_0}^{s} A_{n\varepsilon} \lfloor \ 1 \quad 0 \quad z \quad y \quad -\bar{\omega} \ \rfloor ds \tag{16}$$

Inserting $\mathbf{f}\mathbf{q}_b$ of Eq. (15) and Eq. (1) into Eq. (7), gives the force-displacement relations as follows.

$$\mathbf{F}_b = \mathbf{K}_b \mathbf{q}_b \tag{17}$$

where

$$\mathbf{F}_b = \lfloor \ F_x \quad M_x \quad M_y \quad M_z \quad M_\omega \ \rfloor^{\mathrm{T}} \tag{18}$$

Where, $\mathbf{K}_b$ is the stiffnes matrix which has size of 5×5 that considers beam's behavior at Euler-Bernoulli level for bending and Vlasov level for torsion considering torsional warping restraint effect with elastic couplings.

Timoshenko-Vlasov Level Stiffness Matrix

We can find the relation of the first derivative of beam's generalized strain measures to beam's sectional load by differentiating it as follows.

$$\mathbf{q}_{b,x} = \mathbf{K}_b^{-1} \mathbf{F}_{b,x} = \mathbf{K}_b^{-1} \mathbf{F}_s \tag{19}$$

where

$$\mathbf{F}_s = \lfloor \ 0 \quad 0 \quad F_z \quad F_y \quad 0 \ \rfloor^{\mathrm{T}} \tag{20}$$

With Eqs. (1), (2), (5), (15), and (20), the force-displacement relations considering the transverse shear deformations as follows.

$$\mathbf{F}_t = \mathbf{K}_t \mathbf{q}_t \tag{21}$$

where

$$\mathbf{F}_t = \lfloor \ F_x \quad F_y \quad F_z \quad M_x \quad M_y \quad M_z \quad M_\omega \ \rfloor^{\mathrm{T}} \tag{22}$$

Recovery of stresses and strains

The recovery of stresses is used to check material failure at layerwise level for blade cross-section. The beam's generalized strain measures are computed by inverting cross-sectional stiffness matrix of Eq. (21) as

$$\mathbf{q}_t = \mathbf{K}_t^{-1} \mathbf{F}_t \tag{23}$$

It is advantageous to make use of N_{xs} to compute stress and strain considering material couplings. To use this, F_y and F_z is computed from $\mathbf{q}_t$ by

$$\begin{Bmatrix} F_y \\ F_z \end{Bmatrix} = \mathbf{K}_y \mathbf{q}_t \tag{24}$$

where $\mathbf{K}_y$ is the submatrix with only the transverse shear components of the stiffness matrix $\mathbf{K}_t$. Using Eqs. (14), (24), (3) and (5), and assuming $N_{ss} = 0$, ε_{xx}, ε_{ss} and γ_{xs} are obtained. These strains are transformed into the material coordinate system using the fiber orientation for each layer.

Failure Analysis

In this study, the maximum strain criterion (Ref. 27) is applied to determine the failure of each material layer as did in Bak et al. (Ref. 2). An arbitrary material layer is considered to be safe if it satisfies the following conditions:

$$X_C \leq \varepsilon_{11} \leq X_T \tag{25}$$

$$Y_C \leq \varepsilon_{22} \leq Y_T \tag{26}$$

$$|\gamma_{12}| \leq S \tag{27}$$

where, X_c and X_T are the maximum values of compressive and tensile strain in the fiber direction. Y_c and Y_T are the maximum values of compressive and tensile strains in the transverse direction, respectively and S is the shear strength of the fiber.

Mass Matrix

The mass matrix is computed from the external virtual work done by inertial forces and given by

$$\mathbf{M} = \int_C m \mathbf{B}_m^{\mathrm{T}} \mathbf{B}_m t ds \tag{28}$$

where

$$\mathbf{B}_m = \begin{bmatrix} 1 & 0 & 0 & 0 & z & -y \\ 0 & 1 & 0 & -z & 0 & 0 \\ 0 & 0 & 1 & y & 0 & 0 \end{bmatrix} \tag{29}$$

Where, m is the mass per unit length of laminate at each contour location.

Cross-Sectional Offsets

The centroidal offsets are defined in this section. The tension center is defined on a cross-section as the location where the extensional and bending deformations of the beam are decoupled. The coordinates of tension center are given by

$$y_{tc} = -\frac{K_{16}}{K_{11}} \tag{30}$$

$$z_{tc} = \frac{K_{15}}{K_{11}} \tag{31}$$

The shear center is defined as the point where bending due to shear loads occurs without inducing cross-sectional twist. The locations of shear center are given by

$$y_{sc} = \int_C (N_{xs})_z r_n ds \tag{32}$$

$$z_{sc} = -\int_C (N_{xs})_y r_n ds \tag{33}$$

where $(N_{xs})_z$ is the shear flow along the contour coordinates under the case when F_z is applied and $(N_{xs})_y$ is the shear flow distribution when F_y is applied to the cross-section. The principal bending axes are located such that flap and chordwise bending deformations are decoupled to each other at the reference origin. The orientation angle for the axes is given by

$$\theta_B = \frac{1}{2}\tan^{-1}\left(\frac{2K_{56}}{K_{55}-K_{66}}\right) \tag{34}$$

The mass center is defined in similar manner as tension center by

$$y_{mc} = -\frac{M_{16}}{M_{11}} \tag{35}$$

$$z_{mc} = \frac{M_{15}}{M_{11}} \tag{36}$$

OPTIMIZATION FRAMEWORK

Baseline Configuration

For optimization, a 10 MW reference wind turbine blade presented by Technical University of Denmark (DTU) (Ref. 2) was used. It was made as a reference for developing better designs for large scale wind turbines that can be made possible in the future. Comprehensive data on aerodynamic, structural, and aeroelastic analysis are reported in the document (Ref. 2).

For present analysis, 7 cross-sections among all the cross-sections present in the report were chosen for the cross-sectional strength analysis. In the report, 8 loadcases are presented for every cross-sections. Among these cases, single loadcase which shows maximum magnitude of flapwise bending moment was selected for evaluating strength of each cross-section. The table 1 shows radial location of each cross-section chosen for this study. Each cross-section is constructed

Table 1. Radial location of cross-sections (Ref. 2)

Section no.	r (m)	Section no.	r (m)
1	24.3913	5	68.6847
2	43.0459	6	75.0233
3	49.4376	7	81.8249
4	62.2488		

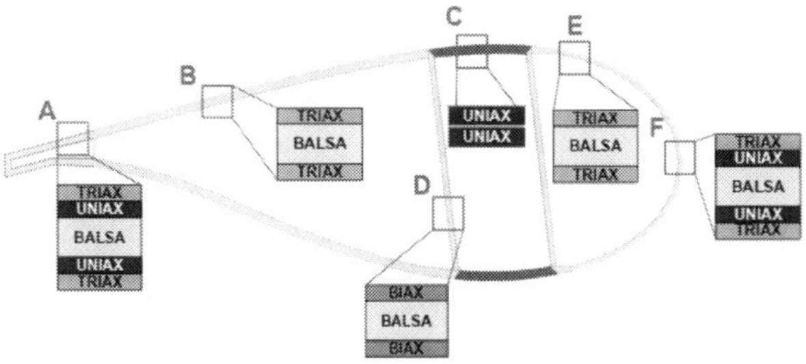

Figure 2. Materials distribution at section no. 2

with uniaxial, biaxial, and triaxial composite laminates with balsa wood substituting honeycomb sandwich core to resist buckling (Ref. 2). An example cross-section for showing how these materials are distributed is depicted in Figure 2.

Optimization Problem

As mentioned previously, mass of the rotor tends to increase proportionally to approximately the cube of the rotor radius based on the historic trends. The increase in blade mass affects the material costs thereby increasing COE. Due to this, many preceding studies focused on reducing mass and increasing annual energy production (AEP). Since AEP requires aerodynamic analysis, and the focus of present structural optimization is limited to the minimization of mass from the baseline configurations.

It is necessary to keep structural strength while minimizing the mass during optimization. Therefore, the constraint on cross-sectional strength was set as the maximum strain value does not exceed the value of the baseline cross-section. For maintaining the stability of rotor, based on the studies of Li et al. (Ref. 12), cross-sectional stiffness for extensional, torsional, and bending stiffnesses were made to change so that the difference between the baseline and optimum cross-sections remain within 5%. Additional constraints were set on the shear and mass center offsets to be greater than or equal to the baseline values. These can be written in mathematical form as follows.

$$\begin{aligned} &\text{Minimize:} && f(\mathbf{x}) = \frac{m}{m_0} \\ &\text{subject to:} && 0.95 \le \left(\frac{EA}{EA_0}, \frac{EI_y}{EI_{y0}}, \frac{EI_z}{EI_{z0}}, \frac{GJ}{GJ_0}\right) \le 1.05, \\ & && y_{sc} \ge y_{sc.0}, \quad y_{mc} \ge y_{mc.0}, \quad R \le R_0 \end{aligned} \tag{37}$$

where $\mathbf{x}$ is the vector containing the design variables, m is the mass per unit length of the cross-section, EA, EI_y, EI_z, GJ

are the extensional, flapwise and chordwise bending, and torsional stiffnesses respectively evaluated at the tension center with coordinate axes oriented along the principal bending axes. y_{sc} and y_{mc} are the coordinates of the shear center and mass center of the cross-section in chordwise direction respectively. R denotes the maximum value of the strain ratio which is defined as the applied strain divided by the maximum strain. For the normal strains, the strain ratio is computed by the applied normal strain divided by the maximum compressive or tensile strain depending on the sign of the value. For the shear strain, it is simply calculated as the ratio of the applied shear strain to the maximum shear strain S in Eq. (27). The variables with the subscript 0 indicate the baseline values. The design variable vector $\mathbf{x}$ contains the variables related to the internal geometry which include the material thickness and the locations of the shear webs of the cross-section. The thickness of balsa wood was varied from 50% to 100% of that of the baseline. The thickness of the other materials, such as uniaxial, biaxial and triaxial plies, are the integer multiples of 0.1 mm with maximum values to be same as that of the baseline. Since the cross-sections are discretized using 1D line segments, after the discretization is complete, points are generated over the segments where each point is denoted by a number. The locations of the two shear webs are determined selecting points located at both pressure and suction parts of the section (see Fig. 3).

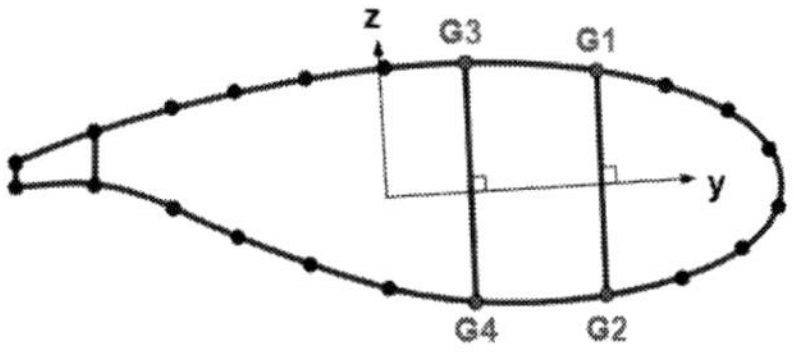

Figure 3. Schematic of design variables for changing shear web locations

Optimization Procedure

The optimization procedure is depicted in Figure 4. Genetic algorithm (Ref. 26) is used in the present problem since it is able to find the global optimum without requiring the gradient of the objective function during the optimization process. The optimization of each cross-section starts with the initialization of the cross-sectional geometry. The internal geometry of each cross-section is changed by the design variables supplied by the optimizer. An input file is generated based on the information of the newly generated cross-section. The cross-section analysis program then computes the cross-sectional properties and performs the stress recovery and failure analysis based on the cross-sectional loads. These results are used to check for the constraint violations. If the constraints are satisfied, the objective function is evaluated. If the change in the objective function values for all solutions in a single generation is less than the termination condition specified, the process is terminated, else the optimization loop is continued.

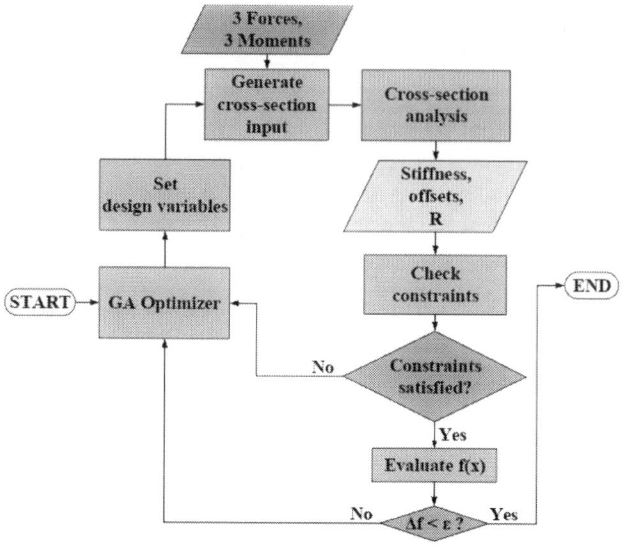

Figure 4. Cross-sectional optimization procedure

NUMERICAL RESULTS

Vaildation of Beam Sectional Analysis

A composite box beam cross-section (Ref. 28) is used to validate the cross-sectional analysis. The geometry and layup of the cross-section are shown in Figure 5. The material properties are listed in Table 2. This cross-section has the circumferentially uniform stiffness (CUS) layup, which means the same stacking sequence is applied around the cross-sectional walls. Therefore, when looked in globally, top-bottom, and left-right walls show anti-symmetry with respect to the horizontal and vertical lines through the center of the cross-section. This induces the coupled behavior of extension-shear and bending-torsion in gross-behavior of the beam.

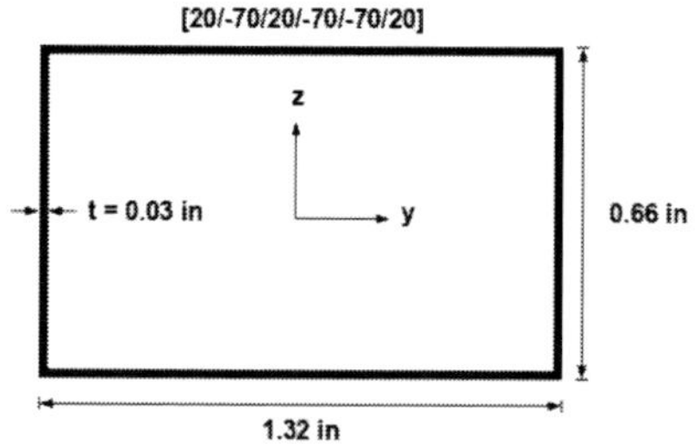

Figure 5. Benchmark cross-section for present cross-sectional analysis

Table 2. Material properties of T300/5208 (Ref. 29)

Parameter	Value	Unit
E_{11}	21.3×10^6	psi
E_{22}	1.6×10^6	psi
ν_{12}	0.28	-
G_{12}	0.9×10^6	psi

Table 3 compares the cross-sectional constants computed from NABSA, which is a cross-sectional analysis tool based on Giavotto et al. (Ref. 17), Hodges et al. (Ref. 28), and the present analysis. NABSA is a 2D FE-based cross-sectional analysis tool considering all the cross-sectional warpings. The results from Hodges et al. are based on the theory of Rehfield et al. (Ref. 21), which is a thin-walled beam theory that computes cross-sectional constants from the displacement-based formulation. Present theory sets the beam's displacements and transverse shear stress as the unknowns together to compute the accurate values of the shear-related cross-sectional constants which is a different feature from the theory of Rehfied et al. (Ref. 21). From Table 3, the results from Hodges et al. show the larger errors in the direct chordwise and flapwise shear stiffness, K_{22} and K_{33} than the present analysis. This is because the direct shear stiffnesses are affected by the shear warping distribution (Ref. 28). On the other hand, the present analysis, which also does not consider the shear warping displacements as in Rehfield et al. (Ref. 28), shows good agreement through all the cross-sectional constants with NABSA, having the maximum difference of 6.3% at the chordwise shear stiffness. This is because the present formulation directly finds exact distribution of shear flow to compute shear-related direct stiffness components. The another benchmark example is used to validate

Table 3. Comparison of stiffness constants of box beam no.2

Stiffness	NABSA (Ref. 28)	TAIL (Ref. 28) (%)*	Present (%)*
$K_{11} \times 10^6$	1.137	1.140 (0.3)	1.139 (0.2)
$K_{22} \times 10^5$	1.741	1.930 (10.9)	1.770 (1.7)
$K_{33} \times 10^4$	6.645	9.404 (41.5)	6.225 (-6.3)
$K_{14} \times 10^5$	-1.522	-1.514 (-0.5)	-1.514 (-0.5)
$K_{25} \times 10^4$	7.370	7.572 (2.7)	7.615 (3.3)
$K_{36} \times 10^4$	8.025	7.572 (-5.6)	7.964 (-0.8)
$K_{44} \times 10^4$	5.241	5.102 (-2.7)	5.102 (-2.6)
$K_{55} \times 10^4$	8.564	8.754 (2.2)	8.576 (0.1)
$K_{66} \times 10^5$	2.597	2.603 (0.2)	2.595 (-0.1)

* Percentage difference with respect to NABSA defined as (method - NABSA)/NABSA × 100.

the predictions of the cross-sectional and inertial properties of the present formulation through the free-vibration analysis of a rotating composite beam (Ref. 15). The cross-sectional geometry and material layup are shown in Figure 6 and the material properties applied to the cross-section are listed in Table 4. Same equation of motion as used in Jung et al. (Ref. 30), is applied in this study to perform the analysis.

Table 5 shows free-vibration analysis results of present formulation compared with Chandra and Chopra (Ref. 31). The present analysis results are close to both the experiment and analyis results of Chandra and Chopra (Ref. 31) with the maximum difference to the experimental value less than 5%. The composite box beam used for free vibration analysis

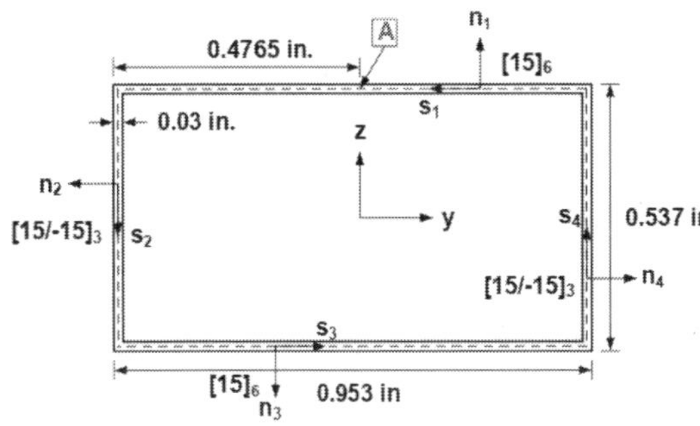

Figure 6. Benchmark cross-section for free-vibration analysis

Table 4. Material properties of AS4/3501-6 (Ref. 31)

Parameter	Value	Unit
E_{11}	20.59×10^6	psi
E_{22}	1.42×10^6	psi
ν_{12}	0.42	-
G_{12}	0.89×10^6	psi
ρ	0.1352×10^{-3}	lbs²/in.⁴

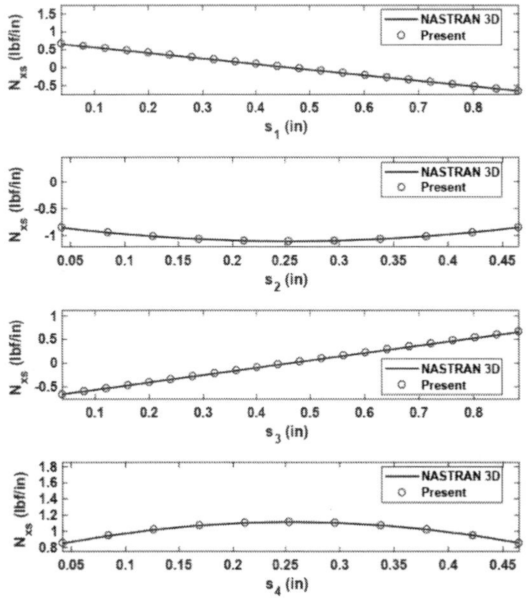

Figure 7. Comparison of shear flow (N_{xs}) distribution of the composite box beam

(Figure 6) is again used to validate the stress computation. A cantilevered beam having the length of 30 in. is used. An unit shear force along z direction is applied at the tip and the stress values at around the half-span (x=14.98in) are extracted. To validate the stresses in the composite beam, a NASTRAN 3D FE model is used. Total of 333,000 linear, 8-noded hexagonal solid elements (CHEXA) are used to model the beam. For the comparison, the axial stress at the center of top flange (point A in Figure 6) and shear flow values along the midplane contour coordinates with respect to the global beam coordinate system are used. For σ_{xx} values, the NASTRAN 3D model and present analysis show -790.24 psi and -793.50 psi respectively showing difference of 0.41%, which shows good correlation.

Table 5. Comparison of rotating frequencies at rotor speed of 1000 RPM

Mode	Chandra and Chopra (Ref. 31)		Present
	Exp.	Analysis	(%)*
1F	35.20	35.40	35.53 (0.94)
1L	53.80	56.00	56.37 (4.78)
2F	188.00	194.00	193.10 (2.71)

* Percentage difference with respect to Experiment defined as (method - Exp.)/Exp.× 100.

Figure 7 shows the shear flow values computed from both the NASTRAN 3D model and present cross-section analysis. The values at all the corners are excluded because the values at these corners showed sharp change in the 3D FE model. The results between the both cases show excellent agreement. This results are coming from the fact that the present formulation computes the exact distribution of the shear flow values along the contour coordinates.

Cross-sectional Optimization

Table 6 summarizes optimization process of all the cross-sections. It takes about 2 seconds in average to evaluate the objective function for all the sections. It is noted that this computation time includes the time taken by the cross-section analysis code from reading the inputs to writing the outputs at every iterations with the optimizer and the other functions necessary to run the GA internally. After the optimization, none of the cases violated the constraints. Figure 8 shows the baseline and optimum cross-sections for section no. 1 and 7, respectively. These figures also show the locations of cross-sectional offsets. These offsets have moved towards the leading edge of the sections because the uniaxial plies are occupying much wider regions in the spar caps and the first shear webs of both sections have moved close to the leading edge. Figure 9 shows the change in the mass with the mass ratio which is defined as the mass per unit length of the optimum devided by the basline. It is observed that at all the cross-sections, the mass were reduced. The maximum reduction in mass occurs at the case of section no. 1, which shows about 19% decrease in the mass. The minimum reduction in the mass is shown at section no. 5 which shows about 5% decrease in the mass. Therefore, all the cross-sections are reduced in the mass more than 5%. Figure 10 shows the change in the maximum value of the strain ratio of each cross-section. The 6 cross-sections show reductions in the maximum value of the strain ratios. Section no. 3 remained almost the same as the baseline cross-sections. Therefore, during the optimization, It is found that those baseline configurations can be improved to the lighter designs having the same or better resistance to the external loads.

CONCLUDING REMARKS

In this study, an analytical composite beam model has been developed to include transverse shear deformation and retraint torsion warping using Reissner's mixed variational theorem. The beam model computes the stiffness constans in a Timoshenko-Vlasov level. The model is validated for thin-walled composite box-section beams with elastic couplings. The maximum deviation in cross-sectional constants is about 5% in the direct chordwise shear stiffness as compared with other detailed finite element-based analysis results. Comparison of predicted rotating natural frequencies exhibits a good correlation in comparison with experimental data. As an application, the cross-sections of a large-scale wind turbine rotor blade are optimized with GA in reducing weights of each cross-section. Without violating constraints, all cross-sections reach reductions in weights by 5 to 20%. Based on the results, the present optimization framework can readily be applied to any structural optimization of wind turbine blades particularly in the preliminary design stage.

ACKNOWLEDGMENTS

This work was conducted at High-Speed Compound Unmanned Rotorcraft (HCUR) research laboratory with the support of Agency for Defense Development (ADD). This research was supported by the EDISON Program through the National Research Foundation of Korea (NRF) funded by the Ministry of Science, ICT & Future Planning (2014M3C1A6038856).

REFERENCES

[1]Moccia, J., Wilkes, J., Pineda, I., Corbetta, G., Casey, Z., Ros, C., and Gonzalez, A., "Wind Energy Scenarios for 2020," Technical report, European Wind Energy Association, 2014.

[2]Bak, C., Zahle, F., Bitsche, R., Kim, T., Yde, A., Henriksen, L., Andersen, P. B., Natarajan, A., and Hansen, M. H., "Design and Performance of a 10 MW Wind Turbine," *Wind Energy*, Vol. 124, 2013.

[3]Griffith, D. T. and Ashwill, T. D., "The Sandia 100-meter All-Glass Baseline Wind Turbine Blade: SNL100-00," Technical report, Sandia National Laboratories, Albuquerque, Report No. SAND2011-3779, 2011.

[4]Griffin, D. A., "Windpact Turbine Design Scaling Studies Technical Area 1-Composite Blades for 80-to 120-meter Rotor," Technical report, National Renewable Energy Lab., Golden, CO (US), 2001.

[5]Veers, P. S., Ashwill, T. D., Sutherland, H. J., Laird, D. L., Lobitz, D. W., Griffin, D. A., Mandell, J. F., Musial, W. D., Jackson, K., Zuteck, M., *et al.*, "Trends in the design, manufacture and evaluation of wind turbine blades," *Wind Energy*, Vol. 6, (3), 2003, pp. 245–259.

Table 6. Optimization summary

Section no.	1	2	3	4	5	6	7
Number of function evaluations	10050	7350	8500	7350	9750	8800	10050
Computation time (sec)	4199	3854	4619	3939	5211	4648	5492
Computation time per evaluation (sec)	2.4	1.9	1.8	1.9	1.9	1.9	1.8
Constraint violation	No	No	No	No	No	No	No

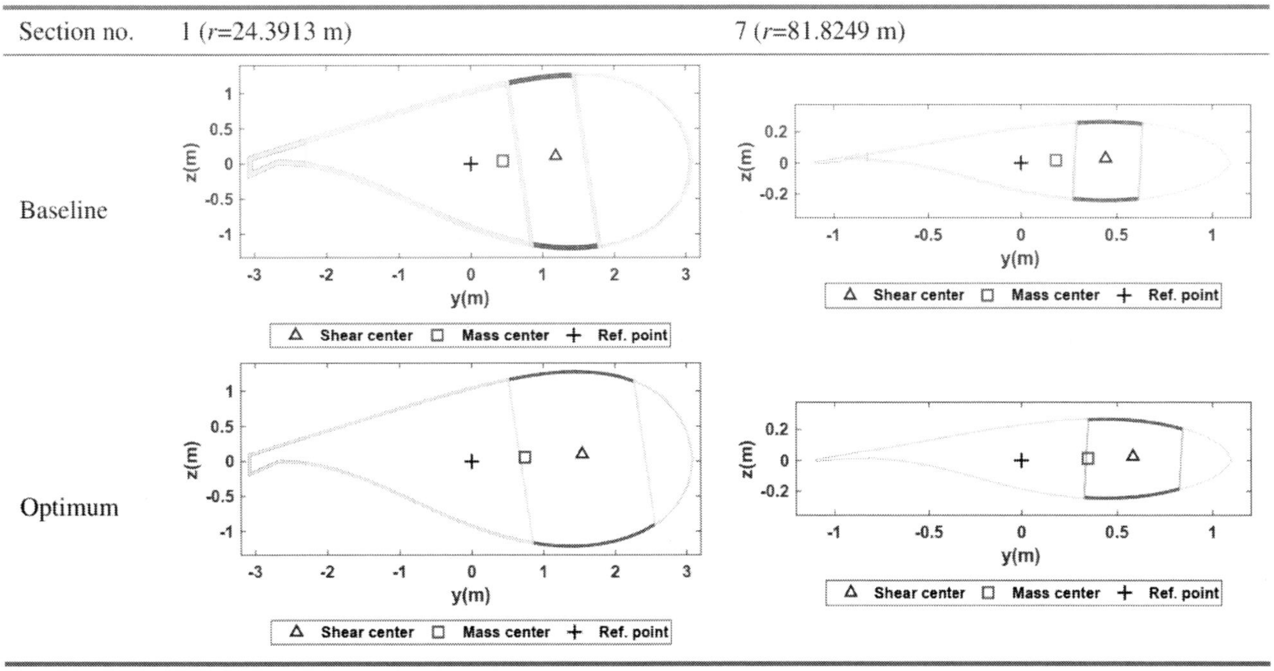

Figure 8. Baseline and optimum cross-sections

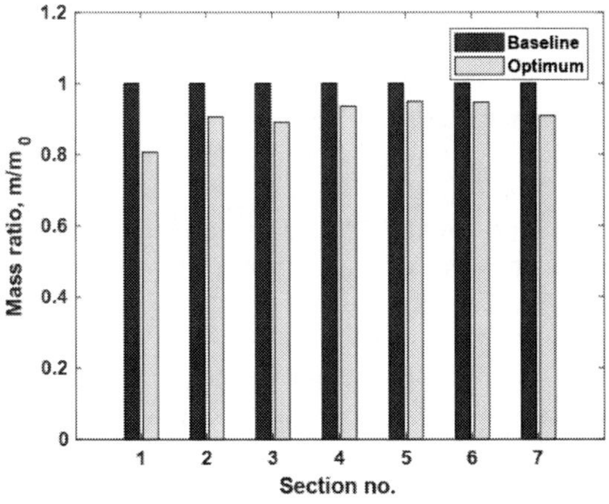

Figure 9. Comparison of mass per unit length between the baseline and optimum cross-sections

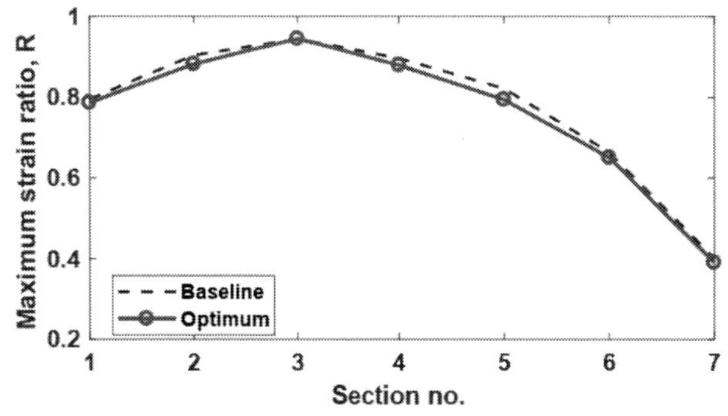

Figure 10. Comparison of maximum strain ratio between the baseline and optimum cross-sections

[6]Pavese, C., Tibaldi, C., Zahle, F., and Kim, T., "Aeroelastic multidisciplinary design optimization of a swept wind turbine blade," *Wind Energy*, we.2131, pp. 1–13. doi: 10.1002/we.2131

[7]Richards, P. W., Griffith, D. T., and Hodges, D. H., "Aeroelastic Design of Large Wind Turbine Blades Considering Damage Tolerance," *Wind Energy*, Vol. 20, (1), 2017, pp. 159–170.

[8]Barlas, A. K., Tibaldi, C., Zahle, F., and Madsen, H., "Aeroelastic Optimization of a 10 MW Wind Turbine Blade with Active Trailing Edge Flaps," Paper AIAA 2016-1262, 34th Wind Energy Symposium, AIAA SciTech Forum, 2016.

[9]Zahle, F., Tibaldi, C., Verelst, D. R., Bak, C., Bitsche, R., and Blasques, J., "Aero-elastic Optimization of a 10 mw Wind Turbine," Paper AIAA 2015-0491, 33rd Wind Energy

Symposium, AIAA SciTech Forum, 2015.

[10]Visweswaraiah, S. B., Ghiasi, H., Pasini, D., and Lessard, L., "Multi-objective Optimization of a Composite Rotor Blade Cross-section," *Composite Structures*, Vol. 96, 2013, pp. 75–81.

[11]Bottasso, C. L., Campagnolo, F., and Croce, A., "Multi-disciplinary Constrained Optimization of Wind Turbines," *Multibody System Dynamics*, Vol. 27, (1), 2012, pp. 21–53.

[12]Li, L., Volovoi, V. V., and Hodges, D. H., "Cross-Sectional Design of Composite Rotor Blades," *Journal of the American Helicopter society*, Vol. 53, (3), 2008, pp. 240–251.

[13]Kenway, G. and Martins, J., "Aerostructural Shape Optimization of Wind Turbine Blades Considering Site-specific Winds," 12th AIAA/ISSMO Multidisciplinary Analysis and Optimization Conference, 2008.

[14]Chen, K.-N. and Pin-Yung, C., "Structural Optimization of 3 MW Wind Turbine Blades Using a Two-step Procedure," *International Journal for Simulation and Multidisciplinary Design Optimization*, Vol. 4, (3-4), 2010.

[15]Chandra, R. and Chopra, I., "Structural Response of Composite Beams and Blades with Elastic Couplings," *Composites Engineering*, Vol. 2, (5–7), 1992, pp. 347–374.
doi: 10.1016/0961-9526(92)90032-2

[16]Jung, S. N., Nagaraj, V. T., and Chopra, I., "Refined Structural Model for Thin- and Thick-walled Composite Rotor Blades," *AIAA Journal*, Vol. 40, (1), 2002, pp. 105–116.
doi: 10.2514/2.1619

[17]Giavotto, V., Borri, M., Mantegazza, P., Ghiringhelli, G., Carmaschi, V., Maffioli, G. C., and Mussi, F., "Anisotropic Beam Theory and Applications," *Computers and Structures*, Vol. 16, (1–4), 1983, pp. 403–413.
doi: 10.1016/0045-7949(83)90179-7

[18]Yu, W., Hodges, D. H., and Ho, J. C., "Variational Asymptotic Beam Sectional Analysis - an Updated Version," *International Journal of Engineering Science*, Vol. 59, 2012, pp. 40–64.
doi: 10.1016/j.ijengsci.2012.03.006

[19]Sung, N. J. and Il-Ju, P., "A simple Mixed-based Approach for Thin-walled Composite Blades with Two-cell Sections," *Journal of mechanical science and technology*, Vol. 19, (11), 2005, pp. 2016.

[20]Cesnik, C. E. and Hodges, D. H., "VABS: A New Concept for Composite Rotor Blade Cross-sectional Modeling," *Journal of the American Helicopter Society*, Vol. 42, (1), 1997, pp. 27–38.

[21]Rehfield, L. W., Atilgan, A. R., and Hodges, D. H., "Non-classical Behavior of Thin-walled Composite Beams with Closed Cross Sections," *Journal of the American Helicopter Society*, Vol. 35, (2), 1990, pp. 42–50.
doi: 10.4050/JAHS.35.42

[22]Smith, E. C. and Chopra, I., "Formulation and Evaluation of an Analytical Model for Composite Box-Beams," *Journal of the American Helicopter Society*, Vol. 36, (3), 1991, pp. 23–35.
doi: 10.4050/JAHS.36.23

[23]Volovoi, V. V. and Hodges, D. H., "Theory of Anisotropic Thin-walled Beams," *Journal of Applied Mechanics*, Vol. 67, (3), 2000, pp. 453–459.
doi: 10.1115/1.1312806

[24]Yu, W., Hodges, D. H., and Ho, J. C., "Variational Asymptotic Beam Sectional Analysis–An Updated Version," *International Journal of Engineering Science*, Vol. 59, 2012, pp. 40–64.

[25]Reissner, E., "On Mixed Variational Formulations in Finite Elasticity," *Acta Mechanica*, Vol. 56, (3-4), 1985, pp. 117–125.
doi: 10.1007/BF01177113

[26]Holland, J. H., *Adaptation in Natural and Artificial Systems: An Introductory Analysis with Applications to Biology, Control, and Artificial Intelligence*, MIT press, 1992.

[27]Gibson, R. F., *Principles of Composite Material Mechanics*, CRC press, 2016.

[28]Hodges, D. H., Atilgan, A. R., Fulton, M. V., and Rehfield, L. W., "Free-Vibration Analysis of Composite Beams," *Journal of the American Helicopter Society*, Vol. 36, (3), 1991, pp. 36–47.

[29]Stemple, A. and Lee, S., "Finite-element Model for Composite Beams with Arbitrary Cross-Sectional Warping," *AIAA journal*, Vol. 26, (12), 1988, pp. 1512–1520.

[30]Jung, S. N., Nagaraj, V., and Chopra, I., "Refined Structural Dynamics Model for Composite Rotor Blades," *AIAA journal*, Vol. 39, (2), 2001.

[31]Chandra, R. and Chopra, I., "Experimental-Theoretical Investigation of the Vibration Characteristics of Rotating Composite Box Beams," *Journal of Aircraft*, Vol. 29, (4), 1992, pp. 657–664.

Performance Analysis of Variable Speed Tail Rotors with Gurney Flaps

Chen Dong
Research assistant
Nanjing University of Aeronautics and Astronautics
Nanjing, China

Dong Han*
Professor
Nanjing University of Aeronautics and Astronautics
Nanjing, China

Lei Yu
Research assistant
Nanjing University of Aeronautics and Astronautics
Nanjing, China

ABSTRACT

Gurney flaps are used for improving the performance of variable speed tail rotors. A validated analytical helicopter model able to predict the main and tail rotor power is utilized. The fixed height Gurney flap has substantially small influence on the tail rotor power, except significant power reduction in high speed flight. This ability can be enhanced by the reduction of the tail rotor speed. The collective pitch of the tail rotor decreases by the deployment of Gurney flap, and the maximum tail rotor thrust can be increased. The Gurney flap can compensate the thrust reduction by the speed decrease of the tail rotor. 5% GF height can nearly remedy 50% thrust reduction introduced by 10% speed reduction. With increasing Gurney flap height, the maximum thrust generated by the tail rotor increases. Too large Gurney flap height can cause the power increase of the tail rotor in hover and medium forward flight. The retractable Gurney can obtain more power savings than the fixed height Gurney flap. However, the benefit is substantially small even in high speed flight. Considering the penalties introduced by the active Gurney flap, the fixed height Gurney flap may be more preferable. The mechanism for the retractable Gurney flap to generate more tail rotor thrust is to increase the lift in the advancing side due to the high dynamic pressure.

INTRODUCTION

Varying the helicopter main rotor speed has been well studied to reduce the main rotor power required and improve the performance in hover and forward flight [1-6]. Changing the output shaft speed of the engine is one of the primary effective means of changing the main rotor speed, which cannot avoid the simultaneous variation of the tail rotor speed. The reduction of the tail rotor speed can not only reduce the tail rotor power [7], but also the tail rotor noise [8]. However, the decrease of the tail rotor speed causes the decrease of the maximum tail rotor thrust due to the decrease of the dynamic pressure, which can deteriorate the capability of balancing the main rotor torque and implementing the yaw control. This can adversely limit the application of variable speed tail rotor. In high speed flight, reducing the tail rotor speed can cause the increase of the tail rotor power required [7]. It is necessary to compensate the accompanying increase in the power and decrease in the maximum thrust to maintain acceptable fuel consumption and handling quality.

The Gurney Flap (GF) is a lift enhancement device fixed on the trailing edge of an airfoil, invented by the race car driver Dan Gurney in 1960s [9]. GFs attracted much attention not only due to its efficiency and simplicity, but also due to its lower power consumption [10]. GFs has been extensively explored to improve helicopter main rotor performance [11-14]. These investigations pointed out that GFs could be used to improve helicopter main rotor performance, especially in high thrust, high speed and/or high altitude. Few researchers investigated the effect of GFs on the performance of tail rotors. Baslamisli et al. applied GF in a tail rotor for the increase of the tail rotor thrust [15]. The fixed height of 5% chord GF was selected for the consideration of safety in operation and easiness in production. This height could decrease the lift to drag ratio, but it was effective to increase the thrust generated by the tail rotor.

In this work, GF is investigated to explore the potential in reducing the tail rotor power and increasing the maximum thrust associated with the decrease of the tail rotor speed. To evaluate and compare the power savings and thrust increase by GFs, a helicopter model able to predict the main and tail rotor power is used, which includes a rigid blade model, aerofoil table look-up method, the Pitt-Peters inflow model, a rigid fuselage model, a tail rotor model and a propulsive trim method. The tail rotor power and corresponding thrust for the fixed and variable height GFs are analyzed to investigate the benefit of GFs in improving the performance of variable speed tail rotors.

MODELING

A helicopter power prediction model is used in this work. The main rotor blade model is based on a rigid beam with a hinge offset and a hinge spring, which are used to match the fundamental flap-wise blade frequency. Look-up table aerofoil aerodynamics is used to calculate the lift and drag coefficients of blade elements according to the local resultant Mach number and angle of attack. The induced velocity over the rotor disk is predicted by the Pitt-Peters inflow model , which captures the first harmonic azimuthal

Presented at the 6th Asian/Austrailian Rotorcraft Forum / Heli Japan 2017, Kanazawa, Japan, November 7-9, 2017.

variation of the induced velocity. The hub forces and moments of the main rotor are derived from the resultant root forces and moments of rotor blades by the blade element theory. The fuselage is treated as a rigid body with aerodynamic forces and moments. These forces and moments acting on the main rotor, tail rotor and fuselage contribute to the equilibrium equations of the helicopter, which are solved to obtain the trimmed pitch controls and rotor attitude angles. The required tail rotor thrust to counter the main rotor torque is determined by the torque divided by the distance from the hub center of the tail rotor to the main rotor shaft. The tail rotor thrust and power are obtained by performing a numerical integration over the blade elements along the blade radius and azimuth with uniform induced velocity. This model has been validated by using the UH-60A flight test data, and applied in analyzing the helicopter performance improvement by dynamic blade twist and variable tail rotor speeds [7, 16]. The flowchart of the performance prediction is shown in Figure 1.

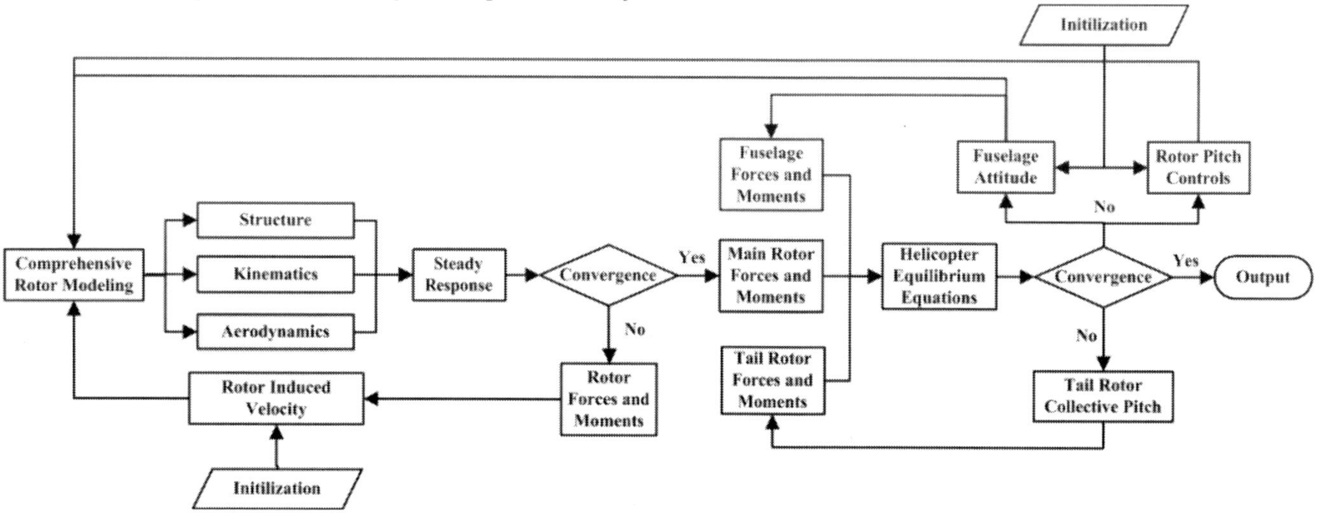

Figure 1. Flowchart of performance prediction.

The parameters of an airfoil with GF are shown in Figure 2. In the figure, d is the height of GF, and c is the chord length. For a retractable GF, the mounting angle β is 90°, and the height is given as a prescribed value. In the following analysis, the maximum height ratio d/c is limited to be less than 5.0% and the GF extends from 70% to 90% of the rotor radius. The methodology in Ref. 11 to capture the change of the aerodynamic characteristics of NACA 0012 airfoil is utilized. The comparisons of the lift coefficients between the test data in Ref. 17 and the prediction based the C81 airfoil table are shown in Figure 3, which shows the prediction is generally in good agreement with the test data.

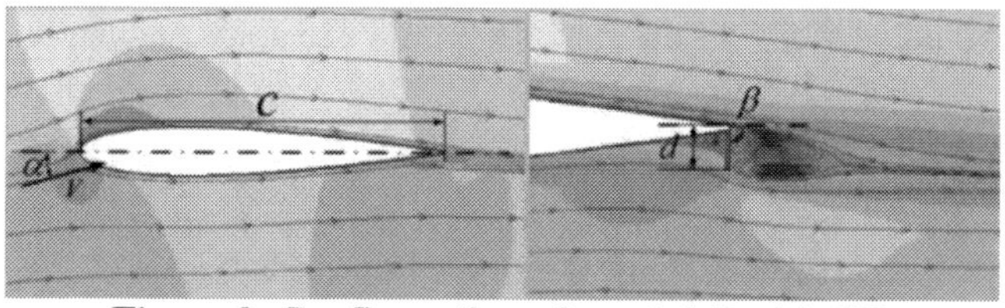

Figure 2. Configuration of an airfoil with GF.

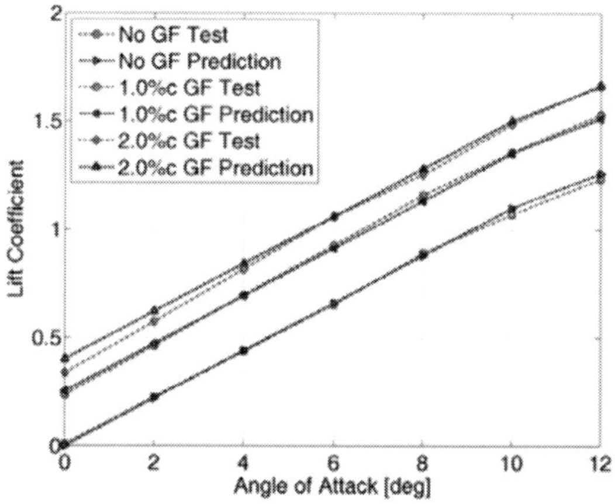

Figure 3. Comparison between test and prediction data.

The power reduction ratio is defined to determine the benefit in tail rotor power savings as

$$\eta = (1 - P/P_b) \times 100\% \quad (1)$$

where, P is the tail rotor power to be compared with, and P_b is the baseline tail rotor power. The baseline power is defined as the power at sea level and 100%Ω (nominal tail rotor speed) without any GF. In the following analysis, the baseline helicopter weight is 8322.3kg, and the corresponding weight coefficient is 0.0065.

The parameters of the baseline main and tail rotors similar to the UH-60A helicopter are listed in Tables 1 and 2 [18-21]. For the performance analysis, only the aerodynamic drag force is considered in the fuselage model. The distance from the hub center of tail rotor to the rotor shaft is 9.93 m.

The vertical distance from the mass center of the helicopter to the rotor hub is 1.78 m.

Table 1: Main rotor parameters

Parameter	Value
Main Rotor Radius	8.18m
Main Rotor Speed (100%)	27.0 rad/s
Blade Chord Length	0.527m
Blade Twist	Nonlinear
Blade Airfoil	SC1095/SC1094R8
Number of Blades	4
Flap Hinge Offset	0.381 m
Blade Mass per Unit Length	13.9 kg/m
Longitudinal Shaft Tilt	3°

Table 2: Tail rotor parameters

Parameter	Value
Tail Rotor Radius	1.68 m
Tail Rotor Blade Chord	0.247 m
Tail Rotor Speed	124.6 rad/s
Tail Rotor Blade Twist	-18°
Blade Airfoil	NACA0012
Number of Blades	4

Fixed Height GF

Figures 4 and 5 show the tail rotor power and the corresponding power reduction with and without the 2%c fixed height GF for different forward speeds. In hover, the reduction of the rotor speed has substantially small influence on the tail rotor power. 10% and 20% reduction of the speed can reduce the power by 2.1% and 3.1%. With the deployment of the 2%c GF, the power reductions change to 2.3% and 3.6%. The increments by the GF are 0.2% and 0.5%, which indicates that the speed reduction dominates the power change and the deployment of GF has very small effect. At a speed of 150km/h, reducing the tail rotor speed can significantly reduce the power and 20% speed reduction can obtain 25.1% power reduction. With the GF, the difference is not distinguishable. At a high speed of 290km/h, the power reductions introduced by the GF for the 0%, 10% and 20% speed reductions are 7.2%, 9.7% and 21.5%, respectively. The GF shows significant power saving ability in high speed flight, especially in lower tail rotor speeds. For the performance improvement, reducing tail rotor speed is more suitable for medium forward flight, and the deployment of GF is suitable for high speed flight.

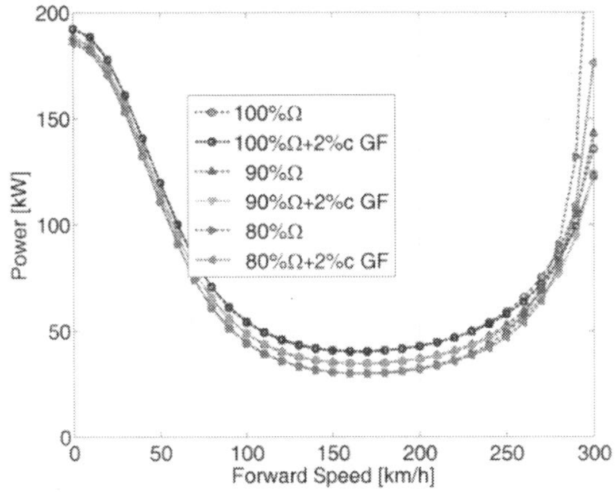

Figure 4. Tail rotor power for different rotor speeds and heights of GF.

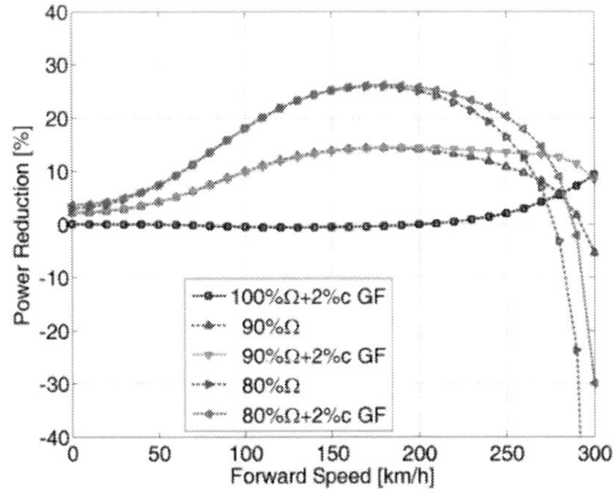

Figure 5. Tail rotor power reductions for different rotor speeds and heights of GF.

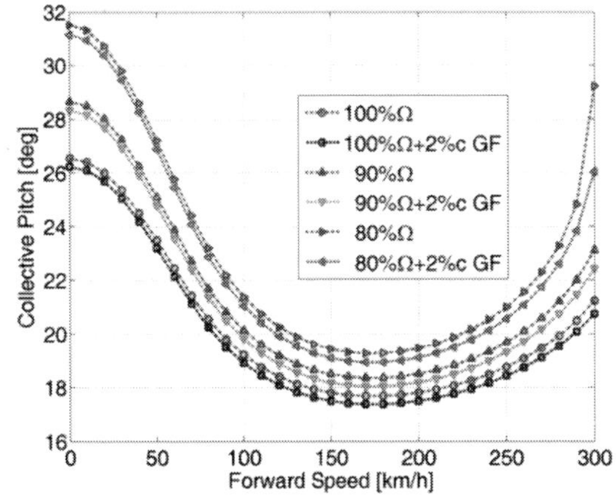

Figure 6. Collective pitch for different rotor speeds and heights of GF.

Figure 6 shows the collective pitch of the tail rotor with the forward speed. The reduction of the rotor speed increases the collective pitch. With the deployment of 2%c height GF, the collective pitch decreases. The change of the collective pitch introduced by the rotor speed is generally larger than that introduced by the GF. Figure 7 shows the distribution of angle of attack (AoA) for different rotor speeds with and without the 2%c height GF at a speed of 300km/h. The distribution of AoA in the advancing side is distinctly larger than that in the retreating side, which is essentially different from that of main rotors. Since tail rotors do not employ cyclic pitch controls and the lateral moment balance are not available, the advancing side contributes much more thrust than the retreating side. The deployment of the fixed height GF decreases the AoA due to the increase of lift coefficient. The speed reduction increases the AoA due to the decease of the dynamic pressure. GF can compensate the AoA increase due to the speed reduction.

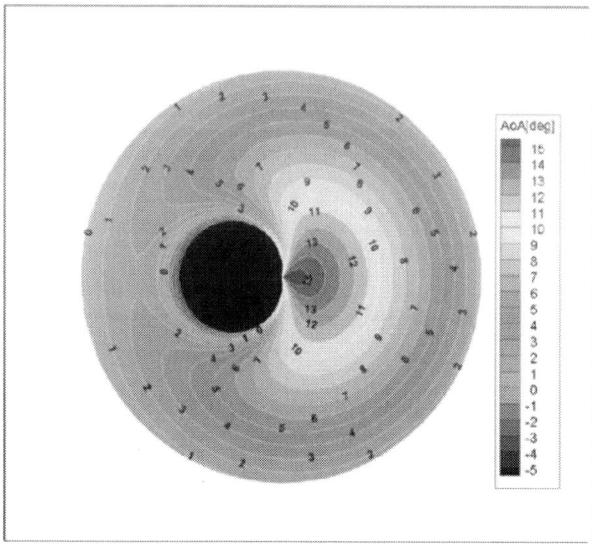

a) 100%Ω without GF

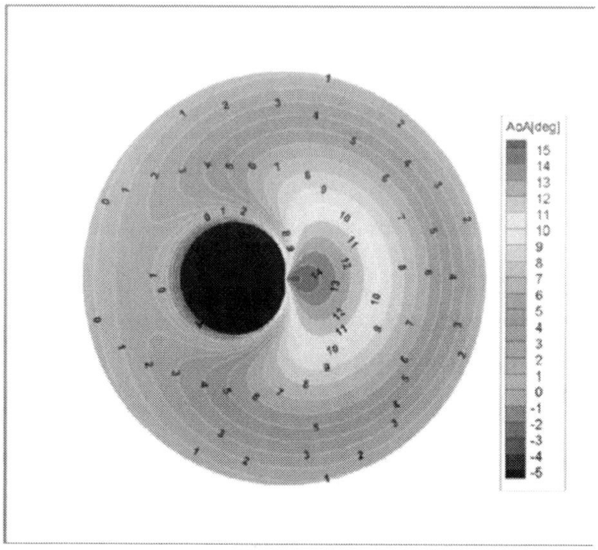

b) 100%Ω with GF

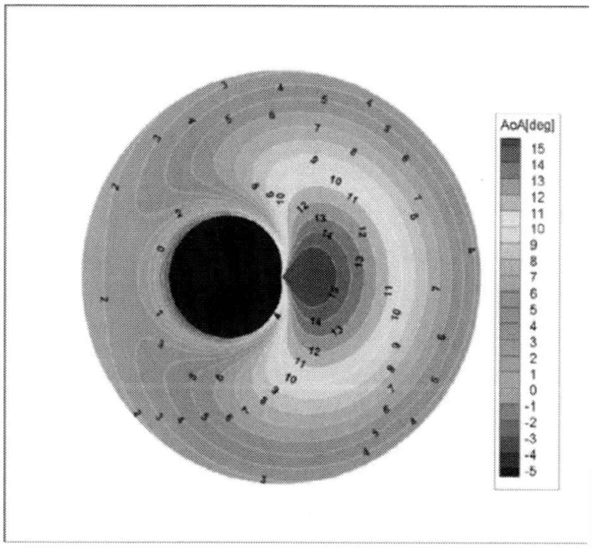

c) 90%Ω without GF

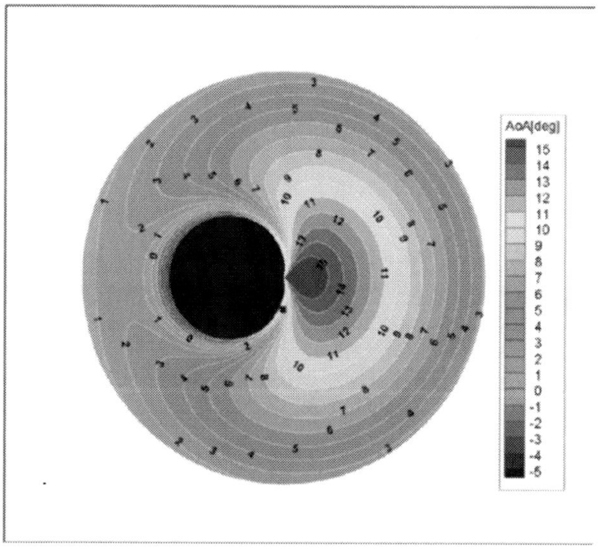

d) 90%Ω with GF

Figure 7. Distribution of angle of attack for different rotor speeds with and without GF.

Figure 8 shows the thrust required to balance the main rotor torque, and the maximum thrust can be generated, for different tail rotor speeds with and without GF. The thrust required is very large in hover, decreases with increasing forward speed, and then increases in medium to high speed flight. The decrease of the tail rotor speed decreases the maximum thrust. At a speed of 300km/h, 20% reduction of the speed almost makes the thrust inadequately to implement the yaw control. In high speed flight, it is necessary to compensate the thrust decease introduced by the speed reduction. With the deployment of GF, the maximum thrust increases, and the amplitude increment is moderate. The 2%c height GF has not the potential in remedying the thrust decrease caused by the 10% speed reduction.

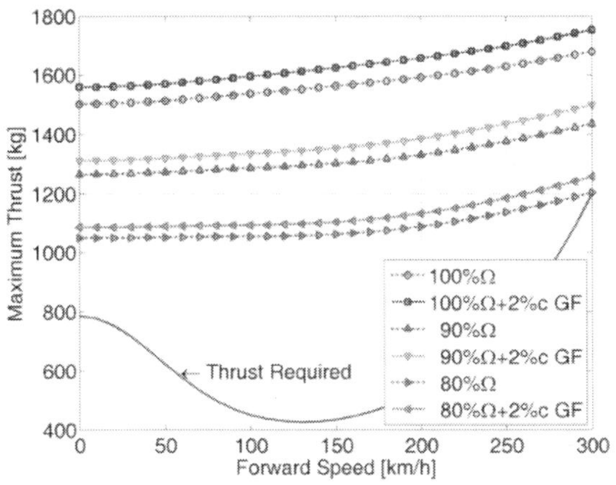

Figure 8. Maximum thrust for different rotor speeds and heights of GF.

Figure 9 shows the power reductions for different heights of GF at 90%Ω. In hover and low to medium forward speed flight, it is obvious that the speed reduction dominates the power savings. The power reduction increases first and then decreases with increasing GF height. Large height is not preferred. At a speed of 150km/h, the power reduces by 13.9% with 10% speed reduction. With the 5%c height GF, that reduction changes to 11.8%. The 5%c height GF increases the power by 2.1%. At a speed of 290km/h, the 2%c and 5%c height GFs can further reduce the power by 9.7% and 13.5%. In high speed flight, the larger height GF is preferred, and it can significantly achieve more power savings than the smaller one.

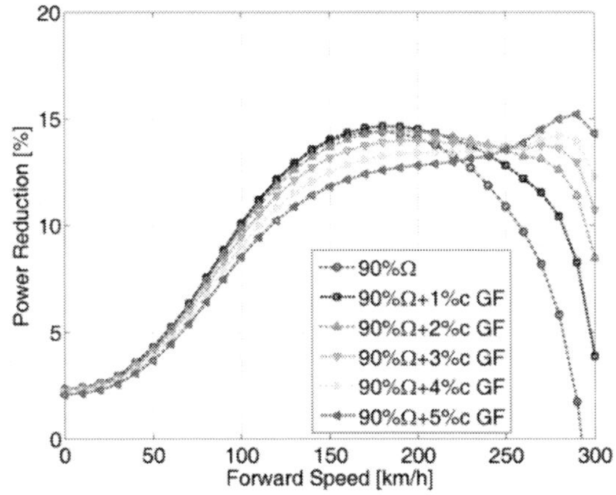

Figure 9. Power reduction for different heights of GF with 90%Ω.

Figure 10 shows the maximum thrust for different GF heights and rotor speeds. It is obvious that 10% reduction of tail rotor speed can significantly reduce the maximum thrust, which can significantly deteriorate the yaw control ability. The maximum thrust increases with the GF height. The 5% GF height can remedy near 50% thrust reduction introduced by the 10% speed reduction. This value can be increased by increasing the GF height. However, too large height GF is not preferred for the consideration of the performance penalties in hover and medium speed forward flight.

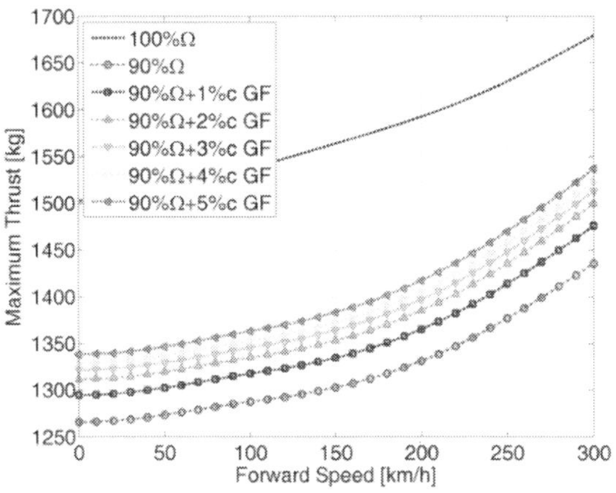

Figure 10. Maximum thrust for different GF heights and rotor speeds.

Retractable GF

The previous analysis is based on fixed height GF, which is simple and easy for production and maintenance. With the change of helicopter flight state (speed, altitude, or weight), the GF height corresponding to the best performance improvement changes. The active GF emerges as a better candidate for performance improvement. In the following analysis, the height of a retractable GF is prescribed as

$$h = A[1 + \sin(n\Omega t + \phi)] \tag{2}$$

where, A is the average GF height, n is the harmonic number, Ω is the tail rotor speed, and ϕ is the phase of the harmonic input.

Figure 11 shows the effect of the phase on the tail rotor power for different harmonic inputs with 2%c average GF height at a speed of 250km/h and. With the increase of harmonic number, the amplitude of the power change decreases. The 1/rev input can cause much larger power change and reduction than the other harmonics. The 3/rev and 4/rev inputs have little effect on the tail rotor power. Lower harmonic input is preferred for the power reduction. When the phase of the 1/rev input shifts to 0°, it can obtain the maximum power reduction. For the 2/rev input, the corresponding phase is 270°. Figure 12 shows the GF height with the azimuth. For the 1/rev input, the GF has the maximum height in the advancing side and the minimum value in the retreating side. For the 2/rev input, the GF has the maximum value at both the advancing and retreating sides. To generate more thrust, the tail rotor sufficiently takes advantage of the high dynamic pressure in the advancing side. The mechanism for the GF to improve the tail rotor performance is completely different from that of the main rotor [12].

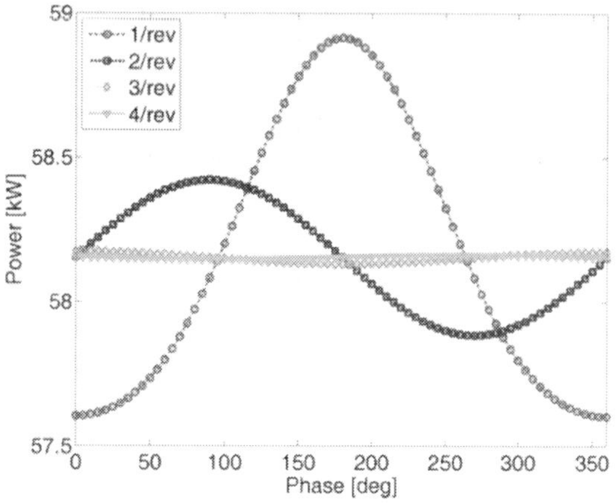

Figure 11. The effect of phase on the power for different harmonic input (250km/h, 100%Ω, 2%c).

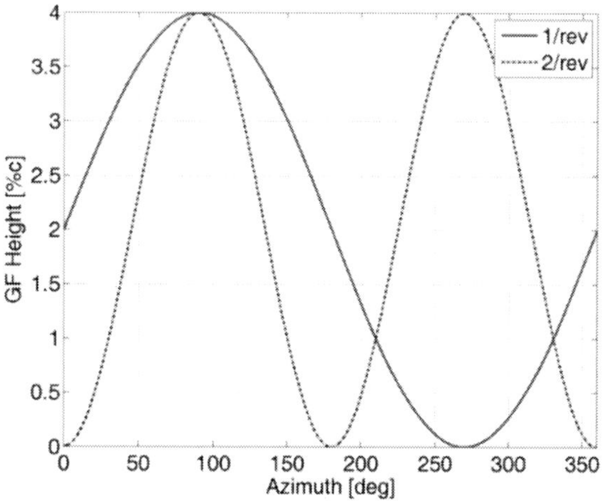

Figure 12. GF height with azimuth for 1/rev and 2/rev inputs.

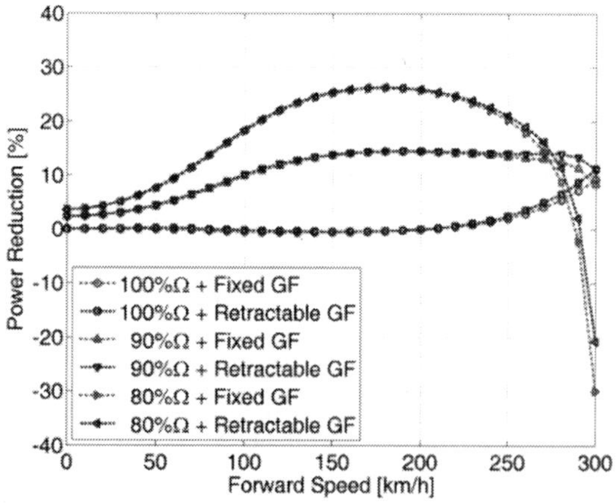

Figure 13. Comparisons of the power reductions by the fixed height and 1/rev retractable GF.

Figure13 shows the comparison of the power reductions by the fixed and 1/rev retractable GFs. The average height of both are 2%c, and the 1/rev GF is shifted to the phase of 0°. Generally, the difference is substantially small except in very high speed flight. At 100%, 90% and 80%Ω, the differences between the fixed height GF and the retractable GF are 1.9%, 2.8% and 9.2%. It obvious that the retractable GF does not have significant advantages over the fixed height GF. If consider the penalties introduced by the active GF, for example weight, complexity and power consumption, the fixed height GF can be more feasible and preferable.

CONCLUSIONS

A validated helicopter model able to predict the main and tail rotor powers was used to explore the performance improvement of variable speed tail rotors by GFs. The analyses yielded the following conclusions:

1) GF has substantially small influence on the tail rotor power in hover and low to medium speed forward flight. In high speed flight, it can significantly decrease the tail rotor power. This benefit increases with the reduction of the tail rotor speed.

2) The collective pitch of the tail rotor decreases by the deployment of GF, and the maximum tail rotor thrust increases.

3) In high speed flight, larger height GF is preferred, and it can achieve much more power savings than the smaller one. At a speed of 290km/h, the 2%c and 5%c height GFs can further reduce the power at 90%Ω by 9.7% and 13.5%.

4) GF can compensate the decrease of the maximum tail rotor thrust. The 5% GF height can nearly remedy 50% thrust reduction introduced by the 10% speed reduction.

5) The retractable GF can obtain more power savings than the fixed height GF. However, the benefit is substantially small even in high speed flight. Considering the penalties caused by the active GF, the fixed height GF may be more feasible and preferable.

6) The mechanism for the retractable GF generating more tail rotor thrust is to increase the lift in the advancing side due to the high dynamic pressure.

Author contact: Chen Dong dongchen@nuaa.edu.cn
Dong Han donghan@nuaa.edu.cn

ACKNOWLEDGMENTS

This work was supported from the National Natural Science Foundation of China (11472129), Science and Technology on Rotorcraft Aeromechanics Laboratory Foundation (61422200504162200002), Foundation of Graduate Innovation Center in NUAA (KFJJ20170102) and Fundamental Research Funds for the Central Universities.

REFERENCES

1. Karem, A. E., "Optimum Speed Rotor," U.S. Patent No. 6007298. 1999.
2. Prouty, R. W., "Should We Consider Variable Rotor Speeds? " *Vertiflite*, Vol. 50, No. 4, 2004, pp. 24-27.
3. Kang, H., Saberi, H., and Grandhi, F., "Dynamic blade shape for improved helicopter rotor performance," *Journal of the American Helicopter Society*, Vol.55, No. 3, 2010, pp. 0320081-03200811.
4. Mistry, M., and Gandhi, F., "Helicopter Performance Improvement with Variable Rotor Radius and RPM," *Journal of the American Helicopter Society*, Vol. 59, No. 4, 2014, pp. 17-35.
5. Misté, G. A., Benini, E., Garavello, A., and Gonzalez-Alcoy, M., "A Methodology for Determining the Optimal Rotational Speed of a Variable RPM Main Rotor/Turboshaft Engine System," *Journal of the American Helicopter Society*, Vol. 60, No. 3, 2015, pp. 0320091-03200911.
6. Han, D., Pastrikakis, V., and Barakos, G. N., "Helicopter Performance Improvement by Variable Rotor Speed and Variable Blade Twist," *Aerospace Science and Technology*, Vol. 54, 2016, pp.164-173.
7. Han, D. and Barakos, G., "Variable Tail Rotor Speeds for Helicopters with Variable Main Rotor Speeds," *The Aeronautical Journal*, Vol. 121, No. 1238, 2017, pp. 433-448.
8. George, A. R., "Helicopter Noise: State-of-the-Art," *Journal of Aircraft*, Vol. 15, No. 11, 1978, pp. 707-715.
9. Wang, J. J., Li, Y. C., Choi, K.-S., "Gurney Flap - Lift Enhancement, Mechanisms and Applications," *Progress in Aerospace Sciences*, Vol. 44, No. 1, 2008, pp. 22-47.
10. Palacios, J., Kinzel, M., and Overmeyer, A, "Active Gurney Flaps: Their Application in a Rotor Blade Centrifugal Field," *Journal of Aircraft*, Vol. 51, No. 2, 2014, pp. 473-489.
11. Kentfield, J. A. C., "The Potential of Gurney Flaps for Improving the Aerodynamic Performance of Helicopter Rotors," AIAA Paper 1993-4883, AIAA International Power Lift Conference, Santa Clara, CA, December 1-3, 1993.
12. Kinzel, M. P., Maughmer, M. D., and Lesieutre, G. A., "Miniature Trailing-Edge Effectors for Rotorcraft Performance Enhancement," *Journal of the American Helicopter Society*, Vol. 52, No, 2, 2007, pp. 146-158.
13. Pastrikakis, V. A., Steijl, R., Barakos G. N., and Małecki, J., "Computational Aeroelastic Analysis of a Hovering W3 Sokol Blade with Gurney Flap," *Journal of Fluids and Structures*, Vol. 53, 2015, pp. 96-111.
14. Pastrikakis, V. A., Steijl, R., and Barakos G. N., "Effect of Active Gurney Flaps on Overall Helicopter Flight Envelope," *The Aeronautical Journal*, Vol. 120, No. 1230, 2016, pp. 1230-1261.
15. Baslamisli, U., Ayan, E., and Yücekayali, A., "Tail Rotor Aerodynamic Performance Improvement with a Gurney Flap Design," the 40th European Rotorcraft Forum, Southampton, UK, September 2-5, 2014.
16. Han, D., Pastrikakis, V. and Barakos, G., "Helicopter Performance Improvement by Dynamic Blade Twist," *Aerospace Science and Technology*, Vol. 58, 2016, pp. 445-452.
17. Li, Y., Wang, J., and Zhang, P., "Effects of Gurney Flaps on a NACA0012 Airfoil," Flow, *Turbulence and Combustion*, Vol. 68, No. 1, 2002, pp. 27-39.
18. Hilbert, K. B., "A Mathematical Model of the UH-60 Helicopter," Technical Report NASA-TM-85890, 1984.
19. DavisS. J., "Pedesign Study for a Modern 4-Bladed Rotor for the RSRA," Technical Report NASA-CR-166155, 1981.
20. Buckanin, R. M., Herbst, M. K., Lockwood, R. A., Skinner, G. L. and Sullivan, P. J., "Airworthiness and Flight Characteristics Test of a Sixth Year Production UH-60A," Final Report, USAAEFA Project No. 83-24, June1985.
21. Nagata, J. I., Piotrowski, J. L., Young, C. J., Lewis, W. D., Losier, P. W. and Lyle, J. A., "Baseline Performance Verification of the 12th Year Production UH-60A Black Hawk Helicopter," Final Report, USAAEFA Project No. 87-32, January 1989.

Analysis on the influence of the variable speed and variable diameter strategies on aerodynamic performance of the tiltrotor

Bo Wang — Doctor
Peng Li — Doctor
Qi-Jun Zhao — Professor
Si-Yu Chen — Ph D Student

National Key Laboratory of Science and Technology on Rotorcraft Aeromechanics, Nanjing University of Aeronautics and Astronautics
Nanjing, People's Republic of China

ABSTRACT

Both VST and VDT concepts provide easy control over the tip speed of the tiltrotor to ameliorate the tiltrotor design, so an investigation is introduced to analyze VST and VDT on the basis of CFD. Firstly, a new efficacious embedded grid method is introduced to realize detailed information transmission between the blade and background grids with a considerable cost. Then, based on the momentum theory and blade element theory, a trimming methodology is established for the tiltrotor trimming of the hovering and cruising flights. A new comprehensive index is proposed to evaluate the hover performance and the cruise performance of the tiltrotor simultaneously. At last, several blades with different rotation speed and radius are investigated respectively to obtain regularities for the tiltrotor design.

NOTATION

c = blade chord
D = drag force
E = total energy
$\boldsymbol{F}$ = unsteady momentum source terms
G = rotational fluxer
H = total enthalpy
L = lift force
N = number of blades
$\boldsymbol{n}$ = normal vector
p = pressure
Q_C = convective inviscid fluxes
Q_V = convective viscid fluxes
Re = Reynolds number
r = distance from the center of the rotor hub
S_Δ = area of actuator disk cell
U = velocity
V_x = X component of the absolute velocity vector
V_y = Y component of the absolute velocity vector
V_z = Z component of the absolute velocity vector
V_b = grid rotating velocity vector
W = vector of conserved flow variables
α_g = geometric angle of attack of a blade section
θ_t = the geometric twist angle
$\tilde{\upsilon}$ = eddy viscosity
ρ = density
ϕ = inflow angle
Ω = control volume
$\partial\Omega$ = boundary of a control volume
ω = angular velocity
$\Delta\psi$ = width along the circumferential direction
i, j, k = grid indices

INTRODUCTION

With a combination of the wing of fixed-wing aircraft and the rotor of helicopters, tiltrotor aircraft possess a distinct advantage of integrating vertical take-off and landing (VTOL) with high-speed cruise. In hover, tiltrotor is needed to support the weight of the aircraft. Consequently, larger rotor diameter and higher tip speed are required to reduce the disk loading. However, the wings generate the major portion of lift needed, and the rotor thrust serves to balance the drag of the aircraft during cruise flight. Therefore, smaller rotor diameter and lower tip speed are adopted for the purpose of a better propulsive efficiency. As a result, in the design of a tiltrotor, a balance should be struck between the aerodynamic performance requirements

Presented at the 6th Asian/Austrailian Rotorcraft Forum / Heli Japan 2017, Kanazawa, Japan, November 7-9, 2017.

of both hovering process and high-speed cruise flight, which inevitably leads to some compromises in the tiltrotor design.

To ameliorate the compromises and simplify the complex designing, variable speed tiltrotor (VST) and variable diameter tiltrotor (VDT) concepts are proposed (Ref. 1-2). In VST design, a gearbox is introduced to control the rotation speed of the fixed-diameter rotor. In VDT method, a retraction system is introduced while rotation speed remains constant. By varying rotor speeds or the diameter of rotor, VST and VDT concepts both provide an easy control over the tip speed of the tiltrotor, significantly reduce disk loading in hover and delay the onset of compressibility effect in high-speed forward flight. Over the past few decades, large numbers of experiments on VST have been done by the Bell Helicopter company (Ref. 3-4) to improve the tiltrotor performance, and Karem (Ref. 5) developed the optimum speed tiltrotor that optimizes aerodynamic efficiency over the flight envelope with continuously variable rotor speeds. In VDT, Fradenburgh (Ref. 6) carried out large quantities of tests and Wang (Ref. 7) have demonstrated that VDT can enhance the aerodynamic performance of civil tiltrotors. Most of the above-mentioned studies were performed via experimental methods (Ref. 8-11) or simplified aerodynamic models (Ref. 11-14), whose application is relatively constrained by limitations of experiment conditions as well as large time-consumption, and the simplified conventional aerodynamic models also proven to be deleterious to the expansion of application. However, CFD method has now become an increasingly prevalent tool in analyzing aerodynamic characteristics of the tiltrotor.

In this paper, a preliminary investigation is introduced to analyze VST and VDT on the basis of CFD. Firstly, a new efficacious embedded grid method is introduced to realize detailed information transmission between the blade and background grids with a considerable cost. Secondly, the three-dimensional quasi-steady Navier-Stokes equations are discretized by means of Jameson's second-order cell-centered finite volume scheme, and an implicit LU-SGS scheme is adopted for time advance. At last, the aerodynamic forces and flowfield of the VST and VDT in different flight modes are attained, and several conclusions are obtained.

MOVING-EMBEDDED GRID METHODOLOGY

Grid System

For the symmetric rotation of tiltrotor rotors in hover and cruise, the grid around rotor can be simplified to a cylindrical grid system (Fig. 1) around the blade (Ref. 15). For a given number of grids, the cylindrical grid system is the most effective configuration for simulation. The interpolation accuracy between the blade and background grids is also improved to some extent.

In this paper, the cylindrical grid system is divided into two parts: the C-O type body-fitted grid around the blade and the background grids. The former can be used to effectively obtain chordwise and spanwise boundary layer flow on the blade. The latter is adaptively adjusted to ensure the acquisition of the flowfield of the tiltrotor.

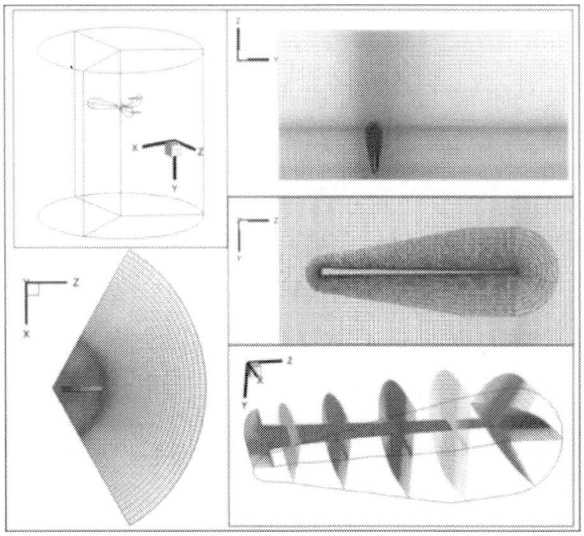

Fig.1 Grid system

Embedded Grid method

The collective pitch angle varies from 0°to 30°in hover, while the variation range extends from 30°to 60°in cruise. The embedded grid method should cover the whole spectrum of the variation. Thus a new embedded grid method is introduced, and it consists of the following three steps.

1. Identifying fringe cells around the blade wall. The fringe cells in the background grid around the blade can be determined on the basis of desired distances, and the relationship between them is simultaneously recorded.

2. Marking hole cells. With the known fringe cells, hole cells can be marked out by collapsing the hole boundary.

3. Determining suitable outer boundary elements. Suitable outer boundary elements can be searched out from a certain distance in radial direction of the blade surface.

Figure 2 illustrates the hole cells and fringe cells of the blade in different collective angles. As can be seen from the diagram, a desired distance away from the blade wall has been fulfilled in a wide range of the collective pitch angles.

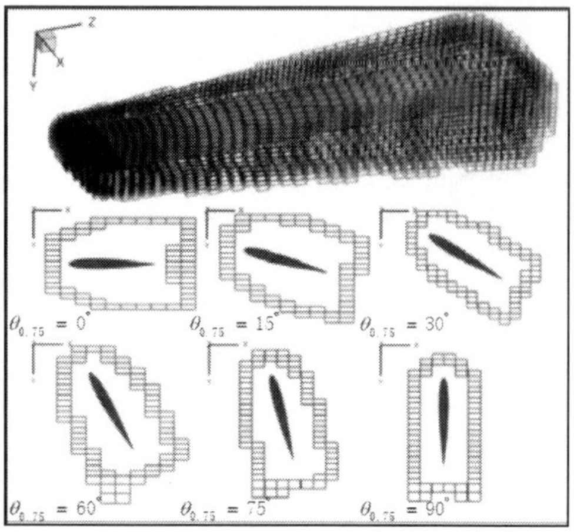

Fig.2 The embedded results of different collective angles

NUMERICAL METHODS

Governing equations

The 3D compressible governing equations are written in a reference frame rotating with a constant angular velocity. In the reference frame, the equations are written in finite integral form as:

$$\frac{\partial}{\partial t}\int_V W\mathrm{d}V + \oint_{\partial V}\left(\boldsymbol{F}_c - \boldsymbol{F}_v\right)\bullet \boldsymbol{n}\mathrm{d}s = \oint_{\partial V} \boldsymbol{J}\bullet \boldsymbol{n}\mathrm{d}s \quad (1)$$

Where, $\boldsymbol{F}_c = \begin{bmatrix} \rho(\boldsymbol{q}-\boldsymbol{q}_r) \\ \rho u(\boldsymbol{q}-\boldsymbol{q}_r)+p\boldsymbol{n}_x \\ \rho v(\boldsymbol{q}-\boldsymbol{q}_r)+p\boldsymbol{n}_y \\ \rho w(\boldsymbol{q}-\boldsymbol{q}_r)+p\boldsymbol{n}_z \\ \rho H(\boldsymbol{q}-\boldsymbol{q}_r)+p\boldsymbol{q}_r \end{bmatrix}$,

$$\boldsymbol{F}_v = \begin{bmatrix} 0 \\ \tau_{xx}\boldsymbol{n}_x+\tau_{xy}\boldsymbol{n}_y+\tau_{xz}\boldsymbol{n}_z \\ \tau_{yx}\boldsymbol{n}_x+\tau_{yy}\boldsymbol{n}_y+\tau_{yz}\boldsymbol{n}_z \\ \tau_{zx}\boldsymbol{n}_x+\tau_{zy}\boldsymbol{n}_y+\tau_{zz}\boldsymbol{n}_z \\ \Phi_x\boldsymbol{n}_x+\Phi_y\boldsymbol{n}_y+\Phi_z\boldsymbol{n}_z \end{bmatrix}, \quad W = \begin{bmatrix} \rho \\ \rho u \\ \rho v \\ \rho w \\ \rho E \end{bmatrix},$$

$\boldsymbol{J} = \begin{bmatrix} 0 \\ -\rho w\Omega \\ 0 \\ \rho u\Omega \\ 0 \end{bmatrix}$, V, ∂V, $\boldsymbol{n}$, Ω, ρ, t, k, T, E and H represents the volume, area, surface vector, angular velocity, density, time, thermal conductivity coefficient, absolute temperature, total energy and total enthalpy respectively. $\boldsymbol{q} = (\mathrm{u}, \mathrm{v}, \mathrm{w})^T$ is the absolute velocity and $\boldsymbol{q}_r$ is the relative velocity. One equation Spalart-Allmaras turbulence model (Ref. 16) is used in the RANS equations. The cell-centered finite volume method with the Jameson's second-order flux difference splitting scheme is employed. The implicit lower-upper symmetric-Gauss-Seidel method scheme is used as the time integration method.

Performance parameter

As the tiltrotor aircraft combine the vertical hover capacities of helicopter and the high-speed of proprotor aircraft, two sets of standards are proposed to evaluate the hover performance and the propulsion efficiency respectively. However, such complicated definitions of parameters significantly increase the complexity of performance analysis. Consequently, a simplification of parameters is introduced in this paper.

$$\begin{cases} \mu = V_{t/v} / (\Omega R) \\ C_T = T / (\rho\pi\Omega^2 R^4) \\ C_Q = P / (\rho\pi\Omega^3 R^5) \\ FM = C_T \times \sqrt{C_T / 2.0} / C_Q \\ \eta = \mu \times C_T / C_Q \end{cases} \quad (2)$$

Rotor trimming

To carry out performance comparison with equivalent thrusts, initial control strategies should be determined under different flight conditions. In this paper, a trimming method to determine initial control strategies for the tiltrotor is established. The inflow angle of a given airfoil should be obtained in order to determine the airfoil lift coefficient under different flight conditions. In most cases, the presupposition of uniform flow could satisfy the actual need for accuracy and precision. A simplification of inflow model for various flight conditions is proposed, which is shown as follows

$$C_T = 2\frac{V_v + V_i}{(\Omega R)}\frac{V_i}{(\Omega R)}k_{ind} \quad (3)$$

Thus the flow circumstance can be derived as

$$\begin{cases} \phi = \tan^{-1}\frac{V_v}{V_t}, \alpha = \theta_0 + \theta_{twist} - \phi \\ Ma = V_a / V_{sound} = \sqrt{(V_v + V_i)^2 + V_t^2} / V_{sound} \end{cases} \quad (4)$$

Where, θ_0 is the collective pitch, θ_{twist} represents the torsion angle.

On the basis of the blade element theory, the thrust coefficient of the spanwise element in hover and cruise can be described as follows

$$\begin{cases} C_L = Airfoil_{C_L}(Ma,\alpha), C_D = Airfoil_{C_D}(Ma,\alpha) \\ dL = \frac{1}{2}\rho V_a^2 c C_L dr, dD = \frac{1}{2}\rho V_a^2 c C_D dr \\ dT = dL\cos\phi - dD\sin\phi \end{cases} \tag{5}$$

Due to the circumferential symmetry of rotor in hover and cruise

$$dC_T = N\frac{dT}{\rho(\Omega R)^2 \bullet (\pi R^2)} = \frac{1}{2}s\sqrt{\xi^2+\lambda^2}\left(C_L\xi - C_D\lambda\right)dr \tag{6}$$

$$C_T = \int_{r_0}^{1} dC_T = s\int_{r_0}^{1}\frac{1}{2}\sqrt{\xi^2+\lambda^2}\left(C_L\xi - C_D\lambda\right)dr \tag{7}$$

Where, axial inflow factor $\lambda = (V_v + V_i)/(\Omega R)$ and the azimuthal flow ratio $\xi = V_t/(\Omega R)$, r_0 is the root cut ratio.

To solve for the collective pitch angle θ_0, the blade element expressions for the incremental thrust in eq. (7) are equated to the one from momentum theory in eq. (3).

RESULTS AND DISCUSSIONS

Caradona-Tung rotor at different rotation speeds

The Caradona-Tung rotor with two blades is taken as the numerical example to verify the the validity of the numerical method. Pitch angle is 8°, and the grid size is 215×49×58. Parameters of calculation states for Caradona-Tung rotor in hover are given in Table 1.

Table 1 Parameters of calculation states for Caradona-Tung rotor in hover

Case	C1	C2	C3	C4	C5	C6
Ma	0.439	0.526	0.612	0.727	0.794	0.877

Fig.3 is a comparison between predicted pressure coefficients and experimental data (Ref. 17) at the typical spanwise position, which demonstrates good agreements between the calculated results and experimental data. It shows that the proposed numerical simulation method is reasonable and feasible to simulate the aerodynamic force of rotor at different rotation speeds.

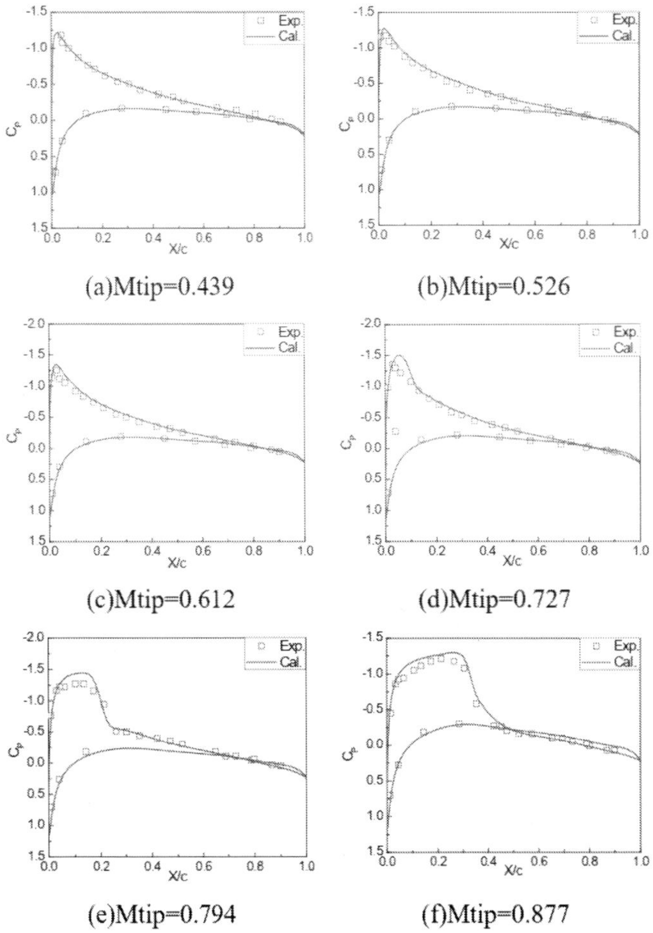

Fig.3 Pressure coefficient distributions on the rotor blade at r/R=0.89

Figure 4 compares the calculated lift coefficients along the blade in different states, and a similar tendency in distribution is observed. For 70% of the blade near the root, the variance remains slight, while the difference of the rest 30% blade proves to be obvious. It suggests that Mach number exerts a more salient impact on the rotor tip than on the root. For some potentially latent effects of transonic speed on the blade tip, the growth rate of the lift coefficient decreases as the rotation speed increases.

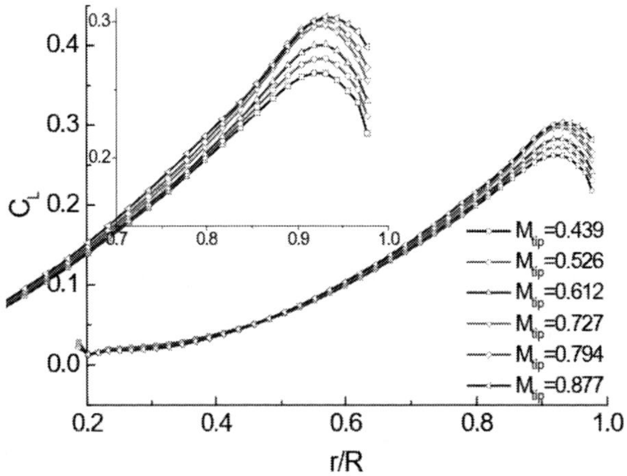

Fig.4 Prediction of the spanwise lift distribution along the blade at different rotation speeds

Figure 5 demonstrates the aerodynamic performance of the Caradona-Tung rotor changing along with different rotation speeds. The lift coefficient increases linearly with the rotation speed, while the torque coefficient soars nearly in an exponential manner. The diagram also indicates that the Figure of Merit first increases with rotation speed and then decreases drastically. Therefore, the optimum rotation speed falls within the range of Mach number from 0.612 to 0.727 at the given collective angle.

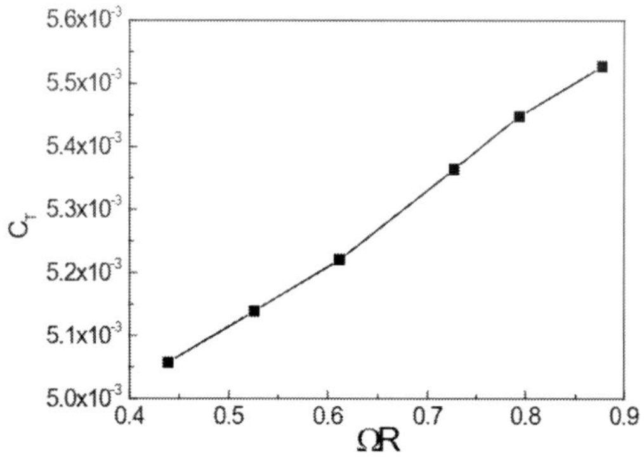

(a) Lift coefficient changes with rotate speeds

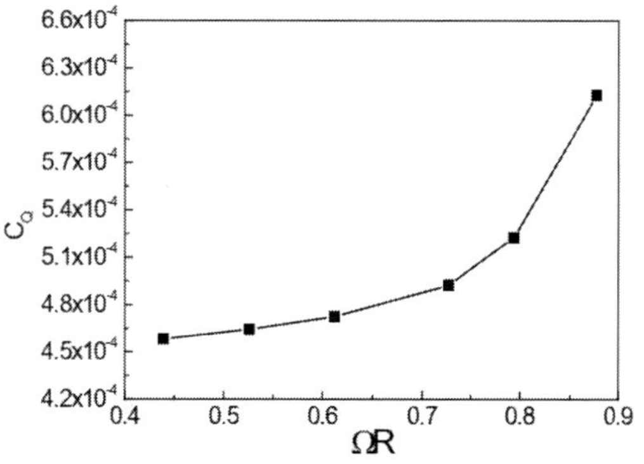

(b) Torque coefficient changes with rotate speeds

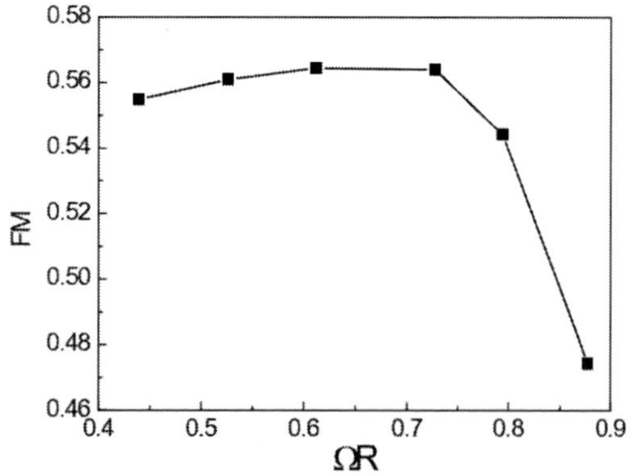

(c) Figure of Merit changes with rotation speeds

Fig.5 Aerodynamic performance of the Caradona-Tung rotor changes with rotate speeds

Figure 6 illustrates the streamlines on the Caradona-Tung blade in different states to show the impact of rotation speed in hover. As the rotation speed increases, the low pressure region expands on the top surface of the blade tip. When the Mach number reaches 0.877, an obvious flow separation can be observed at the blade tip. This demonstrates that the rotor can be kept under an optimum condition by adjusting rotation speed to avoid flow separation.

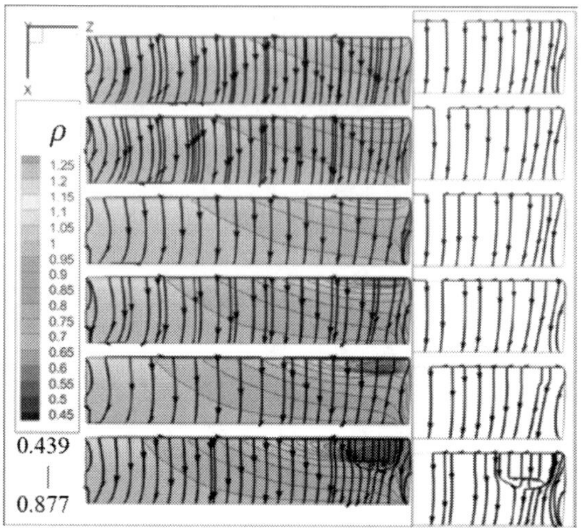

Fig.6 Streamlines on the Caradona-Tung blade

The influence of different rotation speeds on hover performance

Figure 7 presents a comparison of FM at the standard speed against experimental data. The consequence of simulation is relatively smaller than the experimental data, for the root possesses a larger undercut in actual simulation. However, the simulation accuracy proves to be adequate in evaluating aerodynamic performance of the tiltrotor.

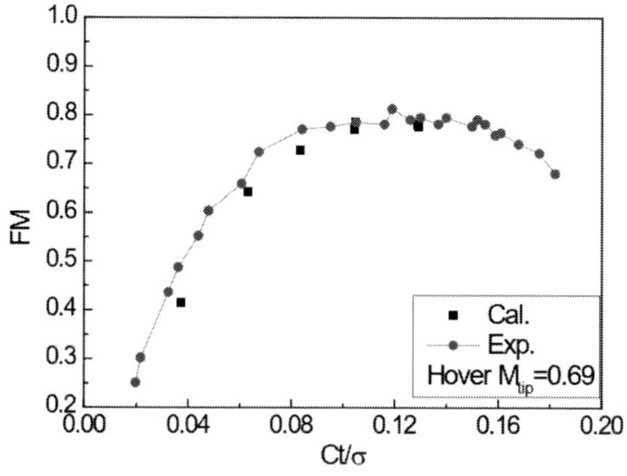

Fig.7 FM of the XV-15

Figure 8 illustrates the FM distribution of XV-15 tiltrotor changing along with the thrust. In the case of small thrust, lower rotation speed leads to a high efficiency, while with intense thrust, high efficiency results from a higher rotation speed. Besides, at a relatively higher rotation speed, the FM remains efficient within a large thrust range.

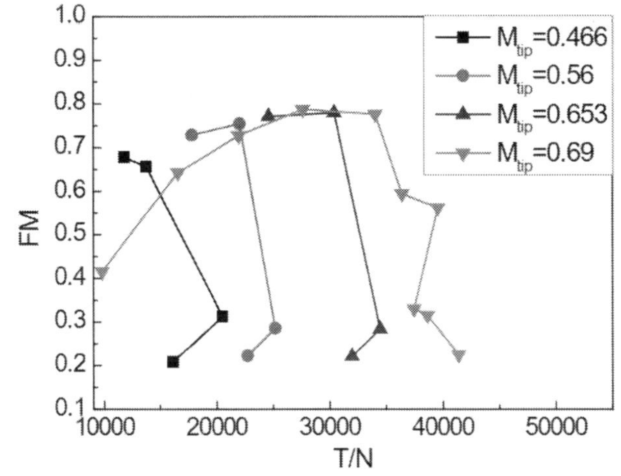

Fig.8 Distribution of the FM changing with thrust of the XV-15

Figure 9 illustrates the streamline distribution on the blade surface at different rotation speeds with maximum thrust. The blade surface pressure and streamline distribution at the certain speed is given in Figure (9a-9c). The geometric angle of attack is mainly due to the large root pitch. As a result, the actual angle of attack near the blade root is larger than the stall angle of the airfoil, which causes an obvious flow separation on blade surface. As the speed increases, the flow separation apparently weakens. Figure (c) and (d) provide approximate high thrust condition. In Figure 9d, the flow separation appears in the rear of the airfoil, which demonstrates that rotor with high rotation speed can work at an ideal range of angles of attack and maintain a high efficiency under the precondition of the same thrust.

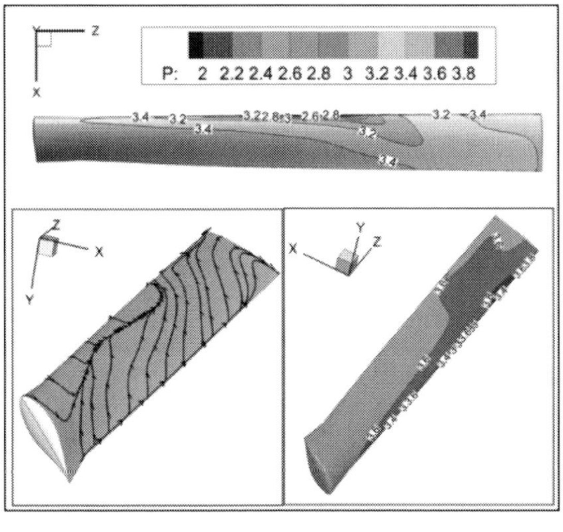

(a)Mtip = 0.466, θ=25°

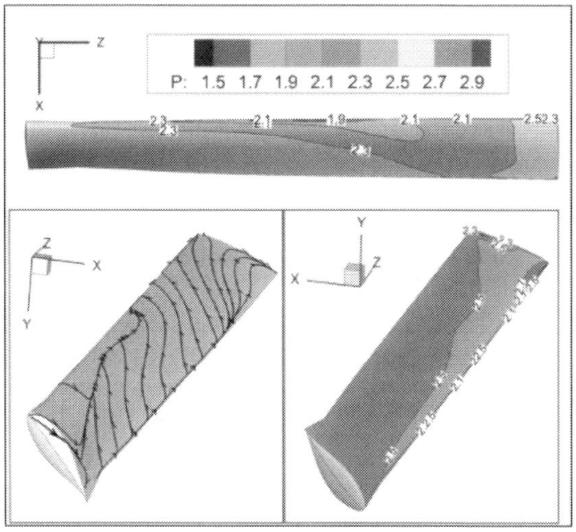

(b)Mtip = 0.56, θ=25°

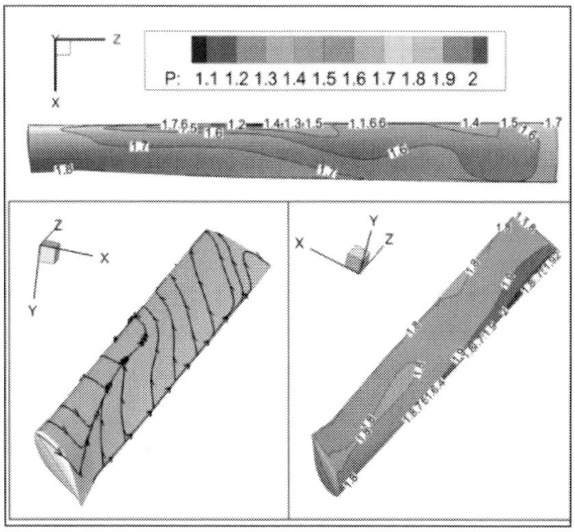

(c)Mtip = 0.653, θ=25°

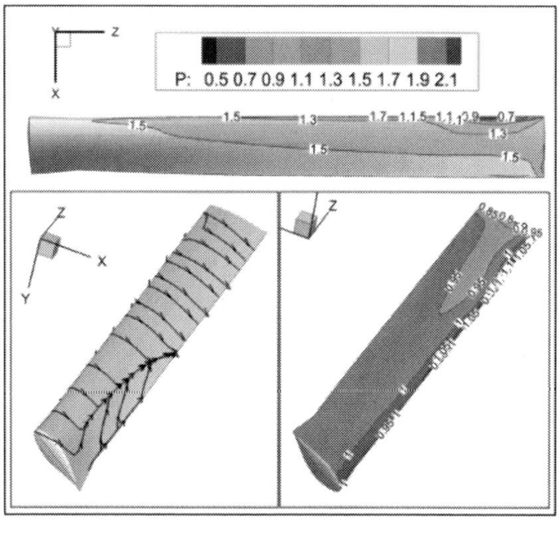

(d)Mtip = 0.69, θ=12.5°

Fig.9 Flow characteristics of the XV-15 at different rotation speeds

The influence of VST and VDT on cruise performance

In order to further analyze the impact of VDT on the cruise performance, two variable diameter strategies are adopted. These two types of rotors can be obtained by scale down the blade tip or blade root on the original configuration. Parameters of VST and VDT configurations are given in Table 2.

Table 2 Parameters of VST and VDT configurations

Configurations	Case	M_{tip}	Radius
Baseline Configuration	A	0.56	R
VST	B	0.69	R
	C	0.653	R
	D	0.466	R
	E	0.621	Tip0.9
	F	0.552	Tip0.8
VDT	G	0.483	Tip0.7
	H	0.621	Root0.9
	I	0.552	Root0.8
	J	0.483	Root0.7

Figure 10 shows the configurations of these two types of blades. The larger twist characteristics of blade root are retained as the blade tip changes (named as Tip class), while the higher lift airfoil of blade tip is retained as the blade root changes (named as Root class).

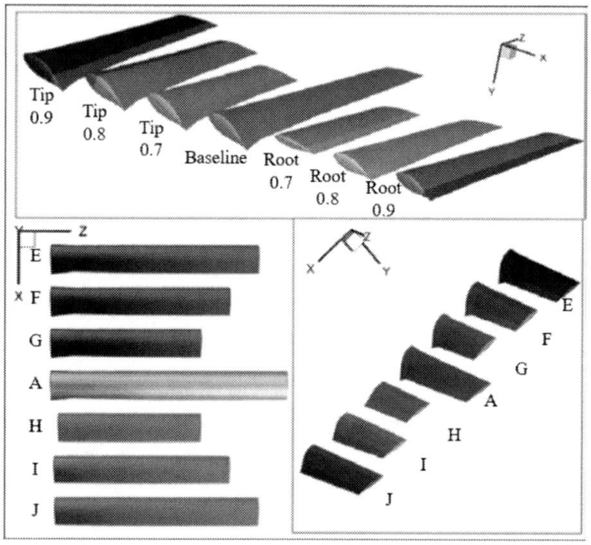

Fig 10 Comparison of different blade configurations

Figure 11 demonstrates the comparisons of propulsive efficiency among rotors of constant diameter, variable diameter, constant rotation speed and variable rotation speed at forward flight speed of 300 knots. The diagram indicates that, compared with the given speed strategy, rotors with variable rotation speed exhibits a relatively better performance. With a thrust of 13,000N, the propulsion efficiencies at the rotation speed of 500 fps, 600 fps, and with the configuration of root-0.8, tip-0.8, are close to each other. However, it should be noted that, a wide range of high propulsive efficiency can be obtained within a large thrust range at the rotation speed of 600 rpm, which further explains why the XV-15 tiltrotor aircraft has adopted 600 rpm as its cruise speed.

For such a design, variable rotation speed strategies are taken to directly alter the tangential inflow speed on blade sections, and variable diameter measures are taken to enhance the solidity of rotors. Both the above-mentioned strategies have realized noticeable decrease of the blade tip speed. To obtain equivalent thrust, pitch angle needs to be trimmed. However, influenced by both the inflow and pitch angle, blade sections with different configurations at the given forward flight speed have already deviated from the optimum working condition, thus leading to the decline of the tiltrotor performance to some extent.

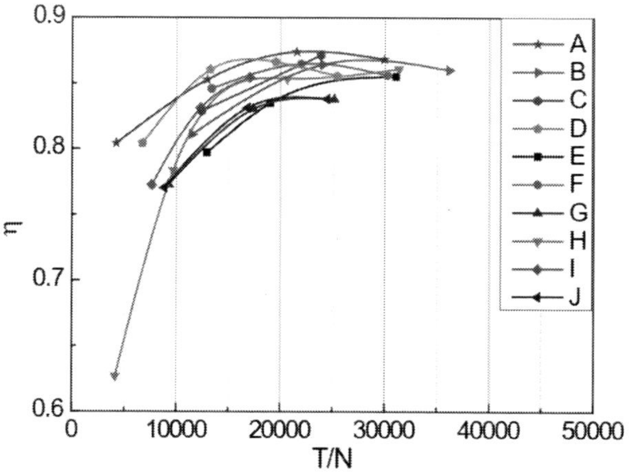

Fig 11 Comparison of aerodynamic characteristics of tiltrotor at cruise speed of 300 knot

To further explore the reason for the decrease of the aerodynamic performance, the nondimensional thrust coefficient for different configurations under the given cruise condition is presented in Fig 12. The diagram illustrates that both the configuration of the original rotation speed (740/700 fps) and the configuration of root-0.9 generate a relatively large negative lift at the inner side of the blade, and these two configurations also generate a relatively high thrust level at the tip compared with the configuration of rotation speed of 600 fps. Besides, it can also be noticed from Fig. 12 that all the other configurations exhibit a similar tendency in thrust distribution that gradually increases from the root to the tip. These conclusions indicate that the distribution of thrust coefficient along the blade tends to be smoother if various measures have been taken to alter the tip speed.

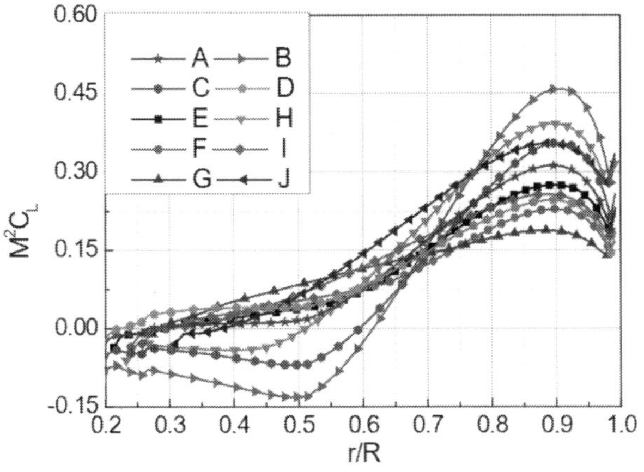

Fig 12 The distribution of thrust coefficient at cruise speed of 300 knots

Figure 13 illustrates the comparison of density distributions on the blade surface under the given cruise condition. The effect of the compressibility on the tip tends to grow with the increase of rotation speed, and the tip class enjoys a great advantage over the root class with respect to compressibility, thus leading to the improvement of the overall efficiency indicated in Fig 12.

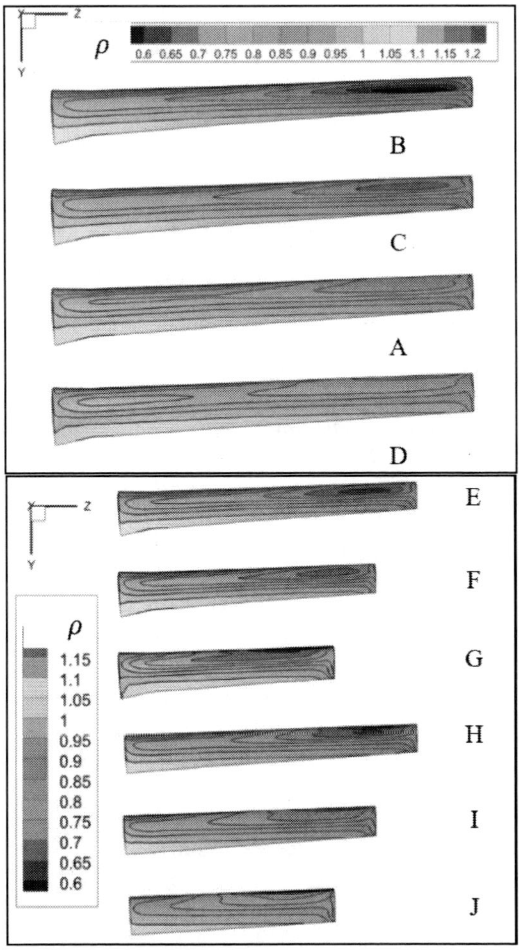

Fig 13 The distribution of the density of the blade in cruise flight

Figure 14 shows the propulsive efficiency comparison between variable speed rotor with constant diameter and variable diameter rotor in constant rotation speed at the speed of 230 knots in forward flight. When the thrust is 13,000 N, the optimum cruise performance can be obtained at the tip speed of 500 fps. Under such circumstances, the performances of variable rotation speed configurations transcend the performances of variable diameter configurations with constant rotation speed, which indicates that under small advance ratio, variable speed configurations perform better than the variable diameter configurations in low speed forward flight. At the given flight speed, the original configuration with a constant rotation speed possesses a wider range of high propulsion efficiency.

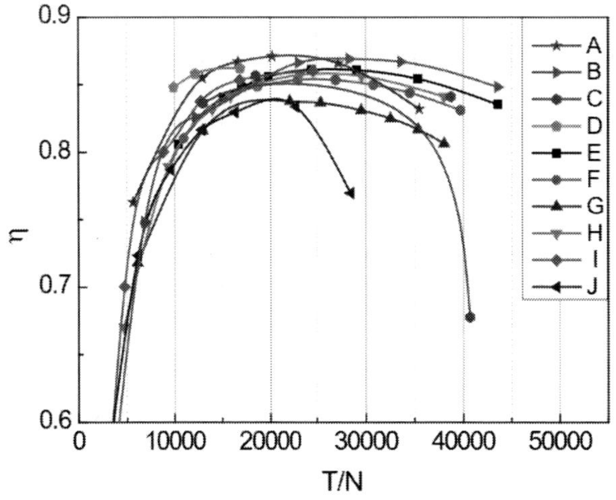

Fig 14 Comparison of the tiltrotor performances at cruise speed of 230 knots

Figure 15 shows the propulsive efficiency comparison between variable speed rotor with original diameter and variable diameter rotor in constant rotation speed at cruise speed of 400 knots. The figure shows that variable diameter configuration performs better, and the optimal configuration of blade turns out to be root-0.8 with a constant rotation speed at the given forward flight speed. It shows that root-0.8 can improve the tiltrotor performance at high cruise flight speed. With the increase of the forward speed, the drag of fuselage will increase significantly. So, the standard thrust is set to 20,000 N. It can be learnt from the figure that, compared with the original constant speed configurations, the propulsion efficiency of variable diameter configurations is significantly improved and they maintain high propulsion efficiency in a wide thrust range. Besides, the Root class performs better than the Tip class, for the Root class can work at a better angle of attack than the Tip class.

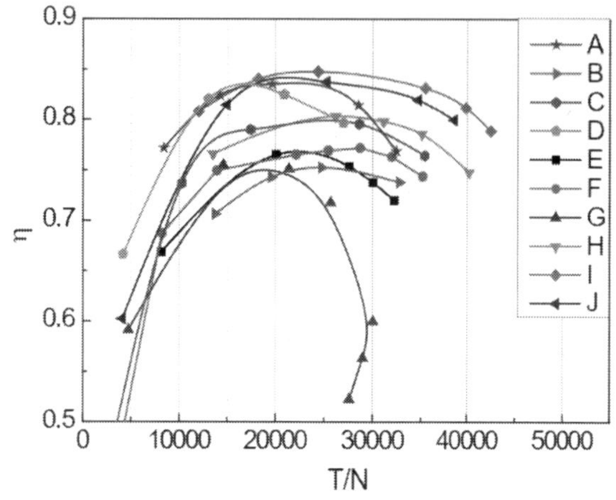

Fig 15 Comparison of the tiltrotor performances at cruise speed of 400 knots

Figure 16 demonstrates the comparison of aerodynamic performance at different cruise speeds with a given thrust. At low cruise speed, variable rotation speed configurations with a constant diameter exhibit a better performance than the others. However, with the increase of the cruise speed, both the variable rotation speed configurations and the variable diameter configurations undergo an obvious decline in efficiency. At the speed of 400 knots, the configuration of root-0.8 outperforms the others. Therefore, the Root class configuration is an advisable choice for the high cruise speed.

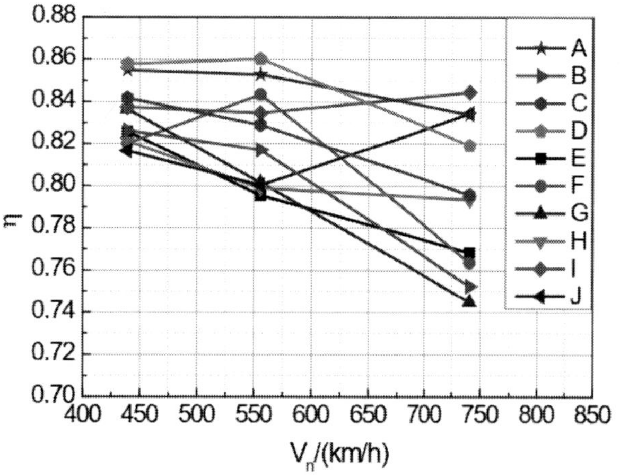

Fig 16 Comparison of the tiltrotor aerodynamic performance at different cruise speeds

CONCLUSIONS

A large amount of numerical analyses about aerodynamic performance of VST and VDT are introduced in hover and cruise flight. From the analyses, several conclusions are drawn.

1. The embedded grid system introduced in this paper can be used to analyze the aerodynamic performance of rotor at variable speed and in variable diameter.

2. The comparison between the simulation analyses and the experimental data indicates that the simulation analysis method established in the paper can satisfy the need of performance analysis at variable speed and in variable diameter.

3. In hover, for a small thrust, low blade rotation speed outperforms high blade rotation speed in terms of hover performance.

4. During low forward speed cruise, variable rotation speed configurations outperforms variable diameter configurations, and the Root class enjoys an obvious performance advantage over the Tip class. During high forward speed cruise, the Root class still maintains a performance advantage over the Tip class. Therefore, it can be concluded that in blade design of variable diameter and variable speed, retain high lift blade properties of the tip can improve tiltrotor performance in high speed cruise.

5. By combining variable rotation speed and variable diameter configurations, the aerodynamic performance of tiltrotor in high speed cruise can be further improved.

Author contact: Bo Wang wangbo@nuaa.edu.cn

Peng Li 494692237@qq.com

Qi-Jun Zhao zhaoqijun@nuaa.edu.cn

Si-Yu Chen chensiyu@nuaa.edu.cn

REFERENCES

[1]Fradenburgh, E., "Improving tiltrotor aircraft performance with variable diameter rotors," Fourteenth European Rotorcraft Forum. Italy, Sept. 1988.

[2]Fradenburgh, E., "Merging two ends of the V/STOL Spectrum," Eighteenth European Rotorcraft Forum, Avignon, France, Sept. 1992.

[3]EUROFAR-The European Future Advanced Rotorcraft Conference Proceedings, Royal Aeronautical Society, April 11, 1989.

[4]Pollack, M.,Warbuton, F. and Curtiss, P., "Asimulation study of tiltrotor vertical takeoff procedures using conventional and variable diameter rotor systems," Sikorsky Aircraft and Princeton University, Paper No. 91-26, Seventeen European Rotorcraft Forum, Berlin, Sept. 1991.

[5]Karem A.E., Optimum Speed Tiltrotor, U.S. Patent Number5, 541365, Novermber 4, 2003.

[6]Fradenburgh E., and Matuska D., "Advancing Tiltrotor State-of-the-Art with Variable Diameter Rotors," American

Helicopter Society 48th Annual Forum Proceedings, Washington, D.C., June 3–5, 1992.

[7]Wang J.M., Jones C.T., and Nixon M.W., "A variable dimeter short-haul civil tiltrotor," American Helicopter Society 55th Annual Forum Proceedings, Montreal, Canada, May 25-27, 1999.

[8]Lamon s., XV-15 Advanced Technology Blade, Ultimate Stres Analysis, Boeing Report Number-D210-12345-1, 1985.

[9]Sobieszczanski-Sobieski j., "Recent experiences in multidisciplinary analysis and optimization," NASA CP-2327, 1984.

[10]Dollyhigh S.M. and Sobieszcazanski-Sobieski J., "Recent experiences with multidisciplinary analysis and optimization in advanced aircraft design," Proc. 3th Air Force/NASA Symp. On Recent Advances in Multidisciplinary Analysis and Optimization, Sept. 24-26, 1990, San Francisco, CA.

[11]Weick F.E., *Aircraft Propeller Design*, McGraw-Hill Book Company, Inc., York, PA, 1930.

[12]Froude, W., "On the elementary relation between pitch, slip and propulsive efficiency," Transctions of the Institute of Naval Architects, 19, 1878, pp.47-57.

[13]Gessow, A., "Effect of rotor0blade twist and plan-form taper on helicopter hovering performance," NACA Technical Note 1542, 1948.

[14]Gessow, A., and Myers, G. C., *Aerodynamics of the Helicopter*, MacMillan Co., New York, NY, 1952, pp. 73-75.

[15]Zhao, Q. J., Xu G. H., Zhao J. G., "New hybrid method for predicting the flowfields of helicopter rotors," *Journal of Aircraft*, 2006, Vol. 43, (2), pp. 132-140.

[16]Spalart, P. R., Allmaras S. R., "A one equation turbulence model for aerodynamic flows," 30th Aerospace Sciences Meeting&Exhibit, 1992.

[17]Caradonna F. X., Laub G. H., Tung C., "An Experimental Investigation of the Parallel Blade-Vortex Interaction." NASA TM-86005, 1984.

[18]Felker F.F., Maisel M.D., and Betzina M.D., "Full scale tiltrotor hover performance," Journal of American Helicopter Society, 1985.

Affective Human Factors in Aviation

Çiğdem GÜNEŞ
Human Factors Specialist, MSc
Turkish Aerospace Industries, Inc. (TAI)
Ankara, TURKEY

Gülşen TÖRE YARGIN
Assist. Prof. Dr.
Middle East Technical University
Ankara, TURKEY

ABSTRACT

Conventional Human Factors/Ergonomics (HF/E) approaches in aviation concentrate on the product's or system's utility and underestimates the power of the users' feelings. By a majority, HF/E emphasizes on ease of access, human performance, human error, workload and situation awareness, which are related with the functional features of the system, without emotional interactions. However, it is now clear that the success of the product in the market comes from its aesthetic appeal and the affect it creates on the users; the interaction between the user and the designed artifact is impressed by emotions. There has been a growing interest in research of industrial design, engineering design, psychology, neuroscience, and human factors engineering to affect and pleasure. There is a shift from cognitive perspective to affective, which concerns promoting pleasure instead of preventing pain (workload, error, etc.) resulting by the design. There are many studies in the literature focused on the consumer products. However, complex systems like aviation industry where performance and safety is the main point, is still impassive to affective design, the theoretical constructs related to affective design have been poorly defined. Affective considerations might provide a new and different perspective in aviation human factors. This study will present the importance of affective human factors in aviation and purposes some suggestions to reflect affective design to cockpit design.

INTRODUCTION

Flying an aircraft is one of the most challenging activities for a human in terms of required cognitive capability to control the aircraft, to trace various displays and information, and to communicate simultaneously. Pilots interact with the aircraft, other aircrafts, ground stations, and airport with control grips, displays, and control panels. Pilots' primary responsibility is to 'aviate', 'navigate', and 'communicate' for a safe flight. During these activities, pilot(s) are in the flight deck in extremely long hours in high workload. Therefore, performance and speed is a key factor for flight deck design. The flight deck and related pilot vehicle systems are designed to enhance the pilots' performance through the human factors considerations. By a majority, Human Factors/Ergonomics (HF/E) regards human performance, human error, workload and situation awareness, which are related with the functional features of the system without emotional interactions (Ref. 35; Ref. 50; Ref. 12). Traditional human factors cognitively formulate the pilots and does not take pilots' emotions account. However, it is known that human can easily impressed by emotions; therefore, design should arouse proper feelings (fun for a mobile phone, excitement or safe for a sport car, etc.) for the designed task. Before proceeding further, it will be worth to clarify some definitions. The term 'affect' is widely used to define the feelings that a human experienced.

Presented at the 6th Asian/Australian Rotorcraft Forum / Heli Japan 2017, Kanazawa, Japan, November 7-9, 2017.

Affect means human's psychological response to the perceptual design details of a product (Ref. 6). In literature, there are several terminologies for affective design like *Affective Human Factors* (Ref. 20; Ref. 18; Ref. 27), *Hedonomics* (Ref. 16; Ref. 38; Ref. 41; Ref. 42), *Engineering Aesthetics* (Ref. 36; Ref. 10), *Kansei Engineering* (Ref. 39; Ref. 43), *Flight Deck Aesthetic* (Ref. 12), Funology (Ref. 1; Ref. 3), *Positive Design* (Ref. 7; Ref. 17), Citarasa (Ref. 29; Ref. 22) and similar other terminologies. Some of them refer to design field and some of them refer to engineering. All of the concepts are not equivalent but they are similar concepts depend on the integration of affective qualities to designed object.

All these concepts discuss sometimes similar, sometimes intersecting, and sometimes complementary issues of design, which collectively mention the emotional effect of the design. In this paper, "Affective Human Factors" terminology will be used, as it is found more comprehensive. By this terminology, both aviation, which is familiar to human factors, and affect, which is representative of industrial design field, is covered. The term is able to reference human factors and give a clue to the reader about the concept.

Affective considerations may provide a new and different perspective in aviation human factors and design. Affective design focusses on emotional responses of users to design and expects the users to attach feelings, such as attractiveness, desirability, or pleasantness, to products or services. Flight deck designers, generally engineers, focus on the function of the design and underestimate the importance of form, whereas in the field of design, it is accepted that form and function cannot be separated; "form and function are one" (Ref. 49, p. 20) and "attractive things work better" (Ref. 40, p. 42).

AFFECTIVE HUMAN FACTORS

Affect, Emotion and Cognition

Affect and cognition are the components of the human information-processing system (Ref. 40, Ref. 41) that has different mechanisms, functions, and outcome. Cognition is a decision-making process, which is analytical and rational and utilizes from thinking, reasoning, and problem-solving processes. Affective system is the judgmental system that decides which things in the environment is good, bad, safe, or dangerous, etc. On the other hand, emotion is the conscious experience of the affect (Ref. 40, Ref. 41). That means affect is the feeling that you could not explore the source, while emotion appeared because of something definable. Norman (Ref. 41) claims that cognition and affect are in mutual effect, each system influenced by each other. However, aviation human factors generally focuses on the cognitive part of this interaction.

Affect in Aviation

A product or service does not sell just for its utility anymore; something it has must attract the users. The key factor that effects the success of a new product is to touch the user's senses and emotions. People have affective reactions to the objects internally that makes some objects more attractive for them that can be caused by the visual appearance of the object, or tactile sense of touching it, or a previous experience. Design activity cannot be considered as just shaping a product physically; a designer also decides consciously or unconsciously how to use the product, how the product will interact with the user and what kind of affect the product will have on the user while designing the product.

On the contrary, the objectives of HF/E are increasing the productivity and the efficiency of work, improve safety, and increase the comfort (Ref. 48). Whilst doing this, the capabilities, limitations, characteristics, behavior, and motivation of the human should be considered. Traditionally HF/E is problem oriented (Ref. 9), calls for solutions to the problems of human interaction with a system, and aviation HF/E has still been based on possible problems. HF/E considers usability, effectiveness, and efficiency and does not focus much on the effect of visual appearance of the product, the feelings it creates on human and the potential effect on performance. In spite of the remarkable background of HF/E research dates back to the Second World War, HF/E has not given much attention to visual aesthetic until mid-1990s (Ref. 47). Affect is new in aviation HF/E emerging in the last decade; aesthetics and emotion in aviation is poorly researched and their interaction with crew's performance is even less (Ref. 12).

Design on the other hand has long been using the power of affect. Designers (not engineers) are aware of the power of affect and researches and industrial applications are studied in consumer product design (Ref. 14), game design (Ref. 26) or automobile design (Ref. 8; Ref. 4; Ref. 21). Although their perspectives towards considering affective qualities in design is different, still HF/E and industrial design has concurrent perspectives (Ref. 15) in terms of designing efficient systems/products for the users and can find common ground on the field of "Affective Human Factors Design" (Ref. 27). Khalid (Ref. 27) declares that by this association both industrial design by integrating assessment and prediction methods and HF/E by considering affective issues will have the opportunity to advance their fields.

In aviation, HF/E considers negative effects that is related to human performance (time and human error) and related to physical or psychological pain (muscle pain and workload). However, the focus has been changing with the concept of "Hedonomics" (a term purposed for defining affect by Hancock) from performance and pain to pleasure. (Ref. 18). Affective human factors is an emerging field but much more research is needed especially for performance-required systems like aviation. Although there is much research on consumer products regarding affective human factors, it has not been considered much for systems like aircrafts, combat vessels, and battle tanks where high safety and performance is required in a dense, clustered operation area. Moreover, not only integration of affect into these systems' design, it is also required to integrate affect into test and evaluation activities. In aviation, test and evaluation (T&E) phases are essential part of the system design and should take place throughout the design process. T&E phases guide system design to improve and eliminate the design defects. Inputs of HF/E through the medium of its own methods and evaluation scales throughout the design phase are critical. HF/E specialists contribute to design process by evaluating the interaction between the designed artifact (both physical and soft) and the operators (both flight crew and flight engineers), the impact of design on human performance and workload, system's tendency to human error and user comfort and satisfaction (Ref. 5). Although subjective outputs of user comfort and satisfaction are the responsibility of HF/E specialist, evaluation is still limited with physical ergonomic values and neither the hedonic values nor attitudes of the pilots are not concerned.

Although affective design is not limited with aesthetic, there is a strong link between the aesthetic and affective design; aesthetic is a facet of affective design. It is obvious that aesthetic design makes a positive contribution to competitive advantage and increase customers' willingness to buy. Firms give more emphasis on the use of aesthetic design are more likely to gain more economic value and competitive advantage than firms ignore its advantage (Ref. 44). Willingness to buy depends on customers' expectations and their perceived utility provided by aesthetic design compared to perceived utility provided by functionality. Khalifa (Ref. 30) explains this phenomenon by grouping the features as basic, expected, and innovative. Basic features are the taken for granted properties that absence of the property will dissatisfy the user. Expected features are the needs of the customer and customer's satisfaction depends on how effectively this need is provided. Innovative features are the properties that are nice to have and absence of the property will not dissatisfy the user. If a product does not achieve basic features it does not matter how aesthetic it is. However, preference of the two products both meet the basic features would depend on the perceived utility provided by aesthetic or functionality.

Research on affect generally focused on the consumer products. Helander and Khalid (Ref. 19) stated that by enhancing the positive emotions and minimizing the negative emotions simultaneously makes the product successful and increases the sales. For example, video games, which generate the feeling of "flow", are more successful indeed the feeling of mastery makes these games addicting (Ref. 26). Johnson and Wiles (Ref. 26) also explored that when the video games played on a console dedicated for that game the effect of affect is more instead of using a personal computer. However, performance-required systems or safety-critical systems, which are the case in aviation, are different from consumer products in terms of needs and expectations. Performance-required systems are more complex than consumer products and mainly include various components, which are also complex itself. These systems are prone to error and usually error cause disaster. Therefore, there is high stress on the operators of these systems compared to the users of consumer products who may feel annoyances because of a failure. In this systems system effectiveness depends on system's ability to accomplish the mission, while in consumer products system effectiveness comes from user satisfaction.

Affect is also important for cognition and perception process where decision-making and human information processing occur. Affective human factors is not just entirely related with fun, there is also much more exalted mission about engaging the user into complex systems. Khalid (Ref. 28) declares that negative emotions cause users to avoid a product while positive emotions make them engaged. Engagement is important for aviation because rejecting the use of automation may also result in human error (Ref. 13).

Isen (Ref. 24) also claims that positive affect provides user to better organized which is essential for clustered work environments like cockpits, more creative and more flexible decision making and problem-solving ability which is essential for time critical systems where the situation may change instantly. In addition, Sondregger and Sauer (Ref. 46) asserted that positive aesthetic qualities are more effective than ergonomic qualities in terms of acceptance of the design. According to their findings, aesthetically appealing design resulted in better performance and performance was not affected after a negative user experience. Bushnell (Ref. 2) prepared a report about the industrial design effect in aerospace and claimed that aesthetic appearance affects human both psychologically and physiologically thus affects human efficiency and attitude. This provides the enhancement of overall productivity and enjoyment.

Performance and Affect

Lindgaard and Whitfield (Ref. 34) exposed that aesthetic has an explicit linkage with performance. Rasmussen (Ref. 45) distinguishes performance to three levels as skill-based, rule-based, and knowledge-based and Lehto (Ref. 32) adds judgment-based as fourth level (Ref. 33) as given in the Figure 1. Skill-based performance is the lowest level of performance and generated automatically such as bicycle riding. Rule-based performance involves the routine steps in a familiar situation such as approach procedure of an aircraft. Distinction between these two levels is not quite distinct. While in the rule-based level procedure can be explicitly explained, and requires conscious attention, in the skill-based level explanation is implicit and without conscious attention. Knowledge-based level is activated when an unfamiliar situation occurs. At this level, the decision maker developed plans based on the mental model of current situation (Ref. 45). Lehto (Ref. 32) placed judgement-based level of performance over the skill-based, rule-based, and knowledge-based and claimed that judgement-based level of performance occurs when affective reactions of a decision-maker effects the goal, decisions or priorities. According to Lehto (Ref. 32), these four levels have distinctive type of errors. At the skill-based level, errors are derived from fail to receive the signal, motor variability, fail to create schema for the situation or create wrong schema. At the rule-based level, errors are derived from applying faulty rules, skipping a step or inability to perceive the state. At the knowledge-based level, errors are derived from cognitive limitations of problem solving and inability to formulate the correct intentions. At judgment-based levels, errors are derived from inappropriate affective reactions (Lehto, 1991). Human error is a crucial problem in aviation as the majority of the accidents are based on in some way, to some form of human error. Major goal of HF/E is to eliminate the human error by reducing workload and supporting human performance and it seems that judgement-based level of performance effected by affective reactions, which makes the interaction more prone to human error.

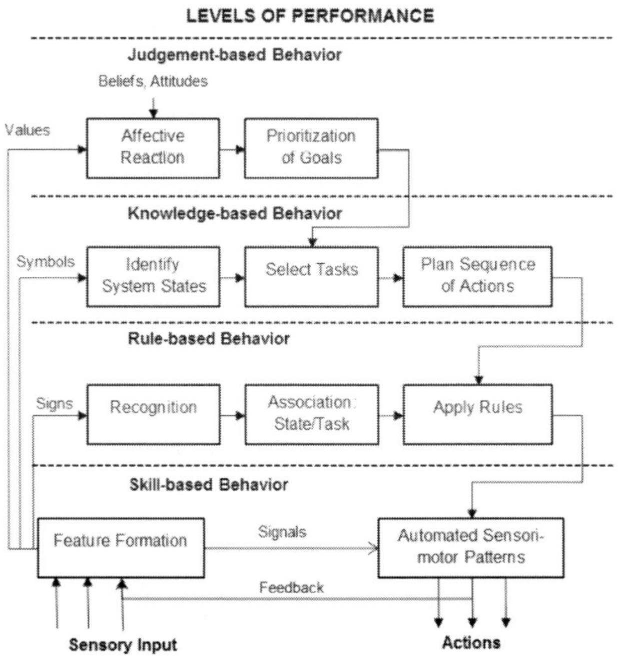

Figure 1: Levels of Performance (Ref. 32, pp. 600)

According to Isen and Means (Ref. 25) positive affect influences the efficiency of decision-making process positively. In their experiment, participants who have good mood are prone to ignore unimportant information and the participants less likely to review the information they had already seen. These participants are faster in each predetermined task and they have better performance.

Positive affect also enhances the cognitive flexibility; the ability to think multiple issues simultaneously (Ref. 38). Murphy et al. (Ref. 38) found that aesthetically pleasing design increase the perceived usability when accompanied with good usability but when the usability has weaknesses these deficiencies gets bigger and perceived usability is decreased. That means aesthetic makes people more discerning about usability.

Affect influences the overall human information processing style. Compared to negative, positive moods induce heuristic processing and encouraging people to give faster decisions and to pay little attention to details. While positive moods avoid systematic thinking, and make reliance on preexisting knowledge, negative moods support systematic processing, make human prone to focus on details and cause little reliance on preexisting knowledge (Ref. 31).

Affective state of the users remarkably influences the performance especially in complex systems where decision support systems have a crucial role (Ref. 23). Moshagen et al. (Ref. 37) also explored that completion time of the tasks in highly aesthetic websites is significantly shorter than in unaesthetic websites. Likewise, Sonderegger and Sauer (Ref. 46) acquired similar results with cell phones that visually appealing design enables users to reduce the completion time of the tasks with less error.

Gannon (Ref. 11) conducted an experiment about the correlation between visual aesthetics and pilot workload, and performance with 24 pilots. He found that overall workload has decreased when the interface is found attractive by the pilots even though only the aesthetics of the PFD (Primary Flight Display) interface graphic design (only the colors and shapes) was manipulated.

CONCLUSIONS

The affective qualities of design have been given more and more consideration in design and human factors fields and there have already been many statements that propound that affect has an explicit effect on the users and their performance. There is not much work in aviation industry with affective qualities and if something may an effect on performance, human factors specialists have to scrutinize it.

Further research is required in order to measure this effect on pilots. Such a study will provide how affective qualities would influence pilot performance and how to manipulate design in order to enhance the performance.

Author contact:

Çiğdem GÜNEŞ cigdem.gunes@tai.com.tr
Gülşen TÖRE YARGIN tore@metu.edu.tr

REFERENCES

[1]Blythe, M., Overbeeke, K., Monk, A., and Wright, P., *Funology: From Usability to Enjoyment*, Kluwer, Boston, 2003.

[2]Bushnell, D. M. "Industrial Design in Aerospace/Role of Aesthetics". NASA/TM-2006-214498 Hampton, Virginia: NASA Langley Research Center, 2006.

[3]Carroll, J. M., "Beyond Fun," *Interactions*, Vol. 11 (4), 2004, pp. 38-40.

[4]Chan, K.-Y., and Chang, S.-C. "A Preliminary Study on the Integration of Engineering and Aesthetics Measures via the Design of Vehicle Silhouettes," *Journal of Mechanical Engineering Science*, Vol. 229 (12), 2015, pp. 2221-2230.

[5]Czaja, S. J., and Nair, S. N., "Human Factors Engineering and Systems Design," edited by G. Salvendy, *Handbook of Human Factors and Ergonomics*, Fourth Edition, John Wiley & Sons Inc., New Jersey, 2012, pp. 38-56.

[6]Demirbilek, O., and Şener, B., "Product Design, Semantics and Emotional Response," *Ergonomics*, Vol. 46 (13-14), 2003, pp.1346-1360.

[7]Desmet, P. M., and Pohlmeyer, A. E., "Positive Design: An Introduction to Design for Subjective Well-Being," *International Journal of Design*, Vol. 7 (3), 2013, pp.5-19.

[8]Desmet, P., Hekkert, P., and Hillen, M., "Values and Emotions, an Empirical Investigation in the Relationship between Emotional Responses to Products and Human Values," 5th European Academy of Design Conference, Barcelona, Spain, 2004.

[9]Edwards, E., "Introductory Overview," edited by E. L. Wiener, and D. C. Nagel, *Human Factors in Aviation*, Academic Press, San Diego, 1988, pp. 3-24.

[10]Faste, R. A., "The Role of Aesthetics in Engineering," *Japan Society of Mechanical Engineers (JSME) Journal*, Vol. 28, 1995, pp. 385-391.

[11]Gannon, A. J., "The Effects of Flightdeck Display Interface Aesthetics on Pilot Performance and Workload," Presented in Partial Fulfillment of the Requirements for the Degree Doctor of Philosophy in the Graduate School of the Arizona State University, 2005.

[12]Gannon, A. J., "Flight Deck Aesthetics and Pilot Performance: New Uncharted Seas," edited by J. A. Wise, V. D. Hopkin, and D. J. Garland, *Handbook of Aviation Human Factors-Human Factors in Transportation*, CRC Press, Baco Ratton, 2010, pp. 382-402.

[13]Güneş, Ç. (2010). "Human Factors Issues of Glass Cockpit Automation," Presented in Partial Fulfillment of the Requirements for the Master of Science in the Graduate School of Natural end Applied Sciences, Middle East Technical University, Ankara, 2010.

[14]Han, S. H., Kim, K. J., Yun, M. H., Hong, S. W., and Kim, J., "Identifying Mobile Phone Design Features Critical tu User Satisfaction," *Human Factors and Ergonomics in Manufacturing & Service Industries*, Vol. 14(1), 2004, pp.15-29.

[15]Hancock, P. A. "Human Factors/Ergonomics," edited by V. S. Ramachandran, *Encyclopedia of Human Behavior*, Elsevier, Oxford, 2012, pp. 358-363.

[16]Hancock, P. A., Pepe, A., and Murphy, L. L., "Hedonomics_The Power of Positive and Pleasurable Ergonomics," *Ergonomics in Design*, 2005, pp.8-14.

[17]Hassenzahl, M., and Tractinsky, N., "User Experience-A Research Agenda," *Behaviour and Information Technology*, Vol. 25(2), 2006, pp.91-97.

[18]Helander, M. G., "Hedonomics - Affective Human Factors Design," Proceedings of Human Factors and Ergonomics Society 46th Annual Meeting, 2002, pp. 978-982.

[19]Helander, M. G., and Khalid, H. M., "Affective and Pleasurable Design," edited by G. Salvendy, *Handbook of Human Factors and Ergonomics*, Third Edition, John Wiley&Sons Inc., New Jersey, 2006, pp. 543-572.

[20]Helander, M. G., and Tham, M. P., "Editorial: Hedonomics-Affective Human Factors Design," *Ergonomics*, Vol.46(13/14), 2003, pp.1269-1272.

[21]Helander, M. G., Khalid, H. M., Lim, T. Y., Peng, H., and Yang, X., "Emotional Needs of Car Buyers and Emotional Intent of Car Designers," *Theoretical Issues in Ergonomics Science*, Vol. 14(5), 2013, pp.455-474.

[22]Helander, M., Khalid, H., Hancock, P., Jeon, M., Seva, R., and Bruder, R., "Affective Design and EQUID: Emotional and Ergonomic Quality in Product Design and Development," Proceedings 19th Triennial Congress of the IEA, Melbourne, 2015.

[23]Hudlicka, E., and McNeese, M. D., "Assessment of User Affective and Belief States for Interface Adaptation: Application to an Air Force Pilot Task," *User Modeling and User-Adapted Interaction*, Vol.12(1), 2002, pp.1-47.

[24]Isen, A. M., "An Influence of Positive Affect on Decision Making in Complex Situations," *Journal of Consumer Psychology*, Vol. 11 (2), 2001, pp.75-85.

[25]Isen, A. M., and Means, B., "The Influence of Positive Affect on Decision-making Strategy," *Social Cognition*, Vol.2 (1), 1983, pp.18-31.

[26]Johnson, D. M., and Wiles, J., "Effective Affective User Interface Design in Games," *Ergonomics*, Vol. 46 (13-14), 2003, pp.1332-1345.

[27]Khalid, H. M., "Guest Editorial: Conceptualizing Affective Human Factors Design," *Theoretical Issues in Ergonomics Science,* Vol. 5 (1), 2004, pp.1-3.

[28]Khalid, H. M., "Embracing Diversity in User Needs for Affective Design," *Applied Ergonomics*, Vol. 37, 2006, pp. 409-418.

[29]Khalid, H. M., "Citarasa Engineering for Modeling Affective Product Design," edited by H. Khalid, A. Hedge, and T. Z. Ahram, *Advances in Ergonomics Modeling and Usability Evaluation*, CRC Press, Boca Raton, 2010, pp. 269–279.

[30]Khalifa, A. S., "Customer Value: A Review of Recent Literature and Integrative Configuration," *Management Decision,* Vol. 42 (5), 2004, pp. 645-666.

[31]Lee, J. D., "Affect, Attention, and Automation," Edited by A. Kramer, D. Wiegmann, and A. Kirlik, *Attention: From theory to Practice*, Oxford University Press, New York, 2006, pp. 73-89

[32]Lehto, M. R. "A Proposed Conceptual Model of Human Behavior and Its Implications for Design of Warnings," *Perceptual and Motor Skills*, Vol. 73, 1991, pp. 595-611.

[33]Lehto, M. R., Nah, F. F.-H., and Yi, J. S., "Decision-Making Models, Decision Support, and Problem Solving," edited by G. Salvendy, *Handbook of Human Factors and Ergonomics*, Fourth Edition, John Wiley & Sons, Inc, New Jersey, 2012, pp. 191-242.

[34]Lindgaard, G., and Whitfield, A., "Usability, Aesthetics and the User Experience: A Theoretical Proposal," Proceedings International Conference on Affective Human Factors Design, Singapore, 2001, pp. 373-378.

[35]Lindgaard, G., and Whitfield, T. W., "Integrating aesthetics within an evolutionary and psychological," *Theoretical Issues in Ergonomics Science*, Vol. 5(1), 2004, pp. 73-90.

[36]Liu, Y. "The Aesthetic and Ethic Dimensions of Human Factors and Design," *Ergonomics*, Vol. 46(13/14), 2003, pp. 1293-1305.

[37]Moshagen, M., Musch, J., and Göritz, A. S., "A blessing, not a curse: Experimental evidence for beneficial effects of visual aesthetics on performance," *Ergonomics*, Vol. 52(10), 2009, pp. 1311-1320.

[38]Murphy, L. L., Stanney, K., and Hancock, P. A., "The Effect of Affect: A Hedonomic Evaluation of Human-Computer Interaction," Proceedings of the 47th Annual Meeting of the Human Factors and Ergonomics Society, 2003, pp. 764-768.

[39]Nagamachi, M., "Kansei Engineering: An Ergonomic Technology for a Product Development," Proceedings of IEA, 1994, pp. 1220-1222.

[40]Norman, D. A., "Emotion and Design: Attractive Things Work Better," *Interactions Magazine*, Vol. 9 (4), 2002, pp. 36-42.

[41]Norman, D. A., *Emotional design: Why do We Love (or Hate) Everyday Things*, Basic Books, New York, 2004.

[42]Oron-Gilad, T., and Hancock, P. A., "From Ergonomics to Hedonomics: Trends in Human Factors and Technology-The Role of Hedonomics Revisited", edited by M. Jeon, *Emotions and Affect in Human Factors and Human-Computer Interaction*, Chapter 7, Academic Press, San Diego, 2017, pp. 185-194.

[43]Perez Mata, M., Ahmed-Kristensen, S., and Yanagisawa, H., "Perception of Aesthetics in Consumer Products," Proceedings of the 19th International Conference on Engineering Design (ICED13), 2013, pp. 527-536.

[44]Porter, M. E., *Competitive Advantage*, The Free Press, New York, 1985.

[45]Rasmussen, J., "Skills, Rules, Knowledge: Signals, Signs, and Symbols and Other Distinctions in Human Performance Models," *IEEE Transactions on Systems, Man, and Cybernetics*, Vol. 13 (3), 1983, pp. 257-267.

[46]Sonderegger, A., and Sauer, J., "The Influence of Design Aesthetics in Usability Testing: Effects on User Performance and Perceived Usability," *Applied Ergonomics*, Vol. 41, 2010, pp. 403-410.

[47]Tractinsky, N., "Aesthetics and Apparent Usability: Empirically Assessing Cultural and Methodological Issues," CHI Conference on Human Factors in Computing Systems, ACM, New York, 1997, pp. 115-122.

[48]Wickens, C. D., Gordon, S. E. and Yili, L., *An Introduction to Human Factors Engineering*, Addison-Wesley Longman, Inc., New York, 1998.

[49]Wright, F. L., *The Natural House*, Horizon Press, New Yok, 1954.

[50]Zukowsky, J., "Design for the jet age: Charles Butler and Uwe Schneider," *Design Issues*, Vol. 13 (3), 1997, pp. 66-81.

Generalized Timoshenko Theory for Functionally Graded Beams with Three-dimensional Warping

Manoj Kumar Dhadwal
Postdoctoral Researcher
Konkuk University
Seoul, Republic of Korea

Sung Nam Jung
Professor
Konkuk University
Seoul, Republic of Korea

ABSTRACT

Functionally graded materials have been gaining considerable attention in the aerospace field since they offer tailored material properties with smooth variations. Although various theories have been proposed to model transverse shear effects at Timoshenko level for functionally graded beams (FGBs), a generalized formulation which includes three-dimensional (3D) warping and elastic couplings is lacked in the earlier works. The present work deals with the development of a generalized multifield variational formulation for FGBs using Reissner's mixed principle considering both displacements and cross-sectional stresses as unknowns. A generalized Timoshenko model for cross-sectional stiffness constants is derived including the generic treatment of elastic couplings through 3D (in-plane and out-of-plane) warping displacements. The resulting formulation thus inherently includes transverse shear and Poisson effects, and associated couplings. The cross-sectional formulation is implemented in a two-dimensional (2D) finite element (FE) code. The elastostatic response from the present analysis is substantiated against 3D FE solutions for various FGBs indicating an excellent correlation. The continuous distributions of sectional stresses for FGBs are also demonstrated, which are computed directly in the present analysis.

INTRODUCTION

The slender beam-like structures offer a wide variety of applications such as in helicopter or wind turbine rotor blades, aircraft wings, and propellers. Generally, laminated composite materials have seen an extensive usage for the structural design due to their low weight and high strength along with the advantage of tailored properties. However, the laminated composites usually required layerwise layup definitions for manufacturing and pose additional damage problems due to abrupt change in the stresses across the layer interfaces. Recently, FGMs are gaining attention due to their continuously varying properties along desired directions which results in a smooth variation of the stresses (Ref. 1). FGM can also be useful as a transition material to smooth out the stress gaps between highly dissimilar materials.

A plethora of research is available on the elastostatic and/or free vibration analysis of FGM beams. A brief overview of a few of the major works is presented next. Horgan and Chan (Ref. 2) proposed analytical solutions for torsional behavior of functionally graded isotropic bars. Sankar (Ref. 3) presented analytical solutions based on Euler-Bernoulli beam theory for orthotropic beams made of FGM. Chakraborty et al. (Ref. 4) used a first-order shear deformation theory for the analysis of isotropic FGBs. Zhong and Yu (Ref. 5) derived analytical solutions for orthotropic cantilevered FGBs. Ke et al. (Ref. 6) performed nonlinear free vibration analysis of functionally graded carbon nanotube-reinforced composites based on Timoshenko beam theory. Giunta et al. (Ref. 7) proposed a hierarchical beam theory for the free vibration analysis of FGBs assuming the kinematical relation using a generic polynomial approximation. Filippi et al. (Ref. 8) performed static analysis of FGM beams using the Carrera unified formulation (Ref. 9). Nguyen et al. (Ref. 10) presented an analytical thin-walled beam theory for open-section FGM beams neglecting in-plane warping. Kim and Lee (Ref. 11) presented exact solutions using a thin-walled approach for the analysis of FGM sandwich beams.

Most of the studies mentioned previously rely on *ad-hoc* assumptions by neglecting the in-plane cross-sectional deformation and/or ignoring some of the stress components. Also, analytical approaches cannot accurately model the arbitrary geometry of beam cross-sections. Because of the material distribution and geometric shape, various elastic couplings can appear which are required for an accurate prediction of the global behavior. An efficient approach for the analysis of beam structures is to decompose 3D beam analysis into local 2D sectional analysis and global one-dimensional (1D) beam analysis (Ref. 12). The local sectional analysis is performed to compute the sectional elastic and inertial properties considering the classical elastic couplings and nonclassical effects due to 3D warping displacements. The computed sectional properties are then used to perform 1D beam static or dynamic analyses. The proper recovery of 3D displacements, strains, and stresses is strongly dependent on the 2D sectional and 1D

Presented at the 6th Asian/Australian ,Rotorcraft Forum / Heli Japan 2017, Kanazawa, Japan, November 7-9, 2017.

beam analyses.

There are numerous analytical as well as computational studies on the composite beams and blades in the literature. Most beam theories are developed from the displacement-based formulations, such as in (Refs. 13–16), which consider the displacements as the only unknown variables. The strains and stresses obtained from the displacement derivatives are therefore discontinuous across element boundaries. A fine FE mesh is usually needed with a large number of elements to achieve the converged stresses. The mixed or multifield theories (Ref. 17) can overcome this drawback which involves modeling of some or all stresses as unknowns which may result in an accurate prediction with continuity across elements using only fewer elements.

The present formulation is developed using the Reissner's multifield variational principle (Ref. 17) for continuous stress distribution in FGM beams. The proposed formulation is implemented into a FE code called multifield variational sectional analysis code (MVSAC) which allows arbitrary geometric shapes and material distributions. The key points of the present formulation are listed as follows: (a) the 3D warping displacements and transverse sectional stresses (one normal and two transverse components, called as *reactive stresses* as defined in Ref. 17) are modeled as unknowns. The remaining nondominant stress components are computed using the material constitutive relations and are expressed in terms of displacement derivatives through the strain relations; and (b) the 3D warping deformations (in- and out-of-plane) and transverse sectional stresses are approximated as linear functions of generalized stress resultants (due to extension, shear, torsion, and bending) at the global 1D beam level, and using FE shape functions at the local 2D sectional level. This leads to a generic nonlinear distribution of both the 3D warping and reactive stress fields over the beam cross-section. The present formulation also considers the classical elastic couplings along with the rigorous treatment of nonclassical couplings due to transverse shear and Poisson deformations. A 6×6 generalized Timoshenko like stiffness matrix is subsequently derived which incorporates Timoshenko like model for transverse shear. The elastostatic responses from the present analysis are validated against the 3D FE solutions for several FGM beams. The distributions of cross-sectional warping and reactive stresses are also illustrated in comparison to conventional homogeneous beams.

CROSS-SECTIONAL ANALYSIS

A brief description of the present multifield formulation is discussed next. The beam is considered to be straight and prismatic. The schematic of the beam section made of FGM with continuously varying properties is shown in Fig. 1. The constitutive relations for a linear elastic FGM are defined using generalized Hooke's law as

$$\sigma = \mathbf{C}(\xi_2, \xi_3)\varepsilon_m \tag{1}$$

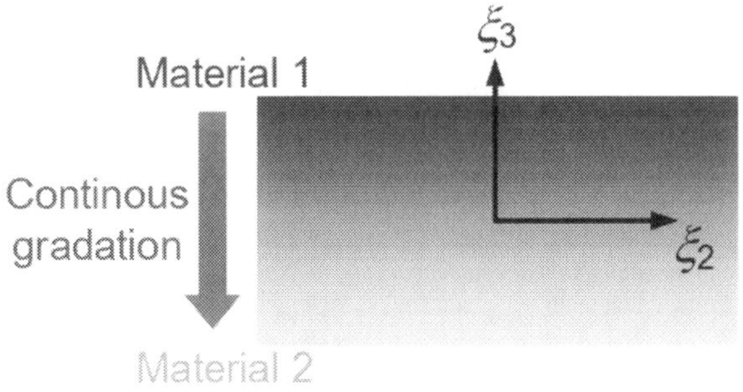

Figure 1. Beam section made of functionally graded material.

where the matrix $\mathbf{C}$ varies continuously with the cross-sectional location.

In the present study, the variation of any FGM property p is assumed to follow the power law as given by (Ref. 7)

$$p = (p_1 - p_2)(\alpha_2\xi_2 + \beta_2)^{k_2}(\alpha_3\xi_3 + \beta_3)^{k_3} + p_2 \tag{2}$$

In case of linear variation along ξ_2 and ξ_3, the exponents k_2 and k_3 will assume the value of unity. Note that the variation of properties can be applied to any material type such as isotropic, orthotropic or anisotropic materials.

Semi-inverted Material Constitutive Relations

For the application of Reissner's multifield variational principle, the stresses and strains are decomposed into sectional stresses acting on the beam section (normal and transverse shear stresses) and stresses acting on the planes normal to the beam section (normal and in-plane shear stresses) as

$$\begin{aligned}
\sigma_s &= \lfloor\, \sigma_{11} \;\; \sigma_{12} \;\; \sigma_{13} \,\rfloor^T, \quad \sigma_n = \lfloor\, \sigma_{22} \;\; \sigma_{33} \;\; \sigma_{23} \,\rfloor^T \\
\varepsilon_s &= \lfloor\, \varepsilon_{11} \;\; \varepsilon_{12} \;\; \varepsilon_{13} \,\rfloor^T, \quad \varepsilon_n = \lfloor\, \varepsilon_{22} \;\; \varepsilon_{33} \;\; \varepsilon_{23} \,\rfloor^T \\
\sigma &= \lfloor\, (\sigma_s)^T \;\; (\sigma_n)^T \,\rfloor^T, \quad \varepsilon = \lfloor\, (\varepsilon_s)^T \;\; (\varepsilon_n)^T \,\rfloor^T
\end{aligned} \tag{3}$$

where the subscript s indicates the sectional stresses, and subscript n represents the stresses on the planes normal to the beam section.

The constitutive relations in the beam coordinate system can then be expressed as

$$\begin{Bmatrix} \sigma_s^a \\ \sigma_n^a \end{Bmatrix} = \begin{bmatrix} \mathbf{C}_{ss} & \mathbf{C}_{sn} \\ \mathbf{C}_{sn}^T & \mathbf{C}_{nn} \end{bmatrix} \begin{Bmatrix} \varepsilon_s^a \\ \varepsilon_n^a \end{Bmatrix} \tag{4}$$

Here, the superscript a indicates the active components computed directly which implies that the stresses are computed using Hooke's law and the strains are computed from kinematical relations.

The semi-inverted form of the constitutive relations can be obtained as

$$\begin{Bmatrix} \varepsilon_s^r \\ \sigma_n^a \end{Bmatrix} = \begin{bmatrix} \overline{\mathbf{C}}_{ss} & \overline{\mathbf{C}}_{sn} \\ -\overline{\mathbf{C}}_{sn}^T & \overline{\mathbf{C}}_{nn} \end{bmatrix} \begin{Bmatrix} \sigma_s^r \\ \varepsilon_n^a \end{Bmatrix} \tag{5}$$

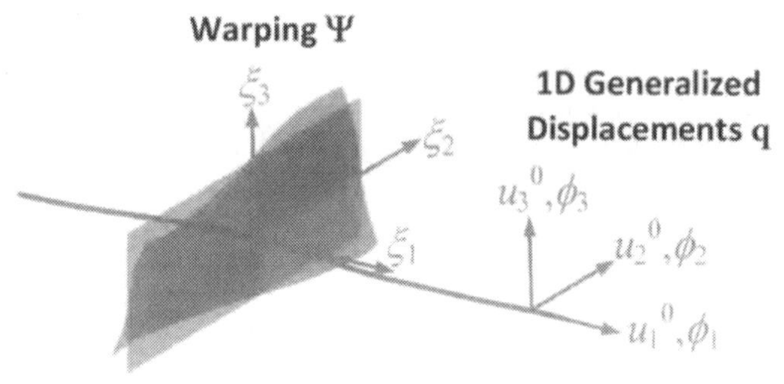

Figure 2. Kinematics of beam section.

where the superscript r indicates the reactive components (reactive sectional strains ε_s^r and stresses σ_s^r). The modified constitutive matrices $\overline{\mathbf{C}}_{ss}$, $\overline{\mathbf{C}}_{sn}$, and $\overline{\mathbf{C}}_{nn}$ are defined as

$$\overline{\mathbf{C}}_{ss} = \mathbf{C}_{ss}^{-1}, \quad \overline{\mathbf{C}}_{sn} = -\mathbf{C}_{ss}^{-1}\mathbf{C}_{sn}$$
$$\overline{\mathbf{C}}_{nn} = \mathbf{C}_{nn} - \mathbf{C}_{ns}\mathbf{C}_{ss}^{-1}\mathbf{C}_{sn} \tag{6}$$

Kinematics

The position vector of an arbitrary material point on an undeformed beam is denoted by $\mathbf{x} = \lfloor\ \xi_1 \quad \xi_2 \quad \xi_3\ \rfloor^T$.

The position vector $\mathbf{X}$ of an arbitrary material point located on the section of a deformed beam is defined as

$$\mathbf{X} = \mathbf{x} + \mathbf{u} \tag{7}$$

where $\mathbf{u} = \lfloor\ u_1 \quad u_2 \quad u_3\ \rfloor^T$ represents the displacement of an arbitrary material point on the section, and is defined as the sum of the rigid rototranslational displacements $\mathbf{u}_b$ and the sectional warping displacements Ψ (see Fig. 2), as given by

$$\mathbf{u} = \mathbf{u}_b + \Psi \tag{8}$$

where

$$\mathbf{u}_b = \mathbf{B}\mathbf{q}, \quad \Psi = \lfloor\ \psi_1 \quad \psi_2 \quad \psi_3\ \rfloor^T, \quad \mathbf{q} = \lfloor\ \mathbf{u}_0^T \quad \Phi^T\ \rfloor^T$$
$$\mathbf{B} = \begin{bmatrix} 1 & 0 & 0 & 0 & \xi_3 & -\xi_2 \\ 0 & 1 & 0 & -\xi_3 & 0 & 0 \\ 0 & 0 & 1 & \xi_2 & 0 & 0 \end{bmatrix} \tag{9}$$

where $\mathbf{q}$ represents the generalized displacements consisting of the translations $\mathbf{u}_0 = \lfloor\ u_1^0 \quad u_2^0 \quad u_3^0\ \rfloor^T$ and rotations $\mathbf{\Phi} = \lfloor\ \phi_1 \quad \phi_2 \quad \phi_3\ \rfloor^T$ of the beam section. These translations and rotations are determined from the elastic analysis of 1D beam, and can be viewed as rigid deformations of the beam section.

Warping Constraints

The warping field is six times redundant as defined in Eq. (8). It can be conveniently considered that the 1D beam displacement field describes an average behavior of the cross-section in terms of rigid translations and rotations. The 1D displacement vector $\mathbf{u}_0$ thus represents the average deformation of the cross-section. The constraints can be stated in compact form as

$$\int_A \mathscr{D}_w \Psi \, dA = \mathbf{0} \tag{10}$$

where

$$\mathscr{D}_w = \begin{bmatrix} 1 & 0 & 0 \\ 0 & 1 & 0 \\ 0 & 0 & 1 \\ 0 & -\partial_3 & \partial_2 \\ \partial_3 & 0 & -\partial_1 \\ -\partial_2 & \partial_1 & 0 \end{bmatrix} \tag{11}$$

where $\partial_1, \partial_2, \partial_3$ represent the derivatives with respect to ξ_1, ξ_2 and ξ_3, respectively.

Strains

With the assumptions of small strains and small local rotations, the 3D strains ε can be expressed in terms of displacements as

$$\varepsilon = \mathscr{L}_s \mathbf{u} + \mathscr{L}_b \mathbf{u}' \tag{12}$$

where $(\cdot)'$ indicates derivative with respect to ξ_1, and $\mathscr{L}_s$ and $\mathscr{L}_b$ are the operator matrices given by

$$\mathscr{L}_s = \begin{bmatrix} 0 & 0 & 0 \\ \partial_2 & 0 & 0 \\ \partial_3 & 0 & 0 \\ 0 & \partial_2 & 0 \\ 0 & 0 & \partial_3 \\ 0 & \partial_3 & \partial_2 \end{bmatrix}, \quad \mathscr{L}_b = \begin{bmatrix} 1 & 0 & 0 \\ 0 & 1 & 0 \\ 0 & 0 & 1 \\ 0 & 0 & 0 \\ 0 & 0 & 0 \\ 0 & 0 & 0 \end{bmatrix} \tag{13}$$

Substitution of displacements from Eqs. (8) and (9) leads to

$$\varepsilon = \mathscr{L}_b \mathbf{B}\Gamma + \mathscr{L}_s \Psi + \mathscr{L}_b \Psi' \tag{14}$$

where Γ represents the generalized strain measures vector defined in terms of $\mathbf{q}$ as

$$\Gamma = \lfloor\ \gamma_1 \quad \gamma_2 \quad \gamma_3 \quad \kappa_1 \quad \kappa_2 \quad \kappa_3\ \rfloor^T = \mathscr{L}_q \mathbf{q} + \mathbf{q}' \tag{15}$$

Here, γ_1 is the extensional strain, γ_2 and γ_3 are the transverse shear strains, κ_1 is the rate of twist (or twist curvature), and κ_2 and κ_3 are the bending curvatures. The matrix $\mathscr{L}_q$ is given by

$$\mathscr{L}_q = \begin{bmatrix} 0 & 0 & 0 & 0 & 0 & 0 \\ 0 & 0 & 0 & 0 & 0 & -1 \\ 0 & 0 & 0 & 0 & 1 & 0 \\ 0 & 0 & 0 & 0 & 0 & 0 \\ 0 & 0 & 0 & 0 & 0 & 0 \\ 0 & 0 & 0 & 0 & 0 & 0 \end{bmatrix} \tag{16}$$

Decomposition of Strains and Strain Compatibility

To be consistent with the semi-inverted form of material constitutive relations defined in Eq. (5), the 3D strains from Eq. (14) can be expressed in decomposed form as

$$\varepsilon_s^a = \mathbf{B}\Gamma + \mathscr{L}_s^s \Psi + \Psi', \quad \varepsilon_n^a = \mathscr{L}_s^n \Psi \tag{17}$$

where

$$\mathscr{L}_s^s = \begin{bmatrix} 0 & 0 & 0 \\ \partial_2 & 0 & 0 \\ \partial_3 & 0 & 0 \end{bmatrix}, \quad \mathscr{L}_s^n = \begin{bmatrix} 0 & \partial_2 & 0 \\ 0 & 0 & \partial_3 \\ 0 & \partial_3 & \partial_2 \end{bmatrix} \quad (18)$$

The compatibility of the reactive strains obtained using Hooke's law ε_s^r and active strains computed from the kinematics ε_s^a is given by the following equation.

$$\varepsilon_s^r = \varepsilon_s^a \quad (19)$$

The above compatibility relation is satisfied in a weak sense in the present multifield formulation.

Discretization of warping and reactive stresses

The warping and reactive stress fields can be discretized using isoparametric shape functions as

$$\begin{aligned} \Psi(\xi_1,\xi_2,\xi_3) &= \mathbf{N}_\psi(\xi_2,\xi_3)\Lambda(\xi_1) \\ \sigma_s^r(\xi_1,\xi_2,\xi_3) &= \mathbf{N}_\sigma(\xi_2,\xi_3)\Upsilon(\xi_1) \end{aligned} \quad (20)$$

where $\mathbf{N}_\psi(\xi_2,\xi_3)$ and $\mathbf{N}_\sigma(\xi_2,\xi_3)$ represent the FE shape function matrices for warping and reactive stress fields, respectively. $\Lambda(\xi_1)$ and $\Upsilon(\xi_1)$ indicate the nodal values of warping and reactive stress fields, respectively.

The warping constraints in Eq. (10) can be discretized using Eq. (20) as

$$\int_A \mathscr{D}_w \mathbf{N}_\psi \Lambda dA = \left(\int_A \mathscr{D}_w \mathbf{N}_\psi dA\right)\Lambda = \mathbf{D}_\psi \Lambda = \mathbf{0} \quad (21)$$

where $\mathbf{D}_\psi$ denotes the discretized warping constraints matrix.

Multifield Variational Governing Equations

The present formulation assumes the sectional stresses to be unknowns in addition to the displacements and uses the variational principle to establish the governing equations, hence the name *multifield variational formulation*. The total energy per unit length of the beam Π_R can be written as

$$\Pi_R = U_s - W_s \quad (22)$$

where U_s is the sectional strain energy and W_s is the external work on the beam section due to applied loads.

The variational energy principle can then be defined as

$$\delta\Pi_R = \delta U_s - \delta W_s = 0 \quad (23)$$

where $\delta\Pi_R$ is the variation of total energy.

The Reissner's semi-complimentary energy functional Φ_R (Ref. 17) can be defined in a generalized form using the active and reactive components of strains and stresses, as given by

$$\Phi_R = \frac{1}{2}\left((\varepsilon_n^a)^T \sigma_n^a - (\sigma_s^r)^T \varepsilon_s^r\right) \quad (24)$$

Beam sectional strain energy The strain energy per unit length of the beam (sectional strain energy) U_s can be obtained in terms of semi-complimentary energy functional Φ_R from Eq. (24) as

$$U_s = \int_A \left[\Phi_R + (\varepsilon_s^r)^T \sigma_s^r\right] dA \quad (25)$$

Substituting the expression for Φ_R from Eq. (24) and introducing the strain compatibility condition from Eq. (19), the first variation of the sectional strain energy can be obtained as

$$\delta U_s = \int_A \left[(\delta\varepsilon_n^a)^T \sigma_n^a + (\delta\varepsilon_s^a)^T \sigma_s^r + (\delta\sigma_s^r)^T (\varepsilon_s^a - \varepsilon_s^r)\right] dA \quad (26)$$

The third term in the above equation is due to the strain compatibility condition, where $\delta\sigma_s^r$ can be viewed as Lagrange multipliers. Note that this condition is a byproduct of the partially mixed formulation.

Substituting semi-inverted constitutive relations, strain-displacement relations, and discretized warping and reactive stress fields from Eqs. (5), (17), and (20), respectively, the strain energy variation becomes

$$\delta U_s = \begin{Bmatrix} \delta\Lambda' \\ \delta\Upsilon \\ \delta\Lambda \\ \delta\Gamma \end{Bmatrix}^T \begin{bmatrix} \mathbf{0} & \mathbf{A} & \mathbf{0} & \mathbf{0} \\ \mathbf{A}^T & -\mathbf{H} & \mathbf{G}^T & \mathbf{R} \\ \mathbf{0} & \mathbf{G} & \mathbf{E} & \mathbf{0} \\ \mathbf{0} & \mathbf{R}^T & \mathbf{0} & \mathbf{0} \end{bmatrix} \begin{Bmatrix} \Lambda' \\ \Upsilon \\ \Lambda \\ \Gamma \end{Bmatrix} \quad (27)$$

where

$$\begin{aligned} \mathbf{A} &= \int_A \mathbf{N}_\psi^T \mathbf{N}_\sigma dA, \quad \mathbf{E} = \int_A (\mathscr{L}_s^n \mathbf{N}_\psi)^T \overline{\mathbf{C}}_{nn} \mathscr{L}_s^n \mathbf{N}_\psi dA \\ \mathbf{G} &= \int_A \left(\mathscr{L}_s^s \mathbf{N}_\psi - \overline{\mathbf{C}}_{sn} \mathscr{L}_s^n \mathbf{N}_\psi\right)^T \mathbf{N}_\sigma dA \\ \mathbf{H} &= \int_A \mathbf{N}_\sigma^T \overline{\mathbf{C}}_{ss} \mathbf{N}_\sigma dA, \quad \mathbf{R} = \int_A \mathbf{N}_\sigma^T \mathbf{B} dA \end{aligned} \quad (28)$$

The matrices **A**, **E**, **G**, **H**, and **R** include the effects of geometric and material couplings of the beam section. The scale of magnitude of the elements of these matrices is significantly different from each other due to the semi-inverted form of material constitutive relations.

External work due to applied loads The sectional stress resultants **F** can be obtained from the tractions **p** $(= \sigma_s)$ acting on the section as given by

$$\mathbf{F} = \int_A \mathbf{B}^T \mathbf{p} dA = \int_A \mathbf{B}^T \sigma_s dA \quad (29)$$

with

$$\mathbf{F} = \lfloor F_1 \quad F_2 \quad F_3 \quad M_1 \quad M_2 \quad M_3 \rfloor^T \quad (30)$$

where F_1 is the extensional force, F_2 and F_3 are the transverse shear forces, M_1 is the torsional moment, and M_2 and M_3 are the bending moments.

Neglecting the surface and body forces, the external work per unit length of the beam W_s is given by

$$W_s = \int_A (\mathbf{u}^T \sigma_s)' dA \quad (31)$$

Using Eqs. (8), (9), and (29), and the definition of Γ from Eq. (15), the variation of external work δW_s is obtained as

$$\delta W_s = \left[\delta\Gamma^T \mathbf{F} + \delta\mathbf{q}^T \left(\mathbf{F}' - \mathscr{L}_q^T \mathbf{F}\right)\right] + \int_A \left[\delta\Psi'^T \sigma_s + \delta\Psi^T (\sigma_s)'\right] dA \tag{32}$$

The first bracketed term in the above equation is independent of warping displacements and depends only on the rigid displacements of the section.

Substituting the discretized warping and reactive stress fields from Eq. (20), δW_s can be expressed as

$$\delta W_s = \begin{Bmatrix} \delta\Lambda' \\ \delta\Upsilon \\ \delta\Lambda \\ \delta\Gamma \end{Bmatrix}^T \begin{Bmatrix} \mathbf{P} \\ \mathbf{0} \\ \mathbf{P}' \\ \mathbf{F} \end{Bmatrix} + \delta\mathbf{q}^T \left(\mathbf{F}' - \mathscr{L}_q^T \mathbf{F}\right) \tag{33}$$

where

$$\mathbf{P} = \int_A \mathbf{N}_\psi^T \sigma_s \, dA, \qquad \mathbf{P}' = \int_A \mathbf{N}_\psi^T (\sigma_s)' \, dA \tag{34}$$

Equilibrium equations The variational equilibrium equation can be obtained by substituting the expressions for strain energy and external variations from Eqs. (27) and (33), respectively, into Eq. (23), which results in

$$\begin{Bmatrix} \delta\Lambda' \\ \delta\Upsilon \\ \delta\Lambda \\ \delta\Gamma \end{Bmatrix}^T \begin{bmatrix} \mathbf{0} & \mathbf{A} & \mathbf{0} & \mathbf{0} \\ \mathbf{A}^T & -\mathbf{H} & \mathbf{G}^T & \mathbf{R} \\ \mathbf{0} & \mathbf{G} & \mathbf{E} & \mathbf{0} \\ \mathbf{0} & \mathbf{R}^T & \mathbf{0} & \mathbf{0} \end{bmatrix} \begin{Bmatrix} \Lambda' \\ \Upsilon \\ \Lambda \\ \Gamma \end{Bmatrix} - \begin{Bmatrix} \delta\Lambda' \\ \delta\Upsilon \\ \delta\Lambda \\ \delta\Gamma \end{Bmatrix}^T \begin{Bmatrix} \mathbf{P} \\ \mathbf{0} \\ \mathbf{P}' \\ \mathbf{F} \end{Bmatrix} = \delta\mathbf{q}^T \left(\mathbf{F}' - \mathscr{L}_q^T \mathbf{F}\right) \tag{35}$$

For arbitrary $\delta\Lambda'$, $\delta\Lambda$, $\delta\Upsilon$, $\delta\Gamma$, and $\delta\mathbf{q}$, and assuming the double derivatives of warping and reactive stresses to be negligible, the following sets of equilibrium equations are obtained as given by

$$\begin{bmatrix} -\mathbf{H} & \mathbf{G}^T & \mathbf{R} \\ \mathbf{G} & \mathbf{E} & \mathbf{0} \\ \mathbf{R}^T & \mathbf{0} & \mathbf{0} \end{bmatrix} \begin{Bmatrix} \Upsilon' \\ \Lambda' \\ \Gamma' \end{Bmatrix} = \begin{Bmatrix} \mathbf{0} \\ \mathbf{0} \\ \mathscr{L}_q^T \mathbf{F} \end{Bmatrix} \tag{36a}$$

$$\begin{bmatrix} -\mathbf{H} & \mathbf{G}^T & \mathbf{R} \\ \mathbf{G} & \mathbf{E} & \mathbf{0} \\ \mathbf{R}^T & \mathbf{0} & \mathbf{0} \end{bmatrix} \begin{Bmatrix} \Upsilon \\ \Lambda \\ \Gamma \end{Bmatrix} = \begin{bmatrix} \mathbf{0} & -\mathbf{A}^T & \mathbf{0} \\ \mathbf{A} & \mathbf{0} & \mathbf{0} \\ \mathbf{0} & \mathbf{0} & \mathbf{0} \end{bmatrix} \begin{Bmatrix} \Upsilon' \\ \Lambda' \\ \Gamma' \end{Bmatrix} + \begin{Bmatrix} \mathbf{0} \\ \mathbf{0} \\ \mathbf{F} \end{Bmatrix} \tag{36b}$$

Note that in the displacement based formulations, the stresses are computed from the derivatives of warping displacements resulting in lower accuracy. In the present multifield formulation, the sectional stresses are assumed as unknowns and computed directly. The reactive sectional stresses can be modeled using the same shape functions as that of the warping field leading to an accurate recovery of the sectional stresses.

Solution for the Warping and Reactive Stresses

Since the 3D beam analysis is decomposed into 2D sectional analysis and 1D beam analysis, the cross-sectional constitutive relations must be expressed either in terms of 1D generalized strains or generalized stress resultants. To resolve the warping and reactive stress fields, they are expressed as linear functions of generalized stress resultants and their derivatives. These relations are computed at the SV level including generalized Timoshenko theory for shear deformations while the boundary effects are neglected. The warping and reactive stress fields are therefore approximated as

$$\Lambda = \widetilde{\Lambda}\mathbf{F}, \quad \Lambda' = \widetilde{\Lambda}'\mathbf{F}, \quad \Upsilon = \widetilde{\Upsilon}\mathbf{F}, \quad \Upsilon' = \widetilde{\Upsilon}'\mathbf{F}$$
$$\Gamma = \widetilde{\Gamma}\mathbf{F}, \quad \Gamma' = \widetilde{\Gamma}'\mathbf{F} \tag{37}$$

where $\widetilde{\Lambda}$ and $\widetilde{\Upsilon}$ are the warping and reactive stress coefficient matrices with size $3n \times 6$ (n represents the total number of nodes in the discretized model), and $\widetilde{\Gamma}$ is a 6×6 matrix which is constant over the section. The terms with $(\cdot)'$ indicate the coefficient matrices corresponding to the derivatives of generalized strain measures present in the sectional stress resultants. The warping displacements (in- and out-of-plane) and reactive stresses are discretized using general FEs leading to a nonlinear distribution over the beam section. The distribution of reactive stresses can be directly computed by multiplying coefficient matrices with the given sectional loads with the same accuracy as that of the displacements.

The approximations of warping displacements and reactive stresses from Eq. (37) along with the warping constraints from Eq. (21) can be substituted into the equilibrium equations obtained in Eq. (36), which leads to

$$\begin{bmatrix} -\mathbf{H} & \mathbf{G}^T & \mathbf{R} & \mathbf{0} \\ \mathbf{G} & \mathbf{E} & \mathbf{0} & \mathbf{D}_\psi^T \\ \mathbf{R}^T & \mathbf{0} & \mathbf{0} & \mathbf{0} \\ \mathbf{0} & \mathbf{D}_\psi & \mathbf{0} & \mathbf{0} \end{bmatrix} \begin{bmatrix} \widetilde{\Upsilon}' \\ \widetilde{\Lambda}' \\ \widetilde{\Gamma}' \\ \Theta'_\psi \end{bmatrix} = \begin{bmatrix} \mathbf{0} \\ \mathbf{0} \\ \mathscr{L}_q^T \\ \mathbf{0} \end{bmatrix} \tag{38a}$$

$$\begin{bmatrix} -\mathbf{H} & \mathbf{G}^T & \mathbf{R} & \mathbf{0} \\ \mathbf{G} & \mathbf{E} & \mathbf{0} & \mathbf{D}_\psi^T \\ \mathbf{R}^T & \mathbf{0} & \mathbf{0} & \mathbf{0} \\ \mathbf{0} & \mathbf{D}_\psi & \mathbf{0} & \mathbf{0} \end{bmatrix} \begin{bmatrix} \widetilde{\Upsilon} \\ \widetilde{\Lambda} \\ \widetilde{\Gamma} \\ \Theta_\psi \end{bmatrix} = \begin{bmatrix} \mathbf{0} & -\mathbf{A}^T & \mathbf{0} & \mathbf{0} \\ \mathbf{A} & \mathbf{0} & \mathbf{0} & \mathbf{0} \\ \mathbf{0} & \mathbf{0} & \mathbf{0} & \mathbf{0} \\ \mathbf{0} & \mathbf{0} & \mathbf{0} & \mathbf{0} \end{bmatrix} \begin{bmatrix} \widetilde{\Upsilon}' \\ \widetilde{\Lambda}' \\ \widetilde{\Gamma}' \\ \Theta'_\psi \end{bmatrix} + \begin{bmatrix} \mathbf{0} \\ \mathbf{0} \\ \mathbf{I} \\ \mathbf{0} \end{bmatrix} \tag{38b}$$

where Θ'_ψ and Θ_ψ are the Lagrange multipliers corresponding to the warping constraints. The structure of the matrix on the left hand side is sparse and symmetric which can be solved using any available sparse direct or iterative solver.

Generalized Timoshenko Like Sectional Stiffness Matrix

The generalized Timoshenko like 6×6 stiffness matrix is derived first using the SV warping solution which is assumed as uniform along the beam axis. With the known warping and reactive stresses from Eq. (38), the strain energy variation (δU_s) from Eq. (27) can be restated as

$$\delta U_s = \delta \mathbf{F}^T \begin{bmatrix} \widetilde{\Lambda}' \\ \widetilde{\Upsilon} \\ \widetilde{\Lambda} \\ \widetilde{\Gamma} \end{bmatrix}^T \begin{bmatrix} \mathbf{0} & \mathbf{A} & \mathbf{0} & \mathbf{0} \\ \mathbf{A}^T & -\mathbf{H} & \mathbf{G}^T & \mathbf{R} \\ \mathbf{0} & \mathbf{G} & \mathbf{E} & \mathbf{0} \\ \mathbf{0} & \mathbf{R}^T & \mathbf{0} & \mathbf{0} \end{bmatrix} \begin{bmatrix} \widetilde{\Lambda}' \\ \widetilde{\Upsilon} \\ \widetilde{\Lambda} \\ \widetilde{\Gamma} \end{bmatrix} \mathbf{F} \tag{39}$$

The variation of the external work can be defined in terms of Timoshenko like sectional flexibility matrix $\mathbf{S}$ as

$$\delta W_s = \delta \Gamma^T \mathbf{F} = \delta \mathbf{F}^T \mathbf{S} \mathbf{F} \tag{40}$$

where $\mathbf{S}$ can be determined using energy principle defined in Eq. (23), which results in

$$\mathbf{S} = \begin{bmatrix} \widetilde{\Lambda}' \\ \widetilde{\Upsilon} \\ \widetilde{\Lambda} \\ \widetilde{\Gamma} \end{bmatrix}^T \begin{bmatrix} \mathbf{0} & \mathbf{A} & \mathbf{0} & \mathbf{0} \\ \mathbf{A}^T & -\mathbf{H} & \mathbf{G}^T & \mathbf{R} \\ \mathbf{0} & \mathbf{G} & \mathbf{E} & \mathbf{0} \\ \mathbf{0} & \mathbf{R}^T & \mathbf{0} & \mathbf{0} \end{bmatrix} \begin{bmatrix} \widetilde{\Lambda}' \\ \widetilde{\Upsilon} \\ \widetilde{\Lambda} \\ \widetilde{\Gamma} \end{bmatrix} \tag{41}$$

The generalized Timoshenko like stiffness matrix $\mathbf{K}$ can be computed by taking the inverse of the flexibility matrix as given by

$$\mathbf{K} = \mathbf{S}^{-1} \tag{42}$$

The above stiffness matrix $\mathbf{K}$ takes into account the effects of elastic couplings, and transverse shear and Poisson deformations. For the case of general anisotropic beams, the 6×6 stiffness matrix may be fully populated.

FE Implementation

The present multifield formulation is implemented in a FE code called MVSAC. The discretized beam cross-section and the material properties are provided as input to the analysis. The C^0-continuous isoparametric Lagrange shape functions are used for the FE discretization of warping and reactive stress fields. Four types of FEs can be used for the discretization: three- and six-noded triangular elements, and four- and eight-noded quadrilateral elements. The 3D warping and reactive stress coefficients are computed using Eq. (38). Using the known warping and reactive stress coefficients, the generalized 6×6 Timoshenko like stiffness matrix is computed. The output summary of the sectional constants is printed in a text file format. ParaView (Ref. 18) is used as a postprocessor for the visualization of 3D warping and reactive stress coefficients.

RESULTS AND DISCUSSION

Several examples are presented to demonstrate the application of the present multifield sectional analysis for FGBs. The FGM properties are assumed to follow linear distribution for the present analyses. The warping displacement modes and the distribution of reactive stress coefficients are illustrated for homogeneous isotropic beams and FGBs. The 1D elastostatic responses computed using the present analysis are validated against the 3D FE solutions.

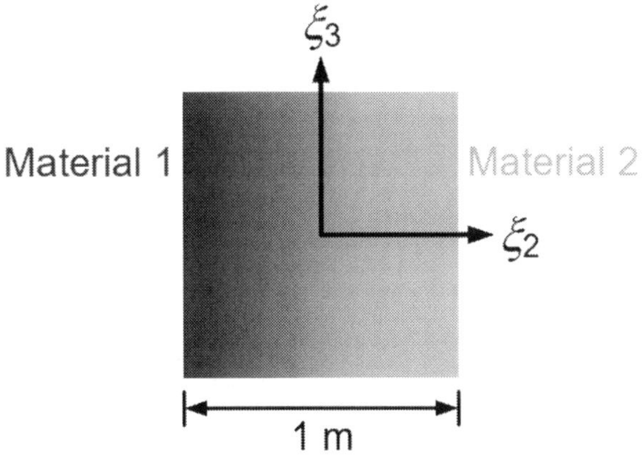

Figure 3. Functionally graded square section beam.

Square Section

First example is a square section (Ref. 7) as shown in Fig. 3. The material 1 and 2 are respectively alumina and steel with properties as (Ref. 7): $E_1 = 3.9 \times 10^5$ MPa, $\nu_1 = 0.25$, $E_2 = 2.1 \times 10^5$, and $\nu_2 = 0.31$. The Young's modulus E and Poisson's ratio ν vary linearly along ξ_2 axis. The section exhibits extension-bending and little shear-torsion couplings due to asymmetric material distribution. The cross-section is discretized using 900 eight-node quadrilateral elements and 2,821 nodes giving a total of 16,926 degrees of freedom. For the present analysis, homogeneous properties are assumed in each element using 30 discrete layers along ξ_2 axis.

The warping deformation modes for the homogeneous alumina and FGM square sections are compared in Fig. 4. For the homogeneous case, the extension (F_1) and bending (M_3) modes indicate in-plane deformation while the shear (F_3) and torsion (M_1) modes indicate out-of-plane deformation. The extension (F_1) mode shows additional in-plane distortion because of coupling with the bending (M_3) mode. Note that the FGM section indicates higher magnitude of warping displacements than those of the homogeneous section. It should be remarked that these warping displacements are essential for accurate prediction of elastically coupled stiffness constants.

In order to validate the present multifield analysis, the elastostatic responses of a cantilevered FGM square section beam are computed under torsional and shear loads. The beam length L is considered to be 10 m. The present results are compared with those obtained from 3D FE analysis using the commercial software MSC.Nastran (www.mscsoftware.com/nastran). The 3D FE model consists of 180,000 20-node brick elements and 759,221 nodes resulting in a total of 2,269,200 degrees of free-

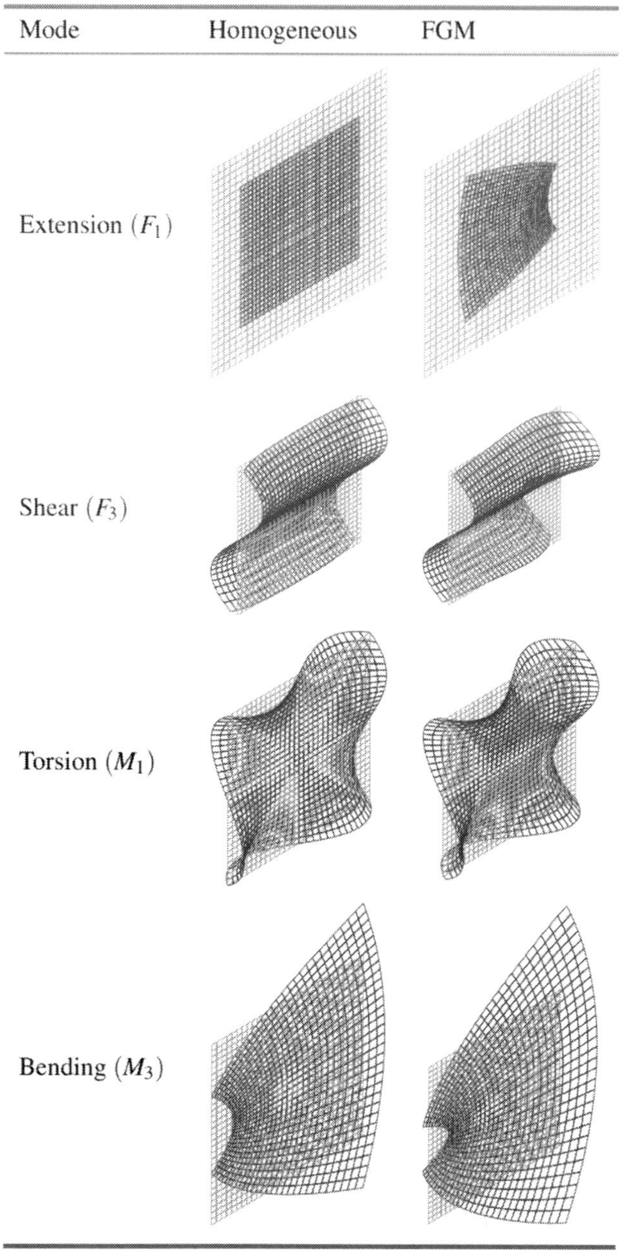

Figure 4. Comparison of warping displacement modes for homogeneous and functionally graded square section beams (exaggerated).

dom. The CPU time consumed by 3D FE analysis is 12,023 seconds whereas the present analysis is computed within 13 seconds indicating a speedup of 925. For the 3D analysis, the torsional moment (M_1) is applied through a shear force couple (F_3) at the tip end. The twist response of FGM square section beam under a tip torsional moment (M_1) of 1 N m is presented in Fig. 5. The present values are identical to those of the Nastran 3D FE solution. Figure 6 shows the bending displacement and bending rotation under a tip shear force $F_3 = 1$ N. The present results match well with those of the 3D FE solutions. The induced twist response under tip shear force is presented in Fig. 7. Note that the magnitude of induced twist is small which results in larger differences due to local boundary effect near the clamped root end and tip loading end. Overall, a good

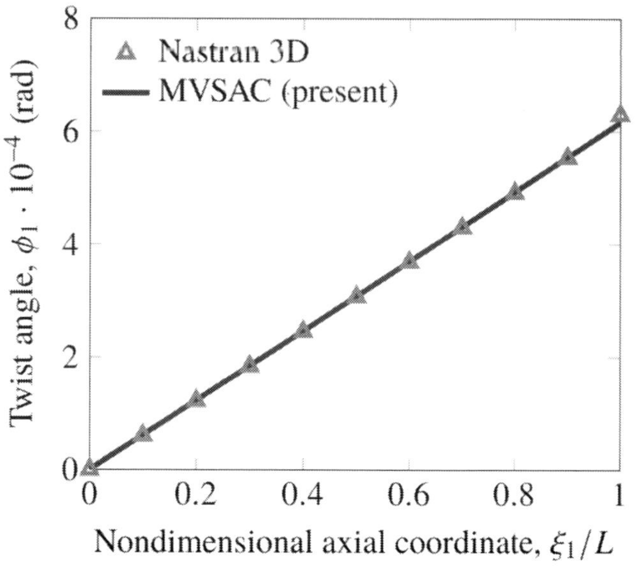

Figure 5. Twist response of functionally graded square section beam under a tip torsional moment.

correlation is achieved by the present multifield analysis in comparison to 3D FE results.

The reactive stress distributions under unit loads are presented in Fig. 8. The stress distributions for homogeneous alumina and FGM square sections are compared. The variation in the distribution of transverse shear stresses σ_{12} and σ_{13} under unit shear force F_3 is clearly visible in the figure. The normal stress σ_{11} under unit bending moment M_2 shows higher concentrations near the left end (alumina) of FGM section. Under the unit torsional moment M_1, the transverse shear stress σ_{12} shows similar distribution for both homogeneous and FGM sections whereas the σ_{13} component shows reduction in the maximum values for FGM section.

Box Section

Next example is a FGM box section as shown in Fig. 9. The material properties are the same as those of the FGM square section. The properties vary linearly along the thickness with 10 discrete layers in each wall. The section does not exhibit any elastic couplings. The section is discretized with 1,200 eight-node quadrilateral elements and 3,840 nodes leading to a total of 23,040 degrees of freedom.

Figure 10 presents the warping deformations of the homogeneous alumina and FGM box sections. The deformation profiles for both sections are similar because of the absence of any couplings with only difference being higher magnitudes for the FGM section. The extension (F_1) and bending (M_3) modes indicate in-plane distortions while the shear (F_3) and torsion (M_1) modes indicate out-of-plane deformations.

The 1D elastostatic response of a cantilevered FGM box beam is discussed next under tip torsional and shear loads.

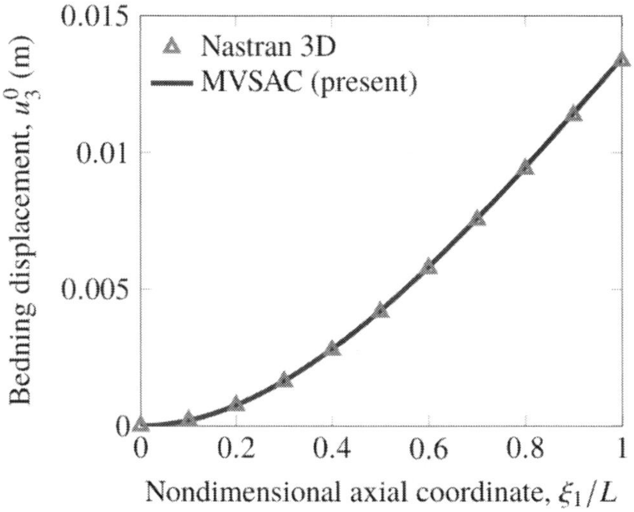

(a)

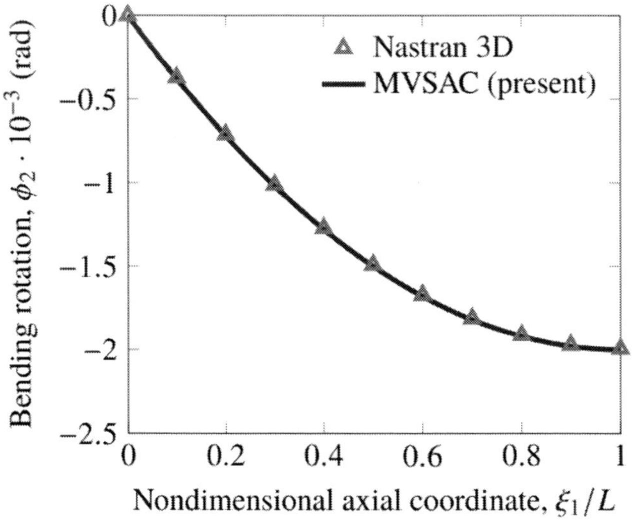

(b)

Figure 6. (a) Bending displacement and (b) bending rotation of functionally graded square section beam under a tip shear force.

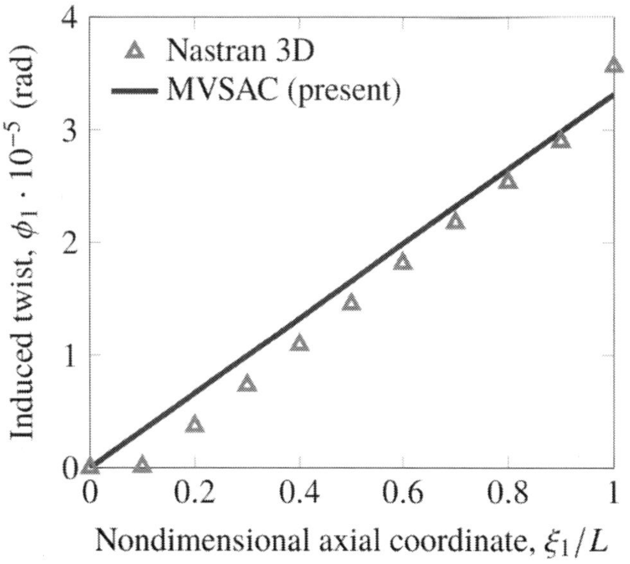

Figure 7. Induced twist response of functionally graded square section beam under a tip shear force.

The beam length L is 10 m. The present results are validated with those obtained using Nastran 3D FE analysis. The 3D FE model is discretized using 240,000 20-node brick elements and 1,035,840 nodes leading to a total of 3,096,000 degrees of freedom. The Nastran 3D required 15,041 seconds of CPU time whereas the present analysis is performed within 13 seconds which is 1,157 faster than Nastran 3D. The torsional moment (M_1) of 1 N m for the Nastran analysis is applied through a shear force (F_3) couple at the tip end. The twist response computed by the present multifield MVSAC is compared with that of Nastran 3D in Fig. 11. The present shows an excellent correlation with Nastran 3D FE solution except near the tip end due to local free boundary effects from concentrated tip shear forces. The bending displacement and bending rotation under a tip shear force (F_3) of 1 N are indicated in Fig. 12. The present results are identical to those computed from 3D FE analysis. The present investigations therefore indicate the capability of the proposed formulation with reasonable accuracy and higher efficiency in comparison to the 3D FE analysis.

Figure 13 illustrates the reactive stress distributions for FGM box section under unit shear, bending, and torsional loads. The stress concentrations at the inner corners are reduced for FGM as compared to the homogeneous (alumina) section. The distribution of stresses is much smoother along the wall thickness for FGM section than that of homogeneous section due to the gradual distribution of internal loads. The smooth variation in the stress distribution can therefore be conveniently exploited using FGMs to avoid material failure.

CONCLUDING REMARKS

A multifield variational sectional analysis is developed for the FGM beams with continuously varying material properties. The 3D warping displacements and sectional stresses are modeled with generic nonlinear distribution over the beam cross-section and are solved simultaneously using the multifield formulation. A generalized Timoshenko like 6×6 stiffness matrix is derived as a part of the analysis consisting of elastic couplings, and transverse shear and Poisson deformation effects. The present results are validated for several FGM beams with elastic couplings. The 1D elastostatic responses computed using the present analysis are shown to be nearly identical to 3D FE solutions. The warping deformations are demonstrated for homogeneous and FGM beams which indicate softening of the section due to the continuous distribution of FGM. The reactive sectional stresses are illustrated for FGM beams which have continuous and smooth distribution over the beam cross-section. The present analysis clearly demonstrates the application for FGM beams with elastic couplings which follow a continuous stress profile that can be conveniently computed from the present multifield analysis without requiring

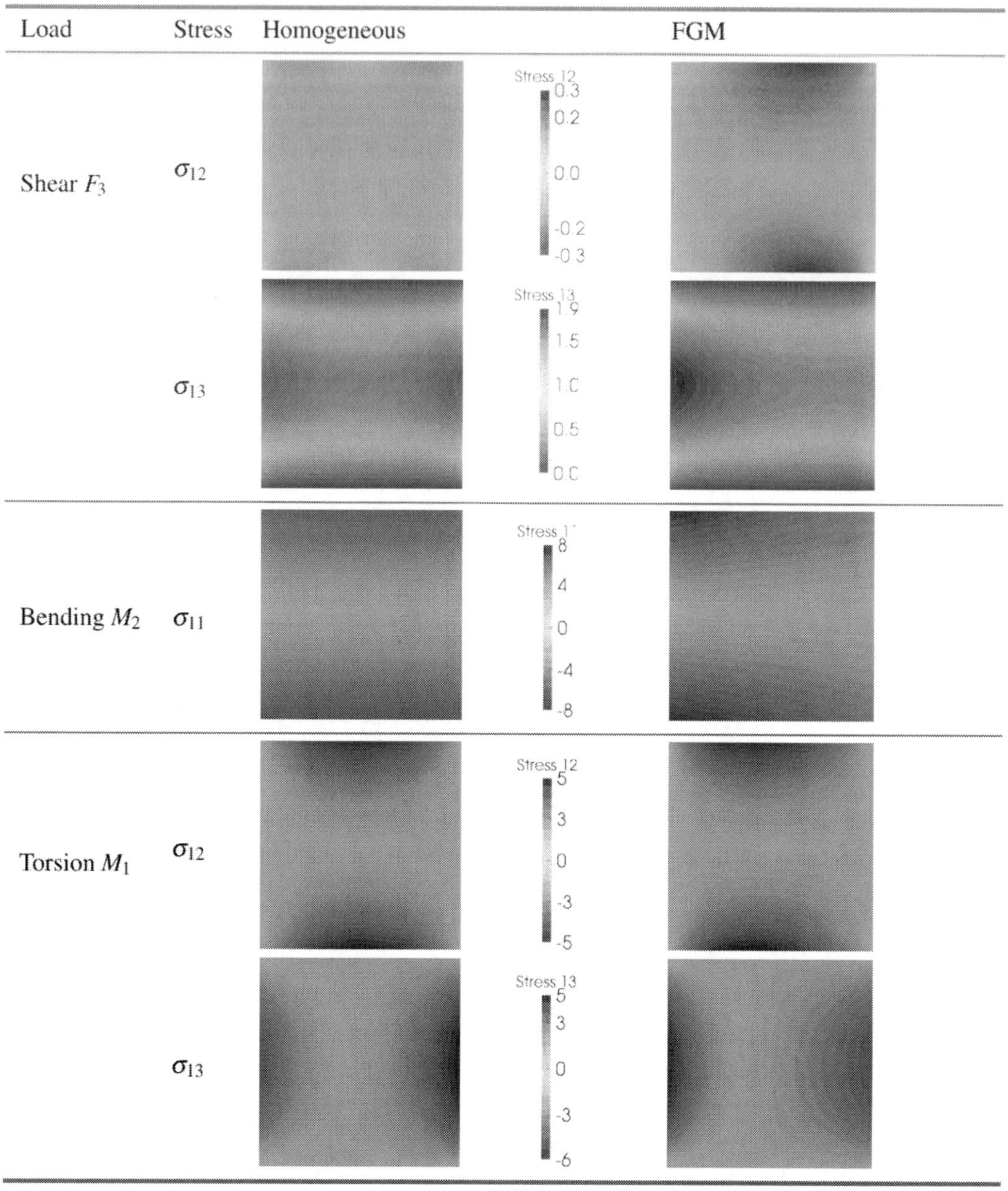

Figure 8. Comparison of reactive stress distributions for homogeneous and functionally graded square section beams (exaggerated).

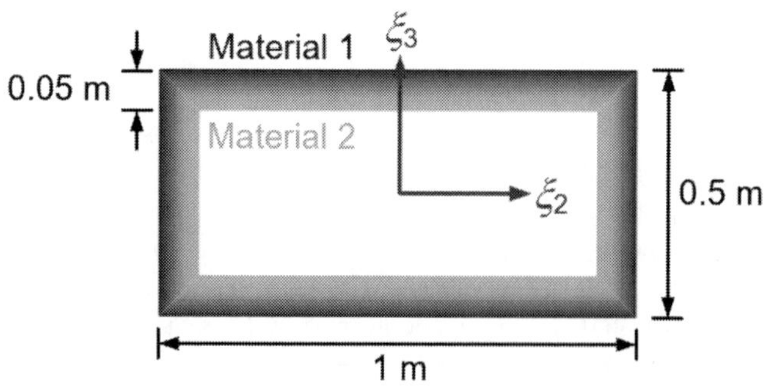

Figure 9. Functionally graded box beam.

displacement derivatives. The future work includes the validation of stresses with the 3D analysis and the implementation of continuous material property variation at the integration points within FEs.

Author contact: Manoj Kumar Dhadwal, dhadwal.manojkumar@gmail.com; Sung Nam Jung, snjung@konkuk.ac.kr

ACKNOWLEDGMENTS

This research was supported by the EDISON Program through the National Research Foundation of Korea (NRF) funded by the Ministry of Science, ICT & Future Planning (2014M3C1A6038856). This research was supported by Ba-

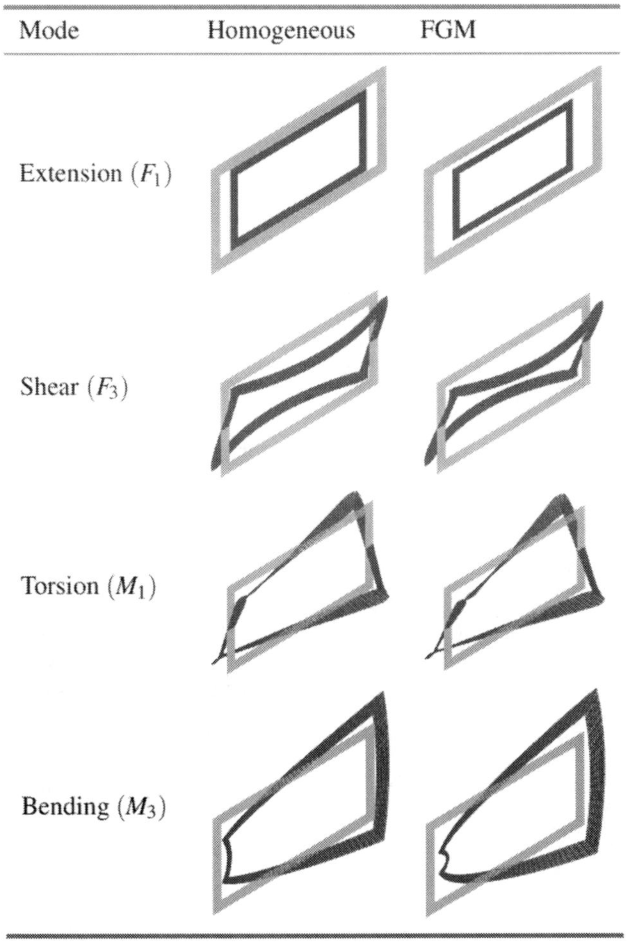

Figure 10. Comparison of warping displacement modes for homogeneous and functionally graded box beams (exaggerated).

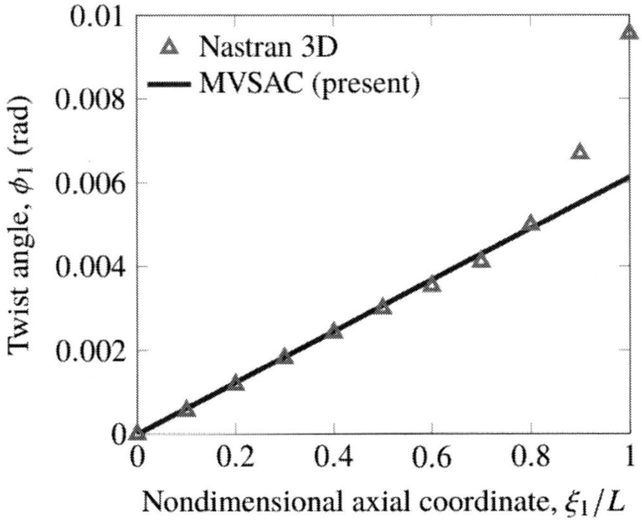

Figure 11. Twist response of functionally graded box beam under a tip torsional moment.

sic Science Research Program through the National Research Foundation of Korea (NRF) funded by the Ministry of Education (2017R1D1A1A09000590).

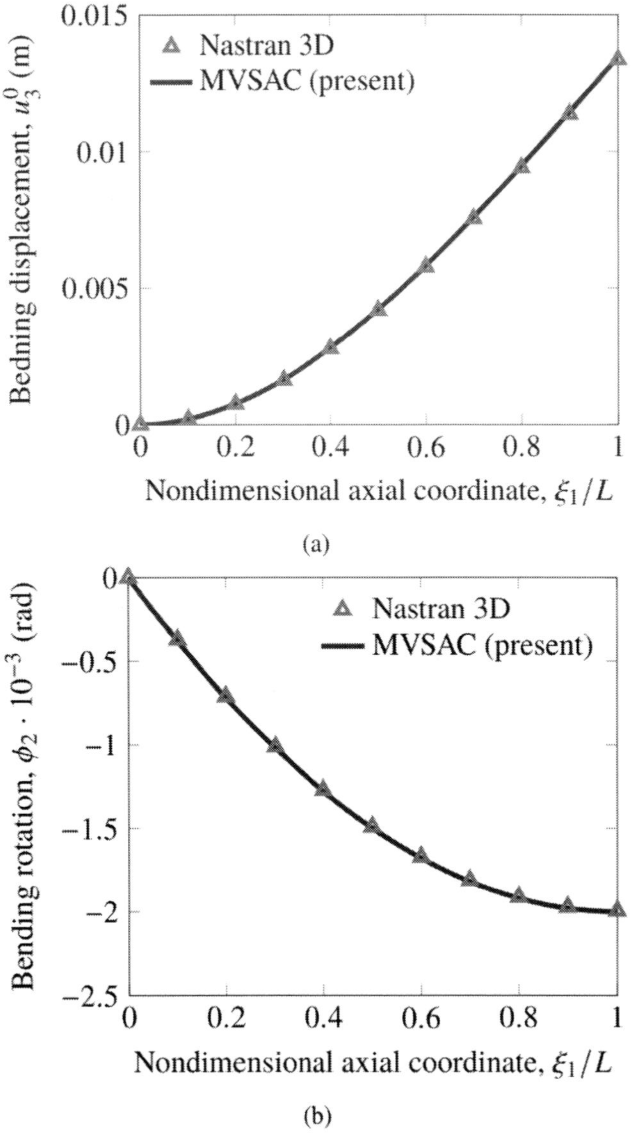

Figure 12. (a) Bending displacement and (b) bending rotation of functionally graded box beam under a tip shear force.

REFERENCES

[1]Miyamoto, Y., Kaysser, W. A., Rabin, B. H., Kawasaki, A., and Ford, R. G., *Functionally Graded Materials: Design, Processing and Applications*, Springer US, Boston, MA, 1999. doi: 10.1007/978-1-4615-5301-4

[2]Horgan, C. and Chan, A., "Torsion of Functionally Graded Isotropic Linearly Elastic Bars," *Journal of Elasticity*, Vol. 52, (2), Aug 1998, pp. 181–199. doi: 10.1023/A:1007544011803

[3]Sankar, B., "An elasticity solution for functionally graded beams," *Composites Science and Technology*, Vol. 61, (5), 2001, pp. 689–696. doi: 10.1016/S0266-3538(01)00007-0

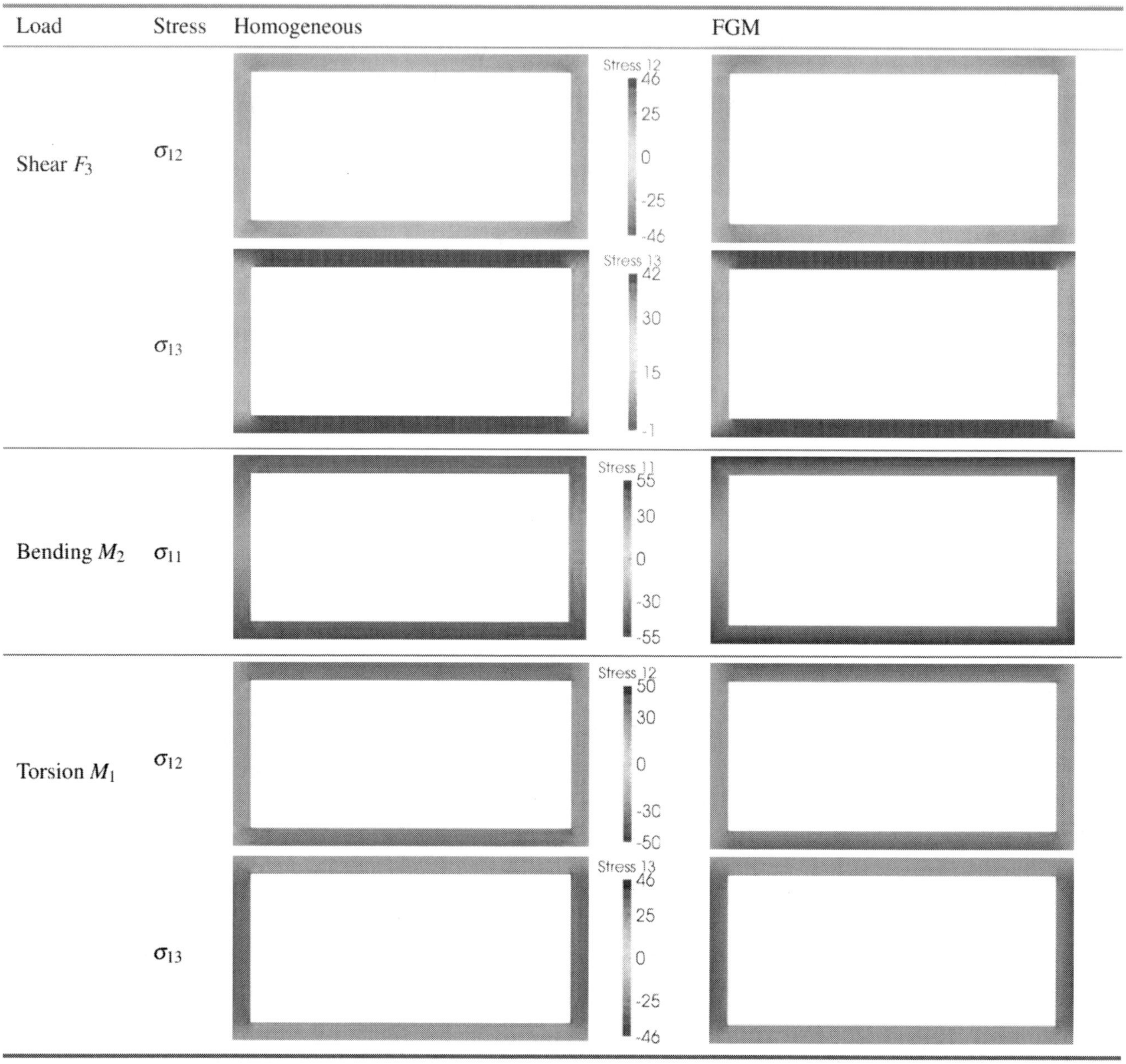

Figure 13. Comparison of reactive stress distributions for homogeneous and functionally graded box beams (exaggerated).

[4]Chakraborty, A., Gopalakrishnan, S., and Reddy, J., "A new beam finite element for the analysis of functionally graded materials," *International Journal of Mechanical Sciences*, Vol. 45, (3), 2003, pp. 519–539.
doi: 10.1016/S0020-7403(03)00058-4

[5]Zhong, Z. and Yu, T., "Analytical solution of a cantilever functionally graded beam," *Composites Science and Technology*, Vol. 67, (3), 2007, pp. 481–488.
doi: 10.1016/j.compscitech.2006.08.023

[6]Ke, L.-L., Yang, J., and Kitipornchai, S., "Nonlinear free vibration of functionally graded carbon nanotube-reinforced composite beams," *Composite Structures*, Vol. 92, (3), 2010, pp. 676–683.
doi: 10.1016/j.compstruct.2009.09.024

[7]Giunta, G., Crisafulli, D., Belouettar, S., and Carrera, E., "Hierarchical theories for the free vibration analysis of functionally graded beams," *Composite Structures*, Vol. 94, (1), 2011, pp. 68–74.
doi: 10.1016/j.compstruct.2011.07.016

[8]Filippi, M., Carrera, E., and Zenkour, A., "Static analyses of FGM beams by various theories and finite elements," *Composites Part B: Engineering*, Vol. 72, (Supplement C), 2015, pp. 1–9.
doi: 10.1016/j.compositesb.2014.12.004

[9]Carrera, E., Giunta, G., and Petrolo, M., *Beam Structures*, John Wiley & Sons, Ltd, 2011.
doi: 10.1002/9781119978565

[10]Nguyen, T.-T., Kim, N.-I., and Lee, J., "Analysis of thin-walled open-section beams with functionally graded materials," *Composite Structures*, Vol. 138, 2016, pp. 75–83.
doi: 10.1016/j.compstruct.2015.11.052

[11]Kim, N.-I. and Lee, J., "Exact solutions for coupled responses of thin-walled {FG} sandwich beams with non-symmetric cross-sections," *Composites Part B: Engineering*, Vol. 122, 2017, pp. 121–135.
doi: 10.1016/j.compositesb.2017.04.016

[12]Hodges, D. H., *Nonlinear composite beam theory*, AIAA, Washington, DC, 2006.
doi: 10.2514/4.866821

[13]Giavotto, V., Borri, M., Mantegazza, P., Ghiringhelli, G., Carmaschi, V., Maffioli, G. C., and Mussi, F., "Anisotropic beam theory and applications," *Computers and Structures*, Vol. 16, (1–4), 1983, pp. 403–413.
doi: 10.1016/0045-7949(83)90179-7

[14]Cesnik, C. E. S. and Hodges, D. H., "VABS: a new concept for composite rotor blade cross-sectional modeling," *Journal of the American Helicopter Society*, Vol. 42, (1), 1997, pp. 27–38.
doi: 10.4050/JAHS.42.27

[15]Yu, W., Volovoi, V. V., Hodges, D. H., and Hong, X., "Validation of the variational asymptotic beam sectional analysis," *AIAA Journal*, Vol. 40, (10), 2002, pp. 2105–2112.
doi: 10.2514/2.1545

[16]Morandini, M., Chierichetti, M., and Mantegazza, P., "Characteristic behavior of prismatic anisotropic beam via generalized eigenvectors," *International Journal of Solids and Structures*, Vol. 47, 2010, pp. 1327–1337.
doi: 10.1016/j.ijsolstr.2010.01.017

[17]Reissner, E., "On mixed variational formulations in finite elasticity," *Acta Mechanica*, Vol. 56, (3-4), 1985, pp. 117–125.
doi: 10.1007/BF01177113

[18]Henderson, A., Ahrens, J., and Law, C., *The ParaView Guide*, Clifton Park, NY, 2004.

Significant Effects Obtained from Health and Usage Monitoring System (HUMS) on JASDF Rescue Helicopter

Kosuke Watanabe
Product Support Department
Mitsubishi Heavy Industries, Ltd.
Nagoya, Japan

Hiroshi Saito
Product Support Department
Mitsubishi Heavy Industries, Ltd.
Nagoya, Japan

Shinsuke Yamazaki
Product Support Department
Mitsubishi Heavy Industries, Ltd.
Nagoya, Japan

Daisuke Matsuura
Product Support Department
Mitsubishi Heavy Industries, Ltd.
Nagoya, Japan

ABSTRACT

In recent years, Health and Usage Monitoring System (HUMS) that records, collects and analyzes operational data to monitor the condition of airframes and equipment has been installed on both military and civil helicopters around the world. Mitsubishi Heavy Industries, Ltd. (MHI) has developed the original HUMS for Japan Air Self-Defense Force (JASDF) UH-60J Rescue Helicopter, which has been in service since 2015. The HUMS has an ability to clarify unusual airframe vibration trouble and detect early faults in the powertrains by using vibration data collected from accelerometers on the airframes and powertrains, respectively. In addition, the HUMS records flight parameters and equipment operational data for troubleshooting of aircraft systems such as engines, flight control and avionics systems. MHI has not only developed on-board devices but also provided operational data management and analysis. This paper describes the overview of UH-60J HUMS, the benefits obtained from HUMS and future plans to expand the benefits.

INTRODUCTION

Within a conventional time-based maintenance paradigm for helicopters, maintainers inspect components at predetermined time intervals and take maintenance actions based on the condition of each component. However, maintainers often tend not to find faults within the components. Therefore, it is possible that maintainers conduct more maintenance inspections than necessary during the scheduled maintenance periods. On the other hand, Condition Based Maintenance (CBM) is a cost-effective maintenance approach that monitors the health condition of the airframes and components by analyzing data collected from sensors throughout the airframes and components. The concept of CBM is shown in Figure 1. Maintainers can accurately make maintenance decisions based on actual condition. There are four goals in the implementation of CBM (Ref.1):

- Reducing burdensome maintenance tasks currently required to assure continued airworthiness
- Increasing aircraft availability
- Improving flight safety
- Reducing sustainment costs

Presented at the 6th Asian/Australian Rotorcraft Forum / Heli Japan 2017, Kanazawa, Japan, November 7-9, 2017.

Health and Usage Monitoring System (HUMS) that records, collects and analyzes a large volume and variety of operational data has been installed on both military and civil helicopters around the world and helped users engage CBM approaches.

The U.S. Army has installed HUMS on most of their helicopters. The main objective of the HUMS installation for U.S. Army is transition from a conventional time-based maintenance paradigm to a more cost-effective CBM paradigm. The U.S Army has reported benefits obtained from the CBM program as shown in Table 1 (Ref.2).

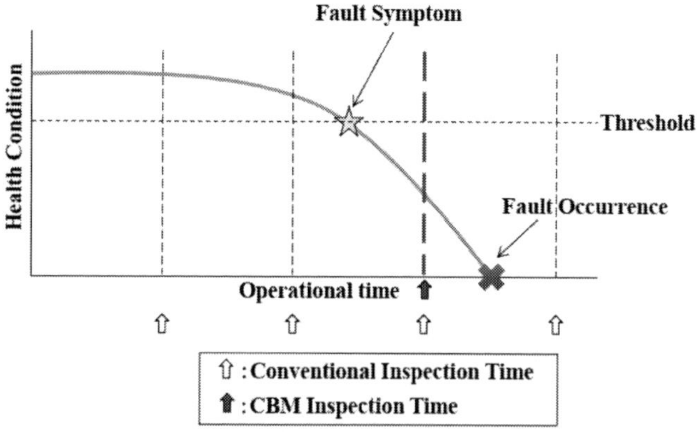

Figure 1. The concept of Condition Based Maintenance.

In order to introduce CBM approaches to Japan Air Self-Defense Force (JASDF) helicopter maintenance program, Mitsubishi Heavy Industries, Ltd. (MHI) has developed the original HUMS for JASDF UH-60J Rescue Helicopter as shown in Figure 2, which has been in service since 2015. The HUMS has an ability to clarify unusual vibration trouble and detect early faults in the powertrains by using vibration data from accelerometers on the airframes and powertrains, respectively. In addition, the HUMS enables more efficient real-time troubleshooting by using HUMS-recorded operational data streaming in MIL-STD-1553B data bus. MHI has not only developed on-board devices but also provided operational data management and analysis. In this paper, the overview of UH-60J HUMS including on-board and ground system, the benefits obtained from HUMS and future plans to expand HUMS benefits are described.

Table1. Example of U.S. Army CBM benefits.

CBM Program Objectives	Benefits (Example)
Reduce Soldier Burden	･343,278 MMHs/Year Avoidance With Over 4,958 Maintenance Events Eliminated ･127 Improved Maintenance Procedures
Increase Platform Availability and Readiness	･Experienced 2.2%-11.3%NMCM[a] Reduction
Enhance Safety	･Avoided 4 Class A Mishaps
Reduce/Control Total Life Cycle Operation & Support Costs	･TBO Extensions on 22 CBM Monitored Parts ･MTF[b] Cost Avoidance of $25.7 M/Year

[a]NMCM: Not Mission Capable Maintenance
[b]MTF: Maintenance Test Flights

Figure 2. UH-60J Rescue Helicopter.

UH-60J HUMS OVERVIEW

UH-60J HUMS developed by MHI continuously monitors the condition of components in order to enhance flight safety, reduce maintenance cost and increase operational readiness.

The HUMS has a range of features including:

- Collecting vibration data from accelerometers on the airframes and powertrains, allows early fault detection of each component.
- Recording flight parameters and equipment operational data streaming in MIL-STD-1553B data bus, enables to make troubleshooting more efficient and faster.
- Monitoring flight parameters and alerting if they exceed their thresholds during the flights, enhances flight safety and makes maintenance decisions faster.
- Automatically collecting and processing data related to engine health conditions, reduces crew workloads.

System Configuration

UH-60J HUMS includes accelerometers on the airframes and powertrains as shown in Figure 3 and a tachometer within the rotor systems. Then, the Vibration Data Collector (VDC) acquires the vibration and tracking data. In addition, the VDC connects to MIL-STD-1553B data bus, enables to record a large volume and variety of aircraft operational data including flight parameters and equipment operational data. The HUMS data acquired by the VDC are stored in the Removable Memory Unit (RMU) including Solid State Drive (SSD) through the Data Transfer Unit (DTU). After each flight, the RMU recorded data are downloaded to the Ground Support Equipment (GSE) and simply displayed on the GSE. UH-60J HUMS system block diagram is shown in Figure 4.

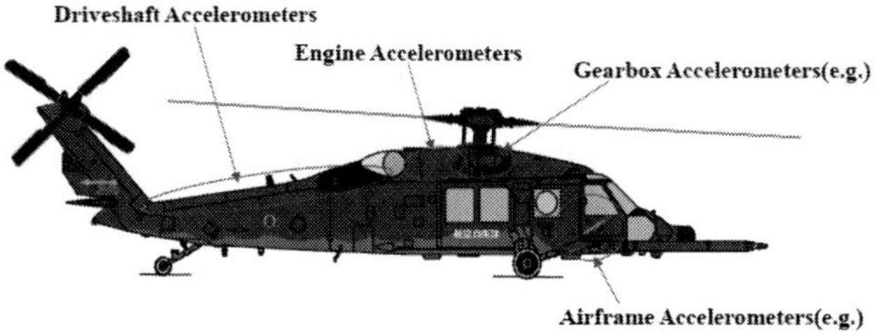

Figure 3. Accelerometers for UH-60J HUMS.

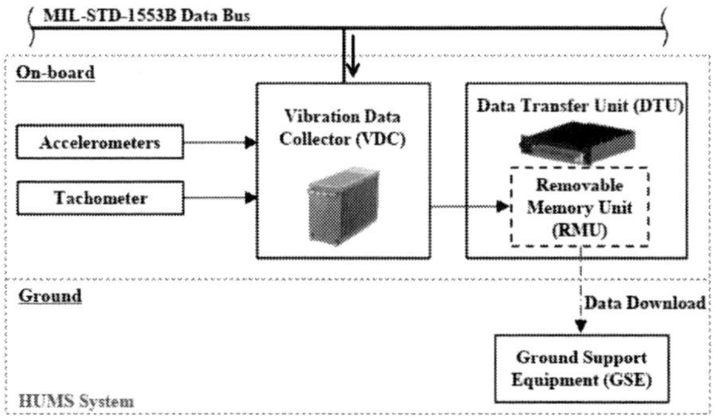

Figure 4. UH-60J HUMS system block diagram.

HUMS Ground Software

HUMS Data Manager (HDM) designed by MHI is installed in UH-60J GSE in order to handle the HUMS data more efficiently.

Figure 5 shows the example of the HDM software displays. The main features of HDM software are:

- Download RMU recorded data.
- Providing graph and table views of vibration data, flight parameters and equipment operational data.
- Display the results of engine performance diagnosis, exceedance monitoring and HUMS self-diagnosis.
- Input maintenance records data.

These features make maintenance decision making and troubleshooting more efficient and faster. In addition, The HDM software can be updated as required by users. For example, exceedance monitoring threshold values are optimized based on actual operation records.

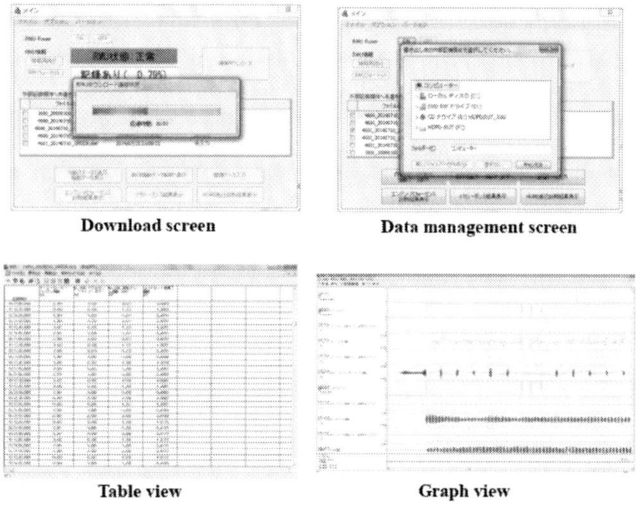

Figure 5. Example of HDM display.

In addition, HUMS data Analysis and Report Program (HARP) is installed in MHI HUMS data analysis working station as shown in Figure 6. The HARP has an ability to automatically calculate and display health parameters which indicate the condition of each component by using vibration data collected from accelerometers on the powertrains. In addition, the HARP enables data analysis in order to precisely diagnose the condition of the components with aircraft parameters synchronized perfectly. Finally, MHI makes reports about the results of diagnosis for an action to take.

Figure 6. MHI HUMS data analysis working station.

HUMS Data Flow

UH-60J HUMS data flow between JASDF and MHI is shown in Figure 7. In the Maintenance Division of helicopter fleet, HUMS data recorded in RMU are downloaded, simply displayed and checked by using HDM software. Then, the collected HUMS data are accumulated in MHI's HUMS database server and managed by MHI HUMS support division. In addition, MHI precisely analyzes the data to monitor the condition of each aircraft. By using accumulated HUMS data, MHI conducts technical investigations in order to reduce scheduled preventive maintenance inspections and increase operational readiness. The results of HUMS data analysis are reported to the Management Division of helicopter fleet. The Management Division considers making maintenance plans, revising maintenance manuals and upgrading the HUMS software based on the reported results. This data flow will make unit operations more efficient and increase aircraft availability.

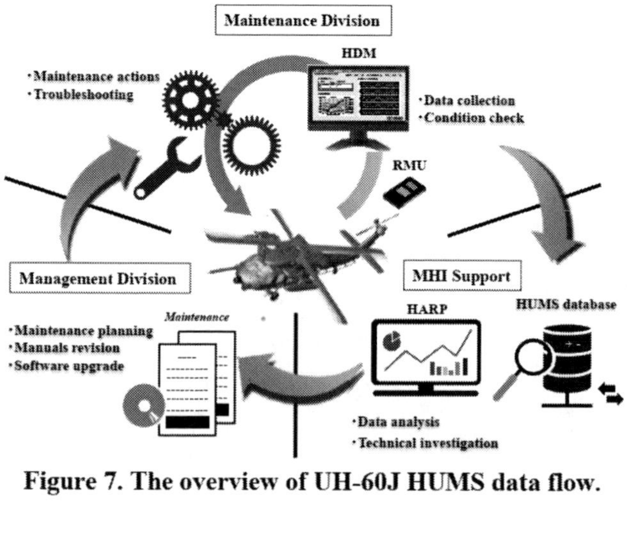

Figure 7. The overview of UH-60J HUMS data flow.

HUMS BENEFITS

There have been benefits obtained from HUMS since the HUMS installation on UH-60J. For example, the HUMS has provided users with more efficient troubleshooting. On the other hand, there have not yet been sufficient achievements of early fault detection of the powertrains.

More Efficient Troubleshooting

UH-60J HUMS records a large volume and variety of flight parameters and equipment operational data streaming in MIL-STD-1553B data bus, which allows users to conduct more efficient real-time troubleshooting than conventional troubleshooting. The difference between conventional troubleshooting and HUMS-based troubleshooting is shown in Figure 8. Regarding the conventional troubleshooting, in case of temporary or non-reproducible equipment faults, maintainers are forced to depend on vague information from crew interviews. Therefore, it is difficult to isolate fault equipment and it is often happened to spend a longer period to resolve the trouble. On the other hand, by using HUMS data, operational status of aircraft systems such as engines, flight control and avionics systems can be tracked. Therefore, users can isolate the fault equipment and replace it correctly and quickly. As a result, users can make troubleshooting faster and recover the aircraft earlier.

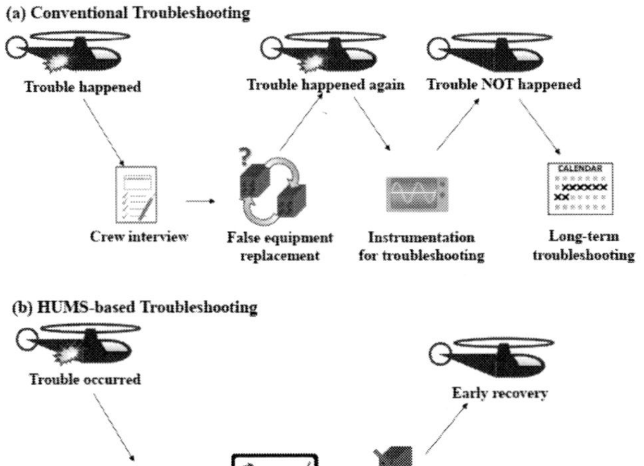

Figure 8. The difference between conventional troubleshooting and HUMS-based troubleshooting.

Early Fault Detection

MHI has provided users with health monitoring and fault detection of the powertrains by analyzing vibration data collected from accelerometers throughout the systems. Mahalanobis Distance (MD) is automatically calculated and displayed by MHI's vibration analysis algorithm installed in the HARP that integrates Condition Indicators (CIs) using Mahalanobis Taguchi Method (MTS) (Ref.3). MHI quickly checks whether MD exceeds its threshold value based on fleet data. This simple diagnosis enables early detection of fault symptom. If the MD exceeds its threshold, MHI analyzes collected data including raw/spectrum vibration data, CIs trends data and maintenance records data in order to precisely diagnose the condition of the components and determine an action to take. This will enable not only to avoid unscheduled maintenance because of component failure but also to increase reliability of the powertrains.

FUTURE WORK

MHI is planning to expand benefits obtained from HUMS by reducing scheduled maintenance inspections and extending component service lives.

Scheduled Maintenance Reduction

Monitoring the condition of components such as the oil cooler and engine drive shaft by analyzing vibration data and maintenance records data enables to eliminate scheduled maintenance inspections of the components. Regarding the oil cooler and engine drive shaft, vibration inspection and tuning with measurement equipment instrumentation and rotor running are conducted every fixed flight hours, can be replaced with quickly HUMS data check after each flight.

In addition, the HUMS can be used for airframe vibration monitoring and tuning. Conventional airframe vibration tuning action is required depending on pilot feeling as shown in Figure 9. In order to acquire vibration data and conduct vibration tuning, measurement equipment instrumentation and maintenance test flights are needed. It costs a lot of maintenance man-hours and aircraft fuels. Therefore, MHI is planning to introduce HUMS-based vibration tuning methodology called Small Rotor Adjustment (SRA), which combines airframe vibration monitoring with a small amount of vibration tuning. If main rotor 1/Rev vibration exceeds SRA threshold value as shown in Figure 10, maintainers only have to slightly adjust the rotor components without unscheduled maintenance test flights (Ref.4).

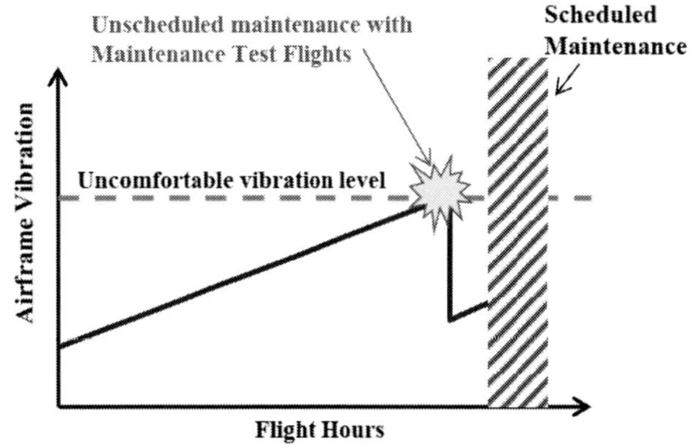

Figure 9. Implementation timing of conventional airframe vibration tuning.

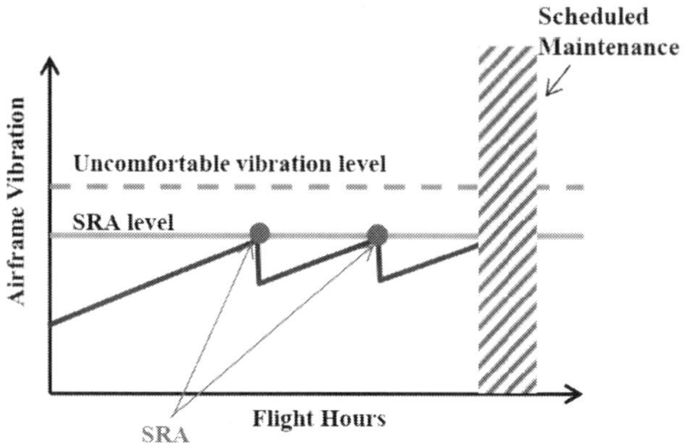

Figure 10. Implementation timing of Small Rotor Adjustment.

These procedures will enable to reduce maintenance cost and increase operational readiness. In order to replace conventional maintenance procedures with HUMS-based maintenance procedures, MHI will continue to collect vibration data and conduct technical investigations to determine appropriate monitoring criteria.

Extending Component Service Lives

The service lives of components are calculated by using conservative design spectrum based on composite worst case (CWC) usage. Therefore, it is possible to keep significantly high reliability of each component. However, in fact, actual usage of each aircraft may be different from CWC usage and individual component service life is probably overestimated.

The concept of HUMS-based fatigue life prediction methodology is shown in Figure 11. The actual aircraft usage can be tracked by processing HUMS-recorded flight parameters (e.g., airspeed, outside air temperature, pressure altitude, radar altitude, pitch rate, roll rate, yaw rate, engine torque and rotor speed…), which enables to accurately analyze component fatigue damage and predict component service lives. By monitoring individual aircraft usage based on HUMS data, it is possible to improve flight spectrum and update component service lives and Time Between Overhaul (TBO), which allows users to reduce aircraft maintenance and life cycle cost. In order to manage and improve individual aircraft component service lives based on HUMS data, MHI is planning to conduct technical investigations about regime recognition algorithms and load/strain prediction methodology.

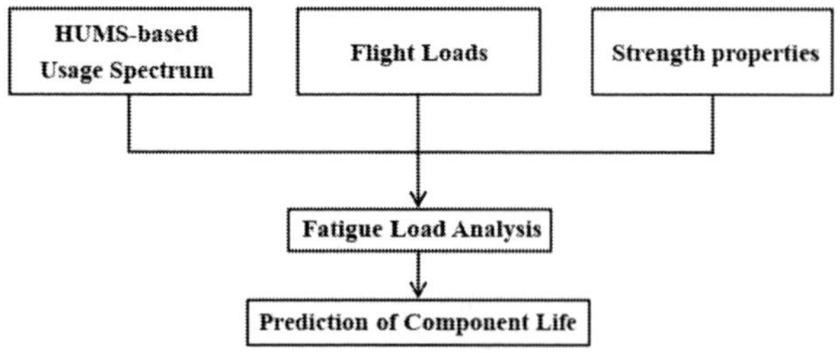

Figure 11. The concept of HUMS-based fatigue life prediction methodology

CONCLUSIONS

UH-60J HUMS developed by MHI has an ability to monitor the condition of the airframes and powertrains by analyzing vibration data collected from accelerometers throughout the aircraft. This enables to avoid unscheduled maintenance as well as improve aircraft reliability. In addition, by using HUMS-recorded operational data streaming in MIL-STD-1553B data bus, users can perform more efficient real-time troubleshooting, which allows not only to reduce maintenance cost but also to increase operational readiness.

MHI is planning to expand the benefits obtained from HUMS. Scheduled/Unscheduled vibration inspection/tuning including maintenance test flights can be reduced by monitoring vibration of the airframes and components such as the oil cooler and engine drive shaft. Moreover, monitoring individual aircraft actual usage based on HUMS data allows users to accurately manage component service lives of each aircraft. MHI will continue to collect HUMS data and conduct technical investigations in order to achieve these plans, which enable to reduce operational cost and improve aircraft availability.

In the U.S., in order to reduce scheduled inspections and preventive maintenance while enhancing safety, Aviation Development Directorate (ADD) and Sikorsky Aircraft Corporation (SAC) aims for integration of health management technologies, which is called Capability-Based Operations and Sustainment Technology-Aviation (COST-A) program (Ref.5). In Japan, in order to obtain significant benefits from HUMS as well, it is necessary to apply next-generation health management technologies to JASDF helicopters. This technology acquisition needs many kinds of demonstrations in a laboratory and flight tests. Therefore, MHI believes that it is important to actively collaborate with the Japan Ministry of Defense.

Author contact:
Kosuke Watanabe kosuke_watanabe@mhi.co.jp ,
Hiroshi Saito hiro4_saito@mhi.co.jp,
Shinsuke Yamazaki shinsuke_yamazaki@mhi.co.jp,
Daisuke Matsuura daisuke_matsuura@mhi.co.jp

REFERENCES

[1]AMCOM Standardization Technical Data Management Division ADS-79D-HDBK, "Condition Based Maintenance for US Army Aircraft Systems", Redstone Arsenal, AL, 2013.

[2]Kennedy, J., "Army Aviation CBM+ and Cost-Wise Readiness", 2013 AHS Airworthiness, CBM, and HUMS Specialists' meeting, Huntsville, AL, February, 2013.

[3]Daisuke, M., Tokihiko, T., "Helicopter Transmissions Gear Fault Detection Technique Using Vibration Monitoring", Paper T112-3, Heli Japan 2010, Saitama, Japan, November 2010.

[4]Hiroshi, S., "Small Rotor Adjustment for SH-60K Helicopter", Paper T122-2, Heli Japan 2010, Saitama, Japan, November 2010.

[5]Brookhart, A., Davis, M., Bates, P., Cycon, J., Lyman, C., Pantelis, P., Baker, T., Bordick, N., "Flight Testing of Health Management Technologies", American Helicopter Society 73rd Annual Forum Proceedings, Ft. Worth, TX, May 2017.

Non-Linear Analysis of Pretwisted Anisotropic Beam with Multiple Delaminations

Prasad S Puvvala
Student
Indian Institute of Science
Bangalore, Karnataka, India

Rama Sateesh K.V
Student
Indian Institute of Science
Bangalore, Karnataka, India

Dineshkumar Harursampath
Professor
Indian Institute of Science
Bangalore, Karnataka, India

ABSTRACT

Fiber reinforced laminated composites are widely used as primary load-carrying members (typically open-section flexbeams) in helicopter rotor blades because of their high specific strengths and stiffnesses (in desired directions) as well as fatigue life. The appearance of delaminations in such composite rotorcraft flexbeams can lead to deterioration of the mechanical properties and, in turn, the helicopter performance as well as safety. Understanding and predicting the influence of free edge delaminations on the overall behavior of the laminates will provide quantitative measures of the extent of the damage and help ensure their damage tolerance.

The present work is aimed at the development of an efficient mathematical model to assess the degradation in the stiffness properties of anisotropic strips (with small pretwist, typical of helicopter flexbeams) in the presence of multiple edge delaminations with the motivation to capture those nonlinear effects that arise by virtue of the highly flexible strip-like beam geometry of rotor blades, amplified by the presence of delamination(s). Variational Asymptotic Method (VAM) is used as a tool for asymptotically-correct dimensional reduction of laminated shell theory to a nonlinear one-dimensional (1-D) theory. A study is performed to investigate the influence of the locations, sizes and combinations of delaminations on the trapeze effect, which is important in rotor blades. The developed model has a potential application in structural health monitoring of rotorcraft flexbeams.

INTRODUCTION

Flexbeams made of Fiber reinforced composites (FRC) are being increasingly used in bearing less helicopter rotor as well as tail rotor blades because of their high specific strengths and stiffnesses. These rotor blades rotating in high speed are the primary load carrying members in a helicopter. The appearance of cracks and delaminations in composite rotorcraft flexbeams, which affects flapwise and lagwise performance of a helicopter, may lead to catastrophic results. Hence, there is a strong need and requirement for prediction of influence of delamination on the strength of the composite flexbeams and develop online damage detection and health monitoring. The longitudinal cracks in helicopter tail rotor flexbeams that are predominantly made of 0^0 layup, causes very less reduction in stiffness. Hence the use of laminated composites in rotorcraft flexbeams requires assessment of their damage such that even longitudinal cracks causing very small reduction in stiffness can be detected. This is a primary motivation for exploring nonlinear nonclassical effects, which are far more sensitive to stiffness changes.

Delamination in laminated fiber reinforced composites is a mode of material damage that occurs mainly due to the presence of interlaminar stresses associated with the anisotropic interaction of the laminating layers. Particularly, if the interlaminar normal stress is tensile, crack may be induced separating two adjacent layers. Also at the free edges of a laminate, the interlaminar normal stresses are very high and would therefore cause the debonding or what is generally known as edge delamination.

A significant amount of research work has been done to predict the interlaminar stresses in the presence of delaminations in composite structures using various methodologies. Layer-wise theory by Barbero and Reddy (Ref. 1) although predicts the interlaminar stresses in the presence of delamination in composite laminates very accurately, it is computationally more expensive than the conventional laminate theories. Kim et al. (Ref. 2) developed

Presented at the 6th Asian/Australian Rotorcraft Forum / Heli Japan 2017, Kanazawa, Japan, November 7-9, 2017.

computationally efficient layerwise composite laminate theory with delamination by reducing the number of structural variables through enforcing continuity of the transverse shear stresses and the in-plane displacement at ply interfaces. The above-mentioned methodologies are finite element based and are useful during the final design stage of the structures. But efficient analytical tools, accurately predicting the behavior of damaged structures, are needed at the preliminary design stage in order to ensure cost-effective performance.

Makeev and Armanios (Ref. 3) developed an approximate elasticity solution for stiffness parameters in damaged laminated beams, exhibiting a three-dimensional (3D) state of strain independent of the longitudinal direction. This method has been used for predicting interlaminar stresses in symmetric laminates under axial extension and torsion. They also observed that an accurate solution for the crack tip stresses is not necessary for an accurate prediction of the laminate stiffness. Makeev and Armanios (Ref. 4) obtained a closed form solution, which takes into account the influence of delamination on the nonlinear extension-twist coupling (trapeze effect) behavior of antisymmetric laminates. The closed form solution that they obtained was in the presence of internal and symmetric edge delamination. But their methodology is specific for the problem of obtaining a closed-form solution to the trapeze effect.

In the present paper, generalized theory for composite strips with a small pretwist in the presence of multiple edge delaminations, without any symmetry restrictions is considered. Moreover, the influence of damage on all inherent non-linear effects, including but not restricted to trapeze effect, is asymptotically captured. Numerical examples are presented to validate the present theory with the results obtained from the established 3-D ANSYS FE solutions. The same example are also used to demonstrate the computational efficiency of the current approach.

PRETWISTED STRIP WITH DELAMINATIONS

In the present analysis, the original 3D problem is simplified by the decomposition into two simpler problems: a 2D problem, which provides in a compact form the cross-sectional properties using a mathematical technique called the Variational Asymptotic Method(VAM), and a non-linear one-dimensional problem along the length of the beam. This is possible because the wavelength of deformation of the strip along the beam length is very large as compared to the cross-sectional dimensions of the strip. This also gives rise to geometric small parameters which are inherent in the structure. The kinematics of the strip is derived in terms of these small parameters. Using these kinematic relations the 2D strain energy of the strip is expressed in terms of small parameters. VAM is then used as a mathematical tool to carry out the dimensional reduction of the 2D problem to 1D problem.

Kinematics

Consider a pretwisted thin composite strip with anisotropic lay-up having multiple free-edge delamination from both ends in different planes offset from mid surface as shown in Fig. 1. Accordingly, the strip is divided into M+N+3 sublaminate domains, groups of plies above and below the delamination interface, denoted by indices $L_1, L_2, L_3, .., L_{M-1}$, L_M and L_{M+1} for the left portion of the strip, $R_1, R_2, R_3, ..., R_{N-1}$, R_N and R_{N+1} for the right portion of the strip and h for the mid portion of the strip respectively. The Cartesian coordinate measures x_i, associated with each of the sublaminate is directed along the length (l), width (b) and thickness (h) of the strip for i=1,2 and 3 respectively, and $h \leq b \leq l$. Hence the natural small parameters are $\delta_h = \frac{h}{b}$; $\delta_b = \frac{b}{l}$ and $\delta_t = bk_1$ (k_1 is the derivative of the pretwist angle with respect to the length along the strip). The domain of the sublaminates is such that $0 \leq x_1 \leq l, -\frac{b}{2} \leq x_2 \leq \frac{b}{2}$ and $-\frac{h_{L_{m+1}}-h_{L_m}}{2} \leq x_3 \leq \frac{h_{L_{m+1}}-h_{L_m}}{2}, 0 \leq m \leq M$ for the left sublaminates and $-\frac{h_{R_{n+1}}-h_{R_n}}{2} \leq x_3 \leq \frac{h_{R_{n+1}}-h_{R_n}}{2}, 0 \leq n \leq N$ for the right sublaminates. Since the strip has a pretwist rate of k_1, the unit vectors associated with the cross-sectional plane b_2 and b_3 are functions of x_1.

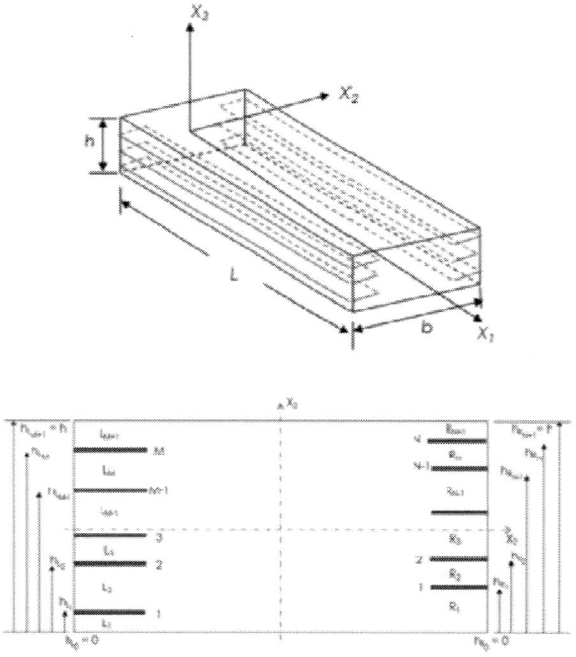

Figure 1. Pretwisted Strip with multiple delaminations

The position vector ($\boldsymbol{r}$) of an arbitrary material point in each of the sublaminate region can be written as

$$r^{(L_m)} = x_1\boldsymbol{b}_1 + x_2\boldsymbol{b}_2(x_1) + \left(x_3 + \frac{h_{L_{m+1}}+h_{L_m}-h}{2}\right)\boldsymbol{b}_3(x_1)$$

$$r^{(R_n)} = x_1\boldsymbol{b}_1 + x_2\boldsymbol{b}_2(x_1) + \left(x_3 + \frac{h_{R_{n+1}}+h_{R_n}-h}{2}\right)\boldsymbol{b}_3(x_1) \qquad -(1)$$

where the superscript associated with $\boldsymbol{r}$ refers to each of the sublaminates in the undeformed state and $\boldsymbol{b}_i$ are the unit vectors.

The position vector ($\boldsymbol{R}$) of an arbitrary material point in the deformed condition

$$\boldsymbol{R}^{(L_m)} = x_1\boldsymbol{b}_1 + u_i\boldsymbol{b}_i + x_2\boldsymbol{b}_2 + w_i^{(L_m)}\boldsymbol{b}_i + \left(x_3 + \frac{h_{L_{m+1}}+h_{L_m}-h}{2}\right)\left(\phi_1^{(L_m)}\boldsymbol{B}_i + \phi_2^{(L_m)}\boldsymbol{B}_3\right) + \Delta_i^{(L_m)}\boldsymbol{B}_i$$

$$\boldsymbol{R}^{(R_n)} = x_1\boldsymbol{b}_1 + u_i\boldsymbol{b}_i + x_2\boldsymbol{b}_2 + w_i^{(R_n)}\boldsymbol{b}_i + \left(x_3 + \frac{h_{R_{n+1}}+h_{R_n}-h}{2}\right)\left(\phi_1^{(R_n)}\boldsymbol{B}_i + \phi_2^{(R_n)}\boldsymbol{B}_3\right) + \Delta_i^{(R_n)}\boldsymbol{B}_i \qquad -(2)$$

where $w_i^{(L_m)}, w_i^{(R_n)}$'s are average warpings (i=1,2,3); $\phi_\alpha^{(L_m)}, \phi_\alpha^{(R_n)}$'s are average local rotations (α= 1,2); u_i are rigid-body displacements; $\boldsymbol{B}_i$ are orthogonal unit vectors introduced by global rotations.

The kinematics of the problem is derived by following the methodology as developed by Hodges et al (Ref.5). The expressions for membrane strains and curvatures are

$$\varepsilon_{11}^{(L_m)} \approx \gamma_{11} - x_2\kappa_3 + k_1x_2^2\kappa_1 + \frac{h_{L_{m+1}}+h_{L_m}-h}{2}\kappa_2 + \underline{\frac{x_2^2\kappa_1^2}{2} + w_3^{(L_m)}\kappa_2}$$

$$\varepsilon_{11}^{(R_n)} \approx \gamma_{11} - x_2\kappa_3 + k_1x_2^2\kappa_1 + \frac{h_{R_{n+1}}+h_{R_n}-h}{2}\kappa_2 + \underline{\frac{x_2^2\kappa_1^2}{2} + w_3^{(R_n)}\kappa_2}$$

$$\varepsilon_{22}^{(L_m)} \approx w_{2,2}^{(L_m)} + \frac{h_{L_{m+1}}+h_{L_m}-h}{2}w_{3,22}^{(L_m)} + \underline{\frac{{w_{3,2}^{(L_m)}}^2}{2}}$$

$$\varepsilon_{22}^{(R_n)} \approx w_{2,2}^{(R_n)} + \frac{h_{R_{n+1}}+h_{R_n}-h}{2}w_{3,22}^{(R_n)} + \underline{\frac{{w_{3,2}^{(R_n)}}^2}{2}}$$

$$2\varepsilon_{12}^{(L_m)} \approx w_{1,2}^{(L_m)} + k_1\left(x_2w_{3,2}^{(L_m)} - w_3^{(L_m)}\right) + \underline{\kappa_1\left(x_2w_{3,2}^{(L_m)} - w_3^{(L_m)}\right)}$$

$$2\varepsilon_{12}^{(R_n)} \approx w_{1,2}^{(R_n)} + k_1\left(x_2w_{3,2}^{(R_n)} - w_3^{(R_n)}\right) + \underline{\kappa_1\left(x_2w_{3,2}^{(R_n)} - w_3^{(R_n)}\right)}$$

$$\rho_{11}^{(L_m)} \approx \kappa_2;\ \rho_{11}^{(R_n)} \approx \kappa_2$$

$$\rho_{22}^{(L_m)} \approx -w_{3,22}^{(L_m)};\quad \rho_{22}^{(R_n)} \approx -w_{3,22}^{(R_n)}$$

$$2\rho_{12}^{(L_m)} \approx -2\kappa_1;\ 2\rho_{12}^{(R_n)} \approx -2\kappa_1 \qquad -(3)$$

In the above equations, the underlined terms are non-linear and arise due to moderate local rotation. The non-underlined terms are lower order terms, hence are more significant. In the zeroth-order approximation only those terms, which are not underlined, are included. But the first order approximation should include all the above terms. Also, it is to be noted that, since $\rho_{11}^{(L_i,R_i)}$ and $2\rho_{12}^{(L_i,R_i)}$ depend only on 1D strain terms they are equal in all the sublaminate.

Let $\varepsilon^{(L_i,R_i)} = \left[\varepsilon_{11}^{(L_i,R_i)} \quad \varepsilon_{22}^{(L_i,R_i)} \quad 2\varepsilon_{12}^{(L_i,R_i)} \quad \rho_{11}^{(L_i,R_i)} \quad \rho_{22}^{(L_i,R_i)} \quad 2\rho_{12}^{(L_i,R_i)}\right]$

The laminate stiffness matrix of the healthy part of the cross-section is given by

$$K_h = \begin{bmatrix} A_h & B_h \\ B_h & D_h \end{bmatrix}$$

and that of the delaminated sublaminate is given by

$$K_{(L,R)}^i = \begin{bmatrix} A_{(L,R)}^i & B_{(L,R)}^i \\ B_{(L,R)}^i & D_{(L,R)}^i \end{bmatrix}$$

The stiffness parameters are defined as

$$\left(A_{h_{ij}}, B_{h_{ij}}, D_{h_{ij}}\right) = \int_{-\frac{h}{2}}^{\frac{h}{2}}\left(\overline{Q}_{ij}, \overline{Q}_{ij}z, \overline{Q}_{ij}z^2\right)dz$$

$$\left(A_{L_{ij}}^i, B_{L_{ij}}^i, D_{L_{ij}}^i\right) = \int_{-\frac{h}{2}}^{\frac{h}{2}}\left(\overline{Q}_{L_{ij}}, \overline{Q}_{L_{ij}}z, \overline{Q}_{L_{ij}}z^2\right)dz$$

$$\left(A_{R_{ij}}^i, B_{R_{ij}}^i, D_{R_{ij}}^i\right) = \int_{-\frac{h}{2}}^{\frac{h}{2}}\left(\overline{Q}_{R_{ij}}, \overline{Q}_{R_{ij}}z, \overline{Q}_{R_{ij}}z^2\right)dz \qquad -(5)$$

The 2D strain energy density of the strip is then given by

$$U_{1D} = \sum_{i=1}^{M+1}\int_{-b/2}^{-m}\varepsilon_L^{i^T}K_{L_d}^i\varepsilon_L^i dx_2 + \int_{-m}^{n}\varepsilon_h^T K_h\varepsilon_h dx_2 + \sum_{i=1}^{N+1}\int_n^{b/2}\varepsilon_R^{i^T}K_{R_d}^i\varepsilon_R^i dx_2 \qquad -(6)$$

where $\varepsilon_{\alpha\beta}$ and $\rho_{\alpha\beta}$ are the strain measure defined in Eqs. (3) and (12), 'm' and 'n' are the location of the damage from the origin. The strain terms to be used in the healthy part of the cross-section reduces to the equations as given by Hodges et.al (Ref.5).

Zeroth – order approximation

Variational Asymptotical Method calls for the minimization of the strain energy expression with respect to the warpings. As mentioned before, the zeroth order approximation should contain all terms up to $O(\varepsilon)$ while the first – order approximation should contain all the relevant terms up to $O(\varepsilon\Delta^2)$ where, Δ denotes the ratio $\epsilon/\delta h^2$.

Closed form expressions for warpings are determined by solving the Euler-Lagrange differential equations which are obtained by the minimization of the 2D strain energy density of the strip, subject to the resulting essential and natural boundary conditions (all zeroth order variables will be expressed denoted by a superscript 0). In addition to natural and essential boundary conditions, the warpings are also required to satisfy the following global constraints

$$\sum_{p=1}^{M+1}\int_{-b/2}^{-m} w_{L_i}^{p} dx_2 + \int_{-m}^{n} w_{h_i} dx_2 + \sum_{p=1}^{N+1}\int_{n}^{b/2} w_{R_i}^{p} dx_2 = 0$$

$$\sum_{p=1}^{m+1}\int_{-b/2}^{-m} w_{L_{3,2}}^{p} dx_2 + \int_{-a}^{a} w_{h_{3,2}} dx_2 + \sum_{p=1}^{N+1}\int_{n}^{b/2} w_{R_{3,2}}^{p} dx_2 = 0 \qquad -(7)$$

At the sublaminate interface between the damaged and undamaged regions, continuity of the warpings should be ensured. In mathematical form this means

$$w_i^{(L,R)} = w_i^{h}$$

$$w_{3,2}^{(L,R)} = w_{3,2}^{h} \qquad -(8)$$

Closed form solutions are now obtained for the warping functions. Once out-of-shell warping expressions are determined in the damaged and undamaged part of the cross-section, in-plane warpings can be readily obtained. The expressions for w_i are omitted here due to the paucity of space. The detailed procedure to determine the out-of-shell and in-plane warpings is provided by Hodges et.al (Ref.5).

First order approximation

The first order approximation to the 2D strains should contain all the terms up to $O(\varepsilon\Delta^2)$. This consists of both the underlined and non-underlined terms in Eq. (3). First order variables will be denoted by a superscript I. The warping displacements are given by $w_i^I = w_i^0 + \tilde{w}_i$, where the perturbation quantities $\tilde{w}_i$ are assumed to be of an order higher than the corresponding zeroth-order quantities, w_i^0. The strain measures in Eq. (3) can then be rewritten as

$$\varepsilon_{(L,R)_{11}}^{(m,n)} = \varepsilon_{(L,R)_{11}}^{0(m,n)} + \tilde{\varepsilon}_{(L,R)_{11}}^{(m,n)}$$

$$\varepsilon_{(L,R)_{22}}^{(m,n)} = \varepsilon_{(L,R)_{22}}^{0(m,n)} + \tilde{\varepsilon}_{(L,R)_{22}}^{(m,n)}$$

$$\varepsilon_{(L,R)_{12}}^{(m,n)} = \varepsilon_{(L,R)_{12}}^{0(m,n)} + \tilde{\varepsilon}_{(L,R)_{12}}^{(m,n)}$$

$$\rho_{(L,R)_{11}}^{(m,n)} = \kappa_2$$

$$\rho_{(L,R)_{22}}^{(m,n)} = \rho_{(L,R)_{22}}^{0(m,n)} + \tilde{\rho}_{(L,R)_{22}}^{(m,n)}$$

$$\rho_{(L,R)_{12}}^{(m,n)} = -2\kappa_1 \qquad -(9)$$

After substitution of these equations into the strain energy expression, the resulting functional is minimized with respect to $\tilde{w}_i$, subject to appropriate boundary conditions. Closed form solutions are then obtained for the perturbations to the zeroth order warping displacements $\tilde{w}_i$. The expressions for $\tilde{w}_i$ are omitted here due to the paucity of space.

Stiffness Coefficients

After obtaining the unknown warping functions we can readily carry out the integration of U_{2D} along the beam width to obtain the beam stain energy density. In simplified form, the beam strain energy density can be represented by,

$$U_{1D} = \frac{1}{2}\varepsilon_L{}^T[S_L]\varepsilon_L + \varepsilon_L^T[S_{LN}]\varepsilon_N + \frac{1}{2}\varepsilon_N^T[S_N]\varepsilon_N \qquad -(10)$$

where the linear and non-linear 1D strain vectors, ε_L and ε_N respectively are defined as follows:

$$\varepsilon_L = \{\gamma_{11}, \kappa_1, \kappa_2, \kappa_3\}^T$$

$$\varepsilon_N = \{\kappa_1{}^2, \kappa_2{}^2, \kappa_2\gamma_{11}, \kappa_2\kappa_3, \kappa_2\kappa_1\}^T \qquad -(11)$$

and the matrices $[S_L]$, $[S_{LN}]$, and $[S_N]$ can be thought of as partitions of a 9 X 9 matrix [S]. The terms of the linear stiffness matrix are presented in Appendix-A. Nonlinear stiffness terms are omitted here due to the paucity of space.

It should be noted that in the case of a healthy cross-section, the above-determined terms of the stiffness matrix reduces to the terms of the stiffness of a healthy cross-section developed by Hodges et.al (Ref.5) by putting the length of the delamination as 0 i.e. by substituting m= n=b/2.

3-D stress field

The solution obtained for warping fields can now be substituted into the 2D strain expression in Eq. (3). The 3D Green-strains can now be derived from the 2D strains using the following relation:

$$\Gamma_{\alpha\beta}^{(1,2)} = \varepsilon_{\alpha\beta}^{(1,2)} + x_3\rho_{\alpha\beta}^{(1,2)} \qquad -(12)$$

The plane stress field at any material point in the strip is obtained from the stress strain relationship in terms of the ply stiffness coefficients, $\bar{Q}$'s which are functions α, the orientations of the fibers in the ply with respect to the strip axis.

$$\begin{Bmatrix}\sigma_{11}\\ \sigma_{22}\\ \sigma_{12}\end{Bmatrix} = \begin{bmatrix}\bar{Q}_{11} & \bar{Q}_{12} & \bar{Q}_{16}\\ \bar{Q}_{12} & \bar{Q}_{22} & \bar{Q}_{26}\\ \bar{Q}_{16} & \bar{Q}_{26} & \bar{Q}_{66}\end{bmatrix}\begin{Bmatrix}\Gamma_{11}\\ \Gamma_{22}\\ \Gamma_{12}\end{Bmatrix} \qquad -(13)$$

NUMERICAL RESULTS AND DISCUSSIONS.

The implementation of the developed composite beam model is validated by comparing with 3-D finite element results for several numerical examples. End twist and distribution of 3D stresses along the thickness of a typical beam under axial load is obtained, whose dimensions are typical of helicopter rotor blades. They are correlated with established 3-D (8-noded hexagonal) Finite Elements of ANSYS commercial FE software.

Correlation of the results with 3-D FE Solutions

To compare the results from the present theory with the solutions of well-established 3-D finite elements (8Noded Hex. element) of ANSYS software, a typical composite beam whose geometry is comparable to that of any helicopter rotor blade is chosen. A beam having dimensions 254mm length x 26mm width x 1.2mm thickness is considered.

The pretwisted composite strip is modeled using ANSYS 1D beam elements and the cross sectional stiffness matrix derived using VAM approach is fed into ANSYS 1D beam model. The stiffness matrix is entered as pre-integrated beam stiffness defined in section property for the beam in ANSYS. The ANSYS pre-integrated stiffness matrix requires additional terms for shear strains in Y and Z. These terms are assumed to be of the same order of magnitude as that of axial stiffness. ANSYS model also requires stiffness term for bi-moment which is assumed to be of similar to torsional stiffness times the square of the thickness of the beam.

Material T300/954-3 graphite/cyanate is considered with properties shown in the Table 1. The hygrothermally stable anti-symmetric stacking sequences, $[\alpha_2/(\alpha - 90)_4/ \alpha_2/-\alpha_2/(90 - \alpha)_4/-\alpha_2]$, are considered.

Table 1. Properties of T300/954-3 graphite/cyanate material system

E_{11} = 135.6 GPa, E_{22} = 9.9 GPa
G_{12} = G_{13} = 4.2 GPa, G_{23} = 2.5 GPa
ν_{12} = 0.3

The stress analysis of the composite beam is carried out for and axial load of 1000N applied at one end and fixed in all degrees of freedom on the other end.

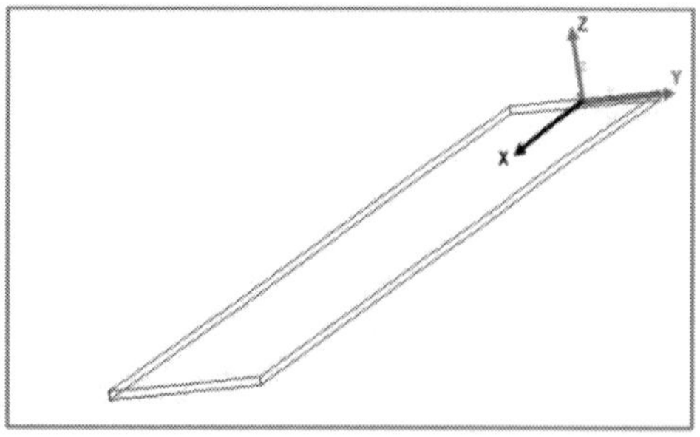

Figure 2. Geometry and coordinate system of the strip

Analyses carried out for healthy beam and delaminated beam models. Geometry and coordinate system for the composite strip and the delamination details are shown in Figures 2 and 3. On one side a delamination of width 6.5mm is modeled throughout the length of the beam at Z=0.0mm. On other side of the strip delaminations are modeled at two locations each at Z=+/- 0.3mm throughout the length of the strip.

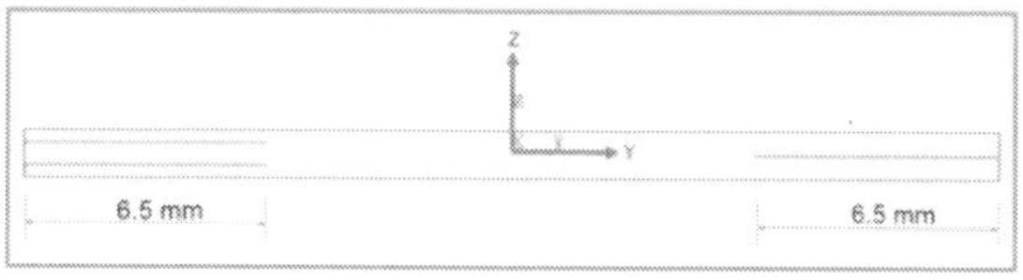

Figure 3. Delamination dimension

The computational advantage of the current approach is also demonstrated in Table 2 where solution time is directly proportional to the degree of freedom (DoF) of the model.

Table 2. Comparison of Degree of Freedom between 3D and 1D models

Model Description	3DSolid Element FE Model	VAM Based 1D FE Model
Healthy Beam	398,478DoF	30 DoF
Delaminated Beam	407,133DoF	30 DoF

Analysis results

Geometrical non-linear analysis was carried out for axial load of 1000N for both undamaged strip and delaminated strip. The axial load and twist relationship is plotted in the Figure 4. The axial-twist relationship is nonlinear and the difference between the end twist between 3D model and the 1D beam model is less than 4%.

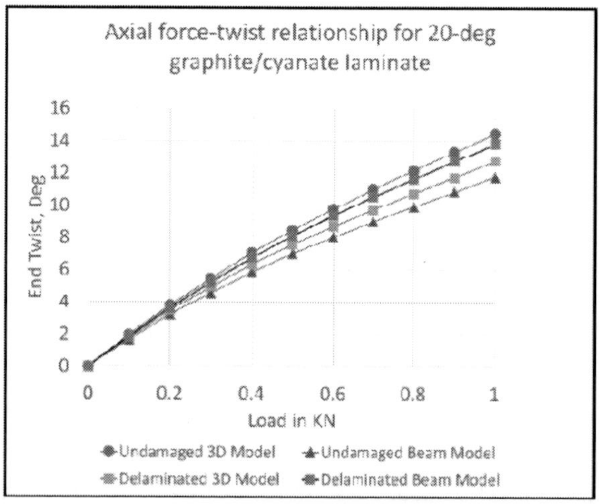

Figure 4. Axial force and end twist relationship

Axial, lateral and shear stresses at the bottom, middle and top layers of the strip for both 3D and 1D undamaged models are presented in the Figures 5,6 and 7.

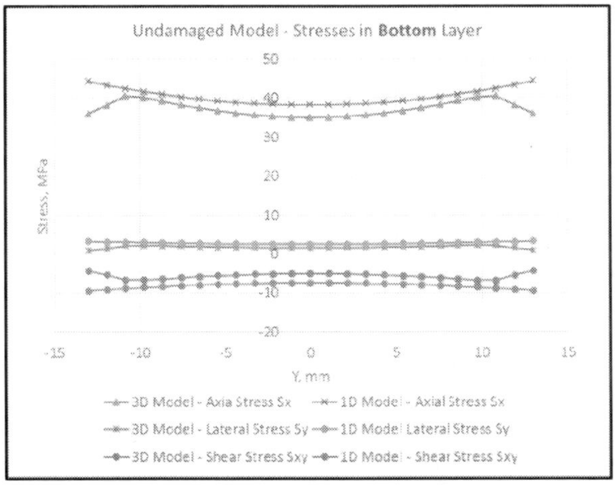

Figure 5. Bottom layer stresses for undamaged model

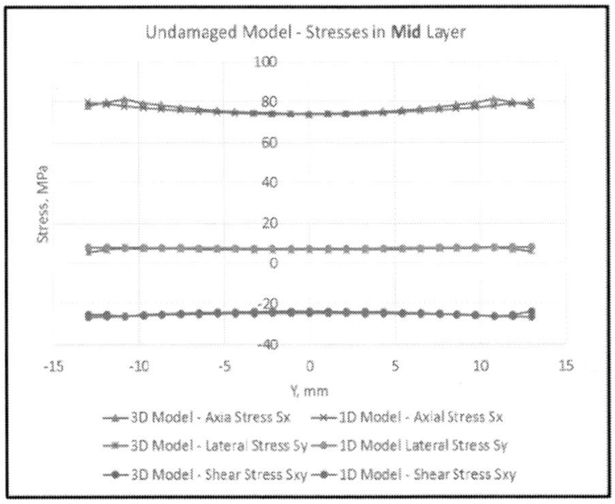

Figure 6. Mid layer stresses for undamaged model

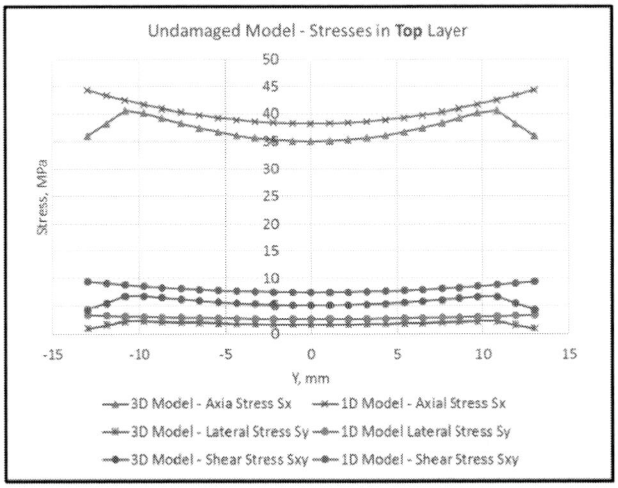

Figure 7. Top layer stresses for undamaged model

Figures 8,9 and 10 shows the axial, lateral and shear stresses at the middle layer for both 3D and 1D delaminated beam models.

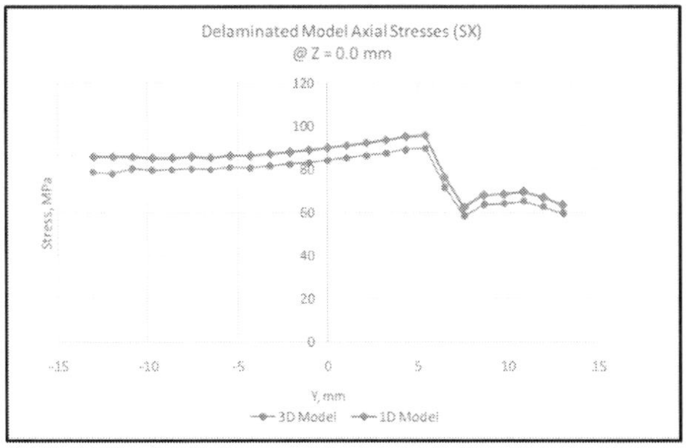

Figure 8. Axial stress distribution across the width at Z=0.0 for Delaminated model

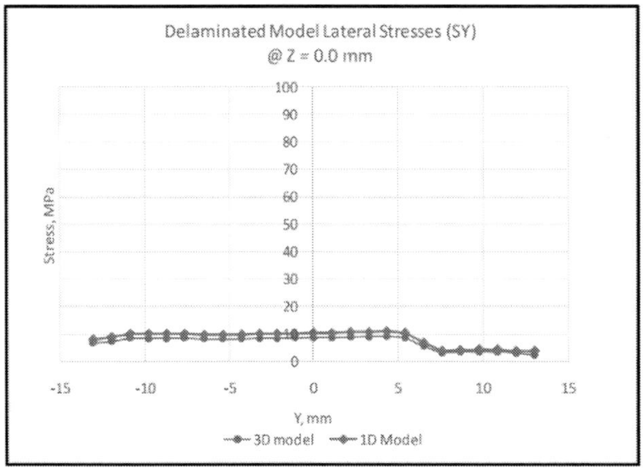

Figure 9. Lateral stress distribution across the width at Z=0.0 for Delaminated model

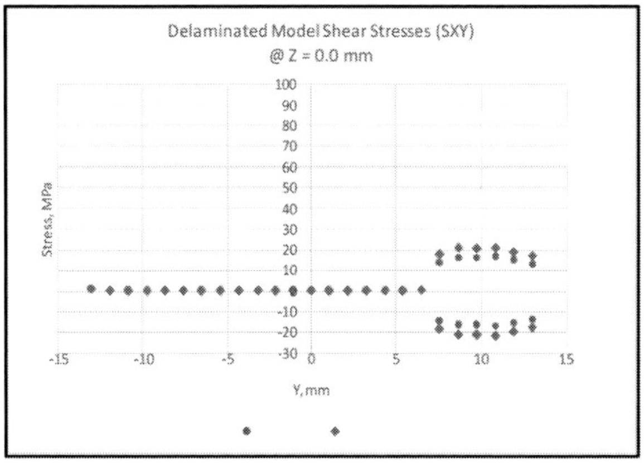

Figure 10. Shear stress distribution across the width at Z=0.0 for Delaminated model

CONCLUSIONS

The degradation in flapwise and lagwise performance of rotor blades due to the presence of delaminations in rotorcraft flexbeams necessitates a study on damage in these structures. The available models to deal with delaminations in composite strips are computationally expensive. Developed a mathematical model to assess the degradation in the stiffness properties of anisotropic strips with small pretwist, typical of helicopter flexbeams in the presence of multiple edge delaminations. The axial force-twist behavior is investigated for a class of antisymmetric hygrothermally stable laminate with multiple free edge delaminations. The accuracy of the developed model is assessed through comparison of axial force-twist and stresses across the thickness with 3D finite elements of ANSYS software.

Author contact: Prasad S Puvvala

prasadps0303070@gmail.com

Appendix A

The stiffness variables used in the paper are defined as follows

$$\overline{AIi_{11}} = AIi_{11} + \frac{AIi_{16}^2\, AIi_{22} - 2\, AIi_{12}\, AIi_{16}\, AIi_{26} + AIi_{12}^2\, AIi_{66}}{AIi_{26}^2 - AIi_{22}\, AIi_{66}}$$

$$\overline{BIi_{11}} = BIi_{11} + \frac{(AIi_{12}\, AIi_{66}\, BIi_{12} + AIi_{16}\, AIi_{22}\, BIi_{16} - AIi_{26}\, (AIi_{16}\, BIi_{1}}{(AIi_{26}^2 - AIi_{22}\, AIi_{66})}$$

$$\overline{BIi_{12}} = BIi_{12} + \frac{(AIi_{12}\, AIi_{66}\, BIi_{22} + AIi_{16}\, AIi_{22}\, BIi_{26} - AIi_{26}\, (AIi_{16}\, BIi_{2}}{(AIi_{26}^2 - AIi_{22}\, AIi_{66})}$$

$$\overline{BIi_{16}} = BIi_{16} + \frac{(AIi_{12}\, AIi_{66}\, BIi_{26} + AIi_{16}\, AIi_{22}\, BIi_{66} - AIi_{26}\, (AIi_{16}\, BIi_{2}}{(AIi_{26}^2 - AIi_{22}\, AIi_{66})}$$

$$\overline{DIi_{11}} = DIi_{11} + \frac{AIi_{66}\, BIi_{12}^2 - 2\, AIi_{26}\, BIi_{12}\, BIi_{16} + AIi_{22}\, BIi_{16}^2}{AIi_{26}^2 - AIi_{22}\, AIi_{66}}$$

$$\overline{DIi_{12}} = DIi_{12} + \frac{(AIi_{66}\, BIi_{12}\, BIi_{22} + AIi_{22}\, BIi_{16}\, BIi_{26} - AIi_{26}\, (BIi_{16}\, BIi_{2}}{(AIi_{26}^2 - AIi_{22}\, AIi_{66})}$$

$$\overline{DIi_{16}} = DIi_{16} + \frac{(AIi_{66}\, BIi_{12}\, BIi_{26} + AIi_{22}\, BIi_{16}\, BIi_{66} - AIi_{26}\, (BIi_{16}\, BIi_{2}}{(AIi_{26}^2 - AIi_{22}\, AIi_{66})}$$

$$\overline{DIi_{22}} = DIi_{22} + \frac{AIi_{66}\, BIi_{22}^2 - 2\, AIi_{26}\, BIi_{22}\, BIi_{26} + AIi_{22}\, BIi_{26}^2}{AIi_{26}^2 - AIi_{22}\, AIi_{66}}$$

$$\overline{ALi_{11}} = \overline{ALi_{11}} - \frac{\overline{BLi_{12}}^2}{\overline{DLi_{22}}};\ \overline{ARi_{11}} = \overline{ARi_{11}} - \frac{\overline{BRi_{12}}^2}{\overline{DRi_{22}}};\ \overline{AH_{11}} = \overline{Ah_{11}} - \frac{\overline{Bh_{12}}^2}{\overline{Dh_{22}}};$$

$$\overline{ALD_{11}} = \sum_{i=1}^{M+1} \overline{ALi_{11}};\ \overline{ARD_{11}} = \sum_{i=1}^{N+1} \overline{ARi_{11}};$$

$$\overline{BLi_{16}} = \overline{BLi_{16}} - \frac{\overline{BLi_{12}}\,\overline{DLi_{26}}}{\overline{DLi_{22}}};\ \overline{BRi_{16}} = \overline{BRi_{16}} - \frac{\overline{BRi_{12}}\,\overline{DRi_{26}}}{\overline{DRi_{22}}};\ \overline{BH_{16}} = \overline{B}$$

$$\overline{BLD_{16}} = \sum_{i=1}^{M+1} \overline{BLi_{16}};\ \overline{BRD_{16}} = \sum_{i=1}^{N+1} \overline{BRi_{16}};$$

$$\overline{BLi_{11}} = \overline{BLi_{11}} - \frac{\overline{BLi_{12}}\,\overline{DLi_{12}}}{\overline{DLi_{22}}};\ \overline{BRi_{11}} = \overline{BRi_{11}} - \frac{\overline{BRi_{12}}\,\overline{DRi_{12}}}{\overline{DRi_{22}}};\ \overline{BH_{11}} = \overline{Bh}$$

$$\overline{BLD_{11}} = \sum_{i=1}^{M+1} \overline{BLi_{11}};\ \overline{BRD_{16}} = \sum_{i=1}^{N+1} \overline{BRi_{11}};$$

$$\overline{DLi_{66}} = \overline{DLi_{66}} - \frac{\overline{DLi_{26}}^2}{\overline{DLi_{22}}};\ \overline{DRi_{66}} = \overline{DRi_{66}} - \frac{\overline{DRi_{26}}^2}{\overline{DRi_{22}}};\ \overline{DH_{66}} = \overline{Dh_{66}} - \frac{\overline{Dh_{26}}^2}{\overline{Dh_{22}}};$$

$$\overline{DLD_{66}} = \sum_{i=1}^{M+1} \overline{DLi_{66}};\ \overline{DRD_{66}} = \sum_{i=1}^{N+1} \overline{DRi_{66}};$$

$$\overline{DLi_{16}} = \overline{DLi_{16}} - \frac{\overline{DLi_{12}}\,\overline{DLi_{26}}}{\overline{DLi_{22}}};\ \overline{DRi_{16}} = \overline{DRi_{16}} - \frac{\overline{DRi_{12}}\,\overline{DRi_{26}}}{\overline{DRi_{22}}};\ \overline{DH_{16}} = \overline{Dh}$$

$$\overline{DLD_{16}} = \sum^{M+1} \overline{DLi_{16}};\ \overline{DRD_{16}} = \sum^{N+1} \overline{DRi_{16}};$$

The terms of the Linear Stiffness Matrix S_L are

$$S11 = m\,(\overline{AH_{11}} - \overline{ALD_{11}}) + n\,(\overline{AH_{11}} - \overline{ARD_{11}}) + \frac{1}{2}\, b\,(\overline{ALD_{11}} + \overline{ARD_{11}})$$

$$S12 = m\,(-2\,\overline{BH_{16}} + 2\,\overline{BLD_{16}}) + b\,(-\overline{BLD_{16}} - \overline{BRD_{16}}) + n\,(-2\,\overline{BH_{16}} + 2\,\overline{BRD}$$

$$\frac{1}{3}\, m^3\,(\overline{AH_{11}} - \overline{ALD_{11}})\, k_1 + \frac{1}{3}\, n^3\,(\overline{AH_{11}} - \overline{ARD_{11}})\, k_1 + \frac{1}{24}\, b^3\,(\overline{ALD_{11}} + \overline{A}$$

$$S13 = \frac{1}{2}\, m\,\left(2\,\overline{BH_{11}} - 2\,\overline{BLD_{11}} + \sum_{i=1}^{M+1} \overline{ALi_{11}}\,(-h + h_{L,i-1} + h_{L,i})\right) +$$

$$\frac{1}{2}\, n\,\left(2\,\overline{BH_{11}} - 2\,\overline{BRD_{11}} + \sum_{i=1}^{N+1} \overline{ARi_{11}}\,(-h + h_{R,i-1} + h_{R,i})\right) +$$

$$\frac{1}{}\, b\, (2\,\overline{BLD_{11}} + 2\,\overline{BRD_{11}} + \sum \overline{ALi_{11}}\,(h - h_{L,i-1} - h_{Li}) + \sum \overline{ARi_{11}}\,(h -$$

$$S22 = 4\, m\,(\overline{DH_{66}} - \overline{DLD_{66}}) + 4\, n\,(\overline{DH_{66}} - \overline{DRD_{66}}) + 2\, b\,(\overline{DLD_{66}} + \overline{DRD_{66}}) -$$

$$\frac{4}{3}\, m^3\,(\overline{BH_{16}} - \overline{BLD_{16}})\, k_1 - \frac{4}{3}\, n^3\,(\overline{BH_{16}} - \overline{BRD_{16}})\, k_1 - \frac{1}{6}\, b^3\,(\overline{BLD_{16}} + \overline{BRD_{16}})$$

$$\frac{1}{5}\, m^5\,(\overline{AH_{11}} - \overline{ALD_{11}})\, k_1^2 + \frac{1}{5}\, n^5\,(\overline{AH_{11}} - \overline{ARD_{11}})\, k_1^2 + \frac{1}{160}\, b^5\,(\overline{ALD_{11}} + \overline{ARD_{1}}$$

$$S23 = m\,\left(-2\,\overline{DH_{16}} + 2\,\overline{DLD_{16}} + \sum_{i=1}^{M+1} \overline{BLi_{16}}\,(h - h_{L,i-1} - h_{L,i})\right) +$$

$$n\,\left(-2\,\overline{DH_{16}} + 2\,\overline{DRD_{16}} + \sum_{i=1}^{N+1} \overline{BRi_{16}}\,(h - h_{R,i-1} - h_{R,i})\right) +$$

$$\frac{1}{2}\, b\, (-2\,\overline{DLD_{16}} - 2\,\overline{DRD_{16}} + \sum_{i=1}^{M+1} \overline{BLi_{16}}\,(-h + h_{L,i-1} + h_{L,i}) + \sum_{i=1}^{N+1} \overline{BRi_{16}}\,(-h$$

$$\frac{1}{6}\, m^3\,\left(2\,\overline{BH_{11}} - 2\,\overline{BLD_{11}} + \sum_{i=1}^{M+1} \overline{ALi_{11}}\,(-h + h_{L,i-1} + h_{L,i})\right) +$$

1 [illegible]

$$S24 = m^2\left(-\overline{BH_{16}} + \overline{BLD_{16}}\right) + n^2\left(\overline{BH_{16}} - \overline{BRD_{16}}\right) + \frac{1}{4} b^2\left(-\overline{BLD_{16}} + \overline{BRD_{16}}\right) +$$

$$\frac{1}{4} m^4\left(\overline{AH_{11}} - \overline{ALD_{11}}\right) k_1 - \frac{1}{4} n^4\left(\overline{AH_{11}} - \overline{ARD_{11}}\right) k_1 + \frac{1}{64} b^4\left(\overline{ALD_{11}} - \overline{ARD_{11}}\right)$$

$$S33 = \frac{1}{4} m\left(4\,\overline{DH_{11}} - 4\,\overline{DLD_{11}} + 4\sum_{i=1}^{M+1}\overline{BLi_{11}}\,(-h + h_{L,i-1} + h_{Li}) - \sum_{i=1}^{M+1}\overline{ALi_{11}}\,(-h\right.$$

$$\frac{1}{4} n\left(4\,\overline{DH_{11}} - 4\,\overline{DRD_{11}} + 4\sum_{i=1}^{N+1}\overline{BRi_{11}}\,(-h + h_{R,i-1} + h_{Ri}) - \sum_{i=1}^{N+1}\overline{ARi_{11}}\,(-h + h_R\right.$$

$$\frac{1}{8} b\left(4\,\overline{DLD_{11}} + 4\,\overline{DRD_{11}} + 4\sum_{i=1}^{M+1}\overline{BLi_{11}}\,(h - h_{L,i-1} - h_{Li}) + \sum_{i=1}^{M+1}\overline{ALi_{11}}\,(-h + h_L\right.$$

$$\left.4\sum_{i=1}^{N+1}\overline{BRi_{11}}\,(h - h_{R,i-1} - h_{Ri}) + \sum_{i=1}^{N+1}\overline{ARi_{11}}\,(-h + h_{R,i-1} + h_{Ri})^2\right)$$

$$S34 = \frac{1}{4} m^2\left(2\,\overline{BH_{11}} - 2\,\overline{BLD_{11}} + \sum_{i=1}^{M+1}\overline{ALi_{11}}\,(-h + h_{L,i-1} + h_{L,i})\right) +$$

$$\frac{1}{4} n^2\left(-2\,\overline{BH_{11}} + 2\,\overline{BRD_{11}} + \sum_{i=1}^{N+1}\overline{ARi_{11}}\,(-h + h_{R,i-1} + h_{R,i})\right) +$$

REFERENCES

[1]Barbero, E. J., and Reddy, J. N., "Modeling of Delamination in Composite Laminates Using A Layer-Wise Plate Theory," International Journal of Solids and Structures, Vol.28, No. 3, 1991, pp. 373-388.

[2]Kim, H. S., Aditi Chattopadhyay and AnindyaGhosal, "Dynamic Analysis of Cross-ply Laminates with Embedded Multiple Delamination," Proc., 44th AIAA/ASME/ASCE/AHS/ASC Structures, Structural Dynamics and Materials Conference, Norfolk, Virginia, April 2003.

[3]Andrew Makeev and Erian, A. Armanios., "Influence of Damage on the Elastic Behavior of Composite Beams," presented at the 55th Annual Forum of the American Helicopter Society, Montreal, Canada, May 1999.

[4]Andrew Makeev and Erian A. Armanios., "A Geometrically Nonlinear Model for Laminated Composite Strips with Extension-Twist Coupling in the Presence of Delamination," Journal of Applied Mechanics, Vol. 65, 1998, pp 685-693.

[5]Hodges, D. H. Dineshkumar, H. Vitali, V. V. and Cesnik, C. E. S,"NonclassicalEffects in Non-linear Analysis of Pretwisted Anisotropic Strips", InternationalJournal of Non-linear Mechanics, Vol. 34, 1999, pp. 259-277.

Helicopter-Ship Interface Testing
Innovative SHOL-X predictive software tool

Alrik Hoencamp
Founder and CEO
AeroMath | Aeronautical Engineering Services
The Hague, Netherlands

It has occasionally happened that either due to prevailing weather conditions, ship availability or helicopter availability, the limits of a helicopter-ship combination could not be fully explored, thereby restricting the operational capability. The goal of this paper is to present an innovative predictive software tool that can be used for optimizing cost and time efficiency of helicopter-ship qualification testing without reducing safety. The main advantage of the presented predictive tool, in conjunction with the "SHOL-X" test methodology, is that the operator can perform early evaluation of safety limits for helicopters operating on ships in a wide range of in-service conditions. In this way the qualification process is less dependent on the successful outcome of solely qualitative assessed test points during dedicated sea trials. As such, the predictive tool can be used to allow a well-considered assessment of the gap between the safe flight envelope, as determined by the helicopter manufacturer, and the user-defined operational flight envelope for a particular helicopter-ship combination. The innovative predictive software tool has, amongst other occasions, successfully been applied since 2011 during the helicopter-ship qualification process of the NH90 NFH across the entire Dutch fleet.

Figure 1; NH90 NFH during shipboard operations

NOTATION

AEO	All Engine Operatives
CFE	Candidate Flight Envelope
NFH	NATO Frigate Helicopter
OAT	Outside Air Temperature
OPV	Ocean Patrol Vessel
PPI	Power Performance Index
SHOL	Ship Helicopter Operational Limitation
TFCP	Trimmed Flight Control Position

Presented at the 6th Asian/Australian Rotorcraft Forum / Heli Japan 2017, Kanazawa, Japan, November 7-9, 2017.

INTRODUCTION

The helicopter limitations for land-based operations determined after extensive factory testing, are usually based on a rigid and unobstructed landing site. On the other hand, the limitations for ship-borne operations are to be based on an obstructed landing site (i.e. flight deck) which may show random oscillatory movement and where strong turbulent wind conditions can prevail. Sometimes an almost marginal facility is provided for take-off, landing and deck handling. Yet, helicopter operations may be required in a wide range of operational conditions (day, night, sea-state, wind, visibility, etc.) with the highest possible payload. Nowadays, in line with the increasing importance of helicopter/ship operations the helicopter manufacturer sometimes provides limitations of a general nature for helicopter-ship operations. However, because of the unique characteristics of each helicopter-type/class-of-ship combination and the innumerable combinations possible it is understandable that usually no (extensive) testing has been carried out by the helicopter manufacturer for all combinations that may be of interest for each operator.

It follows that the limitations given, if any, must be considered as general guidelines, with large safety margins with respect to the helicopter capabilities and pilot ability to control the helicopter, and thus do not provide the maximum operational capability of the helicopter on board the ship. To determine the limitations that allow maximum operational capability of the helicopter within the constraints of safety, a dedicated helicopter-ship qualification programme is to be executed. An innovative three-phase approach for establishing Ship Helicopter Operational Limitations (SHOL), aided by a predictive software tool is presented in this paper named "SHOL-X" as shown in Figure 2 [1].

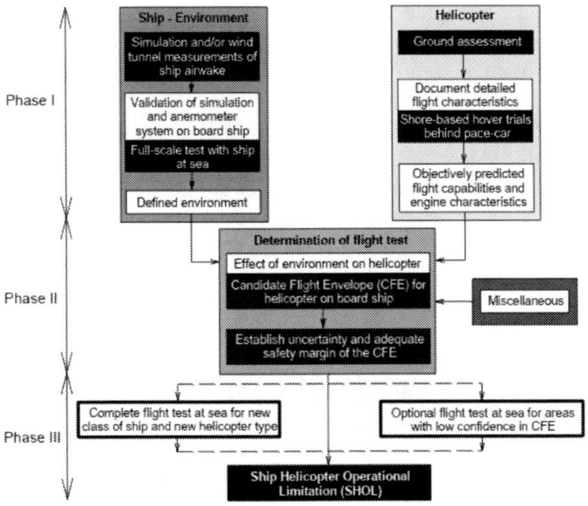

Figure 2; Innovative flow chart SHOL development

"SHOL-X" TEST METHODOLOGY

The proposed approach eliminates subjective elements by determining unambiguous operational envelopes by merging the huge amount of data gathered in the phase prior to the embarked sea trials, whilst making the merging process transparent for everybody involved in the test campaign and the authorizing authorities. In *Phase I* the ship-environment (green box in figure) in which the helicopter will operate is determined by conducting wind tunnel measurements of the airflow in the take-off and landing paths of the ship. This is (still) followed by a validation with full-scale measurements of the airflow above the flight deck of the ship. For the helicopter (yellow box in figure), a ground assessment and shore-based hover trials with a pace-car are carried out to verify precisely the helicopter limitations, including aspects such as e.g. pilot workload in cross-wind conditions, engine performance and control margins resulting in objectively predicted flight capabilities and engine characteristics. Thereafter, in *phase II* (blue box in figure), the so-called "Candidate Flight Envelope" (CFE) is obtained by the predictive tool "SHOL-X", which has been developed by AeroMath, combining the behaviour of the isolated helicopter, the specific environmental conditions for a particular ship type and other miscellaneous items. The CFE is a diagram giving the likely combinations of indicated wind speed (in radial coordinates) and direction (in angular coordinates) for safe take-offs and landings from a particular ship type. The miscellaneous items (red box in figure) could be added to optimize the CFE, for example experience from previous test campaigns with either the helicopter or the ship under test. Once the CFE is developed, the associated uncertainty and adequate safety margins are established prior commencing sea trials. Finally, in *phase III*, based on the CFE, a (partial) flight test campaign on board the ship is conducted preferably in a range of weather conditions by day and by night (black outlined white boxes in figure). This is to determine for the particular helicopter-ship combination the effects on the pilot workload from, for example, visual references, ship motion and turbulence.

The finally determined operational limitations for safe take-off and landing from a particular ship type, known as SHOL, are presented as a set of diagrams (i.e. polar plot) giving the combination of ship motion with indicated wind speed (in radial coordinates) and direction (in angular coordinates) as shown for example in Figure 3. Note that for security reasons, the axes from the remaining polar plots presented in this book have been removed. There are separate SHOLs developed for a range of weights of the helicopter, varying take-off and landing procedures and day or night conditions. The boundaries of these operational envelopes are either defined when approaching and/or exceeding quantitative helicopter limitations, or when the pilot's workload becomes too high to consistently make safe take-off and landings. The proper merging of the quantitative and qualitative assessment is what now defines the operational capacities for a particular helicopter-ship combination. For example, a helicopter with a low thrust margin or inadequate vertical agility is in danger of unintentional contact with the flight deck in high sea states. Alternatively, in strong cross-winds, the limit of tail rotor thrust can be reached such that the pilot cannot maintain heading. Therefore, the SHOL envelopes are equally reliant on an acceptable envelope based on objective helicopter limitations, as on acceptable piloting workload to perform the task. During operational scenarios the combination of ship motion, wind speed and wind direction must be within the relevant SHOL envelope. If they are not, an alternative take-off and landing procedure or different helicopter weight category must be chosen, or the ship must be forced to change course and speed to achieve the environmental conditions within the defined envelope.

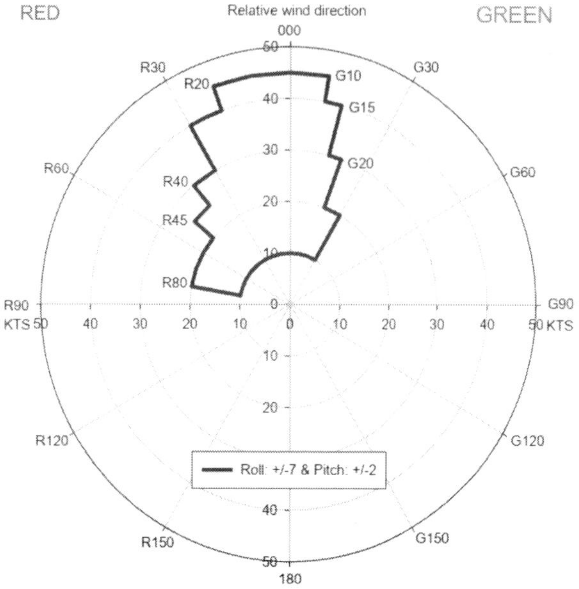

Figure 3; Example SHOL envelope

PHASE I, PRELIMINARY INVESTIGATION OF SHIP AND HELICOPTER

The novelty of the preliminary investigation is the systematic approach to the assembly of the database for the construction of the CFE in the next phase. As such any issues can be properly investigated before being embarked on board the ship whilst under huge pressure to provide the widest operational capabilities within the least amount of time. The preliminary investigation includes an assessment of the expected environment conditions near the ship, and the determination of the helicopter flight characteristics in an omni-directional relative wind envelope.

Ship Environment Aspects

The superstructure of a naval ship is cluttered with constructions of different sizes and shapes (e.g. hangar, antennas and exhaust stakes) due to which the state of the wind after passing through these differs from the undisturbed relative wind. The undisturbed relative wind is the vector summation of the actual wind with the course and speed of the ship. However, the airwake encountered by the helicopter near or over the flight deck during take-off and landing may strongly deviate in speed and direction from it, and could be highly complex with both steady (low frequency) and unsteady (high frequency) airflow effects interacting. In addition, high up-flow and down-flow caused by the ship's hull may be encountered. In this section, a novel approach is demonstrated how wind tunnel measurements are used for an objective assessment of the ship environment prior to conducting sea trials.

The helicopter flight characteristics determined by shore-based hover trials are based on the relative wind conditions encountered by either hovering in natural wind conditions or by using a pace-car. However, near and above the flight deck the relative wind is disturbed by the large ship's superstructure. This disturbed wind is what the helicopter encounters when operating from the ship and is known as local wind. The local wind conditions are determined by wind tunnel measurements at different locations and heights above the flight deck and in the approach and departure paths of the helicopter. Unfortunately, the undisturbed relative wind and local wind conditions are unknown for the operational crew after the test campaign, as the indicated wind by the ship anemometers is their only reference source. The latter is unreliable since by mounting anemometers on a ship with a bluff body, the local air flow at the anemometer location also deviates from the undisturbed relative wind conditions. For this reason, the undisturbed relative wind should be known to the test team, and applicable corrections made to the indicated wind speed on board the ship in order to construct a usable CFE.

Wind tunnel tests on ship models are carried out to determine the airflow characteristics (i.e. airflow deviations with respect to the undisturbed oncoming relative wind) above the flight deck and in the possible approach and departure paths of the helicopter to the ship as function of the relative wind. In addition, the position error of the ship's anemometers are determined which is needed to establish the relationship between the undisturbed relative wind conditions. The wind tunnel measurements were performed, on scale models of the ships in service within the entire Dutch fleet, in the German Dutch Wind Tunnels Large Low-speed Facility (DNW-LLF) as shown for example for the Ocean Patrol Vessel (OPV) in Figure 4 [2].

Figure 4; Example SHOL envelope

Once the data reduction is performed, the data is presented in graphical format. In these graphs the port anemometer is used for red wind conditions and the starboard anemometer is used for green wind conditions. This red and green terminology is used throughout this paper to refer to winds coming from either the port or starboard side of the ship, respectively. Some examples are shown for wind speed coefficient C_v and horizontal flow deviation χ at different heights above the landing spot as a function of the indicated wind direction in Figure 5 and Figure 6, respectively.

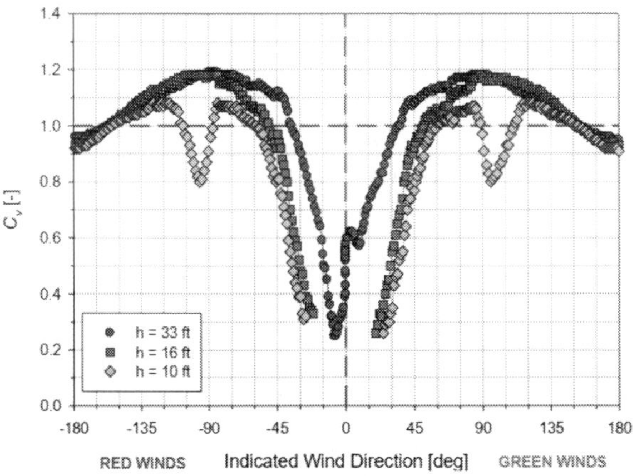

Figure 5; Example wind speed coefficient C_v [1]

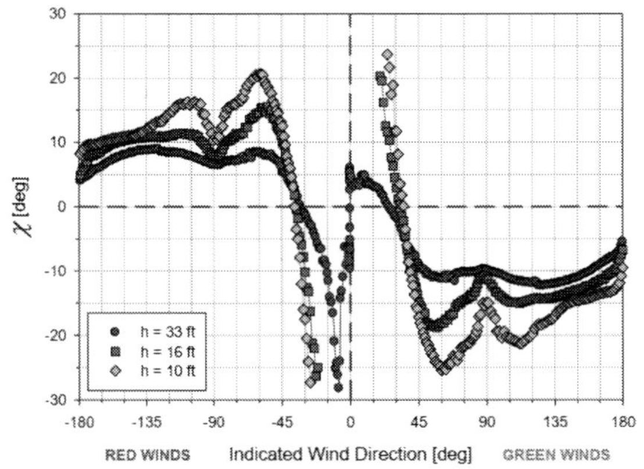

Figure 6; Example horizontal flow deviation χ [1]

In case $C_v < 1$, the indicated wind speed is higher than the local wind speed which the helicopter encounters when operating from the flight deck, and vice versa. The helicopter encounters the local airspeed and wind direction, and thus the flight characteristics in the CFE should be based on the local wind conditions instead of on the indicated wind conditions by the anemometers. The decrease in local wind speed between red 45° and green 45° is caused by the air bubble generated by the hangar in front of the landing spot. There was no reliable data measured in the wind tunnel for winds in the sector between red 30° and green 15° at 16 ft (purple line in figures) and 10 ft (grey line in figures) and thus not presented. In general, a symmetrical airflow pattern is observed. However, the position of an electronic warfare jammer on the starboard side of the ship caused anti-symmetrical airflow at 33 ft above the flight deck (green line in figures). In this case, the electronic warfare jammer has a positive effect on the airflow for winds in the sector between dead-ahead and green 15°, i.e. higher wind speeds and smaller deviations in flow direction.

Both for green and red wind conditions, the local winds have a larger angle when seen from the bow of the ship as indicated by the anemometers (i.e. the wind moves inboard behind the hangar). Once the relative wind is more from abeam and aft, other factors might influence the local wind direction. In the sector between 30° and 150° (both for red and green winds) an over-speed (up to 20 %) occurs. This over-speed is caused by the flow being deflected over the ship's hull; above the hull the streamlines will be compressed, leading to higher wind speeds. For a height of 16 ft the flow for head and stern winds is more or less identical to the flow pattern at 33 ft. For a height of 10 ft, the airflow in the sector 80° to 110° (both for red and green winds) is strongly influenced by vortices generated by the ship's hull. There is a sharp gradient between red 45° and red 15° and between green 15° and green 45° with possibly unpredictable wind speed and wind direction in case the helicopter flies within either of these regions.

Helicopter Aspects

For the determination of the helicopter flight characteristics, shore-based hover trials are conducted. A distinction is made between rejection criteria and dependencies, to be less dependent on dedicated sea trials [1]. More specifically, the subdivision of test data in rejection criteria and dependencies enables (1) assessing other possible conditions for in-service operations and (2) accurate exclusion of test points. A distinction is made between subjective, performance, control position, and aircraft attitude related issues. The flight test data is converted into referred parameters, so that it can be used to produce information relevant to atmospheric conditions and helicopter masses different from those actually tested. In any case, the precise limiting factor(s) for the specified operational conditions should be known, for example, in order to compare test conditions flown during different missions of the test campaign with each other. Since the steady-state helicopter flight characteristics are valid for trimmed conditions without any difference between land- or sea-based operations, the measured helicopter data from the shore-based hover trials can be saved into lookup tables and used for future helicopter-ship qualification trials.

The purpose of the shore-based hover trials is to determine the helicopter flight characteristics, for example, pilot workload, engine performance, Trimmed Flight Control Positions (TFCP), controllability limits and helicopter attitudes in an omni-directional relative wind envelope. This is done in order to complement the flight manual information, which only provides torque required for dead-ahead conditions. These tests are executed within the limitations for land-based operations as given by the helicopter manufacturer. The flight test data obtained from the shore-based hover trials indicates - within the low speed hover envelope - which regions exist where safety margins between available and required helicopter rejection criteria are marginal or even exceeded. This is required for safety reasons, as in these regions, limitations are likely to be exceeded by the operational aircrew during shipboard operations. The flight tests are performed at the required values of referred weights, $W/\sigma\omega^2$, where W is helicopter weight, σ is relative density (relative to international standard atmosphere), and ω is the relative rotorspeed (relative to a standard rotor speed value). The targeted referred weights are set as the operational weight bands for shipboard operations.

A dedicated pace-car, as shown in Figure 7, is used to set up the required relative wind conditions in addition to the actual wind conditions encountered. Note that a pace-car or alternative low-speed measuring device is required as the airspeed indications of a helicopter are generally only reliable above 40 knots with winds on the nose of the helicopter. The pace-car is equipped with a calibrated speed measurement system, a display on top of the dashboard to present the relative wind to the driver, and a wind vane to provide the pilot with a visual reference of the relative wind direction.

Figure 7; Pace-car used for shore-based hover trials

An example of the NH90 NFH helicopter for torque required at 20 knots, as a function of the relative wind direction are shown in Figure 8 [3,4]. The torque required increases for higher referred weights, and depends on the relative wind direction. The increase in torque required between 11 000 kg and 11 750 kg referred weight, is caused by variations in OAT and air pressure, as the maximum take-off weight of the helicopter is limited to 11 000 kg. It shows that the effect for increased take-off weight of the helicopter has a larger effect on torque required, than increasing the Outside Air Temperature (OAT) whilst maintaining the same take-off weight. As the main rotor turns counter-clockwise, when seen from above, more tail rotor thrust and thus torque required, is necessary to prevent the nose to turn into the wind in the sector between dead-ahead and green 90°. The tendency for the nose of the helicopter to turn into the wind is known as the 'weather-cock effect'. For this reason the torque required is highest at relative winds from the direction green 60° to green 120°. An opposite effect occurs for red wind conditions in the sector between red 90° and dead-ahead, with a slightly less or equal demand for tail rotor thrust and thus torque required. Once past the abeam position for either red or green wind conditions, the same effects occur although in opposite direction as the nose of the helicopter wants to turn now in the other direction. The areas that exceed the applicable safety margin for All Engine Operatives (AEO) continuous power are to be excluded from the CFE, which in this case is the sector between green 30° and green 150° for 11 000 kg and 11 750 kg referred weight. Although not shown, for the lower relative wind speeds at 11 000 kg and 11 750 kg referred weight, the torque required exceeds the safety margin and/or the maximum continuous power limitation of 104% torque. Therefore, these low speed relative wind conditions are removed from the respective CFE's.

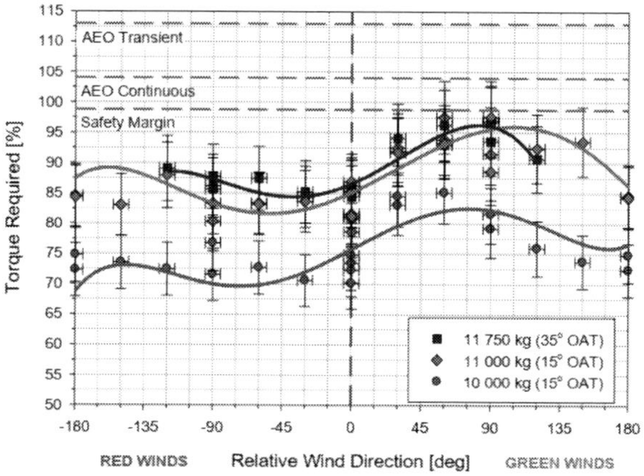

Figure 8; Torque required at 20 knots [1]

In general, the following effects are noted of variations in the environmental conditions to the helicopter flight characteristics in a world-wide theatre. An increase in referred weight results in an increase in torque required, engine gas generator speed N_g, engine power turbine inlet temperature T_{46} and tail rotor thrust. As a result, for higher referred weights, an associated minimum relative wind speed is required to perform flight operations. An increase in referred weight has negligible influence on pedal position, longitudinal cyclic, lateral cyclic, pitch attitude and roll attitude.

PHASE II, THE PREDICTION OF THE CANDIDATE FLIGHT ENVELOPE

The presented predictive tool for the CFE's in this section, is considered original, and can be seen as the most important novelty of this work [1]. There are various inputs required in the predictive tool, i.e. helicopter flight characteristics and limitations, helicopter parameters and ship parameters, atmospheric conditions and the ship environment as shown in Figure 9. There are sub-models for the atmospheric boundary layer, ship airwake, subjective pilot ratings, helicopter performance, helicopter control positions and helicopter attitudes. Furthermore, the algorithm processes the input data such that all relevant rejection criteria and dependencies interact with each other. For this reason, maximum and minimum values of local wind conditions near and above the flight deck are applied to helicopter rejection criteria, stored in lookup tables, for example to ensure that there remains sufficient power and control margin available to perform safe shipboard operations throughout various operational scenarios.

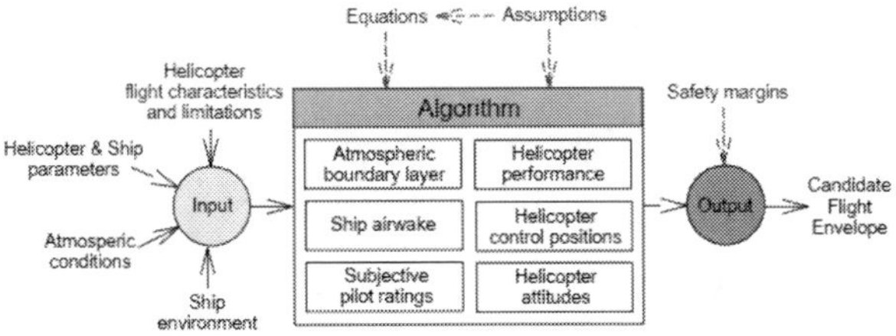

Figure 9; Overview elements predictive tool

The algorithm processes the input data such that all relevant rejection criteria and dependencies interact with each other. For this reason, maximum and minimum values of local wind conditions near and above the flight deck are applied to helicopter rejection criteria, stored in lookup tables, for example, to ensure that there remains sufficient power and control margin available to perform safe shipboard operations throughout various operational scenarios. The output from the predictive tool results in the CFE, based on the indicated wind as presented by the ship's anemometers, which can likely be released to in-service operations for each particular helicopter-ship combination, whilst taking adequate safety margins into account. Unfortunately, there are uncertainties associated with the measurements used as input for the predictive tool, and uncertainties associated with the algorithm itself.

The CFE is used to increase trial effectiveness as it functions as the starting point for sea trials. The aim of the CFE is to achieve the smallest possible number of flying hours that does not affect the quality of the results and/or safety standards. Subsequent flight tests are then concentrated at and around the potential boundaries, reducing flight test time, and for example easing the difficulties of ship availability and the vagaries of the weather, whilst increasing the effectiveness of sea trials. In fact, the CFE is the preliminary envelope, whilst the SHOL is the result of sea trials and is the ultimate version defining the safe operational limitations.

Model Calibration of the Predictive Tool

Based on the sub-models calibration results, it might be expected that the connected total model gives approximately the same agreement between computation and experimental data. This is however not necessarily the case due to the propagation of errors. The calibration phase is used to improve agreement between the computational results and the experimental data obtained by actual sea trials. By adjusting uncertain parameters of the predictive tool, the level of agreement can often be improved significantly. It should however be kept in mind that short-sighted use of model calibration (or fine-tuning of the safety margins) may result in the right model output for the wrong reason. Too often, model calibration is (unintentionally) used to cover up the weaknesses of the underlying conceptual (sub) models.

The only noteworthy calibration performed for the predictive tool is an increase in the minimum wind speed as determined during the wind-tunnel tests, and applied towards the helicopter flight characteristics. Possible causes for the discrepancy between the predictive tool and the experimental data could be either:

- The absence of ground effect in the model;
- The absence in the model for the various blade sections of the rotor experiencing a spatially distributed flow field at each instant of time (as seen inside airwake of a ship), compared to the various blade sections of the same rotor experiencing a basically spatially homogenous flow field (as seen in shore-based hover trials);
- Too large safety margins for the rejection criteria;
- Different piloting techniques in case the helicopter is power limited;
- The application of minimum and maximum wind values throughout the whole approach and departure paths whilst certain phases thereof may be more relevant;
- Any combination of the above items.

In addition, any variance in Power Performance Index (PPI) may influence the engine performance. After the simple calibration of the predictive tool by adjusting the minimum wind speed, the agreement seems to be acceptable from a qualitative point of view. Of course, this does not mean that with the simplifications made in the model (e.g. the absence of a spatially distributed flow field for the various blade sections in the ship airwake), it is assumed that the flow fields encountered during the shore-based hover trials and sea trials are very similar from an aerodynamic perspective. However, the model calibration has demonstrated that the effects on e.g. torque required and flight controls is small enough to be captured by the safety margins.

Construction of the Candidate Flight Envelope

To demonstrate the process for the construction of the CFE, an example that was not used during model calibration, is given for the fore-aft take-off and landing procedure at 10 000 kg, 11 000 kg and 11 600 kg referred weight with an aft-CG. Note that the shore-based hover trials were performed up to 11 750 kg referred weight, whilst the CFE are only developed up to 11 600 kg referred weight. A somewhat lower maximum referred weight is chosen for the Hot & Heavy conditions on board ships, to allow flight operations to be conducted within the speed bracket of the ship's cruise engines in calm wind conditions. Otherwise, the ship's high speed gas turbines would always have to be available in calm wind conditions to operate the helicopter, which unless operational scenarios dictate is not desirable. The first set of examples for a particular ship type is shown for the relevant referred weights in Figure 10, Figure 11 and Figure 12, respectively. Note that for clarity only the rejection criteria are indicated in the sector between red 110° to green 110° above 10 knots (i.e. not in the aft sector). The following rejection criteria were relevant for construction of the CFE for the NH90 NFH [1]:

- Torque required for the lower relative wind speeds;
- It was noted that right pedal position exceeds the 10% safety margin in red wind conditions;
- Pitch attitudes around green 30° at 30 knots;
- Roll attitude in red wind conditions;
- From previous sea trials it was noted that torque required exceeds the rejection criteria in the red 90° azimuth (input miscellaneous items);
- The boundaries for a Hot & Heavy envelope have an additional safety margin due to the more sluggish helicopter response (input miscellaneous items).

For the construction of the CFE the indicated rejection criteria should be excluded as much as possible. In both examples, for all three referred weights, the rejection criteria for pedal position (red wind conditions), pitch attitude (green wind conditions) and roll attitude (red wind conditions) are restricting the CFE (brown dash-dotted line in figure) in similar relative wind conditions. This is because, as demonstrated by the shore-based hover trials, an increase in referred weight has negligible influence on pedal position, longitudinal cyclic, lateral cyclic, pitch attitude and roll attitude. The differences between the CFE's in referred weight are mainly in the lower speed regions, due to an increase in torque required for the higher referred weights.

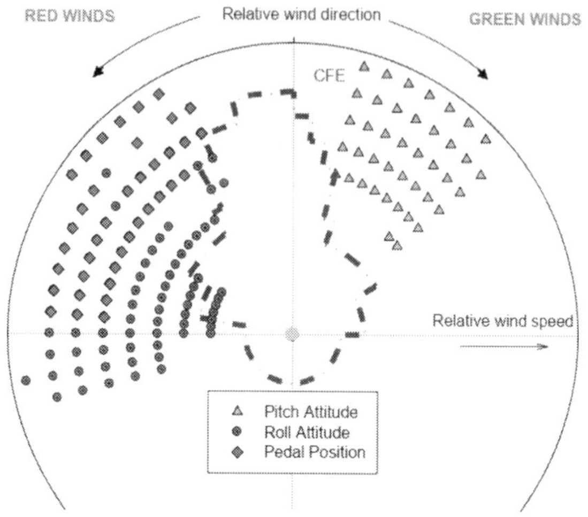

Figure 10; CFE 10 000 kg referred weight [1]

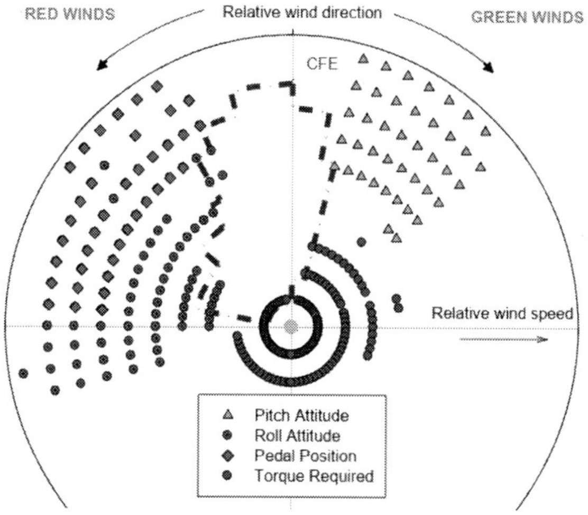

Figure 11; CFE 11 000 kg referred weight [1]

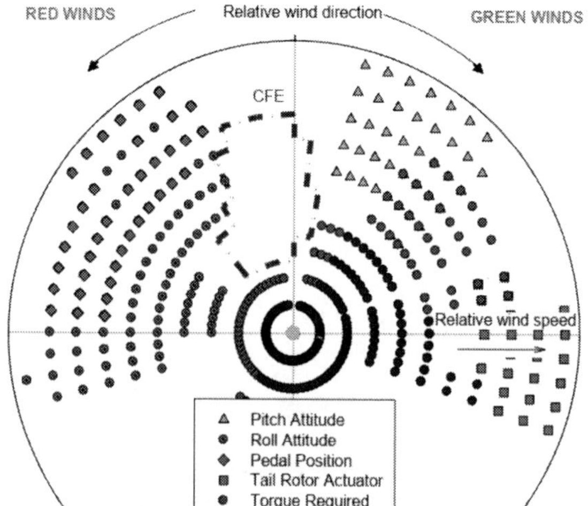

Figure 12; CFE 11 600 kg referred weight [1]

This is because, as demonstrated by the shore-based hover trials, increases in referred weight results in an increase in torque required, engine gas generator speed N_g, engine power turbine inlet temperature T_{46} and tail rotor thrust. As a result, for higher referred weights, an associated minimum relative wind speed is required to perform flight operations.

Turbulence level

The wind tunnel measurements used to predict the ship's airwake are average values, which do not take into account the variety in wind speed or wind direction. However, in reality there is an associated turbulence level for a particular relative wind condition for each ship type. Generally, the turbulence level increases with increasing relative wind speed, and vice versa. The turbulence level can increase to such a high level that the helicopter responds violently in pitch and roll attitude, which causes corresponding torque spikes, increased control activity and higher pilot workload. Typically, these high turbulence levels resulting in violent helicopter responses also define the upper boundaries of the SHOL envelope. There is still some engineering judgement required to set the upper boundaries for the CFE of a particular ship type correctly. Unless adequate safety margins are applied, the upper boundaries are the main focus of the sea trials. Further research should thus focus on identifying areas with increased turbulence, including the associated response to helicopter flight characteristics, such that this could be included in the predictions for the CFE.

Night operations

Different envelopes are produced for use by day and by night. The closure rate to the ship during the approach to the ship is more difficult to judge at night, which may result in a last-minute flare to arrest closure rates and potentially leading to an exceedance of power limitations and/or overshooting the approach. This is undesirable and for safety reasons no tail winds are allowed during night operations. In general, this is the main difference between day and night wind envelopes, including some differences that might occur in the upper boundaries due to turbulence intensity.

PHASE III, SEA TRIALS AND VALIDATION

The "SHOL-X" approach seeks to replace some or much of the experimentation at sea with less expensive simulation results that are validated with experimental data. The rejection criteria are used to accurately define the boundaries of the operational limitations and to assess whether actual limits are approached or whether there is still room for expansion of the envelope. The construction of a SHOL envelope can now be achieved by two different options. The first and most common option is using dedicated sea trials in which the potential boundaries for the various take-off and landing procedures are validated. Preferably, during the sea trials the complete CFE is explored, such that all the evaluated parts can be released to service. By doing so, it also provides a validation of the predictive tool and gives the associated level of confidence for that particular helicopter-ship combination to be used as baseline for successive trials. The second option is the construction of the Hot & Heavy envelope by desk-top analysis alone. This option is new and only feasible in case the predictive tool is applied as described in this paper. The construction of the operational limitation for Hot & Heavy condition are based on the data gathered during the shore-based hover trials, and the flight test results for the other referred weights on board the same ship type. The construction of operational limitations by desk-top analysis alone is a novel approach, and can be seen as the most important achievement of this work.

Uncertainties of the Predictive Tool

There are various inputs required in the predictive tool, i.e. helicopter flight characteristics and limitations, helicopter parameters and ship parameters, atmospheric conditions and the ship environment as shown in Figure 13. There are sub-models for the atmospheric boundary layer, ship airwake, subjective pilot ratings, helicopter performance, helicopter control positions and helicopter attitudes. Unfortunately, there are uncertainties associated with the measurements used as input for the predictive tool, and uncertainties associated with the algorithm itself. There are both parameter / variable uncertainties and measurement uncertainties associated with inputs required for the model. In addition, as the predictive tool is validated against experimental data, the associated measurement uncertainties should be taken into account. Finally, there are uncertainties in the underlying fundamental physical principles of the model caused by the applied assumptions and utilized equations belonging to either the theory uncertainty or model uncertainty.

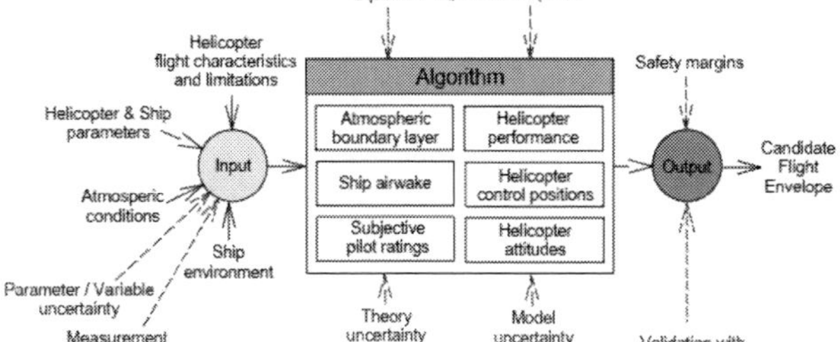

Figure 13; Overview elements predictive tool

The goal of the predictive tool is to make predictions of real-life system behaviour. Since the predictive tool, or any other simulation model, can never fully capture the complex physical phenomena that contribute to real-life system behaviour, it never perfectly agrees with reality. Nevertheless the predictive tool can still be used, as long as the user has some idea of how well the model predictions are for the intended application of the tool – but to do this with confidence – one must know how good the predictions are at the set points selected for validation, and also how good the predictions are in the interpolated or extrapolated regions of the application domain for which there are no experimental results. Therefore, a proper understanding of the predictive tool with the potential sources of uncertainties is essential during the validation process.

Validation of the Predictive Tool

The estimation of a range within which the simulation modelling errors lie is a primary objective of the validation process and is accomplished by comparing a simulation result (solution) with an appropriate experimental result (data) for specified validation variables at a specified set of conditions. The output of the predictive tool is validated to determine the degree of representation of the reality from the perspective of constructing the CFE as a starting point for the sea trials. Validation takes place by quantification of the agreement between experimental and computational data. It is common practice in all fields of engineering and science for comparisons between computational results and experimental data to be made graphically. In this case, the computational constructed CFE, and the experimental data from the sea trials. If the computational results generally agree with the experimental data, the computational model is declared as validated. The credibility of a helicopter-specific model's capability to make predictions about reality must be evaluated against independent data.

During the introduction of the NH90 NFH, there were seven helicopter capable ship types in-service within the Dutch fleet that all have been certified for helicopter-ship operations [5,6,7,8,9,10,11]. The sea trials were conducted in the North Sea up to 11 000 kg referred weight (at approximately 15 °C) In addition, for two ship types sea trials were conducted up to 11 600 kg referred weight, which was achieved by 11 000 kg actual weight at an OAT equivalent to 30.7 °C at sea level, in either the Indian ocean or Caribbean Sea [12,13].

The validation of the CFE's consists of both a qualitative and a quantitative part. This is required as only the most conservative outcome of the subjective pilot ratings and objective flight test data can be released to service. The qualitative part consists of the pilot workload, with acceptable test points (green colour in figure) and unacceptable test points (red colour in figure) as shown for example in Figure 14 to Figure 16, respectively. There is a distinction made between test points flown during day and night (symbol for day is a square and for night is a circle). In case a quantitative safety margin is exceeded, the respective rejection criteria are indicated in a graph as shown for example in Figure 17 to Figure 19, respectively. As torque is the dominant factor within the atmospheric conditions tested, other performance parameters are not shown for clarity. The CFE is represented as the brown dotted line in these figures. Note that the rejection criteria are only indicated if a safety margin is exceeded. The safety margins for the rejection criteria are initially determined during the shore-based hover trials, and can be fine-tuned as more experience is gained during successive sea trials.

A "wedding cake" strategy has been applied in which the results for the higher referred weight are also valid for the lower referred weight and the latter do not have to be tested again. Therefore, unless the test points are not achievable for 11 000 kg referred weight, the required test points are flown for 10 000 kg referred weight. For the examples shown, roll attitude of the helicopter is the main limitation in red wind conditions (which may also result in more torque to manoeuvre the helicopter). The torque utilised is the main limitation for take-offs towards the leeward side of the ship and the lower boundaries of the envelope. Torque utilised, instead of torque required is mentioned, as the pilot may have used temporarily more torque than what would be required. The upper boundaries are mainly defined by the large pitch and roll attitude deviations of the helicopter due to the turbulence resulting from the ship's superstructure, with a corresponding increase in pilot workload. The CFE was somewhat conservative at 11 000 kg referred weight in red wind conditions with regard to roll attitude, as shown in Figure 18, and this was qualitatively confirmed by Figure 15. In all other scenarios, both for the qualitative and quantitative validation, there has proved to be a good agreement between the experimental and computational data. Note that the upper boundaries for 11 600 kg referred weight were not tested due to a lack of sufficient wind and thus cannot be validated with flight test data.

The objective rejection criteria are generally more consistent and restrictive in defining the boundaries of the operational envelopes determined during the sea trials than the subjective ratings given by the test pilots involved (whilst taking into account the skills of the "average" fleet pilot). Although there is no straightforward explanation, this is an important conclusion as it supports the philosophy to use objective rejection criteria to accurately predict the likely outcome of the sea trials. In addition, it enables to minimize the amount of test pilots involved in the sea trials, as it makes the qualitative ratings transparent to the test crew (i.e. to determine whether the test pilots were restrictive or conservative in their decisions). The "SHOL-X" approach aims to replace some or much of the experimentation at sea with less expensive simulation results that are validated with experimental results at selected set points.

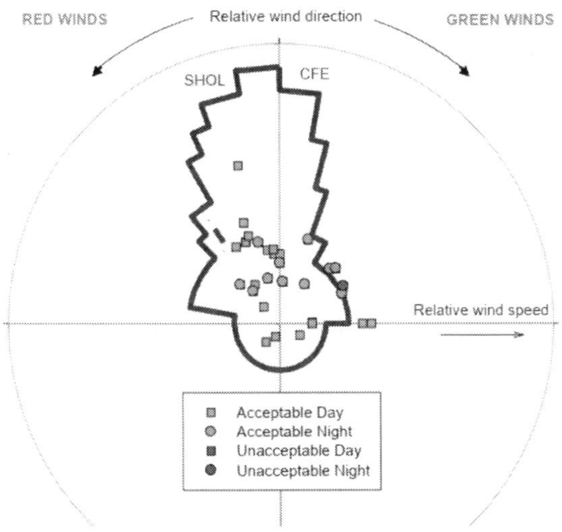

Figure 14; Subjective 10 000 kg ref weight [1]

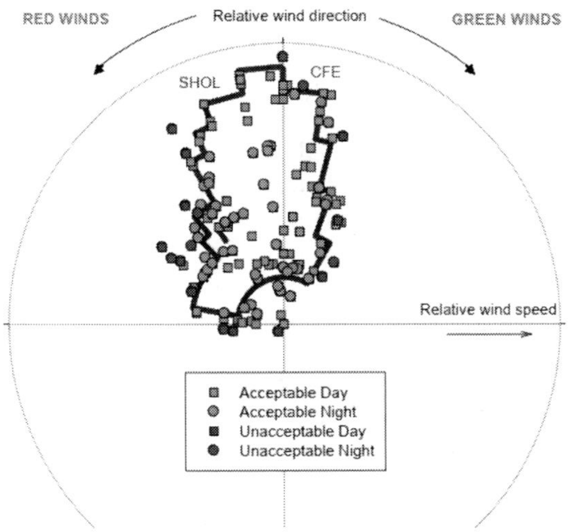

Figure 15; Subjective 11 000 kg ref weight [1]

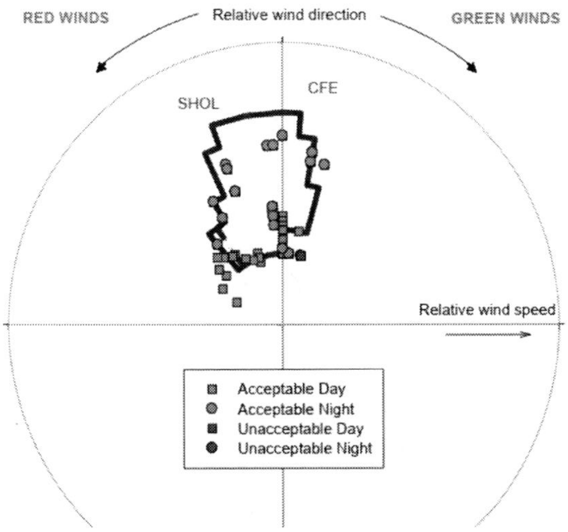

Figure 16; Subjective 11 600 kg ref weight [1]

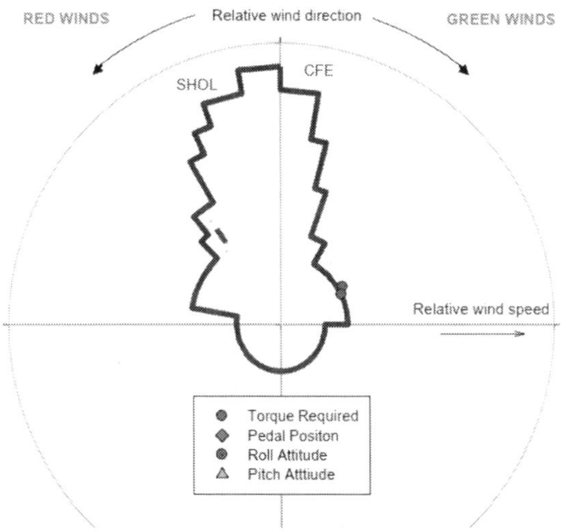

Figure 17; Objective 10 000 kg ref weight [1]

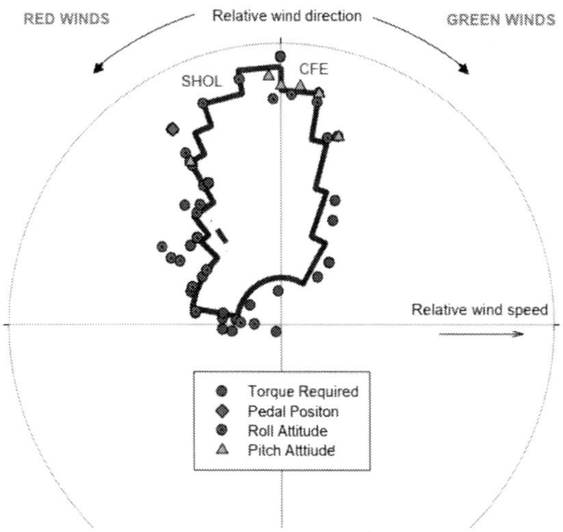

Figure 18; Objective 11 000 kg ref weight [1]

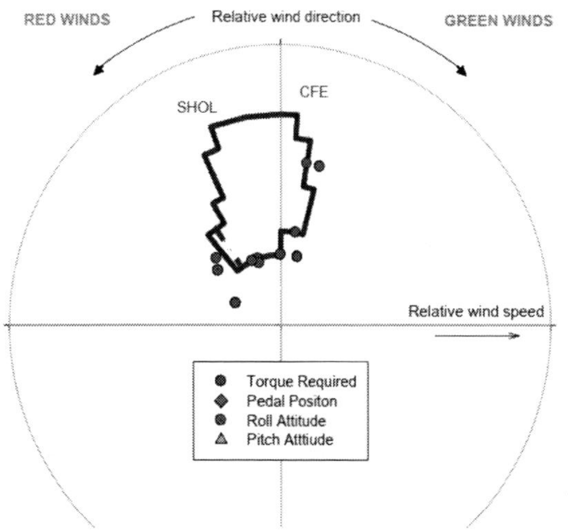

Figure 19; Objective 11 600 kg ref weight [1]

In particular, the objective rejection criteria are used to accurately define the boundaries of the potential operational limitations and by an additional feature of the tool during the validation after the sea trials to assess whether actual limits are approached or whether there is still room for expansion of the envelope. The software tool "SHOL-X" thereby assists in making fair comparisons between various predictions with any helicopter type for diversity in ship types. This does not mean that expert opinion does not count because quantitative data do not capture all intricacies of model validation. The user of the predictive tool should be aware of both the advantages and shortcomings to use the predictions optimally such that savings in time and expenses of sea trials are realized without reducing safety. With the appropriate mindset, that the predictive tool does not provide a perfect solution but only the likely outcome of the sea trials, the graphical validation confirms the predictions made by the predictive tool and it is validated. However, this implicates a minimum knowledge level and skill set for the operator of the predictive tool, which should be provided by dedicated training to ensure all prior phases to construct the CFE are executed correctly.

The fleet remains required for testing of some relative wind conditions and ship motion to assess the pilot workload and visual references acceptability. Therefore, with the novel test methodology applied, 10-15 flight hours are recommended as a bare minimum for sea trials in order to evaluate a particular helicopter-ship combination. The corresponding savings for the sea trials, compared with other current test methodologies, in time required and flight hours consumed for it are roughly 70% whilst at least the same operational capability can be achieved. This is equivalent to a reduction in time required for the sea trials from at least a three week period towards a seven day period, and a reduction from at least 45 flight hours towards a maximum of 15 flight hours required.

CONCLUSIONS

The newly developed predictive tool for the CFE's presented in this paper, is considered original, and can be seen as the most important novelty of this work. There are various inputs required in the predictive tool, i.e. helicopter flight characteristics and limitations, helicopter parameters and ship parameters, atmospheric conditions and the ship environment. There are sub-models for the atmospheric boundary layer, ship airwake, subjective pilot ratings, helicopter performance, helicopter control positions and helicopter attitudes. Furthermore, the algorithm processes the input data such that all relevant rejection criteria and dependencies interact with each other. For this reason, maximum and minimum values of local wind conditions near and above the flight deck are applied to helicopter rejection criteria, stored in lookup tables, for example to ensure that there remains sufficient power and control margin available to perform safe shipboard operations throughout various operational scenarios.

The output from the predictive tool results in the CFE, based on the indicated wind as presented by the ship's anemometers, which can likely be released to in-service operations for each particular helicopter-ship combination, whilst taking adequate safety margins into account. The CFE is used to increase trial effectiveness as it functions as the starting point for sea trials. The aim of the CFE is to achieve the smallest possible number of flying hours that does not affect the quality of the results and/or safety standards. Subsequent flight tests are then concentrated at and around the potential boundaries, reducing flight test time, and for example easing the difficulties of ship availability and the vagaries of the weather, whilst increasing the effectiveness of sea trials. In fact, the CFE is the preliminary envelope, whilst the SHOL is the result of sea trials and is the ultimate version defining the safe operational limitations.

REFERENCES

[1]Hoencamp, A., *Helicopter-Ship Qualification Testing*, "SHOL-X" Test Methodology, AeroMath books, 2016, ISBN 978-94-6259-565-1.

[2]Booij, P.J.A., Airflow measurements on a 1:70 scale model of the Royal Netherlands Navy Patrol Vessel, 2001, NLR contract report CR-2011-588.

[3]Hoencamp, A., Report CLSK Directive 11-32, Part 1; NH90 NFH Shore-Based Hover Trials, 2012.

[4]Hoencamp, A., Report CLSK Directive 11-32, Part 6; NH90 NFH Shore-Based Hover Trials, 2012.

[5]Hoencamp, A., Sicking, J.P., and van Kralingen, R.A., Report CLSK Directive 11-32, Part 2; NH90 NFH sea trials LPD1 "Hr.Ms. Rotterdam", 2012.

[6]Hoencamp, A., Sicking, J.P., and van Kralingen, R.A., Report CLSK Directive 11-32, Part 4; NH90 NFH sea trials LCF "Hr.Ms. De Ruyter", 2012.

[7]Hoencamp, A., Sicking, J.P., and van Kralingen, R.A., Report CLSK Directive 11-32, Part 7; NH90 NFH sea trials OPV "Hr.Ms. Holland", 2012.

[8]Hoencamp, A., Sicking, J.P., and van Kralingen, R.A., Report CLSK Directive 11-32, Part 8; NH90 NFH sea trials M-Frigate "Hr.Ms. van Speijk", 2013.

[9]Hoencamp, A., Sicking, J.P., and van Kralingen, R.A., Report CLSK Directive 11-32, Part 10; NH90 NFH sea trials AOR "Zr.Ms. Amsterdam", 2013.

[10]Hoencamp, A., and van Kralingen, R.A., Report CLSK Directive 11-32, Part 12; NH90 NFH sea trials LPD2 "Zr.Ms. Johan de Witt", 2014.

[11]Hoencamp, A., van Kralingen, R.A., and Campfens, J.A., Report CLSK Directive 11-32, Part 15; NH90 NFH sea trials JSS "Zr.Ms. Karel Doorman", 2015.

[12]Hoencamp, A., and Sicking, J.P., Report CLSK Directive 11-32, Part 11; NH90 NFH sea trials Hot & Heavy AOR "Zr.Ms. Amsterdam", 2013.

[13]Hoencamp, A., Sicking, J.P., and van Kralingen, R.A., Report CLSK Directive 11-32, Part 9; NH90 NFH sea trials Hot & Heavy LCF "Hr.Ms. De Ruyter", 2013.

Deformation Measurement and Modal Identification of an Extremely Flexible Rotor Blade

Daiju Uehara — Graduate Research Assistant
Christopher Cameron — Postdoctoral Fellow
Jayant Sirohi — Associate Professor
Department of Aerospace Engineering and Engineerig Mechanics
The University of Texas at Austin
Austin, Texas 78712

ABSTRACT

This present study describes the deformation measurement and modal identification of an extremely flexible rotor blade in hover. Deformation measurement were performed on two-bladed rotors at four different root pitch angles and rotational speeds by means of a time-resolved digital image correlation technique (DIC). The rotor blade deformations were then processed with the natural excitation technique (NExT) and the eigensystem realization algorithm (ERA) to determine the dynamic characteristics of the flexible blade, including natural frequencies, mode shapes, and damping ratios. The static equilibrium displacements along the quarter-chord were extracted from the 3D displacement distributions of an entire blade span for the flap, lead-lag, and torsional deformations. The modal identification algorithms NExT/ERA successfully extracted modal parameters from the recorded deformation data for the first and second modes. The stability boundary of the flexible rotor blade was found to be dependent on its root pitch angle and rotational speed.

NOTATION

A	State transition matrix in discrete time
A_c	State space matrix in continuous time
B	Controllability matrix
C	Damping matrix or obserbility matrix
C_m	Obserbility matrix in modal coordinates
F	Force vector
$H(n)$	Hankel matrix
i, j	Indices of physical channels
I	Identity matrix
k, n	Indices in frequency and time domain
K	Stiffness matrix
M	Mass matrix
N	The number of channels
O	Null matrix
p	The number of column of data block
P	The ratio of magnitudes of mode shapes
q	The number of row of data block
Q	Weighting factor
R	Radius
R_{ab}	Cross-correlation function of a and b
S_{ab}	Cross-spectral density function of a and b
S_{xx}, S_{yy}, S_{xy}	variance and covariance of x and y
$y, \dot{y}, \ddot{y}$	Displacement, velocity, and acceleration vector
$Y(n)$	Free response matrix to an impulse excitaiton at nth time step
λ	Eigenvalue
Λ	The diagonal matrix of eigenvalues of A
ψ	Eigenvalues matrix of A
ϕ	Mode shape
θ_0	Root pitch angle
α	Phase angle
Ω	Rotor rotational velocity
$\Delta\theta$	Pitch angle displacement
ζ	Damping ratio
Δt	Sampling interval

Presented at the 6th Asian/Australian Rotorcraft Forum/ Heli Japan 2017, Kanazawa, Japan, November 7-9, 2017.

INTRODUCTION

An extremely flexible rotor blade concept has been proposed to increase the weight efficiency and survivability of helicopter rotors. In this concept, the structural stiffness of the rotor blade is negligible, and centrifugal force on the blades during rotation provides stiffness as well as stability. Because the elimination of a structural stiffness requirement results in reduced weight, such rotor blades have been proposed for heavy lift helicopters. Recently, Sicard and Sirohi (Ref. 1) investigated an 18" (45.7 cm) diameter rotor with extremely flexible blades for application in rotary-wing micro aerial vehicles (MAVs). The objective was to change the rotor diameter in-flight to increase hover endurance or access confined spaces as needed. In addition, these flexible rotor blades were found to be damage tolerant. The biggest challenge in implementation of such a concept is to avoid aeroelastic instabilities. This requires a detailed analysis as well as experimental data on the dynamic behavior of the rotor blades during rotation.

Investigating the dynamic characteristics, such as the rotating natural frequencies, damping ratios, and mode shapes, of the flexible blade is more challenging than that of the conventional rigid rotor blades for several reasons. First, modal analysis is not possible on the non-rotating rotor blade because the stiffness mainly relies on the centrifugal force, and the characteristics of the blade may be significantly different from its rotating state. Thus, modal analysis on the flexible blade must be performed when spinning at certain rotational speed. Additionally, classical modal analysis requires frequency response functions or impulse response functions obtained from measurements of both the input force and the output response of the structure. For a conventional rigid rotor blade, the modal analysis in rotating state is typically based on measurements using on-blade instrumentation, including strain gages, accelerometers, or pressure transducers, in order to obtain the input and output information (Ref. 2). The conventional method, however, is not practical for the flexible rotor blade because the presence of the instrumentation may cause a significant change in the blade characteristics.

To overcome these limitations, an optical deformation measurement technique and an output-only operational modal analysis (OMA) and are introduced in the current study. In contrast to classical modal analysis, the output-only operational modal analysis (OMA) only requires the resultant responses of the structure at a certain operating condition. The OMA techniques were first developed in civil engineering along with damage detection and evaluation of structures, such as the large scale bridges and buildings (Refs. 3–5). There are a number of OMA approaches available in the literature, which are well-summarized by Peeters and Roeck (Ref. 6). OMA has also been applied to rotating machinery, for example, James III et al. (Ref. 7) performed operational modal testing on a wind turbine and Qi et al. (Ref. 8) validated their OMA method on an operating linear rotor system, by comparing the experimental results to a finite element model. In this study, the natural excitation technique (NExT) using cross-correlation functions, and a time domain modal identification algorithm called the Eigensystem Realization Algorithm (ERA), were selected among several available methods. This is because the NExT/ERA method has been used as a very powerful tool for the modal analysis of structures undergoing natural vibrations due to ambient excitation in their operating environment (Refs. 9, 10).

Although the present OMA approach does not require the measurement of actual excitation on the flexible blade, there is still a need to measure deformations of the blade without using the conventional on-blade sensors. In the past, several non-contact optical measurement techniques have been used to measure rotating blade deformation, including Projection Moiré Interferometry (Refs. 11, 12), stereo photogrammetry (Ref. 13), and Digital Image Correlation (DIC). In the current study, the stereoscopic DIC technique was selected to measure the flexible rotor blade deformations. This is because the DIC technique is able to measure the three-dimensional displacement over the entire blade span at a high spatial resolution with minimal preparation of the test article, i.e., by only painting the blade surface with random speckle patterns. In previous work, Sirohi and Lawson (Ref. 14) explored the application of the stereoscopic DIC for measuring deformations on rotor blades under rotation. Sicard and Sirohi (Ref. 15) utilized the DIC technique to measure the deformation of an extremely flexible rotor blade. Cameron et al. performed (Ref. 16) full span blade deformation measurement via DIC for a lower rotor blade of a model-scale coaxial rotor system. Thus, DIC has proven to be a useful tool for full-field flexible rotor blade deformation measurement.

Recently, Sergio and Sirohi (Ref. 17) used a combination of DIC and an OMA algorithm called the Ibrahim Time Domain (Ref. 18) method to extract the modal properties of a conventional rotor blade under rotation. This approach was also used by Cameron et al. (Ref. 19) to measure the rotating modal properties of a rigid blade in a model-scale coaxial helicopter rotor. However, these studies were not able to apply the modal identification using NExT/ERA because the field of view in DIC measurement was limited to one quarter of the rotor disk, resulting in an incomplete time history of the response.

The main goal of the present study is to perform three-dimensional, full field deformation measurement on an extremely flexible rotor blade using the DIC technique, and extract the modal parameters of the blade using NExT/ERA for further understanding of the unique characteristics of the blade. The present study extends the range of DIC measurement over the entire rotor disk so that the deformations in continuous time history could be used for the NExT/ERA modal analysis. The time-resolved DIC measurement was performed at four different root pitch angles and rotational speeds in order to assess the dependence of the blade dynamics on pitch angles and rotational speeds.

EXPERIMENTAL SETUP

All the experiments were performed on a hover test stand at the University of Texas at Austin, shown in Figure 1. In this section, the rotor test stand and rotor blade designs are introduced, as well as the DIC measurement set up including specifications of high speed digital cameras, specimen preparation, and light source.

Rotor configuration

The two-bladed flexible rotor consists of two flexible blades and a fixed-pitch rigid hub. The blade structure is made of a ±45° AS4/3501-6 prepreg and a tungsten rod held at the blade tip in a chordwise direction to provide centrifugal stiffening and stabilization. The blade is nominally untwisted with a constant thin circular arc airfoil profile; the thin, open section profile results in a low bending stiffness and a negligible torsional stiffness. The rotor parameters are summarized in Table 1. The rotor is driven by a brushless DC motor Hacker A50-16S and the rotational speed is measured by an optical incremental encoder US DIGITAL E5.

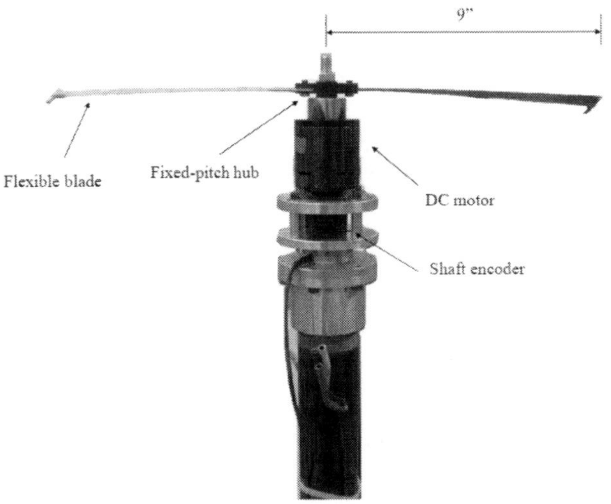

Fig. 1. Side view of the rotor stand.

Table 1. Rotor system parameters.

Rotor diameter (Radius)	406.4 (203.2) [mm]
Root cutout	11 [% R]
Blade airfoil	Thin circular arc
Camber	9%
Chord	26.2 [mm]
Thickness to chord ratio	1.5%
Number of blades	2

DIC measurement set up

The blade deformation measurement was performed by means of the digital image correlation (DIC) technique in the test chamber as shown in Figure 2. The present study uses stereoscopic DIC with two high speed cameras to obtain full three-dimensional displacement fields on the flexible rotor blade. Knowing the positions of two cameras relative to each other and the magnifications of lenses, the stereoscopic DIC algorithm can calculate the absolute three-dimensional coordinates of any surface point in space.

In the current study, two cameras (Phantom Miro M310 with Nikon NIKKOR f/1.8D 35 mm lenses) are mounted on linear camera rails above the rotor plane with their axes oriented approximately 45° with respect to the rotor disk. The locations of two cameras are determined in a way such that the field of view of each camera captures the entire rotor disk, in order to perform time-resolved DIC measurement.

A key step in the implementation of the DIC measurement is the preparation of the test article, which in this case is the flexible rotor blade. The surface of the test article must have a stochastic intensity distribution, which deforms together with the surfaces. As can be seen in Figure 3, fluorescent paint is applied on the blade surface and random black dots are painted on it. A Nd:YLF pulsed laser with wavelength 527 nm routed through an engineered diffuser is used to illuminate the entire rotor disk at any desired instant of time (as shown in Figure 2), and images are captured on the high-speed cameras by using low pass filters to allow only the light emitted by the flourescent paint. An optical encoder on the rotor shaft is used to generate a 16 per-revolution pulse train, which is used as a trigger for laser strobing and image acquisition at 16 evenly-spaced azimuthal locations. Undeformed images for DIC calculations are captured at the same azimuthal locations by rotating the rotor shaft slowly by hand, i.e. at a negligibly small rotational speed, thereby eliminating any aerodynamic forcing. Figure 4 shows the unprocessed images acquired by the high-speed camera, overlaid on each other to indicate the complete rotor disk.

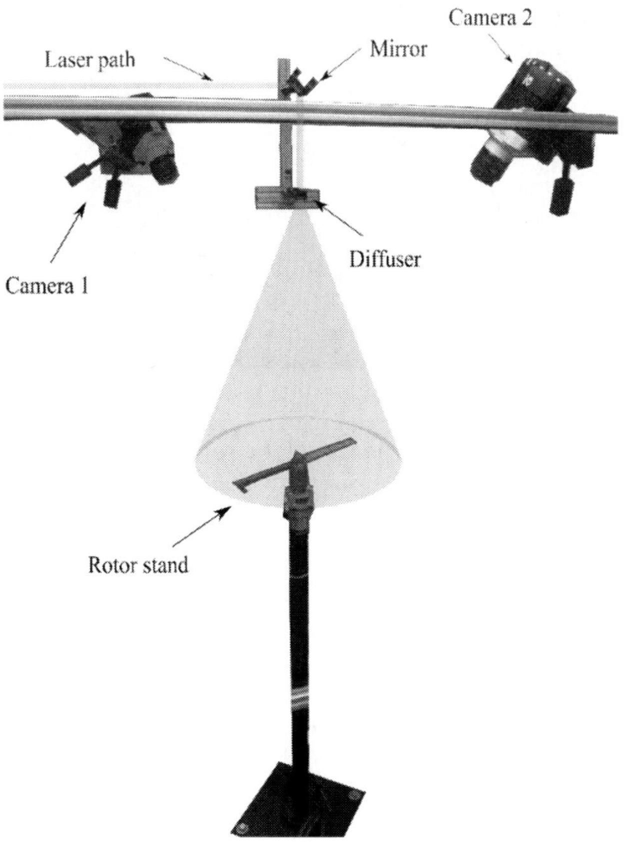

Fig. 2. DIC measurement setup.

EXPERIMENTAL PROCEDURE

Test envelope

Two-bladed rotors were tested at four root pitch angles: 0°, 10°, 20°, and 30°, by using four fixed-pitch hubs. Additionally, each rotor was spun at four different rotational speeds to understand the dynamic behavior of the blade with respect to the rotational speed. For 10°, 20°, and 30° cases, each rotor was spun at 600, 900, 1200, and 1500 RPM, whereas for 0° case the rotor was spun at 600, 900, 1000, and 1200 RPM. This is because an aeroelastic instability occured above 1200

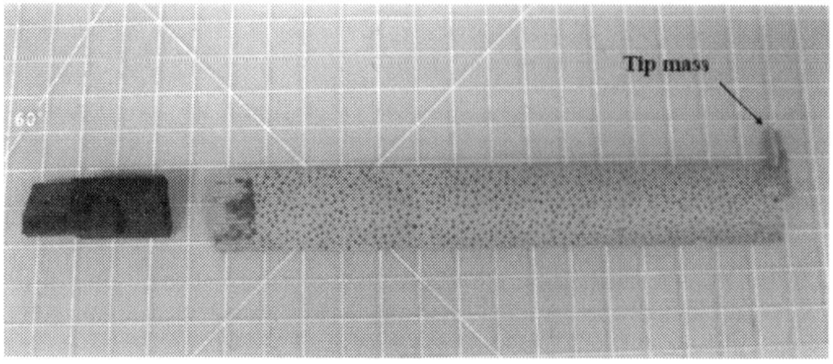

Fig. 3. The hub and the flexible blade with the fluorescent paint and random speckle pattern.

Fig. 4. Unprocessed images of the rotor strobed at 16 azimuthal locations, showing complete rotor disk at 1200 RPM.

RPM for the 0° root pitch rotor. The complete test matrix is shown in Table 2.

Table 2. Summary of the test envelope

Data set	Root pitch angle	Rotational speed (RPM)
1	$\theta_0 = 0°$	600 ,900, 1000, 1200
2	$\theta_0 = 10°$	
3	$\theta_0 = 20°$	600, 900, 1200, 150
4	$\theta_0 = 30°$	

Procedure

The collective pitch angle was first set by selecting the fixed-pitch hub, and the two-bladed rotor with the flexible blades was spun at a target rotational speed. Then the two high-speed cameras and the laser were triggered by the pulse train obtained from the encoder, and 6400 images were continously taken at 16 azimuthal locations at each rotational speed. This procedure was repeated for different collective pitch angles. After the collective pitch and rotational speed sweep, undeformed images were taken by rotating the motor by hand very slowly (approximately 60 RPM). All the images were then post-processed for the DIC deformation calculations and modal identification.

MODAL IDENTIFICATION PROCESS

Modal identification process in this paper mainly consists of three steps. First, time-resolved three-dimensional blade deformations are calculated by DIC algorithm. Second, the response data excited in the operating environment is processed with NExT in order to calculate correlation functions that contain the modal parameter information. Third, ERA is used as the time-domain modal identification scheme to extract modal frequencies, damping, and mode shapes from the correlation functions.

In practice, however, some modal parameters picked up by NExT/ERA include non-physical modes so that one has to introduce a certain procedure to filter out these spurious or fictitious modes. In this study, Consistent Mode Indicator (CMI) and spectrum analysis of recorded responses are used to filter out some fictitious modes found along this modal identification process.

DIC processing

After the images were captured by a stereo pair of high-speed cameras, the DIC software (LaVision DaVis Strain Manager 3D) calculated the deformations of the blade. The complete image, whose size is 1280 x 800 pixels, was divided into a number of interrogation windows and one velocity vector was calculated for each interrogation window. The size of the window was selected as 19 x 19 pixels and the window shift was set to be 5 pixels for mapping the interrogation windows to the whole image, resulting in 54 % overlapping area. Camera scaling factor is approximately 2.24 pixel/mm so that the deformations at total 988 locations (13 chordwise and 76 spanwise locations) were calculated. The sample processed result is shown in Figure 5. To recover the three-dimensional deformation field, corresponding interrogation windows in two different camera images were found by a correlation process, which identifies the same speckle pattern inside both images. With the help of mapping functions calculated during the camera calibration process, the two two-dimensional vectors from each camera were combined to yield a three-dimensional vector. The accuracy of the vectors depends on the size of interrogation window and is estimated to be 0.01%R, while the spatial resolution of the displacement field is 0.2%R.

Natural Excitation Technique (NExT)

Natural Excitation Technique (NExT) is a methodology of modal testing that allows us to identify structural modal parameters under the operating condition. The theoretical development (Ref. 7) for application to this study is presented as follows.

Consider the differential equation of motion for a multi-degree-of-freedom structure

$$\mathbf{M}\ddot{\mathbf{y}}(t) + \mathbf{C}\dot{\mathbf{y}}(t) + \mathbf{K}\mathbf{y}(t) = \mathbf{F}(t) \tag{1}$$

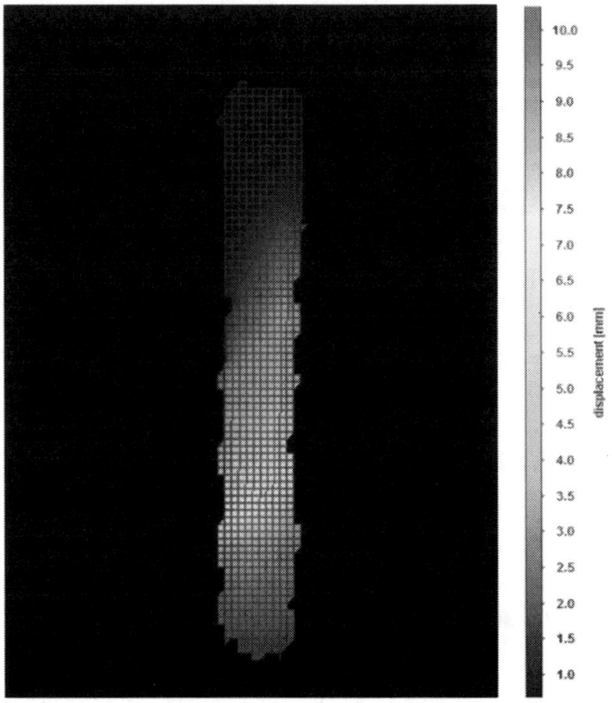

Fig. 5. Vector map processed by DIC.

where M, C, and K are the mass, damping, and stiffness matrices respectively, $F(t)$ is the force vector, and $\ddot{y}(t)$, $\dot{y}(t)$, and $y(t)$ are the acceleration, velocity, and displacement vectors at time t. With the assumption that the structure is excited by a white noise, the equation can be expressed as

$$\mathbf{M}\ddot{\mathbf{R}}_{y_iy_j}(\tau)+\mathbf{C}\dot{\mathbf{R}}_{y_iy_j}(\tau)+\mathbf{K}\mathbf{R}_{y_iy_j}(\tau)=0 \qquad (2)$$

where $R_{y_iy_j}(\tau)$ is the cross-correlation function of the responses $y_i(t)$ and $y_j(t)$ at two different locations i and j on the blade computed by DIC. The cross-correlation function is equivalent to the impulse response of the structure in time domain, as described by James III et al. (Ref. 7). Thus, the modal parameters of the structure can be extracted using the cross-correlation function.

Computing cross-correlation functions is the key step of the NExT algorithm. Based on the fact that the cross-spectral density (frequency domain) is equivalent to the cross-correlation function (time doman), $R_{y_iy_j}(\tau)$ can be calculated using the inverse Fourier transform of the cross-spectral density of $y_i(t)$ and $y_j(t)$ (Ref. 20).

First, the responses $y_i(t)$ and $y_j(t)$ are divided into data blocks, possibly with overlapping. The cross-spectral density function is then computed with ensemble averaging and Hanning window of the Fourier Transform of each data block in order to improve the accuracy of the computation. Then the cross-correlation function between the two channels i and j is obtained using

$$\mathbf{R}_{y_iy_j}(n)=\frac{1}{N}\sum_{k=0}^{N-1}\mathbf{S}_{y_iy_j}(k)exp\left(i\frac{2\pi kn}{N}\right), \quad n=0,\ldots,N-1 \qquad (3)$$

where k and n represent indices in frequency and time domain, respectively, and $S_{y_iy_j}(k)$ is the discrete cross-spectral density function between the two channels i and j.

It should be noted that the assumption made for the theoretical development is not necessarily met for modal analysis on a real structure. The operational modal analysis with NExT has been performed and successful on real structures in past studies (Refs. 7, 21, 22) without satisfying the assumption, i.e. the excitation to the structures is not a white noise; it is typically bandwidth limited and does not have equal power at all frequencies. For appropriate use of this method, it is also essential to mention that the sampling frequency of the cross-correlation functions $R_{y_iy_j}(\tau)$ and the original responses $y_i(t)$ and $y_j(t)$ are equal, and its number of data points are the same as used for the FFT computation. Overall, the analyst needs to determine the number of samples used for FFT, the number of data blocks, the number of overlapping data points, the length of the data blocks, shape of window, and reference channels for calculating cross-correlation functions (Ref. 23).

Eigensystem Realization Algorithm (ERA)

A number of techniques have been invented for identifying the modal parameters from the free response or impulse response data. In the present study, the Eigensystem Realization Algorithm (ERA) (Ref. 24) is applied to the pseudo-impulse response data derived from NExT. Consider the state-space equations in continuous time

$$\begin{aligned}\dot{\mathbf{x}}(t)&=\mathbf{A}_c\mathbf{x}(t)+\mathbf{B}\mathbf{u}(t)\\ \mathbf{y}(t)&=\mathbf{C}\mathbf{x}(t)\end{aligned} \qquad (4)$$

where $x(t)$ is the state vector, $u(t)$ is the excitation vector, $y(n)$ is the response vector to the excitation, and A_c, B, and C are state space matrices in continuous time. Assuming that a unit-impulse excitation is applied to the system, a solution to these equation has the exponential form

$$\mathbf{Y}(t)=\mathbf{C}e^{\mathbf{A}_ct}\mathbf{B} \qquad (5)$$

where $Y(t)$ consists of the free responses of the system to a impulse excitation. In discrete time, the solution is represented as

$$\mathbf{Y}(n)=\mathbf{C}\mathbf{A}^{n-1}\mathbf{B} \qquad (6)$$

where $Y(n)$ is the response at nth time step, $A=e^{A_c\Delta t}$ is the discrete-time state transition matrix and Δt is the time step of system discretization. Thus, the goal of ERA is to estimate the constant matrices A, B, and C from the impulse response data $Y(n)$ acquired from measurement. The system modal parameters are then computed from the eigenvalues and eigenvectors of the constructed state matrices. It is important to mention that the matrices B cannot be estimated for the present modal identification since the NExT is output-only modal analysis technique and the real excitation to the structure is unknown. Regardless of the fact, the modal parameters are obtained from the matrices A and C.

System realization starts from building the generalized Hankel matrix

$$\mathbf{H}(n-1) = \begin{bmatrix} Y(n) & Y(n+1) \dots & Y(n+p) \\ Y(n+1) & \ddots & \vdots \\ \vdots & & \\ Y(n+q) & \dots & Y(n+p+q) \end{bmatrix} \quad (7)$$

where $Y(n)$ is in this case the N x 1 vector which consists of the pseudo-impulse responses computed by NExT at the nth time step, and N is the number of channels corresponding to the number of interrogation windows on DIC processing. The parameters p and q correspond to the size of the Hankel matrix. As a rule of thumb for selecting p and q, the analyst should consider four times the number of expected modes for the number of columns, and the number of points available in the cross-spectral density function for the number of rows in the Hankel matrix. Normally, the Hankel matrix is formed with block rows and block columns data shifted in time by one sample from the previous block row and column. The final block row and column are exceptionally shifted by a large time sample in order to calculate the extended modal amplitude coherence (EMAC), which will be discussed later.

The Hankel matrix for n = 1 can be decomposed by using the singular value decomposition as

$$\mathbf{H}(0) = \begin{bmatrix} CB & CAB \dots & CA^{p-1}B \\ CAB & \ddots & \vdots \\ \vdots & & \\ CA^{q-1}B & \dots & CA^{p-1+q-1}B \end{bmatrix} = \mathbf{R}\Sigma\mathbf{S}^T. \quad (8)$$

Then the nth order system realization is calculated as

$$\begin{aligned} \mathbf{A} &= \Sigma_n^{1/2}\,\mathbf{R}_n^{1/2}\,\mathbf{H}(1)\,\mathbf{S}_n\,\Sigma_n^{-1/2} \\ \mathbf{C} &= \mathbf{E}^T\,\mathbf{R}_n\,\Sigma_n^{1/2} \end{aligned} \quad (9)$$

where $H(1)$ is the Hankel matrix for n = 2 and $E = [I \;\; O]$ with an identity matrix I and a null matrix O of appropriate order. These matrices are then transformed to modal coordinates by multiplying with the eigenvector matrix of A

$$\begin{aligned} \Lambda &= \Psi^{-1}\mathbf{A}\Psi \\ \mathbf{C}_m &= \mathbf{C}\Psi \end{aligned} \quad (10)$$

where Λ is the diagonal matrix of eigenvalues of A and Ψ is the matrix of eigenvectors of A. After transforming back to the continuous time domain, one can obtain the modal damping ratios and damped natural frequencies of the system from the real and imaginary parts of the eigenvalues, respectively. The column vectors of C_m correspond to the mode shapes of each mode.

Consistent Mode Indicator (CMI)

The modal parameters identified with NExT and ERA usually contain some fictitious modes due to the presence of structural nonlinearity and unavoidable noise in measured data. Consistent Mode Indicator (CMI) (Ref. 25) is a primary metric that can filter out the fictitious modes. CMI is calculated as a product of two parameters, Extended Modal Amplitude Coherence (EMAC) and Modal Phase Collinearity as

$$CMI = EMAC \cdot MPC. \quad (11)$$

These two parameters are described in the following sections.

EMAC EMAC consists of two parts, one is calculated based on observability matrix C, and the other is calculated from controllability matrix B. Recalling that the modal analysis along with NExT cannot determine the matrix B, the EMAC from the observability matrix C, named as EMAC^O, is used in the present study.

EMAC^O quantifies a predictability of mode shapes constructed by the identified modal parameters as an indicator of the temporal consistency of selected modes. The actual mode shape at time step T_0 corresponding to the final data block of the Hankel matrix is computed as

$$\Phi(T_0) = \widehat{\mathbf{R}_n}\,\Sigma_n^{1/2}\,\Psi \quad (12)$$

where $\widehat{R_n}$ is the last row block of R_n. Knowing that the final data blocks of the Hankel matrix are shifted by at least 10 times samples from the previous data block, a predicted mode shape at T_0 is calculated as

$$\widetilde{\Phi}(T_0) = \mathbf{C}\mathbf{A}^{q-2+10}\,\Psi = \mathbf{C}_m\Lambda^{q-2+10} \quad (13)$$

The actual and predicted magnitudes of mode shapes are then compared using the ratio of the magnitudes

$$\begin{aligned} P &= |\Phi(T_0)|/|\widetilde{\Phi}(T_0)| \quad for \quad |\Phi(T_0)| \le |\widetilde{\Phi}(T_0)| \\ &= |\widetilde{\Phi}(T_0)|/|\Phi(T_0)| \quad otherwise \end{aligned} \quad (14)$$

Additionally, the actual and predicted phase angles are compared. Define the phase angle α as

$$\alpha = \arctan(\Phi(T_0)/\widetilde{\Phi}(T_0)), \quad -\pi \le \alpha \le \pi \quad (15)$$

Then a weighting factor Q is calculated as

$$\begin{aligned} Q &= 1.0 - (|\alpha|/(\pi/4)) \quad for \quad |\alpha| \le \pi/4 \\ &= 0 \quad otherwise \end{aligned} \quad (16)$$

Finally, EMAC^O is computed as

$$EMAC^O = P \cdot Q \quad (17)$$

MPC MPC, on the other hand, quantifies the spatial consistency of the identified modes by measuring the complexity of an identified mode shape. Calculating the variance and covariance of the real and imaginary parts of the identified mode shapes

$$S_{xx} = \mathrm{Re}(\Phi)^T \mathrm{Re}(\Phi)$$
$$S_{yy} = \mathrm{Im}(\Phi)^T \mathrm{Im}(\Phi) \quad (18)$$
$$S_{xy} = \mathrm{Re}(\Phi)^T \mathrm{Im}(\Phi)$$

The eigenvalues $\lambda_{1,2}$ of the variance-covariance matrix can be computed as

$$\eta = \frac{S_{yy} - S_{xx}}{2S_{xy}} \quad (19)$$

$$\lambda_{1,2} = \frac{S_{xx} + S_{yy}}{2} \pm S_{xy}\sqrt{\eta^2 + 1} \quad (20)$$

Finally, the MPC is defined as

$$MPC = \left(\frac{\lambda_1 - \lambda_2}{\lambda_1 + \lambda_2}\right)^2 \quad (21)$$

Spectrum analysis

Even after filtering out some fictitious modes using the CMI, there are still cases where one needs to identify the pysical mode from several choices. In this study, the modal frequencies picked up from peaks in a spectrum plot of measured responses are used as a reference of the physical mode.

The main step of this method is simply to apply the Fast Fourier Transformation algorithm to deformation data at the arbitrary point on the blade, obtained from DIC processing. Under the conditions of well-separated modal frequencies, the analyst should be able to pick up the resonance peaks in a spectrum plot. It should be noted that measured data in the present study inherently contains harmonics of rotating frequencies components, thus, the analyst need to take it into account during this process.

RESULTS AND DISCUSSION

In this section, the static equilibrium displacements of the flexible blade calculated using DIC algorithm are first presented. Results of modal identification process are then discussed in terms of modal frequencies, mode shapes, and damping rates. Stability boundary of the flexible rotor blade are also examined in this section. The flexible rotor was operated with four types of hubs, whose root pitch angles are 0°, 10°, 20°, and 30°. Since data were collected at four different rotational frequencies, total 16 data sets were processed.

Steady-state deformation

In advance of investigating the dynamic behaviors, the blade deformations in trim state was examined for a range of different root pitch angles. The mean of the time history of spanwise and chordwise variations of deformations was calculated using recorded data. The spanwise deformations of each degree of freedom along the quarter-chord axis were then extracted. It should be noted that the flap and lead-lag bending were measured from non-rotating reference state of the blades and torsional deformations were obtained by calculating slopes of the blade chord between deformed and undeformed images at each spanwise channel location.

Figure 6 to 8 shows the spanwise deformations of each degree of freedom of the blade at different root pitch angles when spinning at 1200 RPM. The lag bending was almost negligible at θ_0 = 10° while the lag bending at θ_0 = 20° and θ_0 = 30° can be clearly seen in Figure 6. The flap bending has the same trend as the lag bending, that is, magnitude of the bending becomes greater as the root pitch angles increases as shown in Figure 7. This is due to the influence of an increase in drag with increasing the local pitch angle. The torsional displacement, however, does not follow the trend as can be seen in Figure 8. The pitch angle displacement is maximum at θ_0 = 10°. The other two operating conditions have approximately identical twist distribution, although the absolute local pitch angles are different due to the difference of root pitch angle. For further understanding of the blade dynamics, an analytical model of the flexible rotor must be developed and validated with the experimental results. Note that θ_0 = 0° case was excluded from the plots because the severe instability was observed during the test at this operating speed.

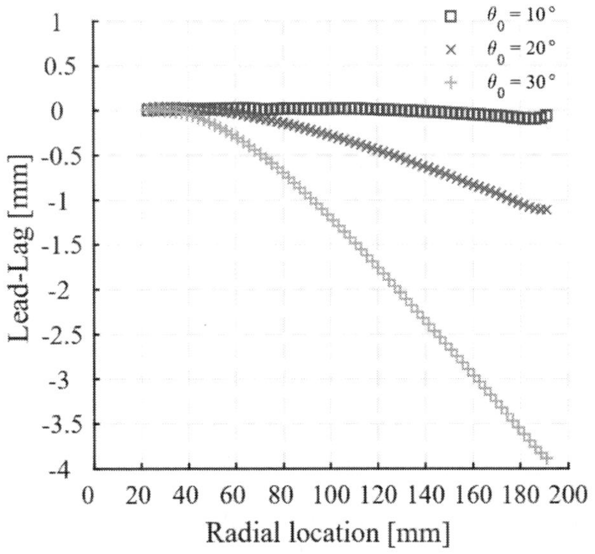

Fig. 6. Lead-lag bending deformation for different root pitch angles, rotating at 1200 RPM.

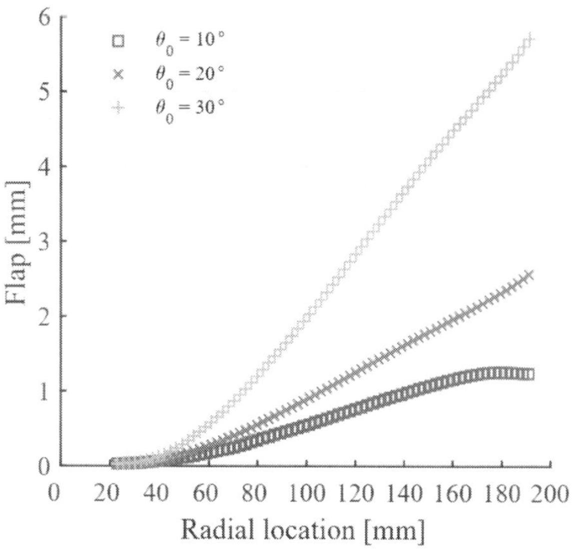

Fig. 7. Flap bending deformation for different root pitch angles, rotating at 1200 RPM.

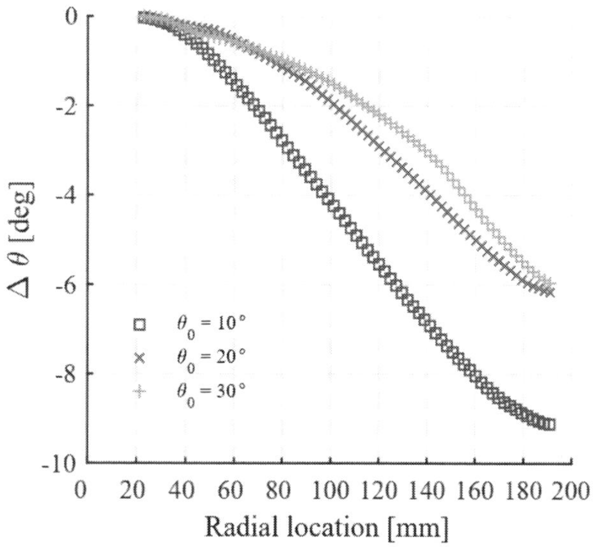

Fig. 8. Torsion bending deformation for different root pitch angles, rotating at 1200 RPM.

Frequencies

The modal frequencies were identified in two different ways, one is the combination of NExT and ERA with CMI, and the other is the spectrum analysis using FFT technique. Figure 9 shows the spectrum plot of local pitch angle displacement at the tip of the blade for θ_0 = 30° case, rotating at 20 Hz. It can be seen that in this operating condition the spectrum peaks are clear enough to identify which peaks are associated with operating frequencies or modal frequencies. The first peak right next to the 1/rev and second peak between 2/rev and 3/rev harmonics correspond to the first and second modal frequencies as shown in Figure 9.

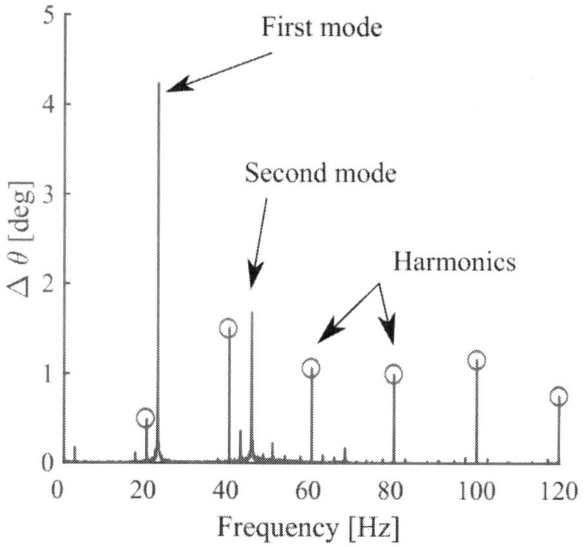

Fig. 9. Spectrum plot of the pitch angle displacement at the tip for θ_0 = 30°.

Fanplots for θ_0 = 0° and θ_0 = 30° are shown in Figure 10 and 11 and the results from the spectrum analysis and NExT/ERA processing at 600 RPM and 1200 RPM are summarized in Table 3 and 4. There is excellent agreement up to the second mode between the spectrum analysis and NExT/ERA modal identifications for all the test cases, except 20° root pitch angle. This is because the peaks corresponding to the modal frequencies were not well identified from the spectrum plot for θ_0 = 20° case as mentioned earlier. It is worth mentioning that all the modes of vibration on the flexible rotor blade are coupled due to the structural characteristics of the blade. To identify which degree of freedom is the dominant component at each mode, corresponding mode shapes are plotted and discussed next.

Note that the peaks for the case when θ_0 = 20° were not as visible as other root pitch angle cases in a spectrum plot, although the NExt/ERA analysis successfully identified the modal frequencies for θ_0 = 20°. A related point to consider is that the modes higher than the third mode were not well identified from spectrum plots for any root pitch angles and rotational velocities. This could be caused by the limitation of measuring the smaller deformation at higher frequencies by means of DIC.

Table 3. Summary of modal frequencies [/rev] for different operating conditions at 600 RPM

Root pitch	θ_0 = 0°	θ_0 = 10°	θ_0 = 20°	θ_0 = 30°
FFT	1.470	1.413	1.401	1.370
NExT/ERA	1.478	1.392	1.410	1.365

Table 4. Summary of modal frequencies [/rev] for different operating conditions at 1200 RPM

Root pitch	$\theta_0 = 0°$	$\theta_0 = 10°$	$\theta_0 = 20°$	$\theta_0 = 30°$
FFT	1.135	1.108	N/A	1.135
NExT/ERA	1.176	1.119	1.015	1.134

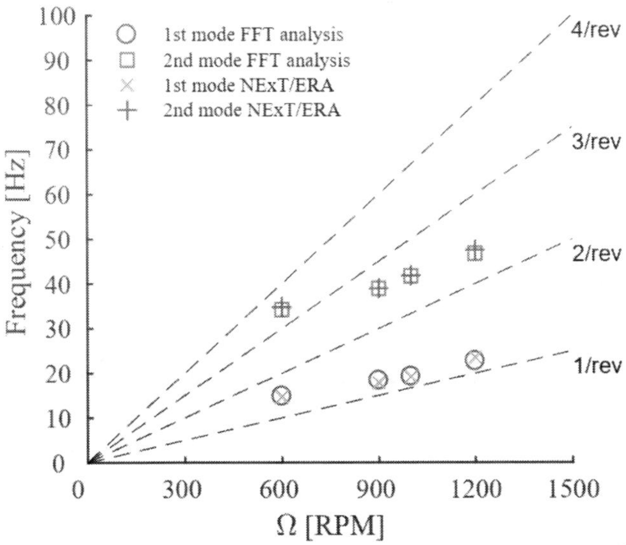

Fig. 10. Fanplot with the spectrum analysis and NExT/ERA process for $\theta_0 = 0°$.

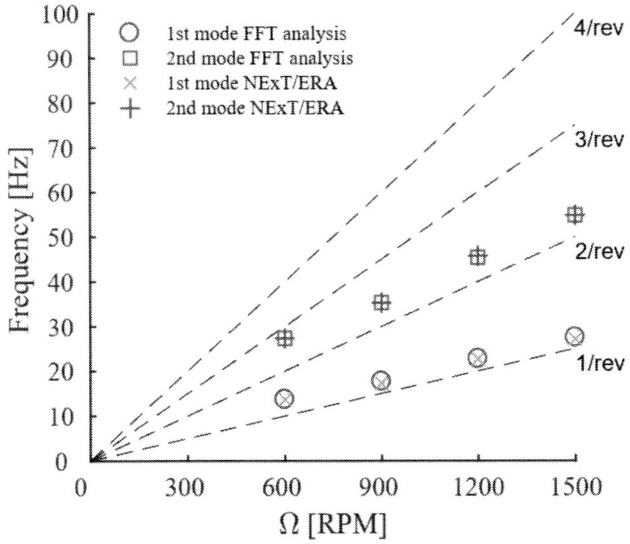

Fig. 11. Fanplot with the spectrum analysis and NExT/ERA process for $\theta_0 = 30°$.

Mode shapes

Mode shapes of the first and second mode along the blade quarter-chord line for $\theta_0 = 0°$ and 30° operated at 1200 RPM are plotted in Figure 12 and 13. Note that the amplitudes of the mode shapes are normalized by the maximum value among the flap, lead-lag and torsional degrees of freedom at each mode.

As can be seen in Figure 12, the first eigenmode of lead-lag degree of freedom contribute much less than that of flap and torsion at $\theta_0 = 0°$, implying that the mode could be coupled with the flap and torsional degree of freedom. The lead-lag mode for the second eigenmode is also incomparable to other two modes, which concludes that the second eigenmode mainly couples the flap and torsion deformations. For $\theta_0 = 30°$, it can be seen in Figure 13 that all of the lead-lag, flap, and torsional deformations for both the first and second modes are fully coupled, although the largest contribution is still the torsional modes. Hence, the results can lead to the conclusion that the first and second eigenmodes for $\theta_0 = 0°$ couple the flap and torsional modes, while the modes for other operating conditions fully couple the flap, lead-lag, and torsional displacements.

It is important to mention that the first torsional mode shape at $\theta_0 = 30°$ appears to be the second mode in terms of the shape itself. The characteristics can also be observed in the shape of the first flap bending mode. One possible reason is that the fictitious mode and real mode obtained from NExT/ERA analysis are so close to each other that the CMI cannot identify the real mode at this root pitch angle. The identification process repeated with different parameters, such as the number of rows and columns in Hankel matrix and reference channels, may be required for better results at this root pitch angle.

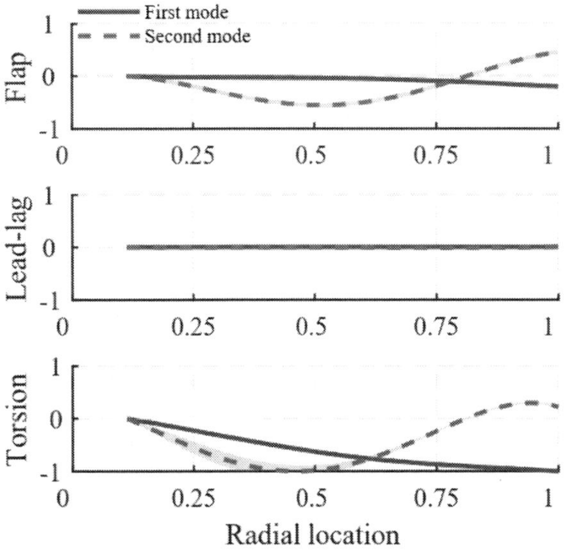

Fig. 12. Measured mode shapes of the first and second modes for each degree of freedom at $\theta_0 = 0°$.

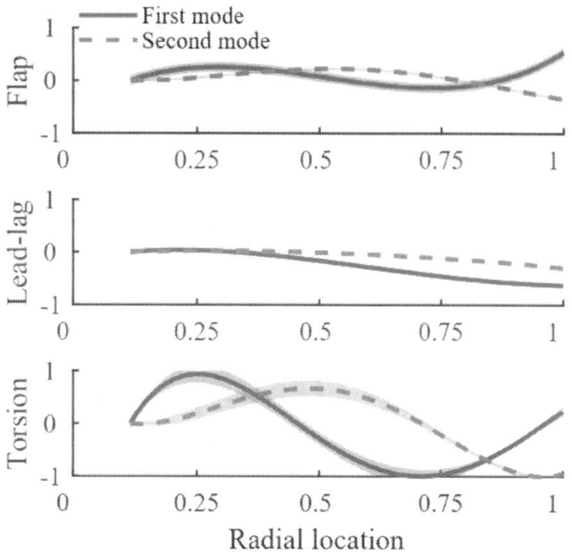

Fig. 13. Measured mode shapes of the first and second modes for each degree of freedom at $\theta_0 = 30°$.

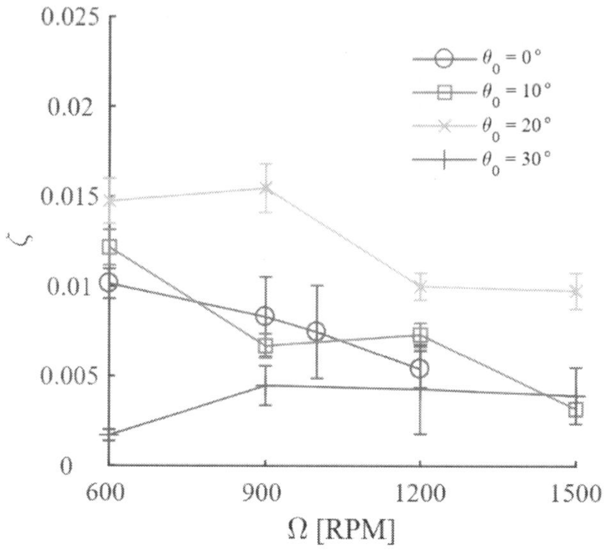

Fig. 14. Measured damping ratios variations of the first mode with respect to diffetent operationd conditions.

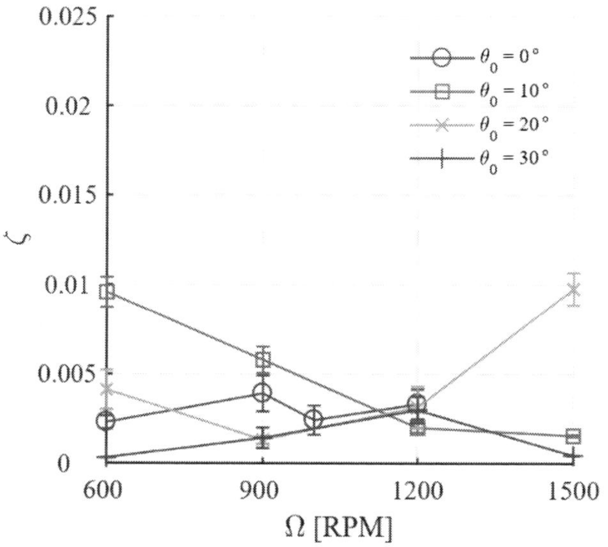

Fig. 15. Measured damping ratios variations of the second mode with respect to diffetent operationd conditions.

Damping ratios

In this section, the identified damping ratios are examined in more detail. Damping ratios were calculated from the real part of eigenvalues of the state matrix A. Figure 14 and 15 summarize the relationship among the damping ratios, pitch angle, and rotational speeds for the first and second modes.

The plot in Figure 14 shows that the modal damping ratios for the first mode at $\theta_0 = 0°$. 10°, and 20° generally decrease as opposed to an increase in blade rotational speeds. At $\theta_0 =$ 30°, it is hard to find a trend from the plot because of the uncertainties bounds. Note that the uncertainty bounds are based on the standard deviation calculated with multiple NExT/ERA analysis. The multiple modal identification processes are performed by taking different reference channels at each process. The damping ratios at $\theta_0 = 20°$ are greater than others over a range of four different rotational frequencies, whereas that θ_0 = 30° are generally the lowest.

The damping ratios of the second mode for different operating conditions are compared in Figure 15. At $\theta_0 = 0°$, the values do not vary much between 600 RPM and 1200 RPM and the magnitudes are less than 0.5%. The damping ratios for $\theta_0 = 30°$ are also less than 0.5% and are generally the lowest over a range of target rotational speeds. The same trend as the first mode can be found at $\theta_0 = 10°$, although the overall magnitudes of the damping became lower than the first mode. At $\theta_0 = 20°$, there is a sudden increase at 1500 RPM, which could be due to modal coupling to a less damped mode.

These results imply that the values of damping ratios correlate with the root pitch angles and rotational speeds, however, some refinement must be made to the modal identification process, in order to identify the trend of damping with respect to operating conditions.

Stability boundary

Unstable behaviors have been observed at some operating conditions during DIC measurement. It is essential to assess the stability boundary of the flexible rotor blade because it could lead to catastrophic damages. By plotting the pitch angle displacement at the tip and examining the slow-motion videos of the blade behavior, the stability regimes were determined for all operating conditions.

Figure 16 and 17 show the pitch angle displacement at the tip of the blade at different operating conditions. At 600 RPM, the pitch angle displacements at $\theta_0 = 0°$ and $30°$ seem to be much greater than other two root pitch angle cases. The result could lead to the fact that 0° and 30° are much more unstable, compared to 10° and 20° root pitch angles at this rotational speed. This conclusion agrees with the experimental observation of the instability regime of the blade.

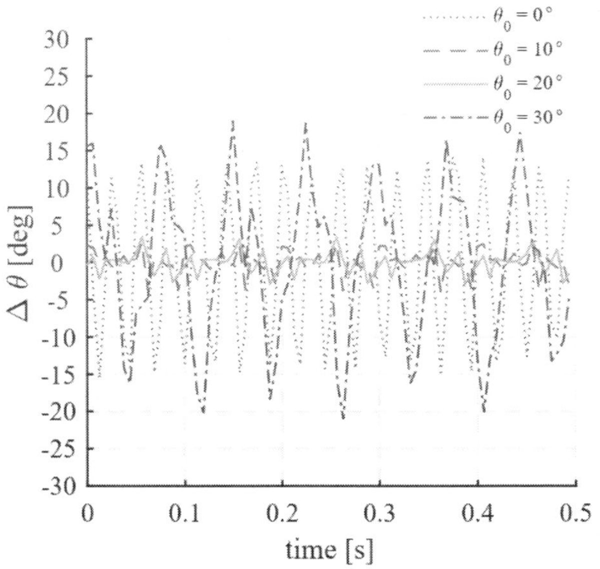

Fig. 16. Time history of the pitch angle displacement at the tip for different root pitch angles, spinning at 600 RPM .

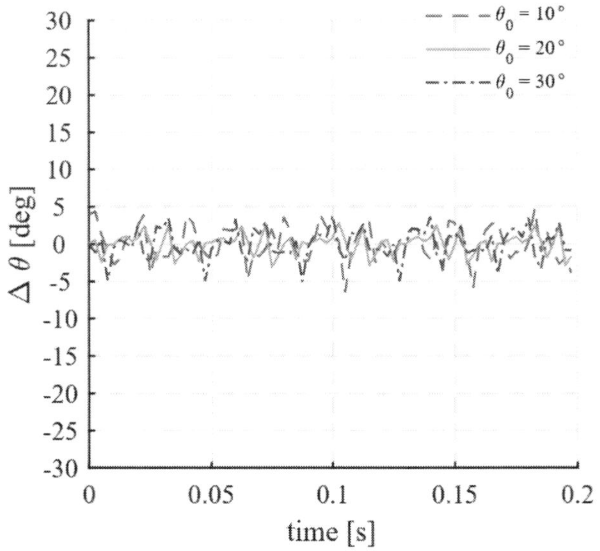

Fig. 17. Time history of the pitch angle displacement at the tip for different root pitch angles, spinning at 1500 RPM .

Based on the tip pitch angle displacement analysis and correlation with the experimental observations, the instability boundary was determined as shown in Figure 18. The flexible rotor blade at $\theta_0 = 0°$ was unstable at any rotational speeds while the blade at $\theta_0 = 10°$ and $20°$ were stable at any rotational speeds. The blade for $\theta_0 = 30°$ are found to be unstable when spinning at 600 and 900 RPM, and become stable at 1200 and 1500 RPM. In conclusion, it is found that the stability boundaries of the flexible rotor blade depend on both the rotational speeds and the root pitch angles of the rotor.

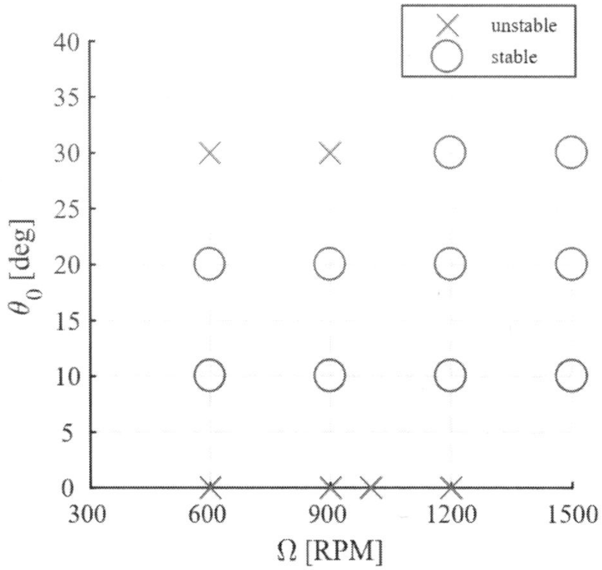

Fig. 18. Stability boundaries of the flexible blade.

CONCLUSIONS

The blade deformations of an extremely flexible rotor blade were measured by means of the time-resolved DIC measurement. The recorded response data were then used to identify the modal natural frequencies, damping ratios, and mode shapes along with NExT and ERA operational modal analysis. Experiments and modal identification processing were performed at four different root pitch angels and rotational speeds.

First, the blade static displacement of each degree of freedom were examined by averaging the recorded data. It was found that the blade deformation in trim state depend on the root pitch angles and rotational speed. Second, the dynamic characteristics of the flexible rotor blade were investigated in terms of the modal frequencies, mode shapes, and damping ratios. There is an excellent agreement between the modal frequencies identified by NExT/ERA and spectrum peaking-picking analysis up to second mode. Damping ratios for the first mode generally decrease as opposed to an increase in blade rotational speeds for all four different root pitch angles. For the second mode, the trends of damping ratios with respect to rotating speeds were different for different root pitch angles and the magnitudes were overall less than 0.5%. Identification of damping ratios appear to have a room for improvement. Mode shapes of lead lag bending, flap bending and torsional displacement along the blade quarter-chord were also well extracted by the identification process. By comparing the amplitudes of mode shapes, the dominant degree of freedom for each eigenmode were determined. The stability boundary were also identified by investigating the time history of pitch angle deformation at the tip of the blade for all the operating conditions. The results were compared with the experimental observation.

As a future work, these experimental results can be used for validation and development of an aeroelastic analysis tool. Since only the first and second modes were identified in this study, additional the NExT/ERA modal analysis with different parameters, for example the higher sampling rates or additional excitation to the blade, may be worth trying to better capture the higher modes.

ACKNOWLEDGMENTS

The work presented was supported by the Vertical Research Center of Excellence, Award number W911W6-17-2-0004, with Dr. Mahendra Bhagwat as Program Manager. The authors would like to thank the lab colleague Mr. Young-Joon Choi for his technical assistance on the experiments.

REFERENCES

[1]Sicard, J., and Sirohi, J., "Experimental Study of an Extremely Flexible Rotor for Microhelicopters," *Journal of Aircraft*, Vol. 49, No.5, 2012, pp. 1306–1314.

[2]Wilkie, W. K., Mirick, P. H., and Langston, C. W. "Rotating Shake Test and Modal Analysis of a Model Helicopter Rotor Blade," NASA TM-4760, 1997, NASA Langley Research Center, Hampton, VA

[3]Peeters, B., and Roeck, D. G., "Reference-based stochastic subspace identification for outpu-only modal analysis," *Mechanical systems and signal processing*, Vol. 13, Issue 6, November 1999, pp. 855–878.

[4]Ndambi, J. M., Vantomme, J., and Harri, K., "Damage Assessment in Reinforced Concrete Beams Using Eigenfrequencies and Mode Shape Derivatives," *Engineering Structures*, Vol. 24, Issue 4, April 2002, pp. 501–515.

[5]Hermans, H., and Van Der Auweraer, H., "Modal Testing and Analysis of Structure under Operational Conditions: Industrial Applications ," *Mechanical Systems and Signal Processing*, Vol. 13, Issue 2, March 1999, pp. 193–216.

[6]Peeters, B., and Roeck, D. G., "Stochastic System Identification for Operational Modal Analysis: A Review," *Journal of Dynamic Systems, Measurement, and Control*, Vol. 123, Issue 4, Febrauary 2001, pp. 659–667.

[7]James, G. H. III., Carne, T. G., and Lauffer, J. P., "The Natural Excitation Technique (NExT) for Modal Parameter Extraction from Operating Wind Turbines," Sandia Technical Report SAND 92-1666, 1993, Sandia National Labs, Albuquerque, NM.

[8]Qi, K., He, Z., Li, Z., Zi, Y., and Chen, X., "Vibration Based Operational Modal Analysis of Rotor System," *Measurement*, Vol. 41, Issue 7, August 2008, pp. 810–816.

[9]Caicedo, J. M., Dyke, S. J., and Johnson, E. A., "Natural Excitation Technique and Eigensystem Realization Algorithm for Phase I of the IASC-ASCE Benchmark Problem," *Journal of Engineering Mechanics*, Vol. 130, Issue 1, January 2004, pp. 49–60.

[10]Shen, F., Zheng, M., Shi, D. F., and Xu, F., "Using the Cross-Correlation Technique to Extract Modal Parameters on Response-Only Data," *Journal of Sound and Vibration*, Vol. 259, Issue 5, January 2003, pp. 1163–1179.

[11]Fleming, G. A., and Gorton, S. A., "Measurement of Rotorcraft Blade Deformation Using Projection Moiré Interferometry," *Shock and Vibration*, Vol. 7, (3), 2000, pp. 149–165.

[12]Sekula, M., "The Development and Hover Test Application of a Projection Moiré Interferometry Blade Displacement Measurement System," American Helicopter Society 68th Annual Forum Proceedings, Fort Worth, TX, 2012.

[13]Abrego, A. L., Olson, L. E., Romander, E. A., Barrows, D. A., and Burner, A. W., "Blade Displacement Measurement Technique Applied to a Full-Scale Rotor Test," American Helicopter Society 68th Annual Forum Proceedings, Fort Worth, TX, 2012.

[14]Sirohi, J., and Lawson, M. S., "Measurement of Helicopter Rotor Blade Deformation Using Digital Image Correlation," *Optical Engineering*, Vol. 51, (4), April 2012, pp. 043603.

[15]Sicard, J., and Sirohi, J., "Measurement of the Deformation of an Extremely Flexible Rotor Blade Using Digital Image Correlation," *Measurement Science and Technology*, Vol. 24, No.6, May 2013, pp. 065203.

[16]Cameron, G. C., Uehara, D., and Sirohi, J., "Transient Hub Loads and Blade Deformation of a Mach-Scale Coaxial Rotor in Hover," 56th AIAA/ASCE/AHS/ASC Structures, Structural Dynamics, and Materials Conference, Kissmmee, Florida, 2015

[17]Rizo-Patron, S. S., and Sirohi, J., "Operational Modal Analysis of a Rotating Cantilever Beam Using High-Speed Digital Image Correlation," 57th AIAA/ASCE/AHS/ASC Structures, Structural Dynamics, and Materials Conference, San Diego, CA, 2016

[18]Ibrahim, S. R., and Milkulcik, E. C., "The Experimental Determination of Vibration Test Parameters from Time Responses," *The Shock and vibration Bulletin*, Vol. 46, No.5, 1976, pp. 187–196.

[19]Cameron, C., Sirohi, J., Feil, R., and Rauleder, J., "Measurement of Transient Loads and Blade Deformation in a Coaxial Counter-Rotating Rotor," American Helicopter Society 73rd Annual Forum Proceedings, Fort Worth, TX, 2017.

[20]Welch, P. D., "The Use of Fast Fourier Transform for the Estimation of Power Spectra: A Method Based on Time Averaging Over Short, Modified Periodograms," *IEEE Transactions on Audio and Electroacoustics*, Vol. 15, Issue 2, June 1967, pp. 70–73.

[21]Farrar, C. R., and James, G. H. III., "System Identification from Ambient Vibration Measurements on a Bridge," *Journal of Sound and Vibration*, Vol. 205, Issue 1, August 1997, pp. 1–18.

[22]Farrar, C. R., and Jauregui, D. A., "Comparative Study of Damage Identification Algorithms Applied to a Bridge II: Numerical Study.," *Smart Materials and Structures*, Vol. 7, No.5, 1998, pp. 721–731.

[23]Caicedo, J. M., "Practical Guidelines for the Natural Excitation Technique (NExT) and the Eigensystem Realization Algorithm (ERA) for Modal Identification Using Ambient Vibration," *Experimental Techinques*, Vol. 35, Issue 4, July/August 2011, pp. 52–58.

[24]Juang, J-N., and Pappa, R. S., "An Eigensystem Realization Algorithm for Modal Parameter Identification and Model Reduction," *Journal of Guidance, Control, and Dynamics*, Vol. 8, No.5, 1985, pp. 620–627.

[25]Pappa, R. S., and Elliott, K. B., "Consistent-Mode Indicator for the Eigensystem Realization Algorithm," *Journal of Guidance, Control, and Dynamics*, Vol. 16, No.5, September-October 1993, pp. 852–858.

Experimental Investigation of the Rotor-Wing Aerodynamic Interaction on a Compound Helicopter in Hover

Keiichiro Makino
Senior
Kanazawa Institute of Technology
Hakusan, Ishikawa, Japan

Takeshi Akasaka
Asistant Professor
Kanazawa Institute of Technology
Hakusan, Ishikawa, Japan

Yusuke Hamamoto
Senior
Kanazawa Institute of Technology
Hakusan, Ishikawa, Japan

Masatoshi Inaba
Engineer
Subaru Corporation
Ota, Gunma, Japan

Yasutada Tanabe
Associate Senior Researcher
Japan Aerospace Exploration Agency
Mitaka, Tokyo, Japan

ABSTRACT

Compound helicopters have been developed to increase the flight speed of the conventional helicopters. In this paper, hovering experiments are performed using a compound helicopter model to elucidate the rotor-wing aerodynamic interaction. The rotor with wings has larger thrust coefficient and total torque coefficient than the rotor without wings. The figure of merit of the rotor with wings is larger compared to the rotor without wings. In addition, the figure of merit of the rotor with wings increases as the distance between rotor and wing decreases. The ratio of wing download to rotor thrust is increased as the distance between rotor and wings is decreased.

NOTATION

C_Q : Coefficient of torque

C_T : Coefficient of thrust

h : Distance between rotor and wing, m

FM : Figure of merit

Q : Torque of rotor, Nm

R : Rotor radius, m

S : Rotor disc area, m^2

T : Rotor thrust, N

σ : Solidity of rotor

Ω : Rotational speed of rotor, rad/s

INTRODUCTION

The conventional helicopter is inferior to the fixed wing aircraft because its lower flight speed and shorter cruising range that is discussed in Ref. 1. Though tilt-rotor aircraft such as the V-22 Osprey have been successfully developed, the mechanism tends to be complicated and there are difficulties in the flight controls especially during the transition between the helicopter and fixed wing modes that is discussed in Ref. 2.

On the other hand, compound helicopters are proposed as helicopters with wings and propellers. A flight test of a compound helicopter model was performed by Japan Aerospace Exploration Agency (JAXA) (Ref. 3). It was found that the compound helicopter was more stable than a conventional helicopter in forward flight. However, aerodynamic interaction between rotor and wings has not been sufficiently studied especially for the compound helicopter configurations and only a few references can be found (Ref. 4 - 6).

In this paper, hovering experiments is carried out using a 1/14 scaled compound helicopter model to elucidate aerodynamic interaction between the rotor and the wings in hover. The rotor performance is investigated by means of force measurements with six-component balances in three positions; i.e. the rotor and both side wings at various distance between the rotor and the wings..

EXPERIMENTAL DEVICES AND METHOD

Rotor model

The baseline experimental helicopter model is a radio controlled helicopter "Innovator MD530" manufactured by Taya Engineering Co., Ltd. This model is modified for present study. Although the model has a tail boom, the tail rotor and tail stabilizers are removed as shown in Figure 1. Main design parameters of the experimental model are shown in Table 1.

Presented at the 6th Asian/Australian Rotorcraft Forum / Heli Japan 2017, Kanazawa, Japan, November 7-9, 2017.

Table 1. Design parameters of Rotor Model.

Rotor radius [m]	R	0.356
Blade chord [m]	c	0.032
Blade material	-	CFRP
Blade airfoil	-	NACA0012
Number of blades	b	2
Solidity of rotor	σ	0.0572
Model length [m]	-	0.53
Model width [m]	-	0.12
Model height [m]	-	0.21
Distance between Rotor and Sensor	-	0.18
Rotational direction (Bird's-eye view)	-	Clockwise

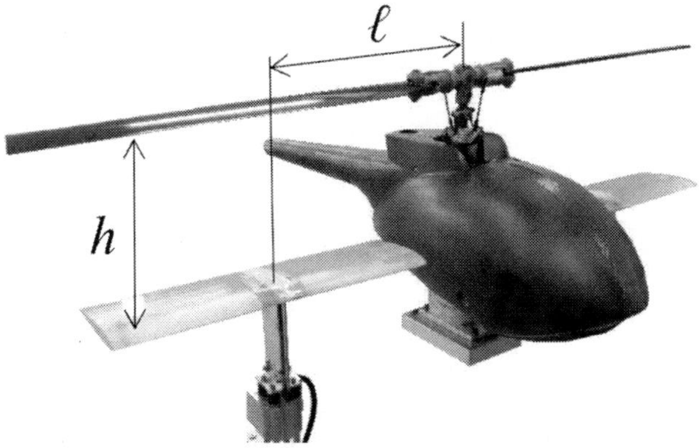

Figure 1. Compound helicopter model.

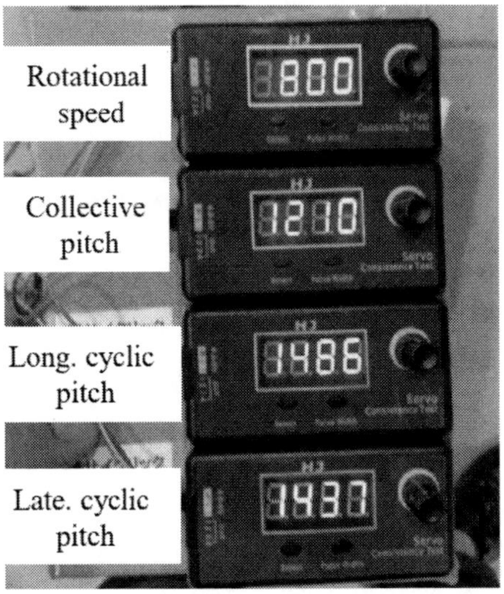

Figure 2. Servo controller for rotor model operation

The rotor model is operated by a servo controller as shown in Figure 2. The collective pitch, longitudinal cyclic pitch, lateral cyclic pitch, and rotational speed of rotor can be controlled independently.

Wing model

The airfoil of wings are NACA2412, and the material of the wings is light curing acrylic resin as shown in Figure 3. The design parameters of the wings are summarized in Table 2.

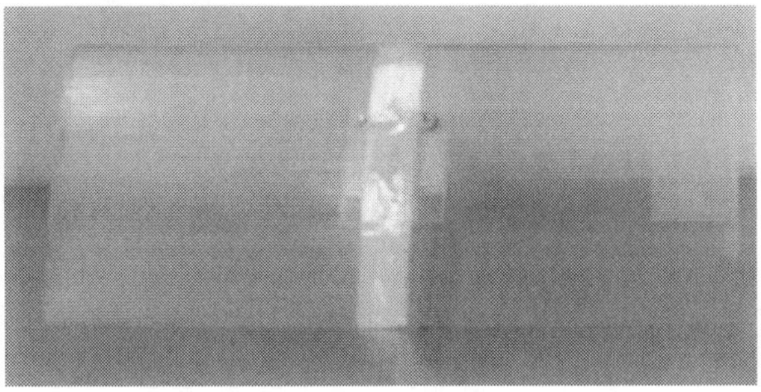

Figure 3. One Wing overview (top view).

Table 2. Design parameters of the wings.

Span [m]	-	0.24 (One Wing)
Span [m]	b_w	0.63 (added to model)
Chord [m]	c_w	0.1
Aspect Ratio	AR	6.3
Airfoil	-	NACA2412
Material	-	Light curing acrylic resin

(a) h/R=0.197

(b) h/R=0.281

(c) h/R=0.365

Figure 4. Mounting position of wing.

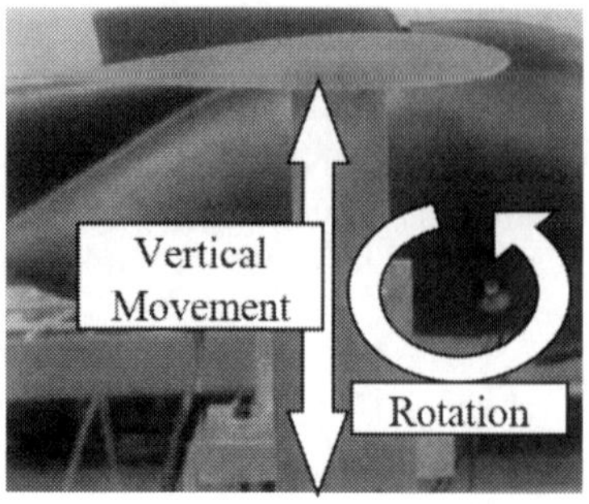

Figure 5. Wing rotational and vertical movement point

The height of the wings can be adjusted to such as h/R = 0.197, 0.281, 0.365 as shown in Figure 4. Wings can rotate around a fixing bolt and its angle can be adjusted as shown in Figure 5

Measuring Equipment

The forces and moments of the experimental model are measured by six-component balances of "CFS034CA301U", "CFS034CA101U" and designated data logger of "LGR101U" manufactured by Leptrino Co., Ltd. as shown in Figure 6.

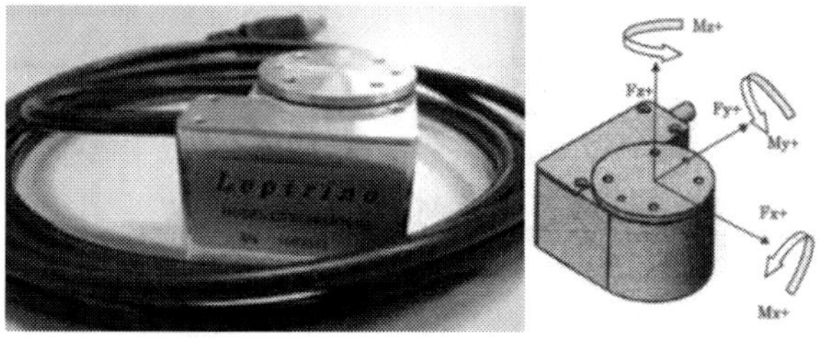

Figure 6. Six-component balance.

Experimantal method

The forces and moments $(F_X,F_Y,F_Z,M_X,M_Y,M_Z)$ of the fuselage model including the rotor are measured by a six-component balance with vibration-proof rubbers underneath the model. Also, both side wings are connected to another two six-component balances with aluminum plates and vibration-proof rubbers. The rotational speed of the rotor is measured by a magnetic tachometer.

The experimental conditions are listed in Table 3. The rotational speed of rotor is 1800 rpm.

Table 3. Experimental conditions

Rotational speed of rotor [rpm]	Ω	1800
Rotor blade pitch angle [deg]	θ	8
Wing incident angle [deg]	i	4
Distance between rotor and wing [m]	h	0.07, 0.1, 0.13
Distance between rotor and wing / R	h/R	0.197, 0.281, 0.365

The longitudinal cyclic pitch and lateral cyclic pitch are controlled to adjust the rolling moment and pitching moment of the rotor to zero.

The signs of the side force and pitching moment of the six-component balances are inverted according to the definition of coordinate axes as shown in Figure 7. Because the axis-X is facing to front and axis-Z is upward, axis-Y is facing to left.

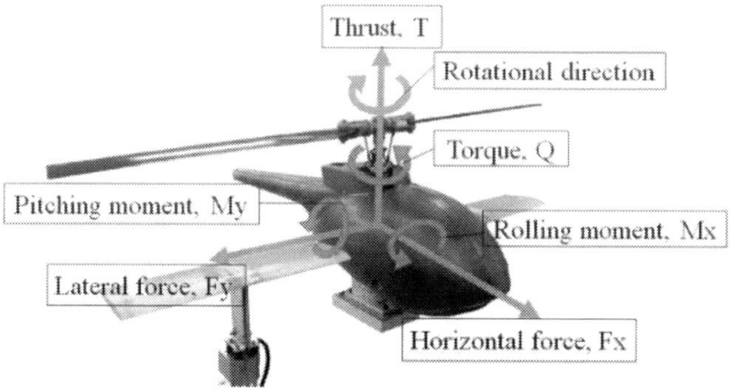

Figure 7. Definition of coordinate axes

During the experiments, the distance between rotor and wing h [m] is changed. Nominal rotational speed of the rotor is 1800 [rpm], with nominal blade pitch angle of 8.0 [deg]. In measurement, sampling rate is 1200 [Hz], and each sampling time is 6.0 [sec].

Equations

Forces (horizontal firce F_X [N], side force F_Y[N], rotor thrust T [N] or wing down load $-F_Z$ [N]) and moments (rotor and wing rolling moment M_X [Nm] and wing pitching moment M_Y [Nm], rotor torque Q [Nm]) are non-dimensionalized as follows: where the air density ρ [kg/m^3], rotor disc area S [m^2], rotor radius R [m], rotor solidity, σ, and rotational speed of rotor Ω [rad/s] are used.

$$C_{F_X}/\sigma = \frac{F_X}{\rho S(R\Omega)^2\sigma}$$

$$C_{F_Y}/\sigma = \frac{F_Y}{\rho S(R\Omega)^2\sigma}$$

$$C_T/\sigma = \frac{T\ or\ F_Z}{\rho S(R\Omega)^2\sigma}$$

$$C_{M_X}/\sigma = \frac{M_X}{\rho S(R\Omega)^2 R\sigma}$$

$$C_{M_Y}/\sigma = \frac{M_Y}{\rho S(R\Omega)^2 R\sigma}$$

$$C_Q/\sigma = \frac{M_z}{\rho S(R\Omega)^2 R\sigma}$$

The figure of merit, *FM* as a measure of efficiency of rotor in hover is calculated as follows with coefficient of thrust C_T and coefficient of torque C_Q.

$$FM = \frac{1/\sqrt{2}\, C_T^{3/2}}{C_Q}$$

RESULTS AND DISCUSSION

Figure 8(a) shows the distribution of thrust coefficients of rotor with wings, total of rotor and wings, right wing, left wing compared to the Baseline (without wings).

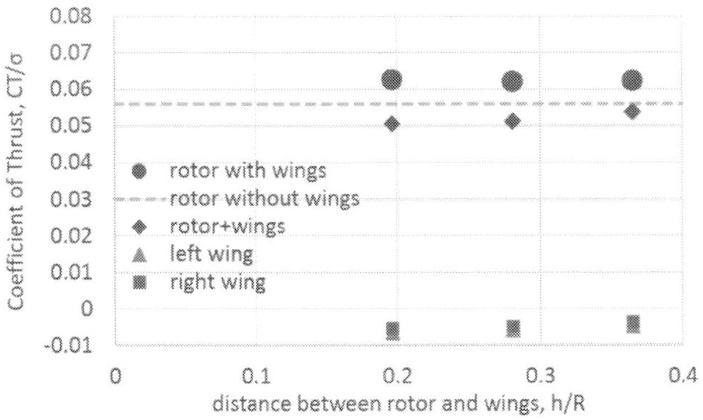

(a) Coefficient of thrust.

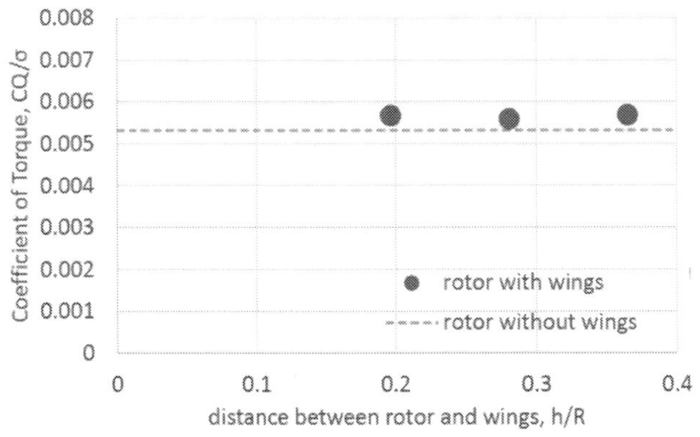

(b) Coefficient of torque.

Figure 8. Characteristics of thrust and torque for distance between rotor and wings.

It is found that the thrust coefficient of rotor with wings is larger than Baseline (rotor without wings). However, the thrust coefficient of the total of rotor and wings is lower than Baseline.

The thrust coefficient of rotor with wings increased as the distance between rotor and wings is decreased. However, the thrust coefficient of total of rotor and wings is decreased as the distance between rotor and wings is decreased. The thrust coefficient of left wing is slightly larger than that of right wing. They decreased as the distance between rotor and wings is increased.

Figure 8(b) shows the torque coefficient of rotor with wings is larger than Baseline (without wings). It is considered that the torque is increased, because the thrust is increased

Characteristics of horizontal and side force coefficient with distance between rotor and wing

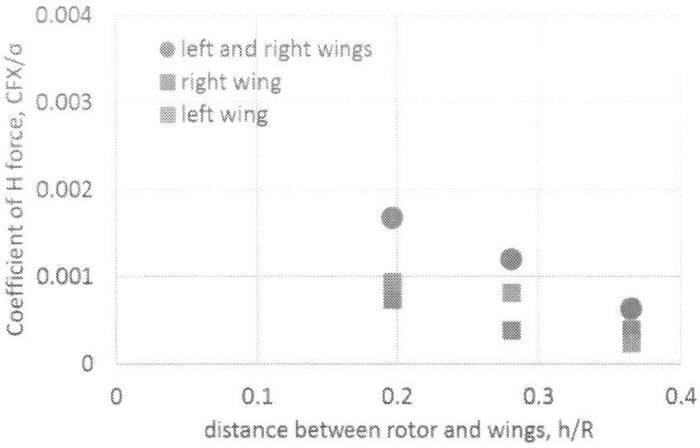

(a) Coefficient of horizontal force.

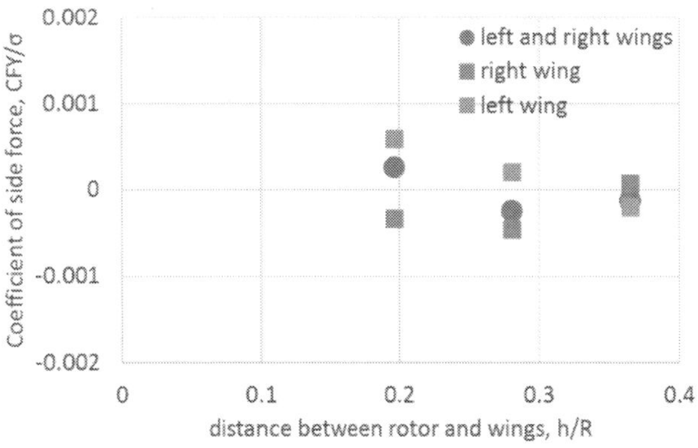

(b) Coefficient of side force.

Figure 9. Characteristics of horizontal force and side force for distance between rotor and wings.

The coefficients of horizontal force and side force are shown in Figure 9. It is found that the horizontal force is generated as the distance between rotor and wings is close. The coefficient of side force of left and right wings is increased toward the fuselage respectively as the distance between rotor and wings is decreased. Because the flow is toward fuselage on the upper surface of wing that is shown by visualized test with tuft (Ref. 5), the side force is generated.

Figure of Merit

Figure 10 shows the figure of merit of rotor with wings is larger than Baseline (without wings). The ground effect of rotor downwash is generated by wing. It is found that the figure of merit is increased as the distance between rotor and wings is decreased.

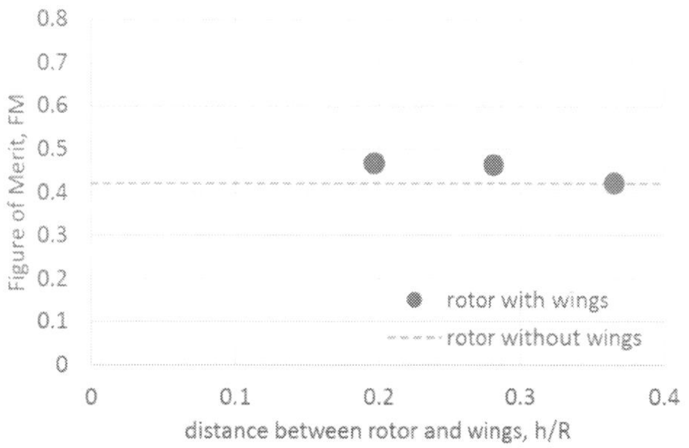

Figure 10. Comparison figure of merit of rotor with wings and rotor without wings.

Wing Download

Figure 11 shows the wing download is increased as the distance between rotor and wings is decreased. It is considered that the wings is affected by rotor downwash as the distance between rotor and wings is close. The wing download – rotor thrust ratio is about 0.2 as the distance between rotor and wings is decreased.

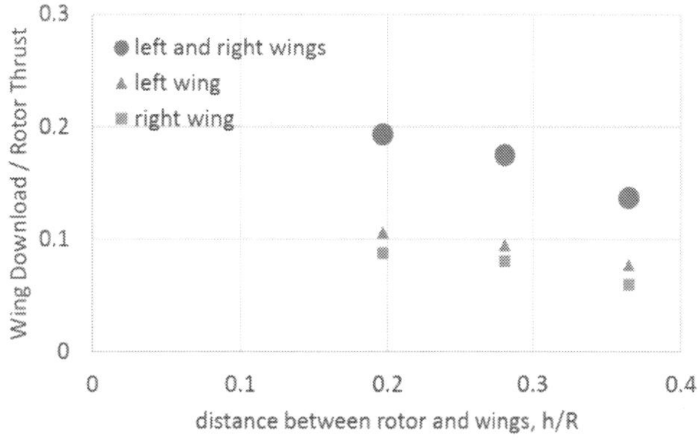

Figure 11. Wing Download / Rotor Thrust for distance between rotor and wings

CONCLUSIONS

We conclude as follows.

1. The thrust and torque of the rotor with wings is larger than the rotor without wings. However, the thrust of the total of rotor with wings and wings is lower than the rotor without wings.

2. The horizontal force of the wings is generated as the distance between rotor and wings is close. The side force of left and right wings is generated toward the fuselage respectively as the distance between rotor and wings is close.

3. The figure of merit of the rotor with wings is larger than the rotor without wings. The figure of merit of the rotor with wings is increases as the distance between the rotor and the wings decreases.

4. The wing download – rotor thrust ratio is increased as the distance between rotor and wings is decreased.

Author contact:
Keiichiro Makino b1411700@planet.kanazawa-it.ac.jp,
Takeshi Akasaka akasaka@neptune.kanazawa-it.ac.jp,
Yusuke Hamamoto b1412022@planet.kanazawa-it.ac.jp,
Masatoshi Inaba inaba.masatoshi@subaru.co.jp,
Yasutada Tanabe tan@chofu.jaxa.jp.

REFERENCES

[1]Yeo, H., and Johnson, W., "Optimum Design of a Compound Helicopter", the Heli Japan 2006 AHS International Meeting on Advanced Rotorcraft Technology and Lift Saving Activities, Nagoya; Japan, Nov. 17, 2006.

[2]Aoki, K., "Thorough verification V-22 Osprey", Softbank creative, Oct. 2012, Chap. 6 Safety of the Osprey, pp-142-166.

[3]Sugiura, M., Tanabe, Y., Kobiki, N., Aoyama, T., "First Report of High Speed Model Helicopter Flight Test", 47th JSASS Annual Meeting, Apr. 2016.

[4]Felker, F. F. and Light, J. S., "Aerodynamic Interaction between a Rotor and Wing in Hover", Journal of the American Helicopter Society, April, 1988.

[5]Inaba, M., Akasaka, T., Furuta, T., Kobiki, N., Tanabe, Y., Aoyama, T., Sugiura, M., "Measurement of the Rotor-Wing Aerodynamic Interference on a Compound Helicopter Model", 54th Aircraft Symposium, JSASS, Oct. 2016.

[6]Tanabe, Y., Sugawara, H., "Numerical Simulation of Aerodynamic Interaction between a Rotor and a Wing", Asia-Australian Rotorcraft Forum, Nov. 2016.

Experimental Investigation of the Rotor-Wing Aerodynamic Interaction on a Compound Helicopter in High Advance Ratio

Yusuke Hamamoto
4th grade Student
Kanazawa Institute of Technology
Nonoich, Ishikawa, Japan

Takeshi Akasaka
Assistant Professor
Kanazawa Institute of Technology
Hakusan, Ishikawa, Japan

Yasutada Tanabe
Associate Senior Researcher
Japan Aerospace Exploration Agency
Mitaka, Tokyo, Japan

ABSTRACT

Compound helicopter based on a single main rotor need to add a wing under the main rotor to generate lift at high speed. An investigation of the rotor/wing aerodynamic interaction in forward flight for advance ratio of 0.7 with various distances between the rotor and the wing in a wind tunnel is described. The results show that decreasing the distance between the rotor and wing decreases the lift coefficient of the wing and increases the drag coefficient and rolling moment coefficient of the wing. Also Y-force of the wing has been observed.

NOTATION

Symbol	Definition	Units
AR	Aspect ratio	[-]
b	Number of blades	[-]
b_w	Wing span	[m]
c_b	Blade chord length	[m]
c_w	Wing chord length	[m]
C_D	Drag coefficient of the wing	[-]
C_L	Lift coefficient of the wing	[-]
C_{Rw}	Rolling moment coefficient of the wing	[-]
C_{Yw}	Y-force coefficient of the wing	[-]
R	Rotor radius	[m]
S_w	Wing area	[m^2]
V	Flow velocity	[m/s]
z	Distance between the rotor and wing	[m]
α	Angle of attack of wing	[deg]
θ_0	Collective pitch	[deg]
θ_{1c}	Longitudinal cyclic pitch	[deg]
μ	Advance ration	[-]
ρ	Air density	[kg/m^3]
σ	Rotor solidity	[-]
Ω	Rotor angular velocity	[rad/s]

INTRODUCTION

Conventional helicopters are operated for disaster-relief activities and emergency rescues because they possess VTOL and hovering performance. However, the conventional helicopter is inferior to fixed wing aircraft in flight speed. Therefore, JAXA proposed compound helicopter of single rotor type (Ref.1) to achieve high flight speed equivalent to the propeller driven fixed wing aircraft.

Compound helicopter is attached with a wing under the main rotor. The rotor downwash will interact with the wing. Felker has carried out a series of testing using simplified model consisting of an isolated rotor and an isolated wing. It reported the relationship of the aerodynamic wing download with flap angle of wing and with the distance between rotor and wing (Ref.2). Also, Inaba indicated that the figure of merit of the rotor is varied by wing chord length and distance between rotor and wing (Ref.3).

There are several researches focusing on the interaction between rotor and wing during hovering. However, there are no researches for high speed flight being reported yet. The high speed compound helicopter can be designed to fly in high advance ratio of 0.6 to 0.7 though advance ratio of the conventional helicopters is 0.3 to 0.4 at highest. Quackenbush indicated that thrust coefficient and H-force coefficient is influenced by the advance ratio through wind tunnel testing (Ref.4). However, rotor/wing aerodynamic interaction was not investigated in their studies.

This paper describes an experimental investigation of rotor / wing aerodynamic interactions in high advance ratio. This test was designed to evaluate the effect of changes in the distance between rotor and wing on the aerodynamic character in advance ratio of 0.7.

EXPERIMENTAL DEVICES AND METHOD

Experimental devices

The rotor used in this experiment is modified from a RC model helicopter Horizon Hobby's Blade 180 CFX. The tail boom and the tail rotor are removed. And the swashplate and throttle is controlled as shown in Figure 1. Command values are sent from the PC, three servomotors are driven through Arduino UNO and the swash plate are moved. The rotation speed of the rotor is controlled by the servo controller.

Presented at the 6th Asian/Australian Rotorcraft Forum & Heli Japan 2017, Kanazawa, Japan, November 7-9, 2017.

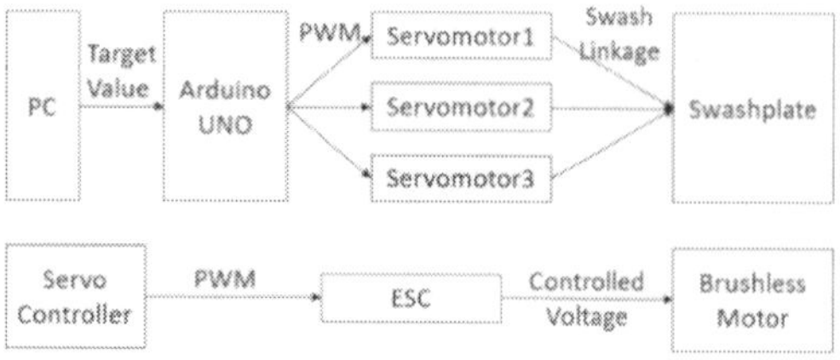

Figure 1. System of swash plate and rotor rotation control.

The design parameters of the rotor are shown in Table 1. And appearance of the rotor is shown in Figure 2.

Table 1. Design parameter of rotor.

Blade Chord[m]	c_b	0.0195
Rotor Radius[m]	R	0.180
Blade Airfoil		NACA0012
Number of Blades	b	2
Solidity of Rotor	σ	0.0690
Angular Velocity[rad/s]	Ω	278
Rotational Direction		CounterClockWise

Figure 2. Rotor overview.

The fixed wing is produced using a Stratasys 3D printer at Kanazawa Institute of Technology. The material of the wing is photocurable resin VeroClear RGD810. The design parameters of the wing is shown in Table 2 and the appearance of the wing is shown in Figure 3. As shown in Table 1 and Table 2, the wing span is 0.7 diameter of the rotor.

Table 2. Design parameters of the wing.

Wing Span[m]	b_w	0.252 (=1.4R)
Wing Chord[m]	c_w	0.042
Wing Airfoil		NACA4412
Aspect Ratio	AR	6.00

Figure 3. Wing overview.

The wind tunnel is shown in Figure 4. The wind tunnel is a small low-speed wind tunnel in Kanazawa Institute of Technology. The experimental section is 0.5 m × 0.5 m and the maximum flow velocity of the wind tunnel is 35 m/s. In addition, a manual triaxial traverse device is mounted on the wind tunnel as shown in Figure 5.

Figure 4. Wind tunnel overview.

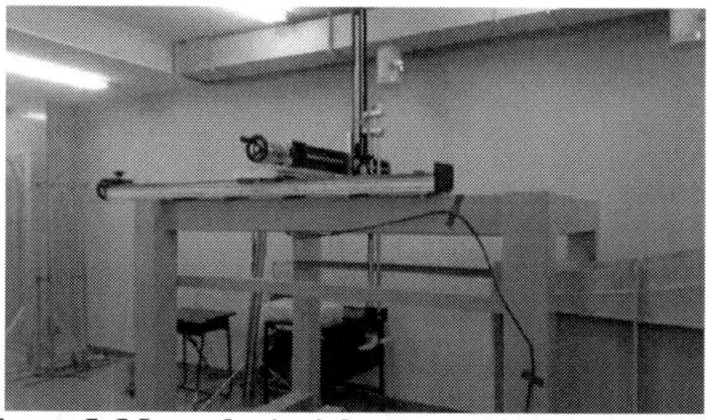

Figure 5. Manual triaxial traverse device overview.

The force and moments of the rotor and wing are measured using balances. The balance for the rotor is "CFS034CA101U" as shown in Figure 6, the balance for the wing is "SFS080F200M1R0U6" as shown in Figure 7 and they are manufactured by Leptrino Co.,Ltd.

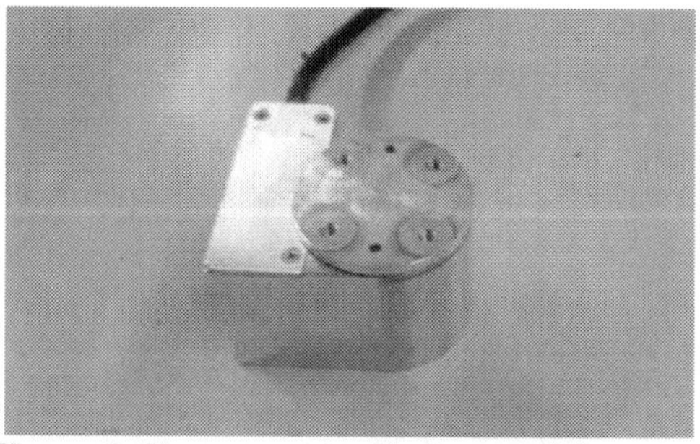

Figure 6. Six-component balance of the rotor.

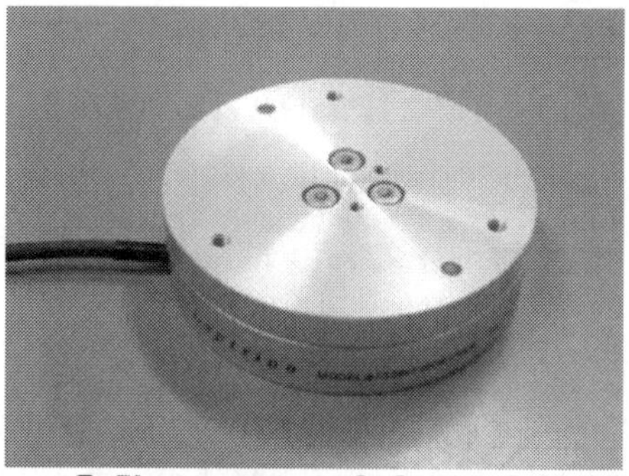

Figure 7. Six-component balance of the wing.

The wing is supported through a strut and attached to a six-component balance. The appearance of this strut and the cover are shown in Figure 8.

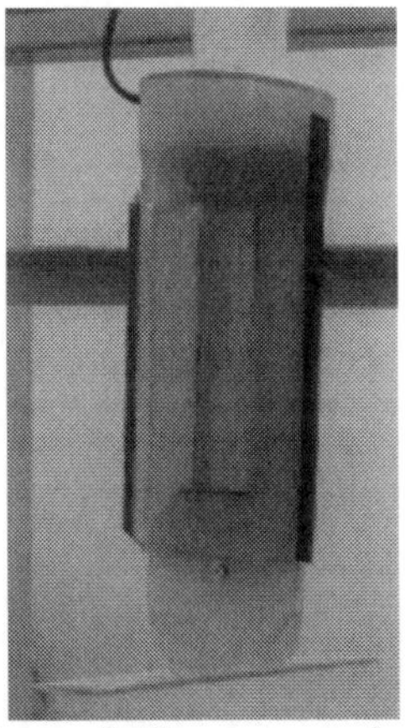

Figure 8. Cover overview.

Experimental Method

The advance ratio is obtained by non-dimensionalizing the flow velocity with the blade tip speed. In this experiment, the rotor shaft angle of attack is 0 deg. The rotor angular velocity is 278 rad/s to produce advance ratio of 0.7. The experimental overview is shown in Figure 9. The wing is attached to the manual triaxial traverse device equipped with the wind tunnel in order of jigs, balance, strut and wing. The balance and the strut is covered with a cover. Also, upper surface of the wing is toward the ground direction. The angle of attack of the wing is set to 2 deg at this study.

The rotor is attached to the wind tunnel in order of jigs, balance and rotor. The rotor downwash is upward. The collective pitch is -5 deg. In order to simulate actual flight condition, the experiment is desirable to perform condition that the total lift of the rotor and the wing are constant. However, the pitch angle of the rotor is constant this time because this experiment is the initial test. The longitudinal cyclic pitch is 5.19 deg to trim the tip pass plane parallel to uniform flow in flow velocity of 35 m/s. The lateral cyclic pitch is 0 deg because it is considered that the inclination of the tip pass plane to lateral distance is small.

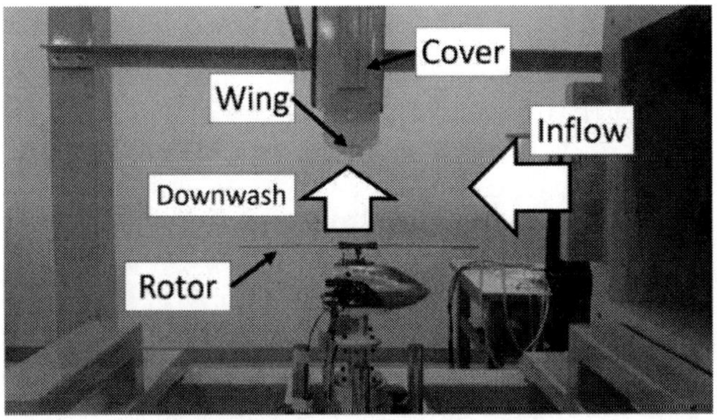

Figure 9. Experiment overview.

The flow velocity is 35 m/s and distance between the rotor and wing is 0.2 to 0.6 radius of the rotor. The experimental condition of the distance between the rotor and the wing is shown in Table 3.

Table 3. Experimental condition.

Distance between Rotor and Wing, z[m]	z/R[-]
0.036	0.20
0.054	0.30
0.072	0.40
0.090	0.50
0.108	0.60

The coordinate axis of the wing is defined as shown in Figure 10 and the coordinate of rotor is defined as shown in Figure 11.

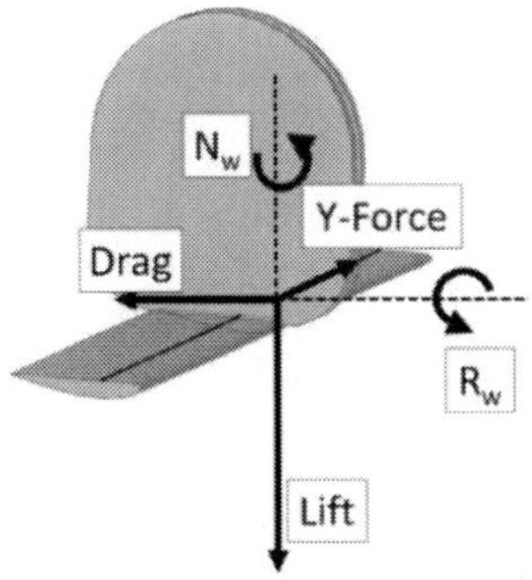

Figure 10. Coordinate axes of the wing.

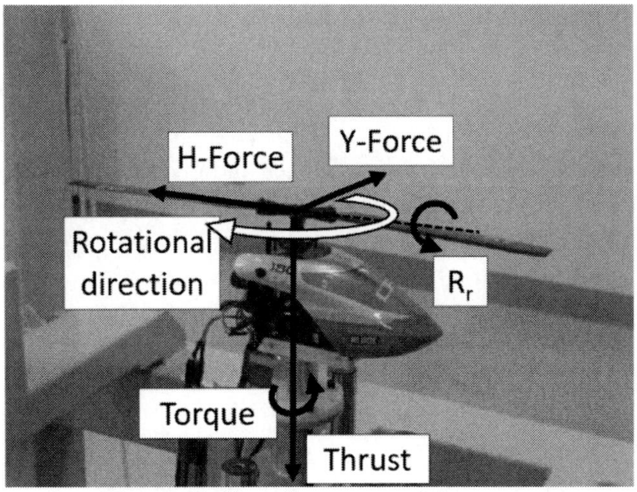

Figure 11. Coordinate axes of the rotor.

The lift, drag and Y-force of the wing are non-dimensionalized by the following equation. Where the lift, drag and Y-force are F_w.

$$C_{F_w} = \frac{F_w}{0.5\rho V^2 S_w}$$

The rolling moment of the wing is non-dimensionalized by the following equation.

$$C_{N_w} = \frac{N_w}{0.5\rho V^2 S_w c_w}$$

RESULTS AND DISCUSSION

Figure 12 shows the reverse flow region in advance ration of 0.7. It can be seen that the left side of the wing is under the reverse flow region of the rotor.

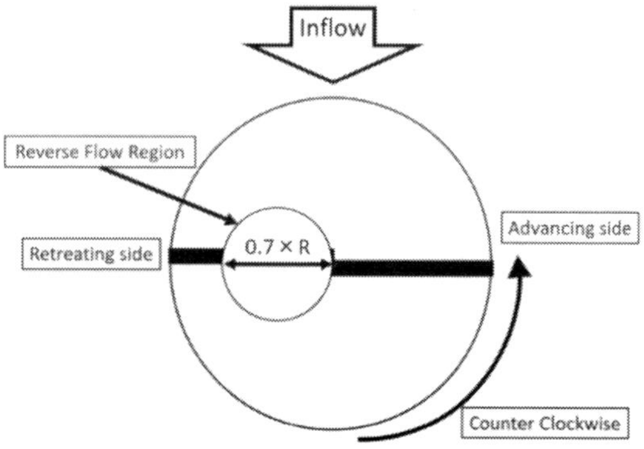

Figure 12. Reverse flow region overview (μ = 0.7).

Figures 13, 14 and 15 show the lift coefficient, the drag coefficient and the rolling moment coefficient of the wing respectively in the interaction test that vary with the distance between the rotor and wing. In Figures 13 and 14, decreasing the distance between the rotor and wing decreases the lift coefficient of the wing and increases the drag coefficient of the wing. Figure 15 shows that decreasing the distance between the rotor and wing increases rolling moment coefficient of wing. Increasing the rolling moment coefficient of the wing means to increase rolling of the wing to the right direction. The presence of the downwash of the rotor cause the angle of attack of the wing decreases and the flow velocity around the wing increases. Also, the downwash of the rotor is considered to decreases with decreasing distance between the rotor and the wing. The downwash of the rotor is deflected by the uniform flow of advance ratio of 0.7.

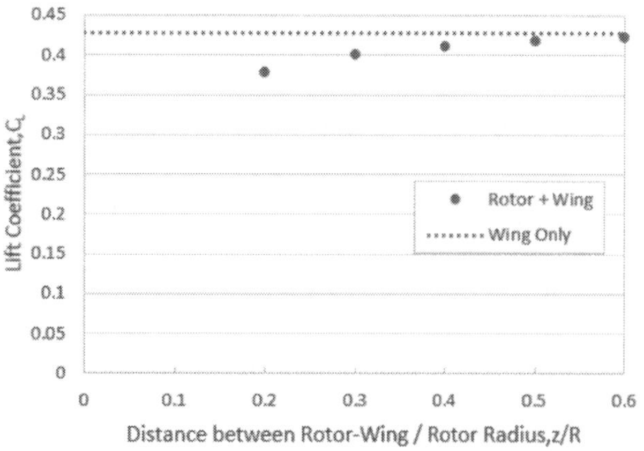

Figure 13. Lift coefficient of the wing in aerodynamic interaction.

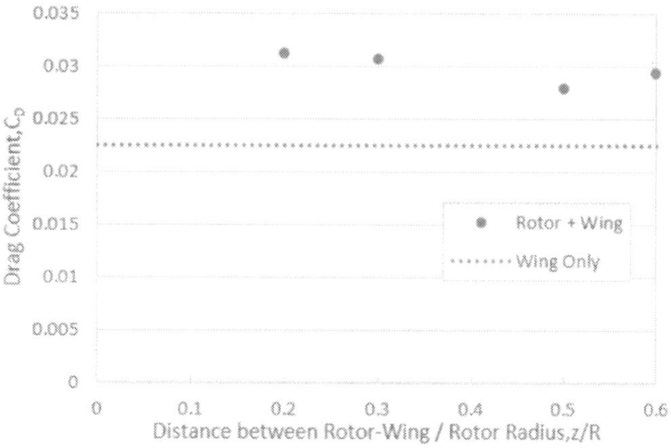

Figure 14. Drag coefficient of the wing in aerodynamic interaction.

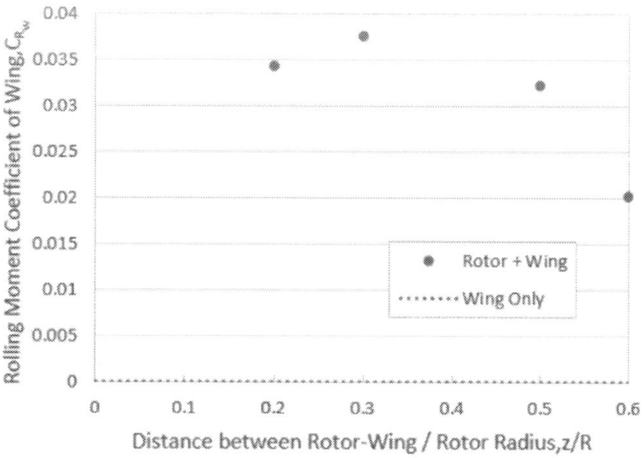

Figure 15. Rolling moment coefficient of the wing in aerodynamic interaction.

Figure 16 shows that the Y-force coefficient of the wing in the interaction test varies with the distance between the rotor and wing. The Y-force coefficient of wing in aerodynamic interaction is observed. The Y-force coefficient of the wing in aerodynamic interaction takes negative values. Also, Reference 5 shown that the rotor downwash on the upper surface of the wing under the rotor in hover flow to the fuselage as shown in Figure 17. Therefore, we consider that the rotor downwash of only the advancing side blade causes side slip for the wing in aerodynamic interaction to right.

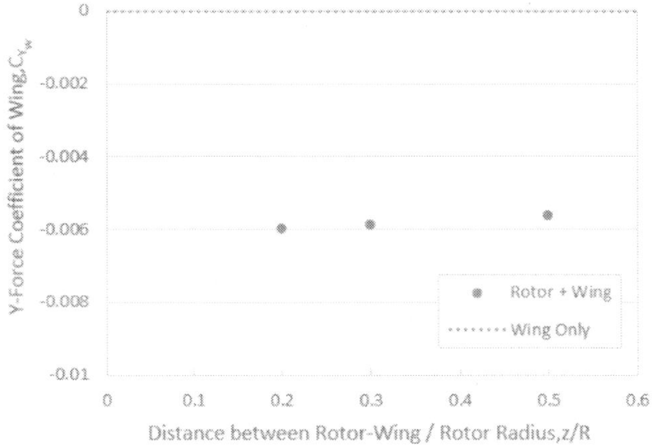

Figure 16. Y-force coefficient of the wing in aerodynamic interaction.

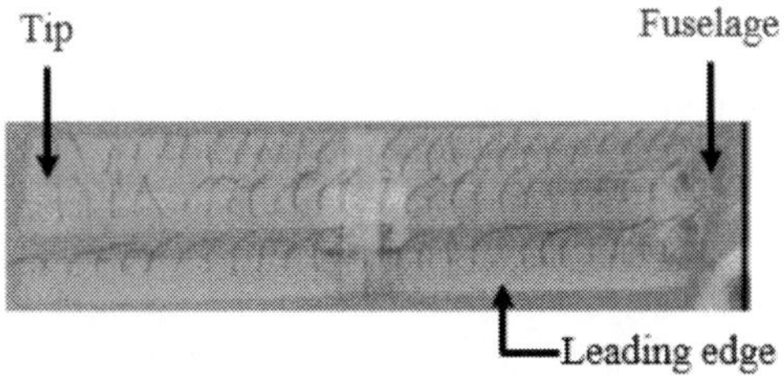

(a) Under retreating side flow (Author retouched).

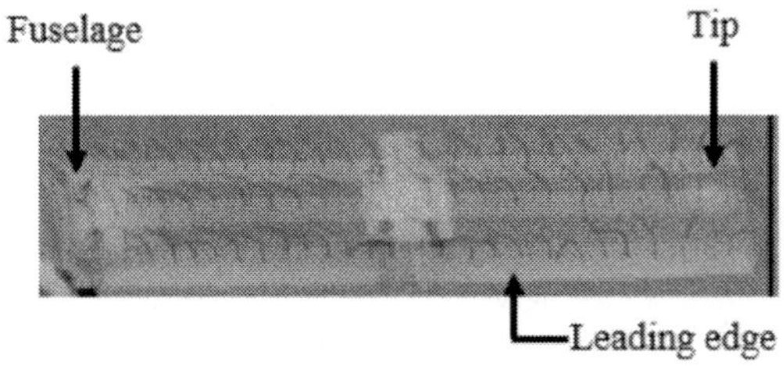

(b) Under advancing side flow (Author retouched).

Figure 17. Visualized flow of the wing in hover (Ref.5).

CONCLUSIONS

In this wind tunnel experiment at advance ratio of 0.7, the following results are obtained by varying the distance between the rotor and wing in aerodynamic interaction.

1. The lift coefficient of the wing decreases with decreasing distance between the rotor and the wing.

2. The drag coefficient of the wing increases with decreasing distance between the rotor and the wing.

3. The rolling moment coefficient of the wing increases with decreasing distance between the rotor and wing.

4. The Y-force of the wing is observed because of the aerodynamic interaction.

As future works, following issues will be considered.

1. Aerodynamic characteristics of the rotor at high advance ratio of 0.7.

2. In order to simulate actual flight conditions, experiments with the total lift of rotor and wing being constant.

3. Clarification of the mechanism that wing performance at high advance ratio decrease with decreasing distance between rotor and wing by visualizing flow between rotor and wing.

Yusuke Hamamoto b1412022@planet.kanazawa-it.ac.jp
Takeshi Akasaka akasaka@neptune.kanazawa-it.ac.jp
Yasutada Tanabe tan@chofu.jaxa.jp

REFERENCES

[1]Tanabe, Y., Kobiki, N., Sugiura, M., Aoyama, T., Sugawara, H., "Design of a Scale-Down Model of High Speed Compound Helicopter", The 54th Aircraft Symposium, Japan, Toyama, Oct 2016.

[2]Light, F. F., and Jeffrey, S. F., "Rotor/Wing Aerodynamic Interaction in Hover," NACA TN88255, 1986.

[3]Inaba, M.., Akasaka, T., Furuta, T., Kobiki, N., Tanabe, Y., "Experimental Investigation of Rotor-Wing Aerodynamic Interaction on a Compound Helicopter in Hover", 5th Asian-Australian Rotorcraft Forum, Japan, Toyama, Nov 2016.

[4]Quackenbush, T. R., Wachspress, D. A., McKillp, Jr, R. M., Sibilia, M. J.,"Aerodynamic studies of High Advance Ratio Rotor System", American Helicopter Society 67th Annual Forum, Virginia, Virginia Beach, May 2011.

[5]Inaba, M.., Akasaka, T., Furuta, T., Kobiki, N., Tanabe, Y., Aoyama, T., Sugiura, M., "Measurement of the Rotor-Wing Aerodynamic Interference on a Compound Helicopter Model", The 54th Aircraft Symposium, Toyama, Oct 2016.

From ERATO Basic Research to the Blue Edge™ Rotor Blade

Berend G. van der Wall
Senior Scientist
DLR, Germany

Christoph Kessler
Rotorcraft Branch Head
DLR, Germany

Yves Delrieux
Research Engineer
ONERA, France

Philippe Beaumier
Deputy Director
Applied Aerod. Dept.
ONERA, France

Marc Gervais
Head Aerodynamics
& Performance
Airbus Helicopters, France

Jean-François Hirsch
Overall Blade Design
Airbus Helicopters, France

Kurt Pengel
Project Engineer
DNW, Netherlands

Pascal Crozier
Project Engineer
ONERA, France

Abstract

In 2015, Airbus Helicopters unveiled the secrecy around its Dauphin successor and presented the all new H160 helicopter. A special feature immediately attracting attention is its unusual and revolutionary fore-aft swept main rotor blade. This design, aiming at significantly reducing the blade-vortex interaction noise signature and also reducing fast forward flight power requirements, finds its origins far back in the 1990s, when DLR and ONERA formed a joint team to acoustically optimize a rectangular reference rotor blade. Based on state-of-the-art comprehensive rotor codes and a 50-50 work share, the ERATO rotor blade design was developed, patented worldwide and tested on a rotor test rig and in the wind tunnel. Airbus Helicopters (then: Eurocopter) took up that design, optimized hover and forward flight high lift performance and prepared it for serial production, until it finally made it as the Blue Edge™ rotor blade on the Airbus Helicopters H160 helicopter. The paper covers the history and technical achievements, wind tunnel test results and flight tests.

NOMENCLATURE

a_∞	Speed of sound, m/s
bpf	Blade passage frequency, multiples of N_b/rev
c	Airfoil chord, m
C_n	Normal force coeff., $C_n = 2L'/(\rho_\infty c V^2)$
C_p	Pressure coefficient, $C_p = 2(p - p_\infty)/(\rho_\infty V^2)$
C_P	Power coefficient, $C_P = P/(\rho_\infty \pi R^2 (\Omega R)^3)$
C_P^*	Power coefficient, based on Ω^*
C_T	Thrust coefficient, $C_T = T/(\rho_\infty \pi R^2 (\Omega R)^2)$
C_W	Weight coefficient, $C_W = W/(\rho_\infty \pi R^2 (\Omega R)^2)$
C_W^*	Weight coefficient, based on Ω^*
L'	Section lift per unit span, N/m
M	Mach number, $M = V/a_\infty$
M_h	Hover tip Mach number, $M_h = \Omega R/a_\infty$
N_b	Number of blades
p_∞	Steady pressure
P	Power, kW
P_b^*	Power coeff., based on Ω^*, $P_b^* = 200\, C_P^*/\sigma$
R	Rotor radius, m
T	Rotor thrust, N
V	Relative velocity at a blade element, m/s
V_∞	Wind speed, true air speed, m/s
W	Virtual model weight force, N
α_S	Shaft angle of attack, deg
γ	Flight path angle, deg
Γ_b	Blade element circulation, $\Gamma_b = L'/(\rho V)$, m²/s
μ	Advance ratio, $\mu = V_\infty \cos\alpha/(\Omega R)$
μ^*	Advance ratio, based on Ω^*
ρ_∞	Air density, kg/m³
σ	Rotor solidity, $\sigma = N_b c/(\pi R)$
ψ	Rotor blade azimuth angle, deg
Ω	Rotor rotational speed, rad/s
Ω^*	Reference rotational speed, 100 rad/s
AH	Airbus Helicopters
AHS	American Helicopter Society International
BVI	Blade Vortex Interaction
DGAC	Direction Générale de l'Aviation Civile
DLR	Deutsches Zentrum für Luft- und Raumfahrt e.V. (German Aerospace Center)
DNW	Deutsch-Niederländische Windkanäle (German Dutch Wind Tunnels)
EASA	European Aviation Safety Agency

Originally presented at the AHS 72nd Annual Forum, West Palm Beach, Florida, May 17–19, 2016.

ERATO Etude d´un Rotor Aéroacoustique Technologiquement Optimisé (Aeroacoustically optimized rotor)
EMS Emergency Medical Service
EPN Effective Perceived Noise Level
FCM Fringe Correlation Method
ICAO International Civil Aviation Organization
LLS Laser Light Sheet
MBB Messerschmitt-Bölkow-Blohm
MTOW Maximum Take-Off Weight
ONERA Office National d'Etudes et Recherches Aérospatiales (The French Aerospace Lab)
PIV Particle Image Velocimetry
SPA Strain Pattern Analysis

1. INTRODUCTION

Since the early developments of rotary wing aircraft in the late 19th century, the first flight in 1907, and a period of maturation during 1920 to 1945, helicopters have made tremendous progress in performance, handling qualities, comfort, reliability, and efficiency. Some additional features make helicopters especially useful for many missions, which to date cannot be performed by any other contemporary series production aircraft. These include their capabilities to hover, to climb or sink vertically or almost vertically, to fly slowly in any horizontal direction (even backwards), and still maintain good handling qualities and manoeuvrability. This is why helicopters have conquered their niche in the aircraft market.

These advantages, but also their relatively small outer dimensions (i.e. their footprint), allow helicopters to fly at low altitudes within an obstacle backdrop and land almost everywhere even in confined areas. Despite the undoubtedly increased maturity of helicopters, some challenges still remain like high noise levels, high vibration levels, high demands on hover figure of merit and high speed forward flight, and hence limited capabilities in terms of maximum speed and range. While most of the above mentioned drawbacks of helicopters are of concern for the crew (e.g. high vibration level) and operators (e.g. limited speed and range), this is different for the noise radiated by helicopters.

Helicopters often operate at low altitudes and close to populated areas, e.g EMS helicopters which land on or next to hospitals in cities, see [1]. This fact and the high noise level themselves have a direct impact on the annoyance of the population on the ground, [1], [2], and might hinder acceptance or even increase of helicopter operation in proximity to populated areas even for well accepted EMS missions. Mooreman has documented the impact of increasing helicopter operations and the decreasing acceptance of the public due to noise issues for large cities like the Los Angeles County [3], Long Island, and Chicago [4], but also for US national parks [5].

Therefore, low noise emissions of helicopters are mandatory [1], [6]. This need is also documented by the initiation of the Noise Initiative by AHS International[1]. The following noise sources contribute to the total helicopter noise [6]:

- main rotor:
  - thickness and loading noise,
  - blade vortex interaction (BVI),
  - high speed impulsive noise,
  - blade wake interaction,
  - trailing edge noise,
- tail rotor (same as for main rotor and in addition interaction with body and main rotor wakes),
- engines, (compressor, turbine, combustion),
- and airframe (fuselage, skids).

While BVI is of more concern during descent or maneuvering flight [7], [8], thickness and loading noise are of general importance during all flight segments. Both, BVI noise and high speed impulsive noise show an impulsive character at high levels [8]. The impulsive noise is very annoying, since it dominates the noise emission in the mid-frequency range to which human subjective response is very sensitive [7]. For BVI occurrence in descent, this is all the worse, since in this flight segment the helicopter gets closer to the ground and hence closer to the population [1].

The BVI phenomenon is caused by unsteady pressure fluctuations on the rotor blades [8], [9]. These fluctuations result from interactions of a blade with vortices generated by preceding blades. The responsible surface pressure fluctuations are concentrated mainly at the blade's leading edge [8] and at the blade tip. Experiments have shown that two BVI locations occur in descent flight condition: one on the advancing side in the first quadrant of the rotor disc, the other one on the retreating side in the fourth quadrant [7], [10].

The noise has a strong directivity characteristics, which is mostly forward and below the rotor plane. Due to this, pilots are in most cases unaware of the noise emission and the considerable community annoyance they are causing [8]. Maximum noise levels of helicopters are defined by the International Civil Aviation Organization (ICAO) in [11], Annex 16, and most of the helicopters certified by the European Aviation Safety Agency (EASA) are certified in takeoff, overflight (i.e. horizontal flight), and approach according to Chapter 8 of[11], see [12]. Allowed levels are shown in Figure 1. A summary on

---

[1] https://vtol.org/what-we-do/noise-initiative

helicopter related noise issues and their relevance to a community is given by Leverton and Pike [13].

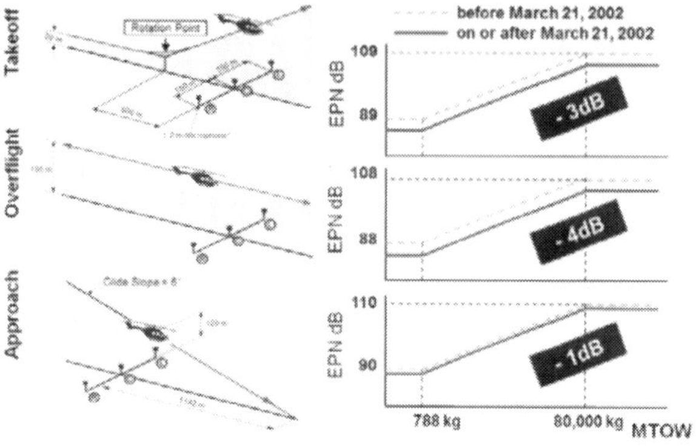

Figure 1: ICAO noise certification limits of Annex 16, Volume I, Chapter 8.

According to this, the reaction of the public to helicopters dependent on several factors, some of which are completely unrelated to the maximum helicopter noise level. The subjective character of the sound is equally or more important than the absolute noise level itself, which seems to contradict the stipulation of maximum noise levels. Improvements should not only focus on the reduction of the noise level itself, but also on reducing the impulsive character of the noise. This would lead to greater acceptance in the public.

Recommendations given in [13] include a reduction of main and tail rotor tip speeds, usage of thin blade profile sections, low noise blade tips, increased number of blades, etc. For older helicopter designs it is being recommended to fly at higher heights above the ground and use noise abatement procedures for normal operations. Also, these helicopters should not fly at very high speeds in order to minimize high-speed impulsive noise intensity and exposure time.

Two mechanisms are addressed in [13] to reduce the noise emissions of helicopters: the first one is related to passive helicopter design aspects, the second one to helicopter operations. Some of these noise related helicopter design aspects are already used at modern helicopters. How well passive improvements can reduce noise emissions is shown in Figure 2. For selected turbine powered helicopters which have been certified according to chapter 8 and which have a MTOW of 1 to 6 to this figure shows noise values in approach in comparison to the permissible noise levels. Both limits of Figure 1 have been included in the figure. In contrast to those shown in Figure 1, they are no straight lines, since a linear MTOW-axis has been used for Figure 2 to allow for a better separation of the data points. For the data points of Figure 2 it has not been distinguished whether helicopters have to fulfil the strong new limits or the somewhat relaxed older ones. This difference is for residents in the vicinity of helicopter operations actually of minor importance. Noteworthy is the No Tail Rotor System (NOTAR) design installed on the MD902 [14], but noise radiation could also be reduced by flight procedures as demonstrated on DLR's EC135 test aircraft, [15].

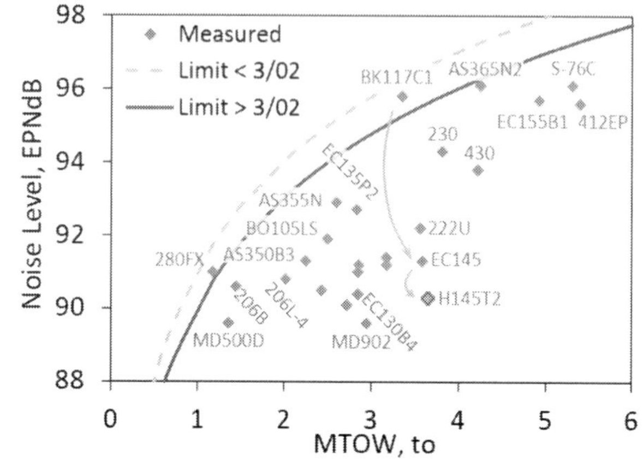

Figure 2: Measured noise levels and ICAO limits of various helicopters during approach, [12].

In addition to the above mentioned passive noise reduction, active rotor control can contribute to a significant noise reduction (5 to 6dB), but would increase system complexity of helicopters. It can additionally reduce vibrations or power consumption in high speed forward flight. A survey on various technologies has been given e.g. in [6], [16], and [17]. ONERA experience in active flap technology can be found in [18] and [19].

In the future, even stronger noise certification requirements are foreseeable. This and the in general negative aspect of noise emission by aircraft in the public both motivate research establishments and manufacturers to address further noise reduction technologies. An overview on current noise reduction activities at US and European helicopter manufacturers is given in [14]. An overview on various blade design programs with the intention to reduce BVI noise is given in [8]. However, it is being mentioned that not all projects focusing on the blade tip design were successful. Some could simply not reduce BVI noise, others generated high control loads or even showed power penalties.

How well proper blade design can reduce noise while also showing power benefits has been demonstrated in a joint ONERA-DLR research program called ERATO (Etude d´un Rotor Aéroacoustique Technologiquement Optimisé), see [20]. This research project especially has targeted the BVI noise reduction and hence has addressed a noise source with very impulsive character which has been mentioned above. The excellent results triggered a follow-on project in order to further raise the maturity level of this technology, in which Airbus Helicopters and ONERA designed and tested a full-scale variant of the ERATO concept. Airbus Helicopters has first shown this modified ERATO blade as its Blue Edge™ design at the Heli-Expo 2010 in Houston. Now, the Blue Edge™ blade has found its way to the H160 helicopter which is currently under flight test evaluation. This paper gives an overview on both the ONERA-DLR research project ERATO and its results as well as the development activities at Airbus Helicopters.

2. THE ONERA-DLR ERATO PROGRAM

In 1991 ONERA and DLR formed a joint team to design a rotor blade with reduced BVI noise signature purely by passive design, leading to the ERATO rotor blade design [21]-[27]. Quarterly meetings took place to exchange results of parameter variations and define the actions for the next steps. Industry representatives from Aerospatiale and MBB (from 1992 on as Eurocopter; renamed in 2014 to Airbus Helicopters) joined the meetings as industrial observers and contributing the design optimization with their expertise. Since the rotor was to be tested in both the German Dutch Wind tunnels Large Low speed Facility DNW-LLF on DLR's rotor test rig (December 1998) as well as in ONERA's S1MA large transonic wind tunnel (March 1998) the radius was chosen as 2.1 m and the rotational sense counter-clockwise as seen from above.

The modeling and simulation codes employed were mainly based on Blade Element Theory: the S4 isolated high resolution rotor code on DLR side, and the Airbus Helicopters R85 code further improved by ONERA, which later formed the basis of the HOST code. DLR used a prescribed wake method based on the Beddoes model while ONERA used their MESIR free-wake. The reference flight condition was a representative 6° descent condition at about 70 kts (35 m/s) flight speed and scaled-down loading. With the boundary to keep the thrust weighted solidity constant the following parameters were varied in work share with the overall goal to reduce BVI noise radiation:

- chord distribution (taper)
- quarter chord line distribution (blade sweep)
- twist distribution
- airfoil distribution
- rotor rotational speed (tip Mach number)

The starting point of the optimization process was the ONERA 7AD rotor blade geometry. The ERATO blade design freeze happened in 1996 and a Mach scaled model rotor was jointly built [21]. Figure 3 shows details of the ERATO blade design, compared to the 7AD rotor reference blade.

The design features of the ERATO aeroacoustically optimized blade are:

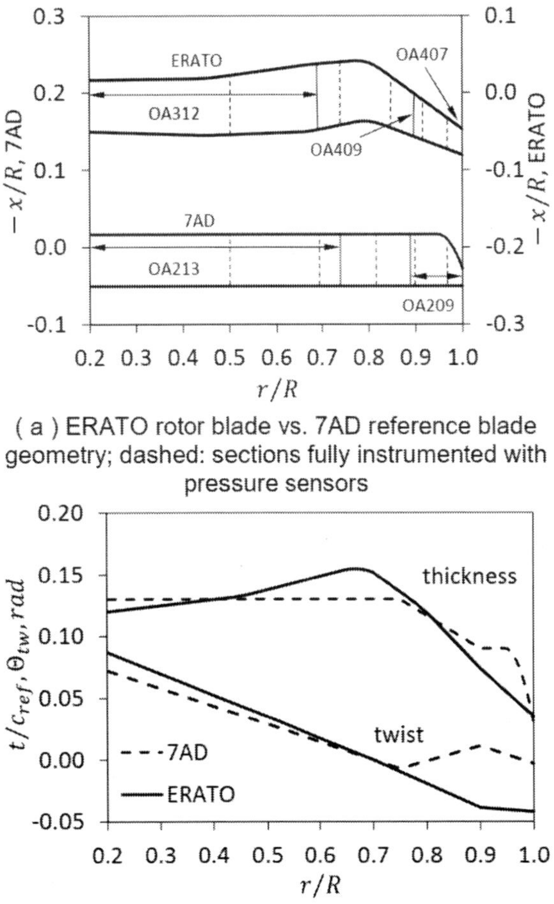

(a) ERATO rotor blade vs. 7AD reference blade geometry; dashed: sections fully instrumented with pressure sensors

(b) ERATO and 7AD blade twist and absolute thickness

Figure 3: ERATO aeroacoustically optimized rotor blade design compared to the 7AD reference rotor blade.

- the largest chord is located around 65% radius (leading to a strong taper towards the tip) with the goal to redistribute the lift from the tip towards more inboard, aiming at reducing the tip vortex strength;
- a combination of fore-aft sweep with the goal to distribute BVI interaction topology from a quasi-parallel BVI (7AD) to azimuthally/radially distributed BVI (ERATO);
- as a side effect, sweep reduces the effective Mach number on the blade and thus alleviates compressibility effects, leading to less power drag in high speed flight and – though no design goal – reducing rotor power consumed;
- usage of modern next generation rotorcraft airfoils with high lift capability in the inner regions up to 70% radius; combined with linear strakes to thinner and finally transonic airfoils at the blade tip with reduced shock characteristics at high subsonic Mach numbers (again reducing drag and power);

- usage of higher twisted blades compared to the reference rotor which again aims at redistribution of lift towards more inboard locations.

Of course, the fore-aft swept blade design was suspected to generate aero-elastic problems due to strong couplings of especially flap and torsion modes, compared to a straight blade such as the 7AD. The structural design was therefore carefully chosen at least to avoid strong coupling at the nominal rotational speed. Any flap-torsion mode proximity was requested not to be close to a blade harmonic number, i.e., 4/rev or 8/rev (as the number of blades was 4) to avoid excessive vibration transmitted to the non-rotating frame. The computed fan diagram of both model-scale rotors, using a finite element beam method based on the Houbold-Brooks equations is given in Figure 4.

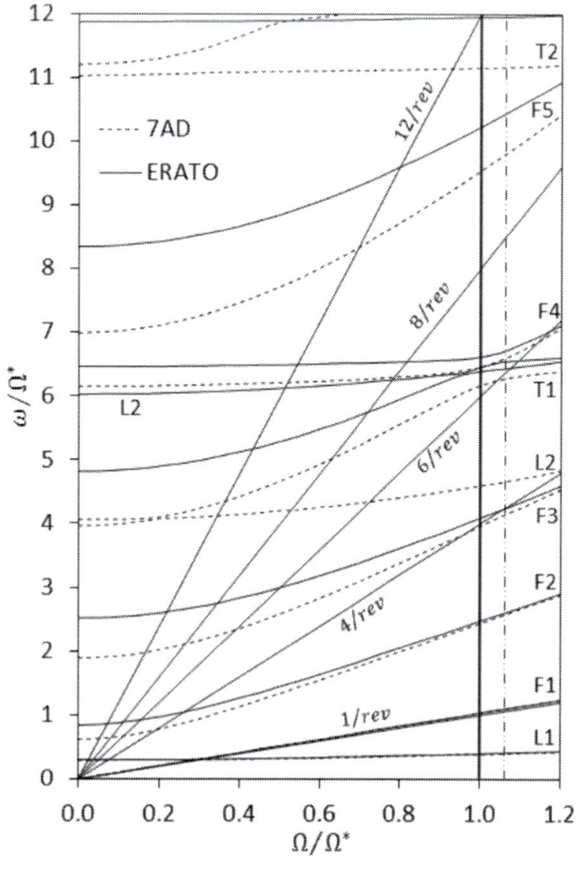

Figure 4: Fan diagram of the 7AD and ERATO blades.

While the 7AD rotor shows a proximity of the fourth flap mode to the first torsion mode close to 6/rev with the second lag mode well separated from it, the ERATO rotor has all three of them in close proximity around 6.5/rev at its nominal speed of rotation, but far away from the blade harmonics 4 and 8/rev.

2.1 Wind Tunnel Testing in DNW-LLF and S1MA

In order to get maximum use of wind tunnel tests and to understand the underlying physics, one rotor blade was heavily instrumented with 125 absolute pressure sensors and 32 strain gauges. The other blades got dummies for blade dynamics reasons and some strain gauges as well. All blades were equipped with flap, lag and pitch sensors at the articulation and pitch bearing.

The DNW-LLF tests, where in the open jet configuration about 78 m/s maximum wind speed could be achieved (Figure 5), focused on rotor acoustics, vibrations, and performance in hover and at low to moderate speeds. The S1MA tests (Figure 10) aimed at high speed flight with respect to rotor performance and noise, high lift (including stall) and vibration. DNW testing [20], [22]-[23] capabilities employed in its low back-ground noise anechoic environment included a microphone traverse for measurement of the noise carpet below the rotor (centered in the lower half of Figure 5). Optical blade deformation measurements were performed by a fringe correlation method (FCM, complementary to the strain gauges readings within the blades), [24], that was provided and operated by DNW. With this method it is possible to measure flapwise and torsional deflections along the blade span at selected azimuthal angles. This technique is based on the fact that a grid, which is projected on the surface of a structure, will change its pattern when the structure is deformed or displaced.

Figure 5. The ONERA-DLR ERATO rotor mounted in the DNW LLF 8 m by 6 m anechoic open test section.

For this purpose the projection of the grid has to be executed from the opposite direction of the recording device, which was a fast motion video camera in the ERATO test campaign. By comparing the image of the non-deformed and non-displaced blade with one that is deformed and deflected, the relative deformation and displacement of the structure can be determined. The result of FCM is thus the difference of the total deflections of two rotor-conditions. Figure 6 represents a fringe pattern projected onto an ERATO blade.

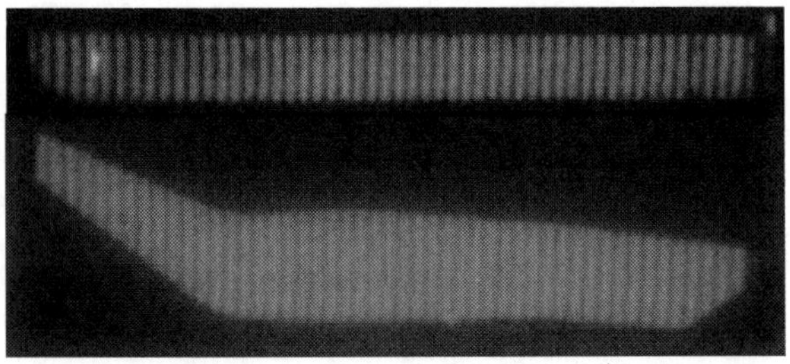

Figure 6. A Fringe Correlation pattern projected onto one of the 7AD and ERATO rotor blades.

An example result from the reference operating condition representing a 6° descent flight is shown in Figure 7. The FCM provides information of local deflections from all visible surface points of the blade, while the strain gauge measurements provide the local bending moments at the point of the sensor, and deflections of the entire blade are computed using beam bending theory with many simplifying assumptions included.

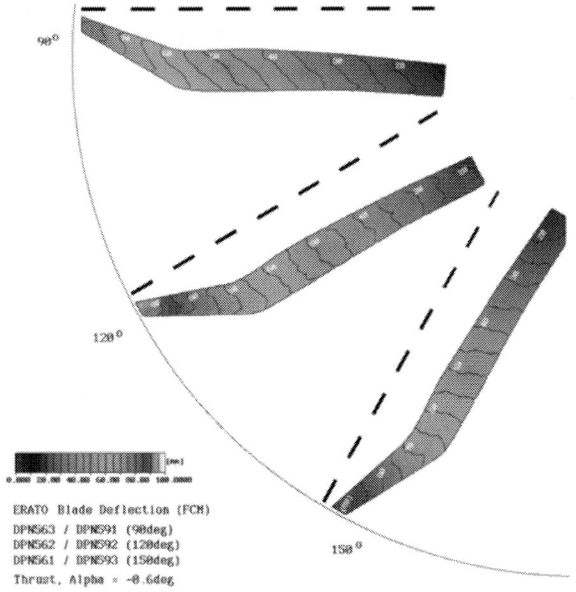

Figure 7. Vertical deflection of the ERATO blade measured at different azimuth during rotation.

In order to determine tip vortex parameters as vortex core diameter, swirl velocity and circulation, DNW provided their Particle Image Velocimetry (PIV) system for 2-D flow field measurements at pre-selected locations in the vicinity of the BVI area. DNW conducted the measurements and also performed the data analysis. Figure 8 represents two vector maps of a tip vortex cross section: one instantaneous and one averaged vector map. Tip vortex line locations and actual rotor blade locations have been measured by means of DNW's Laser Light Sheet (LLS) flow field visualization method. The results were used in order to calculate blade-vortex miss distances at rotor blade azimuths of strong BVI. Furthermore, these flow field visualization measurements have been done for getting an impression of the tip vortex trajectories while traveling downstream the rotor disk for a better understanding of the BVI noise phenomena. Figure 9 represents results gained by means of the LLS method.

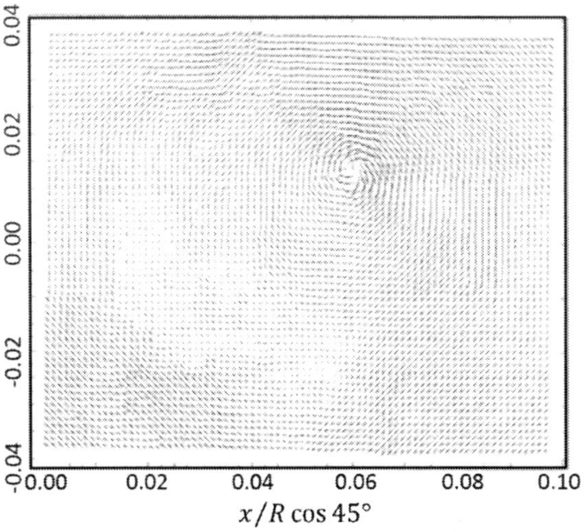

(a) instantaneous velocity vector field

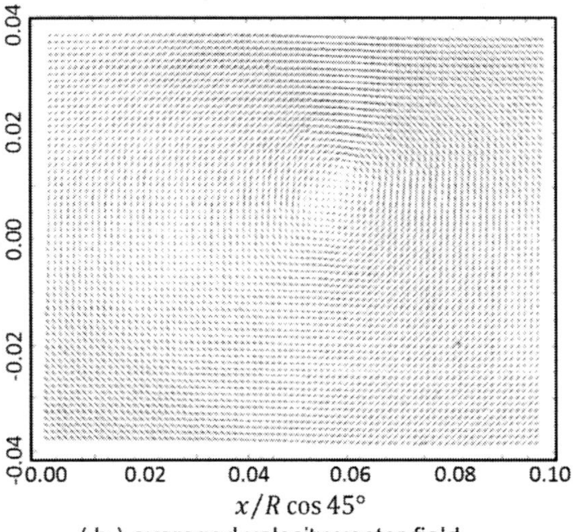

(b) averaged velocity vector field

Figure 8. Vector maps of a tip vortex cross section measured by means of PIV.

Blade surface pressure measurements during the DNW test were performed by DLR Göttingen. 125 absolute pressure sensors were installed on the ERATO reference blade, distributed among 5 radial sections on upper and lower surface as indicated in Figure 3 (a) with additional ones located between these sections along the leading edge. The 7AD rotor had a similar number of pressure sensors, but distributed among the blades. A 6-component rotor balance was employed in all tests as part of DLR's rotor test rig with data acquired by DLR and some identical operational conditions were performed in both DNW and S1MA wind tunnels to cross-check and ensure the similarity of results.

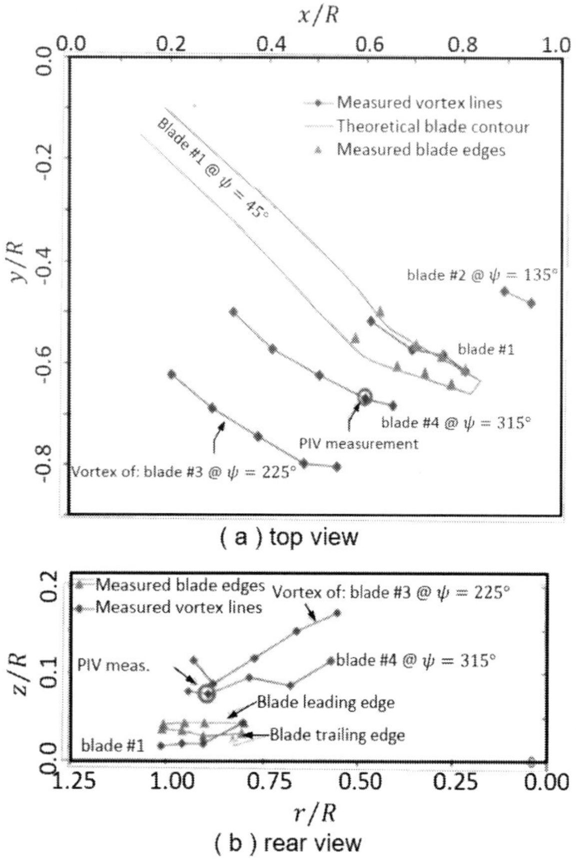

Figure 9. Tip vortex and rotor blade location measured with the LLS method. $V_\infty = 35\,m/s, \gamma = 6°$.

In S1MA, the test section was equipped with its "low speed" acoustic liners (black surfaces in Figure 10) which can withstand Mach Numbers up to 0.32. Acoustic measurements were realized using 10 microphones located especially in the area affected by BVI noise. Tests included also load measurements with a 6-component rotor balance located inside the test rig, blade deformation measurements using the Strain Pattern Analysis method (SPA), pressure measurement on the blades to establish correlation between noise levels and Mach number distribution on the blades. All data were acquired by S1MA wind tunnel operators and delivered to ONERA and DLR scientists for analysis after the test.

2.2 Wind tunnel test results

Rotor Noise Radiation

Typically for helicopters, the maximum BVI noise radiation occurs in landing approach, i.e., at flight speeds of about 70 kts (35 m/s) and 6° descent angle. In DNW-LLF a descent angle sweep from climb to steep descent was carried out to identify the maximum BVI noise radiation condition. This is shown in Figure 11 for various rotor loadings.

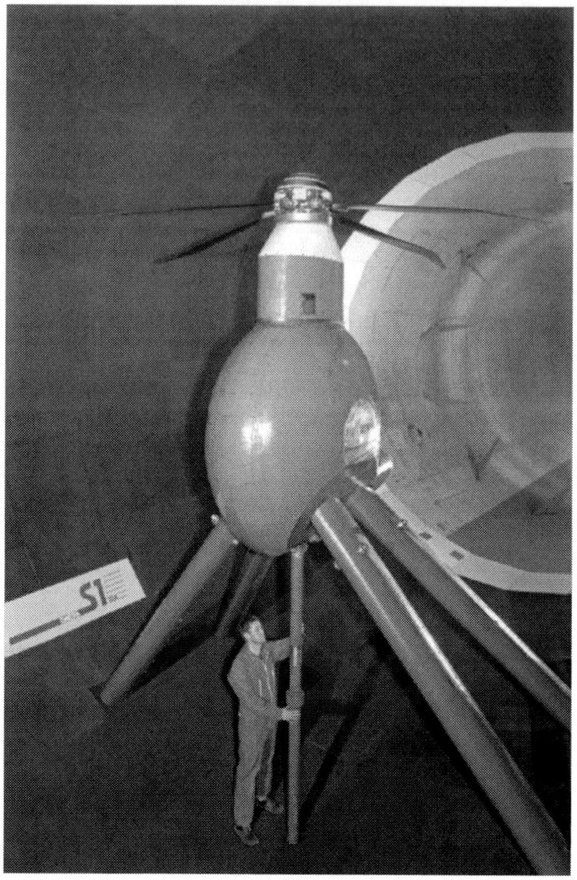

Figure 10. The ONERA-DLR ERATO rotor mounted in the S1MA 8 m Ø closed test section.

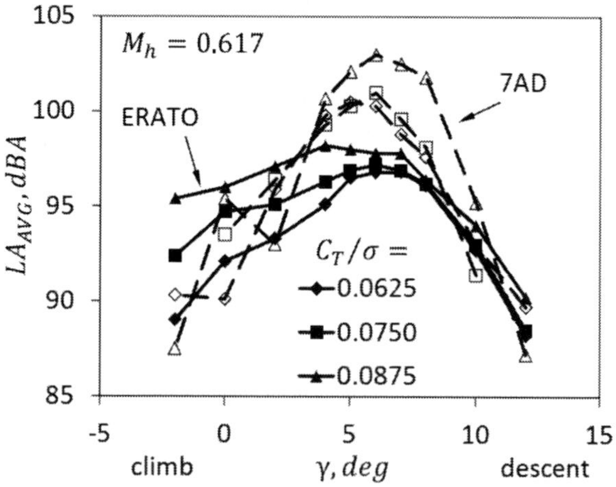

Figure 11: BVI noise reduction potential of the ERATO rotor at $V_\infty = 35\,m/s$.

At the reference 35 m/s, 6° descent condition a significant BVI noise reduction of roughly 4 dBA was demonstrated, verifying the numerical predictions, Figure 12. One of the major results of these tests was that noise could be reduced for a wide range of flight conditions, demonstrating a significant robustness of acoustic gains with respect to variations of descent angle, advancing velocity, trim conditions etc.

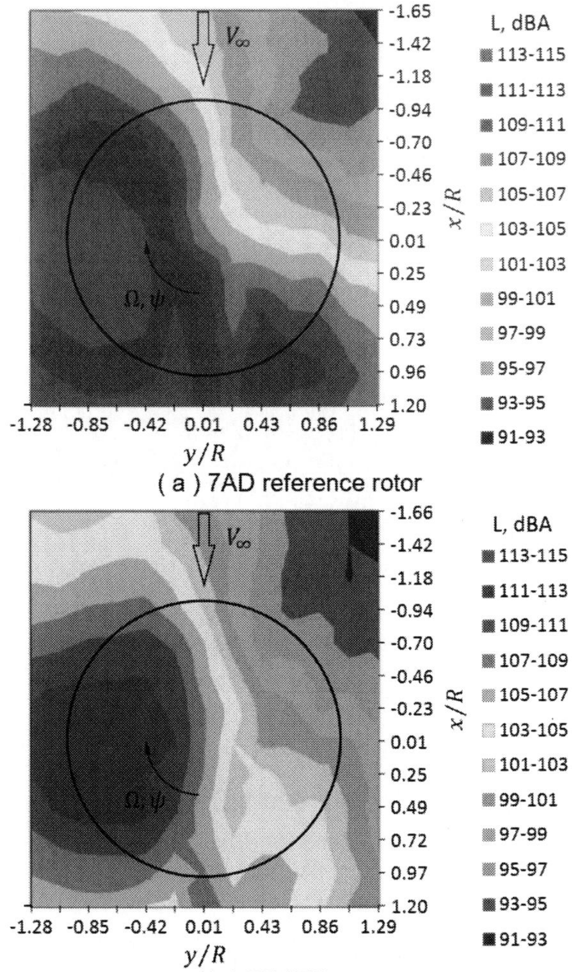

(a) 7AD reference rotor

(b) ERATO rotor

Figure 12: Acoustic results of DNW-LLF tests showing the ERATO rotor with significantly less BVI noise signature compared to the straight reference blade.
$V_\infty = 35\, m/s, C_T/\sigma = 0.0625, \gamma = 6°$.

The reason of BVI noise reduction was mainly seen in a different BVI topology due to the blade planform. The double sweep causes the BVI not to happen over a large radial portion simultaneously as is the case with a rectangular blade; rather it distributes the BVI event in time and space. Figure 13 (a) shows the high-frequency, BVI-related content of leading edge pressure distributions (high-pass filtered at 10/rev) as source of the noise and in (b) the same is shown for the ERATO rotor.

It can be observed in this figure in a qualitative manner that the blade pressure waves in the first and fourth quadrant of the 7AD rotor appear more intense than those of the ERATO rotor. Accordingly, the noise radiation intensity of the ERATO rotor is found significantly lower.

The source of BVI noise are high frequency loading oscillations in the first and fourth quadrant of the rotor disk, where the tip vortex axis and the blade leading edge are almost parallel to each other and simultaneously interact on a long radial part of the blade. Any noise reduction therefore is partly due to a reduction of these BVI. This is substantiated in Figure 13 for both rotors in the maximum BVI noise radiation operational condition.

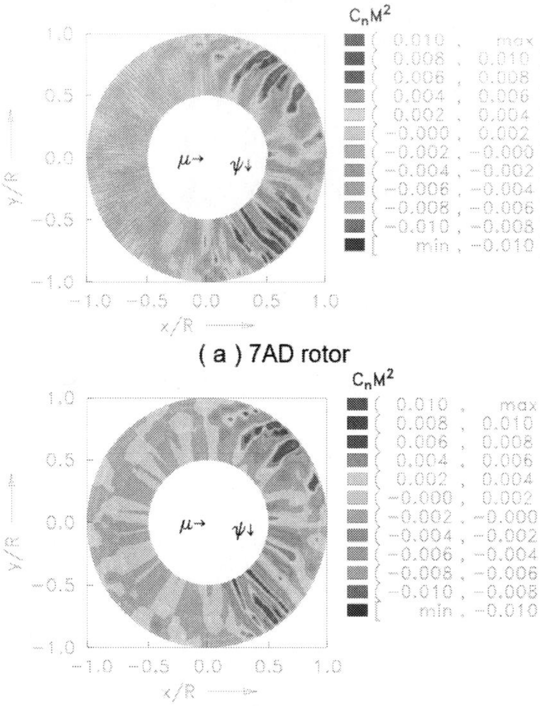

(a) 7AD rotor

(b) ERATO rotor

Figure 13: Rotor blade loading, high-pass filtered at 10/rev, showing BVI locations and intensity.
$V_\infty = 35\, m/s, C_T/\sigma = 0.0625, \gamma = 6°$.

It can be seen that the ERATO rotor has less amount of BVI in terms of number of BVI events and intensity on both the advancing side and the retreating side of the rotor. This was quantitatively confirmed by the post-treatment of PIV measurements which emphasized the reduction of the vortex intensity emitted by the ERATO blades (Figure 14). A further proof of this observation is given in acoustically relevant time derivative of the section loading at a representative radial station close to 85% radius (ERATO) and 81.5% radius (7AD) in Figure 15 (a) for the advancing side of the rotor and in (b) for the retreating side. In this case the time derivative is based on the 6-40 bpf of the data.

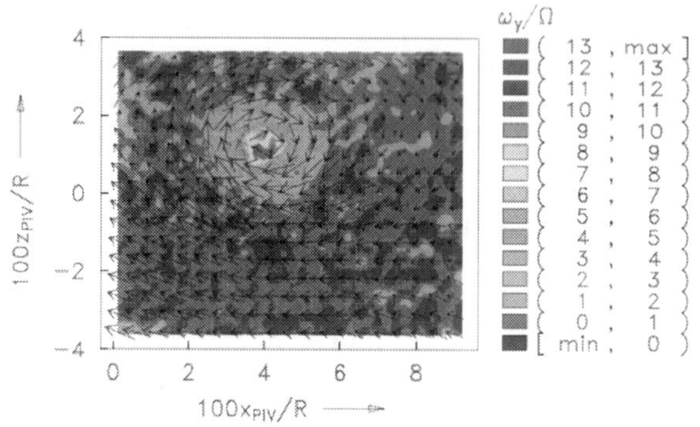

(a) 7AD rotor, $\Gamma = 4.23\,m^2/s$

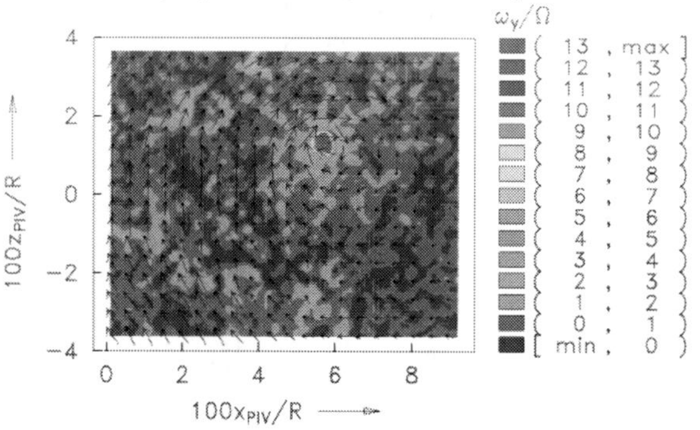

(b) ERATO rotor, $\Gamma = 3.56\,m^2/s$

Figure 14: Velocity field and vorticity ω_y derived from it around a vortex as measured by PIV. $V_\infty = 35\,m/s, C_T/\sigma = 0.0875, M_h = 0.617, \gamma = 6°$.

The grey shaded areas are those of parallelism of the vortex axis with the blade leading edge for a straight rotor blade. It is obvious that the 7AD rotor has stronger BVI interactions and a longer range of azimuth where the intensity is relatively large, compared to the ERATO rotor, i.e., the vortices are stronger and appear to pass the rotor blade's path in a closer manner than in case of the ERATO rotor. While the first can be explained with the blade loading at the origin of the vortex, i.e., around $\psi = 135°$ azimuth on the advancing side and around 225° on the retreating side, the latter can only be explained with a different vortex trajectory while travelling downstream across the rotor disk.

The rotor blade circulation at 135° azimuth is shown next in Figure 16, computed from the section loading. It can clearly be seen that the gradient of circulation towards the blade tip, which feeds the tip vortex, is much larger in case of the 7AD rotor than in case of the ERATO rotor. The second assumption, a different tip vortex trajectory while travelling downstream across the rotor disk, can only be due to a higher dynamic low frequency loading content in case of the ERATO rotor, compared to the 7AD. This phenomenon is known from the HART test, where this was actively controlled by means of HHC. Based on the concept of momentum theory, any additional local lift (or download) is associated with an additional downwash (or upwash) induced by it. A vortex travelling this local down- or upwash gets additional trajectory deflections.

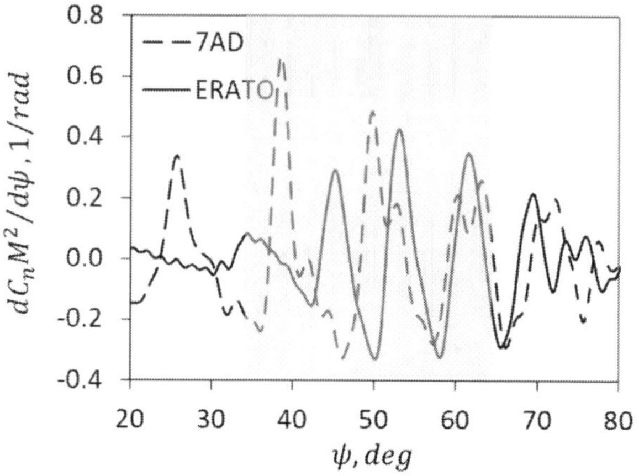

(a) advancing side

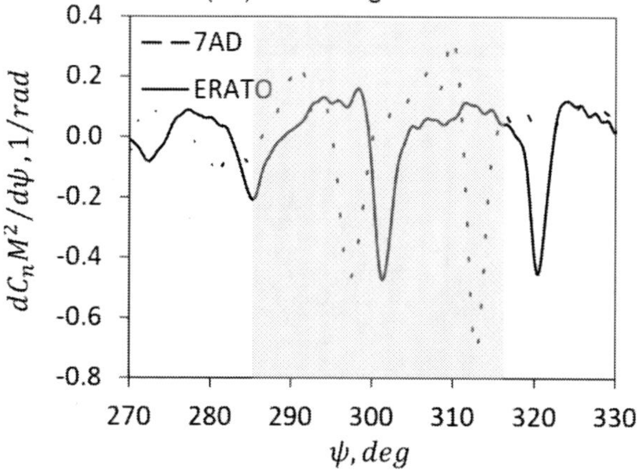

(b) retreating side

Figure 15: Rotor blade loading, high-pass filtered at 10/rev, showing BVI locations and intensity. $V_\infty = 35\,m/s, C_T/\sigma = 0.0625, \gamma = 6°$.

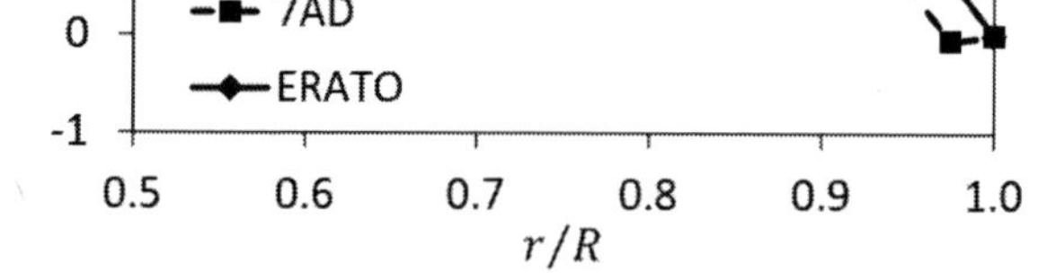

Figure 16: Instantaneous rotor blade circulation. $V_\infty = 35\,m/s, \gamma = 6°$.

The low frequency content of rotor blade loading (0-10/rev) is shown in Figure 17, making proof of the assumption. While the 7AD rotor has a relatively smooth loading distribution with small to moderate harmonic content, the ERATO rotor has stronger dynamic fluctuations, quite similar to what has been observed in the HART test using HHC active rotor blade control.

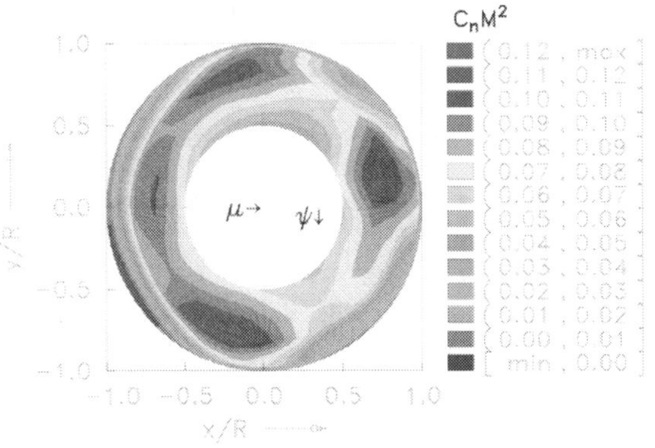

(a) 7AD rotor

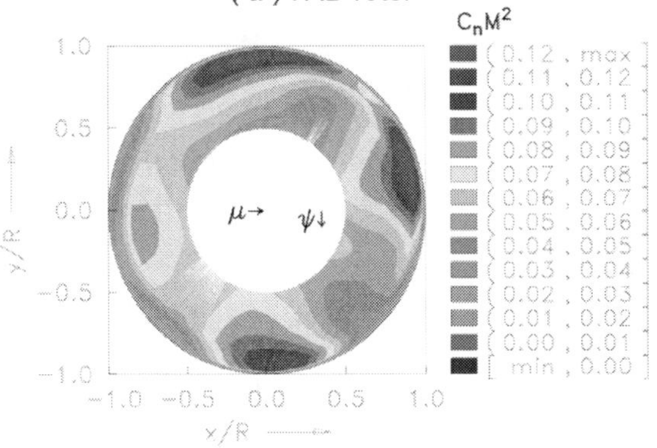

(b) ERATO rotor

Figure 17: Rotor blade loading, low-pass filtered at 10/rev; $V_\infty = 100\ km/h, \gamma = 6°$.

This was expected since due to the fore-aft swept blade shape of the ERATO blade, more flap-torsion couplings were expected from the numerical simulations during the design phase, leading to a larger dynamic content of the elastic blade deformation and thus in the lift distribution, as confirmed by the measurement of elastic deformations using the Strain Pattern Analysis (SPA) method (Figure 18).

The experiments therefore confirm a larger rotor blade dynamic motion content, leading to a larger dynamic loading content, and finally to modified tip vortex trajectories that help reducing BVI noise.

The gains in loading, thickness and transonic (high speed impulsive) noise levels were measured in S1MA wind tunnel for level flight conditions and are shown in Figure 19 where the difference between ERATO and 7AD rotor for identical operational conditions are shown. The 7AD rotor was always operated at a tip Mach number of 0.661, while the ERATO rotor was operated with different tip Mach numbers. In any condition, the rotor trim ensured the same overall lift and propulsive force for both rotors. For the same tip Mach numbers, gains are between 2 and 3 dBA, but are significantly larger with reduced tip Mach number.

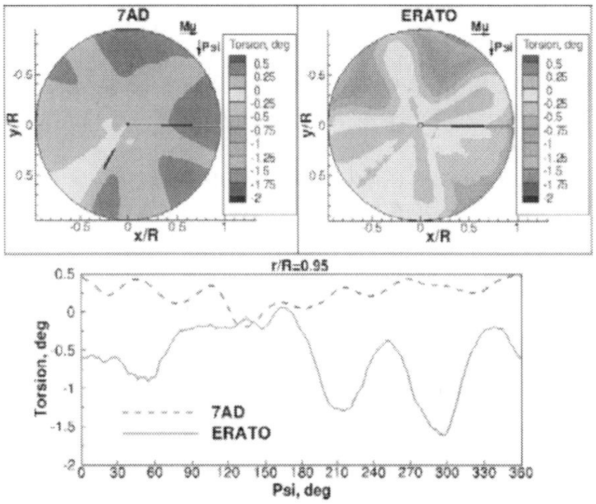

Figure 18: Torsion deformation measured by SPA method in S1MA. $M_h = 0.617, \mu = 0.283, C_T/\sigma = 0.075$.

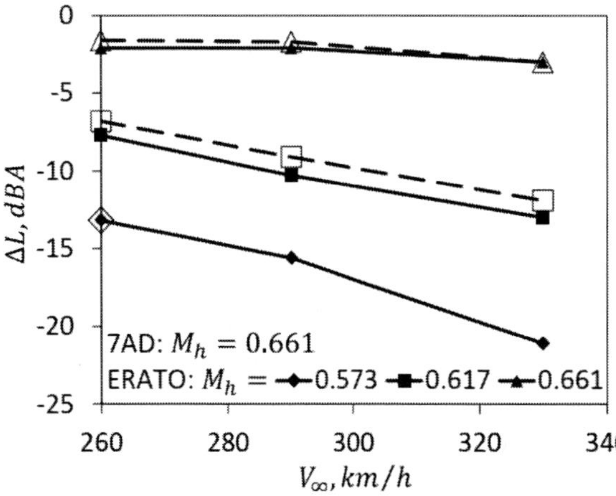

Figure 19: Gains in maximum noise levels of the ERATO rotor relative to the 7AD reference rotor in level flight; solid: $C_T/\sigma = 0.0625$, dashed: 0.075.

The ERATO rotor is not equipped with a specifically optimized blade tip to reduce the High Speed Impulsive (HIS) noise. Nevertheless, thanks to the reduced thickness and thanks to the backward sweep, the tests have shown that the delocalization phenomenon was delayed to higher advancing speeds with the ERATO rotor (explaining the 3.5 dB reduction at 330 km/h (183 kts), Figure 19).

The influence of the rotor loading on the acoustic radiation has been analyzed. At ICAO certification flight conditions (120 and 140 kts = 216 and 252 km/h), the ERATO rotor provides gains whatever the lift coefficient, up to 7 dB at $C_T/\sigma = 0.0875$ which corresponds to a strong BVI case. These tendencies were confirmed by the DNW tests as shown in the previous section. Even if the flight conditions simulated in S1MA where not those for which the ERATO design was optimized, noise reductions have been demonstrated for 75% of the measured points.

Rotor Power Consumption in forward flight

The high speed and high load tests at S1MA wind tunnel of ONERA (Refs. [20], [25]) confirmed a smaller power consumption of the ERATO rotor throughout the entire range of operational conditions of up to 9% compared to the reference rotor and 5% in average, see Figure 20.

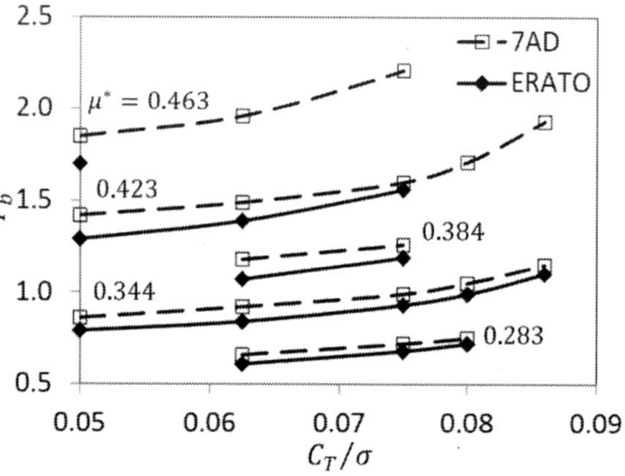

Figure 20: Power gains of the ERATO rotor.

This is mainly due to the use of new generation airfoils specifically designed for high subsonic speed conditions (OA4-OA3 series on the ERATO blade instead of OA2 series on the reference 7AD rotor blades) and to a lesser extend to the back sweep of the blade tip, reducing the compressibility effects and hence rotor power required.

The good behavior of the ERATO blade in transonic conditions is illustrated by the contours of local Mach numbers in Figure 21, extracted on-line during the S1MA tests from the measurement of the pressure distributions.

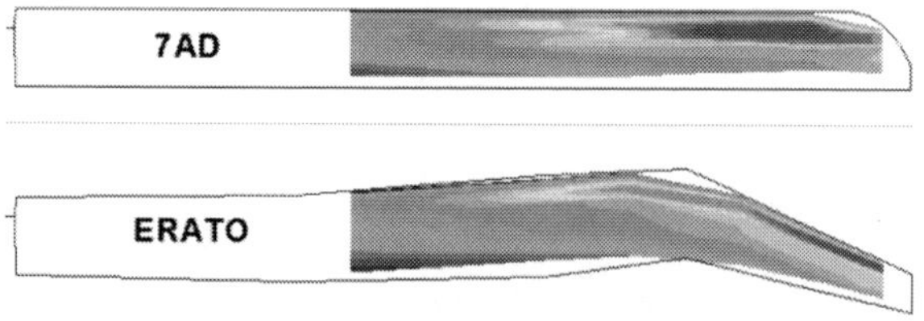

Figure 21: Measured contours of local Mach numbers on the blade on the advancing side ($\psi = 90°$) for high speed conditions in S1MA. Red: $M > 1$.

Efficiency in Hover

Although not part of the specifications given for the optimization of the rotor, hover performance assessment was done by Airbus Helicopters. The same 7AD and ERATO rotor models as used in DNW-LLF and S1MA were installed on the Marignane test rig, and the Figure of Merit (FM) of the two rotors was measured for different tip Mach numbers. A sample result of these measurements is illustrated in Figure 22. Both rotors show similar performance for thrust coefficients $C_T/\sigma < 0.087$, with even slightly higher efficiency for the ERATO rotor than for the 7AD rotor.

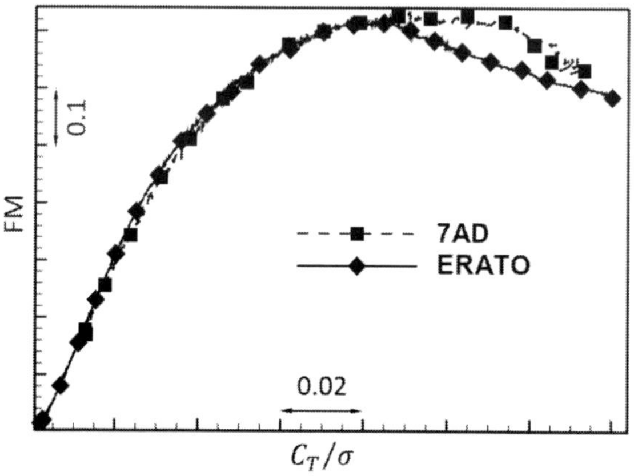

Figure 22: Measured hover performance of 7AD and ERATO rotors, $M_h = 0.617$.

However, beyond $C_T/\sigma \approx 0.087$, there is a sudden loss of efficiency of the ERATO rotor, whereas the 7AD reference rotor continues to maintain efficiency higher than 0.71 up to $C_T/\sigma \approx 0.115$. This poor performance at high thrust of the ERATO rotor clearly indicated a premature stall of the blade which was not anticipated during the design phase, but was identified as a major topic to be addressed for future improvements of the rotor performance.

3. SHANEL AND SHANEL-L (2009-2012) CALCULATIONS AND DEVELOPMENTS ON ERATO CASES

The optimization process used for the ERATO design was based on sets of comprehensive codes. In parallel, the CFD-CAA capabilities available at ONERA have been developed and improved. In particular, the CFD capabilities to well capture the emitted vortices and more challenging to be able to properly model their characteristics from their emission up to the azimuths of interaction have been developed in the frame of the CHANCE (2001-2006) and SHANEL and SHANEL-L (2009-2012) programs.

In the framework of the French-German CHANCE cooperation, Chimera techniques have been developed and used for automatic mesh generation and adaptation in the elsA solver. In 2004, these methods have been applied to the HART Baseline testcase, but no BVI was simulated due to too coarse blades and background grids. In the framework of SHANEL, the objectives were to capture the blade-vortex interaction with fine reference meshes and to improve the efficiency of the computations by applying recent numerical methods (high-order schemes, vorticity confinement, etc.). The simulations, using a loose CFD-CSD coupling were then applied to the complex geometry of the ERATO blades. The CFD results were used as input data for the noise analysis and compared to the HART experimental measurements.

One of the objectives was to test the efficiency of the Vorticity Confinement (VC) technique previously validated using the HART II database. The CFD were performed with the elsA code (weakly coupled with the HOST code) to solve the 3D URANS (Unsteady Reynolds-Averaged Navier-Stokes) and the acoustic predictions were made using the Ffowcs Williams-Hawkings KIM code [32]. As shown in Figure 23, the impulsive nature of BVI events is much better reproduced by the CFD simulations using the Vorticity Confinement (VC) technique, because it better preserves the vorticity of the vortices from their emission up to the interaction azimuths with a moderate number of mesh points (24 M).

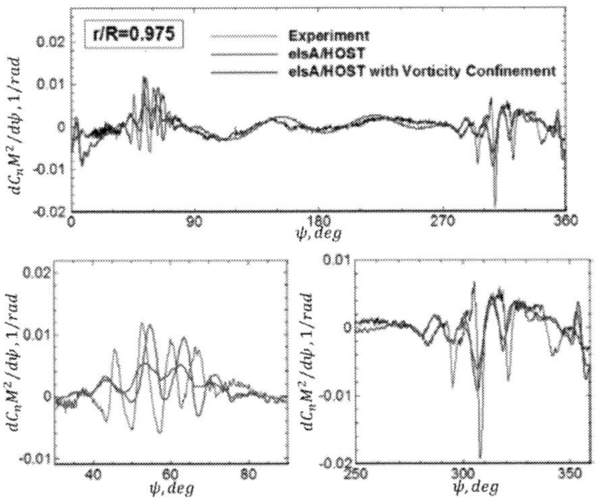

Figure 23: Measured azimuthal derivatives of section loads compared to numerical simulation.

As a consequence, noise predictions using as an input the VC results are in much better agreement with measurements, as shown by Figure 24.

4. FULL SCALE ADAPTATION AND DESIGN: FROM ERATO TO BLUE EDGETM

Following the ONERA-DLR ERATO project, a new project was launched in year 2000 on the French side with Airbus Helicopters and ONERA as acting partners in order to take benefit of the acoustically optimized ERATO blades with the objective of developing a full-scale prototype rotor. The project was subdivided into four major phases, providing clear progression and knowledge capitalization during each phase, [31]. The four phases were:

- Phase 1: Selection of candidates. Starting from lessons learnt with ERATO project, a design phase was launched in order to significantly improve the hover performance and some observed dynamic behaviour. Numerical simulations in aerodynamics, acoustics and dynamics were mainly used in this optimization loop in order to select the candidate which offered the best compromise between all these significant objectives.
- Phase 2: Technology and structural design. After the selection of a candidate, the team was fully in a position to launch the structural design, dynamics optimization, strength assessment, detailed design and manufacturing tools.

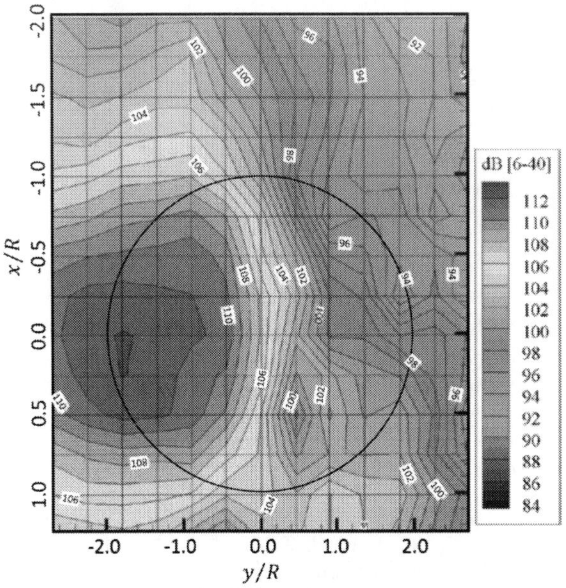

(a) experiment

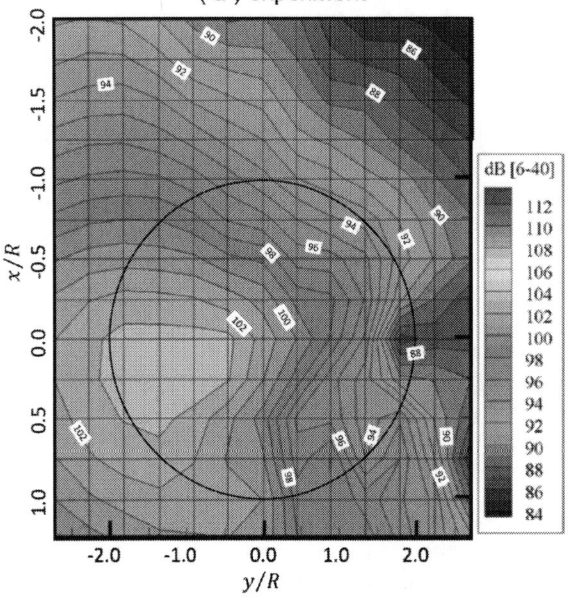

(b) *els*A KIM without VC

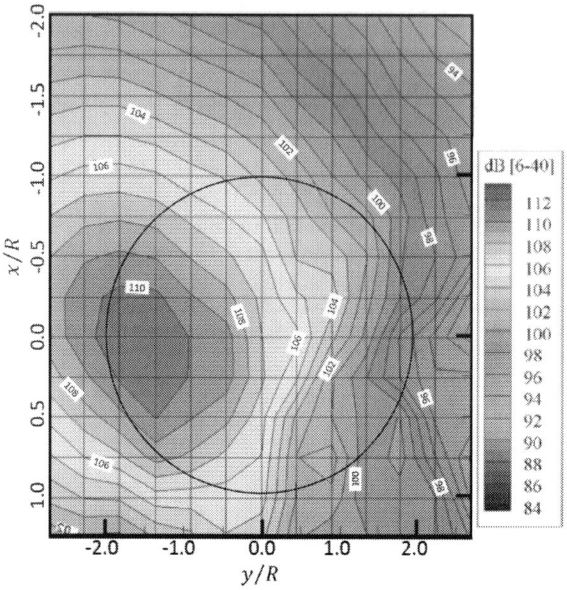

(c) *els*A KIM with VC

Figure 24: Measured BVI noise contours compared to numerical simulation.

- Phase 3: Validation of the design. Manufacturing began in this phase, followed by ground demonstration tests in order to validate the design and to obtain the flight clearance for the rotor.
- Phase 4: Flight demonstration on a H155 demonstrator in order to validate the aerodynamic, acoustic and dynamic characteristics and performance.

A dedicated team was created with representatives from each key area through which the blade would ultimately be designed, manufactured and tested. This multi-disciplinary approach gave wider joint ownership to the team and stimulated a creative spirit, which was necessary during certain critical phases of such an innovative design. The project was also the right platform to implement a range of simulation and design tools which could later be applied to other applications.

During the first phase of this project, the challenge at the beginning of the design phase was to propose blade shapes featuring the fore-aft sweep concept of the ERATO rotor promising for acoustics, while significantly improving the hover performance in terms of Figure of Merit. Thanks to the use of CFD for the first time in a helicopter rotor design exercise, the poor behavior in hover of the ERATO rotor could be explained and attributed to premature stall on the outboard blade sections located in the aft part of the blade, where the chord length of the aerodynamic sections were too small (Figure 25). CFD could reproduce the impact of the premature stall on the rotor performance, as demonstrated by Figure 26.

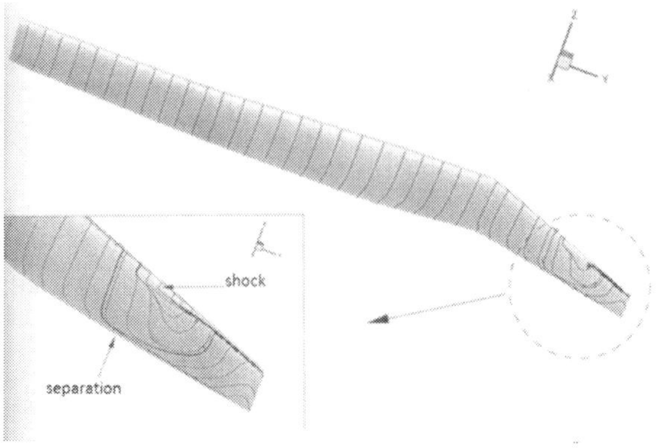

Figure 25: ERATO rotor stall in hover (CFD prediction).

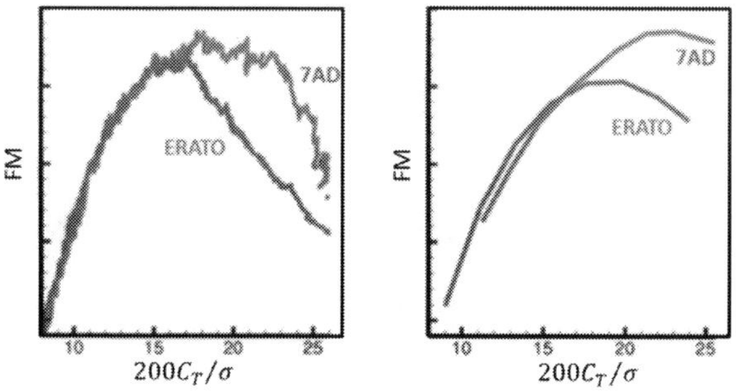

(a) experiment (b) CFD

Figure 26: Validation of CFD method for predicting the hover performance of 7AD and ERATO rotors.

This validation gave confidence in the available numerical tools to achieve the objectives of the project. As a first step, five candidate rotors were proposed, all with increased chord values on the outer part of the blade, and numerically evaluated in hover conditions, using as a reference the C0 rotor designed by Airbus Helicopters, Figure 27.

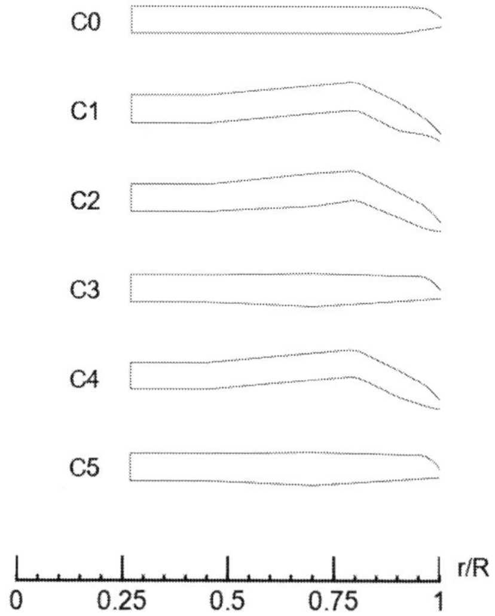

Figure 27: Initial blade shapes evaluated in hover conditions (C0 = reference).

The result of these computations, as is illustrated in Figure 28, was that all proposed candidates showed improved performance compared to the C0, with a maximum FM improvement by approximately 3 counts.The detailed investigation of the origin of the gains concluded that FM improvement came from the increased twist distribution and from the anhedral put at the blade tip which was absent on the C0 blade. Following this first step, the C4 candidate was selected and an optimization of its tip shape was carried out. It has to be highlighted here that this was the first time that a CFD solver was introduced by ONERA and Airbus Helicopters in an optimization loop to design a rotor blade.

The optimizer could modify the twist and anhedral law of the blade tip ($r > 0.9R$), using as a cost function the maximum FM. Several designs came out from the optimization runs, each featuring improved hover performance, the selection being finally made outside the optimization loop based on the noise radiated by the rotor in typical descent flight conditions, using the ONERA 'HMMAP' aero-acoustics chain [27]: this led to the so-called C4P-OPT candidate.

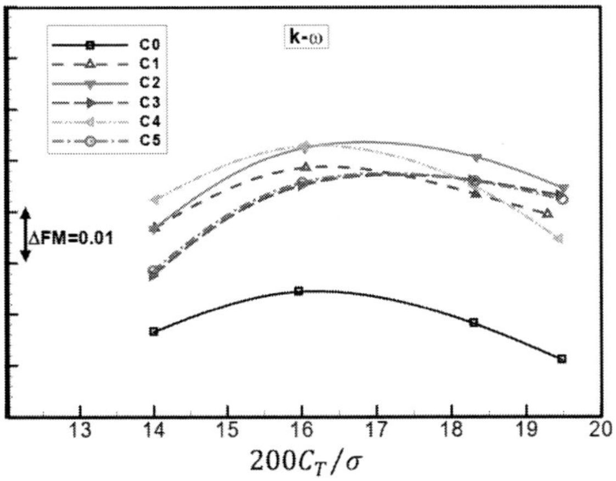

Figure 28: CFD evaluation of initial blade designs in hover conditions.

Since all the optimizations carried out in this first phase of the design were done using a rigid blade assumption, the next phase was to define blade structural data and repeat the aerodynamic and acoustic simulations using the soft blade assumption. Taking into account the elastic blade deformations did not significantly modify the good hover performance of the optimized blade, but led to reduced acoustic gains (but still significant for descent angles higher than 6°, see Figure 29).

At that step of the project, the main issue was related to torsion deformations and control loads, which were found to be too high in high-speed forward flight. The best way to reduce the torsion deformations - which were mainly due to bending-torsion coupling resulting from the fore-aft sweep concept - was to try to modify the sweep angles of the blade. Since the radiated noise in descent flight is highly sensitive to this parameter, reaching a good compromise between acceptable torsion deformations and promising noise reduction was a real challenge.

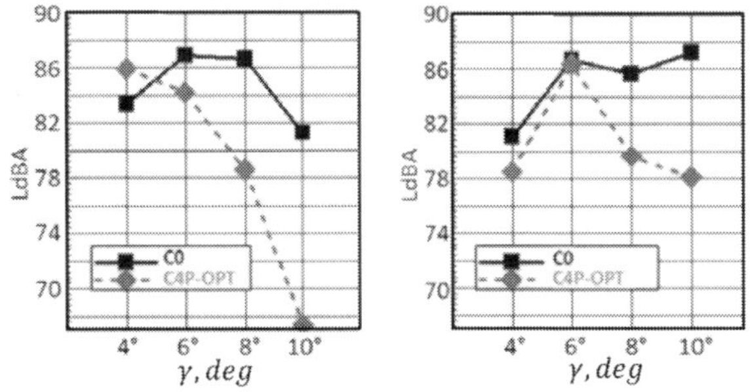

(a) rigid blade (b) soft blade

Figure 29: predicted noise levels in descent flight: influence of blade deformations, $V_\infty = 40\,m/s$.

In this phase, the main modifications brought to the design were the following: the sweep angles were changed on the final Blue Edge™ design and the spanwise extension of the backward swept part of the blade was reduced. This led to a good aerodynamic-acoustic-dynamic compromise (Figure 30), with high expected FM in hover, and promising noise reductions in low-speed descent flight.

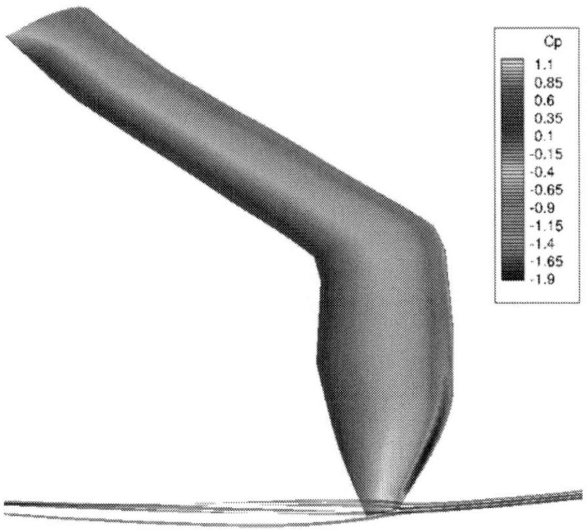

Figure 30: Blue Edge™ optimized blade with surface pressure coefficient distribution in hover.

An assessment of the flight conditions for which the Blue Edge™ design was acoustically efficient was numerically done and is illustrated in Figure 31: it was expected that the most favorable conditions would be the low speed – high descent angle conditions (typically 60 knots with descent angle higher than 6°).

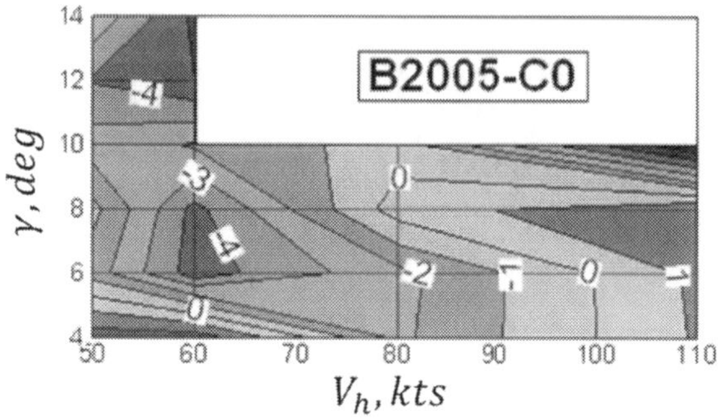

Figure 31: Final evaluation of the Blue Edge™. Contour plot of noise gains with respect to the C0 reference blade.

The feasibility of ERATO blade on the H155 helicopter was studied as soon as the interest of the fore-aft swept shape was confirmed by the results of the wind tunnel tests. The first studies showed instability of the blade for rotational speeds higher than the nominal one. The aeroelastic optimization of this kind of blades was difficult because of important backward offset. Indeed, it is of common usage to adapt the stiffness's and polar inertias with an adequate stacking of fabrics composing the structure of the blade. However, with this kind of geometry, most of the matter is found behind the lift center, therefore the centers of gravity, elastic and torsion are in a position ready to generate important couplings between the flapping and torsion modes. Although efforts were made to decrease the polar inertia of the backward swept area and to minimize the offsets of these centers compared to the lift center, this was not sufficient to have an acceptable margin with the phenomenon of instability. For the project of Blue Edge™ blade, the design of this zone was fundamental to improve this situation.

The requirements of design set out again from the same needs: to minimize the polar inertia of the backward offset by maintaining the torsion mode above 4/rev. Obtaining a product more mature and ready with the serial production, it was asked to make a blade design with the techniques of the state of the art, to minimize the increase of weight and the manufacturing costs. Consequently, the studied concept was to optimize the position of the center of gravity of the backward offset, to be closest to the total lift center. The way chosen was to avoid adding ballast while centering by counterweights but rather reducing to the maximum the backward area by reorganizing the placement of the components of the blade tip.

Usually, the spar of the blade, located in the leading edge, makes it possible to center the structure around the lift center while ensuring the hold of the centrifugal load and to bring an important bending stiffness. Preserving this principle in the zone of offset did not make sense because the angle of the swept edge is not aligned with the centrifugal force. It is of use to place the balancing system in the tip of the blade for a better effectiveness. By putting it in the zone behind the leading edge, it enables also to center this part of the blade correctly. To sustain in an optimal way the centrifugal load, the unidirectional glass fiber rovings of the spar were divided into two distinct ways at the beginning of the inboard swept zone. A part of the spar preserved its usual path while remaining in the leading edge and second was prolonged in a rectilinear way up to the balancing system. This structure with double spar, reinforced by carbon fabrics reinforcement, concentrates the matter at the good place with providing an optimal path for the efforts, Figure 32.

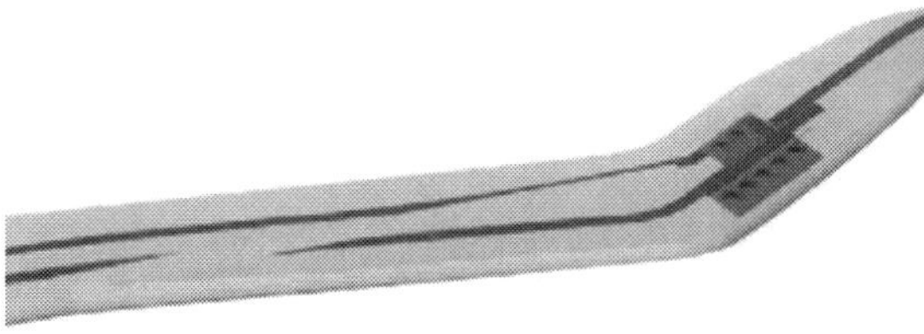

Figure 32: Blade structure with double spar.

This structural design is well visible during the manufacturing of the blades, Figure 33. To validate this solution, a section of this part of the blade was tested in fatigue, see Figure 34.

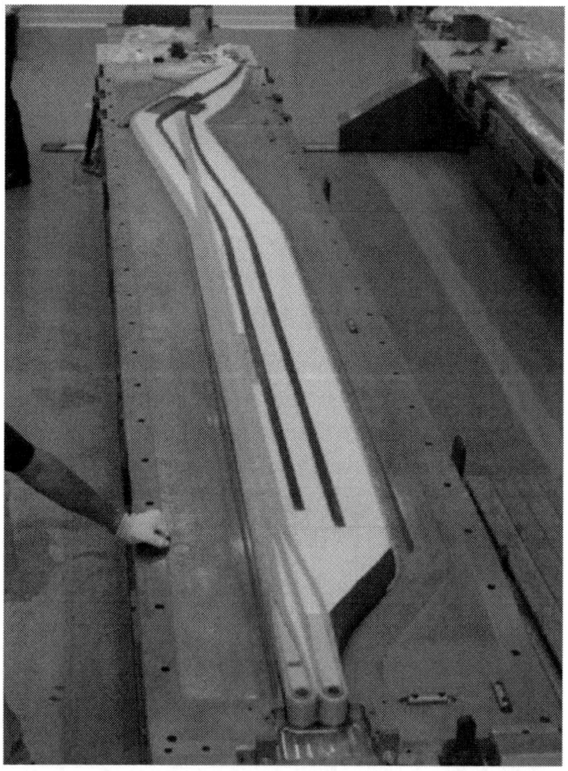

Figure 33: Molded structure of the experimental Blue Edge™ blade.

Figure 34: Fatigue test on double swept area.

Solicited by multi-axial efforts, it passed the levels of vibratory fatigue successfully. Measurements from flight tests confirmed the good distribution of the constraints in the structure and confirmed the margins of resistance. The fore-aft swept shape was a challenge for the technological design of the blade. This technological definition has been patented [33] and became a reference in the state of the art of the design of Blue Edge™-type blades, from which H160's main rotor blades were derived.

5. VALIDATION OF CONCEPT AT FULL-SCALE

The newly designed full-scale blade first flew on an H155 Dauphin demonstrator aircraft in Summer 2007, Figure 35. Detailed results of the extensive flight test campaign were previously highlighted in [31] and [34]. During the more than 75 flight hours conducted within the test program, the full flight envelope was explored and specific attention was paid to performance, loads, airframe vibrations, and in particular to a thorough acoustics assessment.

The inflight validation of the concept was acquired by first confirming that the aircraft performances were in line with expectations (gains in HOGE and same power in level flight compared to the serial H155 straight blade). This was an important confirmation step since the original ERATO blade design had been modified in order to improve hover performance.

Figure 35: Flight Tests of H155 Demonstrator with the Blue Edge™ Optimized Rotor.

A large part of the validation also involved assessing the loads and vibrations. Pitch rod static loads were found to be lowered with the new blade and dynamic loads proved to be slightly increased at low reduced weight and slightly increased at high reduced weight. Airframe vibrations were also confirmed to be slightly lower with the new blades.

Since improved acoustics in approach was the main driver for the new blade, a strong focus was given to validating this aspect of the design and gaining a deeper understanding of the physical mechanisms at play for reducing BVI noise through a double-swept planform. Indeed, a significant portion of the flight test program was dedicated to assessing noise using advanced measurement means.

A first assessment was done using near-field microphones attached to the airframe. This allowed covering the whole flight envelope in a reduced time in order to obtain comparative results of the two blade shapes in the same atmospheric conditions. The expected noise reduction was thus quickly confirmed in a qualitative manner, and it was noted that impulsive noise was mostly eliminated from a large portion of the flight envelope (Figure 36).

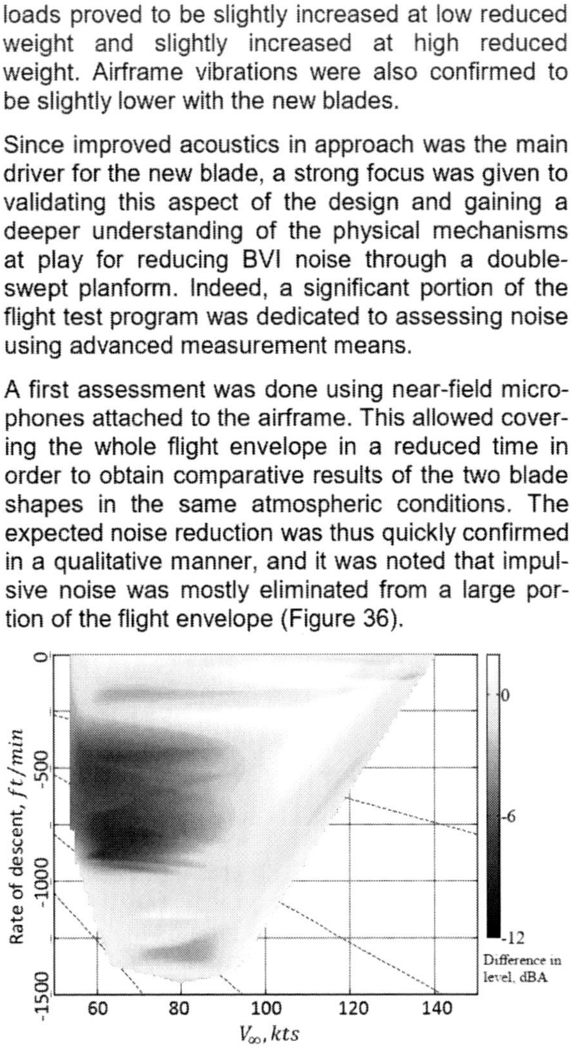

Figure 36: Difference between near-field A-weighted sound pressure levels for the two rotors (negative values where the reference rotor is louder), from [34].

This feature offers greater capability for developing noise abatement procedures which use moderate descent angles and deceleration rates.

Following the near-field tests, an extensive flight testing campaign was performed using a seven-microphone transversal ground array and focusing on providing a deeper analysis of the noteworthy features previously measured. The flights were of course performed with both the reference and Blue Edge™ rotors during successive days, ensuring again consistent ambient conditions. The tests with ground microphones not only allowed quantifying the operational noise reductions, but also were used to assess the directivity patterns associated with the two rotors.

The analysis focused on the flight conditions which showed the most differences between the straight blades and the Blue Edge™ blades. This was a descent at 60 kts with an 8° glideslope. In this condition, noise reductions of close to 5 EPNdB were measured close to the centerline, Figure 37. The analysis showed that this noise reduction was directly linked to a reduction of BVI noise. Interestingly, retreating side noise reductions were more significant than on the advancing side.

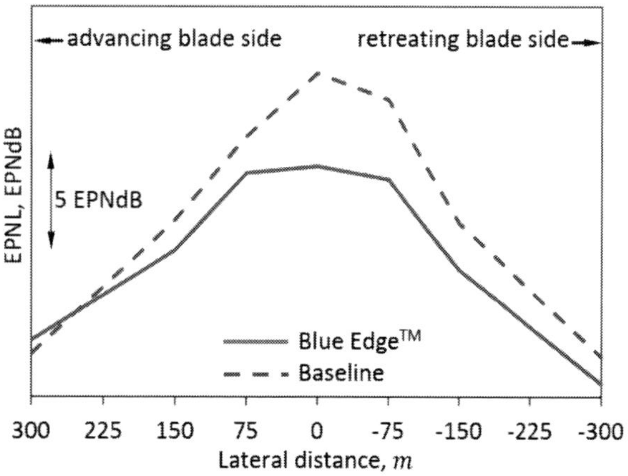

Figure 37: EPNL as a function of lateral distance for the baseline and the Blue Edge™ rotor, [34].

Further investigations of the measurements and additional computations using an acoustic solver coupled to a freewake code allowed to confirm that the blade's backward sweep was very efficient in shifting the directivities more towards the advancing side of the blade, to observers directly to the side of the aircraft. These observer positions are quite beneficial to noise reductions as highlighted in [34].

Noise footprint assessments also shown in [34] illustrate the fact that the significant gains measured on the centerline and retreating side of the aircraft provide significant reductions in the surface area exposed to the loudest noise contours, which highlights the high potential of this rotor for community noise mitigation.

6. CONCLUSIONS

The Blue Edge™ rotor blade is an excellent example of technology transfer from fundamental research to industrial application. After numerical optimization of the blade geometry using state-of-the-art comprehensive codes and structural design optimization to address aeroelastic issues, Mach-scaled model rotor blades were built and tested in large-scale wind tunnels. The results validated numerical predictions and industry took over the design for adaptation and further optimization with respect to industrial application. The major conclusions are:

- State-of-the-art comprehensive codes completed by high-order aerodynamic models (CFD) can successfully be employed for aero-acoustical optimization of rotor blades with respect to BVI noise reduction.
- Mach-scale models are invaluable to make a reliable proof of the design and generate data for a deep understanding of the physics involved.
- Based on fundamental work performed by ONERA and DLR within the ERATO project, Airbus Helicopters, together with ONERA, continued research and development with the aim of industrial application.
- Special attention was first given to further aerodynamic refinements of the blade tip for improvement of hover performance with respect to figure of merit, modifying the twist distribution and introducing anhedral.
- Special attention was given to the aeroelastic tailoring of the blade to assure structural integrity, leading to a double spar design.
- The results were confirmed by industry during flight tests, validating the transferability of model-scale wind tunnel tests to full-scale application.

The ERATO research project is a good example of efficient collaboration between the research and industrial worlds. Indeed, this very innovative blade design was initially conceived and tested using a wide array of the most advanced means, and then successfully adapted and flight tested at full-scale on a Dauphin demonstrator aircraft. This research has allowed demonstrating the maturity of the technology in time to allow the prototype of the Dauphin successor (the H160, Figure 38), to successfully launch its flight test campaign with the Blue Edge™ blades.

ACKNOWLEDGMENTS

The following people are deeply acknowledged for their scientific contribution at different steps of this research: Jean Prieur, Patrick Gardarein, Arnaud Le Pape (all from ONERA), Wolf Splettstößer, Markus Raffel and the rotor test team (all from DLR), Reinert Müller (from FIBUS for support regarding the FCM technology and data evaluation). The test teams of DNW and S1MA wind tunnels are acknowledged for their support during the tests. From Airbus Helicopters, Antoine Baud, Paul Cranga, Patrice Rauch, and Vincent Gareton deserve special thanks, as well as the flight test crew and ground support. The authors also wish to thank the French DGAC for their financial support to the ERATO, Blue Edge and Shanel projects.

Figure 38: Flight testing trials of an Airbus Helicopters H160 prototype (©T. Rostang).

Author contact:
Berend G. van der Wall berend.vanderwall@dlr.de
Christoph Kessler christoph.kessler@dlr.de
Yves Delrieux yves.delrieux@onera.fr
Philippe Beaumier philippe.beaumier@onera.fr
Marc Gervais marc.gervais@airbus.com
Jean-François Hirsch jean-francois.hirsch@airbus.com
Kurt Pengel kurt.pengel@dnw.aero
Pascal Crozier pascal.crozier@onera.fr

7. REFERENCES

[1] Kessler, Ch., "Helicopter emergency medical service: motivation for focused research," *CEAS Aeronautical Journal*, **6**(3), 2015, pp. 337-394.

[2] Hirschberg, M., "Investing in Tomorrow's Civil Rotorcraft," *Vertiflite*, **60**(4), 2014, pp. 4-5.

[3] Mooreman, R. W., "Helicopter Noise: The People's Perspective – Los Angeles County," *Vertiflite*, **60**(6), 2014, pp. 14-18.

[4] Mooreman, R. W., "Helicopter Noise: The People's Perspective – Long Island and Chicago," *Vertiflite*, **61**(1), 2015, pp. 28-30.

[5] Mooreman, R. W., "Noise in the Parks," *Vertiflite*, **61**(2), 2015, pp. 34-38.

[6] Kessler, Ch., "Active Rotor Control for Helicopters: Motivation and Survey on Higher Harmonic Control," *CEAS Aeronautical Journal*. **1**(1-4), 2011, pp. 3-22.

[7] Splettstoesser, W. R., Kube, R., Wagner, W., Seelhorst, U., Boutier, A., Micheli, F., Mercker, E., and Pengel, K., "Key Results From a Higher Harmonic Control Aeroacoustic Rotor Test (HART)," *Journal of the American Helicopter Society*, **42**(1), 1997, pp. 58-78.

[8] Yu, Y. H., "Rotor Blade-Vortex Interaction Noise," *Progress in Aerospace Sciences*, **36**(2), 2000, pp. 97-115.

[9] Spiegel, P, and Rahier, G., "Theoritical study and prediction of BVI noise including close interactions," American Helicopter Society Technical Specialists' Meeting on Rotorcraft Acoustics and Fluid Dynamics, Philadelphia, PA, Oct. 15-17, 1991.

[10] van der Wall, B. G., Burley, C. L., Yu, Y. H., Richard, H., Pengel, K., and Beaumier, P., "The HART II Test-Measurement of Helicopter Rotor Wakes," *Aerospace Science and Technology*, **8**(4), 2004, pp. 273–284.

[11] N.N., "Environmental Protection, Volume I Aircraft Noise," Annex 16 to the Convention on International Civil Aviation, 6th Edition, ISBN-13: 978-92-9231-837-6, July 2011.

[12] Web-Site of EASA on Noise Type Certificate-Approved Noise Levels: www.easa.europa.eu/document-library/noise-type-certificates-approved-noise-levels.

[13] Leverton, J. W. and Pike, A. C., "Helicopter Noise – What is Important from a Community Prospective," American Helicopter Society 63rd Annual Forum Proceedings, Virginia Beach, VA, May 1-3, 2007.

[14] Hampson, M., "OEM Perspectives on Noise," *Vertiflite*, **60**(1), 2014, pp. 18-23.

[15] Spiegel, P., Guntzer, F., Le Duc, A., and Buchholz, H., "Aeroacoustic Flight Test Data Analysis and Guidelines for Noise-Abatement-Procedure Design and Piloting," 34th European Rotorcraft Forum Proceedings, Liverpool, UK, Sept. 16-19, 2008.

[16] Kessler, Ch., "Active Rotor Control for Helicopters: Individual Blade Control and Swashplateless Rotor Designs," *CEAS Aeronautical Journal*. **1**(1-4), 2011, pp. 23-54.

[17] Friedmann, P. P., "On-Blade Control of Rotor Vibration, Noise, and Performance: Just Around the Corner?" The 33rd Alexander Nikolsky Honorary Lecture. *Journal of the American Helicopter Society*, **59**(4), 2014, pp. 1-37.

[18] Crozier, P., Leconte, P., Delrieux, Y., Gimonet, B., Le Pape, A., and Mercier des Rochettes, H., "Wind Tunnel Tests of a Helicopter Rotor with

Active Flaps," 32nd European Rotorcraft Forum, Maastricht, Netherlands, Sept. 12-14, 2006.

[19] Delrieux, Y., Le Pape, A., Leconte, P., Crozier, P., Gimonet, B., and Mercier des Rochettes, H., "Wind-Tunnel Assessment of the Concept of Active Flaps on a Helicopter Rotor Model", American Helicopter Society 63rd Annual Forum Proceedings, Virginia Beach, VA, May 1-3, 2007.

[20] Prieur, J. and Splettstoesser, W. R., "ERATO: an ONERA DLR Cooperative Programme on Aeroacoustic Rotor Optimisation," 25th European Rotorcraft Forum, Rome, Italy, Sept. 14-16, 1999.

[21] Splettstoesser, W. R., Prieur, J., Pahlke, K., Schultz, K.-J., van der Wall, B. G., Delrieux, Y., Gardarein, P., Geoffroy, P., and Leconte, P., "Main Phase of the ERATO Cooperation on Aeroacoustic Rotor Optimization." DLR Internal Report, DLR-IB 129-97/10, Braunschweig, Germany, 1997.

[22] Splettstoesser, W. R., Schultz, K.-J., Buchholz, H., van der Wall, B. G., Junker, B., Wagner, W., Mercker, E., and Pengel, K., "ERATO Rotor Validation Test in the DNW - Documentation and Representative Results," DLR Internal Report, DLR-IB 129-99/30, Braunschweig, Germany, 1999.

[23] Splettstoesser, W. R., van der Wall, B. G., Junker, B., Schultz, K.-J., Beaumier, P., Delrieux, Y., Leconte, P., and Crozier, P., "Wind Tunnel Test and Proof of Design for an Aeroacoustically Optimized Rotor," 25th European Rotorcraft Forum, Rome, Italy, Sept. 14-16, 1999.

[24] Müller, R.H.G., Pengel, K., and van der Wall, B. G., "Blade Deflection measurement at the Low Noise ERATO Rotor," 26th European Rotorcraft Forum, The Hague, Netherlands, Sept. 26-29, 2000.

[25] Delrieux, Y., Prieur, J., Costes, M., Gardarein, P., Beaumier, P., des Rochettes, H.M., Leconte, P., Crozier, P., Splettstoesser, W. R., van der Wall, B. G., Junker, B., Schultz, K.-J., Mercker, E., Pengel, K., Philippe, J. J., and Gmelin, B., "The ONERA-DLR Aeroacoustic Rotor Optimisation Programme ERATO: Methodology and Achievements," American Helicopter Society Aerodynamics, Acoustics, and Test and Evaluation Technical Specialists Meeting, San Francisco, CA, Jan. 23-25, 2002.

[26] Beaumier, P., and Bousquet, J.-M., "Applied CFD for Analysing Aerodynamic Flows Around Helicopters, " 24th International Congress of the Aeronautical Sciences (ICAS2004), Yokohama, Japan, Aug. 29-Sept. 3, 2004.

[27] Beaumier, P., and Delrieux, Y., "Description and validation of the ONERA computational method for the prediction of blade-vortex interaction noise," *Aerospace Science and Technology*, **9**(1), 2005, pp. 31–43.

[28] Le Pape, A., and Beaumier, P., "Numerical optimization of helicopter rotor aerodynamic performance in hover," *Aerospace Science and Technology*, **9**(3), 2005, pp. 191–201.

[29] Le Pape, A., "Numerical Aerodynamic Optimization of Helicopter Rotors: Multi-Objective Optimization in Hover and Forward Flight Conditions," 31st European Rotorcraft Forum, Florence, Italy, Sept. 13-15, 2005.

[30] León, E.R., Le Pape, A., Désidéri, J.-A., Alfano, D., and Costes, M., "Concurrent Aerodynamic Optimization of Rotor Blades Using a Nash Game Method," 69th Annual Forum of the American Helicopter Society Proceedings, Phoenix, AZ, May 21-23, 2013.

[31] Rauch, P., Gervais, M., Cranga, P., Baud, A., Hirsch, J.-F., Walter, A., and Beaumier, P., "Blue Edge™: The Design, Development and Testing of a New Blade Concept," 67th Annual Forum of the American Helicopter Society Proceedings, Virginia Beach, VA, May 3-5, 2011.

[32] Rahier, G., and Prieur, J., "An Efficient Kirchhoff Integration Method for Rotor Noise Prediction Starting Indifferently from Subsonically or Supersonically Rotating Meshes." American Helicopter Society 53rd Annual Forum Proceedings, Virginia Beach, April 29-May 1, 1997.

[33] Hirsch, J.-F., and Rochegude, B., "Rotor blade provided with radial section and at least one arrowed section before and/or after," European Patent EP1961658, 2008.

[34] Gervais, M., and Gareton, V., "Analysis of Main Rotor Noise Reduction Due to Novel Planform Design – The Blue Edge™ Blade," 37th European Rotorcraft Forum, Gallarate, Italy, Sept. 13-15, 2011.

NUMERICAL ANALYSIS ON DUCTED FAN TAIL ROTOR IN FORWARD FLIGHT CONDITION

Nahyeon Roh
Ph.D. Candidate
Pusan National University
Pusan National Univ.,
Jangjeon 2-dong,
Geumjeong-gu,
Busan, Korea

Donghun Park
Assistant Professor
Pusan National University
Pusan National Univ.,
Jangjeon 2-dong,
Geumjeong-gu,
Busan, Korea

Chelhun Woo
Research Engineer
Korea Aerospace Industries, Ltd.,
Sanam-myeon, Sacheon-si,
Gyeongsangnam-do,
Republic of Korea, Korea

Sejong Oh
Professor
Pusan National University
Pusan National Univ.,
Jangjeon 2-dong,
Geumjeong-gu,
Busan, Korea

ABSTRACT

During forward flight, the ducted fan tail rotor experience loss of rotor thrust and improvement of vertical fin thrust. This result in different thrust contribution of each component compare to hover flight condition. In this study, numerical research on ducted fan tail rotor is conducted to investigate the flow characteristics and aerodynamic performance in forward flight. The collective pitch angle is varied along forward speed depending on change of main rotor torque. The rotor thrust is weakened as forward speed increases. Whereas effect of vertical fin is improved since the vertical fin acts like fixed wing. As a result of thrust compensate by vertical fin, total thrust is increased after 60knot, despite the loss of rotor thrust. Total thrust is increased up to 2times from minimum point for 60knot to maximum point for 140knot. At 140knot, total power loading is about 4.2times than rotor power loading.

NOTATION

r	: Rotor radius
T	: Thrust
P	: Power
PL	: Power Loading

INTRODUCTION

Typical single rotor helicopter requires a tail rotor to compensate the torque of main rotor and to ensure lateral controllability and stability. In general, the types of tail rotors can be classified into three types: common tail rotor exposed to the outside, ducted fan tail rotor which has the shroud around the rotor[1], and NOTAR[2] utilizing Coanda effect.

The aerodynamic performance of common tail rotor is affected by several flow phenomena such as momentum flow of main rotor, main rotor trailing vortex, vortex ring from tail rotor and freestream wind(Figure 1). Even the vertical fin and rotor are adjacent, the thrust generating by vertical fin is suffered from tail rotor wake.

On the other hand, ducted fan tail rotor reduces the possibilities of the influence of any other wakes, because the rotor is shrouded. In addition, since the section of vertical fin and rotor are separated, it offers advantages in generating additional thrust through the vertical fin. Therefore, for the ducted fan tail rotor, less power will be required to produce a given total thrust. At the same time, boundary layer separation occurs at duct inlet in forward flight. Since the separation increases flow unsteadiness and loss of shroud thrust, it leads to vibration of aerodynamic performance.

Unlike hover flight where efficiency of rotor and shroud are important, forward flight condition could increase the thrust contribution of any other components. Therefore, analysis on the component thrust is needed to suggest the main parameters in forward flight for the detail design. The most of previous research was conducted in a limited forward speed and thrust changes of each component such as shroud, vertical fin was not analyzed.

In this study, numerical simulations are conducted to investigate the aerodynamic performance and flow

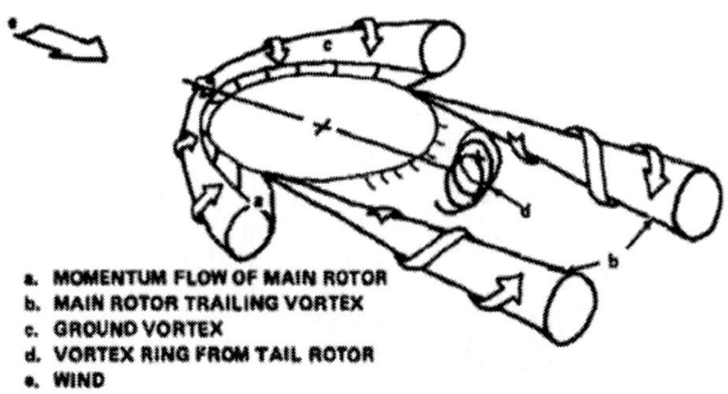

Figure 1. Flow environment of tail rotor [3]

Presented at the 6th Asian/Austrailian Rotorcraft Forum / Heli Japan 2017, Kanazawa, Japan, November 7-9, 2017.

(a) Full geometry **(b) Shroud** **(c) Outer Shroud** **(d) Vertical Fin**

Figure 2. Ducted Fan Tail Rotor Major Component

characteristics of ducted fan tail rotor in forward flight. The collective pitch angle is varied to balance the main rotor torque at each forward speed condition. Proper collective pitch angle along forward speed is suggested by tests on full scale helicopter. Tests is carried out by Airbus Helicopter. Based on the flow analysis, aerodynamiac performance changes of each component is presented.

NUMERICAL METHODS

Solution Procedures

In the present work, a series of computations are carried out using open source code, OpenFOAM. (Full Navier-stokes equation). The PIMPLE algorithm that combines characteristics of SIMPLE (Semi Implicit Method for Pressure Linked Equations) algorithm with PISO (Pressure Implicit with Splitting of Operator) algorithm has been used. The Spalart-Allmaras turbulence model was used.

Furthermore, to overcome the limitations of full CFD analysis that require a complex grid and huge computational resource, CFD-Actuator Disk Method(ADM) hybrid solver was developed and applied. In ADM, the rotor was modelled as an actuator disk. Therefore, the presence of the blades had been represented as momentum source terms which required two-dimensional airfoil lift coefficient and drag coefficient

as a function of Mach number. Thus, the whole domain was simulated with the Navier-Stokes except from the fan which was considered as an actuator disk.

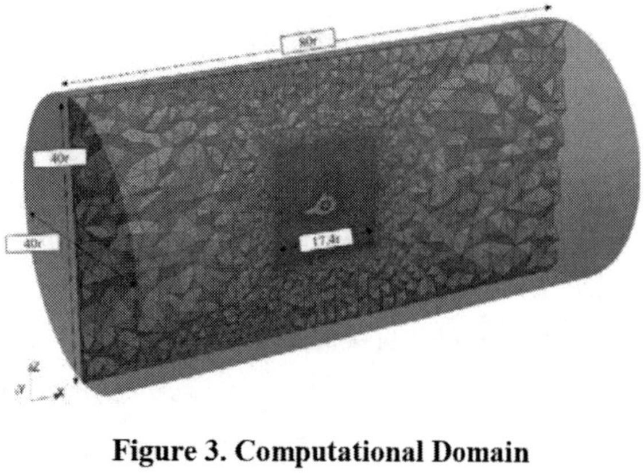

Figure 3. Computational Domain

Analysis Shape and Computational Domain

The isolated ducted fan tail rotor of EC155b1[4] is analyzed. The region and name of selected major component is shown in Figure2. The rotor has 10 blades with even spacing. Each blade consists of 5 airfoils along span. In fact, actual rotor has 10 blades with unequal blade spacing. Though it is known that the spacing between the blades only affects the noise and does not affect the aerodynamic performance[5,6]. So in order to adopting advantages of ADM, even blade spacing is assumed.

As shown in Figure 3, the complete domain encompasses the region of $-20 \leq x/r \leq 20$, $-40 \leq y/r \leq 40$ and $-20 \leq z/r \leq 20$. Total number of cell is 1.5×10^7

RESULTS

To investigate the effect of forward speed, forward speed is varied from knot to 140knot, with interval of 20knot. The collective pitch angle is determined by referring the tests results of full scale helicopter. With the increasing of forward speed, the required collective pitch angle is decreased from 14.4degree to 2.3degree.

Flow Characteristics

The streamline around the tail rotor and surface pressure are illustrated in Figure 4. The color on streamline means velocity in flow direction. As can be seen in Figure 4, rotor wake is weakened as forward speed increases. At 20knot, effect of rotor suction force is clear even around vertical fin. While effect of suction force on vertical fin is diminished at 60knot. Therefore, it can be expected that the rotor thrust will be decreased with increasing of forward speed.

The fin surface, can be seen in the figure, is the upper surface of airfoil. Therefore, figure indicates pressure decrease on upper surface. It means that the vertical fin induces similar results to fixed wing with increasing freestream velocity. Because the airfoil of vertical fin has

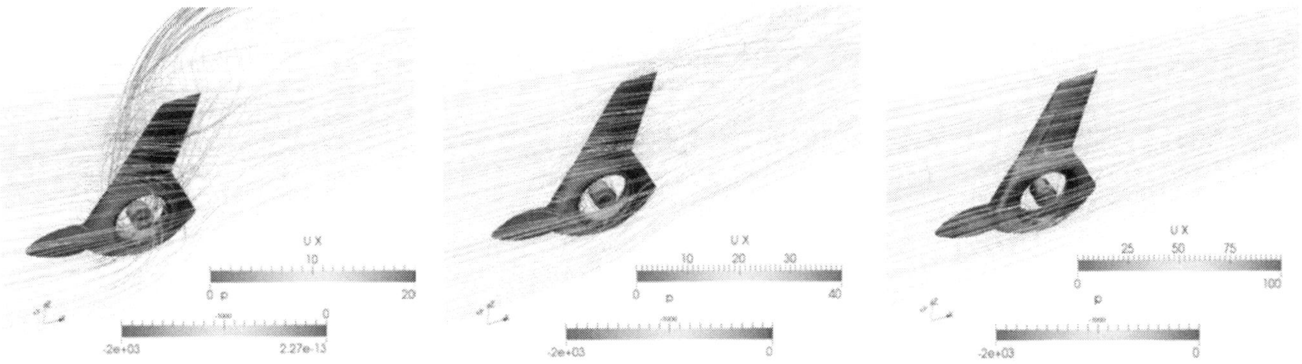

(a) 20knot (b) 40knot (c) 100knot

Figure 4. Streamline with Pressure Distribution on Surface

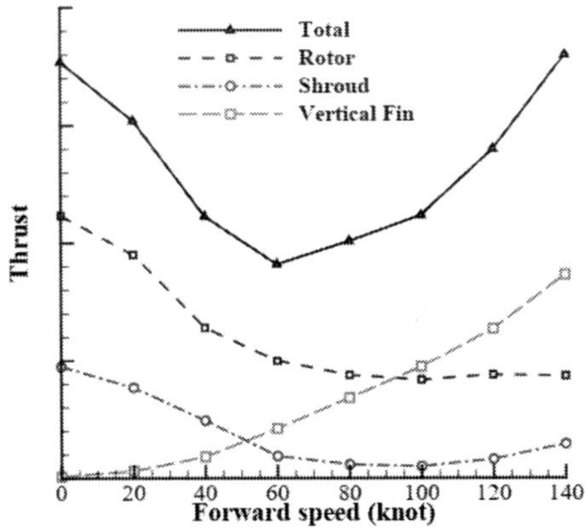

Figure 5. Thrust along with forward speed

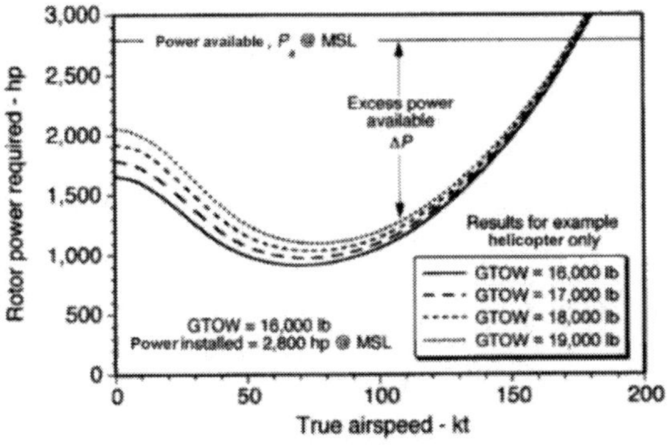

Figure 6. Main rotor power at different GTOW [7]

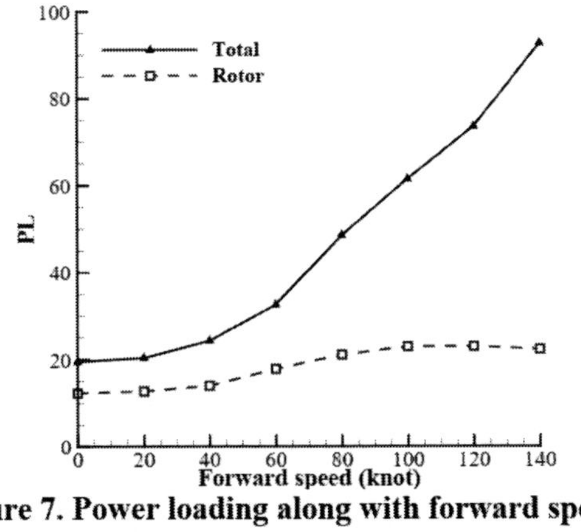

Figure 7. Power loading along with forward speed

camber, increasing of thrust is predictable. Pressure distributions on outer shroud

and hub look alike that of vertical fin, though they are relatively narrow.

Aerodynamic Performance

To evaluate the aerodynamic performance of ducted fan tail rotor, thrust and power loading are examined. Thrust of each component indicates separately. Power loading is calculated in 2 ways, using total thrust and rotor thrust.

As can be seen in Figure 5, rotor thrust is steadily decreased. It is consistent result with flow analysis. This results from both decreasing of collective pitch angle and losing of suction momentum. Thrust of shroud is determined by not only duct inlet pressure drop but also efflux of flow along diffuser. Quantity of momentum gain through rotor influences efflux efficiency. Therefore, decreasing of rotor thrust leads to loss of shroud thrust. On the contrary, since the vertical fin acts like a fixed wing, thrust of vertical fin is increased. The vertical fin thrust occupies about 50% of total thrust at 140knot.

Thrust of complete ducted fan tail rotor is changed due to the each component thrust, decreased until 60knot and increased after that. Total thrust is increased up to 2times from minimum point for 60knot to maximum point for 140knot. Trends of total thrust shows good agreement with general main rotor torque(Figure 6) in forward flight.

Figure 7 shows power loading which means thrust to power ratio. To calculated power loading, total thrust and rotor thrust is applied as can be seen in equation(1) and equation(2).

$$PL_{total} = \frac{T_{total}}{P} \tag{1}$$

$$PL_{rotor} = \frac{T_{rotor}}{P} \tag{2}$$

From Figure 7, it could be known that the total power loading is obviously larger than rotor power loading. At 140knot, total power loading is about 4.2times of rotor power loading.

Despite comparably small difference, performance of rotor is increased. This is results of the lower-tip loss effect associated with a shrouded design, so further reduction in induced power requirement is offered.

For detail design, it is expected that the research on the shape and area of the vertical fin will be valuable to improving the forward flight performance.

CONCLUSIONS

In this study, numerical simulations are conducted to investigate the aerodynamic performance and flow characteristics of ducted fan tail rotor in forward flight. For the numerical simulations, CFD-Actuator Disk Method hybrid solver was developed and applied.

The collective pitch angle of tail rotor is varied along forward speed, since the torque of main rotor is changed. The collective pitch angle is determined by referring the tests results of full scale helicopter.

Through flow characteristics analysis, it is confirmed that the rotor wake is weakened as forward speed increases. Also by comparing surface pressure along forward speed, it is can be known that the vertical fin induces similar results to fixed wing with increasing freestream velocity.

As a result of thrust compensate by vertical fin, total thrust is increased after 60knot, despite the loss of rotor thrust. Total power loading is increased steadily. At 140knot, maximum velocity among analysis condition, total power loading is larger than that of rotor only, about 4.2times. Importance of vertical fin design is confirmed, since vertical fin occupies about 50% of total thrust.

Author contact: Nahyeon Roh nahyeon7782@pusan.ac.kr,
Donghun Park parkdh@pusan.ac.kr,
Sejong Oh tazo@pusan.ac.kr,

ACKNOWLEDGMENTS

This work was supported by the KEIT Research Grant of 2016 ('10053155')

REFERENCES

[1]Mouille, Rene. The "Fenestron," Shrouded Tail Rotor of the SA. 341 Gazelle.: Journal of the American Helicopter Society 15.4, 1970, pp. 31-37.

[2]VanHorn, James R.: Circulation control slots in helicopter yaw control system: U.S. Patent No. 4,948,068. 14 Aug. 1990

[3]Johnson, W., *Helicopter Theory*, Princeton University Press, Princeton, NJ, 1980, pp. 808–813.

[4]www.eurocopter.com

[5]Vuillet, A., and F. Morelli. "New aerodynamic design of the Fenestron for improved performance." (1986).

[6]Mouille, R., and F. D. Ambra. "The 'fenestron'–a shrouded tail rotor concept for helicopters." 42nd AHS Forum, Washington, DC. 1986.

[7]Leishman, J. G., *Principles of Helicopter Aerodynamics*, Cambridge University Press, New York, NY, 2000, Chapter 5.

CFD/CSD coupling investigations on aeroelastic characteristics of rotor with different blade-tip planforms

Ma Li	**Zhao Qijun**
Doctoral student	Professor
Nanjing University of Aeronautics and Astronautics	Nanjing University of Aeronautics and Astronautics

ABSTRACT

For accurate aeroelastic analysis, the unsteady rotor flowfield is solved by Computational Fluid Dynamics (CFD) module based on RANS equations and moving-embedded grid system, while Computational Structural Dynamics (CSD) module is introduced to handle blade flexibility. In CFD module, dual time-stepping algorithm is employed in temporal discretization, JST scheme is adopted in spatial discretization and S-A turbulent model is used to introduce the viscous effect. The CSD module is developed based on Hamilton's variational principles and moderate deflection beam theory. Grid deformation is implemented using algebraic method through coordinate transformations to achieve deflections with high quality and efficiency. A CFD/CSD loose coupling strategy is developed to transfer information between fluid and structure. The CFD and CSD module is validated, then loose coupling is adopted in UH-60A rotor aeroelastic loads prediction under high speed forward flight condition and the calculated results are compared with test data which show good agreement. Finally, effects of blade-tip geometry are investigated through parametric study of aeroelastic characteristics on rotors. The results are evaluated through aeroelastic loads variation and elastic response, and some meaningful conclusions about rotor aeroelastic characteristics are drawn.

1. Introduction

The rotor is one of the most critical components of the helicopter. Compared with the conventional blade tip, the advanced helicopter rotor with new blade tip has greater advantages in aerodynamic characteristics. The blade tip region of rotor is not only the key part of rotor aerodynamic load, but also the formation and escape of blade tip vortex (Ref. 1, Ref. 2). On the other hand, rotor blades are highly flexible, structural deformations have significant influences on rotor aerodynamic characteristics. Therefore, the aerodynamic loads and structural motion responses of rotor blades are coupled with each other. In order to estimate the aeroelastic load of helicopter rotor accurately, aerodynamics, structural dynamics and numerical calculation methods are needed to construct a rational aeroelastic synthetic analysis method for helicopter rotor blades

Nowadays, the research on the shape of rotor blade-tip is still in constant improvement and development in the world. In 1978, the rectangle, taper and swept shape of the rotor were studied by Stroub et al in Ref. 3, which swept taper has better aerodynamic characteristics. In 1984, Desopper et al. (Ref. 4) studied the aerodynamic performance of PF2 blade tip in unsteady transonic flow, and obtained better rotor aerodynamic performance. In 1989, Perry in Ref. 5 studied on blade with parabollic swept tip and BERP blade, the results show that the blade tip can effectively weaken the influence of the shock on the blade, significantly improve the aerodynamic performance of the rotor. In 2013, the "CLOR" shape of the blade (Ref. 6) tips were designed by Zhao Qijun, Wang Bo et al., the results of "CLOR" tip aerodynamic characteristics test show that the "CLOR" tip shape has good aerodynamic performance and low aerodynamic noise level. In 2016, Zhao Guoqing (Ref. 7) studied on the influence of a blade with new tip on the transonic characteristics of rotors, but did not consider the effect of elastic blade. Up to now, unconventional rotor blade design increases the difficulty in aerodynamics characteristics analysis and makes CFD/CSD coupling a necessary tool in high-fidelity airloads prediction. CFD/CSD coupling can also provide better understanding for the flowfield around advanced blade-tip.

The subject of this paper is to apply the advanced blade-tip CSD module to coupling and investigate the UH-60A rotor flowfield in high speed forward flight condition using CFD/CSD loose coupling with different tip swept angles.

2. Numerical Methodology

2.1 CFD Methodology

2.1.1 Grid Generation Method

In the present work, the C-O body-fitting blade grid is used in spatial discretization. Firstly, the 2D grid (Ref. 8) around rotor airfoil is generated, then, geometric twist distributions, chord length variations and quarter-chord positions are imposed with the tip approximated to be a bevel tip; finally, the grid is folded at root to cover the rootcut portion and tip region to capture the tip vortex generated at this region, see Fig.1.

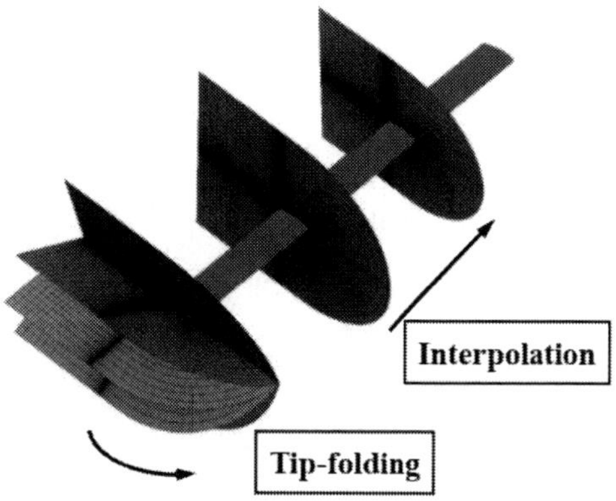

Figure 1 C-O body-fitting blade grid

When the forward flight is calculated, the rotor moves forward and the vortex moves out of the back. Then the Cartesian mesh is adopted. In order to accurately simulate the development of the rotor tip vortex, the rotor wake region is encrypted for different calculation states (as shown in Figure 2), which is helpful to effectively calculate the velocity, spatial distribution and other information of the rotor. In the calculation of rotor CFD in this paper, the blade body fitted grid size is 237×45×61(chordwise, spanwise, and normal), and the background mesh size is 160×164×152(longitudinal, vertical, and lateral).

2.1.2 Moving Embedded Grid Method

In the flowfield calculation considering the elastic deformation, the hole boundary needs to be determined at every azimuth since the relative positions between blade grid cells and background grid cells are varying. In this paper, an efficient "disturbance diffraction" (Ref. 9) method is used to rapidly determine the hole boundary of the blade in the background mesh. The minimum distance method (Ref.10) is used to search the contribution of the artificial boundary of the background mesh, and then the flow field information between the rotor mesh and the background mesh is completed and the moving embedded grid systemis shown in Fig.2.

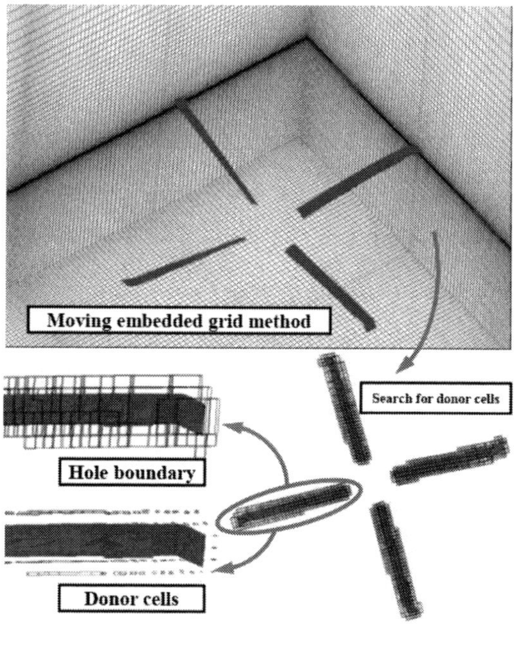

Figure 2 Moving embedded grid system

2.1.3 Grid Deformation Method

The blade deformation is carried out based on the algebraic method (Ref. 11).The elastic displacement of each blade azimuth and profile was calculated through the CSD method, we would get the coordinate transformation matrix T_{DC} and linear deformation matrix x_{lin}. The deformed grid coordinates are given by:

$$\begin{bmatrix} x' \\ y' \\ z' \end{bmatrix} = T_{DC} \begin{bmatrix} x \\ y \\ z \end{bmatrix} + x_{lin} \quad (1)$$

The blade deformations are shown in Fig.3, it can be seen that the deform algorithm achieves grid deformation and maintain good grid quality at the same time.

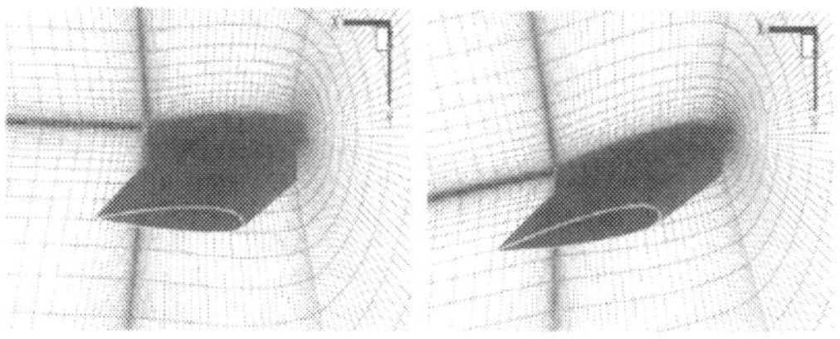

Figure 3 Deformable blade grid

2.1.4 Flowfield Solver

In order to simulate the unsteady flow characteristics of rotor accurately and capture the shock position and flow separation of blade surface, the integral RANS equation (Ref. 12) is used as the governing equation for the flow field of a rotor body fitted grid:

$$\frac{\partial}{\partial t}\int_{\Omega} W d\Omega + \oint_{\partial\Omega} (F_c^M - F_v) \cdot n dS = 0 \qquad (2)$$

$$W = \begin{bmatrix} \rho \\ \rho u \\ \rho v \\ \rho w \\ \rho E \end{bmatrix}, F_c^M = \begin{bmatrix} \rho V_r \\ \rho u V_r + n_x p \\ \rho v V_r + n_y p \\ \rho w V_r + n_z p \\ \rho H V_r + V_t p \end{bmatrix}, F_v = \begin{bmatrix} 0 \\ n_x \tau_{xx} + n_y \tau_{xy} + n_z \tau_{xz} \\ n_x \tau_{yx} + n_y \tau_{yy} + n_z \tau_{yz} \\ n_x \tau_{zx} + n_y \tau_{zy} + n_z \tau_{zz} \\ n_x \theta_x + n_y \theta_y + n_z \theta_z \end{bmatrix}$$

where Ω is the control volume at any spatial station, $\partial\Omega$ is the boundary of Ω, dS is a face element on $\partial\Omega$. The S-A (Ref. 13) turbulence model was adopted which has a small amount of computation and is robust to complex flow calculations. In the unsteady flow field of rotor, the two time method is used to solve the problem. Dual time-stepping algorithm is adopted to simulate the unsteady flowfield. In order to improve the efficiency of the flow field solution, the pseudo time stepping method adopts implicit LU-SGS scheme. The time derivatives of the governing equations are discretized:

$$V \cdot \frac{W^{m+1} - W^m}{\Delta t} + \oiint_{\partial V} (F_c^{n,m+1} - F_c^{n,m}) \mathrm{d}s - \oiint_{\partial V} (F_v^{n,m+1} - F_v^{n,m}) \mathrm{d}s - V \cdot (Q^{m+1} - Q^m) = Res^m \qquad (3)$$

Where the Res is the residual value.

2.2 CSD Methodology

The CSD module is developed based on Hamilton's variational principle (Ref. 14), see Eq.(4):

$$\int_{t_1}^{t_2} (\delta U - \delta T - \delta W) dt = 0 \qquad (4)$$

The finite element model has 14 degrees of freedom (DOFs) without warping and transverse shear, and the end nodes have six DOFs while the middle node has two DOFs (see Fig.4) where u, v and w is linear axial, lag and flap deformation, $v_{,x}$ and $w_{,x}$ is lagging and flapping derivatives while ϕ is elastic torsion.

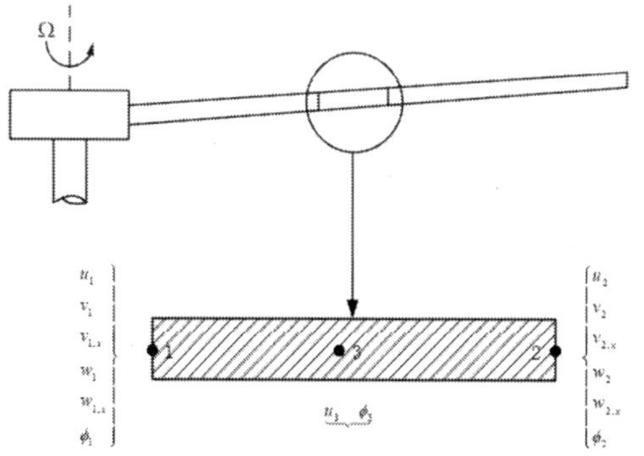

Figure 4 Nodal degrees of freedom

For the blades employed in this study, the sectional properties are obtained from experimental data. Then, these sectional properties are introduced to mass, damping and stiffness matrix assembly procedure. The assembled dynamic equations are solved iteratively by using Newmark-beta method.

2.2.1 Advanced Blade-Tip Modification

To analyze advanced rotor blade-tip with swept or anhedral or combined tip variations, see Fig.5. the CSD module is modified to include complex blade-tip (Ref. 15).

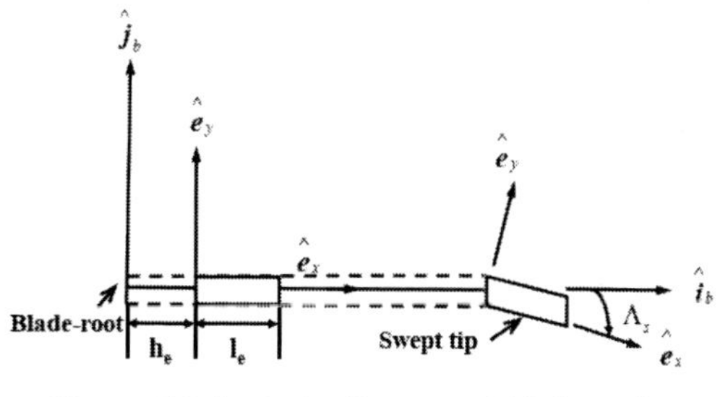

Figure 5 Schematic diagram of blade with swept tip

For advanced blade-tip analysis with combined forward-backward swept such as BERP and highly nonlinear twist like UH-60A, several tip elements rather than one are required to simulate tip configuration details.

The modified tip beam element could be expressed as:

$$\begin{Bmatrix} \mathbf{q} \\ \dot{\mathbf{q}} \\ \ddot{\mathbf{q}} \end{Bmatrix}^{n-1} = [\Lambda]^{n-1} \begin{Bmatrix} \mathbf{q} \\ \dot{\mathbf{q}} \\ \ddot{\mathbf{q}} \end{Bmatrix}^{n-i} \tag{5}$$

where $\mathbf{q}$, $\dot{\mathbf{q}}$, $\ddot{\mathbf{q}}$ is the vector of displacement, velocity and acceleration of beam element, respectively. $[\Lambda]^{n-1}$ is the nonlinear transformation matrix related to tip geometric configuration. And the transformation from local to global coordinate system is similar and can be expressed as Eq.(6).

$$\begin{aligned} &[\Lambda^L][M_i][\Lambda^{LC}]\ddot{q}^G + \\ &[\Lambda^L]([M_i][\Lambda^M]+[C_i][\Lambda^{LC}])\dot{q}^G + \\ &[\Lambda^L][K_i][\Lambda^{LK}]q^G + F_i^G = 0 \end{aligned} \tag{6}$$

2.3 CFD/CSD Coupling Strategy

After the CFD and CSD module are established, coupling strategy is required to transfer fluid and structural information between the two modules. Loose coupling (LC) exchanges information on a per revolutionary basis and is adopted in airloads prediction in level steady flight condition (Ref. 16). In this paper, LC is carried out in rotor airloads prediction and investigated in details, and the flowchart of LC is shown in Fig.6.

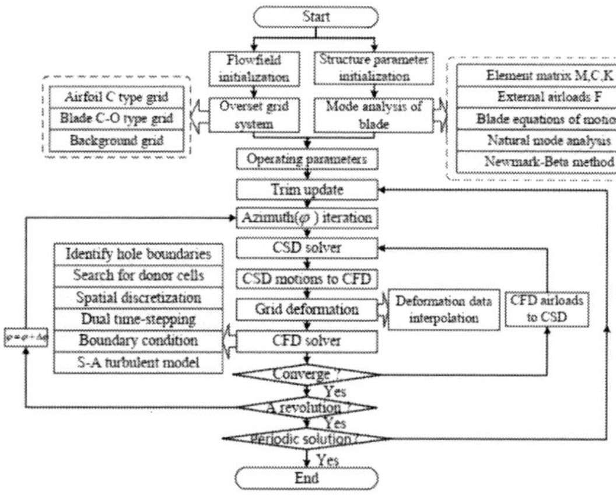

Figure 6 The CFD/CSD loose coupling flowchart

3. Modular Validation

3.1 CFD Module Validation

The CFD module is validated by calculating the pressure distributions of the Caradom-Tung rotor (Ref. 17) using rigid blade assumption. The operating condition in forward flight is $\mu = 0.2$ and $M_{tip} = 0.8$, and the results are as shown in Fig.7(a) and (b). It can be seen that the CFD module employed is capable of capturing accurate airloads and can be used in CFD/CSD coupling.

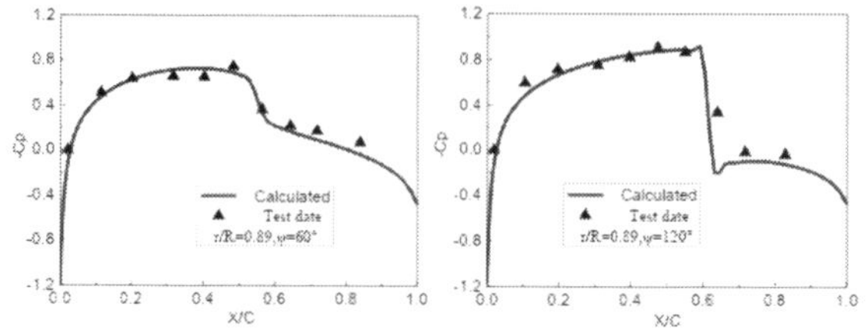

(a) $r/R = 0.89, \psi = 60^\circ$ (b) $r/R = 0.89, \psi = 120^\circ$

Figure 7 C-T rotor pressure distributions

3.2 CSD Module Validation

The CSD module is validated through two different blade models: the Maryland rotating beam (Ref. 18) and the UH-60A (Ref. 19) rotor.

The Maryland rotating beam is analyzed and compared with experimental data in Fig 8. In this analysis, 20 elements are used for the straight portion while 6 elements are employed to simulate the swept tip. The calculated frequencies agree well with the test data. Both flapping and torsional frequencies are accurately computed and it demonstrates the validity of the modified advanced blade-tip ability over a wide range of tip angle variation.

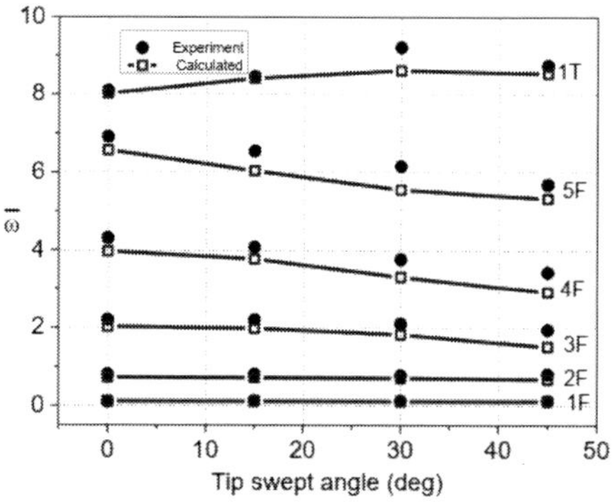

Figure 5 Validation for swept angle variations

The UH-60A natural frequencies in Fig.9 show that the developed advanced blade-tip modification is able to analyze highly nonlinear blade-tip shapes and can obtain better results compared with single-tip element. The solid line is calculated by using UMARC (Ref. 20) with first hand sectional properties, therefore it is taken as reference values for comparison.

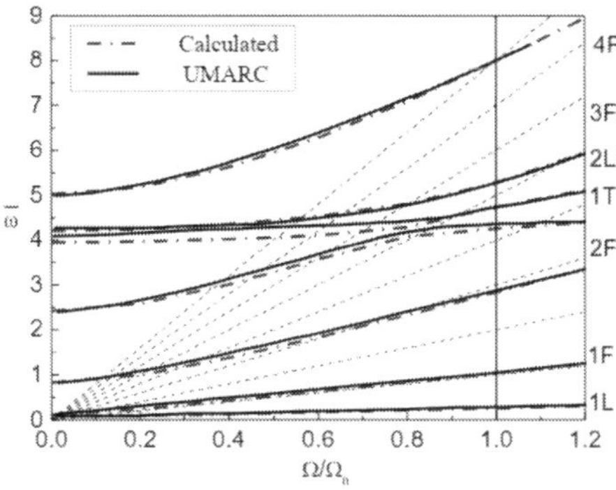

Figure 6 UH-60A blade natural frequencies

Therefore, the modified CSD module is proven to be reliable for CFD/CSD coupling and is used in the present work.

4. CFD/CSD Coupling Results

Results for UH-60A rotor are presented for high speed level steady flight.

4.1 Baseline Validation

The normal force distribution of UH-60A rotor in C8534 (advance ration 0.368, tip Mach number 0.642) (Ref. 21) condition is calculate using rigid rotor CFD method and CFD/CSD coupling method. The corresponding collective and cyclic pitch angles are $\theta_0 = 14.32$, $\theta_{1s} = -9.54$ and $\theta_{1c} = 3.26$, Figure 7 shows the vorticity magnitude. As can be seen, the tip vortex trajectory develops backward as the blade moves forward. In the course of development, the tip vortex can not keep on the rotor trajectory, and gradually dissipate and merge, and the other vortex is formed on both sides of the rotor. From the point of view of the overall blade tip vortex, the CFD/CSD method can capture a complete vortex.

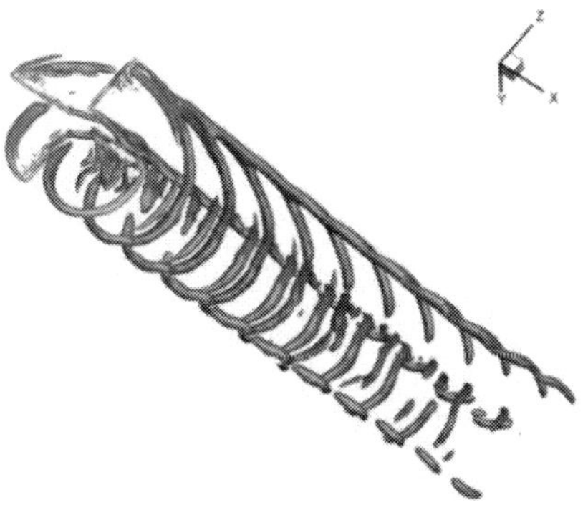

Figure 10 UH-60A rotor vorticity contour

Figure 11 shows the corresponding data for airloads. CFD/CSD coupling can provide better airloads prediction accuracy compare to the rigid CFD method and exhibits a pronounced shock interaction on the advancing side of the blade.

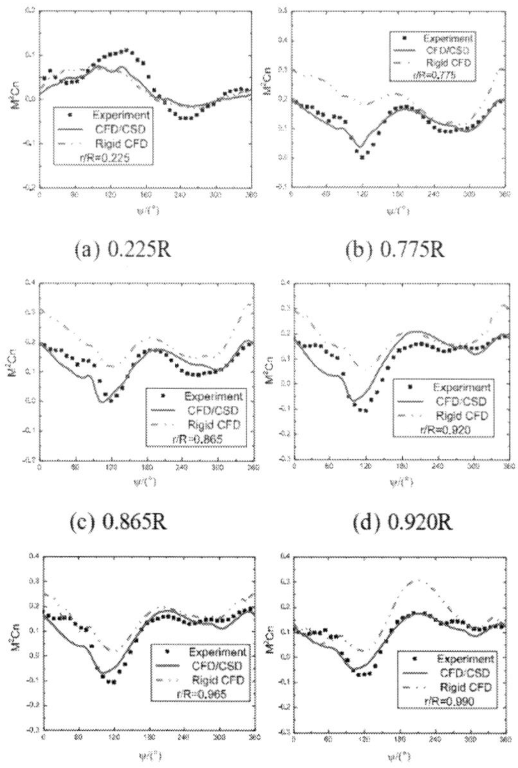

(a) 0.225R (b) 0.775R

(c) 0.865R (d) 0.920R

(e) 0.965R (f) 0.99R

Figure 7 Normal force distribution of UH-60A rotor

Figure 12 shows the surface pressure distribution of blade. The legend on the right hand side is pressure in Pascal. As can be seen in the elastic effect, pressure distribution of blade tip region has been significantly improved, shock of advancing side strength is weakened, which is mainly due to the elastic deformation caused by airloads, thus relaxing the shock strength of the blade tip。

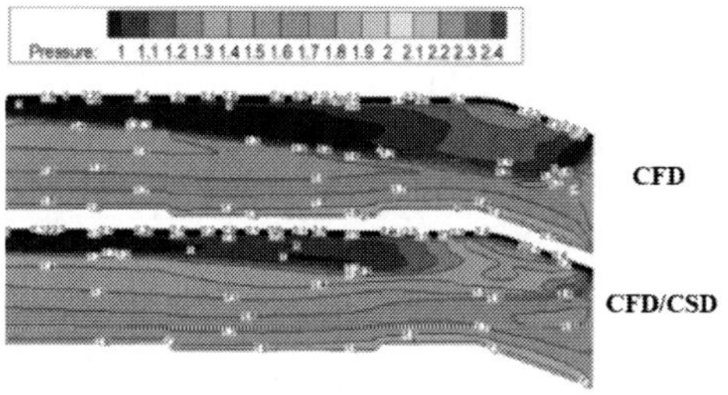

(a) 60°azimuth angle

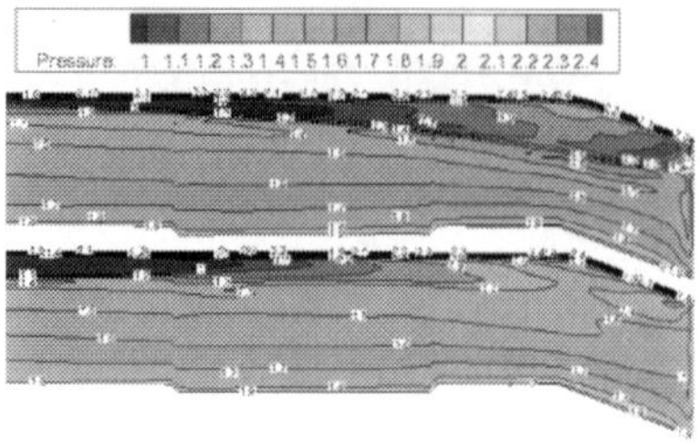

(b) 90° azimuth angle

Figure 12 Surface pressure distribution of blade at different azimuth angles

Figure 13 shows the comparisons of computed and measured root motion (Ref. 22) for the condition. The pitch angle measured is in best agreement with the computation.

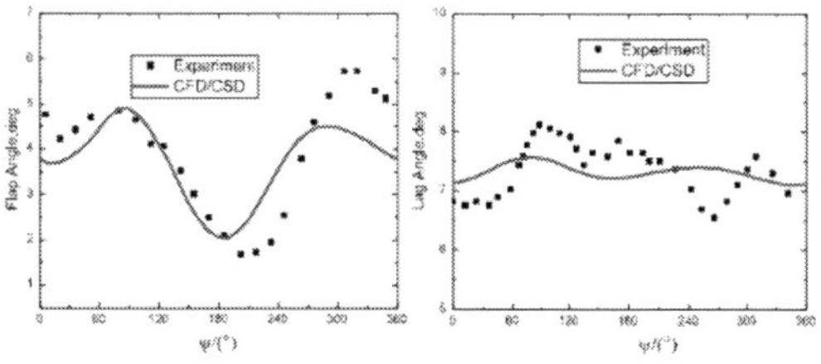

(a) Flap (b) Lag

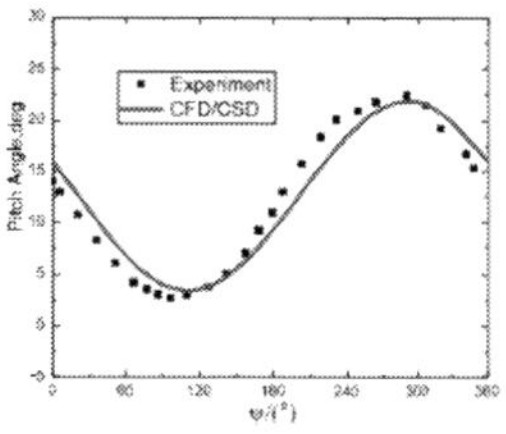

(c) Pitch

Figure 13 Comparison of predicted and measured blade root motion

Figure 13 shows the comparisons of computed and measured elastic bending and elastic twist of blade-tip.

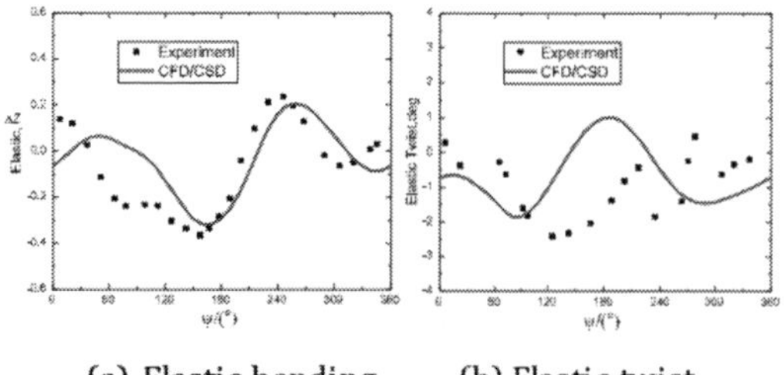

(a) Elastic bending (b) Elastic twist

Figure 14 Comparison of predicted and measured elastic bending and elastic twist of blade-tip

4.2 Parametric Investigations of Blade-tip

To investigate the aerodynamic characteristics of rotor with different tip swept angles, rotor blade-tip swept angle are varied from 10 degree forward to 30 degree backward. For different tip swept angles, SF10 means swept forward tip of 10 degree and SB0, SB10, SB20 and SB30 represent swept backward tip of 0 degree, 10 degree, 20 degree and 30 degree, respectively.

4.2.1 Effect on Airloads Prediction

Figure 15 shows the pressure distributions on blade tips with different swept angles on the advancing side. It can be seen from the figure, with the sweep angle increases, the negative pressure center moves to the tip position, then the shock wave / boundary layer interaction flow separation caused by shifts toward the tip. The reference blade (SB20) has the most gentle gradient of pressure at the blade-tip. The forward swept tip causes negative pressure to move forward. The negative pressure zone is close to the blade-tip when blade with 30 degrees swept (SB30) and the flow separation at the outermost end of the blade-tip is intensified.

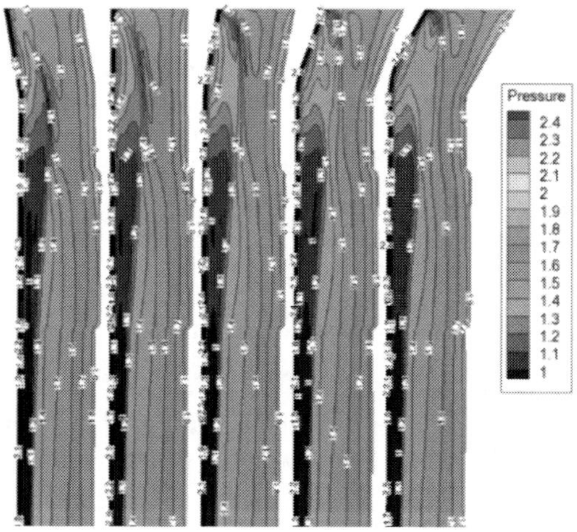

Figure 15 Comparisons of pressure distributions of blades with different swept angles at 90° azimuth angle

The normal force coefficient distributions calculated by using CFD/CSD coupling method for blades with different tip swept angles are shown in Figure 16. At the advancing side, the peaks of negative lift at 120 degree azimuth angle reduce with the increase of tip swept angle, but SB20 blade has minimum negative lift peak indicates that swept tip could reduce the relative inflow Mach number normal to the blade leading-edge and moderate shockwave at the blade-tip region of advancing side.

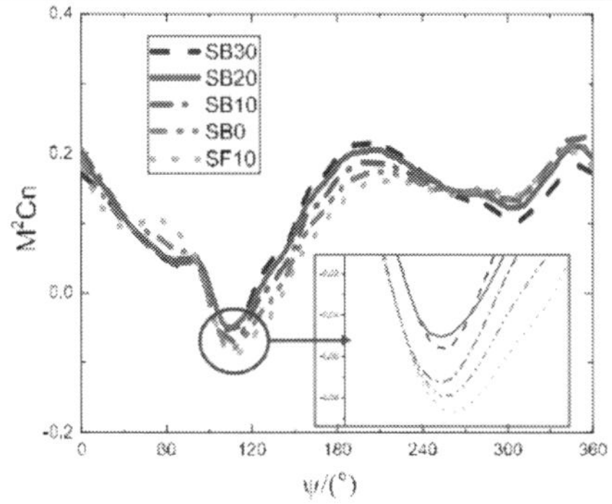

Figure 16 Normal force coefficient distributions of blades with different swept angles at 0.865R

4.2.2 Effect on Structural Dynamic characteristics

Natural frequencies with different swept angles are shown in Figure 17. The change of tip swept angle has almost no effect on low order flapping and lagging frequency, but the torsion frequency increase with the increase of tip swept angle. Besides the SB10 and SF10 have the same torsion frequency which means the same angle of sweepback and sweep forward will not change the natural frequency. Elastic twist of blade-tip with different swept angles are shown in Figure 18. With the increase of sweep angle, torsional deformation at on the advancing side decreases, while increase on the retreating side, the SB30 blade has the maximum value of the tip deformation in a cycle.

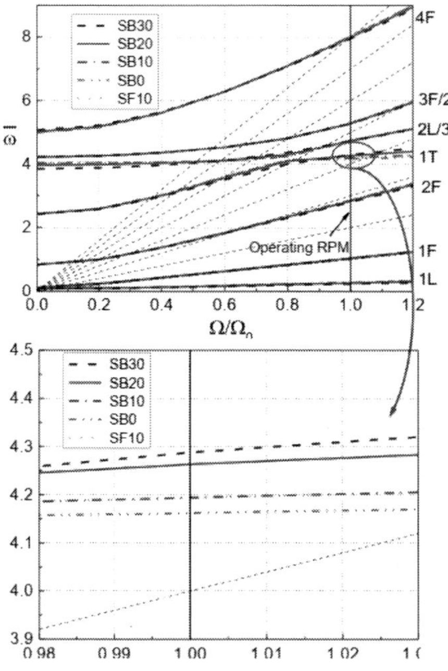

Figure 17 natural frequencies with different swept angles

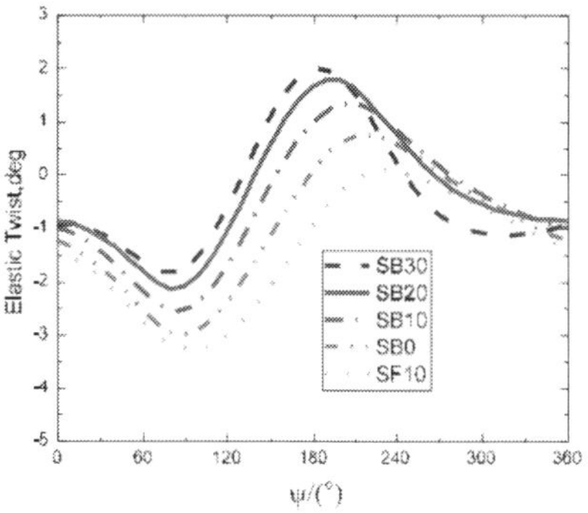

Figure 18 elastic twist of blade-tip with different swept angles

5. Conclusions

An unsteady Navier-Stokes CFD solver for structured grids is loosely coupled with a CSD solver based on moderate deformation beam theory and used to simulate high speed forward flight condition of UH-60A rotor. Parametric investigations on the effects of tip swept angle are carried out. The following concluding remarks are obtained:

(1) The CFD/CSD coupling method established in the present work is able to simulate the unsteady flowfield and aerodynamic airloads as well as the natural frequency of rotor with advanced blade-tip shape and blade displacement in flap, lag and twist. The correlation between computation and measurement was generally good.

(2) Backward swept tip could make the negative pressure center moves to the tip position. The reference blade (SB20) has the gentle gradient of pressure at the blade-tip and minimum negative lift peak at the advancing side. The forward swept tip causes negative pressure to move forward and an area of strong shock waves appears at the corner of swept.

(3) Backward swept tip could increase the torsion frequency, and has almost no effect on flapping and lagging frequency. With the increase of sweep angle, torsional deformation on the advancing side decreases, while increase on the retreating side.

(4) Better aeroelastic characteristics could be obtained through varying blade-tip planform, which could be beneficial to further study of aerodynamic optimization using CFD/CSD method.

4 REFERENCES

[1]Tung, C., Caradonna, F.X., Johnson W, The prediction of transonic flows on an advancing rotor, *Journal of the American Helicopter Society*, vol.32, no.3, p.4-9, 1986.

[2]Conlisk A T. Modern helicopter rotor aerodynamics. *Progress in Aerospace Science*, vol. 37, no.5, p.419-476, 2001.

[3]Stroub R H. Full-scale wind tunnel test of a modern helicopter main rotor-Investigation of tip Mach number effects and comparisons of four tip shapes. The 34th Annual Forum of the American Helicopter Society, 1978.

[4]Desopper A. Study of the unsteady transonic flow on rotor blade with different tip shapes. European Rotorcraft Forum, 10th, The Hague, Netherlands, Aug. 28-31, 1984.

[5]Perry F J. The contribution of planform area to the performance of the BERP rotor. *Journal of the American Helicopter Society*, vol. 34, no.1, p.64-65, 1989.

[6]WangBo, Zhao Qi-jun, Zhao Guo-qing. Experimental research and numerical analysis on aerodynamic characteristics of rotors with improved CLOR blade-tip. *Acta Aeronautica et Astronautica Sinica*, vol. 34, no.2, p.235-245, 2013.

[7]Zhao Guo-qing, Zhao Qi-jun, Wu Qi. Analyses on inhibiting of transonic characteristics of rotor with new type blade-tip. *Journal of Aerospace Power*, vol. 31, no.1, p.143-152, 2016.

[8]HSU K, LEE S L. A numerical technique for two-dimensional grid generation with grid control at all of the boundaries. *Journal of Computational Physics*, vol. 96, no.2, p.451-469, 1991.

[9]Zhao Guo-qing, Zhao Qi-jun, Wu Qi. A universal moving-embedded grid method for CFD simulation of unsteady aerodynamic characteristics of rotor. *Journal of Aerospace Power*, vol. 42, no.3, p.546-554, 2015. (in Chinese)

[10] Zhao Guo-qing, Zhao Qi-jun, Wang Bo. Simulations and parametric analyses on unsteady dynamic stall characteristics of rotor airfoil based on CFD method. *Acta Aerodynamica Sinica*, vol. 37, no.1, p.72-81, 2015. (in Chinese)

[11]Datta A, Sitarman J, Choppa I, et al. CFD/CSD prediction of rotor vibratory loads in high-speed flight. *Journal of Aircraft*, vol. 43, no.6, p.1698-1709, 2006.

[12]Blazek, J., Computational fluid dynamics: principles and applications, second edition, Elsevier Ltd, 2005.

[13]Spalart P R, Allmaras S R. A one-equation turbulence model for aerodynamic flows: AIAA-1991-0439. Reston: AIAA, 1992.

[14]Yuan, K.A., Friedmann, P.P., Aeroelasticity and structural optimization of composite helicopter rotor blades with swept tips, NASA CR 4665, 1995.

[15]Panda B. Assembly of moderate-rotation finite elements used in helicopter rotor dynamics. *Journal of the American Helicopter Society*, vol. 32, no.4, p.63-69, 1987.

[16]Johnson, W., Milestones in rotorcraft aeromechanics, NASA TP 2011-215971, 2011.

[17]Caradonna F X, Laub G H, Tung C. An experimental investigation of the parallel blade-vortex interaction. *NASA TM-86005*, 1984.

[18]Davis, S.J., Predesign study for a modern 4-bladed rotor for the RSRA, *NASA CR 166155*, 1981.

[19]Abhishek, A., Datta, A., Chopra, I., Prediction of UH-60A structural loads using multibody analysis and swashplate dynamics. *Journal of Aircraft*, vol.46, no.2, p.474-490, 2009.

[20]Hamade, K.S., Kufeld, R.M., Modal analysis of UH-60A instrumented rotor blades, *NASA TM 4239*, 1990.

[21]Sitaraman, J., Baeder, J.D., Chopra, I., Validation of UH-60A rotor blade aerodynamic characteristics using CFD. *American Helicopter Society 59th Annual Forum*, Phoenix, Arizona, May 6-8, 2003.

[22]Abrego, A. I., Olson, L. E., Romander, E. A., Barrow, D. A., and Burner, A. W., "Blade Displacement Measurement Technique Applied to a Full-Scale Rotor Test," *American Helicopter Society 68th Annual Forum Proceedings*, 2012.

Aerodynamic Performance Analysis for Rotor with Ice Accretion in Forward Flight

Chen Xi
Ph. D student
Nanjing University of Aeronautics and Astronautics
Nanjing, China

Zhao Qi-Jun
Professor
Nanjing University of Aeronautics and Astronautics
Nanjing, China

Chen Hao
Master student
Nanjing University of Aeronautics and Astronautics
Nanjing, China

ABSTRACT

Based on a 3D icing simulation method, the aerodynamic performance of iced rotor in forward flight is calculated and analyzed. The ice amount is calculated using a coupling method of 3-D CFD and 3-D icing model. Firstly, the orthogonal and body-fitted grids are generated around rotor blade by interpolating and folding approach of section airfoil grid. Then, the CFD simulation method for 3-D flowfield around the blade is established by solving the unsteady RANS equations. Based on the Eulerian approach, the conservation equations of water droplet mass and momentum are solved to obtain the droplet flowfield properties on the same grid used in CFD simulations. Followed that, combined with the characters of rotor blade motion, a 3-D icing model considering the influences of the centrifugal force, the movement of water film and the variation of the azimuthal angle is developed. The comparisons between the calculated results and experimental data about the ice accretion on rotor in forward flight indicate that the present simulation method is effective. At last, the variation of aerodynamic performance of iced rotor and the influence of icing on aerodynamic performance are calculated and analyzed. Additionally, the temperature in forward flight on the rotor performance degradation are also calculated, and some new conclusions are obtained. The influence of icing on aerodynamic characteristic is bigger in the advancing blade than that in the retreating blade. At the moderate temperature, the blade lift coefficient decrease mostly due to the formation of the typical glaze ice.

NOTATION

W = conservative variable

F = convective fluxes

F_v = viscous fluxes

m = mass in the mass balances, $kg \cdot s^{-1}$

q = heat in the thermal balances, $W \cdot s^{-1}$

ρ_d = apparent density of water droplet, $kg \cdot m^{-3}$

α = droplet volume fraction

ρ_w = density of water, $kg \cdot m^{-3}$

Re = Reynolds number

S = shadow zone variable

R = radius of rotor

Subscript

d = water droplet

f = water film

imp = water impinged on the surface

out = water flowing out of the cell

in = water flowing into the cell from adjacent cells

so = water freezing on the surface

evp = water evaporated and sublimated in the cell

INTRODUCTION

The ice accretion has been a significant factor in limiting the safe flight regimes of various aircraft (Ref. 1). When helicopters fly in the icing condition, the ice accretion may occur on blades of rotors, and the designed aerodynamic shapes of rotors may be modified and degraded, result in some serious air crash accidents (Ref. 2). Therefore, the research of ice accretion for helicopter, especially on rotors, is one of the important issues in helicopter safe flight.

Initial research work was mainly concerned with the rotor-blade airfoil icing, having experience with experimental and theoretical studies on airplane and propeller system in the icing condition. Korkan (Ref. 3, 4) studied the aerodynamic problems on rotors of helicopter model with simulated ice shape by experimental method, and these simulated ice shapes are derived from a

---

Presented at the 6th Asian/Austrailian Rotorcraft Forum / Heli Japan 2017, Kanazawa, Japan, November 7-9, 2017.

predetermined natural ice. Lee studied the iced rotor in a wind tunnel test program in full scale (Ref. 5), and this iced rotor was also a replica of the main rotor of a UH-1H helicopter in hover in an artificial cloud (Ref. 6). However, due to limitations of the experimental conditions and security considerations, the real time measurement of aerodynamic performance for rotors in ice accretion process is difficult and expensive. As a result, a few papers have been found on the aerodynamic performance of rotor by experiments in ice accretion process.

With the development of computers, numerical simulation method provides the possibility to analysis aerodynamic performance of iced rotor in ice accretion process. In recent years, many numerical methods have been developed successfully to simulate ice accretion on aircraft, especially for the fixed-wing aircraft, such as: LEWICE (Ref. 7), ONERA (Ref. 8), FENSAP-ICE (Ref. 9), etc. Tran (Ref. 10) developed a numerical method to predict ice accretion on airfoils including thermodynamic effect. Cao (Ref. 11) presented a numerical simulation method to predict the ice accretions on 3-D fixed wings. In the researches of icing models, Blackmore (Ref. 12, 13) introduced an icing model that describes spongy ice formation in atmospheric icing, and a dendritic-growth layer is assumed to relate the microscopic ice growth with the macroscopic mass/energy conservation. Kong (Ref. 14) developed an aircraft supercooled icing model, in which the influences of the flow velocity on ice growth are taken into account.

However, these methods and icing models are mainly applied for airfoil or fixed-wing aircraft, and rotor icing is more difficult due to the complexity of the rotor movement. Rotors are operated in strong non-linear flowfield containing 3-D flow features, unsteady flow and centrifugal force which are unique for rotors (Ref. 15, 16). These features may not be considered in the numerical methods above, and they are also the difficulties of the numerical simulation for ice accretion on rotors, especially for the forward flight state.

In order to solver the complexity of the simulation method for predicting ice accretion on rotors, some simplified methods are also developed. Cao (Ref. 17) investigated the trim and flight characteristics of the CH-47B tandem twin rotor helicopter under an icing condition. Rajmohan (Ref. 18) developed a quasi-steady numerical method for ice accretion on helicopter rotors in forward flight using the LEWICE3D. Bain (Ref. 19) used the same approach to predict the ice accretion and ice shedding on rotors in forward flight. In these methods, they extracted representative 2-D airfoil conditions for blade sections at different radial and azimuthal locations, and predicted ice buildup on the rotor accounting for the diverse operating environments. Ice accretion on rotors could be conveniently achieved by using these methods, but it is difficult to obtain accurate ice shapes with 3-D characters. For other hand, since the accreted ice has different shapes according to different flight conditions, such as different advance ratios in forward flight, the aerodynamic performance of iced rotors are also different even in a same icing condition. An aerodynamic performance analysis method based on a simplified icing model, such as ignoring the 3-D characters of rotors, may not get an accurate result. Therefore, the icing model for rotors may need to be further developed, and some features unique for ice accretion on rotors also should be considered, such as the centrifugal force.

The objective of the present work is to develop a new 3-D icing model for simulating the ice accretion on rotors, and to analyze the aerodynamic performance of iced rotors. The ice amount is calculated using a coupling method of 3-D flowfield solver and the 3-D icing model for rotors. The influences of the centrifugal force and the effects of water film motion on blade surface are considered in the present 3-D icing model, and the influence of azimuthal angle variation in forward flight is taken into consideration. The present 3-D icing model and numerical method is validated in this paper by the experimental data of Spinning Rotor Blade (SRB) developed at Anti-icing Material International Laboratory (Ref. 20). Additionally, the effects of the temperature on the rotor performance degradation in forward flight are analyzed, and some new conclusions are obtained.

NUMERICAL METHODS

The numerical simulation approach for ice accretion on rotors consists of several modules, including: (1) CFD solver module for predicting the flowfield, (2) Eulerian method for obtaining the droplet impingement on the rotor, (3) 3-D icing model for simulating the ice accretion on the rotor, and (4) Grid regeneration for the iced blade. If the variation of surface shape is large enough to affect the flowfield of the rotor, the CFD solver should be invoked to compute the new flowfield around the iced rotor, and the functional relationship of icing information with the azimuthal angle should be reestablished in forward flight. The main analytical procedure of the ice accretion on a rotor is illustrated in Fig. 1.

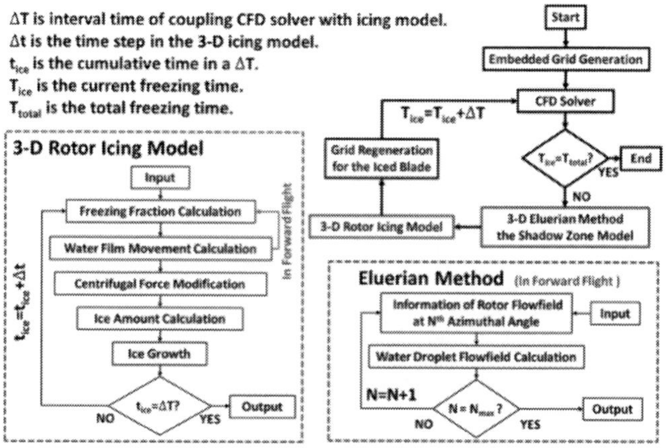

Figure 1. Overview of the numerical simulation process for ice accretion on rotor.

The hybrid method (Ref. 21) was employed as a base of flowfield solver to conduct efficient predictions of the 3-D viscous flowfield of a helicopter rotor under forward flight conditions.

Grids around rotor blade are generated by interpolating and folding of airfoil section grids constructed by solving Poisson equations. Structured Cartesian grids are used as the background grid system. Fig. 2 shows the moving-embedded grid system of a rotor in hover and forward flight.

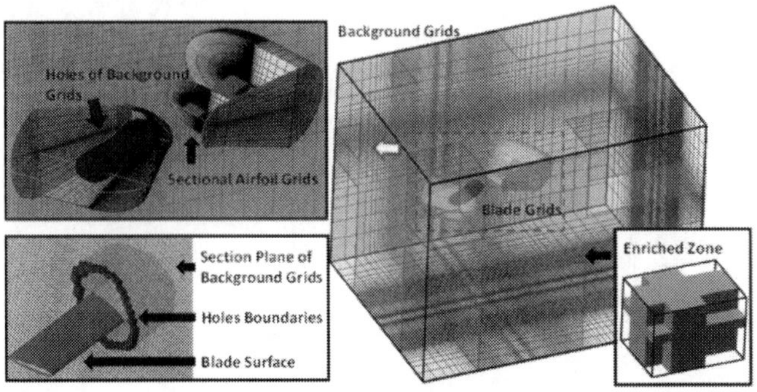

Figure 2. Overall schematic of a moving-embedded grid system for the rotor.

The CFD methods based on the Reynolds-averaged Navier-Stokes equations are employed, and the governing equations in integral form for predicting the flowfield of the rotor are described as:

$$\frac{\partial}{\partial t}\iiint \vec{W} d\Omega + \iint_{S} (\vec{F} - \vec{F}_v)\cdot \vec{n} ds = 0 \tag{1}$$

Considering the general situation that the flowfield of the helicopter rotor is characterized by unsteady phenomena, and the blade is in relative motion, the Eulerian method is applied for the simulation of ice accretion on rotors.

After introducing the concept of water droplet volume fraction, which indicates the volume of the water droplet per volume of the air, and the apparent density of water droplet is constructed:

$$\rho_d = \alpha \cdot \rho_w \tag{2}$$

The continuity and momentum equations for droplets in 3-D applications on rotors could be simplified as:

$$\frac{\partial}{\partial t}\iiint_{\Omega} \vec{W}_d d\Omega + \iint_{S} \vec{F}_d \cdot \vec{n} ds = \iiint_{\Omega} \vec{R} + \vec{R}_d d\Omega \tag{3}$$

The physics of the rotor icing process is extremely complex, the 3-D rotor icing model (Ref. 22) is developed to predict ice accretion considering the 3-D flow effects of rotor, the water film movement and the influence of the centrifugal force due to rotor rotation. In this model, water droplets which impinge on a unit come into being a thin, continuous film of water which covers the blade and iced surface. Some freeze immediately at the point of impact, and some run back along the surface into adjacent cells caused by the action of the airflow. Figure 3 gives the water film movement on the surface of blade.

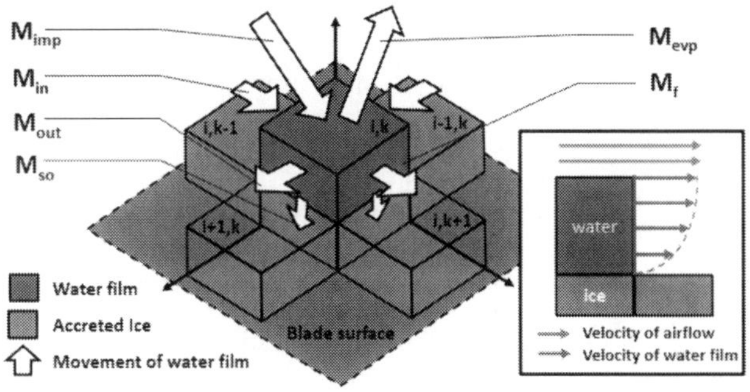

Figure 3. Water film movement on the surface of blade.

The mass and thermal balances are performed in each control cell:

$$m_{imp} + m_f(t) + \sum m_{in} = m_f(t+\Delta t) + m_{so} + \sum m_{out} + m_{evp} \tag{4}$$

$$q_{imp} + q_f(t) + \sum q_{in} = q_{evp} + q_{so} + \sum q_{out} + q_{hc} + q_{cnd} + q_f(t+\Delta t) \tag{5}$$

CALCULATED RESULTS AND ANALYSES

Aerodynamic performance on iced airfoil

The experimental data of iced NACA23012 airfoil at the NASA Langley Low Turbulence Pressure Tunnel (LTPT) is selected to validate the accuracy of the numerical method for aerodynamic performance of iced airfoil (Ref. 23). The LTPT measurements were made at a Mach number of 0.208 and a Reynolds number of approximately 2×10^6. The modified NACA23012 section and iced shape are shown in the figure 4.

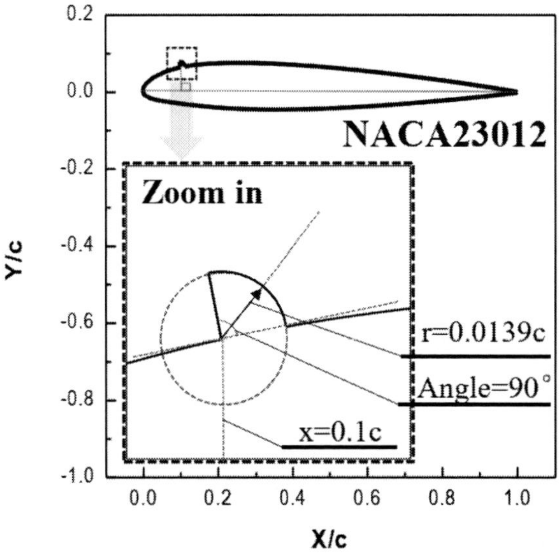

Figure 4. Modified NACA 23012 section and detail view of ice shape.

Figure 5 shows pressure coefficient distributions of airfoil surface with and without ice at AoA = 0°. As shown in the figure, numerical results show a very good level of agreement with LTPT data, and it indicates that the present CFD method is reliable to simulate the aerodynamic performance of iced airfoil.

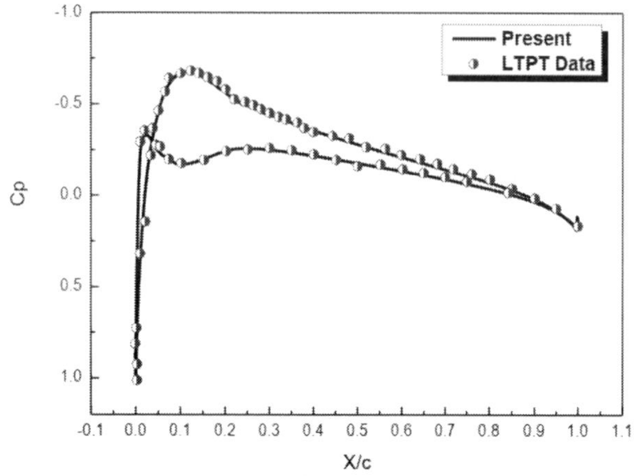

(a) clean surface

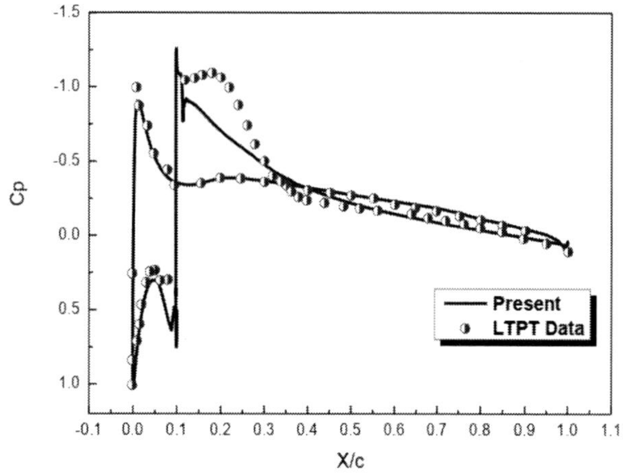

(b) iced surface

Figure 5. NACA 23012 surface pressure coefficient comparison.

Ice accretion on rotors in forward flight

The experimental data of Spinning Rotor Blade (SRB) developed at Anti-icing Material International Laboratory (Ref. 20) is selected to validate the accuracy of the numerical method for predicting ice accretion on rotor in forward flight. The rotor is a 1/18-scale model of a small helicopter. The rotor diameter is 0.78 m, and the blade-tip speed is 130 m/s. The forward flight speed of the rotor was 15 m/s. The blades are untwisted, and made of NACA 0012 sections with a constant chord 0.07 m and a root cut out of 0.075 m. The liquid water content was 0.84 g/m^3, and the median diameter of the water drops was 27 μm. The ambient temperature was -15°C. In this experiment, the rotor was operated at a fixed collective pitch of 6° with zero cyclic pitch.

Figure 6 shows 3-D iced blade and a sample of the correlation at some radial sections. As can be seen, the ice amount (the area of the ice shape) increases gradually from blade root to blade tip. The ice shape represents the characteristics of rime ice near the blade root (r<0.4R), and then represents the characteristics of glaze ice.

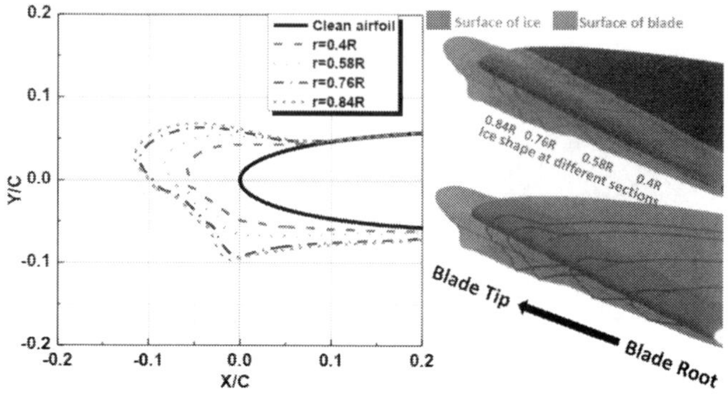

Figure 6. Results of ice accretion on the blade in forward flight.

Figure 7 shows the comparisons of the predicted ice thickness at the tip of blade section with calculated results from the reference (Ref. 19) and experimental data (Ref. 20). As can be seen, the variation tendency of ice amount obtained by calculated methods in Ref. 19 are in good agreements with the experimental data, but there is a noticeable deviation in the value of ice amount. By comparison, the ice amount calculated by the proposed method is closer to the experimental data, and it indicates that the new 3-D icing model is reliable to simulate the ice accretion on rotor blade in forward flight.

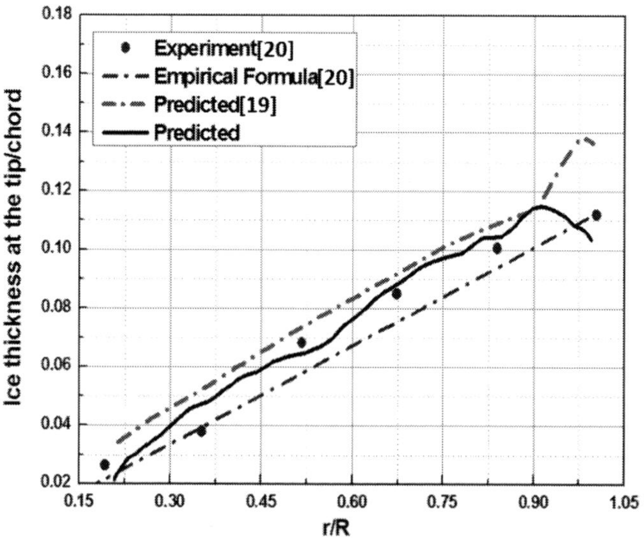

Figure 7. Ice thickness along the blade spanwise with the experimental data.

Aerodynamic performance on iced rotors in forward flight

Figure 8 shows the distributions of lift coefficient (C_L) of the clean and iced blade in forward flight. In the advancing blade, the lift is affected in the range from 0° to 110° azimuthal angle, and it decreases maximum by 18% at 70° azimuthal angle. In the retreating blade, the lift is affected in the range from 250° to 330° azimuthal angle, and it decreases maximum by 9% at 280° azimuthal angle. Figure 9 shows the distributions of required power (C_Q) of clean and iced blade, and there is a large increase when ice forms on the blade surface.

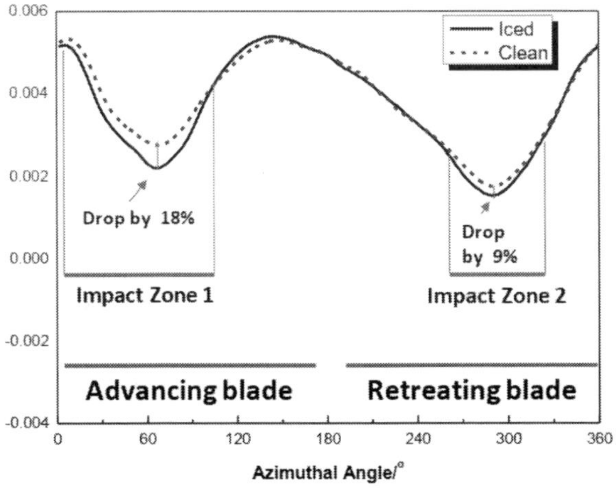

Figure 8. Lift coefficient distributions of the clean and iced blade.

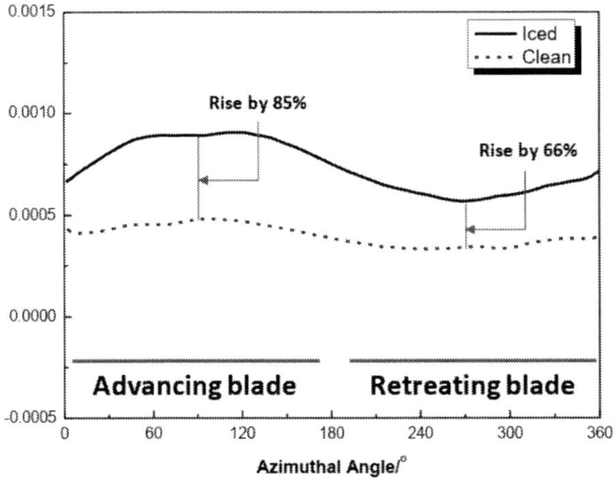

Figure 9. Required power distributions of clean and iced blade.

In order to reveal the reason of the decrease of lift and the increase of required power, figure 10 gives the pressure distributions of clean and iced blade surface at 270° azimuthal angle. Due to the ice accretion at the leading edge of blade, there is significant change in the movement of airflow over the blade surface, result in low-pressure zones emerging at the lower surface. Compared with the clean blade, the low-pressure zone at the lower surface of iced blade is much bigger.

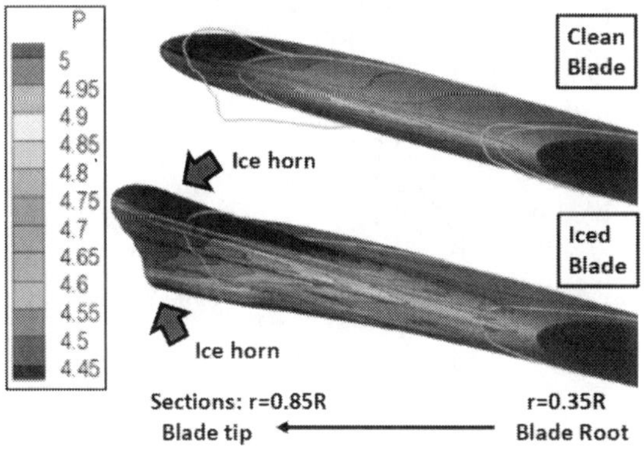

(a) blade surface

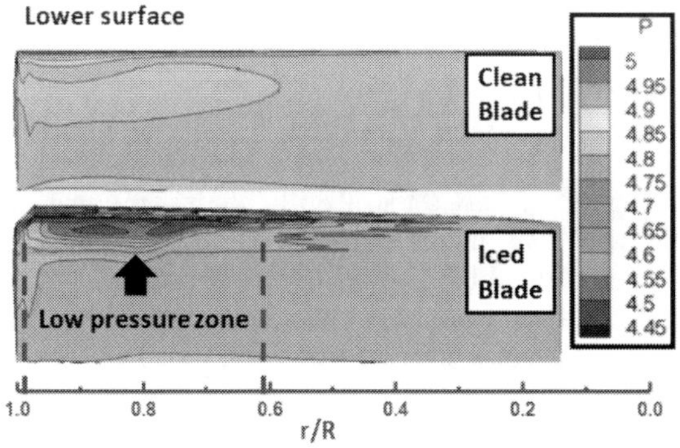

(b) lower surface

Figure 10. Pressure distributions of clean and iced blade surface at 270° azimuthal angle.

Figure 11 gives the sectional pressure distribution of clean and iced blades at 270° azimuthal angle. At the 0.35R blade section, there is no significant variation in the pressure distribution due to the small ice amount. At the 0.85R blade section, the ice amount is bigger, and there is obvious horn-shaped ice at the leading edge. The movement of air flow is affected by the lower ice horn, and there even be a flow separation on the lower surface, and the low-pressure zones emerges as seen in the figure.

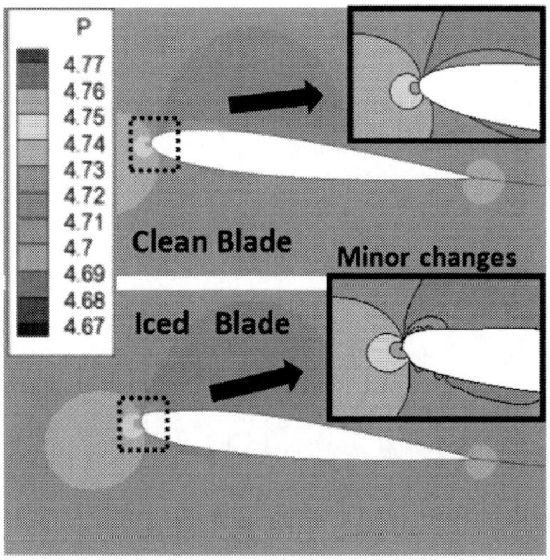

(a) 0.35R section

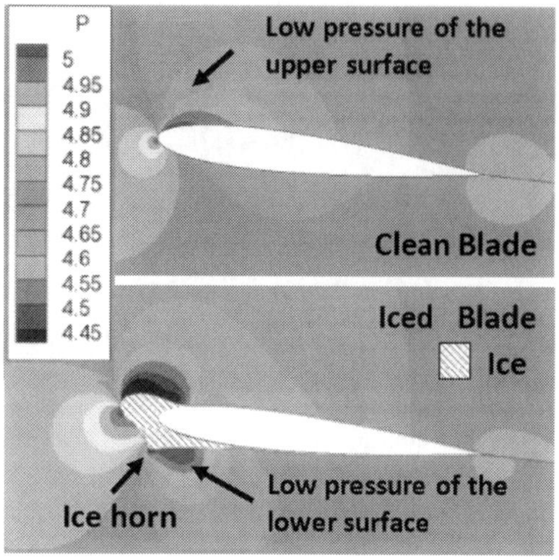

(b) 0.85R section

Figure 11. Sectional pressure distribution of clean and iced blade at 270° azimuthal angle.

Figure 12 give the 0.35R blade sectional pressure coefficient distribution at two azimuthal angles. Compared with the clean blade, there are a decrease in the lower surface and an increase in the upper surface at r=0.35R, causing a loss of lift force. However, the chord is longer by the accreted ice at the leading edge of blade, and the lift force might be increased. As a result, there might not be a significant variation in the sectional lift force near the blade root, and there might even be a small increase of the lift force in some icy conditions.

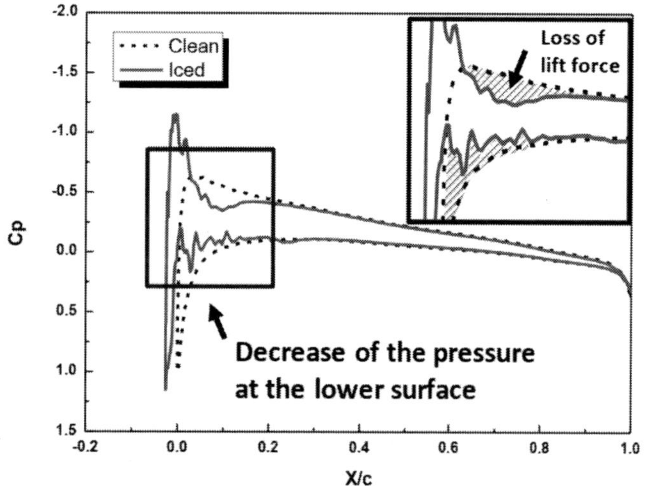

(a) 90° azimuthal angle

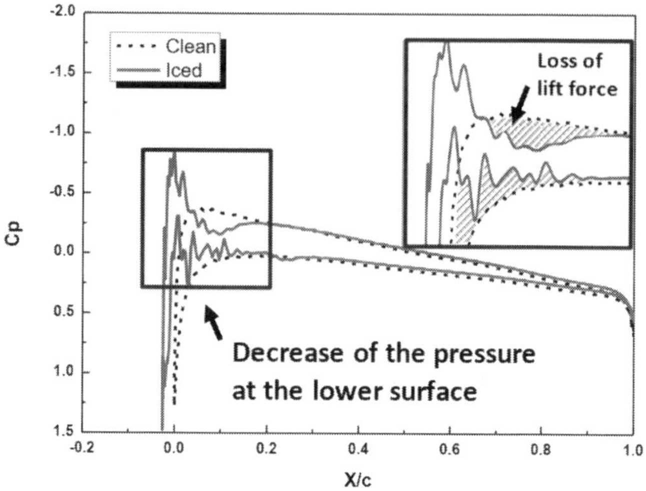

(b) 270° azimuthal angle

Figure 12. Pressure coefficient distribution on the surface of clean blade and iced blade at r=0.35R section.

Figure 13 give the 0.85R blade sectional pressure coefficient distribution at two azimuthal angles. Near the blade tip, there are an increase of lift force at the leading edge as in the r=0.35R section mentioned earlier. Unlike the region near the blade root, there is a sharp reduction in the pressure coefficient of the lower surface, and this can create a zone of negative lift force.

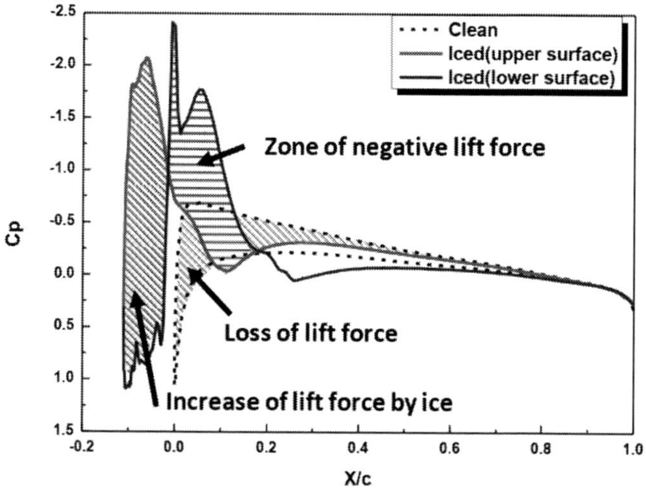

(a) 90° azimuthal angle

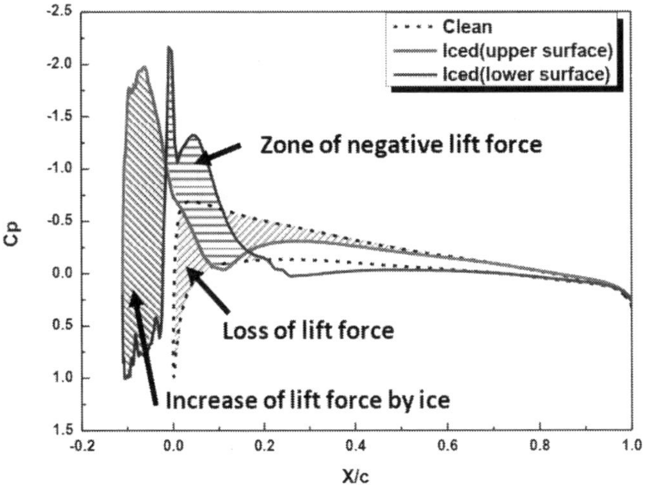

(b) 270° azimuthal angle

Figure 13. Pressure coefficient distribution on the surface of clean blade and iced blade at r=0.85R section

The aerodynamic performance of iced rotors at different temperature in forward flight

The ice accretions on rotor are calculated at different temperature, the other conditions are the same as the experiment of Spinning Rotor Blade. Figure 14 shows the ice accretion area along the spanwise direction of the blade at different temperatures in forward flight. As can be seen, the ice area on section gradually increases as the temperature decreases.

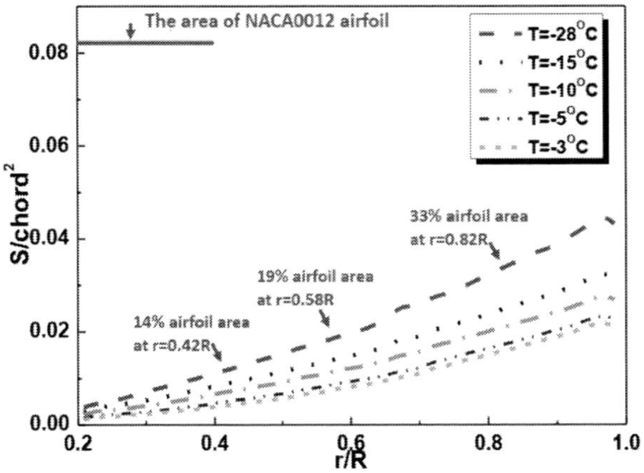

Figure 14. Ice area on different blade sections along span in forward flight at different temperatures.

Figure 15 shows the distributions of the lift coefficient (C_L) and required power (C_Q) of the blade at different temperatures. In the advancing blade (near the 60° azimuthal angle), the ice formed on the blade at T=-10°C had the biggest effects on the lift force. Since there is a smaller ice amount at the higher temperature (T=-5°C), the lift force of blade is less affected. Although the ice amount at the lower temperature (t=-28°C) is much more than others, the loss of lift is not great. It is because that the rime ice at lower temperature have a similar streamline shape. In terms of the required power, there are a smaller increase at the lower temperature and a larger increase at the higher temperature.

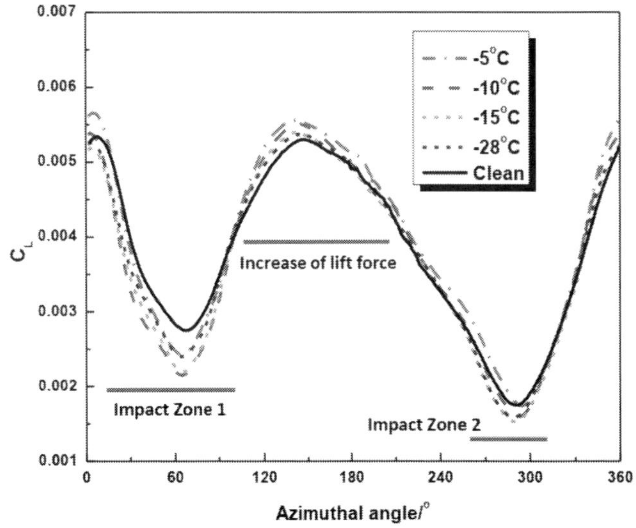

(a) lift coefficient

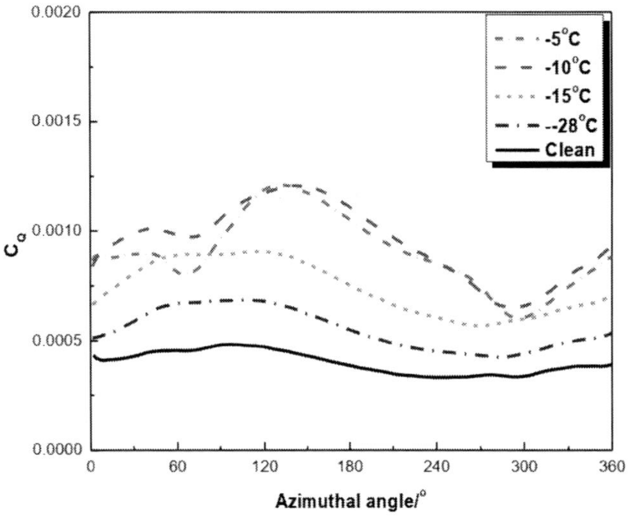

(b) required power

Figure 15. Distributions of the lift coefficient and required power of the blade at different temperatures.

Figure 16 shows the distributions of the pressure coefficient (C_P) of the blade sections at -5°C and -28°C temperatures. At the higher temperature (T=-5°C), the pressure on the lower surface markedly decreases, and it is smaller than that of the upper surface at some zone. Therefore, there is a larger zone of negative lift force at the blade section, especially near the blade tip, ant it is the main reason of lift loss. At the lower temperature (T=-28°C), although the pressure on the lower surface decreases, the area of negative lift is smaller compared with that of higher temperature. In addition, this phenomenon is not obvious near the blade root, since ice amount is smaller.

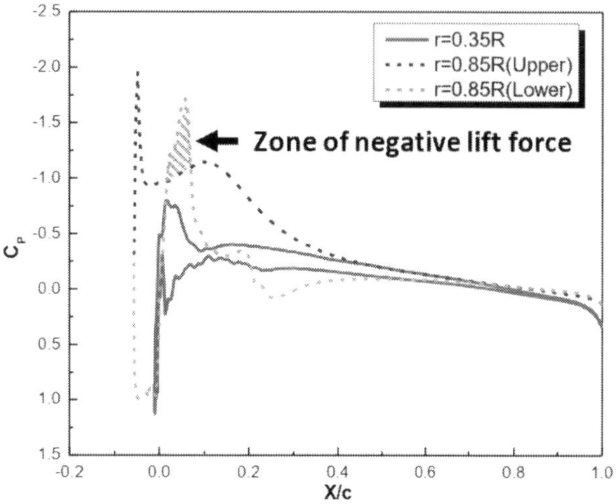

(a) T=-5°C

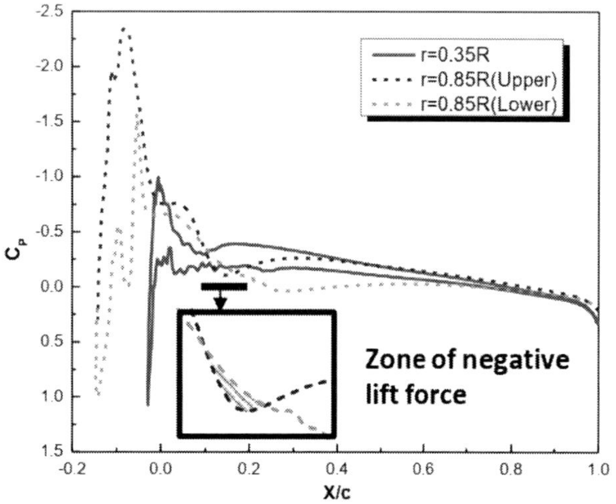

(a)T=-28°C

Figure 16. Distributions of the sectional pressure coefficient at different temperatures.

CONCLUSIONS

Based on 3-D icing simulation method for rotor, aerodynamic performance of iced rotors is calculated, and a corresponding numerical method for calculating aerodynamic performance of rotors in forward flight during ice accretion process is developed. Main conclusions are obtained from the calculated results as follows:

1) In the forward flight, the influence of icing on blade lift coefficient is bigger in the advancing blade than that in the retreating blade. There is a large increase in the required power when ice forms on the blade surface.

2) In a blade section of iced rotor, the pressure coefficient often has a decrease in the lower surface and an increase in the upper surface compared to clean blade, result in a loss of lift force. However, the chord is longer by the accreted ice at the leading edge of blade, and sectional lift force may increase in some case. Due to the increase of ice amount along the spanwise direction, the influence of icing on aerodynamic performance of blade section gradually increase.

3) At the higher temperature, the pressure on the lower surface markedly decreases, and it is smaller than that of the upper surface in some area. There is a larger zone of negative lift force at the blade section, especially near the blade tip, ant it is the main reason of lift loss. At the lower temperature, although the pressure on the lower surface decreases, the area of negative lift is smaller compared with that of higher temperature.

Author contact:

Xi Chen chenxicc@nuaa.edu.cn

Qijun Zhao (Corresponding author) zhaoqijun@nuaa.edu.cn

Hao Chen chenhao@nuaa.edu.cn

REFERENCES

[1]Flemming, R., and David, A., "High Speed Ice Accretion on Rotorcraft Airfoils," NASA CR 3910, 1985.

[2]Hartman, P., "Prediction of Ice Accumulation and Airfoil Performance Degradation: A Boeing-CIRA Research Collaboration," AHS International 62nd Annual Forum, 2006.

[3]Korkan, K. D., Cross, E. J., and Miller, T. L., "Performance Degradation of a Model Helicopter Rotor with a Generic Ice Shape," Journal of Aircraft, Vol. 21, No. 10, 1984, pp. 823-830.

[4]Korkan, K. D., Dadone, L., and Shaw, J. R., "Performance Degradation of Helicopter Rotor in Forward Flight due to Ice," Journal of Aircraft, Vol. 22, No. 8, 1985, pp. 713-718.

[5]Lee, J., "Aerodynamic Evaluation of a Helicopter Rotor Blade with Ice Accretion in Hover," AIAA paper 1984-0608, 1984.

[6]John, D. L., "Documentation of Ice Shapes on the Main Rotor of a UH-1H Helicopter in Hover," NASA 168332, 1984.

[7]Wright, W., "Further Refinement of the LEWICE SLD Model," AIAA paper 2006-0464, 2006.

[8]Hedde, T., Guffond, D., "ONERA three-dimensional Icing Model," AIAA Journal, Vol. 33(6), 1995, pp. 1038-1044.

[9]Nilamdeen, S., and Habashi, W., "Fensap-ice: Modeling of Water Droplets and Ice Crystals," AIAA paper 2009-4128, 2009.

[10]Tran, P., Brahimi, M. T., Paraschivoiu, I., Pueyo, A., and Tezok F., "Ice Accretion on Aircraft Wings with Thermodynamic Effects," Journal of Aircraft, Vol. 32, No. 2, 1995, pp. 444-446.

[11]Cao, Y. H., Ma, C., Zhang, Q., and Sheridan, J., "Numerical Simulation of Ice Accretions on an Aircraft Wing," Aerospace Science & Technology, Vol. 23, No. 1, 2012, pp.296-304.

[12]Blackmore, R. Z., and Lozowski, E. P., "A Theoretical Spongy Spray Icing Model with Surficial Structure," Atmospheric Research, Vol. 49, No. 4, 1998, pp. 267-288.

[13]Blackmore, R. Z., Makkonen, L., and Lozowski, E. P., "A New Model of Spongy Icing from First Principles," Journal of Geophysical Research, Vol. 107, No. D21, 2002, pp. 9-1–9-15.

[14]Kong, W., and Liu, H., "Development and Theoretical Analysis of an Aircraft Supercooled Icing Model," Journal of Aircraft, Vol. 51, No.3, 2014, pp. 975-986.

[15]Srinivasan, G. R., Baeder, J. D., Obayashi, S., and Mccroskey, W. J., "Flowfield of a Lifting Rotor in Hover - A Navier-Stokes Simulation," AIAA Journal, Vol. 30, No. 10, 1992, pp. 2371-2378.

[16]Rajagopalan, R. G., Mathur, S. R., "Three Dimensional Analysis of a Rotor in Forward Flight," Journal of the American Helicopter Society, Vol. 38, No. 3, 1993, pp. 14-25.

[17]Cao, Y. H., Li, G., and Zhong, G., "Tandem Helicopter Trim and Flight Characteristics in the Icing Condition," Journal of Aircraft, Vol. 47, No.5, 2010, pp. 1559-1569.

[18]Rajmohan, N., Bain, J., Nucci, M., and Kreeger, R., "Icing Studies for the UH-60A Rotor in Forward Flight," Proceedings of the 2010 AHS Aeromechanics Specialists' Conference, American Helicopter Society, San Francisco, CA, Jan 20-22, 2010, pp. 261-276.

[19]Bain, J., Cajigas, J., Sankar, L., Flemming, R. J., and Aubert, R., "Prediction of Rotor Blade Ice Shedding Using Empirical Methods," AIAA paper 2010-7985, 2010.

[20]Fortin, G., and Perron, J., "Spinning Rotor Blade Tests in Icing Wind Tunnel," AIAA paper 2009-4260, 2009.

[21]Zhao, Q. J., Xu, G. H., and Zhao, J. G., "New Hybrid Method for Predicting the Flowfields of Helicopter Rotors," Journal of Aircraft, Vol. 43, No. 2, 2006, pp. 372-380.

[22]Chen, X., Zhao, Q. J., "Numerical Simulations for Ice Accretion on Rotors Using New Three-Dimensional Icing Model," Journal of Aircraft, Vol. 54, No. 4, 2017, pp. 1428-1442.

[23]Broeren, A. P., Lee, S., Lamarre, C. and Bragg, M. B., "Effect of airfoil geometry on performance with simulated ice accretions volume 1: experimental investigation," DOT/FAA/AR-03/64, August 2003.

Chen Xi is a Ph.D. student in the College of Aerospace Engineering at Nanjing University of Aeronautics and Astronautics, and his research interests are ice simulation of rotor, helicopter CFD. Email: chenxicc@nuaa.edu.cn.

Zhao Qi-jun is a professor and Ph.D. advisor in the College of Aerospace Engineering at Nanjing University of Aeronautics and Astronautics. His main research interests are helicopter CFD, helicopter aerodynamics, rotor aerodynamic shape design, and active flow control on aerodynamic characteristics of rotor et.al. Email: zhaoqijun@nuaa.edu.cn.

CFD Analysis on Aerodynamic Characteristics of Helicopter Ducted Tail Rotor Based on Sliding Mesh

Cao Chen-Kai
PhD student
Nanjing University of Aeronautics and Astronautics
Nanjing, China

Zhao Qi-Jun
Professor
Nanjing University of Aeronautics and Astronautics
Nanjing, China

Ni Tong-Bing
PhD student
Nanjing University of Aeronautics and Astronautics
Nanjing, China

ABSTRACT

In order to calculate the aerodynamic characteristics of the ducted tail rotor, a computational fluid dynamics (CFD) method is established to solve the flowfield of the FANTAIL ducted tail rotor under different flight conditions. A sliding mesh method is used to transfer data between the computational domains of duct and tail rotor. A body-fitted grid technology is presented based on grid generation methodology by solving Poisson equations. Based on the grid generation technology, the unsteady Reynolds-averaged Navier-Stokes (RANS) equations are chosen as the governing equations for predicting the flowfield of the ducted tail rotor and the highly-efficient implicit scheme of lower-upper symmetric Gauss-Seidel (LU-SGS) is adopted. In order to capture the aerodynamic characteristics of the ducted tail rotor more accurately, the Spalart-Allmaras turbulence model is employed to close the RANS equations. The flowfield and aerodynamic characteristics of the FANTAIL ducted tail rotor under different flight conditions are calculated respectively and compared with the available experimental data. By analyses and comparisons on the results of unsteady aerodynamic of the ducted tail rotor under different flight conditions, some meaningful conclusions are obtained, and these investigation results could provide a good foundation for ducted tail rotor design in the future.

INTRODUCTION

Tail rotor is a very important component of helicopter, the flow field and the characteristics of tail rotor have important influences on the performance, flight quality and noise characteristics of helicopter. The analysis of tail rotor has been one of the research hotspots. The tail rotor is divided into conventional tail rotor and ducted tail rotor. Compared with the conventional tail rotor, the ducted tail rotor has higher safety and better aerodynamic performance, less noise and other advantages. Therefore, the ducted tail rotor has been used widely. Due to the existence of duct, the slip form and the aerodynamic characteristics of ducted tail rotor have changed greatly. Furthermore, the aerodynamic characteristics of ducted tail rotor are closely related to its flight state. Therefore, to carry out the research on the aerodynamic characteristics of ducted tail rotor have great value for the design and use of helicopter.

To analyse the aerodynamic characteristics of the ducted tail rotor, a series of methods have been developed including experimental method (Ref. [1] to [2]), theoretical analysis method (Ref. [3]) and the computational fluid dynamics method, which were presented in Ref. [4] to [7]. The experimental method is direct and reliable, but the cost of the experimental method is higher and the cycle is longer, which limits the application of the experimental method. Compared to the experimental method, the theoretical analysis method has shorter cycle and lower cost, so the theoretical analysis method has been widely used. However, it's difficult to accurately capture the local characteristics of the flow field around the ducted tail rotor and the influences of the change of structure parameters. The CFD method can capture the characteristics of the flow field, especially the details of the transonic flow field.

The characteristics of the flow field and the performance of FANTAIL in hover, side flight and forward flight have been studied by Alpman (Ref. [3]), which were carried out by using Euler equation and disk theory. However, the effects of viscous were not considered. More detailed aerodynamic characteristics of the flow field of FANTAIL in hover have been analysed by Ruzicka (Ref. [5]) which were based on Navier-Stokes equation and overset grid technology. But, the state of side and forward flight have not been studied.

In order to further investigate the flow field and aerodynamic characteristics of the tail rotor, it is necessary to develop a more accurate, more general and practical CFD method. In this paper, a CFD analysis method and a sliding mesh system are established to solve the flowfield of the FANTAIL ducted tail rotor under different flight conditions. The validity and reliability of the method are verified by a

Presented at the 6th Asian/Austrailian Rotorcraft Forum / Heli Japan 2017, Kanazawa, Japan, November 7-9, 2017.

variety of examples. On the basis of this, the aerodynamic characteristics of the FANTAIL ducted tail rotor in different states (hover, forward flight, side flight, etc.) are studied and analyzed.

METHODLOGY AND MODEL

Geometric model

The FANTAIL ducted tail rotor is employed and its basic parameters are shown in Table 1.

Table 1. Parameters of FANTAIL ducted tail rotor.

Variable	Values
Rotor chord	0.1676 m
Rotor radius (R)	0.685m
Center body radius (R_0)	40%R
Blade negative twist	-7°
Tip speed	205 m/s
Blade number	8
Spread angle	5°
Lip radius	15%R

The geometric model has been simplified. The FANTAIL ducted tail rotor model does not include vertical tail, flat tail and the thickness of the outer channel along the circumference is unchanged. Figure 1 shows a schematic diagram of the FANTAIL ducted tail rotor model.

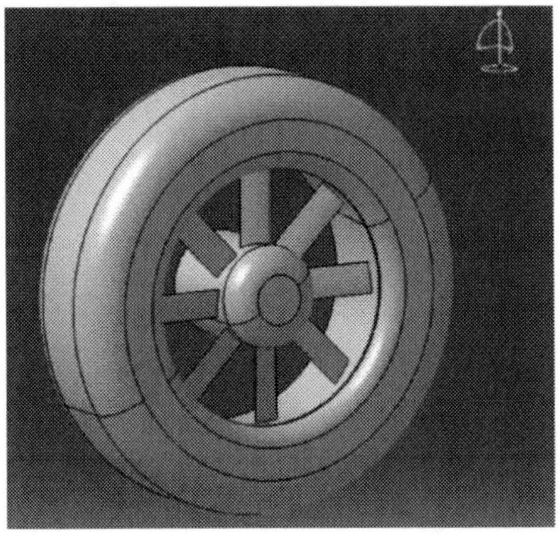

Figure 1. FANTAIL ducted tail rotor model.

The sliding mesh method allows relative sliding between adjacent grids. The sliding mesh method has the characteristics of fast and efficient, and it is not necessary to deform, re-generate or mesh digging the grid during the relative movement of different grids, so it is easy to simulate the flow field with rotation period.

Sliding grid system

The sliding mesh method allows relative sliding between adjacent grids. The sliding mesh method has the characteristics of fast and efficient, and it is not necessary to deform, re-generate or mesh digging the grid during the relative movement of different grids, so it is easy to simulate the flow field with rotation period.

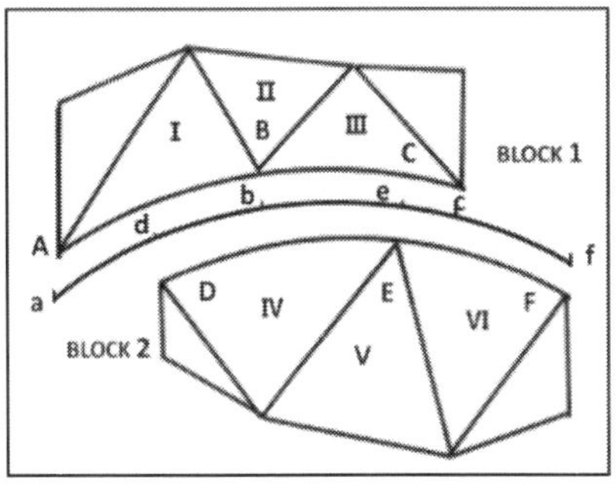

Figure 2. Two - dimensional slip interface.

In Figure 2, the interface is composed of A-B and B-C and D-E and E-F, and a-d, d-b and b-e are generated at the intersection. The d-b plane, the b-e plane and the e-c surface are formed at the overlapping of the two cell regions to form the inner region, and the remaining a-d plane and the c-f plane form a periodic region. When calculating the flow of the interface into the IV unit, use the d-b and b-e faces instead of the D-E surface and transfer information from I and III units to the IV unit.

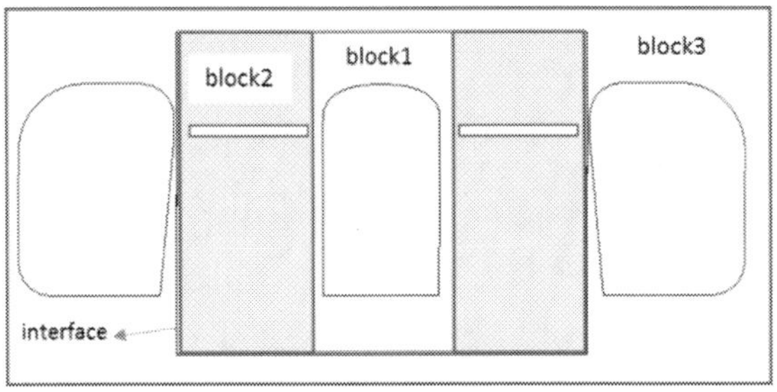

Figure 3. Block layouts for the ducted tail rotor.

In order to establish a sliding grid system, the flow field near the ducted tail rotor flow field is divided into four parts. The part containing the central body is block 1, and the part containing the duct is block 3. The rotation area containing the tail rotor in the middle of block 3 and block 1 is block 2.

Figure 4 shows the o-type body grid around the central body and the outer duct. The grids near the wall have been encrypted.

(a) Grids around the central body

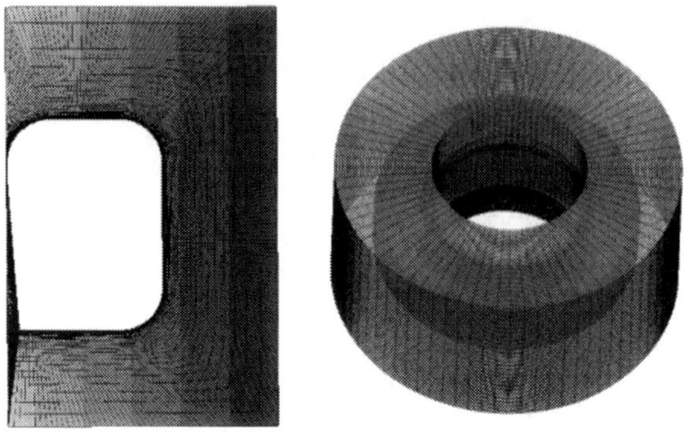

(b) Grids around the duct

Figure 4. Grids around the Static parts.

As the blades rotate around the central body, in order to adapt to the overall grid distribution and the sliding mesh method, the outer shape of the blade grid is made into a circle. In order to solve the problem that the gap between the blade tip and the duct wall is too small, the multi-block grids have been employed around the blade grid, and Fig. 5 shows the distribution of the blade grid.

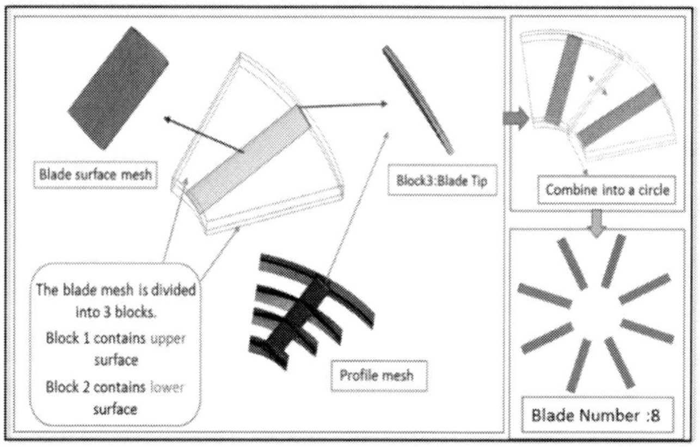

Figure 5. Multi-block grids around the Rotating parts.

Because of the tip-shroud clearance gap between the tip of the blade and the wall of the duct, the traditional nested grid method is difficult to carry out the digging process, and the sliding mesh method with the interface sliding relative can solve the problem.

CFD solver

The unsteady Reynolds-averaged Navier-Stokes (RANS) equations are employed as the governing equations.

$$\frac{\partial}{\partial t}\int_{\Omega} \boldsymbol{W} \mathrm{d}\Omega + \int_{\partial\Omega} \left(\boldsymbol{F}_c - \boldsymbol{F}_v\right) \mathrm{d}\boldsymbol{S} = 0 \tag{1}$$

The Spalart-Allmaras (Ref. [8]) turbulence model is used for the RANS closure. The S-A model is established by comprehensively analyzing the experimental results of two-dimensional mixed layer, laminar flow and plate boundary layer. The simulation ability of the separated flow is strong and the calculation results are satisfactory.

The second-order lower-upper symmetric Gauss-Seidel (LU-SGS) scheme are selected to improve calculation accuracy.

$$\begin{cases} (D+L)D^{-1}(D+U)W^{n} = -R^{*n} \\ (D+L)W^{(1)} = -R^{*n} \\ (D+U)W^{n} = -DW^{(1)} \end{cases} \tag{2}$$

RESULTS AND DISCUSSIONS

Hover

In order to verify the effectiveness of the numerical method of aerodynamic characteristics of the ducted tail rotor based on the sliding mesh and the N-S equation, the thrust of tail rotor and the total thrust of ducted tail rotor at different collective pitch have been calucated, and compared with the test data. It can be seen from figure 6 that the calculated results in this paper are in good agreement with the experimental result. The thrust and the total thrust increase with the increase of the collective pitch, which is consistent the test result.

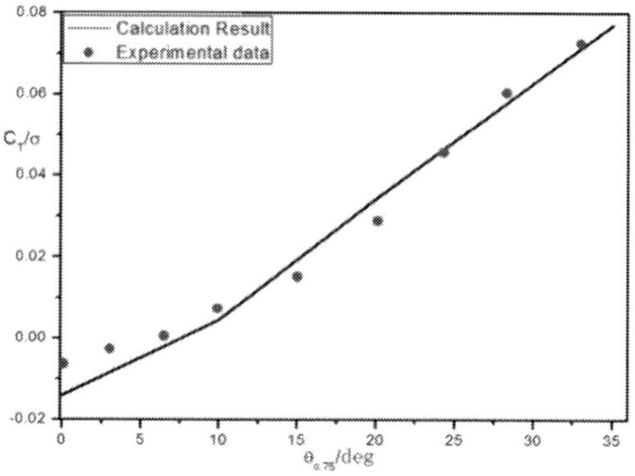

(a) The thrust of tail rotor

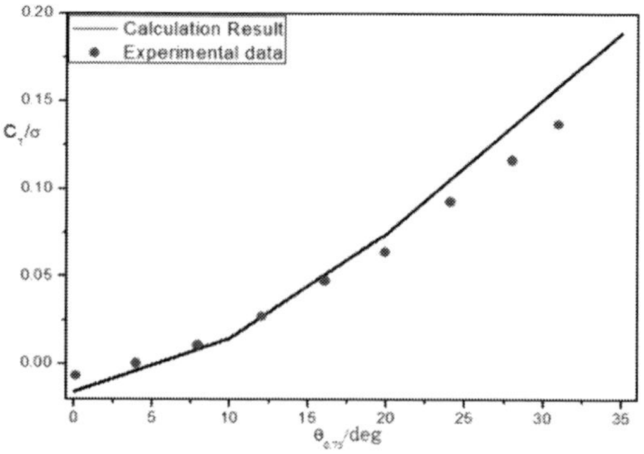

(b) The total thrust of ducted tail rotor

Figure 6. Variation of rotor thrust and total thrust at different collective pitch in hover.

It can be concluded from figure 7 that the airflow is separated at the lip of the duct and produces a recirculation zone, which also appears at the aeras below the duct wall and the central body, and with the increase of collective pitch, the induced speed is also increased, the recirculation zone becomes larger.

(a) collective pitch :10°

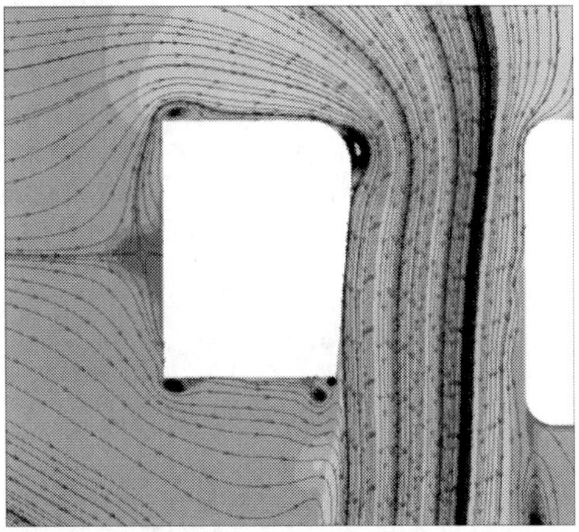

(b) collective pitch :20°

(c) collective pitch :35°

Figure 7. Flow field near the duct wall at different collective pitch in hover.

Forward flight

Compared with the state of hover, the flow field of the ducted tail rotor at forward flight is more complicated. Figure 8 shows the spanwise distribution of the resultant velocity of rotor disk axial induced velocity and forward velocity at different collective pitch at the forward speed of 102m/s. It can be seen from the figure that the axial induced velocity is basically the same at different positions when r/R is less than 0.9, which is related to the negative torsion of the tail rotor blade. At the tip of the blade, the axial induced velocity decreases rapidly due to the influence of the duct wall, which is similar to that in hover. It can be concluded that the axial induced velocity of the same spanwise position increases with the increase of the collective pitch in forward flight.

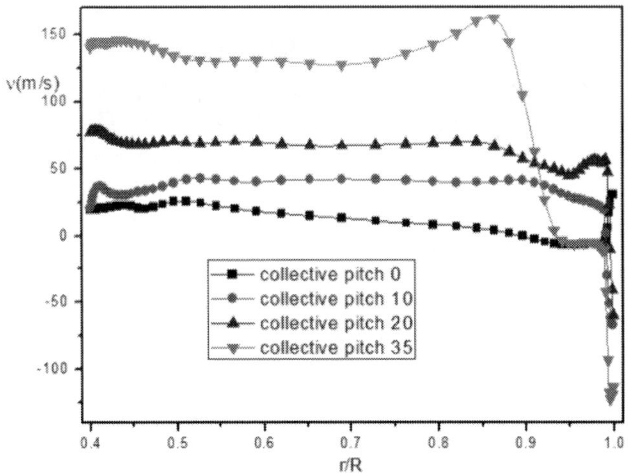

Figure 8. The velocity distribution in rotor disk at different collective pitch.

Figure 9 shows the flow field near the duct wall of the ducted tail rotor. As can be seen from the figure, the airflow separates from the outer edge of the duct, and as the collective pitch increases, separation is more obvious. When the airflow cross through the duct lip, the airflow reattach in the lip smooth transition zone, and further flow into the duct. When the downwash flows out of the duct, the airflow separation occurs further at the outlet of the duct, and as the collective pitch increases, the separation area is also expanded.

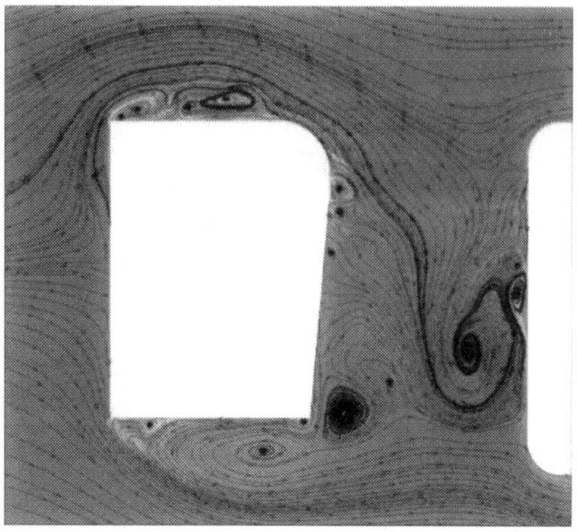

(a) collective pitch :10°

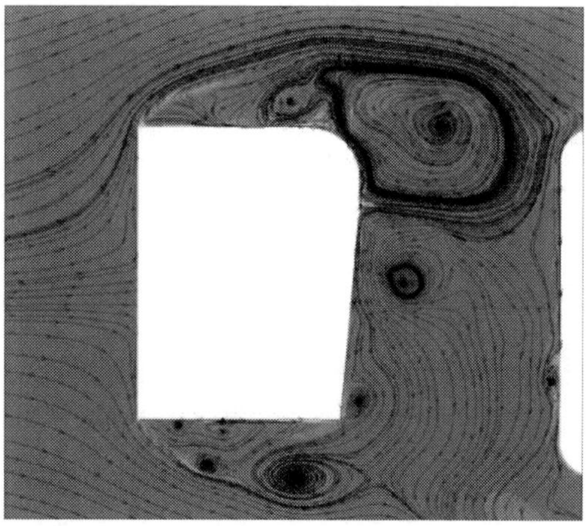

(b) collective pitch :20°

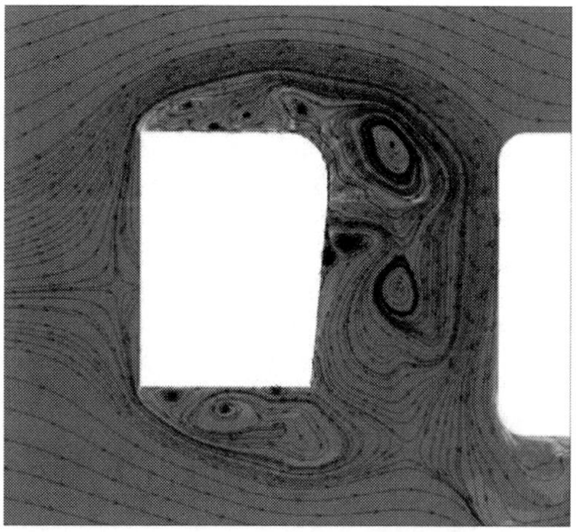

(c) collective pitch :35°

Figure 9. Flow chart near the surface of the duct wall at different collective pitch at the forward speed of 100 m/s.

Figure 10 shows the vorticity distribution of the ducted tail rotor at the different collective pitch at the forward speed of 102 m/s. It can be seen from the comparison that with the increase of the collective pitch, the vortices produced by the tail rotor are obviously enhanced, because the induction of the tail rotor is enhanced with the increase of the collective pitch. As a result, air separation occurs easily when airflow cross through the outer edge of the duct wall.

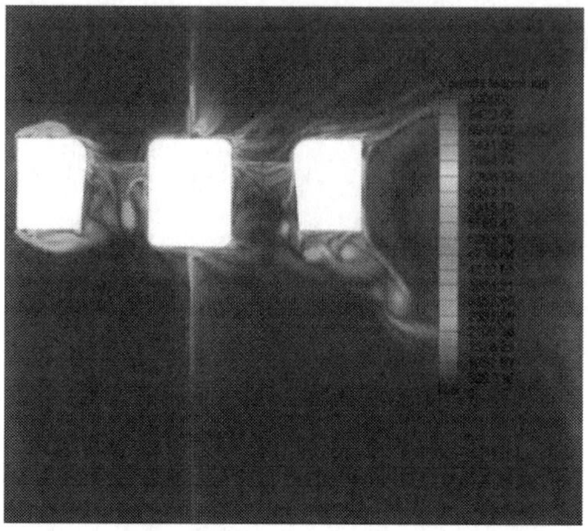

(a) collective pitch :10°

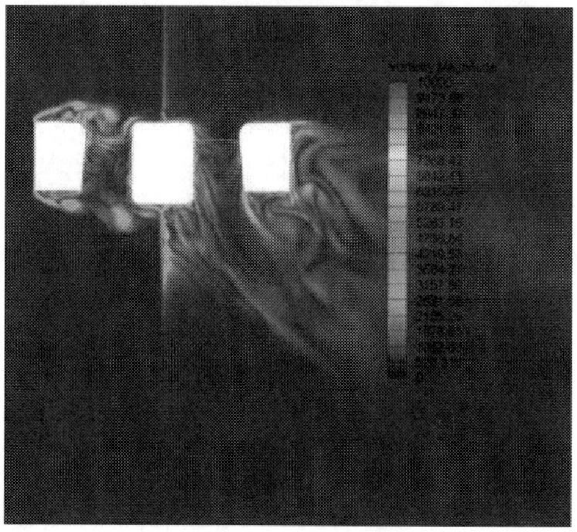

(b) collective pitch :20°

(c) collective pitch :35°

Figure 10. The axial vorticity distribution of Y = 0 profile at the forward speed of 100 m/s.

Sideward flight

Figure 11 shows the spanwise distribution of the resultant velocity of rotor disk axial induced velocity and lateral velocity at different collective pitch at the left-sideward speed of 43 m/s. It can be seen from the figure that with the increase of the collective pitch, the induced velocity in rotor disk increases, and the spanwise variation is not obvious, which is similar to the results of the forward flight and hover.

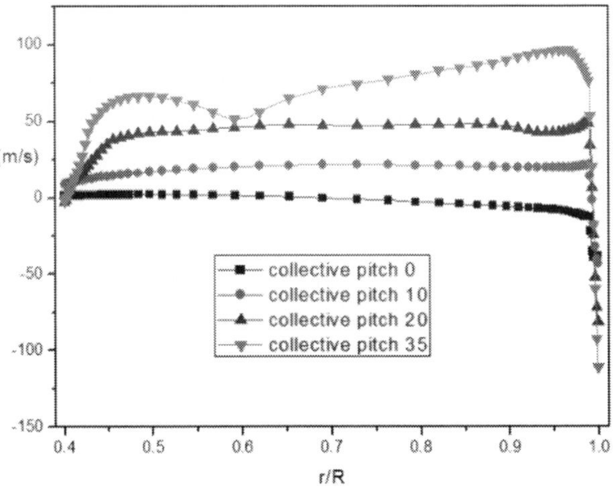

Figure 11. The velocity distribution in rotor disk at different collective pitch.

Figure 12 shows the flow field of the duct wall at different collective pitch. As can be seen from the figure, , the downwash of the tail rotor is affected by the sideward flow, resulting in an airflow separation around the outlet of the duct and around the duct wall along the outside of the duct. When the airflow cross through the upper edge of the duct, the separation intensifies, which will lead to a significant reduction of the thrust or even negative thrust. With the increase of the collective pitch, the axial induced velocity becomes larger. The interference of the sideward speed decreases and the separation around the duct is weakened. As a result, the thrust generated by the duct wall increases.

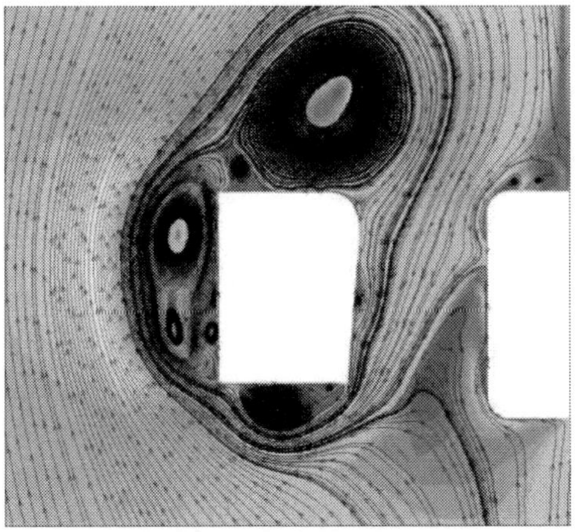

(a) collective pitch :10°

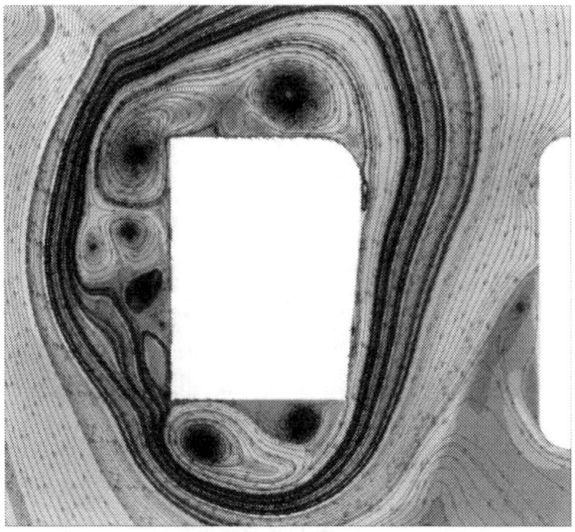

(b) collective pitch :20°

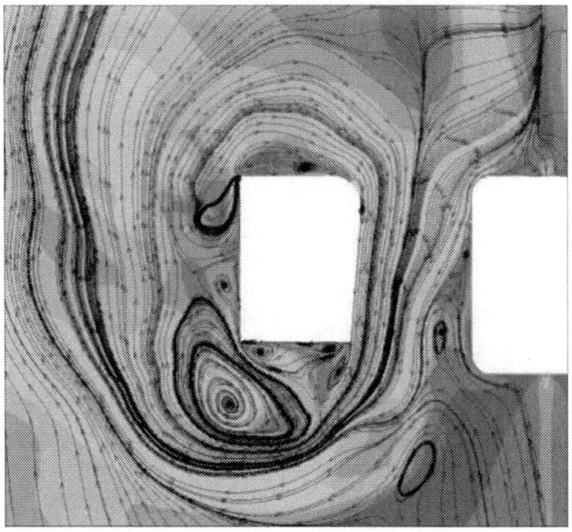

(c) collective pitch :35°

Figure 12. Flow field near the duct wall with different collective pitch at the left-sideward speed of 43 m/s.

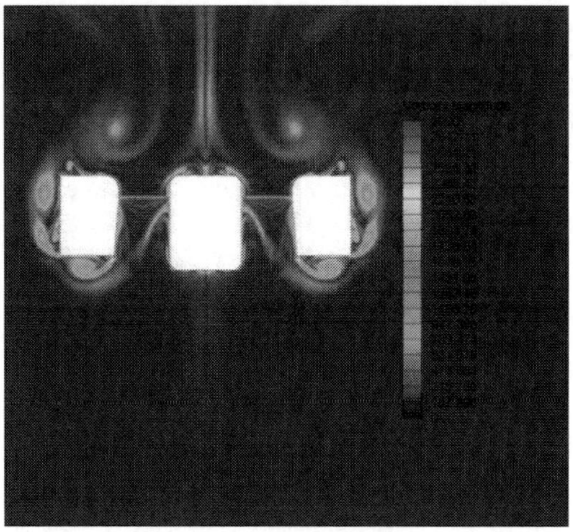

(a) collective pitch :10°

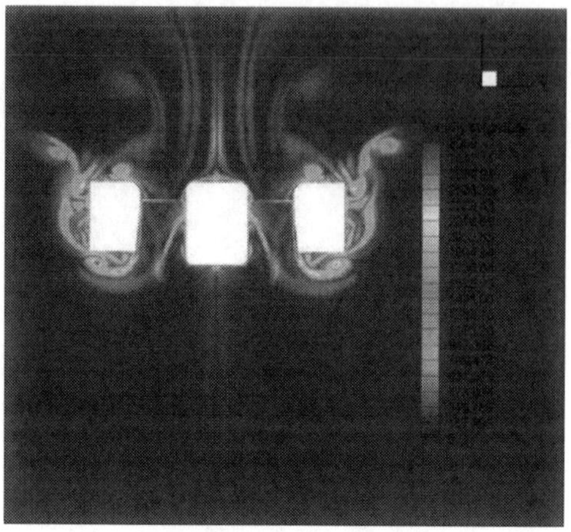

(b) collective pitch :20°

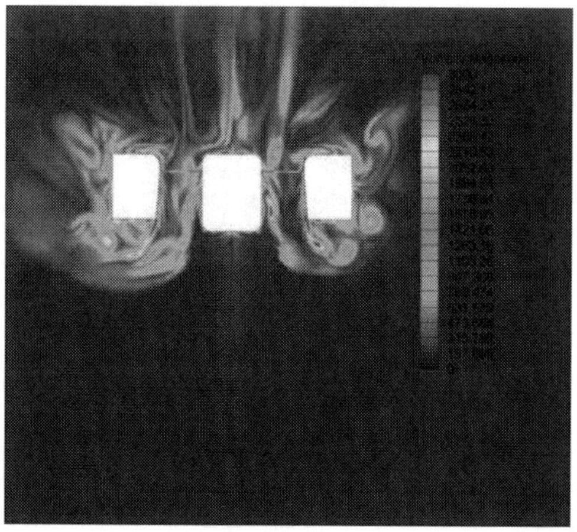

(c) collective pitch :35°

Figure 13. The axial vorticity distribution of Y = 0 profile at the left-sideward speed of 34 m/s.

It can be seen from figure 13 that when the lateral velocity is large and the collective pitch is relatively small, the airflow cross through the duct outlet will rapidly rise, and the air separation will occur near the outer duct surface.

CONCLUSIONS

A body-fitted grid technology and a sliding mesh method are presented based upon the structured grid generation methodology. A CFD method is established to solve the flowfield and aerodynamic characteristics of the ducted tail rotor under different flight conditions by solving N-S equations. Some instructive conclusions are obtained:

(1) The established sliding mesh method is effective to solve the small tip-shroud clearance gap and simulate the flow field of the ducted tail rotor.

(2) The ducted tail rotor disk thrust accounted for about 50% of the total thrust; The thrust of rotor disk and total thrust increase with the increase of the collective pitch

(3) The axial induced velocity increase with the increase of the collective pitch under different flight conditions in different flight conditions.

REFERENCES

[1]KEYS C, SHEFFLER M, WEINER S, et al. LH wind tunnel testing: key to advanced aerodynamic design [A]. 47th American Helicopter Society Annual Forum [C], Phoenix, 1991.

[2]AKTURKL A, SHAVALIKUL A, CAMCI C. PIV measurements and computational study of a 5 inch ducted fan for VSTOL UAV Applacations [R].AIAA-2009-332.

[3]E.Alpman, L. N. Long. B. D. Kothmann. Understanding Ducted-Rotor Antitorque and Directional Control Characteristics Part I. Steady-State Simulation. Journal of Aircraft, Vol.41, No.5, September-October 2004.

[4]R.G. Rajagopalan, C. N. Keys. Detailed Aerodynamic Analysis of the RAH-66 FANTAILTM Using CFD. Journal of the American Helicopter Society, 1997, Vol.42 (No.4):310-320.

[5]G. C. Ruzicka, R. C. Strawn. Edward T. Meadowcroft, Discrete Blade CFD Analysis of Ducted Tail Fan Flow. AIAA 2004-0048.

[6]I-Chung Chang, R.G. Rajagopalan. CFD Analysis for Ducted Fans with Validation. AIAA 2003-4079.

[7]H.D. Lee, O.J. Kwon, J. Joo. Aerodynamic Performance Analysis of a Helicopter Shrouded Tail Rotor Using an Unstructured Mesh Flow Solver. The 5th Asian Computational Fluid Dynamics Forum, 2003.

[8]SPALART P R, ALLMARAS S R. An one-equation turbulence model for aerodynamic flows [R]. AIAA-92-0439.

[9]D. J. Mavriplis, A. Jameson, L. Martinelli. Multigrid Solution of the Navier-Stokes Equations on Triangular Meshes. AIAA Paper 89-0194, 1989.

[10]T. A. Nygaard, A. C. Dimanlig, E. T. Meadowcroft. Application of a Momentum Source Model to the RAH-66 Comanche FANTAIL, AHS 4th Decennial Specialist's Conference on Aeromechanics, San Francisco, CA, Jan 21-23,2004.

[11]M.G. Clemmons, "Antitorque Safety and the RAH-66 FANTAIL™, 48th AHS Forum, Washington,DC,June 1992.

[12]A. Jameson, D. J. Mavriplis. Finite Volume Solution of the Two Dimensional Euler Equations on A Regular Triangular Meshes, AIAA Journal, Vol.24, No.4, 1986, pp: 6ll-618.

[13]D. J. Mavriplis, A. Jameson, L. Martinelli. Multigrid Solution of the Navier-Stokes Equations on Triangular Meshes. AIAA Paper 89-0194, 1989.

Numerical research on the aerodynamic impact of ship superstructures for helicopter operations

Xu Yi
Master student
Nanjing University of Aeronautics and Astronautics
Nanjing, China

Shi Yong-jie
Associate professor
Nanjing University of Aeronautics and Astronautics
Nanjing, China

Xu Guo-hua
Professor
Nanjing University of Aeronautics and Astronautics
Nanjing, China

ABSTRACT

A coupling numerical methodology has been developed using Navier-Stokes solver for ship/rotor flowfield simulation during the helicopter shipboard landing, where the momentum source approach is employed to account for rotor operations. In this method, the second-order upwind scheme is used in the spatial discretization, and the k-ω turbulence model is selected to improve the simulation precision for flow structure. The goal of this paper is to investigate the potential benefits of some aerodynamic modifications to the simplified landing platform dock-17 (LPD-17) ship. Qualitative and quantitative analysis of ship airwake between the baseline and modified ship geometries highlight the changes of aerodynamic environment over the flight deck. Root-mean-square (RMS) forces and moments have been obtained from the time histories at five important locations through the landing approach for different starboard wind-over-deck (WOD) angles. Compared with the baseline ship geometry, the notch and side-ramp modifications both show the encouraging reduction in RMS loadings, but the improvement effect is limited in the large WOD condition

NOTATION

b	Ship beam (m)
h	Height of hangar (m)
l	Length of deck (m)
L	Length of ship (m)
Ma	Mach number
Ma_{tip}	Rotor tip mach number
θ	Collective pitch (°)
φ	Rotor shaft angle (°)

INTRODUCTION

The shipboard landing of the helicopter in a single-spot ship is one of the most challenging task. The random six degree of freedom motion of small-sized deck and the ship airwake caused by the bluff-body flow separation both have adverse effects on the helicopter operations. Therefore, the safe helicopter landing depends heavily on the specific Ship-Helicopter Operation Limits (SHOLs) which are derived from First of Class Flight Trials (FOCFTs). However, the determination of SHOLs is a very hard and complex task, which is highly susceptible to the wind speed, WOD and sea conditions. To improve the efficiency and reduce the cost, a series of methods have been developed for the research of ship/helicopter coupling flowfield, and the Computational fluid dynamics (CFD) technology is widely used, which were presented in Ref. 1， Ref. 2 and Ref. 3.

During the approaching to the landing spot, the ship airwake with large velocity gradient and unsteady characteristics disturb the aerodynamic environment around the helicopter and results in a sudden change of local aerodynamic force, which greatly increase the pilot workload. An alternative method that can reduce the adverse effects of ship airwake on the helicopter is to modify the ship superstructure, or in other words, the flow control of ship airwake. The wind tunnel test and CFD simulation of flow control of ship airwake have been conducted in recent years. Shafer (Ref. 4), Greenwell (Ref. 5), Kääriä (Ref. 6) and Czerwiec (Ref. 7) all shows that the flow control devices have a strong influence on the spatial and temporal characteristics of ship airwake, but not exactly the improvement.

Most of the previous studies only focus on the flow control of ship airwake in the case of without considering the influence of helicopter, which should be researched further, as Crozon (Ref. 8) shows that the ship/helicopter coupling flowfield is not just the superposition of ship airwake and rotor induced wake, instead, there is the complex interactional effect in two wakes. Due to the more serious aerodynamic interaction in the crosswind, as compared with that in the headwind, the researches of flow control derived from the modification of ship superstructure have been performed in this paper at different starboard crosswind conditions.

Presented at the 6th Asian/Austrailian Rotorcraft Forum / Heli Japan 2017, Kanazawa, Japan, November 7-9, 2017.

METHODLOGY AND MODEL

CFD solver

The unsteady Reynolds-averaged Navier-Stokes (RANS) equations are employed as the governing equations and the k-ω turbulence model is used for the RANS closure. The second-order upwind Roe scheme and the second-order lower-upper symmetric Gauss-Seidel (LU-SGS) scheme are selected to improve calculation accuracy. The momentum source approach is used to model the aerodynamic effect of the rotor.

Numerical setup

As the scale of the ship and velocity of the flow determine the convective time scale (CTS). The CTS can be defined as the approximate time it takes for a fluid particle to pass along the ship. A sufficiently long simulation time in order to arrive at those independent samples amounts easily to at least 8 CTS, or even significantly more when higher accuracy in the mean flow is required, which is discussed in Ref. 9. Thus the 1/20-scale landing platform dock-17 (LPD-17) ship is chosen to conduct the numerical simulation. For the convenience of calculation, the ship hull and the superstructure of the ship model are simplified, as shown in Figure 1. The length of deck (l) and the width of ship beam (b) are 3.2m and 1.6m, respectively, with a hangar (h) of 0.8-meter height.

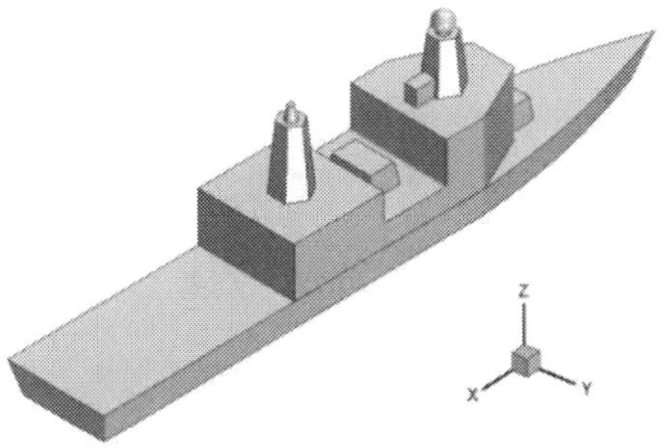

Figure 1. Sketch of the LPD-17 ship.

The 'Dolphin' helicopter rotor (Ref. 10) is employed and its basic parameters are shown in Table 1. To match the size of ship deck, a 1/10-scale rotor is employed to perform the simulation, but it maintains the tip speed constantly. The effect of fuselage is ignored as it barely influence the main flow characteristics of helicopter/ship coupling flowfield due to its narrow configuration (Ref. 11). The numerical simulations in this paper have been conducted at three conditions: starboard 15°, 30° and 45° WOD, with a 15m/s wind.

Table 1. Parameters of "Dolphin" helicopter rotor.

Variable	Values
Rotor chord	0.385 m
Rotor diameter	11.93 m
Blade negative twist	-10°
Tip speed	208 m/s
Blade number	4

In the case of crosswind, the shedding vortex and the shear layer that separated from the top and right-side edge of the hangar would fall down to the deck, which has adverse effects on the shipboard operations. Therefore, to improve the aerodynamic environment of rotor in the starboard crosswind condition, the modifications are conducted along the right-side surface of the hangar, as shown in Figure 2.

The flow control method 1 is the notch modification at the right side of the hangar and the cross-sectional area of the notch modification in the x direction is 0.2×0.2 m at the scale. While the flow control method 2 is the side-ramp modification and its width is 0.135 m (1/6 of hangar height), which is set at an angle of 30° to the longitudinal symmetry plane.

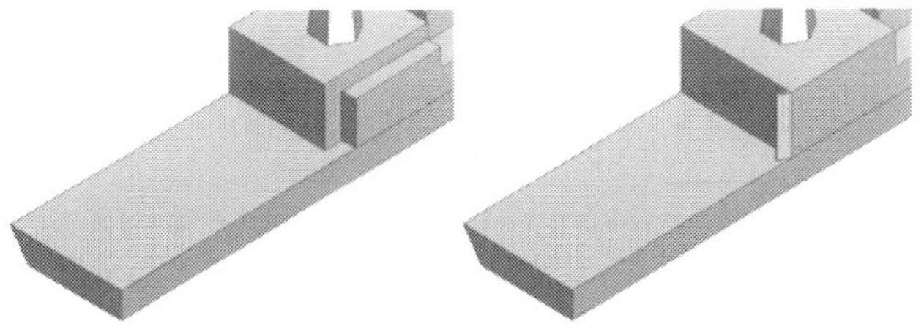

(a) Notch modification (b) Side-ramp modification

Figure 2. Maps of vertical hangar edge modifications.

A rectangular computational domain (8L×6L×6L, where L=10m is the overall ship length) was employed, and the sea and ship surface were specified as no-slip walls. The front and right-side surfaces are set as velocity-inlet boundary condition, while other surfaces are set as pressure-outlet boundary condition. The grid (Figure 3) was refined in the vicinity of the ship hull and the first wall unit value was set as 2 mm in order to meet the requirement of standard wall function in turbulence model calculation. The actuator disk was located in the refined grid area above the flight deck and the overall grid size was 6.8 million cells.

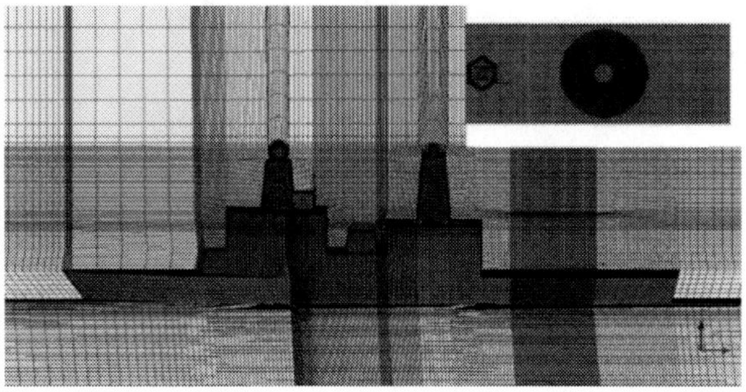

Figure 3. Views of the grid used for the ship/rotor simulation.

Numerical method validation

For lacking of the experiment data of coupling rotor/ship flow, the simulations of simple frigate shape (SFS2) and rotor/fuselage interaction are performed to demonstrate the validation of the CFD method, which are discussed in Ref. 12 and 13, and the specific model size are described in detail in the references.

Figure 4 presents the horizontal velocity at two monitoring lines in the headwind condition, with a 12.87 m/s of wind. The two monitoring lines are located 10.66 m above the deck, 1/4 and 3/4 deck length away from the hangar, respectively. There are some small differences between simulated results and test data, however the velocity distribution of the two is quite consistent

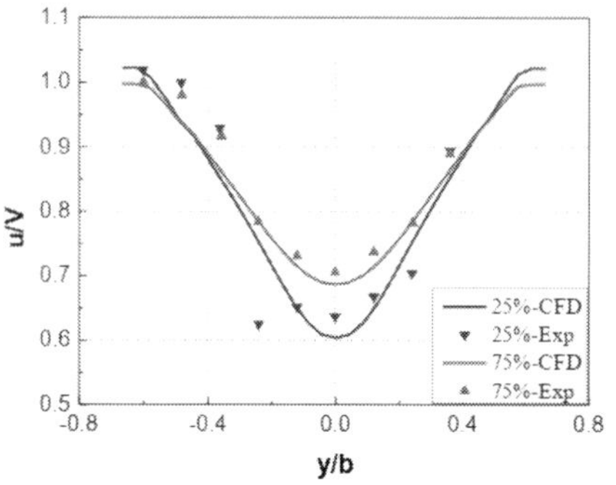

Figure 4. Comparison of horizontal velocity on two lateral lines over the flight deck.

Then, a numerical case of a coupling rotor/fuselage interaction is presented, which operates at a low speed flight case, with Ma_{tip}=0.295, Ma=0.0295, and rotor shaft angle φ -6.0 degrees (positive tilted forward). Figure 5 shows the predicted time-averaged pressure distribution with test data at the top lines on fuselage. As shown, two peaks of pressure coefficient due to the aerodynamic interaction between the rotor induced downwash and the fuselage could be predicted by the CFD solver, and the magnitudes and locations of which are close to the experiment data, indicating the developed method is suitable for solving complex aerodynamic interference problem of rotor machine

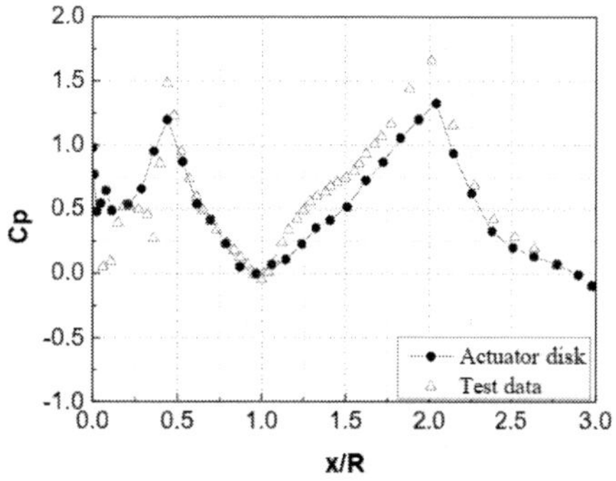

Figure 5. Comparison of the predicted pressure distribution with test data on the top of the fuselage.

RESULTS AND DISCUSSIONS

Isolated ship

Figure 6 shows the contours of velocity in the horizontal plane (0.8m above the deck) at starboard 30° WOD. The distribution of velocity vector is also shown in the figure. Due to the blocking effect of the ship tower and the hangar, two low-velocity region with vortex are generated at the both side of the flight deck, as seen in Figure 6 (a). The large velocity gradient above the deck would result in a sudden change of local aerodynamic force, which adversely influence the stability and controllability of the helicopter. Compared to the baseline, the notch modification (Method 1) makes the low-velocity region at the right side of the deck deviate to the deck center, which leads to the improvement of velocity distribution. However, the increase of the turbulent level of velocity vector above the deck center brings about the adverse effect on shipboard operation. For the side-ramp modification (Method 2), the desired effect is to relieve the strong flow separation due to the sharp edges and reduce the turbulent intensity of ship airwake. As shown in Figure 6 (c), the low-velocity region at the right side of deck is not generated as the inflow velocity deflects with the side-ramp modification, which also results in the reduction of turbulence level of velocity field, as compared with the baseline.

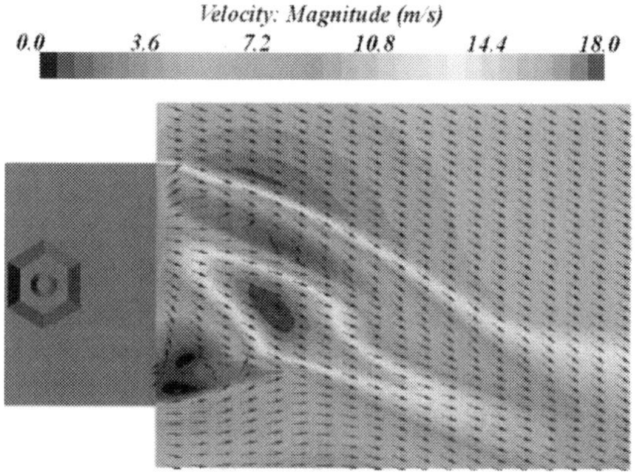

(a) Baseline

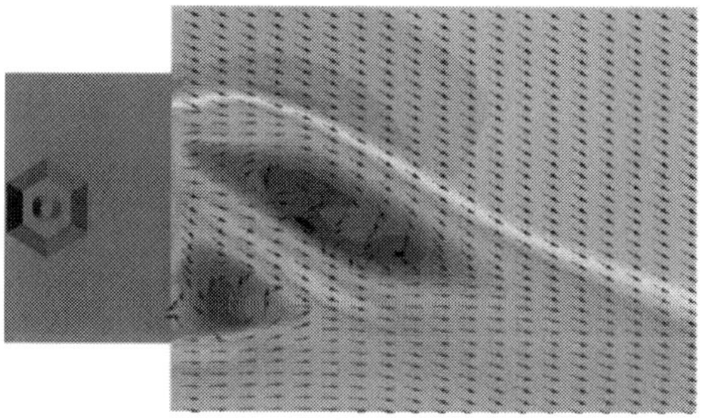

(b) Notch modification

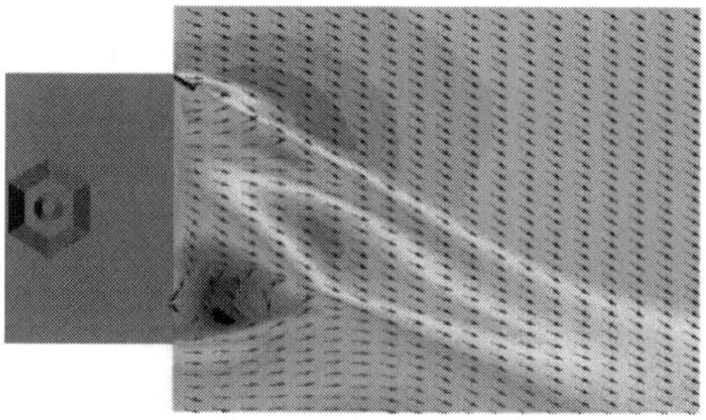

(c) Side-ramp modification

Figure 6. Velocity contours in the horizontal plane for baseline, notch and side-ramp modification at starboard 30° WOD.

Figure 7 presents the comparison of vertical velocity at a monitoring line (0.8 m above the deck, 1.6 m away from the hangar) at 30° WOD. For the helicopter, the increased vertical velocity reduce the effective angle of attack of the main rotor blades and hence the thrust force produced by the main rotor, which further reduces the available control margin of helicopter operations. The minimum of vertical velocity is clearly seen at the left side of deck ($-0.5 \leq y/b \leq 0$), and this can be explained by that the shedding vortex separated from the top of ship tower contains the large vertical velocity component. As seen, there is an improvement in the vertical velocity distribution of the notch modification, while the side-ramp modification has little effect.

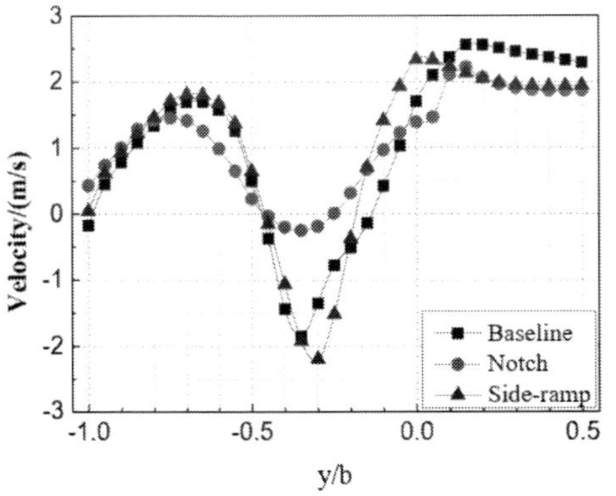

Figure 7. Comparison of vertical velocity between baseline, notch modification and side-ramp.

Shipboard landing

Not only the spatial characteristics but also the temporal characteristics of ship airwake have a great influence on the shipboard operation. To study the effect of flow control devices on the unsteady rotor loadings, the approach technique that a lateral translation from a port side approach position to a hover over the deck center is developed, keeping the relative altitude between the rotor and the deck constant, and five monitoring positions are selected in the flight envelope (Figure 8). To ensure the efficiency and accuracy of numerical simulation, 15-second time histories of rotor loadings are recorded in every monitoring position for the analysis of the magnitude of the unsteady disturbance.

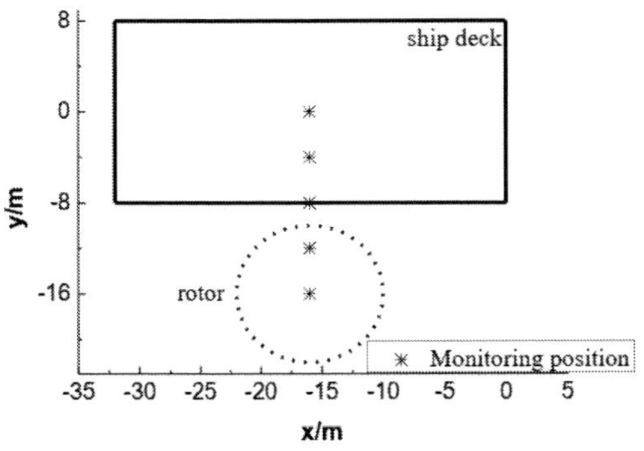

Figure 8. Distribution map of monitoring locations.

During the approaching, the high frequency disturbances do not cause the pilot much difficulty because of the aircraft's inertia. While the low frequency disturbances might result in the large amplitude displacement, but the pilot have enough time to input correct control command. Generally, the unsteady aerodynamic loadings having frequencies in the range 0.2~2 Hz bring the most significant control challenges for the pilot.

Therefore, to correctly measure the magnitude of the unsteady disturbance from within the frequency bandwidth (0.2~2 Hz), the Power Spectral Density (PSD) plots have been conducted according to the force and moment time-histories and the square root of the integral between 0.2~2 Hz in PSD has been taken as a measure of the unsteady loading, which is referred to as the root-mean-square (RMS) loading (e.g. RMS thrust).

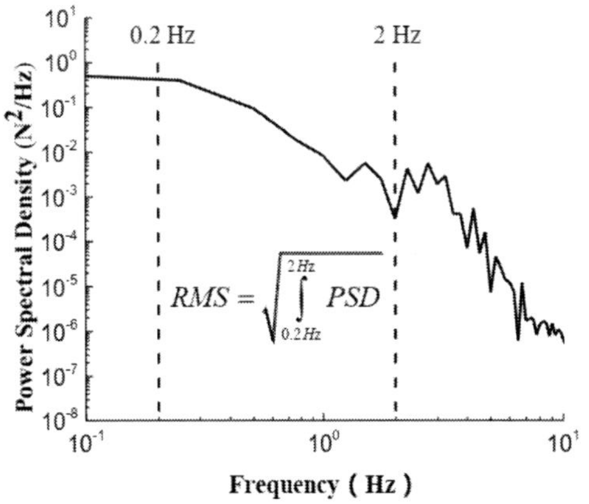

Figure 9. The RMS loadings are calculated from the within the frequency bandwidth 0.2~2 Hz.

Figure 10 shows the RMS forces and moments for baseline, notch modification and side-ramp modification at five monitoring positions at starboard 30° WOD, where the rotor operates at the condition of $Ma_{tip}=0.643, \theta=12.82°$. The locations y/b=0 and y/b= -0.5 represent the deck center and left-side deck edge, respectively. As can be seen in Figure 10, the peak of RMS loadings is located around y/b= -0.5, which is can be explained by the rotor is wholly immersed in the ship airwake at y/b= -0.5, thus the shear layer and shedding vortex separated from the ship superstructure pose severe turbulent fluctuation around the rotor. While the rotor is located at y/b=0, part of the rotor is exposed to the freestream with low velocity fluctuation because of the crosswind condition.

The notch modification is equally as effective as the side-ramp modification over the landing spot (y/b=0) with reduction of 10%-25%, compared with the baseline, in all six RMS loadings. For the notch modification, there is improvement of 15%-35% between $-1 \leq y/b \leq -0.25$, in RMS drag and roll. Although an increase in RMS pitch occurs at y/b= -0.5, this point is an isolated increase and the general effect of the notch modification is still positive. In RMS thrust and side-force, the notch modification has reduced the peak levels, which occur at y/b= -0.25, by 49% and 33%, respectively. The reduction of RMS loadings for the side-ramp modification can also be seen between -1≤y/b≤-0.25, where there are very encouraging reductions in RMS drag, side force and thrust of up to 48%, 33% and 50%, respectively. While the improvements of RMS pitch and roll are not very significant, by average 13% and 17% between -1≤y/b≤-0.25, respectively.

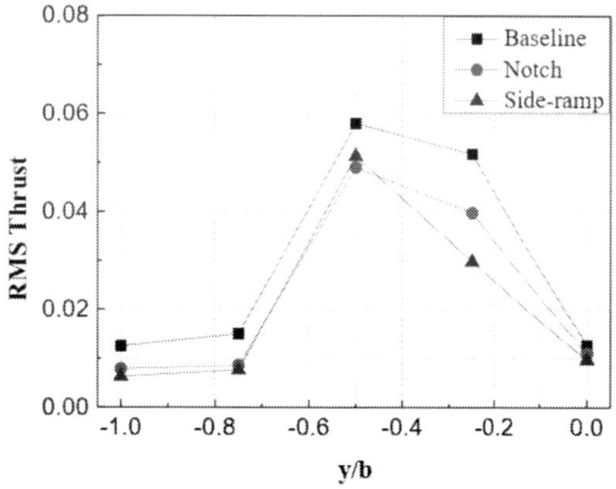

(a) **RMS Thrust**

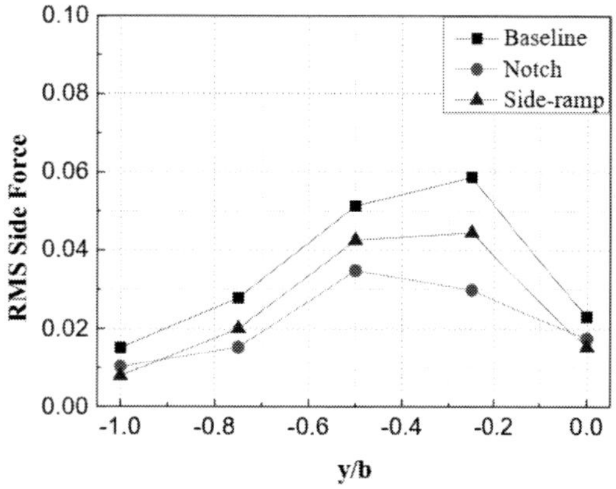

(b) **RMS Side Force**

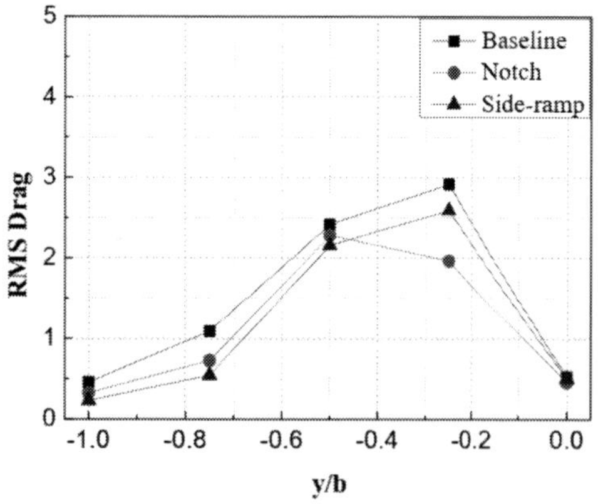

(c) **RMS Drag**

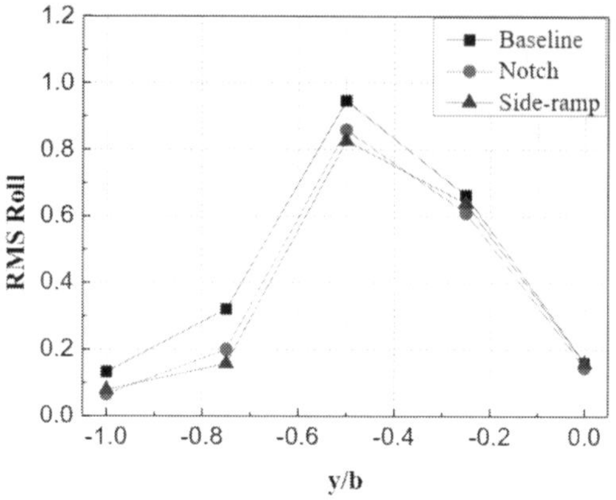

(d) **RMS Roll**

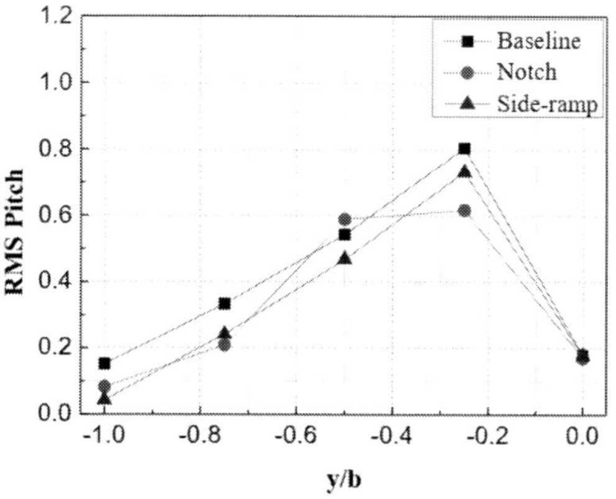

(e) **RMS Pitch**

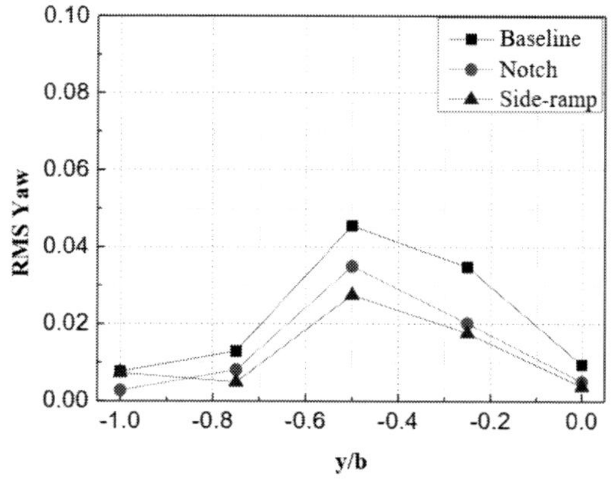

(f) **RMS Yaw**

Figure 10. RMS loadings for baseline, notch and side-ramp modifications in starboard 30° WOD

The RMS side force and roll at starboard 15° and 45° WOD are subsequently presented to compare with the starboard 30° WOD in terms of the improvement effect of flow control devices on the lateral unsteady disturbance.

Figure 11, which maps the RMS side force and roll for baseline, notch modification and side-ramp modification at starboard 15° WOD, illustrates that the distribution of RMS loadings is quite different from that at starboard 30° WOD.

From the port to the deck center, the RMS loadings is gradually increased, because the region with the maximum turbulence intensity still locates around the deck center in the small WOD condition. The notch modification has reduced RMS side force by 20-30% between $-1 \leq y/b \leq 0$, compared with the baseline. While the side-ramp modification has reduced by as much as 45% and 42% for RMS side force at y/b= -0.25 and -0.75, respectively. For the RMS roll, the notch modification has reduced the peak value by 42% at y/b= -0.25, while the reduction of the side-ramp modification is only up to 30%. Generally, the improvement at starboard 15° WOD is about 8% better than that at starboard 30° WOD.

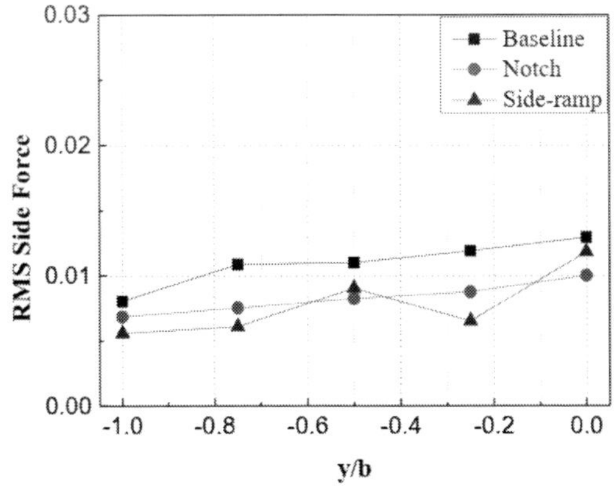

(a) RMS Side Force

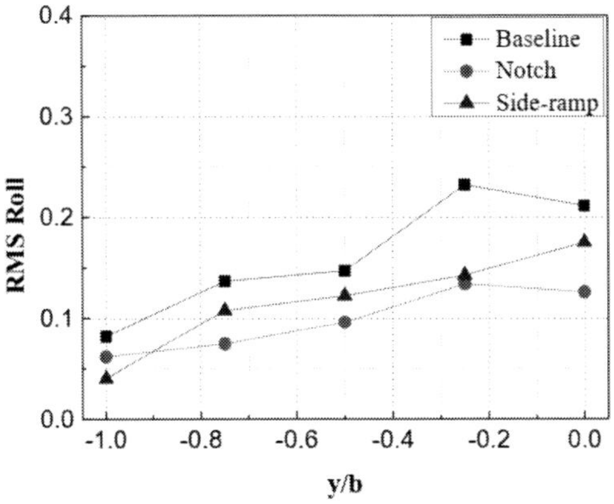

(a) RMS Roll

Figure 11. RMS loadings for baseline, notch and side-ramp modifications in starboard 15° WOD

At the starboard 45° WOD condition, the peak value of RMS loadings shift towards to the deck edge due to the obvious deflection of ship airwake, as shown in Figure 12. It is clearly found that the notch and side-ramp modification have increased the RMS side force by 40% and 30% between $-0.75 \leq y/b \leq -0.25$, respectively. The discouraging increase in RMS roll for the notch modification is up to 35% at y/b= -0.5, although there is improvement of 10% between $-1 \leq y/b \leq -0.75$. Overall, the improvement of flow control devices is not ideal at starboard 45° WOD. It can be deduced that the improvement effects of the notch and side-ramp modification are limited in large WOD condition because of the fixed installation.

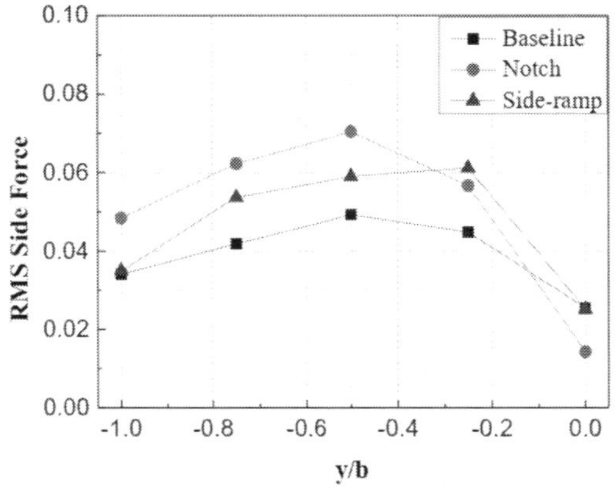

(a) RMS side force

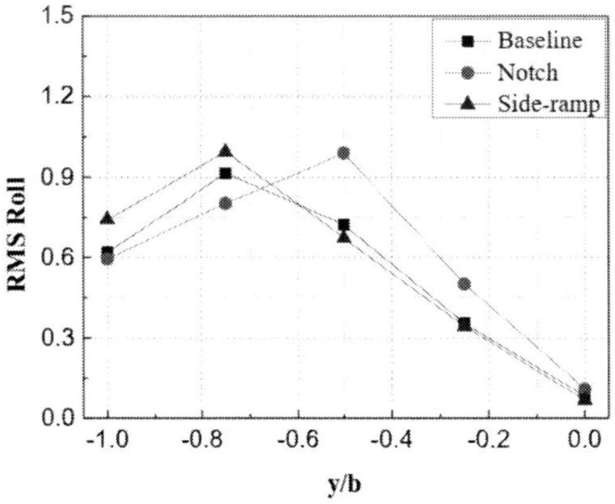

(b) RMS roll

Figure 12. RMS loadings for baseline, notch and side-ramp modifications in starboard 45° WOD.

CONCLUSIONS

(1) For the notch modification, the deviation of the windward deck flowfield is mainly responsible for the reduction of the vertical velocity and the improvement of the velocity gradient over the landing spot. The notch modification is also proved to be useful for the improvement of RMS loadings.

(2) For the side-ramp modification, the control of flow separation at the windward edge of the superstructure improves the velocity spatial distribution and the turbulent intensity over the flight deck, which is contributed to the reduction of the unsteady disturbance of ship airwake on the helicopter.

(3) The improvement of the flow control devices is easily influenced by the change of WOD as they are fixed in constant dimensions and angle. As a result, the improvement effects of notch and side-ramp modification are limited in the large WOD condition.

ACKNOWLEDGMENTS

This work was supported by National Natural Science Foundation of China: [Grant Number 11302103].

REFERENCES

[1]Syms, G. F., "Numerical simulation of frigate airwakes," *International Journal of Computational Fluid Dynamics*, Vol. 18, (2), 2004, pp. 199-207.

[2]Forrest, J. S., and Owen, I., "An investigation of ship airwakes using Detached-Eddy Simulation," *Computers & Fluids*, Vol. 39, (4), 2010, pp. 656-673.

[3]Shi, Y. J., Xu, Y., Zong, K., and Xu, G. H., "An investigation of coupling ship/rotor flowfield using steady and unsteady rotor methods," *Engineering Applications of Computational Fluid Mechanics*, Vol. 11, (1), 2017, pp. 417-434.

[4]Shafer, D. M., "Active and passive flow control over the flight deck of small naval vessels;" Virginia Polytechnic Institute, April 2005.

[5]Greenwell, D. I., and Barrett, R. V., "Inclined screens for control of ship airwake," AIAA 2006-3502, 3rd AIAA Flow Control Conference, San Francisco, California, June 5-8, 2006.

[6]Kääriä, C. H., Wang, Y., White, M. D., and Owen, I., "An experimental technique for evaluating the aerodynamic impact of ship superstructures on helicopter operations," *Ocean Engineering*, Vol. 61, (6), 2013, pp. 97-108.

[7]Czerwiec, R. M., and Polshy, S. A., "LHA Airwake Wind Tunnel and CFD Comparison with and Without Bow Flap," 22nd Applied Aerodynamics Conference and Exhibit, Providence, Rhode Island, 16-19 August, 2004.

[8]Crozon, C., Steijl, R., and Barakos, G. N., "Numerical Study of Helicopter Rotors in a Ship Airwake," *Journal of Aircraft*, Vol. 51, (6), 2014, pp. 1813-1832.

[9]Muijden, J. V., Boelens, O., Vorst, J. V. D., and Gooden, J., "Computational ship airwake determination to support helicopter-ship dynamic interface assessment," Aiaa Computational Fluid Dynamics Conference, 2013.

[10]Le, P. A., Gatard, J., and Monnier, J. C., "Experimental investigations of rotor-fuselage aerodynamic interactions using a helicopter powered mode," ONERA-TP-04-176, 2004

[11]Su, D. C., Shi, Y. J., Xu, G. H. and Zong, K., "Numerical study of the coupled flowfield of ship/helicopter," *Acta Aeronautica et Astronautica Sinica*. Vol. 38, (7), 2017. (in Chinese).

[12]Zhang, F., Xu, H., and Ball, N. G., "Numerical simulation of unsteady flow over SFS2 ship model," 47th AIAA Aerospace Sciences Meeting, Orlando, Florida. 2009.

[13]Park, Y. M., Nam, H. J., and Kwon, O. J., "Simulation of unsteady rotor-fuselage interactions using unstructured adaptive meshes," American Helicopter Society 59th Annual Forum.

Xu Yi is a master student in aircraft design at Nanjing University of Aeronautics and Astronautics, and his research interests are helicopter CFD, aerodynamic interaction of ship/helicopter. Email: xuyi7057@nuaa.edu.cn

Shi Yong-jie is an associate professor in the College of Aerospace Engineering at Nanjing University of Aeronautics and Astronautics. His main research interests are helicopter CFD, helicopter aerodynamics, aerodynamic interaction of ship/helicopter, aerodynamic shape design of rotor blade and rotor aeroacoustics. Email: shiyongjie@nuaa.edu.cn

Development of a Ducted Rotor for Multicopters

Koichi Yonezawa
Project researcher
Central Research Institute
of Electric Power Industry
Abiko, Japan

Hironori Matsumoto
Graduate Student
Osaka University
Toyonaka, Japan

Kazuyasu Sugiyama
Professor
Osaka University
Toyonaka, Japan

Hiroshi Tokutake
Associate Professor
Kanazawa University
Kanazawa, Japan

Yasutada Tanabe
Associate senior Researcher
Japan Aerospace Exploration Agency
Tokyo, Japan

Shigeru Sunada
Professor
Nagoya University
Nagoya, Japan

ABSTRACT

Electric powered multicopters are utilized in various fields. The multicopters have become the most successful small-scale unmanned aerial vehicles (UAVs) thanks to the simple mechanisms. One of the biggest problems of the electric powered small-scale UAVs is the flight distance and the flight duration. In order to extend the distance and the duration of the flight, it is important to improve the aerodynamic efficiency of the rotor. In the present paper, aerodynamic characteristics of ducted rotors are examined. Experiments are carried out with the various duct contours and the relationship between the contours and the aerodynamic characteristics are discussed.

NOTATION

C_T: Thrust coefficient
C_Q: Torque coefficient
CP: Rotor collective pitch angle
A_E: Duct outlet area
A_I: Duct inlet area
A_R: Rotor rotation plane area
L_E: Duct outlet section length
L_I: Duct intlet section length
L: Duct length

INTRODUCTION

When a large-scale disaster occurs, such as a big earthquake, a heavy rain, and a tsunami, it is very important to collect information of the disaster site in order to make suitable plans of the rescue, the evacuation and the recovery. A small-scale unmanned aerial vehicle (UAV) represented by a multicopter is one of the most useful devices to observe the situation. The multicopters are required to be capable of stable, safety and long-time flights in hard weather conditions[1]. The authors are carrying out investigations to improve the propulsion device of the multicopter to enhance the agility, the safety and the efficiency[2]. A multicopter with advanced mechanisms has been developed through the research activities as reported previously[3]. The rotor with a variable collective pitch mechanism has been adopted as shown in Fig. 1. Using the variable pitch rotors, the agility of the multirotor is enhanced because the duration of the thrust control becomes shorter than that of the fixed blade rotor. In addition, the downward thrust can be produced using the variable pitch rotor. The improvement of the agility enhances the flight stability in turbulent condition and consequently the safety of the multicopter is enhanced. To improve the efficiency of the rotor, a ducted rotor has been examined and it has been confirmed that both the maximum thrust and the figure of merit of the rotor can be increased. As another advantage, the ducted rotor improves the safety of the multicopter because the duct surrounds the rotating rotor blades and prevents the rotor blades from the smashing persons and structures.

In the present study, the aerodynamic characteristics of the ducted rotor are examined in detail. Various kinds of the duct contours are compared experimentally.

Presented at the 6th Asian/Austrailian Rotorcraft Forum / Heli Japan 2017, Kanazawa, Japan, November 7-9, 2017.

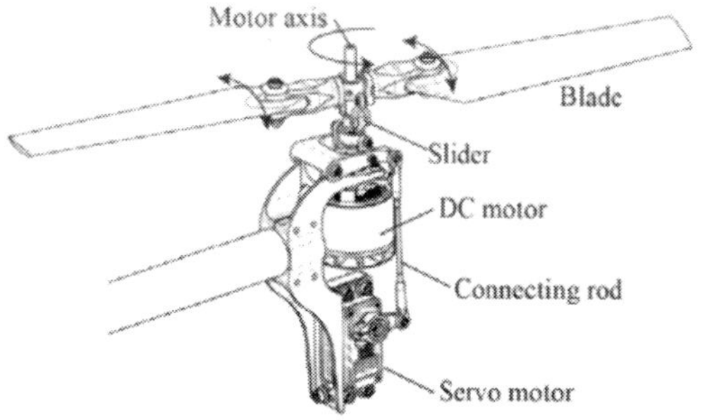

Fig. 1 The first prototype of multicopter (top) and the rotor with variable collective pitch angle.

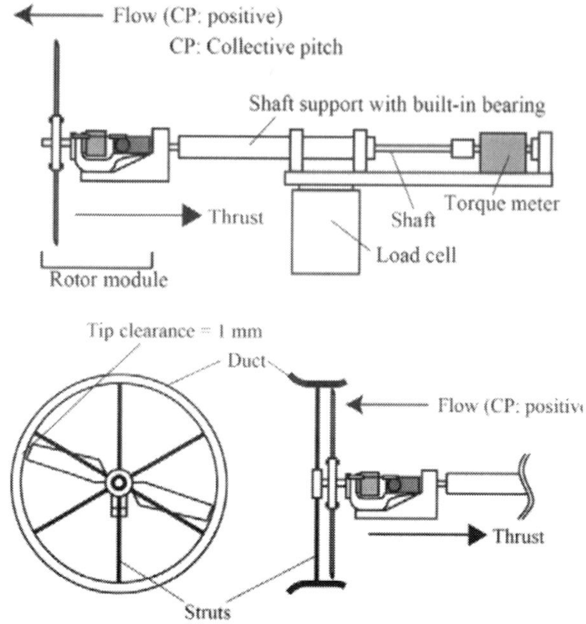

Fig. 2 Experimental apparatus. Thrust and axial torque measurement setup (top) and equipment of the rotor duct (bottom).

Table 1 Specification of rotor

Rotor diameter	0.33 m
Number of blades per rotor	2
Blade twist angle	0
Blade airfoil	NACA0009
Blade planform	Tapered, taper ratio = 0.6
Rotational angular speed	600 rad/s (5800 rpm)
Reynolds number	2.0×10^5

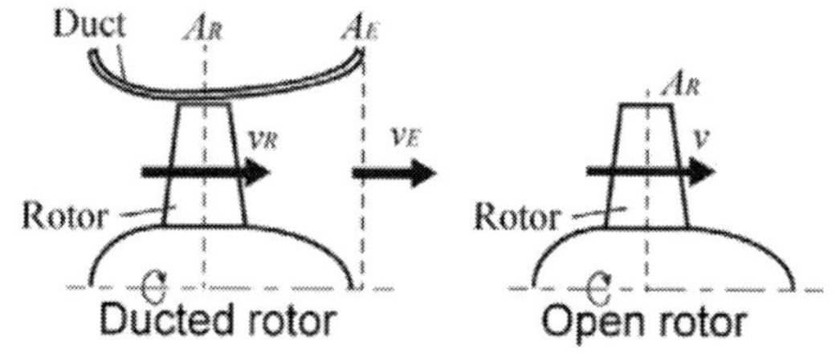

Fig. 3 Flow in ducted and open rotors

EXPERIMENTAL SETUP

Specification of the rotor module is shown in Table 1. The rotor blade is driven by the DC motor and the collective pitch is determined using a servo motor. The DC motor and the servomotor are controlled by the pulse-width modulation (PWM).

The thrust and the axial torque of the rotor was measured to evaluate the aerodynamic performance of the rotor. The experimental apparatus is shown in Fig. 2. The rotor thrust was measured using a load cell (Nissho electric works, LMC-61322, rated force capacity is 10 N) and the axial torque was measured using a torque meter (UNIPULSE corp., UTM-II 1Nm, rated torque is 1 Nm). The duct rotor was supported using six struts with the diameter of 3 mm. The hub of the struts was installed at the motor axis with a bearing. Therefore, the assembly of the rotor duct was supported by the motor. A clearance between the blade tip and the inner wall of the duct was 1 mm with the assemble error of less than 0.3 mm.

The thrust and the axial torque were measured at various collective pitch angles, *CPs*.

DUCT DESIGN

Aerodynamics of the ducted rotor can be evaluated using equation of momentum theory and Bernoulli's principle. Areas of the flow path and flow velocities through the rotor rotational plane and the duct outlet are defined as shown in Fig. 3. Here, pressures at far field, at the inlet of the rotor rotational plane, and at outlet of the rotor rotational plane are defined as p_0, p_1 and p_2, respectively. Then the thrust of the rotor blade is calculated as follows:

$$T_R = A_R\left(p_1 - p_2\right) = \frac{1}{2}\rho A_R v_E^2 \quad (1).$$

On the other hand, the total thrust of the ducted rotor is calculated considering the conservation of the flow momentum as follows:

$$T_H = \left(\text{mass flow rate}\right)\cdot v_E = \rho A_R v_R v_E \quad (2).$$

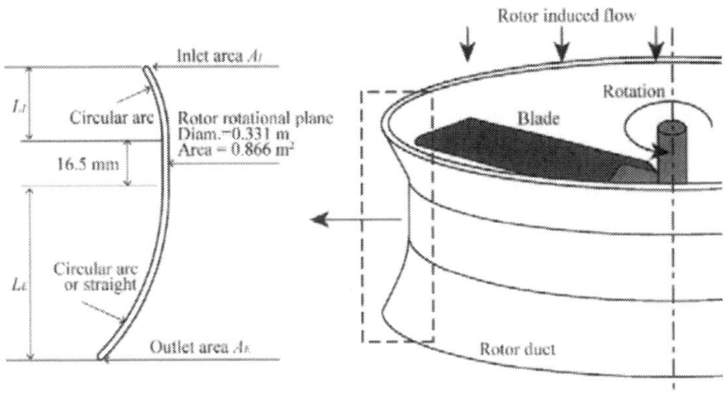

Fig. 4 Definition of duct dimensions

Table 2 Dimensions of rotor ducts

	A	B	C	D	E	F	G	H
L_I(m)	0.0250	0.0250	0.0250	0.0250	0.0250	0.0250	0.0250	0.0250
L_E(m)	0.06	0.06	0.06	0.06	0.060	0.06	0.0210	0.0210
L (m)	0.102	0.102	0.102	0.102	0.102	0.102	0.0625	0.0625
A_I(m^2)	0.101	0.101	0.101	0.101	0.0937	0.118	0.101	0.118
A_E(m^2)	0.0866	0.0888	0.0910	0.104	0.104	0.104	0.0866	0.866
Geometry	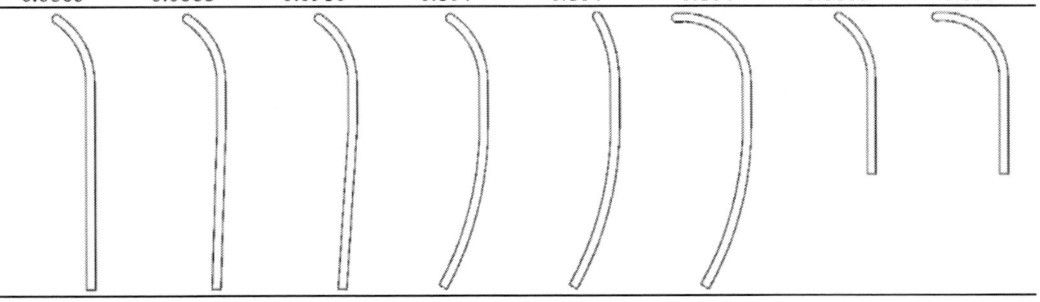							

The power required to drive the ducted rotor is:

$$P_H = T_R v_R = \frac{1}{2}\rho A_R v_E^2 v_R \quad (3).$$

For the open rotor, the thrust and the power are respectively given as:

$$T_{Hu} = 2\rho A_R v^2 \quad (4),$$

$$P_{Hu} = T_{Hu} v = 2\rho A v^3 \quad (5).$$

When the power is identical between the ducted and the open rotors, $P_H = P_{Hu}$, the relationship of the thrust is:

$$\frac{T_H}{T_{Hu}} = \left(\frac{2v_R}{v_E}\right)^{1/3} = \left(\frac{2A_E}{A_R}\right)^{1/3} \quad (6).$$

Equation (6) means the thrust of the ducted can increase when $A_E/A_R > 1/2$. When the thrust is identical between the ducted and the open rotors, $T_H = T_{Hu}$, the relationship of the power is:

$$\frac{P_H}{P_{Hu}} = \left(\frac{v_E}{2v_R}\right)^{1/2} = \left(\frac{A_R}{2A_E}\right)^{1/2} \quad (7).$$

Equation (7) means the power of the ducted can be saved when $A_E/A_R > 1/2$.

According to these theoretical suggestions, it is better to increase the duct outlet area rather than to reduce. Although an influence of the inlet area does not appear, the inlet contour is expected to affect the loss and the suction force. Then the influence of the inlet contour has been examined experimentally.

Figure 4 shows schematic of the ducted rotor. The rotor duct contour consists of the inlet section, the shroud section, and the outlet section. The dimensions of the shroud section are fixed as shown in Fig. 4. Geometries of the inlet and the outlet sections are varied as shown in Table 2. Using the ducts A, B, C, and D, the influence of the outlet contour is discussed. Using the ducts D, E, and F the influence of the outlet contour is discussed. The ducts G and F have short outlet sections. The ducts, which were used for the measurements, were manufactured using a 3-D printer (Tiertime Technology Co., Ltd., UPBOX) and were made from ABS resin.

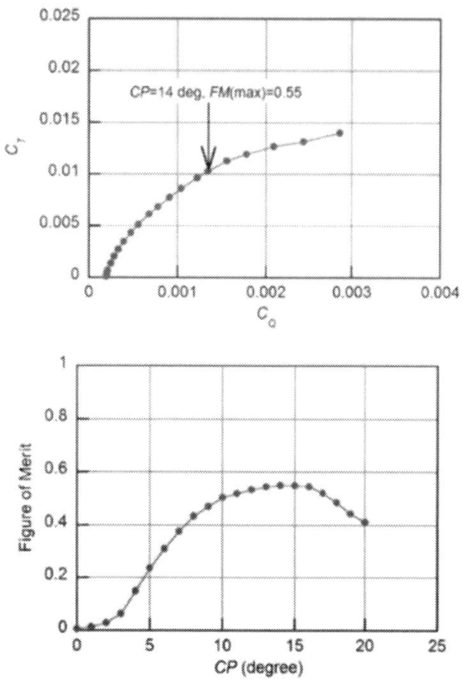

Fig. 5 Measurement of aerodynamic characteristics of open rotor

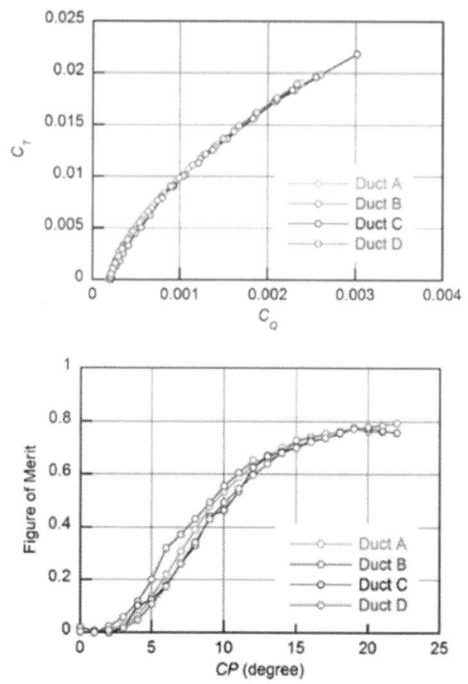

Fig. 6 Measurement of aerodynamic characteristics of ducted rotor A, B, C and D. The influence of the outlet geometry can be examined.

RESULTS AND DISCUSSIONS

The hovering thrust and the axial torque were measured at various collective pitch angles of the rotor, *CP*. The measured values of the thrust and the torque are normalized and the thrust coefficient, C_T and the torque coefficient C_Q are defined as follows:

$$C_T = \frac{Thrust}{\rho (R\Omega)^2 \pi R^2} \quad \cdots (1),$$

$$C_Q = \frac{Torque}{\rho (R\Omega)^2 \pi R^3} \quad \cdots (2).$$

In addition, the rotor efficiency is evaluated as the figure of merit defined as follows:

$$\text{Figure of merit} = \frac{C_T^{3/2}}{\sqrt{2}C_Q} \quad \cdots (3).$$

The aerodynamic characteristics of the open rotor is shown in Fig. 5. C_T and C_Q monotonically increase as *CP* increases. The figure of merit shows a maximum value of 0.55 at $CP = 14°$. According to these results, the stall occurs at $CP > 15°$.

Aerodynamic characteristics of the ducted rotors with the duct A, B, C and D are shown in Fig. 6. Comparing the results shown in Fig. 6, an influence of the outlet geometry on the aerodynamic performance can be examined. The characteristics shown as the C_T - C_Q curves are almost identical to each other. Comparing with the result of the open rotor, C_T becomes larger at the same C_Q conditions. The values of the figure of merit are also almost identical to each other and larger than that of the open rotor. Consequently, the influence of the outlet geometry is not significant. These results are different from the theoretical suggestion mentioned in the previous section. Then numerical simulations were carried out using the validated CFD software for the flow around rotorcraft, rFlow3D. Velocity vector fields obtained numerically are shown in Fig. 7. The numerical results show that the boundary layer separation occurs in the outlet section of the duct D. Consequently, the effect of the increase of the cross-section area does not appear. Similar flow pattern may occur in the experiment of the ducts B and C though the wall angles of the outlet section are smaller than the duct D. One of the reason why the outlet geometry does not affect the properties may be that the flow Reynolds number is small.

Aerodynamic characteristics of the ducted rotors with the duct D, E and F are shown in Fig. 8. Comparing the results shown in Fig. 8, an influence of the inlet geometry on the aerodynamic performance can be examined. Comparing C_T - C_Q curves, obvious differences can be observed. As the inlet area increases, C_T becomes larger at the same C_Q condition. The values of the figure of merit show the same tendencies.

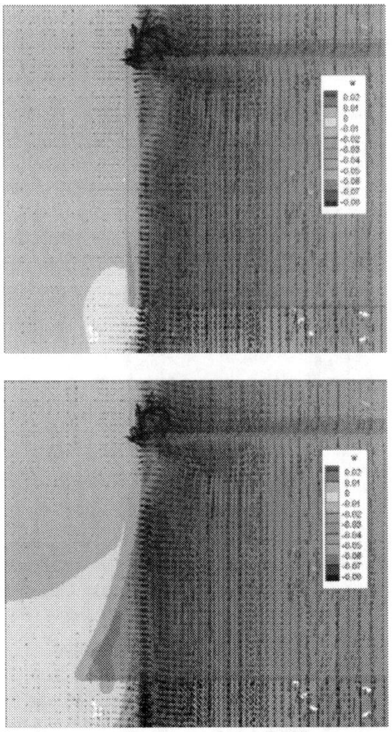

Fig. 7 Instantanous velocity vector fields around the duct outlet of duct A (top) and duct D (bottom). Unsteady CFD results using rFlow3D

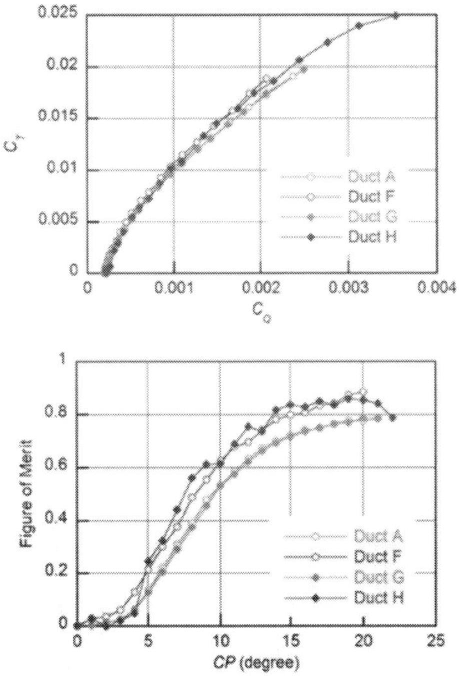

Fig. 9 Measurement of aerodynamic characteristics of ducted rotor D, E and E. The influence of the length of outtlet section can be examined.

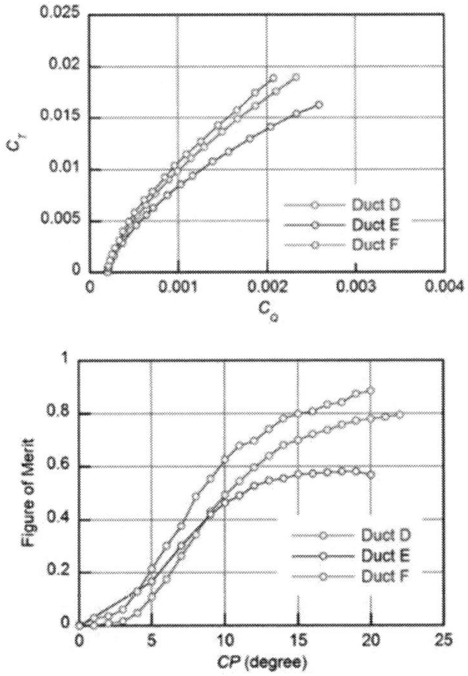

Fig. 8 Measurement of aerodynamic characteristics of ducted rotor D, E and E. The influence of the intlet geometry can be examined.

The maximum figure of merit of the ducted rotor F is 0.89, which is much larger than that of the open rotor, 0.55.

As shown in Figs. 6 and 8, the results suggest that the inlet section is important but the outlet section does not affect the rotor performance. Then the outlet section of the duct A and G were cut and examined their performances. Figure 9 shows the comparison of the aerodynamic characteristics of longer and shorter ducted rotors. As a result, it can be confirmed that the outlet section does not affect the aerodynamic characteristics. Values of the figure of merit of the duct A and F are almost the same and those of the duct G and H are also almost the same

According to the measurement, the duct H is the best choice of the eight ducts examined in the present study because the figure of merit is the highest and the duct weight is smallest. The duct G can be suitable choice when the distance between each rotor is not large enough to implement the duct H.

Finally, the performance of the ducted rotor was examined using our first prototype of multicopter with the ducts G [4]. The test apparatus is shown in Fig. 10. In this experiment, the thrust was measured using the load cell but the axial torque could not be measured. Then the figure of merit was evaluated from the electric power consumption though a loss of the motor was included.

The thrust coefficient and the figure of merit are shown in Fig. 11. As observed in the experiments using the single rotor module, C_T and the figure of merit of the ducted rotor are larger than that of the open rotor. Since the loss of the motor

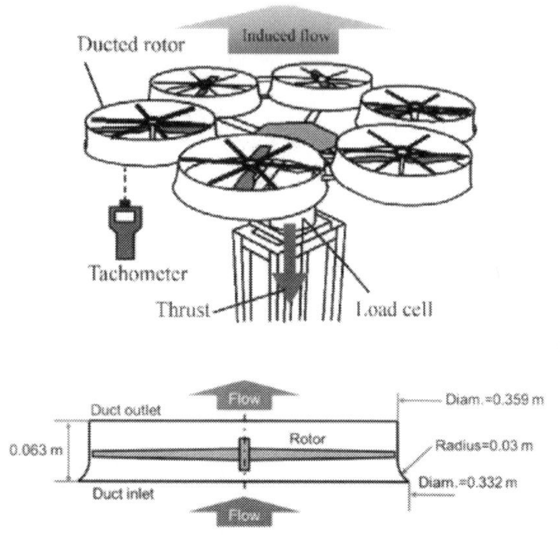

Fig. 10 Schematic aerodynamic test apparatus (top) and duct dimensions (bottom). Test stand is located upstream of rotors to prevent aerodynamic interactions[4].

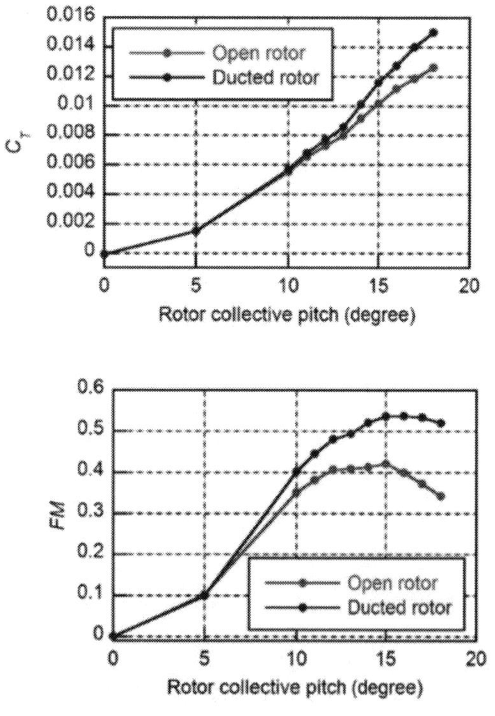

Fig. 11 Thrust coefficient and figure of merit versus rotor collective pitch angle with ducted rotor G [4].

included, the figure of merit of both open and the ducted rotor becomes lower than of those obtained in the experiments using the single rotor module.

CONCLUSION

The ducted rotors with various duct contours have been developed and the aerodynamic characteristics during hovering have been examined. As a result of the experimental study, it has been clarified that the duct geometry of the inlet affects the rotor performance significantly but the duct geometry of the outlet does not affect it. It is also confirmed that the larger thrust can obtained and the figure of merit increases using the ducted rotor. Further investigations of ducted rotor are under way to develop the reliable multicopter, such as the influence of the lateral wind, aerodynamics near a wall.

This research was supported by ImPACT, Cabinet Office, Government of Japan.

REFERENCES

1. http://www.jst.go.jp/impact/en/program/07.html

2. Sunada, S., Tanabe, Y., Yonezawa, K., and Tokutake, H., "Improvement of flight performance of Minisurveyor by using pitch control and decreasing the number of rotors," Proceedings of 2016 Asia-Pacific International Symposium on Aerospace Technology (APISAT-2016), Toyama, Japan, 2016.

3. Yonezawa, K. Yoshida, N., Sugiyama, K., Tokutake, H., Tanabe, Y., Sunada, S., "Development of a Multicopter with Ducted and Variable Pitch Rotors," 5th Asian Australian Rotorcraft Forum, Singapore, November, 2016.

4. Yonezawa, K., Matsumoto, H., Sugiyama, K., Tokutake, H., Tanabe, Y., Sunada, S., "Development of High-Performance Multi-Rotor Drone," 14th International Conference on Flow Dynamics, Sendai, November 1-3, 2017.

Modeling of Aerodynamic Characteristics of Drone and Improvement of Gust Response

Raijo Torita	Takuji Kishi	Hiroshi Tokutake
Graduate student	Graduate student	Associate Professor
Kanazawa University	Kanazawa University	Kanazawa University
Kanazawa, Japan	Kanazawa, Japan	Kanazawa, Japan

Shigeru Sunada	Yasutada Tanabe	Koichi Yonezawa
Professor	Associate senior Researcher	Project researcher
Nagoya University	Japan Aerospace Exploration Agency	Central Research Institute of Electric Power Industry
Nagoya, Japan	Tokyo, Japan	Abiko, Japan

ABSTRACT

Gust disturbances cause drones to crash. Gusts cause variations in dynamics, which affect the aerodynamic forces and moments generated by rotors. Gross modeling of rotor behavior and its application to flight control during a gust disturbance are necessary for safe flight. In this study, a gust response alleviation system based upon real-time modeling of the drone's dynamics is proposed. The gust input is used to construct the drone dynamics. In addition to the use of feedback control of motion, an appropriate control input to cancel the aerodynamic force and moment variation due to the gust input is proposed. This control input is derived analytically without iteration by using inverse dynamics of the drone. The proposed system is validated with numerical simulations.

NOTATION

A_d	: rotor-disk area, m^2
a	: lift slope
α	: angle of attack, rad
b	: number of blade
C_d	: drag coefficient
c	: chord length of blade, m
dr	: blade element radius, m
dL, dD	: propeller section lift and drag, N
e	: estimation error
g	: acceleration of gravity, m/s^2
h_r	: distance between rotor-disk plane and C.G. plane, m
θ_t	: pitch angle(blade element), rad
I_{xx}, I_{yy}, I_{zz}, I_{xz}	: moment of inertia, kg･m^2
K_{PP}, K_{PQ}, K_{PR}	: proportional gain of PID controller I
K_{PL}, K_{PM}, K_{PN}	: proportional gain of PID controller II
L_o	: observer gain
L, M, N	: moment acting on drone, N･m
L_c, M_c, N_c	: moment command (sum of output from PID controller I and II), N･m

L_f, M_f	: output from PID controller II, N·m
l	: length between C.G. and rotor, m
m	: gross weight of drone, kg
n_l, n_m	: noise of disturbance moment, N·m
P, Q, R	: angular velocity of drone, rad/s
p, q	: angular velocity of drone (difference with trim point), rad/s
ρ	: density of air, kg/m^3
R_r	: rotor radius, m
T	: rotor thrust, N
T_{DP}, T_{DQ}, T_{DR}	: derivative time constant of PID controller I
T_{DL}, T_{DM}, T_{DN}	: derivative time constant of PID controller II
T_{IP}, T_{IQ}, T_{IR}	: integration time constant of PID controller I
T_{IL}, T_{IM}, T_{IN}	: integration time constant of PID controller II
U, V, W	: translation velocity of drone, m/s
U_T	: tangential component of relative velocity with rotor blade and atmosphere, m/s
U_P	: perpendicular component of relative velocity with rotor blade and atmosphere, m/s
Vc	: upward velocity of drone, m/s
Vr	: composition of relative velocity with rotor blade and atmosphere, m/s
v_i	: induced velocity, m/s
x_p	: state variable of drone
x_d	: disturbance moment input, N·m
z	: state variable expanded to disturbance moment input
Φ, Θ, Ψ	: Euler angle of drone, rad
φ	: inflow angle, rad
X_a, Y_a, Z_a	: aerodynamic force, N
Ψ_r	: azimuth angle of rotor blade, rad
Ω	: rotor rotational velocity, rad/s

Subscript

c	: command
d	: disturbance
j	: rotor number
0	: trim point (hovering)

INTRODUCTION

The multirotor unmanned aerial vehicle or drone has become very popular in recent years because of its hovering capability, vertical take-off and landing capability, and simple structure[1]. However, many crash incidents have been caused by a gust disturbance[2]. A gust generates a large dynamics response for a small drone along with aerodynamic interaction between the rotors[3]. This results in a dynamics variation from the nominal model used for conventional controller design. Many previous studies about the control of a drone are based on the simple linearized dynamics model[4)5)]. A more precise dynamics model may aid the development of improved controller design.

In this study, the aerodynamic forces of the rotors were formulated analytically with blade element theory and momentum theory. The model developed by us enabled real-time calculation of the rotational velocity commands of rotors by realizing the desired control input in actual flight conditions. The proposed flight control system is constructed from the PID controller, which increases drone stability by using feedback from both the angular velocity and the disturbance observer. The estimated disturbances are fed back to the rotational velocity commands. This reduces the variation of aerodynamic forces at the rotors caused by gust disturbances. The performance of the proposed method is validated through the numerical simulations.

MODELING OF DRONE DYNAMICS

Equation of motion

In this study, a quadrotor drone is analyzed. The rotational direction of its rotors is shown in Figure. 1. Equations of

Presented at the 6th Asian/Australian Rotorcraft Forum / Heli Japan 2017, Kanazawa, Japan, November 7-9, 2017.

motion expressed in body axis are given by Eqs. (1) to (6). Relationships between Euler angle and angular velocity are given by Eqs. (7) to (9). The aerodynamic forces and moments acting on the drone are calculated by combining blade element theory and momentum theory.

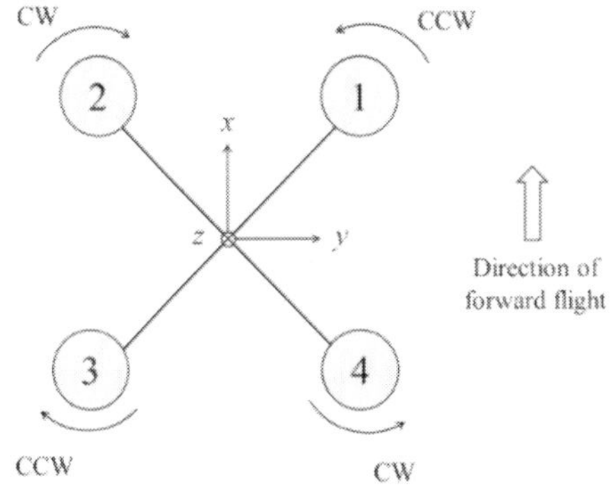

Figure 1. Rotor configuration

$$m(\dot{U}+QW-RV)=-mg\sin\Theta+X_a \tag{1}$$

$$m(\dot{V}+RU-PW)=-mg\cos\Theta\sin\Phi+Y_a \tag{2}$$

$$m(\dot{W}+PV-QU)=mg\cos\Theta\cos\Phi+Z_a \tag{3}$$

$$I_{xx}\dot{P}-I_{xz}\dot{R}+(I_{zz}-I_{yy})QR-I_{xz}PQ=L \tag{4}$$

$$I_{yy}\dot{Q}+(I_{xx}-I_{zz})RP+I_{xz}(P^2-R^2)=M \tag{5}$$

$$-I_{xz}\dot{P}+I_{xz}\dot{R}+(I_{yy}-I_{xx})PQ+I_{xz}QR=N \tag{6}$$

$$\dot{\Phi}=P+Q\sin\Phi\tan\Theta+R\cos\Phi\tan\Theta \tag{7}$$

$$\dot{\Theta}=Q\cos\Phi-R\sin\Phi \tag{8}$$

$$\dot{\psi}=Q\sin\Phi\sec\Theta+R\cos\Phi\sec\Theta \tag{9}$$

Blade element theory

Each rotating blade is divided into smaller blade elements. Lift and drag forces acting on a blade element are shown in Figure 2 and described by Eqs. (10) and (11). Aerodynamic forces acting on the rotor is obtained by integration of *dL* and *dD* to span direction. [6]

$$dL_j=\frac{1}{2}\rho\left(U_{Tj}^2+U_{Pj}^2\right)a\left(\theta_t-\frac{U_{Pj}}{U_{Tj}}\right)cdr \tag{10}$$

$$dD_j=\frac{1}{2}\rho c_d\left(U_{Tj}^2+U_{Pj}^2\right)cdr \tag{11}$$

The relative velocity of a rotor blade to the atmosphere in tip-path plane, U_{Tj}, is the sum of rotor velocity caused by rotation and velocity caused by drone motion. The perpendicular component to tip-path plane of relative velocity, U_{Pj}, is the sum of induced velocity and velocity caused by drone motion.

$$\begin{aligned}U_{Tj}&=r\Omega+U\sin\psi_r+V\cos\psi_r\\&+lR\cos\left(\frac{\pi}{2}j+\frac{3}{4}\pi\right)(\sin\psi_r+\cos\psi_r)\end{aligned} \tag{12}$$

$$\begin{aligned}U_{Pj}&=v_{ij}-W\\&+\frac{l}{\sqrt{2}}\left(P\cos\left(\frac{\pi}{2}j+\frac{3}{4}\pi\right)-Q\cos\left(\frac{\pi}{2}j+\frac{\pi}{4}\right)\right)\end{aligned} \tag{13}$$

Aerodynamic forces acting on a single rotor are obtained by integrating Eqs. (10) and (11) to blade span direction and azimuth angle direction, and then multiplying by number of blades (Eqs. (14) to (16)).

$$X_{aj}=\frac{\rho bcR_r}{2}\left\{\left(a\theta_t U_{pj}+C_d R_0\Omega_j\right)\left(\frac{U}{2}+lR\cos(\frac{\pi}{2}j+\frac{3}{4}\pi)\right)\right\} \tag{14}$$

$$Y_{aj}=-\frac{\rho bcR_r}{2}\left\{\left(a\theta_t U_{Pj}+c_d R_0\Omega_j\right)\left(\frac{V}{2}+lR\cos\left(\frac{\pi}{2}j+\frac{3}{4}\pi\right)\right)\right\} \tag{15}$$

$$\begin{aligned}Z_{aj}=\frac{\rho bc}{2}\Bigg\{&a\theta\Big(\frac{R_r{}^3\Omega_j^2}{3}+\frac{R_r}{2}\left(U^2+V^2\right)\\&+\frac{lR}{\sqrt{2}}lR_r\cos\left(\frac{\pi}{2}j+\frac{3}{4}\pi\right)(U+V)+\frac{l^2R_r}{2}R^2\Big)\\&-U_{Pj}\cdot\frac{R_r{}^2\Omega_j}{2}(a+c_d)\Bigg\}\end{aligned} \tag{16}$$

Aerodynamic forces and moments acting on the drone are calculated from the aerodynamic forces of all rotors.

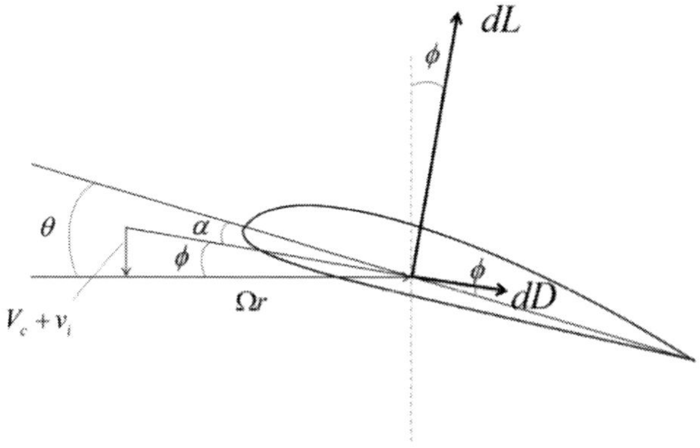

Figure 2. Aerodynamic forces at a blade element

Momentum Theory

The formulated aerodynamic force derived from blade element theory contains induced velocity, which cannot be determined. Therefore, momentum theory is included in the model to calculate the induced velocity. In momentum theory, the thrust acting on a rotor is expressed as[7]

$$T = 2\rho A_d v_i^2 \tag{17}$$

By equating Eqs. (16) and (17), the aerodynamic forces and induced velocity are obtained explicitly from rotor rotational velocity.

A flow chart of steps used to calculate aerodynamic forces by combining blade element theory and momentum theory is shown in Figure 3.

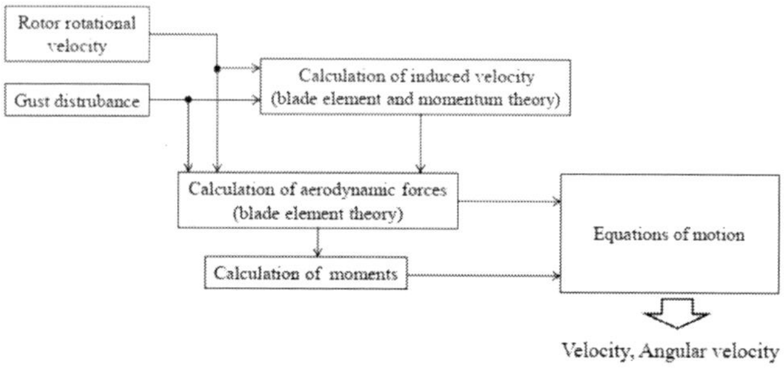

Figure 3. Flow chart of dynamics model

CONTROL SYSTEM

The flight control system designed for this study is shown in Figure 4. The sensed angular velocity and the estimated disturbance are fed back through PID controllers I and II, respectively. The sum of outputs from the PID controllers is the moment command. This command is transformed to the rotor rotational velocity by calculating the inverse dynamics of rotor dynamics model in real-time. Because the rotor dynamics model is formulated theoretically as a function of rotor rotational velocity and flight conditions, a precise control input can be realized. In this study, angular velocity and rotor rotational velocity are measured and provided for PID controller I and the observer. A motor model was assumed as a first-order lag with a time constant of 0.5.

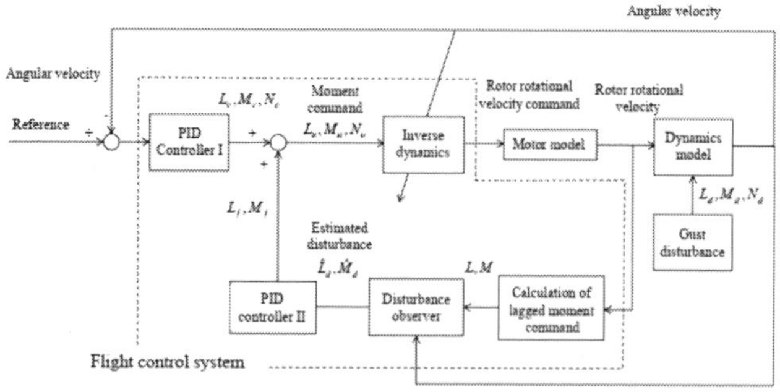

Figure 4. Flight control system

PID Controller

The PID controller of the angular velocity and disturbance feedback are described as Eqs. (18) and (19), respectively.

$$L_c = K_{PP}\left(-P - \frac{1}{T_{IP}}\int P dt - T_{DP}\frac{dP}{dt}\right) \tag{18.a}$$

$$M_c = -K_{PQ}\left(Q + \frac{1}{T_{IQ}}\int Q dt + T_{DQ}\frac{dQ}{dt}\right) \tag{18.b}$$

$$N_c = -K_{PR}\left(R + \frac{1}{T_{IR}}\int R dt + T_{DR}\frac{dR}{dt}\right) \tag{18.c}$$

$$L_f = K_{PL}\left(\hat{L}_d + \frac{1}{T_{IL}}\int \hat{L}_d dt + T_{DL}\frac{d\hat{L}_d}{dt}\right) \tag{19.a}$$

$$M_f = K_{PM}\left(\hat{M}_d + \frac{1}{T_{IM}}\int \hat{M}_d dt + T_{DM}\frac{d\hat{M}_d}{dt}\right) \tag{19.b}$$

The roll moment and pitch moment commands are determined as follows:

$$L_u = L_c + L_f \tag{20.a}$$

$$M_u = M_c + M_f \tag{20.b}$$

Disturbance Observer

The linearized dynamics of the drone and the assumed disturbance dynamics are shown in Eqs. (21) to (23), respectively.

$$\dot{x}_p = Ax_p + B_{d1}x_d + B_p u \tag{21}$$

$$y = Cx_p \tag{22}$$

$$\dot{x}_d = A_d x_d + B_{d2} u + B_w w \tag{23}$$

Combining the state equations of drone and disturbance, the extended state equation is obtained.

$$\dot{z} = \tilde{A}z + \tilde{B}_p u + \tilde{B}_w w \tag{24}$$

$$y = \tilde{C}z \tag{25}$$

$$z = \begin{bmatrix} x_p & x_d \end{bmatrix}^T \tag{26}$$

The observer for the extended state equation is constructed as follows.

$$\dot{\hat{z}} = A\hat{z} + \tilde{B}_p u + L(y - \hat{y}) \tag{27}$$

The dynamics of the estimated error $e = z - \hat{z}$ becomes

$$\dot{e} = \left(\tilde{A} - L_o\tilde{C}\right)e + B_w w \qquad (28)$$

We allow the transfer function from w to z become $G(s)$. The observer gain L which minimizes $\|G\|_\infty$ can be obtained solving the linear matrix inequality of the bounded real lemma.[8)]

NUMERICAL SIMULATIONS

The proposed method was applied to the testbed drone developed by our laboratory (Figure 5). The configuration of the testbed drone is listed in Table 1.

Figure 5. Testbed drone

Table 1. Testbed drone Characteristics

Characteristic	English	Metric
Rotor diameter	0.070 ft	0.23 m
Arm Length between Center of Rotors	0.14 ft	0.46 m
Gross Weight	0.50 lb	1.1 kg

The state matrices and state vector of drone dynamics are follows.

$$x_p = [p \quad q]^T,\ x_d = [L_d \quad M_d]^T\ u = [L \quad M]^T \qquad (29)$$

$$A = \begin{bmatrix} -0.072 & 0 \\ 0 & -0.072 \end{bmatrix} \qquad (30)$$

$$B_p = B_{d1} = \begin{bmatrix} 1.5\times10^2 & 0 \\ 0 & 1.5\times10^2 \end{bmatrix} \qquad (31)$$

$$C = \begin{bmatrix} 1 & 0 \\ 0 & 1 \end{bmatrix} \qquad (32)$$

$$w = [n_l \quad n_m]^T \qquad (33)$$

$$A_d = \begin{bmatrix} 0 & 0 \\ 0 & 0 \end{bmatrix} B_{d2} = \begin{bmatrix} 0 & 0 \\ 0 & 0 \end{bmatrix}, B_w = \begin{bmatrix} 1 & 0 \\ 0 & 1 \end{bmatrix} \qquad (34)$$

$$\tilde{A} = \begin{bmatrix} A & B_{d1} \\ \mathbf{0} & A_d \end{bmatrix}, \tilde{B}_p = \begin{bmatrix} B_p \\ B_{d2} \end{bmatrix}, \tilde{B}_w = \begin{bmatrix} \mathbf{0} \\ I \end{bmatrix} \qquad (35)$$

$$\tilde{C} = \begin{bmatrix} 1 & 0 & 0 & 0 \\ 0 & 1 & 0 & 0 \end{bmatrix}^T \qquad (36)$$

The observer gain was determined from these dynamics as follows:

$$L_o = 10^8 \times \begin{bmatrix} 0.017 & 0 & 3.0 & 0 \\ 0 & 0.0014 & 0 & 2.5 \end{bmatrix}^T \qquad (37)$$

The determined PID gains are shown in Table 2. These gains are determined to minimize the peak values of angular velocity response to the disturbance moment input. In this simulation, minimizing of roll and pitch rate, in which estimated disturbance was fed back was given priority.

Table 2. PID gains of PID controllers

K_{PP}	T_{IP}	T_{DP}	K_{PQ}	T_{IQ}
2.0	0.045	0.45	2.0	0.045
T_{DQ}	K_{PR}	T_{IR}	T_{DR}	K_{PL}
0.45	10	0.050	0.50	1.0
T_{IL}	T_{DL}	K_{PM}	T_{IM}	T_{DM}
1.0×10^4	1.0×10^{-6}	1.0	1.0×10^4	1.0×10^{-6}

The designed flight control system was validated through numerical simulations. The responses for the moment disturbances resulting from only angular velocity feedback as well as both angular velocity and estimated disturbance feedback were analyzed. Figure 6 shows the moment input and estimated values by the designed observer. The maximum value of moment input is about 0.009 Nm, which is the moment generated by a gust of 4.3 m/s. It can be seen that the observer precisely estimates moment input. However, the continuous oscillation, which is caused by the high gain of the observer, remains.

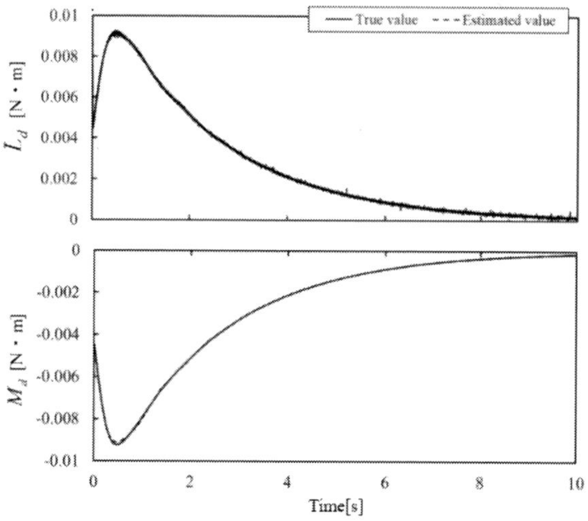

Figure 6. Estimation results of the disturbance

The time histories of angular velocities, Euler angles, and moment commands are shown in Figures 7, 8, and 9, respectively. Responses in rolling and pitching are suppressed by disturbance feedback. However, control inputs from proposed controller were almost same as these from only angular velocity feedback. The structure of flight control system using estimated disturbance would be improved.

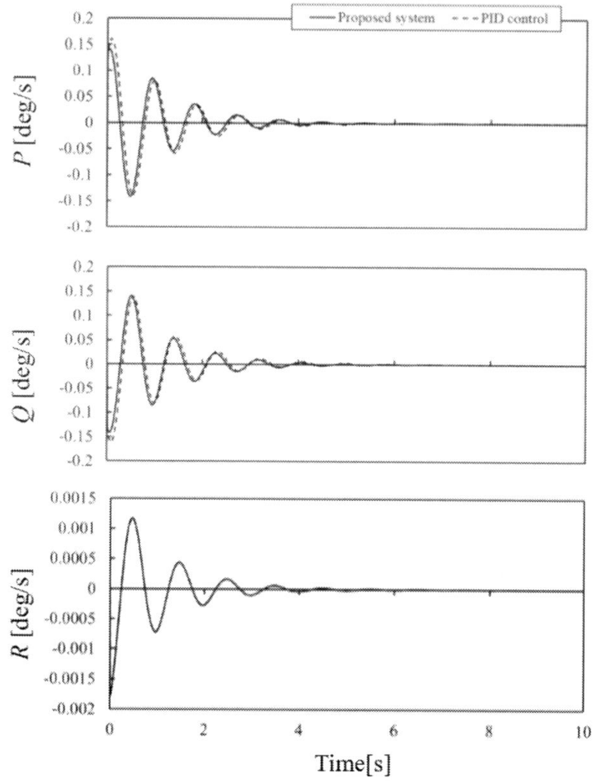

Figure 7. Angular velocity responses

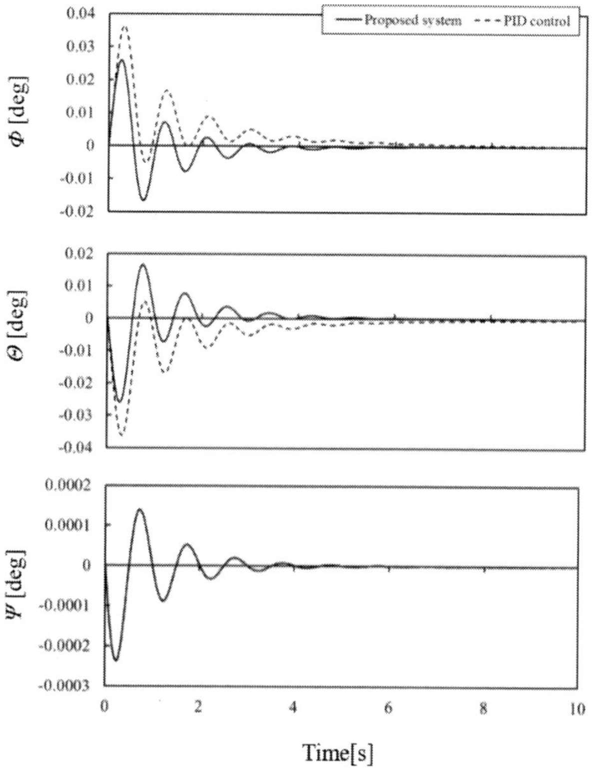

Figure 8. Euler angle responses

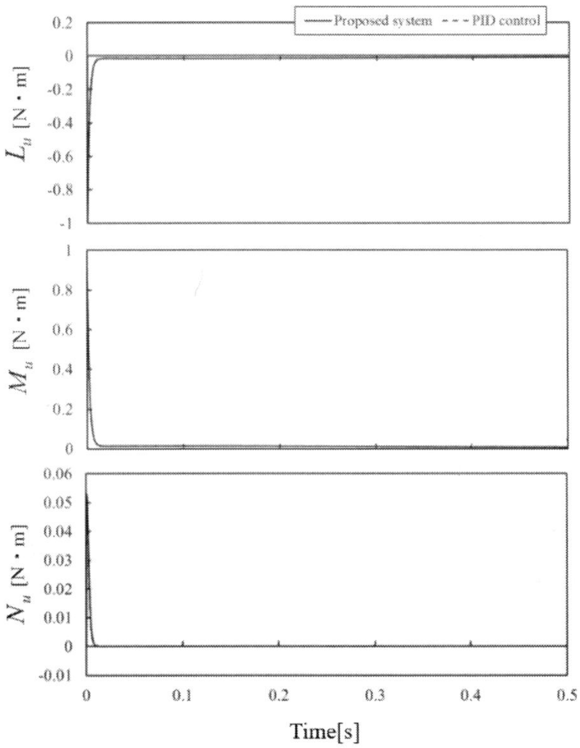

Figure 9. Control inputs

FLIGHT TESTING

Flight testing using the testbed drone (shown in Figure 5) was conducted. Steady hovering flight with angular velocity feedback was performed, and the disturbance input was analyzed off-line from the logged data. Figure 10 shows the estimated disturbance moment. These results are reasonable orders of magnitude. However, estimation accuracy could not be determined because the disturbance moment was could not be measured in flight test. In future work, implementation of on-line disturbance observer and proposed control system will be worked on.

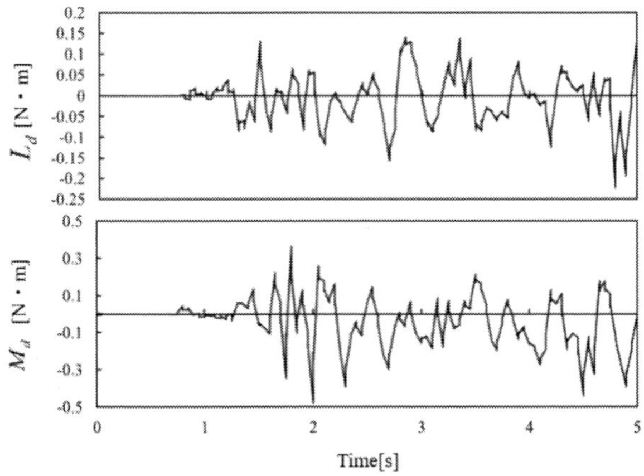

Figure 10. Estimation results of disturbance moment

CONCLUSIONS

A gust alleviation control system based upon real-time calculation of rotor dynamics and a disturbance observer was developed and tested. The angular velocities and estimated disturbance input are used as feedback to decrease the drone responses. The appropriate rotational velocity of rotors was calculated by the theoretical model. The performance of the proposed system was analyzed by numerical simulation. Additionally, flight tests for the validation of the disturbance observer were conducted.

Variable gust disturbances exist in many flight fields where drones are used. These gusts are often found close to structures. The proposed system could be effective for flight missions to observe buildings or bridges.

Author contact: Raijo Torita r.torita@stu.kanazawa-u.ac.jp
Hiroshi Tokutake tokutake@se.kanazawa-u.ac.jp

ACKNOWLEDGMENTS

This research was supported by ImPACT, Cabinet Office, Government of Japan.

REFERENCES

[1]Tawara, M., Nonami, K., "General Airframe Design and Implementation with Low-Cost for Multi-Rotor Type Helicopters," Transactions of the Japan Society of Mechanical Engineers, Series C, vol. 78, No.787, 2012-3, pp.872-882(in Japanese)

[2]Ministry of Land, Infrastructure, Transport and Tourism, "The regulations of flight of UAV," 2017, http://www.mlit.go.jp/koku/koku_tk10_000003.html

[3]Hwang, J. Y., Jung, M. K., Kwon, O. J., "Numerical Study of Aerodynamic Performance of a Multirotor Unmanned-Aerial-Vehicle Configuration," *Journal of Aircraft*, Vol. 52, No. 3, May-June 2015, pp.839-846

[4]Hoffmann, G. M., Huang, H., Waslander, S. L., Tomlin, C. J., "Quadrotor Helicopter Flight Dynamics and Control: Theory and Experiment," AIAA Guidance, Navigation and Control Conference and Exhibit, Hilton Head, South Carolina, 20-23 August 2007

[5]Altug, E., Ostrowski, J. P., Taylor, C. J., "Control of a Quadrotor Helicopter Using Dual Camera Visual Feedback," *The International Journal of Robotics Research*, Vol. 24, No. 5, May 2005, pp.329-341

[6]Prouty, R. W., *Helicopter performance, Stability, and Control*, Krieger Publishing Company, Inc., Malabar, Florida, 1986, pp. 10–32.

[7]Bramwell, A. R. S., Done, G., Balmford, D., *Bramwell's Helicopter Dynamics*, American Institute of Aeronautics and Astronautics, Inc., Reston, Virginia and Butterworth-Heinemann, Oxford, Great Britain, 2001, pp. 34–42.

[8]Tokutake, H., Okada, S., Sunada S., "Disturbance Preview Controller and Its Application to a Small UAV," *Transactions of the Japan Society for Aeronautical and Space Sciences*, Vol.55, No.1, January 2012, pp76-78

Development of High-Fidelity Flight Dynamic Models for Unmanned Rotorcrafts and Multi-copters with Applications

Seong Han Lee
PhD Student
Konkuk University
Seoul, Republic of Korea

Do Hyeon Lee
PhD Student
Konkuk University
Seoul, Republic of Korea

Sung Wook Hur
PhD Student
Konkuk University
Seoul, Republic of Korea

Sang Ho Kang
Master Student
Konkuk University
Seoul, Republic of Korea

Chang-Joo Kim
Professor
Konkuk University
Seoul, Republic of Korea

ABSTRACT

This study treats techniques of high-fidelity flight dynamics models for unmanned rotorcraft, multi-copters, and the propeller-driven aircrafts which have the wing components with control surfaces. The paper tries to provide a unified approach to the flight-dynamic modeling to handle the general configuration. For this purpose, the high-fidelity rotor models used in the HETLAS (Helicopter Trim, Linearization, and Simulation) program are adopted to handle the propellers with the RPM or pitch variations with the clockwise or counterclockwise directions of rotation. The rotor models include the dynamic inflow and rotor dynamic models for flap and lead-lag motions and it can handle the cyclic pitch controls for the rotorcrafts. The aerodynamic wings with control surfaces are modeled by combining the strip theory with the lifting line theory. The effects of the control surface deflection on wing aerodynamic forces and moments are modeled with various empirical formulas and the best candidate among them is selected by comparing relative accuracy with the experimental data. The proposed rotor and wing component models are integrated to build example aircraft configurations and tested to validate the present approaches. After rigorous validation works, the resultant flight dynamic models are applied to the flight dynamic analysis of the multi-copters with tilted rotors and wings with control surfaces to show its applicability of trim, linearization, and simulation analyses.

NOTATION

AC	: Aerodynamic Center
ABC	: Advancing Blade Concept
CG	: Center of Gravity
CW/CCW	: Clockwise/Counter Clockwise
L	: Coordinate Transformation Matrix
u, v, w	: Linear velocity (m/s)
α	: Angle of Attack (rad)
σ	: Rotating Direction
ϕ, θ	: Euler Angles or Attachment Angle(rad)
AZ	: Azimuth Angle
FWD	: Forward
HC/WC	: From CG to Hub Fixed or to Wing Coordinate
n/t	: Normal/Tangential Component
ST	: Starboard Side Tilt Angle
S/P	: Starboard side/Port side

INTRODUCTION

A Typical fixed wing aircraft requires a certain distance of runway to take off and has the disadvantage that it cannot fly hovering flight. And rotorcrafts have shorter range than the fixed wing aircraft, but it cannot fly at high speed. As the use of fixed wing aircraft or rotorcraft is restricted due to missions, the necessity of development of composite rotorcraft utilizing the advantages of fixed wing and rotary wing aircraft which can be applied to various missions has been developed, and various types of unmanned aircraft/multi-copter have been proposed. The proposed aircrafts have a variety of shapes with a combination of rotor, propeller and wing with control surfaces. It is difficult to develop a flight dynamic model for each aircraft, therefore, a generalized aerodynamic model is necessary to be applied whatever shape is provided.

This paper shows the general approach to building the flight dynamic model to deal with any aircraft configuration. There are two aerodynamic models, one is a generalized rotor model and the other one is a generalized wing model. The rotor model using HETLAS program which is discussed in Ref. 1 and Ref. 2 handles the propeller with the RPM or pitch variations and the rotor with rotor dynamics such as flap and lead-lag motion and with cyclic pitch control. It considers rotating directions of propeller or rotor. The installation of each rotor or propeller is defined by the forward and lateral tilt angles and the location of the hub center relative to the CG of the aircraft. The generalized wing model with control surface is modeled by combining the strip theory with the lifting line theory which is discussed in Ref 3. The effect on the control surface deflection on wing aerodynamic forces and moments are modeled with various empirical formulas and the best candidate among them is selected by comparing relative accuracy with the experimental data. The empirical formulas

Presented at the 6th Asian/Austrailian Rotorcraft Forum / Heli Japan 2017, Kanazawa, Japan, November 7-9, 2017.

are discussed in Ref. 4. The installation of each wing is defined by the elevation angle, attachment angle, airfoil configuration and its location relative to the aircraft CG. To validate the proposed models, they are integrated to build example aircraft configurations.

GENERALIZED AERODYNAMIC MODEL

This chapter discusses to build each model that which assumption are applied and which techniques are used.

Generalized Rotor Component Model

This section discusses the generalization methodology of rotor and propeller. The rotor and propeller generally differs the control input, location and dynamics as shown in the table 1. The rotor can distinguish among a conventional main rotor, a tail rotor and ABC rotor through the application of dynamics such as flap, lead-lag dynamics. The rotor generally controls the collective pitch and 2 types of cyclic pitches with constant RPM to produce the thrust. On the other hand, the propeller, is different with rotor, controls RPM with fixed blade pitch to produce thrust. Furthermore, the location and orientation of the propeller and the rotor are different depending on the purpose of use. The unmanned aircraft or multi-copter can use lots of rotors and propellers. Therefore, the generalized rotor model considers following 4 steps to distinguish a rotor or propeller. 1) Appliance rotor dynamics, 2) The way to control, 3) The mounting location, 4) The orientation.

Table 1 Definition of Rotor and Propeller

Type	Dynamics	Control	Location	Orientation
Propeller	No	COLL/ RPM	Front/ Rear	90 deg FWD tilt
Main Rotor	Flap/Lag/ RPM	COLL &2CP	Top/ Center	Vertical(Reference) with small FWD tilt
Tail Rotor	Flap/ RPM	COLL	Rear	±90 deg sideward tilt with small cant angle
Teetering Rotor	Flap	COLL &2CP	Top/ Center	Vertical with small FWD tilt
ABC Rotor	No	COLL& 2CP	Top/ Center	Vertical with small FWD tilt
Ducted Fan	No	COLL (Thrust vectoring)	As specified	As specified

COLL.: Collective, CP: Cyclic Pitch

The equations of motion of the rotor can be derived into various formulations, but it is difficult to judge whether the induced equation of motion can be used without any limitation. In the case of the nonlinear induced equations of motion, it is fixed according to the sequence of the hinges mounted on the rotor, and when adding and removing the hinges, the equation of motion must be re-derived. Therefore, it is necessary to be able to easily modify and verify the equations of motion even if the sequence of the hinge is changed or added or removed according to the operational state and purpose. To suggest solving this problem, the relation kinematics, velocity, and acceleration between two arbitrary coordinate systems are generalized, and flapping, lead-lag, and torque equations of the rotor are derived. The equations of motion of the rotor or propeller are generalized using relative motion equation between two arbitrary coordinate systems. There is detail derivation in Ref. 1. The direction of aerodynamic forces and moments in lead-lag and torque equations differ the rotating direction of rotor, the rotating direction, σ , is applied to the equations of motion in equation 2~4 of Ref. 2. The equations of motion are defined when the rotor rotates CCW direction. And if the rotor rotates CW direction, the parameter σ is equal to negative 1. In the case of propeller, flap and lead-lag equations are not considered, only torque equations are applied. The aerodynamic forces and moments are calculated using nonlinear aerodynamic coefficients. Those are defined as relative velocity of wind and angle of attack at a blade section. And those are estimated using table lookup method in the aerodynamic coefficient data of the airfoil. The inflow velocity over the rotor is calculated using various theories such as uniform inflow model or peter-he inflow model.

In general, a main rotor is attached perpendicular to the fuselage. And a tail rotor and a propeller are attached different directions with the main rotor. Furthermore, the unmanned aircraft and multi-copter can have lots of rotors or propellers. Therefore, it is important to define the location and orientation of each rotor or propeller. Firstly, to define the location of them, the reference position should be defined. The aerodynamic forces and moments of rotors and propellers act on the CG of aircraft. Therefore, the reference position should set as the CG position. Then those hub center should be defined relative position vector to the reference position or specified in the aircraft datum. Secondly, the reference orientation direction should be defined. Before setting the reference direction, the coordinate transfer matrices around x, y, z-axis are defined as shown in equation from 1 to 3, respectively.

$$\mathbf{L}_1(x) = \begin{bmatrix} 1 & 0 & 0 \\ 0 & \cos x & \sin x \\ 0 & -\sin x & \cos x \end{bmatrix} \tag{1}$$

$$\mathbf{L}_2(y) = \begin{bmatrix} \cos y & 0 & -\sin y \\ 0 & 1 & 0 \\ \sin y & 0 & \cos y \end{bmatrix} \quad (2)$$

$$\mathbf{L}_3(z) = \begin{bmatrix} \cos z & \sin z & 0 \\ -\sin z & \cos z & 0 \\ 0 & 0 & 1 \end{bmatrix} \quad (3)$$

The positive direction of x-axis is defined as the direction of the rear fuselage. The positive direction of y-axis and z-axis are defined as the direction of the starboard side and opposite direction of the Earth's center. The orientation is specified starboard side tilt angle (θ_{ST}), forward tilt angle (θ_{FWD}) and azimuth rotation angle (θ_{AZ}) which rotate around x, y, z- axis, respectively. Therefore, the coordinate transfer matrix from CG to the hub fixed coordinate is defined as shown in equation 4. For example, the main rotor of convention helicopter is defined as $(\theta_{ST}, \theta_{FWD}, \theta_{AZ}) = (0,0,0)$. In addition, the tail rotor attached to the port side of empennage is defined as $(\theta_{ST}, \theta_{FWD}, \theta_{AZ}) = (-\pi/2, 0, 0)$.

$$\mathbf{L}_{HC} = \mathbf{L}_1(\theta_{azim})\mathbf{L}_2(-\theta_{right})\mathbf{L}_3(\pi - \theta_{fwd}) \quad (4)$$

The rotor and propeller configuration are analyzed using the developed generalized rotor model. They are rotating CCW direction and orientation are shown in equation 5.

$$(\theta_{ST}, \theta_{FWD}, \theta_{AZ}) = (0,0,0)_{\text{Rotor}}, (0, \pi/2, 0)_{\text{Prop}} \quad (5)$$

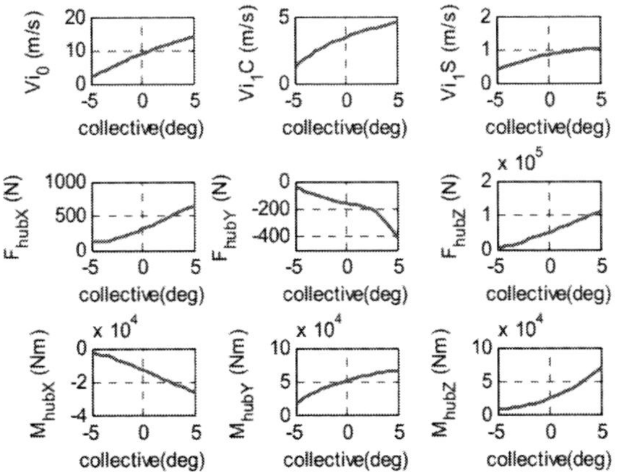

Figure 1 Analysis of Rotor Configuration

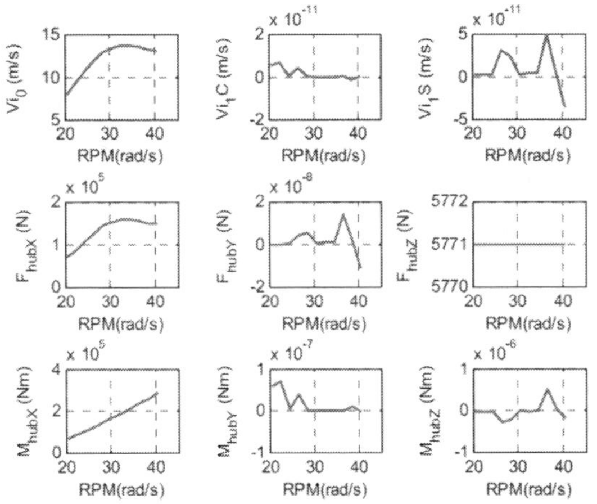

Figure 2 Analysis of Propeller Configuration

Figure 1 and figure 2 show the inflow velocities and the aerodynamic forces and moments. Figure 1 is the result of rotor configuration with respect to collective sweep. Also figure 2 is the result of propeller configuration with respect to RPM variation. The blade pitch angle is 10 degrees. Those attachment angles are the same with equation 5. This analysis conducts at the velocity $u = 20.0knots$.

Generalized Wing Component Model

The generalization methodology of wings with control surface is discussed in this section. In general, the wings in a conventional fixed-wing aircraft are classified main wings and horizontal and vertical stabilizers as shown in table 2. It describes the control surfaces, locations and orientations of various wing types. The unmanned aircraft can have lots of wings and control surfaces, regardless of the conventional fixed-wing aircraft configuration. Therefore, the method considers the following 4 steps to define each wing. 1) The wing configuration, 2) Application of Control Surface, 3) The wing location. 4) The wing Orientation.

Table 2 Definition of Wings

Type	Control surface	Location	Orientation
MW	Aileron/Flap	Fuselage Center	Horizontal with dihedral, sweep, and twist distribution along the span
HS	Elevator	Rear Fuselage	The same with MW, the airfoil may be upside down orientation
VS	Rudder	Rear Fuselage	±90 deg upward tilt, the airfoil may be upside down orientation
Others	As specified	As specified	As specified

MW: Main Wing, HS/VS: Horizontal/Vertical Stabilizer

The calculation of aerodynamic forces is discussed before defining the wing configuration. The aerodynamic wing model are used to estimate aerodynamic coefficient of the wing element using nonlinear 2-D airfoil coefficient data by table lookup method. The aerodynamic force of the wing element can be calculated by calculating the vortex intensity through the lift-line theory discussed in Ref. 3 which is applied considering the calculation time and applying the predicted induction velocity to the strip theory through the Biot-Savart law. Then, the induced velocity is calculated by considering influence of peripheral aerodynamic surfaces. In addition, the aerodynamic forces of wing with control surface are calculated by applying lift increment estimation, which is empirical formula, in the process to calculate aerodynamic forces using strip theory.

To define wing configuration, the reference position on the aircraft should be define. The CG at the aircraft is specified as a reference position for expressing the position of the wing on the aircraft as the same with the generalized rotor model. Firstly, the position of wing is expressed using the relative position vector between CG and AC of the root chord of a wing. Because the wing can mount with various angle, two angles are set based on the starboard main wing to apply the same wing aerodynamic model to the aerodynamic calculation of individual wings. The elevation angle and the attachment angle are defined for wing geometry. The positive elevation angle direction is assumed CW direction in x-axis of CG coordinate and the positive attachment angle direction is assumed CCW direction in y-axis of CG coordinate. For example, to be deal with the vertical stabilizer, the angles are defined as $(\theta_{elevation}, \theta_{attachement}) = (\pi/2, 0)$. The coordinate transformation matrix from CG to wing coordinate is expressed in equation 6.

$$\mathbf{L}_{WC} = \mathbf{L}_1(\phi_{elevation})\mathbf{L}_2(\theta_{attachment}) \tag{6}$$

Secondly, the information along the wing span direction such as an airfoil, chord length, anhedral angle, sweep angle, airfoil upside/down and application of control surface is defined to determine the wing configuration. The airfoil upside/down is important to calculate the aerodynamic forces. The positive lift direction of horizontal stabilizer is given in the opposite direction to the main wing for moment balance. In other words, the aerodynamics of horizontal stabilizer is a y-axis symmetrical relationship with the main wing aerodynamics. Therefore, if the airfoil is attached in the opposite direction, the coordinate transformation matrix should be the form of equation 7.

$$\mathbf{L}_{WC} = \mathbf{L}_1(\phi_{elevation})\mathbf{L}_2(\theta_{attachment})\mathbf{L}_2(\pi) \tag{7}$$

The aerodynamic forces of wing with control surface is calculated by applying strip theory with lift increment estimation due to control surface deflection. In case of main wing, it uses more than 2 types of control surface such as flap and aileron. The flap of starboard side wing moves in the same direction as the flap of port wing. However, the aileron doesn't move the same direction. Therefore, individual control surface is numbered as convenience to the application of the strip theory. In the other words, this allows to specify the number of control surfaces and to provide the convenience of interfacing with control surfaces attached to the same wing or other wings.

To generalize the various shape of control surface such as flap, elevator and rudder, plain flap shape is considered. The lift increment estimation is a technique for calculating the lift increment by the deflection of the control surface deflection. The comparison of the lift coefficient increments of the wing element using the three of estimation methods and the lift coefficient obtained from the wind tunnel test is discussed in Ref. 4. As a result of the comparison, the proposed method in DATCOM which is Ref. 5 is the most similar to the wind tunnel test result, therefore it is applied to the generalized aerodynamic wing model. To use the proposed method, Reynolds number, Mach number, airfoil thickness ratio, chord ratio between airfoil and control surface, and the thickness ratio at 90% and 99% chord are needed. The lift coefficient of wing element with control surface is estimated as shown in equation 8.

$$C_l = C_{l,0} + C_{l_\alpha}\alpha + C_{l,surface}\delta_{surface} \tag{8}$$

As it is mentioned earlier, the wing can be mounted various angles on unmanned aircrafts. The angles of attack with respect to relative velocity on wing elements are required to calculate the aerodynamic forces and they are different for each wing. For example, when the aerodynamic wing model is applied to the horizontal and vertical stabilizers, the angle of attack is calculated differently even if the same wind is applied to both wings. Therefore, it is important to generalize the calculation of the angle of attack on each wing.

The angle of attack is generalized using following procedure. Firstly, the unit vector of wing section airfoil is calculated using equation 6 or 7. Then, the tangential and normal components of the relative wind in the airfoil are estimated using equation 9.

$$\begin{aligned}\mathbf{V}_{ac} &= \mathbf{V}_\infty - (\mathbf{V}_\infty \cdot \mathbf{e}_y)\mathbf{e}_y \leftarrow \mathbf{e}_y = -(\mathbf{u}_c \times \mathbf{u}_n) \\ &= V_c\mathbf{u}_c + V_n\mathbf{u}_n\end{aligned} \tag{9}$$

The angle of attack can be calculated using the velocity magnitudes of the tangential and normal component of relative wind velocity as shown in equation from 10 to 13.

$$\alpha = \tan^{-1}(|V_n/V_c|), \qquad where\ V_c > 0, V_n > 0 \tag{10}$$

$$\alpha = -\tan^{-1}(|V_n/V_c|), \qquad where\ V_c > 0, V_n < 0 \tag{11}$$

$$\alpha = \pi/2 + \tan^{-1}(|V_n/V_c|),\ where\ V_c < 0, V_n > 0 \quad (12)$$

$$\alpha = -\pi + \tan^{-1}(|V_n/V_c|),\ where\ V_c < 0, V_n < 0 \quad (13)$$

The aerodynamic coefficient of the tapered wing with control surface partly attached was compared with the wind tunnel test result of the Ref. 6 to verify the generalized aerodynamic model. The configuration of wing is the same with the figure 14 in Ref. 6. The figure 3 compares the lift coefficients with respect to angle of attack of the generalized aerodynamic model and the wind tunnel test results. The figure 4 describes the inflow distribution over the wing with various angle of attack.

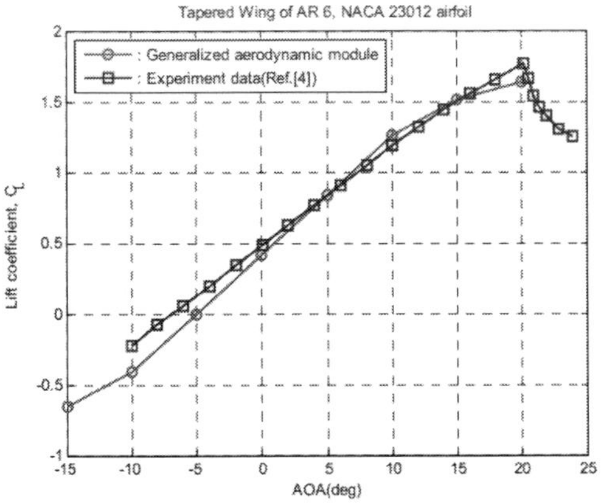

Figure 3 Lift Coefficients of Tapered Wing

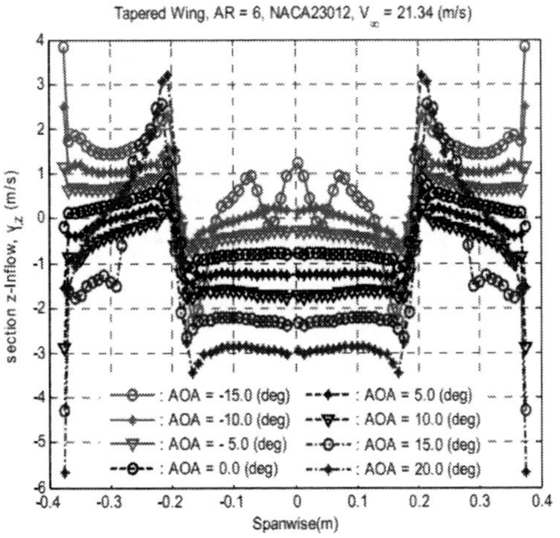

Figure 4 Inflow Distribution of Tapered Wing

Furthermore, the data of the main wing and the horizontal and vertical stabilizer shown in the table 3 are applied to the model. The table 3 describes the elevation angle, Airfoil upside/down and the wing root AC position vector. The positive value in airfoil upside/down defines that the airfoil faces upside, on the other hand, the negative value means the airfoil faces down. The location defines the position vector between CG and wing root AC point. Also, the figure 5 describes the aerodynamic forces when the data of table 3 is applied to the model.

Table 3 Characteristics of Each Wing

	MW(S)	MW(P)	HS(S)	HS(P)	VS
Elevation Angle	0	0	0	0	$\frac{\pi}{2}$
Airfoil Upside/down	1	1	-1	-1	-1
Location (m)	$\begin{pmatrix}0\\0\\0\end{pmatrix}$	$\begin{pmatrix}0\\0\\0\end{pmatrix}$	$\begin{pmatrix}3.8\\0\\-0.5\end{pmatrix}$	$\begin{pmatrix}3.8\\0\\-0.5\end{pmatrix}$	$\begin{pmatrix}3.4\\0\\-0.7\end{pmatrix}$

MW: Main Wing, HS/VS: Horizontal/Vertical Stabilizer

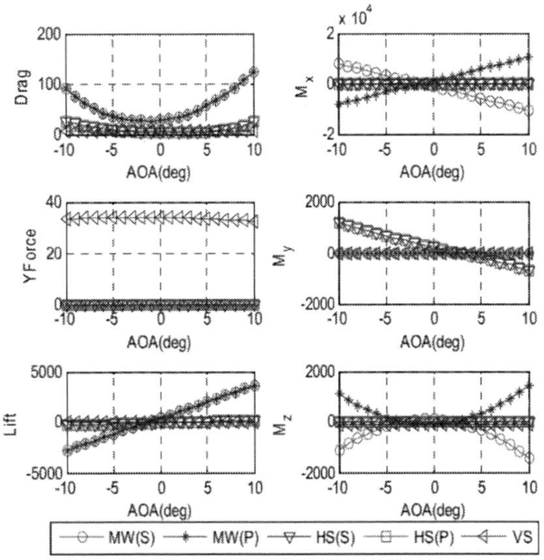

Figure 5 Aerodynamic Forces of Each Wing

APPLICATION

The proposed rotor model and aerodynamic model are integrated to build an example rotorcraft model. The forward level flight trim analysis is conducted using the example model. The example model is a 600Kg class aircraft, and the rotor diameter is about 8 meters. The figure 6 describes the trim analysis results with respect to the relative wind velocity. In addition, the figure 7 describes the simulation result from trimmed condition at 60 knots forward level flight using doublet control with 1.0 deg amplitude to collective pitch.

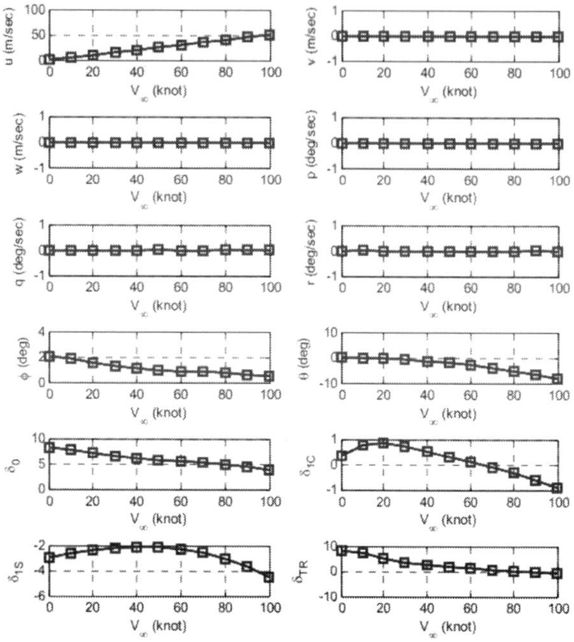

Figure 6 Trim Results of Unmanned Aircraft

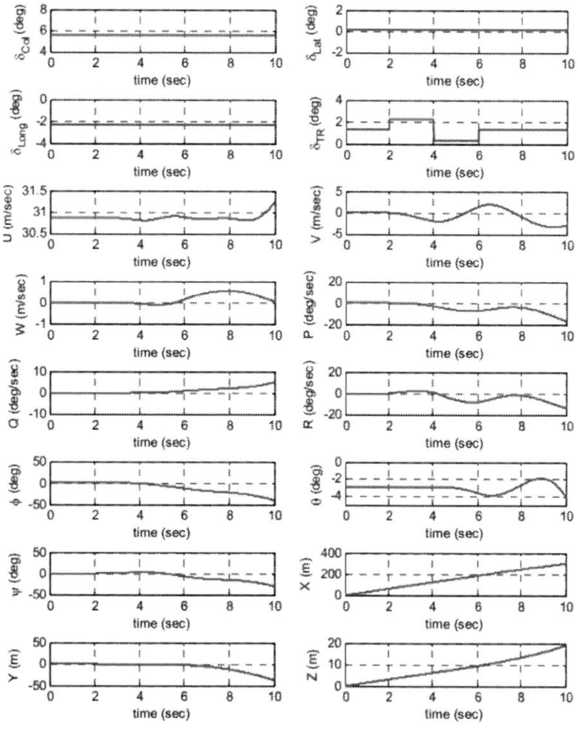

Figure 7 Simulation Result using Doublet Input to Collective Pitch

CONCLUSIONS

The generalized rotor and aerodynamic models are developed to handle any configuration of unmanned aircraft and multi-copter. The rotor model handles the propeller with RPM variation and includes the rotor dynamics of flap and lead-lag motions and it can handle the collective and cyclic pitch controls. Also, it uses various mounting angles to define a rotor or a propeller location. The aerodynamic models are built by combining strip theory with lifting line theory and with empirical formulas for the effect of control surface deflection on a wing. Many parameters are defined in the span-wise direction to describe a wing configuration. The example aircraft model is built by integrating the proposed models and is applied the flight dynamic analysis. And the model shows applicability of flight dynamic analysis. In conclusion, the proposed modeling approach shows handling any aircraft configuration with easiness to change, to add and to remove any component. In addition, it shows the applicability of trim, linearization and simulation analysis.

Author contact:

Seong Han Lee dldltjdgks@naver.com

Do Hyeon Lee ldh2620@konkuk.ac.kr

Sung Wook Hur sungwookhur@konkuk.ac.kr

Sang Ho Kang funho5@naver.com

Chang-Joo Kim cjkim@konkuk.ac.kr

ACKNOWLEDGMENTS

This work was conducted at High-Speed Compound Unmanned Rotorcraft (HCUR) research laboratory with the support of Agency for Defense Development (ADD).

REFERENCES

[1]Kim, C. J, "Implicit Formulation of Rotor Aeromechanic Equations for Helicopter Flight Simulation," *Journal of the Korean Society for Aeronautical & Space Science*, Vol. 30, (3), April. 2002, pp. 8-16.

[2]Yun, Y. H., Yang, C. D., Kim, C. J, Cho, I. J., "Flight Dynamic Analysis program, HETLAS for Development of Helicopter FBW," *KSAS 2012 Spring Conference*, April. 2012, pp.1270-1275

[3]Phillips, W. F., Snyder, D. O., "Modern Adaptation of Prandtl's Classic Lifting-Line Theory," Journal of Aircraft, Vol. 37, No. 4, July-August. 2000.

[4]Lee, S.H., Lee, D. H., Hur, S. H., Kang, S. H., Kim, C. J., "General Methods of Rotor and Wing Aerodynamic Modeling for Drone/UAV Flight Dynamic Model," *KSAS 2017 Spring Conference*, April. 2017, pp. 666-667.

[5]Fink, R., "USAF Stability and Control DATCOM," AFWAL-TR-83-3048, 1978

[6]Pearson, H.A., Anderson, R. F., "Calculation of the Aerodynamic Characteristics of Tapered Wings with Partial-Span Flaps," *NACA Report No. 665*.

Low Speed Performance Analysis of a Compound Rotorcraft with Tandem Ducted Fan and Pusher

Wonik Seo
Ph.D student
INHA University
Incheon, South Korea

Keeyoung Choi
Professor
INHA University
Incheon, South Korea

ABSTRACT

Most VTOL aircraft have the advantage of taking off vertically, but they are typically slow compared to fixed-wing aircraft. To overcome the speed limit, a new compound rotorcraft with wing, tandem ducted fan and a pusher is proposed. This paper presents initial results of performance characteristics of the new design using FLIGHTLAB. The aircraft features differential vane for roll and yaw control. Low speed ducted fan aircraft tilts fuselage and generate forward component of the total rotor thrust for forward flight. However, this may increase the fuselage drag significantly at high speed. The new design has a pusher that generates forward thrust that compensates drag without increasing the frontal area. This extra mechanism adds extra weight and complexity of the system. For low speed flight only, the pusher may be replaced with slight body tilt. As a result, it is important to predict the performance characteristics as function of combination of the various component for optimal design and operation of the aircraft. This paper presents performance characteristics of the compound rotorcraft for various flight status, which will be helpful in further detailed optimal design.

NOTATION

A	Disk Area
c	Blade chord
c_l	Lift coefficient
c_d	Drag coefficient
r	Annulus radius
T	Thrust
U_T	Tangential induced speed
U_P	Axial induced speed
X_a	Lateral stick position
X_b	Longitudinal stick position
X_c	Collective stick position
X_p	Pedal stick position
X_t	Pusher collective stick position
α	Angle of attack
ρ	Air density
ϕ	Roll attitude
Φ	Inflow angle

INTRODUCTION

Rotorcraft, which can hover and take-off and land vertically, is highly utilized compared to fixed-wing aircraft. However, the conventional helicopter has a disadvantage in that the advancing side of the main rotor falls into stall due to the compressive effect during forward flight and the speed of flight is limited due to the rapid increase of drag. Because of this, the range of the cruising distance is shortened, and the operating range is limited. As a result, interest in developing more efficient VTOL aircraft globally has been growing.

The US Defense Advanced Research Projects Agency (DARPA) has attempted to overcome the limitations of current VTOL aircraft by using the innovative combination of fixed-wing aircraft and rotorcraft through the VTOL X-plane project. Boeing, Sikorsky, Karem Aircraft, and Aurora Flight Science participated in this project. By the end of 2015, the initial designs of the four companies were evaluated and the shape of the final development was Aurora Flight Science's Lightening Strike VTOL X-Plane (XV-24A). Aurora Flight Science, a military drone manufacturer, will develop an aircraft capable of carrying out test flights by 2018 and complete commercialization by 2030 to secure a technology that can achieve vertical take-off and landing and similar speed compared to fixed-wing aircraft (Ref. 1). In the VTOL X-plane project, Boeing suggested a tilting ducted fan aircraft called phantom swift with two fans in front of and behind in the fuselage and wingtip fans to create forward thrust (Ref. 2). Ducted fan has disadvantages such as pitching moment and drag problem in flight, and low efficiency of control vane in certain flight conditions. However, compared to general rotor system, it has advantages such as high efficiency, safety problem and noise reduction effect at low speed flight. Thus, the research on this type of aircraft has been carried out continuously from the past.

The first vertical take-off and landing aircraft with a ducted fan was the Piasecki VZ-8 Airgeep in the late 1950s, a ducted fan aircraft in a tandem configuration. Although these early aircrafts were not actively developed due to failure to satisfy the FAA safety standards for their performance and

Presented at the 6th Asian/Australian Rotorcraft Forum / Heli Japan 2017, Kanazawa, Japan, November 7-9, 2017.

maneuverability, technological advances are now being actively researched for VTOL aircraft with Ducted fan system such as Urban Aeronautics Airmule (Ref. 3).

Eric L. Tobias (Ref. 3) developed the Tandem Ducted Fan Aircraft Simulation Model with MATLAB and compared it with experimental data of FLIGHTLAB and 29-inch single ducted fan UAV. Simulation of Steady-state, Trim, and Maneuver states and vane control of ducted fan were conducted. Ozdemir (Ref. 4) applied a dynamic inversion controller to the Tandem Ducted fan aircraft using Tobias' model and conducted a study to verify the effectiveness of the vane in lateral / longitudinal maneuvering using a combination of cyclic pitch and vanes for the aircraft.

Since a typical tandem ducted fan aircraft tilts its fuselage to generate forward thrust, it has a low cruising speed due to a sudden increase in drag as the forward speed increases. In order to overcome these limitations, this study proposes a shape with pusher type propeller, wing and differential control vanes to control roll and yaw as shown in figure 1. This type of aircraft produces forward thrust by using a pusher in low speed flight. However, there is a possibility that the slight body tilt effectively affects performance. For this reason, we modeled the Tandem Ducted Fan with Pusher type aircraft using FLIGHTLAB and conducted the simulation under various flight conditions to give the results of flight. In future studies, it will be necessary to analyze the efficiency of the control surface in terms of stability and maneuverability in this section.

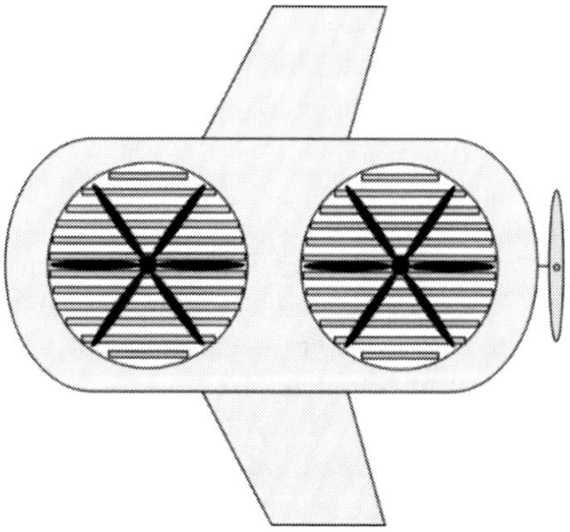

Figure 1. Concept of the compound rotorcraft with tandem ducted fan with pusher

CFD ANALYSIS

In the previous research, CFD analysis was performed by applying Actuator Disk Method and Actuator Surface Method to the 3 - dimensional Euler equation, which is the governing equation of inviscid compressive flow, to obtain aerodynamic data of tandem ducted fan aircraft (Ref. 5). In this study, we used this result to analyze the performance analysis of low - speed Tandem ducted fan aircraft.

MODELING

Ducted Fan

The ducted fan modeled in a blade element ducted fan model. FLIGHTLAB's blade element ducted fan model uses nonlinear and nonuniform airloads to calculate the airload according to the angle of attack and Mach number for each segment of the blade in the inner loop and analyzes the fan wake model using the enhanced finite state dynamic wake in the outer loop (Ref. 6). And the effects on the aerodynamic interaction between duct, fan, and fuselage can be considered. It also includes the ability to calculate duct lift using the duct lift augmentation factor (Ref. 7). Since this duct model cannot simulate the duct characteristics and duct interference model accurately, FLIGHTLAB has developed a model that can simulate maneuvering flight under unsteady flow condition using ring vortex method. However, the Tandem ducted fan shape proposed in this study does not satisfy the independent duct shape for the Ring vortex method because it has two ducts inside the blunt fuselage. Therefore, Quasi-Steady Duct Airloads method using empirical data table with insufficient fidelity compared to ring vortex method was used for duct model. The airfoil used for ducted fan modeling in this study is NACA0012. And the finite state interference model is used as the interference model of the fan to set the aerodynamic force to act on the control vane in the nonuniform and unsteady rotor wake flow field. The configuration of the modeled fan is shown in table 1.

Table 1. Fan configuration

Characteristic	English	Metric
Number of fan blades	6	6
Fan Radius	3.28084ft	1m
Rotor rotational speed	2000rpm	2000rpm
Blade tip loss factor	0.98	0.98

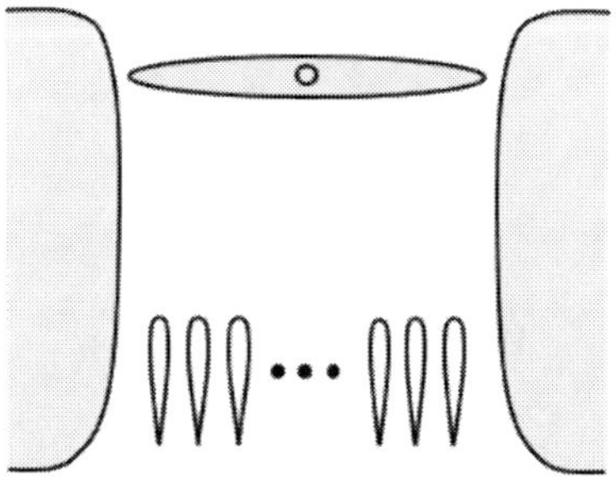

Figure 2. Ducted fan structure

Pusher Type Propeller

The pusher type propeller uses the ducted fan tail rotor model among the rotor model options of FLIGHTLAB. Ducted fan rotor model was derived by combining blade element analysis and unsteady momentum theory. The apparent mass parameter for unsteady inflow dynamics is calculated from the Peters / He finite state induced flow model in FLIGHTLAB (Ref. 7). Since the model used in this study is a pusher model without duct, duct interference is not affected and the fan wake contraction factor is set to 1.0, which is modeled as a general propeller.

Table 2. Propeller configuration

Characteristic	English	Metric
Number of rotor blades	5	5
Rotor Radius	3.28084ft	1m
Rotor rotational speed	2000rpm	2000rpm
Blade linear twist	- 30deg	-30deg
Blade chord	0.328084ft	0.1m
Blade root cut	0.2	0.2
Lift curve slope	5	5

Control Vanes

The control vane is located at the exit of the duct and acts in the wake of the rotor to generate control forces and control moments. The ducted fan control vanes were modeled as a total of 24 control vanes, 12 each for the front and rear, and NACA0012 was used for airfoil. The span of control vanes is set differently for each position and set to operate in the lateral direction as shown in figure 3.

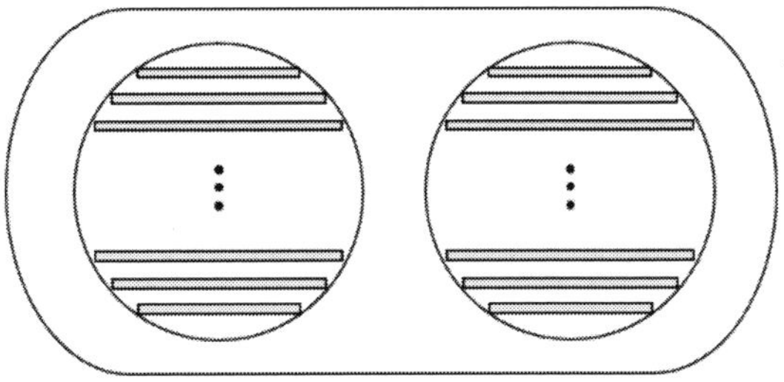

Figure 3. Control vanes arrangement

Control Logic

A typical tandem ducted fan aircraft controls pitch, roll, yaw, and heave using the front and rear fan thrust and control vanes in the duct. In this study, heave is controlled by using collective and pitch is controlled by using thrust difference of front and rear fans by using longitudinal stick. Differential vanes control the roll and yaw attitude. The front control vane is modeled as vane 1 to vane 12, and the rear control vane is modeled as vane 13 to vane 24. The lateral sticks were used to control roll direction of the front and rear control vanes in the same angle and direction, while yaw direction control was controlled by varying the angles or directions of the front and rear control vanes. Finally, a pusher collective stick was added to increase the thrust using a pusher propeller so that the collective angle of the pusher propeller can be controlled. table 3 and figure 4 below show the control allocation and the control logic configured in CSGE of FLIGHTLAB.

Table 3. Control Allocation

Attitude Control	Control Deflection
Heave	Collective pitch
Roll	Symmetric control vane
Pitch	Differential collective pitch
Yaw	Asymmetric control vane

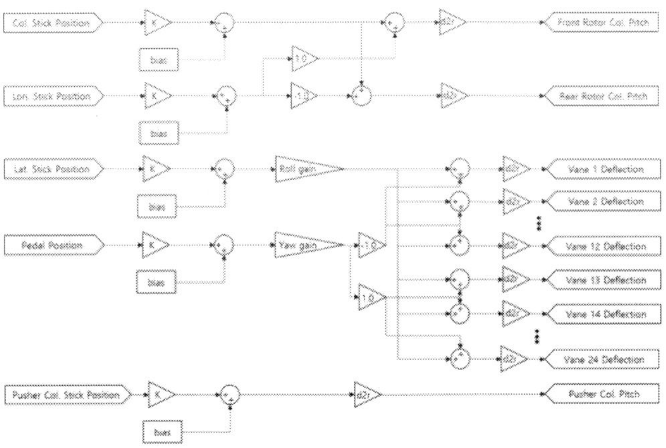

Figure 4. Control Logic

TRIM ANALYSIS

Trim condition

In order to analyze the trim of the aircraft, it is necessary to find the trim parameters of the aircraft to make the six acceleration components (3-axis acceleration, 3-axis rotational angular acceleration) zero. In this study, the trim parameters such as the basic control inputs (X_c , X_a , X_b , X_p), the pusher control input X_t , and the aircraft roll attitude angle are determined by changing body tilt angle(θ) of the aircraft to -5deg, 0, 5deg.

Trim simulation results

As the forward speed increases, the pitching moment generated by the fuselage and the fan. Thus, the aircraft tends to keep the nose up with positive pitching moment. Therefore, by pushing the longitudinal control input forward, the thrust of the front fan is reduced and the thrust of the rear fan is increased to generate a negative pitching moment to maintain the trim. The roll moment generated by the fuselage is controlled by moving the control vanes in the same direction in the front and rear using the lateral control input. However, when the fuselage angle is 5 degrees, the side force on the fuselage occurs in the opposite direction unlike the rest of conditions and the yaw moment occurs as the roll attitude angle increases. Therefore, when the body tilt angle is 5 degrees, the control vanes move in opposite directions through the pedal input while the forward speed increases, and the roll and yaw moments are simultaneously controlled. In addition, it can be confirmed that it is most efficient to perform the forward flight using only the thrust of the pusher by setting the body tilt angle to zero from the viewpoint of the total required power.

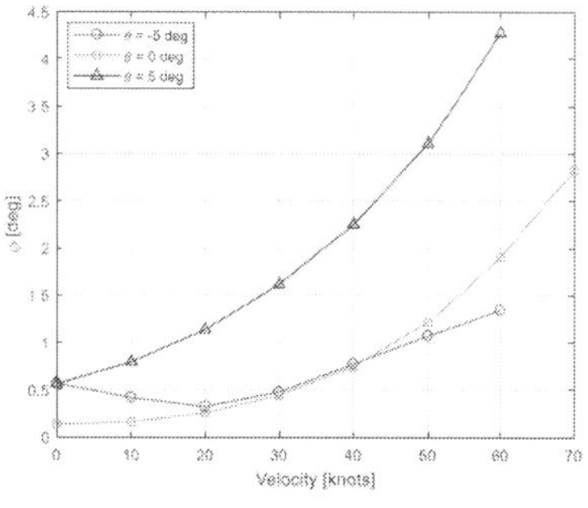

Figure 5. Roll attitude

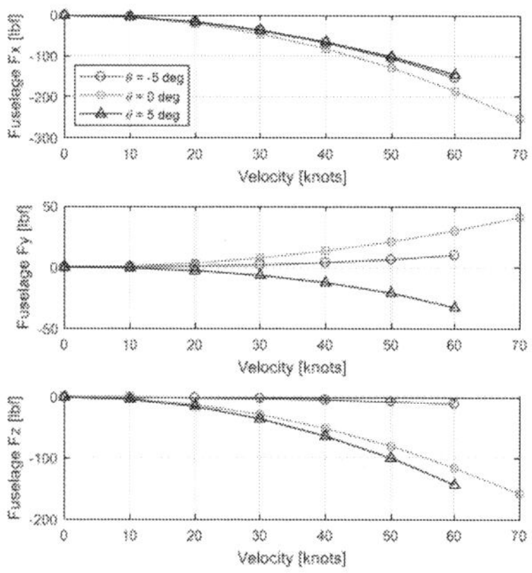

Figure 6. Fuselage force

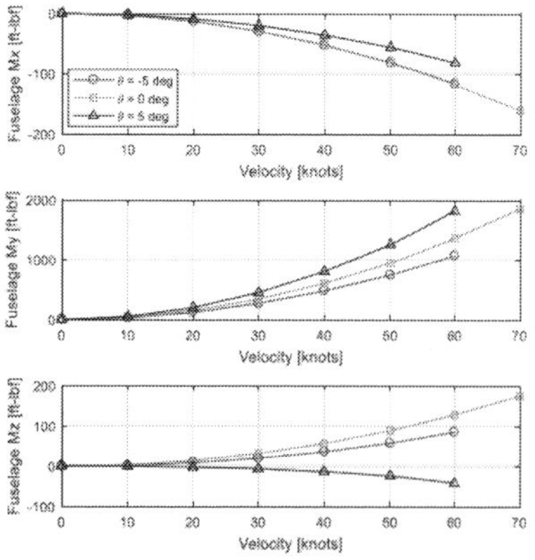

Figure 7. Fuselage moment

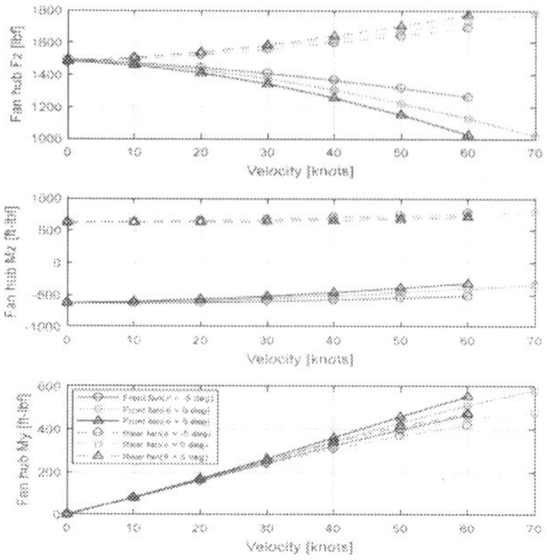

Figure 8. Fan hub force and moment

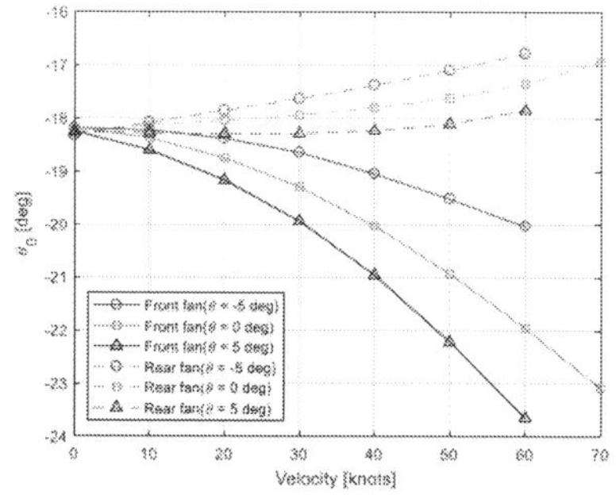

Figure 9. Fan collective angle

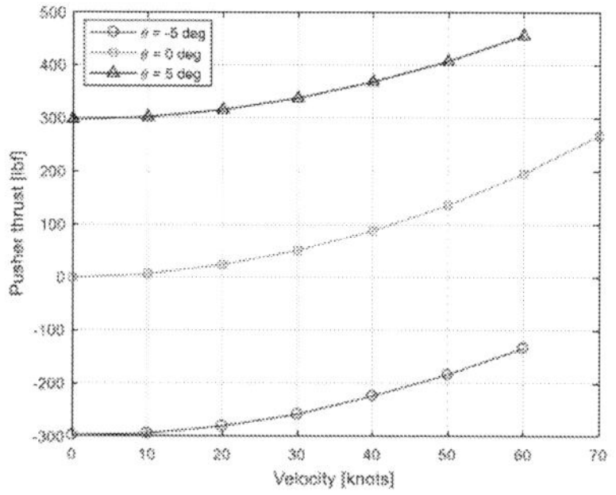

Figure 10. Pusher thrust

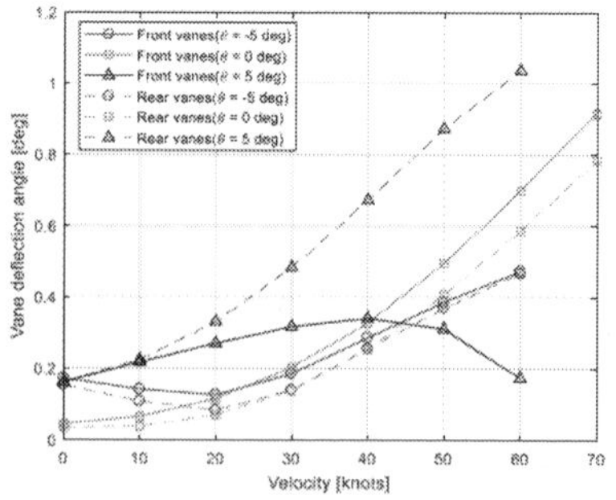

Figure 11. Vane deflection angle

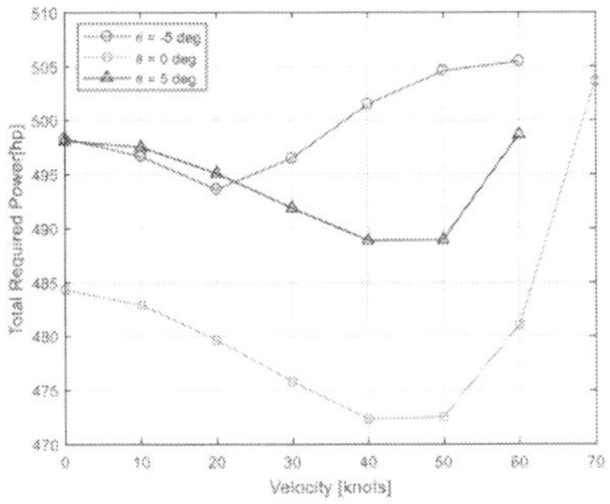

Figure 12. Total required power

CONCLUSIONS

Forward thrust and lift increase are essential for extending the flight distance of conventional tandem ducted fan aircraft. Therefore, performance analysis of a tandem ducted fan with a pusher and wing was conducted. Based on aerodynamic data of CFD analysis, duct, control vane and pusher are modeled using FLIGHTLAB and trim condition analysis is performed and analyzed for body tilt angles in low speed.

The results are summarized as follows.

1. As the forward flight speed increases, thrust of the rear fan increases to compensate the pitching moment generated at the fan hub.

2. It is possible to control the roll and yaw attitude using differential control vanes in the lateral direction.

3. From the point of view of required power, it is effective to fly by controlling the angle of the fuselage horizontally.

Based on these results, we conducted a basic study on the low speed flight of the compound rotorcraft for overcoming the maximum speed limitation and extending the operating range of the tandem ducted fan type aircraft. In the future, it is necessary to study the concept of flying with the fan closed similar to fixed-wing aircraft at high speed flight. Therefore, the results of this study can be used as an initial research to select an efficient transition zone between low speed and high speed flight of tandem ducted fan aircraft.

Author contact: Wonik Seo swigreen@gmail.com Keeyoung Choi kchoi@inha.ac.kr

ACKNOWLEDGMENTS

This work was conducted at High-Speed Compound Unmanned Rotorcraft (HCUR) research laboratory with the support of Agency for Defense Development (ADD).

REFERENCES

[1]https://www.flightglobal.com/news/articles/video-auroras-lightningstrike-wins-darpa-vtol-x-pl-422720/

[2]https://www.darpa.mil

[3]Tobias, E.L., Horn, J.F., "Simulation Analysis of the Controllability of a Tandem Ducted Fan Aircraft," AIAA Atmospheric Flight Conference and Exhibit, Honolulu, HI, August 2008.

[4]Ozdemir, G.T, "Flight Control Design of Tandem Ducted Fan Aircraft using Redundant Control Effectors," Master's Thesis, Pennsylvania State University, Pennsylvania, PA, May 2010.

[5]Seoungyu Choi, Seungsoo Lee, "Aerodynamic Analysis of Fan-In-Body Unmanned Air Vehicle," Korean Society for Aeronautical & Space Sciences Spring Conference Proceedings, Samcheok, April 2017.

[6]Chengjian, H., Hong, X., "An Unsteady Ducted Fan Model for Rotorcraft Flight Simulation," American Helicopter Society 62nd Annual Forum Proceedings, Phoenix, AZ, May 2006.

[7]Advanced Rotorcraft Technology Inc., "FLIGHTLAB Theory Manual," Vol. 1.

Design, Manufacturing, and Experimental Investigation of Composite Rotor Blades for Small-Scale Unmanned Helicopters with Flight Test Verification

D. R. Abhiram
PhD Student
Indian Institute of Science
Bangalore, India.

Ranjan Ganguli
Professor
Indian Institute of Science
Bangalore, India.

Dineshkumar Harursampath
Assistant Professor
Indian Institute of Science
Bangalore, India.

ABSTRACT

The main rotor is a critical component of the helicopter as it controls its position and attitude, and also governs its behaviour in hover, vertical flight, and forward flight. This work presents a design, analysis and fabrication study of main rotor blades made of composite materials for small-scale unmanned helicopters. In the present scenario in India, composite rotor blades for small unmanned helicopters have to be imported. Additionally, the design specifications of the commercially available composite rotor blades are limited in terms of blade pre-twist, profile, plan-form variation. To address this situation, a detailed study on available commercial composite rotor blades is being carried out. This study presents initial work carried out in this direction. Experimental investigations are conducted on commercial rotor blades to determine its mass properties, material properties and structural properties. Based on the data obtained from these experimental studies, identical composite rotor blades are fabricated which have better stiffness and lower weight. Flight test verifications are conducted using a small unmanned composite helicopter. The unmanned helicopter was flown successfully using the fabricated composite rotor blades validating their applicability.

INTRODUCTION

Interest in development of unmanned aerial vehicles (UAVs) has been on the increase ever since its inception. Developed initially for military purposes, UAVs have come to be used for many civilian purposes as well. They have been employed for a variety of military purposes such as for surveying of sensitive areas, use as armed interceptors, etc. In the civilian sector, they have been used for taking ground images from high altitudes for use in classification of farm lands, surveying of forests, monitoring the growth of cities, etc.

The AIAA defines an UAV as An aircraft which is designed or modified, not to carry a human pilot and is operated through electronic input initiated by the flight controller or by an on board autonomous flight management control system that does not require flight controller intervention. (Ref. 1).

UAVs can be broadly classified based on their mechanism of lift generation as: fixed wing UAVs; rotary wing UAVs; and flapping wing UAVs (Ref. 2).

Fixed wing UAVs are aircrafts with fixed wings. They are driven by propellers and powered either by batteries or engines. Predator is an example of fixed wing UAV used for military purposes.

Flapping wing UAVs generate lift and propulsive forces by mimicking the flapping motion of birds and insects. Flapping wing UAVs have generated a lot of interest in the academic research field and many different models are being pursued across various research organizations. Development of flapping wing UAVs for practical use is still in the initial stages.

Rotary wing UAVs use rotating wings to generate thrust and propulsive forces. In addition to tasks that can be performed by fixed-wing UAVs, rotary wing UAVs, or unmanned rotorcraft, can use their unique abilities such as hovering, slow speed flight in forward, backward and side-ways directions, and vertical climb and descent, to perform tasks in a variety of situations. This versatility of unmanned rotorcraft makes them suitable for performing tasks such as inspection of power lines, scanning for damage on dam walls, monitoring of volcanoes, and tsunami affected regions, spraying of agrochemicals over farms, surveillance aid to security personnel, continuous monitoring of a certain fixed area, etc. (Ref. 3). Unmanned rotorcraft can be classified broadly into three categories based on their size and characteristics as: Type-I, Type-II and Type-III unmanned rotorcraft (Ref. 2).

Type-I unmanned rotorcraft are relatively large in size and rotor span. They are usually powered by internal combustion engines and therefore, have relatively more power. This makes them suitable for missions that require longer endurance and heavier payload, compared to the other two types. Type-I unmanned rotorcraft are dominated by the traditional main rotor- tail rotor type helicopters. Examples include the Northrop Grumman MQ-8 Fire Scout (Ref. 4), the Yamaha R-MAX and R-50 (Ref. 5), the X-Copter (Ref. 6), etc.

Type-II unmanned rotorcraft are smaller than Type-I ro-

Presented at the 6th Asian/Austrailian Rotorcraft Forum / Heli Japan 2017, Kanazawa, Japan, November 7 – 9, 2017.

torcraft. They are powered by either on-board batteries or small internal combustion engines. Therefore, their payload carrying capacity and endurance is limited by the available power. However, several Type-II unmanned helicopters are being used in civilian applications; examples include the autonomous unmanned helicopter being developed at IIT Kanpur (Ref. 7), India, Dauphin, Maxi-Joker-2, etc. (Ref. 2).

Type-III unmanned rotorcraft are the smallest. Besides single rotor designs such as Trex-450, they are also available in multiple-rotor designs such as the co-axial rotor Lama-V4, quad-rotor - Dragonflyer, etc. (Ref. 2). Being very small in size they have very low endurance and almost no payload carrying capacity.

While Type-III unmanned rotorcraft are being used in academic researches apart from hobby purposes, the small-scale helicopters that can be categorized as Type-I and Type-II have plenty of practical applications, as previously mentioned.

Some small unmanned helicopters have already been put to use. The MQ-8 Fire Scout of Northrop Grumman, an unmanned autonomous helicopter with attached gunships, is used for naval reconnaissance (Ref. 4). Yamaha R-50 and R-MAX have been used in Japan for crop-dusting (Ref. 5). Similar agricultural helicopter, called X-Copter, is being designed by Korea's Skytech (Ref. 6).

Their versatility, relatively larger power availability and payload carrying capacity, and relatively longer endurance makes such small unmanned helicopters very suitable for a large number of practical applications.

In spite of their huge potential to be gainfully employed for a large number of practical applications, the use of small unmanned helicopters has been very limited. This is largely due to lack of available literature related to design and analysis of small unmanned helicopters and also due to the high costs involved in their operations (Ref. 6). The results from research on full-scale helicopters are not directly applicable to the small-scale helicopters due to aeroelastic scaling effects (Ref. 8). Research on rotary wing micro air vehicles (MAVs) has focussed on their design, analysis and hover performance (Ref. 9), unconventional MAV configurations (Ref. 10) and airfoil design for ultra-low Reynolds number flight (Ref. 11). (Ref. 3) is one of the research works focussed on structural design and fabrication of rotor blades for unmanned rotorcraft. This however, addresses the rotor blade design and fabrication for small-scale fixed pitch rotors of a quad-rotor. Research on small unmanned rotorcraft has largely focussed on its autonomous control and guidance and navigation aspects (Ref. 12). Thus, literature on design and fabrication of composite rotor blades for small-scale unmanned helicopters is sparse.

The present work focuses on the design and fabrication of composite main rotor blades for a small-scale unmanned helicopter of Type-II category. Experimental investigations are conducted on commercially available composite rotor blades to determine its structural and mass properties as well as material characterization. With sufficient information from the experimental work, composite rotor blades are fabricated which are geometrically identical to the commercial rotor blades. The objective of composite blade fabrication is to understand the feasibility of manufacturing of such rotor blades for small-scale unmanned helicopters and to develop rotor blades which have superior structural properties with reduced weight.

EXPERIMENTAL INVESTIGATIONS

A commercially available rotor blade of a small-scale unmanned helicopter is selected as the baseline rotor blade. This serves as a starting point for the design and also as an effective benchmark. Radix YEI-YB-710-V2 rotor blades, shown in Figure 1, manufactured by Curtis Youngblood has been selected as the baseline rotor blade (Ref. 13). These rotor blades are made of carbon fiber composite material and are used as main rotor blades in the 700-class unmanned helicopters which come under the bracket of Type-II unmanned rotorcraft. These are also used as the main rotor blades for the unmanned helicopter at IIT Kanpur (Ref. 7).

Morphological Study

CAD Model of Rotor Blade

A 3D CAD model of the Radix blade is obtained by means of a 3D scanner. The cloud points from the 3D scan is shown in Figure 2 The extracted surface model and the final CAD model of the rotor blade are shown in Figure 3 and Figure 4, respectively. The CAD model is useful for determining the Radix blade geometrical details, such as the cross-section at different radial stations.

Mass Properties

The Radix blade is cut through the cross-section into seven parts of equal lengths to measure its mass and to determine the inertial properties. The weights of each cut section was measured using a precision electronic weighing machine. Figure 5 shows the second cut-section from the tip of the rotor blade. A steel wire of approximately 2mm diameter is present near the leading edge throughout the blade length. The cross-section of the steel wire increases towards the tip of the blade. Therefore, the cut sections at the tip have almost twice the weight of the remaining sections. The root section also has more weight as it has more thickness due to additional layers of composite at the blade root region.

Material Characterization

The Radix blade cut sections are subjected to a material characterization test using scanning electron microscope (SEM) with energy-dispersive X-ray spectroscopy (EDS). It is a useful tool for material elemental characterization (Ref. 14). The SEM with EDS facility is used to obtain information on the materials used in blade. The material characterization study revealed that the metallic insert present along the leading edge of the blade is of Steel. Figure 6 depicts a magnified view of the rotor blade cross-section under study in the SEM facility.

Structural Characterization

The Radix blade is subjected to a 3-point bending test to determine its flexural stiffness in the out-of-plane bending mode (flap-mode). Three different specimens are used in the test. The specimens are subjected to gradually increasing load in a micro-UTM until failure. Figures 7, 8, and 9 show the plots of applied load against the mid-point deflection for the specimen used in the UTM. The initial linear regions are considered to evaluate the stiffness. This simple experiment for stiffness characterization provides accurate results as proven in similar studies on helicopter rotor blade properties measurement (Ref. 15) and (Ref. 16). The mean flexural stiffness of the Radix blade is given in Table 1.

COMPOSITE ROTOR BLADE FABRICATION AND FLIGHT TEST

The preliminary experiments provide sufficient information for fabrication of composite rotor blades. The 3D CAD model of the Radix blade is used to obtain an accurate mould for the blade fabrication, shown in Figure 10. Several rotor blades are fabricated to obtain a rotor blade identical to the Radix blades.

Blade 1: AGB-1
The first blade is fabricated using glass-fiber with epoxy resin. Though the use of glass-fibers increases the weight and decrease the stiffness, it is used to check for feasibility of manufacturing and for any practical challenges that may arise during the fabrication process. Blade AGB-1 is shown in Figure 11 (bottom) against the Radix blade (top).

Blade 2: AGB-2
The second blade is fabricated with carbon fiber composites. The top-sheet is made of layers of ± 45 carbon-fiber fabric to account for bending and torsional stiffness, and the interior is filled with polyurethane foam (PUF). The fabricated blade is light in weight but not sufficiently stiff. The use of only the top-sheet to provide bending stiffness is the reason for lack of sufficient stiffness.

Blade 3: AGB-3
The third blade is fabricated using a tapering thickness for top-sheet (with thickness decreasing from the root towards the tip) and with two spars along the blade length. The spars are made of carbon fibers and of I-section, located along the leading and trailing edge of the blade. The use of a tapered top-sheet and two spars makes the blade extremely stiff when compared to the Radix blade but involves a considerable weight penalty of about 150 g. The increased weight would adversely affect the performance of the helicopter.

Blade 4: AGB-4
The use of two spars is identified as the cause for increased weight in the AGB-3. The fourth blade is fabricated with layers of ± 45 woven carbon fiber fabrics as the top sheet. A single spar made of carbon fiber with I-section is used. This is placed next to the leading edge, close to the quarter-chord position. Experiments are conducted to determine the flexural rigidity of the AGB-4 blade and the blade is found to have better flexural stiffness than the Radix blades. The flexural rigidity of the Radix blades and AGB-4 blades is given in Table 1 for comparison. The AGB-4 blade is also 33% lighter than the Radix blade which amounts to an overall blade weight reduction of 132 g. The AGB-4 blades are shown in Figure 12.

Flight Test using AGB-4 Blades
A pair of AGB-4 fabricated composite blades are used to fly a Rave Ballistic helicopter - a small-scale remote controlled unmanned Type-II helicopter. The flight test is successful with the stable flight of the helicopter with the AGB-4 rotor blades. A video captured during the flight test will be shown during the presentation.

Fig. 1. Radix V2 main rotor blade from Curtis Youngblood

Fig. 2. Cloud points from 3D scan

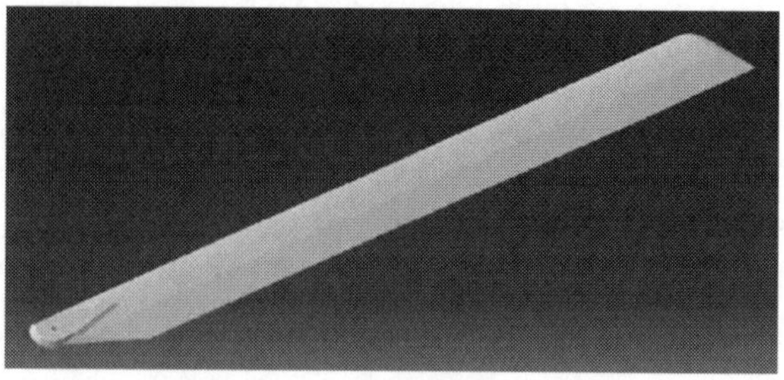

Fig. 3. Surface model of Radix blade

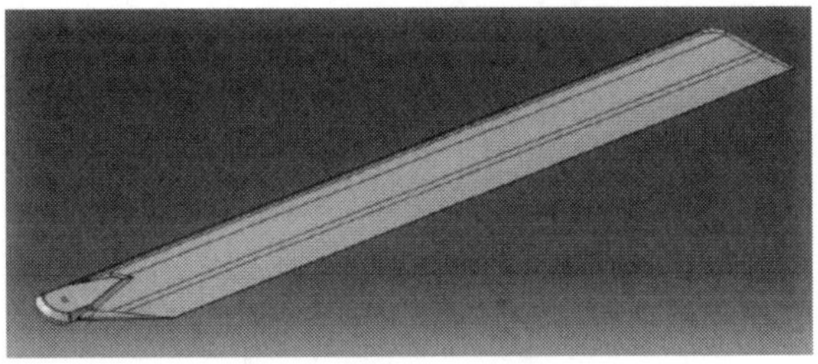

Fig. 4. CAD model of Radix blade

Fig. 5. Cut-section of rotor blade near tip

Table 1. Radix blades and AGB-4 blades comparison

Blade	Mass (g)	Flap Stiffness (N-mm^2)
Radix	196	3.49 X 10^7
AGB-4	130	4.22 X 10^7

CONCLUSIONS

This work focuses on the design and fabrication of composite main rotor blades for a small-scale unmanned helicopter. The fabricated rotor blades are based on commercially available composite rotor blades, Radix blades, designed for small-scale unmanned helicopters of Type-II category. Experimental investigations are conducted on the Radix rotor blade to determine its mass properties, material characterization and structural characterization. Morphological study is carried out to give a better understanding of the Radix blade cross sectional design. A CAD model of the Radix blade is developed to accurately extract the geometric features.

Using the data from the experimental investigation, four composite rotor blades are fabricated. These are used to demonstrate the feasibility of manufacturing of composite he-

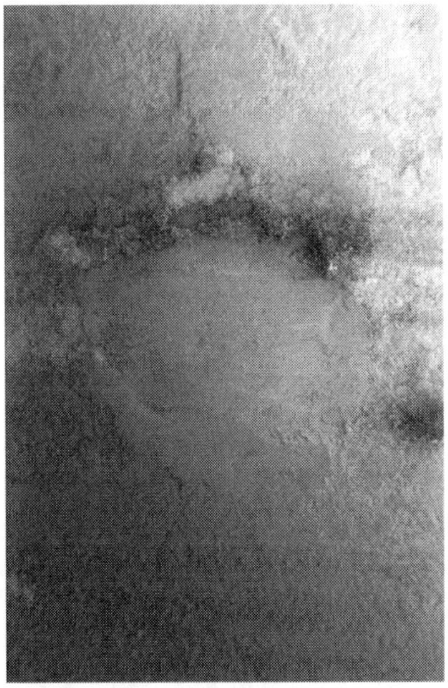

Fig. 6. Magnified image of the steel wire insert in the Radix blade

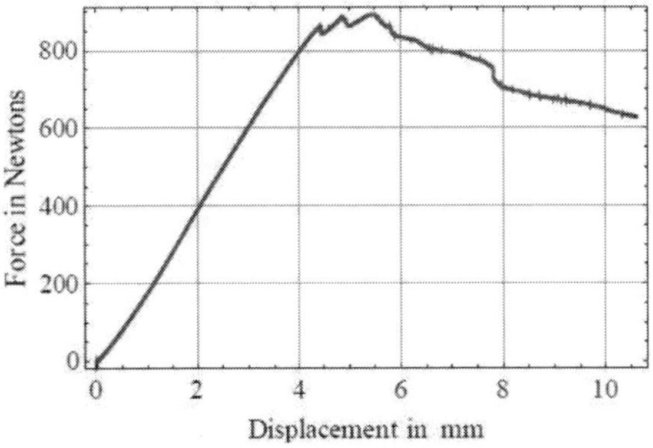

Fig. 7. 3-point bending test of rotor blade specimen 1

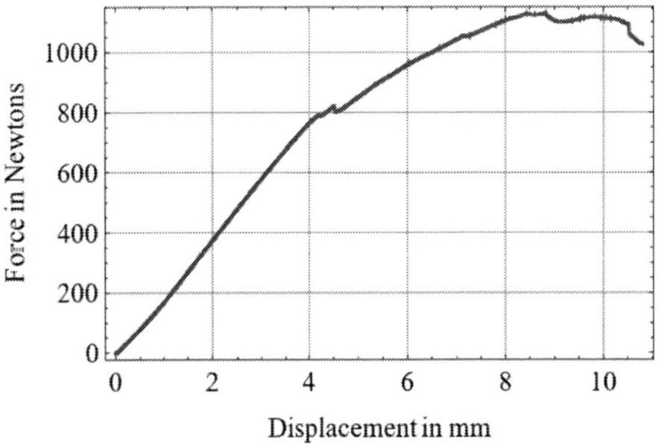

Fig. 8. 3-point bending test of rotor blade specimen 2

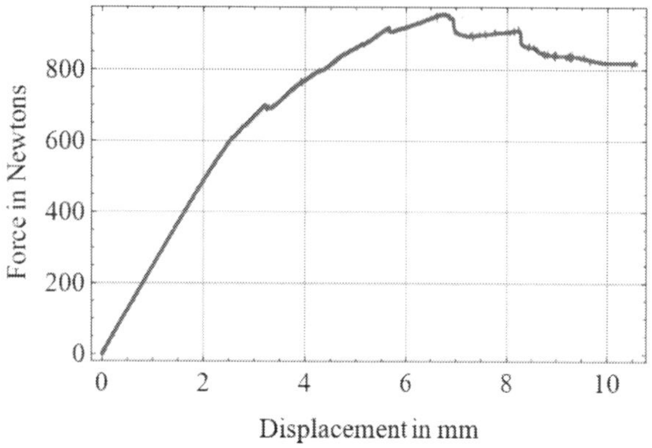

Fig. 9. **3-point bending test of rotor blade specimen 3**

Fig. 10. Mould used for rotor blade fabrication

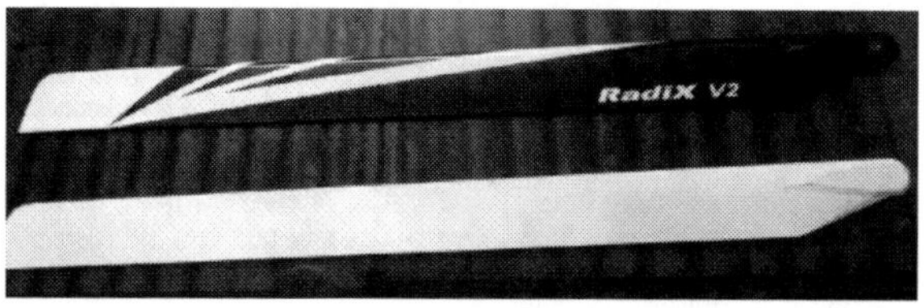

Fig. 11. Fabricated blade, AGB-1 (bottom) and Radix blade (top)

Fig. 12. Fabricated blade, AGB-4 (bottom) and Radix blade (top)

licopter main rotor blades. The final design of the fabricated rotor blade, AGB-4 shows promising results with experiments demonstrating improved flexural stiffness compared to the baseline Radix blade with a weight reduction of about 32%. Successful flight test were conducted using the AGB-4 rotor blades as the main rotor blades on a Type-II unmanned helicopter.

ACKNOWLEDGMENTS

The authors would like to thank the Aeronautical Research and Development Board (ARDB), Ministry of Defence, India for financially supporting this work. The authors would also like to thank Mr. Kalaiselvam, Mr. Ravi Kumar, and the Reinforced Plastics Industry, Bangalore for their support in fabrication and Mr. Gautham Anand for his support in flight test.

REFERENCES

[1]Nonami, K., "Prospect and recent research & development for civil use autonomous unmanned aircraft as UAV and MAV," *Journal of System Design and Dynamics*, Vol. 1, (2), 2007, pp. 120–128.

[2]Cai, G., Chen, B. M., and Lee, T. H., "An overview on development of miniature unmanned rotorcraft systems," *Frontiers of Electrical and Electronic Engineering in China*, Vol. 5, (1), 2010, pp. 1–14.

[3]Pounds, P., Mahony, R., and Corke, P., "Small-scale aeroelastic rotor simulation, design and fabrication," Proceedings of the Australasian Conference on Robotics and Automation, 2005.

[4]Grumman, N., "MQ-8B Fire Scout," Accessed on: 19-08-2014, 2011.

[5]Sato, A., "The RMAX helicopter UAV," Technical report, DTIC Document, 2003.

[6]Shim, D. H., Han, J.-S., and Yeo, H.-T., "A development of unmanned helicopters for industrial applications," *Unmanned Aircraft Systems*, Springer, 2009, pp. 407–421.

[7]Venkatesan, C., Swaroop, B., Haritha, P., and Gupta, R., "Development of autonomous mini helicopter: Challenges faced," *Journal of Aerospace Sciences and Technologies*, Vol. 65, (1A), 2013.

[8]Friedmann, P., "Aeroelastic scaling for rotary-wing aircraft with applications," *Journal of Fluids and Structures*, Vol. 19, (5), 2004, pp. 635–650.

[9]Bohorquez, F., Samuel, P., Sirohi, J., Pines, D., Rudd, L., and Perel, R., "Design, analysis and hover performance of a rotary wing micro air vehicle," *Journal of the American Helicopter Society*, Vol. 48, (2), 2003, pp. 80–90.

[10]Sirohi, J., Parsons, E., and Chopra, I., "Hover performance of a cycloidal rotor for a micro air vehicle," *Journal of the American Helicopter Society*, Vol. 52, (3), 2007, pp. 263–279.

[11]Kunz, P. J., *Aerodynamics and design for ultra-low Reynolds number flight*, Ph.D. thesis, Stanford University, 2003.

[12]Wendel, J., Meister, O., Schlaile, C., and Trommer, G. F., "An integrated GPS/MEMS-IMU navigation system for an autonomous helicopter," *Aerospace Science and Technology*, Vol. 10, (6), 2006, pp. 527–533.

[13]Inc., C. Y. E., "Radix V2 main rotor blade," Accessed on: 30-10-2015, 2015.

[14]Goldstein, J., Newbury, D. E., Echlin, P., Joy, D. C., Romig Jr, A. D., Lyman, C. E., Fiori, C., and Lifshin, E., *Scanning electron microscopy and X-ray microanalysis: a text for biologists, materials scientists, and geologists*, Springer Science & Business Media, 2012.

[15]Jung, S. N. and Lau, B. H., "Determination of HART I blade structural properties by laboratory testing," Technical report, Tech. Rep., NASA-CR-2012-216039, NASA Ames Research Center, 2012.

[16]Jung, S. N., You, Y. H., Dhadwal, M. K., Riemenschneider, J., and Hagerty, B. P., "Study on Blade Property Measurement and Its Influence on Air/Structural Loads," *AIAA Journal*, Vol. 53, (11), 2015, pp. 3221–3232.

Numerical Investigations on Aerodynamic and Acoustic Characteristics of a Rigid Coaxial Rotor in Hover

Zhu Zheng
Ph.D. student

Zhao Qijun
Professor

Yuan Xin
Graduate student

National Key Laboratory of Science and Technology on Rotorcraft Aeromechanics,
Nanjing University of Aeronautics and Astronautics
Nanjing, China

ABSTRACT

A numerical method which combins CFD (Computational fluid dynamics) method and FW-H (Ffowcs Williams-Hawkings) equations is established to simulate the acoustic characteristics of the coaxial rigid rotor in hover. In order to accurately simulate the vortex interaction and unsteady characteristics, based on the moving-embedded grid technology and a dual-time method, the simulation method is developed by solving the compressible Reynolds-averaged Navier-Stokes (RANS) equations with Baldwin-Lomax turbulence model. The acoustic characteristics of the coaxial rotor can be predicted by formulation Farassat 1A, and the motion law and airloads of sound sources can be directly obtained from blade surface girds. Then, the sound pressure time history and noise spectrum characteristics on the thickness, loading and total noise of the coaxial rotor have been analyzed in detail. The simulated results demonstrate that the feature that two axially aligned rotors rotating in opposite directions has significant effects on the sound pressure wave, and the phases of peak value in different observation direction show differences, and some phenomenon such as Venturi effect, blade-vortex interaction, downwash effect make the blade loads show obvious unsteady characteristics, which cause the strong loading noise radiation. The total noise is widely dominated by thickness noise in the low frequency band and by loading noise in the high frequency band.

INTRODUCTION

Helicopter provides significant advantages over fixed-wing aircrafts by the capabilities of hovering, loitering, vertical take-off and landing. But conventional helicopters with a single main rotor and a tail rotor do not have the ability of high-speed flight mainly due to the shock of advancing blade and stalling of retreating blade. Also, the additional power requirements associated with the tail rotor and transmission system may be reallocated for additional payload capability. These above disadvantages restrict the application prospect of helicopter. The helicopter with coaxial rotor configuration offers potential technological solutions for improving the power utilization efficiency due to its counter-rotating rotors and removal of the tail rotor. By offloading the retreating blades and reducing rotor rotational speed, the lift potential of the advancing blades can be more optimally exploited, which means the helicopter with coaxial rigid rotors have the ability of high-speed flight. Therefore, the helicopter with coaxial rigid rotors has become new research hotspot in the recent international high-speed helicopter field.

Unlike the quasi-steady flowfield of a single rotor, the flowfield of coaxial rotors shows obvious unsteady characteristics under hovering condition. The coaxial rotor configuration produces a more complicated flowfield than a single rotor system due to various blade-vortex and vortex-vortex interactions, which brings serious challenges about studying the interactional aerodynamic and acoustic characteristics of coaxial rotors.

So far, some research programs have been carried out to study the aerodynamic and acoustic characteristics of the coaxial rigid rotors. Leishman developed the blade element momentum theory (BEMT) suitable for coaxial rigid rotors[1]. Andrew[2], Bagai[3] and Huang[4] studied the aerodynamic characteristics of coaxial rigid rotors by free wake vortex model. The vorticity transport model was applied to the coaxial rotors analysis by Kim[5]. Tong[6] used the CFD method with momentum source terms to simulate the effects of coaxial rotor blades on its flowfield. All these methods above can solve the flowfield quickly, but can hardly obtain the unsteady airloads of blade surface accurately. Lakshminarayan[7] combining the sliding and embedded mesh, Xv[8] and Ye[9] applying the unstructured mesh conducted the numerical simulations of the flowfield and aerodynamic characteristics of coaxial rotors and get the detailed features of flowfield, but the acoustics characteristics of coaxial rotors were not involved. Wachspress[10] using the vortex filament model and WOPWOP program, and Kim[11-12] using vorticity transport model and Farassat1A equations did some research on the acoustics characteristics of coaxial rotors. But these methods cannot accurately obtain the accurate unsteady loads and the sound pressure of coaxial rotors cannot be solved. To sum up, most of these investigates are mainly focused on the flowfield and aerodynamic characteristics of coaxial rotors, only a limited amount of work has been carried out on its

Presented at the 6th Asian/Austrailian Rotorcraft Forum / Heli Japan 2017, Kanazawa, Japan, November 7-9, 2017.

acoustic characteristics, and the methods for predicting blade loads that necessary for noise calculation were not accurate enough.

In this paper, the unsteady flowfield of coaxial rotors has been simulated by CFD method based on the moving-embedded gird technology, the dual time-stepping approach and Baldwin-Lomax model, thus the accurate noise source information can be obtained; then the numerical method based on acoustic analogy is established to calculate the noise characteristics of coaxial rotors; the sound pressure time history, noise spectrum characteristics and directivity patterns on the thickness, loading and total noise of the coaxial rotor have been analyzed in detail. The comparisons of noise characteristics between coaxial rotors and single rotor have been conducted, and some meaningful conclusions are given.

PREDICTION METHOD FOR THE ACOUSTICS OF COAXIAL ROTORS

CFD method

The unsteady flowfield of coaxial rotor is calculated by solving three dimensional RANS equations with Baldwin-Lomax turbulence model based on moving-embedded grid system. The governing equations could be written in finite-volume forms as

$$\frac{\partial}{\partial t}\iiint_V \boldsymbol{W}\mathrm{d}V + \iint_S \boldsymbol{F}\mathrm{d}S = \iint_S \boldsymbol{F}_v \mathrm{d}S \tag{1}$$

Where $\boldsymbol{W}$ is the vector of conservative variables, $\boldsymbol{F}$ is the convective flux, and $\boldsymbol{F}_v$ is the viscous flux.

The coefficients of viscosity are calculated by Baldwin-Lomax turbulence model. The nonslip boundary condition is adopted for blade surfaces and nonreflecting condition is adopted for the outermost boundary of both blade grids and background grids. A dual time-stepping approach is adopted, and four-step Rung-Kutta scheme is used in pseudo time step. A finite-volume method is used for spatial discretization. The three-order ROE-MUSCL scheme and central difference scheme are applied to calculate inviscid flux and viscid flux, respectively.

Grid is the basis of CFD technology. The upper rotor and lower rotor rotate in opposite directions, thus the single grid cannot simulate the flowfield of coaxial rotors, and the moving-embedded grids are applied in this paper. The improved methods of grid generation, identifying the hole boundary and searching donor elements can be found in Ref.13 and Ref.14 in detail. Fig.1 gives the schematic of moving-embedded grid system. The local grids located at the rotor plane and blade-tip are refined. The processes of grid generation and CFD flowfield simulation are completed in Intel Visual Fortran compiler. The detailed features of the unsteady flowfield of coaxial rotors can be obtained by CFD method, and then be used as the information of noise sources for acoustics calculation.

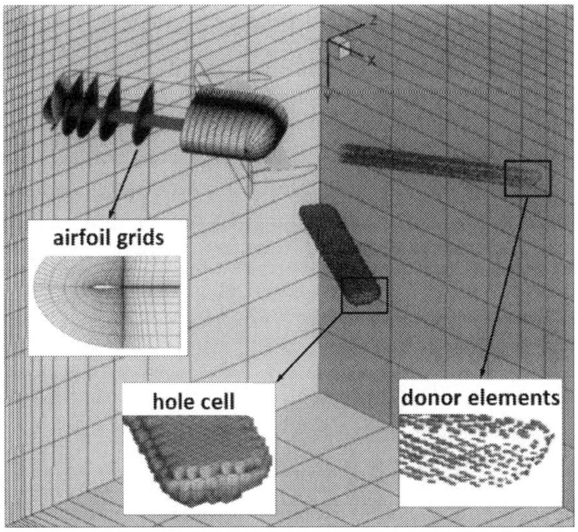

Fig.1 Schematic of moving-embedded grid system

Acoustic analogy

In terms of the study about the prediction of rotor acoustics, the famous FW-H equations[15] have been proposed by Ffowcs Williams and Hawkings based on the Lighthill analogy, and it considered the effects of moving solid boundary on fluid for the first time. The FW-H equations divide the noise sources into three terms, i.e. the monopoles caused by blade geometry and motion, the dipoles caused by surface pressure distributions and the quadrupoles caused by nonlinear phenomena in turbulent transonic flow. The quadrupole noise has been ignored as the blade-tip Mach number in hover is small.

Farassat et al. [16] put forward the solution formula (Farassat 1A) for FW-H equations in time domain. The formulation Farassat 1A can be written as

$$p' = p'_T + p'_L \tag{2}$$

Where,

$$4\pi p'_T = \int \left[\frac{\rho_0(\dot{\boldsymbol{U}}\cdot\hat{\boldsymbol{n}} + \boldsymbol{U}\cdot\dot{\hat{\boldsymbol{n}}})}{r(1-\boldsymbol{M}\cdot\hat{\boldsymbol{r}})^2} \right]_{\mathrm{ret}} \mathrm{d}s + \int \left[\frac{\rho_0 \boldsymbol{U}\cdot\hat{\boldsymbol{n}}(r\dot{\boldsymbol{M}}\cdot\hat{\boldsymbol{r}} + a_0(\boldsymbol{M}\cdot\hat{\boldsymbol{r}} - \boldsymbol{M}\cdot\boldsymbol{M}))}{r^2(1-\boldsymbol{M}\cdot\hat{\boldsymbol{r}})^3} \right]_{\mathrm{ret}} \mathrm{d}s,$$

$$4\pi p'_T = \frac{1}{a_0}\int \left[\frac{\dot{\boldsymbol{L}}\cdot\hat{\boldsymbol{r}}}{r(1-\boldsymbol{M}\cdot\hat{\boldsymbol{r}})^2} \right]_{\mathrm{ret}} \mathrm{d}s + \int \left[\frac{\boldsymbol{L}\cdot\hat{\boldsymbol{r}} - \boldsymbol{L}\cdot\boldsymbol{M}}{r(1-\boldsymbol{M}\cdot\hat{\boldsymbol{r}})^2} \right]_{\mathrm{ret}} \mathrm{d}s + \frac{1}{a_0}\int \left[\frac{\boldsymbol{L}\cdot\hat{\boldsymbol{r}}(r\dot{\boldsymbol{M}}\cdot\hat{\boldsymbol{r}} + a_0(\boldsymbol{M}\cdot\hat{\boldsymbol{r}} - \boldsymbol{M}\cdot\boldsymbol{M}))}{r^2(1-\boldsymbol{M}\cdot\hat{\boldsymbol{r}})^3} \right]_{\mathrm{ret}} \mathrm{d}s$$

Where, p' is the total sound pressure, p'_T is the monopole sound pressure, p'_L is the dipole sound pressure. a_0 and ρ_0 is the speed and density of sound in the undisturbed medium. The superscript "·" means derivative of time. The subscript "ret" means the quantity evaluated at retarded time. $\hat{\boldsymbol{r}}$ is the unit vector in the radiation direction, $\hat{\boldsymbol{n}}$ is the unit outward normal vector to surface, $\boldsymbol{M}$ is the local Mach number vector of source. r is the distance between observer and source and s is the surface of the acoustic source integration. $\boldsymbol{U}$ is the local velocity vector of source surface. $\boldsymbol{L}=l\cdot\hat{\boldsymbol{n}}$, where l is the load of source surface.

For the given observer position and time t, each source corresponds to a retarded time τ. In the integral calculation, the equation of source retarded time will be solved many times, and the Newton iterative method is used. The equation for solving source retarded time is as follows:

$$\tau=t-\frac{r}{a_0} \tag{3}$$

A numerical program has been written in Intel Visual Fortran compiler based on formula Farassat 1A for calculating the coaxial rotor noise. The sound pressure of coaxial rotors can be obtained by the linear superposition of that of upper rotor and lower rotor. The two axially aligned rotors rotate in opposite directions, which makes the motion laws of sound sources of coaxial rotor different from that of single rotor. The motion and position parameters of the integral surface have significant effects on the thickness noise. The complicated unsteady flowfield can lead to the unsteady airloads of noise source, which have significant effects on the loading noise. Thus, in order to achieve the accurate sound pressure, the blade surfaces are chosen as the integral surfaces of noise sources. The motion law and airloads of sound sources can be obtained from flowfield by CFD method directly, which can avoid the interpolation error and improve precision. Fig.2 gives the schematic of blade grids and noise calculation process of coaxial rotors.

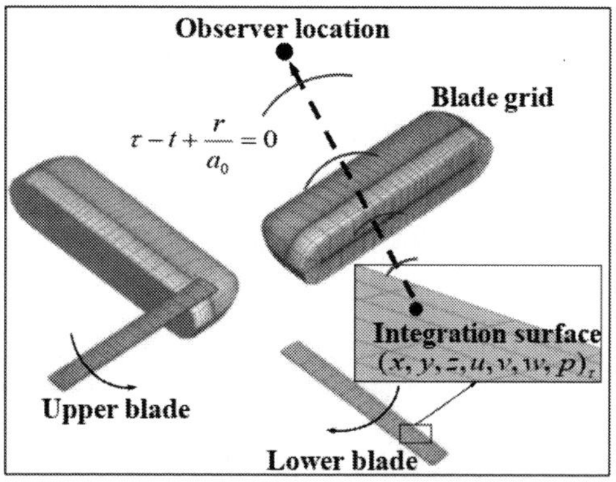

Fig.2 Schematic of blade grids and noise calculation process of coaxial rotors

Method validations

Fig.3 gives the comparisons of calculated aerodynamic performance with experimental data[17], where C_T is the thrust coefficient, C_Q is the torque coefficient. The calculated results are consistent with the experimental data. Fig.4 gives comparisons of calculated sound pressure with reference data[18] for the UH-1H rotor model in hover. Fig.5(a), (b) give the comparisons of calculated sound pressure with experimental data[19] for the AH-1 rotor model in descent flight, and serious blade-vortex interaction occurs, further details of rotor and motion parameters are given in Ref.19. It can be seen that the hovering state of a single rotor can be regarded as a quasi-steady state and the sound pressure waveform varies smoothly. Oppositely, for the unsteady state with blade-vortex interaction, the sound pressure waveform varies intensely [20] [21]. In general, the numerical method used in this paper is capable to predict the noise characteristics well both in the quasi-steady state and unsteady state, which offers a basic analysis method for the prediction of acoustic characteristics of coaxial rotors.

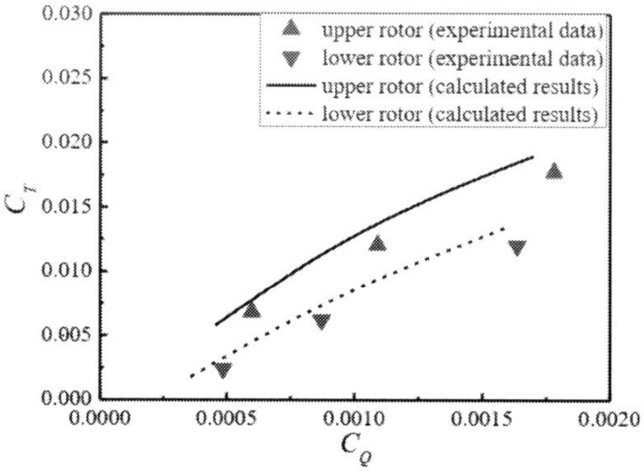

Fig.3 Comparisons of calculated aerodynamic performance with experimental data

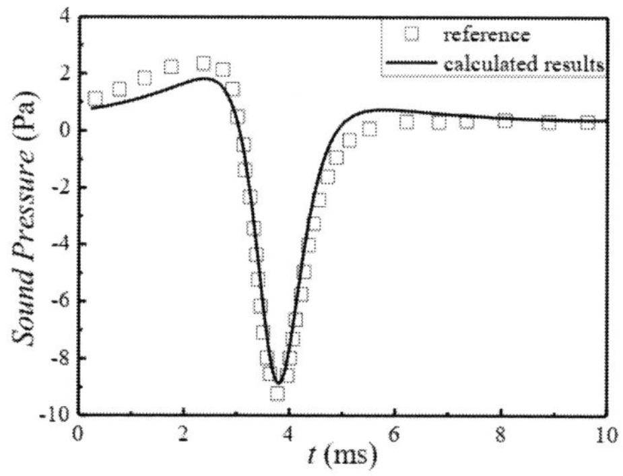

Fig.4 Comparisons of calculated sound pressure with reference data for the UH-1H rotor model

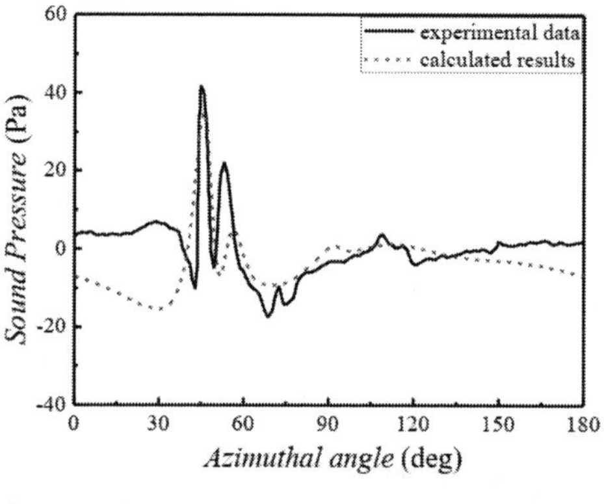

Fig.5 Comparisons of calculated sound pressure with experimental data for the AH-1 rotor model

RESULT ANALYSIS

The model rotor applied in this paper consists of two two-bladed rigid rotors arranged to form a coaxial system. The blades have a rectangular planform with untwisted NACA0012 airfoil and have an aspect ratio of 10. The radius and vertical spacing are 1m and 0.15m respectively, and the cutoff position is 0.2m. The maximum thickness of blade is 0.012m. In order to avoid the large computation cost in trimming process based on CFD method, the collective angles of the upper rotor and lower rotor are fixed to 8 deg. The blade-tip Mach number is 0.65, namely, the rotational speed is 1941rpm, which matches the actual rotational speed of helicopter. When viewed from above, the upper rotor rotates in an anticlockwise direction and the lower rotor rotates clockwise. The initial azimuthal angles of upper rotor and lower rotor are both 0 deg.

Unsteady aerodynamic characteristics of coaxial rotors

Fig. 6 gives the iso-surfaces of vorticity magnitude when the upper rotor and lower rotor are aligned (vorticity magnitude = 0.25). As seen in Fig.6, complicated blade-vortex and vortex-vortex interactions occur in the flowfield of coaxial rotors. Close vortex-surface interactions and impingement phenomenon have been observed among the vortexes from upper rotor and lower blades, which make the motion pattern and position of blade-tip vortex change suddenly, especially the strength decrease rapidly. The interfered blade-tip vortexes shed from upper blades may be affected by the blade-tip vortexes shed from lower blades again, and the blade-tip vortexes become distorted and then disappeared after interaction.

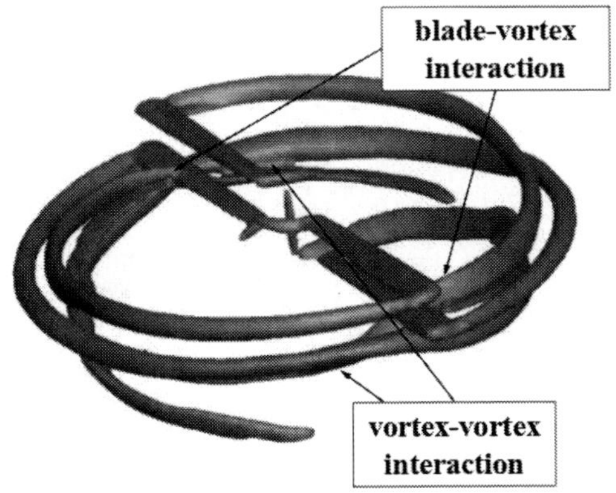

Fig. 6 Iso-surfaces of vorticity magnitude when the upper rotor and lower rotor are aligned

Fig. 7 gives the azimuthal variation of the lift coefficient. Fig. 8 gives the azimuthal variation of the normal force coefficient for different sections. The azimuthal angles when the upper blade encounters the lower blade are 0°, 90°, 180° and 270°, thus each rotation period consists of four loading period. For the single rotor, the hovering state can be regarded as the quasi-steady state, the blade surface airloads that have significant effects on dipole noise change a little with azimuthal angles, thus the loading noise waveform varies smoothly. As seen in Fig.7 and Fig.8, the aerodynamic forces of the upper rotor and lower rotor vary predictably due to the unsteady aerodynamic characteristics of the flowfield of coaxial rotors. The Venturi effect caused by the opposite rotation directions of upper and lower rotors can leads to a reduction in pressure between the rotors[7]. In addition, the lower rotor is seriously affected by the downwash of upper rotor, which decreases the actual angle of attack of lower rotor and leads to the thrust loss of lower rotor.

The induction between the upper rotor and lower rotor can also have influence on the rotor thrust. As the blades of the upper and lower rotors approach each other, the bound circulation[7] caused by blade loads induces an upwash on the other blade, which offsets the effect that downwash on the lower rotor. With the increasing of upwash, the thrust increases as the angle of attack increases. When the upper and lower blades are far away from each other, the upwash caused by bound circulation changes sign, and acts as a downwash, which leads to the reduction of angle of attack and rotor thrust.

In general, the airloads of coaxial rotors and single rotor are different obviously. Also, there exist obvious differences between the amplitude and change law of upper and lower rotors, which has significant effects on the loading noise of coaxial rotors.

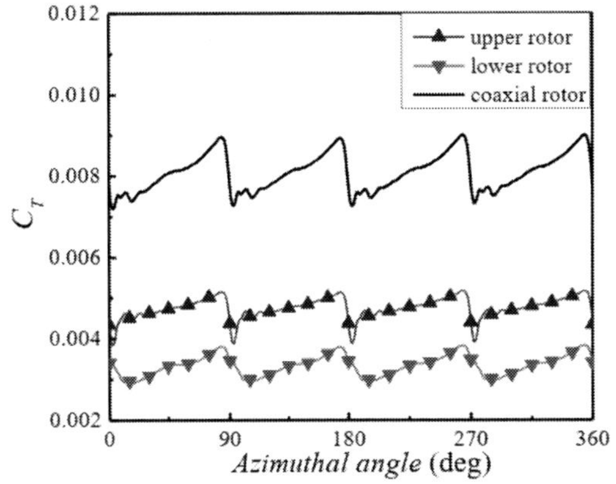

Fig. 7 Azimuthal variation of the lift coefficient

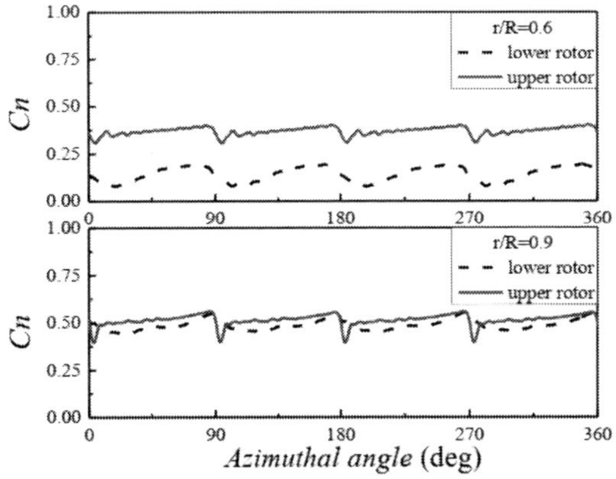

Fig. 8 Azimuthal variation of the normal force coefficient for different sections

Acoustic characteristics of coaxial rotors

The observer locations are shown in Fig.9, all these observers translate with the same forward flight speed as the rotor hub, such that the distance between the rotation axis and the observers is fixed at r/R=5, where R is the radius. The observer #1 is located in the rotor plane with azimuthal angle of 180 deg. The observer #1, #2 and #3 are located in the same horizontal plane. The angle between the observer #2, #3 and the downstream direction is 15 deg, 45 deg respectively.

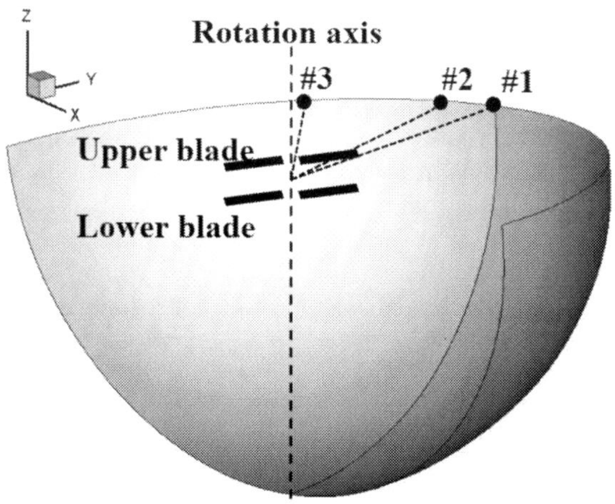

Fig. 9 Schematic of observer locations

Fig.10(a), (b) and (c) give the thickness, loading and total noise sound pressure time history of coaxial rotors at three observers located in horizontal plane. There exist large differences among the sound pressure time histories of coaxial rotors at different observers. For the conventional single rotor, though the positive or negative peak values of observers with different observation angles in the rotor plane occur in different azimuthal angles, the phase differences between two adjacent positive or negative peaks are also the same. For the coaxial rotors system, the upper rotor and lower rotor rotate in opposite directions, which cause the phase differences between adjacent positive or negative peaks of observers in the same rotor plane show obvious differences.

Fig.11 (a), (b) and (c) give the thickness and loading noise sound pressure time history at observer #1,#2 and #3. The effects of observation angles on sound pressure waveform can be known by analyzing the sound pressure time history of thickness noise. For both the upper rotor and lower rotor, the peak values occur two times in one rotational period.

For the observer #1, as the initial azimuthal angles of upper and lower rotors are both 0 deg, the phases of peak values of the upper rotor and lower rotor are the same. The phases of negative peak of the thickness sound pressure waveform are the same as that of a single two-bladed rotor. The negative peak values at observer #1 are larger than those at other observers in rotor plane. The sound pressure time history of coaxial rotors at observer #1 shows the similar characteristics as that of a single two-bladed rotor.

For the observer #2, the angle between the observer #2 and the downstream direction is 15 deg, that is the reason why the phase difference between two adjacent negative peaks of thickness sound pressure at observer #2 is 30 deg. In addition, the sound pressure time history of coaxial rotors has similar features with scissors tail rotor[22].

For the observer #3, the phase difference between two adjacent negative peaks of sound pressure of coaxial rotor is exactly 90 deg, which is the same as that of a single four-bladed rotor. It can be seen that the negative peak values of the upper rotor and lower rotor are almost identical. It is because that the upper rotor and lower rotor have the same blade shape and operate under the same blade-tip Mach number, and the thickness noise is mainly depended on the position and velocity of acoustic source that located in blade surface.

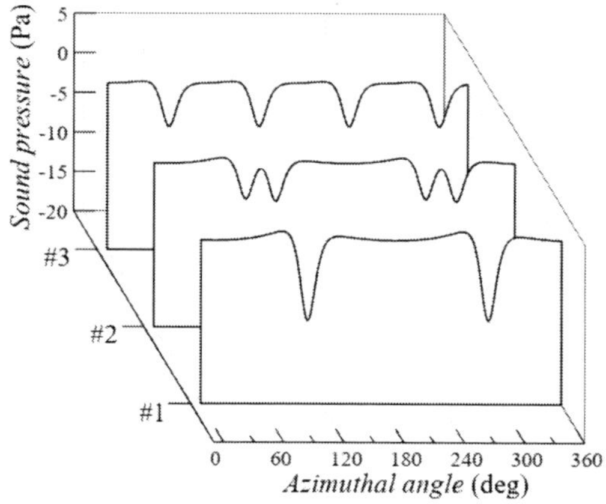

(a) thickness noise

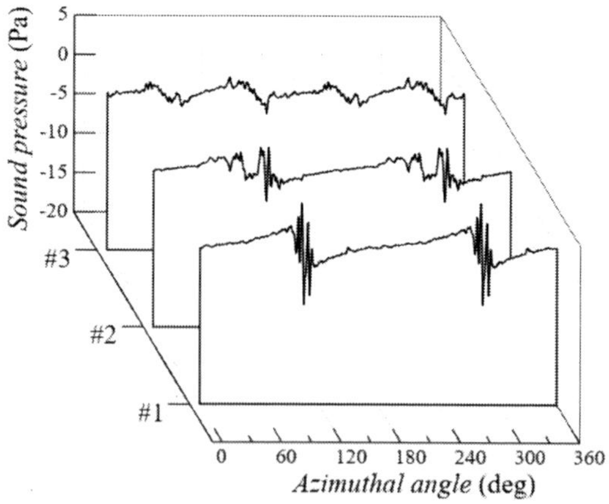

(b) loading noise

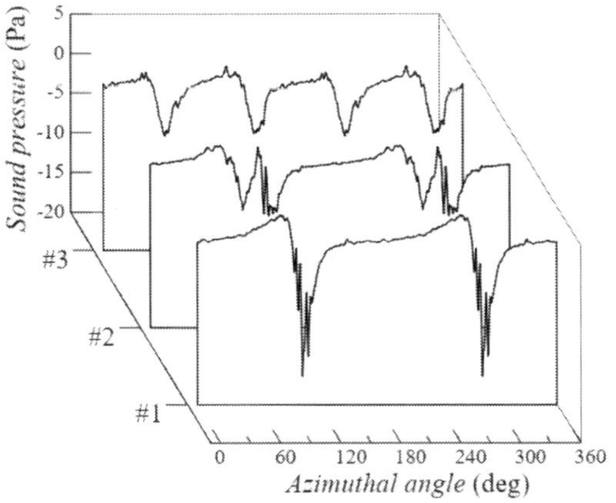

(c) total noise

Fig. 10 Sound pressure time history of coaxial rotors

As shown in Fig.11, there exist obvious differences between the negative peak values of upper rotor and lower rotor. Compared with the smooth waveform of thickness noise, the loading noise sound pressure waveform varies more violently. The tip vortexes shed from the upper rotor cause severe aerodynamic interference to the lower rotor, coming with complicated blade-vortex interaction noise. The downwash from upper rotor can also lead to periodic variation of the airloads of lower blade. Then the strong loading noise radiation occurs due to the unsteady loads. For the upper rotor, the tip vortex and downwash from lower rotor can hardly interfere the upper rotor. But, the loads of the upper blades show obvious unsteady characteristics due to the Venturi effect, which can also cause the serious loading noise radiation of upper rotor.

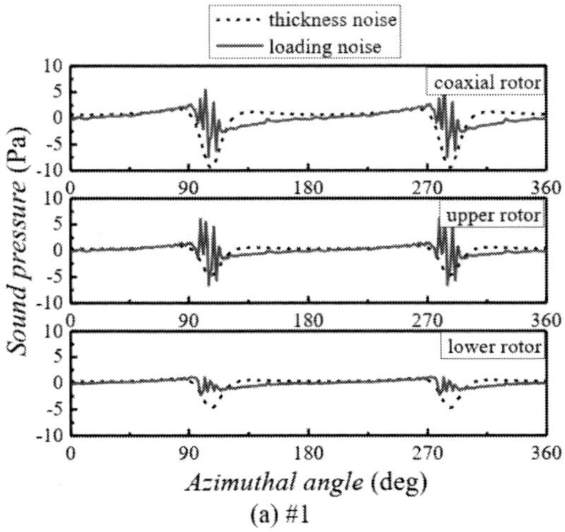

(a) #1

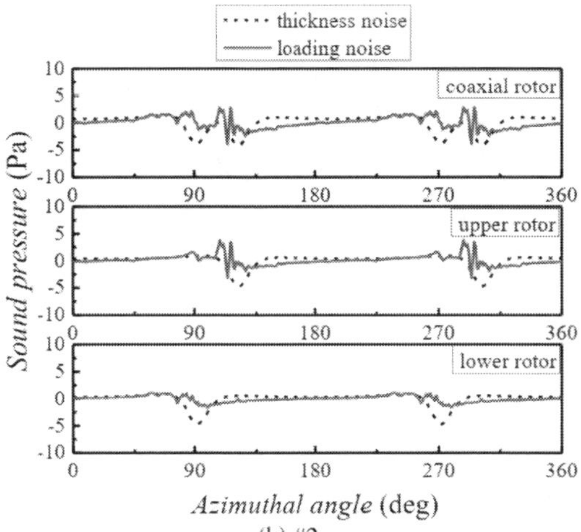

(b) #2

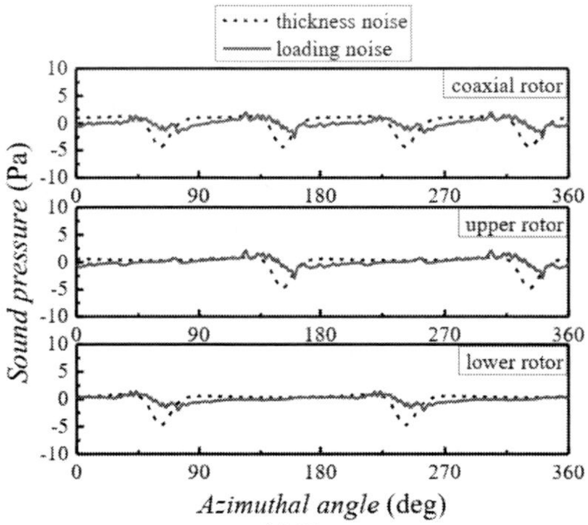

(c) #3

Fig. 11 Sound pressure time history of upper rotor and lower rotor

Fig.12 (a) and (b) give the spectrum frequency of thickness noise, loading noise and total noise of coaxial rotors at observers #1 and #3 respectively (BPF=35.7Hz). It can be seen that there exist significant differences between the spectrum characteristics at these observers with different observation angles in rotor plane. In general, the amplitude of thickness noise is higher than that of loading noise at lower frequencies, but lower than that of loading noise at higher frequencies while the thickness noise decays rapidly with increasing of frequency. The total noise is widely dominated by thickness noise at lower frequencies and the amplitude of total noise decreases rapidly. With the increase of frequency, the total noise is widely dominated by loading noise gradually at higher frequencies and the decay rate decreases.

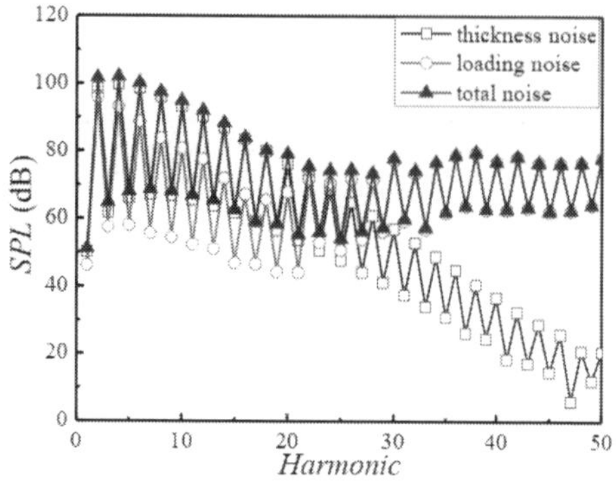

(a) #1

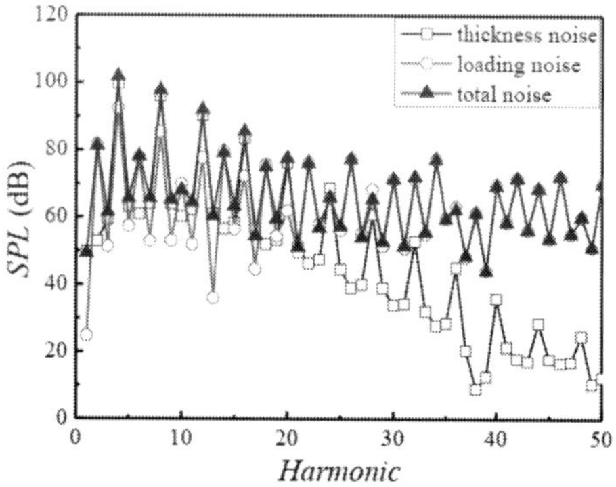

(b) #3

Fig. 12 spectrum frequency

To gain insight into the propagation characteristics, Fig.13(a) and (b) give the comparisons of noise directivity patterns in rotor plane (the distance between rotation axis and the observers is fixed at r/R=5.0) between coaxial rotors and a single four-bladed rotor. For the conventional single rotor, the hovering state can be regarded as a steady state, thus the loading noise radiation intensity from different observer directions with the same distance is also the same. It can be seen that the thickness noise level of coaxial rotors is smaller than that of single rotor in some directions. It is because the opposite rotation directions of upper and lower rotors produce the phenomenon similar to "modulated effect" of uneven blades[23], which may reduce the noise radiation intensity to a certain extent. The loading noise level of coaxial rotors is smaller than that of single rotor in some directions. Though the unsteady aerodynamic characteristics of upper and lower blades lead to the strong loading noise radiation, the modulated effect weakens the radiation intensity in some direction. The radiation intensity of loading noise is significantly lower than that of thickness noise, therefore the total noise level is mainly dominated by thickness noise level.

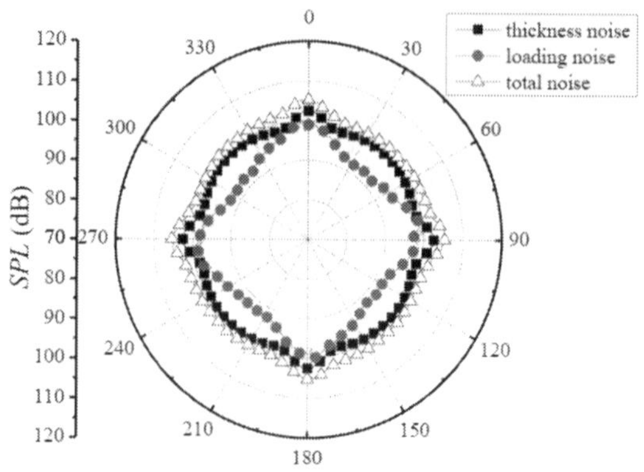

(a) coaxial rotors

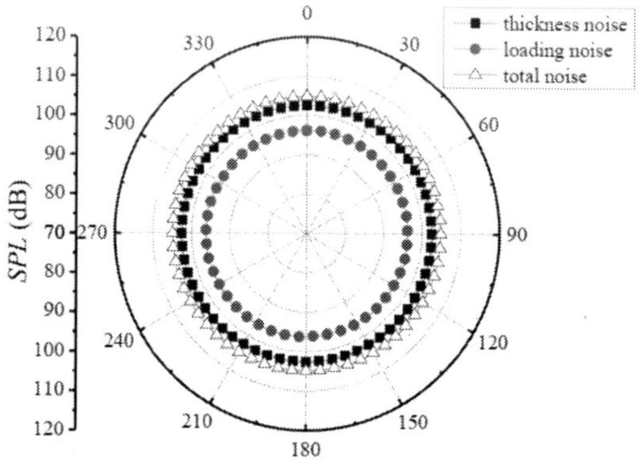

(b) single rotor

Fig.13 comparisons of noise directivity patterns in rotor plane

CONCLUSIONS

A solver based on CFD/FW-H method is established to simulate the acoustic characteristics of the coaxial rigid rotors in hover. The sound pressure time history and noise spectrum characteristics on the thickness, loading and total noise of the coaxial rotor have been analyzed in detail. The conclusions are as follows:

(1) The coaxial rotor configuration produces a more complicated flowfield than a single rotor system under hovering condition due to various blade-vortex and vortex-vortex interactions, and the airloads of coaxial rotors shows obvious unsteady characteristics;

(2) The upper rotor and lower rotor rotate in opposite directions, therefore the phase differences between two adjacent positive or two negative peaks of observers in the same rotor plane are not identical. In the same horizontal plane, the thickness noise level is always larger than the loading noise level; the total noise is widely dominated by thickness noise in the low frequency band, and by loading noise in the high frequency band respectively;

(3) The loading noise of coaxial rotors is stronger than that of the single rotor due to the unsteady characteristics of airloads of coaxial rotors, therefore the serious BVI noise radiation occurs.

REFERENCES

[1] Leishman J G, Syal M., "Figure of merit definition for coaxial rotors," *Journal of the American Helicopter Society*, Vol. 53, (3), July. 2008, pp. 290–300.

[2] Bagai A, Leishman J G., "Coaxial rotor aerodynamics in hover," *Vertica*, Vol. 5, (2), Sep. 1981, pp. 163–172.

[3] Leishman J G, Syal M., "Free-wake analysis of tandem, tilt-rotor and coaxial rotor configurations," *Journal of the American Helicopter Society*, Vol. 41, (3), Num. 1996, pp. 196–207.

[4] Huang S L, Xu G H, Li C H., "Flow field analysis of coaxial twin rotor based on free wake," *Journal of Nanjing University of Aeronautics and Astronautics*, Vol. 40, (6), Dec. 2008, pp. 721–726. (in Chinese)

[5] Kim H W, Brown R E., "A rational approach to comparing the performance of coaxial and conventional rotors," *Journal of the American Helicopter Society*, Vol. 55, (1), Jan. 2010, pp. 12003(1)-12003(9).

[6] Tong Z L, Sun M., "Navier-Stokes calculations of coaxial rotor aerodynamic," *Acta Aeronautica et Astronautica Sinica*, Vol. 19, (1), Jan. 1998, pp. 1-5. (in Chinese)

[7] Lakshminarayan V K, Baeder J D., "Computational investigation of micro-scale coaxial rotor aerodynamics in hover," *Journal of the Aircraft*, Vol. 47, (3). 2010, pp. 940–955.

[8] Xu H Y, Ye Z Y., "Coaxial rotor helicopter in hover based on unstructured dynamic overset grids," *Journal of Aircraft*, Vol. 47, (5). 2010, pp. 1820–1824.

[9] Ye L, Xu G H., "Calculation on flow field and aerodynamic force of coaxial rotors in hover with CFD method," *Acta Aerodynamic Sinica*, Vol. 30, (4), Aug. 2012, pp. 437-442. (in Chinese)

[10] Wachspress D A, Quackenbush T R., "Impact of rotor design on coaxial rotor performance, wake geometry and noise," American Helicopter Society 62nd Annual Forum Proceedings, Phoenix, America, May 2006.

[11] Kim H W, Kenyon A R, Brown R E, Duraisamy K., "Interactional aerodynamics and acoustics of a hingeless coaxial helicopter with an auxiliary propeller in forward flight," *Aeronautical Journal*, Vol. 113, (1140), July. 2009, pp. 65-78.

[12] Kim H W, Duraisamy K, Brown R E., "Effect of rotor stiffness and lift offset on the aeroacoustics of a coaxial rotor in level flight," American Helicopter Society 65th Annual Forum Proceedings, Grapevine, TX, May 2009.

[13] Zhao Q J, Xu G H, Zhao J G., "New hybrid method for predicting the flowfields of helicopter rotors," *Journal of the Aircraft*, Vol. 43, (2). 2006, pp. 372-380.

[14] Wang B, Zhao Q J, Xu G., "Simulation for unsteady flow-field of forward rotor based upon a new moving-embedded grid method," *Acta Aerodynamic Sinica*, Vol. 30, (1), Feb. 2012, pp. 14-21. (in Chinese)

[15] Williams J E F, Hawkings D L., "Sound generation by turbulence and surfaces in arbitrary motion," *Philosophical Transactions of the Royal Society of London. Series A, Mathematical and Physical Sciences*, Vol. 264, (1151), 1969, pp. 321-342.

[16] Farassat F., "Linear acoustic formulas for calculation of rotating blade noise," *AIAA Journal*, Vol. 19, (9), 1981, pp. 1122–1130.

[17] Coleman C P., "A survey of theoretical and experimental coaxial rotor aerodynamic research," NACA TP 3675, 1997.

[18] Baeder J D, Gallman J M, Yu Y H., "A computational study of the aeroacoustics of rotors in hover," *Journal of the American Helicopter Society*, Vol. 42, (1), Jan. 1997, pp. 39–53.

[19] Boxwell D A, Schmitz F H, Splettstoesser W R., "Helicopter Model Rotor Blade Vortex Interaction Impulsive Noise: Scalability and Parametric Variations," *Journal of the American Helicopter Society*, Vol. 32, (1), Jan. 1987, pp. 3–12.

[20] Zhang Q, Wang H M, Hu Z W., "Analysis for helicopter noise signal based on wavelet transform," *Acta Acustica*, Vol. 26, (5), Sep. 2001, pp. 450-454. (in Chinese)

[21] Wang H M, Zhang Q, Hu Z W., "An experimental study of AS350B2 helicopter noise," *Acta Acustica*, Vol. 26, (5), Mar. 2003, pp. 177-181. (in Chinese)

[22] Zhu Z, Zhao Q J., "Numerical analyses for aerodynamic and noise characteristics of helicopter scissors tail rotor," American Helicopter Society 70th Annual Forum Proceedings, Quebec, Canada, May 2014.

[21] Li D H, Yang Y C., "Analysis on thickness noise generated from helicopter rotor with uneven blade spacing," *Acta Acustica*, Vol. 36, (5), Sep. 2011, pp. 513-519. (in Chinese)

Triple Swept Rotor Design for Helicopter BVI Noise Reduction

Sanghyun Chae
Senior Researcher
Korea Aerospace Research Institute
Daejeon, Republic of Korea

Jinyoung Mo
Researcher
Hyundai Motor Company
Hwaseong, Gyeonggi-do, Republic of Korea

Kihoon Jeong
Principal Researcher
Korea Aerospace Research Institute
Daejeon, Republic of Korea

Seongyong Wie
Senior Researcher
Korea Aerospace Research Institute
Daejeon, Republic of Korea

Deogkwan Kim
Principal Researcher, Head
Korea Aerospace Research Institute
Daejeon, Republic of Korea

Kwanjung Yee
Associate Professor
Seoul National University
Seoul, Republic of Korea

ABSTRACT

Triple swept rotor design has been conducted with the main objective being to reduce the blade vortex interaction(BVI) noise and to improve aerodynamic performance on hover and high speed forward flight. The triple swept blade inherits the BVI noise reduction from BlueEdge, and this shape is devised to take into account of the reduction of the pitching moment. Derivative shapes of the triple swept blade have high possibility to cause structural problem, a constant thrust weighted solidity and a pitching moment are set to design constraints. For the efficient design optimization, the aerodynamic analysis was performed by hybrid method coupling computational fluid dynamics and freewake model, and efficient global optimization(EGO) based on Kriging model was used to search the optimal points. Farasat1A method predicts the aeroacoustic pressure on the far-field for the evaluation of the BVI noise. Design results are presented that show that the designed blades have improved aeroacoustic performances. ANOVA and SOM explain the relation between design variables and objectives (BVI noise reduction, hovering and forward flight performances).

INTRODUCTION

Noise issues have been a serious impediment to the public acceptance of helicopters, as well as its affect on passenger comfort. The thickness and loading of the main and tail rotor blades accelerate air flows and generates acoustic waves. The compressible these waves radiate noise to an observer far from the noise source. Additionally, the helicopter has other various noise sources resulting from the aerodynamic interactions between the rotor wakes and other solid surfaces. The detailed explanation of helicopter noises and noise reduction technologies are presented in Ref. 1.

Among them, a significant impulsive noise is caused by unsteady pressure fluctuations occurring on the lifting rotor blades. When the helicopter is operating in descending or decelerating flight, the rotor blades pass in close proximity to their previously shed or trailed tip vortices, and have sharp changes in pressure on their surfaces. Thus, this impulsive noise is called blade vortex interaction (BVI) noise.

Recently, "Eurocopter's Blue Edge Blade" achieved a significant reduction of BVI noise. The leading edge changes of this blade can be used to reduce BVI noise at chosen observer locations (Ref. 2). The forward backward sweep of the "Blue Edge Blade" encourages the BVI to be more oblique, which reduces the strength of the acoustic pulses while fostering cancellation at chosen observer locations.

Figure 1. EC155 equipped with Blue Edge blades. (Ref. 2)

Presented at the 6th Asian/Australian Rotorcraft Forum / Heli Japan 2017, Kanazawa, Japan, November 7-9, 2017.

However, a double swept planform such as "Blue Edge Blade" may have a considerable nose-up pitch moment. The forward backward sweep makes a protruding area which are located at fore 1/4 chord line of the blade root; the high lift on this area is connected to the nose-up pitching moment. If pitching moments exceed a certain level, stall can occur on the blade.

To prevent a pitching moment problem and retain a BVI noise reduction, a triple swept planform is devised. The backward forward backward sweep of the triple swept blade has "Blue Edge Blade" like sweep shape on outboard, so this planform inherits the noise reduction characteristics of "Blue Edge Blade". The triple swept blade region on the backward inboard generate nose-down pitching moments which may reduce the overall nose-up pitching moments of the blade.

In this study, an optimization design has been conducted with the main objective being to reduce the BVI noise and to improve aerodynamic performance of a triple swept planform blade. To this end, a high fidelity aero-acoustic analysis tool and efficient optimization techniques are employed, and the design variables are carefully designated to describe an arbitrary planform shape.

NUMERICAL METHODLOGY

Hybrid Computational Fluid Dynamics Methodology

The hybrid method is an efficient computational tool to simulate the complex interaction between blades and wakes. This method consists of a computational fluid dynamics method and a freewake method. The computational fluid dynamics predicts flow fields near the blade only, and the freewake method models the wake convection in the far field.

The hybrid method is a finite volume based three-dimensional unsteady viscous compressible flow solver. The flow is modeled by first principles using the Navier-Stokes equations. The Navier-Stokes equations are solved in the transformed body-fitted coordinate system using a time-accurate, finite volume scheme. A third-order spatially accurate Roe scheme is used for computing the inviscid fluxes, and second-order central differencing scheme is used for viscous terms. The Navier-Stokes equations are integrated in time by means of an approximate lowever upper symmetric Gauss Seidel implicit time-marching scheme. The flow is assumed to be turbulent everywhere, and hence no transition model is currently used. The Spalart-Allmaras turbulence model is used.

In the hybrid CFD methodology, the flowfield near the blade is resolved through the Navier-Stokes solution, whereas the influence of the other blades and of the trailing and shed vorticity in the far-field wake are accounted for by modeling them as a collection of piecewise linear bound and wake vortex elements, as shown in Fig. 2. The near wake is captured inherently in the Navier-Stokes analysis. The use of such a hybrid Navier-Stokes/vortex modeling of the wake in the far field. The vortex model is based on a Lagrangian wake approach, where a collection of vortex elements is released from the rotor blade trailing edge and is convected downstream by a combination of the freestream velocity and the bound and wake vortex element self-induced velocities. The strength of the vortex elements is based on the radial and temporal gradients of the bound circulation and the number of spanwise wake elements and wake time step increment chosen by the user. The influence of these vorticities on the blade aerodynamics from the wake model is computed by appropriately specifying the vortex-induced velocities at the far-field boundary of the Navier-Stokes domain, neglecting the contribution of the elements within the CFD volume grid released from the blade. The multiple trailer wake model is based on Prandtl's lifting-line theory. For a three-dimensional blade, the bound vorticity, located at the quarter-chord line of the blade is trailed into the wake from the blade tip and root. Vorticity is also shed from the blade midspan regions because of radial changes in the bound circulation. Therefore, the single tip vortex is replaced by user-specified number of multiple vortex segments trailed from all of the blades. The trailers are equally distributed along the blade span. The strength of the vortex elements is based on radial gradient of bound circulation and number of wake trailers chosen by the user.

In the present computations, the hybrid CFD solution accounts for the effects of the trailing wake as well as the shed wake. A multitrailer wake model is used. The effects of four revolutions of shed wake on the Navier-Stokes domain are included. The time step used in the CFD solution is 0.1 degree azimuth. The effects of turbulent mixing are modeled using the SA turbulence model. No transition model has been used in the current simulation. The induced velocity due to shed and trailing wake outside the Navier- Stokes domain is computed once every 5 degree of azimuth and imposed as velocity boundary conditions.

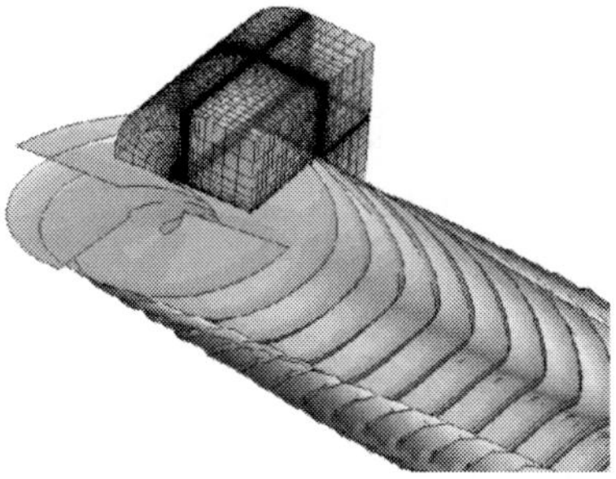

Figure 1. Concept of hybrid CFD methodology.

Computational Grid

Figure 3 shows the overall perspective and the top views of the C-H type of grid system that have been used for the computations where the baseline blade has a rectangular planform, and is further modified during the shape optimization process. The grid consists of 829,350 points

(171×96×50: chordwise × spanwise × normal), and 9,625 points (125×77: chordwise × spanwise) are distributed along the blade surface. The size of computational domain and boundary conditions are presented in Fig. 3.

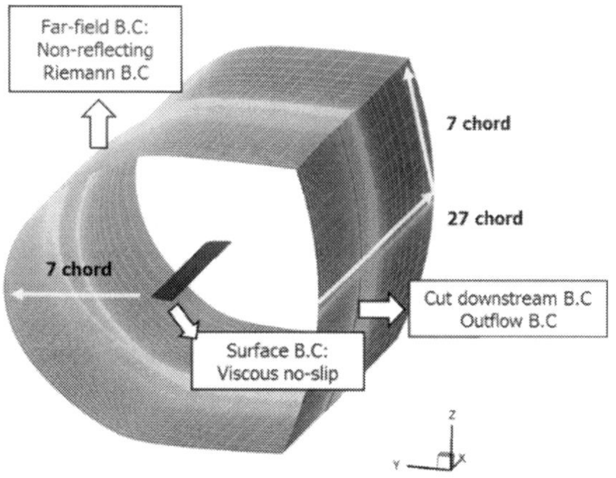

Figure 2. Grid system.

Acoustics

For rotor noise prediction, the Ffowcs-Wiliams Hawkings (FW-H) equation(Ref. 3) is used.

$$\frac{1}{c_0^2}\frac{\partial^2 p'}{\partial t^2} - \nabla^2 p' = \frac{\partial}{\partial t}\left[\rho_0 v_n |\nabla f| \delta(f)\right] - \frac{\partial}{\partial x_i}\left[l_i |\nabla f| \delta(f)\right] - \frac{\partial^2}{\partial x_i \partial x_j}\left[T_{ij} H(f)\right] \quad , (1)$$

In the present study, the Farassat formula is implemented using monopole and dipole terms. The Farassat formula(Ref. 4 and 5) is shown below,

$$4\pi p_T'(\vec{x},t) = \int_{f=0}\left[\frac{\rho_0 \dot{v}_n}{r(1-M_r)^2}\right]_{ret} dS + \int_{f=0}\left[\frac{\rho_0 v_n (r\dot{M}_i \hat{r}_i + c_0 M_r - c_0 M^2)}{r^2(1-M_r)^3}\right]_{ret} dS \quad , (2)$$

$$4\pi p_L'(\vec{x},t) = \frac{1}{c_0}\int_{f=0}\left[\frac{\dot{l}_i \hat{r}_i}{r(1-M_r)^2}\right]_{ret} dS + \int_{f=0}\left[\frac{l_r - l_i M_i}{r(1-M_r)^2}\right]_{ret} dS + \frac{1}{c_0}\int_{f=0}\left[\frac{l_r (r\dot{M}_i \hat{r}_i + c_0 M_r - c_0 M^2)}{r^2(1-M_r)^3}\right]_{ret} dS \quad , (3)$$

In the Farassat formula, the quadruple term is not considered, because noise induced by the flow shear stress effect is not dominant. The monopole and dipole terms represent the thickness and loading noise. In particular, loading noise represented by the dipole term is related to the BVI noise, since the vortex interaction induces changed aerodynamic loading on the blade surface.

OPTIMIZATION METHOD

For the modification of the blade shape to improve the acoustic and aerodynamic performances, an optimization method should be combined with the CFD flow solver in some manner. In the present study, the CFD solver is weakly coupled with the optimization method through surrogate models – namely the kriging model and the optimization method is selected for its computational efficiency.

In this study the MEDOC(Ref. 6), which has been developed for the kriging model based EGO (efficient global optimization) algorithm(Ref. 7) and post-processing tools, has been chosen as it has been used rotor blade shape design applications for improvement of aeroacoustic performance.

Overall procedure of MEDOC(EGO algorithm)

The overall procedure of the optimization shown in Fig. 4 is as follows:

1. Initial sample points are selected by an 'Improved Distributed Sampling' (Ref. 8).

2. The CFD solver evaluates the sample points.

3. The kriging model is constructed with the sample points.

4. The GA starts: Initially 512 populations are generated and evaluated using the kriging model.

5. The parents are selected.

6. Crossover and mutations transpire.

7. The fitness of the new generation is evaluated.

8. If the number of generations does not exceed 100, repeat the procedure from Step 4.

9. Additional sample points may be selected using the EI by the user.

10. Repeat procedure from Step 2 until the optima of the kriging model are not updated over 5 iterations.

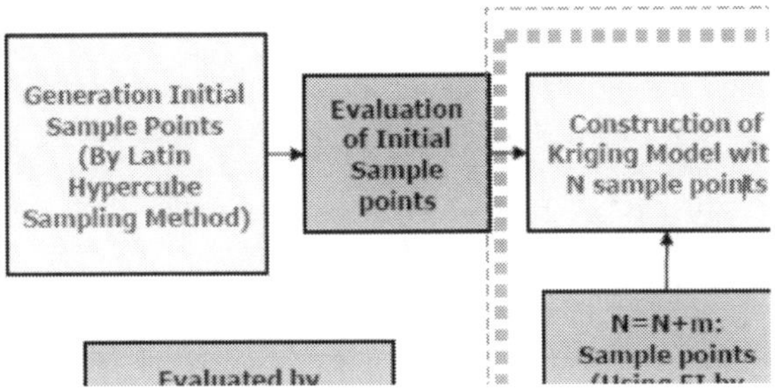

Figure 3. Flowchart of EGO

DEFINITION OF DESIGN PROBLEM

In this research, well documented and publicly available SA365C Dauphin rotor is chosen as the baseline. SA365C was originally developed by Aérospatiale in 1970's, and has been in continuous production for more than 40 years. (currently produced by Airbus Helicopters) This helicopter has a simple rectangular planform of 15.5 aspect ratio, a single symmetric airfoil (NACA0012) and 4 blades. Maximum take-off weight and tip Mach number of SA365 are 3,000 kg and 0.644, respectively. However, any twist angle is not applied to the baseline and designed planforms to focus on the design effects of triple swept planform. The specific values of the baseline rotor are presented in Table 1.

Table 1. Baseline rotor blade configuration

Description	Value
Number of blade	4
Aspect ratio	15.5
Twist	0°
Airfoil	NACA0012
Maximum take-off weight (MTOW)	3,000 kg
Tip Mach number	0.644
Planform shape	Rectangular

The objective functions aim to reduce the BVI noise characteristics and to improve aerodynamic performances of hovering and high speed forward flight. The multi objectives are defined as given in Eqs. (4)-(6).

$$OASPL_{non-dimensional} = \frac{|OASPL_{Baseline} - OASPL_{Design}|}{OASPL_{Baseline}}, \quad (4)$$

$$C_{P_{non-dimensional}} = \frac{|C_{P_{Baseline}} - C_{P_{Design}}|}{C_{P_{Baseline}}}, \quad (5)$$

$$FM_{non-dimensional} = \frac{|FM_{Baseline} - FM_{Design}|}{FM_{Baseline}}, \quad (6)$$

Figure 5 shows the definition of the design 13 variables to define the desired blade planform. 8 design variables of SA1 ~ SA4, R1 ~ R4 define the backward forward backward sweep of blade. The other design variables are model a chord length, thickness and anhedral respectively.

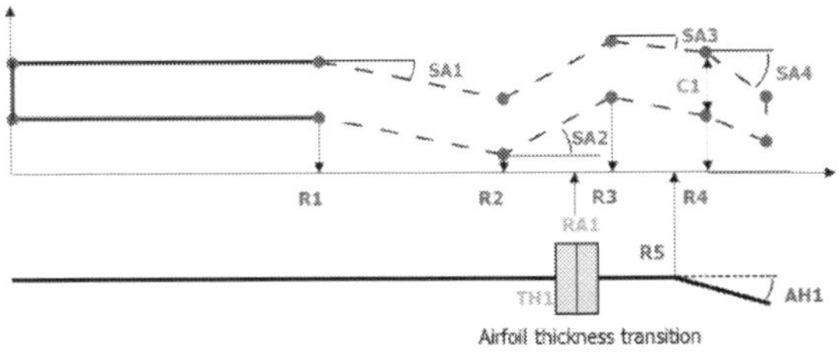

Figure 4. Diagram of swept blade planform with 13 design variables.

Randomly generated blade planforms may have strange or unacceptable shapes, with extremely small chord lengths or excessive curvatures, as demonstrated in Fig.6. Geometric constraints are applied to avoid these impractical blade shapes during the optimization process. (Fig. 7)

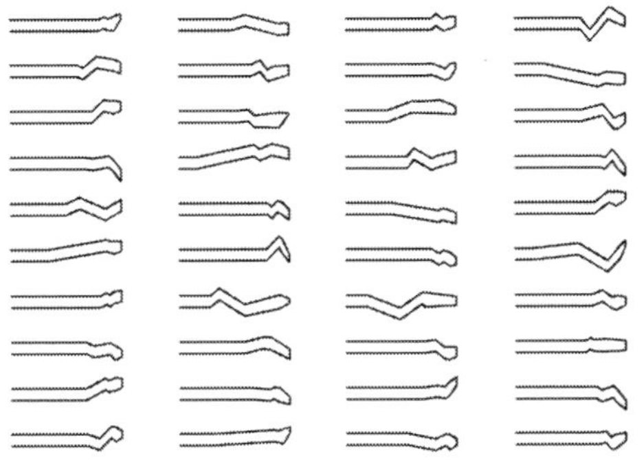

Figure 6. Planform shapes in the initial design space.

Figure 7. Example of discarded blade shapes.

DESIGN RESULTS AND DISCUSSION

Throughout the optimization process, 2~4 sample points are added to sequentially improve the kriging model every iteration using the EI. With the inclusion of the sample points, the accuracy of the kriging model is also enhanced while at the same time, the EIs of sample points are decreased because of the uncertainty of the surrogate model and the probability to become an optimum are reduced.

The theoretical convergence can be achieved when EI becomes less than 1% of the current best function value (Ref. 8), but this is an extremely difficult criterion to achieve in practice, especially for this kind of multi-objective design. Thus, the convergence to the minimum of the objective at each iteration step is used as a practical criterion and Fig. 8 shows the convergence of the minimum and maximum values of three objective functions and it can be seen that the minimum values are converged for both objectives from about the 10th iteration.

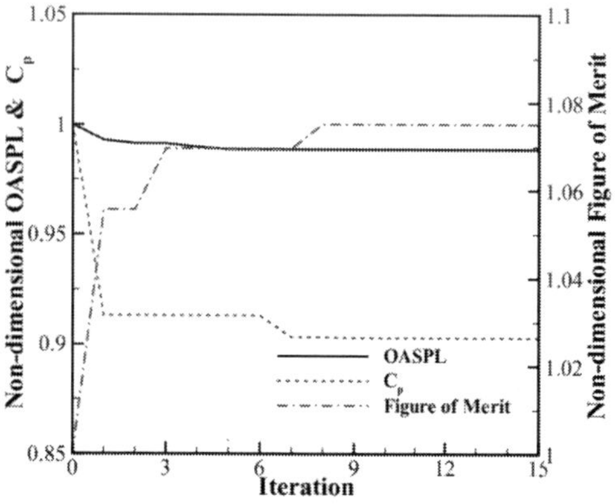

Figure 8. Change of objective values

The objective function values of the designed points are shown in Table. 8.. The values of the reference planform shapes, the baseline is also presented for comparison.

Table 2. Objective values of each blade shape.

	Blade shape	figure of merit	Power coefficient	OASPL[dB]
Baseline		0.6391	0.3512	103.459
Design1		0.6405 (+0.29%)	0.345 (-1.76%)	102.307 (-1.152dB)
Design2		0.6660 (+4.21%)	0.317 (-9.96%)	103.378 (-0.081dB)
Design3		0.6871 (+7.51%)	0.329 (-6.31%)	103.377 (-0.082dB)
Design4		0.6746 (+5.56%)	0.321 (-8.67%)	102.852 (-0.607dB)

The figure 9 describes the planform shapes and specific values of shape features. Triple swept shapes are clearly presented in Fig. 9 (a) and (b), and these shapes(Design 1 and 2) has lower value of OASPL than other designed shapes. The maximum decrease of OASPL(BVI noise reduction) is obtained by the design 1 shape, which has large region of backward forward backward triple swept. The best figure merit (hovering) value is achieved by the design 3, which has the highest anhedral angle. The lowest thickness of the design 2 makes the lowest power coefficient at high speed forward flight. The design 4 shows compromised performances of BVI noise reduction, high performance on hovering and forward flight.

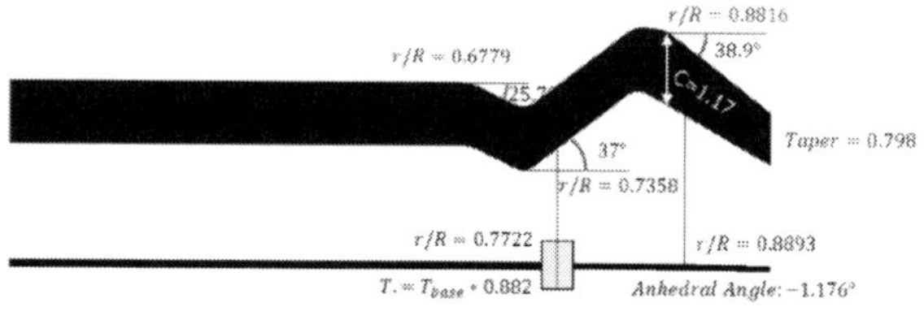

(a) Design 1.

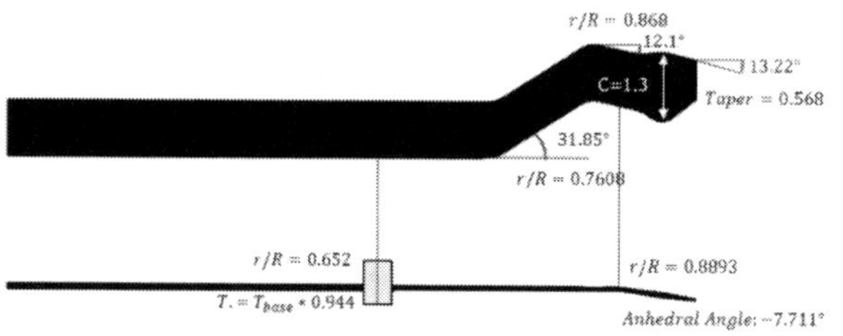

(b) Design 2.

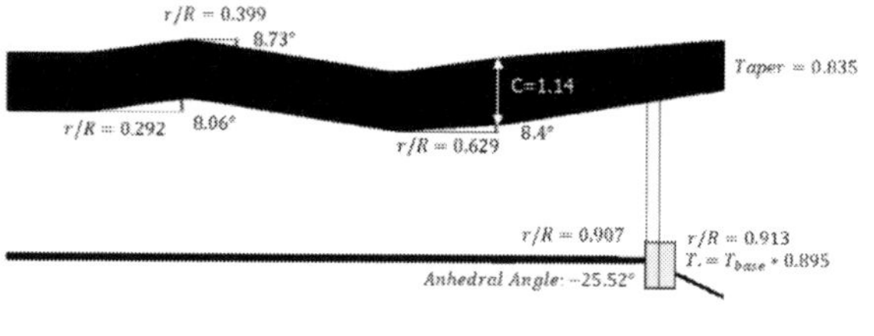

(c) Design 3.

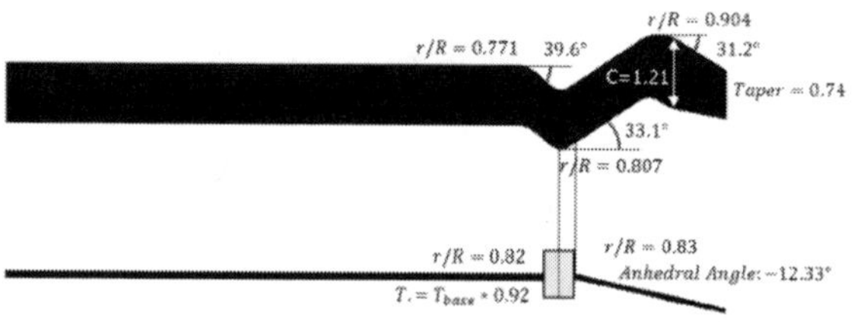

(d) Design 4.

Figure 9. Planform shapes of designed blades.

CONCLUDING REMARKS

The paper has presented a study to obtain improved helicopter blade shapes using a triple swept concept, which can simultaneously reduce the BVI noise and improve aerodynamic performances. By performing a thorough analysis of the design process and the results presented, the following conclusions are reached. Triple swept planform has a significant effect on the reduction of BVI noise. As known well, the anhedral and the thin airfoil thickness on the blade tip improve hovering and forward flight performance respectively.

More detail and specific analysis for designed shapes and triple swept concepts will be discussed on the presentation.

Author contact: Sanghyun Chae nyugnas@kari.re.kr, Jinyoung Mo jintaylor99@snu.ac.kr, Kihoon Jeong khchung@kari.re.kr, Seongyoung Wie wie@kari.re.kr, Deogkwan Kim shine@kari.re.kr, Kwanung Yee kjyee@snu.ac.kr

ACKNOWLEDGMENTS

This research was supported by the 'Conceptual Study of Advanced Concept VTOL Unmanned Aerial Vehicle'(FR17160) research program of the Korea Aerospace Research Instititude.

REFERENCES

[1]Schmitz, F. H., "The Challenges and Possibilities of a Truly Quiet Helicopter, 29th Alexander A. Nikolsky Honorary Lecture," *Journal of the American Helicopter Society*, Vol. 61, (4), Oct. 2016, pp. 1–33. Doi: 10.4050/JAHS.61.041001

[2]Gervais, M., and Gareton, V., "Analysis of Main Rotor Noise Re- duction due to Novel Planform Design—The Blue Edge™ Blade," 37th European Rotorcraft Forum, Gallarate, Italy, September 13–15, 2011.

[3]Ernsthausen, W., "The source of propeller noise," Luftfahrtforschung 8:433–440, 1936. [translated as NACA TM 825, 1937].

[4]Farassat, F., Succi, G. P., "The prediction of helicopter discrete frequency noise," *Vertica* Vol. 7, (4) pp. 309-20, 1983.

[5]Brentner, K. S., "Prediction of helicopter discrete frequency rotor noise—a computer program incorporating realistic blade motions and advanced acoustic formulation," NASA TM 87721, 1986.

[6]Chae, S., Yee, K., Obayashi, S., and Jeong S., "Helicopter Rotor Blade Planform Design Using a Newly Developed MDO Platform – MEDOC," Heli Japan 2010, Sitama, Japan, November 1-3, 2010.

[7]Donald, R. J., Matthias, S., and William, J. W., "Efficient Global Optimization of Expensive Black-Box Function," *Journal of Global Optimization*, Vol. 13, No. 4, 1998, pp. 455–492. doi:10.1023/A:1008306431147

[8]Beachkofski, B. K., and Grandhi, R. V., "Improved Distributed Hypercube Sampling," 43rd AIAA/ASME/ASCE/AHS/ASC Structures, Structural Dynamics, and Materials Conference, AIAA Paper 2002-1274, April 2002.

Development and validation of a hybrid method for predicting helicopter rotor impulsive noise

Liangquan Wang Ph.D. Candidate

Guohua Xu Professor

Yongjie Shi Associate Professor

National Key Laboratory of Science and Technology on Rotorcraft Aeromechanics,
Nanjing University of Aeronautics and Astronautics
Nanjing, China

ABSTRACT

Prediction of helicopter rotor impulsive noise is very challenging. This paper describes a new approach to predict rotor impulsive noise of both high-speed impulsive (HSI) noise and blade-vortex interaction (BVI) noise. This approach has been developed by combining the advantages of three different methods: 1) a Computational Fluid Dynamics (CFD) method based on Reynolds-Averaged Navier-Stokes (RANS) equations to account for the viscous and compressible effects near the blade; 2) a Vorticity Transport Model (VTM) to predict rotor wake system without artificial dissipation; and 3) an acoustic calculation method, based on Ffowcs-Williams Hawkings (FW-H) equation implemented to a permeable data surface, to effectively compute both the HSI noise and BVI noise. The developed approach is validated through available test data, for the cases of UH-1H model rotor, AH-1/OLS rotor and Helishape 7A rotor operated at different flight conditions. Peak sound pressure level of HSI noise is accurately predicted with relative errors less than 7%. Additionally, acoustic waveform of BVI noise is well captured though it is sensitive to small changes in aerodynamic load. It is suggested that present hybrid method is versatile for the prediction of rotor impulsive noise with moderate computational cost.

NOTATION

c	blade chord length
C_{Mp}	pitching moment around the rotor hub
C_{Mr}	rolling moment around the rotor hub
C_T	rotor thrust coefficient
E	total energy per unit mass
$\boldsymbol{F}_c$	inviscid flux
$\boldsymbol{F}_v$	viscous flux
M	Mach number
p	pressure of the flow
q_j	heat flux
R	rotor radius
S	surface of the grid cell
$\boldsymbol{S}$	vorticity source
$\boldsymbol{u}_b$	blade segment velocity relative to the air
W	conservative variables
$\boldsymbol{x}$	position vector
α	angle of attack
$\boldsymbol{\Gamma}_b$	blade bound circulation
θ_0	collective pitch
θ_{1c}	lateral cyclic pitch
θ_{1s}	longitudinal cyclic pitch
μ	rotor advance ratio
ν	air kinematic viscosity coefficient
ρ	fluid density
τ_{ij}	stress tensor
ψ	blade azimuth angle
$\boldsymbol{\omega}$	vorticity
Ω	control volume

INTRODUCTION

Helicopter rotor aeroacoustics has become an important issue since the 1960s. High-speed impulsive (HSI) noise and blade-vortex interaction (BVI) noise, known together as impulsive noise, are the most annoying components and getting the most attention. HSI noise source is generated with rotor moving at high tip speeds and focused on a narrow transonic flow area around the blade tip. While BVI noise normally occurs in low speed forward and descent flight. Blade-vortex interaction is associated with a shed tip vortex in rotor wake intersected by following blades on the advancing and retreating side, and the resulting unsteady airload and noise are very sensitive to the miss distance between the tip vortex and blade. The key point of accurately predicting HSI and BVI noise lies in the modeling of transonic flow and vortical rotor wake, receptively. Moreover, typical propagation mechanisms for those two types of noise are also distinct. Peak sound pressure of HSI noise locates in the rotor disk plane whereas BVI noise mainly radiates below the rotor disk. For these reasons, a versatile approach for accurately predicting both HSI and BVI noise is quite promising in the design process of

---

Presented at the 6th Asian/Austrailian Rotorcraft Forum / Heli Japan 2017, Kanazawa, Japan, November 7-9, 2017.

modern quiet helicopters. To date, such an approach has yet to be developed because this work is very challenging.

Computational fluid dynamics methods, based on nonlinear Euler or Navier-Stokes equations, are exclusive to simulate the rotor HSI noise with transonic flow involved at high tip speeds. Many studies have been carried out on this topic, representative work can be found in Refs.1-4. In those studies, overset grid systems are implemented, and the overall flowfield is discretized into two sets of grids: a blade body-fitted grid and an off-body (or background) Cartesian gird. They are used to solve the flow near and far from the blade, respectively. Body-fitted grid provides enough flexibility in dealing with geometrically complex blades. However, in order to capture the rotor wake which plays a vital role in BVI phenomenon, the number of background gird points used is far more than the body-fitted grid. This greatly increase the requirement of computation resources and time (Ref.5-6). Even so, it is still not at a level acceptable for BVI prediction because of inherent numerical dissipation.

Classical Lagrangian wake models (e.g., free wake model (Ref.7), prescribed wake model (Ref.8)), based on the potential flow assumption, can avoid the problem of excessive numerical dissipation that CFD encountered and are of low computation cost. Some scholars coupled them with CFD method to investigate rotor aerodynamics and the resulting noise (Ref.9-13). The major deficiency of those wake models is that they rely heavily on empirical parameters such as the vortex decay factor and vortex core size. Due to the progress in wake modeling techniques, a novel type of Lagrangian model, which is referred as Vorticity Transport Model (VTM), can explicitly preserve the vorticity of rotor wake without introducing any empirical parameters. It has been successfully used in modeling the complicate rotor wake in recent years (Ref.14-16).

Ffowcs Williams-Hawkings (FW-H) equation (Ref.17) is the foundation in rotor acoustics research. Rotor noise is expressed as monopole, dipole, and quadruple sources in the FW-H equation. Direct volume integral of the problematic quadrupole term is difficult and time consuming (Ref.18). For a long period of time, the most commonly used form among the various solutions (Ref.19-24) of this equation is Formulation 1A. In original Formulation 1A, blade surface is chosen as control surface, and rotor thickness noise, loading noise (including BVI noise) can be calculated effectively by this formulation. Nevertheless, the quadrupole noise source in the transonic flow region is neglected. Hence, the original Formulation 1A is incapable of predicting HSI noise. To address this problem, a permeable data surface (Ref.25,26) which encloses all the nonlinear flow region has been recently used for surface integral. In this way, the quadrupole noise source is accounted for while the demand of computational cost is much less than direct volume integral. This formulation is therefore abbreviated as FW-Hpds.

The main objective of this study is to develop a robust and efficient tool for predicting helicopter rotor HSI and BVI impulsive noise. The advantages of Computational Fluid Dynamics model, Vorticity Transport Model, FW-Hpds method have been combined together in the proposed simulation tool. The rest of this paper is organized as follows: Section 2 presents the above-mentioned three methods in some detail, as well as a trimming procedure to determine the rotor control angles in forward flight. In Section 3, HSI noise of a hovering UH-1H rotor and an advancing AH-1/OLS rotor are investigated. In Section 4, BVI aerodynamic and acoustic results of the AH-1/OLS rotor and a Helishape 7A rotor are presented. Finally, the conclusions are summarized in Section 5.

COMPUTATIONAL METHODOLOGY

Aerodynamic and Flowfield Solver

In this research, the entire rotor flowfield is decomposed into two domains (Figure 1): the near-body Eulerain domain which covers a relatively small region of the computational zone and the off-body Lagrangian domain. CFD is applied to a C-O topology gird (C airfoil section and O tip cap, see Figure 2) to predict the flow in the vicinity of the rotating blades. On the other hand, the rest of the flowfield is resolved by VTM, which employs a set of particles to model the rotor wake.

The structured, body-fitted blade grid in C-O topology is generated by an algebraic technique. It moves with the blade to consider blade rigid motions such as rotation and pitching (Ref.27). To ensure that all the transonic flow in high-speed flight is enclosed, the outer boundary of C-O topology gird extends $3c$ away from the blade surface, where c is the blade chord length. The normal grid spacing at the blade surface is $1\times10^{-5}\,c$ for viscous calculation. The grid is clustered at the leading and trailing edges of the blade and near the blade tip to resolve the high gradients flow. Radial grid spacing near blade tip is about $0.02c$.

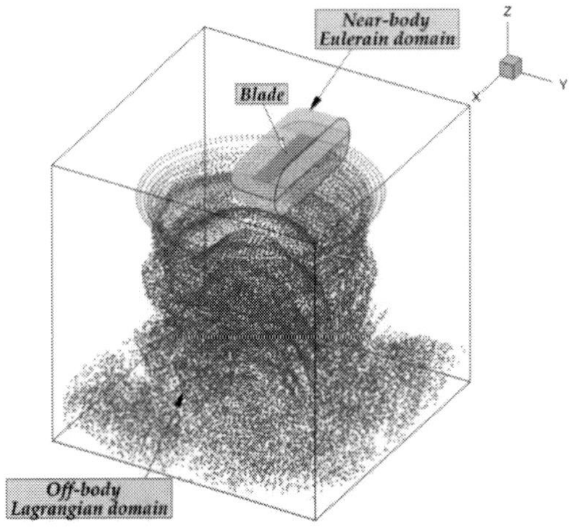

Figure 1. Schematic of the computational zone (only one blade is shown).

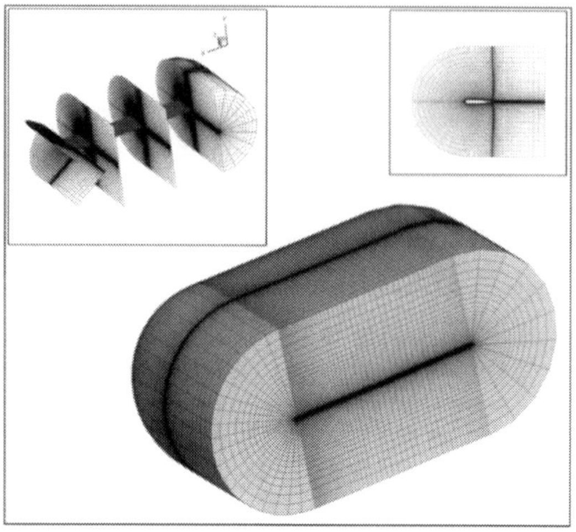

Figure 2. Near-body C-O topology grid of the rotor blade.

CFD method is employed to solve compressible Navier-Stokes (N-S) equations. The conservative form of mass, momentum and energy equations are as follows

$$\frac{\partial \rho}{\partial t}+\frac{\partial \rho u_j}{\partial x_j}=0 \quad (1)$$

$$\frac{\partial \rho u_i}{\partial t}+\frac{\partial \rho u_i u_j}{\partial x_j}=-\frac{\partial p}{\partial x_i}+\frac{\partial \tau_{ij}}{\partial x_j} \quad (2)$$

$$\frac{\partial \rho E}{\partial t}+\frac{\partial (\rho E+p)u_j}{\partial x_j}=\frac{\partial u_i \tau_{ij}}{x_j}-\frac{\partial q_j}{\partial x_j} \quad (3)$$

where ρ is the fluid density, u_j the velocity vector, p the pressure, τ_{ij} the stress tensor, E the total energy per unit mass, and q_j is the heat flux.

In our CFD code, the above N-S equations are solved in Reynolds-Averaged form using a finite volume algorithm. For any grid cell, equations (1-3) can be rewritten as

$$\frac{\partial}{\partial t}\int_{\Omega} W d\Omega+\oint_{\partial\Omega}(F_c-F_v)\, dS=0 \quad (4)$$

where Ω is the control volume, S denotes the surface of the grid cell. W is the conservative variables, and F_c, F_v are inviscid and viscous fluxes, respectively. Inviscid fluxes are calculated by Roe's flux-difference splitting scheme (Ref.28), while turbulence closure is achieved by Spalart-Allmaras one-equation model (Ref.29). In addition, Implicit Lower-Upper Symmetric Gauss-Seidel (LU-SGS) scheme (Ref.30) is chosen for temporal discretization.

The off-body flowfield is solved by the Vorticity Transport Model (Ref.31), which is derived from incompressible Navier-Stokes equations. VTM follows Lagrangian description and is written in vorticity-velocity form, and the governing equations can be expressed as

$$\frac{d\mathbf{x}}{dt}=\mathbf{u} \quad (5)$$

$$\frac{d\boldsymbol{\omega}}{dt}=\boldsymbol{\omega}\cdot\nabla\mathbf{u}+\nu\nabla^2\boldsymbol{\omega}+\mathbf{S} \quad (6)$$

where $\mathbf{x}$ is the position vector of the vortex filaments in the flow, $\boldsymbol{\omega}=\nabla\times\mathbf{u}$ is the vorticity, and ν is the air kinematic viscosity coefficient. The first term on the right-hand side of equation (6) is called vortex stretching term, which describes the dilation and rotation effects of the vortex filaments. The second term $\nu\nabla^2\boldsymbol{\omega}$ is referred as viscous diffusion term, which represents the effects of air viscosity. And the last term $\mathbf{S}$ is the vorticity source.

Instead of the dense background grid implemented in most traditional full CFD analysis, a set of particles (Ref.32) are spread into the rotor far flowfield to model the transport of rotor wake. Those particles are able to track the wake in a natural way. The vortex stretching term in equation (6) is solved by multiplying the velocity gradient with the particle vorticity, and the particle strength exchange (PSE) algorithm (Ref.33) is adopted to solve the viscous diffusion term. Because of the conservative property of PSE algorithm, VTM can explicitly preserve the vorticity of rotor wake without any artificial dissipation. The source term $\mathbf{S}$ is the vorticity shed and trailed from blade segment, and it is determined by

$$\mathbf{S}=-\frac{d\boldsymbol{\Gamma}_b}{dt}+\mathbf{u}_b\nabla\cdot\boldsymbol{\Gamma}_b \quad (7)$$

where $\boldsymbol{\Gamma}_b$ is the bound circulation of the blade segment, and $\mathbf{u}_b$ is the local relative velocity to the air.

Position and strength of the ith particle at a new time step are computed through the following time marching scheme

Predictor step:

$$\mathbf{x}_i^{n+1}{}_{(p)}=\mathbf{x}_i^n+\frac{\Delta t}{2}\left(3\mathbf{u}_i{}^n-\mathbf{u}_i{}^{n-1}\right) \quad (8)$$

$$\boldsymbol{\omega}_i{}^{n+1}{}_{(p)}=\boldsymbol{\omega}_i{}^n+\frac{\Delta t}{2}\left(3\frac{d\boldsymbol{\omega}_i{}^n}{dt}-\frac{d\boldsymbol{\omega}_i{}^{n-1}}{dt}\right) \quad (9)$$

Corrector step:

$$x_i^{n+1}{}_{(c)} = x_i^n + \frac{\Delta t}{2}\left(u_i^{n+1}{}_{(p)} + u_i^n\right) \quad (10)$$

$$\omega_i^{n+1}{}_{(c)} = \omega_i^n + \frac{\Delta t}{2}\left(\frac{d\omega_i^{n+1}{}_{(p)}}{dt} + \frac{d\omega_i^n}{dt}\right) \quad (11)$$

The blade pitch angle in forward flight can be described as the Fourier series

$$\theta(t) = \theta_0 + \theta_{1c}\cos\psi + \theta_{1s}\sin\psi \quad (12)$$

where θ_0, θ_{1c}, θ_{1s} denote the rotor collective pitch, lateral and longitudinal cyclic pitch angle, respectively. ψ is the blade azimuth angle.

The rotor is trimmed to match the experimental thrust coefficient CT and to eliminate the pitching and rolling moments (i.e., $C_{Mp} = 0$, $C_{Mr} = 0$) around the rotor hub. To launch the trimming procedure, initial guess of control settings are determined by blade-element theory. The relationship between (CT, C_{Mp}, C_{Mr}) and (θ_0, θ_{1c}, θ_{1s}) is nonlinear, in order to obtain the correction angles for each rotor trimming cycle, the following Newton-Raphson iteration is utilized

$$\begin{pmatrix}\Delta\theta_0\\ \Delta\theta_{1c}\\ \Delta\theta_{1s}\end{pmatrix} = \begin{bmatrix}\frac{\partial C_T}{\partial\theta_0} & \frac{\partial C_T}{\partial\theta_{1c}} & \frac{\partial C_T}{\partial\theta_{1s}}\\ \frac{\partial C_{Mp}}{\partial\theta_0} & \frac{\partial C_{Mp}}{\partial\theta_{1c}} & \frac{\partial C_{Mp}}{\partial\theta_{1s}}\\ \frac{\partial C_{Mr}}{\partial\theta_0} & \frac{\partial C_{Mr}}{\partial\theta_{1c}} & \frac{\partial C_{Mr}}{\partial\theta_{1s}}\end{bmatrix}^{-1} \times \begin{pmatrix}C_T^{exp} - C_T\\ -C_{Mp}\\ -C_{Mr}\end{pmatrix} \quad (13)$$

where the matrix in equation (13) is knowns as Jacobi matrix. The Newton-Raphson iteration continues until the trim solution is converged. Specifically, it requires 3 or 4 trimming cycles before convergence. New control settings for each cycle are given by

$$\begin{pmatrix}\theta_0\\ \theta_{1c}\\ \theta_{1s}\end{pmatrix}_{new} = \begin{pmatrix}\theta_0\\ \theta_{1c}\\ \theta_{1s}\end{pmatrix}_{old} + \begin{pmatrix}\Delta\theta_0\\ \Delta\theta_{1c}\\ \Delta\theta_{1s}\end{pmatrix} \quad (14)$$

Acoustic Solver

The Lighthill acoustic analogy (Ref.34), as formulated in integral representation by FW-H equation, is the basis for predicting rotor noise generated by complex blade motions and aerodynamics. In FW-H equation, total rotor noise is expressed as monopole and dipole sources distributing on blade surface, and quadruple source in the volume surrounding the blade. The monopole term is used to model rotor thickness noise and the dipole term is used to model loading noise (including the impulsive BVI noise). The quadrupole term is negligible when rotor rotates at low blade tip Mach numbers, and the computational demand for direct volume integral of this term is very high. For rotor acoustic studies in early years, the quadrupole term in FW-H equation is usually neglected and the integral surface has often been placed on the solid (or impermeable) blade body. However, quadrupole noise source becomes important when transonic flow is involved, as identified by Hanson and Fink (Ref.35). In the FW-Hpds formulation, the integral surface is moved away from the solid blade to a virtual or permeable data surface[23]. In this way, the transonic-associated contributions to overall rotor noise in high-speed flight can also be included. More information about this permeable data surface formulation can be found in Ref.18.

Primitive flow variables on the permeable data surface required for acoustic prediction are exported from converged solution on the C-O topology CFD grid. Figure 3 is 2-D illustration of a NACA0012 airfoil at 4.0° angle of attack and Mach number $M = 0.8$. As revealed in the figure, all the transonic flow regions are enclosed in the permeable data surface. For the case of 3D rotor acoustic prediction, a small and a large sample surface are shown in Figure 4.

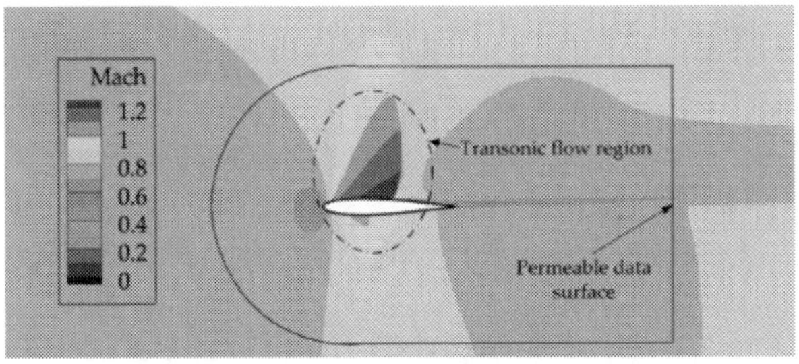

Figure 3. Mach contours around a NACA0012 airfoil at $\alpha = 4.0°$and $M = 0.8$, and permeable data surface is chosen to enclose all the transonic regions.

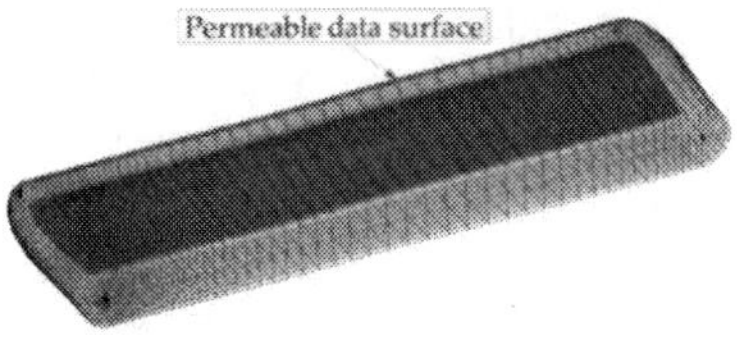

(a)

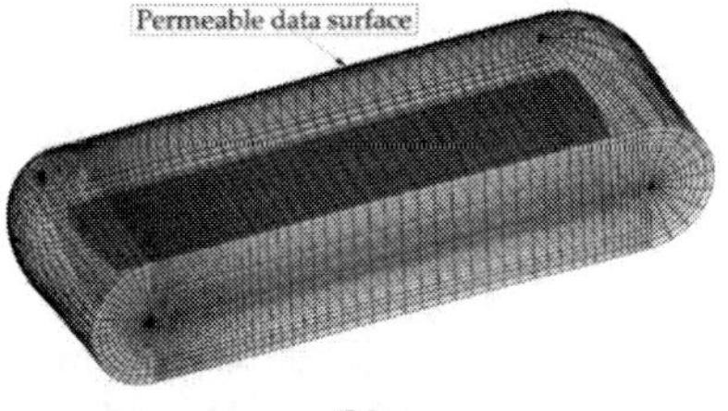

(b)

Figure 4. Samples of the permeable surface extracted from C-O topology grid for rotor acoustic prediction. (a) small surface; (b) large surface.

Coupling Framework

The flow chart of CFD, VTM and FW-Hpds coupling strategy is shown in Figure 5. First the initial control angles are computed by blade-element theory, following that CFD/VTM simulates the rotor aerodynamics and flowfield. At every time step, CFD predicted vortex sources are transmitted into VTM, and VTM predicted flow variables are fed back to CFD at the outer boundary of the C-O topology grid. In our research, a time step of 1° is used for time marching in HSI case, and 0.25° azimuth increment is selected for BVI case which requires a smaller step to capture the unsteady airload. When one rotor revolution of simulation is accomplished, trimming procedure is performed to obtain new control settings. This process is repeated until converged aerodynamic results are obtained. Then permeable surface grid and associated flow data are exported to acoustic solver, to compute sound pressure time history at specific observer location. Note that trimming procedure is not necessary in hover simulations.

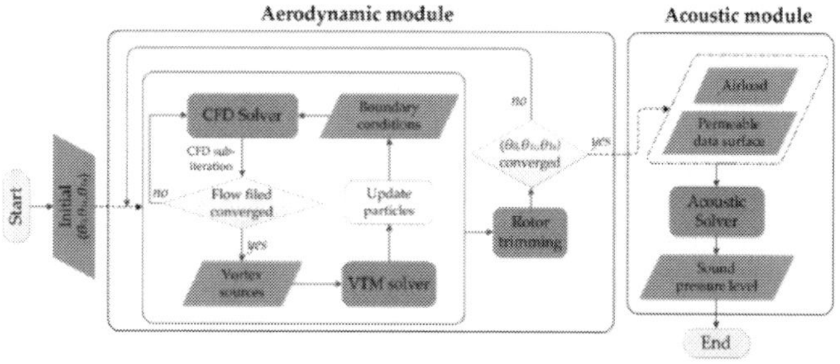

Figure 5. Flow chart of the coupling methodology.

RESULTS AND DISCUSSION

HSI Noise Prediction

In this section, we concentrate on HSI noise prediction. A one-seventh scale model of the two-bladed UH-1H main rotor is investigated firstly. The model rotor blades which utilize NACA 0012 airfoil section are rectangular and untwisted, with a radius of 1.045m and a chord length of 0.076m. The rotor is operated at non-lifting hover condition (i.e., θ_0 = 0) with several tip Mach numbers. In order to check the CFD grid sensitivity, a baseline grid and an additional fine grid are generated. The baseline grid consists of 225×43×74 grid points (225 points are used in the chordwise direction with 161 points distributed on the airfoil surface, 43 points are used in the normal direction, and 74 points in the spanwise direction), while the fine grid consists of 310×58×116 grid points. The number of vortex particles used in the off-body flowfiled simulation remains the same, and the cutoff size of VTM zone is 3R×3R×3R for hovering simulations.

The predicted acoustic pressure time histories are compared with the experimental data conducted by Boxwell, Yu and Schmitz (Ref.36). Figure 6 shows the comparison for tip Mach numbers M = 0.70 and 0.85. The microphone locates in the rotor disk plane (where HSI peak negative pressure appears) and is 3.09 radii away from the hub center. Predicted acoustic signals (large permeable data surface is used here) match well with measured data in the experiment, though peak negative pressure at M = 0.85 is slightly underpredicted. The fine grid obtains just slightly better results than the baseline grid, however, the computational cost of fine grid (2.08 million total grid points) is much higher than the baseline one (0.71 million total grid points and only approximately 1/3 of the fine grid). Figure 7 compares the acoustic results of applying different permeable data surfaces. The large data surface corresponds to the 25th grid line normal to blade surface, and the small one corresponds to the 13th grid line. It is demonstrated that there is nearly no difference between them. In the following numerical cases, we only perform acoustic computations on the baseline grid with large permeable data surface. Acoustic results at three different observer locations for M = 0.88 are shown in Figure 8. Again, the agreement between predicted and measured results is reasonably good, and the relative errors are less than 7%.

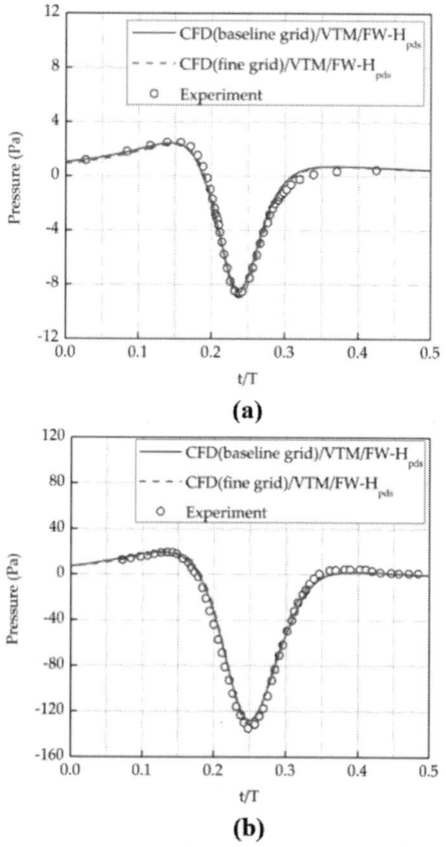

Figure 6. Comparison of acoustic pressure time history for different CFD gird resolution, and microphone locates 3.09 radii away from rotor hub. (a) M = 0.70; (b) M = 0.85.

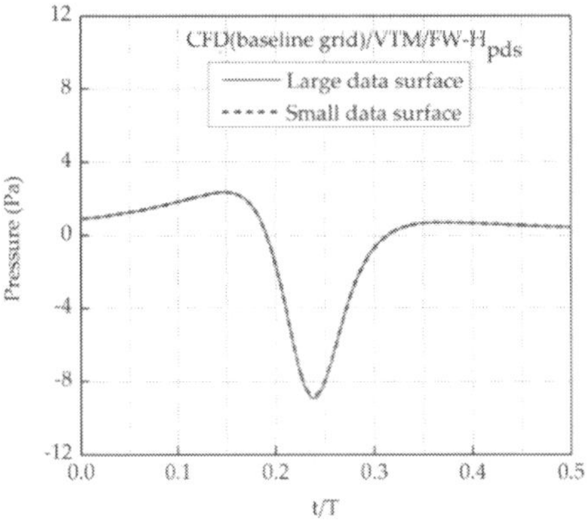

(a)

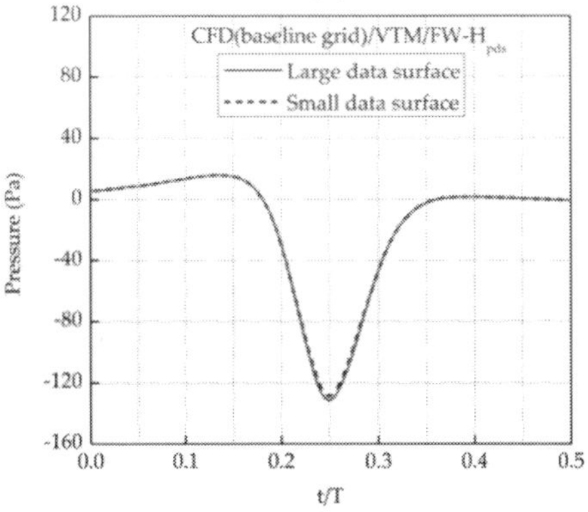

(b)

Figure 7. Comparison of acoustic pressure time history for different permeable data surfaces. (a) $M = 0.70$; (b) $M = 0.85$.

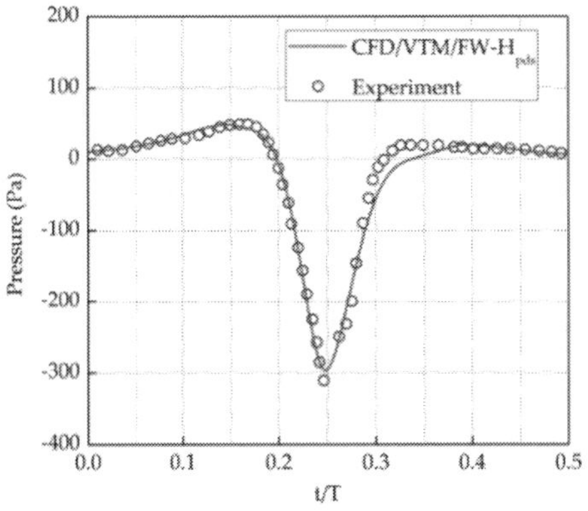

(a)

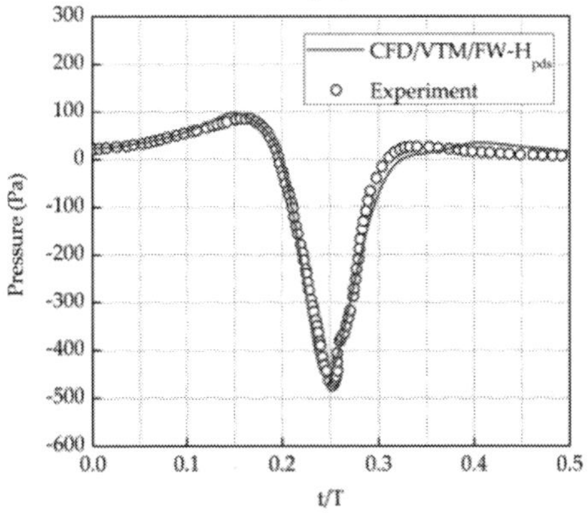

(b)

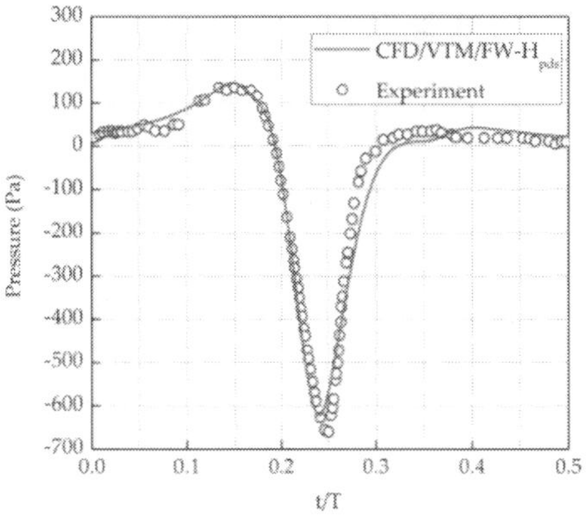

(c)

Figure 8. Comparison of acoustic pressure time history for different microphone locations, and the rotor is operated at $M = 0.88$. (a) 3.09 R; (b) 2.18 R; (c) 1.78 R.

Non-dimensional fluid density (ρ/ρ_0) distributions around the airfoil at a spanwise blade section of r/R = 0.98 are depicted in Figure 9. For $M = 0.70$, the entire flow filed is subcritical and quadruple noise source is not involved, hence, the peak negative pressure remains at a low level (approximately 9 Pa). For a tip Mach number of 0.85, the air is severely compressed at the leading edge and shock wave emerges at the middle chord (Figure 9(b)), forming a supersonic pocket on the surface of blade tip (shown in Figure 10(a)). At this tip Mach number, the contribution of quadrupole noise source becomes significant, and the peak negative pressure jumps to about 135 Pa. In Figure 10(b), the supersonic pocket further expands to larger region when M reaches 0.88, resulting in a more intense HSI noise (about 300 Pa at 3.09 R observer location).

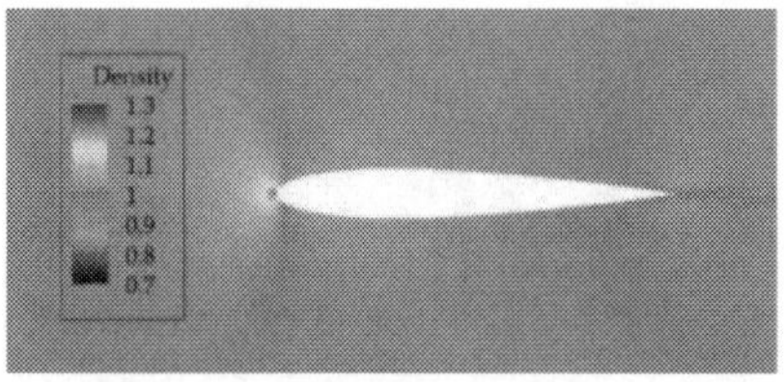

(a)

(b)

(c)

Figure 9. Non-dimensional fluid density around the airfoil, r/R = 0.98. (a) $M = 0.70$; (b) $M = 0.85$; (c) $M = 0.88$.

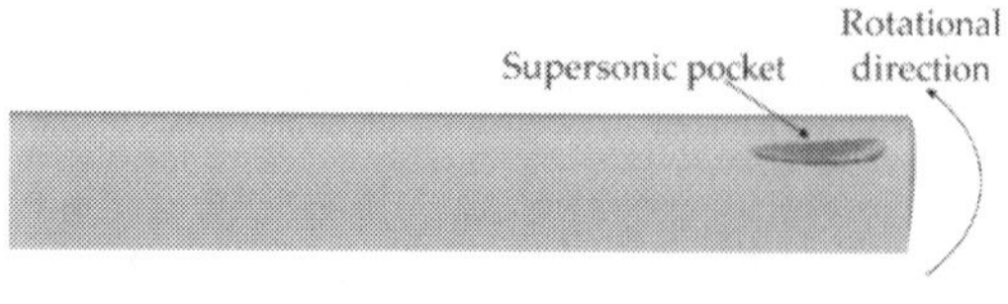

(a)

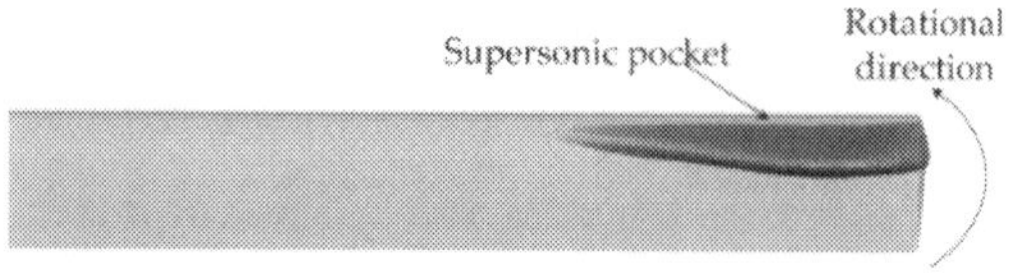

(b)

Figure 10. Supersonic pocket near the blade tip of UH-1H model rotor. (a) $M = 0.85$; (b) $M = 0.88$.

The second numerical case is an advancing AH-1 Operational Loads Survey (AH-1/OLS) rotor. This rotor has two rectangular blades. The blade has a radius of 0.958m and is 0.1039m in chord length. A modified BHT-540 profile is chosen as airfoil section with 9.71% thickness. Linear twist from blade root to tip is -8.2°. Comprehensive acoustic test of this rotor was performed in German-Dutch Wind tunnel (DNW), for both high-speed forward flight (Ref.37) and low-speed blade-vortex interaction flight (see Ref.38, and will be discussed in section 3.2.1) conditions. The locations of the microphones in the wind tunnel are shown in Figure 11, and the corresponding coordinates are tabulated in Table 1. As shown, Mic #2, 6, 8 are positioned in rotor disk plane to measure HSI noise, while Mic #3, 7, 9 are placed forward and down the rotor to gather BVI noise information.

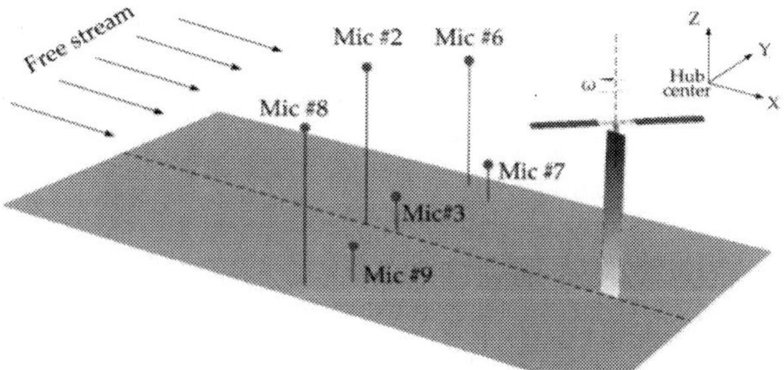

Figure 11. Location of the microphones in DNW for AH-1/OLS acoustic test.

Table 1. Coordinates of the microphones in DNW.

Mic number	X (m)	Y (m)	Z (m)
#2	-3.30	0	0
#6	-2.85	1.65	0
#8	-2.85	-1.65	0
#3	-2.85	0	-1.65
#7	-2.47	1.43	-1.65
#9	-2.47	-1.43	-1.65

The test case selected for HSI noise validation corresponds to an advance ratio of $\mu = 0.27$, advancing tip Mach number $M_{AT} = 0.84$, and a time-averaged thrust coefficient $C_T = 0.0054$. Table 2 presents the collective pitch (θ_0), lateral (θ_{1c}) and longitudinal (θ_{1s}) cyclic pitch angles after trimming solution. For forward flight simulation, the wake cutoff distance in the free stream direction extends to 5R. The acoustic signals for the three in-plane microphones are shown in Figure 12. It can be observed that predicted acoustic waveforms demonstrate good correlation with the measurement. For the microphones located on the advancing side (Mic #6) and centerline (Mic #2), the amplitudes of acoustic pressure are close (approximately 70 Pa), while it decreases rapidly to an amplitude of 38 Pa at the retreating microphone (Mic #8).

Table 2. Control settings for AH-1/OLS rotor in high speed forward flight, $\mu = 0.27$.

θ_0	θ_{1c}	θ_{1s}
7.83°	4.64°	-3.97°

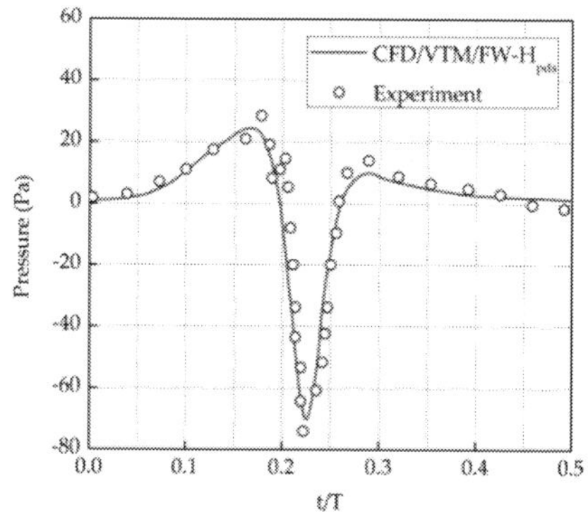

(a)

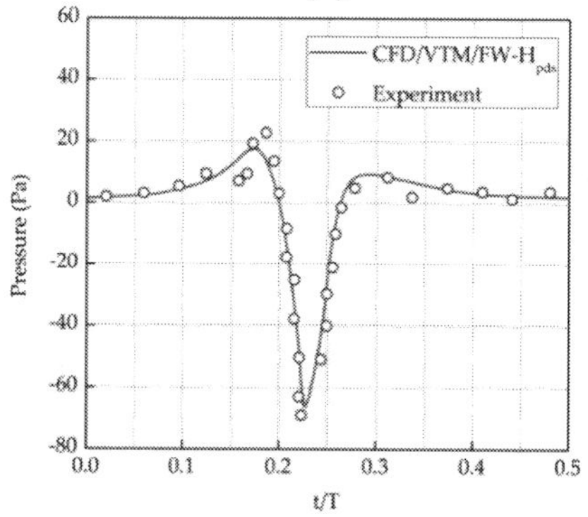

(b)

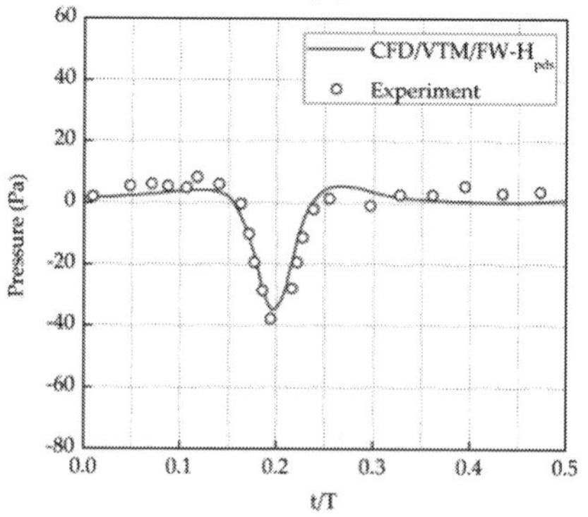

(c)

Figure 12. Comparison of in-plane HSI noise for AH-1/OLS rotor. (a) Mic #6, advancing side microphone; (b) Mic #2, centerline microphone; (c) Mic #8, retreating side microphone.

BVI Noise Prediction

The acoustic signal for HSI noise is characterized by a single peak negative pressure. However, the acoustic waveform in blade-vortex interaction dominated flight condition is more complicated. Unlike HSI noise mainly generated by air compressibility, BVI noise is closely related to the highly unsteady airload in low speed forward and descent flight. Rotor wake plays an important role to BVI noise signature. When a tip vortex shed from one rotor blade and interacts with following blades, the local airloads would fluctuate dramatically. Vorticity preservation of rotor wake and accurate prediction of blade airloads are prerequisites for relaible BVI acoustic analysis. In what follows, we continue to disscuss the acoustics of AH-1/OLS rotor but for a BVI-dominated flight case. Afterwards, the Helishape 7A rotor is going to be investigated in detail. Predicted rotor wake, blade surface pressure and airload, as well as the resulting BVI noise will be shown.

The calculated BVI case for AH-1/OLS rotor corresponds to an advance ratio of μ = 0.146. Trimmed control settings are listed in Table 3. Comparisons between measurement (Ref.38) and this study at Mic #3, 7, 9 are shown in Figure 13. It can be seen that for all three microphones located below the rotor plane, acoustic waveforms in shape and amplitude are predicted well by present CFD/VTM/FW-Hpds method. At Mic #3 (on the centerline) and Mic #9 (on the retreating side), a strong BVI pulse is prominent in the acoustic pressure time history and is accompanied by several moderate pulses. As for the microphone located on the advancing side (Mic #7), the magnitudes of the positive and negative spikes are comparable.

Table 3. Control settings for AH-1/OLS rotor in BVI flight case, μ = 0.146.

θ_0	θ_{1c}	θ_{1s}
5.65°	2.39°	-2.08°

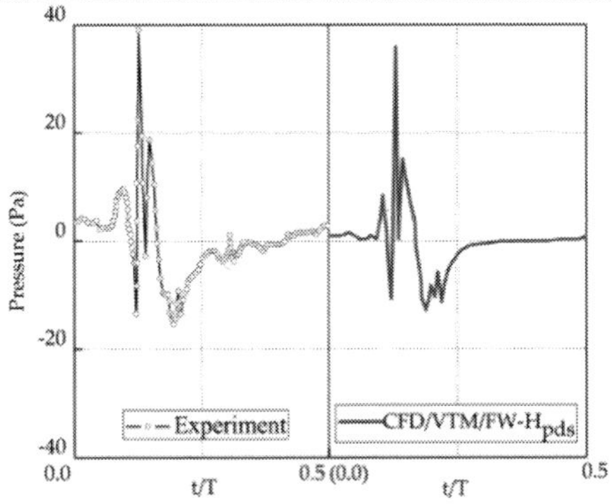

(a)

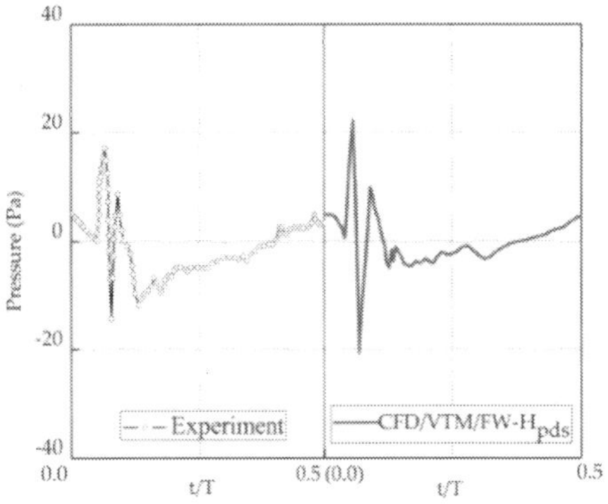

(b)

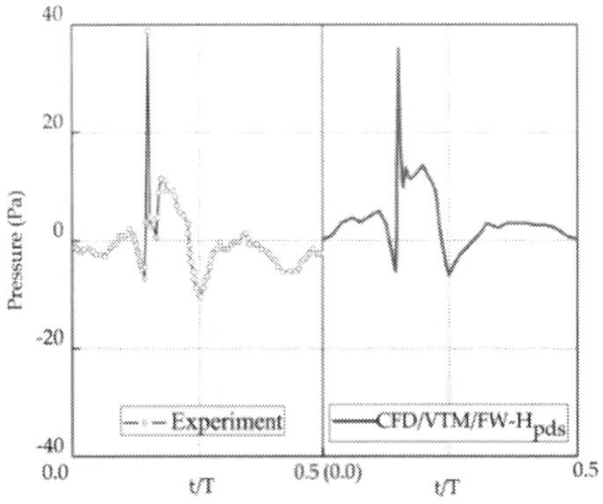

(c)

Figure 13. Comparison of measured and predicted BVI noise signal for AH-1/OLS rotor. (a)Mic #3; (b) Mic #7; (c) Mic #9.

The four-bladed Helishape 7A rotor utilizes modern ONERA OA213 and OA209 airfoil sections, with unconventional blade geometric twist. Figure 14 shows the planform of 7A rotor blade. It has a large aspect ratio of 15, and the blade chord is 0.14m. Advance ratio, thrust coefficient, and advancing tip Mach number of the considered case are μ = 0.167, CT = 0.007 and M_{AT} = 0.783, respectively. Table 4 gives the trimmed control settings.

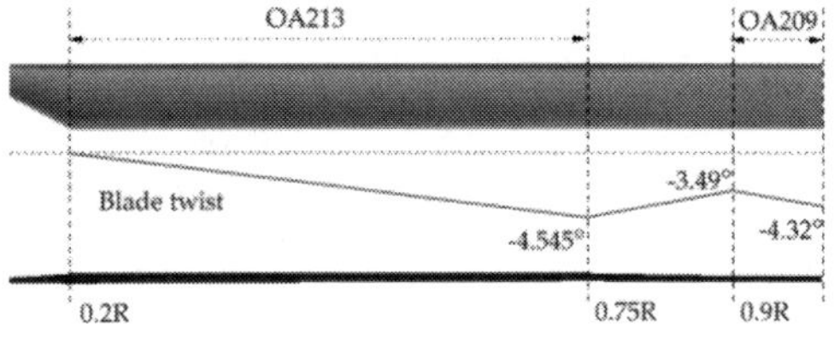

Figure 14. Planform of 7A rotor blade.

Table 4. Control settings for Helishape 7A rotor in BVI flight case, μ = 0.167.

θ_0	θ_{1c}	θ_{1s}
4.76°	2.97°	-2.81°

Figure 15 presents the 3D wake visualization solved by Vortex Transport Model for the Helishape 7A rotor. It is plotted by vorticity iso-surface. From this figure, we can clearly see the rollup process of tip vortices at the outboard region of advancing and retreating blade sides, and the wake structure is persevered quite well even after six rotor revolutions, which is the key point to accurate BVI airload and acoustic prediction. Compared with full CFD method that usually track the vortical rotor wake less than two rotor revolutions (Ref.5), VTM shows its capability in rotor wake capture. In addition, the vortex filaments that interact parallel with the blade in the first and fourth quadrant are labeled in Figure 15(b). Wake contour at different downstream locations are depicted in Figure 16. As seen in the figure, concentrated tip vortex on the retreating side is stronger than on the advancing side. Furthermore, a root vortex structure is gradually formed along the centerline of the wake.

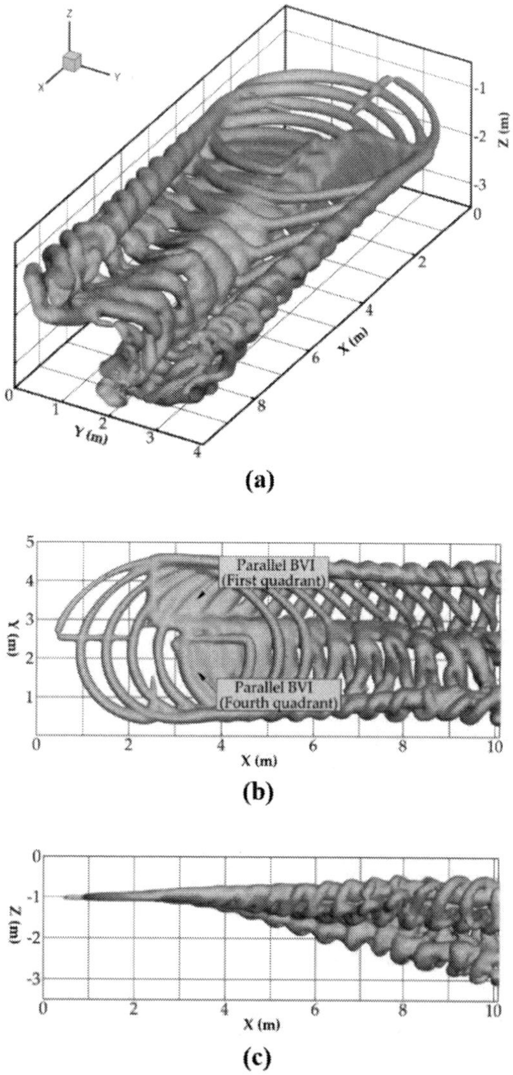

(a)

(b)

(c)

Figure 15. Wake structure for the Helishape 7A rotor in BVI flight condition. (a) isometric view; (b) top view; (c) side view.

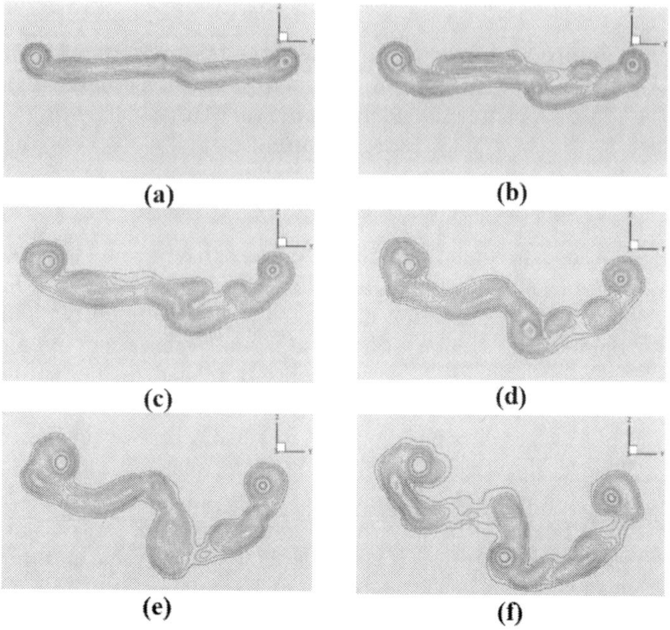

(a) (b) (c) (d) (e) (f)

Figure 16. Wake contour at different downstream locations for Helishape 7A rotor in BVI flight condition. (a) x/R=0.5; (b) x/R=1.0; (c) x/R=1.5; (d) x/R=2.0; (e) x/R=2.5; (f) x/R=3.0.

Since BVI noise prediction is highly depend on the unsteady blade surface aerodynamics, blade pressure distributions at two spanwise locations: r/R = 0.82 and 0.92 are plotted in Figure 17 for azimuth angles of ψ = 0°, 90°, 180°, 270°. For the most part, the agreement between prediction and experimental data (Ref.39) is fairly good. Figure 18 further shows the regarded blade sectional lift, which is obtained by pressure integral along the chordwise direction. As shown, the sectional lift fluctuates dramatically around 50° azimuth angle in the first quadrant and around 300° in the fourth quadrant. Despite the challenges in BVI airload prediction, the gross features of computed results for those two local BVI events are satisfactory. The high-frequency fluctuations of lift are captured well compared with the measurements, with some differences in amplitude and phase. This may be attributed to aeroelastic effects, which become significant for rotor blades with large aspect ratio in BVI flight. In current study, blade elasticity is not modeled in the hybrid solver. Figure 19 gives the comparison of pressure time history at two microphones positioned forward and below the rotor disk plane. The acoustic waveforms correlate well with the experimental data overall, and the underprediction of the peak pressure may be due to the lack of blade elasticity model as aforementioned.

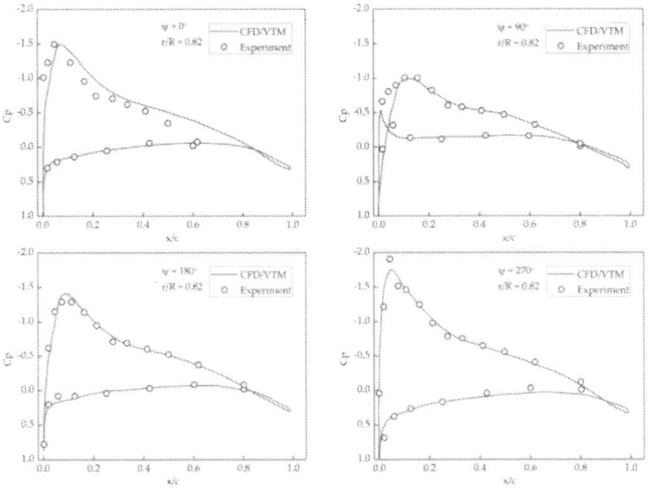

(a)

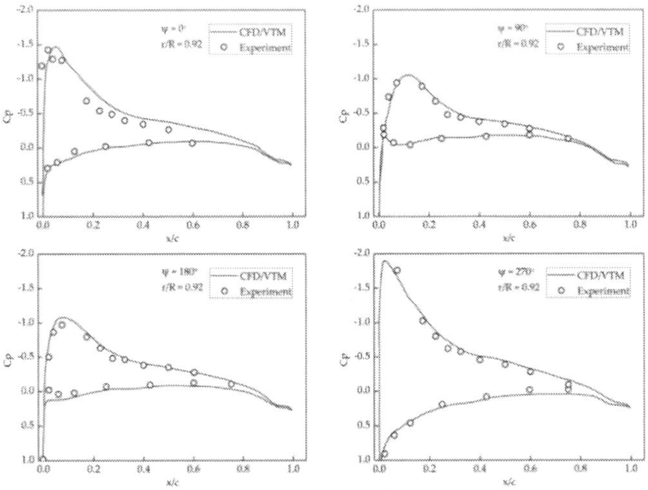

(b)

Figure 17. Blade pressure coefficient for Helishape 7A rotor at various azimuth angles. (a) r/R=0.82; (b) r/R=0.92.

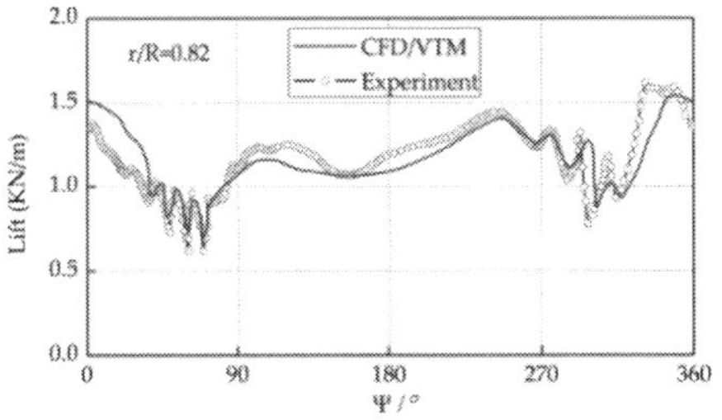

(a)

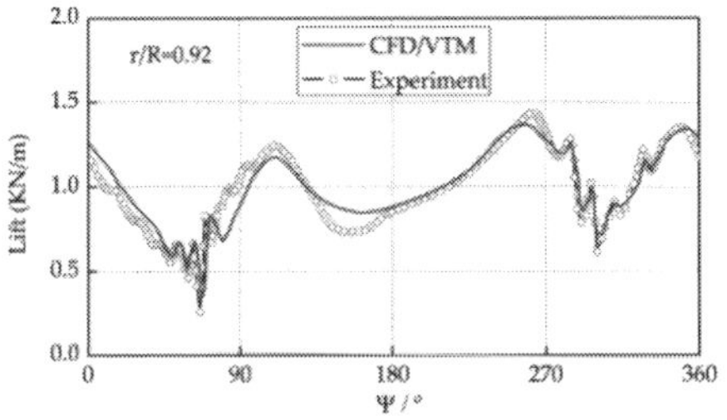

(b)

Figure 18. Time history of blade sectional airload for Helishape 7A rotor in one revolution. (a) r/R=0.82; (b) r/R=0.92.

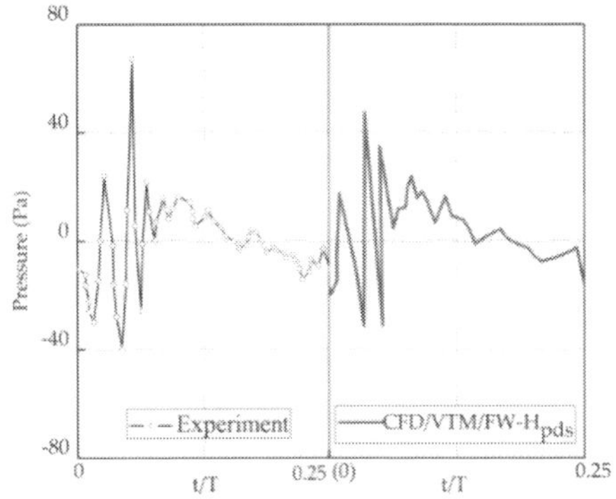

(a)

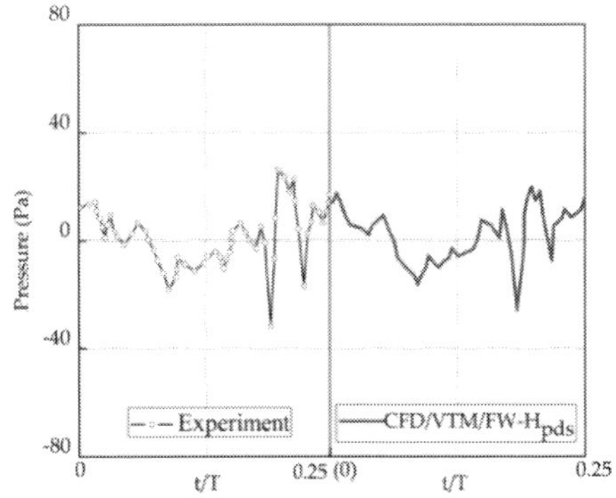

(b)

Figure 19. Comparison of measured and predicted BVI noise signal for Helishape 7A rotor. (a) advancing side microphone, (-0.51, 1.61, -2.29) m; (b) retreating side microphone, (-0.51, -1.61, -2.29) m.

Computational Cost of the Hybrid Solver

All the simulations in this paper were performed on an eight-core processor with CPU speeds of 4.0 GHz. The aerodynamic solver (CFD/VTM) code was parallelized during the computation, and aeroacoustic code was running in serial mode. As described above, HSI case and BVI case correspond to 1° and 0.25° timestep, respectively. In addition, the number of vortex particles used in the off-body flowfield is about 23,000 for HSI case, and 57,000 particles are used for BVI case to fully model the rotor wake. CPU running time in forward flight aerodynamic simulation (CFD/VTM) for a two-bladed rotor is documented in Table 5. Typically, a HSI case occupies 2.4 hours of total wall clock time, and a BVI case needs 10.6 hours of total wall clock time due to the demand for accurately capturing the unsteady airload. VTM solver module shows a higher computational efficiency when compared to CFD solver module. Furthermore, acoustic analysis by FW-Hpds at each observer location requires only approximately 37 seconds. In short, the developed hybrid solver is well suited for routine use in helicopter rotor impulsive noise prediction.

Table 5. CPU wall clock time in aerodynamic simulation for a two-bladed rotor.

Solver	CFD (baseline grid)	VTM	Total
HSI case	1.9 hour	0.5 hour	2.4 hour
BVI case	8.1 hour	2.5 hour	10.6 hour

CONCLUSIONS

In this study, Computational Fluid Dynamics method, Vorticity Transport Model and FW-H equation with permeable data surface have been combined together to predict both the high-speed impulsive noise and blade-vortex interaction noise for helicopter rotors. Several numerical simulations have been performed and validated through available experimental data. The following conclusions can be drawn from this paper:

(1) Proposed hybrid method is robust and efficient. The aerodynamic solver module is not only capable of accurately predicting the transonic flow in high-speed flight, but also model the vortical rotor wake and highly unsteady airload at BVI flight condition effectively. Since the dense background grid is no longer needed, it is of moderate computational cost compared with traditional full CFD method. Moreover, the acoustic formulation that applied a permeable data surface is suitable for both BVI and HSI noise prediction. It is suggested that this hybrid solver provides a promising tool for helicopter rotor acoustic analysis.
(2) From the hovering UH-1H model rotor case, it can be seen that the peak sound pressure level of HSI noise is insensitive to the choice of permeable data surface, as long as the transonic flow region is enclosed.
(3) The vortex particles utilized by the VTM can capture the rotor wake without any artificial numerical dissipation. The capability of VTM for vorticity preservation is demonstrated by simulation, and wake structure of Helishape 7A rotor can be clearly identified more than six rotor revolutions with about 57,000 particles spread in the off-body flowfield. Though BVI acoustic characteristics are difficult to predict, we still obtain favorable results due to the application of Vorticity Transport Model.
(4) The Helishape 7A case shows that aeroelastic effects might be significant for rotor blades with large aspect ratio in BVI flight. It is helpful to incorporate blade elastic model into present hybrid solver for further improving the BVI acoustic predicting accuracy in future work.

Author contact:

Liangquan Wang wangliangquan@nuaa.edu.cn
Guohua Xu ghxu@nuaa.edu.cn
Yongjie Shi shiyongjie@nuaa.edu.cn

ACKNOWLEDGMENTS

If used, this should be placed at the end of the paper, before the references. If used, one or more acknowledgments to contributors, a sponsor or technical monitor should be placed at the end of the paper, before the references section. Also include any research or project support or funding sources here.

REFERENCES

[1]Strawn, R. C., Biswas, R., and Garceau, M., "Unstructured adaptive mesh computations of rotorcraft high-speed impulsive noise," *Journal of Aircraft*, 1995, Vol. 32, (4), pp. 754-760.

[2]Datta, A., Sitaraman, J., Chopra, I., and Baeder, J. D., "CFD/CSD prediction of rotor vibratory loads in high-speed flight," *Journal of Aircraft*, 2006, Vol. 43, (6), pp. 1698-1709.

[3]Antoniadis, A. F., Drikakis, D., and Zhong, B., "Assessment of CFD methods against experimental flow measurements for helicopter flows," *Aerospace Science & Technology*, 2012, Vol. 19, (1), pp. 86-100.

[4]Pahlke, K., and Wall, B. G., "Chimera simulations of multibladed rotors in high-speed forward flight with weak fluid-structure-coupling," *Aerospace Science & Technology*, 2005, Vol. 9, (5), pp. 379-389.

[5]Pomin, H., and Wagner, S., "Navier-Stokes analysis of helicopter rotor aerodynamics in hover and forward flight," *Journal of Aircraft*, 2002, Vol. 39, (5), pp. 813-821.

[6]Jarkowski, M., Woodgate, M. A., and Barakos, G. N., "Towards consistent hybrid overset mesh methods for rotorcraft CFD," *International Journal for Numerical Methods in Fluids*, 2014, Vol. 74, (8), pp. 543-576.

[7]Miller, W. O., and Bliss, D. B., "Direct periodic solutions of rotor free wake calculations," *Journal of the American Helicopter Society*, 1993, Vol. 38, (2), pp. 53-60.

[8]Wall, B. G., "Helicopter rotor BVI airloads computation using advanced prescribed wake modeling," 29th AIAA Applied Aerodynamics Conference, Honolulu, USA, June 27-30, 2011.

[9]Biava, M., Bindolino, G., and Vigevano, L., "Single blade computations of helicopter rotors in forward flight," Proceedings of the 41st Aerospace Sciences Meeting & Exhibit, Reno, USA, January 6-9, 2003.

[10]Wie, S. Y., Dong, K. I., Kwon, J. H., and Lee D. J. "Numerical simulation of rotor using coupled computational fluid dynamics and free wake," *Journal of Aircraft*, 2010, Vol. 47, (4), pp. 1167-1177.

[11]Sugiura, M., Tanabe, Y., and Sugawara, H., "Development of a hybrid method of CFD and prescribed wake model for helicopter BVI noise prediction," *Transactions of the Japan Society for Aeronautical & Space*

Sciences, 2013, Vol. 56, (6), pp. 343-350.

[12]Algermissen, G., and Wagner, S., "Computation of helicopter BVI noise by coupling free-wake, Euler and Kirchhoff method," 4th AIAA/CEAS Aeroacoustics Conference, Toulouse, France, June 2-4, 1998.

[13]Cao, Y. H., Yu, Z. Q., Su, Y., and Kang K., "Combined free wake/CFD methodology for predicting transonic rotor flow in hover," *Chinese Journal of Aeronautics*, 2002, Vol. 15, (2), pp. 65-71.

[14]Fletcher, T. M., and Brown, R. E., "Helicopter tail rotor thrust and main rotor wake coupling in crosswind flight," *Journal of Aircraft*, 2010, Vol. 47, (6), pp. 2136-2148.

[15]He, C. J., and Zhao, J. G., "Modeling rotor wake dynamics with viscous vortex particle method," *AIAA Journal*, 2012, Vol. 47, (4), pp. 902-915.

[16]Shi, Y. J., Xu, G. H. and Wei P., "Rotor wake and flow analysis using a coupled Eulerian-Lagrangian method," *Engineering Applications of Computational Fluid Mechanics*, 2016, Vol. 10, (1), pp. 384-402.

[17]Williams, J. E. F., and Hawkings, D. L., "Sound generated by turbulence and surfaces in arbitrary motion," *Philosophical Transactions of the Royal Society A*, 1969, Vol. 264, pp. 321-342.

[18]Morgans, A. S., Karabasov, S. A., Dowling, A. P., and Hynes, T. P., "Transonic helicopter noise," *AIAA Journal*, 2012, Vol. 43, (7), pp. 1512-1524.

[19]Farassat, F., "Linear acoustic formulas for calculation of rotating blade noise," *AIAA Journal*, 1981, Vol. 19, (9), pp. 1122-1130.

[20]Brentner, K. S., "Prediction of helicopter discrete frequency rotor noise - A computer program incorporating realistic blade motions and advanced formulation," NASA TM-87721, 1986.

[21]Strawn, R. C., Biswas, R., and Lyrintzis, A. S., "Helicopter noise predictions using Kirchhoff methods," *Journal of Computational Acoustics*, 2011, Vol. 4, (3), pp. 321-339.

[22]Brentner, K. S., "An efficient and robust method for predicting helicopter rotor high-speed impulsive noise," *Journal of Sound & Vibration*, 1997, Vol. 203, (1), pp. 87-100.

[23]Francescantonio, P. D., "A new boundary integral formulation for the prediction of sound radiation," *Journal of Sound & Vibration*, 1997, Vol. 202, (4), pp. 491-509.

[24]Brentner, K. S., and Farassat, F., "Analytical comparison of the acoustic analogy and Kirchhoff formulation for moving surfaces," *AIAA Journal*, 1998, Vol. 36, (8), pp. 1379-1386.

[25]Koushik, S., and Schmitz, F. H., "Understanding in-plane helicopter Blade-Vortex Interaction (BVI) noise," American Helicopter Society 68th Annual Forum, Fort Worth, USA, May1-3, 2012.

[26]Ikeda, T., Enomoto, S., and Yamamoto, K., "Quadrupole effects in the Ffowcs Williams-Hawkings equation using permeable control surface," 18th AIAA/CEAS Aeroacoustics Conference, Colorado Springs, USA, June 4-6, 2012.

[27]Shi, Y. J., Zhao, Q. J., Xu, G. H., "An analytical study of parametric effects on rotor–vortex interaction noise," *Proceedings of the Institution of Mechanical Engineers Part G Journal of Aerospace Engineering*, 2011, Vol. 225, (3), pp. 259–268.

[28]Roe, P. L., "Approximate Riemann solvers, parameter vectors, and difference schemes," *Journal of Computational Physics*, 1981, Vol. 43, pp. 357-372.

[29]Spalart, P. R., and Allmaras S. R., "A one-equation turbulence model for aerodynamic flows," 30th Aerospace Sciences Meeting and Exhibit, Reno, USA, January 6-9, 1992.

[30]Jameson, A., and Yoon, S., "Lower-Upper implicit schemes with multiple grids for the Euler equations," *AIAA Journal*, 1987, Vol. 25, (7), pp. 929-935.

[31]Brown, R. E., and Line, A. J., "Efficient high-resolution wake modeling using the vorticity transport equation," *AIAA Journal*, 2005, Vol. 43, (7), pp. 1434-1443.

[32]Zhao, J. G., and He, C. J., "Enhancements of viscous vortex particle method for fundamental rotor wake dynamics simulation," 50th AIAA Aerospace Sciences Meeting and Exhibit, Nashville, USA, January 9-12, 2012.

[33]Eldredge, J. D., Leonard, A., and Colonius, T., "A general deterministic treatment of derivatives in particle methods," *Journal of Computational Physics*, 2002, Vol. 180, (2), pp. 686-709.

[34]Lighthill, M. J., "On sound generated aerodynamically. I: general theory," *Proceedings of the Royal Society of London*, 1952, Vol. 211, pp. 564-587.

[35]Hanson, D. B., and Fink, M. R., "The importance of quadrupole sources in prediction of transonic tip speed propeller noise.," *Journal of Sound & Vibration*, 1979, Vol. 62, pp. 19-38.

[36]Boxwell, D. A., Yu, Y. H., and Schmitz F. H., "Hovering impulsive noise - Some measured and calculated results," *Vertica*, 1978, Vol. 3, pp. 35-45.

[37]Boxwell, D. A., Schmitz, F. H., Splettstoesser, W. R., and Schultz K. J., "Model helicopter rotor high-speed impulsive noise: measured acoustics and blade pressures," NASA TM-85850, 1983.

[38]Boxwell, D. A., Schmitz, F. H., Splettstoesser, W. R., and Schultz K. J., "Helicopter model rotor-blade vortex interaction impulsive noise: scalability and parametric variations," *Journal of the American Helicopter Society*, 1987, Vol. 32, (1), pp. 3-12.

[39]Splettstoesser, W. R., Junker, B., and Wagner, W., "A parametric wind tunnel test on rotorcraft aerodynamics and aeroacoustics (Helishape)-test procedures and representative results," *Aeronautical Journal*, 1997, Vol. 101, pp. 143-154.

Suppression on Rotor BVI Noise by IBC Active Control Methodology

Ni Tong-Bing *Zhao Qi-Jun* *Ma Li*
Ph. D student Professor Ph. D studet
ntb@nuaa.edu.cn zhaoqijun@nuaa.edu.cn lim@nuaa.edu.cn
(National Key Laboratory of Rotorcraft Aeromechanics, Nanjing University of Aeronautics and Astronautics, Nanjing, 210016)

Abstract

A CFD/CSD/FW-Hpds integrated method is established to investigate the individual blade control (IBC) technique for reducing the rotor blade-vortex interaction (BVI) noise. The Navier-Stokes equations are adopted to simulate the rotor flowfield for the noise computation. To improve computational accuracy and efficiency of the present numerical methods, Roe-MUSCL scheme, S-A turbulence model and implicit LU-SGS scheme are employed. Combined with CFD solver, a CFD/CSD coupling strategy is developed to calculate viscous flow around the rotor and elastic deformation of blades. The CSD module is developed based on Hamilton's variational principles. Based upon the simulated flowfield by CFD/CSD coupling method, numerical calculations on aeroacoustic characteristics of rotor are conducted by the FW-Hpds equations with the penetrable integral surface. The aeroacoustic characteristics of the UH-60A rotor in oblique descending flight with and without IBC control are calculated with different control parameters under different flight conditions. Comparisons of wavelet time-scale graphs with and without IBC control show that the characters of the BVI noise signal is obvious without IBC control and the rotor noise hotspots can be weakened with IBC control while the rotor BVI noise signal and energy can be effectively attenuated. Comparisons on the effects of different control parameters such as phase angle and frequency show that the BVI noise sound pressure level (SPL) can be reduced more than 5.5dB by selecting reasonable IBC control inputs and the BVI noise reduction effect of second-order harmonic control is better than that of the fourth-order harmonic control for a four-bladed rotor. The IBC control effect shows that the BVI noise can be effectively reduced under different flight conditions.

Key words: rotor; blade-vortex interaction noise; individual blade active control; CFD/CSD coupling method; parameter influence

1 Introduction

BVI noise is a very high amplitude impulsive sound, which will dominate the helicopter noise when it happens [1-2]. The accurate prediction of rotor BVI noise and the effective noise suppression method to reduce rotor BVI noise have become the main research content in the field of rotor acoustic investigation.

The mechanism of BVI noise has many influence factors such as the vortex strength, blade loads, miss-distance and the interaction angle between the vortex and the blade. Several passive noise reduction concepts have been investigated for diffusing tip vortices and reducing the blade loads to reduce BVI noise [3-5], which show that these tip design concepts could reduce BVI noise. However, the aerodynamic shape and interior structure of blade are greatly changed. Sometimes, the blade tip design method has not reduced BVI noise SPL without rotor performance degradation. At the same time, active blade control techniques such as the individual blade control (IBC) have been investigated for noise and vibration reduction by modifying the rotor wake and the blade motion and deflections, which would not change the shape of helicopter blade.

The flight tests [6-10] show that in the oblique descending flight and landing, the helicopter blade root pitch active control technology is effective to reduce BVI noise. The experiment of an IBC system in the 40×80 foot NASA Ames wind tunnel showed that IBC can reduce BVI noise greatly. The BO-105 helicopter flight tests [11-12] and the full-scale tests in the NASA Ames wind tunnel [13] have shown a noise benefit exceeding to 6dB. Splettstoesser et al. [14] conducted a collaborative flight test to reduce the external noise of the helicopter in oblique descending flight with IBC method, the result showed that the noise reduction benefit was obtained significantly and the SPL can be reduced by 5dB. However, all of the researches were conducted by experimental methods. Wind tunnel measurements of performance, loads, and vibration of a full-scale UH-60A main rotor with the IBC system are compared with the calculations using the coupled CAMRAD II/OVERFLOW 2 analyses by Yeo [15], which showed that IBC technique can be conducted effectively by analysis methods, but the noise reduction analyses were not conducted. In addition, it is controversy whether the second-order harmonic or the fourth-order harmonic control is better to reduce the BVI noise for a four-bladed rotor by IBC method. Stephen et al. [16] claims that the second-order harmonic control is better and the BVI noise can be reduced by up to 12 dB using 3.0° of 2/rev IBC while Feng [17] claims that the fourth-order harmonic control is the most obvious control in each harmonic control, which need to be analyzed and verified specially.

In view of the complex of noise prediction method by IBC active control technology with too many influence factors, an integrated numerical simulation method based on CFD/CSD/FW-Hpds equations is adopted. On the present numerical method basis, the effect investigations of different IBC control parameters such as different phase angle show that the BVI noise could be effectively reduced under different flight conditions with reasonable IBC control, which provide a theoretical basis for IBC control application to reduce helicopter rotor BVI noise.

2 CFD/CSD/FW-Hpds Integrated Method

2.1 Grid Generation

Considering the periodic motions of blade, a moving-embedded grid generation method is required to obtain accurate airloads and noise sources information of rotor. The grid system mainly consists of two components: the body-fitted and body-orthogonal C-O grids around the blade and the background grids.

Considering the complex geometry of rotor blade, a parameterized body-fitted grid generation method is established for the blade while the body-fitted grids around some typical airfoils are generated by solving Poisson equations (Equations 1), and the grid generation quality is improved by employing source term correction approach.

$$\begin{cases} \alpha x_{\xi\xi} - 2\beta x_{\xi\eta} + \gamma x_{\eta\eta} = -J^2(x_{\xi}p + x_{\eta}q) \\ \alpha y_{\xi\xi} - 2\beta y_{\xi\eta} + \gamma y_{\eta\eta} = -J^2(y_{\xi}p + y_{\eta}q) \end{cases} \quad (1)$$

Where p, q are the equations source terms to control the grid density and the orthogonality at the wall and the outer boundary in the grid generation process, and

$$\begin{cases} \alpha = x_{\eta}^2 + y_{\eta}^2 \\ \beta = x_{\xi}x_{\eta} + y_{\xi}y_{\eta} \\ \gamma = x_{\xi}^2 + y_{\xi}^2 \end{cases} .$$

Then, the grids around the blade are generated by stretching between the blade typical sections and folding at the blade tip and root respectively. Finally, grid refinement is quickly fulfilled by solving Poisson equations considering the chord length, quarter-chord line and twist distributions of blade. The grids are properly refined within and near the boundary layer.

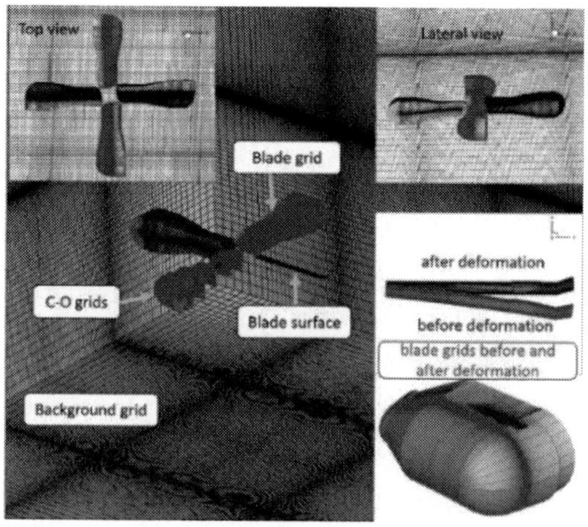

Fig.1 Overall schematic of a moving-embedded grid system

The background grids are the Cartesian grids. Fig.2 shows the overall schematic of the moving-embedded grid system. Considering the elastic blade, the comparison of blade grids before and after the deformation are given in this figure.

The moving-embedded grid generation method is established to adapt the motion characters of blades including flapping, pitching and rotation with elastic deformation, the hole-map method is simplified in vertical direction to increase the efficiency by identifying the hole cell and hole boundary in background grids [18-19]. When the azimuthal angle is updated, the hole cell can be re-recognized and re-marked. The inverse map is established in the similar algorithm.

2.2 CFD/CSD Coupling Method

In order to obtain accurate airloads to preciously estimate the BVI noise sources information, the unsteady, three-dimensional flowfield details are required. And rotor blades are highly flexible, structural deformations have significant influences on rotor aerodynamic characteristics. Therefore, the CFD code and CSD code interact with each other in the prediction of rotor aerodynamic characteristics.

The Navier-Stokes governing equations in the conservative form are:

$$\frac{d}{dt}\iiint_V \vec{Q}dv + \iint_S \vec{F}\cdot\vec{n}ds - \iint_S \vec{F}_v\cdot\vec{n}ds = 0 \qquad (2)$$

Where

$$\vec{Q}=\begin{bmatrix}\rho\\ \rho u\\ \rho v\\ \rho w\\ \rho E\end{bmatrix} \quad \vec{F}=\begin{bmatrix}\rho V_r\\ \rho u V_r+n_x p\\ \rho v V_r+n_y p\\ \rho w V_r+n_z p\\ \rho H V_r+V_t p\end{bmatrix} \quad \vec{F}_v=\begin{bmatrix}0\\ n_x\tau_{xx}+n_y\tau_{xy}+n_z\tau_{xz}\\ n_x\tau_{yx}+n_y\tau_{yy}+n_z\tau_{yz}\\ n_x\tau_{zx}+n_y\tau_{zy}+n_z\tau_{zz}\\ n_x\varphi_x+n_y\varphi_y+n_z\varphi_z\end{bmatrix}.$$

$\vec{Q}$ is the conservation variable, $\vec{F}$ is the modified-convective flux , $\vec{F}_v$ is the viscous flux, V is the volume of a grid cell, and S is the boundary of a grid cell. $V_r = n_x u + n_y v + n_z w - V_t$, $V_t = n_x\partial x/\partial t + n_y\partial y/\partial t + n_z\partial z/\partial t$ is the contravariant velocity of cell surface, n_x , n_y and n_z are the unit normal vector of dS respectively, the viscosity terms are $\tau_{xx} = 2\mu u_x - (2/3)\mu\nabla q$, $\tau_{xy} = \mu(u_y + v_x)$, $\varphi_x = u\tau_{xx} + v\tau_{xy} + w\tau_{xz} + K\partial T/\partial x$.

In spatial discretization, the third-order MUSCL interpolation scheme and flux-difference splitting Roe scheme [20-21] are adopted, the monotone upwind interpolation method is used to calculate the surface around the variable unit:

$$\begin{aligned} U_R &= U_{i+1} - \frac{1}{4}[(1+\kappa)\Delta_- + (1-\kappa)\Delta_+]U_{i+1} \\ U_L &= U_i + \frac{1}{4}[(1+\kappa)\Delta_+ + (1-\kappa)\Delta_-]U_i \end{aligned} \qquad (3)$$

Where $\Delta_+ = \dfrac{U_{i+1}-U_i}{U_{i+1}+(U_{i+1}-U_i)/2}, \Delta_- = \dfrac{U_i - U_{i+1}}{U_i-(U_i-U_{i+1})/2}$,

K is a free parameter, take $\kappa = \dfrac{1}{3}$ corresponding to the third-order upwind scheme.

The dual-time stepping method is adopted while the implicit LU-SGS scheme is employed to the pseudo-time step. The viscous region of blade is calculated by S-A turbulent model. The inner boundary is no-slip condition for Navier-stokes equations, and the farfield boundary condition is determined by Riemann invariant. To further enhance the computational efficiency, an appropriate parallel strategy is employed.

The CSD module is developed based on Hamilton's variational principle:

$$\int_{t_1}^{t_2}(\delta U - \delta T - \delta W)dt = 0 \qquad (4)$$

Where δU is the variation of strain energy, δT is the variation of kinematic energy, and δW is the virtual work done by external forces which are aerodynamics loads in this case.

When the CFD method and the CSD method are established respectively, a loose CFD/CSD coupling strategy is developed to transfer information between fluid and structure. When the moving-embedded grid system is generated, after the initial calculation, the CSD module is calculated and the rotor blade deformation is obtained firstly. The CFD blade grids are deformed based upon the blade deformation results, and the deformed blade grids are input into the CFD calculation module secondly. Then the CFD calculation results are returned to the CSD module, and the iterative calculation of above steps is implemented until the rotor blades structural deformations and airloads converge.

2.3 FW-Hpds Method

The aeroacoustic characteristics of rotor can be predicted by the Ffowcs Williams-Hawkings equations with penetrable data surface (FW-Hpds equations) in the time domain, which is valid even if the integration surface is in the nonlinear region.

$$p'(x,t) = p'_T(x,t) + p'_L(x,t) \tag{5}$$

$$\begin{aligned} 4\pi p'_T(x,t) &= \int_f \left[\frac{\rho_0(\dot{U}_n + U_{\dot{n}})}{r(1-M_r)^2}\right]_{ret} ds \\ &+ \int_f \left[\frac{\rho_0 U_n (r\dot{M}_r + c_0(M_r - M^2))}{r^2(1-M_r)^3}\right]_{ret} ds \end{aligned} \tag{6}$$

$$\begin{aligned} 4\pi p'_L(x,t) &= \frac{1}{c_0}\int_f \left[\frac{\dot{L}_r}{r(1-M_r)^2}\right]_{ret} ds \\ &+ \int_f \left[\frac{L_r - L_i M_i}{r(1-M_r)^2}\right]_{ret} ds \\ &+ \frac{1}{c_0}\int_f \left[\frac{L_r(r\dot{M}_r + c_0(M_r - M^2))}{r^2(1-M_r)^3}\right]_{ret} ds \end{aligned} \tag{7}$$

$$U_i = [1-(\rho/\rho_0)]v_i + (\rho u_i/\rho_0) \tag{8}$$

$$L_i = P_{ij} n_j + \rho u_i (u_n - v_n) \tag{9}$$

Where p'_T is the sound pressure of thickness noise, p'_L is the sound pressure of loading noise, ρ_0 is the air density, c_0 is the sound velocity, r is the distance from the source point to the observation point, Ma is Mach number, Ma_r is the component of Mach number in r-direction, v_i is the moving velocity of integral surface, u_i is the velocity of fluid on the integral surface, ρ is the density of fluid on the integral surface, P_{ij} is the compressive stress tensor, L_r is the component of the blade surface load in r-direction, $(\dot{\ })$ represents the time derivative and the subscript ret indicates retarded time.

2.4 Validation of the Method

2.4.1 Validation of the CFD/CSD Coupling Method

In order to verify the validity of the rotor CFD/CSD coupling method for the aerodynamic performance simulation of the rotor in forward flight, the UH-60A helicopter rotor which has wind tunnel test data and specific structural parameters is selected as a numerical example. The calculation status are as follow: advance ratio $\mu = 0.368$, pitch angle $\theta(t) = 13.55^\circ + 3.39^\circ \cos\psi(t) - 9.62^\circ \sin\psi(t)$, where $\psi(t)$ stands for azimuthal angle. According to the structural characteristics of the UH-60A blade, 50 finite element units are divided along the spanwise direction.

Fig.2 shows the comparisons of the predicted and measured blade surface pressure coefficient distributions of the UH-60A rotor in forward flight [22], while the rotor CFD/CSD coupling predicted data (considering the elastic deformation) and the rigid rotor CFD predicted data are also compared. In Fig.2, r/R represents the relative radius of the blade section. As can be seen from the figure, compared to the rigid rotor CFD predicted data, the CFD/CSD coupling predicted data considering the elastic deformation have better agreements with the experimental data at different azimuthal angles, which indicates that the CFD/CSD coupling method developed in this paper has high precision and can be used to simulate the aerodynamic characteristics of the rotor in forward flight. The CFD/CSD coupling method can provide reliable noise source information for simulating the BVI noise characteristics of the rotor effectively.

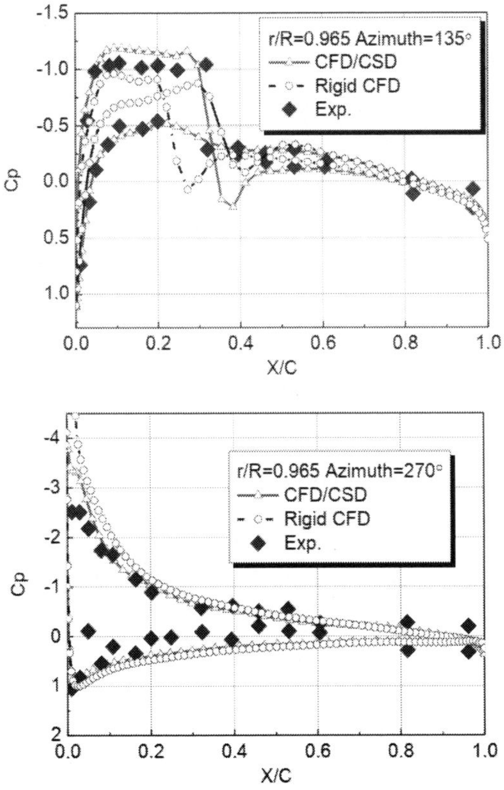

Fig.2 Comparisons of the predicted and measured blade surface pressure coefficient distributions of the UH-60A rotor in forward flight

2.4.2 Validation of the CFD/CSD/FW-Hpds Integrated Method

The noise characteristics of rotor under a BVI test condition 10014 state [23] of the AH-1G/OLS model rotor in forward flight have been simulated. The calculation status are: the blade tip Mach number $Ma_{tip}=0.644$, advance ratio $\mu=0.164$, tip-path-plane angle of attack $\alpha_{TPP}=1^{\circ}$, lift coefficient $C_T=0.0054$, Reynolds number $\mathrm{Re}=1.6\times10^6$. In this state, the rotor not only has periodic variable pitch motion, but also has periodic flapping motion. The pitch and flapping laws of rotor blades are as follows: pitch angle $\theta(t)=6.14^{\circ}+0.9^{\circ}\cos\psi(t)-1.39^{\circ}\sin\psi(t)$, flapping angle $\beta(t)=0.5^{\circ}-1.0^{\circ}\cos\psi(t)$.

In Fig.3, the comparisons of the predicted and measured time histories of sound pressure have been shown, while the position MIC#3 is in the direction of propagation with large BVI noise. The numerical results are identical to the experimental results. The positive peak sound pressure at approximately 45° azimuthal angle is calculated accurately in amplitude and phase, and this is the typical characteristic of the BVI state. It is clearly shown that the method has been proved to be accurate and effective to simulate the aerodynamic and aeroacoustic characteristics, especially the BVI noise characteristics.

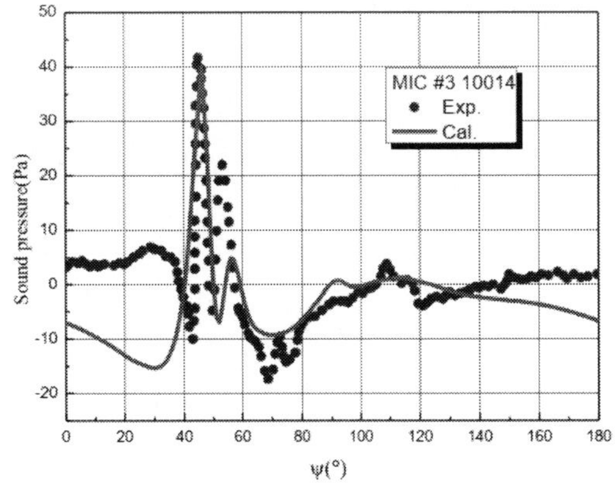

Fig.3 Comparisons of the predicted and measured time histories of sound pressure of the AH-1G/OLS model rotor in forward flight

3 Parameter Influence of IBC Active Control

The aeroacoustic characteristics of the UH-60A rotor in oblique descending flight are calculated with IBC control and without IBC control. The calculation status are: $Ma_{tip}=0.642$, $\mu=0.164$, $\alpha_s=-4^{\circ}$. The trace angle is 10°. The blade pitch angle including IBC input is defined as follows:

$$\theta=\theta_0+\theta_{1C}\cos\psi+\theta_{1S}\sin\psi+\sum_{k=1}^{n}\theta_{IBCk}\cos(k_{IBCk}\psi+\psi_{IBCk}) \quad (10)$$

Where n is the number of harmonics of different frequencies, θ_0 is the collective pitch angle, θ_{1C} is the lateral cyclic pitch angle and θ_{1S} is the longitudinal cyclic pitch angle. If the single frequency harmonic input is used, the formula is simplified as: $\theta=\theta_0+\theta_{1C}\cos\psi+\theta_{1S}\sin\psi+\theta_{IBC}\cos(k_{IBC}\psi+\psi_{IBC})$, where θ_{IBC} is the amplitude input, k_{IBC} is the frequency input, and ψ_{IBC} is the phase angle input.

3.1 Phase influence

Table 1 shows the comparisons of C_T (thrust coefficient) and C_Q (torque coefficient) with and without IBC phase sweeping control, the amplitude is 2° while the frequency order is 2/rev. It can be seen that the thrust coefficient and the torque coefficient remain stable with IBC control, which proves that the aerodynamic performance of the rotor can be guaranteed while the BVI noise is reduced.

Table 1. Comparisons of C_T and C_Q with phase sweeping

Phase angle	C_T (E-03)	C_Q (E-04)
Without IBC	**6.2**	**3.27**
0°	**6.3**	**3.34**
45°	**6.2**	**3.32**
90°	**6.2**	**3.26**
135°	**6.3**	**3.20**
180°	**6.3**	**3.21**
225°	**6.4**	**3.26**

The comparisons of noise sound pressure level under polar coordinate with IBC control (amplitude 2°, phase angle 135°, the frequency order 2/rev) and without IBC control are given in Fig.4. The observer positions are 3.09R away from the disk center and 30° below the rotor disc. It is clearly that the noise radiation is toward to the forward flight direction. With the IBC control, the BVI noise of the rotor in this observation plane is obviously reduced (up to 4dB), especially in the region where the BVI noise radiation is relatively large in front of the rotor.

The wavelet time-scale graphs with and without IBC control are given in Fig.5 while the amplitude is 2° and the frequency order is 2/rev. The abscissa is in one rotation period and the ordinate is the size of the scale. The sixth-order wavelet decomposition is adopted for helicopter rotor noise while the scale a=24. In Fig.5, the characters of the BVI noise signal is obvious without IBC control, the energy is strong and corresponding to the time domain impulse. As can be seen, the BVI noise signal and energy is obviously weakened with the IBC control.

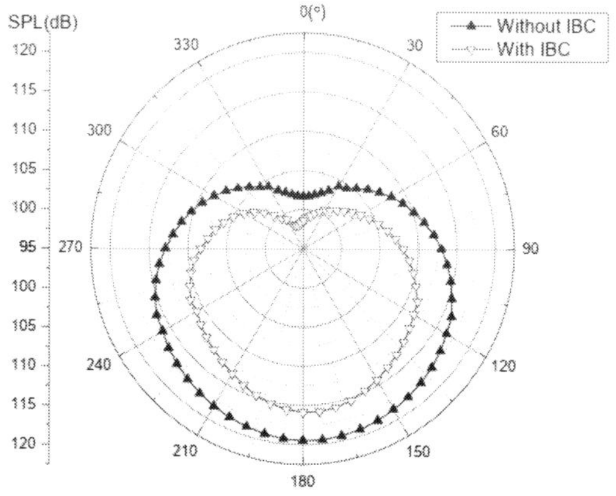

Fig.4 Comparisons of noise sound pressure level under polar coordinate with and without IBC control

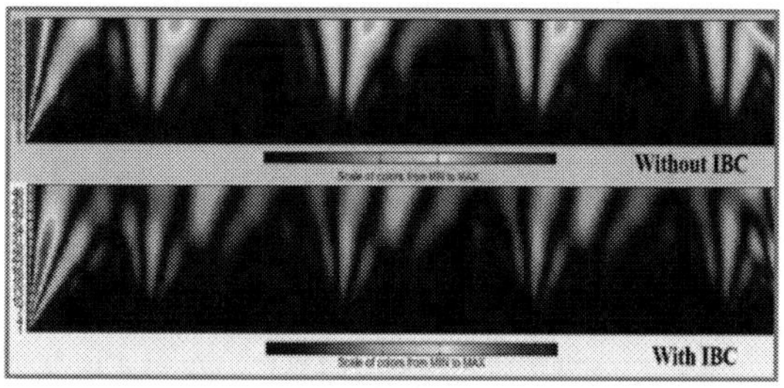

Fig.5 Comparisons of wavelet time-scale graphs with and without IBC control

Fig.6 shows the comparisons of noise SPL with different phase angles. The IBC active control amplitude is 2°, the frequency order is 2/rev, and the phase angles are swept from 0° to 360°. As can be seen from Fig.8, the BVI noise in the advancing side can be greatly reduced while selecting reasonable phase angles (such as from 100° to 200°). It is obvious that a significant reduction of the predicted BVI noise level is achieved with a minimum value at the phase of 135° and 150° in the advancing side, where the noise reduction benefit of about 5.5 dB is obtained. At the same time, it can be seen that the BVI noise can also be reduced in the retreating side while selecting reasonable phase angles, when the phase angle reaches to 210°, the BVI noise can be reduced by 2dB.

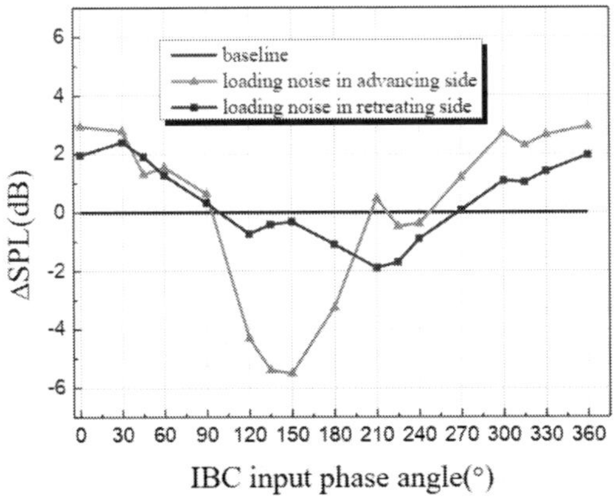

Fig.6 Comparisons of noise SPL with different IBC phase angle inputs

The comparisons of time-variation C_T with different phase sweeping are given in Fig.7, which show that the large fluctuations of time-variation C_T tend to be stable with IBC phase angle control from 90° to 180° while the noise SPL reduced and the average values remain constant. The periodic load of rotor fluctuates moderately and the vibration decreases while the vibration reduction is one of the main purposes of IBC methodology, which show that the BVI noise and the vibration can be decreased simultaneously by reasonable IBC control.

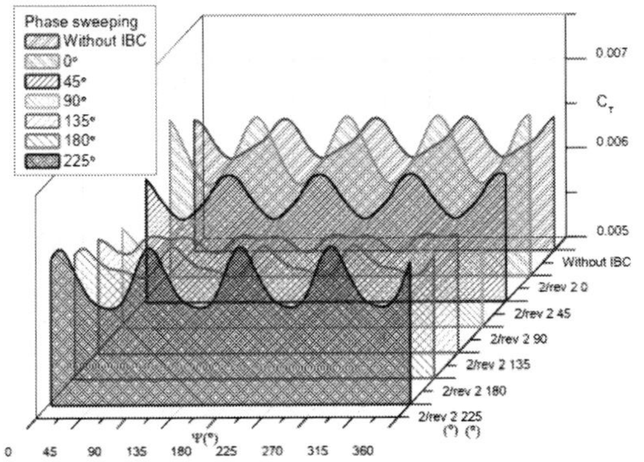

Fig.7 Comparisons of C_T with phase sweeping

3.2 Frequency influence

It is controversy whether the second-order harmonic or the fourth-order harmonic control is better to reduce the BVI noise for a four-bladed rotor [22-23]. In order to solve this question, Fig.8 shows the rotor noise SPL comparisons of 2/rev and 4/rev with different phase angle controls. The results show that the BVI noise reduction effect of second-order harmonic control is better than that of the fourth-order harmonic control for a four-bladed rotor, especially in the retreating side. With 2/rev frequency control, the noise can be reduced in large range of phase angles and BVI noise can be reduced more.

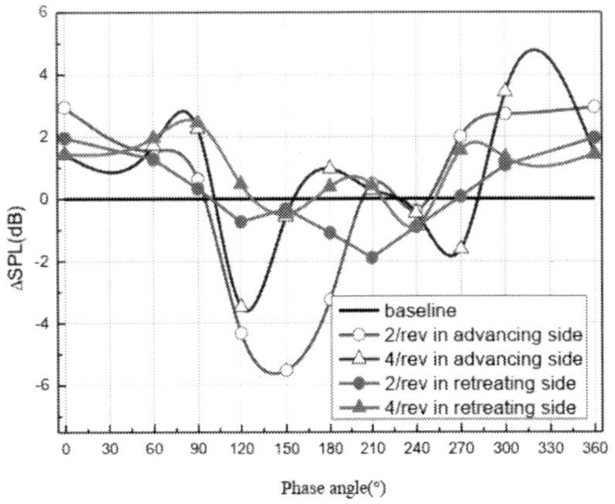

Fig.8 Comparisons of noise SPL with 2/rev and 4/rev frequency inputs

3.3 Descent Angle Influence

Fig.9 shows the comparisons of noise SPL with different descent angle with and without IBC control. The IBC control parameters are: amplitude 2°, phase angle 135°, frequency order 2/rev, and the advance ratio 0.164. The oblique descent flight condition is one of the most easily conditions to generate helicopter BVI noise. As can be seen from the figure, the rotor BVI noise SPL increases with the increase of descent angle without IBC control. When the descent angle reaches to10°, the BVI noise SPL arrives at the maximum value. Then, the BVI noise gradually decreases with the descent angle continues to increase. With IBC control, BVI noise can be effectively reduced under different flight trajectory conditions, especially when the track angle is 10° while the BVI noise SPL arrives at the maximum value. As can be seen, the IBC control method can effectively reduce the BVI noise in the oblique descending flight state.

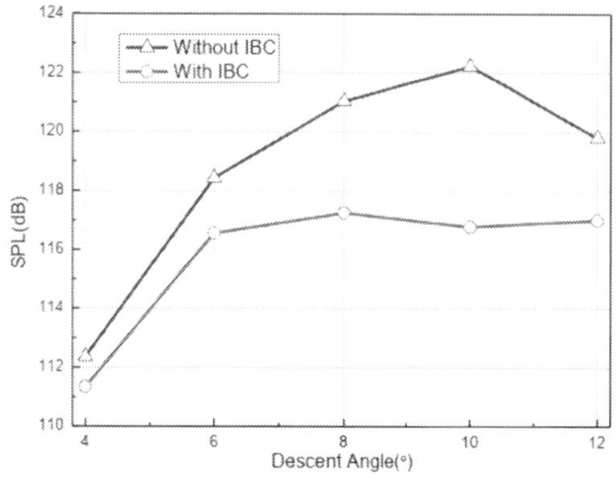

Fig.9 Comparisons of noise SPL with different descent angles

Fig.10 shows the wavelet analysis results of different descent angle states with and without IBC control, the scale a=32. The noise signal and energy at the 10° descent angle are obviously stronger than that of the 4° descent angle. In different descent angle states, the rotor noise hotspots have been weakened with IBC control and the rotor BVI noise signal and energy have been effectively attenuated.

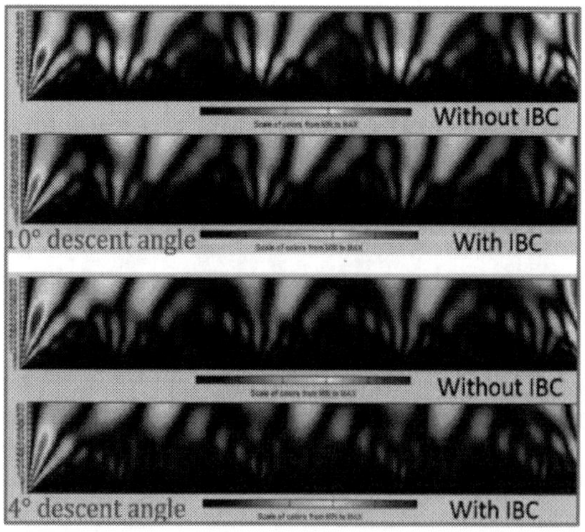

Fig.10 Comparisons of wavelet time-scale graphs with and without IBC control with different descent angles

3.4 Advance Ratio Influence

Fig.11 shows the comparisons of noise SPL with different advance ratios with and without IBC control. The IBC control parameters are: amplitude 2°, phase angle 135°, frequency order 2/rev, and the trace angle is 10°. As can be seen from the figure, the BVI noise increases with the increase of advance ratio without IBC control; when the advance ratio reaches to 0.20, the BVI noise arrives at the maximum value; then, when the advance ratio continues to increase, the BVI noise gradually decreases. That is because when the advance ratio is further increased, the blade tip vortex generated by the preceding blade is rapidly shed from the rotor disk, which reduces the BVI noise SPL. It can be seen that the noise reduction in the advancing side is obvious with IBC control, while the advancing side is the strongest radiation direction of the BVI noise. The BVI noise SPL can be reduced by 7dB at the advance ratio of 0.25. However, the noise in the retreating side is not strong while the noise reduction effect is not obvious. As can been seen from the figure, the noise SPL can be reduced by 2.2dB at the advance ratio of 0.25 in the retreating side.

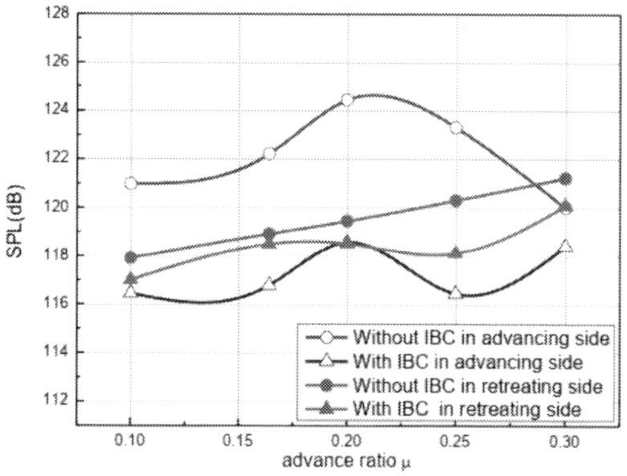

Fig.11 Comparisons of noise SPL with different advance ratios

The comparisons of wavelet time-scale graphs without IBC control with different advance ratios are given in Fig.12 while the scale a=32. It can be seen that the characters of the BVI noise signal are obvious without IBC control, especially when the advance ratio is 0.20, where the energy is strong and the BVI noise is the strongest.

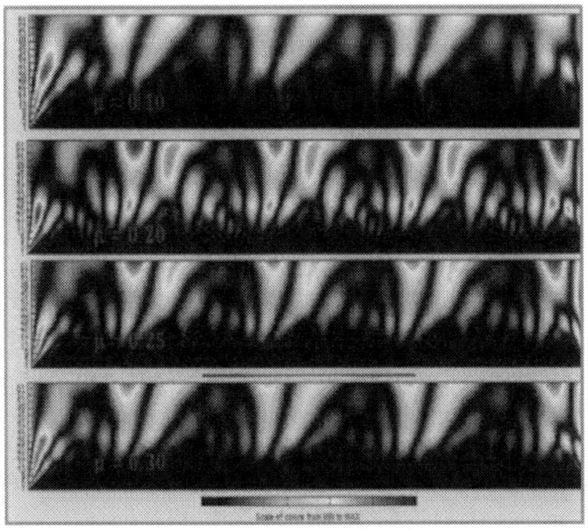

Fig.12 Comparisons of wavelet time-scale graphs without IBC control with different advance ratios

4 Conclusions

(1) The present CFD/SCD/FW-Hpds integrated method has been proved to be effective to simulate the unsteady aerodynamic and acoustic characteristics of rotor in forward flight.

(2) It can be seen from the calculated results that the rotor BVI noise SPL can be reduced more than 5.5dB by selecting reasonable IBC control.

(3) The BVI noise reduction effect of second-order harmonic control is better than that of the fourth-order harmonic control for a four-bladed rotor.

(4) The control effects show that BVI noise SPL can be effectively reduced by IBC control method under different descent angle and different advance ratio.

(5) Comparisons of wavelet time-scale graphs show that the characters of the BVI noise signal is obvious without IBC control in oblique descending flight. The rotor noise hotspots can be obviously weakened with IBC control while the BVI noise signal and energy can be effectively attenuated.

Reference

[1]Megan S M. Helicopter Blade-Vortex Interaction Noise with Comparisons to CFD Calculations [C]. NASA Technical Memorandum, 110423, December 1996.

[2]Yung H Y. Rotor Blade-Vortex Interaction Noise [J]. Progress in Aerospace Science, 2000, 36: 97-115.

[3]Zhao Q J, Xu G H. A Study on Aerodynamic and Acoustic Characteristics of Advanced Tip-shape Rotors [J]. Journal of the American Helicopter Society, 2007, 52(3): 201-213.

[4]Shi Y J，Zhao Q J，Xu G H. An Analytical Study of Parametric Effects on Rotor-Vortex Interaction Noise [J]. Proceedings of the Institution of Mechanical Engineers Part G Journal of Aerospace Engineering, 2011, 225(3): 259-268.

[5]Hu H, Jr L A J, Baeder J D. Computational Simulation and Analysis of Double-swept Blade in BVI Noise Reduction [J]. Advances in Engineering Software, 2005, 36(2): 67-76.

[6]Splettstoesser W R, Schultz K J, Kube R, et al. A Higher Harmonic Control Test in the DNW to Reduce Impulsive BVI Noise [J]. Journal of the American Helicopter Society, 1994, 39(4):3-13.

[7]Splettstoesser W R, Kube R, Wagner W, et al. Key Results from a Higher Harmonic Control Aeroacoustic Rotor Test (HART) in the German-Dutch Wind Tunnel [J]. Journal of the American Helicopter Society, 1997, 42(1):58-78.

[8]Pengel K, Merjer E, Lehmann G. The Higher-Harmonic Control Aeroacoustic Rotor Test (HART) in the DNW-LLF [R]. AD-N68171-99-M-6269, November, 2000.

[9]Rendy P C, Mark B T, Roberto C. A High-Order, Linear Time-Invariant Model for Application to Higher Harmonic Control and Flight Control System Interaction [R]. NASA/TP-2006-213460, 2006.

[10]Norman T R, Theodore C R, Shinoda P, Fuerst D, Arnold U T P, Makinen S, et al. Full-Scale Wind Tunnel Test of a UH-60 Individual Blade Control System for Performance Improvement and Vibration, Loads and Noise Control [C]. Proceedings of the 65th Annual Forum of American Helicopter Society, Grapevine, TX, May 27-29, 2009.

[11]Niesl G, Swanson S M, Jacklin S A, et al. Effect of Individual Blade Control on Noise Radiation [C]. AGARD Aeroacoustics Conference, Berlin, Germany,

1994.

[12]Richter P, Eisbrecher H D, Kloeppel V. Design and First Tests of Individual Blade Control Actuators [C]. Proceedings of the 16th Annual Forum of European Rotorcraft, Glasgow, U K, 1990.

[13]Swanson S M, Jacklin S A, Blaas A, Kube R, Niesl G. Individual Blade Control Effects on Blade-Vortex Interaction Noise [C]. Proceedings of the 50th Annual Forum of American Helicopter Society, 1994.

[14]Splettstoesser W R, Schultz K J, Wall B, Buchholz H. Helicopter Noise Reduction by Individual Blade Control (IBC)-Selected Flight Test and Simulation Results [C]. RTA-AVT Symposium, Braunschweig, Germany, 8-12 May 2000.

[15]Yeo H, Romander E A, Norman T R. Investigation of Rotor Performance and Loads of a UH-60A Individual Blade Control System [C]. Proceedings of the 66th Annual Forum of American Helicopter Society, 2010.

[16]Stephen A J, Axel H, Gary D S, Thomas R N, Patrick S. Full Scale Wind Tunnel Test of an Individual Blade Control System for a UH-60 Helicopter [C]. Proceedings of the 58th Annual Forum of American Helicopter Society, 2002.

[17]Feng J B, Lu Y, Xu J F, et al. Research on the Effect of Open-loop Active Blade-pitch Control on Rotor BVI Noise Alleviation [J]. Acta Aeronautica Sinica, 2014, 35(11): 2901-2909. (in Chinese).

[18]Aoyama A, Yang C, Saito S. Numerical Analysis of Active Flap for Noise Reduction Using Moving Overlapped Grid Method. Journal of the American Helicopter Society, 2005, 52(3): 189-200.

[19]Li P, Zhao Q J, Zhu Q X. CFD Calculations on the Unsteady Aerodynamic Characteristics of a Tilt-rotor in a Conversion Mode. Chinese Journal of Aeronautics, 2015, 28(6): 1593-1605.

[20]Zhao Q J, Xu G H, Zhao J G. New Hybrid Method for Predicting the Flowfields of Helicopter Rotors [J]. Journal of Aircraft, 2006, 43(2): 372-380.

[21]Benedict M, Jarugumilli T, Lakshminarayan V, Chopra I. Effect of Flow Curvature on Forward Flight Performance of a Micro-Air-Vehicle-Scale Cycloidal-Rotor. AIAA Journal, 2014, 52(6): 1159-1169.

[22]Potsdam M, Yeo H, Johnson W. Rotor Airloads Prediction Using Loose Aerodynamic/Structural Coupling [R]. AD-A523729, 2004.

[23]Yu Y H, Tung C, Gallman J, et al. Aerodynamics and Acoustics of Rotor Blade-Vortex Interactions [J]. Journal of Aircraft, 1995, 32(5): 970-977.

Ni Tong-Bing is a Ph.D. student in aircraft design at Nanjing University of Aeronautics and Astronautics, and his research interests are helicopter CFD, helicopter aerodynamics, and rotor aeroacoustics. **E-mail:** ntb@nuaa.edu.cn

Zhao Qi-Jun is a professor and Ph.D. advisor in the College of Aerospace Engineering at Nanjing University of Aeronautics and Astronautics, where he received his Ph.D. degree in aircraft design. His main research interests are helicopter CFD, helicopter aerodynamics, aerodynamic shape design of rotor blade, active flow control on aerodynamic characteristics of rotor, and rotor aeroacoustics. **E-mail:** zhaoqijun@nuaa.edu.cn

Ma Li is a Ph.D. student in aircraft design at Nanjing University of Aeronautics and Astronautics, and his research interests are helicopter CFD and helicopter aerodynamics. **E-mail:** lim@nuaa.edu.cn

6th Asian-Australian Rotorcraft Forum (ARF) and Heli Japan 2017

Kanazawa, Japan

November 7 – 9, 2017

U.S. ARMY RDECOM

Recent Advances in High Fidelity Rotorcraft Aeromechanics Analyses

AMRDEC

TECHNOLOGY DRIVEN. **WARFIGHTER FOCUSED.**

Dr. Hyeonsoo Yeo

US Army Aviation Development Directorate

Aviation & Missile Research, Development, and Engineering Center

Research, Development, and Engineering Command

Ames Research Center, Moffett Field, CA

Overview

- Introduction of Aviation Development Directorate (ADD)
- Rotorcraft aeromechanics challenges
- Aeromechanics analysis tools
- Current status and capabilities
  - Rotor loads
  - Vibratory hub loads
  - Aeroelastic stability
- Ongoing and future research
- Concluding remarks

2

US Army Aviation Development Directorate (ADD)

AMRDEC

- Army's Aviation S&T (Science and Technology) organization
- Manage and conduct basic research (6.1), applied research (6.2), and advanced technology development (6.3)
- Discover, mature and demonstrate technologies that support desired aviation capabilities

ADD
Aviation Development Directorate

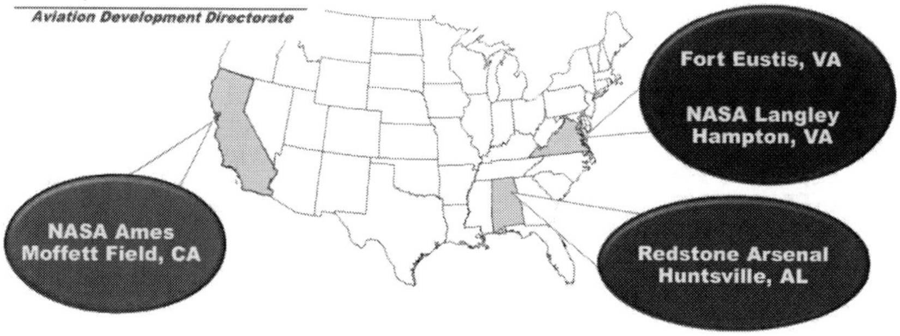

3

Research Activities of Aviation Development Directorate (ADD)

AMRDEC

Wind Tunnel Testing

Flight Testing

Flight Simulation

Concept Design and Analysis

Computational Aeromechanics

4

Rotorcraft Aeromechanics Challenges

Complex, highly interactive, multi-disciplinary phenomena

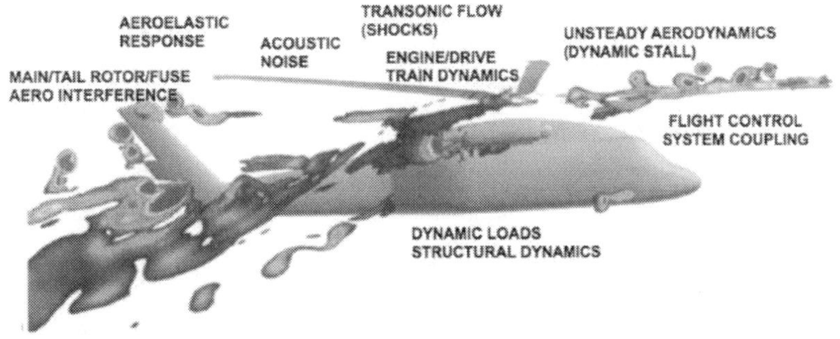

- Rotorcraft aeromechanics have significant influence on performance, loads, vibration, noise, aeroelastic stability of a vehicle
- Accurate prediction of rotorcraft aeromechanics is critical for successful design of future advanced rotorcraft

5

Computational Aeromechanics Software Development

RCAS (Rotorcraft Comprehensive Analysis System)

(HELIcopter Overset Simulations)

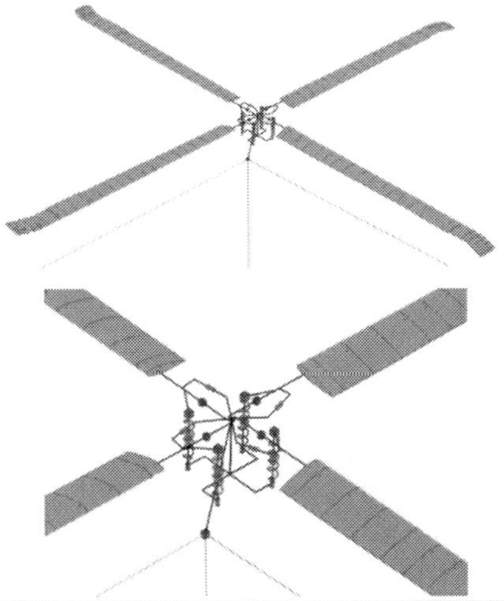

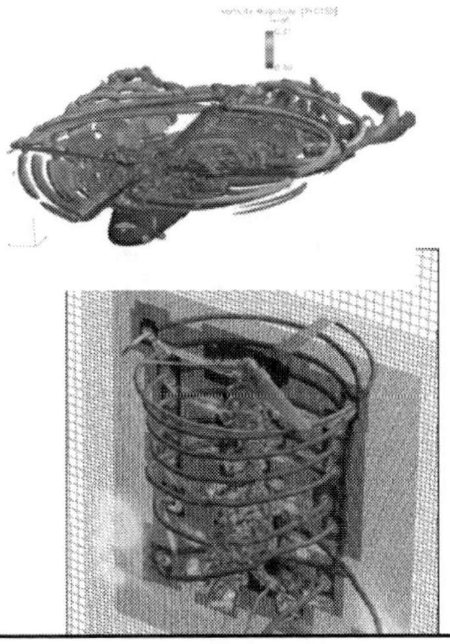

6

RCAS

- Comprehensive code developed and maintained by US Army ADD and Advanced Rotorcraft Technology, Inc. (ART)
- Multi-disciplinary software system capable of modeling a wide range of rotorcraft configurations
- Structural model: finite element, multibody dynamics formulation
  Aerodynamics: a lifting line approach combined with various inflow model
- Continually enhanced and validated
- Extensive usage by US government, industry and academia

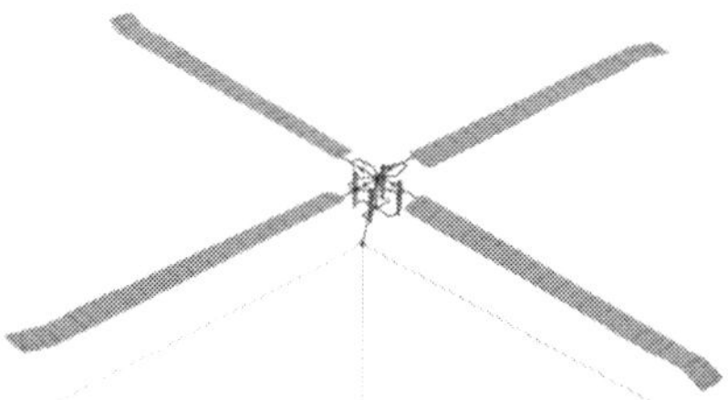

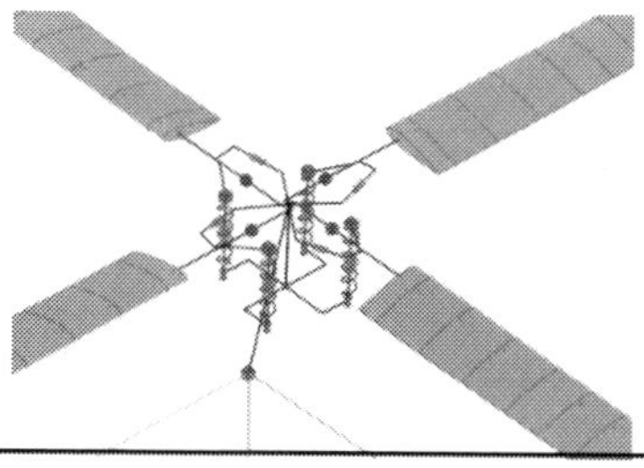

7

Helios

- Rotary-wing product of the CREATE™-AV(air vehicles) program sponsored by the AMRDEC and Department of Defense High Performance Computing (HPC) Modernization Office
- State-of-the-art, accurate, efficient, and easy-to-use modeling and simulation tool
- Continually enhanced and validated

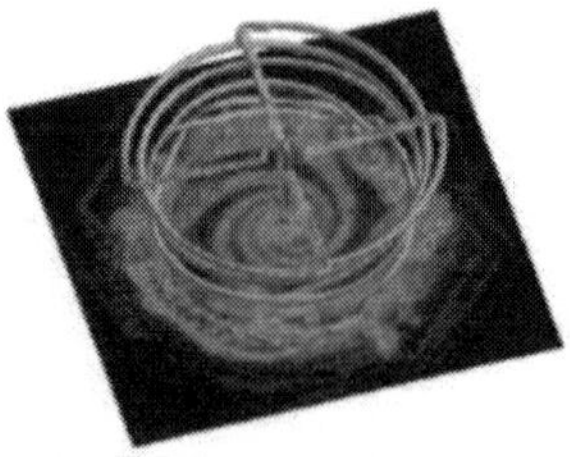

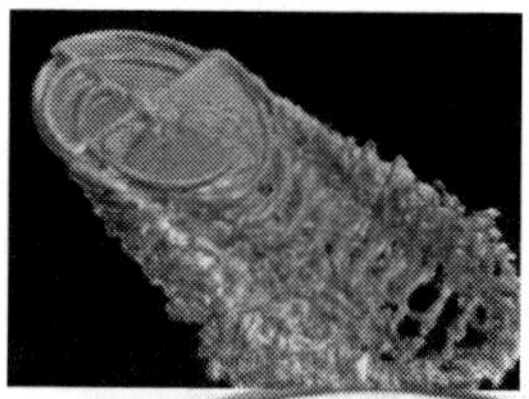

8

Helios Computational Model

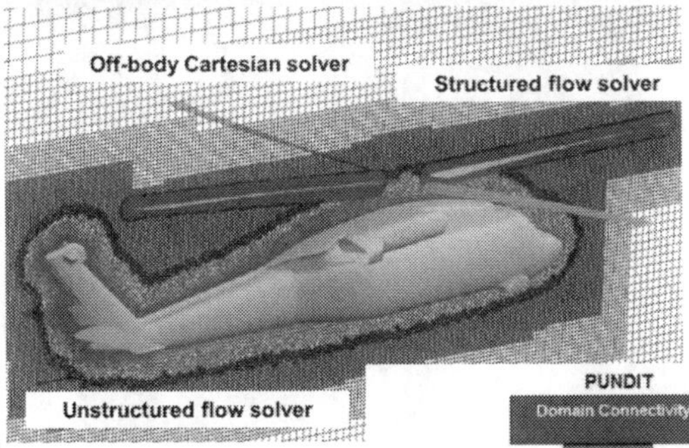

Multi-mesh / multi-solver paradigm

- Near-body solver for complex geometry
- Off-body solver for speed and adaptive mesh refinement
- Automated and efficient interpolation schemes between grids

Component modular architecture

Framework that facilitates rapid introduction of new technology

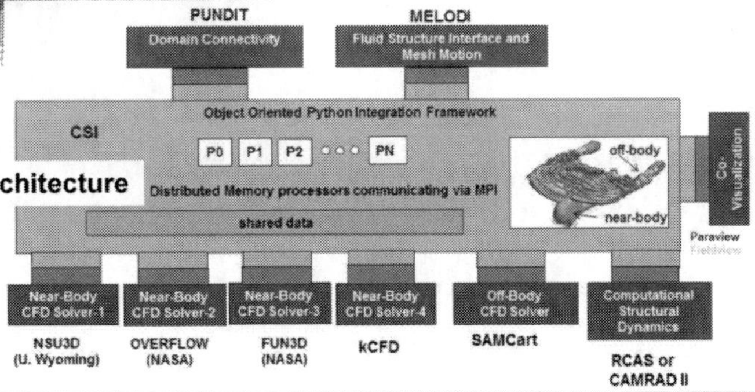

9

CFD/CA Coupling Methods

- Coupled CFD/CA (computational fluid dynamics / comprehensive analysis)
  - CFD solves aerodynamic loads
  - CA solves blade motion and trim

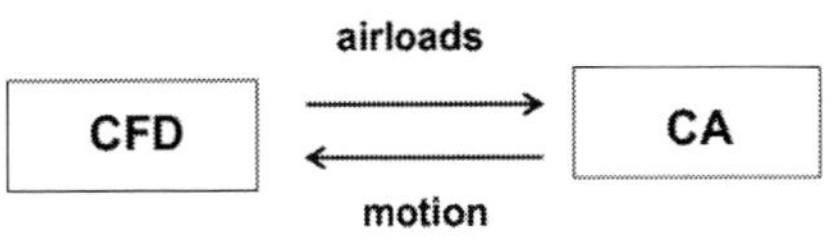

- Loose or "delta" coupling
  - Steady flight
  - Coupling exchanges done on a per revolution, periodic basis
- Tight coupling
  - Maneuver or transient analyses
  - Coupling exchanges at each time step

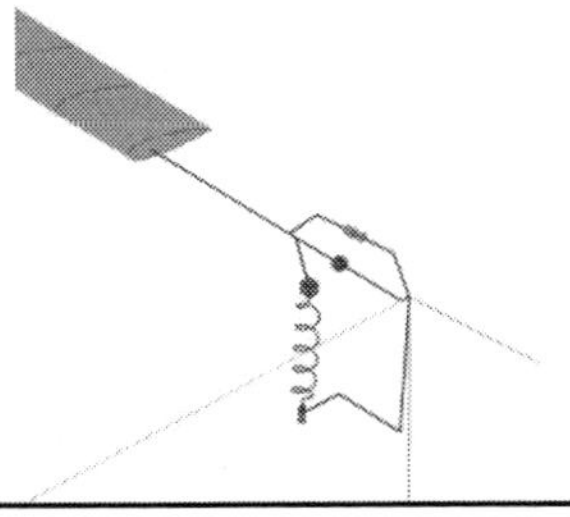

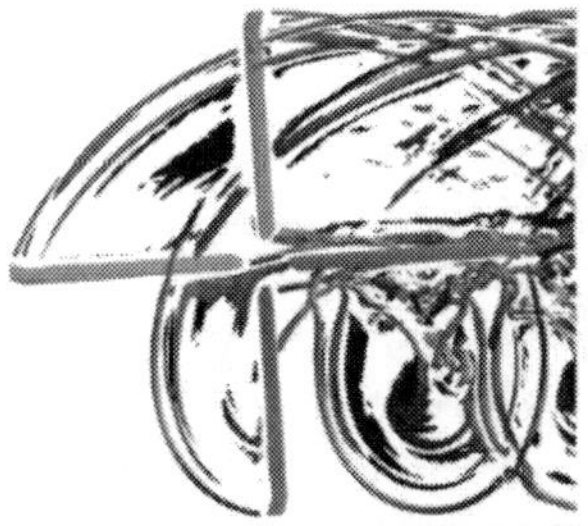

10

Test Data Used for Validation

AMRDEC

- Rotor airloads and structural loads
  - UH-60A flight test
  - ONERA 7A rotor wind tunnel test

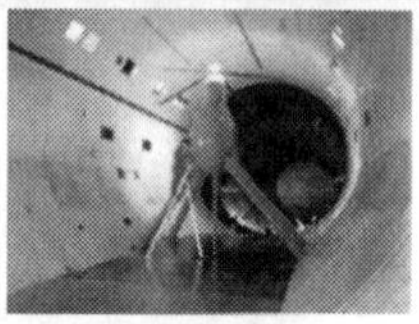

- Vibratory hub loads
  - UH-60A with IBC (individual blade control) wind tunnel test

- Aeroelastic stability
  - Advanced dynamics model blade (straight and swept tip) wind tunnel test

High-quality research data sets are critical for code development/validation

11

NASA/Army UH-60A Airloads Program

AMRDEC

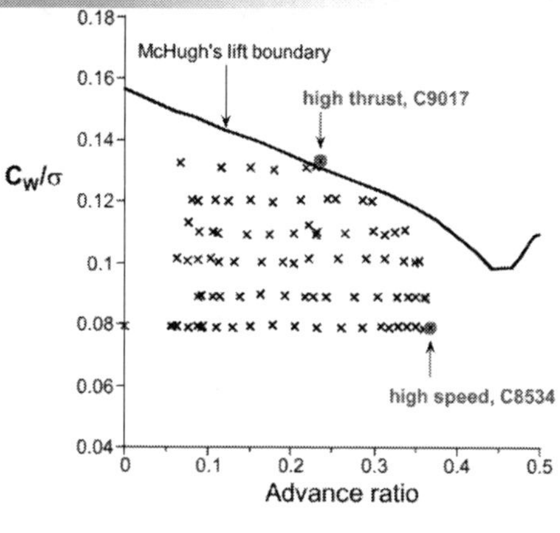

- High Speed (C8534)
  - $\mu = 0.37$, $C_W/\sigma = 0.08$
- High thrust (C9017)
  - $\mu = 0.24$, $C_W/\sigma = 0.13$

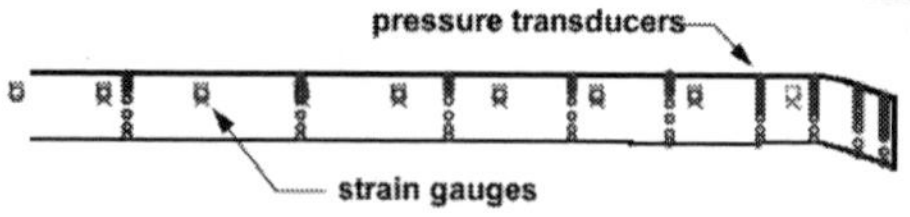

12

Blade Section Airloads (UH-60A, @ 92%R)

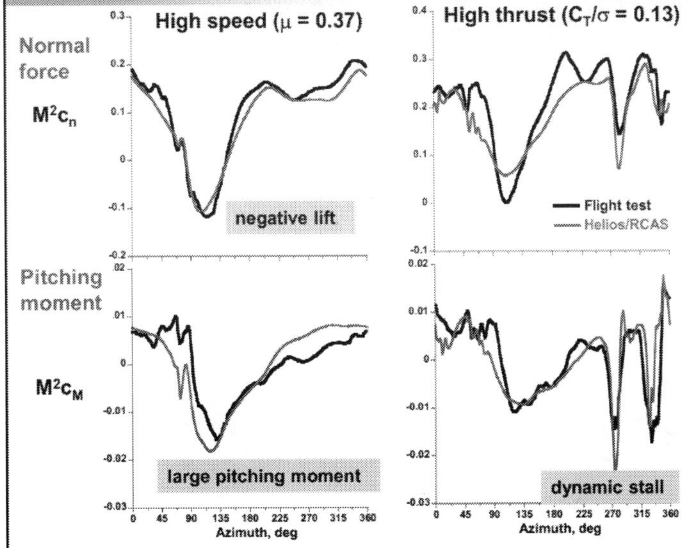

- Coupled analysis captures all the important flow physics
- Coupled analysis shows good agreement with test data

13

Structural Loads (UH-60A)

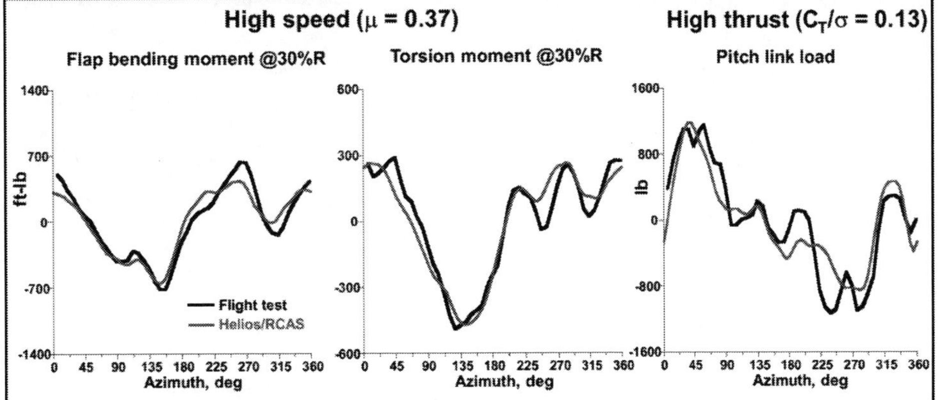

- Accurate prediction of structural loads is important for rotor dynamic component design
- Both amplitude and phase are well predicted

14

Chord Bending Moment (UH-60A)

AMRDEC

High speed (μ = 0.37)

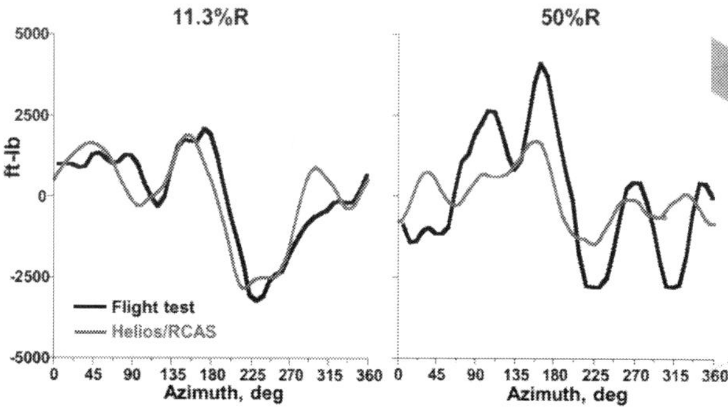

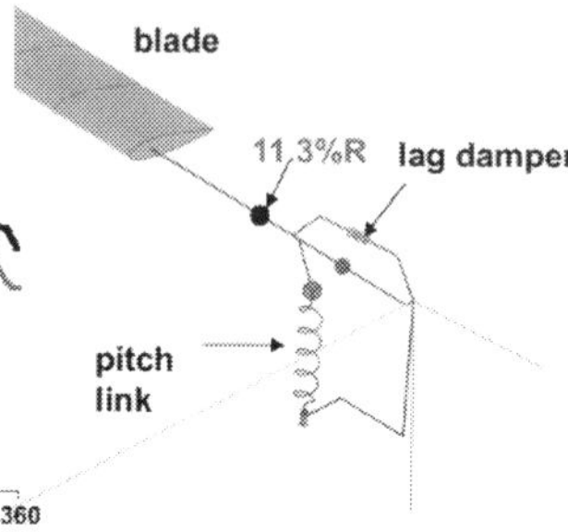

- CBM at 11.3%R is reasonably well predicted with a nonlinear lag damper model
- CBM at 50%R is significantly underpredicted
- Drivetrain dynamics and hub impedance need to be included

15

ONERA 7A Rotor

AMRDEC

- Cooperative research on rotor loads
  - United States/France Project Agreement on Rotorcraft Aeromechanics and Human Factors Integration Research
  - US Army ADD and French ONERA investigated ONERA 7A rotor airloads and structural loads at high-speed and high-thrust wind tunnel test conditions

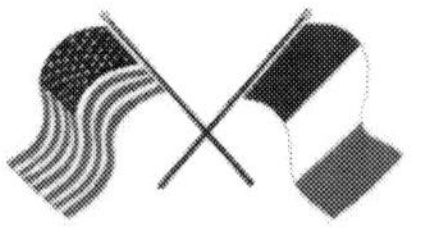

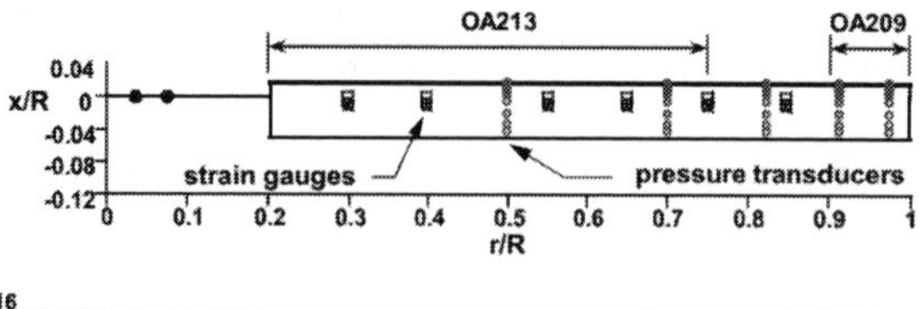

ONERA S1MA transonic wind tunnel

16

Chord Bending Moment (7A @ 40%R)

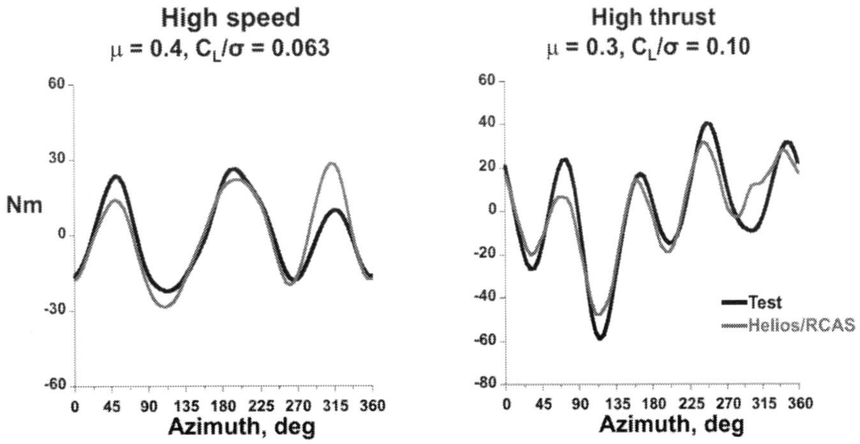

- Very good correlation with test data
- Both waveform and phase are well captured

17

UH-60A with Individual Blade Control (IBC)

- IBC to improve rotor performance and reduce vibration and noise
- German ZFL-designed IBC actuators replaced existing pitch links
- Provides unique vibratory hub loads data

18

18

4/rev Vibratory Hub Loads (0.9 deg 3/rev IBC, μ = 0.35)

AMRDEC

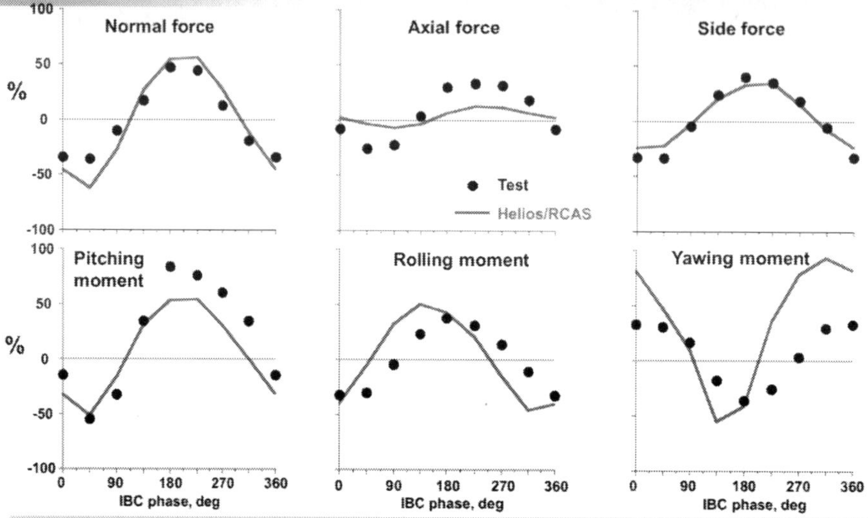

- IBC actuation $\theta = A \cos (N \psi - \phi)$, A:amplitude, N: frequency, ϕ : phase
- Vibratory hub load reduction: 26% (axial force) to 54% (pitching moment)
- Good agreement, considering difficulties in accurate vibratory loads prediction

19

UH-60A UTTAS Pull-up Maneuver

AMRDEC

- Many rotor components and fixed-system controls are sized by the loads in maneuvers
- Rotor loads analysis with prescribed hub motions and blade control angle
- Trim analysis (loose coupling) followed by maneuver analysis (tight coupling)

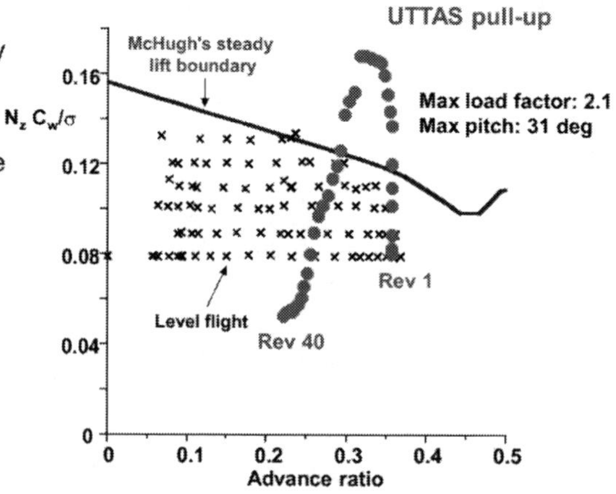

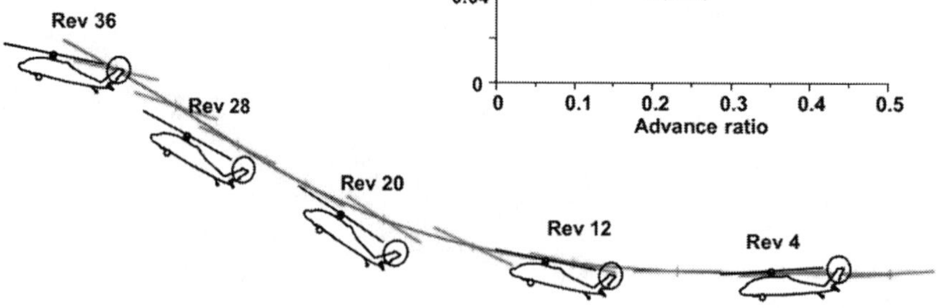

20

UTTAS Maneuver

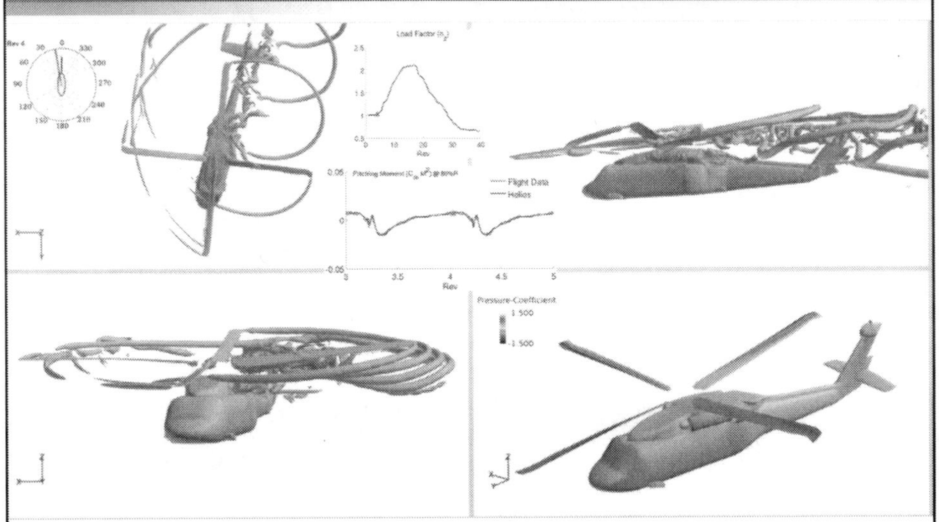

21

UH-60A Pitch Link Load (UTTAS Maneuver)

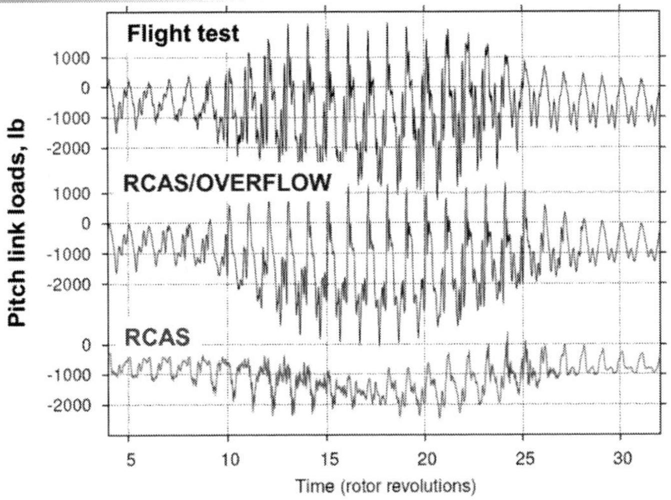

- Dramatic improvement in pitch link load prediction by coupled analysis
- High-fidelity analysis is needed for high-g maneuvers

22

Aeroelastic Stability

AMRDEC

- Aeroelastic instability poses a critical risk in a new rotorcraft development program
- Advanced Dynamic Model (ADM) wind tunnel testing
  - Mach-scaled, soft-inplane hingeless rotor (7.5-ft dia)
  - Regressing lag mode damping measured, μ = hover to 0.45
  - Torsion frequencies:
    2.59/rev for straight blades
    2.75/rev for swept-tip blades

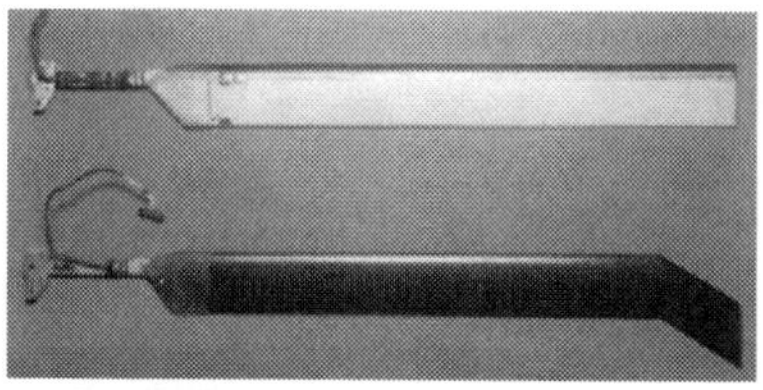

- Stability testing procedure
  - Trim the rotor for a specified operating condition
  - Excite regressing lag mode through swashplate
  - Post-process transient decay chordwise bending moment @ 12%R via moving-block method for lag mode damping

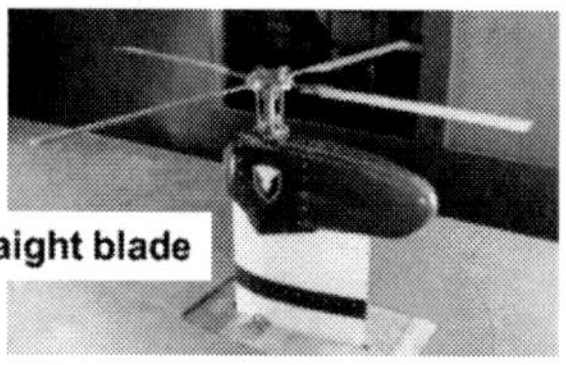

23

Coupled Stability Analysis

AMRDEC

- Coupled CFD/CA approach
  - Trim (loose coupling) ⇨ Excitation (tight coupling) ⇨ Transient response analysis (tight coupling)
  - Identify damping from transient decay response via moving block method

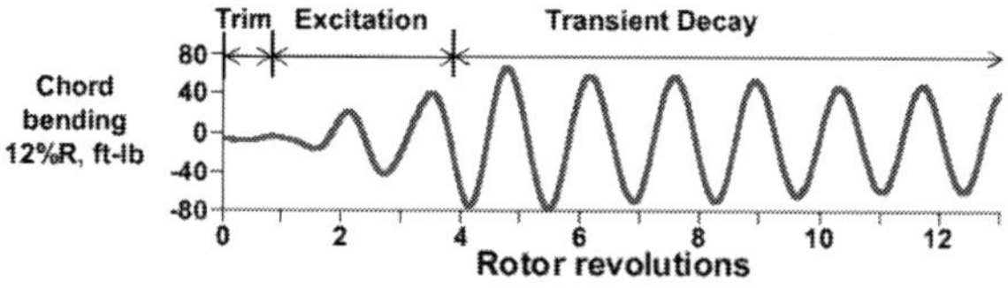

- Challenges of CFD and aeroelastic stability
  - Minimize CFD/CA computational burden for TC transient time histories
    - Effective excitation methods
    - Pure transient decay response
  - Sensitive, lightly damped, hingeless rotor lag mode

24

Cyclic Pitch & Tip Force Excitation

AMRDEC

- Cyclic pitch excitation
  - For wind tunnel test
  - Indirectly generate lead-lag response
  - Small lead-lag response
- Tip force excitation
  - For efficient simulation
  - Apply inplane force to the tip of each rotor blade
  - Large lead-lag response
- 18 revolutions
  - Trim: 1 – 5 rev
  - No excitation: 5 – 6 rev
  - Excitation: 6 – 9.5 rev
  - Decay: 9.5 – 18 rev
- Perturbation motions in rotating frame

Cyclic pitch excitation

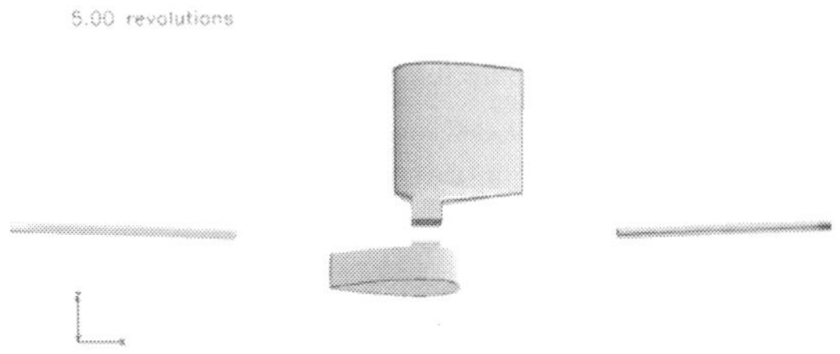

Blade-tip force excitation

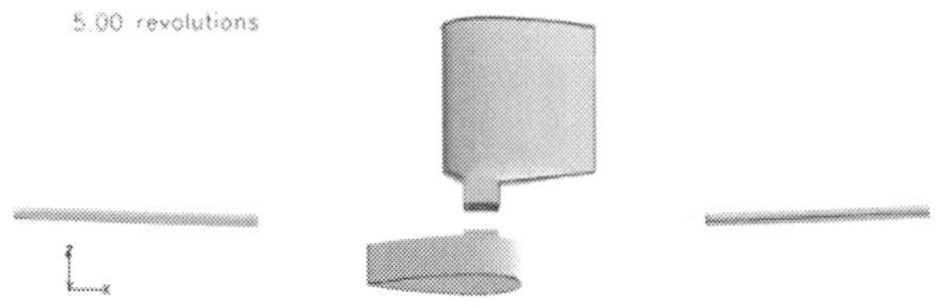

25

Forward Flight Stability
ADM Rotor Blades

AMRDEC

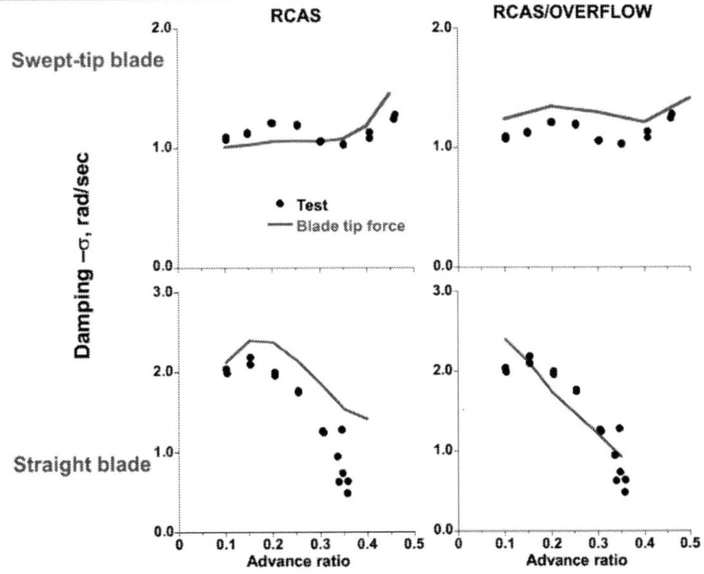

Coupled analysis better captures "up-down-up" damping trend

Coupled analysis better captures "decreasing damping" trend

26

3-D Rotor Structural Dynamics

AMRDEC

Objective

- Develop 3-D FEM and multibody dynamics tool for rotor structural analysis and couple with CFD

Technical challenges

- Efficient 3-D structural mesh generation directly from CAD
- Scalable 3-D structural dynamics solvers
- Consistent and energy conserving fluid-structure interface

On going work

- Collaboration between US Army and University of Maryland

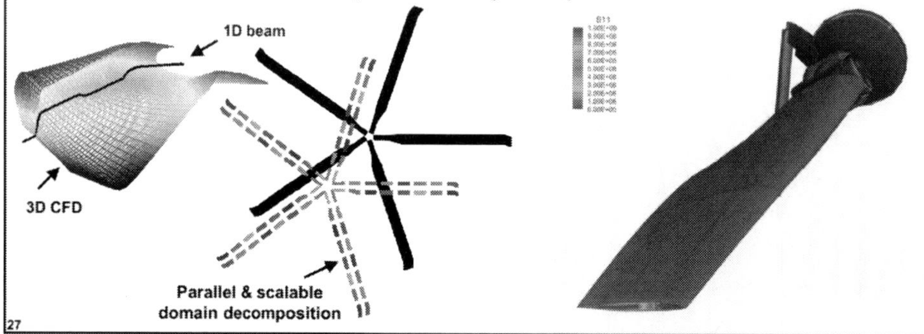

27

Elastic Fuselage Coupling with CFD

AMRDEC

Objective

- Develop coupling methodology for computational structural dynamics (CSD) and computational fluid dynamics (CFD) models of flexible rotorcraft fuselage structure to predict interactional buffet airloads and tail shake

Technical challenges

- Rigorous, consistent and energy conserving fluid-structure interface
- Able to handle incompatible geometries and meshes between CFD and CSD
- Easy-to-use interface with minimal effort from the analyst

On-going work

- Advanced Rotorcraft Technology Inc. and ATA Engineering, Inc. under SBIR

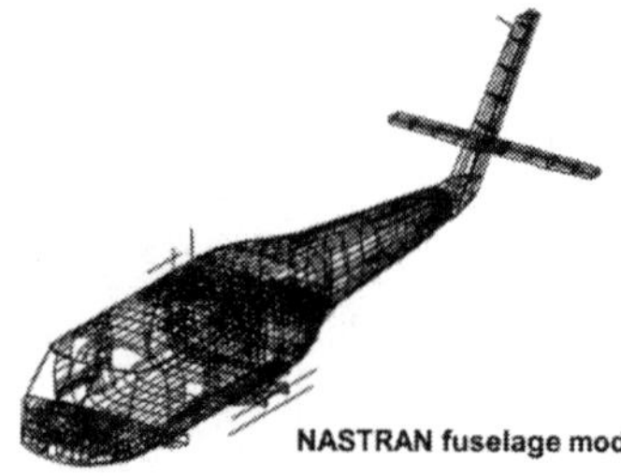

NASTRAN fuselage model

CFD fuselage model

28

Tiltrotor Whirl Flutter

Objective

- Develop high-fidelity coupled rotor/wing model to predict tiltrotor whirl flutter speed

Technical challenges

- Modeling of flexible wing as well as rotor with arbitrary motion

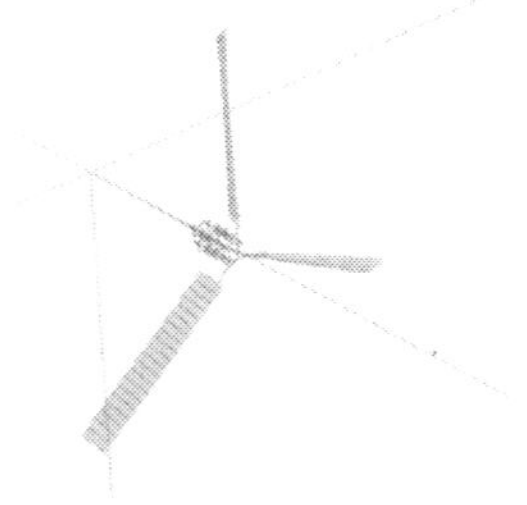

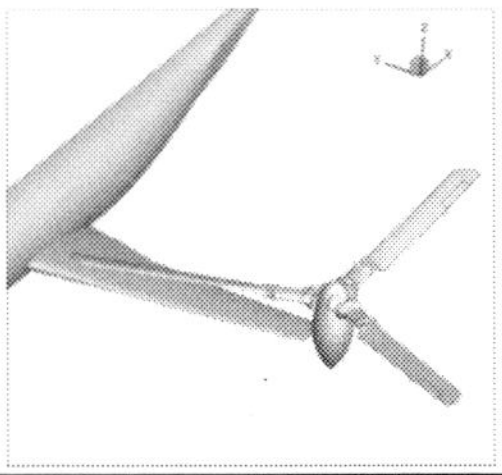

29

Tiltrotor Transition Flight

Objective

- Develop fully coupled CFD/CA tiltrotor model to predict rotor and wing loads and tail interactional loads at transition flight

Technical challenges

- Expensive computational cost
- Realistic conversion schedule

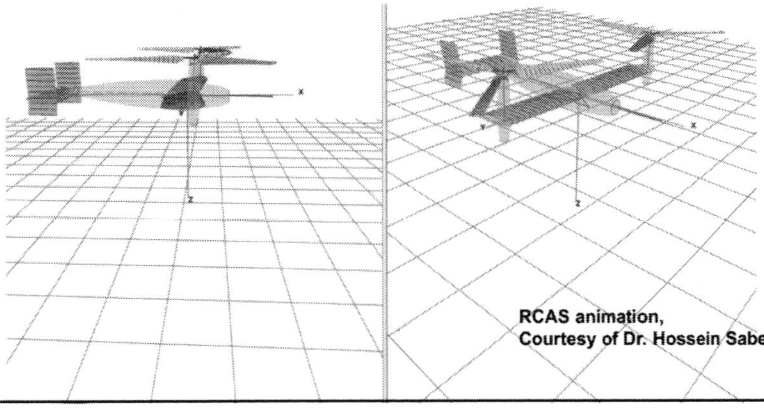

RCAS animation, Courtesy of Dr. Hossein Saberi

30

Concluding Remarks

- Significant progress in rotorcraft aeromechanics prediction capability has recently been made
- Efficiency, accuracy, and automation of our analysis tools need to be continually improved
- Validations need to be performed for a wide range of rotorcraft
- Ultimate goal is rotorcraft design using fully-automated high-fidelity analysis including optimization

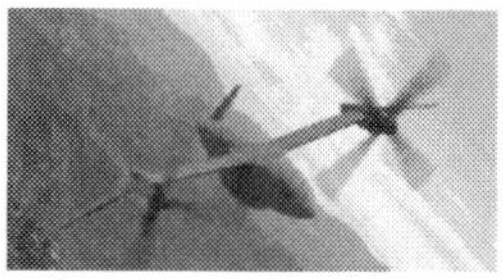

31

Questions?

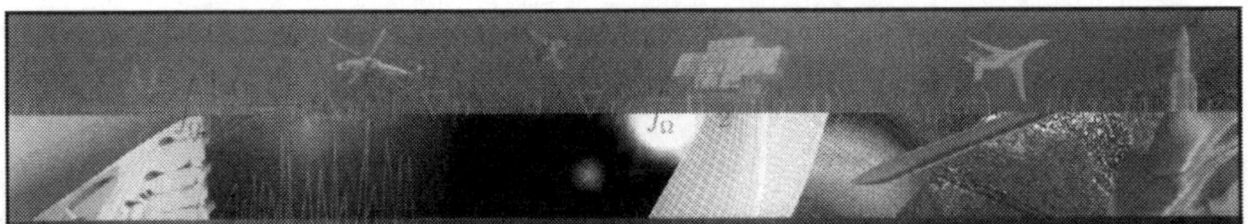

ONERA Rotorcraft Highlights

Presentation to 6th ARF & HELI JAPAN 2017
Nov. 7-9, 2017
Blanche Demaret
Program Director for Rotorcraft

ONERA
THE FRENCH AEROSPACE LAB

Summary

- **ONERA's identity**
- **Rotorcraft Highlights**
  - Aerodynamics
  - Green Rotorcraft (Clean Sky)
  - Flight Safety
  - Racer (Clean Sky 2)
- **Future challenges**

2 B. Demaret_6th ARF&Heli Japan 2017_ Nov. 7-9, 2017

ONERA

The French Aerospace Research Center
• A public entity created in 1946
• Reporting to the French Ministry of Defense
• 1985 employees, 8 centers
• Including 238 doctoral students and PhD
• 235 M€ budget
• 55% Contract based budget
• Recognized internationally as one of the first Research Centers
• Largest fleet of wind tunnels in Europe
3
B. Demaret_6th ARF&Heli Japan 2017_ Nov. 7-9, 2017
ONERA

ONERA Mission: Develop and direct research in the aerospace field
Prepare Tomorrow Defence Systems
Provide Expertise to Governement
Meet the Aerospace Challenges for the Future
Contribute to the Competitiveness of the Aerospace Industry
4
B. Demaret_6th ARF&Heli Japan 2017_ Nov. 7-9, 2017
ONERA

ONERA International Cooperations

- Bilateral or multilateral government agreements : NATO/STO, DGA bilateral agreements
- Scientific and technical associations and universities world wide
- Agreements with research organizations or government agencies : Multilateral (IFAR) or Bilateral :

USA (NASA, USARMY)

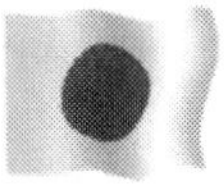
Japan (JAXA)

China (CAE)

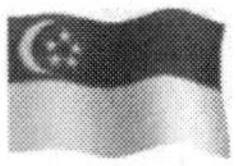
Singapore (DSO)

Russia (CIAM & TsAGI)

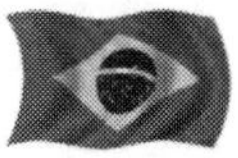
Brazil (DCTA IPEV)

ONERA shapes European aerospace research

CLORA ESRE EREA GARTEUR

EREA
- 5831 employees in aviation
- 426 M€
- ONERA Chair (2016-2017)

ESRE
- 3800 employees
- 430 M€

CLORA
- 34 members
- ONERA chair (2016)

GARTEUR
7 countries, 13 active « action group »
French chairmanship (2013-2014-2015)

European Union
- Horizon 2020 (aeronautics, security, space, etc.), Clean Sky 2, SESAR 2020
- 12 million € volume (European projects)
- ACARE (Advisory Council for Aeronautics Research in Europe)

French-German partnership : ONERA/DLR
- **Integrated helicopter research program (2000)**
- ONERA-DLR-Airbus program (2005)

A specific organization in accordance with the mission of ONERA

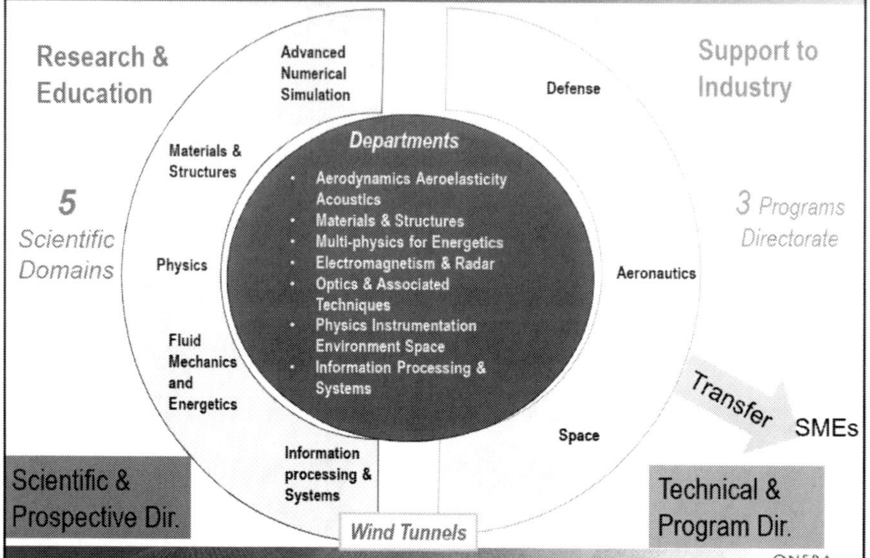

Highlights - Aeronautics

- All rotor designs for Airbus Helicopters are based on profiles defined by ONERA esp. double swept rotorblade.
- ONERA's elsA code for aerodynamics is used on a daily basis by AIRBUS and SAFRAN R&T and design teams.
- The ONERA single-crystal superalloy AM1 co-developed with Safran Aircraft Engines, ARMINES and Tecphy SA was introduced in the monocrystalline blades of the high pressure turbine of the Silvercrest engine.

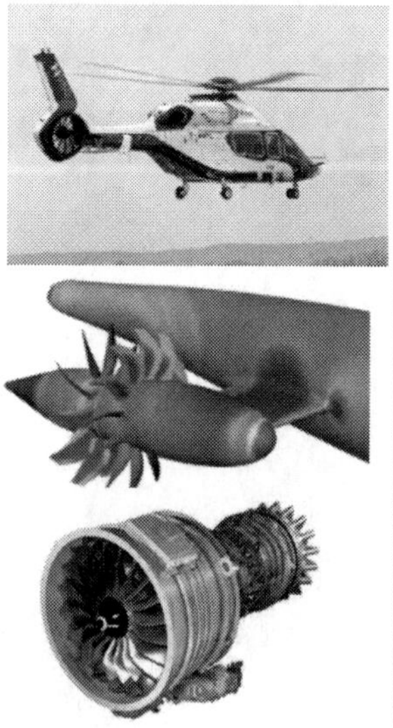

Highlights - Aeronautics

- Real-time airspeed measurement: development of an airborne lidar (AIM2).

- Turbulence detection: fibre laser technology available for wind measurements beyond 10 km.

- Ground vibration tests (GVT) on all aircrafts of the AIRBUS family (with DLR). These tests are routinely performed before the first flight of each aircraft.

Addressing all range of R&T and R&D

Up to new blade shapes

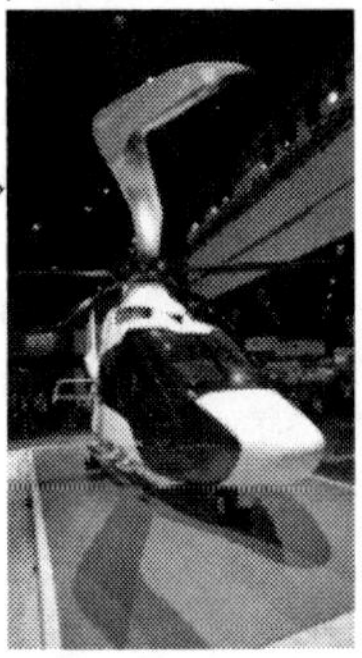

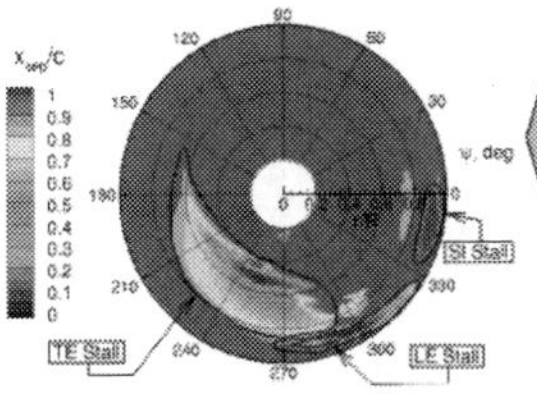

Prospective and physical knowledge:
Analysis of various 3D dynamic stall events

Tools for simulation

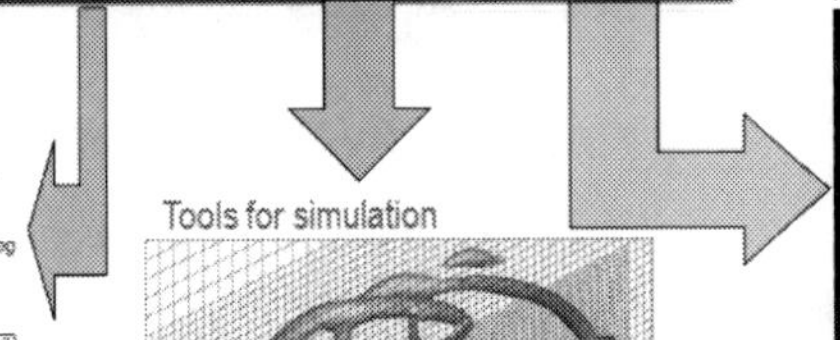

ONERA Rotorcraft Research Facilities

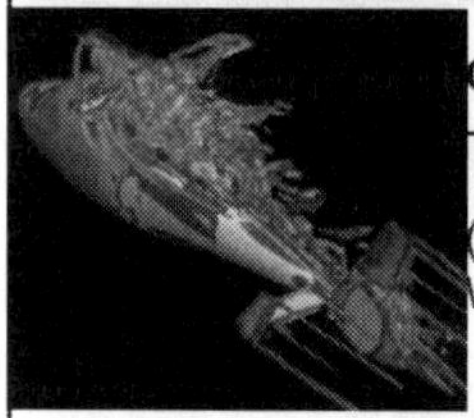

CFD tools : elsA
-Simulation of fuselage/rotor-head drag reduction in CleanSky/GRC2 (~6% drag reduction confirmed by WTT)

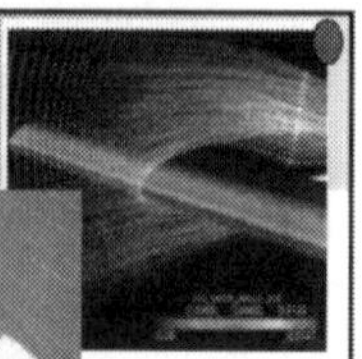

ICING:
- 2D and 3D numerical tools
- Physical knowledge of accretion and ice behavior in flying conditions.

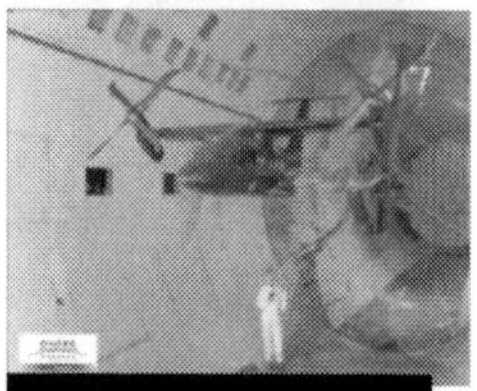

S1 MODANE Wind tunnel:
High speed rotorcraft configuration
(NICETRIP WTT- AHS71 F. Lebrun paper)

L1 Research WT:
- Rotorcraft ship interactions
- Motorized model for multirotors platform

PycsHel:
Prototyping and Design of Systems for Helicopters using ministicks, visual cueing

Common Research Programme for Rotorcraft

DLR ONERA THE FRENCH AEROSPACE LAB

- ONERA and DLR signed a cooperative agreement in 1998,
- A common research programme for the rotorcraft domain was installed in 2000,
- Joint teams, cooperative activities, exchanges, benchmarks
- Harmonized activities supported by French and German governmental agencies
- Common proposals were successfully contracted in Clean Sky and Clean Sky 2 European Research Programs.

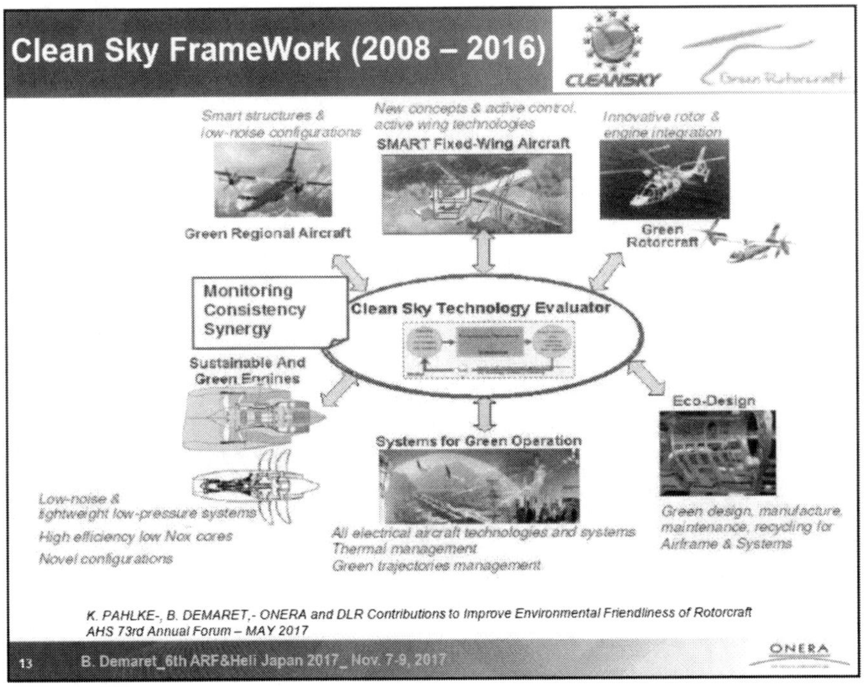

Clean Sky FrameWork (2008 – 2016)
CLEANSKY
Smart structures & low-noise configurations
New concepts & active control, active wing technologies
SMART Fixed-Wing Aircraft
Innovative rotor & engine integration
Green Regional Aircraft
Green Rotorcraft
Monitoring
Consistency
Synergy
Clean Sky Technology Evaluator
Sustainable And Green Engines
Eco-Design
Systems for Green Operation
Low-noise & lightweight low-pressure systems
High efficiency low Nox cores
Novel configurations
All electrical aircraft technologies and systems
Thermal management
Green trajectories management
Green design, manufacture, maintenance, recycling for Airframe & Systems
K. PAHLKE-, B. DEMARET,- ONERA and DLR Contributions to Improve Environmental Friendliness of Rotorcraft
AHS 73rd Annual Forum – MAY 2017
13
B. Demaret_6th ARF&Heli Japan 2017_ Nov. 7-9, 2017
ONERA

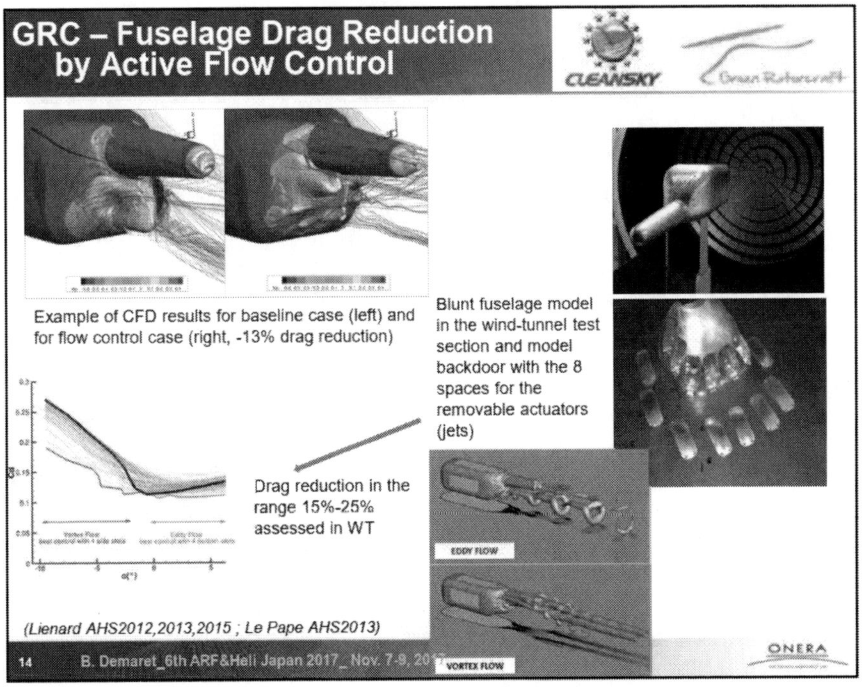

GRC – Fuselage Drag Reduction by Active Flow Control
CLEANSKY
Example of CFD results for baseline case (left) and for flow control case (right, -13% drag reduction)
Blunt fuselage model in the wind-tunnel test section and model backdoor with the 8 spaces for the removable actuators (jets)
Drag reduction in the range 15%-25% assessed in WT
EDDY FLOW
VORTEX FLOW
(Lienard AHS2012,2013,2015 ; Le Pape AHS2013)
14
B. Demaret_6th ARF&Heli Japan 2017_ Nov. 7-9, 2017
ONERA

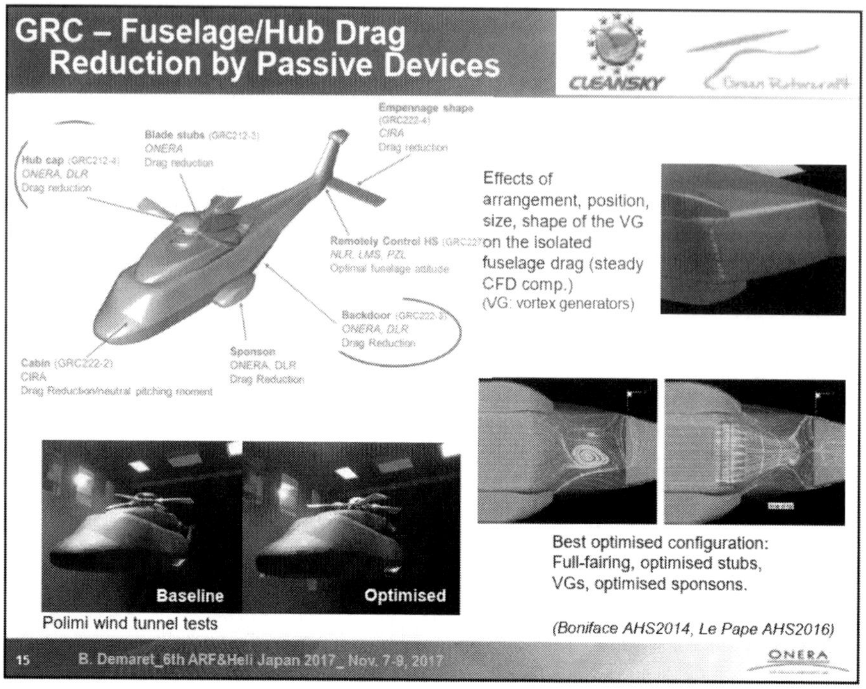
GRC – Fuselage/Hub Drag Reduction by Passive Devices
CLEANSKY
Hub cap (GRC212-4)
ONERA, DLR
Drag reduction
Blade stubs (GRC212-3)
ONERA
Drag reduction
Empennage shape (GRC222-4)
CIRA
Drag reduction
Remotely Control HS
NLR, LMS, PZL
Optimal fuselage attitude
Backdoor (GRC222-3)
ONERA, DLR
Drag Reduction
Sponson
ONERA, DLR
Drag Reduction
Cabin (GRC222-2)
CIRA
Drag Reduction/neutral pitching moment
Effects of arrangement, position, size, shape of the VG on the isolated fuselage drag (steady CFD comp.)
(VG: vortex generators)
Baseline
Optimised
Polimi wind tunnel tests
Best optimised configuration: Full-fairing, optimised stubs, VGs, optimised sponsons.
(Boniface AHS2014, Le Pape AHS2016)
15
B. Demaret_6th ARF&Heli Japan 2017_ Nov. 7-9, 2017
ONERA

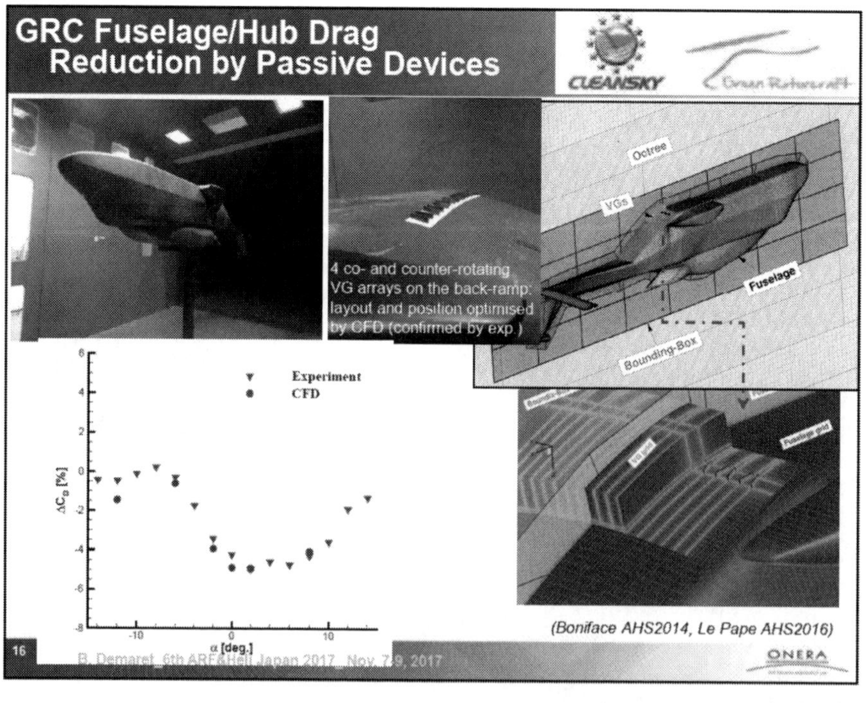
GRC Fuselage/Hub Drag Reduction by Passive Devices
CLEANSKY
4 co- and counter-rotating VG arrays on the back-ramp: layout and position optimised by CFD (confirmed by exp.)
Octree
VGs
Fuselage
Bounding-Box
Experiment
CFD
ΔC_D [%]
α [deg.]
(Boniface AHS2014, Le Pape AHS2016)
16
B. Demaret_6th ARF&Heli Japan 2017_ Nov. 7-9, 2017
ONERA

GRC Fuselage/Hub Drag Reduction

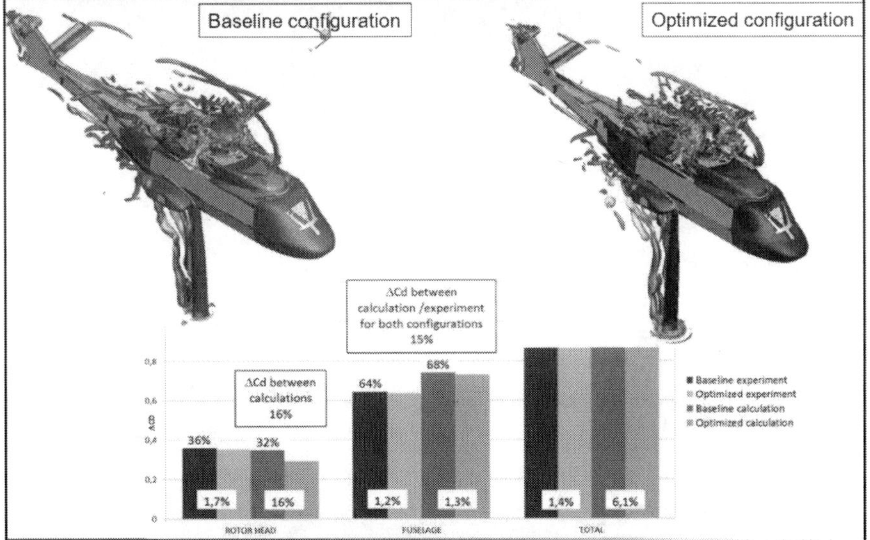

Obstacle Avoidance in Flight using Haptic Feedback

Developments of haptic obstacle avoidance feedbacks for the active cyclic sidestick :

- Relative position of the obstacles w.r.t. the helicopter supposed ideally known through perfect sensors (no detection or estimation).
- 2D avoidance only (within horizontal plane): obstacles were extended vertically for piloting convenience.
- 2 different techniques for the computation of the haptic feedback applied to the stick:

A -Virtual Force Fields (VFFs);

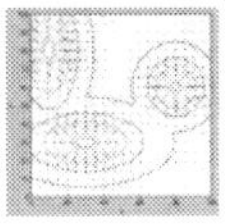

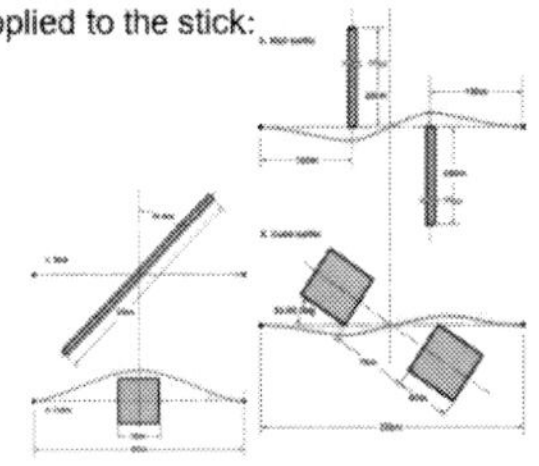

B- Avoidance by geometric approach derived time to contact

RESULTS:

→ Preliminary qualitative evaluation of the benefits of haptics feedback in different situations.

→ Both techniques were promising. Virtual force fields adapted to low speeds, geometric approach adapted to moderate/high speeds.

"Using Haptic Feedbacks for Obstacle Avoidance in Helicopter Flight", L. Binet, T. Rakotomamonjy, 6th EUCASS European Conference for Aero-Space Sciences, Cracovie, 2015

Tactile Cueing for Reactive Obstacle Avoidance for Low Speed Maneuvers

- ONERA and DLR developed haptic obstacle avoidance systems for the active cyclic sidestick, which:
  - indicate obstacle direction
  - increase intensity with decreasing distance or time-to-collision
  - provide different tactile cueing patterns
- Pilots generally accepted a tactile cue for obstacle
- Situational awareness improved
- Ticker, i.e. repetitive pulses away from the obstacle are a promising concept

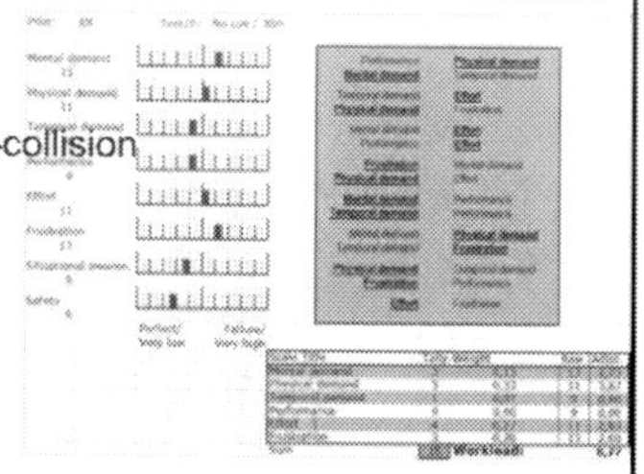

"ONERA-DLR Joint Research On Tactile Cueing For Reactive Obstacle Avoidance Dedicated To Low Speed Helicopter Maneuvers", T. Rakatomamonjy, L. Binet, M. Müllhäuser, ERF Forum, Lille, France, 5-8 September 2016

19 B. Demaret_6th ARF&Heli Japan 2017_ Nov. 7-9, 2017 ONERA

Assistance Systems for Helicopter Ship Landing including visual servoing techniques

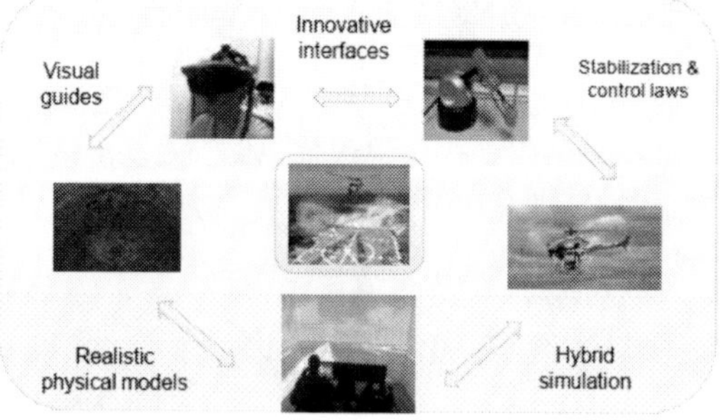

Main research axes:

- Model & integrate the dynamics of the platform and the aerology in the control loops;
- Find an optimal piloting strategy through high-order control laws;
- Use haptics interfaces and synthetic / or augmented vision systems to give a feedback to the pilot

- Flight tests with the participation of CEPA (French Navy Center for Practical Experimentation and Reception)
- Collect data for model validation
- Define methods for models calibration and validation

CONTAHCT - ONERA funded, 3-year duration (2016-18)

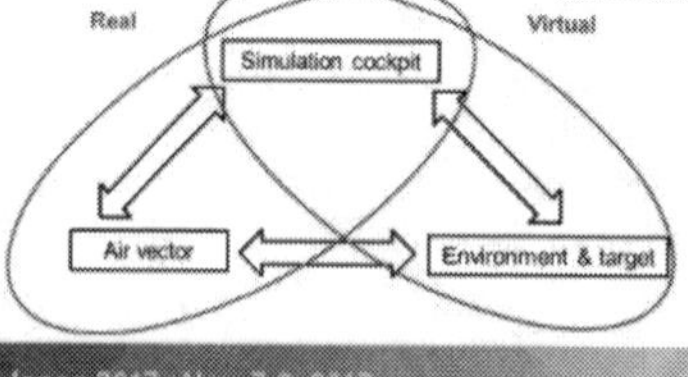

20 B. Demaret_6th ARF&Heli Japan 2017_ Nov. 7-9, 2017 ONERA

Vision Based Control for Helicopter Ship Landing Assistance

Objective: To help pilot decision

Model improvement and data management

- Increase Performances of the ship movements simulation (time >10s).
- Provide data for ship/helo respective position through sensors (on the ship and on the helo).

Then develop a strategy:

- landing phases sequence : (rendez-vous, approach, wait and final approach)
- Criteria for the deck landing point
- Include in the criteria the environment constraints

Image features, extracted from camera image (current image vs reference image)

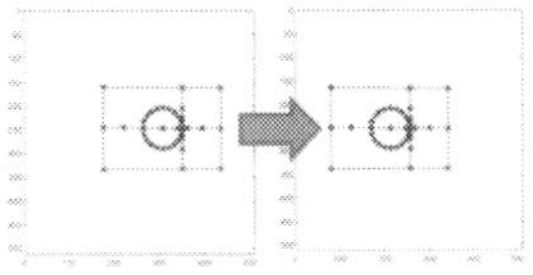

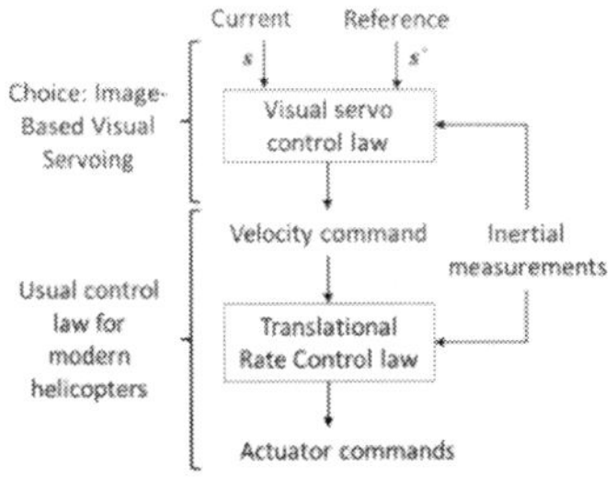

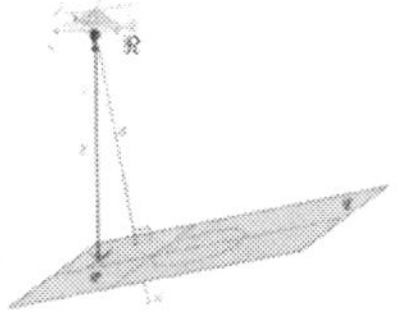

Vision-based control for helicopter ship landing with handling qualities constraints, Truong, Rakotomamonjy, Taghizad, Biannic, 20th IFAC ACA, Sherbrooke, 2016

Visually controlled helicopter: a control scheme for helicopter landing, Truong, Rakotomamonjy, Biannic,, 42nd ERF, Lille, 2016

Simulation facilities: LABSIM/PycsHel

VESA : Visual Environment for Simulations in Aeronautics

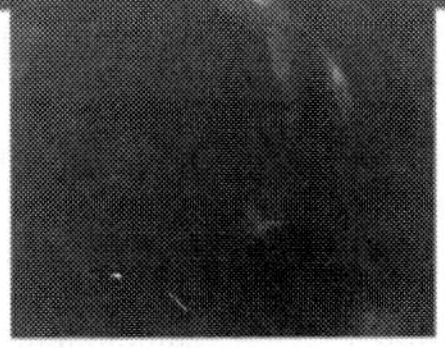

- Four-sided immersive CAVE projection
- Orthographic height-field based terrain
- Day/night conditions, procedural weather (clouds, precipitations, fog,...)
- Realistic 3D toolkit for sea state & ship wakes (Triton by Sundog^{TM} Soft.)
- Pseudo-random procedural real-time building generation

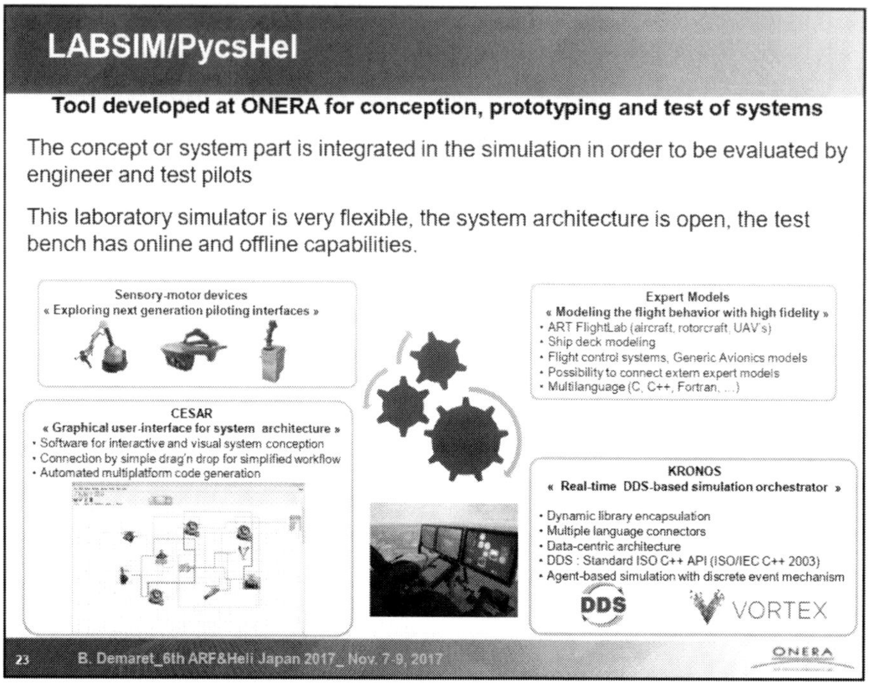
LABSIM/PycsHel
Tool developed at ONERA for conception, prototyping and test of systems
The concept or system part is integrated in the simulation in order to be evaluated by engineer and test pilots
This laboratory simulator is very flexible, the system architecture is open, the test bench has online and offline capabilities.
Sensory-motor devices
« Exploring next generation piloting interfaces »
Expert Models
« Modeling the flight behavior with high fidelity »
• ART FlightLab (aircraft, rotorcraft, UAV's)
• Ship deck modeling
• Flight control systems, Generic Avionics models
• Possibility to connect extern expert models
• Multilanguage (C, C++, Fortran, ...)
CESAR
« Graphical user-interface for system architecture »
• Software for interactive and visual system conception
• Connection by simple drag'n drop for simplified workflow
• Automated multiplatform code generation
KRONOS
« Real-time DDS-based simulation orchestrator »
• Dynamic library encapsulation
• Multiple language connectors
• Data-centric architecture
• DDS : Standard ISO C++ API (ISO/IEC C++ 2003)
• Agent-based simulation with discrete event mechanism
DDS
VORTEX
23
B. Demaret_6th ARF&Heli Japan 2017_ Nov. 7-9, 2017
ONERA

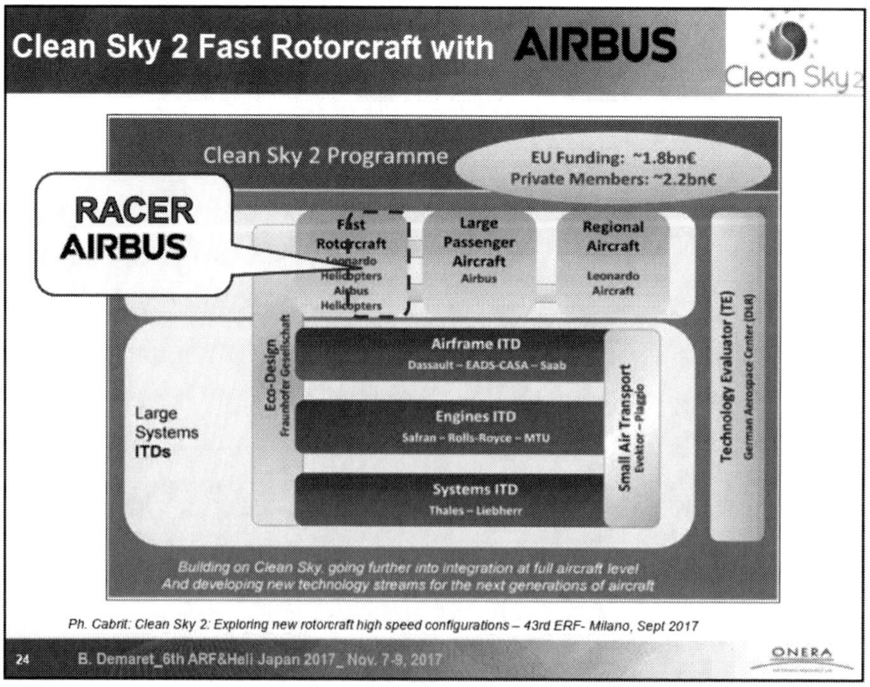
Clean Sky 2 Fast Rotorcraft with AIRBUS
Clean Sky 2
Clean Sky 2 Programme
EU Funding: ~1.8bn€
Private Members: ~2.2bn€
RACER
AIRBUS
Fast Rotorcraft
Leonardo Helicopters
Airbus Helicopters
Large Passenger Aircraft
Airbus
Regional Aircraft
Leonardo Aircraft
Eco-Design
Fraunhofer Gesellschaft
Airframe ITD
Dassault – EADS-CASA – Saab
Engines ITD
Safran – Rolls-Royce – MTU
Systems ITD
Thales – Liebherr
Small Air Transport
Evektor – Piaggio
Technology Evaluator (TE)
German Aerospace Center (DLR)
Large Systems ITDs
Building on Clean Sky, going further into integration at full aircraft level
And developing new technology streams for the next generations of aircraft
Ph. Cabrit: Clean Sky 2: Exploring new rotorcraft high speed configurations – 43rd ERF- Milano, Sept 2017
24
B. Demaret_6th ARF&Heli Japan 2017_ Nov. 7-9, 2017
ONERA

Aerodynamic and Aeroacoustic Design of Airbus Helicopter's Racer Compound Rotorcraft

Clean Sky 2

Clean Sky 2-project in AIRFRAME ITD

Common activity of ONERA (lead) and DLR

Activity started in July 2015, main objectives :

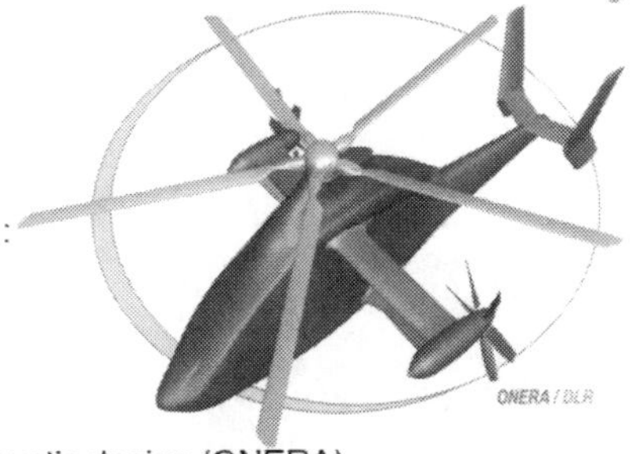

- Wing aerodynamic design (DLR)
- Lateral Rotors aerodynamic and aero-acoustic design (ONERA)
- Global acoustic evaluation (DLR & ONERA)
- Tail parts aerodynamic design : horizontal empennage (DLR), vertical fin (ONERA)

→ Taking into account **aerodynamic interactions** between components

→ Complying with demonstrator **development schedule**

Press release at Pairs International Airshow June 2017

ONERA DLR ONERA

Aerodynamic and Aeroacoustic Design of Airbus Helicopter's Racer" Compound Rotorcraft

Clean Sky 2

Lateral Rotors provided (ONERA)

→ Two different designs for left and right

➔ significant performance gains in hover and forward flight,

➔ noise reduced

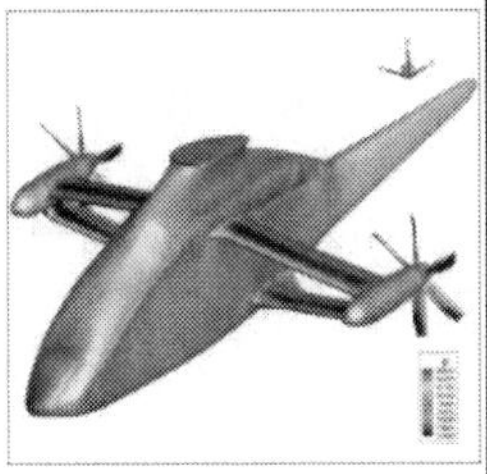

Tail parts design provided (ONERA)

➔ includes detailed analysis of interactions with main rotor and lateral rotors (ONERA& DLR)

➔ design specifications met

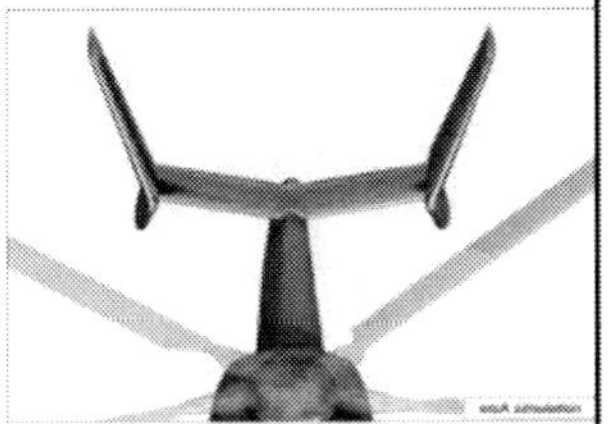

AIRBUS ONERA DLR Clean Sky 2 ONERA

Aerodynamic and Aeroacoustic Design of Airbus Helicopter's Racer Compound Rotorcraft

Clean Sky 2

- Global acoustic assessment for various approaches, climbs, hover, and fly-overs (DLR&ONERA)
  → for preliminary and updated flight conditions, helicopter and airplane modes

DLR UPM simulation

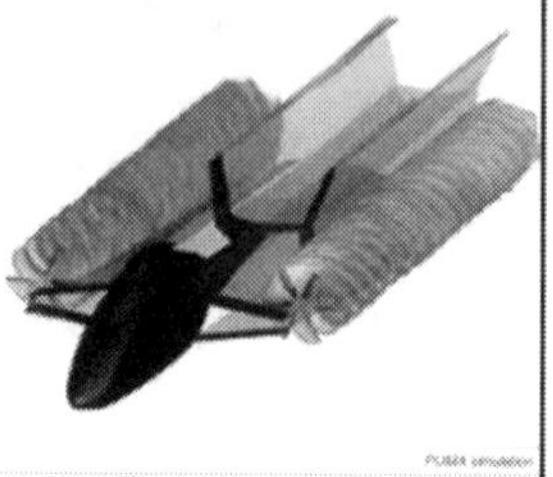

ONERA elsA PUMA KIM simulation

→ include fuselage noise scattering effects

Potential publications at AHS-May 2018 and ERF2018

AIRBUS ONERA DLR ONERA

Key Objectives & Future Challenges

Provide the tools to make virtual engineering a reality

- Multi-fidelity multi-disciplinary design and analysis tools
- Multi-fidelity multi-discipl. multi-objective optimization tools
- Systemic use of test facilities for generation of validation data in all relevant disciplines including human factors (pilot or operator or remote pilot for autonomous systems)

To enable

- Reduction of non-recurring costs (identify and mitigate risks early in the development)
- New propulsion solutions, new concepts for new missions
- Optionally/fully autonomous vehicles
- Innovative technologies answering the societal needs (mobility incl. quality of life and environmental aspects)

Acknowledgements

The author thanks the rotorcraft teams from ONERA (and DLR) for their major contribution to rotorcraft progress

In each topic, the relevant published paper is referenced.

Acknowledgements for the support of the Clean Sky Joint Technology Initiative (grant number CSJU-GAM-GRC-2008-001) – which is part of the European Union 7th Framework Program (FP7/2007-2013).

Acknowledgements for the support of the Clean Sky 2 Joint Technology Initiative (Amendment N° to Grant Agreement CSJU-CS2-GAM-AIR-2014-15-01 Annex 1 - Issue B) ITD AIRFRAME which is part of the European Union Framework Program.

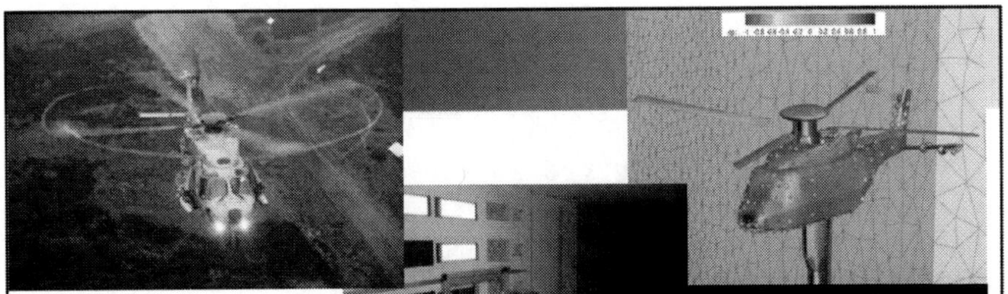

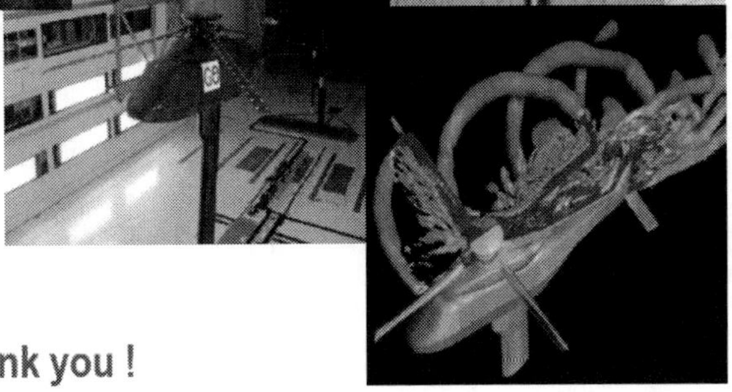

Blanche Demaret
Rotorcraft Programme Director
bdemaret@onera.fr

Thank you !

ENVELOPE PROTECTION FOR CAREFREE MANEUVERING

Dr. J.V.R. Prasad
Professor and VLRCOE Associate Director
School of Aerospace Engineering
Georgia Institute of Technology
Atlanta, GA 30332

6th Asian-Australian Rotorcraft Forum
Kanazawa, Japan
November 7-8, 2017

Acknowledgment

- Dr. Joe Horn (PhD, 1999)
- Dr. Ilkay Yavrucuk (PhD, 2003)
- Dr. Geoffrey Jeram (PhD, 2004)
- Dr. Suraj Unnikrishnan (PhD, 2006)
- Dr. JongKi Moon (PhD, 2009)
- Dr. Keeryun Kang (PhD, 2012)

Outline

- Motivation
- Limit Detection and Avoidance
- Envelope Protection in Manned Aircraft
- Envelope Protection in Unmanned Aircraft
- Current Research

Significance of Envelope Protection

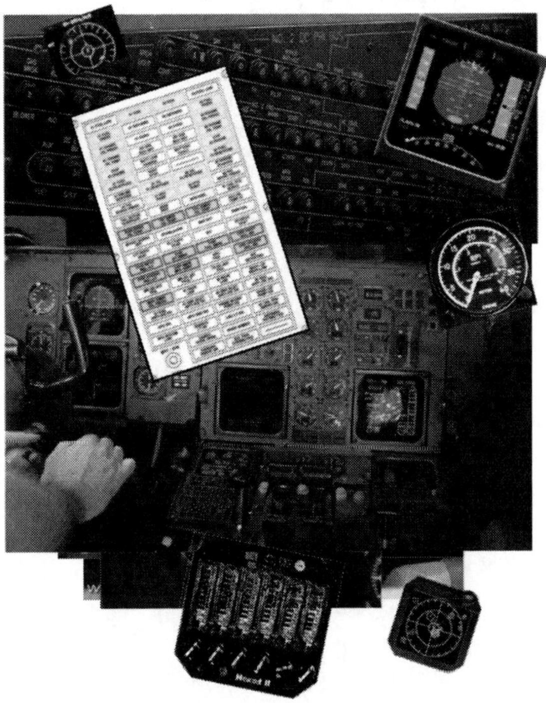

- Maneuver Envelope
  - Enhancement
  - Maximization
- Rotorcraft challenges:
  - Multiple, simultaneous, highly coupled, and nonlinear Limits
  - Extensive pilot training
  - Visual / Audio Saturation
  - Extreme cockpit workload
  - Compromise: Safety vs. Performance

Envelope Limiting Systems on Fixed Wing Aircraft

- Fixed wing aircraft have relatively simple flight envelopes.
- Simple devices are historically used to increase pilot awareness.
- Envelope protection systems are installed into FBW systems.

Limit Avoidance

Leave It to the Pilot:

- Increased Workload
- Simplified Operational Limits Result in Sub-Optimal Performance
- Difficult with FBW systems

Automatic Limit Avoidance with FBW Control System:

- Can Not Override
- Changes Aircraft Response
- Pilot Not Aware of Limit

Pilot Cueing System:

- Visual
- Aural
- Tactile

Dynamic Limit Margin Prediction

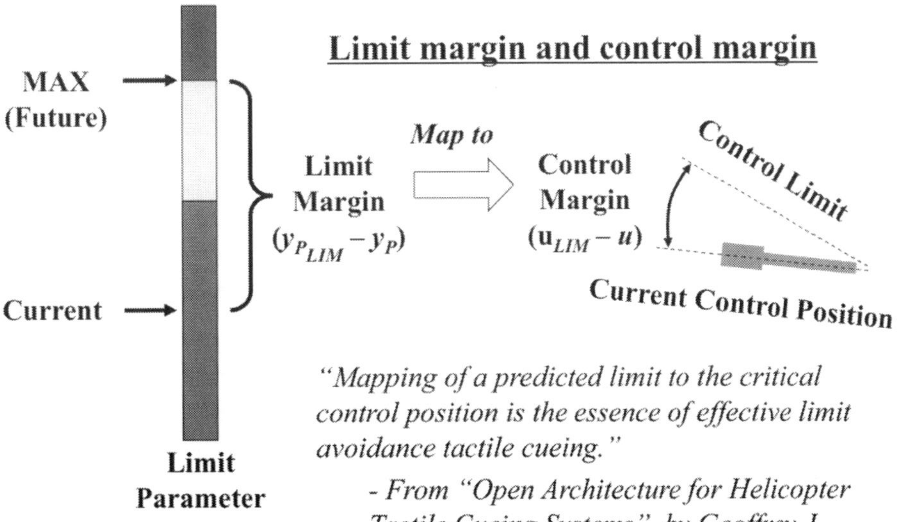

"Mapping of a predicted limit to the critical control position is the essence of effective limit avoidance tactile cueing."

- From "Open Architecture for Helicopter Tactile Cueing Systems", by Geoffrey J. Jeram & J. V. R. Prasad

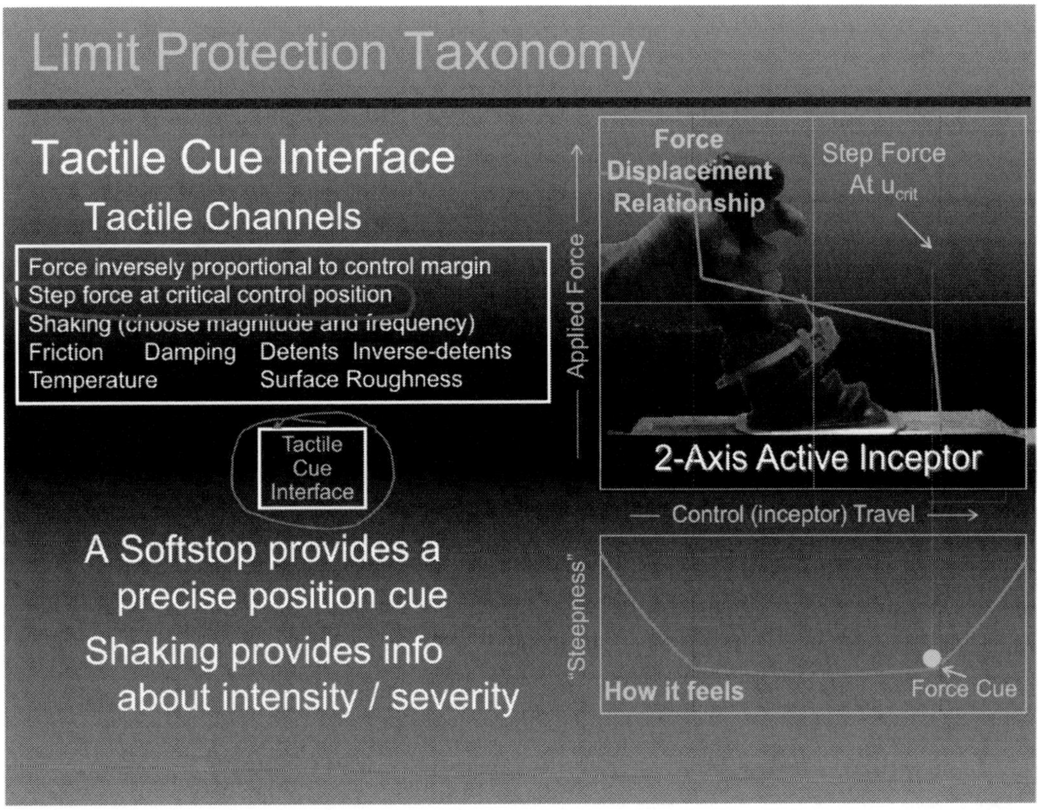

Examples of Envelope Limits

- V-n Envelope - Maximum load factor, stall limits, maximum airspeed
- Rotor Flapping Limit - Mechanical interference, fatigue loading limits, droop stops
- Component Structural Limit - (e.g. pitch link load)
- Torque Limits - Maximum transmission torque or maximum engine torque
- Control Margin - Unexpected saturation of control actuators

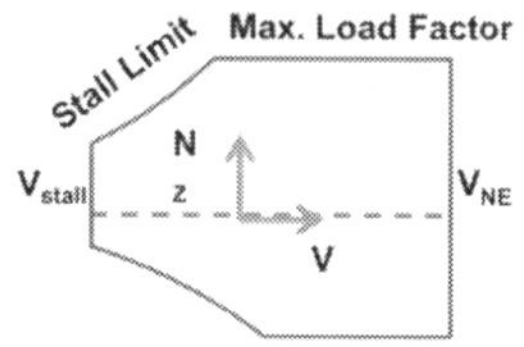

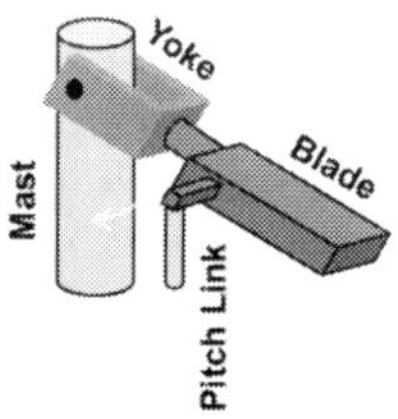

Limit Response Types

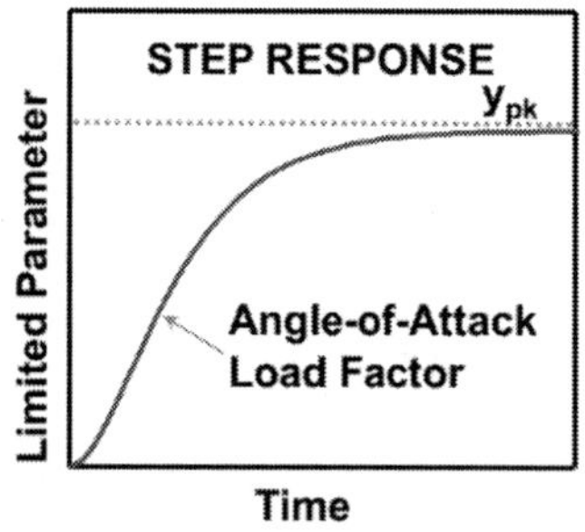

Steady-State Response Critical

LIMIT DETECTION - Estimate value of limited parameter in quasi-steady-state (dynamic trim)

LIMIT AVOIDANCE - Find control deflection that causes the limited parameter to reach envelope limit in dynamic trim

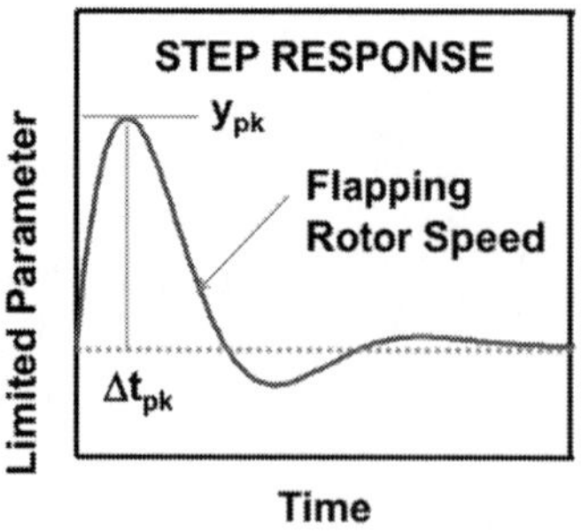

Peak Response Critical

LIMIT DETECTION - Estimate peak response of limited parameter to control step input

LIMIT AVOIDANCE - Find control deflection that causes the limited parameter to peak at envelope limit

LIMIT PREDICTION USING DYNAMIC TRIM FOR STEADY STATE CRITICAL LIMITS

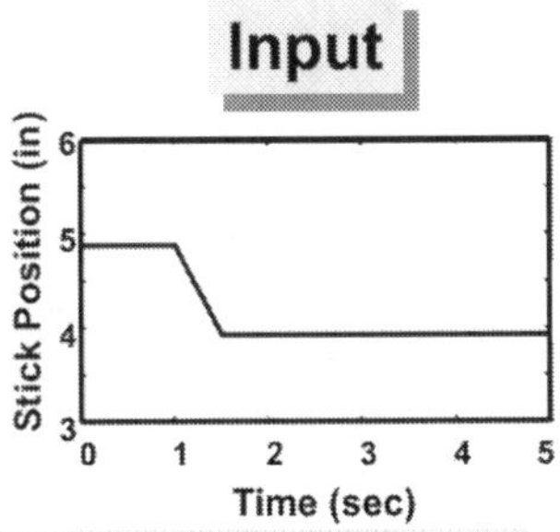

$$\mathbf{x} = \begin{bmatrix} \mathbf{x}_{\text{slow}}^{\text{T}} & \mathbf{x}_{\text{fast}}^{\text{T}} \end{bmatrix}^{\text{T}}$$
$$\mathbf{x}_{\text{slow}} = \begin{bmatrix} V & \gamma & \phi \end{bmatrix}^{\text{T}}$$
$$\mathbf{x}_{\text{fast}} = \begin{bmatrix} q & \alpha & p & r & \beta \end{bmatrix}^{\text{T}}$$
$$\dot{\mathbf{x}}_{\text{fast}} = \mathbf{0}$$
$$\dot{q} = \dot{\alpha} = \dot{p} = \dot{r} = \dot{\beta} = 0$$

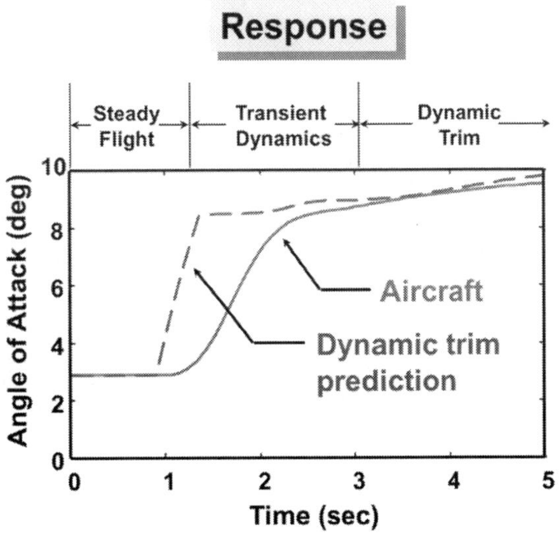

V-22 Limit Avoidance System

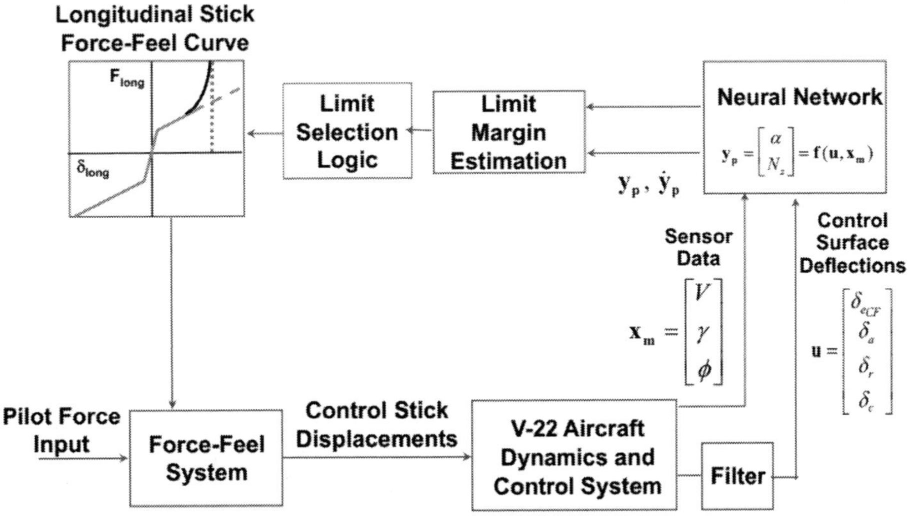

Envelope Expansion Experiment*

High Speed, Wind-Up Turn Maneuver

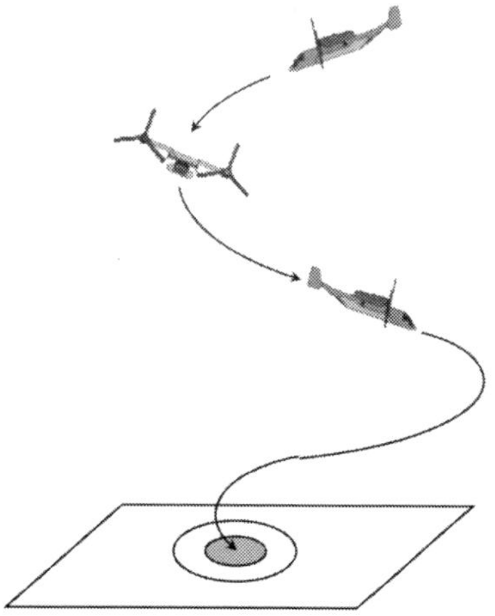

- Descend Rapidly from 15,000 ft
- Hold a Desired Dive Speed of 250, 270, 290 or 310 knots
- Perform a High-G Banked Turn in Order to Maintain Airspeed and Position over Ground
- Pull-Up to Level Flight Below 2000 ft
- Avoid Exceeding AOA Buffet Limit and Maximum Load Factor Limit
- Perform with and without Tactile Cueing System
- Simulated Structural Failure Based on Time Integration of Limit Exceedence
- Pilot Training without Cueing System

* Conducted in Boeing simulator

Typical Results

Wind-Up Turn, 290 kts, Pilot B

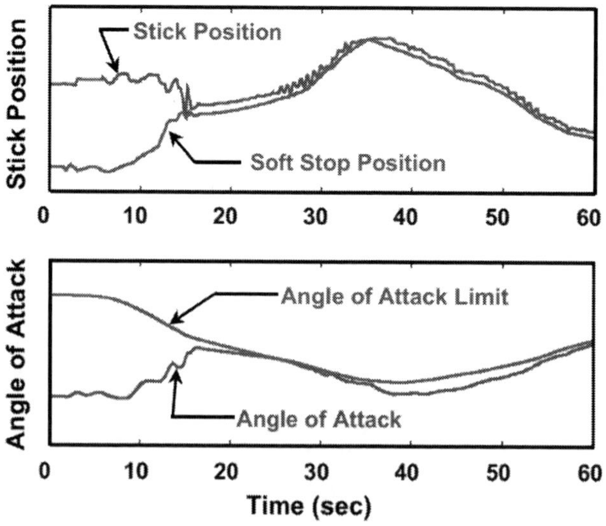

Safe Envelope Expansion
High Speed Wind-Up Turn Maneuver

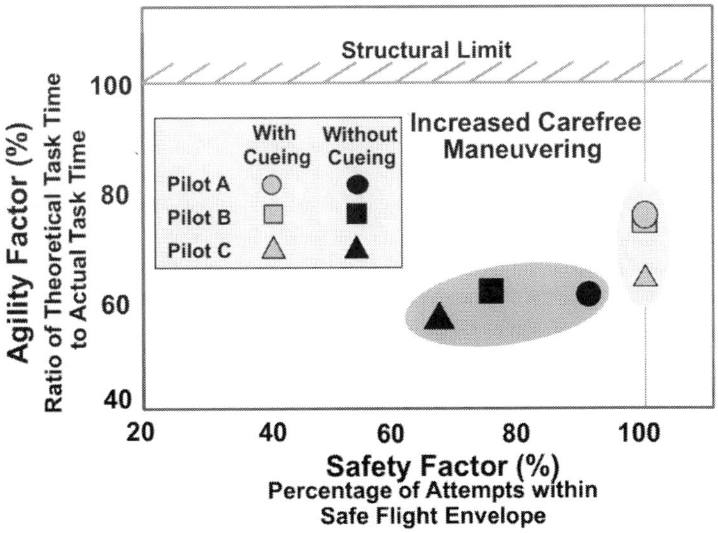

LIMIT PREDICTION USING AN OPTIMAL CONTROL FORMULATION

- Dynamic trim method can work for steady state critical limit parameters.
- Need a method that can work for all limit parameter types.
- Find a very aggressive *optimal control profile* that takes limit parameter response to the limit boundary.
- Use this control profile to estimate control limit. How?
- Such a methodology will be applicable to all limit parameter types.

Optimal Control Problem Formulation

Find an optimal control profile that minimizes the following objective function → $J = \int_{t_0}^{t_f} (1 + Wu^2)dt$

Taking the limit parameter response to the limit boundary. → $y_p(t_f) = y_{\lim}$

- Final time t_f is free.
- Objective function minimizes time to reach limit boundary.
- The control weighting w is related to pilot aggressiveness.
- Increasing the value of control weighting will result in increasing the time taken to reach the limit boundary.

Optimal Control Solution to Control Limit

Define- Area norm of a control signal u_p between the time interval $[t_0, t_f]$ as

$$u_{AN}(u_p(.),[t_0\ t_f]) = \sqrt{\int_{t_0}^{t_f} u_p^2} \,/\, \sqrt{t_f - t_0}$$

Proposition- *Area norm of the optimal control profile is an estimate of the control limit.*

- Let (u_p^*, t_f^*) denote the optimal solution.
- The area norm of the optimal control solution u_{AN}^* is the *smallest step control input* that would take the limit parameter response to the limit boundary in time t_f^*
- Any other step control input with a lower value of critical time will have a larger value than the area norm of optimal solution.

Simulation Test Maneuver - Swoop

- Designed upon ADS-33 specifications.
- Used to evaluate envelope protection for both steady and transient pitch related limits.

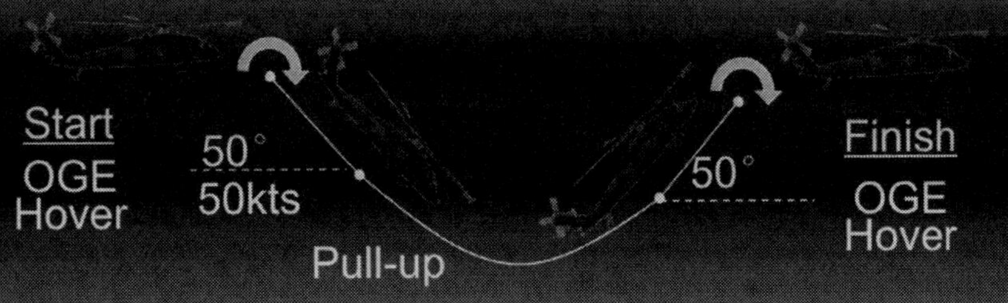

Configuration	Weight (lbs)	X axis CG location
Nominal	16825	357.3 inches
Heavy	22000	357.3 inches

RIPTIDE Simulation Demo

OPLP 2LIM Demo.avi

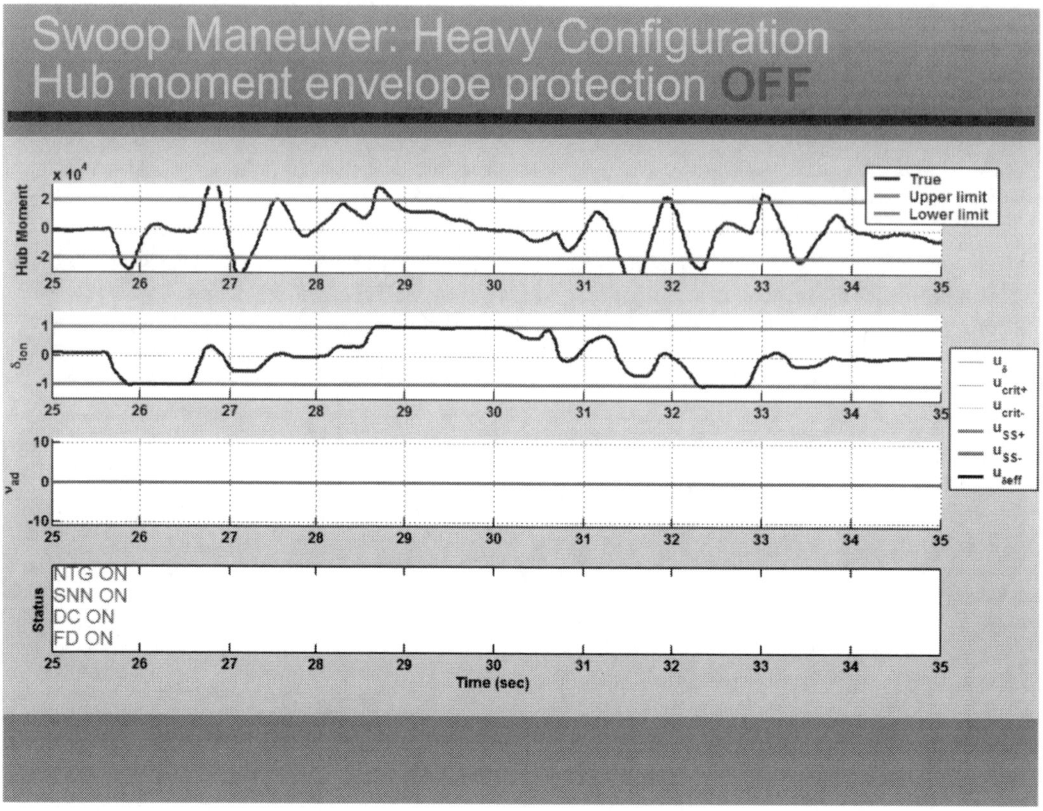
Swoop Maneuver: Heavy Configuration
Hub moment envelope protection OFF
x 10^4
Hub Moment
True
Upper limit
Lower limit
δlon
vad
Status
NTG ON
SNN ON
DC ON
FD ON
Time (sec)

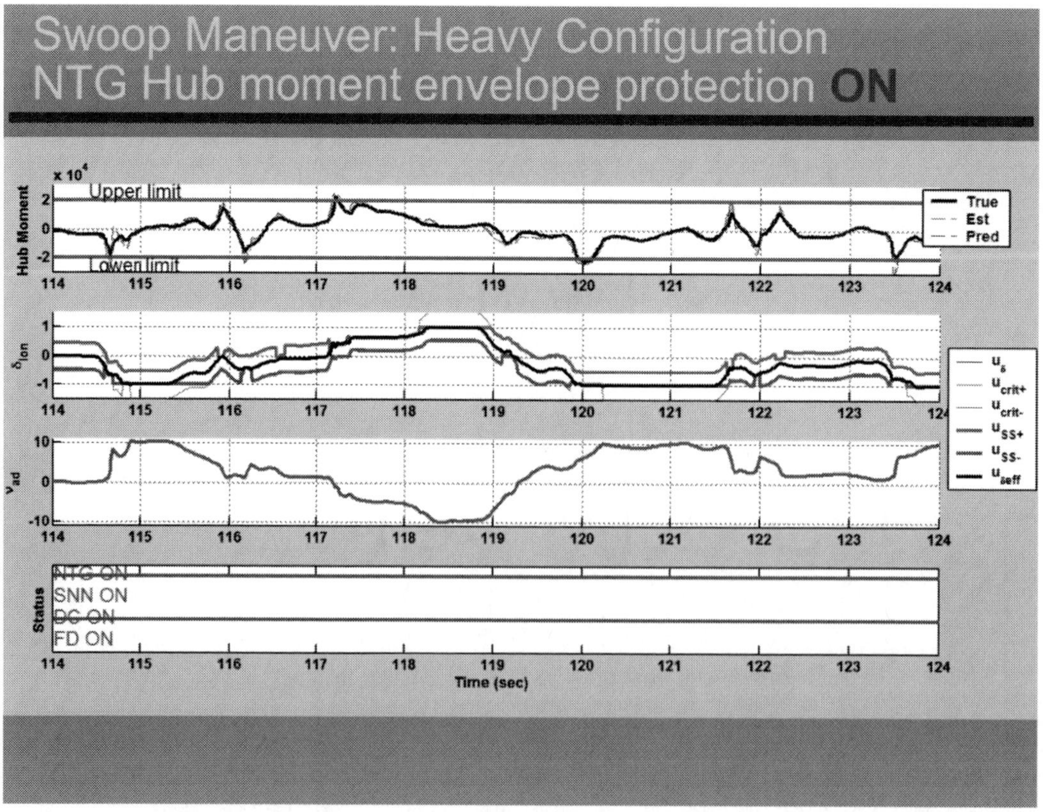
Swoop Maneuver: Heavy Configuration
NTG Hub moment envelope protection ON
x 10^4
Upper limit
Lower limit
Hub Moment
True
Est
Pred
δlon
vad
Status
NTG ON
SNN ON
DC ON
FD ON
Time (sec)

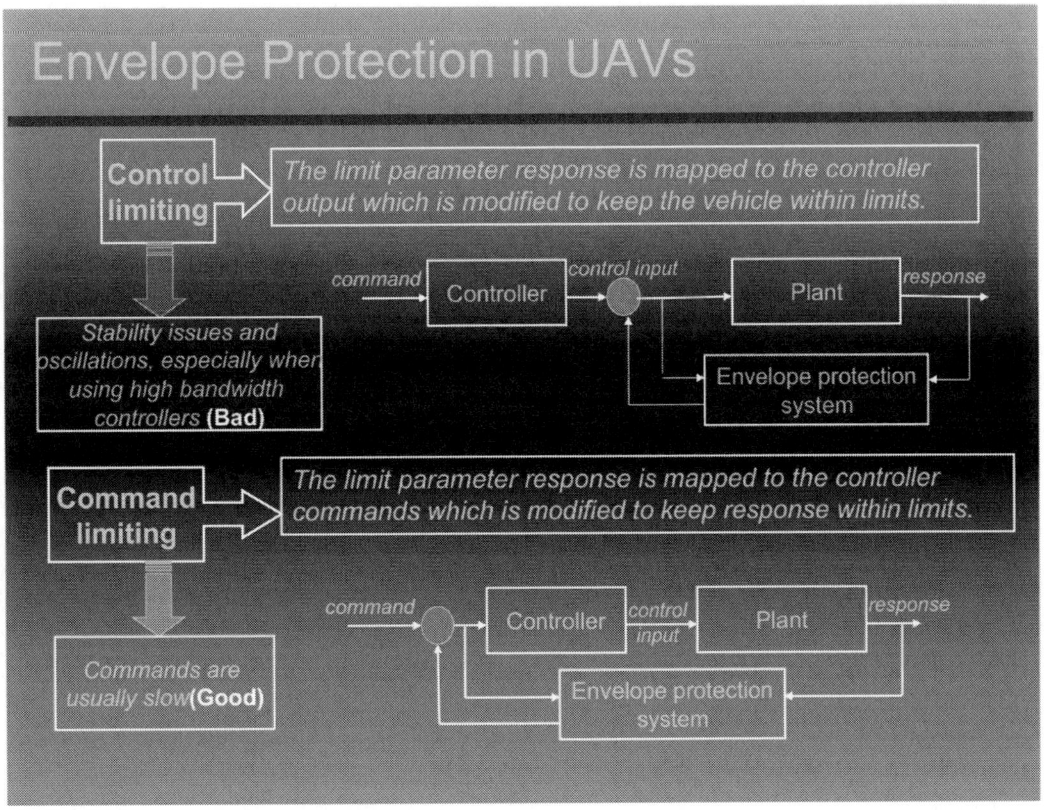
Envelope Protection in UAVs
Control limiting
The limit parameter response is mapped to the controller output which is modified to keep the vehicle within limits.
Stability issues and oscillations, especially when using high bandwidth controllers (Bad)
command
Controller
control input
Plant
response
Envelope protection system
Command limiting
The limit parameter response is mapped to the controller commands which is modified to keep response within limits.
Commands are usually slow(Good)
command
Controller
control input
Plant
response
Envelope protection system

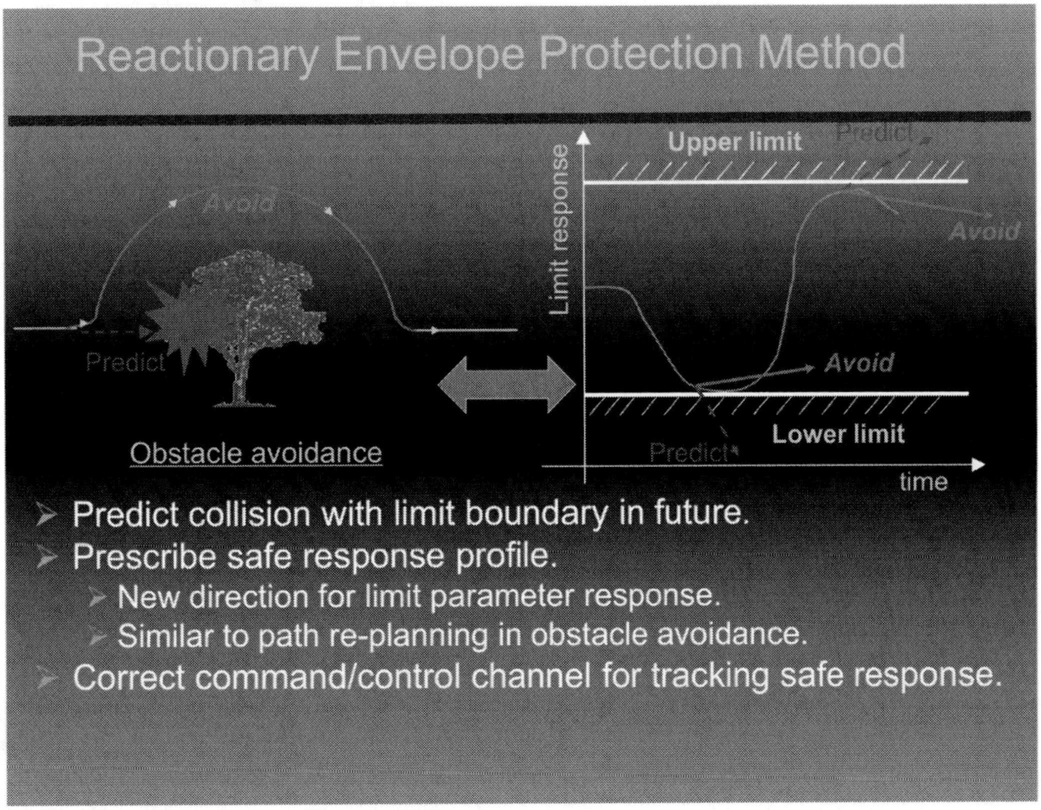
Reactionary Envelope Protection Method
Avoid
Predict
Obstacle avoidance
Limit response
Upper limit
Predict
Avoid
Avoid
Lower limit
Predict
time
Predict collision with limit boundary in future.
Prescribe safe response profile.
New direction for limit parameter response.
Similar to path re-planning in obstacle avoidance.
Correct command/control channel for tracking safe response.

Test maneuver- Aggressive eturn

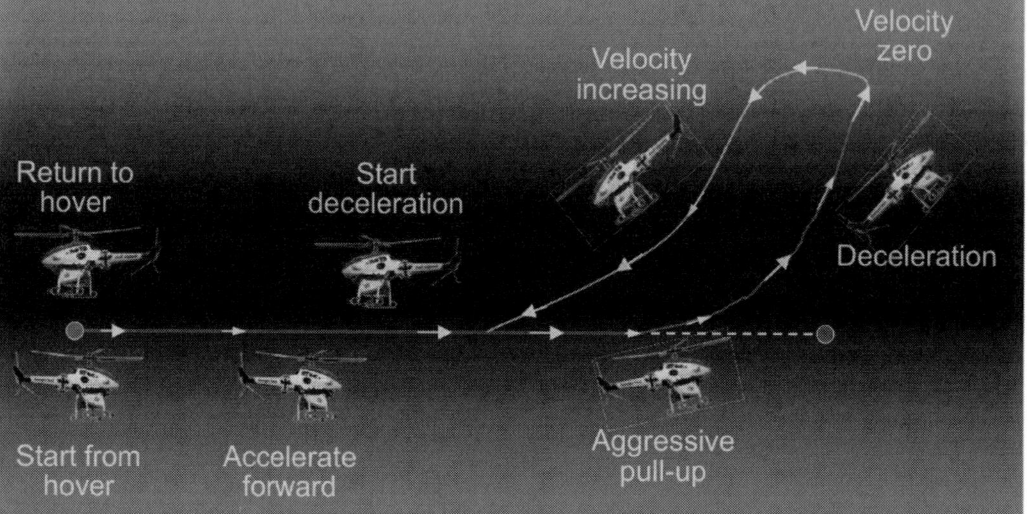

GTMax Avionics Hardware

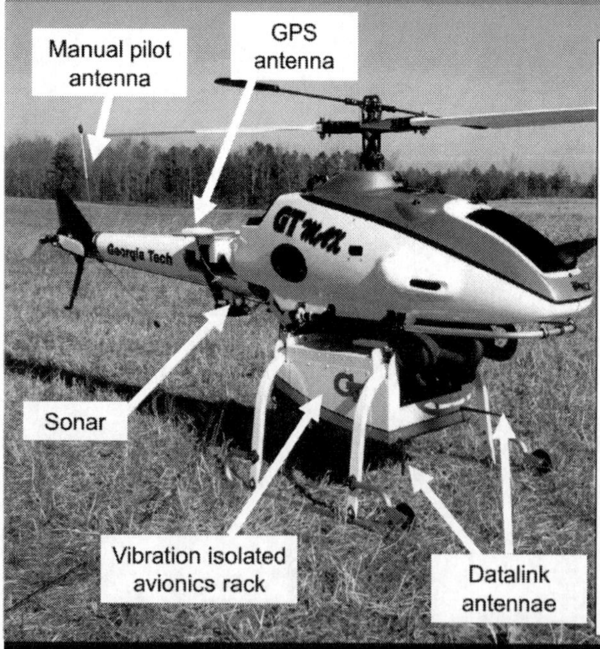

- Flight Computers
  - 266MHz & 800 MHz embedded PCs, Ethernet, flash drives
- Sensors
  - Inertial Measurement Units (x2)
  - Differential GPS
  - Magnetometer, sonar altimeter
  - RPM, voltage/fuel warnings
  - Various camera systems
- Data Links
  - 11 Mbps Ethernet data link
  - RS-232 serial data link

FLIGHT TESTING USING GTMAX

Flight test results for load factor limit protection-Load factor response comparison

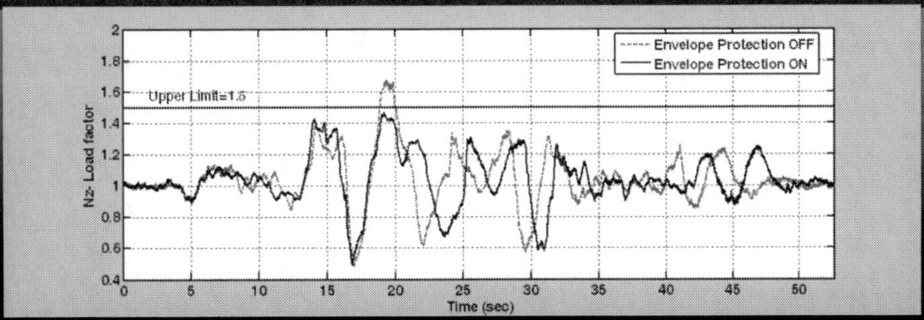

- Load factor response maintained within the prescribed limit.
- Maneuver time increases as expected.

Flight test results for load factor limit protection- eturn trajectory comparison

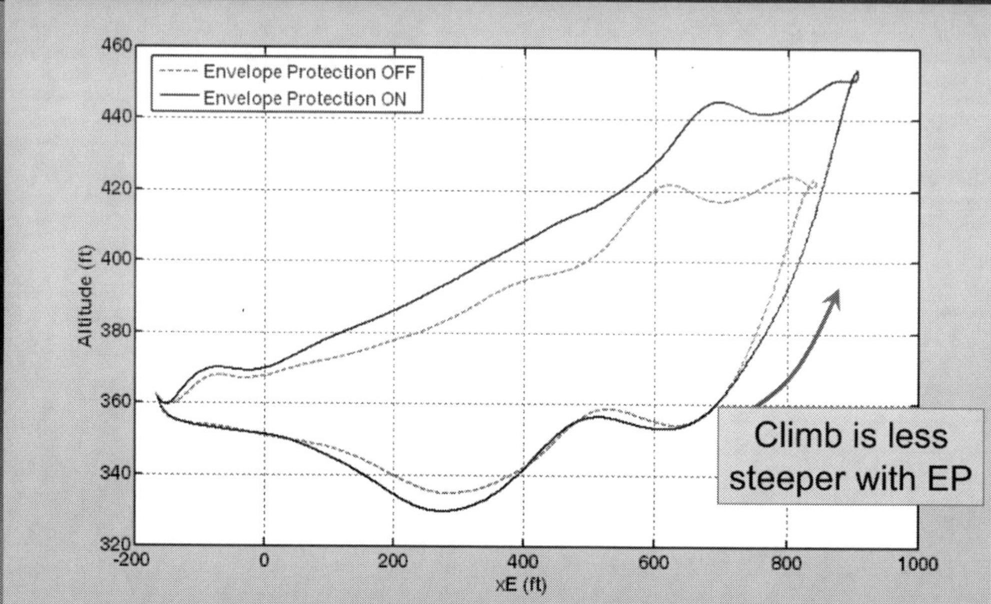

Integration of Envelope Protection in Autonomous UAVs

- Objective
  - Determine guidance commands for accomplishing the overall mission while simultaneously taking into account obstacle avoidance (external constraints) and envelope protection (internal constraints) in the presence of uncertainty.

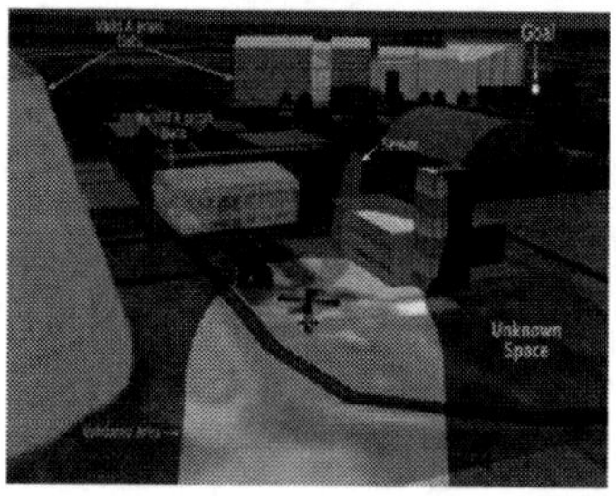

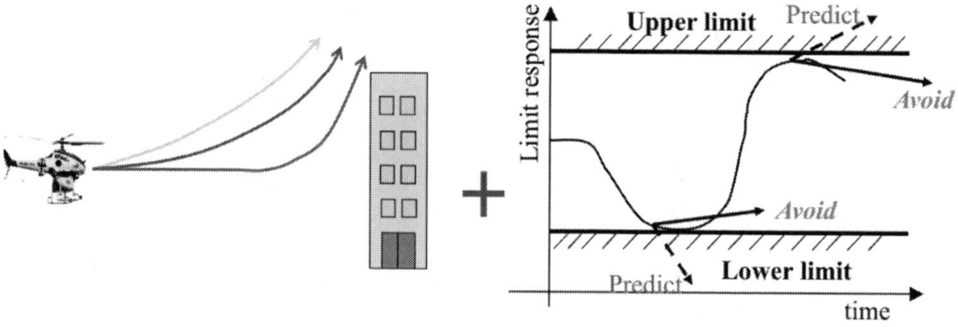

Obstacle Avoidance with Envelope Protection

Generic Concern on OA: **How to avoid an obstacle without violating a vehicle operational limit boundary?**

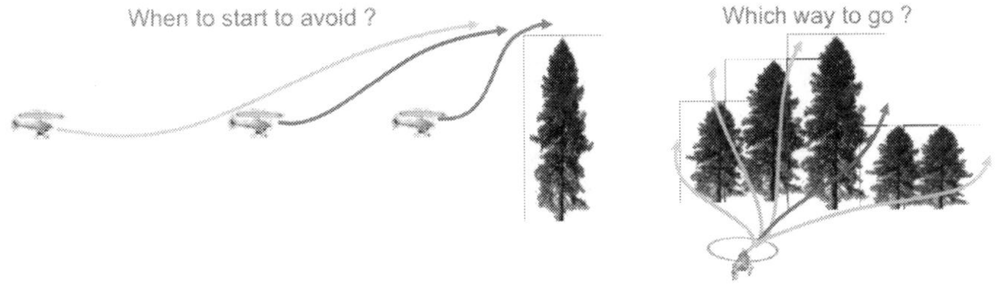

- The typical decision making in obstacle avoidance problem
  - How to determine obstacle avoidance maneuvers when vehicle is moving through an unknown obstacle-rich area?
  - How to reduce excess excursions above obstacles?
  - How to decide whether to pull-up or go around obstacles?
  - How to account for multiple limits on vehicle performance and motion?
  - How to carry out the decisions and the trajectory generation in real time?

Integrated Approach for Obstacle Avoidance and Envelope Protection

- An optimal control formulation in which
  - Obstacles are treated as external path constraints
  - Envelope limits are treated as internal path constraints

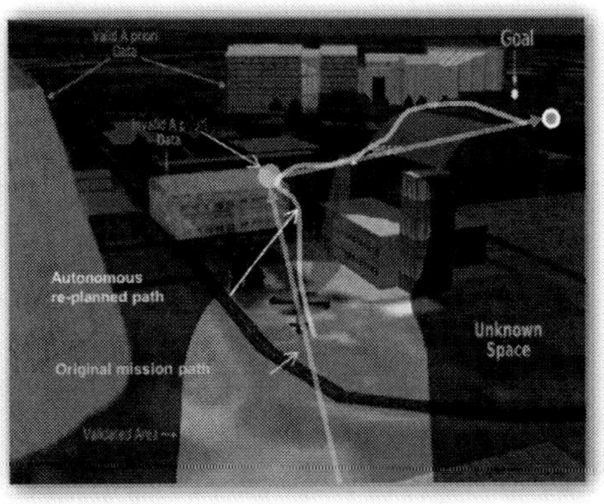

Trajectory Optimization

❋ Overall procedure for trajectory optimization

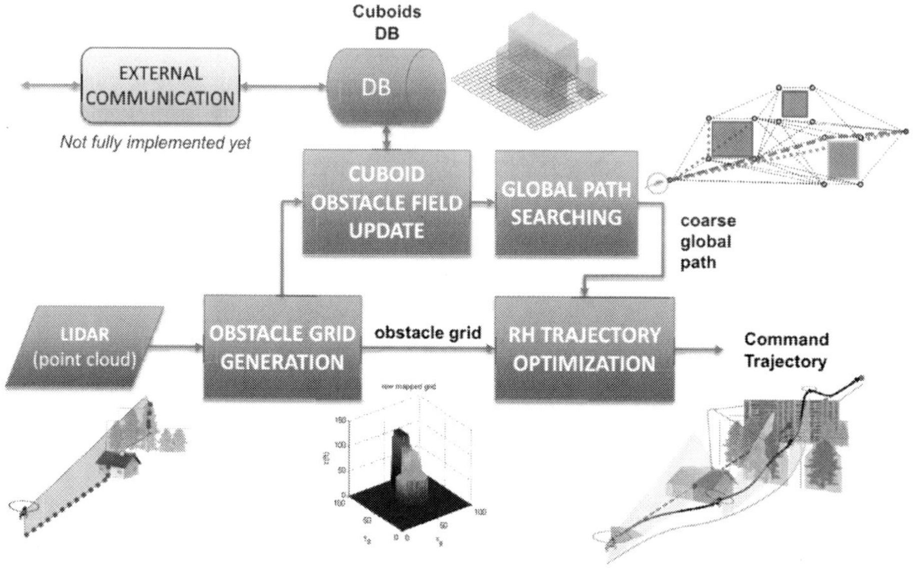

Trajectory Optimization

❋ Local(receding horizon) trajectory optimization formulation

❖ The scope of optimization : **find open-loop optimal trajectory within the sensor range**

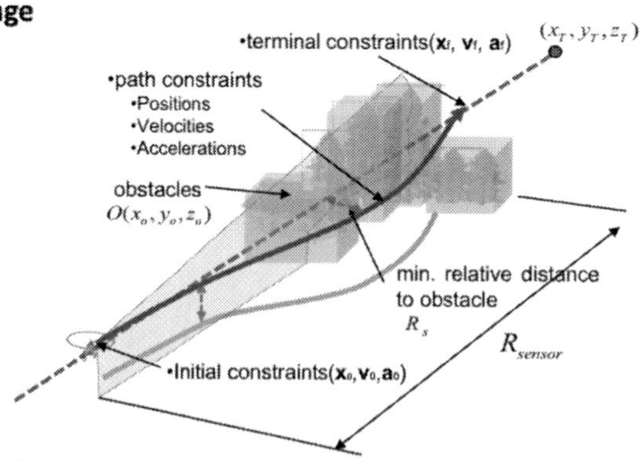

❖ The objective : **Minimum path deviation from the original mission path.**

➢ Performance Index : Quadratic distance from the target position

$$J = \int_{t}^{t_{free}} \left[\frac{(x_T - x(\tau))^2}{R_x^2} + \frac{(y_T - y(\tau))^2}{R_y^2} + \frac{(z_T - z(\tau))^2}{R_z^2} \right] d\tau, \quad R_x, R_y, R_z \text{ are selected constants}$$

Framework Implementation

❋ Simulation and onboard software implementation

- ❖ Simulation : GUST(Georgia Tech UAV Simulation Tool)
- ❖ Onboard SW: Implemented in Onboard2

Sensor: Sick LD-MRS HD
- FOV : Max. 110°
- Range : 0.5 ~ 250m
- Scan Freq : up to 50Hz
- Mounted on airframe to scan vertically

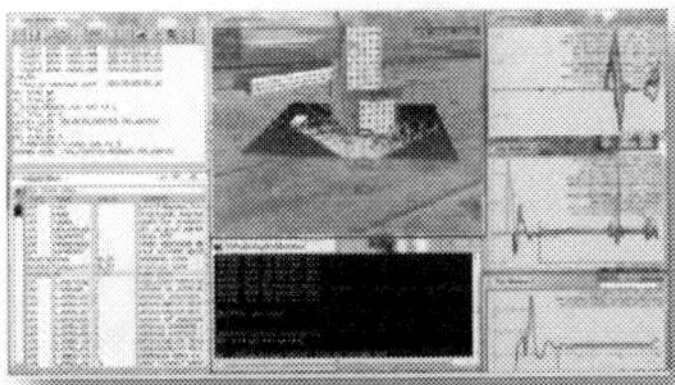

GUST simulation environment
- C/C++ programming
- Real-time simulation
- Provides Software-in-the-loop and Hardware-in-the-loop environment

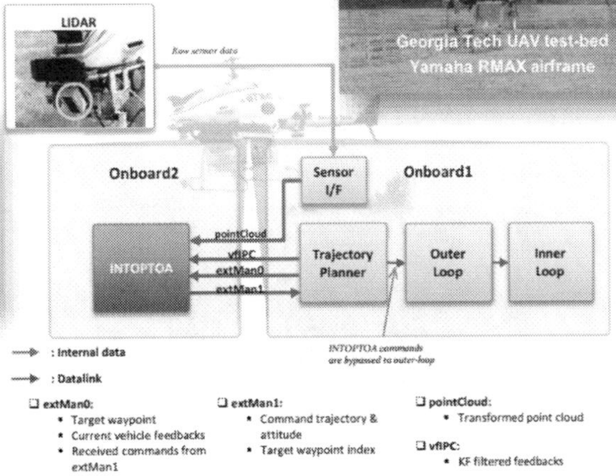

Framework Implementation

❋ Simulation Demonstration

Flight Test Evaluations

❋ Benchmark Case 6

Simulation

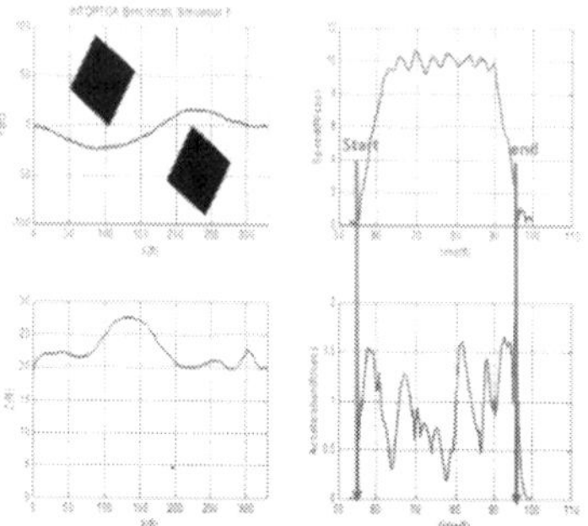

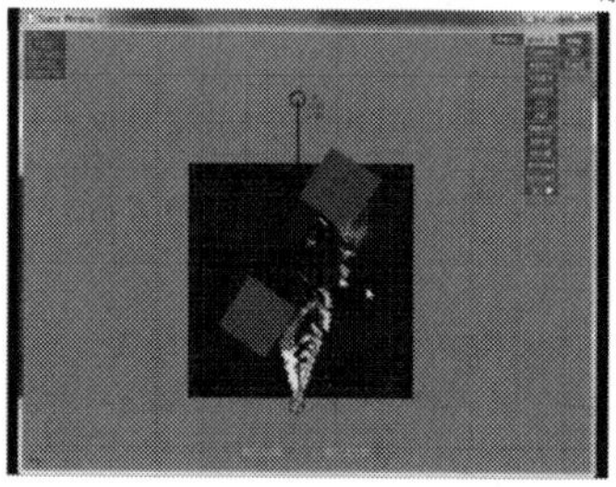

FT

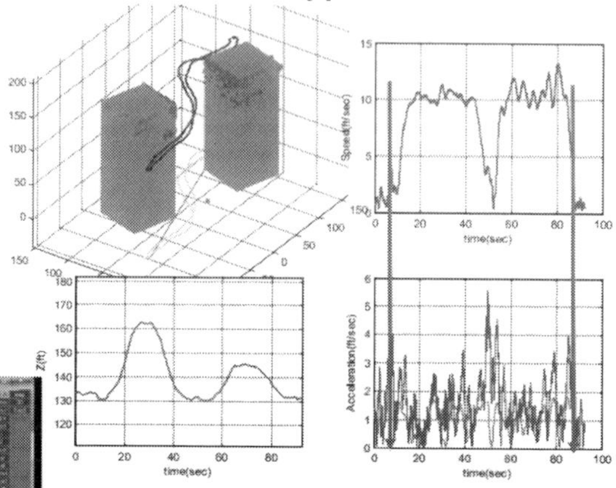

Comparisons of Time for Trajectory

Baseline	OFN of Ref.	Sim.	FT
39.8	51.9	**40.7**	40

Flight test evaluations

❋Flight demonstration

- ❑ **Date : 2012.01.24**
- ❑ **Test Summary**
  - ❖ **McKenna Mout Site, Fort Benning, GA**
  - ❖ **Flight Speed: 15ft/sec, 25ft/sec**
  - ❖ **Clearance conditions:**
    - o Min. Height : 60ft
    - o Clearance : 50ft

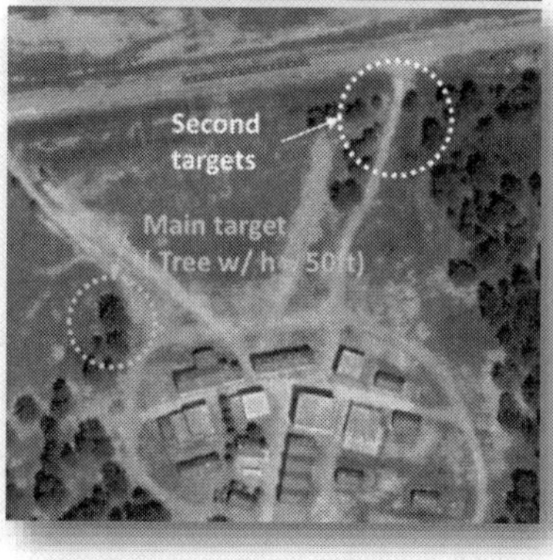

Flight test evaluations

❋ Flight demonstration

- ❖ Overall trajectory coincided with the simulation
- ❖ INTOPTOA successfully conducted avoidance on the main target
- ❖ On second target, LIDAR was interfered by the Sun

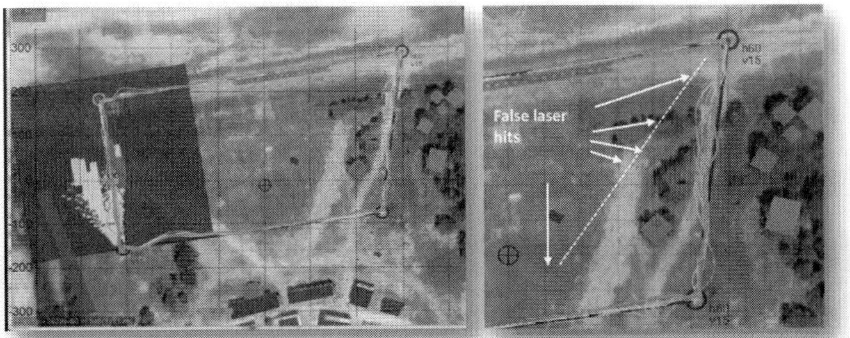

Overlapped flight test trajectory(green) and M/C simulation (Yellow)

Current Research

➢ Active Load Alleviation and Component Life Extension through Pilot Cueing

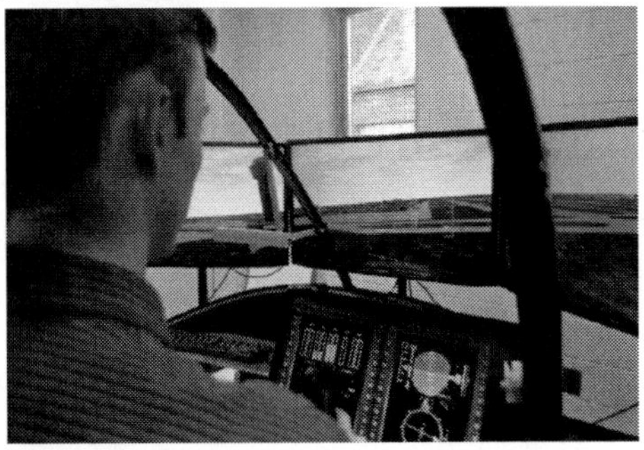

➢ Pilot Guidance for Shipboard Landing Envelope Expansion

ENVELOPE PROTECTION FOR
CAREFREE MANEUVERING
Georgia Tech
Thank You!